Handbook of Fluorescent Probes and Research Chemicals

Sixth Edition

Molecular Probes

by Richard P. Haugland, Ph.D.

Michelle T.Z. Spence, Ph.D., Editor

Iain D. Johnson, Ph.D., Technical Information Coordinator

Design, Layout and Graphics
Lynda J. Gansel, Shannon L. Peters and Isamu Sato

Publications Assistants
Ching-Ying Cheung, Gretchen A. Klaubert, Cathy Patton, Stephanie Perkins, Lisa J. Petersen and David Roomes

Product Administration
Leslie Bayer; Martha McReynolds Jr.; Ann O. Skaugset, Ph.D.; Dawn Swank; and Mary Wisegarver

Contributors
Stephen H. Chamberlain, Ph.D.; Ian Clements; Zhenjun Diwu, Ph.D.; Angélique Dreijer, Ph.D.; Kyle Gee, Ph.D.; David Hagen, Ph.D.; Rosaria P. Haugland, Ph.D.; Iain D. Johnson, Ph.D.; Hee Chol Kang, Ph.D.; Dieter H. Klaubert, Ph.D.; Michael Kuhn; Tom Landon, Ph.D.; Karen D. Larison; Nabi Malekzadeh; Fei Mao, Ph.D.; Paul J. Millard, Ph.D.; David J. Phelps, Ph.D.; Martin Poot, Ph.D.; Victoria L. Singer, Ph.D.; Diane Ryan; Richard M. Schubert, Ph.D.; Michelle T.Z. Spence, Ph.D.; Josh Stahl; Onno van de Stolpe; Gerald A. Thomas, Ph.D.; Donald A. Upson, Ph.D.; K. Sam Wells, Ph.D.; Stephen Yue, Ph.D.; Cailan Zhang; Yu-Zhong Zhang, Ph.D.; and Mingjie Zhou, Ph.D.

This Handbook would not have been possible without the help of many others in the Accounting, Analytical Services, Biosciences, Custom and Bulk Sales, Customer Service, Facilities, Human Resources, Information Systems, Legal, Organic Chemistry, Packaging, Purchasing, Quality Assurance, Receiving, Safety/Regulatory, Shipping and Technical Assistance Departments.

Molecular Probes, Inc., Eugene, OR 97402; Founded in 1975.
Molecular Probes Europe BV, Leiden, The Netherlands; Founded in 1995.

First Edition published 1978. Second Edition 1981. Third Edition 1985.
Fourth Edition 1989. Fifth Edition 1992. Sixth Edition 1996.

ISBN 0-9652240-0-7 (US dollar version)
ISBN 0-9652240-1-5 (Deutsche mark version)
ISBN 0-9652240-2-3 (Unpriced version)

Printed in the United States of America

Customer Service and Technical Assistance

Molecular Probes has offices in Eugene, Oregon, USA and in Leiden, The Netherlands. **Molecular Probes, Inc.** in Eugene, Oregon serves customers from North America, Central America, South America, Asia, Australia, New Zealand, Pakistan and islands in the Pacific. **Molecular Probes Europe BV** in Leiden, The Netherlands handles all accounts in Europe, Africa, the Middle East, Russia and the Commonwealth of Independent States; orders from these areas must be placed with the Leiden office.

Molecular Probes, Inc.
4849 Pitchford Avenue
Eugene, OR 97402-9165 USA

PO Box 22010
Eugene, OR 97402-0469 USA
Phone: (541) 465-8300
Fax: (541) 344-6504
Web: http://www.probes.com

Customer Service Department
Hours: 7:00 am to 5:00 pm (Pacific Time)
Phone: (541) 465-8338
Fax: (541) 344-6504
E-mail: order@probes.com

For US and Canada
Toll Free Order Phone: (800) 438-2209
Toll Free Order Fax: (800) 438-0228

Technical Assistance Department
Hours: 8:00 am to 4:00 pm
Phone: (541) 465-8353
Fax: (541) 465-4593
E-mail: tech@probes.com

Custom and Bulk Sales
Phone: (541) 465-8390
Fax: (541) 984-5658
E-mail: custom@probes.com

Molecular Probes Europe BV
PoortGebouw
Rijnsburgerweg 10
2333 AA Leiden
The Netherlands
Phone: +31-71-5233378
Fax: +31-71-5233419
Web: http://www.probes.com

Customer Service Department
Hours: 9:00 to 16:30 (Central European Time)
Phone: +31-71-5236850
Fax: +31-71-5233419
E-mail: eurorder@probes.nl

For Germany, Finland, France and Switzerland
Toll Free Order Phone: +31-800-5550
Toll Free Order Fax: +31-800-5551

Technical Assistance Department
Hours: 9:00 to 16:30
Phone: +31-71-5233431
Fax: +31-71-5233419
E-mail: eurotech@probes.nl

Custom and Bulk Sales
Phone: +31-71-5236850
Fax: +31-71-5233419
E-mail: eurocustom@probes.nl

Ordering Information

Molecular Probes' *Handbook of Fluorescent Probes and Research Chemicals*

Ordering from Molecular Probes, Inc. and Molecular Probes Europe BV

Placing an Order

Molecular Probes has offices in Eugene, Oregon, USA and in Leiden, The Netherlands. **Molecular Probes, Inc.** in Eugene, Oregon serves customers from North America, Central America, South America, Asia, Australia, New Zealand, Pakistan and islands in the Pacific. **Molecular Probes Europe BV** in Leiden, The Netherlands handles all accounts in Europe, Africa, the Middle East, Russia and the Commonwealth of Independent States; orders from these areas *must* be placed with the Leiden office.

Orders can be placed with our Customer Service Department at the addresses or telephone numbers indicated on the front inside cover or through our Web Site (http://www.probes.com). When placing an order, please provide the following information:

- Contact name, institution, department and telephone number
- Account number — If you do not know your account number, we can provide it for you
- Address for the invoice
- Delivery address
- Purchase order number (if applicable)
- Catalog number(s) and product description(s)
- Order quantity and unit size of product(s)
- Value-added tax (VAT) number — for customers from EU member states only

All orders are subject to acceptance by Molecular Probes, and each product comes with a license for research use only. We require that an account be established or that an order be prepaid prior to shipment. All orders for Molecular Probes' products placed through our distributors are subject to the prices, terms and conditions set by the individual distributor; these may differ from those of Molecular Probes.

Acknowledgements, Cancellations and Confirming Orders

It is the policy of Molecular Probes to ship orders within 24 hours of receipt whenever possible; perishable products that must not be shipped over a weekend are an exception. Please inform us when placing your order if you require either accelerated or delayed delivery. We take pride in the fact that virtually all of our products are continuously in stock in both the Eugene and Leiden offices and almost all orders are shipped complete on the same day they are received. We do not send order acknowledgements. Unless required by your institution, *do not send confirming orders*. Customers will be responsible for payment of invoices for any orders placed by mail, telephone, fax or e-mail or through our Web site that result from duplicate orders not indicated as "CONFIRMING." Back-ordered items are so indicated on the invoice. If the delivery time on back-ordered items exceeds 30 days, an approximate delivery date, which may be subject to revision, will be communicated. Cancellations must be received prior to shipment of the material. We reserve the right to refuse cancellation of orders.

Prices and Payment Terms

This Handbook is printed in three versions:

- **US dollar version** (ISBN 0-9652240-0-7) quotes prices in US dollars for orders placed with **Molecular Probes, Inc.** for delivery in the United States, Canada and Mexico only.
- **Deutsche mark (DM) version** (ISBN 0-9652240-1-5) quotes prices in Deutsche marks for orders placed with **Molecular Probes Europe BV** for delivery in Europe, Africa, the Middle East, Russia and the Commonwealth of Independent States. Molecular Probes Europe BV uses a conversion factor that is periodically adjusted to make quotations in several other European currencies.
- **Unpriced version** (ISBN 0-9652240-2-3) does not list prices and is used by Molecular Probes, Inc., Molecular Probes Europe BV and most of our distributors. If you have received this unpriced version from Molecular Probes, Inc. or Molecular Probes Europe BV, it will be accompanied by a separate price list. If it is received from one of our distributors, then you should contact that distributor for pricing information.

Prices shown in this Handbook, or in an accompanying price list, are valid at the time of publication and are subject to correction or change without notice. If you require pricing information prior to placing your order, the current prices of virtually all products ordered directly from Molecular Probes are available for all countries through our Web site (http://www.probes.com) or by contacting our Customer Service Department. The prices in effect at the time of shipment will be charged, and shipment will be made promptly even if prices have been nominally increased. When placing your order, please supply the quoted prices or reference our *pro forma* invoice number. Any governmental or banking fees that are levied on products or invoices — taxes, assessments, duties, customs charges, money transfer charges — are the responsibility of the customer.

Quantity Discounts

On most products, we offer a 20% discount off the single-unit price when 5–24 units of the same item catalog number are purchased. Additional discounts may be available for larger numbers of units or standing orders. See Custom, OEM and Bulk Sales below for further information.

Handling Charges

A handling fee will be charged to those institutions with special invoicing and shipping requirements. We also charge a fee for filling special packaging requests and for all orders that require an Export Certificate for Animal Products. To cover extra paperwork, a small fee is charged for shipments from Molecular Probes, Inc. to institutions outside of the United States.

Payment Terms

Invoices are sent within 24 hours of shipping the order. Terms of payment are net 30 days from the date of shipment. Payment must be *received* within this period to ensure eligibility for future nonprepaid shipment. The amount received must be sufficient to cover both the invoiced amount and *any* bank charges that may be incurred. Late charges may be added to invoices not paid within the 30-day time period. Late charges must be paid before subsequent orders can be shipped. Shipments will be suspended if payments are not received within the specified time, and subsequent orders may be held until all overdue payments from the entire institution from which the order was received are paid in full.

Payments for all shipments from **Molecular Probes, Inc.** in Eugene, Oregon, USA must be made by a check in US dollars or by bank transfer to the addresses on the invoice. Both offices of Molecular Probes accept Visa or MasterCard. Include the full name and account number of the card holder, and expiration date of card. **Molecular Probes Europe BV** in Leiden, The Netherlands currently invoices and accepts payments in 12 different European currencies; see the invoice for the proper address and account numbers to which to submit payments. For the countries listed in the table below, we accept payment in local currency after applying the appropriate exchange rates. We reserve the right to adjust prices and to change exchange rates to cover currency fluctuations.

Exchange rates for converting Deutsche mark (DM) prices into local currencies.

Country	Currency	Multiply DM Price By *
Austria	Austrian schilling	7
Belgium	Belgian franc	20.66
Denmark	Danish krone	4
Finland	Finnish mark	3.1
France	French franc	3.53
Italy	Italian lira	1150
The Netherlands	Dutch guilder	1.09
Spain	Spanish peseta	90
Sweden	Swedish krona	4.5
Switzerland	Swiss franc	0.8
United Kingdom	British pound	0.42

* Due to the uncertainty of the currency market, these conversion factors are subject to change without notice. The exchange rates in effect at the time of shipment will be charged.

Shipping, Storage and Returns

Method of Shipment

Shipping is FCA, Free Carrier (Eugene, Oregon, USA or Leiden, The Netherlands) according to INCOTERMS 1990. Orders are sent to customers by the best method both for protecting the products during transit and for meeting commitments on delivery dates. Products, even those for which freezer storage is recommended, are stable for the period of shipment. Some products must be shipped cold without freezing by the fastest practical method. A fee will be added for items requiring dry ice or extra ice packs necessitated by long delivery time. Unless other means are specified, shipment is via courier service. Customer requests for special packaging and alternative methods of shipment will be accepted and followed within our abilities to meet the specifications; any extra charges are billed to the customer. At all times, Molecular Probes will adhere to government regulations for packaging and shipping.

Shipping Charges

All shipping charges billed are the responsibility of the customer and are normally prepaid by Molecular Probes and added to the invoice. Partial shipments of available items are made when another item is backordered; in this case we will pay the shipping charge on the backordered item(s). All invoices for partial shipments are due on the date specified on the invoice, without regard to subsequent deliveries. Two separate packages may be required if the order includes products with different shipping or incompatible storage requirements. In these cases, appropriate shipping charges for both shipments will be added to the invoice. In most cases, shipment can be made freight collect if desired.

Return Goods Policy

Please inspect your packages upon receipt. If the goods have been damaged in transit, our Customer Service Department can assist you in filing a claim with the carrier. Please inform us immediately of any shortage or shipment errors. Any claims for such errors must be made within 10 business days. If it is our error, we will do whatever is necessary to ship the correct products as soon as possible. Any returns must be authorized by our Customer Service Department. WE WILL NOT ACCEPT return shipments unless we have given you prior written permission and shipping instructions. In most cases, items that you order in error cannot be returned because of the sensitive nature of many of our products and the difficulty and expense of requalifying returned items. If we decide to accept returns caused by your error, these must be in new, unopened, unused and undamaged condition, and you will be charged a per-unit 20% restocking charge. If items are accepted for return, Molecular Probes will advise the method of packing, labeling and shipping. To ensure proper credit, each return must include the customer name, address, purchase order number, invoice number and catalog number of all returned items, along with **Molecular Probes' return authorization number** and a written statement describing the reason for the return.

Storage Conditions

Recommended storage for each item is normally indicated by codes on the product's label; these storage codes are defined on the Product Support Sheet that accompanies each shipment (see also page vii). Storage codes for most products can also be found in the data tables within each section of the Handbook and through our Web site (http://www.probes.com). Labels are placed on the outside of packages to alert receiving personnel to open and store immediately; shipments requiring freezer or refrigerator storage are labeled accordingly. Some kits have mixed storage requirements that require separate storage of their components.

Custom, OEM and Bulk Sales and Contract Research

Molecular Probes, Inc. manufactures over 90% of the products it sells and can provide many of these in bulk for manufacturers in the diagnostics, pharmaceutical and biotechnology industries (however, see TSCA below). Molecular Probes is currently seeking ISO 9001 certification and has experience working with companies in regulated industries. Molecular Probes also has extensive experience working with academic and industrial researchers who need solutions to problems that involve chemical synthesis, assay development or contract research. Resources of the company include over 75 research scientists, manufacturing facilities and an extensive patent portfolio. For information and quotations on listed products, please contact our Custom, OEM and Bulk Sales Department.

Research and Development Use Only and Toxic Substances Control Act (TSCA)

Except as provided by written contracts, products ordered from Molecular Probes are sold for research and development use only. They are not intended for food, drug, household, agricultural or cosmetic use. Molecular Probes reserves the right to refuse orders to any customer who plans to use our products for any other purposes. Products to be used for purposes other than research and development, including additional products not listed in this Handbook, are available only under license from Molecular Probes and may require prior approval from the Environmental Protection Agency (EPA) or inclusion on the TSCA inventory. The buyer assumes responsibility to ensure that the products purchased are approved for use under TSCA, if applicable.

Many of Molecular Probes' products are covered by patents and trademarks owned by Molecular Probes or other companies (see also page 679). No patented product from Molecular Probes is offered or sold for incorporation into any product for resale or for use in providing services for which a fee is charged without specific written permission and, in some cases, a license from Molecular Probes.

Use of Molecular Probes' products must be supervised by a technically qualified individual. Products defined as hazardous by the OSHA Hazard Communication Standard are accompanied by a Material Safety Data Sheet (MSDS). Other products are accompanied by a letter stating that no MSDS is required.

Disclaimer of Warranty, Remedy and Liability

The products shipped by Molecular Probes are warranted to conform to the chemical descriptions as provided in this Handbook or our other publications. However, *Molecular Probes does not warrant or guarantee that its products are merchantable or satisfactory for any particular purpose, nor free from any claim of foreign or domestic patent infringement by a third party, and there are no warranties, express or implied, to such effect. Molecular Probes will not be liable for any incidental, consequential or contingent damages involving their use. Our responsibility is limited only to replacement of items ordered.*

All information in this Handbook, at our Web site and provided by our Technical Assistance Department is believed to be accurate, but it remains the responsibility of the investigator to confirm all technical aspects of the applications. We appreciate receiving any additions, corrections, or updates to information in this Handbook and will include these in the upcoming electronic version of the Handbook at our Web site. We reserve the right to discontinue or change specifications or prices and to correct any errors or omissions at any time without incurring obligations. Molecular Probes will not be responsible for claims made by any of its distributors that exceed those made in Molecular Probes' own publications.

Technical Assistance at Our Web Site (http://www.probes.com)

At Molecular Probes' Web site, we are developing an electronic version of this Handbook and other databases that should prove extremely useful for the researcher. In addition to containing all of the text from this Handbook, our Web site provides:

- **Product searches** by product name or catalog number
- **Bibliographies** for all products for which we have references
- **Keyword searches** of our entire bibliography of over 25,000 references
- **Product information sheets** for many kits and reagents
- **Technical bulletins**, including *BioProbes* newsletters and other product literature
- **Chemical structure**, **technical data** and **material safety and data sheets**
- **Color photomicrographs** that show our products in action

Visit our Web site often for new additions to our bibliography, as well as new products and upgraded search capabilities. Also look for special sales and introductory specials on some important products.

If you do not have access to the Internet or you need assistance that is not available at that site, further information on the scientific and technical background of our products can be obtained by contacting our Technical Assistance Department at the numbers listed on the inside front cover.

How to Use the Handbook

Molecular Probes' *Handbook of Fluorescent Probes and Research Chemicals*

The *Handbook of Fluorescent Probes and Research Chemicals* organizes Molecular Probes' products according to their major applications, thereby facilitating discussion of probe usage. This organization also encourages exploration of cutting-edge reagents and techniques that may be unfamiliar to many researchers or overshadowed by traditional methodology. We have provided several mechanisms that will point you to the preferred products for your research, be it in cell biology, biochemistry, biophysics, microbiology, molecular biology or neuroscience.

Tools for Navigating Through the Handbook

Product Names and Catalog Numbers

Many products have alternative names that are either abbreviations of the principal name or synonyms that have gained widespread acceptance in the scientific literature. In the section price lists, alternative names are shown in parentheses following the principal name — Principal name (Alternative name). In the Master Price List/Product Index, reverse listings — Alternative name (Principal name) — are also given for many products to help you find a given product regardless of whether you know the principal or the alternative name.

Our alpha-numeric catalog numbers comprise the first significant letter of the principal name followed by a unique number. Numerals, *cis*- and *trans*- (in isomer mixtures), *N*- (for nitrogen), *p*- (for *para*), α, β and other similar descriptors are not considered to be significant for catalog number assignment. These characters are also not used when alphabetizing product names in the price lists.

Master Price List/Product Index

The Master Price List/Product Index at the back of the Handbook includes all products available from Molecular Probes at the time of printing, with the exception of dextrans (Section 15.5), FluoSpheres® polystyrene microspheres (Section 6.2) and optical filters (Section 26.5). As in the section price lists, products are arranged in alphabetical order of name. In the Master Price List/Product Index, many products that have alternative names (see above) are listed under both the principal and alternative name to aid in locating a product of interest. The last column of the Master Price List/Product Index serves as a product index by indicating the Handbook section(s) in which technical information on the product can be found.

Keyword Index

The Keyword Index follows the Master Price List/Product Index and provides selected keywords, which are alphabetically listed, along with the page numbers on which they appear. These keywords were chosen to help you quickly locate the Handbook section of interest.

Finding the Product of Interest

This Handbook includes products released by Molecular Probes before September 1996. Additional products and search capabilities are available through our Web site (http://www.probes.com). All information, including price and availability, is subject to change without notice.

- **If you know the name of the product you want:** Look up the product by name in the **Master Price List/Product Index** at the back of the book. The Master Price List/Product Index is arranged alphabetically, by both product name and alternative name, using the alphabetization rules outlined above. This index will refer you directly to all the Handbook sections that discuss the product.

- **If you DO NOT know the name of the product you want:** Search the **Keyword Index** for the words that best describe the applications in which you are interested. The Keyword Index will refer you directly to page numbers on which the listed topics are discussed. Alternatively, search the **Table of Contents** for the chapter or section title that most closely relates to your application.

- **If you only know the alpha-numeric catalog number of the product you want but not its name:** Look for the catalog number in the **Master Price List/Product Index** at the back of the book. First locate the group of catalog numbers with the same alpha identifier as the catalog number of interest and then scan this group for the correct numeric identifier. The Master Price List/Product Index will refer you directly to all the Handbook sections that discuss this product. Alternatively, you may use the search features available through our Web site (http://www.probes.com) to obtain the product name of any of our current products given only the catalog number (or numeric identifier); see below for more details.

- **If you only know part of a product name or are searching for a product introduced after publication of this Handbook:** Search Molecular Probes' up-to-date product list through our **Web site (http://www.probes.com)**. The product search feature on our home page allows you to search our entire product list, including products that have been added since the completion of this Handbook, using a partial product name, complete product name or catalog number. Product-specific bibliographies can also be obtained from our Web site once the product of interest has been located.

- **If you still cannot find the product that you are looking for or are seeking advice in choosing a product:** Please contact our Technical Assistance Department by phone, e-mail, fax or mail.

Chapter Organization

The Handbook is divided into **Chapters**, numbered 1 through 27. Chapters 1 through 26 each deal with a product group with common properties or applications. When a product has multiple applications, it appears in more than one chapter. Each chapter is further subdivided into a variable number of **Sections** to further group products with a more narrowly defined set of properties or applications. Sections are numbered sequentially within each chapter; e.g., sections of Chapter 1 are numbered 1.1, 1.2, 1.3, etc. Chapter and section numbers and titles are listed in the **Table of Contents**. In addition, the chapter and section number and the section title appear at the bottom of each right-hand page throughout the Handbook.

Chapter 27 consists of two sections: new products and sale products. The new products in Section 27.1 were added to our inventory too late for inclusion into the appropriate Handbook chapter; the sections in which they belong are indicated for each product. The sale products in Section 27.2 are listed according to the section of Chapters 1–26 to which they relate. Therefore, if you are looking for products for a particular application by consulting the relevant section of Chapters 1–26, there may be additional products of interest in Chapter 27.

Each Handbook section contains several elements in the following order: (1) **text**, tables and figures describing the products allocated to the section, (2) **bibliography** of literature cited in the text, (3) technical **data table**, (4) drawings of **chemical structures** and (5) section **price list** for products discussed in the section. Shaded boxes dispersed throughout the Handbook contain product highlights, which feature products that are not otherwise grouped together within the section, and technical notes, which focus on specific techniques for using a group of products. More detailed descriptions of these elements can be found below.

Text

The text describes the applications and properties of products. Where possible, we have included literature citations for particular applications. In some cases, we have speculated on possible applications for new or existing products. Although the information given is believed to be reliable, the original literature should be consulted for experimental details. More extensive references for many products are available through our Web site (http://www.probes.com).

Figures and tables are numbered sequentially within each chapter, without reference to the specific section number in which they appear. Except where attributed otherwise, figures are based on data collected at Molecular Probes. We will usually grant authorization upon request for reproduction of figures in scientific and technical literature, where such use does not conflict with Molecular Probes' commercial interests. Please submit requests to the Publications Department at Molecular Probes by e-mail, fax or mail. We can also provide camera-ready enlargements of figures for a small service charge.

Bibliography

Literature cited in the text is indicated by superscripted numerals that refer to the bibliography list immediately following the section text. Unlike previous editions, we have not included a master bibliography at the end of the Handbook because our total product bibliography has grown to over 25,000 references, with new references being added daily. Furthermore, an up-to-date product bibliography is now accessible through our World Wide Web site (http://www.probes.com), providing what we believe is a much more effective means of disseminating this information. Product references can be accessed from our Web site using either the catalog number of our product or the partial or complete product name. Visit our Web site often for new additions to our bibliography, as well as new products and upgraded search capabilities. Product-specific bibliographies are also available through our Technical Assistance Department upon request, as are *free* keyword searches of the titles in our bibliography.

Data Tables

Most sections of this Handbook include a table of technical data, listing chemical and spectroscopic properties of products described in the section. Definitions of the contents of these tables are as follows:

Cat #: The alpha-numeric catalog number for a product described in the section. Data tables are organized firstly in alphabetical order using the alpha identifier of the catalog number, and secondly in numerical order using the numeric identifier of the catalog number (e.g., A-36, A-37, B-153, B-162, C-10, C-194, etc.). Protein conjugates, dextran conjugates, FluoSpheres polystyrene microspheres and kits are not generally listed in the data tables.

Structure: The location for the chemical structure drawing. Abbreviations are as follows:

✓	Shown in the current section
C	Not shown due to complexity
P	Proprietary information.
Sn.n	Shown in section number n.n
T	Available upon request from our Technical Assistance Department or through our Web site (http://www.probes.com)

MW: Molecular weight (MW) for the anhydrous compound, except in a few cases for which the exact degree of hydration has been determined and is specified in the name of the product. We caution that products may not be completely anhydrous upon receipt, even if they are represented in this form by the chemical structure drawing and MW value. For compounds isolated and sold in salt form, MW is inclusive of counterions unless annotated otherwise. In some cases (e.g., nucleotide derivatives, some peptides and proteins), the exact salt form has not been established, and consequently the MW value is approximate, denoted by a preceding ~ symbol. For these products, we recommend measuring the absorbance of a solution and calculating concentration with the extinction coefficient.

Storage: Recommended storage conditions for products. Abbreviations are as follows:

A	Material may be air sensitive
AA	Air sensitive, use under a N_2 or Ar atmosphere
D	Desiccation *recommended*
DD	Desiccation *required*
F	-20°C freezer storage *recommended*
FF	-20°C freezer storage *required* upon arrival

L	Protect material from light, *especially in solution*
LL	Protect material from light *at all times*
MIXED	Contains components with incompatible long-term storage requirements; store components separately as advised
NC	Storage temperature not critical; may be stored frozen, refrigerated or at room temperature
RO	4°C refrigeration *required*; for longer storage freeze in small aliquots
RR	4°C refrigeration *required*, DO NOT FREEZE
UF	-70°C ultra-low freezer storage *required*

Many products are stable under less stringent conditions for short periods and can be shipped accordingly without any deleterious effects. Irrespective of the conditions upon receipt, long-term storage should always be in accordance with the storage code. Storage codes are also printed on the labels of all product containers.

Soluble: Recommended solvent for preparing stock solutions of at least 1 mM. Recommendations are based on the best knowledge and experience of our technical staff but have not necessarily been tested. If a pH is indicated, the product generally ionizes and should be dissolved in aqueous buffer in the specified pH range. We appreciate receiving suggestions for alternative solvents and corrections to this information based on practical experience. Abbreviations for solvents are defined below (see **Solvent**).

Abs: The longest-wavelength (unless noted otherwise) absorption maximum (in nanometers) determined in the solvent listed in the column headed **Solvent**.

$\{\epsilon \times 10^{-3}\}$: Molar extinction coefficient determined at the wavelength listed in the column headed **Abs**. Listed values should be multiplied by 1000 to convert to the conventional units of $cm^{-1}M^{-1}$; for example, {75} = 75,000 $cm^{-1}M^{-1}$. Values above 10,000 $cm^{-1}M^{-1}$ are usually rounded to the nearest 1000. In most cases, extinction coefficients have not been rigorously determined, and the values may vary somewhat among production lots.

Em: Fluorescence emission maximum (in nanometers) in the solvent listed in the column headed **Solvent**. Em values are generally not corrected for instrument response characteristics, resulting in small variations when compared to measurements in other laboratories. Considerable environment-dependent variation of Em occurs for some products; when known this is indicated with a footnote.

Solvent: Solvent used for acquisition of spectroscopic data, including Abs, $\{\epsilon \times 10^{-3}\}$ and Em. Solubility of the product in this solvent is not necessarily any greater than is required to obtain an absorption spectrum (i.e., 10 µM or less). In some cases, the concentrated stock solution used to prepare the working solution for measuring spectra may require a different solvent. Refer to the column headed **Soluble** for recommended solvents for preparing stock solutions. This stock solution is then diluted into the indicated **Solvent** for acquisition of spectroscopic data. Abbreviations are as follows:

$CHCl_3$	Chloroform
DMF	Dimethylformamide
DMSO	Dimethylsulfoxide
EtOAc	Ethyl acetate
EtOH	Ethanol
H_2O	Unbuffered water
pH 7	pH 7 aqueous buffer
H_2O/DNA	Spectra measured for probes bound to dsDNA in aqueous solution
MeCN	Acetonitrile
MeOH	Methanol
THF	Tetrahydrofuran

Data for ion indicators are usually listed for aqueous solutions both with and without the target ion. The indicator dissociation constant for the target ion, in most cases determined in Molecular Probes' laboratories, is listed in a separate column headed $\mathbf{K_d}$.

Notes: Extensions, identified by numbers (<1>, etc.) or symbols (*, †, ‡, etc.), to the listed data. For example, in cases where a product exists in two forms (e.g., substrate/enzymatic product, free/bound, unreacted/reacted), data for one form is listed in the table and for the other form in a footnote. Other spectroscopic parameters reported in some footnotes include fluorescence quantum yields (QY) and excited state lifetimes (τ; units = seconds × 10^{-9} = nsec). We do not routinely determine QY and τ for our products but may include this data in the footnotes when it is known.

Chemical Structures

All of the chemical structures have been drawn using the ChemWindow® program (SoftShell® International, Ltd.). They are arranged following the same ordering scheme used in the technical data tables. With very few exceptions, each product structure is shown in only one section of the Handbook. For products included in multiple sections, crossreferences on the structure page and in the **Structure** column of the data table give the section number location of the structure drawing. Similar structures within a section are usually combined into a single drawing. Instances where this results in a product structure being located out of alphabetical sequence (e.g., C-400 shown under D-399) are indicated by placeholders on the structure page. Some structures are not shown for proprietary reasons. Most structures of products in the catalog number range A-96001 to D-99028 are not shown in the Handbook but are available through our Web site (http://www.probes.com) or from our Technical Assistance Department upon request. Structures for most chemicals will soon be available through our Web site. We will provide camera-ready enlargements of chemical structure drawings for a small service charge.

Section Price Lists

Most Handbook sections conclude with a price list giving the catalog number, name, unit size and unit price for the products covered in the section. Also listed is the 20% discounted unit price applicable to most products when 5 or more units of the same item are purchased at one time. Additional discounts may be available on purchases of larger quantities of a product. Products are arranged in alphabetical order of name and marked **New** when they have been introduced since the publication of the Fifth Edition of this Handbook in 1992.

Table of Contents

Molecular Probes' *Handbook of Fluorescent Probes and Research Chemicals*

Introduction to Fluorescence Techniques

by Iain D. Johnson

Fluorescent probes enable researchers to detect particular components of complex biomolecular assemblies, including live cells, with exquisite sensitivity and selectivity. The purpose of this introduction is to briefly outline fluorescence techniques for newcomers to the field.

The Fluorescence Process

Fluorescence is the result of a three-stage process that occurs in certain molecules (generally polyaromatic hydrocarbons or heterocycles) called fluorophores or fluorescent dyes. A fluorescent probe is a fluorophore designed to localize within a specific region of a biological specimen or to respond to a specific stimulus. The process responsible for the fluorescence of fluorescent probes and other fluorophores is illustrated by a simple electronic-state diagram (Jablonski diagram) shown in Figure 1.

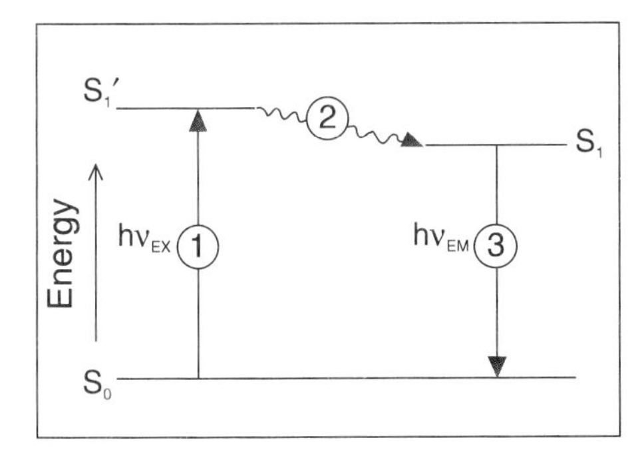

Figure 1 Jablonski diagram illustrating the processes involved in the creation of an excited electronic singlet state by optical absorption and subsequent emission of fluorescence. The labeled stages 1, 2, 3 are referred to in the text.

Stage (1): Excitation

A photon of energy $h\nu_{EX}$ is supplied by an external source such as an incandescent lamp or a laser and absorbed by the fluorophore, creating an excited electronic singlet state (S_1'). This process distinguishes fluorescence from chemiluminescence, in which the excited state is created by a chemical reaction.

Stage (2): Excited-State Lifetime

The excited state exists for a finite time (typically 1–10 × 10^{-9} seconds). During this time, the fluorophore undergoes conformational changes and is also subject to a multitude of possible interactions with its molecular environment. These processes have two important consequences. First, the energy of S_1' is partially dissipated, yielding a relaxed singlet excited state (S_1) from which fluorescence emission originates. Second, not all the molecules initially excited by absorption (Stage 1) return to the ground state (S_0) by fluorescence emission. Other processes such as collisional quenching, fluorescence energy transfer and intersystem crossing (see below) may also depopulate S_1. The fluorescence quantum yield, which is the ratio of the number of fluorescence photons emitted (Stage 3) to the number of photons absorbed (Stage 1), is a measure of the relative extent to which these processes occur.

Stage (3): Fluorescence Emission

A photon of energy $h\nu_{EM}$ is emitted, returning the fluorophore to its ground state S_0. Due to energy dissipation during the excited-state lifetime, the energy of this photon is lower, and therefore of longer wavelength, than the excitation photon $h\nu_{EX}$. The difference in energy or wavelength represented by ($h\nu_{EX}$–$h\nu_{EM}$) is called the Stokes shift. The Stokes shift is fundamental to the sensitivity of fluorescence techniques because it allows emission photons to be detected against a low background, isolated from excitation photons. In contrast, absorption spectrophotometry requires measurement of transmitted light relative to high incident light levels at the same wavelength.

Fluorescence Spectra

The entire fluorescence process is cyclical. Unless the fluorophore is irreversibly destroyed in the excited state (an important phenomenon known as photobleaching, see below), the same fluorophore can be repeatedly excited and detected. For polyatomic molecules in solution, the discrete electronic transitions represented by $h\nu_{EX}$ and $h\nu_{EM}$ in Figure 1 are replaced by rather broad energy spectra called the fluorescence excitation spectrum and fluorescence emission spectrum, respectively. The bandwidths of these spectra are parameters of particular importance for applications in which two or more different fluorophores are simultaneously detected (see below). With few exceptions, the fluorescence excita-

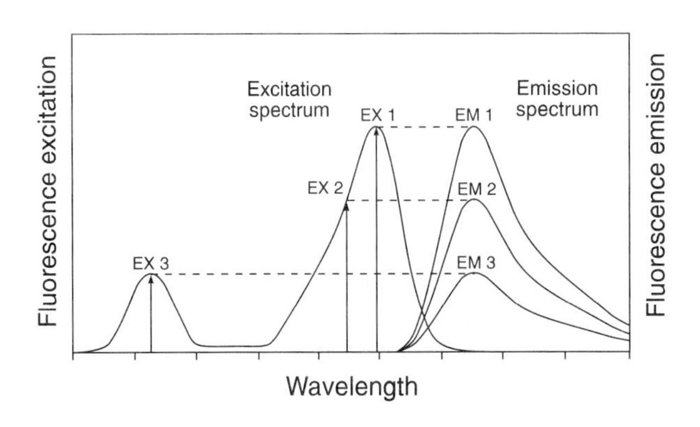

Figure 2 Excitation of a fluorophore at three different wavelengths (EX 1, EX 2, EX 3) does not change the emission profile but does produce variations in fluorescence emission intensity (EM 1, EM 2, EM 3) that correspond to the amplitude of the excitation spectrum.

tion spectrum of a single fluorophore species in dilute solution is identical to its absorption spectrum. Under the same conditions, the fluorescence emission spectrum is independent of the excitation wavelength, due to the partial dissipation of excitation energy during the excited-state lifetime as illustrated in Figure 1. The emission intensity is proportional to the amplitude of the fluorescence excitation spectrum at the excitation wavelength (Figure 2).

Fluorescence Detection

Fluorescence Instrumentation

Four essential elements of fluorescence detection systems can be identified from the preceding discussion: (1) an excitation source, (2) a fluorophore, (3) wavelength filters to isolate emission photons from excitation photons and (4) a detector that registers emission photons and produces a recordable output, usually as an electrical signal or a photographic image. Regardless of the application, compatibility of these four elements is essential for optimizing fluorescence detection.

Fluorescence instruments are primarily of three types, each providing distinctly different information:

- A **spectrofluorometer** measures the *average* properties of bulk (µL to mL) samples.
- A **fluorescence microscope** resolves fluorescence as a function of spatial coordinates in two or three dimensions.
- A **flow cytometer** measures fluorescence per cell in a flowing stream, allowing subpopulations within a large sample to be identified and quantitated.

Each type of instrument produces different measurement artifacts and makes different demands on the fluorescent probe. For example, although photobleaching is often a significant problem in fluorescence microscopy, it is not a major impediment in flow cytometry because the dwell time of individual cells in the excitation beam is short.

Fluorescence Signals

Fluorescence intensity is quantitatively dependent on the same parameters as absorbance — defined by the Beer–Lambert law as the product of the molar extinction coefficient, optical pathlength and solute concentration — as well as on the fluorescence quantum yield of the dye and the excitation source intensity and fluorescence collection efficiency of the instrument. In dilute solutions or suspensions, fluorescence intensity is linearly proportional to these parameters. When sample absorbance exceeds about 0.05 in a 1 cm pathlength, the relationship becomes nonlinear and measurements may be distorted by artifacts such as self-absorption and the inner-filter effect.[1] Because fluorescence quantitation is dependent on the instrument, fluorescent reference standards are essential for calibrating measurements made at different times or using different instrument configurations.[2-4] To meet these requirements, Molecular Probes offers high-precision fluorescent microsphere reference standards for fluorescence microscopy and flow cytometry (see Sections 26.1 and 26.2, respectively).

A spectrofluorometer is extremely flexible, providing continuous ranges of excitation and emission wavelengths. Laser-scanning microscopes and flow cytometers, however, require probes that are excitable at a single fixed wavelength. In contemporary instruments, the excitation source is usually the 488 nm spectral line of the argon-ion laser. As shown in Figure 3, separation of the fluorescence emission signal (S1) from Rayleigh-scattered excitation light (EX) is facilitated by a large fluorescence Stokes shift (i.e., separation of A1 and E1). Biological samples labeled with fluorescent probes typically contain more than one fluorescent species, making signal-isolation issues more complex. Additional optical signals, represented in Figure 3 as S2, may be due to background fluorescence or to a second fluorescent probe.

Background Fluorescence

Fluorescence detection sensitivity is severely compromised by background signals, which may originate from endogenous sample constituents (referred to as autofluorescence) or from unbound or nonspecifically bound probes (referred to as reagent background). Detection of autofluorescence can be minimized either by selecting filters that reduce the transmission of E2 relative to E1 or by selecting probes that absorb and emit at longer wavelengths. Although narrowing the fluorescence detection bandwidth increases the resolution of E1 and E2, it also compromises the overall fluorescence intensity detected. Signal distortion caused by autofluorescence of cells, tissues and biological fluids is most readily minimized by using probes that can be excited at >500 nm. Furthermore, at longer wavelengths, light scattering by dense media such as tissues is much reduced, resulting in greater penetration of the excitation light.

Multicolor Labeling Experiments

A multicolor labeling experiment entails the deliberate introduction of two or more probes to simultaneously monitor different biochemical functions. This technique has major applications in flow cytometry,[5,6] DNA sequencing,[7,8] fluorescence *in situ* hybridization[9,10] and fluorescence microscopy.[11,12] Signal isolation and data analysis are facilitated by maximizing the spectral separation of the multiple emissions (E1 and E2 in Figure 3). Consequently, fluorophores with narrow spectral bandwidths, such as Molecular Probes' BODIPY® dyes (see Section 1.2), are particularly useful in multicolor applications.[7] An ideal combination of dyes for multicolor labeling would exhibit strong absorption at a coincident excitation wavelength and well-separated emission spectra (Figure 3). Unfortunately, it is not easy to find dyes with the requisite combination of a large extinction coefficient for absorption and a large Stokes shift[13] (see the Technical Note "Limitations of Low Molecular Weight Dyes" on page 109).

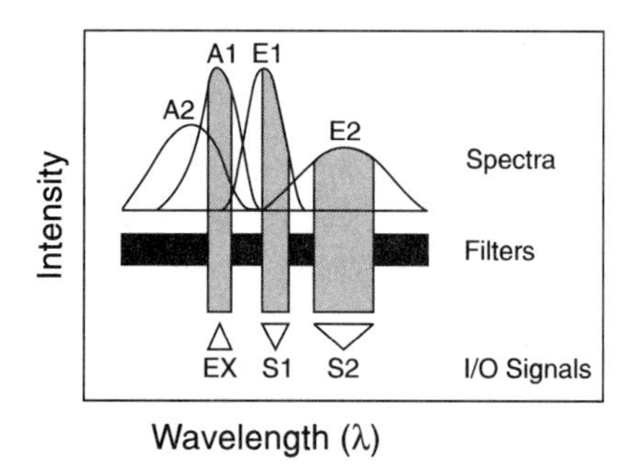

Figure 3 *Fluorescence detection of mixed species. Excitation (EX) in overlapping absorption bands A1 and A2 produces two fluorescent species with spectra E1 and E2. Optical filters isolate quantitative emission signals S1 and S2.*

Ratiometric Measurements

In some cases, for example the calcium indicators fura-2 and indo-1 (see Section 22.2) and the pH indicators BCECF, SNARF® and SNAFL® (see Section 23.2), the free and ion-bound forms of fluorescent ion indicators have different emission or excitation spectra. With this type of indicator, the ratio of the optical signals (S1 and S2 in Figure 3) can be used to monitor the association equilibrium and to calculate ion concentrations. Ratiometric measurements eliminate distortions of data caused by photobleaching and variations in probe loading and retention, as well as by instrumental factors such as illumination stability. For a thorough discussion of ratiometric techniques, see the Technical Note "Loading and Calibration of Intracellular Ion Indicators" on page 549.

Fluorescence Output of Fluorophores

Comparing Different Dyes

Fluorophores currently used as fluorescent probes offer sufficient permutations of wavelength range, Stokes shift and spectral bandwidth to meet requirements imposed by instrumentation (e.g., 488 nm excitation), while allowing flexibility in the design of multicolor labeling experiments (Figure 4). The fluorescence output of a given dye depends on the efficiency with which it absorbs and emits photons, and its ability to undergo repeated excitation/emission cycles. Absorption and emission efficiencies are most usefully quantified in terms of the molar extinction coefficient (ϵ) for absorption and the quantum yield (QY) for fluorescence. Both are constants under specific environmental conditions. The value of ϵ is specified at a single wavelength (usually the absorption maximum), whereas QY is a measure of the total photon emission over the entire fluorescence spectral profile. Fluorescence intensity per dye molecule is proportional to the product of ϵ and QY, the measure used to produce the rank ordering of the fluorophores in Figure 4. The range of these parameters among fluorophores of current practical importance is approximately 5000 to 200,000 $cm^{-1}M^{-1}$ for ϵ and 0.05 to 1.0 for QY. Phycobiliproteins such as R-phycoerythrin (see Section 6.4) have multiple fluorophores on each protein and consequently have much larger extinction coefficients (on the order of 2×10^6 $cm^{-1}M^{-1}$) than low molecular weight fluorophores.

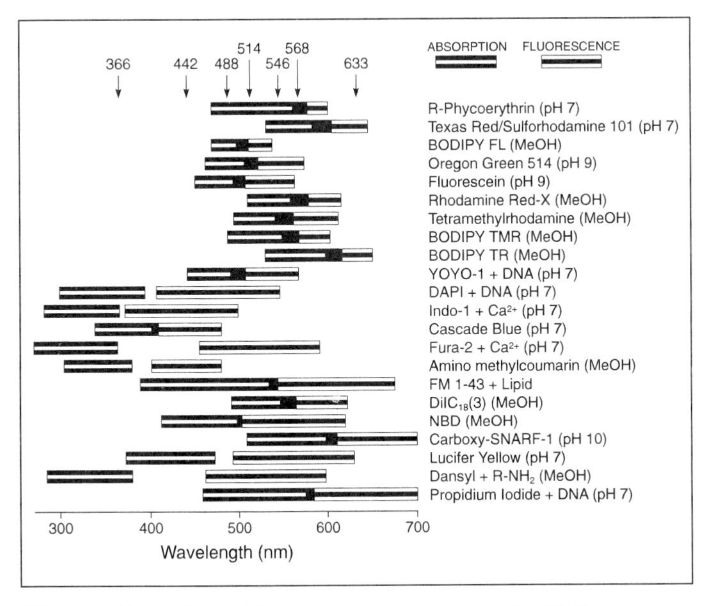

Figure 4 *Absorption and fluorescence spectral ranges for 22 fluorophores of current practical importance. The range encompasses only those values of the absorbance or the fluorescence emission that are >25% of the maximum value. Fluorophores are arranged vertically in rank order of the product* $\epsilon_{max} \times QY$*, in either methanol or aqueous buffer as specified. Some important excitation source lines are indicated on the upper horizontal axis.*

Photobleaching

Under high-intensity illumination conditions, the irreversible destruction or photobleaching of the excited fluorophore becomes the factor limiting fluorescence detectability. Recent investigations have provided a detailed description of the multiple photochemical reaction pathways responsible for photobleaching of fluorescein.[14,15] Some pathways include reactions between adjacent dye molecules, making the process considerably more complex in labeled biological specimens than in dilute solutions of free dye. In all cases, photobleaching originates from the triplet excited state, which is created from the singlet state (S_1, Figure 1) via an excited-state process called intersystem crossing.

The most effective remedy for photobleaching is to maximize detection sensitivity, which allows the excitation intensity to be reduced. Detection sensitivity is enhanced by low-light detection devices such as CCD cameras, as well as by high–numerical aperture objectives and the widest emission bandpass filters compatible with satisfactory signal isolation. Alternatively, a less photolabile fluorophore may be substituted in the experiment. Molecular Probes' Oregon Green™ dyes are important new fluorescein substitutes that provide greater photostability (see Figure 1.6 in Chapter 1, Figure 7.1 in Chapter 7 and Figure 11.1 in Chapter 11) than fluorescein, yet are compatible with standard fluorescein optical filters. Antifade reagents such as Molecular Probes' *SlowFade*™ and ProLong™ products (see Section 26.1) can also be applied to reduce photobleaching; however, they are usually incompatible with live cells. In general, it is difficult to predict the necessity for and effectiveness of such countermeasures because photobleaching rates are dependent on the fluorophore's environment.[15-17]

Signal Amplification

The most straightforward way to enhance fluorescence signals is to increase the number of fluorophores available for detection. Fluorescent signals can be amplified using 1) avidin–biotin or antibody–hapten secondary detection techniques, 2) enzyme-labeled secondary detection reagents in conjunction with fluorogenic substrates [18,19] or 3) probes that contain multiple fluorophores such as phycobiliproteins and Molecular Probes' FluoSpheres® fluorescent microspheres; all three strategies are discussed in Chapter 6.

Simply increasing the probe concentration can be counterproductive and often produces marked changes in the probe's chemical and optical characteristics. It is important to note that the effective intracellular concentration of probes loaded by bulk permeabilization methods (see the Technical Note "Loading and Calibration of Intracellular Ion Indicators" on page 549) is usually much higher (>10-fold) than the extracellular incubation concentration. Also, increased labeling of proteins or membranes ultimately leads to precipitation of the protein or gross changes in membrane permeability. Antibodies labeled with more than four to six fluorophores per protein may exhibit reduced specificity and reduced binding affinity. Furthermore, at high degrees of substitution, the extra fluorescence obtained per added fluorophore typically decreases due to self-quenching (see Figure 1.5 in Chapter 1).

Environmental Sensitivity of Fluorescence

Fluorescence spectra and quantum yields are generally more dependent on the environment than absorption spectra and extinction coefficients. For example, coupling a single fluorescein label to a protein reduces fluorescein's QY ~60% but only decreases its ϵ by ~10%. Interactions either between two adjacent fluorophores or between a fluorophore and other species in the surrounding environment can produce environment-sensitive fluorescence.

Fluorophore–Fluorophore Interactions

Fluorescence quenching can be defined as a bimolecular process that reduces the fluorescence quantum yield without changing the fluorescence emission spectrum; it can result from transient excited-state interactions (collisional quenching) or from formation of nonfluorescent ground-state species. Self-quenching is the quenching of one fluorophore by another; it therefore tends to occur when high loading concentrations or labeling densities are used (see Figures 1.5 and 1.11 in Chapter 1). The substrates in several of Molecular Probes' EnzChek™ Assay Kits (see Sections 10.2 and 10.4) are heavily labeled and therefore highly quenched biopolymers that exhibit dramatic fluorescence enhancement upon enzymatic cleavage. Studies of the self-quenching of carboxyfluorescein show that the mechanism involves energy transfer to nonfluorescent dimers.[20]

Fluorescence resonance energy transfer (FRET) is a strongly distance-dependent excited-state interaction in which emission of one fluorophore is coupled to the excitation of another. See the Technical Note "Fluorescence Resonance Energy Transfer" on page 46 for a discussion of FRET applications.

Some excited fluorophores interact to form excimers, which are excited-state dimers that exhibit altered emission spectra. Excimer formation by the polyaromatic hydrocarbon pyrene is described in Section 13.2 (see especially Figure 13.6 in Chapter 13).

Because they all depend on the interaction of adjacent fluorophores, self-quenching, FRET and excimer formation can be exploited for monitoring a wide array of molecular assembly or fragmentation processes such as membrane fusion (see the Technical Note "Assays of Volume Change, Membrane Fusion and Permeability" on page 364), nucleic acid hybridization, ligand–receptor binding and polypeptide hydrolysis.

Other Environmental Factors

Many other environmental factors exert influences on fluorescence properties. The three most common are:

- Solvent polarity (solvent in this context includes interior regions of cells, proteins, membranes and other biomolecular structures)
- Proximity and concentrations of quenching species
- pH of the aqueous medium

Fluorescence spectra may be strongly dependent on solvent. This characteristic is most usually observed with fluorophores that have large excited-state dipole moments, resulting in fluorescence spectral shifts to longer wavelengths in polar solvents. Representative fluorophores include the aminonaphthalenes such as prodan (see Figure 14.3 in Chapter 14), badan (see Figure 2.2 in Chapter 2) and dansyl, which are effective probes of environmental polarity in, for example, a protein's interior.[21]

Binding of a probe to its target can dramatically affect its fluorescence quantum yield. Probes that have a high fluorescence quantum yield when bound to a particular target but are otherwise effectively nonfluorescent yield extremely low reagent background signals (see above). Molecular Probes' ultrasensitive SYBR®, SYTO®, PicoGreen® and OliGreen™ nucleic acid stains (see Section 8.1) are prime examples of this strategy. Similarly, fluorogenic enzyme substrates, which are nonfluorescent until they are converted to fluorescent products by enzymatic cleavage (see below), allow sensitive detection of enzymatic activity.

Extrinsic quenchers, the most ubiquitous of which are paramagnetic species such as O_2 and heavy atoms such as iodide, reduce fluorescence quantum yields in a concentration-dependent manner. If quenching is caused by collisional interactions, as is usually the case, information on the proximity of the fluorophore and quencher and their mutual diffusion rate can be derived. Many fluorophores are also quenched by proteins. Examples are NBD, fluorescein and BODIPY dyes, in which the effect is apparently due to charge-transfer interactions with aromatic amino acid residues.[22-24] Consequently, antibodies raised against these fluorophores are effective and highly specific fluorescence quenchers (see Section 7.3).

Fluorophores that have strongly pH-dependent absorption and fluorescence characteristics can be used as physiological pH indicators; for example, see the pH-dependent spectra of BCECF, SNARF and SNAFL dyes in Section 23.2. Fluorescein and hydroxycoumarins (umbelliferones) are further examples of this type of fluorophore. Structurally, pH sensitivity is due to a reconfiguration of the fluorophores's π-electron system that occurs upon protonation. Molecular Probes' BODIPY fluorophore, which lacks protolytically ionizable substituents, provides a spectrally equivalent alternative to fluorescein for applications requiring a pH-*in*sensitive probe (see Figure 1.7 in Chapter 1).

Modifying Environmental Sensitivity of a Fluorophore

The environmental sensitivity of a fluorophore can be transformed by structural modifications to achieve a desired probe specificity. For example, conversion of the prototropic 3'- and 6'-hydroxyl groups of fluorescein to acetate esters yields fluorescein diacetate. This derivatization causes fluorescein to adopt the nonfluorescent lactone configuration that is also prevalent at low pH[25] (see Figure 23.1 in Chapter 23); cleavage of the acetates by esterases under appropriate pH conditions releases fluorescent anionic fluorescein. Fluorogenic substrates for other hydrolytic enzymes can be created by replacing acetates with other appropriate functional groups (Sections 10.2 and 10.3). Furthermore, unlike fluorescein, fluorescein diacetate is uncharged and therefore somewhat membrane permeant. This property forms the basis of an important noninvasive method for loading polar fluorescent indicators into cells in the form of membrane-permeant precursors that can be activated by intracellular esterases[26] (see the Technical Note "Loading and Calibration of Intracellular Ion Indicators" on page 549).

Selected Books and Articles on Fluorescence Techniques

The preceding discussion has introduced some general principles to consider when selecting a fluorescent probe. Application-specific details are addressed in subsequent chapters of this Handbook. For in-depth treatments of fluorescence techniques and their biological applications, the reader is referred to the many excellent books and review articles listed below.

Fluorescence: Fundamentals and Applications

Bright, F.V., "Bioanalytical Applications of Fluorescence Spectroscopy," Anal Chem 60, 1031 (1988).

Cantor, C.R. and Schimmel, P.R., *Biophysical Chemistry Part 2*, W.H. Freeman (1980) pp. 433–465.

Dewey, T.G., Ed., *Biophysical and Biochemical Aspects of Fluorescence Spectroscopy*, Plenum Publishing (1991).

Goldberg, M.C., Ed., *Luminescence Applications in Biological, Chemical, Environmental and Hydrological Sciences (ACS Symposium Series 383)*, American Chemical Society (1989).

Guilbault, G.G., Ed., *Practical Fluorescence, Second Edition*, Marcel Dekker (1990).

Harris, D.A. and Bashford, C.L., Eds., *Spectrophotometry and Spectrofluorimetry: A Practical Approach*, IRL Press (1987).

Hemmilä, I.A., *Applications of Fluorescence in Immunoassays*, John Wiley and Sons (1991).

Jameson, D.M. and Reinhart, G.D., Eds., *Fluorescent Biomolecules*, Plenum Publishing (1989).

Lakowicz, J.R., Ed., *Topics in Fluorescence Spectroscopy: Techniques (Volume 1); Principles (Volume 2); Biochemical Applications (Volume 3)*, Plenum Publishing (1991).

Lakowicz, J.R., Ed., *Principles of Fluorescence Spectroscopy*, Plenum Publishing (1983).

Mathies, R.A., Peck, K. and Stryer, L., "Optimization of High-Sensitivity Fluorescence Detection," Anal Chem 62, 1786 (1990).

McGown, L.B. and Warner, I.M., "Molecular Fluorescence, Phosphorescence, and Chemiluminescence Spectrometry," Anal Chem 66, 428R (1994).

Parker, C.A., *Photoluminescence of Solutions*, Elsevier (1968).

Royer, C.A., "Approaches to Teaching Fluorescence Spectroscopy," Biophys J 68, 1191 (1995).

Taylor, D.L. *et al.*, Eds., *Applications of Fluorescence in the Biomedical Sciences*, A.R. Liss (1986).

Fluorescence Microscopy

Herman, B. and Jacobson, K., Eds., *Optical Microscopy for Biology*, Wiley-Liss (1990).

Herman, B. and LeMasters, J.J., Eds., *Optical Microscopy: Emerging Methods and Applications*, Academic Press (1992).

Inoue, S., *Video Microscopy*, Plenum Press (1986).

Jovin, T.M. and Arndt-Jovin, D.J., "Luminescence Digital Imaging Microscopy," Ann Rev Biophys Biophys Chem 18, 271 (1989).

Matsumoto, B., Ed., *Cell Biological Applications of Confocal Microscopy (Methods in Cell Biology, Volume 38)*, Academic Press (1993).

Pawley, J.B., Ed., *Handbook of Biological Confocal Microscopy, Second Edition*, Plenum Publishing (1995).

Ploem, J.S. and Tanke, H.J., *Introduction to Fluorescence Microscopy*, Oxford University Press (1987).

Rost, F.W.D., *Fluorescence Microscopy, Volumes 1 and 2*, Cambridge University Press (1992).

Rost, F.W.D., *Quantitative Fluorescence Microscopy*, Cambridge University Press (1991).

Rubbi, C.P., *Light Microscopy: Essential Data*, John Wiley and Sons (1994).

Stevens, J.K., Mills, L.R. and Trogadis, J.E., Eds., *Three-Dimensional Confocal Microscopy: Volume Investigation of Biological Systems*, Academic Press (1994).

Taylor, D.L. and Wang, Y.L., Eds., *Fluorescence Microscopy of Living Cells in Culture, Parts A and B (Methods in Cell Biology, Volumes 29 and 30)*, Academic Press (1989).

Tsien, R.Y., "Fluorescence Imaging Creates a Window on the Cell," Chem Eng News 72, 34 (1994).

Wang, X.F. and Herman, B., Eds., *Fluorescence Imaging Spectroscopy and Microscopy*, John Wiley and Sons (1996).

Flow Cytometry

Darzynkiewicz, Z., Robinson, J.P. and Crissman, H.A., Eds., *Flow Cytometry, Parts A and B (Methods in Cell Biology, Volumes 41 and 42)*, Academic Press (1994). **Available from Molecular Probes (M-7887, M-7888; see Section 26.6)**.

Gilman-Sachs, A., "Flow Cytometry," Anal Chem 66, 700A (1994).

Girvan, A.L., *Flow Cytometry: First Principles*, Wiley-Liss (1992).

Grogan, W.M. and Collins, J.M., *Guide to Flow Cytometry Methods*, Marcel Dekker (1990).

Jensen, B.D. and Horan, P.K., "Flow Cytometry: Rapid Isolation and Analysis of Single Cells," Meth Enzymol 171, 549 (1989).

Lloyd, D., Ed., *Flow Cytometry in Microbiology*, Springer-Verlag (1993).

Melamed, M.R., Lindmo, T. and Mendelsohn, M.L., Eds., *Flow Cytometry and Sorting, Second Edition*, Wiley-Liss (1990).

Ormerod, M.G., Ed., *Flow Cytometry: A Practical Approach, Second Edition*, IRL Press (1994).

Robinson, J.P., Ed., *Handbook of Flow Cytometry Methods*, Wiley-Liss (1993).

Shapiro, H.M., *Practical Flow Cytometry, Third Edition*, Wiley-Liss (1994).

Watson, J.V., Ed., *Introduction to Flow Cytometry*, Cambridge University Press (1991).

Fluorophores and Fluorescent Probes

Berlman, I.B., *Handbook of Fluorescence Spectra of Aromatic Molecules, Second Edition*, Academic Press (1971).

Czarnik, A.W., Ed., *Fluorescent Chemosensors for Ion and Molecule Recognition (ACS Symposium Series 538)*, American Chemical Society (1993).

Drexhage, K.H., "Structure and Properties of Laser Dyes" in *Dye Lasers, Third Edition*, F.P. Schäfer, Ed., Springer-Verlag, (1990) pp. 155–200.

Giuiliano, K.A. *et al.*, "Fluorescent Protein Biosensors: Measurement of Molecular Dynamics in Living Cells", Ann Rev Biophys Biomol Struct 24, 405 (1995).

Griffiths, J., *Colour and Constitution of Organic Molecules*, Academic Press (1976).

Haugland, R.P., "Coupling of Monoclonal Antibodies with Fluorophores" in *Monoclonal Antibody Protocols (Methods in Molecular Biology, Volume 45)*, W.C. Davis, Ed., Humana Press (1995) pp. 205–221.

Haugland, R.P., "Spectra of Fluorescent Dyes Used in Flow Cytometry," Meth Cell Biol 42, 641 (1994).

Hermanson, G.T., *Bioconjugate Techniques*, Academic Press (1996). **Available from Molecular Probes (B-7884, see Section 26.6)**.

Johnson, I.D., Ryan, D. and Haugland, R.P., "Comparing Fluorescent Organic Dyes for Biomolecular Labeling" in *Methods in Nonradioactive Detection*, G.C. Howard, Ed., Appleton and Lange (1993) pp. 47–68.

Kasten, F.H., "Introduction to Fluorescent Probes: Properties, History and Applications" in *Fluorescent and Luminescent Probes for Biological Activity*, W.T. Mason, Ed., Academic Press (1993) pp. 12–33.

Krasovitskii, B.M. and Bolotin, B.M., *Organic Luminescent Materials*, VCH Publishers (1988).

Lakowicz, J.R., Ed., *Topics in Fluorescence Spectroscopy: Probe Design and Chemical Sensing (Volume 4)*, Plenum Publishing (1994).

Mason, W.T., Ed., *Fluorescent and Luminescent Probes for Biological Activity*, Academic Press (1993). **Available from Molecular Probes (F-7889, see Section 26.6)**.

Tsien, R.Y., "Fluorescent Probes of Cell Signaling," Ann Rev Neurosci 12, 227 (1989).

Waggoner, A.S., "Fluorescent Probes for Cytometry" in *Flow Cytometry and Sorting, Second Edition*, M.R. Melamed, T. Lindmo and M.L. Mendelsohn, Eds., Wiley-Liss (1990) pp. 209–225.

Wells, S. and Johnson, I., "Fluorescent Labels for Confocal Microscopy" in *Three-Dimensional Confocal Microscopy: Volume Investigation of Biological Systems*, J.K. Stevens, L.R. Mills and J.E. Trogadis, Eds., Academic Press (1994) pp. 101–129.

1. Analyst 119, 417 (1994); **2.** Meth Cell Biol 42, 605 (1994); **3.** Meth Cell Biol 30, 113 (1989); **4.** *Luminescence Applications in Biological, Chemical, Environmental and Hydrological Sciences (ACS Symposium Series 383)*, M.C. Goldberg, Ed., American Chemical Society (1989) pp. 98–126; **5.** Meth Cell Biol 41, 61 (1994); **6.** Methods: Comp Meth Enzymol 2, 192 (1991); **7.** Science 271, 1420 (1996); **8.** Anal Biochem 223, 39 (1994); **9.** Proc Natl Acad Sci USA 89, 1388 (1992); **10.** Cytometry 11, 126 (1990); **11.** Meth Cell Biol 38, 97 (1993); **12.** Meth Cell Biol 30, 449 (1989); **13.** Haugland, R.P. in *Optical Microscopy for Biology*, B. Herman and K. Jacobson, Eds., Wiley-Liss (1990) pp. 143–157; **14.** Biophys J 70, 2959 (1996); **15.** Biophys J 68, 2588 (1995); **16.** J Cell Biol 100, 1309 (1985); **17.** J Org Chem 38, 1057 (1973); **18.** Cytometry 23, 48 (1996); **19.** J Histochem Cytochem 43, 77 (1995); **20.** Anal Biochem 172, 61 (1988); **21.** Nature 319, 70 (1986); **22.** Biophys J 69, 716 (1995); **23.** Biochemistry 16, 5150 (1977); **24.** Immunochemistry 14, 533 (1977); **25.** Spectrochim Acta A 51, L7 (1995); **26.** Proc Natl Acad Sci USA 55, 134 (1966).

Books from Molecular Probes

To augment our selection of fluorescent probes, we are pleased to offer several valuable reference books for scientists utilizing fluorescence techniques for their research applications. We recommend these recent editions to anyone interested in up-to-date guides on fluorescence methods.

Cat #	Product Name	Unit Size	Price Per Unit ($) 1–4 Units	5–24 Units
B-7884 ***New***	Bioconjugate Techniques. G.T. Hermanson, Academic Press (1996); 785 pages, soft cover	each	60.00	60.00
F-7889 ***New***	Fluorescent and Luminescent Probes for Biological Activity. A Practical Guide to Technology for Quantitative Real-Time Analysis. W.T. Mason, ed. Academic Press (1993); 433 pages, comb bound	each	65.00	65.00
H-8999 ***New***	Handbook of Fluorescent Probes and Research Chemicals, Sixth Edition. R.P. Haugland, Molecular Probes, Inc. (1996); 700 pages, soft cover	each	25.00	20.00
M-7890 ***New***	Methods in Cell Biology, Volume 40: A Practical Guide to the Study of Calcium in Living Cells. R. Nuccitelli, ed. Academic Press (1994); 342 pages, comb bound	each	45.00	45.00
M-7887 ***New***	Methods in Cell Biology, Volume 41: Flow Cytometry, Part A. Z. Darzynkiewicz, J.P. Robinson, H.A. Crissman, eds. Academic Press (1994); 591 pages, comb bound	each	59.00	59.00
M-7888 ***New***	Methods in Cell Biology, Volume 42: Flow Cytometry, Part B. Z. Darzynkiewicz, J.P. Robinson, H.A. Crissman, eds. Academic Press (1994); 697 pages, comb bound	each	59.00	59.00

Chapter 1

Fluorophores and Their Amine-Reactive Derivatives

Contents

1.1 Introduction to Amine Modification

Molecular Probes puts at your command a full spectrum of fluorophores and haptens for labeling biopolymers and derivatizing low molecular weight molecules. Chapters 1–5 describe the chemical and spectral properties of the reactive reagents we offer, whereas the remainder of this Handbook is primarily devoted to our diverse collection of fluorescent probes and their applications in cell biology, biochemistry, biophysics, microbiology, molecular biology, neuroscience and other areas.

Common Applications for Amine-Reactive Probes

Labeling Biopolymers

Amine-reactive probes are widely used to modify proteins, peptides, ligands, synthetic oligonucleotides and other biomolecules. In contrast to our thiol-reactive reagents (see Chapter 2), which frequently serve as probes of protein structure and function, amine-reactive dyes are most often used to prepare bioconjugates for immunochemistry, fluorescence *in situ* hybridization, cell tracing, receptor labeling and fluorescent analog cytochemistry.[1] In these applications, the stability of the chemical bond between the dye and biomolecule is particularly important because the conjugate is typically stored and used repeatedly over a relatively long period of time. Moreover, these conjugates are often subjected to rigorous hybridization and washing steps that demand a strong dye–biomolecule linkage.

Our selection of amine-reactive fluorophores for modifying biomolecules covers the entire visible spectrum (Table 1.1). An up-to-date bibliography is available upon request from our Technical Assistance Department or through our Web site (http://www.probes.com) for every amine-reactive probe for which we have references. Our Technical Assistance Department can also provide you with product-specific bibliographies, as well as keyword searches of the over 25,000 literature references in our extensive bibliography database. Included in Chapter 1 are discussions of the properties of Molecular Probes' most important proprietary fluorophores, including:

Table 1.1 Approximate spectral properties of the conjugates of amine-reactive probes described in Sections 1.2–1.7.

Fluorophore	COOH *	SE *	Other *	Abs (nm)	Em (nm)	Notes
AMCA-S		A-6120		346	442	• Higher fluorescence per attached dye than AMCA
AMCA		A-6118 (X)		349	448	• Widely used blue fluorescent labeling dye • Compact structure
BODIPY 493/503	D-2190	D-2191		500	506	• Higher 488 nm absorptivity than the BODIPY FL fluorophore
BODIPY FL	D-2183 [1] D-6101 (X) D-3834 (C_5)	D-2184 D-6140 (SSE) D-6141 (CASE) D-6102 (X) D-6184 (C_5)		505	513	• pH insensitive • Narrow spectral bandwidth • Succinimidyl ester derivative (BODIPY FL, SE; D-2184) is useful for automated DNA sequencing [7] • Succinimidyl ester with a cysteic acid spacer (BODIPY FL, CASE; D-6141) is the preferred reactive BODIPY FL for protein conjugation
BODIPY FL Br_2	D-6143	D-6144		533	548	• Useful for DAB photoconversion
BODIPY R6G		D-6180		528	550	• BODIPY substitute for rhodamine 6G • Useful for automated DNA sequencing [7]
BODIPY 530/550	D-2186	D-2187		534	554	• pH insensitive • Narrow spectral bandwidth
BODIPY TMR		D-6117 (X)		542	574	• BODIPY substitute for TMR
BODIPY 558/568	D-2218	D-2219		558	569	• pH insensitive • Narrow spectral bandwidth
BODIPY 564/570	D-2221	D-2222		565	571	• pH insensitive • Narrow spectral bandwidth • Useful for automated DNA sequencing [7]
BODIPY 576/589	D-2224	D-2225		576	590	• pH insensitive • Narrow spectral bandwidth
BODIPY 581/591	D-2227	D-2228		584	592	• pH insensitive • Narrow spectral bandwidth • Useful for automated DNA sequencing [7]
BODIPY TR		D-6116 (X)		589	617	• BODIPY substitute for Texas Red fluorophore
Cascade Blue			C-2284 (AA)	400	420	• Resistant to quenching upon protein conjugation • Water soluble
Cl-NERF	C-6831			518	544	• pH sensitive between pH 3–5
Dansyl		D-6104 (X)	D-21 (SC) D-1302 (ITC)	340	520 [7]	• Environmentally sensitive • Large Stokes shift
Dialkylaminocoumarin	D-126 D-1421	D-374 D-1412	D-347 (ITC)	375 435	470 [3] 475 [4]	• Longer wavelength alternatives to AMCA
4′,5′-Dichloro-2′,7′-dimethoxyfluorescein	C-6170	C-6171		522	550	• Succinimidyl ester derivative (6-JOE, SE; C-6171) is widely used for automated DNA sequencing [8-10]
2′,7′-Dichlorofluorescein	C-368 †		D-6078 (ITC)	510	532	• pH insensitive at pH >6.
DM-NERF	D-6830			515	542	• pH sensitive between pH 4.5–6.5
Eosin	C-301 †		E-18 (ITC)	524	544	• Useful for DAB photoconversion • Phosphorescent
Eosin F_3S	C-6167			535	542	• Photostable eosin derivative useful for DAB photoconversion
Erythrosin			E-332 (ITC)	530	555	• Phosphorescent
Fluorescein	C-1359 C-1360 C-1904 †	C-2210 C-6164 C-1311 † F-6106 (X) F-2181 (X) † F-6129 (X) † F-6130 (EX)	D-16 (DTA) D-17 (DTA) F-143 (ITC) F-144 (ITC) F-1906 (ITC) F-1907 (ITC) F-1919 (ITC)	494	518	• Most widely used green fluorescent labeling dye • Absorption overlaps the 488 nm spectral line of the argon-ion laser • Prone to photobleaching • pH sensitive between pH 5–8 • Succinimidyl ester derivative (5-FAM, SE; C-2210) is widely used for automated DNA sequencing [8-10] • Fluorescein-5-EX succinimidyl ester (F-6130) is the preferred reactive fluorescein for protein conjugation

Table 1.1, continued Approximate spectral properties of the conjugates of amine-reactive probes described in Sections 1.2–1.7.

Fluorophore	COOH *	SE *	Other *	Abs (nm)	Em (nm)	Notes
Hydroxycoumarin	H-185 H-1428	H-1193 H-1411		385 360	445 [5] 455 [6]	• pH sensitive • Compact structure
Isosulfan blue			I-6204 (SC)	650	none	• Nonfluorescent photosensitizer
Lissamine rhodamine B			L-20 (SC) † L-1908 (SC) †	570	590	• Optimal for 568 nm excitation • Photostable
Malachite green			M-689 (ITC)	630	none	• Nonfluorescent photosensitizer
Methoxycoumarin	M-1420	M-1410		340	405	• pH-insensitive alternative to hydroxycoumarin
Naphthofluorescein	C-652 †	C-653 †		605	675	• Very long-wavelength excitation and emission • pH sensitive
NBD		S-1167 (X)	C-10 (AH) F-486 (AH)	465	535	• Environmentally sensitive • Compact structure
Oregon Green 488	O-6146 O-6148	O-6147 O-6149	O-6080 (ITC) †	496	524	• Photostable fluorescein substitute • pH insensitive at pH >6
Oregon Green 500	O-6135	O-6136		503	522	• Photostable fluorescein substitute • pH insensitive at pH >6
Oregon Green 514	O-6138	O-6139		511	530	• Exceptionally photostable • pH insensitive at pH >6
PyMPO		S-6110	I-6076 (ITC)	415	570	• Large Stokes shift
Pyrene	P-164	P-6115 P-130 P-6114 (CASE)	P-24 (SC) P-331 (ITC)	345	378	• Long excited-state lifetime • Spectral shifts due to excimer emission
Rhodamine 6G	C-6109 C-2213	C-6127 C-6128 C-6157 †		525	555	• Absorption matched for 514 nm excitation • Spectra intermediate between those of fluorescein and TMR
Rhodamine Green	R-6150 †	R-6107 † R-6112 † R-6113 (X) †		502	527	• Photostable fluorescein substitute • pH insensitive
Rhodamine Red		R-6160 (X) †		570	590	• Rhodamine Red-X succinimidyl ester generally yields higher fluorescence per attached dye than Lissamine rhodamine B sulfonyl chloride and is more stable in H_2O
Rhodol Green	R-6152 †	R-6108 † R-6111 †		499	525	• Photostable fluorescein substitute • pH sensitive between 4.5–6.5
2′,4′,5′,7′-Tetrabromo-sulfonefluorescein	C-6165	C-6166		528	544	• Eosin derivative useful for DAB photoconversion
Tetramethylrhodamine (TMR)	C-6121 C-6122 C-300 †	C-2211 C-6123 C-1171 † T-6105 (X) †	T-1480 (ITC) T-1481 (ITC) T-490 (ITC) †	555	580	• pH insensitive • Photostable • Prone to aggregation • Succinimidyl ester derivative (6-TAMRA, SE; C-6123) is widely used for automated DNA sequencing [8-10]
Texas Red		T-6134 (X) †	T-353 (SC) † T-1905 (SC) †	595	615	• Good spectral separation from fluorescein • Texas Red-X succinimidyl ester typically yields higher fluorescence per attached dye than Texas Red sulfonyl chloride and is more stable in H_2O
X-rhodamine	C-6124 C-6156 C-1308 †	C-6125 C-6126 C-1309 †	X-491 (ITC) †	580	605	• Succinimidyl ester derivative (6-ROX, SE; C-6126) is widely used for automated nucleic acid sequencing [8-10]

The absorption (Abs) and fluorescence emission (Em) maxima listed in this table are for the goat anti-mouse IgG or dextran conjugates in aqueous buffer.
* COOH = carboxylic acid; SE = succinimidyl ester; (AA) = acetyl azide, (AH) = aryl halide, (C_5) = pentanoic acid, (CASE) = cysteic acid, succinimidyl ester, (DTA) = dichlorotriazine, (EX) = seven-atom spacer that is more hydrophilic than X, (ITC) = isothiocyanate, (SC) = sulfonyl chloride, (SSE) = sulfosuccinimidyl ester, (X) = an aminohexanoyl spacer between the dye and SE. † mixed isomers. **1.** Dipropionic acid (D-6103) is also available; **2.** Emission spectra of dansylated proteins may vary considerably depending on the dye attachment site and the degree of labeling; **3.** Spectral maxima for D-374; **4.** Spectral maxima for D-1412; **5.** Spectral maxima for H-1193; **6.** Spectral maxima for H-1411; **7.** Science 271, 1420 (1996); **8.** Anal Biochem 223, 39 (1994); **9.** Nucleic Acids Res 20, 2471 (1992); **10.** Proc Natl Acad Sci USA 86, 9178 (1989).

- BODIPY dyes (see Section 1.2)
- Oregon Green 488, Oregon Green 500, Oregon Green 514, Rhodol Green and Rhodamine Green dyes, our exceptional new fluorescein substitutes (see Sections 1.4 and 1.5)
- Rhodamine Red-X succinimidyl ester, a new stable reactive form of Lissamine rhodamine B (see Section 1.6)
- Texas Red sulfonyl chloride and the new Texas Red-X succinimidyl ester (see Section 1.6)
- UV-excitable Cascade Blue fluorophore (see Section 1.7)
- AMCA-X and AMCA-S succinimidyl esters, our new reactive aminocoumarins (see Section 1.7)

Amine-reactive reagents with similar spectra, rather than the same reactive group, are generally discussed together in the following sections. The notable exception to this organization is the description of our wide selection of BODIPY fluorophores, which are available with fluorescein-, tetramethylrhodamine- and Texas Red–like spectral properties, as well as with emissions that fall in the gaps between the spectra of these common dyes. BODIPY dyes are considered as a group in Section 1.2.

Preparing the Optimal Bioconjugate

The preferred bioconjugate usually has a high fluorescence yield (or suitable degree of substitution in the case of a haptenylated conjugate) yet retains the critical parameters of the unlabeled biomolecule, such as selective binding to a receptor or nucleic acid, activation or inhibition of a particular enzyme or ability to incorporate into a biological membrane. Frequently, conjugates that have the highest degree of substitution precipitate or bind nonspecifically. It may therefore be necessary to have a less-than-maximal fluorescence yield to preserve function or binding specificity. Although conjugating dyes to biomolecules is usually rather easy, preparing the *optimal* conjugate may require extensive experimentation. Thus for the most critical assays, we recommend that you consider preparing and optimizing your own conjugates. We provide a free technical bulletin (request MP 0143) that describes how to use several of our amine-reactive dyes for labeling biomolecules. The procedure is fairly simple and requires no special equipment. Following conjugation, it is very important to remove as much unconjugated dye as possible, usually by gel filtration, dialysis or a combination of these techniques. Presence of free dye, particularly if it remains chemically reactive, can greatly complicate subsequent experiments with the bioconjugate.

With the exception of the phycobiliproteins, all the dyes used to prepare Molecular Probes' fluorescent bioconjugates are amine-reactive reagents and are described in this chapter. We have also developed some very useful FluoReporter® Kits for labeling proteins or oligonucleotides with several of our most important dyes, as well as with biotin or the dinitrophenyl hapten; see the Product Highlight "FluoReporter Protein and Oligonucleotide Labeling Kits" on page 18. Each of these kits contains the reactive dye, along with a detailed protocol for preparing and purifying the conjugates; the FluoReporter Protein Labeling Kits also contain separation media. Conjugations with fluorescent polystyrene microspheres and phycobiliproteins require unique procedures that are described in Sections 6.2 and 6.4, respectively.

Derivatizing Low Molecular Weight Molecules

Some amine-reactive probes described in this chapter are also important reagents for various bioanalytical applications, including amine quantitation, protein and nucleic acid sequencing and chromatographic and electrophoretic analysis of low molecular weight molecules. Those reagents particularly useful for derivatizing low molecular weight amines — including fluorescamine, *o*-phthaldialdehyde, our ATTO-TAG reagents, NBD chloride and dansyl chloride — are discussed in Section 1.8. However, many of the reactive dyes described in Sections 1.2 through 1.7 can also be used as derivatization reagents; likewise, some of the derivatization reagents in Section 1.8 can be utilized for biomolecule conjugation.

Reactivity of Amino Groups

Amine-reactive probes described in this chapter are mostly acylating reagents that form carboxamides, sulfonamides, ureas or thioureas upon reaction with amines. The kinetics of the reaction depend on the reactivity and concentration of both the acylating reagent and the amine. Of course, buffers that contain free amines such as Tris must be avoided when using *any* amine-reactive probe. Reagents for direct alkylation and reductive alkylation are described in Chapters 2 and 3.

The most significant factors affecting the amine's reactivity are its class and its basicity. Virtually all proteins have lysine residues, and most have a free amine at the N-terminus. Aliphatic amines such as lysine's ϵ-amino group are moderately basic and reactive with most acylating reagents. However, the concentration of the free base form of aliphatic amines below pH 8 is very low; thus, the kinetics of acylation reactions of amines by isothiocyanates, succinimidyl esters and other reagents are strongly pH dependent. A pH of 8.5–9.5 is usually optimal for modifying lysine residues. In contrast, the α-amino group at a protein's N-terminus usually has a pK_a of ~7, so it can sometimes be selectively modified by reaction at near neutral pH. Furthermore, although amine acylation should usually be carried out above pH 8.5, the acylation reagents tend to degrade in the presence of water, with the rate increasing as the pH increases. Protein modification by succinimidyl esters can typically be done at pH 8.5, whereas isothiocyanates usually require a pH >9 for optimal conjugations; this high pH may be a factor when working with base-sensitive proteins.

Aromatic amines, which are uncommon in biomolecules, are very weak bases and thus unprotonated at pH 7. Modification of aromatic amines requires a highly reactive reagent, such as an isothiocyanate, sulfonyl chloride or acid halide, but can be done at any pH above ~4. A tyrosine residue can be selectively modified to form an *o*-aminotyrosine aromatic amine, which can then be reacted at a relatively low pH with certain amine-reactive probes (see Section 3.1).

In aqueous solution, acylating reagents are virtually unreactive with the amide group of peptide bonds and the side-chain amides of glutamine and asparagine residues, the guanidinium group of arginine, the imidazolium group of histidine and the nonbasic amines found in nucleotides such as adenosine or guanosine.

Isothiocyanates

Molecular Probes does not sell any isocyanate (R–NCO) reagents because they are very susceptible to deterioration during storage. However, some acyl azides (see Section 3.1) are readily converted to isocyanates (see Figure 3.1 in Chapter 3). As an alter-

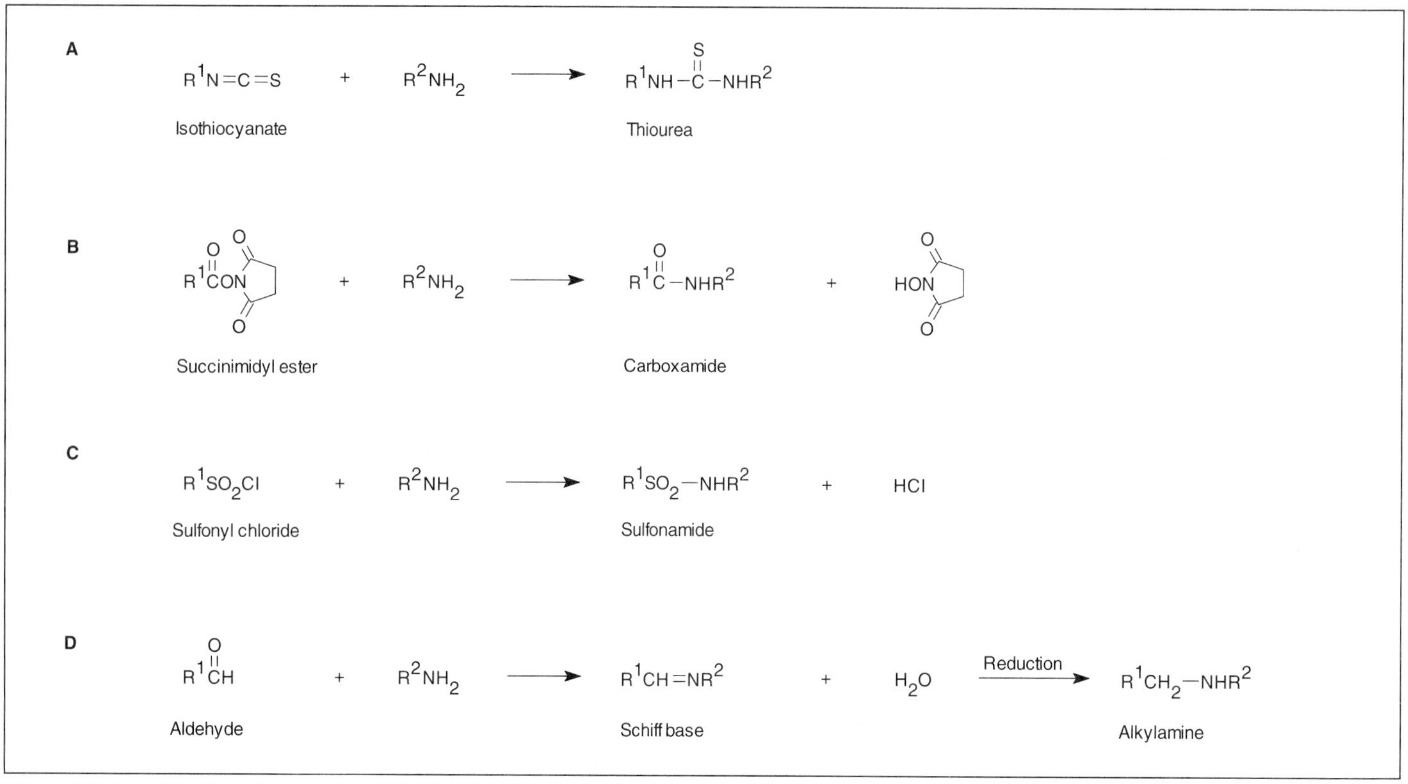

A

$$R^1N{=}C{=}S + R^2NH_2 \longrightarrow R^1NH{-}\overset{S}{\overset{\parallel}{C}}{-}NHR^2$$

Isothiocyanate — Thiourea

B

$$R^1\overset{O}{\overset{\parallel}{C}}ON(\text{succinimide}) + R^2NH_2 \longrightarrow R^1\overset{O}{\overset{\parallel}{C}}{-}NHR^2 + HON(\text{succinimide})$$

Succinimidyl ester — Carboxamide

C

$$R^1SO_2Cl + R^2NH_2 \longrightarrow R^1SO_2{-}NHR^2 + HCl$$

Sulfonyl chloride — Sulfonamide

D

$$R^1\overset{O}{\overset{\parallel}{C}}H + R^2NH_2 \longrightarrow R^1CH{=}NR^2 + H_2O \xrightarrow{\text{Reduction}} R^1CH_2{-}NHR^2$$

Aldehyde — Schiff base — Alkylamine

Figure 1.1 *Reaction of a primary amine with A) an isothiocyanate, B) a succinimidyl ester, C) a sulfonyl chloride and D) an aldehyde.*

native to the unstable isocyanates, we offer a large selection of iso*thio*cyanates (R–NCS), which are moderately reactive but quite stable in water and most solvents. Isothiocyanates form thioureas upon reaction with amines (Figure 1.1A). Although the thiourea product is reasonably stable, it has been reported that antibody conjugates prepared from fluorescent isothiocyanates deteriorate over time, prompting us to use fluorescent succinimidyl esters and sulfonyl halides almost exclusively for synthesizing our bioconjugates. Despite the growing number of choices in amine-reactive fluorophores, fluorescein isothiocyanate (FITC) and tetramethylrhodamine isothiocyanate (TRITC) remain the most widely used reactive fluorescent dyes for preparing fluorescent antibody conjugates.

Succinimidyl Esters and Carboxylic Acids

Succinimidyl esters are excellent reagents for amine modification because the amide bonds they form (Figure 1.1B) are as stable as peptide bonds. These reagents are generally stable during storage if well desiccated and show good reactivity with aliphatic amines and very low reactivity with aromatic amines, alcohols, phenols (including tyrosine) and histidine. Succinimidyl esters will also react with thiols in organic solvents to form thioesters. Succinimidyl ester hydrolysis can compete with conjugation, but this side reaction is usually slow below pH 9. Some succinimidyl esters may not be compatible with a specific application because they can be quite insoluble in aqueous solution. To overcome this limitation, Molecular Probes also offers carboxylic acid derivatives of many of its fluorophores, which can be converted into *sulfo*succinimidyl esters. These sulfonated reagents are more water soluble than their succinimidyl ester counterparts and sometimes eliminate the need for organic solvents in the conjugation reaction. However, they are also more polar, making them less likely to react with buried amines in proteins or to penetrate cell membranes. Because of their combination of reactivity and polarity, sulfosuccinimidyl esters are not easily purified by chromatographic means and thus only a few are currently available from Molecular Probes; contact our Custom and Bulk Sales Department for information about availability or custom synthesis of other sulfosuccinimidyl esters. Sulfosuccinimidyl esters can generally be prepared *in situ* simply by dissolving the carboxylic acid dye in a buffer that contains *N*-hydroxysulfosuccinimide (NHSS, H-2249; see Section 3.3) and 1-ethyl-3-(3-dimethylaminopropyl)carbodiimide (EDAC, E-2247; see Section 3.3). Addition of NHSS to the buffer has been shown to enhance the yield of carbodiimide-mediated conjugations.[2] The carboxylic acid derivatives may also be useful for preparing acid chlorides and anhydrides, which, unlike succinimidyl esters, can be used to modify aromatic amines and alcohols.

Sulfonyl Halides

Sulfonyl chlorides, including the dansyl, Lissamine rhodamine B and Texas Red derivatives, are highly reactive. These reagents are quite unstable in water, especially at the higher pH required for reaction with aliphatic amines. For example, we have determined that Texas Red sulfonyl chloride is totally hydrolyzed within 2–3 minutes in pH 8.5 aqueous solution at room temperature. Protein modification by this reagent is best done at low temperature. Once conjugated, however, the sulfonamides that are formed (Figure 1.1C) are extremely stable; they even survive complete protein hydrolysis (for example, dansyl end-group analysis). Sulfonyl chlorides can also react with phenols (including tyrosine), aliphatic alcohols (including polysaccharides), thiols (such as cysteine) and imidazoles (such as histidine), but these reactions are not common in proteins or in aqueous solution. Sulfonyl chloride conjugates of thiols and imidazoles are generally unstable, and conjugates of aliphatic alcohols are subject to nucleophilic displacement.[3] Note that sulfonyl chlorides are unstable in dimethylsulfoxide (DMSO) and should never be used in that solvent.[4] In contrast to sulfonyl chlorides, sulfonyl fluorides have very low reactivity, making them useful for the selective modification of active-site serine residues of some enzymes (see Section 10.9).

Other Amine-Reactive Reagents

Aldehydes and ketones react with amines to form Schiff bases (Figure 1.1D). Notable aldehyde-containing reagents include fluorescamine, *o*-phthaldialdehyde (OPA), naphthalenedicarboxaldehyde (NDA) and the 3-acylquinolinecarboxaldehyde (ATTO-TAG) reagents of Novotny and collaborators.[5,6] All these reagents are useful for the sensitive quantitation of amines in solution, as well as by HPLC and capillary electrophoresis. In addition, certain arylating reagents such as NBD chloride, NBD fluoride and dichlorotriazines react with both amines and thiols, forming bonds with amines that are particularly stable.

1. Meth Cell Biol 29, 1 (1989); **2.** Anal Biochem 156, 220 (1986); **3.** J Phys Chem 83, 3305 (1979); **4.** J Org Chem 31, 3880 (1966); **5.** Anal Chem 63, 408 (1991); **6.** J Chromatography 499, 579 (1990).

1.2 BODIPY Dyes Spanning the Visible Spectrum

Overview of Our BODIPY Fluorophores

Our patented BODIPY fluorophores have spectral characteristics that are often superior to those of fluorescein, tetramethylrhodamine, Texas Red and longer-wavelength dyes and may be substituted for these dyes in most applications. With derivatives that span the visible spectrum (Figure 1.2), BODIPY dyes are proving to be extremely versatile. We use them to generate fluorescent conjugates of proteins, nucleotides, oligonucleotides and dextrans, as well as to prepare fluorescent enzyme substrates, fatty acids, phospholipids, receptor ligands and polystyrene microspheres. Amine-reactive BODIPY dyes are discussed below; thiol-reactive BODIPY dyes are included in Section 2.2 and other reactive BODIPY dyes useful for derivatizing aldehydes, ketones and carboxylic acids are described in Sections 3.2 and 3.3. Because of potential traces of amines in dimethylformamide, we now recommend that this solvent not be used with any BODIPY dyes, whether amine-reactive or not. We generally use dimethylsulfoxide (DMSO) in our labeling protocols.

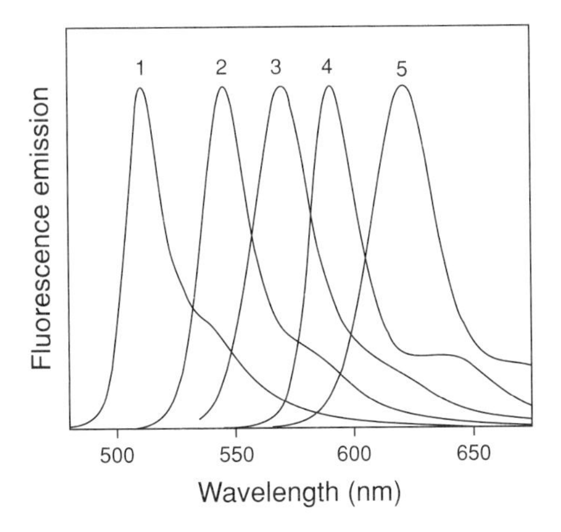

Figure 1.2 *Normalized fluorescence emission spectra of 1) BODIPY FL, 2) BODIPY R6G, 3) BODIPY TMR, 4) BODIPY 581/591 and 5) BODIPY TR fluorophores in methanol.*

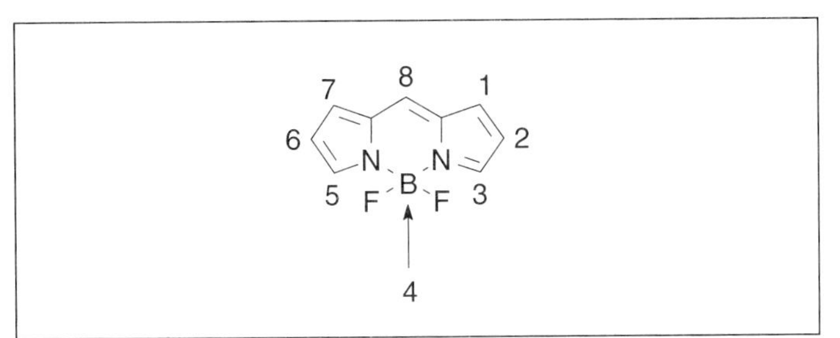

Figure 1.3 *The structure and numbering of the BODIPY fluorophore, 4,4-difluoro-4-bora-3a,4a-diaza-s-indacene.*

The basic structure of the BODIPY fluorophore is shown in Figure 1.3. Solutions of the alkyl-substituted derivatives have a green, fluorescein-like fluorescence. However, when substituents that yield additional conjugation are added to the parent molecule, both the absorption and emission spectra of the resulting derivatives can shift to significantly longer wavelengths, with emission maxima of greater than 650 nm possible. Our goal is to develop BODIPY dyes that are optimal for the widely used excitation sources and that match the common optical filter sets. Accordingly, our best BODIPY substitutes for the fluorescein, rhodamine 6G, tetramethylrhodamine and Texas Red fluorophores are named BODIPY FL, BODIPY R6G, BODIPY TMR and BODIPY TR, respectively (Figure 1.4). Because we have so many BODIPY dyes, we have had to develop a systematic strategy for naming them. Except for BODIPY FL, BODIPY R6G, BODIPY TMR and BODIPY TR, we now identify these dyes with the registered trademark BODIPY followed by the approximate absorption/emission maxima in nm (determined in methanol); for example, BODIPY 581/591.

The BODIPY fluorophores, reactive dyes and conjugates are covered by U.S. Patent Nos. 4,774,339; 5,187,288; 5,248,782; 5,274,113; 5,433,896; 5,451,663 and other U.S. and foreign patents pending. These products are offered for research purposes only. Molecular Probes welcomes inquiries about licensing these products for resale or other commercial uses.

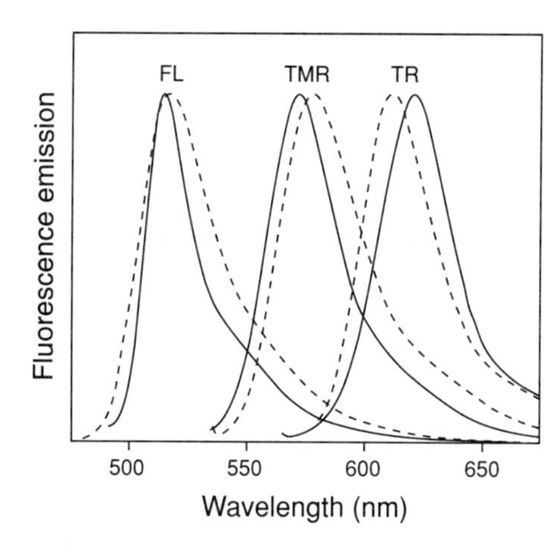

Figure 1.4 Normalized fluorescence emission spectra of goat anti–mouse IgG conjugates of fluorescein (FL), tetramethylrhodamine (TMR) and Texas Red (TR), shown by dashed lines (---), compared to BODIPY FL, BODIPY TMR and BODIPY TR conjugates shown by solid lines (—).

BODIPY FL: A Fluorescein Substitute

With the most fluorescein-like spectra of the BODIPY dyes, the green fluorescent BODIPY FL fluorophore (excitation/emission maxima ~503/512 nm) has several characteristics [1,2] that make it potentially superior to fluorescein. These include:

- High extinction coefficient (ϵ >80,000 $cm^{-1}M^{-1}$)
- High fluorescence quantum yield (often approaching 1.0, even in water)
- Spectra that are insensitive to solvent polarity and pH [1]
- Narrow emission bandwidth (Figure 1.4), resulting in a higher peak intensity than that of fluorescein
- Little or no spectral overlap with longer wavelength dyes such as tetramethylrhodamine and Texas Red (Figure 1.4), making BODIPY FL one of the preferred green fluorescent dyes for multicolor applications [3]
- Greater photostability than fluorescein in some environments [4] (see Figure 1.6 in Section 1.4)
- Lack of ionic charge

Longer-Wavelength BODIPY Dyes

We have found that we can synthesize new BODIPY fluorophores by simply changing the substituents on the parent molecule. This discovery has led to the creation of a series of longer-wavelength BODIPY dyes with fluorescence spectra that span the visible spectrum (Figure 1.2). The BODIPY R6G (excitation/emission maxima ~528/547 nm), BODIPY TMR (excitation/emission maxima ~543/569 nm) and BODIPY TR (excitation/emission maxima ~592/618 nm) fluorophores are spectrally similar to the rhodamine 6G (R-634, see Section 12.2), tetramethylrhodamine and Texas Red fluorophores, respectively, and thus compatible with standard optical filter sets designed for these important dyes. Significantly, the absorption and emission properties of these longer-wavelength BODIPY derivatives retain most of the advantages of the BODIPY FL fluorophore, including narrow bandwidths, high extinction coefficients and good fluorescence quantum yields. Oligonucleotide conjugates of several of our BODIPY dyes have recently been reported to be useful for DNA sequencing,[5,6] in part because the dye exhibits minimal effect on the mobility of the fragment during electrophoresis. Like the BODIPY FL fluorophore, however, most of these dyes have a small Stokes shift, which may require that they be excited or detected at suboptimal wavelengths. Nevertheless, even when suboptimal excitation is required, the BODIPY dyes are among the most intensely fluorescent dyes available.

BODIPY Succinimidyl Esters and Carboxylic Acids

BODIPY Succinimidyl Esters

Except for the new BODIPY FL cysteic acid succinimidyl ester and sulfosuccinimidyl ester derivatives (see below), most of the amine-reactive BODIPY dyes in this section are succinimidyl esters of BODIPY propionic acids. We recently introduced the reactive succinimidyl ester derivatives of BODIPY FL-X, BODIPY TMR-X and BODIPY TR-X (D-6102, D-6117, D-6116), which contain an additional seven-atom aminohexanoyl spacer ("X") between the fluorophore and the succinimidyl ester group. This spacer helps to separate the fluorophore from its point of attachment, potentially reducing the interaction of the fluorophore with the biomolecule to which it is conjugated and making it more accessible to secondary detection reagents such as anti-dye antibodies.[7-9] For amplifying BODIPY FL's signal or converting it into an electron-dense signal, we offer an unlabeled anti–BODIPY FL polyclonal antibody, as well as its biotin and fluorescein conjugates (A-5770, A-5771, A-6420; see Section 7.3). In addition, we now have a polyclonal anti–BODIPY TR antibody (A-6414, see Section 7.3) available. These antibodies crossreact with other BODIPY dyes to some extent, but not with other fluorophores. They should therefore not be used for simultaneous detection of more than one BODIPY-based dye.

The BODIPY propionic acid succinimidyl esters (D-2184, D-2187, D-2191, D-2219, D-2222, D-2225, D-2228, D-6180) and pentanoic acid succinimidyl ester (BODIPY FL C_5, D-6184) are particularly useful for preparing conjugates of peptides, nucleotides, drugs, toxins, sphingolipids and other low molecular weight ligands that contain amines. Several probes for receptors and ion channels are already available from Molecular Probes (see Chapters 17 and 18), and BODIPY dye–labeled nucleotides are described in Section 8.2. Like eosin dyes (see Section 1.5), the succinimidyl ester of the brominated BODIPY FL propionic acid (BODIPY FL Br_2, SE; D-6144) may be useful as a triplet probe or for preparing conjugates that are used in photoconversion studies (see the Technical Note "Fluorescent Probes for Photoconversion of Diaminobenzidine" on page 264).

BODIPY dyes are unusual in that they are relatively nonpolar and electrically neutral. We have found these properties to sometimes enhance the affinity of their ligand conjugates for receptors, as long as the overall conjugate is not too lipophilic. BODIPY conjugates also tend to be more permeant to live cells than are conjugates of charged fluorophores. In addition, with their high peak intensity, reactive BODIPY dyes are among the most detectable amine-derivatization reagents available for HPLC and capillary electrophoresis.[10]

Water-Soluble BODIPY FL Succinimidyl Esters

As is common with fluorescent dyes, conjugation of BODIPY dyes to proteins is sometimes accompanied by significant fluorescence quenching. Because of this potential problem, we do *not* recommend using the simple BODIPY propionic acid succinimidyl esters discussed above for preparing most protein conjugates, although peptides labeled with a single BODIPY dye can be quite fluorescent.[11] To reduce quenching in our protein conjugates, Molecular Probes usually uses the succinimidyl ester of BODIPY FL cysteic acid (BODIPY FL, CASE; D-6141), which contains a sulfonated spacer that appears to decrease the interaction between the fluorophore and the protein. Both this cysteic acid derivative and our new *sulfo*succinimidyl ester of BODIPY FL (BODIPY FL, SSE; D-6140) are quite water soluble and potentially useful for preparing conjugates of most proteins and other biomolecules. Sulfosuccinimidyl esters of biotin (B-6352, B-6353; see Section 4.2) are sometimes employed as cell-impermeant probes for selectively labeling the outer membrane of cells in topological studies [12-14] (see Section 9.4); these sulfonated BODIPY FL succinimidyl esters may be similarly useful.

BODIPY Carboxylic Acids

BODIPY FL (D-2183) and other BODIPY propionic acids (D-2186, D-2190, D-2218, D-2221, D-2224, D-2227), BODIPY dipropionic acid (D-6103), BODIPY pentanoic acid (BODIPY FL C_5, D-3834), brominated BODIPY FL (BODIPY FL Br_2, D-6143) and BODIPY FL-X (D-6101) are also available. These carboxylic acid derivatives can be used to prepare the water-soluble, amine-reactive sulfosuccinimidyl esters [15] or activated with other reagents to form anhydrides or acid halides.

BODIPY Conjugates and BODIPY Labeling Kits

BODIPY Nucleotide and Oligonucleotide Conjugates

Fluorescence quenching is usually not a problem if the BODIPY derivative is conjugated to nucleotides, oligonucleotides or low molecular weight amines in which the stoichiometry of modification is 1:1. In fact, we have actually observed fluorescence enhancement in some cases, making the BODIPY oligonucleotide conjugates generally the brightest available for DNA sequencing and other applications. BODIPY FL–labeled oligonucleotide primers have also been shown to have lower photodestruction rates than fluorescein-labeled primers, improving the detectability of labeled DNA in sequencing gels.[4] Oligonucleotide conjugates of several of our BODIPY dyes have recently been reported to be useful for DNA sequencing,[5,6] in part because the dye exhibits minimal effect on the mobility of the fragment during electrophoresis. Molecular Probes offers FluoroTide™ M13pUC primers labeled with the BODIPY FL, BODIPY R6G, BODIPY TMR and BODIPY TR dyes (see Section 8.2). In addition, six BODIPY-labeled ChromaTide™ nucleotides — both UTP and dUTP derivatives — are described in Section 8.2. Among these is the BODIPY FL-14-dUTP conjugate (C-7614, see Section 8.2), which has been used as a terminal deoxynucleotidyl transferase (TdT) substrate in the detection of apoptotic cells.[16]

BODIPY Oligonucleotide Labeling Kits

Molecular Probes has developed FluoReporter Oligonucleotide Amine Labeling Kits for preparing BODIPY conjugates from amine-derivatized oligonucleotides; see the Product Highlight "FluoReporter Protein and Oligonucleotide Labeling Kits" on page 18 for more details. Three of these FluoReporter Kits (F-6082, F-6083, F-6084) contain reactive BODIPY succinimidyl esters — BODIPY FL-X, BODIPY TMR-X or BODIPY TR-X succinimidyl esters — with spectra similar to those of fluorescein, tetramethylrhodamine and Texas Red fluorophores, respectively (Figure 1.4). We also offer four FluoReporter kits containing BODIPY FL, BODIPY R6G, BODIPY 564/570 or BODIPY 581/591 succinimidyl esters (F-6079, F-6092, F-6093, F-6094), which contain the reactive BODIPY dyes found to be useful for automated DNA sequencing.[5] In addition to these amine labeling kits, we offer two FluoReporter Oligonucleotide Phosphate Labeling Kits (F-6097, F-6098; see Section 8.2), which provide cadaverine derivatives of the BODIPY FL or BODIPY TMR dyes for preparing fluorescent phosphoramidates from 5'-phosphate–labeled (and potentially 3'-phosphate–labeled) oligonucleotides.

1. J Am Chem Soc 116, 7801 (1994); **2.** Haugland, R.P. in *Optical Microscopy for Biology*, B. Herman and K. Jacobson, Eds., Wiley-Liss (1990) pp. 143–157; **3.** J Microscopy 168, 219 (1992); **4.** Electrophoresis 13, 542 (1992); **5.** Science 271, 1420 (1996); **6.** Nucleic Acids Res 20, 2471 (1992); **7.** Biochim Biophys Acta 1104, 9 (1992); **8.** Biochim Biophys Acta 776, 217 (1984); **9.** Biochemistry 21, 978 (1982); **10.** Anal Chem 67, 139 (1994); **11.** Lett Pept Sci 1, 235 (1995); **12.** BioTechniques 18, 55 (1994); **13.** J Cell Biol 127, 2081 (1994); **14.** J Cell Biol 127, 2021 (1994); **15.** Anal Biochem 156, 220 (1986); **16.** Cytometry 20, 172 (1995).

1.2 Data Table *BODIPY Dyes Spanning the Visible Spectrum*

Cat #	Structure	MW	Storage	Soluble	Abs	{ε × 10⁻³}	Em	Solvent	Notes
D-2183	✓	292	F,L	MeCN, DMSO	505	{91}	511	MeOH	<1>
D-2184	✓	389	F,D,L	MeCN, DMSO	502	{82}	510	MeOH	<1>
D-2186	✓	416	F,L	MeCN, DMSO	535	{62}	552	MeOH	<1>
D-2187	✓	513	F,D,L	MeCN, DMSO	534	{77}	551	MeOH	<1>
D-2190	✓	320	F,L	MeCN, DMSO	495	{87}	503	MeOH	<1>
D-2191	✓	417	F,D,L	MeCN, DMSO	500	{79}	509	MeOH	<1>
D-2218	✓	346	F,L	MeCN, DMSO	559	{84}	569	MeOH	<1>
D-2219	✓	443	F,D,L	MeCN, DMSO	559	{97}	568	MeOH	<1>
D-2221	✓	366	F,L	MeCN, DMSO	564	{143}	570	MeOH	<1>
D-2222	✓	463	F,D,L	MeCN, DMSO	563	{142}	569	MeOH	<1>
D-2224	✓	329	F,L	MeCN, DMSO	578	{96}	591	MeOH	<1>
D-2225	✓	426	F,D,L	MeCN, DMSO	575	{83}	588	MeOH	<1>
D-2227	✓	392	F,L	MeCN, DMSO	582	{154}	591	MeOH	<1>
D-2228	✓	489	F,D,L	MeCN, DMSO	581	{136}	591	MeOH	<1>
D-3834	✓	320	F,L	MeCN, DMSO	505	{96}	511	MeOH	<1>

Cat #	Structure	MW	Storage	Soluble	Abs	{ε × 10⁻³}	Em	Solvent	Notes
D-6101	✓	407	F,L	MeCN, DMSO	504	{82}	510	MeOH	<1>
D-6102	✓	502	F,D,L	MeCN, DMSO	504	{85}	510	MeOH	<1>
D-6103	✓	336	F,L	MeCN, DMSO	509	{94}	515	MeOH	<1>
D-6116	✓	634	F,D,L	MeCN, DMSO	588	{68}	616	MeOH	<1>
D-6117	✓	608	F,D,L	MeCN, DMSO	544	{56}	570	MeOH	<1>
D-6140	✓	491	F,D,L	H_2O*, DMSO	502	{75}	510	MeOH	<1>
D-6141	✓	641	F,D,L	H_2O*, DMSO	504	{82}	511	MeOH	<1>
D-6143	✓	450	F,L	MeCN, DMSO	531	{75}	545	MeOH	<1>
D-6144	✓	547	F,D,L	MeCN, DMSO	530	{64}	545	MeOH	<1>
D-6180	✓	437	F,D,L	MeCN, DMSO	528	{70}	547	MeOH	<1>
D-6184	✓	417	F,D,L	MeCN, DMSO	504	{86}	511	MeOH	<1>

For definitions of the contents of this data table, see "How to Use the Handbook" on page vi.
Structure: Chemical structure drawing: ✓ = shown in this section; Sn.n = shown in Section number n.n.

* Sulfosuccinimidyl esters and cysteic acid succinimidyl esters are water soluble and may be dissolved in buffer at pH ~8 for reaction with amines. Long-term storage in water is NOT recommended due to hydrolysis.

<1> The absorption and fluorescence spectra of BODIPY derivatives are relatively insensitive to the solvent.

1.2 Structures *BODIPY Dyes Spanning the Visible Spectrum*

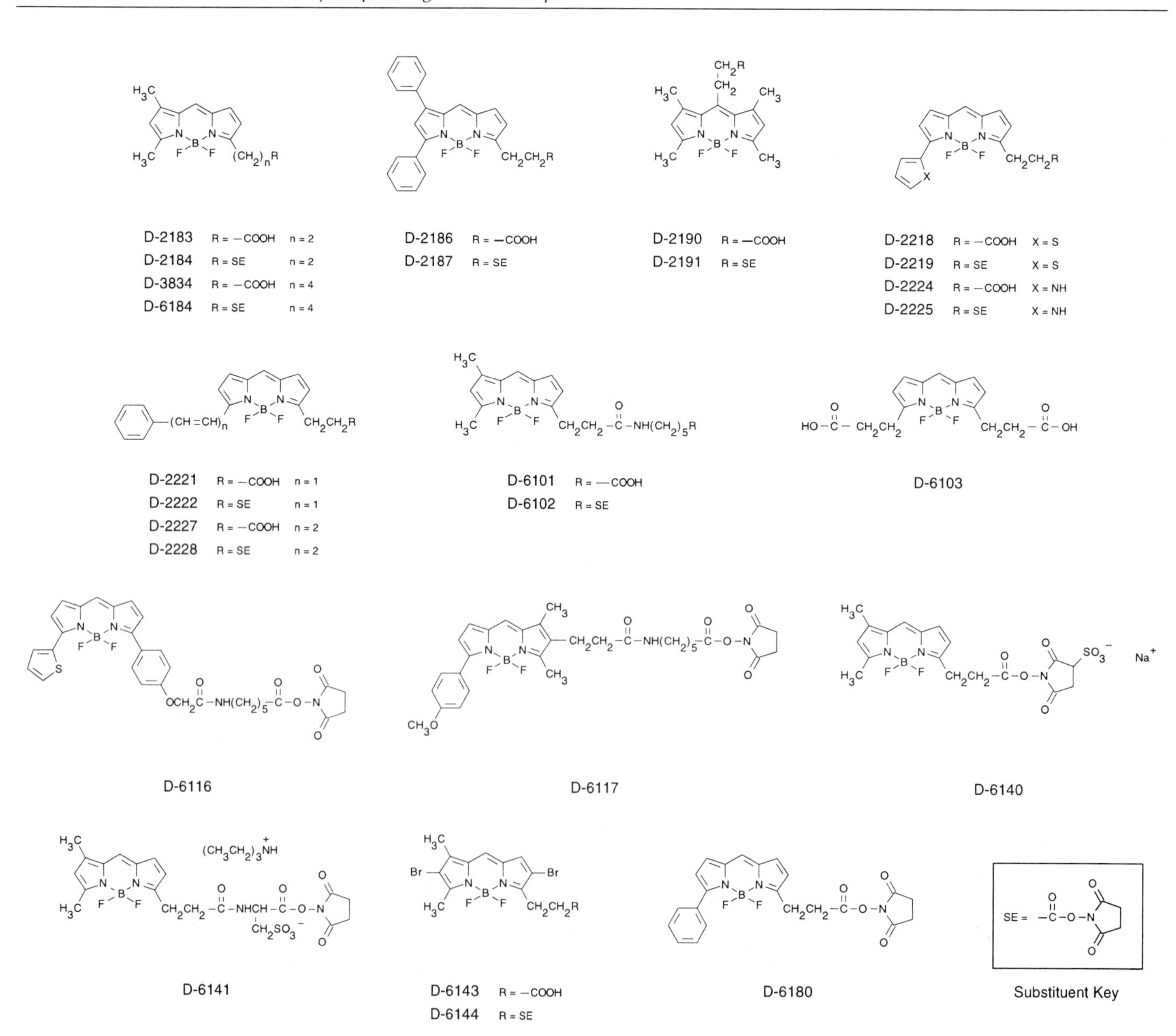

Cat #		Product Name	Unit Size	Price Per Unit ($) 1–4 Units	5–24 Units
D-6143	**New**	2,6-dibromo-4,4-difluoro-5,7-dimethyl-4-bora-3a,4a-diaza-s-indacene-3-propionic acid (BODIPY® FL Br_2)	5 mg	95.00	76.00
D-6144	**New**	2,6-dibromo-4,4-difluoro-5,7-dimethyl-4-bora-3a,4a-diaza-s-indacene-3-propionic acid, succinimidyl ester (BODIPY® FL Br_2, SE)	5 mg	145.00	116.00
D-6103	**New**	4,4-difluoro-4-bora-3a,4a-diaza-s-indacene-3,5-dipropionic acid (BODIPY® 500/510)	5 mg	95.00	76.00
D-3834		4,4-difluoro-5,7-dimethyl-4-bora-3a,4a-diaza-s-indacene-3-pentanoic acid (BODIPY® FL C_5)	1 mg	38.00	30.40
D-6184	**New**	4,4-difluoro-5,7-dimethyl-4-bora-3a,4a-diaza-s-indacene-3-pentanoic acid, succinimidyl ester (BODIPY® FL C_5, SE)	5 mg	145.00	116.00
D-2183		4,4-difluoro-5,7-dimethyl-4-bora-3a,4a-diaza-s-indacene-3-propionic acid (BODIPY® FL)	5 mg	95.00	76.00
D-2184		4,4-difluoro-5,7-dimethyl-4-bora-3a,4a-diaza-s-indacene-3-propionic acid, succinimidyl ester (BODIPY® FL, SE)	5 mg	145.00	116.00
D-6140	**New**	4,4-difluoro-5,7-dimethyl-4-bora-3a,4a-diaza-s-indacene-3-propionic acid, sulfosuccinimidyl ester, sodium salt (BODIPY® FL, SSE)	5 mg	145.00	116.00
D-6101	**New**	6-((4,4-difluoro-5,7-dimethyl-4-bora-3a,4a-diaza-s-indacene-3-propionyl)amino)hexanoic acid (BODIPY® FL-X)	5 mg	95.00	76.00
D-6102	**New**	6-((4,4-difluoro-5,7-dimethyl-4-bora-3a,4a-diaza-s-indacene-3-propionyl)amino)hexanoic acid, succinimidyl ester (BODIPY® FL-X, SE)	5 mg	145.00	116.00
D-6141	**New**	N-(4,4-difluoro-5,7-dimethyl-4-bora-3a,4a-diaza-s-indacene-3-propionyl)cysteic acid, succinimidyl ester, triethylammonium salt (BODIPY® FL, CASE)	5 mg	125.00	100.00
D-6117	**New**	6-((4,4-difluoro-1,3-dimethyl-5-(4-methoxyphenyl)-4-bora-3a,4a-diaza-s-indacene-2-propionyl)amino) hexanoic acid, succinimidyl ester (BODIPY® TMR-X, SE)	5 mg	145.00	116.00
D-2186		4,4-difluoro-5,7-diphenyl-4-bora-3a,4a-diaza-s-indacene-3-propionic acid (BODIPY® 530/550)	5 mg	95.00	76.00
D-2187		4,4-difluoro-5,7-diphenyl-4-bora-3a,4a-diaza-s-indacene-3-propionic acid, succinimidyl ester (BODIPY® 530/550, SE)	5 mg	145.00	116.00
D-6180	**New**	4,4-difluoro-5-phenyl-4-bora-3a,4a-diaza-s-indacene-3-propionic acid, succinimidyl ester (BODIPY® R6G, SE)	5 mg	145.00	116.00
D-2227		4,4-difluoro-5-(4-phenyl-1,3-butadienyl)-4-bora-3a,4a-diaza-s-indacene-3-propionic acid (BODIPY® 581/591)	5 mg	95.00	76.00
D-2228		4,4-difluoro-5-(4-phenyl-1,3-butadienyl)-4-bora-3a,4a-diaza-s-indacene-3-propionic acid, succinimidyl ester (BODIPY® 581/591, SE)	5 mg	145.00	116.00
D-2224		4,4-difluoro-5-(2-pyrrolyl)-4-bora-3a,4a-diaza-s-indacene-3-propionic acid (BODIPY® 576/589)	5 mg	95.00	76.00
D-2225		4,4-difluoro-5-(2-pyrrolyl)-4-bora-3a,4a-diaza-s-indacene-3-propionic acid, succinimidyl ester (BODIPY® 576/589, SE)	5 mg	145.00	116.00
D-2221		4,4-difluoro-5-styryl-4-bora-3a,4a-diaza-s-indacene-3-propionic acid (BODIPY® 564/570)	5 mg	95.00	76.00
D-2222		4,4-difluoro-5-styryl-4-bora-3a,4a-diaza-s-indacene-3-propionic acid, succinimidyl ester (BODIPY® 564/570, SE)	5 mg	145.00	116.00
D-2190		4,4-difluoro-1,3,5,7-tetramethyl-4-bora-3a,4a-diaza-s-indacene-8-propionic acid (BODIPY® 493/503)	5 mg	95.00	76.00
D-2191		4,4-difluoro-1,3,5,7-tetramethyl-4-bora-3a,4a-diaza-s-indacene-8-propionic acid, succinimidyl ester (BODIPY® 493/503, SE)	5 mg	145.00	116.00
D-2218		4,4-difluoro-5-(2-thienyl)-4-bora-3a,4a-diaza-s-indacene-3-propionic acid (BODIPY® 558/568)	5 mg	95.00	76.00
D-2219		4,4-difluoro-5-(2-thienyl)-4-bora-3a,4a-diaza-s-indacene-3-propionic acid, succinimidyl ester (BODIPY® 558/568, SE)	5 mg	145.00	116.00
D-6116	**New**	6-(((4-(4,4-difluoro-5-(2-thienyl)-4-bora-3a,4a-diaza-s-indacene-3-yl)phenoxy)acetyl)amino)-hexanoic acid, succinimidyl ester (BODIPY® TR-X, SE)	5 mg	145.00	116.00
F-6079	**New**	FluoReporter® BODIPY® FL Oligonucleotide Amine Labeling Kit *5 labelings*	1 kit	175.00	140.00
F-6082	**New**	FluoReporter® BODIPY® FL-X Oligonucleotide Amine Labeling Kit *5 labelings*	1 kit	175.00	140.00
F-6093	**New**	FluoReporter® BODIPY® 564/570 Oligonucleotide Amine Labeling Kit *5 labelings*	1 kit	175.00	140.00
F-6094	**New**	FluoReporter® BODIPY® 581/591 Oligonucleotide Amine Labeling Kit *5 labelings*	1 kit	175.00	140.00
F-6092	**New**	FluoReporter® BODIPY® R6G Oligonucleotide Amine Labeling Kit *5 labelings*	1 kit	175.00	140.00
F-6083	**New**	FluoReporter® BODIPY® TMR-X Oligonucleotide Amine Labeling Kit *5 labelings*	1 kit	175.00	140.00
F-6084	**New**	FluoReporter® BODIPY® TR-X Oligonucleotide Amine Labeling Kit *5 labelings*	1 kit	175.00	140.00

FluoReporter® Protein and Oligonucleotide Labeling Kits

Product (Catalog Number)	Features	Contents
FluoReporter Protein Labeling Kits • FITC (F-6434) • Fluorescein-EX (F-6433) • Oregon Green™ 488 (F-6153) • Oregon Green™ 500 (F-6154) • Oregon Green™ 514 (F-6155) • Rhodamine Red™-X (F-6161) • Tetramethylrhodamine (F-6163) • Texas Red®-X (F-6162)	The FluoReporter Protein Labeling Kits facilitate research-scale preparation of protein conjugates labeled with some of our best dyes. Typically, labeling and purifying conjugates with the FluoReporter Protein Labeling Kits can be completed in under three hours, with very little hands-on time. Each FluoReporter Protein Labeling Kit provides sufficient reagents for 5 to 10 labeling reactions of 0.2 to 2 mg of protein each.	• Five vials of the amine-reactive dye • Anhydrous DMSO • Reaction tubes, each containing a stir bar • Stop reagent • Ten spin columns • Collection tubes
FluoReporter Biotin-XX Protein Labeling Kit (F-2610)	This kit is designed for five biotinylation reactions, each with 5 to 20 mg of protein; up to 100 mg of protein may be labeled. A gel filtration column is provided for purifying the labeled proteins from excess biotin reagent. Once purified, the degree of biotinylation can be determined using the included avidin–biotin displacement assay; biotinylated goat IgG is provided as a control.	• Biotin-XX, succinimidyl ester • Anhydrous DMSO • A gel filtration column • Avidin–HABA complex • Biotinylated goat IgG
FluoReporter Mini-Biotin-XX Protein Labeling Kit (F-6347)	This kit permits efficient biotinylation of small amounts of antibodies or other proteins. The water-soluble biotin-XX sulfosuccinimidyl ester has a 14-atom spacer that enhances the binding of biotin derivatives to avidin's relatively deep binding sites. The ready-to-use spin columns provide a convenient method of purifying the biotinylated protein from excess reagents. Sufficient reagents are provided for five biotinylation reactions of 0.1 to 3 mg each.	• Biotin-XX, sulfosuccinimidyl ester • Reaction tubes, each containing a stir bar • Five spin columns • Collection tubes • Dialysis tubing
FluoReporter Biotin/DNP Protein Labeling Kit (F-6348)	The degree of biotinylation of proteins labeled with DNP-X–biocytin-X succinimidyl ester can be assessed from the optical absorbance of DNP (ϵ = 15,000 $cm^{-1}M^{-1}$ at ~360 nm). The conjugates are recognized by both avidin derivatives and anti-DNP antibodies, permitting a choice of detection techniques. Sufficient reagents are supplied for 5 to 10 labeling reactions of 0.2 to 2 mg of protein each.	• DNP-X–biocytin-X, succinimidyl ester • Anhydrous DMSO • Reaction tubes • Stop reagent • Ten spin columns • Collection tubes
FluoReporter Oligonucleotide Amine Labeling Kits • Biotin-XX (F-6081) • BODIPY® FL (F-6079) • BODIPY® FL-X (F-6082) • BODIPY® R6G (F-6092) • BODIPY® TMR-X (F-6083) • BODIPY® 564/570 (F-6093) • BODIPY® 581/591 (F-6094) • BODIPY® TR-X (F-6084) • DNP-X (F-6085) • Fluorescein-X (F-6086) • Oregon Green™ 488 (F-6087) • Rhodamine Green™-X (F-6088) • Rhodamine Red™-X (F-6089) • Tetramethylrhodamine (F-6090) • Texas Red®-X (F-6091)	The FluoReporter Oligonucleotide Amine Labeling Kits permit the easy preparation of labeled oligonucleotides by reacting amine-derivatized oligonucleotides with a wide selection of our amine-reactive succinimidyl esters. The amine-reactive haptens and fluorophores in most of our 15 different FluoReporter Oligonucleotide Amine Labeling Kits contain aminohexanoic spacers ("X") to reduce the label's interaction with the oligonucleotide and enhance its accessibility to secondary detection reagents. The protocol has been optimized for labeling 5′-amine–modified oligonucleotides, 18 to 24 bases in length. Shorter or longer oligonucleotides may be labeled by the same procedure; however, adjustments to the protocol may be necessary. Sufficient reagents are provided in each kit for five complete labeling reactions of 100 µg of oligonucleotide each. The conjugates can be purified with our FluoReporter Labeled Oligonucleotide Purification Kit (F-6100, see below).	• Five vials of the amine-reactive label • Anhydrous DMSO • Labeling buffer • A detailed protocol for labeling
FluoReporter Oligonucleotide Phosphate Labeling Kits • Biotin-X-C_5 (F-6095) • BODIPY® FL-C_5 (F-6096) • BODIPY® TMR-C_5 (F-6097) • Rhodamine Red™-C_2 (F-6098) • Texas Red®-C_5 (F-6099)	These kits use proprietary coupling technology to link aliphatic amines to 5′-phosphate–terminated oligonucleotides to form phosphoramidate adducts. Unphosphorylated oligonucleotides can be enzymatically phosphorylated using T4 polynucleotide kinase prior to use of these labeling kits. Sufficient reagents are provided in each kit for five complete labeling reactions of 100 µg oligonucleotide each. The conjugates can be purified with our FluoReporter Labeled Oligonucleotide Purification Kit (F-6100, see below).	• Five vials of the phosphate-reactive label • Anhydrous DMSO • Labeling buffer • A detailed protocol for labeling
FluoReporter Labeled Oligonucleotide Purification Kit (F-6100)	The crude, labeled oligonucleotide is simply precipitated with ethanol to remove the excess reactive reagent, adsorbed on a spin column, washed to remove any unconjugated oligonucleotide and then eluted with an elution buffer to yield the conjugate. Isolated yields for the combined conjugation and purification steps are usually >50%, and the products are typically 90% pure as determined by HPLC. Conjugates can be used for most procedures without additional purification.	• Five spin columns • Buffers for column equilibration, loading, washing and elution • A detailed protocol

1.3 Fluorescein: The Predominant Green Fluorophore

Spectral Properties of Fluorescein

The amine-reactive fluorescein derivatives are probably the most common fluorescent derivatization reagents for covalently labeling proteins. In addition to its relatively high absorptivity, excellent fluorescence quantum yield and good water solubility, fluorescein (F-1300, excitation/emission maxima ~494/520 nm) has an excitation maximum that closely matches the 488 nm spectral line of the argon-ion laser, making it the predominant fluorophore for confocal laser scanning microscopy [1] and flow cytometry applications. In addition, fluorescein's protein conjugates are not inordinately susceptible to precipitation. However, fluorescein-based dyes and their conjugates have several drawbacks, including a relatively high rate of photobleaching [2] (Figure 1.6 below; see also Figure 7.1 in Chapter 7 and Figure 11.1 in Chapter 11), as well as pH-sensitive fluorescence [3] (pK_a ~6.4) that is significantly reduced below pH 7 (see Figure 1.7 in Section 1.4). Also, the fluorescence of fluorescein (and most other dyes) is typically quenched >50% upon conjugation to biopolymers. Conjugating more fluorescein molecules to the protein or nucleic acid will not necessarily improve the fluorescence because high degrees of dye conjugation result in even greater quenching [4,5] (Figure 1.5). And finally, the fluorescence emission spectrum of fluorescein is relatively broad (Figure 1.4), thereby limiting its utility in some multicolor applications. Although dyes with improved fluorescence properties have been difficult to find, Molecular Probes has developed some potentially outstanding fluorescein substitutes that are described in Sections 1.4 and 1.5.

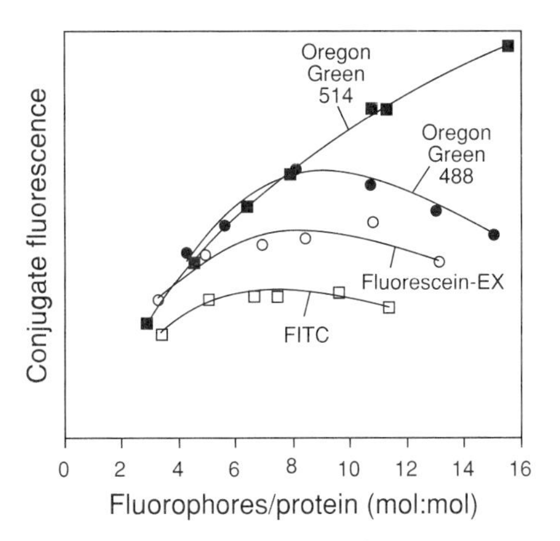

Figure 1.5 *Comparison of relative fluorescence as a function of the number of fluorophores attached per protein for goat anti–mouse IgG conjugates prepared using Oregon Green 514 succinimidyl ester (■), Oregon Green 488 succinimidyl ester (●) fluorescein-5-EX succinimidyl ester (○) and fluorescein isothiocyanate (FITC, □). Conjugate fluorescence is determined by measuring the fluorescence quantum yield of the conjugated dye relative to that of the free dye and multiplying by the number of fluorophores per protein.*

Fluorescein Isothiocyanates (FITC)

Single-Isomer FITC Preparations

Despite the availability of alternative amine-reactive fluorescein derivatives that yield conjugates with superior stability and comparable fluorescence, fluorescein isothiocyanate (FITC) remains one of the most popular fluorescent labeling reagents. The synthesis of fluorescein isothiocyanate, carboxyfluorescein (FAM, see below) and similar fluorescein-derived reagents yields a mixture of isomers at the 5- and 6-positions of fluorescein's "bottom" ring (see the chemical structures at the end of this section). Spectra of the two isomers are almost indistinguishable in both wavelength and intensity. However, the isomers may differ in the geometry of their binding to proteins, and the conjugates may elute under different chromatographic conditions or migrate differently in an electrophoretic gel when the dyes are used for high-resolution DNA sequencing. Thus, certain applications may require the single-isomer preparations. Many fluorescein (and rhodamine) probes are available from Molecular Probes either as a mixture of isomers or as purified single isomers.

The 5-isomer or "isomer I" of FITC (F-143) is the most widely used FITC isomer, probably because it is easier to isolate in pure form, but we also offer the 6-isomer or "isomer II" (F-144). Because isothiocyanates may deteriorate during storage, we recommend purchasing the 5-isomer of FITC specially packaged in individual vials (F-1906, F-1907). This FITC isomer is also available adsorbed on Celite® diatomaceous earth (F-1919), which makes it easier to weigh small quantities of the reactive dye. Adsorption of FITC on Celite has been reported to accelerate its labeling of proteins,[6] although we have not been able to confirm this. In addition, Celite provides a hydrophobic surface for those reagents that are insoluble in water, eliminating the need for organic solvents. FITC is, however, readily soluble in aqueous solutions with pH above 6. FITC is also available in our FluoReporter FITC Protein Labeling Kit (F-6434). We offer the similar FluoReporter Fluorescein-EX Protein Labeling Kit (F-6433) containing our unique fluorescein-5-EX succinimidyl ester (F-6130), the reactive fluorescein that we use to prepare most of our protein conjugates. These kits and their components are described in the Product Highlight "FluoReporter Protein and Oligonucleotide Labeling Kits" on page 18.

Applications for FITC

In addition to its widespread use for preparing immunoreagents, FITC has a multitude of other applications. Oligonucleotide conjugates of FITC are frequently employed as hybridization probes.[7] Peptide conjugates of FITC and other fluorescent isothiocyanates are susceptible to Edman degradation, making them useful for high-sensitivity amino acid sequencing;[8] FITC-labeled amino acids and peptides have been separated by capillary electrophoresis with a detection limit of fewer than 1000 molecules.[9,10] FITC has also been used to detect proteins in gels [11-13] and on nitrocellulose membranes [14-16] and is a selective inhibitor of several membrane ATPases.[17-19] Furthermore, fluorescein-to-fluorescein excited-state energy transfer leads to self-quenching; see the Technical Note "Fluorescence Resonance Energy Transfer" on page 46. This self-quenching has allowed scientists to follow the assembly of fluorescein-labeled C9 complement protein from its subunits.[20,21]

Succinimidyl Esters of Carboxyfluorescein (FAM)

Mixed-Isomer and Single-Isomer Preparations of Carboxyfluorescein Succinimidyl Ester

Although many other companies prepare their fluorescein bioconjugates with FITC, Molecular Probes prefers to use amine-reactive succinimidyl esters of carboxyfluorescein (commonly called FAM), which yield carboxamides that are more resistant to hydrolysis. We offer both mixed-isomer and single-isomer preparations of FAM (C-1904, C-1359, C-1360) and FAM succinimidyl esters (C-1311, C-2210, C-6164). FAM is an important dye for oligonucleotide labeling and automated DNA sequencing applications.[22-24] A study comparing the relative conjugation rate of several reactive fluorescein derivatives with a protein or L-lysine and the stability of the resulting conjugates concluded that carboxyfluorescein succinimidyl ester showed superior performance, followed by fluorescein dichlorotriazine (DTAF, see below). FITC was both the slowest and yielded the least stable conjugates;[25] however, the degree of substitution was most easily controlled with FITC.[25]

Succinimidyl Esters of Fluorescein Containing Spacer Groups

We also prepare succinimidyl esters of fluorescein that contain aliphatic spacers between the fluorophore and the reactive group. These include a mixed-isomer (F-2181, F-6129) and a single-isomer (F-6106) preparation of fluorescein-X succinimidyl ester (SFX), which contains a seven-atom aminohexanoyl spacer ("X") between the FAM fluorophore and the succinimidyl ester. We also offer fluorescein-5-EX succinimidyl ester (F-6130), which contains a seven-atom spacer that is somewhat more hydrophilic than the spacer in SFX. These spacers separate the fluorophore from the biomolecule to which it is conjugated, potentially reducing the quenching that typically occurs upon conjugation. We have determined that conjugates of some proteins prepared with fluorescein-5-EX succinimidyl ester are up to twice as fluorescent as the corresponding conjugates labeled with FITC at the same degree of dye substitution (Figure 1.5). Consequently, we now recommend this fluorescein derivative as the preferred dye for preparing most fluoresceinated proteins. Fluorescein-5-EX succinimidyl ester is also available in our convenient FluoReporter Fluorescein-EX Protein Labeling Kit (F-6433) and 5-SFX is used in our FluoReporter Fluorescein-X Oligonucleotide Amine Labeling Kit (F-6086); see the Product Highlight "FluoReporter Protein and Oligonucleotide Labeling Kits" on page 18 for more details on these labeling kits.

The spacers in our SFX and fluorescein-5-EX succinimidyl esters may also make the fluorophore more accessible to secondary detection reagents.[26-28] For example, the spacers should make the fluorescein moiety more available for quenching by our polyclonal and monoclonal anti-fluorescein antibodies, a technique used to determine the accessibility of the fluorophore in proteins, membranes and cells.[27,29] Fluorescein is also frequently used as a hapten on a primary detection reagent that can be either amplified or converted into a longer-wavelength or electron-dense signal with the appropriate secondary detection reagent. Section 7.3 describes our extensive selection of antibodies to fluorescein and other dyes.

Fluorescein Dichlorotriazine (DTAF)

The isomeric dichlorotriazines of fluorescein (5-DTAF, D-16; 6-DTAF, D-17) are highly reactive with proteins [30,31] and are commonly used to prepare biologically active fluorescein tubulin.[32] Unlike other reactive fluoresceins, the DTAF dyes also react directly with polysaccharides and other alcohols in aqueous solution at pH above 9 but cannot be used to modify alcohols in the presence of better nucleophiles such as amines or thiols.[33]

1. Wells, S. and Johnson, I. in *Three-Dimensional Confocal Microscopy: Volume Investigation of Biological Specimens*, J.K. Stevens, L.R. Mills and J.E. Trogadis, Eds., Academic Press (1994) pp. 101–129; **2.** Biophys J 68, 2588 (1995); **3.** Spectrochim Acta A 51, L7 (1995); **4.** Anal Biochem 173, 59 (1988); **5.** Clin Chem 25, 1554 (1979); **6.** Experientia 16, 430 (1966); **7.** J Histochem Cytochem 38, 467 (1990); **8.** Biosci Biotech Biochem 58, 300 (1994); **9.** J Chromatography 480, 141 (1989); **10.** Science 242, 562 (1988); **11.** Anal Biochem 174, 38 (1988); **12.** Anal Biochem 132, 334 (1983); **13.** Agr Biol Chem 41, 2059 (1977); **14.** Anal Biochem 177, 263 (1989); **15.** Anal Biochem 164, 303 (1987); **16.** Electrophoresis 8, 25 (1987); **17.** J Biol Chem 259, 9532 (1984); **18.** Biochim Biophys Acta 731, 9 (1983); **19.** Biochim Biophys Acta 626, 255 (1980); **20.** Biochemistry 23, 3260 (1984); **21.** Biochemistry 23, 3248 (1984); **22.** Anal Biochem 223, 39 (1994); **23.** Nucleic Acids Res 20, 2471 (1992); **24.** Proc Natl Acad Sci USA 86, 9178 (1989); **25.** Bioconjugate Chem 6, 447 (1995); **26.** Biochim Biophys Acta 1104, 9 (1992); **27.** Biochim Biophys Acta 776, 217 (1984); **28.** Biochemistry 21, 978 (1982); **29.** Biochemistry 30, 1692 (1991); **30.** J Immunol Meth 17, 361 (1977); **31.** J Immunol Meth 13, 305 (1976); **32.** Meth Enzymol 134, 519 (1986); **33.** Carbohydrate Res 30, 375 (1973).

1.3 Data Table *Fluorescein: The Predominant Green Fluorophore*

Cat #	Structure	MW	Storage	Soluble	Abs	{ε × 10⁻³}	Em	Solvent	Notes
C-1311	✓	473	F,D,L	DMF, DMSO	495	{74}	519	pH 9	<1>
C-1359	✓	376	L	pH>6, DMF	492	{79}	518	pH 9	<1>
C-1360	✓	376	L	pH>6, DMF	492	{81}	515	pH 9	<1>
C-1904	✓	376	L	pH>6, DMF	492	{78}	517	pH 9	<1>
C-2210	✓	473	F,D,L	DMF, DMSO	494	{78}	520	pH 9	<1>
C-6164	✓	473	F,D,L	DMF, DMSO	496	{83}	516	pH 9	<1>
D-16	✓	495	F,D,L	pH>6*, DMF	492	{70}	516	pH 9	<1>
D-17	✓	495	F,D,L	pH>6*, DMF	493	{72}	518	pH 9	<1>
F-143	✓	389	F,DD,L	pH>6†, DMF	494	{73}	519	pH 9	<1, 2>
F-144	✓	389	F,DD,L	pH>6†, DMF	494	{80}	519	pH 9	<1, 2>
F-1300	S10.1	332	L	pH>6, DMF	490	{88}	514	pH 9	<1>
F-2181	✓	587	F,D,L	DMF, DMSO	494	{73}	520	pH 9	<1>

Cat #	Structure	MW	Storage	Soluble	Abs	{ε x 10^-3}	Em	Solvent	Notes
F-6106	✓	587	F,D,L	DMF, DMSO	494	{73}	521	pH 9	<1>
F-6130	✓	591	F,D,L	DMF, DMSO	491	{86}	515	pH 9	<1>

F-1906, F-1907, F-1919 see F-143; F-6129 see F-2181

For definitions of the contents of this data table, see "How to Use the Handbook" on page vi.
Structure: Chemical structure drawing: ✓ = shown in this section; Sn.n = shown in Section number n.n.

* Unstable in water. Use immediately.
† Isothiocyanates are unstable in water and should not be stored in aqueous solution.
<1> Absorption and fluorescence of fluorescein derivatives are pH dependent. See data for F-1300 and C-1904 in Section 23.2.
<2> F-143 and F-144 extinction coefficients decrease about 10% upon protein conjugation and are pH dependent [J Immunol Meth 5, 103 (1974)]. F-143 fluorescence lifetime is 3.8 nsec.

1.3 Structures *Fluorescein: The Predominant Green Fluorophore*

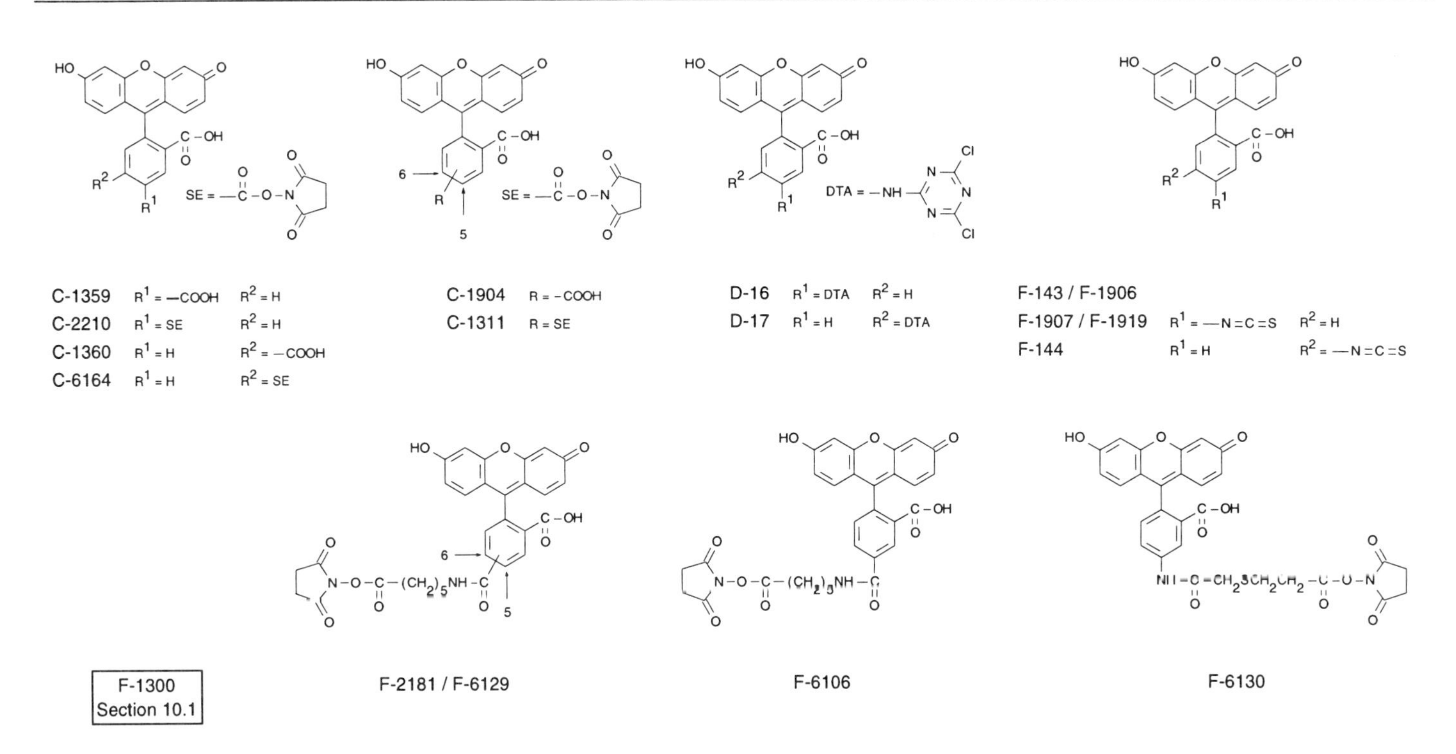

1.3 Price List *Fluorescein: The Predominant Green Fluorophore*

Cat #		Product Name	Unit Size	Price Per Unit ($) 1–4 Units	5–24 Units
C-1359		5-carboxyfluorescein (5-FAM) *single isomer*	100 mg	55.00	44.00
C-1360		6-carboxyfluorescein (6-FAM) *single isomer*	100 mg	55.00	44.00
C-1904		5-(and-6)-carboxyfluorescein (5(6)-FAM) *high purity* *mixed isomers*	100 mg	55.00	44.00
C-2210		5-carboxyfluorescein, succinimidyl ester (5-FAM, SE) *single isomer*	10 mg	95.00	76.00
C-6164	**New**	6-carboxyfluorescein, succinimidyl ester (6-FAM, SE) *single isomer*	10 mg	95.00	76.00
C-1311		5-(and-6)-carboxyfluorescein, succinimidyl ester (5(6)-FAM, SE) *mixed isomers*	100 mg	75.00	60.00
D-16		5-(4,6-dichlorotriazinyl)aminofluorescein (5-DTAF) *single isomer*	100 mg	75.00	60.00
D-17		6-(4,6-dichlorotriazinyl)aminofluorescein (6-DTAF) *single isomer*	100 mg	75.00	60.00
F-6434	**New**	FluoReporter® FITC Protein Labeling Kit *5-10 labelings*	1 kit	195.00	156.00
F-6433	**New**	FluoReporter® Fluorescein-EX Protein Labeling Kit *5-10 labelings*	1 kit	195.00	156.00
F-6086	**New**	FluoReporter® Fluorescein-X Oligonucleotide Amine Labeling Kit *5 labelings*	1 kit	175.00	140.00
F-1300		fluorescein *reference standard*	1 g	40.00	32.00
F-6106	**New**	6-(fluorescein-5-carboxamido)hexanoic acid, succinimidyl ester (5-SFX) *single isomer*	5 mg	65.00	52.00
F-2181		6-(fluorescein-5-(and-6)-carboxamido)hexanoic acid, succinimidyl ester (5(6)-SFX) *mixed isomers*	10 mg	65.00	52.00
F-6129	**New**	6-(fluorescein-5-(and-6)-carboxamido)hexanoic acid, succinimidyl ester (5(6)-SFX) *mixed isomers* *special packaging*	10x1 mg	78.00	62.40

Cat #		Product Name	Unit Size	Price Per Unit ($) 1–4 Units	5–24 Units
F-6130	New	fluorescein-5-EX, succinimidyl ester *single isomer*	10 mg	65.00	52.00
F-143		fluorescein-5-isothiocyanate (FITC 'Isomer I') *≥95% by HPLC*	1 g	98.00	78.40
F-1906		fluorescein-5-isothiocyanate (FITC 'Isomer I') *≥95% by HPLC* *special packaging*	10x10 mg	48.00	38.40
F-1907		fluorescein-5-isothiocyanate (FITC 'Isomer I') *≥95% by HPLC* *special packaging*	10x100 mg	125.00	100.00
F-1919		fluorescein-5-isothiocyanate *10% adsorbed on Celite®*	1 g	45.00	36.00
F-144		fluorescein-6-isothiocyanate (FITC 'Isomer II')	1 g	98.00	78.40

1.4 Fluorescein Substitutes

Limitations of Fluorescein

As mentioned in Section 1.3, the photobleaching and pH sensitivity of unsubstituted fluorescein make quantitative measurements with this fluorophore problematic. Furthermore, fluorescein's relatively high photobleaching rate limits the sensitivity that can be obtained, a significant disadvantage for applications requiring ultrasensitive detection such as DNA sequencing (see Section 8.3), fluorescence *in situ* hybridization (see Section 8.4) and localization of low-abundance receptors. These limitations have encouraged the development of alternative fluorophores. However, because of the widespread availability of optical filter sets designed to efficiently excite and detect fluorescein's fluorescence (see Section 26.5) and the near-optimal match of fluorescein dyes to the 488 nm spectral line of the argon-ion laser, useful fluorescein substitutes must closely replicate fluorescein's spectra.

There are no new dyes available that completely solve fluorescein's photobleaching problems, but Molecular Probes has developed some novel fluorescein-like dyes that are more photostable than fluorescein and have less or no pH sensitivity in the physiological pH range. When compared with fluorescein, all of these dyes exhibit the same or slightly longer-wavelength spectra (absorption ~490–500 nm) and comparably high fluorescence quantum yields. Alternatively, where they can be used, our FluoSpheres® fluorescent microspheres (see Section 6.2) provide a means of preparing bioconjugates that have a combination of fluorescence intensity and photostability far superior to that of any simple dye conjugate.

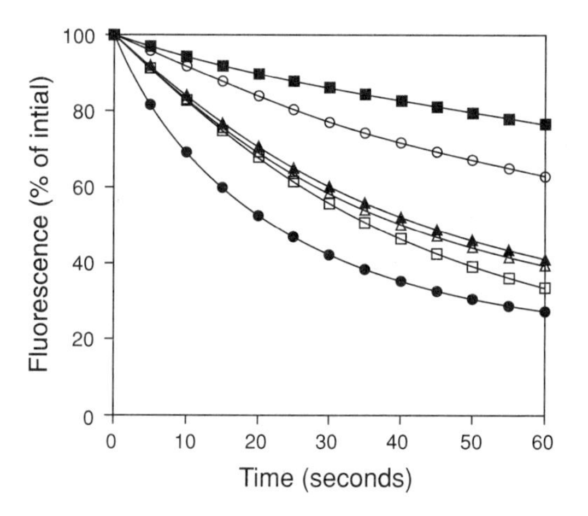

Figure 1.6 Comparison of photostability of green fluorescent antibody conjugates. The following fluorescent goat anti–mouse IgG conjugates (see Section 7.2) were used to detect mouse anti-human antibody labeling of HEp-2 cells on prefixed test slides (INOVA Diagnostics Corp.): Oregon Green 514 (O-6383, ■), Rhodol Green (R-6412, ○), Oregon Green 500 (O-6386, ▲), BODIPY FL (B-2752, △), Oregon Green 488 (O-6380, □) or fluorescein (F-2761, ●). Samples were continuously illuminated and viewed on a fluorescence microscope using an Omega® Optical fluorescein longpass filter set (O-5717, see Section 26.5). Images were acquired every 5 seconds using a Star 1™ CCD camera (Photometrics) and Image-1® software (Universal Imaging Corp.). For each conjugate, three data sets, representing different fields of view, were averaged and then normalized to the same initial fluorescence intensity value to facilitate comparison.

Oregon Green 488 Dye: A Perfect Match to Fluorescein

Reactive Oregon Green 488 Dyes

We are pleased to report that we have created a dye — Oregon Green 488 (2',7'-difluorofluorescein, D-6145) — with absorption and emission spectra that are a perfect match to those of fluorescein. Not only are the excitation and emission spectra of our new Oregon Green 488 dye virtually superimposable on those of fluorescein, but bioconjugates prepared from the reactive Oregon Green 488 dyes have important advantages when directly compared with conjugates of fluorescein. These include:

- Greater photostability (Figure 1.6)
- A lower pK_a (pK_a = 4.7 versus 6.4 for fluorescein) (Figure 1.7)
- Higher fluorescence and less quenching at comparable degrees of substitution on proteins (see Figure 1.5 in Section 1.3)

We have prepared a variety of reactive derivatives to enable researchers to take advantage of the excellent spectral properties of the Oregon Green 488 dye. These include the FITC analog Oregon Green 488 isothiocyanate (F_2FITC, O-6080), the single-isomer preparations of Oregon Green 488 carboxylic acid (O-6146, O-6148) and their corresponding succinimidyl esters (O-6147, O-6149). Oregon Green 488 iodoacetamide, which is useful for thiol labeling, is described in Section 2.2.

Oregon Green 488 Conjugates and Oregon Green 488 Labeling Kits

When compared directly to their fluorescein analogs, Oregon Green 488 conjugates typically yield higher fluorescence yields and

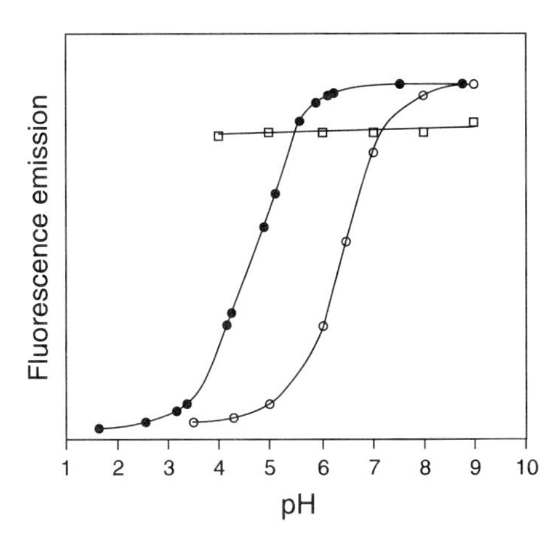

Figure 1.7 *Comparison of pH-dependent fluorescence of Oregon Green 488 (●), carboxyfluorescein (○) and Rhodamine Green (□). Fluorescence intensities were measured for equal concentrations of the three dyes using excitation/emission at 490/520 nm (●, ○) or 500/530 nm (□).*

greater resistance to photobleaching. We have therefore used the succinimidyl ester of Oregon Green 488 carboxylic acid to prepare conjugates of antibodies and protein A (see Section 7.2), streptavidin and NeutraLite™ avidin (see Section 7.5), lectins (see Section 7.6), the FluoroTide M13/pUC oligonucleotide primer and ChromaTide dUTP (see Section 8.2), phalloidin and DNase I (see Section 11.1), dextrans (see Section 15.5), fibrinogen (see Section 16.3) and the calcium chelator BAPTA (see Sections 22.3 and 22.4). Our new FluoReporter Oregon Green 488 Protein Labeling Kit (F-6153) and FluoReporter Oregon Green 488 Oligonucleotide Amine Labeling Kit (F-6087) — which contain the 5-isomer of Oregon Green 488 succinimidyl ester — make conjugations with the Oregon Green 488 dye just as easy as preparing fluorescein conjugates with the corresponding FluoReporter Fluorescein Labeling Kits (F-6433, F-6086; see Section 1.3). See the Product Highlight "FluoReporter Protein and Oligonucleotide Labeling Kits" on page 18 for further information on these kits.

The Oregon Green fluorophores, reactive dyes and conjugates are the subject of patent applications filed by Molecular Probes, Inc. and are offered for research purposes only. Molecular Probes welcomes inquiries about licensing these products for resale or other commercial uses.

Rhodol Green and Rhodamine Green Dyes

Rhodol Green and Rhodamine Green are our tradenames for carboxyrhodol and carboxyrhodamine 110, respectively. Reactive versions of both of these fluorophores were originally developed by Molecular Probes for possible use in DNA sequencing and other applications.[1,2] Like the Oregon Green dyes, Rhodol Green and Rhodamine Green dyes offer a combination of desirable properties, including good photostability (Figure 1.6), high extinction coefficients (ϵ >75,000 $cm^{-1}M^{-1}$) and high fluorescence quantum yields, particularly in their nucleotide and nucleic acid conjugates.[3] The Rhodamine Green fluorophore was recently used to label the peptide gastrin;[4] however, in general, Rhodol Green and Rhodamine Green fluorophores are less suitable for protein conjugations than the new Oregon Green dyes.

Reactive Rhodol Green Dyes and Their Conjugates

Our patented Rhodol Green reactive dyes[5] and their conjugates combine high photostability (Figure 1.6) with low pH sensitivity in the physiological range[1] (the pK_a of Rhodol Green is about 5.6; see Figure 23.13 in Chapter 23). Spectra of Rhodol Green conjugates are slightly red-shifted from those of fluorescein, but the same optical filter sets can be used for detecting both fluorophores. Two different Rhodol Green succinimidyl esters are available: the new succinimidyl ester of Rhodol Green carboxylic acid (5(6)-CRF 492, SE; R-6108) reacts with amines to directly yield the fluorescent Rhodol Green conjugate, whereas the succinimidyl ester of the trifluoroacetylated Rhodol Green carboxylic acid (5(6)-CRF 492 TFA, SE; R-6111) requires removal of the TFA groups following conjugation by treatment with a base such as hydroxylamine for protein conjugates or ammonia for oligonucleotide conjugates (Figure 1.8). A protocol for this deprotection step is included with the product.

In addition to the amine-reactive Rhodol Green dyes, Molecular Probes offers Rhodol Green carboxylic acid (5(6)-CRF 492, R-6152), as well as conjugates of secondary antibodies (see Section 7.2), streptavidin (see Section 7.5) and dextrans (see Section 15.5).

Reactive Rhodamine Green Dyes and Their Conjugates

The Rhodamine Green fluorophore,[2] (Figure 1.6) is even more photostable than the Rhodol Green and Oregon Green 488 dyes and about equivalent in photostability to the Oregon Green 514 dye. Moreover, the fluorescence of its conjugates is completely insensitive to pH between 4 and 9 (Figure 1.7). The absorption and fluorescence emission maxima of Rhodamine Green conjugates are shifted about 7 nm toward the red from that of fluorescein; however, like the Rhodol Green conjugates, they remain compatible with standard fluorescein optical filter sets. Conjugates of the Rhodamine Green fluorophore with amines can be prepared either directly from its succinimidyl ester (5(6)-CR 110, SE; R-6107) or indirectly from its TFA-protected derivative (5(6) CR 110 TFA, SE; R-6112). The succinimidyl ester of Rhodamine Green-X (R-6113) has an additional seven-atom aminohexanoyl spacer ("X") to potentially reduce interaction of the fluorophore and its reaction site. Rhodamine Green carboxylic acid (5(6)-CR 110, R-6150) can be reacted with EDAC and NHSS (E-2247, H-2249; see Section 3.3) to form the *sulfo*succinimidyl ester,[6] which is more water soluble than its succinimidyl ester counterpart and sometimes eliminates the need for organic solvents in the conjugation reaction.

Although the Rhodamine Green dye is one of the most photostable of the fluorescein substitutes, its fluorescence when conjugated to proteins is often quenched and these conjugates tend to

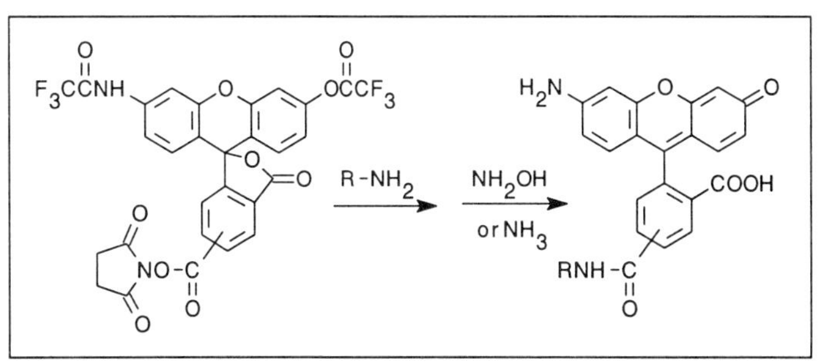

Figure 1.8 *Conjugation of Rhodol Green TFA, succinimidyl ester (R-6111) to an amine, followed by deprotection of the fluorophore with either hydroxylamine or ammonia.*

precipitate from solution. Therefore, we do not recommend any of the Rhodamine Green succinimidyl esters for preparing protein conjugates. However, when conjugated to dextrans, nucleotides and oligonucleotides, the Rhodamine Green fluorophore remains quite fluorescent. Molecular Probes currently has available the Rhodamine Green dextran conjugates (see Section 15.5) and ChromaTide Rhodamine Green UTP and dUTP (C-7628, C-7629; see Section 8.2), as well as the FluoReporter Rhodamine Green-X Oligonucleotide Amine Labeling Kit (F-6088; see the Product Highlight "FluoReporter Protein and Oligonucleotide Labeling Kits" on page 18).

1. Anal Biochem 207, 267 (1992); **2.** Haugland, R.P. in *Optical Microscopy for Biology*, Herman, B. and Jacobson, K., Eds., Wiley-Liss (1990) pp. 143–157; **3.** U.S. Patent No. 5,366,860; **4.** Lett Pept Sci 1, 235 (1995); **5.** U.S. Patent Nos. 5,227,487; 5,442,045 and other patents pending; **6.** Anal Biochem 156, 220 (1986).

1.4 Data Table *Fluorescein Substitutes*

Cat #	Structure	MW	Storage	Soluble	Abs	{ε × 10⁻³}	Em	Solvent	Notes
D-6145	✓	368	L	pH>6, DMF	490	{87}	514	pH 9	<1>
O-6080	✓	425	F,DD,L	DMF, DMSO	493	{78}	520	pH 9	<1, 2>
O-6146	✓	412	L	pH>6, DMF	492	{85}	518	pH 9	<1>
O-6147	✓	509	F,D,L	DMF, DMSO	495	{76}	521	pH 9	<1>
O-6148	✓	412	L	pH>6, DMF	492	{88}	517	pH 9	<1>
O-6149	✓	509	F,D,L	DMF, DMSO	496	{82}	516	pH 9	<1>
R-6107	✓	508	F,D,L	DMF, DMSO	504	{78}	532	MeOH	
R-6108	✓	472	F,D,L	DMF, DMSO	496	{63}	523	pH 9	<3>
R-6111	✓	664	F,DD,L	DMF, DMSO	<300		none		<4>
R-6112	✓	663	F,DD,L	DMF, DMSO	<300		none		<5>
R-6113	✓	621	F,D,L	DMF, DMSO	503	{74}	528	MeOH	
R-6150	✓	411	L	pH>6, DMF	502	{89}	524	MeOH	
R-6152	✓	412	DD,L	pH>6, DMF	494	{73}	521	pH 9	<3>

For definitions of the contents of this data table, see "How to Use the Handbook" on page vi.
Structure: Chemical structure drawing: ✓ = shown in this section.

<1> Absorption and fluorescence of Oregon Green 488 derivatives are pH dependent. See data for O-6146 and O-6148 in Section 23.3.
<2> Isothiocyanates are unstable in water and should not be stored in aqueous solution.
<3> Absorption and fluorescence of Rhodol Green derivatives are pH dependent. See data for R-6152 in Section 23.3.
<4> R-6111 is converted to R-6108 by cleavage of the trifluoroacetyl-protecting groups.
<5> R-6112 is converted to R-6107 by cleavage of the trifluoroacetyl-protecting groups.

1.4 Structures *Fluorescein Substitutes*

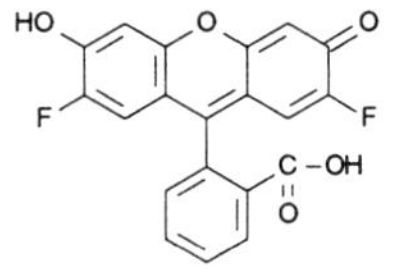

D-6145

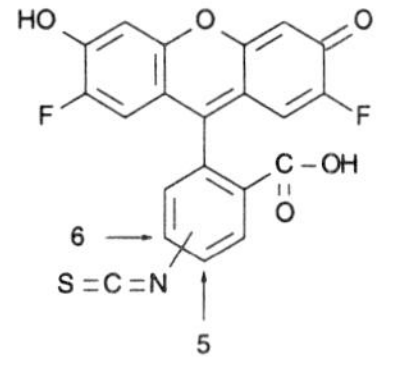

O-6080

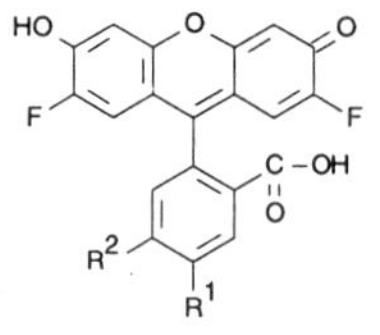

O-6146 R^1 = —COOH R^2 = H
O-6147 R^1 = SE R^2 = H
O-6148 R^1 = H R^2 = —COOH
O-6149 R^1 = H R^2 = SE

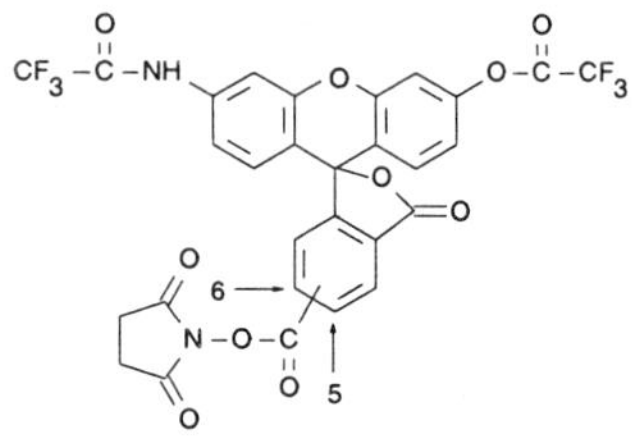

R-6111

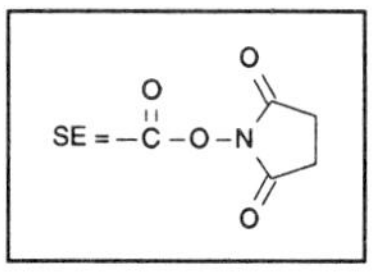

Substituent Key

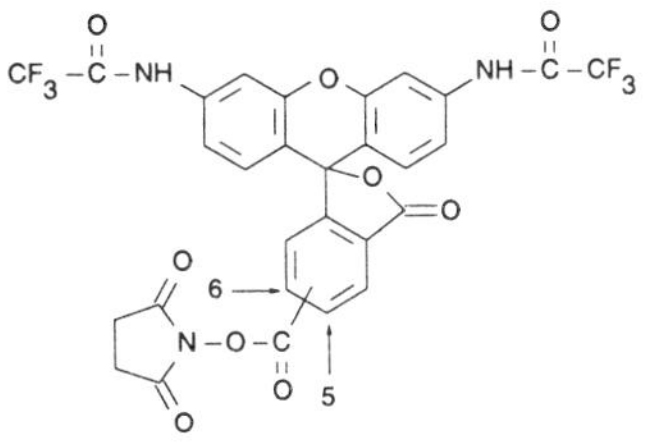

R-6112

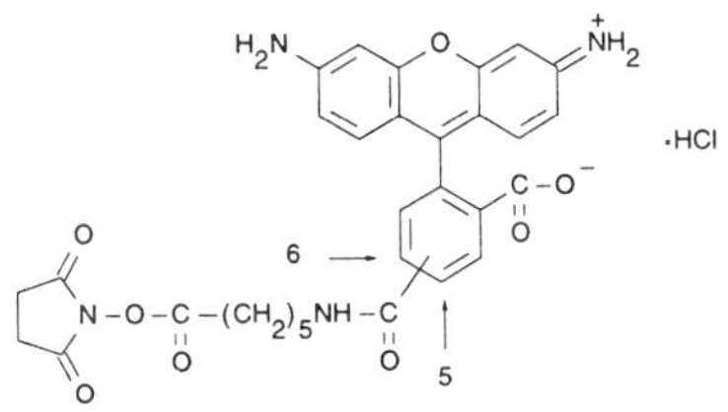

R-6113

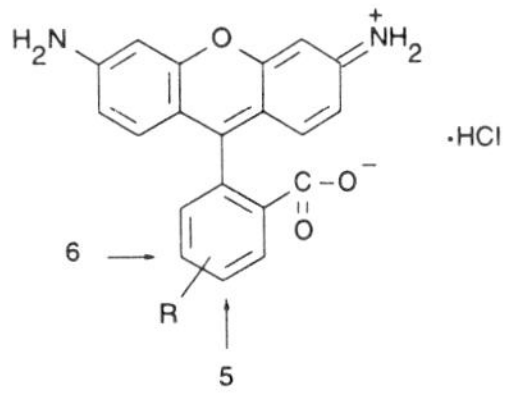

R-6150 R = —COOH
R-6107 R = SE

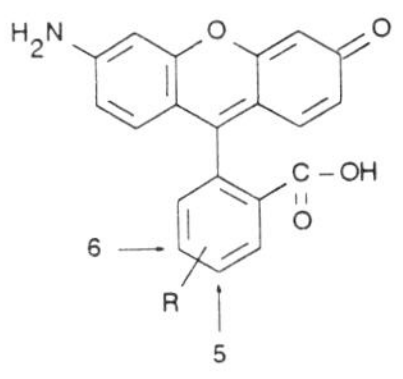

R-6152 R = —COOH ·HCl
R-6108 R = SE

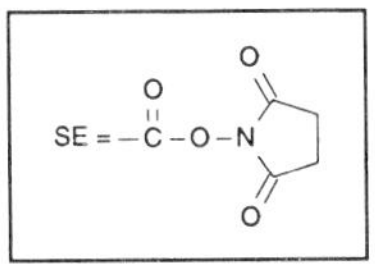

Substituent Key

1.4 Price List *Fluorescein Substitutes*

Cat #		Product Name	Unit Size	Price Per Unit ($) 1–4 Units	5–24 Units
D-6145	***New***	2′,7′-difluorofluorescein (Oregon Green™ 488)	10 mg	20.00	16.00
F-6087	***New***	FluoReporter® Oregon Green™ 488 Oligonucleotide Amine Labeling Kit *5 labelings*	1 kit	175.00	140.00
F-6153	***New***	FluoReporter® Oregon Green™ 488 Protein Labeling Kit *5-10 labelings*	1 kit	195.00	156.00
F-6088	***New***	FluoReporter® Rhodamine Green™-X Oligonucleotide Amine Labeling Kit *5 labelings*	1 kit	175.00	140.00
O-6146	***New***	Oregon Green™ 488 carboxylic acid *5-isomer*	5 mg	75.00	60.00
O-6148	***New***	Oregon Green™ 488 carboxylic acid *6-isomer*	5 mg	75.00	60.00
O-6147	***New***	Oregon Green™ 488 carboxylic acid, succinimidyl ester *5-isomer*	5 mg	95.00	76.00
O-6149	***New***	Oregon Green™ 488 carboxylic acid, succinimidyl ester *6-isomer*	5 mg	95.00	76.00
O-6080	***New***	Oregon Green™ 488 isothiocyanate (F_2FITC) *mixed isomers*	5 mg	95.00	76.00
R-6150	***New***	Rhodamine Green™ carboxylic acid, hydrochloride (5(6)-CR 110) *mixed isomers*	5 mg	95.00	76.00
R-6107	***New***	Rhodamine Green™ carboxylic acid, succinimidyl ester, hydrochloride (5(6)-CR 110, SE) *mixed isomers*	5 mg	135.00	108.00
R-6112	***New***	Rhodamine Green™ carboxylic acid, trifluoroacetamide, succinimidyl ester (5(6)-CR 110 TFA, SE) *mixed isomers*	5 mg	135.00	108.00
R-6113	***New***	Rhodamine Green™-X, succinimidyl ester, hydrochloride *mixed isomers*	5 mg	135.00	108.00
R-6111	***New***	Rhodol Green™ carboxylic acid, N,O-bis-(trifluoroacetyl), succinimidyl ester (5(6)-CRF 492 TFA, SE) *mixed isomers*	5 mg	135.00	108.00
R-6152	***New***	Rhodol Green™ carboxylic acid, hydrochloride (5(6)-CRF 492) *mixed isomers*	5 mg	75.00	60.00
R-6108	***New***	Rhodol Green™ carboxylic acid, succinimidyl ester (5(6)-CRF 492, SE) *mixed isomers*	5 mg	145.00	116.00

1.5 Dyes with Absorption Maxima Between 500 and 540 nm

Molecular Probes has also developed some excellent new fluorophores with absorption maxima between 500 and 540 nm, the approximate spectral gap between the absorption maxima of fluorescein and tetramethylrhodamine. Some of these dyes are designed to be optimally excited with the 514 nm spectral line of the argon-ion laser or with the 532 nm spectral line of the frequency-doubled Nd-YAG laser. Moreover, they are likely to assume increasing importance in multicolor applications, such as DNA sequencing and fluorescence *in situ* hybridization, which demand a greater number of fluorophores with distinct spectra.

Oregon Green 500 and Oregon Green 514 Dyes

Reactive Oregon Green 500 Dyes

Replacement of the carboxylic acid of the Oregon Green 488 dye by a sulfonic acid yields a dye that we call Oregon Green 500. The single-isomer carboxylic acid of the Oregon Green 500 dye (O-6135) can be activated to the succinimidyl ester (O-6136), which is readily conjugated to amines. The Oregon Green 500 dye has spectra only slightly shifted from that of Oregon Green 488 or fluorescein fluorophores (excitation/emission maxima ~497/517 nm in pH 9 buffer) with similar quantum yields and slightly better photostability (Figure 1.6). Furthermore, even at high degrees of substitution, Oregon Green 500 protein conjugates do not exhibit significant fluorescence quenching. We have prepared Oregon Green 500 conjugates of antibodies (see Section 7.2) and avidins (see Section 7.5).

Reactive Oregon Green 514 Dyes

In addition to our new Oregon Green 488 and Oregon Green 500 difluorinated dyes, we have introduced two pentafluorofluorescein derivatives, Oregon Green 514 carboxylic acid (O-6138) and its succinimidyl ester (O-6139). We have discovered that a combination of fluorination at the 2' and 7' positions of fluorescein and fluorination of the "bottom" ring of fluorescein produces a dye (2',4,5,6,7,7'-hexafluorofluorescein, HFF; H-7018; see Section 23.3) that exhibits a moderate shift in its fluorescence spectra of about 15 nm relative to those of the fluorescein or Oregon Green 488 dyes. The fluorine atoms at the 4- and 6-positions of HFF are susceptible to displacement by nucleophiles, a property we exploited to synthesize Oregon Green 514 carboxylic acid.

Because of the near match of its absorption maximum on proteins (~512 nm) to the 514 nm spectral line of the argon-ion laser, the Oregon Green 514 fluorophore is an important dye for both confocal laser scanning microscopy and flow cytometry applications. As with Oregon Green 500 conjugates, the fluorescence of protein conjugates prepared from the Oregon Green 514 succinimidyl ester (O-6139) is not appreciably quenched, even at high degrees of substitution (see Figure 1.5 in Section 1.3). Significantly, conjugates of the Oregon Green 514 fluorophore have greater photostability than those of any other green fluorescent dye that is useful for protein conjugation (Figure 1.6; see also Figure 7.1 in Chapter 7 and Figure 11.1 in Chapter 11). Oregon Green 514's superior photostability permits the acquisition of many more photons before the photodestruction of the dye, making this fluorophore the preferred fluorescein substitute for all fluorescence imaging applications.

Oregon Green Conjugates and Oregon Green Protein Labeling Kits

Because of their favorable spectral properties, we have used the Oregon Green 500 and Oregon Green 514 dyes to prepare conjugates of antibodies (see Section 7.2), avidins (see Section 7.5), concanavalin A (see Section 7.6), phalloidin (see Section 11.1) and α-bungarotoxin (see Section 18.2). Furthermore, to facilitate the preparation of protein conjugates from Oregon Green 500 or Oregon Green 514, we offer FluoReporter Oregon Green 500 and Oregon Green 514 Protein Labeling Kits (F-6154, F-6155), which essentially match our FluoReporter Oregon Green 488 Protein Labeling Kit (F-6153, see Section 1.4). See the Product Highlight "FluoReporter Protein and Oligonucleotide Labeling Kits" on page 18 for further information.

The Oregon Green fluorophores, reactive dyes and conjugates are the subject of patent applications filed by Molecular Probes, Inc. and are offered for research purposes only. Molecular Probes welcomes inquiries about licensing these products for resale or other commercial uses.

2′,7′-Dichlorofluorescein

Unlike fluorination of fluorescein, as in the Oregon Green 488 dye (see Section 1.4), chlorination of fluorescein to yield 2',7'-dichlorofluorescein is accompanied by about a 10 nm red shift in the absorption and emission spectra. Protein conjugates prepared with the mixed 5- and 6-isomers of carboxy-2',7'-dichlorofluorescein (C-368) and with the new 2',7'-dichlorofluorescein-5-isothiocyanate (Cl_2FITC, D-6078) exhibit a lower pK_a than those prepared with reactive unsubstituted fluorescein derivatives. Furthermore, Cl_2FITC-labeled oligonucleotide primers are reported to give improved performance over the corresponding FITC-labeled primers in DNA sequencing.[1]

Rhodol Dyes

Substitution of the Rhodol Green dye (see Section 1.4) by various groups shifts the spectra to longer wavelengths.[2] One of these dyes, Cl-NERF (C-6831) retains Rhodol Green's high photostability and exhibits an absorption maximum in aqueous solution at exactly 514 nm, a spectral line of the argon-ion laser. Although it is not readily isolated in pure form, the succinimidyl ester of Cl-NERF is useful for preparing bioconjugates, including pH-sensitive tracers for analysis of phagocytosis into acidic organelles (see Chapter 17). Another carboxylated rhodol dye, DM-NERF (D-6830), has similar spectral properties and a higher pK_a (5.4 for DM-NERF versus 3.8 for Cl-NERF; see Section 23.3). Both carboxylic acids [3] can be activated to form water-soluble, amine-reactive sulfosuccinimidyl esters.[4]

Carboxyrhodamine 6G and JOE

Carboxyrhodamine 6G

The pure 5- and 6-isomers of carboxyrhodamine 6G (C-6109, C-2213) have spectral characteristics that are intermediate between those of fluorescein and tetramethylrhodamine (see Figure 1.10 in Section 1.6). With a peak absorption at ~520 nm, the isomers of carboxyrhodamine 6G (CR 6G) are an excellent match to the 514 nm spectral line of the argon-ion laser. Conjugates prepared from the mixed-isomer (C-6157) or single-isomer (C-6127, C-6128) preparations of CR 6G succinimidyl esters or from the more water-soluble sulfosuccinimidyl esters (prepared from the carboxylic acids [4]) tend to have a higher fluorescence quantum yield than tetramethylrhodamine conjugates as well as excellent photostability. As with the Rhodamine Green dyes, the carboxyrhodamine 6G dyes are more suitable for preparing nucleotide and oligonucleotide conjugates than protein conjugates.[5] The 5- and 6-isomers of CR 6G dye (C-6128) have been reported to have spectroscopic and electrophoretic properties that are superior to the JOE dye that is often used for DNA sequencing.[6,7] One of our reactive BODIPY dyes (BODIPY R6G, see Section 1.2) has spectra similar to carboxyrhodamine 6G but with narrower absorption and emission spectra (see Figure 1.2 in Section 1.2), which may be advantageous for multicolor applications.

JOE

We also offer a single-isomer preparation of 6-carboxy-4',5'-dichloro-2',7'-dimethoxyfluorescein (6-JOE, C-6170) and its succinimidyl ester (6-JOE, SE; C-6171). This JOE isomer is an important dye for oligonucleotide labeling and automated DNA sequencing applications.[7-10] Its intermediate-wavelength spectra (absorption/emission maxima 520/548 nm), high quantum yield (~0.75) and low pH sensitivity (pK_a ~4.3) may make this a useful dye for protein conjugation too.

Eosins and Erythrosin

Eosin and Erythrosin

The reactive eosin and erythrosin dyes are usually not chosen for their fluorescence properties — the fluorescence quantum yield of eosin is typically only about 10–20% that of fluorescein, and erythrosin is even less fluorescent — but rather for their ability to act as phosphorescent probes or as photosensitizers.[11] With its high quantum yield (~0.57) for singlet oxygen generation,[12,13] eosin can be used as an effective photooxidizer of diaminobenzidine (DAB) in high-resolution electron microscopy studies;[14,15] see the Technical Note "Fluorescent Probes for Photoconversion of Diaminobenzidine" on page 264. Eosin conjugates of antibodies (see Section 7.2), streptavidin (see Section 7.5), phalloidin (see Section 11.1), α-bungarotoxin (see Section 18.2), dextrans (see Section 15.5) and a lipid (H-193, see Section 14.3) are available for this technique.

Like their thiol-reactive counterparts in Section 2.2, eosin and erythrosin isothiocyanates (E-18, E-332) are particularly useful as phosphorescent probes for measuring rotational properties of proteins, virus particles and other biomolecules in solution and in membranes.[16-21] In addition, they are employed for fluorescence resonance energy transfer studies[22-24] and for fluorescence recovery after photobleaching (FRAP) measurements of lateral diffusion.[25] We also offer 5-(and-6)-carboxyeosin (C-301), which can potentially be converted to a water-soluble, amine-reactive sulfosuccinimidyl ester.[4]

New Eosin Analogs

In our new 5-carboxy-2',4',5',7'-tetrabromosulfonefluorescein (C-6165), the carboxylic acid usually found in eosin dyes is replaced by a sulfonic acid. The resulting dye is somewhat more photostable than eosin but is likely to have a similar triplet yield. Because the ability to generate singlet oxygen is lost when a dye bleaches, it is possible that conjugates prepared from the succinimidyl ester of this dye (C-6166) will produce singlet oxygen for longer periods, making them more useful than eosin conjugates for photoconversion studies. We have determined that the brominated analog of Oregon Green 514 (eosin F_3S, C-6167) photobleaches more slowly than eosin, which may make eosin F_3S the best new dye for use as a phosphorescent probe and as a photosensitizer for singlet oxygen production in photoconversion studies. Its spectra (excitation/emission maxima ~535/552 nm in pH 9 buffer) closely resemble those of tetramethylrhodamine.

1. Electrophoresis 13, 542 (1992); **2.** Anal Biochem 207, 267 (1992); **3.** U.S. Patent Nos. 5,227,487; 5,442,045 and other patents pending; **4.** Anal Biochem 156, 220 (1986); **5.** U.S. Patent No. 5,366,860; **6.** Nature Med 2, 246 (1996); **7.** Anal Biochem 238, 165 (1996); **8.** Anal Biochem 223, 39 (1994); **9.** Nucleic Acids Res 20, 2471 (1992); **10.** Proc Natl Acad Sci USA 86, 9178 (1989); **11.** J Gen Microbiol 139, 841 (1993); **12.** Photochem Photobiol 37, 271 (1983); **13.** J Am Chem Soc 99, 4306 (1977); **14.** J Cell Biol 126, 901 (1994); **15.** J Cell Biol 126, 877 (1994); **16.** Spectroscopy 5, 20 (1990); **17.** Biochemistry 26, 2207 (1987); **18.** FEBS Lett 175, 329 (1984); **19.** J Mol Biol 179, 55 (1984); **20.** Biochemistry 22, 3082 (1983); **21.** Biochim Biophys Acta 732, 347 (1983); **22.** Eur J Biochem 173, 561 (1988); **23.** FEBS Lett 230, 109 (1988); **24.** Biophys J 51, 205 (1987); **25.** Biophys J 58, 1321 (1990).

1.5 Data Table *Dyes with Absorption Maxima Between 500 and 540 nm*

Cat #	Structure	MW	Storage	Soluble	Abs	$\{\epsilon \times 10^{-3}\}$	Em	Solvent	Notes
C-301	✓	692	L	pH>6, DMF	519	{100}	542	pH 9	
C-368	✓	445	L	pH>6, DMF	504	{90}	529	pH 8	<1>
C-2213	✓	495	D,L	pH>6, DMF	518	{88}	543	MeOH	
C-6109	✓	496	D,L	pH>6, DMF	520	{101}	546	MeOH	
C-6127	✓	556	F,D,L	pH>6, DMF	524	{108}	557	MeOH	
C-6128	✓	556	F,D,L	DMF, DMSO	524	{102}	550	MeOH	
C-6157	✓	556	F,D,L	DMF, DMSO	524	{84}	552	MeOH	
C-6165	✓	728	L	pH>6, DMF	526	{98}	544	pH 9	
C-6166	✓	1084	F,D,L	DMF, DMSO	529	{89}	544	pH 9	
C-6167	✓	792	L	pH>6, DMF	535	{99}	552	pH 9	
C-6170	✓	505	L	pH>6, DMF	520	{71}	548	pH 9	
C-6171	✓	602	F,D,L	DMF, DMSO	520	{71}	548	pH 9	
C-6831	S23.3	452	D,L	pH>6	514	{84}	540	pH 9	<2>
D-6078	✓	458	F,DD,L	pH>6*, DMF	506	{81}	527	pH 9	<1, 3>
D-6830	S23.3	431	D,L	pH>6	510	{76}	536	pH 9	<2>
E-18	✓	705	F,DD,L	pH>6*, DMF	521	{95}	544	pH 9	<3>
E-332	✓	893	F,DD,L	pH>6*, DMF	529	{90}	553	pH 9	<3>
O-6135	✓	448	L	pH>6, DMF	497	{84}	517	pH 9	<4>
O-6136	✓	647	F,D,L	DMF, DMSO	499	{78}	519	pH 9	<4>
O-6138	✓	512	L	pH>6, DMF	506	{88}	526	pH 9	<4>
O-6139	✓	609	F,D,L	DMF, DMSO	506	{85}	526	pH 9	<4>

For definitions of the contents of this data table, see "How to Use the Handbook" on page vi.
Structure: Chemical structure drawing: ✓ = shown in this section; Sn.n = shown in Section number n.n.

* Isothiocyanates are unstable in water and should not be stored in aqueous solution.

<1> Absorption and fluorescence of dichlorofluorescein derivatives are pH dependent. See data for C-368 in Section 23.3.

<2> Absorption and fluorescence of NERF derivatives are pH dependent. See data for D-6830 and C-6831 in Section 23.3.

<3> E-18 and E-332 also exhibit phosphorescence with an emission maximum at ~680 nm. The phosphorescence lifetime is ~1 msec for E-18 and 0.5 msec for E-332 [Biochem J 183, 561 (1979); Spectroscopy 5, 20 (1990)]. The fluorescence lifetimes are 1.4 nsec (QY = 0.2) for E-18 and 0.1 nsec (QY = 0.02) for E-332 [J Am Chem Soc 99, 4306 (1977)].

<4> Absorption and fluorescence of Oregon Green 500 and Oregon Green 514 derivatives are pH dependent. See data for O-6135 and O-6138 in Section 23.3.

1.5 Structures *Dyes with Absorption Maxima Between 500 and 540 nm*

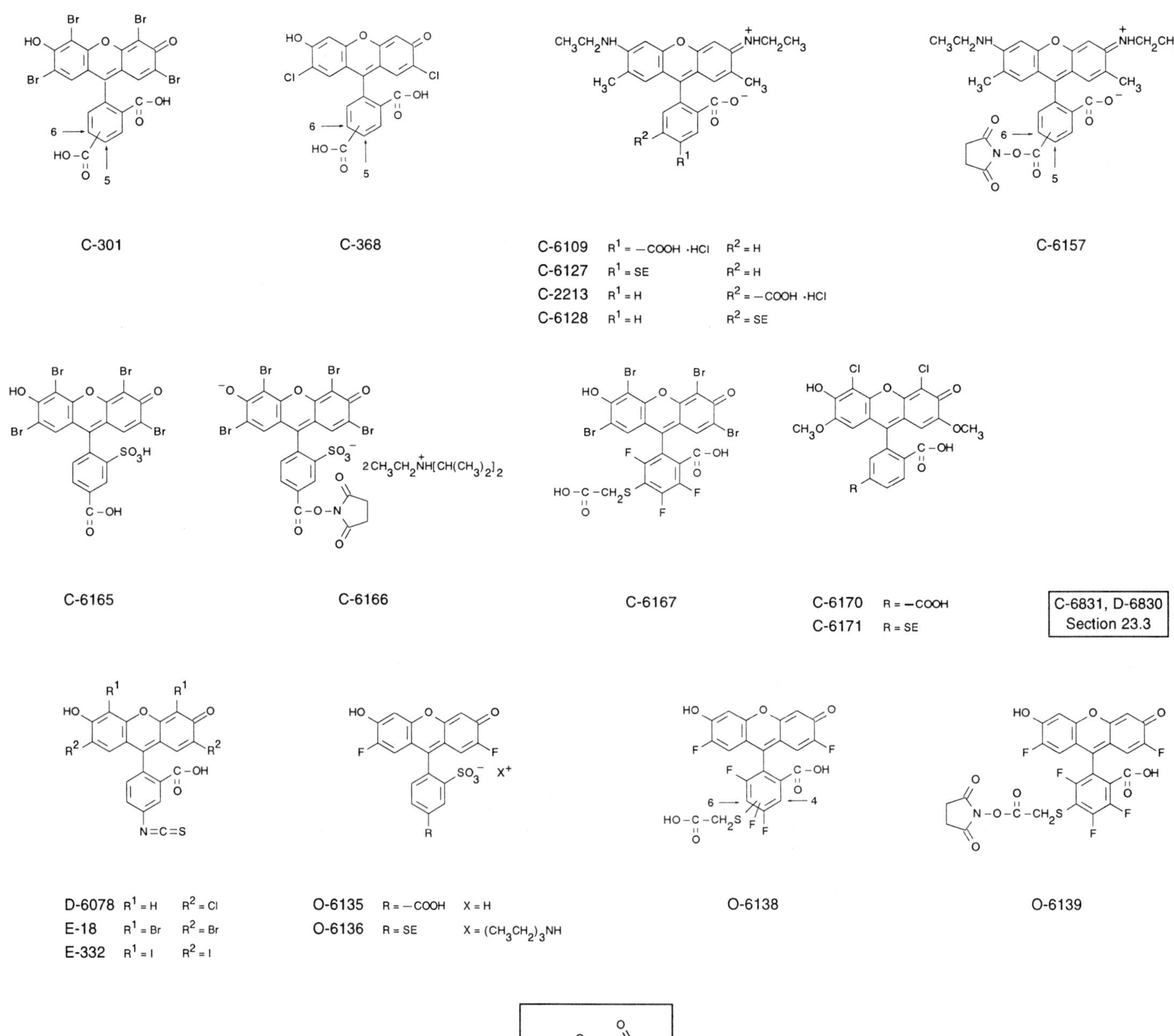

Substituent Key

1.5 Price List *Dyes with Absorption Maxima Between 500 and 540 nm*

Cat #		Product Name	Unit Size	Price Per Unit ($) 1–4 Units	5–24 Units
C-6170	*New*	6-carboxy-4′,5′-dichloro-2′,7′-dimethoxyfluorescein (6-JOE)	5 mg	75.00	60.00
C-6171	*New*	6-carboxy-4′,5′-dichloro-2′,7′-dimethoxyfluorescein, succinimidyl ester (6-JOE, SE)	5 mg	125.00	100.00
C-368		5-(and-6)-carboxy-2′,7′-dichlorofluorescein *mixed isomers*	100 mg	48.00	38.40
C-301		5-(and-6)-carboxyeosin *mixed isomers*	100 mg	38.00	30.40
C-6167	*New*	6-carboxymethylthio-2′,4′,5′,7′-tetrabromo-4,5,7-trifluorofluorescein (eosin F_3S)	5 mg	75.00	60.00
C-6109	*New*	5-carboxyrhodamine 6G, hydrochloride (5-CR 6G) *single isomer*	5 mg	75.00	60.00
C-2213		6-carboxyrhodamine 6G, hydrochloride (6-CR 6G) *single isomer*	5 mg	75.00	60.00
C-6127	*New*	5-carboxyrhodamine 6G, succinimidyl ester (5-CR 6G, SE) *single isomer*	5 mg	125.00	100.00
C-6128	*New*	6-carboxyrhodamine 6G, succinimidyl ester (6-CR 6G, SE) *single isomer*	5 mg	125.00	100.00
C-6157	*New*	5-(and-6)-carboxyrhodamine 6G, succinimidyl ester (5(6)-CR 6G, SE) *mixed isomers*	5 mg	75.00	60.00
C-6165	*New*	5-carboxy-2′,4′,5′,7′-tetrabromosulfonefluorescein	5 mg	75.00	60.00
C-6166	*New*	5-carboxy-2′,4′,5′,7′-tetrabromosulfonefluorescein, succinimidyl ester, bis-(diisopropylethylammonium) salt	5 mg	95.00	76.00
C-6831	*New*	Cl-NERF	5 mg	95.00	76.00
D-6078	*New*	2′,7′-dichlorofluorescein-5-isothiocyanate (Cl_2FITC)	25 mg	55.00	44.00
D-6830	*New*	DM-NERF	5 mg	95.00	76.00
E-18		eosin-5-isothiocyanate	100 mg	98.00	78.40
E-332		erythrosin-5-isothiocyanate	25 mg	185.00	148.00
F-6154	*New*	FluoReporter® Oregon Green™ 500 Protein Labeling Kit *5-10 labelings*	1 kit	195.00	156.00
F-6155	*New*	FluoReporter® Oregon Green™ 514 Protein Labeling Kit *5-10 labelings*	1 kit	195.00	156.00
O-6135	*New*	Oregon Green™ 500 carboxylic acid *5-isomer*	5 mg	75.00	60.00
O-6136	*New*	Oregon Green™ 500 carboxylic acid, succinimidyl ester, triethylammonium salt *5-isomer*	5 mg	95.00	76.00
O-6138	*New*	Oregon Green™ 514 carboxylic acid	5 mg	75.00	60.00
O-6139	*New*	Oregon Green™ 514 carboxylic acid, succinimidyl ester	5 mg	95.00	76.00

1.6 Long-Wavelength Dyes

The long-wavelength rhodamines, which can be considered Rhodamine Green derivatives with substituents on the nitrogens, are among the most photostable fluorescent labeling reagents available. Moreover, their spectra are not affected by changes in pH between 4 and 10, an important advantage over the fluoresceins for many biological applications. The most common members of this group are the tetramethylrhodamines — including the reactive isothiocyanate TRITC and carboxylic acid TAMRA derivatives — as well as the X-rhodamines. The X prefix of the X-rhodamines, which include Texas Red derivatives, refers to the fluorophore's extra julolidine rings. These rings prevent rotation about the nitrogen atoms, resulting in a shift in the fluorophore's spectra to longer wavelengths and usually an increase in its fluorescence quantum yield. The long-wavelength BODIPY dyes — BODIPY TMR and BODIPY TR — with spectra similar to tetramethylrhodamine and Texas Red dyes, respectively, are described above in Section 1.2.

Tetramethylrhodamine

Tetramethylrhodamine (TMR) is an important fluorophore for preparing protein conjugates, especially fluorescent antibody and avidin derivatives used in immunochemistry. Under the popular name TAMRA, the carboxylic acid of TMR has also achieved prominence as a dye for oligonucleotide labeling and automated DNA sequencing applications.[1-3] The detection limit of TMR-labeled amino acids by capillary electrophoresis has been reported to be ~600 molecules.[4] The fluorescence quantum yield of TMR conjugates is usually only about one fourth that of fluorescein conjugates. However, because TMR is readily excited by the intense 546 nm spectral line from mercury-arc lamps used in most fluorescence microscopes and is intrinsically more photostable than fluorescein, TMR conjugates often appear to be brighter than the corresponding fluorescein conjugates. TMR is also efficiently excited by the 543 nm spectral line of the green He-Ne laser, which is increasingly being used for analytical instrumentation. TMR conjugates are not well excited by the 568 nm line of the Ar-Kr mixed gas laser used in many confocal laser scanning microscopes.

The absorption spectrum of TMR-labeled proteins is frequently complex (Figure 1.9), usually splitting into two absorption peaks at about 520 and 550 nm,[5] so that the actual degree of substitution is difficult to determine. Excitation at wavelengths in the range of the short-wavelength peak fails to yield the expected amount of fluorescence, indicating that it arises from a nonfluorescent dye aggregate. Furthermore, when the TMR-labeled protein conjugate is denatured by guanidine hydrochloride, the long-wavelength absorption increases, the short-wavelength peak mostly disappears and the fluorescence yield almost doubles (Figure 1.9). This change in the absorption spectrum indicates that there is, most likely, a decrease in the extinction coefficient of TMR on conjugation to proteins. The absorption spectra of TMR-labeled nucleotides and of other probes such as our rhodamine phalloidin (R-415, see Section 11.1) do *not* split into two peaks, indicating a degree of substitution of

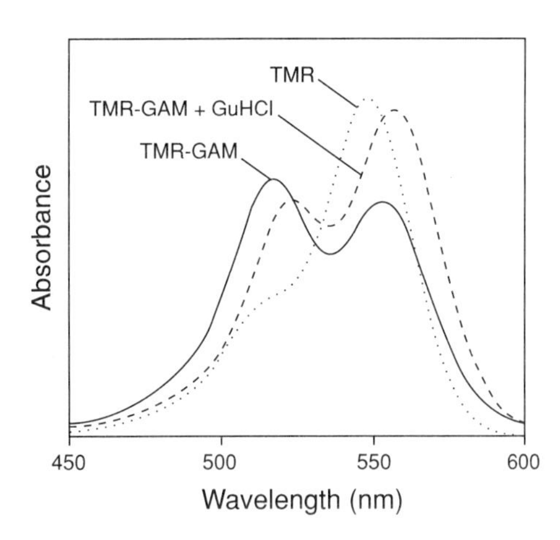

Figure 1.9 *Effect of protein conjugation on tetramethylrhodamine's absorption spectrum. The absorption spectrum of tetramethylrhodamine conjugated to goat anti–mouse IgG (TMR-GAM) shows an additional peak at about 520 nm when compared to the spectrum of the same concentration of free dye (TMR). Partial unfolding of the protein in the presence of 4.8 M guanidine hydrochloride (TMR-GAM + GuHCl) results in a spectrum more similar to that of the free dye.*

one dye molecule per biomolecule. The emission spectrum of TMR conjugates does not vary much with the degree of substitution.[5]

Mixed-Isomer and Single-Isomer TRITC Preparations

Our tetramethylrhodamine isothiocyanate (TRITC) is of the highest quality available from any commercial source. Both our mixed-isomer (T-490) and high-purity single-isomer (T-1480, T-1481) TRITC preparations typically have extinction coefficients above 80,000 $cm^{-1}M^{-1}$, whereas competitive TRITCs have extinction coefficients reported to be below 50,000 $cm^{-1}M^{-1}$. TRITC is widely used by other companies to prepare most of their so-called "rhodamine" immunoconjugates; however, they also often employ reactive versions of rhodamine B or Lissamine rhodamine B, which have somewhat different spectra, resulting in some confusion in matching the product name to the correct fluorophore.

Succinimidyl Esters of Carboxytetramethylrhodamine (TAMRA)

Almost all of Molecular Probes' TMR conjugates are prepared using succinimidyl esters of carboxytetramethylrhodamine (TAMRA), rather than TRITC, because bioconjugates from succinimidyl esters are more stable and often more fluorescent. We offer the mixed-isomer (C-300) and single-isomer (C-6121, C-6122) preparations of TAMRA, as well as the corresponding mixed-isomer (C-1171) and single-isomer (C-2211, C-6123) succinimidyl esters. The single-isomer preparations of TAMRA are most important for high-resolution techniques such as DNA sequencing[2] and separation of TAMRA-labeled carbohydrates by capillary electrophoresis.[6] The convenient FluoReporter Tetramethylrhodamine (5-TAMRA) Oligonucleotide Amine Labeling Kit (F-6090) provides the 5-isomer of TAMRA succinimidyl ester for preparing TMR-labeled oligonucleotides, whereas our FluoReporter Tetramethylrhodamine Protein Labeling Kit (F-6163) supplies the mixed-isomer 5(6)-TAMRA succinimidyl ester for preparing TMR-labeled proteins. See the Product Highlight "FluoReporter Protein and Oligonucleotide Labeling Kits" on page 18 for further details on the kits.

We have also prepared the mixed isomer TAMRA-X succinimidyl ester (5(6)-TAMRA-X, SE; T-6105), which contains a seven-atom aminohexanoyl spacer ("X") between the reactive group and the fluorophore. This spacer helps to separate the fluorophore from its point of attachment, potentially reducing the interaction of the fluorophore with the biomolecule to which it is conjugated and making it more accessible to secondary detection reagents.[7-9] Polyclonal anti-tetramethylrhodamine and anti–Texas Red antibodies that recognize tetramethylrhodamine, Rhodamine Red, X-rhodamine and Texas Red fluorophores are now available (see Section 7.3).

Lissamine Rhodamine B and Rhodamine Red-X Dyes

Lissamine Rhodamine B Sulfonyl Chloride

Lissamine rhodamine B sulfonyl chloride (L-20, L-1908) is much less expensive than Texas Red sulfonyl chloride (see below) and the fluorescence emission spectra of its protein conjugates lie between those of tetramethylrhodamine and Texas Red conjugates (Figure 1.10). Although the absorption spectral shift relative to tetramethylrhodamine is not large, it is sufficient to permit conjugates of Lissamine rhodamine B to be excited by the 568 nm line of the Ar-Kr mixed gas laser used in many confocal laser scanning microscopes. Furthermore, the protein conjugates of Lissamine rhodamine B are easier to purify and more chemically stable than those of tetramethylrhodamine. Like Texas Red sulfonyl chloride, Lissamine rhodamine B sulfonyl chloride is actually a mixture of isomeric sulfonyl chlorides.

Rhodamine Red-X Succinimidyl Ester

Lissamine rhodamine B sulfonyl chloride is unstable, particularly in aqueous solution, making it somewhat difficult to achieve reproducible conjugations using this dye. As with the new succinimidyl ester of Texas Red-X (see below), we have prepared the succinimidyl ester of Lissamine rhodamine B–labeled aminohexanoic acid, which we have named Rhodamine Red-X succinimidyl ester (R-6160). Rhodamine Red-X succinimidyl ester is resistant to hydrolysis at the pH used for conjugation and provides a spacer between the fluorophore and the reactive site. Moreover, we have

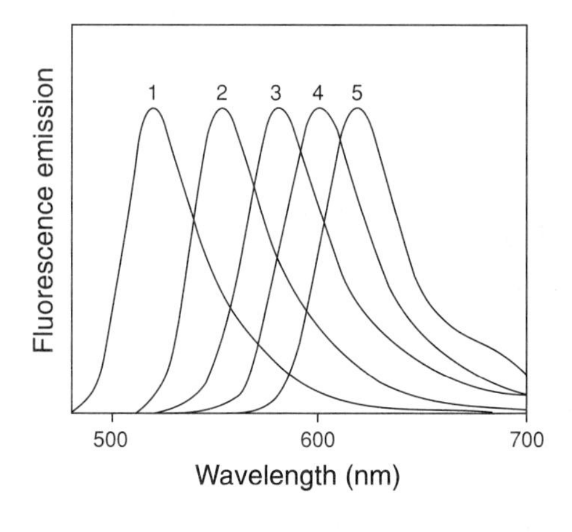

Figure 1.10 *Normalized fluorescence emission spectra of goat anti–mouse IgG conjugates of 1) fluorescein, 2) rhodamine 6G, 3) tetramethylrhodamine, 4) Lissamine rhodamine B and 5) Texas Red dyes.*

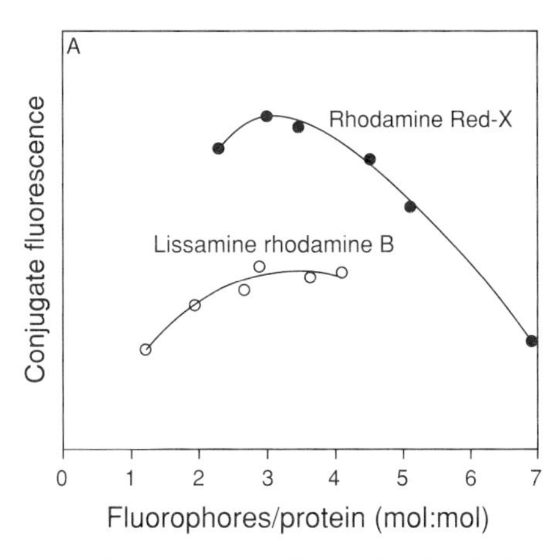

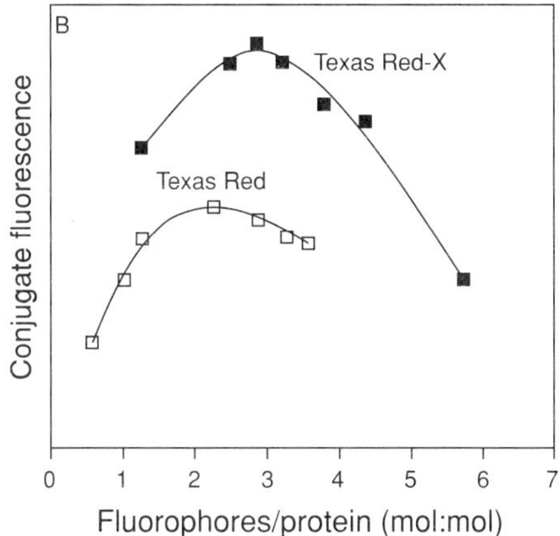

Figure 1.11 *Comparison of relative fluorescence of goat anti–mouse IgG conjugates of A) Rhodamine Red-X succinimidyl ester (●) and Lissamine rhodamine B sulfonyl chloride (○), and B) Texas Red-X succinimidyl ester (■) and Texas Red sulfonyl chloride (□). Conjugate fluorescence is determined by measuring the fluorescence quantum yield of the conjugated dye relative to that of the free dye and multiplying by the number of fluorophores per protein. Higher numbers of fluorophores attached per protein are attainable with Rhodamine Red-X and Texas Red-X due to the lesser tendency of these dyes to induce protein precipitation.*

found that protein conjugates of Rhodamine Red-X are frequently brighter than those of Lissamine rhodamine B (Figure 1.11) and more resistant to precipitation during storage.[10] Rhodamine Red-X succinimidyl ester is used in our FluoReporter Rhodamine Red-X Protein Labeling Kit (F-6161) and FluoReporter Rhodamine Red-X Oligonucleotide Amine Labeling Kit (F-6089). See the Product Highlight "FluoReporter Protein and Oligonucleotide Labeling Kits" on page 18 for further information on both kits.

Rhodamine Red-X succinimidyl ester and its conjugates are offered for research purposes only. Molecular Probes welcomes inquiries about licensing these products for resale or other commercial uses.

X-Rhodamine

The derivatives of carboxy-X-rhodamine (ROX) — a dye that was originally developed at Molecular Probes — are widely used for oligonucleotide labeling and automated DNA sequencing applications.[1-3] Conjugates of this dye and of the similar isothiocyanate (5(6)-XRITC, X-491) have longer-wavelength spectra than those of Lissamine rhodamine B but somewhat shorter-wavelength spectra than those of Texas Red conjugates. The mixed 5- and 6-isomers (C-1308) and pure 5-isomer (C-6124) and 6-isomer (C-6156) of ROX are available, as are mixed-isomer (C-1309) and single-isomer (C-6125, C-6126) preparations of the succinimidyl ester of ROX.

Texas Red and Texas Red-X Dyes

The Texas Red fluorophore emits at a longer wavelength than either tetramethylrhodamine or Lissamine rhodamine B, making Texas Red conjugates among the most commonly used long-wavelength "third labels" in fluorescence microscopy and flow cytometry. Unlike the other rhodamines, the Texas Red fluorophore exhibits very little spectral overlap with fluorescein (Figure 1.10), and its fluorescence can be distinguished from that of phycoerythrin.[11-13] Moreover, the fluorescence quantum yield of Texas Red conjugates is usually higher than that of tetramethylrhodamine or Lissamine rhodamine B conjugates. When the correct optical filter sets are used (see Section 26.5), Texas Red conjugates appear brighter and have lower background than conjugates of the other commonly used red fluorescent dyes. Texas Red conjugates are particularly well suited for the 568 nm spectral line of the Ar-Kr mixed gas laser now used in many confocal laser scanning microscopes or the 594 nm spectral line of the orange He-Ne laser.

Texas Red Sulfonyl Chloride

Texas Red sulfonyl chloride is Molecular Probes' trademarked mixture of isomeric sulfonyl chlorides of sulforhodamine 101.[14] This reagent is quite unstable in water, especially at the higher pH required for reaction with aliphatic amines. For example, dilute solutions of Texas Red sulfonyl chloride are totally hydrolyzed within 2–3 minutes in pH 8.5 aqueous solution at room temperature. Protein modification by this reagent is best done at low temperature. Once conjugated, however, the sulfonamides that are formed (see Figure 1.1C in Section 1.1) are extremely stable; they even survive complete protein hydrolysis. Because Texas Red sulfonyl chloride degrades upon exposure to moisture, Molecular Probes offers this reactive dye specially packaged as a set of 10 vials (T-1905), each containing approximately one mg of Texas Red sulfonyl chloride for small-scale conjugations. We also offer the 10 mg unit size packaged in a single vial (T-353) for larger-scale conjugations. Each milligram of Texas Red sulfonyl chloride modifies approximately 8–10 mg of protein. Polyclonal anti-tetramethylrhodamine and anti–Texas Red antibodies that recognize tetramethylrhodamine, Rhodamine Red, X-rhodamine and Texas Red fluorophores are now available (see Section 7.3).

Texas Red-X Succinimidyl Esters

Texas Red sulfonyl chloride's susceptibility to hydrolysis and low solubility in water may complicate its conjugation to some biomolecules. To overcome this difficulty, Molecular Probes is pleased to introduce its new Texas Red-X succinimidyl ester, which contains an additional seven-atom aminohexanoyl spacer ("X") between the fluorophore and its reactive group [10] (T-6134). New thiol-reactive Texas Red derivatives based on a similar synthetic approach are described in Section 2.2. Texas Red-X succinimidyl ester offers significant advantages over Texas Red sulfonyl chloride for the preparation of bioconjugates:

- In the absence of protein, greater than 80% of Texas Red-X succinimidyl ester's reactivity is retained in pH 8.5 solution after one hour at room temperature.
- Much less Texas Red-X succinimidyl ester (usually half of the amount of Texas Red sulfonyl chloride) is required to yield the same degree of substitution.
- Conjugations with Texas Red-X succinimidyl ester are more reproducible.
- Unlike Texas Red sulfonyl chloride, which can form unstable products with tyrosine, histidine, cysteine and other residues in proteins, the Texas Red-X succinimidyl ester reacts almost exclusively with amines.
- Protein conjugates prepared with Texas Red-X succinimidyl ester have a higher fluorescence yield than those with the same degree of dye substitution prepared with Texas Red sulfonyl chloride (Figure 1.11).
- We have noted a decreased tendency of Texas Red-X protein conjugates to precipitate during the reaction or upon storage.

Texas Red-X Conjugates and Texas Red-X Labeling Kits

Because of these advantages, we are now converting some of our Texas Red protein and other conjugates to the new-and-improved Texas Red-X conjugates. Consequently, Texas Red-X conjugates of proteins (see Chapter 7), dUTP and a FluoroTide M13/pUC oligonucleotide primer (see Section 8.2), phalloidin (see Section 11.1), α-bungarotoxin (see Section 18.2) and a phospholipid (T-7710, see Section 13.4) are already available and additional Texas Red-X conjugates are being developed. Protein and oligonucleotide conjugates of Texas Red-X are readily prepared using our FluoReporter Texas Red-X Protein Labeling Kit (F-6162) and FluoReporter Texas Red-X Oligonucleotide Amine Labeling Kit (F-6091), respectively. See the Product Highlight "FluoReporter Protein and Oligonucleotide Labeling Kits" on page 18 for further information on these kits. Polyclonal anti-tetramethylrhodamine and anti–Texas Red antibodies that recognize tetramethylrhodamine, Rhodamine Red, X-rhodamine and Texas Red fluorophores are now available (see Section 7.3).

Texas Red-X succinimidyl ester and its conjugates are the subject of patent applications filed by Molecular Probes, Inc. and are offered for research purposes only. We welcome inquiries about licensing these products for resale or other commercial uses.

Naphthofluorescein

Naphthofluorescein carboxylic acid and its succinimidyl ester (C-652, C-653) have emission maxima of approximately 660 nm in aqueous solution at pH 10, making them the longest-wavelength reactive probes currently available from Molecular Probes. However, their fluorescence is pH dependent (pK_a ~7.6), requiring a relatively alkaline pH for maximal fluorescence.

Nonfluorescent Malachite Green and Isosulfan Blue

Malachite green is a nonfluorescent photosensitizer that absorbs at long wavelengths (~630 nm). Its photosensitizing action can be targeted to particular cellular sites by conjugating malachite green isothiocyanate (M-689) to specific antibodies. Enzymes and other proteins within ~10 Å of the binding site of the malachite green–labeled antibody can then be selectively destroyed upon irradiation with long-wavelength light. Recent studies by Jay and colleagues have demonstrated that this photoinduced destruction of enzymes in the immediate vicinity of the chromophore is apparently the result of localized production of hydroxyl radicals, which have short lifetimes that limit their diffusion from the site of their generation.[15] Earlier studies had supported a thermal mechanism of action.[16-18]

Our new reactive sulfonyl chloride of isosulfan blue (I-6204) is potentially a lower cost alternative to malachite green isothiocyanate for the preparation of photosensitizing bioconjugates. Although the structure of its chromophore and its absorption properties are virtually identical to those of malachite green, we are not aware of its use as a photosensitizer. As with Texas Red and Lissamine rhodamine B sulfonyl chlorides, isosulfan blue sulfonyl chloride is a mixture of isomeric sulfonyl chlorides.

1. Anal Biochem 223, 39 (1994); **2.** Nucleic Acids Res 20, 2471 (1992); **3.** Proc Natl Acad Sci USA 86, 9178 (1989); **4.** J Chromatography 608, 117 (1992); **5.** Anal Biochem 80, 585 (1977); **6.** J Chromatography B 675, 307 (1994); **7.** Biochim Biophys Acta 1104, 9 (1992); **8.** Biochim Biophys Acta 776, 217 (1984); **9.** Biochemistry 21, 978 (1982); **10.** Bioconjugate Chem 7, 482 (1996); **11.** Acta Histochem 89, 85 (1990); **12.** Proc Natl Acad Sci USA 85, 3546 (1988); **13.** Cell Biophys 7, 129 (1985); **14.** J Immunol Meth 50, 193 (1982); **15.** Proc Natl Acad Sci USA 91, 2659 (1994); **16.** Biophys J 61, 956 (1992); **17.** Biophys J 61, 631 (1992); **18.** Proc Natl Acad Sci USA 85, 5454 (1988).

1.6 Data Table *Long-Wavelength Dyes*

Cat #	Structure	MW	Storage	Soluble	Abs	{ε x 10⁻³}	Em	Solvent	Notes
C-300	✓	430	L	DMF, DMSO	540	{95}	565	MeOH	<1>
C-652	✓	476	L	pH>6, DMF	598	{49}	668	pH 10	<2>
C-653	✓	574	F,D,L	DMF, DMSO	602	{42}	672	pH 10	<2>
C-1171	✓	528	F,D,L	DMF, DMSO	546	{95}	576	MeOH	<1>
C-1308	✓	535	D,L	pH>6, DMF	568	{98}	595	MeOH	<1>
C-1309	✓	632	F,D,L	DMF, DMSO	576	{80}	601	MeOH	<1>
C-2211	✓	528	F,D,L	DMF, DMSO	546	{89}	579	MeOH	<1>
C-6121	✓	430	L	pH>6, DMF	542	{91}	568	MeOH	<1>
C-6122	✓	430	L	pH>6, DMF	540	{103}	564	MeOH	<1>
C-6123	✓	528	F,D,L	DMF, DMSO	547	{91}	573	MeOH	<1>

Cat #	Structure	MW	Storage	Soluble	Abs	{ϵ × 10⁻³}	Em	Solvent	Notes
C-6124	✓	535	L	pH>6, DMF	567	{92}	591	MeOH	<1>
C-6125	✓	632	F,D,L	DMF, DMSO	574	{78}	602	MeOH	<1>
C-6126	✓	632	F,D,L	DMF, DMSO	575	{82}	602	MeOH	<1>
C-6156	✓	535	L	pH>6, DMF	570	{113}	590	MeOH	<1>
I-6204	S21.2	563	F,DD,L	DMF, MeCN	646	{87}	none	MeOH	
L-20	✓	577	F,DD,L	DMF, MeCN*	568	{88}	583	MeOH	
M-689	✓	486	F,DD,L	DMF, DMSO	628	{76}	none	MeOH	<3>
R-6160	✓	769	F,D,L	DMF, DMSO	560	{129}	580	MeOH	
T-353	✓	625	F,DD,L	DMF, MeCN*	587	{85}	602	$CHCl_3$	
T-490	✓	444	F,DD,L	DMF, DMSO	544	{83}	572	MeOH	<3,4>
T-1480	✓	444	F,DD,L	DMF, DMSO	543	{93}	571	MeOH	<3,4>
T-1481	✓	444	F,DD,L	DMF, DMSO	544	{90}	572	MeOH	<3,4>
T-6105	✓	641	F,D,L	DMF, DMSO	543	{90}	571	MeOH	
T-6134	✓	817	F,D,L	DMF, DMSO	583	{116}	603	MeOH	
X-491	✓	548	F,DD,L	DMF, DMSO	572	{92}	596	MeOH	<3>

L-1908 see L-20; T-1905 see T-353

For definitions of the contents of this data table, see "How to Use the Handbook" on page vi.
Structure: Chemical structure drawing: ✓ = shown in this section; Sn.n = shown in Section number n.n.

* Do NOT dissolve in DMSO.
<1> Abs and Em for TAMRA and ROX dyes in pH 8 buffer are red-shifted approximately 8 nm compared to Abs and Em in MeOH, with ϵ lower by ~10%.
<2> Absorption and fluorescence of naphthofluorescein derivatives are pH dependent. See data for C-652 in Section 23.2. Fluorescence quantum yield ~0.14 at pH 9.5 [Cytometry 10, 151 (1989)].
<3> Isothiocyanates are unstable in water and should not be stored in aqueous solution.
<4> Protein conjugates of T-490, T-1480 and T-1481 often exhibit two absorption peaks at about 520 and 545 nm. The 520 nm peak is due to nonfluorescent dye aggregates.

1.6 Structures *Long-Wavelength Dyes*

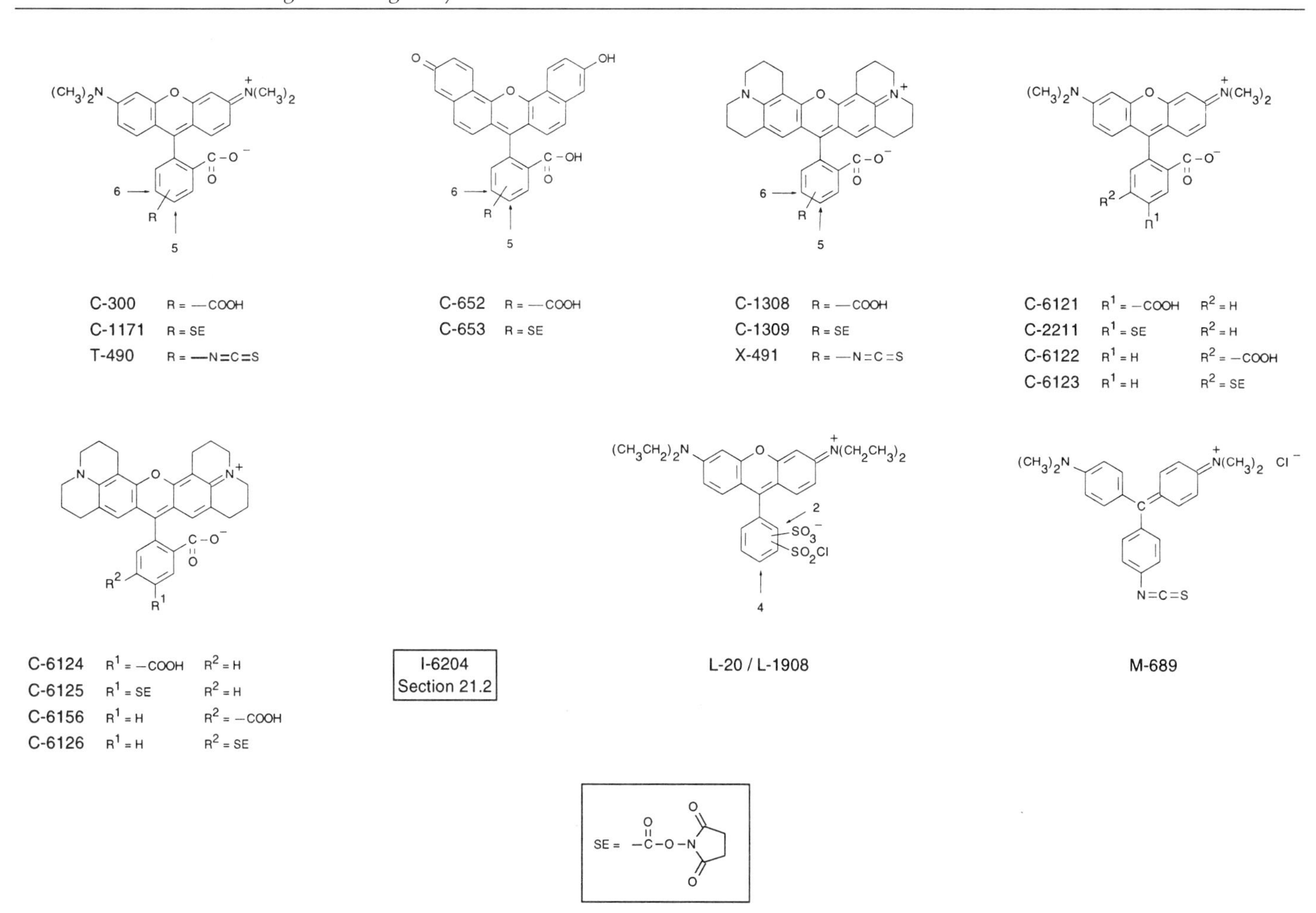

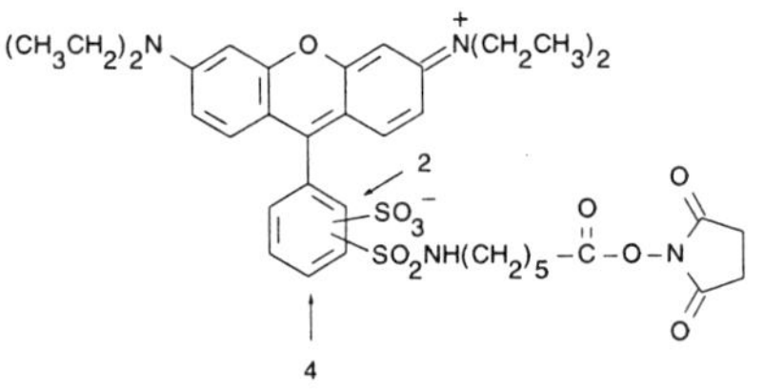

R-6160

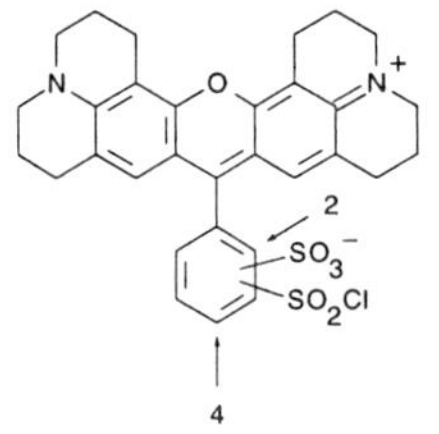

T-353 / T-1905

T-490 see C-300

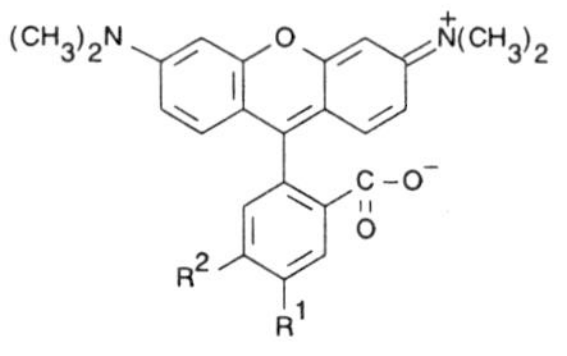

T-1480 R^1 = —N=C=S R^2 = H
T-1481 R^1 = H R^2 = —N=C=S

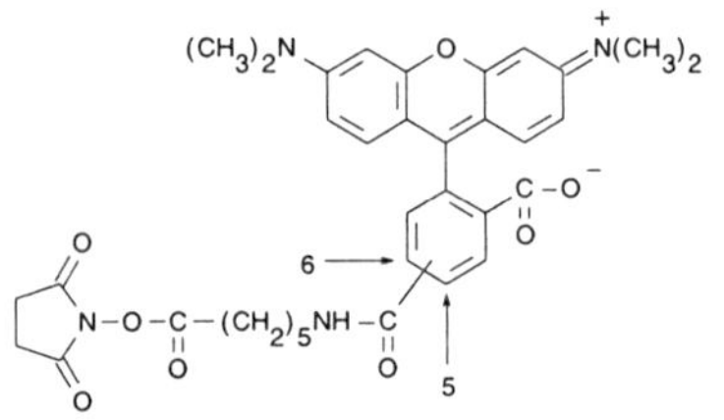

T-6105

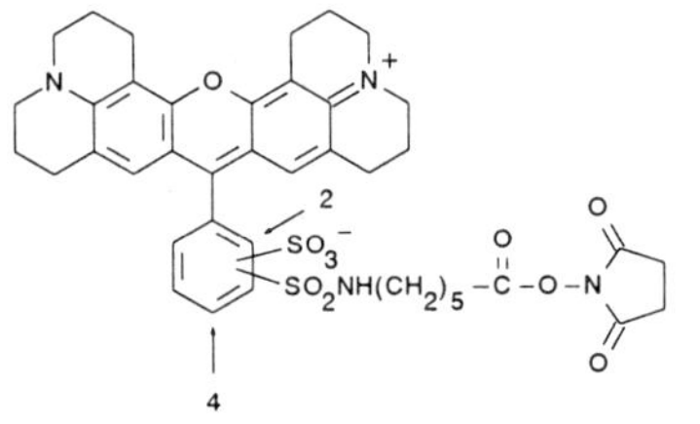

T-6134

X-491 see C-1308

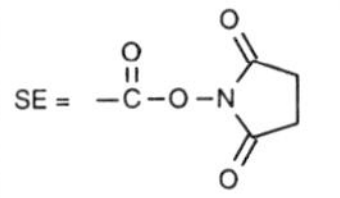

SE = —C(=O)—O—N (succinimidyl)

Substituent Key

1.6 Price List *Long-Wavelength Dyes*

Cat #		Product Name	Unit Size	Price Per Unit ($) 1–4 Units	5–24 Units
C-652		5-(and-6)-carboxynaphthofluorescein *mixed isomers*	100 mg	45.00	36.00
C-653		5-(and-6)-carboxynaphthofluorescein, succinimidyl ester *mixed isomers*	25 mg	48.00	38.40
C-6121	**New**	5-carboxytetramethylrhodamine (5-TAMRA) *single isomer*	10 mg	95.00	76.00
C-6122	**New**	6-carboxytetramethylrhodamine (6-TAMRA) *single isomer*	10 mg	95.00	76.00
C-300		5-(and-6)-carboxytetramethylrhodamine (5(6)-TAMRA) *mixed isomers*	100 mg	80.00	64.00
C-2211		5-carboxytetramethylrhodamine, succinimidyl ester (5-TAMRA, SE) *single isomer*	5 mg	148.00	118.40
C-6123	**New**	6-carboxytetramethylrhodamine, succinimidyl ester (6-TAMRA, SE) *single isomer*	5 mg	135.00	108.00
C-1171		5-(and-6)-carboxytetramethylrhodamine, succinimidyl ester (5(6)-TAMRA, SE) *mixed isomers*	25 mg	118.00	94.40
C-6124	**New**	5-carboxy-X-rhodamine (5-ROX) *single isomer*	10 mg	95.00	76.00
C-6156	**New**	6-carboxy-X-rhodamine (6-ROX) *single isomer*	10 mg	95.00	76.00
C-1308		5-(and-6)-carboxy-X-rhodamine (5(6)-ROX) *mixed isomers*	25 mg	60.00	48.00
C-6125	**New**	5-carboxy-X-rhodamine, succinimidyl ester (5-ROX, SE) *single isomer*	5 mg	135.00	108.00
C-6126	**New**	6-carboxy-X-rhodamine, succinimidyl ester (6-ROX, SE) *single isomer*	5 mg	135.00	108.00
C-1309		5-(and-6)-carboxy-X-rhodamine, succinimidyl ester (5(6)-ROX, SE) *mixed isomers*	25 mg	118.00	94.40
F-6089	**New**	FluoReporter® Rhodamine Red™-X Oligonucleotide Amine Labeling Kit *5 labelings*	1 kit	175.00	140.00
F-6161	**New**	FluoReporter® Rhodamine Red™-X Protein Labeling Kit *5-10 labelings*	1 kit	195.00	156.00
F-6163	**New**	FluoReporter® Tetramethylrhodamine Protein Labeling Kit *5-10 labelings*	1 kit	195.00	156.00
F-6090	**New**	FluoReporter® Tetramethylrhodamine (5-TAMRA) Oligonucleotide Amine Labeling Kit *5 labelings*	1 kit	175.00	140.00
F-6091	**New**	FluoReporter® Texas Red®-X Oligonucleotide Amine Labeling Kit *5 labelings*	1 kit	175.00	140.00
F-6162	**New**	FluoReporter® Texas Red®-X Protein Labeling Kit *5-10 labelings*	1 kit	195.00	156.00
I-6204	**New**	isosulfan blue sulfonyl chloride *mixed isomers*	5 mg	78.00	62.40
L-20		Lissamine™ rhodamine B sulfonyl chloride *mixed isomers*	1 g	135.00	108.00
L-1908		Lissamine™ rhodamine B sulfonyl chloride *mixed isomers* *special packaging*	10x10 mg	65.00	52.00
M-689		malachite green isothiocyanate	10 mg	135.00	108.00
R-6160	**New**	Rhodamine Red™-X, succinimidyl ester *mixed isomers*	5 mg	75.00	60.00
T-6105	**New**	6-(tetramethylrhodamine-5-(and-6)-carboxamido)hexanoic acid, succinimidyl ester (5(6)-TAMRA-X, SE) *mixed isomers*	10 mg	118.00	94.40

Cat #		Product Name	Unit Size	Price Per Unit ($) 1–4 Units	5–24 Units
T-1480		tetramethylrhodamine-5-isothiocyanate (5-TRITC; G isomer)	5 mg	75.00	60.00
T-1481		tetramethylrhodamine-6-isothiocyanate (6-TRITC; R isomer)	5 mg	75.00	60.00
T-490		tetramethylrhodamine-5-(and-6)-isothiocyanate (5(6)-TRITC) *mixed isomers*	10 mg	85.00	68.00
T-353		Texas Red® sulfonyl chloride *mixed isomers*	10 mg	98.00	78.40
T-1905		Texas Red® sulfonyl chloride *mixed isomers* *special packaging*	10x~1 mg	118.00	94.40
T-6134	**New**	Texas Red®-X, succinimidyl ester *mixed isomers*	5 mg	95.00	76.00
X-491		X-rhodamine-5-(and-6)-isothiocyanate (5(6)-XRITC) *mixed isomers*	10 mg	95.00	76.00

1.7 Fluorophores Excited with Ultraviolet Light

Shorter-wavelength amine-reactive fluorophores are infrequently used for preparing bioconjugates because dyes excited with longer wavelengths, and therefore lower energy, are widely available and less likely to cause photodamage to labeled biomolecules. Moreover, many cells and tissues autofluoresce when excited with UV light, thereby precluding the use of blue fluorescent conjugates in a number of applications. However, for certain multicolor fluorescence applications, including immunohistochemistry, *in situ* hybridization and neuronal tracing, a blue fluorescent probe provides a contrasting color that is easily distinguished from the green, yellow, orange or red fluorescence of the longer-wavelength probes. The short-wavelength reactive dyes that we recommend for preparing the brightest blue fluorescent bioconjugates are Cascade Blue and various amino- or hydroxycoumarin derivatives. The amine-reactive naphthalene and pyrene derivatives are important for the production of environmentally sensitive probes for protein structural studies; their thiol-reactive counterparts are discussed in Section 2.3. Many of our UV-excitable reactive dyes are more commonly employed for such bioanalytical techniques as HPLC derivatization, amino acid sequencing and protein determination and are therefore discussed below in Section 1.8.

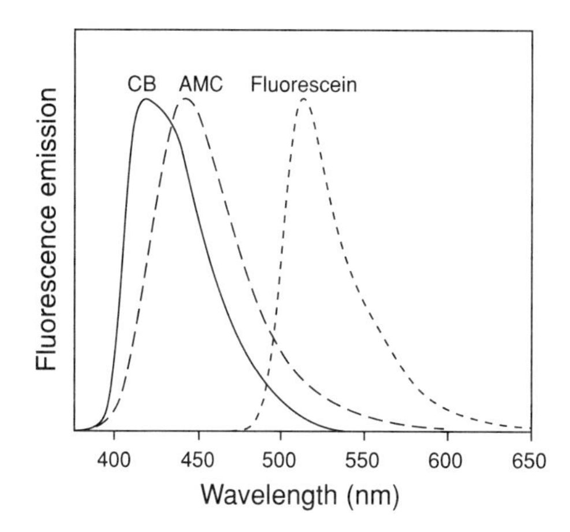

Figure 1.12 *Normalized fluorescence emission spectra of Cascade Blue (CB), aminomethylcoumarin (AMC) and fluorescein in aqueous solutions.*

Cascade Blue Dye

Cascade Blue acetyl azide is the amine-reactive derivative of the trademarked and patented sulfonated pyrene [1,2] that Molecular Probes uses to prepare its Cascade Blue–labeled proteins. This reagent, however, is difficult to purify. To meet requests for a reactive Cascade Blue derivative, we offer a Cascade Blue acetyl azide preparation (C-2284) that is ~60% pure and packaged according to the net weight of the reactive dye. The remaining impurities are inorganic salts or unreactive forms of the dye that can readily be removed following conjugation. The water-soluble Cascade Blue acetyl azide is the recommended reactive dye for preparing blue fluorescent bioconjugates. As compared to the aminocoumarin derivatives, the Cascade Blue fluorophore shows less spectral overlap with fluorescein (Figure 1.12), an important advantage for multicolor applications. In addition, this reactive Cascade Blue derivative has high absorptivity, is highly fluorescent and, unlike most dyes, resists quenching upon protein conjugation (see Figure 7.3 in Chapter 7). Even at low degrees of substitution, Cascade Blue conjugates are significantly more fluorescent than the more familiar 7-amino-4-methylcoumarin-3-acetic acid (AMCA) conjugates.[1] Although a standard DAPI/Hoechst optical filter set can be used with Cascade Blue conjugates, the fluorescence will be brighter when viewed through optical filters optimized for Cascade Blue's spectral properties (O-5703, O-5707, O-5708; see Section 26.5).

In addition to its use for preparing protein conjugates, the membrane-impermeant Cascade Blue acetyl azide may be useful for identifying proteins located on extracellular cell surfaces (see Section 9.4). Molecular Probes offers fixable Cascade Blue derivatives for use as polar tracers (see Section 15.3) and Cascade Blue conjugates of a variety of proteins, nucleotides and dextrans, as well as antibodies directed against the Cascade Blue fluorophore (see Section 7.3).

Cascade Blue acetyl azide and its conjugates are covered by U.S. Patent No. 5,132,432 and are offered for research purposes only. Molecular Probes welcomes inquiries about licensing these products for resale or other commercial uses.

Coumarin Derivatives

Aminocoumarin Derivatives

Although Cascade Blue conjugates generally have a higher absorptivity and fluorescence quantum yield than the corresponding aminocoumarin conjugates, the longer emission wavelength of the aminocoumarin conjugates makes them appear visibly brighter when viewed through a microscope because of the limited spectral sensitivity of the human eye to Cascade Blue's shorter-wavelength fluorescence. Derivatives of various 7-aminocoumarins are the most extensively utilized labeling reagents for preparing these conjugates.[3] For preparing brightly blue fluorescent conjugates of proteins and nucleic acids, we offer AMCA-X succinimidyl ester (AMCA-X, SE; A-6118), which contains a seven-atom aminohexanoyl spacer ("X") between the fluorophore and the reactive group. Slightly longer-wavelength conjugates can be prepared from the succinimidyl esters (D-374, D-1412) or free acids (D-126, D-1421) of 7-dialkylaminocoumarins.[4] Also quite fluorescent and more strongly absorbing is the coumarin isothiocyanate, CPI [5] (D-347).

Some of our newest conjugates of 7-aminocoumarin are prepared from a sulfonated coumarin derivative that we call AMCA-S (see Chapter 7). Protein conjugates of AMCA-S have bright blue fluorescence emission at wavelengths slightly shorter than AMCA-X conjugates (442 nm versus 448 nm), which reduces the dye's spectral overlap with that of fluorescein. Conjugates can be prepared using the new succinimidyl ester of AMCA-S (AMCA-S, SE; A-6120), which is more water soluble than other reactive AMCA derivatives.

Hydroxy- and Alkoxycoumarin Derivatives

The hydroxycoumarins (H-185, H-1193, H-1411, H-1428) exhibit pH-sensitive spectral properties, whereas the methoxycoumarins (M-1410, M-1420) are not sensitive to pH.[6,7] Hydroxycoumarins are often used to prepare reactive intermediates for the synthesis of radioiodinated materials.[7,8] The spectral properties of the hydroxycoumarins allow their quantitation prior to iodination.[8] Most of these coumarins are also very useful as intermediates for derivatizing amines and synthesizing new probes.

Naphthalenes, Including Dansyl Chloride

The naphthalene-based probes tend to have emission spectra that are sensitive to the environment. Environmentally sensitive probes respond to perturbations in the local environment by undergoing spectral changes. For example, changes in solvation that occur because of ligand binding, protein assembly or protein denaturation can often evoke changes in the fluorescence properties of these probes. This property has made dansyl chloride (5-dimethylaminonaphthalene-1-sulfonyl chloride, D-21: see Section 1.8) and other naphthalene dyes important tools for protein structural studies.

Dansyl chloride is nonfluorescent until it reacts with amines. The resulting dansyl amides have environmentally sensitive fluorescence quantum yields and emission maxima along with large Stokes shifts. Despite the weak absorptivity (ϵ ~4000 $cm^{-1}M^{-1}$ at 330–340 nm) and moderate fluorescence quantum yield of dansyl sulfonamides, dansyl chloride is also widely used as a derivatization reagent for end-group analysis of proteins, amino acid analysis and HPLC detection (see Section 1.8). The succinimidyl ester of dansylaminohexanoic acid (dansyl-X, SE; D-6104) contains a seven-atom spacer ("X") that places the dansyl fluorophore further from its reaction site, potentially reducing the interaction of the fluorophore with the biomolecule to which it is conjugated and making it more accessible to antibody binding.[10-12] A polyclonal antibody to the 1,5-dansyl fluorophore that significantly enhances the dye's fluorescence is described in Section 7.3.

Protein conjugates of the two isomers of dansyl chloride (2,5-dansyl chloride, D-22; 2,6-dansyl chloride, D-23) have smaller Stokes shifts and appreciably longer fluorescence lifetimes (up to ~30 nanoseconds) than those of 1,5-dansyl chloride, making these isomers among the best available probes for fluorescence depolarization studies.[13] Conjugates of mansyl chloride (M-405) and TNS chloride (T-406) are virtually nonfluorescent unless bound to hydrophobic sites.[14] These naphthalene derivatives may be particularly useful for preparing fluorescent drug or ligand analogs that are expected to bind to hydrophobic sites in proteins or membranes. The lipophilicity of these reagents may also facilitate the labeling of sites within the membrane-spanning portions of cellular proteins. Other reactive naphthalene derivatives include 4-((5-dimethylaminonaphthalene-1-sulfonyl)amino)phenylisothiocyanate (D-1302, see Section 1.8), which is used as a fluorescent Edman degradation reagent,[15,16] and 4-dimethylaminonaphthyl-1-isothiocyanate (D-427, see Section 1.8).

Pyrenes

Conjugates of the pyrene succinimidyl esters (P-130, P-6114, P-6115) have exceptionally long fluorescence lifetimes (sometimes >100 nanoseconds) and relatively short-wavelength spectra. They have primarily been used for labeling polynucleotides [17-19] and for studying protein conformation;[20,21] pyrene derivatives have also been utilized as oxygen sensors.[22,23] The long fluorescence lifetime of pyrenebutyric acid permits time-gating of the fluorescence, which is a useful technique for discriminating between the dye signal and sample autofluorescence [24] and has been exploited for fluorescence immunoassays.[25] For preparing pyrene conjugates with long fluorescence lifetimes, we now recommend our new, more water-soluble succinimidyl ester of *N*-(1-pyrenebutanoyl)-cysteic acid (P-6114). The amine-reactive 1-pyreneisothiocyanate (P-331) and 1-pyrenesulfonyl chloride (P-24) are described in Section 1.8.

Pyrene dyes frequently form excited-state dimers (excimers) with fluorescence emission at longer wavelengths than the monomers (see Figure 13.6 in Chapter 13). This property has been used to detect protein interactions (see Section 9.4), nucleic acid hybridization,[26] phospholipase activity (see Section 20.5) and transition metal ions.[27]

Pyridyloxazole Derivatives

The new pyridyloxazole derivatives, including the succinimidyl ester (PyMPO, SE; S-6110) and isothiocyanate (PyMPO-ITC, I-6076) as well as the thiol-reactive maleimide and epoxide (M-6026, E-6051; see Section 2.2), fill the spectral gap between UV-excited dyes and the fluoresceins. These derivatives of the laser dye PyMPO exhibit absorption maxima near 415 nm and unusually high Stokes shifts with emission at 560–580 nm.[28] Like the naphthalene dyes, the oxazole dyes exhibit environment-sensitive fluorescence spectra.

1. Anal Biochem 198, 119 (1991); **2.** U.S. Patent No. 5,132,432; **3.** Histochem J 18, 497 (1986); **4.** Biochemistry 27, 8889 (1988); **5.** Biochemistry 29, 10734 (1990); **6.** Anal Chem 40, 803 (1968); **7.** J Chem Soc S 12 (1949); **8.** FEBS Lett 182, 185 (1985); **9.** J Immunol Meth 81, 123 (1985); **10.** Biochim Biophys Acta 1104, 9 (1992); **11.** Biochim Biophys Acta 776, 217 (1984); **12.** Biochemistry 21, 978 (1982); **13.** J Biochem Biophys Meth 5, 1 (1981); **14.** Biochemistry 29, 5856 (1990); **15.** Biol Chem Hoppe-Seyler 367, 1259 (1986); **16.** FEBS Lett 198, 150 (1986); **17.** Nucleic Acids Res 21, 5085 (1993); **18.** EMBO J 11, 3777 (1992); **19.** Anal Biochem 183, 231 (1989); **20.** Biochemistry 30, 4298 (1991); **21.** Biochemistry 29, 3082 (1990); **22.** Anal Chem 59, 279 (1987); **23.** Mikrochim Acta 1, 153 (1984); **24.** J Biochem Biophys Meth 29, 157 (1994); **25.** Anal Biochem 174, 101 (1988); **26.** Photochem Photobiol 62, 836 (1995); **27.** Tetrahedron Lett 36, 1451 (1995); **28.** IEEE J Quantum Electronics 16, 777 (1980).

1.7 Data Table *Fluorophores Excited with Ultraviolet Light*

Cat #	Structure	MW	Storage	Soluble	Abs	{ε × 10^{-3}}.	Em	Solvent	Notes
A-6118	✓	443	F,D,L	DMF, DMSO	353	{19}	442	MeOH	
A-6120	✓	410	F,DD,L	DMSO, H_2O*	353	{19}	437	MeOH	
C-2284	✓	607	F,D,LL	H_2O†, MeOH	396	{29}	410	MeOH	<1>
D-22	✓	270	F,DD,L	DMF, MeCN‡	403	{2.9}	none	MeOH	<2>
D-23	✓	270	F,DD,L	DMF, MeCN‡	380	{16}	none	$CHCl_3$	<2>
D-126	✓	247	L	pH>6, DMF	370	{22}	459	MeOH	
D-347	✓	364	F,DD,L	DMF, MeCN	388	{32}	475	MeOH	<3>
D-374	✓	344	F,D,L	DMF, MeCN	376	{22}	468	MeOH	
D-1412	✓	358	F,D,L	DMF, MeCN	432	{56}	472	MeOH	
D-1421	✓	261	L	pH>6, DMF	409	{33}	473	pH 9	
D-6104	✓	462	F,DD,L	DMF, MeCN	335	{4.2}	518	MeOH	
H-185	✓	206	L	pH>6, DMF	386	{29}	448	pH 10	<4>
H-1193	✓	303	F,D,L	DMF, MeCN	419	{36}	447	MeOH	
H-1411	✓	331	F,D,L	DMF, MeCN	328	{15}	386	$CHCl_3$	
H-1428	✓	234	L	pH>6, DMF	360	{19}	455	pH 10	
I-6076	✓	480	F,DD,L	DMF, DMSO	416	{27}	571	MeOH	<3, 5>
M-405	✓	332	F,DD,L	DMF, MeCN‡	373	{19}	none	MeOH	<6>
M-1410	✓	317	F,D,L	DMF, MeCN	358	{26}	410	MeOH	
M-1420	✓	220	L	pH>6, DMF	336	{20}	402	pH 9	
P-130	✓	385	F,D,L	DMF, DMSO	340	{43}	376	MeOH	<7>
P-6114	✓	559	F,DD,L	H_2O*, DMSO	341	{38}	376	MeOH	<7>
P-6115	✓	357	F,D,L	DMF, DMSO	340	{44}	376	MeOH	<7>
S-6110	✓	582	F,D,L	DMF, DMSO	415	{26}	570	MeOH	<5>
T-406	✓	332	F,DD,L	DMF, MeCN‡	381	{18}	none	MeOH	<6>

For definitions of the contents of this data table, see "How to Use the Handbook" on page vi.
Structure: Chemical structure drawing: ✓ = shown in this section.

* This succinimidyl ester derivative is water soluble and may be dissolved in buffer at ~pH 8 for reaction with amines. Long-term storage in water is NOT recommended due to hydrolysis.
† Unstable in water. Use immediately.
‡ Do NOT dissolve in DMSO.
<1> Cascade Blue dyes have a second absorption peak at about 376 nm with ε ~80% of the 395–400 nm peak.
<2> D-22 butylamine conjugate: Abs = 375 nm (ε = {3.1}), Em = 470 nm in MeOH. D-23 butylamine conjugate: Abs = 375 nm (ε = {13}), Em = 419 nm in $CHCl_3$.
<3> Isothiocyanates are unstable in water and should not be stored in aqueous solution.
<4> Abs = 339 nm (ε = {19}), Em = 448 nm at pH 4.
<5> Spectra shift to shorter wavelengths in nonpolar solvents. Fluorescence quantum yields of propylamine adducts in water are ~0.25.
<6> M-405 butylamine conjugate: Abs = 324 nm (ε = {21}), Em = 480 nm in MeOH. T-406 butylamine conjugate: Abs = 327 nm (ε = {25}), Em = 450 nm in MeOH.
<7> Pyrene derivatives exhibit structured spectra. The absorption maximum is usually about 340 nm with a subsidiary peak at about 325 nm. There are also strong absorption peaks below 300 nm. The emission maximum is usually about 376 nm with a subsidiary peak at 396 nm. Excimer emission at about 470 nm may be observed at high concentrations.

1.7 Structures *Fluorophores Excited with Ultraviolet Light*

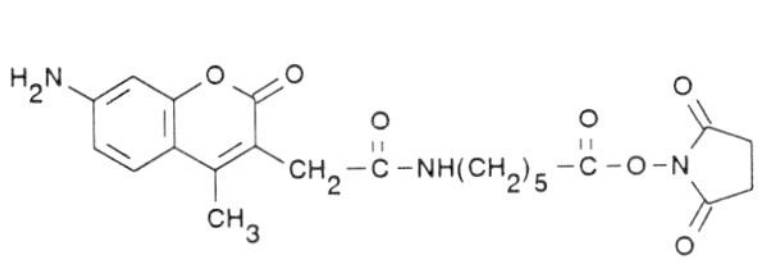

A-6118

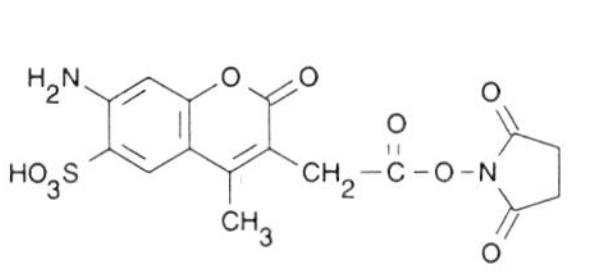

A-6120

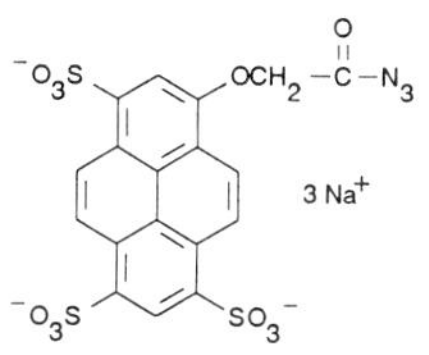

C-2284

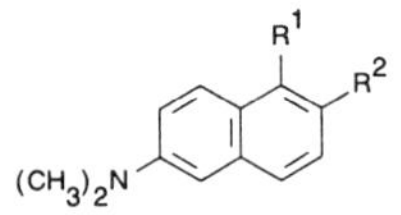

D-22 R^1 = $-SO_2Cl$ R^2 = H
D-23 R^1 = H R^2 = $-SO_2Cl$

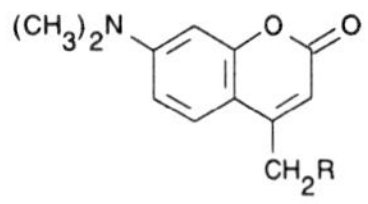

D-126 R = —COOH
D-374 R = SE

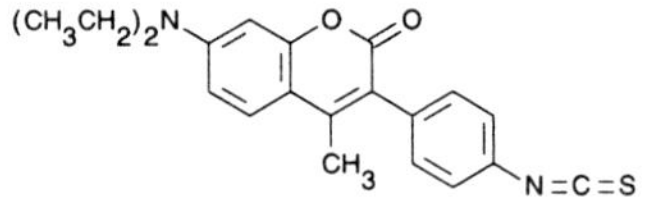

D-347

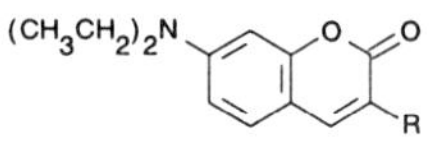

D-1421 R = —COOH
D-1412 R = SE

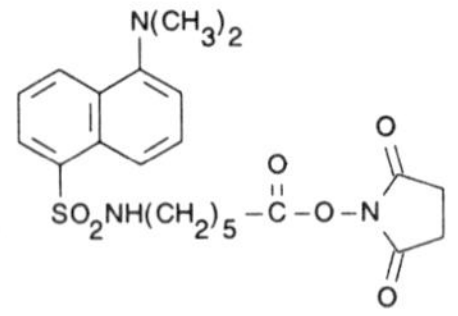

D-6104

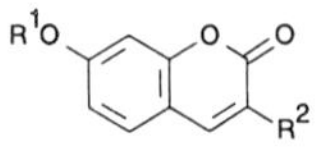

H-185 R^1 = H R^2 = —COOH
H-1193 R^1 = H R^2 = SE
M-1420 R^1 = $-CH_3$ R^2 = —COOH
M-1410 R^1 = $-CH_3$ R^2 = SE

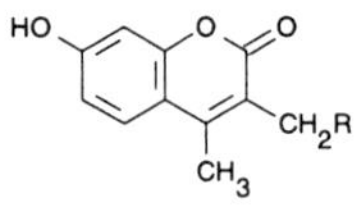

H-1428 R = —COOH
H-1411 R = SE

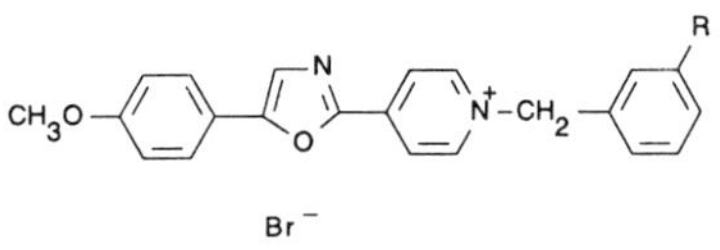

I-6076 R = —N=C=S
S-6110 R = SE

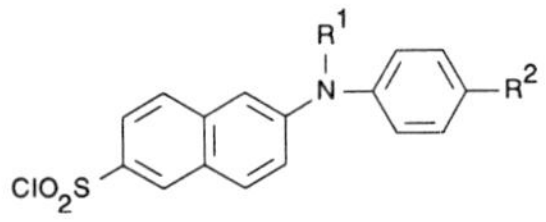

M-405 R^1 = $-CH_3$ R^2 = H
T-406 R^1 = H R^2 = $-CH_3$

M-1410, M-1420 see H-185

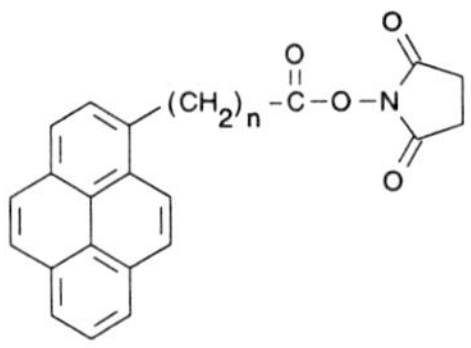

P-130 n = 3
P-6115 n = 1

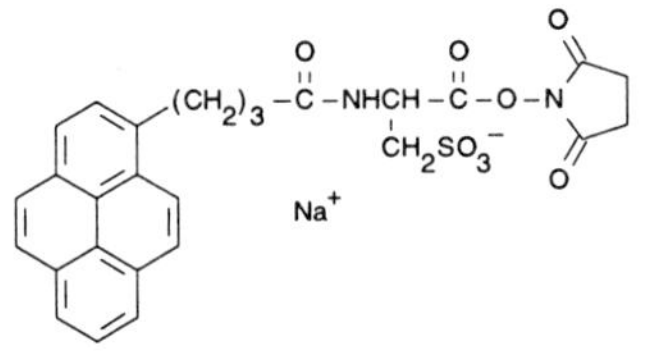

P-6114

S-6110 see I-6076

T-406 see M-405

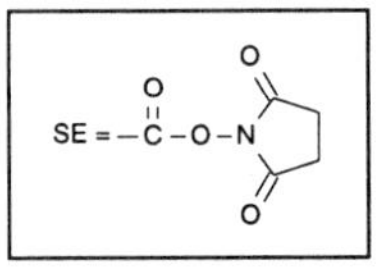

Substituent Key

1.7 Price List *Fluorophores Excited with Ultraviolet Light*

Cat #		Product Name	Unit Size	Price Per Unit ($) 1–4 Units	5–24 Units
A-6118	**New**	6-((7-amino-4-methylcoumarin-3-acetyl)amino)hexanoic acid, succinimidyl ester (AMCA-X, SE)	10 mg	98.00	78.40
A-6120	**New**	7-amino-3-((((succinimidyl)oxy)carbonyl)methyl)-4-methylcoumarin-6-sulfonic acid (AMCA-S, SE)	5 mg	95.00	76.00
C-2284		Cascade Blue® acetyl azide, trisodium salt	5 mg	125.00	100.00
D-1421		7-diethylaminocoumarin-3-carboxylic acid	100 mg	38.00	30.40
D-1412		7-diethylaminocoumarin-3-carboxylic acid, succinimidyl ester	25 mg	75.00	60.00
D-347		7-diethylamino-3-(4′-isothiocyanatophenyl)-4-methylcoumarin (CPI)	25 mg	98.00	78.40
D-126		7-dimethylaminocoumarin-4-acetic acid (DMACA)	100 mg	88.00	70.40
D-374		7-dimethylaminocoumarin-4-acetic acid, succinimidyl ester (DMACA, SE)	25 mg	115.00	92.00
D-6104	**New**	6-((5-dimethylaminonaphthalene-1-sulfonyl)amino)hexanoic acid, succinimidyl ester (dansyl-X, SE)	25 mg	48.00	38.40
D-22		2-dimethylaminonaphthalene-5-sulfonyl chloride	100 mg	125.00	100.00
D-23		2-dimethylaminonaphthalene-6-sulfonyl chloride	100 mg	125.00	100.00
H-185		7-hydroxycoumarin-3-carboxylic acid *reference standard*	100 mg	95.00	76.00
H-1193		7-hydroxycoumarin-3-carboxylic acid, succinimidyl ester	25 mg	78.00	62.40
H-1428		7-hydroxy-4-methylcoumarin-3-acetic acid	100 mg	60.00	48.00
H-1411		7-hydroxy-4-methylcoumarin-3-acetic acid, succinimidyl ester	25 mg	78.00	62.40
I-6076	**New**	1-(3-isothiocyanatophenyl)-4-(5-(4-methoxyphenyl)oxazol-2-yl)pyridinium bromide (PyMPO-ITC)	5 mg	98.00	78.40
M-1420		7-methoxycoumarin-3-carboxylic acid	100 mg	48.00	38.40

Cat #		Product Name	Unit Size	Price Per Unit ($) 1–4 Units	5–24 Units
M-1410		7-methoxycoumarin-3-carboxylic acid, succinimidyl ester	25 mg	65.00	52.00
M-405		6-(N-methylanilino)naphthalene-2-sulfonyl chloride (mansyl chloride)	25 mg	135.00	108.00
P-6115	New	1-pyreneacetic acid, succinimidyl ester	25 mg	48.00	38.40
P-130		1-pyrenebutanoic acid, succinimidyl ester	100 mg	115.00	92.00
P-6114	New	N-(1-pyrenebutanoyl)cysteic acid, succinimidyl ester, sodium salt	5 mg	75.00	60.00
S-6110	New	1-(3-(succinimidyloxycarbonyl)benzyl)-4-(5-(4-methoxyphenyl)oxazol-2-yl)pyridinium bromide (PyMPO, SE)	5 mg	98.00	78.40
T-406		6-(p-toluidinyl)naphthalene-2-sulfonyl chloride (TNS chloride)	25 mg	135.00	108.00

1.8 Reagents for Analysis of Low Molecular Weight Amines

Not only are low molecular weight amines abundantly distributed in nature, but numerous drugs, synthetic probes and other molecules of interest contain amino groups. The sensitive detection, identification and quantitation of amines is an important application of many of the reactive fluorophores in this section. Some of these reagents have also been used to indirectly detect other analytes such as carbohydrates, carboxylic acids, cyanide and thiols.

It is usually difficult to compare the sensitivity for amine detection of the different reagents because it depends heavily on the equipment and detection technology used. However, many of the assays are rapid, reliable and adaptable to a variety of different sample types and instrumentation.

Fluorescamine

Fluorescamine (F-2332) is intrinsically nonfluorescent but reacts in milliseconds with primary aliphatic amines, including peptides and proteins, to yield a fluorescent derivative (Figure 1.13A). Modifications to the reaction protocol permit fluorescamine to be used to detect those amino acids containing secondary amines,[1] such as proline. Excess reagent is rapidly converted to a nonfluorescent product by reaction with water,[2-5] making fluorescamine useful for determining protein concentrations of solutions.[6,7] Fluorescamine can also be used to detect proteins in gels and to analyze low molecular weight amines by TLC, HPLC and capillary electrophoresis.[8] For chemiluminescence detection methods, fluorescamine adducts can be treated with bis-trichlorophenyl oxalate[9] (TCPO, B-1174).

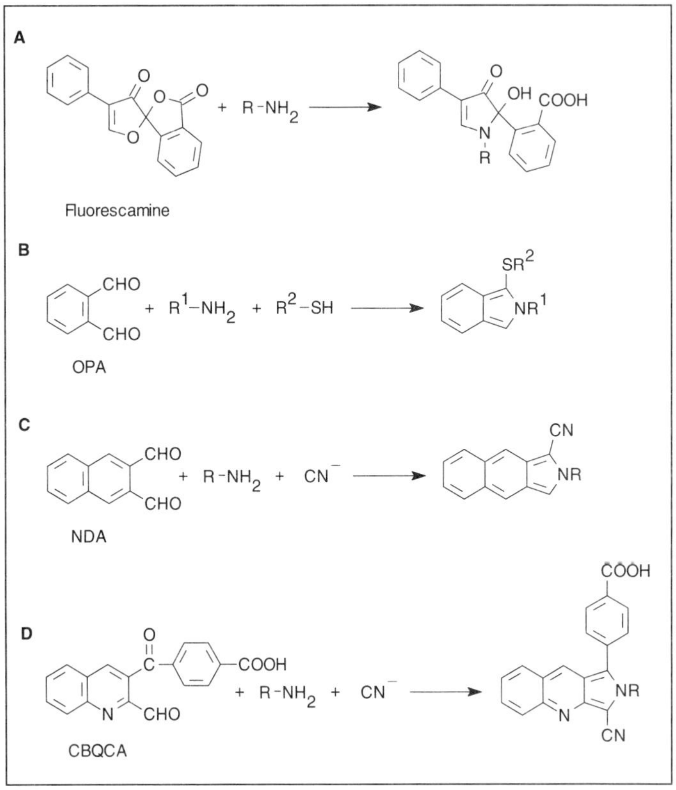

A

Fluorescamine + R-NH$_2$ →

B

OPA + R^1–NH$_2$ + R^2–SH →

C

NDA + R-NH$_2$ + CN$^-$ →

D

CBQCA + R-NH$_2$ + CN$^-$ →

Figure 1.13 Fluorogenic amine-derivatization reactions of A) fluorescamine (F-2332), B) o-phthaldialdehyde (OPA, P-2331), C) naphthalene-2,3-dicarboxaldehyde (NDA, N-1138) and D) 3-(4-carboxybenzoyl)-quinoline-2-carboxaldehyde (CBQCA, A-6222).

Dialdehydes: OPA, NDA and ADA

Analyte Detection with OPA, NDA and ADA

The homologous aromatic dialdehydes, *o*-phthaldialdehyde[10] (OPA, P-2331), naphthalene-2,3-dicarboxaldehyde[11] (NDA, N-1138) and anthracene-2,3-dicarboxaldehyde (ADA, A-1139) are essentially nonfluorescent until reacted with a primary amine in the presence of excess cyanide or a thiol, such as 2-mercaptoethanol, 3-mercaptopropionic acid or the less obnoxious sulfite,[12] to yield a fluorescent isoindole (Figures 1.13B and 1.13C). Modified protocols that use an excess of an amine and limiting amounts of other nucleophiles permit the determination of carboxylic acids[13] and thiols,[14] as well as of cyanide in blood, urine and other samples.[15-18] Without an additional nucleophile, NDA forms fluorescent adducts with both hydrazine and methylated hydrazines (excitation/emission maxima ~403/500 nm), whereas ADA selectively detects submicrogram levels of hydrazine (excitation/emission maxima ~476/549 nm) but does not form fluorescent products upon reaction with methylated hydrazines.[19]

Sensitivity of NDA and ADA

Amine adducts of NDA and ADA have longer-wavelength spectral characteristics and greater sensitivity than the amine adducts of OPA. The stability and detectability of the amine derivatives of NDA and ADA are also superior;[20,21] the detection of glycine with NDA and cyanide is reported to be 50-fold more sensitive than with OPA and β-mercaptoethanol.[11] The limit for electrochemical detection of the NDA adduct of asparagine has been determined to be as low as 36 attomoles [22,23] (36×10^{-18} moles). The NDA adduct of a geminal bisphosphonate reportedly has a fluorescence quantum yield of 0.82.[24] Although the fluorescent adducts of NDA and ADA can be detected directly, the use of the chemiluminescent reagent bis-(2,4,6-trichlorophenyl) oxalate (TCPO, B-1174) significantly enhances the sensitivity of their detection.[25-27]

Applications for OPA, NDA and ADA

OPA and NDA are widely used for both pre- and postcolumn derivatization of amines (and thiols) separated by HPLC [28] or by capillary electrophoresis.[29-33] Recently, the amines in a single cell have been analyzed by capillary electrophoresis using a sequence of on-capillary lysis, derivatization with NDA and cyanide and laser-excited detection.[34,35] Derivatives of NDA and ADA can be excited by the He-Cd laser.[29] An extensive bibliography is available for both OPA and NDA from our Technical Assistance Department or through our Web site (http://www.probes.com).

ATTO-TAG Reagents

Sensitivity of ATTO-TAG CBQCA and ATTO-TAG FQ

Molecular Probes exclusively offers ATTO-TAG CBQCA (A-6222, A-2333) and ATTO-TAG FQ (A-2334) for ultrasensitive detection of primary amines, including those in peptides and glycoproteins.[36] These reagents combine high sensitivity, visible-wavelength excitation and freedom from background fluorescence, making them useful for research, analytical and forensic applications. Developed by Novotny and collaborators, the ATTO-TAG reagents are similar to OPA and NDA in that they react with amines in the presence of cyanide or thiols to form highly fluorescent isoindoles [37-44] (Figure 1.13D).

ATTO-TAG CBQCA reacts specifically with amines to form charged conjugates that can be analyzed by electrophoresis techniques. Using ATTO-TAG CBQCA, researchers analyzed the amines in human cerebrospinal fluid by capillary electrophoresis.[37] Carbohydrates, including polysaccharides, that lack amines can be detected following reductive amination with ammonia and $NaCNBH_3$.[43] ATTO-TAG CBQCA conjugates are maximally excited at ~456 nm or by the 442 nm spectral line of the He-Cd laser, with peak emission at ~550 nm, whereas ATTO-TAG FQ conjugates are maximally excited at ~480 nm or by the 488 nm spectral line of the argon-ion laser, with peak emission at ~590 nm. In capillary zone electrophoresis, the sensitivity of detection of the laser-induced fluorescence should be in the subattomole range ($<10^{-18}$ moles) for ATTO-TAG CBQCA and subfemtomole range ($<10^{-15}$ moles) for ATTO-TAG FQ. Sensitivity for detection of reductively aminated glucose using ATTO-TAG CBQCA is reported to be 75 zeptomoles [45] (75×10^{-21} moles). ATTO-TAG reagents can, of course, be used in HPLC and other modes of chromatography with either absorption or fluorescence detection. Moreover, it may be possible to detect amine adducts of ATTO-TAG reagents by enhancement with the chemiluminescent reagent bis-(2,4,6-trichlorophenyl) oxalate [25-27] (TCPO, B-1174). A very sensitive assay that uses ATTO-TAG CBQCA for rapid quantitation on protein amines (C-6667) is described in Section 9.1.

ATTO-TAG Reagents and Kits

Cyclodextrins have been reported to amplify the signal from ATTO-TAG CBQCA conjugates up to 10-fold [44,46] so we include β-cyclodextrin in our ATTO-TAG Amine Derivatization Kits (A-2333, A-2334). The kits include:

- 5 mg ATTO-TAG CBQCA *or* ATTO-TAG FQ
- Potassium cyanide
- β-Cyclodextrin
- Protocol for amine modification

The ATTO-TAG CBQCA and FQ Amine Derivatization Kits contain sufficient reagents for derivatizing approximately 150 and 100 samples, respectively, depending on the amine concentration and sample volume. For your convenience, we also offer the ATTO-TAG CBQCA derivatization reagent separately (CBQCA, A-6222).

7-Nitrobenz-2-Oxa-1,3-Diazole (NBD) Derivatives

NBD chloride (C-10) was first introduced in 1968 as a fluorogenic derivatization reagent for amines.[47] It also reacts with thiols and alcohols, although these adducts absorb and emit at shorter wavelengths and are less fluorescent than amine derivatives.[48] NBD fluoride (F-486) usually yields the same products as NBD chloride but is much more reactive;[49] for example, the reaction of NBD fluoride with glycine is reported to be 500 times faster than the reaction of NBD chloride with glycine.[50] Unlike OPA and fluorescamine, both NBD chloride and NBD fluoride react with secondary amines and are therefore capable of derivatizing proline and hydroxyproline.[49,51] NBD chloride and NBD fluoride are extensively used as derivatization reagents for chromatographic analysis of amino acids [52] and other low molecular weight amines.[53] NBD fluoride has been used for the enantiomeric separation of D,L-amino acids on a chiral column.[54]

The absorption and fluorescence emission spectra, quantum yields and extinction coefficients of NBD conjugates are all markedly dependent on solvent;[55,56] in particular, the fluorescence quantum yield in water of NBD-amine adducts is very low (<0.01). Fluorescence of lysine-modified NBD-actin is sensitive to polymerization.[57] Inactivation of ATPases by NBD chloride apparently involves a tyrosine modification followed by intramolecular migration of the label to a lysine residue.[58,59] NBD is also a functional analog of the dinitrophenyl hapten and is quenched upon binding to anti-dinitrophenyl antibodies [56] (see Section 7.3). NBD aminohexanoic acid (NBD-X, N-316) and its succinimidyl ester (NBD-X, SE; S-1167) are precursors to NBD-labeled phospholipids (see Sections 13.2 and 13.4), NBD C_6-ceramide (N-1154, see Section 12.4) and other probes.

Dansyl Chloride and Other Sulfonyl Chlorides

Many of the blue and blue-green fluorescent probes already described in Section 1.7, including dansyl chloride, are particularly useful as chromatographic derivatization reagents. In addition, they are generally good acceptors for fluorescence resonance energy transfer from tryptophan, as well as good donors to longer-wavelength dyes such as dabsyl chloride (D-1537).

Dansyl Chloride

Since its development by Weber in 1951,[60] dansyl chloride (D-21) has been used extensively to determine the N-terminal amino acid residue of proteins and to prepare fluorescent derivatives of drugs, amino acids, oligonucleotides and proteins for detection by numerous chromatographic methods.[61] Dansyl chloride, which is nonfluorescent, reacts with amines to form fluorescent dansyl amides that exhibit large Stokes shifts along with environmentally sensitive fluorescence quantum yields and emission maxima.

Pyrene and Anthracene Sulfonyl Chlorides

The absorptivity (and therefore ultimate fluorescence output) of dansyl derivatives is weak compared with that of the more strongly UV light–absorbing fluorophores such as pyrene. Thus, 1-pyrenesulfonyl chloride (P-24) should have greater sensitivity for detection of amines. The fluorescence lifetime of pyrenesulfonamides can also be relatively long (up to ~30 nanoseconds), making them useful for fluorescence anisotropy measurements.[62] The reagent 2-anthracenesulfonyl chloride (A-448) is a very good amine-derivatizing reagent, though underutilized in this application. Reactive anthracenes are especially useful as HPLC derivatization reagents because they exhibit a strong UV absorption with an extinction coefficient typically >50,000 $cm^{-1}M^{-1}$ near 250 nm.

Chromophoric Sulfonyl Chloride

Dabsyl chloride (D-1537) is a common amine-derivatization reagent for detecting proteins by HPLC,[63,64] as well as by gel and capillary electrophoresis.[65-67] Conjugates of dabsyl chloride have broad and intense visible absorption, making them useful as acceptors in fluorescence resonance energy transfer applications.

FITC and Other Isothiocyanates

FITC

Isothiocyanates for preparing bioconjugates have been described in several sections of this chapter. However, FITC (F-143, F-144, F-1906, F-1907, F-1919; see Section 1.3) and some other isothiocyanates can also be used for derivatizing low molecular weight amines and, like phenyl isothiocyanate, for microsequencing of peptides as their thiohydantoins. A unique method for specific derivatization of the N-terminus of peptides by FITC has recently been described.[68] FITC-labeled amino acids and peptides have been separated by capillary electrophoresis with a detection limit of fewer than 1000 molecules.[69,70]

Other Fluorescent and Chemiluminescent Isothiocyanates

The dansyl isothiocyanate (D-1302) is a fluorescent Edman degradation reagent.[71,72] However, other isothiocyanates such as the acridine (A-417), anthracene (A-409), benzofuran (D-1332), coumarin (D-347; see Section 1.7), naphthalene (D-427), pyrene (P-331) and pyridyloxazole (I-6076, see Section 1.7) generally have higher detectability and may also be useful in Edman degradation or amine labeling. The coumarin[73] and benzofuran derivatives probably have the most intense blue fluorescence, whereas the pyridyloxazole derivative has a huge Stokes shift of >150 nm.

Isoluminol isothiocyanate (ILITC, I-1195) confers chemiluminescent properties to amines and proteins and serves as a precolumn HPLC derivatization reagent for amino acids.[74]

Succinimidyl Esters and Carboxylic Acids

Succinimidyl esters have a high selectivity for reaction with aliphatic amines. Most of the succinimidyl reagents described elsewhere in this chapter can be used to derivatize low molecular weight amines for separation by chromatography or capillary electrophoresis. The BODIPY and fluorescein derivatives usually yield the greatest sensitivity, particularly when the conjugate is detected following laser excitation. Analysis by capillary electrophoresis shows that carboxyfluorescein succinimidyl ester reacts faster and yields more stable amine conjugates than FITC and DTAF.[75] The long-wavelength succinimidyl esters of carboxyfluorescein (FAM), carboxytetramethylrhodamine (TAMRA) and carboxy-X-rhodamine (ROX) are widely used for preparing probes for oligonucleotide labeling, fluorescence *in situ* hybridization and DNA sequencing.[76-78] BODIPY dyes (see Section 1.2) have significant potential for DNA sequencing.[77,79] The new succinimidyl esters of Texas Red-X and Rhodamine Red-X (see Section 1.6) are likely to have similar utility. The UV-excitable coumarins described in Section 1.7 have good absorptivity at ~320–380 nm with purple to bright blue emission at 400–500 nm. Other succinimidyl esters and acids that have good absorptivity, particularly in the UV, include the acridine (9-AC, SE; A-1127), anthracene (A-176, S-1144) and pyrene (P-164) derivatives.

Succinimidyl 4-*O*-(4,4′-dimethoxytrityl)oxybutyrate (S-2182) has been employed to estimate the degree of amine substitution of polymers.[80] It may also be useful as an amine-detection reagent in HPLC where the absorption of the trityl cation can be detected by acidification in a postcolumn reactor.

The Smallest Reactive Fluorophores

The unusually stable benzotriazolyl derivatives of anthranilic acid (B-680) and 4-aminobenzoic acid[81] (B-681), *N*-methylisatoic anhydride (M-25) and the succinimidyl ester of *N*-methylanthranilic acid (S-128) are all useful precursors for preparing bioconjugates with small fluorophores. The small size of these fluorophores should reduce the likelihood that they will interfere with the function of the biomolecule, an important advantage when designing site-selective probes. They are often used to prepare fluorescent derivatives of biologically active peptides and toxins[82-85] and, in combination with a quencher, to prepare fluorogenic endoprotease substrates.[86,87] *N*-methylisatoic anhydride also reacts with ribonucleotides to yield fluorescent MANT nucleotide analogs.[88-90]

Chromophoric Succinimidyl Ester

Dabcyl has a broad and intense visible absorption but no fluorescence, making this dye useful as an acceptor in fluorescence resonance energy transfer applications. Biomolecules double-labeled with dabcyl and the appropriate fluorophore can be used to monitor proteolytic cleavage, conformational changes and other dynamic spatial movements. Dabcyl succinimidyl ester (dabcyl, SE; D-2245) is particularly useful for preparation of quenched fluorogenic substrates for proteases, including our HIV protease and renin substrates[91-93] (H-2930, R-2931; see Section 10.4), papain,[94,95] Alzheimer's disease–associated proteases[96] and others.[97,98] Fluorogenic substrates using this quenching group have also recently been prepared for interleukin-1β–converting enzyme[99] (ICE), a cysteine protease that is proposed to function in the onset of apoptosis.[100]

Amine-Reactive Spin Label

The pyrrolidine spin label (S-520) is the most common amine-reactive spin label. It is useful for preparing spin-labeled proteins,[101] peptides (EGF [102]), tRNA [103,104] and various low molecular weight probes.[105]

Bolton–Hunter Reagent

The Bolton–Hunter reagent (H-1586) is a succinimidyl ester that is used to introduce a phenolic residue into proteins, peptides and other amines that can be subsequently radioiodinated.[106,107]

Probe for Surface Modification

The new succinimidyl ester of 3,6-dioxaheptanoic acid (D-6137) can potentially be used to reduce the basicity of proteins and amine-containing polymers while retaining their hydrophilic character.

Crosslinking and Thiolation Reagents

Molecular Probes has available a number of succinimidyl esters that are useful for crosslinking biomolecules and thiolation of amines. These products are described in Chapter 5.

Detection of Amines by Ion-Pairing Chromatography

Moderately basic amines, including most nonchromophoric aliphatic amines, can be detected as ion pairs by addition of an appropriate chromophoric or fluorescent strong acid. Sulfonic acids are preferred because they form strong ion pairs. One of the best sulfonic acid derivatives available from Molecular Probes for this purpose is 9,10-dimethoxyanthracene-2-sulfonic acid (DAS, D-6948).[108-110]

Detection of Arginine in Proteins

Glyoxals such as phenylglyoxal (P-6221) and the chromophoric 4-nitrophenylglyoxal (N-1559) react with both arginine and guanosine. Arginine residues in proteins have been quantitatively determined with 4-nitrophenylglyoxal.[111] Like DIDS (D-337, see Section 18.5), aryl glyoxals are also powerful inhibitors of anion transport in red blood cells; see Section 18.5 for further discussion.[112,113]

1. Biochem Biophys Res Comm 50, 352 (1973); **2.** J Immunol Meth 112, 121 (1988); **3.** Arch Biochem Biophys 163, 400 (1974); **4.** Arch Biochem Biophys 163, 390 (1974); **5.** Science 178, 871 (1972); **6.** Clin Chim Acta 157, 73 (1986); **7.** J Lipid Res 27, 792 (1986); **8.** J Chromatography 502, 247 (1990); **9.** Biomed Chromatography 8, 207 (1994); **10.** Proc Natl Acad Sci USA 72, 619 (1975); **11.** Anal Chem 59, 1096 (1987); **12.** J Chromatography A 668, 323 (1994); **13.** Anal Biochem 189, 122 (1990); **14.** J Chromatography 564, 258 (1991); **15.** J Chromatography 582, 131 (1992); **16.** Anal Chim Acta 225, 351 (1989); **17.** Biomed Chromatography 3, 209 (1989); **18.** Anal Sci 2, 491 (1986); **19.** Analyst 119, 1907 (1994); **20.** Anal Chem 59, 411 (1987); **21.** J Org Chem 51, 3978 (1986); **22.** Anal Biochem 178, 202 (1989); **23.** Anal Chem 61, 432 (1989); **24.** J Chromatography 534, 139 (1990); **25.** J Chromatography 511, 155 (1990); **26.** J Pharm Biomed Anal 8, 477 (1990); **27.** J Chromatography 464, 343 (1989); **28.** Anal Biochem 180, 279 (1989); **29.** Anal Meth Instrument 2, 133 (1995); **30.** Anal Chem 63, 417 (1991); **31.** J Chromatography 540, 343 (1991); **32.** Anal Chem 62, 2189 (1990); **33.** Science 242, 224 (1988); **34.** Anal Chem 67, 58 (1995); **35.** Science 246, 57 (1989); **36.** ATTO-TAG CBQCA is licensed to Molecular Probes under U.S. Patent No. 5,459,272; **37.** Anal Chem 66, 3512 (1994); **38.** Anal Chem 66, 3477 (1994); **39.** Electrophoresis 14, 373 (1993); **40.** Anal Chem 63, 413 (1991); **41.** Anal Chem 63, 408 (1991); **42.** J Chromatography 559, 223 (1991); **43.** Proc Natl Acad Sci USA 88, 2302 (1991); **44.** J Chromatography 499, 579 (1990); **45.** J Chromatography A 716, 221 (1995); **46.** J Chromatography 519, 189 (1990); **47.** Biochem J 108, 155 (1968); **48.** FEBS Lett 6, 346 (1970); **49.** Anal Chim Acta 130, 377 (1981); **50.** Anal Chim Acta 170, 81 (1985); **51.** J Chromatography 278, 167 (1983); **52.** Anal Biochem 116, 471 (1981); **53.** Anal Chim Acta 290, 3 (1994); **54.** Biomed Chromatography 9, 10 (1995); **55.** Photochem Photobiol 54, 361 (1991); **56.** Biochemistry 16, 5150 (1977); **57.** J Biol Chem 269, 3829 (1994); **58.** Eur J Biochem 142, 387 (1984); **59.** J Biol Chem 259, 14378 (1984); **60.** Biochem J 51, 155 (1951); **61.** J Liquid Chromatography 12, 2733 (1989); **62.** J Colloid Interface Sci 135, 435 (1990); **63.** J Chromatography 553, 123 (1991); **64.** Meth Enzymol 91, 41 (1983); **65.** Anal Chem 62, 2193 (1990); **66.** Anal Biochem 141, 121 (1984); **67.** Anal Biochem 128, 412 (1983); **68.** J Chromatography 239, 608 (1992); **69.** J Chromatography 480, 141 (1989); **70.** Science 242, 562 (1988); **71.** Biol Chem Hoppe-Seyler 367, 1259 (1986); **72.** FEBS Lett 198, 150 (1986); **73.** Biochemistry 29, 10734 (1990); **74.** Anal Lett 19, 2277 (1986); **75.** Bioconjugate Chem 6, 447 (1995); **76.** Anal Biochem 223, 39 (1994); **77.** Nucleic Acids Res 20, 2471 (1992); **78.** Proc Natl Acad Sci USA 86, 9178 (1989); **79.** Science 271, 1420 (1996); **80.** Anal Biochem 180, 253 (1989); **81.** Aust J Chem 36, 1629 (1983); **82.** Peptides 13, 663 (1992); **83.** J Neurosci Meth 13, 119 (1985); **84.** J Biol Chem 259, 6117 (1984); **85.** J Biol Chem 258, 11948 (1983); **86.** Anal Biochem 212, 58 (1993); **87.** Anal Biochem 162, 213 (1987); **88.** Biochemistry 30, 422 (1991); **89.** Biochemistry 29, 3309 (1990); **90.** J Biol Chem 257, 13354 (1982); **91.** Anal Biochem 210, 351 (1993); **92.** Science 247, 954 (1990); **93.** Tetrahedron Lett 31, 6493 (1990); **94.** Arch Biochem Biophys 306, 304 (1993); **95.** FEBS Lett 297, 100 (1992); **96.** Bioorg Medicinal Chem Lett 2, 1555 (1992); **97.** Anal Biochem 204, 96 (1992); **98.** J Med Chem 35, 3727 (1992); **99.** Peptide Res 7, 72 (1994); **100.** Science 267, 1445 (1995); **101.** Biochemistry 16, 3746 (1982); **102.** Biochemistry 30, 8976 (1991); **103.** Biochemistry 9, 2526 (1970); **104.** Proc Natl Acad Sci USA 62, 1195 (1969); **105.** Arch Biochem Biophys 232, 477 (1984); **106.** Clin Chim Acta 66, 97 (1976); **107.** Biochem J 133, 529 (1973); **108.** Anal Chem 52, 700 (1980); **109.** J Chromatography 172, 141 (1979); **110.** Anal Chim Acta 67, 89 (1973); **111.** Anal Biochem 111, 220 (1981); **112.** Biochim Biophys Acta 1026, 43 (1990); **113.** J Biosciences 15, 179 (1990).

1.8 Data Table *Reagents for Analysis of Low Molecular Weight Amines*

Cat #	Structure	MW	Storage	Soluble	Abs	{ε x 10^{-3}}	Em	Solvent	Notes
A-176	S13.3	250	L	DMSO	366	{8.9}	414	MeOH	
A-409	✓	235	F,DD,L	DMF, MeCN	358	{5.8}	453	MeOH	<1>
A-417	✓	236	F,DD,L	DMF, MeCN	392	{12}	455	MeOH	
A-448	✓	277	F,DD,L	DMF, MeCN*	382	{4.0}	421	MeOH	<1>
A-1127	✓	320	F,D,L	DMF, DMSO	362	{11}	462	MeOH	
A-1139	✓	234	L	EtOH	546	ND	570	MeOH	<2, 3>
A-2334	✓	251	F,D,L	EtOH	486	ND	591	MeOH	<4>
A-6222	✓	305	F,D,L	MeOH	465	ND	560	MeOH	<3, 5, 6>
B-680	✓	254	F,D,L	DMF, MeCN	358	{7.1}	415	MeOH	
B-681	✓	254	F,D,L	DMF, MeCN	312	{34}	none	MeOH	
B-1174	S24.3	449	F,D	DMF, MeCN	296	{3.8}	none	MeOH	<7>
C-10	✓	200	F,D,L	DMF, MeCN	336	{9.8}	none	MeOH	<8, 9>

Cat #	Structure	MW	Storage	Soluble	Abs	{ε × 10^-3}	Em	Solvent	Notes
D-21	✓	270	F,DD,L	DMF, MeCN*	372	{3.9}	none	$CHCl_3$	<10>
D-427	✓	228	F,DD,L	DMF, MeCN	343	{16}	435	MeOH	
D-1302	✓	383	F,DD,L	DMF, MeCN	335	{4.2}	536	MeOH	
D-1332	✓	294	F,DD,L	DMF, MeCN	348	{38}	425	MeOH	<11>
D-1537	✓	324	F,DD,L	DMF, MeCN*	466	{33}	none	MeOH	<12>
D-2245	✓	366	F,D,L	DMF, DMSO	453	{32}	none	MeOH	<12>
D-6137	✓	231	F,D	DMF, DMSO	<300		none		
D-6948	S21.3	340	L	H_2O, DMF	340	{6.3}	464	pH 7	
F-486	✓	183	F,DD,L	MeCN, $CHCl_3$	328	{8.0}	none	MeOH	<8, 9>
F-2332	✓	278	F,DD,L	MeCN	381	{7.6}	470	MeCN	<13>
H-1586	✓	263	F,D	DMF, MeCN	278	{1.3}	none	MeOH	
I-1195	✓	219	F,DD	DMF, MeCN	280	{21}	none	MeOH	<14>
M-25	✓	177	D	DMF, DMSO	316	{3.5}	386	MeOH	<15>
N-316	S13.2	294	L	DMSO	467	{23}	539	MeOH	<9>
N-1138	✓	184	L	DMF, MeCN	462	ND	520	MeOH	<3, 16>
N-1559	✓	179	F	DMF, MeCN	265	{8.7}	none	MeOH	
P-24	✓	301	F,DD,L	DMF, MeCN*	350	{28}	380	MeOH	<1>
P-164	S13.2	246	L	DMF, DMSO	342	{30}	382	MeOH	
P-331	✓	259	F,DD,L	DMF, MeCN	341	{33}	395	MeOH	<1>
P-2331	✓	134	L	EtOH	334	{5.7}	455	pH 9	<17>
P-6221	✓	152†	F	EtOH	247	{13}	none	MeOH	
S-128	✓	248	F,D,L	DMF, MeCN	368	{6.5}	437	MeOH	
S-520	✓	281	F,D	DMF, MeCN	<300		none		
S-1144	✓	347	F,D,L	DMF, DMSO	366	{7.6}	412	MeOH	
S-1167	✓	391	F,D,L	DMF, DMSO	466	{22}	535	MeOH	<9>
S-2182	✓	504	F,DD	MeCN	275	{3.0}	307	MeOH	

A-2333 see A-6222

For definitions of the contents of this data table, see "How to Use the Handbook" on page vi.
Structure: Chemical structure drawing: ✓ = shown in this section; Sn.n = shown in Section number n.n..

* Do NOT dissolve in DMSO.
† MW is for the hydrated form of this product.
<1> Spectra of reaction product with butylamine.
<2> Spectra of A-1139 with butylamine + cyanide. Absorption spectrum has multiple peaks. Unreacted reagent in MeOH: Abs = 361 nm (ε = {5.4}), Em = 410 nm.
<3> ND = not determined.
<4> Spectra of A-2334 with glycine + cyanide. Unreacted reagent in MeOH: Abs = 282 nm (ε = {21}), nonfluorescent.
<5> Spectra of A-6222 with glycine + cyanide. Unreacted reagent in MeOH: Abs = 254 nm (ε = {46}), nonfluorescent.
<6> Solubility in methanol is improved by addition of base (e.g., 1–5% (v/v) 0.2 M KOH).
<7> B-1174 is most effective at pH 6–8 [Anal Chim Acta 177, 103 (1985)].
<8> Spectra for primary aliphatic amine adduct of C-10 in MeOH: Abs = 465 nm (ε = {22}), Em = 535 nm (QY = 0.3). Spectra for secondary aliphatic amine adduct of C-10 in MeOH: Abs = 485 nm (ε = {25}), Em = 540 nm (QY <0.1). Aromatic amine adducts are nonfluorescent. F-486 gives the same derivatives as C-10 but is more reactive.
<9> NBD amine derivatives are almost nonfluorescent in water; emission spectra and quantum yields in other solvents are variable [Biochemistry 16, 5150 (1977)].
<10> D-21 butylamine conjugate has Abs = 337 nm (ε = {5.3}), Em = 492 nm in $CHCl_3$. Em and QY are highly solvent dependent: Em = 496 nm (QY = 0.45) in dioxane, 536 nm (QY = 0.28) in MeOH and 557 nm (QY = 0.03) in H_2O [Biochemistry 6, 3408 (1967)]. ε typically decreases upon conjugation to proteins (ε = {3.4} at 340 nm) [Anal Biochem 25, 412 (1968)]. Fluorescence lifetimes (τ) of protein conjugates are typically 12–20 nsec [Arch Biochem Biophys 133, 263 (1969); Arch Biochem Biophys 128, 163 (1968)].
<11> Spectra of this compound are in methanol containing a trace of KOH.
<12> D-1537 reaction product with butylamine: Abs = 435 nm (ε = {31}), nonfluorescent in MeOH. D-2245 reaction product with butylamine: Abs = 428 nm (ε = {32}), nonfluorescent in MeOH.
<13> F-2332 spectra are for reaction product with butylamine. Fluorescence quantum yield/lifetime of adduct are 0.23/7.5 nsec in EtOH [Arch Biochem Biophys 163, 390 (1974)]. Unreacted reagent in MeCN: Abs = 234 nm (ε = {28}), nonfluorescent.
<14> I-1195 emits chemiluminescence (Em ~ 425 nm) upon oxidation in basic aqueous solutions.
<15> M-25 amide reaction product with butylamine has Abs = 353 nm (ε = {5.9}), Em = 426 nm in methanol.
<16> Spectra of N-1138 with glycine + cyanide. Unreacted reagent in MeOH: Abs = 279 nm (ε = {5.5}), Em = 330 nm.
<17> Spectral data are for the reaction product of P-2331 with alanine and 2-mercaptoethanol. The spectra and the stability of the adduct depend on the amine and thiol reactants [Biochim Biophys Acta 576, 440 (1979)]. Unreacted reagent in H_2O: Abs = 257 nm (ε = {1.0}).

1.8 Structures *Reagents for Analysis of Low Molecular Weight Amines*

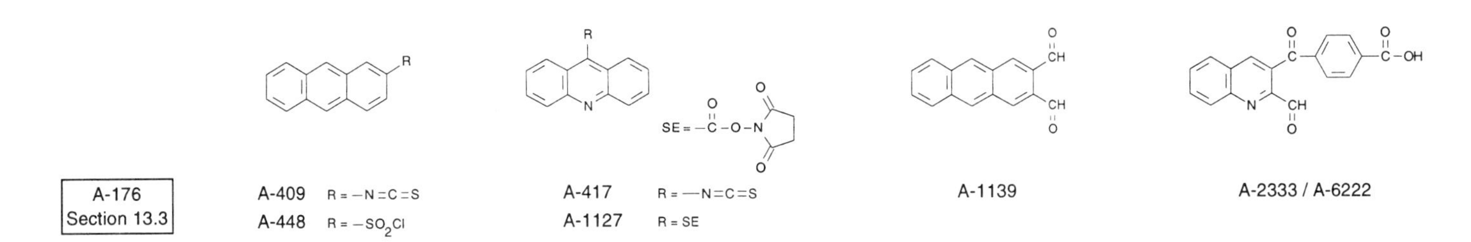

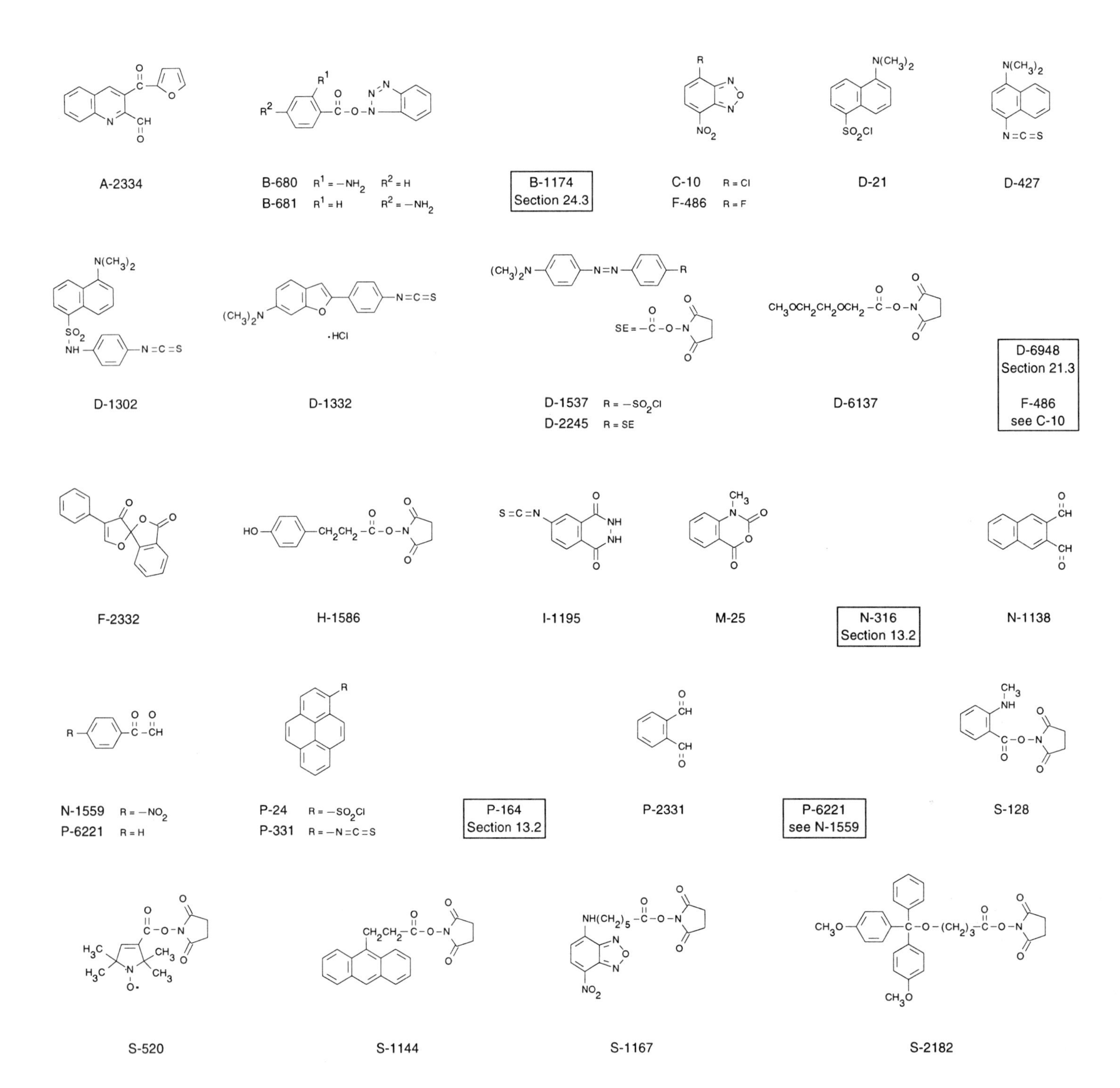

1.8 Price List *Reagents for Analysis of Low Molecular Weight Amines*

Cat #	Product Name	Unit Size	Price Per Unit ($) 1–4 Units	5–24 Units
A-1127	acridine-9-carboxylic acid, succinimidyl ester (9-AC, SE)	100 mg	78.00	62.40
A-417	9-acridineisothiocyanate	100 mg	35.00	28.00
A-1139	anthracene-2,3-dicarboxaldehyde (ADA)	25 mg	40.00	32.00
A-409	2-anthraceneisothiocyanate	100 mg	95.00	76.00
A-176	9-anthracenepropionic acid	100 mg	38.00	30.40
A-448	2-anthracenesulfonyl chloride	100 mg	48.00	38.40
A-2333	ATTO-TAG™ CBQCA Amine Derivatization Kit	1 kit	98.00	78.40
A-6222 ***New***	ATTO-TAG™ CBQCA derivatization reagent (CBQCA; 3-(4-carboxybenzoyl)quinoline-2-carboxaldehyde)	10 mg	110.00	88.00

Cat #		Product Name	Unit Size	Price Per Unit ($) 1–4 Units	5–24 Units
A-2334		ATTO-TAG™ FQ Amine Derivatization Kit	1 kit	98.00	78.40
B-681		benzotriazol-1-yl 4-aminobenzoate	100 mg	48.00	38.40
B-680		benzotriazol-1-yl anthranilate	100 mg	48.00	38.40
B-1174		bis-(2,4,6-trichlorophenyl) oxalate (TCPO)	1 g	25.00	20.00
C-10		4-chloro-7-nitrobenz-2-oxa-1,3-diazole (NBD chloride; 4-chloro-7-nitrobenzofurazan)	1 g	24.00	19.20
D-6948	***New***	9,10-dimethoxyanthracene-2-sulfonic acid, sodium salt (DAS)	100 mg	25.00	20.00
D-1537		4-dimethylaminoazobenzene-4′-sulfonyl chloride (dabsyl chloride)	100 mg	24.00	19.20
D-1332		N-(4-(6-dimethylamino-2-benzofuranyl)phenylisothiocyanate	25 mg	48.00	38.40
D-1302		4-((5-dimethylaminonaphthalene-1-sulfonyl)amino)phenylisothiocyanate	100 mg	38.00	30.40
D-21		5-dimethylaminonaphthalene-1-sulfonyl chloride (dansyl chloride)	1 g	40.00	32.00
D-427		4-dimethylaminonaphthyl-1-isothiocyanate	100 mg	28.00	22.40
D-2245		4-((4-(dimethylamino)phenyl)azo)benzoic acid, succinimidyl ester (dabcyl, SE)	100 mg	68.00	54.40
D-6137	***New***	3,6-dioxaheptanoic acid, succinimidyl ester	25 mg	48.00	38.40
F-2332		fluorescamine	100 mg	20.00	16.00
F-486		4-fluoro-7-nitrobenz-2-oxa-1,3-diazole (NBD fluoride; 4-fluoro-7-nitrobenzofurazan)	25 mg	95.00	76.00
H-1586		3-(4-hydroxyphenyl)propionic acid, succinimidyl ester (Bolton-Hunter reagent)	1 g	38.00	30.40
I-1195		isoluminol isothiocyanate (ILITC)	100 mg	48.00	38.40
M-25		N-methylisatoic anhydride *high purity*	1 g	75.00	60.00
N-1138		naphthalene-2,3-dicarboxaldehyde (NDA)	100 mg	65.00	52.00
N-316		6-(N-(7-nitrobenz-2-oxa-1,3-diazol-4-yl)amino)hexanoic acid (NBD-X)	100 mg	50.00	40.00
N-1559		4-nitrophenylglyoxal	100 mg	48.00	38.40
P-6221	***New***	phenylglyoxal, monohydrate	1 g	8.00	6.40
P-2331		o-phthaldialdehyde (OPA) *high purity*	1 g	32.00	25.60
P-164		1-pyrenecarboxylic acid	100 mg	48.00	38.40
P-331		1-pyreneisothiocyanate	100 mg	95.00	76.00
P-24		1-pyrenesulfonyl chloride	100 mg	78.00	62.40
S-1144		succinimidyl 9-anthracenepropionate	100 mg	38.00	30.40
S-2182		succinimidyl 4-O-(4,4′-dimethoxytrityl)oxybutyrate	25 mg	45.00	36.00
S-1167		succinimidyl 6-(7-nitrobenz-2-oxa-1,3-diazol-4-yl)aminohexanoate (NBD-X, SE)	25 mg	98.00	78.40
S-128		succinimidyl N-methylanthranilate	100 mg	98.00	78.40
S-520		succinimidyl 2,2,5,5-tetramethyl-3-pyrroline-1-oxyl-3-carboxylate	100 mg	55.00	44.00

Technical Assistance at Our Web Site (http://www.probes.com)

At Molecular Probes' Web site, we are developing an electronic version of this Handbook and other databases that should prove extremely useful for the researcher. In addition to containing all of the text from this Handbook, our Web site provides:

- **Product searches** by product name or catalog number
- **Bibliographies** for all products for which we have references
- **Keyword searches** of our entire bibliography of over 25,000 references
- **Product information sheets** for many kits and reagents
- **Technical bulletins**, including *BioProbes* newsletters and other product literature
- **Chemical structure**, **technical data** and **material safety and data sheets**
- **Color photomicrographs** that show our products in action

Visit our Web site often for new additions to our bibliography, as well as new products and upgraded search capabilities. Also look for special sales and introductory specials on some important products.

If you do not have access to the Internet or you need assistance that is not available at that site, further information on the scientific and technical background of our products can be obtained by contacting our Technical Assistance Department at the numbers listed on the inside front cover.

Fluorescence Resonance Energy Transfer

Fluorescence resonance energy transfer (FRET) is a distance-dependent interaction between the electronic excited states of two dye molecules in which excitation is transferred from a donor molecule to an acceptor molecule *without emission of a photon.* FRET is dependent on the inverse sixth power of the intermolecular separation,[1] making it useful over distances comparable with the dimensions of biological macromolecules. Thus, FRET is an important technique for investigating a variety of biological phenomena that produce changes in molecular proximity.[2-10]

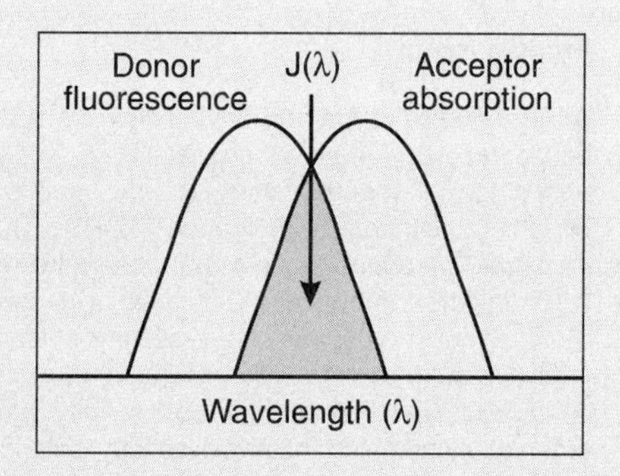

Schematic representation of the FRET spectral overlap integral.

Primary Conditions for FRET

- Donor and acceptor molecules must be in close proximity (typically 10–100 Å).
- Absorption spectrum of the acceptor must overlap fluorescence emission spectrum of the donor (see figure).
- Donor and acceptor transition dipole orientations must be approximately parallel.

Förster Radius

The distance at which energy transfer is 50% efficient (i.e., 50% of excited donors are deactivated by FRET) is defined by the Förster radius (R_o). The magnitude of R_o is dependent on the spectral properties of the donor and acceptor dyes:

$$R_O = [8.8 \times 10^{23} \cdot \kappa^2 \cdot n^{-4} \cdot QY_D \cdot J(\lambda)]^{1/6}\ \text{Å}$$

where κ^2 = dipole orientation factor (range 0 to 4, $\kappa^2 = 2/3$ for randomly oriented donors and acceptors)
QY_D = fluorescence quantum yield of the donor in the absence of the acceptor
n = refractive index
$J(\lambda)$ = spectral overlap integral (see figure)

$$= \int \epsilon_A(\lambda) \cdot F_D(\lambda) \cdot \lambda^4 d\lambda\ \text{cm}^3\text{M}^{-1}$$

where ϵ_A = extinction coefficient of acceptor
F_D = fluorescence emission intensity of donor as a fraction of the total integrated intensity

Donor/Acceptor Pairs

In most applications, the donor and acceptor dyes are different, in which case FRET can be detected by the appearance of sensitized fluorescence of the acceptor or by quenching of donor fluorescence. When the donor and acceptor are the same, FRET can be detected by the resulting fluorescence depolarization.[11] Some typical values of R_o are listed in the table above. Note that because the component factors of R_o (see above) are dependent on environment, the actual value observed in a specific experimental situation is somewhat variable. Extensive compilations of R_o values can be found in the literature.[4,5,7,10]

Typical values of R_o.

Donor	Acceptor	R_o (Å)
Fluorescein	Tetramethylrhodamine	55
IAEDANS	Fluorescein	46
EDANS	DABCYL	33
Fluorescein	Fluorescein	44
BODIPY FL	BODIPY FL	57

Selected Applications of FRET

- Structure and conformation of proteins [12-17]
- Spatial distribution and assembly of protein complexes [18-22]
- Receptor/ligand interactions [23-26]
- Immunoassays [27,28]
- Structure and conformation of nucleic acids [29-34]
- Detection of nucleic acid hybridization [35-39]
- Automated DNA sequencing [40,41]
- Distribution and transport of lipids [42-45]
- Membrane fusion assays [46-49]
- Membrane potential sensing [50]
- Fluorogenic protease substrates [51-55]
- Indicators for cyclic AMP [56] and zinc [57]

1. Proc Natl Acad Sci USA 58, 719 (1967); **2.** J Struct Biol 115, 175 (1995); **3.** Meth Enzymol 246, 300 (1995); **4.** Anal Biochem 218, 1 (1994); **5.** Van der Meer, B.W. *et al., Resonance Energy Transfer Theory and Data* VCH Publishers (1994); **6.** Meth Cell Biol 30, 219 (1989); **7.** J Muscle Res Cell Motility 8, 97 (1987); **8.** Photochem Photobiol 38, 487 (1983); **9.** Ann Rev Biochem 47, 819 (1978); **10.** Meth Enzymol 48, 347 (1978); **11.** Biophys J 69, 1569 (1995); **12.** Biochemistry 35, 4795 (1996); **13.** Biochemistry 34, 8693 (1995); **14.** Biochemistry 34, 6475 (1995); **15.** Biochemistry 33, 11900 (1994); **16.** Biochemistry 33, 10171 (1994); **17.** J Biol Chem 268, 15588 (1993); **18.** Biochemistry 34, 7904 (1995); **19.** Biochemistry 33, 13102 (1994); **20.** Biochemistry 33, 5539 (1994); **21.** J Photochem Photobiol B 12, 323 (1992); **22.** J Biol Chem 264, 8699 (1989); **23.** Biochemistry 33, 11875 (1994); **24.** J Cell Physiol 159, 176 (1994); **25.** Biophys J 60, 307 (1991); **26.** J Biol Chem 259, 5717 (1984); **27.** Anal Biochem 174, 101 (1988); **28.** Anal Biochem 108, 156 (1980); **29.** Anal Biochem 221, 306 (1994); **30.** Biophys J 66, 99 (1994); **31.** Nucleic Acids Res 22, 920 (1994); **32.** Science 266, 785 (1994); **33.** Biochemistry 32, 13852 (1993); **34.** Proc Natl Acad Sci USA 90, 2994 (1993); **35.** Biochemistry 34, 285 (1995); **36.** Nucleic Acids Res 22, 662 (1994); **37.** Morrison, L.E. in *Nonisotopic DNA Probe Techniques*, L.J. Kricka, Ed., Academic Press (1992) pp. 311–352; **38.** Anal Biochem 183, 231 (1989); **39.** Proc Natl Acad Sci USA 85, 8790 (1988); **40.** Anal Chem 67, 3676 (1995); **41.** Proc Natl Acad Sci USA 92, 4347 (1995); **42.** Biochemistry 34, 4846 (1995); **43.** Biochemistry 31, 2865 (1992); **44.** Meth Cell Biol 29, 75 (1989); **45.** J Biol Chem 258, 5368 (1983); **46.** Biochim Biophys Acta 1189, 175 (1994); **47.** Biophys J 67, 1117 (1994); **48.** Meth Enzymol 221, 239 (1993); **49.** Biochemistry 20, 4093 (1981); **50.** Biophys J 69, 1272 (1995); **51.** Contillo, L.G. *et al.* in *Techniques in Protein Chemistry, Volume 5*, J.W. Crabb, Ed., Academic Press (1994) pp. 493–500; **52.** Anal Biochem 210, 351 (1993); **53.** Bioconjugate Chem 4, 537 (1993); **54.** Anal Biochem 197, 347 (1991); **55.** Science 247, 954 (1990); **56.** Nature 349, 694 (1991); **57.** J Am Chem Soc 118, 6514 (1996).

Chapter 2

Thiol-Reactive Probes

Contents

Technical Notes and Product Highlights

Related Chapters

2.1 Introduction to Thiol Modification

Common Applications for Thiol-Reactive Probes

Labeling Biopolymers

In contrast to the amine-reactive reagents described in Chapter 1, thiol-reactive dyes are principally used to prepare fluorescent peptides, proteins and oligonucleotides for probing biological structure, function and interactions. Because the thiol functional group can be labeled with high selectivity, thiol modification is often employed to estimate both inter- and intramolecular distances using excited-state energy transfer; see the Technical Note "Fluorescence Resonance Energy Transfer" on page 46. Table 2.1 lists selected thiol-reactive iodoacetamide and maleimide derivatives that have been used to investigate specific biomolecules. Thiol-reactive dyes can also be reacted with thiolated primers for DNA sequencing,[1,2] with thiouridine-modified tRNA for studying its association with protein synthesis machinery [3,4] and with thiol-containing proteins to facilitate electrophoretic detection.[5] In addition, some of the polar fluorescent probes described in Section 2.4 may be useful for determining the topography of thiol-containing proteins and polypeptides in biological membranes.

Derivatizing Low Molecular Weight Molecules

Several of the thiol-reactive probes described in this chapter are also useful for derivatizing low molecular weight thiols for various analytical assays that employ chromatographic and electrophoretic separation. An extensive review by Shimada and Mitamura has recently been published that describes the use of several of our thiol-reactive reagents for derivatizing thiol-containing compounds.[6]

Table 2.1 *Selected examples of iodoacetamide and maleimide derivatives used to investigate specific biomolecules. An up-to-date bibliography is available from our Web site for every thiol-reactive probe for which we have references. Our Technical Assistance Department can also provide you with product-specific bibliographies, as well as with keyword searches of the over 25,000 literature references in our extensive bibliography database.*

- **acetylcholine receptor** I-9, P-28
- **actin cys-10** I-3, I-14
- **actin cys-374** B-1508, D-1521, E-99, E-333, F-150, I-3, I-9, I-14, I-15, P-28, P-29, T-6006
- **actin-depolymerizing factor** E-118
- **α-actinin** I-3, T-6006
- **adrenodoxin** I-3, I-14
- **apoprotein(a) (LDL)** L-1338
- **bacteriophage T4** F-6053
- **bacteriorhodopsin** B-1355
- **band 3 protein (erythrocyte)** E-118
- **blood coagulation factor 5a** T-6006
- **bovine serum albumin** F-150, I-3, I-14, T-6006
- **brush-border membrane proteins** P-28
- **α-bungarotoxin** E-99
- **caldesmon** I-3, P-28, T-6006
- **calmodulin** B-1508, D-346, D-1521, I-3, I-7, I-14, M-8
- **chloroplast-coupling factor** D-82, D-346, P-28
- **colicin** I-14
- **creatine kinase** D-346, I-3, I-7, I-9, I-106
- **crystallins** I-14, M-8, P-28
- **cyclic AMP–dependent protein kinase** I-9, I-14
- **cytochrome c oxidase** I-9
- **desmin** T-6006
- **DNA** I-3
- **enolase** F-6053
- **F_1-ATPase** I-3, I-7
- **fatty acid synthase** D-346
- **ferrochelatase** M-8
- **fibronectin** D-346, F-150, P-28
- **glucagon (on tryptophan thiol)** I-3, I-9, T-6027, T-6028
- **glutamate dehydrogenase** I-106
- **glutamine synthetase** D-1522, I-14
- **glutathione *S*-transferase** P-28
- **hemoglobin** I-3
- **hemolysin** A-484
- **histones H3 and H4** I-3, I-14, I-15, P-28, T-6006
- **immunoglobulins G and E** D-346, D-1522, I-14
- **isocitrate dehydrogenase** I-106
- **kinesin** I-3
- **lac repressor protein** I-14
- **α-lactalbumin (on methionine)** I-3
- **maltose binding protein** A-433, I-9
- **microtubule-associated proteins** I-3, I-15
- **myosin** D-1521, E-99, E-118, F-150, I-3, I-7, I-9, I-14, I-15, M-8, P-29, T-6006, T-6027, T-6028
- **myosin light chains** A-433, B-1508, B-2103, D-1521, D-2006, I-3, I-9, I-14, T-6006
- **nucleosomal protein** I-3, I-14
- **neurofilaments** D-346, F-150, I-3, M-1602
- **oligonucleotides** I-3, I-14
- **papain** I-7, I-9
- **phenylalanine hydroxylase** I-3
- **plasminogen activator inhibitor 1** D-2004
- **phosphatase inhibitor 2** D-346
- **phosphoenolpyruvate carboxykinase** D-346, I-14, P-28
- **phosphotransferase (bacterial)** P-28
- **plasminogen activator inhibitor** D-2005
- **prealbumin** I-14
- **protein phosphatase-1 and -2A** P-28
- **pyruvate dehydrogenase** E-118
- **pyruvate oxidase** M-8
- **rhodopsin** D-346, E-333, P-28
- **ribonuclease** I-14
- **ribosomal proteins** B-1508, D-346, I-3, I-14, P-28
- **RNA polymerase** D-346, F-150, P-29
- **sarcoplasmic reticulum ATPase** D-1521, E-99, E-333, F-150, I-3, I-14, M-8, P-28
- **Na^+/K^+-ATPase** F-6053, I-3, M-8, P-28
- **spectrin** E-118, E-333, I-3, I-14
- **thiolated oligonucleotides** D-346, E-118
- **thiophosphorylated RNA** D-346
- **thrombin** I-3, I-7, T-6006
- **transducin** I-3, M-8
- **tRNA** D-346, D-404, F-150, I-3, I-14
- **tRNA synthetase** I-14
- **tropomyosin** B-1508, I-14 (on lysine), P-28, P-29
- **troponin C** B-1508, D-346, D-1521, E-99, I-7, I-9, I-14
- **troponin I** D-346, D-1521, I-3, I-9, I-14
- **tryptophan synthetase** I-14
- **tubulin** T-6027, T-6028
- **ubiquinone** I-9
- **UDPG dehydrogenase** E-99, I-9, T-6027, T-6028

Reactivity of Thiol Groups

In proteins, thiol groups (also called mercaptans or sulfhydryls) are present in cysteine residues. Thiols can also be generated by selectively reducing cystine disulfides with reagents such as DTT [7] (dithiothreitol, D-1532), DTE [8] (1,4-dithioerythritol, D-8452) or 2-mercaptoethanol, each of which must then be removed by dialysis or gel filtration before reaction with the thiol-reactive probe.[9] Unfortunately, removal of DTT, DTE or 2-mercaptoethanol is sometimes accompanied by air oxidation of the thiols back to the disulfides. Reformation of the disulfide bond can be avoided by using the recently developed reducing reagent TCEP [10] (tris-(2-carboxyethyl)phosphine, T-2556), which does not need to be removed prior to thiol modification because it does not contain thiols. Our pH-insensitive and less polar phosphine derivative tris-(2-cyanoethyl)phosphine (T-6052) may show greater reactivity with buried disulfides. Several reagents have also been developed for introducing thiols into proteins, nucleic acids and lipids. Because these reagents are particularly important for crosslinking two biomolecules, they are discussed in Chapter 5. See also the Technical Note "Interconversion of Functional Groups" below for reagents that convert a thiol into another functional group.

The thiol-reactive functional groups are primarily alkylating reagents, including iodoacetamides, maleimides, benzylic halides and bromomethylketones. Arylating reagents such as NBD halides react with thiols or amines by a similar substitution of the aromatic halide. Reaction of any of these functional groups with thiols usually proceeds rapidly at or below room temperature in the physiological pH range (pH 6.5–8.0) to yield chemically stable thioethers.

Thiols also react with many of the amine-reactive reagents described in Chapter 1, including isothiocyanates and succinimidyl esters. However, the products appear to be insufficiently stable to be useful for modifying thiols in proteins. Although the thiol–isothiocyanate product (a dithiocarbamate) can react with an adjacent amine to yield a thiourea, the dithiocarbamate is more likely to react with water, consuming the reactive reagent without forming a covalent adduct.

Iodoacetamides

Iodoacetamides readily react with all thiols, including those found in peptides, proteins and thiolated polynucleotides, to form thioethers (Figure 2.1A); they are somewhat more reactive than bromoacetamides. However, when a protein's cysteine residues are blocked or absent, iodoacetamides can sometimes react with methionine residues.[11,12] They may also react with histidine [13] or potentially tyrosine, but generally only if free thiols are absent. Although iodoacetamides can react with the free base form of amines, most aliphatic amines, except the α-amino group at a protein's N-terminus, are protonated and thus relatively unreactive below pH 8. In addition, iodoacetamides react with thiolated oligonucleotide primers, as well as with thiophosphates and thiouridine residues present in certain nucleic acids,[1,3,4,14] but not usually with the common nucleotides.

Iodoacetamides are intrinsically unstable in light, especially in solution; reactions should therefore be carried out under subdued light. Adding cysteine, glutathione or mercaptosuccinic acid to the reaction mixture will quench the reaction of thiol-reactive probes, forming highly water-soluble adducts that are easily removed by dialysis or gel filtration. Although the thioether bond formed when an iodoacetamide reacts with a protein thiol is very stable, the bioconjugate loses its fluorophore during amino acid hydrolysis to yield *S*-carboxymethylcysteine.

Maleimides

Maleimides are excellent reagents for thiol-selective modification, quantitation and analysis. The reaction involves addition of the thiol across the double bond of the maleimide to yield a thioether (Figure 2.1B). The applications of these fluorescent and chromophoric analogs of *N*-ethylmaleimide (NEM) strongly overlap those of iodoacetamides, although maleimides apparently do not react with methionine, histidine or tyrosine. Reaction of maleimides with amines usually requires a higher pH than reaction of maleimides with thiols. Hydrolysis of maleimides to a nonreact-

Interconversion of Functional Groups

Several reagents can be used to convert one functional group to another, thereby providing greater flexibility in selecting reactive probes for functional group modification. These reagents are discussed in Chapter 5, except where otherwise noted. Among the reagents are the following:

- **Aldehyde (Reducing Sugar) or Ketone to Amine**: NH_3 + $NaCNBH_3$ (widely available)
- **Aldehyde or Ketone to Thiol**: AMBH (A-1642)
- **Amine to Aldehyde**: SFB (S-1580, see Chapter 3)
- **Amine to Carboxylic Acid**: succinic anhydride (widely available)
- **Amine to Iodoacetamide**: succinimidyl iodoacetate (I-6001); 4-nitrophenyl iodoacetate (N-1505); SIAX (S-1666); SIAXX (S-1668)
- **Amine to Maleimide**: SMCC (S-1534); EMCS (S-1563); SMB (S-1535)
- **Amine to Thiol**: iminothiolane (IT, I-2553); SPDP (S-1531); SATA (S-1553); SATP (S-1573)
- **Disulfide to Thiol**: DTT (D-1532); DTE (D-8452); TCEP (T-2556); tris-(2-cyanoethyl)phosphine (T-6052)
- **Thiol to Aldehyde**: chloroacetaldehyde (widely available)
- **Thiol to Amine**: *N*-(2-iodoethyl)trifluoroacetamide (I-1536, see Chapter 2)
- **Thiol to Carboxylic Acid**: iodoacetic acid; 3-bromopropionic acid (widely available)
- **Thiol to Ketone**: 3-bromopyruvic acid (B-6058, see Chapter 2)

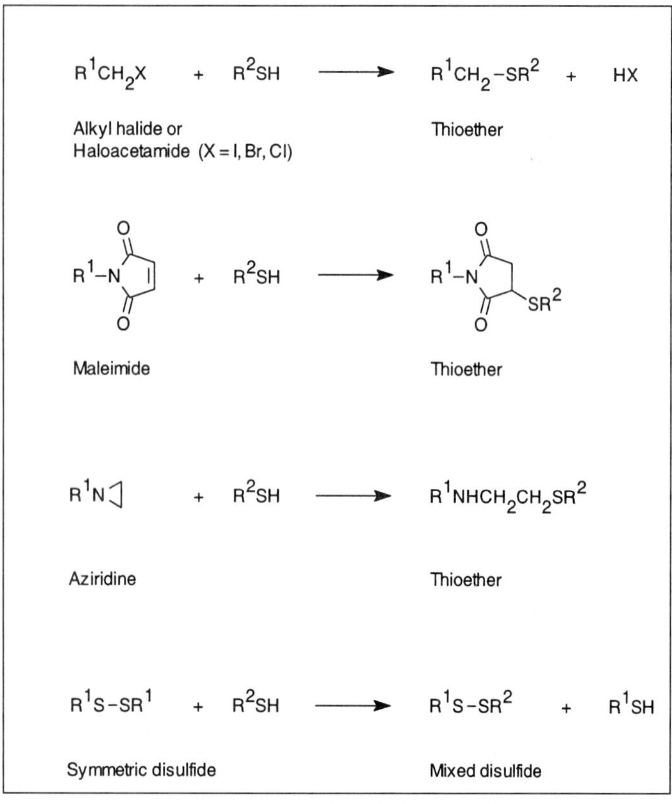

Figure 2.1 *Reaction of a thiol with A) an iodoacetamide, B) a maleimide, C) an aziridine and D) a symmetric disulfide (e.g., didansyl-L-cystine).*

ive maleamic acid can compete significantly with thiol modification, particularly above pH 8. Furthermore, maleimide adducts can hydrolyze, which may cause a significant change in the fluorescence properties of the conjugate,[15] or they can ring-open by nucleophilic reaction with an adjacent amine to yield crosslinked products.[16] This latter reaction can potentially be enhanced by raising the pH above 9 after conjugation. Several maleimides, including the coumarin, pyrene and stilbene derivatives, are not appreciably fluorescent until after conjugation with thiols and may therefore be useful for thiol quantitation.

Other Thiol-Reactive Reagents

A variety of other thiol-reactive probes are available. These include alkylating agents (alkyl halides, aziridines, epoxides) and arylating agents (NBD chloride, ABD fluoride). Aziridines and epoxides react by ring-opening to form very stable thioethers (Figure 2.1C). We also offer a fluorescent disulfide, didansyl-L-cystine (D-146, see Section 2.3). Symmetric disulfides undergo a thiol–disulfide interchange reaction to yield a new asymmetric disulfide (Figure 2.1D), a reaction that is freely reversible and is thiol-*specific*. The adducts from any of these thiol-reactive probes are, in general, more resistant to hydrolysis than those from iodoacetamides or maleimides, which are somewhat sensitive to hydrolysis at the amide linkage connecting the fluorophore to its reactive group.

1. Nucleic Acids Res 16, 2203 (1988); **2.** Nucleic Acids Res 15, 4593 (1987); **3.** Biochemistry 24, 692 (1985); **4.** J Mol Biol 156, 113 (1982); **5.** Anal Biochem 160, 376 (1987); **6.** J Chromatography B 659, 227 (1994); **7.** Eur J Biochem 168, 169 (1987); **8.** BioTechniques 8, 512 (1990); **9.** Meth Enzymol 143, 246 (1987); **10.** J Org Chem 56, 2648 (1991); **11.** Biochemistry 25, 5036 (1986); **12.** Biochemistry 25, 4887 (1986); **13.** Biochemistry 20, 7021 (1981); **14.** Biochemistry 22, 1208 (1983); **15.** Biophys J 50, 75 (1986); **16.** Biochemistry 15, 2863 (1976).

2.1 Data Table *Introduction to Thiol Modification*

Cat #	Structure	MW	Storage	Soluble	Abs	Em
D-1532	✓	154	D	H_2O	<300	none
D-8452	✓	154	D	H_2O	<300	none
T-2556	✓	287	D	pH>5	<300	none
T-6052	✓	193	D	MeCN	<300	none

For definitions of the contents of this data table, see "How to Use the Handbook" on page vi.
Structure: Chemical structure drawing: ✓ = shown in this section.

2.1 Structures *Introduction to Thiol Modification*

$HSCH_2C(OH)(H)$–$C(H)(OH)CH_2SH$
D-1532

$HSCH_2C(HO)(H)$–$C(OH)(H)CH_2SH$
D-8452

$P(CH_2CH_2$–$C(=O)$–$OH)_3 \cdot HCl$
T-2556

$P(CH_2CH_2CN)_3$
T-6052

2.1 Price List *Introduction to Thiol Modification*

Cat #		Product Name	Unit Size	Price Per Unit ($) 1–4 Units	5–24 Units
D-8452	New	1,4-dithioerythritol (DTE)	1 g	21.00	16.80
D-1532		dithiothreitol (DTT)	1 g	18.00	14.40
T-2556		tris-(2-carboxyethyl)phosphine, hydrochloride (TCEP)	1 g	25.00	20.00
T-6052	New	tris-(2-cyanoethyl)phosphine	1 g	24.00	19.20

2.2 Thiol-Reactive Probes Excited with Visible Light

Among the thiol-reactive probes, the BODIPY, fluorescein, Oregon Green, tetramethylrhodamine and Texas Red derivatives have the strongest absorptivity and highest fluorescence quantum yields. This combination of attributes makes these compounds the preferred reagents for preparing protein conjugates to study the structural properties, diffusion and interactions of proteins using techniques such as fluorescence recovery after photobleaching (FRAP) or excited-state energy transfer (FRET) (see the Technical Note "Fluorescence Resonance Energy Transfer" on page 46). In the following sections, thiol-reactive reagents with similar spectra, rather than the same reactive group, are generally discussed together. The exception to this organization is the description of our reactive BODIPY fluorophores, which are available in several choices of excitation and emission wavelengths and are considered as a group. Table 2.2 summarizes the iodoacetamide and maleimide probes listed in this section.

Table 2.2 *Spectral properties of the iodoacetamides and maleimides described in Section 2.2.*

Derivative	Iodoacetamide	Maleimide
BODIPY FL	D-2005, D-6003	
BODIPY 507/545	D-6004	
BODIPY 530/550	D-2006	
Diethylaminocoumarin	D-404	D-346
Eosin	E-99	E-118
Erythrosin	E-333	
Fluorescein	I-3 *, I-15 †	F-150
Oregon Green 488	O-6010	
PyMPO		M-6026
Tetramethylrhodamine	T-6006 *	T-6027 *, T-6028 †
Texas Red	T-6009 ‡	T-6008

* 5-isomer. † 6-isomer. ‡ bromoacetamide.

BODIPY Derivatives

The BODIPY iodoacetamides and the bromomethyl BODIPY probe yield thiol adducts with several important properties:

- High extinction coefficients (ϵ >50,000 $cm^{-1}M^{-1}$)
- High fluorescence quantum yields, often approaching 1.0 even in water
- Narrow emission bandwidths (see Figure 1.4 in Chapter 1)
- Good photostability
- Spectra that are insensitive to solvent polarity and pH [1]
- Lack of ionic charge, which is especially useful when preparing membrane probes and cell-permeant reagents

Our selection of thiol-reactive BODIPY reagents includes two iodoacetamides of the fluorescein-like BODIPY FL fluorophore (excitation/emission ~503/512 nm; D-2005, D-6003), the new BODIPY 507/545 iodoacetamide (D-6004), the longer-wavelength BODIPY 530/550 iodoacetamide (D-2006) and BODIPY 493/503 methyl bromide (B-2103). These thiol-reactive BODIPY probes are suitable for labeling thiolated oligonucleotides and for detecting thiol conjugates separated by HPLC and capillary electrophoresis using ultrasensitive laser-scanning techniques.[1] The BODIPY probes are chemically stable between about pH 3 and 10, although they are less stable to extremes of pH than are fluorescein derivatives.

Fluorescein Derivatives, Including Thiol-Reactive Oregon Green Dyes

Fluorescein Derivatives

The excellent water solubility of fluorescein iodoacetamide single isomers (I-3, I-15) and fluorescein maleimide [2] (F-150) at pH 7 makes it easy to prepare green fluorescent thiol conjugates of biomolecules. When compared with these iodoacetamide and maleimide derivatives, 5-(bromomethyl)fluorescein (B-1355) reacts more slowly with thiols of peptides, proteins and thiolated nucleic acids but forms stronger thioether bonds that are expected to remain stable under conditions required for complete amino acid analysis. 5-Bromomethylfluorescein has perhaps the highest intrinsic detectability of all thiol-reactive probes, particularly for instrumentation that uses the 488 nm spectral line of the argon-ion laser. Furthermore, its negative charges should make capillary electrophoretic separation of 5-bromomethylfluorescein adducts possible.

New Oregon Green Derivative

One of our new fluorescein substitutes — Oregon Green 488 (2',7'-difluorofluorescein, O-6145; see Section 1.4) — has absorption and emission spectra that are a perfect match to those of fluorescein. In addition to the Oregon Green 488 isothiocyanate, carboxylic acid and succinimidyl ester derivatives (see Section 1.4), we have recently synthesized Oregon Green 488 iodoacetamide (O-6010). This thiol-reactive probe yields conjugates that have several important advantages when directly compared with conjugates of fluorescein. These include:

- Greater photostability (see Figure 1.6 in Chapter 1)
- A lower pK_a (pK_a for 2',7'-difluorofluorescein = 4.8; pK_a for fluorescein = 6.4) (see Figure 1.7 in Chapter 1)
- Higher fluorescence and less quenching at comparable degrees of substitution (see Figure 1.5 in Chapter 1)

Eosin and Erythrosin Derivatives

Although eosin and erythrosin iodoacetamides (E-99, E-333) and eosin maleimide [2] (E-118) are much less fluorescent than the corresponding fluorescein derivatives, they are more phosphorescent and better photosensitizers.[3] With its high quantum yield of 0.57 for singlet oxygen generation,[4-6] eosin can be used as an effective photooxidizer of diaminobenzidine (DAB) in high-resolution electron microscopy studies;[7,8] see the Technical Note "Fluorescent Probes for Photoconversion of Diaminobenzidine" on page 264. Eosin conjugates of antibodies (see Section 7.2), streptavidin (see Section 7.5), phalloidin (see Section 11.1), a lipid probe (see Section 14.3), dextrans (see Section 15.5) and α-bungarotoxin (see Section 18.2) are available for this technique.

Presently, the principal application of eosin and erythrosin conjugates, including those prepared with the thiol-reactive derivatives in this chapter and the amine-reactive derivatives in Section 1.5, is to follow localized rotational motions in proteins, protein assemblies and proteins in membranes using phosphorescence anisotropy.[9-11] Eosin (excitation/emission maxima ~519/540 nm) and erythrosin (excitation/emission maxima ~532/555 nm) derivatives efficiently absorb the fluorescence from fluorescein and other fluorophores such as BODIPY, dansyl and coumarins, making them good acceptors in fluorescence resonance energy transfer techniques. Although usually selectively reactive with thiols, eosin maleimide reportedly also reacts with a specific lysine residue of the band-3 protein in human erythrocytes, inhibiting anion exchange in these cells.[12,13]

Rhodamine Derivatives, Including Thiol-Reactive Texas Red Dyes

Tetramethylrhodamine Derivatives

Tetramethylrhodamine iodoacetamide (TMRIA, T-6006) and tetramethylrhodamine maleimide (T-6027, T-6028) yield photostable, pH-insensitive, red-orange fluorescent thiol conjugates.[14,15] However, the iodoacetamide and maleimide derivatives are difficult to prepare in pure form and different batches of our mixed-isomer products have contained variable mixtures of the 5- and 6-isomers. Apparently certain cytoskeletal proteins preferentially react with individual isomers, leading to complications in the interpretation of labeling results.[16,17] Consequently, we now prepare the 5-isomer of TMRIA (T-6006) and the 5-isomer (T-6027) and 6-isomer (T-6028) of tetramethylrhodamine maleimide and have discontinued sale of the mixed-isomer products (formerly T-488 and T-489). Literature references for the mixed isomers are sent with the single-isomer products.

Texas Red Derivatives

We are pleased to introduce the new bromoacetamide and maleimide derivatives of our Texas Red fluorophore (T-6009, T-6008), the longest-wavelength thiol-reactive dyes available. Unlike tetramethylrhodamine conjugates, Texas Red conjugates exhibit very little spectral overlap with fluorescein conjugates and can be excited by the 568 nm spectral line of the Ar-Kr mixed gas laser used in some confocal microscopes (see Figure 1.10 in Chapter 1). Thus, Texas Red conjugates are ideal as a second label in multicolor applications. Bromoacetamides are only slightly less reactive with thiols than are iodoacetamides. Like most Texas Red derivatives, the bromoacetamide and maleimide are mixtures of the isomeric sulfonamides.

Coumarin Derivatives

The coumarin iodoacetamide (DCIA, D-404) and maleimide (CPM, D-346) are probably the best UV-excitable fluorescent thiol reagents available. Not only do these reagents cost significantly less than the similar coumarin maleimide DACM, but they yield conjugates that are more fluorescent than the corresponding DACM adducts. The coumarin fluorophore is an excellent fluorescence energy acceptor from tryptophan and a good donor to fluorescein, eosin and selected BODIPY dyes,[18] making DCIA and CPM especially valuable for protein structure studies and for study of protein–membrane interactions (Table 2.1). Nucleolar protein staining by CPM has been used to distinguish highly proliferating cancer cells in a flow cytometry assay.[19] In addition to modifying proteins, CPM reacts with thiophosphorylated RNA [20] and thiolated oligonucleotides.[21,22] Fluorescence emission of the coumarins is moderately sensitive to environment.

The maleimide derivative of coumarin (and several other fluorophores) is essentially nonfluorescent until it react with thiols, permitting thiol quantitation without a separation step. CPM has been used to follow the release of picomoles of thiols from acetylthiocholine by acetylcholinesterase [23,24] and to determine cystamine, cysteamine [25] and thiol content of proteins, cells and plasma.[26-29] CPM can also be used to quantitate thiols using a fluorescence microplate reader.[30]

Pyridyloxazole Derivatives

Our recently introduced pyridyloxazole dyes — including the thiol-reactive PyMPO maleimide (M-6026) and PyMPO epoxide (E-6051), as well as the amine-reactive isothiocyanate and succinimidyl ester derivatives (I-6076, S-6110; see Section 1.7) — fill a spectral gap between UV-excitable dyes and the fluoresceins. These derivatives of the laser dye PyMPO exhibit absorption maxima between 410–420 nm (ϵ >20,000 $cm^{-1}M^{-1}$) and unusually high Stokes shifts with emission at 560–570 nm.[31]

1. J Am Chem Soc 116, 7801 (1994); **2.** U.S. Patent No. 4,213,904; **3.** J Gen Microbiol 139, 841 (1993); **4.** Adv Photochem 18, 315 (1993); **5.** Photochem Photobiol 37, 271 (1983); **6.** J Am Chem Soc 99, 4306 (1977); **7.** J Cell Biol 126, 901 (1994); **8.** J Cell Biol 126, 877 (1994); **9.** Biochemistry 30, 3538 (1991); **10.** Biochemistry 29, 10023 (1990); **11.** Spectroscopy 5, 20 (1990); **12.** Biochemistry 29, 8283 (1990); **13.** Biochim Biophys Acta 1025, 199 (1990); **14.** Biophys J 65, 113 (1993); **15.** Biochemistry 28, 2204 (1989); **16.** Biophys J 68, 78s (1995); **17.** Biochemistry 31, 12431 (1992); **18.** Biochem Biophys Res Comm 207, 508 (1995); **19.** J Histochem Cytochem 41, 1413 (1993); **20.** Biochemistry 30, 4821 (1991); **21.** Anal Biochem 191, 295 (1990); **22.** Proc Natl Acad Sci USA 87, 1744 (1990); **23.** Biochemistry 29, 10640 (1990); **24.** Anal Biochem 133, 450 (1983); **25.** Anal Biochem 170, 432 (1988); **26.** Anal Biochem 154, 186 (1986); **27.** Cytometry 3, 349 (1982); **28.** J Histochem Cytochem 29, 1377 (1981); **29.** J Histochem Cytochem 29, 314 (1981); **30.** Toxicol Appl Pharmacol 100, 485 (1989); **31.** IEEE J Quantum Electronics 16, 777 (1980).

2.2 Data Table *Thiol-Reactive Probes Excited with Visible Light*

Cat #	Structure	MW	Storage	Soluble	Abs	{ε × 10⁻³}	Em	Solvent	Notes
B-1355	✓	425	F,D,L	pH>6, DMF	492	{81}	515	pH 9	
B-2103	S3.4	341	F,D,L	DMSO, MeCN	515	{55}	525	MeOH	<1>
D-346	✓	402	F,D,L	DMSO	384	{33}	469	MeOH	<2>
D-404	✓	490	F,D,L	DMSO	384	{31}	470	MeOH	<2, 3>
D-2005	✓	502	F,D,L	DMSO, MeCN	504	{81}	511	MeOH	<2, 3>
D-2006	✓	626	F,D,L	DMSO, MeCN	534	{69}	552	MeOH	<2, 3>
D-6003	✓	417	F,D,L	DMSO, MeCN	502	{76}	510	MeOH	<2, 3>
D-6004	✓	431	F,D,L	DMSO, MeCN	508	{69}	543	MeOH	<2, 3>
E-99	✓	831	F,D,L	pH>6, DMF	519	{100}	540	pH 9.5	<2, 3>
E-118	✓	743	F,D,L	pH>6, DMF	524	{103}	545	MeOH	<2>
E-333	✓	1020	F,D,L	pH>6, DMF	532	{97}	555	MeOH	<2, 3>
E-6051	✓	458	F,D,L	DMSO	412	{26}	564	MeOH	<4>
F-150	✓	427	F,D,L	pH>6, DMF	492	{83}	515	pH 9	<2>
I-3	✓	515	F,D,L	pH>6, DMF	492	{75}	515	pH 9	<2, 3>
I-15	✓	515	F,D,L	pH>6, DMF	492	{81}	516	pH 9	<2, 3>
M-6026	✓	471	F,D,L	DMSO	412	{23}	561	MeOH	<2, 4>
O-6010	✓	551	F,D,L	pH>6, DMF	491	{68}	516	pH 9	<2, 3>
T-6006	✓	569	F,D,L	DMSO	543	{87}	567	MeOH	<2, 3>
T-6008	✓	729	F,D,L	DMSO	582	{108}	600	MeOH	<2>
T-6009	✓	812	F,D,L	DMSO	583	{113}	603	MeOH	<2>
T-6027	✓	482	F,D,L	DMSO	541	{91}	567	MeOH	<2>
T-6028	✓	482	F,D,L	DMSO	541	{91}	567	MeOH	<2>

For definitons of the contents of this data table, see "How to Use the Handbook" on page vi.
Structure: Chemical structure drawing: ✓ = shown in this section; Sn.n = shown in section number n.n.

<1> B-2103 spectra are for the unreacted reagent. The thiol adduct has Abs = 493 nm, Em = 503 nm in MeOH.
<2> Spectral data of the 2-mercaptoethanol adduct.
<3> Iodoacetamides in solution undergo rapid photodecomposition to unreactive products. Minimize exposure to light prior to reaction.
<4> Spectra shift to shorter wavelengths in nonpolar solvents. Fluorescence quantum yields of 1-propanethiol adducts in water are ~0.3.

2.2 Structures *Thiol-Reactive Probes Excited with Visible Light*

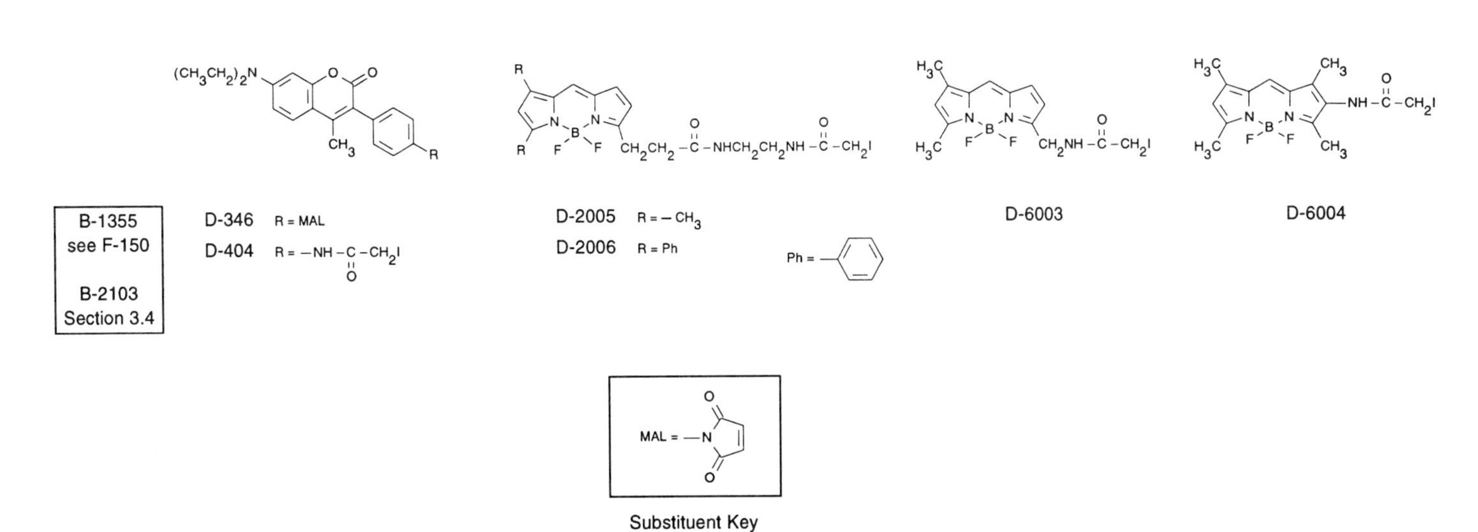

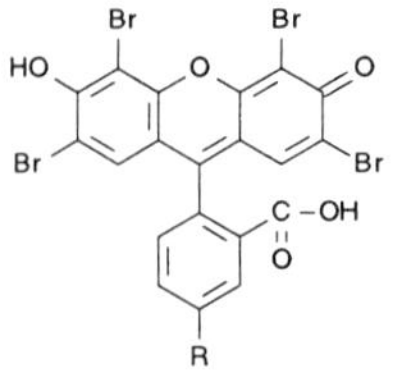

E-99 R = —NH—C(=O)—CH_2I
E-118 R = MAL

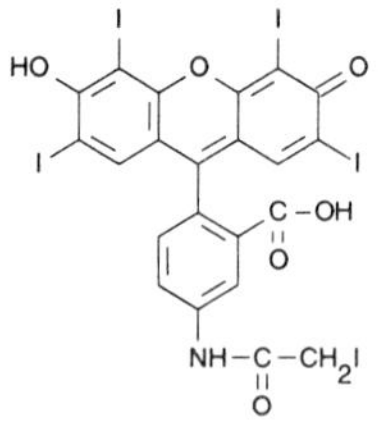

E-333

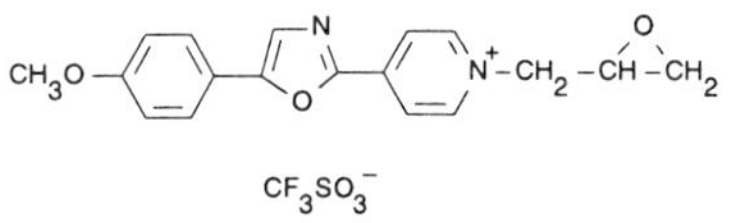

E-6051

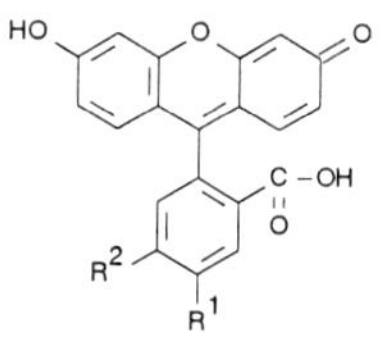

F-150 R^1 = MAL R^2 = H
I-3 R^1 = —NH—C(=O)—CH_2I R^2 = H
I-15 R^1 = H R^2 = —NH—C(=O)—CH_2I
B-1355 R^1 = —CH_2Br R^2 = H

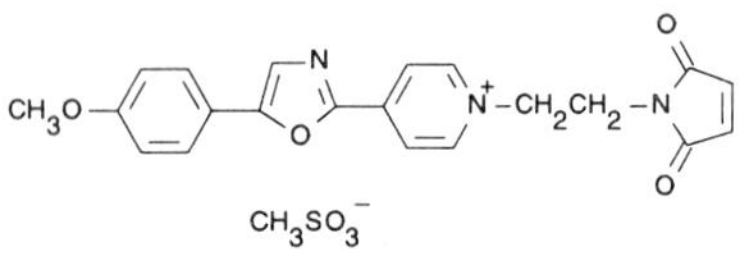

M-6026

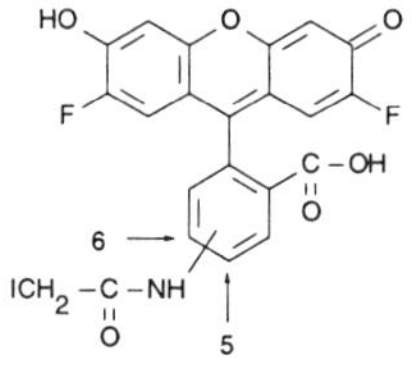

O-6010

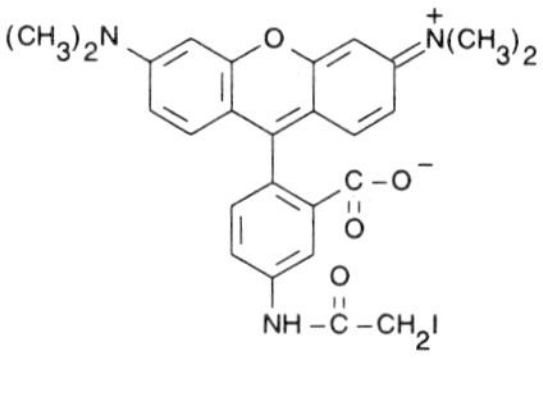

T-6006

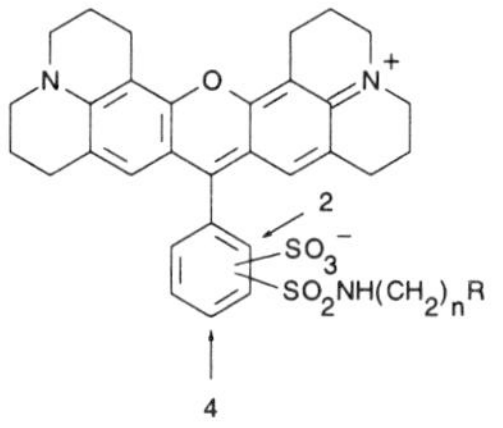

T-6008 R = MAL n = 2
T-6009 R = —NH—C(=O)—CH_2Br n = 5

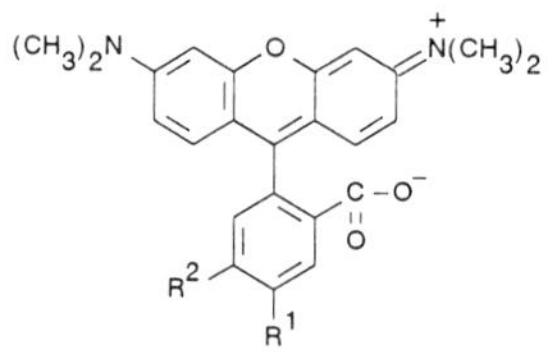

T-6027 R^1 = MAL R^2 = H
T-6028 R^1 = H R^2 = MAL

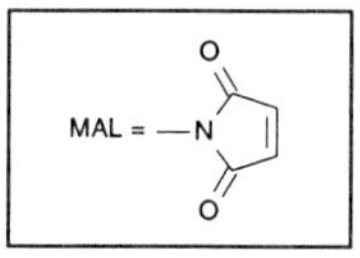

MAL =

Substituent Key

2.2 Price List *Thiol-Reactive Probes Excited with Visible Light*

Cat #		Product Name	Unit Size	Price Per Unit ($) 1–4 Units	Price Per Unit ($) 5–24 Units
B-2103		8-bromomethyl-4,4-difluoro-1,3,5,7-tetramethyl-4-bora-3a,4a-diaza-s-indacene (BODIPY® 493/503 methyl bromide)	5 mg	95.00	76.00
B-1355		5-(bromomethyl)fluorescein	10 mg	125.00	100.00
D-404		7-diethylamino-3-((4′-(iodoacetyl)amino)phenyl)-4-methylcoumarin (DCIA)	25 mg	158.00	126.40
D-346		7-diethylamino-3-(4′-maleimidylphenyl)-4-methylcoumarin (CPM)	25 mg	158.00	126.40
D-2005		N-(4,4-difluoro-5,7-dimethyl-4-bora-3a,4a-diaza-s-indacene-3-propionyl)-N′-iodoacetyl ethylenediamine (BODIPY® FL IA)	5 mg	95.00	76.00
D-6003	***New***	N-(4,4-difluoro-5,7-dimethyl-4-bora-3a,4a-diaza-s-indacene-3-yl)methyl)iodoacetamide (BODIPY® FL C_1 IA)	5 mg	95.00	76.00
D-2006		N-(4,4-difluoro-5,7-diphenyl-4-bora-3a,4a-diaza-s-indacene-3-propionyl)-N′-iodoacetylethylenediamine (BODIPY® 530/550 IA)	5 mg	95.00	76.00
D-6004	***New***	N-(4,4-difluoro-1,3,5,7-tetramethyl-4-bora-3a,4a-diaza-s-indacene-2-yl)iodoacetamide (BODIPY® 507/545 IA)	5 mg	95.00	76.00
E-99		eosin-5-iodoacetamide	100 mg	155.00	124.00
E-118		eosin-5-maleimide	25 mg	155.00	124.00

Cat #		Product Name	Unit Size	Price Per Unit ($) 1–4 Units	5–24 Units
E-6051	**New**	1-(2,3-epoxypropyl)-4-(5-(4-methoxyphenyl)oxazol-2-yl)pyridinium trifluoromethanesulfonate (PyMPO epoxide)	5 mg	98.00	78.40
E-333		erythrosin-5-iodoacetamide	25 mg	155.00	124.00
F-150		fluorescein-5-maleimide	25 mg	155.00	124.00
I-3		5-iodoacetamidofluorescein (5-IAF)	100 mg	138.00	110.40
I-15		6-iodoacetamidofluorescein (6-IAF)	100 mg	138.00	110.40
M-6026	**New**	1-(2-maleimidylethyl)-4-(5-(4-methoxyphenyl)oxazol-2-yl)pyridinium methanesulfonate (PyMPO maleimide)	5 mg	98.00	78.40
O-6010	**New**	Oregon Green™ 488 iodoacetamide *mixed isomers*	5 mg	95.00	76.00
T-6006	**New**	tetramethylrhodamine-5-iodoacetamide (5-TMRIA) *single isomer*	5 mg	175.00	140.00
T-6027	**New**	tetramethylrhodamine-5-maleimide *single isomer*	5 mg	125.00	100.00
T-6028	**New**	tetramethylrhodamine-6-maleimide *single isomer*	5 mg	125.00	100.00
T-6009	**New**	Texas Red® C_5 bromoacetamide	5 mg	95.00	76.00
T-6008	**New**	Texas Red® C_2 maleimide	5 mg	95.00	76.00

2.3 Environment- and Conformation-Sensitive Probes

The spectra of certain dyes tend to be particularly sensitive to ligand and metal binding, protein association and chaotropic reagents. When protein conjugates of these dyes are denatured or undergo a change in conformation, a decrease in fluorescence intensity and a shift in emission to longer wavelengths is often observed. Sensitivity to conformation and environment makes these probes especially useful for investigating protein structure and assembly, following protein transport through membranes and studying ligand binding to receptors. Coumarin (see Section 2.2), benzoxadiazole (NBD and SBD) and aminonaphthalene (e.g., dansyl) fluorophores are particularly susceptible to solvent-induced spectral shifts.

Benzoxadiazole Derivatives, Including NBD Probes

NBD Chloride and NBD Fluoride

Benz-2-oxa-1,3-diazoles (also called benzofurazans) are a diverse group of reactive dyes that include both nitrated derivatives of the NBD series and sulfonated analogs of the SBD series. NBD chloride (C-10) and the more reactive NBD fluoride (F-486) are common reagents for amine modification (see Section 1.8). However, they also react with thiols [1-3] and cysteine in several proteins [4-8] to yield thioethers (Table 2.1). NBD conjugates of thiols usually have much shorter-wavelength absorption and weaker fluorescence than do NBD conjugates of amines.[3] Selective modification of cysteines in the presence of reactive lysines and tyrosines is promoted by carrying out the reaction at pH <7;[9,10] however, NBD-thiol conjugates are often unstable, resulting in time-dependent label migration to adjacent lysine residues.[3,10]

SBD Probes

Thiol conjugates of benz-2-oxa-1,3-diazole-4-sulfonamides [11,12] (ABD-F, F-6053) and morpholinosulfonamide (C-6055) are much more stable in aqueous solution than are the thiol conjugates prepared from NBD chloride or NBD fluoride.[11] These SBD probes are nonfluorescent until reacted with thiols and therefore can be used to quantitate thiols in solution,[13] as well as thiols separated by HPLC [14,15] or TLC.[16] Subpicomole sensitivity has been reported for determination of cysteine, glutathione and other thiols separated by reversed-phase HPLC.[17,18] ABD-cysteine conjugates are very stable to acid hydrolysis but labeling is partially reversed in basic solution containing DTT [19] (D-1532, see Section 2.1). ABD-F can also be combined with tributylphosphine for the determination of disulfides in peptides and proteins.[20,21] As possible alternatives to the malodorous tributylphosphine, we recommend TCEP (tris-(2-carboxyethyl)phosphine, T-2556; see Section 2.1) or the less polar tris-(2-cyanoethyl)phosphine (T-6052, see Section 2.1). Due to its high polarity, TCEP may selectively reduce only those disulfides that are located on protein surfaces.

IANBD Ester and IANBD Amide

When conjugating the NBD fluorophore to thiols located in hydrophobic sites of proteins, we recommend using the iodoacetate ester (IANBD ester, I-9) or, preferably, the more hydrolytically stable iodoacetamide [22] (IANBD amide, D-2004). These reactive reagents exhibit appreciable fluorescence only after reaction with buried or unsolvated thiols, and this fluorescence is highly sensitive to changes in the solvation level of the fluorophore. Site-selective labeling of myosin in glycerinated muscle fibers with IANBD ester has been reported.[23]

Naphthalene Derivatives, Including Thiol-Specific Didansyl Cystine

Acrylodan and Its Bromoacetyl Analog

Although acrylodan (A-433) and the new 6-bromoacetyl-2-dimethylaminonaphthalene (badan, B-6057) generally react with thiols more slowly than do iodoacetamides or maleimides, they form very strong thioether bonds that are expected to remain stable under conditions required for complete amino acid analysis. The fluorescence emission peak and intensity of these adducts (Figure 2.2) are particularly sensitive to conformational changes or ligand binding, making these dyes some of the most useful thiol-reactive probes for protein structure studies (Table 2.1). For example, the

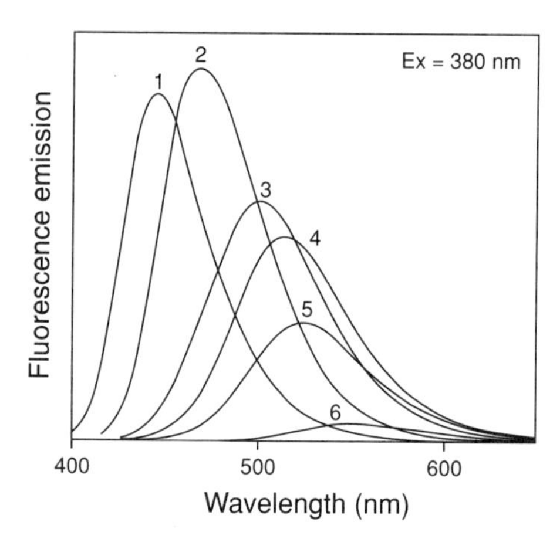

Figure 2.2 *Fluorescence emission spectra of 2-mercaptoethanol adducts of badan (B-6057) in: 1) toluene, 2) chloroform, 3) acetonitrile, 4) ethanol, 5) methanol and 6) water. Each solution contains the same concentration of adduct.*

acrylodan conjugate of an intestinal fatty acid–binding protein, ADIFAB (A-3880, see Section 20.6), is a sensor for free fatty acids.[24] Also, myosin regulatory light chain labeled with acrylodan shows a spectral response upon phosphorylation by myosin light chain kinase.[25] The environment-sensitive spectral shifts of acrylodan and badan conjugates may make these probes useful for distinguishing thiols that are located in membranes versus those exposed to aqueous solvation in cells.

IAANS and MIANS

To develop appreciable fluorescence, both the reactive anilinonaphthalenesulfonate iodoacetamide (IAANS, I-7) and maleimide (MIANS, M-8) must be reacted with thiols that are located in hydrophobic sites. Often, however, buried unsolvated thiol residues are exceptionally reactive, allowing these sites to be selectively modified by these reagents. Like most other maleimides, MIANS (also called Mal-ANS) is nonfluorescent until it has reacted with a thiol. Several cytoskeletal proteins, membrane ATPases and other proteins have been investigated using these probes (Table 2.1).

IAEDANS

The fluorescence of IAEDANS (I-14) is quite dependent upon environment; its conjugates frequently respond to ligand binding by undergoing spectral shifts and changes in fluorescence intensity that are determined by the degree of aqueous solvation. Advantages of this reagent include high water solubility above pH 4 and a relatively long fluorescence lifetime (sometimes >20 nanoseconds, although commonly 10–15 nanoseconds), making the conjugates useful for polarization and rotational studies.[26-29] In addition, because it has a large Stokes shift [30] and an emission that overlaps well with the absorption of fluorescein and BODIPY FL dyes, IAEDANS is an excellent reagent for fluorescence resonance energy transfer measurements of proximity up to about 60 Å.[31,32] IAEDANS usually reacts with thiols; however, it has been reported to react with a lysine residue in tropomyosin.[33]

Dansyl Aziridine

Another probe with environmentally sensitive spectral properties, dansyl aziridine [34] (D-151), forms very strong thioether bonds that remain stable under conditions required for complete amino acid analysis. Although it reacts with a methionine residue in troponin C,[35,36] dansyl aziridine is primarily a thiol-reactive reagent (Table 2.1). The fluorescence of the calmodulin and troponin C conjugates of dansyl aziridine is sensitive to Ca^{2+} binding.[36-41]

Didansyl-L-Cystine

Disulfides such as didansyl-L-cystine (D-146) are the only type of fluorescent thiol-*specific* reagents available. Disulfide derivatives undergo a thiol–disulfide interchange reaction to form mixed disulfides (Figure 2.1D).[42] The fluorescent disulfide that is initially formed, however, can subsequently transfer its fluorophore to neighboring thiols. The disulfide linkage formed by didansyl cystine can also be cleaved with reagents such as dithiothreitol (D-1532, see Section 2.1).

Pyrene Derivatives

Pyrene Maleimide

Not only is pyrene maleimide (P-28) essentially nonfluorescent until it has reacted with thiols, but once excited, pyrene-thiol conjugates can interact to form excited-state dimers (excimers) that emit at longer wavelengths than the lone excited fluorophore. Pyrene maleimide conjugates have very long fluorescence lifetimes (>100 nanoseconds), giving proximal pyrene rings within 6–10 Å of each other ample time to form the spectrally altered excimer (see Figure 13.6 in Chapter 13). Excimers may form between labeled sites in a single protein, as it does in tropomyosin,[43-46] lens crystallins [47] and sarcoplasmic reticulum ATPase,[48,49] or between sites in interacting biomolecules. Excimer formation can be used to monitor diffusion or to define interacting molecules within a functional unit of assembled biomolecules. Despite its low solubility, pyrene maleimide has been conjugated to several proteins (Table 2.1) and used as an HPLC derivatization reagent for thiols and reduced disulfides.[50] In several papers, *N*-(1-pyrene)maleimide (P-28) has been incorrectly named *N*-(3-pyrene)maleimide or variants of that nomenclature.

Pyrene Iodoacetyl Derivatives

The fluorescence of the actin monomer labeled with pyrene iodoacetamide (P-29) has been demonstrated to change upon polymerization, making this probe an excellent tool for following the kinetics of actin polymerization.[51-53] Using pyrene iodoacetamide, researchers can investigate the influence of several actin-binding proteins and of cytochalasin on the rate of actin polymerization. Conjugates from pyrenemethyl iodoacetate (P-4) and the more hydrolytically stable pyrenemethyl-, pyreneethyl- and pyrenepropyliodoacetamides (P-2007, P-6030, P-6031) have the longest excited-state fluorescence lifetimes (>100 nanoseconds) of all reported thiol-reactive probes. These pyrene-derived dyes may also form excited-state dimers with altered fluorescence emission.

1. Anal Chem 57, 1864 (1985); **2.** Anal Chem 55, 1786 (1983); **3.** FEBS Lett 6, 346 (1970); **4.** Arch Biochem Biophys 281, 6 (1990); **5.** Biochemistry 29, 10613 (1990); **6.** Biochemistry 29, 7309 (1990); **7.** J Biochem 107, 562 (1990); **8.** Biochim Biophys Acta 956, 217 (1988); **9.** J Biol Chem 266, 13777 (1991); **10.** J Biol Chem 258, 5419 (1983); **11.** Anal Chim Acta 290, 3 (1994); **12.** Anal Chem 56, 2461 (1984); **13.** Chem Pharm Bull 38, 2290 (1990); **14.** J Chromatography 553, 433 (1991); **15.** J Chromatography 514, 189 (1990); **16.** J Chromatography 502, 230 (1990); **17.** J Biochem 97, 1811 (1985); **18.** J Chromatography 282, 495 (1983); **19.** Techniques in Protein Chemistry V, 189 (1994); **20.** Anal Biochem 214, 128 (1993); **21.** J Biol Chem 264, 7185 (1989); **22.** J Biol Chem 270, 5395 (1995); **23.** Biochemistry 28, 2204 (1989); **24.** J Biol Chem 267, 23495 (1992); **25.** J Biol Chem 269, 12880 (1994); **26.** Biochemistry 24, 3731 (1985); **27.** Biochim Biophys Acta 773, 321 (1984); **28.** Proc Natl Acad Sci USA 73, 133 (1976); **29.** Biochemistry 12, 2250 (1973); **30.** Biochemistry 12, 4154 (1973); **31.** Biopolymers 27, 821 (1988); **32.** J Muscle Res Cell Motil 8, 97 (1987); **33.** Eur J Biochem 187, 155 (1990); **34.** Biochim Biophys Acta 336, 421 (1974); **35.** Biochemistry 28, 6751 (1989); **36.** J Biol Chem 253, 6451 (1978); **37.** Biochemistry 30, 5238 (1991); **38.** Biopolymers 26, 1189 (1987); **39.** J Muscle Res Cell Motil 8, 428 (1987); **40.** Meth Enzymol 102, 148 (1983); **41.** Biochemistry 21, 5135 (1982); **42.** Science 185, 1176 (1974); **43.** Biochemistry 26, 4922 (1987); **44.** Biochemistry 24, 6631 (1985); **45.** J Biol Chem 255, 11296 (1980); **46.** J Biol Chem 253, 3757 (1978); **47.** J Biol Chem 265, 14277 (1990); **48.** Biophys J 51, 513 (1987); **49.** Eur J Biochem 130, 5 (1983); **50.** J Chromatography 564, 258 (1991); **51.** J Biochem 116, 236 (1994); **52.** J Muscle Res Cell Motil 4, 235 (1983); **53.** Eur J Biochem 114, 33 (1981).

2.3 Data Table *Environment- and Conformation-Sensitive Probes*

Cat #	Structure	MW	Storage	Soluble	Abs	{ϵ x 10^{-3}}	Em	Solvent	Notes
A-433	✓	225	L	DMF, MeCN	391	{20}	500	MeOH	<1>
B-6057	✓	292	F,L	DMF, MeCN	387	{21}	520	MeOH	<2>
C-10	S1.8	200	F,D,L	DMF, MeCN	336	{9.8}	none	MeOH	<3>
C-6055	✓	304	F,D,L	DMF, DMSO	329	{7.9}	none	MeOH	<4>
D-146	✓	707	L	pH>6	328	{8.4}	563	pH 7	<5>
D-151	✓	276	F,D,L	DMF, MeCN	340	{4.1}	543	MeOH	<6>
D-2004	✓	419	F,D,L	DMF, DMSO	478	{25}	541	MeOH	<7, 8>
F-486	S1.8	183	F,DD,L	MeCN, $CHCl_3$	328	{8.0}	none	MeOH	<3>
F-6053	✓	217	F,D,L	DMF, DMSO	320	{4.8}	none	MeOH	<4>
I-7	✓	504	F,D,L	DMF	327	{26}	463	MeOH	<7, 8>
I-9	✓	406	F,D,L	DMF, MeCN	472	{23}	536	MeOH	<7, 8>
I-14	✓	434	F,D,L	pH>6	336	{5.7}	490	pH 8	<8, 9>
M-8	✓	416	F,D,L	DMF, MeOH	313	{14}	420	MeOH	<7>
P-4	✓	400	F,D,L	DMF, MeCN	341	{43}	376	MeOH	<7, 8, 10>
P-28	✓	297	F,D,L	DMF, DMSO	338	{40}	375	MeOH	<7, 10>
P-29	✓	385	F,D,L	DMF, DMSO	339	{26}	384	MeOH	<7, 8>
P-2007	✓	399	F,D,L	DMSO	341	{41}	377	MeOH	<7, 8, 10>
P-6030	✓	413	F,D,L	DMSO	341	{46}	376	MeOH	<7, 8, 10>
P-6031	✓	427	F,D,L	DMSO	341	{43}	376	MeOH	<7, 8, 10>

For definitons of the contents of this data table, see "How to Use the Handbook" on page vi.
Structure: Chemical structure drawing: ✓ = shown in this section; Sn.n = shown in section number n.n.

<1> Fluorescence of A-433 is weak, increasing markedly upon reaction with thiols. Em (QY) for the 2-mercaptoethanol adduct are: 540 nm (0.18) in H_2O, 513 nm (0.57) in MeOH, 502 nm (0.79) in EtOH, 468 nm (0.78) in MeCN, 435 nm (0.83) in dioxane [J Biol Chem 258, 7541 (1983)].

<2> Em for 2-mercaptoethanol adduct of B-6057: 550 nm in H_2O (pH 7), 523 nm in MeOH, 514 nm in EtOH, 502 nm in MeCN, 469 nm in $CHCl_3$, 457 nm in dioxane, 445 nm in toluene. Abs is relatively independent of solvent.

<3> Spectra of 2-mercaptoethanol adduct of C-10 in MeOH: Abs = 425 nm (ϵ = {13}), Em = 520 nm. Emission spectra of adducts are highly environment dependent, and all NBD products are almost nonfluorescent in water. F-486 gives the same derivatives as C-10 but is more reactive.

<4> Spectra of reaction products with dimethylaminoethanethiol in methanol: for F-6053, Abs = 376 nm (ϵ ~{8.0}), Em ~510 nm; for C-6055, Abs = 386 nm (ϵ ~{9.0}), Em ~485 nm.

<5> D-146 thiol conjugate has ϵ = {4.0}. Em = 527 nm on G-actin, shifting to 520 nm upon polymerization [Arch Biochem Biophys 142, 333 (1971)].

<6> D-151 conjugated to calmodulin with and without Ca^{2+} has Em = 532 nm and 550 nm, respectively [Meth Enzymol 102, 148 (1984)]. On troponin I, Em = 500 nm (with Ca^{2+}) and 550 nm (without Ca^{2+}) [Biochemistry 21, 5669 (1982)]. Fluorescence lifetimes of conjugates are 14 to 20 nsec (Ca^{2+}-dependent).

<7> Spectral data of the 2-mercaptoethanol adduct.

<8> Iodoacetamides in solution undergo rapid photodecomposition to unreactive products. Minimize exposure to light prior to reaction.

<9> 2-Mercaptoethanol derivative of I-14 has essentially similar spectral characteristics in aqueous solution [Biochemistry 12, 4154 (1973)]. Fluorescence lifetime = 21 nsec when conjugated to myosin subfragment-1 [Biochemistry 12, 2250 (1973)].

<10> Pyrene derivatives exhibit structured spectra. The absorption maximum is usually about 340 nm with a subsidiary peak at about 325 nm. There are also strong absorption peaks below 300 nm. The emission maximum is usually about 376 nm with a subsidiary peak at 396 nm. Excimer emission at about 470 nm may be observed at high concentrations.

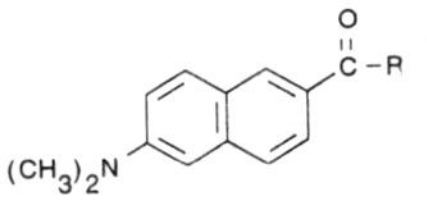

2.3 Structures *Environment- and Conformation-Sensitive Probes*

A-433 R = —CH=CH$_2$
B-6057 R = —CH$_2$Br

C-10
Section 1.8

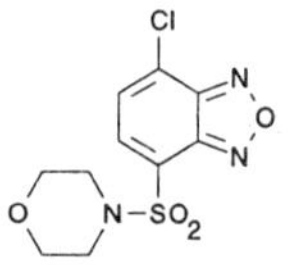

C-6055

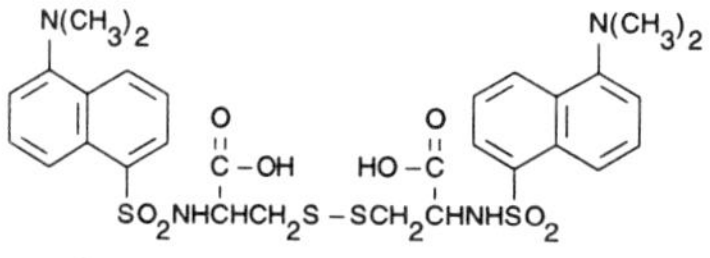

D-146

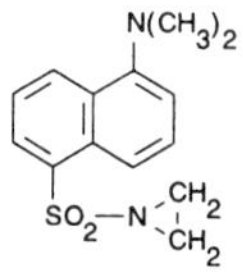

D-151

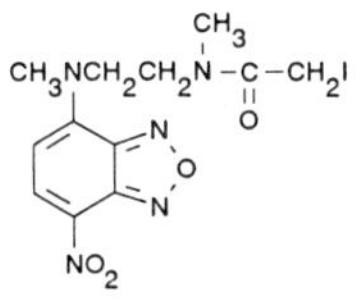

D-2004

F-486
Section 1.8

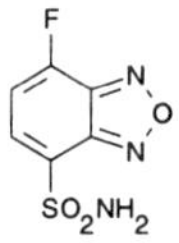

F-6053

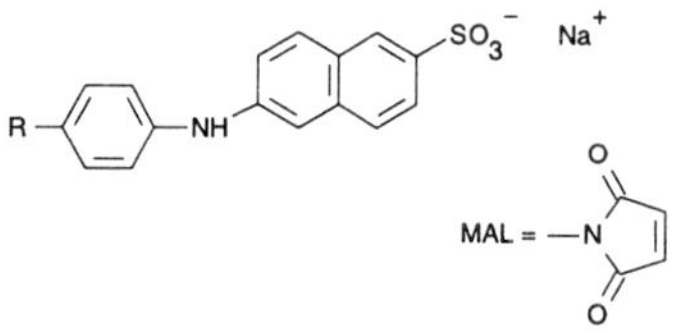

I-7 R = —NH—C(=O)—CH$_2$I
M-8 R = MAL

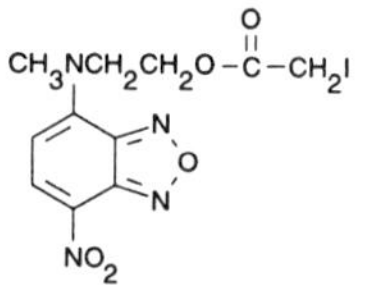

I-9

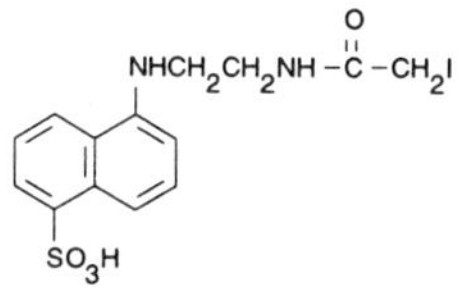

I-14

M-8
see I-7

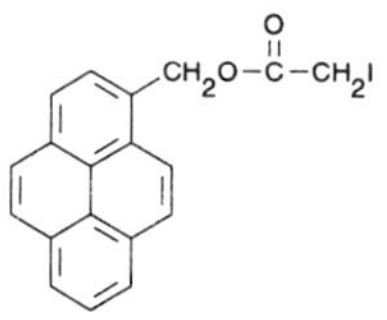

P-4

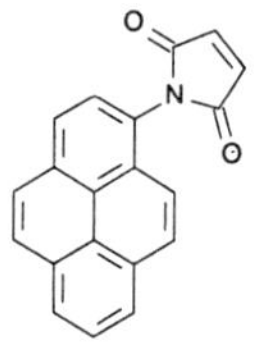

P-28

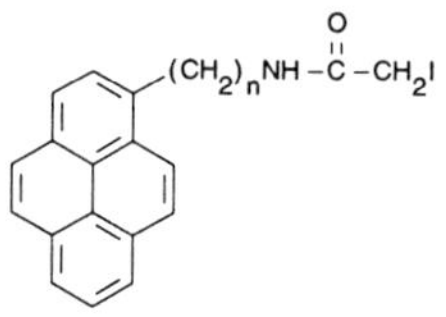

P-29 n = 0
P-2007 n = 1
P-6030 n = 2
P-6031 n = 3

2.3 Price List *Environment- and Conformation-Sensitive Probes*

Cat #		Product Name	Unit Size	Price Per Unit ($) 1–4 Units	5–24 Units
A-433		6-acryloyl-2-dimethylaminonaphthalene (acrylodan)	25 mg	155.00	124.00
B-6057	**New**	6-bromoacetyl-2-dimethylaminonaphthalene (badan)	10 mg	75.00	60.00
C-6055	**New**	N-(7-chlorobenz-2-oxa-1,3-diazol-4-yl)sulfonyl morpholine	10 mg	45.00	36.00
C-10		4-chloro-7-nitrobenz-2-oxa-1,3-diazole (NBD chloride; 4-chloro-7-nitrobenzofurazan)	1 g	24.00	19.20
D-146		didansyl-L-cystine	100 mg	48.00	38.40
D-151		5-dimethylaminonaphthalene-1-sulfonyl aziridine (dansyl aziridine)	100 mg	68.00	54.40
D-2004		N,N′-dimethyl-N-(iodoacetyl)-N′-(7-nitrobenz-2-oxa-1,3-diazol-4-yl)ethylenediamine (IANBD amide)	25 mg	95.00	76.00
F-6053	**New**	7-fluorobenz-2-oxa-1,3-diazole-4-sulfonamide (ABD-F)	10 mg	55.00	44.00
F-486		4-fluoro-7-nitrobenz-2-oxa-1,3-diazole (NBD fluoride; 4-fluoro-7-nitrobenzofurazan)	25 mg	95.00	76.00
I-7		2-(4′-(iodoacetamido)anilino)naphthalene-6-sulfonic acid, sodium salt (IAANS)	100 mg	95.00	76.00
I-9		N-((2-(iodoacetoxy)ethyl)-N-methyl)amino-7-nitrobenz-2-oxa-1,3-diazole (IANBD ester)	100 mg	95.00	76.00
I-14		5-((((2-iodoacetyl)amino)ethyl)amino)naphthalene-1-sulfonic acid (1,5-IAEDANS)	100 mg	35.00	28.00
M-8		2-(4′-maleimidylanilino)naphthalene-6-sulfonic acid, sodium salt (MIANS)	100 mg	95.00	76.00
P-6030	**New**	N-(1-pyreneethyl)iodoacetamide	5 mg	48.00	38.40
P-29		N-(1-pyrene)iodoacetamide	100 mg	115.00	92.00
P-28		N-(1-pyrene)maleimide	100 mg	75.00	60.00
P-2007		N-(1-pyrenemethyl)iodoacetamide (PMIA amide)	25 mg	95.00	76.00
P-4		1-pyrenemethyl iodoacetate (PMIA ester)	100 mg	95.00	76.00
P-6031	**New**	N-(1-pyrenepropyl)iodoacetamide	10 mg	65.00	52.00

2.4 Other Thiol-Reactive Reagents

Bimanes for Thiol Derivatization

The monobromobimanes (M-1378, M-1380), which are essentially nonfluorescent until conjugated, readily react with several low molecular weight thiols including glutathione,[1-3] *N*-acetylcysteine,[4] mercaptopurine,[5] peptides[6] and plasma thiols,[7] as well as with carboxylic acids (see Section 3.4). Although monobromobimane (M-1378) is the most extensively used reagent, its derivative monobromotrimethylammoniobimane (M-1380) contains a positive charge, thus permitting separation of its conjugates by electrophoresis[8] or cation-exchange chromatography.[9] The membrane permeability of this charged bimane as well as its ability to access sites within a protein may differ from that of the uncharged bimane probes.[10,11] These reagents, which were originally described by Kosower and colleagues,[12,13] are also useful for detecting the distribution of protein thiols in cells before and after chemical reduction of disulfides.[14] Both monobromobimane and the more thiol-selective monochlorobimane (M-1381) have been extensively used for detecting glutathione in live cells (see Section 16.3).

Dibromobimane (D-1379) is an interesting crosslinking reagent for proteins because it is unlikely to fluoresce until *both* of its alkylating groups have reacted. It has been used to crosslink thiols in myosin,[15-17] hemoglobin[18] and mitochondrial ATPase.[19]

Polar Reagents for Determining Thiol Accessibility

Like IAEDANS mentioned above, the iodoacetamide and maleimide derivatives of stilbene and lucifer yellow (A-484, A-485 and L-1338) have high water solubility and are readily conjugated to thiols. Their combination of high polarity and membrane impermeability may make these polysulfonates useful for determining whether thiol-containing proteins and polypeptide chains are exposed at the extracellular or cytoplasmic membrane surface. Their protein adducts are charged and can be detected by gel or capillary electrophoresis. Lucifer yellow iodoacetamide (L-1338) reacts with the single exposed thiol of low-density lipoproteins (LDL) without reacting with the two buried thiols of the lipoprotein(a) component, whereas acrylodan (A-433, see Section 2.3) reacts with all three thiols.[20] Lucifer yellow iodoacetamide has also been used in a flow cytometry assay of accessible thiols on the cell surface.[21] The sulfonated stilbene iodoacetamide (A-484) was used to surface label key single-cysteine mutants of staphylococcal α-hemolysin in order to determine structural changes that occur during oligomerization and pore formation.[22]

Salicylic Acid Derivatives for Time-Resolved Assays

Proteins derivatized with the salicylic acid iodoacetamide (I-106) or maleimide (M-84) complex with terbium (T-1247, see Section 9.2), sensitizing its luminescence. Because the excited-state lifetime of terbium is typically 0.4 milliseconds, these complexes can be used for time-resolved assays, particularly in liquid chromatography.[23-25] The iodoacetamide derivative is a specific inhibitor of glutamate dehydrogenase and other enzymes.[26] The maleimide derivative, which is water soluble at neutral pH, has low fluorescence until reacted with thiols, making it potentially useful for thiol quantitation.

1,10-Phenanthroline Haloacetamides for Preparing Metal-Binding Conjugates

Conjugation of the bromoacetamide or iodoacetamide of 1,10-phenanthroline (P-6878, P-6879) to thiol-containing ligands confers the metal-binding properties of this important complexing agent on the ligand. For example, the covalent copper–phenanthroline complex of oligonucleotides or nucleic acid–binding molecules in combination with hydrogen peroxide acts as a chemical nuclease to selectively cleave DNA or RNA (see Section 8.3).[27-32]

Reagents for Quantitating Thiols

Thiol and Sulfide Quantitation Kit

Ultrasensitive colorimetric quantitation of both protein and nonprotein thiols is now possible using our Thiol and Sulfide Quantitation Kit (T-6060). In this assay, which is based on a method reported by Singh,[33,34] thiols or sulfides reduce a disulfide-inhibited derivative of papain, stoichiometrically releasing the active enzyme. Activity of the enzyme is then measured using the chromogenic papain substrate L-BAPNA. Although thiols and inorganic sulfides can also be quantitated using 5,5'-dithiobis-(2-nitrobenzoic acid) (DTNB or Ellman's reagent, D-8451), the enzymatic amplification step in our quantitation kit yields a sensitivity for detection of thiols or sulfides of approximately 0.2 nanomoles, or about 100-fold better than that obtained with DTNB. Thiols in proteins and potentially other high molecular weight molecules can be detected indirectly by incorporating the disulfide cystamine into the reaction mixture. Cystamine undergoes an exchange reaction with protein thiols, yielding 2-mercaptoethylamine (cysteamine), which then releases active papain. The Thiol and Sulfide Quantitation Kit contains:

- Papain-$SSCH_3$, the disulfide-inhibited papain derivative
- L-BAPNA, a chromogenic papain substrate
- DTNB (Ellman's reagent), for calibrating the assay
- Cystamine
- L-Cysteine, a thiol standard
- Reaction buffer
- Protocols for measuring thiols, inorganic sulfides and maleimides

Sufficient reagents are provided for approximately 50 assays using standard 1 mL cuvettes or 250 assays using a microplate format. This kit can also be used to detect phosphines, sulfites and cyanides with detection limits of about 0.5, 1 and 5 nanomoles, respectively (see Section 24.3).

Ellman's Reagent (DTNB) for Quantitating Thiols

Ellman's reagent (5,5'-dithiobis-(2-nitrobenzoic acid) or DTNB, D-8451), is an important reagent for quantitating thiols in proteins, cells and plasma by absorption measurements.[35] It readily forms a mixed disulfide with thiols, liberating the chromophore 5-mercapto-2-nitrobenzoic acid (absorption maximum 410 nm, ϵ ~13,600 $cm^{-1}M^{-1}$.[36] Only protein thiols that are accessible to this water-soluble reagent are modified.[37,38] Inaccessible thiols can usually be quantitated by carrying out the titration in the presence of 6 M guanidinium chloride. DTNB conjugates of glutathione and other thiols have been separated by HPLC.[39] A new cleavable crosslinking reagent based on DTNB is described in Section 5.2.

Reagents for Protein Sequencing Applications

N-(2-Iodoethyl)trifluoroacetamide (I-1536) can be used to convert thiols to amines for modification by amine-reactive reagents. Furthermore, modification of cysteine residues in proteins by *N*-(2-iodoethyl)trifluoroacetamide makes these sites susceptible to cleavage by trypsin, thus providing a means for selectively preparing additional fragments for protein sequencing.[40]

N-Isopropyl iodoacetamide (NIPIA, I-6002) has been recommended for use in the phenylthiohydantoin (PTH) microsequencing of reductively alkylated proteins. The NIPIA-alkylated cysteine PTH can be detected as a sharp peak in standard reverse-phase HPLC analysis of PTH amino acids.[41]

Other Fluorescent and Nonfluorescent Thiol-Reactive Reagents

Maleimides for HPLC Derivatization

The fluorogenic maleimide *N*-(4-(6-dimethylamino-2-benzofuranyl)phenyl)maleimide (D-1333) and the chromophoric maleimide DABMI (D-1521) are principally used to derivatize thiols for HPLC detection.[42,43] The broad visible absorption of DABMI conjugates makes DABMI one of the best nonfluorescent thiol-reactive acceptors for fluorescence resonance energy transfer studies.[44,45]

Methyl Methanethiolsulfonate

Methyl methanethiolsulfonate (MMTS, M-6539) is a relatively compact disulfide that reversibly blocks thiol groups of proteins. Researchers have used this disulfide to modify a highly reactive thiol of creatine kinase and demonstrated that this thiol is not essential for the enzyme's activity.[46] The disulfide-inhibited papain in Molecular Probes' Thiol and Sulfide Quantitation Kit (T-6060, see above) has been blocked with MMTS and can be reactivated by thiols, thus forming the basis of our ultrasensitive thiol quantitation assay.

Vinyl Sulfone

Vinyl sulfones are extensively used as a gelatin hardener in the photographic film industry. These functional groups form very stable thioether adducts.[47] UniBlue A vinyl sulfone (U-6059) is an inexpensive water-soluble reagent for modifying thiols or amines with a long-wavelength, nonfluorescent chromophore.

3-Bromopyruvic Acid

The reagent 3-bromopyruvic acid (B-6058) converts thiols to a ketone derivative. Interconversion of organic functional groups in this manner greatly expands the available probes because the ketone derivative can potentially be coupled to any of the various hydrazines and amines listed in Chapter 3.

TEMPO Iodoacetamide

The spin-labeled TEMPO iodoacetamide (I-591) has been used extensively in protein structure studies.[48] It can also act as a short-range quencher of fluorescent probes.

1. J Biol Chem 263, 14107 (1988); **2.** J Am Chem Soc 108, 4527 (1986); **3.** J Pharm Pharmacol 35, 384 (1983); **4.** Anal Lett 21, 741 (1988); **5.** J Chromatography 309, 409 (1984); **6.** Histochemistry 86, 281 (1987); **7.** J Chromatography 424, 141 (1988); **8.** Anal Biochem 107, 1 (1980); **9.** Anal Biochem 111, 357 (1981); **10.** Arch Biochem Biophys 282, 309 (1990); **11.** J Leukocyte Biol 45, 177 (1989); **12.** Meth Enzymol 143, 76 (1987); **13.** Proc Natl Acad Sci USA 76, 3382 (1979); **14.** Mol Reprod Devel 37, 318 (1994); **15.** Biochem Biophys Res Comm 152, 1 (1988); **16.** Biochemistry 26, 1889 (1987); **17.** Proc Natl Acad Sci USA 82, 1658 (1985); **18.** Biochim Biophys Acta 622, 201 (1980); **19.** FEBS Lett 150, 207 (1982); **20.** Biochemistry 30, 11245 (1991); **21.** Biochem Soc Trans 23, 85 (1995); **22.** FEBS Lett 356, 66 (1994); **23.** FEBS Lett 333, 96 (1993); **24.** J Chromatography 552, 625 (1991); **25.** Anal Chem 62, 2051 (1990); **26.** Biochemistry 23, 3789 (1984); **27.** Nucleic Acids Res 22, 4789 (1994); **28.** Bioconjugate Chem 4, 69 (1993); **29.** Meth Enzymol 208, 414 (1991); **30.** Ann Rev Biochem 59, 207 (1990); **31.** J Am Chem Soc 111, 4941 (1989); **32.** Proc Natl Acad Sci USA 86, 9702 (1989); **33.** Bioconjugate Chem 5, 348 (1994); **34.** Anal Biochem 213, 49 (1993); **35.** Biochem J 89, 296 (1963); **36.** Meth Enzymol 233, 380 (1994); **37.** Meth Enzymol 143, 45 (1987); **38.** Meth Enzymol 91, 49 (1983); **39.** Anal Biochem 138, 95 (1984); **40.** Anal Biochem 106, 43 (1980); **41.** Anal Biochem 209, 109 (1993); **42.** Chromatography 414, 11 (1987); **43.** Biochem J 211, 163 (1983); **44.** Biochemistry 34, 6475 (1995); **45.** Biochemistry 33, 13102 (1994); **46.** Biochemistry 14, 766 (1975); **47.** J Am Chem Soc 103, 7615 (1981); **48.** Biochemistry 19, 1219 (1980).

Product-Specific Bibliographies and Keyword Searches

Our Technical Assistance Department can provide you with product-specific bibliographies, as well as keyword searches of the over 25,000 literature references in our extensive bibliography database. Our bibliography database is also searchable through our Web site (http://www.probes.com).

2.4 Data Table *Other Thiol-Reactive Reagents*

Cat #	Structure	MW	Storage	Soluble	Abs	{ε × 10⁻³}	Em	Solvent	Notes
A-484	✓	624	F,D,L	H_2O	329	{39}	408	pH 8	<1, 2>
A-485	✓	536	F,D	H_2O	322	{29}	411	pH 8	<1>
B-6058	S3.2	167	F,D	MeOH	<300		none		
D-1333	✓	332	F,D,L	DMF, MeCN	357	{33}	444	MeOH	<1>
D-1379	✓	350	L	DMF, MeCN	391	{6.1}	<3>	MeOH	
D-1521	✓	320	F,D,L	DMF, MeCN	419	{34}	none	MeOH	<1>
D-8451	S5.2	396	D	pH>6	324	{11}	none	pH 8	<4>
I-106	✓	321	F,D,L	pH>6, DMF	304	{9.5}	402	pH 9	<1, 2>
I-591	✓	338	F,D,L	EtOH	<300		none		<2>
I-1536	✓	267	F	DMF, MeCN	<300		none		
I-6002	✓	227	F,D	DMF, MeCN	<300		none		<2>
L-1338	✓	649	F,D,L	H_2O	426	{11}	531	pH 7	<1, 2>
M-84	✓	233	F,D	pH>6, DMSO	307	{3.4}	408	MeOH	<1>
M-1378	✓	271	F,L	DMF, MeCN	398	{5.0}	<3>	pH 7	
M-1380	✓	409	L	H_2O*	378	{5.5}	<3>	pH 7	
M-1381	✓	227	F,L	DMSO	380	{6.0}	<3>	MeOH	
M-6539	✓	126	F,D	EtOH†	<300		none		
P-6878	✓	316	F,D,L	DMSO	270	{29}	none	$CHCl_3$	
P-6879	✓	363	F,D,L	DMSO	270	{28}	none	$CHCl_3$	<2>
U-6059	✓	506	F,D,L	H_2O	617	{11}	none	MeOH	

For definition of the contents of this data table, see "How to Use the Handbook" on page vi.
Structure: Chemical structure drawing: ✓ = shown in this section; Sn.n = shown in section number n.n.

* Unstable in water. Use immediately.
† This product is intrinsically a liquid at room temperature.
<1> Spectral data of the 2-mercaptoethanol adduct.
<2> Iodoacetamides in solution undergo rapid photodecomposition to unreactive products. Minimize exposure to light prior to reaction.
<3> D-1379, M-1378, M-1380 and M-1381 are almost nonfluorescent until reacted with thiols. Em = 475–485 nm for thiol adducts (QY ~0.1–0.3) [Meth Enzymol 143, 76 (1987)].
<4> D-8451 reaction product with thiols has Abs = 410 nm (ε = {14}) [Meth Enzymol 233, 380 (1994)].

2.4 Structures *Other Thiol-Reactive Reagents*

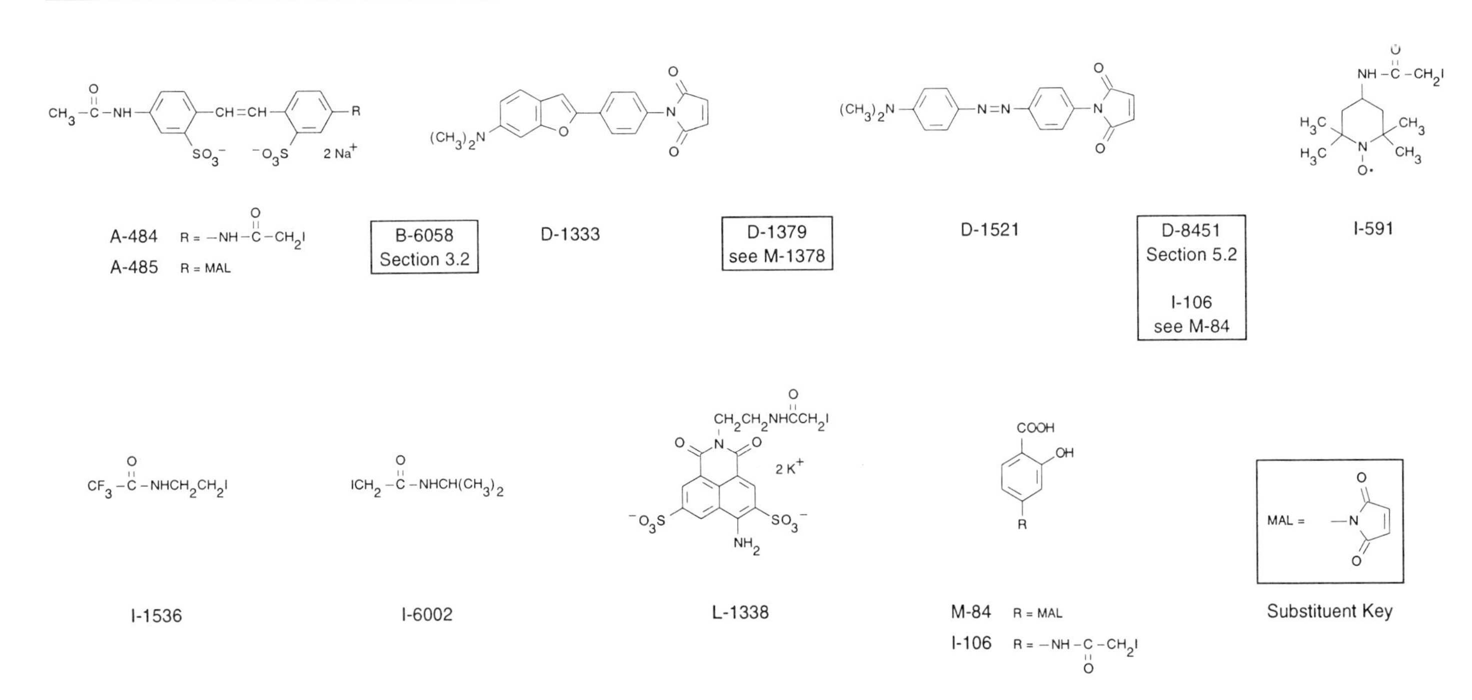

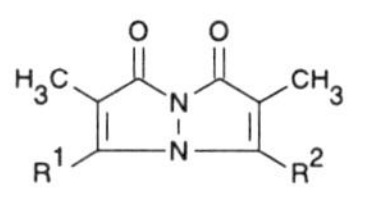

M-1378 $R^1 = -CH_3$ $R^2 = -CH_2Br$
D-1379 $R^1 = -CH_2Br$ $R^2 = -CH_2Br$
M-1380 $R^1 = -CH_3$ $R^2 = -CH_2\overset{+}{N}(CH_3)_3\ Br^-$
M-1381 $R^1 = -CH_2Br$ $R^2 = -CH_2Cl$

M-6539 $CH_3SO_2SCH_3$

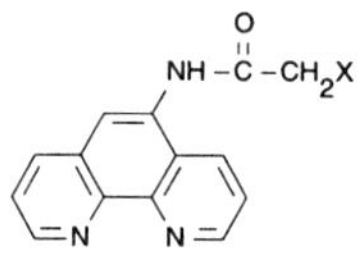

P-6878 X = Br
P-6879 X = I

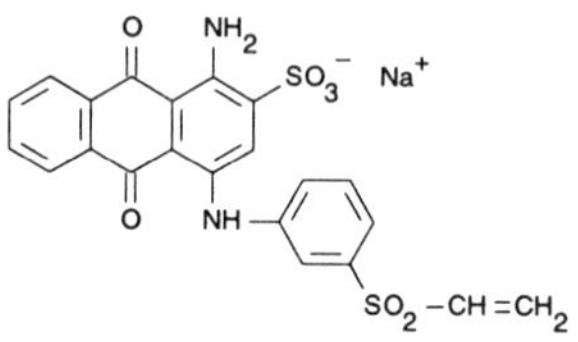

U-6059

2.4 Price List *Other Thiol-Reactive Reagents*

Cat #		Product Name	Unit Size	Price Per Unit ($) 1–4 Units	5–24 Units
A-484		4-acetamido-4′-((iodoacetyl)amino)stilbene-2,2′-disulfonic acid, disodium salt	25 mg	95.00	76.00
A-485		4-acetamido-4′-maleimidylstilbene-2,2′-disulfonic acid, disodium salt	25 mg	95.00	76.00
B-6058	***New***	3-bromopyruvic acid	1 g	8.00	6.40
D-1379		dibromobimane (bBBr)	25 mg	35.00	28.00
D-1333		N-(4-(6-dimethylamino-2-benzofuranyl)phenyl)maleimide	25 mg	38.00	30.40
D-1521		4-dimethylaminophenylazophenyl-4′-maleimide (DABMI)	100 mg	55.00	44.00
D-8451	***New***	5,5′-dithiobis-(2-nitrobenzoic acid) (DTNB; Ellman's reagent)	10 g	45.00	36.00
I-106		4-iodoacetamidosalicylic acid	100 mg	48.00	38.40
I-591		N-(4-(iodoacetyl)amino)-2,2,6,6-tetramethylpiperidinyl-1-oxyl (TEMPO iodoacetamide)	25 mg	55.00	44.00
I-1536		N-(2-iodoethyl)trifluoroacetamide	1 g	35.00	28.00
I-6002	***New***	N-isopropyl iodoacetamide (NIPIA)	100 mg	45.00	36.00
L-1338		lucifer yellow iodoacetamide, dipotassium salt	25 mg	95.00	76.00
M-84		4-maleimidylsalicylic acid	100 mg	48.00	38.40
M-6539	***New***	methyl methanethiolsulfonate (MMTS) *special packaging*	10x10 mg	35.00	28.00
M-1378		monobromobimane (mBBr)	25 mg	43.00	34.40
M-1380		monobromotrimethylammoniobimane bromide (qBBr)	25 mg	38.00	30.40
M-1381		monochlorobimane (mBCl)	25 mg	78.00	62.40
P-6878	***New***	N-(1,10-phenanthrolin-5-yl)bromoacetamide	5 mg	75.00	60.00
P-6879	***New***	N-(1,10-phenanthrolin-5-yl)iodoacetamide	5 mg	75.00	60.00
T-6060	***New***	Thiol and Sulfide Quantitation Kit *50-250 assays*	1 kit	95.00	76.00
U-6059	***New***	UniBlue A vinyl sulfone, sodium salt *>80% by HPLC*	100 mg	18.00	14.40

Thiol-Reactive Probes Discussed in Other Chapters of the Handbook

Molecular Probes offers a number of other thiol-reactive probes with speciality applications. For example, our DetectaGene™ β-galactosidase substrates and our CellTracker™ and MitoTracker™ probes are mildly thiol-reactive chloromethyl derivatives that show superior cellular retention when compared with their nonchloromethyl counterparts. These and other thiol-reactive probes are discussed in the chapter appropriate for their application.

Chapter 4 Biotins and Haptens — Including thiol-reactive biotinylation and haptenylation reagents.
Chapter 5 Crosslinking and Photoreactive Reagents — Including several reagents that contain iodoacetamide or maleimide groups for crosslinking biomolecules or synthesizing thiol-reactive probes, as well as reagents for introducing thiols into biopolymers.
Chapter 10 Enzymes, Enzyme Substrates and Enzyme Inhibitors — Including fluorogenic substrates for glycosidases, proteases and microsomal dealkylase that contain mildly thiol-reactive chloromethyl groups.
Chapter 12 Probes for Organelles — Including thiol-reactive MitoTracker mitochondrion-selective probes.
Chapter 15 Fluorescent Tracers of Cell Morphology and Fluid Flow — Including thiol-reactive CellTracker probes.
Chapter 17 Probes for Following Endocytosis and Exocytosis — Including a cell-permeant, thiol-reactive derivative of dichlorodihydrofluorescein.
Chapter 23 pH Indicators — Including a thiol-reactive, cell-permeant derivative of the carboxy SNARF®-1 pH indicator.

Chapter 3

Reagents for Modifying Groups Other Than Thiols or Amines

Contents

3.1 Reagents for Modifying Alcohols

Alcohols in Proteins: Serine, Threonine and Tyrosine Residues

Although alcohols (including phenols such as tyrosine and the alcohol groups in carbohydrates) are very abundant in biomolecules, their chemical reactivity in aqueous solution is low. Few reagents are selective for alcohols in aqueous solution, especially in the presence of more reactive nucleophiles such as thiols and amines. It is therefore difficult to selectively modify serine, threonine and tyrosine residues in proteins except when they exhibit unusual reactivity by residing at an enzyme's active site.

Serine and Threonine Residues

Several sulfonyl fluorides, including phenylmethanesulfonyl fluoride (PMSF) and acylating reagents, inhibit proteases by modifying their essential serine residues; these probes are discussed in detail in Section 10.6. Recent reports have described the apparent labeling of certain serine and threonine residues in myosin by 9-anthroylnitrile [1-4] (A-1440). As an alternative approach, nonacylated N-terminal serine and threonine residues in proteins can be oxidized with periodate to yield aldehydes [5-7] that can be subsequently modified with a variety of hydrazine derivatives (see Section 3.2).

Tyrosine Residues

Modification of tyrosine residues is sometimes a side reaction when proteins are reacted with sulfonyl chlorides, iodoacetamides or other reactive dyes described in Chapters 1 and 2. For example, NBD chloride (C-10, see Section 1.8) reacts with an active-site tyrosine in monoamine oxidase enzymes, causing strong inhibition.[8] NBD chloride also reacts with the tyrosine residues in glutathione *S*-transferase,[9] bacterial F_1-ATPase,[10] mitochondrial F_1-ATPase [11,12] and chloroplast H^+-ATPase.[13]

A tyrosine residue in a protein can be selectively modified by first nitrating the *ortho* position of its phenolic ring using tetranitromethane, and then reducing the *o*-nitrotyrosine with sodium dithionite to form an *o*-aminotyrosine. The *o*-aminotyrosine can react with most amine-reactive reagents (except succinimidyl esters) at pH ≤7.[14] This indirect method has been used to selectively modify tyrosine residues in actin with dansyl chloride and dabsyl chloride [15,16] (D-21, D-1537; see Section 1.8).

A second method for modifying a tyrosine residue is to react it with an aromatic diazonium salt such as the one prepared from the biotinylated aromatic amine N^{α}-(4-aminobenzoyl) biocytin [17] (A-1604, see Section 4.2), to yield a colored azo dye adduct. Although usually not fluorescent, this azo dye adduct can be reduced to an *o*-aminotyrosine using sodium dithionite and then reacted with amine-reactive reagents as described above.

A third method for modifying tyrosine groups in peptides is to convert the phenol group in tyrosine residues to a salicylaldehyde derivative, and then to react the salicylaldehyde with 1,2-diamino-4,5-dimethoxybenzene (D-1463, see Section 3.2) to form a fluorescent benzimidazole.[18-20] The salicylaldehyde generated in this Reimer–Tiemann reaction can also potentially react with hydrazine derivatives (see Section 3.2) to yield fluorescent derivatives, though this reaction has apparently not yet been reported.

Alcohols in Carbohydrates

As with derivatization of alcohols in proteins, it is difficult to selectively modify most carbohydrates in aqueous solution because of their low reactivity and the competing hydrolysis of the reactive reagents. However, several reagents are available for derivatizing reducing sugars (which contain a low equilibrium concentration of the reactive aldehyde function), as well as for modifying aldehydes and ketones obtained by periodate oxidation of various carbohydrates. To pursue this labeling approach, see Section 3.2 for a description of our aldehyde- and ketone-reactive reagents.

DTAF

The isomeric fluorescein dichlorotriazines (5-DTAF, D-16; 6-DTAF, D-17), which readily modify amines in proteins (see Section 1.3), are among the few reagents that are reported to react directly with polysaccharides and other alcohols in aqueous solution at pH >9.[21]

N-Methylisatoic Anhydride

In the absence of other reactive functions, *N*-methylisatoic anhydride (M-25) will convert ribonucleotides and probably other carbohydrates to fluorescent esters with excitation/emission maxima of ~350/446 nm in mildly basic aqueous solution.[22-26] The small size and moderate environmental sensitivity of this fluorophore, which is a synthetic precursor to brightly blue fluorescent *N*-methylanthranilic acid amides and esters,[25] may be of use for preparing site-selective probes. Low molecular weight alcohols are better derivatized by this reagent in aprotic organic solvents.[27,28]

Dansyl Boronic Acid

m-Dansylaminophenylboronic acid (D-2281) reacts with vicinal diols (hydroxyl groups on adjacent carbon atoms) and certain amino alcohols to form cyclic complexes [29] that have a fluorescence intensity and peak emission dependent on the environment of the dansyl fluorophore. This interesting reagent binds reversibly to cell-wall carbohydrates and probably glycoproteins,[29] as well as to glycosylated (but not deglycosylated) human serum albumin.[30] The boronic acid specifically inhibits certain serine hydrolases that have adjacent histidine residues, including subtilisin,[31] lipoprotein lipase,[32] human milk lipase [33] and the β-lactamase from *Enterobacter cloacae* P99.[34] Dansylaminophenylboronic acid is also used as an HPLC derivatization reagent for vicinal diols.[35]

BMPM Chloride

BMPM chloride (B-2282) is a pyrene analog of the common alcohol-protecting reagent trityl chloride. This unique reagent has been used as a fluorescent 5'-protecting group for oligonucleotide synthesis.[36] It can be removed and quantitated using mild acid hydrolysis.

Fluoral P

4-Amino-3-penten-2-one (fluoral P, A-6250; see Section 3.2) is reported to selectively react with formaldehyde to form a fluorescent dihydropyridine product (excitation/emission maxima ~410/510 nm).[37] Formaldehyde can be quantitatively formed by

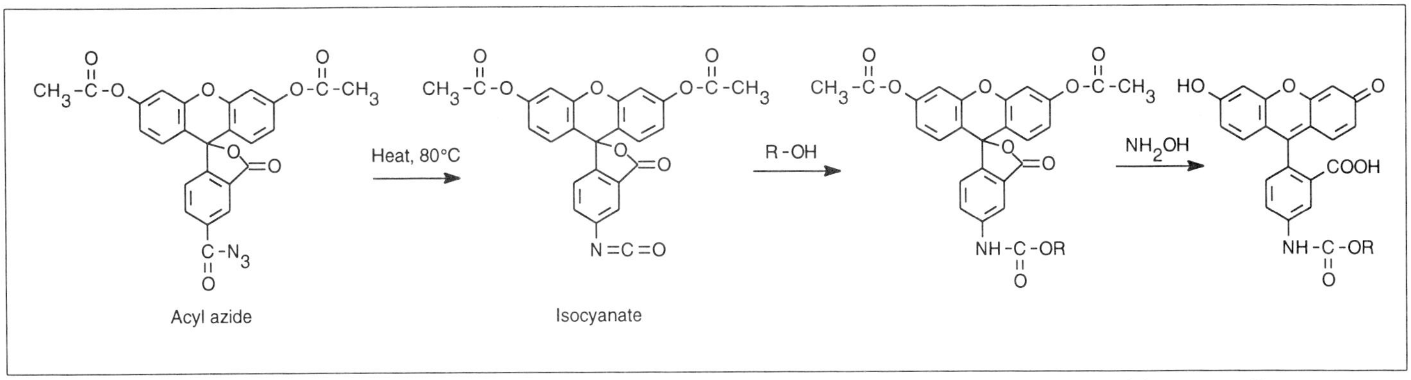

Heat, 80°C

R-OH

NH_2OH

Acyl azide

Isocyanate

Figure 3.1 *Derivatization of an alcohol using the diacetate of fluorescein-5-carbonyl azide (F-6218). This process consists of three steps: 1) rearrangement of the acyl azide to an isocyanate; 2) reaction of the isocyanate with an alcohol; and 3) deprotection of the nonfluorescent alcohol derivative using hydroxylamine.*

periodate oxidation of a wide variety of vicinal diols, including ethylene glycol, glycerol and certain carbohydrates, as well as serine (but not threonine). Therefore, fluoral P may also serve as a useful analytical reagent for selective determination of these molecules. Triglycerides and phospholipids have been successfully detected with fluoral P after saponification to glycerol or phosphoglycerol, followed by periodate oxidation to formaldehyde.[38]

Alcohol Modification in Organic Solvents

Three functional groups — acyl azides, acyl nitriles and acid chlorides — react directly with aliphatic amines to yield the same products as do the corresponding succinimidyl esters. However, when reacted in organic solvents, these reagents can also form derivatives of alcohols and phenols, making them extremely useful for sensitive analysis of alcohols by HPLC or capillary electrophoresis.

Isocyanates Prepared from Acyl Azides

Alcohols are much easier to modify in anhydrous organic solvents than in aqueous solution. Perhaps the most effective reagents are isocyanates, which are much more reactive with alcohols (and amines) than are isothiocyanates but are not sufficiently stable to make their preparation commercially practical. Fortunately, isocyanates can often be prepared by Curtius rearrangement of acyl azides (Figure 3.1). When an acyl azide and alcohol are heated together in an organic solvent such as toluene, dioxane or DMF at 80°C, the acyl azide will rearrange to form an isocyanate that then reacts with the alcohol to form a stable urethane. As little as 50 femtograms of the urethane conjugates prepared from the coumarin derivatives 7-methoxycoumarin-3-carbonyl azide (M-1445) and 7-diethylaminocoumarin-3-carbonyl azide (D-1446) can be detected using an HPLC fluorescence detector.[39] Alcohol conjugates prepared from the new carbonyl azides of fluorescein diacetate (F-6218) and tetramethylrhodamine (T-6219) may provide even higher sensitivity, particularly with instruments that employ the argon-ion laser. Following rearrangement and alcohol conjugation, the acetates of the fluorescein derivative can be removed by hydrolysis at pH 9–10.

Acyl Nitriles

9-Anthroylnitrile (A-1440) reacts with alcohols, such as steroids and acylglycerols, in organic solvents to yield carboxylate esters that are useful for HPLC.[40-43] Alcohol derivatizations with acyl nitriles generally must be performed in organic solvents using the catalyst quinuclidine (Q-1590).

Acid Chlorides

Preparation of other simple carboxylate esters of alcohols usually requires synthesis of a highly reactive acid chloride, which is derived from a carboxylic acid in combination with a reagent such as thionyl chloride or oxalyl chloride. Several fluorescent carboxylic acids are listed in Chapter 1. However, the carboxylate esters produced with acid chlorides or acyl nitriles are usually not as chemically stable as the urethanes produced with acyl azides.

1. Biochemistry 33, 6867 (1994); **2.** J Biol Chem 265, 18791 (1990); **3.** J Biol Chem 265, 18786 (1990); **4.** J Biol Chem 264, 18188 (1989); **5.** Biochem J 108, 883 (1968); **6.** Biochem J 95, 180 (1965); **7.** Biochem J 94, 17 (1965); **8.** J Neurochem 55, 813 (1990); **9.** J Biol Chem 266, 13777 (1991); **10.** J Biol Chem 265, 2483 (1990); **11.** Biochim Biophys Acta 1057, 208 (1991); **12.** J Biol Chem 264, 1361 (1989); **13.** Z Naturforsch 49c, 204 (1994); **14.** Eur J Biochem 60, 67 (1975); **15.** Biochem Int'l 22, 125 (1990); **16.** Biochemistry 18, 3589 (1979); **17.** Biochem Biophys Res Comm 138, 872 (1986); **18.** J Chromatography 430, 271 (1988); **19.** J Chromatography 356, 171 (1986); **20.** J Chromatography 344, 267 (1985); **21.** Carbohydrate Res 30, 375 (1973); **22.** Biochemistry 30, 422 (1991); **23.** Biochemistry 29, 3309 (1990); **24.** Biochim Biophys Acta 742, 496 (1983); **25.** J Biol Chem 257, 13354 (1982); **26.** Arch Biochim Biophys 155, 70 (1973); **27.** Synthesis 39, 266 (1982); **28.** J Org Chem 24, 1214 (1959); **29.** Biochem Biophys Res Comm 96, 157 (1980); **30.** Clin Chim Acta 149, 13 (1985); **31.** FEBS Lett 133, 36 (1981); **32.** Biochim Biophys Acta 746, 217 (1983); **33.** J Mol Catalysis 52, 317 (1989); **34.** Biochemistry 28, 6875 (1989); **35.** Anal Chim Acta 228, 101 (1990); **36.** Tetrahedron Lett 28, 5157 (1987); **37.** Anal Chim Acta 119, 349 (1980); **38.** Anal Chim Acta 142, 13 (1982); **39.** Chem Pharm Bull 33, 1164 (1985); **40.** J Chromatography A 667, 113 (1994); **41.** J Chromatography 491, 37 (1989); **42.** Anal Chim Acta 147, 397 (1983); **43.** J Chromatography 276, 289 (1983).

3.1 Data Table *Reagents for Modifying Alcohols*

Cat #	Structure	MW	Storage	Soluble	Abs	{ε x 10^{-3}}	Em	Solvent	Notes
A-1440	✓	231	F,DD,L	DMF, MeCN	361	{7.5}	470	MeOH	<1>
B-2282	✓	463	F,DD,L	MeCN	348	{47}	397	$CHCl_3$	<2>
D-16	S1.3	495	F,D,L	pH>6, DMF	492	{70}	516	pH 9	<3>
D-17	S1.3	495	F,D,L	pH>6, DMF	493	{72}	518	pH 9	<3>
D-1446	✓	286	F,DD,L	DMF, MeCN	436	{57}	478	MeOH	
D-2281	✓	370	D,L	DMF, DMSO	337	{4.6}	517	MeOH	<4>
F-6218	✓	485	FF,D	DMF, MeCN	<300		none		<5>
M-25	S1.8	177	D	DMF, DMSO	316	{3.5}	386	MeOH	<6>
M-1445	✓	245	FF,DD,L	DMF, MeCN	360	{25}	415	MeOH	
Q-1590	✓	111	D,A	pH<8, DMF	<300		none		
T-6219	✓	455	FF,D,L	DMF, MeCN	545	{86}	578	MeOH	

For definitions of the contents of this data table, see "How to Use the Handbook" on page vi.
Structure: Chemical structure drawing: ✓ = shown in this section; Sn.n = shown in Section number n.n.

<1> Absorption spectrum of A-1440 has subsidiary peaks at 380 nm and 344 nm. Emission spectrum is unstructured. Ester derivatives formed by reaction with alcohols have essentially similar spectra.

<2> The thymidine ether derivative of B-2282 has essentially the same spectral characteristics. The nonfluorescent carbonium ion formed in trifluoroacetic acid has Abs = 636 nm (ε = {25}) and 515 nm (ε = {72}).

<3> Unstable in water. Use immediately.

<4> Fluorescence of D-2281 when bound to proteins is typically blue shifted (Em ~490 nm).

<5> Spectra after acetate hydrolysis are similar to C-1359 (Abs = 492 nm (ε = {79}), Em = 518 nm at pH 9).

<6> Ester reaction products of M-25 with alcohols have Abs = 350 nm (ε = {5.7}), Em = 446 nm in water (pH 8).

3.1 Structures *Reagents for Modifying Alcohols*

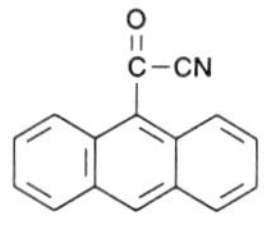

A-1440

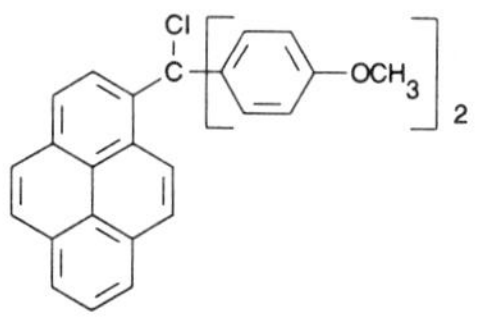

B-2282

D-16, D-17
Section 1.3

D-1446
see M-1445

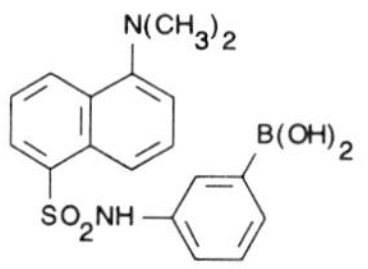

D-2281

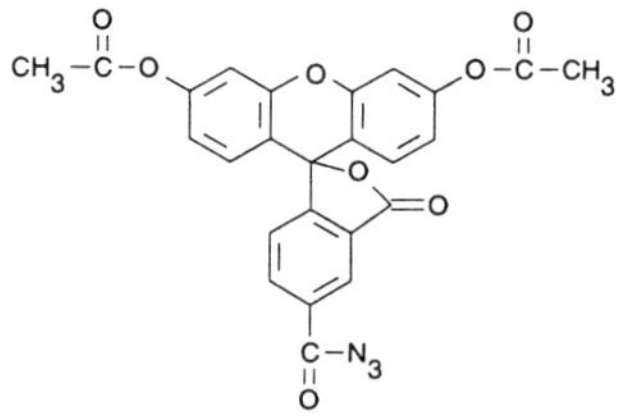

F-6218

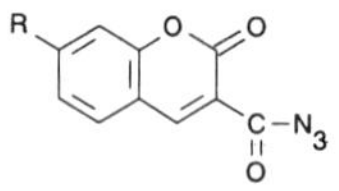

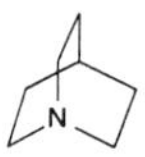

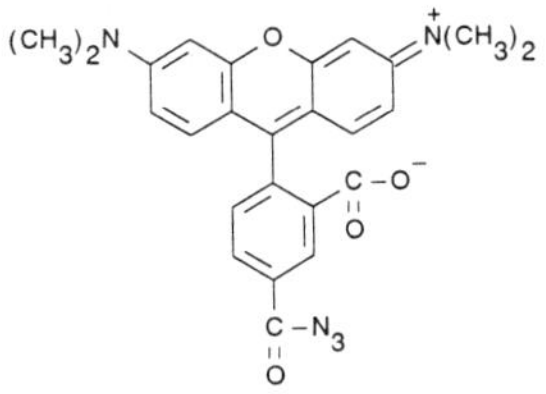

M-25
Section 1.8

M-1445 R = —OCH_3
D-1446 R = —$N(CH_2CH_3)_2$

Q-1590

T-6219

3.1 Price List *Reagents for Modifying Alcohols*

Cat #		Product Name	Unit Size	Price Per Unit ($) 1–4 Units	5–24 Units
A-1440		9-anthroylnitrile	25 mg	55.00	44.00
B-2282		bis-(4-methoxyphenyl)-1-(pyrenyl)methyl chloride (BMPM chloride)	25 mg	35.00	28.00
D-2281		m-dansylaminophenylboronic acid	100 mg	48.00	38.40
D-16		5-(4,6-dichlorotriazinyl)aminofluorescein (5-DTAF) *single isomer*	100 mg	75.00	60.00
D-17		6-(4,6-dichlorotriazinyl)aminofluorescein (6-DTAF) *single isomer*	100 mg	75.00	60.00
D-1446		7-diethylaminocoumarin-3-carbonyl azide	25 mg	88.00	70.40
F-6218	*New*	fluorescein-5-carbonyl azide, diacetate	10 mg	78.00	62.40
M-1445		7-methoxycoumarin-3-carbonyl azide	25 mg	88.00	70.40
M-25		N-methylisatoic anhydride *high purity*	1 g	75.00	60.00
Q-1590		quinuclidine	100 mg	8.00	6.40
T-6219	*New*	tetramethylrhodamine-5-carbonyl azide	5 mg	98.00	78.40

3.2 Hydrazines and Aromatic Amines for Modifying Aldehydes, Ketones and Vicinal Diols

Aldehydes and ketones are present in a number of low molecular weight molecules such as drugs, steroid hormones, reducing sugars and some metabolic intermediates such as pyruvate and α-ketoglutarate. Except for polysaccharides containing free reducing sugars, biopolymers generally lack aldehyde and ketone groups. Even those aldehydes and ketones that are found in the open-ring form of simple carbohydrates are usually in equilibrium with the closed-ring form of the sugar. The infrequent occurrence of aldehydes and ketones has stimulated the development of techniques to selectively introduce these functional groups into certain biomolecules, thus providing unique sites for chemical modification and greatly extending the applications of the probes found in this section. See the Technical Note "Interconversion of Functional Groups" on page 80 for reagents that convert amines and thiols into aldehydes and ketones.

Introducing Aldehydes and Ketones into Biomolecules

Periodate Oxidation

The most common method for introducing aldehydes and ketones into polysaccharides and glycoproteins (including antibodies) is periodate oxidation of vicinal diols. Periodate will also oxidize certain β-aminoethanol derivatives, as well as methionine (to its sulfoxide) and certain thiols (usually to disulfides). These reactions, however, usually occur at a slower rate than oxidation of vicinal diols. Periodate oxidation of the 3'-terminal ribose provides one of the few methods of selectively modifying RNA.[1,2] Periodate-oxidized ribonucleotides can subsequently be converted to fluorescent nucleotide probes by reaction with fluorescent hydrazines and amines.[3] In addition, N-terminal serine and threonine residues of proteins can be selectively oxidized to aldehyde groups, thus allowing highly selective modification of certain proteins such as corticotrophin.[4,5] Moreover, because antibodies are glycosylated at sites distant from the antigen-binding region, modification of periodate-oxidized antibodies by hydrazines usually does not inactivate the antibody as sometimes occurs with FITC, TRITC and Texas Red® sulfonyl chloride labeling. Researchers have also used the hydrazine derivatives in this section to detect periodate-oxidized glycoproteins in gels.[6]

Enzyme-Mediated Oxidation with Galactose Oxidase

A second specific method for introducing aldehydes into biomolecules is through the use of galactose oxidase, an enzyme that oxidizes galactose residues to aldehydes, particularly in glycoproteins.[7-11] This method was used to label live corn and rose protoplasts with hydrazines and aromatic amines.[12] The introduction of galactose residues can be especially advantageous for structural studies because it provides a means of selectively labeling specific sites on biomolecules. For example, galactose has been specifically inserted into the carbohydrate moiety of rhodopsin using a galactosyl transferase.[11]

Conversion of Amines and Thiols

Aldehydes and ketones can be introduced into biomolecules at existing amine or thiol functions. Aldehydes can be introduced at amine sites using succinimidyl 4-formylbenzoate (SFB, S-1580) and ketones can be introduced at thiol groups with the water-soluble reagent 3-bromopyruvic acid [13] (B-6058). Aldehyde-modified alkaline phosphatase has been coupled to a hydrazide-modified oligonucleotide using SFB.[14]

Coupling Hydrazines and Amines to Amine-Containing Biomolecules without Introducing Aldehydes and Ketones

Common tissue fixatives such as formaldehyde and glutaraldehyde can be used to couple hydrazine and amine derivatives to proteins and other amine-containing polymers. For example, lucifer yellow CH (L-453) can be conjugated to surrounding biomolecules by common aldehyde fixatives in order to preserve the dye's staining pattern during subsequent tissue manipulations.[15] Glutaraldehyde has also been used to couple biotin hydrazides (see Section 4.2) directly to nucleic acids,[16] a reaction that is potentially useful for conjugating fluorescent hydrazine derivatives to DNA.

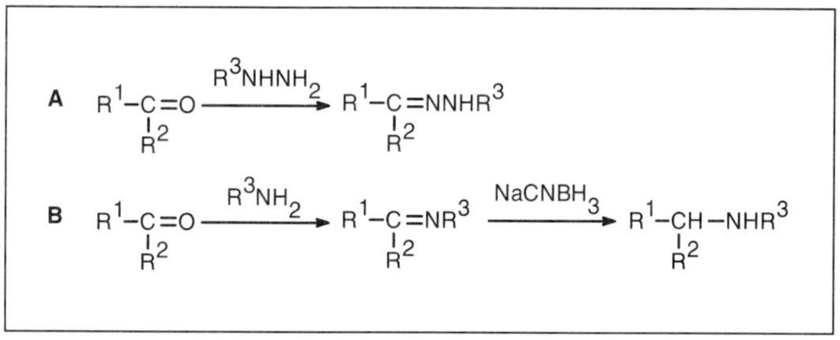

A $R^1-\underset{|\,R^2}{C}=O \xrightarrow{R^3NHNH_2} R^1-\underset{|\,R^2}{C}=NNHR^3$

B $R^1-\underset{|\,R^2}{C}=O \xrightarrow{R^3NH_2} R^1-\underset{|\,R^2}{C}=NR^3 \xrightarrow{NaCNBH_3} R^1-\underset{|\,R^2}{CH}-NHR^3$

Figure 3.2 Modifying aldehydes and ketones with A) hydrazine derivatives and B) amine derivatives.

Hydrazines

Reactivity of Hydrazine Derivatives

Although certain aromatic amines such as 8-aminonaphthalene-1,3,6-trisulfonic acid (ANTS, A-350) and 2-aminopyridine (A-6252) have been extensively utilized to modify reducing sugars for analysis and sequencing, the most reactive reagents for forming stable conjugates of aldehydes and ketones are usually hydrazine derivatives, including hydrazides, semicarbazides and carbohydrazides. Hydrazine derivatives react with ketones to yield relatively stable hydrazones (Figure 3.2A), and with aldehydes to yield hydrazones that are somewhat less stable though they may be formed faster. These hydrazones can be reduced with sodium borohydride to further increase the stability of the linkage. Hydrazine derivatives also have amine-like reactivity and, in some cases, can be coupled to water-soluble carbodiimide-activated carboxylic acid groups in drugs, peptides and proteins;[17-19] see Section 3.3 for more details. Because the fluorescence of the hydrazide-containing reagents usually does not change significantly upon reaction with aldehydes or ketones, especially after chemical reduction, the excess reagent required to drive the equilibrium reaction must be separated from the conjugate by washing, gel filtration, dialysis or other separation techniques.

Fluorescent Hydrazine Derivatives Excited with Visible Light

Molecular Probes offers a large number of fluorescent hydrazine derivatives for reaction with aldehydes or ketones (Table 3.1). The most extensively used visible light–excitable fluorescent hydrazine derivative, fluorescein-5-thiosemicarbazide (F-121), has been coupled to a wide variety of biomolecules, including:

- Enzyme-oxidized live plant protoplasts [12]
- Immunoglobulins [20]
- Na^+/K^+-ATPase glycoprotein [21]
- Periodate-oxidized glycoproteins in gels [6]
- Periodate-oxidized RNA [1,2,22-25]
- Thrombin and antithrombin [26]

Probably the brightest hydrazides, however, are the fluorescein hydrazide with the extra spacer between the fluorophore and the reactive group (C-356), the BODIPY® hydrazides (D-2371, D-2372) and Texas Red hydrazide (T-6256), which is >90% single isomer by HPLC. Because they are more photostable than the fluorescein derivatives, the Texas Red and BODIPY hydrazides should be the best reagents for detecting aldehydes and ketones in laser-excited chromatographic methods.

Fluorescent Hydrazine Derivatives Excited with UV Light

Dansyl hydrazine (D-100) has been by far the most widely used UV-excitable hydrazine probe for derivatizing aldehydes and ketones for chromatographic analysis. It has been used to modify:

- Bile acids [27-29]
- Glycoproteins [30-32]
- Human chorionic gonadotropin [33]
- Oligosaccharides [34]
- Reducing sugars [35-39]
- Rhodopsin [11,17]
- RNA [40-42]
- Steroids [29,43-46]

Table 3.1 Molecular Probes' hydrazine and amine derivatives.

Derivative	Hydrazines	Amines *
Acridone		A-6289
Biotin †	B-1549, B-1600, B-1603, B-2600, B-6350	A-1593, A-1594, A-1604, B-1592, B-1596, B-1597, B-2605, N-6356
t-BOC		M-6246, M-6247, M-6248
BODIPY FL	D-2371	D-2390
BODIPY 530/550	D-2372	D-2391
BODIPY TR		D-6251
Cascade Blue	C-687	C-621, C-622
Coumarin	D-355	A-191
Dansyl	D-100	D-112, D-113
Dinitrophenyl		D-1552
Eosin		A-117, A-434
Fluorescein	C-356, F-121	A-456, A-1351, A-1353, A-1363, A-1364, A-6255
FMOC	F-6290	
Lucifer yellow	L-453	A-1339, A-1340
Naphthalene		A-91, A-350
PEG		P-6281, P-6282, P-6283, P-6285
Pyrene	P-101	P-2421, P-6253, P-6254
Pyridine		A-6252
Quinoline		A-497
Rhodamine		A-1318, L-2424
Texas Red	T-6256	T-2425

* Amine derivatives are listed in Section 3.3, except for D-1552, which appears in Section 12.3. † Biotin derivatives are listed in Section 4.2.

A unique application that has been reported for dansyl hydrazine, but which is likely a general reaction of hydrazine derivatives, is the detection of *N*-acetylated or *N*-formylated proteins through transfer of the acyl group to the fluorescent hydrazide.[47,48] The higher absorptivity and fluorescence of the coumarin and pyrene hydrazine derivatives (D-355, P-101) should make their conjugates about 10 times more detectable than those of dansyl hydrazine. We also offer the shorter-wavelength derivatization reagent, FMOC hydrazine (F-6290), which was employed in the HPLC determination of neutral monosaccharides in glycoproteins. In this study, the FMOC-derivatized sugar hydrazones were detected by fluorescence (excitation/emission wavelengths 270/320 nm) with a detection limit of 0.05–0.4 picomoles and by UV absorption (at 263 nm) with a detection limit of 1–3 picomoles.[49]

Lucifer Yellow and Cascade Blue Hydrazides

Lucifer yellow CH (L-453) is most commonly used as an aldhyde-fixable neuronal tracer. This membrane-impermeant hydrazide also reacts with periodate-oxidized cell-surface glycoproteins,[21,50] oxidized ribonucleotides[51] and gangliosides.[52] Cascade Blue hydrazide[53] (C-687) exhibits high absorptivity (ϵ >28,000 $cm^{-1}M^{-1}$), fluorescence quantum yield (0.54) and water solubility (~1%). This trisulfonated pyrene derivative should have applications similar to those of lucifer yellow CH. See Section 15.3 for a discussion of the use of these probes as fixable polar tracers.

Nonfluorescent Hydrazine Derivatives

In addition to the fluorescent hydrazine derivatives, we offer several nonfluorescent biotin hydrazides (see Section 4.2) that can be detected using fluorescent or enzyme-labeled avidin or streptavidin (see Section 7.5), as well as the nonfluorescent thiol-containing AMBH (A-1642), which can be used to convert aldehydes and ketones to the more reactive thiols for modification with any thiol-reactive reagent.

Aromatic Amines

Primary aliphatic and aromatic amines couple reversibly with aldehydes and ketones to form hydrolytically unstable Schiff bases[7] (Figure 3.2B). The reversibility of this modification makes reagents that contain amines less desirable unless the Schiff base is reduced to a stable amine derivative by sodium borohydride[54,55] or sodium cyanoborohydride.[33] Chemical reduction also retains the amine's original charge. Sequencing of carbohydrate polymers using fluorescent derivatives has usually relied on derivatization of the reducing end of the polymer with a fluorescent amine.[56] Certain aromatic amines and amidines have been extensively utilized for coupling to aldehydes, ketones, monosaccharides and the reducing end of carbohydrate polymers:

- 2-Aminoacridone (A-6289) forms conjugates that can be separated as their borate complexes by polyacrylamide gel electrophoresis.[56-58] It is also used to prepare fluorogenic substrates for proteases.[59]
- 2-Aminopyridine (A-6252) is extensively used for carbohydrate derivatization in both manual and automated procedures.[60-63]
- ANTS (A-350) has a high charge that permits electrophoretic separation of the degradation products of complex oligosaccharides.[56,64-66]
- Benzamidine (B-6546) is a sensitive postcolumn HPLC derivatization reagent for detecting picomoles of reducing sugars in alkaline medium.[67,68] Recently, its utility has been extended to nonreducing sugars.[69]

The aromatic diamine DDB (D-1463), which forms heterocyclic compounds with certain aldehydes and ketones, has been used to selectively detect aromatic aldehydes in the presence of aliphatic aldehydes, including carbohydrates.[70] It has also been employed for fluorometric determination of ascorbic acid,[71,72] lactic acid,[73] α-keto acids[74-76] and *N*-acetylneuraminic acids[77] by HPLC. Endogenous leucine-enkephalin and other tyrosine-containing peptides can be detected by HPLC at concentrations as low as 5.6 picomoles per gram of rat brain tissue using a Reimer–Tiemann reaction followed by derivatization with DDB.[78]

Alternatively, aldehydes and ketones can be transformed into primary aliphatic amines by reaction with ammonia, ethylenediamine or other nonfluorescent diamines.[79] This chemistry is particularly useful because the products can then be coupled with any of the amine-reactive reagents described in Chapter 1 such as the succinimidyl esters of TAMRA[80] (C-1171, C-6121, C-6122; see Section 1.6).

Fluoral P for Formaldehyde Detection

4-Amino-3-pentene-2-one (fluoral P, A-6250) is reported to selectively react with formaldehyde to form a fluorescent dihydropyridine product (excitation/emission maxima ~410/510 nm).[81] When other aldehydes, ketones, saccharides and freshly distilled solvents were tested at concentrations 2000 times that of detectable formaldehyde, no fluorescent product was detected.[82] Fluoral P has been used for the fluorometric determination of aspartame[83] and hydrogen peroxide[84] after these compounds were enzymatically converted to formaldehyde by the α-chymotrypsin–alcohol oxidase system and the catalase–methanol system, respectively.

1. Biochemistry 30, 4821 (1991); **2.** Biochemistry 19, 5947 (1980); **3.** Bioconjugate Chem 5, 436 (1994); **4.** Biochem J 94, 17 (1965); **5.** Biochem J 83, 91 (1962); **6.** Anal Biochem 161, 245 (1987); **7.** Meth Enzymol 247, 30 (1994); **8.** Anal Biochem 170, 271 (1988); **9.** Meth Enzymol 138, 429 (1987); **10.** Biochem Biophys Res Comm 92, 1215 (1980); **11.** J Supramol Struct 6, 291 (1977); **12.** Protoplasma 139, 117 (1987); **13.** FEBS Lett 166, 194 (1984); **14.** Anal Biochem 178, 43 (1989); **15.** Nature 292, 17 (1981); **16.** Chem Pharm Bull 37, 1831 (1989); **17.** Biochim Biophys Acta 897, 384 (1987); **18.** Annals NY Acad Sci 463, 214 (1984); **19.** Biochemistry 22, 5 (1983); **20.** J Immunol Meth 109, 289 (1988); **21.** Biochemistry 24, 322 (1985); **22.** J Mol Biol 221, 441 (1991); **23.** Eur Biophys J 16, 45 (1988); **24.** Biochemistry 25, 5298 (1986); **25.** Eur J Biochem 142, 261 (1984); **26.** Biochim Biophys Acta 785, 1 (1984); **27.** J Chromatography 272, 261 (1983); **28.** J Chromatography 260, 115 (1983); **29.** J Chromatography 217, 349 (1981); **30.** Anal Biochem 157, 100 (1986); **31.** Anal Biochem 96, 208 (1979); **32.** Anal Biochem 73, 192 (1976); **33.** Biochim Biophys Acta 670, 181 (1981); **34.** Anal Biochem 146, 143 (1985); **35.** J Chromatography 464, 343 (1989); **36.** J Chromatography 333, 123 (1985); **37.** J Chromatography 256, 27 (1983); **38.** Anal Biochem 114, 153 (1981); **39.** J Chromatography 139, 343 (1977); **40.** Biochemistry 30, 5238 (1991); **41.** Eur J Biochem 129, 211 (1982); **42.** Nucleic Acids Res 8, 3229 (1980); **43.** Chromatographia 19, 452 (1984); **44.** J Chromatography 233, 61 (1982); **45.** J Chromatography 232, 1 (1982); **46.** J Chromatography 226, 1 (1981); **47.** J Cell Biol 106, 1607 (1988); **48.** Anal Biochem 29, 186 (1969); **49.** J Chromatography 646, 45 (1993); **50.** Biochem Biophys Res Comm 112, 872 (1983); **51.** Biochemistry 27, 6039 (1988); **52.** J Cell Biol 100, 721 (1985); **53.** Anal Biochem 198, 119 (1991); **54.** Biochemistry 26, 2162 (1987); **55.** Biochim Biophys Acta 597, 285 (1980); **56.** Anal Biochem 222, 270 (1994); **57.** Anal Biochem 216, 243 (1994); **58.** Anal

Biochem 196, 238 (1991); **59.** Anal Biochem 171, 393 (1988); **60.** J Chromatography 587, 177 (1991); **61.** Anal Biochem 145, 245 (1985); **62.** J Biochem 95, 197 (1984); **63.** J Biochem 90, 407 (1981); **64.** Meth Enzymol 230, 150 (1993); **65.** Electrophoresis 12, 94 (1991); **66.** Biochem J 270, 705 (1990); **67.** Anal Chim Acta 252, 173 (1991); **68.** J Chromatography 553, 255 (1991); **69.** Chromatographia 38, 12 (1994); **70.** Anal Chim Acta 134, 39 (1982); **71.** Chem Pharm Bull 33, 3499 (1985); **72.** J Chromatography 344, 351 (1985); **73.** J Chromatography 566, 1 (1991); **74.** Chem Pharm Bull 35, 687 (1987); **75.** Anal Chim Acta 172, 167 (1985); **76.** J Chromatography 344, 33 (1985); **77.** Anal Biochem 164, 138 (1987); **78.** J Chromatography 430, 271 (1988); **79.** Proc Natl Acad Sci USA 88, 2302 (1991); **80.** J Chromatography B 657, 307 (1994); **81.** Anal Chim Acta 119, 349 (1980); **82.** Analyst 119, 1413 (1994); **83.** Analyst 115, 435 (1990); **84.** J Chromatography 411, 423 (1987).

3.2 Data Table *Hydrazines and Aromatic Amines for Modifying Aldehydes, Ketones and Vicinal Diols*

Cat #	Structure	MW	Storage	Soluble	Abs	{ε × 10^{-3}}	Em	Solvent	Notes
A-350	S15.3	427	L	H_2O	353	{7.2}	520	H_2O	
A-1642	S5.2	191	D,A	EtOH	<300		none		
A-6250	✓	99	D	DMSO, MeOH	298	{18}	none	MeOH	
A-6252	✓	94	D	H_2O	295	{4.2}	352	MeOH	<1>
A-6289	✓	247	D,L	DMF, DMSO	425	{5.2}	531	MeOH	<2>
B-6058	✓	167	F,D	MeOH	<300		none		
B-6546	S10.6†	175†	D	H_2O, MeOH	<300		none		<3>
C-356	✓	493	L	pH>7, DMF	492	{78}	516	pH 8	
C-687	S15.3	596	L	H_2O*	399	{30}	421	H_2O	<4>
D-100	✓	265	L	EtOH	336	{4.4}	534	MeOH	
D-355	✓	275	D,L	MeCN, DMF	420	{46}	468	MeOH	
D-1463	✓	241	D,L	EtOH	298	{3.1}	359	MeOH	
D-2371	✓	306	F,D,L	MeOH, MeCN	503	{71}	510	MeOH	<5>
D-2372	✓	430	F,D,L	DMF, MeOH	534	{79}	551	MeOH	<5>
F-121	✓	421	D,L	pH>7, DMF	492	{85}	516	pH 9	
F-6290	✓	254	F,D	DMF, DMSO	299	{5.8}	309	MeOH	<6>
L-453	S15.3	457	L	H_2O*	428	{12}	536	H_2O	
P-101	✓	302	D,L	MeCN, DMF	341	{43}	376	MeOH	<7>
S-1580	✓	247	F,D	DMF, MeCN	<300		none		
T-6256	✓	621	L	DMF	584	{94}	605	MeOH	

For definitions of the contents of this data table, see "How to Use the Handbook" on page vi.
Structure: Chemical structure drawing: ✓ = shown in this section; Sn.n = shown in Section number n.n.

* Maximum solubility in water is ~1% for C-687 and ~8% for L-453.
† MW is for the hydrated form of this product.
<1> Spectral characteristics of A-6252 in water are pH dependent. Protonated form (pK_a = 6.9) is more fluorescent than the neutral form [J Phys Chem 72, 1982 (1968)].
<2> Spectra of this compound are in methanol containing a trace of KOH.
<3> B-6546 reacts with reducing sugars in 1 M KOH at 100°C, yielding fluorescent products with Em = 470 nm (excitation at 360 nm) [J Chromatography 533, 255 (1991)].
<4> Cascade Blue dyes have a second absorption peak at about 376 nm with ε ~80% of the 395–400 nm peak.
<5> The absorption and fluorescence spectra of BODIPY derivatives are relatively insensitive to the solvent.
<6> F-6290 has stronger absorption at 264 nm (ε = {20}). This peak is usually used for excitation in fluorescence-detected HPLC or capillary electrophoresis.
<7> Pyrene derivatives exhibit structured spectra. The absorption maximum is usually about 340 nm with a subsidiary peak at about 325 nm. There are also strong absorption peaks below 300 nm. The emission maximum is usually about 376 nm with a subsidiary peak at 396 nm. Excimer emission at about 470 nm may be observed at high concentrations.

3.2 Structures *Hydrazines and Aromatic Amines for Modifying Aldehydes, Ketones and Vicinal Diols*

A-350 Section 15.3

A-1642 Section 5.2

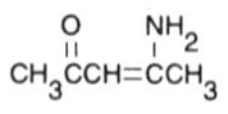

A-6250

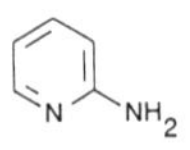

A-6252

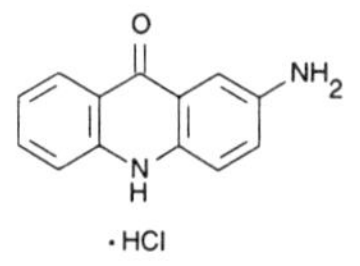

A-6289

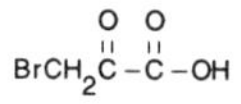

B-6058

B-6546 Section 10.6

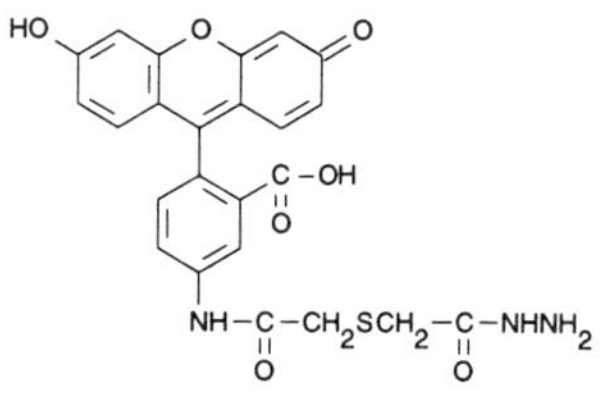

C-356

C-687 Section 15.3 | D-100 | D-355 | D-1463 | D-2371 R = $-CH_3$, D-2372 R = Ph

F-121 | F-6290 | L-453 Section 15.3 | P-101 | S-1580 | T-6256

3.2 Price List *Hydrazines and Aromatic Amines for Modifying Aldehydes, Ketones and Vicinal Diols*

Cat #		Product Name	Unit Size	Price Per Unit ($) 1–4 Units	5–24 Units
A-1642		2-acetamido-4-mercaptobutyric acid hydrazide (AMBH)	100 mg	40.00	32.00
A-6289	**New**	2-aminoacridone, hydrochloride	25 mg	75.00	60.00
A-350		8-aminonaphthalene-1,3,6-trisulfonic acid, disodium salt (ANTS)	1 g	88.00	70.40
A-6250	**New**	4-amino-3-pentene-2-one (fluoral P)	1 g	38.00	30.40
A-6252	**New**	2-aminopyridine	5 g	16.00	12.80
B-6546	**New**	benzamidine, hydrochloride, monohydrate	5 g	18.00	14.40
B-6058	**New**	3-bromopyruvic acid	1 g	8.00	6.40
C-356		5-(((2-(carbohydrazino)methyl)thio)acetyl)aminofluorescein	25 mg	138.00	110.40
C-687		Cascade Blue® hydrazide, trisodium salt	10 mg	95.00	76.00
D-1463		1,2-diamino-4,5-dimethoxybenzene, dihydrochloride (DDB)	100 mg	73.00	58.40
D-355		7-diethylaminocoumarin-3-carboxylic acid, hydrazide (DCCH)	25 mg	90.00	72.00
D-2371		4,4-difluoro-5,7-dimethyl-4-bora-3a,4a-diaza-s-indacene-3-propionic acid, hydrazide (BODIPY® FL hydrazide)	5 mg	95.00	76.00
D-2372		4,4-difluoro-5,7-diphenyl-4-bora-3a,4a-diaza-s-indacene-3-propionic acid, hydrazide (BODIPY® 530/550 hydrazide)	5 mg	95.00	76.00
D-100		5-dimethylaminonaphthalene-1-sulfonyl hydrazine (dansyl hydrazine)	100 mg	27.00	21.60
F-6290	**New**	N-(9-fluorenylmethoxycarbonyl) hydrazine (FMOC hydrazine)	25 mg	68.00	54.40
F-121		fluorescein-5-thiosemicarbazide	100 mg	95.00	76.00
L-453		lucifer yellow CH, lithium salt	25 mg	48.00	38.40
P-101		1-pyrenebutanoic acid hydrazide	100 mg	95.00	76.00
S-1580		succinimidyl 4-formylbenzoate (SFB)	100 mg	20.00	16.00
T-6256	**New**	Texas Red® hydrazide *>90% single isomer*	5 mg	95.00	76.00

3.3 Amidation Reagents for Carboxylic Acids and Glutamine

In general, carboxylic acids can be converted to either amide derivatives or acylhydrazides, which are discussed in this section, or to esters, which are discussed in the following section. Alternatively, the half-protected *tert*-butyloxycarbonyl (*t*-BOC) ethylenediamine, propylenediamine and cadaverine derivatives (M-6247, M-6248, M-6246) are useful for converting organic solvent–soluble carboxylic acids into aliphatic amines (Figure 3.3). Following coupling of the amine to an activated carboxylic acid, the *t*-BOC group can be quantitatively removed from most compounds with trifluoroacetic acid (Figure 3.3). The resultant aliphatic amines can then be modified with any of the amine-reactive reagents described in Chapter 1 or coupled to solid-phase matrices for affinity chromatography.

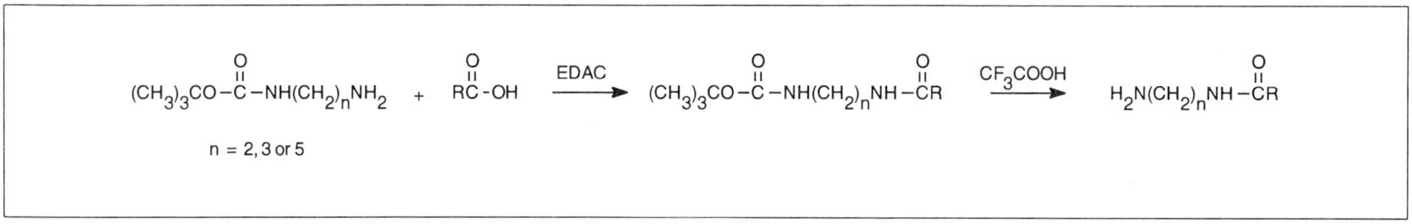

$$(CH_3)_3CO-\overset{O}{\overset{\|}{C}}-NH(CH_2)_nNH_2 + R\overset{O}{\overset{\|}{C}}-OH \xrightarrow{EDAC} (CH_3)_3CO-\overset{O}{\overset{\|}{C}}-NH(CH_2)_nNH-\overset{O}{\overset{\|}{C}}R \xrightarrow{CF_3COOH} H_2N(CH_2)_nNH-\overset{O}{\overset{\|}{C}}R$$

n = 2, 3 or 5

Figure 3.3 *Conversion of a carboxylic acid group into an aliphatic amine. The activated carboxylic acid is derivatized with a half-protected aliphatic diamine (M-6246, M-6247, M-6248), followed by removal of the t-BOC–protecting group with trifluoroacetic acid.*

Coupling Hydrazines and Amines to Carboxylic Acids

Modification In Aqueous Solutions

The carboxylic acids of water-soluble biopolymers such as proteins have been coupled to hydrazines (see Section 3.2) and amines in aqueous solution using water-soluble carbodiimides such as 1-ethyl-3-(3-dimethylaminopropyl)carbodiimide (EDAC, E-2247). Including *N*-hydroxysulfosuccinimide (H-2249) in the reaction mixture has been shown to improve the coupling efficiency of EDAC-mediated protein–carboxylic acid conjugations.[1] As an alternative to EDAC, the hydrophobic reagent EEDQ (E-6288) may be useful for coupling the more hydrophobic amine labels to a protein's carboxylic acids,[2,3] potentially including those buried in a cell's membrane. Like EDAC, EEDQ also forms intramolecular protein–protein crosslinks, for example in myosin.[4] To reduce intra- and interprotein coupling to lysine residues, which is a common side reaction, carbodiimide-mediated coupling should be performed in a concentrated protein solution, at a low pH, using a large excess of the nucleophile.[5-8]

The fluorescein glycine amides (A-1363, A-1364) and various hydrazines may be the best probes for this application because they are more likely to remain reactive at a lower pH than are other aliphatic amines.[9] Fluoresceinyl glycine amide has been coupled to the carboxylic acid of a cyclosporin derivative by EDAC.[10] Quantitative analysis of carboxylic acids, including sugar carboxylates, in aqueous solution using 1-naphthylethylenediamine and *o*-phthaldialdehyde (P-2331, see Section 1.8) has also been reported.[11]

Modification in Organic Solvents

Peptide synthesis research has led to the development of numerous reagents for coupling carboxylic acids to amines in organic solution. Among these are the conversion of carboxylic acids to succinimidyl esters or mixed anhydrides. Dicyclohexylcarbodiimide or diisopropylcarbodiimide are widely used to promote amide formation in organic solution. Another recommended derivatization method for coupling organic solvent–soluble carboxylic acids, including peptides, to aliphatic amines without racemization is the combination of 2,2'-dipyridyldisulfide and triphenylphosphine.[12,13]

Unlike fluorescent aliphatic amines, fluorescent aromatic amines such as those derived from 7-amino-4-methylcoumarin (A-191), 7-amino-4-trifluoromethylcoumarin (AFC, A-6249), 6-aminoquinoline (6-AQ, A-497) and 2-aminoacridone (A-6289, see Section 3.2) exhibit a shift in their absorption and emission (if any) to much shorter wavelengths upon forming carboxamides. This property makes these aromatic amines ideal reagents for preparing peptidase substrates (see Section 10.4). Aromatic amines can generally be coupled to acid halides and anhydrides, with organic solvents usually required for efficient reaction. 5-Aminofluorescein (A-6255) and 5-aminoeosin (A-117) are the key precursors to a wide variety of fluorescein and eosin probes, respectively. The spectrum of 5-aminofluorescein does not shift upon conjugation to carboxylic acids; however, its quantum yield of <0.02 increases considerably upon acylation.

Hydrazine and Aliphatic Amine Derivatives

Molecular Probes provides a wide selection of carboxylic acid–reactive reagents, including several different BODIPY, fluorescein, eosin, rhodamine and Texas Red hydrazine and amine derivatives (Table 3.1), all of which are particularly useful for drug analog synthesis, as well as probes for fluorescence polarization immunoassays.[14-16] Some of the more important probes and their potential applications include:

- BODIPY aliphatic amines (D-2390, D-2391, D-6251), for preparing pH-insensitive probes from carboxylic acid derivatives
- Isomeric aminomethyl fluoresceins (A-1351, A-1353), which are more reactive than 5-aminofluorescein (A-6255) and are readily coupled to activated carboxylic acids [14,17]
- Dansyl ethylenediamine (D-112) and dansyl cadaverine (D-113), for carboxylic acid derivatization [18] and glutamine transamidation reactions [19]
- EDANS (A-91), for preparing radioactive IAEDANS,[16,20] energy transfer–quenched substrates for endopeptidases (see Section 10.4), labeled sugar carboxylates,[21] and an ATP substrate for DNA-dependent RNA polymerase [22]
- 1-Pyrenemethylamine (P-2421), 1-pyreneethylamine (P-6253) and 1-pyrenepropylamine (P-6254), for synthesizing new probes that have excited-state lifetimes of ~100 nanoseconds and also for preparing derivatives of carboxylic acids for chromatographic analysis
- Hydrazine (L-453, C-687; see Section 3.2) and amine (A-1339, A-1340, C-621, C-622) derivatives of lucifer yellow and Cascade Blue, which are precursors of highly fluorescent, water-soluble probes
- Hydrazine and amine derivatives of biotin (see Section 4.2), which are versatile intermediates for synthesizing biotin-containing probes [5,23]

Kits for Coupling Amine Probes to Nucleotides and Oligonucleotides

Our FluoReporter® Oligonucleotide Phosphate Labeling Kits (see Section 8.3) provide the reagents needed to couple aliphatic amines to the 5'-phosphate group of oligonucleotides. The phosphoramidates that are formed have reasonable stability and can be used as hybridization probes for Northern and Southern blots, as well as for fluorescence *in situ* hybridization. Nucleotide phosphoramidates formed by first coupling ethylenediamine to the terminal

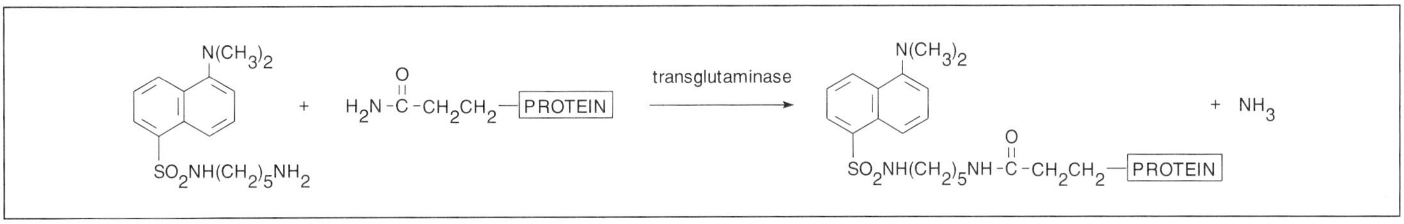

$N(CH_3)_2$ (naphthalene) $SO_2NH(CH_2)_5NH_2$ + $H_2N\text{-}\overset{O}{\overset{\|}{C}}\text{-}CH_2CH_2$—PROTEIN —transglutaminase→ $N(CH_3)_2$ (naphthalene) $SO_2NH(CH_2)_5NH\text{-}\overset{O}{\overset{\|}{C}}\text{-}CH_2CH_2$—PROTEIN + NH_3

Figure 3.4 *Transglutaminase-mediated labeling of a protein using dansyl cadaverine (D-113).*

phosphate followed by reaction with an amine-reactive dye have been separated by capillary electrophoresis.[24] Our convenient FluoReporter Kits may be similarly useful for preparing derivatives of nucleotides, oligonucleotides and other phosphates for subsequent chromatographic or electrophoretic separation.

Transamidation of Glutamine Residues in Proteins

A special enzyme-catalyzed transamidation reaction of glutamine residues in some proteins and peptides — including actin,[25] melittin (M-98403, see Section 20.3),[26] rhodopsin [27] and factor XIII [19,28] — enables their selective modification by amine-containing probes. The NH_2 group of certain glutamine residues can be replaced with an aliphatic amine to form a labeled glutamine amide, a reaction that can be catalyzed by a transglutaminase enzyme.[26,27] (Figure 3.4). This unique method for selective protein modification requires the formation of a complex consisting of the glutamine residue, the aliphatic amine probe and the enzyme. It has been found that a short aliphatic spacer in the amine probe enhances the reaction. The cadaverine ($–NH(CH_2)_5NH–$) spacer is usually optimal. Although dansyl cadaverine (D-113) is probably the most widely used, fluorescein cadaverine [29] (A-456), eosin cadaverine (A-434), tetramethylrhodamine cadaverine [30] (A-1318), Texas Red cadaverine (T-2425) and BODIPY TR cadaverine (D-6251) are the most fluorescent transglutaminase substrates available. The intrinsic transglutaminase activity in sea urchin eggs has been used to covalently incorporate dansyl cadaverine during embryonic development.[31] Two biotin cadaverines (A-1594, B-1596; see Section 4.2) are also available for transglutaminase reactions.[32,33]

Modification of Proteins with Fluorescent Carbodiimides

When carboxylic acids are reacted with carbodiimides in the absence of a nucleophile, they may rearrange to form a stable *N*-acylurea (Figure 3.5). If the carbodiimide contains a fluorophore (such as C-428, C-442), then the fluorophore will be specifically incorporated into the protein. This reaction has been used to label:

- Chloroplast-coupling factor [34,35]
- Mitochondrial proton-channel protein [36]
- Plant tonoplast ATPase [37]
- Proteins in the sarcoplasmic reticulum [38-40]

A similar mechanism of labeling may occur in some dicyclohexylcarbodiimide (DCC)–inhibited proteins, in which DCC appears to react with a carboxyl residue within a very hydrophobic sequence of the protein.[41]

Amine-Terminated Poly(Ethylene Glycol) Methyl Ethers

Amine-terminated poly(ethylene glycol) methyl ethers that have average molecular weights of 550, 750, 2000 and 5000 (P-6283, P-6282, P-6285, P-6281) are available for coupling to carboxylic acids. These hydrophilic polymers can be used to solubilize and stabilize acidic drugs, dyes, proteins and other biomolecules,[42] as well as to prepare new polar tracers.

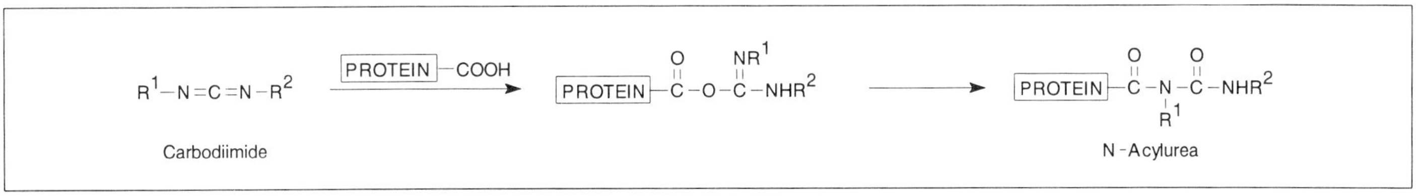

$R^1–N=C=N–R^2$ (Carbodiimide) —PROTEIN—COOH→ PROTEIN—$\overset{O}{\overset{\|}{C}}\text{-}O\text{-}\overset{NR^1}{\overset{\|}{C}}\text{-}NHR^2$ → PROTEIN—$\overset{O}{\overset{\|}{C}}\text{-}\underset{R^1}{N}\text{-}\overset{O}{\overset{\|}{C}}\text{-}NHR^2$ (N-Acylurea)

Figure 3.5 *Carbodiimide modification of a carboxylic acid group in a protein, followed by rearrangement to yield a stable N-acylurea.*

1. Anal Biochem 156, 220 (1986); **2.** Biochem Soc Trans 12, 405 (1984); **3.** J Am Chem Soc 90, 1651 (1968); **4.** Biochemistry 33, 6867 (1994); **5.** J Histochem Cytochem 38, 377 (1990); **6.** J Planar Chromatography 2, 65 (1989); **7.** Biochim Biophys Acta 897, 384 (1987); **8.** Biochemistry 22, 5 (1983); **9.** Meth Enzymol 25B, 616 (1972); **10.** Bioconjugate Chem 3, 32 (1992); **11.** Anal Biochem 219, 189 (1994); **12.** J Chromatography 645, 75 (1993); **13.** Tetrahedron Lett 22, 1901 (1970); **14.** Anal Biochem 162, 89 (1987); **15.** Clin Chem 31, 1193 (1985); **16.** Biochemistry 12, 4154 (1973); **17.** Bioconjugate Chem 5, 459, (1994); **18.** Biochim Biophys Acta 1085, 223 (1991); **19.** Anal Biochem 44, 221 (1971); **20.** J Labelled Compounds Radiopharmaceut 20, 1265 (1983); **21.** Biosci Biotech Biochem 56, 186 (1992); **22.** Arch Biochem Biophys 246, 564 (1986); **23.** Annals NY Acad Sci 463, 214 (1984); **24.** J Chromatography 608, 171 (1992); **25.** Biochemistry 27, 938 (1988); **26.** FEBS Lett 278, 51 (1991); **27.** Biochemistry 17, 2163 (1978); **28.** Anal Biochem 131, 419 (1983); **29.** Biochemistry 27, 3483 (1988); **30.** Biochemistry 27, 4512 (1988); **31.** Biochemistry 29, 5103 (1990); **32.** J Biol Chem 269, 24596 (1994); **33.** Anal Biochem 205, 166 (1992); **34.** Biochemistry 29, 9879 (1990); **35.** Biochemistry 28, 3063 (1989); **36.** Biochemistry 24, 7366 (1985); **37.** Plant Physiol 95, 707 (1991); **38.** Biochim Biophys Acta 979, 113 (1989); **39.** Biochim Biophys Acta 827, 419 (1985); **40.** Biochim Biophys Acta 730, 201 (1983); **41.** Trends Biochem Sci 9, 310 (1984); **42.** Anal Biochem 131, 25 (1983).

3.3 Data Table *Amidation Reagents for Carboxylic Acids and Glutamine*

Cat #	Structure	MW	Storage	Soluble	Abs	{ε × 10^{-3}}	Em	Solvent	Notes
A-91	✓	288	L	pH>10, DMF	335	{5.9}	493	pH 8	
A-117	✓	663	L	pH>6, DMF	523	{109}	542	MeOH	
A-191	S10.1	175	L	DMF, DMSO	351	{18}	430	MeOH	
A-434	✓	807	D,L	pH>6, DMF	515	{93}	540	pH 8	
A-456	✓	492	D,L	pH>6, DMF	493	{81}	517	pH 9	
A-497	S10.1	144	L	DMF, EtOH	354	{3.7}	445	MeOH	
A-1318	✓	515	D,L	DMF, EtOH	544	{78}	571	MeOH	
A-1339	✓	492	L	H_2O	425	{12}	532	H_2O	
A-1340	✓	534	L	H_2O	426	{11}	531	H_2O	
A-1351	✓	398	L	pH>6, DMF	492	{80}	516	pH 9	
A-1353	✓	398	L	pH>6, DMF	492	{68}	516	pH 9	
A-1363	✓	404	L	pH>6, DMF	491	{80}	515	pH 9	
A-1364	✓	418	L	pH>6, DMF	494	{81}	517	pH 9	
A-6249	✓	229	D,L	DMF, DMSO	376	{18}	480	MeOH	
A-6255	✓	347	L	pH>6	488	{86}	518	pH 9	<1>
C-428	✓	292	F,D	DMF, MeCN	333	{8.9}	414	$CHCl_3$	<2>
C-442	✓	323	F,D	DMF, MeCN	362	{40}	398	$CHCl_3$	<2>
C-621	✓	624	L	H_2O	399	{30}	423	H_2O	<3>
C-622	✓	667	L	H_2O	399	{29}	422	H_2O	<3>
D-112	✓	293	L	EtOH, DMF	335	{4.6}	526	MeOH	
D-113	✓	335	L	EtOH, DMF	335	{4.6}	518	MeOH	
D-2390	✓	371	F,D,L	DMSO, MeCN	503	{76}	510	MeOH	<4>
D-2391	✓	495	F,D,L	DMSO, MeCN	534	{60}	551	MeOH	<4>
D-6251	✓	545	F,D,L	DMSO, MeCN	588	{64}	616	MeOH	<4>
E-2247	✓	192	F,D	H_2O	<300		none		
E-6288	✓	246	D	DMSO, MeCN	261	{8.0}	none	MeOH	
H-2249	✓	217	D	H_2O	<300		none		
L-2424	✓	601	L	DMF, DMSO	561	{126}	581	MeOH	
M-6246	✓	202	D,A	DMF, MeCN	<300		none		<5>
M-6247	✓	160	D,A	DMF, MeCN	<300		none		<5>
M-6248	✓	174	D,A	DMF, MeCN	<300		none		
P-2421	✓	268	L	DMF, DMSO	340	{39}	376	MeOH	<6>
P-6253	✓	282	L	DMF, DMSO	339	{44}	376	MeOH	<6>
P-6254	✓	296	L	DMF, DMSO	340	{41}	376	MeOH	<6>
P-6281	✓	~5000	NC	H_2O, MeOH	<300		none		
P-6282	✓	~750	NC	H_2O, MeOH	<300		none		
P-6283	✓	~550	NC	H_2O, MeOH	<300		none		
P-6285	✓	~2000	NC	H_2O, MeOH	<300		none		
T-2425	✓	691	L	DMF	591	{81}	612	pH 9	

For definitions of the contents of this data table, see "How to Use the Handbook" on page vi.
Structure: Chemical structure drawing: ✓ = shown in this section; Sn.n = shown in Section number n.n.

<1> Fluorescence of A-6255 is weak (QY ~0.02).
<2> Spectra are for the reaction product with acetic acid.
<3> Cascade Blue dyes have a second absorption peak at about 376 nm with ε ~80% of the 395–400 nm peak.
<4> The absorption and fluorescence spectra of BODIPY derivatives are relatively insensitive to the solvent.
<5> This product is intrinsically a liquid or an oil at room temperature.
<6> Pyrene derivatives exhibit structured spectra. The absorption maximum is usually about 340 nm with a subsidiary peak at about 325 nm. There are also strong absorption peaks below 300 nm. The emission maximum is usually about 376 nm with a subsidiary peak at 396 nm. Excimer emission at about 470 nm may be observed at high concentrations.

Licensing and OEM

The BODIPY®, Cascade Blue®, Oregon Green™, Rhodamine Green™, Rhodol Green™, SNAFL®, SNARF® and Texas Red® fluorophores were developed by Molecular Probes' scientists, and most are patented or have patents pending. Molecular Probes welcomes inquiries about licensing these dyes, as well as their reactive versions and conjugates, for resale or other commercial purposes. Molecular Probes also manufactures many other fluorescent dyes (e.g., FITC and TRITC; succinimidyl esters of FAM, JOE, ROX and TAMRA; reactive phycobiliproteins), crosslinking reagents (e.g., SMCC and SPDP) and biotinylation reagents, which are extensively used by other companies to prepare conjugates. We offer special discounts on almost all of our products when they are purchased in bulk quantities. Please contact our Custom and Bulk Sales Department for more specific information.

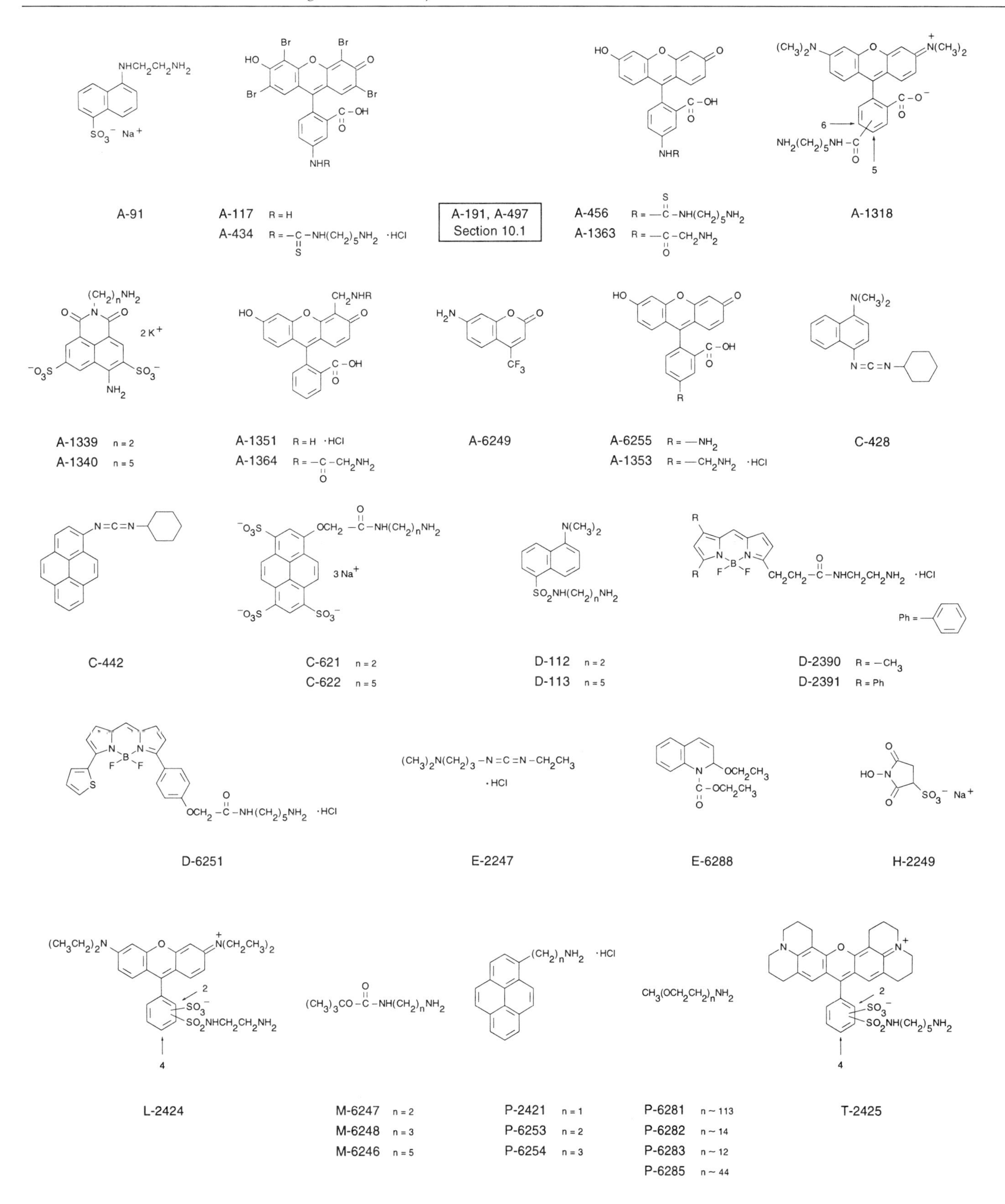
A-91
A-117 R = H
A-434 R = —C(=S)—NH(CH2)5NH2 ·HCl
A-191, A-497 Section 10.1
A-456 R = —C(=S)—NH(CH2)5NH2
A-1363 R = —C(=O)—CH2NH2
A-1318
A-1339 n = 2
A-1340 n = 5
A-1351 R = H ·HCl
A-1364 R = —C(=O)—CH2NH2
A-6249
A-6255 R = —NH2
A-1353 R = —CH2NH2 ·HCl
C-428
C-442
C-621 n = 2
C-622 n = 5
D-112 n = 2
D-113 n = 5
D-2390 R = —CH3
D-2391 R = Ph
D-6251
E-2247
E-6288
H-2249
L-2424
M-6247 n = 2
M-6248 n = 3
M-6246 n = 5
P-2421 n = 1
P-6253 n = 2
P-6254 n = 3
P-6281 n ~ 113
P-6282 n ~ 14
P-6283 n ~ 12
P-6285 n ~ 44
T-2425

Cat #		Product Name	Unit Size	Price Per Unit ($) 1–4 Units	5–24 Units
A-1363		5-(aminoacetamido)fluorescein (fluoresceinyl glycine amide)	10 mg	98.00	78.40
A-1364		4′-((aminoacetamido)methyl)fluorescein	10 mg	98.00	78.40
A-117		5-aminoeosin	100 mg	70.00	56.00
A-1339		N-(2-aminoethyl)-4-amino-3,6-disulfo-1,8-naphthalimide, dipotassium salt (lucifer yellow ethylenediamine)	25 mg	38.00	30.40
A-91		5-((2-aminoethyl)amino)naphthalene-1-sulfonic acid, sodium salt (EDANS)	1 g	75.00	60.00
A-6255	*New*	5-aminofluorescein *single isomer*	100 mg	18.00	14.40
A-191		7-amino-4-methylcoumarin *reference standard*	100 mg	45.00	36.00
A-1351		4′-(aminomethyl)fluorescein, hydrochloride	25 mg	98.00	78.40
A-1353		5-(aminomethyl)fluorescein, hydrochloride	10 mg	98.00	78.40
A-1318		5-(and-6)-((N-(5-aminopentyl)amino)carbonyl)tetramethylrhodamine (tetramethylrhodamine cadaverine) *mixed isomers*	10 mg	98.00	78.40
A-1340		N-(5-aminopentyl)-4-amino-3,6-disulfo-1,8-naphthalimide, dipotassium salt (lucifer yellow cadaverine)	25 mg	70.00	56.00
A-434		5-((5-aminopentyl)thioureidyl)eosin, hydrochloride (eosin cadaverine)	25 mg	48.00	38.40
A-456		5-((5-aminopentyl)thioureidyl)fluorescein (fluorescein cadaverine)	25 mg	138.00	110.40
A-497		6-aminoquinoline (6-AQ) *reference standard*	1 g	24.00	19.20
A-6249	*New*	7-amino-4-trifluoromethylcoumarin (AFC) *reference standard*	100 mg	18.00	14.40
C-622		Cascade Blue® cadaverine, trisodium salt	10 mg	98.00	78.40
C-621		Cascade Blue® ethylenediamine, trisodium salt	10 mg	98.00	78.40
C-428		N-cyclohexyl-N′-(4-(dimethylamino)naphthyl)carbodiimide (NCD-4)	25 mg	138.00	110.40
C-442		N-cyclohexyl-N′-(1-pyrenyl)carbodiimide	25 mg	65.00	52.00
D-2390		4,4-difluoro-5,7-dimethyl-4-bora-3a,4a-diaza-s-indacene-3-propionyl ethylenediamine, hydrochloride (BODIPY® FL EDA)	5 mg	145.00	116.00
D-2391		4,4-difluoro-5,7-diphenyl-4-bora-3a,4a-diaza-s-indacene-3-propionyl ethylenediamine, hydrochloride (BODIPY® 530/550 EDA)	5 mg	95.00	76.00
D-6251	*New*	5-(((4-(4,4-difluoro-5-(2-thienyl)-4-bora-3a,4a-diaza-s-indacene-3-yl)phenoxy)acetyl)amino)-pentylamine, hydrochloride (BODIPY® TR cadaverine)	5 mg	95.00	76.00
D-112		5-dimethylaminonaphthalene-1-(N-(2-aminoethyl))sulfonamide (dansyl ethylenediamine)	100 mg	78.00	62.40
D-113		5-dimethylaminonaphthalene-1-(N-(5-aminopentyl))sulfonamide (dansyl cadaverine)	100 mg	78.00	62.40
E-6288	*New*	2-ethoxy-1-ethoxycarbonyl-1,2-dihydroquinoline (EEDQ)	5 g	16.00	12.80
E-2247		1-ethyl-3-(3-dimethylaminopropyl)carbodiimide, hydrochloride (EDAC)	100 mg	15.00	12.00
H-2249		N-hydroxysulfosuccinimide, sodium salt (NHSS)	100 mg	25.00	20.00
L-2424		Lissamine™ rhodamine B ethylenediamine	10 mg	115.00	92.00
M-6246	*New*	mono-N-(t-BOC)-cadaverine	100 mg	35.00	28.00
M-6247	*New*	mono-N-(t-BOC)-ethylenediamine	1 g	35.00	28.00
M-6248	*New*	mono-N-(t-BOC)-propylenediamine	1 g	35.00	28.00
P-6283	*New*	poly(ethylene glycol) methyl ether, amine-terminated, average MW 550	100 mg	38.00	30.40
P-6282	*New*	poly(ethylene glycol) methyl ether, amine-terminated, average MW 750	100 mg	38.00	30.40
P-6285	*New*	poly(ethylene glycol) methyl ether, amine-terminated, average MW 2000	1 g	95.00	76.00
P-6281	*New*	poly(ethylene glycol) methyl ether, amine-terminated, average MW 5000	1 g	95.00	76.00
P-6253	*New*	1-pyreneethylamine, hydrochloride	5 mg	48.00	38.40
P-2421		1-pyrenemethylamine, hydrochloride (1-aminomethylpyrene, hydrochloride)	100 mg	58.00	46.40
P-6254	*New*	1-pyrenepropylamine, hydrochloride	5 mg	48.00	38.40
T-2425		Texas Red® cadaverine (Texas Red® C_5)	5 mg	125.00	100.00

3.4 Esterification Reagents for Carboxylic Acids

Esterification of carboxylic acids in aqueous solution is usually not possible, and esters tend to be unstable in water. Biologically important molecules, especially the nonchromophoric fatty acids, bile acids and prostaglandins, are typically modified by carboxylic acid–reactive reagents in organic solvents. Recent reviews [1,2] should be consulted for extensive discussions of fluorescent derivatization reagents for biomedical chromatography.

Fluorescent Diazoalkanes

HPLC derivatization reagents for carboxylic acids include several fluorescent analogs of the common esterification reagent diazomethane. Diazoalkanes react without the addition of catalysts and may be useful for direct carboxylic acid modification of proteins and synthetic polymers. Direct and selective modification of a glutamic acid residue in a membrane Na^+/K^+-ATPase with a cou-

Table 3.2 *Selected examples of carboxylic acid–reactive probes used to investigate specific biomolecules. An up-to-date bibliography is available from our Web site (http://www.probes.com) for every reactive probe for which we have references. Our Technical Assistance Department can also provide you with product-specific bibliographies, as well as with keyword searches of the over 25,000 literature references in our extensive bibliography database.*

amino acids A-1400,[1] P-1405[2] **arachidonic acid** A-1400[3] **bile acids** A-1400,[4] B-344,[5] B-1115[6,7] **biotin** B-344,[8] P-1405[9] **carnitine** A-1400[10] **eicosapentaenoic acid** A-1400[11]	**fatty acids** A-1400,[12-16] B-344,[17-20] D-1401 and D-1402[21,22] **5-fluorouracil** B-344,[23] B-1450[17] **glycyrrhizin** B-344[24] **herbicides** A-1400[25] **imidazole acetic acid** B-344[26]	**okadaic acid, shellfish toxins** A-1400[27] **oxalic acid** A-1400[28] **phosphoinositides (on phosphate)** A-1400[29] **phosphonic acids** A-1400[30] **polyether antibiotics,** **including monensin and laslocid** A-1400 and B-1115[31]	**prostaglandins** A-1122,[32-35] A-1400,[36,37] B-344,[38] P-1405[22] **sorbic acid** B-1456[39] **steroid acids** P-1405[40] **undecylenic acid** B-344[41] **valproic acid** B-344[42]

1. J Chromatography 348, 425 (1985); **2.** Tetrahedron Lett 28, 679 (1987); **3.** J Chromatography 357, 199 (1986); **4.** J Chromatography 413, 247 (1987); **5.** J Liquid Chromatography 12, 491 (1989); **6.** J Chromatography 400, 149 (1987); **7.** J Chromatography 272, 29 (1983); **8.** Anal Biochem 128, 359 (1983); **9.** J Chromatography 456, 421 (1988); **10.** J Chromatography 445, 175 (1988); **11.** J Chromatography 295, 463 (1984); **12.** Biochim Biophys Acta 1046, 277 (1990); **13.** J Chromatography 526, 331 (1990); **14.** J Biol Chem 263, 5724 (1988); **15.** Fresenius Z Anal Chem 329, 47 (1987); **16.** Anal Biochem 107, 116 (1980); **17.** J Chromatography 502, 423 (1990); **18.** Anal Biochem 171, 150 (1988); **19.** J Chromatography 411, 297 (1987); **20.** Anal Chem 49, 442 (1977); **21.** Anal Chem 60, 2067 (1988); **22.** Chem Pharm Bull 31, 3014 (1983); **23.** J Chromatography 530, 57 (1990); **24.** J Chromatography 567, 151 (1991); **25.** J Chromatography 541, 359 (1991); **26.** J Chromatography 416, 63 (1987); **27.** Toxicon 29, 21 (1991); **28.** Anal Biochem 128, 459 (1983); **29.** Anal Biochem 179, 127 (1989); **30.** Anal Chem 59, 1056 (1987); **31.** J Chromatography 396, 261 (1987); **32.** J Chromatography 427, 209 (1988); **33.** Anal Biochem 165, 220 (1987); **34.** Prostaglandins 32, 301 (1986); **35.** Anal Chem 56, 1866 (1984); **36.** J Liquid Chromatography 11, 1273 (1988); **37.** J Chromatography 253, 271 (1982); **38.** Proc Natl Acad Sci USA 82, 8315 (1985); **39.** J Chromatography 543, 69 (1991); **40.** J Chromatography 564, 27 (1991); **41.** J Chromatography 456, 201 (1988); **42.** J Chromatography 487, 496 (1989).

marin diazomethane (D-1402) has been reported.[3] Fluorescent diazoalkanes also react with phosphates[4] and potentially with lipid-associated carboxylic acids in membrane-bound proteins or with free fatty acids.

The most common fluorescent diazomethyl derivative has been 9-anthryldiazomethane (ADAM, A-1400), a reagent that has been used to derivatize several types of compounds (Table 3.2). Unfortunately, ADAM is not very stable and may decompose during storage. The 1-pyrenyldiazomethane[5-7] (PDAM, P-1405) and coumarin diazomethanes[5,6,8] (D-1401, D-1402) are recommended as replacements for ADAM because they have much better chemical stability. Moreover, the detection limit for PDAM conjugates is reported to be about 20–30 femtomoles, which is five times better than that for ADAM conjugates.[5] ADAM and PDAM have been used to detect several types of acids (Table 3.2) and to measure plasma fatty acids.[9] In addition, fatty acids derivatized with these reagents have been used to measure phospholipase A_2 activity.[10] It has been reported that photolysis of pyrenemethyl esters liberates the free carboxylic acid,[7] making PDAM a potential protecting group for carboxylic acids.

Fluorescent alcohols such as 1-pyrenebutanol (P-244) and the new BODIPY FL propanol (D-6300) react with acid halides and isocyanates to yield aliphatic esters and urethanes. Like diazoalkanes, these reagents are primarily useful as synthetic precursors.

Fluorescent Alkyl Halides

The low nucleophilicity of carboxylic acids requires that they be converted to anions (typically cesium or quaternary ammonium are used as counterions) before they can be esterified with alkyl halides in organic solvents. The alkyl halides 4-bromomethyl-7-methoxycoumarin (BMB, B-344), 4-bromomethyl-6,7-dimethoxycoumarin (B-1456) and 3-bromomethyl-6,7-dimethoxy-1-methyl-2(1*H*)-quinoxazolinone (B-1458) are commonly used to derivatize prostaglandins,[13-15] fatty acids[16] and biotin,[17] and they also react with phosphonic acids.[18] 1-(Bromoacetyl)pyrene (B-1115) is a versatile reagent that has been used to derivatize antibiotics[19,20] and bile acids.[21-23] Some of the other reagents listed in Table 3.2, however, may have advantages over BMB and panacyl bromide. In particular, the esters derived from the coumarin diazomethane (B-344, D-1401) can be excited by the 325 nm spectral line of the He-Cd laser, which is used for laser-induced fluorescence in some analytical instrumentation. Conjugates prepared with 3-bromoacetyl-7-diethylaminocoumarin (B-1450) have relatively strong visible absorption at the 442 nm spectral line of the He-Cd laser ($\epsilon_{442\,nm}$ = 53,000 $cm^{-1}M^{-1}$). Conjugates of 6-bromoacetyl-2-dimethylaminonaphthalene (badan, B-6057) have a high Stokes shift, as well as spectral properties that are very environmentally sensitive. 5-(Bromomethyl)fluorescein (B-1355) and BODIPY 493/503 methyl bromide (B-2103) have the strongest absorptivity and fluorescence of the currently available carboxylic acid–derivatization reagents.[24] Molecular Probes' BODIPY 493/503 methyl bromide reacts with anions of carboxylic acids during heating in an organic solvent such as methanol or acetonitrile. The high absorptivity, electrical neutrality and intense green fluorescence of these conjugates may make this BODIPY methyl bromide the preferred reagent for carboxylic acid determinations.

All of the alkyl halides in this section also react with thiol groups, including those in proteins. Although more commonly used as thiol-reactive reagents, the monobromobimanes (M-1378, M-1380; see Section 2.4) have been reported to react with carboxylic acids in organic solvents.[25] The coumarin iodoacetamide DCIA (D-404, see Section 2.2) has also been used to derivatize carboxylic acids;[26] other iodoacetamides in Chapter 2 will probably react similarly.

Trifluoromethanesulfonates

The naphthalimide sulfonate ester (N-2461) reacts rapidly with the anions of carboxylic acids in acetonitrile to give adducts that are reported to be detectable by absorption at 259 nm down to 100 femtomoles and by fluorescence at 394 nm down to 4 femtomoles.[27] This reagent will likely react with other nucleophiles too, including thiols, amines, phenols (tyrosine) and probably histidine. The nonfluorescent reagents 4-bromophenacyl bromide (B-6286) and 4-bromophenacyl trifluoromethanesulfonate (B-6287) each yield the same chromophoric esters for HPLC detection of carboxylic acids;[28] however, carboxylic acids in dilute solutions are reported to be much easier to modify with trifluoromethanesulfonate derivatives.[29-31] The 4-bromophenacyl bromide is a potent inhibitor of phospholipase A_2 and carboxypeptidase C_n and is reported to modify histidine residues in proteins.[32-35]

1. J Chromatography B 659, 139 (1994); **2.** J Chromatography B 659, 85 (1994); **3.** J Biol Chem 269, 6892 (1994); **4.** Anal Biochem 179, 127 (1989); **5.** Anal Chem 60, 2067 (1988); **6.** J Chromatography 456, 421 (1988); **7.** Tetrahedron Lett 28, 679 (1987); **8.** Chem Pharm Bull 31, 3014 (1983); **9.** J Chromatography 380, 247 (1986); **10.** J Biol Chem 263, 5724 (1988); **11.** Anal Sci 1, 295 (1985); **12.** J Chromatography 269, 81 (1983); **13.** Prostaglandins 42, 355 (1991); **14.** J Chromatography 427, 209 (1988); **15.** Anal Biochem 165, 220 (1987); **16.** J Liquid Chromatography 16, 2915 (1993); **17.** Anal Biochem 200, 85 (1992); **18.** Anal Chem 59, 1056 (1987); **19.** J Chromatography A 657, 349 (1993); **20.** J Chromatography 396, 261 (1987); **21.** J Liquid Chromatography 14, 605 (1991); **22.** J Chromatography 400, 149 (1987); **23.** J Chromatography 272, 29 (1983); **24.** Anal Chem 68, 327 (1996); **25.** J Org Chem 46, 1666 (1981); **26.** Anal Chem. 59, 1203 (1987); **27.** J Chromatography 508, 133 (1990); **28.** Anal Chem 47, 1797 (1975); **29.** J Chromatography 613, 203 (1993); **30.** Anal Biochem 205, 14 (1992); **31.** Clin Chem Acta 212, 55 (1992); **32.** Biochemistry 19, 743 (1980); **33.** Eur J Biochem 94, 531 (1979); **34.** J Biol Chem 252, 2405 (1977); **35.** J Biochem 76, 375 (1974).

3.4 Data Table *Esterification Reagents for Carboxylic Acids*

Cat #	Structure	MW	Storage	Soluble	Abs	{ε × 10⁻³}	Em	Solvent	Notes
A-1122	✓	419	F,D,L	DMF, MeCN	362	{8.5}	494	MeOH	<1, 2>
A-1400	✓	218	FF,D,L	DMF, MeCN	364	{6.1}	411	MeOH	<1, 3>
B-344	✓	269	L	DMF, MeCN	330	{13}	none	MeOH	<4>
B-1115	✓	323	F,L	DMF, MeCN	360	{18}	387	MeOH	
B-1355	S2.2	425	F,D,L	pH>6, DMF	492	{81}	515	pH 9	
B-1450	✓	338	F,L	DMF, MeCN	442	{56}	470	$CHCl_3$	
B-1456	✓	299	F,L	DMF, MeCN	354	{10}	435	MeOH	<4>
B-1458	✓	313	F,L	$CHCl_3$	384	{13}	454	MeOH	
B-2103	✓	341	F,D,L	DMSO, MeCN	515	{55}	525	MeOH	
B-6057	S2.3	292	F,L	DMF, MeCN	387	{21}	520	MeOH	
B-6286	✓	278	F,L	DMF, DMSO	261	{16}	none	MeOH	
B-6287	✓	347	F,DD	DMF, DMSO	258	{14}	none	MeOH	
D-1401	✓	216	FF,D,L	DMF, MeCN	320	{27}	393	MeOH	<1, 5>
D-1402	✓	257	FF,D,L	DMF, MeCN	395	{19}	477	MeOH	<1, 5>
D-6300	✓	278	F,L	DMSO, MeCN	504	{96}	510	MeOH	
N-2461	✓	373	FF,DD,L	DMF, $CHCl_3$	260	{59}	395	MeOH	
P-244	S14.6	274	L	DMF, $CHCl_3$	341	{41}	376	MeOH	<6>
P-1405	✓	242	FF,L	DMF, MeCN	340	{41}	375	MeOH	<1, 6, 7>

For definitions of the contents of this data table, see "How to Use the Handbook" on page vi.
Structure: Chemical structure drawing: ✓ = shown in this section; Sn.n = shown in Section number n.n.

<1> Spectra are for the reaction product with acetic acid.
<2> A-1122 also has a much stronger absorption peak at 253 nm (ε = {174}) [Anal Biochem 125, 30 (1982)].
<3> A-1400 acetate ester derivative also has Abs = 256 nm (ε = {63}); however, fluorescence excitation at 360 nm gives highest detection sensitivity [Anal Biochem 107, 116 (1980)]. For unreacted reagent, Abs = 399 nm (ε = {7.4}).
<4> Spectra for the reaction product of B-344 with acetic acid: Abs = 322 nm (ε = {15}), Em = 395 nm (QY = 0.08) in MeOH [J Chromatography 178, 249 (1979)]. Spectra for reaction product of B-1456 with acetic acid: Abs = 340 nm (ε = {13}), Em = 425 nm (QY = 0.42) in MeOH [J Chromatography 269, 81 (1983)].
<5> Acetate ester derivatives of D-1401 and D-1402 in EtOH have fluorescence quantum yields of 0.03 and 0.48, respectively. Unreacted D-1401 is nonfluorescent; unreacted D-1402 quantum yield is 0.05 [Chem Pharm Bull 31, 3014 (1983)].
<6> Pyrene derivatives exhibit structured spectra. The absorption maximum is usually about 340 nm with a subsidiary peak at about 325 nm. There are also strong absorption peaks below 300 nm. The emission maximum is usually about 376 nm with a subsidiary peak at 396 nm. Excimer emission at about 470 nm may be observed at high concentrations.
<7> Unreacted P-1405 is nonfluorescent with Abs = 383 nm (ε = {31}) in MeOH.

3.4 Structures *Esterification Reagents for Carboxylic Acids*

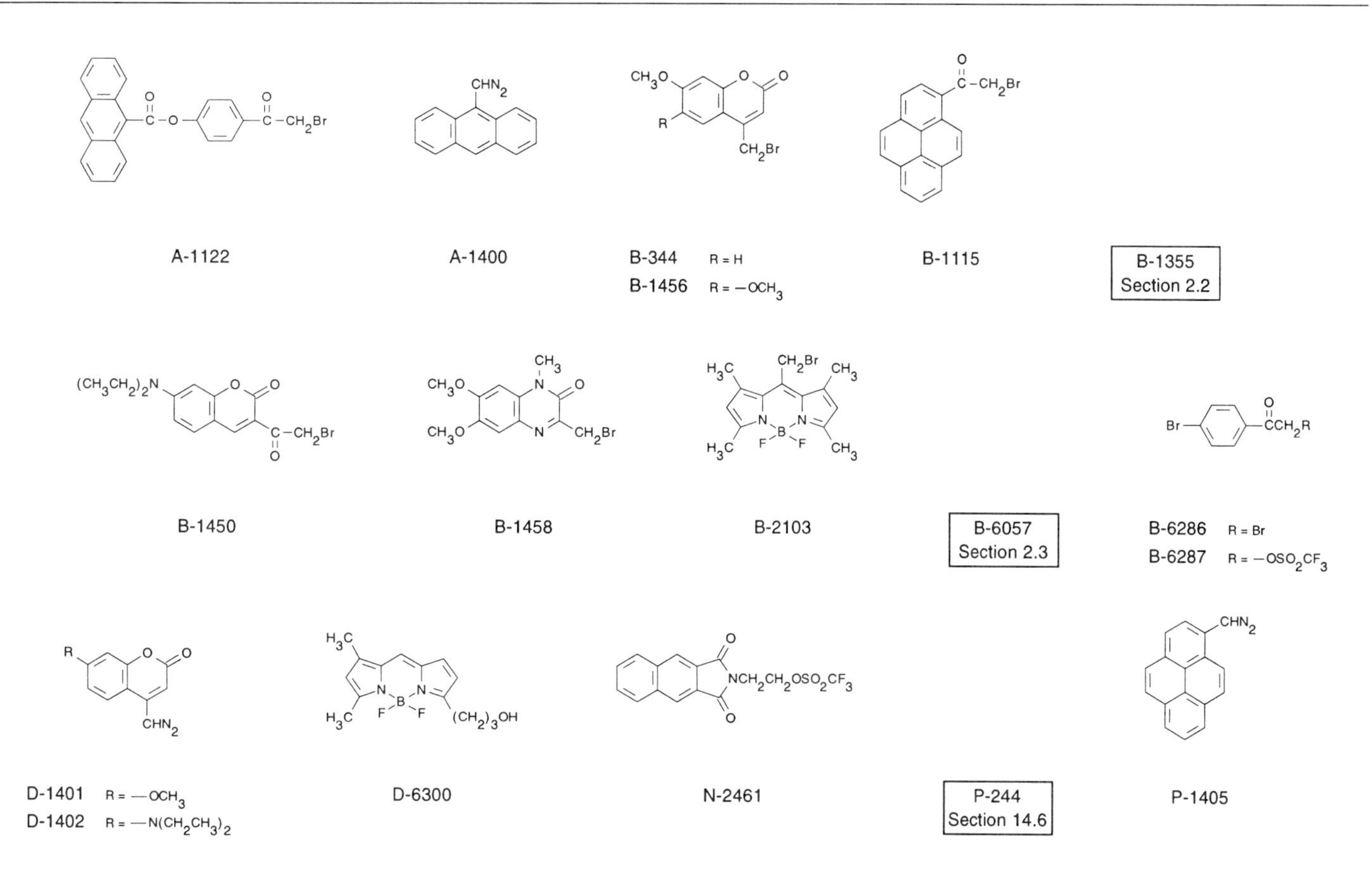

3.4 Price List *Esterification Reagents for Carboxylic Acids*

Cat #		Product Name	Unit Size	Price Per Unit ($) 1–4 Units	5–24 Units
A-1122		4-(9-anthroyloxy)phenacyl bromide (panacyl bromide)	100 mg	118.00	94.40
A-1400		9-anthryldiazomethane (ADAM)	25 mg	118.00	94.40
B-1450		3-bromoacetyl-7-diethylaminocoumarin	25 mg	60.00	48.00
B-6057	**New**	6-bromoacetyl-2-dimethylaminonaphthalene (badan)	10 mg	75.00	60.00
B-1115		1-(bromoacetyl)pyrene	100 mg	48.00	38.40
B-2103		8-bromomethyl-4,4-difluoro-1,3,5,7-tetramethyl-4-bora-3a,4a-diaza-s-indacene (BODIPY® 493/503 methyl bromide)	5 mg	95.00	76.00
B-1456		4-bromomethyl-6,7-dimethoxycoumarin	100 mg	24.00	19.20
B-1458		3-bromomethyl-6,7-dimethoxy-1-methyl-2(1H)-quinoxazolinone	25 mg	48.00	38.40
B-1355		5-(bromomethyl)fluorescein	10 mg	125.00	100.00
B-344		4-bromomethyl-7-methoxycoumarin (BMB)	100 mg	38.00	30.40
B-6286	**New**	4-bromophenacyl bromide (2,4′-dibromoacetophenone)	5 g	18.00	14.40
B-6287	**New**	4-bromophenacyl trifluoromethanesulfonate	25 mg	48.00	38.40
D-1402		4-diazomethyl-7-diethylaminocoumarin	10 mg	88.00	70.40
D-1401		4-diazomethyl-7-methoxycoumarin	25 mg	88.00	70.40
D-6300	**New**	4,4-difluoro-5,7-dimethyl-4-bora-3a,4a-diaza-s-indacene-3-propanol (BODIPY® FL propanol)	10 mg	95.00	76.00
N-2461		2-(2,3-naphthalimino)ethyl trifluoromethanesulfonate	100 mg	53.00	42.40
P-244		1-pyrenebutanol	100 mg	65.00	52.00
P-1405		1-pyrenyldiazomethane (PDAM)	25 mg	88.00	70.40

Interconversion of Functional Groups

Several reagents can be used to convert one functional group to another, thereby providing greater flexibility in selecting reactive probes for functional group modification. These reagents are discussed in Chapter 5, except where otherwise noted. Among the reagents are the following:

- **Aldehyde (Reducing Sugar) or Ketone to Amine**: NH_3 + $NaCNBH_3$ (widely available)
- **Aldehyde or Ketone to Thiol**: AMBH (A-1642)
- **Amine to Aldehyde**: SFB (S-1580, see Chapter 3)
- **Amine to Carboxylic Acid**: succinic anhydride (widely available)
- **Amine to Iodoacetamide**: succinimidyl iodoacetate (I-6001); 4-nitrophenyl iodoacetate (N-1505); SIAX (S-1666); SIAXX (S-1668)
- **Amine to Maleimide**: SMCC (S-1534); EMCS (S-1563); SMB (S-1535)
- **Amine to Thiol**: SPDP (S-1531); SATA (S-1553); SATP (S-1573)
- **Disulfide to Thiol**: DTT (D-1532); DTE (D-8452); TCEP (T-2556); tris-(2-cyanoethyl)phosphine (T-6052)
- **Thiol to Aldehyde**: chloroacetaldehyde (widely available)
- **Thiol to Amine**: *N*-(2-iodoethyl)trifluoroacetamide (I-1536, see Chapter 2)
- **Thiol to Carboxylic Acid**: iodoacetic acid; 3-bromopropionic acid (widely available)
- **Thiol to Ketone**: 3-bromopyruvic acid (B-6058, see Chapter 2)

Chapter 4

Biotins and Haptens

Contents

Technical Notes and Product Highlights

Related Chapters

4.1 Introduction to Avidin–Biotin and Antibody–Hapten Techniques

The high affinity and specificity of avidin–biotin and antibody–hapten interactions have been exploited for diverse applications in immunology, histochemistry, *in situ* hybridization, affinity chromatography and many other areas.[1,2] Biotinylation and haptenylation reagents provide the "tag" that transforms poorly detectable molecules into probes that can be recognized by a labeled detection reagent. Once tagged with biotin or a hapten, a biomolecule of interest — such as an antibody, lectin, drug, polynucleotide, polysaccharide or receptor ligand — can be used to probe cells, tissues or complex solutions. This tagged molecule can then be detected with the appropriate avidin or anti-hapten antibody conjugate, which has been labeled with a fluorophore, fluorescent microsphere, enzyme, chromophore or colloidal gold. Avidin–biotin and antibody–hapten techniques are compatible with flow cytometry and light, electron and fluorescence microscopy, as well as with solution-based methods such as enzyme-linked immunosorbent assays (ELISAs). Moreover, by judicious choice of detection reagents and sandwich protocols, these techniques can be employed to amplify the signal from low-abundance analytes. For example, avidin or streptavidin can serve as a bridge between two biotinylated biomolecules, which is a common immunohistochemical technique for signal amplification and improved tissue penetration. Sections 7.2 and 7.5 describes our assortment of labeled antibody and avidin probes, respectively; we offer the largest selection available from any commercial source. Where they can be used, our avidin- and biotin-coated FluoSpheres® fluorescent microspheres (see Section 6.2) provide an alternative detection technology that offers a combination of fluorescence intensity and photostability far superior to that of any simple dye conjugate.

1. Meth Enzymol 184 (1990); **2.** Anal Biochem 171, 1 (1988).

4.2 Biotinylation and Haptenylation Reagents

Molecular Probes is the primary manufacturer of a diverse array of biotinylation and haptenylation reagents for labeling biomolecules. Reviews of the methods that we use to prepare biotin [1] and fluorescent [2] conjugates of antibodies have recently been published. To make the labeling reactions particularly easy, we have developed some very useful FluoReporter Kits for labeling proteins and oligonucleotides with biotin, DNP (see below), or a choice of several different fluorophores. Each of these kits contains the preferred reactive dye — many of which have spacers to reduce interactions between the label and the biomolecule — along with a detailed protocol for preparing the conjugates. The protein labeling kits also provide the separation media for purifying labeled protein conjugates from the reaction mixture. Our FluoReporter Labeled Oligonucleotide Purification Kit provides a simple spin-column method for isolating labeled oligonucleotides from the reaction mixture. See the Product Highlight "FluoReporter Protein and Oligonucleotide Labeling Kits" on page 124 for a complete description.

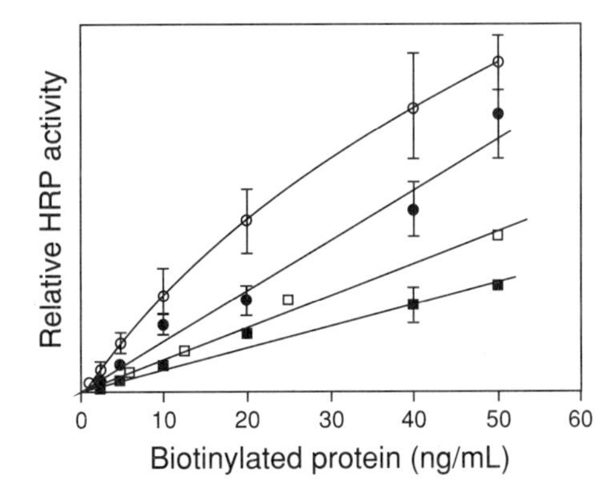

Figure 4.1 *ELISA-type assay comparing the binding capacity of bovine serum albumin (BSA) and goat anti-mouse antibody (GAM) biotinylated with either biotin-X or biotin-XX. The assay was developed using streptavidin–horseradish peroxidase conjugate (0.2 µg/mL) and o-phenylenediamine dihydrochloride (OPD). The number of biotin/mole was: 4.0 biotin-X/GAM (●), 4.4 biotin-XX/GAM (○), 6.7 biotin-X/BSA (■) and 6.2 biotin-XX/GAM (□). Error bars on some data points have been omitted for clarity. Reprinted with permission from Meth Mol Biol 45, 223 (1995).*

Amine-Reactive Biotinylation Reagents

The primary building blocks for preparing biotinylation reagents are biotin (B-1595), biotin-X (B-1598) and biotin-XX, where "X" represents a seven-atom aminohexanoyl spacer between biotin and the reactive carboxylic acid. This spacer helps to separate the biotin moiety from its point of attachment, potentially reducing the interaction of biotin with the biomolecule to which it is conjugated and enhancing its ability to bind to the relatively deep biotin binding sites in avidin [3] (Figure 4.1). When these biotin derivatives are converted to reactive acid chlorides or mixed anhydrides, they can be used to synthesize biotin conjugates of biomolecules that contain poorly reactive functional groups such as alcohols and aromatic amines. We also offer biocytin (ϵ-biotinoyl-L-lysine, B-1592) and biocytin-X (B-2605), each of which contain a primary amine that allows them to be fixed in cells with aldehyde fixatives, facilitating subsequent detection with labeled avidin and streptavidin conjugates. Biocytin derivatives are commonly employed as microinjectable cell tracers and are discussed in Section 15.3.

Amine-Reactive Biotinylation Reagents

Although Molecular Probes' biotin succinimidyl ester (B-1513) can be used to biotinylate amines in peptides, proteins and other biomolecules,[4] we recommend the biotin-X and biotin-XX succinimidyl esters (B-1582, B-1606), and especially the new water-soluble biotin-X and biotin-XX sulfosuccinimidyl esters (B-6353, B-6352), because their additional 7- and 14-atom spacers greatly facilitate avidin binding (Figure 4.1). Molecular Probes is licensed to manufacture and sell biotin-X succinimidyl ester [5] and uses this or the biotin-XX derivative to prepare all its biotinylated protein and dextran conjugates.

Amine-Reactive Chromophoric Biotin Derivative

Determining a protein's degree of biotinylation is difficult because of the lack of visible absorbance by the biotin molecule. To facilitate this determination, Molecular Probes offers an amine-reactive chromophoric derivative, biotin-X 2,4-dinitrophenyl-X-L-lysine succinimidyl ester [6] (DNP-X–biocytin-X, SE; B-2604). Following protein conjugation, the extent of biotinylation can be determined from the absorbance of the DNP chromophore ($\epsilon_{360\ nm}$ = 15,000 $cm^{-1}M^{-1}$). Incorporation of the DNP moiety into the biotinylating reagent does not affect its complexation with avidin. We have also introduced a FluoReporter Biotin/DNP Protein Labeling Kit (see below), which contains sufficient DNP-X–biocytin-X succinimidyl ester for 5 to 10 labeling reactions of 0.2–2 mg each.

The DNP-X–biocytin-X succinimidyl ester is a unique amine-reactive reagent with versatile applications. Because this reagent comprises both DNP and biotin moieties, molecules labeled with it may be probed with Molecular Probes' anti-DNP antibodies (see Section 7.3) or with our avidin conjugates (see Section 7.5), facilitating correlated fluorescence and electron microscopy studies. This chromophoric biotin is also very useful for preparing conjugates for Molecular Devices' patented Threshold® Immuno-Ligand Assay System.[7] The Threshold System is designed to quantitate DNA and protein impurities in biopharmaceuticals by a urease-mediated signal amplification system that employs a silicon sensor to convert the chemical signal into an electronic signal.[7]

FluoReporter Kits for Biotinylating Proteins and Oligonucleotides

FluoReporter Mini-Biotin-XX Protein Labeling Kit

Molecular Probes' new FluoReporter Mini-Biotin-XX Protein Labeling Kit (F-6347) provides a method for efficiently biotinylating small amounts of antibodies or other proteins. The water-soluble biotin-XX sulfosuccinimidyl ester contained in this kit readily reacts with a protein's amines to yield a biotin moiety that is linked via two aminohexanoyl chains ("XX"). This 14-atom spacer has been shown to enhance the binding of biotin derivatives to avidin's relatively deep binding sites (Figure 4.1). Each FluoReporter Mini-Biotin-XX Labeling Kit contains:

- Biotin-XX, sulfosuccinimidyl ester
- Reaction tubes, each containing a magnetic spin bar
- Spin columns plus collection tubes
- Dialysis tubing (molecular weight cut-off ~12,000–14,000)
- Protocol for preparing and purifying the biotinylated protein

The ready-to-use spin columns provide an extremely convenient method of purifying the biotinylated protein from excess biotinylation reagents. Alternatively, the researcher may choose to remove excess reagents by dialysis, thereby avoiding further dilution of the biotinylated protein. The FluoReporter Mini-Biotin-XX Protein Labeling Kit contains sufficient reagents for five biotinylation reactions of 0.1–3 mg protein each.

FluoReporter Biotin-XX Protein Labeling Kit

We also offer the FluoReporter Biotin-XX Protein Labeling Kit (F-2610) for larger-scale biotinylation reactions. Once the labeled protein is purified from excess biotin reagent, its degree of biotinylation can be determined using an avidin–biotin displacement assay;[8,9] biotinylated goat IgG is provided as a control. The FluoReporter Biotin-XX Protein Labeling Kit supplies:

- Biotin-XX, succinimidyl ester
- Dimethylsulfoxide (DMSO)
- Gel filtration column
- Avidin–HABA complex
- Biotinylated goat IgG
- Protocol for preparing and purifying the biotinylated protein, as well as for quantitating the degree of labeling

The FluoReporter Biotin-XX Protein Labeling Kit provides sufficient reagents for five labeling reactions of 5–20 mg protein each.

FluoReporter Biotin/DNP Protein Labeling Kit

The new FluoReporter Biotin/DNP Protein Labeling Kit (F-6348) is similar to our other FluoReporter Protein Labeling Kits, except that it contains DNP-X–biocytin-X succinimidyl ester as the reactive label. When proteins are labeled with this chromophoric biotin derivative, the degree of biotinylation can be readily assessed from the extinction coefficient of DNP ($\epsilon_{360\ nm}$ = 15,000 $cm^{-1}M^{-1}$). An additional feature of the conjugates labeled with DNP-X–biocytin-X succinimidyl ester is that they can be recognized by both avidin derivatives and anti-DNP antibodies, enabling researchers to choose among several detection techniques suitable for fluorescence and electron microscopy. Each FluoReporter Biotin/DNP Protein Labeling Kit contains:

- DNP-X–biocytin-X, succinimidyl ester
- Dimethylsulfoxide (DMSO)
- Reaction tubes, each containing a magnetic spin bar
- Spin columns plus collection tubes
- Protocol for preparing and purifying the protein conjugate, as well as for quantitating the degree of labeling

The FluoReporter Biotin/DNP Protein Labeling Kit supplies sufficient reagents for 5 to 10 labeling reactions of 0.2–2 mg protein each.

FluoReporter Kits for Labeling Oligonucleotides with Biotin or DNP

Molecular Probes' FluoReporter Oligonucleotide Labeling Kits provide the researcher with a simple and convenient method for efficiently labeling oligonucleotides with a wide variety of our fluorophores and haptens. Labeling is not only economical, it is easy and very reproducible. Labeled oligonucleotides prepared with these FluoReporter Kits can serve as primers for DNA sequencing and amplification and as probes for Northern and Southern blots, colony and plaque lifts and mRNA *in situ* hybridization experiments.

In addition to kits for labeling oligonucleotides with different fluorophores (see Section 8.2 and the Product Highlight "Fluo-

Reporter Protein and Oligonucleotide Labeling Kits" on page 124), we offer the FluoReporter Biotin-XX (F-6081) and FluoReporter DNP-X (F-6085) Oligonucleotide Amine Labeling Kits, which provide stable biotin-XX and DNP-X succinimidyl esters, respectively, for labeling amine-modified oligonucleotides. For directly labeling 3'- or 5'-phosphate–terminated oligonucleotides, we also offer the FluoReporter Biotin-X-C_5 Oligonucleotide Phosphate Labeling Kit (F-6095). Labeled oligonucleotides prepared with any of these kits can be purified by HPLC or gel electrophoresis, or simply by using the spin columns and protocol provided in our FluoReporter Labeled Oligonucleotide Purification Kit (F-6100, see below).

Our FluoReporter Kits have been optimized for labeling oligonucleotides containing 18 to 25 bases but may be useful for labeling either shorter or longer oligonucleotides.[10] The FluoReporter Biotin-XX, Biotin-X-C_5 and DNP-X Oligonucleotide Labeling Kits provide sufficient reagents for five complete labeling reactions. Each kit contains:

- Five vials of biotin-XX succinimidyl ester, biotin-X-cadaverine *or* DNP-X succinimidyl ester, each sufficient for labeling 100 µg of oligonucleotide
- Anhydrous dimethylsulfoxide (DMSO) for dissolving the reactive reagent
- Labeling buffer
- Detailed protocol for oligonucleotide labeling

Our FluoReporter Labeled Oligonucleotide Purification Kit (F-6100), which can be used with any of our FluoReporter Oligonucleotide Labeling Kits, contains:

- Five spin columns
- Separate buffers for column loading, equilibration, washing and elution
- Detailed protocol that has been tested with all of our FluoReporter Oligonucleotide Labeling Kits

Sufficient columns and buffers are provided for purification of five labeling reactions of 100 µg oligonucleotide each. Isolated yields for the combined conjugation and purification steps are usually >60%, and the products are typically >90% pure as determined by HPLC. This kit may be useful for purifying conjugates of many other reactive dyes and haptens too. Oligonucleotide conjugates can be used for most procedures without additional purification.

Other Biotinylation Reagents

Thiol-Reactive Biotinylation Reagents

Although amine-reactive reagents are more commonly employed, thiol-reactive biotin iodoacetamide [11] (B-1591) and biotin maleimide [12-14] (M-1602) can also be used to label proteins and thiol-modified oligonucleotides. Electrophoretically separated thiolated proteins treated with biotin maleimide have been detected in Western blots by an avidin–alkaline phosphatase conjugate.[15] Biotin iodoacetamide was recently used as an enzyme substrate in an unusual chemical reaction catalyzed by a ribozyme.[16] Biotin maleimide is reported to be membrane impermeant, making it useful for determining topology of thiol groups in cell membranes.[17]

Biotinylation of Tyrosine, Histidine and Guanosine Residues

The aromatic amino group of N^{α}-(4-aminobenzoylbiocytin) (A-1604) is converted by diazotization to a phenol-reactive diazonium salt that is readily coupled to a protein's tyrosine and histidine residues, yielding a chromophoric conjugate.[18,19] Direct biotinylation of guanosine residues in DNA by this reactive diazonium salt also promises to be an important technique for preparing *in situ* hybridization probes.[20] This chemistry represents one of the few methods for direct modification of tyrosine, histidine and guanosine residues.

Biotinylation of Nucleic Acids and Carbohydrates

As described in Chapter 3, aldehydes generated by periodate oxidation of vicinal diols in glycoproteins, polysaccharides and RNA or of N-terminal serine and threonine residues in proteins can be biotinylated using biotin hydrazide (B-1549) and its biotin-X (B-1600) and biotin-XX (B-2600) analogs.[21,22] Biocytin hydrazide (B-1603) and biocytin-X hydrazide (B-6350) react similarly and may be preferred because of their higher water solubility.[13,23] Biotin hydrazides have been used to quantitate periodate-oxidized glycoproteins on electroblots,[24] as well as to biotinylate:

- Antibodies [22,25]
- Calmodulin [18]
- Cytidine residues [26]
- Erythropoietin [21]
- Glucagon [27]
- Glycoproteins [28,29], sialic acid and carbohydrates [30]
- Low-density lipoproteins [31] (LDL)
- Nucleic acid hybridization probes [32,33]
- Steroids [34]

Modification of Abasic Sites in Nucleic Acids

The biotinylated hydroxylamine ARP (A-6346) was recently used to modify abasic sites in DNA — those apurinic sites and apyrimidinic lesions thought to be important intermediates in carcinogenesis.[35-37] Once the aldehyde group in an abasic site is modified with ARP, the resulting biotinylated DNA can be quantitated with labeled avidin conjugates (see Section 7.5).

Biotinylation of Carboxylic Acids

The biotin amines and hydrazides can be coupled to chemically activated carboxylic acids. The amine-containing biotin derivatives (A-1593, A-1594, B-1596, B-1597, N-6356) are versatile intermediates for coupling biotin to DNA, carboxylic acids [38,39] and other biomolecules.[11] The biotin cadaverines (A-1594, B-1596) and potentially our unique norbiotinamine [40] (N-6356) are useful for transglutaminase-mediated modification of glutamine residues in certain proteins [41] (see Section 3.3) and for the microplate-based assay of transglutaminase activity.[42-44]

Synthesis of Biotinylated Peptides

Biocytin is a naturally occurring amino acid derivative (ε-D-biotinoyl-L-lysine). The *t*-BOC derivative of biocytin (B-6349) can be used in automated synthesis of biotinylated peptides. It has also been attached to the synthesis resin as the first residue to provide for automated synthesis of C-terminal biotinylated peptides.[45] Other reagents for automated synthesis of labeled peptides are described in Section 9.3. Alternatively, biotin iodoacetamide (B-1591) and biotin maleimide (M-1602) can be used to couple biotin to cysteine residues of peptides and proteins.

Table 4.1 Selected haptenylation reagents and their corresponding anti-hapten antibodies.

Cat #	Reactive Hapten	Unlabeled and Labeled Anti-Hapten Antibodies * (Cat #)
A-2952	3-Amino-3-deoxydigoxigenin hemisuccinamide, SE †	Available from other suppliers
D-6102	BODIPY FL-X, SE	Anti–BODIPY FL dye (A-5770, A-5771, A-6420) ‡
D-6116	BODIPY TR-X, SE	Anti–BODIPY TR dye (A-6414) ‡
C-2284	Cascade Blue acetyl azide	Anti–Cascade Blue dye (A-5760, A-5761)
D-6104	Dansyl-X, SE	Anti-dansyl (A-6398)
D-2248	DNP-X, SE	Anti-DNP (A-6423, A-6424, A-6425, A-6430, A-6435)
B-2604	DNP-X–biocytin-X, SE	Anti-DNP (A-6423, A-6424, A-6425, A-6430, A-6435) or avidin/streptavidin conjugates
F-2181	5(6)-SFX	Anti-fluorescein (A-889, A-981, A-982, A-6413, A-6418, A-6421)
F-6130	Fluorescein-EX, SE	
L-1338	Lucifer yellow iodoacetamide	Anti–lucifer yellow (A-5750, A-5751)
T-6105	5(6)-TAMRA-X, SE	Anti-tetramethylrhodamine (A-6397); anti–Texas Red (A-6399) ‡
R-6160	Rhodamine Red-X, SE	
T-6134	Texas Red-X, SE	

* See Section 7.3 for a description of these anti-hapten antibodies. † This reagent is not licensed for use in the practice of U.S. Patent Nos. 5,344,757; 5,354,657 or other patents owned by Boehringer Mannheim GmbH. ‡ Both the anti-tetramethylrhodamine and the anti–Texas Red antibodies crossreact with tetramethylrhodamine, Lissamine rhodamine, Rhodamine Red and Texas Red fluorophores. Thus, these fluorophores should not be used simultaneously to generate separate signals in a multicolor experiment. Similarly, the anti–BODIPY dye antibodies crossreact with other BODIPY dyes and therefore should not be used for simultaneous detection of more than one dye based on the BODIPY fluorophore.

Haptenylation Reagents

A prerequisite for multicolor applications such as fluorescence *in situ* hybridization is the availability of multiple hapten molecules along with their complementary binding proteins. The avidin–biotin system can provide only single-color detection, whereas antibody–hapten methods can generate a number of unique signals, limited only by the specificity of the antibody–hapten detection and the ability to distinguish the signals from different antibodies. The characteristics of a suitable hapten include a unique chemical structure that is not commonly found in cells, a high degree of antigenicity so as to elicit good antibody production and a means for incorporating the hapten into the detection system. Molecular Probes' ever-increasing selection of haptenylation reagents enables researchers to covalently attach haptens to nucleotides, proteins, enzymes and other biomolecules.

In addition to our wide range of biotinylation reagents discussed above, Molecular Probes provides some unique haptenylation reagents, including reactive versions of digoxigenin [46] (A-2952, A-2953), dinitrophenyl-X [47] (DNP-X, SE; D-2248) and several fluorophores (Table 4.1). The succinimidyl ester and iodoacetamide derivatives of digoxigenin have recently been shown to inhibit the Na^+/K^+-ATPase by binding to the cardiac steroid receptor site.[48] We usually recommend haptenylation reagents that contain spacers between the hapten and the reactive groups to reduce potential interactions with the biomolecule to which it is conjugated and to make the hapten maximally available to secondary detection reagents. Most of the preferred haptenylation reagents in Table 4.1 possess this feature.

DNP-X–biocytin-X succinimidyl ester (B-2604) can be used to simultaneously incorporate both a hapten and biotin into a biomolecule, enabling researchers to readily determine the degree of substitution from the absorbance of the DNP chromophore. Likewise, using a fluorophore as a hapten offers significant advantages for determining the degree of labeling. Fluorescein has been found to be an excellent hapten for *in situ* hybridization because it binds with high affinity to its anti-fluorescein antibody.[49-51] With the recent addition of anti-dansyl, anti-tetramethylrhodamine and anti–Texas Red® antibodies to our line of detection reagents, we have greatly expanded the number of potential haptens. Because the anti-tetramethylrhodamine and anti–Texas Red antibodies crossreact with the tetramethylrhodamine, Lissamine™ rhodamine, Rhodamine Red™ and Texas Red fluorophores, these antibody–fluorophore combinations should not be used simultaneously to generate separate signals in a multicolor experiment. Similarly, the anti–BODIPY® dye antibodies crossreact with other BODIPY dyes and therefore should not be used for simultaneous detection of more than one dye based on the BODIPY fluorophore.

1. Meth Mol Biol 45, 223 (1995); **2.** Meth Mol Biol 45, 205 (1995); **3.** Biochemistry 21, 978 (1982); **4.** Proc Natl Acad Sci USA 71, 3537 (1974); **5.** Biotin-X, SE is licensed to Molecular Probes under U.S. Patent No. 4,656,252; **6.** DNP-X–biocytin-X, SE is licensed to Molecular Probes under U.S. Patent No. 5,180,828; **7.** Anal Chem 63, 850 (1991); **8.** FEBS Lett 328, 165 (1993); **9.** Meth Enzymol 18, 418 (1970); **10.** FASEB J 8, A1445 (1994); **11.** Biochem J 251, 935 (1988); **12.** Anal Biochem 161, 262 (1987); **13.** Biochem Biophys Res Comm 136, 80 (1986); **14.** Anal Biochem 149, 529 (1985); **15.** Radiation Res 117, 326 (1989); **16.** Nature 374, 777 (1995); **17.** J Biol Chem 270, 843 (1995); **18.** Biochem J 275, 733 (1991); **19.** Biochem Biophys Res Comm 138, 872 (1986); **20.** Nucleic Acids Res 16, 7197

(1988); **21.** Blood 74, 952 (1989); **22.** Immunol Lett 8, 273 (1984); **23.** J Biochem 5, 357 (1986); **24.** Anal Biochem 163, 204 (1987); **25.** J Immunol Meth 168, 209 (1994); **26.** Biochem Biophys Res Comm 142, 519 (1987); **27.** Biochim Biophys Acta 631, 49 (1980); **28.** J Cell Biol 111, 2909 (1990); **29.** Meth Enzymol 138, 429 (1987); **30.** Anal Biochem 170, 271 (1988); **31.** Biochem J 229, 785 (1985); **32.** Chem Pharm Bull 37, 1831 (1989); **33.** Nucleic Acids Res 14, 6227 (1986); **34.** J Steroid Biochem 35, 633 (1990); **35.** Biochemistry 32, 8276 (1993); **36.** Biochemistry 31, 3703 (1992); **37.** Biochemistry 11, 3610 (1972); **38.** J Histochem Cytochem 38, 377 (1990); **39.** Annals NY Acad Sci 463, 214 (1984); **40.** Bioconjugate Chem 7, 271 (1996) **41.** J Biol Chem 269, 24596 (1994); **42.** Anal Biochem 223, 88 (1994); **43.** J Biol Chem 269, 28309 (1994); **44.** Anal Biochem 205, 166 (1992); **45.** Anal Biochem 202, 68 (1992); **46.** This reagent is not licensed for use in the practice of U.S. Patent Nos. 5,344,757; 5,354,657 or other patents owned by Boehringer Mannheim GmbH; **47.** Eur J Cell Biol 56, 223 (1991); **48.** Eur J Biochem 227, 61 (1995); **49.** Nucleic Acids Res 19, 3237 (1991); **50.** J Histochem Cytochem 38, 467 (1990); **51.** Voss, Jr., E.W. in *Fluorescein Hapten: An Immunological Probe*, E.W. Voss, Jr., Ed., CRC Press (1984) pp. 3–14.

4.2 Data Table *Biotinylation and Haptenylation Reagents*

Cat #	Structure	MW	Storage	Soluble	Abs	{ε x 10⁻³}	Em	Solvent	Notes
A-1593	✓	367	NC	DMF, DMSO	<300		none		
A-1594	✓	328	NC	DMF, DMSO	<300		none		
A-1604	✓	606	NC	DMF, DMSO	<300		none		
A-2952	✓	587	F,D	DMF, DMSO	<300		none		
A-6346	✓	331	F,D	DMF, DMSO	<300		none		
B-1513	✓	341	F,D	DMF, DMSO	<300		none		
B-1549	✓	258	D	DMF, DMSO	<300		none		
B-1582	✓	455	F,D	DMF, DMSO	<300		none		
B-1591	✓	454	F,D	DMF, DMSO	<300		none		<1>
B-1592	S15.3	372	NC	H_2O	<300		none		
B-1595	✓	244	NC	pH>6, DMF	<300		none		
B-1596	✓	556	NC	DMF, DMSO	<300		none		
B-1597	✓	400	NC	DMF, DMSO	<300		none		
B-1598	✓	357	NC	pH>6, DMF	<300		none		
B-1600	✓	371	D	DMF, DMSO	<300		none		
B-1603	S15.3	387	D	pH>6, DMF	<300		none		
B-1606	✓	568	F,D	DMF, DMSO	<300		none		
B-2600	✓	485	D	DMF, DMSO	<300		none		
B-2604	✓	862	FF,DD,L	DMF	362	{15}	none	pH 8	
B-2605	S15.3	486	NC	pH>6, DMF	<300		none		
B-6349	S9.3	473	F,D	DMF, MeCN	<300		none		
B-6350	S15.3	500	D	pH>6, DMF	<300		none		
B-6352	✓	670	F,D	DMF, pH>6*	<300		none		
B-6353	✓	557	F,D	DMF, pH>6*	<300		none		
C-2284	S1.7	607	F,D,LL	H_2O†, MeOH	396	{29}	410	MeOH	<2>
D-6116	S1.2	634	F,D,L	MeCN, DMSO	588	{68}	616	MeOH	<3>
D-2248	✓	394	F,D,L	DMF, DMSO	348	{18}	none	MeOH	
D-6102	S1.2	502	F,D,L	MeCN, DMSO	504	{85}	510	MeOH	<3>
D-6104	S1.7	462	F,DD,L	DMF, MeCN	335	{4.2}	518	MeOH	
F-2181	S1.3	587	F,D,L	DMF, DMSO	494	{73}	520	pH 9	<4>
F-6130	S1.3	591	F,D,L	DMF, DMSO	491	{86}	515	pH 9	<4>
I-2953	✓	557	F,D	DMF, DMSO	<300		none		<1>
L-1338	S2.4	649	F,D,L	H_2O	426	{11}	531	pH 7	<1, 5>
M-1602	✓	524	F,D	pH>6, DMF	<300		none		
N-6356	✓	252	D	DMF, pH<6	<300		none		
R-6160	S1.6	769	F,D,L	DMF, DMSO	560	{129}	580	MeOH	
T-6105	S1.6	641	F,D,L	DMF, DMSO	543	{90}	571	MeOH	
T-6134	S1.6	817	F,D,L	DMF, DMSO	583	{116}	603	MeOH	

For definitions of the contents of this data table, see “How to Use the Handbook” on page vi.
Structure: Chemical structure drawing: ✓ = shown in this section; Sn.n = shown in Section number n.n

* Sulfosuccinimidyl esters and cysteic acid succinimdyl esters are water soluble and may be dissolved in buffer at ~pH 8 for reaction with amines. Long-term storage in water is not recommended due to hydrolysis.
† Unstable in water. Use immediately.
<1> Iodoacetamides in solution undergo rapid photodecomposition to unreactive products. Minimize exposure to light prior to reaction.
<2> Cascade Blue dyes have a second absorption peak at about 376 nm with ε ~80% of the 395–400 nm peak.
<3> The absorption and fluorescence spectra of BODIPY derivatives are relatively insensitive to the solvent.
<4> Absorption and fluorescence of fluorescein derivatives are pH dependent. See data for F-1300 and C-1904 in Section 23.2.
<5> Spectral data of the 2-mercaptoethanol adduct.

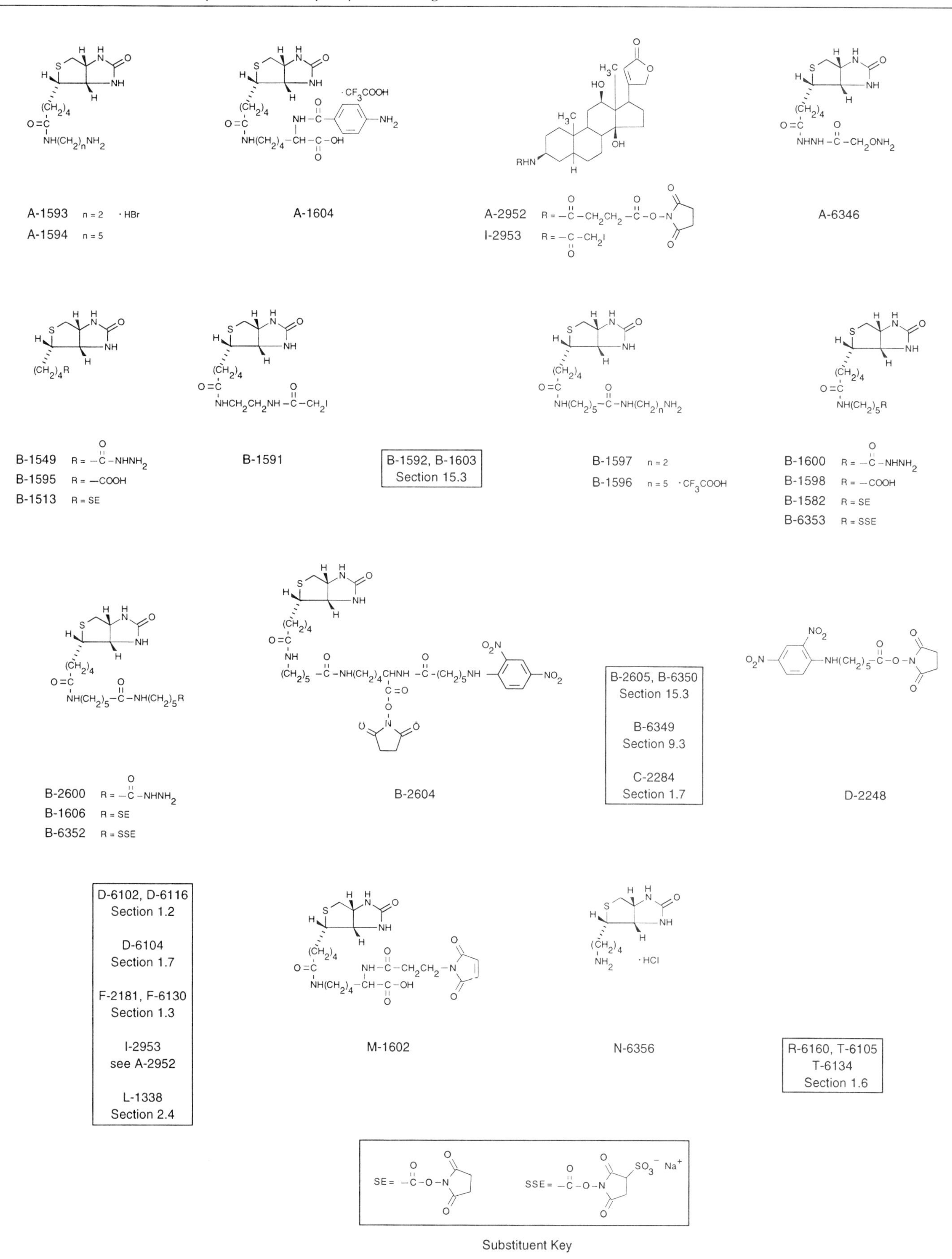
A-1593 n = 2 · HBr
A-1594 n = 5
A-1604
· CF3COOH
A-2952 R = —C—CH2CH2—C—O—N
I-2953 R = —C—CH2I
A-6346
B-1549 R = —C—NHNH2
B-1595 R = —COOH
B-1513 R = SE
B-1591
B-1592, B-1603
Section 15.3
B-1597 n = 2
B-1596 n = 5 · CF3COOH
B-1600 R = —C—NHNH2
B-1598 R = —COOH
B-1582 R = SE
B-6353 R = SSE
B-2600 R = —C—NHNH2
B-1606 R = SE
B-6352 R = SSE
B-2604
B-2605, B-6350
Section 15.3
B-6349
Section 9.3
C-2284
Section 1.7
D-2248
D-6102, D-6116
Section 1.2
D-6104
Section 1.7
F-2181, F-6130
Section 1.3
I-2953
see A-2952
L-1338
Section 2.4
M-1602
N-6356
· HCl
R-6160, T-6105
T-6134
Section 1.6
SE = —C—O—N
SSE = —C—O—N
SO3⁻ Na⁺
Substituent Key

Cat #		Product Name	Unit Size	Price Per Unit ($) 1–4 Units	5–24 Units
A-1604		N^{α}-(4-aminobenzoyl)biocytin, trifluoroacetate	25 mg	85.00	68.00
A-2952		3-amino-3-deoxydigoxigenin hemisuccinamide, succinimidyl ester	5 mg	95.00	76.00
A-1593		N-(2-aminoethyl)biotinamide, hydrobromide (biotin ethylenediamine)	25 mg	60.00	48.00
A-6346	**New**	N-(aminooxyacetyl)-N′-(D-biotinoyl) hydrazine (ARP)	10 mg	98.00	78.40
A-1594		N-(5-aminopentyl)biotinamide (biotin cadaverine)	25 mg	85.00	68.00
B-1592		biocytin (ε-biotinoyl-L-lysine)	100 mg	60.00	48.00
B-1603		biocytin hydrazide	25 mg	85.00	68.00
B-1595		D(+)-biotin	1 g	32.00	25.60
B-1549		biotin hydrazide	100 mg	24.00	19.20
B-1598		6-((biotinoyl)amino)hexanoic acid (biotin-X)	100 mg	48.00	38.40
B-1600		6-((biotinoyl)amino)hexanoic acid, hydrazide (biotin-X hydrazide; BACH)	50 mg	48.00	38.40
B-1582		6-((biotinoyl)amino)hexanoic acid, succinimidyl ester (biotin-X, SE; biotinamidocaproate, N-hydroxysuccinimidyl ester)	100 mg	80.00	64.00
B-6353	**New**	6-((biotinoyl)amino)hexanoic acid, sulfosuccinimidyl ester, sodium salt (Sulfo-NHS-LC-Biotin; biotin-X, SSE)	25 mg	75.00	60.00
B-1597		2-(((N-(biotinoyl)amino)hexanoyl)amino)ethylamine (biotin-X ethylenediamine)	10 mg	75.00	60.00
B-2600		6-((6-((biotinoyl)amino)hexanoyl)amino)hexanoic acid, hydrazide (biotin-XX hydrazide)	25 mg	80.00	64.00
B-1606		6-((6-((biotinoyl)amino)hexanoyl)amino)hexanoic acid, succinimidyl ester (biotin-XX, SE)	100 mg	135.00	108.00
B-6352	**New**	6-((6-((biotinoyl)amino)hexanoyl)amino)hexanoic acid, sulfosuccinimidyl ester, sodium salt (biotin-XX, SSE)	25 mg	75.00	60.00
B-1596		5-(((N-(biotinoyl)amino)hexanoyl)amino)pentylamine trifluoroacetate salt (biotin-X cadaverine)	10 mg	85.00	68.00
B-2605		ε-(6-(biotinoyl)amino)hexanoyl-L-lysine (biocytin-X)	10 mg	60.00	48.00
B-6350	**New**	ε-(6-(biotinoyl)amino)hexanoyl-L-lysine, hydrazide (biocytin-X hydrazide)	25 mg	90.00	72.00
B-6349	**New**	ε-biotinoyl-α-*tert*-butoxycarbonyl-L-lysine (α-(t-BOC) biocytin)	100 mg	95.00	76.00
B-1591		N-(biotinoyl)-N′-(iodoacetyl)ethylenediamine	25 mg	88.00	70.40
B-1513		D-biotin, succinimidyl ester (succinimidyl D-biotin)	100 mg	53.00	42.40
B-2604		biotin-X 2,4-dinitrophenyl-X-L-lysine, succinimidyl ester (DNP-X-biocytin-X, SE)	5 mg	145.00	116.00
C-2284		Cascade Blue® acetyl azide, trisodium salt	5 mg	125.00	100.00
D-6102	**New**	6-((4,4-difluoro-5,7-dimethyl-4-bora-3a,4a-diaza-s-indacene-3-propionyl)amino)hexanoic acid, succinimidyl ester (BODIPY® FL-X, SE)	5 mg	145.00	116.00
D-6116	**New**	6-(((4-(4,4-difluoro-5-(2-thienyl)-4-bora-3a,4a-diaza-s-indacene-3-yl)phenoxy)acetyl)amino)-hexanoic acid, succinimidyl ester (BODIPY® TR-X, SE)	5 mg	145.00	116.00
D-6104	**New**	6-((5-dimethylaminonaphthalene-1-sulfonyl)amino)hexanoic acid, succinimidyl ester (dansyl-X, SE)	25 mg	48.00	38.40
D-2248		6-(2,4-dinitrophenyl)aminohexanoic acid, succinimidyl ester (DNP-X, SE)	25 mg	75.00	60.00
F-6348	**New**	FluoReporter® Biotin/DNP Protein Labeling Kit *5-10 labelings*	1 kit	245.00	196.00
F-6095	**New**	FluoReporter® Biotin-X-C_5 Oligonucleotide Phosphate Labeling Kit *5 labelings*	1 kit	175.00	140.00
F-6081	**New**	FluoReporter® Biotin-XX Oligonucleotide Amine Labeling Kit *5 labelings*	1 kit	175.00	140.00
F-2610		FluoReporter® Biotin-XX Protein Labeling Kit *5 labelings of 5-20 mg protein each*	1 kit	165.00	132.00
F-6085	**New**	FluoReporter® Dinitrophenyl-X (DNP-X) Oligonucleotide Amine Labeling Kit *5 labelings*	1 kit	175.00	140.00
F-6100	**New**	FluoReporter Labeled Oligonucleotide Purification Kit *five spin columns plus buffers*	1 kit	50.00	40.00
F-6347	**New**	FluoReporter® Mini-biotin-XX Protein Labeling Kit *5 labelings of 0.1-3 mg protein each*	1 kit	165.00	132.00
F-2181		6-(fluorescein-5-(and-6)-carboxamido)hexanoic acid, succinimidyl ester (5(6)-SFX) *mixed isomers*	10 mg	65.00	52.00
F-6130	**New**	fluorescein-5-EX, succinimidyl ester *single isomer*	10 mg	65.00	52.00
I-2953		3-iodoacetylamino-3-deoxydigoxigenin	5 mg	95.00	76.00
L-1338		lucifer yellow iodoacetamide, dipotassium salt	25 mg	95.00	76.00
M-1602		3-(N-maleimidylpropionyl)biocytin	25 mg	68.00	54.40
N-6356	**New**	norbiotinamine, hydrochloride	10 mg	65.00	52.00
R-6160	**New**	Rhodamine Red™-X, succinimidyl ester *mixed isomers*	5 mg	75.00	60.00
T-6105	**New**	6-(tetramethylrhodamine-5-(and-6)-carboxamido)hexanoic acid, succinimidyl ester (5(6)-TAMRA-X, SE) *mixed isomers*	10 mg	118.00	94.40
T-6134	**New**	Texas Red®-X, succinimidyl ester *mixed isomers*	5 mg	95.00	76.00

4.3 Biotin Conjugates

Molecular Probes prepares a wide array of biotin conjugates, all of which are included in this section's product list. We will also custom-conjugate biotin, fluorophores or other haptens to proteins, oligonucleotides or biomolecules of interest. See the Technical Note "Custom Immunogen Preparation" on page 92 for more information or contact our Custom and Bulk Sales Department to request a quotation.

Quantitation of Biotin and Avidin

The amount of free biotin in solution has been measured by several methods, including a microplate reader assay in which free biotin competes with biotinylated bovine serum albumin (BSA) for binding avidin–β-galactosidase,[1] as well as a sensitive fluorometric displacement assay for the biotin–avidin interaction that employs 2,6-ANS [2] (A-50, see Section 14.5). Probably the most sensitive technique is a homogeneous competitive binding assay that uses biotinylated recombinant aequorin (A-6786) to detect as little as 4 attomoles of biotin.[3] Biotin and low molecular weight biotin derivatives have also been quantitated by their fluorescence enhancement of fluorescein streptavidin (S-869, see Section 7.5), a technique employed in an HPLC-based binding assay with a reported sensitivity of 97 pg and 149 pg for biotin and biocytin, respectively.[4] In addition, the degree of protein biotinylation has been determined using fluorescein biotin (B-1370) in a fluorescence polarization assay that can detect about 2–20 nM biotinylated BSA.[5]

1. BioTechniques 13, 543 (1992); **2.** Anal Biochem 151, 178 (1985); **3.** Anal Chem 66, 1837 (1994); **4.** Anal Chem 67, 1014 (1995); **5.** Clin Chem 40, 2112 (1994).

Fluorescein Biotin and Other Biotin Derivatives

Fluorescein Biotin

Fluorescein biotin (B-1370) was developed by Molecular Probes as an alternative to radioactive biotin for detecting and quantitating biotin binding sites by either fluorescence or absorbance. A fluorescence polarization assay that employs competitive binding of fluorescein biotin to assess the degree of protein biotinylation has recently been reported.[1] A similar derivative was used for determining avidin and biotin concentrations by fluorescence depolarization.[2] See the Technical Note "Quantitation of Biotin and Avidin" above for other quantitation methods.

Other Fluorescent Biotin Derivatives

For cell-tracing experiments, we offer lucifer yellow cadaverine biotin-X [3] (L-2601), lucifer yellow biocytin (L-6950) and Cascade Blue® biocytin (L-6949) — reagents that incorporate a fluorophore and biotin moiety in a single molecule. It was recently reported that our lucifer yellow cadaverine biotin-X is well retained in aldehyde-fixed tissues, even after sectioning, extraction with detergents and several washes.[4] Because the biocytin conjugates contain free primary amines, they should be more efficiently fixed with aldehydes. Once these probes are fixed, the researcher can choose to detect them directly by fluorescence or indirectly with either labeled avidin conjugates (see Section 7.5) or labeled anti-fluorophore antibodies (see Section 7.3).

LysoTracker™ Biotin and LysoTracker™ Biotin/DNP

The LysoTracker family of probes is a new group of acidotropic reagents for labeling and tracing acidic organelles in live cells.[5,6] These weakly basic amines selectively accumulate in cellular compartments with low internal pH and can be used to investigate the biosynthesis and pathogenesis of lysosomes. LysoTracker Biotin (L-7537) and LysoTracker Biotin/DNP (L-7538), which consist of a biotin moiety linked to a weak base that is only partially protonated at neutral pH, are permeant to cell membranes and typically concentrate in spherical organelles. Their mechanism of retention has not been firmly established but is likely to involve protonation and retention in the organelles' membranes, although staining is generally not reversed by subsequent treatment of the cells with weakly basic cell-permeant compounds. Following aldehyde fixation, the biotin moiety can be detected using one of our many avidin, streptavidin or NeutraLite™ avidin conjugates. In addition to the biotin moiety, the LysoTracker Biotin/DNP probe contains the hapten DNP and should therefore also be detectable with our anti-DNP antibodies. To complement these biotin LysoTracker probes, we offer four fluorescent derivatives — LysoTracker Blue DND-22, LysoTracker Green DND-26, LysoTracker Yellow DND-68 and LysoTracker Red DND-99 — that can be detected by fluorescence microscopy without cell fixation; see Section 12.3 for more information on these acidic organelle stains.

Biotinylated Dextrans

In addition to the low molecular weight biotinylated tracers described above, Molecular Probes prepares biotinylated versions of a wide variety of dextrans (Figure 4.2), including dextrans that are double-labeled with fluorophores and biotin moieties for correlated fluorescence and electron microscopy studies. See Section 15.5 for discussion of the applications of these reagents, particularly as cell tracers.

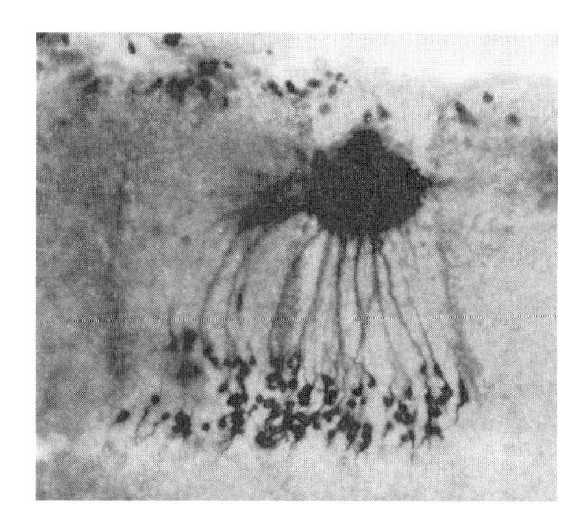

Figure 4.2 *Motor neuron in a three-day chick embryo labeled with lysine-fixable biotinylated 3000 MW dextran (BDA-3000, D-7135). Filled neurons were detected with biotinylated horseradish peroxidase and diaminobenzidine using standard avidin/streptavidin bridging techniques. Reprinted with permission from J Neurosci Meth 50, 95 (1993).*

Biotinylated Proteins

Our biotinylated primary and secondary antibodies, phycobiliproteins and enzymes are invaluable detection reagents for a broad assortment of assays (see discussions in Chapters 6 and 7). Biotinylated R-phycoerythrin (P-811) can be used in combination with an avidin or streptavidin bridge to considerably amplify the detection of biotinylated targets. Our enzyme conjugates include biotin-XX derivatives of horseradish peroxidase (P-917) and alkaline phosphatase (A-927). The biotin conjugate of AquaLite® recombinant aequorin (A-6786) is a unique reagent that has been used in an ultrasensitive assay for biotin.[7]

Biotinylated Microspheres

Biotinylated FluoSpheres microspheres have significant potential for signal amplification techniques (see discussion in Section 6.2). Like biotinylated R-phycoerythrin, they can be used with bridging techniques to detect biotinylated targets. Our intensely fluorescent FluoSpheres microspheres are much brighter and more photostable than low molecular weight dye conjugates.

Biotinylated Site-Selective Probes

Biotin conjugates of low molecular weight ligands permit amplified detection of ligand binding. They may also be useful for immobilizing receptor ligands on streptavidin agarose (S-951, see Section 7.5) for affinity isolation of receptors. Our selection of biotinylated ligands include:

- Biotin-XX conjugate of α-bungarotoxin[8,9] (B-1196, see Section 18.2), for labeling the α-subunit of the acetylcholine receptor
- Biotin-XX conjugate of epidermal growth factor[10-13] (EGF) as well as biotinylated EGF complexed with Texas Red streptavidin (E-3477, E-3480; see Section 17.3), for labeling EGF receptors
- Biotin-XX conjugate of phalloidin[14,15] (B-7474, see Section 11.1), for labeling F-actin

Biotinylated Lipids

Like the reactive phospholipids in Chapter 5, phospholipid derivatives of biotin and biotin-X (B-1550, B-1616) can be used to prepare liposomes that retain high affinity for avidin conjugates.[16] 12-(Biotinoyl)aminododecanoic acid (B-2602) may be useful for preparing phospholipids and sphingolipids; see the Technical Note "Synthetic Methods for Preparing New Phospholipid Probes" on page 303. Its amine-reactive succinimidyl ester[17] (B-2603) provides a means of preparing long nonpolar intermolecular crosslinks.

1. Clin Chem 40, 2112 (1994); **2.** Anal Chem 60, 853 (1988); **3.** J Neurosci Meth 53, 23 (1994); **4.** J Neurosci Meth 45, 59 (1993); **5.** Cytometry Suppl 7, 77, abstract #426B (1994); **6.** Mol Biol of the Cell 5, 113a, abstract #653 (1994); **7.** Anal Chem 66, 1837 (1994); **8.** J Cell Biol 124, 661 (1994); **9.** J Biol Chem 268, 25108 (1993); **10.** J Histochem Cytochem 42, 307 (1994); **11.** J Histochem Cytochem 41, 313 (1993); **12.** J Histochem Cytochem 40, 1353 (1992); **13.** Anal Biochem 188, 97 (1990); **14.** J Cell Biol 130, 591 (1995); **15.** Anal Biochem 200, 199 (1992); **16.** Meth Enzymol 149, 119 (1987); **17.** 12-((Biotinoyl)amino)dodecanoic acid, succinimidyl ester is licensed to Molecular Probes under U.S. Patent No. 4,656,252.

4.3 Data Table *Biotin Conjugates*

Cat #	Structure	MW	Storage	Soluble	Abs	{ε x 10^-3}	Em	Solvent	Notes
B-1196	C	~8400	FF,D	H_2O	<300		none		<1>
B-1370	✓	831	L	DMF, pH>6	494	{75}	518	pH 9	
B-1550	S13.4	1019	FF,D	<2>	<300		none		
B-1616	S13.4	1133	FF,D	<2>	<300		none		
B-2602	S13.3	442	NC	DMSO	<300		none		
B-2603	S13.3	539	F,D	DMSO	<300		none		
B-7474	C	~1300	F	MeOH, H_2O*	<300		none		<1>
C-6949	✓	957	D,L	H_2O	400	{31}	417	pH 8	<3>
E-3477	C	~6600	FF,D	H_2O	<300		none		<1>
E-3480	C	<4>	FF,D,L	H_2O	596	ND	612	pH 7	<5>
L-2601	S15.3	873	D,L	H_2O	428	{11}	531	H_2O	
L-6950	✓	850	D,L	H_2O	428	{11}	532	pH 7	
L-7537	S12.3	512	D	DMSO†	<300		none		
L-7538	S12.3	835	D,L	DMSO†	363	{16}	none	pH 8	

For definitions of the contents of this data table, see "How to Use the Handbook" on page vi.
Structure: Chemical structure drawing: ✓ = shown in this section; Sn.n = shown in Section number n.n; C = not shown due to complexity.
MW: ~ indicates an approximate value.

* Although phallotoxins are soluble in water, their storage in water is not recommended, particularly in dilute solution.

† This product is packaged as a solution in the solvent indicated in "Soluble."

<1> α-Bungarotoxin, phallotoxin and EGF conjugates have approximately 1 label per mole.

<2> Chloroform is the most generally useful solvent for preparing stock solutions of phospholipids. Solutions of glycerophosphoethanolamines in ethanol up to 1–2 mg/mL should be obtainable, using sonication to aid dispersion if necessary. Labeling of aqueous samples with nonmiscible phospholipid stock solutions can be accomplished by evaporating the organic solvent, followed by hydration and sonication, yielding a suspension of liposomes. Information on solubility of natural phospholipids can be found in Marsh, D., *CRC Handbook of Lipid Bilayers,* CRC Press (1990) pp. 71–80.

<3> Cascade Blue dyes have a second absorption peak at about 376 nm with ϵ ~80% of the 395–400 nm peak.
<4> E-3480 is a complex of E-3477 with Texas Red streptavidin (S-872, see Section 7.5), which typically incorporates ~3 dyes/streptavidin (MW ~60,000 daltons).
<5> ND = not determined.

4.3 Structures *Biotin Conjugates*

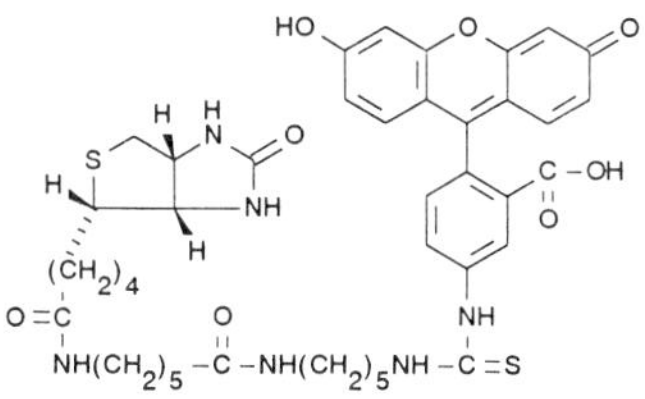

B-1370

B-1550, B-1616
Section 13.4

B-2602, B-2603
Section 13.3

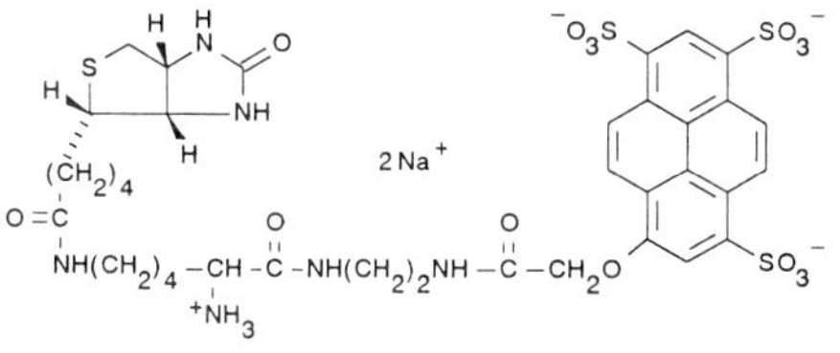

C-6949

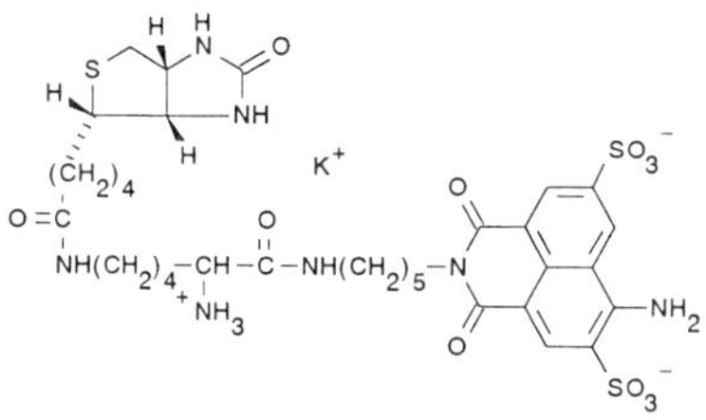

L-6950

L-2601
Section 15.3

L-7537, L-7538
Section 12.3

4.3 Price List *Biotin Conjugates*

Fluorescein Biotin and other Biotin Derivatives *(see Chapters 12 and 15)*

Cat #		Product Name	Unit Size	Price Per Unit ($) 1–4 Units	5–24 Units
B-1370		5-((N-(5-(N-(6-(biotinoyl)amino)hexanoyl)amino)pentyl)thioureidyl)fluorescein (fluorescein biotin)	5 mg	98.00	78.40
C-6949	**New**	Cascade Blue® biocytin, disodium salt, fixable (biocytin Cascade Blue®)	5 mg	95.00	76.00
L-6950	**New**	lucifer yellow biocytin, potassium salt, fixable (biocytin lucifer yellow)	5 mg	95.00	76.00
L-2601		lucifer yellow cadaverine biotin-X, dipotassium salt	10 mg	95.00	76.00
L-7537	**New**	LysoTracker™ biotin *4 mM solution in DMSO* *special packaging*	10x50 µL	125.00	100.00
L-7538	**New**	LysoTracker™ biotin/DNP *4 mM solution in DMSO* *special packaging*	10x50 µL	125.00	100.00

Biotinylated Dextrans *(see Chapter 15)*

Cat #		Product Name	Unit Size	1–4 Units	5–24 Units
D-7135	**New**	dextran, biotin, 3000 MW, lysine fixable (BDA-3000)	10 mg	125.00	100.00
D-7134	**New**	dextran, biotin, 3000 MW, neutral	10 mg	110.00	88.00
D-1956		dextran, biotin, 10,000 MW, lysine fixable (BDA-10,000)	25 mg	95.00	76.00
D-1856		dextran, biotin, 10,000 MW, neutral	100 mg	80.00	64.00
D-1957		dextran, biotin, 70,000 MW, lysine fixable (BDA-70,000)	25 mg	95.00	76.00
D-7142	**New**	dextran, biotin, 500,000 MW, lysine fixable (BDA-500,000)	10 mg	110.00	88.00
D-7143	**New**	dextran, biotin, 2,000,000 MW, lysine fixable (BDA-2,000,000)	10 mg	110.00	88.00
D-7146	**New**	dextran, DMNB-caged fluorescein and biotin, 10,000 MW, lysine fixable	5 mg	150.00	120.00
D-7147	**New**	dextran, DMNB-caged fluorescein and biotin, 70,000 MW, lysine fixable	5 mg	150.00	120.00
D-7156	**New**	dextran, fluorescein and biotin, 3000 MW, anionic, lysine fixable (micro-emerald)	5 mg	150.00	120.00
D-7178	**New**	dextran, fluorescein and biotin, 10,000 MW, anionic, lysine fixable (mini-emerald)	10 mg	150.00	120.00
D-7165	**New**	dextran, fluorescein and biotin, 70,000 MW, anionic, lysine fixable	10 mg	150.00	120.00
D-7162	**New**	dextran, tetramethylrhodamine and biotin, 3000 MW, lysine fixable (micro-ruby)	5 mg	150.00	120.00
D-3312		dextran, tetramethylrhodamine and biotin, 10,000 MW, lysine fixable (mini-ruby)	10 mg	150.00	120.00
D-7133	**New**	dextran, tetramethylrhodamine and biotin, 40,000 MW, lysine fixable	10 mg	150.00	120.00

Biotinylated Lipids *(see Chapter 13)*

Cat #		Product Name	Unit Size	1–4 Units	5–24 Units
B-2602		12-((biotinoyl)amino)dodecanoic acid	25 mg	53.00	42.40
B-2603		12-((biotinoyl)amino)dodecanoic acid, succinimidyl ester	25 mg	48.00	38.40
B-1616		N-((6-(biotinoyl)amino)hexanoyl)-1,2-dihexadecanoyl-*sn*-glycero-3-phosphoethanolamine, triethylammonium salt (biotin-X DHPE)	5 mg	115.00	92.00
B-1550		N-(biotinoyl)-1,2-dihexadecanoyl-*sn*-glycero-3-phosphoethanolamine, triethylammonium salt (biotin DHPE)	10 mg	115.00	92.00

Biotinylated Site-Selective Probes *(see Chapter 18)*

Cat #		Product Name	Unit Size	Price Per Unit ($) 1–4 Units	5–24 Units
B-7474	***New***	biotin-XX phalloidin	50 U	198.00	158.40
B-1196		α-bungarotoxin, biotin-XX conjugate	500 µg	175.00	140.00
E-3477		epidermal growth factor, biotin-XX conjugate (biotin EGF)	20 µg	160.00	128.00
E-3480		epidermal growth factor, biotinylated, complexed to Texas Red® streptavidin	100 µg	160.00	128.00

Biotinylated Proteins *(see Chapter 7)*

Cat #		Product Name	Unit Size	1–4 Units	5–24 Units
A-927		alkaline phosphatase, biotin-XX conjugate *5 mg/mL*	200 µL	95.00	76.00
A-5771		anti-BODIPY® FL, rabbit IgG fraction, biotin-XX conjugate *≥ 1 mg/mL*	0.5 mL	148.00	118.40
A-5761		anti-Cascade Blue®, rabbit IgG fraction, biotin-XX conjugate *≥ 1 mg/mL*	0.5 mL	148.00	118.40
A-6435	***New***	anti-dinitrophenyl-KLH, rabbit IgG fraction, biotin-XX conjugate *2 mg/mL*	0.5 mL	148.00	118.40
A-982		anti-fluorescein, rabbit IgG fraction, biotin-XX conjugate *≥ 1 mg/mL*	0.5 mL	148.00	118.40
A-5751		anti-lucifer yellow, rabbit IgG fraction, biotin-XX conjugate *≥ 1 mg/mL*	0.5 mL	148.00	118.40
A-6786	***New***	AquaLite® aequorin, biotinylated *recombinant*	25 µg	175.00	140.00
B-2763		biotin-XX goat anti-mouse IgG (H+L) conjugate *2 mg/mL*	0.5 mL	75.00	60.00
B-2770		biotin-XX goat anti-rabbit IgG (H+L) conjugate *2 mg/mL*	0.5 mL	75.00	60.00
P-917		peroxidase from horseradish, biotin-XX conjugate	10 mg	75.00	60.00
P-811		R-phycoerythrin, biotin-XX conjugate *4 mg/mL*	0.5 mL	135.00	108.00
P-2757		protein A, biotin-XX conjugate	1 mg	75.00	60.00

Biotinylated Microspheres *(see Chapter 6)*

Cat #		Product Name	Unit Size	1–4 Units	5–24 Units
F-8766		FluoSpheres® biotin-labeled microspheres, 0.04 µm, yellow-green fluorescent (505/515) *1% solids*	0.4 mL	165.00	132.00
F-8767		FluoSpheres® biotin-labeled microspheres, 0.2 µm, yellow-green fluorescent (505/515) *1% solids*	0.4 mL	165.00	132.00
F-8769		FluoSpheres® biotin-labeled microspheres, 1.0 µm, nonfluorescent *1% solids*	0.4 mL	165.00	132.00
F-8768		FluoSpheres® biotin-labeled microspheres, 1.0 µm, yellow-green fluorescent (505/515) *1% solids*	0.4 mL	165.00	132.00

Custom Immunogen Preparation

Low molecular weight molecules (<2000 daltons) or haptens generally will not elicit an immune response unless conjugated to a carrier protein such as bovine serum albumin (BSA) or keyhole limpet hemocyanin (KLH). Preparing these immunogens often requires introducing reactive groups into the haptens through chemical synthesis. Molecular Probes has considerable experience synthesizing reactive chemical species, including reactive forms of drugs, natural products and herbicides. In addition to their use for preparing immunogens, these reactive haptens can be used to generate new detection reagents and site-selective probes, as well as affinity matrices for isolating antibodies and receptors. We provide our custom services on an exclusive or nondisclosure basis when requested. Please contact our Custom and Bulk Sales Department for further information.

Chapter 5

Crosslinking and Photoreactive Reagents

Contents

Related Chapters

5.1 Introduction to Crosslinking Reagents

Bifunctional "crosslinking" reagents contain two reactive groups, thereby providing a means of covalently linking two target groups. The reactive groups in a chemical crosslinking reagent (see Section 5.2) typically belong to the classes of functional groups — including succinimidyl esters, maleimides and iodoacetamides — described in Chapters 1–3. In contrast, one or both of the reactive groups in a photoreactive crosslinking reagent (see Section 5.3) require light activation before reacting with a target group. Crosslinking a biopolymer (such as an antibody, enzyme, avidin or nucleic acid) with a low molecular weight molecule (such as a drug, toxin, peptide or oligonucleotide) or another biopolymer yields a stable heteroconjugate that can serve as a detection reagent in a wide variety of research and diagnostic assays (see Section 6.1) or as an immunogen designed to elicit antibody production. Crosslinking reagents are also useful for probing the spatial relationships and interactions within and between biomolecules.

In **homobifunctional crosslinking reagents**, the reactive groups are identical. These reagents couple like target groups such as two thiols or two amines and are predominantly used to form intramolecular crosslinks. When used to conjugate two different biomolecules, for example an enzyme to an antibody, these relatively nonspecific reagents tend to yield high molecular weight aggregates.

In **heterobifunctional crosslinking reagents**, the reactive groups have dissimilar chemistry, allowing the formation of crosslinks between unlike functional groups. As with homobifunctional crosslinking reagents, heterobifunctional crosslinking reagents can still form multiple intermolecular crosslinks to yield high molecular weight aggregates, but their conjugation reactions are more easily controlled by optimizing the stoichiometry of the target molecules. Thus, heterobifunctional crosslinking reagents are very useful for preparing conjugates between two different biomolecules.

An additional variation is the **"zero-length" crosslinking reagent** — a reagent that forms a chemical bond between two groups without itself being incorporated into the product. The water-soluble carbodiimide EDAC and the more hydrophobic EEDQ (E-2247, E-6288; see Section 5.2), both of which are used to couple carboxylic acids to amines, are examples of zero-length crosslinking reagents.

A noncovalent interaction between two molecules that is of sufficient affinity to prevent the conjugate from falling apart can also function as a crosslink. For example, reactive derivatives of phospholipids can be used to link the liposomes or cell membranes in which they are incorporated to antibodies or enzymes. Biotinylation and haptenylation reagents (see Chapter 4) can also be thought of as heterobifunctional crosslinking reagents because they comprise a chemically reactive group as well as a biotin or hapten moiety that binds with high affinity to avidin or an anti-hapten antibody, respectively. Similarly, avidin, streptavidin and NeutraLite™ avidin can tightly bind up to four biotinylated molecules, and antibodies can bind up to two haptens.

5.2 Chemical Crosslinking Reagents

The most common schemes for forming a heteroconjugate involve the indirect coupling of an amine group on one biomolecule to a thiol group on a second biomolecule, usually by a two- or three-step reaction sequence. The high reactivity of thiols (see Chapter 2) and — with the exception of a few proteins such as β-galactosidase — their relative rarity in most biomolecules make thiol groups ideal targets for controlled chemical crosslinking. If neither molecule contains a thiol group, then one or more can be introduced using one of several thiolation methods. The thiol-containing biomolecule is then reacted with an amine-containing biomolecule using a heterobifunctional crosslinking reagent such as one of those described in Amine–Thiol Crosslinking below.

Thiolation of Biomolecules

Introducing Thiol Groups

Several methods are available for introducing thiols into biomolecules, including the reduction of intrinsic disulfides, as well as the conversion of amine, aldehyde or carboxylic acid groups to thiol groups:

- Cystines in proteins can be reduced to cysteine residues by dithiothreitol (DTT, D-1532), dithioerythritol (DTE, D-8452), tris-(2-carboxyethyl)phosphine (TCEP, T-2556) or tris-(2-cyanoethyl)phosphine (T-6052). However, reduction may result in loss of protein activity or specificity. Excess DTT or DTE must be carefully removed under conditions that prevent reformation of the disulfide,[1] whereas excess TCEP usually does not need to be removed before carrying out the crosslinking reaction.
- Amines can be indirectly thiolated by reaction with succinimidyl 3-(2-pyridyldithio)propionate[2] (SPDP, S-1531), followed by reduction of the pyridyldithiopropionyl conjugate with DTT or TCEP (Figure 5.1). Reduction releases the chromophore 2-pyridinethione, which can be used to determine the degree of thiolation.
- Amines can be indirectly thiolated by reaction with succinimidyl acetylthioacetate[3] (SATA, S-1553) or succinimidyl acetylthiopropionate[4] (SATP, S-1573), followed by removal of the acetyl group with 50 mM hydroxylamine or hydrazine. These reagents are most useful when disulfides are essential for activity, as is the case for some peptide toxins.
- Amines can be directly thiolated using high concentrations of 2-iminothiolane[5,6] (IT, I-2553) (Figure 5.2).
- Thiols can be incorporated at carboxylic acid groups by an EDAC-mediated reaction with cystamine, followed by reduction of the disulfide with DTT or TCEP;[7,8] see Amine–Carboxylic Acid Crosslinking below.

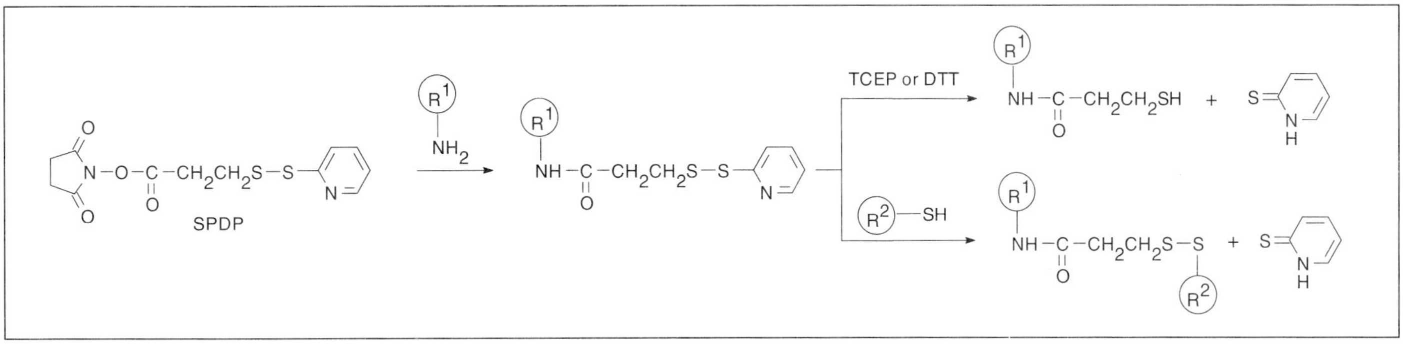

Figure 5.1 *SPDP derivatization reactions. SPDP (S-1531) reacts with an amine-containing biomolecule at pH 7 to 9, yielding a mixed disulfide. The mixed disulfide can then be reacted with a reducing agent such as TCEP or DTT to release a pyridyldithiopropionyl conjugate or with a thiol-containing biomolecule to form a disulfide-linked biomolecule pair. Either reaction can be quantitated by measuring the amount of 2-pyridinethione chromophore released during the reaction.*

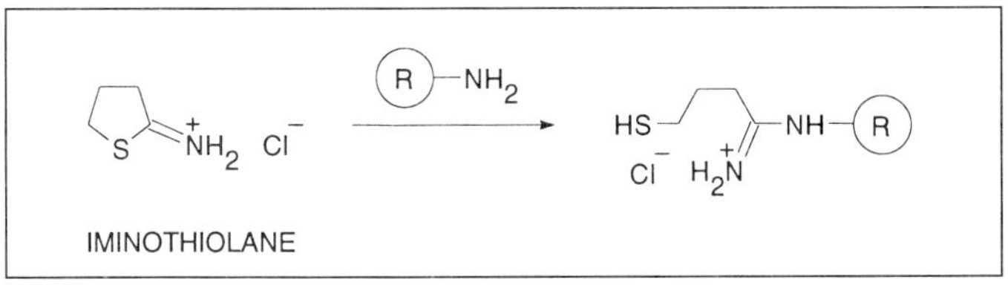

Figure 5.2 *Direct thiolation of an amine-containing biomolecule using 2-iminothiolane (IT, I-2553).*

- Periodate-oxidized carbohydrates, glycoproteins (including antibodies and avidin) or RNA can be directly thiolated using 2-acetamido-4-mercaptobutanoic acid hydrazide [9] (AMBH, A-1642); see Amine–Carbohydrate and Thiol–Carbohydrate Crosslinking below.
- Tryptophan residues in thiol-free proteins can be oxidized to mercaptotryptophan residues, which can then be modified by iodoacetamides or maleimides.[10-12]

Our preferred reagent combination for protein thiolation is SPDP/DTT or SPDP/TCEP.[13] Molecular Probes uses SPDP to prepare a reactive R-phycoerythrin derivative (P-806, see Section 6.4), providing researchers with the optimal number of pyridyldisulfide groups for crosslinking the phycobiliprotein to thiolated antibodies, enzymes and other biomolecules through disulfide linkages.[14] More commonly, the pyridyldisulfide groups are first reduced to thiols, which are then reacted with maleimide- or iodoacetamide-derivatized proteins (Figure 5.1). SPDP can also be used to thiolate oligonucleotides [15] and, like all of the thiolation reagents in this section, to introduce the highly reactive thiol group into peptides, onto cell surfaces, or onto affinity matrices for subsequent reaction with fluorescent, enzyme-coupled or other thiol-reactive reagents. In addition, because the pyridyldithiopropionyl conjugate releases the 2-pyridinethione chromophore upon reduction, SPDP is useful for quantitating the number of reactive amines in an affinity matrix.[16]

Measuring Thiolation of Biomolecules

To ensure success in forming heterocrosslinks, it is important to know that a molecule has the proper degree of thiolation. We generally find that two to three thiol residues per protein are optimal. Following removal of excess reagents, the degree of thiolation in proteins or other molecules thiolated with SPDP can be directly determined by measuring release of the chromophore 2-pyridinethione [2] ($\epsilon_{343\ nm}$ ~8000 $cm^{-1}M^{-1}$).

Alternatively, the degree of thiolation and presence of residual thiols in a solution can be assessed using 5,5'-dithiobis-(2-nitrobenzoic acid) (DTNB, Ellman's reagent; D-8451), which stoichiometrically yields the chromophore 5-mercapto-2-nitrobenzoic acid ($\epsilon_{410\ nm}$ ~13,600 $cm^{-1}M^{-1}$) upon reaction with a thiol group.[17,18] DTNB can also be used to quantitate residual phosphines, including TCEP;[19] in this case, two molecules of 5-mercapto-2-nitrobenzoic acid are formed per reaction with one molecule of phosphine.

Thiol and Sulfide Quantitation Kit

Ultrasensitive colorimetric quantitation of both protein and nonprotein thiols is now possible using our Thiol and Sulfide Quantitation Kit (T-6060). In this assay, which is based on a method reported by Singh,[20,21] thiols reduce a disulfide-inhibited derivative of papain, stoichiometrically releasing the active enzyme. Activity of the enzyme is then measured using the chromogenic papain substrate L-BAPNA. Although thiols can also be quantitated using DTNB (Ellman's reagent, see above), the enzymatic amplification step in our quantitation kit enables researchers to detect as little as 0.2 nanomoles of thiols — a sensitivity that is about 100-fold better than that achieved with DTNB. Thiols in proteins and potentially other high molecular weight molecules can be detected indirectly by incorporating the disulfide cystamine into the solution. Cystamine undergoes an exchange reaction with protein thiols, yielding 2-mercaptoethylamine (cysteamine), which then releases active papain. All traces of reducing agents must be removed before determining free thiols in proteins. The Thiol and Sulfide Quantitation Kit contains:

- Papain-$SSCH_3$, the disulfide-inhibited papain derivative
- L-BAPNA, a chromogenic papain substrate
- DTNB (Ellman's reagent), for calibrating the assay
- Cystamine
- L-Cysteine, a thiol standard
- Buffer
- Protocols for measuring thiols, inorganic sulfides and maleimides

Sufficient reagents are provided for approximately 50 assays using standard 1 mL assay volumes and standard cuvettes or 250 assays using a microplate format. This kit can also be used to detect phosphines, sulfites and cyanides with detection limits of about 0.5, 1 and 5 nanomoles, respectively (see Section 24.3).

Thiol–Thiol Crosslinking

Oxidation

Thiol residues in close proximity can be oxidized to disulfides by either an intra- and intermolecular reaction. Except in certain circumstances, however, this oxidation reaction is reversible and difficult to control.

Dibromobimane

Dibromobimane (bBBr, D-1379) is a thiol–thiol crosslinking reagent that generally exhibits fluorescence only after *both* of its alkylating groups have reacted. This homobifunctional reagent has been used to crosslink thiols in myosin subfragment-1,[22-24] hemoglobin [25] and mitochondrial ATPase.[26] Both *inter*molecular crosslinking between protein subunits [27] and *intra*molecular crosslinking between cysteine residues or thiols of reduced cystine residues are possible.

Amine–Amine Crosslinking

The scientific literature contains numerous references to reagents that form crosslinks between amines of biopolymers. These materials include glutaraldehyde, bis(imido esters), bis(succinimidyl esters), diisocyanates and diacid chlorides.[28] However, these homobifunctional crosslinking reagents tend to yield high molecular weight aggregates, making them unsuitable for preparing conjugates between two different amine-containing biomolecules. Such conjugates are more commonly prepared by thiolating one or more amines on one of the biomolecules and converting one or more amines on the second biomolecule to a thiol-reactive functional group such as a maleimide or iodoacetamide, as described below in Amine–Thiol Crosslinking.

Formaldehyde and Glutaraldehyde

Direct amine–amine crosslinking routinely occurs during fixation of proteins, cells and tissues with formaldehyde or glutaraldehyde. These common aldehyde fixatives are also used to crosslink amine and hydrazine derivatives to proteins and other amine-containing polymers. For example, lucifer yellow CH (L-453, see Section 15.3) is nonspecifically conjugated to surrounding biomolecules by aldehyde fixatives in order to preserve the dye's staining pattern during subsequent tissue manipulations.[29] Also, biotin hydrazides (see Section 4.2) have been directly coupled to nucleic acids,[30] a reaction that is potentially useful for conjugating fluorescent hydrazides to DNA.

Glutaraldehyde is still used by some companies and research laboratories to couple horseradish peroxidase, which has only six lysine residues,[31] to proteins with a larger numbers of lysine residues. However, this practice can result in variable molecular weights and batch-to-batch inconsistency. Consequently, we prepare our horseradish peroxidase conjugates (see Section 7.5) using SPDP- or SMCC-mediated reactions (Figures 5.1 and 5.3).

Bis(Succinimidyl Esters)

The bis(succinimidyl esters) of 5,5'-dithiobis-(2-nitrobenzoic acid) (DTNB, SE; D-6316) and ethylene glycol bis-(succinic acid) (EGS, E-6306) can link amine groups within or between biomolecules by forming stable amide bonds. Moreover, each reagent contains a linkage that can be severed to permit uncoupling of the biomolecules. Crosslinking by the amine-reactive derivative of DTNB (Ellman's reagent) is reversed by disulfide-reducing agents, including DTT and TCEP. The resulting yellow-colored reduction products enable the researcher to visualize the uncoupled thiol-containing biomolecules. The ester bonds within the EGS reagent are readily cleaved by either hydroxylamine or hydrazine.[32] Reversible crosslinking by EGS and presumably the succinimidyl ester of DTNB provides a means of recovering ligands adsorbed to an affinity matrix [32,33] and of determining the proximity of protein subunits or interacting proteins.[34,35]

Amine–Thiol Crosslinking

Indirect crosslinking of the amines in one biomolecule to the thiols in a second biomolecule is the predominant method for forming a heteroconjugate. If one of the biomolecules does not already contain one or more thiol groups, it is necessary to introduce them using one of the thiolation procedures described above in Thiolation of Biomolecules. Thiol-reactive groups such as maleimides or iodoacetamides are typically introduced into the second biomolecule by modifying a few of its amines with a heterobifunctional crosslinker containing both a succinimidyl ester and either a maleimide or an iodoacetamide. The maleimide- or iodoacetamide-modified biomolecule is then reacted with the thiol-containing biomolecule to form a stable thioether crosslink. Chromatographic methods are usually employed to separate the higher molecular weight heteroconjugate from the unconjugated biomolecules.

Introducing Maleimides or Iodoacetamides at Amines

Heterobifunctional crosslinking reagents for introducing thiol-reactive groups at amine sites include the following amine-reactive maleimide and iodoacetamide derivatives:

- Succinimidyl *trans*-4-(maleimidylmethyl)cyclohexane-1-carboxylate [36] (SMCC, S-1534), our preferred reagent because of the superior chemical stability of its maleimide [37] (Figure 5.3)

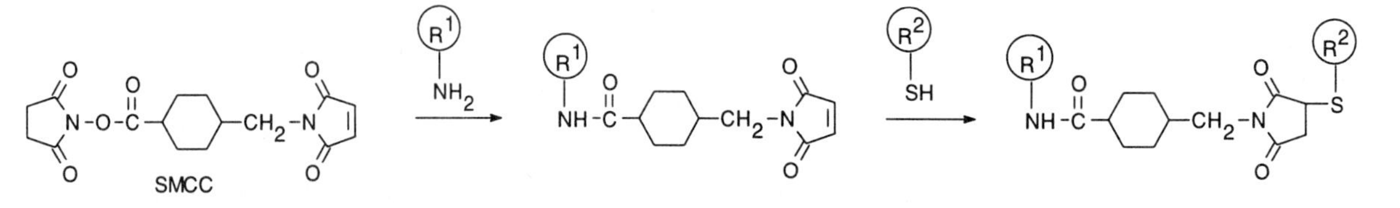

Figure 5.3 *Two-step reaction sequence for crosslinking biomolecules using the heterobifunctional crosslinker SMCC (S-1534).*

- Succinimidyl 3-maleimidylbenzoate (SMB or MBS, S-1535), which is used to selectively form actin–myosin crosslinks,[38-40] as well as to generate immunogens consisting of a peptide or other low molecular weight antigen conjugated to a carrier protein such as serum albumin or keyhole limpet hemocyanin [41,42] (KLH)
- Succinimidyl 6-maleimidylhexanoate [43] (EMCS, S-1563), which is reported to be less immunogenic than SMCC and SMB [44]
- 4-Nitrophenyl iodoacetate [45,46] (NPIA, N-1505) as well as succinimidyl ester iodoacetamides such as succinimidyl iodoacetate [47,48] (I-6001), succinimidyl 6-((iodoacetyl)amino)-hexanoate [49] (SIAX, S-1666) and succinimidyl 6-(6-(((iodoacetyl)amino)hexanoyl)amino)hexanoate (SIAXX, S-1668), which are short-range linkers that are also useful for synthesizing new iodoacetamide probes [50] and for crosslinking amines and thiols within proteins [45]

Introducing Disulfides at Amines

As discussed above, our preferred method for preparing heteroconjugates employs the thiolation reagent SPDP (S-1531). The pyridyldisulfide intermediate that is initially formed by reaction of SPDP with amines can form an unsymmetrical disulfide through reaction with a second thiol-containing molecule [2,51] (Figure 5.1). The thiol-containing target can be a molecule such as β-galactosidase that contains intrinsic thiols or a molecule in which thiols have been introduced using one of the thiolation procedures described above in Thiolation of Biomolecules. In either case, it is essential that all reducing agents, including both DTT and TCEP, are absent. The heteroconjugate's disulfide bond is about as stable and resistant to reduction as those found in proteins; it can be reduced with DTT or TCEP to generate two thiol-containing biomolecules.

Protein–Protein Crosslinking Kit

Our new Protein–Protein Crosslinking Kit (P-6305) provides all the reagents and purification media required to perform three protein–protein conjugations when neither protein contains thiol residues. The chemistry used to thiolate the amines of one of the proteins with SPDP and to convert the amines of the second protein to thiol-reactive maleimides with SMCC is shown in Figures 5.1 and 5.3, respectively. Included in the kit are:

- SPDP, for thiolating amines
- SMCC, for converting amines to thiol-reactive maleimides
- TCEP, for reducing the pyridyldisulfide intermediate
- *N*-ethylmaleimide (NEM), for capping residual thiols
- Six reaction tubes, each containing a magnetic stir bar
- Spin columns plus collection tubes
- Dimethylsulfoxide (DMSO)
- Detailed crosslinking protocol

The Protein–Protein Crosslinking Kit was designed to prepare and purify protein–protein conjugates; however, it can be readily modified for generating peptide–protein or enzyme–nucleic acid conjugates or for conjugating biomolecules to affinity matrices.

Molecular Probes has considerable experience in preparing protein–protein conjugates and will apply this expertise to your particular application through our custom synthesis service. We provide custom conjugation services on an exclusive or nondisclosure basis when requested. For more information or a quotation, please contact our Custom and Bulk Sales Department.

Assaying Maleimide- and Iodoacetamide-Modified Biomolecules

The potential instability of maleimide derivatives and the photosensitivity of iodoacetamide derivatives may make it advisable to assay the modified biomolecule for thiol reactivity before conjugation with a thiol-containing biomolecule. Fluorescein SAMSA (A-685), which is our only fluorescent reagent that can generate a free thiol group, was designed for assaying whether or not a biomolecule is adequately labeled with a heterobifunctional maleimide or iodoacetamide crosslinker. Brief treatment of fluorescein SAMSA with NaOH, pH 10 liberates a free thiol. By adding base-treated fluorescein SAMSA to a small aliquot of the crosslinker-modified biomolecule, the researcher can check to see whether the biomolecule has been sufficiently labeled before proceeding to the next step. The degree of modification can be approximated from either the absorbance or fluorescence of the conjugate following quick purification on a gel filtration column.

Alternatively, thiol reactivity of the modified biomolecule can be assayed using the reagents provided in our Thiol and Sulfide Quantitation Kit (T-6060) described above.[52,21] Once unconjugated reagents have been removed, a small aliquot of the maleimide- or iodoacetamide-modified biomolecule can be reacted with excess cysteine. Thiol-reactive groups can then be quantitated by determining the amount of cysteine consumed in this reaction using the Thiol and Sulfide Quantitation Kit.

Fluorescent Amine–Thiol Crosslinking Reagent

Both chloro groups of 7-chlorobenz-2-oxa-1,3-diazole-4-sulfonyl chloride (C-6056) can be displaced by nucleophiles. However, the sulfonyl chloride is expected to react with an amine to form a sulfonamide and the aromatic 7-chloro atom is more likely to be displaced by a thiol, resulting in a fluorescent crosslink. The group that is displaced first may depend on the pH of the medium.

Amine–Carboxylic Acid Crosslinking

Two reagents available from Molecular Probes — 1-ethyl-3-(3-dimethylaminopropyl)carbodiimide (EDAC, E-2247) and 2-ethoxy-1-ethoxycarbonyl-1,2-dihydroquinoline (EEDQ, E-6288) — can react with biomolecules to form "zero-length" crosslinks, usually within a molecule or between subunits of a protein complex. In this chemistry, the crosslinking reagent is not incorporated into the final product. The water-soluble carbodiimide EDAC crosslinks a specific amine and carboxylic acid between subunits of allophycocyanin, thereby stabilizing its assembly.[53] Molecular Probes uses EDAC to stabilize allophycocyanin in some of its allophycocyanin conjugates (see Section 6.4). EDAC has also been used to form intramolecular crosslinks in myosin subfragment-1,[54] intermolecular crosslinks in actomyosin [55] and crosslinks between proteins and DNA.[56] Both EDAC and the more hydrophobic EEDQ were used to crosslink proteins in a mitochondrial ATP synthase complex.[57] Addition of *N*-hydroxysuccinimide or *N*-hydroxysulfosuccinimide (NHSS, H-2249) is reported to enhance the yield of carbodiimide-mediated conjugations,[58] indicating the *in situ* formation of a succinimidyl ester–activated protein.

Reaction of carboxylic acids with cystamine and EDAC followed by reduction with DTT results in thiolation at carboxylic acids.[8] This indirect route to amine–carboxylic acid coupling is particularly suited to acidic proteins with few amines, carbohydrate polymers,[7] heparin, poly(glutamic acid) and synthetic polymers lacking amines. The thiolated biomolecules can also be reacted with any of the probes described in Chapter 2.

Amine–Carbohydrate and Thiol–Carbohydrate Crosslinking

As noted in Chapter 3, carbohydrates are usually unreactive in aqueous solution. An important method for their modification and subsequent crosslinking employs 2-acetamido-4-mercaptobutanoic acid hydrazide (AMBH, A-1642), which reacts with periodate-oxidized carbohydrates, glycoproteins (including antibodies and avidin but not streptavidin) or RNA to form a stable thiol-containing hydrazone.[59] This derivative can then be coupled to maleimide-, iodoacetamide- or SPDP-modified biomolecules.

Crosslinking Liposomes and Cell Membranes to Biomolecules

Unlike the chemical crosslinkers described above, all of which form covalent bonds with their targets, reagents used to crosslink liposomes, cell membranes and potentially other lipid assemblies to biomolecules typically comprise a phospholipid derivative to anchor one end of the crosslink in the lipid layer and a reactive group to attach the membrane assembly to the target biomolecule.

Chemically Reactive and Biotinylated Phospholipids

Phosphatidylethanolamines such as our high purity dihexadecanoylphosphoethanolamine (DPPE, D-7705) serve as building blocks for synthesis of phospholipid crosslinking agents; see the Technical Note "Synthetic Methods for Preparing New Phospholipid Probes" on page 303. Molecular Probes offers a maleimide-containing phospholipid (MMCC DHPE, M-1618), as well as an SPDP-derived phospholipid (PDP DHPE, P-1619). As with other pyridyldisulfide derivatives, PDP DHPE can be coupled to thiolated biomolecules via a disulfide link or reduced to the free thiol with DTT or TCEP and then coupled to iodoacetamide- or maleimide-derivatized biomolecules. Both MMCC DHPE and PDP DHPE can be incorporated into liposomes prepared from DPPC (D-7704) or other phospholipids, then coupled to thiolated antibodies,[60] streptavidin,[61] lectins [62] and other proteins.[63,64] Similarly, our phospholipid derivatives of biotin and biotin-X (B-1550, B-1616) can be used to prepare liposomes that have high affinity for avidin conjugates.[65]

Applications for Liposome Bioconjugates

Liposome bioconjugates are versatile reagents that can serve as a means of targeted delivery — either of the contents of the liposome's aqueous cavity or of the components in its lipid membrane — to a particular site recognized by its biomolecule tag. Representative applications include:

- Following receptor-mediated endocytosis of liposomes by flow cytometry [66]
- Loading liposomes with fluorescent dyes, including any of the polar tracers described in Section 15.3, for amplified detection in imaging and flow cytometry [67-70]
- Measuring anti-protein antibody using antigen-bearing liposomes in a liposome immune lysis assay [71] (LILA);
- Studying lateral and structural organization at aqueous interfaces [72-75]
- Targeting delivery of enzyme inhibitors [76] and oligodeoxyribonucleotides [77] into cells

1. Meth Enzymol 143, 246 (1987); **2.** Biochem J 173, 723 (1978); **3.** Anal Biochem 132, 68 (1983); **4.** Chem Pharm Bull 33, 362 (1985); **5.** Glycobiology 3, 279 (1993); **6.** Biochemistry 17, 5399 (1978); **7.** Biosci Biotech Biochem 56, 186 (1992); **8.** Biochim Biophys Acta 1038, 382 (1990); **9.** Biochem Int'l 1, 353 (1980); **10.** Biochim Biophys Acta 971, 307 (1988); **11.** Biochim Biophys Acta 971, 298 (1988); **12.** J Biol Chem 255, 10884 (1980); **13.** Meth Mol Biol 45, 235 (1995); **14.** J Cell Biol 93, 981 (1982); **15.** Nucleic Acids Res 17, 4404 (1989); **16.** J Biochem Biophys Meth 12, 349 (1986); **17.** Meth Enzymol 233, 380 (1994); **18.** Meth Enzymol 91, 49 (1983); **19.** Anal Biochem 220, 5 (1994); **20.** Bioconjugate Chem 5, 348 (1994); **21.** Anal Biochem 213, 49 (1993); **22.** Biochem Biophys Res Comm 152, 1 (1988); **23.** Biochemistry 26, 1889 (1987); **24.** Proc Natl Acad Sci USA 82, 1658 (1985); **25.** Biochim Biophys Acta 622, 201 (1980); **26.** FEBS Lett 150, 207 (1982); **27.** Anal Biochem 225, 174 (1995); **28.** Meth Enzymol 172, 584 (1989); **29.** Nature 292, 17 (1981); **30.** Chem Pharm Bull 37, 1831 (1989); **31.** Eur J Biochem 96, 483 (1979); **32.** Biochem Biophys Res Comm 87, 734 (1979); **33.** BioTechniques 8, 276 (1990); **34.** J Biol Chem 270, 2053 (1995); **35.** Biochemistry 19, 2260 (1980); **36.** Eur J Biochem 101, 395 (1979); **37.** Anal Biochem 198, 75 (1991); **38.** FEBS Lett 345, 113 (1994); **39.** Biochemistry 31, 10070 (1992); **40.** Biochemistry 31, 389 (1992); **41.** J Immunol Meth 75, 383 (1984); **42.** J Biochem 84, 491 (1978); **43.** J Immunol Meth 45, 195 (1981); **44.** J Immunol Meth 120, 133 (1989); **45.** Biochemistry 27, 2964 (1988); **46.** Biochemistry 26, 3168 (1987); **47.** Biochemistry 33, 375 (1994); **48.** Eur J Biochem 140, 63 (1984); **49.** Biochemistry 33, 10607 (1994); **50.** Biochemistry 12, 4154 (1973); **51.** J Cell Biol 93, 981 (1982); **52.** Bioconjugate Chem 5, 348 (1994); **53.** Cytometry 8, 91 (1987); **54.** Biochemistry 33, 6867 (1994); **55.** Biophys Chem 68, 35s (1995); **56.** J Mol Biol 123, 149 (1978); **57.** J Biol Chem 270, 2053 (1995); **58.** Anal Biochem 156, 220 (1986); **59.** Biochem Int'l 1, 353 (1980); **60.** Biol of the Cell 47, 111 (1983); **61.** Anal Biochem 207, 341 (1992); **62.** Cell Biol Int'l Rep 9, 1123 (1985); **63.** J Immunol Meth 132, 25 (1990); **64.** Meth Enzymol 149, 111 (1987); **65.** Meth Enzymol 149, 119 (1987); **66.** Biochem J 214, 189 (1983); **67.** J Fluorescence 3, 33 (1993); **68.** J Immunol Meth 121, 1 (1989); **69.** Cytometry 8, 562 (1987); **70.** J Immunol Meth 100, 59 (1987); **71.** J Immunol Meth 75, 351 (1984); **72.** Biophys J 68, 312 (1995); **73.** Biophys J 66, 305 (1994); **74.** Biophys J 65, 2160 (1993); **75.** J Membrane Biol 135, 83 (1993); **76.** J Cell Biol 102, 1630 (1986); **77.** Proc Natl Acad Sci USA 87, 2448 (1990).

5.2 Data Table *Chemical Crosslinking Reagents*

Cat #	Structure	MW	Storage	Soluble	Abs	{ε × 10⁻³}	Em	Solvent	Notes
A-685	✓	522	F,D,L	pH>6, DMF	491	{78}	515	pH 9	
A-1642	✓	191	D,A	EtOH	<300		none		
B-1550	S13.4	1019	FF,D	<1>	<300		none		
B-1616	S13.4	1133	FF,D	<1>	<300		none		<1>
C-6056	✓	253	F,DD	DMF, MeCN	322	{9.0}	none	MeOH	<2>
D-1379	S2.4	350	L	DMF, MeCN	391	{6.1}	<3>	MeOH	
D-1532	S2.1	154	D	H_2O	<300		none		
D-6316	✓	590	F,D	DMF, DMSO	316	{15}	none	MeOH	
D-7704	S13.4	734	FF,D	<1>	<300		none		
D-7705	S13.4	692	FF,D	<1>	<300		none		
D-8451	✓	396	D	pH>6	324	{11}	none	pH 8	<4>
D-8452	S2.1	154	D	H_2O	<300		none		
E-2247	S3.3	192	F,D	H_2O	<300		none		
E-6288	S3.3	246	D	DMSO, MeCN	261	{8.0}	none	MeOH	
E-6306	✓	456	F,D	DMF, DMSO	<300		none		
H-2249	S3.3	217	D	H_2O	<300		none		
I-2553	✓	138	F,DD	pH<10, MeCN	<300		none		<5>
I-6001	✓	283	F,DD,LL	DMSO, MeCN	<300		none		
M-1618	S13.4	1012	FF,D	<1>	<300		none		
N-1505	✓	307	F,DD	DMF, MeCN	268	{10}	none	MeCN	<6>
P-1619	S13.4	990	FF,D	<1>	281	{4.9}	none	MeOH	
S-1531	✓	312	F,D	DMF, MeCN	282	{4.7}	none	MeOH	<7>
S-1534	✓	334	F,D	DMF, MeCN	<300		none		
S-1535	✓	313	F,D	DMF, MeCN	<300		none		
S-1553	✓	231	F,D	DMF, MeCN	<300		none		
S-1563	✓	308	F,D	DMF, MeCN	<300		none		
S-1573	✓	245	F,D	DMF, MeCN	<300		none		
S-1666	✓	396	F,DD,LL	DMF, DMSO	<300		none		
S-1668	✓	509	F,DD,LL	DMF, DMSO	<300		none		
T-2556	S2.1	287	D	pH>5	<300		none		
T-6052	S2.1	193	D	MeCN	<300		none		

For definitions of the contents of this data table, see "How to Use the Handbook" on page vi.
Structure: Chemical structure drawing: ✓ = shown in this section; Sn.n = shown in Section number n.n

<1> Chloroform is the most generally useful solvent for preparing stock solutions of phospholipids. Glycerophosphocholines are usually freely soluble in ethanol. Solutions of glycerophosphoethanolamines in ethanol up to 1–2 mg/mL should be obtainable, using sonication to aid dispersion if necessary. Labeling of aqueous samples with nonmiscible phospholipid stock solutions can be accomplished by evaporating the organic solvent followed by hydration and sonication, yielding a suspension of liposomes. Information on solubility of natural phospholipids can be found in Marsh, D., *CRC Handbook of Lipid Bilayers*, CRC Press (1990) pp. 71–80.

<2> Thioether + sulfonamide reaction product of C-6056 should have spectra similar to those of the thiol adduct of C-6055 (see Section 2.3), which exhibits Abs ~386 nm, Em ~485 nm in methanol.

<3> D-1379 is almost nonfluorescent until reacted with thiols. Em = 475–485 nm for thiol adducts (QY ~0.1–0.3) [Meth Enzymol 143, 76 (1987)].

<4> D-8451 reaction product with thiols has Abs = 410 nm (ε = {14}) [Meth Enzymol 233, 380 (1994)].

<5> Loss of I-2553 absorption at 248 nm (ε ~{8.8} in 0.1 M HCl) occurs upon basic hydrolysis (pH 10) or upon reaction with amines [Biochemistry 17, 5399 (1978)].

<6> Upon basic hydrolysis, N-1505 forms *p*-nitrophenol (N-6477, see Section 10.1) with Abs = 399 nm (ε = {18}) in pH 9 buffer.

<7> After conjugation of S-1531, the degree of substitution can be determined by measuring the amount of 2-pyridinethione formed by treatment with DTT (D-1532) or TCEP (T-2556) using its absorbance at 343 nm (ε = {8.0}) [Biochem J 173, 723 (1978)].

5.2 Structures *Chemical Crosslinking Reagents*

$R^1 = H \quad R^2 = -SC(O)CH_3$ or $R^1 = -SC(O)CH_3 \quad R^2 = H$

A-685

A-1642

B-1550, B-1616
Section 13.4

C-6056

D-1379
Section 2.4

D-1532
Section 2.1

D-7704, D-7705
Section 13.4

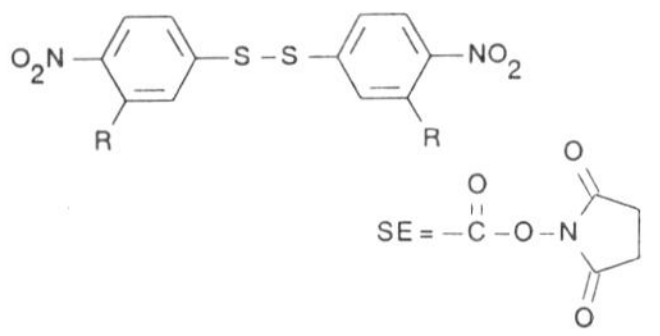

D-8451 R = —COOH
D-6316 R = SE

D-8452 Section 2.1

E-2247, E-6288 Section 3.3

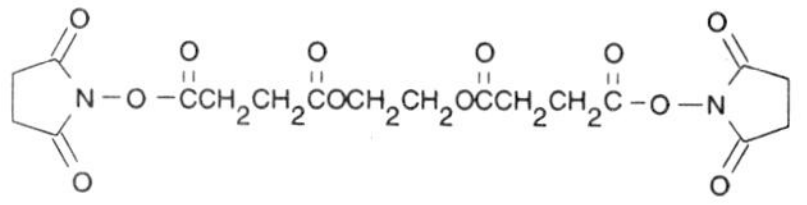

E-6306

H-2249 Section 3.3

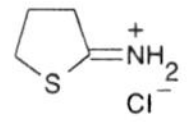

I-2553

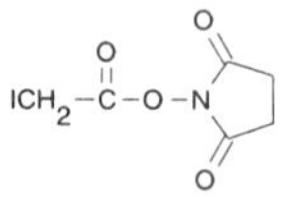

I-6001

M-1618 Section 13.4

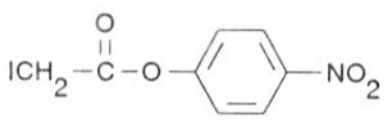

N-1505

P-1619 Section 13.4

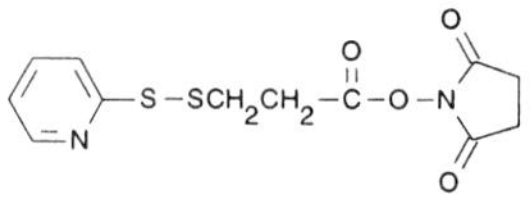

S-1531

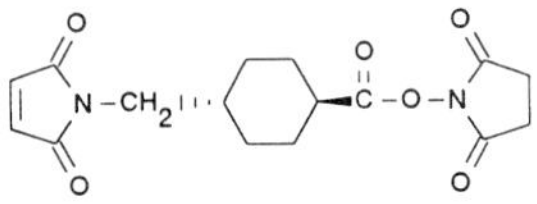

S-1534

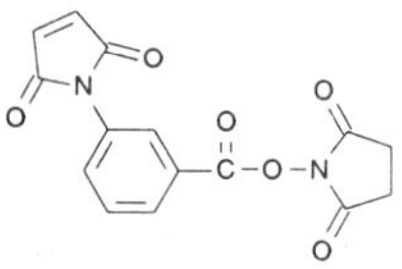

S-1535

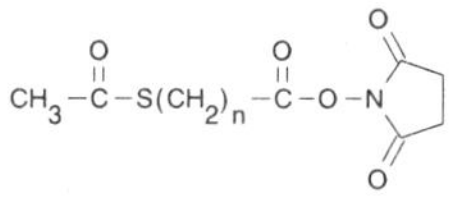

S-1553 n = 1
S-1573 n = 2

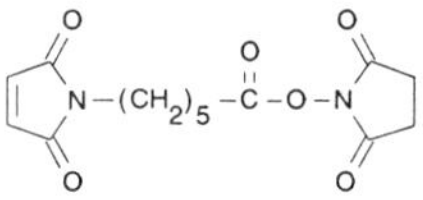

S-1563

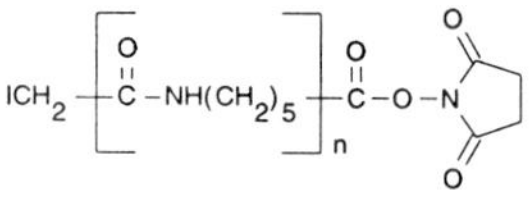

S-1666 n = 1
S-1668 n = 2

T-2556, T-6052 Section 2.1

5.2 Price List *Chemical Crosslinking Reagents*

Cat #		Product Name	Unit Size	Price Per Unit ($) 1–4 Units	5–24 Units
A-1642		2-acetamido-4-mercaptobutanoic acid hydrazide (AMBH)	100 mg	40.00	32.00
A-685		5-((2-(and-3)-S-(acetylmercapto)succinoyl)amino)fluorescein (SAMSA fluorescein) *mixed isomers*	25 mg	95.00	76.00
B-1616		N-((6-(biotinoyl)amino)hexanoyl)-1,2-dihexadecanoyl-*sn*-glycero-3-phosphoethanolamine, triethylammonium salt (biotin-X DHPE)	5 mg	115.00	92.00
B-1550		N-(biotinoyl)-1,2-dihexadecanoyl-*sn*-glycero-3-phosphoethanolamine, triethylammonium salt (biotin DHPE)	10 mg	115.00	92.00
C-6056	**New**	7-chlorobenz-2-oxa-1,3-diazole-4-sulfonyl chloride	25 mg	45.00	36.00
D-1379		dibromobimane (bBBr)	25 mg	35.00	28.00
D-7704	**New**	1,2-dihexadecanoyl-*sn*-glycero-3-phosphocholine (DPPC)	1 g	95.00	76.00
D-7705	**New**	1,2-dihexadecanoyl-*sn*-glycero-3-phosphoethanolamine (DPPE)	1 g	95.00	76.00
D-8451	**New**	5,5′-dithiobis-(2-nitrobenzoic acid) (DTNB; Ellman's reagent)	10 g	45.00	36.00
D-6316	**New**	5,5′-dithiobis-(2-nitrobenzoic acid, succinimidyl ester) (DTNB, SE)	25 mg	65.00	52.00
D-8452	**New**	1,4-dithioerythritol (DTE)	1 g	21.00	16.80
D-1532		dithiothreitol (DTT)	1 g	18.00	14.40
E-6288	**New**	2-ethoxy-1-ethoxycarbonyl-1,2-dihydroquinoline (EEDQ)	5 g	16.00	12.80
E-2247		1-ethyl-3-(3-dimethylaminopropyl)carbodiimide, hydrochloride (EDAC)	100 mg	15.00	12.00
E-6306	**New**	ethylene glycol bis-(succinic acid), bis-(succinimidyl ester) (EGS)	100 mg	18.00	14.40
H-2249		N-hydroxysulfosuccinimide, sodium salt (NHSS)	100 mg	25.00	20.00
I-2553		2-iminothiolane, hydrochloride	100 mg	15.00	12.00
I-6001	**New**	iodoacetic acid, succinimidyl ester (succinimidyl iodoacetate)	100 mg	65.00	52.00
M-1618		N-((4-maleimidylmethyl)cyclohexane-1-carbonyl)-1,2-dihexadecanoyl-*sn*-glycero-3-phosphoethanolamine, triethylammonium salt (MMCC DHPE)	5 mg	135.00	108.00
N-1505		4-nitrophenyl iodoacetate (NPIA)	1 g	48.00	38.40
P-6305	**New**	Protein-Protein Crosslinking Kit *3 conjugations*	1 kit	175.00	140.00
P-1619		N-((2-pyridyldithio)propionyl)-1,2-dihexadecanoyl-*sn*-glycero-3-phosphoethanolamine, triethylammonium salt (PDP DHPE)	5 mg	135.00	108.00
S-1553		succinimidyl acetylthioacetate (SATA)	100 mg	48.00	38.40
S-1573		succinimidyl acetylthiopropionate (SATP)	100 mg	48.00	38.40
S-1666		succinimidyl 6-((iodoacetyl)amino)hexanoate (SIAX)	100 mg	85.00	68.00

Cat #		Product Name	Unit Size	Price Per Unit ($) 1–4 Units	5–24 Units
S-1668		succinimidyl 6-(6-(((iodoacetyl)amino)hexanoyl)amino)hexanoate (SIAXX)	25 mg	85.00	68.00
S-1535		succinimidyl 3-maleimidylbenzoate (SMB)	100 mg	42.00	33.60
S-1563		succinimidyl 6-maleimidylhexanoate (EMCS)	25 mg	48.00	38.40
S-1534		succinimidyl *trans*-4-(maleimidylmethyl)cyclohexane-1-carboxylate (SMCC)	100 mg	98.00	78.40
S-1531		succinimidyl 3-(2-pyridyldithio)propionate (SPDP)	100 mg	115.00	92.00
T-6060	*New*	Thiol and Sulfide Quantitation Kit *50-250 assays*	1 kit	95.00	76.00
T-2556		tris-(2-carboxyethyl)phosphine, hydrochloride (TCEP)	1 g	25.00	20.00
T-6052	*New*	tris-(2-cyanoethyl)phosphine	1 g	24.00	19.20

5.3 Photoreactive Crosslinking and Labeling Reagents

Nonfluorescent Photoreactive Crosslinking Reagents

In contrast to chemical crosslinking reagents, which are often used to prepare bioconjugates, photoreactive crosslinking reagents are important tools for determining the proximity of two sites. Thus, these probes can be employed to define relationships between two reactive groups on a protein, on a ligand and its receptor or on separate biomolecules within an assembly. In the latter case, photoreactive crosslinking reagents can potentially reveal interactions among proteins, nucleic acids and membranes in live cells. The general scheme for defining spatial relationships usually involves photoreactive crosslinking reagents that contain a chemically reactive group as well as a photoreactive group. These crosslinkers are first chemically reacted with one molecule, for example a receptor ligand, and then this modified molecule is coupled to a second molecule, for example the ligand's receptor, using UV illumination. Depending on the reactive properties of the chemical and photoreactive groups, these crosslinkers can be used to couple like or unlike functional groups.

Molecular Probes offers four types of photoreactive reagents for covalent labeling:

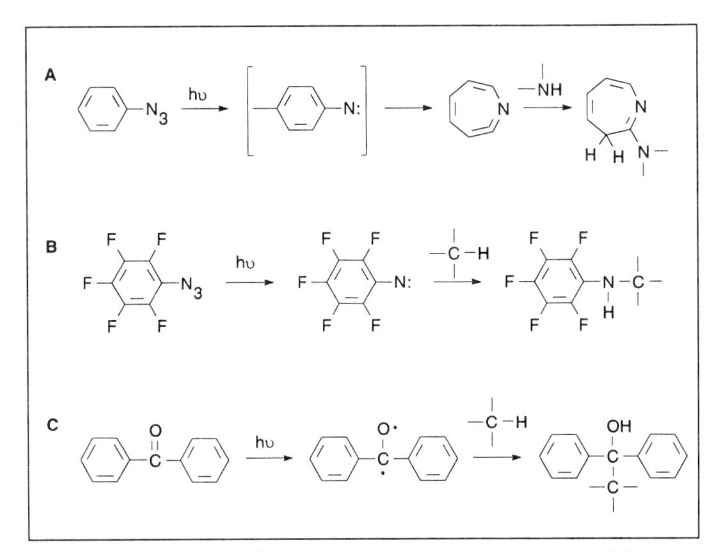

Figure 5.4 *Three types of photoreactive crosslinking reagents and their reactions: A) simple aryl azide, B) fluorinated aryl azide, C) benzophenone derivative.*

- **Simple aryl azides** that upon illumination (usually at <360 nm) generate reactive intermediates that form bonds with nucleophilic groups (Figure 5.4A)
- **Fluorinated aryl azides** that generate reactive nitrenes, thereby producing more C–H insertion products than the simple aryl azides (Figure 5.4B)
- **Benzophenone derivatives** that can be repeatedly excited at <360 nm without loss of reactivity (Figure 5.4C)
- **Diazopyruvate derivative** that upon photolysis generates a chemically reactive acylating agent (Figure 5.5)

Simple Aryl Azides

Azide-containing photoreactive crosslinking reagents are important both for modifying biomolecules and for synthesizing new photoreactive probes. FNPA (F-1501), the original photoreactive reagent described by Fleet and Porter,[1] has been reacted with amines, phenols, thiols and other nucleophiles in several biologically active molecules to yield new aryl azides. The FNPA-modified molecules can then be photoactivated with UV light to form chromophoric crosslinked derivatives.

In addition, we offer the succinimidyl ester derivative of 4-azidosalicylic acid (A-6307), which can be iodinated to a radioactive derivative,[2-4] and of 4-azidobenzoic acid (A-1548), which is a versatile amine-reactive crosslinker for preparing protein conjugates[5-7] and for synthesizing new photoreactive probes.[8,9]

4-Azidophenyl disulfide[10] (4,4'-dithiobis-(phenylazide), A-1506) and *N*-(4-azidophenylthio)phthalimide[11,12] (A-6312) react specifically with thiols to form mixed disulfides. The relative hydrophobicity of these thiol-reactive reagents makes them useful for crosslinking proteins from within the membrane core.[10] These reversible crosslinkers can be cleaved with reducing agents such as DTT or TCEP (D-1532, T-2556; see Section 5.2) prior to the second stage of two-dimensional gel electrophoresis, allowing the oligomerization state of crosslinked protein complexes to be ascertained.[11,13,14]

In addition to these simple aryl azides, we offer the "transferable" aryl azide *N*-((2-pyridyldithio)ethyl)-4-azidosalicylamide (PEAS, P-6317). This new aryl azide undergoes disulfide–thiol interchange of its pyridyldisulfide groups with the thiol groups of biomolecules to form mixed disulfides in the same way as SPDP.[15]

Fluorinated Aryl Azides: True Nitrene-Generating Reagents

Although the simple aryl azides may be initially photolyzed to electron-deficient aryl nitrenes, it has been shown that these rapidly

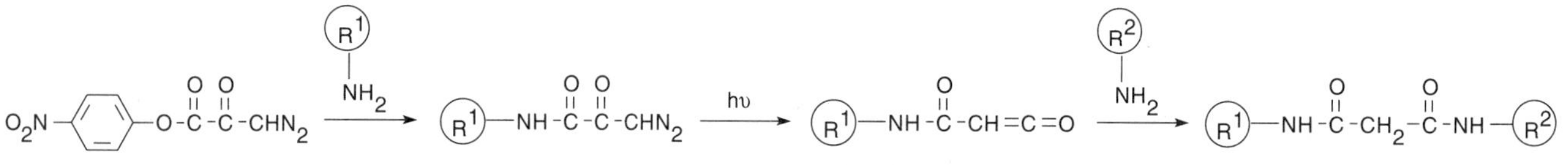

Figure 5.5 4-Nitrophenyl 3-diazopyruvate (N-2551), which generates a chemically reactive acylating agent upon photolysis.

ring-expand to form dehydroazepines — molecules that tend to react with nucleophiles rather than form C–H insertion products.[16,17] In contrast, Keana and Cai have shown that the photolysis products of the fluorinated aryl azides are clearly aryl nitrenes[18] and undergo characteristic nitrene reactions such as C–H bond insertion with high efficiency. Moreover, 4-azido-2,3,5,6-tetrafluorobenzoic acid (A-2521) and conjugates prepared from its amine-reactive succinimidyl ester (A-2522) may have quantum yields for formation of photocrosslinked products that are superior to those of the nonfluorinated aryl azides. An important application of the succinimidyl ester of 4-azido-2,3,5,6-tetrafluorobenzoic acid is the photofunctionalization of polymer surfaces.[19,20] 4-Azido-2,3,5,6-tetrafluorobenzyl amine (A-6308) is a useful building block for other fluorinated photoreactive reagents.

The polyfluorinated maleimide azides — TFPAM-3 (A-6309) and the longer-chain TFPAM-6 (A-6315) — are perhaps the most efficient reagents for photocrosslinking thiols to adjacent residues. TFPAM-3 and TFPAM-6 have been used to identify the interacting subunits of F_1F_0-ATPase involved in conformational changes associated with ATP synthesis.[21] Another polyfluorinated maleimide, TFPAM-SS1 (A-6311), undergoes thiol reaction and photolytic nitrene insertion to form reversible crosslinks that can be cleaved by disulfide reduction. It is therefore functionally similar to the previously discussed homobifunctional aryl azide crosslinker 4-azidophenyl disulfide. The internal disulfide in the crosslink can be reduced with DTT or TCEP (see Section 5.2) to yield free thiols (Figure 5.6), which can then be derivatized with fluorescent or radioisotopic labeling reagents to allow identification of the crosslinked products.[22]

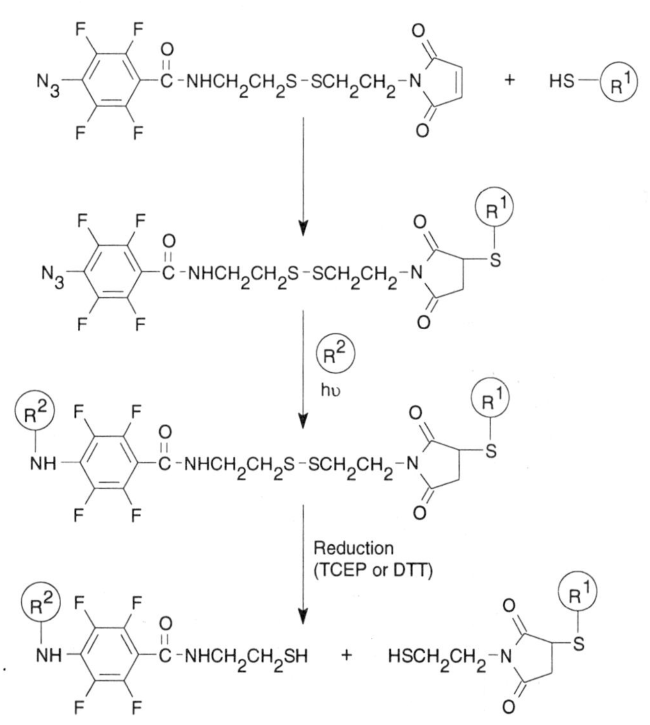

Figure 5.6 Reversible photocrosslinking using TFPAM-SS1 (A-6311).

Benzophenone-Based Photoreactive Reagents

The benzophenone reagents generally have higher crosslinking yields than the aryl azide photoreactive reagents.[23] Molecular Probes offers four thiol- or amine-reactive benzophenones (B-1508, B-1509, B-1526, B-1577) for efficient irreversible protein crosslinking of actin,[24] calmodulin,[25,26] myosin,[27,28] tropomyosin,[29] troponin[30-32] and other proteins. They are also useful for synthesizing new benzophenone-modified peptides,[33,34] carbohydrates[35] and other photoreactive probes.

We also offer the benzophenone phenylalanine analog D,L-Bpa (B-1584). This probe has been shown to interact directly with calmodulin[36] and has been used to synthesize photoreactive peptides.[36-39]

Diazopyruvate Derivative: A Photoreactive Acylating Agent

4-Nitrophenyl 3-diazopyruvate (N-2551) reacts with amines to yield diazopyruvic acid amides that undergo UV photolysis to produce ketene amides. These in turn efficiently acylate nucleophilic species to form malonic acid amide derivatives.[40] Photolyzed diazopyruvate-modified proteins efficiently acylate nucleophilic species, including both amines and alcohols, to form intra- or interprotein crosslinks (Figure 5.5). 4-Nitrophenyl 3-diazopyruvate has been used to couple calmodulin to calmodulin-sensitive adenylate cyclase,[41,42] to crosslink aldolase[40] and to synthesize photoreactive peptides.[43]

Fluorescent Photoreactive Reagents

Photoreactive "fluorescent" reagents are, in some cases, nonfluorescent until they are photolyzed with UV light. Because they are aryl azides, these photoreactive reagents may form insertion or addition products with nearby residues and solvents. These photoproducts, which may have relatively undefined structures, can usually be detected and quantitated by their fluorescence. Applications for the photoreactive fluorescent reagents include crosslinking receptors and other cell membrane components, labeling nucleic acids and synthesizing new photoreactive probes.

Photoreactive Fluorescent Crosslinking Reagents

5-Azidonaphthalene-1-sulfonyl chloride[44-46] (A-6310) is a versatile amine-reactive reagent that resembles dansyl chloride (D-21, see Section 1.8) in structure. Its amine conjugates are not fluorescent until photolyzed. This reagent is useful for crosslinking within and between biomolecules and for preparing new fluorogenic photoreactive probes.

The very polar reagent 4,4'-diazidostilbene-2,2'-disulfonic acid (DASD, D-6313) has the potential to form either intra- or interprotein crosslinks that may be fluorescent. It is expected to be membrane impermeant and may therefore be useful for topological studies. DASD has been used to prepare an enzyme–polymer film[47] and to study the topology of microsomal glucose-6-phosphatase.[48] Like other stilbenedisulfonates (see Section 18.5), DASD may also be an anion channel blocker.

Photoreactive Fluorescent Labeling of Nucleic Acids

Ethidium monoazide (E-1374) and ethidium diazide (E-3561) can be photolyzed in the presence of DNA or RNA to yield fluorescently labeled nucleic acids, both in solution and in cells.[49-52] The efficiency of the irreversible photolytic coupling of ethidium monoazide, which intercalates into nucleic acids like ethidium bromide, is unusually high[53] (>40%). Likewise, photolyzed ethidium diazide has been reported to covalently crosslink strands of DNA with an efficiency of about 30%.[54] Photolabeling of DNA can be used to follow its transport,[55] phase transitions[56] and diffusion.[57] In addition to its utility for studying DNA dynamics, the membrane-impermeant ethidium monoazide is reported to label only those cells with compromised membranes and can therefore serve as a fixable cell viability probe. A mixed population of live and dead cells labeled with this reagent retains its staining pattern after aldehyde fixation,[58,59] thereby reducing exposure to potentially pathogenic cells during cell viability analysis. Also, leukocyte phagocytosis was investigated by flow cytometry using ethidium monoazide–labeled *Candida albicans*.[60]

Polar Photoreactive Fluorescent Reagents for Labeling Proteins and Cell Surfaces

Researchers have used Molecular Probes' polar lucifer yellow and Cascade Blue® azide derivatives (A-629, C-625) to photolabel Ca^{2+}-ATPase in sarcoplasmic reticulum vesicles.[61] As expected, photoactivation of the dyes resulted primarily in the labeling of the cytoplasmic domain of this protein. In this study, antibodies to lucifer yellow and Cascade Blue were used to detect fluorophore-labeled proteins in Western blots; for more details on these antibodies, see the Product Highlight "Anti–Lucifer Yellow and Anti–Cascade Blue Antibodies" on page 333. It may also be possible to visualize labeled proteins directly in gels or on blots using UV illumination. Another polar azide, 5-azidonaphthalene-1-sulfonic acid (A-1156), may be similarly useful.[62,63]

A by-product of our development of the new Oregon Green™ dyes (see Chapter 1) is the novel polyfluorinated photoreactive reagent 6-azido-4,5,7-trifluorofluorescein (6-ATFF, A-6320). This probe is expected to generate green fluorescent adducts with proteins or other biomolecules and to retain the advantages of fluorinated aryl azides over nonfluorinated aryl azides.

1. Biochem J 128, 499 (1972); **2.** Biochemistry 33, 12092 (1994); **3.** Anal Chem 121, 286 (1992); **4.** Anal Biochem 121, 286 (1982); **5.** Peptides 14, 325 (1993); **6.** J Biol Chem 265, 9595 (1990); **7.** J Biol Chem 253, 1743 (1978); **8.** J Lipid Res 35, 45 (1994); **9.** J Neurochem 63, 1544 (1994); **10.** J Biol Chem 251, 6413 (1976); **11.** Anal Biochem 121, 321 (1982); **12.** J Biol Chem 262, 8524 (1977); **13.** Biochemistry 20, 6754 (1981); **14.** J Biol Chem 251, 7413 (1976); **15.** Traut, R.R. *et al.*, *Protein Function: A Practical Approach*, IRL Press (1989) p. 101; **16.** Ann Rev Biochem 62, 483 (1993); **17.** Adv Photochem 17, 69 (1992); **18.** J Org Chem 55, 3640 (1990); **19.** Bioconjugate Chem 5, 151 (1994); **20.** J Am Chem Soc 115, 814 (1993); **21.** Biochemistry 31, 2956 (1992); **22.** J Biol Chem 268, 20831 (1993); **23.** Biochemistry 33, 5661 (1994); **24.** Arch Biochem Biophys 240, 627 (1985); **25.** Biochemistry 33, 518 (1994); **26.** J Biol Chem 263, 542 (1988); **27.** Arch Biochem Biophys 288, 584 (1991); **28.** J Biol Chem 266, 2272 (1991); **29.** Biochemistry 25, 7633 (1986); **30.** Science 247, 1339 (1990); **31.** Biochemistry 27, 6893 (1988); **32.** Biochemistry 26, 7042 (1987); **33.** Biochemistry 32, 2741 (1993); **34.** J Protein Chem 3, 479 (1985); **35.** Carbohydrate Res 78, C4 (1980); **36.** J Biol Chem 261, 10695 (1986); **37.** Biochemistry 34, 2130 (1995); **38.** J Biol Chem 270, 10125 (1995); **39.** J Biol Chem 270, 1213 (1995); **40.** Biochemistry 28, 6346 (1989); **41.** J Biol Chem 267, 5847 (1992); **42.** Biochemistry 28, 6023 (1989); **43.** J Biol Chem 269, 27618 (1994); **44.** Biochem Biophys Res Comm 179, 259 (1991); **45.** Agr Biol Chem 52, 547 (1988); **46.** Agr Biol Chem 48, 2695 (1984); **47.** Anal Biochem 97, 320 (1979); **48.** Arch Biochem Biophys 275, 202 (1989); **49.** Biochemistry 30, 5644 (1991); **50.** Photochem Photobiol 43, 7 (1986); **51.** J Biol Chem 259, 11090 (1984); **52.** Photochem Photobiol 36, 31 (1982); **53.** Biochemistry 20, 1887 (1981); **54.** Biochemistry 22, 3226 (1983); **55.** J Biol Chem 269, 4910 (1994); **56.** Biochemistry 30, 10931 (1991); **57.** Macromolecules 22, 4550 (1989); **58.** Cytometry 19, 243 (1995); **59.** Cytometry 12, 133 (1991); **60.** Cytometry 11, 610 (1990); **61.** Biochim Biophys Acta 1068, 27 (1991); **62.** Biochemistry 22, 3954 (1983); **63.** Anal Biochem 103, 26 (1980).

5.3 Data Table *Photoreactive Crosslinking and Labeling Reagents*

Cat #	Structure	MW	Storage	Soluble	Abs	{ε × 10^{-3}}	Em	Solvent	Notes
A-629	✓	637	F,LL	H_2O	426	{11}	534	H_2O	
A-1156	✓	271	F,LL	pH>5	307	{6.6}	none	pH 7	<1>
A-1506	✓	300	F,LL	DMF, MeCN	265	{32}	none	MeOH	<2>
A-1548	✓	260	F,D,LL	DMF, MeCN	280	{25}	none	MeOH	<2>
A-2521	✓	235	F,LL	MeOH	258	{17}	none	MeOH	<2>
A-2522	✓	332	F,D,LL	DMF	273	{23}	none	EtOH	<2>
A-6307	✓	276	F,D,LL	DMF, DMSO	278	{21}	none	MeOH	<3>
A-6308	✓	220	F,LL	DMF	250	{17}	none	MeOH	<2>
A-6309	✓	371	F,D,LL	DMSO	250	{14}	none	MeOH	<2>
A-6310	✓	268	F,DD,LL	DMF	338	{7.5}	none	MeOH	<1>
A-6311	✓	449	F,D,LL	DMSO	256	{20}	none	MeOH	<2>
A-6312	✓	296	F,D,LL	DMSO	270	{19}	none	MeOH	<2>
A-6315	✓	413	F,D,LL	DMSO	250	{13}	none	MeOH	<2>

Cat #	Structure	MW	Storage	Soluble	Abs	{ε × 10⁻³}	Em	Solvent	Notes
A-6320	✓	427	D,LL	pH>6	506	{91}	526	pH 9	
B-1508	✓	277	F,D	DMF, MeCN	282	{9}	none	MeOH	<2, 4>
B-1509	✓	365	F,D	DMF, MeCN	300	{24}	none	MeOH	<2, 4>
B-1526	✓	239	F,DD	DMF, MeCN	300	{26}	none	MeOH	<2>
B-1577	✓	323	F,D	DMF, MeCN	256	{27}	none	MeOH	<2>
B-1584	S9.3	269	NC	pH>6, DMF	260	{19}	none	MeOH	
C-625	✓	770	F,LL	H_2O, MeOH	400	{29}	421	H_2O	<5>
D-6313	✓	538†	F,LL	H_2O	334	{34}	none	pH 7	<6>
E-1374	S8.1	420	F,LL	DMF, EtOH	464	{5.8}	625	pH 3	<7>
E-3561	S8.1	402	F,LL	pH 3	432	{5.7}	496	pH 3	<8>
F-1501	✓	182	F,D,LL	DMF, MeCN	244	{17}	none	EtOH	<9>
N-2551	✓	235	FF,D,LL	DMF, MeCN	400	{18}	none	pH 9	<10>
P-6317	✓	347	F,D,LL	DMSO	271	{24}	none	MeOH	<3>

For definitions of the contents of this data table, see "How to Use the Handbook" on page vi.
Structure: Chemical structure drawing: ✓ = shown in this section; Sn.n = shown in Section number n.n

* Sulfosuccinimidyl esters are water soluble and may be dissolved in buffer at pH ~8 for reaction with amines. Long-term storage in water is NOT recommended due to hydrolysis.

† MW is for the hydrated form of this product.

<1> Nonfluorescent until photolyzed. Photoproduct of A-1156 has Abs = 350 nm and Em = 545–578 nm depending on environment [Biochemistry 22, 3954 (1983)].

<2> This compound has weaker visible absorption at >300 nm but no discernible absorption peaks in this region.

<3> The absorption spectra of these compounds include shoulders at longer wavelengths: A-6307, Abs = 312 nm (ε = {9.2}); P-6317, Abs = 306 nm (ε = {10}).

<4> Spectral data of the 2-mercaptoethanol adduct.

<5> Cascade Blue dyes have a second absorption peak at about 376 nm with ε ~80% of the 395–400 nm peak.

<6> Changes color upon exposure to ultraviolet light.

<7> E-1374 spectral data are for the free dye. Fluorescence is weak, but intensity increases ~15-fold on binding to DNA. After photocrosslinking to DNA, Abs = 504 nm (ε ~{4.0}), Em = 600 nm [Biochemistry 19, 3221 (1980); Nucleic Acids Res 5, 4891 (1978)].

<8> E-3561 spectral data are for the free dye. Fluorescence is weak and decreases in intensity upon binding to DNA [Biochemistry 19, 3221 (1980)].

<9> F-1501 spectra are for the unreacted reagent. Much weaker absorption occurs at longer wavelengths (Abs ~450 nm, ε ~{0.7}). The reaction product with amines has Abs = 456 nm (ε = {5.0}) and is nonfluorescent.

<10> On basic hydrolysis, N-2551 forms *p*-nitrophenol (N-6477, see Section 10.1), Abs = 399 nm (ε = {18}) in pH 9 buffer.

5.3 Structures *Photoreactive Crosslinking and Labeling Reagents*

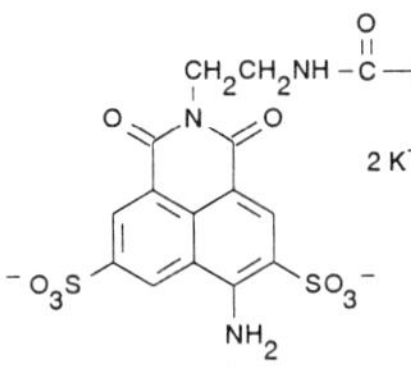

A-629

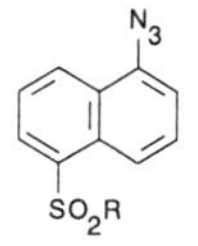

A-1156 R = O^- Na^+
A-6310 R = Cl

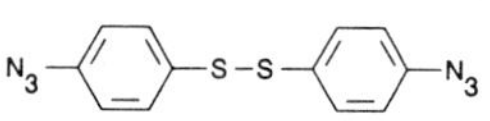

A-1506

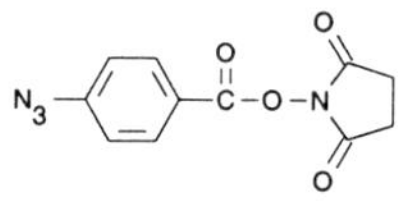

A-1548

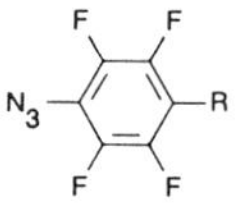

A-2521 R = —COOH
A-2522 R = SE
A-6308 R = $—CH_2NH_2$

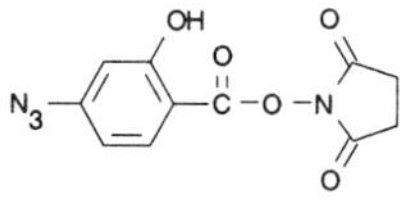

A-6307

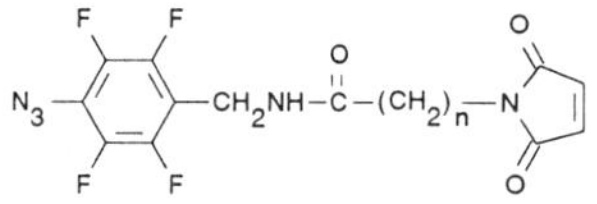

A-6309 n = 2
A-6315 n = 5

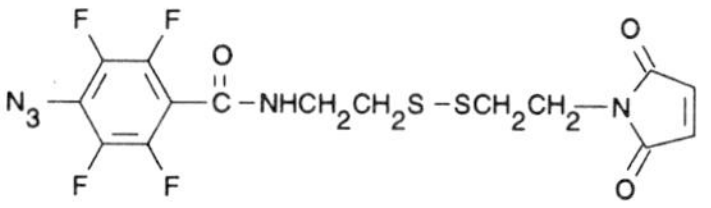

A-6311

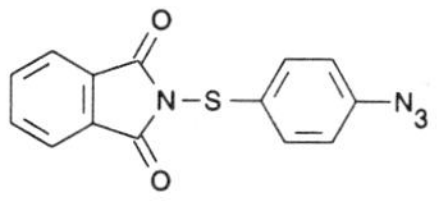

A-6312

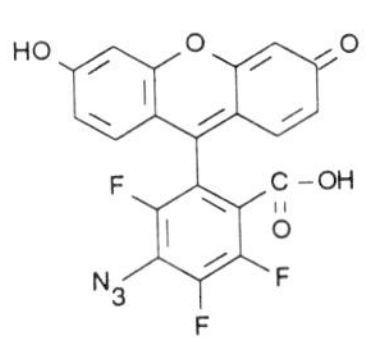

A-6320

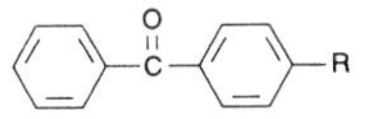

B-1509 R = $-NH-\overset{O}{\overset{\|}{C}}-CH_2I$
B-1508 R = MAL
B-1526 R = $-N=C=S$
B-1577 R = SE

B-1584
Section 9.3

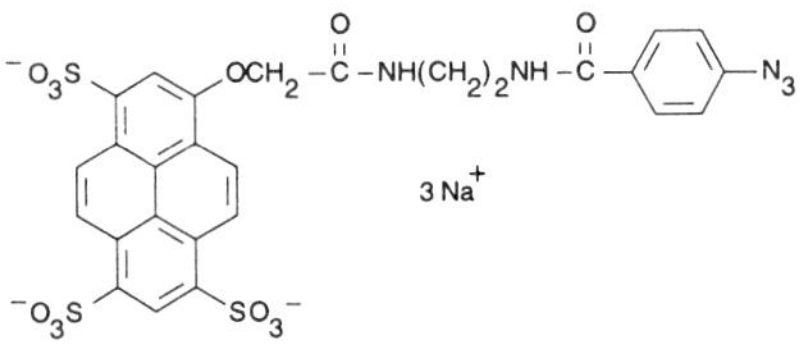

C-625

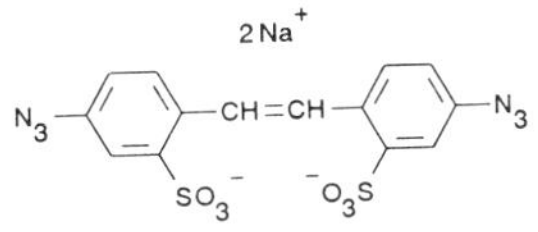

D-6313

E-1374, E-3561
Section 8.1

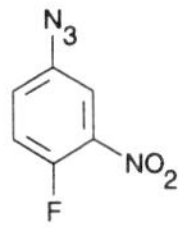

F-1501

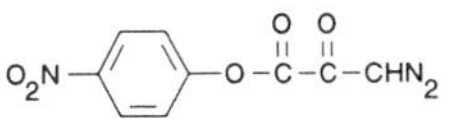

N-2551

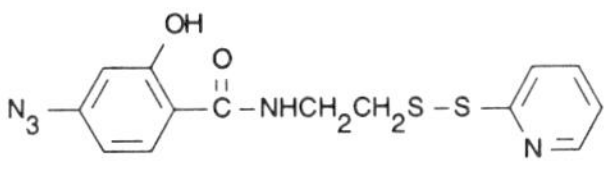

P-6317

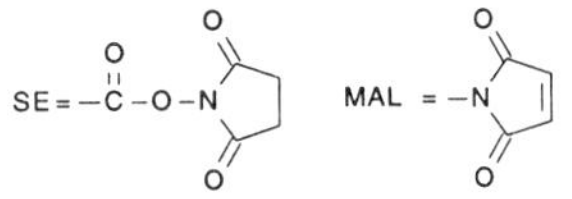

Substituent Key

5.3 Price List *Photoreactive Crosslinking and Labeling Reagents*

Cat #		Product Name	Unit Size	Price Per Unit ($) 1–4 Units	5–24 Units
A-1548		4-azidobenzoic acid, succinimidyl ester	100 mg	38.00	30.40
A-629		N-(((4-azidobenzoyl)amino)ethyl)-4-amino-3,6-disulfo-1,8-naphthalimide, dipotassium salt (lucifer yellow AB)	10 mg	95.00	76.00
A-1156		5-azidonaphthalene-1-sulfonic acid, sodium salt	100 mg	48.00	38.40
A-6310	*New*	5-azidonaphthalene-1-sulfonyl chloride	10 mg	68.00	54.40
A-1506		4-azidophenyl disulfide (4,4′-dithiobis-(phenylazide))	100 mg	45.00	36.00
A-6312	*New*	N-(4-azidophenylthio)phthalimide	25 mg	28.00	22.40
A-6307	*New*	4-azidosalicylic acid, succinimidyl ester	25 mg	38.00	30.40
A-2521		4-azido-2,3,5,6-tetrafluorobenzoic acid	100 mg	53.00	42.40
A-2522		4-azido-2,3,5,6-tetrafluorobenzoic acid, succinimidyl ester	25 mg	85.00	68.00
A-6311	*New*	N-(2-((2-(((4-azido-2,3,5,6-tetrafluoro)benzoyl)amino)ethyl)dithio)ethyl)maleimide (TFPAM-SS1)	5 mg	98.00	78.40
A-6308	*New*	4-azido-2,3,5,6-tetrafluorobenzyl amine	25 mg	85.00	68.00
A-6315	*New*	N-(4-azido-2,3,5,6-tetrafluorobenzyl)-6-maleimidohexanamide (TFPAM-6)	5 mg	95.00	76.00
A-6309	*New*	N-(4-azido-2,3,5,6-tetrafluorobenzyl)-3-maleimidopropionamide (TFPAM-3)	5 mg	95.00	76.00
A-6320	*New*	6-azido-4,5,7-trifluorofluorescein (6-ATFF)	10 mg	95.00	76.00
B-1509		benzophenone-4-iodoacetamide	100 mg	85.00	68.00
B-1526		benzophenone-4-isothiocyanate	100 mg	85.00	68.00
B-1508		benzophenone-4-maleimide	100 mg	85.00	68.00
B-1577		4-benzoylbenzoic acid, succinimidyl ester	100 mg	35.00	28.00
B-1584		p-benzoyl-DL-phenylalanine (DL-Bpa)	25 mg	24.00	19.20
C-625		Cascade Blue® aminoethyl 4-azidobenzamide, trisodium salt	10 mg	95.00	76.00
D-6313	*New*	4,4′-diazidostilbene-2,2′-disulfonic acid, disodium salt, tetrahydrate (DASD)	1 g	35.00	28.00
E-3561		ethidium diazide chloride	5 mg	95.00	76.00
E-1374		ethidium monoazide bromide	5 mg	115.00	92.00
F-1501		4-fluoro-3-nitrophenyl azide (FNPA)	100 mg	24.00	19.20
N-2551		4-nitrophenyl 3-diazopyruvate	25 mg	45.00	36.00
P-6317	*New*	N-((2-pyridyldithio)ethyl)-4-azidosalicylamide (PEAS; AET)	10 mg	65.00	52.00

A Sampling of Sampler Kits

Most of Molecular Probes' products are used in minute quantities, making "sample sizes" impractical. However, we have put together a number of Sampler Kits containing a set of application-specific probes, sometimes in smaller quantities and always at lower cost than the corresponding components sold separately. These Sampler Kits enable you to test our products and find the optimal probe for your particular application. Look in the designated Handbook chapter, contact our Technical Assistance Department or visit our Web site (http://www.probes.com) for more information on any of these kits.

Cat #		Product Name	Unit Size	Price Per Unit ($) 1–4 Units	5–24 Units
Chapter 7 *Protein Conjugates for Biological Detection*					
W-7024	***New***	Wheat Germ Agglutinin Sampler Kit *four fluorescent conjugates, 1 mg each*	1 kit	98.00	78.40
Chapter 8 *Nucleic Acid Detection*					
N-7565	***New***	Nucleic Acid Stains Dimer Sampler Kit	1 kit	125.00	100.00
N-7566	***New***	Nucleic Acid Stains Monomer Sampler Kit	1 kit	65.00	52.00
S-7580	***New***	SYBR® Green Nucleic Acid Gel Stain Starter Kit	1 kit	55.00	55.00
S-6655	***New***	SYPRO® Protein Gel Stain Starter Kit	1 kit	55.00	55.00
S-7572	***New***	SYTO® Live-Cell Nucleic Acid Stain Sampler Kit #1 *SYTO® dyes 11-16* *50 µL each*	1 kit	135.00	108.00
S-7554	***New***	SYTO® Live-Cell Nucleic Acid Stain Sampler Kit #2 *SYTO® dyes 20-25* *50 µL each*	1 kit	135.00	108.00
Chapter 10 *Enzymes, Enzyme Substrates and Enzyme Inhibitors*					
I-6614	***New***	Indolyl β-D-Galactopyranoside Sampler Kit *contains 5 mg each of B-1690, B-8407, I-8414, I-8420, M-8421*	1 kit	65.00	52.00
P-6548	***New***	Protease Inhibitor Sampler Kit	1 kit	195.00	156.00
R-6564	***New***	Resorufin Ether Sampler Kit	1 kit	98.00	78.40
T-6500	***New***	Tetrazolium Salt Sampler Kit *100 mg of T-6490, X-6493, M-6494, N-6495, I-6496, N-6498*	1 kit	125.00	100.00
Chapter 12 *Probes for Organelles*					
Y-7530	***New***	Yeast Mitochondrial Stain Sampler Kit	1 kit	95.00	76.00
Y-7531	***New***	Yeast Vacuole Marker Sampler Kit	1 kit	95.00	76.00
Chapter 15 *Fluorescent Tracers of Cell Morphology and Fluid Flow*					
L-7781	***New***	Lipophilic Tracer Sampler Kit	1 kit	150.00	120.00
Chapter 18 *Probes for Receptors and Ion Channels*					
B-6850	***New***	Brevetoxin Sampler Kit *5 µg of B-6851, B-6852, B-6853, B-6854, B-6855*	1 kit	135.00	108.00
Chapter 21 *Probes For Reactive Oxygen Species,Including Nitric Oxide*					
N-7925	***New***	Nitric Oxide Synthase (NOS) Inhibitor Kit	1 kit	195.00	156.00
Chapter 22 *Indicators for Ca^{2+}, Mg^{2+}, Zn^{2+} and Other Metals*					
B-6767	***New***	BAPTA Acetoxymethyl Ester Sampler Kit *2x1 mg each of B-1205, D-1207, D-1209, D-1213*	1 kit	75.00	60.00
C-6777	***New***	Coelenterazine Sampler Kit *contains 25 µg of C-2944, C-6776, C-6779, C-6780, C-6781*	1 kit	255.00	204.00
B-6768	***New***	BAPTA Buffer Kit *10 mg each of B-1204, D-1206, D-1208, D-1211*	1 kit	75.00	60.00
Chapter 26 *Tools for Fluorescence Applications*					
F-7321	***New***	Flow Cytometry Alignment Standards Sampler Kit, 2.5 µm	1 kit	55.00	44.00
F-7322	***New***	Flow Cytometry Alignment Standards Sampler Kit, 6.0 µm	1 kit	55.00	44.00
T-7284	***New***	TetraSpeck™ Fluorescent Microspheres Sampler Kit	1 kit	95.00	76.00

Chapter 6

Fluorescence Detection Methods, Including FluoSpheres® and ELF® Technologies

Contents

6.1 Introduction to Detection Methods

Fluorophore- and hapten-labeled proteins, nucleic acids, polysaccharides and lipids are important reagents for both research and diagnostic applications because they are amenable to sensitive detection techniques. Sometimes the labeled biomolecules are used simply as diffusible tracers for defining the physical delimitations of the cell or tissue. More commonly, however, the labeled biomolecule selectively binds a particular antigen, carbohydrate, nucleic acid sequence or previously bound hapten, thus providing a means of detecting these biological targets. This section introduces the reagents and methods commonly used in fluorescence detection; the remaining sections in this chapter focus on three novel approaches for further amplifying the signal in these methods. Section 6.2 describes Molecular Probes' unique selection of intensely fluorescent microspheres, including beads labeled with biotin, streptavidin, NeutraLite™ avidin or protein A. Section 6.3 discusses our proprietary Enzyme-Labeled Fluorescence (ELF) substrates, which form a highly fluorescent precipitate at the site of enzymatic activity. Section 6.4 focuses on phycobiliproteins, a family of highly fluorescent proteins widely used in multicolor applications. Our large selection of fluorescent and haptenylated proteins — including antibodies, enzymes, avidins, protein A and lectins — are listed in Chapter 7. Applications of our many fluorescent detection reagents for fluorescence *in situ* hybridization (FISH) are described in Section 8.4.

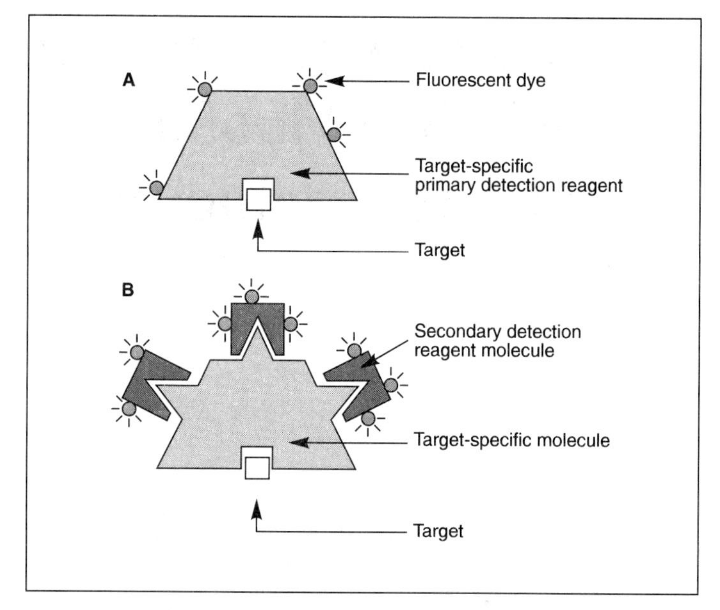

Figure 6.1 *Schematic diagram of primary and secondary detection reagents. A) In primary detection methods, the target-specific molecule includes one or more detectable moieties, shown here as radiant orbs. B) In secondary detection methods, the target-specific molecule contains binding sites or haptens that can be selectively recognized by secondary detection reagents. For example, these sites might be antigenic epitopes that bind antibodies. Alternatively, the target-specific molecule might be conjugated to either biotin or fluorescent dyes, thereby creating a molecule that can be detected with any of our avidin and streptavidin conjugates or our anti–fluorescent dye antibodies (see Chapter 7). As shown here, the target-specific molecule may contain multiple sites for binding the secondary detection reagent, thereby providing a simple system for amplifying the signal. Techniques in which multiple layers of secondary reagents are applied, thereby creating elaborate detection complexes, are widely used to achieve further signal amplification. The detectable moieties on the secondary reagents may be either fluorescent molecules or enzymes that can be developed for analysis by adding the appropriate substrate (see Chapter 10).*

Definitions: Detection Reagents

Primary Detection Reagents

Any easily detectable molecule that binds directly to a specific target is defined as a primary detection reagent. Such reagents are detected either by their intrinsic fluorescence, absorption, electron spin or electron density, or by virtue of a tightly associated label that confers one of these properties on the molecule. Many of the site-selective products offered by Molecular Probes can be considered primary detection reagents, including our fluorescent lectins (see Section 7.6), nucleic acid stains (see Chapter 8), phallotoxins (see Section 11.1), organelle probes (see Chapter 12) and various drug and toxin analogs (see Chapter 18). These primary detection reagents can typically be detected by methods such as fluorescence microscopy, fluorometry and flow cytometry.

Secondary Detection Reagents

Although many biomolecules such as antibodies and lectins bind selectively to a biological target, they usually must first be chemically modified before they can be detected. Often the biomolecule is conjugated to a fluorescent or chromophoric dye, a spin label or a heavy atom complex such as colloidal gold or ferritin. In many cases, however, the researcher may wish to avoid the time and expense required for these conjugations, choosing instead to employ a secondary detection reagent, defined as any easily detectable molecule that can be indirectly linked to the molecule of interest. Often, secondary detection reagents recognize a particular class of molecules. For example, Molecular Probes' goat anti-mouse antibodies — available conjugated to a wide range of fluorophores and haptens — can be used to localize a tremendous variety of target-specific mouse monoclonal antibodies.

Figure 6.1 compares the type of complexes formed in primary and secondary detection methods. Primary and secondary detection methods are used to detect specific molecules in cells and tissues, as well as on blots and in solution. Although secondary detection methods usually require additional steps and may sometimes yield unsatisfactory signal-to-noise ratios, they produce a signal that is usually amplified over that obtained with primary detection reagents. The scheme for secondary detection shown in Figure 6.1B represents the simplest method of secondary detection. Techniques in which multiple secondary reagents are applied to create elaborate detection complexes are widely used to achieve amplification.[1,2]

Common Experimental Protocols Using Primary and Secondary Detection Reagents

The types of experiments commonly performed in the laboratory with primary and secondary detection methods include:

- A target-specific polyclonal or monoclonal primary antibody is first bound to its target and then detected by a fluorescently labeled polyclonal secondary antibody raised against the immunoglobulin of the primary antibody. The properties of the amine-reactive dyes used by Molecular Probes to prepare these conjugates are described in detail in Chapter 1. The secondary antibody conjugates themselves and their applications are discussed in Section 7.2.

- A target-specific molecule such as an antibody, lectin, nucleic acid hybridization probe, drug or toxin is first conjugated to a low molecular weight hapten such as a fluorophore, digoxigenin, dinitrophenyl (DNP) or biotin; see Section 4.2 for a discussion of our haptenylation and biotinylation reagents. After binding its target, this haptenylated biomolecule is then detected using a labeled protein that binds selectively to the hapten. Commonly used secondary detection reagents include antibodies, avidins and streptavidins labeled with a fluorescent dye or other detectable marker (see Sections 7.2 and 7.5).

- In nucleic acid detection, including fluorescence *in situ* hybridization (FISH), the hybridization probe may incorporate a fluorescent label, thereby providing a straightforward method of primary detection. Alternatively, the hybridization probe may be modified to include either a hapten that can be detected with a secondary detection reagent or an enzyme that can be detected with a fluorogenic or chromogenic substrate. Our FluoReporter® Labeling Kits are designed to simplify the chemical labeling of oligonucleotides (see Section 8.2).

- In an enzyme-linked immunosorbent assay (ELISA), the primary or secondary detection reagent is usually conjugated to an enzyme. The signal is then developed by adding a suitable fluorogenic or chromogenic enzyme substrate. Catalytic turnover of substrate by the enzyme often results in significant amplification of the signal. Thus, enzyme-conjugated detection reagents often provide greater sensitivity than that achieved with direct dye conjugates. The substrates used to detect the enzymatic activity — and indirectly the amount of the target — typically yield a soluble fluorophore or chromophore. Molecular Probes' extensive assortment of substrates for use in ELISAs is described in Chapter 10.

- Enzyme-conjugated detection reagents can also be used to localize a signal in cells or tissues. In this case, however, it is essential to use an enzyme substrate that directly yields a fluorescent, colored or electron-dense precipitate at the site of the enzyme complex. Molecular Probes' new ELF technology, which is described in Section 6.3, provides a greater degree of resolution than many conventional enzyme-mediated fluorescence staining methods.

1. J Histochem Cytochem 43, 31 (1995); **2.** Histochem J 20, 75 (1988).

Limitations of Low Molecular Weight Dyes

The argon-ion laser is the excitation source in many flow cytometers and confocal laser scanning microscopes, as well as in certain laser scanners. In these instruments, the wavelengths used to excite green, yellow, orange and red fluorescent dyes are limited primarily to the laser's 488 and 514 nm spectral lines, which severely restricts simultaneous multicolor detection. For example, when excited at 488 or 514 nm, the Texas Red® fluorophore has a particularly low fluorescence output (see figure) that is easily obscured by the more intense fluorescein fluorescence in a double-labeling experiment, even when detected past 600 nm. For these applications, it would be useful to have a red fluorescent dye with an absorption maximum closer to the spectral lines of the argon-ion laser. Unfortunately, very few low molecular weight dyes have a combination of a large Stokes shift — defined as the separation of the absorption and emission maxima — and high fluorescence output.[1] In addition to this Stokes shift limitation, low molecular weight dyes may be impractical for some applications because they do not provide a bright enough fluorescent signal. Typically, only a limited number of dyes can be attached to a biomolecule without interfering with its binding specificity or causing it to precipitate. It may therefore be necessary to prepare a bioconjugate with a less-than-maximal fluorescence yield in order to preserve its important properties.

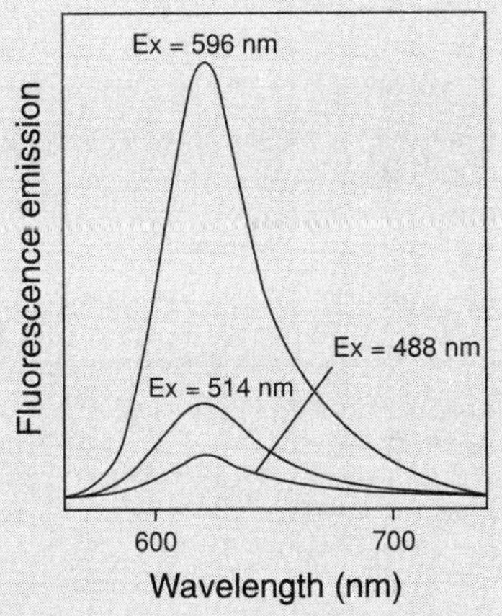

Fluorescence intensity of the Texas Red bovine serum albumin conjugate (A-824, see Section 15.7) when excited at 488, 514 and 596 nm.

Molecular Probes offers two product lines that specifically address the limitations of low molecular weight dyes. Our intensely fluorescent TransFluoSpheres® polystyrene microspheres (see Section 6.2) have been especially designed to provide both a large Stokes shift and a high fluorescence yield. For instruments with argon-ion laser excitation sources, we have developed a series of five TransFluoSpheres beads all of which have an excitation maximum at 488 nm but emit at different wavelengths: 560 nm, 605 nm, 645 nm, 685 nm or 720 nm (see Figure 6.4 in Chapter 6). This set of TransFluoSpheres beads enables researchers to detect five experimental parameters simultaneously with a single excitation wavelength. We also offer phycobiliproteins, a family of highly soluble red fluorescent proteins with high quantum yields and large extinction coefficients, as well as phycobiliprotein-labeled secondary detection reagents (see Section 6.4). In addition, Molecular Probes can custom-label antibodies or other target-specific proteins with our fluorescent microspheres or phycobiliproteins to provide detection reagents that match your needs; for more information or a quotation, please contact our Custom and Bulk Sales Department.

1. Haugland, R.P. in *Optical Microscopy for Biology*, B. Herman and K. Jacobson, Eds., Wiley-Liss (1990) pp. 143–157.

6.2 FluoSpheres and TransFluoSpheres Fluorescent Microspheres

Although low molecular weight reactive dyes are versatile and easy to use, they are not without limitations; see the Technical Note "Limitations of Low Molecular Weight Dyes" on page 109. For example, the fluorescence output of the dye–biomolecule conjugate is often limited by the number of dyes that can be attached to the biomolecule without disrupting its function. Our highly fluorescent microspheres — both the FluoSpheres and our new TransFluoSpheres beads (see Color Plate 1 in "Visual Reality") — provide a means of overcoming this limitation. Moreover, our TransFluoSpheres beads are designed to facilitate multicolor detection, particularly in applications that use lasers, with their inherent limited number of excitation wavelengths. TransFluoSpheres beads contain a series of two or more proprietary dyes that have been carefully chosen to ensure excited-state energy transfer between the dyes. This unique strategy enables us to fine-tune both the excitation and emission wavelengths of our microspheres so that they match a particular instrument's excitation source and detection sensitivity and complement the spectra of other fluorophores in a multicolor experiment.

Properties of Our Fluorescent Microspheres

Molecular Probes' intensely fluorescent FluoSpheres and TransFluoSpheres beads are manufactured using high-quality, ultraclean polystyrene microspheres. These microspheres are loaded with Molecular Probes' proprietary dyes, making them the brightest fluorescent microspheres available (Table 6.1). We employ methods to ensure that each bead is heavily loaded with dye. The protective environment within the bead shields the dye from many of the environmental effects that cause photobleaching of exposed fluorophores.

The stability, uniformity and reproducibility of fluorescent microspheres, as well as the extensive selection of colors available, make them the preferred standards for research and diagnostic assays that use fluorescence. We have introduced several new microsphere-based products for calibrating and aligning fluorescence microscopes and flow cytometers (see Sections 26.1 and 26.2, respectively). By carefully selecting dyes that can be incorporated within the microspheres, we are able to duplicate the spectra of widely used fluorophores such as fluorescein or R-phycoerythrin that are not soluble in polystyrene beads. Not only are our yellow-green fluorescent beads more photostable, but their emission spectra are not affected by changes in pH, as are conventional fluorescein-labeled microspheres.

We also offer fluorescent microspheres labeled with biotin, streptavidin, NeutraLite avidin and protein A, which are described below. We would be happy to consider discounts for large-volume orders, as well as custom preparation of microspheres with other colors, sizes and surface coatings; please contact our Custom and Bulk Sales Department for more information. The FluoSpheres and TransFluoSpheres beads and most of our microsphere-based standards for fluorescence microscopy and flow cytometry are covered by U.S. Patent No. 5,326,692 and other U.S. and foreign patents pending. These products are offered for research purposes only. Molecular Probes welcomes inquiries about licensing these products for resale or other commercial uses.

Table 6.1 *Comparison of the fluorescence intensity of Molecular Probes' yellow-green fluorescent FluoSpheres beads with other commercially available yellow-green fluorescent microspheres.*

Supplier	Size (µm)	Fluorescence Intensity *	CV † for Intensity
Molecular Probes (F-8852)	1.02	1998	4.40%
Molecular Probes (F-8853)	2.07	8998	3.26%
Company A	0.84	3.7	30.28%
Company A	1.55	5.2	11.69%
Company B	1.01	12.6	2.49%
Company B	1.94	595	2.91%
Company C	0.93	116	4.62%
Company C	1.48	434	1.92%
Company D	0.85	17	5.19%
Company D	1.84	119	3.08%

* Median value for fluorescence intensity (in arbitrary units), measured for 10,000 individual beads per sample excited at 488 nm using flow cytometry. † Coefficient of variation.

Common Applications for Fluorescent Microspheres

Fluorescent microspheres have been used as retrograde neuronal tracers [1-3], microinjectable cell tracers [4,5] (see Section 15.6) and standardization reagents for flow cytometry [6] (see Section 26.5). Furthermore, they have been employed to:

- Determine blood flow in tissues [7-10] (see Section 15.6)
- Investigate phagocytic processes [11-13] (see Section 17.1)
- Detect low-abundance receptors [14-16]
- Follow the fate of transplanted cells [17] such as donor erythrocytes in patients who had received allogenic bone marrow transplants [18]
- Identify the exchangeable GTP-binding site on the plus-end of taxol-stabilized fluorescent microtubules [19]
- Image three different *Candida albicans* antigens simultaneously [20]
- Investigate binding mechanisms of neural cell adhesion molecules [21,22]
- Track the lateral mobility of GPI-anchored proteins in supported bilayers,[23] Thy1 molecules in the plasmalemma of live fibroblasts [24] and surface receptors during cell division [25]

- Detect amines on the surface of self-assembled monolayers of a microfabricated device [26]
- Make kinesin force measurements with optical tweezers [27]

In addition, fluorescent microspheres are potentially more sensitive than colorimetric methods in most, if not all, of the major microsphere-based diagnostic test systems presently in use, including latex-agglutination tests, filter-separation tests, particle-capture ELISA methods and two-particle sandwich techniques.

FluoSpheres Fluorescent Microspheres

A Wide Array of Fluorescent Colors

Molecular Probes' FluoSpheres fluorescent microspheres contain dyes with excitation and emission wavelengths that cover the entire spectrum from the near ultraviolet to the near infrared. Figure 6.2 shows the normalized emission spectra for ten of our eleven fluorescent colors of FluoSpheres beads. Because long-wavelength (>680 nm) light can penetrate tissues, it may be possible to perform experiments using our new far red and infrared fluorescent microspheres that were not previously possible with beads that emit at shorter wavelengths. We would like to highlight the following FluoSpheres products:

- Our new **blue fluorescent FluoSpheres beads** with excitation/emission maxima of 350/440 nm contain an improved blue fluorescent dye that provides superior brightness and a longer shelf life. We also offer blue fluorescent FluoSpheres beads with slightly shorter-wavelength fluorescence spectra (excitation/emission maxima = 365/415 nm).
- Our **yellow-green fluorescent FluoSpheres beads** are excited very efficiently using the 488 nm spectral line of the argon-ion laser and have exceptionally intense fluorescence (Table 6.1).
- Although the **red fluorescent FluoSpheres beads** are maximally excited at 580 nm, the excitation band is broad enough so that the beads emit well at 605 nm, even when excited at 488 nm. Our orange and red/orange fluorescent FluoSpheres beads have excitation maxima of 540 and 565 nm, respectively.
- Our **nile red fluorescent FluoSpheres beads** have broad excitation and emission bandwidths, making them compatible with fluorescein, rhodamine and Texas Red® optical filter sets.
- Our **crimson and dark red fluorescent FluoSpheres beads** are efficiently excited by the 633 nm spectral line of the He-Ne laser. Although the dark red fluorescent particles are significantly less fluorescent than the crimson fluorescent particles, they fluoresce at wavelengths that are longer and clearly distinguishable from those of the crimson fluorescent particles.
- Our **far red fluorescent FluoSpheres beads** with excitation/emission maxima of 690/720 nm are compatible with diode lasers — inexpensive excitation sources that are increasingly being used in fluorescence instrumentation.[28] These far red fluorescent beads may also prove useful for making direct fluorescence measurements in autofluorescent materials such as blood, plant tissues and marine organisms.
- Our **infrared fluorescent FluoSpheres beads** with excitation/emission maxima of 715/755 nm are the longest-wavelength fluorescent microspheres currently available. These beads absorb and emit at wavelengths at which most tissues are almost optically transparent.

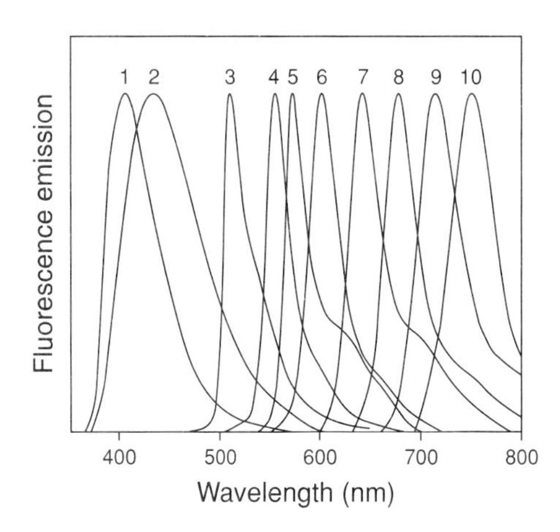

Figure 6.2 *Normalized fluorescence emission spectra of our FluoSpheres beads, named according to their excitation/emission maxima (nm): 1) blue (365/415), 2) blue (350/440), 3) yellow-green (505/515), 4) orange (540/560), 5) red-orange (565/580), 6) red (580/605), 7) crimson (625/645), 8) dark red (660/680), 9) far red (690/720) and 10) infrared (715/755) FluoSpheres beads.*

We have found our FluoSpheres beads to be many times brighter than fluorescent particles from other companies (Table 6.1). Table 6.2 shows the approximate number of unquenched fluorescein equivalents in our yellow-green fluorescent FluoSpheres beads. The intensity of the beads is sufficient to allow visualization of single particles, even for our smallest microspheres, which appear as point sources; see the description of our PS-Speck™ Microscope Point Source Kit (P-7220) in Section 26.1. Moreover, aqueous suspensions of FluoSpheres beads do not fade significantly when illuminated by a 250-watt xenon-arc lamp for 30 minutes. Indeed, most of our FluoSpheres beads show little or no photobleaching, even when excited with the intense illumination required for fluorescence microscopy.

Although some of our FluoSpheres beads are available in limited sizes and surface functions, we will prepare custom orders upon request. FluoSpheres beads can also be prepared with intensities

Table 6.2 *Fluorescein equivalents in our yellow-green fluorescent FluoSpheres beads.*

Microsphere Diameter (µm)	Fluorescein Equivalents per Microsphere
0.02	1.8×10^2
0.04	3.5×10^2
0.1	7.4×10^3
0.2	1.1×10^5
0.5	2.0×10^6
1.0	1.3×10^7
2.0	3.1×10^7
10	1.1×10^{10}
15	3.7×10^{10}

Table 6.3 *Summary of Molecular Probes' FluoSpheres fluorescent microspheres.**

Microspheres	0.02 µm	0.04 µm	0.1 µm	0.2 µm	0.5 µm	1.0 µm	2.0 µm	4.0 µm
Carboxylate-Modified Microspheres								
Blue (365/415)	F-8781 10 mL		F-8796 10 mL	F-8805 10 mL		F-8814 10 mL	F-8824 2 mL	
Blue (350/440)			F-8797 10 mL			F-8815 10 mL		
Yellow-green (505/515)	F-8787 10 mL	F-8795 1 mL	F-8803 10 mL	F-8811 10 mL	F-8813 10 mL	F-8823 10 mL	F-8827 2 mL	
Nile red (535/575)	F-8784 10 mL			F-8808 10 mL		F-8819 10 mL	F-8825 2 mL	
Orange (540/560)	F-8785 10 mL	F-8792 1 mL	F-8800 10 mL	F-8809 10 mL		F-8820 10 mL		
Red-orange (565/580)		F-8794 1 mL	F-8802 10 mL			F-8822 10 mL		
Red (580/605)	F-8786 10 mL	F-8793 1 mL	F-8801 10 mL	F-8810 10 mL	F-8812 10 mL	F-8821 10 mL	F-8826 2 mL	
Crimson (625/645)	F-8782 2 mL	F-8788 1 mL		F-8806 2 mL		F-8816 2 mL		
Dark red (660/680)	F-8783 2 mL	F-8789 1 mL		F-8807 2 mL				
Far red (690/720)		F-8790 0.4 mL	F-8798 1 mL			F-8817 1 mL		
Infrared (715/755)		F-8791 0.4 mL	F-8799 1 mL			F-8818 1 mL		
Sulfate Microspheres								
Blue (365/415)						F-8849 10 mL		F-8854 2 mL
Yellow-green (505/515)	F-8845 10 mL		F-8847 10 mL	F-8848 10 mL		F-8852 10 mL	F-8853 2 mL	F-8859 2 mL
Orange (540/560)			F-8846 10 mL			F-8850 10 mL		F-8857 2 mL
Red (580/605)						F-8851 10 mL		F-8858 2 mL
Crimson (625/645)								F-8855 2 mL
Dark Red (660/680)								F-8856 2 mL
Aldehyde-Sulfate Microspheres								
Yellow-green (505/515)	F-8760 10 mL				F-8761 10 mL	F-8762 10 mL		
Amine-Modified Microspheres								
Yellow-green (505/515)				F-8764 5 mL		F-8765 5 mL		
Red (580/605)				F-8763 5 mL				

All products listed in this table are $135 per unit; we offer a 20% discount when 5–24 units of a particular FluoSpheres product are purchased at one time.
* FluoSpheres beads are supplied as aqueous suspensions containing 2% solids, except for the 0.04 µm microspheres, which are supplied as aqueous suspensions containing 5% solids. All sizes fall within a narrow range as discussed on page 113. Sizes indicated in the above tables are nominal and may vary from batch to batch. Actual sizes, as determined by electron microscopy, are specified on the product labels.

that are *lower* than those of our regular selection, a desirable feature in some multicolor applications. FluoSpheres beads with calibrated intensities are already offered in our InSpeck™ Microscope Intensity Calibration Kits (see Section 26.1) and LinearFlow™ Flow Cytometry Intensity Calibration Kits (see Section 26.2), which are each available in several fluorescent colors.

A Wide Range of Sizes

To meet the diverse needs of our customers, we offer FluoSpheres beads in a variety of sizes (Table 6.3). The smallest microspheres are currently about 0.02 µm in diameter, with a coefficient of variation (CV) of about 20%, as determined by electron microscopy. The size uniformity improves with increasing size, with the CV decreasing from 5% for 0.1 µm FluoSpheres beads to ~1% for those with 10–15 µm diameters. The sizes specified in the product names are nominal bead diameters; because of batch variation in the undyed microspheres, the actual mean diameters shown on the product labels may differ from the nominal diameters, especially for the smaller microspheres. Because of their small size, 0.02–0.04 µm microspheres are transparent to light in aqueous suspensions and behave very much like true solutions.

Five Different Surface Functional Groups

We prepare FluoSpheres beads with five different surface functional groups, making them compatible with a variety of conjugation strategies. Our fluorescent dyes have negligible effect on the surface properties of the polystyrene beads or on their protein adsorption. We caution, however, that the surface properties have an important role in the functional utility of the microspheres; we cannot guarantee the suitability of a particular bead type for all applications.

- **Carboxylate-modified FluoSpheres beads** have pendent carboxylic acids, making them suitable for covalent coupling of proteins and other amine-containing biomolecules using water-soluble carbodiimide reagents such as EDAC (E-2247, see Section 3.3). In order to both decrease nonspecific binding and provide additional functional groups for conjugation, we are now using carboxylate-modified beads with a much larger number of carboxylic acids on their surface. The approximate surface charge is specified on the product information sheet that accompanies each vial of FluoSpheres beads.
- **Sulfate FluoSpheres beads** are relatively hydrophobic particles that will passively adsorb almost any protein, including BSA, IgG, avidin and streptavidin.
- **Aldehyde-sulfate FluoSpheres beads**, which are sulfate microspheres that have been modified to add surface aldehyde groups, are designed to react with proteins and other amines under very mild conditions.
- **Amine-modified FluoSpheres beads** can be coupled to a wide variety of amine-reactive molecules, including succinimidyl esters and isothiocyanates of haptens and drugs or carboxylic acids of proteins, using a water-soluble carbodiimide. The amine surface groups can also be reacted with SPDP (S-1531, see Section 5.2) to yield (after reduction) microspheres with sulfhydryl groups.

Biotin-, Avidin- and Protein A–Labeled Microspheres

Molecular Probes offers yellow-green fluorescent, red fluorescent and nonfluorescent microspheres labeled with biotin, streptavidin, NeutraLite avidin or protein A (Table 6.4). NeutraLite avidin has been specially processed to remove carbohydrates and lower the isoelectric point, resulting in a near-neutral protein that has significantly less nonspecific binding than conventional avidin. These covalently labeled microspheres provide our customers with valuable tools for improving the sensitivity of flow cytometry applications and immunodiagnostic assays. They may also be useful as tracers that can be detected with standard enzyme-mediated avidin/streptavidin methods. Additional sizes and colors of these labeled microspheres can be custom-ordered through our Custom and Bulk Sales Department.

Table 6.4 *Summary of biotin-, streptavidin, NeutraLite avidin– and protein A–labeled FluoSpheres microspheres.**

Microspheres	0.04 µm	0.2 µm	1.0 µm
Biotin-Labeled Microspheres			
Yellow-green (505/515)	F-8766 0.4 mL	F-8767 0.4 mL	F-8768 0.4 mL
Nonfluorescent			F-8769 0.4 mL
Streptavidin-Labeled Microspheres			
Yellow-green (505/515)	F-8780 0.4 mL		
NeutraLite Avidin–Labeled Microspheres			
Yellow-green (505/515)	F-8771 0.4 mL	F-8774 0.4 mL	F-8776 0.4 mL
Red (580/605)	F-8770 0.4 mL	F-8773 0.4 mL	F-8775 0.4 mL
Nonfluorescent	F-8772 0.4 mL		F-8777 0.4 mL
Protein A–Labeled Microspheres			
Yellow-green (505/515)	F-8778 0.4 mL		F-8779 0.4 mL

All products listed in this table are $165 per unit; we offer a 20% discount when 5–24 units of a particular FluoSpheres product are purchased at one time.

* Biotin- and NeutraLite avidin–labeled FluoSpheres beads are supplied as aqueous suspensions containing 1% solids and 0.02% Tween®; the streptavidin- and protein A–labeled FluoSpheres beads are supplied as aqueous suspensions containing 0.5% solids without surfactant. All sizes fall within a narrow range as discussed on page 113. Sizes indicated in the above tables are nominal and may vary from batch to batch. Actual sizes, as determined by electron microscopy, are specified on the product label.

Fluorescent Microsphere Starter Kits

For first-time users, we offer four types of fluorescent microsphere starter kits:

- **FluoSpheres Fluorescent Color Kit** (F-8889) consists of 1 mL samples of yellow-green, orange, red, crimson and dark red fluorescent carboxylate-modified 0.04 µm FluoSpheres beads packaged as high-density, azide-free suspensions for microinjection.
- **FluoSpheres Size Kits** contain 1 mL samples of carboxylate-modified FluoSpheres beads in 0.02, 0.1, 0.2, 0.5, 1.0 and

2.0 µm sizes and are available in yellow-green (F-8888) or red (F-8887) fluorescent colors.

- **FluoSpheres Blood Flow Determination Fluorescent Color Kits** provide several different fluorescent colors of our 10 µm (F-8890) or 15 µm (F-8891, F-8892, F-8893) FluoSpheres polystyrene microspheres (see Section 15.6).

TransFluoSpheres Fluorescent Microspheres — A Breakthrough in Microsphere Technology

Advantages of TransFluoSpheres Fluorescent Microspheres

Molecular Probes' new TransFluoSpheres fluorescent microspheres [29] are especially designed to overcome the limitations imposed by modern fluorescence instrumentation (Figure 6.3). Many flow cytometers, confocal laser scanning microscopes and laser scanners incorporate the argon-ion laser as the excitation source, thereby limiting the available excitation wavelengths to the laser's 488 and 514 nm spectral lines and severely restricting simultaneous multicolor detection. Ideally, it would be useful to have a series of fluorescent dyes with absorption maxima close to the argon-ion laser's spectral lines but with emission maxima at a variety of longer wavelengths. This approach would require that some of the dyes exhibit large Stokes shifts — defined as the separation of the absorption and emission maxima. Unfortunately, very few low molecular weight dyes have a combination of a large Stokes shift and high fluorescence output (see the Technical Note "Limitations of Low Molecular Weight Dyes" on page 109). For example, the Texas Red fluorophore — often used in combination with fluorescein — has a particularly low absorption at 488 and 514 nm. In applications that employ the argon-ion laser as an excitation source,

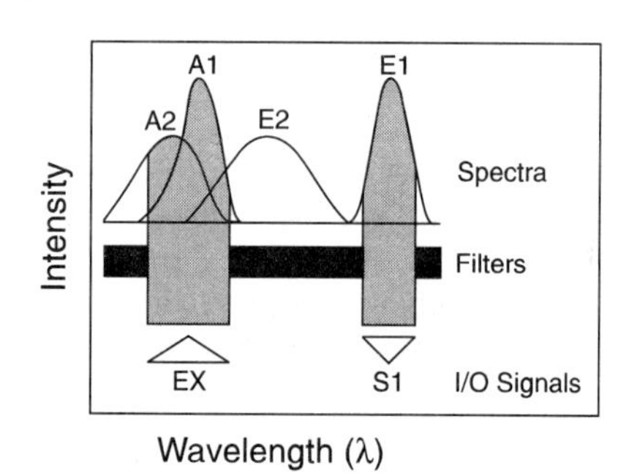

Figure 6.3 Schematic diagram of the advantages of the large Stokes shift exhibited by our TransFluoSpheres beads. A1 and E1 represent the absorption and emission bands of a typical TransFluoSpheres bead. The large separation of the absorption and emission maxima (Stokes shift) is characteristic of our TransFluoSpheres beads. Unlike most low molecular fluorescent dyes, which show considerable overlap of their absorption and emission spectra, the TransFluoSpheres beads can be excited (EX) across the entire absorption band A1 and the resulting fluorescence can be detected across the full emission band E1, thereby allowing the researcher to maximize the signal (S1). Moreover, because of the large Stokes shifts of the TransFluoSpheres beads, researchers can often avoid problems associated with autofluorescence. The absorption and emission bands of a typical autofluorescent component are represented in this figure by A2 and E2. Although the endogenous fluorescent species will be excited simultaneously with the TransFluoSpheres beads, the resulting emission (E2) does not coincide with E1 and is therefore readily rejected by suitably chosen optical filters.

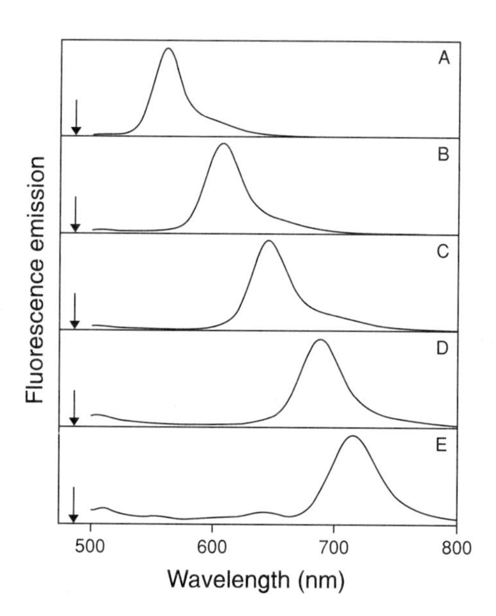

Figure 6.4 Fluorescence emission spectra of our new series of five 488 nm–excitable TransFluoSpheres beads, named according to their excitation/emission maxima (nm): A) 488/560, B) 488/605, C) 488/645, D) 488/685 and E) 488/720. The arrow in each spectrum represents the 488 nm spectral line of the argon-ion laser.

Texas Red conjugates have a low fluorescence output that is easily obscured by the more intense fluorescein fluorescence, even when detected past 600 nm.

Our TransFluoSpheres beads, which incorporate two or more fluorescent dyes that undergo excited-state energy transfer, exhibit Stokes shifts that can be extremely large. Each microsphere contains a dye with an excitation peak that maximally overlaps the spectral output of a commonly used excitation source (for example, the 488 nm spectral line of the argon-ion laser, Figure 6.4). In addition, the microsphere contains one or more longer-wavelength dyes that are carefully chosen to create a relay series that can efficiently transfer the energy from the initially excited dye to the longest-wavelength acceptor dye. The proprietary dyes used in the TransFluoSpheres beads are optimally loaded to ensure that the excitation energy is efficiently transferred from dye to dye so that essentially only the longest-wavelength dye in the series exhibits significant fluorescence. Because these TransFluoSpheres beads fluoresce at a wavelength that is considerably longer than the excitation wavelength, they provide a signal that can be detected in samples with significant Rayleigh or Raman scattering or with endogenous fluorescent compounds such as bilins, flavins and certain drugs. Also, the large Stokes shifts exhibited by the TransFluoSpheres beads allow the use of broadband filters, both to excite the sample and to detect the emission, resulting in a greater fluorescent signal (Figure 6.3).

TransFluoSpheres Beads to Match Your Excitation Source

Molecular Probes currently offers TransFluoSpheres beads that are compatible with a variety of excitation sources:

- A series of five argon-ion laser–excitable TransFluospheres beads, all of which have an excitation maximum at 488 nm but emit at different wavelengths: 560 nm, 605 nm, 645 nm, 685 nm or 720 nm (Figure 6.4)

- Red He-Ne laser–excitable TransFluoSpheres beads with excitation/emission maxima of 633/720 nm
- Green He-Ne laser–excitable TransFluoSpheres beads with excitation/emission maxima of 543/620 nm; these TransFluoSpheres beads are also efficiently excited by a mercury-arc lamp
- UV-excitable TransFluoSpheres beads with excitation/emission maxima of 365/515 nm

The series of TransFluoSpheres beads with 488 nm excitation enables researchers to detect five experimental parameters simultaneously. TransFluoSpheres beads can also be combined with our more traditional FluoSpheres beads or with low molecular weight dyes for multicolor detection. Using carbodiimide reagents such as EDAC (E-2247, see Section 3.3), researchers can couple protein or other amine-containing molecules to our carboxylate-modified TransFluoSpheres beads, making these microspheres suitable for a wide range of applications. TransFluoSpheres beads can be used in the major microsphere-based diagnostic test systems, as well as in experiments that currently employ standard fluorescent microspheres to detect cell-surface antigens, trace neurons and study phagocytosis.[2,30-33]

TransFluoSpheres fluorescent microspheres are available with nominal bead diameters of 0.04 µm, 0.1 µm and 1.0 µm (Table 6.5). In addition, we offer NeutraLite avidin–labeled TransFluoSpheres beads with excitation/emission maxima of 488/605 nm (see Section 7.5). If your wavelength or bead-size requirements are not met by our current selection, we invite you to inquire about custom synthesis by contacting our Custom and Bulk Sales Department. Molecular Probes can also fine-tune the excitation and emission of our microspheres to match your needs. In addition, we can covalently label our TransFluoSpheres beads with other target-specific proteins to provide detection reagents that potentially have greater sensitivity in flow cytometry applications and immunodiagnostic assays.

Table 6.5 *Summary of Molecular Probes' TransFluoSpheres fluorescent microspheres.**

Excitation/Emission (nm)	Size		
	0.04 µm	**0.1 µm**	**1.0 µm**
365/515	T-8863 0.5 mL	T-8871 0.5 mL	T-8879 0.5 mL
488/560	T-8864 0.5 mL	T-8872 0.5 mL	T-8880 0.5 mL
488/605	T-8865 0.5 mL	T-8873 0.5 mL	T-8881 0.5 mL
488/605 NeutraLite avidin–labeled	T-8860 0.4 mL	T-8861 0.4 mL	T-8862 0.4 mL
488/645	T-8867 0.5 mL	T-8875 0.5 mL	T-8883 0.5 mL
488/685	T-8868 0.5 mL	T-8876 0.5 mL	T-8884 0.5 mL
488/720	T-8869 0.5 mL	T-8877 0.5 mL	T-8885 0.5 mL
543/620	T-8866 0.5 mL	T-8874 0.5 mL	T-8882 0.5 mL
633/720	T-8870 0.5 mL	T-8878 0.5 mL	T-8886 0.5 mL

All products listed in this table are $175 per unit; we offer a 20% discount when 5–24 units of a particular TransFluoSpheres product are purchased at one time.
* All TransFluoSpheres beads are based on carboxylate-modified polystyrene microspheres and are supplied as aqueous suspensions containing 2% solids, except the NeutraLite avidin–labeled TransFluoSpheres beads, which are supplied as aqueous suspensions containing 1% solids.

1. J Neurosci 14, 6621 (1994); **2.** J Neurosci 13, 5082 (1993); **3.** Nature 310, 498 (1984); **4.** Cell Motil Cytoskeleton 8, 293 (1987); **5.** Devel Growth Differ 28, 461 (1986); **6.** Clin Immunol Immunopathol 55, 173 (1990); **7.** Circulatory Shock 41, 156 (1993); **8.** J Appl Physiol 74, 2585 (1993); **9.** J Cereb Blood Flow Metab 13, 359 (1993); **10.** J Autonomic Nervous Sys 30, 159 (1990); **11.** Cell Immunol 156, 508 (1994); **12.** J Leukocyte Biol 43, 148 (1988); **13.** Science 215, 64 (1982); **14.** Biochem Biophys Res Comm 181, 1223 (1991); **15.** Lea, T. *et al.*, *Flow Cytometry and Sorting*, Wiley-Liss (1990) pp. 367–380; **16.** Am J Physiol 255, C452 (1988); **17.** J Histochem Cytochem 41, 1579 (1993); **18.** Br J Haematol 72, 239 (1989); **19.** Science 261, 1044 (1993); **20.** J Immunol Meth 116, 213 (1989); **21.** Devel Biol 149, 213 (1992); **22.** J Cell Biol 118, 937 (1992); **23.** J Membrane Biol 135, 83 (1993); **24.** J Membrane Biol 144, 231 (1995); **25.** J Cell Biol 127, 963 (1994); **26.** Science 268, 272 (1995); **27.** Science 260, 232 (1993); **28.** Cytometry 15, 267 (1994); **29.** U.S. Patent No. 5,326,692 and other patents pending; **30.** Biophys J 65, 2396 (1993); **31.** Brain Res 630, 115 (1993); **32.** J Membrane Biol 135, 82 (1993); **33.** Cytometry 13, 423 (1992).

6.2 Price List *FluoSpheres and TransFluoSpheres Fluorescent Microspheres*

Cat #	Product Name	Unit Size	Price Per Unit ($) 1–4 Units	5–24 Units
F-8889	FluoSpheres® Fluorescent Color Kit, carboxylate-modified microspheres, 0.04 µm *five colors, 1 mL each* *5% solids, azide free*	1 kit	295.00	236.00
F-8887	FluoSpheres® Size Kit #1, carboxylate-modified microspheres, red fluorescent (580/605) *six sizes, 1 mL each* *2% solids*	1 kit	295.00	236.00
F-8888	FluoSpheres® Size Kit #2, carboxylate-modified microspheres, yellow-green fluorescent (505/515) *six sizes, 1 mL each* *2% solids*	1 kit	295.00	236.00

For prices of carboxylate-modified, sulfate, aldehyde-sulfate and amine-modified FluoSpheres® microspheres, see Table 6.3.
For prices of biotin-, streptavidin-, NeutraLite avidin– and Protein A–labeled FluoSpheres® microspheres, see Table 6.4.
For prices of TransFluoSpheres® fluorescent microspheres, see Table 6.5.

FluoSpheres® Fluorescent Microspheres for Blood Flow Determination

For nearly 30 years, radioactively labeled microspheres have been used to estimate regional organ perfusion. In order to minimize radiation exposure and to reduce expensive storage and disposal of radioactive materials, researchers are beginning to replace the radiolabeled beads in these studies with fluorescent microspheres. Molecular Probes' intensely fluorescent FluoSpheres beads for blood flow determination are especially designed for these applications. Radioactive (^{125}I) and fluorescent FluoSpheres beads yielded comparable results when used to assess coronary blood flow in dogs,[1] as well as to measure perfusion of dog lung, pig heart and pig kidney.[2] A complete description of our FluoSpheres fluorescent microspheres for blood flow determination can be found in Section 15.6.

Additional technical support, including a detailed applications manual on the use of fluorescent microspheres for blood flow determination, is available from the Fluorescent Microsphere Resource Center (FMRC) at the University of Washington, Seattle. The FMRC can be contacted by phone (206) 685–9479 or fax (206) 685–8673; information can also be obtained through a filter transfer protocol (FTP) server (fmrc.pulmcc.washington.edu), the internet (glenny@pele.pulmcc.washington.edu) or their Web site (http://fmrc.pulmcc.washington.edu/fmrc/fmrc.html).

1. Circulatory Shock 41, 156 (1993); **2.** J Appl Physiol 74, 2585 (1993).

FluoSpheres Beads for Blood Flow Determination

Cat #	Product Name	Unit Size	Price Per Unit ($) 1–4 Units	Price Per Unit ($) 5–24 Units
F-8829	FluoSpheres® polystyrene microspheres, 10 µm, blue fluorescent (365/415) *for blood flow determination* *3.6x10^6 beads/mL*	10 mL	150.00	120.00
F-8830 ***New***	FluoSpheres® polystyrene microspheres, 10 µm, blue-green fluorescent (430/465) *for blood flow determination* *3.6x10^6 beads/mL*	10 mL	150.00	120.00
F-8831	FluoSpheres® polystyrene microspheres, 10 µm, crimson fluorescent (625/645) *for blood flow determination* *3.6x10^6 beads/mL*	10 mL	150.00	120.00
F-8833	FluoSpheres® polystyrene microspheres, 10 µm, orange fluorescent (540/560) *for blood flow determination* *3.6x10^6 beads/mL*	10 mL	150.00	120.00
F-8834	FluoSpheres® polystyrene microspheres, 10 µm, red fluorescent (580/605) *for blood flow determination* *3.6x10^6 beads/mL*	10 mL	150.00	120.00
F-8835 ***New***	FluoSpheres® polystyrene microspheres, 10 µm, scarlet fluorescent (645/680) *for blood flow determination* *3.6x10^6 beads/mL*	10 mL	150.00	120.00
F-8836	FluoSpheres® polystyrene microspheres, 10 µm, yellow-green fluorescent (505/515) *for blood flow determination* *3.6x10^6 beads/mL*	10 mL	150.00	120.00
F-8837	FluoSpheres® polystyrene microspheres, 15 µm, blue fluorescent (365/415) *for blood flow determination* *1.0x10^6 beads/mL*	10 mL	150.00	120.00
F-8838 ***New***	FluoSpheres® polystyrene microspheres, 15 µm, blue-green fluorescent (430/465) *for blood flow determination* *1.0x10^6 beads/mL*	10 mL	150.00	120.00
F-8839	FluoSpheres® polystyrene microspheres, 15 µm, crimson fluorescent (625/645) *for blood flow determination* *1.0x10^6 beads/mL*	10 mL	150.00	120.00
F-8841	FluoSpheres® polystyrene microspheres, 15 µm, orange fluorescent (540/560) *for blood flow determination* *1.0x10^6 beads/mL*	10 mL	150.00	120.00
F-8842	FluoSpheres® polystyrene microspheres, 15 µm, red fluorescent (580/605) *for blood flow determination* *1.0x10^6 beads/mL*	10 mL	150.00	120.00
F-8843 ***New***	FluoSpheres® polystyrene microspheres, 15 µm, scarlet fluorescent (645/680) *for blood flow determination* *1.0x10^6 beads/mL*	10 mL	150.00	120.00
F-8844	FluoSpheres® polystyrene microspheres, 15 µm, yellow-green fluorescent (505/515) *for blood flow determination* *1.0x10^6 beads/mL*	10 mL	150.00	120.00

Fluorescent Color Kits for Regional Blood Flow Determination

Cat #	Product Name	Unit Size	Price Per Unit ($) 1–4 Units	Price Per Unit ($) 5–24 Units
F-8890	FluoSpheres® Blood Flow Determination Fluorescent Color Kit #1, polystyrene microspheres, 10 µm *seven colors, 10 mL each* *3.6x10^6 beads/mL*	1 kit	775.00	620.00
F-8891	FluoSpheres® Blood Flow Determination Fluorescent Color Kit #2, polystyrene microspheres, 15 µm *seven colors, 10 mL each* *1.0x10^6 beads/mL*	1 kit	775.00	620.00
F-8892 ***New***	FluoSpheres® Blood Flow Determination Fluorescent Color Kit #3, polystyrene microspheres, 15 µm *five colors, 10 mL each* *1.0x10^6 beads/mL*	1 kit	600.00	480.00
F-8893 ***New***	FluoSpheres® Blood Flow Determination Fluorescent Color Kit #4, polystyrene microspheres, 15 µm *five colors, 2 mL each* *1.0x10^6 beads/mL*	1 kit	165.00	132.00

6.3 Enzyme-Labeled Fluorescence (ELF) Signal Amplification Technology

When detecting specific biomolecules in cells and tissues, enzyme-mediated detection methods can provide significant signal amplification due to the catalytic turnover of the fluorogenic or chromogenic substrate. NBT is frequently combined with the phosphatase substrate BCIP (N-6495, B-6492; see Section 9.2) to produce a colored precipitate at the site of phosphatase activity in histochemical assays, *in situ* hybridization techniques and Western, Northern and Southern blot analyses. Because fluorometric methods are potentially more sensitive than colorimetric methods, Molecular Probes has been actively engaged in research to develop fluorogenic substrates for enzyme-mediated detection.

Our newly patented ELF-97 (**E**nzyme-**L**abeled **F**luorescence) phosphate is an alkaline phosphatase substrate with several unique properties that make it superior to many of the existing reagents for these applications. Upon enzymatic cleavage, this weakly blue fluorescent substrate yields a bright yellow-green fluorescent precipitate that exhibits an unusually large Stokes shift and excellent photostability.[1] The ELF-97 phosphatase substrate has proven to be a particularly powerful tool for mRNA *in situ* hybridization methods. Unlike the radio-active signal produced by conventional methods, ELF mRNA detection signals can be developed in minutes or even seconds and can be clearly distinguished from sample pigmentation, which often obscures both radioactive and colorimetric signals. Moreover, in this application, the yellow-green fluorescent ELF-97 alcohol precipitate produces a signal that is up to 40 times brighter than that achieved when using either directly labeled fluorescent hybridization probes or fluorescent secondary detection methods.[2-4]

To encourage research using the ELF technology, we have introduced additional ELF-97 substrates for esterase, β-glucuronidase, guanidinobenzoatase, lipase, microsomal dealkylase and sulfatase, all of which are described in Chapter 10. ELF substrates that yield precipitates with contrasting colors and ELF substrates for other enzymes are under development.

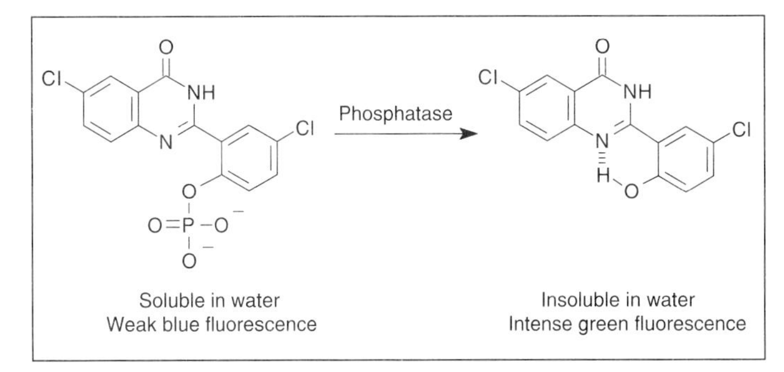

Figure 6.5 Principle of enzyme-mediated formation of the fluorescent ELF-97 alcohol precipitate from the ELF-97 phosphatase substrate.

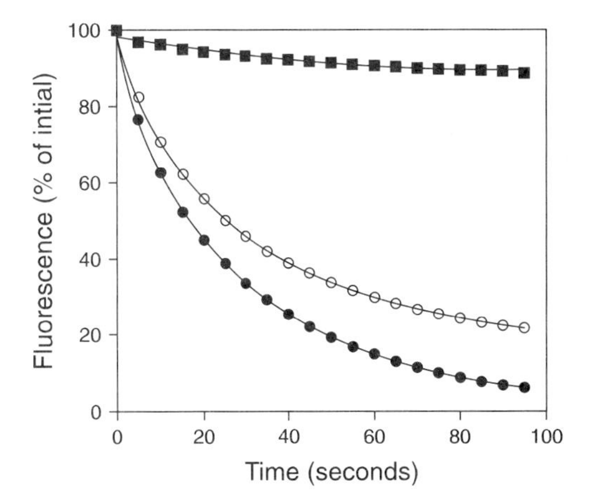

Figure 6.6 Photostability comparison for ELF- and fluorescein-labeled tubulin preparations. Tubulin in acetone-fixed CRE BAG 2 mouse fibroblasts was labeled with anti–β-tubulin monoclonal antibody (Boehringer Mannheim Corp.) and then detected using biotin-XX goat anti–mouse IgG (B-2763, see Section 7.2) in conjunction with either our ELF-97 Cytological Labeling Kit #2 (E-6603, ■) or fluorescein streptavidin (S-869, see Section 7.5; ●). Alternatively, anti-tubulin labeling was detected directly using fluorescein goat anti–mouse IgG (F-2761, see Section 7.2; ○). Photostablility of labeling produced by the three methods was compared by continously illuminating stained samples on a fluorescence microscope using Omega® Optical longpass optical filter sets O-5717 for fluorescein or O-5705 for ELF-97 signals (see Section 20.5). Images were acquired every 5 seconds using a Star 1™ CCD camera (Photometrics); the average fluorescence intensity in the field of view was calculated with Image-1® software (Universal Imaging Corp.) and expressed as a fraction of the initial intensity. Three data sets, representing different fields of view, were averaged for each conjugate to obtain the plotted time courses.

Spectral Characteristics of the ELF-97 Signal

Figure 6.5 shows the conversion of the water-soluble ELF-97 phosphatase substrate (E-6588, E-6589) into its hydrolysis product, the ELF-97 alcohol (E-6578). Under our reaction conditions, this highly insoluble planar molecule forms an intense yellow-green fluorescent precipitate at the site of phosphatase activity. Like other crystalline fluorescent molecules containing an intramolecular hydrogen bond, the ELF-97 alcohol precipitate provides a fluorescent signal that is not only extremely photostable but also has a very large Stokes shift.[5,6] In some applications, we have found the ELF-97 signal to be orders-of-magnitude more photostable than direct or indirect detection with fluorescein conjugates[7] (Figure 6.6), thereby allowing ample time to focus and photograph ELF-labeled tissue under high magnification. For example, in tissue samples stained with the ELF-97 Immunohistochemistry Kit, we have observed that about 40% of the ELF-97 signal remains after two hours of constant illumination from an epifluorescence microscope. Also, because the fluorescence emission of the ELF-97 alcohol precipitate is separated from its excitation maximum by over 100 nm (Figure 6.7), the ELF-97 signal can be clearly distinguished from most cell and tissue autofluorescence, which generally has a Stokes shift of much less than 100 nm.[7]

The extremely high Stokes shift of the ELF-97 alcohol precipitates makes the ELF-97 substrates ideal for use in multicolor appli-

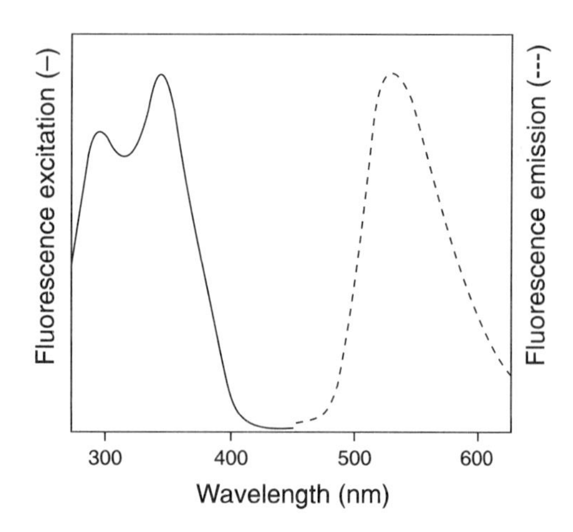

Figure 6.7 The normalized excitation and emission spectra of the ELF-97 alcohol precipitate (E-6578), which is generated by enzymatic cleavage of the soluble ELF-97 phosphatase substrate (E-6588, E-6589).

cations. The ELF-97 signal can be visualized simultaneously with blue fluorescent probes (such as Cascade Blue®–labeled secondary reagents or the DAPI and Hoechst counterstains) using a fluorescence microscope fitted with a standard DAPI/Hoechst longpass optical filter set. Because the ELF-97 alcohol precipitate and the blue fluorescent label have distinct emission spectra, the two signals can be easily distinguished.[7] In addition, the excitation spectra of tetramethylrhodamine (TMR) and Texas Red (TR) dyes are well separated from that of the ELF-97 alcohol precipitate. Thus, ELF-97 signals and the red fluorescence of TMR- or TR-labeled secondary reagents (or a red fluorescent counterstain such as propidium iodide) can be visualized sequentially with the appropriate optical filter sets without bleed-through.[7] Also, ELF-97 signals can be distinguished from fluorescein signals because their excitation wavelengths do not significantly overlap. For those researchers wishing to isolate the signal, we offer an ELF-97 optical filter set (O-5705; excitation = 365 ± 12.5 nm, emission ≥ 515 nm); see Section 26.5 for our complete selection of Omega Optical filters.

ELF-97 Kits for a Wide Variety of Applications

We are pleased to offer six ELF-97 Kits — five application-specific kits for secondary detection and one kit designed for detecting endogenous phosphatase activity in cells and tissues. A general scheme for the secondary detection methods used to develop the ELF-97 signal for *in situ* hybridization, cytological labeling and immunohistochemistry is shown in Figure 6.8. Please note that the components of these ELF-97 Kits are not interchangeable. Through the course of our product development, we have found that unique substrate formulations, buffers and protocols are required for optimal sensitivity in each type of biological application. We strongly recommend that each kit be used only for the applications for which it was developed. However, we also offer the ELF-97 phosphate (E-6588, E-6589) as well as the ELF-97 alcohol cleavage product (E-6578) as separate products for those researchers who want to develop their own applications. We have found that addition of 1–5 µM ELF-97 alcohol to the enzyme detection medium usually improves the quality of precipitation by reducing the crystal size. Addition of other components to the buffers may also be required to maximize the signal intensity and localize the signal to the target.

ELF-97 mRNA *In Situ* Hybridization Kits

The optimized reagents and protocols in our ELF-97 mRNA *In Situ* Hybridization Kits provide a rapid and sensitive assay for detecting mRNA *in situ* hybridization signal in cells and tissue sections. In conventional mRNA *in situ* hybridization, radioactively labeled DNA or RNA probes are hybridized to the experimental sample and then detected by applying a photosensitive emulsion to the microscope slides.[8] Typically, the emulsion is exposed for days to weeks before it is developed and photographed using white-light microscopy. In contrast, ELF-97 signals develop in minutes, producing a fluorescent precipitate that is significantly brighter than signals achieved either with directly labeled fluorescent nucleic acid probes or with hapten-labeled probes in combination with fluorophore-labeled secondary detection reagents.[2-4] Moreover, the fluorescent ELF-97 signals can be clearly distinguished from sample pigmentation, which often obscures the dark silver grains in the emulsion.

In addition to a detailed protocol, the ELF-97 mRNA *In Situ* Hybridization Kits include:

- Streptavidin–alkaline phosphatase conjugate (Kit #2 only)
- ELF wash, blocking and developing buffers
- Application-specific ELF-97 phosphatase substrate solution
- Hoechst 33342 nucleic acid counterstain
- ELF mounting medium
- 50 plastic coverslips

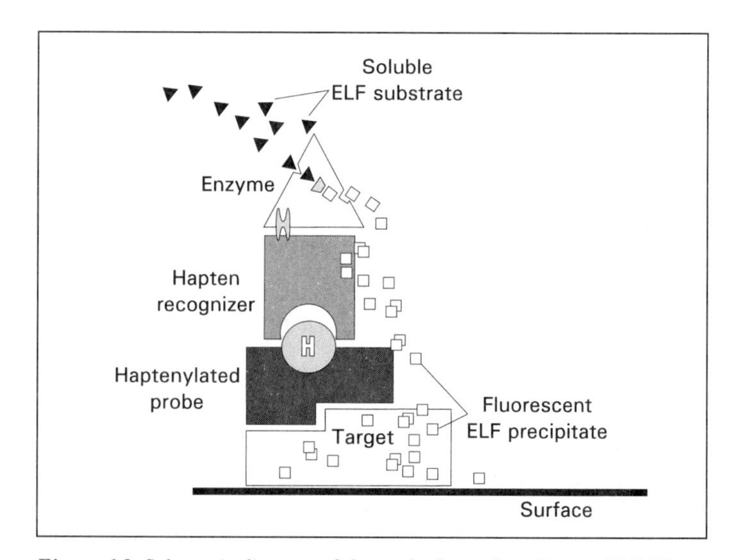

Figure 6.8 Schematic diagram of the methods employed in our ELF-97 mRNA In Situ Hybridization, Cytological Labeling and Immunohistochemistry Kits. Samples are probed with haptenylated or biotinylated target-specific probes such as antibodies or hybridization probes. Next, alkaline phosphatase conjugates of streptavidin or the hapten-specific probe are applied. Alternatively, a biotinylated antibody and biotinylated alkaline phosphatase can be used with standard bridging methods to increase the penetration in tissue, a method that is employed in our ELF-97 Immunohistochemistry Kit. The sample is then incubated with the ELF-97 phosphatase substrate, which forms an intense yellow-green fluorescent ELF-97 alcohol precipitate at the site of alkaline phosphatase activity.

The ELF-97 mRNA *In Situ* Hybridization Kit #2 (E-6605), which contains a streptavidin–alkaline phosphatase conjugate, can be used to detect biotinylated DNA or RNA probes. The ELF-97 mRNA *In Situ* Hybridization Kit #1 (E-6604), which does not include the streptavidin–alkaline phosphatase conjugate, is designed for use with other alkaline phosphatase conjugates — such as an anti-fluorescein–alkaline phosphatase conjugate — that have been applied to detect DNA or RNA probes conjugated to other haptens such as fluorescein. Each kit contains sufficient reagents for 50 slides or coverslips; some kit components are also available separately (see below or contact our Technical Assistance Department for details). Other useful reagents for *in situ* hybridization are described in Section 8.4.

ELF-97 Cytological Labeling Kits

Our ELF-97 Cytological Labeling Kits facilitate the detection of a broad range of cellular targets. Molecular Probes' researchers have used these ELF-97 Kits to stain both actin filaments and microtubules (see Color Plate 1 in "Visual Reality") and have found that the resolution of the ELF-97 signal is approximately equivalent to that achieved with direct fluorophore-labeled probes and secondary reagents but the intensity of the ELF-97 signal is about an order-of-magnitude greater. Labeling cell-surface receptors with the ELF-97 Cytological Labeling Kits also results in signals that are many times brighter and more photostable than those produced by conventional methods.[9] These versatile kits can potentially be used to detect any subcellular structure that can be selectively labeled with a biotinylated or haptenylated ligand (Figure 6.9). In addition to a detailed protocol, the ELF-97 Cytological Labeling Kits include:

- Streptavidin–alkaline phosphatase conjugate (Kit #2 only)
- ELF wash, blocking and developing buffers
- Application-specific ELF-97 phosphatase substrate solution plus additives
- ELF mounting medium

Like the ELF-97 mRNA *In Situ* Hybridization Kits, the ELF-97 Cytological Labeling Kits are available with (E-6603) or without (E-6602) the streptavidin–alkaline phosphatase conjugate. Enough reagents are included for 50 slides or coverslips; some kit components are also available separately (see below or contact our Technical Assistance Department for details).

ELF-97 Immunohistochemistry Kit

The ELF-97 substrate in our ELF-97 Immunohistochemistry Kit (E-6600) has been specially formulated to reduce nonspecific staining in immunohistochemical preparations. In addition to a detailed protocol, the kit contains the key reagents for detecting antigens in tissue sections, including streptavidin and biotinylated alkaline phosphatase. The streptavidin provided in the kit is used to link the biotinylated alkaline phosphatase with a biotinylated secondary antibody — a common immunohistochemical technique for optimizing tissue penetration. We have used the ELF-97 Immunohistochemistry Kit to characterize a number of antibodies generated against the zebrafish retina (see Color Plate 1 in "Visual Reality") and found that the ELF-97 staining pattern was identical to that seen with fluorophore-conjugated secondary reagents.[7] The ELF-97 signal could easily be visualized despite this tissue's considerable autofluorescence. Moreover, the staining was approximately

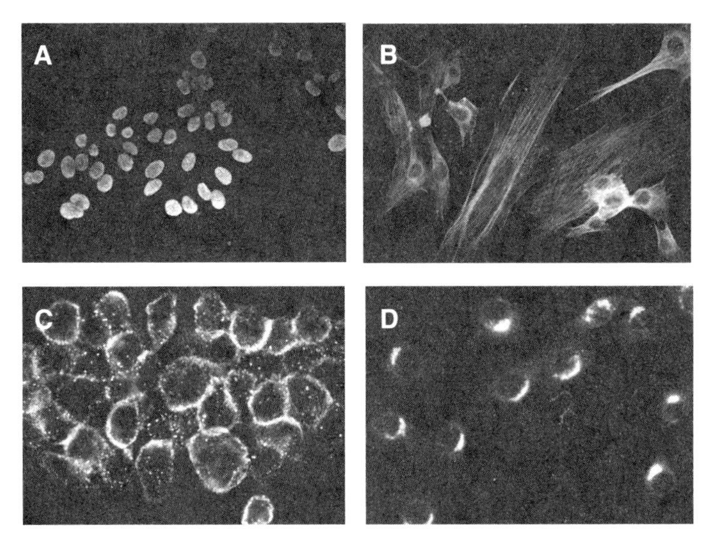

Figure 6.9 *Cellular targets developed for visualization with the reagents in our new ELF-97 Cytological Labeling Kits using the methods described in Figure 6.8. A) Nuclei in a commercial preparation of human epithelial cells (HEp-2) that have been labeled with human anti-nuclear antibodies and then incubated with biotin-XX goat anti–human IgG (B-6452, see Section 7.2). B) Acetone-fixed mouse fibroblast (CRE BAG 2) cells that have been treated with Triton® X-100 and then incubated with biotin-XX phalloidin (B-7474, see Section 11.1), a probe specific for actin stress fibers. C) Formaldehyde-fixed human carcinoma cells that have been incubated with biotin-XX epidermal growth factor (E-3477, see Section 17.3). D) Mouse fibroblast cells that have been probed with antibodies directed against rat medial Golgi cisternae (a gift from Vivek Malhotra; University of California, San Diego) and then incubated with biotin-XX goat anti–rabbit IgG (B-2770, see Section 7.2). In each case, the biotinylated probe was detected with a streptavidin–alkaline phosphatase conjugate followed by incubation with the ELF-97 phosphatase substrate.*

500 times more photostable than that produced by fluorescein-labeled secondary reagents.[7]

The ELF-97 Immunohistochemistry Kit provides enough reagents to prepare 50 mL of detection solution, which is sufficient for staining 250 to 1000 sections. Included in the ELF-97 Immunohistochemistry Kit are:

- Streptavidin
- Biotinylated alkaline phosphatase
- Application-specific ELF-97 phosphatase substrate solution
- ELF reaction buffer
- ELF mounting medium

Some kit components are also available separately (see below or contact our Technical Assistance Department for details). We also offer biotinylated secondary antibodies, including goat anti-mouse (B-2763), anti-rabbit (B-2770) and anti-human (B-6452) antibodies and donkey anti-sheep (B-6453) antibodies for use in this application; see Section 7.2 for a description of these products.

ELF-97 Endogenous Phosphatase Detection Kit

Molecular Probes' scientists have used the ELF-97 phosphatase substrate to develop a novel fluorescence-based assay for detecting phosphatases in tissue sections;[9] see Color Plate 1 in "Visual Reality." Phosphatases have been commonly used as enzyme markers, allowing researchers to identify primordial germ cells,[10] to distinguish subpopulations of bone marrow stromal cells [11] and to investigate *in vitro* differentiation in carcinoma cell lines.[12-14] In contrast to the time-consuming Gomori technique,[15] the simple protocol that accompanies our ELF-97 Endogenous Phosphatase Detection Kit (E-6601) takes only minutes to perform and does not require unpleasant reagents such as ammonium sulfide. We have successfully used the ELF-97 Endogenous Phosphatase Detection Kit to localize phosphatase activity in several different zebrafish tissues. The patterns of ELF-97 staining in intestine, kidney, ovary and gills were essentially identical to the patterns of black precipitate produced by the conventional Gomori technique. Although optimally excited with UV light, the ELF-97 precipitate can also reportedly be excited with the 488 nm spectral line of the argon-ion laser, making it compatible with flow cytometry and confocal laser scanning microscopy.[16] Using ELF-97 phosphate and confocal laser scanning microscopy, researchers have developed a semi-automated method for analyzing the position within a regenerating newt limb of transfected cells expressing the alkaline phosphatase reporter gene.[16] Some of the new ELF-97 substrates that are described in Chapter 10 may have similar utility for detecting enzymatic activity in live or fixed cells or in nondenaturing gels. In addition to a simple protocol, the ELF-97 Endogenous Phosphatase Detection Kit provides enough reagents to stain 50 to 250 tissue sections, including:

- Application-specific ELF-97 phosphatase substrate solution
- ELF detection buffer
- ELF mounting medium

Some kit components are also available separately (see below or contact our Technical Assistance Department for details).

Accessories for the ELF-97 Kits

ELF Spin Filters

Like many enzyme substrates, the ELF-97 developing solution must be filtered before use for optimal staining. Our ELF spin filters (E-6606) — spin-filtration devices with a pore size of 0.2 µm — are both convenient and efficient, allowing a very small volume to be filtered without significant loss of sample. These spin filters are recommended for use with all of our ELF-97 Kits.

ELF-97 Optical Filter Set

ELF-97 staining can be visualized using a fluorescence microscope equipped with a standard Hoechst/DAPI longpass optical filter set or with the ELF-97 optical filter set (excitation = 365 ± 12.5 nm, emission ≥515 nm; O-5705). Molecular Probes offers high-quality Omega Optical filter sets that are optimized to match the spectral properties of our dyes. For further information on our complete line of Omega Optical filter sets, see Section 26.5.

ELF-97 Kit Components

Our ELF-97 Kits provide sufficient amounts of each component to carry out the extensively tested protocols for which they were designed. However, it is possible that extra buffers will be required when modifying or optimizing our procedures for a particular application. Therefore, we also offer many of the ELF-97 Kit components individually for those researchers who need to replenish their supply. Please contact our Technical Assistance Department for availability. The solutions of pure ELF-97 phosphate (E-6588, E-6589) cannot be substituted for the substrate reagent contained in the kits because these reagents contain additional additives that are essential to each application.

We also offer the streptavidin–alkaline phosphatase conjugate (S-921) used in our ELF-97 Cytological Labeling and mRNA *In Situ* Hybridization Kits and the biotin-XX–alkaline phosphatase conjugate (A-927) used in the ELF-97 Immunohistochemistry Kit. However, because these enzyme conjugates are not supplied at the same concentration or specific activity as those included in the kits and have not been tested with the ELF-97 phosphatase substrate, the amount required to achieve optimal ELF-97 staining must be determined empirically.

1. U.S. Patent Nos. 5,316,906 and 5,443,986; EP 0 641 351; and patents pending; **2.** Am J Human Genet Suppl 55, A271, abstract #1585 (1994); **3.** FASEB J 8, A1444, abstract #1081 (1994); **4.** Mol Biol of the Cell Suppl 4, 226a, abstract #1313 (1993); **5.** Optics Comm 64, 457 (1987); **6.** J Phys Chem 74, 4473 (1970); **7.** J Histochem Cytochem 43, 77 (1995); **8.** Pardue, M.L. in *Nucleic Acid Hybridization: A Practical Approach*, B.D. Hames and S.J. Higgens, Eds., IRL Press (1985) pp. 179–202; **9.** Mol Biol of the Cell Suppl 4, 226a, abstract #1312 (1993); **10.** Anatomical Record 118, 135 (1954); **11.** J Histochem Cytochem 40, 1059 (1992); **12.** Devel Biol 88, 279 (1981); **13.** Cell 5, 229 (1975); **14.** Proc Natl Acad Sci USA 70, 3899 (1973); **15.** Pearse, A.G.E. in *Histochemistry: Theoretical and Applied, Third Edition*, Baltimore, Williams & Wilkins (1968); **16.** J Histochem Cytochem 44, 559 (1996).

6.3 Data Table *Enzyme-Labeled Fluorescence (ELF) Signal Amplification Technology*

Cat #	Structure	MW	Storage	Soluble	Abs	{ε × 10⁻³}	Em	Solvent	Notes
E-6578	S10.3	307	L	DMSO*	345	ND	530	pH 8	<1, 2>
E-6588	S10.3	431	F,D,L	H_2O*	289	{12}	<3>	pH 10	<4>

E-6589 see E-6588

For definitions of the contents of this data table see "How to Use the Handbook" on page vi.
Structure: Chemical structure drawings are shown in Figure 6.5 and in Section 10.3.

* This product is packaged as a solution in the solvent indicated in "Soluble."
<1> ND = not determined.
<2> ELF-97 alcohol is essentially insoluble in water. Spectral maxima listed are for an aqueous suspension; for this reason the value of {ε} cannot be determined.
<3> Fluorescence of the unhydrolyzed substrate is very weak.
<4> Enzymatic cleavage of the phosphate group yields E-6578.

6.3 Price List *Enzyme-Labeled Fluorescence (ELF) Signal Amplification Technology*

Cat #		Product Name	Unit Size	Price Per Unit ($) 1–4 Units	5–24 Units
A-927		alkaline phosphatase, biotin-XX conjugate *5 mg/mL*	200 µL	95.00	76.00
E-6578	**New**	ELF®-97 alcohol *1 mM solution in DMSO*	1 mL	24.00	19.20
E-6602	**New**	ELF®-97 Cytological Labeling Kit #1 *50 assays*	1 kit	148.00	118.40
E-6603	**New**	ELF®-97 Cytological Labeling Kit #2 *with streptavidin, alkaline phosphatase conjugate* *50 assays*	1 kit	198.00	158.40
E-6601	**New**	ELF®-97 Endogenous Phosphatase Detection Kit	1 kit	135.00	108.00
E-6600	**New**	ELF®-97 Immunohistochemistry Kit	1 kit	198.00	158.40
E-6604	**New**	ELF®-97 mRNA *In Situ* Hybridization Kit #1 *50 assays*	1 kit	175.00	140.00
E-6605	**New**	ELF®-97 mRNA *In Situ* Hybridization Kit #2 *with streptavidin, alkaline phosphatase conjugate* *50 assays*	1 kit	225.00	180.00
E-6589	**New**	ELF®-97 phosphatase substrate (ELF®-97 phosphate) *5 mM solution in water* *contains 2 mM azide*	1 mL	150.00	120.00
E-6588	**New**	ELF®-97 phosphatase substrate (ELF®-97 phosphate) *5 mM solution in water* *sterile filtered*	1 mL	160.00	128.00
E-6606	**New**	ELF® spin filters *20 filters*	1 box	48.00	38.40
S-921		streptavidin, alkaline phosphatase conjugate *2 mg/mL*	0.5 mL	155.00	124.00

6.4 Phycobiliproteins

Phycobiliproteins are a family of reasonably stable and highly soluble fluorescent proteins derived from cyanobacteria and eukaryotic algae. These proteins contain covalently linked tetrapyrrole groups that play a biological role in collecting light and, through fluorescence resonance energy transfer, conveying it to a special pair of chlorophyll molecules located in the photosynthetic reaction center.[1] Because of their role in light collection, phycobiliproteins have evolved to maximize both absorption and fluorescence and to minimize the quenching caused either by internal energy transfer or by external factors such as changes in pH or ionic composition.[2,3]

Spectral Characteristics of Phycobiliprotein

The phycobiliproteins B-phycoerythrin (B-PE), R-phycoerythrin (R-PE) and allophycocyanin (APC) are among the best dyes currently available for applications that require either high sensitivity or simultaneous multicolor detection.[4-7] Quantum yields of up to 0.98 and extinction coefficients of up to 2.4 million have been reported for these fluorescent proteins (Table 6.6). On a molar basis, the fluorescence yield is equivalent to at least 30 unquenched fluorescein or 100 rhodamine molecules at comparable wavelengths. The fluorescence of a single molecule of B-PE has been detected.[8,9] In practical applications such as flow cytometry and immunoassays,[10,11] the sensitivity of B-PE– and R-PE–conjugated antibodies is five to 10 times greater than that of the corresponding fluorescein conjugate.[12,13] Using R-PE–conjugated streptavidin, researchers have detected fewer than 100 receptor-bound biotinylated antibodies per cell by flow cytometry.[14]

Figures 6.10 and 6.11 show the spectra for B-PE, R-PE and APC. R-PE can be excited efficiently at 488 nm either with an argon-ion laser or with a broadband illumination source (xenon- or mercury-arc lamps) and a standard fluorescein optical filter set. With the proper emission filters, fluorescein (or any of the fluorescein substitutes described in Section 1.4) and R-PE can be simultaneously detected at approximately 520 nm and at greater than 575 nm, respectively, making R-PE conjugates ideal for multicolor flow cytometry applications. One of the fluorescent microsphere suspensions in our new CompenFlow™ Flow Cytometry Compensation Kit (C-7301, see Section 26.2) has spectra that almost exactly match those of the phycoerythrins. This microsphere suspension is designed to help flow cytometry operators set up compensation circuits that properly remove unwanted phycoerythrin signals from secondary channels. Conjugates prepared from APC are ideal for use with the 633 nm spectral line of the red He-Ne laser,[10,15] or the 647 nm spectral line of the krypton-ion laser.

Table 6.6 *Absorption spectra for B-PE, R-PE and APC.*

Cat #	Phycobiliprotein	Molecular Weight	Absorption Max (nm)	ϵ (cm^{-1}M^{-1})	Emission Max (nm)	Fluorescence QY
P-800	B-Phycoerythrin	240,000	546, 565	2,410,000	575	0.98
P-801	R-Phycoerythrin	240,000	480, 546, 565	1,960,000	578	0.82
A-803, A-819	Allophycocyanin	104,000	650	700,000	660	0.68

ϵ = extinction coefficient. QY = quantum yield.

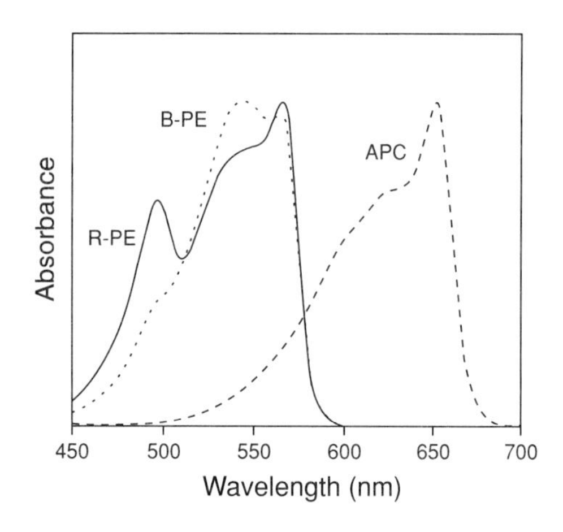

Figure 6.10 *Absorption spectra for B-PE, R-PE and APC.*

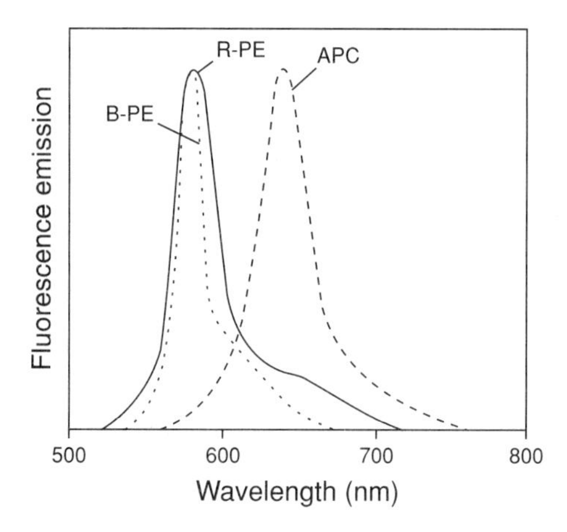

Figure 6.11 *Normalized emission spectra for B-PE, R-PE and APC.*

Pure Phycobiliproteins

Molecular Probes was the first company to make the phycobiliproteins available for research, and we continue to supply many of the phycobiliproteins used in commercial applications.[16] Bulk quantities of B-PE (P-800), R-PE (P-801), APC (A-803), chemically crosslinked APC (A-819) or their conjugates are available at a considerable discount.

Phycobiliproteins undergo some loss of fluorescence upon freezing. The pure proteins are shipped in ammonium sulfate suspension and are stable for at least one year when stored at 4°C. The conjugates and modified derivatives are shipped in solutions containing sodium azide to inhibit bacterial growth and typically have a useful life of more than six months. *All phycobiliproteins and their derivatives should be stored refrigerated, never frozen.*

Phycobiliprotein Conjugates

Reactive Phycobiliprotein Derivative

Conjugates of R-phycoerythrin with other proteins are generally prepared with the pyridyldisulfide derivative of R-PE (P-806). This derivative can be directly reacted with thiolated antibodies, enzymes and other biomolecules to form a disulfide linkage. More commonly, the pyridyldisulfide groups in this derivative are first reduced to thiols, which are then reacted with maleimide- or iodoacetamide-derivatized proteins (see Figure 5.1 in Chapter 5). Because the pyridyldisulfide derivative of R-PE is somewhat unstable, we recommend using it within three months of receipt. Phycobiliproteins can be conveniently crosslinked to other proteins using the reagents and protocol provided in our Protein–Protein Crosslinking Kit (P-6305, see Section 5.2), which is described in detail in Chapter 5.

Phycobiliprotein-Labeled Secondary Detection Reagents

Molecular Probes regularly prepares R-PE conjugates of goat anti-mouse (P-852) and goat anti-rabbit (P-2771) antibodies and NeutraLite avidin (A-2660), as well as both B-PE and R-PE conjugates of streptavidin (S-867, S-866). Because allophycocyanin tends to dissociate into subunits when highly diluted, Molecular Probes prepares its APC conjugates of goat anti-mouse antibody (A-865) and streptavidin (S-868) from chemically crosslinked APC (A-819), a protein complex that does not dissociate, even in strongly chaotropic salts.[17] In addition, biotinylated R-PE (P-811) can be used with standard avidin/streptavidin bridging techniques to detect biotinylated molecules.[18]

Custom Phycobiliprotein Conjugates

Molecular Probes has carried out hundreds of successful conjugations with phycobiliproteins, beginning soon after their use was disclosed in 1982.[7] We are experts in doing custom conjugations of phycobiliproteins to antibodies and other proteins and welcome your inquiries for specific conjugates. For more information or a quotation, please contact Molecular Probes' Custom and Bulk Sales Department.

1. J Fluorescence 1, 135 (1991); **2.** J Biol Chem 264, 1 (1989); **3.** Meth Enzymol 167, 291 (1988); **4.** J Immunol Meth 112, 279 (1988); **5.** Proc Natl Acad Sci USA 85, 7312 (1988); **6.** Trends Biochem Sci 9, 423 (1984); **7.** J Cell Biol 93, 981 (1982); **8.** Proc Natl Acad Sci USA 86, 4087 (1989); **9.** Anal Chem 59, 2158 (1987); **10.** Anal Lett 24, 1075 (1991); **11.** J Histochem Cytochem 39, 921 (1991); **12.** Eur Biophys J 15, 141 (1987); **13.** Clin Chem 29, 1582 (1983); **14.** J Immunol Meth 135, 247 (1990); **15.** Cytometry 15, 267 (1994); **16.** Molecular Probes is licensed to produce phycobiliproteins and their conjugates under patents owned by Stanford University, including U.S. Patent Nos. 4,520,110; 4,859,582; and 5,055,556. Phycobiliproteins and their conjugates are not available for resale or for incorporation into products for sale unless the purchaser is a licensee of these patents; **17.** Cytometry 8, 91 (1987); **18.** J Biol Chem 265, 15776 (1990).

6.4 Price List *Phycobiliproteins*

Cat #	Product Name	Unit Size	Price Per Unit ($) 1–4 Units	5–24 Units
A-803	allophycocyanin *4 mg/mL*	0.5 mL	95.00	76.00
A-819	allophycocyanin, crosslinked (APC-XL) *4 mg/mL*	250 µL	198.00	158.40
A-865	allophycocyanin, crosslinked, goat anti-mouse IgG (H+L) conjugate *1 mg/mL*	0.5 mL	195.00	156.00
A-2660	avidin, NeutraLite™, R-phycoerythrin conjugate *1 mg/mL*	1 mL	145.00	116.00
P-800	B-phycoerythrin *4 mg/mL*	0.5 mL	95.00	76.00
P-801	R-phycoerythrin *4 mg/mL*	0.5 mL	95.00	76.00
P-811	R-phycoerythrin, biotin-XX conjugate *4 mg/mL*	0.5 mL	135.00	108.00
P-852	R-phycoerythrin goat anti-mouse IgG (H+L) conjugate *1 mg/mL*	1 mL	165.00	132.00
P-2771	R-phycoerythrin goat anti-rabbit IgG (H+L) conjugate *1 mg/mL*	0.5 mL	95.00	76.00
P-806	R-phycoerythrin, pyridyldisulfide derivative *2 mg/mL*	1 mL	135.00	108.00
S-868	streptavidin, allophycocyanin, crosslinked, conjugate *1 mg/mL*	0.5 mL	285.00	228.00
S-867	streptavidin, B-phycoerythrin conjugate *1 mg/mL*	1 mL	145.00	116.00
S-866	streptavidin, R-phycoerythrin conjugate *1 mg/mL*	1 mL	145.00	116.00

FluoReporter® Protein and Oligonucleotide Labeling Kits

Product (Catalog Number)	Features	Contents
FluoReporter Protein Labeling Kits • FITC (F-6434) • Fluorescein-EX (F-6433) • Oregon Green™ 488 (F-6153) • Oregon Green™ 500 (F-6154) • Oregon Green™ 514 (F-6155) • Rhodamine Red™-X (F-6161) • Tetramethylrhodamine (F-6163) • Texas Red®-X (F-6162)	The FluoReporter Protein Labeling Kits facilitate research-scale preparation of protein conjugates labeled with some of our best dyes. Typically, labeling and purifying conjugates with the FluoReporter Protein Labeling Kits can be completed in under three hours, with very little hands-on time. Each FluoReporter Protein Labeling Kit provides sufficient reagents for 5 to 10 labeling reactions of 0.2 to 2 mg of protein each.	• Five vials of the amine-reactive dye • Anhydrous DMSO • Reaction tubes, each containing a stir bar • Stop reagent • Ten spin columns • Collection tubes
FluoReporter Biotin-XX Protein Labeling Kit (F-2610)	This kit is designed for five biotinylation reactions, each with 5 to 20 mg of protein; up to 100 mg of protein may be labeled. A gel filtration column is provided for purifying the labeled proteins from excess biotin reagent. Once purified, the degree of biotinylation can be determined using the included avidin–biotin displacement assay; biotinylated goat IgG is provided as a control.	• Biotin-XX, succinimidyl ester • Anhydrous DMSO • A gel filtration column • Avidin–HABA complex • Biotinylated goat IgG
FluoReporter Mini-Biotin-XX Protein Labeling Kit (F-6347)	This kit permits efficient biotinylation of small amounts of antibodies or other proteins. The water-soluble biotin-XX sulfosuccinimidyl ester has a 14-atom spacer that enhances the binding of biotin derivatives to avidin's relatively deep binding sites. The ready-to-use spin columns provide a convenient method of purifying the biotinylated protein from excess reagents. Sufficient reagents are provided for five biotinylation reactions of 0.1 to 3 mg each.	• Biotin-XX, sulfosuccinimidyl ester • Reaction tubes, each containing a stir bar • Five spin columns • Collection tubes • Dialysis tubing
FluoReporter Biotin/DNP Protein Labeling Kit (F-6348)	The degree of biotinylation of proteins labeled with DNP-X–biocytin-X succinimidyl ester can be assessed from the optical absorbance of DNP (ϵ = 15,000 $cm^{-1}M^{-1}$ at ~360 nm). The conjugates are recognized by both avidin derivatives and anti-DNP antibodies, permitting a choice of detection techniques. Sufficient reagents are supplied for 5 to 10 labeling reactions of 0.2 to 2 mg of protein each.	• DNP-X–biocytin-X, succinimidyl ester • Anhydrous DMSO • Reaction tubes • Stop reagent • Ten spin columns • Collection tubes
FluoReporter Oligonucleotide Amine Labeling Kits • Biotin-XX (F-6081) • BODIPY® FL (F-6079) • BODIPY® FL-X (F-6082) • BODIPY® R6G (F-6092) • BODIPY® TMR-X (F-6083) • BODIPY® 564/570 (F-6093) • BODIPY® 581/591 (F-6094) • BODIPY® TR-X (F-6084) • DNP-X (F-6085) • Fluorescein-X (F-6086) • Oregon Green™ 488 (F-6087) • Rhodamine Green™-X (F-6088) • Rhodamine Red™-X (F-6089) • Tetramethylrhodamine (F-6090) • Texas Red®-X (F-6091)	The FluoReporter Oligonucleotide Amine Labeling Kits permit the easy preparation of labeled oligonucleotides by reacting amine-derivatized oligonucleotides with a wide selection of our amine-reactive succinimidyl esters. The amine-reactive haptens and fluorophores in most of our 15 different FluoReporter Oligonucleotide Amine Labeling Kits contain aminohexanoic spacers ("X") to reduce the label's interaction with the oligonucleotide and enhance its accessibility to secondary detection reagents. The protocol has been optimized for labeling 5′-amine–modified oligonucleotides, 18 to 24 bases in length. Shorter or longer oligonucleotides may be labeled by the same procedure; however, adjustments to the protocol may be necessary. Sufficient reagents are provided in each kit for five complete labeling reactions of 100 µg of oligonucleotide each. The conjugates can be purified with our FluoReporter Labeled Oligonucleotide Purification Kit (F-6100, see below).	• Five vials of the amine-reactive label • Anhydrous DMSO • Labeling buffer • A detailed protocol for labeling
FluoReporter Oligonucleotide Phosphate Labeling Kits • Biotin-X-C_5 (F-6095) • BODIPY® FL-C_5 (F-6096) • BODIPY® TMR-C_5 (F-6097) • Rhodamine Red™-C_2 (F-6098) • Texas Red®-C_5 (F-6099)	These kits use proprietary coupling technology to link aliphatic amines to 5′-phosphate–terminated oligonucleotides to form phosphoramidate adducts. Unphosphorylated oligonucleotides can be enzymatically phosphorylated using T4 polynucleotide kinase prior to use of these labeling kits. Sufficient reagents are provided in each kit for five complete labeling reactions of 100 µg oligonucleotide each. The conjugates can be purified with our FluoReporter Labeled Oligonucleotide Purification Kit (F-6100, see below).	• Five vials of the phosphate-reactive label • Anhydrous DMSO • Labeling buffer • A detailed protocol for labeling
FluoReporter Labeled Oligonucleotide Purification Kit (F-6100)	The crude, labeled oligonucleotide is simply precipitated with ethanol to remove the excess reactive reagent, adsorbed on a spin column, washed to remove any unconjugated oligonucleotide then eluted with an elution buffer to yield the conjugate. Isolated yields for the combined conjugation and purification steps are usually >50% and the products are typically 90% pure as determined by HPLC. Conjugates can be used for most procedures without additional purification.	• Five spin columns • Buffers for column equilibration, loading, washing and elution • A detailed protocol

Chapter 7

Protein Conjugates for Biological Detection

Contents

Technical Notes and Product Highlights

Related Chapters

7.1 A Wide Variety of Protein Conjugates

The quality of the conjugate depends to a large degree on the quality of the protein from which it is made as well as the spectral properties of the fluorophore. Molecular Probes endeavors to use the highest-quality proteins in its conjugates. In addition, our dyes and conjugation methods yield conjugates that are typically brighter than other commercially available conjugates, yet have low background and better spectral resolution. Moreover, all our fluorescent secondary antibody, avidin and streptavidin conjugates are tested on real cell samples to ensure low nonspecific binding and high specific staining. Tables 7.1 and 7.2 list our current offerings of fluorescent and biotinylated secondary immunoreagents and avidins. We also offer anti-dye antibodies as well as biotinylated enzymes for use in diverse detection schemes.

Molecular Probes prepares protein conjugates of a wide variety of fluorophores, ranging from the blue fluorescent Cascade Blue® and new AMCA-S dyes to the red fluorescent Texas Red® dye and phycobiliproteins. Properties of the low molecular weight dyes that we use to prepare our conjugates are described in detail in Chapter 1. In particular we would like to highlight our:

- **Oregon Green™ conjugates.** Our new green fluorescent Oregon Green conjugates may soon replace fluorescein-labeled secondary reagents as the standard tools in fluorescence-based immunoassays and *in situ* hybridization applications. The Oregon Green 488 dye has excitation and emission spectra virtually identical to those of fluorescein yet offers greater photostability and a fluorescence signal that is essentially independent of pH above pH 6. The Oregon Green 500 and Oregon Green 514 dyes are even more photostable (Figure 7.1, see also Figure 1.6 in Chapter 1 and Figure 11.1 in Chapter 11). Furthermore, the decreased tendency of the Oregon Green dyes to quench their fluorescence upon protein conjugation allows us to prepare conjugates that are more fluorescent than fluorescein conjugates (see Figure 1.5 in Chapter 1).

- **Rhodol Green™ conjugates.** The Rhodol Green fluorophore is efficiently excited by the 488 nm spectral line of the argon-ion laser, and its conjugates have greater photostability and lower pH sensitivity than fluorescein conjugates (see Figure 23.3 in Chapter 23).

- **BODIPY® conjugates.** We prepare a large number of reactive BODIPY dyes (see Section 1.2) and have chosen two of these — BODIPY FL and BODIPY TMR — as preferred substitutes for fluorescein and tetramethylrhodamine, respectively. Unlike fluorescein's fluorescence, the green fluorescence of BODIPY FL conjugates is pH independent. BODIPY TMR conjugates are typically more fluorescent than tetramethylrhodamine conjugates. Both dyes have narrow emission spectra, making them particularly useful for multicolor applications (see Figure 1.4 in Chapter 1).

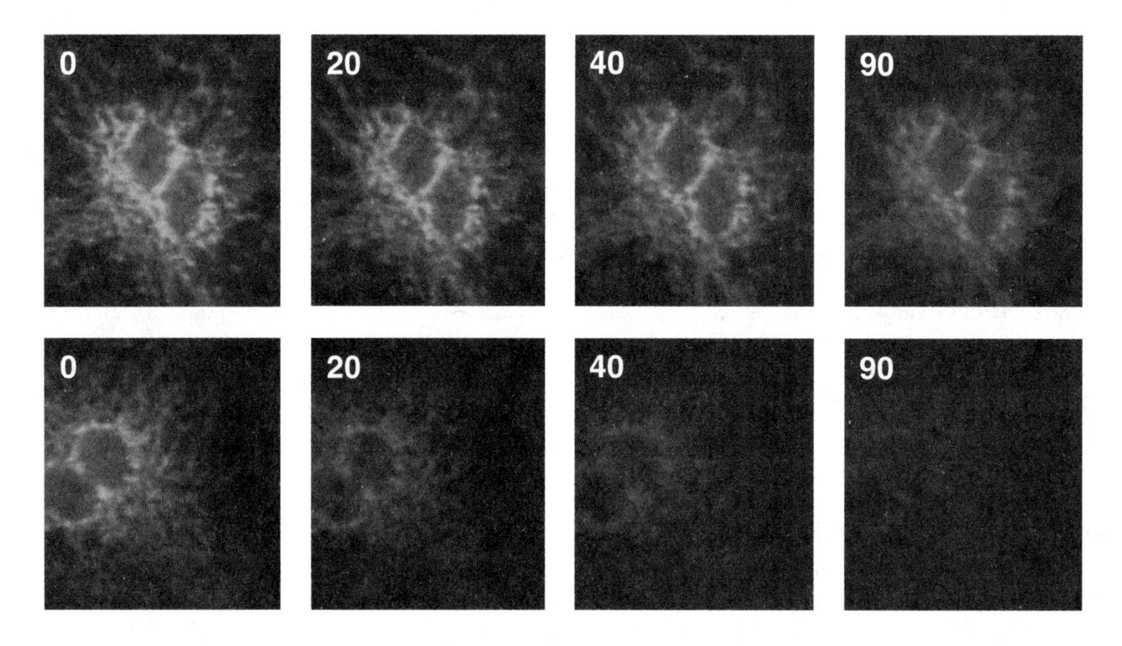

Figure 7.1 *Comparison of photostability of immunofluorescent staining by Oregon Green 514 goat anti–mouse IgG (O-6383, upper series) and by fluorescein goat anti–mouse IgG (F-2761, lower series); see Section 7.2 for a complete list of fluorescent secondary antibodies. Bovine pulmonary arterial endothelial cells (BPAEC) were fixed with formaldehyde and permeabilized in cold acetone. Following blocking in 1% BSA, 1% normal goat serum, 0.1% Tween® 20 in PBS, samples were incubated for one hour with 60 µg/mL mouse monoclonal anti–human cytochrome oxidase subunit I antibody (A-6403, see Section 7.4), after which they were rinsed and incubated with fluorescent anti-mouse secondary antibodies at 10 µg/mL for 30 minutes. Samples were continuously illuminated and viewed on a fluorescence microscope using an Omega® Optical fluorescein longpass filter set (O-5717, see Section 26.5), a Star 1™ CCD camera (Photometrics) and Image-1® software (Universal Imaging Corp.). Images acquired 0, 20, 40 and 90 seconds after the start of illumination (as indicated in the top left hand corner of each panel) clearly demonstrate the superior photostability of the Oregon Green 514 conjugate.*

- **Fluorescein conjugates.** Although we feel that our new Oregon Green dyes will provide superior performance in most applications, we continue to provide high-quality fluorescein conjugates for those researchers who prefer using fluorescein in their applications. Molecular Probes has developed a reactive fluorescein that typically yields conjugates with significantly greater fluorescence than other commercially available fluorescein-labeled proteins. Figure 1.5 in Chapter 1 shows the fluorescence intensity of IgG labeled in the traditional manner using FITC, compared with that of IgG labeled using Molecular Probes' unique fluorescein-EX succinimidyl ester (F-6130, see Section 1.3). As can be seen, labeling with the fluorescein-EX reagent ensures that a greater signal is obtained for each IgG-bound fluorescein. Protein conjugates prepared from succinimidyl esters of fluorescein also have higher chemical stability than those prepared from the isothiocyanate FITC.

- **Eosin-labeled secondary reagents.** Our yellow-green fluorescent eosin-labeled secondary reagents provide researchers with an efficient probe for singlet oxygen–mediated photooxidation of diaminobenzidine (DAB), thereby generating a signal that can be detected by light or electron microscopy; see the Technical Note "Fluorescent Probes for Photoconversion of Diaminobenzidine" on page 264.

- **Red fluorescent Rhodamine Red™-X and Texas Red-X conjugates.** Molecular Probes uses the succinimidyl esters of our proprietary Rhodamine Red-X and Texas Red-X fluorophores to prepare several new detection reagents. The aminohexanoyl spacer ("X") apparently lessens the quenching that sometimes occurs when fluorescent dyes are conjugated to proteins. We have found that some of our new Rhodamine Red-X and Texas Red-X protein conjugates are about twice as fluorescent as the corresponding conjugates prepared from Lissamine™ rhodamine B sulfonyl chloride and Texas Red sulfonyl chloride (see Figure 1.11 in Chapter 1), thus providing a better signal-to-noise ratio. We continue to supply most of our original Texas Red conjugates for those customers who have developed protocols using these products. Texas Red and Texas Red-X conjugates emit at wavelengths that have little overlap with the fluorescence of fluorescein or phycoerythrin and are particularly useful for multicolor applications.[1-4] Rhodamine Red-X conjugates have maximal absorption at ~570 nm, making them the preferred probes for excitation by the 568 nm spectral line of the Ar-Kr laser used in some confocal laser scanning microscopes.

- **Phycobiliprotein conjugates.** In addition to our selection of immunoreagents labeled with organic dyes, Molecular Probes prepares phycobiliprotein-labeled secondary reagents. The fluorescence yield of these red fluorescent proteins is theoretically equivalent to at least 30 fluorescein or 100 rhodamine molecules at comparable wavelengths. Because of their exceptional fluorescence, phycobiliprotein-labeled detection reagents have been used extensively in flow cytometry to detect cell-specific expression of surface antigens.[5-7] Researchers have used a phycobiliprotein-conjugated antibody to detect interleukin-4 in a microplate assay and found that it was the only tested fluorophore that produced adequate signal.[8] We also offer allophycocyanin-labeled streptavidin — one of the few probes that can be excited by the 633 nm spectral line of the He-Ne laser.[9] Section 6.4 discusses the spectral properties of phycobiliproteins in more detail.

- **Cascade Blue and AMCA-S conjugates.** Although less frequently used because of their spectral overlap with sample autofluorescence and their generally lower fluorescence yields, blue fluorescent fluorophores remain important for multicolor applications such as fluorescence *in situ* hybridization. The brightest UV-excitable dyes are Molecular Probes' patented Cascade Blue dyes [10] and 7-amino-4-methylcoumarin-3-acetic acid (AMCA). We have found that protein conjugates of our new sulfonated AMCA derivative, which we call AMCA-S, are

Superior Antifade Reagents for a Multitude of Applications

Molecular Probes offers a variety of antifade reagents to meet the diverse needs of the research community. The antifade formulation in our original *SlowFade*™ Antifade Kit (S-2828) reduces the fading rate of fluorescein to almost zero, making this antifade reagent especially useful for quantitative measurements. However, this *SlowFade* formulation does initially quench fluorescein's fluorescence. Our new *SlowFade Light* Antifade Kit (S-7461) contains an antifade formulation that slows fluorescein's photobleaching by about fivefold with little or no quenching of the fluorescent signal, thereby dramatically increasing the signal-to-noise ratio in photomicroscopy. The latest addition to our antifade arsenal — ProLong™ Antifade Kit (P-7481) — outperforms all other commercially available reagents, with little or no quenching of the fluorescent signal. In addition to inhibiting the fading of fluorescein, tetramethylrhodamine and Texas Red® dyes, the ProLong antifade reagent retards the fading of DNA-bound nucleic dyes such as DAPI, propidium iodide and YOYO®-1. Its compatibility with a multitude of dyes and dye complexes makes the Prolong Antifade Kit an especially valuable tool for multiparameter analyses such as fluorescence *in situ* hybridization (FISH). For more information on these antifade reagents, see Section 26.1.

Cat #		Product Name	Unit Size	Price Per Unit ($) 1–4 Units	5–24 Units
P-7481	*New*	ProLong™ Antifade Kit	1 kit	98.00	78.40
S-2828		*SlowFade*™ Antifade Kit	1 kit	55.00	44.00
S-7461	*New*	*SlowFade*™ *Light* Antifade Kit	1 kit	55.00	44.00

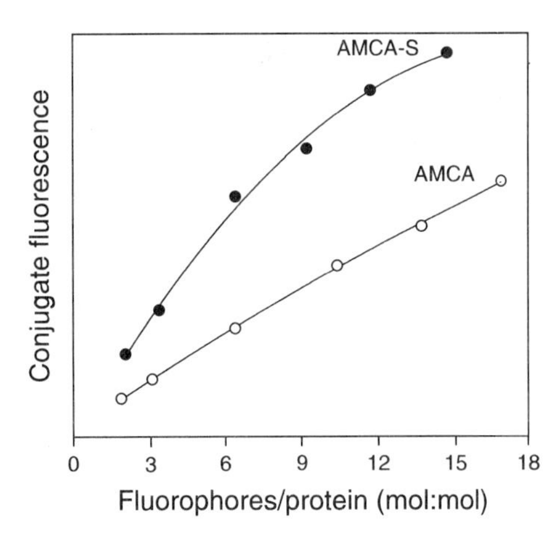

Figure 7.2 *Comparison of relative fluorescence of streptavidin conjugates of 7-amino-4-methylcoumarin-3-acetic acid, succinimidyl ester (AMCA, ○) and the sulfonated derivative AMCA-S (●). Conjugate fluorescence is determined by measuring the fluorescence quantum yield of the conjugated dye relative to that of the free dye and multiplying by the number of fluorophores per protein.*

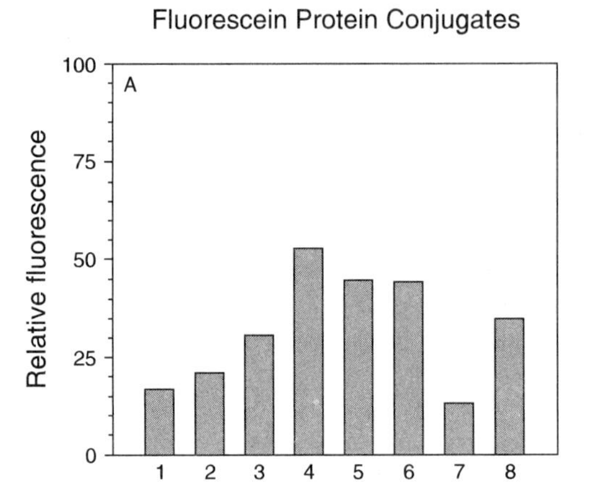

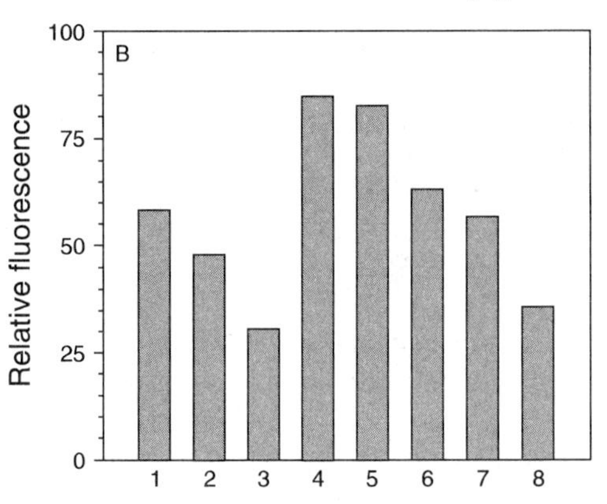

Figure 7.3 *Histograms showing the fluorescence per fluorophore for A) fluorescein and B) Cascade Blue conjugated to various proteins, relative to the fluorescence of the free dye in aqueous solution, represented by 100 on the Y-axis. The proteins represented are: 1) avidin, 2) bovine serum albumin, 3) concanavalin A, 4) goat IgG, 5) ovalbumin, 6) protein A, 7) streptavidin and 8) wheat germ agglutinin.*

typically twice as fluorescent as AMCA conjugates (Figure 7.2). Furthermore, AMCA-S conjugates have slightly shorter-wavelength emission than AMCA conjugates (~442 nm versus 448 nm), thus yielding better separation of their emission from that of fluorescein or BODIPY FL dyes. Cascade Blue conjugates are intrinsically brighter than AMCA [11] or AMCA-S conjugates and have improved spectral resolution from emission of fluorescein (see Figure 1.12 in Chapter 1), an important advantage for multicolor applications. However, the longer emission wavelength of the AMCA-S conjugates makes them appear visibly brighter because of the limited spectral sensitivity of the human eye to the shorter-wavelength fluorescence of the Cascade Blue dye. In addition, the Cascade Blue dye resists quenching upon protein conjugation, unlike other fluorophores such as fluorescein (Figure 7.3).

- **Fluorescent microspheres.** Where they can be used, conjugates of our FluoSpheres® and TransFluoSpheres® polystyrene microspheres provide the greatest versatility in selection of wavelengths and other properties [12] (see Section 6.2). Not only are they intensely fluorescent, but TransFluoSpheres beads also have extremely large Stokes shifts (see Figure 6.3 in Chapter 6). Fluorescent microspheres labeled with biotin, streptavidin, NeutraLite avidin and protein A are described below.

- **Enzyme-Labeled Fluorescence (ELF®) technology.** Our ELF detection reagents, which are described in detail in Section 6.3, can be used to enhance the detection of biotinylated antibodies and other haptenylated probes. When used in combination with alkaline phosphatase–streptavidin conjugates or alkaline phosphatase–labeled primary detection reagents, the ELF-97 phosphatase substrate yields a fine yellow-green fluorescent precipitate at the site of enzymatic activity that is much more photostable than any simple dye-labeled antibody conjugates (see Figure 6.6 in Chapter 6).

- **Enzyme conjugates.** Molecular Probes offers streptavidin and NeutraLite avidin conjugates of alkaline phosphatase, horseradish peroxidase and β-galactosidase, which are all prepared by methods that result in an approximate 1:1 ratio of enzyme to carrier protein, thereby ensuring the retention of both carrier-protein binding and enzymatic activity. These conjugates, as well as our biotin conjugates of alkaline phosphatase and horseradish peroxidase, are described in Section 7.5.

1. Stewart, C.C., *Multiparameter Flow Cytometry*, CRC Press (1994); **2.** Acta Histochem 89, 85 (1990); **3.** Proc Natl Acad Sci USA 85, 3546 (1988); **4.** Cell Biophys 7, 129 (1985); **5.** J Cell Biol 116, 1291 (1992); **6.** J Immunol Meth 149, 159 (1992); **7.** Proc Natl Acad Sci USA 85, 4672 (1988); **8.** J Immunol Meth 128, 109 (1990); **9.** Cytometry 15, 267 (1994); **10.** U.S. Patent No. 5,132,432; **11.** Anal Biochem 198, 119 (1991); **12.** Lea, T. *et al.* in *Flow Cytometry and Sorting, Second Edition*, M.R. Melamed, T. Lindmo and M.L. Mendelsohn, Eds., Wiley-Liss (1990) pp. 367–380.

7.2 Secondary Immunoreagents

Molecular Probes provides research scientists with one of the widest possible selections of secondary immunoreagents for use in fluorescence microscopy and flow cytometry. In addition to offering an extensive line of species-specific anti-IgG antibodies, we prepare a variety of fluorescent conjugates of protein A, a bacterial protein that binds with high affinity to the Fc portion of various classes and subclasses of immunoglobulins from many species.

Species-Specific Secondary Antibodies

Our anti–human, –mouse, –rabbit, –rat and –sheep IgG antibodies are conjugated to a wide variety of fluorophores, including our new green fluorescent Oregon Green (see Figure 7.1 in Section 7.1), Rhodol Green and BODIPY FL fluorophores and red fluorescent Rhodamine Red-X and Texas Red-X fluorophores (Table 7.1). Unlabeled goat antibodies to mouse IgG, rabbit IgG and rat IgG are also available.

These species-specific antibodies, which are raised against IgG heavy and light chains, are affinity-purified and absorbed against the sera of a number of species to minimize crossreactivity. Because the Fc portion of the light chain does not vary among the antibody classes, these secondary antibodies will detect other immunoglobulin classes besides IgG, although the extent of non-IgG immunoglobulin binding has not been tested and may vary from lot to lot.

We also offer biotin conjugates of alkaline phosphatase and horseradish peroxidase (A-927, P-917; see Section 7.5). By using an avidin, streptavidin or NeutraLite avidin bridge, researchers can link our biotinylated secondary antibodies to these biotinylated enzymes — a method that is often preferred because it reduces nonspecific staining.

Protein A and Protein G

Protein A and protein G are bacterial proteins that bind with high affinity to the Fc portion of various classes and subclasses of immunoglobulins from a variety of species. In particular, protein A binds to human IgG_1, IgG_2, IgG_4, IgM, IgA and IgE subclasses and all mouse subclasses except IgG_1, as well as to cat, cow, dog, goat, guinea pig, pig and rabbit antibodies. Unlike protein A, protein G binds to human IgG_3 and mouse IgG_1 subclasses and horse, rat and sheep antibodies but not to human IgM, IgA or IgE subclasses or cat antibodies. Protein A and protein G both fail to recognize the human IgD subclass or chicken antibodies.

We offer protein A conjugated to the Cascade Blue (P-963), Oregon Green 488 (P-6395), fluorescein (P-846), Texas Red (P-826) fluorophores and to biotin-XX (P-2757). Our unlabeled recombinant protein A (P-6445) and protein G (P-6444) can be custom-conjugated to meet the researcher's specific needs. We also offer yellow-green fluorescent protein A–labeled polystyrene FluoSpheres microspheres with 0.04 µm and 1 µm diameters (F-8778, F-8779). Additional colors and sizes can be custom-ordered.

Table 7.1 *Molecular Probes' selection of secondary antibodies, which are raised against the IgG heavy and light chains from mouse, rabbit and other species.*

Label (Abs/Em Maxima) *	Goat IgG Antibodies to:		
	Mouse IgG	Rabbit IgG	Other IgG †
AMCA-S (346/442)	A-6471	A-6472	
AMCA-X (349/448)	A-6440	A-6441	
Cascade Blue (400/420)	C-962	C-2764	C-2794 (rat)
Fluorescein (494/518)	F-2761	F-2765	F-2780 (human) F-2810 (sheep) ‡
Oregon Green 488 (496/524)	O-6380	O-6381	O-6382 (rat)
Rhodol Green (499/525)	R-6412	R-6470	
Oregon Green 500 (503/522)	O-6386	O-6387	
BODIPY FL (505/513)	B-2752	B-2766	
Oregon Green 514 (511/530)	O-6383	O-6384	
Eosin (524/544)	E-6436	E-6437	
BODIPY TMR-X (542/574)	B-6450		
Tetramethylrhodamine (555/580)	T-2762	T-2769	
R-Phycoerythrin (565/578)	P-852	P-2771	
Rhodamine Red-X (570/590)	R-6393	R-6394	
Texas Red (595/615)	T-862	T-2767	
Texas Red-X (595/615)	T-6390	T-6391	T-6392 (rat)
Allophycocyanin (650/660)	A-865		
Biotin (<300/none)	B-2763	B-2770	B-6452 (human) B-6453 (sheep) ‡
Unlabeled	G-6448	G-6446	G-6447 (rat)

With the exception of the BODIPY FL and BODIPY TMR antibody conjugates, which are lyophilized for improved stability, all of our antibody conjugates are packaged in buffered solutions containing 2 mM sodium azide as a preservative. At the time of preparation, the products are certified to be free of unconjugated dyes and are tested in a cytological experiment for low nonspecific staining.
* Approximate absorption and emission maxima in nm for conjugates. † "Other IgG" column shows the IgG species that these antibodies are raised against in parentheses. ‡ These are donkey IgG antibodies, in contrast to the others in the table, which are goat IgG antibodies.

Because single fluorescent microspheres can be detected, FluoSpheres beads have significant potential for ultrasensitive flow cytometry applications and immunodiagnostic assays [1] (see Section 6.2).

1. Lea, T. *et al.* in *Flow Cytometry and Sorting, Second Edition*, M.R. Melamed, T. Lindmo and M.L. Mendelsohn, Eds., Wiley-Liss (1990) pp. 367–380.

Cat #		Product Name	Unit Size	Price Per Unit ($) 1–4 Units	5–24 Units
A-865		allophycocyanin, crosslinked, goat anti-mouse IgG (H+L) conjugate *1 mg/mL*	0.5 mL	195.00	156.00
A-6471	**New**	AMCA-S goat anti-mouse IgG (H+L) conjugate *2 mg/mL*	0.5 mL	75.00	60.00
A-6472	**New**	AMCA-S goat anti-rabbit IgG (H+L) conjugate *2 mg/mL*	0.5 mL	75.00	60.00
A-6440	**New**	AMCA-X goat anti-mouse IgG (H+L) conjugate *2 mg/mL*	0.5 mL	75.00	60.00
A-6441	**New**	AMCA-X goat anti-rabbit IgG (H+L) conjugate *2 mg/mL*	0.5 mL	75.00	60.00
B-6453	**New**	biotin-XX donkey anti-sheep IgG (H+L) conjugate *2 mg/mL*	0.5 mL	75.00	60.00
B-6452	**New**	biotin-XX goat anti-human IgG (H+L) conjugate *2 mg/mL*	0.5 mL	75.00	60.00
B-2763		biotin-XX goat anti-mouse IgG (H+L) conjugate *2 mg/mL*	0.5 mL	75.00	60.00
B-2770		biotin-XX goat anti-rabbit IgG (H+L) conjugate *2 mg/mL*	0.5 mL	75.00	60.00
B-2752		BODIPY® FL goat anti-mouse IgG (H+L) conjugate	1 mg	75.00	60.00
B-2766		BODIPY® FL goat anti-rabbit IgG (H+L) conjugate	1 mg	75.00	60.00
B-6450	**New**	BODIPY® TMR-X goat anti-mouse IgG (H+L) conjugate	1 mg	75.00	60.00
C-962		Cascade Blue® goat anti-mouse IgG (H+L) conjugate *2 mg/mL*	0.5 mL	75.00	60.00
C-2764		Cascade Blue® goat anti-rabbit IgG (H+L) conjugate *2 mg/mL*	0.5 mL	75.00	60.00
C-2794		Cascade Blue® goat anti-rat IgG (H+L) conjugate *2 mg/mL*	0.5 mL	75.00	60.00
E-6436	**New**	eosin goat anti-mouse IgG (H+L) conjugate *2 mg/mL*	0.5 mL	75.00	60.00
E-6437	**New**	eosin goat anti-rabbit IgG (H+L) conjugate *2 mg/mL*	0.5 mL	75.00	60.00
F-2810		fluorescein donkey anti-sheep IgG (H+L) conjugate *2 mg/mL*	0.5 mL	75.00	60.00
F-2780		fluorescein goat anti-human IgG (H+L) conjugate *2 mg/mL*	0.5 mL	75.00	60.00
F-2761		fluorescein goat anti-mouse IgG (H+L) conjugate *2 mg/mL*	0.5 mL	75.00	60.00
F-2765		fluorescein goat anti-rabbit IgG (H+L) conjugate *2 mg/mL*	0.5 mL	75.00	60.00
F-8778	**New**	FluoSpheres® protein A-labeled microspheres, 0.04 µm, yellow-green fluorescent (505/515) *0.5% solids*	0.4 mL	165.00	132.00
F-8779	**New**	FluoSpheres® protein A-labeled microspheres, 1.0 µm, yellow-green fluorescent (505/515) *0.5% solids*	0.4 mL	165.00	132.00
G-6448	**New**	goat anti-mouse IgG (H+L) *affinity adsorbed against human IgG and human serum* *5 mg/mL*	1 mL	75.00	60.00
G-6446	**New**	goat anti-rabbit IgG (H+L) *affinity adsorbed against human, mouse and bovine serum* *5 mg/mL*	1 mL	95.00	76.00
G-6447	**New**	goat anti-rat IgG (H+L) *affinity adsorbed against mouse IgG and mouse serum* *5 mg/mL*	1 mL	125.00	100.00
O-6380	**New**	Oregon Green™ 488 goat anti-mouse IgG (H+L) conjugate *2 mg/mL*	0.5 mL	75.00	60.00
O-6381	**New**	Oregon Green™ 488 goat anti-rabbit IgG (H+L) conjugate *2 mg/mL*	0.5 mL	75.00	60.00
O-6382	**New**	Oregon Green™ 488 goat anti-rat IgG (H+L) conjugate *2 mg/mL*	0.5 mL	75.00	60.00
O-6386	**New**	Oregon Green™ 500 goat anti-mouse IgG (H+L) conjugate *2 mg/mL*	0.5 mL	75.00	60.00
O-6387	**New**	Oregon Green™ 500 goat anti-rabbit IgG (H+L) conjugate *2 mg/mL*	0.5 mL	75.00	60.00
O-6383	**New**	Oregon Green™ 514 goat anti-mouse IgG (H+L) conjugate *2 mg/mL*	0.5 mL	75.00	60.00
O-6384	**New**	Oregon Green™ 514 goat anti-rabbit IgG (H+L) conjugate *2 mg/mL*	0.5 mL	75.00	60.00
P-852		R-phycoerythrin goat anti-mouse IgG (H+L) conjugate *1 mg/mL*	1 mL	165.00	132.00
P-2771		R-phycoerythrin goat anti-rabbit IgG (H+L) conjugate *1 mg/mL*	0.5 mL	95.00	76.00
P-6445	**New**	protein A *recombinant*	5 mg	45.00	36.00
P-2757		protein A, biotin-XX conjugate	1 mg	75.00	60.00
P-963		protein A, Cascade Blue® conjugate	1 mg	75.00	60.00
P-846		protein A, fluorescein conjugate	1 mg	75.00	60.00
P-6395	**New**	protein A, Oregon Green™ 488 conjugate	1 mg	75.00	60.00
P-826		protein A, Texas Red® conjugate	1 mg	75.00	60.00
P-6444	**New**	protein G *recombinant*	5 mg	45.00	36.00
R-6393	**New**	Rhodamine Red™-X goat anti-mouse IgG (H+L) conjugate *2 mg/mL*	0.5 mL	75.00	60.00
R-6394	**New**	Rhodamine Red™-X goat anti-rabbit IgG (H+L) conjugate *2 mg/mL*	0.5 mL	75.00	60.00
R-6412	**New**	Rhodol Green™ goat anti-mouse IgG (H+L) conjugate *2 mg/mL*	0.5 mL	75.00	60.00
R-6470	**New**	Rhodol Green™ goat anti-rabbit IgG (H+L) conjugate *2 mg/mL*	0.5 mL	75.00	60.00
T-2762		tetramethylrhodamine goat anti-mouse IgG (H+L) conjugate *2 mg/mL*	0.5 mL	75.00	60.00
T-2769		tetramethylrhodamine goat anti-rabbit IgG (H+L) conjugate *2 mg/mL*	0.5 mL	75.00	60.00
T-862		Texas Red® goat anti-mouse IgG (H+L) conjugate *2 mg/mL*	0.5 mL	75.00	60.00
T-2767		Texas Red® goat anti-rabbit IgG (H+L) conjugate *2 mg/mL*	0.5 mL	75.00	60.00
T-6390	**New**	Texas Red®-X goat anti-mouse IgG (H+L) conjugate *2 mg/mL*	0.5 mL	75.00	60.00
T-6391	**New**	Texas Red®-X goat anti-rabbit IgG (H+L) conjugate *2 mg/mL*	0.5 mL	75.00	60.00
T-6392	**New**	Texas Red®-X goat anti-rat IgG (H+L) conjugate *2 mg/mL*	0.5 mL	75.00	60.00

7.3 Anti-Dye Antibodies

Anti-Fluorophore Antibodies

In addition to being useful for direct optical detection, some fluorescent and nonfluorescent dyes make excellent haptens that can be recognized by secondary detection reagents in applications such as *in situ* hybridization. Antibodies to dyes provide unique opportunities both for signal enhancement and for correlated fluorescence and electron microscopy studies. Essentially all of the methods that use biotin and avidin reagents (see Section 7.5) are also possible using dyes as haptens as long as the corresponding anti-fluorophore antibody is also available (see Table 4.1 in Chapter 4). One advantage of using fluorescent dyes as haptens instead of biotin-based techniques is that the hapten signal is usually visible preceding the secondary detection step. Availability of noncross-reacting antibodies to a variety of haptens is essential for any multicolor application. Molecular Probes provides the largest assortment of anti-dye antibodies commercially available, including rabbit polyclonal IgG antibodies to the fluorescein, tetramethylrhodamine, Texas Red, dansyl, BODIPY FL, BODIPY TR, lucifer yellow and Cascade Blue fluorophores and to the dinitrophenyl hapten. Biotin- and fluorophore-conjugated versions of some of these antibodies are also available.

Antibodies to Fluorescein and Oregon Green Dyes

The high affinity of anti-fluorescein antibodies (A-889, A-6413, A-6421) makes fluorescein an ideal hapten for various detection schemes.[1,2] Researchers have found fluorescein–anti-fluorescein enzyme-linked immunosorbent assays (ELISAs) to be similar in sensitivity to biotin–streptavidin methods [3] and to display extremely low nonspecific binding.[4] Our biotin-XX–labeled anti-fluorescein antibody (A-982) is an excellent reagent for converting a fluorescence-based detection into an enzyme-amplified light or electron microscopy technique. BODIPY FL–labeled anti-fluorescein antibody (A-6418) can be used to enhance the green fluorescent signal of the fluorescein hapten without changing its fluorescence color. Texas Red–labeled anti-fluorescein antibody (A-981) can be used to convert the green fluorescence of fluorescein conjugates into red fluorescence, or potentially to amplify the signal from fluorescein conjugates. As expected, all of our anti-fluorescein antibodies also strongly crossreact with the structurally similar Oregon Green dyes.

In addition to our anti-fluorescein rabbit polyclonal antibody (A-889), Molecular Probes offers an anti-fluorescein monoclonal antibody and a polyclonal anti-fluorescein Fab fragment. Our high-affinity anti-fluorescein mouse monoclonal 4-4-20 [5,6] (A-6421) may reduce nonspecific binding in ELISAs and other second-step detection assays. Our polyclonal anti-fluorescein Fab fragment (A-6413) provides researchers with a probe that more efficiently penetrates immunohistochemical preparations. Furthermore, because the Fab fragment no longer contains the Fc portion, nonspecific interactions with Fc receptor–bearing cells are eliminated.

Some of the more important applications for anti-fluorescein antibodies — almost all of which could also be carried out with our other anti-dye antibodies and their complementary dyes — include:

- Amplification of the signal from a fluorescein-labeled phenol that had been deposited in an enzyme-mediated reaction [7]
- Detection of fluorescein-labeled primary antibodies [8,9]
- Development of fluorescein-labeled cell preparations for electron microscopy [10]
- Investigation of the uptake of fluorescein dextran in kidney proximal tubules [11]
- Localization of mRNA sequences in a double *in situ* hybridization experiment in which both fluorescein- and biotin-labeled oligonucleotides were used [12]
- Preparation of an anti-fluorescein affinity matrix, which was used to immobilize a fluoresceinated protein in order to study its protein–protein interactions *in vitro* [13]

Antibodies to Tetramethylrhodamine, Rhodamine Red and Texas Red Dyes

As with the anti-fluorescein antibodies, our new rabbit polyclonal antibodies to the tetramethylrhodamine and Texas Red fluorophores (A-6397, A-6399) are effective reagents for binding these dye haptens and quenching their fluorescence. However, these antibodies strongly crossreact with all rhodamines, including tetramethylrhodamine, Rhodamine Red and Texas Red dyes, and cannot be used for simultaneous detection of more than one rhodamine-based dye. These anti-tetramethylrhodamine and anti–Texas Red antibodies do not appear to crossreact with fluorescein, and our anti-fluorescein antibodies do not crossreact with tetramethylrhodamine, Rhodamine Red or Texas Red dyes.

Antibodies to Lucifer Yellow and Cascade Blue Dyes

Lucifer yellow CH and Cascade Blue hydrazide are frequently employed as polar tracers for neuronal cell labeling (see Section 15.3). Our unconjugated (A-5750, A-5760) and biotinylated (A-5751, A-5761) rabbit polyclonal antibodies to these dyes are useful in standard enzyme-mediated immunohistochemical methods for permanently labeling neuronal tissue.[14-18] Anti–lucifer yellow antibody (A-5750) has also been used to follow dye-coupling in smooth muscle cells by electron microscopy.[19,20] Anti–Cascade Blue antibody (A-5760) has been employed in Western blot analysis to identify cytoplasmic and luminal domains of the sarcoplasmic reticulum Ca^{2+}-ATPase, which had been photolabeled with Cascade Blue aminoethyl 4-azidobenzamide [21] (C-625, see Section 5.3).

Antibodies to BODIPY FL and BODIPY TR Dyes

Our unlabeled and biotinylated rabbit polyclonal antibodies to the BODIPY FL fluorophore (A-5770, A-5771) crossreact with some other BODIPY dyes but do not crossreact appreciably with any of the other fluorophores. In solution assays, we have found that anti–BODIPY FL antibody effectively quenches the fluorescence of BODIPY FL, but quenches BODIPY TR to a lesser degree and does not significantly quench BODIPY TMR. Our fluorescein-labeled anti–BODIPY FL antibody (A-6420) can be used to enhance the green fluorescent signal of the BODIPY FL hapten without changing its fluorescence color. Like the anti–BODIPY FL antibody, our rabbit polyclonal antibody to the BODIPY TR fluorophore (A-6414) also crossreacts with the other BODIPY dyes to some extent but not with other fluorophores. These anti-BODIPY antibodies should therefore not be used for simultaneous detection of more than one BODIPY-based dye.

Antibody to Dansyl Dye

In contrast to the other anti-fluorophore antibodies, which quench the fluorescence of the dye to which they bind, our rabbit polyclonal anti-dansyl antibody (A-6398) typically *enhances* the fluorescence of dansyl amides by greater than 10-fold. Binding of the anti-dansyl antibody also blue-shifts the emission spectrum of the fluorophore in water from ~520 nm to ~450 nm. These properties, combined with the high Stokes shift of the dansyl dye, make this antibody particularly useful for determining the topography of dansyl-labeled probes, including that of dansyl-labeled phospholipids (see Section 13.4) in cell and artificial membranes.[22] The dansyl hapten is preferably incorporated into biomolecules using the succinimidyl ester of dansyl-X (D-6104, see Section 4.2) because its aminohexanoyl spacer ("X") reduces the interaction of the fluorophore with the biomolecule to which it is conjugated and makes it more accessible to anti-dansyl antibodies.[22-24]

Efficient Quenching by Anti-Fluorophore Antibodies

Quenching Efficiencies

Except for the anti-dansyl antibody, which enhances dansyl fluorescence, all of our anti-fluorophore antibodies strongly quench the fluorescence of their complementary dyes in free solution. For example, anti-fluorescein antibody effects up to 95% quenching of the fluorescence of both fluorescein and Oregon Green 488 fluorophores. Anti-fluorescein antibody also quenches other fluorescein derivatives such as carboxyfluorescein, Calcium Green™-1 and BCECF, making this antibody useful for reducing background fluorescence caused by leakage of these dyes from the cell.[25] However, anti-fluorescein's quenching of our fluorescein-based Ca^{2+} indicators is apparently dependent on whether or not Ca^{2+} is bound; Calcium Green-1 is quenched 89% in the presence of 5 µM Ca^{2+}, whereas it is quenched only 46% in the presence of 10 mM EGTA. Maximal quenching efficiencies for fluorescein analogs (all at 5 nM dye) are as follows (values may vary somewhat from batch to batch):

- Oregon Green 488 dye, 95%
- Oregon Green 500 dye, 90%
- Oregon Green 514 dye, 92%
- carboxyfluorescein, 93%
- Calcium Green-1 (in the presence of 5 µM Ca^{2+}), 89%
- Calcium Green-1 (in the presence of 10 mM EGTA), 46%
- BCECF, 43%
- fluo-3 (in the presence of 5 µM Ca^{2+}), 32%
- Rhodamine Green dye, 9%
- calcein, <5%
- tetramethylrhodamine, <5%

Our initial preparations of anti-tetramethylrhodamine and anti–Texas Red antibodies are somewhat less effective as fluorescence quenchers of their complementary fluorophores, with maximal quenching efficiencies of ~80% and ~60% respectively. Anti–BODIPY FL antibody typically quenches the BODIPY FL fluorescence by ~85%. It also quenches BODIPY TR fluorescence by ~45% but does not significantly quench BODIPY TMR fluorescence. Likewise, the anti–BODIPY TR antibody quenches BODIPY TR fluorescence by ~70%, and BODIPY FL and BODIPY TMR fluorescence to a lesser extent. The antibodies to lucifer yellow and Cascade Blue quench the fluorescence of their complementary fluorophores by ~85% and ~80%, respectively. In addition, anti-DNP antibodies have been reported to quench aminonitrobenzoxadiazoles [26] (NBD amines).

Quenching Assay

Molecular Probes uses quenching assays to determine the concentration of specific anti-dye antibody in these purified IgG fractions. As supplied, 10 µL of the antibody solution corresponds to the amount needed to obtain 50% of the maximal fluorescence quenching (or enhancement, in the case of anti-dansyl antibody) of 1 mL of a 50 nM solution of the corresponding dye, assayed in 100 mM sodium phosphate, pH 8.0. It is then assumed that 20 µL will produce maximal quenching of this amount of dye, which is confirmed by experiment. All maximal quenching values are determined using free fluorophore; the maximal quenching of fluorophore covalently bound to protein is often significantly less due to steric hindrance. Because the percentage of dye-specific IgG differs from lot to lot, we vary the concentration of total purified IgG to ensure that all vials contain similar quantities of anti-dye IgG or Fab fragment. Thus, first-time users must initially determine the correct antibody dilution for their application, but subsequent lots of the same product can be used at the same dilution.

This quenching assay cannot be applied to our BODIPY FL–labeled anti-fluorescein antibody or our fluorescein-labeled anti–BODIPY FL antibody because these antibodies are conjugated with a fluorophore spectrally similar to the fluorophore that they quench. Both antibodies are provided at a fixed concentration of 2 mg/mL. Also, the anti-fluorescein monoclonal antibody is not calibrated by the quenching assay; however, when it is reconstituted at 1 mg/mL, as described above, the quenching potency is similar to that of the polyclonal anti-fluorescein antibody.

Applications for Fluorescence Quenching by Anti-Fluorophore Antibodies

Fluorescence quenching of dye haptens by anti-dye antibodies provides a useful measure of topography in cells, proteins and membranes. For example, researchers have used anti-fluorescein quenching assays to determine the accessibility of a fluorescein-labeled ATP-binding site in both Na^+/K^+-ATPase and Ca^{2+}/Mg^{2+}-ATPase.[27,28] In addition, anti-fluorophore antibodies have been used as cell-impermeant probes for determining whether fluorescent dye–conjugated ligands, proteins, bacteria or other biomolecules have been internalized by endocytic or pinocytic processes [29-32] (see Section 17.3). Anti-fluorophore antibodies also permit background-free observation of fusion events in an assay designed to monitor the fusion of membrane vesicles *in vitro*.[33] However, as noted above, these antibodies may quench dye-labeled proteins less effectively than they quench free dyes.

Anti-Dinitrophenyl Antibody

Because of its high affinity for the dinitrophenyl hapten [34,35] (DNP), our anti-DNP polyclonal rabbit antibodies are excellent reagents for probing DNP-labeled molecules.[36] It has been reported that human chromosomes can be probed with equal sensitivities using either biotinylated, DNP-modified or digoxigenin-modified cosmid probes.[37] Anti-DNP antibodies have been used to localize a DNP-labeled DNA probe in HIV-infected cells.[38] Researchers have also reported using anti-DNP antibodies to probe for DNP-labeled IgG as a method for detecting sparse antigens [39] and, in conjunction with DNP-labeled BSA (A-843, see Section 15.7), to study the Fc receptor–mediated endocytosis of IgG complexes.[40,41]

In addition to unlabeled anti-DNP antibody (A-6430), Molecular Probes offers anti-DNP antibody conjugates of biotin-XX (A-6435), fluorescein (A-6423), tetramethylrhodamine (A-6424) and Texas Red-X (A-6396). These anti-DNP antibodies are prepared against DNP–keyhole limpet hemocyanin (DNP-KLH) and thus do not crossreact with bovine serum albumin (BSA), a common blocking reagent in hybridization applications.

For use in conjunction with anti-DNP antibody, Molecular Probes offers the DNP-X succinimidyl ester (D-2248, see Section 4.2) for labeling proteins and amine-modified DNA. The literature also describes methods for incorporating DNP into DNA using DNP-labeled primers.[42] In addition to recognizing DNP, anti-DNP antibody crossreacts with trinitrobenzenesulfonic acid, making this antibody useful both for localizing and for isolating cell-surface molecules labeled with either DNP or TNP.[42] Furthermore, anti-DNP antibody has been reported to quench aminonitrobenzoxadiazoles [26] (NBD amines).

Auxiliary Reagents for Use with Anti-Dye Antibodies

Use of the anti-fluorophore or anti-dinitrophenyl antibodies requires a means for the selective incorporation of their haptens into proteins, nucleic acids, cells and other biomolecules. Hapten-labeled probes are particularly important for localizing the target in cells and organelles, on blots and in other applications. Detection with an anti-dye antibody permits amplification significantly beyond what is possible with the dye itself. This method can also be used for correlated fluorescence and electron microscopy studies. Molecular Probes has an assortment of reactive dye haptens and hapten-labeled probes for these applications, including:

- Chemically reactive haptenylation reagents complementary to each of our anti-dye antibodies (see Table 4.1 in Chapter 4)
- FluoReporter® Protein Labeling Kits for incorporating biotin, biotin/DNP or a wide variety of fluorophores (see the Product Highlight "FluoReporter Protein and Oligonucleotide Labeling Kits" on page 124)
- FluoReporter® Oligonucleotide Amine or Phosphate Labeling Kits for chemical labeling of oligonucleotides with biotin, DNP or a large selection of fluorophores (see the Product Highlight "FluoReporter Protein and Oligonucleotide Labeling Kits" on page 124)
- ChromaTide™ nucleotides — dUTP or UTP labeled with fluorescein, Oregon Green 488, BODIPY FL, tetramethylrhodamine, Texas Red, Texas Red-X or Cascade Blue fluorophores (see Section 8.2)
- CellTracker™ Green CMFDA, CellTracker Orange CMTMR and CellTracker Green BODIPY, derived from fluorescein, tetramethylrhodamine and BODIPY FL dyes, respectively, for cell tracing (see Section 15.2)
- LysoTracker™ Green DND–26 (L-7526), which contains the BODIPY FL fluorophore, and LysoTracker Biotin/DNP (L-7538) and DAMP (D-1552), which contain a dinitrophenyl moiety, probes for selective staining and ultrastructural localization of intracellular compartments with low pH (see Section 12.3)
- MitoTracker™ Orange CMTMRos and MitoTracker Red CMXRos, both of which are recognized by the anti-tetramethylrhodamine and anti–Texas Red antibodies, for selective mitochondrial staining (see Section 12.2)

1. J Histochem Cytochem 38, 467 (1991); **2.** Voss Jr., E.W., *Fluorescein Hapten: An Immunological Probe*, CRC Press (1984); **3.** J Immunol Meth 122, 115 (1989); **4.** J Chemilumin Biolumin 2, 215 (1988); **5.** J Immunol Meth 169, 35 (1994); **6.** Mol Immunol 22, 871 (1985); **7.** J Immunol Meth 137, 103 (1991); **8.** J Histochem Cytochem 38, 325 (1990); **9.** J Immunol Meth 117, 45 (1989); **10.** J Cell Biol 111, 249 (1990); **11.** Am J Physiol 258, C309 (1990); **12.** J Histochem Cytochem 38, 467 (1991); **13.** J Biol Chem 257, 13095 (1982); **14.** Cell 81, 631 (1995); **15.** J Neurosci 15, 4851 (1995); **16.** Lübke, J. in *Neuroscience Protocols*, F.G. Wouterlood, Ed., Elsevier Science Publishers (1993) 90-050-06-01–13; **17.** J Neurosci Meth 41, 45 (1992); **18.** Devel Biol 94, 391 (1982); **19.** Circulation Res 70, 49 (1992); **20.** Beny, J.L. in *Endothelium-Derived Relaxing Factors*, G.M. Rubanyi and P.M. Vanhoutte, Eds., Karger (1990) pp. 117–123; **21.** Biochim Biophys Acta 1068, 27 (1991); **22.** Biochim Biophys Acta 1104, 9 (1992); **23.** Biochim Biophys Acta 776, 217 (1984); **24.** Biochemistry 21, 978 (1982); **25.** J Biol Chem 266, 24540 (1991); **26.** Biochemistry 16, 5150 (1977); **27.** Biochemistry 30, 1692 (1991); **28.** FEBS Lett 253, 273 (1989); **29.** Biochemistry 30, 2888 (1991); **30.** Biochim Biophys Acta 817, 238 (1985); **31.** Biochim Biophys Acta 778, 612 (1984); **32.** J Biol Chem 259, 5661 (1984); **33.** FEBS Lett 197, 274 (1986); **34.** J Exp Med 145, 931 (1977); **35.** Adv Immunol 2, 1 (1962); **36.** J Immunol Meth 150, 193 (1992); **37.** Science 247, 64 (1990); **38.** BioTechniques 9, 186 (1990); **39.** J Histochem Cytochem 38, 69 (1990); **40.** Cell 58, 317 (1989); **41.** J Cell Biol 98, 1170 (1984); **42.** Nucleic Acids Res 18, 3175 (1990).

7.3 Price List *Anti-Dye Antibodies*

Cat #		Product Name	Unit Size	Price Per Unit ($) 1–4 Units	5–24 Units
A-5770		anti-BODIPY® FL, rabbit IgG fraction *≥ 1 mg/mL*	0.5 mL	148.00	118.40
A-5771		anti-BODIPY® FL, rabbit IgG fraction, biotin-XX conjugate *≥ 1 mg/mL*	0.5 mL	148.00	118.40
A-6420	New	anti-BODIPY® FL, rabbit IgG fraction, fluorescein conjugate *2 mg/mL*	0.5 mL	148.00	118.40
A-6414	New	anti-BODIPY® TR, rabbit IgG fraction *≥ 1 mg/mL*	0.5 mL	148.00	118.40
A-5760		anti-Cascade Blue®, rabbit IgG fraction *≥ 1 mg/mL*	0.5 mL	148.00	118.40
A-5761		anti-Cascade Blue®, rabbit IgG fraction, biotin-XX conjugate *≥ 1 mg/mL*	0.5 mL	148.00	118.40
A-6398	New	anti-dansyl, rabbit IgG fraction *≥ 1 mg/mL*	0.5 mL	148.00	118.40
A-6430	New	anti-dinitrophenyl-KLH, rabbit IgG fraction	1 mg	148.00	118.40
A-6435	New	anti-dinitrophenyl-KLH, rabbit IgG fraction, biotin-XX conjugate *2 mg/mL*	0.5 mL	148.00	118.40
A-6423	New	anti-dinitrophenyl-KLH, rabbit IgG fraction, fluorescein conjugate *2 mg/mL*	0.5 mL	148.00	118.40
A-6424	New	anti-dinitrophenyl-KLH, rabbit IgG fraction, tetramethylrhodamine conjugate *2 mg/mL*	0.5 mL	148.00	118.40

Cat #		Product Name	Unit Size	Price Per Unit ($) 1–4 Units	5–24 Units
A-6396	New	anti-dinitrophenyl-KLH, rabbit IgG fraction, Texas Red®-X conjugate *2 mg/mL*	0.5 mL	148.00	118.40
A-6421	New	anti-fluorescein, mouse monoclonal 4-4-20	0.5 mg	195.00	156.00
A-6413	New	anti-fluorescein, rabbit IgG Fab fragments *≥ 1 mg/mL*	0.5 mL	195.00	156.00
A-889		anti-fluorescein, rabbit IgG fraction *≥ 1 mg/mL*	0.5 mL	148.00	118.40
A-982		anti-fluorescein, rabbit IgG fraction, biotin-XX conjugate *≥ 1 mg/mL*	0.5 mL	148.00	118.40
A-6418	New	anti-fluorescein, rabbit IgG fraction, BODIPY® FL-X conjugate *2 mg/mL*	0.5 mL	148.00	118.40
A-981		anti-fluorescein, rabbit IgG fraction, Texas Red® conjugate *≥ 1 mg/mL*	0.5 mL	148.00	118.40
A-5750		anti-lucifer yellow, rabbit IgG fraction *≥ 1 mg/mL*	0.5 mL	148.00	118.40
A-5751		anti-lucifer yellow, rabbit IgG fraction, biotin-XX conjugate *≥ 1 mg/mL*	0.5 mL	148.00	118.40
A-6397	New	anti-tetramethylrhodamine, rabbit IgG fraction *≥ 1 mg/mL*	0.5 mL	148.00	118.40
A-6399	New	anti-Texas Red®, rabbit IgG fraction *≥ 1 mg/mL*	0.5 mL	148.00	118.40

7.4 Primary Antibodies for Diverse Applications

Our polyclonal and monoclonal primary antibodies are directed toward some important reporter gene products as well as toward proteins found in synaptic vesicles, mitochondria and yeast. Many of these antibodies are discussed in greater detail in other chapters.

Polyclonal Antibodies Against Reporter Gene Products

Anti–Glutathione *S*-Transferase Antibody

One common partner in protein fusions is glutathione *S*-transferase (GST), a protein whose natural binding specificity can be exploited to facilitate its purification.[1] Because the GST portion of the fusion protein retains its affinity for glutathione, the fusion protein can be conveniently purified from the cell lysate in a single step by affinity chromatography on glutathione agarose[2-7] (see Figure 9.5 in Chapter 9). For the purification of GST fusion proteins, Molecular Probes offers glutathione linked via the sulfur atom to crosslinked beaded agarose (G-2879, G-6664, G-6665; see Section 9.1). We also prepare anti-GST antibody (A-5800), which has been generated against a 260–amino acid N-terminal fragment of the *Schistosoma japonica* enzyme expressed in *E. coli*, for identifying GST fusion proteins on Western blots. This anti-GST antibody may be useful for detecting glutathione transferase distribution in cells.[8]

Anti–*Aequorea Victoria* Green Fluorescent Protein Rabbit Serum

Expression of the intrinsically fluorescent green fluorescent protein (GFP) from *Aequorea victoria* has become a popular method for following gene expression and protein localization.[9,10] Molecular Probes offers a polyclonal rabbit antibody that is raised against GFP purified directly from the jellyfish *A. victoria*. The anti-GFP antibody (A-6455) facilitates the detection of native GFP, recombinant GFP and GFP fusion proteins by Western blot analysis and immunoprecipitation.

Anti–β-Glucuronidase Antibody

The *E. coli* β-glucuronidase (GUS) gene is a popular reporter gene in plants.[11-14] For Western blot and immunohistochemical analysis of transformed plant tissue,[15,16] Molecular Probes offers unlabeled rabbit anti–β-glucuronidase (A-5790) raised against *E. coli* type X-A β-glucuronidase. Our fluorogenic and chromogenic β-glucuronidase substrates are described in Section 10.2.

Other Polyclonal Antibodies

Anti–Synapsin I Antibody

Synapsin I is an actin-binding protein that is localized exclusively to synaptic vesicles and thus serves as an excellent marker for synapses in brain and other neuronal tissues.[17,18] Synapsin I inhibits neurotransmitter release, an effect that is abolished upon its phosphorylation by Ca^{2+}/calmodulin–dependent protein kinase II (CaM kinase II, C-6462; see Section 20.3).[19] Antibodies directed against synapsin I have proven valuable in molecular and neurobiology research, for example, to estimate synaptic density and to follow synaptogenesis.[20-22]

Molecular Probes offers anti–bovine synapsin I rabbit polyclonal antibody in two forms: as affinity-purified antibody (A-6442) or as unfractionated serum (A-6443). This antibody was isolated from rabbits immunized against bovine brain synapsin I but is also active against human, rat and mouse forms of the antigen; it has little or no activity against synapsin II. The affinity-purified antibody was fractionated from the serum using column chromatography in which bovine synapsin I was covalently bound to the column matrix. Affinity-purified anti–synapsin I is suitable for immunohistochemistry, Western blots, enzyme-linked immunoadsorbent assays and immunoprecipitations. Unfractionated rabbit serum is provided as an economical alternative for Western blots and immunoprecipitations.

Anti–NMDA Receptor Antibodies

N-methyl-D-aspartate (NMDA) receptors constitute cation channels of the central nervous system that are gated by the excitatory neurotransmitter L-glutamate.[23,24] Activation of NMDA receptors is essential for inducing long-term potentiation (LTP), a form of activity-dependent synaptic plasticity that is implicated in the learning process in animal behavioral models.[25] The biophysical properties of NMDA receptor channels contributing to LTP include Ca^{2+} permeability, voltage-dependent Mg^{2+} block and slow-gating kinetics.[26-29] NMDA receptor channel activities play a role in neuronal development and in disorders such as epilepsy and ischemic neuronal cell death. As targets for ethanol, NMDA receptors may also function in the pathology of alcoholism.[30,31]

In vitro reconstitution experiments with the cloned NMDA receptor subunit 1 and any one of the four NMDA receptor subunits 2A, 2B, 2C and 2D revealed that the physical properties of the heteromeric NMDA receptor channel appear to be imparted by the particular NMDA receptor subunit 2.[32] NMDA receptor subunits 2A and 2B are detected predominantly in the hippocampus and cortex, whereas 2C is found mainly in the cerebellum. Thus, cellular expression profiles of the NMDA receptor subunits 2A, 2B, 2C and 2D may contribute to the biophysical properties of NMDA receptors in specific central neurons.

For neurobiologists, Molecular Probes offers rabbit polyclonal antibodies to NMDA receptor subunits 2A, 2B and 2C in two forms: as affinity-purified antibodies or as unfractionated sera. The anti–NMDA receptor subunit 2A and 2B antibodies were generated against fusion proteins containing amino acid residues 1253–1391 of subunit 2A and 984–1104 of subunit 2B, respectively. These two antibodies are active against mouse, rat and human forms of the antigens and are specific for the subunit against which they were generated. In contrast, the anti–NMDA receptor subunit 2C antibody was generated against amino acid residues 25–130 of subunit 2C and recognizes the 140,000-dalton subunit 2C, as well as the 180,000-dalton subunit 2A and subunit 2B from mouse, rat and human. The affinity-purified antibodies were fractionated from sera by affinity chromatography in which NMDA receptor subunit fusion proteins were bound to a column matrix. Our affinity-purified anti–NMDA receptor subunit 2A (A-6473), 2B (A-6474) and 2C (A-6475) antibodies are suitable for immunohistochemistry, Western blots, enzyme-linked immunosorbent assays (ELISAs) and immunoprecipitations. Unfractionated rabbit sera for subunits 2A (A-6467), 2B (A-6468) and 2C (A-6469) are also available as economical alternatives for Western blot analysis and immunoprecipitation.

Anti–Phosphatidylinositol-Specific Phospholipase C Antibody

Phosphatidylinositol-specific phospholipase C (PI-PLC) from *Bacillus cereus* (EC 3.1.4.10) cleaves the phospholipid phosphatidylinositol (PI) into two molecules: water-soluble *myo*-inositol 1:2-cyclic phosphate and lipid-soluble diacylglycerol[33-35] (DAG). *B. cereus* PI-PLC cleaves PI, lyso-PI and glycosylphosphatidylinositol (GPI)-containing structures in a Ca^{2+}-independent manner.[36] This enzyme is a member of a large class of ubiquitous PI-PLCs. The smaller PI-PLCs (~35,000 daltons), including *B. cereus* PI-PLC and the nearly identical *B. thuringiensis* PI-PLC, are secreted by bacteria.[36] The larger PI-PLCs exist in eukaryotes and generate second messengers in the PI signal transduction pathway.[37]

Molecular Probes now offers an inhibitory mouse monoclonal antibody (A-6400) that is suitable for immunological detection of *B. cereus* PI-PLC in Western blots or ELISAs, as well as for structure–function studies of the active enzyme. This antibody, monoclonal A72-24, recognizes both *B. cereus* PI-PLC and *B. thuringiensis* PI-PLC.[38] The binding site of the antibody has been mapped by tryptic digests to an 8000-dalton stretch of the *B. cereus* enzyme that has been proposed to contain at least part of the catalytic site.[38] Monoclonal A72-24 inhibits the activity of both the *B. cereus* and *B. thuringiensis* PI-PLC in assays based on PI hydrolysis and the release of the GPI-linked enzyme acetylcholinesterase from bovine erythrocytes.[35] To complement this antibody, Molecular Probes also offers a highly purified *B. cereus* PI-PLC (P-6466, see Section 20.5) suitable for both enzymology and cell biology applications.

Monoclonal Antibodies Against Cytochrome Oxidase

Cytochrome oxidase (COX) catalyzes the transfer of electrons from reduced cytochrome *c* to molecular oxygen, with a concomitant translocation of protons across the mitochondrial inner membrane.[39] This mitochondrial membrane–bound enzyme is composed of both mitochondrial-encoded subunits (subunits I, II and III) and nuclear-encoded subunits (all others), with a total of 11 subunits for yeast COX and 13 subunits for mammalian COX.

Molecular Probes' monoclonal antibodies against specific subunits of yeast, bovine and human COX can be used to investigate the structure and biogenesis of this important mitochondrial protein complex. The binding specificity exhibited by our anti-COX monoclonal antibody preparations allows researchers to investigate the regulation, assembly and orientation of COX subunits from a variety of organisms[40-44] (see Tables 12.4 and 12.5 in Chapter 12). The anti–human COX antibodies (A-6402, A-6403, A-6404) are potentially valuable tools for the analysis of mitochondrial myopathies and related disorders.[45]

These monoclonal antibodies were selected for their ability to detect native COX by solid-phase binding assays such as particle-concentration fluorescence immunoassay (PCFIA) and enzyme-linked immunosorbent assay (ELISA). In addition, these antibodies recognize the corresponding denatured COX subunit by Western blot analysis and may be employed to test other subcellular preparations for mitochondrial contamination. Several of these antibodies can also be used to visualize mitochondria in fixed tissue using standard immunohistochemical techniques (see Tables 12.4 and 12.5 in Chapter 12). Detailed information regarding the IgG isotype and recommended working concentration is provided with each product. These monoclonal antibodies are described in more detail with our other organelle probes in Section 12.5.

Monoclonal Antibodies for Yeast Cell Biology

Molecular Probes provides an array of immunoreagents for researchers studying aspects of cell biology with the yeast *Saccharomyces cerevisiae* (Table 12.6). In particular, we offer monoclonal antibodies that recognize:

- Vacuolar membrane proteins — integral membrane proteins, which include alkaline phosphatase (A-6458) and the 100,000-dalton subunit of the yeast vacuolar H^+-ATPase (V-ATPase, A-6426), as well as peripheral membrane proteins, which include the 60,000- and 69,000-dalton subunits of V-ATPase (A-6427, A-6422)
- Vacuolar lumen protein — carboxypeptidase Y (A-6428)
- Mitochondrial membrane proteins — porin (A-6449) and four subunits of yeast COX (A-6405, A-6407, A-6408, A-6432)
- Endoplasmic reticulum membrane protein — dolichol phosphate mannose synthase (Dol-P-Man synthase, Dpm1p; A-6429)
- Cytosolic protein — 3-phosphoglycerate kinase (PGK, A-6457)

We have selected this set of monoclonal antibodies because they are compatible with both Western blotting of denatured proteins and protein immunolocalization in fixed yeast cells. Other potential uses of these antibodies include the development of ELISAs to determine either the level of enrichment of a particular yeast organelle or the level at which the organelle contaminates a subcellular fraction. Detailed information regarding the IgG isotype and recommended working concentration is provided with each product. These monoclonal antibodies are described along with our other organelle-selective probes in Section 12.5.

1. Proc Natl Acad Sci USA 83, 8703 (1986); **2.** Meth Mol Genet 1 (Part A), 267 (1993); **3.** BioTechniques 13, 856 (1992); **4.** BioTechniques 10, 178 (1991); **5.** Nucleic Acids Res 19, 4005 (1991); **6.** Science 252, 712 (1991); **7.** Gene 67, 31 (1988); **8.** Cytometry 20, 134 (1995); **9.** Nature 369, 400 (1994); **10.** Science 263, 802 (1994); **11.** Mol Gen Genet 216, 321 (1989); **12.** Mol Gen Genet 215, 38 (1988); **13.** Plant Mol Biol 1, 387 (1987); **14.** Proc Natl Acad Sci USA 84, 8447 (1986); **15.** J Biol Chem 269, 17635 (1994); **16.** Plant Mol Biol 15, 821 (1990); **17.** Science 226, 1209 (1984); **18.** J Cell Biol 96, 1337 (1983); **19.** Greengard, P., Benfenati, F. and Valtorta, F. in *Molecular and Cellular Mechanisms of Neurotransmitter Release*, L. Stjärne, *et al.*, Eds. Raven Press (1994) pp. 31–45; **20.** J Neurosci 12, 1736 (1992); **21.** J Neurosci 11, 1617 (1991); **22.** J Neurosci 9, 2151 (1989); **23.** Neuron 12, 529 (1994); **24.** Nature 354, 31 (1991); **25.** J Neurosci 9, 3040 (1989); **26.** Nature 346, 565 (1990); **27.** Nature 325, 529 (1987); **28.** Nature 321, 519 (1986); **29.** Nature 307, 462 (1984); **30.** Mol Pharmacol 45, 324 (1994); **31.** Neurosci Lett 152, 13 (1993); **32.** Science 256, 1217 (1992); **33.** Meth Enzymol 197, 493 (1991); **34.** Biochemistry 29, 8056 (1990); **35.** Meth Enzymol 71, 731 (1981); **36.** Bioorg Medicinal Chem Lett 2, 49 (1994); **37.** Science 244, 546 (1989); **38.** Biochim Biophys Acta 1047, 47 (1990); **39.** Ann Rev Biochem 59, 569 (1990); **40.** Meth Enzymol 260, 117 (1995); **41.** J Cell Biol 126, 1375 (1994); **42.** J Biol Chem 268, 18754 (1993); **43.** Biochemistry 30, 3674 (1991); **44.** J Biol Chem 266, 7688 (1991); **45.** Pediatric Res 28, 529 (1990).

7.4 Price List *Primary Antibodies for Diverse Applications*

Cat #		Product Name	Unit Size	Price Per Unit ($) 1–4 Units	5–24 Units
A-6455	***New***	anti-*Aequorea victoria* green fluorescent protein, rabbit serum (anti-GFP)	100 µL	98.00	78.40
A-6400	***New***	anti-*Bacillus cereus* phosphatidylinositol-specific phospholipase C, mouse monoclonal A72-24	100 µg	98.00	78.40
A-6409	***New***	anti-bovine cytochrome oxidase subunit IV, mouse monoclonal 10G8-D12-C12 *specificity: human, bovine, chicken*	250 µg	195.00	156.00
A-6431	***New***	anti-bovine cytochrome oxidase subunit IV, mouse monoclonal 20E8-C12 *specificity: human, bovine, rat, mouse*	250 µg	195.00	156.00
A-6456	***New***	anti-bovine cytochrome oxidase subunit Vb, mouse monoclonal 16H12-H9 *specificity: human, bovine, rat, mouse, chicken*	250 µg	195.00	156.00
A-6410	***New***	anti-bovine cytochrome oxidase subunit VIa-H, mouse monoclonal 4H2-A5	250 µg	195.00	156.00
A-6411	***New***	anti-bovine cytochrome oxidase subunit VIa-L, mouse monoclonal 14A3-AD2-BH4	250 µg	195.00	156.00
A-6401	***New***	anti-bovine cytochrome oxidase subunit VIc, mouse monoclonal 3G5-F7-G3 *specificity: human, bovine, rat*	100 µg	195.00	156.00
A-6442	***New***	anti-bovine synapsin I, rabbit IgG fraction *affinity purified* *specificity: human, bovine, rat, mouse*	10 µg	120.00	96.00
A-6443	***New***	anti-bovine synapsin I, rabbit serum *specificity: human, bovine, rat, mouse*	100 µL	98.00	78.40
A-5790		anti-β-glucuronidase, rabbit IgG fraction *2 mg/mL*	0.5 mL	148.00	118.40
A-5800		anti-glutathione S-transferase, rabbit IgG fraction *3 mg/mL*	0.5 mL	155.00	124.00
A-6403	***New***	anti-human cytochrome oxidase subunit I, mouse monoclonal 1D6-E1-A8	100 µg	195.00	156.00
A-6402	***New***	anti-human cytochrome oxidase subunit I, mouse monoclonal 5D4-F5	100 µg	195.00	156.00
A-6404	***New***	anti-human cytochrome oxidase subunit II, mouse monoclonal 12C4-F12	100 µg	195.00	156.00
A-6473	***New***	anti-rat NMDA receptor, subunit 2A, rabbit IgG fraction *affinity purified* *specificity: human, rat, mouse*	10 µg	240.00	192.00
A-6467	***New***	anti-rat NMDA receptor, subunit 2A, rabbit serum *specificity: human, rat, mouse*	50 µL	196.00	156.80
A-6474	***New***	anti-rat NMDA receptor, subunit 2B, rabbit IgG fraction *affinity purified* *specificity: human, rat, mouse*	10 µg	240.00	192.00
A-6468	***New***	anti-rat NMDA receptor, subunit 2B, rabbit serum *specificity: human, rat, mouse*	50 µL	196.00	156.80
A-6475	***New***	anti-rat NMDA receptor, subunit 2C, rabbit IgG fraction *affinity purified* *specificity: human, rat, mouse*	10 µg	240.00	192.00
A-6469	***New***	anti-rat NMDA receptor, subunit 2C, rabbit serum *specificity: human, rat, mouse*	50 µL	176.00	140.80
A-6405	***New***	anti-yeast cytochrome oxidase subunit I, mouse monoclonal 11D8-B7	100 µg	195.00	156.00
A-6407	***New***	anti-yeast cytochrome oxidase subunit II, mouse monoclonal 4B12-A5	250 µg	195.00	156.00
A-6408	***New***	anti-yeast cytochrome oxidase subunit III, mouse monoclonal DA5	250 µg	195.00	156.00
A-6432	***New***	anti-yeast cytochrome oxidase subunit IV, mouse monoclonal 1A12-A12	250 µg	195.00	156.00
A-6429	***New***	anti-yeast dolichol phosphate mannose synthase, mouse monoclonal 5C5-A7	250 µg	195.00	156.00
A-6449	***New***	anti-yeast mitochondrial porin, mouse monoclonal 16G9-E6	250 µg	195.00	156.00
A-6457	***New***	anti-yeast 3-phosphoglycerate kinase (PGK), mouse monoclonal 22C5-D8	250 µg	195.00	156.00
A-6458	***New***	anti-yeast vacuolar alkaline phosphatase, mouse monoclonal 1D3-A10	250 µg	195.00	156.00
A-6428	***New***	anti-yeast vacuolar carboxypeptidase Y (CPY), mouse monoclonal 10A5-B5	250 µg	195.00	156.00
A-6427	***New***	anti-yeast vacuolar H^+-ATPase 60 kDa subunit, mouse monoclonal 13D11-B2	250 µg	195.00	156.00
A-6422	***New***	anti-yeast vacuolar H^+-ATPase 69 kDa subunit, mouse monoclonal 8B1-F3	250 µg	195.00	156.00
A-6426	***New***	anti-yeast vacuolar H^+-ATPase 100 kDa subunit, mouse monoclonal 10D7-A7-B2	250 µg	195.00	156.00

7.5 Avidin, Streptavidin and NeutraLite Avidin

The high affinity of avidin for biotin was first exploited in histochemical applications in the mid-1970s.[1,2] This egg-white protein and its bacterial counterpart, streptavidin, have since become standard reagents for diverse detection schemes.[3] In their simplest form, such methods entail applying a biotinylated probe to the sample and then detecting the bound probe with labeled avidin or streptavidin. These techniques are commonly used to localize antigens in cells and tissues [4,5] and to detect biomolecules in immunoassays and DNA hybridization techniques [6-9] (see Section 8.4). Chapter 4 contains a complete list of all of our biotin-containing probes, including biotinylation reagents, biotin tracers and biotinylated site-selective probes.

Binding Characteristics of Avidin, Streptavidin and NeutraLite Avidin

Avidin, streptavidin and NeutraLite avidin bind four biotins per molecule with high affinity and specificity. Their multiple binding sites permit a number of techniques in which unlabeled avidin, streptavidin or NeutraLite avidin can be used to bridge two biotinylated reagents. This bridging method, which is commonly used to link a biotinylated probe to a biotinylated enzyme in enzyme-linked immunohistochemical applications, often eliminates the background problems that can occur when using direct avidin– or streptavidin–enzyme conjugates. High-purity unlabeled avidin (A-887), streptavidin (S-888) and NeutraLite avidin (A-2666) are available in bulk from Molecular Probes at reasonable prices. We also offer avidin and streptavidin specially packaged in smaller unit sizes for extra convenience (A-2667, S-2669). Our avidin, streptavidin and deglycosylated NeutraLite avidin all bind >12 µg of biotin per mg protein.

Avidin

Avidin is a positively charged 66,000-dalton glycoprotein [10] with an isoelectric point of about 10.5. It is thought that avidin's positively charged residues and its oligosaccharide component (heterogeneous structures composed largely of mannose and *N*-acetylglucosamine) can interact nonspecifically with negatively charged cell surfaces and nucleic acids, sometimes causing background problems in histochemical applications and flow cytometry. Methods have been developed to suppress this nonspecific avidin binding.[11] Avidin and its conjugates selectively bind to a component in mast cell granules and can be used to identify mast cells in normal and diseased human tissue.[12,13]

Streptavidin

Streptavidin, a nonglycosylated 60,000-dalton protein with a near-neutral isoelectric point, reportedly exhibits less nonspecific binding than avidin. However, streptavidin contains a tripeptide sequence Arg–Tyr–Asp (RYD) that apparently mimics the Arg–Gly–Asp (RGD) binding sequence of fibronectin, a component of the extracellular matrix that specifically promotes cellular adhesion.[14] This universal recognition sequence binds integrins and related cell-surface molecules. Background problems sometimes associated with streptavidin may be attributable to this tripeptide.

NeutraLite Avidin

Molecular Probes now provides an alternative to the commonly used avidin and streptavidin. Our conjugates of NeutraLite avidin — a protein that has been processed to remove the carbohydrate and lower its isoelectric point — can sometimes reduce background staining. The methods used to deglycosylate the avidin reportedly retain both its specific binding [15] and its complement of amine-conjugation sites. NeutraLite avidin conjugates have been shown to provide improved detection of single-copy genes in metaphase chromosome spreads.[16]

Secondary Detection with Avidins

Avidin, streptavidin and NeutraLite avidin conjugates are used as secondary detection reagents in histochemical applications, flow cytometry,[17,18] blot analysis and immunoassays. These reagents can also be employed to localize biocytin, biocytin-X, biotin ethylenediamine, Cascade Blue biocytin or lucifer yellow biocytin — derivatives of biotin that are used as neuroanatomical tracers [19,20] (see Section 15.3). The following are commonly used methods for employing avidin, streptavidin and NeutraLite avidin as secondary detection reagents:

- **Direct procedure.** A biotinylated primary probe — such as an antibody, single-stranded nucleic acid probe or lectin — is bound to tissues, cells or other surfaces. Excess protein is removed by washing, and detection is mediated by reagents such as our fluorescent avidins, streptavidins or NeutraLite avidins or our enzyme-conjugated streptavidins plus substrate.

- **Bridging methods.** A biotinylated antibody or oligonucleotide is used to probe a tissue, cell or other surface. This preparation is then treated with unlabeled avidin, streptavidin or NeutraLite avidin. Excess reagents are removed by washing, and detection is mediated by a biotinylated detection reagent such as fluorescein biotin (B-1370, see Section 4.3), biotinylated R-phycoerythrin (P-811, see Section 6.4), biotinylated FluoSpheres microspheres (see Section 6.2), or a biotinylated enzyme (P-917, A-927) plus substrate.

- **Indirect procedure.** An unlabeled primary antibody is bound to a cell followed by a biotinylated species-specific secondary antibody. After washing, the complex is detected by one of the two procedures described above.

Fluorescent Conjugates of Avidin, Streptavidin and NeutraLite Avidin

Fluorophore-Labeled Avidins

Fluorescent avidin and streptavidin are extensively used in DNA hybridization techniques,[21,22] immunohistochemistry and multicolor flow cytometry.[23-25] Molecular Probes' selection of avidin, streptavidin and NeutraLite avidin conjugates keeps growing as we introduce improved fluorophores. We continue to provide some avidin, NeutraLite avidin and streptavidin conjugates of fluorescein, Lissamine Rhodamine B and Texas Red dyes. However, we strongly recommend your evaluation of our new green fluorescent Oregon Green, red-orange fluorescent Rhodamine Red-X and red fluorescent Texas Red-X conjugates, which exhibit excitation and emission maxima nearly identical to these traditional reagents but are generally brighter. Moreover, conjugates of the Oregon Green dyes, in particular those of Oregon Green 514, are much more photostable and less pH sensitive than fluorescein conjugates (Figure 7.1, see also Figures 1.6 and 1.7 in Chapter 1 and Figure 11.1 in Chapter 11). For blue fluorescent labeling, we now offer an improved AMCA-S streptavidin (S-6364), which shows superior brightness in side-by-side testing with AMCA streptavidin (see Figure 7.2 in Section 7.1). Our phycoerythrin streptavidins (S-866, S-867) have the most intense fluorescence of all avidin conjugates and are important labels for multicolor flow cytometry. Additionally, we offer allophycocyanin streptavidin — one of the few probes that can be excited by the 633 nm spectral line of the He-Ne laser.[26] A complete list of our current offerings of fluorophore- and enzyme-labeled avidins, streptavidins and NeutraLite avidins can be found in Table 7.2.

Biotin-, Streptavidin- and NeutraLite Avidin–Labeled Fluorescent Microspheres

Molecular Probes also offers biotin, streptavidin and NeutraLite avidin conjugates of the intensely fluorescent FluoSpheres and TransFluoSpheres polystyrene microspheres in a variety of colors and sizes (see the product list at the end of this section). Because single fluorescent microspheres can be detected, FluoSpheres and TransFluoSpheres beads have significant potential for ultrasensitive flow cytometry applications and immunodiagnostic assays.[27] They may also be useful as tracers that can be detected with standard enzyme-mediated histochemical methods. Additional sizes and colors of these labeled microspheres can be custom-ordered through our Custom and Bulk Sales Department.

Enzyme Conjugates of Biotin, Streptavidin and NeutraLite Avidin

Enzyme conjugates are used in enzyme-linked immunosorbent assays[28] (ELISAs), blotting techniques,[29] *in situ* hybridization[30] and cytochemistry and histochemistry.[31] Enzyme-mediated *in situ* techniques using these conjugates provide better resolution and are safer, more sensitive and faster than radioactive methods. Most frequently, the enzymes of choice are horseradish peroxidase, alkaline phosphatase and *E. coli* β-galactosidase because of their high turnover rate, stability, ease of conjugation and relatively low cost. Molecular Probes has prepared highly active enzyme conjugates of streptavidin and NeutraLite avidin, as well as of biotin-XX. Fluorogenic substrates for ELISAs are often much more sensitive than chromogenic substrates in these important assays. Our fluorogenic and chromogenic substrates for these assays are described in Chapter 10; our patented fluorogenic ELF-97 phosphatase substrate, which is proving superior to direct fluorophore conjugates in many applications, is discussed in Section 6.3.

Our enzyme conjugates of streptavidin and NeutraLite avidin are prepared by techniques that yield an approximate 1:1 ratio of enzyme to avidin analog, thus ensuring maximum retention of activity of both enzyme and carrier protein. We offer streptavidin conjugates of alkaline phosphatase, horseradish peroxidase and β-galactosidase (S-921, S-911, S-931) and the NeutraLite avidin conjugate of peroxidase[32] (A-2664). To decrease background problems, researchers often prefer to use the biotin-XX conjugates of peroxidase (P-917) and alkaline phosphatase (A-927) in conjunction with an avidin or streptavidin bridge for indirect detection of a

Table 7.2 *Molecular Probes' selection of avidin, streptavidin and NeutraLite avidin conjugates.*

Label (Abs/Em Maxima) *	Avidins		
	Avidin	Streptavidin	NeutraLite Avidin
Fluorescent Conjugates			
AMCA-S (346/442)		S-6364	A-6379
Cascade Blue (400/420)			A-2663
Lucifer yellow (428/532)		S-884	
Fluorescein (494/518)	A-821	S-869	A-2662
Oregon Green 488 (496/524)		S-6368	A-6374
Rhodol Green (499/525)		S-6371	
Oregon Green 500 (503/522)		S-6367	A-6376
BODIPY FL (505/513)	A-2641	S-2642	A-2661
Oregon Green 514 (511/530)		S-6369	A-6375
Eosin (524/544)		S-6372	
Tetramethylrhodamine (555/580)		S-870	A-6373
B-Phycoerythrin (565/575)		S-867	
R-Phycoerythrin (565/578)		S-866	A-2660
Rhodamine B (570/590)		S-871	
Rhodamine Red-X (570/590)		S-6366	A-6378
Texas Red (595/615)	A-820	S-872	A-2665
Texas Red-X (595/615)		S-6370	
Allophycocyanin (650/660)		S-868	
Aequorin, Agarose and Enzyme Conjugates			
Aequorin		A-6787	
Agarose		S-951	
Alkaline phosphatase		S-921	
β-Galactosidase		S-931	
Horseradish peroxidase		S-911	A-2664
Unlabeled Avidins			
Unlabeled	A-887 A-2667	S-888 S-2669 †	A-2666

* Approximate absorption and emission maxima in nm for conjugates. † Special packaging.

wide array of biotinylated biomolecules. Our biotinylated peroxidase and alkaline phosphatase conjugates are prepared with a reactive biotin-XX derivative, which contains the longest available spacer and allows high avidin affinity.

Aequorin Conjugates for Sensitive Bioluminescent Assays

Molecular Probes' biotin and streptavidin conjugates of AquaLite® recombinant aequorin (A-6786, A-6787) are extremely versatile bioluminescent probes for detecting antigens, nucleic acids and other biological targets.[33] The aequorin used in these conjugates consists of a 22,000-dalton recombinant apoaequorin protein from the jellyfish *Aequorea victoria*, along with molecular oxygen and the prosthetic luminophore coelenterazine. Once aequorin binds Ca^{2+}, the complex emits a flash of blue light (emission maximum 466 nm) that persists for about two seconds. Because excitation light sources are not needed for this reaction, the background noise and sample autofluorescence sometimes associated with fluorescence techniques are eliminated. Our new coelenterazine analogs, which are described in Section 22.5, may provide even greater chemiluminescence when reconstituted into apoaequorin than is possible using coelenterazine. Please contact our Technical Assistance Department about the possible availability of apoaequorin complexed with other coelenterazines.

The aequorin conjugates can be employed in an ELISA-type assay using commercially available luminometers. Researchers have found that these aequorin-based bioluminescence immunoassays (BLIA) are five- to six-fold more sensitive than those employing chemiluminescent detection and at least 50-fold more sensitive than the most sensitive colorimetric methods.[34] Moreover, these aequorin-based assays exhibit a remarkably large linear dynamic range, spanning six orders of magnitude.[35] Using the BLIA microplate assay, researchers have been able to detect about 40 femtomoles of Forssman glycolipid, a sensitivity that is an order-of-magnitude better than achieved with a radioactive overlay technique on thin-layer chromatographic plates.[35] Researchers have also been able to adapt the aequorin-based BLIA microplate assay to determine the kinetic constants of solution-phase glycosyltransferase reactions [36] and to detect attomole quantities of biotin.[37]

Alternatively, the aequorin signal can be detected on blots with instant photographic and X-ray films. In these assays, it is not necessary to expose films for long periods of time as is required with radioactive probes. Using biotinylated aequorin, researchers have been able to detect 6 pg of human transferrin on dot blots, 10 ng of biotinylated protein standards on Western blots and 8 ng of target DNA on Southern blots.[35]

Streptavidin Agarose

Molecular Probes also offers streptavidin conjugated to 4% beaded crosslinked agarose (S-951) — a matrix that can be used to isolate biotinylated proteins.[38] In addition, biotinylated antibodies can be bound to streptavidin agarose to generate affinity matrices for the large-scale isolation of antigens.[38] In a recent report, staurosporine-treated myotubules were incubated with biotinylated α-bungarotoxin (B-1196, see Section 18.2) in order to isolate the acetylcholine receptors (AChRs) on streptavidin agarose and assess staurosporine's effect on the degree of phosphorylation of this receptor.[39] Streptavidin agarose has also been used to investigate the turnover of cell-surface proteins that had previously been derivatized with an amine-reactive biotin [40] (B-1582, see Section 4.2).

Auxiliary Reagents for Use with Avidins

In addition to the direct conjugates of avidins, Molecular Probes offers an extensive selection of biotinylated products for use in conjunction with avidins; see Chapter 4 for a complete list. Combining one of our biotinylated antibodies (see Table 7.1) with a fluorescent- or enzyme-labeled avidin provides a facile method for indirect detection of antibodies from various animal sources. Biotinylated R-phycoerythrin (P-811, see Section 6.4) can be used with an avidin or streptavidin bridge to detect biotinylated biomolecules;[41] this bridging technique may substantially reduce the nonspecific staining that is commonly seen when using phycobiliproteins for immunohistochemical applications.

Biotinylated liposomes can be prepared using Molecular Probes' biotin conjugates of phospoethanolamine (B 1550, B 1616; see Section 4.3). Avidin has been used to form a bridge between a biotinylated liposome loaded with fluorescent dyes and a target-specific biotinylated detection reagent. Biotinylated liposomes containing carboxyfluorescein have been employed in an immunoassay that was reported to be both faster and 100-fold more sensitive than the comparable peroxidase-based ELISA.[42]

Molecular Probes offers a broad selection of biotinylating reagents, including FluoReporter Biotin-XX and Biotin/DNP Protein Labeling Kits (F-2610, F-6347, F-6348; see Section 4.2) and FluoReporter Biotin Oligonucleotide Labeling Kits (F-6081, F-6095; see Section 8.2).

1. Proc Natl Acad Sci USA 71, 3537 (1974); **2.** Biochim Biophys Acta 264, 165 (1972); **3.** Meth Enzymol 184 (1990); **4.** J Cell Biol 111, 1183 (1990); **5.** Physiol Plantarum 79, 231 (1990); **6.** Cytometry 11, 126 (1990); **7.** Proc Natl Acad Sci USA 87, 6223 (1990); **8.** Science 249, 928 (1990); **9.** Anal Biochem 171, 1 (1988); **10.** Adv Protein Chem 29, 85 (1975); **11.** J Histochem Cytochem 29, 1196 (1981); **12.** J Histochem Cytochem 33, 27 (1985); **13.** J Invest Dermatol 83, 214 (1984); **14.** Biochem Biophys Res Comm 170, 1236 (1990); **15.** Biochem J 248, 167 (1987); **16.** Trends Genet 9, 71 (1993); **17.** J Microbiol Meth 12, 1 (1990); **18.** Biochemistry 16, 5150 (1977); **19.** J Neurosci 10, 3421 (1990); **20.** Brain Res 497, 361 (1989); **21.** Histochemistry 85, 4 (1986); **22.** Proc Natl Acad Sci USA 83, 2934 (1986); **23.** J Immunology 137, 1486 (1986); **24.** Meth Enzymol 108, 197 (1984); **25.** J Immunology 129, 532 (1982); **26.** Cytometry 15, 267 (1994); **27.** Lea, T. *et al.* in *Flow Cytometry and Sorting, Second Edition*, M.R. Melamed, T. Lindmo and M.L. Mendelsohn, Eds., Wiley-Liss (1990) pp. 367–380; **28.** Harlow, E. and Lane, D., *Antibodies: A Laboratory Manual*, Cold Spring Harbor Press (1988) pp. 553–612; **29.** *Short Protocols in Molecular Biology*, F.M. Ausubel *et al.*, Eds. Wiley-Interscience (1989) pp. 304–306; **30.** *Molecular Neuroanatomy*, F.W. van Leeuwen, Ed., Elsevier Science Publisher (1988); **31.** J Histochem Cytochem 27, 1131 (1979); **32.** Histochemistry 84, 333 (1986); **33.** Stults, N.L. *et al.* in *Bioluminescence and Chemiluminescence: Current Status*, P. Stanly and L. Kricka, Eds., Wiley (1991) pp. 533–536; **34.** Smith, D.F. *et al.* in *Bioluminescence and Chemiluminescence: Current Status*, P. Stanley and L. Kricka, Eds., John Wiley & Sons (1991) pp. 529–532; **35.** Biochemistry 31, 1433 (1992); **36.** Anal Biochem 199, 286 (1991); **37.** Anal Chem 66, 1837 (1994); **38.** J Chromatography 510, 3 (1990); **39.** J Cell Biol 125, 661 (1994); **40.** Biochemistry 28, 574 (1989); **41.** J Biol Chem 265, 15776 (1990); **42.** Anal Biochem 176, 420 (1989).

Cat #		Product Name	Unit Size	Price Per Unit ($) 1–4 Units	5–24 Units
A-927		alkaline phosphatase, biotin-XX conjugate *5 mg/mL*	200 µL	95.00	76.00
A-6786	**New**	AquaLite® aequorin, biotinylated *recombinant*	25 µg	175.00	140.00
A-6787	**New**	AquaLite® aequorin, streptavidin *recombinant*	1 unit	240.00	192.00
A-2641		avidin, BODIPY® FL conjugate	5 mg	75.00	60.00
A-887		avidin, egg white	100 mg	195.00	156.00
A-2667		avidin, egg white	5 mg	18.00	14.40
A-821		avidin, fluorescein conjugate	5 mg	75.00	60.00
A-2666		avidin, NeutraLite™	5 mg	78.00	62.40
A-6379	**New**	avidin, NeutraLite™, AMCA-S conjugate	1 mg	65.00	52.00
A-2661		avidin, NeutraLite™, BODIPY® FL conjugate	1 mg	65.00	52.00
A-2663		avidin, NeutraLite™, Cascade Blue® conjugate	1 mg	65.00	52.00
A-2662		avidin, NeutraLite™, fluorescein conjugate	1 mg	65.00	52.00
A-2664		avidin, NeutraLite™, horseradish peroxidase conjugate	1 mg	65.00	52.00
A-6374	**New**	avidin, NeutraLite™, Oregon Green™ 488 conjugate	1 mg	65.00	52.00
A-6376	**New**	avidin, NeutraLite™, Oregon Green™ 500 conjugate	1 mg	65.00	52.00
A-6375	**New**	avidin, NeutraLite™, Oregon Green™ 514 conjugate	1 mg	65.00	52.00
A-6378	**New**	avidin, NeutraLite™, Rhodamine Red™-X conjugate	1 mg	65.00	52.00
A-2660		avidin, NeutraLite™, R-phycoerythrin conjugate *1 mg/mL*	1 mL	145.00	116.00
A-6373	**New**	avidin, NeutraLite™, tetramethylrhodamine conjugate	1 mg	65.00	52.00
A-2665		avidin, NeutraLite™, Texas Red® conjugate	1 mg	65.00	52.00
A-820		avidin, Texas Red® conjugate	5 mg	75.00	60.00
F-8766		FluoSpheres® biotin-labeled microspheres, 0.04 µm, yellow-green fluorescent (505/515) *1% solids*	0.4 mL	165.00	132.00
F-8767		FluoSpheres® biotin-labeled microspheres, 0.2 µm, yellow-green fluorescent (505/515) *1% solids*	0.4 mL	165.00	132.00
F-8769		FluoSpheres® biotin-labeled microspheres, 1.0 µm, nonfluorescent *1% solids*	0.4 mL	165.00	132.00
F-8768		FluoSpheres® biotin-labeled microspheres, 1.0 µm, yellow-green fluorescent (505/515) *1% solids*	0.4 mL	165.00	132.00
F-8772		FluoSpheres® NeutraLite™ avidin-labeled microspheres, 0.04 µm, nonfluorescent *1% solids*	0.4 mL	165.00	132.00
F-8770		FluoSpheres® NeutraLite™ avidin-labeled microspheres, 0.04 µm, red fluorescent (580/605) *1% solids*	0.4 mL	165.00	132.00
F-8771		FluoSpheres® NeutraLite™ avidin-labeled microspheres, 0.04 µm, yellow-green fluorescent (505/515) *1% solids*	0.4 mL	165.00	132.00
F-8773		FluoSpheres® NeutraLite™ avidin-labeled microspheres, 0.2 µm, red fluorescent (580/605) *1% solids*	0.4 mL	165.00	132.00
F-8774		FluoSpheres® NeutraLite™ avidin-labeled microspheres, 0.2 µm, yellow-green fluorescent (505/515) *1% solids*	0.4 mL	165.00	132.00
F-8777		FluoSpheres® NeutraLite™ avidin-labeled microspheres, 1.0 µm, nonfluorescent *1% solids*	0.4 mL	165.00	132.00
F-8775		FluoSpheres® NeutraLite™ avidin-labeled microspheres, 1.0 µm, red fluorescent (580/605) *1% solids*	0.4 mL	165.00	132.00
F-8776		FluoSpheres® NeutraLite™ avidin-labeled microspheres, 1.0 µm, yellow-green fluorescent (505/515) *1% solids*	0.4 mL	165.00	132.00
F-8780	**New**	FluoSpheres® streptavidin-labeled microspheres, 0.04 µm, yellow-green fluorescent (505/515) *0.5% solids*	0.4 mL	165.00	132.00
P-917		peroxidase from horseradish, biotin-XX conjugate	10 mg	75.00	60.00
S-888		streptavidin	5 mg	135.00	108.00
S-2669		streptavidin *special packaging*	10x100 µg	48.00	38.40
S-951		streptavidin agarose *sedimented bead suspension*	5 mL	125.00	100.00
S-921		streptavidin, alkaline phosphatase conjugate *2 mg/mL*	0.5 mL	155.00	124.00
S-868		streptavidin, allophycocyanin, crosslinked, conjugate *1 mg/mL*	0.5 mL	285.00	228.00
S-6364	**New**	streptavidin, AMCA-S conjugate	1 mg	95.00	76.00
S-2642		streptavidin, BODIPY® FL conjugate	1 mg	95.00	76.00
S-6372	**New**	streptavidin, eosin conjugate	1 mg	95.00	76.00
S-869		streptavidin, fluorescein conjugate	1 mg	95.00	76.00
S-931		streptavidin, β-galactosidase conjugate	1 mg	125.00	100.00
S-884		streptavidin, lucifer yellow conjugate	1 mg	95.00	76.00
S-6368	**New**	streptavidin, Oregon Green™ 488 conjugate	1 mg	95.00	76.00
S-6367	**New**	streptavidin, Oregon Green™ 500 conjugate	1 mg	95.00	76.00
S-6369	**New**	streptavidin, Oregon Green™ 514 conjugate	1 mg	95.00	76.00
S-911		streptavidin, peroxidase from horseradish conjugate	1 mg	95.00	76.00
S-867		streptavidin, B-phycoerythrin conjugate *1 mg/mL*	1 mL	145.00	116.00
S-866		streptavidin, R-phycoerythrin conjugate *1 mg/mL*	1 mL	145.00	116.00
S-871		streptavidin, rhodamine B conjugate	1 mg	95.00	76.00
S-6366	**New**	streptavidin, Rhodamine Red™-X conjugate	1 mg	95.00	76.00
S-6371	**New**	streptavidin, Rhodol Green™ conjugate	1 mg	95.00	76.00
S-870		streptavidin, tetramethylrhodamine conjugate	1 mg	95.00	76.00

Cat #		Product Name	Unit Size	Price Per Unit ($) 1–4 Units	5–24 Units
S-872		streptavidin, Texas Red® conjugate	1 mg	95.00	76.00
S-6370	*New*	streptavidin, Texas Red®-X conjugate	1 mg	95.00	76.00
T-8860	*New*	TransFluoSpheres® NeutraLite™ avidin-labeled microspheres, 0.04 µm (488/605) *1% solids*	0.4 mL	175.00	140.00
T-8861	*New*	TransFluoSpheres® NeutraLite™ avidin-labeled microspheres, 0.1 µm (488/605) *1% solids*	0.4 mL	175.00	140.00
T-8862	*New*	TransFluoSpheres® NeutraLite™ avidin-labeled microspheres, 1.0 µm (488/605) *1% solids*	0.4 mL	175.00	140.00

Add Free Biotin to Obtain Brighter Signals from Fluorescent Avidin Conjugates

Fluorophores conjugated to avidin and streptavidin may be quenched significantly, apparently because the dyes interact with amino acid residues in the biotin-binding pocket. Exceptions include Cascade Blue®– and phycobiliprotein-labeled avidin and streptavidin; the dyes in these conjugates are not quenched because they do not interact with the biotin-binding site. A significant recovery of the avidin or streptavidin conjugate's fluorescence can be obtained if biotin (B-1595, see Chapter 4.2) is added as a final incubation step in the staining procedure (see figure). Fluorescence enhancement of avidin conjugates by biotin has been shown to occur in <100 milliseconds.[1] Biotin apparently blocks the interaction of the fluorophore with residues in the biotin-binding pocket that quench the fluorescence, enhancing the fluorescence of the stained tissue, often severalfold.

1. Biophys J 69, 716 (1995).

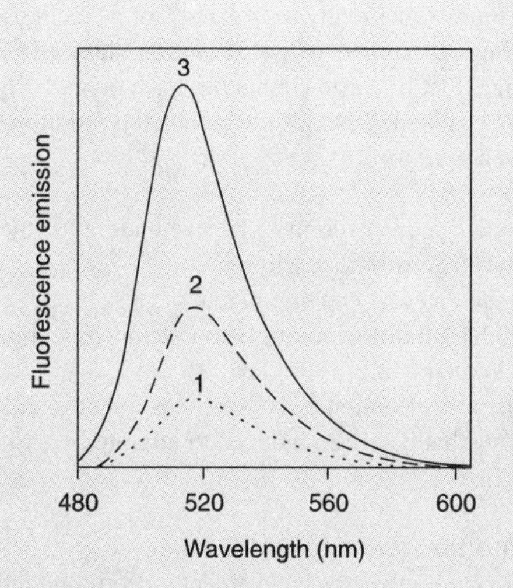

Spectra showing the fluorescence of 1) fluorescein-labeled avidin, 2) fluorescein-labeled avidin after addition of 10 µM biotin and 3) free fluorescein at the same concentration as the fluorescein label in the avidin conjugate.

7.6 Lectin Conjugates

Cellular proteoglycans, glycoproteins and glycolipids may contain any of a wide variety of oligosaccharides. Although most abundant on the cell surface, oligosaccharide residues are sometimes also found covalently attached to the constituents found within the cell. Often, specific oligosaccharides are associated with a specific cell type or organelle. Lectins, which bind to specific configurations of sugar molecules, can thus serve to identify cell types or cellular components, making them versatile primary detection reagents in histochemical applications and flow cytometry.[1]

Fluorescent Lectin Conjugates

Molecular Probes offers several fluorescent conjugates of concanavalin A (Con A) and wheat germ agglutinin (WGA) — two of the most commonly used lectins in cell biology. The 104,000-dalton tetrameric Con A selectively binds to α-mannopyranosyl and α-glucopyranosyl residues. The 36,000-dalton dimeric WGA (normally cationic) binds to sialic acid and *N*-acetylglucosaminyl residues. In addition, WGA undergoes axonal transport[2,3] and has been shown to cross from axonal endings into adjacent neurons.[4]

Our newest lectin conjugates include Con A and WGA conjugates of our green fluorescent Oregon Green 488 dye (C-6741, W-6748), which is more fluorescent and more photostable than fluorescein, and a Con A conjugate of our Oregon Green 514 dye (C-6742), which is our most photostable green fluorophore for protein labeling. In addition, we have tested our new Texas Red-X conjugate of WGA (W-6746) and the previously offered Texas Red conjugate (former catalog number W-831) side-by-side and found the Texas Red-X conjugate produced much brighter staining with less background. Similar testing has shown that our new blue fluorescent AMCA-S Con A (C-6749) and AMCA-S WGA (W-6745) are significantly more fluorescent than the corresponding AMCA conjugates (former catalog numbers C-878 and W-879). For researchers interested in testing our fluorescent WGA conjugates in their application, we also now offer a Wheat Germ Agglutinin Stain Sampler Kit (W-7024) which contains 1 mg quantities each of WGA conjugates of the AMCA-S, Oregon Green 488, tetramethylrhodamine and Texas Red-X dyes.

Applications for Fluorescent Con A and WGA Conjugates

Applications for Fluorescent Con A

In most cell types, Con A binds to the endoplasmic reticulum, whereas WGA binds to the Golgi apparatus,[5,6] making these fluorescent lectins ideal reagents for facilitating the immunolocalization of oncogene products,[7] specific intracellular enzymes,[8,9] viral proteins [10] and components of the cytoskeleton.[11] However, Con A reportedly binds specifically to isolated Golgi fractions from rat liver, enabling researchers to use fluorescein-labeled Con A to examine the effect of chronic ethanol intake on carbohydrate content in these organelles using flow cytometry.[12] Fluorescent Con A has also been used to:

- Determine if human sperm cells have undergone the progesterone-induced acrosome reaction [13]
- Investigate receptor capping in leukocytes [14]
- Measure lateral diffusion of glycoproteins, glycolipids and viruses in membranes [15,16]
- Show the redistribution of cell-surface glycoproteins in murine fibroblasts that had been induced to migrate by exposure to an electric field [17]

Applications for Fluorescent WGA

Nuclear core complexes have recently been found to contain several proteins with *O*-linked *N*-acetylglucosaminyl residues.[18-22] In a study of nuclear protein transport, nuclei isolated from monkey kidney epithelial cells were demonstrated to be intact by their bright staining with fluorescein WGA (W-834); fluorescein Con A (C-827), which binds to residues accessible only in nuclei with compromised membranes, was used as a negative control for intact nuclei.[23] Fluorescent WGA has also been employed to monitor the reconstitution of the nuclear core complex in *Xenopus* egg extracts.[24]

Fluorescent lectins are also useful in microbiology applications. Fluorescent WGA conjugates stain chitin in fungal cell walls,[25] as well as gram-positive but not gram-negative bacteria.[26] Fluorescent WGA conjugates are used in our ViaGram™ Red+ Bacterial Gram Stain and Viability Kit to differentiate gram-positive and gram-negative bacteria [27] (see Section 16.2). Fluorescent WGA has also been shown to bind to sheathed microfilariae and has been used to detect filarial infection in blood smears.[28]

In addition to these nuclear core and microbiology studies, fluorescent WGA has been used to:

- Bind the sarcolemma of rat and dog cardiac myocytes, even within the intercalated discs and transverse tubules, allowing researchers to map the distribution of gap junctions in these cell types [29]
- Determine the intracellular distribution of altered lysosomal proteins, enabling researchers to define the sequence requirements for proper cell sorting [30]
- Identify the differentiation state of Madin–Darby canine kidney (MDCK) cells [31]
- Investigate plant hemicelluloses [32]
- Measure cell membrane potential in combination with potential-sensitive membrane probes [33]

1. Science 246, 227 (1989); **2.** J Neurosci Meth 9, 185 (1983); **3.** J Neurosci 2, 647 (1982); **4.** Brain Res 344, 41 (1985); **5.** Cytometry 19, 112 (1995); **6.** J Cell Biol 85, 429 (1980); **7.** J Cell Biol 111, 3097 (1990); **8.** J Biol Chem 269, 1727 (1994); **9.** J Cell Biol 111, 2851 (1990); **10.** J Cell Biol 110, 625 (1990); **11.** J Cell Biol 111, 1929 (1990); **12.** Exp Cell Res 207, 136 (1993); **13.** Mol Cell Endocrinol 101, 221 (1994); **14.** J Virol Meth 29, 257 (1990); **15.** FEBS Lett 246, 65 (1989); **16.** Cell 23, 423 (1981); **17.** J Cell Biol 127, 117 (1994); **18.** Ann Rev Cell Biol 8, 495 (1992); **19.** Science 258, 942 (1992); **20.** Biochim Biophys Acta 1071, 83 (1991); **21.** Cell 64, 489 (1991); **22.** Physiol Rev 71, 909 (1991); **23.** J Biol Chem 269, 4910 (1994); **24.** J Biol Chem 269, 9289 (1994); **25.** Investigative Ophthalmol Visual Sci 27, 500 (1986); **26.** Appl Environ Microbiol 56, 2245 (1990); **27.** Purchase of the ViaGram Red+ Bacterial Gram Stain and Viability Kit is accompanied by a research license under U.S. Patent No. 5,137,810 for determination of bacterial gram sign with fluorescent lectins; **28.** Int'l J Parasitol 20, 1099 (1990); **29.** J Mol Cell Cardiol 24, 1443 (1992); **30.** J Cell Biol 111, 955 (1990); **31.** Cell Physiol Biochem 3, 42 (1993); **32.** Protoplasma 156, 67 (1990); **33.** Biophys J 69, 1272 (1995).

7.6 Price List *Lectin Conjugates*

Cat #		Product Name	Unit Size	Price Per Unit ($) 1–4 Units	5–24 Units
C-6749	*New*	concanavalin A, AMCA-S conjugate	5 mg	48.00	38.40
C-827		concanavalin A, fluorescein conjugate	10 mg	48.00	38.40
C-6741	*New*	concanavalin A, Oregon Green™ 488 conjugate	5 mg	48.00	38.40
C-6742	*New*	concanavalin A, Oregon Green™ 514 conjugate	5 mg	48.00	38.40
C-860		concanavalin A, tetramethylrhodamine conjugate	10 mg	48.00	38.40
C-825		concanavalin A, Texas Red® conjugate	10 mg	48.00	38.40
W-6745	*New*	wheat germ agglutinin, AMCA-S conjugate	5 mg	98.00	78.40
W-966		wheat germ agglutinin, Cascade Blue® conjugate	5 mg	98.00	78.40
W-834		wheat germ agglutinin, fluorescein conjugate	5 mg	98.00	78.40
W-6748	*New*	wheat germ agglutinin, Oregon Green™ 488 conjugate	5 mg	98.00	78.40
W-7024	*New*	Wheat Germ Agglutinin Sampler Kit *four fluorescent conjugates, 1 mg each*	1 kit	98.00	78.40
W-849		wheat germ agglutinin, tetramethylrhodamine conjugate	5 mg	98.00	78.40
W-6746	*New*	wheat germ agglutinin, Texas Red®-X conjugate	5 mg	98.00	78.40

Chapter 8

Nucleic Acid Detection

Contents

Technical Notes and Product Highlights

Related Chapters

8.1 Nucleic Acid Stains

Molecular Probes prepares the most extensive assortment of nucleic acid stains commercially available, many of which have been developed in our research laboratories. Table 8.1 serves as a quick reference guide to our nucleic acid stains. The properties of each chemical dye class are discussed below.

Cyanine Dyes

During the past five years, Molecular Probes has invented several new cyanine dyes that share some unique properties:

- High molar absorptivity, with extinction coefficients usually >50,000 $cm^{-1}M^{-1}$ at visible wavelengths
- Very low intrinsic fluorescence, with quantum yields usually <0.01 when not bound to nucleic acids
- Large fluorescence enhancements (often over 1000-fold) upon binding to nucleic acid, with increases in quantum yields to as high as 0.9
- Moderate to very high affinity for nucleic acids, with little or no staining of other biopolymers

Furthermore, representatives of this new class of nucleic acid stains have fluorescence excitations and emissions that span the visible spectrum from blue to near infrared (Figure 8.1 and Table 8.1) with additional absorption peaks in the UV, making them compatible with many different types of instrumentation.

Although these cyanine dyes share common optical properties, chemical differences allow us to assign them to distinct subclasses. The **TOTO series** of dyes includes the TOTO®, YOYO®, BOBO™ and POPO™ analogs, which are symmetric dimers of cyanine dyes [1] with four positive charges and very high affinity for nucleic acids. These dyes are generally considered to be cell impermeant,[2] although their use to stain reticulocytes permeabilized by 5% DMSO has been reported.[3] YOYO-1 has been microinjected into cells in order to follow mitotic chromosomes through at least six cell cycles in fertilized sea urchin eggs.[4]

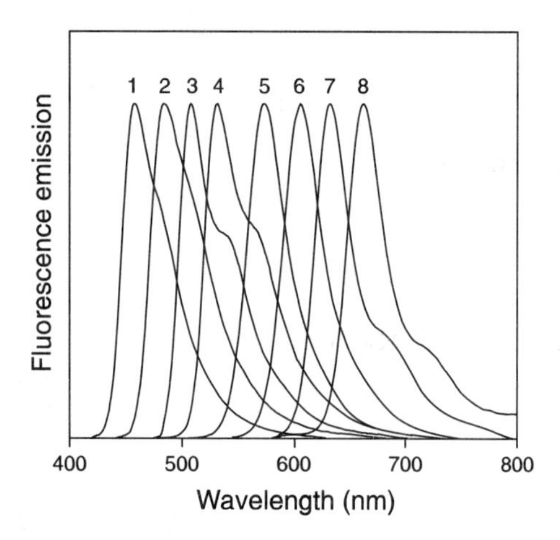

Figure 8.1 *Normalized fluorescence emission spectra of DNA-bound dimeric cyanine dimers: 1) POPO-1 (P-3580), 2) BOBO-1 (B-3582), 3) YOYO-1 (Y-3601), 4) TOTO-1 (T-3600), 5) POPO-3 (P-3584), 6) BOBO-3 (B-3586), 7) YOYO-3 (Y-3606) and 8) TOTO-3 (T-3604).*

Our patented **TO-PRO series** of dyes includes TO-PRO™, YO-PRO™, BO-PRO™ and PO-PRO™ analogs, which comprise a single cyanine dye and a cationic side chain.[5] The dyes in the TO-PRO series are spectrally analogous to the corresponding dimeric cyanine dyes; however, with only two positive charges, they exhibit reduced affinity for nucleic acids relative to the dyes in the TOTO series.[5] Like their dimeric counterparts, these monomeric cyanine dyes are typically impermeant to cells.[6]

The **SYTO® dyes** are lower-affinity nucleic acid stains that passively diffuse through cell membranes.[7] These cell-permeant, visible light–excitable dyes can be used to stain RNA and DNA in both live and dead eukaryotic cells, as well as in gram-positive and gram-negative bacteria.

In addition to synthesizing new dyes, we have developed several important bioanalytical applications that use a variety of nucleic acid stains, some of which are cyanine dyes:

- **SYBR® Green I and SYBR® Green II nucleic acid gel stains** are ultrasensitive gel stains that surpass conventional dyes, including ethidium bromide, by more than an order of magnitude in nucleic acid sensitivity (see Section 8.3). Furthermore, Ames testing by an independent laboratory has shown that SYBR Green I stain is significantly less mutagenic than ethidium bromide (see Figure 8.7 in Section 8.3).
- **SYBR DX DNA blot stain** allows the direct detection of DNA on filter membranes after Southern transfer, with sensitivity equivalent to that achieved with silver-enhanced gold staining (see Section 8.3).
- **PicoGreen® and OliGreen™ quantitation reagents** set a new benchmark for the detection and quantitation of DNA and oligonucleotides in solution, with extremely simple and rapid protocols as well as linear ranges that span four orders of magnitude in nucleic acid concentration (see Section 8.3).
- **CyQUANT™ GR dye** is a new reagent for quantitating cell proliferation that can reliably detect the nucleic acids in fewer than 50 cells (see Sections 16.2 and 16.3).
- **SYTOX® Green nucleic acid stain** is an extremely versatile nucleic acid stain for identifying cells with compromised membranes, counterstaining fixed cells and chromosomes, enumerating bacteria, performing flow cytometric cell-cycle analysis and, in conjunction with methyl green, producing chromosome banding patterns (see Sections 8.4 and 16.3).
- **SYTO 9, SYTO 10, SYBR 14, FUN-1™ and DEAD Red™ dyes** are specialty reagents that we have utilized for viability assays in bacteria, yeast, spermatozoa and other cells (see Sections 16.2 and 16.3).

Table 8.1 Properties of Molecular Probes' nucleic acid stains.

Nucleic Acid Stain (Cat #)	Ex/Em (nm/nm) *	Properties and Uses
Acridine homodimer (A-666)	431/498	• Impermeant • AT selective • High-affinity DNA labeling
Acridine orange (A-1301, A-3568 †)	500/526 (DNA) 460/650 (RNA)	• Permeant • RNA/DNA discrimination measurements • Lysosome labeling • Flow cytometry
Actinomycin D (A-7592)	nonfluorescent	• GC selective • Used with DAPI or Hoechst 33258 for chromosome banding
7-Aminoactinomycin D (7-AAD) (A-1310)	546/647	• Weakly permeant • GC selective • Flow cytometry
9-Amino-6-chloro-2-methoxyacridine (ACMA) (A-1324)	419/483	• AT selective • Alternative to quinacrine for chromosome Q banding • Membrane phenomena
BOBO-1 (B-3582)	462/481	• Impermeant • High-affinity DNA labeling • Dead-cell stain • Chromosome and cell counterstain • Electrophoresis prestain
BOBO-3 (B-3586)	570/604	• Impermeant • High-affinity DNA labeling • Dead-cell stain • Chromosome and cell counterstain • Electrophoresis prestain
BO-PRO-1 (B-3583)	462/481	• Impermeant • Dead-cell stain • Electrophoresis
BO-PRO-3 (B-3587)	575/599	• Impermeant • Dead-cell stain • Electrophoresis
4′,6-Diamidino-2-phenylindole (DAPI) (D-1306, D-3571)	358/461	• Semi-permeant • AT selective • Cell-cycle studies • Mycoplasma detection • Chromosome and nuclei counterstain
Dihydroethidium (D-1168)	518/605 ‡	• Permeant • Redox indicator • Blue fluorescent until oxidized to ethidium
4′,6-(Diimidazolin-2-yl)-2-phenylindole (DIPI) (D-7594)	364/455	• AT selective • Used with chromomycin for chromosome banding
Ethidium–acridine heterodimer (E-667)	455 and 532/619	• Impermeant • AT selective • High-affinity DNA labeling
Ethidium bromide (E-1305, E-3565 †)	518/605	• Impermeant • dsDNA intercalator • Dead-cell stain • Chromosome counterstain • Electrophoresis • Flow cytometry • Argon-ion laser excitable
Ethidium diazide (E-3561)	432/496 § (unbound)	• Impermeant • Photocrosslinks DNA
Ethidium homodimer-1 (EthD-1) (E-1169)	528/617	• Impermeant • High-affinity DNA labeling • Dead-cell stain • Electrophoresis prestain • Argon-ion and green He-Ne laser excitable
Ethidium homodimer-2 (EthD-2) (E-3599)	535/624	• Impermeant • Very high-affinity DNA labeling • Electrophoresis prestain
Ethidium monoazide (E-1374)	464/625 § (unbound)	• Impermeant • Photoaffinity label • Compatible with fixation procedures
Hexidium iodide (H-7593)	518/600	• Permeant • Stains nucleic and cytoplasm of eukaryotes and some bacteria
Hoechst 33258 (bis-benzimide) (H-1398, H-3569 †)	352/461	• Permeant • AT selective • Minor-groove binding • dsDNA-selective binding • Live-cell stain • Cell-cycle studies • Chromosome and cell counterstain
Hoechst 33342 (H-1399, H-3570 †)	350/461	• Permeant • AT selective • Minor-groove binding • Live-cell stain • Cell-cycle studies • Chromosome and cell counterstain
Hydroxystilbamidine, methanesulfonate (H-7599)	385/emission varies with nucleic acid	• AT selective • Spectra dependent on secondary structure and sequence • RNA/DNA discrimination • Nuclear stain in tissue
LDS 751 (L-7595)	543/712	• Permeant • High Stokes shift • Long-wavelength spectra • Flow cytometry
OliGreen ssDNA quantitation reagent (O-7582)	498/518 (random sequence 40-mer)	• Ultrasensitive reagent for solution quantitation of ssDNA and oligonucleotides
PicoGreen dsDNA quantitation reagent (P-7581)	502/523	• Ultrasensitive reagent for solution quantitation of dsDNA
POPO-1 (P-3580)	434/456	• Impermeant • High-affinity DNA labeling • Dead-cell stain • Chromosome and cell counterstain • Electrophoresis prestain
POPO-3 (P-3584)	534/570	• Impermeant • High-affinity DNA labeling • Dead-cell stain • Chromosome and cell counterstain • Electrophoresis prestain

Table 8.1, continued *Properties of Molecular Probes' nucleic acid stains.*

Nucleic Acid Stain (Cat #)	Ex/Em (nm/nm) *	Properties and Uses
PO-PRO-1 (P-3581)	435/455	• Impermeant • Dead-cell stain • Electrophoresis
PO-PRO-3 (P-3585)	539/567	• Impermeant • Dead-cell stain • Electrophoresis
Propidium iodide (P-1304, P-3566 †)	535/617	• Impermeant • Dead-cell stain • Chromosome and cell counterstain
SYBR Green I nucleic acid gel stain (S-7567)	494/521	• Ultrasensitive gel stain for DNA and oligonucleotides post-electrophoresis
SYBR Green II RNA gel stain (S-7568)	492/513	• Sensitive stain for RNA and ssDNA post-electrophoresis
SYTO 11 live-cell nucleic acid stain (S-7573)	508/527	• Permeant ¶ • Green fluorescent live-cell stain • Argon-ion laser excitable
SYTO 12 live-cell nucleic acid stain (S-7574)	500/522	• Permeant ¶ • Green fluorescent live-cell stain • Argon-ion laser excitable
SYTO 13 live-cell nucleic acid stain (S-7575)	488/509	• Permeant ¶ • Green fluorescent live-cell stain • Argon-ion laser excitable
SYTO 14 live-cell nucleic acid stain (S-7576)	517/549	• Permeant ¶ • Yellow-green fluorescent live-cell stain • Argon-ion laser excitable
SYTO 15 live-cell nucleic acid stain (S-7577)	516/546	• Permeant ¶ • Yellow-green fluorescent live-cell stain • Argon-ion laser excitable
SYTO 16 live-cell nucleic acid stain (S-7578)	488/518	• Permeant ¶ • Green fluorescent live-cell stain • Argon-ion laser excitable
SYTO 20 live-cell nucleic acid stain (S-7555)	512/530	• Permeant ¶ • Green fluorescent live-cell stain • Argon-ion laser excitable
SYTO 21 live-cell nucleic acid stain (S-7556)	494/517	• Permeant ¶ • Green fluorescent live-cell stain • Argon-ion laser excitable
SYTO 22 live-cell nucleic acid stain (S-7557)	515/535	• Permeant ¶ • Green fluorescent live-cell stain • Argon-ion laser excitable
SYTO 23 live-cell nucleic acid stain (S-7558)	499/520	• Permeant ¶ • Green fluorescent live-cell stain • Argon-ion laser excitable
SYTO 24 live-cell nucleic acid stain (S-7559)	490/515	• Permeant ¶ • Green fluorescent live-cell stain • Argon-ion laser excitable
SYTO 25 live-cell nucleic acid stain (S-7560)	521/556	• Permeant ¶ • Yellow-green fluorescent live-cell stain • Argon-ion laser excitable
SYTO 17 red fluorescent nucleic acid stain (S-7579)	621/634	• Permeant ¶ • Stains live and dead bacteria and eukaryotic cells • Primarily cytoplasmic labeling • Long-wavelength spectra
SYTOX Green nucleic acid stain (S-7020)	504/523	• Impermeant • High-affinity dead-cell stain • Chromosome and cell counterstain • Used with methyl green for chromosome banding • Flow cytometry
TO-PRO-1 (T-3602)	515/531	• Impermeant • Dead-cell stain • Electrophoresis • Argon-ion laser excitable
TO-PRO-3 (T-3605)	642/661	• Impermeant • Dead-cell stain • Electrophoresis • Long-wavelength spectra
TO-PRO-5 (T-7596)	747/770	• Impermeant • Dead-cell stain • Electrophoresis • Very long-wavelength spectra
TOTO-1 (T-3600)	514/533	• Impermeant • High-affinity DNA labeling • Dead-cell stain • Chromosome and cell counterstain • Electrophoresis prestain • Argon-ion laser excitable
TOTO-3 (T-3604)	642/660	• Impermeant • High-affinity DNA labeling • Dead-cell stain • Chromosome and cell counterstain • Electrophoresis prestain • Long-wavelength spectra
YO-PRO-1 (Y-3603)	491/509	• Impermeant • Dead-cell stain • Electrophoresis • Argon-ion laser excitable
YO-PRO-3 (Y-3607)	612/631	• Impermeant • Dead-cell stain • Electrophoresis • Orange He-Ne laser excitable
YOYO-1 (Y-3601)	491/509	• Impermeant • Ultrasensitive high-affinity DNA labeling • Single molecule labeling • Dead-cell stain • Chromosome and cell counterstain • Used with methyl green for chromosome banding • Electrophoresis prestain
YOYO-3 (Y-3606)	612/631	• Impermeant • High-affinity DNA labeling • Dead-cell stain • Chromosome and cell counterstain • Electrophoresis prestain • Orange He-Ne laser excitable

* All excitation and emission maxima were determined for dyes bound to double-stranded calf thymus DNA in aqueous solution, unless otherwise indicated. † Available in aqueous solution for those wishing to avoid potentially hazardous and mutagenic powders. ‡ After oxidation to ethidium. § Prior to photolysis; after photolysis the spectra of the dye/DNA complexes are similar to those of ethidium bromide/DNA complexes. ¶ All SYTO stains label both nuclei and cytoplasm; some SYTO dyes label mitochondria as well.

TOTO, YOYO, BOBO and POPO Dyes

Appropriately designed dimers of nucleic acid–binding dyes have been shown to have nucleic acid binding affinities several orders-of-magnitude greater than those of their parent compounds.[8-10] For example, the intrinsic DNA binding affinity constants of ethidium bromide (E-1305) and ethidium homodimer-1 (E-1169) are 1.5×10^5 and 2×10^8 M^{-1}, respectively, in 0.2 M Na^+.[11] In light of these observations, we have developed an important set of dimeric cyanine dyes — the TOTO, YOYO, BOBO and POPO dyes — that are among the most sensitive and highest affinity fluorescent probes available for nucleic acid staining. Their sensitivity is sufficient for detecting single molecules of labeled nucleic acids by optical imaging and flow cytometry (see Section 8.3 and photo on back cover) and for tracking labeled virus particles in microbial communities by fluorescence microscopy.[12]

In the TOTO-1 dimeric cyanine dye (T-3600), the positively charged side chains of two TO-PRO-1 monomeric cyanine dyes (T-3602, see below) are covalently linked to form a molecule with four positive charges that has greatly enhanced affinity for nucleic acids — over 100-times greater than that of the TO-PRO-1 monomer. TOTO-1 exhibits a higher affinity for double-stranded DNA (dsDNA) than even the ethidium homodimers (see below) and also binds to both single-stranded DNA (ssDNA) and RNA. The extraordinary stability of TOTO-1/nucleic acid complexes [9,13,14] ensures that the dye/DNA association remains stable, even during electrophoresis; thus, samples can be prestained with nanomolar dye concentrations *prior to* electrophoresis,[15,16] thereby reducing the hazards inherent in handling large volumes of ethidium bromide staining solutions.[9,14,17] In contrast, the binding of thiazole orange, the parent compound of TOTO-1 and TO-PRO-1, is rapidly reversible, limiting the dye's sensitivity and rendering its nucleic acid complex unstable to electrophoresis.[17]

In addition to their superior binding properties, TOTO-1 and the other cyanine dimers are essentially nonfluorescent in the absence of nucleic acids and exhibit significant fluorescence enhancements upon DNA binding [13,18] (100- to 1000-fold), which compares favorably with the fluorescence enhancement of thiazole orange upon DNA binding [19] (~3000-fold). Furthermore, fluorescence quantum yields of the cyanine dimers bound to DNA are high (generally between 0.2 and 0.6), and their extinction coefficients are an order-of-magnitude greater than those of the ethidium homodimers.[13]

Simply by changing the aromatic rings and the number of carbon atoms linking the cyanine monomers (see Structures in Section 8.1), we were able to synthesize an extended series of these dyes. These structural changes produce dramatic shifts in the molecules' absorption and emission spectra and reduce the quantum yields of the bound dyes but cause little or no change in their high affinity for DNA. The names of the dyes reflect their basic structure and spectral characteristics. For example, YOYO-1 (491/509) has one carbon atom bridging the aromatic rings of the oxacyanine dye and exhibits absorption/emission maxima of 491/509 nm when bound to dsDNA. The dsDNA complex of YOYO-3 (612/631) — which differs from YOYO-1 only in the number of bridging carbon atoms (three) — has absorption/emission maxima of 612/631 nm when bound to dsDNA. Fluorescence spectra for POPO, BOBO, YOYO and TOTO dyes bound to dsDNA are shown in Figure 8.1. The spectra of these dyes at dye:base ratios of less than 1:1 are essentially the same for the corresponding dye/ssDNA and dye/RNA complexes. At higher dye:base ratios, ssDNA and RNA complexes of all of the monomethine ("-1") dyes of the TOTO series and TO-PRO series have red-shifted emissions, whereas corresponding complexes of the trimethine ("-3") analogs do not.

The 491 nm absorption maximum of YOYO-1 when bound to dsDNA makes it the preferred dye for excitation by the 488 nm spectral line of the argon-ion laser or by a fluorescence microscope equipped with a standard fluorescein optical filter set. Nucleic acid–bound TOTO-1 absorbs maximally near 514 nm, which is ideal for excitation by the 514 nm spectral line of the argon-ion laser. YOYO-3 (612/631) and TOTO-3 (642/660) are longer-wavelength analogs of YOYO-1 and TOTO-1. TOTO-3 can be excited by both the red He-Ne laser at 633 nm and the 647 nm spectral line of the Ar-Kr laser used in some confocal laser scanning microscopes. TOTO-3 (and TO-PRO-3) is an effective dye for detecting DNA using a flow cytometer equipped with an inexpensive 3 mW visible-wavelength diode laser that provides excitation at 635 nm.[20] BOBO-1 (462/481) and POPO-1 (434/456) are high-affinity, blue fluorescent nucleic acid stains, whereas BOBO-3 (570/602) and POPO-3 (534/570) emit at wavelengths between those of TOTO-1 and YOYO-3 (Table 8.1). Some of these dyes are better suited for other laser lines such as those of the He-Cd (442 nm), green He-Ne (543 nm), Kr (568 nm and 647 nm) and orange He-Ne (594 nm) excitation sources.

Recent work has focused on the nucleic acid binding modes of these cyanine dimers. YOYO-1 was found to exhibit at least two distinct binding modes: at low dye:base pair ratios, the binding mode appears to consist primarily of intercalation; at high dye:base pair ratios, a second mode involving external binding begins to contribute.[21,22] Circular dichroism measurements also indicate a possible difference in the binding modes of YOYO-1 to ssDNA and dsDNA.[23] These data are consistent with our own results, including the observation that YOYO-1's fluorescence emission shifts to longer wavelengths at high dye:base ratios upon binding to single-stranded nucleic acids and that the salt, ethanol and sodium dodecyl sulfate sensitivity of YOYO-1 binding to DNA is a function of the dye:base pair ratio.[24]

The binding modes of the other members of the TOTO series have also been characterized. TOTO-1 and ethidium homodimer-1 [25,26] (E-1169, see below) are capable of bis-intercalation, although they are reported to interact with dsDNA and ssDNA with similar high affinity.[8] Electrophoresis and fluorescence lifetime measurements have shown that YOYO-3 also appears to intercalate into DNA.[27] During application development, we have determined that staining of nucleic acids by BOBO-1 and POPO-1 is much faster (occurring within minutes) than staining by YOYO-1 and TOTO-1 (which can take as long as 2 hours to reach equilibrium under the same experimental conditions), indicating possible differences in their binding mechanisms.

Fluorescence yield and lifetime measurements have been used to assess the base selectivity of an extensive series of these dyes.[18] NMR studies of TOTO-1 interactions with a double-stranded 8-mer indicated that it is a bis-intercalator with interactions in the minor groove and that it distorts the helix by unwinding it where the dye is bound.[28] These same researchers also reported that TOTO-1 exhibits strong sequence selectivity for the site CTAG.[29,30]

All dyes of the TOTO series are supplied as 1 mM solutions in dimethylsulfoxide (DMSO). These cationic dyes appear to be readily adsorbed out of aqueous solutions onto surfaces (particularly glass) but are very stable once complexed to nucleic acids. A number of applications of these dyes for staining nucleic acids in solutions, gels and cells are described in Sections 8.3, 8.4 and 16.3.

TO-PRO, YO-PRO, BO-PRO and PO-PRO Dyes

Our nine monomeric nucleic acid stains of the TO-PRO series have all the exceptional spectral properties of the dimeric cyanine dyes listed above. The absorption and emission spectra of these monomeric cyanine dyes cover the visible and near-IR spectrum (Table 8.1). They also have relatively narrow emission bandwidths, thus facilitating multicolor applications in imaging and flow cytometry. YO-PRO-1 (491/509) and TO-PRO-1 (515/531) are optimally excited by the argon-ion laser. TO-PRO-3 (642/661) (and TOTO-3) is an effective dye for detecting DNA using a flow cytometer equipped with an inexpensive 3 mW visible-wavelength diode laser that provides excitation at 635 nm.[20] In flow cytometric analyses, TO-PRO-3 has also been excited directly by the red He-Ne laser[31] and indirectly by the argon-ion laser using fluorescence resonance energy transfer from propidium iodide.[32] Although the DNA-induced fluorescence enhancement of TO-PRO-5 (T-7596) is not as large as that observed with our other cyanine dyes, its spectral characteristics (excitation/emission maxima ~745/770 nm) provide a unique alternative for multicolor applications and specialized instrumentation.

Binding affinity of the TO-PRO series of dyes to dsDNA is lower than that of the TOTO series of dyes but is still very high, with dissociation constants in the micromolar range.[33] They also bind to RNA and ssDNA, although with somewhat lower fluorescence yields.

All dyes of the TO-PRO series are supplied as 1 mM solutions in DMSO. Various applications of the TO-PRO series of dyes for staining nucleic acids are described in Sections 8.3, 8.4 and 16.3.

Nucleic Acid Stain Sampler Kits

Molecular Probes offers the Nucleic Acid Stains Dimer and Monomer Sampler Kits, containing spectrally distinct analogs of the dimeric and monomeric cyanine dyes, respectively. The Nucleic Acid Stains Dimer Sampler Kit (N-7565) contains 10 µL of a 1 mM DMSO solution of each of our eight dimeric dyes of the TOTO series. The Nucleic Acid Stains Monomer Sampler Kit (N-7566) contains 50 µL of a 1 mM DMSO solution of each of eight monomeric dyes of the TO-PRO series (the ninth dye in this series — TO-PRO-5 — is not included). These sampler kits are ideal for researchers who wish to test the full spectral series.

Green Fluorescent Cell-Permeant SYTO Nucleic Acid Stains

Molecular Probes has recently introduced a set of nucleic acid stains capable of penetrating intact live cells. The SYTO live-cell nucleic acid stains have several characteristics in common:

- Permeability to virtually all cell membranes, including mammalian cells and bacteria (see Chapter 16)
- High molar absorptivity, with extinction coefficients >50,000 $cm^{-1}M^{-1}$ at visible wavelengths
- Extremely low intrinsic fluorescence, with quantum yields typically <0.01 when not bound to nucleic acids
- Quantum yields when bound to nucleic acids that are typically >0.4
- Excitation and emission spectra similar to those of fluorescein (except for the SYTO 17 red fluorescent nucleic acid stain, see below)

These novel SYTO stains provide researchers with visible light–excitable dyes for labeling DNA and RNA in live cells (see Color Plate 4 in "Visual Reality" and photo on front cover). The SYTO dyes may also be useful for nucleic acid detection in solution, in electrophoretic gels, on blots and in other assays. SYTO dyes differ from each other in one or more characteristics — including cell permeability, fluorescence enhancement upon binding nucleic acids, excitation and emission spectra (Table 8.1) and DNA/RNA selectivity and binding affinity. They are compatible with a variety of fluorescence instruments that use either a monochromatic source such as the argon-ion laser or a conventional broadband illumination source (e.g., mercury- and xenon-arc lamps).

The SYTO dyes can stain both DNA and RNA. In most cases the fluorescence wavelengths and emission intensities are similar for solution measurements of DNA or RNA binding. Exceptions are SYTO 12 and SYTO 14, which are about twice as fluorescent on RNA than DNA, and SYTO 16, which is about twice as fluorescent on DNA than RNA. Consequently, the SYTO dyes do not act exclusively as nuclear stains in live cells and should therefore not be equated in this regard with compounds such as Hoechst 33258 or Hoechst 33342 that, because of their DNA selectivity, readily stain cell nuclei at low concentrations in most cells. Stained eukaryotic cells will generally show diffuse cytoplasmic staining as well as nuclear staining. Particularly intense staining of intranuclear bodies is frequently observed. Because these dyes are cell permeant and most contain a net positive charge at neutral pH, they may also stain mitochondria. In addition, SYTO dyes will stain both gram-positive and gram-negative bacterial cells. Dead yeast cells are brightly stained with the SYTO dyes, and live yeast cells may exhibit mitochondrial or nuclear staining. Recent reports demonstrate the utility of some of the SYTO dyes for detecting apoptosis,[34-37] and dyes structurally similar to the SYTO dyes have been used to detect multidrug-resistant cells[38] (see Section 16.3). We anticipate that many more applications will be found for these unique, visible light–excitable, live-cell nucleic acid stains.

Twelve different green fluorescent SYTO live-cell nucleic acid stains are currently available as individual reagents (S-7573 through S-7578, S-7555 through S-7560). These dyes are packaged as 250 µL of a 5 mM solution in DMSO (except SYTO 16 and SYTO 20, which have a concentration of 1 mM). In addition, our two SYTO Live-Cell Nucleic Acid Stain Sampler Kits permit evaluation of an extensive range of SYTO dyes. Kit #1 (S-7572) contains 50 µL samples of SYTO 11 to SYTO 16 and Kit #2 (S-7554) contains 50 µL samples of SYTO 20 to SYTO 25, with all dyes provided at the same concentrations as in the individual products. To assist the researcher in designing staining protocols, we include a detailed product information sheet describing the spectral properties of the dyes with each purchase of a sampler kit or individual reagent. Recommended dye concentrations for cell staining depends on the assay and may vary widely: 1–20 µM for bacteria, 1–100 µM for yeast and 10 nM–5 µM for other eukaryotes.

Two other green fluorescent SYTO dyes — SYTO 9 and SYTO 10 — are available in our LIVE/DEAD® *Bac*Light Kits (see Section 16.2); the dye SYBR 14 in our LIVE/DEAD Sperm Viability Kit (L-7011, see Section 16.2) is also in the SYTO family of dyes. SYTO 18 (S-7529) selectively stains yeast mitochondria and is discussed with other mitochondrial probes in Section 12.2.

Red Fluorescent Cell-Permeant SYTO 17 Nucleic Acid Stain

Molecular Probes has recently introduced another member to our growing family of nucleic acid stains for live cells. The cell-permeant SYTO 17 red fluorescent nucleic acid stain (S-7579) shares the nucleic acid–binding properties of the green fluorescent SYTO dyes described above. However, unlike the fluorescein-like spectra of the green fluorescent SYTO dyes bound to nucleic acids, the excitation and emission spectra of the SYTO 17/nucleic acid complex (excitation/emission maxima ~621/634 nm) are quite similar to those of the commonly used Texas Red® fluorophore. We have found that our SYTO 17 dye brightly stains both live and dead bacteria of every species tested, making it a useful tool for determining total number of bacteria in a population with either fluorometry or fluorescence microscopy. In addition to staining bacteria, the SYTO 17 dye labels live and dead eukaryotic cells, primarily in the extranuclear region. This red fluorescent cell-permeant nucleic acid stain may also prove useful as a counterstain when used in combination with green fluorescent antibodies, lectins or the cell-impermeant SYTOX Green nucleic acid stain (see Section 16.3).

Green Fluorescent Cell-Impermeant SYTOX Green Nucleic Acid Stain

SYTOX Green nucleic acid stain (S-7020) is a high-affinity nucleic acid stain that easily penetrates cells with compromised plasma membranes and yet will not cross the membranes of live cells. It is especially useful for staining both gram-positive and gram-negative bacteria — and probably virus particles [12,39] — where an exceptionally bright signal is required. After brief incubation with SYTOX Green stain, the nucleic acids of dead cells fluoresce bright green when excited with the 488 nm spectral line of the argon-ion laser or with any other 450–500 nm source. Unlike DAPI or Hoechst dyes, SYTOX Green nucleic acid stain shows little base selectivity. These properties, combined with its 1000-fold fluorescence enhancement upon nucleic acid binding and high quantum yield, make our new SYTOX Green stain a simple and quantitative single-step dead-cell indicator for use with epifluorescence and confocal laser scanning microscopes, fluorometers, fluorescence microplate readers and flow cytometers (see Figure 16.11 in Chapter 16).

SYTOX Green nucleic acid stain may be used with blue and red fluorescent surface labels for multiparameter analyses (see Color Plate 2 in "Visual Reality"). It is also possible to combine SYTOX Green nucleic acid stain with SYTO 17 red fluorescent nucleic acid stain for two-color visualization of dead and live cells (see Section 16.3). Because SYTOX Green nucleic acid stain is an excellent DNA counterstain for chromosome labeling and for fixed cells and tissues (see photo on back cover), we have incorporated it into our FISH Counterstain Kit #2 (F-7588), Chromosome Banding Kit #2 (C-7587) and Cytological Nuclear Counterstain Kit (C-7590), which are discussed in Section 8.4.

Phenanthridines and Acridines

Ethidium Bromide and Propidium Iodide

Ethidium bromide (EtBr; E-1305, E-3565) and propidium iodide (PI; P-1304, P-3566) are structurally similar phenanthridinium intercalators. PI is more water soluble and less membrane permeant than EtBr, although both dyes are generally excluded from viable cells. EtBr and PI can be excited with mercury- or xenon-arc lamps or with the argon-ion laser, making them suitable for fluorescence microscopy, confocal laser scanning microscopy (see photo on back cover), flow cytometry and fluorometry. These dyes bind with little or no sequence preference at a stoichiometry of one dye per 4–5 base pairs of DNA.[40] Excitation of the EtBr/DNA complex may result in photobleaching of the dye and single-stranded breaks.[41] Both EtBr and PI also bind to RNA, necessitating treatment with nucleases to distinguish between RNA and DNA. Once these dyes are bound to nucleic acids, their fluorescence is enhanced 20- to 30-fold, their excitation maxima are shifted ~30–40 nm to the red and their emission maxima are shifted ~15 nm to the blue [42] (Figure 8.2 and Table 8.1). Although their molar absorptivities (extinction coefficients) are relatively low, EtBr and PI exhibit sufficiently large Stokes shifts to allow simultaneous detection of nuclear DNA and fluorescein-labeled antibodies, provided the proper optical filters are used.

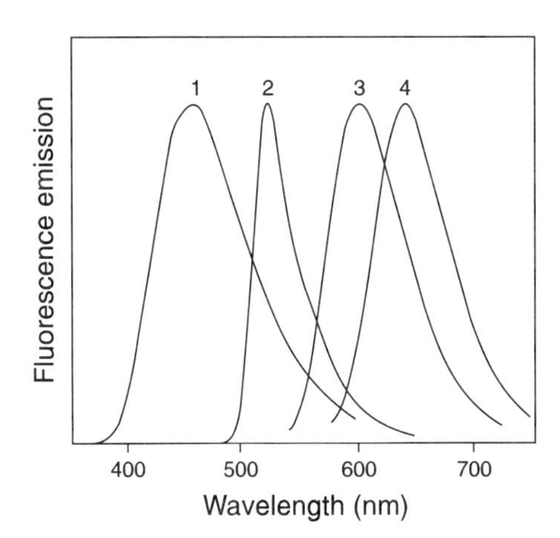

Figure 8.2 Normalized fluorescence emission spectra of DNA-bound 1) Hoechst 33258 (H-1398), 2) acridine orange (A-1301), 3) ethidium bromide (E-1305) and 4) 7-aminoactinomycin D (A-1310).

EtBr is currently the most commonly used general nucleic acid gel stain. However, our new SYBR Green nucleic acid gel stains is not only much more sensitive, but has been shown to be significantly less mutagenic than EtBr by Ames testing (see Section 8.3). EtBr and PI are known to be potent mutagens and must be handled with extreme care. Solutions containing EtBr or PI can be decontaminated by filtration with activated charcoal, which is then incinerated, thus providing an economical decontamination procedure.[43] Alternatively, the dyes can be completely degraded in buffer by reaction with sodium nitrite and hypophosphorous acid.[44] EtBr and PI are offered as solids (E-1305, P-1304) as well as in aqueous solution (E-3565, P-3566), enabling researchers to avoid contact with the mutagenic powders.

Hexidium Iodide

Molecular Probes' newly patented hexidium iodide (H-7593) is a moderately lipophilic phenanthridinium dye that is permeant to mammalian cells and selectively stains gram-positive bacteria in the presence of gram-negative bacteria.[45] Hexidium iodide yields slightly shorter-wavelength spectra upon DNA binding than our ethidium or propidium dyes. Generally, both the cytoplasm and nuclei of eukaryotic cells show staining with hexidium iodide; however, mitochondria and nucleoli can also be stained.

Dihydroethidium

Dihydroethidium (D-1168, also see Sections 17.1 and 21.4), which is structurally identical to Hydroethidine™ (a trademark of Prescott Labs), is a chemically reduced ethidium derivative that is permeant to live cells. Dihydroethidium exhibits blue fluorescence in the cytoplasm. Many viable cells dehydrogenate the probe to ethidium, which then fluoresces red upon DNA intercalation.[46-48]

Ethidium Homodimers

Ethidium homodimer-1 (EthD-1, E-1169), ethidium homodimer-2 (EthD-2, E-3599) and ethidium–acridine heterodimer (E-667, see below) strongly bind to dsDNA, ssDNA, RNA and oligonucleotides with a large fluorescence enhancement (>30-fold). EthD-1 also binds with high affinity to triplex nucleic acid structures.[25] One molecule of EthD-1 binds per four base pairs in dsDNA,[11] and the dye's intercalation is not sequence selective.[49] It was originally reported that only one of the two phenanthridinium rings of EthD-1 is bound at a time;[11] subsequent reports indicate that bis-intercalation appears to be involved in staining both double-stranded and triplex nucleic acids.[25,26]

The spectra and other properties of the EthD-1 and EthD-2 dimers are almost identical. However, the DNA affinity of EthD-2 is about twice that of EthD-1 and is close to that of our highest-affinity nucleic acid stain, TOTO-1.[33] EthD-2 is also about twice as fluorescent bound to dsDNA than to RNA. Because both EthD-1 and EthD-2 can be excited with UV light or by the 488 nm spectral line of the argon-ion laser, either dye can be used with TOTO-1, YOYO-1 or SYTOX Green nucleic acid stains for multicolor experiments, yielding emissions that are reasonably distinct (Figure 8.3). The ethidium homodimers are impermeant to cells with intact membranes, a property that makes EthD-1 useful as a viability indicator in our LIVE/DEAD Viability/Cytotoxicity Kit (L-3224, see Section 16.2). These dyes have also been used to detect DNA in solution,[49] although they are not as sensitive as our new PicoGreen dsDNA quantitation reagent (P-7581, P-7589; see Section 8.3).

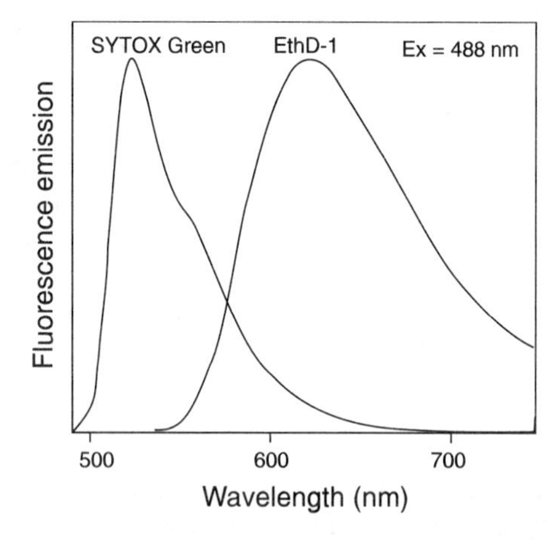

Figure 8.3 *Normalized fluorescence emission spectra of DNA-bound SYTOX Green nucleic acid stain (S-7020) and ethidium homodimer-1 (EthD-1, E-1169). Both spectra were obtained using excitation at 488 nm.*

Ethidium Monoazide and Ethidium Diazide

Nucleic acids can be covalently photolabeled by various DNA intercalators. Ethidium monoazide (E-1374) is a fluorescent photoaffinity label that, after photolysis, binds covalently to nucleic acids both in solution and in cells with compromised membranes.[50-54] The quantum yield for covalent photolabeling by ethidium monoazide is unusually high (>0.4). Photolysis of ethidium diazide (E-3561) has been reported to covalently crosslink strands of DNA with an efficiency of about 30%.[55,56]

The membrane-impermeant ethidium monoazide is reported to only label dead cells and is therefore useful for assaying the viability of pathogenic cells. A mixed population of live and dead cells incubated with this reagent can be illuminated with a visible-light source, washed, fixed and then analyzed in order to determine the viability of the cells at the time of photolysis.[57] This method not only reduces some of the hazards inherent in working with pathogenic cells, but also is compatible with immunocytochemical analyses requiring fixation. We have developed an alternative two-color fluorescence-based assay for determining the original viability of fixed samples that employs our cell-permeant, green fluorescent SYTO 10 and cell-impermeant, red fluorescent DEAD Red nucleic acid stains; the LIVE/DEAD Reduced Biohazard Viability/Cytotoxicity Kit (L-7013) is described in Section 16.2.

In addition to its utility as a viability indicator, ethidium monoazide has been used to irreversibly label the DNA of *Candida albicans* in order to investigate phagocytic capacity of leukocytes.[58] Ethidium monoazide has also been employed to "footprint" drug-binding sites on DNA[59] and to probe for ethidium-binding sites in DNA[60] and transfer RNA[61] (tRNA).

Acridine Orange

Molecular Probes offers highly purified, flow cytometry–grade acridine orange (A-1301, A-3568), a dye that interacts with DNA and RNA by intercalation or electrostatic attractions. In condensed chromatin, however, the bulk of DNA is packed in a way that does not allow efficient acridine orange intercalation.[62] This cationic dye has green fluorescence with an emission maximum at 525 nm when bound to DNA. Upon association with RNA, its emission is shifted to ~650 nm (red fluorescence).

Acridine orange is available as a solid (A-1301) and, for ease of handling, as a 10 mg/mL aqueous solution (A-3568).

Acridine Homodimer and Ethidium–Acridine Heterodimer

The water-soluble acridine homodimer — bis-(6-chloro-2-methoxy-9-acridinyl)spermine (A-666) — is one of several acridine dimers that have been described in the literature. This dye has extremely high affinity for AT-rich regions of nucleic acids, making it particularly useful for chromosome banding.[63,64] Acridine homodimer emits a blue-green fluorescence when bound to DNA, yielding fluorescence that is proportional to the fourth power of the AT base-pair content.[65] Acridine homodimer has recently been recommended as an alternative to quinacrine for Q banding because of its greater brightness and higher photostability.[63]

Molecular Probes also prepares ethidium–acridine heterodimer (E-667). The efficiency of fluorescence resonance energy transfer from the acridine to the phenanthridine portion of this heterodimer is related to the square of the AT base-pair content of the DNA.[11] Thus, the fluorescence excitation spectrum of this heterodimer can be used to characterize AT content in fixed single cells and chromosomes by ratio imaging techniques.[11]

9-Amino-6-Chloro-2-Methoxyacridine

ACMA (9-amino-6-chloro-2-methoxyacridine; A-1324) is a DNA intercalator that binds to poly(d(A-T)) with a binding affinity constant of 2×10^5 M^{-1} at pH 7.4.[66,67] Excitation of the ACMA/DNA complex (excitation/emission maxima ~419/483 nm) is possible with most UV light sources, making it compatible for use with both shorter- and longer-wavelength dyes. ACMA also apparently binds to membranes in the energized state and becomes quenched if a pH gradient forms.[68] It has been employed to follow cation and anion movement across membranes [68-71] and to study the proton-pumping activity of various membrane ATPases [72,73] (see Section 23.3).

Indoles and Imidazoles

Hoechst 33258 and Hoechst 33342

The bisbenzimide dyes — Hoechst 33258 and Hoechst 33342 — are cell-permeant, minor groove–binding DNA stains that fluoresce bright blue upon binding to DNA. Hoechst 33342 has slightly higher membrane permeability than Hoechst 33258,[42] but both dyes are water soluble (up to 2% solutions can be prepared) and relatively nontoxic. Hoechst dyes, which can be excited with the UV lines of the argon-ion laser and most conventional fluorescence excitation sources, exhibit relatively large Stokes shifts (excitation/emission maxima ~350/460 nm, Figure 8.2 and Table 8.1), making them suitable for multicolor labeling experiments. These dyes appear to show a wide spectrum of sequence-dependent DNA affinities and bind with sufficient strength to poly(d(A-T)) sequences that they can displace several DNA intercalators.[74] They also exhibit multiple binding modes and distinct fluorescence emission spectra that are dependent on dye:base pair ratios.[75]

Hoechst 33258 and Hoechst 33342 are available as solids (H-1398, H-1399) and, for ease of handling, as 10 mg/mL aqueous solutions (H-3569, H-3570).

DAPI and DIPI

DAPI (4',6-diamidino-2-phenylindole; D-1306, D-3571) and DIPI [76] (4',6-(diimidazolin-2-yl)-2-phenylindole, D-7594) can be excited with a mercury-arc lamp or with the UV lines of the argon-ion laser. Like the Hoechst dyes, the blue fluorescent DAPI apparently associates with the minor groove of dsDNA, preferentially binding to AT clusters;[77] there is evidence that DAPI also binds to DNA sequences containing as few as two consecutive AT base pairs, perhaps employing a different binding mode.[78-80] DAPI is thought to employ an intercalating binding mode with RNA that is AU selective.[81]

The selectivity of DAPI for DNA over RNA is reported to be greater than that displayed by ethidium bromide and propidium iodide.[82] Furthermore, the DAPI/RNA complex exhibits a longer-wavelength fluorescence emission maximum than the DAPI/dsDNA complex (~500 nm versus ~460 nm) and a quantum yield that is only about 20% as high.[83]

Binding of DAPI to dsDNA produces an ~20-fold fluorescence enhancement, apparently due to the displacement of water molecules from both DAPI and the minor groove.[84] Although the Hoechst dyes may be somewhat brighter in some applications, their photostability when bound to dsDNA is less than that of DAPI. In the presence of appropriate salt concentrations, DAPI does not usually exhibit fluorescence enhancement upon binding to ssDNA or GC base pairs.[85] However, the fluorescence of DAPI does increase significantly upon binding to detergents,[86] dextran sulfate,[87] polyphosphates and other polyanions.[88] A recent review by Kapuscinski discusses the mechanisms of DAPI binding to nucleic acids, its spectral properties and its uses in flow cytometry and for chromosome staining.[89]

DIPI, an analog of DAPI that also exhibits AT selectivity,[90] is a new addition to our selection of nucleic acid stains. DIPI has been used with several DNA-binding drugs to produce specific chromosomal banding patterns.[91-93]

Other Nucleic Acid Stains

7-Aminoactinomycin D and Actinomycin D

7-AAD (7-aminoactinomycin D, A-1310) is a fluorescent intercalator that undergoes a spectral shift upon association with DNA. 7-AAD/DNA complexes can be excited by the argon-ion laser and emit beyond 610 nm (Figure 8.2 and Table 8.1), making this nucleic acid stain useful for multicolor fluorescence microscopy, confocal laser scanning microscopy (see photo on front cover) and flow cytometry.[95-98] 7-AAD appears to be generally excluded from live cells, although it has been reported to label the nuclear region of live cultured mouse L cells and salivary gland polytene chromosomes of *Chironomus thummi thummi* larvae.[99] 7-AAD binds selectively to GC regions of DNA,[100] yielding a distinct banding pattern in polytene chromosomes and chromatin.[99,101] This sequence selectivity has been exploited for chromosome banding studies [102,103] (see Section 8.4).

Actinomycin D (A-7592) is a nonfluorescent intercalator that exhibits high GC selectivity and causes distortion at its binding site.[104] Binding of the nonfluorescent actinomycin D to nucleic acids results in absorbance changes of the dye.[105] Like 7-AAD, actinomycin D has also been used for chromosome banding studies.[106] Binding of actinomycin D to ssDNA is reported to inhibit reverse transcriptase and other polymerases.[107]

Hydroxystilbamidine

Hydroxystilbamidine (H-7599) — a trypanocidal drug that has previously been sold for research use as a neuronal tracer [108,109] under the trademark Fluoro-Gold™ (a trademark of FluoroChrome, Inc.) — is an interesting probe of nucleic acid conformation with nucleic acid staining properties that were first described in 1973.[110] Hydroxystilbamidine, a nonintercalating dye, exhibits AT-selective binding that is reported to favor regions with secondary structure. The interaction between hydroxystilbamidine and DNA has been investigated using binding isotherms [111] and temperature-jump relaxation studies.[112]

Hydroxystilbamidine has some unique spectral properties upon binding nucleic acids. At pH 5, the free dye exhibits UV excitation maxima at ~330 nm and ~390 nm, with dual emission at ~450 nm and ~600 nm. Although the red fluorescent component remains present when bound to DNA, it is never observed when the dye is bound to RNA, allowing potential discrimination to be made between these two types of polynucleotides. The enhancement of its metachromatic fluorescence upon binding to DNA is proportional to the square of the AT base-pair content. Hydroxystilbamidine is reported to have red fluorescence when bound to calf thymus DNA and T5 DNA, orange fluorescence with *M. lysodekticus* DNA and blue-violet fluorescence on poly(d(A-T)).[110]

Because hydroxystilbamidine has been unavailable commercially, or its identity has been obscured by a trademark, its use as a nucleic acid stain in cellular applications has not been extensively tested. However, Murgatroyd described use of its metachromatic fluorescence properties for the selective permanent staining of DNA (with yellow fluorescence), mucosubstances and elastic fibers in paraffin sections.[113] He also reported that hydroxystilbamidine (as its isethionate salt [114)] is nonmutagenic in *Salmonella typhimurium* by the Ames test.[113]

LDS 751

We now offer LDS 751 (L-7595), a cell-permeant nucleic acid stain that has been used to discriminate intact nucleated cells from nonnucleated and damaged nucleated cells [115,116] as well as to identify distinct cell types in mixed populations of neutrophils, leukocytes and monocytes by flow cytometry.[117] LDS 751, which has its peak excitation at ~543 nm on dsDNA, can be excited by the argon-ion laser at 488 nm and is useful in multicolor analyses due to its long-wavelength emission maximum (~712 nm). Binding of LDS 751 to dsDNA results in an ~20-fold fluorescence enhancement. When LDS 751 binds to RNA, we have observed a significant red shift in its excitation maximum to 590 nm and blue shift in its emission maxima to 607 nm, which may permit its use to discriminate DNA and RNA in cells. A recent report has ascribed the name LDS 751 to another dye called styryl 8;[118] however, their chemical structures are not the same.

Psoralens

Psoralens (furocoumarins) are photoreactive nucleic acids stains that are useful for probing DNA and RNA structure. The cell-permeant 4,5',8-trimethylpsoralen (trioxsalen, T-1365), psoralen (P-96045) and angelicine (isopsoralen, A-96003) intercalate into dsDNA where, upon illumination at <400 nm, they covalently bind to pyrimidines;[119] their intrinsic blue fluorescence is usually quenched upon DNA binding. Interstrand DNA crosslinking and inhibition of DNA and RNA synthesis by trioxsalen and psoralen can result if pyrimidines are on opposite strands,[120-122] leading to heavy crosslinking of naked or readily accessible DNA. However, in eukaryotic chromatin, such regions are interspersed with stretches of 50 or more base pairs that show no photocrosslinking due to protection of the DNA by bound proteins.[123] Trioxsalen is the most effective inhibitor of nucleic acid synthesis among the psoralens, and supercoiled DNA photocrosslinks faster with trioxsalen than linear DNA.[124] Angelicine is a monofunctional DNA photobinding reagent that does not crosslink DNA.

When illuminated with UV light, the psoralens are known to generate singlet oxygen.[125] They are commonly used in dermatology as photosensitizing agents to treat psoriasis and other skin diseases.[126] The triplet state quantum yield of angelicine (0.33) is the highest reported among the psoralens.[127]

1. U.S. Patent No. 5,410,030; CA 2,119,126; and patents pending; **2.** Anal Biochem 221, 78 (1994); **3.** European Patent Application No. EP 0 634 640 A1; **4.** M. Terasaki and L. Jaffe, personal communication; **5.** U.S. Patent No. 5,321,130; **6.** Appl Environ Microbiol 60, 3284 (1994); **7.** U.S. Patent Nos. 5,436,134; 5,445,946 and patents pending; **8.** Nucleic Acids Res 23, 1215 (1995); **9.** Nucleic Acids Res 19, 327 (1991); **10.** Biochemistry 17, 5071 (1978); **11.** Biochemistry 17, 5078 (1978); **12.** Appl Environ Microbiol 61, 3623 (1995); **13.** Nucleic Acids Res 20, 2803 (1992); **14.** Proc Natl Acad Sci USA 87, 3851 (1990); **15.** Nucleic Acids Res 21, 5720 (1993); **16.** Nature 359, 859 (1992); **17.** BioTechniques 10, 616 (1991); **18.** J Phys Chem 99, 17936 (1995); **19.** Cytometry 7, 508 (1986); **20.** Cytometry 15, 267 (1994); **21.** J Am Chem Soc 116, 8459 (1994); **22.** J Phys Chem 98, 10313 (1994); **23.** Anal Chem 67, 663A (1995); **24.** FASEB J 7, A1087 (1993); **25.** Bioorg Med Chem 3, 701 (1995); **26.** Nucleic Acids Res 23, 2413 (1995); **27.** Biochem Mol Biol Int'l 34, 1189 (1994); **28.** Biochemistry 34, 8542 (1995); **29.** Nucleic Acids Res 24, 859 (1996); **30.** Nucleic Acids Res 23, 753 (1995); **31.** Cytometry 17, 185 (1994); **32.** Cytometry 17, 310 (1994); **33.** Haugland, R.P., *Molecular Probes' Handbook of Fluorescent Probes and Research Chemicals, Fifth Edition*, (1992–1994) p. 224; **34.** Cytometry 21, 265 (1995); **35.** Mol Biol of the Cell 6 (suppl), 444a, abstract 1805 (1995); **36.** Nature 377, 20 (1995); **37.** Neuron 15, 961 (1995); **38.** Cytometry 20, 218 (1995); **39.** Limnol Oceanogr 40, 1050 (1995); **40.** J Mol Biol 13, 269 (1965); **41.** Biochemistry 29, 981 (1990); **42.** Meth Cell Biol 30, 417 (1989); **43.** Chromatographia 29, 167 (1990); **44.** Anal Biochem 162, 453 (1987); **45.** U.S. Patent No. 5,437,980; **46.** J Immunol Meth 170, 117 (1994); **47.** FEMS Microbiol Lett 101, 173 (1992); **48.** J Histochem Cytochem 34, 1109 (1986); **49.** Anal Biochem 94, 259 (1979); **50.** Biochemistry 30, 5644 (1991); **51.** Photochem Photobiol 43, 7 (1986); **52.** J Biol Chem 259, 11090 (1984); **53.** Photochem Photobiol 36, 31 (1982); **54.** J Mol Biol 92, 319 (1975); **55.** Biochemistry 22, 3226 (1983); **56.** Biochemistry 20, 1887 (1981); **57.** Cytometry 12, 133 (1991); **58.** Cytometry 11, 610 (1990); **59.** Eur J Biochem 182, 437 (1989); **60.** Biochemistry 19, 3221 (1980); **61.** Photochem Photobiol 36, 31 (1982); **62.** Exp Cell Res 194, 147 (1991); **63.** Meth Mol Biol 29, 83 (1994); **64.** Biochemistry 18, 3354 (1979); **65.** Proc Natl Acad Sci USA 72, 2915 (1975); **66.** Eur J Biochem 180, 359 (1989); **67.** J Biomol Struct Dynamics 5, 361 (1987); **68.** Biochim Biophys Acta 722, 107 (1983); **69.** Biochim Biophys Acta 1143, 215 (1993); **70.** Eur Biophys J 19, 189 (1991); **71.** Biochemistry 19, 1922 (1980); **72.** J Biol Chem 269, 10221 (1994); **73.** Biochim Biophys Acta 1183, 161 (1993); **74.** Biochemistry 29, 9029 (1990); **75.** J Histochem Cytochem 33, 333 (1985); **76.** Annalen Chemie 749, 68 (1971); **77.** Biochemistry 26, 4545 (1987); **78.** Biochemistry 32, 2987 (1993); **79.** J Biol Chem 268, 3944 (1993); **80.** Biochemistry 29, 8452 (1990); **81.** Biochemistry 31, 3103 (1992); **82.** Nucleic Acids Res 6, 3535 (1979); **83.** J Histochem Cytochem 38, 1323 (1990); **84.** Biochem Biophys Res Comm 170, 270 (1990); **85.** Nucleic Acids Res 6, 3519 (1979); **86.** Nucleic Acids Res 5, 3775 (1978); **87.** Can J Microbiol 26, 912 (1980); **88.** Biochim Biophys Acta 721, 394 (1982); **89.** Biotechnic Histochem 70, 220 (1995); **90.** Stain Technol 60, 7 (1985); **91.** Human Genet 57, 1 (1987); **92.** Science 225, 57 (1984); **93.** Eur J Cell Biol 20, 290 (1980); **94.** Cytometry 12, 570 (1991); **95.** Cytometry 12, 221 (1991); **96.** Cytometry 12, 172 (1991); **97.** J Immunology 136, 2769 (1986); **98.** J Histochem Cytochem 23, 793 (1975); **99.** Histochem J 17, 131 (1985); **100.** Biopolymers 18, 1749 (1979); **101.** Cytometry 20, 296 (1995); **102.** Proc Natl Acad Sci USA 75, 5655 (1978); **103.** Chromosoma 68, 287 (1978); **104.** J Mol Biol 225, 445 (1992); **105.** Biochemistry 32, 5881 (1993); **106.** Cancer Genet Cytogenet 1, 187 (1980); **107.** Biochemistry 35, 3525 (1996); **108.** U.S. Patent No. 4,716,905; **109.** J Neurocytol 18, 333 (1989); **110.** Biochemistry 12, 4827 (1973); **111.** Biochim Biophys Acta 407, 24 (1975); **112.** Biochim Biophys Acta 407, 43 (1975); **113.** Histochemistry 74, 107 (1982); **114.** Not available from Molecular Probes; **115.** J Immunol Meth 123, 103 (1989); **116.** Cytometry 9, 477 (1988); **117.** J Immunol Meth 163, 155 (1993); **118.** J Photochem Photobiol A: Chem 84, 45 (1994); **119.** Annals NY Acad Sci 346, 355 (1980); **120.** J Virol 27, 127 (1978); **121.** Biochim Biophys Acta 254, 30 (1971); **122.** FEBS Lett 9, 121 (1970); **123.** Science 193, 62 (1976); **124.** Biochem Biophys Acta 521, 529 (1978); **125.** Photochem Photobiol 30, 331 (1979); **126.** Photochem Photobiol 32, 813 (1980); **127.** Photochem Photobiol 27, 273 (1978).

8.1 Data Table *Nucleic Acid Stains*

Cat #	Structure	MW	Storage	Soluble	Abs	{$\epsilon \times 10^{-3}$}	Em	Solvent	Notes
A-666	✓	686	L	DMSO, DMF	431	ND	498	<1>	<2, 3>
A-1301	✓	302	L	EtOH, DMF	500	{53}	526	<1>	<4>
A-1310	✓	1270	F,L	DMF, DMSO	546	{25}	647	<1>	
A-1324	✓	259	L	DMF, DMSO	412	{8.2}	471	MeOH	<5>
A-7592	✓	1255	F,L	DMF, DMSO	442	{23}	none	MeOH	
A-96003	✓	186	L	$CHCl_3$, DMSO	295	{10}	440	MeOH	<6>
B-3582	✓	1203	F,D,L	DMSO*	462	{114}	481	<1>	<7>
B-3583	✓	595	F,D,L	DMSO*	462	{58}	481	<1>	<7>
B-3586	✓	1255	F,D,L	DMSO*	570	{148}	604	<1>	<7>
B-3587	✓	621	F,D,L	DMSO*	575	{81}	599	<1>	<7>
D-1168	✓	315	FF,L,AA	DMF, DMSO	355	{14}	<8>	MeCN	<9>
D-1306	✓	350	L	H_2O, DMF	358	{21}	461	<1>	
D-3571	✓	457	L	H_2O, MeOH	358	{20}	461	<1>	
D-7594	✓	402	D,L	H_2O, DMF	364	{25}	455	<1>	
E-667	✓	844	F,D,L	H_2O, DMF	532	ND	619	<1>	<2, 10>
E-1169	✓	857	F,D,L	DMSO	528	{7.0}	617	<1>	<11>
E-1305	✓	394	L	H_2O, DMSO	518	{5.2}	605	<1>	<12>
E-1374	✓	420	F,LL	DMF, EtOH	464	{5.8}	625	pH 3	<13>
E-3561	✓	402	F,LL	pH 3	432	{5.7}	496	pH 3	<14>
E-3599	✓	1293	F,D,L	DMSO*	535	{8.0}	624	<1>	<11>
H-1398	✓	624†	L	H_2O, DMF	352	{40}	461	<1>	
H-1399	✓	616†	L	H_2O, DMF	350	{45}	461	<1>	
H-7593	✓	497	L	DMSO	518	{3.9}	600	<1>	<15>
H-7599	✓	473	F,D,L	H_2O	385	{15}	536	<1>	<16>
L-7595	✓	472	L	DMSO, EtOH	543	{46}	712	<1>	
O-7582	P	<17>	F,D,LL	DMSO*	498	<17>	518	<1>	<7>
P-1304	✓	668	L	H_2O, DMF	535	{5.4}	617	<1>	<18>
P-3580	✓	1171	F,D,L	DMSO*	434	{92}	456	<1>	<7>
P-3581	✓	579	F,D,L	DMSO*	435	{50}	455	<1>	<7>
P-3584	✓	1223	F,D,L	DMSO*	534	{146}	570	<1>	<7>
P-3585	✓	605	F,D,L	DMSO*	539	{88}	567	<1>	<7>
P-7581	P	<17>	F,D,LL	DMSO*	502	<17>	523	<1>	<7>
P-96045	✓	186	L	$CHCl_3$, DMSO	295	{10}	443	MeOH	<6>
S-7020	P	~600	F,D,L	DMSO*	504	{67}	523	<1>	<7>
S-7555	P	~450	F,D,L	DMSO*	512	{64}	530	<1>	<7>
S-7556	P	~500	F,D,L	DMSO*	494	{43}	517	<1>	<7>
S-7557	P	~350	F,D,L	DMSO*	515	{43}	535	<1>	<7>
S-7558	P	~400	F,D,L	DMSO*	499	{46}	520	<1>	<7>
S-7559	P	~550	F,D,L	DMSO*	490	{58}	515	<1>	<7>
S-7560	P	~450	F,D,L	DMSO*	521	{57}	556	<1>	<7>
S-7573	P	~400	F,D,L	DMSO*	508	{75}	527	<1>	<7>
S-7574	P	~300	F,D,L	DMSO*	500	{54}	522	<1>	<7>
S-7575	P	~400	F,D,L	DMSO*	488	{74}	509	<1>	<7>
S-7576	P	~500	F,D,L	DMSO*	517	{60}	549	<1>	<7>
S-7577	P	~400	F,D,L	DMSO*	516	{55}	546	<1>	<7>
S-7578	P	~450	F,D,L	DMSO*	488	{42}	518	<1>	<7>
S-7579	P	~650	F,D,L	DMSO*	621	{88}	634	<1>	<7>
T-1365	✓	228	L	$CHCl_3$, DMSO	296	{11}	445	MeOH	<6>
T-3600	✓	1303	F,D,L	DMSO*	514	{117}	533	<1>	<7>
T-3602	✓	645	F,D,L	DMSO*	515	{63}	531	<1>	<7>
T-3604	✓	1355	F,D,L	DMSO*	642	{154}	660	<1>	<7>
T-3605	✓	671	F,D,L	DMSO*	642	{102}	661	<1>	<7>
T-7596	✓	697	F,D,L	DMSO*	747	{108}	770	<1>	<7>
Y-3601	✓	1271	F,D,L	DMSO*	491	{99}	509	<1>	<7>
Y-3603	✓	629	F,D,L	DMSO*	491	{52}	509	<1>	<7>
Y-3606	✓	1323	F,D,L	DMSO*	612	{167}	631	<1>	<7>
Y-3607	✓	655	F,D,L	DMSO*	612	{100}	631	<1>	<7>

A-3568 see A-1301; E-3565 see E-1305; H-3569 see H-1398; H-3570 see H-1399; P-3566 see P-1304; P-7589 see P-7581.

For definitions of the contents of this data table, see "How to Use the Handbook" on page vi.
Structure: Chemical structure drawing: ✓ = shown in this section; P = proprietary information.
MW: ~ indicates an approximate value, not including counterions.

* This product is packaged as a solution in the solvent indicated in "Soluble."
† MW is for the hydrated form of this product.
<1> Spectra represent aqueous solutions of DNA-bound dye. ϵ values are derived by comparing the absorbance of the DNA-bound dye with that of free dye in a reference solvent (H_2O or MeOH).
<2> ND = not determined.
<3> A-666 in MeOH: Abs = 418 nm (ϵ = {12}), Em = 500 nm.

<4> A-1301 bound to RNA has Em ~650 nm [Cytometry 2, 201 (1982)].

<5> Spectra of this compound are in methanol acidified with a trace of HCl.

<6> Absorption spectra of psoralens have a longer wavelength peak at ~340 nm. In fluorescence excitation spectra, this peak is usually more intense than the ~300 nm peak. It is also the activation maximum for photoaddition reactions of psoralens. Em for DNA-bound psoralens is ~455 nm [Ann Rev Biochem 54, 1151 (1985)].

<7> This product is essentially nonfluorescent, except when bound to DNA or RNA.

<8> D-1168 has blue fluorescence (Em ~420 nm) until oxidized to ethidium (E-1305). The reduced dye does not bind to nucleic acids [FEBS Lett 26, 169 (1972)].

<9> This compound is susceptible to oxidation, especially in solution. Store solutions under argon or nitrogen. Oxidation appears to be catalyzed by illumination.

<10> E-667 has stronger absorption at shorter wavelengths; Abs = 455 nm (bound to DNA). In MeOH, Abs = 447 nm (ϵ = {8.8}) and 527 nm (ϵ = {4.9}), Em = 635 nm.

<11> E-1169 in H_2O: Abs = 493 nm (ϵ = {9.1}). E-3599 in H_2O: Abs = 498 nm (ϵ = {10.8}). Both compounds are very weakly fluorescent in H_2O. The fluorescence quantum yield increases >40-fold on binding to DNA.

<12> E-1305 in H_2O: Abs = 480 nm (ϵ = {5.6}), Em = 620 nm (weakly fluorescent). Fluorescence is enhanced >10-fold on binding to DNA.

<13> E-1374 spectral data are for the free dye. Fluorescence is weak, but intensity increases ~15-fold on binding to DNA. After photocrosslinking to DNA, Abs = 504 nm (ϵ ~{4.0}), Em = 600 nm [Nucleic Acids Res 5, 4891 (1978); Biochemistry 19, 3221 (1980)].

<14> E-3561 spectral data are for the free dye. Fluorescence is weak and decreases in intensity on binding to DNA [Biochemistry 19, 3221 (1980)].

<15> H-7593 in H_2O: Abs = 482 nm (ϵ = {5.5}), Em = 625 nm (very weak fluorescence).

<16> Fluorescence enhancement of H-7599 upon binding polydeoxyribonucleotides increases linearly with the square of the AT base content. Emission on the red edge of the spectrum (~600 nm) is often appreciable on DNA but is absent on RNA [Biochemistry 12, 4827 (1973)].

<17> The active ingredients of P-7581 and O-7582 are organic dyes with MW <1000. The exact MW values and extinction coefficients of these dyes are proprietary.

<18> P-1304 in H_2O: Abs = 493 nm (ϵ = {5.9}), Em = 636 nm (weakly fluorescent). Fluorescence is enhanced >10-fold on binding to DNA.

8.1 Structures *Nucleic Acid Stains*

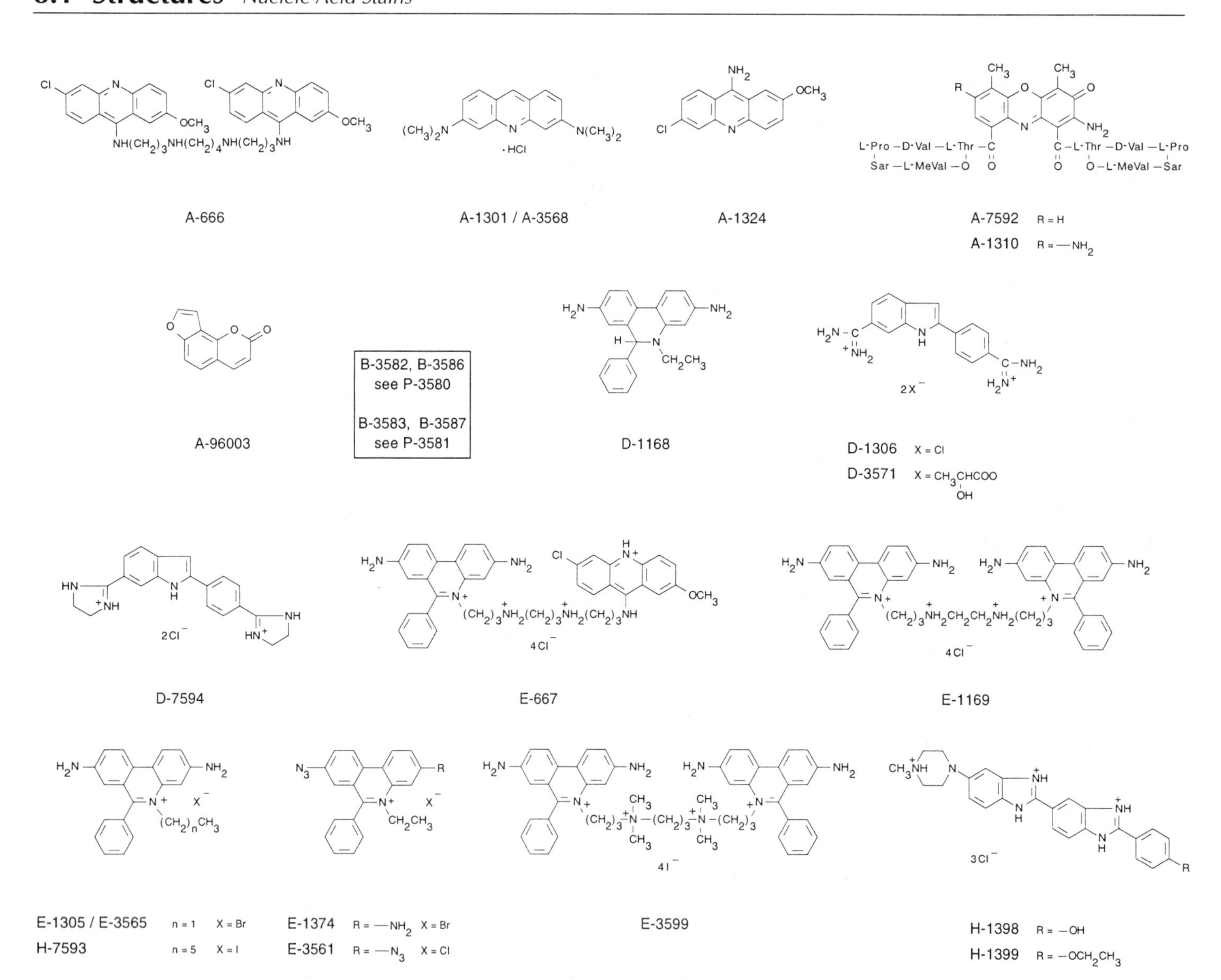

H-7593
see E-1305

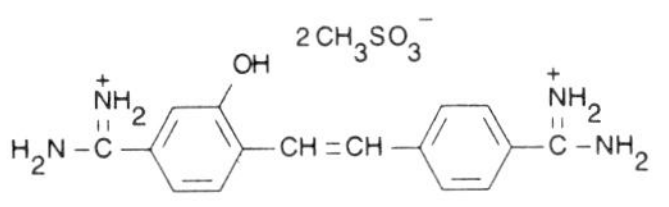

H-7599

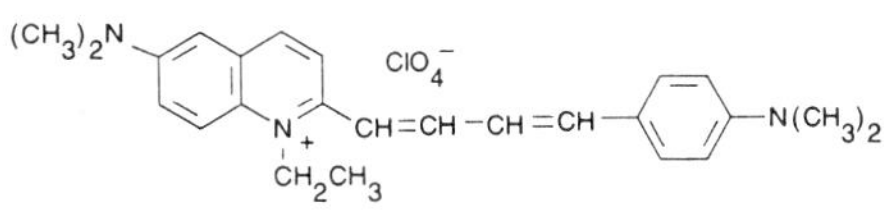

L-7595

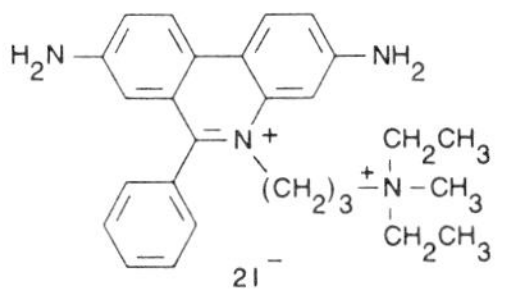

P-1304 / P-3566

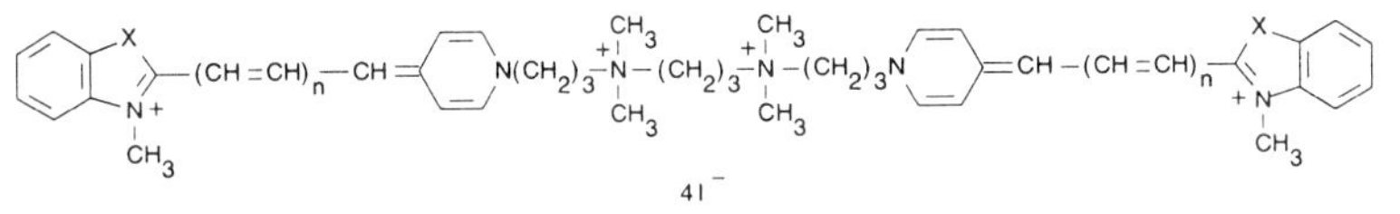

	X	n
P-3580	X = O	n = 0
B-3582	X = S	n = 0
P-3584	X = O	n = 1
B-3586	X = S	n = 1

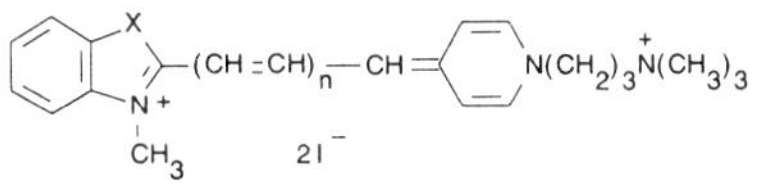

	X	n
P-3581	X = O	n = 0
B-3583	X = S	n = 0
P-3585	X = O	n = 1
B-3587	X = S	n = 1

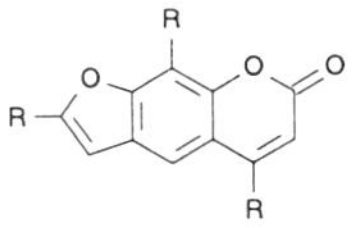

	R
P-96045	R = H
T-1365	R = $-CH_3$

	X	n
T-3600	X = S	n = 0
Y-3601	X = O	n = 0
T-3604	X = S	n = 1
Y-3606	X = O	n = 1

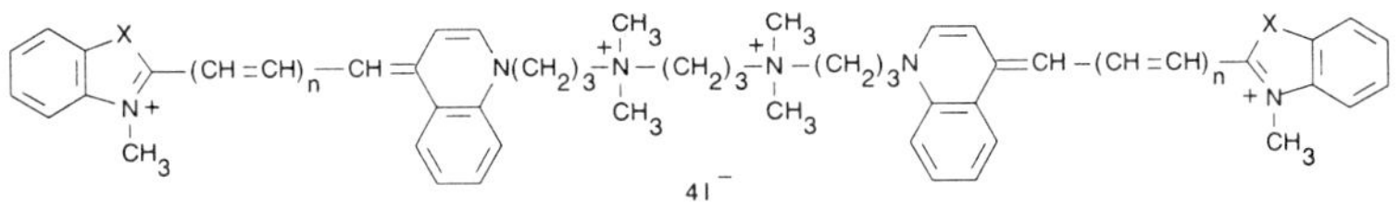

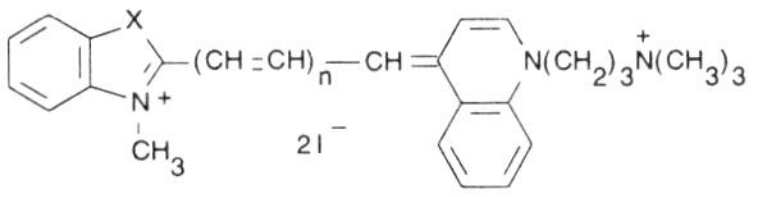

	X	n
T-3602	X = S	n = 0
Y-3603	X = O	n = 0
T-3605	X = S	n = 1
Y-3607	X = O	n = 1
T-7596	X = S	n = 2

8.1 Price List *Nucleic Acid Stains*

Cat #		Product Name	Unit Size	Price Per Unit ($) 1–4 Units	5–24 Units
A-666		acridine homodimer (bis-(6-chloro-2-methoxy-9-acridinyl)spermine)	10 mg	98.00	78.40
A-1301		acridine orange *high purity*	1 g	48.00	38.40
A-3568		acridine orange *10 mg/mL solution in water*	10 mL	28.00	22.40
A-7592	**New**	actinomycin D *≥98% by HPLC*	10 mg	55.00	44.00
A-1310		7-aminoactinomycin D (7-AAD)	1 mg	50.00	40.00
A-1324		9-amino-6-chloro-2-methoxyacridine (ACMA)	100 mg	78.00	62.40
A-96003	**New**	angelicine (isopsoralen)	20 mg	55.00	44.00
B-3582		BOBO™-1 iodide (462/481) *1 mM solution in DMSO*	200 µL	225.00	180.00
B-3586		BOBO™-3 iodide (570/602) *1 mM solution in DMSO*	200 µL	225.00	180.00
B-3583		BO-PRO™-1 iodide (462/481) *1 mM solution in DMSO*	1 mL	135.00	108.00
B-3587		BO-PRO™-3 iodide (575/599) *1 mM solution in DMSO*	1 mL	135.00	108.00
D-1306		4′,6-diamidino-2-phenylindole, dihydrochloride (DAPI)	10 mg	37.00	29.60
D-3571		4′,6-diamidino-2-phenylindole, dilactate (DAPI, dilactate)	10 mg	37.00	29.60
D-1168		dihydroethidium (also called Hydroethidine - a trademark of Prescott Labs)	25 mg	98.00	78.40
D-7594	**New**	4′,6-(diimidazolin-2-yl)-2-phenylindole, dihydrochloride (DIPI)	10 mg	65.00	52.00
E-667		ethidium-acridine heterodimer	1 mg	95.00	76.00
E-1305		ethidium bromide	1 g	12.00	9.60
E-3565		ethidium bromide *10 mg/mL solution in water*	10 mL	28.00	22.40
E-3561		ethidium diazide chloride	5 mg	95.00	76.00
E-1169		ethidium homodimer-1 (EthD-1)	1 mg	155.00	124.00
E-3599		ethidium homodimer-2 (EthD-2) *1 mM solution in DMSO*	200 µL	138.00	110.40
E-1374		ethidium monoazide bromide	5 mg	115.00	92.00
H-7593	**New**	hexidium iodide	5 mg	95.00	76.00
H-3569		Hoechst 33258 (bis-benzimide) *10 mg/mL solution in water*	10 mL	40.00	32.00
H-1398		Hoechst 33258, pentahydrate (bis-benzimide)	100 mg	20.00	16.00
H-3570		Hoechst 33342 *10 mg/mL solution in water*	10 mL	40.00	32.00

Cat #		Product Name	Unit Size	Price Per Unit ($) 1–4 Units	5–24 Units
H-1399		Hoechst 33342, trihydrochloride, trihydrate	100 mg	20.00	16.00
H-7599	**New**	hydroxystilbamidine, methanesulfonate	10 mg	75.00	60.00
L-7595	**New**	LDS 751	10 mg	65.00	52.00
N-7565	**New**	Nucleic Acid Stains Dimer Sampler Kit	1 kit	125.00	100.00
N-7566	**New**	Nucleic Acid Stains Monomer Sampler Kit	1 kit	65.00	52.00
O-7582	**New**	OliGreen™ ssDNA quantitation reagent *for ssDNA and oligonucleotides* *200-2000 assays*	1 mL	175.00	140.00
P-7589	**New**	PicoGreen® dsDNA Quantitation Kit *200-2000 assays*	1 kit	175.00	140.00
P-7581	**New**	PicoGreen® dsDNA quantitation reagent *200-2000 assays*	1 mL	150.00	120.00
P-3580		POPO™-1 iodide (434/456) *1 mM solution in DMSO*	200 µL	225.00	180.00
P-3584		POPO™-3 iodide (534/570) *1 mM solution in DMSO*	200 µL	225.00	180.00
P-3581		PO-PRO™-1 iodide (435/455) *1 mM solution in DMSO*	1 mL	135.00	108.00
P-3585		PO-PRO™-3 iodide (539/567) *1 mM solution in DMSO*	1 mL	135.00	108.00
P-1304		propidium iodide	100 mg	48.00	38.40
P-3566		propidium iodide *1.0 mg/mL solution in water*	10 mL	28.00	22.40
P-96045	**New**	psoralen	100 mg	95.00	76.00
S-7550	**New**	SYBR® DX DNA blot stain *1000X concentrate in DMSO*	1 mL	175.00	140.00
S-7563	**New**	SYBR® Green I nucleic acid gel stain *10,000X concentrate in DMSO*	500 µL	125.00	100.00
S-7567	**New**	SYBR® Green I nucleic acid gel stain *10,000X concentrate in DMSO*	1 mL	235.00	188.00
S-7585	**New**	SYBR® Green I nucleic acid gel stain *10,000X concentrate in DMSO* *special packaging*	20x50 µL	265.00	212.00
S-7564	**New**	SYBR® Green II RNA gel stain *10,000X concentrate in DMSO*	500 µL	125.00	100.00
S-7568	**New**	SYBR® Green II RNA gel stain *10,000X concentrate in DMSO*	1 mL	235.00	188.00
S-7586	**New**	SYBR® Green II RNA gel stain *10,000X concentrate in DMSO* *special packaging*	20x50 µL	265.00	212.00
S-7580	**New**	SYBR® Green Nucleic Acid Gel Stain Starter Kit	1 kit	55.00	55.00
S-7573	**New**	SYTO® 11 live-cell nucleic acid stain *5 mM solution in DMSO*	250 µL	98.00	78.40
S-7574	**New**	SYTO® 12 live-cell nucleic acid stain *5 mM solution in DMSO*	250 µL	98.00	78.40
S-7575	**New**	SYTO® 13 live-cell nucleic acid stain *5 mM solution in DMSO*	250 µL	98.00	78.40
S-7576	**New**	SYTO® 14 live-cell nucleic acid stain *5 mM solution in DMSO*	250 µL	98.00	78.40
S-7577	**New**	SYTO® 15 live-cell nucleic acid stain *5 mM solution in DMSO*	250 µL	98.00	78.40
S-7578	**New**	SYTO® 16 live-cell nucleic acid stain *1 mM solution in DMSO*	250 µL	98.00	78.40
S-7555	**New**	SYTO® 20 live-cell nucleic acid stain *1 mM solution in DMSO*	250 µL	98.00	78.40
S-7556	**New**	SYTO® 21 live-cell nucleic acid stain *5 mM solution in DMSO*	250 µL	98.00	78.40
S-7557	**New**	SYTO® 22 live-cell nucleic acid stain *5 mM solution in DMSO*	250 µL	98.00	78.40
S-7558	**New**	SYTO® 23 live-cell nucleic acid stain *5 mM solution in DMSO*	250 µL	98.00	78.40
S-7559	**New**	SYTO® 24 live-cell nucleic acid stain *5 mM solution in DMSO*	250 µL	98.00	78.40
S-7560	**New**	SYTO® 25 live-cell nucleic acid stain *5 mM solution in DMSO*	250 µL	98.00	78.40
S-7572	**New**	SYTO® Live-Cell Nucleic Acid Stain Sampler Kit #1 *SYTO® dyes 11-16* *50 µL each*	1 kit	135.00	108.00
S-7554	**New**	SYTO® Live-Cell Nucleic Acid Stain Sampler Kit #2 *SYTO® dyes 20-25* *50 µL each*	1 kit	135.00	108.00
S-7579	**New**	SYTO® 17 red fluorescent nucleic acid stain *5 mM solution in DMSO*	250 µL	98.00	78.40
S-7020	**New**	SYTOX® Green nucleic acid stain *5 mM solution in DMSO*	250 µL	98.00	78.40
T-3602		TO-PRO™-1 iodide (515/531) *1 mM solution in DMSO*	1 mL	135.00	108.00
T-3605		TO-PRO™-3 iodide (642/661) *1 mM solution in DMSO*	1 mL	135.00	108.00
T-7596	**New**	TO-PRO™-5 iodide (745/770) *1 mM solution in DMSO*	1 mL	135.00	108.00
T-3600		TOTO®-1 iodide (514/533) *1 mM solution in DMSO*	200 µL	225.00	180.00
T-3604		TOTO®-3 iodide (642/660) *1 mM solution in DMSO*	200 µL	225.00	180.00
T-1365		4,5′,8-trimethylpsoralen (trioxsalen)	100 mg	45.00	36.00
Y-3603		YO-PRO™-1 iodide (491/509) *1 mM solution in DMSO*	1 mL	135.00	108.00
Y-3607		YO-PRO™-3 iodide (612/631) *1 mM solution in DMSO*	1 mL	135.00	108.00
Y-3601		YOYO®-1 iodide (491/509) *1 mM solution in DMSO*	200 µL	225.00	180.00
Y-3606		YOYO®-3 iodide (612/631) *1 mM solution in DMSO*	200 µL	225.00	180.00

Product-Specific Bibliographies and Keyword Searches

Our Technical Assistance Department can provide you with product-specific bibliographies, as well as keyword searches of the over 25,000 literature references in our extensive bibliography database. Our bibliography database is also searchable through our Web site (http://www.probes.com).

8.2 Chemically Modified Nucleotides, Oligonucleotides and Nucleic Acids

Chemical modification of nucleic acids is not as straightforward as protein modification. The nonbasic "amines" of adenosine and guanosine are virtually unreactive with the amine-reactive reagents described in Chapter 1, and the other major reactive groups found in proteins — thiols, carboxylic acids and alcohols — are usually absent or not abundant in natural nucleic acids. Consequently, only a few techniques have been used for the direct labeling of DNA and RNA.[1,2] Generally, nucleotides or oligonucleotides are labeled during chemical synthesis, then enzymatically converted into labeled nucleic acid polymers or directly used as primers or hybridization probes.[3-9] For example, incorporation of amines or thiols during their synthesis permits nucleotides and oligonucleotides to be modified using the reagents in Chapters 1 and 2.

In addition to producing the widest assortment of nucleic acid stains (see Section 8.1), Molecular Probes supplies many of the most important dyes for nucleotide and oligonucleotide labeling, nucleic acid sequencing and direct or indirect nucleic acid modification. Our ChromaTide™ and FluoroTide™ products provide researchers with labeled nucleotides and oligonucleotides for enzymatic incorporation into nucleic acids, and our FluoReporter® Oligonucleotide Labeling and Purification Kits supply the reagents needed to reliably label and purify synthetic oligonucleotides without expensive equipment.

ChromaTide Labeled Nucleotides

Molecular Probes offers a series of uridine and deoxyuridine triphosphates, each conjugated to one of ten different fluorophores. The spectral diversity of the ChromaTide nucleotides (Table 8.2) gives researchers a great deal of flexibility in choosing a label that is compatible with a particular optical detection system or multicolor experiment. Our ChromaTide nucleotides[9] contain a unique aminoalkynyl linker[10] between the fluorophore and the nucleotide that is designed to reduce the interaction of the fluorophore with the nucleic acid and to make the hapten more accessible to secondary reagents (Figure 8.4). In addition to this four-atom bridge, several of these nucleotides contain a seven- to ten-atom spacer, which further separates the dye from the base.

Our newest ChromaTide nucleotides are the Oregon Green™ 488, Rhodamine Green™ and Texas Red-X conjugates of dUTP (C-7630, C-7629, C-7631) and the Rhodamine Green conjugate of UTP (C-7628). As compared to the corresponding fluorescein conjugates, the Oregon Green 488 and Rhodamine Green conjugates have similar fluorescence spectra but superior photostability (see Section 1.4). The Texas Red-12-dUTP (C-7631) has an emission spectrum in solution that is narrower and about 25% more intense than that of Texas Red-5-dUTP (C-7608).

Preliminary experiments have shown that the ChromaTide nucleotides are functional with a variety of nucleic acid modifying enzymes:

Table 8.2 *Fluorophore labels for ChromaTide nucleotides.*

ChromaTide Nucleotide Cat #	Fluorophore Label (Cat #, see Section n.n) *
C-7613, C-7614	BODIPY FL-X dye (D-6102, see Section 1.2)
C-7615, C-7616	BODIPY TMR-X dye (D-6117, see Section 1.2)
C-7617, C-7618	BODIPY TR-X dye (D-6116, see Section 1.2)
C-7611, C-7612	Cascade Blue dye (C-2284, see Section 1.7)
C-7603, C-7604	Fluorescein-X (F-2181, see Section 1.3)
C-7630	Oregon Green 488 dye (O-6147, see Section 1.4)
C-7628, C-7629	Rhodamine Green dye (R-6107, see Section 1.4)
C-7605, C-7606	Tetramethylrhodamine (T-1480 †, see Section 1.6)
C-7607, C-7608	Texas Red dye (T-353 ‡, see Section 1.6)
C-7631	Texas Red-X dye (T-6134, see Section 1.6)

* The catalog number for the labeling group and the Handbook Section number that contains its structure are indicated for each ChromaTide nucleotide. Fluorophores are attached to the terminal amine of the alkynyl spacer via a carboxamide linkage (arrow A in Figure 8.4) except for the following: † thiourea linkage, ‡ sulfonamide linkage.

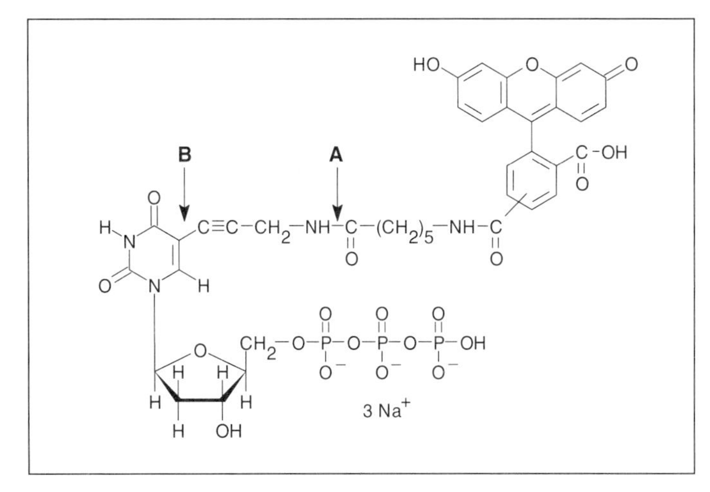

Figure 8.4 *Structure of ChromaTide fluorescein-12-dUTP (C-7604). This structure is representative of our other ChromaTide labeled nucleotides. Fluorophore labels are attached via a four-atom aminoalkynyl spacer (between arrows A and B) to either deoxyuridinetriphosphate (dUTP) or uridinetriphosphate (UTP). Fluorophore labels for other ChromaTide nucleotides are indicated in Table 8.2.*

- Taq polymerase in polymerase chain reaction (PCR) assays [11]
- DNA polymerase I in primer extension assays
- Klenow polymerase in random primer labeling
- Terminal deoxynucleotidyl transferase for 3'-end labeling (see Color Plate 4 in "Visual Reality")
- SP6 RNA polymerase, T3 RNA polymerase and T7 RNA polymerase for *in vitro* transcription

ChromaTide nucleotides can also potentially be used as substrates in DNA sequencing reactions. Nucleic acids labeled with ChromaTide nucleotides can serve as probes for chromosome and mRNA FISH experiments, as well as for Southern, Northern, colony and plaque hybridization.

Labeled Oligonucleotides

Dyes for Nucleic Acid Sequencing

Molecular Probes manufactures most of the dyes that are used in nucleic acid sequencing and provides these in reactive forms for preparing conjugates. Because the electrophoretic separation step during sequencing is highly sensitive to the chemical structure of the fragments, the use of single-isomer labels is essential. In addition to providing high-purity reactive succinimidyl esters of the common FAM, JOE, TAMRA and ROX dyes,[12-14] Molecular Probes prepares amine-reactive single isomers of carboxyrhodamine 6G (CR 6G) (Table 8.3). The 6-isomer of the CR 6G dye has been reported to have spectroscopic and electrophoretic properties that are superior to the JOE dye often used for automated DNA sequencing.[15] Also, oligonucleotide conjugates of several of our BODIPY® dyes (see Section 1.2) have recently been reported to be very useful for DNA sequencing, in part because the dyes have a minimal effect on the mobility of the fragment during electrophoresis and also exhibit well-resolved spectra with narrow bandwidths [16] (see Figure 1.2 in Chapter 1). The BODIPY dyes [17] are all high-purity, pH-insensitive single isomers. Oligonucleotides labeled with multiple dyes that form excited-state energy transfer pairs have been shown to enhance the detection in sequencing applications that depend on the argon-ion laser for excitation.[15,16,18,19]

Contact our Custom and Bulk Sales Department for information about custom synthesis of amine-reactive single isomers of our Rhodamine Green, Rhodol Green™ or other dyes [20,21] or about availability of any of our reactive dyes in bulk.

Table 8.3 Amine-reactive dyes for nucleic acid sequencing.

Cat #	Reactive Dye *	Handbook Location
C-2210	5-FAM, SE †	Section 1.3
C-6164	6-FAM, SE	Section 1.3
C-6127	5-CR 6G, SE	Section 1.5
C-6128	6-CR 6G, SE	Section 1.5
C-6171	6-JOE, SE †	Section 1.5
C-2211	5-TAMRA, SE	Section 1.6
C-6123	6-TAMRA, SE †	Section 1.6
C-6125	5-ROX, SE	Section 1.6
C-6126	6-ROX, SE †	Section 1.6
D-2184	BODIPY FL, SE ‡	Section 1.2
D-6140	BODIPY FL, SSE	Section 1.2
D-6102	BODIPY FL-X, SE	Section 1.2
D-6180	BODIPY R6G, SE ‡	Section 1.2
D-6117	BODIPY TMR-X, SE	Section 1.2
D-2222	BODIPY 564/570, SE ‡	Section 1.2
D-2228	BODIPY 581/591, SE ‡	Section 1.2
D-6116	BODIPY TR-X, SE	Section 1.2

* FAM = carboxyfluorescein; CR 6G = carboxyrhodamine 6G; JOE = carboxy-4′,5′-dichloro-2′,7′-dimethoxyfluorescein; TAMRA = carboxytetramethylrhodamine; ROX = carboxy-X-rhodamine; BODIPY = a substituted 4,4-difluoro-4-bora-3a,4a-diaza-*s*-indacene derivative (see Figure 1.3 in Chapter 1); SE = succinimidyl ester; SSE = water-soluble sulfosuccinimidyl ester. † These are the most widely used isomers for DNA sequencing [Anal Biochem 223, 39 (1994); Nucleic Acids Res 20, 2471 (1992); Proc Natl Acad Sci USA 86, 9178 (1989)]. ‡ These BODIPY derivatives were recently reported to be useful for automated DNA sequencing, in part because the dyes have a minimal effect on the mobility of the fragment during electrophoresis and also exhibit well-resolved spectra with narrow bandwidths [Science 271, 1420 (1996)].

FluoReporter Oligonucleotide Labeling Kits

Molecular Probes' FluoReporter Oligonucleotide Labeling Kits [9] provide researchers with a convenient method for efficiently labeling oligonucleotides with a wide variety of our fluorophores and haptens. Labeling is not only economical, but easy and very reproducible. Our FluoReporter Labeled Oligonucleotide Purification Kit (F-6100) provides a simple way to purify the conjugates without resorting to HPLC or gel electrophoresis for purification.[22]

We offer two types of FluoReporter Oligonucleotide Labeling Kits. The FluoReporter Oligonucleotide Amine Labeling Kits use stable succinimidyl esters to label amine-modified synthetic oligonucleotides, whereas the FluoReporter Oligonucleotide Phosphate Labeling Kits use proprietary coupling technology to conjugate aliphatic amines to 3'- or 5'-phosphate–terminated oligonucleotides in a single step. For unphosphorylated oligonucleotides, a 5'-phosphate can be added enzymatically with T4 polynucleotide kinase, prior to use of the FluoReporter Oligonucleotide Phosphate Labeling Kits.

The amine-reactive haptens and fluorophores in most of our fifteen different **FluoReporter Oligonucleotide Amine Labeling Kits** contain aminohexanoyl spacers ("X") to reduce the label's interaction with the oligonucleotide and enhance its accessibility to secondary detection reagents. Our BODIPY FL-X, BODIPY TMR-X and BODIPY TR-X Kits (F-6082, F-6083, F-6084) contain reactive versions of our patented BODIPY fluorophores with emission properties similar to those of fluorescein, rhodamine 6G, tetramethylrhodamine and Texas Red dyes, respectively (see Figure 1.4 in Chapter 1). We also offer BODIPY FL, BODIPY R6G, BODIPY 564/570 and BODIPY 581/591 Kits (F-6079, F-6092, F-6093, F-6094), which contain reactive BODIPY dyes found to be useful for automated DNA sequencing.[16] The BODIPY fluorophores exhibit high extinction coefficients, excellent quantum yields and a fluorescence emission that is quite photostable and insensitive to pH. The narrow absorption and emission bandwidths of these BODIPY fluorophores make them particularly well suited to multicolor applications. The FluoReporter Oregon Green 488, Rhodamine Green-X, Rhodamine Red™-X and Texas Red-X Oligonucleotide Amine Labeling Kits (F-6087, F-6088, F-6089, F-6091) contain some of our newest and most photostable dyes (see Sections 1.4 and 1.6). In addition to these kits, we offer FluoReporter Kits for labeling amines with fluorescein-X, tetramethylrhodamine, biotin-XX and DNP-X (F-6086, F-6090, F-6081, F-6085). The reactive dyes in some of the kits contain mixed isomers, and their oligonucleotide conjugates may resolve into two peaks in very high resolution separation techniques.

Conventional methods for modifying terminal phosphate groups require a multistep synthesis.[23-25] In contrast, the **FluoReporter Oligonucleotide Phosphate Labeling Kits** permit the single-step covalent labeling of 3'- or 5'-phosphate–terminated oligonucleotides with cadaverine derivatives of the BODIPY FL, BODIPY TMR, or Texas Red fluorophores or biotin ligand (F-6096, F-6097, F-6099, F-6095) or with the ethylenediamine derivative of our Rhodamine Red dye (F-6098). The resulting phosphoramidate adducts have reasonable chemical stability, particularly in neutral solution. These kits can also be used to double-label radioactively labeled oligonucleotides or, in combination with T4 polynucleotide kinase, to fluorescently label oligonucleotides lacking a 5'-phosphate. In addition, we have found that the simple method provided in our FluoReporter Oligonucleotide Phosphate Labeling Kits can be used to label linear DNA restriction fragments, although such conjugates may require an alternative purification method.

The FluoReporter Oligonucleotide Amine Labeling Kits and FluoReporter Oligonucleotide Phosphate Labeling Kits provide sufficient reagents for five complete labeling reactions. Each FluoReporter Kit contains:

- Five vials of the amine-reactive *or* phosphate-reactive label, each sufficient for labeling 100 µg of amine-derivatized oligonucleotide
- Anhydrous dimethylsulfoxide (DMSO) for dissolving the reactive reagent
- Labeling buffer
- Detailed protocol for oligonucleotide labeling

Our FluoReporter Kits have been optimized for labeling oligonucleotides containing 18 to 25 bases but may be useful for labeling either shorter or longer oligonucleotides. Fluorescent, biotinylated or DNP-labeled oligonucleotides can be purified from the reaction mixture with our FluoReporter Labeled Oligonucleotide Purification Kit (see below) or with standard HPLC or gel electrophoresis methods. After purification, labeled oligonucleotides can serve as primers for DNA sequencing, DNA amplification or cDNA preparation and as probes for Northern and Southern blots, colony and plaque lifts and mRNA *in situ* hybridization experiments. Fluorescence anisotropy measurements can detect hybridization of fluorescent oligonucleotides in homogeneous solution.[26] RNA oligonucleotides are useful for probing RNA secondary structure, in combination with the dsRNA-specific RNase H and the ssRNA-specific RNase A and RNase T1.

FluoReporter Labeled Oligonucleotide Purification Kit

Purification of fluorescent, biotinylated or DNP-labeled oligonucleotides is made easy with our new FluoReporter Labeled Oligonucleotide Purification Kit (F-6100). The crude, labeled oligonucleotide is simply precipitated with ethanol to remove the excess reactive reagent, adsorbed on the spin column, washed to remove any unconjugated oligonucleotide and then eluted with an elution buffer to yield the conjugate. Isolated yields for the combined conjugation and purification steps are usually >60%, and the products are typically >90% pure as determined by HPLC. This kit may be useful for purifying oligonucleotide conjugates of many of our other reactive dyes and haptens too. Oligonucleotide conjugates can be used for most procedures without additional purification. Each FluoReporter Oligonucleotide Purification Kit contains:

- Five spin columns
- Separate buffers for column equilibration, washing and elution
- Detailed protocol that has been tested with all of our FluoReporter Oligonucleotide Labeling Kits

Sufficient columns and buffers are provided for purification of five labeling reactions of 100 µg oligonucleotide each.

FluoroTide Oligonucleotide Primers

Molecular Probes offers FluoroTide oligonucleotide M13/pUC (-21) primers[9] conjugated to fluorescein, Oregon Green 488, Texas Red-X, BODIPY FL, BODIPY R6G, BODIPY TMR and BODIPY TR dyes (F-3621, F-6677, F-6676, F-3622, F-7632, F-7633, F-7634). Our primer conjugates are prepared by attaching the dye to the 5'-(6-aminohexyl)–modified oligonucleotide, purified by preparative HPLC and packaged by optical density units measured at 260 nm. These primers are useful for automated single- or double-stranded sequencing of DNA fragments cloned into M13mp vectors and pUC or pUC-related plasmids.[27] They can also be used to synthesize hybridization sequences for probing blots, chromosome squashes, plaques, colonies and mRNA, as well as to generate a variety of probes for forensic and diagnostic applications.[28] The four BODIPY dye–labeled primers have well-resolved spectra with narrow bandwidths (see Figure 1.4 in Chapter 1), making them spectrally distinct from other fluorescently labeled primers and probes. Similar BODIPY dye–labeled primers have recently been reported to be very useful for DNA sequencing because the dyes do not produce the mobility artifacts exhibited by other dyes commonly used for DNA sequencing.[16]

Chemical Modification of Nucleic Acid Polymers

DNA and RNA are unreactive with most common chemical reagents, and special methods are necessary for their modification. Only a few general methods are available for modifying nucleic acid polymers.

Cytidine Residues

DNA and RNA can be modified by reacting their cytidine residues with sodium bisulfite to form sulfonate intermediates that can then be directly coupled to hydrazides or aliphatic amines.[29,30] For example, biotin hydrazides (see Section 4.2) have been used in a bisulfite-mediated reaction to couple biotin to cytidine residues in oligonucleotides.[31] Virtually any of the fluorescent, biotin or other hydrazides or aliphatic amines in Chapters 3 and 4, except possibly the BODIPY derivatives, can potentially be used in this reaction. The bisulfite-activated cytidylic acid can also be coupled to aliphatic diamines such as ethylenediamine.[32] The amine-modified

DNA or RNA can then be modified with any of the amine-reactive dyes described in Chapter 1.

Phosphate Groups

Our FluoReporter Oligonucleotide Phosphate Labeling Kits provide the reagents and a protocol for the single-step modification of terminal phosphate residues of oligonucleotides or restriction fragments. Although phosphate groups of nucleotides and oligonucleotides are not very reactive in aqueous solution, their terminal phosphate groups can react with carbodiimides and similar reagents in combination with nucleophiles to yield labeled phosphodiesters, phosphoramidates and phosphorothioates.[33] For example, it has been reported that DNA can be reacted quantitatively with carbonyl diimidazole and a diamine such as ethylenediamine to yield a phosphoramidate that has a free primary amine and that this amine can then be modified with amine-reactive reagents of the type described in Chapter 1.[23-25,34] Fluorescent or biotinylated amines have been coupled to the 5'-phosphate of tRNA using dithiodipyridine and triphenylphosphine.[35] Wang and Giese have reported an apparently general method for labeling phosphates, including nucleotides, for capillary electrophoresis applications that employs an imidazole derivative prepared from our BODIPY FL hydrazide[36] (D-2371, see Section 3.2).

Abasic Sites

The biotinylated hydroxylamine ARP (A-6346) has been recently used to modify abasic sites in DNA — those apurinic sites and apyrimidinic lesions thought to be important intermediates in carcinogenesis.[37-39] Once the aldehyde group in an abasic site is modified with ARP, the resulting biotinylated DNA can be detected with avidin conjugates (see Section 7.5).

Terminal Ribose Group of RNA

Selective oxidation of the 3'-end of RNA by periodate yields a dialdehyde. This dialdehyde can then be coupled with a fluorescent or biotin hydrazide reagent[40-42] (see Sections 3.2 and 4.2).

Specialized Methods

A few other specialized methods have been developed for nucleic acid modification. These include:

- Synthesis of DNA using fluorescent 2'- or 3'-acyl derivatives of uridine triphosphate and terminal deoxyribonucleotide transferase[43]
- Use of a fluorescent iodoacetamide or maleimide, along with T4 polynucleotide kinase and ATP-γ-*S* (ATP with a sulfur in the terminal phosphate) to introduce a thiophosphate at the 5'-terminus of 5'-dephosphorylated RNA[41] or DNA
- Introduction of 4-thiouridine at the 3'-terminus of DNA using calf thymus terminal deoxynucleotidyl transferase followed by treatment with ribonuclease and reaction with thiol-reactive probes[44,45]
- Direct reaction of thiol-reactive reagents with 4-thiouridine residues in nucleic acids[8,9,35,38,46,47]
- Direct reaction of amine- or thiol-reactive reagents with aminoacyl tRNA or thioacetylated aminoacyl tRNA[35,48,49]
- Reaction of the X-base of tRNA with isothiocyanates[50] or replacement of other uncommon bases in tRNA by fluorophores[51-53]

1. Kessler, C. in *Nonisotopic Probing, Blotting, and Sequencing*, L.J. Kricka, Ed., Academic Press (1995) pp. 41–109; **2.** Kricka, L.K. in *Nonisotopic Probing, Blotting, and Sequencing*, L.J. Kricka, Ed., Academic Press (1995) pp. 3–40; **3.** Histochem J 27, 94 (1995); **4.** Proc Natl Acad Sci USA 89, 9509 (1992); **5.** Electrophoresis 13, 542 (1992); **6.** Proc Natl Acad Sci USA 86, 9178 (1989); **7.** Nucleic Acids Res 16, 2203 (1988); **8.** Anal Biochem 131, 419 (1983); **9.** ChromaTide labeled nucleotides, FluoroTide labeled oligonucleotides and oligonucleotides prepared using our FluoReporter Oligonucleotide Labeling Kits are intended for research use only; use of labeled nucleotides and oligonucleotides for any other purposes or in any patented application may require licenses from Molecular Probes and other companies; **10.** U.S. Patent No. 5,047,519 owned by E.I. DuPont de Nemours and Co.; **11.** The PCR process is covered by patents owned by Hoffmann–LaRoche, Inc. Purchase of these products does not convey a license under these patents. Information about licenses for PCR can be obtained from Perkin–Elmer Corp. or Roche Molecular Systems, Inc.; **12.** Anal Biochem 223, 39 (1994); **13.** Nucleic Acids Res 20, 2471 (1992); **14.** Proc Natl Acad Sci USA 86, 9178 (1989); **15.** Nature Med 2, 246 (1996); **16.** Science 271, 1420 (1996); **17.** U.S. Patent Nos. 4,774,339; 5,187,288; 5,248,782; 5,274,113; 5,433,896; 5,451,663 and other U.S. and foreign patents pending; **18.** Proc Natl Acad Sci USA 92, 4347 (1995); **19.** Anal Biochem 231, 131 (1995); **20.** Anal Biochem 207, 267 (1992); **21.** Haugland, R.P. in *Optical Microscopy for Biology*, Herman, B. and Jacobson, K., Eds., Wiley-Liss (1990) pp. 143–157; **22.** FASEB J 8, A1445 (1994); **23.** Anal Biochem 218, 444 (1994); **24.** Biochem Biophys Res Comm 200, 1239 (1994); **25.** Meth Mol Biol 26, 145 (1994); **26.** Anal Chem 67, 3945 (1995); **27.** Meth Enzymol 101, 20 (1983); **28.** FEBS Lett 351, 231 (1994); **29.** J Clin Microbiol 23, 311 (1986); **30.** Biochemistry 19, 1774 (1980); **31.** Biochem Biophys Res Comm 142, 519 (1987); **32.** Biochem J 108, 883 (1968); **33.** Nucleic Acids Res 22, 920 (1994); **34.** J Chromatography 608, 171 (1992); **35.** Biochemistry 29, 10734 (1990); **36.** Anal Chem 65, 3518 (1993); **37.** Biochemistry 32, 8276 (1993); **38.** Biochemistry 31, 3703 (1992); **39.** Biochemistry 11, 3610 (1972); **40.** Bioconjugate Chem 5, 436 (1994); **41.** Biochemistry 30, 4821 (1991); **42.** Biochemistry 19, 5947 (1980); **43.** Molekulyarnaya Biologiya 11, 598 (1977); **44.** Anal Biochem 170, 271 (1988); **45.** Nucleic Acids Res 7, 1485 (1979); **46.** Biochemistry 24, 692 (1985); **47.** J Mol Biol 156, 113 (1982); **48.** J Am Chem Soc 113, 2722 (1991); **49.** Eur J Biochem 172, 663 (1988); **50.** Eur Biophys J 16, 45 (1988); **51.** Eur J Biochem 98, 465 (1979); **52.** Meth Enzymol 29, 667 (1974); **53.** FEBS Lett 18, 214 (1971).

Cat #		Product Name	Unit Size	Price Per Unit ($) 1–4 Units	5–24 Units
A-6346	*New*	N-(aminooxyacetyl)-N′-(D-biotinoyl) hydrazine (ARP)	10 mg	98.00	78.40
C-7614	*New*	ChromaTide™ BODIPY® FL-14-dUTP (BODIPY® FL-14-dUTP) *1 mM in buffer*	25 µL	175.00	175.00
C-7613	*New*	ChromaTide™ BODIPY® FL-14-UTP (BODIPY® FL-14-UTP) *1 mM in buffer*	25 µL	175.00	175.00
C-7616	*New*	ChromaTide™ BODIPY® TMR-14-dUTP (BODIPY® TMR-14-dUTP) *1 mM in buffer*	25 µL	175.00	175.00
C-7615	*New*	ChromaTide™ BODIPY® TMR-14-UTP (BODIPY® TMR-14-UTP) *1 mM in buffer*	25 µL	175.00	175.00
C-7618	*New*	ChromaTide™ BODIPY® TR-14-dUTP (BODIPY® TR-14-dUTP) *1 mM in buffer*	25 µL	175.00	175.00
C-7617	*New*	ChromaTide™ BODIPY® TR-14-UTP (BODIPY® TR-14-UTP) *1 mM in buffer*	25 µL	175.00	175.00
C-7612	*New*	ChromaTide™ Cascade Blue®-7-dUTP (Cascade Blue®-7-dUTP) *1 mM in buffer*	25 µL	175.00	175.00
C-7611	*New*	ChromaTide™ Cascade Blue®-7-UTP (Cascade Blue®-7-UTP) *1 mM in buffer*	25 µL	175.00	175.00
C-7604	*New*	ChromaTide™ fluorescein-12-dUTP (fluorescein-12-dUTP) *1 mM in buffer*	25 µL	175.00	175.00
C-7603	*New*	ChromaTide™ fluorescein-12-UTP (fluorescein-12-UTP) *1 mM in buffer*	25 µL	175.00	175.00
C-7630	*New*	ChromaTide™ Oregon Green™ 488-5-dUTP (Oregon Green™ 488-5-dUTP) *1 mM in buffer*	25 µL	175.00	175.00
C-7629	*New*	ChromaTide™ Rhodamine Green™-5-dUTP (Rhodamine Green™-5-dUTP) *1 mM in buffer*	25 µL	175.00	175.00
C-7628	*New*	ChromaTide™ Rhodamine Green™-5-UTP (Rhodamine Green™-5-UTP) *1 mM in buffer*	25 µL	175.00	175.00
C-7606	*New*	ChromaTide™ tetramethylrhodamine-5-dUTP (TMR-5-dUTP) *1 mM in buffer*	25 µL	175.00	175.00
C-7605	*New*	ChromaTide™ tetramethylrhodamine-5-UTP (TMR-5-UTP) *1 mM in buffer*	25 µL	175.00	175.00
C-7608	*New*	ChromaTide™ Texas Red®-5-dUTP (Texas Red®-5-dUTP) *1 mM in buffer*	25 µL	175.00	175.00
C-7631	*New*	ChromaTide™ Texas Red®-12-dUTP (Texas Red®-12-dUTP) *1 mM in buffer*	25 µL	175.00	175.00
C-7607	*New*	ChromaTide™ Texas Red®-5-UTP (Texas Red®-5-UTP) *1 mM in buffer*	25 µL	175.00	175.00
F-6095	*New*	FluoReporter® Biotin-X-C_5 Oligonucleotide Phosphate Labeling Kit *5 labelings*	1 kit	175.00	140.00
F-6081	*New*	FluoReporter® Biotin-XX Oligonucleotide Amine Labeling Kit *5 labelings*	1 kit	175.00	140.00
F-6096	*New*	FluoReporter® BODIPY® FL-C_5 Oligonucleotide Phosphate Labeling Kit *5 labelings*	1 kit	175.00	140.00
F-6079	*New*	FluoReporter® BODIPY® FL Oligonucleotide Amine Labeling Kit *5 labelings*	1 kit	175.00	140.00
F-6082	*New*	FluoReporter® BODIPY® FL-X Oligonucleotide Amine Labeling Kit *5 labelings*	1 kit	175.00	140.00
F-6093	*New*	FluoReporter® BODIPY® 564/570 Oligonucleotide Amine Labeling Kit *5 labelings*	1 kit	175.00	140.00
F-6094	*New*	FluoReporter® BODIPY® 581/591 Oligonucleotide Amine Labeling Kit *5 labelings*	1 kit	175.00	140.00
F-6092	*New*	FluoReporter® BODIPY® R6G Oligonucleotide Amine Labeling Kit *5 labelings*	1 kit	175.00	140.00
F-6097	*New*	FluoReporter® BODIPY® TMR-C_5 Oligonucleotide Phosphate Labeling Kit *5 labelings*	1 kit	175.00	140.00
F-6083	*New*	FluoReporter® BODIPY® TMR-X Oligonucleotide Amine Labeling Kit *5 labelings*	1 kit	175.00	140.00
F-6084	*New*	FluoReporter® BODIPY® TR-X Oligonucleotide Amine Labeling Kit *5 labelings*	1 kit	175.00	140.00
F-6085	*New*	FluoReporter® Dinitrophenyl-X (DNP-X) Oligonucleotide Amine Labeling Kit *5 labelings*	1 kit	175.00	140.00
F-6086	*New*	FluoReporter® Fluorescein-X Oligonucleotide Amine Labeling Kit *5 labelings*	1 kit	175.00	140.00
F-6100	*New*	FluoReporter® Labeled Oligonucleotide Purification Kit *five spin columns plus buffers*	1 kit	50.00	40.00
F-6087	*New*	FluoReporter® Oregon Green™ 488 Oligonucleotide Amine Labeling Kit *5 labelings*	1 kit	175.00	140.00
F-6088	*New*	FluoReporter® Rhodamine Green™-X Oligonucleotide Amine Labeling Kit *5 labelings*	1 kit	175.00	140.00
F-6098	*New*	FluoReporter® Rhodamine Red™-C_2 Oligonucleotide Phosphate Labeling Kit *5 labelings*	1 kit	175.00	140.00
F-6089	*New*	FluoReporter® Rhodamine Red™-X Oligonucleotide Amine Labeling Kit *5 labelings*	1 kit	175.00	140.00
F-6090	*New*	FluoReporter® Tetramethylrhodamine (5-TAMRA) Oligonucleotide Amine Labeling Kit *5 labelings*	1 kit	175.00	140.00
F-6099	*New*	FluoReporter® Texas Red®-C_5 Oligonucleotide Phosphate Labeling Kit *5 labelings*	1 kit	175.00	140.00
F-6091	*New*	FluoReporter® Texas Red®-X Oligonucleotide Amine Labeling Kit *5 labelings*	1 kit	175.00	140.00
F-3622		FluoroTide™ M13/pUC (-21) primer, 5′-BODIPY® FL labeled, X d(TGTAAAACGACGGCCAGT)	0.1 OD	95.00	76.00
F-7632	*New*	FluoroTide™ M13/pUC (-21) primer, 5′-BODIPY® R6G labeled, X d(TGTAAAACGACGGCCAGT)	0.1 OD	95.00	76.00
F-7633	*New*	FluoroTide™ M13/pUC (-21) primer, 5′-BODIPY® TMR labeled, X d(TGTAAAACGACGGCCAGT)	0.1 OD	95.00	76.00
F-7634	*New*	FluoroTide™ M13/pUC (-21) primer, 5′-BODIPY® TR labeled, X d(TGTAAAACGACGGCCAGT)	0.1 OD	95.00	76.00
F-3621		FluoroTide™ M13/pUC (-21) primer, 5′-fluorescein labeled, X d(TGTAAAACGACGGCCAGT)	0.1 OD	95.00	76.00
F-6677	*New*	FluoroTide™ M13/pUC (-21) primer, 5′-Oregon Green™ 488 labeled, X d(TGTAAAACGACGGCCAGT)	0.1 OD	95.00	76.00
F-6676	*New*	FluoroTide™ M13/pUC (-21) primer, 5′-Texas Red®-X labeled, X d(TGTAAAACGACGGCCAGT)	0.1 OD	95.00	76.00

8.3 *In Vitro* Applications for Nucleic Acid Stains and Probes

This section describes reagents and kits for making quantitative measurements and qualitative observations of nucleic acids in solution, in electrophoretic gels and on blots, as well as for detecting the activity of nucleases. Most of the nucleic acid stains described in this section are listed in the tables and product list associated with Section 8.1, unless otherwise indicated.

Nucleic Acid Quantitation in Solution

Molecular Probes has developed some exceptional new products — the PicoGreen dsDNA quantitation reagent and the OliGreen ssDNA quantitation reagent — for quantitating nucleic acids in solutions. These new reagents provide valuable alternatives to conventional Hoechst dyes and ethidium bromide, which have much lower sensitivity [1,2] or poor linearity.[1] Ultrasensitive nucleic acid detection in solution will facilitate:

- Accurate measurements of cDNA yields before cloning
- Detection of nucleic acids in drug preparations
- Determination of plasmid yields from mini- or large-scale preps
- Normalization of amounts of templates used per reaction in electrophoretic mobility–shift (bandshift) assays, DNA footprinting or filter binding assays, without depleting template supplies
- Quantitation of dsDNA products of S1 nuclease assays
- Quantitation of polymerase chain reaction (PCR) products [3] from low–cycle number or low–target number reactions

PicoGreen dsDNA Quantitation Reagent

Through our intensive research efforts in both chemical synthesis and bioassay development, Molecular Probes' scientists are pleased to announce a major advance in nucleic acid detection — ultrasensitive DNA quantitation with our new PicoGreen dsDNA quantitation reagent (P-7581) and kit (P-7589). Due to its high affinity for double-stranded DNA (dsDNA) and large fluorescence enhancement upon binding nucleic acids, the PicoGreen reagent has proven exceptionally sensitive for quantitating dsDNA in solution. The PicoGreen dsDNA assay has several important features:

- **Sensitivity.** With the PicoGreen reagent, we can reliably detect as little as 25 pg/mL dsDNA (50 pg dsDNA in a 2 mL assay volume) using fluorescein excitation and emission wavelengths and a standard fluorometer or 250 pg/mL (50 pg dsDNA in a 200 µL assay volume) using a fluorescence microplate reader. The sensitivity of the PicoGreen dsDNA assay surpasses that of conventional absorbance measurements at 260 nm (an A_{260} of 0.1 corresponds to a 5 µg/mL dsDNA solution), and even exceeds that of YO-PRO-1– and YOYO-1–based assays, which have reported detection limits of approximately 2.5 ng/mL [4] and 0.5 ng/mL,[5] respectively.
- **Linearity.** When used in a standard fluorometer, the PicoGreen dsDNA assay is linear over four orders of magnitude — from 25 pg/mL to 1 µg/mL (Figure 8.5) — with a single dye concentration, whereas YOYO-1–based assays exhibit a limited linear range of 0.5–100 ng/mL.[5]
- **Superior to conventional Hoechst 33258–based DNA assays, as well as to blot and dipstick assays.** The PicoGreen dsDNA assay is not only more sensitive than Hoechst 33258–based DNA assays (~10 ng/mL) by greater than 400-fold, but also faster, much less expensive and more precise than dipstick or blot assays.
- **Minimal interference by ssDNA, RNA and other common contaminants.** Using the recommended PicoGreen dsDNA assay protocol, we have found that single-stranded DNA (ssDNA) and RNA do not significantly contribute to the fluorescence signal, even when present at equimolar concentrations. Also, our testing shows that this linearity is maintained in the presence of several compounds commonly found in nucleic acid preparations, including salts, urea, ethanol, chloroform, detergents, proteins and agarose.

These exceptional properties make the PicoGreen dsDNA assay the most sensitive and user-friendly reagent available for quantitating dsDNA in solution. We anticipate that the PicoGreen reagent will become the standard reagent for nucleic acid quantitation, allowing an accurate assessment of dsDNA yields in a host of molecular biology procedures such as DNA amplification reactions,[6] DNA minipreps, cDNA synthesis and nuclease protection assays — without wasting precious sample. For example, it has recently been used to quantitate dsDNA samples before and after PCR amplification,[3,7,8] as well as to determine PCR amplification yields in a method for direct cycle sequencing of PCR products.[3,9] In addition, use of the PicoGreen dsDNA assay should provide a facile method for detecting nucleic acid contamination in compounds prepared from biological sources.

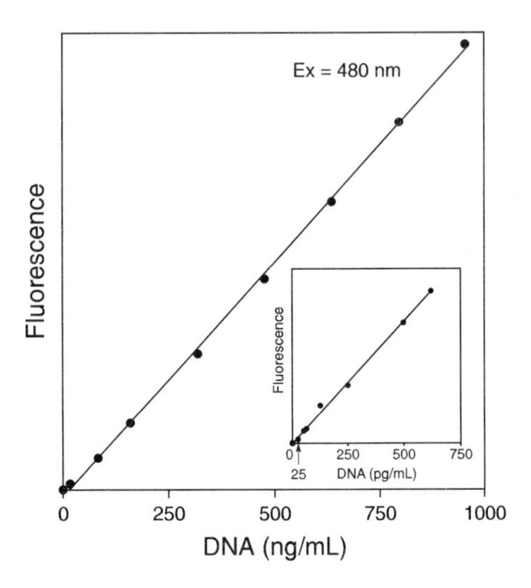

Figure 8.5 *Linear quantitation of calf thymus DNA from 25 pg/mL to 1000 ng/mL using the PicoGreen dsDNA Quantitation Reagent (P-7581, P-7589). Samples in 10 mm × 10 mm cuvettes were excited at 480 nm. The fluorescence emission intensity was measured at 520 nm using a spectrofluorometer and plotted as a function of DNA concentration. The inset shows an enlargement of the results obtained with DNA concentrations between zero and 750 pg/mL.*

Each vial of PicoGreen dsDNA quantitation reagent (P-7581) contains sufficient dye for at least 200 assays using a 2 mL assay volume and a standard fluorometer, or 2000 assays using a 200 µL assay volume and a fluorescence microplate reader, and is accompanied by a simple protocol that ensures linear and reproducible quantitation of dsDNA. We also now provide the PicoGreen reagent in a new kit form — the PicoGreen dsDNA Quantitation Kit (P-7589) — which contains:

- PicoGreen dsDNA quantitation reagent
- Low-fluorescence, nucleic acid–free, concentrated assay buffer
- dsDNA standard solution for assay calibration
- Detailed protocol

Like the individual reagent, this kit provides sufficient material for 200 assays using a 2 mL assay volume and a standard fluorometer or 2000 assays using a 200 µL assay volume and a fluorescence microplate reader. Use of an assay buffer that has low fluorescence, such as that provided in the PicoGreen dsDNA Quantitation Kit, is essential for obtaining the maximal sensitivity from this sensitive assay.

OliGreen ssDNA Quantitation Reagent

For researchers working with oligonucleotides as well as companies that synthesize oligonucleotides, Molecular Probes is pleased to offer the OliGreen ssDNA quantitation reagent (O-7582) for quantitating small amounts of ssDNA and oligonucleotides in solution. Short, synthetic oligonucleotides are used in a number of molecular biology techniques, including DNA sequencing, site-directed mutagenesis, DNA amplification, antisense gene suppression and *in situ* and blot hybridization. Unfortunately, the classic methods for quantitating oligonucleotides are not very sensitive, often requiring highly concentrated samples. The most commonly used technique for measuring oligonucleotide and ssDNA concentrations is the determination of absorbance at 260 nm (A_{260}). The major disadvantages of the absorbance method are the large relative contribution of nucleotides to the signal, the interference caused by contaminants commonly found in nucleic acid preparations and the relative insensitivity of the assay (an A_{260} of 0.1 corresponds to ~3 µg/mL solution of a synthetic 24-mer M13 sequencing primer).

The remarkable properties of our OliGreen ssDNA quantitation reagent offers researchers a valuable alternative for ssDNA and oligonucleotide quantitation:

- **Sensitivity.** Our new OliGreen ssDNA quantitation reagent enables researchers to quantitate as little as 100 pg/mL ssDNA or oligonucleotide (200 pg in a 2 mL assay volume) with a standard fluorometer and fluorescein excitation and emission wavelengths. Thus, quantitation with the OliGreen reagent is 10,000 times more sensitive than quantitation oligonucleotides with absorbance methods and 500 times more sensitive than detecting oligonucleotides on electrophoretic gels stained with ethidium bromide. Using a fluorescence microplate reader, we routinely obtain detection limits of 1 ng/mL ssDNA or oligonucleotide (200 pg in a 200 µL assay volume).
- **Versatility.** We have quantitated oligonucleotides ranging from 10 to 50 nucleotides in length, as well as several sources of ssDNA such as M13 and ϕX174 viral DNA and denatured calf thymus DNA, and obtained similar sensitivity.
- **Linearity.** The linear detection range of the OliGreen assay in a standard fluorometer extends over four orders of magnitude — from 100 pg/mL to 1 µg/mL — with a single dye concentration (Figure 8.6).
- **Minimal interference by common contaminants.** We have shown that the linearity of the OliGreen assay is maintained in the presence of several compounds commonly found to contaminate nucleic acid preparations, including salts, urea, ethanol, chloroform, detergents, proteins, ATP and agarose; however, many of these compounds do affect the signal intensity, so standard curves should be generated using solutions that closely mimic those of the samples. The OliGreen assay can even be performed using samples as complex as whole blood or serum.[10] Nucleotides and short oligonucleotides of six bases or less do not interfere in the quantitation assay, but the OliGreen ssDNA quantitation reagent does exhibit fluorescence enhancement when bound to dsDNA and RNA.

Our recent experiments with homopolymers have demonstrated that the OliGreen reagent exhibits significant base selectivity. The OliGreen reagent shows a large fluorescence enhancement when bound to poly(dT) but only a relatively small fluorescence enhancement when bound to poly(dG) and little signal with poly(dA) and poly(dC). Thus, it is important to use an oligonucleotide with similar base composition when generating the standard curve.

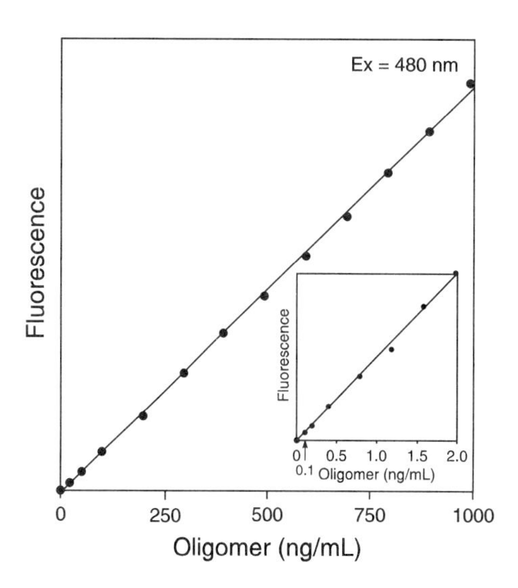

Figure 8.6 *Linear quantitation of a synthetic 24-mer (an M13 sequencing primer) from 0.1 to 1000 ng/mL using the OliGreen ssDNA Quantitation Reagent (O-7582). Samples in 10 mm × 10 mm cuvettes were excited at 480 nm. The fluorescence emission intensity was measured at 520 nm using a spectrofluorometer and plotted as a function of oligonucleotide concentration. The inset shows an enlargement of the results obtained with oligonucleotide concentrations between zero and 2.0 ng/mL.*

The OliGreen ssDNA quantitation reagent comes with a simple protocol and enough dye for 200 assays using a 2 mL assay volume and a standard fluorometer or up to 2000 assays using a 200 µL assay volume and a fluorescence microplate reader.

SYBR Green Dyes for Nucleic Acid Quantitation in Solution

Specific detection of RNA in the presence of DNA is not yet possible either in solution or in cells. However, a recent publication describes the use of the SYBR Green II RNA gel stain (S-7568) for quantitating RNA in solution with nanogram sensitivity.[11] Fluorescence of the SYBR Green II/RNA complex is reported to be linear from 5 to 1000 ng/mL RNA or from 2 to 100 ng/mL, using two different dye concentrations. This level of sensitivity is at least 100-fold greater than that of ethidium bromide and 1000-fold better than spectrophotometric measurements. Furthermore, the selectivity for RNA over DNA of the SYBR Green II dye is better than that of other common nucleic acid stains. Because there are no dyes available that exhibit fluorescence *specifically* enhanced by RNA, quantitation of RNA in solution is best conducted following pretreatment of samples with DNase.

Molecular biologists frequently need a quick estimate of nucleic acid concentration of solutions. We have found that the SYBR Green dyes make excellent detection reagents for DNA and RNA spotted onto plastic wrap or paraffin sheets (e.g., Parafilm®) in drop-spot tests. See the Technical Note "SYBR Green I and SYBR Green II Dyes for Staining Nucleic Acids on Plastic Wrap or Paraffin Sheets" on page 164 for further information.

Other Stains for Nucleic Acid Quantitation in Solution

The dimeric cyanine dyes TOTO-1 and YOYO-1 are also useful for sensitive fluorometric measurement of dsDNA, ssDNA and RNA in solution.[5] The linear range of DNA detection for these assays encompasses over two orders of magnitude, with a sensitivity limit of about 0.5 ng/mL. TOTO-1, YOYO-1 and YO-PRO-1 have been used to quantitate PCR amplification products in a homogeneous human leukocyte antigen (HLA) typing method that requires no transfer or washing steps, thus minimizing the risk of sample contamination.[3,12,13] Other applications of our dyes for nucleic acid quantitation in solution include:

- YOYO-1 for solution quantitation of oligonucleotides,[14] PCR products [3,15] and nuclear run-on assays [16]
- YO-PRO-1 for quantitating dsDNA in a fluorescence microplate reader, with a reported sensitivity limit of about 2.5 ng/mL [4]
- YO-PRO-1 for quantitating RNA isolated from *Xenopus* embryos [17]
- YO-PRO-1 for direct counting of viruses in marine and freshwater environments,[18] an application in which our SYTOX Green nucleic acid stain may prove superior because it typically has more intense fluorescence than YO-PRO-1
- PO-PRO-3 for quantitating DNA in a fluorescence microplate reader,[19] with fluorescence measurements reported to be independent of base-pair composition
- Ethidium bromide for measuring the yield of PCR products [3,20]
- Ethidium bromide and acridine homodimer for quantitating covalently closed, circular DNA and for measuring the activity of polymerases, deoxynucleotidyl transferases, ligases, gyrases, topoisomerases and nucleases [1,21,22]

Dyes such as the ethidium homodimers and our dimeric cyanine dyes — the TOTO, YOYO, BOBO and POPO dyes — exhibit a high affinity for double-stranded nucleic acids but label small single-stranded oligonucleotides less well. This characteristic of ethidium homodimer-1 was exploited to assay short self-annealing oligonucleotides for their ability to hybridize.[23] Because our dimeric cyanine dyes and PicoGreen dsDNA quantitation reagent have extremely low intrinsic fluorescence in the absence of DNA, high fluorescence enhancements upon binding, higher quantum yields and larger extinction coefficients than ethidium homodimer-1,[4,24] they should prove superior in this application.

The high affinity for nucleic acids and general membrane impermeability of the YOYO-1 and TOTO-1 stains may permit their use to follow nucleic acid incorporation into cells by ballistic, implantation (see photo on front cover) and other permeabilization methods. These dyes have already been used to follow cell-to-cell trafficking of nucleic acids in plants [25,26] and virus-particle adsorption by host cells.[27]

Single Molecule Nucleic Acid Detection

Some of our new cyanine dyes are so bright that they can be used to directly visualize single nucleic acid molecules in the fluorescence microscope. YOYO-1 has also been used to follow the making and breaking of single chemical bonds.[28] A number of laboratories have taken advantage of the high sensitivity of these dyes to study biopolymer behavior:

- Video microscopy has been used to observe relaxation of YOYO-1–stained phage lambda DNA multimers, after stretching in a fluid flow [29] or with optical tweezers [30] (see photo on back cover). TOTO-1 has also been used in this application.[31]
- Individual YOYO-1/ssDNA molecular complexes have been imaged in solution by fluorescence video microscopy.[32]
- Molecular combing, a technique which uses a receding fluid interface to elongate DNA molecules for optical mapping of genetic loci, was developed using YOYO-1.[33]

SYBR® Green I and SYBR® Green II Dyes for Staining Nucleic Acids on Plastic Wrap or Paraffin Sheets

Molecular biologists frequently need a quick estimate of nucleic acid concentration of solutions. We have found that the SYBR Green dyes make excellent detection reagents for DNA and RNA spotted onto plastic wrap or paraffin sheets (e.g., Parafilm® brand). Extremely small quantities of the dyes are required for this assay. Using the following simple procedure, we have obtained 10- to 25-fold better sensitivity with SYBR Green dyes than with 1 µM ethidium bromide, depending on the nucleic acid tested.

Step 1. Dilute an aliquot of the SYBR Green I (S-7563, S-7567, S-7585) or SYBR Green II (S-7564, S-7568, S-7586) stock DMSO solutions 1:10,000 in TE (10 mM Tris-HCl, 1 mM EDTA, pH 8).

Step 2. Mix 5 µL of a nucleic acid–containing sample with 5 µL of the 1:10,000 dilution of SYBR Green I dye or SYBR Green II dye prepared in step 1.

Step 3. Spot the mixture onto plastic wrap or a paraffin sheet that is directly placed on a transilluminator.

Step 4. Illuminate with 300 nm transillumination and photograph through a SYBR Green gel stain photographic filter (S-7569) using Polaroid® 667 black-and-white print film. Photography is essential for obtaining the highest sensitivity.

Step 5. For a semiquantitative assay, repeat steps 2 through 4 with known amounts of nucleic acid and compare signals with those from experimental samples.

Detection limits per spot of phage lambda DNA are approximately 3 ng and 30 ng with SYBR Green I dye and ethidium bromide, respectively. Visual detection limits for ribosomal RNA under the same conditions are 3 ng, 9 ng and 80 ng with SYBR Green II dye, SYBR Green I dye and ethidium bromide, respectively.

- Highly sensitive sheath flow techniques have also been developed for detecting and discriminating the size of single TOTO-1/DNA molecular complexes.[34-36]
- Large DNA fragments stained with TOTO-1 have been sorted by flow cytometry. This extremely rapid analytical method yields a linear fluorescence intensity–to-size relationship over a 10–50 kilobase pair range.[37]
- TOTO-1, YOYO-1 and SYBR Green I dyes have been used to visualize lambda DNA that has been stretched between beads with optical tweezers.[38]

DAPI has also been employed to detect a single DNA molecule in solution [39] and by fluorescence microscopy [40] and to detect femtograms of DNA in single cells and chloroplasts.[41]

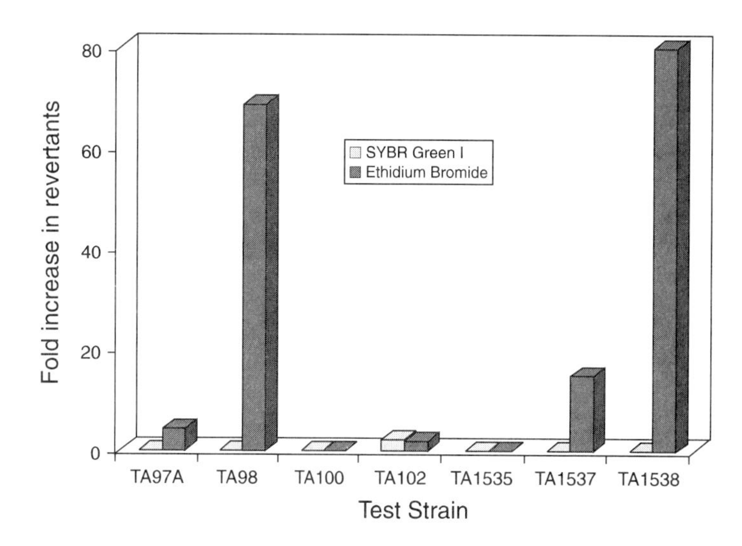

Figure 8.7 Comparison of the mutagenicity of SYBR Green I nucleic acid gel stain and ethidium bromide (EtBr) as assessed by the Ames mammalian microsome reverse mutation assay [Mutation Res 113, 173 (1983); Proc Natl Acad Sci USA 70, 2281 (1973)]. The mean fold increase in the number of revertants per plate is shown for seven Salmonella typhimurium test strains grown in the presence of mammalian microsomal enzymes from PCB-induced rat liver and either EtBr or SYBR Green I nucleic acid gel stain.

Nucleic Acid Detection in Gels

Molecular Probes' new generation of green fluorescent nucleic acid gel stains — SYBR Green I and SYBR Green II — are well on their way to becoming the preferred reagents for staining DNA and RNA in electrophoretic gels (see Color Plate 2 in "Visual Reality" and photo on back cover). The detection limits provided by each of these two gel stains are significantly lower than those obtained with the conventional gel stain ethidium bromide. In addition, Ames testing performed by an independent laboratory has shown that SYBR Green I stain is significantly less mutagenic than ethidium bromide [42] (Figure 8.7).

SYBR Green I Nucleic Acid Gel Stain

Molecular Probes' SYBR Green I nucleic acid gel stain is the most sensitive stain available for detecting nucleic acids in agarose and polyacrylamide gels. This remarkable sensitivity can be attributed to a combination of unique dye characteristics. SYBR Green I stain exhibits exceptional affinity for DNA and a large fluorescence enhancement upon binding to DNA — at least an order-of-magnitude greater than that of ethidium bromide. Also, the fluorescence quantum yield of the SYBR Green I/DNA complex (~0.8) is over five times greater than that of the ethidium bromide/DNA complex (~0.15). Some other important features of SYBR Green I nucleic acid gel stain include:

- **Sensitivity for dsDNA detection.** Less than 20 pg per band of dsDNA can be detected in a SYBR Green I–stained gel using 254 nm epi-illumination, Polaroid® 667 black-and-white print film and a SYBR Green gel stain photographic filter (S-7569) — a sensitivity that is at least 25 times greater than that achieved with ethidium bromide and standard 300 nm transillumination. Even with 300 nm transillumination, as little as 60 pg dsDNA per band can be detected with SYBR Green I stain.[43-45] (Figure 8.8)

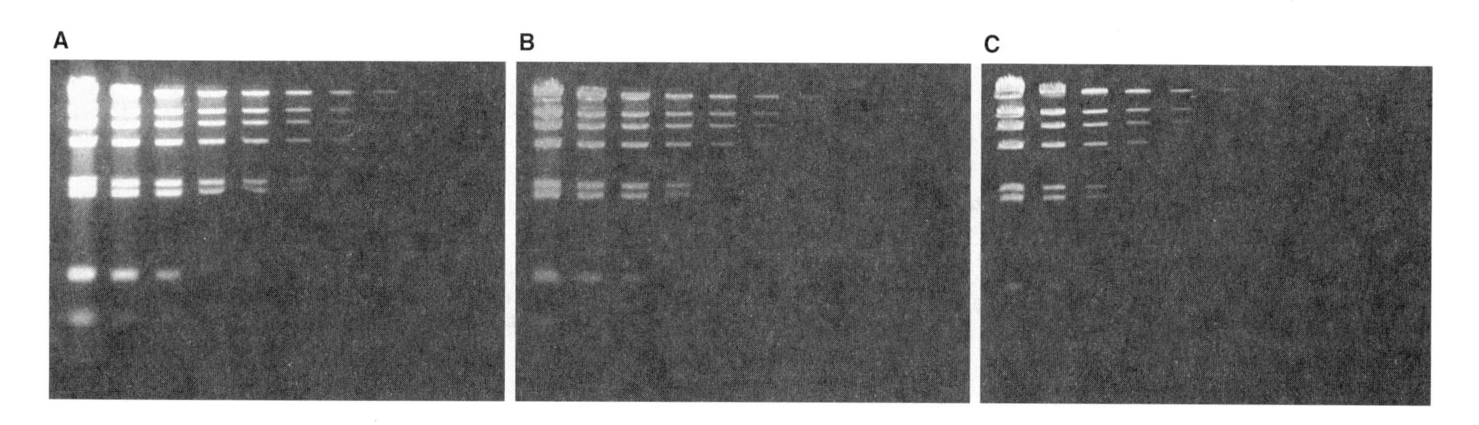

Figure 8.8 Comparison of dsDNA detection in native gels using SYBR Green I nucleic acid gel stain and ethidium bromide. Identical threefold dilutions of bacteriophage lambda DNA digested with Hind III restriction endonuclease were electrophoresed on 1% agarose gels. Gels were stained for 30 minutes with a 1:10,000 dilution of SYBR Green I nucleic acid gel stain (S-7563, S-7567, S-7585) and not destained (panels A and B) or with 5 µg/mL ethidium bromide (E-1305) for 30 minutes and destained for a further 30 minutes in water (panel C). Gel staining was visualized using 254 nm epi-illumination (panel A) or 300 nm transillumination (panels B and C) and then photographed using Polaroid 667 black-and-white print film and a SYBR Green gel stain photographic filter (S-7569, panels A and B) or an ethidium bromide gel stain photographic filter (E-7591, panel C).

- **Sensitivity for oligonucleotide detection.** We have discovered that our SYBR Green I nucleic acid gel stain is nearly two orders-of-magnitude more sensitive than ethidium bromide for staining oligonucleotides in gels. Using 254 nm epi-illumination, we can detect 1–2 ng of a synthetic 24-mer on 5% polyacrylamide gels (Figure 8.9). This surprising result only adds to SYBR Green I dye's distinction as the most sensitive stain available for detecting dsDNA in agarose or polyacrylamide gels.
- **Ease of use.** Because the SYBR Green I dye rapidly penetrates the gel matrix and has negligible background fluorescence in the absence of DNA, the staining procedure can be completed in 10–40 minutes with no destaining step required prior to photography. The staining time may have to be increased slightly with thicker precast gels. Each mL of SYBR Green I nucleic acid gel stain contains enough reagent to stain at least 100 agarose or polyacrylamide minigels. Reuse of the reagent can increase the number of gels stained per vial.
- **Compatibility with in-gel subcloning protocols.** Presence of typical staining concentrations of the SYBR Green I dye does not significantly inhibit the ability of several restriction endonucleases, including *Hin*d III and *Eco*R I to cleave DNA.[43] This property makes SYBR Green I staining potentially compatible with in-gel subcloning protocols.[46] SYBR Green I stain is also easily removed from dsDNA by simple ethanol precipitation; see the Technical Note "Removal of SYBR Green I Nucleic Acid Gel Stain from Double-Stranded DNA" on page 168.

Because the nucleic acid–bound SYBR Green dyes exhibit spectral characteristics (excitation/emission maxima ~497/520 nm) very similar to those of fluorescein along with secondary excitation peaks in the UV range (Figure 8.10), they are compatible with a

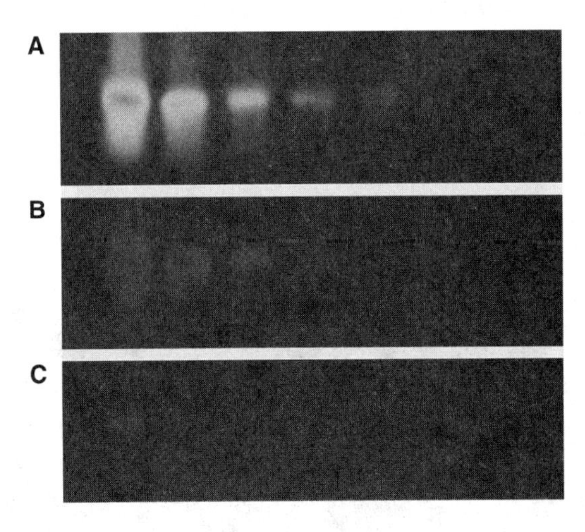

Figure 8.9 Comparison of single-stranded oligonucleotide detection using SYBR Green I nucleic acid gel stain and ethidium bromide. Identical three-fold dilutions of a synthetic, single-stranded 24-mer were electrophoresed on 10% polyacrylamide gels. Gels were stained for 30 minutes with a 1:10,000 dilution of SYBR Green I nucleic acid gel stain (S-7563, S-7567, S-7585) and not destained (panels A and B) or with 5 µg/mL ethidium bromide (E-1305) for 30 minutes and destained for a further 30 minutes in water (panel C). Gel staining was visualized using 254 nm epi-illumination (panel A) or 300 nm transillumination (panels B and C) and then photographed using Polaroid 667 black-and-white print film and a SYBR Green gel stain photographic filter (S-7569, panels A and B) or an ethidium bromide gel stain photographic filter (E-7591, panel C).

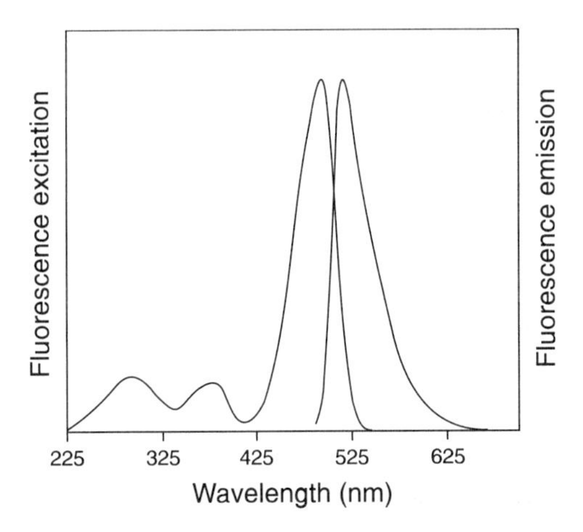

Figure 8.10 Fluorescence excitation and emission spectra of dsDNA-bound SYBR Green I nucleic acid gel stain.

wide variety of instrumentation, ranging from UV epi- and trans-illuminators to mercury-arc lamp and argon-ion laser excitation gel scanners. For optimal sensitivity with black-and-white print film and UV illumination, SYBR Green dye–stained gels should be photographed through Molecular Probes' SYBR Green gel stain photographic filter (S-7569). A gel photographic filter holder is also available (G-6657) to protect the filter and keep it flat during photography.

The ultrasensitivity of the SYBR Green I dye makes it useful for detecting the products of DNA and RNA amplification reactions by gel electrophoresis,[6] restriction mapping small amounts of DNA and detecting the products of bandshift and nuclease protection assays. Two independent laboratories found that use of SYBR Green I stain allows the detection of DNA amplification products from low–target number PCR.[3,47,48] Amplification products that were at the limit of detection using ethidium bromide were easily detected using the SYBR Green I dye. Reverse transcription PCR reaction products were also detected with high sensitivity following gel electrophoresis and SYBR Green I staining, allowing the cycle number to be lowered, which reduces heteroduplex formation during amplification.[49]

In other gel-based techniques, SYBR Green I nucleic acid gel stain has enabled researchers to eliminate silver staining and sometimes even radioactivity from their protocols. SYBR Green I staining was shown to be as sensitive as silver staining — as well as more rapid and less expensive — in a nonradioactive detection method devised for detection of hypervariable simple sequence repeats in electrophoretic gels.[50] Likewise, in a gel-based assay for detection of telomerase activity (telomeric repeat amplification protocol or TRAP) in human cells and tumors, SYBR Green I staining was found to be more sensitive than silver staining.[51] Moreover, unlike silver stains, SYBR Green I stain did not label proteins carried over from the reaction mixture. Using commercially available image analysis instruments, these researchers also reported that the results obtained with SYBR Green I staining were comparable to those achieved with a radioisotope-based TRAP assay.

SYBR DNA Molecular Weight Standards and DNA Gel Loading Solutions

For the convenience of researchers, we now provide DNA molecular weight standards prelabeled with a green fluorescent dye

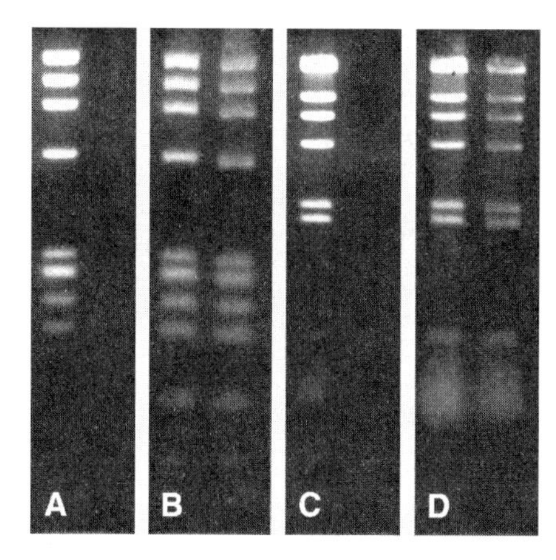

Figure 8.11 Molecular Probes' SYBR DNA molecular weight standards: A) low-range standards (S-6680, first lane) and unlabeled control (second lane) electrophoresed on a 2% agarose gel in 89 mM Tris base, 89 mM boric acid, 1 mM EDTA, pH 8.0 (TBE); B) same gel as in panel A, poststained with SYBR Green I nucleic acid gel stain; C) high-range standards (S-6681, first lane) and unlabeled control (second lane) electrophoresed on a 1% agarose gel in TBE; D) same gel as in panel C, poststained with SYBR Green I nucleic acid gel stain. The gels were visualized using 254 nm epi-illumination and then photographed through a SYBR Green gel stain photographic filter (S-7569) using Polaroid 667 black-and-white print film.

that has spectra similar to the SYBR Green I stain. The SYBR DNA molecular weight standards (S-6680, S-6681) are prepared at a dye:base pair ratio that does not affect their electrophoretic mobility, allowing scientists to monitor DNA migration in real time during electrophoresis, as well as after staining to determine the size of bands in other lanes (Figure 8.11). Enough of either high-range standards (phage lambda DNA digested with *Hin*d III) (S-6681) or low-range standards (phage ϕX174 RF DNA digested with *Hin*f I) (S-6680) are provided for loading approximately 100 minigel lanes each.

We also provide convenient 6X concentrated DNA gel loading solutions for adding to samples prior to electrophoresis. These are available with either 15% Ficoll® (D-7562) or 40% sucrose (D-7561); both contain bromophenol blue and xylene cyanol marker dyes.

SYBR Green II RNA Gel Stain

SYBR Green II RNA gel stain is a highly sensitive dye for detecting RNA or ssDNA in agarose or polyacrylamide gels (Figure 8.12). Outstanding features of the SYBR Green II RNA gel stain include its high binding affinity for RNA and large fluorescence enhancement and quantum yield upon binding RNA. Although it is not specific for RNA, the SYBR Green II dye exhibits a larger fluorescence quantum yield when bound to RNA (~0.54) than to dsDNA (~0.36). This property is unusual among nucleic stains; most show far greater quantum yields and fluorescence enhancements when bound to double-stranded nucleic acids. Moreover, the fluorescence quantum yield of the SYBR Green II/RNA complex is over seven times greater than that of the ethidium bromide/RNA complex [52] (~0.07). The affinity of SYBR Green II RNA gel stain for RNA is also higher than that of ethidium bromide, and its fluorescence enhancement upon binding to RNA is well over an order-of-magnitude greater. Other important properties of SYBR Green II RNA gel stain include:

- **Sensitivity.** Using 254 nm epi-illumination, Polaroid 667 black-and-white print film and a SYBR Green gel stain photographic filter (S-7569), we have been able to detect as little as 100 pg ribosomal RNA (rRNA) per band on native 1% agarose gels and <1 ng rRNA per band on 5% polyacrylamide gels stained with SYBR Green II RNA gel stain. The detection limit of SYBR Green II–stained native gels excited with 300 nm transillumination is approximately 500 pg per band, as compared to about 1.5 ng for ethidium bromide–stained gels [45] (Figure 8.12).
- **Ease of Use.** Like SYBR Green I stain, the SYBR Green II RNA gel stain has a very low intrinsic fluorescence, eliminating the need to destain gels.
- **Compatibility with urea and formaldehyde gels.** The fluorescence of SYBR Green II/RNA complexes does not appear to be quenched in the presence of urea or formaldehyde, so denaturing gels do not have to be washed free of the denaturant prior to staining.
- **Compatibility with Northern blots.** Recent work at Molecular Probes has shown that SYBR Green II staining is also compatible with agarose/formaldehyde gels. The formaldehyde does not have to be removed prior to staining, and the sensitivity of SYBR Green II staining is 5–10 times better than that of

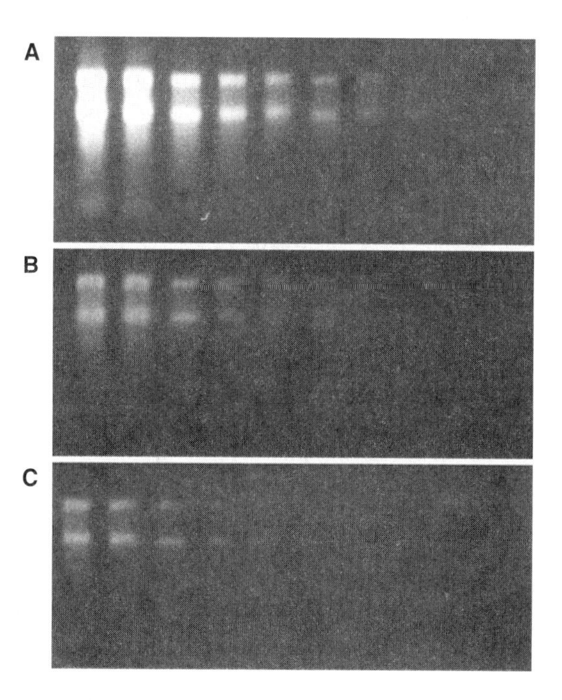

Figure 8.12 Comparison of RNA detection in nondenaturing gels using SYBR Green II RNA gel stain and ethidium bromide. Identical twofold dilutions of E. coli ribosomal RNA were electrophoresed on 1% agarose gels using Tris-borate buffer. Gels were stained for 20 minutes with a 1:10,000 dilution of SYBR Green II RNA gel stain (S-7564, S-7568, S-7586) and not destained (panels A and B) or with 5 µg/mL ethidium bromide (E-1305) for 20 minutes and destained for a further 20 minutes in water (panel C). Gel staining was visualized using 254 nm epi-illumination (panel A) or 300 nm transillumination (panels B and C) and then photographed using Polaroid 667 black-and-white print film and a SYBR Green gel stain photographic filter (S-7569, panels A and B) or an ethidium bromide gel stain photographic filter (E-7591, panel C).

ethidium bromide on these gels. In addition, staining formaldehyde gels with the SYBR Green II dye does not interfere with transfer of the RNA to filters or subsequent hybridization in Northern blot analysis as long as 0.1% to 0.3% SDS is included in prehybridization and hybridization buffers.[53]

SYBR Green II RNA gel stain facilitates the detection of viroid RNAs and other multicopy cellular RNA species. This gel stain has recently been used to visualize the migration behavior of 5S rRNA species after electrophoresis through a denaturing gradient gel, a method that was used to discriminate different acidophile species in a mixed culture.[54] SYBR Green II RNA gel stain should also improve the analysis of small aliquots from an RNA preparation, leaving the researcher with more material to carry out the primary experiment, be it Northern blotting, start-site mapping or cDNA preparation.

In addition to its use for detecting RNA, SYBR Green II RNA gel stain is proving useful for single-strand conformation polymorphism (SSCP) analysis,[55] which demands extremely sensitive detection techniques.[56] Many of the nonradioisotopic SSCP methods currently in use require long, complex procedures such as silver staining or chemiluminescence-mediated signal amplification.[57-60] Ethidium bromide has recently been employed to detect SSCP in precast polyacrylamide minigels.[61] Not only is this ethidium bromide–based SSCP technique simple, rapid and reproducible, but it allows precise temperature control — an important parameter in SSCP analysis. Our SYBR Green II RNA gel stain and potentially our SYBR Green I nucleic acid gel stain should be more sensitive than ethidium bromide in this application. SYBR Green II RNA gel stain was recently used to detect Ki-*ras* mutants by SSCP analysis and was reported to yield tenfold better sensitivity than standard silver staining techniques.[55]

SYBR Green Nucleic Acid Gel Stains: Special Packagings and Starter Kit

In addition to providing the SYBR Green nucleic acid gel stains packaged as 500 µL or 1 mL stock solutions in DMSO (S-7563, S-7564, S-7567, S-7568), we make them available as a set of 20 individual vials, each containing 50 µL of the DMSO stock solution (S-7585, S-7586). This convenient packaging makes it easy to supply each member of the lab with an aliquot of stock solution, or to share stock with other laboratories. Special packaging also minimizes potential losses due to contamination, spills and light exposure. Each milliliter of gel stain provides enough reagent to prepare 10 liters of staining solution. Although best results are obtained with freshly diluted dye, properly prepared staining solution can be stored for up to a week, if kept refrigerated and protected from light, and can be reused 2–3 times with little loss of signal.

Our SYBR Green Nucleic Acid Gel Stain Starter Kit (S-7580) is designed for laboratories that want to sample these new products. The kit includes single 50 µL vials of both SYBR Green I and SYBR Green II stains and a SYBR Green gel stain photographic filter, along with complete directions for their use and photography.

SYBR dyes are covered by U.S. Patent No. 5,436,134 and other U.S. and foreign patents pending. These products are offered for research purposes only. Molecular Probes welcomes inquiries about licensing these products for resale or other commercial uses.

Ethidium Bromide

Ethidium bromide (EtBr, E-1305) is the most commonly used dye for DNA and RNA detection in gels.[62] It binds to single-, double- and triple-stranded DNA.[63,64] Ethidium bromide has also been used to detect protein–DNA complex formation in bandshift assays [65] and to observe single DNA molecules undergoing gel electrophoresis.[66,67] In addition to the solid form (E-1305), we supply ethidium bromide in a 10 mg/mL concentrated stock solution (E-3565) for those wishing to avoid contact with mutagenic powders. The ethidium bromide gel stain photographic filter (E-7591) and a filter holder (G-6657) are available for photographing ethidium bromide–stained gels.

Cyanine Monomers for Staining DNA After Electrophoresis

Although the SYBR Green dyes are now the preferred gel stains, four of the monomeric cyanine dyes — TO-PRO-1, YO-PRO-1, BO-PRO-1 and PO-PRO-1 — are also sensitive reagents for staining gels after electrophoresis and are compatible with UV trans- or epi-illumination or with laser-excited gel scanners. Their range of absorption maxima may make them superior to the SYBR Green dyes when using some lasers as excitation sources. The limit of detection of dsDNA with these dyes is about 60 pg/band, using 254 nm epi-illumination and Polaroid 667 black-and-white print film photography.

Removal of SYBR® Green I Nucleic Acid Gel Stain from Double-Stranded DNA

Removal of nucleic acid stains from double-stranded DNA (dsDNA) may be important for some molecular biology protocols. SYBR Green I nucleic acid gel stain (S-7563, S-7567, S-7585) can be easily extracted from a dsDNA fragment isolated from a SYBR Green I–stained gel by simple ethanol precipitation. We have found that at least 99.9% of SYBR Green I stain is removed from dsDNA using the following protocol:

Step 1. Add NaCl to the SYBR Green I–stained DNA to a final concentration of 100 mM.
Step 2. Add 2½ volumes of absolute or 95% ethanol.
Step 3. Incubate the mixture for at least 20 minutes on ice.
Step 4. Centrifuge for at least 10 minutes in a microcentrifuge at 4°C.
Step 5. Remove the ethanol and wash the pellet once with 70–95% ethanol.
Step 6. Dry the sample and resuspend the dsDNA in an appropriate buffer such as TE (10 mM Tris-HCl, 1 mM EDTA, pH 8).

Cyanine and Ethidium Dimers for Staining DNA Prior to Electrophoresis

The extraordinary stability of the cyanine and ethidium dimer–nucleic acid complexes [24,68,69] ensures that the dye/DNA association remains stable during electrophoresis; thus, samples can be prestained with nanomolar dye concentrations *before* electrophoresis,[70,71] thereby reducing the hazards inherent in handling large volumes of ethidium bromide staining solutions.[68,69,72] The fluorescence intensities of both EthD-1/DNA and TOTO-1/DNA complexes are directly proportional to the amount of DNA in a band; however, TOTO-1 has a smaller effect on the electrophoretic mobility of DNA fragments than EthD-1. Furthermore, unlike EthD-1–labeled DNA, in which up to two-thirds of the bound dye can be transferred to excess unlabeled DNA, TOTO-1's transfer to unlabeled DNA is reported to be only about 15–20%, even when the TOTO-1/DNA complexes are incubated for up to 10 hours with a 100-fold excess of uncomplexed dsDNA.[24,73,74] This property is valuable for multiplex electrophoretic separations,[24] particularly since our cyanine nucleic acid stains are available in so many visually distinct colors. If two DNA populations are stained with spectrally distinct cyanine dimer dyes and run in the same lane, simultaneous two-color detection can potentially eliminate errors caused by lane-to-lane variations in electrophoretic mobility. Binding of TOTO-1, YOYO-1 and ethidium homodimer-1 to DNA initially results in inhomogeneous binding that yields double bands in DNA gel electrophoresis.[75] These double bands can be avoided by incubating complexes for times long enough to allow binding to come to equilibrium or by heating samples to 50°C for two hours.

An extremely sensitive confocal laser gel scanner has been exploited in multiplex electrophoretic separations to detect as little as four picograms per band of TOTO-1– and YOYO-1–stained dsDNA;[24,71,74,76] although sophisticated equipment is required for achieving these low detection limits, such equipment is *not essential* for detecting larger quantities of nucleic acid/dye complexes. TOTO-1 has been used to label DNA prior to electrophoresis in order to detect cystic fibrosis mutant alleles with a laser-excited fluorescence gel scanner,[77] as well as to detect DNA amplification products on agarose gels with standard UV transillumination.[78] TOTO-1 has also been used to label nine DNA fragments of the dystrophin gene that were simultaneously generated using the polymerase chain reaction.[79] The resolution obtained by gel electrophoresis of these labeled fragments compared favorably to that observed using fluorophore-labeled primers. TOTO-3 and POPO-3 have been similarly used to analyze DNA with a xenon lamp–based luminescence analyzer.[80]

Ethidium homodimer-1 (EthD-1) has been used for fluorescence detection of 30–60 picograms DNA per band on polyacrylamide gels using a confocal laser fluorescence gel scanner.[68,69,72] Ethidium homodimer-2 (EthD-2) has higher affinity for nucleic acids than EthD-1 and may transfer less rapidly between nucleic acids polymers.[81]

Electrophoretic Mobility–Shift (Bandshift) Assays

EthD-1, YOYO-1 and TOTO-1 have been shown by several laboratories to be useful tools for labeling DNA prior to electrophoresis in electrophoretic mobility–shift (bandshift) assays. EthD-1 and TOTO-1 were used to examine interactions between the binding domain of the *Kluyveromyces lactis* heat shock transcription factor and its specific binding site,[82] and YOYO-1 has been used to study the association of *E. coli* RNA polymerase with DNA templates.[83] All eight of our spectrally distinct, high-affinity dimeric cyanine dyes (Figure 8.1) and the ethidium homodimers are potentially useful for multicomponent analysis in this application.

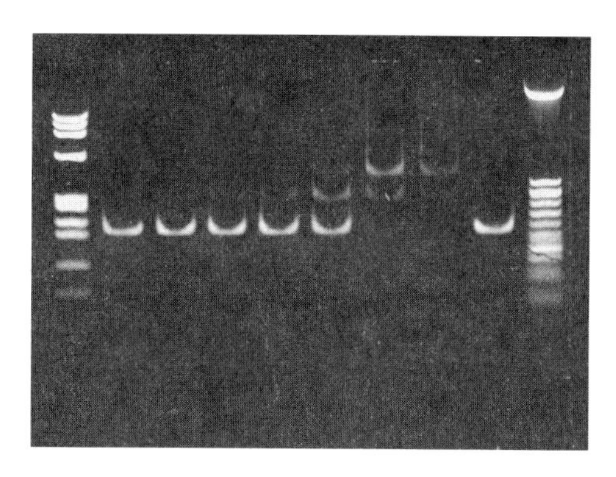

Figure 8.13 *Electrophoretic bandshift assays using SYBR Green I nucleic acid gel stain (S-7563, S-7567, S-7585). The association between a 208 bp fragment purified from Ava I-digested plasmid p5S208-12 and a mutant restriction endonuclease (EcoR I/Gln 111) was analyzed on a 4% native polyacrylamide gel. Samples containing approximately 50 ng total fragment and various amounts of mutant enzyme were subjected to electrophoresis and stained with SYBR Green I nucleic acid gel stain. Gel staining was visualized using 254 nm epi-illumination and then photographed using Polaroid 667 black-and-white print film and a SYBR Green gel stain photographic filter (S-7569). Lane 1 contains Hae III-digested ϕX174 RF DNA markers; lanes 2 through 9 contain 0, 0.05, 0.1, 0.2, 0.4, 0.6, 0.8 and 0 µM EcoR I/Gln 111; lane 10 contains Hha I-digested plasmid p5S208-12 as a size standard.*

The SYBR Green I nucleic acid gel stain can also be used as a sensitive reagent to detect bandshift products. SYBR Green I dye is much more sensitive than ethidium bromide and, when used as a poststain, obviates the need for radioactive or fluorescent labeling of DNA templates prior to carrying out the bandshift assay [84] (Figure 8.13).

Phenanthroline-Labeled Probes

The thiol-reactive bromoacetamide and iodoacetamide of 1,10-phenanthroline (P-6878, P-6879; see Section 2.4) are useful adjunct reagents for bandshift assays. Conjugation to thiol-containing ligands confers the metal-binding properties of this important complexing agent on the ligand. For example, the covalent copper–phenanthroline complex of oligonucleotides or nucleic acid–binding molecules in combination with hydrogen peroxide acts as a chemical nuclease to selectively cleave DNA or RNA.[85-91] The reagents can also be conjugated to proteins to detect nucleic acid binding and targeted cleavage.[92] Metal complexes of *N,N'*-bis-(salicylidene)ethylenediamine (B-6875, see Section 22.6) are reported to selectively cleave right-handed dsDNA.[93]

Other Nucleic Acid Stains for Gel Staining Applications

DAPI reportedly provides a significantly more sensitive means of detecting dsDNA in agarose gels than ethidium bromide.[94] Selective detection of dsDNA in the presence of dsRNA in gels with DAPI has been reported.[95] Likewise, the Hoechst dyes have been used to detect DNA in the presence of RNA in agarose gels.[96] DNA conformational changes during gel electrophoresis have been investigated with acridine orange.[97,98]

Capillary Electrophoresis

Researchers are using several of our high-sensitivity nucleic acid stains in capillary electrophoresis, a rapid and sensitive technique that is superior to slab-gel electrophoresis for resolving similar-length DNA fragments:[99]

- SYBR Green I nucleic acid gel stain has been reported to exhibit a large linear detection range and high resolution of DNA fragments from 100 to 1000 base pairs in length.
- The OliGreen reagent has been employed to detect short single-stranded oligonucleotides using capillary gel electrophoresis with laser-induced fluorescence detection; formation of the fluorescent oligonucleotide complexes is accomplished on the column.[100] In this study, the specificity of the OliGreen reagent enabled researchers to detect oligonucleotides in plasma without prior sample handling.
- Both TOTO-1 and ethidium bromide have been used with capillary-array electrophoresis for high-speed, high-throughput parallel separation of DNA fragments.[101,102] This technique may prove useful for DNA sequencing or for analysis of DNA amplification products.
- Capillary electrophoresis has been used to quantitate DNA complexes with YOYO-1 in polymerase chain reaction (PCR) mixtures.[3,103,104]
- Our dimeric cyanine dyes may also prove useful with a new high-speed sequencing method developed for use with capillary electrophoresis.[105]

In these capillary electrophoresis applications, fragments stained with PO-PRO-3 or POPO-3 can be excited with the 543 nm He-Ne laser, whereas those stained with the TOTO-1, TO-PRO-1, YO-PRO-1, YOYO-1, SYBR Green I, PicoGreen and OliGreen reagents are suitable for instruments employing the argon-ion laser.[104,106-110]

As an alternative labeling method, pre- or post-separation chemical derivatization of thiol- or amine-containing oligonucleotides is possible with many of the dyes described in Chapters 1 and 2. The thiol-reactive BODIPY, fluorescein and Oregon Green dyes are particularly suitable for labeling thiolated oligonucleotides and for applications that use ultrasensitive laser-scanning techniques.[111] Several papers have been published on separation of fluorescent oligonucleotides by capillary electrophoresis.[112-115]

Nucleic Acid Sequencing

Molecular Probes prepares a variety of products that can be used with dideoxy sequencing methods:

- Single isomers of the reactive succinimidyl esters of FAM, JOE, TAMRA and ROX are widely used to label oligonucleotide primers, deoxynucleotides and dideoxynucleotides.[116-118] (Table 8.3)
- Several other dyes in Chapter 1, including BODIPY FL, BODIPY R6G, BODIPY 564/570, BODIPY 581/591,[119] carboxydichlorofluorescein,[120] carboxyrhodamine 6G, Rhodamine Green [121] and Texas Red [122] derivatives, have been used to label primers, deoxynucleotides or dideoxynucleotides for nucleic acid sequencing (Table 8.3 and also Table 1.1 in Chapter 1).
- ChromaTide UTP and dUTP derivatives labeled with a variety of dyes are useful as substrates for DNA or RNA amplification experiments, primer extension assays, random primer labeling and potentially DNA sequencing (see Section 8.2).
- FluoroTide M13/pUC (-21) sequencing primers are available labeled with a variety of dyes or haptens (see Section 8.2). Products of sequencing reactions primed with the FluoroTide primers can be directly detected using appropriate laser scanners or automated sequencers or indirectly detected on filter membranes using our anti-dye antibodies (see Section 7.3).
- FluoReporter Oligonucleotide Labeling Kits provide a convenient method for fluorescently labeling oligonucleotide primers (see Section 8.2).
- Ultrasensitive SYBR Green I and SYBR Green II nucleic acid gel stains can potentially be used in combination with fluorescence laser scanners, automated sequencers and capillary electrophoresis systems [105] for analyzing DNA sequencing products. The SYBR Green gel stains may be particularly useful for detecting the products of cycle-sequencing reactions, in which the DNA fragments are more abundant. It is not necessary to remove urea from these gels prior to staining, as staining is not perturbed by its presence under normal conditions.
- PicoGreen dsDNA quantitation reagent and OliGreen ssDNA quantitation reagent enable researchers to determine the concentration of DNA and oligonucleotides prior to sequencing.[9]
- Alkaline phosphatase and β-galactosidase conjugates of streptavidin (S-921, S-931; see Section 7.5) can be used in combination with commercially available chemiluminescent substrates for detection of sequencing products on filter membranes.[123,124]

Nucleic Acid Detection on Filter Membranes, Plastic Wrap and Paraffin Sheets

SYBR DX DNA Blot Stain for Ultrasensitive DNA Detection on Filter Membranes

The most commonly used methods for detecting denatured DNA or RNA after blotting on filter membranes are ethidium bromide staining, which gives rise to extremely high background fluorescence, and methylene blue staining, which has low sensitivity and also yields high background signals. Silver staining or gold staining followed by silver enhancement provides 10- to 100-fold better sensitivity than ethidium bromide but is very expensive, time-consuming and tedious. Also, because of the affinity of gold for sulfur, only agarose gels containing less than 0.1% sulfate are suitable for use with gold staining; higher amounts of sulfate will result in unacceptably high background signals.

We have found that our new fluorescent SYBR DX DNA blot stain [125,126] (S-7550) provides a rapid and easy method for staining denatured DNA and other nucleic acids on filters, with sensitivity superior to that of any of our other nucleic acid stains. Direct staining of immobilized nucleic acids following Southern blotting gives an indication of the efficiency of transfer from the gel to the membrane; thus, there is no need to re-examine the gel for residual DNA. After Southern transfer, DNA is fixed to the membrane by UV crosslinking or baking in a vacuum oven as usual. Then, the blot is simply incubated with diluted SYBR DX stain for as little as 10 minutes and photographed. Unlike silver-enhanced gold labeling, no blocking steps are required. We have found that we can

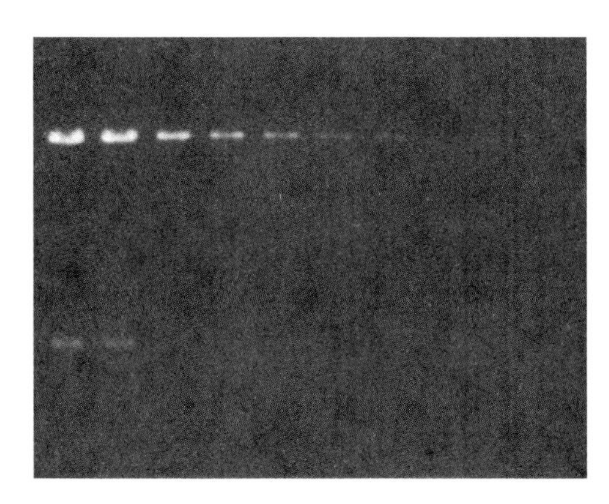

Figure 8.14 DNA stained with SYBR DX DNA blot stain (S-7550) following Southern transfer. M13mp19 RF DNA was digested with Hind III and Bgl II restriction enzymes, creating fragments of 6548 bp and 701 bp. Ten samples of the digested DNA were prepared by serial twofold dilution and applied to an agarose gel for separation by electrophoresis. The samples ranged in total DNA content from 100 ng to 0.2 ng. The DNA from the gel was then transferred to a Hybond®-N+ positively charged nylon membrane (Amersham Life Science, Inc.) by Southern blotting, stained with SYBR DX DNA blot stain, visualized with 254 nm epi-illumination and then photographed through a SYBR Green gel stain photographic filter (S-7569) using Polaroid 667 black-and-white print film. Not all of the bands visible in the original photograph are visible in this reproduction.

detect as little as 200 pg/band of denatured M13 RF DNA following Southern transfer to a nylon membrane, using SYBR DX DNA blot stain with 254 nm epi-illumination, a SYBR Green gel stain photographic filter (S-7569) and Polaroid 667 black-and-white print film. The sensitivity was reduced using 300 nm transillumination due to light scattering through the nylon membrane. However, in both cases, the sensitivity was at least 10-fold better than that obtained with ethidium bromide in parallel experiments (Figure 8.14). SYBR DX staining can also be analyzed with commercially available CCD documentation systems and laser scanners equipped with appropriate filters.

Both neutral and positively charged nylon membranes can be used with the SYBR DX DNA blot stain, and staining is fully compatible with subsequent hybridization and colorimetric detection of alkaline phosphatase–coupled probes using the chromogenic enzyme substrates nitro blue tetrazolium (NBT) and 5-bromo-4-chloro-3-indolyl phosphate (BCIP) (N-6495, B-6492, B-8405, N-6547; see Section 9.2). Blots that have been hybridized and detected by radioactive, chemiluminescent or colorimetric reagents can also be stained with the SYBR DX dye, provided that the filter is not blocked with ssDNA. Staining allows the positions of bands detected by hybridization to be easily compared with markers or other distinct species on the blot.

SYBR DX DNA blot stain (S-7550) is provided as a 1000-fold concentrate in DMSO. This amount of dye is sufficient for staining about 50–100 blots, and the staining solution can be reused at least three times with little loss in staining intensity.

Secondary Detection Reagents for Detecting Nucleic Acids on Blots

Molecular Probes sells a full line of fluorophore- and hapten-labeled secondary antibodies (see Section 7.2) and avidins (see Section 7.5), as well as an extensive selection of anti-dye antibodies (see Section 7.3). Besides their use in standard detection methods, many of these products are compatible with commercially available laser scanners or CCD documentation systems for nonradioactive detection of hybridization signals on blots. In addition, we supply streptavidin conjugates of alkaline phosphatase, β-galactosidase and horseradish peroxidase (see Section 7.5) that can be used in combination with our many chromogenic enzyme substrates (see Chapter 10) or with commercially available chemiluminescent substrates for detecting nucleic acids on blots.

ELF®-97 Phosphatase Substrate

Our ELF-97 alkaline phosphatase substrate (E-6588, E-6589; see Section 6.3) can also be used for detection of hybridization on blots, but we do not currently supply a kit designed for this application. We have found that the sensitivity for nucleic acid detection using the ELF reagent is at least 10-fold better than can be achieved with colorimetric substrates.[127] In addition, when singly labeled oligonucleotide probes are used, we have been able to attain sensitivities comparable to those achieved with singly radiolabeled oligonucleotides.

Bioluminescent Aequorin Conjugates and Chemiluminescent Luminol

The biotinylated and streptavidin-conjugated forms of the calcium-sensitive photoprotein aequorin (see Section 7.5) are versatile bioluminescent reagents for detecting DNA on Southern blots.[128,129] Using biotinylated aequorin, researchers have been able to detect 8 ng of target DNA on Southern blots.[128]

The aequorin used in these conjugates consists of a 22,000-dalton recombinant apoaequorin protein from the jellyfish *Aequorea victoria*, along with molecular oxygen and the prosthetic luminophore coelenterazine. Once aequorin binds Ca^{2+}, the complex emits a flash of blue light (emission maximum 466 nm) that persists for about two seconds. The aequorin signal can be detected on blots with instant photographic and X-ray films; it is not necessary to expose films for long periods of time, as is required with radioactive probes. Moreover, because excitation light sources are not needed for this reaction, the background noise and sample autofluorescence sometimes associated with fluorescence techniques are eliminated. Our new coelenterazine analogs, which are described in Section 22.5, may provide even greater chemiluminescence when reconstituted into apoaequorin than is possible using coelenterazine. Please contact our Technical Assistance Department about the possible availability of apoaequorin complexed with other coelenterazines.

Chemiluminescence generated with luminol (L-8455, see Section 10.5) has been used to detect horseradish peroxidase- and fluorescein-labeled nucleic acid probes on Southern blots [130] and to quantitate <150 pg human genomic DNA immobilized on a nylon membrane.[131]

Nuclease Detection

Nucleases catalyze the degradation of intact nucleic acids into smaller fragments that can be detected by their altered electrophoretic pattern using nucleic acid gel stains or by their decreased number of dye-binding sites using solution assays. Likewise, topoisomerases, gyrases, ligases, splicing enzymes and other DNA-

or RNA-binding proteins yield altered electrophoretic mobility patterns, either due to their effect on DNA conformation or due to DNA/protein complex formation, and thus should also be detectable using our sensitive gel stains.

EnzChek™ DNase Gel Assay and EnzChek DNase Solution Assay Kits

Contaminating DNases are often responsible for poor resolution of DNA fragments, degradation of samples and nicking of supercoiled plasmids. Conventional DNase assays detect DNase activity by monitoring the increase in UV absorbance that occurs when the base pairs unstack as the DNA is degraded. This absorbance method, however, is intrinsically insensitive as it requires large sample volumes and relies on small changes in absorbance. To meet the demand for a more sensitive DNase assay, Molecular Probes now introduces two EnzChek DNase Assay Kits — a gel-based assay for highest sensitivity and a solution-based assay for ease of use and for potentially detecting nicking of supercoiled DNA.

EnzChek DNase Gel Assay

By taking advantage of our ultrasensitive SYBR Green I nucleic acid gel stain, we have developed a gel-based DNase assay that provides extremely low detection limits.[132] Test DNA supplied in our EnzChek DNase Gel Assay Kit (E-6636) is simply incubated with a sample suspected of containing DNase, electrophoresed through an agarose gel and then stained with SYBR Green I dye. This EnzChek Kit allows researchers to detect less than 10^{-5} Kunitz units of DNase activity (approximately 5 pg of DNase I with a specific activity of 2000 Kunitz units/mg) in an 8 µL aliquot. Furthermore, the assay can be completed in one hour, from adding DNA to your unknown to visualizing the gel staining pattern, and will detect the activity of both DNA exonucleases and DNA endonucleases. We anticipate that our new EnzChek DNase gel assay will prove useful for researchers concerned about the integrity of their DNA samples and also in industrial settings in which reagents must be certified "nuclease free." Each EnzChek DNase Gel Assay Kit contains:

- SYBR Green I nucleic acid gel stain
- *Pst* I–linearized ϕX174 RF DNA
- Concentrated digestion buffer
- Stop buffer
- 15% Ficoll solution
- DNA molecular size markers
- Detailed protocol

Sufficient reagents are provided to perform 100 assays. Analysis of the gel by UV illumination and black-and-white print film requires the use of the SYBR Green gel stain photographic filter (S-7569), which is available separately.

EnzChek DNase Solution Assay

Our new EnzChek DNase Solution Assay Kit (E-6637) is an extremely simple and rapid solution-based assay for endonuclease activity. This kit contains supercoiled test DNA, as well as our novel monomeric cyanine nucleic acid stain YO-PRO-1, which exhibits spectral characteristics similar to those of fluorescein when bound to DNA. In the presence of DNase I or other endonucleases, the YO-PRO-1/DNA complex exhibits an immediate decrease in fluorescence that serves as a sensitive and convenient indicator of nuclease contamination. Using this kit, researchers can detect less than 4×10^{-3} Kunitz units of DNase I (approximately 2 ng of active enzyme) in a 2 to 50 µL aliquot. Moreover, this assay can be completed in about 30 minutes. Each EnzChek DNase Solution Assay Kit contains:

- YO-PRO-1 nucleic acid stain
- ϕX174 RF DNA
- Concentrated digestion buffer
- Detailed protocol

Sufficient reagents are provided to perform approximately 25 assays using standard cuvettes in a fluorometer.

EnzChek RNase Gel Assay

Contaminating RNases cause degradation of mRNA templates, making cDNA preparation or reverse transcription PCR assays difficult, as well as causing poor band resolution on Northern blots or gels of transcription products. The current methods used to detect RNases include absorbance methods similar to those employed for DNase quantitation,[133,134] radioactive procedures,[135] or gel assays based on ethidium bromide staining. The absorption-based methods are intrinsically insensitive, as they require large sample volumes and rely on small changes in absorbance. Methods that use radioactive tracers have the inherent problems of user safety and expensive waste disposal. Ethidium bromide–based staining methods are limited in sensitivity by the weak interaction of that dye with RNA.

Our new EnzChek RNase gel assay employs SYBR Green II RNA gel stain and an MS2 single-stranded RNA phage template in a simple agarose gel assay that takes about 1 hour to perform, including gel electrophoresis and staining.[132] With our EnzChek RNase Gel Assay Kit (E-6641), researchers can detect as little as 10^{-8} Kunitz units [134] of RNase A activity (single picogram levels of RNase A with a specific activity of 50 Kunitz units/mg), or as little as 5×10^{-3} Egami units [133] of RNase T1 (approximately 14 pg of RNase T1 with a specific activity of 350,000 Egami units/mg) in 8 µL samples. A sample suspected of containing RNase is incubated for five minutes with a MS2 single-stranded RNA phage substrate; the sample is then subjected to agarose gel electrophoresis followed by staining with SYBR Green II RNA dye. For even greater sensitivity, the suspect sample can be incubated with the substrate for longer times. The EnzChek RNase gel assay provides the speed and sensitivity needed both by researchers concerned about the integrity of their RNA samples and by industrial users who must certify reagents as "RNase-free." Each EnzChek RNase Gel Assay Kit contains:

- SYBR Green II RNA gel stain
- MS2 RNA
- Concentrated digestion buffer
- Stop buffer
- Loading buffer
- DNA molecular weight standards
- RNase-free H_2O
- Detailed protocol

Sufficient reagents are provided to perform 100 assays. Analysis of the gel by UV illumination and black-and-white print film requires the use of the SYBR Green gel stain photographic filter (S-7569), which is available separately.

YOYO-1–Based Nuclease Assay

YOYO-1 has also been used in a fluorescence-based microplate assay for nuclease activity.[136] This assay takes advantage of the large fluorescence enhancement of YOYO-1 binding nucleic acids and corresponding lack of fluorescence in the presence of released nucleotides and very small nucleic acid fragments. Other dyes — in particular the PicoGreen dsDNA quantitation reagent — may be more suitable for this assay.

α-Amanitin: An RNA Polymerase Inhibitor

The mushroom toxin, α-amanitin (A-6920) from *Amanita phalloides*, is a potent inhibitor of RNA polymerase II with activity *in vitro* at nanomolar concentrations, thus blocking RNA synthesis. It is used extensively for *in vitro* transcription experiments with eukaryote transcription factors isolated from cellular extracts. It does not inhibit mammalian RNA polymerases I and III even at concentrations of 1 µM.[137] Yeast RNA polymerase II also shows some sensitivity to α-amanitin but only at micromolar concentrations; yeast RNA polymerase I requires 300-fold higher concentrations for effective inhibition.[138] α-Amanitin is taken up by hepatocytes and can be used to selectively prevent mRNA synthesis *in vivo*, while allowing cells to produce rRNA and tRNAs.

Nucleic Acid Conformational Analysis

A number of conventional dyes have been used for analysis of nucleic acid conformation *in vitro* and *in vivo*:

- Acridine orange is one of the most popular and versatile fluorescent stains for histochemistry and cytochemistry and can provide a wide variety of information about the *in situ* content, molecular structure, conformation and environment of many nucleic acid–containing cell constituents.[139]
- Fluorescence photobleaching of ethidium monoazide–labeled DNA permits measurement of slow reorientational motions.[140]
- The fluorescence intensity and binding affinity of Hoechst dyes appear to be highly dependent on the sequence and conformation of the DNA base pairs.[141-143] For example, Hoechst 33258 staining can discriminate parallel and antiparallel stem regions in hairpin DNA conformations.[144]
- Psoralen has been used to assay DNA conformations such as cruciforms [145] and Z-DNA regions in supercoiled B-DNA.[146]

We also anticipate that several of our new cyanine dyes — in particular the SYTO dyes — may be useful in these applications because many of these stains appear to yield environment-sensitive metachromatic shifts upon binding to nucleic acids. Fluorescence of the TOTO-1, YOYO-1, BOBO-1 and POPO-1 dyes is dependent on nucleic acid secondary structure; a shift to longer-wavelength emission and a concomitant drop in quantum yield are observed upon binding of these dyes to single-stranded nucleic acids at high dye:base ratios.[147] Most of our unsymmetrical cyanine dyes show this spectral shift, and some show sequence-selectivity in their fluorescence intensity as well.

1. Anal Biochem 230, 353 (1995); **2.** Anal Biochem 191, 31 (1990); **3.** The PCR process is covered by patents owned by Hoffmann–LaRoche, Inc. Purchase of these products does not convey a license under these patents. Information about licenses for PCR can be obtained from Perkin–Elmer Corp. or Roche Molecular Systems, Inc.; **4.** Biophys J 61, A314 (1992); **5.** Anal Biochem 208, 144 (1993); **6.** Mol Cell Probes 9, 145 (1995); **7.** BioTechniques 21, 372 (1996); **8.** Nucleic Acids Res 24, 2623 (1996); **9.** BioTechniques 20, 676 (1996); **10.** FASEB J 9, A1422 (1995); **11.** Anal Biochem 232, 144 (1995); **12.** Anal Biochem 221, 340 (1994); **13.** Human Immunol 39, 1 (1994); **14.** BioTechniques 16, 1032 (1994); **15.** Anal Biochem 218, 458 (1994); **16.** Anal Biochem 221, 202 (1994); **17.** Neuron 14, 865 (1995); **18.** Limnol Oceanogr 40, 1050 (1995); **19.** BioTechniques 18, 136 (1995); **20.** BioTechniques 9, 310 (1990); **21.** Nucleic Acids Res 7, 567 (1979); **22.** Nucleic Acids Res 7, 571 (1979); **23.** BioTechniques 15, 1060 (1993); **24.** Nucleic Acids Res 20, 2803 (1992); **25.** Cell 76, 925 (1994); **26.** Plant Cell 5, 1783 (1993); **27.** Appl Environ Microbiol 61, 3623 (1995); **28.** Proc Natl Acad Sci USA 92, 2278 (1995); **29.** Science 264, 822 (1994); **30.** Science 264, 819 (1994); **31.** Science 268, 83 (1995); **32.** CR Acad Sc Paris 316, 459 (1993); **33.** Science 265, 2096 (1994); **34.** Anal Chem 67, 1755 (1995); **35.** Anal Chem 65, 849 (1993); **36.** Ber Bunsenges Phys Chem 97, 1535 (1993); **37.** Nucleic Acids Res 21, 803 (1993); **38.** Paul Matsudaira, MIT Whitehead Institute, personal communication; **39.** J Biochem 89, 693 (1981); **40.** J Mol Biol 152, 501 (1981); **41.** J Histochem Cytochem 34, 761 (1986); **42.** Singer, V.L., Lawlor, T.E. and Yue, S.T., manuscript in preparation; **43.** Biomedical Products 19, 68 (1994); **44.** Biophys J 66, A159 (1994); **45.** FASEB J 8, A1266 (1994); **46.** BioTechniques 3, 452 (1985); **47.** Felix Baker, Stanford University, personal communication; **48.** Andreas Oberhauser, Mayo Clinic, personal communication; **49.** PCR Meth Appl 4, 234 (1995); **50.** BioTechniques 19, 223 (1995); **51.** Meth Cell Sci 17, 1 (1995); **52.** Cytometry 7, 508 (1986); **53.** J Chin Biochem Soc 23, 37 (1994); **54.** Appl Environ Microbiol 62, 1969 (1996); **55.** Diagnostic Molec Pathol 4 (1996) in press; **56.** Genomics 5, 874 (1989); **57.** Trends Genet 8, 49 (1992); **58.** Nucleic Acids Res 19, 3154 (1991); **59.** Nucleic Acids Res 19, 2500 (1991); **60.** Nucleic Acids Res 19, 405 (1991); **61.** Nucleic Acids Res 21, 3637 (1993); **62.** Anal Chem 63, 2038 (1991); **63.** J Biol Chem 266, 5417 (1991); **64.** Nucleic Acids Res 19, 1521 (1991); **65.** Anal Biochem 190, 331 (1990); **66.** Biopolymers 28, 1491 (1989); **67.** Science 243, 203 (1989); **68.** Nucleic Acids Res 19, 327 (1991); **69.** Proc Natl Acad Sci USA 87, 3851 (1990); **70.** Nucleic Acids Res 21, 5720 (1993); **71.** Nature 359, 859 (1992); **72.** BioTechniques 10, 616 (1991); **73.** Nucleic Acids Res 23, 1215 (1995); **74.** Meth Enzymol 217, 414 (1993); **75.** Nucleic Acids Res 23, 2413 (1995); **76.** Rev Sci Instrum 65, 807 (1994); **77.** Mol Cell Probes 8, 245 (1994); **78.** Modern Pathology 7, 385 (1994); **79.** BioTechniques 15, 274 (1993); **80.** BioTechniques 20, 708 (1996); **81.** Haugland, R.P., *Molecular Probes' Handbook of Fluorescent Probes and Research Chemicals, Fifth Edition* (1992–1994) p. 224; **82.** J Biol Chem 268, 25229 (1993); **83.** Proc Natl Acad Sci USA 91, 6870 (1994); **84.** FASEB J 10, A1128, abstract #751 (1996); **85.** Nucleic Acids Res 22, 4789 (1994); **86.** Proc Natl Acad Sci USA 91, 1721 (1994); **87.** Biochemistry 33, 3848 (1994); **88.** Bioconjugate Chem 4, 69 (1993); **89.** Meth Enzymol 208, 414 (1991); **90.** J Am Chem Soc 111, 4941 (1989); **91.** Proc Natl Acad Sci USA 86, 9702 (1989); **92.** Science 265, 959 (1994); **93.** J Org Chem 58, 820 (1993); **94.** J Biochem Biophys Meth 6, 95 (1982); **95.** Nucleic Acids Res 6, 3535 (1979); **96.** Nucleic Acids Res 15, 10589 (1987); **97.** Ann Rev Biophys Chem 20, 415 (1991); **98.** Biochemistry 29, 3396 (1990); **99.** Anal Chem 64, 1737 (1992); **100.** L. Reyderman and S. Stavchansky, University of Texas at Austin, personal communication; **101.** Anal Chem 66, 1424 (1994); **102.** Anal Biochem 215, 163 (1993); **103.** Anal Biochem 224, 140 (1995); **104.** J Chromatography B 658, 271 (1994); **105.** Anal Chem 67, 1913 (1995); **106.** Anal Biochem 231, 359 (1995); **107.** FASEB J 9, A1423 (1995); **108.** J Chromatography A 669, 205 (1994); **109.** Appl Theoret Electrophoresis 3, 235 (1993); **110.** J Microcolumn Separation 5, 275 (1993); **111.** J Am Chem Soc 116, 7801 (1994); **112.** Anal Chem 66, 1941 (1994); **113.** Anal Chem 65, 3518 (1993); **114.** J Chromatography A 652, 83 (1993); **115.** J Chromatography A 652, 75 (1993); **116.** Anal Chem 67, 1197 (1995); **117.** Anal Biochem 223, 39 (1994); **118.** U.S. Patent No. 4,855,225; **119.** Science 271, 1420 (1996); **120.** Electrophoresis 13, 542 (1992); **121.** Nucleic Acids Res 20, 2471 (1992); **122.** Anal Biochem 224, 117 (1995); **123.** Martin, C.S. and Bronstein, I. in *Nonisotopic Probing, Blotting, and Sequencing*, L.J. Kricka, Ed., Academic Press (1995) pp. 493–512; **124.** Flick, P.K. in *Nonisotopic Probing, Blotting, and Sequencing*, L.J. Kricka, Ed., Academic Press (1995) pp. 475–492; **125.** FASEB J 10, A1129, abstract #752 (1996); **126.** Patents pending; **127.** Paragas, V.B. *et al.*, manuscript submitted; **128.** Biochemistry 31, 1433 (1992); **129.** Stults, N.L., Stocks, N.A., Cummings, R.D., Cormier, M.J. and Smith, D.F. in *Bioluminescence and Chemiluminescence: Current Status*, P. Stanley and L. Kricka, Eds., John Wiley and Sons, Chichester, U.K. (1991) pp. 533–536; **130.** Durrant, I. in *Nonisotopic Probing, Blotting and Sequencing, Second Edition*, L.J. Kricka, Ed., Academic Press, San Diego (1995) pp. 195–215; **131.** Nucleic Acids Res 20, 5061 (1992); **132.** FASEB J 9, A1400 (1995); **133.** Prog Nucleic Acids Res and Mol Biol 3, 59 (1964); **134.** J Gen Physiol 24, 15 (1940); **135.** Nucleic Acids Res 10, 833 (1982); **136.** BioTechniques 18, 231 (1995); **137.** Science 170, 447 (1970); **138.** Sentenac, A. and Hall, B. in *The Molecular Biology of the Yeast Saccharomyces, Metabolism and Gene Expression*, J.N. Strathern, E.W. Jones and J.R. Broach, Eds. Cold Spring Harbor Laboratory Press (1982) pp. 561–606; **139.** Darzynkiewicz, Z. and Kapuscinski, J. in *Flow Cytometry and Sorting, Second Edition*, M.R. Melamed, *et al.*, Eds., Wiley-Liss (1990) pp. 291–314; **140.** Biophys J 53, 215 (1988); **141.** Biochemistry 30, 182 (1991); **142.** Biochemistry 29, 10181 (1990); **143.** Nucleic Acids Res 18, 3753 (1990); **144.** Science 241, 551 (1988); **145.** Proc Natl Acad Sci USA 80, 1797 (1983); **146.** Biochemistry 26, 1343 (1987); **147.** FASEB J 7, A1087 (1993).

8.3 Price List *In Vitro Applications for Nucleic Acid Stains and Probes*

Cat #		Product Name	Unit Size	Price Per Unit ($) 1–4 Units	5–24 Units
A-6920	*New*	α-amanitin	1 mg	90.00	72.00
D-7562	*New*	DNA gel-loading solution with 15% Ficoll® *6X concentrate*	1 mL	24.00	19.20
D-7561	*New*	DNA gel-loading solution with 40% sucrose *6X concentrate*	1 mL	24.00	19.20
E-6636	*New*	EnzChek™ DNase Gel Assay Kit *100 assays*	1 kit	135.00	108.00
E-6637	*New*	EnzChek™ DNase Solution Assay Kit *25 assays*	1 kit	135.00	108.00
E-6641	*New*	EnzChek™ RNase Gel Assay Kit *100 assays*	1 kit	135.00	108.00
E-7591	*New*	ethidium bromide gel stain photographic filter	each	29.00	29.00
G-6657	*New*	gel photographic filter holder	each	18.00	18.00
O-6682	*New*	oligonucleotide molecular weight standards *50 gel lanes*	500 µL	50.00	40.00
O-7582	*New*	OliGreen™ ssDNA quantitation reagent *for ssDNA and oligonucleotides* *200-2000 assays*	1 mL	175.00	140.00
P-7589	*New*	PicoGreen® dsDNA Quantitation Kit *200-2000 assays*	1 kit	175.00	140.00
P-7581	*New*	PicoGreen® dsDNA quantitation reagent *200-2000 assays*	1 mL	150.00	120.00
S-6680	*New*	SYBR® DNA molecular weight standards *low range* *100 gel lanes*	1 mL	50.00	40.00
S-6681	*New*	SYBR® DNA molecular weight standards *high range* *100 gel lanes*	1 mL	50.00	40.00
S-7550	*New*	SYBR® DX DNA blot stain *1000X concentrate in DMSO*	1 mL	175.00	140.00
S-7563	*New*	SYBR® Green I nucleic acid gel stain *10,000X concentrate in DMSO*	500 µL	125.00	100.00
S-7567	*New*	SYBR® Green I nucleic acid gel stain *10,000X concentrate in DMSO*	1 mL	235.00	188.00
S-7585	*New*	SYBR® Green I nucleic acid gel stain *10,000X concentrate in DMSO* *special packaging*	20x50 µL	265.00	212.00
S-7564	*New*	SYBR® Green II RNA gel stain *10,000X concentrate in DMSO*	500 µL	125.00	100.00
S-7568	*New*	SYBR® Green II RNA gel stain *10,000X concentrate in DMSO*	1 mL	235.00	188.00
S-7586	*New*	SYBR® Green II RNA gel stain *10,000X concentrate in DMSO* *special packaging*	20x50 µL	265.00	212.00
S-7569	*New*	SYBR® Green gel stain photographic filter	each	29.00	29.00
S-7580	*New*	SYBR® Green Nucleic Acid Gel Stain Starter Kit	1 kit	55.00	55.00

8.4 Chromosome Banding and Fluorescence *In Situ* Hybridization

Applications of many of our nucleic acid stains for chromosome banding and fluorescence *in situ* hybridization are described below. The use of nucleic acids stains and related products for staining live cells, analyzing cell cycle, measuring cell proliferation and detecting apoptotic and dead cells is discussed in Chapter 16. Most of the nucleic acid stains described in this section are listed in the tables and product list associated with Section 8.1, unless otherwise indicated.

Chromosome Banding and Analysis

Chromosome Banding Kits

Counterstain-enhanced chromosome banding is a widely used technique for karyotype analysis and chromosome structure studies.[1] In general, the staining pattern is accomplished by combining an AT- or GC-selective DNA-binding dye with a nucleic acid counterstain. We offer two Chromosome Banding Kits (C-7584, C-7587), which contain the same DNA-binding dye but different green fluorescent counterstains. Each Chromosome Banding Kit contains:

- Concentrated YOYO-1 (Kit #1, C-7584) *or* SYTOX Green nucleic acid stain (Kit #2, C-7587) in DMSO
- Methyl Green
- Detailed chromosome banding protocol

Sufficient reagents are supplied for staining approximately 200 slide preparations or up to 600 coverslip preparations, when using the concentrations recommended in the protocol.

These Chromosome Banding Kits provide sharp, clear chromosome bands that can be resolved in the fluorescence microscope. The green fluorescent YOYO-1 and SYTOX Green dyes are both efficiently excited with the argon-ion laser, allowing analysis of chromosome structure by confocal laser scanning microscopy as well. YOYO-1 bound to DNA has somewhat narrower and slightly shorter-wavelength spectra than SYTOX Green nucleic acid stain bound to DNA, whereas the use of SYTOX Green dye eliminates the need for RNase treatment of slides. Methyl Green is a major-groove binding dye that binds selectively to AT sequences along the chromosome;[2] its binding quenches the fluorescence of the YOYO-1 or SYTOX Green stains bound to DNA, giving rise to a banding pattern that indicates the location of AT-rich regions (see Color Plate 2 in "Visual Reality"). This phenomenon has been exploited to examine metaphase chromatin structure [3] and represents an extremely simple, rapid, fluorescence-based banding method that may prove useful for general karyotype analysis.

Cyanine Dimers

Staining with our TOTO-1 and YOYO-1 nucleic acid stains has been shown to allow extremely sensitive flow cytometric analysis of nuclei and isolated human chromosomes.[4] In this study, YOYO-1 produced more than 1000 times the fluorescence signal obtained with mithramycin, and histograms of both TOTO-1 and YOYO-1 on RNase-treated nuclei provided coefficients of variation that were at least as low as those found with propidium iodide or mithramycin. These researchers also found that when nuclei were simultaneously stained with YOYO-1 and Hoechst 33258, the ratio of the fluorescence of these two dyes varied as a function of cell cycle. This observation suggests that the cyanine dyes might be useful for examining cell cycle–dependent changes that occur in chromatin structure. Dual-wavelength excitation of either TOTO-1– or YOYO-1–stained chromosomes reportedly permits specific chromosomes to be identified and sorted.[5]

Acridine Homodimer and Heterodimer

The water-soluble acridine homodimer has extremely high affinity for AT-rich regions of nucleic acids, making it particularly useful for chromosome banding.[6,7] Acridine homodimer emits a blue-green fluorescence when bound to DNA, yielding fluorescence that is proportional to the fourth power of the AT base-pair content.[8] Acridine homodimer has recently been recommended as an alternative to quinacrine for Q banding because of its greater brightness and higher photostability.[6]

Molecular Probes also prepares ethidium–acridine heterodimer. The efficiency of fluorescence resonance energy transfer from the acridine to the phenanthridine portion of this heterodimer is related to the square of the AT base-pair content of the DNA.[9] Thus, the fluorescence excitation spectrum of this heterodimer can be used to characterize AT content in fixed single cells and chromosomes by ratio imaging techniques.[9]

Other Dyes and Chromosome Banding Reagents

A wide variety of fluorescent nucleic acid stains have been used for chromosome banding:[1,6,10,11]

- Hoechst dyes have been used in chromosome sorting, multivariate analysis and karyotyping.[12]
- High-resolution flow karyotyping has also been carried out with DAPI.[13,14]
- Combinations of DAPI, DIPI or Hoechst 33258 with nonfluorescent DNA-binding drugs have been used for chromosome binding studies.[15]
- DIPI and Hoechst dyes have been employed in combination with chromomycin and a high-resolution, dual-laser method to sort 21 unique human chromosome types onto nitrocellulose filters, followed by hybridization to gene-specific probes.[16]
- 7-Aminoactinomycin D (7-AAD) binds selectively to GC regions of DNA,[17] yielding a distinct banding pattern in polytene chromosomes and chromatin.[18,19]
- 9-Amino-6-chloro-2-methoxyacridine (ACMA) fluoresces with greater intensity in AT-rich regions on chromosomes,[20] yielding a staining pattern similar to the Q-banding pattern produced with quinacrine.

Fluorescence *In Situ* Hybridization (FISH)

Fluorescence *in situ* hybridization (FISH) offers many advantages over radioactive methods for localizing and determining the relative abundance of specific nucleic acid sequences in cells, tissue, interphase nuclei and metaphase chromosomes. Not only are fluorescence techniques fast and precise, they allow the simultaneous analysis of multiple probes.[21-26] Potentially, researchers can visualize over 40 signals simultaneously by using single probes that contain two or more haptens (in varying proportions) in conjunction with highly discriminating optical filters.[27-29] Multicolor fluorescent DNA and RNA hybridization techniques are important in clinical diagnostics [30] and gene mapping.[26]

Molecular Probes offers a wide variety of reagents for the detection of *in situ* hybridization signals. These include tools for preparing labeled hybridization probes, second-step reagents for detecting labeled probes, kits and reagents for amplifying signals and dyes for counterstaining nuclei or chromosome spreads.

Probe Preparation for FISH

Our ChromaTide dUTP derivatives (see Section 8.2) can be used for nick translation with *Escherichia coli* DNA polymerase I, random-primer labeling of DNA probes with Klenow polymerase or 3'-end labeling with terminal deoxynucleotidyl transferase. Our ChromaTide UTP derivatives can be used to prepare RNA probes with T3, T7 and SP6 RNA polymerases. Protocols for enzymatic labeling are available through our Technical Assistance Department. In addition, we provide 16 different FluoReporter Oligonucleotide Labeling Kits for chemically labeling either the amine moieties of amine-modified synthetic oligonucleotides or the 3'- or 5'-phosphates of conventional oligonucleotides (see Section 8.2).

Secondary Detection Reagents for FISH

Tables 7.1 and 7.2 of Chapter 7 list a wide variety of antibodies, avidin, streptavidin and NeutraLite™ avidins labeled with fluorophores, haptens or enzymes. These secondary detection reagents are important for multicolor fluorescence *in situ* hybridization applications. NeutraLite avidin conjugates have been shown to provide improved detection of single-copy genes in metaphase chromosome spreads.[31]

We also supply various labeled and unlabeled rabbit antibodies to fluorescein, tetramethylrhodamine, Texas Red, BODIPY FL, BODIPY TR, Cascade Blue®, lucifer yellow and dansyl fluorophores and the dinitrophenyl hapten (see Section 7.3) that can be used to amplify signals from probes containing those labels, to restore fluorescence of partially bleached samples and to discriminate among probes labeled with different haptens. Preferred haptenylation reagents and their corresponding anti-hapten antibodies are listed in Table 4.1 in Chapter 4. Our fluorescent goat anti-rabbit conjugates and avidin conjugates (see Tables 7.1 and 7.2 in Chapter 7) can be used to further amplify signals from our anti-dye antibodies.

Anti-fluorescein antibodies have been employed to simultaneously detect two different mRNA sequences in double *in situ* hybridizations using fluorescein- or biotin-labeled oligonucleotides.[24] The high affinity of anti-fluorescein antibodies makes fluorescein an excellent hapten for *in situ* hybridizations and other second-step detections.[24,32] Researchers have found fluorescein–anti-fluorescein ELISA techniques to be similar in sensitivity to biotin–streptavidin methods [33] and to display extremely low nonspecific binding.[34] Our anti–BODIPY FL preparation has been shown to bind specifically to BODIPY FL–labeled oligonucleotides, where it has been detected with an alkaline phosphatase–conjugated anti–rabbit IgG.[35] We also now offer a fluorescein conjugate of anti–BODIPY FL antibody (A-6420, see Section 7.3) for amplifying the green fluorescence of a BODIPY FL–labeled probe.

Our biotin-XX (A-6435), fluorescein (A-6423), tetramethylrhodamine (A-6424) and Texas Red-X (A-6396) conjugates of anti-dinitrophenyl (DNP) antibodies (see Section 7.3) are especially suitable for detecting hybridization probes labeled with the DNP hapten.[26,36] These anti-DNP antibodies are prepared against the DNP–keyhole limpet hemocyanin (KLH) conjugate and thus do not crossreact with bovine serum albumin (BSA), which is commonly used as a blocking or carrier molecule in hybridization applications. Anti-DNP antibodies have been used to localize a DNP-labeled DNA probe in HIV-infected cells.[37] It has also been reported that human chromosomes can be probed with equal sensitivities by biotinylated, DNP-modified and digoxigenin-modified cosmid probes.[26]

ELF-97 Signal Amplification

Our Enzyme-Labeled Fluorescence (ELF) technology, which is discussed in greater detail in Section 6.3, promises to become an important development for amplifying fluorescence *in situ* hybridization detection. Upon enzymatic cleavage, the weakly blue fluorescent ELF-97 phosphatase substrate yields a bright yellow-green fluorescent precipitate that exhibits an unusually large Stokes shift and excellent photostability.[38] We have developed and optimized procedures for using this ELF-97 phosphatase substrate to detect mRNA *in situ* hybridization in cells and tissue sections (Figure 8.15). We now provide the key reagents for this application in our new ELF-97 mRNA *In Situ* Hybridization Kits (E-6604, E-6605).

In conventional mRNA *in situ* hybridization, radioactively labeled DNA or RNA probes are hybridized to the experimental sample and then detected by applying a photosensitive emulsion to the microscope slides.[39-41] Typically, the emulsion is exposed for days to weeks before it is developed and photographed using white-light microscopy.

In contrast, ELF signals develop in minutes, producing a green fluorescent precipitate of the ELF-97 alcohol that is significantly brighter than signals achieved either with directly labeled fluorescent nucleic acid probes or with hapten-labeled probes in combination with fluorophore-labeled secondary detection reagents.[42-44] Only a short oligonucleotide with as few as 18 bases and a single hapten molecule is required as the probe. Fluorescence of the precipitated dye is exceptionally stable to illumination in the microscope (see Figure 6.6 in Chapter 6) and has an extremely high

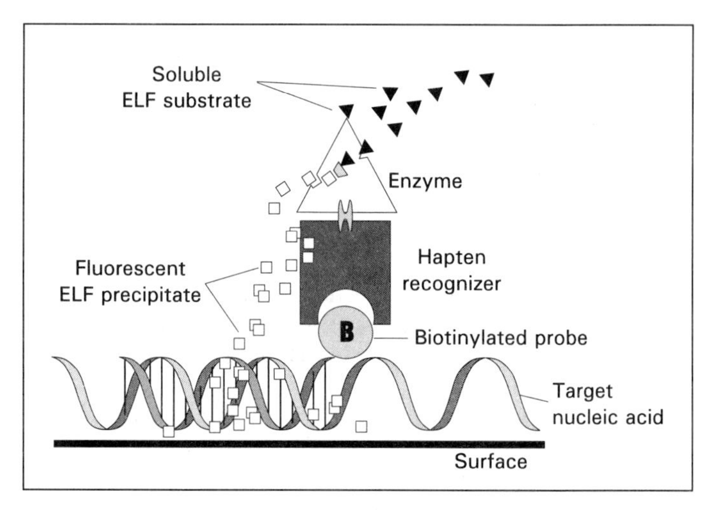

Figure 8.15 *Schematic representation of mRNA in situ hybridization detection using the enzyme labeled fluorescence (ELF) process.*

Stokes shift — greater than 140 nm (see Figure 6.7 in Chapter 6). Moreover, the fluorescent ELF signals can be clearly distinguished from sample pigmentation, which often obscures the dark silver grains in the emulsion. ELF amplification technology can be combined with a wide variety of haptens to increase the versatility of the reagents or can be used with directly labeled alkaline phosphatase–nucleic acid probes.

In addition to a detailed protocol, the ELF-97 mRNA *In Situ* Hybridization Kit #2 (E-6605) supplies the reagents for using the ELF-97 phosphatase substrate to detect hybridization of biotinylated DNA or RNA probes in cells or tissue sections, including a streptavidin–alkaline phosphatase conjugate:

- Streptavidin–alkaline phosphatase conjugate (Kit #2 only)
- ELF wash, blocking and developing buffers
- Application-specific ELF-97 phosphatase substrate solution plus additives
- Nucleic acid counterstain, Hoechst 33342
- ELF mounting medium
- 50 plastic coverslips

This Kit #2 is also useful for detecting hybridization probes labeled with other haptens when it is used in combination with our biotinylated anti-hapten or anti-dye antibodies or secondary antibodies (see Table 7.1 in Chapter 7). The ELF-97 mRNA *In Situ* Hybridization Kit #1 (E-6604) is identical to Kit #2 except that the streptavidin–alkaline phosphatase conjugate is not included. The ELF-97 mRNA *In Situ* Hybridization Kit #1 (E-6604) is designed for use with other alkaline phosphatase conjugates — such as an anti-fluorescein–alkaline phosphatase conjugate — that have been applied to detect DNA or RNA probes conjugated to other haptens such as fluorescein. Each kit contains sufficient reagents for 50 slides or coverslips. These kits' protocols require that the ELF developing solution be filtered before use. Our ELF spin filters (E-6606) — spin-filtration devices with a pore size of 0.2 µm — are both convenient and efficient, allowing a very small volume to be filtered without significant loss of sample.

Colorimetric Signal Amplification

We also offer the colorimetric enzyme substrates NBT/BCIP (N-6495, B-6492, B-8405, N-6547, see Section 9.2), X-Gal

(B-1690, see Section 10.2) and tetramethylbenzidine (TMB, T-8463, T-8464, see Section 10.5), which can be used to detect hapten-labeled probes with our conjugates of alkaline phosphatase (S-921), β-galactosidase (S-931) and horseradish peroxidase (S-911), respectively [45] (see Section 7.5).

FISH Counterstain Kits

Molecular Probes provides a spectrum of dyes for counterstaining hybridized metaphase or interphase chromosomes in FISH assays. The most commonly used chromosome counterstains are the red fluorescent propidium iodide and the blue fluorescent DAPI. Our two new FISH Counterstain Kits #1 and #2 (F-7583, F-7588) contain these conventional dyes (see Color Plate 2 in "Visual Reality"), as well as one of our proprietary green fluorescent dyes:

- Concentrated YOYO-1 (Kit #1, F-7583) *or* SYTOX Green nucleic acid stain (Kit #2, F-7588) in DMSO
- Concentrated propidium iodide in water
- Concentrated DAPI in water
- Detailed counterstaining protocol

Each of the nucleic acid stains is provided as a convenient 300X solution, ready for dilution and staining. Sufficient reagents are supplied for staining 100 slides with each counterstain.

Molecular Probes' researchers have found that both YOYO-1 and SYTOX Green dyes provide simple and reliable green fluorescent counterstains for FISH analysis, though they differ somewhat in their properties and applications. YOYO-1 staining requires RNase treatment for optimal background reduction, whereas SYTOX Green staining does not. In addition, SYTOX Green counterstaining is more rapid than YOYO-1 counterstaining. With the SYTOX Green dye, staining is complete in a few minutes; with YOYO-1, a longer incubation is required for stable binding. The spectral properties of the two green fluorescent dyes also differ. The emission spectrum for YOYO-1/DNA complexes is much narrower than that of SYTOX Green/DNA complexes and may therefore be better suited to multicolor applications. Nucleic acids counterstained with either of these green fluorescent dyes can be efficiently excited with the mercury-arc lamp or argon-ion laser and can be visualized using standard fluorescein optical filter sets.

In addition to DAPI [46-48] and propidium iodide,[49] Hoechst 33258 and ethidium bromide and a number of our dimeric cyanine stains have been used as chromosome counterstains for fluorescence *in situ* hybridization.[50] POPO-1 can be used as either a blue or green fluorescent counterstain, depending on the optical filter chosen for observation. The monomeric cyanine ("TO-PRO series") dyes can also be used in this application.

Cytological Nuclear Counterstain Kit

For RNA *in situ* hybridization with tissue culture cells, we recommend our Cytological Nuclear Counterstain Kit (C-7590), which contains three spectrally distinct fluorescent dyes for staining nuclei in fixed-cell preparations:

- Concentrated SYTOX Green nucleic acid stain in DMSO
- Concentrated propidium iodide in water
- Concentrated DAPI in water
- Detailed protocols for fixations and staining

Each of the nucleic acid stains is provided as a convenient 300X solution, ready for dilution and staining. Sufficient reagents are supplied for staining 300 slides with each counterstain.

The Cytological Nuclear Counterstain Kit is especially valuable for multicolor applications, in which an appropriate counterstain can be selected to contrast spectrally with other fluorescent probes in the sample. When used according to our protocols, the counterstains in this kit stain nuclei specifically, with little or no cytoplasmic labeling. Observed by fluorescence microscopy, the nuclei stand out in vivid contrast to fluorescent probes of other cell structures. With slight protocol modifications, the SYTOX Green dye or propidium iodide can be used to label both RNA and DNA in order to visualize the cytoplasm as well as the nucleus. The counterstain protocols are compatible with a wide range of cytological labeling techniques, including direct or indirect antibody-based detection methods, mRNA *in situ* hybridization or detection of specific cellular structures with fluorescent probes such as our mitochondrion-selective MitoTracker™ (see Section 12.2) and F-actin–selective phalloidin (see Section 11.1) probes. These counterstains can also serve to fluorescently label cells for analysis in multicolor flow cytometry experiments.

1. Human Genet 57, 1 (1981); **2.** FEBS Lett 315, 61 (1993); **3.** Cell 76, 609 (1994); **4.** Cytometry 15, 129 (1994); **5.** U.S. Patent No. 5,418,169; **6.** Meth Mol Biol 29, 83 (1994); **7.** Biochemistry 18, 3354 (1979); **8.** Proc Natl Acad Sci USA 72, 2915 (1975); **9.** Biochemistry 17, 5078 (1978); **10.** BioEssays 15, 349 (1993); **11.** Am J Human Genet 51, 17 (1992); **12.** Cytometry 11, 80 (1990); **13.** Cytometry 11, 184 (1990); **14.** Stain Technol 60, 7 (1985); **15.** Eur J Cell Biol 20, 290 (1980); **16.** Science 225, 57 (1984); **17.** Biopolymers 18, 1749 (1979); **18.** Cytometry 20, 296 (1995); **19.** Histochem J 17, 131 (1985); **20.** Exp Cell Res 117, 451 (1978); **21.** Histochem J 27, 4 (1995); **22.** Exp Cell Res 194, 310 (1991); **23.** Nucleic Acids Res 19, 3237 (1991); **24.** J Histochem Cytochem 38, 467 (1990); **25.** Nature 345, 93 (1990); **26.** Science 247, 64 (1990); **27.** J Histochem Cytochem 41, 1755 (1993); **28.** Proc Natl Acad Sci USA 89, 1388 (1992); **29.** Cytometry 11, 126 (1990); **30.** Science 250, 559 (1990); **31.** Trends Genet 9, 71 (1993); **32.** E.W. Voss Jr., *Fluorescein Hapten: An Immunological Probe*, CRC Press (1984); **33.** J Immunol Meth 122, 115 (1989); **34.** J Chemilumin Biolumin 2, 215 (1988); **35.** G.S. Schneider, Tropix, Inc., personal communication; **36.** Nucleic Acids Res 18, 3175 (1990); **37.** BioTechniques 9, 186 (1990); **38.** U.S. Patent Nos. 5,316,906 and 5,443,986; **39.** Angerer, L.M. and Angerer, R. in *In Situ Hybridization: A Practical Approach*, D.G. Wilkinson, Ed., IRL Press (1992) pp. 15–32; **40.** Wilkinson, D.G. in *In Situ Hybridization: A Practical Approach*, D.G. Wilkinson, Ed., IRL Press (1992) pp. 1–14; **41.** Pardue, M.L. in *Nucleic Acid Hybridization: A Practical Approach*, B.D. Hames and S.J. Higgens, Eds., IRL Press (1985) pp. 179–202; **42.** Am J Human Genet Suppl 55, A271, abstract #1585 (1994); **43.** FASEB J 8, A1444, abstract #1081 (1994); **44.** Mol Biol of the Cell Suppl 4, 226a, abstract #1313 (1993); **45.** Naturwissenschaften 73, 553 (1986); **46.** J Microscopy 157, 73 (1990); **47.** Proc Natl Acad Sci USA 87, 9358 (1990); **48.** Proc Natl Acad Sci USA 87, 6634 (1990); **49.** BioTechniques 16, 441 (1994); **50.** Cytometry, Suppl 6, A427B (1993).

Cat #		Product Name	Unit Size	Price Per Unit ($) 1–4 Units	5–24 Units
C-7584	**New**	Chromosome Banding Kit #1 *YOYO®-1/Methyl Green* *200 slides*	1 kit	95.00	76.00
C-7587	**New**	Chromosome Banding Kit #2 *SYTOX® Green/Methyl Green* *200 slides*	1 kit	95.00	76.00
C-7590	**New**	Cytological Nuclear Counterstain Kit *DAPI/SYTOX® Green/PI* *300 slides*	1 kit	75.00	60.00
E-6604	**New**	ELF®-97 mRNA *In Situ* Hybridization Kit #1 *50 assays*	1 kit	175.00	140.00
E-6605	**New**	ELF®-97 mRNA *In Situ* Hybridization Kit #2 *with streptavidin, alkaline phosphatase conjugate* *50 assays*	1 kit	225.00	180.00
E-6606	**New**	ELF® spin filters *20 filters*	1 box	48.00	38.40
F-7583	**New**	FISH Counterstain Kit #1 *DAPI/YOYO®-1/PI* *100 slides*	1 kit	95.00	76.00
F-7588	**New**	FISH Counterstain Kit #2 *DAPI/SYTOX® Green/PI* *100 slides*	1 kit	95.00	76.00

Hints for Finding the Product of Interest

This Handbook includes products released by Molecular Probes before September 1996. Additional products and search capabilities are available through our Web site (http://www.probes.com). All information, including price and availability, is subject to change without notice.

- **If you know the name of the product you want:** Look up the product by name in the **Master Price List/Product Index** at the back of the book. The Master Price List/Product Index is arranged alphabetically, by both product name and alternative name, using the alphabetization rules outlined on page vi. This index will refer you directly to all the Handbook sections that discuss the product.

- **If you DO NOT know the name of the product you want:** Search the **Keyword Index** for the words that best describe the applications in which you are interested. The Keyword Index will refer you directly to page numbers on which the listed topics are discussed. Alternatively, search the **Table of Contents** for the chapter or section title that most closely relates to your application.

- **If you only know the alpha-numeric catalog number of the product you want but not its name:** Look for the catalog number in the **Master Price List/Product Index** at the back of the book. First locate the group of catalog numbers with the same alpha identifier as the catalog number of interest and then scan this group for the correct numeric identifier. The Master Price List/Product Index will refer you directly to all the Handbook sections that discuss this product. Alternatively, you may use the search features available through our Web site (http://www.probes.com) to obtain the product name of any of our current products given only the catalog number (or numeric identifier).

- **If you only know part of a product name or are searching for a product introduced after publication of this Handbook:** Search Molecular Probes' up-to-date product list through our **Web site (http://www.probes.com)**. The product search feature on our home page allows you to search our entire product list, including products that have been added since the completion of this Handbook, using a partial product name, complete product name or catalog number. Product-specific bibliographies can also be obtained from our Web site once the product of interest has been located.

- **If you still cannot find the product that you are looking for or are seeking advice in choosing a product:** Please contact our Technical Assistance Department by phone, e-mail, fax or mail.

Chapter 9

Peptide and Protein Detection, Analysis and Synthesis

Contents

Related Chapters

9.1 Detection and Quantitation of Proteins in Solution

Several colorimetric methods have been described for quantitating proteins in solution, including the widely used Bradford [1] and Lowry [2] assays, as well as an assay described by Smith that uses bicinchoninic acid [3] (BCA). However, because they rely on absorption-based measurements, these methods are limited in sensitivity and effective range. Molecular Probes has now developed two unique fluorometric methods for quantitating proteins in solution — NanoOrange Protein Quantitation Kit (N-6666) and CBQCA Protein Quantitation Kit (C-6667) — which outperform *all* existing methods (Table 9.1). We also offer several other fluorescent reagents for protein determination.

NanoOrange Protein Quantitation Kit

Our new NanoOrange Protein Quantitation Kit (N-6666) provides an ultrasensitive assay for measuring the concentration of proteins in solution.[4] The NanoOrange reagent complements our PicoGreen® dsDNA quantitation reagent and OliGreen™ ssDNA quantitation reagent (P-7581, P-7589, O-7582; see Section 8.3) — all premier dyes for fluorometric quantitation of biopolymers in solution.

***Table 9.1** A comparison of reagents for detecting and quantitating proteins in solution.*

Assay	Detection Wavelengths (nm) †	Sensitivity and Effective Range	Mechanism of Action	Notes
NanoOrange protein quantitation assay (N-6666)	485/590	10 ng/mL to 10 µg/mL	Binds to detergent coating on proteins and hydrophobic regions of proteins; unbound dye is nonfluorescent	• High sensitivity • Little protein-to-protein variation • Rapid and accurate assay with a simple procedure • Compatible with reducing agents
Bradford assay [1] (Coomassie Brilliant Blue)	595	1 µg/mL to 1.5 mg/mL	Directly binds specific amino acids and protein tertiary structures; dye color changes from brown to blue	• High protein-to-protein variation • Not compatible with detergents • Rapid assay • Useful when accuracy is not crucial
BCA method [2] (bicinchoninic acid)	562	0.5 µg/mL to 1.2 mg/mL	Cu^{2+} is reduced to Cu^{+} in the presence of proteins at high pH; the BCA chelates Cu^{+} ions, forming purple-colored complexes	• Compatible with detergents, chaotropes and organic solvents • Not compatible with reducing agents • Must read samples within 10 minutes
Lowry assay [3] (biuret reagent plus Folin–Ciocalteu reagent)	750	1 µg/mL to 1.5 mg/mL	Cu^{2+} is reduced to Cu^{+} in the presence of proteins at high pH; the biuret reagent chelates the Cu^{+} ion, then the Folin–Ciocalteu reagent enhances the blue color	• Lengthy procedure with carefully timed steps • Not compatible with detergents or reducing agents
CBQCA protein quantitation assay [4-8] (C-6667)	450/550	10 ng/mL to 150 µg/mL	Reacts with primary amine groups on proteins in the presence of cyanide or thiols; unbound dye is nonfluorescent	• Sensitivity depends on number of amines present • Reacts with other primary amines • Not compatible with buffers containing amines or thiols
Fluorescamine [9-12] (F-2332)	390/475	0.3 µg/mL to 13 µg/mL	Reacts with primary amine groups on proteins; unbound dye is nonfluorescent	• Sensitivity depends on number of amines present • Reagent is unstable • Not compatible with Tris or glycine buffers
OPA [13-15] (*o*-phthaldialdehyde) (P-2331)	340/455	0.2 µg/mL to 25 µg/mL	Reacts with primary amine groups on proteins in the presence of β-mercaptoethanol; unbound dye is nonfluorescent	• Sensitivity depends on number of amines present • Not compatible with Tris or glycine buffers • Low cost
UV absorption [16]	205 280	10 µg/mL to 50 µg/mL 50 µg/mL to 2 mg/mL	Peptide bond absorption Tryptophan and tyrosine absorption	• Sensitivity depends on number of aromatic amino acid residues present • Nondestructive • Low cost

† Excitation/emission wavelength maxima or absorbance wavelength maximum in nm.

1. Anal Biochem 72, 248 (1976); **2.** Anal Biochem 150, 76 (1985); **3.** J Biol Chem 193, 265 (1951); **4.** Anal Chem 63, 408 (1991); **5.** Anal Chem 63, 413 (1991); **6.** Proc Natl Acad Sci USA 88, 2302 (1991); **7.** J Chromatography 559, 223 (1991); **8.** J Chromatography 519, 189 (1990); **9.** Science 178, 871 (1972); **10.** Clin Chim Acta 157, 73 (1986); **11.** J Lipid Res 27, 792 (1986); **12.** Anal Biochem 214, 346 (1993); **13.** Anal Biochem 115, 203 (1981); **14.** BioTechniques 4, 130 (1986); **15.** J Immunol Meth 172, 141 (1994); **16.** Scopes, R.K., *Protein Purification: Principles and Practice, Second Edition*, Springer-Verlag (1987) pp. 280–283.

Features of the NanoOrange Protein Quantitation Kit

The NanoOrange Protein Quantitation Kit has several important features:

- **Ease of use.** The NanoOrange assay protocol is much easier to perform than the Lowry method. Protein samples are simply added to the diluted NanoOrange reagent, and the mixtures are heated at 95°C for 10 minutes. After cooling the mixtures to room temperature, their fluorescence emissions are measured directly. The interaction of proteins with the NanoOrange reagent produces a large fluorescence enhancement that may be used to generate a standard curve for protein determination; fluorescence of the reagent in the absence of proteins is negligible.
- **Sensitivity and effective range.** The NanoOrange assay can detect protein at a final concentration as low as 10 ng/mL when a standard spectrofluorometer or filter minifluorometer is used. A single protocol is suitable for quantitating protein concentrations between 10 ng/mL and 10 µg/mL — an effective range of three orders of magnitude (Figure 9.1).
- **Stability.** The NanoOrange reagent and its protein complex have high chemical stability. In contrast to the Bradford and BCA assays, readings can be taken up to six hours after sample preparation with no loss in signal, provided that samples are protected from light.
- **Little protein-to-protein variability** (Figure 9.2). The NanoOrange assay is not only more sensitive but shows less protein-to-protein variability than Bradford assays.
- **Insensitivity to sample contaminants.** Unlike the Lowry and BCA assays, the NanoOrange assay is compatible with the presence of reducing agents. Furthermore, the high sensitivity of the assay and stability of the protein–dye complex make it possible to dilute out most potential contaminants, including detergents and salts (Table 9.2). Nucleic acids do not interfere with protein quantitation using the NanoOrange reagent.

Our NanoOrange protein quantitation reagent is suitable for use with a variety of instrumentation. Fluorescence is measured using instrument settings or filters that provide excitation/emission at ~485/590 nm, commonly available for both spectrofluorometers and microplate readers. A spectrofluorometer — either a standard fluorometer or a minifluorometer — offers the greatest effective range and lowest detection limits for this assay. The Turner Designs TD-700 minifluorometer (see Section 26.3), which we used to develop the assay, is the most economical instrument for protein quantitation with the NanoOrange reagent, as well as for nucleic acid quantitation with our PicoGreen or OliGreen reagents. The effective range on this minifluorometer extends over the full range of the NanoOrange assay, from 10 ng/mL to 10 µg/mL. Furthermore, with this filter minifluorometer, we have obtained the same high sensitivity when the fluorescence of the sample is measured using inexpensive disposable glass test tubes and an appropriate sample adaptor. With microplate-based fluorescence readers, the NanoOrange assay is useful over a somewhat narrower range, from 100 ng/mL to 10 µg/mL in final protein concentration. Performing the assay in microplates permits smaller assay volumes, and therefore less sample consumed than when using standard cuvettes.

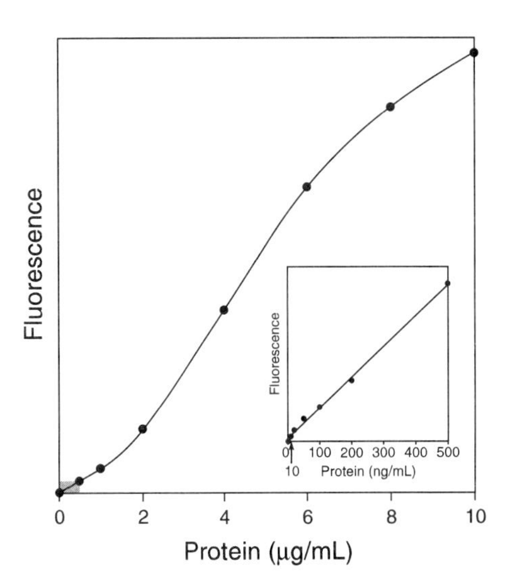

Figure 9.1 Quantitative analysis of bovine serum albumin (BSA) using the NanoOrange Protein Quantitation Kit (N-6666). Fluorescence measurements were carried out on an SLM SPF-500™C fluorometer using excitation/emission wavelengths of 485/590 nm. The inset corresponds to the shaded area in the lower left corner of the plot (0 to 500 ng protein per mL) and illustrates the detection limit of 10 ng/mL.

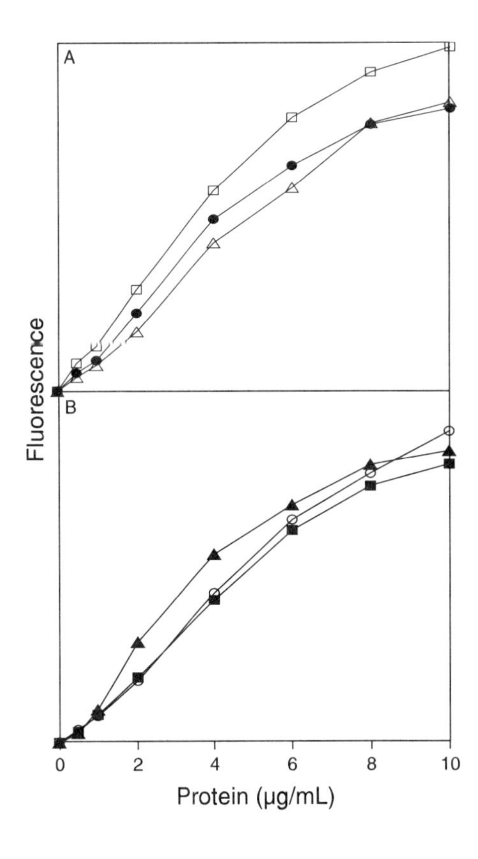

Figure 9.2 Quantitative analysis of six different proteins using the NanoOrange Protein Quantitation Kit (N-6666). Proteins represented are: Panel A) bovine serum albumin (BSA, □), trypsin (●) and carbonic anhydrase (△); Panel B) IgG (■), streptavidin (□) and RNase A (▲). The y-axis fluorescence intensity scale is the same in both panels, illustrating the minimal protein-to-protein staining variation of the NanoOrange assay. Each point is the average of six sample readings, with background fluorescence (zero protein) subtracted. Data were collected on a CytoFluor® 2350 microplate reader (PerSeptive Biosystems, Inc.) with excitation/emission wavelengths set at 485 ± 20/590 ± 35 nm.

Table 9.2 Tolerance levels for contaminants in the NanoOrange protein quantitation assay.

Contaminating Compound	Maximum Tolerable Concentration *
Glycerol	10% by volume
Poly(ethylene glycol) (PEG)	1% by volume
Urea	1 M
Dithiothreitol (DTT), β-mercaptoethanol	100 mM
KCl, NaCl, sodium acetate, sodium phosphate	20 mM
Ammonium sulfate, ascorbic acid, HEPES buffer, HCl, NaOH, sodium azide, sucrose	10 mM
EDTA	5 mM
Calcium chloride, magnesium chloride	1 mM
Amino acids	100 µg/mL
DNA	100 ng/mL
Sodium dodecyl sulfate (SDS)	0.01%
Tween® 20, Triton® X-100	0.001%

* Compounds present in the final assay solution at or below the indicated concentrations do not appreciably interfere with the NanoOrange protein quantitation assay. Whenever feasible, the blank and protein standards should be prepared in a solution closely matching that of the experimental samples.

NanoOrange Kit Contents

Our NanoOrange Protein Quantitation Kit supplies:

- 500X concentrated NanoOrange reagent in dimethylsulfoxide (DMSO)
- 10X concentrated NanoOrange diluent
- Bovine serum albumin (BSA) protein reference standard
- Detailed protocol for protein quantitation

The amount of dye supplied in the kit is sufficient for ~200 assays using a 2 mL assay volume and a standard or minifluorometer, or ~2000 assays using a 200 µL assay volume and a fluorescence microplate reader.

Other Reagents for Protein Quantitation in Solution

Most other fluorescent reagents for general protein quantitation in solution detect accessible primary amines. The sensitivity of assays based on these reagents therefore depends on the number of amines available — a function of both the protein's structure and its amino acid composition. For example, horseradish peroxidase (~40,000 daltons), which has only six lysine residues,[5] will be detected less efficiently than egg white avidin (~66,000 daltons), which has 36 lysine residues,[6,7] and bovine serum albumin (~66,000 daltons), which has 59 lysine residues.[8] However, the assays are generally rapid and easy to conduct, particularly in minifluorometer and fluorescence microplate reader formats.

Certain dyes that detect primary aliphatic amines, including ATTO-TAG™ CBQCA (A-6222), fluorescamine (F-2332) and *o*-phthaldialdehyde (OPA, P-2331), have been the predominant reagents for fluorometric determination of proteins in solution (Table 9.1). These same reagents, and others such as naphthalene-2,3-dicarboxaldehyde [9,10] (NDA, N-1138; see Section 1.8), have frequently been utilized for amino acid analysis of hydrolyzed proteins (see Section 1.8).

CBQCA Protein Quantitation Kit

The ATTO-TAG CBQCA reagent was originally developed as a chromatographic derivatization reagent for amines [11,12] (see Section 1.8). However, this reagent is also useful for quantitating amines in solution, including the number of accessible amines in proteins. ATTO-TAG CBQCA is more water soluble than either fluorescamine or *o*-phthaldialdehyde and much more stable in aqueous solution than fluorescamine. Moreover, ATTO-TAG CBQCA provides greater sensitivity for protein quantitation in solution than either fluorescamine or *o*-phthaldialdehyde (Figure 9.3). Molecular Probes has recently developed the CBQCA Protein Quantitation Kit (C-6667), which employs the ATTO-TAG CBQCA reagent for rapid and sensitive protein quantitation in solution [13] (Table 9.1). Protein quantitation with this kit has a detection limit of about 10 ng using 100–200 µL assay volumes and a fluorescence microplate reader or using 1–2 mL assay volumes and a fluorometer; the effective range is 10 ng to 150 µg. Each CBQCA Protein Quantitation Kit contains:

- ATTO-TAG CBQCA detection reagent
- Potassium cyanide
- Bovine serum albumin (BSA) protein reference standard
- Detailed protocol for protein quantitation

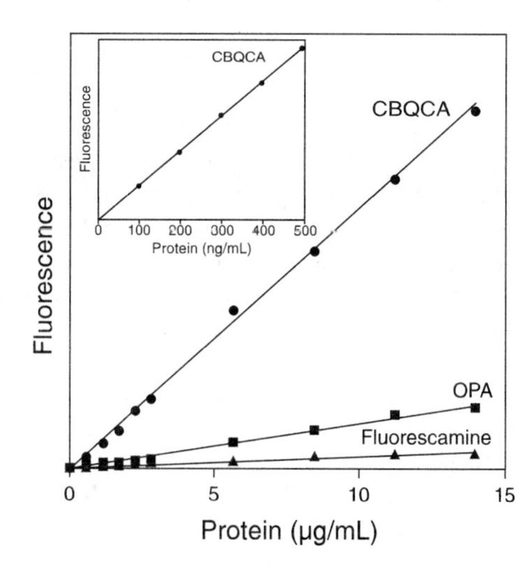

Figure 9.3 Comparison of fluorometric quantitation of bovine serum albumin (BSA) using ATTO-TAG CBQCA, which is supplied in the CBQCA Protein Quantitation Kit (C-6667), OPA (P-2331) and fluorescamine (F-2332). BSA samples were derivatized using large molar excesses of the fluorogenic reagents and were analyzed on a CytoFluor 2350 microplate reader (PerSeptive Biosystems, Inc.). Excitation/emission wavelengths were 360/460 nm for OPA and fluorescamine and 485/530 nm for ATTO-TAG CBQCA. The inset shows an enlargement of the results obtained using CBQCA to assay protein concentrations between zero and 500 ng/mL.

The CBQCA Protein Quantitation Kit provides sufficient reagents for 300–800 assays using a standard fluorometer, minifluorometer or fluorescence microplate reader.

Fluorescamine

Fluorescamine (F-2332) is intrinsically nonfluorescent but reacts in milliseconds with primary aliphatic amines, including peptides and proteins, to yield a fluorescent derivative [14] (see Figure 1.13A in Chapter 1). This amine-reactive reagent has been shown to be useful for determining protein concentrations of aqueous solutions [15,16] and for measuring the number of accessible lysine residues in proteins.[8] Protein quantitation with fluorescamine is particularly well suited to a minifluorometer or fluorescence microplate reader.[17] Fluorescamine can also be used to detect proteins in gels and to analyze low molecular weight amines by TLC, HPLC and capillary electrophoresis.[18]

o-Phthaldialdehyde

The combination of *o*-phthaldialdehyde (OPA, P-2331) and 2-mercaptoethanol (see Figure 1.13B in Chapter 1) provides a rapid and simple method of determining protein concentration in the range of 0.2 µg/mL to 25 mg/mL.[19] As compared to fluorescamine, OPA is both more soluble and stable in aqueous buffers and its sensitivity for detection of peptides is 5–10 times better.[20] The OPA assay for lysine content is reasonably reliable over a broad range of proteins.[8] OPA (and likely the ATTO-TAG CBQCA reagent) can also be used to detect *increases* in the concentration of free amines that result from protease-catalyzed protein hydrolysis.[21]

FluoReporter® Cell Protein Assay Kits

Our FluoReporter Colorimetric Cell Protein Assay Kit is based on an electrostatic dye-binding method for measuring protein content of trichloroacetic acid (TCA)–fixed cells, a method used at the National Cancer Institute (NCI) for high-throughput screening of new chemotherapeutic reagents.[22-24] This kit enables researchers to rapidly quantitate numbers of adherent or nonadherent cells based on protein content. All manipulations are carried out in microplate wells and can be completed in less than four hours. The FluoReporter Colorimetric Cell Protein Assay Kit (F-2961) contains the anionic xanthene dye sulforhodamine B, which forms an electrostatically stabilized complex with basic amino acid residues under moderately acidic conditions. The protein–dye complex (absorption maximum ~565 nm) can then be detected spectrophotometrically after removal of unbound dye from TCA-fixed cells. This FluoReporter assay is linear for determining protein content of 2000 to 200,000 mouse myeloma P3X cells. Because sulforhodamine B is fluorescent, the protein–dye complex can also be measured with a fluorescence microplate reader using excitation/emission maxima of ~485/590 nm, which provides suboptimal excitation but produces a linear relationship between fluorescence intensity and cell number. The FluoReporter Colorimetric Cell Protein Assay Kit contains:

- Sulforhodamine B
- Concentrated solubilization solution
- Protocol for determining cell numbers

Each kit provides sufficient reagents for ~2500 microplate-well assays.

Based on similar principles, our FluoReporter Fluorometric Cell Protein Assay Kit (F-2960) contains the red fluorescent anionic dye sulforhodamine 101 (excitation/emission maxima ~586/602 nm) and enables researchers to detect 100 to 100,000 mouse myeloma P3X cells. The FluoReporter Fluorometric Cell Protein Assay Kit contains:

- Sulforhodamine 101 in 1% acetic acid
- Concentrated counterstain solution in 1% acetic acid
- Concentrated solubilization solution
- Protocol for determining cell numbers

Each kit provides sufficient reagents for ~2000 microplate-well assays. Electrostatic dye binding properties of sulforhodamine 101 (S-359, see Section 15.3) have also been exploited to quantitate proteins in cells by flow cytometry.[25-28]

Selective Protein Detection and Quantitation in Solution

Albumin Blue 580 For Nonimmunological Assay of Serum and Urinary Albumins

The new reagent Albumin Blue 580 (A-6663) is a unique dye for quantitating trace quantities of albumin in biological fluids, including human serum and urine.[29] Elevated urinary albumin is highly predictive for renal damage, making it an important parameter to measure for early diagnosis of the disease. Albumin Blue 580 is similar in use to Albumin Blue 670,[30,31] but exhibits enhanced chemical stability. Free Albumin Blue 580 is effectively nonfluorescent in aqueous solution; however, upon binding to albumin, Albumin Blue 580 exhibits a significant fluorescence enhancement (Figure 9.4) and strong red fluorescence (excitation/emission maxima ~590/620 nm). The long wavelengths for fluorescence excitation and detection make it possible to detect albumin with

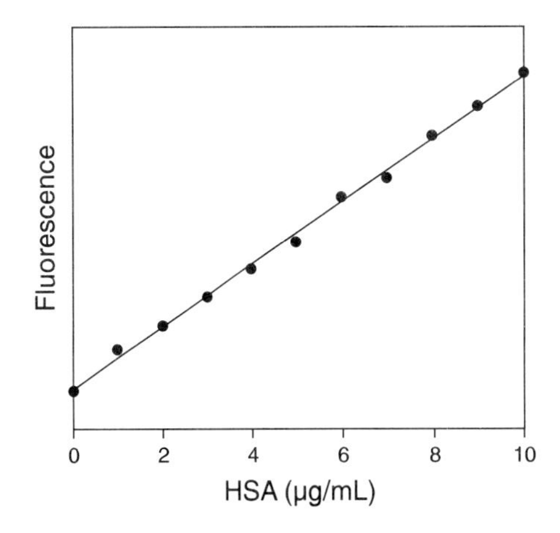

Figure 9.4 *Fluorometric quantitation of human serum albumin (HSA) using Albumin Blue 580 (A-6663). Samples of HSA in phosphate-buffered saline were incubated with 0.67 µM Albumin Blue 580. Fluorescence measurements were performed on a CytoFluor 2350 microplate reader (PerSeptive Biosystems, Inc.) with excitation/emission wavelengths set at 590/645 nm. Under these conditions, the fluorescence response becomes nonlinear at HSA concentrations >10 µg/mL and saturates completely at >30 µg/mL.*

little interference from other fluorescent components in urine and blood. Moreover, this fluorescence response is quite selective for albumin; fluorescence enhancement by other proteins in blood or urine such as transferrin and γ globulin is much weaker. Detection limits for human serum albumin (HSA) of about 0.2 µg/mL can be attained, making this assay competitive with more complex immunoassay methods for detecting elevated urinary albumin excretion (microalbuminuria).[30,31]

Glutathione Agarose for Glutathione *S*-Transferase Fusion Protein Purification

In protein fusion techniques, the coding sequence of one protein is fused in-frame with another so that the expressed hybrid protein possesses desirable properties of both parent proteins. One common partner in these engineered products is glutathione *S*-transferase (GST), a protein whose natural binding specificity can be exploited to facilitate its purification.[32] Because the GST portion of the fusion protein retains its affinity for glutathione, the fusion protein can be conveniently purified from the cell lysate in a single step by affinity chromatography on glutathione agarose[33-38] (Figure 9.5). For purification of GST fusion proteins, Molecular Probes offers glutathione linked via the sulfur atom to crosslinked beaded agarose. This reagent is available as:

- A sedimented bead suspension in bulk (10 mL, G-2879)
- A set of eight prepacked columns, each containing 1 mL of glutathione agarose (G-6664)
- A set of five prepacked columns, each containing 2 mL of glutathione agarose (G-6665)

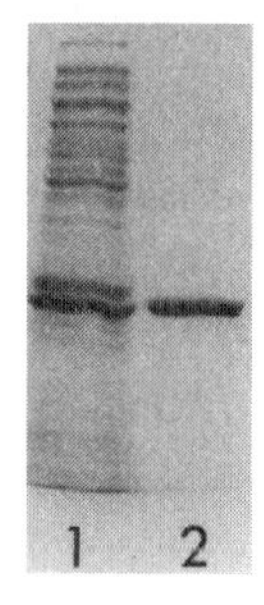

Figure 9.5 Coomassie® Brilliant Blue–stained SDS-polyacrylamide gel demonstrating the purification of a glutathione S-transferase (GST) fusion protein using glutathione agarose (G-2879). Lane 1 contains crude supernatant from an Escherichia coli lysate and lane 2 contains the affinity-purified GST fusion protein.

Each mL of gel can bind approximately 5–6 mg bovine liver GST. Adding excess free glutathione liberates the GST fragment from the matrix, which can then be regenerated by washing with a high-salt buffer. Our high-quality rabbit anti–glutathione *S*-transferase antibody (A-5800) is available for detection of GST fusion proteins on Western blots.

Following purification, the fusion protein can serve as an immunogen for antibody production[39,40] or its properties can be compared with those of native GST to provide insights on the normal function of the polypeptide of interest. Such methods have been used to investigate biological properties of many proteins. Recent examples include the cleavage of the capsid assembly protein ICP35 by the herpes simplex virus type 1 protease,[41] the role of the Rho GTP-binding protein in *lbc* oncogene function[42] and the association of v-Src with cortactin in Rous sarcoma virus–transformed cells.[43] In fact, the Ca^{2+}-binding properties of a protein kinase C–GST fusion protein were examined while the GST fusion protein was still bound to the glutathione agarose.[44] Likewise, the DNA protein–protein interactions of a DNA binding protein–GST fusion protein can be assessed using an affinity column consisting of the fusion protein bound to glutathione agarose.[33] Alternatively, the GST fusion expression vector can be engineered to encode a recognition sequence for a site-specific protease, such as thrombin or factor Xa, between the GST structural gene and gene of interest.[45-48] Once the fusion protein is bound to the affinity matrix, the site-specific enzyme can be added to release the protein.

Phosphoryl Choline for C-Reactive Protein Purification

The concentration of C-reactive protein, an acute-phase reactant in serum, rapidly increases by up to 1000-fold following infection or trauma or as a result of pancreatic cancer.[49,50] Elevated levels of C-reactive protein are also a sensitive indicator of pneumonia.[51] C-reactive protein has high, Ca^{2+}-dependent affinity for various phosphoryl cholines,[52] chromatin and nucleosome core particles.[53,54] *p*-Aminophenyl phosphoryl choline (A-7494) is a useful reagent for preparing an affinity matrix that can be used to isolate C-reactive protein from serum.[55]

Streptavidin Agarose

Biotinylated proteins, oligonucleotides, carbohydrates and other molecules readily bind to streptavidin agarose (S-951). Streptavidin agarose can also be used to generate an affinity matrix by binding a target-specific primary antibody in conjunction with the appropriate biotinylated, species-selective secondary antibody (for a complete listing of Molecular Probes' biotinylated secondary antibodies, see Section 7.2). The binding capacity of the streptavidin–agarose conjugate is measured in an assay using fluorescein biotin (B-1370, see Section 4.3). Typically, the conjugate binds 15–20 µg (18–24 nanomoles) fluorescein biotin per mL of sedimented gel.

EnzChek™ Assay Kits

Molecular Probes prepares numerous chromogenic and fluorogenic substrates that are useful for quantitating enzymes in experimental samples. In addition, we have developed several EnzChek Assay Kits especially designed for detecting protease, dextranase, 5'-nucleotidase, DNase and RNase activity (see Chapter 10). Most of these products are described in Chapter 10.

1. Anal Biochem 72, 248 (1976); **2.** J Biol Chem 193, 265 (1951); **3.** Anal Biochem 150, 76 (1985); **4.** FASEB J 10, A1512, abstract #2954 (1996); **5.** Eur J Biochem 96, 483 (1979); **6.** Adv Protein Chem 29, 85 (1975); **7.** J Biol Chem 246, 698 (1971); **8.** Anal Biochem 115, 203 (1981); **9.** Anal Chem 62, 1580 (1990); **10.** Anal Chem 62, 1577 (1990); **11.** Anal Chem 63, 413 (1991); **12.** Anal Chem 63, 408 (1991); **13.** You, W.W. *et al.*, Anal Biochem, in press; **14.** Science 178, 871 (1972); **15.** Clin Chim Acta 157, 73 (1986); **16.** J Lipid Res 27, 792 (1986); **17.** Anal Biochem 214, 346 (1993); **18.** J Chromatography 502, 247 (1990); **19.** J Immunol Meth 172, 141 (1994); **20.** Proc Natl Acad Sci USA 72, 619 (1975); **21.** Anal Biochem 123, 41 (1982); **22.** Eur J Cancer 29A, 395 (1993); **23.** J Natl Cancer Inst 82, 1113 (1990); **24.** J Natl Cancer Inst 82, 1107 (1990); **25.** Biotech Histochem 67, 73 (1992); **26.** Cytometry 12, 68 (1991); **27.** Histochemistry 75, 353 (1982); **28.** Stain Technol 53, 205 (1979); **29.** Albumin Blue 580 is licensed to Molecular Probes under U.S. Patent No. 5,182,214; **30.** Anal Biochem 200, 254 (1992); **31.** Clin Chem 38, 2089 (1992); **32.** Proc Natl Acad Sci USA 83, 8703 (1986); **33.** Meth Mol Genet 1 (Part A), 267 (1993); **34.** BioTechniques 13, 856 (1992); **35.** BioTechniques 10, 178 (1991); **36.** Nucleic Acids Res 19, 4005 (1991); **37.** Science 252, 712 (1991); **38.** Gene 67, 31 (1988); **39.** J Cell Biol 131, 1003 (1995); **40.** J Cell Biol 130, 651 (1995); **41.** J Biol Chem 270, 30168 (1995); **42.** J Biol Chem 270, 9031 (1995); **43.** J Biol Chem 270, 26613 (1995); **44.** J Biol Chem 268, 3715 (1993); **45.** J

3715 (1993); **45.** J Biol Chem 270, 24525 (1995); **46.** J Cell Biol 129, 189 (1995); **47.** Mol Biol of the Cell 6, 247 (1995); **48.** Biochemistry 31, 5841 (1992); **49.** Cancer 75, 2077 (1995); **50.** J Immunology 143, 2553 (1989); **51.** Chest 107, 1028 (1995); **52.** J Immunology 126, 856 (1981); **53.** J Immunology 141, 4266 (1988); **54.** J Biol Chem 259, 7311 (1984); **55.** Biochemistry 11, 766 (1972).

9.1 Data Table *Detection and Quantitation of Proteins in Solution*

Cat #	Structure	MW	Storage	Soluble	Abs	{ε × 10⁻³}	Em	Solvent	Notes
A-6222	S1.8	305	F,D,L	MeOH*	465	ND	560	MeOH	<1, 2>
A-6663	✓	307	F,D,L	DMSO	580	{108}	<3>	pH 7	
A-7494	✓	274	D	H_2O	293	{1.7}	none	MeOH	
F-2332	S1.8	278	F,DD,L	MeCN	381	{7.6}	470	MeCN	<4>
P-2331	S1.8	134	L	EtOH	334	{5.7}	455	pH 9	<5>

For definitions of the contents of this data table, see "How to Use the Handbook" on page vi.
Structure: Chemical structure drawing: ✓ = shown in this section; Sn.n = shown in Section number n.n.

* Solubility in methanol is improved by addition of base (e.g., 1–5% (v/v) 0.2 M KOH).
<1> Spectra of A-6222 with glycine + cyanide. Unreacted reagent in MeOH: Abs = 254 nm (ε = {46}), nonfluorescent.
<2> ND = not determined.
<3> Albumin Blue 580 is essentially nonfluorescent in free solution. Fluorescence is strongly enhanced upon binding to human serum albumin (Abs = 590 nm, ε = {108}, Em = 620 nm).
<4> F-2332 spectra are for reaction product with butylamine. Fluorescence quantum yield/lifetime of adduct are 0.23/7.5 nsec in EtOH [Arch Biochem Biophys 163, 390 (1974)]. Unreacted reagent in MeCN: Abs = 234 nm (ε = {28}), nonfluorescent.
<5> Spectral data are for the reaction product of P-2331 with alanine and 2-mercaptoethanol. The spectra and the stability of the adduct depend on the amine and thiol reactants [Biochim Biophys Acta 576, 440 (1979)]. Unreacted reagent in H_2O: Abs = 257 nm (ε = {1.0}).

9.1 Structures *Detection and Quantitation of Proteins in Solution*

A-6222
Section 1.8

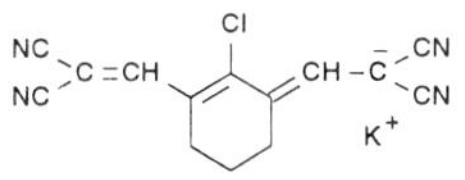

A-6663

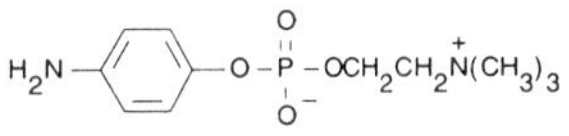

A-7494

F-2332, P-2331
Section 1.8

9.1 Price List *Detection and Quantitation of Proteins in Solution*

Cat #		Product Name	Unit Size	Price Per Unit ($) 1–4 Units	5–24 Units
A-6663	***New***	Albumin Blue 580, potassium salt	10 mg	90.00	72.00
A-7494	***New***	p-aminophenyl phosphoryl choline	100 mg	18.00	14.40
A-5800		anti-glutathione S-transferase, rabbit IgG fraction *3 mg/mL*	0.5 mL	155.00	124.00
A-6222	***New***	ATTO-TAG™ CBQCA derivatization reagent (CBQCA; 3-(4-carboxybenzoyl)quinoline-2-carboxaldehyde)	10 mg	110.00	88.00
C-6667	***New***	CBQCA Protein Quantitation Kit *300-800 assays*	1 kit	98.00	78.40
F-2961		FluoReporter® Colorimetric Cell Protein Assay Kit	1 kit	100.00	80.00
F-2960		FluoReporter® Fluorometric Cell Protein Assay Kit	1 kit	135.00	108.00
F-2332		fluorescamine	100 mg	20.00	16.00
G-2879		glutathione agarose, linked through sulfur *sedimented bead suspension*	10 mL	90.00	72.00
G-6664	***New***	glutathione agarose, linked through sulfur *eight 1.0 mL prepacked columns*	1 set	125.00	100.00
G-6665	***New***	glutathione agarose, linked through sulfur *five 2.0 mL prepacked columns*	1 set	125.00	100.00
N-6666	***New***	NanoOrange™ Protein Quantitation Kit *200-2000 assays*	1 kit	135.00	108.00
P-2331		o-phthaldialdehyde (OPA) *high purity*	1 g	32.00	25.60
S-951		streptavidin agarose *sedimented bead suspension*	5 mL	125.00	100.00

9.2 Detection of Proteins in Gels and on Blots

Studies of protein identification, structure and function often rely on the sensitive detection of proteins in gels and on blots. Some proteins — such as cytochrome *c*, myoglobin, hemoglobin and ferritin — can be visualized directly in gels due to their intrinsic color or fluorescence.[1,2] However, most proteins require staining reagents for their detection.

SYPRO Orange and SYPRO Red Protein Gel Stains

Our recently introduced SYBR® Green nucleic acid gel stains have proven to be extremely sensitive and easy-to-use reagents for fluorescence-based detection of DNA and RNA in gels (see Section 8.3). Now, with the development of our new SYPRO Orange and SYPRO Red protein gel stains,[3,4] Molecular Probes is revolutionizing the detection of proteins in gels. Not only do the SYPRO protein gel stains provide a fluorescence-based method that is much faster and easier to perform than conventional colorimetric staining methods, but they produce ultrasensitive protein detection (Figure 9.6). Furthermore, the simple one-step SYPRO staining procedure is compatible with Western blot analysis because it does not interfere with protein transfer to filter membranes or subsequent detection by immunoreagents.

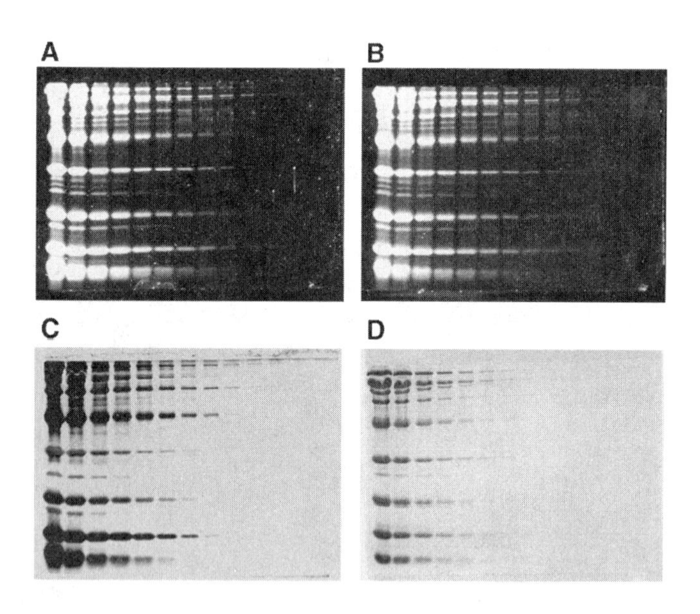

Figure 9.6 *Comparison of sensitivity achieved with SYPRO, silver and Coomassie Brilliant Blue staining methods. Identical SDS-polyacrylamide gels were stained with A) SYPRO Orange protein gel stain (S-6650), B) SYPRO Red protein gel stain (S-6653), C) silver stain and D) Coomassie Brilliant Blue stain, according to standard protocols. The SYPRO-stained gels were photographed using 300 nm transillumination, a SYPRO Orange/Red protein gel stain photographic filter (S-6656) and Polaroid® 667 black-and-white print film. The silver- and Coomassie Brilliant Blue–stained gels were photographed with transmitted white light and Polaroid 667 black-and-white print film; no photographic filter was used.*

Features of the SYPRO Protein Gel Stains

Currently the most common methods for staining protein gels are Coomassie Brilliant Blue staining[5] and silver staining.[6] Although Coomassie Brilliant Blue is an inexpensive reagent, its staining is relatively insensitive, requires long processing times and consumes large volumes of organic solvents, resulting in high hidden costs. Silver staining may be up to 100 times more sensitive than Coomassie Brilliant Blue staining but is expensive and entails several labor-intensive steps as well as solutions that are not reusable. Moreover, silver staining exhibits a high degree of protein-to-protein variability; staining intensity and color are very dependent on polypeptide sequence and degree of glycosylation, and bands are often detectable only as negatively stained patches.

Molecular Probes' new SYPRO Orange (S-6650, S-6651) and SYPRO Red (S-6653, S-6654) protein gel stains provide a fluorescence-based alternative for protein detection in sodium dodecyl sulfate (SDS)–polyacrylamide gels that is as sensitive as silver staining.[3,7] In the presence of excess SDS, nonpolar regions of polypeptides are coated with detergent molecules, forming a micelle-like structure with a nearly constant SDS:protein ratio (1.4 g SDS:1.0 g protein); this constant charge per mass ratio is the basis of molecular weight determination by SDS-polyacrylamide gel electrophoresis.[8] Our SYPRO dyes appear to bind to the SDS coat surrounding proteins in SDS-polyacrylamide gels. Thus, SYPRO staining exhibits little protein-to-protein variation and is linearly related to protein mass (Figure 9.7). Some other important features of the SYPRO stains include:

- **Ease of use.** Following electrophoresis, the gel is stained for as little as 10–60 minutes and then briefly rinsed. No hazardous solvents are required for SYPRO staining, and no separate fixation or destaining steps are required. After staining, the gel is immediately ready for photography, or it can be stored in or out of the staining solution for days.
- **Sensitivity and effective range.** The SYPRO protein gel stains provide a sensitivity level of 1–2 ng per band in SDS-polyacrylamide minigels when visualized with standard 300 nm transillumination and photographed using Polaroid 667 black-and-white print film and a SYPRO Orange/Red protein gel stain photographic filter (S-6656) (Figure 9.6). Polaroid photography enhances the sensitivity of SYPRO staining by 25- to 50-fold over visible detection because the film is much more efficient than the human eye at capturing the orange and red signals and can integrate the signal throughout the duration of the exposure. Furthermore, using commercially available scanners, we have found that protein detection in gels stained with either of the SYPRO dyes is linear over three orders of magnitude in protein quantity[3] (Figure 9.7).
- **Versatility.** The SYPRO protein gel stains are compatible with SDS, urea/SDS, two-dimensional and native polyacrylamide gels, as well as with agarose gels. Staining proteins in native gels results in more protein-to-protein variation and lower sensitivity than staining SDS-denatured proteins, due to variations in hydrophobicity of the target polypeptides. However, sensitivity of SYPRO staining in native gels can be improved if gels are soaked in SDS solution after electrophoresis but prior to staining.

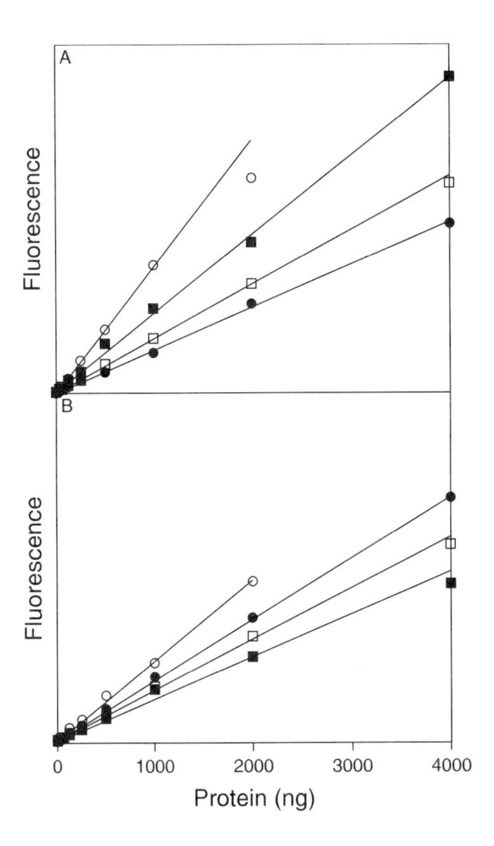

Figure 9.7 *Linear quantitation of proteins on an electrophoretic gel using SYPRO Orange protein gel stain (S-6650). Standard protein mixtures (commercially available broad-range markers) were serially diluted and electrophoresed on a 15% SDS-polyacrylamide gel and then stained with SYPRO Orange protein gel stain according to the standard protocol. The gel was then scanned using a Molecular Dynamics' Storm™ gel and blot analysis system (excitation/emission 488/>520 nm) and analyzed to yield the fluorescence intensities of the stained bands. The y-axis fluorescence intensity scale is the same in both panels, illustrating the minimal protein-to-protein staining variation of the SYPRO Orange gel stain. Detection limits are between 2 and 16 ng of protein; the linear detection ranges are approximately 1000-fold. Proteins represented are: Panel A) β-galactosidase (○), lysozyme (■), bovine serum albumin (BSA, □) and phosphorylase B (●); Panel B) myosin (○), soybean trypsin inhibitor (■), ovalbumin (□) and carbonic anhydrase (●).*

- **Uniform protein staining.** Unlike silver staining,[9] the SYPRO dyes exhibit relatively low protein-to-protein variability in SDS-polyacrylamide gels (Figure 9.7) and do not stain nucleic acids, which are sometimes found in protein mixtures from cell or tissue extracts. In addition, the SYPRO dyes only weakly stain lipopolysaccharides in bacterial lysates, whereas these biopolymers are strongly detected by silver staining. Glycoproteins (such as the IgG variable subunit) and proteins with prosthetic groups (such as bovine cytochrome oxidase) are also efficiently stained with the SYPRO dyes.
- **Photostability.** Gels stained with the SYPRO dyes are relatively photostable, enabling the researcher to acquire multiple photographic images and to use long film exposure times (2–8 seconds). SYPRO-stained gels are readily photographed using Polaroid 667 black-and-white print film, a SYPRO Orange/Red protein gel stain photographic filter (S-6656) and standard 300 nm transillumination. The SYPRO stains are also suitable for use with laser scanners and CCD image acquisition systems. Gels that have been partially photobleached can be restained with little loss of sensitivity.
- **Chemical stability.** The SYPRO gel stains are chemically stable; fluorescence of the stained gel is stable overnight or for several days, and staining solutions can be stored for months.
- **Compatible with blotting and sequencing.** Because the SYPRO dyes do not covalently bind to proteins, antigenic sites are unmodified and, following transfer to filter membranes, are available for antibody binding. Proteins may therefore be visualized prior to Western blot analysis, using either nitrocellulose or nylon membranes, without interfering with subsequent immunodetection [3,10] (Figure 9.8). Furthermore, after transfer to PDVF membranes, SYPRO Orange–stained proteins can be microsequenced through at least their initial five or six residues.[10]
- **Economy.** The SYPRO gel stains are not only much less expensive than silver staining kits but faster and less laborious to use. Use of the SYPRO dyes avoids the costs of purchase and disposal of large amounts of organic solvents that are required for Coomassie Brilliant Blue staining. Furthermore, the SYPRO staining solutions can be reused up to four times, with little loss in sensitivity.

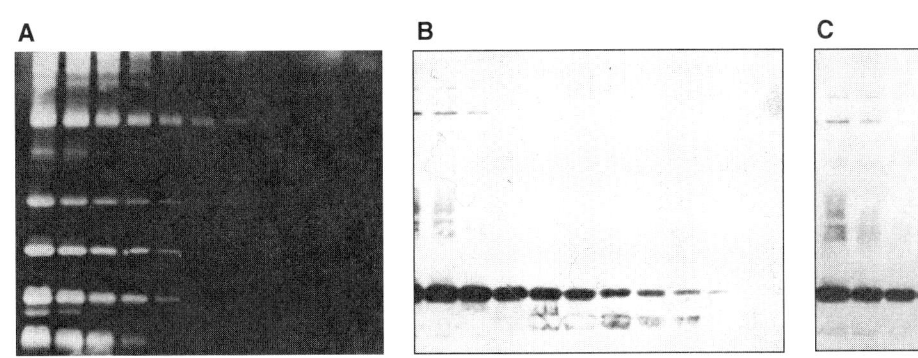

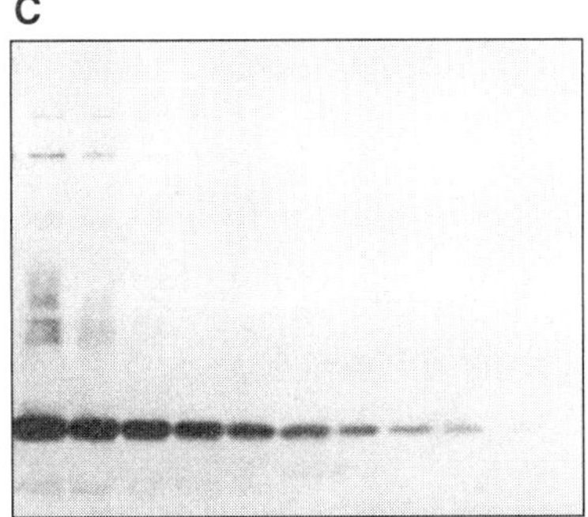

Figure 9.8 *Compatibility of SYPRO staining with Western blot analysis. A two-fold dilution series of bovine cytochrome oxidase (COX) was prepared in loading buffer and electrophoresed on two identical SDS-polyacrylamide gels. One gel, shown in Panel A, was then stained with SYPRO Orange stain in transfer buffer (25 mM Tris, 192 mM glycine, 20% methanol) and photographed using 300 nm transillumination, a SYPRO Orange/Red protein gel stain photographic filter (S-6656) and Polaroid 667 black-and-white print film., whereas the second gel was not stained. Both gels were then electroblotted onto nitrocellulose and sequentially probed with polyclonal antibodies directed against the VIc subunit of COX, biotinylated secondary antibodies (see Section 7.2) and streptavidin–alkaline phosphatase conjugate (S-921, see Section 7.5), followed by detection with the NBT/BCIP Reagent Kit (N-6547). The sensitivity for detection of the COX VIc subunit was identical in both blots, showing that SYPRO staining does not interfere with subsequent Western blot analysis.*

The SYPRO Orange and SYPRO Red stains have similar staining properties and are equally suitable for use in most applications. We have observed that SYPRO Orange staining is slightly brighter, whereas SYPRO Red staining has a lower background fluorescence. Although their peak absorption is in the visible range (Figure 9.9), both SYPRO dyes are also efficiently excited by UV or broadband illumination, including standard 300 nm transillumination. The SYPRO Orange stain produces bright orange fluorescence (see Color Plate 2 in "Visual Reality") that is even visible using standard room light for excitation. The intensity of fluorescence using white-light excitation is less than that seen with UV illumination; however, this property should make the SYPRO Orange stain particularly useful for laboratories that want to avoid, or do not have access to, UV illumination. We recommend the SYPRO Orange protein gel stain for gel scanners that employ argon-ion lasers and the SYPRO Red protein gel stain for laser-scanning instruments that employ green He-Ne or Nd:YAG lasers. Protein detection in gels stained with SYPRO Orange or SYPRO Red dyes has been shown to be linear over three orders of magnitude in protein quantity using the Molecular Dynamics' Storm™ gel and blot analysis system (Figure 9.7) or Hitachi Software's FMBIO® 100 fluorescent imaging device, respectively, with detection limits in the single nanogram range.[9]

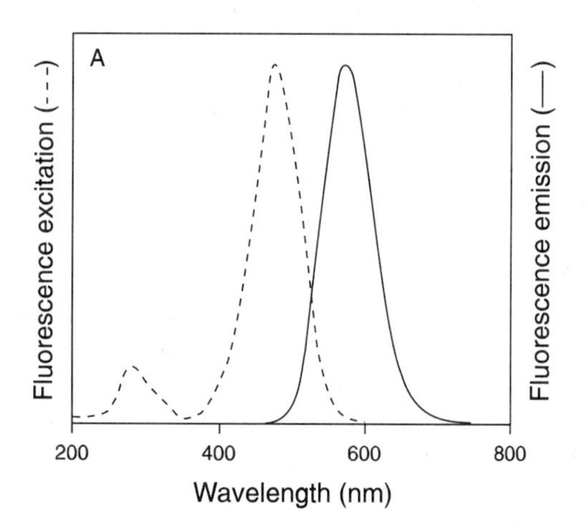

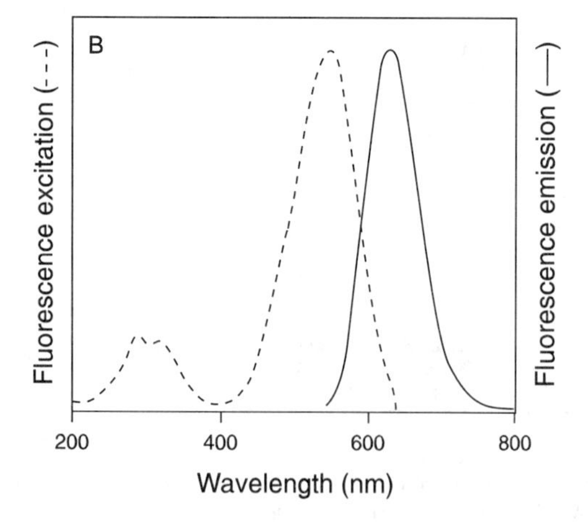

Figure 9.9 *The fluorescence excitation and emission spectra of A) SYPRO Orange (S-6650) and B) SYPRO Red (S-6653) protein gel stains diluted 1:10,000 in water containing 0.05% SDS and 150 µg/mL bovine serum albumin (BSA).*

SYPRO Protein Gel Stains and Sampler Kit

The SYPRO Orange and SYPRO Red protein gel stains (patents pending) are available as 500 µL stock solutions in dimethylsulfoxide (DMSO), either in a single vial (S-6650, S-6653) or specially packaged as a set of 10 vials, each containing 50 µL (S-6651, S-6654). The reagents are supplied as 5000-fold concentrates so that 500 µL yields 2.5 L of staining solution, and this staining solution may be reused 3–4 times with little loss in sensitivity. Photography of proteins in gels requires use of the SYPRO Orange/Red protein gel stain photographic filter (S-6656). A gel photographic filter holder (G-6657), which is useful for protecting the SYPRO Orange/Red filter, is also available.

Our SYPRO Orange and SYPRO Red protein gel stains are now available in a SYPRO Protein Gel Stain Starter Kit (S-6655) for scientists who would like to try these stains in their particular applications. In addition to providing both SYPRO dyes, the kit includes a SYPRO Orange/Red gel photographic filter and a sample of our broad range protein molecular weight standards (see below), to aid in their evaluation. Enough of the SYPRO Orange and SYPRO Red dyes are supplied in the starter kit to stain about 5–20 minigels each.

Protein Molecular Weight Standards

Molecular Probes offers three different protein mixtures for use as molecular weight markers in SDS-polyacrylamide gel electrophoresis (Figure 9.10):

- High-range markers: 8 polypeptides from 30,000 to 205,000 daltons (P-6647)
- Low-range markers: 8 polypeptides from 6500 to 55,000 daltons (P-6648)
- Broad-range markers: 11 polypeptides from 6500 to 205,000 daltons (P-6649)

These mixtures are balanced formulations of purified proteins that give rise to sharp, well-separated bands when the gel is stained with SYPRO Orange or SYPRO Red protein gel stains. Each mixture provides sufficient protein for loading about 200 gel lanes. A

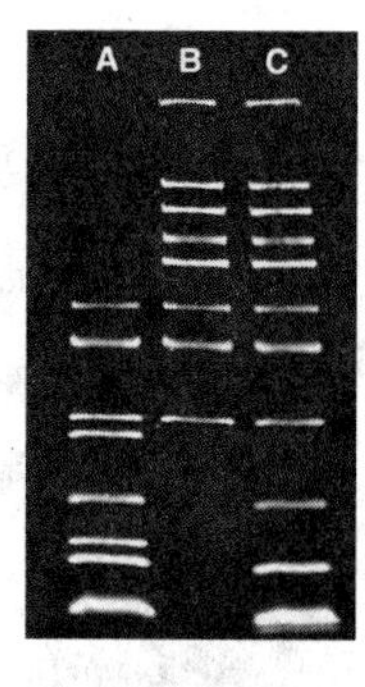

Figure 9.10 *Molecular Probes' protein molecular weight standards: A) low-range (P-6648), B) high-range (P-6647) and C) broad-range (P-6649) mixtures. A 2 µL sample of each standard mixture was mixed with 8 µL loading buffer, heated at 95°C for 4 minutes and electrophoresed on an SDS-polyacrylamide gradient gel (4–20% polyacrylamide). After electrophoresis, the gel was stained with SYPRO Orange protein gel stain and photographed using 300 nm transillumination, a SYPRO Orange/Red protein gel stain photographic filter (S-6656) and Polaroid 667 black-and-white print film.*

sample of the broad-range markers is included in our SYPRO Protein Gel Stain Starter Kit (S-6655, see above).

Selective Protein Detection in Gels

Detection of Glycoproteins in Gels

Fluorescent and biotinylated lectins (see Section 7.6) have been used to locate glycoproteins in gels and blots.[11-13] Preceding electrophoresis, glycoproteins in a protein mixture can also be selectively labeled using fluorescent hydrazides (see Section 3.2) or biotin hydrazides [14,15] (see Section 4.2). The biotin hydrazide–labeled glycoproteins are then detected using fluorophore- or enzyme-labeled conjugates of avidin, streptavidin or NeutraLite™ avidin (see Section 7.5).

Detection of Calcium-Binding Proteins in Gels

The luminescent lanthanides terbium and europium, which are available from Molecular Probes as the chloride salts (Tb^{3+} from $TbCl_3$, T-1247; Eu^{3+} from $EuCl_3$, E-6960), selectively stain calcium-binding proteins in SDS-polyacrylamide gels.[16] With some modifications to the staining protocol, these lanthanides can also be used to detect all protein bands.[16]

Detection of Enzymes in Nondenaturing Gels

A wide variety of enzymes have been detected in nondenaturing gels by using various chromogenic substrates, including X-Gal (B-1690), X-GlcU (B-1691) and NBT/BCIP [17] (N-6495, B-6492). In unpublished experiments, we have shown that our ELF®-97 phosphatase substrate (E-6589, see Section 10.3), which forms a highly fluorescent precipitate at the site of enzymatic activity, can detect both acid and alkaline phosphatase in nondenaturing polyacrylamide gels. We have also demonstrated that our ELF-97 β-D-glucuronidase substrate (E-6587, see Section 10.2) has similar utility for detecting β-glucuronidase in native gels, with a detection limit of less than 1–5 ng of the enzyme [18] (Figure 9.11).

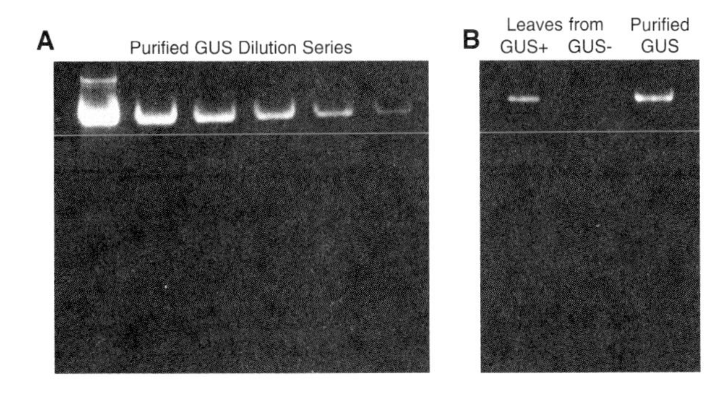

Figure 9.11 *In situ gel assay of β-D-glucuronidase (GUS) with ELF-97 β-D-glucuronide (E-6587, see Section 10.2). A) Twofold dilutions of purified GUS enzyme or B) cell extracts from single leaves from GUS-positive and -negative Arabidopsis plants were electrophoresed on a native 7.5% polyacrylamide gel. Following electrophoresis, the gel was washed with 0.1 M sodium phosphate, pH 7.0 containing 0.2% Triton® X-100 at room temperature for 60 minutes and then incubated with 15 µM ELF-97 β-D-glucuronide in 0.1 M sodium phosphate, pH 7.0 at 37°C for 30–60 minutes. The gel was photographed using 300 nm transillumination, a SYBR Green gel stain photographic filter (S-7569, see Section 8.3) and Polaroid 667 black-and-white print film.*

Protein Detection on Blots

Proteins are frequently separated by electrophoresis, then blotted onto nylon or nitrocellulose filter membranes for subsequent Western blot analysis.[19] Our SYPRO dyes (see above) do not interfere with either the transfer of proteins from SDS-polyacrylamide gels to blots or their subsequent detection or sequencing. Alternatively, proteins can be directly spotted onto a membrane for detection by an immunoreagent, adsorbed onto a plastic surface for enzyme-linked immunosorbent assays (ELISAs) or immunoprecipitated. Molecular Probes has available a variety of primary antibodies that are useful for detecting specific proteins on filter membranes or by immunoprecipitation. These include antibodies directed against:

- Cytochrome oxidase (COX) subunits (see Section 12.5)
- β-Glucuronidase (see Section 10.1)
- Glutathione *S*-transferase (GST) (see Section 7.4)
- Green fluorescent protein (GFP) (see Section 7.4)
- Synapsin I (see Section 20.3)
- NMDA receptor subunits (see Section 20.3)
- PIP_2-specific phospholipase C (see Section 20.5)
- Yeast vacuolar proteins (see Section 12.5)
- Fluorescent dyes and the dinitrophenyl (DNP) hapten (see Section 7.3)

The anti-GST and anti-GFP antibodies are particularly useful for detecting fusion proteins constructed with the GST and GFP reporter proteins, respectively. The anti-dye and anti-DNP antibodies can be employed for the selective detection of labeled proteins on blots; Table 4.1 in Chapter 4 lists our recommended haptenylation reagents and the corresponding anti-hapten antibodies. For example, aldehydes generated by periodate oxidation of glycoproteins have been modified by 2,4-dinitrophenylhydrazine or biotin hydrazides (see Section 4.2) and then detected using anti-DNP antibodies [20] or avidin derivatives,[15] respectively. Likewise, thiol-containing proteins can be labeled with (maleimidylpropionyl)-biocytin (M-1602, see Section 4.2) and subsequently separated by electrophoresis, blotted and detected with fluorescent- or enzyme-labeled avidin derivatives.[21] Biotinylated lectins (see Section 7.6) can also be used as general detection reagents for glycoproteins on blots.[12,13]

Sulfonated Dyes

Various sulfonated dyes, including Coomassie Brilliant Blue, selectively adsorb to the positive charges of proteins on blots when stained from dilute acetic acid or trichloroacetic acid solutions. Sulforhodamine B (S-1307) and benzoxanthene yellow (Hoechst 2495, B-6947) have also been used for quantitative protein detection on blots. When applied to proteins electroblotted onto PVDF membranes, sulforhodamine B produces a red fluorescent signal with minimal nonspecific staining and a reported protein detection limit of ~10 ng.[22] The intensity of benzoxanthene yellow staining can be quantitated densitometrically or following extraction of the dye from the blot using a weakly basic solution; protein detection is reported to be linear in the range of 10 ng to 20 µg.[23] The original benzoxanthene yellow procedure has been modified for use in a dot blot assay to quantitate proteins in the presence of carrier ampholytes and other potentially interfering compounds,[24] making it compatible with isoelectric focusing (IEF) and two-dimensional gel electrophoresis sample buffers.

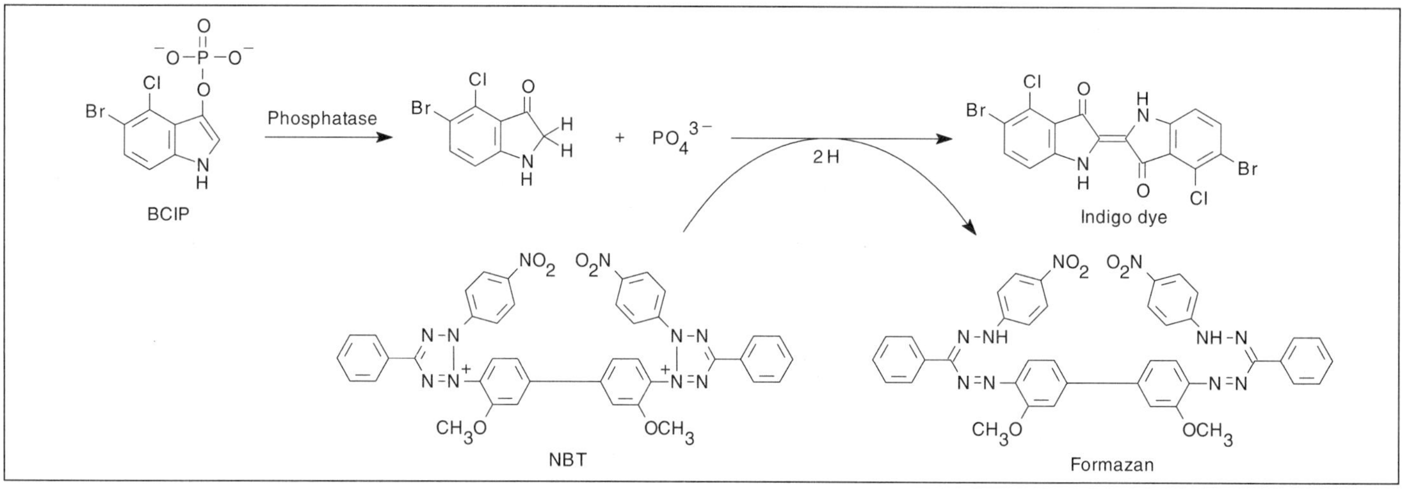

Figure 9.12 *Principle of enzyme-linked detection using our NBT/BCIP (N-6547) Reagent Kit. Phosphatase hydrolysis of BCIP is coupled to reduction of NBT, yielding a formazan and an indigo dye that together form a black-purple colored precipitate.*

DiOC$_5$(3)

Dipentyloxacarbocyanine (DiOC$_5$(3), D-272) has been recommended as the preferred reagent for detecting proteins electroblotted onto activated glass filter paper after SDS-polyacrylamide gel electrophoresis prior to microsequencing.[25] In room light, bands containing 1 µg or more of protein were visible as orange spots. Upon 254 nm illumination, detection limits improved to about 20–50 ng. Destaining was not required for visualization or subsequent sequencing, as the dye was separated from the protein during the first ethyl acetate or butyl chloride wash.

Enzyme-Amplified Protein Detection on Blots

Protein detection on blots is frequently amplified by use of a biotinylated primary or secondary antibody, in conjunction with an enzyme-conjugated secondary detection reagent, such as streptavidin conjugates of alkaline phosphatase (S-921), β-galactosidase (S-931) or horseradish peroxidase (S-911), and chromogenic substrates that yield insoluble precipitates. Suitable substrates include:

- The combination of nitro blue tetrazolium (NBT, N-6495) and 5-bromo-4-chloro-3-indolyl phosphate (BCIP; B-6492, B-8405), also available in our new NBT/BCIP Reagent Kit (N-6547), for alkaline phosphatase[26,27] (Figure 9.12)
- 5-Bromo-4-chloro-3-indolyl galactoside (X-Gal, B-1690), as well as analogs of X-Gal that yield blue, blue-green, green, magenta or pink colored precipitates (see Section 10.2), for β-galactosidase[28,29]
- Tetramethylbenzidine (TMB; T-8463, T-8464) for peroxidase[30,31]

4-Methylumbelliferyl phosphate (M-6491, M-8425; see Section 10.3) has also been used to detect proteins on blots,[27] but the blue fluorescent product rapidly diffuses from the reaction site.

1. Anal Biochem 111, 235 (1981); **2.** Annals NY Acad Sci 39, 209 (1939); **3.** Anal Biochem 239, 238 (1996); **4.** Anal Biochem 239, 223 (1996); **5.** Anal Biochem 89, 264 (1978); **6.** Meth Enzymol 182, 477 (1990); **7.** J NIH Res 7, 82 (1995); **8.** Andrews, A. T., *Electrophoresis: Theory, Techniques, and Biochemical and Clinical Applications*, Oxford University Press (1986) pp. 117–147; **9.** Anal Biochem 165, 33 (1987); **10.** Am Biotech Lab 14, 12 (1996); **11.** Anal Biochem 170, 491 (1988); **12.** Meth Enzymol 104, 447 (1984); **13.** Anal Biochem 96, 208 (1979); **14.** Anal Biochem 161, 245 (1987); **15.** Anal Biochem 163, 204 (1987); **16.** Anal Biochem 216, 439 (1994); **17.** Anal Biochem 203, 1 (1992); **18.** Zhou, M. *et al.*, J Biochem Biophys Meth (1996) in press; **19.** Proc Natl Acad Sci USA 76, 4350 (1979); **20.** Meth Enzymol 233, 346 (1994); **21.** Radiation Res 117, 326 (1989); **22.** J Protein Chem 9, 259 (1990); **23.** Hoppe–Seyler Z Physiol Chem 360, 1657 (1979); **24.** Anal Biochem 196, 99 (1991); **25.** J Biol Chem 261, 4229 (1986); **26.** Meth Enzymol 121, 497 (1986); **27.** Anal Biochem 136, 180 (1984); **28.** BioTechniques 12, 656 (1992); **29.** Clin Chem 40, 1893 (1992); **30.** Anal Biochem 209, 323 (1993); **31.** J Clin Microbiol 29, 1836 (1991).

9.2 Data Table *Detection of Proteins in Gels and on Blots*

Cat #	Structure	MW	Storage	Soluble	Abs	{ε x 10^{-3}}	Em	Solvent	Notes
B-1690	S10.2	409	F,D	DMSO	290	{5.0}	none	H_2O	<1>
B-1691	S10.2	522	F,D	pH>6	292	{4.8}	none	MeOH	<1>
B-6492	S10.3	370	F,D	DMSO, H_2O	292	{3.8}	none	H_2O	<1>
B-6947	✓	616	D,L	DMSO, H_2O	442	{19}	474	MeOH	
B-8405	S10.3	434	F,D	DMSO, H_2O	290	{5.9}	none	H_2O	<1>
D-272	S25.3	544	D,L	DMSO	484	{155}	500	MeOH	
E-6960	S15.3	366*	D	H_2O	337	{17}	613	H_2O	<2>
N-6495	S21.4	818	D,L	H_2O, DMSO	256	{64}	none	MeOH	<3>
S-1307	S15.3	559	L	H_2O	565	{84}	586	H_2O	
T-1247	S15.3	373*	D	H_2O	270	{4.7}	545	H_2O	<4>

Cat #	Structure	MW	Storage	Soluble	Abs	{ε × 10^{-3}}	Em	Solvent	Notes
T-8463	S10.5	240	D	DMSO	287	{25}	none	MeOH	<5>
T-8464	S10.5	313	D	DMSO, H_2O	292	{19}	none	MeOH	<5>

For definitions of the contents of this data table, see "How to Use the Handbook" on page vi.
Structure: Chemical structure drawing: ✓ = shown in this section; Sn.n = shown in Section number n.n.

* MW is for the hydrated form of this product.
<1> Enzymatic hydrolysis of this substrate yields a water-insoluble blue colored indigo dye (Abs ~615 nm).
<2> Absorption and fluorescence of E-6960 are extremely weak unless it is chelated. Data are for detergent-solubilized NTA (T-411, see Section 15.3) chelate.
<3> Enzymatic reduction product of N-6495 is a water-insoluble formazan, Abs = 605 nm after solubilization in DMSO or DMF. See Histochemistry 76, 381 (1982) and Prog Histochem Cytochem 9, 1 (1977) for further information.
<4> Absorption and fluorescence of T-1247 are extremely weak unless it is chelated. Data are for dipicolinic acid (D-1248, see Section 15.3) chelate. Fluorescence spectrum has secondary peak at 490 nm.
<5> Enzymatic oxidation product of T-8463 and T-8464 is blue colored (Abs = 650 nm with a second peak at 370 nm) in basic solutions, or yellow colored (Abs = 450 nm) in acidic solutions.

9.2 Structures *Detection of Proteins in Gels and on Blots*

B-1690, B-1691
Section 10.2

B-6492, B-8405
Section 10.3

B-6947

D-272
Section 25.3

E-6960, S-1307
T-1247
Section 15.3

N-6495
Section 21.4

T-8463, T-8464
Section 10.5

9.2 Price List *Detection of Proteins in Gels and on Blots*

Cat #		Product Name	Unit Size	Price Per Unit ($) 1–4 Units	5–24 Units
B-6947	**New**	benzoxanthene yellow (Hoechst 2495)	100 mg	18.00	14.40
B-1690		5-bromo-4-chloro-3-indolyl β-D-galactopyranoside (X-Gal)	1 g	55.00	44.00
B-1691		5-bromo-4-chloro-3-indolyl β-D-glucuronide, cyclohexylammonium salt (X-GlcU, CHA)	100 mg	115.00	92.00
B-6492	**New**	5-bromo-4-chloro-3-indolyl phosphate, disodium salt (BCIP, Na)	1 g	75.00	60.00
B-8405	**New**	5-bromo-4-chloro-3-indolyl phosphate, p-toluidine salt (BCIP, toluidine)	1 g	110.00	88.00
D-272		3,3′-dipentyloxacarbocyanine iodide ($DiOC_5(3)$)	100 mg	85.00	68.00
E-6960	**New**	europium(III) chloride, hexahydrate *99.99%*	1 g	20.00	16.00
G-6657	**New**	gel photographic filter holder	each	18.00	18.00
N-6547	**New**	NBT/BCIP Reagent Kit	1 kit	135.00	108.00
N-6495	**New**	nitro blue tetrazolium chloride (NBT)	1 g	65.00	52.00
P-6649	**New**	protein molecular weight standards *broad range* *200 gel lanes*	400 µL	65.00	52.00
P-6647	**New**	protein molecular weight standards *high range* *200 gel lanes*	400 µL	45.00	36.00
P-6648	**New**	protein molecular weight standards *low range* *200 gel lanes*	400 µL	45.00	36.00
S-921		streptavidin, alkaline phosphatase conjugate *2 mg/mL*	0.5 mL	155.00	124.00
S-931		streptavidin, β-galactosidase conjugate	1 mg	125.00	100.00
S-911		streptavidin, peroxidase from horseradish conjugate	1 mg	95.00	76.00
S-1307		sulforhodamine B	5 g	78.00	62.40
S-6650	**New**	SYPRO™ Orange protein gel stain *5000X concentrate in DMSO*	500 µL	95.00	76.00
S-6651	**New**	SYPRO™ Orange protein gel stain *5000X concentrate in DMSO* *special packaging*	10x50 µL	110.00	88.00
S-6656	**New**	SYPRO™ Orange/Red protein gel stain photographic filter	each	35.00	35.00
S-6655	**New**	SYPRO™ Protein Gel Stain Starter Kit	1 kit	55.00	55.00
S-6653	**New**	SYPRO™ Red protein gel stain *5000X concentrate in DMSO*	500 µL	95.00	76.00
S-6654	**New**	SYPRO™ Red protein gel stain *5000X concentrate in DMSO* *special packaging*	10x50 µL	110.00	88.00
T-1247		terbium(III) chloride, hexahydrate	1 g	20.00	16.00
T-8464	**New**	3,3′,5,5′-tetramethylbenzidine, dihydrochloride (TMB, HCl)	5 g	95.00	76.00
T-8463	**New**	3,3′,5,5′-tetramethylbenzidine, free base (TMB)	5 g	75.00	60.00

9.3 Reagents for Peptide Analysis, Sequencing and Synthesis

The dominant chemistry for sequencing peptides employs the nonfluorescent reagent, phenyl isothiocyanate, which forms phenylthiohydantoins (PTH) in the sequencing reaction. Several of our fluorescent probes and research chemicals have been used for N-terminal amino acid analysis and peptide sequencing, as well as for protein fragment modification prior to PTH sequencing.

N-Terminal Amino Acid Analysis

Except when it is already blocked by formylation, acetylation, pyroglutamic acid formation or other chemistry, the N-terminal amino acid of proteins can be labeled with a variety of fluorescent and chromophoric reagents from Chapter 1; however, only those functional groups that survive complete protein hydrolysis, such as sulfonamides, are useful for N-terminal amino acid analysis. Dansyl chloride (D-21) and dabsyl chloride (D-1537) are the most commonly employed reagents for such analyses.[1-3]

Nonacylated N-terminal serine and threonine residues of proteins can be periodate-oxidized to aldehydes[4-6] that can then be modified by a variety of hydrazine derivatives listed in Chapter 3. Only peptides and proteins that contain these two terminal amino acids become fluorescent, although oxidation of the carbohydrate portion of glycoproteins to aldehydes may cause interference in this analysis.

N-Acetylated or *N*-formylated proteins have been detected by transfer of the acyl group to dansyl hydrazine (D-100) and subsequent chromatographic separation of the fluorescent product.[7,8] The sensitivity of this method can likely be improved by the use of other fluorescent hydrazides described in Section 3.2.

Peptide Sequencing

Fluorescent Isothiocyanates

As analogs of phenyl isothiocyanate, the peptide conjugates of fluorescein isothiocyanate (FITC; F-143, F-1906, F-1907), dansyl isothiocyanate (D-1302) and other fluorescent isothiocyanates in Chapter 1 are susceptible to Edman degradation via their thiohydantoins. Thus, these fluorescent reagents are potentially useful for ultrasensitive amino acid sequencing.[9-12]

Reagents for Modified Sequencing

Modification of cysteine residues in proteins by *N*-(2-iodoethyl)-trifluoroacetamide (I-1536) makes these sites susceptible to cleavage by trypsin, thus providing a means for selectively preparing additional fragments for protein sequencing.[13] *N*-Isopropyl iodoacetamide (NIPIA, I-6002) has been recommended for use in the phenylthiohydantoin (PTH) microsequencing of reductively alkylated proteins. The NIPIA-alkylated cysteine PTH can be detected as a sharp peak in standard reverse-phase HPLC analysis of PTH amino acids.[14]

Peptide Synthesis

Peptides specifically labeled with fluorescent dyes, haptens, photoactive groups or radioisotopes are important both as probes for receptors and as substrates for enzymes (see Section 10.4). Labeled peptides can be prepared by modifying isolated peptides or by incorporating the label during solid-phase synthesis.

Labeling Peptides in Solution

Appropriately substituted synthetic peptides can be labeled in solution by almost any of the reactive probes in Chapters 1–5. Many peptides contain multiple residues that can be modified, potentially leading to complex mixtures of products, some of which may be biologically inactive. Modification of the peptide's thiol group by one of the thiol-reactive reagents described in Chapter 2 is usually facile and very efficient. If the peptide is synthetic, or can be modified by site-directed mutagenesis, then incorporation of a cysteine residue at the desired site of labeling is recommended. The N-terminus of peptides, which has a lower pK_a than the ε-amino group of lysine residues, can sometimes be labeled in the presence of other amines if the pH is kept near neutral. Conversion of tyrosine residues to *o*-aminotyrosines (see Section 3.1) can be used to provide selective sites for peptide modification unless the tyrosine residues are essential for the biological activity of the peptide.

Radiolabeling of Peptides and Proteins

Although Molecular Probes does not perform radiochemical syntheses, we provide some important reagents for peptide and protein iodination. The Bolton–Hunter reagent[15] (H-1586) is a widely used amine-reactive succinimidyl ester for introducing phenolic groups into peptides and proteins. These phenolic groups can then be readily radioiodinated using published procedures.[16,17] Incorporation of the 7-hydroxy-4-methylcoumarin-3-acetic acid residue, which also contains a phenolic group that can be radioiodinated, into peptides or proteins via its succinimidyl ester (H-1411) can be quantitated by the conjugate's absorbance or fluorescence.[18,19]

Solid-Phase Synthesis of Labeled Peptides

Because specific labeling of peptides in solution is problematic, it may be more convenient to conjugate the fluorophore to the N-terminus of a resin-bound peptide *before* removal of other protecting groups and release of the labeled peptide from the resin. About five equivalents of an amine-reactive fluorophore are usually used per amine of the immobilized peptide. Fluorescein, eosin, Oregon Green™, Rhodamine Green™, Rhodol Green™, tetramethylrhodamine, Rhodamine Red™, Texas Red®, coumarin and NBD fluorophores, the dabcyl chromophore and biotin are all reasonably stable to hydrogen fluoride (HF), as well as to most other acids.[20-25] With the possible exception of the coumarins, these fluorophores are also stable to reagents used for deprotection of peptides synthesized using FMOC chemistry.[26] The BODIPY® fluorophore may be unstable to the conditions used to remove some protecting groups.

Molecular Probes has recently prepared some unique reagents for automated synthesis of peptides that are specifically labeled with fluorophores, chromophores and haptens. Use of these precursors permits the incorporation of these groups at specific sites in the peptide's sequence. The *t*-BOC and α-FMOC derivatives of ε-dabcyl-L-lysine (B-6217, D-6216) can be used to incorporate the dabcyl chromophore at selected sites in the peptide sequence. The dabcyl chromophore, which has broad visible absorption, has been extensively used as a quenching group in the automated synthesis of HIV protease substrate (H-2930, see Section 10.4), renin substrate (R-2931, see Section 10.4) and other fluorogenic peptidase substrates [27-30] (see Figure 10.9 in Chapter 10). The dabcyl group can also be incorporated at the N-terminus by using dabcyl succinimidyl ester [25,31] (D-2245).

EDANS is the most common fluorophore for pairing with the dabcyl quencher in fluorescence resonance energy transfer (FRET) experiments. This fluorophore is conveniently introduced during automated synthesis of peptides by using 5-((2-(*t*-BOC)-γ-glutamylaminoethyl)amino)naphthalene-1-sulfonic acid (B-6215).[25,27] We also offer α-(*t*-BOC)-ε-dansyl-L-lysine (B-6142) for incorporation of the dansyl fluorophore into peptides during synthesis. Like EDANS, its fluorescence overlaps the absorption of the dabcyl dye, which may make the combination of dansyl and dabcyl labels useful for rapid synthesis of a variety of fluorogenic substrates for endopeptidases.[32]

Site-specific biotinylation of peptides can be achieved using the *t*-BOC–protected derivative of biocytin (B-6349).[24] The racemic benzophenone phenylalanine analog (DL-Bpa, B-1584) can be incorporated into peptides following its *t*-BOC or FMOC protection.[33-35] Resolution of the diastereomers is usually accomplished during HPLC purification of the products; the unprotected benzophenone can also be resolved using published enzymatic methods.[35]

Other protected amino acids for automated peptide synthesis are being developed or can be prepared by custom synthesis. Molecular Probes also performs custom syntheses of fluorophore-, chromophore-, hapten- and biotin-labeled peptides and sometimes undertakes these projects as research collaborations. Please contact our Technical Assistance Department for details.

Synthesis of Peptides that Contain Caged Amino Acids

Our caged amino acids are designed to yield peptides that are biologically *inactive* until they are photolyzed with UV light; see Chapter 19 for more information on caged probes. Those caged on the ε-amino group of lysine (B-7099), the phenol of tyrosine (B-7077), the γ-carboxylic acid of glutamic acid (B-7078) or the thiol of cysteine (B-7097, D-7098) can be used for the specific incorporation of caged amino acids in the sequence. The amino acids alanine, glycine, leucine, isoleucine, methionine, phenylalanine, tryptophan and valine that are caged on the α-amine with the NVOC caging group can be used to prepare peptides that are caged on the N-terminus or caged intermediates that can be selectively photolyzed to yield the active amino acid either in a polymer or in solution.[36] Certain other caged amino acids, which are principally used as caged neurotransmitters, are described in Section 18.3.

1. J Chromatography 553, 123 (1991); **2.** Anal Biochem 174, 38 (1988); **3.** Anal Biochem 170, 542 (1988); **4.** Biochem J 108, 883 (1968); **5.** Biochem J 95, 180 (1965); **6.** Biochem J 94, 17 (1965); **7.** J Cell Biol 106, 1607 (1988); **8.** Anal Biochem 29, 186 (1969); **9.** Biosci Biotech Biochem 58, 300 (1994); **10.** Biol Chem Hoppe–Seyler 367, 1259 (1986); **11.** FEBS Lett 198, 150 (1986); **12.** Anal Biochem 141, 446 (1984); **13.** Anal Biochem 106, 43 (1980); **14.** Anal Biochem 209, 109 (1993); **15.** Biochem J 133, 529 (1973); **16.** Anal Biochem 125, 427 (1982); **17.** Biochem Biophys Res Comm 80, 849 (1978); **18.** FEBS Lett 182, 185 (1985); **19.** J Immunol Meth 81, 123 (1985); **20.** Biochemistry 33, 7211 (1994); **21.** Biochemistry 33, 6966 (1994); **22.** J Biol Chem 269, 15124 (1994); **23.** Techniques in Protein Chemistry V, 493 (1994); **24.** Anal Biochem 202, 68 (1992); **25.** J Med Chem 35, 3727 (1992); **26.** Biochemistry 33, 10951 (1994); **27.** Bioorg Medicinal Chem Lett 2, 1665 (1992); **28.** J Protein Chem 9, 663 (1990); **29.** Science 247, 954 (1990); **30.** Tetrahedron Lett 31, 6493 (1990); **31.** FEBS Lett 297, 100 (1992); **32.** Arch Biochem Biophys 306, 304 (1993); **33.** Int'l J Peptide Protein Res 45, 106 (1995); **34.** Biochemistry 30, 336 (1991); **35.** J Biol Chem 261, 10695 (1986); **36.** J Am Chem Soc 92, 6333 (1970).

9.3 Data Table *Reagents for Peptide Analysis, Sequencing and Synthesis*

Cat #	Structure	MW	Storage	Soluble	Abs	{ε × 10⁻³}	Em	Solvent	Notes
B-1584	✓	269	NC	pH>6, DMF	260	{19}	none	MeOH	
B-6142	✓	480	F,D,L	DMF, MeCN	327	{4.6}	556	pH 8	
B-6215	✓	496	F,D,L	DMF	341	{5.4}	470	MeOH	
B-6217	✓	498	F,D,L	DMF, MeCN	428	{30}	none	MeOH	
B-6349	✓	473	F,D	DMF, MeCN	<300		none		
B-7077	✓	476	F,D,LL	DMSO, MeCN	340	{6.0}	none	MeOH	<1>
B-7078	✓	442	F,D,LL	DMSO, MeCN	340	{4.8}	none	MeOH	<1>
B-7097	✓	416	F,D,LL	DMSO, MeCN	346	{4.8}	none	MeOH	<1>
B-7099	✓	485	F,D,LL	DMSO, MeCN	340	{7.9}	none	MeOH	<1>
D-21	S1.8	270	F,DD,L	DMF, MeCN*	372	{3.9}	none	$CHCl_3$	<2>
D-100	S3.2	265	L	EtOH	336	{4.4}	534	MeOH	
D-1302	S1.8	383	F,DD,L	DMF, MeCN	335	{4.2}	536	MeOH	
D-1537	S1.8	324	F,DD,L	DMF, MeCN*	466	{33}	none	MeOH	<3>
D-2245	S1.8	366	F,D,L	DMF, DMSO	453	{32}	none	MeOH	<3>
D-6216	✓	620	F,D,L	DMF, MeCN	427	{30}	none	MeOH	
D-7086	✓	328	F,D,LL	DMSO, MeCN	348	{6.4}	none	pH 7	<1>
D-7087	✓	314	F,D,LL	DMSO, MeCN	347	{6.4}	none	pH 7	<1>
D-7088	✓	370	F,D,LL	DMSO, MeCN	344	{6.9}	none	$CHCl_3$	<1>
D-7089	✓	370	F,D,LL	DMSO, MeCN	348	{6.5}	none	pH 7	<1>
D-7092	✓	388	F,D,LL	DMSO, MeCN	347	{6.2}	none	pH 7	<1>
D-7093	✓	404	F,D,LL	DMSO, MeCN	348	{6.0}	none	pH 7	<1>
D-7094	✓	443	F,D,LL	DMSO, MeCN	348	{6.1}	none	pH 7	<1>

Cat #	Structure	MW	Storage	Soluble	Abs	{ε x 10^{-3}}	Em	Solvent	Notes
D-7095	✓	416	F,D,LL	DMSO, MeCN	347	{6.4}	none	pH 7	<1>
D-7098	✓	539	F,D,LL	DMSO,MeCN	347	{5.1}	none	MeOH	<1>
F-143	S1.3	389	F,DD,L	pH>6†, DMF	494	{73}	519	pH 9	<4, 5>
H-1411	S1.7	331	F,D,L	DMF, MeCN	328	{15}	386	$CHCl_3$	
H-1586	S1.8	263	F,D	DMF, MeCN	278	{1.3}	none	MeOH	
I-1536	S2.4	267	F	DMF, MeCN	<300		none		
I-6002	S2.4	227	F,D	DMF, MeCN	<300		none		<6>

F-1906, F-1907 see F-143

For definitions of the contents of this data table, see "How to Use the Handbook" on page vi.
Structure: Chemical structure drawing: ✓ = shown in this section; Sn.n = shown in Section number n.n.

* Do NOT dissolve in DMSO.
† Isothiocyanates are unstable in water and should not be stored in aqueous solution.
<1> All caged products are sensitive to light. They should be protected from illumination, except when photolysis is intended.
<2> D-21 butylamine conjugate has Abs = 337 nm (ε = {5.3}), Em = 492 nm in $CHCl_3$. Em and QY are highly solvent dependent: Em = 496 nm (QY = 0.45) in dioxane, 536 nm (QY = 0.28) in MeOH and 557 nm (QY = 0.03) in H_2O [Biochemistry 6, 3408 (1967)]. ε typically decreases upon conjugation to proteins (ε = {3.4} at 340 nm) [Anal Biochem 25, 412 (1968)].
<3> D-1537 reaction product with butylamine: Abs = 435 nm (ε = {31}), nonfluorescent in MeOH. D-2245 reaction product with butylamine Abs = 428 nm (ε = {32}), nonfluorescent in MeOH.
<4> F-143 extinction coefficient decreases about 10% on protein conjugation and is pH dependent [J Immunol Meth 5, 103 (1974)].
<5> Absorption and fluorescence of fluorescein derivatives are pH dependent. See data for F-1300 in Section 23.2.
<6> Iodoacetamides in solution undergo rapid photodecomposition to unreactive products. Minimize exposure to light prior to reaction.

9.3 Structures *Reagents for Peptide Analysis, Sequencing and Synthesis*

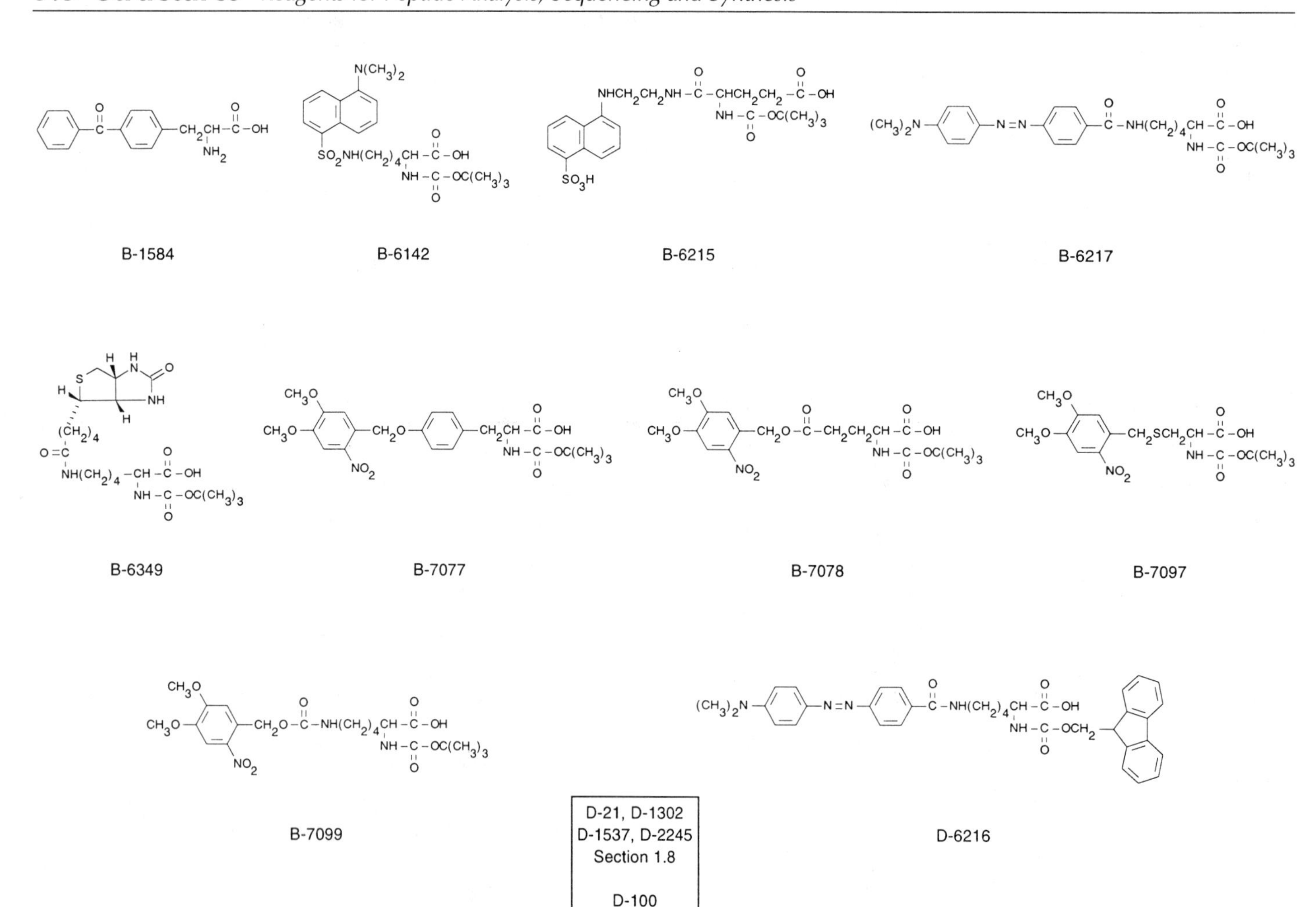

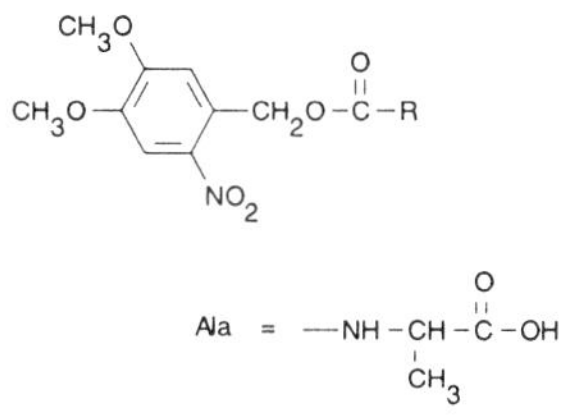

D-7086 R = L-Ala
D-7087 R = Gly
D-7088 R = L-Ile
D-7089 R = L-Leu
D-7092 R = L-Met
D-7093 R = L-Phe
D-7094 R = L-Trp
D-7095 R = L-Val

Ala = —NH–CH(CH$_3$)–C(=O)–OH

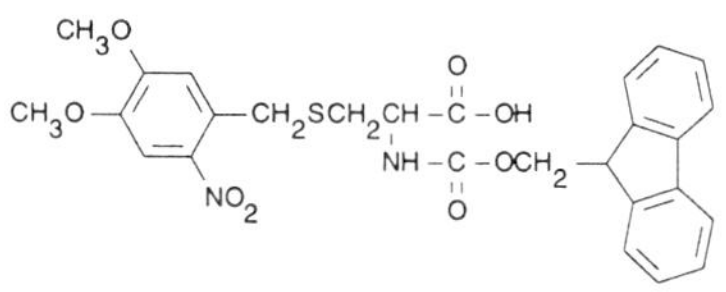

D-7098

F-143 Section 1.3

H-1411 Section 1.7

H-1586 Section 1.8

I-1536, I-6002 Section 2.4

9.3 Price List *Reagents for Peptide Analysis, Sequencing and Synthesis*

Cat #		Product Name	Unit Size	Price Per Unit ($) 1–4 Units	5–24 Units
B-1584		p-benzoyl-DL-phenylalanine (DL-Bpa)	25 mg	24.00	19.20
B-6349	**New**	ε-biotinoyl-α-*tert*-butoxycarbonyl-L-lysine (α-(t-BOC) biocytin)	100 mg	95.00	76.00
B-7097	**New**	N-(t-BOC)-S-(4,5-dimethoxy-2-nitrobenzyl)-L-cysteine (N-(t-BOC)-S-(DMNB-caged) L-cysteine)	100 mg	175.00	140.00
B-7099	**New**	α-(t-BOC)-ε-(N-(((4,5-dimethoxy-2-nitrobenzyl)oxy)carbonyl))-L-lysine (α-(t-BOC)-ε-(NVOC-caged) L-lysine)	100 mg	175.00	140.00
B-7077	**New**	N-(t-BOC)-O-(4,5-dimethoxy-2-nitrobenzyl)-L-tyrosine (N-(t-BOC)-O-(DMNB-caged) L-tyrosine)	100 mg	175.00	140.00
B-6142	**New**	α-(t-BOC)-ε-(5-dimethylaminonaphthalene-1-sulfonyl)-L-lysine (α-(t-BOC)-ε-dansyl-L-lysine)	100 mg	175.00	140.00
B-6217	**New**	α-(t-BOC)-ε-(4-dimethylaminophenylazobenzoyl)-L-lysine (α-(t-BOC)-ε-dabcyl-L-lysine)	100 mg	175.00	140.00
B-7078	**New**	N-(t-BOC)-L-glutamic acid γ-(4,5-dimethoxy-2-nitrobenzyl ester) (N-(t-BOC)-γ-(DMNB-caged) L-glutamic acid)	100 mg	175.00	140.00
B-6215	**New**	5-((2-(t-BOC)-γ-glutamylaminoethyl)amino)naphthalene-1-sulfonic acid	100 mg	175.00	140.00
D-7098	**New**	S-(4,5-dimethoxy-2-nitrobenzyl)-N-FMOC-L-cysteine (S-(DMNB-caged) N-FMOC-L-cysteine)	100 mg	175.00	140.00
D-7086	**New**	N-(((4,5-dimethoxy-2-nitrobenzyl)oxy)carbonyl)-L-alanine (N-(NVOC-caged) L-alanine)	25 mg	35.00	28.00
D-7087	**New**	N-(((4,5-dimethoxy-2-nitrobenzyl)oxy)carbonyl) glycine (N-(NVOC-caged) glycine)	25 mg	35.00	28.00
D-7088	**New**	N-(((4,5-dimethoxy-2-nitrobenzyl)oxy)carbonyl)-L-isoleucine (N-(NVOC-caged) L-isoleucine)	25 mg	35.00	28.00
D-7089	**New**	N-(((4,5-dimethoxy-2-nitrobenzyl)oxy)carbonyl)-L-leucine (N-(NVOC-caged) L-leucine)	25 mg	35.00	28.00
D-7092	**New**	N-(((4,5-dimethoxy-2-nitrobenzyl)oxy)carbonyl)-L-methionine (N-(NVOC-caged) L-methionine)	25 mg	35.00	28.00
D-7093	**New**	N-(((4,5-dimethoxy-2-nitrobenzyl)oxy)carbonyl)-L-phenylalanine (N-(NVOC-caged) L-phenylalanine)	25 mg	35.00	28.00
D-7094	**New**	N-(((4,5-dimethoxy-2-nitrobenzyl)oxy)carbonyl)-L-tryptophan (N-(NVOC-caged) L-tryptophan)	25 mg	35.00	28.00
D-7095	**New**	N-(((4,5-dimethoxy-2-nitrobenzyl)oxy)carbonyl)-L-valine (N-(NVOC-caged) L-valine)	25 mg	35.00	28.00
D-1537		4-dimethylaminoazobenzene-4′-sulfonyl chloride (dabsyl chloride)	100 mg	24.00	19.20
D-1302		4-((5-dimethylaminonaphthalene-1-sulfonyl)amino)phenylisothiocyanate	100 mg	38.00	30.40
D-21		5-dimethylaminonaphthalene-1-sulfonyl chloride (dansyl chloride)	1 g	40.00	32.00
D-100		5-dimethylaminonaphthalene-1-sulfonyl hydrazine (dansyl hydrazine)	100 mg	27.00	21.60
D-2245		4-((4-(dimethylamino)phenyl)azo)benzoic acid, succinimidyl ester (dabcyl, SE)	100 mg	68.00	54.40
D-6216	**New**	ε-(4-((4-(dimethylamino)phenyl)azo)benzoyl)-α-9-fluorenylmethoxycarbonyl-L-lysine (ε-dabcyl-α-FMOC-L-lysine)	100 mg	175.00	140.00
F-143		fluorescein-5-isothiocyanate (FITC 'Isomer I') *≥95% by HPLC*	1 g	98.00	78.40
F-1906		fluorescein-5-isothiocyanate (FITC 'Isomer I') *≥95% by HPLC* *special packaging*	10x10 mg	48.00	38.40
F-1907		fluorescein-5-isothiocyanate (FITC 'Isomer I') *≥95% by HPLC* *special packaging*	10x100 mg	125.00	100.00
H-1411		7-hydroxy-4-methylcoumarin-3-acetic acid, succinimidyl ester	25 mg	78.00	62.40
H-1586		3-(4-hydroxyphenyl)propionic acid, succinimidyl ester (Bolton-Hunter reagent)	1 g	38.00	30.40
I-1536		N-(2-iodoethyl)trifluoroacetamide	1 g	35.00	28.00
I-6002	**New**	N-isopropyl iodoacetamide (NIPIA)	100 mg	45.00	36.00

9.4 Probes for Protein Topology and Interactions

Labeling Cell-Surface Proteins

The membrane of intact cells provides an effective barrier to the entry of polar dyes. Thus, a polar reactive dye will usually only label functional groups on the cell's outer surface. After cell disruption, the proteins can be separated by electrophoresis and the labeled proteins detected by their fluorescence, by use of anti-dye antibodies or by other means. This technique provides information about which proteins are exposed in biological membranes and, in particular, about the topology and conformational changes of membrane-bound proteins. Polar reactive dyes may also be useful for the study of protein folding and of protein topology in reconstituted vesicles.

Labeling Cell-Surface Thiols and Disulfides

Recommended reagents for labeling cell-surface thiols include the stilbene iodoacetamide and maleimide (A-484, A-485),[1-5] lucifer yellow iodoacetamide (L-1338),[6] 5-iodoacetamidofluorescein (5-IAF, I-3), fluorescein-5-maleimide (F-150), tetramethylrhodamine maleimide[7] (T-6027, T-6028), 1,5-IAEDANS (I-14) and possibly the cationic monobromotrimethylammoniobimane (qBBr, M-1380). The cationic bimane is essentially nonfluorescent until conjugated to a thiol; its ability to penetrate to sites within a protein or cell differs from that of the uncharged bimane probes (see Section 2.4), as does its membrane permeability.[8,9] The membrane-impermeant tetramethylrhodamine maleimide was recently used to investigate conformational changes of the *Shaker* K^+ channel associated with channel gating.[7] Our antibodies to lucifer yellow, fluorescein and tetramethylrhodamine (see Section 7.3) may facilitate isolation and detection of proteins labeled with the thiol-reactive lucifer yellow, fluorescein and tetramethylrhodamine derivatives, respectively.

Disulfides in peptides and proteins can be reduced to thiols by a variety of reagents (see Section 5.2). The three negative charges of the phosphine TCEP (T-2556) are likely to render it membrane impermeant, which may make it particularly useful for selective reduction of disulfides on the cell's outer membrane. The less polar tris-(2-cyanoethyl)phosphine (T-6052) may be a useful complemetary reagent for reducing intramembrane and intracellular disulfides. Unlike the dithiols DTT and DTE (see Section 5.2), phosphines do not introduce thiol groups and therefore do not usually need to be removed before reaction with thiol-reactive reagents.

Labeling Cell-Surface Amines

The sulfosuccinimidyl esters of biotin (B-6352, B-6353) have frequently been used to label proteins in the outer membrane surface,[10-14] yielding labeled proteins that can be separated by electrophoresis and then detected with a fluorophore- or enzyme-conjugated avidin derivative (see Section 7.5). A recent abstract describes the labeling and specific detection of tight junctions in developing embryos by the impermeant sulfosuccinimidyl ester of biotin-X (B-6353).[15] Recommended polar fluorescent reagents from Chapter 1 for labeling cell surface amines include Cascade Blue® acetyl azide (C-2284), the succinimidyl ester of BODIPY FL cysteic acid (BODIPY FL, CASE; D-6141) and the sulfosuccinimidyl ester of BODIPY FL (BODIPY FL, SSE; D-6140). Antibodies to the BODIPY FL and Cascade Blue fluorophores (see Section 7.3) may facilitate the isolation and detection of labeled proteins. In addition, fluorescamine (F-2332, see Section 9.1) is reported to be useful for labeling proteins on the outside of cells without penetrating through the membrane; excess unreacted fluorescamine is rapidly hydrolyzed.[16]

Labeling Other Functional Groups on the Cell Surface

The following approaches may be useful for labeling functional groups other than thiols or amines on the cell's surface, though in most cases they have not been tested for this application:

- Periodate- or galactose oxidase–mediated oxidation of cell-surface glycoproteins can be used to introduce aldehyde residues on the cell's surface; these can then be reacted with a membrane-impermeant fluorescent hydrazide such as lucifer yellow CH (L-453), Cascade Blue hydrazide (C-687) or one of several biotin hydrazides listed in Section 4.2. This method has been used to label plant protoplasts with lucifer yellow CH and Texas Red hydrazide[17] (T-6256); cell monolayers with biotin-X hydrazide[18,19] (BACH, B-1600); and erythrocyte ghosts with fluorescein-5-thiosemicarbazide (F-121), lucifer yellow CH (L-453), 1-pyrenebutanoic acid hydrazide (P-101) and several other dyes.[20] Our anti-dye antibodies and avidin derivatives (see Sections 7.3 and 7.5) may facilitate isolation and detection of the modified residues.

- Tyrosine resides can be coupled with an aromatic diazonium salt, such as the one prepared from N^{α}-(4-aminobenzoyl)-biocytin (A-1604), to form a colored azo dye that can be detected on blots using avidin or streptavidin conjugates[21] (see Section 7.5).

- Photolysis of the membrane-impermeant lucifer yellow and Cascade Blue azide derivatives (A-629, C-625) will indiscriminately label cell-surface proteins and lipids. These probes have been shown to photolabel cytoplasmic and luminal domains of Ca^{2+}-ATPase in sarcoplasmic reticulum vesicles.[22] The labeled proteins and peptides can be detected on blots using our antibodies to the Cascade Blue or lucifer yellow fluorophores[22] (see Section 7.3).

- Transamidation of cell-surface glutamine residues by the combination of a transglutaminase enzyme and a fluorescent or biotinylated aliphatic amine can form stable amides[23] (see Section 3.3). Impermeability of the enzyme restricts this reaction to a limited number of proteins on the cell's surface. This technique was used to selectively label erythrocyte band 3 protein with dansyl cadaverine (D-113). Following protease treatment, the dansylated peptides were isolated using an anti-dansyl affinity column.[24]

Labeling Intramembrane and Intracellular Proteins

A number of reactive reagents from Molecular Probes have very low water solubility, making them potentially useful for labeling functional groups from within the cell's membrane. However, the high membrane permeability of these reagents also permits their entry into cells, where they may react with a wide variety of proteins, glutathione or other cell constituents.

Labeling Intracellular and Intramembrane Thiols and Disulfides

Thiols, either intrinsic or prepared from disulfides by chemical reduction with a neutral reagent like tris-(2-cyanoethyl)phosphine (T-6052), are the most likely nucleophiles for labeling from within the membrane because they are more readily accommodated in the membrane than are amines. Particularly recommended for intramembrane or intracellular thiol modification are *N*-(1-pyrene)-maleimide (P-28), the coumarin maleimide CPM (D-346) and monobromobimane (mBBr, M-1378), all of which are essentially nonfluorescent until reacted with thiols. CPM has already been used to determine thiol content of proteins, cells and plasma.[25-28] CPM is reported to selectively stain thiols in the nucleoli.[29] The bimanes, including monobromobimane, have been shown to be useful for detecting the distribution of protein thiols in cells before and after chemical reduction of disulfides.[30]

Labeling Intracellular and Intramembrane Carboxylic Acids

The diazomethane derivative, 4-(diazomethyl)-7-diethylaminocoumarin[31] (D-1402) and fluorescent carbodiimides such as NCD-4[32-34] (C-428) selectively modify buried carboxylic acids of membrane-bound ATPases and other proteins. The carboxylic acid–reactive 1-pyrenyldiazomethane (PDAM, P-1405) is a particularly lipophilic probe that may be useful for intramembrane labeling.

Other Techniques for Labeling Membrane-Bound and Intracellular Proteins

Another relevant technique for studying membrane-bound protein structure uses trypsin to hydrolyze exposed proteins on the membrane's surface. The protein remaining in the membrane is labeled with fluorescein-5-maleimide (F-150), and the fluorescent products are analyzed by gel electrophoresis.[35]

The low intrinsic reactivity and high selectivity of sulfonyl *fluorides* for inhibiting serine proteases (see Section 10.9) may permit localization of serine proteases in live cells. The fluorescent BODIPY FL sulfonyl fluoride (BODIPY FL AEBSF, D-6201) is recommended for this possible application because of its membrane permeability, high fluorescence yield and long-wavelength excitation and emission spectra. Dansyl fluoride (D-11), which has much weaker UV-excited fluorescence, has been extensively used to selectively modify the active site of serine proteases in solution[36,37] and may have similar utility in cells.

The selective modification of a single protein in a tissue by a simple reactive dye is not common. However, the enhanced reactivity of the thiol groups of myosin in glycerinated muscle fibers apparently permits the site-selective labeling using tetramethylrhodamine-5-iodoacetamide (5-TMRIA, T-6006) and IANBD ester[38-41] (I-9).

Detecting Protein–Protein and Protein–Nucleic Acid Interactions

Several techniques are available for the study of protein–protein and protein–nucleic acid interactions that use fluorescent or other probes. Some methods are of potential general utility, and others are specific for particular proteins.

Fluorescence Resonance Energy Transfer

When two sites are labeled, one with a fluorescent donor and the other with a fluorescent or nonfluorescent acceptor, their close proximity to each other (10–100 Å) can be detected by excited-state energy transfer (see the Technical Note "Fluorescence Resonance Energy Transfer" on page 46). For example, energy transfer has been observed between labeled myosin and actin[42,43] and between fluorescent oligonucleotides hybridized to adjacent nucleic acid sequences.[44] In addition, fluorescein-to-fluorescein excited-state energy transfer leads to self-quenching. This self-quenching has allowed scientists to follow the assembly of fluorescein-labeled C9 complement protein from its subunits.[45,46]

Environment- and Conformation-Sensitive Probes

Interaction of two protein molecules, both of which are labeled with pyrene maleimide (P-28), during the long excited-state lifetime of the dye may be monitored by fluorescence if the sites diffuse to within ~10 Å of each other. The close proximity of the pyrenes results in the formation of excited-state pyrene dimers (excimers), which exhibit a distinctive red-shifted emission (see Figure 13.6 in Chapter 13). Excimers can form between adjacent labeled sites in a single protein, as occurs in tropomyosin,[47-50] lens crystallins[51] and sarcoplasmic reticulum ATPase,[52,53] or between interacting sites of two different labeled biomolecules.

Fluorescence of the actin monomer (G-actin) labeled with pyrene iodoacetamide (P-29) has been demonstrated to change upon polymerization, making this probe an excellent tool for following the kinetics of actin polymerization.[54-56] Using this probe, researchers can investigate the influence of several actin-binding proteins and of cytochalasin on the rate of actin polymerization.

Chemical and Photoreactive Crosslinking Reagents

Chemical or light-induced crosslinking is a useful tool for detecting *intra*- or *inter*molecular interactions of proteins (see Section 5.3). Cleavable crosslinking reagents such as the bis(succinimidyl ester)s of 5,5'-dithiobis-(2-nitrobenzoic acid) (DTNB, SE; D-6316) and ethylene glycol bis-(succinic acid) (EGS, E-6306) can link amine groups within or between proteins by forming stable amide bonds. Each reagent also contains a linkage that can be severed to permit uncoupling to permit analysis of the crosslinked products by peptide mapping or gel electrophoresis. Crosslinking by the amine-reactive derivative of DTNB (Ellman's reagent) is reversed by disulfide-reducing agents, including DTT and TCEP (see Section 5.2). The ester bonds within the EGS reagent are readily cleaved by either hydroxylamine or hydrazine.[57,58]

Similarly, the polyfluorinated maleimide azide TFPAM-SS1 (A-6311) is a heterobifunctional crosslinker that undergoes both thiol reaction and photolytic nitrene insertion (Figure 5.5). The internal disulfide in the crosslink can be reduced with DTT or TCEP to yield free thiols, which can then be derivatized with fluorescent or radioisotopic labeling reagents to allow identification of the crosslinked products.[59]

The photoreactive crosslinker *N*-((2-pyridyldithio)ethyl)-4-azidosalicylamide (PEAS, AET; P-6317) is a new aryl azide that undergoes disulfide–thiol interchange of its pyridyldisulfide groups with the thiol groups of biomolecules to form mixed disulfides in the same way as SPDP[60] (see Figure 5.1 in Chapter 5). The phenolic group of PEAS can also be radioiodinated prior to conjugation.

Affinity Chromatography

Streptavidin agarose (S-951, see Section 9.1) as well as immobilized antibodies to fluorophores or other haptens (see Section 7.3) have been used to bind biotinylated or haptenylated proteins to affinity matrices. Mixtures containing proteins or nucleic acids with which the immobilized protein are expected to interact can be passed through the matrix and, following denaturation, detected by electrophoresis or other methods. For example, an anti-fluorescein affinity matrix has been used to immobilize a fluoresceinated protein, thus facilitating the investigation of its *in vitro* protein–protein interactions.[61] Alternatively, the protein–protein interactions of a DNA binding protein–GST fusion protein can be assessed using an affinity column consisting of the fusion protein bound to glutathione agarose[62] (G-2879, G-6664, G-6665; see Section 9.1).

Bandshift Assays

Protein–nucleic acid interactions are most conveniently monitored using electrophoretic mobility–shift (bandshift) assays (see Section 8.3), although many of the techniques mentioned above for detecting protein–protein interactions may also be effective. EthD-1, YOYO®-1 and TOTO®-1 have been shown by several laboratories to be useful tools for labeling DNA in electrophoretic mobility-shift (bandshift) assays. EthD-1 and TOTO-1 (E-1169, T-3600; see Sections 8.1 and 8.3) were used to examine interactions between the binding domain of the *Kluyveromyces lactis* heat shock transcription factor and its specific binding site,[63] and YOYO-1 (Y-3601, see Sections 8.1 and 8.3) has been used to study the association of *Escherichia coli* RNA polymerase with DNA templates.[64] All eight of our high-affinity dimeric cyanine dyes (see Figure 8.1 in Chapter 8), and the ethidium homodimers are potentially useful in this application.

Bandshift assays can also be detected by staining the gels after electrophoresis with SYBR Green I nucleic acid gel stain (S-7567, see Section 8.3), which is much more sensitive than ethidium bromide and obviates the need for radioactive labeling of DNA templates for bandshift assays (see Figure 8.13 in Chapter 8).

1. J Biol Chem 270, 23065 (1995); **2.** J Biol Chem 270, 16167 (1995); **3.** J Biol Chem 270, 843 (1995); **4.** FEBS Lett 356, 66 (1994); **5.** Proc Natl Acad Sci USA 91, 12828 (1994); **6.** Biochem Soc Trans 23, 38S (1995); **7.** Science 271, 213 (1996); **8.** Arch Biochem Biophys 282, 309 (1990); **9.** J Leukocyte Biol 45, 177 (1989); **10.** J Biol Chem 270, 27228 (1995); **11.** BioTechniques 18, 55 (1994); **12.** J Cell Biol 127, 2081 (1994); **13.** J Cell Biol 127, 2021 (1994); **14.** Cell 73, 1435 (1993); **15.** Mol Biol of the Cell Suppl 6, 193a, abstract #1119 (1995); **16.** Biochem Biophys Res Comm 67, 760 (1975); **17.** Protoplasma 139, 117 (1987); **18.** J Biol Chem 270, 29607 (1995); **19.** J Cell Biol 109, 2117 (1989); **20.** Biochemistry 24, 322 (1985); **21.** Biochem Biophys Res Comm 138, 872 (1986); **22.** Biochim Biophys Acta 1068, 27 (1991); **23.** Biochemistry 17, 2163 (1978); **24.** J Biol Chem 269, 22907 (1994); **25.** Anal Biochem 154, 186 (1986); **26.** Cytometry 3, 349 (1982); **27.** J Histochem Cytochem 29, 1377 (1981); **28.** J Histochem Cytochem 29, 314 (1981); **29.** J Histochem Cytochem 41, 1413 (1993); **30.** Mol Reprod Devel 37, 318 (1994); **31.** J Biol Chem 269, 6892 (1994); **32.** Biochemistry 32, 9586 (1993); **33.** Biochemistry 24, 7366 (1985); **34.** Biochim Biophys Acta 730, 201 (1983); **35.** J Biol Chem 269, 22533 (1994); **36.** Thrombosis Res 12, 15 (1977); **37.** Biochim Biophys Acta 439, 194 (1976); **38.** Biophys J 68, 81s (1995); **39.** Biophys J 68, 78s (1995); **40.** J Mol Biol 223, 185 (1992); **41.** Biochemistry 28, 2204 (1989); **42.** J Muscle Res Cell Motil 8, 97 (1987); **43.** Biochemistry 30, 3189 (1991); **44.** Nucleic Acids Res 22, 920 (1994); **45.** Biochemistry 23, 3260 (1984); **46.** Biochemistry 23, 3248 (1984); **47.** Biochemistry 26, 4922 (1987); **48.** Biochemistry 24, 6631 (1985); **49.** J Biol Chem 255, 11296 (1980); **50.** J Biol Chem 253, 3757 (1978); **51.** J Biol Chem 265, 14277 (1990); **52.** Biophys J 51, 513 (1987); **53.** Eur J Biochem 130, 5 (1983); **54.** J Biochem 116, 236 (1994); **55.** J Muscle Res Cell Motil 4, 235 (1983); **56.** Eur J Biochem 114, 33 (1981); **57.** J Biol Chem 270, 2053 (1995); **58.** Biochem Biophys Res Comm 87, 734 (1979); **59.** J Biol Chem 268, 20831 (1993); **60.** Traut, R.R. *et al., Protein Function: A Practical Approach*, IRL Press (1989) p. 101; **61.** J Biol Chem 257, 13095 (1982); **62.** Meth Mol Genet 1 (Part A), 267 (1993); **63.** J Biol Chem 268, 25229 (1993); **64.** Proc Natl Acad Sci USA 91, 6870 (1994).

9.4 Data Table *Probes for Protein Topology and Interactions*

Cat #	Structure	MW	Storage	Soluble	Abs	{ε x 10^{-3}}	Em	Solvent	Notes
A-484	S2.4	624	F,D,L	H_2O	329	{39}	408	pH 8	<1, 2>
A-485	S2.4	536	F,D	H_2O	322	{29}	411	pH 8	<1>
A-629	S5.3	637	F,LL	H_2O	426	{11}	534	H_2O	
A-1604	S4.2	606	NC	DMF, DMSO	<300		none		
A-6311	S5.3	449	F,D,LL	DMSO	256	{20}	none	MeOH	<3>
B-1600	S4.2	371	D	DMF, DMSO	<300		none		
B-6352	S4.2	670	F,D	DMF, pH>6*	<300		none		
B-6353	S4.2	557	F,D	DMF, pH>6*	<300		none		
C-428	S3.3	292	F,D	DMF, MeCN	333	{8.9}	414	$CHCl_3$	<4>
C-625	S5.3	770	F,LL	H_2O, MeOH	400	{29}	421	H_2O	<5>
C-687	S15.3	596	L	H_2O†	399	{30}	421	H_2O	<5>
C-2284	S1.7	607	F,D,LL	H_2O‡, MeOH	396	{29}	410	MeOH	<5>
D-11	S10.6	253	F,D,L	DMF, MeCN	356	{3.6}	none	MeOH	<6>
D-113	S3.3	335	L	EtOH, DMF	335	{4.6}	518	MeOH	
D-346	S2.2	402	F,D,L	DMSO	384	{33}	469	MeOH	<1>
D-1402	S3.4	257	FF,D,L	DMF, MeCN	395	{19}	477	MeOH	<4, 7>
D-6140	S1.2	491	F,D,L	H_2O*, DMSO	502	{75}	510	MeOH	<8>
D-6141	S1.2	641	F,D,L	H_2O*, DMSO	504	{82}	511	MeOH	<8>
D-6201	S10.6	477	F,D,L	DMF, MeCN	503	{84}	510	MeOH	
D-6316	S5.2	590	F,D	DMF, DMSO	316	{15}	none	MeOH	

Cat #	Structure	MW	Storage	Soluble	Abs	{ε × 10^{-3}}	Em	Solvent	Notes
E-6306	S5.2	456	F,D	DMF, DMSO	<300		none		
F-121	S3.2	421	D,L	pH>7, DMF	492	{85}	516	pH 9	
F-150	S2.2	427	F,D,L	pH>6, DMF	492	{83}	515	pH 9	<1>
I-3	S2.2	515	F,D,L	pH>6, DMF	492	{75}	515	pH 9	<1, 2>
I-9	S2.3	406	F,D,L	DMF, MeCN	472	{23}	536	MeOH	<1, 2>
I-14	S2.3	434	F,D,L	pH>6	336	{5.7}	490	pH 8	<2, 9>
L-453	S15.3	457	L	H_2O§	428	{12}	536	H_2O	
L-1338	S2.4	649	F,D,L	H_2O	426	{11}	531	pH 7	<1, 2>
M-1378	S2.4	271	F,L	DMF, MeCN	398	{5.0}	<10>	pH 7	
M-1380	S2.4	409	L	H_2O‡	378	{5.5}	<10>	pH 7	
P-28	S2.3	297	F,D,L	DMF, DMSO	338	{40}	375	MeOH	<1, 11>
P-29	S2.3	385	F,D,L	DMF, DMSO	339	{26}	384	MeOH	<1, 2>
P-101	S3.2	302	D,L	MeCN, DMF	341	{43}	376	MeOH	<11>
P-1405	S3.4	242	FF,L	DMF, MeCN	340	{41}	375	MeOH	<4, 11, 12>
P-6317	S5.3	347	F,D,LL	DMSO	271	{24}	none	MeOH	<13>
T-2556	S2.1	287	D	pH>5	<300		none		
T-6006	S2.2	569	F,D,L	DMSO	543	{87}	567	MeOH	<1, 2>
T-6027	S2.2	482	F,D,L	DMSO	541	{91}	567	MeOH	<1>
T-6028	S2.2	482	F,D,L	DMSO	541	{91}	567	MeOH	<1>
T-6052	S2.1	193	D	MeCN	<300		none		
T-6256	S3.2	621	L	DMF	584	{94}	605	MeOH	

For definitions of the contents of this data table, see "How to Use the Handbook" on page vi.
Structure: Chemical structure drawing: Sn.n = shown in Section number n.n.

* Sulfosuccinimidyl esters and cysteic acid succinimidyl esters are water soluble and may be dissolved in buffer at ~pH 8 for reaction with amines. Long-term storage in water is NOT recommended due to hydrolysis.
† Maximum solubility in water is ~1% for C-687.
‡ Unstable in water. Use immediately.
§ Maximum solubility in water is ~8% for L-453.
<1> Spectral data of the 2-mercaptoethanol adduct.
<2> Iodoacetamides in solution undergo rapid photodecomposition to unreactive products. Minimize exposure to light prior to reaction.
<3> This compound has weaker visible absorption at >300 nm but no discernible absorption peaks in this region.
<4> Spectra are for the reaction product with acetic acid.
<5> Cascade Blue dyes have a second absorption peak at about 376 nm with ε ~80% of the 395–400 nm peak.
<6> D-11 serine ester formed via reaction with chymotrypsin has Abs = 355 nm (ε = {4.2}), Em = 535 nm [Biochim Biophys Acta 439, 194 (1976)].
<7> Acetate ester derivatives of D-1402 in EtOH have fluorescence quantum yields = 0.48. Unreacted D-1402 quantum yield is 0.05 [Chem Pharm Bull 31, 3014 (1983)].
<8> The absorption and fluorescence spectra of BODIPY derivatives are relatively insensitive to the solvent.
<9> 2-Mercaptoethanol adduct of I-14 has essentially similar spectral characteristics in aqueous solution [Biochemistry 12, 4154 (1973)].
<10> M-1378 and M-1380 are almost nonfluorescent until reacted with thiols. Em = 475–485 nm for thiol adducts (QY ~0.1–0.3) [Meth Enzymol 143, 76 (1987)].
<11> Pyrene derivatives exhibit structured spectra. The absorption maximum is usually about 340 nm with a subsidiary peak at about 325 nm. There are also strong absorption peaks below 300 nm. The emission maximum is usually about 376 nm with a subsidiary peak at 396 nm. Excimer emission at about 470 nm may be observed at high concentrations.
<12> Unreacted P-1405 is nonfluorescent, Abs = 383 nm (ε = {31}) in MeOH.
<13> The absorption spectra of this compound includes a shoulder at longer wavelengths: Abs = 306 nm (ε = {10}).

9.4 Price List *Probes for Protein Topology and Interactions*

Cat #	Product Name	Unit Size	Price Per Unit ($) 1–4 Units	Price Per Unit ($) 5–24 Units
A-484	4-acetamido-4′-((iodoacetyl)amino)stilbene-2,2′-disulfonic acid, disodium salt	25 mg	95.00	76.00
A-485	4-acetamido-4′-maleimidylstilbene-2,2′-disulfonic acid, disodium salt	25 mg	95.00	76.00
A-1604	N^{α}-(4-aminobenzoyl)biocytin, trifluoroacetate	25 mg	85.00	68.00
A-629	N-(((4-azidobenzoyl)amino)ethyl)-4-amino-3,6-disulfo-1,8-naphthalimide, dipotassium salt (lucifer yellow AB)	10 mg	95.00	76.00
A-6311 ***New***	N-(2-((2-(((4-azido-2,3,5,6-tetrafluoro)benzoyl)amino)ethyl)dithio)ethyl)maleimide (TFPAM-SS1)	5 mg	98.00	78.40
B-1600	6-((biotinoyl)amino)hexanoic acid hydrazide (biotin-X hydrazide; BACH)	50 mg	48.00	38.40
B-6353 ***New***	6-((biotinoyl)amino)hexanoic acid, sulfosuccinimidyl ester, sodium salt (Sulfo-NHS-LC-Biotin; biotin-X, SSE)	25 mg	75.00	60.00
B-6352 ***New***	6-((6-((biotinoyl)amino)hexanoyl)amino)hexanoic acid, sulfosuccinimidyl ester, sodium salt (biotin-XX, SSE)	25 mg	75.00	60.00
C-2284	Cascade Blue® acetyl azide, trisodium salt	5 mg	125.00	100.00
C-625	Cascade Blue® aminoethyl 4-azidobenzamide, trisodium salt	10 mg	95.00	76.00
C-687	Cascade Blue® hydrazide, trisodium salt	10 mg	95.00	76.00
C-428	N-cyclohexyl-N′-(4-(dimethylamino)naphthyl)carbodiimide (NCD-4)	25 mg	138.00	110.40
D-1402	4-diazomethyl-7-diethylaminocoumarin	10 mg	88.00	70.40
D-346	7-diethylamino-3-(4′-maleimidylphenyl)-4-methylcoumarin (CPM)	25 mg	158.00	126.40
D-6140 ***New***	4,4-difluoro-5,7-dimethyl-4-bora-3a,4a-diaza-s-indacene-3-propionic acid, sulfosuccinimidyl ester, sodium salt (BODIPY® FL, SSE)	5 mg	145.00	116.00

Cat #		Product Name	Unit Size	Price Per Unit ($) 1–4 Units	5–24 Units
D-6201	**New**	4-(2-(4,4-difluoro-5,7-dimethyl-4-bora-3a,4a-diaza-s-indacene-3-propionyl)aminoethyl)benzenesulfonyl fluoride (BODIPY® FL AEBSF)	1 mg	48.00	38.40
D-6141	**New**	N-(4,4-difluoro-5,7-dimethyl-4-bora-3a,4a-diaza-s-indacene-3-propionyl)cysteic acid, succinimidyl ester, triethylammonium salt (BODIPY® FL, CASE)	5 mg	125.00	100.00
D-113		5-dimethylaminonaphthalene-1-(N-(5-aminopentyl))sulfonamide (dansyl cadaverine)	100 mg	78.00	62.40
D-11		5-dimethylaminonaphthalene-1-sulfonyl fluoride (dansyl fluoride)	100 mg	18.00	14.40
D-6316	**New**	5,5′-dithiobis-(2-nitrobenzoic acid, succinimidyl ester) (DTNB, SE)	25 mg	65.00	52.00
E-6306	**New**	ethylene glycol bis-(succinic acid), bis-(succinimidyl ester) (EGS)	100 mg	18.00	14.40
F-150		fluorescein-5-maleimide	25 mg	155.00	124.00
F-121		fluorescein-5-thiosemicarbazide	100 mg	95.00	76.00
I-3		5-iodoacetamidofluorescein (5-IAF)	100 mg	138.00	110.40
I-9		N-((2-(iodoacetoxy)ethyl)-N-methyl)amino-7-nitrobenz-2-oxa-1,3-diazole (IANBD ester)	100 mg	95.00	76.00
I-14		5-((((2-iodoacetyl)amino)ethyl)amino)naphthalene-1-sulfonic acid (1,5-IAEDANS)	100 mg	35.00	28.00
L-453		lucifer yellow CH, lithium salt	25 mg	48.00	38.40
L-1338		lucifer yellow iodoacetamide, dipotassium salt	25 mg	95.00	76.00
M-1378		monobromobimane (mBBr)	25 mg	43.00	34.40
M-1380		monobromotrimethylammoniobimane bromide (qBBr)	25 mg	38.00	30.40
P-101		1-pyrenebutanoic acid hydrazide	100 mg	95.00	76.00
P-29		N-(1-pyrene)iodoacetamide	100 mg	115.00	92.00
P-28		N-(1-pyrene)maleimide	100 mg	75.00	60.00
P-1405		1-pyrenyldiazomethane (PDAM)	25 mg	88.00	70.40
P-6317	**New**	N-((2-pyridyldithio)ethyl)-4-azidosalicylamide (PEAS; AET)	10 mg	65.00	52.00
T-6006	**New**	tetramethylrhodamine-5-iodoacetamide (5-TMRIA) *single isomer*	5 mg	175.00	140.00
T-6027	**New**	tetramethylrhodamine-5-maleimide *single isomer*	5 mg	125.00	100.00
T-6028	**New**	tetramethylrhodamine-6-maleimide *single isomer*	5 mg	125.00	100.00
T-6256	**New**	Texas Red® hydrazide *>90% single isomer*	5 mg	95.00	76.00
T-2556		tris-(2-carboxyethyl)phosphine, hydrochloride (TCEP)	1 g	25.00	20.00
T-6052	**New**	tris-(2-cyanoethyl)phosphine	1 g	24.00	19.20

Chapter 10

Enzymes, Enzyme Substrates and Enzyme Inhibitors

Contents

10.1 Introduction to Enzyme Substrates and Their Reference Standards

Molecular Probes offers a large assortment of both common and uncommon fluorogenic and chromogenic enzyme substrates. We prepare substrates for detecting very low levels of enzymatic activity in fixed cells, cell extracts and purified preparations, as well as substrates for enzyme-linked immunosorbent assays (ELISAs). We have also developed effective methods for detecting enzymes in live cells. In this section, we describe the characteristics of our enzyme substrates and the fluorophores and chromophores from which they are derived, focusing primarily on the suitability of these substrates for different types of enzyme assays. Fluorophores and chromophores available as reference standards can be found in the data table, structure drawings and product list at the end of this section. Substrates for specific enzymes are described in subsequent sections.

Substrates Yielding Soluble Fluorescent Products

Solution assays designed to quantitate enzymatic activity in cell extracts or other biological fluids typically employ substrates that yield highly fluorescent or intensely absorbing water-soluble products. ELISAs also rely on these substrates for indirect quantitation of analytes.[1-3] An ideal substrate for fluorescence-based solution assays yields a highly fluorescent, water-soluble product with optical properties significantly different from those of the substrate. If the fluorescence spectra of the substrate and product overlap significantly, analysis will likely require a separation step, especially when using excess substrate to obtain pseudo–first-order kinetics. Fortunately, many substrates are metabolized to products that have longer-wavelength excitation or emission spectra (Figure 10.1). These fluorescent products can typically be quantitated in the presence of unreacted substrate using a fluorometer or a fluorescence microplate reader, which facilitates high-throughput analysis and requires relatively small assay volumes. Moreover, the front-face optics in many microplate readers allows researchers to use more concentrated solutions, which may both improve the linearity of the kinetics and reduce inner-filter effects.

When the spectral characteristics of the substrate and its metabolic product are similar, techniques such as thin-layer chromatography (TLC), high-performance liquid chromatography (HPLC), capillary electrophoresis, solvent extraction or ion exchange can be used to separate the product from unconsumed substrate prior to analysis. For example, our *FAST* CAT® Kits (F-2900, F-6616, F-6617; see Section 10.5) utilize TLC to separate the fluorescent chloramphenicol substrate from its acetylation products, whereas our new EnzChek™ 5'-Nucleotidase Assay Kit (E-6643, see Section 10.3) employs a simple precipitation procedure to separate the BODIPY® FL–AMP substrate from its BODIPY FL–adenosine product.

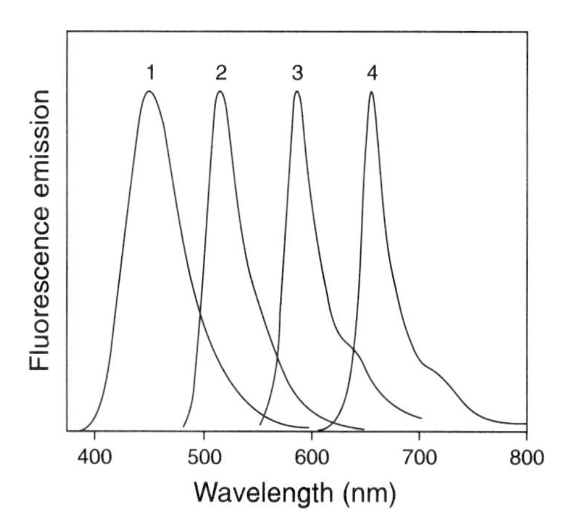

Figure 10.1 *Normalized emission spectra of 1) 7-hydroxy-4-methylcoumarin (H-189), 2) fluorescein (F-1300), 3) resorufin (R-363) and 4) DDAO (H-6482) in aqueous solution at pH 9. These fluorophores correspond to the hydrolysis products of several of our fluorogenic enzyme substrates.*

Substrates Derived from Water-Soluble Blue Fluorophores

Hydroxy- and amino-substituted coumarins have been the most widely used fluorophores for preparing fluorogenic substrates. Coumarin-based substrates produce highly soluble, intensely blue fluorescent products. Phenolic dyes with high pK_as — such as 7-hydroxycoumarin (often called umbelliferone) and the more common 7-hydroxy-4-methylcoumarin (β-methylumbelliferone, H-189) — are not fully deprotonated and therefore not fully fluorescent unless the pH of the reaction mixture is raised to above pH ~10. Thus, substrates derived from these fluorophores are not often used for continuous measurement of enzymatic activity in solution or live cells. The similar 3-cyano-7-hydroxycoumarin (C-183) and β-trifluoromethylumbelliferone (7-hydroxy-4-trifluoromethylcoumarin, T-659) and our new 6,8-difluoro-7-hydroxy-4-methylcoumarin (D-6566) have lower pK_as, making them suitable for a broader range of applications. Moreover, the β-trifluoromethylumbelliferone product has longer-wavelength spectra and stronger absorbance than β-methylumbelliferone. Ether and ester substrates derived from these phenolic dyes may be fluorescent but exhibit shorter-wavelength absorption and emission spectra.

Aromatic amines, including the commonly used 7-amino-4-methylcoumarin (AMC, A-191), 7-amino-4-trifluoromethylcoumarin (AFC, A-6249) and 6-aminoquinoline (6-AQ, A-497), are partially protonated at low pH (<~5) but fully deprotonated at physiological pH. Thus, their fluorescence spectra are not subject to variability due to pH-dependent protonation/deprotonation when assayed near or above physiological pH. These dyes are used to prepare peptidase substrates in which the amide has shorter-wavelength absorption and emission spectra than the amine hydrolysis product.

Substrates Derived from Water-Soluble Green Fluorophores

Compared to coumarin- or naphthalene-based substrates, substrates derived from fluoresceins, rhodamines, resorufins and other dyes often provide significantly greater sensitivity in fluorescence-based enzyme assays. In addition, most of these longer-wavelength dyes have extinction coefficients that are 5 to 25 times that of coumarins, nitrophenols or nitroanilines, making them potentially useful as sensitive chromogenic substrates.

Substrates based on the derivatives of fluorescein or rhodamine 110 (R110, R-6479) usually incorporate two moieties, each of which serves as a substrate for the enzyme. Consequently, they are cleaved first to the monosubstituted analog and then to the free fluorophore. Because the monosubstituted analog often absorbs and emits light at the same wavelengths as the ultimate hydrolysis product, this initial hydrolysis complicates the interpretation of hydrolysis kinetics.[4] However, when highly purified, the disubstituted fluorescein- and rhodamine 110–based substrates have virtually no visible-wavelength absorbance or background fluorescence, making them extremely sensitive detection reagents. For example, researchers have reported that the activity of as few as 1.6 molecules of β-galactosidase can be detected with fluorescein di-β-D-galactopyranoside and capillary electrophoresis.[5]

Substrates Derived from Water-Soluble Red Fluorophores

Long-wavelength fluorophores are often preferred because background absorbance and autofluorescence are generally lower when longer excitation wavelengths are used. Substrates derived from the red fluorescent resorufin (R-363) and the dimethylacridinone derivative 7-hydroxy-9*H*-(1,3-dichloro-9,9-dimethylacridin-2-one) (DDAO, H-6482) contain only a single hydrolysis-sensitive moiety, thereby avoiding the biphasic kinetics of both fluorescein- and rhodamine-based substrates.[6]

Resorufin is used to prepare several substrates for glycosidases, hydrolytic enzymes and dealkylases. In most cases, the relatively low pK_a of resorufin (~6.0) permits continuous measurement of enzymatic activity. Thiols such as DTT or mercaptoethanol should be avoided in assays utilizing resorufin-based substrates.

Substrates derived from DDAO, a red He-Ne laser–excitable fluorophore, generally exhibit good water solubility, low K_ms and high turnover rates.[7] In addition, the difference between the excitation maximum of the DDAO-based substrates and that of the product is greater than 150 nm (see Figure 10.2 in Section 10.2), which allows the two species to be easily distinguished.

Substrates for Live-Cell Enzyme Assays

Molecular Probes has developed a number of innovative strategies for investigating enzymatic activity in live cells.[8,9] For example, we offer a diverse set of probes that can passively enter the cell, yet once inside are processed by intracellular enzymes to generate products with improved cell retention. Molecular Probes also offers kits and reagents for detecting the expression of several common reporter genes in cells and cell extracts. These include substrates for β-galactosidase (Section 10.2), β-glucuronidase (Section 10.2), alkaline phosphatase (Section 10.3), chloramphenicol acetyltransferase (CAT, see Section 10.5) and luciferase (Section 10.5).

Thiol-Reactive Fluorophores

Molecular Probes prepares a number of enzyme substrates for live-cells assays that incorporate a mildly thiol-reactive chloromethyl moiety. Once inside the cell, this chloromethyl group undergoes what is believed to be a glutathione *S*-transferase–mediated reaction to produce a membrane-impermeant glutathione–fluorescent dye adduct, although our experiments suggest that they may also react with other intracellular components. Regardless of the mechanism, many cell types loaded with these chloromethylated substrates are both fluorescent and viable for at least 24 hours after loading and often through several cell divisions. Furthermore, unlike the free dye, the peptide–fluorescent dye adducts contain amino groups and can therefore be covalently linked to surrounding biomolecules by fixation with formaldehyde or glutaraldehyde. This property permits long-term storage of the labeled cells or tissue and, in cases where the anti-dye antibody is available (see Section 7.3), amplification of the conjugate by standard immunohistochemical techniques. Chloromethyl analogs of fluorogenic substrates for glycosidases (our DetectaGene™ products, see Section 10.2), peptidases, dealkylases, peroxidases and esterases are available. Molecular Probes' MitoTracker™ and CellTracker™ probes are also based on this principle (see Sections 12.2 and 15.2, respectively).

Lipophilic Fluorophores

Lipophilic analogs of fluorescein and resorufin exhibit many of the same properties as the water-soluble fluorophores, including relatively high extinction coefficients and good quantum yields. In most cases, however, substrates based on these lipophilic analogs load more readily into cells, and their fluorescent products are better retained after cleavage than their water-soluble counterparts. Lipophilic substrates and their products probably also distribute differently in cells and may tend to associate with lipid regions of the cell. When passive cell loading or enhanced dye retention are critical parameters of the experiment, we recommend using our lipophilic substrates for glycosidases (our ImaGene™ products, see Section 10.2), peroxidases and dealkylases.

Substrates Yielding Insoluble Fluorescent Products

Alkaline phosphatase, β-galactosidase and horseradish peroxidase conjugates are widely used as secondary detection reagents for immunohistochemical analysis and *in situ* hybridization, as well as for protein and nucleic acid detection by Western, Southern and Northern blots. Also, various methods such as chromatography, isoelectric focusing and gel electrophoresis are commonly employed to separate enzymes preceding their detection. A recent review by Weder and Kaiser discusses the use of a wide variety of fluorogenic substrates for the detection of electrophoretically separated hydrolases.[10]

In order to precisely localize enzymatic activity in a tissue or cell, on a blot or in a gel, the substrate must yield a product that immediately precipitates at the site of enzymatic activity. In addition to the commonly used chromogenic substrates, including X-Gal, BCIP and NBT, Molecular Probes has developed fluorogenic ELF substrates for alkaline phosphatase and several other hydrolytic enzymes (see also Section 6.3). Our ELF substrates fluoresce only weakly in the blue range. However, upon enzymatic cleavage, these substrates form the intensely fluorescent yellow-green ELF-97 alcohol (E-6578), which precipitates immediately at the site of enzymatic activity (see photos on front and back covers and Color Plate 1 in "Visual Reality"). The fluorescent ELF precipitate is exceptionally photostable and has a high Stokes shift (see Figures 6.6 and 6.7 in Chapter 6). We offer ELF substrates for several different enzymes, as well as several ELF Kits based on our ELF-97 phosphatase substrate. See Section 6.3 for a complete discussion of our ELF technology.

Substrates Based on Excited-State Energy Transfer

The principle of excited-state energy transfer (see the Technical Note "Fluorescence Resonance Energy Transfer" on page 46) can also be used to generate fluorogenic substrates. For example, the EDANS fluorophore in our HIV protease and renin substrates is effectively quenched by a nearby DABCYL acceptor chromophore (see Figure 10.10 in Section 10.4). This chromophore has been carefully chosen for maximal overlap of its absorbance with the fluorophore's fluorescence, thus ensuring that the fluorescence is quenched through excited-state energy transfer. Proteolytic cleavage of the substrate results in the spatial separation of the fluorophore and the acceptor chromophore, thereby restoring the fluorophore's fluorescence.[11-14] Many of the dyes described in Chapter 1 have been used to form energy-transfer pairs, some of which can be introduced during automated synthesis of peptides using modified amino acids described in Section 9.3.

The protease substrates in two of our EnzChek Protease Assay Kits (E-6638, E-6639; see Section 10.4) are heavily labeled casein conjugates; the close proximity of dye molecules results in considerable self-quenching. Hydrolysis of the protein to smaller fragments is accompanied by a dramatic increase in fluorescence, which forms the basis of a simple and sensitive continuous assay for a variety of proteases. Our EnzChek Dextranase Assay Kit (E-6644, see Section 10.2) also utilizes this strategy to measure dextranase activity. In addition, we offer a phospholipase A substrate (bis-BODIPY FL C_{11}-PC, B-7701; see Section 20.5) that contains a BODIPY FL fluorophore on each phospholipid acyl chain. The proximity of the BODIPY FL fluorophores on adjacent phospholipid acyl chains causes fluorescence self-quenching that is relieved only when the fluorophores are separated by phospholipase A–mediated cleavage.

Fluorescent Derivatization Reagents for Discontinuous Enzyme Assays

The mechanism of some enzymes makes it difficult to obtain a continuous optical change during reaction with an enzyme substrate. However, a discontinuous assay can often be developed by derivatizing the reaction products with one of the reagents described in Chapters 1–3, usually followed by a separation step in order to generate a product-specific fluorescent signal. For example, fluorescamine (F-2332, see Section 1.8) or *o*-phthaldialdehyde (OPA, P-2331; see Section 1.8) can detect the rate of *any* peptidase reaction by measuring the increase in the concentration of free amines in solution.[15,16] The activity of enzymes that produce free coenzyme A from its esters can be detected using thiol-reactive reagents such as 5,5'-dithiobis-(2-nitrobenzoic acid) (DTNB, D-8451; see Section 5.2) or 7-fluorobenz-2-oxa-1,3-diazole-4-sulfonamide [17] (ABD-F, F-6053; see Section 2.3). The products of enzymes that metabolize low molecular weight substrates can frequently be detected by chromatographic or electrophoretic analysis. HPLC or capillary zone electrophoresis can also be used to enhance the sensitivity of reactions that yield fluorescent products.[18]

Chromogenic Substrates

Soluble UV-Absorbing Chromophores

Several purely chromogenic substrates for hydrolytic enzymes are derived from 2-nitrophenol (*o*-nitrophenol, N-6476) or 4-nitrophenol (*p*-nitrophenol, N-6477). Hydrolysis of substrates based on these chromophores releases highly soluble, UV-absorbing products. Moreover, the strong electron-withdrawing properties of the nitro group in these substrates often accelerates enzymatic hydrolysis rates.

Insoluble Chromophoric Products

A number of chromogenic substrates for hydrolytic enzymes are derived from indolyl chromophores. These initially form a colorless — and sometimes blue fluorescent — 3-hydroxyindole ("indoxyl"), which spontaneously, or through mediation of an oxidizing agent such as nitro blue tetrazolium (NBT, N-6495; see Section 21.4) or potassium ferricyanide,[19] is converted to an intensely colored indigo dye that typically precipitates from the medium (see Figure 9.12 in Chapter 9). Indoxyl β-D-glucuronide (I-8416, see Section 10.2) has also been reported to yield chemiluminescence emission.[20] Halogenated indolyl derivatives are generally preferred because they produce finer precipitates that are less likely to diffuse from the site of formation, making them especially useful for detecting enzymatic activity in cells and tissues, on blots and in gels.

1. Meth Mol Biol 32, 461 (1994); **2.** J Immunol Meth 150, 5 (1992); **3.** Meth Enzymol 70, 419 (1980; **4.** Biochemistry 30, 8535 (1991); **5.** Anal Biochem 226, 147 (1995); **6.** Angew Chem Int Ed Engl 30, 1646 (1991); **7.** DDAO-based substrates are licensed to Molecular Probes under U.S. Patent No. 4,810,636 and its foreign equivalents; **8.** Biotech Histochem 70, 243 (1995); **9.** J Fluorescence 3, 119 (1993); **10.** J Chromatography A 698, 181 (1995); **11.** Bioorg Medicinal Chem Lett 2, 1665 (1992); **12.** J Protein Chem 9, 663 (1990); **13.** Science 247, 954 (1990); **14.** Tetrahedron Lett 31, 6493 (1990); **15.** Biochemistry 29, 6670 (1990); **16.** Anal Biochem 123, 41 (1982); **17.** Chem Pharm Bull 38, 2290 (1990); **18.** Anal Biochem 115, 177 (1981); **19.** Histochemie 23, 266 (1970); **20.** Clin Chem 40, 1580 (1994).

10.1 Data Table *Introduction to Enzyme Substrates and Their Reference Standards*

Cat #	Structure	MW	Storage	Soluble	Abs	{ε x 10^{-3}}	Em	Solvent	Notes
A-191	✓	175	L	DMF, DMSO	351	{18}	430	MeOH	<1>
A-497	✓	144	L	DMF, EtOH	354	{3.7}	445	MeOH	<2>
A-6249	S3.3	229	D,L	DMF, DMSO	376	{18}	480	MeOH	
C-183	✓	187	L	pH>8, DMF	408	{43}	450	pH 9	
C-2110	S15.2	210	F,D,L	DMSO	353	{14}	466	pH 9	
D-6566	✓	212	L	pH>6, DMF	358	{18}	452	pH 9	<3>
E-6578	✓	307	L	DMSO*	345	ND	530	pH 8	<4, 5>
F-1300	✓	332	L	pH>6, DMF	490	{88}	514	pH 9	<6>
H-185	✓	206	L	pH>6, DMF	386	{29}	448	pH 10	<3>
H-189	✓	176	L	pH>9, MeOH	360	{19}	449	pH 9	<3>
H-6482	✓	308	L	DMF	646	{41}	659	pH 10	
N-6476	✓	139	L	DMSO, DMF	420	{21}	none	pH 10	
N-6477	✓	139	L	DMSO, DMF	399	{18}	none	pH 9	
R-363	✓	235	L	pH>7, DMF	571	{54}	585	pH 9	<7, 8>
R-6479	✓	367	L	DMSO	499	{88}	521	MeOH	<9>
T-659	✓	230	L	MeOH	385	{16}	502	pH 10	<3>

For definitions of the contents of this data table, see "How to Use the Handbook" on page vi.
Structure: Chemical structure drawing: ✓ = shown in this section; Sn.n = shown in Section number n.n.

* This product is packaged as a solution in the solvent indicated in "Soluble."
<1> A-191 in aqueous solution (pH 7.0): Abs = 342 nm (ε = {16}), Em = 441 nm.
<2> A-497 in aqueous solution (pH 7.0): Abs = 339 nm (ε = {3.6}), Em = 550 nm.
<3> Spectra of these hydroxycoumarins are pH dependent. Below the pK_a, Abs shifts to shorter wavelengths (325–340 nm) and fluorescence intensity decreases. Approximate pK_a values are: 7.8 (H-189, C-2111), 7.5 (H-185), 7.3 (T-659) and 4.9 (D-6566).
<4> ND = not determined.
<5> ELF-97 alcohol is insoluble in water. Spectral maxima listed are for an aqueous suspension; for this reason, the value of {ε} cannot be determined.
<6> Absorption and fluorescence of fluorescein are pH dependent. See data for F-1300 in Section 23.2.
<7> Absorption and fluorescence of resorufin are pH dependent. Below the pK_a (~6.0), Abs shifts to ~480 nm and both ε and fluorescence quantum yield are markedly lower.
<8> R-363 is unstable in the presence of thiols such as dithiothreitol (DTT) and β-mercaptoethanol.
<9> R-6479 in aqueous solution (pH 7.0): Abs = 496 nm (ε = {80}), Em = 520 nm.

10.1 Structures *Introduction to Enzyme Substrates and Their Reference Standards*

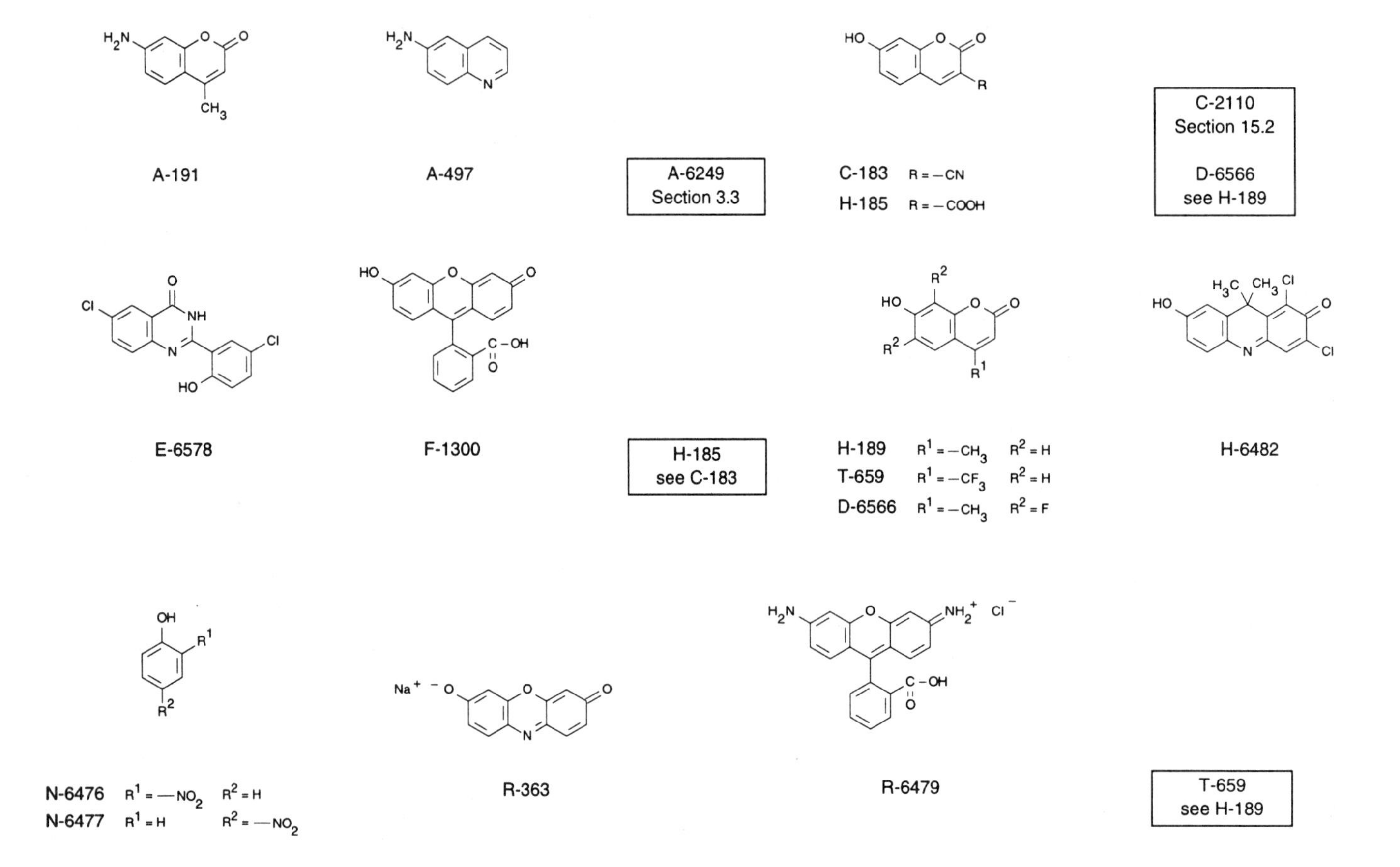

10.1 Price List *Introduction to Enzyme Substrates and Their Reference Standards*

Cat #		Product Name	Unit Size	Price Per Unit ($) 1–4 Units	5–24 Units
A-191		7-amino-4-methylcoumarin *reference standard*	100 mg	45.00	36.00
A-497		6-aminoquinoline (6-AQ) *reference standard*	1 g	24.00	19.20
A-6249	***New***	7-amino-4-trifluoromethylcoumarin (AFC) *reference standard*	100 mg	18.00	14.40
C-2110		CellTracker™ Blue CMAC (7-amino-4-chloromethylcoumarin)	5 mg	95.00	76.00
C-183		3-cyano-7-hydroxycoumarin	100 mg	35.00	28.00
D-6566	***New***	6,8-difluoro-7-hydroxy-4-methylcoumarin *reference standard*	10 mg	45.00	36.00
E-6578	***New***	ELF®-97 alcohol *1 mM solution in DMSO*	1 mL	24.00	19.20
F-1300		fluorescein *reference standard*	1 g	40.00	32.00
H-185		7-hydroxycoumarin-3-carboxylic acid *reference standard*	100 mg	95.00	76.00
H-6482	***New***	7-hydroxy-9H-(1,3-dichloro-9,9-dimethylacridin-2-one) (DDAO) *reference standard*	10 mg	45.00	36.00
H-189		7-hydroxy-4-methylcoumarin *reference standard*	1 g	12.00	9.60
N-6476	***New***	2-nitrophenol (o-nitrophenol) *reference standard*	1 g	8.00	6.40
N-6477	***New***	4-nitrophenol (p-nitrophenol) *reference standard*	1 g	8.00	6.40
R-363		resorufin, sodium salt *reference standard*	100 mg	18.00	14.40
R-6479	***New***	rhodamine 110 (R110) *reference standard*	25 mg	12.00	9.60
T-659		β-trifluoromethylumbelliferone (7-hydroxy-4-trifluoromethylcoumarin) *reference standard*	100 mg	18.00	14.40

10.2 Detecting Glycosidases

Glycosidase enzymes exhibit very high selectivity for hydrolysis of their preferred sugars. For example, β-galactosidase (G-6463) rapidly hydrolyzes β-D-galactopyranosides but usually does not hydrolyze either the anomeric α-D-galactopyranosides or the isomeric β-D-glucopyranosides. Endogenous glycosidase activity is frequently used to characterize strains of microorganisms [1-3] and to selectively label organelles of mammalian cells;[4-6] defects in glycosidase activity are characteristic of several diseases.[7,8]

In addition, glycosidases are important reporter gene markers. Specifically, *lacZ*, which encodes β-galactosidase, is extensively used as a reporter gene in animals and yeast, whereas the β-glucuronidase (GUS) gene is a popular reporter gene in plants.[9-13] Glycosidase substrates are also used in conjunction with glycosidase-conjugated secondary detection reagents in immunohistochemical techniques and enzyme-linked immunosorbent assays [14] (ELISAs).

Molecular Probes' complete line of fluorogenic and chromogenic glycosidase substrates are listed in Table 10.1. The common inducer of β-galactosidase activity, isopropyl β-D-thiogalactopyranoside (IPTG, I-6621), and other inducers and inhibitors of glycosidase activity are discussed in Section 10.6.

Table 10.1 *Glycosidase enzymes and their fluorogenic and chromogenic substrates.*

Carbohydrate (Enzyme)	Fluorophore Label	Chromophore Label	Notes on Enzyme Activity
N-Acetyl-β-D-galactosaminide (Neuraminidase, E.C. 3.2.1.18)		• 5-Bromo-4-chloro-3-indolyl [1] (B-8400) • 4-Nitrophenyl [2-5] (N-8426)	• Useful for differentiating *Candida albicans* from *Candida parapsilosis* [2,6]
N-Acetyl-β-D-glucosaminide (*N*-Acetyl-β-D-glucosaminidase, E.C. 3.2.1.30; Chitinase, E.C. 3.2.1.14)	• 6′-(*O*-(Carboxymethyl))-2′,7′-dichlorofluorescein [7] (C-6541) • Resorufin [8] (R-6540)	• 5-Bromo-4-chloro-3-indolyl [1,9] (B-8401) • 4-Nitrophenyl [4,5,9-11] (N-8427)	• Degrades chitin, which is present in diatoms, fungi, protozoans, arachnids, insects, crustaceans, nematodes and other invertebrates [8,9] • Found in many mammalian tissues and fluids, particularly in human kidney, lung and liver lysosomes [12] • Elevated activity in human urine may predict early renal disease [13-15] • Useful as a lysosomal marker [16,17] • Useful for detecting *Candida albicans* [18]
β-D-Cellobioside (Cellulase, E.C. 3.2.1.4)		• 4-Nitrophenyl [4,19,20] (N-8428)	• Degrades cellulose and xylan
β-D-Fucopyranoside (β-Fucosidase, E.C. 3.2.1.38)		• 4-Nitrophenyl [4,21-23] (N-8429)	
α-D-Galactopyranoside (α-Galactosidase, E.C. 3.2.1.22)		• 4-Nitrophenyl [4,11,24] (N-8430)	• Deficiency characterizes Fabry's disease [25,26] • Useful for identifying streptococci and enterococci [27]
β-D-Galactopyranoside (β-Galactosidase, E.C. 3.2.1.23)	• C_8FDG [28] (O-2892) • C_{12}FDG [29-31] (D-2893, I-2904) • 3-Carboxyumbelliferyl [32,33] (C-1488) • 4-Chloromethylcoumarin (D-2921) • 5-Chloromethylfluorescein [34] (D-2920) • DDAO [35] (D-6488) • 2-Dodecylresorufin [36] (I-2906) • Fluorescein [37-39] (FDG, F-1179) • 4-Methylumbelliferyl [4,40] (M-1489) • Resorufin [41] (R-1159) • 4-Trifluoromethylumbelliferyl [42,43] (T-657)	• 5-Bromo-4-chloro-3-indolyl [44] (X-gal, B-1690) • 5-Bromo-6-chloro-3-indolyl (B-8407) • 6-Chloro-3-indolyl [45,46] (C-8411) • Indoxyl (I-8414) • 5-Iodo-3-indolyl (I-8420) • *N*-Methyl-3-indolyl (M-8421) • 2-Nitrophenyl [47] (ONPG, N-8431) • 4-Nitrophenyl [3,4,48] (N-8432) • Phenyl [49,50] (P-8458)	• Useful as a reporter gene marker [37-39,45] • Useful for ELISAs [33,51-53] • Useful for enumerating coliforms from the family *Enterobacteriaceae* [11,54,55] • Useful for classifying mycobacteria [56]
α-D-Glucopyranoside (α-D-Glucosidase, E.C. 3.2.1.20)	• 4-Methylumbelliferyl [4,56-63] (M-8422)	• 4-Nitrophenyl [4,22,64-66] (N-8433)	• Glycoprotein found in endoplasmic reticulum and post-Golgi apparatus structures [57] • Deficiency characterizes glycogen storage disease type II (Pompe's disease), an autosomal recessive disorder [67]

Table 10.1, continued *Glycosidase enzymes and their fluorogenic and chromogenic substrates.*

Carbohydrate (Enzyme)	Fluorophore Label	Chromophore Label	Notes on Enzyme Activity
β-D-Glucopyranoside (β-Glucosidase, E.C. 3.2.1.31)	• Fluorescein (F-2881) • 4-Methylumbelliferyl [4,68-70] (M-8423) • Resorufin [71] (R-1160)	• 5-Bromo-4-chloro-3-indolyl [1,11,72] (B-1689) • Indoxyl [73] (I-8415) • 4-Nitrophenyl [4,20,74-76] (N-8434)	• Cleaves glucocerebrosides to glucose and ceramide [77,78] • Deficiency in acid β-glucosidase, which leads to abnormal lysosomal storage, characterizes Gaucher's disease [70,71,79] • Useful as an endoplasmic reticulum marker [16] • Useful for bacterial differentiation [3,76]
β-D-Glucuronide (β-Glucuronidase, GUS; E.C. 3.2.1.31)	• 3-Carboxyumbelliferyl (C-1492) • C_{12}FDGlcU (I-2908) • DDAO (D-6486) • 2-Dodecylresorufin (I-2910) • ELF-97 (E-6587) • Fluorescein (F-2915) • 4-Methylumbelliferyl [4,80-84] (M-1490) • Resorufin (R-1161) • 4-Trifluoromethylumbelliferyl [42] (T-658)	• 5-Bromo-4-chloro-3-indolyl [80,85] (X-GlcU; B-1691, B-8404) • 5-Bromo-6-chloro-3-indolyl [86] (B-8408) • 6-Chloro-3-indolyl (C-8412) • Indoxyl [87,88] (I-8416) • 4-Nitrophenyl [4,89,90] (N-8435) • Phenyl [91] (P-8459)	• Useful as a reporter gene marker [92,93] • Useful as a lysosomal marker [82,84] • Useful for detecting *E. coli*; 94–96% of *E. coli* contain this enzyme, but it is less common in *Shigella* (44–58%), *Salmonella* (20–29%) and *Yersina* strains [11,94-96]
α-D-Mannopyranoside (α-Mannosidase, E.C. 3.2.1.24)		• 4-Nitrophenyl [4,97-99] (N-8436)	• Found in liver, fibroblasts and other tissues • Deficiency characterizes mannosidosis, a lysosomal storage disease [100-102] • Mannosidase II is a marker for the Golgi complex [16]

1. Histochemistry 58, 203 (1978); **2.** J Clin Microbiol 28, 614 (1990); **3.** Meth Microbiol 19, 105 (1987); **4.** Anal Biochem 104, 182 (1980); **5.** J Biol Chem 245, 5153 (1970); **6.** J Clin Microbiol 32, 3034 (1994); **7.** Chem Pharm Bull 41, 1513 (1993); **8.** Anal Sci 8, 161 (1992); **9.** Anal Biochem 208, 74 (1993); **10.** J Cell Biol 119, 259 (1992); **11.** Microbiol Rev 55, 335 (1991); **12.** Anal Biochem 148, 50 (1985); **13.** J Clin Lab Anal 5, 1 (1991); **14.** Clin Chem 92, 459 (1983); **15.** Toxicol 23, 99 (1982); **16.** Nature 369, 113 (1994); **17.** Cell 52, 73 (1988); **18.** J Clin Microbiol 25, 2424 (1987); **19.** Biochemistry 33, 6371 (1994); **20.** Anal Biochem 138, 481 (1984); **21.** Appl Environ Microbiol 57, 1644 (1991); **22.** J Membrane Biol 4, 113 (1971); **23.** J Biol Chem 243, 103 (1968); **24.** Proc Natl Acad Sci USA 77, 6319 (1980); **25.** Human Gene Ther 6, 905 (1995); **26.** Pediatric Annals 5, 313 (1976); **27.** Appl Environ Microbiol 45, 622 (1988); **28.** Cytometry 20, 324 (1995); **29.** Appl Environ Microbiol 60, 4638 (1994); **30.** Proc Natl Acad Sci USA 89, 10681 (1992); **31.** FASEB J 5, 3108 (1991); **32.** Infection and Immunity 61, 5231 (1993); **33.** Anal Biochem 146, 211 (1985); **34.** J Neurosci 15, 1025 (1995); **35.** U.S. Patent No. 4,810,636; **36.** Biotech Bioeng 42, 1113 (1993); **37.** Cytometry 17, 216 (1994); **38.** Devel Biol 161, 77 (1994); **39.** Proc Natl Acad Sci USA 85, 2603 (1988); **40.** Anal Biochem 215, 24 (1993); **41.** Anal Chim Acta 163, 67 (1984); **42.** Biochem Int'l 24, 1135 (1991); **43.** Anal Lett 21, 193 (1988); **44.** BioTechniques 7, 576 (1989); **45.** BioTechniques 18, 434 (1995); **46.** Am J Pathol 141, 1331 (1992); **47.** J Immunol Meth 150, 23 (1992); **48.** J Appl Biochem 2, 390 (1980); **49.** Appl Environ Microbiol 64, 1497 (1995); **50.** J Biol Chem 267, 2737 (1992); **51.** Exp Parasitol 73, 440 (1991); **52.** J Immunol Meth 54, 297 (1982); **53.** J Virol Meth 3, 155 (1981); **54.** J Appl Bacteriol 64, 65 (1988); **55.** Appl Environ Microbiol 35, 136 (1978); **56.** Zbl Bakt 280, 476 (1994); **57.** J Cell Biol 110, 309 (1990); **58.** DNA 17, 99 (1988); **59.** Histochem J 6, 491 (1974); **60.** Science 181, 352 (1973); **61.** FEBS Lett 27, 161 (1972); **62.** Biochem Biophys Res Comm 43, 913 (1971); **63.** Science 167, 1268 (1970); **64.** J Insect Physiol 15, 2273 (1969); **65.** Biochem J 65, 389 (1957); **66.** Nature 179, 1190 (1957); **67.** Biochem Biophys Res Comm 208, 886 (1995); **68.** J Clin Microbiol 34, 376 (1996); **69.** Phytochemistry 11, 1947 (1972); **70.** Proc Natl Acad Sci USA 68, 2810 (1971); **71.** Cell Biochem Function 11, 167 (1993); **72.** Biochemistry 34, 14547 (1995); **73.** Zbl Bakt (Naturwiss) I Abt Orig 202, 97 (1967); **74.** J Biol Chem 268, 9337 (1993); **75.** J Clin Microbiol 27, 1719 (1989); **76.** Zbl Bakt 272, 191 (1989); **77.** J Biol Chem 269, 2283 (1994); **78.** Biochem J 63, 39 (1956); **79.** Crit Rev Biochem Mol Biol 25, 385 (1990); **80.** J Appl Bacteriol 74, 223 (1993); **81.** Arch Biochem Biophys 286, 394 (1991); **82.** Cell Signalling 3, 625 (1991); **83.** Plant Sci 78, 73 (1991); **84.** J Immunol Meth 100, 211 (1987); **85.** EMBO J 6, 3901 (1987); **86.** BioTechniques 19, 352 (1995); **87.** Appl Environ Microbiol 59, 3534 (1993); **88.** Can J Microbiol 34, 690 (1988); **89.** Biochemistry 17, 385 (1978); **90.** Chem Pharm Bull 8, 239 (1960); **91.** J Chromatography 527, 59 (1990); **92.** BioTechniques 8, 39 (1990); **93.** Plant Mol Biol Reporter 5, 387 (1987); **94.** Appl Environ Microbiol 59, 4378 (1993); **95.** Appl Environ Microbiol 50, 1383 (1985); **96.** J Clin Microbiol 13, 483 (1981); **97.** Int'l J Biochem 19, 395 (1987); **98.** Clin Chim Acta 48, 335 (1973); **99.** Clin Chim Acta 47, 9 (1973); **100.** Vet Pathol 15, 141 (1978); **101.** Arch Neurol 34, 45 (1977); **102.** Res Comm Chem Pathol Pharmacol 12, 499 (1975).

Some Common Fluorogenic β-Galactosidase Substrates

Fluorescein Digalactoside

Probably the most sensitive substrate for detecting β-galactosidase is fluorescein di-β-D-galactopyranoside (FDG, F-1179). Nonfluorescent FDG is sequentially hydrolyzed by β-galactosidase, first to fluorescein monogalactoside (FMG) and then to highly fluorescent fluorescein. Enzyme-mediated hydrolysis of FDG can be followed by the increase in either absorbance or fluorescence. Although the turnover rates of FDG and its analogs are considerably slower than that of the common spectrophotometric galactosidase substrate, 2-nitrophenyl β-D-galactopyranoside [15,16] (ONPG, N-8431), the absorbance of fluorescein is about fivefold greater than that of *o*-nitrophenol. Moreover, fluorescence-based measurements can be several orders-of-magnitude more sensitive than absorbance-based measurements. Fluorescence assays employing FDG are also reported to be 100- to 1000-fold more sensitive than radioisotope-based ELISAs.[17]

In addition to their use in ELISAs, the FDG substrate has proven very effective for identifying *lacZ*-positive cells with fluorescence microscopy [18-21] and flow cytometry.[22-27] FDG has been employed to identify cells infected with recombinant herpesvirus,[28] to detect unique patterns of β-galactosidase expression in live transgenic zebrafish embryos [18] and to monitor β-galactosidase expression in bacteria.[29-31] The purity of FDG and its analogs is very important because an extremely low fluorescence background of the reagent is necessary for most applications. Our stringent quality control ensures that the fluorescent contamination is less than 50 ppm.

The FluoReporter® *lacZ* Flow Cytometry Kits (F-1930, 50-test kit; F-1931, 250-test kit) provide materials and protocols for quantitating β-galactosidase activity with FDG in single cells using flow cytometry. These kits are accompanied by a license to practice patented techniques for loading FDG by hypotonic shock and improving retention of fluorescein in *lacZ*-positive cells.[32] In addition to a detailed protocol, each FluoReporter *lacZ* Flow Cytometry Kit contains convenient premixed solutions of:

- FDG
- Phenylethyl β-D-thiogalactopyranoside (PETG; also available separately as a solid, see P-1692 in Section 10.6), a broad-spectrum galactosidase inhibitor for stopping the reaction [33]
- Chloroquine diphosphate for inhibiting acidic hydrolysis of the substrate
- Propidium iodide for detecting dead cells

Sufficient reagents are provided for 50 (F-1930) or 250 (F-1931) flow cytometry assays. This flow cytometry assay enables researchers to detect heterogeneous expression patterns and to sort and clone individual cells expressing known quantities of β-galactosidase. Practical reviews on using FDG for flow cytometric analysis and sorting of *lacZ*-positive cells are available.[34,35]

FDG's fluorescent hydrolysis product, fluorescein (F-1300, see Section 10.1), rapidly leaks from cells under physiological conditions, making FDG's use problematic for prolonged studies. Our DetectaGene and ImaGene substrates have been specifically designed to improve retention of the fluorescent products in cells (see below).

Resorufin Galactoside

Unlike FDG, resorufin β-D-galactopyranoside (R-1159) requires only a single-step hydrolysis reaction to attain full fluorescence.[36] This substrate is especially useful for sensitive enzyme measurements in ELISAs.[14,37] The relatively low pK_a (~6.0) of its hydrolysis product, resorufin (R-363, see Section 10.1), permits its use for continuous measurement of enzymatic activity at physiological pH. Resorufin galactoside has also been used to quantitate β-galactosidase activity in single yeast cells by flow cytometry [38] and to detect immobilized β-galactosidase activity in bioreactors.[39,40]

DDAO Galactoside

Although DDAO-based substrates are intrinsically fluorescent (excitation/emission ~460/610 nm), β-galactosidase–catalyzed hydrolysis of DDAO galactoside [41] (D-6488) liberates the much longer-wavelength DDAO fluorophore (excitation/emission ~645/660 nm) (Figure 10.2). Not only can DDAO (H-6482, see Section 10.1) be excited without interference from the substrate, but its fluorescence emission is detected at wavelengths that are well beyond the autofluorescence exhibited by most biological samples. The low pK_a of DDAO (~5.5) permits continuous monitoring of β-galactosidase activity at physiological pH.

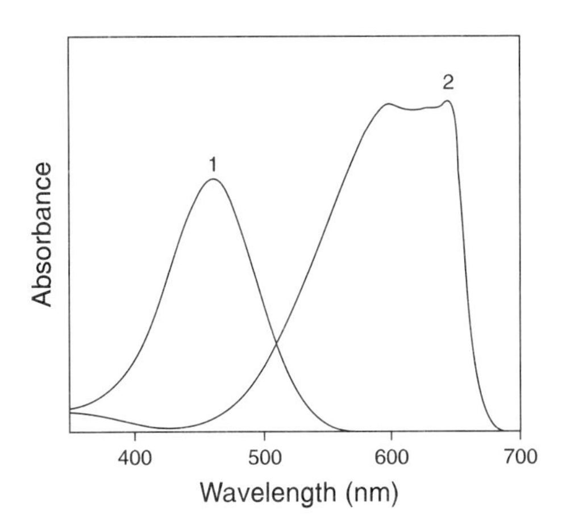

Figure 10.2 *Absorption spectra of 1) DDAO galactoside (D-6488) and 2) DDAO (H-6482) at equal concentrations in pH 9 aqueous buffer. These spectra show the large spectral shift accompanying enzymatic cleavage of DDAO-based substrates.*

Methylumbelliferyl Galactoside and Its Trifluoromethylated Analog

The fluorogenic β-galactosidase substrate β-methylumbelliferyl β-D-galactopyranoside (MUG, M-1489) is commonly used to detect β-galactosidase activity in cell extracts,[42-44] lysosomes [45] and human blood serum.[46] However, MUG's hydrolysis product, 7-hydroxy-4-methylcoumarin (β-methylumbelliferone, H-189, see Section 10.1), has a relatively high pK_a (~7.8), precluding its use for continuous measurement of enzymatic activity. The lower pK_a of the hydrolysis product of β-trifluoromethylumbelliferyl β-D-galactopyranoside (T-657), β-trifluoromethylumbelliferone (T-659, see Section 10.1), allows its detection at physiological pH.[47,48] β-Trifluoromethylumbelliferone also has longer-wavelength excitation and emission spectra and stronger absorbance than β-methylumbelliferone, which may be advantageous when detecting β-galactosidase activity in cell extracts.[49]

Carboxyumbelliferyl Galactoside and the FluoReporter *lacZ*/Galactosidase Quantitation Kit

Hydrolysis of 3-carboxyumbelliferyl β-D-galactopyranoside (CUG, C-1488) by β-galactosidase yields 7-hydroxycoumarin-3-carboxylic acid [50] (H-185, see Section 10.1). 7-Hydroxycoumarin has a pK_a below the pH at which the turnover rate is optimal, facilitating the use of CUG for continuous measurements of β-galactosidase activity. In addition, CUG is very water soluble and can be used over a wide range of concentrations in enzymatic activity measurements.[51,52] Our FluoReporter *lacZ*/Galactosidase Quantitation Kit (F-2905) provides a CUG-based method for quantitating β-galactosidase activity in ELISAs or *lacZ*-positive cell extracts. Each kit contains:

- CUG [50]
- 7-Hydroxycoumarin-3-carboxylic acid
- Detailed protocol suitable for any fluorescence microplate reader

Sufficient reagents are provided for 300 β-galactosidase assays. We have demonstrated a practical detection limit of ~0.5 pg of β-galactosidase using this kit and a CytoFluor® 2300 (PerSeptive Biosystems, Inc.) fluorescence microplate reader.

Modified Fluorogenic β-Galactosidase Substrates with Improved Cellular Retention

The primary problems associated with detecting *lacZ* expression in live cells using fluorogenic substrates are:

- Difficulty in loading the substrates under physiological conditions
- Leakage of the fluorescent product from live cells
- High levels of endogenous β-galactosidase activity in many cells

Our DetectaGene and ImaGene Kits are designed to improve the sensitivity of β-galactosidase assays by yielding products that are better retained in viable cells and, in the case of the ImaGene Kits, by providing substrates that can be passively loaded into live cells. High levels of endogenous β-galactosidase activity remain an obstacle when detecting low levels of *lacZ* expression.

DetectaGene *lacZ* Gene Expression Kits

The substrates in our DetectaGene Green and Blue *lacZ* Gene Expression Kits (D-2920, D-2921) — 5-chloromethylfluorescein di-β-D-galactopyranoside (CMFDG) and 4-chloromethylcoumarin-7-yl β-D-galactopyranoside (CMCG), respectively — are galactose derivatives that have been chemically modified to include a mildly thiol-reactive chloromethyl group (Figure 10.3). Once loaded into the cell by microinjection, hypotonic shock or other technique (see Table 15.1 in Chapter 15), the DetectaGene substrate undergoes two reactions: 1) its galactose moieties (two per molecule for CMFDG, one for CMCG) are cleaved by intracellular β-galactosidase and 2) either simultaneously or sequentially, its chloromethyl moiety reacts with glutathione and possibly other intracellular thiols to form a membrane-impermeant peptide–fluorescent dye adduct [53] (Figure 10.3). Because peptides do not readily cross the plasma membrane, the resulting fluorescent adduct is much better retained than the free dye, even in cells that have been kept at 37°C. We have found that *lacZ*-positive cells loaded from medium containing 50 µM of CMFDG are as fluorescent as those loaded with 2 mM FDG. Furthermore, unlike the free dye, the peptide–fluorescent dye adducts contain amino groups and can therefore be covalently linked to surrounding biomolecules by fixation with formaldehyde or glutaraldehyde. This property permits long-term storage of the labeled cells or tissue and, in cases where the anti-dye antibody is available (see Section 7.3), amplification of the conjugate by standard immunohistochemical techniques.

The CMFDG substrate in our DetectaGene Green *lacZ* Gene Expression Kit was used to stain *lacZ*-expressing floor plate cells in tissue dissected from a developing mouse embryo [54] and to identify *lacZ*-enhancer–trapped *Drosophila* neurons in culture. In the latter study, the neurons' fluorescence could still be visualized 24 hours after dye loading, and the fluorescent CMFDG-loaded neurons exhibited a normal pattern and time course of axonal outgrowth and branching.[55] CMFDG also has been microinjected into primary hepatocytes, fibroblasts and glioma cells to detect β-galactosidase activity [56] and has been incorporated into an electrophysiological recording pipette to confirm the identity of neurons cotransfected with the *lacZ* gene and a second gene encoding Ca^{2+}/calmodulin-dependent protein kinase II [57] (CaM kinase II).

The DetectaGene Green CMFDG and DetectaGene Blue CMCG *lacZ* Gene Expression Kits contain:

- DetectaGene Green CMFDG *or* DetectaGene Blue CMCG (Figure 10.3)
- Phenylethyl β-D-thiogalactopyranoside (PETG; also available separately as a solid, see P-1692 in Section 10.6), a broad-spectrum galactosidase inhibitor for stopping the reaction [33]
- Verapamil for inhibiting product efflux [58] (DetectaGene Green CMFDG Kit only)
- Chloroquine diphosphate for inhibiting acidic hydrolysis of the substrate
- Propidium iodide for detecting dead cells
- Detailed protocol for detecting β-galactosidase activity

When used at the recommended dilutions, sufficient reagents are provided for approximately 200 flow cytometry tests with the DetectaGene Green CMFDG Kit or 50 flow cytometry tests with the DetectaGene Blue CMCG Kit. Verapamil is a recent addition to the DetectaGene Green CMFDG *lacZ* Gene Expression Kit because we have observed that cell retention of the fluorescent dye–peptide adduct can be considerably improved in many cell types by adding verapamil to the medium.[58]

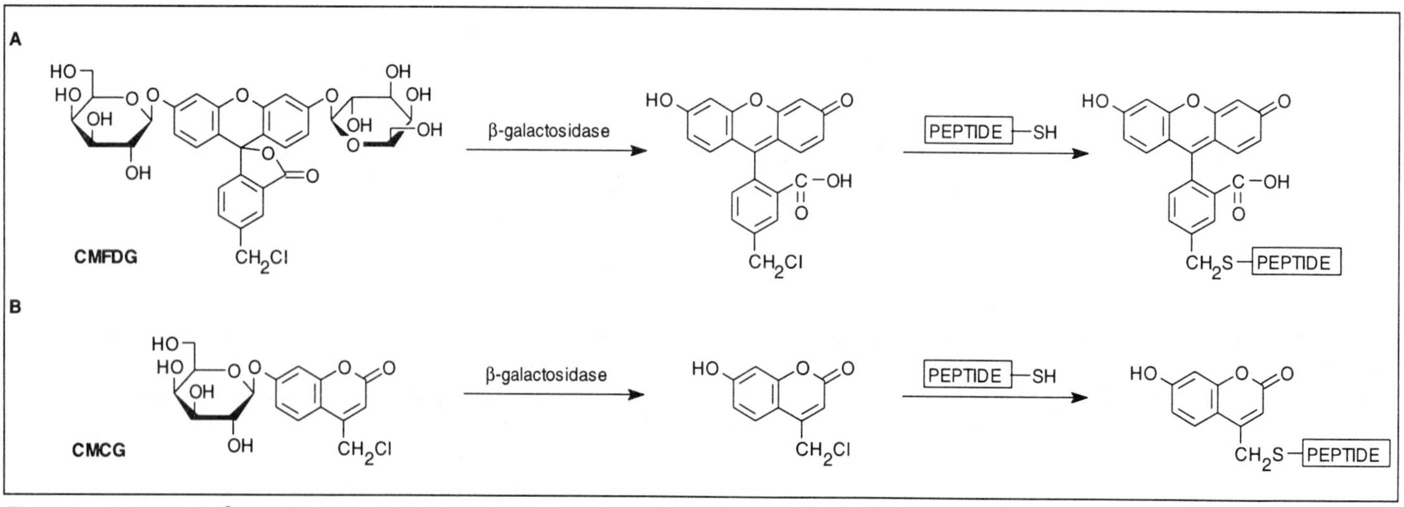

Figure 10.3 *Sequential β-galactosidase hydrolysis and peptide conjugate formation of A) CMFDG (D-2920) and B) CMCG (D-2921).*

ImaGene *lacZ* Gene Expression Kits

The fluorescein- and resorufin-based galactosidase substrates in our ImaGene Green™ and ImaGene Red™ *lacZ* Gene Expression Kits (I-2904, I-2906) have been covalently modified to include a 12-carbon lipophilic moiety. Unlike FDG or CMFDG, these lipophilic fluorescein- and resorufin-based substrates — abbreviated C_{12}FDG and C_{12}RG for the ImaGene Green and ImaGene Red substrate, respectively — can be loaded simply by adding the substrate to the aqueous medium in which the cells or organisms are growing, either at ambient temperatures or at 37°C. Once inside the cell, the substrates are cleaved by β-galactosidase, producing fluorescent products that are retained by the cells, probably by incorporation of their lipophilic tails within the cellular membranes. Mammalian NIH 3T3 *lacZ*-positive cells grown for several days in medium containing 60 µM C_{12}FDG appear morphologically normal, continue to undergo cell division and remain fluorescent for up to three cell divisions after replacement with substrate-free medium.[53,59]

A recent report concludes that the C_{12}FDG substrate in our ImaGene Green *lacZ* Expression Kits is superior to FDG for flow cytometric detection of β-galactosidase activity in live mammalian cells.[60] Using C_{12}FDG with flow cytometric methods, researchers have:

- Assessed levels of *lacZ* gene expression in recombinant Chinese hamster ovary (CHO) cells throughout the cell cycle, which was monitored with Hoechst 33342 [61] (H-1399, see Section 8.1)
- Identified endocrine cell precursors in dissociated fetal pancreatic tissue based on their high levels of endogenous acid β-galactosidase [62]
- Measured β-galactosidase activity in single recombinant *E. coli* bacteria [63]
- Sorted β-galactosidase–expressing mouse sperm cells [64] and insect cells that harbor recombinant baculovirus [65,66]

The C_{12}FDG substrate has also proven useful for a fluorescence microscopy study of the expression of a *lacZ* reporter gene under the control of a mammalian homeobox gene promoter in mosaic transgenic zebrafish.[67] In addition, lipophilic β-galactosidase substrates have been employed to diagnose the deficiency in β-galactocerebrosidase activity that typifies Krabbe disease in human patients.[7,68] In some cell types, C_{12}FDG has been reported to produce high levels of background fluorescence that may prohibit its use in assaying low β-galactosidase expression.[20]

Molecular Probes' ImaGene Green C_{12}FDG or ImaGene Red C_{12}RG *lacZ* Gene Expression Kits contain:

- ImaGene Green C_{12}FDG *or* ImaGene Red C_{12}RG
- Phenylethyl β-D-thiogalactopyranoside (PETG; also available separately as a solid, see P-1692 in Section 10.6), a broad-spectrum galactosidase inhibitor for stopping the reaction [33]
- Chloroquine diphosphate for inhibiting acidic hydrolysis of the substrate
- Detailed protocol for detecting β-galactosidase activity

Sufficient reagents are provided for 100–200 assays, depending on the volume used for each experiment.

We also offer the somewhat less lipophilic 5-octanoylaminofluorescein di-β-D-galactopyranoside (C_8FDG, O-2892). The C_8FDG analog was recently shown to be optimal for investigating the expression of *lacZ* fusion genes in sporulating cultures of *Bacillus subtilis*.[69]

Fluorogenic β-Glucuronidase Substrates

The substrate 4-methylumbelliferyl β-D-glucuronide (MUGlcU, M-1490) is probably the most commonly used fluorogenic reagent for identifying *E. coli* contamination and for detecting reporter GUS gene expression in plants and plant extracts.[70,71] However, β-glucuronidase substrates based on fluorescein, resorufin and carboxyumbelliferone may be much more sensitive and yield products that are fluorescent at physiological pH, making them useful for continuous monitoring of enzymatic activity. In addition, we offer a fluorogenic ELF-97 β-D-glucuronidase substrate that produces an intensely green fluorescent precipitate at the site of enzymatic activity that can clearly be distinguished from most autofluorescence.

Fluorescein Diglucuronide

Fluorescein di-β-D-glucuronide (FDGlcU, F-2915) is colorless and nonfluorescent until it is hydrolyzed to the monoglucuronide and then to the highly fluorescent fluorescein (F-1300, see Section 10.1). FDGlcU has been used to detect β-glucuronidase activity in plant extracts containing the GUS reporter gene [71] and may also be useful for assaying lysosomal enzyme release from neutrophils.[72,73]

Resorufin Glucuronide

β-Glucuronidase–catalyzed hydrolysis of resorufin β-D-glucuronide (R-1161) yields the visible-wavelength excitable resorufin fluorophore [74] (R-363, see Section 10.1). Unlike the fluorescein-based substrates, hydrolysis of resorufin glucuronide requires only a single-step hydrolysis reaction to attain full fluorescence.

DDAO Glucuronide

DDAO β-D-glucuronide [41] (DDAO GlcU, D-6486) forms the longest-wavelength fluorescent product of all the β-glucuronidase substrates, a property that may facilitate GUS detection in highly autofluorescent samples such as plant extracts. As with resorufin-based substrates, only a single hydrolysis step is required to generate the highly fluorescent product.

Coumarin Glucuronides

The glucuronide 4-methylumbelliferyl β-D-glucuronide (MUGlcU, M-1490) has been used extensively to detect *E. coli* in food,[75,76] water,[77] urine [78] and environmental samples.[79] MUGlcU is stable to the conditions required for sterilization of media. A fluorogenic bioassay using MUGlcU has been developed to assess the detrimental effect of Li^+, Al^{3+}, Cr^{6+} and Hg^{2+} on the proliferation of *E. coli*.[80]

MUGlcU is also commonly used to identify plant tissue expressing the GUS reporter gene,[74,81,82] including nondestructive assays that allow propagation of the transformed plant lines.[84,85] In addition, MUGlcU has served as a sensitive substrate for lysosomal enzyme release from neutrophils.[72,73]

β-Trifluoromethylumbelliferyl β-D-glucuronide (T-658) and 3-carboxyumbelliferyl β-D-glucuronide (CUGlcU, C-1492) yield products that have lower pK_as than the product of MUGlcU hydrolysis,[47,48,74] allowing continuous measurement of enzymatic activity at physiological pH. The products of these substrates also exhibit longer-wavelength excitation and emission spectra, which can be advantageous for cells that have high levels of endogenous fluorescence such as plant cells.

ImaGene Green and Red β-D-Glucuronidase Substrates

Molecular Probes also offers lipophilic analogs of fluorescein di-β-D-glucuronide and resorufin β-D-glucuronide in our ImaGene Green and ImaGene Red GUS Gene Expression Kits (I-2908, I-2910), respectively. We have shown that these lipophilic substrates freely diffuse across the membranes of viable cultured tobacco leaf cells or protoplasts under physiological conditions. Furthermore, the fluorescent cleavage products are retained by the plant cell for hours to days, facilitating long-term measurements of GUS expression. Molecular Probes' ImaGene Green C_{12}FDGlcU and ImaGene Red C_{12}RGlcU GUS Gene Expression Kits contain:

- ImaGene Green C_{12}FDGlcU *or* ImaGene Red C_{12}RGlcU
- D-Glucaric acid-1,4-lactone, a β-glucuronidase inhibitor for stopping the reaction
- Detailed protocol for detecting β-glucuronidase activity

Sufficient reagents are provided for approximately 100 tests, depending on the volume used for each experiment.

ELF-97 β-D-Glucuronide

Molecular Probes' ELF-97 β-D-glucuronidase substrate (ELF-97 β-D-glucuronide, E-6587) may be the ideal substrate for analyzing GUS expression in transgenic plants. Upon hydrolysis, this fluorogenic substrate produces a bright yellow-green fluorescent precipitate at the site of enzymatic activity. This fluorescent precipitate has some unique spectral characteristics that make it easily distinguishable from the endogenous fluorescent components commonly found in plants, including an extremely large Stokes shift (see Section 6.3 for a description of our patented ELF technology). We have been using this important new substrate to detect GUS expression in *Arabidopsis* and have found that, after only four hours incubation, signal can be detected in whole-leaf cuttings from GUS-positive plants.[85] Homogenization of a small portion of a leaf from a GUS-positive *Arabidopsis* plant followed by separation on a nondenaturing gel yields a discrete band corresponding to the glucuronidase enzyme (Figure 10.4). We have been able to detect as little as 0.5 ng of purified β-glucuronidase in a nondenaturing gel incubated with ELF-97 glucuronide. This substrate may also be useful for detecting GUS fusion proteins in gels, for identifying *E. coli* in agarose-containing medium [86] and for assaying lysosomal enzyme release from neutrophils.[72,73]

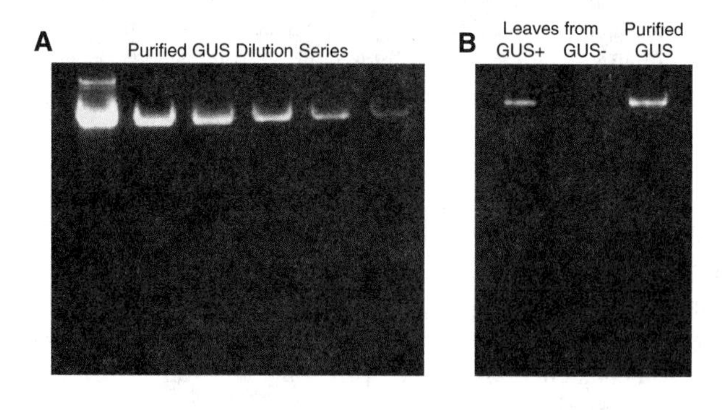

Figure 10.4 *In situ gel assay of β-D-glucuronidase (GUS) with ELF-97 β-D-glucuronide. A) Twofold dilutions of purified GUS enzyme or B) cell extracts from single leaves from GUS-positive and -negative Arabidopsis plants were electrophoresed through a native 7.5% polyacrylamide gel. Following electrophoresis, the gel was washed with 0.1 M sodium phosphate, pH 7.0 containing 0.2% Triton® X-100 at room temperature for 60 minutes and then incubated with 15 µM ELF-97 β-D-glucuronide in 0.1 M sodium phosphate, pH 7.0 at 37°C for 30–60 minutes. The gel was photographed using 300 nm transillumination, a SYBR Green gel stain photographic filter (S-7569, see Section 8.3) and Polaroid® 667 black-and-white print film.*

Fluorogenic β-Glucosidase Substrates

β-Glucosidase, which is a marker for the endoplasmic reticulum, is present in nearly all species. Its natural substrate is a glucosylceramide (see Section 20.6). People with Gaucher's disease have mutations in the acid β-glucosidase gene that result in abnormal lysosomal storage.[87,88] Plant β-glucosidases are implicated in a variety of key metabolic events and growth-related responses.[89]

Fluorescein Diglucoside

As with the other fluorescein diglycosides, Molecular Probes' fluorogenic fluorescein di-β-D-glucopyranoside (FDGlu, F-2881) is likely to yield the greatest sensitivity for detecting β-glucosidase activity in both cells [87] and cell extracts. This substrate has been used to demonstrate the utility of *Saccharomyces cerevisiae* and *Candida albicans* exo-1,3-β-glucanase genes as reporter genes.[90] Because these reporter genes encode secreted proteins, assays for reporter gene expression do not require cell permeabilization.

Resorufin Glucoside

Resorufin β-D-glucopyranoside (R-1160) has served as a sensitive fluorogenic substrate for ELISAs and for the study of β-glucosidase deficiency in Gaucher's disease.[87]

Methylumbelliferyl Glucoside

The fluorogenic β-glucosidase substrate 4-methylumbelliferyl β-D-glucopyranoside (MUGlu, M-8423) has been used to differentiate bacteria,[2,91,92] including streptococci.[93] Like resorufin glucoside, this substrate has also been employed in the study of Gaucher's disease.[94,95]

Substrates for *N*-Acetylglucosaminidase, Neuraminidase, Chitinase and Dextranase

The activity of urinary *N*-acetyl-β-D-glucosaminidase provides a sensitive measure of renal health.[96,97] *N*-acetyl-β-D-glucosaminidase assays have also been used to detect the *Candida albicans* yeast pathogen.[98]

We now offer 6'-(*O*-(carboxymethyl))-2',7'-dichlorofluorescein 3'-(*O*-(*N*-acetyl-β-D-glucosaminide) (CM-DCF-NAG, C-6541), a fluorogenic substrate that is soluble at the acidic pH required for optimal activity. Even with absorbance measurements, CM-DCF-NAG is reported to be 2.7 times more sensitive than the conventional chromogenic substrate 2-chloro-4-nitrophenyl *N*-acetyl-β-D-glucosaminide.[99] Unlike the fluorescein diglycosides, this fluorescein-based substrate requires only a single hydrolytic step to yield maximal fluorescence. CM-DCF-NAG may also be useful for assaying chitinase; chitin is a linear polymer of β-(1→4)-*N*-acetylglucosamine present in diatoms, fungi, protozoans, arachnids, insects, crustaceans, nematodes and other invertebrates. Resorufin *N*-acetyl-β-D-glucosaminide (RNAG, R-6540) is a similarly useful *N*-acetylglucosaminidase or chitinase substrate [100] that yields the longer-wavelength dye resorufin (R-363, see Section 10.1).

EnzChek Dextranase Assay Kit

Dextranase (α-1,6-glucan-6-glucanohydrolase; EC 3.2.1.11) is an enzyme produced by bacteria [101] and fungi [102,103] that hydrolyzes the α-1,6-glycosidic linkages of the hydrophilic polysaccharide dextran, also known as glucan. This enzyme appears to be involved in the ability of oral streptococci to participate in the production of dental plaque and subsequent formation of caries.[101] The genes from *S. salivarius, S. sobrinus* and *Arthrobacter globiformis* have been cloned and sequenced.[101,104-106] A bacterial gene encoding an inhibitor of dextranase activity has also been cloned, and the role of this inhibitor in cariogenicity is under investigation.[106-108]

Current methods for detecting dextranase activity, such as an assay for the hydrolysis of Blue Dextran 2000 [109] or the use of dinitrosalicylic acid color reagent,[110] require large volumes of reagents and cumbersome manipulations. Our EnzChek Dextranase Assay Kit (E-6644) provides a simple and sensitive continuous assay for dextranase activity. This kit contains a dextran derivative that is heavily labeled with the pH-insensitive green fluorescent BODIPY FL dye, resulting in almost total quenching of the conjugate's fluorescence (<5% of the fluorescence of the corresponding free dyes). Dextranase-catalyzed hydrolysis of the BODIPY FL–dextran substrate releases highly fluorescent BODIPY FL–labeled fragments. The accompanying increase in fluorescence, which can be measured using a spectrofluorometer, minifluorometer or fluorescence microplate reader and standard fluorescein excitation/emission wavelengths, serves as a convenient measure of dextranase activity. Each EnzChek Dextranase Assay Kit contains:

- BODIPY FL–labeled 70,000 MW dextran
- Concentrated reaction buffer
- Dextranase for calibrating a standard curve
- Protocol for quantitating dextranase activity

This kit provides sufficient reagents for ~100 assays using 2 mL assay volumes and standard fluorescence cuvettes or ~1000 assays using 200 µL assay volumes and 96-well microplates. This single-step assay, which we developed using a dextranase from *Penicillium*, allows researchers to detect as few as 10^{-4} U/mL of dextranase activity and is potentially useful for screening dextranase inhibitors.

Chromogenic Glycosidase Substrates

Molecular Probes offers a wide variety of chromogenic glycosidase substrates, including those derived from the yellow colored 4-nitrophenol chromophore (N-6477, see Section 10.1) or 5-bromo-4-chloro-3-indolyl derivatives (Table 10.1), all available at very competitive prices.

β-Galactosidase Substrates

The widely used β-galactosidase substrate — 5-bromo-4-chloro-3-indolyl β-D-galactopyranoside (X-Gal, B-1690) — yields a dark blue precipitate at the site of enzymatic activity. X-Gal is useful for numerous histochemical and molecular biology applications, including detection of *lacZ* activity in cells and tissues. In addition to X-Gal, Molecular Probes offers 5-bromo-6-chloro-3-indolyl (B-8407), 6-chloro-3-indolyl [111,112] (C-8411), 5-iodoindolyl (I-8420), *N*-methyl-3-indolyl (M-8421) and indoxyl (I-8414) β-D-galactopyranosides. These form magenta, pink, purple, green and blue colored precipitates, respectively. Our Indolyl β-D-Galactopyranoside Sampler Kit (I-6614) contains 5 mg samples of four of these special indolyl galactosides (C-8411 is not included) plus X-Gal (B-1690).

β-Glucuronidase Substrates

The chromogenic substrate 5-bromo-4-chloro-3-indolyl β-D-glucuronic acid (X-GlcU), which is available as either its cyclohexylammonium or sodium salt (B-1691, B-8404), forms a dark blue precipitate. X-GlcU is routinely used to detect GUS expression in transformed plant cells and tissues.[9,113-119] However, because it is relatively difficult to differentiate the blue color of the product of X-GlcU against the dark-green chloroplasts,[120] we also offer the isomeric 5-bromo-6-chloro-3-indolyl β-D-glucuronide [121] (B-8408), 6-chloro-3-indolyl β-D-glucuronide (C-8412) and indoxyl β-D-glucuronide [122] (I-8416), which form magenta, pink and blue colored precipitates, respectively. X-GlcU can also be used to detect *E. coli* contamination in food and water.[86,123]

Auxiliary Products for Glycosidase Research

Phenylethyl β-D-Thiogalactopyranoside (PETG)

Phenylethyl β-D-thiogalactopyranoside (PETG, P-1692; see Section 10.6) is a cell-permeant inhibitor of β-galactosidase activity.[33,124] We provide PETG in our FluoReporter, DetectaGene and ImaGene *lacZ* Gene Expression Kits for stopping the enzymatic reaction.

Isopropyl β-D-Thiogalactopyranoside (IPTG)

Isopropyl β-D-thiogalactopyranoside (IPTG, I-6621; see Section 10.6) is an inhibitor of the *lac* repressor, thereby producing enhanced expression from the *Escherichia coli* promoter.[125,126] It may be useful in conjunction with X-Gal or one of our diverse fluorogenic β-galactosidase substrates for reporter gene detection.

β-Galactosidase and Its Streptavidin Conjugate

Molecular Probes also offers β-galactosidase (G-6463) and its streptavidin conjugate (S-931), a reagent used in a variety of ELISAs.[127] Streptavidin β-D-galactosidase reportedly provided enhanced sensitivity over that obtained with the avidin horseradish peroxidase in the detection of a variety of mammalian interleukins and their receptors by ELISA.[128] This reagent has also been used in fluorometric-reverse (IgE-capture) [129] and fluorescence-sandwich [130] ELISAs.

Anti–β-Glucuronidase Antibody

In combination with a fluorophore- or enzyme-labeled anti–rabbit IgG secondary antibody, our anti–β-glucuronidase antibody (A-5790, see Section 7.4) can be used to detect the GUS enzyme in transformed plant tissue using Western blotting or immunohistochemical techniques.[131,132] Furthermore, this antibody, which is raised in rabbits against *E. coli* type X-A β-glucuronidase, can be immobilized in microplate wells in order to capture GUS enzyme from cell lysates.[133] The enzymatic activity can subsequently be determined using any of our fluorogenic or chromogenic β-glucuronidase substrates.[134]

1. J Clin Microbiol 30, 1402 (1992); **2.** Microbiol Rev 55, 335 (1991); **3.** Meth Microbiol 19, 105 (1987); **4.** Nature 369, 113 (1994); **5.** J Cell Biol 110, 309 (1990); **6.** Anal Biochem 148, 50 (1985); **7.** Clin Chim Acta 205, 87 (1992); **8.** Crit Rev Biochem Mol Biol 25, 385 (1990); **9.** *GUS Protocols: Using the GUS Gene as a Reporter of Gene Expression*, S.R. Gallagher, Ed., Academic Press (1992); **10.** Mol Gen Genet 216, 321 (1989); **11.** Mol Gen Genet 215, 38 (1988); **12.** Plant Mol Biol 1, 387 (1987); **13.** Proc Natl Acad Sci USA 83, 8447 (1986); **14.** J Immunol Meth 150, 5 (1992); **15.** J Immunol Meth 150, 23 (1992); **16.** J Immunol Meth 48, 133 (1982); **17.** Exp Parasitol 73, 440 (1991); **18.** Devel Biol 161, 77 (1994); **19.** Mol Microbiology 13, 655 (1994); **20.** J Neurosci 13, 1418 (1993); **21.** Eur J Immunol 19, 1619 (1989); **22.** Cytometry 17, 216 (1994); **23.** Eur J Cell Biol 62, 324 (1993); **24.** J Biol Chem 268, 9762 (1993); **25.** Neuron 9, 1117 (1992); **26.** Science 251, 81 (1991); **27.** Proc Natl Acad Sci USA 85, 2603 (1988); **28.** J Virol Meth 44, 99 (1993); **29.** Mol Microbiol 13, 655 (1994); **30.** Proc Natl Acad Sci USA 90, 8194 (1993); **31.** Appl Environ Microbiol. 56, 3861 (1990); **32.** Purchase of the FluoReporter *lacZ* Flow Cytometry Kits is accompanied by a research license under U.S. Patent No. 5,070,012; **33.** Carbohydrate Res 56, 153 (1977); **34.** Methods: Companion to Meth Enzymol 2, 261 (1991); **35.** Methods: Companion to Meth Enzymol 2, 248 (1991); **36.** Anal Chim Acta 163, 67 (1984); **37.** Oncogene 10, 2323 (1995); **38.** Cytometry 9, 394 (1988); **39.** Annals NY Acad Sci 613, 333 (1990); **40.** Anal Chim Acta 213, 245 (1988); **41.** DDAO-based substrates are licensed to Molecular Probes under U.S. Patent No. 4,810,636 and its foreign equivalents; **42.** Anal Biochem 215, 24 (1993); **43.** Neuron 10, 427 (1993); **44.** Proc Natl Acad Sci USA 84, 156 (1987); **45.** J Histochem Cytochem 33, 965 (1985); **46.** Clin Chim Acta 12, 647 (1965); **47.** Anal Lett 21, 193 (1988); **48.** J Histochem Cytochem 34, 585 (1986); **49.** Biochem Int'l 24, 1135 (1991); **50.** CUG is licensed to Molecular Probes under U.S. Patent No. 4,226,978; **51.** Infection and Immunity 61, 5231 (1993); **52.** Anal Biochem 146, 211 (1985); **53.** J Fluorescence 3, 119 (1993); **54.** Development 119, 1217 (1993); **55.** J Neurosci 15, 1025 (1995); **56.** Exp Cell Res 219, 372 (1995); **57.** Science 266, 1881 (1994); **58.** Poot, M. and Arttamangkul, S., submitted; **59.** FASEB J 5, 3108 (1991); **60.** Appl Environ Microbiol 60, 4638 (1994); **61.** Biotech Bioeng 42, 1113 (1993); **62.** J Clin Endocrinol Metab 78, 1232 (1994); **63.** Biotech Bioeng 42, 708 (1993); **64.** Proc Natl Acad Sci USA 89, 10681 (1992); **65.** Meth Cell Biol 42, 563 (1994); **66.** BioTechniques 14, 274 (1993); **67.** Genes Dev 6, 591 (1992); **68.** J Lab Clin Med 110, 740 (1987); **69.** Cytometry 20, 324 (1995); **70.** BioTechniques 8, 39 (1990); **71.** Plant Mol Biol Rep 5, 387 (1987); **72.** Cell Signalling 3, 625 (1991); **73.** J Immunol Meth 100, 211 (1987); **74.** Naleway, J.J. in *GUS Protocols: Using the GUS Gene as a Reporter of Gene Expression*, S.R. Gallagher, Ed., Academic Press (1992) pp. 61–76; **75.** J Assoc Official Anal Chem 71, 589 (1988); **76.** Appl Environ Microbiol 50, 1383 (1985); **77.** Can J Microbiol 39, 1066 (1993); **78.** J Microbiol Meth 12, 51 (1990); **79.** J Microbiol Meth 12, 235 (1990); **80.** BioTechniques 16, 888 (1994); **81.** Plant Sci 78, 73 (1991); **82.** Plant Mol Biol 15, 527 (1990); **83.** Meth Enzymol 216, 357 (1992); **84.** Plant Mol Biol Rep 10, 37 (1992); **85.** Zhou, M. *et al.*, J Biochem Biophys Meth (1996) in press; **86.** J Appl Bacteriol 74, 223 (1993); **87.** Cell Biochem Function 11, 167 (1993); **88.** Crit Rev Biochem Mol Biol 25, 385 (1990); **89.** *β-Glucosidases. Biochemistry and Molecular Biology (ACS Symposium Series Volume 533)*, A. Esen, Ed., American Chemical Society (1993); **90.** Yeast 10, 747 (1994); **91.** Zbl Bakt 280, 476 (1994); **92.** J Clin Pathol 28, 686 (1975); **93.** J Clin Microbiol 13, 483 (1981); **94.** Proc Natl Acad Sci USA 68, 2810 (1971); **95.** Eur J Biochem 7, 34 (1968); **96.** J Clin Lab Anal 5, 1 (1991); **97.** Toxicology 23, 99 (1982); **98.** J Clin Microbiol 25, 2424 (1987); **99.** Chem Pharm Bull 41, 1513 (1993); **100.** Anal Sci 8, 161 (1992); **101.** J Bacteriol 176, 3839 (1994); **102.** Biochim Biophys Acta 3094, 357 (1973); **103.** Appl Microbiol 20, 421 (1970); **104.** Gene 156, 93 (1995); **105.** J Bacteriol 176, 7730 (1994); **106.** J Bacteriol 176, 7213 (1994); **107.** J Bacteriol 177, 1703 (1995); **108.** J Bacteriol 176, 7206 (1994); **109.** Can J Biochem 48, 226 (1970); **110.** Anal Biochem 2, 127 (1960); **111.** BioTechniques 18, 434 (1995); **112.** Currently available only from Molecular Probes Europe BV, Leiden, The Netherlands; inquire about availability from Molecular Probes, Inc., Eugene, Oregon; **113.** BioTechniques 19, 106 (1995); **114.** Bio/Technology 8, 833 (1990); **115.** Plant Cell Physiol 31, 805 (1990); **116.** Science 249, 1285 (1990); **117.** Science 248, 471 (1990); **118.** Nature 342, 837 (1990); **119.** EMBO J 6, 3901 (1987); **120.** BioTechniques 7, 922 (1989); **121.** BioTechniques 19, 352 (1995); **122.** Can J Microbiol 34, 690 (1988); **123.** Lett Appl Microbiol 13, 212 (1991); **124.** Anal Biochem 199, 119 (1991); **125.** Environ Mol Mutagen 26, 16 (1995); **126.** Barkley, M.D. and Bourgeois, S. in *The Operon*, J. Miller, Ed., Cold Spring Harbor Laboratory (1978) pp. 177–220; **127.** J Immunol Meth 125, 279 (1989); **128.** Biochemistry 26, 4922 (1987); **129.** J Immunol Meth 116, 181 (1989); **130.** J Immunol Meth 110, 129 (1988); **131.** J Biol Chem 269, 17635 (1994); **132.** Plant Mol Biol 15, 821 (1990); **133.** J Clin Microbiol 32, 1444 (1994); **134.** Appl Environ Microbiol 53, 1073 (1987).

10.2 Data Table *Detecting Glycosidases*

Cat #	Structure	MW	Storage	Soluble	Abs	{ε × 10⁻³}	Em	Solvent	Product*
B-1689	✓	409	F,D	DMSO	290	{4.9}	none	H_2O	<1>
B-1690	✓	409	F,D	DMSO	290	{5.0}	none	H_2O	<1>
B-1691	✓	522	F,D	pH>6	292	{4.8}	none	MeOH	<1>
B-8400	✓	450	F,D	DMSO	290	{4.4}	none	pH 7	<1>
B-8401	✓	450	F,D	DMSO	290	{4.5}	none	pH 7	<1>
B-8404	✓	445	F,D	pH>6	290	{4.5}	none	pH 7	<1>
B-8407	✓	409	F,D	DMSO	294	{4.8}	none	pH 7	<2>
B-8408	✓	522	F,D	pH>6	294	{4.6}	none	pH 7	<2>
C-1488	✓	368	F,D	pH>6, DMSO	330	{16}	396†	pH 8	H-185‡
C-1492	✓	458	F,D	pH>6	329	{18}	395†	pH 7	H-185‡
C-6541	✓	701	F,D	pH>5, DMSO	282	{8.7}	none	MeOH	<3>
C-8411	✓	330	F,D	DMSO	285	{4.9}	none	pH 7	<4>
C-8412	✓	443	F,D	pH>6	285	{4.9}	none	pH 7	<4>
D-2893	✓	854	F,D	DMSO	289	{6.0}	none	MeOH	D-109§
D-2920	✓	705	F,D,L	<5>	273	{4.8}	none	MeOH	<6>
D-2921	✓	373	F,D,L	<5>	321	{12}	394†	MeOH	C-2111¶
D-6486	✓	484	F,D,L	DMSO	463	{26}	607	MeOH	H-6482‡
D-6488	✓	470	F,D,L	DMSO	465	{24}	608	pH 7	H-6482‡
E-6587	✓	483	F,D,L	DMSO, H_2O	302	{12}	ND†	MeOH	E-6578‡
F-1179	✓	657	F,D	DMSO#	273	{6.4}	none	MeOH	F-1300‡
F-2881	✓	657	F,D	DMSO	272	{6.2}	none	MeOH	F-1300‡
F-2915	✓	685	F,D	pH>6, DMSO	272	{5.7}	none	MeOH	F-1300‡
I-2906	✓	544	F,D,L	<7>	448	{20}	none	MeOH	D-7786§
I-2908	✓	889	F,D	<5>	290	{5.4}	none	MeOH	D-109§
I-2910	✓	558	F,D	<5>	459	ND	none	H_2O	D-7786§
I-8414	✓	295	F,D	DMSO	282	{4.3}	none	MeOH	<8>
I-8415	✓	349⊕	F,D	H_2O	282	{5.2}	none	MeOH	<8>
I-8416	✓	408	F,D	pH>6	281	{5.1}	none	MeOH	<8>
I-8420	✓	421	F,D	DMSO	290	{4.6}	none	MeOH	<9>
M-1489	✓	338	D	DMSO, H_2O	316	{14}	376†	pH 9	H-189‡

Cat #	Structure	MW	Storage	Soluble	Abs	{ε × 10^{-3}}	Em	Solvent	Product*
M-1490	✓	352	F,D	pH>6	316	{12}	375†	pH 9	H-189‡
M-8421	✓	309	F,D	DMSO	293	{4.7}	none	MeOH	<10>
M-8422	✓	338	F,D	DMSO	317	{14}	374†	pH 9	H-189‡
M-8423	✓	338	F,D	DMSO	316	{15}	372†	pH 9	H-189‡
N-8426	✓	342	F,D	DMSO	295	{11}	none	MeOH	N-6477‡
N-8427	✓	342	F,D	H_2O	295	{11}	none	MeOH	N-6477‡
N-8428	✓	463	F,D	H_2O	296	{11}	none	MeOH	N-6477‡
N-8429	✓	285	F,D	H_2O	297	{12}	none	MeOH	N-6477‡
N-8430	✓	301	F,D	H_2O	297	{12}	none	MeOH	N-6477‡
N-8431	✓	301	F,D	H_2O	302	{3.2}	none	MeOH	N-6476‡
N-8432	✓	301	F,D	H_2O	296	{12}	none	MeOH	N-6477‡
N-8433	✓	301	F,D	H_2O	296	{13}	none	MeOH	N-6477‡
N-8434	✓	301	F,D	H_2O	296	{11}	none	MeOH	N-6477‡
N-8435	✓	315	F,D	H_2O	294	{11}	none	MeOH	N-6477‡
N-8436	✓	301	F,D	DMSO	296	{12}	none	MeOH	N-6477‡
O-2892	✓	798	F,D	DMSO	289	{5.5}	none	MeOH	<11>
P-8458	✓	256	D	H_2O	266	{1.0}	none	MeOH	<12>
P-8459	✓	288⊕	D	H_2O	266	{1.0}	none	MeOH	<12>
R-1159	✓	375	F,D,L	DMSO	469	{19}	none	pH 9	R-363‡
R-1160	✓	375	F,D,L	DMSO	468	{18}	none	pH 9	R-363‡
R-1161	✓	427	F,D,L	pH>6	479	{17}	none	pH 8	R-363‡
R-6540	✓	416	F,D,L	DMSO	468	{17}	none	pH 9	R-363‡
T-657	✓	392	F,D	DMSO	325	{13}	410†	pH 9	T-659‡
T-658	✓	406	F,D	pH>6	325	{11}	410†	pH 8	T-659‡

I-2904 see D-2893

For definitions of the contents of this data table, see "How to Use the Handbook" on page vi.
Structure: Chemical structure drawing: ✓ = shown in this section.
ND = not determined.

* Catalog number or footnote description of the enzymatic cleavage product.
† Fluorescence of the unhydrolyzed substrate is very weak.
‡ See Section 10.1.
§ See Section 14.3.
¶ See Section 15.2.
\# F-1179 is soluble at 1 mM in water, but it is best to prepare a stock solution in DMSO.
⊕ MW is for the hydrated form of this product.
<1> Enzymatic cleavage of this substrate yields a water-insoluble, blue colored indigo dye (Abs ~615 nm).
<2> Enzymatic cleavage of this substrate yields a water-insoluble, magenta colored indigo dye (Abs ~565 nm).
<3> Product of *N*-acetyl-β-D-glucosaminidase hydrolysis is 6′-(*O*-carboxymethyl)-2′,7′-dichlorofluorescein, Abs = 489 nm (ε = {5.6}) in MeOH [Chem Pharm Bull 41, 1513 (1993)].
<4> Enzymatic cleavage of this substrate yields a water-insoluble, pink colored indigo dye (Abs ~540 nm).
<5> This product is packaged as a solution in 1:1 (v/v) DMSO/H_2O.
<6> Enzymatic cleavage of this substrate yields 5-chloromethylfluorescein, with spectroscopic properties similar to F-1300 (see Section 10.1).
<7> This product is packaged as a solution in 7:3 (v/v) DMSO/EtOH.
<8> Enzymatic cleavage of this substrate yields a water-insoluble, blue colored indigo dye (Abs ~680 nm).
<9> Enzymatic cleavage of this substrate yields a water-insoluble, purple colored indigo dye.
<10> Enzymatic cleavage of this substrate yields a water-insoluble, green colored indigo dye.
<11> Enzymatic cleavage of this substrate yields a 5-acylaminofluorescein derivative with spectroscopic properties similar to D-109 (see Section 14.3).
<12> Enzymatic cleavage of this substrate yields phenol.

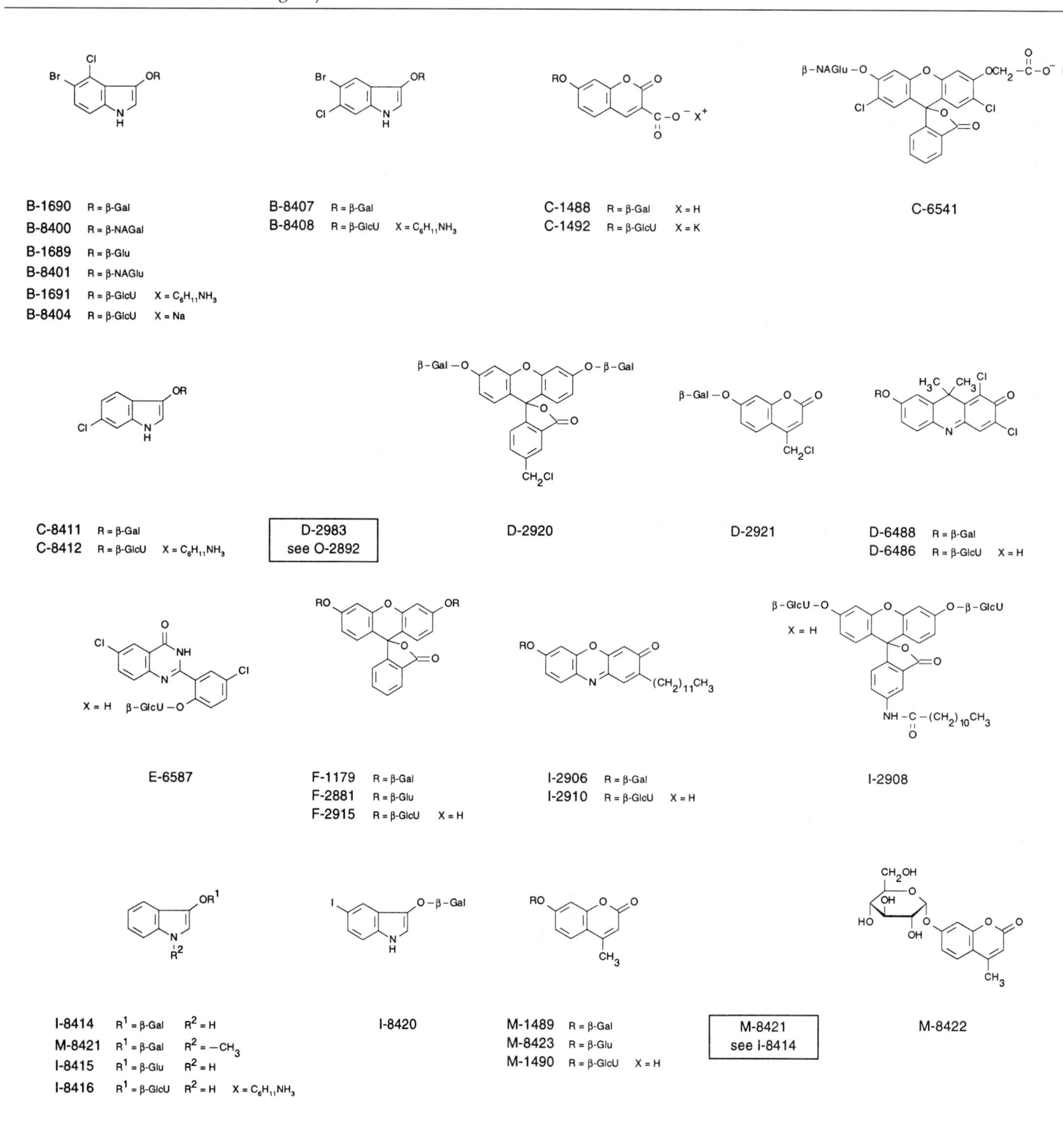

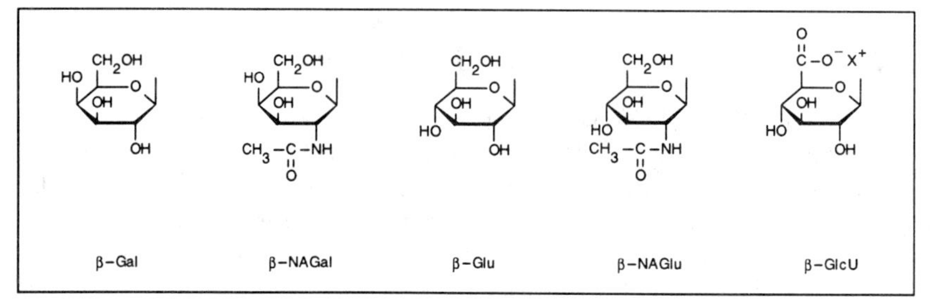

Substituent Key

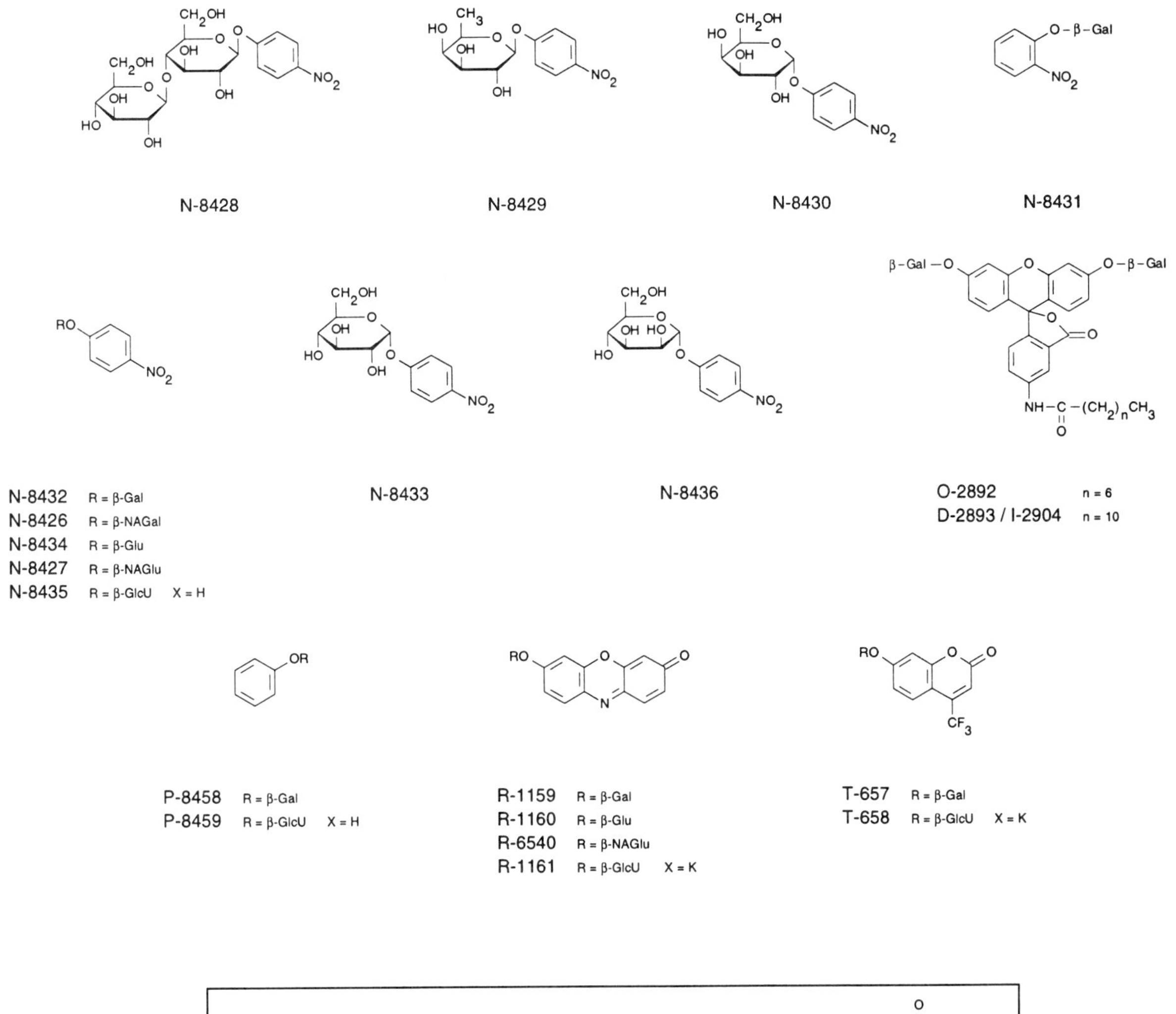

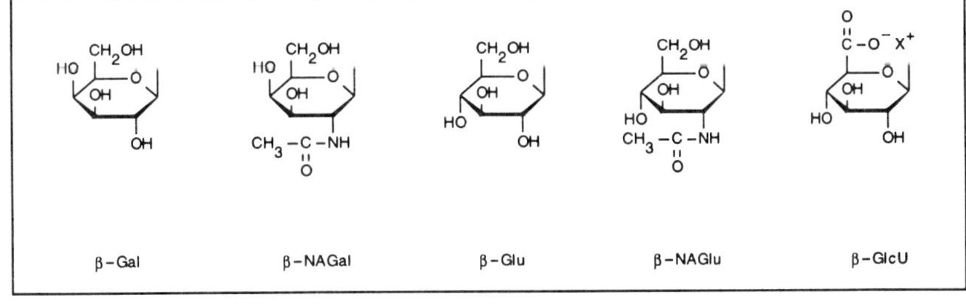

Substituent Key

10.2 Price List *Detecting Glycosidases*

Cat #		Product Name	Unit Size	Price Per Unit ($) 1–4 Units	5–24 Units
A-5790		anti-β-glucuronidase, rabbit IgG fraction *2 mg/mL*	0.5 mL	148.00	118.40
B-8400	***New***	5-bromo-4-chloro-3-indolyl N-acetyl-β-D-galactosaminide	25 mg	95.00	76.00
B-8401	***New***	5-bromo-4-chloro-3-indolyl N-acetyl-β-D-glucosaminide	25 mg	48.00	38.40
B-1690		5-bromo-4-chloro-3-indolyl β-D-galactopyranoside (X-Gal)	1 g	55.00	44.00
B-1689		5-bromo-4-chloro-3-indolyl β-D-glucopyranoside (X-Glu)	100 mg	98.00	78.40
B-1691		5-bromo-4-chloro-3-indolyl β-D-glucuronide, cyclohexylammonium salt (X-GlcU, CHA)	100 mg	115.00	92.00
B-8404	***New***	5-bromo-4-chloro-3-indolyl β-D-glucuronide, sodium salt (X-GlcU, Na)	50 mg	120.00	96.00
B-8407	***New***	5-bromo-6-chloro-3-indolyl β-D-galactopyranoside	100 mg	65.00	52.00
B-8408	***New***	5-bromo-6-chloro-3-indolyl β-D-glucuronide, cyclohexylammonium salt	25 mg	95.00	76.00
C-6541	***New***	6′-(O-(carboxymethyl))-2′,7′-dichlorofluorescein 3′-(O-(N-acetyl-β-D-glucosaminide)) (CM-DCF-NAG)	5 mg	95.00	76.00
C-1488		3-carboxyumbelliferyl β-D-galactopyranoside (CUG)	10 mg	98.00	78.40
C-1492		3-carboxyumbelliferyl β-D-glucuronide (CUGlcU)	10 mg	48.00	38.40

Cat #		Product Name	Unit Size	Price Per Unit ($) 1–4 Units	5–24 Units
C-8411	***New***	6-chloro-3-indolyl β-D-galactopyranoside	100 mg	inquire	inquire
C-8412	***New***	6-chloro-3-indolyl β-D-glucuronide, cyclohexylammonium salt	25 mg	inquire	inquire
D-2921		DetectaGene™ Blue CMCG *lacZ* Gene Expression Kit	1 kit	148.00	118.40
D-2920		DetectaGene™ Green CMFDG *lacZ* Gene Expression Kit	1 kit	295.00	236.00
D-6488	***New***	9H-(1,3-dichloro-9,9-dimethylacridin-2-one-7-yl) β-D-galactopyranoside (DDAO galactoside)	5 mg	135.00	108.00
D-6486	***New***	9H-(1,3-dichloro-9,9-dimethylacridin-2-one-7-yl) β-D-glucuronide (DDAO GlcU)	5 mg	135.00	108.00
D-2893		5-dodecanoylaminofluorescein di-β-D-galactopyranoside (C_{12}FDG)	5 mg	195.00	156.00
E-6587	***New***	ELF®-97 β-D-glucuronidase substrate (ELF®-97 β-D-glucuronide)	5 mg	175.00	140.00
E-6644	***New***	EnzChek™ Dextranase Assay Kit *100-1000 assays*	1 kit	135.00	108.00
F-1930		FluoReporter® *lacZ* Flow Cytometry Kit *50 assays*	1 kit	178.00	142.40
F-1931		FluoReporter® *lacZ* Flow Cytometry Kit *250 assays*	1 kit	725.00	580.00
F-2905		FluoReporter® *lacZ*/Galactosidase Quantitation Kit *300 assays*	1 kit	98.00	78.40
F-1179		fluorescein di-β-D-galactopyranoside (FDG)	5 mg	115.00	92.00
F-2881		fluorescein di-β-D-glucopyranoside (FDGlu)	5 mg	70.00	56.00
F-2915		fluorescein di-β-D-glucuronide (FDGlcU)	5 mg	148.00	118.40
G-6463	***New***	β-galactosidase *from *Escherichia coli**	5 mg	95.00	76.00
I-2904		ImaGene Green™ C_{12}FDG *lacZ* Gene Expression Kit	1 kit	225.00	180.00
I-2908		ImaGene Green™ C_{12}FDGlcU GUS Gene Expression Kit	1 kit	225.00	180.00
I-2906		ImaGene Red™ C_{12}RG *lacZ* Gene Expression Kit	1 kit	195.00	156.00
I-2910		ImaGene Red™ C_{12}RGlcU GUS Gene Expression Kit	1 kit	195.00	156.00
I-6614	***New***	Indolyl β-D-Galactopyranoside Sampler Kit *contains 5 mg each of B-1690, B-8407, I-8414, I-8420, M-8421*	1 kit	65.00	52.00
I-8414	***New***	3-indoxyl β-D-galactopyranoside	100 mg	85.00	68.00
I-8415	***New***	3-indoxyl β-D-glucopyranoside, trihydrate	100 mg	35.00	28.00
I-8416	***New***	3-indoxyl β-D-glucuronide, cyclohexylammonium salt	25 mg	38.00	30.40
I-8420	***New***	5-iodo-3-indolyl β-D-galactopyranoside	25 mg	48.00	38.40
M-8421	***New***	N-methyl-3-indolyl β-D-galactopyranoside	25 mg	48.00	38.40
M-1489		4-methylumbelliferyl β-D-galactopyranoside (MUG)	1 g	25.00	20.00
M-8422	***New***	4-methylumbelliferyl α-D-glucopyranoside	100 mg	38.00	30.40
M-8423	***New***	4-methylumbelliferyl β-D-glucopyranoside (MUGlu)	1 g	38.00	30.40
M-1490		4-methylumbelliferyl β-D-glucuronide (MUGlcU)	100 mg	48.00	38.40
N-8426	***New***	4-nitrophenyl N-acetyl-β-D-galactosaminide	100 mg	35.00	28.00
N-8427	***New***	4-nitrophenyl N-acetyl-β-D-glucosaminide	1 g	75.00	60.00
N-8428	***New***	4-nitrophenyl β-D-cellobioside	100 mg	48.00	38.40
N-8429	***New***	4-nitrophenyl β-D-fucopyranoside	250 mg	55.00	44.00
N-8430	***New***	4-nitrophenyl α-D-galactopyranoside	1 g	35.00	28.00
N-8431	***New***	2-nitrophenyl β-D-galactopyranoside (ONPG)	5 g	35.00	28.00
N-8432	***New***	4-nitrophenyl β-D-galactopyranoside	1 g	24.00	19.20
N-8433	***New***	4-nitrophenyl α-D-glucopyranoside	1 g	20.00	16.00
N-8434	***New***	4-nitrophenyl β-D-glucopyranoside	1 g	35.00	28.00
N-8435	***New***	4-nitrophenyl β-D-glucuronide	100 mg	25.00	20.00
N-8436	***New***	4-nitrophenyl α-D-mannopyranoside	1 g	48.00	38.40
O-2892		5-octanoylaminofluorescein di-β-D-galactopyranoside (C_8FDG)	5 mg	135.00	108.00
P-8458	***New***	phenyl β-D-galactopyranoside	1 g	22.00	17.60
P-8459	***New***	phenyl β-D-glucuronide, monohydrate	100 mg	35.00	28.00
R-6540	***New***	resorufin N-acetyl-β-D-glucosaminide (RNAG)	5 mg	128.00	102.40
R-1159		resorufin β-D-galactopyranoside	25 mg	103.00	82.40
R-1160		resorufin β-D-glucopyranoside	25 mg	103.00	82.40
R-1161		resorufin β-D-glucuronide, potassium salt	5 mg	128.00	102.40
S-931		streptavidin, β-galactosidase conjugate	1 mg	125.00	100.00
T-657		β-trifluoromethylumbelliferyl β-D-galactopyranoside	100 mg	68.00	54.40
T-658		β-trifluoromethylumbelliferyl β-D-glucuronide	25 mg	125.00	100.00

10.3 Detecting Enzymes That Metabolize Phosphates and Polyphosphates

Cells utilize a wide variety of phosphate and polyphosphate esters as enzyme substrates, second messengers, membrane structural components and vital energy reservoirs. This section includes an assortment of reagents and methods for detecting the metabolism of phosphate esters. Our diverse array of fluorogenic and chromogenic substrates include substrates for phosphatases, as well as reagents to follow enzymes such as ATPase, 5'-nucleotidase and DNA and RNA polymerases. In addition, we have several substrates for phospholipases and phosphodiesterases that are described in Section 20.5.

By far the largest group of chromogenic and fluorogenic substrates for phosphate-ester metabolizing enzymes are those for simple phosphatases such as alkaline and acid phosphatase, both of which hydrolyze phosphate monoesters to an alcohol and inorganic phosphate. Conjugates of calf intestinal alkaline phosphatase are widely used as secondary detection reagents in ELISAs,[1] immunohistochemical techniques[2] and Northern, Southern and Western blot analyses (see Sections 8.3 and 9.2). In addition, phosphatases serve as enzyme markers, allowing researchers to identify primordial germ cells,[3] to distinguish subpopulations of bone marrow stromal cells[4] and to investigate *in vitro* differentiation in carcinoma cell lines.[5-7] *P ALP-1*, the gene for human placental alkaline phosphatase, has been used as a eukaryotic reporter gene and was found superior to *lacZ* for lineage studies in murine retina.[8,9] This gene has also been engineered to produce a secreted alkaline phosphatase, allowing quantitation of gene expression without disrupting the cells.[10]

Phosphatase Substrates Yielding Soluble Fluorescent Products

Fluorescein Diphosphate

First described in 1963,[11] fluorescein diphosphate (FDP, F-2999) is perhaps the most sensitive fluorogenic phosphatase substrate available. The nonfluorescent and colorless FDP is hydrolyzed to fluorescein (F-1300, see Section 10.1), which exhibits superior spectral properties (ϵ ~90,000 $cm^{-1}M^{-1}$, quantum yield ~0.92). We have succeeded in preparing a highly purified FDP and find it to be an excellent substrate for alkaline phosphatase in ELISAs, providing detection limits 50 times lower than those obtained with the chromogenic 4-nitrophenyl phosphate.[12] The high pH required to monitor alkaline phosphatase activity is advantageous because it also enhances fluorescein's fluorescence. FDP has been used to measure endogenous phosphatase in a new assay for cell adhesion and migration that is reported to be as sensitive as ^{51}Cr-release assays.[13]

Dimethylacridinone (DDAO) Phosphate

Our new dimethylacridinone (DDAO) phosphatase substrate[14] (D-6487) yields a hydrolysis product that is efficiently excited by the 633 nm spectral line of the He-Ne laser (ϵ ~78,000 $cm^{-1}M^{-1}$) to produce bright red fluorescence with absorption/emission maxima of ~634/665 nm. Although the substrate itself is fluorescent, the difference between the substrate's excitation maximum and that of the product is over 200 nm (see Figure 10.2 in Section 10.2), allowing the two species to be easily distinguished. Like other DDAO substrates, our DDAO phosphate has good water solubility, a low K_m and a high turnover rate.

Methylumbelliferyl Phosphate and Our New Difluorinated Methylumbelliferyl Phosphate

We offer 4-methylumbelliferyl phosphate (MUP), the most widely used fluorogenic substrate for phosphatase detection, as either its free acid (M-6491) or dicyclohexylammonium salt (M-8425). MUP has been used for a variety of ELISA protocols[15,16] in which the relatively high pH optimum of alkaline phosphatase permits continuous detection of the rate of formation of 4-methylumbelliferone (7-hydroxy-4-methylcoumarin, H-189; see Section 10.1). MUP has also been used to count cells based on their alkaline phosphatase activity,[17] to detect PCR amplification products[18,19] and to identify and characterize bacteria.[20,21]

We are also pleased to introduce 6,8-difluoro-4-methylumbelliferyl phosphate (DiFMUP, D-6567), which exhibits extraordinary spectral properties that are proving advantageous for the assay of both acid and alkaline phosphatase activity. The hydrolysis product of DiFMUP — 6,8-difluoro-4-methylumbelliferone (6,8-difluoro-7-hydroxycoumarin, D-6566; see Section 10.1) — exhibits both a lower pK_a (4.9 versus 7.8) and a higher fluorescence quantum yield (0.89 versus 0.63) than the hydrolysis product of MUP. The low pK_a of its hydrolysis product makes DiFMUP a sensitive substrate for the continuous assay of acid phosphatases, which is not possible with MUP because its fluorescence must be measured at alkaline pH. Furthermore, with its high fluorescence quantum yield, DiFMUP increases the sensitivity of both acid and alkaline phosphatase measurements. As with our fluorinated fluorescein derivatives (Oregon Green™ dyes, see Sections 1.4 and 1.5), fluorination renders the methylumbelliferone fluorophore much less susceptible to photobleaching yet does not significantly affect the extinction coefficient or excitation/emission maxima.

MANT-cGMP

2'-(*N*-Methylanthraniloyl)guanosine 3',5'-cyclicmonophosphate (MANT-cGMP, M-1035) is a substrate for cyclic nucleotide phosphodiesterases.[22-24] The decrease of its fluorescence upon hydrolysis can be used for continuous activity measurements, or the hydrolysis product can be quantitated by HPLC.

Cyclic GDP Ribose

Cyclic ADP ribose (cADP-ribose, C-7619; see Section 20.2) is generated from NAD^+ by ADP-ribosyl cyclase, an enzyme found in a variety of cell types, including rat pituitary,[25] rabbit liver, brain, spleen, kidney and heart.[26] ADP-ribosyl cyclase also mediates the formation of the fluorescent nucleotide cyclic GDP ribose (cGDP-ribose, C-7620) from nicotinamide guanine dinucleotide (NGD^+), providing a convenient fluorescence assay for cyclase activity in tissue extracts or pure enzyme preparations.[27,28]

ELF-97 Phosphate — A Phosphatase Substrate That Yields a Fluorescent Precipitate

Molecular Probes' scientists have made a major breakthrough in the development of substrates that yield fluorescent precipitates at the site of enzymatic activity — a process we call **E**nzyme-**L**abeled **F**luorescence[29] (ELF). Our first product in this line was the ELF-97 phosphatase substrate (ELF-97 phosphate; E-6588, E-6589). Upon enzymatic cleavage, this weakly blue fluorescent substrate yields a green fluorescent precipitate that is up to 40 times brighter than the signal achieved when using either directly labeled fluorescent hybridization probes or fluorescent secondary detection methods in comparable applications.[30] ELF-97 phosphate can be used over a wide pH range to selectively detect either acid- or alkaline-phosphatase activity. Although optimally excited with UV light, the ELF-97 precipitate can also reportedly be excited with the 488 nm spectral line of the argon-ion laser, making it compatible with flow cytometry and confocal laser scanning microscopy.[31] Using ELF-97 phosphate and confocal laser scanning microscopy, researchers have developed a semi-automated method for analyzing the position within a regenerating newt limb of transfected cells expressing the alkaline phosphatase reporter gene.[31] Kits based on our ELF-97 phosphate include:

- ELF-97 mRNA *In Situ* Hybridization Kits (E-6604, E-6605)
- ELF-97 Cytological Labeling Kits (E-6602, E-6603)
- ELF-97 Immunohistochemistry Kit (E-6600)
- ELF-97 Endogenous Phosphatase Detection Kit (E-6601)

These kits and their contents are described in more detail in Section 6.3. ELF-97 phosphate is also available separately as a 5 mM solution in water containing sodium azide (E-6589) or in sterile solution (E-6588). This substrate can be used to detect endogenous phosphatase activity or in combination with streptavidin alkaline phosphatase (S-921) or other alkaline phosphatase conjugates for signal amplification. Filtration of ELF-97 phosphate through an ELF spin filter (E-6606) is recommended before use. We have found that enzyme-catalyzed precipitation is often improved by incorporating a trace amount of the hydrolysis product — the yellow-green fluorescent ELF-97 alcohol — in the detection buffer. The ELF-97 alcohol is available in a concentrated solution in DMSO (E-6578, see Section 10.1).

Our ELF technology is covered by U.S. Patent Nos. 5,316,906 and 5,443,986; EP 0 641 351; and patents pending. These products are offered for research purposes only. Molecular Probes welcomes inquiries about licensing these products for resale or other commercial uses.

Chromogenic and Nonchromogenic Phosphatase Substrates

BCIP and Other Indolyl Phosphates

5-Bromo-4-chloro-3-indolyl phosphate (BCIP) is commonly used with a number of different chromogens in various histological and molecular biology techniques (see Sections 8.3 and 9.2). Hydrolysis of this indolyl phosphate, followed by oxidation, produces a blue colored precipitate at the site of enzymatic activity. BCIP is available as either its sodium (B-6492) or *p*-toluidine (B-8405) salt. We also offer 5-bromo-6-chloro-3-indolyl phosphate[32] (B-8409), 6-chloro-3-indolyl phosphate[33] (C-8413) and three different salts of 3-indoxyl phosphate (I-8417, I-8418, I-8419), yielding magenta, pink and blue colored precipitates, respectively. These substrates, which produce products that are insoluble in water, ethanol and xylene, can be used with hematoxylin stains.[32]

NBT and INT Co-Precipitants for the BCIP Reaction

Nitro blue tetrazolium (NBT, N-6495; see Section 21.4) is the most commonly used electron-transfer agent and co-precipitant for the BCIP reaction, forming a dark blue, precisely localized precipitate in the presence of alkaline phosphatase.[34,35] Our BCIP and NBT reagents are already priced very competitively, but our NBT/BCIP Reagent Kit (N-6547) offers 1-gram samples of both the sodium salt of BCIP and NBT at an even lower price.

Like NBT, 2-(4-iodophenyl)-3-(4-nitrophenyl)-5-phenyltetrazolium chloride (INT, I-6496; see Section 21.4) can be co-precipitated with BCIP to form a brick-red signal.[36] The BCIP/INT system yields a precipitate that is distinct against the blue background of hematoxylin-stained tissue sections.

p-Nitrophenol–Based Substrates

4-Nitrophenyl phosphate (PNPP, N-6613) is probably the most widely used chromogenic phosphatase substrate that yields a soluble product.[37] Hydrolysis of PNPP liberates yellow colored 4-nitrophenol (N-6477, see Section 10.1).

Creatine Phosphate

Creatine phosphate (phosphocreatine, C-8448) is an essential high-energy phosphate that serves as a substrate for creatine phosphokinase, which is particularly abundant in skeletal muscle. Creatine phosphate is often used as a component of *in vitro* ATP-regenerating systems.[38]

Kits for Detecting Phosphatases, Polymerases and Nucleases

Molecular Probes is pleased to introduce several unique new products for following the activity of phosphatases, polymerases and nucleases. Two of our most versatile new products are the EnzChek Phosphate and EnzChek Pyrophosphate Assay Kits, which provide colorimetric assays for inorganic phosphate and pyrophosphate, respectively. The EnzChek phosphate assay permits the continuous measurement of the activity of ATPases, GTPases, phosphatases, nucleotidases, kinases and a number of enzymes that produce or consume inorganic phosphate. In the EnzChek pyrophosphate assay, this colorimetric phosphate assay is coupled with the enzyme pyrophosphatase in order to monitor the activity of pyrophosphate-producing enzymes such as DNA- and RNA-polymerase and adenylate cyclase. Other kits in the EnzChek product line designed to detect phosphate-metabolizing enzymes include the EnzChek 5'-Nucleotidase Assay Kit, as well as the EnzChek DNase and RNase Assay Kits.

EnzChek Phosphate Assay Kit

Continuous assay of many phosphate ester–metabolizing enzymes is difficult because suitable substrates are not available. It usually has been necessary to determine inorganic phosphate release using tedious colorimetric assays or radioisotope-based methods.

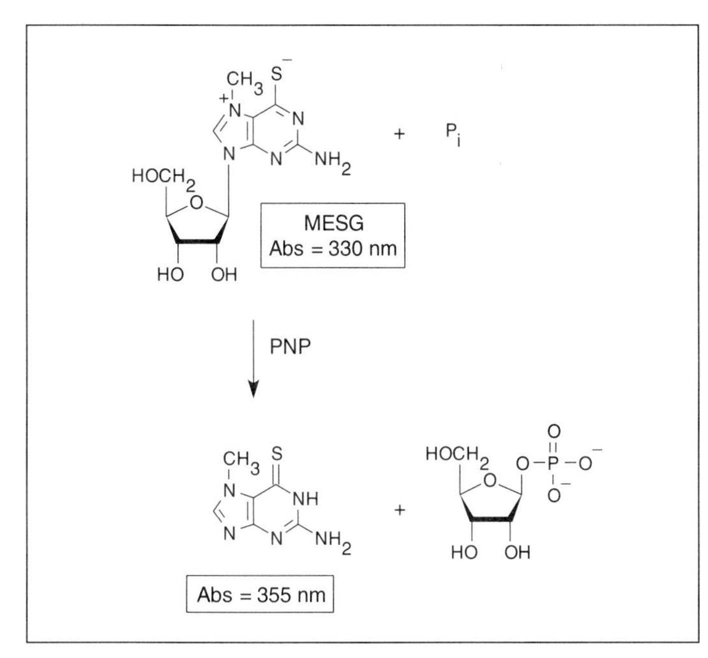

Figure 10.5 *Enzymatic conversion of 2-amino-6-mercapto-7-methylpurine ribonucleoside (MESG) to ribose-1-phosphate and 2-amino-6-mercapto-7-methylpurine by purine nucleoside phosphorylase (PNP), reagents supplied in our EnzChek Phosphate Assay Kit (E-6646). The accompanying change in the absorption maximum (Abs) allows quantitation of inorganic phosphate (P_i) consumed in the reaction.*

Our EnzChek Phosphate Assay Kit (E-6646), which is based on a method originally described by Webb,[39,40] provides a rapid and highly sensitive enzymatic assay for detecting inorganic phosphate through formation of a chromophoric product (Figure 10.5). This new spectrophotometric technique permits the continuous assay of ATPase activity, and potentially the activity of many other enzymes such as GTPases and phosphatases that produce inorganic phosphate. Each EnzChek Phosphate Assay Kit contains:

- 2-Amino-6-mercapto-7-methylpurine ribonucleoside (MESG)
- Purine nucleoside phosphorylase (PNP)
- Concentrated reaction buffer
- KH_2PO_4 standard
- Protocol for detecting and quantitating phosphate

In the presence of inorganic phosphate, MESG is enzymatically converted by PNP to ribose-1-phosphate and the chromophoric product 2-amino-6-mercapto-7-methylpurine. Although the MESG reagent is somewhat unstable above pH 7, the reaction can be performed over the pH range of 6.5 to 8.5 with the proper controls.[40] This kit contains sufficient reagents for about 100 phosphate assays using 1 mL assay volumes and standard cuvettes.

The substrate MESG and the enzyme PNP included in our EnzChek Phosphate Assay Kit have already been adapted for monitoring the kinetics of phosphate release by:

- Actin-activated myosin ATPase [40]
- Aspartate transcarbamylase [39]
- Dethiobiotin synthetase [41]
- Glycerol kinase [40]
- Glycogen phosphorylase [42]
- GTPases [43,44]
- *myo*-Inositol monophosphatase [45]
- Phospholysine and phosphohistidine phosphatases [46,47]
- Phosphorylase *a* phosphatase [48]
- Phosphorylase kinase [49]
- Serine phosphatase [50]

Moreover, the EnzChek phosphate assay is sufficiently fast and quantitative to permit stopped-flow kinetic experiments on enzymes that produce phosphate, an important development for mechanistic enzyme studies.[40]

Although this kit is usually used to determine the inorganic phosphate produced by enzymes such as ATPases, it can also be used to specifically quantitate inorganic phosphate contamination of reagents and solution, with a detection limit of ~2 µM (~0.2 µg/mL) and an effective range between 2 and 150 µM inorganic phosphate (between 2 and 150 nanomoles phosphate in a 1 mL volume). For example, the assay has been used for the rapid assay of inorganic phosphate in the presence of high concentrations of acid-labile phosphates using a microplate reader.[51] The kit's reagents can also be used as a phosphate "mop" to remove all inorganic phosphate from a protein solution.[52]

EnzChek Pyrophosphatase Assay Kit

We have adapted the method provided in our EnzChek Phosphate Assay Kit to permit the sensitive spectrophotometric detection of pyrophosphate, which is converted by the enzyme pyrophosphatase to inorganic phosphate. Because two moles of phosphate are released per mole of pyrophosphate consumed, the sensitivity limit of the EnzChek Pyrophosphatase Assay Kit is 1 µM (~0.2 µg/mL). This assay can potentially be adapted to continuously detect enzymes that liberate pyrophosphate such as aminoacyl-tRNA synthetase,[53] luciferase and acetyl-CoA synthetase [54] and potentially DNA and RNA polymerases and adenylate cyclase. Each EnzChek Pyrophosphate Assay Kit (E-6645) contains:

- Inorganic pyrophosphatase
- 2-Amino-6-mercapto-7-methylpurine ribonucleoside (MESG)
- Purine nucleoside phosphorylase (PNP)
- Concentrated reaction buffer
- $Na_2P_2O_7$ standard
- Protocol for detecting and quantitating pyrophosphate

The kit contains sufficient reagents for about 100 pyrophosphate assays using 1 mL assay volumes and standard cuvettes.

EnzChek 5′-Nucleotidase Assay Kit

5'-Nucleotidase (5'-ribonucleotide phosphohydrolase; EC 3.1.3.5) catalyzes the dephosphorylation of purine and pyrimidine ribo- and deoxyribonucleoside 5'-monophosphates. This enzyme is present as both cytoplasmic and membrane-associated forms in bacterial, plant and vertebrate cells.[55] Extracellular production of adenosine by membrane-associated 5'-nucleotidase is linked to a number of adenosine receptor–mediated events, including vasodilation,[56-60] glomerular filtration rate,[61] neurotransmitter release,[62,63] inflammatory response [64,65] and lipolysis.[66] Cytoplasmic 5'-nucleotidase is involved in regulating intracellular levels of nucleotides, as well as in controlling the secretion of adenosine.[55] Deficiencies in 5'-nucleotidase activity have been studied in several disease states, including hemolytic anemia,[67] multiple myeloma [68] and hairy cell, chronic B lymphocytic and acute lymphoblastic leukemias.[69-71] Because of their central role in maintaining intracellular levels of nucleoside 5'-monophosphates and extracellular levels of adenosine, cytoplasmic and membrane-associated 5'-nucleotidases have been important targets of pharmaceutical research.

Molecular Probes' new EnzChek 5'-Nucleotidase Assay Kit (E-6643) provides a quick and sensitive fluorescence-based method for assaying 5'-nucleotidase activity. Conventional assays for 5'-nucleotidase activity have relied on colorimetric determination of the released inorganic phosphate,[72] absorbance-based measurements of the conversion of the product adenosine to inosine by adenosine deaminase [73] or a separation-based radioisotopic assay using ^{3}H-AMP as a substrate.[74] In the EnzChek assay, the green fluorescent BODIPY FL–AMP substrate is hydrolyzed by 5'-nucleotidase to a BODIPY FL–labeled adenosine analog, which is easily separated from the unreacted substrate by precipitation. After a brief centrifugation step to pellet the precipitated substrate, the fluorescence of the supernatant, which contains the product, is measured in a fluorometer or fluorescence microplate reader using standard fluorescein settings or filters. Each EnzChek 5'-Nucleotidase Assay Kit contains:

- BODIPY FL AMP
- Reaction buffer
- Barium hydroxide for stopping the reaction
- Zinc sulfate for precipitating unreacted substrate
- Protocol for quantitating 5'-nucleotidase activity

Sufficient reagents are provided for 100 assays. This sensitive assay can be used to detect as few as 10^{-4} U of 5'-nucleotidase activity; however, phosphatase contamination of experimental samples must be rigorously avoided. If present, alkaline phosphatase and other phosphatases can catalyze the hydrolysis of BODIPY FL AMP and create spurious results with the EnzChek 5'-nucleotidase assay. Our EnzChek Phosphate Assay Kit (E-6646, see above) can also be used to assay 5'-nucleotidase by measuring the release of inorganic phosphate.

EnzChek DNase Gel Assay Kit

The EnzChek DNase and RNase Assay Kits, which are designed primarily for detecting nuclease contamination in laboratory reagents, provide ultrasensitive fluorometric detection of DNase or RNase activity (see also Section 8.3). The EnzChek DNase Gel Assay Kit (E-6636) allows researchers to detect less than 10^{-5} Kunitz units of DNase activity (approximately 5 pg of DNase I with a specific activity of 2000 Kunitz units/mg) in an 8 µL aliquot. Furthermore, the assay can be completed in one hour, from adding DNA to your unknown to visualizing the gel staining pattern, and will detect the activity of both DNA exonucleases and DNA endonucleases. Each EnzChek DNase Gel Assay Kit contains:

- SYBR® Green I nucleic acid gel stain
- *Pst* I–linearized ϕX174 RF DNA
- Concentrated digestion buffer
- Stop buffer
- 15% Ficoll® solution
- DNA molecular size markers
- Detailed protocol

Sufficient reagents are provided to perform 100 assays. Analysis of the gel by UV illumination and black-and-white print film requires the use of the SYBR Green gel stain photographic filter (S-7569, see Section 8.3), which is available separately.

EnzChek RNase Gel Assay Kit

Similarly, the EnzChek RNase gel assay employs SYBR Green II RNA gel stain and an MS2 single-stranded RNA phage template in a simple agarose gel assay that takes about 1 hour to perform, including gel electrophoresis and staining.[75] With our EnzChek RNase Gel Assay Kit (E-6641), researchers can detect as little as 10^{-8} Kunitz units [76] of RNase A activity (single picogram levels of RNase A with a specific activity of 50 Kunitz units/mg), or as little as 5×10^{-3} Egami units [77] of RNase T1 (approximately 14 pg of RNase T1 with a specific activity of 350,000 Egami units/mg) in 8 µL samples. Each EnzChek RNase Gel Assay Kit contains:

- SYBR Green II RNA gel stain
- MS2 RNA
- Concentrated digestion buffer
- Stop buffer
- Loading buffer
- DNA molecular weight standards
- RNase-free H_2O
- Detailed protocol

Sufficient reagents are provided to perform 100 assays. Analysis of the gel by UV illumination and black-and-white print film requires the use of the SYBR Green gel stain photographic filter (S-7569, see Section 8.3), which is available separately.

EnzChek DNase Solution Assay Kit

In addition to these gel-based nuclease assay kits, we also offer the EnzChek DNase Solution Assay Kit (E-6637), which is an extremely simple and rapid solution-based assay for endonuclease activity. Using this kit, researchers can detect less than 4×10^{-3} Kunitz units DNase I (approximately 2 ng of active enzyme) in a 2 to 50 µL aliquot in about 30 minutes. Each EnzChek DNase Solution Assay Kit contains:

- YO-PRO™-1 nucleic acid stain
- ϕX174 RF DNA
- Concentrated digestion buffer
- Detailed protocol

Sufficient reagents are provided to perform approximately 25 assays using standard cuvettes in a fluorometer.

ATP Determination Kit

Molecular Probes now offers a convenient ATP Determination Kit (A-6608) for the sensitive bioluminescence-based detection of ATP with recombinant firefly luciferase and its substrate, luciferin. This assay is based on luciferase's requirement for ATP to produce light. In the presence of Mg^{2+}, luciferase catalyzes the reaction of luciferin, ATP and O_2 to form oxyluciferin, AMP, CO_2, pyrophosphate and ~560 nm light (Figure 16.9).

The luciferin–luciferase bioluminescence assay is extremely sensitive; most luminometers can detect as little as 1 picomole of pre-existing ATP or ATP as it is generated in kinetic systems. This sensitivity has led to its widespread use for detecting ATP in various enzymatic reactions, as well as for measuring viable cell number [78] and for detecting low-level bacterial contamination in samples such as blood, milk, urine, soil and sludge.[79-83] The luciferin–luciferase bioluminescence assay has also been used to determine cell proliferation and cytotoxicity in both bacterial[84,85] and mammalian cells,[86,87] and to distinguish cytostatic versus cytocidal potential of anticancer drugs on malignant cell growth.[88]

Each ATP Determination Kit contains:

- Luciferin
- Luciferase
- Dithiothreitol (DTT)
- ATP
- 20X concentrated reaction buffer
- Protocol for ATP quantitation

Unlike many other commercially available ATP detection kits, our ATP Determination Kit provides the luciferase and luciferin packaged separately, which enables researchers to optimize the reaction conditions for their particular instruments and samples. Each ATP Determination Kit provides sufficient reagents to perform 200 ATP assays using 0.5 mL sample volumes or 500 ATP assays using 0.2 mL sample volumes.

1. Meth Mol Biol 32, 461 (1994); **2.** J Clin Microbiol 19, 230 (1984); **3.** Anatomical Record 118, 135 (1954); **4.** J Histochem Cytochem 40, 1059 (1992); **5.** Devel Biol 88, 279 (1981); **6.** Cell 5, 229 (1975); **7.** Proc Natl Acad Sci USA 70, 3899 (1973); **8.** BioTechniques 14, 818 (1993); **9.** Proc Natl Acad Sci USA 89, 693 (1992); **10.** Meth Enzymol 216, 362 (1992); **11.** Proc Natl Acad Sci USA 50, 1 (1963); **12.** J Immunol Meth 149, 261 (1992); **13.** J Immunol Meth 192, 165 (1996); **14.** DDAO-based substrates are licensed to Molecular Probes under U.S. Patent No. 4,810,636 and its foreign equivalents; **15.** Anal Chem 65, 1147 (1993); **16.** J Immunol Meth 150, 23 (1992); **17.** In Vitro Cell Devel Biol 25, 105 (1989); **18.** Anal Biochem 205, 1 (1992); **19.** Mol Cell Probes 6, 489 (1992); **20.** Zbl Bakt 280, 476 (1994); **21.** Microbiol Rev 55, 335 (1991); **22.** J Biol Chem 268, 22895 (1993); **23.** Anal Biochem 162, 291 (1987); **24.** J Biol Chem 257, 13354 (1982); **25.** J Biol Chem 264, 11725 (1989); **26.** J Biol Chem 266, 16985 (1991); **27.** J Biol Chem 270, 30327 (1995); **28.** J Biol Chem 269, 30260 (1994); **29.** U.S. Patent Nos. 5,316,906 and 5,443,986; EP 0 641 351; and patents pending; **30.** Am Soc Biochem Mol Biol, abstract #968 (1994); **31.** J Histochem Cytochem 44, 559 (1996); **32.** J Histochem Cytochem 42, 551 (1994); **33.** Currently available only from Molecular Probes Europe BV, Leiden, The Netherlands; inquire about availability from Molecular Probes, Inc., Eugene, Oregon; **34.** BioTechniques 12, 656 (1992); **35.** Histochemistry 58, 203 (1978); **36.** Am J Pathol 141, 1331 (1992); **37.** J Immunol Meth 143, 49 (1991); **38.** Ann Rev Biochem 54, 831 (1985); **39.** Anal Biochem 218, 449 (1994); **40.** Proc Natl Acad Sci USA 89, 4884 (1992); **41.** Biochemistry 34, 10976 (1995); **42.** Anal Biochem 221, 348 (1994); **43.** Biochemistry 34, 15592 (1995); **44.** Biochem J 287, 555 (1992); **45.** Biochem J 307, 585 (1995); **46.** Anal Biochem 222, 14 (1994); **47.** Biochem J 296, 293 (1993); **48.** Anal Biochem 226, 68 (1995); **49.** Anal Biochem 230, 55 (1995); **50.** Biochemistry 33, 2380 (1994); **51.** Anal Biochem 230, 173 (1995); **52.** Biochemistry 33, 8262 (1994); **53.** Nucleic Acids Res 23, 2886 (1995); **54.** Upson, R.H., Haugland, R.P., Malekzadeh, N. and Haugland, R.P., Anal Biochem, in press; **55.** Biochem J 285, 345 (1992); **56.** J Clin Invest 95, 1747 (1995); **57.** J Clin Invest 95, 285 (1995); **58.** J Biol Chem 262, 6296 (1987); **59.** Biochemistry 20, 5188 (1981); **60.** J Biol Chem 253, 1938 (1978); **61.** Am J Physiol 263, 991 (1992); **62.** Neuron 15, 909 (1995); **63.** Proc Natl Acad Sci USA 89, 8586 (1992); **64.** J Immunol 153, 4159 (1994); **65.** J Immunol 151, 5716 (1993); **66.** Biochem Biophys Acta 1221, 315 (1994); **67.** J Inherit Metab Dis 13, 701 (1990); **68.** Acta Haematol 78, 41 (1987); **69.** Br J Haematol 62, 545 (1986); **70.** Cancer 58, 96 (1986); **71.** Cancer Res 41, 4821 (1981); **72.** Enzyme 45, 194 (1991); **73.** J Biol Chem 245, 6274 (1970); **74.** Cell 17, 143 (1979); **75.** FASEB J 9, A1400 (1995); **76.** J Gen Physiol 24, 15 (1940); **77.** Prog in Nucleic Acids Res and Mol Biol 3, 59 (1964); **78.** J Biolumin Chemilumin 10, 29 (1995); **79.** Anal Biochem 175, 14 (1988); **80.** Bio/Technology 6, 634 (1988); **81.** J Clin Microbiol 20, 644 (1984); **82.** J Clin Microbiol 18, 521 (1983); **83.** Meth Enzymol 57, 3 (1978); **84.** Biotech Bioeng 42, 30 (1993); **85.** J Biolumin Chemilumin 6, 193 (1991); **86.** Biochem J 295, 165 (1993); **87.** J Immunol Meth 160, 81 (1993); **88.** J Natl Cancer Inst 77, 1039 (1986).

10.3 Data Table *Detecting Enzymes That Metabolize Phosphates and Polyphosphates*

Cat #	Structure	MW	Storage	Soluble	Abs	{ε × 10^{-3}}	Em	Solvent	Product*
B-6492	✓	370	F,D	DMSO, H_2O	292	{3.8}	none	H_2O	<1>
B-8405	✓	434	F,D	DMSO, H_2O	290	{5.9}	none	H_2O	<1>
B-8409	✓	434	F,D	DMSO, H_2O	292	{5.4}	none	pH 7	<2>
C-7620	S20.2	557	FF,D	H_2O	284	{5.8}	396	pH 7	NA
C-8413	✓	355	F,D	DMSO, H_2O	286	{6.3}	none	pH 7	<3>
C-8448	✓	327†	FF,DD	H_2O	<300		none		NA
D-6487	✓	422	F,D,L	DMSO, H_2O	478	{26}	628	pH 7	H-6482‡
D-6567	✓	326	F,D	DMSO, H_2O	320	{14}	385§	pH 9	D-6566‡
E-6588	✓	431	F,D,L	H_2O¶	289	{12}	ND§	pH 10	E-6578‡
F-2999	✓	560	F,D	H_2O	272	{5.0}	none	MeOH	F-1300‡
I-8417	✓	423	F,D	H_2O	282	{5.0}	none	MeOH	<4>
I-8418	✓	257	F,D	H_2O	282	{4.8}	none	MeOH	<4>
I-8419	✓	320	F,D	H_2O	282	{5.8}	none	MeOH	<4>
M-1035	S20.4	~500	F,D,L	H_2O	359	{5.2}	448	pH 7	<5>
M-6491	✓	256	F,D	DMSO, H_2O	319	{15}	383§	pH 9	H-189‡
M-8425	✓	509†	F,D	H_2O	318	{14}	385§	pH 9	H-189‡
N-6613	✓	371†	F,D,L	H_2O	310	{11}	none	pH 7	N-6477‡

E-6589 see E-6588.

For definitions of the contents of this data table, see "How to Use the Handbook" on page vi.
Structure: Chemical structure drawing: ✓ = shown in this section; Sn.n = shown in Section number n.n.
MW: ~ indicates that the molecular weight of this product is approximate because the exact degree of hydration and salt form has not been conclusively established.
ND = not determined.
NA = not applicable.

* Catalog number or footnote description of the enzymatic cleavage product.
† MW is for the hydrated form of this product.
‡ See Section 10.1.

§ Fluorescence of the unhydrolyzed substrate is very weak.
¶ This product is packaged as a solution in the solvent indicated in "Soluble."
<1> Enzymatic cleavage of this substrate yields a water-insoluble, blue colored indigo dye (Abs ~615 nm).
<2> Enzymatic cleavage of this substrate yields a water-insoluble, magenta colored indigo dye (Abs ~565 nm).
<3> Enzymatic cleavage of this substrate yields a water-insoluble, pink colored indigo dye (Abs ~540 nm).
<4> Enzymatic cleavage of this substrate yields a water-insoluble, blue colored indigo dye (Abs ~680 nm).
<5> Cleavage of M-1035 by cyclic nucleotide phosphodiesterase produces MANT-GMP, which has similar spectra but is less fluorescent [Anal Biochem 162, 291 (1987)].

10.3 Structures *Detecting Enzymes that Metabolize Phosphates and Polyphosphates*

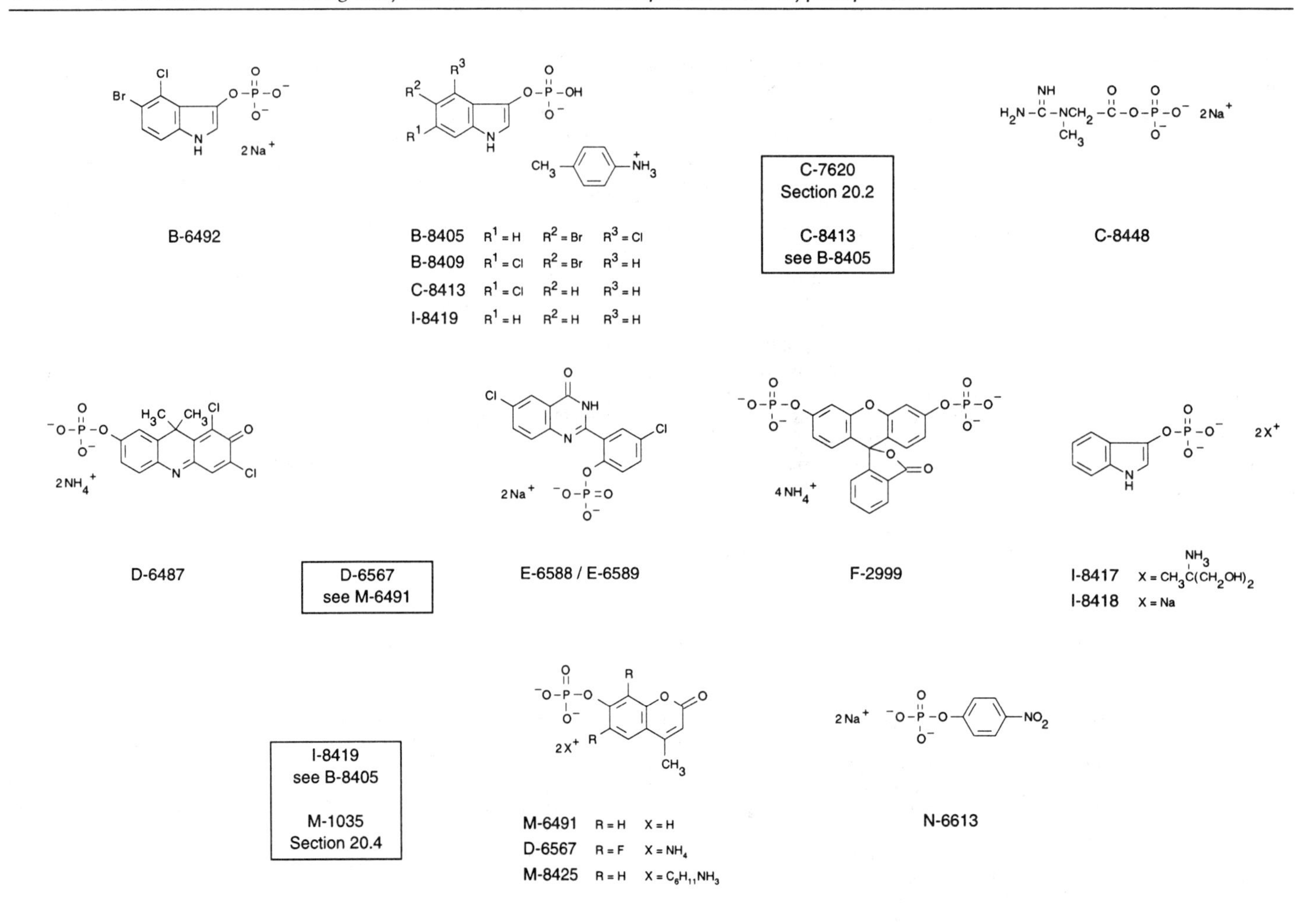

10.3 Price List *Detecting Enzymes That Metabolize Phosphates and Polyphosphates*

Cat #		Product Name	Unit Size	Price Per Unit ($) 1–4 Units	5–24 Units
A-6608	***New***	ATP Determination Kit *200-1000 assays*	1 kit	95.00	76.00
B-6492	***New***	5-bromo-4-chloro-3-indolyl phosphate, disodium salt (BCIP, Na)	1 g	75.00	60.00
B-8405	***New***	5-bromo-4-chloro-3-indolyl phosphate, p-toluidine salt (BCIP, toluidine)	1 g	110.00	88.00
B-8409	***New***	5-bromo-6-chloro-3-indolyl phosphate, p-toluidine salt	100 mg	48.00	38.40
C-8413	***New***	6-chloro-3-indolyl phosphate, p-toluidine salt	100 mg	inquire	inquire
C-8448	***New***	creatine phosphate, disodium salt, tetrahydrate (phosphocreatine)	5 g	44.00	35.20
C-7620	***New***	cyclic guanosine 5′-diphosphate ribose (cGDP-ribose)	100 µg	100.00	80.00
D-6487	***New***	9H-(1,3-dichloro-9,9-dimethylacridin-2-one-7-yl) phosphate, diammonium salt (DDAO phosphate)	5 mg	135.00	108.00
D-6567	***New***	6,8-difluoro-4-methylumbelliferyl phosphate, ammonium salt (DiFMUP)	5 mg	38.00	30.40
E-6602	***New***	ELF®-97 Cytological Labeling Kit #1 *50 assays*	1 kit	148.00	118.40
E-6603	***New***	ELF®-97 Cytological Labeling Kit #2 *with streptavidin, alkaline phosphatase conjugate* *50 assays*	1 kit	198.00	158.40
E-6601	***New***	ELF®-97 Endogenous Phosphatase Detection Kit	1 kit	135.00	108.00
E-6600	***New***	ELF®-97 Immunohistochemistry Kit	1 kit	198.00	158.40
E-6604	***New***	ELF®-97 mRNA *In Situ* Hybridization Kit #1 *50 assays*	1 kit	175.00	140.00

Cat #		Product Name	Unit Size	Price Per Unit ($) 1–4 Units	5–24 Units
E-6605	*New*	ELF®-97 mRNA *In Situ* Hybridization Kit #2 *with streptavidin, alkaline phosphatase conjugate* *50 assays*	1 kit	225.00	180.00
E-6589	*New*	ELF®-97 phosphatase substrate (ELF®-97 phosphate) *5 mM solution in water* *contains 2 mM azide*	1 mL	150.00	120.00
E-6588	*New*	ELF®-97 phosphatase substrate (ELF®-97 phosphate) *5 mM solution in water* *sterile filtered*	1 mL	160.00	128.00
E-6606	*New*	ELF® spin filters *20 filters*	1 box	48.00	38.40
E-6636	*New*	EnzChek™ DNase Gel Assay Kit *100 assays*	1 kit	135.00	108.00
E-6637	*New*	EnzChek™ DNase Solution Assay Kit *25 assays*	1 kit	135.00	108.00
E-6643	*New*	EnzChek™ 5′-Nucleotidase Assay Kit *100 assays*	1 kit	95.00	76.00
E-6646	*New*	EnzChek™ Phosphate Assay Kit *100 assays*	1 kit	135.00	108.00
E-6645	*New*	EnzChek™ Pyrophosphate Assay Kit *100 assays*	1 kit	155.00	124.00
E-6641	*New*	EnzChek™ RNase Gel Assay Kit *100 assays*	1 kit	135.00	108.00
F-2999		fluorescein diphosphate, tetraammonium salt (FDP)	5 mg	165.00	132.00
I-8417	*New*	3-indoxyl phosphate, di(2-amino-2-methyl-1,3-propanediol) salt	1 g	75.00	60.00
I-8418	*New*	3-indoxyl phosphate, disodium salt	1 g	75.00	60.00
I-8419	*New*	3-indoxyl phosphate, p-toluidine salt	1 g	75.00	60.00
M-1035		2′-(N-methylanthraniloyl)guanosine 3′,5′-cyclicmonophosphate, sodium salt (MANT-cGMP)	10 mg	95.00	76.00
M-8425	*New*	4-methylumbelliferyl phosphate, dicyclohexylammonium salt, trihydrate	1 g	50.00	40.00
M-6491	*New*	4-methylumbelliferyl phosphate, free acid	1 g	38.00	30.40
N-6547	*New*	NBT/BCIP Reagent Kit	1 kit	135.00	108.00
N-6613	*New*	4-nitrophenyl phosphate, disodium salt, hexahydrate (PNPP)	5 g	75.00	60.00
S-921		streptavidin, alkaline phosphatase conjugate *2 mg/mL*	0.5 mL	155.00	124.00

10.4 Detecting Peptidases and Proteases

Peptidases and proteases play essential roles in protein activation, cell regulation and signaling, as well as in the generation of amino acids for protein synthesis or utilization in other metabolic pathways. In general, peptidases cleave shorter peptide bonds in peptides, and proteases cleave longer peptides and proteins. Depending on their site of cleavage, peptidases can be classified as **exopeptidases** if they preferentially hydrolyze amino acid residues from the terminus of a peptide or **endopeptidases** if they cleave internal peptide bonds.[1] Exopeptidases are further divided into aminopeptidases and carboxypeptidases depending on whether they hydrolyze residues from the amine or carboxy terminus.

Although the spectral properties of fluorogenic peptidase and protease substrates and their hydrolysis products are easily predictable, the utility of a given substrate for an enzyme depends on the kinetics of hydrolysis by the enzyme, which, in turn, depends on the substrate's concentration and amino acid sequence, as well as on the pH, temperature and presence of cofactors in the medium. For measurements in live cells, the suitability of a particular substrate also hinges on its accessibility to the enzyme and the cellular retention of the hydrolysis product. In addition to these factors, the chromophore or fluorophore conjugated to the substrate can influence its hydrolysis rate and specificity, as well as the permeability of the substrate and its hydrolysis product. Several of our newer substrates are based on sequences that have been shown to be effective when conjugated with other dyes, typically 7-amino-4-methylcoumarin or *p*-nitroaniline, but have not yet been tested for specificity.

Peptidase Substrates

Fluorophores and Chromophores for Peptidase Substrates

The carboxy terminus of single amino acids and short peptides can be conjugated to certain amine-containing fluorophores to create fluorogenic peptidase substrates. The dyes used to make these substrates are fluorescent at physiological pH. However, when the dyes are coupled in an amide linkage to peptides, the absorbance maxima are usually shortened significantly. The resulting substrates are sometimes fluorescent but with relatively short-wavelength emission spectra. In an extreme case such as that of rhodamine 110–based substrates, detectable absorbance and fluorescence are completely eliminated by amide formation. Peptidase activity releases the fluorophore, restoring its free-dye fluorescence.

Molecular Probes currently uses eight different amine-containing dyes in its peptidase substrates (Figure 10.6 and Table 10.2):

- 7-Amino-4-methylcoumarin (AMC, A-191; see Section 10.1), a blue fluorescent dye used extensively to label substrates for detecting enzymatic activity in cells, homogenates and solutions and on blots
- 7-Amino-4-chloromethylcoumarin (CMAC, C-2110; see Section 10.1), a mildly thiol-reactive analog of AMC; CMAC-based substrates yield fluorescent peptidase products with improved retention in live cells
- 6-Aminoquinoline (6-AQ, A-497; see Section 10.1), a dye with a large Stokes shift (absorption/emission maximum

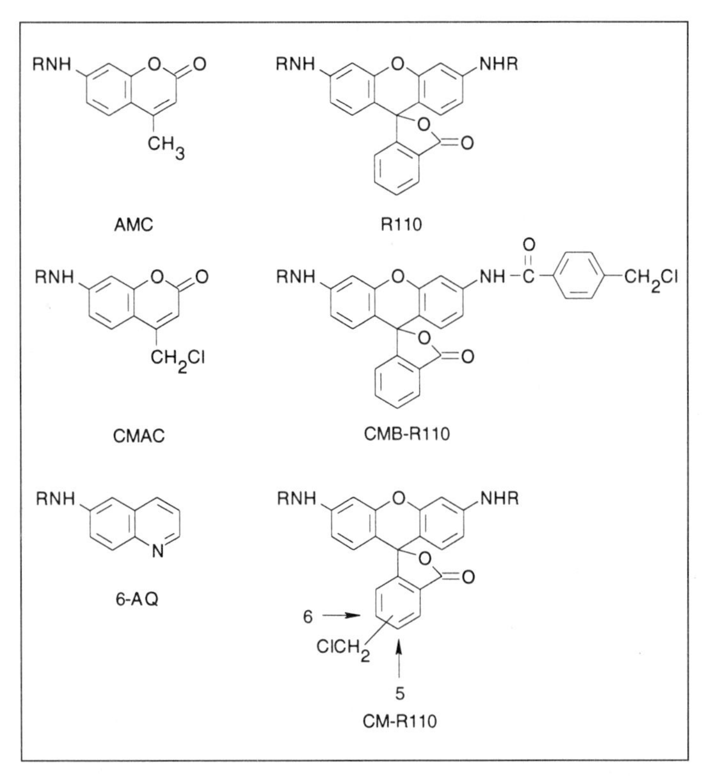

Figure 10.6 *Generic structures of peptidase substrates derived from 7-amino-4-methylcoumarin (AMC), 7-amino-4-chloromethylcoumarin (CMAC), 6-aminoquinoline (6-AQ), rhodamine 110 (R110), N-(4-chloromethyl)benzoyl rhodamine 110 (CMB-R110) and 5-(and-6)-chloromethyl-rhodamine 110 (CM-R110). The carboxy terminal–linked peptide substituents (R) for specific substrates are defined in Tables 10.3 and 10.4.*

~345/550 nm), yielding a fluorescent signal that can be clearly distinguished from the fluorescence of the substrate, as well as from that of most autofluorescent components.

- Rhodamine 110 (R110, R-6479; see Section 10.1), a visible light–excitable dye with stronger absorbance than either AMC or 6-AQ; R110-based substrates [2] comprise two identical amino acids or peptides attached to a single fluorophore
- *N*-(4-Chloromethyl)benzoyl rhodamine 110 (CMB-R110) — a version of R110 specifically designed to assay enzymatic activity in live cells or tissues;[3] unlike R110-based substrates, CMB-R110–based substrates are *mono*amides of amino acids and peptides, thus simplifying reaction kinetics
- 5-(and-6)-Chloromethylrhodamine 110 (CM-R110) — a thiol-reactive analog of rhodamine 110; CM-R110–based substrates are bisamides that may yield substrates with improved retention in live cells
- 6-Amino-6-deoxyluciferin (ADL, A-6611; see Section 10.5), an alternative substrate for luciferase that can be detected by chemiluminescence
- *p*-Nitroaniline, the base of important chromophoric peptidase substrates

Most of our single-amino acid substrates are primarily useful for detecting aminopeptidase activity (Table 10.3). Our peptidase substrates containing multiple amino acids or chemically blocked amino acids are useful for detecting various endopeptidases and dipeptidyl peptidases (Table 10.4). Where they are known, references for those peptidase substrates based on 7-amino-4-methyl-coumarin are listed in these tables, even if the substrates are not available from Molecular Probes. In several cases the utility of substrates based on other fluorophores has not been published. As noted in Table 10.4, a single substrate may be useful for assaying multiple enzymes.

Substrates Based on AMC, 6-AQ and *p*-Nitroaniline

Substrates based on the AMC and 6-AQ fluorophores and the *p*-nitroaniline chromophore are commonly used for peptidase and protease assays in cells, homogenates and solutions. Molecular Probes offers a selection of these substrates and provides references for their use in a range of enzyme assays (Tables 10.3 and 10.4).

Although absorbance of 6-AQ is quite weak (ϵ ~3000 $cm^{-1}M^{-1}$), its fluorescence spectra do not overlap with either the substrate's fluorescence or sample autofluorescence, thereby providing improved assay sensitivity.[4-6] The various aminoquinoline substrates have been utilized for bacterial profiling based on differential activity of aminopeptidases.[7]

Table 10.2 *Spectral properties of coumarin-, quinoline- and rhodamine 110–based peptidase substrates and their products.*

Dye	Substrate*			Product			Solvent
	Abs †	ε ‡	Em §	Abs †	ε ‡	Em §	
AMC	324	16,000	390 ¶	342	16,000	441	pH 7
CMAC	328	14,000	405 ¶	353	14,000	466	pH 7
6-AQ	315	3500	370	339	3600	550	pH 7
R110 and CM-R110	232	57,000	none	499 #	88,000 #	521 #	MeOH
CMB-R110	232	57,000	none	492 ⊕	24,000 ⊕	529 ⊕	MeOH

* Peptidase substrates should be stored frozen (-20°C) and desiccated until required for use. The recommended solvent for preparation of stock solutions is DMSO. Molecular weights of individual substrates are indicated on the container label and are also available through our Web site (http://www.probes.com). † Absorption maximum (nm). ‡ Molar extinction coefficient at Abs (cm^{-1} M^{-1}). § Fluorescence emission maximum (nm). ¶ Fluorescence of the unhydrolyzed substrate is very weak. # Abs = 496 nm (ϵ = 80,000 $cm^{-1}M^{-1}$), Em = 520 nm in pH 7 buffer. ⊕ The cleavage product is a 4-(chloromethyl)benzoyl amide derivative of rhodamine 110. Spectral properties estimated based on data reported by Leytus and co-workers [Biochem J 209, 299 (1983)].

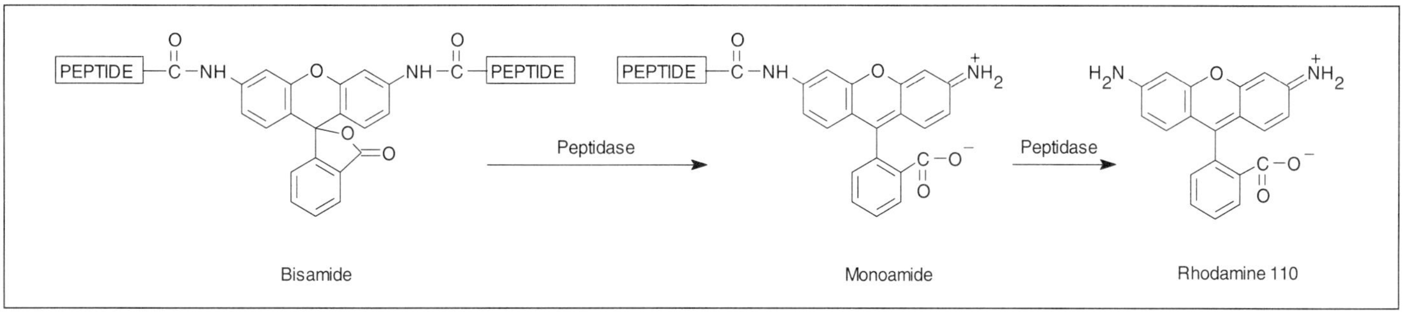

Figure 10.7 Sequential peptidase cleavage of a rhodamine 110–based substrate. The nonfluorescent bisamide substrate is first converted to the fluorescent monoamide and then to the highly fluorescent rhodamine 110.

The *p*-nitroaniline–based BAPNA (B-6552, B-6553) is the most common chromogenic substrate for assaying the amidase activity of trypsin and papain.[8-10] The optically active form (L-BAPNA, B-6552) is more water-soluble than the DL-form of BAPNA (B-6553).

Table 10.3 Aminopeptidase substrates containing a single amino acid.

Amino Acid	AMC	CMAC	CMB-110	6-AQ
L-Ala	A-8437 [1-6]	A-6527	R-6559	A-6591 [7,8]
L-Arg	NA [2,4,9-14]	A-6528		
L-Citrulline	A-8438 [15]			
Gly	A-8439 [2,8,10]	A-6526	R-6556	A-6590
L-Hydroxy-proline				A-6597 [8]
L-Ile	NA [10]			A-6595
L-Leu	A-8440 [2-4,16,17]	A-6525	R-6557 R-6509* [18]	
L-Lys	NA [1,2,11]	A-6570		A-6598 [8]
L-Met	A-8441 [2,3]			A-6594
L-Phe	A-8442 [2,4,19]		R-6577 *	A-6593 [8]
L-Pro	NA [2,20]			A-6596
L-Pyro-glutamic	A-8443 [10,21,22]			
L-Val	NA [10]			A-6592

Abbreviations: AMC = 7-amino-4-methylcoumarin; CMAC = 7-amino-4-methylcoumarin; CMB-R110 = 4-(chloromethyl)benzoyl amide of rhodamine 110; 6-AQ = 6-aminoquinoline; NA = not available from Molecular Probes.
* Rhodamine 110–based aminopeptidase substrate.

1. J Biol Chem 269, 13651 (1994); **2.** J Biol Chem 269, 13644 (1994); **3.** Zbl Bakt 280, 476 (1994); **4.** Exp Parasitol 76, 127 (1993); **5.** J Appl Bacteriol 69, 822 (1990); **6.** Biochem J 211, 567 (1983); **7.** Am J Physiol 265, C129 (1993); **8.** Appl Spectroscopy 44, 400 (1990); **9.** Arch Biochem Biophys 299, 334 (1992); **10.** FEMS Microbiol Lett 44, 349 (1987); **11.** Eur J Biochem 147, 307 (1985); **12.** Meth Enzymol 80, 535 (1981); **13.** Biochem J 191, 487 (1980); **14.** Biochem J 187, 909 (1980); **15.** Eur J Biochem 210, 759 (1992); **16.** Anal Biochem 200, 352 (1992); **17.** Chem Pharm Bull 25, 362 (1977); **18.** Cytometry 20, 334 (1995); **19.** Comp Biochem Physiol 104C, 13 (1993); **20.** J Clin Microbiol 34, 376 (1996); **21.** J Biochem 84, 467 (1978); **22.** J Biochem 83, 1145 (1978).

Visible Light–Excitable Substrates Based on Rhodamine 110

Molecular Probes' bisamide derivatives of rhodamine 110 (R110, R-6479; see Section 10.1) are sensitive and selective substrates for assaying protease activity in solution or inside live cells.[2] Originally developed by Walter F. Mangel and colleagues, these substrates contain an amino acid or peptide covalently linked to each of R110's amino groups, thereby suppressing both its visible absorption and fluorescence.[11,12] Upon enzymatic cleavage, the nonfluorescent bisamide substrate is converted in a two-step process first to the fluorescent monoamide and then to the even more fluorescent R110 (Figure 10.7). The fluorescence intensities of the monoamide and of R110 are constant from pH 3–9. Both of these hydrolysis products exhibit spectral properties similar to those of fluorescein, with peak excitation and emission wavelengths of 496 nm and 520 nm, respectively, making them compatible with flow cytometers and other argon-ion laser–based instrumentation. Substrates based on R110 may also be useful for sensitive absorptimetric assays because they yield an intensely absorbing dye ($\epsilon_{496\,nm}$ ~80,000 $cm^{-1}M^{-1}$ in pH 6 solution).

Bis-(CBZ-Arg)-R110 (BZAR, R-6501) is a general substrate for serine proteases that has proven to be 50- to 300-fold more sensitive than the analogous AMC–based substrate.[11-13] This enhanced sensitivity can be attributed both to the greater fluorescence of the enzymatic product and to the enhanced reactivity of the cleavage site. In addition, BZAR inhibits guanidinobenzoatase activity in tumor cells.[13,14] The tripeptide derivative bis-(CBZ-Ile-Pro-Arg)-R110 (BZiPAR, R-6505) allows direct and continuous monitoring of enzyme turnover, making it useful for determining individual kinetic constants of fast-acting, irreversible trypsin inhibitors.[15] R110 was also derivatized with the consensus sequence for a human adenovirus protease (bis-(CBZ-Leu-Arg-Gly-Gly)-R110, R-6511) and then employed to identify cofactors required for adenovirus virion protease activity.[16,17]

Several peptide derivatives of R110 have been used to assay protease activity in live cells. BZiPAR has been shown to enter intact cells where it is cleaved by lysosomal proteases.[18] Bis-(CBZ-Phe-Arg)-R110 (R-6502) has been employed for flow cytometric analysis of the cysteine proteases cathepsin B and L in human monocytes and rat macrophages.[19-21] In similar experiments, bis-(CBZ-Ala-Ala)-R110 (R-6504) was used to measure the activity of the lysosomal serine protease elastase.[20]

A Chemiluminescent Peptidase Substrate

6-Amino-6-deoxyluciferin (ADL, A-6611; see Section 10.5) is an alternative substrate for luciferase. Blocking the amine by amide formation completely blocks the chemiluminescent reaction, thus

Table 10.4 *Peptidase and protease substrates.*

Peptide	Enzyme(s)	AMC	CMAC	R110	CMB-R110
L-Ala-L-Pro	Dipeptidylpeptidase IV [1-4] (CD26)		A-6524		R-6555
t-BOC-L-Leu-Gly-L-Arg	Plasminogen activator	NA [5,6]		R-6515	
t-BOC-L-Leu-L-Met	Calpain		A-6520 [7]	R-6513	
CBZ-L-Ala-L-Ala	Elastase			R-6504 [8]	C-6569 *
CBZ-L-Ala-L-Ala-L-Ala-L-Ala	Elastase			R-6506 [9]	
CBZ-L-Ala-L-Arg	Elastase, trypsin			R-6508 [10,11]	
CBZ-L-Ala-L-Arg-L-Arg	Cathepsin B, cathepsin L [12]			NA [8]	
CBZ-L-Arg	Trypsin	NA [13,14]	A-6575	R-6501 [11,15-17]	R-6560
CBZ-L-Arg-L-Arg	Cathepsin B, cathepsin L, cathepsin O	A-6519 [18-21]		NA [19]	
CBZ-Gly-Gly-L-Arg	Plasminogen activator, urokinase [3,22]	NA [23-25]		R-6514	
CBZ-Gly-L-Pro-L-Arg	Granzyme A, prostatin, thrombin, trypase	NA [26-30]		R-6517	
CBZ-L-Ile-L-Pro-L-Arg	Trypsin		A-6576	R-6505 [15,16,31]	
CBZ-L-Phe-L-Arg	Cathepsin B, cathepsin C, cathepsin L, cathepsin O, follipsin, kallikrein, plasmin, prohormone thiol protease	A-6521 [6,18,20,21,32-37]	A-6522	R-6502 [8,19,38]	
CBZ-L-Pro-L-Arg	Thrombin, trypsin	NA [6]		R-6507	
CBZ-L-Val-L-Leu-L-Lys	Plasmin, prohormone thiol protease	NA [33,37,39]	A-6523		
CBZ-L-Val-L-Lys-L-Met	Amyloid A4-generating enzyme	NA [40]		R-6512	
CBZ-L-Val-L-Pro-L-Arg	Thrombin	NA [6]		R-6516	
(Glutaryl)-Gly-L-Arg	Urokinase	NA [6,37]		R-6510	
Gly-L-Arg	Dipeptidylpeptidase I, urokinase [41,42]	NA [6,43]	A-6573		
Gly-L-Pro	Dipeptidylpeptidase IV [4,44-47] (CD26)	NA [43,48-52]	A-6574		R-6561
L-Leu-L-Arg-Gly-Gly	Human adenovirus proteinase			R-6511 [53,54]	
L-Phe-L-Pro	Dipeptidylpeptidase II, dipeptidylpeptidase IV [2,47,55] (CD26)		A-6571		R-6558
L-Pro-L-Phe-L-Arg	Cathepsin C, kallikrein	NA [6,34,36,37,56,57]		R-6518	

Abbreviations: AMC = 7-amino-4-methylcoumarin; CMAC = 7-amino-4-methylcoumarin; Rh-110 = rhodamine 110 bis(peptide amide); CMB-R110 = 4-(chloromethyl)benzoyl amide of rhodamine 110; CBZ = benzyloxycarbonyl; t-BOC = *tert*-butoxycarbonyl; NA = not available from Molecular Probes. * 5-Chloromethylrhodamine 110 bis(peptide amide).

1. Clin Exp Immunol 89, 192 (1992); **2.** Histochem J 24, 637 (1992); **3.** J Periodontal Res 24, 353 (1989); **4.** Anal Biochem 74, 466 (1976); **5.** J Immunol 126, 1963 (1981); **6.** J Biochem 82, 1495 (1977); **7.** J Biol Chem 268, 23593 (1993); **8.** Biol Chem Hoppe-Seyler 373, 547 (1992); **9.** Anal Chem 65, 2352 (1993); **10.** J Immunol 154, 5376 (1995); **11.** Biochem J 215, 253 (1983); **12.** J Biol Chem 253, 4319 (1978); **13.** Biochemistry 33, 3252 (1994); **14.** Clin Chim Acta 138, 221 (1984); **15.** Biochem Biophys Res Comm 146, 107 (1987); **16.** Biochim Biophys Acta 788, 74 (1984); **17.** Biochem J 209, 299 (1983); **18.** FEBS Lett 341, 197 (1994); **19.** Biol Chem Hoppe-Seyler 373, 433 (1992); **20.** Arch Biochem Biophys 259, 131 (1987); **21.** Biochem J 201, 367 (1982); **22.** Clin Chem 27, 256 (1981); **23.** Biochem J 183, 555 (1979); **24.** Proc Natl Acad Sci USA 76, 4225 (1979); **25.** Proc Natl Acad Sci USA 75, 750 (1978); **26.** Biochemistry 34, 5164 (1995); **27.** J Biol Chem 269, 27650 (1994); **28.** J Biol Chem 269, 18843 (1994); **29.** Biochim Biophys Acta 956, 133 (1988); **30.** FEBS Lett 157, 265 (1983); **31.** Photochem Photobiol 44, 461 (1986); **32.** J Biol Chem 270, 558 (1995); **33.** Arch Biochem Biophys 314, 171 (1994); **34.** J Biochem 113, 441 (1993); **35.** FEBS Lett 257, 388 (1989); **36.** Biochem J 193, 187 (1981); **37.** Meth Enzymol 80, 341 (1981); **38.** Glia 7, 183 (1993); **39.** J Biochem 88, 183 (1980); **40.** Neurosci Lett 115, 329 (1990); **41.** Arch Biochem Biophys 295, 280 (1992); **42.** Anal Biochem 83, 143 (1977); **43.** Histochem 79, 87 (1983); **44.** Eur J Biochem 210, 161 (1992); **45.** Histochemistry 89, 151 (1988); **46.** Histochemistry 81, 167 (1984); **47.** Histochemistry 59, 153 (1979); **48.** J Biochem 116, 1182 (1994); **49.** J Biochem 111, 770 (1992); **50.** Biochemie 71, 757 (1989); **51.** Clin Chem 33, 1463 (1987); **52.** Biochem Med 19, 351 (1978); **53.** J Biol Chem 271, 536 (1996); **54.** Nature 361, 274 (1993); **55.** Biochim Biophys Acta 1069, 14 (1991); **56.** J Biol Chem 269, 16890 (1994); **57.** J Biol Chem 269, 13644 (1994).

permitting aminodeoxyluciferin to be used to prepare peptidase substrates. Currently, Molecular Probes has available a single substrate — 6-(*N*-acetyl-L-phenylalanyl)amino-6-deoxyluciferin (A-6609) — that has been used in conjunction with luciferase to selectively detect levels of α-chymotrypsin as low as 0.3 ng.[22] Please contact our Technical Assistance Department with requests for additional peptidase substrates based on 6-amino-6-deoxy-luciferin.

Tosylarginine Methyl Ester

N^{α}-*p*-Tosyl-L-arginine methyl ester (TAME, T-6536) is a nonchromogenic substrate for trypsin, thrombin, plasmin, kallikrein and other proteases.[23-25] Protease activity is usually determined by measuring the acid produced during hydrolysis of TAME[24] or by spectrophotometric detection of methanol or the residual ester.[23,26]

Peptidase Substrates Designed for Live-Cell Assays

The problems with detecting peptidase activity in live cells using fluorogenic substrates are similar to those for detecting glycosidase activity in cells (see Section 10.2):

- Difficulty in loading the substrates under physiological conditions
- Leakage of the fluorescent product from live cells
- Lack of specificity of the synthetic substrates

Although substrates based on the 7-amino-4-methylcoumarin (AMC) and 6-aminoquinoline (6-AQ) fluorophores are moderately permeant to most cell membranes, their fluorescent products are very poorly retained in live cells (Figure 10.8). Molecular Probes has developed chloromethylated versions of the AMC and rhodamine 110 (R110) fluorophore to potentially improve the cellular retention of fluorescent products. Note that many of these substrates have been developed by analogy with other fluorophore-labeled substrates. In most cases, the kinetics, specificity and utility of these new chloromethylated substrates have not yet been published. Bibliographies sent with the products will be updated as we receive publications on their uses.

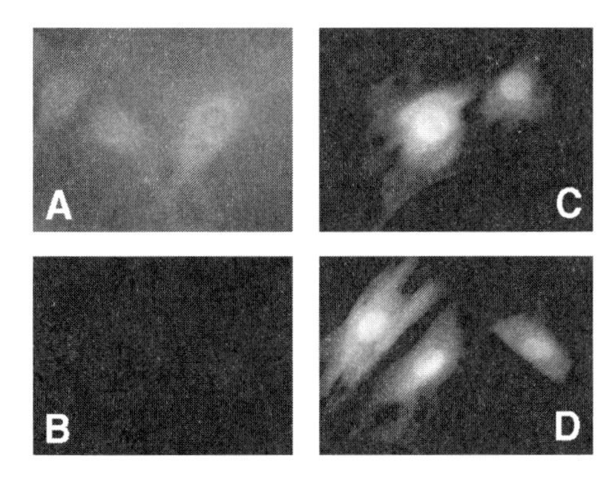

Figure 10.8 *Comparison of cellular retention of the enzymatic product from AMC-based and CMAC-based substrates. After incubating in 25 µM Leu-AMC for 30 minutes, NIH 3T3 cells showed dim blue fluorescence (panel A) that disappeared once the cells were transferred to fresh medium and allowed to grow for 4 hours (panel B). In contrast, after incubating in Molecular Probes' new Leu-CMAC (A-6525) at 37°C for 30 minutes, NIH 3T3 cells showed bright blue fluorescence (panel C) that persisted once the cells were transferred to fresh medium and allowed to grow for 40 hours (panel D).*

Aminochloromethylcoumarin (CMAC) Peptidase Substrates

Peptidase substrates based on 7-amino-4-chloromethylcoumarin (CellTracker Blue CMAC, C-2110; see Sections 10.1 and 15.2) passively diffuse into several types of cells, where the thiol-reactive chloromethyl group is enzymatically conjugated to glutathione by intracellular glutathione *S*-transferase or reacts with protein thiols, thus transforming the substrate into a membrane-impermeant probe.[27] Subsequent peptidase cleavage results in a blue fluorescent glutathione conjugate. When tested side-by-side with the traditional protease substrates, the hydrolysis products of our chloromethyl derivatives show superior retention in live cells (Figure 10.8), and staining persists even through cell division. Care must be taken to excite the fluorescence of the hydrolysis products in cells beyond ~380 nm because the substrate–peptide adduct absorbs below this wavelength, and its weak fluorescence may contribute to the overall fluorescence of the cell.

Most of our CMAC-based substrates (see Tables 10.3 and 10.4) are so new that their use in cells has not yet been reported. However, the fluorogenic *t*-BOC-Leu-Met-CMAC substrate (A-6520) has been used to measure calpain activity in hepatocytes following the addition of extracellular ATP.[28] When the Lys-CMAC substrate (A-6570) is conjugated to glutathione, its product (A-6565) becomes membrane impermeant. The glutathione conjugate of Lys-CMAC has proven useful for characterizing the topology of lysyl peptidase in cells.[29]

Chloromethyl Derivatives of Rhodamine 110–Based Peptidase Substrates

Several R110-based substrates have been shown to readily diffuse into live cells, making them useful for analyzing intracellular protease activity of cathepsins, aminopeptidases and other proteases.[18-20,30] As with the glutathione-reactive substrates based on CMAC (Figure 10.6), we have recently prepared chloromethyl derivatives of our R110-based substrates that will potentially improve the retention of fluorescent products in live cells. With these asymmetric CMB rhodamine 110 (CMB-R110) substrates, which contain a single peptide amide and a thiol-reactive (4-chloromethyl)benzamide (Figure 10.6), a single hydrolysis step and simultaneous or sequential reaction with intracellular thiols such as glutathione yield green fluorescent adducts that should be well retained in live cells. In cells lacking the peptidase activity, the substrate is expected to react with intracellular thiols, but these adducts will remain essentially nonfluorescent. The relatively lipophilic CMB group may also facilitate uptake of the substrate relative to the corresponding R110-based substrate.

We have prepared several CMB-R110–based substrates that are analogs of known AMC- or R110-based peptidase substrates; their utility for detecting enzymatic activity in cells has not yet been reported (Tables 10.3 and 10.4). These new substrates should be useful for detecting stimulation or inhibition of peptidase activity in cells by external agents and for analyzing or sorting cells by flow cytometry based on their peptidase activity. In contrast to the asymmetric monoamides, one of our new thiol-reactive R110-based substrates (C-6569) is a bisamide containing a chloromethyl moiety attached to the "bottom" ring of rhodamine 110 (CM-R110) along with two peptide groups (Figure 10.6).

Substrates for HIV Protease and Renin

Alternative strategies have been employed to create substrates specifically for endopeptidases. Our HIV protease and renin substrates (H-2930, R-2931) utilize fluorescence resonance energy transfer (see the Technical Note "Fluorescence Resonance Energy Transfer" on page 46) to generate a spectroscopic response to protease cleavage. In this type of substrate, both an acceptor molecule and a fluorescent molecule are attached to the peptide or protein. The acceptor molecule is carefully chosen so that its absorbance overlaps with the fluorophore's excited-state fluorescence (Figure 10.9), thus ensuring that the fluorescence is quenched through resonance energy transfer. Enzyme hydrolysis of the substrate results in spatial separation of the fluorophore and the acceptor molecule, thereby restoring the fluorophore's fluorescence (Figure 10.10). See Section 9.3 for a discussion of our reagents for synthesizing labeled peptides and peptidase substrates.

Substrate for Detecting HIV Protease Activity

Our HIV protease substrate 1 (H-2930) is a peptide that includes the HIV protease cleavage site, along with two covalently modified amino acid residues — one that has been linked to EDANS and the other to DABCYL.[31,32] Proteolytic cleavage releases a fragment containing only the EDANS fluorophore, thus liberating it from the quenching effect of the nearby DABCYL chromophore (Figure 10.10). HIV protease activity can be followed by exciting the sample at 340 nm and measuring the resulting fluorescence at 490 nm. This HIV protease substrate has been used to analyze the effects of solvent composition, incubation time and enzyme concentration on HIV-1 protease activity[33] and to investigate a newly designed inhibitor of the enzyme.[34] One milligram of HIV protease substrate 1 is sufficient for approximately 120 enzyme assays using 2 mL assay volumes and standard fluorescence cuvettes or ~1600 assays using 150 µL assay volumes and microcuvettes.

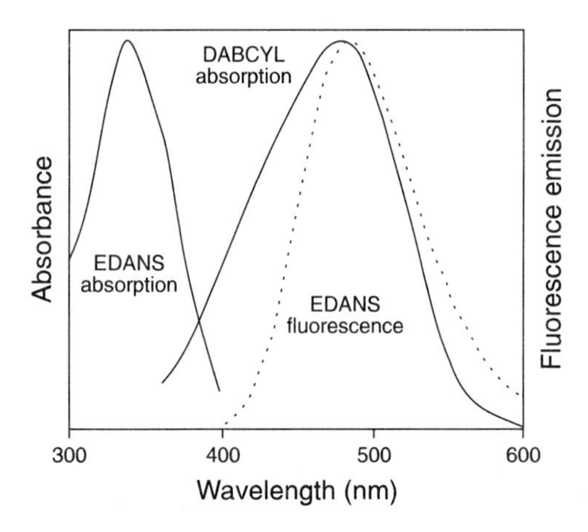

Figure 10.9 Spectral overlap between EDANS fluorescence and DABCYL absorption, which is required for efficient quenching of EDANS fluorescence by resonance energy transfer to nonfluorescent DABCYL.

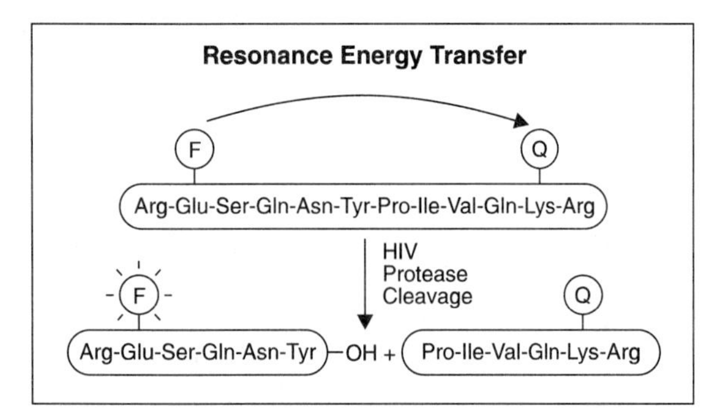

Figure 10.10 Principle of the fluorogenic response to protease cleavage exhibited by HIV protease substrate 1 (H-2930). Quenching of the EDANS fluorophore (F) by distance-dependent resonance energy transfer to the DABCYL quencher (Q) is eliminated upon cleavage of the intervening peptide linker.

Human Renin Substrate

Assaying renin activity with the renin substrate 1 (R-2931) is analogous to assaying HIV protease activity with the HIV protease substrate described above. Renin, an aspartic protease, plays an important role in blood-pressure regulation and is therefore a target for anti-hypertension therapeutics. Using renin substrate 1, researchers have discovered a stable, partially active conformational variant of recombinant human prorenin.[35] This substrate has also been used to investigate the kinetics and pH stability of recombinant human renin.[36] A fluorogenic substrate similar to our renin substrate 1 was used to develop a microplate-based assay for screening renin inhibitors.[37] One milligram of renin substrate 1 is sufficient for approximately 100 enzyme assays using 2 mL volumes and standard fluorescence cuvettes or ~1400 assays using 150 µL assay volumes and microcuvettes.

Kits for Detecting Protease Activity

Often it is necessary to have fluorogenic substrates for the assay of purified enzymes with unknown specificity or for which there are no known useful substrates. Assay for contamination of biological preparations by unknown proteases requires substrates that can detect a variety of enzymes. Our four new EnzChek Assay Kits provide exceptionally fast, simple and direct fluorescence assays for detecting a wide variety of metallo-, serine, acid and thiol proteases. Our two EnzChek Protease Assay Kits (E-6638, E-6639) measure the increase in fluorescence intensity that results from protease hydrolysis of a heavily labeled casein derivative, whereas our two EnzChek Polarization Assay Kits for Proteases (E-6658, E-6659) monitor fluorescence polarization changes that occur during protease hydrolysis of a lightly labeled fluorescent casein derivative. Although the detection principles of these assays are quite different, no separation steps are required and both assays are rapid, sensitive and can detect a wide variety of proteases.

EnzChek Protease Assay Kits for Fluorescence Intensity Measurements

Our two EnzChek Protease Assay Kits (E-6638, E-6639) contain a casein derivative that is heavily labeled with the green fluorescent BODIPY FL or red fluorescent BODIPY TR-X dyes, resulting in almost total quenching of the conjugate's fluorescence; they typically exhibit <3% of the fluorescence of the corresponding free dyes. Protease-catalyzed hydrolysis relieves this quenching, yielding brightly fluorescent BODIPY FL– or BODIPY TR-X–labeled peptides (Figure 10.11). The increase in fluorescence, which can be measured with a spectrofluorometer, minifluorometer or fluorescence microplate reader, is proportional to protease activity.

In contrast to the conventional fluorescein thiocarbamoyl (FTC)–casein protease assay, these EnzChek assays do not involve any separation steps and, consequently, can be used to continuously measure the kinetics of a variety of exopeptidases and endopeptidases over a wide pH range. They can also be used to measure the total substrate turnover at a fixed time following addition of the enzyme. We have found that these protease assays are over 100-times more sensitive and much easier to perform than the labor-intensive FTC–casein assay. Detection limits for fluorescence intensity measurements with these kits are given in Table 10.5.

In addition to their utility for detecting protease contamination in culture medium and other experimental samples, BODIPY FL casein and BODIPY TR-X casein appear to have significant potential as general nontoxic, pH-insensitive markers for phagocytic cells in culture (see Section 17.1). We have shown that uptake of these quenched conjugates by neutrophils is accompanied by hydrolysis of the labeled proteins by intracellular proteases and the generation of fluorescent products that are well retained in cells.[38] This phagocytosis assay is readily performed in a fluorescence microplate reader or a flow cytometer; localization of the fluorescent products can be determined by fluorescence microscopy.

BODIPY FL and BODIPY TR-X casein can be used interchangeably, depending on whether green or red fluorescence is desired. The peptide hydrolysis products of BODIPY FL casein exhibit green fluorescence that is optimally excited by the argon-ion laser, permitting flow sorting of the cells. The red fluorescent BODIPY TR-X–labeled peptides, with excitation and emission spectra similar to those of the Texas Red® fluorophore, should be useful for multilabeling experiments or measurements in the presence of green autofluorescence. Each EnzChek Protease Assay Kit includes:

- BODIPY FL casein *or* BODIPY TR-X casein
- 20X digestion buffer
- Detailed protocol

Sufficient reagents are supplied for ~100 assays using 2 mL assay volumes and standard fluorescence cuvettes or ~1000 assays using 200 µL assay volumes and 96-well microplates.

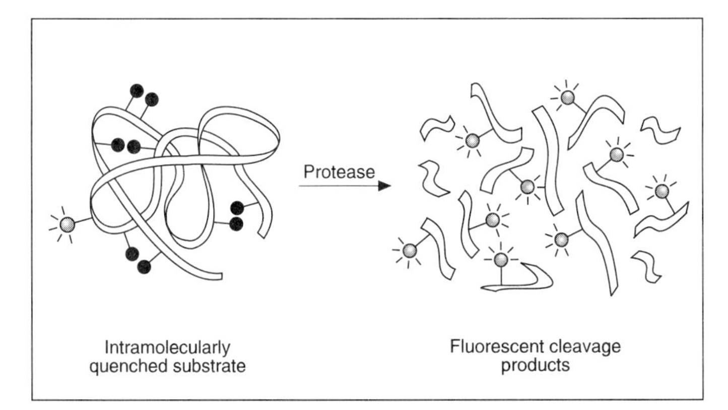

Figure 10.11 *Principle of protease detection via the disruption of intramolecular self-quenching. Protease-catalyzed hydrolysis of the heavily labeled and almost totally quenched casein substrates provided in our EnzChek Protease Assay Kits (E-6638, E-6639) relieves the intramolecular self-quenching, yielding brightly fluorescent peptides.*

Table 10.5 *Detection limits of the EnzChek Protease Assay Kits (E-6638, E-6639).*

Enzyme (Source)	Class	Detection Limit (Units) *	Buffer Conditions
Elastase, Type IV (porcine pancreas)	Serine protease	2.2×10^{-3}	10 mM Tris-HCl, pH 8.8
Chymotrypsin, Type II (bovine pancreas)	Serine protease	5.0×10^{-5}	10 mM Tris-HCl, pH 7.8
Thermolysin (*Bacillus thermoproteolyticus* Rokko)	Acid protease	4.4×10^{-5}	10 mM Tris-HCl, pH 7.8
Trypsin, Type IX (porcine pancreas)	Serine protease	1.3×10^{-2}	10 mM Tris-HCl, pH 7.8
Papain (papaya latex)	Cysteine protease	2.1×10^{-4}	10 mM MES, pH 6.2
Pepsin (porcine stomach mucosa)	Acid protease	2.1×10^{-3}	10 mM HCl, pH 2.0
Elastase (*Pseudomonas aeruginosa*)	Metalloprotease	1.0×10^{-3}	20 mM sodium phosphate, pH 8.0
Cathepsin D	Acid protease	2.0×10^{-4}	20 mM sodium citrate, pH 5.0
Elastase (human leukocyte)	Serine protease	1.0×10^{-3}	10 mM Tris-HCl pH 7.5

* The detection limit is defined as the amount of enzyme required to cause a 10–20% change in fluorescence compared to the control sample at 22°C. Enzyme unit definitions are standard definitions for each individual enzyme. Detection limits were determined with BODIPY FL casein and with BODIPY TR-X casein; both substrates yielded similar results. Detection limits may vary with instrumentation.

EnzChek Protease Assay Kits for Fluorescence Polarization Measurements

When a fluorescent molecule tethered to a protein is excited by polarized fluorescent light, the polarization of fluorescence emission is dependent on the rate of molecular tumbling. Upon proteolytic cleavage of the fluorescently labeled protein, the resulting smaller peptides tumble faster, and the emitted light is depolarized relative to that measured with the intact protein. We have introduced two EnzChek Polarization Assay Kits for Proteases (E-6658, E-6659) that contain green fluorescent BODIPY FL casein or red fluorescent BODIPY TR-X casein with an optimal degree of substitution for fluorescence polarization–based protease assays.[39,40] Fluorescence polarization technology is more sensitive than many nonradioactive protease assays and allows measurements to be taken in real time, permitting the collection of kinetic data. Our BODIPY dyes have adequate fluorescence lifetimes and pH-insensitive fluorescence — two prerequisites for successful measurement of protease activity by fluorescence polarization. Each EnzChek Polarization Assay Kit contains:

- BODIPY FL casein *or* BODIPY TR-X casein
- 5X digestion buffer
- Detailed protocol

Sufficient reagents are supplied for ~100 assays using 2 mL assay volumes and standard fluorescence cuvettes or ~1000 assays using 200 µL assay volumes and 96-well microplates. With the advent of affordable instrumentation, these kits provide an important tool for high-throughput screening of proteases and their inhibitors in research laboratories.

Fluorescein Casein

We also offer fluorescein casein (C-2990) for assaying protease activity. In this assay, unhydrolyzed fluorescein casein must be precipitated with trichloroacetic acid, separated by centrifugation, transferred for measurement and then pH-adjusted for fluorescein signal enhancement.[41-45] Fluorescein casein may be useful for a continuous assay if monitored by fluorescence polarization.[40] Fluorescein casein has been reported to be rapidly degraded by *Bacteroides gingivalis* but only slowly by streptococci.[46]

Alternative Methods for Detecting Protease Activity

Peptidases typically liberate a free amine for each hydrolysis step. Thus, fluorogenic amine detection reagents such as fluorescamine (F-2332, see Section 1.8) and *o*-phthaldialdehyde (P-2331, see Section 1.8) have been used to detect the rate of amine production by peptidases.[47-50]

Protease assays conducted in highly autofluorescent or strongly light-scattering solutions (such as crude cell and tissue extracts) can sometimes be improved by extracting the fluorescent hydrolysis product from the assay mixture with an organic solvent such as toluene, chloroform or ethyl acetate.[51] Most unhydrolyzed peptidase substrates will remain in the aqueous layer.

Endopeptidase substrates that are singly labeled at the amine terminus with a fluorophore usually do not undergo a fluorescence change upon hydrolysis of internal peptide bonds; however, fluorescence (or absorbance) of the fluorophore that remains attached to the cleaved peptide can be used to detect the hydrolysis product during separation by HPLC or capillary electrophoresis.

1. Meth Enzymol 244, 1 (1994); **2.** Symmetric rhodamine 110–based substrates are licensed to Molecular Probes under U.S. Patent Nos. 4,557,862 and 4,640,893; **3.** Patent pending; **4.** J Med Chem 27, 1166 (1984); **5.** Anal Biochem 126, 447 (1982); **6.** Anal Biochem 116, 408 (1981); **7.** Appl Spectroscopy 44, 400 (1990); **8.** Meth Enzymol 19, 226 (1970); **9.** Biochemistry 3, 180 (1964); **10.** Arch Biochem Biophys 95, 271 (1961); **11.** Biochem J 215, 253 (1983); **12.** Biochem J 209, 299 (1983); **13.** Anticancer Res 8, 1179 (1988); **14.** J Enzym Inhib 2, 209 (1988); **15.** Biochim Biophys Acta 788, 74 (1984); **16.** J Biol Chem 271, 536 (1996); **17.** Nature 361, 274 (1993); **18.** Photochem Photobiol 44, 461 (1986); **19.** Glia 7, 183 (1993); **20.** Biol Chem Hoppe-Seyler 373, 547 (1992); **21.** Biol Chem Hoppe-Seyler 373, 433 (1992); **22.** Anal Biochem 221, 329 (1994); **23.** Chem Pharm Bull 30, 2512 (1982); **24.** Meth Enzymol 45, 273 (1976); **25.** Can J Biochem Physiol 37, 1393 (1959); **26.** Arch Biochem Biophys 97, 159 (1962); **27.** U.S. Patent No. 5,362,628 and patents pending; **28.** J Biol Chem 268, 23593 (1993); **29.** J Biol Chem 271, 3328 (1996); **30.** Cytometry 20, 334 (1995); **31.** Science 247, 954 (1990); **32.** Tetrahedron Lett 31, 6493 (1990); **33.** J Biol Chem 267, 20028 (1992); **34.** Science 249, 527 (1990); **35.** J Protein Chem 9, 663 (1990); **36.** J Protein Chem 10, 553 (1991); **37.** Anal Biochem 210, 351 (1993); **38.** Zhou, M. *et al.*, manuscript in preparation; **39.** J Biomol Screening 1, 33 (1996); **40.** BioTechniques 17, 585 (1994); **41.** Cytometry 15, 213 (1994); **42.** Anal Biochem 197, 347 (1991); **43.** Anal Biochem 191, 133 (1990); **44.** Anal Biochem 143, 30 (1984); **45.** Anal Biochem 121, 290 (1982); **46.** FEMS Microbiol Lett 67, 257 (1990); **47.** Biochemistry 29, 6670 (1990); **48.** Anal Biochem 123, 41 (1982); **49.** Anal Biochem 87, 556 (1978); **50.** Biochem Biophys Res Comm 53, 75 (1973); **51.** Anal Biochem 87, 257 (1978).

10.4 Data Table *Detecting Peptidases and Proteases*

Cat #	Structure	MW	Storage	Soluble	Abs	{ε x 10^{-3}}	Em	Solvent	Product*
A-6609	✓	469	F,D,L,A	H_2O	321	{17}	450	pH 8	A-6611†
B-6552	✓	435	F,D	DMSO, H_2O	312	{16}	none	MeOH	<1>
B-6553	✓	435	F,D	DMSO	312	{16}	none	MeOH	<1>
C-2990	C	~26000‡	F,D,L	H_2O	496	ND	519	pH 8	<2>
H-2930	C	~2016	F,D,L	DMF, H_2O	430	{23}	none	MeOH	<3>
R-2931	C	~2281	F,D,L	DMF, H_2O	460	{13}	none	H_2O	<3>
T-6536	✓	379	F,D	H_2O	227	{13}	none	MeOH	NA

Properties of the following groups of peptidase substrates are summarized in Table 10.2:

AMC (A-6519, A-6521, A-8437 through A-8443)
CMAC (A-6520, A-6522 through A-6528, A-6565, A-6570, A-6571, A-6573 through A-6576)
6-AQ (A-6590 through A-6598)
CM-R110 (C-6569)
CMB-R110 (R-6555 through R-6561)
R110 (R-6501, R-6502, R-6504 through R-6518, R-6577)

For definitions of the contents of this data table, see "How to Use the Handbook" on page vi.
Structure: Chemical structure drawing: ✓ = shown in this section; C = not shown due to complexity.
MW: ~ indicates an approximate value.
ND = Not determined.
NA = Not applicable.

* Catalog number or footnote description of the enzymatic cleavage product.
† See Section 10.5.
‡ This product is a dye-protein conjugate. The number of dyes attached per protein is indicated on the product label.
<1> Enzymatic hydrolysis of this substrate yields 4-nitroaniline, Abs = 410 nm (ε = {8.8}).
<2> Protease hydrolysis results in fluorescence enhancement without a spectral shift.
<3> Fluorescence of this substrate is >99% quenched. The proteolytic cleavage products fluoresce at 500 nm (excitation at 335 nm).

10.4 Structures *Detecting Peptidases and Proteases*

A-6609

B-6552

B-6553

T-6536

Structures representing the following groups of peptidase substrates are shown in Figure 10.6:

AMC (A-6519, A-6521, A-8437 through A-8443)
CMAC (A-6520, A-6522 through A-6528, A-6565, A-6570, A-6571, A-6573 through A-6576)
6-AQ (A-6590 through A-6598)
CM-R110 (C-6569)
CMB-R110 (R-6555 through R-6561)
R110 (R-6501, R-6502, R-6504 through R-6518, R-6577)

10.4 Price List *Detecting Peptidases and Proteases*

Cat #		Product Name	Unit Size	Price Per Unit ($) 1–4 Units	5–24 Units
A-6609	New	6-(N-acetyl-L-phenylalanyl)amino-6-deoxyluciferin	1 mg	95.00	76.00
A-6527	New	7-amino-4-chloromethylcoumarin, L-alanine amide, hydrochloride (CMAC, Ala)	5 mg	95.00	76.00
A-6524	New	7-amino-4-chloromethylcoumarin, L-alanyl-L-proline amide, hydrochloride (CMAC, Ala-Pro)	5 mg	95.00	76.00
A-6528	New	7-amino-4-chloromethylcoumarin, L-arginine amide, dihydrochloride (CMAC, Arg)	5 mg	95.00	76.00
A-6520	New	7-amino-4-chloromethylcoumarin, t-BOC-L-leucyl-L-methionine amide (CMAC, t-BOC-Leu-Met)	5 mg	95.00	76.00
A-6575	New	7-amino-4-chloromethylcoumarin, CBZ-L-arginine amide, hydrochloride (CMAC, CBZ-Arg)	5 mg	95.00	76.00
A-6576	New	7-amino-4-chloromethylcoumarin, CBZ-L-isoleucyl-L-prolyl-L-arginine amide, hydrochloride (CMAC, CBZ-Ile-Pro-Arg)	5 mg	95.00	76.00
A-6522	New	7-amino-4-chloromethylcoumarin, CBZ-L-phenylalanyl-L-arginine amide, hydrochloride (CMAC, CBZ-Phe-Arg)	5 mg	95.00	76.00
A-6523	New	7-amino-4-chloromethylcoumarin, CBZ-L-valyl-L-leucyl-L-lysine amide, hydrochloride (CMAC, CBZ-Val-Leu-Lys)	5 mg	95.00	76.00
A-6526	New	7-amino-4-chloromethylcoumarin, glycine amide, hydrochloride (CMAC, Gly)	5 mg	95.00	76.00
A-6573	New	7-amino-4-chloromethylcoumarin, glycyl-L-arginine amide, dihydrochloride (CMAC, Gly-Arg)	5 mg	95.00	76.00
A-6574	New	7-amino-4-chloromethylcoumarin, glycyl-L-proline amide, hydrochloride (CMAC, Gly-Pro)	5 mg	95.00	76.00
A-6525	New	7-amino-4-chloromethylcoumarin, L-leucine amide, hydrochloride (CMAC, Leu)	5 mg	95.00	76.00
A-6570	New	7-amino-4-chloromethylcoumarin, L-lysine amide, dihydrochloride (CMAC, Lys)	5 mg	95.00	76.00
A-6571	New	7-amino-4-chloromethylcoumarin, L-phenylalanyl-L-proline amide, hydrochloride (CMAC, Phe-Pro)	5 mg	95.00	76.00
A-6565	New	7-amino-4-(S-glutathionyl)methylcoumarin, L-lysine amide	5 mg	135.00	108.00
A-8437	New	7-amino-4-methylcoumarin, L-alanine amide, trifluoroacetate salt	100 mg	65.00	52.00
A-6519	New	7-amino-4-methylcoumarin, CBZ-L-arginyl-L-arginine amide, dihydrochloride	25 mg	95.00	76.00
A-6521	New	7-amino-4-methylcoumarin, CBZ-L-phenylalanyl-L-arginine amide, hydrochloride	25 mg	95.00	76.00
A-8438	New	7-amino-4-methylcoumarin, L-citrulline amide, hydrobromide	10 mg	75.00	60.00
A-8439	New	7-amino-4-methylcoumarin, glycine amide, hydrobromide	100 mg	75.00	60.00
A-8440	New	7-amino-4-methylcoumarin, L-leucine amide, hydrochloride	25 mg	38.00	30.40
A-8441	New	7-amino-4-methylcoumarin, L-methionine amide, trifluoroacetate salt	25 mg	38.00	30.40
A-8442	New	7-amino-4-methylcoumarin, L-phenylalanine amide, trifluoroacetate salt	25 mg	38.00	30.40
A-8443	New	7-amino-4-methylcoumarin, L-pyroglutamic acid amide	5 mg	38.00	30.40
A-6591	New	6-aminoquinoline, L-alanine amide, di(trifluoroacetate) salt	10 mg	45.00	36.00
A-6590	New	6-aminoquinoline, L-glycine amide, di(trifluoroacetate) salt	10 mg	45.00	36.00
A-6597	New	6-aminoquinoline, L-hydroxyproline amide, di(trifluoroacetate) salt	10 mg	45.00	36.00
A-6595	New	6-aminoquinoline, L-isoleucine amide, di(trifluoroacetate) salt	10 mg	45.00	36.00
A-6598	New	6-aminoquinoline, L-lysine amide, tri(trifluoroacetate) salt	10 mg	45.00	36.00
A-6594	New	6-aminoquinoline, L-methionine amide, di(trifluoroacetate) salt	10 mg	45.00	36.00
A-6593	New	6-aminoquinoline, L-phenylalanine amide, di(trifluoroacetate) salt	10 mg	45.00	36.00
A-6596	New	6-aminoquinoline, L-proline amide, di(trifluoroacetate) salt	10 mg	45.00	36.00
A-6592	New	6-aminoquinoline, L-valine amide, di(trifluoroacetate) salt	10 mg	45.00	36.00
B-6553	New	N-benzoyl-DL-arginine, p-nitroanilide, hydrochloride (DL-BAPNA)	1 g	16.00	12.80
B-6552	New	N-benzoyl-L-arginine, p-nitroanilide, hydrochloride (L-BAPNA)	100 mg	32.00	25.60
C-2990		casein, fluorescein conjugate	25 mg	68.00	54.40
C-6569	New	5-(and-6)-chloromethylrhodamine 110, bis-(CBZ-L-alanyl-L-alanine amide) (CM-R110, $(CBZ\text{-}Ala\text{-}Ala)_2$) *mixed isomers*	1 mg	75.00	60.00
E-6658	New	EnzChek™ Polarization Assay Kit for Proteases *green fluorescence* *100-1000 assays*	1 kit	135.00	108.00

Cat #		Product Name	Unit Size	Price Per Unit ($) 1–4 Units	5–24 Units
E-6659	New	EnzChek™ Polarization Assay Kit for Proteases *red fluorescence* *100-1000 assays*	1 kit	135.00	108.00
E-6638	New	EnzChek™ Protease Assay Kit *green fluorescence* *100-1000 assays*	1 kit	135.00	108.00
E-6639	New	EnzChek™ Protease Assay Kit *red fluorescence* *100-1000 assays*	1 kit	135.00	108.00
H-2930		HIV Protease Substrate 1 (Arg-Glu(EDANS)-Ser-Gln-Asn-Tyr-Pro-Ile-Val-Gln-Lys(DABCYL)-Arg)	1 mg	150.00	120.00
R-2931		Renin Substrate 1 (Arg-Glu(EDANS)-Ile-His-Pro-Phe-His-Leu-Val-Ile-His-Thr-Lys(DABCYL)-Arg)	1 mg	150.00	120.00
R-6515	New	rhodamine 110, bis-(t-BOC-L-leucylglycyl-L-arginine amide), dihydrochloride	5 mg	75.00	60.00
R-6513	New	rhodamine 110, bis-(t-BOC-L-leucyl-L-methionine amide)	5 mg	75.00	60.00
R-6504	New	rhodamine 110, bis-(CBZ-L-alanyl-L-alanine amide)	5 mg	75.00	60.00
R-6506	New	rhodamine 110, bis-(CBZ-L-alanyl-L-alanyl-L-alanyl-L-alanine amide)	5 mg	75.00	60.00
R-6508	New	rhodamine 110, bis-(CBZ-L-alanyl-L-arginine amide), dihydrochloride	5 mg	75.00	60.00
R-6501	New	rhodamine 110, bis-(CBZ-L-arginine amide), dihydrochloride (BZAR)	5 mg	75.00	60.00
R-6514	New	rhodamine 110, bis-(CBZ-glycylglycyl-L-arginine amide), dihydrochloride	5 mg	75.00	60.00
R-6517	New	rhodamine 110, bis-(CBZ-glycyl-L-prolyl-L-arginine amide), dihydrochloride	5 mg	75.00	60.00
R-6505	New	rhodamine 110, bis-(CBZ-L-isoleucyl-L-prolyl-L-arginine amide), dihydrochloride (BZiPAR)	5 mg	75.00	60.00
R-6502	New	rhodamine 110, bis-(CBZ-L-phenylalanyl-L-arginine amide), dihydrochloride	5 mg	75.00	60.00
R-6507	New	rhodamine 110, bis-(CBZ-L-prolyl-L-arginine amide), dihydrochloride	5 mg	75.00	60.00
R-6512	New	rhodamine 110, bis-(CBZ-L-valyl-L-lysyl-L-methionine amide), dihydrochloride	5 mg	75.00	60.00
R-6516	New	rhodamine 110, bis-(CBZ-L-valyl-L-prolyl-L-arginine amide), dihydrochloride	5 mg	75.00	60.00
R-6510	New	rhodamine 110, bis-(glutarylglycyl-L-arginine amide), dihydrochloride	5 mg	75.00	60.00
R-6509	New	rhodamine 110, bis-(L-leucine amide), dihydrochloride	5 mg	75.00	60.00
R-6511	New	rhodamine 110, bis-(L-leucyl-L-arginylglycylglycine amide), tetrahydrochloride	5 mg	75.00	60.00
R-6577	New	rhodamine 110, bis-(L-phenylalanine amide), di(trifluoroacetic acid) salt	5 mg	75.00	60.00
R-6518	New	rhodamine 110, bis-(L-prolyl-L-phenylalanyl-L-arginine amide), tetrahydrochloride	5 mg	75.00	60.00
R-6559	New	rhodamine 110, 4-(chloromethyl)benzoyl amide, L-alanine amide, hydrochloride (CMB-R110-Ala)	1 mg	75.00	60.00
R-6555	New	rhodamine 110, 4-(chloromethyl)benzoyl amide, L-alanyl-L-proline amide, hydrochloride (CMB-R110-Ala-Pro)	1 mg	75.00	60.00
R-6560	New	rhodamine 110, 4-(chloromethyl)benzoyl amide, CBZ-L-arginine amide, hydrochloride (CMB-R110-Arg)	1 mg	75.00	60.00
R-6556	New	rhodamine 110, 4-(chloromethyl)benzoyl amide, glycine amide, hydrochloride (CMB-R110-Gly)	1 mg	75.00	60.00
R-6561	New	rhodamine 110, 4-(chloromethyl)benzoyl amide, glycyl-L-proline amide, hydrochloride (CMB-R110-Gly-Pro)	1 mg	75.00	60.00
R-6557	New	rhodamine 110, 4-(chloromethyl)benzoyl amide, L-leucine amide, hydrochloride (CMB-R110-Leu)	1 mg	75.00	60.00
R-6558	New	rhodamine 110, 4-(chloromethyl)benzoyl amide, L-phenylalanyl-L-proline amide, hydrochloride (CMB-R110-Phe-Pro)	1 mg	75.00	60.00
T-6536	New	N^{α}-p-tosyl-L-arginine methyl ester, hydrochloride (TAME)	5 g	18.00	14.40

ELF®-97 Substrates for a Variety of Enzymes and Applications *(see Section 6.3)*

Cat #		Product Name	Unit Size	Price Per Unit ($) 1–4 Units	5–24 Units
E-6578	New	ELF®-97 alcohol *1 mM solution in DMSO*	1 mL	24.00	19.20
E-6602	New	ELF®-97 Cytological Labeling Kit #1 *50 assays*	1 kit	148.00	118.40
E-6603	New	ELF®-97 Cytological Labeling Kit #2 *with streptavidin, alkaline phosphatase conjugate* *50 assays*	1 kit	198.00	158.40
E-6601	New	ELF®-97 Endogenous Phosphatase Detection Kit	1 kit	135.00	108.00
E-6580	New	ELF®-97 esterase substrate (ELF®-97 acetate)	5 mg	95.00	76.00
E-6587	New	ELF®-97 β-D-glucuronidase substrate (ELF®-97 β-D-glucuronide)	5 mg	175.00	140.00
E-6585	New	ELF®-97 guanidinobenzoatase substrate (ELF®-97 guanidinobenzoate) *>70%*	5 mg	125.00	100.00
E-6600	New	ELF®-97 Immunohistochemistry Kit	1 kit	198.00	158.40
E-6583	New	ELF®-97 lipase substrate (ELF®-97 palmitate)	5 mg	95.00	76.00
E-6581	New	ELF®-97 microsomal dealkylase substrate (ELF®-97 ethyl ether)	5 mg	95.00	76.00
E-6604	New	ELF®-97 mRNA *In Situ* Hybridization Kit #1 *50 assays*	1 kit	175.00	140.00
E-6605	New	ELF®-97 mRNA *In Situ* Hybridization Kit #2 *with streptavidin, alkaline phosphatase conjugate* *50 assays*	1 kit	225.00	180.00
E-6589	New	ELF®-97 phosphatase substrate (ELF®-97 phosphate) *5 mM solution in water* *contains 2 mM azide*	1 mL	150.00	120.00
E-6588	New	ELF®-97 phosphatase substrate (ELF®-97 phosphate) *5 mM solution in water* *sterile filtered*	1 mL	160.00	128.00
E-6579	New	ELF®-97 sulfatase substrate (ELF®-97 sulfate)	5 mg	125.00	100.00
E-6606	New	ELF® spin filters *20 filters*	1 box	48.00	38.40

10.5 Substrates for Miscellaneous Enzymes

Fluorogenic substrates that detect glycosidases (see Section 10.2) and phosphatases (see Section 10.3) have been, by far, the dominant probes for measuring enzymatic activity. Exactly the same fluorophores and chromophores — fluorescein, resorufin, umbelliferones (7-hydroxycoumarins), 4-nitrophenol, and indolyl dyes — can be used to prepare substrates for other hydrolytic enzymes (for example, lipases, cholinesterases, sulfatases) or the ether-metabolizing microsomal dealkylase (cytochrome) enzymes. We have also applied our **E**nzyme-**L**abeled **F**luorescence (ELF) technology to produce unique substrates that yield photostable fluorescent precipitates at the site of enzymatic activity.

In addition, we offer substrates for chloramphenicol acetyltransferase (CAT) and luciferase, which are usually *not* widely expressed in cells. These substrates are important tools for detecting cells transfected with reporter genes that encode these enzymes.[1,2]

Esterases

Esterases include cholinesterase, lipase and nonspecific intracellular esterases. Carboxyfluorescein diacetates (C-195, C-369; see Section 16.3) may be preferred as general esterase substrates because they are inexpensive and quite water soluble at neutral pH. Our ELF-97 esterase substrate (ELF-97 acetate, E-6580) is unique in that it yields a fluorescent yellow-green precipitate at the site of enzymatic activity, making it useful for assaying esterase activity in agarose-containing medium or potentially in live cells. Care must be taken, however, to control spontaneous hydrolysis of any ester, which increases with increasing pH.

The substrate 7-acetoxy-*N*-methylquinolinium iodide (A-1434) has been used to determine acetylcholinesterase activity.[3-5] Acetylcholinesterase also hydrolyzes *S*-acetylthiocholine, forming acetate and thiocholine. The maleimide CPM (D-346, see Section 2.2), which forms a fluorescent product when reacted with thiocholine, has been used in an ultrasensitive assay for acetylcholinesterase.[6,7] Molecular Probes also offers several cholinesterase inhibitors (see Section 10.6).

Lipases generally include cholesterol esterases and glycerol ester hydrolases. The fluorogenic 4-methylumbelliferyl palmitate substrate[8,9] (M-8424) has been used to characterize lipase activity in mycobacteria[10] and brain,[11] as well as in the diagnosis of Wolman's disease, a genetic disease characterized by low lysosomal acid lipase activity.[12,13] Upon hydrolysis, glycerol tris-(1-pyrenebutyrate) (G-7702) releases a monomeric pyrene, and a concomitant shift in fluorescence from ~470 nm to ~400 nm is observed (see Figure 13.6 in Chapter 13); this substrate has been used to assay lipases from human stomach and gastric juices.[14] Our ELF-97 lipase substrate (ELF-97 palmitate, E-6583) offers the added advantage of forming an insoluble, fluorescent precipitate, making it potentially useful for *in situ* assays. We also offer the chromogenic 5-bromo-4-chloro-3-indolyl butyrate and caprylate substrates (B-8402, B-8403), which yield blue colored precipitates and are useful for histochemical demonstration of carboxyesterase activity.[15,16] For a discussion of our extensive selection of substrates and other probes for phospholipases, see Section 20.5.

Guanidinobenzoatase

Fluorescein mono-(4-guanidinobenzoate) (FMGB, F-6489) and fluorescein di-(4-guanidinobenzoate) (FDGB, F-2890) are substrates for guanidinobenzoatase, an enzyme bound to the surface of actively migrating tumor cells.[17] This enzyme has been shown to proteolytically degrade the Gly-Arg-Gly-Asp sequence of fibronectin.[18] Hydrolysis of our ELF-97 guanidinobenzoatase substrate (ELF-97 guanidinobenzoate, E-6585) yields a fluorescent yellow-green precipitate, which may prove useful for identifying tumor cells in a mixed-cell population or for following tumor cell invasion and metastasis.[19] The fluorescein guanidinobenzoate substrates are also used as active-site titrants for serine proteases (see Section 10.6).

Sulfatases

The sulfatases belong to a highly conserved gene family; considerable sequence similarity exists among sulfatases of both prokaryotic and eukaryotic origins.[20-23] Nine different sulfatases have been characterized biochemically from lysosomal or microsomal membrane compartments. Assessment of arylsulfatase A activity in human leukocytes has been shown to be an important diagnostic tool for an autosomal recessive lysosomal storage disease called metachromatic leukodystrophy.[24,25] Lung carcinoma cells are reported to have enhanced arylsulfatase activity.[26] Another lysosomal storage disorder, multiple sulfatase deficiency, is characterized by a decreased activity of all known sulfatases.[27]

Our products for assaying arylsulfatase activity include the fluorogenic ELF-97 sulfatase substrate (ELF-97 sulfate, E-6579) and the chromogenic indolyl substrates (B-8406, B-8410). ELF-97 sulfate is expected to yield a photostable yellow-green fluorescent precipitate, whereas the indolyl sulfates (B-8406, B-8410) produce dark blue and magenta precipitates, respectively.

Microsomal Dealkylases (Cytochromes)

Metabolic oxidation of chemical compounds, including many pollutants, is the function of the cytochrome-mediated monooxygenase or mixed-function oxidase system. Several enzymes are involved, including cytochrome P-448 monooxygenase (aryl hydrocarbon hydroxylase), which is induced by carcinogenic polyaromatic hydrocarbons. Cytochrome P-450 is a useful marker of endoplasmic reticulum membranes.[28] The very low turnover rate of these enzymes can be followed using various fluorogenic alkyl ether derivatives of coumarin,[29] resorufin[30] and fluorescein.[31]

Resorufin-Based Microsomal Dealkylase Substrates

The four resorufin ether substrates (R-351, R-352, R-441, R-1147), which all yield fluorescent resorufin (excitation/emission maxima = 571/585 nm), have been extensively used to differentiate isozymes of cytochrome P-450.[30,32-37] A fluorescence microplate assay has been developed for the simultaneous determination of ethoxyresorufin *O*-de-ethylase and total protein concentration using ethoxyresorufin (R-352) and fluorescamine [38] (F-2332, see Section 9.1). Because the cytochrome isozymes exhibit differential activity on the various resorufin ethers, we offer the Resorufin Ether Sampler Kit (R-6564), which contains 1 mg samples of our four resorufin ethers, 10 mg of resorufin (R-363, see Section 10.1) and an extensive bibliography on these derivatives.

Alternatively, cytochrome P-450 activity can be assayed with phenoxazone (P-6549). Cytochrome P-450 adds a hydroxyl group to phenoxazone to produce the highly fluorescent resorufin.[39]

Coumarin-Based Microsomal Dealkylase Substrates

Possibly the most extensively used substrate for detecting microsomal dealkylase has been 7-ethoxycoumarin (E-186), which upon dealkylation yields a blue fluorescent product.[40,41] Fluorescence detection of the de-ethylation of 3-cyano-7-ethoxycoumarin (C-684) is reported to be 50–100 times more sensitive than that of resorufin ethyl ether (ethoxyresorufin), primarily because of the faster turnover rate of 3-cyano-7-ethoxycoumarin;[42,43] however, ethoxyresorufin exhibits lower fluorescence background due to its more favorable spectral shifts. The de-ethylase product of 3-cyano-7-ethoxycoumarin, 3-cyano-7-hydroxycoumarin (C-183, see Section 10.1), has a lower pK_a than that of 7-ethoxycoumarin,[40,41] allowing continuous measurements of enzyme activity at pH 7.

The cytochrome P-450 substrate 7-ethoxy-4-trifluoromethylcoumarin (E-2882) yields a product with a fluorescence emission that is distinct from that of the substrate and of NADPH, making this substrate useful for the direct measurement of enzymatic activity.[44,45] Researchers have shown that this substrate is cleaved by at least the 1A2, 2E1 and 2B1 isozymes of cytochrome P-450.[44,46]

Although we are not aware of its use as a substrate for aryl hydrocarbon hydroxylase, the succinimidyl ester of coumarin-3-carboxylic acid (SECCA, C-7933) can potentially be used to prepare conjugates of biomolecules that will be hydroxylated to highly fluorescent 7-hydroxycoumarin derivatives.[47-49]

New Microsomal Dealkylase Substrates

The fluorescent products of most microsomal dealkylase substrates rapidly leak from live cells, making them ineffective for measuring intracellular enzymatic activity by imaging or flow cytometric analysis. Using techniques for product retention that proved successful for our patented DetectaGene and ImaGene glycosidase substrates, we have developed unique new substrates that can potentially be used to detect dealkylase activity in single cells.

Like the DetectaGene β-galactosidase substrates, the mildly thiol-reactive chloromethyl moiety in our new 4-chloromethyl-7- ethoxycoumarin (C-6532) and 5-chloromethylfluorescein diethyl ether (C-6533) should react with glutathione or other intracellular thiols to produce a product that is retained in cells through cell division. Fluorescein ethers are known microsomal dealkylase substrates.[43] Adding a chloromethyl moiety to our glycosidase and peptidase substrates (see Sections 10.2 and 10.4 and Figures 10.3 and 10.6) has enabled researchers to identify cells with enzymatic activity 24 hours after loading the substrate.[50]

Patterned after our ImaGene β-galactosidase substrates, the new lipophilic C_{12}-ethoxyresorufin (D-6530), 5-dodecanoylaminofluorescein diethyl ether (D-6534) and 7-ethoxy-4-heptadecylcoumarin (E-6531) substrates are essentially nonfluorescent until enzymatically dealkylated to red, green or blue fluorescent lipophilic products, respectively. We anticipate that the lipophilic tails of these substrates will incorporate into cell membranes, dramatically improving cell retention of their enzymatic products.

Our patented ELF detection technology (see Chapter 6) permits the specific localization of enzymatic activity in cells. Our new ELF-97 microsomal dealkylase substrate (ELF-97 ethyl ether, E-6581), which yields a fluorescent yellow-green precipitate at the site of enzymatic activity, is likely to have a specificity similar to that of resorufin ethyl ether (R-352); however, its use for detection of microsomal dealkylase activity has not yet been reported.

Molecular Probes also offers resorufin 3β-hydroxy-22,23-bisnor-5-cholenyl ether [51] (R-6535), which is cleaved by cytochrome $P\text{-}450_{scc}$ to yield a red fluorescent product that can be detected in single cells by flow cytometry. This fluorogenic substrate has been used to study the regulation of steroid biosynthesis in live cells.[52,53]

Chemiluminescent Microsomal Dealkylase Substrate

The various isozymes of microsomal dealkylase are known to cleave simple ethers of a wide variety of phenols. Luciferin methyl ether (L-6619) is inactive as a luciferase substrate until its ether is cleaved to release free luciferin. Although we are not aware of its use as a microsomal dealkylase substrate, luciferin methyl ether can potentially provide an ultrasensitive assay for dealkylase activity if it is turned over by the enzyme.

Peroxidase

Current Strategies for Fluorescent Peroxidase Assays

Although horseradish peroxidase is an important enzyme for both histochemistry and ELISAs, fluorogenic peroxidase substrates have not been extensively used for its detection. Fluorogenic peroxidase substrates such as the dihydrofluoresceins (also known as fluorescins; D-399, C-400, D-1194), dihydrorhodamines (D-632, D-633, D-638; see Section 21.4) and dihydroethidium (D-1168, see Section 21.4) are converted to highly fluorescent products in the presence of the enzyme and hydrogen peroxide. Because these substrates are insufficiently stable for routine use in ELISA assays, Molecular Probes has converted the dihydrofluoresceins to diacetates. When used in intracellular applications, the acetates are cleaved by endogenous esterases, releasing the intact substrate. However, when used for *in vitro* assays, esterase or mild base must first be added to cleave the acetates, releasing the substrate. The dihydrofluoresceins have been used to measure peroxidase activity [54] and to detect hydroperoxide formation.[55-58]

Currently our best fluorogenic substrate for peroxidase is 10-acetyl-3,7-dihydroxyphenoxazine (A-6550), which reacts with hydrogen peroxide to form red fluorescent resorufin (R-363, see Section 10.1). This reagent is at least as sensitive as the common peroxidase substrate *o*-phenylenediamine but is much more stable in solution. The new reagent 10-acetyl-3,7-dihydroxy-2-dodecylphenoxazine (A-6551) is based on the lipophilic resorufin derivative 2-dodecylresorufin (D-7786; see Section 14.3), which also forms the basis of our ImaGene Red glycosidase substrates (see Section 10.2). Therefore, we anticipate that oxidation of this new peroxidase substrate by reactive oxygen species, including hydroperoxides, will form a fluorescent product that is well retained in living cells by virtue of its lipophilic tail.

Luminol: A Chemiluminescent Peroxidase Substrate

Nonisotopic immunoassays utilizing peroxidase conjugates and the chemiluminescent horseradish peroxidase substrate luminol (L-8455) have provided a rapid and sensitive method for quantitating a wide variety of analytes, including cholesterol,[59] digoxin[60] and acetylcholine.[61] Addition of trace amounts of luciferin (L-2911, L-2912, L-2916; see below) has been shown to considerably enhance the sensitivity in the assay of thyroxine, digoxin, α-fetoprotein and other analytes.[62] A method that employs luminol has been developed for the quantitation of very limiting samples of human DNA from single hairs, saliva, small blood stains and paraffin-embedded and fixed tissue sections. Using a biotinylated oligodeoxynucleotide probe to membrane-immobilized DNA, a streptavidin–horseradish peroxidase conjugate and luminol, researchers have detected 150 pg of human DNA.[63]

Chromogenic Peroxidase Substrates

The noncarcinogenic peroxidase substrate 3,3',5,5'-tetramethylbenzidine (TMB) is a commonly used chromogenic substrate for detecting horseradish peroxidase in ELISA applications[64] and in immunohistochemical and immunoblot techniques.[65] The chemical stability of TMB solutions is reported to be higher than that of diaminobenzidine solutions.[65] TMB has been used to detect peroxidase activity in neutrophils[66] and spermatozoa.[67] We offer both the TMB dihydrochloride salt (T-8464) and free base (T-8463).

2,2'-Azino-bis-(3-ethylbenzothiazine-6-sulfonic acid (also called ABTS™, a trademark of Boehringer Mannheim GmbH, A-6499) is a very sensitive peroxidase substrate for ELISAs and yields a water-soluble blue-green colored product.[68,69] It is also a useful chromogenic substrate for determination of monoamine oxidase activity in tissues.[70] In addition, ABTS can be oxidized in the presence of hydrogen peroxide by cytochrome *c*.[71]

In addition to peroxidase substrates, Molecular Probes offers horseradish peroxidase (HRP) conjugates of streptavidin and biotin (S-911, P-917; see Section 7.5), as well as fluorescent conjugates of peroxidase for use as tracers (see Section 15.7).

Glucose Oxidase

Hydrogen peroxide produced in glucose oxidase-mediated reactions has been detected with 2',7'-dichlorodihydrofluorescein diacetate[57] (D-399). We have shown that the hydrogen peroxide thus formed can be continuously determined in a coupled reaction with peroxidase that uses 10-acetyl-3,7-dihydroxyphenoxazine (A-6550) as a substrate.

Chloramphenicol Acetyltransferase (CAT)

Because of the close correlation between its transcript levels and enzymatic activity and the excellent sensitivity of the enzyme assay, the chloramphenicol acetyltransferase (CAT) reporter gene system has proven to be a powerful tool for investigating transcriptional elements in animal[72,73] and plant cells.[1,2] Most conventional CAT assays require incubation of cell extracts with radioactive substrates, typically [^{14}C]-chloramphenicol or [^{14}C]-acetyl CoA, followed by organic extraction and autoradiography or scintillation counting.[74-76] Molecular Probes' *FAST* CAT Chloramphenicol Acetyltransferase Assay Kits[77] contain unique fluorescent BODIPY chloramphenicol substrates that take advantage of the exquisite sensitivity of fluorescence techniques, thus eliminating the need for hazardous radiochemicals, film, fluors, scintillation counters and expensive radioactive waste disposal.[78,79] The original *FAST* CAT Kit and our new *FAST* CAT Green and Yellow (deoxy) Kits provide detection limits similar to those achieved with conventional radioactive methods and yield results that are easily visualized using a hand-held UV lamp.

FAST CAT Chloramphenicol Acetyltransferase Assay Kit

The green fluorescent BODIPY FL chloramphenicol substrate in our original *FAST* CAT Chloramphenicol Acetyltransferase Assay Kit (F-2900) has a K_m for purified CAT of 7.4 µM and a V_{max} of 375 pmole/unit/min, values that are similar to those of ^{14}C-labeled chloramphenicol[79,80] (K_m = 12 µM and V_{max} = 120 pmole/unit/min). Cell extracts are simply incubated with BODIPY FL chloramphenicol and acetyl CoA at 37°C. After a suitable incubation period, ice-cold ethyl acetate is added both to terminate the reaction and to extract the fluorescent substrate and its acetylated derivatives, which are then resolved by thin-layer chromatography (TLC). Brightly fluorescent, well-separated spots can easily be visualized by illuminating the TLC plate with a hand-held UV lamp; when CAT activity is sufficiently high, the TLC plates can even be monitored by visual inspection of the colored substrate and product. To quantitate CAT activity, each of the TLC spots is scraped into a microfuge tube, extracted with methanol and measured using either a fluorometer or spectrophotometer. HPLC analysis of the fluorescent products has also been used to further enhance the assay's sensitivity.[81]

These attributes have enabled researchers to use this *FAST* CAT substrate to measure CAT activity in crude cellular extracts of transfected ovarian granulosa cells.[80] Our *FAST* CAT Kit has also been employed to study hormonal regulation of prodynorphin gene expression[82,83] and to measure the rate of hair growth in single follicles of transgenic mice.[84] In the latter study, the BODIPY FL chloramphenicol substrate was used in conjunction with fluorescence-detected HPLC to provide 1000-fold greater sensitivity than traditional HPLC-UV detection of CAT products.

Each *FAST* CAT Chloramphenicol Acetyltransferase Assay Kit is available in a 100-test size and includes:

- BODIPY FL chloramphenicol substrate
- Mixture of the 1- and 3-acetyl and 1,3-diacetyl BODIPY FL derivatives, which serve as a reference standard for the fluorescent products
- Detailed protocol

The *FAST* CAT reagents are guaranteed to be stable for at least one year under recommended storage conditions.

FAST CAT (Deoxy) Chloramphenicol Acetyltransferase Assay Kits

Molecular Probes' two new *FAST* CAT (deoxy) Chloramphenicol Acetyltransferase Assay Kits (F-6616, F-6617) contain substrates that greatly simplify the quantitation of chloramphenicol acetyltransferase (CAT) activity and extend the linear detection range of Molecular Probes' original *FAST* CAT assay.[85] Our original BODIPY FL chloramphenicol *FAST* CAT substrate is a BODIPY FL derivative of the CAT enzyme's natural substrate, chloramphenicol. Chloramphenicol contains two acetylation sites,

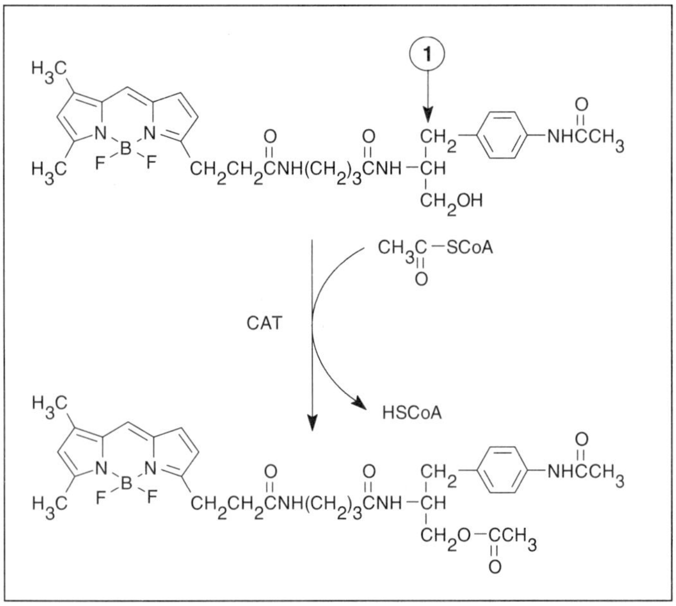

Figure 10.12 The green fluorescent BODIPY FL 1-deoxychloramphenicol substrate in our FAST CAT Green (deoxy) Chloramphenicol Acetyltransferase Assay Kit (F-6616). CAT-mediated acetylation of this substrate, or the BODIPY TMR 1-deoxychloramphenicol in our FAST CAT Yellow (deoxy) Chloramphenicol Acetyltransferase Assay Kit (F-6617), results in single fluorescent products because these substrates contain only one hydroxyl group that can be acetylated. In contrast, the BODIPY FL chloramphenicol substrate in our original FAST CAT Kit (F-2900) contains a second hydroxyl group at the 1-position (indicated by the labeled arrow). This hydroxyl group undergoes a nonenzymatic transacetylation step, restoring the original hydroxyl for a second acetylation. CAT-mediated acetylation of this chloramphenicol substrate produces three fluorescent products, thus complicating the analysis.

only one of which is acetylated by the CAT enzyme. Once the CAT enzyme adds an acetyl group to this position, the acetyl group is nonenzymatically transferred to the second site, leaving the original position open for another enzymatic acetylation.[86-88] Therefore, enzyme acetylation of our original BODIPY FL *FAST* CAT substrate produces three products — one diacetylated and two monoacetylated chloramphenicols — thus complicating the quantitative analysis of CAT gene activity. More importantly, because the nonenzymatic transacetylation is the rate-limiting step, the rate of product accumulation may not accurately reflect CAT activity.[86,89,90]

To overcome this limitation, we have modified the original *FAST* CAT substrate, producing reagents that undergo a single acetylation reaction (Figure 10.12). The green fluorescent BODIPY FL *deoxy*-chloramphenicol and yellow fluorescent BODIPY 543/569 *deoxy*-chloramphenicol substrates in our *FAST* CAT Green and *FAST* CAT Yellow (deoxy) Chloramphenicol Acetyltransferase Assay Kits (F-6616, F-6617) are acetylated at a single position, yielding only one fluorescent product [85,90] (Figure 10.12). Thus, the rate of acetylation is a direct function of enzymatic activity. This simplified reaction scheme provides a straightforward and reliable measure of CAT activity and extends the linear detection range of our original *FAST* CAT assay.

Each *FAST* CAT Green or Yellow (deoxy) Chloramphenicol Acetyltransferase Assay Kit is available in a 100-test size and includes:

- BODIPY FL *or* BODIPY TMR 1-deoxychloramphenicol substrate
- 3-Acetyl BODIPY FL or BODIPY TMR derivative, which serves as a reference standard for the fluorescent product
- Detailed protocol

The FAST CAT reagents are guaranteed to be stable for at least one year under recommended storage conditions. The CAT substrate in our *FAST* CAT Green (deoxy) Chloramphenicol Acetyltransferase Assay Kit (F-6616) is spectrally identical to the green fluorescent BODIPY FL chloramphenicol substrate in our original *FAST* CAT Kit. The FAST CAT Yellow (deoxy) Chloramphenicol Acetyltransferase Assay Kit (F-6617) contains a red-orange fluorescent BODIPY TMR derivative, which can be excited by the 543 nm spectral line of the green He-Ne laser. The availability of two spectrally distinct CAT substrates gives researchers the flexibility to choose the fluorophore compatible with a particular excitation source or multicolor labeling experiment.

Deoxychloramphenicol

For researchers who prefer to use [^{14}C]-acetyl CoA but would like to take advantage of the increased linear detection range and simplified reaction kinetics provided by our new CAT substrates, Molecular Probes also offers unlabeled 1-deoxychloramphenicol (D-6620), which is converted to a singly acetylated product by chloramphenicol acetyltransferase.[90]

Luciferase

Firefly luciferase (*Photinus*-luciferin:oxygen 4-oxidoreductase or luciferin 4-monooxygenase, EC 1.13.12.7) produces light by the ATP-dependent oxidation of luciferin (see Figure 16.9 in Chapter 16). The 560 nm chemiluminescence from this reaction peaks within seconds, with light output that is proportional to luciferase concentration when substrates are present in excess.[91] The *luc* gene, which encodes the 62,000-dalton firefly luciferase, is a popular reporter gene for plants,[1,2,92-95] bacteria [96,97] and mammalian cells [98,99] and for monitoring baculovirus gene expression in insects.[100,101] Chemiluminescent techniques are virtually background-free, making the *luc* reporter gene ideal for detecting low-level gene expression.[102]

Luciferin

The substrate for firefly luciferase, D-(-)-2-(6'-hydroxy-2'-benzothiazolyl)thiazoline-4-carboxylic acid, commonly known as luciferin, was first isolated by Bitler and McElroy (9 mg from approximately 15,000 fireflies!).[103] Molecular Probes is a primary manufacturer of synthetic luciferin (L-2911) and its water-soluble sodium and potassium salts (L-2912, L-2916). The physical properties of these derivatives are identical to those of the natural compound. Our prices for luciferin are significantly lower than those of many other suppliers; additional discounts are available for bulk purchases or standing orders.

Typically, luciferase expression is measured by adding the substrates ATP and luciferin to cell lysates and then analyzing light production with a luminometer. As little as 0.02 pg (250,000 molecules) of luciferase can be reliably measured using a standard scintillation counter.[104] Moreover, a CCD-imaging method of detecting *luc* gene expression in single cells has been developed.[105]

Caged Luciferin

Although luciferase activity is sometimes measured in living cells,[105,106] *in vivo* quantitation appears to be limited by the difficulty in delivering luciferin into intact cells.[107] Molecular Probes' DMNPE-caged luciferin (L-7085) readily crosses cell membranes. Once the caged luciferin is inside the cell, active luciferin can be released either instantaneously by a flash of UV light or continuously by the action of endogenous intracellular esterases, which are found in many cell types. This probe should facilitate *in vivo* luciferase assays in two important ways. First, caged luciferin improves the sensitivity and quantitative analysis of these assays by allowing more efficient delivery of luciferin into intact cells. Second, hydrolysis by intracellular esterases provides a continuous supply of active luciferin, permitting long-term measurements and reducing the need for rapid mixing protocols and costly injection devices. Moreover, DMNPE-caged luciferin may make it easier to follow dynamic changes in gene expression in live cells. Molecular Probes also offers DMNPE-caged ATP (A-1049, see Section 19.2), which can be used in conjunction with DMNPE-caged luciferin (L-7085) for this *in vivo* luciferase assay.[108]

6-Amino-6-Deoxyluciferin

6-Amino-6-deoxyluciferin (ADL, A-6611) is an alternative substrate for luciferase.[109] Conjugation of an amino acid or peptide to ADL's amine completely blocks its chemiluminescent reaction; thus ADL is also useful for preparing peptidase substrates (see Section 10.4).

Luciferin Methyl Ether

The methyl ether of luciferin (L-6619) is an inhibitor of the luciferase–luciferin light-generating system with a K_i of 1.0 µM.[110]

Recombinant Luciferase

Molecular Probes offers highly purified, recombinant firefly luciferase (L-7105) for *in vitro* assays and cellular microinjection procedures, as well as for generating standard activity curves. Purified firefly luciferase has been microinjected into single isolated rat cardiomyocytes [111] and hepatocytes [112] to monitor cytosolic ATP and other cytosolic parameters of metabolically compromised cells.

Luciferin–Luciferase Assays for ATP, Anesthetics and Hormones

Luciferin has been used in an exquisitely sensitive ATP assay,[113,114] which allows the detection of as little as femtomolar quantities of ATP.[115] This bioluminescent ATP assay has been employed to determine cell proliferation and cytotoxicity in both bacteria [116,117] and mammalian cells.[112,118] Molecular Probes now provides all the reagents needed for this important assay in its ATP Determination Kit (A-6608), which is described in detail in Section 10.3.

Researchers have also adapted the luciferin–luciferase ATP assay system for detecting single base changes in a solid-phase DNA sequencing method.[119] In addition, amphipathic and hydrophobic substances, including certain anesthetics and hormones, compete with luciferin for the hydrophobic site on the luciferase molecule, providing a convenient method to assay subnanomolar concentrations of these substances.[120] A protein A–luciferase fusion protein has been developed that can be used in bioluminescence-based immunoassays.[121,122]

1. Meth Cell Biol 50, 425 (1995); **2.** Meth Mol Biol 55, 147 (1995); **3.** J Am Chem Soc 113, 5467 (1991); **4.** Arch Biochem Biophys 113, 195 (1966); **5.** Biochem Pharmacol 15, 411 (1966); **6.** Biochemistry 29, 10640 (1990); **7.** Anal Biochem 133, 450 (1983); **8.** Anal Chem 40, 459R (1968); **9.** Anal Biochem 21, 279 (1967); **10.** Zbl Bakt 280, 476 (1994); **11.** J Neurochem 46, 140 (1986); **12.** J Biol Chem 270, 27766 (1995); **13.** Biochim Biophys Acta 794, 89 (1984); **14.** Biochim Biophys Acta 963, 340 (1988); **15.** J Clin Microbiol 27, 1390 (1989); **16.** Biochem Genet 25, 335 (1987); **17.** Anticancer Res 9, 247 (1989); **18.** Biochem Soc Trans 14, 742 (1986); **19.** J Biol Chem 269, 14666 (1994); **20.** Biochem Biophys Res Comm 181, 677 (1991); **21.** J Bacteriol 172, 2131 (1990); **22.** J Biol Chem 265, 3374 (1990); **23.** Devel Biol 135, 53 (1989); **24.** Mol Chem Neuropathol 24, 43 (1995); **25.** Human Mutation 4, 233 (1994); **26.** Cancer Res 40, 3804 (1980); **27.** Cell 82, 271 (1995); **28.** J Biol Chem 270, 24327 (1995); **29.** Anal Biochem 191, 354 (1990); **30.** Biochem J 240, 27 (1986); **31.** Anal Biochem 133, 46 (1983); **32.** Biochem Pharmacol 47, 893 (1994); **33.** Biochem Pharmacol 46, 933 (1993); **34.** Anal Biochem 188, 317 (1990); **35.** Biochem Pharmacol 40, 2145 (1990); **36.** Eur J Immunol 16, 829 (1986); **37.** Biochem Pharmacol 34, 3337 (1985); **38.** Anal Biochem 222, 217 (1994); **39.** Chem-Biol Interact 45, 243 (1983); **40.** Anal Biochem 115, 177 (1981); **41.** Hoppe-Seyler Z Physiol Chem 353, 1171 (1972); **42.** Anal Biochem 172, 304 (1988); **43.** Biochem J 247, 23 (1987); **44.** Biochem Pharmacol 46, 1577 (1993); **45.** Biochem Pharmacol 37, 1731 (1988); **46.** Arch Biochem Biophys 323, 303 (1995); **47.** Free Rad Biol Med 18, 669 (1995); **48.** Radiation Res 138, 177 (1994); **49.** Int'l J Radiation Biol 63, 445 (1993); **50.** J Neurosci 15, 1025 (1995); **51.** Resorufin 3β-hydroxy-22,23-bisnor-5-cholenyl ether is licensed to Molecular Probes under U.S. Patent No. 5,110,725. Molecular Probes, Inc. notifies the public and all users that MediGene, Inc. has the exclusive right to make, use and sell this probe in the field of isolating, separating and recovering fetal cells. Molecular Probes, Inc. and seller of this compound are obligated to prominently display this notice with any sale or use of this compound. Purchaser represents that it will not use this compound in the field of isolating, separating and recovering fetal cells. Purchase of this probe does not convey any license by implication, estoppel or otherwise with respect to such use; **52.** J Steroid Biochem Mol Biol 43, 479 (1992); **53.** Endocrinology 128, 2654 (1991); **54.** Anal Biochem 11, 6 (1965); **55.** J Clin Invest 87, 711 (1991); **56.** J Lab Clin Med 117, 291 (1991); **57.** Anal Biochem 187, 129 (1990); **58.** Anal Biochem 134, 111 (1983); **59.** Biochim Biophys Acta 1210, 151 (1994); **60.** Clin Chem 31, 1335 (1985); **61.** J Neurochem 39, 248 (1982); **62.** Nature 305, 158 (1983); **63.** Nucleic Acids Res 20, 5061 (1992); **64.** J Immunol Meth 143, 49 (1991); **65.** BioTechniques 8, 58 (1990); **66.** Anal Biochem 132, 345 (1983); **67.** J Cell Physiol 151, 466 (1992); **68.** Meth Enzymol 70, 419 (1980); **69.** Biochem J 145, 93 (1975); **70.** Anal Biochem 138, 86 (1984); **71.** Arch Biochem Biophys 288, 112 (1991); **72.** Development 109, 577 (1990); **73.** Proc Natl Acad Sci USA 87, 6848 (1990); **74.** Gene 67, 271 (1988); **75.** J Virol 62, 297 (1988); **76.** Anal Biochem 156, 251 (1986); **77.** U.S. Patent Nos. 5,262,545 and 5,364,764; **78.** Meth Enzymol 216, 369 (1992); **79.** BioTechniques 8, 170 (1990); **80.** Anal Biochem 197, 401 (1991); **81.** J Biol Chem 270, 28392 (1995); **82.** Mol Cell Neurosci 3, 278 (1992); **83.** Mol Endocrinol 6, 2244 (1992); **84.** Eur J Clin Chem Clin Biochem 31, 41 (1993); **85.** BioTechniques 19, 488 (1995); **86.** Biochemistry 30, 3763 (1991); **87.** Biochemistry 30, 3758 (1991); **88.** Biochemistry 29, 2075 (1990); **89.** Ann Rev Biophys Chem 20, 363 (1991); **90.** Nucleic Acids Res 19, 6648 (1991); **91.** Mol Cell Biol 7, 725 (1987); **92.** J Biolumin Chemilumin 8, 267 (1993); **93.** Meth Enzymol 216, 397 (1992); **94.** Devel Genetics 11, 224 (1990); **95.** J Biolumin Chemilumin 5, 141 (1990); **96.** J Gen Microbiol 138, 1289 (1992); **97.** Meth Mol Cell Biol 1, 107 (1989); **98.** Meth Mol Biol 7, 237 (1991); **99.** BioTechniques 7, 1116 (1989); **100.** FEBS Lett 274, 23 (1990); **101.** Gene 91, 135 (1990); **102.** Anal Biochem 176, 28 (1989); **103.** Arch Biochem Biophys 72, 358 (1957); **104.** Anal Biochem 171, 404 (1988); **105.** J Biolumin Chemilumin 5, 123 (1990); **106.** Bio/Technology 10, 565 (1992); **107.** Biochem J 276, 637 (1991); **108.** BioTechniques 15, 848 (1993); **109.** Anal Biochem 221, 329 (1994); **110.** Arch Biochem Biophys 134, 381 (1969); **111.** J Biolumin Chemilumin 9, 363 (1994); **112.** Biochem J 295, 165 (1993); **113.** J Appl Biochem 3, 473 (1981); **114.** Anal Biochem 29, 381 (1969); **115.** Lett Appl Microbiol 1, 208 (1990); **116.** Biotech Bioeng 42, 30 (1993); **117.** J Biolumin Chemilumin 6, 193 (1991); **118.** J Immunol Meth 160, 81 (1993); **119.** Anal Biochem 208, 171 (1993); **120.** Anal Biochem 190, 304 (1990); **121.** Anal Biochem 208, 300 (1993); **122.** J Immunol Meth 137, 199 (1991).

Cat #	Structure	MW	Storage	Soluble	Abs	{ϵ x 10⁻³}	Em	Solvent	Product*
A-1434	✓	329	F,D	DMF	317	{7.9}	410	pH 5	<1>
A-6499	✓	549	D,L	H_2O	340	{40}	none	pH 7	<2>
A-6550	S21.4	257	FF,D,AA	DMSO	280	{6.0}	none	pH 8	R-363†
A-6551	S21.4	426	FF,D,AA	DMSO	286	{8.0}	none	MeOH	D-7786‡
A-6611	✓	279	F,D,L,A	pH>6, DMSO	349	{17}	521	pH 7	<3>
B-8402	✓	317	F,D	DMSO	289	{6.1}	none	MeOH	<4>
B-8403	✓	373	F,D	DMSO	289	{6.2}	none	MeOH	<4>
B-8406	✓	365	F,D	H_2O	287	{6.0}	none	pH 7	<4>
B-8410	✓	365	F,D	H_2O	292	{5.0}	none	pH 7	<5>
C-400	S21.4	531	F,D,AA	pH>6, DMSO	290	{5.6}	none	MeCN	<6>
C-684	✓	215	L	DMSO	356	{20}	411	pH 7	C-183†
C-6532	✓	239	L	DMSO	326	{13}	397§	MeOH	C-2111¶
C-6533	✓	437	F,D	DMSO	275	{8.2}	none	MeOH	<7>
C-7933	S21.3	287	F,D,L	DMSO	299	{16}	431	MeOH	<8>
D-399	S21.4	487	F,D,AA	pH>6, DMSO	258	{11}	none	MeOH	<6>
D-1194	✓	418	F,D,AA	pH>6, DMSO	272	{4.6}	none	MeOH	<6>
D-6530	✓	410	L	DMSO	457	{23}	none	MeOH	D-7786‡
D-6534	✓	586	F	DMSO	240	{37}	none	$CHCl_3$	D-109‡
D-6620	✓	307	F,D	DMSO	273	{10}	none	MeOH	<9>
E-186	✓	190	L	DMSO, DMF	324	{11}	390§	pH 7	<10>
E-2882	✓	258	L	DMSO, DMF	333	{14}	415	MeOH	T-659†
E-6531	✓	429	L	DMSO	321	{15}	379§	MeOH	H-181‡
E-6579	✓	409	F,D,L	DMSO, H_2O	287	{14}	ND#	pH 7	E-6578†
E-6580	✓	349	F,D,L	DMSO, DMF	289	{16}	ND#	MeOH	E-6578†
E-6581	✓	335	NC	DMSO, DMF	313	{15}	ND§	MeOH	E-6578†
E-6583	✓	546	F,D,L	DMSO, DMF	288	{14}	ND#	MeOH	E-6578†
E-6585	✓	505	F,D,L	DMSO, DMF	277	{29}	ND#	MeOH	E-6578†
F-2890	✓	728	F,D	DMF	272	{45}	none	MeOH	F-1300†
F-2900	✓	583	F,D,L	MeOH	504	{80}	511	MeOH	<11>
F-6489	✓	530	F,D,L	DMF	452	{15}	519	pH 7	F-1300†
F-6616	✓	567	F,D,L	MeOH	504	{81}	510	MeOH	<9>
F-6617	✓	674	F,D,L	MeOH	545	{60}	570	MeOH	<9>
G-7702	✓	903	FF,D,L	$CHCl_3$	341	{107}	473	MeOH	<12>
L-2911	✓	280	F,D,L,A	pH>6, DMSO	328	{18}	532	pH 7	<13>
L-2912	✓	302	F,D,L,A	pH>6	328	{17}	533	pH 7	<13>
L-2916	✓	318	F,D,L,A	pH>6	328	{18}	533	pH 7	<13>
L-6619	✓	294	F,D,L,A	DMSO	328	{18}	436	pH 7	<14>
L-7085	✓	490	FF,D,LL⊕	DMSO, DMF	334	{22}	none	MeOH	<15>
L-8455	S21.3	177	D,L	DMF	355	{7.5}	411	MeOH	<16>
M-8424	✓	415	F,D	DMF	310	{9.8}	379#	MeOH	H-189†
P-6549	S21.3	125	L	DMSO	445	{10}	none	MeOH	<17>
R-351	✓	227	L	DMSO	463	{23}	none	MeOH	R-363†
R-352	✓	241	L	DMSO	464	{23}	none	MeOH	R-363†
R-441	✓	303	L	DMSO	463	{21}	none	MeOH	R-363†
R-1147	✓	283	L	DMSO	465	{21}	none	MeOH	R-363†
R-6535	✓	528	L	DMSO	466	{22}	none	MeOH	R-363†
T-8463	✓	240	D	DMSO	287	{25}	none	MeOH	<18>
T-8464	✓	313	D	DMSO, H_2O	292	{19}	none	MeOH	<18>

For definitions of the contents of this data table, see "How to Use the Handbook" on page vi.
Structure: Chemical structure drawing: ✓ = shown in this section; Sn.n = shown in Section number n.n.
ND = not determined.

* Catalog number or footnote description of the enzymatic cleavage, addition or oxidation product.
† See Section 10.1.
‡ See Section 14.3.
§ Fluorescence of the alkylated substrate is very weak.
¶ See Section 15.2.
# Fluorescence of the unhydrolyzed substrate is very weak.
⊕ All photoactivatable probes are sensitive to light. They should be protected from illumination, except when photolysis is intended.
<1> A-1434 is hydrolyzed by acetylcholinesterase to 7-hydroxy-*N*-methylquinoline, Abs = 406 nm (ϵ = {9.8}), Em = 506 nm in pH 8 buffer.
<2> Peroxidase oxidation product has Abs = 414 nm (ϵ = {36}) in H_2O [BioTechniques 15, 271 (1993)].
<3> ATP-dependent oxidation of A-6611 by luciferase (L-7105) results in bioluminescence (Em = 605 nm). The bioluminescence spectrum is pH independent [J Am Chem Soc 88, 2015 (1966)].
<4> Enzymatic cleavage of this substrate yields a water-insoluble, blue colored indigo dye (Abs ~615 nm).
<5> Enzymatic cleavage of this substrate yields a water-insoluble, magenta colored indigo dye (Abs ~565 nm).
<6> Dihydrofluorescein diacetates are colorless and nonfluorescent until both the acetates are hydrolyzed and the products are subsequently oxidized to fluorescein derivatives. The materials contain less than 0.1% of oxidized derivative when initially prepared. C-400 and D-399 give 2′,7′-dichlorofluorescein derivatives, which have Abs = 504 nm (ϵ = {90}), Em = 529 nm. D-1194 gives fluorescein (F-1300, see Section 10.1).
<7> Enzymatic cleavage of this substrate yields 5-chloromethylfluorescein, with spectroscopic properties similar to F-1300 (see Section 10.1).

<8> Enzymatic hydroxylation yields a 7-hydroxycoumarin derivative with spectroscopic properties similar to H-185 (see Section 10.1).
<9> Acetylation by chloramphenicol acetyltransferase (CAT) yields a 3-acetyl-1-deoxychloramphenicol derivative with similar spectroscopic properties to the substrate.
<10> Enzymatic cleavage of this substrate yields 7-hydroxycoumarin (umbelliferone), which has similar spectroscopic properties to H-189 (see Section 10.1).
<11> Acetylation by chloramphenicol acetyltransferase (CAT) yields a mixture of 1-acetyl-, 3-acetyl- and 1,3-diacetyl-chloramphenicol derivatives. Spectroscopic properties of these products are similar to those of the substrate.
<12> The intact compound exhibits pyrene excimer fluorescence. Lipase hydrolysis yields P-32 (see Section 13.2).
<13> ATP-dependent oxidation of luciferin by luciferase (L-7105) results in bioluminescence (Em = 560 nm) at neutral and alkaline pH. Bioluminescence is red-shifted (Em = 617 nm) under acidic conditions [J Am Chem Soc 88, 2015 (1966)].
<14> L-6619 is a luciferase inhibitor ($K_i = 2 \times 10^{-6}$ M) [Arch Biochem Biophys 134, 381 (1969)].
<15> L-7085 is converted to bioluminescent luciferin (L-2911) upon ultraviolet photolysis.
<16> L-8455 emits chemiluminescence (Em = 425 nm) upon oxidation in basic aqueous solutions.
<17> Enzymatic hydroxylation yields resorufin (R-363, see Section 10.1).
<18> Enzymatic oxidation product of T-8463 and T-8464 is blue colored (Abs = 650 nm with a second peak at 370 nm) in basic solutions, or yellow colored (Abs = 450 nm) in acidic solutions.

10.5 Structures *Substrates for Miscellaneous Enzymes*

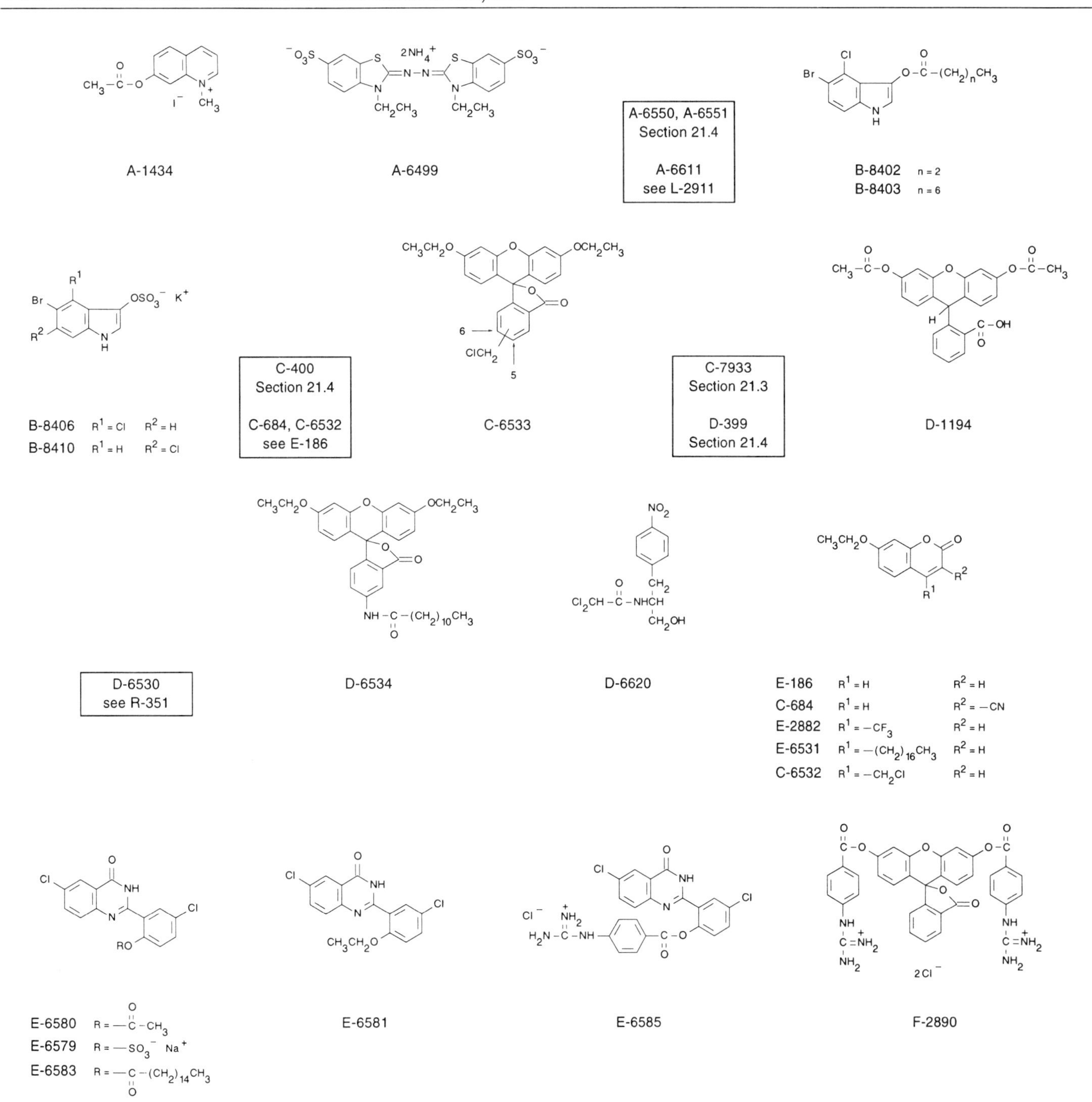

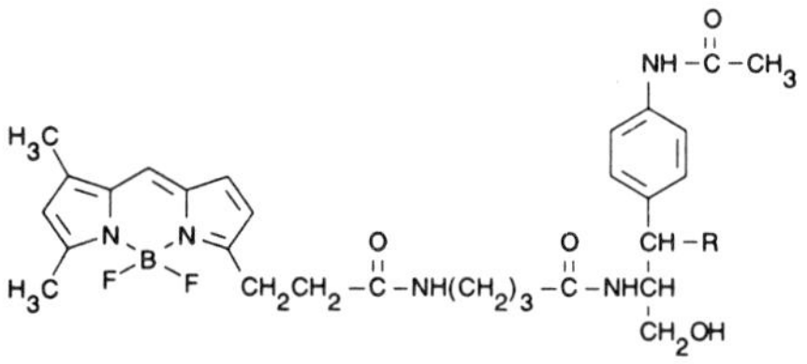

F-2900 R = —OH
F-6616 R = H

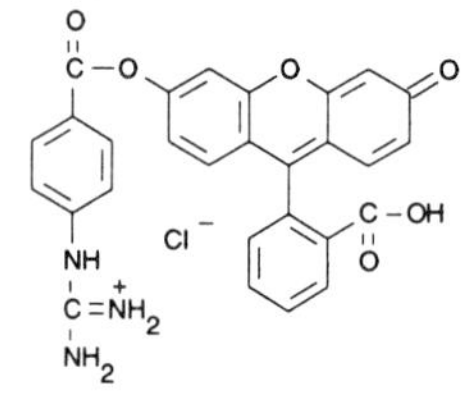

F-6489

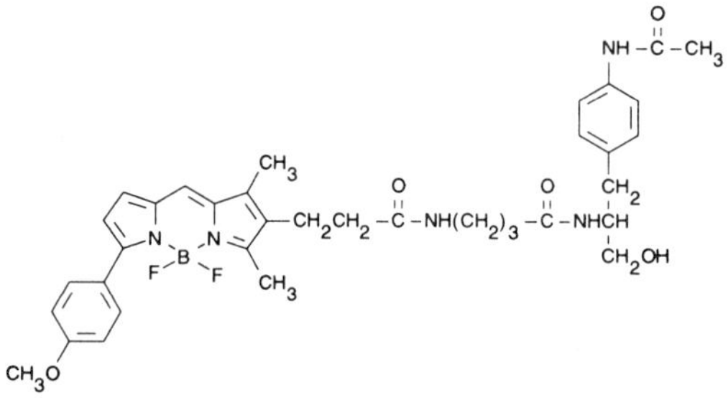

F-6617

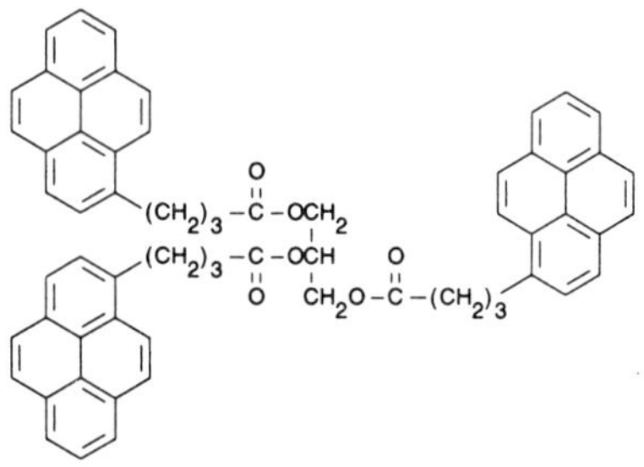

G-7702

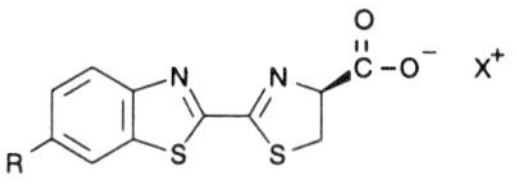

L-2911	R = —OH	X = H
L-2912	R = —OH	X = Na
L-2916	R = —OH	X = K
A-6611	R = $—NH_2$	X = H
L-6619	R = $—OCH_3$	X = H

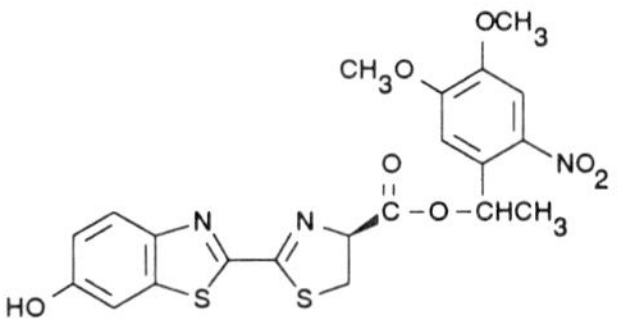

L-7085

L-8455
Section 21.3

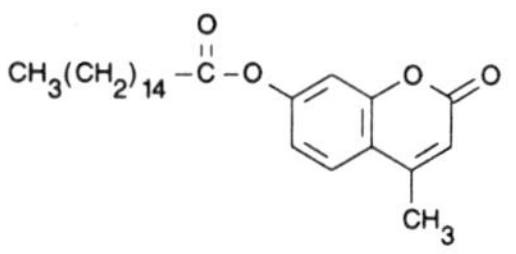

M-8424

P-6549
Section 21.3

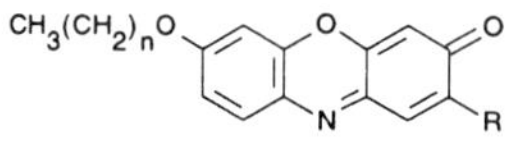

R-351	R = H	n = 0
R-352	R = H	n = 1
R-1147	R = H	n = 4
D-6530	R = $—(CH_2)_{11}CH_3$	n = 1

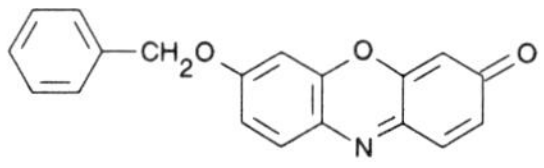

R-441

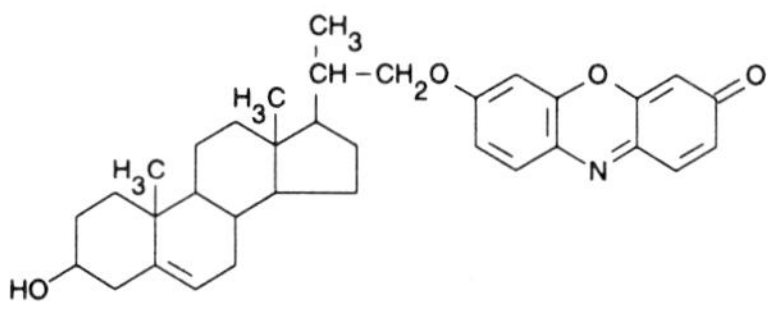

R-6535

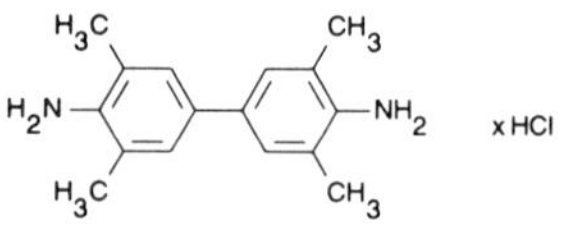

T-8463 x = 0
T-8464 x = 2

10.5 Price List *Substrates for Miscellaneous Enzymes*

Cat #		Product Name	Unit Size	Price Per Unit ($) 1–4 Units	5–24 Units
A-1434		7-acetoxy-N-methylquinolinium iodide	25 mg	35.00	28.00
A-6551	**New**	10-acetyl-3,7-dihydroxy-2-dodecylphenoxazine *special packaging*	10x100 µg	135.00	108.00
A-6550	**New**	10-acetyl-3,7-dihydroxyphenoxazine *special packaging*	10x1 mg	85.00	68.00
A-6611	**New**	6-amino-6-deoxyluciferin (ADL)	5 mg	95.00	76.00
A-6499	**New**	2,2′-azino-bis-(3-ethylbenzothiazine-6-sulfonic acid), diammonium salt	1 g	18.00	14.40
B-8402	**New**	5-bromo-4-chloro-3-indolyl butyrate	25 mg	48.00	38.40
B-8403	**New**	5-bromo-4-chloro-3-indolyl caprylate	25 mg	48.00	38.40
B-8406	**New**	5-bromo-4-chloro-3-indolyl sulfate, potassium salt (X-sulfate)	100 mg	35.00	28.00
B-8410	**New**	5-bromo-6-chloro-3-indolyl sulfate, potassium salt	100 mg	48.00	38.40
C-400		5-(and-6)-carboxy-2′,7′-dichlorodihydrofluorescein diacetate (carboxy-H_2DCFDA) *mixed isomers*	25 mg	88.00	70.40
C-6532	**New**	4-chloromethyl-7-ethoxycoumarin	10 mg	48.00	38.40
C-6533	**New**	5-(and-6)-chloromethylfluorescein diethyl ether *mixed isomers*	5 mg	95.00	76.00
C-7933	**New**	coumarin-3-carboxylic acid, succinimidyl ester (SECCA)	25 mg	65.00	52.00
C-684		3-cyano-7-ethoxycoumarin	10 mg	85.00	68.00
D-6620	**New**	1-deoxychloramphenicol (L-(2-dichloroacetylamino)-3-(4-nitrophenyl)propanol)	5 mg	98.00	78.40
D-399		2′,7′-dichlorodihydrofluorescein diacetate (2′,7′-dichlorofluorescin diacetate; H_2DCFDA)	100 mg	45.00	36.00
D-1194		dihydrofluorescein diacetate (H_2FDA)	1 g	58.00	46.40
D-6534	**New**	5-dodecanoylaminofluorescein diethyl ether	5 mg	95.00	76.00
D-6530	**New**	2-dodecyl-6-ethoxyphenoxazin-3-one (C_{12}-ethoxyresorufin)	5 mg	95.00	76.00
E-6580	**New**	ELF®-97 esterase substrate (ELF®-97 acetate)	5 mg	95.00	76.00
E-6585	**New**	ELF®-97 guanidinobenzoatase substrate (ELF®-97 guanidinobenzoate) *>70%*	5 mg	125.00	100.00
E-6583	**New**	ELF®-97 lipase substrate (ELF®-97 palmitate)	5 mg	95.00	76.00
E-6581	**New**	ELF®-97 microsomal dealkylase substrate (ELF®-97 ethyl ether)	5 mg	95.00	76.00
E-6579	**New**	ELF®-97 sulfatase substrate (ELF®-97 sulfate)	5 mg	125.00	100.00
E-186		7-ethoxycoumarin	100 mg	24.00	19.20
E-6531	**New**	7-ethoxy-4-heptadecylcoumarin	10 mg	55.00	44.00
E-2882		7-ethoxy-4-trifluoromethylcoumarin	25 mg	60.00	48.00
F-2900		*FAST* CAT® Chloramphenicol Acetyltransferase Assay Kit *100 assays*	1 kit	95.00	76.00
F-6616	**New**	*FAST* CAT® Green (deoxy) Chloramphenicol Acetyltransferase Assay Kit *100 assays*	1 kit	125.00	100.00
F-6617	**New**	*FAST* CAT® Yellow (deoxy) Chloramphenicol Acetyltransferase Assay Kit *100 assays*	1 kit	125.00	100.00
F-2890		fluorescein di-(4-guanidinobenzoate), dihydrochloride	5 mg	75.00	60.00
F-6489	**New**	fluorescein mono-(4-guanidinobenzoate), hydrochloride	5 mg	95.00	76.00
G-7702	**New**	glycerol tris-(1-pyrenebutyrate)	5 mg	48.00	38.40
L-7105	**New**	luciferase, firefly *recombinant* *in solution*	1 mg	80.00	64.00
L-7085	**New**	D-luciferin, 1-(4,5-dimethoxy-2-nitrophenyl)ethyl ester (DMNPE-caged luciferin)	5 mg	65.00	52.00
L-2911		D-luciferin, free acid	25 mg	80.00	64.00
L-6619	**New**	luciferin methyl ether	5 mg	85.00	68.00
L-2916		D-luciferin, potassium salt	25 mg	88.00	70.40
L-2912		D-luciferin, sodium salt	25 mg	88.00	70.40
L-8455	**New**	luminol (3-aminophthalhydrazide)	25 g	75.00	60.00
M-8424	**New**	4-methylumbelliferyl palmitate	100 mg	20.00	16.00
P-6549	**New**	phenoxazone	10 mg	98.00	78.40
R-441		resorufin benzyl ether (benzyloxyresorufin)	10 mg	98.00	78.40
R-6564	**New**	Resorufin Ether Sampler Kit	1 kit	98.00	78.40
R-352		resorufin ethyl ether (ethoxyresorufin)	5 mg	98.00	78.40
R-6535	**New**	resorufin 3β-hydroxy-22,23-bisnor-5-cholenyl ether	1 mg	125.00	100.00
R-351		resorufin methyl ether (methoxyresorufin)	5 mg	98.00	78.40
R-1147		resorufin pentyl ether (pentoxyresorufin)	5 mg	98.00	78.40
S-911		streptavidin, peroxidase from horseradish conjugate	1 mg	95.00	76.00
T-8464	**New**	3,3′,5,5′-tetramethylbenzidine, dihydrochloride (TMB, HCl)	5 g	95.00	76.00
T-8463	**New**	3,3′,5,5′-tetramethylbenzidine, free base (TMB)	5 g	75.00	60.00

10.6 Enzyme Inhibitors, Activators and Active-Site Titrants

Enzyme inhibitors are essential reagents for working with cells and protein solutions and for determining enzyme classification and mechanism.[1] Molecular Probes provides a selection of the most effective enzyme inhibitors, as well as some unique natural product inhibitors or activators (Tables 10.6 and 10.7). We also offer several fluorescent and light-activated enzyme inhibitors.

Inhibitors and Active-Site Titrants for Proteases

Sulfonyl Fluorides

The high reactivity of an active-site serine residue in certain enzymes, including α-chymotrypsin, trypsin, thrombin and subtilisin, permits its selective modification by sulfonyl fluorides. The

Table 10.6 *Protease inhibitors.*

Cat #	Probe	Properties
A-6480	4-Aminobenzamidine	• Inhibitor of several arginine-selective proteases, including trypsin, thrombin and kallikrein [1] • Intrinsically fluorescent • Useful as a structural probe of thrombin [2] and other serine proteases [3,4]
A-6202	4-(2-Aminoethyl)benzenesulfonyl fluoride (AEBSF)	• Water-soluble inhibitor of trypsin- and chymotrypsin-like proteases [5,6] • Relatively stable and nontoxic [7]
B-6546	Benzamidine	• Inhibitor of several arginine-selective proteases, including trypsin, thrombin and kallikrein [1]
B-6286	4-Bromophenacyl bromide	• Inhibitor of carboxypeptidase C_n and phospholipase A_2 [8-11]
C-6485	*N*-CBZ-L-leucyl-L-leucyl-L-tyrosine, diazomethylketone	• Potent cell-permeant inhibitor of calpain,[12-14] cathepsin B [15] and cathepsin L [16]
E-6538	1,2-Epoxy-3-(4-nitrophenoxy)propane (EPNP)	• Inhibitor of carboxyl (acid) proteases that contain aspartate residues at their active site, including gastricsin, pepsin and renin [17] • Inhibiter of simian immunodeficiency virus protease [18] (SIV PR)
E-6545	*trans*-Epoxysuccinyl-L-leucylamide-(4-guanidino)butane (E-64)	• Inhibitor of cysteine proteases, including papain, calpain and cathepsins B, H, L and S [19,20]
F-99013	Fisonate	• Synthetic inhibitor of a broad spectrum of serine proteases [21,22]
F-99014	Flodiphoside	• Synthetic irreversible inhibitor of thrombin [23]
L-6543	Leupeptin	• Inhibitor of cathepsin B, clostripain, papain, thrombin and trypsin [24]
M-6539	Methyl methanethiolsulfonate (MMTS)	• Reversible inhibitor of cysteine proteases such as papain [25,26]
P-6542	Pepstatin A	• Microbial inhibitor of carboxyl (acid) proteases that contain aspartate residues at their active site, including cathepsin D, pepsin and renin [27-30]
P-6544	1,10-Phenanthroline	• General metalloprotease inhibitor [31]
P-6203	Phenylmethanesulfonyl fluoride (PMSF)	• Irreversible inhibitor of serine proteases,[32] including α-chymotrypsin, trypsin, dipeptidylpeptidase II, dipeptidylpeptidase IV, kallikrein, subtilisin and thrombin [7,32,33] • Inhibitor of acetylcholinesterase [33]
T-6483	N^{α}-Tosyl-L-lysine chloromethyl ketone (TLCK)	• Inhibitor of trypsin-like serine proteases [34,35] and certain cysteine proteases such as papain [36] • Does not inhibit chymotrypsin
T-6484	*N*-Tosyl-L-phenylalanine chloromethyl ketone (TPCK)	• Inhibitor of chymotrypsin and papain [37,38] • Does not inhibit trypsin

1. J Biol Chem 240, 1579 (1965); **2.** J Biol Chem 257, 14891 (1982); **3.** J Biol Chem 270, 30007 (1995); **4.** Biochemistry 33, 7897 (1994); **5.** J Biol Chem 269, 4539 (1994); **6.** J Biol Chem 268, 27355 (1993); **7.** Anal Biochem 86, 574 (1978); **8.** Biochemistry 19, 743 (1980); **9.** Eur J Biochem 94, 531 (1979); **10.** J Biol Chem 252, 2405 (1977); **11.** J Biochem (Tokyo) 76, 375 (1974); **12.** Biochem Biophys Res Comm 214, 1130 (1995); **13.** Exp Cell Res 215, 164 (1994); **14.** J Biol Chem 268, 23593 (1993); **15.** Biochem J 253, 751 (1988); **16.** Biochem J 274, 497 (1991); **17.** J Biol Chem 246, 4510 (1971); **18.** Biochemistry 32, 12498 (1993); **19.** J Biol Chem 269, 30238 (1994); **20.** Biochem J 201, 189 (1982); **21.** Bioorganic Chem (Russian) 16, 1500 (1990); **22.** Bioorganic Chem (Russian) 15, 987 (1989); **23.** Biochemistry (Moscow) 57, 21 (1992); **24.** Chem Pharm Bull 17, 1896 (1969); **25.** Biochemistry 25, 5595 (1986); **26.** Biochemistry 14, 766 (1975); **27.** J Biol Chem 267, 24725 (1992); **28.** Biochemistry 24, 3165 (1985); **29.** J Antibiotics 29, 97 (1976); **30.** Meth Enzymol 45, 689 (1976); **31.** FASEB J 9, 974 (1995); **32.** Biochem Pharmacol 27, 2693 (1978); **33.** J Pharmacol Exp Ther 167, 98 (1969); **34.** J Biol Chem 246, 4594 (1979); **35.** Biochemistry 4, 2219 (1965); **36.** Arch Biochem Biophys 124, 70 (1968); **37.** FEBS Lett 2, 143 (1969); **38.** Biochemistry 2, 252 (1963).

low intrinsic reactivity of sulfonyl fluorides makes them uniquely useful for selectively labeling proteases in complex mixtures such as cell lysates. The most important nonfluorescent inhibitors of serine proteases are phenylmethanesulfonyl fluoride [2,3] (PMSF, P-6203) and its water-soluble analog 4-(2-aminoethyl)benzenesulfonyl fluoride [4,5] (AEBSF, A-6202), both of which are available from Molecular Probes.

Fluorescent sulfonyl fluorides are useful probes of enzyme structure and localization. Dansyl fluoride (D-11) selectively reacts with serine residues at the reactive sites of chymotrypsin,[6] thrombin,[7] subtilisin,[8] urokinase [9] and other enzymes. Dansyl fluoride was recently used to locate tumor cells in frozen sections of carcinoma tissues by its selective reaction with guanidinobenzoatase.[10]

We have also synthesized a new fluorescent derivative of the serine protease inhibitor AEBSF, BODIPY FL AEBSF (D-6201). This probe exhibits fluorescein-like spectral characteristics and may be useful for labeling proteases in live cells or, as in the example above of dansyl fluoride,[10] to identify tumor cells; however, its specificity and efficacy as an inhibitor has not yet been established.

Protease Inhibitors

Molecular Probes offers a diverse selection of protease inhibitors (Table 10.6). Because protease inhibitors are frequently used in combination, our new Protease Inhibitor Sampler Kit (P-6548) should be particularly useful. Each kit contains eight of the most commonly used inhibitors:

- *trans*-Epoxysuccinyl-L-leucylamido-(4-guanidino)butane (also called "E-64," our catalog number E-6545)
- Leupeptin (L-6543)
- Pepstatin A (P-6542)
- *N*-Tosyl-L-phenylalanine chloromethyl ketone (TPCK, T-6484)
- N^{α}-Tosyl-L-lysine chloromethyl ketone (TLCK, T-6483)
- Phenylmethanesulfonyl fluoride (PMSF, P-6203)
- Benzamidine (B-6546)
- 1,10-Phenanthroline (P-6544)

These protease inhibitors are also offered individually under the catalog numbers listed above. Their selectivities are indicated in Table 10.6.

Photolabile Serine-Protease Inhibitors

Molecular Probes prepares two photolabile serine protease inhibitors — *trans-p*-(diethylamino)-*o*-hydroxy-α-methylcinnamic acid, *p*-nitrophenyl ester (D-7112) and *trans-p*-(diethylamino)-*o*-hydroxy-α-methylcinnamic acid, *p*-amidinophenyl ester (D-7113) — for investigating the active site of chymotrypsin, factor X_a and thrombin.[11] These compounds react with the active-site serine residue to form an inactive enzyme derivative. The enzyme derivatives can then be photoactivated by exposure to 365 nm light, yielding both active enzyme and a fluorescent coumarin derivative. Photolabile inhibitors are important tools for effecting the photo-

Table 10.7 *Glycosidase, phosphatase and miscellaneous enzyme inhibitors and activators.*

Cat #	Probe	Properties
A-6920	α-Amanitin	• RNA polymerase II inhibitor with activity *in vitro* at nanomolar concentrations [1] • Does not inhibit mammalian RNA polymerases I and III, even at concentrations of 1 µM
A-96015	Askendoside D	• Cardiac glycoside Na^+/K^+-ATPase inhibitor [2]
D-99007 D-99008	M1 M2	• Ca^{2+}/Mg^{2+}-ATPase inhibitor
B-99011 D-99009 D-99010	M3 M4 M5	• Ca^{2+}/Mg^{2+}-ATPase activator
E-96021	Erysimoside	• Cardiac glycoside Na^+/K^+-ATPase inhibitor [2]
H-7599	Hydroxystilbamidine	• Inhibitor of cellular ribonucleases [3]
I-6621	Isopropyl β-D-thiogalactopyranoside (IPTG)	• Inhibitor of the *lac* repressor, which regulates β-galactosidase synthesis • Widely used inducer of the *lac* promoter [4,5]
M-98403	Melittin	• Na^+/K^+-ATPase inhibitor [6]
P-96054	Pentazolone	• Monoamine oxidase inhibitor and anti-asthmatic compound [7,8]
P-1692	Phenylethyl β-D-thiogalactopyranoside (PETG)	• β-Galactosidase inhibitor [9-13]
P-8460	Phenyl β-D-thioglucuronide	• Inhibitor of *uidR* repressor, which regulates β-glucuronidase synthesis [14]
S-96205	Strophanthidin 3-acetate (acetylstrophanthidin)	• Membrane-permeant cardiac glycoside Na^+/K^+-ATPase inhibitor [15]
T-6537	Trifluoromethanesulfonamide	• Potent carbonic anhydrase inhibitor (K_i >10^{-9} M) in solution and cells [16,17]

1. Science 170, 447 (1970); **2.** Biochim Biophys Acta 937, 335 (1988); **3.** J Cell Biol 87, 292 (1980); **4.** Nucleic Acids Res 18, 5347 (1990); **5.** Biochem Biophys Res Comm 128, 1268 (1985); **6.** Arch Biochem Biophys 283, 249 (1990); **7.** Indian J Pharmacol 25, 101 (1993); **8.** Indian J Pharmacol 14, 979 (1992); **9.** Eur J Cell Biol 62, 324 (1993); **10.** Proc Natl Acad Sci USA 90, 8194 (1993); **11.** Anal Biochem 199, 119 (1991); **12.** Cytometry 12, 291 (1991); **13.** Carbohydrate Res 56, 153 (1977); **14.** J Bacteriol 127, 418 (1976); **15.** Biochim Biophys Acta 937, 335 (1988); **16.** FEBS Lett 350, 319 (1994); **17.** J Biol Chem 268, 26233 (1993).

release of a biological catalyst and for modeling the light-dependent enzymatic activity observed in many biological systems.

Thiol-Caging Probe for Cysteine Proteases and Other Cysteine-Containing Proteins

The thiol-reactive probe α-bromo-2-nitrophenylacetic acid (BNPA, B-6610) can selectively block the activity of thiols,[12] including cysteine-containing peptides and proteins, as well as cysteine proteases and other enzymes in which the active site is a cysteine residue. UV photolysis of this caging group restores the enzymatic activity. Proteins such as creatine kinase and myosin may also be caged by modification of cysteine residues that are adjacent to the active site or that are otherwise important to the protein function.

Active-Site Titrant: Fluorescein 4-Guanidinobenzoates

Our fluorescein mono-(4-guanidinobenzoate) (FMGB, F-6489; see Section 10.5) and fluorescein di-(4-guanidinobenzoate) (FDGB, F-2890; see Section 10.5) serve as active-site titrants for serine proteases by releasing fluorescein upon formation of the stable acylated enzyme.[13-15] Detection limits for serine proteases using FMGB are reported to be on the order of 1–10 pM.[13] As compared to 4-methylumbelliferyl ester derivatives, FMGB provides much lower detection limits — 20-fold lower for active-site titration of trypsin.[14] A continuous fluorometric assay can detect as little as 20 nM human α-factor X_a with FMGB.[16] FMGB has been used to determine the concentration of active kallikrein formed by cleavage of human high molecular weight kininogen.[17] FDGB has been used as an active-site titrant of the esterase activity of plasmin.[18] FMGB and FDGB are also substrates for guanidinobenzoatase, an enzyme bound to the surface of actively migrating tumor cells [19] (see Section 10.5).

Miscellaneous Enzymes, Enzyme Inhibitors and Enzyme Activators

Molecular Probes has available several other enzyme inhibitors and activators (Tables 10.5 and 10.6). Some of these materials are unique natural products produced for Molecular Probes by Latoxan and have limited published references.

The following enzymes, enzyme inhibitors and enzyme activators have proven useful for the investigation of signal transduction mechanisms and are described in Chapter 20.

Probes for Studying Protein Kinases

- Protein kinase A, protein kinase C and Ca^{2+}/calmodulin kinase II (see Section 20.3)
- Bisindolylmaleimide (GF 109203X, D-7475; see Section 20.3), a highly selective competitive inhibitor of PKC with respect to ATP [20-22] (IC_{50} = 10 nM)
- Fluorescent bisindolylmaleimides, including fim-1, rim-1 and BODIPY FL bisindolylmaleimide (F-7462, R-7454, B-7485; see Section 20.3), probes for localizing PKC in cells [23]
- Cardiotoxin (C-98102, see Section 20.3), a potent PKC-specific inhibitor peptide [24,25] (IC_{50} = 1–3 µM)
- Hypericin (H-7476, see Section 20.3), a light-activated inhibitor of PKC [26-28] (IC_{50} = 3.4 µM)
- Hypocrellins A and B (H-7515, H-7516; see Section 20.3), selective and potent inhibitors of PKC, with IC_{50} values of 3.6 and 9.0 µM, respectively [29,30]
- Staurosporine (antibiotic AM-2282, S-7456; see Section 20.3), a competitive inhibitor of both PKC and PKA,[31-33] with an IC_{50} of 2.7 nM and 8.2 nM, respectively
- 4β-phorbol 12β-myristate 13α-acetate (TPA, PMA; P-3453, P-3454; see Section 20.3) and 1,2-dioctanoyl-sn-glycerol (DiC_8 or DOG, D-1576; see Section 20.3), an effective cell-permeant activators of protein kinase C [34,35] (PKC)

Probes for Studying Protein Phosphatases and Ca^{2+}-ATPase

- Okadaic acid (O-3452, O-7457; see Section 20.3), an effective inhibitor of serine and threonine phosphatases, leading to hyperphosphorylation of numerous cellular proteins [36-38]
- Thapsigargin (T-7458, T-7459; see Section 20.2), a tumor promoter that releases Ca^{2+} from intracellular stores by specifically inhibiting the endoplasmic reticulum Ca^{2+}-ATPase [39] and does not directly affect plasma membrane Ca^{2+}-ATPases, IP_3 production or protein kinase C activity.[40,41]

Probes for Studying Lipid Trafficking

- D 609 (tricyclodecan-9-yl xanthogenate, T-6615; see Section 20.5), a selective inhibitor of phosphatidylcholine-specific phospholipase C [42]
- Phospholipase substrates (see Section 20.5), important probes for signal transduction research
- Fluorescent and nonfluorescent ceramides (see Sections 12.4 and 20.5), probes with a variety of biological activities, including activation of stress-activated protein kinases [43] and protein phosphatase 2A [44,45]
- PDMP and the more potent PPMP (D-7522, H-7523; see Section 20.6), which elevate intracellular ceramide levels by blocking the metabolism of ceramide by glucosylceramide synthase in the Golgi complex [46,47]

Acetylcholinesterase Inhibitors

We also offer several acetylcholinesterase inhibitors, which are discussed in Section 18.2 along with our cholinergic receptor probes. These include:

- Bomin-1, bomin-2 and bomin-3 (E-99019, P-99020, B-99021), which are selective irreversible inhibitors of carboxylesterases
- Cardiotoxin [48] from *Naja nigricollis* (C-98102)
- Desoxypeganine [49] from Zygophyllaceae (D-96009)
- Fasciculin 2 from *Dendroaspis angusticeps* (F-98107), a noncompetitive inhibitor of mammalian and electric eel acetylcholinesterase at picomolar concentrations [50-55] and of avian and insect acetylcholinesterases and butyryl cholinesterase at micromolar to millimolar concentrations [55,56]
- Galanthamine [57,58] from Amaryllidaceae (G-96028)
- Peganole [49,59] from Papaveraceae (P-96043), which inhibits both cholinesterases and monoamine oxidases

Fluorescent β-Lactamase Inhibitor

m-Dansylaminophenylboronic acid (D-2281, see Section 3.1) has been reported to be an active-site–directed competitive inhibitor of β-lactamases with a dissociation constant of about 2 µM.[60-62]

Betaine Aldehyde Chloride

Betaine aldehyde chloride (B-6529) is the first metabolic oxidation product of choline in a reaction by choline oxidase and is itself a substrate for NAD^+-betaine aldehyde dehydrogenase, an enzyme that is important in the response of organisms, particularly plants and bacteria, to osmotic stress.[63,64] The reaction consumes oxygen and produces H_2O_2. Betaine aldehyde, a stable hydrate, functions as an osmoprotectant when added to cultures of *E. coli*.[65]

Miscellaneous Carbohydrates

D-(+)-Galactosamine (G-8453) has been used as an inducing reagent in studies of hepatic tissue injury [66] and as an inhibitor of protein synthesis.[67] In addition to its general use as a substrate for research of metabolic enzymatic activity, α-D-galactose 1-phosphate (G-8454) was used in a study involving the purification of human erythrocyte uridylyl transferase [68] (UDP galactose: α-D-galactose-1-phosphate uridylyl transferase, EC 2.7.7.12), an enzyme that is deficient in galactosemia.[69,70] The glycosides phenyl β-D-thioglucuronide (P-8460) and methyl β-D-glucuronide (M-8456) are potential glucuronidase substrates or inhibitors. When used in combination with mannonic acid, phenyl thioglucuronide induces β-glucuronidase expression in *E. coli*.[71] We also offer *N*-acetylneuraminic acid (sialic acid, A-8444), a product of the hydrolysis of a sialyl glycoside by neuraminidase [72-74] (sialidase, EC 3.2.1.18). Influenza C viruses and human and bovine coronaviruses recognize sialic acid–containing viruses on erythrocytes.[75]

1. Meth Enzymol 244, 1 (1994); **2.** J Am Chem Soc 85, 997 (1963); **3.** J Biol Chem 237, 3245 (1962); **4.** Thrombosis Res 2, 343 (1973); **5.** J Med Chem 14, 119 (1971); **6.** Biochim Biophys Acta 439, 194 (1976); **7.** J Biol Chem 266, 23016 (1991); **8.** Biochem J 238, 923 (1986); **9.** Biochim Biophys Acta 704, 403 (1982); **10.** J Enzym Inhib 10, 3 (1996); **11.** J Am Chem Soc 115, 9371 (1993); **12.** Chem Biol 2, 139 (1995); **13.** Anal Biochem 147, 487 (1985); **14.** J Am Chem Soc 104, 7299 (1982); **15.** Biochemistry 20, 4298 (1981); **16.** Arch Biochem Biophys 273, 375 (1989); **17.** J Biol Chem 269, 16318 (1994); **18.** Proc Natl Acad Sci USA 77, 3796 (1980); **19.** Anticancer Res 9, 247 (1989); **20.** Biochem Biophys Res Comm 206, 119 (1995); **21.** J Med Chem 35, 994 (1992); **22.** J Biol Chem 266, 15771 (1991); **23.** J Biol Chem 268, 15812 (1993); **24.** J Biol Chem 266, 2753 (1991); **25.** Eur J Biochem 174, 103 (1988); **26.** Med Res Rev 15, 111 (1995); **27.** Biochem Biophys Res Comm 165, 1207 (1989); **28.** Proc Natl Acad Sci USA 86, 5963 (1989); **29.** Photochem Photobiol 61, 529 (1995); **30.** Biochem Pharmacol 47, 373 (1994); **31.** J Immunology 149, 3894 (1992); **32.** Biochem Biophys Res Comm 158, 105 (1989); **33.** Biochem Biophys Res Comm 135, 397 (1986); **34.** Cell 81, 917 (1995); **35.** Nature 308, 693 (1984); **36.** J Cell Biol 131, 1291 (1995); **37.** J Neurosci 15, 6475 (1995); **38.** Eur J Biochem 193, 671 (1990); **39.** J Biol Chem 270, 11731 (1995); **40.** Proc Natl Acad Sci USA 87, 2466 (1990); **41.** J Biol Chem 264, 12266 (1989); **42.** Cell 71, 765 (1992); **43.** J Biol Chem 270, 22689 (1995); **44.** J Biol Chem 270, 4088 (1995); **45.** J Biol Chem 268, 15523 (1993); **46.** J Biol Chem 270, 2859 (1995); **47.** Adv Lipid Res 26, 183 (1993); **48.** Biochem J 161, 229 (1977); **49.** Chem Nat Comp 25, 729 (1989); **50.** Cell 83, 503 (1995); **51.** J Biol Chem 270, 20391 (1995); **52.** J Biol Chem 270, 19694 (1995); **53.** J Biol Chem 269, 11233 (1994); **54.** Biochemistry 32, 12074 (1993); **55.** J Physiol (Paris) 79, 232 (1984); **56.** Cervenansky, C. *et al.* in *Snake Toxins*, A.L. Harvey, Ed., Pergammon Press (1991) pp. 303–321; **57.** J Pharmacol Exp Ther 265, 1474 (1993); **58.** Electroencephalography Clin Neurophysiol 82, 445 (1992); **59.** Dokl Akad Nauk SSSR 267, 469 (1982); **60.** Biochemistry 34, 7757 (1995); **61.** Biochemistry 34, 3561 (1995); **62.** Biochemistry 28, 6863 (1989); **63.** J Biol Chem 268, 23818 (1993); **64.** Proc Natl Acad Sci USA 82, 3678 (1985); **65.** J Bacteriol 165, 849 (1986); **66.** Ann Rev Pharmacol Toxicol 35, 655 (1995); **67.** Can J Microbiol 29, 1532 (1983); **68.** Biochim Biophys Acta 657, 374 (1981); **69.** Eur J Pediatr 154, S21 (1995); **70.** Human Gen 93, 167 (1994); **71.** J Bacteriol 127, 418 (1976); **72.** Biochim Biophys Acta 327, 114 (1973); **73.** Anal Biochem 23, 150 (1968); **74.** J Biol Chem 240, 3501 (1965); **75.** Proc Natl Acad Sci USA 85, 4526 (1988).

10.6 Data Table *Enzyme Inhibitors, Activators and Active-Site Titrants*

Cat #	Structure	MW	Storage	Soluble	Abs	{ε x 10^{-3}}	Em	Solvent	Notes
A-6202	✓	240	F,D	pH<8, EtOH	<300		none		
A-6480	✓	208	D	H_2O	<300		none		
A-6920	✓	919	F,D	H_2O	304	{13}	ND	MeOH	
A-8444	✓	309	D	H_2O	<300		none		
A-96015	T	887	D	DMSO	<300		none		
B-6286	S3.4	278	F,L	DMF, DMSO	261	{16}	none	MeOH	
B-6529	✓	156*	F,D	H_2O	<300		none		
B-6546	✓	175*	D	H_2O, MeOH	<300		none		
B-6610	✓	260	F,D	DMSO	<300		none		
B-99011	✓	286	D,L	DMSO, EtOH	<300		none		
C-6485	✓	566	FF,D,L	DMSO	<300		none		
D-11	✓	253	F,D,L	DMF, MeCN	356	{3.6}	none	MeOH	<1>
D-6201	✓	477	F,D,L	DMF, MeCN	503	{84}	510	MeOH	
D-7112	✓	370	F,D,LL	DMSO, MeOH	381	{36}	none	MeOH	
D-7113	✓	540	F,D,LL	DMSO, MeOH	379	{36}	465	MeOH	
D-99007	✓	224	D,L	DMSO, EtOH	<300		none		
D-99008	✓	252	D,L	DMSO, EtOH	<300		none		
D-99009	✓	236	D,L	DMSO, EtOH	<300		none		
D-99010	✓	288	D,L	DMSO, EtOH	<300		none		
E-6538	✓	195	F,D	DMSO	303	{11}	none	MeOH	
E-6545	✓	357	F,DD	DMSO	<300		none		
E-96021	✓	715*	F,D	H_2O	<300		none		
F-99013	✓	387	F,D	DMSO	<300		none		
F-99014	✓	545	F,D	DMSO	<300		none		
G-8453	✓	216	D	H_2O	<300		none		
G-8454	✓	426*	F,D	H_2O	<300		none		
H-7599	S8.1	473	F,D,L	H_2O	361	{17}	536	pH 7	
I-6621	✓	238	D	DMSO, H_2O	<300		none		
L-6543	✓	~475	F,D	H_2O	<300		none		
M-6539	S2.4	126	F,D	EtOH†	<300		none		

Cat #	Structure	MW	Storage	Soluble	Abs	{ε × 10^{-3}}	Em	Solvent	Notes
M-8456	✓	230	D	H_2O	<300		none		
M-98403	C	~2847	FF,D	H_2O	<300		<2>		
O-3452	S20.3	805	F,D,AA	DMSO‡	<300		none		
O-7457	S20.3	827	F,D,AA	H_2O	<300		none		
P-1692	✓	300	F,D	DMSO, H_2O	<300		none		
P-6203	✓	174	F,D	MeOH	<300		none		
P-6542	✓	686	F,D	DMSO, MeOH	<300		none		
P-6544	✓	198*	D	DMSO	<300		none		
P-8460	✓	286	D	DMSO, H_2O	<300		none		
P-96054	✓	251	D	EtOH	<300		none		
S-96205	✓	447	F,D	DMSO	<300		none		
T-6483	✓	369	F,D	MeOH, pH<6	<300		none		
T-6484	✓	352	F,D	MeOH	<300		none		
T-6537	✓	149	D	DMSO	<300		none		

For definitions of the contents of this data table, see "How to Use the Handbook" on page vi.
Structure: Chemical structure drawing: ✓ = shown in this section; Sn.n = shown in Section number n.n; T = available on request from our Technical Assistance Department and also through our Web Site (http://www.probes.com); C = not shown due to complexity.
MW: ~ indicates an approximate value.
ND = not determined.

* MW is for the hydrated form of this product.
† This product is intrinsically a liquid at room temperature.
‡ This product is packaged as a solution in the solvent indicated in "Soluble."
<1> D-11 serine ester formed via reaction with chymotrypsin has Abs = 355 nm (ε = {4.2}), Em = 535 nm [Biochim Biophys Acta 439, 194 (1976)].
<2> This peptide exhibits intrinsic tryptophan fluorescence (Em ~350 nm) when excited at <300 nm.

10.6 Structures *Enzyme Inhibitors, Activators and Active-Site Titrants*

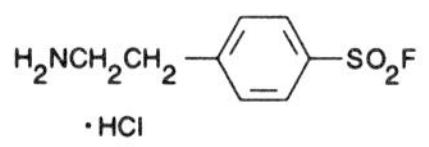

A-6202

A-6480 see B-6546

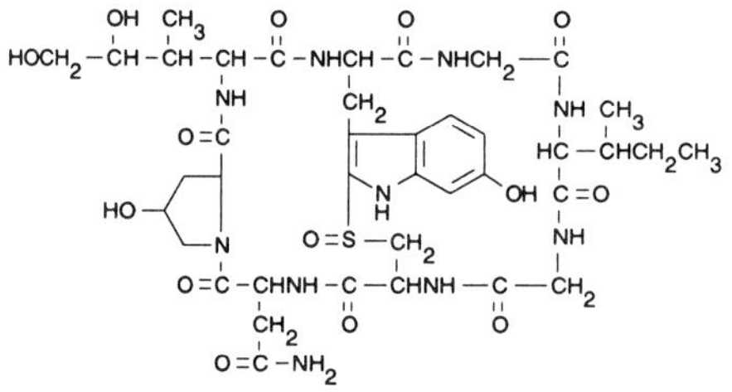

A-6920

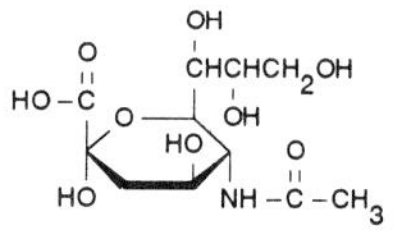

A-8444

B-6286 Section 3.4

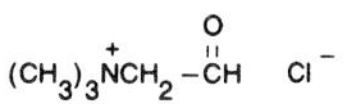

B-6529

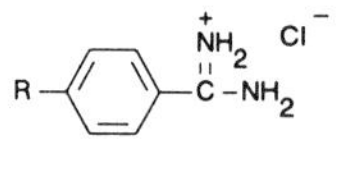

B-6546 R = H
A-6480 R = $-NH_3^+$ Cl^-

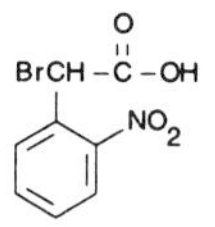

B-6610

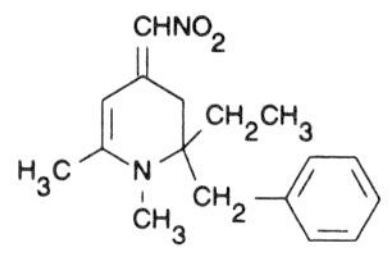

B-99011

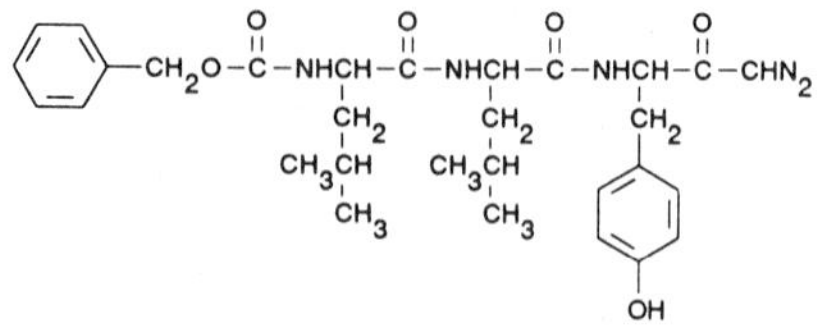

C-6485

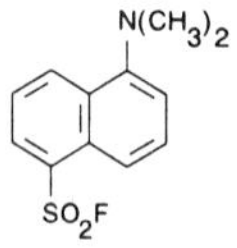

D-11

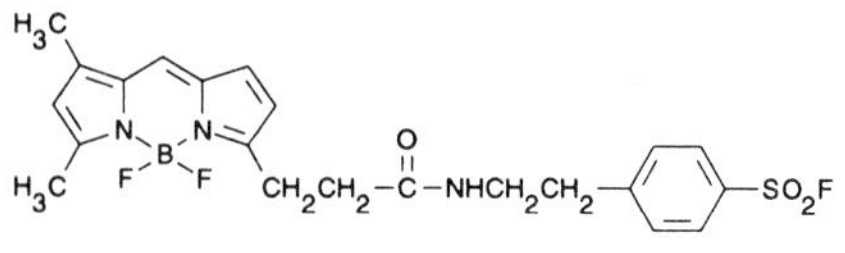

D-6201

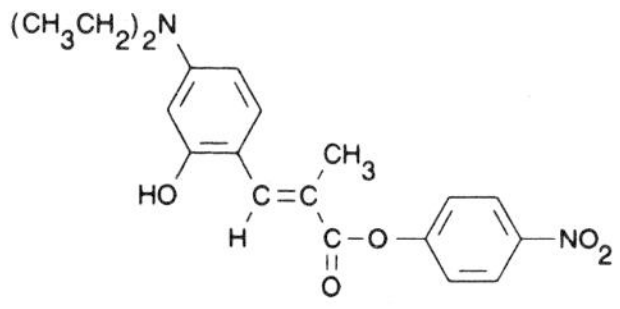

D-7112

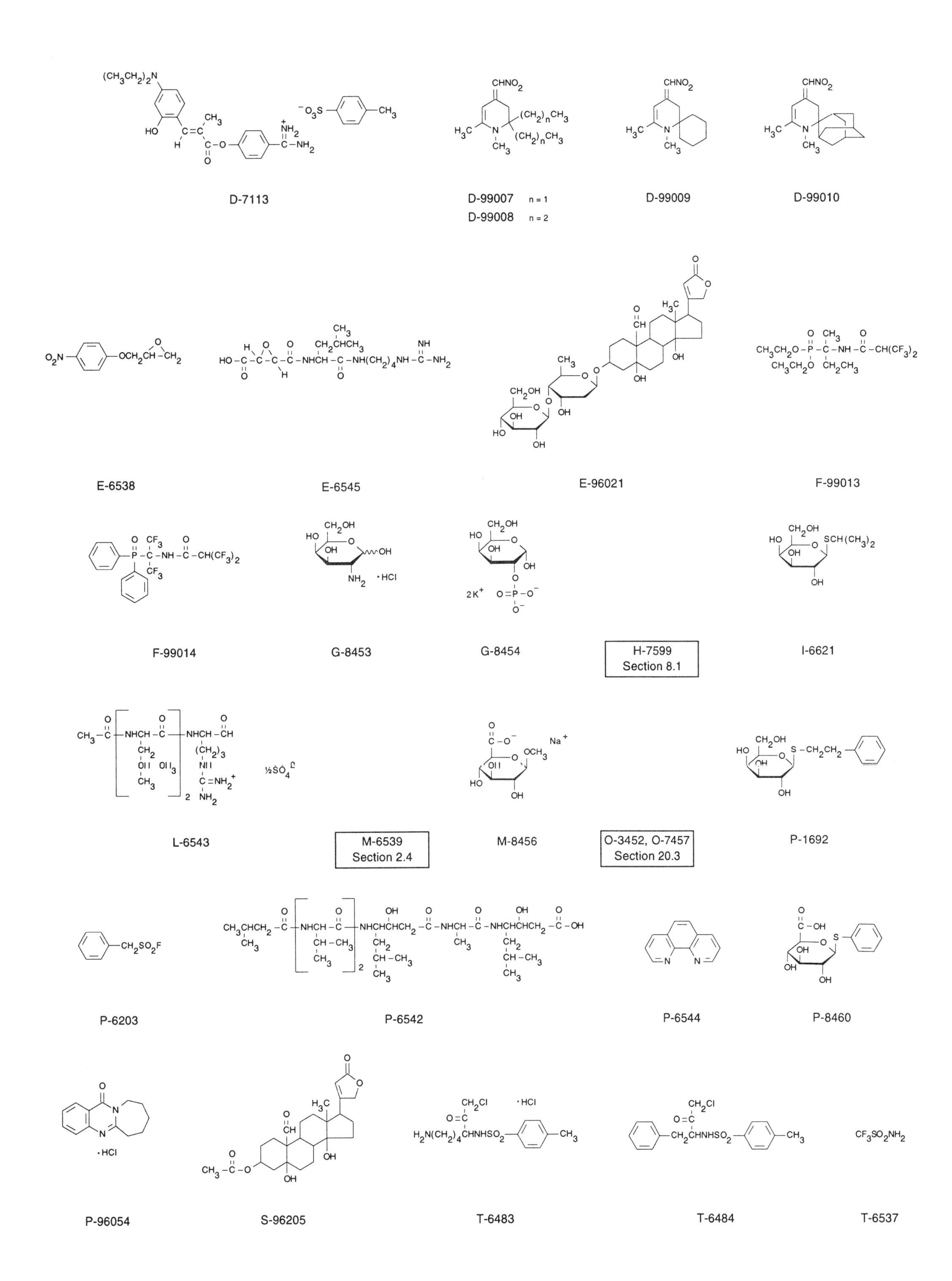
D-7113
D-99007 n = 1
D-99008 n = 2
D-99009
D-99010
E-6538
E-6545
E-96021
F-99013
F-99014
G-8453
G-8454
H-7599
Section 8.1
I-6621
L-6543
M-6539
Section 2.4
M-8456
O-3452, O-7457
Section 20.3
P-1692
P-6203
P-6542
P-6544
P-8460
P-96054
S-96205
T-6483
T-6484
T-6537

Cat #		Product Name	Unit Size	Price Per Unit ($) 1–4 Units	5–24 Units
A-8444	**New**	N-acetylneuraminic acid (sialic acid)	100 mg	42.00	33.60
A-6920	**New**	α-amanitin	1 mg	90.00	72.00
A-6480	**New**	4-aminobenzamidine, dihydrochloride	1 g	18.00	14.40
A-6202	**New**	4-(2-aminoethyl)benzenesulfonyl fluoride, hydrochloride (AEBSF)	25 mg	28.00	22.40
A-96015	**New**	askendoside D	5 mg	68.00	54.40
B-6546	**New**	benzamidine, hydrochloride, monohydrate	5 g	18.00	14.40
B-99011	**New**	6-benzyl-1,2-dimethyl-6-ethyl-4-nitromethinyl-1,4,5,6-tetrahydropyridine (M3)	100 µg	215.00	172.00
B-6529	**New**	betaine aldehyde chloride, monohydrate	100 mg	68.00	54.40
B-6610	**New**	α-bromo-2-nitrophenylacetic acid (BNPA)	5 mg	75.00	60.00
B-6286	**New**	4-bromophenacyl bromide (2,4′-dibromoacetophenone)	5 g	18.00	14.40
C-6485	**New**	N-CBZ-L-leucyl-L-leucyl-L-tyrosine, diazomethyl ketone	10 mg	95.00	76.00
D-7113	**New**	*trans*-p-(diethylamino)-o-hydroxy-α-methylcinnamic acid, p-amidinophenyl ester, p-toluenesulfonic acid salt	5 mg	65.00	52.00
D-7112	**New**	*trans*-p-(diethylamino)-o-hydroxy-α-methylcinnamic acid, p-nitrophenyl ester	5 mg	65.00	52.00
D-99007	**New**	6,6-diethyl-N,2-dimethyl-4-nitromethinyl-1,4,5,6-tetrahydropyridine (M1)	100 µg	95.00	76.00
D-6201	**New**	4-(2-(4,4-difluoro-5,7-dimethyl-4-bora-3a,4a-diaza-s-indacene-3-propionyl)aminoethyl)-benzenesulfonyl fluoride (BODIPY® FL AEBSF)	1 mg	48.00	38.40
D-11		5-dimethylaminonaphthalene-1-sulfonyl fluoride (dansyl fluoride)	100 mg	18.00	14.40
D-99008	**New**	N,2-dimethyl-6,6-dipropyl-4-nitromethinyl-1,4,5,6-tetrahydropyridine (M2)	100 µg	128.00	102.40
D-99010	**New**	1,2-dimethyl-3-nitromethylene-6-spiroadamantyl-5,6-dihydropyridine (M5)	100 µg	270.00	216.00
D-99009	**New**	1,2-dimethyl-3-nitromethylene-5-spirocyclohexyl-5,6-dihydropyridine (M4)	100 µg	215.00	172.00
E-6538	**New**	1,2-epoxy-3-(4-nitrophenoxy)propane (EPNP) *high purity*	100 mg	24.00	19.20
E-6545	**New**	*trans*-epoxysuccinyl-L-leucylamido-(4-guanidino)butane (E-64)	10 mg	65.00	52.00
E-96021	**New**	erysimoside, monohydrate	5 mg	60.00	48.00
F-99013	**New**	fisonate	10 mg	60.00	48.00
F-99014	**New**	flodiphoside	10 mg	60.00	48.00
G-8453	**New**	D-(+)-galactosamine, hydrochloride	1 g	25.00	20.00
G-8454	**New**	α-D-galactose 1-phosphate, dipotassium salt, pentahydrate	100 mg	38.00	30.40
H-7599	**New**	hydroxystilbamidine, methanesulfonate	10 mg	75.00	60.00
I-6621	**New**	isopropyl β-D-thiogalactopyranoside (IPTG) *dioxane free*	1 g	24.00	19.20
L-6543	**New**	leupeptin hemisulfate	10 mg	45.00	36.00
M-98403	**New**	melittin *from *Apis mellifera* bee venom*	1 mg	45.00	36.00
M-8456	**New**	methyl β-D-glucuronide, sodium salt	100 mg	35.00	28.00
M-6539	**New**	methyl methanethiolsulfonate (MMTS) *special packaging*	10x10 mg	35.00	28.00
O-3452		okadaic acid *500 µg/mL solution in DMSO*	50 µL	100.00	80.00
O-7457	**New**	okadaic acid, sodium salt	25 µg	95.00	76.00
P-96054	**New**	pentazolone, hydrochloride	100 mg	38.00	30.40
P-6542	**New**	pepstatin A (iso-valeryl-L-Val-L-Val-Sta-L-Ala-Sta)	25 mg	55.00	44.00
P-6544	**New**	1,10-phenanthroline, monohydrate *≥99%*	5 g	18.00	14.40
P-1692		phenylethyl β-D-thiogalactopyranoside (PETG)	10 mg	43.00	34.40
P-6203	**New**	phenylmethanesulfonyl fluoride (PMSF) *high purity*	5 g	32.00	25.60
P-8460	**New**	phenyl β-D-thioglucuronide	100 mg	35.00	28.00
P-6548	**New**	Protease Inhibitor Sampler Kit	1 kit	195.00	156.00
S-96205	**New**	strophanthidin 3-acetate (acetylstrophanthidin)	5 mg	55.00	44.00
T-6483	**New**	N^{α}-tosyl-L-lysine chloromethyl ketone, hydrochloride (TLCK)	100 mg	25.00	20.00
T-6484	**New**	N-tosyl-L-phenylalanine chloromethyl ketone (TPCK)	1 g	42.00	33.60
T-6537	**New**	trifluoromethanesulfonamide	100 mg	38.00	30.40

Product-Specific Bibliographies and Keyword Searches

Our Technical Assistance Department can provide you with product-specific bibliographies, as well as keyword searches of the over 25,000 literature references in our extensive bibliography database. Our bibliography database is also searchable through our Web site (http://www.probes.com).

Chapter 11

Probes for Actin, Tubulin and Nucleotide-Binding Proteins

Contents

11.1 Probes for Actin

Phalloxins for F-Actin

Molecular Probes prepares several fluorescent and biotinylated derivatives of phalloidin and phallacidin for selectively labeling F-actin. Phallotoxins are bicyclic peptides isolated from the deadly *Amanita phalloides* mushroom. They can be used interchangeably in most applications and bind competitively to the same sites on F-actin. Table 11.1 lists the available phallotoxin derivatives, along with their spectral properties and binding constants for F-actin. A product-specific bibliography and a detailed staining protocol are included with each phallotoxin derivative. One vial of the fluorescent phallotoxins contains sufficient reagent for staining ~300 microscope slide preparations; one vial of biotin-XX phalloidin, which must be used at a higher concentration, contains sufficient reagent for ~50 microscope slide preparations. We also offer unlabeled phalloidin (P-3457) and phallacidin (P-3458) for blocking F-actin staining by labeled phallotoxins or for promoting actin polymerization.

Properties of Phallotoxin Derivatives

The fluorescent and biotinylated phallotoxin derivatives stain F-actin selectively at nanomolar concentrations and are readily water-soluble, thus providing convenient labels for identifying and quantitating actin in tissue sections, cell cultures or cell-free preparations.[1-3] It was recently reported that the F-actin in living neurons can be efficiently labeled using cationic liposomes containing fluorescent phallotoxins, such as BODIPY® FL phallacidin[4] (B-607). This procedure permits the labeling of entire cell cultures with minimum disruption. Labeled phallotoxins have similar affinity for both large and small filaments and bind in a stoichiometric ratio of about one phallotoxin per actin subunit in muscle and nonmuscle cells; they reportedly do not bind to monomeric G-actin, unlike some antibodies against actin.[3,5] Phallotoxins have further advantages over antibodies for actin labeling: 1) their binding properties do not change appreciably with actin from different species, including plants and animals; and 2) their nonspecific staining is negligible; thus, the contrast between stained and unstained areas is high.

Phallotoxins shift actin's monomer/polymer equilibrium toward the polymer, lowering the critical concentration for polymerization up to 30-fold.[6,7] Furthermore, depolymerization of F-actin by cytochalasins, potassium iodide and elevated temperatures is inhibited by phallotoxin binding. Because the phallotoxin derivatives are relatively small, with approximate diameters of 12–15 Å and molecular weights <1500 daltons, a wide variety of actin-binding proteins — including myosin, tropomyosin, troponin and DNase I — can still bind to actin after treatment with fluorescent phallotoxins. Even more significantly, phallotoxin-labeled actin filaments retain certain functional characteristics; labeled glycerinated muscle fibers still contract, and labeled actin filaments still move on solid-phase myosin substrates.[8-10]

Table 11.1 *Spectral characteristics and dissociation constants of our actin-selective probes.*

Cat #	Actin-Selective Probe	Ex/Em *	K_d (nM) †
Probes for F-Actin			
C-606	Coumarin phallacidin	355/443	24
N-354	NBD phallacidin	465/536	18
F-432	Fluorescein phalloidin	496/516 ‡	18
O-7466	Oregon Green 488 phalloidin	496/520 ‡	ND
B-607	BODIPY FL phallacidin	505/512	38
O-7465	Oregon Green 514 phalloidin	511/528 ‡	ND
E-7463	Eosin phalloidin	524/544	ND
B-7491	BODIPY R6G phalloidin	529/547	ND
R-415	Rhodamine phalloidin	554/573 ‡	40
B-3475	BODIPY 558/568 phalloidin	558/569	ND
B-3416	BODIPY 581/591 phalloidin	584/592	13
B-7464	BODIPY TR-X phallacidin	589/617	ND
T-7471	Texas Red-X phalloidin	591/608 ‡	ND
B-7474	Biotin-XX phalloidin	NA	10.5
P-3458	Phallacidin	NA	ND
P-3457	Phalloidin	NA	6.7
J-7473	Jasplakinolide	NA	15
Probes for G-Actin			
D-970	DNase I, fluorescein	494/517	ND
D-7497	DNase I, Oregon Green 488	496/516	ND
D-971	DNase I, tetramethyl-rhodamine	555/580	ND
D-972	DNase I, Texas Red	597/618	ND

* Excitation/emission maxima in nm. Spectra of phallotoxins are either in aqueous buffer, pH 7–9 (denoted ‡) or in methanol; DNase I spectra are in aqueous buffer, pH 7–8. † Rhodamine phalloidin's fluorescence increases upon binding to actin, a phenomenon that allowed Molecular Probes to determine the ligand's binding constant. The binding constants of the other conjugates were determined by competitive binding with rhodamine phalloidin, as described in J Biol Chem 269, 14869 (1994) and Anal Biochem 200, 199 (1992). All binding constants were determined on rabbit skeletal muscle actin. ND = not determined. NA = not applicable.

New Oregon Green™ Phalloidins

Green fluorescent actin stains are popular reagents for labeling F-actin in fixed and permeabilized cells. Unfortunately, the green fluorescent fluorescein phalloidin and NBD phallacidin photobleach rapidly, sometimes making photography difficult. We have used two of our new Oregon Green dyes (see Sections 1.4 and 1.5) to prepare Oregon Green 488 phalloidin (O-7466) and the slightly longer-wavelength Oregon Green 514 phalloidin (O-7465) (see

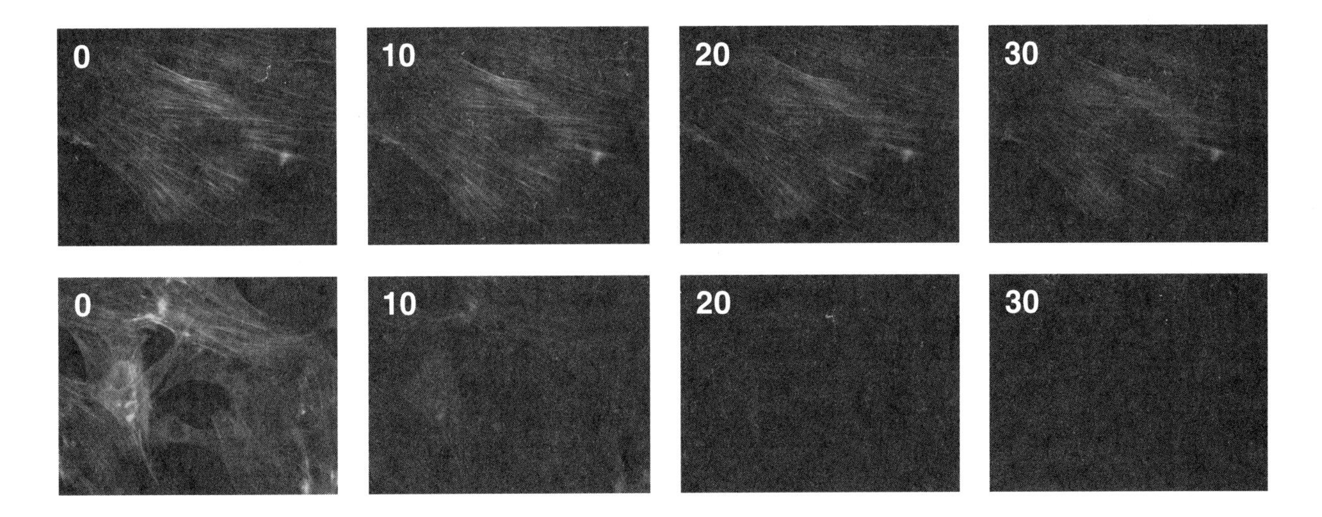

Figure 11.1 *Photostability comparison for Oregon Green 514 phalloidin (O-7465, upper series) and fluorescein phalloidin (F-432, lower series). CRE BAG 2 fibroblasts were fixed with formaldehyde, permeabilized with acetone and then stained with the fluorescent phallotoxins. Samples were continuously illuminated and viewed on a fluorescence microscope using an Omega® Optical fluorescein longpass optical filter set (O-5717, see Section 26.5), a Star 1™ CCD camera Photometrics) and Image-1® software (Universal Imaging Corp.). Images acquired 0, 10, 20 and 30 seconds after the start of illumination (as indicated in the top left hand corner of each panel) clearly demonstrate the superior photostability of Oregon Green 514 phalloidin.*

photo on back cover). The excitation and emission spectra of the Oregon Green 488 dye are virtually superimposable on those of fluorescein, and both Oregon Green 488 and Oregon Green 514 may be viewed with standard fluorescein optical filter sets. As shown in Figures 11.1 and 11.2, Oregon Green 514 phalloidin is by far the most photostable of the green fluorescent stains for F-actin. Particularly when mounted in *SlowFade*™ *Light* antifade reagent (S-7461, see Section 26.1), actin labeled with Oregon Green 514 phalloidin exhibits bright long-lasting fluorescence, making it easy to visualize and photograph.

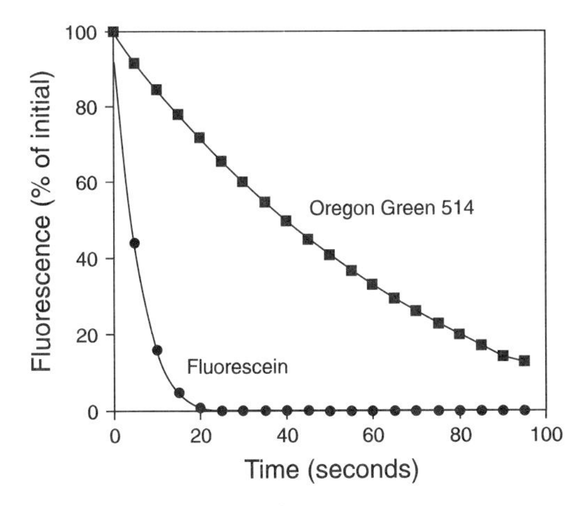

Figure 11.2 *Fluorescence intensity time courses for specimens illustrated in Figure 11.1. Images were acquired every 5 seconds using a Star 1 CCD camera (Photometrics); the average fluorescence intensity in the field of view was calculated with Image-1 software (Universal Imaging Corp.) and expressed as a fraction of the initial intensity. Three data sets, representing different fields of view, were averaged for each labeled phalloidin to obtain the plotted time courses.*

BODIPY Phallotoxins

BODIPY phallotoxin conjugates (B-607, B-3416, B-3475, B-7464, B-7491) have some important advantages over the conventional NBD, fluorescein and rhodamine phallotoxins. The BODIPY FL, BODIPY R6G, BODIPY 558/568 and BODIPY TR-X fluorophores exhibit excitation and emission spectra similar to those of fluorescein, rhodamine 6G, rhodamine B and Texas Red® dyes, respectively, and can be used with standard optical filter sets. Furthermore, BODIPY dyes are more photostable than these traditional fluorophores [11] and have narrower emission bandwidths (see Figure 1.4 in Chapter 1), making them especially useful for double- and triple-labeling experiments. BODIPY FL phallacidin (B-607), which reportedly gives a signal superior to fluorescein phalloidin,[12] has been used for quantitating F-actin and determining its distribution in cells.[13,14] BODIPY 581/591 phalloidin (B-3416), which can also be visualized with standard Texas Red optical filter sets, emits at a slightly shorter wavelength than BODIPY TR-X but may have significant green fluorescence when excited using a standard fluorescein bandpass optical filter. Consequently, we recommend using Texas Red-X phalloidin or BODIPY TR-X phallacidin as the red fluorophore for multicolor applications.

Rhodamine Phalloidin and New Texas Red-X Phalloidin

Rhodamine phalloidin (R-415) remains the standard for red fluorescent phallotoxins, with over 600 citations in our bibliography database. It is well excited by the mercury-arc lamp in most fluorescence microscopes. However, our new Texas Red-X phalloidin (T-7471) will be a welcome replacement for rhodamine phalloidin in many multicolor applications because its emission is better separated from that of fluorescein and of our new green fluorescent Oregon Green dyes (see Figure 1.10 in Chapter 1). Moreover, Texas Red-X conjugates can be excited by the 568 nm spectral line of the Kr-Ar laser used in several confocal scanning laser microscopes, whereas those of the tetramethylrhodamine dye used to prepare rhodamine phalloidin are only poorly excited by this laser.

Other Labeled Phallotoxins, Including a New Eosin Conjugate

The original yellow-green fluorescent NBD phallacidin (N-354) and green fluorescent fluorescein phalloidin (F-432) remain in use despite their relatively poor photostability (Figure 11.2). We recommend the Oregon Green 488, Oregon Green 514 and BODIPY FL phalloidins as the preferred green fluorescent actin stains. Coumarin phallacidin (C-606) is the only blue fluorescent phallotoxin conjugate available for staining actin.[15]

We have recently introduced eosin phalloidin (E-7463), which may be useful for correlated fluorescence and electron microscopic studies; see the Technical Note "Fluorescent Probes for Photoconversion of Diaminobenzidine" on page 264. Deerinck and colleagues have reported that eosin-mediated photooxidation of diaminobenzidine followed by treatment with osmium tetraoxide yields an insoluble, electron-dense DAB oxidation product that can be visualized by either light or electron microscopy.[16] Biotin-XX phalloidin (B-7474) also permits detection of F-actin by electron microscopy and light microscopy techniques.[17] This biotin conjugate can be visualized with fluorophore- or enzyme-labeled avidin and streptavidin (see Section 7.5) or with our novel ELF® signal amplification technology (see Figure 6.9B in Chapter 6).

Jasplakinolide — A Cell-Permeant F-Actin Probe

Molecular Probes now offers jasplakinolide (J-7473), a macrocyclic peptide (Figure 11.3) isolated from the marine sponge *Jaspis johnstoni*.[18,19] Jasplakinolide is a potent inducer of actin polymerization *in vitro* and competes with phalloidin for actin binding (K_d = 15 nM).[20] Moreover, unlike other known actin stabilizers such as phalloidins and virotoxins, jasplakinolide appears to be somewhat cell permeant and therefore can potentially be used to manipulate actin polymerization in live cells. This peptide, which also exhibits fungicidal, insecticidal and antiproliferative activity, should be useful for investigating cell processes mediated by actin polymerization and depolymerization, including cell adhesion, locomotion, endocytosis and vesicle sorting and release.

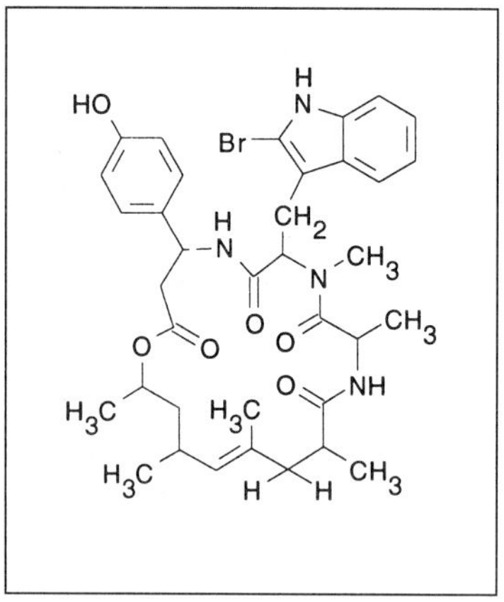

Figure 11.3 *Structure of jasplakinolide (J-7473), a macrocyclic peptide.*

DNase I Conjugates for G-Actin

Molecular Probes' fluorescent conjugates of bovine pancreatic DNase I (~31,000 daltons) selectively label monomeric G-actin. Fluorescein DNase I (D-970) has been used in combination with fluorescently labeled phallotoxins to simultaneously visualize G-actin pools and filamentous F-actin[21-23] and to study the disruption of microfilament organization in live nonmuscle cells.[24] Likewise, a photo on the back cover of this Handbook shows a mouse fibroblast labeled with both Oregon Green 488 phalloidin and Texas Red DNase I, demonstrating the complex network of F-actin and the presence of G-actin throughout the cytoplasm as well as at the cell periphery. Researchers at Molecular Probes triple-labeled endothelial cells with fluorescein DNase I, BODIPY 581/591 phalloidin and a monoclonal anti-actin antibody detected with a Cascade Blue®–labeled secondary antibody.[21] They found that the monoclonal antibody, which binds to both G-actin and F-actin, colocalized with the DNase I and phalloidin conjugates, suggesting that these three probes recognize unique binding sites on the actin molecule. The influence of cytochalasins on actin structure in monocytes has been quantitated by flow cytometry using Texas Red DNase I (D-972) and BODIPY FL phallacidin (B-607) to stain the G-actin and F-actin pools, respectively.[25] Fluorescent DNase I has also been used as a model system to study the interactions of nucleotides, cations and cytochalasin D with monomeric actin.[26] Researchers can choose fluorescein (D-970), Oregon Green 488 (D-7497), tetramethylrhodamine (D-971) or Texas Red (D-972) DNase I conjugates (Table 11.1), depending on their multicolor application and their detection instrumentation.

Assays for Actin Polymerization and Actin-Binding Proteins

FluoReporter® Actin Assay Kit

Molecular Probes' FluoReporter Actin Assay Kit (F-7452) provides a convenient and rapid fluorometric assay for quantitating G-actin and F-actin in cell lysates. The relative abundance of G- and F-actin plays a critical role in a variety of cell functions, including motility, reproduction and growth. Conventionally, the degree of polymerization of DNA has been determined from absorbance measurements.[27] To provide increased accuracy and sensitivity, we have developed an assay that employs one of our novel ultrasensitive nucleic acid stains, YO-PRO™-1. The researcher simply adds both YO-PRO-1–stained DNA and DNase I to a sample containing G-actin and then monitors the fluorescence of the nucleic acid/dye complex. When the sample's G-actin binds to DNase I, nuclease activity is inhibited and the DNA/dye complex remains intact. In the absence of G-actin, DNase I will degrade the DNA/dye complex, resulting in a loss of fluorescence. Thus, the polymerization state of the nucleic acid, as monitored by fluorescence, is a measure of the levels of G-actin in the sample. The FluoReporter Actin Assay Kit includes:

- YO-PRO-1 nucleic acid stain
- Calf thymus DNA
- DNase I
- G-actin
- Cell lysis buffer
- Actin depolymerization buffer
- Reaction buffer
- Detailed protocol for carrying out this fluorometer-based assay

Enough reagents are provided to produce five standard curves and to test 50 G-actin samples.

Other Assays for Quantitating F-Actin

Other quantitative assays for F-actin have used fluorescein phalloidin,[29] rhodamine phalloidin,[30] BODIPY FL phallacidin [14] or NBD phallacidin.[31] Using fluorescein phalloidin to quantitate F-actin, researchers have demonstrated the loss of F-actin from cells during apoptosis.[32] The addition of propidium iodide (P-1304, P-3566; see Section 8.1) to the cell suspensions enabled these researchers to estimate the cell cycle distributions of both the apoptotic and nonapoptotic cell populations.[32] The change in F-actin content in proliferating adherent cells has been quantitated using the ratio of rhodamine phalloidin fluorescence to ethidium bromide fluorescence.[33] The spectral separation of the signals in this assay may be improved by using a green fluorescent stain for F-actin and a high-affinity red fluorescent nucleic acid stain, such as our new Oregon Green 488 phalloidin (O-7466) and ethidium homodimer-1 (E-1169, see Section 8.1). In addition to these phallotoxin-based assays for F-actin, the fluorescence of actin monomers labeled with pyrene iodoacetamide (P-29) has been demonstrated to change upon polymerization, making this probe an excellent tool for following the kinetics of actin polymerization and the effects of actin-binding proteins on polymerization.[34-36]

Assays for Actin-Binding Proteins

Enhancement of the fluorescence of certain phallotoxins upon binding to F-actin can be a useful tool for following the kinetics and extent of binding of specific actin-binding proteins. We have used the change in fluorescence of rhodamine phalloidin to determine the dissociation constant of various phallotoxins [37] (Table 11.1). The enhancement of rhodamine phalloidin's fluorescence upon actin binding has also been used to measure the kinetics and extent of gelsolin severing of actin filaments.[38] In this study, the ion indicator mag-fura-5 (M-3103, see Section 22.2) was employed to determine the dependence of this severing on divalent ion concentrations. The affinity and rate constants for rhodamine phalloidin binding to actin are not affected by saturation of actin with either myosin subfragment-1 or tropomyosin, indicating that these two actin-binding proteins do not bind to the same sites as the phalloidin.[4]

The actin-binding protein synapsin I is localized exclusively to synaptic vesicles and thus serves as an excellent marker for synapses in brain and other neuronal tissues.[39,40] Synapsin I inhibits neurotransmitter release, an effect that is abolished upon its phosphorylation by Ca^{2+}/calmodulin–dependent protein kinase II [41] (CaM kinase II, C-6462, see Section 20.3). For assaying the localization and abundance of synapsin I by Western blot analysis, immunohistochemistry, enzyme-linked immunosorbent assay (ELISA) or immunoprecipitation, Molecular Probes offers rabbit anti-bovine synapsin I both as an affinity-purified IgG (H+L) fraction and as whole serum (A-6442, A-6443; see Section 20.3). Although raised against bovine synapsin I, this antibody also recognizes human, rat and mouse synapsin I; it has little or no activity against synapsin II.

1. Meth Enzymol 194, 729 (1991); **2.** J Muscle Res Cell Motil 9, 370 (1988); **3.** Wieland, T., *Peptides of Poisonous Amanita Mushrooms*, Springer-Verlag (1986); **4.** Neurosci Lett 207, 17 (1996); **5.** Biochemistry 33, 14387 (1994); **6.** Eur J Biochem 165, 125 (1987); **7.** J Cell Biol 105, 1473 (1987); **8.** J Cell Biol 115, 67 (1991); **9.** Nature 326, 805 (1987); **10.** Proc Natl Acad Sci USA 83, 6272 (1986); **11.** J Cell Biol 114, 1179 (1991); **12.** J Cell Biol 127, 1637 (1994); **13.** J Cell Biol 116, 197 (1992); **14.** Histochem J 22, 624 (1990); **15.** J Muscle Res Cell Motil 14, 594 (1993); **16.** J Cell Biol 126, 901 (1994); **17.** J Cell Biol 130, 591 (1995); **18.** J Am Chem Soc 108, 3123 (1986); **19.** Tetrahedron Lett 27, 2797 (1986); **20.** J Biol Chem 269, 14869 (1994); **21.** J Histochem Cytochem 42, 345 (1994); **22.** Biotech Histochem 68, 8 (1993); **23.** J Histochem Cytochem 40, 1605 (1992); **24.** Proc Natl Acad Sci USA 87, 5474 (1990); **25.** J Biol Chem 269, 3159 (1994); **26.** Eur J Biochem 182, 267 (1989); **27.** Cell 15, 935 (1978); **28.** J Cell Sci 100, 187 (1991); **29.** Proc Natl Acad Sci USA 77, 6624 (1980); **30.** J Cell Biol 130, 613 (1995); **31.** J Cell Biol 98, 1265 (1984); **32.** Cytometry 20, 162 (1995); **33.** J Cell Biol 129, 1590 (1995); **34.** J Biol Chem 270, 7125 (1995); **35.** J Muscle Res Cell Motil 4, 235 (1983); **36.** Eur J Biochem 114, 33 (1981); **37.** Anal Biochem 200, 199 (1992); **38.** J Biol Chem 269, 32916 (1994); **39.** Science 226, 1209 (1984); **40.** J Cell Biol 96, 1337 (1983); **41.** P. Greengard, F. Benfenati, F. Valtora in *Molecular and Cellular Mechanisms of Neurotransmitter Release* L. Stjärne *et al.*, Raven Press (1994), pp. 31–45.

Superior Antifade Reagents for a Multitude of Applications

Molecular Probes offers a variety of antifade reagents to meet the diverse needs of the research community. The antifade formulation in our original *SlowFade*™ Antifade Kit (S-2828) reduces the fading rate of fluorescein to almost zero, making this antifade reagent especially useful for quantitative measurements. However, this *SlowFade* formulation does initially quench fluorescein's fluorescence. Our new *SlowFade Light* Antifade Kit (S-7461) contains an antifade formulation that slows fluorescein's photobleaching by about fivefold with little or no quenching of the fluorescent signal, thereby dramatically increasing the signal-to-noise ratio in photomicroscopy. The latest addition to our antifade arsenal — ProLong™ Antifade Kit (P-7481) — outperforms all other commercially available reagents, with little or no quenching of the fluorescent signal. In addition to inhibiting the fading of fluorescein, tetramethylrhodamine and Texas Red® dyes, the ProLong antifade reagent retards the fading of DNA-bound nucleic dyes such as DAPI, propidium iodide and YOYO®-1. Its compatibility with a multitude of dyes and dye complexes makes the Prolong Antifade Kit an especially valuable tool for multiparameter analyses such as fluorescence *in situ* hybridization (FISH). For more information on these antifade reagents, see Section 26.1.

Cat #		Product Name	Unit Size	Price Per Unit ($) 1–4 Units	Price Per Unit ($) 5–24 Units
P-7481	*New*	ProLong™ Antifade Kit	1 kit	98.00	78.40
S-2828		*SlowFade*™ Antifade Kit	1 kit	55.00	44.00
S-7461	*New*	*SlowFade*™ *Light* Antifade Kit	1 kit	55.00	44.00

11.1 Data Table *Probes for Actin*

Cat #	Structure	MW	Storage	Soluble	Abs	{ϵ × 10^{-3}}	Em	Solvent	Notes
B-607	C	~1125	F,L	MeOH, H_2O*	505	{83}†	512	MeOH	<1>
B-3416	C	~1150	F,L	MeOH, H_2O*	584	{145}†	592	MeOH	<1>
B-3475	C	~1115	F,L	MeOH, H_2O*	558	{85}†	569	MeOH	<1>
B-7464	C	~1400	F,L	MeOH	589	{62}†	617	MeOH	<1, 2>
B-7474	C	~1300	F	MeOH, H_2O*	<300		none		<1>
B-7491	C	~1150	F,L	MeOH, H_2O*	529	{73}†	547	MeOH	<1>
C-606	C	~1100	F,L	MeOH, H_2O*	355	{16}†	443	MeOH	<1>
D-970	C	~32,000	F,D,L	H_2O	494	ND	517	pH 8	<3, 4>
D-971	C	~32,000	F,D,L	H_2O	555	ND	580	pH 7	<3, 4>
D-972	C	~32,000	F,D,L	H_2O	597	ND	618	pH 7	<3, 4>
D-7497	C	~32,000	F,D,L	H_2O	496	ND	516	pH 8	<3, 4>
E-7463	C	~1500	F,L	MeOH, H_2O*	524	{100}†	544	MeOH	<1>
F-432	C	~1175	F,L	MeOH, H_2O*	496	{84}†	516	pH 8	<1>
J-7473	✓	710	F,D	MeOH	278	{8.0}	none	MeOH	
N-354	C	~1040	F,L	MeOH, H_2O*	465	{24}†	536	MeOH	<1>
O-7465	C	~1281	F,L	MeOH, H_2O*	511	{85}†	528	pH 9	<1>
O-7466	C	~1180	F,L	MeOH, H_2O*	496	{86}†	520	pH 9	<1>
P-29	S2.3	385	F,D,L	DMF, DMSO	339	{26}	384	MeOH	<5, 6>
P-3457	C	~790	F	MeOH, H_2O*	<300		none		
P-3458	C	~825	F	MeOH, H_2O*	290	{11}	none	MeOH	
R-415	C	~1350	F,L	MeOH, H_2O*	542	{85}†	565	MeOH	<1, 7>
T-7471	C	~1490	F,L	MeOH, H_2O*	583	{95}†	603	MeOH	<1, 7>

For definitions of the contents of this data table, see "How to Use the Handbook" on page vi.
Structure: Chemical structure drawing: ✓ = shown in Figure 11.3; Sn.n = shown in Section number n.n; C = not shown due to complexity.
MW: ~ indicates an approximate value.

* Although this phallotoxin is water soluble, storage in water is not recommended, particularly in dilute solution.
† The value of ϵ listed for this phallotoxin conjugate is for the labeling dye in free solution. Use of this value for the conjugate assumes a 1:1 dye:peptide labeling ratio and no change of ϵ due to dye–peptide interactions.
<1> This phallotoxin conjugate has approximately 1 label per mole.
<2> B-7464 is not directly soluble in H_2O. Aqueous dispersions can be prepared by dilution of a stock solution in MeOH.
<3> ND = not determined.
<4> This protein conjugate contains multiple labels. The number of labels per protein is indicated on the product label.
<5> Spectral data of the 2-mercaptoethanol adduct.
<6> Iodoacetamides in solution undergo rapid photodecomposition to unreactive products. Minimize exposure to light prior to reaction.
<7> In aqueous solutions (pH 7.0), Abs/Em = 554/573 nm for R-415 and 591/608 nm for T-7471.

11.1 Price List *Probes for Actin*

Cat #		Product Name	Unit Size	Price Per Unit ($) 1–4 Units	Price Per Unit ($) 5–24 Units
B-7474	**New**	biotin-XX phalloidin	50 U	198.00	158.40
B-3475		BODIPY® 558/568 phalloidin	300 U	198.00	158.40
B-3416		BODIPY® 581/591 phalloidin	300 U	198.00	158.40
B-607		BODIPY® FL phallacidin	300 U	198.00	158.40
B-7491	**New**	BODIPY® R6G phalloidin	300 U	198.00	158.40
B-7464	**New**	BODIPY® TR-X phallacidin	300 U	198.00	158.40
C-606		coumarin phallacidin	300 U	198.00	158.40
D-970		deoxyribonuclease I, fluorescein conjugate	5 mg	135.00	108.00
D-7497	**New**	deoxyribonuclease I, Oregon Green™ 488 conjugate	5 mg	135.00	108.00
D-971		deoxyribonuclease I, tetramethylrhodamine conjugate	5 mg	135.00	108.00
D-972		deoxyribonuclease I, Texas Red® conjugate	5 mg	135.00	108.00
E-7463	**New**	eosin phalloidin	300 U	198.00	158.40
F-7452	**New**	FluoReporter® Actin Assay Kit	1 kit	195.00	156.00
F-432		fluorescein phalloidin	300 U	198.00	158.40
J-7473	**New**	jasplakinolide	300 U	198.00	158.40
N-354		N-(7-nitrobenz-2-oxa-1,3-diazol-4-yl)phallacidin (NBD phallacidin)	100 µg	198.00	158.40
O-7466	**New**	Oregon Green™ 488 phalloidin	300 U	198.00	158.40
O-7465	**New**	Oregon Green™ 514 phalloidin	300 U	198.00	158.40
P-3458		phallacidin	1 mg	80.00	64.00
P-3457		phalloidin	1 mg	80.00	64.00
P-29		N-(1-pyrene)iodoacetamide	100 mg	115.00	92.00
R-415		rhodamine phalloidin	300 U	198.00	158.40
T-7471	**New**	Texas Red®-X phalloidin	300 U	198.00	158.40

11.2 Probes for Tubulin

Unlabeled and Fluorescent Tubulin

For researchers studying microtubule dynamics *in vitro* and *in vivo*, Molecular Probes offers highly purified tubulin from bovine brain (T-7451) — the standard source of tubulin for research — as well as tetramethylrhodamine-labeled tubulin (T-7460) (Figure 11.4). Our fluorescent tubulin preparation contains approximately four fluorophores per tubulin dimer and is subjected to three cycles of temperature-dependent polymerization/depolymerization to select for functional subunits prior to packaging.[1,2] Tetramethylrhodamine-labeled tubulin has been employed to directly observe cell cycle–dependent microtubule dynamics [3] and mitotic spindle morphogenesis [4] in *Xenopus* oocyte extracts and to follow the loss of localized fluorescence that occurs after cells are treated with anti-microtubule agents.[5] Using tetramethylrhodamine-labeled tubulin and GTP-coated beads prepared with our yellow-green fluorescent FluoSpheres® microspheres (see Section 6.2), Mitchison recently identified an exchangeable GTP-binding site on the plus-end of paclitaxel-stabilized fluorescent microtubules, suggesting that β-tubulin is exposed at the plus-end and α-tubulin at the minus-end of microtubules.[6]

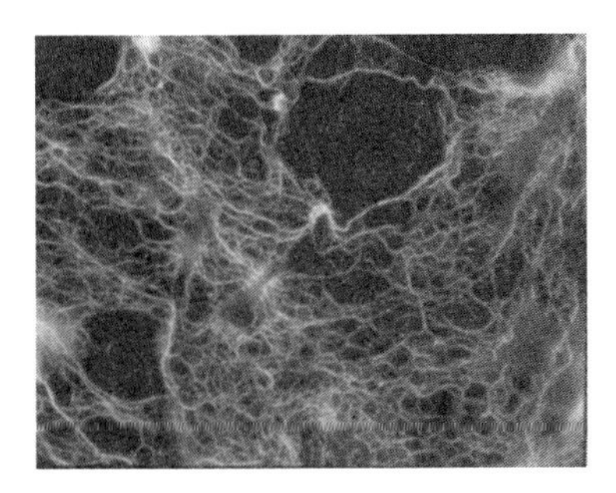

Figure 11.4 *Tetramethylrhodamine-labeled tubulin (T-7460) copolymerized with unlabeled tubulin (T-7451) in G-PEM buffer (80 mM PIPES, 1 mM EGTA, 1 mM $MgCl_2$ and 1 mM GTP) containing 10% DMSO. After 10 minutes at 37°C, the polymerized tubulin was diluted 1:1000 with G-PEM buffer containing 60% glycerol and 0.1% glutaraldehyde and photographed through a tetramethylrhodamine longpass optical filter set.*

Tubulin-Selective Probes

Paclitaxel and Caged Paclitaxel

Molecular Probes offers paclitaxel (P-3456) for *research purposes only* at a purity of >98% by HPLC. Paclitaxel, formerly referred to as taxol in some scientific literature, is the approved generic name for the anticancer pharmaceutical Taxol® (a registered trademark of Bristol-Meyers Squibb Co.). Paclitaxel promotes tubulin assembly, producing aggregates that cannot be depolymerized by dilution, calcium ions, cold or a number of microtubule-disrupting drugs.[7,8] Cultured cells treated with paclitaxel are blocked in G_2 and M phases of the cell cycle.[9]

We have recently synthesized a photoactivatable paclitaxel probe (P-7082) by attaching a (4,5-dimethoxy-2-nitrobenzyl)oxycarbonyl (NVOC) caging group to the side-chain hydroxyl that is known to be required for paclitaxel's tubulin-assembly activity.[10] See Chapter 19 for a discussion of the properties and applications of our many caged probes.

BODIPY Paclitaxel Conjugates

We are pleased to make available two potentially important fluorescent derivatives of paclitaxel. When live cells are incubated with an ~80 nM solution of our new green fluorescent BODIPY FL paclitaxel (P-7500) or red-orange fluorescent BODIPY 564/570 paclitaxel (P-7501), we observe a staining pattern that may correspond to labeled tubulin filaments; however, not all staining by these fluorescent paclitaxel derivatives appears filamentous. Preincubation with 5 µM paclitaxel blocks a significant fraction of the binding and nocodazole alters the distribution of the probe. These fluorescent derivatives also promote polymerization of isolated tubulin at 4°C in a cell-free system. Although localization and inhibition experiments are still underway, we anticipate that these probes will be useful new tools for studying cytoskeletal organization. We appreciate receiving feedback on their performance under diverse experimental conditions.

Inhibitors of Microtubule Assembly

Demecolcine (D-7499), the generic name for the mitotic inhibitor Colcemid® (a registered trademark of Ciba-Geigy AG), is a probe that disrupts microtubule assembly in live cells.[11,12] Desacetyl colchicine (D-1638) has been attached to solid matrices for affinity isolation of tubulin [13] and to a heme nonapeptide that causes tubulin depolymerization.[14] Fluorescein colchicine (C-662) is reported to stain and inhibit assembly of microtubules [15] and to exert the same effects as unlabeled colchicine on mitosis, motility and secretion.[16]

Bis-ANS (B-153) is a potent inhibitor of *in vitro* microtubule assembly.[17] This fluorescent probe binds to the hydrophobic clefts of proteins about 10–100 times better than 1,8-ANS (A-47, see Section 14.5) and exhibits a significant enhancement of fluorescence upon binding. The bis-ANS binding site on tubulin lies near the contact region that is critical for microtubule assembly, but is distinct from the binding sites for the antimitotic drugs colchicine, vinblastine, podophyllotoxin and maytansine.[18-20] Bis-ANS was recently used to investigate structural changes in tubulin monomers during time- and temperature-dependent aging of tubulin.[21]

Other Probes for Tubulin

The nuclear stain DAPI (D-1306) binds tightly to purified tubulin *in vitro* without interfering with microtubule assembly or GTP hydrolysis. DAPI binds to tubulin at sites different from those of paclitaxel, colchicine and vinblastine, and its binding is accompanied by absorption spectral shifts and fluorescence enhancement. The affinity of DAPI for polymeric tubulin is sevenfold greater than for dimeric tubulin, making DAPI a sensitive tool for investigating microtubule assembly kinetics.[22-25]

DCVJ (4-(dicyanovinyl)julolidine; D-3923) has been reported to be a useful probe for following polymerization of tubulin in live cells.[26] DCVJ emits strong green fluorescence upon binding to bovine brain calmodulin.[27] The hydrophobic surfaces of tubulin have also been probed with the environment-sensitive probes nile red [28] (N-1142) and prodan [29] (P-248).

1. Meth Enzymol 196, 478 (1991); **2.** Development 103, 675 (1988); **3.** Cell 62, 579 (1990); **4.** J Cell Biol 112, 925 (1991); **5.** J Cell Biol 130, 639 (1995); **6.** Science 261, 1044 (1993); **7.** J Biol Chem 269, 23399 (1994); **8.** Pharmacol Ther 25, 83 (1984); **9.** Cancer Treat Rep 62, 1219 (1978); **10.** Proc Natl Acad Sci USA 81, 4090 (1984); **11.** Proc Natl Acad Sci USA 87, 1884 (1990); **12.** Biochemistry 25, 5804 (1986); **13.** Chem Pharm Bull 34, 3791 (1986); **14.** Eur J Cell Biol 26, 310 (1982); **15.** Exp Cell Res 141, 211 (1982); **16.** J Cell Biol 76, 619 (1978); **17.** J Biol Chem 259, 14647 (1984); **18.** Biochemistry 33, 11900 (1994); **19.** Biochemistry 33, 11891 (1994); **20.** Biochemistry 25, 3536 (1986); **21.** Biochemistry 34, 13367 (1995); **22.** Acta Histochem 94, 54 (1993); **23.** Arch Biochem Biophys 303, 159 (1993); **24.** Eur J Biochem 165, 613 (1987); **25.** J Biol Chem 260, 2819 (1985); **26.** Biochemistry 28, 6678 (1989); **27.** J Biochem 109, 499 (1991); **28.** J Biol Chem 265, 14899 (1990); **29.** Eur J Biochem 204, 127 (1992).

11.2 Data Table *Probes for Tubulin*

Cat #	Structure	MW	Storage	Soluble	Abs	{ε x 10^{-3}}	Em	Solvent	Notes
B-153	S14.5	673	L	pH>6	395	{23}	500	MeOH	<1, 2>
C-662	✓	747	F,L	pH>6, DMF	492	{85}	517	pH 8	
D-1306	S8.1	350	L	H_2O, DMF	344	{41}	450	pH 7	
D-1638	✓	357	F	pH<8, DMF	351	{17}	none	MeOH	
D-3923	S14.5	249	L	DMF, DMSO	456	{61}	493	MeOH	
D-7499	✓	371	F	DMSO	351	{19}	none	MeOH	
N-1142	S14.5	318	L	DMF, DMSO	552	{45}	636	MeOH	<3>
P-248	S14.5	227	L	DMF, MeCN	361	{18}	498	MeOH	<4>
P-3456	✓	854	F,D	MeOH, DMSO	228	{30}	none	MeOH	
P-7082	✓	1093	FF,D,LL	DMSO	334	{6.4}	none	MeOH	<5>
P-7500	✓	1024	FF,D,L	DMSO	504	{66}	511	MeOH	
P-7501	✓	1099	FF,D,L	DMSO	565	{121}	571	MeOH	
T-7451	C	~110,000	UF	<6>	<300				
T-7460	C	~110,000	UF,L	<6>	555	ND	580	<6>	<7, 8>

For definitions of the contents of this data table, see "How to Use the Handbook" on page vi.
Structure: Chemical structure drawing: ✓ = shown in this section; Sn.n = shown in Section number n.n; C = not shown due to complexity.
MW: ~ indicates an approximate value.

<1> Fluorescence quantum yields of ANS derivatives are environment dependent and are particularly sensitive to the presence of water. Em is also somewhat solvent dependent [Biochim Biophys Acta 694, 1 (1982)].
<2> B-153 is soluble in water at 0.1–1.0 mM after heating.
<3> The absorption and fluorescence spectra and fluorescence quantum yield of N-1142 are highly dependent on solvent [Anal Biochem 167, 228 (1987); J Lipid Res 26, 781 (1985)].
<4> The emission spectrum of P-248 is solvent dependent (see Figure 14.3 in Chapter 14). Em = 401 nm in cyclohexane, 440 nm in $CHCl_3$, 462 nm in MeCN, 496 nm in EtOH and 531 nm in H_2O [Biochemistry 18, 3075 (1979)]. The absorption spectrum is only slightly solvent dependent.
<5> All caged products are sensitive to light. They should be protected from illumination, except when photolysis is intended.
<6> Tubulin and its conjugates are supplied as solutions in G-PEM buffer (80 mM PIPES, 1 mM $MgCl_2$, 1 mM EGTA, 1 mM GTP, 10% glycerol pH 6.8). They should not be exposed to solutions with pH >7.0 or <6.8. Spectra are measured in G-PEM buffer.
<7> ND = not determined.
<8> This product is a multiply labeled protein conjugate. The number of labels per protein is indicated on the product label.

11.2 Structures *Probes for Tubulin*

B-153
Section 14.5

C-662

D-1306
Section 8.1

D-1638 R = H
D-7499 R = $-CH_3$

D-3923, N-1142
P-248
Section 14.5

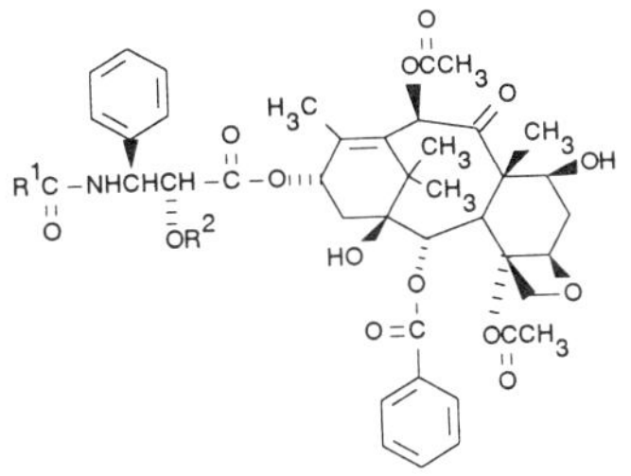

	R^1	R^2
P-3456	R^1 = Ph	R^2 = H
P-7082	R^1 = Ph	R^2 = NVOC
P-7500	R^1 = BODIPY FL	R^2 = H
P-7501	R^1 = BODIPY 564/570	R^2 = H

Ph =

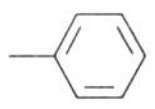

NVOC =

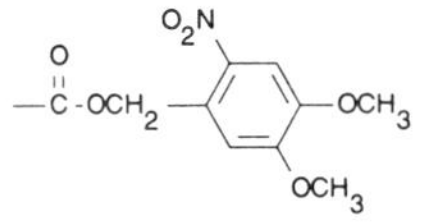

BODIPY FL =

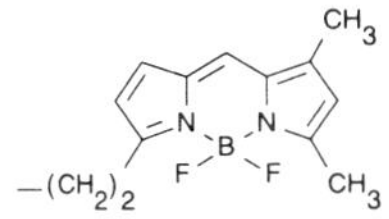

BODIPY 564/570 =

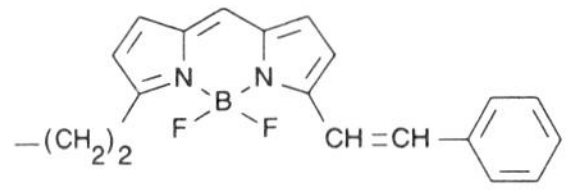

11.2 Price List *Probes for Tubulin*

Cat #		Product Name	Unit Size	Price Per Unit ($) 1–4 Units	5–24 Units
B-153		bis-ANS (4,4′-dianilino-1,1′-binaphthyl-5,5′-disulfonic acid, dipotassium salt)	10 mg	125.00	100.00
C-662		colchicine fluorescein	5 mg	125.00	100.00
D-7499	**New**	demecolcine (Colcemid®; N-desacetyl-N-methylcolchicine)	10 mg	38.00	30.40
D-1638		desacetyl colchicine	10 mg	48.00	38.40
D-1306		4′,6-diamidino-2-phenylindole, dihydrochloride (DAPI)	10 mg	37.00	29.60
D-3923		4-(dicyanovinyl)julolidine (DCVJ)	25 mg	48.00	38.40
N-1142		nile red	25 mg	60.00	48.00
P-3456		paclitaxel (Taxol equivalent) *for use in research only*	5 mg	40.00	32.00
P-7501	**New**	paclitaxel, BODIPY® 564/570 conjugate (BODIPY® 564/570 Taxol)	10 µg	168.00	134.40
P-7500	**New**	paclitaxel, BODIPY® FL conjugate (BODIPY® FL Taxol)	10 µg	168.00	134.40
P-7082	**New**	paclitaxel, 2′-(4,5-dimethoxy-2-nitrobenzyl)carbonate ($O^{2'}$-(NVOC-caged) Taxol)	1 mg	95.00	76.00
P-248		6-propionyl-2-dimethylaminonaphthalene (prodan)	100 mg	135.00	108.00
T-7451	**New**	tubulin from bovine brain *10 mg/mL*	100 µL	125.00	100.00
T-7460	**New**	tubulin from bovine brain, tetramethylrhodamine conjugate *10 mg/mL*	10 µL	145.00	116.00

11.3 Nucleotide Analogs and Phosphate Assays

Nucleotide analogs have been important structural and mechanistic probes for studies of skinned muscle fibers, cytoskeletal proteins and several other types of nucleotide-binding proteins. Most important among the fluorescent nucleotide probes have been etheno and trinitrophenyl (TNP) nucleotides, which are discussed in this section. In addition to these fluorescent analogs, photoactivatable or "caged" nucleotides are powerful tools for studying the kinetics and mechanism of nucleotide-binding proteins because they enable the researcher to control the release of active nucleotide both spatially and temporally.[1,2] We also offer caged and photolabile chelators, which either bind or release Ca^{2+} upon photolysis, as well as caged ionophores, including caged A-23187 and caged nigericin derivatives, which may be useful for studying ion regulation in muscle and other cells. See Chapter 19 for a thorough discussion of our caged probes and their photolysis properties.

Ethenoadenosine Nucleotides

The ethenoadenosine nucleotides, developed in 1972 by Leonard and collaborators,[3,4] bind like endogenous nucleotides to several proteins. The properties and applications of etheno nucleotides have been comprehensively reviewed by Leonard.[5,6] The etheno ATP analog (ϵ-ATP, E-1023) can often mimic ATP in both binding and function. This probe has been used to replace ATP in actin polymerization reactions [7] and is frequently incorporated in place of the tightly bound actin nucleotide.[8] It also supports contraction of actomyosin, facilitates the measurement of nucleotide-exchange kinetics in actin [9] and carbamoyl-phosphate synthetase [10] and serves as a substrate for myosin, which converts it to ϵ-ADP (E-1022).[11] ϵ-ATP is not a substrate for firefly luciferase;[12] however, when added to a reaction mixture containing luciferase, luciferin and

ATP, ϵ-ATP enhances light production, probably by increasing the rate at which the enzyme releases the inhibitory product, oxyluciferin.[13]

The etheno nucleotides usually bind to nucleotide binding sites with high specificity. Because their blue fluorescence is not very environmentally sensitive, protein binding is typically detected by fluorescence anisotropy or circular dichroism. The radiative lifetime of the ethenoadenosines is unusually long (>30 nanoseconds). Binding of etheno nucleotides is frequently accompanied by energy transfer from the protein's tryptophan residues to the nucleotide,[14] and the probes are often used as fluorescence resonance energy transfer donors to longer-wavelength dyes to establish distances within and between cytoskeletal and other proteins.[15-17]

Trinitrophenyl Nucleotides

Unlike the etheno derivatives, the free trinitrophenyl (TNP) nucleotides are essentially nonfluorescent in water. The TNP nucleotides undergo an equilibrium transition to a semiquinoid structure that has relatively long-wavelength spectral properties;[18,19] this form is only fluorescent when bound to the nucleotide-binding site of some proteins. The TNP derivative of ATP frequently exhibits a spectral shift upon protein binding and actually binds with higher affinity than ATP to several proteins. The broad, long-wavelength absorption of TNP nucleotides makes them useful in fluorescence resonance energy transfer studies;[17,20,21] (see the Technical Note "Fluorescence Resonance Energy Transfer" on page 456). The TNP derivatives of ATP (TNP-ATP, T-7602), ADP (TNP-ADP, T-7601) and AMP (TNP-AMP, T-7624) have been used as substrates, inhibitors or structural probes of a wide variety of proteins, including:

- Myosin [18]
- Tubulin [22]
- Na^+/K^+-ATPase [21]
- Gastric H^+/K^+-ATPase [23]
- Erythrocyte Ca^{2+}/Mg^{2+}-ATPase [24]
- F_1-ATPase [25-27]
- Sarcoplasmic reticulum ATPase [28-30]
- Adenylate kinase [31]
- Aspartokinase [32]
- Phosphoglycerate kinase [33]

Molecular Probes also now prepares the TNP analog of GTP (T-7600). This analog has been used as a spectroscopic probe for the GTP inhibitory site of liver glutamate dehydrogenase [34] and as an inhibitor of adenylate cyclase.[35]

We have found that chromatographically purified TNP nucleotides are unstable during lyophilization. Consequently, these derivatives are sold in aqueous solution at 5 mg/mL and should be frozen immediately upon arrival.

Caged Nucleotides

Caged nucleotides are nucleotide analogs in which the terminal phosphate is esterified with a blocking group, rendering the molecule biologically inactive. Photolytic removal of the caging group by UV illumination results in a pulse of the nucleotide — often on a microsecond to millisecond time scale — at the site of illumination. Because photolysis or "uncaging" can be temporally controlled and confined to the area of illumination, the popularity of this technique is growing. Molecular Probes is supporting this development by synthesizing a variety of caged nucleotides, ligands for biological receptors, probes involved in signal transduction and fluorescent dyes. Our current selection of caged nucleotides includes the following probes:

- NPE-caged ATP (A-1048)
- DMNPE-caged ATP (A-1049)
- Desyl-caged ATP (A-7073)
- NPE-caged ADP (A-7056)
- NPE-caged c-AMP (N-1045)
- DMNPE-caged c-AMP (D-1037)
- DMNPE-caged GTP-γ-*S* (G-1053)
- DMNPE-caged GDP-β-*S* (G-3620)

Chapter 19 discusses our selection of caged probes and the properties of the different caging groups that we use (see Table 19.1 in Chapter 19).

Researchers investigating the cytoskeleton have benefited greatly from recent advances in caging technology, primarily originating from the work of Trentham,[36,37] Kaplan [38,39] and their colleagues. NPE-caged ADP (A-7056) is a useful probe for studying the effect of photolytic release of ADP in muscle fibers [40,41] and isolated sarcoplasmic reticulum.[42] It is sometimes difficult to properly abstract papers that describe experiments with caged ATP because they could be referring to either NPE-caged ATP (A-1048), DMNPE-caged ATP (A-1049), desyl-caged ATP (A-7073) or earlier caged versions of this nucleotide. Most researchers have used NPE-caged ATP; however, we have seen a recent trend toward greater use of DMNPE-caged ATP. Caged ATP has been employed in a variety of experimental systems, including tissues, cells and isolated proteins:

- Skinned cardiac and skeletal muscle fibers [40,43-46]
- Submitochondrial particles [47]
- Isolated myosin subfragment-1 [48,49]
- F-actin [50]
- Sarcoplasmic reticulum Ca^{2+}-ATPase [51]
- ATP-regulated K^+ channels of pancreatic β-cells [52] and rat heart cells [53]

Because the caged nucleotides may be added to an experimental system at relatively high concentrations, use of the enzyme apyrase was recommended by Sleep and Burton [54] to eliminate any traces of ATP that may be present in the caged ATP probes.[43,55-57] Once the caged ATP solutions have been preincubated with apyrase, the enzyme can be removed by centrifugal filtration.[43,56]

Although the chemical caging process may confer membrane permeability on the caged ligand, as is the case for caged cAMP,[58] these caged nucleotides are generally cell impermeant and must be microinjected into cells or loaded by other techniques (see Table 15.1 in Chapter 15). Permeabilization of cells with staphylococcal α-toxin or the saponin ester β-escin is reported to make the membrane of smooth muscle cells permeable to low molecular weight (<1000 daltons) molecules, while retaining high molecular weight compounds.[59] α-Toxin permeabilization has permitted the introduction of caged nucleotides, including caged ATP (A-1048) and caged GTP-γ-*S* (G-1053), as well as of 1,4,5-IP_3 (I-3716, see Section 20.2) into smooth muscle cells.[60,61]

Kits for Assaying Phosphate, Pyrophosphate and ATP

EnzChek™ Phosphate Assay Kit

Our EnzChek Phosphate Assay Kit (E-6646), which is based on a method originally described by Webb,[62,63] provides a rapid and highly sensitive enzymatic assay for detecting inorganic phosphate through formation of a chromophoric product (see Figure 10.5 in Chapter 10). This new spectrophotometric technique permits the continuous assay of ATPases activity, and potentially the activity of many other enzymes, such as GTPases and phosphatases, that produce inorganic phosphate. Each EnzChek Phosphate Assay Kit contains:

- 2-Amino-6-mercapto-7-methylpurine ribonucleoside (MESG)
- Purine nucleoside phosphorylase (PNP)
- Concentrated reaction buffer
- KH_2PO_4 standard
- Protocol for detecting and quantitating phosphate

In the presence of inorganic phosphate, MESG is enzymatically converted by PNP to ribose-1-phosphate and the chromophoric product, 2-amino-6-mercapto-7-methylpurine. Although the MESG reagent is somewhat unstable at pH above 7, the reaction can be performed over the pH range of 6.5 to 8.5 with the proper controls.[63] This kit contains sufficient reagents for about 100 phosphate assays using 1 mL assay volumes and standard cuvettes.

The substrate MESG and the enzyme PNP included in our EnzChek Phosphate Assay Kit have already been adapted for monitoring the kinetics of phosphate release by actin-activated myosin ATPase and glycerol kinase,[63] as well as for continuous assay of the enzyme aspartyl transcarbamylase — an enzyme that catalyzes transfer of the carbamoyl group from carbamoyl-phosphate to aspartic acid with liberation of inorganic phosphate.[62] Moreover, the enzymatic assay provided in our EnzChek Phosphate Assay Kit is sufficiently fast and quantitative to permit stopped-flow kinetic experiments on enzymes that produce phosphate, an important development for mechanistic enzyme studies.[63] Although this kit is usually used to determine the inorganic phosphate produced by enzymes such as ATPases, it can also be used to specifically quantitate inorganic phosphate, with a sensitivity of ~2 µM (~0.2 µg/mL).

EnzChek Pyrophosphate Assay Kit

We have adapted the method provided in our EnzChek Phosphate Assay Kit to permit the sensitive spectrophotometric detection of pyrophosphate, which is converted by the enzyme pyrophosphatase to inorganic phosphate.[64] The EnzChek Pyrophosphate Assay Kit (E-6645) enables researchers to detect as little as ~1 µM pyrophosphate (~0.2 µg/mL). This pyrophosphate assay can potentially be adapted to continuously detect enzymes that liberate pyrophosphate such as DNA and RNA polymerases, adenylate cyclase and acyl-CoA synthetase. Each EnzChek Pyrophosphate Assay Kit contains:

- Inorganic pyrophosphatase
- 2-Amino-6-mercapto-7-methylpurine ribonucleoside (MESG)
- Purine nucleoside phosphorylase (PNP)
- Concentrated reaction buffer
- $Na_2P_2O_7$ standard
- Protocol for detecting and quantitating pyrophosphate

The kit contains sufficient reagents for about 100 pyrophosphate assays using 1 mL assay volumes and standard cuvettes.

ATP Determination Kit

Molecular Probes now offers a convenient ATP Determination Kit (A-6608) for the sensitive bioluminescent detection of ATP with firefly luciferase and its substrate luciferin. In the presence of Mg^{2+}, luciferase catalyzes the reaction of luciferin, ATP and O_2 to form oxyluciferin, AMP, CO_2, pyrophosphate and light (Em ~560 nm at pH 7.8). The luciferin–luciferase bioluminescence-based assay is extremely sensitive; most luminometers can detect as little as 1 picomole ATP. This sensitivity has led to its widespread use for detecting ATP produced in enzymatic reactions, as well as for quantitating ATP in various biological samples. The assay has also been used to detect bacterial contamination in milk, soil and sludge because of the ubiquitous presence of ATP in all living matter.[65-67] Each ATP Determination Kit contains:

- Luciferase
- Luciferin
- Dithiothreitol (DTT)
- Reaction buffer
- ATP
- Protocol for quantitating ATP

Unlike many other commercially available ATP detection kits, our ATP Determination Kit provides the luciferase and luciferin packaged separately, which enables the researcher to optimize their reaction conditions for a particular instrument and sample. Each ATP Determination Kit provides sufficient reagents to perform 200 ATP assays using 0.5 mL sample volumes or 500 ATP assays using 0.2 mL sample volumes.

1. Corrie, J.E.T. and Trentham, D.R. in *Bioorganic Photochemistry, Volume 2*, H. Morrison, Ed., J. Wiley (1993) pp. 243–305; **2.** Ann Rev Physiol 55, 755 (1993); **3.** Biochem Biophys Res Comm 46, 597 (1972); **4.** Science 175, 646 (1972); **5.** Chemtracts Biochem Mol Biol 4, 251 (1993); **6.** Chemtracts Biochem Mol Biol 3, 273 (1992); **7.** Biochemistry 27, 3812 (1988); **8.** J Biol Chem 268, 8683 (1993); **9.** J Cell Biol 106, 1553 (1988); **10.** Biochemistry 27, 8050 (1988); **11.** J Biol Chem 259, 11920 (1984); **12.** Proc Natl Acad Sci USA 70, 1664 (1973); **13.** Am J Physiol 261, C1210 (1991); **14.** J Biol Chem 269, 31359 (1994); **15.** Eur J Biochem 217, 737 (1993); **16.** Eur J Biochem 205, 591 (1992); **17.** J Muscle Res Cell Motil 13, 132 (1992); **18.** Biochim Biophys Acta 453, 293 (1976); **19.** Biochim Biophys Acta 320, 635 (1973); **20.** Biochemistry 31, 3930 (1992); **21.** Biophys J 61, 553 (1992); **22.** Biophys Chem 48, 359 (1994); **23.** Biochemistry 29, 3179 (1990); **24.** Clin Chim Acta 156, 165 (1986); **25.** Biochim Biophys Acta 1188, 108 (1994); **26.** J Biol Chem 269, 15431 (1994); **27.** J Biol Chem 268, 6978 (1993); **28.** J Biol Chem 269, 11147 (1994); **29.** Biochemistry 32, 3414 (1993); **30.** J Biol Chem 268, 6917 (1993); **31.** Biochim Biophys Acta 719, 509 (1982); **32.** J Biol Chem 258, 12940 (1983); **33.** Biochem J 301, 885 (1994); **34.** J Biol Chem 260, 4784 (1985); **35.** Arch Biochem Biophys 245, 369 (1986); **36.** J Am Chem Soc 110, 7170 (1988); **37.** J Physiol 352, 575 (1984); **38.** Nature 288, 38 (1980); **39.** Biochemistry 17, 1929 (1978); **40.** Biophys J 68, 78s (1995); **41.** J Mol Biol 223, 185 (1992); **42.** Annals NY Acad Sci 402, 478 (1982); **43.** Biophys J 67, 1933 (1994); **44.** Biophys J 67, 1925 (1994); **45.** Biophys J 67, 1141 (1994); **46.** J Muscle Res Cell Motil 14, 666 (1993);

47. J Biol Chem 268, 25320 (1993); **48.** Biochemistry 33, 6038 (1994); **49.** Biochemistry 30, 11036 (1991); **50.** Biophys J 59, 1235 (1991); **51.** Biochemistry 34, 4864 (1995); **52.** Biochim Biophys Acta 1092, 347 (1991); **53.** Pflügers Arch 415, 510 (1990); **54.** Biophys J 57, 542a (1990); **55.** J Biol Chem 270, 23966 (1995); **56.** Biophys J 66, 115 (1994); **57.** J Biolumin Chemilumin 9, 29 (1994); **58.** Nature 310, 74 (1984); **59.** Meth Cell Biol 31, 63 (1989); **60.** Ann Rev Physiol 52, 857 (1990); **61.** J Biol Chem 264, 5339 (1989); **62.** Anal Biochem 218, 449 (1994); **63.** Proc Natl Acad Sci USA 89, 4884 (1992); **64.** Upson, R.H., Haugland, R.P., Malekzadeh, N. and Haugland, R.P., Anal Biochem, in press; **65.** Anal Biochem 175, 14 (1988); **66.** Bio/Technology 6, 634 (1988); **67.** Meth Enzymol 57, 3 (1978).

11.3 Data Table *Nucleotide Analogs and Phosphate Assays*

Cat #	Structure	MW	Storage	Soluble	Abs	{ε × 10⁻³}	Em	Solvent	Notes
A-1048	✓	~700	FF,D,LL	H_2O	259	{18}	none	MeOH	<1, 2, 3>
A-1049	✓	~760	FF,D,LL	H_2O	351	{4.4}	none	H_2O	<1, 2>
A-7056	✓	~614	FF,D,LL	H_2O	259	{15}	none	MeOH	<1, 2, 3>
A-7073	✓	~753	FF,LL	H_2O*	254	{24}	none	pH 7	<1, 2, 3>
D-1037	S20.4	~524	F,D,LL	DMSO	338	{6.1}	none	MeOH	<1, 2>
E-1022	✓	~517	FF,D	H_2O	265	{5.8}	411	pH 7	<4>
E-1023	✓	~619	FF,D	H_2O	265	{5.0}	411	pH 7	<4>
G-1053	S20.4	~800	FF,D,LL	H_2O	352	{3.7}	none	H_2O	<1, 2>
G-3620	S20.4	~703	FF,D,LL	H_2O	352	{3.1}	none	H_2O	<1, 2>
N-1045	S20.4	~478	FF,D,LL	DMSO	258	{19}	none	MeOH	<1, 2, 3>
T-7600	✓	~799	FF,L	H_2O*	409	{26}	none	pH 8	<5>
T-7601	✓	~681	FF,L	H_2O*	408	{26}	none	pH 8	<5>
T-7602	✓	~782	FF,L	H_2O*	408	{26}	none	pH 8	<5>
T-7624	✓	~579	FF,L	H_2O*	408	{26}	none	pH 8	<5>

For definitions of the contents of this data table, see "How to Use the Handbook" on page vi.
Structure: Chemical structure drawing: ✓ = shown in this section; Sn.n = shown in Section number n.n.
MW: ~ indicates that the molecular weight (MW) of this product is approximate because the exact degree of hydration and salt form has not been conclusively established.

* This product is packaged as a solution in the solvent indicated in "Soluble."
<1> Caged nucleotide esters are free of contaminating free nucleotides when initially prepared. However, some decomposition may occur during storage.
<2> All caged products are sensitive to light. They should be protected from illumination, except when photolysis is intended.
<3> This compound has weaker visible absorption at >300 nm but no discernible absorption peaks in this region.
<4> The absorption spectra of E-1022 and E-1023 have two peaks of almost equal intensity at 265 nm and 275 nm. Absorption extends to approximately 350 nm, but there are no resolved peaks at >300 nm.
<5> Trinitrophenyl nucleotides are, in fact, very weakly fluorescent in water (Em ~560 nm). Fluorescence is blue-shifted and more intense in organic solvents (DMSO, EtOH) and when bound to proteins (Em ~540 nm). Absorption spectrum also has a second, less-intense peak at about 470 nm. Data from Biochim Biophys Acta 719, 509 (1982).

11.3 Structures *Nucleotide Analogs and Phosphate Assays*

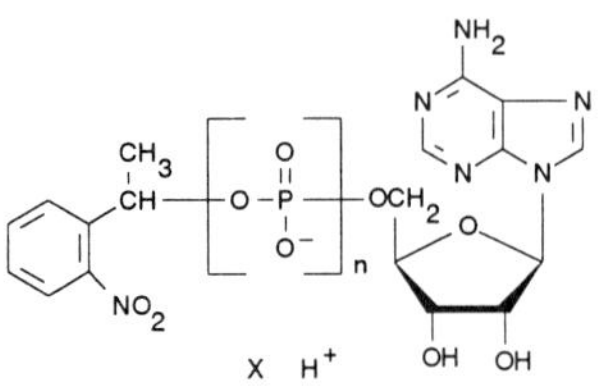

A-1048 n = 3 X = 2 Na^+
A-7056 n = 2 X = K^+

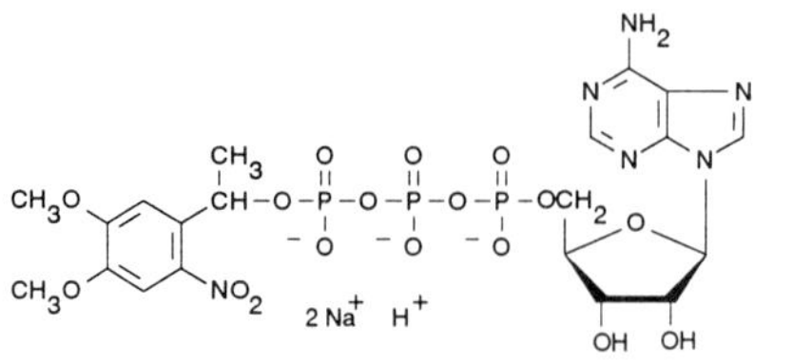

A-1049

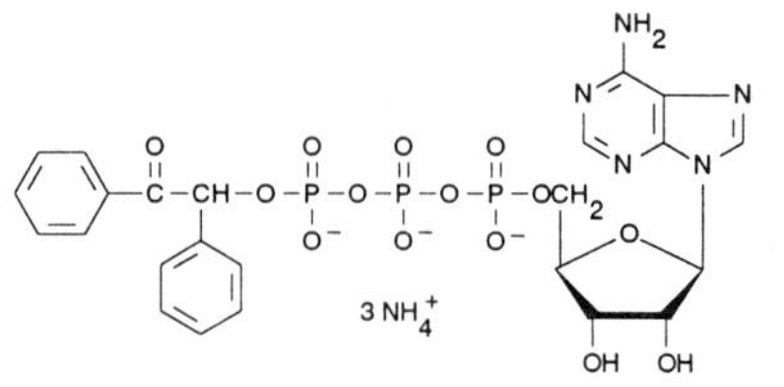

A-7073

Have You Moved?

Updates on our new products and new product applications are provided in our periodic newsletter *BioProbes*. If you have moved, please contact our Customer Service Department so that our technical literature finds its way to you.

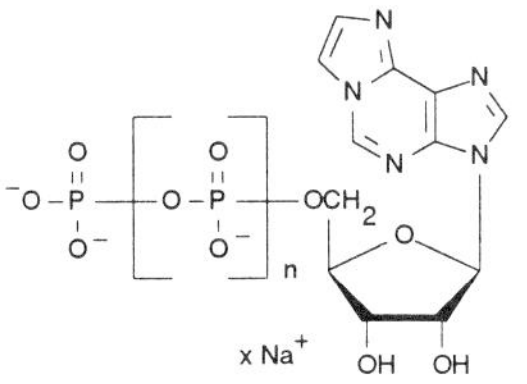

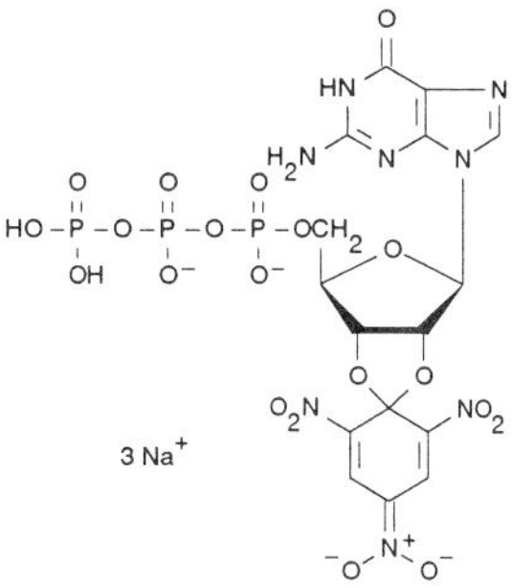

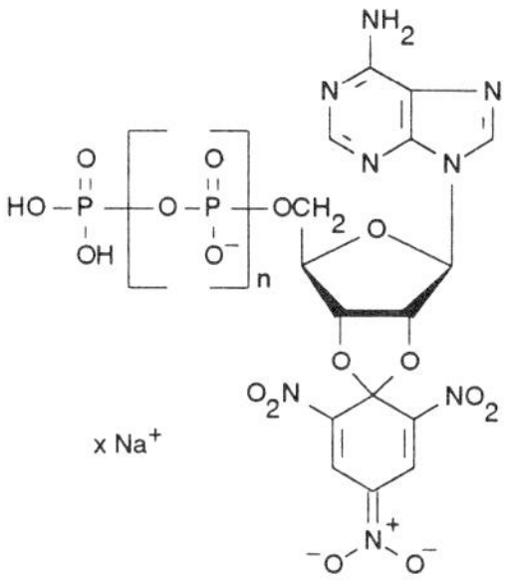

D-1037
Section 20.4

E-1022 n = 1 x = 3
E-1023 n = 2 x = 4

G-1053, G-3620
N-1045
Section 20.4

T-7600

T-7624 n = 0 x = 1
T-7601 n = 1 x = 2
T-7602 n = 2 x = 3

11.3 Price List *Nucleotide Analogs and Phosphate Assays*

Cat #		Product Name	Unit Size	Price Per Unit ($) 1–4 Units	5–24 Units
A-7056	***New***	adenosine 5′-diphosphate, P^2-(1-(2-nitrophenyl)ethyl) ester, monopotassium salt (NPE-caged ADP)	5 mg	103.00	82.40
A-1049		adenosine 5′-triphosphate, P^3-(1-(4,5-dimethoxy-2-nitrophenyl)ethyl) ester, disodium salt (DMNPE-caged ATP)	5 mg	103.00	82.40
A-7073	***New***	adenosine 5′-triphosphate, P^2-(1,2-diphenyl-2-oxo)ethyl ester, ammonium salt (desyl-caged ATP) *5 mM solution in water*	400 µL	145.00	116.00
A-1048		adenosine 5′-triphosphate, P^3-(1-(2-nitrophenyl)ethyl) ester, disodium salt (NPE-caged ATP)	5 mg	103.00	82.40
A-6608	***New***	ATP Determination Kit *200-1000 assays*	1 kit	95.00	76.00
D-1037		4,5-dimethoxy-2-nitrobenzyl adenosine 3′,5′-cyclicmonophosphate (DMNB-caged cAMP)	5 mg	78.00	62.40
E-6646	***New***	EnzChek™ Phosphate Assay Kit *100 assays*	1 kit	135.00	108.00
E-6645	***New***	EnzChek™ Pyrophosphate Assay Kit *100 assays*	1 kit	155.00	124.00
E-1022		1,N^6-ethenoadenosine 5′-diphosphate (ε-ADP)	25 mg	65.00	52.00
E-1023		1,N^6-ethenoadenosine 5′-triphosphate (ε-ATP)	25 mg	118.00	94.40
G-3620		guanosine 5′-O-(2-thiodiphosphate), $P^{2(S)}$-(1-(4,5-dimethoxy-2-nitrophenyl)ethyl) ester, ammonium salt (S-(DMNPE-caged) GDP-β-S)	1 mg	155.00	124.00
G-1053		guanosine 5′-O-(3-thiotriphosphate), $P^{3(S)}$-(1-(4,5-dimethoxy-2-nitrophenyl)ethyl) ester, triammonium salt (S-(DMNPE-caged) GTP-γ-S)	1 mg	155.00	124.00
N-1045		1-(2-nitrophenyl)ethyl adenosine 3′,5′-cyclicmonophosphate (NPE-caged cAMP)	5 mg	78.00	62.40
T-7601	***New***	2′-(or-3′)-O-(trinitrophenyl)adenosine 5′-diphosphate, disodium salt (TNP-ADP) *5 mg/mL in water*	2 mL	160.00	128.00
T-7624	***New***	2′-(or-3′)-O-(trinitrophenyl)adenosine 5′-monophosphate, sodium salt (TNP-AMP) *5 mg/mL in water*	2 mL	48.00	38.40
T-7602	***New***	2′-(or-3′)-O-(trinitrophenyl)adenosine 5′-triphosphate, trisodium salt (TNP-ATP) *5 mg/mL in 0.1 M Tris buffer pH 9*	2 mL	160.00	128.00
T-7600	***New***	2′-(or-3′)-O-(trinitrophenyl)guanosine 5′-triphosphate, trisodium salt (TNP-GTP) *5 mg/mL in water*	1 mL	160.00	128.00

Product-Specific Bibliographies and Keyword Searches

Our Technical Assistance Department can provide you with product-specific bibliographies, as well as keyword searches of the over 25,000 literature references in our extensive bibliography database. Our bibliography database is also searchable through our Web site (http://www.probes.com).

Fluorescent Probes for Photoconversion of Diaminobenzidine

Photoconversion of Diaminobenzidine

Molecular Probes is pleased to offer a variety of fluorescent probes for photoconverting diaminobenzidine (DAB), enabling researchers to take advantage of an important development in correlated fluorescence, transmitted and electron microscopy. In 1982, Maranto first described the use of the fluorophore lucifer yellow for DAB photoconversion.[1] When a fluorophore is exposed to light of an appropriate wavelength, excitation from the electronic ground state to a higher singlet state occurs. Instead of emitting a photon, the excited state of the fluorophore may undergo intersystem crossing to the triplet state. Transfer of energy to ground state triplet oxygen ($^3O_2$) generates the toxic and highly reactive singlet oxygen ($^1O_2$), which is capable of causing damage to lipids, proteins and nucleic acids.[2] However, the reactive potential of $^1O_2$ can also be harnessed to oxidize diaminobenzidine (DAB) into an electron-opaque osmiophilic precipitate within cells. The resulting DAB reaction product exhibits exceptionally uniform, nondiffusible staining properties, making it extremely useful for subsequent electron microscopy investigation of cellular ultrastructure.[3]

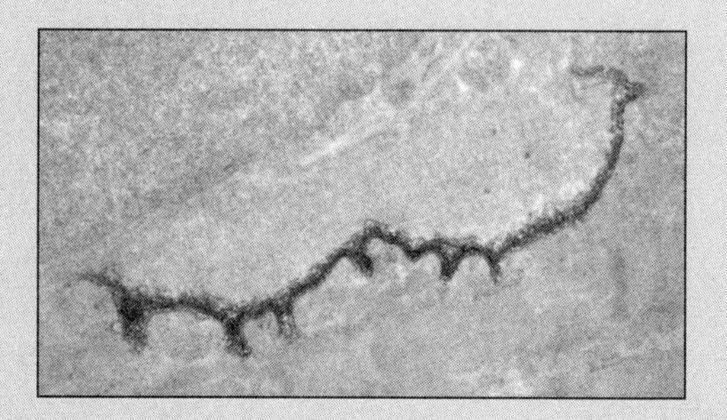

Electron micrograph of a 80 nm–thick section of formaldehyde-fixed rat solius muscle, which was first stained with eosin bungarotoxin (B-7483) and then used to photoconvert DAB into an insoluble osmiophilic polymer. Photo contributed by Thomas J. Deerinck, University of California, San Diego.

Eosin and Other Brominated Probes

In 1994, Deerinck and colleagues reported a simple method for eosin-mediated photoconversion of DAB.[4,5] Halogenated derivatives of fluorescein dyes are known to be effective photosensitizers and singlet oxygen generators.[6] Eosin is a brominated analog of fluorescein that has a $^1O_2$ yield 19 times greater than fluorescein and is an excellent dye for photoconverting DAB.[4,7] Furthermore, the small size of eosin promotes exceptional penetration into tissues resulting in increased resolution for electron microscopy. We offer amine- and thiol-reactive eosin derivatives, as well as several eosin-based conjugates, including secondary antibodies (E-6436, E-6437; see Section 7.2) and streptavidin (S-6372, see Section 7.5), that are known to photoconvert DAB.[4] We have also prepared brominated analogs of several site-selective probes, some of which appear in the product list below. Other fluorescent tracers that have been used to photoconvert DAB include:[8]

- BODIPY® FL C_5-ceramide [9] (D-3521, see Section 12.4)
- DiI [10,11] (D-282, see Section 15.4)
- Fluorescent polystyrene microspheres [12] (see Section 6.2)
- Fluoro-ruby [13] (D-1817, see Section 15.5)
- Lucifer yellow [14] (L-453, see Section 15.3)
- Propidium iodide [15] (P-1304, see Section 8.1)

1. Science 217, 953 (1982); **2.** J Photochem Photobiol 11, 241 (1991); **3.** Science 217, 3 (1982); **4.** J Cell Biol 126, 901 (1994); **5.** J Cell Biol 126, 877 (1994); **6.** Adv Photochem 18, 315 (1993); **7.** Photochem Photobiol 37, 271 (1983); **8.** Lübke, J. in *Neuroscience Protocols*, F.G. Wouterlood, Ed., Elsevier Science Publishers (1993) 93-050-06-01–13; **9.** Cell 73, 1079 (1993); **10.** J Histochem Cytochem 38, 725 (1990); **11.** Neuroscience 28, 3 (1989); **12.** Brain Res 630, 115 (1993); **13.** J Histochem Cytochem 41, 777 (1993); **14.** J Neurosci Meth 36, 309 (1991); **15.** J Neurosci Meth 45, 87 (1992).

Cat #		Product Name	Unit Size	Price Per Unit ($) 1–4 Units	5–24 Units
B-7483	New	α-bungarotoxin, eosin conjugate	500 µg	175.00	140.00
C-6166	New	5-carboxy-2′,4′,5′,7′-tetrabromosulfonefluorescein, succinimidyl ester, bis-(diisopropylethylammonium) salt	5 mg	95.00	76.00
C-2926		CellTracker™ Yellow-Green CMEDA (5-chloromethyleosin diacetate)	1 mg	125.00	100.00
D-7166	New	dextran, eosin, 10,000 MW, anionic, lysine fixable	25 mg	110.00	88.00
D-7546	New	N-(2,6-dibromo-4,4-difluoro-5,7-dimethyl-4-bora-3a,4a-diaza-s-indacene-3-pentanoyl)-sphingosine (BODIPY® FL Br_2C_5-ceramide)	250 µg	98.00	78.40
D-7767	New	6,6′-dibromo-3,3′-dioctadecyloxacarbocyanine perchlorate (Br_2-DiOC_{18}(3))	5 mg	135.00	108.00
D-7766	New	5,5′-dibromo-1,1′-dioctadecyl-3,3,3′,3′-tetramethylindocarbocyanine perchlorate (Br_2-DiIC_{18}(3))	5 mg	135.00	108.00
E-6436	New	eosin goat anti-mouse IgG (H+L) conjugate *2 mg/mL*	0.5 mL	75.00	60.00
E-6437	New	eosin goat anti-rabbit IgG (H+L) conjugate *2 mg/mL*	0.5 mL	75.00	60.00
E-99		eosin-5-iodoacetamide	100 mg	155.00	124.00
E-18		eosin-5-isothiocyanate	100 mg	98.00	78.40
E-118		eosin-5-maleimide	25 mg	155.00	124.00
E-7463	New	eosin phalloidin	300 U	198.00	158.40
L-7542	New	LysoTracker™ Green Br_2 *1 mM solution in DMSO* *special packaging*	20x50 µL	125.00	100.00
S-6372	New	streptavidin, eosin conjugate	1 mg	95.00	76.00
T-7539	New	2′,4′,5′,7′-tetrabromorhodamine 123 bromide	5 mg	48.00	38.40
T-7017	New	2′,4′,5′,7′-tetrabromo-4,5,6,7-tetrafluorofluorescein diacetate (Br_4TFFDA)	5 mg	125.00	100.00

Chapter 12

Probes for Organelles

Contents

12.1 A Diverse Selection of Organelle Probes

Molecular Probes offers a diverse array of cell-permeant fluorescent stains that selectively associate with the mitochondria, lysosomes, endoplasmic reticulum and Golgi apparatus in live cells (Figure 12.1 and Table 12.1). These probes, which are compatible with most fluorescence instrumentation, provide researchers with powerful tools for investigating respiration, mitosis, substrate degradation and detoxification, intracellular transport and sorting and more. Moreover, unlike antibodies, these fluorescent probes can be used to investigate organelle structure and activity in live cells with minimal disruption of cellular function.

We have also recently introduced a collection of organelle-specific monoclonal antibodies for both yeast and mammalian cells that can be used for immunolocalization, immunoprecipitation or Western blot analysis. Cell-permeant and -impermeant fluorescent stains for the nucleus are described in Chapters 8, probes for the cytoskeleton in Chapter 11, and plasma membrane stains in Chapter 14. A variety of probes for phagovacuoles, endosomes and lysosomes — including membrane markers as well as ligands for studying receptor-mediated endocytosis — are discussed in Chapter 17.

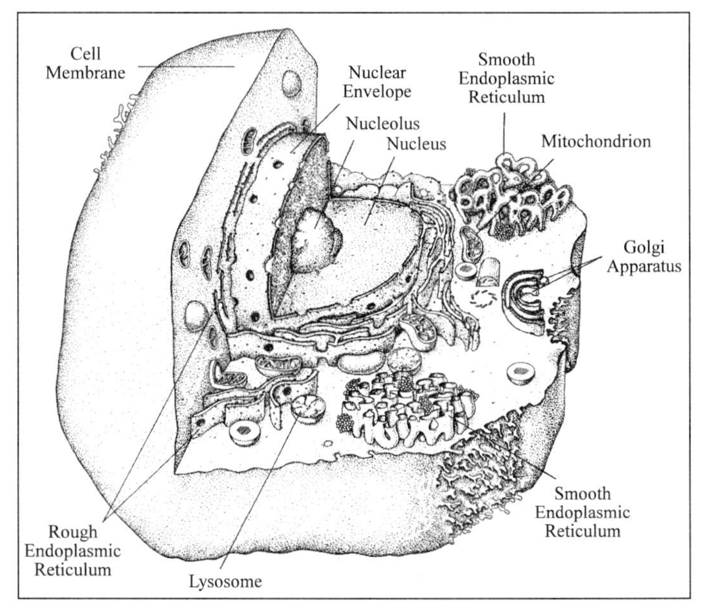

Figure 12.1 An animal cell, as interpreted from electron micrographs. Adapted with permission from Curtis, H. and Barnes, N.S., Biology, Fifth Edition, Worth Publishers (1989).

12.2 Cell-Permeant Probes for Mitochondria

Mitochondria are the intracellular organelles responsible for aerobic metabolism in eukaryotic cells. Their abundance varies with cellular energy level and is a function of cell type, cell-cycle stage and proliferative state. For example, brown adipose tissue cells,[1] hepatocytes [2] and certain renal epithelial cells [3] tend to be rich in active mitochondria, whereas quiescent immune-system progenitor or precursor cells show little staining with mitochondrion-selective dyes.[4] The uptake of most mitochondrion-selective dyes is dependent on mitochondrial membrane potential; nonyl acridine orange and possibly our new MitoTracker Green FM and MitoFluor™ Green probes are notable exceptions. Mitochondrion-selective reagents enable researchers to probe mitochondrial activity, localization and abundance,[5,6] as well as to monitor the effects of some pharmacological agents, such as anesthetics that alter mitochondrial function.[7] Molecular Probes offers a variety of cell-permeant stains for mitochondria described below, as well as monoclonal antibodies directed against the important mitochondrial membrane protein cytochrome oxidase (COX), which are discussed in Section 12.5.

MitoTracker Probes: Fixable Mitochondrion-Selective Probes

Although conventional fluorescent stains for mitochondria, such as rhodamine 123 and tetramethylrosamine, are readily sequestered by functioning mitochondria, they are subsequently washed out of the cells once the mitochondrion's membrane potential is lost. This characteristic limits their use in experiments in which cells must be treated with aldehyde fixatives or other agents that affect the energetic state of the mitochondria. To overcome this limitation, Molecular Probes has developed MitoTracker probes [8] — a series of novel mitochondrion-selective stains that are concentrated by active mitochondria and well retained during cell fixation and permeabilization [9] (Figure 12.2). Because MitoTracker probes are fixable, the sample retains the fluorescent staining pattern characteristic of live cells during subsequent processing steps for immunocytochemistry, *in situ* hybridization or electron microscopy. MitoTracker reagents should also eliminate some of the difficulties of working with pathogenic cells because, once the mitochondria are stained, the cells can be treated with fixatives before the sample is analyzed.

Properties of MitoTracker Probes

MitoTracker probes are cell-permeant mitochondrion-selective dyes that contain a mildly thiol-reactive chloromethyl moiety. The chloromethyl group appears to be responsible for keeping the dye associated with the mitochondria after fixation. To label mitochondria, cells are simply incubated in submicromolar concentrations of the MitoTracker probe, which passively diffuses across the plasma membrane and accumulates in active mitochondria. Once their mitochondria are labeled, the cells can be treated with aldehyde fixatives to allow further processing of the sample; subsequent cold acetone permeabilization does not appear to disturb the MitoTracker staining pattern.

Table 12.1 Molecular Probes' organelle-selective probes.

Blue Fluorescent and Nonfluorescent Probes		Green Fluorescent Probes		Yellow and Orange Fluorescent Probes		Red Fluorescent Probes	
Cat #	Probe	Cat #	Probe	Cat #	Probe	Cat #	Probe
Probes for Mitochondria — see Section 12.2							
L-6868	Lucigenin †	A-1372 D-273 D-303 D-378 M-7502 M-7514 R-302 S-7529 T-3168	Nonyl acridine orange $DiOC_6(3)$ $DiOC_2(5)$ $DiOC_7(3)$ (for plant mitochondria) MitoFluor Green MitoTracker Green FM * Rhodamine 123 SYTO 18 yeast mitochondrial stain JC-1 ‡	D-288 D-308 D-426 M-7510 R-634 R-648 T-639 T-668 T-669	4-Di-1-ASP (DASPMI) 2-Di-1-ASP (DASPMI) DASPEI MitoTracker Orange CMTMRos * Rhodamine 6G Rhodamine B, hexyl ester Tetramethylrosamine Tetramethylrhodamine, methyl ester Tetramethylrhodamine, ethyl ester	M-7512 T-3168	MitoTracker Red CMXRos * JC-1 ‡
Probes for Mitochondria (Probes Requiring Intracellular Oxidation) — see Section 12.2							
		D-632	Dihydrorhodamine 123	D-633 D-638 M-7511	Dihydrorhodamine 6G Dihydrotetramethylrosamine MitoTracker Orange CM-H_2TMRos *	M-7513	MitoTracker Red CM-H_2XRos *
Probes for Acidic Organelles, Including Lysosomes — see Section 12.3							
D-1552 H-7599 L-7525 L-7537 L-7538	DAMP § Hydroxystilbamidine ¶ LysoTracker Blue DND-22 ¶ LysoTracker biotin § LysoTracker biotin/DNP §	L-7526 L-7542	LysoTracker Green DND-26 LysoTracker Green Br_2	D-113 L-7527	Dansyl cadaverine LysoTracker Yellow DND-68	L-7528	LysoTracker Red DND-99
Probes for Acidic Organelles, Including Lysosomes (pH-Sensitive Probes) — see Section 12.3							
L-7532 L-7533 L-7545	LysoSensor Blue DND-192 ¶ LysoSensor Blue DND-167 ¶ LysoSensor Yellow/Blue DND-160 ‡	L-7534 L-7535	LysoSensor Green DND-153 LysoSensor Green DND-189	A-1301 L-7545	Acridine orange LysoSensor Yellow/Blue DND-160 ‡	N-3246	Neutral red
Probes for Endoplasmic Reticulum — see Section 12.4							
		B-7447 D-272 D-273	BODIPY FL brefeldin A $DiOC_5(3)$ $DiOC_6(3)$	B-7449 D-282 D-384 R-648 R-634 T-668 T-669	BODIPY 558/568 brefeldin A $DiIC_{18}(3)$ $DiIC_{16}(3)$ Rhodamine B, hexyl ester Rhodamine 6G Tetramethylrhodamine, methyl ester Tetramethylrhodamine, ethyl ester		
Probes for Golgi Apparatus — see Section 12.4							
B-7450	Brefeldin A	B-7447 D-3521 D-3522 D-7546 N-1154 N-3524	BODIPY FL brefeldin A BODIPY FL C_5-ceramide ‡ BODIPY FL C_5-sphingomyelin BODIPY FL Br_2C_5-ceramide ‡ NBD C_6-ceramide NBD C_6-sphingomyelin	B-7449	BODIPY 558/568 brefeldin A	D-3521 D-7540 D-7546	BODIPY FL C_5-ceramide ‡ BODIPY TR ceramide ‡ BODIPY FL Br_2C_5-ceramide ‡

* Aldehyde-fixable probe. † Chemiluminescent probe. ‡ Dual-emission spectrum. § Nonfluorescent probe. ¶ Blue fluorescent probe.

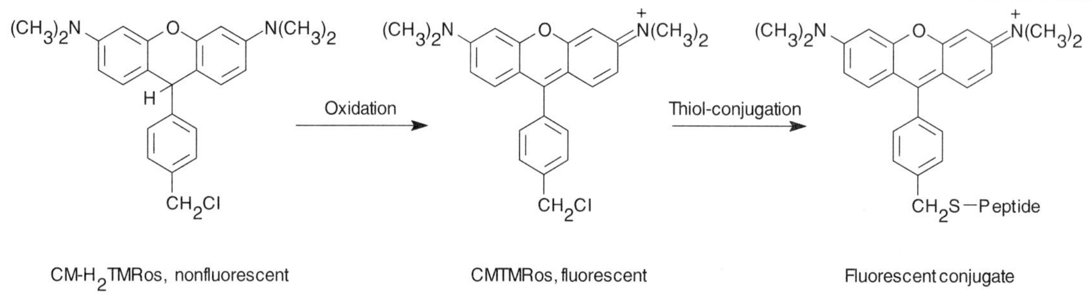

$(CH_3)_2N$ … $N(CH_3)_2$ — Oxidation → — Thiol-conjugation →

CH_2Cl | CH_2Cl | CH_2S—Peptide

CM-H_2TMRos, nonfluorescent | CMTMRos, fluorescent | Fluorescent conjugate

Figure 12.2 *Intracellular reactions of our fixable mitochondrion-selective MitoTracker Orange CM-H_2TMRos (M-7511). When this cell-permeant probe enters an actively respiring cell, it is oxidized to MitoTracker Orange CMTMRos and sequestered in the mitochondria, where it can react with thiols on proteins and peptides to form aldehyde-fixable conjugates.*

Molecular Probes offers five MitoTracker reagents that differ in spectral characteristics, oxidation state and fixability (Table 12.2). MitoTracker probes are provided in specially packaged sets of 20 vials, each containing 50 µg for reconstitution as required.

Table 12.2 *Spectral characteristics of MitoTracker probes.*

Cat #	MitoTracker Probe	Abs* (nm)	Em* (nm)	Oxidation State
M-7514	MitoTracker Green FM †	490	516	NA
M-7510	MitoTracker Orange CMTMRos	551	576	Oxidized
M-7511	MitoTracker Orange CM-H_2TMRos	551 ‡	576 ‡	Reduced
M-7512	MitoTracker Red CMXRos	578	599	Oxidized
M-7513	MitoTracker Red CM-H_2XRos	578 ‡	599 ‡	Reduced

* Absorption (Abs) and fluorescence emission (Em) maxima, determined in methanol; values may vary somewhat in cellular environments. † MitoTracker Green FM is nonfluorescent in aqueous environments. ‡ These reduced MitoTracker Probes are not fluorescent until oxidized.

MitoTracker Rosamine Derivatives

We offer MitoTracker derivatives of the orange fluorescent tetramethylrosamine and its nonfluorescent reduced form, dihydrotetramethylrosamine (MitoTracker Orange CMTMRos, M-7510; MitoTracker Orange CM-H_2TMRos, M-7511), as well as MitoTracker derivatives of the red fluorescent X-rosamine (MitoTracker Red CMXRos, M-7512; MitoTracker Red CM-H_2XRos, M-7513). The X-rosamine derivatives produce a longer-wavelength fluorescence that is well resolved from fluorescein's fluorescence in double-labeling experiments; see Color Plate 2 in "Visual Reality." Unlike MitoTracker Orange CMTMRos and MitoTracker Red CMXRos, the reduced versions of these probes do not fluoresce until they enter an actively respiring cell, where they are oxidized to the fluorescent mitochondrion-selective probe and then sequestered in the mitochondria. MitoTracker Orange CMTMRos has proven useful for:

- Colocalizing a novel kinesin motor protein involved in transport of mitochondria along microtubules [10]
- Studying the localization of mitochondria in fibroblasts transformed with cDNA of wild-type and mutant kinesin heavy chains [11]
- Determining the mechanism by which mitochondrial shape is established and maintained in yeast [12]

Cellular Autofluorescence

Autofluorescence often limits the detectability of fluorescent probes in cells and tissues. The extent and nature of autofluorescence depends on the experimental specimen and preparation. In mammalian cells, autofluorescence is principally due to flavin coenzymes (FAD and FMN, absorption/emission ~450/515 nm) and reduced pyridine nucleotides (NADH, absorption/emission ~340/460 nm).[1,2] In plant cells, lignins are a major source of green autofluorescence, whereas porphyrins such as chlorophyll cause the longer-wavelength red autofluorescence. It has been estimated that the autofluorescence of an average cultured NIH 3T3 fibroblast that has been excited at 488 nm is equivalent to the fluorescence of about 34,000 fluorescein molecules.[3] Several instrumental methods for compensating for autofluorescence in flow cytometry measurements have been developed.[3-5] For fixed cells and tissue, autofluorescence due to cellular components or the aldehyde fixation procedure itself (particularly when glutaraldehyde is used) can be significantly diminished by washing with 0.1% sodium borohydride in phosphate-buffered saline for 30 minutes prior to staining.[6,7]

1. J Histochem Cytochem 27, 44 (1979); **2.** J Histochem Cytochem 27, 36 (1979); **3.** Cytometry 7, 558 (1986); **4.** Cytometry 9, 539 (1988); **5.** Cytometry 7, 566 (1986); **6.** Cytometry 8, 235 (1987); **7.** Bacallao, R., Kiai, K. and Jesaitis, L. in *Handbook of Biological Confocal Microscopy*, J.B. Pawley, Ed., Plenum Press (1995) pp. 311–325.

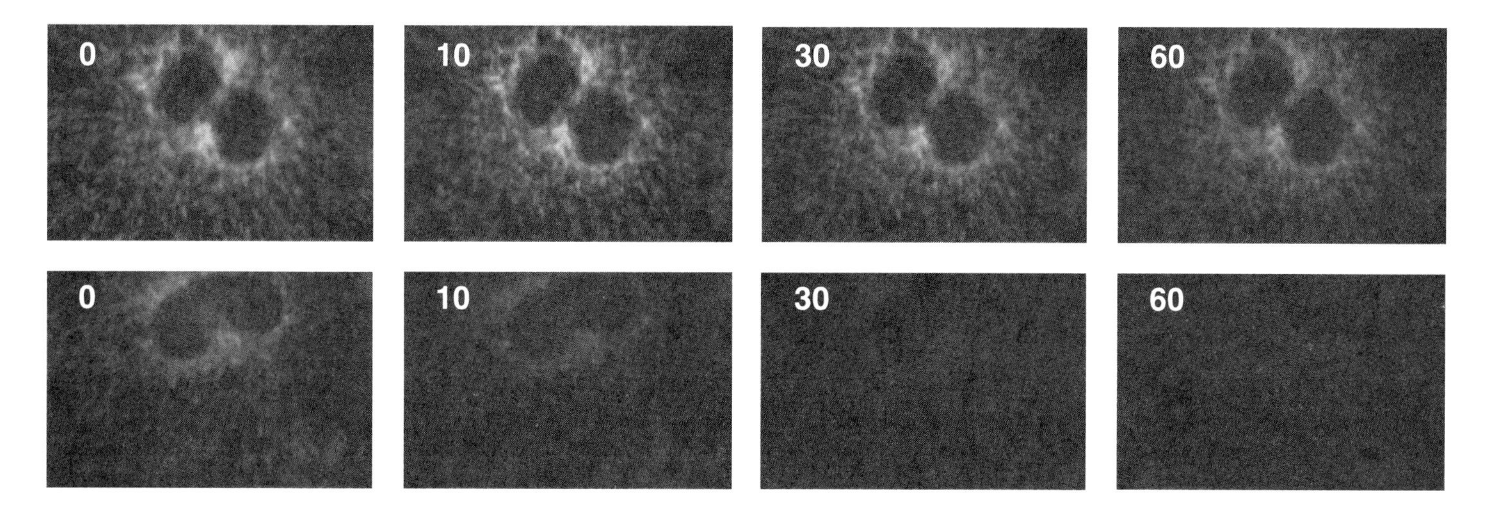

Figure 12.3 *Photostability comparison of mitochondrial staining by MitoFluor Green (M-7502, upper series) and rhodamine 123 (R-302, lower series). HeLa cells were stained with 100 nM MitoFluor Green or 500 nM rhodamine 123 in growth medium for 20 minutes at 37°C. Cells were then rinsed in Hanks' Balanced Salt Solution (HBSS) with 10% calf serum. Samples were continuously illuminated and viewed on a fluorescence microscope using an Omega® Optical fluorescein longpass optical filter set (O-5717, see Section 26.5), a Star 1™ CCD camera (Photometrics) and Image-1® software (Universal Imaging Corp.). Images acquired 0, 10, 30 and 60 seconds after the start of illumination (as indicated in the top left hand corner of each panel) clearly demonstrate the superior photostability of MitoFluor Green.*

MitoTracker Orange CMTMRos and the reduced form CM-H_2TMRos have also been used to investigate the metabolic state of *Pneumocystis carinii* mitochondria.[13] Following fixation, the oxidized forms of the tetramethylrosamine and X-rosamine MitoTracker dyes can be detected directly by fluorescence or indirectly with either anti-tetramethylrhodamine or anti–Texas Red® antibodies (A-6397, A-6399; see Section 7.3).

MitoTracker Green FM and MitoFluor Green

Mitochondria in cells stained with nanomolar concentrations of MitoTracker Green FM (M-7514) exhibit bright green, fluorescein-like fluorescence; see Color Plate 2 in "Visual Reality." MitoTracker Green FM is essentially nonfluorescent in aqueous solutions and only becomes fluorescent once it accumulates in the lipid environment of mitochondria. Background fluorescence is negligible, enabling researchers to clearly visualize mitochondria in live cells immediately following addition of the stain without a wash step. As a companion to MitoTracker Green FM, we have recently introduced MitoFluor Green (M-7502). MitoFluor Green has a structure similar to MitoTracker Green FM but lacks its reactive chloromethyl moieties and is not as well retained following aldehyde fixation.

Like nonyl acridine orange (see below), both MitoTracker Green FM and MitoFluor Green appear to preferentially accumulate in mitochondria regardless of mitochondrial membrane potential, making them important tools for determining mitochondrial mass. Using these probes, we have selectively stained mitochondria in both live and fixed cells. Furthermore, MitoTracker Green FM and MitoFluor Green are substantially more photostable than the widely used fluorescent dye rhodamine 123 (Figure 12.3) and produce a brighter, more mitochondrion-selective signal at lower concentrations. Because their emission maxima are blue-shifted approximately 10 nm relative to the emission maximum of rhodamine 123, MitoTracker Green FM and MitoFluor Green produce a fluorescent staining pattern that should be better resolved from that of red fluorescent probes in double-labeling experiments.

JC-1: A Dual-Emission Potential-Sensitive Probe

The green fluorescent JC-1 (5,5',6,6'-tetrachloro-1,1',3,3'-tetraethylbenzimidazolylcarbocyanine iodide, T-3168) exists as a monomer at low concentrations or at low membrane potential. However, at higher concentrations (aqueous solutions above 0.1 µM) or higher potentials, JC-1 forms red fluorescent "J-aggregates" that exhibit a broad excitation spectrum and an emission maximum at ~590 nm (see Figure 25.1 in Chapter 25). Thus, the emission of this styryl dye can be used as a sensitive measure of mitochondrial membrane potential. Various types of ratio measurements are possible by combining signals from the green fluorescent JC-1 monomer (absorption/emission maxima ~510/527 nm in water) and the J-aggregate (emission maximum 590 nm), which can be effectively excited anywhere between 485 nm and its absorption maximum at 585 nm. Optical filters designed for fluorescein and tetramethylrhodamine (see Section 26.5) can be used to separately visualize the monomer and J-aggregate forms, respectively; see Color Plate 2 in "Visual Reality." Alternatively, both forms can be observed simultaneously using a standard fluorescein longpass optical filter set (see Section 26.5). Chen and colleagues have used JC-1 (T-3168) to investigate mitochondrial potentials in live cells by ratiometric techniques.[14-16] JC-1 has also been used to:

- Analyze the effects of drugs by flow cytometry [17]
- Detect human encephalomyopathy [18]
- Follow mitochondrial changes during apoptosis [19,20]
- Investigate mitochondrial poisoning, uncoupling and anoxia [21]
- Monitor effects of ellipticine on mitochondrial potential [22]

Mitochondrion-Selective Rhodamines and Rosamines

Rhodamine 123

Rhodamine 123 (R-302) is a cell-permeant, cationic, fluorescent dye that is readily sequestered by active mitochondria without inducing cytotoxic effects.[23] Uptake and equilibration of rhodamine 123 is rapid (a few minutes) compared to dyes such as DASPMI, which may take 30 minutes or longer.[6] Viewed through a fluorescein longpass optical filter, the mitochondria of cells stained by rhodamine 123 appear yellow-green. Viewed through a tetramethylrhodamine longpass optical filter, however, these same mitochondria appear red. Unlike the lipophilic rhodamine and carbocyanine dyes, rhodamine 123 does not stain the endoplasmic reticulum.

Rhodamine 123 has been used with a variety of cell types such as presynaptic nerve terminals,[24] live bacteria,[25,26] plants [27,28] and human spermatozoa.[29] Using flow cytometry, researchers used rhodamine 123 to sort respiratory-deficient yeast cells [30,31] and to isolate those lymphocytes that are responsive to mitogen stimulation.[32] Rhodamine 123 has also been used to study:

- Apoptosis [20,33]
- Axoplasmic transport of mitochondria [34]
- Bacterial viability and vitality [25]
- Mitochondrial enzymatic activities [35,36]
- Mitochondrial transmembrane potential and other membrane activities [7,27,37-39]
- Multidrug resistance [40-46] (See Section 16.3)
- Mycobacterial drug susceptibility [47,48]
- Oocyte maturation [49]

Although rhodamine 123 is usually not retained by cells when they are washed, a variety of human carcinoma cell lines (but not sarcomas or leukemic cells) retain the dye for unusually long periods [50] (>24 h), making rhodamine 123 a potential anticancer agent for photodynamic therapy.[51-56] Rhodamine 123 is known to be preferentially taken up and retained by mitochondria of carcinoma cells [57] and to inhibit their proliferation;[58,59] cardiac muscle cells also retain rhodamine 123 for days.[60]

Tetrabromorhodamine 123

The brominated analog of rhodamine 123, tetrabromorhodamine 123 (T-7539), is also potentially useful for photodynamic therapy. Rhodamine 123 is a relatively weak phototoxin [55] with a quantum yield for singlet oxygen ($^1O_2$) generation of <0.01. Continued illumination of rhodamine 123–stained cells causes mitochondrial fragmentation,[61] possibly due to this release of activated oxygen. Tetrabromorhodamine 123 is a much more effective $^1O_2$ generator,[55] with a quantum yield of 0.7. Although it is not localized strictly to mitochondria in cells, tetrabromorhodamine 123 is highly phototoxic to carcinoma cells [55] and its photoproduct is well retained. Tetrabromorhodamine 123 does not stain the nucleus in live cells; however, it binds to DNA in solution, where it has been used as a triplet probe for DNA internal flexibility.[62]

Rosamines and Other Rhodamine Derivatives

Other mitochondrion-selective dyes include tetramethylrosamine (T-639) and 4-dimethylaminotetramethylrosamine (D-7517), which contrast well with fluorescein for multicolor applications, and rhodamine 6G (R-634),[54,63-65] which has an absorption maximum between that of rhodamine 123 and tetramethylrosamine. Although 4-dimethylaminotetramethylrosamine is intrinsically nonfluorescent in solution at pH 7 where it is not protonated, we have observed mitochondrion-selective fluorescent staining by this probe in live cells. Rhodamine 6G has been employed to study microvascular reperfusion injury [66] and the stimulation and inhibition of F_1-ATPase from the thermophilic bacterium PS3.[67]

At low concentrations, lipophilic rhodamine dyes selectively stain mitochondria in live cells.[68] Molecular Probes' researchers have observed that low concentrations of the hexyl ester of rhodamine B (R-648) accumulate selectively in mitochondria and appear to be relatively nontoxic. We have included this probe in our new Yeast Mitochondrial Stain Sampler Kit (Y-7530). At higher concentrations, rhodamine B hexyl ester and rhodamine 6G stain the endoplasmic reticulum of animal cells [68] (see Section 12.4).

The accumulation of tetramethylrhodamine methyl and ethyl esters (TMRM, T-668; TMRE, T-669) in mitochondria and endoplasmic reticulum has also been shown to be driven by membrane potential.[69,70] (see Section 25.3). Moreover, because of their reduced hydrophobic character, these probes exhibit potential-independent binding to cells that is 10 to 20 times lower than that seen with rhodamine 6G.[71] Tetramethylrhodamine ethyl ester has been described as one of the best fluorescent dyes for dynamic and *in situ* quantitative measurements — better than rhodamine 123 — because it is rapidly and reversibly taken up by live cells.[72-74] TMRM and TMRE have been used to measure mitochondrial depolarization related to cytosolic Ca^{2+} transients [75] and to image time-dependent mitochondrial membrane potentials.[73]

Reduced Rhodamines and Rosamines

Inside live cells, the colorless dihydrorhodamines and dihydrotetramethylrosamine are oxidized to fluorescent products that stain mitochondria.[76] However, the oxidation may occur in organelles other than the mitochondria. Dihydrorhodamine 123 (D-632) reacts with hydrogen peroxide in the presence of peroxidases,[77] iron or cytochrome *c* [78] to form rhodamine 123. This reduced rhodamine has been used to monitor reactive oxygen intermediates in rat mast cells [79] and to measure hydrogen peroxide in endothelial cells.[78] Other reduced rhodamines that have been shown to be taken up and oxidized by live cells include dihydrorhodamine 6G (D-633) and dihydrotetramethylrosamine [80,81] (D-638). Chloromethyl derivatives of reduced rosamines (MitoTracker Orange CM-H_2TMRos, M-7511; MitoTracker Red CM-H_2XRos, M-7513), which can be fixed in cells by aldehyde fixation, have been described above.

Other Mitochondrion-Selective Probes

Carbocyanines

Most carbocyanine dyes with short (C_1–C_6) alkyl chains (see Section 25.3) stain mitochondria of live cells when used at low concentrations (~0.5 µM or ~0.1 µg/mL); those with pentyl or hexyl substituents also stain the endoplasmic reticulum when used at higher concentrations (~5–50 µM or ~1–10 µg/mL). DiOC_6(3) (D-273) stains mitochondria in live yeast [12,82,83] and other eukaryotic cells,[65,84] as well as sarcoplasmic reticulum in beating heart cells.[85] It has also been used recently to demonstrate mitochondria moving along microtubules.[10] Photolysis of mitochondrion- or endoplasmic reticulum–bound $DiOC_6(3)$ specifically destroys the microtubules of cells without affecting actin stress fibers, producing a highly localized inhibition of intracellular organelle motility.[86] The red fluorescent dye, $DiOC_2(5)$ (D-303) is reported to be particularly suitable for vital staining of live presynaptic nerve terminals, which are rich in mitochondria, and for visualizing motor nerve terminals in live animals.[24] Several other potential-sensitive carbocyanine probes described in Section 25.3 also stain mitochondria in live cultured cells.[87,88]

The carbocyanine $DiOC_7(3)$ (D-378), which exhibits spectra similar to those of fluorescein, is a versatile dye that has been reported to be a sensitive probe for mitochondria in plant cells.[89] Its other uses include:

- Distinguishing cycling and noncycling fibroblasts [90] and viable and nonviable bacteria [91]
- Following the reorganization of the endoplasmic reticulum during fertilization in the ascidian egg [92]
- Identifying functional vasculature in murine tumors [93,94]
- Studying multidrug resistance [95]
- Visualizing the detailed morphology of neurites of Alzheimer's disease neurons [96]

Styryl Dyes

The isomeric styryl dyes DASPMI (4-Di-1-ASP, D-288; 2-Di-1-ASP, D-308) and DASPEI (D-426) can be used to stain mitochondria in live cells.[6,97] These dyes have large fluorescence Stokes shifts and are taken up relatively slowly as a function of membrane potential.[1] They have been shown to be useful for:

- Determining the distribution of mitochondria in yeast [98,99] and yeast mutants [30]
- Diagnosing mitochondrial disorders [100]
- Long-term imaging of live mammalian nerve cells and their connections [101-103]
- Monitoring the metabolic state of *Pneumocystis carinii* mitochondria [13]
- Screening aberrant mitochondrial distribution and morphology in yeast [104]
- Studying the effects of a sphingolipid synthesis inhibitor on mitochondrial structure and function [105]

Lipophilic Acridine Orange Derivatives

Nonyl acridine orange (A-1372) is well retained in the mitochondria of live HeLa cells for up to 10 days, making it a useful probe for following mitochondria during isolation and after cell fusion.[106-108] The mitochondrial uptake of this metachromatic dye is *not* dependent on membrane potential. It is toxic at high concentrations [109] and apparently binds to all mitochondria, regardless of their energetic state.[110,111] This derivative has been used to analyze mitochondria by flow cytometry,[112] to characterize multidrug resistance [113] and to measure changes in mitochondrial mass during apoptosis in rat thymocytes.[20]

Lucigenin

The well-known chemiluminescent probe lucigenin (L-6868) has recently been shown to accumulate in mitochondria of alveolar macrophages.[114] Relatively high concentrations of the dye (~100 µM) are required to obtain fluorescent staining; however, low concentrations reportedly yield a chemiluminescent response to stimulated superoxide generation within the mitochondria.[114] Lucigenin from Molecular Probes has been highly purified to remove a bright blue fluorescent contaminant that is found in some commercial samples.

Yeast Mitochondrial Stain Sampler Kit

Fluorescence microscopy has been extensively used to study yeast.[12,115] In response to many requests for yeast mitochondrial indicators, Molecular Probes now offers a Yeast Mitochondrial Stain Sampler Kit (Y-7530). This kit contains sample quantities of five different probes that have been found to selectively label yeast mitochondria. Both well-characterized and proprietary mitochondrion-selective probes are provided:

- Rhodamine 123 [31]
- Rhodamine B hexyl ester [68]; see Color Plate 2 in "Visual Reality."
- MitoTracker Green FM
- SYTO® 18 yeast mitochondrial stain
- $DiOC_6(3)$ [12,82,116]

The new mitochondrion-selective nucleic acid stain included in this kit — SYTO 18 yeast mitochondrial stain — exhibits a pronounced fluorescence enhancement upon binding to nucleic acids, resulting in very low background fluorescence even in the presence of dye. SYTO 18 is an effective mitochondrial stain in live yeast but neither penetrates nor stains the mitochondria of higher eukaryotic cells. Each of the components of the Yeast Mitochondrial Stain Sampler Kit is also available separately, including SYTO 18 yeast mitochondrial stain (S-7529).

1. FEBS Lett 170, 181 (1984); **2.** Arch Biochem Biophys 282, 358 (1990); **3.** J Microscopy 132, 143 (1983); **4.** Cytometry 12, 179 (1991); **5.** Microscopy Res Technique 27, 198 (1990); **6.** Int'l Rev Cytol 122, 1 (1990); **7.** Biochem J 271, 269 (1990); **8.** U.S. Patent No. 5,459,268 and other patents pending; **9.** Poot, M. *et al.*, J Histochem Cytochem (1996) in press; **10.** Cell 79, 1209 (1994); **11.** J Cell Biol 131, 1039 (1995); **12.** J Cell Biol 126, 1375 (1994); **13.** J Euk Microbiol 41, S79 (1994); **14.** Chen, L.B. and Smiley, S.T. in *Fluorescent and Luminescent Probes for Biological Activity*, W.T. Mason, Ed., Academic Press (1994) pp. 124–132; **15.** Biochemistry 30, 4480 (1991); **16.** Proc Natl Acad Sci USA 88, 3671 (1991); **17.** Biochem Biophys Res Comm 197, 40 (1993); **18.** Proc Natl Acad Sci USA 92, 729 (1995); **19.** Neuron 15, 961 (1995); **20.** Exp Cell Res 214, 323 (1994); **21.** Cardiovascular Res 27, 1790 (1993); **22.** Biophys J 65, 1767 (1993); **23.** Proc Natl Acad Sci USA 77, 990 (1980); **24.** Nature 310, 53 (1984); **25.** J Appl Bacteriol 72, 410 (1992); **26.** FEMS Microbiol Lett 21, 153 (1984); **27.** Plant Physiol 98, 279 (1992); **28.** Planta 17, 346 (1987); **29.** J Histochem Cytochem 41, 1247 (1993); **30.** J Cell Biol 111, 967 (1990); **31.** Current Genetics 18, 265 (1990); **32.** Proc Natl Acad Sci USA 78, 2383 (1981); **33.** J Cell Biol 123, 1207 (1993); **34.** Brain Res 528, 285 (1990); **35.** J Cell Biol 112, 385 (1991); **36.** Biochim Biophys Acta 975, 377 (1989); **37.** Cytometry 15, 335 (1994); **38.** J Biol Chem 269, 14546 (1994); **39.** Biochem J 288, 207 (1992); **40.** Cytometry 17, 50 (1994); **41.** Eur J Cancer 30A, 1117 (1994); **42.** Mol Pharmacol 45, 1145 (1994); **43.** Proc Natl Acad Sci USA 91, 4654 (1994); **44.** Mol Pharmacol 43, 51 (1993); **45.** Exp Cell Res 190, 69 (1990); **46.** Exp Cell Res 174, 168 (1987); **47.** Biochemistry 33, 7056 (1994); **48.** J Microbiol Meth 7, 139 (1987); **49.** Biol of Reprod 30, 13 (1984); **50.** Annals NY Acad Sci 397, 299 (1982); **51.** Pharmacol Ther 63, 1 (1994); **52.** Exp Cell Res 192, 198 (1991); **53.** Photochem Photobiol 52, 703 (1990); **54.** Biophys J 56, 979 (1989); **55.** Cancer Res 49, 3961 (1989); **56.** Photochem Photobiol 48, 613 (1988); **57.** Cancer Res 45, 6093 (1985); **58.** Science 218, 1117 (1982); **59.** Biochem Biophys Res Comm 118, 717 (1984); **60.** Proc Natl Acad Sci USA 79, 5292 (1982); **61.** Cancer Res 50, 4167 (1990); **62.** Proc Natl Acad Sci USA 79, 5896 (1982); **63.** Histochemistry 94, 303 (1990); **64.** Exp Pathol 31, 47 (1987); **65.** J Cell Biol 88, 526 (1981); **66.** Transplantation 58, 403 (1994); **67.** J Bioenergetics Biomembranes 25, 679 (1993); **68.** J Cell Sci 101, 315 (1992); **69.** Biophys J 56, 1053 (1989); **70.** Biophys J 53, 785 (1988); **71.** J Fluorescence 3, 265 (1993); **72.** Loew, L.M. in *Cell Biology: A Laboratory Handbook, Volume 2*, J.E. Celis, Ed., Academic Press (1994) pp. 399–403; **73.** Biophys J 65, 2396 (1993); **74.** Loew, L.M. *et al.* in *Optical Microscopy for Biology*, B. Herman and K. Jacobson, Eds., Wiley-Liss (1990) pp. 131–143; **75.** Proc Natl Acad Sci USA 91, 12579 (1994); **76.** Chen, L.B. in *Fluorescence Microscopy of Living Cells In Culture, Part A*, D.L. Taylor and Y. Wang, Eds., Academic Press (1989) pp. 103–123; **77.** Eur J Biochem, 217, 973 (1993); **78.** Arch Biochem Biophys 302, 348 (1993); **79.** APMIS 102, 474 (1994); **80.** J Cell Physiol 156, 428 (1993); **81.** Biochem Biophys Res Comm 175, 387 (1991); **82.** Cell Motil Cytoskeleton 25, 111 (1993); **83.** Meth Cell Biol 31, 357 (1989); **84.** Meth Cell Biol 29, 125 (1989); **85.** Exp Cell Res 125, 514 (1980); **86.** Cancer Res 55, 2063 (1995); **87.** Biochem Biophys Res Comm 108, 526 (1982); **88.** J Cell Biol 88, 56 (1981); **89.** Plant Physiol 84, 1385 (1987); **90.** Nature 290, 5807 (1981); **91.** J Appl Bacteriol 78, 309 (1995); **92.** J Cell Biol 120, 1337 (1993); **93.** Br J Cancer 62, 903 (1990); **94.** Br J Cancer 59, 706 (1989); **95.** Biochemistry 34, 3858 (1995); **96.** J Cell Biol 107, 2703 (1988); **97.** Biochim Biophys Acta 423, 1 (1976); **98.** Meth Cell Biol 31, 357 (1991); **99.** J Cell Sci 66, 21 (1984); **100.** J Inher Metab Dis 14, 45 (1991); **101.** J Neurocytol 19, 67 (1990); **102.** J Neurosci 7, 1207 (1987); **103.** Trends Neurosci 10, 398 (1987); **104.** J Cell Biol 126, 1361 (1994); **105.** Biochemistry 31, 3581 (1992); **106.** Histochemistry 82, 51 (1985); **107.** Histochemistry 80, 385 (1984); **108.** Histochemistry 79, 443 (1983); **109.** FEBS Lett 260, 236 (1990); **110.** Eur J Biochem 194, 389 (1990); **111.** Biochem Biophys Res Comm 164, 185 (1989); **112.** Basic Appl Histochem 33, 71 (1989); **113.** Cancer Res 51, 4665 (1991); **114.** Free Rad Biol Med 17, 117 (1994); **115.** Meth Cell Biol 31, 357 (1989); **116.** Biochem Int'l 2, 503 (1981).

12.2 Data Table *Cell-Permeant Probes for Mitochondria*

Cat #	Structure	MW	Storage	Soluble	Abs	{ε × 10⁻³}	Em	Solvent	Notes
A-1372	S14.3	473	L	DMSO, EtOH	495	{84}	519	MeOH	
D-273	S25.3	573	D,L	DMSO	484	{154}	501	MeOH	
D-288	✓	366	L	DMF	475	{45}	605	MeOH	<1>
D-303	S25.3	486	D,L	DMSO	579	{238}	601	MeOH	
D-308	✓	366	L	DMF	461	{38}	585	MeOH	<1>
D-378	S25.3	601	D,L	DMSO	482	{148}	504	MeOH	
D-426	✓	380	L	DMF	461	{39}	589	MeOH	<1>
D-632	S21.4	346	F,D,AA	DMF, DMSO	289	{7.6}	none	MeOH	<2, 3>
D-633	S21.4	445	F,D,AA	DMF, DMSO	296	{11}	none	MeOH	<2, 3>
D-638	S21.4	344	F,D,AA	THF, DMSO	235	{58}	none	MeOH	<2, 3, 4>
D-7517	✓	422	L	DMF, DMSO	556	{91}	585	MeOH	<5>
L-6868	S24.2	511	L	H_2O	455	{7.4}	505	H_2O	<6, 7>
M-7502	✓	603	L	DMSO	489	{112}	517	MeOH	
M-7510	✓	427	F,D,L	DMSO	551	{102}	576	MeOH	
M-7511	✓	393	F,D,AA	DMSO	235	{57}	none	MeOH	<2, 3>
M-7512	✓	532	F,D,L	DMSO	578	{116}	599	MeOH	
M-7513	✓	497	F,D,AA	DMSO	245	{45}	none	MeOH	<2, 3>
M-7514	✓	672	F,D,L	DMSO	490	{111}	516	MeOH	
R-302	✓	381	F,D,L	MeOH, DMF	507	{101}	529	MeOH	
R-634	✓	479	F,D,L	EtOH	528	{100}	551	MeOH	
R-648	✓	563	F,D,L	DMF, DMSO	556	{123}	578	MeOH	
S-7529	P	~450	F,D,L	DMSO*	483	{64}	none	pH 7	<8>
T-639	✓	379	L	DMF, DMSO	550	{87}	574	MeOH	
T-668	S25.3	501	F,D,L	DMSO, MeOH	548	{110}	573	MeOH	
T-669	S25.3	515	F,D,L	DMSO, EtOH	549	{109}	574	MeOH	
T-3168	S25.3	652	D,L	DMSO, EtOH	514	{195}	529	MeOH	<9>
T-7539	✓	741	F,D,L	MeOH, DMF	524	{91}	550	MeOH	

For definitions of the contents of this data table, see "How to Use the Handbook" on page vi.
Structure: Chemical structure drawing: ✓ = shown in this section; Sn.n = shown in Section number n.n; P = proprietary information.
MW: ~ indicates an approximate value, not including counterions.

* This product is packaged as a solution in the solvent indicated in "Soluble."
<1> Abs and Em of styryl dyes are usually at shorter wavelengths in cellular environments than in reference solvents such as methanol.
<2> This compound is susceptible to oxidation, especially in solution. Store solutions under argon or nitrogen. Oxidation appears to be catalyzed by illumination.
<3> These compounds are essentially colorless and nonfluorescent until oxidized. Oxidation products (in parentheses) are as follows: D-632 (R-302); D-633 (R-634); D-638 (T-639); M-7511 (M-7510); M-7513 (M-7512).
<4> Prepare a 4–5 mM stock solution of D-638 by initially dissolving in THF and diluting 1:5 with DMSO [Biochem Biophys Res Comm 175, 387 (1991)]. Long-term storage of THF solutions should be avoided due to tendency for peroxide formation.
<5> D-7517 is nonfluorescent unless protonated. Spectra measured in acidified methanol.
<6> L-6868 has much stronger absorption at 368 nm (ε = {36}).
<7> This compound emits chemiluminescence at 470 nm upon oxidation in basic aqueous solutions.
<8> S-7529 is fluorescent when bound to DNA (Abs = 490 nm, Em = 507 nm).
<9> T-3168 forms J-aggregates with absorption/fluorescence emission maxima at 585/590 nm at concentrations above 0.1 µM in aqueous solutions (pH 8.0) [Biochemistry 30, 4480 (1991)].

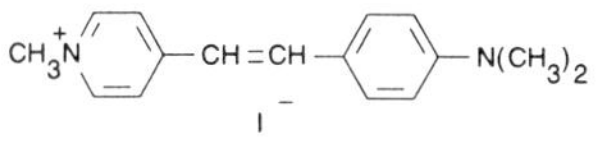

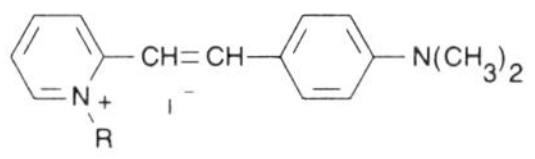

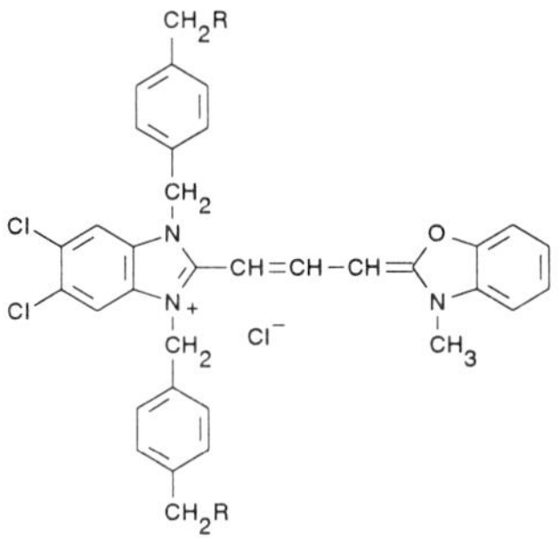

CH_2R
CH_2
Cl
Cl
N
$-CH=CH-CH=$
O
N
N^+
CH_2
Cl^-
CH_3
CH_2R

$CH_3\overset{+}{N}$ — CH = CH — $N(CH_3)_2$
I^-

CH= CH — $N(CH_3)_2$
N^+
R
I^-

A-1372
Section 14.3

D-273, D-303, D-378
Section 25.3

D-288

D-308 R = $-CH_3$
D-426 R = $-CH_2CH_3$

D-632, D-633
D-638
Section 21.4

D-7517
see T-639

L-6868
Section 24.2

M-7502 R = H
M-7514 R = Cl

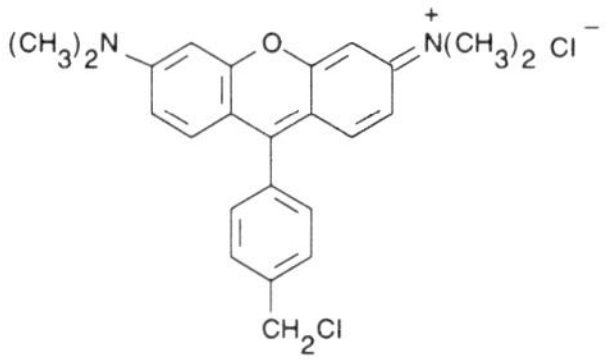

$(CH_3)_2N$ O $\overset{+}{N}(CH_3)_2$ Cl^-
CH_2Cl

M-7510

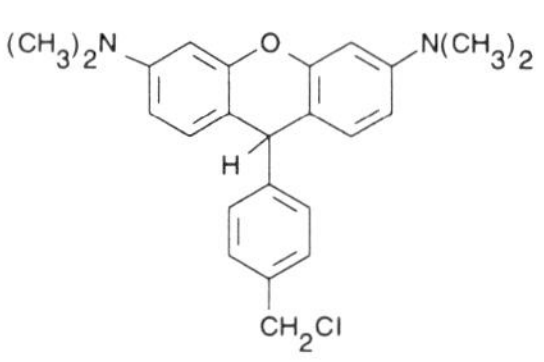

$(CH_3)_2N$ O $N(CH_3)_2$
H
CH_2Cl

M-7511

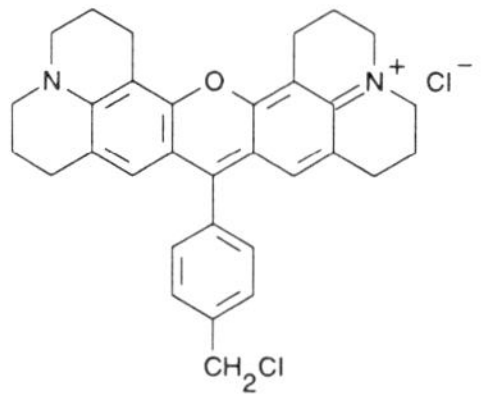

N O N^+ Cl^-
CH_2Cl

M-7512

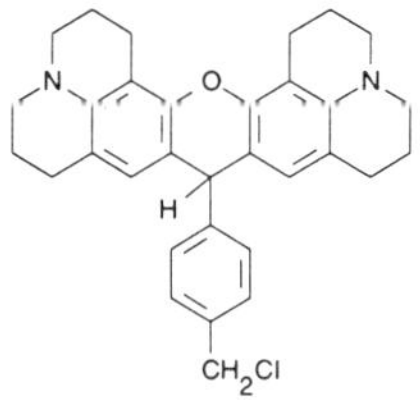

N O N
H
CH_2Cl

M-7513

H_2N R R O $\overset{+}{N}H_2$ X^-
R R
$C-OCH_3$
O

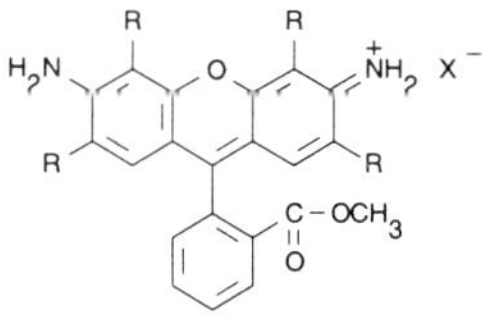

R-302 R = H X = Cl
T-7539 R = Br X = Br

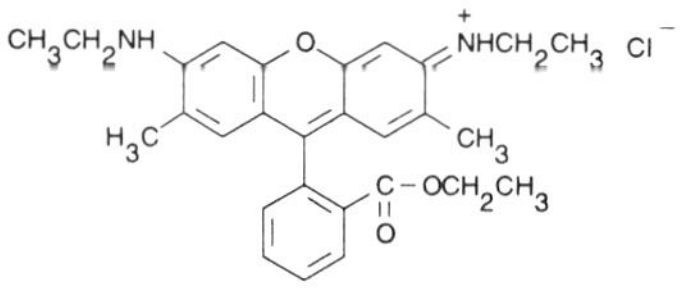

CH_3CH_2NH O $\overset{+}{N}HCH_2CH_3$ Cl^-
H_3C CH_3
$C-OCH_2CH_3$
O

R-634

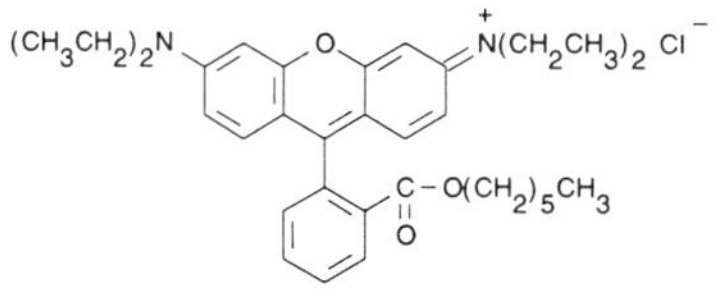

$(CH_3CH_2)_2N$ O $\overset{+}{N}(CH_2CH_3)_2$ Cl^-
$C-O(CH_2)_5CH_3$
O

R-648

$(CH_3)_2N$ O $\overset{+}{N}(CH_3)_2$ Cl^-
R

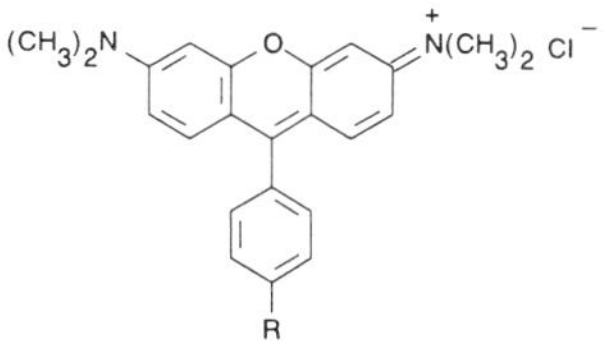

T-639 R = H
D-7517 R = $-N(CH_3)_2$

T-668, T-669
T-3168
Section 25.3

T-7539
see R-302

Cat #		Product Name	Unit Size	Price Per Unit ($) 1–4 Units	5–24 Units
A-1372		acridine orange 10-nonyl bromide (nonyl acridine orange)	100 mg	45.00	36.00
D-303		3,3′-diethyloxadicarbocyanine iodide ($DiOC_2(5)$)	1 g	35.00	28.00
D-378		3,3′-diheptyloxacarbocyanine iodide ($DiOC_7(3)$)	100 mg	68.00	54.40
D-273		3,3′-dihexyloxacarbocyanine iodide ($DiOC_6(3)$)	100 mg	85.00	68.00
D-632		dihydrorhodamine 123	10 mg	98.00	78.40
D-633		dihydrorhodamine 6G	25 mg	98.00	78.40
D-638		dihydrotetramethylrosamine	10 mg	65.00	52.00
D-426		2-(4-(dimethylamino)styryl)-N-ethylpyridinium iodide (DASPEI)	1 g	75.00	60.00
D-288		4-(4-(dimethylamino)styryl)-N-methylpyridinium iodide (4-Di-1-ASP)	1 g	75.00	60.00
D-308		2-(4-(dimethylamino)styryl)-N-methylpyridinium iodide (2-Di-1-ASP)	1 g	75.00	60.00
D-7517	New	4-dimethylaminotetramethylrosamine chloride	5 mg	45.00	36.00
L-6868	New	lucigenin (bis-N-methylacridinium nitrate) *high purity*	10 mg	15.00	12.00
M-7502	New	MitoFluor™ Green	1 mg	98.00	78.40
M-7514	New	MitoTracker™ Green FM *special packaging*	20x50 µg	98.00	78.40
M-7511	New	MitoTracker™ Orange CM-H_2TMRos *special packaging*	20x50 µg	125.00	100.00
M-7510	New	MitoTracker™ Orange CMTMRos *special packaging*	20x50 µg	98.00	78.40
M-7513	New	MitoTracker™ Red CM-H_2XRos *special packaging*	20x50 µg	125.00	100.00
M-7512	New	MitoTracker™ Red CMXRos *special packaging*	20x50 µg	98.00	78.40
R-302		rhodamine 123	25 mg	30.00	24.00
R-634		rhodamine 6G chloride	1 g	18.00	14.40
R-648		rhodamine B, hexyl ester, chloride (R 6)	10 mg	85.00	68.00
S-7529	New	SYTO® 18 yeast mitochondrial stain *5 mM solution in DMSO*	250 µL	98.00	78.40
T-7539	New	2′,4′,5′,7′-tetrabromorhodamine 123 bromide	5 mg	48.00	38.40
T-3168		5,5′,6,6′-tetrachloro-1,1′,3,3′-tetraethylbenzimidazolylcarbocyanine iodide (JC-1; $CBIC_2(3)$)	5 mg	195.00	156.00
T-669		tetramethylrhodamine, ethyl ester, perchlorate (TMRE)	25 mg	72.00	57.60
T-668		tetramethylrhodamine, methyl ester, perchlorate (TMRM)	25 mg	72.00	57.60
T-639		tetramethylrosamine chloride	25 mg	72.00	57.60
Y-7530	New	Yeast Mitochondrial Stain Sampler Kit	1 kit	95.00	76.00

12.3 Cell-Permeant Probes for Lysosomes and Other Acidic Organelles

Molecular Probes' acidotropic reagents can be used to stain lysosomes, as well as several other types of acidic compartments such as trans-Golgi vesicles, endosomes and subpopulations of coated vesicles in fibroblasts, secretory vesicles in insulin-secreting pancreatic β-cells, spermatazoa acrosomes and plant vacuoles.[1] Lysosomes contain glycosidases, acid phosphatases, elastase, cathepsins, carboxypeptidases and a variety of other proteases. Molecular Probes prepares a number of substrates for detecting the activity of these hydrolytic enzymes; see Chapter 10 for more information. Our EnzChek™ 5'-Nucleotidase Assay Kit (E-6643, see Section 10.3) is a new fluorometric assay for 5'-nucleotidase, which is used as a lysosomal marker for isolated lysosomes.[2] 5'-Nucleotidase activity is also found in the Golgi apparatus.[3]

LysoTracker Probes: Acidic Organelle–Selective Probes

Weakly basic amines selectively accumulate in cellular compartments with low internal pH and can be used to investigate the biosynthesis and pathogenesis of lysosomes.[4,5] The most frequently used acidic organelle probe, DAMP (D-1552), is not fluorescent and therefore must be used in conjunction with anti-DNP antibodies conjugated to a fluorophore, enzyme or ferritin in order to visualize the staining pattern.[6] The fluorescent probes neutral red (N-3246) and acridine orange (A-1301, A-3568) are also commonly used for staining acidic organelles, though they lack specificity.[1,7]

These limitations have motivated us to search for alternative acidic organelle–selective probes, both for short-term and long-term tracking studies. The LysoTracker probes are new fluorescent and biotinylated acidotropic probes for labeling and tracing acidic organelles in live cells. These probes have several important features, including high selectivity for acidic organelles and effective labeling of live cells at nanomolar concentrations. Furthermore, the LysoTracker probes are available in several fluorescent colors (Figure 12.4 and Table 12.3), making them especially suitable for multicolor applications.

The LysoTracker probes, which consist of a fluorophore or biotin moiety linked to a weak base that is only partially protonated at neutral pH, are freely permeant to cell membranes and typically concentrate in spherical organelles; see Color Plate 2 in "Visual Reality." We have found that the fluorescent LysoTracker probes must be used at low concentrations, usually about 50 nM, to achieve optimal selectivity. Their mechanism of retention has not

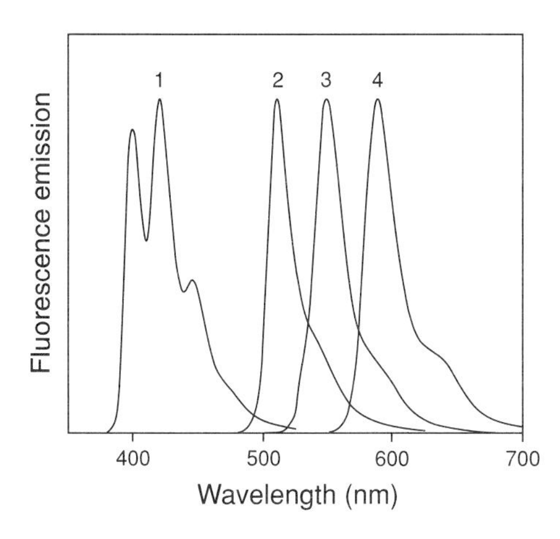

Figure 12.4 *Normalized fluorescence emission spectra of 1) LysoTracker Blue DND-22 (L-7525), 2) LysoTracker Green DND-26 (L-7526), 3) LysoTracker Yellow DND-68 (L-7527) and 4) LysoTracker Red DND-99 (L-7528) in aqueous solutions, pH 6.0.*

Table 12.3 *Summary of our LysoTracker and LysoSensor probes.*

Cat #	Probe	Abs * (nm)	Em * (nm)	pK_a
L-7525	LysoTracker Blue DND-22	373	422	ND
L-7526	LysoTracker Green DND-26	504	511	ND
L-7542	LysoTracker Green Br_2	532	545	ND
L-7527	LysoTracker Yellow DND-68	534	551	ND
L-7528	LysoTracker Red DND-99	577	590	ND
L-7537	LysoTracker Biotin	NA	NA	ND
L-7538	LysoTracker Biotin/DNP	ND	NA	ND
L-7533	LysoSensor Blue DND-167	373	425	5.1
L-7532	LysoSensor Blue DND-192	374	424	7.5
L-7535	LysoSensor Green DND-189	443	505	5.2
L-7534	LysoSensor Green DND-153	442	505	7.5
L-7545	LysoSensor Yellow/Blue DND-160	384 329	540 † 440 ‡	4.2

* Absorption (Abs) and fluorescence emission (Em) maxima, determined in aqueous buffer or methanol; values may vary somewhat in cellular environments. † At pH 3, ‡ at pH 7; this dye has pH-dependent dual-excitation and dual-emission peaks (Figure 12.5). ND = Not determined. NA = Not applicable.

been firmly established but is likely to involve protonation and retention in the organelles' membranes, although staining is generally not reversed by subsequent treatment of the cells with weakly basic cell-permeant compounds. Furthermore, the larger acidic compartments of cells stained with LysoTracker Red DND-99 (L-7528) usually retain their staining pattern following fixation with aldehydes. LysoTracker Green DND-26 was used to localize zinc to acidic compartments in a study of the mechanism of ZnT-2, a membrane protein that facilitates vesicular sequestration of zinc.[8] This LysoTracker probe was also used to identify acidic organelles labeled with rhodamine B in denervated skeletal muscle.[9] Because the lysosomal fluorescence may constitute only a small portion of total cellular fluorescence in LysoTracker dye–stained cells, it is difficult to use these reagents to quantitate the number of lysosomes by flow cytometry or fluorometry.

Like the fluorescent LysoTracker dyes, LysoTracker Biotin (L-7537) and LysoTracker Biotin/DNP (L-7538) accumulate in acidic organelles. Following aldehyde fixation, the biotin moiety of either probe can be detected using one of our many avidin, streptavidin or NeutraLite™ avidin conjugates. LysoTracker Biotin/DNP also contains a dinitrophenyl (DNP) moiety that should be detectable with our anti-DNP antibodies (see Section 7.3). The brominated analog of LysoTracker Green (LysoTracker Green Br_2, L-7542) has potential utility for ultrastructural localization of lysosomes by photoconversion of diaminobenzidine (DAB); see the Technical Note "Fluorescent Probes for Photoconversion of Diaminobenzidine" on page 264.

LysoSensor Probes: Acidic Organelle–Selective pH Indicators

For researchers studying the dynamic aspects of lysosome biogenesis and function in live cells, we introduce our new LysoSensor probes — fluorescent pH indicators that partition into acidic organelles. The LysoSensor dyes are acidotropic probes that appear to accumulate in acidic organelles as the result of protonation. This protonation also relieves the fluorescence quenching of the dye by its weakly basic side chain, resulting in an increase in fluorescence intensity. Thus, the LysoSensor reagents exhibit a pH-dependent increase in fluorescence intensity upon acidification, in contrast to the LysoTracker probes, which exhibit fluorescence that is not substantially enhanced at acidic pH.

We offer five LysoSensor reagents that differ in color and pK_a (Table 12.3). Because these probes may localize in the membranes of organelles, it is probable that the pK_a values listed in Table 12.3 will not be equivalent to those measured in cellular environments and that only qualitative and semiquantitative comparisons of organelle pH will be possible. The blue and green fluorescent LysoSensor probes are available with optimal pH sensitivity in either the acidic or neutral range (pK_a ~5.2 or ~7.5). With their low pK_a values, LysoSensor Blue DND-167 (L-7533) and LysoSensor Green DND-189 (L-7535) are almost nonfluorescent except when inside acidic compartments, whereas LysoSensor Blue DND-192 (L-7532) and LysoSensor Green DND-153 (L-7534) are brightly fluorescent at neutral pH. LysoSensor Yellow/Blue DND-160 (L-7545) is unique in that it exhibits both dual-excitation and dual-emission spectral peaks that are pH-dependent (Table 12.3). In acidic organelles LysoSensor Yellow/Blue DND-160 has predominantly yellow fluorescence, and in less acidic organelles it has blue fluorescence (Figure 12.5). Either dual-excitation or dual-emission measurements may permit ratio imaging of the pH in acidic organelles such as lysosomes.

These probes can be used singly (or potentially in combination) to investigate the acidification of lysosomes and alterations of lysosomal function or trafficking that occur in cells. For example, lysosomes in some tumor cells have a lower pH than normal lysosomes,[10] whereas other tumor cells contain lysosomes with higher pH.[11] In addition, cystic fibrosis and other diseases result in defects in the acidification of some intracellular organelles,[12] and the LysoSensor probes may prove useful in studying these aberrations. As in LysoTracker dye–stained cells, the lysosomal fluorescence in LysoSensor dye–stained cells may constitute only a small portion of

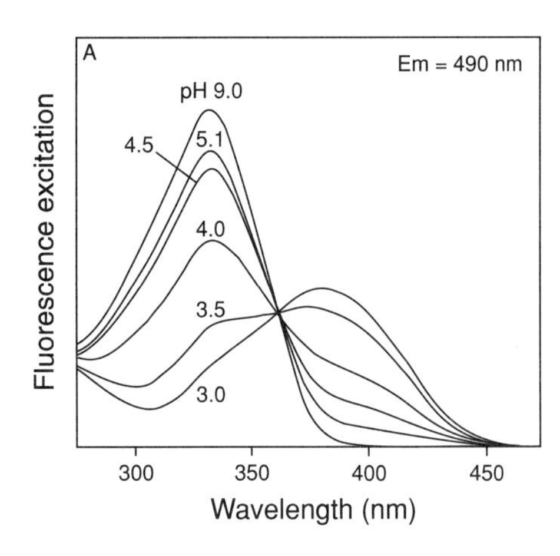

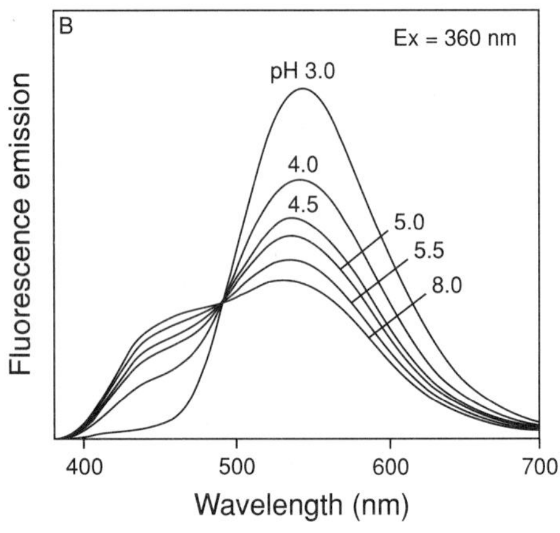

Figure 12.5 *The pH-dependent spectral response of LysoSensor Yellow/Blue DND-160 (L-7545): A) fluorescence excitation spectra and B) fluorescence emission spectra.*

total cellular fluorescence, making it difficult to quantitate the number of lysosomes or their pH by flow cytometry or fluorometry.

DAMP and Other Lysosomotropic Probes

DAMP

The reagent DAMP (*N*-(3-((2,4-dinitrophenyl)amino)propyl)-*N*-(3-aminopropyl)methylamine, dihydrochloride; D-1552), is a weakly basic amine that is taken up in acidic organelles of live cells. This cell-permeant acidotropic reagent can be detected with anti-DNP antibodies (see Section 7.3), including those labeled with fluorochromes, ferritin, colloidal gold or enzymes,[6] making DAMP broadly applicable for detecting acidic organelles by electron and light microscopy. For example, DAMP has been used to investigate:

- Acidification of yolk platelets in sea urchin embryos [13]
- Defective acidification of intracellular organelles in cells from cystic fibrosis patients [12]
- Dependence of pH on the conversion of proinsulin in beta cells [14]
- Development of autophagic vacuoles [15,16]
- Location of intracellular acidic compartments during viral infection [16]

As alternatives to DAMP, our new fluorescent LysoTracker and LysoSensor probes described above have significant potential in many of these applications. Because they can be visualized directly without any secondary detection reagents, the LysoTracker and LysoSensor reagents enable researchers to study acidic organelles and follow their dynamic processes in live cells.

Other Lysosomotropic Probes

Our high-purity neutral red (N-3246) is a common lysosomal probe that stains lysosomes a fluorescent red.[7,18] It has also been used to determine the number of adherent and nonadherent cells in a microplate assay [19] and to stain cells in brain tissue.[20,21]

In addition, dansyl cadaverine [22,23] (D-113), the DNA intercalator acridine orange [7,24] (A-1301, A-3568) and hydroxystilbamidine [25-31] (H-7599) have been reported to be useful lysosomotropic reagents. Recently, dansyl cadaverine was found to selectively label autophagic vacuoles, at least some of which had already fused with lysosomes; it did not, however, accumulate in early or late endosomes.[32]

Ligands for Receptor-Mediated Endocytosis

In addition to these lysosomotropic probes, Molecular Probes prepares a wide variety of low-density lipoproteins (LDL) and fluorescent transferrin conjugates. Once internalized, LDL dissociates from its receptor and ultimately accumulates in lysosomes.[33] The contrasting fluorescence of DiI LDL (L-3482, see Section 17.3) and fluorescein transferrin (T-2871, see Section 17.3) permits their simultaneous use to follow the lysosomally directed pathway and the recycling pathways, respectively.[34] The pH sensitivity of the fluorescein-transferrin conjugate has been exploited to investigate events occurring during endosomal acidification.[35-38] Likewise, dextrans labeled with fluorescent pH indicators (see Sections 15.5 and 23.4) can be used to monitor the uptake and internal processing of exogenous materials in acidic organelles through endocytosis.[39-43]

Yeast Vacuole Markers

The biogenesis of the yeast vacuole has been extensively studied as a model system for eukaryotic organelle assembly.[44-47] Using a combination of genetic and biochemical approaches, researchers have isolated a large collection of yeast vacuolar protein sorting (*vps*) mutants [48] and characterized the vacuolar H^+-ATPase (V-ATPase) responsible for compartment acidification.[49] To facilitate the investigation of yeast vacuole structure and function, Molecular Probes now offers the Yeast Vacuole Marker Sampler Kit (Y-7531), as well as monoclonal antibodies directed against vacuolar proteins, including V-ATPase (see Section 12.5).

Yeast Vacuole Marker Sampler Kit

The Yeast Vacuole Marker Sampler Kit (Y-7531) contains sample quantities of a series of both novel and well-established vacuole marker probes that show promise for the study of yeast cell biology:

- 5-(and-6)-carboxy-2',7'-dichlorofluorescein diacetate (carboxy-DCFDA)
- CellTracker™ Blue CMAC

- Aminopeptidase substrate Arg-CMAC
- Dipeptidyl peptidase substrate Ala-Pro-CMAC
- Yeast vacuole membrane marker MDY-64

Our recent experiments have demonstrated that several cell-permeant derivatives of 7-amino-4-chloromethylcoumarin (CMAC) are largely sequestered within yeast vacuoles. The corresponding 7-amino-4-methylcoumarin derivatives are known to be substrates for yeast vacuolar enzymes.[50-52] This sampler kit's three coumarin-based vacuole markers selectively stain the lumen of the yeast vacuole. To complement the blue fluorescent staining of the lumen, we provide a novel green fluorescent membrane marker MDY-64 for staining the yeast vacuole membrane; see Color Plate 2 in "Visual Reality." Membrane staining can also be accomplished using the red fluorescent probe FM™ 4-64 described below. The commonly used vacuole marker 5-(and-6)-carboxy-2',7'-dichlorofluorescein diacetate (carboxy-DCFDA, D-369; see Section 16.3) is supplied for use as a standard.[44] Each of the components in the Yeast Vacuole Marker Sampler Kit, including the proprietary yeast vacuole membrane marker MDY-64 (Y-7536), is also available separately for those researchers who find that one particular dye is well suited for their application.

FM 4-64

One of our FM dyes, FM 4-64 (T-3166, see also Section 17.2) has recently been reported to selectively stain yeast vacuolar membranes with red fluorescence (excitation/emission maxima ~515/640 nm).[53,54] This lipophilic styryl dye is proving to be an important tool for visualizing vacuolar organelle morphology and dynamics and for studying the endocytic pathway in yeast.

FUN-1™ Cell Stain: An Important New Vital Stain for Yeast

FUN-1 cell stain (F-7030) has unique properties in actively growing yeast. When used at micromolar concentrations, the FUN-1 stain is freely taken up by several species of yeast and fungi, producing an red-orange fluorescent product. Actively growing yeast sequester this product into their vacuoles, generating cylindrical red-orange fluorescent structures as the dye accumulates.[55,56] FUN-1 cell stain is also available in our LIVE/DEAD® Yeast Viability Kit (L-7009, see Section 16.2).

1. J Cell Biol 106, 539 (1988); **2.** Cell 17, 143 (1979); **3.** J Cell Biol 119, 1173 (1992); **4.** Cell 52, 329 (1988); **5.** *Lysosomes in Biology and Pathology, Volume 2*, J.T. Dingle and H.B. Fell, Eds., American Elsevier Publishing (1969); **6.** Proc Natl Acad Sci USA 81, 4838 (1984); **7.** Allison, A.C. and Young, M.R. in *Lysosomes in Biology and Pathology, Volume 2*, J.T. Dingle and H.B. Fell, Eds., American Elsevier Publishing (1969) pp. 600–628; **8.** EMBO J 15, 1784 (1996); **9.** J Histochem Cytochem 44, 267 (1996); **10.** Wilman, D.E.V. and Connors, T.A., *Molecular Aspects of Anticancer Drug Action*, S. Neidle and M.J. Waring, Eds., Macmillian (1983) pp. 233–282; **11.** J Biol Chem 265, 4775 (1990); **12.** Nature 352, 70 (1991); **13.** J Cell Biol 117, 1211 (1992); **14.** J Cell Biol 126, 1149 (1994); **15.** J Cell Biol 111, 329 (1990); **16.** J Cell Biol 110, 1935 (1990); **17.** J Biol Chem 269, 17577 (1994); **18.** In Vitro Toxicol 3, 219 (1990); **19.** Anal Biochem 213, 426 (1993); **20.** Jpn J Physiol 43, S161 (1993); **21.** Brain Res 573, 1 (1992); **22.** Immunology 51, 319 (1984); **23.** J Immunology 131, 125 (1983); **24.** Anal Biochem 192, 316 (1991); **25.** Brain Res 553, 135 (1991); **26.** J Neurocytol 18, 333 (1989); **27.** J Cell Biol 90, 665 (1981); **28.** J Cell Biol 90, 656 (1981); **29.** Infection and Immunity 11, 441 (1975); **30.** Biochem Pharmacol 19, 1251 (1970); **31.** Life Sci 3, 1407 (1964); **32.** Eur J Cell Biol 66, 3 (1995); **33.** J Cell Biol 121, 1257 (1993); **34.** J Cell Sci 107, 2177 (1994); **35.** Biochemistry 31, 5820 (1992); **36.** J Biol Chem 266, 3469 (1991); **37.** J Bioenergetics Biomembranes 23, 147 (1991); **38.** J Biol Chem 265, 6688 (1990); **39.** J Biol Chem 269, 12918 (1994); **40.** Am J Physiol 258, C309 (1990); **41.** J Cell Physiol 131, 200 (1987); **42.** Proc Natl Acad Sci USA 82, 8523 (1985); **43.** Exp Cell Res 150, 36 (1984); **44.** Meth Enzymol 194, 644 (1991); **45.** Trends Biochem Sci 14, 347 (1989); **46.** D.J. Klinonsky *et al.* in *Molecular Biology of Intracellular Protein Sorting and Organelle Assembly*, R. Bradshaw, L. McAllister-Henn and M. Douglas, Eds., Alan R. Liss (1988) pp. 173–186; **47.** J.H. Rothman and T.H. Stevens in *Protein Transfer and Organelle Biogenesis*, R. Das and P. Robbins, Eds., Academic Press (1988) pp. 318–354; **48.** Mol Biol of the Cell 3, 1389 (1992); **49.** J Biol Chem 264, 19236 (1989); **50.** Meth Enzymol 194, 428 (1991); **51.** Arch Biochem Biophys 226, 292 (1983); **52.** FEBS Lett 131, 296 (1981); **53.** Mol Biol of the Cell 7, 985 (1996); **54.** J Cell Biol 128, 779 (1995); **55.** Cytometry Suppl 7, 76, abstract # 417A (1994); **56.** J Cell Biol 126, 1375 (1994).

12.3 Data Table *Cell-Permeant Probes for Lysosomes and Other Acidic Organelles*

Cat #	Structure	MW	Storage	Soluble	Abs	{ε × 10⁻³}	Em	Solvent	Notes
A-1301	S8.1	302	L	EtOH, DMF	489	{67}	520	MeOH	<1>
D-113	S3.3	335	L	EtOH, DMF	335	{4.6}	518	MeOH	
D-1552	✓	384	F,D,L	pH<7, DMF	349	{16}	none	MeOH	
F-7030	✓	529	F,L	DMSO*	508	{71}	none	pH 7	<2>
H-7599	S8.1	473	F,D,L	H_2O	361	{17}	536	pH 7	
L-7525	✓	524	D,L	DMSO*	373	{9.6}	422	pH 7	<3>
L-7526	✓	362	F,D,L	DMSO*	504	{54}	511	MeOH	
L-7527	✓	486	F,D,L	DMSO*	534	{51}	551	MeOH	
L-7528	✓	399	F,D,L	DMSO*	577	{48}	590	MeOH	
L-7532	✓	292	L	DMSO*	374	{11}	424	pH 5	<4>
L-7533	✓	376	L	DMSO*	373	{11}	425	pH 5	<4>
L-7534	✓	356	L	DMSO*	442	{17}	505	pH 5	<4>
L-7535	✓	398	L	DMSO*	443	{16}	505	pH 5	<4>
L-7537	✓	512	D	DMSO*	<300		none		
L-7538	✓	835	D,L	DMSO*	363	{16}	none	pH 8	
L-7542	✓	556	F,D,L	DMSO*	532	{70}	545	MeOH	
L-7545	✓	366	L	DMSO*	384	{21}	540	pH 3	<5>
N-3246	✓	289	D,L	H_2O, EtOH	541	{39}	640	<6>	
T-3166	S17.2	608	D,L	H_2O, DMSO	543	{43}	<7>	MeOH	
Y-7536	✓	384	F,L	DMSO, DMF	456	{27}	505	MeOH	

A-3568 see A-1301

For definitions of the contents of this data table, see "How to Use the Handbook" on page vi.
Structure: Chemical structure drawing: ✓ = shown in this section; Sn.n = shown in Section number n.n.

* This product is packaged as a solution in the solvent indicated in "Soluble."
<1> Abs = 500 nm, Em = 526 nm bound to DNA. See Section 8.1.
<2> F-7030 is fluorescent when bound to DNA (Em = 538 nm). Uptake and processing of the dye by live yeast results in red-shifted fluorescence (Em ~590 nm).
<3> L-7525 has structured spectra, with additional peaks at Abs = 394 nm and Em = 401 nm.
<4> This LysoSensor dye exhibits increasing fluorescence as pH decreases with no spectral shift. L-7532 and L-7533 have additional peaks at Abs = 394 nm and Em = 401 nm.
<5> L-7545 spectra are pH dependent. Abs and Em shift to shorter wavelengths at pH >5 (see Figure 12.5).
<6> Spectra of N-3246 are pH dependent (pK_a ~6.7). Data reported are for 1:1 (v/v) EtOH/1% acetic acid.
<7> Fluorescence of T-3166 in MeOH is very weak. Excitation/emission at 515/640 nm are suitable for detection of yeast vacuole membrane staining with this dye [J Cell Biol 128, 779 (1995)].

12.3 Structures *Cell-Permeant Probes for Lysosomes and Other Acidic Organelles*

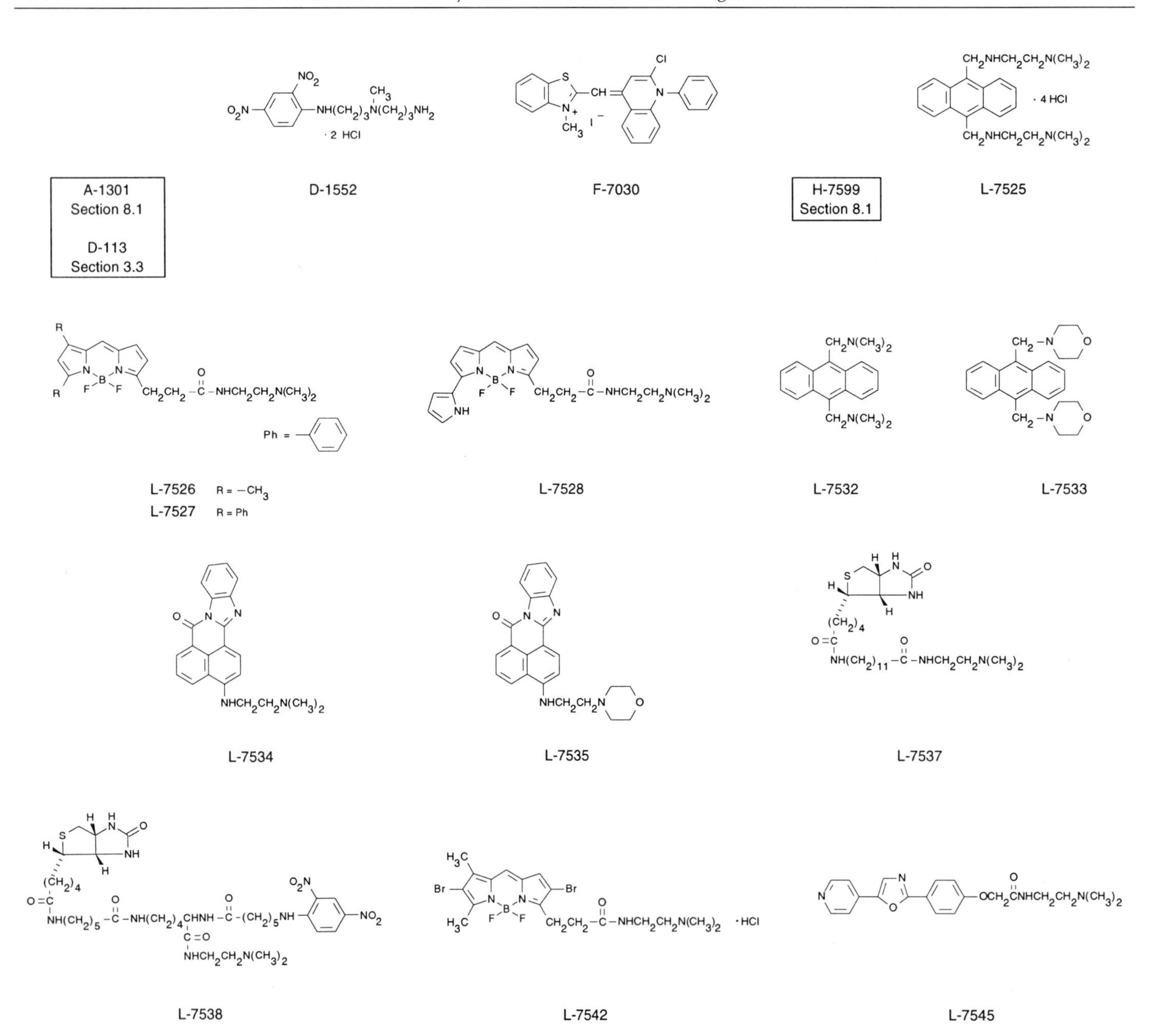

N-3246

T-3166 Section 17.2

Y-7536

12.3 Price List *Cell-Permeant Probes for Lysosomes and Other Acidic Organelles*

Cat #		Product Name	Unit Size	Price Per Unit ($) 1–4 Units	5–24 Units
A-1301		acridine orange *high purity*	1 g	48.00	38.40
A-3568		acridine orange *10 mg/mL solution in water*	10 mL	28.00	22.40
D-113		5-dimethylaminonaphthalene-1-(N-(5-aminopentyl))sulfonamide (dansyl cadaverine)	100 mg	78.00	62.40
D-1552		N-(3-((2,4-dinitrophenyl)amino)propyl)-N-(3-aminopropyl)methylamine, dihydrochloride (DAMP)	100 mg	98.00	78.40
F-7030	*New*	FUN-1™ cell stain *10 mM solution in DMSO*	100 µL	75.00	60.00
H-7599	*New*	hydroxystilbamidine, methanesulfonate	10 mg	75.00	60.00
L-7533	*New*	LysoSensor™ Blue DND-167 *1 mM solution in DMSO* *special packaging*	20x50 µL	125.00	100.00
L-7532	*New*	LysoSensor™ Blue DND-192 *1 mM solution in DMSO* *special packaging*	20x50 µL	125.00	100.00
L-7534	*New*	LysoSensor™ Green DND-153 *1 mM solution in DMSO* *special packaging*	20x50 µL	125.00	100.00
L-7535	*New*	LysoSensor™ Green DND-189 *1 mM solution in DMSO* *special packaging*	20x50 µL	125.00	100.00
L-7545	*New*	LysoSensor™ Yellow/Blue DND-160 *1 mM solution in DMSO* *special packaging*	20x50 µL	125.00	100.00
L-7537	*New*	LysoTracker™ biotin *4 mM solution in DMSO* *special packaging*	10x50 µL	125.00	100.00
L-7538	*New*	LysoTracker™ biotin/DNP *4 mM solution in DMSO* *special packaging*	10x50 µL	125.00	100.00
L-7525	*New*	LysoTracker™ Blue DND-22 *1 mM solution in DMSO* *special packaging*	20x50 µL	125.00	100.00
L-7542	*New*	LysoTracker™ Green Br_2 *1 mM solution in DMSO* *special packaging*	20x50 µL	125.00	100.00
L-7526	*New*	LysoTracker™ Green DND-26 *1 mM solution in DMSO* *special packaging*	20x50 µL	125.00	100.00
L-7528	*New*	LysoTracker™ Red DND-99 *1 mM solution in DMSO* *special packaging*	20x50 µL	125.00	100.00
L-7527	*New*	LysoTracker™ Yellow DND-68 *1 mM solution in DMSO* *special packaging*	20x50 µL	125.00	100.00
N-3246		neutral red *high purity*	25 mg	50.00	40.00
T-3166		N-(3-triethylammoniumpropyl)-4-(6-(4-(diethylamino)phenyl)hexatrienyl)pyridinium dibromide (FM™ 4-64)	1 mg	165.00	132.00
Y-7531	*New*	Yeast Vacuole Marker Sampler Kit	1 kit	95.00	76.00
Y-7536	*New*	yeast vacuole membrane marker MDY-64	1 mg	48.00	38.40

12.4 Cell-Permeant Probes for the Endoplasmic Reticulum and Golgi Apparatus

The endoplasmic reticulum (ER) and Golgi apparatus are primarily responsible for the proper sorting of lipids and proteins in cells. Consequently, most of the probes for these organelles are either lipids or chemicals that affect protein movement. Certain aspects of lipid trafficking through these organelles that involve signal transduction are described in Section 20.6. Enzymes in the ER are also involved in synthesis of cholesterol and membranes and in the detoxification of hydrophobic drugs through the cytochrome P-450 system (see Section 10.5).

In both live and fixed cells, the flattened membranous sacs of the ER and the Golgi apparatus can be stained with a variety of lipophilic probes and then distinguished on the basis of their morphology. Alternatively, the Golgi apparatus can be selectively stained with one of the fluorescent ceramide analogs, which tend to associate preferentially with the trans-Golgi. Molecular Probes also offers a wide selection of fluorescent and biotinylated lectin conjugates (see Section 7.6). In most cell types, concanavalin A (Con A) binds to the ER, whereas wheat germ agglutinin (WGA) binds to the Golgi apparatus,[1,2] making these fluorescent lectins useful reagents for facilitating the immunolocalization of oncogene products,[3] specific intracellular enzymes,[4,5] viral proteins [6] and components of the cytoskeleton.[7] However, fluorescein-labeled Con A is reported to bind specifically to isolated Golgi fractions from rat liver and was used to examine the effect of chronic ethanol intake on carbohydrate content in these organelles using flow cytometry.[8]

Carbocyanine Dyes

Short-Chain Carbocyanine Dyes

Terasaki and co-workers used the short-chain carbocyanine $DiOC_6(3)$ (D-273, see Section 12.2) to visualize the ER in both live and aldehyde-fixed cells.[9,10] This dye and the similar $DiOC_5(3)$

(D-272) have since been used extensively to study structural interactions and dynamics of the ER in neurons,[11,12] yeast,[13] onion epidermis,[14] and to examine the morphological relationships between the ER, mitochondria, intermediate filaments and microtubules in various cell types.[15,16] $DiOC_6(3)$ and $DiOC_5(3)$ pass through the plasma membrane and stain intracellular membranes with a fluorescein-like fluorescence; ER membranes can be easily distinguished by their characteristic morphology.[17] Caution must be exercised, however, in using the carbocyanines as probes for the ER. It has been reported that ER staining with $DiOC_6(3)$ does not occur until the mitochondria round up and lose the fluorochrome.[18] Rhodamine 6G and the hexyl ester of rhodamine B (R-634, R-648; see Section 12.2) appear to stain like $DiOC_6(3)$, except they are apparently less toxic and they fluoresce orange, providing possibilities for multicolor labeling.[17] When used at low concentrations, these slightly lipophilic rhodamine dyes tend to stain only mitochondria of live cells (see also Section 12.2).[19]

Long-Chain Carbocyanine Dyes

Terasaki and Jaffe have used the long-chain carbocyanines $DiIC_{16}(3)$ and $DiIC_{18}(3)$ (D-384, D-282) to label ER membranes. They achieve selective labeling of the ER by microinjecting a saturated solution of DiI in oil into sea urchin eggs.[20,21] As noted in the section on dialkylcarbocyanine and dialkylaminostyryl probes (see Section 14.2), DiI diffuses only in continuous membranes. This method has been successful in several other egg types but was not effective in molluscan or arthropod axons.

Brefeldin A and Fluorescent Analogs

The fungal metabolite brefeldin A (BFA) has proven valuable for dissecting the cellular processes, including vesicle formation [22] and kinesin distribution,[23] involved in exporting newly synthesized proteins.[24] In addition to the natural product isolated from *Penicillium brefeldianum* (B-7450), we now offer green and orange fluorescent BODIPY® BFA derivatives (B-7447, B-7449).

Brefeldin A

BFA (B-7450) has multiple targets in cells.[25] Exposing cells to BFA causes a distortion in intracellular protein traffic from the ER to the Golgi apparatus and the eventual loss of Golgi apparatus morphology; removal of BFA completely reverses these effects.[24,26-30] BFA also alters the morphology of endosomes and lysosomes.[31] BFA has been used to prevent retinoic acid potentiation of immunotoxins,[32] to study translocation of proteins in polarized epithelial cells [33] and to investigate the regulation of ADP-ribosylation factor binding to the Golgi apparatus.[34] BFA action can be monitored using fluorescent endosomal markers such as lucifer yellow CH (L-453, see Section 17.4) [31] and tetramethylrhodamine-labeled transferrin [35] (T-2872, see Section 17.3).

Fluorescent Brefeldin A

Our collaborators in Jonathan Yewdells' laboratory at NIAID have found that the new green fluorescent BODIPY FL and red-orange fluorescent BODIPY 558/568 BFA derivatives (B-7447, B-7449) are selectively localized in the ER and Golgi apparatus in four different cell lines.[36] BODIPY 558/568 BFA may be used in conjunction with NBD C_6-ceramide (N-1154) to investigate the ER and Golgi apparatus simultaneously. However, the biological activity of the fluorescent BFA derivatives may be limited and appears to be dependent on cleavage of the BODIPY fluorophore from BFA by intracellular esterases.[36] Two isomeric esters are isolated in the synthesis of the fluorescent brefeldins; we are selling only "isomer 1" of each product.

Fluorescent Ceramide Analogs

NBD C_6-ceramide (N-1154) and BODIPY FL C_5-ceramide (D-3521), both of which can be used with fluorescein optical filter sets, are selective stains for the Golgi apparatus.[37,38] With spectral properties similar to those of Texas Red, our new BODIPY TR ceramide (D-7540) should be useful for double-labeling experiments, as well as for staining cells and tissues that have substantial amounts of green autofluorescence. In addition, the BODIPY TR fluorophore is ideal for imaging microscopy with CCD cameras or other red-sensitive detectors.

NBD C_6-Ceramide and NBD C_6-Sphingomyelin

NBD C_6-ceramide (N-1154) has been used extensively as a selective stain of the trans-Golgi in both live and fixed cells.[39-46] Researchers have employed NBD C_6-ceramide to investigate:

- Effects of BFA (B-7450) on the transport of proteins from the Golgi apparatus to the ER [26,28,47]
- Farber's disease, a genetically inherited disease of lipid metabolism [48]
- Inhibition of glycoprotein traffic through secretory pathways [49]
- Intracellular trafficking and targeting of thrombin receptors [50]
- Secretory activity during the isolation of secretion mutants by fluorescence-activated cell sorting [51]

Furthermore, NBD C_6-ceramide's fluorescence is apparently sensitive to the cholesterol content of the Golgi apparatus, a phenomenon that is not observed with BODIPY FL C_5-ceramide.[52] If NBD C_6-ceramide–containing cells are starved for cholesterol, the NBD C_6-ceramide that accumulates within the Golgi apparatus appears to be severely photolabile. However, this NBD photobleaching can be reduced by stimulation of cholesterol synthesis. Thus, NBD C_6-ceramide may be useful in monitoring the cholesterol content of the Golgi apparatus in live cells.

NBD C_6-ceramide's conversion to the NBD C_6-glycosyl ceramide and NBD C_6-sphingomyelin (N-3524) has been observed *in vivo*.[53-55] Metabolism of the probe in live Chinese hamster ovary (CHO) fibroblasts has been used to define lipid-transport pathways.[53,56] NBD C_6-ceramide is also reported to be metabolized to NBD C_6-sphingomyelin in *Plasmodium falciparum*-infected erythrocytes, but not in normal erythrocytes.[57,58] Like NBD C_6-ceramide, NBD C_6-sphingomyelin (N-3524) has been used for the study of lipid trafficking between organelles.[56,59,60] Normal fibroblasts hydrolyze NBD C_6-sphingomyelin and process it through the Golgi apparatus.[61] However, in human skin fibroblasts from patients with Niemann–Pick disease, which is characterized by a lack of lysosomal sphingomyelinase activity, NBD C_6-sphingomyelin accumulates in the lysosomes.

BODIPY Ceramides and BODIPY Sphingomyelin

The green fluorescent BODIPY FL C_5-ceramide and red fluorescent BODIPY TR ceramide derivatives (D-3521, D-7540) are somewhat more fade-resistant and brighter than the NBD derivative and can likely be substituted for the NBD C_6-ceramide in many of its applications. During normal resting intracellular transport, the kinetics of dye loading and transport may differ somewhat between the BODIPY and NBD analogs.[62] BODIPY FL C_5-ceramide has proven to be an excellent structural marker for the Golgi apparatus, visualized either by fluorescence microscopy [63,64] or, following diaminobenzidine (DAB) conversion, electron microscopy.[65-67] We have recently prepared the brominated analog of BODIPY FL C_5-ceramide (BODIPY FL Br_2C_5-ceramide, D-7546) for use in dye-mediated photoconversion of diaminobenzidine (DAB). The presence of heavy atoms in site-selective probes has been shown to increase the production of singlet oxygen, making them much more effective photooxidizers of DAB; see Technical Note "Fluorescent Probes for Photoconversion of Diaminobenzidine" on page 264. Thus, this brominated analog may provide enhanced resolution of Golgi apparatus in electron microscopy applications. Photoconversion of DAB using BODIPY FL C_5-ceramide (D-3521) has already been reported.[65,66] BODIPY FL C_5-ceramide has also been used to:

- Delineate the Golgi apparatus in the cytoarchitecture of size-excluding compartments in live cells [68]
- Investigate both the inhibition of glycoprotein transport by ceramides [49] and the possible link between protein secretory pathways and sphingolipid biosynthesis [69]
- Isolate mammalian secretion mutants [51]
- Visualize tubovesicular membranes induced by *Plasmodium falciparum* [57]

BODIPY FL C_5-ceramide exhibits concentration-dependent fluorescence properties that provide additional benefits for imaging the Golgi apparatus. At high concentrations, our nonpolar BODIPY FL fluorophore forms excimers, resulting in a shift of the fluorophore's emission maximum from 515 nm (green) to ~620 nm (red). BODIPY FL C_5-ceramide accumulation is sufficient for excimer formation in the trans-Golgi but not in the surrounding cytoplasm. Red emission longpass optical filters can thus be used to selectively visualize the Golgi apparatus; see Color Plate 2 in "Visual Reality." Moreover, this two-color property can be used to quantitate BODIPY FL C_5-ceramide accumulation by ratio imaging.[37,69,70] Like BODIPY FL C_5-ceramide, the red fluorescent BODIPY TR ceramide appears to form long-wavelength excimers when concentrated in the Golgi apparatus; in this case, however, the excimers exhibit infrared fluorescence.

We also offer BODIPY FL C_5-sphingomyelin (D-3522), the likely metabolic product of BODIPY FL C_5-ceramide,[37] as well as two BODIPY FL C_5- and C_{12}-glucocerebrosides (D-7548, D-7547, see Section 20.6) BODIPY FL C_5-glucocerebroside is reportedly internalized by endocytic and nonendocytic pathways that are quite different from those governing BODIPY FL C_5-sphingomyelin (D-3522) internalization.[71]

1. Cytometry 19, 112 (1995); **2.** J Cell Biol 85, 429 (1980); **3.** J Cell Biol 111, 3097 (1990); **4.** J Biol Chem 269, 1727 (1994); **5.** J Cell Biol 111, 2851 (1990); **6.** J Cell Biol 110, 625 (1990); **7.** J Cell Biol 111, 1929 (1990); **8.** Exp Cell Res 207, 136 (1993); **9.** J Cell Biol 103, 1557 (1986); **10.** Cell 38, 101 (1984); **11.** J Cell Biol 127, 1021 (1994); **12.** Nature 310, 53 (1984); **13.** Cell Motil Cytoskeleton 25, 111 (1993); **14.** Eur J Cell Biol 52, 328 (1990); **15.** Biochem Cell Biol 70, 1174 (1992); **16.** Cell Motil Cytoskeleton 15, 71 (1990); **17.** J Cell Sci 101, 315 (1992); **18.** Microscopy Res Technique 27, 198 (1994); **19.** Histochemistry 94, 303 (1990); **20.** J Cell Biol 120, 1337 (1993); **21.** J Cell Biol 114, 929 (1991); **22.** J Cell Biol 122, 1197 (1993); **23.** J Cell Biol 128, 293 (1995); **24.** J Biol Chem 261, 11398 (1986); **25.** Cell 67, 449 (1991); **26.** J Biol Chem 268, 2341 (1993); **27.** J Cell Biol 116, 1071 (1992); **28.** Proc Natl Acad Sci USA 88, 9818 (1991); **29.** Cell 60, 821 (1990); **30.** J Cell Biol 111, 2295 (1990); **31.** J Cell Biol 119, 273 (1992); **32.** J Cell Biol 125, 743 (1994); **33.** J Cell Biol 124, 83 (1994); **34.** Nature 364, 818 (1993); **35.** J Cell Biol 118, 267 (1992); **36.** J Histochem Cytochem 9, 907 (1995); **37.** J Cell Biol 113, 1267 (1991); **38.** J Cell Biol 109, 2067 (1989); **39.** J Biol Chem 268, 18390 (1993); **40.** Am J Physiol 260, G119 (1991); **41.** Neurochem Res 16, 551 (1991); **42.** Biochem Soc Trans 18, 361 (1990); **43.** Eur J Cell Biol 53, 173 (1990); **44.** Science 229, 1051 (1985); **45.** Science 228, 745 (1985); **46.** Proc Natl Acad Sci USA 80, 2608 (1983); **47.** J Cell Biol 112, 567 (1991); **48.** Arch Biochem Biophys 208, 444 (1981); **49.** J Biol Chem 268, 4577 (1993); **50.** J Biol Chem 269, 27719 (1994); **51.** Mol Biol of the Cell 6, 135 (1995); **52.** Proc Natl Acad Sci USA 90, 2661 (1993); **53.** J Cell Biol 111, 977 (1990); **54.** Biochemistry 27, 6197 (1988); **55.** J Cell Biol 105, 1623 (1987); **56.** J Cell Biol 108, 2169 (1989); **57.** J Cell Biol 124, 449 (1994); **58.** Mol Biochem Parasitol 49, 143 (1991); **59.** J Cell Biol 121, 1257 (1993); **60.** J Cell Biol 117, 259 (1992); **61.** J Cell Biol 111, 429 (1990); **62.** Meth Cell Biol 38, 221 (1993); **63.** Cytometry 14, 251 (1993); **64.** J Cell Biol 120, 399 (1993); **65.** J Cell Biol 127, 29 (1994); **66.** Cell 73, 1079 (1993); **67.** Eur J Cell Biol 58, 214 (1992); **68.** J Cell Sci 106, 565 (1993); **69.** Biochemistry 31, 3581 (1992); **70.** FASEB J 8, 574 (1994); **71.** J Cell Biol 125, 769 (1994).

12.4 Data Table *Cell-Permeant Probes for the Endoplasmic Reticulum and Golgi Apparatus*

Cat #	Structure	MW	Storage	Soluble	Abs	{ε × 10⁻³}	Em	Solvent	Notes
B-7447	✓	554	F,D,L	DMSO, EtOH	503	{83}	510	MeOH	
B-7449	✓	609	F,D,L	DMSO, EtOH	559	{80}	568	MeOH	
B-7450	✓	280	F,D	DMSO, EtOH	<300		none		
D-272	S25.3	544	D,L	DMSO	484	{155}	500	MeOH	
D-282	S15.4	934	L	DMSO, EtOH	549	{148}	565	MeOH	
D-384	S15.4	878	L	DMSO, EtOH	549	{148}	565	MeOH	
D-3521	✓	602	FF,D,L	$CHCl_3$, DMSO	505	{91}	511	MeOH	<1>
D-3522	S13.2	767	FF,D,L	<2>	505	{77}	512	MeOH	
D-7540	✓	706	FF,D,L	$CHCl_3$, DMSO	589	{60}	617	MeOH	
D-7546	✓	759	FF,D,L	$CHCl_3$, DMSO	533	{80}	545	MeOH	
N-1154	✓	576	FF,D,L	$CHCl_3$, DMSO	466	{22}	536	MeOH	<3>
N-3524	S13.2	741	FF,D,L	<2>	466	{22}	536	MeOH	<3>

For definitions of the contents of this data table, see "How to Use the Handbook" on page vi.
Structure: Chemical structure drawing: ✓ = shown in this section; Sn.n = shown in Section number n.n.

<1> Em for D-3521 shifts to ~620 nm when high concentrations of the probe (>5 mol %) are incorporated in lipid mixtures [J Cell Biol 113, 1267 (1991)].

<2> Chloroform is the most generally useful solvent for preparing stock solutions of phospholipids (including sphingomyelins). Information on solubility of natural phospholipids can be found in Marsh, D., *CRC Handbook of Lipid Bilayers*, CRC Press (1990) pp. 71–80.

<3> NBD derivatives are almost nonfluorescent in water but exhibit strong fluorescence enhancement in lipid environments.

12.4 Structures *Cell-Permeant Probes for the Endoplasmic Reticulum and Golgi Apparatus*

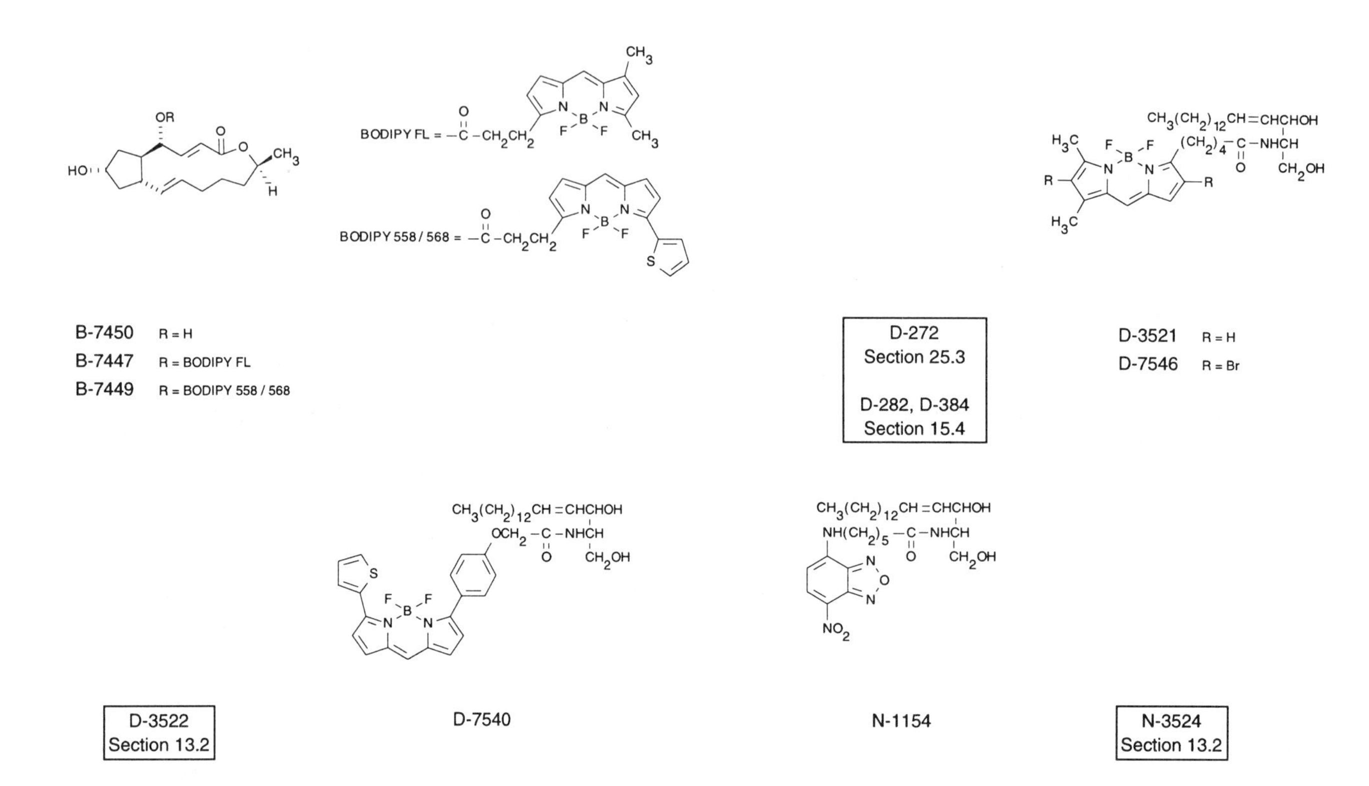

12.4 Price List *Cell-Permeant Probes for the Endoplasmic Reticulum and Golgi Apparatus*

Cat #		Product Name	Unit Size	Price Per Unit ($) 1–4 Units	5–24 Units
B-7450	***New***	brefeldin A *from *Penicillium brefeldianum**	5 mg	42.00	33.60
B-7449	***New***	brefeldin A, BODIPY® 558/568 conjugate *isomer 1*	25 µg	95.00	76.00
B-7447	***New***	brefeldin A, BODIPY® FL conjugate *isomer 1*	25 µg	95.00	76.00
D-7546	***New***	N-(2,6-dibromo-4,4-difluoro-5,7-dimethyl-4-bora-3a,4a-diaza-s-indacene-3-pentanoyl)sphingosine (BODIPY® FL Br_2C_5-ceramide)	250 µg	98.00	78.40
D-3521		N-(4,4-difluoro-5,7-dimethyl-4-bora-3a,4a-diaza-s-indacene-3-pentanoyl)sphingosine (BODIPY® FL C_5-ceramide)	250 µg	125.00	100.00
D-3522		N-(4,4-difluoro-5,7-dimethyl-4-bora-3a,4a-diaza-s-indacene-3-pentanoyl)sphingosyl phosphocholine (BODIPY® FL C_5-sphingomyelin)	250 µg	125.00	100.00
D-7540	***New***	N-((4-(4,4-difluoro-5-(2-thienyl)-4-bora-3a,4a-diaza-s-indacene-3-yl)phenoxy)acetyl)sphingosine (BODIPY® TR ceramide)	250 µg	98.00	78.40
D-384		1,1′-dihexadecyl-3,3,3′,3′-tetramethylindocarbocyanine perchlorate ($DiIC_{16}(3)$)	100 mg	135.00	108.00
D-282		1,1′-dioctadecyl-3,3,3′,3′-tetramethylindocarbocyanine perchlorate ('DiI'; $DiIC_{18}(3)$)	100 mg	135.00	108.00
D-272		3,3′-dipentyloxacarbocyanine iodide ($DiOC_5(3)$)	100 mg	85.00	68.00
N-1154		6-((N-(7-nitrobenz-2-oxa-1,3-diazol-4-yl)amino)hexanoyl)sphingosine (NBD C_6-ceramide)	1 mg	125.00	100.00
N-3524		N-((6-(7-nitrobenz-2-oxa-1,3-diazol-4-yl)amino)hexanoyl)sphingosyl phosphocholine (NBD C_6-sphingomyelin)	1 mg	125.00	100.00

12.5 Organelle-Specific Monoclonal Antibodies

Monoclonal Antibodies Against Yeast, Bovine and Human Cytochrome Oxidase

To facilitate the study of cytochrome oxidase (COX) structure and mitochondrial biogenesis, Molecular Probes offers subunit-specific anti–yeast, anti–bovine and anti–human COX monoclonal antibodies. COX catalyzes the transfer of electrons from reduced cytochrome *c* to molecular oxygen, with a concomitant translocation of protons across the mitochondrial inner membrane.[1] This mitochondrial membrane–bound enzyme is composed of both mitochondrial-encoded subunits (subunits I, II and III) and nuclear-encoded subunits (all others), with a total of 11 subunits for yeast COX and 13 subunits for mammalian COX. The binding specificity exhibited by our anti-COX monoclonal antibody preparations allows researchers to investigate the regulation, assembly and orientation of COX subunits from a variety of organisms[2-6] (Tables 12.4 and 12.5). The anti–human COX antibodies are potentially valuable tools for the analysis of mitochondrial myopathies and related disorders.[7,8] Anti–bovine COX subunit VIa-H and anti–bovine COX subunit VIa-L antibodies may be particularly useful in studies of tissue-specific COX deficiencies because they detect the heart- and liver-specific isoforms, respectively of subunit VIa.[9]

These monoclonal antibodies were selected for their ability to detect native COX by solid-phase binding assays, including as particle-concentration fluorescence immunoassay (PCFIA) and enzyme-linked immunosorbent assay (ELISA). In addition, these antibodies recognize the corresponding denatured COX subunit by Western blot analysis and may be employed to test other subcellular preparations for mitochondrial contamination. Several of these antibodies can also be used to visualize mitochondria in fixed tissue using standard immunohistochemical techniques (Tables 12.4 and 12.5). Detailed information regarding the IgG isotype and recommended working concentration is provided with each product. For detection of these monoclonal antibodies, Molecular Probes offers anti-mouse secondary antibodies labeled with biotin or a wide range of fluorophores (see Section 7.2).

Table 12.4 *Anti–yeast COX monoclonal antibodies.*

Cat #	Yeast COX Subunit Specificity	Mouse Monoclonal	Mouse Isotype	Western Analysis	Immuno fluorescence
A-6405	Subunit I	11D8-B7	$IgG_{2b,k}$	Yes	No
A-6407	Subunit II	4B12-A5	$IgG_{2a,k}$	Yes	No
A-6408	Subunit III *	DA5	$IgG_{2a,k}$	Yes	Yes
A-6432	Subunit IV	1A12-A12	$IgG_{1,k}$	Yes	No

* Monoclonal DA5 has also been shown to crossreact with *Neurospora crassa* COX subunit III.

Monoclonal Antibodies for Yeast Cell Biology

In addition to the anti–yeast COX monoclonal antibodies described above, Molecular Probes offers eight other immunoreagents that have proven useful for studying many aspects of cell biology with the yeast *Saccharomyces cerevisiae* (Table 12.6). We have selected this set of monoclonal antibodies because they are

Table 12.5 *Anti–vertebrate COX monoclonal antibodies.*

Cat #	Vertebrate COX Subunit Specificity	Mouse Monoclonal	Mouse Isotype	Reactivity by Western Blot Analysis *					Immuno-fluorescence
				Human	Bovine	Rat	Mouse	Chicken	
A-6403	Subunit I	1D6-E1-A8	$IgG_{2a,k}$	+	+	ND	+	ND	Yes §
A-6402	Subunit I	5D4-F5	$IgG_{2a,k}$	+	+	ND	+	+	Yes
A-6404	Subunit II	12C4-F12	$IgG_{2a,k}$	+	±	±	±	±	Yes
A-6431	Subunit IV	20E8-C12	$IgG_{2a,k}$	+	+	+	+	-	Yes
A-6409	Subunit IV	10G8-D12-C12	$IgG_{2a,k}$	+	+	-	-	+	Yes
A-6456	Subunit Vb	16H12-H9	$IgG_{2b,k}$	+	+	+	+	+	No
A-6410	Subunit VIa-H †	4H2-A5	$IgG_{2a,k}$	±	+	-	-	-	No
A-6411	Subunit VIa-L ‡	14A3-AD2-BH4	$IgG_{1,k}$	±	+	±	±	-	No
A-6401	Subunit VIc	3G5-F7-G3	$IgG_{2b,k}$	+	+	+	ND	ND	Yes §

* +, strong reactivity; ±, weak reactivity; -, no reactivity. † Heart-specific and ‡ liver-specific isoform of COX subunit VIa. § These monoclonal antibodies yield the strongest and most consistent signals when used for indirect immunofluorescence. ND = not determined.

Table 12.6 Monoclonal antibodies for yeast cell biology.

Cat #	Yeast Antigen Recognized By Antibody	Yeast Subcellular Structure In Which Antigen Resides	Mouse Monoclonal	Mouse Isotype	Western Analysis	Immuno-fluorescence
A-6449	Mitochondrial porin	Mitochondrial outer membrane	16G9-E6	$IgG_{1,\kappa}$	Yes	Yes*
A-6422	V-ATPase 69,000-dalton subunit	Vacuole membrane	8B1-F3	$IgG_{2a,\kappa}$	Yes	Yes
A-6426	V-ATPase 100,000-dalton subunit	Vacuole membrane	10D7-A7-B2	$IgG_{2a,\kappa}$	Yes	Yes
A-6427	V-ATPase 60,000-dalton subunit	Vacuole membrane	13D11-B2	$IgG_{1,\kappa}$	Yes	Yes*
A-6458	Alkaline phosphatase (ALP)	Vacuole membrane	1D3-A10	$IgG_{1,\kappa}$	Yes	Yes*
A-6428	Carboxypeptidase Y (CPY)	Vacuole lumen	10A5-B5	$IgG_{1,\kappa}$	Yes	Yes
A-6429	Dol-P-Man synthase (Dpm1p)	Endoplasmic reticulum membrane	5C5-A7	$IgG_{1,\kappa}$	Yes	Yes*
A-6457	3-Phosphoglycerate kinase (PGK)	Cytosol	22C5-D8	$IgG_{1,\kappa}$	Yes	Yes

* These monoclonal antibodies yield the strongest and most consistent signals when used for indirect immunofluorescence; the anti-ALP antibody is the most reliable monoclonal antibody for detecting yeast vacuolar membranes.

compatible with both Western blotting of denatured proteins and protein immunolocalization in fixed yeast cells. Other potential uses of these antibodies include the development of ELISAs to determine either the level of enrichment of a particular yeast organelle or the level at which the organelle contaminates a subcellular fraction.

- **For the detection of yeast vacuolar membranes,** we offer four monoclonal antibodies directed against integral or peripheral membrane proteins. Two of the antibodies are specific for **integral membrane proteins** of the yeast vacuoles: Monoclonal 1D3-A10 is specific for vacuolar alkaline phosphatase (ALP), the product of the *PHO8* gene; this antibody is ideal for indirect immunofluorescence detection of vacuolar membranes in fixed yeast cells, as well as for detection of ALP by Western blotting.[10] Monoclonal 10D7-A7-B2 recognizes the 100,000-dalton vacuolar H^+-ATPase (V-ATPase) subunit [11] (the product of the *VPH1* gene [12]) and is especially useful for Western blotting. The other two antibodies are directed against **peripheral membrane proteins** of the yeast vacuoles, specifically two subunits of the V-ATPase complex that are associated with the cytosolic face of the yeast vacuolar membrane: Monoclonal 13D11-B2 recognizes the 60,000-dalton subunit [13] (the B-subunit, the product of the *VMA2* or *VAT2* gene), whereas monoclonal 8B1-F3 binds the 69,000-dalton subunit [13,14] (the A-subunit, the product of the *VMA1* or *TFP1* gene). Both of these antibodies are useful for indirect immunofluorescence analysis,[15] as well as for Western blotting.

- **For the detection of yeast vacuolar lumen,** we offer monoclonal 10A5-B5. This antibody is specific for carboxypeptidase Y (CPY), a 61,000-dalton soluble glycoprotein located in the vacuolar lumen.[16] Monoclonal 10A5-B5 is useful for both indirect immunofluorescence, as well as for Western blotting.

- **For the detection of yeast mitochondrial membranes,** the anti–yeast mitochondrial porin monoclonal 16G9-E6 is available. Porin is an abundant ~30,000-dalton integral membrane protein that resides in the outer membrane of yeast mitochondria.[17] We also offer four anti–yeast COX antibodies (described above), which recognize specific subunits of the inner mitochondrial membrane protein, cytochrome oxidase.

- **For the detection of yeast endoplasmic reticulum membranes,** we offer the anti–dolichol phosphate mannose synthase (Dol-P-Man synthase, Dpm1p) monoclonal 5C5-A7. The yeast Dol-P-Man synthase is an ~30,000-dalton integral membrane protein that resides in the endoplasmic reticulum;[18] the monoclonal antibody was prepared against the cytosolic domain of the protein.

- **For the detection of yeast cytosol,** we offer a monoclonal antibody directed against the yeast cytosolic protein phosphoglycerate kinase (PGK). Monoclonal 22C5-D8 can be used for the detection of yeast PGK by Western blotting.

Detailed information regarding the IgG isotype and recommended working concentrations are provided with each product. For detection of these monoclonal antibodies, Molecular Probes offers anti-mouse secondary antibodies labeled with biotin or a wide range of fluorophores (see Section 7.2).

1. Ann Rev Biochem 59, 569 (1990); **2.** Meth Enzymol 260, 117 (1995); **3.** J Cell Biol 126, 1375 (1994); **4.** J Biol Chem 268, 18754 (1993); **5.** Biochemistry 30, 3674 (1991); **6.** J Biol Chem 266, 7688 (1991); **7.** Biochim Biophys Acta 1315, 199 (1996); **8.** Pediatric Res 28, 529 (1990); **9.** Biochim Biophys Acta 1225, 95 (1993); **10.** Mol Cell Biol 16, 2700 (1996); **11.** J Biol Chem 267, 447 (1992); **12.** J Biol Chem 267, 14294 (1992); **13.** J Biol Chem 264, 19236 (1989); **14.** Mol Biol of the Cell 3, 1389 (1992); **15.** Mol Biol of the Cell 7, 985 (1996); **16.** Eur J Cell Biol 65, 305 (1994); **17.** Biochim Biophys Acta 894, 109 (1987); **18.** *Guidebook to the Secretory Pathway*, Sambrook & Tooze Publication, Oxford University Press (1994).

12.5 Price List *Organelle-Specific Monoclonal Antibodies*

Cat #		Product Name	Unit Size	Price Per Unit ($) 1–4 Units	5–24 Units
A-6409	***New***	anti-bovine cytochrome oxidase subunit IV, mouse monoclonal 10G8-D12-C12 *specificity: human, bovine, chicken*	250 µg	195.00	156.00
A-6431	***New***	anti-bovine cytochrome oxidase subunit IV, mouse monoclonal 20E8-C12 *specificity: human, bovine, rat, mouse*	250 µg	195.00	156.00
A-6456	***New***	anti-bovine cytochrome oxidase subunit Vb, mouse monoclonal 16H12-H9 *specificity: human, bovine, rat, mouse, chicken*	250 µg	195.00	156.00
A-6410	***New***	anti-bovine cytochrome oxidase subunit VIa-H, mouse monoclonal 4H2-A5	250 µg	195.00	156.00
A-6411	***New***	anti-bovine cytochrome oxidase subunit VIa-L, mouse monoclonal 14A3-AD2-BH4	250 µg	195.00	156.00
A-6401	***New***	anti-bovine cytochrome oxidase subunit VIc, mouse monoclonal 3G5-F7-G3 *specificity: human, bovine, rat*	100 µg	195.00	156.00
A-6403	***New***	anti-human cytochrome oxidase subunit I, mouse monoclonal 1D6-E1-A8	100 µg	195.00	156.00
A-6402	***New***	anti-human cytochrome oxidase subunit I, mouse monoclonal 5D4-F5	100 µg	195.00	156.00
A-6404	***New***	anti-human cytochrome oxidase subunit II, mouse monoclonal 12C4-F12	100 µg	195.00	156.00
A-6405	***New***	anti-yeast cytochrome oxidase subunit I, mouse monoclonal 11D8-B7	100 µg	195.00	156.00
A-6407	***New***	anti-yeast cytochrome oxidase subunit II, mouse monoclonal 4B12-A5	250 µg	195.00	156.00
A-6408	***New***	anti-yeast cytochrome oxidase subunit III, mouse monoclonal DA5	250 µg	195.00	156.00
A-6432	***New***	anti-yeast cytochrome oxidase subunit IV, mouse monoclonal 1A12-A12	250 µg	195.00	156.00
A-6429	***New***	anti-yeast dolichol phosphate mannose synthase, mouse monoclonal 5C5-A7	250 µg	195.00	156.00
A-6449	***New***	anti-yeast mitochondrial porin, mouse monoclonal 16G9-E6	250 µg	195.00	156.00
A-6457	***New***	anti-yeast 3-phosphoglycerate kinase (PGK), mouse monoclonal 22C5-D8	250 µg	195.00	156.00
A-6458	***New***	anti-yeast vacuolar alkaline phosphatase, mouse monoclonal 1D3-A10	250 µg	195.00	156.00
A-6428	***New***	anti-yeast vacuolar carboxypeptidase Y (CPY), mouse monoclonal 10A5-B5	250 µg	195.00	156.00
A-6427	***New***	anti-yeast vacuolar H^+-ATPase 60 kDa subunit, mouse monoclonal 13D11-B2	250 µg	195.00	156.00
A-6422	***New***	anti-yeast vacuolar H^+-ATPase 69 kDa subunit, mouse monoclonal 8B1-F3	250 µg	195.00	156.00
A-6426	***New***	anti-yeast vacuolar H^+-ATPase 100 kDa subunit, mouse monoclonal 10D7-A7-B2	250 µg	195.00	156.00

Hints for Finding the Product of Interest

This Handbook includes products released by Molecular Probes before September 1996. Additional products and search capabilities are available through our Web site (http://www.probes.com). All information, including price and availability, is subject to change without notice.

- **If you know the name of the product you want:** Look up the product by name in the **Master Price List/Product Index** at the back of the book. The Master Price List/Product Index is arranged alphabetically, by both product name and alternative name, using the alphabetization rules outlined on page vi. This index will refer you directly to all the Handbook sections that discuss the product.

- **If you DO NOT know the name of the product you want:** Search the **Keyword Index** for the words that best describe the applications in which you are interested. The Keyword Index will refer you directly to page numbers on which the listed topics are discussed. Alternatively, search the **Table of Contents** for the chapter or section title that most closely relates to your application.

- **If you only know the alpha-numeric catalog number of the product you want but not its name:** Look for the catalog number in the **Master Price List/Product Index** at the back of the book. First locate the group of catalog numbers with the same alpha identifier as the catalog number of interest and then scan this group for the correct numeric identifier. The Master Price List/Product Index will refer you directly to all the Handbook sections that discuss this product. Alternatively, you may use the search features available through our Web site (http://www.probes.com) to obtain the product name of any of our current products given only the catalog number (or numeric identifier).

- **If you only know part of a product name or are searching for a product introduced after publication of this Handbook:** Search Molecular Probes' up-to-date product list through our **Web site (http://www.probes.com)**. The product search feature on our home page allows you to search our entire product list, including products that have been added since the completion of this Handbook, using a partial product name, complete product name or catalog number. Product-specific bibliographies can also be obtained from our Web site once the product of interest has been located.

- **If you still cannot find the product that you are looking for or are seeking advice in choosing a product:** Please contact our Technical Assistance Department by phone, e-mail, fax or mail.

Lipid Mixing Assays of Membrane Fusion

Fluorometric methods for assaying membrane fusion exploit processes, such as nonradiative energy transfer, fluorescence quenching and pyrene-excimer formation, that are dependent on probe concentration.[1-7] Assays of membrane fusion report either the mixing of membrane lipids (described here) or the mixing of the aqueous contents of the fused entities (see the Technical Note "Assays of Volume Change, Membrane Fusion and Membrane Permeability" on page 364). Chapter 14 describes additional methods for detecting membrane fusion based on image analysis.

For use in conjunction with membrane fusion assays, we offer the fusogenic peptide melittin[43,44] (M-98403, see Section 20.3); high-purity phospholipids (D-7704, D-7705, see Section 13.4) and a variety of cell permeabilization reagents (see Section 16.3).

NBD–Rhodamine Energy Transfer

- **Principle:** Struck, Hoekstra and Pagano introduced lipid mixing assays based on NBD–rhodamine energy transfer.[8] In this method, membranes labeled with a combination of fluorescence energy transfer donor and acceptor lipid probes — typically NBD-PE (N-360, see Section 13.4) and N-Rh-PE (L-1392, see Section 13.4), respectively — are mixed with unlabeled membranes. Fluorescence resonance energy transfer, detected as rhodamine emission at ~585 nm resulting from NBD excitation at ~470 nm, *decreases* when the average spatial separation of the probes is increased upon fusion of labeled membranes with unlabeled membranes. The reverse detection scheme, in which fluorescence resonance energy transfer *increases* upon fusion of membranes separately labeled with donor and acceptor probes, has also proven to be a useful lipid mixing assay.[9]
- **Applications**[10-19]

Octadecyl Rhodamine B Self-Quenching

- **Principle:** Lipid mixing assays based on self-quenching of octadecyl rhodamine B (R 18, O-246; see Section 14.3) were originally described by Hoekstra and co-workers.[20] Octadecyl rhodamine B self-quenching occurs when the probe is incorporated into membrane lipids at concentrations of 1–10 mole percent.[21] Unlike phospholipid analogs, octadecyl rhodamine B can readily be introduced into existing membranes in large amounts. Fusion with unlabeled membranes results in dilution of the probe, which is accompanied by increasing fluorescence[22,23] (excitation/emission maxima 560/590 nm). The assay may be compromised by effects such as spontaneous transfer of the probe to unlabeled membranes, quenching of fluorescence by proteins and probe-related inactivation of viruses; the prevalence of these effects is currently debated.[24-26]
- **Applications** (widely used for detecting virus–cell fusion)[27-38]

Pyrene-Excimer Formation

- **Principle:** Pyrene-labeled fatty acids (e.g., P-31, P-96, P-243; see Section 13.2) can be biosynthetically incorporated into viruses and cells in sufficient quantities to produce the degree of labeling required for long-wavelength pyrene-excimer fluorescence (see Figure 13.6 in Chapter 13). This excimer fluorescence is diminished upon fusion of labeled membranes with unlabeled membranes. Fusion can be monitored by following the increase in the ratio of monomer (~400 nm) to excimer (~470 nm) emission, with excitation at about 340 nm. This method appears to circumvent some of the potential artifacts of the octadecyl rhodamine B self-quenching technique[25] and therefore provides a useful alternative for virus–cell fusion applications.
- **Applications**[25,27,39-42]

1. Meth Enzymol 220, 15 (1993); **2.** Meth Enzymol 220, 3 (1993); **3.** Hepatology 12, 61S (1990); **4.** Ann Rev Biophys Biophys Chem 18, 187 (1989); **5.** Düzgünes, N. and Bentz, J. in *Spectroscopic Membrane Probes, Volume 1*, L.M. Loew, Ed., CRC Press (1988) pp. 117–159; **6.** S.J. Morris *et al.*, *Spectroscopic Membrane Probes, Volume 1*, L.M. Loew, Ed., CRC Press (1988) pp. 161–191; **7.** Biochemistry 26, 8435 (1987); **8.** Biochemistry 20, 4093 (1981); **9.** Meth Enzymol 221, 239 (1993); **10.** Biochemistry 33, 12615 (1994); **11.** Biochemistry 33, 5805 (1994); **12.** Biochemistry 33, 3201 (1994); **13.** Biophys J 67, 1117 (1994); **14.** J Biol Chem 269, 15124 (1994); **15.** J Biol Chem 269, 4050 (1994); **16.** J Biol Chem 268, 1716 (1993); **17.** Biochemistry 31, 2629 (1992); **18.** Biochemistry 30, 5319 (1991); **19.** J Biol Chem 266, 3252 (1991); **20.** Biochemistry 23, 5675 (1984); **21.** J Biol Chem 265, 13533 (1990); **22.** Biophys J 65, 325 (1993); **23.** Biophys J 58, 1157 (1990); **24.** Biochim Biophys Acta 1190, 360 (1994); **25.** Biochemistry 32, 11330 (1993); **26.** Biochemistry 32, 900 (1993); **27.** Biochemistry 33, 9110 (1994); **28.** Biochemistry 33, 1977 (1994); **29.** Biochim Biophys Acta 1191, 375 (1994); **30.** J Biol Chem 269 5467 (1994); **31.** Biochem J 294, 325 (1993); **32.** J Biol Chem 268, 25764 (1993); **33.** J Biol Chem 268, 9267 (1993); **34.** Virology 195, 855 (1993); **35.** Biochemistry 31, 10108 (1992); **36.** Exp Cell Res 195, 137 (1991); **37.** J Virol 65, 4063 (1991); **38.** Biochemistry 29, 4054 (1990); **39.** EMBO J 12, 693 (1993); **40.** J Virol 66, 7309 (1992); **41.** Biochemistry 27, 30 (1988); **42.** Biochim Biophys Acta 860, 301 (1986); **43.** Biochem J 305, 785 (1995); **44.** Biochim Biophys Acta 732, 668 (1983).

Chapter 13

Fluorescent Phospholipids, Fatty Acids and Sterols

Contents

Technical Notes and Product Highlights

Related Chapters

13.1 Introduction to Fluorescent Phospholipids, Fatty Acids and Sterols

Fluorescent analogs of three naturally occurring lipid classes — phospholipids, fatty acids and sterols — are employed primarily as probes of biological membrane structure and as tracers of lipid metabolism and transport. However, in addition to their structural and metabolic functions, lipids can act as mediators in cellular signaling processes. Lipid probes, such as phospholipase substrates, ceramides and platelet-aggregating factor (PAF) analogs, that are specifically used to investigate lipid-mediated signal transduction processes are discussed in Sections 20.5 and 20.6.

Most **phospholipids** are esters of glycerol comprising two fatty acyl residues (nonpolar tails) and a single phosphate ester substituent (polar head group). In sphingomyelins, such as BODIPY FL, NBD and pyrene derivatives (D-3522, N-3524, P-3527; see Section 13.2), the amino alcohol sphingosine replaces glycerol as the structural backbone. Fluorescent phospholipid analogs can be further classified according to where the fluorophore is attached. A fluorophore can be attached either to one of the fatty acyl chains or to the polar head group, and will accordingly be situated either in the nonpolar interior or at the water/lipid interface, respectively, when the fluorescent analog is incorporated into a lipid bilayer membrane (Figure 13.1).

Fluorescent **fatty acids** can often be used interchangeably with the corresponding phospholipids as membrane probes; however, fatty acids transfer more readily between aqueous and lipid phases.[1] Although fatty acids are ionized at neutral pH in water (pK_a ~5), their pK_a is typically about 7 in membranes [2] and thus a significant fraction of membrane-bound fatty acids are neutral species.

Fluorescent **sterols** and **cholesteryl esters** are widely used as structural probes and transport markers for these important lipid constituents of membranes and lipoproteins.

1. Biochemistry 25, 1717 (1986); **2.** Biochim Biophys Acta 728, 159 (1983).

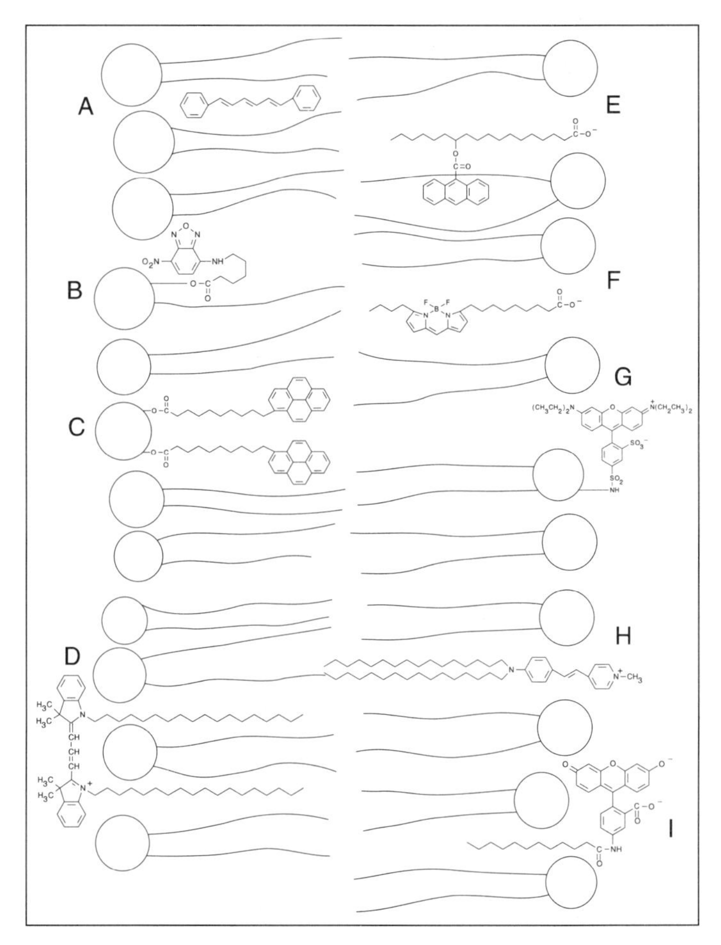

Figure 13.1 Location and orientation of representative fluorescent membrane probes in a phospholipid bilayer: (A) DPH, (B) NBD-C_6-HPC, (C) bis-pyrene-PC, (D) DiI, (E) anthroyloxy-FA, (F) BODIPY 500/510 FA, (G) N-Rh-PE, (H) DiA and (I) C_{12}-fluorescein.

13.2 Acyl Chain–Labeled Phospholipids and Their Corresponding Fluorescent Fatty Acids

Table 13.1 lists the acyl chain–labeled phospholipids and their corresponding fatty acids covered in this section.

Phospholipids with BODIPY Dye–Labeled Acyl Chains

Spectral Properties

The nonpolar BODIPY fluorophore has been used to produce the longest-wavelength fluorescent phospholipids and fatty acids currently available. The spectral coverage of our BODIPY membrane probes currently extends from the BODIPY 500/510 phospholipids and fatty acids to the BODIPY 581/591 analogs; these catalog names include the approximate absorption and fluorescence emission maxima (in nanometers) immediately following the BODIPY registered trademark. The spectral properties of BODIPY phospholipids and fatty acids are summarized in Table 13.2 and are superior in several respects to those of the NBD analogs discussed below (e.g., see Figure 13.2).

Unlike NBD, the BODIPY fluorophore is intrinsically lipophilic and readily localizes in the membrane interior. Fluorescence polarization measurements have been used to show that the lipid/water partition coefficient of BODIPY fatty acids is large, particularly for fluid-phase lipids, and that the angular motion of the BODIPY

Table 13.1 *Acyl chain–labeled phospholipids and their corresponding fluorescent fatty acids.*

Fluorophore *	Phospholipid †	Fatty Acid ‡
BODIPY (500/510)	• B-3794, D-3793, D-3795 (PC)	• B-3824, D-3823, D-3825, D-3826 (C_{13}) §
BODIPY FL (503/512)	• D-3792, D-3803 (PC) • D-3802, D-3804 (PE) • D-3805 (PA) • D-3522 (SM)	• D-3834 (C_5) ¶ • D-3862 (C_{11}) • D-3822 (C_{12}) • D-3821 (C_{16})
BODIPY (530/550)	• D-3812, D-3815 (PC) • D-3813 (PE)	• D-3833 (C_5) ¶ • D-3832 (C_{12})
BODIPY (581/591)	• D-3806 (PC)	• D-3860 (C_5) ¶ • D-3861 (C_{11})
DPH (362/433)	• D-476 (PC)	• P-1342 (C_1) • P-459 (C_3)
NBD (470/530)	• N-3786, N-3787 (PC) • N-3524 (SM)	• N-316 (C_6) • N-678 (C_{12}) • M-1119 (C_{18})
Pyrene (340/376)	• H-361, H-3818 (PC) • H-3784 (PE) • H-3809 (PG) • H-3743, H-3810 (PM) • P-3527 (SM)	• P-164 (C_1) • P-32, P-1903 (C_4) • P-3840 (C_6) • P-31 (C_{10}) • P-96 (C_{12}) • P-243 (C_{16})
Perylene (440/448)	• H-3790 (PC)	• P-646 (C_{12})

* Numbers in parentheses are approximate fluorescence excitation/emission maxima in nm. † All probes are diacylglycerophospholipids with the exception of sphingomyelins (SM). Diacylglycerol headgroup is indicated as follows: PC = phosphocholine, PE = phosphoethanolamine, PG = phosphoglycerol, PM = phosphomethanol, PA = phosphate (phosphatidic acid). Note that we use the stereochemical nomenclature convention for glycerolipid head groups (e.g., *sn*-glycero-3-phosphocholine) instead of the system based on L-α-glycerol phosphate (e.g., L-α-phosphatidylcholine). Systematic names for fatty acids (e.g. hexadecanoic acid, CH_3-$(CH_2)_{14}$-COOH) are used rather than the trivial synonyms (e.g., palmitic acid). ‡ Fatty acid carbon chain length (excluding the fluorophore) is indicated by subscript (C_5 = pentanoic acid). § BODIPY (500/510) fluorophores are inserted in the hydrocarbon chain at varying positions rather than attached at the terminus as in all the other fatty acid probes listed. ¶ Propionic acid (C_3) derivatives of these BODIPY dyes are also available: see Section 1.2).

500/510 fluorophore incorporated in phospholipids increases as its distance from the membrane surface increases.[1] Furthermore, the fluorophore is completely inaccessible to the membrane-impermeant anti–BODIPY FL antibody (A-5770, see Section 7.3), which recognizes the BODIPY 500/510 derivative.

In fluorescence resonance energy transfer (FRET) measurements, the green fluorescent BODIPY 500/510 and BODIPY FL analogs are excellent donors to longer-wavelength BODIPY probes [2] (Figure 13.5) and acceptors from coumarin-labeled phospholipids.[3] These probe combinations offer several alternatives to the widely used NBD–rhodamine fluorophore pair for researchers using FRET techniques to study lipid transfer and membrane fusion.

Applications

BODIPY dye–labeled phospholipids and their corresponding fluorescent fatty acids have been used in a number of recent studies of membrane structure and metabolism:

- Researchers applied BODIPY FL C_5-ceramide (D-3521, see Section 12.4) and BODIPY FL C_5-sphingomyelin (D-3522), its primary metabolic product,[4] to a hepatocyte hybrid cell line in order to identify a barrier to lipid diffusion between the cells' apical and basolateral plasma membrane domains.[5]
- Biosynthetic incorporation of BODIPY fatty acids into BHK cells has been characterized by HPLC analysis.[6] Incorporation levels into glycerophosphocholine were found to be greater than 90% for BODIPY 500/510 dodecanoic acid (D-3823). Microscopic examination of biosynthetically labeled cells revealed localized areas of red-shifted fluorescence. This long-wavelength fluorescence results from the accumulation of BODIPY dye–labeled neutral lipids in cytoplasmic droplets at concentrations sufficient to induce the formation of BODIPY excimers.
- BODIPY FL dodecanoic acid (D-3822) has been employed to examine the co-transfer of lipids and membrane proteins from human neutrophils to the parasite *Schistosoma mansoni*.[7]

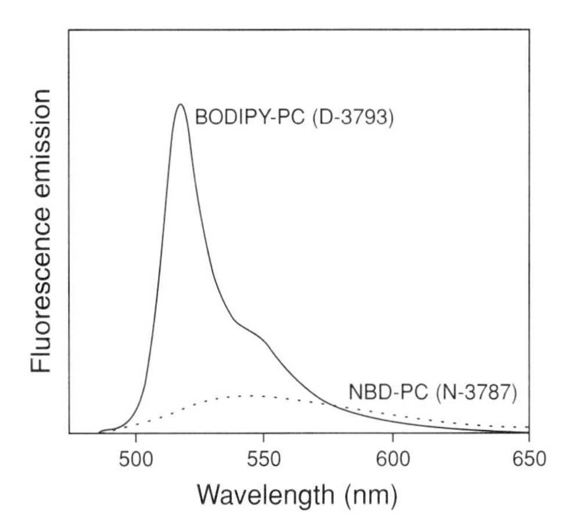

Figure 13.2 *Fluorescence spectra (excitation at 475 nm) of β-BODIPY 500/510 C_{12}-HPC (D-3793) (solid line) and NBD-C_{12}-HPC (N-3787) (dashed line) incorporated in DOPC (dioctadecenoylphosphocholine) liposomes at molar ratios of 1:400 mole:mole (labeled:unlabeled PC). The integrated intensities of the spectra are proportional to the relative fluorescence quantum yields of the two probes.*

Table 13.2 *Spectral properties of BODIPY lipid probes.*

Wavelength Range *	• Fluorescence excitation maxima from 500 nm to 581 nm • Emission maxima from 510 nm to 591 nm
Spectral Bandwidth	• Narrow (Figure 13.3)
Fluorescence Stokes Shift	• Small (Figure 13.3) • Spectral overlap results in homotransfer Förster radius (R_o) = 57 Å for BODIPY FL probes
Fluorescence Quantum Yield	• High • Typically 0.9 in fluid phase lipid bilayers
Molar Absorptivity	• High • ϵ_{max} typically >80,000 $cm^{-1}M^{-1}$
Sensitivity to Environment	• Generally low • Fluorescence quantum yields of fatty acids are not diminished in water • Fluorescence is quenched by collisional interactions with aromatic amino acids
Concentration Dependence	• Long-wavelength excimer emission is detectable at incorporation levels of about 1:10 mole:mole with respect to unlabeled phospholipid in lipid bilayers (Figure 13.4)

* Data in this table are compiled from J Am Chem Soc 116, 7801 (1994) and Anal Biochem 198, 228 (1991).

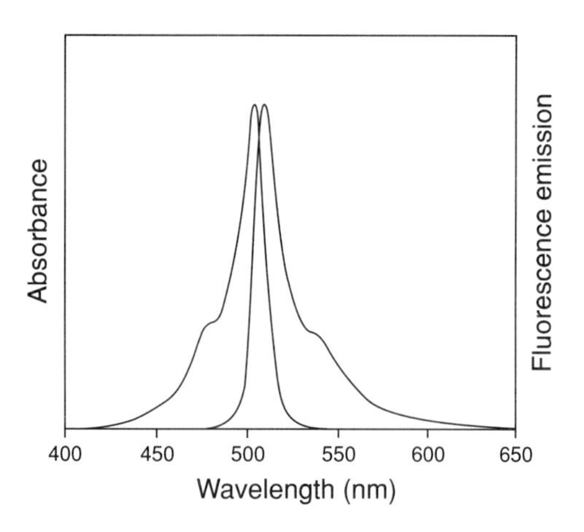

Figure 13.3 *Absorption and corrected fluorescence emission spectra of C_4-BODIPY 500/510 C_9 (B-3824) in methanol, showing the extensive spectral overlap.*

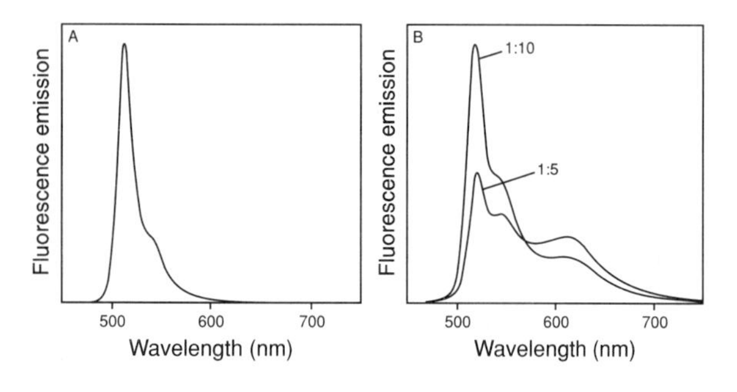

Figure 13.4 *A) Fluorescence spectrum of β-C_8-BODIPY 500/510 C_5-HPC (D-3795) incorporated in DOPC (dioctadecenoylphosphocholine) liposomes at 1:100 mole:mole (labeled:unlabeled PC). B) Fluorescence spectra at high molar incorporation levels: 1:10 mole:mole and 1:5 mole:mole.*

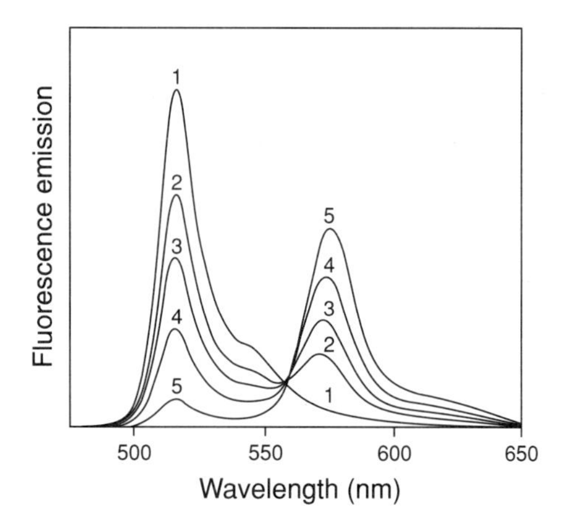

Figure 13.5 *Fluorescence resonance energy transfer from β-BODIPY 500/510 C_{12}-HPC (D-3793) (peak at 516 nm) to BODIPY 558/568 C_{12} (D-3835) (peak at 572 nm) in DOPC (dioctadecenoylphosphocholine) lipid bilayers using 475 nm excitation. Ratio of acceptors to donors is: 1) 0; 2) 0.2; 3) 0.4; 4) 0.8; and 5) 2.0.*

Researchers have also used BODIPY fatty acids and phospholipids to visualize compartmentalization of specific lipid classes in this parasite.[8]

- BODIPY FL undecanoic acid (D-3862) has been biosynthetically incorporated into SupT1 (CD4-positive human lymphoblast) cells and then used to visualize cell–cell fusion induced by the envelope glycoprotein of type 1 human immunodeficiency virus[9] (HIV-1).
- BODIPY FL C_{12}-HPC (D-3792) has been used to examine lipid–protein interactions involved in bacterial protein secretion via FRET measurements.[10]

Phospholipids with DPH-Labeled Acyl Chains

Spectral Properties

Diphenylhexatriene (DPH) carboxylic acid (P-1342), DPH propionic acid (P-459) and the glycerophosphocholine analog β-DPH HPC (D-476) are related to the neutral membrane probe DPH (D-202, see Section 14.4) and the cationic DPH derivatives TMA-DPH and TMAP-DPH (T-204 and P-3900, see Section 14.4). DPH and its derivatives exhibit strong fluorescence enhancement when incorporated into membranes, as well as sensitive fluorescence polarization (anisotropy) responses to lipid ordering. The internal motions of phospholipid acyl chains and their controlling effects on the fluorescence polarization dynamics of DPH and its derivatives continue to stimulate a considerable amount of research.[11-13]

Incorporation of DPH propionic acid and β-DPH HPC into lipid bilayers has been investigated and compared with that of other linear polyene probes.[14] In contrast to DPH, β-DPH HPC shows preferential partitioning into fluid-phase phospholipids relative to coexisting gel phases.[15,16] In plasma membrane preparations, DPH propionic acid appears to distribute primarily in the outer lipid monolayer, whereas the cationic derivative TMA-DPH preferentially localizes in the inner monolayer.[17,18] DPH propionic acid is similar to natural fatty acids in its binding affinity and competitive displacement from bovine serum albumin and hepatic fatty acid–binding protein.[19]

Applications

DPH derivatives are predominately used to investigate the structure and dynamics of the membrane interior by either fluorescence polarization or lifetime measurements. Researchers have used DPH propionic acid and β-DPH HPC as probes for lipid–protein interactions,[20,21] alcohol-induced perturbations of membrane structure,[17,22,23] phospholipid–cholesterol interactions[24] and lipid phase transitions.[25-28] β-DPH HPC was originally devised to improve the localization of DPH in membranes in order to monitor fusion processes.[29] Its fluorescence decay lifetime provides a sensitive index for discriminating poly(ethylene glycol)–mediated fusion from lipid transfer between unilamellar vesicles.[30] In addition to membrane fusion, β-DPH HPC has been used to monitor various other lipid-transfer processes.[31-33]

Phospholipids with NBD-Labeled Acyl Chains

Spectral Properties

Our acyl-modified nitrobenzoxadiazole (NBD) lipid probes include the hexanoyl- and dodecanoyl-NBD glycerophosphocho-

Table 13.3 *Spectral properties of NBD and pyrene lipid probes.*

Spectral Property	NBD	Pyrene
Ex/Em (nm) *	470/530	340/376 †
QY (τ) ‡	0.32 (5–10 nsec) §	0.6 (>100 nsec) ¶
Concentration Dependence	Self-quenched at high concentrations **	Excimer emission at high concentrations †
Environmental Sensitivity	Very weak fluorescence in water	Very sensitive to quenching by oxygen

* Typical fluorescence excitation/emission maxima in lipid environments.
† Monomer emission = 376 nm; excimer emission ~470 nm (Figure 13.6).
‡ QY = fluorescence quantum yield; τ = fluorescence decay lifetime. Excimer emission predominates at probe mole fraction >0.05 in lipid vesicles. § Values, which are solvent dependent, are typical for lipid environments; see Biochemistry 32, 3804 (1993) and Biochemistry 31, 2865 (1992). ¶ Values, which are highly oxygen sensitive, are for monomer emission. ** 98% self-quenched at mole fraction >0.5 in lipid vesicle membranes, as reported in Biochemistry 20, 2783 (1981).

lines (NBD-C_6-HPC and NBD-C_{12}-HPC; N-3786 and N-3787), the corresponding fluorescent fatty acids (N-316, N-678) and NBD C_6-sphingomyelin (N-3524). Table 13.3 compares the spectral properties of these probes with those of the pyrene lipid probes described below. Unlike the BODIPY phospholipids, the location of the relatively polar NBD fluorophore of NBD-C_6-HPC and NBD-C_{12}-HPC in phospholipid bilayers does *not* appear to conform to expectations based on the probe structure. A variety of physical evidence indicates that the NBD moiety "loops back" to the head-group region [34-36] (Figure 13.1). In fact, the fluorophore in this acyl-modified phospholipid appears to probe the same location as does the head group–labeled glycerophosphoethanolamine derivative NBD-PE [37] (N-360, see Section 13.4).

These NBD probes transfer spontaneously between membranes, with NBD-C_6-HPC transferring more rapidly than its more lipophilic C_{12} counterpart.[38,39] NBD-C_6-HPC can be readily removed ("back-exchanged") from the plasma membrane by incubating the labeled cells either with unlabeled lipid vesicles [40] or with bovine serum albumin.[41-43] This property is useful for quantitating lipid transfer and for studying phospholipid distribution asymmetry in lipid bilayers.[37,44-46]

Applications

Acyl-modified NBD phospholipids are used for investigating lipid traffic, either by directly visualizing NBD fluorescence,[47-50] by exploiting NBD self-quenching [51-53] or by fluorescence resonance energy transfer methods.[38,47,54,55] NBD fatty acids are also valuable probes for lipid-transport studies.[56,57] Lateral domains in model monolayers, bilayers and cell membranes have been characterized using NBD phospholipids in conjunction with fluorescence recovery after photobleaching (FRAP),[58-60] NBD fluorescence self-quenching [61-63] and direct microscopy techniques.[64-67] Transmembrane lipid distribution has been assessed using fluorescence resonance energy transfer from NBD-HPC to rhodamine DHPE (L-1392, see Section 13.4) [44,46,68] or alternatively by selective dithionite reduction of NBD phospholipids in the outer membrane monolayer.[69] NBD C_6-sphingomyelin (N-3524), a metabolic product of NBD C_6-ceramide (N-1154, see Section 12.4), is a useful probe for monitoring endocytosis, translocation and intracellular distribution of sphingolipids.[70-75] It is also of particular interest to those studying Niemann–Pick disease, a lysosomal lipid-storage disorder resulting from sphingomyelinase deficiency.[76]

Phospholipids with Pyrene-Labeled Acyl Chains

Spectral Properties

Phospholipid analogs with pyrene-labeled *sn*-2 acyl chains are among the most popular fluorescent membrane probes. Molecular Probes offers pyrene-labeled phospholipids and fatty acids with several different head groups and chain lengths (Table 13.1), including a sphingomyelin analog (P-3527). The spectral properties of the pyrene lipid probes are summarized in Table 13.3. Of primary importance in terms of practical applications is the concentration-dependent formation of excited-state pyrene dimers (excimers), which exhibit a distinctive red-shifted emission (peak ~470 nm) (Figure 13.6). Excimer formation may reflect both diffusion-controlled interactions and nonrandom static distributions of the fluorophore.[77-80] The mechanistic question is of particular significance in relation to the use of pyrene-excimer fluorescence to measure lateral diffusion rates in membranes.[81] Diffusion distances detected by pyrene-excimer fluorescence are dependent on the excited-state lifetime of the monomer (~100 nanoseconds) and are much shorter than those detectable by FRAP techniques.[82]

Applications

The excimer-forming properties of pyrene are well suited to monitoring membrane fusion (see the Technical Note "Lipid Mixing Assays of Membrane Fusion" on page 286) and phospholipid transfer processes.[83-86] The monomer/excimer emission ratio can also be used to characterize membrane structural domains and their dependence on temperature, lipid composition and other external factors.[87-89] For example, by determining the ratio of pyrenedecanoic acid (P-31) monomer emission to excimer emission, researchers have been able to measure membrane fluidity in cells both by flow cytometry [90] and by fluorescence microscopy.[91]

Pyrenedecanoyl glycerophosphocholine (β-py-C_{10}-HPC, H-361) has been used to elucidate the effect of extrinsic species such as Ca^{2+},[92] platelet-activating factor,[93] drugs,[94] membrane-associated proteins,[95-97] ethanol and cholesterol [98,99] on lipid bilayer structure

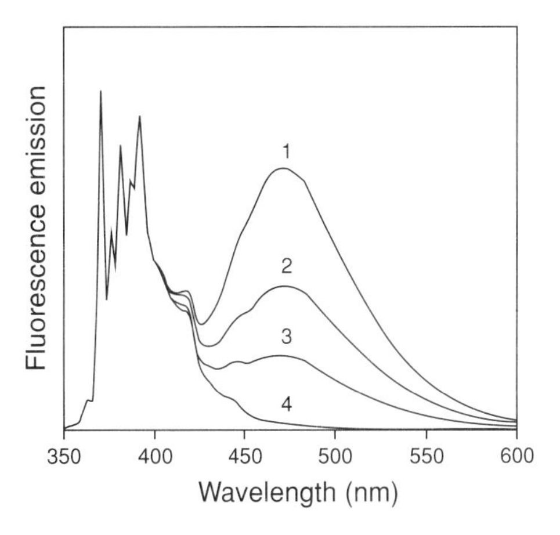

Figure 13.6 *Excimer formation by pyrene in ethanol. Spectra are normalized to the 371.5 nm peak of the monomer. All spectra are identical below 400 nm after normalization. Spectra are as follows: 1) 2 mM pyrene, purged with argon to remove oxygen; 2) 2 mM pyrene, air-equilibrated; 3) 0.5 mM pyrene (argon-purged); and 4) 2 µM pyrene (argon-purged). The monomer-to-excimer ratio (371.5 nm/470 nm) is dependent on both pyrene concentration and the excited-state lifetime, which is variable because of quenching by oxygen.*

and dynamics. Researchers have used liposomes incorporating both py-C_{10}-HPC and lipid-conjugated monoclonal antibodies directed against cell-surface antigens to selectively kill T lymphocytes by photoablation.[100] Pyrenedodecanoic acid (P-96) has also been employed as a cytotoxic photosensitizer.[101-103] The long excited-state lifetime of pyrene (Table 13.3) renders the fluorescence very susceptible to oxygen quenching, and consequently these probes can be used to measure oxygen concentrations in solutions,[104] lipid bilayers [105] and cells.[106,107]

Pyrene fatty acids (particularly pyrenedodecanoic acid) display substantially superior biosynthetic incorporation levels and product distribution than do anthroyloxy, dansyl, DPH and NBD fatty acids or parinaric acid.[86,108] The metabolic products obtained are strongly dependent on the fatty acyl chain length.[109] Because incorporation is accompanied by pyrene fluorescence enhancement, it is unnecessary to wash away the free acid.[110] By measuring pyrenedecanoic acid incorporation, researchers have been able to identify lipid storage defects in fibroblasts and lymphoid cell lines.[111,112]

Phospholipids with Perylene-Labeled Acyl Chains

Perylene is structurally similar to pyrene and also forms excimers, but has longer-wavelength absorption and emission, a shorter excited-state lifetime and a higher fluorescence quantum yield. We offer a glycerophosphocholine (H-3790) derived from perylene dodecanoic acid (P-646). Unlike perylen*oyl* fatty acids,[113,114] in which the fluorophore "loops back" to the membrane surface, our hydrocarbon-linked analogs are expected to position the perylene fluorophore at varying depths within the membrane interior.

1. Anal Biochem 198, 228 (1991); **2.** J Fluorescence 4, 295 (1994); **3.** Biochem Biophys Res Comm 207, 508 (1995); **4.** J Cell Biol 113, 1267 (1991); **5.** J Cell Biol 123, 1761 (1993); **6.** Biochem Biophys Res Comm 187, 1594 (1992); **7.** J Cell Sci 106, 485 (1993); **8.** J Lipid Res 36, 1 (1995); **9.** J Virology 69, 1462 (1995); **10.** Biochemistry 35, 3063 (1996); **11.** Biochem Biophys Res Comm 201, 709 (1994); **12.** Biophys Chem 48, 337 (1994); **13.** Chem Phys Lipids 64, 99 (1993); **14.** J Am Chem Soc 110, 3198 (1988); **15.** Biochemistry 30, 7491 (1991); **16.** Biochemistry 24, 6178 (1985); **17.** Biochim Biophys Acta 1112, 14 (1992); **18.** Biochemistry 29, 7282 (1990); **19.** Biochim Biophys Acta 731, 465 (1983); **20.** Biochim Biophys Acta 1104, 273 (1992); **21.** Biochemistry 29, 6720 (1990); **22.** Biochemistry 27, 9175 (1988); **23.** Biochim Biophys Acta 946, 85 (1988); **24.** Biophys J 56, 1245 (1989); **25.** Biochemistry 31, 3759 (1992); **26.** Biophys J 58, 1527 (1990); **27.** Chem Phys Lipids 56, 149 (1990); **28.** Biophys J 55, 1025 (1989); **29.** Biochim Biophys Acta 692, 196 (1982); **30.** J Fluorescence 4, 153 (1994); **31.** Biochemistry 30, 6780 (1991); **32.** Biochemistry 30, 5555 (1991); **33.** Biochemistry 30, 4193 (1991); **34.** Biochemistry 32, 10826 (1993); **35.** Biochim Biophys Acta 938, 24 (1988); **36.** Biochemistry 26, 39 (1987); **37.** Chem Phys Lipids 53, 1 (1990); **38.** Biochemistry 21, 1720 (1982); **39.** Biochemistry 20, 2783 (1981); **40.** J Biol Chem 255, 5404 (1980); **41.** J Biol Chem 269, 22517 (1994); **42.** Biochemistry 32, 3714 (1993); **43.** Proc Natl Acad Sci USA 86, 9896 (1989); **44.** Biochemistry 33, 6721 (1994); **45.** Kok, J.W. and Hoekstra, D. in *Fluorescent and Luminescent Probes for Biological Activity*, W.T. Mason, Ed., Academic Press (1993) pp. 100–119; **46.** Biochemistry 31, 2865 (1992); **47.** J Cell Biol 123, 1403 (1993); **48.** J Cell Biol 113, 235 (1991); **49.** J Cell Biol 112, 267 (1991); **50.** J Biol Chem 265, 5337 (1990); **51.** Biochim Biophys Acta 1082, 255 (1991); **52.** Biochemistry 27, 1889 (1988); **53.** J Biol Chem 262, 14172 (1987); **54.** Am J Physiol 267, G80 (1994); **55.** J Biol Chem 258, 5368 (1983); **56.** Parasitology 105, 81 (1992); **57.** Am J Physiol 261, G83 (1991); **58.** Biochemistry 33, 8225 (1994); **59.** J Cell Biol 112, 1143 (1991); **60.** J Cell Biol 105, 755 (1987); **61.** Biochemistry 30, 7961 (1991); **62.** Biochemistry 24, 7123 (1985); **63.** Biochemistry 21, 1055 (1982); **64.** Biochemistry 33, 4483 (1994); **65.** Biochemistry 32, 12591 (1993); **66.** Chem Phys Lipids 57, 227 (1991); **67.** Proc Natl Acad Sci USA 88, 1364 (1991); **68.** Chem Phys Lipids 57, 29 (1991); **69.** Chem Phys Lipids 70, 205 (1994); **70.** Biochemistry 33, 9968 (1994); **71.** J Cell Biol 127, 725 (1994); **72.** J Cell Biol 121, 1257 (1993); **73.** J Cell Biol 117, 259 (1992); **74.** J Cell Biol 114, 231 (1991); **75.** J Cell Biol 108, 2169 (1989); **76.** J Cell Biol 111, 429 (1990); **77.** Biophys J 55, 885 (1989); **78.** J Phys Chem 93, 7170 (1989); **79.** Biochim Biophys Acta 858, 221 (1986); **80.** Biochemistry 24, 2773 (1985); **81.** Biochim Biophys Acta 339, 103 (1974); **82.** Biophys J 57, 281 (1990); **83.** Biochemistry 32, 11074 (1993); **84.** Biochemistry 31, 5912 (1992); **85.** Biochemistry 29, 1593 (1990); **86.** Chem Phys Lipids 50, 191 (1989); **87.** Biophys J 66, 2029 (1994); **88.** Biophys J 66, 1981 (1994); **89.** Biophys J 63, 903 (1992); **90.** J Immunol Meth 96, 225 (1987); **91.** Biochemistry 29, 1949 (1990); **92.** Biochemistry 27, 3433 (1988); **93.** Chem Phys Lipids 53, 129 (1990); **94.** J Biol Chem 268, 1074 (1993); **95.** Biochemistry 32, 11711 (1993); **96.** Biochemistry 32, 5411 (1993); **97.** Biochemistry 32, 5373 (1993); **98.** Biophys J 66, 729 (1994); **99.** Biochemistry 30, 2463 (1991); **100.** Proc Natl Acad Sci USA 84, 246 (1987); **101.** J Biol Chem 269, 11734 (1994); **102.** Leukemia Res 16, 453 (1992); **103.** Exp Hematol 18, 89 (1990); **104.** Anal Chem 59, 279 (1987); **105.** Biophys J 47, 813 (1985); **106.** J Cell Physiol 107, 329 (1981); **107.** Biochim Biophys Acta 279, 393 (1972); **108.** Chem Phys Lipids 58, 111 (1991); **109.** J Biol Chem 267, 6563 (1992); **110.** Biochim Biophys Acta 943, 477 (1988); **111.** Biochim Biophys Acta 1005, 130 (1989); **112.** Biochim Biophys Acta 644, 233 (1981); **113.** Chem Phys Lipids 62, 293 (1992); **114.** Chem Phys Lipids 37, 165 (1985).

13.2 Data Table *Acyl Chain–Labeled Phospholipids and Their Corresponding Fluorescent Fatty Acids*

Cat #	Structure	MW	Storage	Soluble	Abs	{ε × 10^{-3}}	Em	Solvent	Notes
B-3794	✓	882	FF,D,L	<1>	510	{92}	515	EtOH	<1, 2>
B-3824	✓	404	F,L	DMSO	509	{101}	515	MeOH	<2>
D-476	✓	782	FF,D,L	<1>	354	{81}	428	MeOH	<1, 3>
D-3522	✓	767	FF,D,L	<1>	505	{77}	512	MeOH	<1, 2>
D-3792	✓	896	FF,D,L	<1>	506	{86}	513	EtOH	<1, 2>
D-3793	✓	882	FF,D,L	<1>	509	{86}	513	EtOH	<1, 2>
D-3795	✓	882	FF,D,L	<1>	508	{89}	516	EtOH	<1, 2>
D-3802	✓	854	FF,D,L	<1>	505	{86}	512	MeOH	<1, 2>
D-3803	✓	798	FF,D,L	<1>	503	{80}	512	MeOH	<1, 2>
D-3804	✓	756	FF,D,L	<1>	504	{83}	510	MeOH	<1, 2>
D-3805	✓	747	FF,D,L	<1>	504	{79}	511	MeOH	<1, 2>
D-3806	✓	898	FF,D,L	<1>	582	{124}	593	MeOH	<1, 2>
D-3812	✓	1020	FF,D,L	<1>	535	{55}	553	EtOH	<1, 2>
D-3813	✓	978	FF,D,L	<1>	534	{66}	551	MeOH	<1, 2>
D-3815	✓	922	FF,D,L	<1>	534	{64}	552	MeOH	<1, 2>
D-3821	✓	474	F,L	DMSO	505	{90}	512	MeOH	<2>
D-3822	✓	418	F,L	DMSO	505	{87}	511	MeOH	<2>

Cat #	Structure	MW	Storage	Soluble	Abs	{ε × 10⁻³}	Em	Solvent	Notes
D-3823	✓	404	F,L	DMSO	508	{97}	514	MeOH	<2>
D-3825	✓	404	F,L	DMSO	509	{100}	515	MeOH	<2>
D-3826	✓	404	F,L	DMSO	509	{102}	516	MeOH	<2>
D-3827	✓	404	F,L	DMSO	494	{100}	504	MeOH	<2>
D-3829	✓	478	F,L	DMSO	505	{111}	512	MeOH	<2>
D-3832	✓	542	F,L	DMSO	534	{76}	552	MeOH	<2>
D-3833	✓	444	F,L	DMSO	534	{74}	552	MeOH	<2>
D-3834	✓	320	F,L	DMSO	505	{96}	511	MeOH	<2>
D-3835	✓	472	F,L	DMSO	559	{91}	568	MeOH	<2>
D-3836	✓	398	F,L	DMSO	542	{81}	563	MeOH	<2>
D-3837	✓	482	F,L	DMSO	542	{72}	563	MeOH	<2>
D-3838	✓	394	F,L	DMSO	564	{148}	570	MeOH	<2>
D-3839	✓	478	F,L	DMSO	564	{145}	570	MeOH	<2>
D-3858	✓	357	F,L	DMSO	578	{94}	590	MeOH	<2>
D-3859	✓	441	F,L	DMSO	578	{100}	590	MeOH	<2>
D-3860	✓	420	F,L	DMSO	582	{140}	591	MeOH	<2>
D-3861	✓	504	F,L	DMSO	582	{140}	591	MeOH	<2>
D-3862	✓	404	F,L	DMSO	505	{92}	510	MeOH	<2>
H-361	✓	850	FF,D,L	<1>	342	{37}	376	MeOH	<1, 4, 5>
H-3743	✓	745	FF,D,L	<1>	341	{37}	377	MeOH	<1, 4, 5>
H-3784	✓	808	FF,D,L	<1>	341	{36}	376	MeOH	<1, 4, 5>
H-3790	✓	928	FF,D,L	<1>	442	{29}	448	EtOH	<1, 6>
H-3809	✓	856	FF,D,L	<1>	341	{38}	376	MeOH	<1, 4, 5>
H-3810	✓	801	FF,D,L	<1>	341	{40}	376	MeOH	<1, 4, 5>
H-3818	✓	794	FF,D,L	<1>	341	{37}	377	MeOH	<1, 4, 5>
M-1119	✓	477	L	DMSO	483	{21}	543	MeOH	<7>
N-316	✓	294	L	DMSO	467	{23}	539	MeOH	<7>
N-678	✓	378	L	DMSO	467	{24}	536	MeOH	<7>
N-3524	✓	741	FF,D,L	<1>	466	{22}	536	MeOH	<7>
N-3786	✓	772	FF,D,L	<1>	465	{21}	533	EtOH	<7>
N-3787	✓	856	FF,D,L	<1>	465	{22}	534	EtOH	<7>
P-31	✓	373	L	DMF, DMSO	341	{43}	377	MeOH	<4, 5>
P-32	✓	288	L	DMF, DMSO	341	{43}	376	MeOH	<4, 5>
P-96	✓	401	L	DMF, DMSO	341	{44}	377	MeOH	<4, 5>
P-164	✓	246	L	DMF, DMSO	342	{30}	382	MeOH	
P-243	✓	457	L	DMF, DMSO	341	{43}	377	MeOH	<4, 5>
P-459	✓	292	L	DMF, DMSO	354	{82}	430	MeOH	<3>
P-646	✓	451	L	DMF, DMSO	440	{34}	448	MeOH	<6>
P-1342	✓	263	L	DMF, DMSO	357	{60}	436	MeOH	<3>
P-1903	✓	288	L	DMF, DMSO	341	{43}	376	MeOH	<4, 5>
P-3527	✓	819	FF,D,L	<1>	341	{39}	377	MeOH	<1, 4, 5>
P-3779	✓	552	FF,D,L	<1>	340	{39}	378	MeOH	<1, 4, 5>
P-3840	✓	316	L	DMF, DMSO	341	{42}	377	MeOH	<4, 5>

For definitions of the contents of this data table, see "How to Use The Handbook" on page vi.
Structure: Chemical structure drawing: ✓ = shown in this section.

<1> Chloroform is the most generally useful solvent for preparing stock solutions of phospholipids (including sphingomyelins). Glycerophosphocholines are usually freely soluble in ethanol. Most other glycerophospholipids (phosphoethanolamines, phosphatidic acids and phosphoglycerols) are less soluble in ethanol, but solutions up to 1–2 mg/mL should be obtainable, using sonication to aid dispersion if necessary. Labeling of aqueous samples with nonmiscible phospholipid stock solutions can be accomplished by evaporating the organic solvent, followed by hydration and sonication, yielding a suspension of liposomes. Information on the solubility of natural phospholipids can be found in Marsh, D., *CRC Handbook of Lipid Bilayers*, CRC Press (1990) pp. 71–80.

<2> The absorption and fluorescence spectra of BODIPY derivatives are relatively insensitive to the solvent.

<3> Diphenylhexatriene (DPH) and its derivatives are essentially nonfluorescent in water. Absorption and emission spectra have multiple peaks. The wavelength, resolution and relative intensity of these peaks are environment dependent. Abs and Em values are for the most intense peak in the solvent specified.

<4> Pyrene derivatives exhibit structured spectra. The absorption maximum is usually about 340 nm with a subsidiary peak at about 325 nm. There are also strong absorption peaks below 300 nm. The emission maximum is usually about 376 nm with a subsidiary peak at 396 nm. Excimer emission at about 470 nm may be observed at high concentrations.

<5> Alkylpyrene fluorescence lifetimes are up to 110 nsec and are very sensitive to oxygen.

<6> Perylene derivatives have fluorescence lifetimes of approximately 5 nsec. Spectra are structured, with subsidiary peaks at 413 nm (absorption) and 477 nm (emission).

<7> NBD derivatives are almost nonfluorescent in water. NBD derivatives of primary amines such as N-316 and N-678 have a much higher quantum yield in methanol than those of secondary amines such as M-1119. The quantum yields of NBD amine derivatives increase upon binding to membranes.

B-3794 see D-3795

B-3824 see D-3826

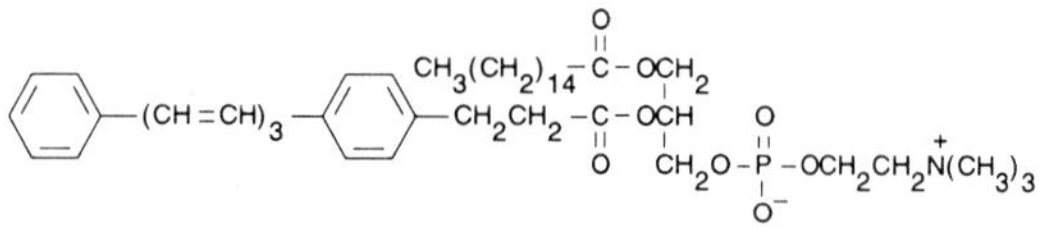

D-476

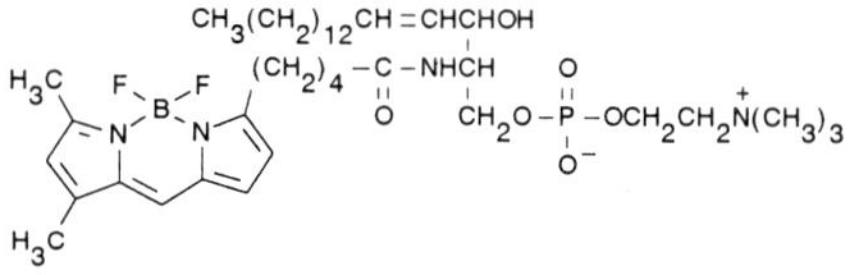

D-3522

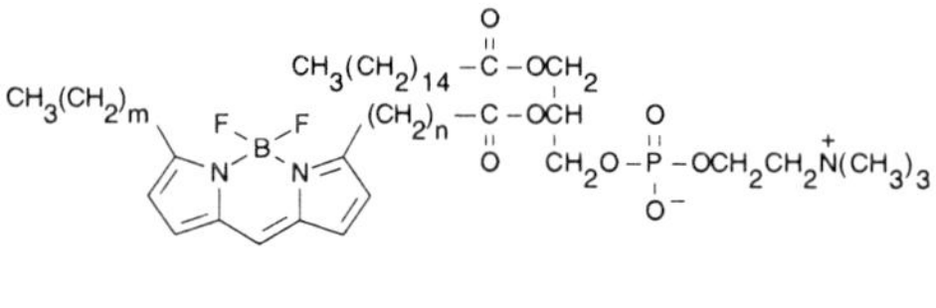

D-3795 n = 4 m = 7
B-3794 n = 8 m = 3
D-3793 n = 11 m = 0

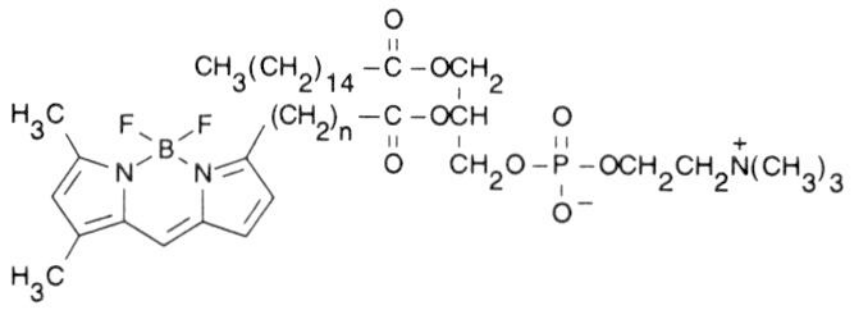

D-3803 n = 4
D-3792 n = 11

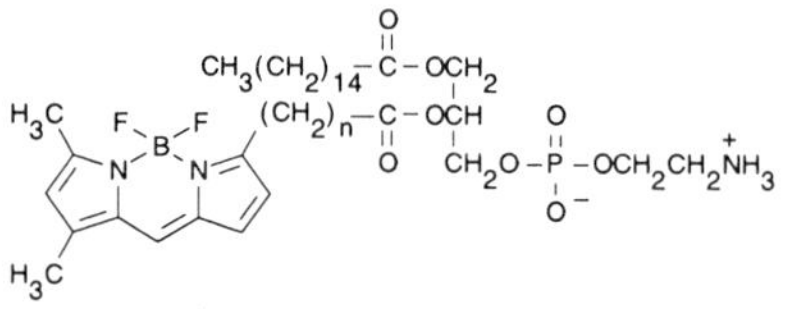

D-3804 n = 4
D-3802 n = 11

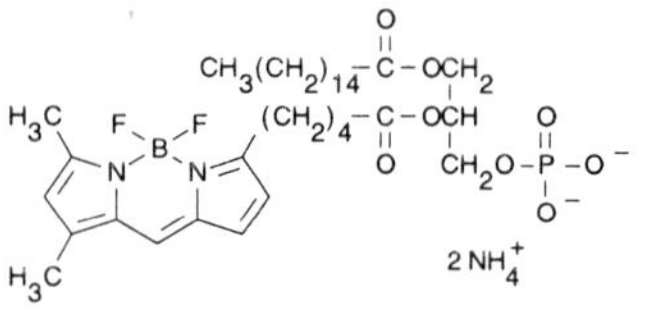

D-3805

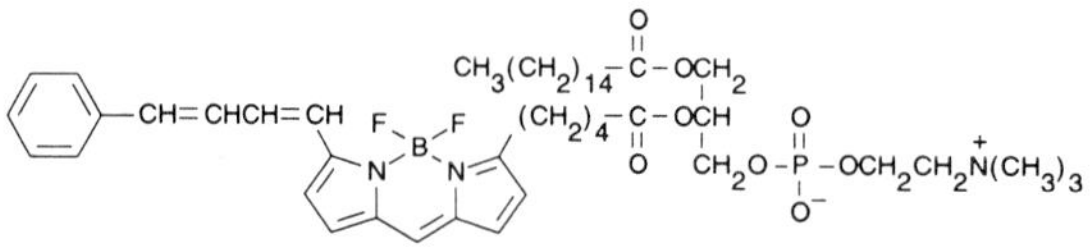

D-3806

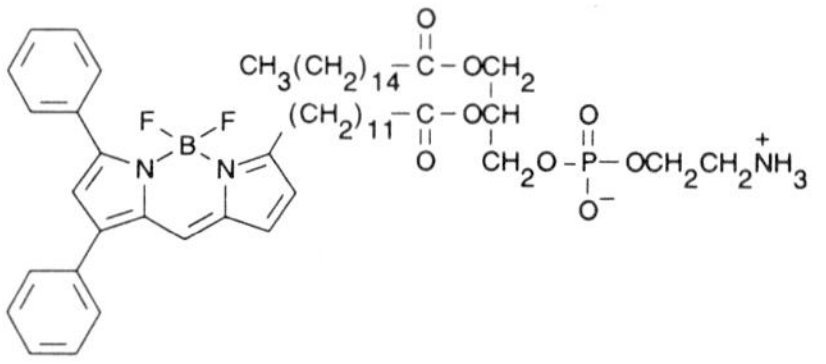

D-3813

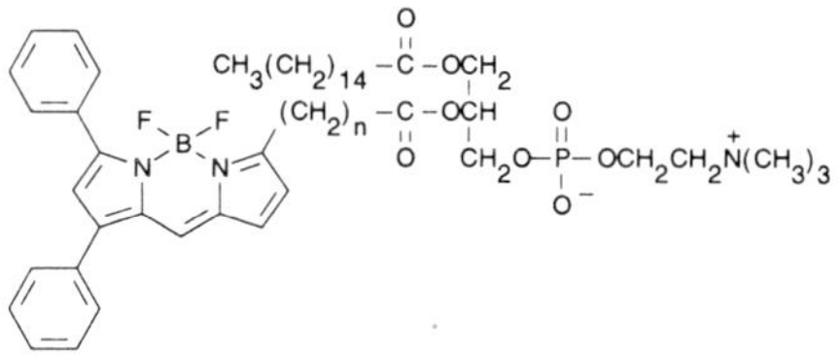

D-3815 n = 4
D-3812 n = 11

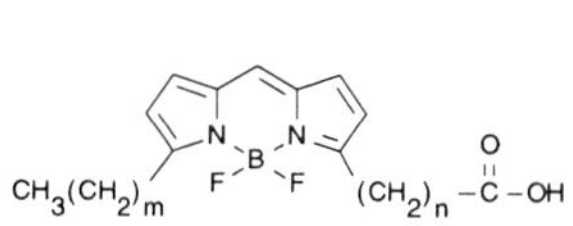

D-3826 n = 2 m = 9
D-3825 n = 4 m = 7
B-3824 n = 8 m = 3
D-3823 n = 11 m = 0

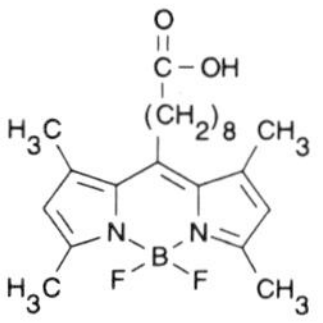

D-3827

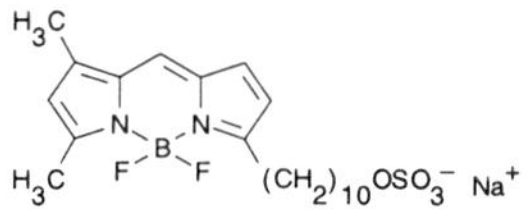

D-3829

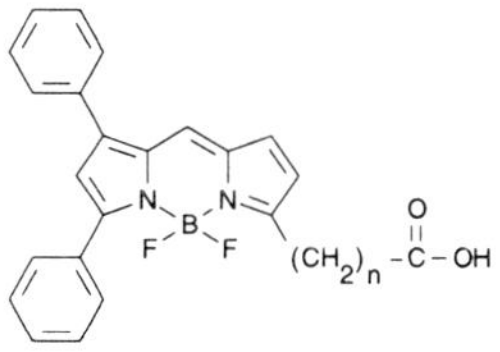

D-3833 n = 4
D-3832 n = 11

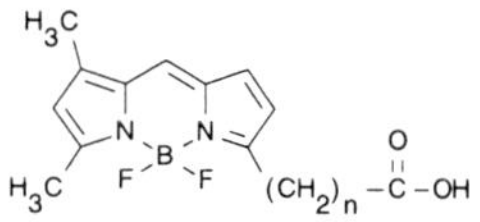

D-3834 n = 4
D-3862 n = 10
D-3822 n = 11
D-3821 n = 15

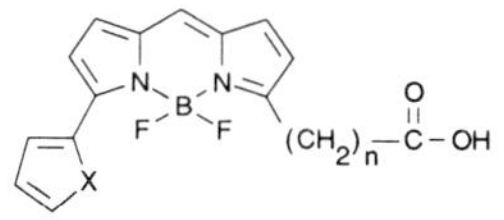

D-3835 n = 11 X = S
D-3858 n = 4 X = NH
D-3859 n = 10 X = NH

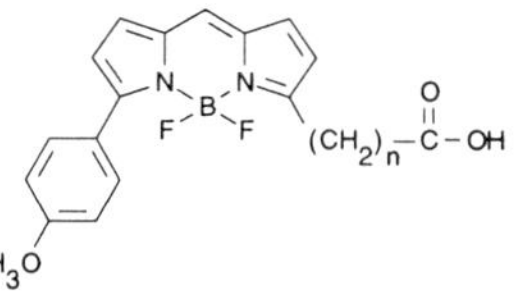

D-3836 n = 4
D-3837 n = 10

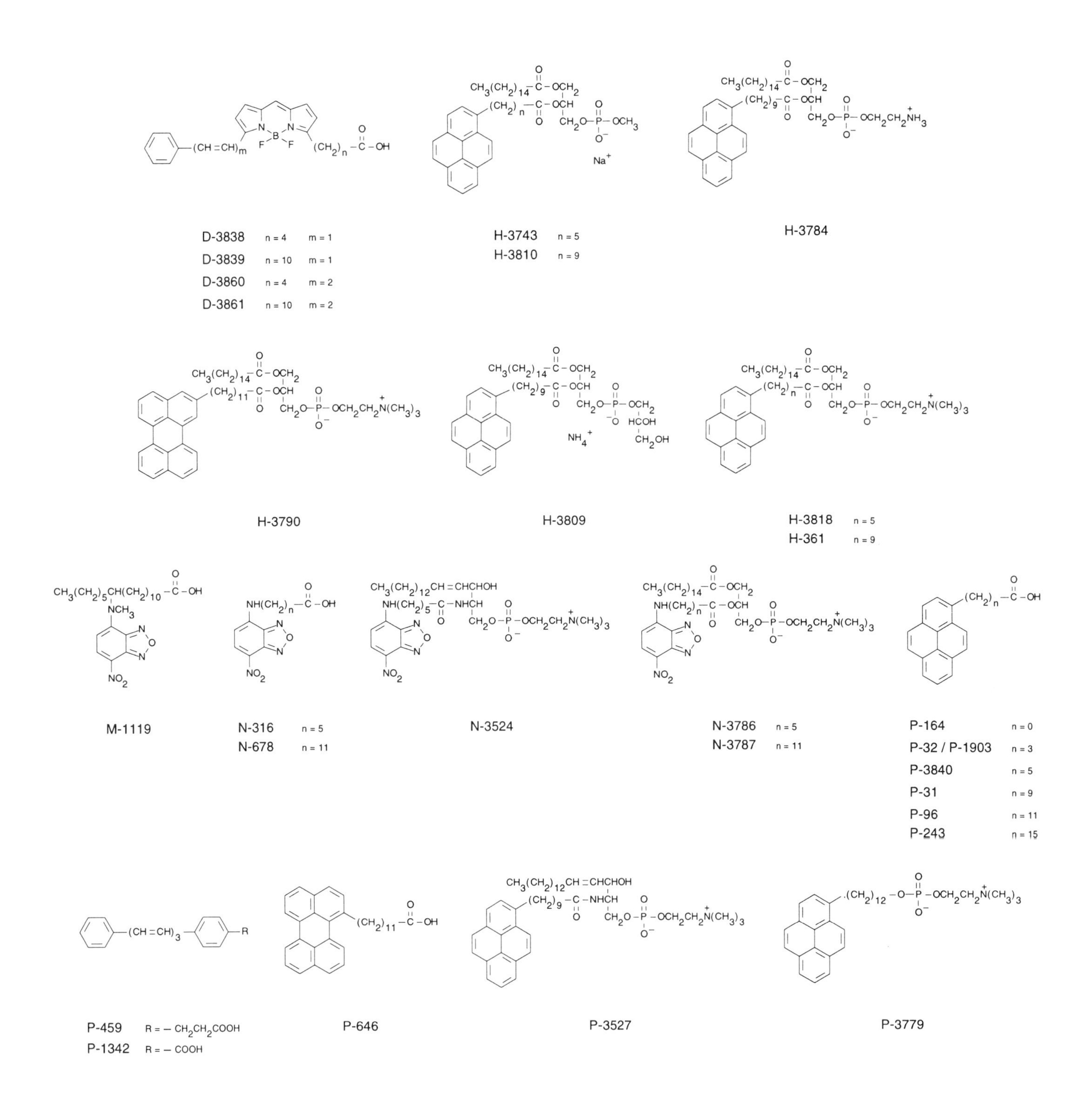

13.2 Price List *Acyl Chain–Labeled Phospholipids and Their Corresponding Fluorescent Fatty Acids*

Cat #	Product Name	Unit Size	Price Per Unit ($) 1–4 Units	5–24 Units
B-3824	5-butyl-4,4-difluoro-4-bora-3a,4a-diaza-s-indacene-3-nonanoic acid (C_4-BODIPY® 500/510 C_9)	1 mg	38.00	30.40
B-3794	2-(5-butyl-4,4-difluoro-4-bora-3a,4a-diaza-s-indacene-3-nonanoyl)-1-hexadecanoyl-*sn*-glycero-3-phosphocholine (β-C_4-BODIPY® 500/510 C_9-HPC)	100 µg	38.00	30.40
D-3826	5-decyl-4,4-difluoro-4-bora-3a,4a-diaza-s-indacene-3-propionic acid (C_{10}-BODIPY® 500/510 C_3)	1 mg	38.00	30.40
D-3829	4,4-difluoro-5,7-dimethyl-4-bora-3a,4a-diaza-s-indacene-3-decyl sulfate, sodium salt (BODIPY® FL C_{10} sulfate)	1 mg	38.00	30.40
D-3822	4,4-difluoro-5,7-dimethyl-4-bora-3a,4a-diaza-s-indacene-3-dodecanoic acid (BODIPY® FL C_{12})	1 mg	38.00	30.40
D-3792	2-(4,4-difluoro-5,7-dimethyl-4-bora-3a,4a-diaza-s-indacene-3-dodecanoyl)-1-hexadecanoyl-*sn*-glycero-3-phosphocholine (β-BODIPY® FL C_{12}-HPC)	100 µg	38.00	30.40

Cat #	Product Name	Unit Size	Price Per Unit ($) 1–4 Units	5–24 Units
D-3802	2-(4,4-difluoro-5,7-dimethyl-4-bora-3a,4a-diaza-s-indacene-3-dodecanoyl)-1-hexadecanoyl-*sn*-glycero-3-phosphoethanolamine (β-BODIPY® FL C_{12}-HPE)	100 µg	38.00	30.40
D-3821	4,4-difluoro-5,7-dimethyl-4-bora-3a,4a-diaza-s-indacene-3-hexadecanoic acid (BODIPY® FL C_{16})	1 mg	38.00	30.40
D-3834	4,4-difluoro-5,7-dimethyl-4-bora-3a,4a-diaza-s-indacene-3-pentanoic acid (BODIPY® FL C_5)	1 mg	38.00	30.40
D-3805	2-(4,4-difluoro-5,7-dimethyl-4-bora-3a,4a-diaza-s-indacene-3-pentanoyl)-1-hexadecanoyl-*sn*-glycero-3-phosphate, diammonium salt (β-BODIPY® FL C_5-HPA)	100 µg	38.00	30.40
D-3803	2-(4,4-difluoro-5,7-dimethyl-4-bora-3a,4a-diaza-s-indacene-3-pentanoyl)-1-hexadecanoyl-*sn*-glycero-3-phosphocholine (β-BODIPY® FL C_5-HPC)	100 µg	38.00	30.40
D-3804	2-(4,4-difluoro-5,7-dimethyl-4-bora-3a,4a-diaza-s-indacene-3-pentanoyl)-1-hexadecanoyl-*sn*-glycero-3-phosphoethanolamine (β-BODIPY® FL C_5-HPE)	100 µg	38.00	30.40
D-3522	N-(4,4-difluoro-5,7-dimethyl-4-bora-3a,4a-diaza-s-indacene-3-pentanoyl)sphingosyl phosphocholine (BODIPY® FL C_5-sphingomyelin)	250 µg	125.00	100.00
D-3862	4,4-difluoro-5,7-dimethyl-4-bora-3a,4a-diaza-s-indacene-3-undecanoic acid (BODIPY® FL C_{11})	1 mg	38.00	30.40
D-3832	4,4-difluoro-5,7-diphenyl-4-bora-3a,4a-diaza-s-indacene-3-dodecanoic acid (BODIPY® 530/550 C_{12})	1 mg	38.00	30.40
D-3812	2-(4,4-difluoro-5,7-diphenyl-4-bora-3a,4a-diaza-s-indacene-3-dodecanoyl)-1-hexadecanoyl-*sn*-glycero-3-phosphocholine (β-BODIPY® 530/550 C_{12}-HPC)	100 µg	38.00	30.40
D-3813	2-(4,4-difluoro-5,7-diphenyl-4-bora-3a,4a-diaza-s-indacene-3-dodecanoyl)-1-hexadecanoyl-*sn*-glycero-3-phosphoethanolamine (β-BODIPY® 530/550 C_{12}-HPE)	100 µg	38.00	30.40
D-3833	4,4-difluoro-5,7-diphenyl-4-bora-3a,4a-diaza-s-indacene-3-pentanoic acid (BODIPY® 530/550 C_5)	1 mg	38.00	30.40
D-3815	2-(4,4-difluoro-5,7-diphenyl-4-bora-3a,4a-diaza-s-indacene-3-pentanoyl)-1-hexadecanoyl-*sn*-glycero-3-phosphocholine (β-BODIPY® 530/550 C_5-HPC)	100 µg	38.00	30.40
D-3836	4,4-difluoro-5-(4-methoxyphenyl)-4-bora-3a,4a-diaza-s-indacene-3-pentanoic acid (BODIPY® 542/563 C_5)	1 mg	38.00	30.40
D-3837	4,4-difluoro-5-(4-methoxyphenyl)-4-bora-3a,4a-diaza-s-indacene-3-undecanoic acid (BODIPY® 542/563 C_{11})	1 mg	38.00	30.40
D-3823	4,4-difluoro-5-methyl-4-bora-3a,4a-diaza-s-indacene-3-dodecanoic acid (C_1-BODIPY® 500/510 C_{12})	1 mg	38.00	30.40
D-3793	2-(4,4-difluoro-5-methyl-4-bora-3a,4a-diaza-s-indacene-3-dodecanoyl)-1-hexadecanoyl-*sn*-glycero-3-phosphocholine (β-BODIPY® 500/510 C_{12}-HPC)	100 µg	38.00	30.40
D-3825	4,4-difluoro-5-octyl-4-bora-3a,4a-diaza-s-indacene-3-pentanoic acid (C_8-BODIPY® 500/510 C_5)	1 mg	38.00	30.40
D-3795	2-(4,4-difluoro-5-octyl-4-bora-3a,4a-diaza-s-indacene-3-pentanoyl)-1-hexadecanoyl-*sn*-glycero-3-phosphocholine (β-C_8-BODIPY® 500/510 C_5-HPC)	100 µg	38.00	30.40
D-3860	4,4-difluoro-5-(4-phenyl-1,3-butadienyl)-4-bora-3a,4a-diaza-s-indacene-3-pentanoic acid (BODIPY® 581/591 C_5)	1 mg	38.00	30.40
D-3806	2-(4,4-difluoro-5-(4-phenyl-1,3-butadienyl)-4-bora-3a,4a-diaza-s-indacene-3-pentanoyl)-1-hexadecanoyl-*sn*-glycero-3-phosphocholine (β-BODIPY® 581/591 C_5-HPC)	100 µg	38.00	30.40
D-3861	4,4-difluoro-5-(4-phenyl-1,3-butadienyl)-4-bora-3a,4a-diaza-s-indacene-3-undecanoic acid (BODIPY® 581/591 C_{11})	1 mg	38.00	30.40
D-3858	4,4-difluoro-5-(2-pyrrolyl)-4-bora-3a,4a-diaza-s-indacene-3-pentanoic acid (BODIPY® 576/589 C_5)	1 mg	38.00	30.40
D-3859	4,4-difluoro-5-(2-pyrrolyl)-4-bora-3a,4a-diaza-s-indacene-3-undecanoic acid (BODIPY® 576/589 C_{11})	1 mg	38.00	30.40
D-3838	4,4-difluoro-5-styryl-4-bora-3a,4a-diaza-s-indacene-3-pentanoic acid (BODIPY® 564/570 C_5)	1 mg	38.00	30.40
D-3839	4,4-difluoro-5-styryl-4-bora-3a,4a-diaza-s-indacene-3-undecanoic acid (BODIPY® 564/570 C_{11})	1 mg	38.00	30.40
D-3827	4,4-difluoro-1,3,5,7-tetramethyl-4-bora-3a,4a-diaza-s-indacene-8-nonanoic acid (BODIPY® 493/503 C_9)	1 mg	38.00	30.40
D-3835	4,4-difluoro-5-(2-thienyl)-4-bora-3a,4a-diaza-s-indacene-3-dodecanoic acid (BODIPY® 558/568 C_{12})	1 mg	38.00	30.40
D-476	2-(3-(diphenylhexatrienyl)propanoyl)-1-hexadecanoyl-*sn*-glycero-3-phosphocholine (β-DPH HPC)	1 mg	48.00	38.40
H-3790	1-hexadecanoyl-2-(3-perylenedodecanoyl)-*sn*-glycero-3-phosphocholine	1 mg	38.00	30.40
H-361	1-hexadecanoyl-2-(1-pyrenedecanoyl)-*sn*-glycero-3-phosphocholine (β-py-C_{10}-HPC)	1 mg	65.00	52.00
H-3784	1-hexadecanoyl-2-(1-pyrenedecanoyl)-*sn*-glycero-3-phosphoethanolamine (β-py-C_{10}-HPE)	1 mg	65.00	52.00
H-3809	1-hexadecanoyl-2-(1-pyrenedecanoyl)-*sn*-glycero-3-phosphoglycerol, ammonium salt (β-py-C_{10}-PG)	1 mg	65.00	52.00
H-3810	1-hexadecanoyl-2-(1-pyrenedecanoyl)-*sn*-glycero-3-phosphomethanol, sodium salt (β-py-C_{10}-HPM)	1 mg	65.00	52.00
H-3818	1-hexadecanoyl-2-(1-pyrenehexanoyl)-*sn*-glycero-3-phosphocholine (β-py-C_6-HPC)	1 mg	65.00	52.00
H-3743	1-hexadecanoyl-2-(1-pyrenehexanoyl)-*sn*-glycero-3-phosphomethanol, sodium salt (β-py-C_6-HPM)	1 mg	65.00	52.00
M-1119	12-(N-methyl)-N-((7-nitrobenz-2-oxa-1,3-diazol-4-yl)amino)octadecanoic acid (12-NBD stearic acid)	5 mg	65.00	52.00
N-678	12-(N-(7-nitrobenz-2-oxa-1,3-diazol-4-yl)amino)dodecanoic acid	100 mg	148.00	118.40
N-3787	2-(12-(7-nitrobenz-2-oxa-1,3-diazol-4-yl)amino)dodecanoyl-1-hexadecanoyl-*sn*-glycero-3-phosphocholine (NBD-C_{12}-HPC)	5 mg	50.00	40.00
N-316	6-(N-(7-nitrobenz-2-oxa-1,3-diazol-4-yl)amino)hexanoic acid (NBD-X)	100 mg	50.00	40.00
N-3786	2-(6-(7-nitrobenz-2-oxa-1,3-diazol-4-yl)amino)hexanoyl-1-hexadecanoyl-*sn*-glycero-3-phosphocholine (NBD-C_6-HPC)	5 mg	50.00	40.00
N-3524	N-((6-(7-nitrobenz-2-oxa-1,3-diazol-4-yl)amino)hexanoyl)sphingosyl phosphocholine (NBD C_6-sphingomyelin)	1 mg	125.00	100.00
P-646	3-perylenedodecanoic acid	10 mg	148.00	118.40

Cat #	Product Name	Unit Size	Price Per Unit ($) 1–4 Units	5–24 Units
P-1342	4-((6-phenyl)-1,3,5-hexatrienyl)benzoic acid (DPH carboxylic acid)	25 mg	38.00	30.40
P-459	3-(4-(6-phenyl)-1,3,5-hexatrienyl)phenylpropionic acid (DPH propionic acid)	25 mg	148.00	118.40
P-32	1-pyrenebutanoic acid (pyrenebutyric acid)	1 g	68.00	54.40
P-1903	1-pyrenebutanoic acid *high purity*	100 mg	68.00	54.40
P-164	1-pyrenecarboxylic acid	100 mg	48.00	38.40
P-31	1-pyrenedecanoic acid	25 mg	68.00	54.40
P-3527	N-(1-pyrenedecanoyl)sphingosylphosphocholine (py-C_{10}-sphingomyelin)	1 mg	115.00	92.00
P-96	1-pyrenedodecanoic acid	25 mg	68.00	54.40
P-3779	12-(1-pyrenedodecyl)phosphocholine	1 mg	48.00	38.40
P-243	1-pyrenehexadecanoic acid	5 mg	98.00	78.40
P-3840	1-pyrenehexanoic acid	100 mg	68.00	54.40

13.3 Other Fluorescent and Spin-Labeled Fatty Acids

Anthroyloxy Fatty Acids

Anthroyloxy fatty acids were first introduced by Waggoner and Stryer,[1] and a large series have since been developed. Molecular Probes offers several isomeric 9-anthroyloxy derivatives of octadecanoic (C_{18}) acid, in addition to derivatives of C_{16} and C_{11} saturated fatty acids and an unsaturated octadecenoic acid (A-40). Reviews of the physical and spectral properties of anthroyloxy fatty acids are available.[2,3] As shown by a variety of fluorescence quenching and energy-transfer measurements, the anthroate fluorophore in the fatty acid analogs localizes at a depth in the membrane prescribed by its position along the alkyl chain.[4-7] As with most probes of this type, anthroyloxy fatty acids (with the exception of A-39) appear to preferentially partition into fluid lipid phases.[8]

The fluorescence emission spectrum of the anthroate fluorophore is sensitive to environmental viscosity because of an excited-state conformational relaxation process.[9] The fluorescence Stokes shifts and lifetimes of the anthroyloxy fatty acids are sensitive to small amounts of polar solvents due to hydrogen-bonding interactions.[2] Anthracenepropionic acid (A-176) lacks the ester linkage of the anthroyloxy fatty acids, resulting in a much more structured emission spectrum. Anthroyloxy fatty acids and anthryl fatty acids undergo bimolecular photodimerization, a process that can be used to measure translational diffusion rates.[10,11] Because the product of the photodimerization is nonfluorescent, continuous or high-power excitation of these probes should be avoided for most purposes. Biosynthetic incorporation of anthroyloxy fatty acids by cells is less efficient than for that of pyrene-labeled fatty acids and is markedly dependent on the position of the anthroate fluorophore relative to the carboxyl group.[12] Selected applications of anthroyloxy-FA are listed in Table 13.4.

Table 13.4 Selected applications of anthroyloxy fatty acids (A-36, A-37, A-39, A-40, A-92, A-171, A-172, A-241, A-242, A-326).

Membrane Structure and Dynamics
- Structural organization of membrane lipids[1-3]
- Interaction of membranes with cholesterol,[4,5] diethylstilbestrol,[6] polyunsaturated fatty acids,[7] dexamethasone,[8,9] hemoglobin,[10] DDT,[11] calcium,[12] surfactants[13] and tocopherol and ubiquinones[14,15]
- Transverse fluidity gradients in model lipid bilayers[16] and cell membranes[17-19]
- Spatial location of protein tryptophan residues relative to the transverse axis of the lipid bilayer by fluorescence energy transfer[20,21]

Fatty Acid Transport
- Intermembrane and protein-membrane fatty acid transfer[22-27]
- Characterization of fatty acid–binding sites of proteins[28-31]

Membrane Physiology
- Lipid peroxidation[32-34]
- Membrane permeability[35-38]

1. Biochim Biophys Acta 1194, 99 (1994); **2.** Biophys Chem 49, 153 (1994); **3.** Biophys J 63, 309 (1992); **4.** Biochim Biophys Acta 736, 137 (1983); **5.** Biochim Biophys Acta 693, 246 (1982); **6.** J Biol Chem 267, 11923 (1992); **7.** Chem Phys Lipids 57, 87 (1991); **8.** Biochem J 253, 401 (1988); **9.** Biochem J 248, 455 (1987); **10.** Biochemistry 27, 6425 (1988); **11.** Biochim Biophys Acta 896, 181 (1987); **12.** Biochim Biophys Acta 812, 473 (1985); **13.** Chem Phys Lipids 46, 107 (1988); **14.** Eur J Biochem 171, 661 (1988); **15.** FEBS Lett 179, 238 (1985); **16.** Chem Phys Lipids 71, 61 (1994); **17.** Biochim Biophys Acta 1067, 171 (1991); **18.** J Cell Physiol 144, 42 (1990); **19.** J Lipid Res 31, 261 (1990); **20.** Biochemistry 30, 5491 (1991); **21.** Biochemistry 24, 1883 (1985); **22.** Biochemistry 32, 11074 (1993); **23.** Biochemistry 32, 8622 (1993); **24.** Biochemistry 32, 2053 (1993); **25.** Biochim Biophys Acta 1168, 307 (1993); **26.** J Biol Chem 267, 77 (1992); **27.** J Biol Chem 266, 13473 (1991); **28.** Biochemistry 33, 5791 (1994); **29.** Biochemistry 33, 5783 (1994); **30.** Biochemistry 32, 3797 (1993); **31.** J Biol Chem 268, 7885 (1993); **32.** Am J Physiol 264, G1009 (1993); **33.** Chem Phys Lipids 62, 123 (1992); **34.** Biochim Biophys Acta 861, 311 (1986); **35.** Biochemistry 33, 7056 (1994); **36.** J Biol Chem 269, 17784 (1994); **37.** Biochim Biophys Acta 984, 158 (1989); **38.** Biochim Biophys Acta 979, 177 (1989).

Dansyl Fatty Acids

Phospholipids incorporating dansyl undecanoic acid (DAUDA, D-94) have been employed in conjunction with fluorescence resonance energy transfer techniques to monitor protein-induced domain formation in membranes.[13,14] The dansyl fluorophore in dansyl undecanoic acid undergoes a 60-fold fluorescence enhancement and large spectral shift on binding to fatty acid–binding proteins.[15] This property has been exploited to analyze fatty acid–binding proteins by HPLC[15,16] and also to develop a fluorometric phospholipase A_2 assay (see Section 20.5) based on competitive fatty acid displacement.[17-19]

cis- and *trans*-Parinaric Acids

The naturally occurring polyunsaturated *cis*-parinaric acid (P-1, P-1901) is converted to the *trans* isomer (P-2, P-1902) by facile photoisomerization. Initially developed as membrane probes by Hudson and co-workers,[20,21] the parinaric acids are the closest structural analogs of intrinsic membrane lipids among currently available fluorescent probes. Their chemical and physical properties are well characterized.[22,23] The lowest absorption band of parinaric acid has two main peaks around 300 nm and 320 nm with large extinction coefficients. Parinaric acid offers several experimentally advantageous optical properties, including a very large fluorescence Stokes shift (~100 nm) and an almost complete lack of fluorescence in water. In addition, the fluorescence decay lifetime of parinaric acid varies from 1 to 40 nanoseconds, depending on the molecular packing density in phospholipid bilayers. Consequently, minutely detailed information on lipid-bilayer dynamics can be obtained.[24] The association between lateral density and orientational order of phospholipids can be directly observed in the time-resolved fluorescence anisotropy decay of *trans*-parinaric acid.[25-27]

Both parinaric acid isomers have large lipid/water partition coefficients.[15] In systems with coexisting gel (solid) and liquid-crystalline (fluid) phase lipids, *trans*-parinaric acid (P-2) partitions preferentially into the solid phase with a partition coefficient of about five, whereas *cis*-parinaric acid (P-1) exhibits an almost equal distribution.[15,28] This property contributes to the unique sensitivity of *trans*-parinaric acid in detecting highly ordered lipid domains. Similar partition behavior is observed for parinaroyl phospholipids.[15,29,30] The competitive displacement of parinaric acid from serum albumin has been used as the basis for assays of free fatty acids in serum or plasma [31] and of lipase activity.[6] Selected applications of the parinaric acids are summarized in Table 13.5.

Parinaric acid's extensive unsaturation makes it quite susceptible to oxidation, a property that can be exploited for detection of lipid peroxidation (Table 13.5). Consequently, we offer both parinaric acid isomers specially packaged in ampules sealed under argon (P-1901 and P-1902). Stock solutions of 1 mg/mL (for P-1) or 0.5 mg/mL (for P-2) should be prepared in ethanol containing a small amount of butylated hydroxytoluene (BHT) and stored protected from light under an inert argon atmosphere at -20°C.[32] With rigorous observance of these conditions, the product should be stable in solution for at least six months. Appearance of yellow coloration in the solid material or lack of solubility in ethanol at the levels indicated above is indicative of oxidative degradation. During experiments, we advise handling parinaric acid samples under inert gas and preparing solutions using degassed buffers and solvents. Parinaric acid is also somewhat photolabile, undergoing photodimerization under intense illumination, resulting in loss of fluorescence.[33]

Table 13.5 Selected applications of parinaric acids (P-1, P-2, P-1901, P-1902).

Application
Membrane Structure
• Lipid–protein interactions [1-3]
• Lipid clustering [4-6]
Lipid Transport Processes
• Structural characterization of lipoproteins [7]
• Mechanism of fatty acid–binding proteins [8,9] and phospholipid-transfer proteins [10,11]
Lipid Peroxidation
• Evaluation of antioxidants [12-15]
• Measurement of peroxidation in lipoproteins [16,17]
• Relationship of peroxidation to cytotoxicity [18,19] and apoptosis [20]

1. Biochemistry 32, 12420 (1993); **2.** Biochemistry 30, 6195 (1991); **3.** Biochemistry 29, 6714 (1990); **4.** Biophys J 65, 2248 (1993); **5.** Biochemistry 31, 1816 (1992); **6.** Eur Biophys J 20, 53 (1991); **7.** Chem Phys Lipids 60, 1 (1991); **8.** Biochemistry 33, 3327 (1994); **9.** J Biol Chem 268, 7885 (1993); **10.** Biochemistry 29, 8548 (1990); **11.** Biochemistry 23, 1505 (1984); **12.** Biochemistry 32, 10692 (1993); **13.** J Biol Chem 268, 10906 (1993); **14.** Chem Phys Lipids 53, 309 (1990); **15.** Anal Biochem 196, 443 (1991); **16.** Proc Natl Acad Sci USA 91, 1183 (1994); **17.** Arch Biochem Biophys 297, 147 (1992); **18.** Cell 77, 817 (1994); **19.** Cytometry 13, 686 (1992); **20.** Cell 75, 241 (1993).

Nitroxide Fatty Acids

The nitroxide-labeled fatty acids are important short-range quenchers of fluorescent probes in membrane interiors.[5,25,34,35] The lipid/water partition coefficient of 5-doxylstearic acid (D-522) is significantly larger than that of 12-doxylstearic acid (D-524) and 16-doxylstearic acid (D-526), a fact that must be considered when determining the effective quencher concentration in the membrane.[36] To avoid this complication, these fatty acid quenchers are often incorporated into phospholipid probes.[4,5] Nitroxide fatty acids have been used to assess the location within lipid bilayers of extrinsic fluorescent probes [37-39] and of intrinsically fluorescent antioxidants and anesthetics.[40,41] These probes also continue to be useful for electron spin resonance (ESR) studies of membrane structure and dynamics.[42,43] ESR provides a particularly sensitive index of water penetration into lipid bilayers.[44] The spin-labeled cationic detergent CAT 16 (D-531), which has a nitroxide moiety attached to the terminal ammonium group, is offered for quenching fluorophores located at the membrane surface.[45]

Biotinylated Fatty Acid

The biotinylated fatty acid (B-2602) may be useful for preparing biotinylated phospholipids and sphingolipids; see the Technical Note "Synthetic Methods for Preparing New Phospholipid Probes" on page 303. Its amine-reactive succinimidyl ester (B-2603) provides a means of preparing long nonpolar intermolecular crosslinks.

Fatty Acid–Binding Protein

Fatty acid–binding proteins are small cytosolic proteins found in a variety of mammalian tissues. Studies of their physiological function frequently employ the use of fluorescent fatty acid probes (Tables 13.4, 13.5). To facilitate these studies, Molecular Probes offers rat intestinal fatty acid–binding protein, a low molecular weight (15,000 daltons) protein with high affinity for free fatty acids [46] (I-FABP, I-3881). ADIFAB (A-3880), a conjugate of I-FABP and the polarity-sensitive acrylodan fluorophore (A-433, see Section 2.3), is a dual-wavelength fluorescent indicator of free fatty acids [47,48] (see Figure 20.5 in Chapter 20). It is designed to provide quantitative monitoring of free fatty acids without resorting to separative biochemical methods. With appropriate precautions, which are described in the product information sheet accompanying this product, ADIFAB can be used to determine free fatty acid concentrations between 1 nM and >20 µM.

1. Proc Natl Acad Sci USA 67, 579 (1970); **2.** Chem Phys Lipids 70, 155 (1994); **3.** Lentz, B.R. in *Spectroscopic Membrane Probes, Volume 1*, L.M. Loew, Ed., CRC Press (1988) pp. 13–41; **4.** Biochemistry 31, 5322 (1992); **5.** Biochemistry 31, 5312 (1992); **6.** Cytometry 12, 247 (1991); **7.** Biochim Biophys Acta 731, 465 (1983); **8.** Biochim Biophys Acta 939, 124 (1988); **9.** J Phys Chem 80, 533 (1976); **10.** Eur J Biochem 171, 669 (1988); **11.** Eur J Biochem 143, 373 (1984); **12.** Chem Phys Lipids 58, 111 (1991); **13.** Biochemistry 33, 4483 (1994); **14.** Biochemistry 27, 2033 (1988); **15.** Biochem J 238, 419 (1986); **16.** Analyst 117, 1859 (1992); **17.** Anal Biochem 212, 65 (1993); **18.** Biochem J 278, 843 (1991); **19.** Biochem J 266, 435 (1990); **20.** Biochemistry 16, 819 (1977); **21.** Biochemistry 16, 813 (1977); **22.** Chem Phys Lipids 64, 99 (1993); **23.** Hudson, B. and Cavalier, S.A. in *Spectroscopic Membrane Probes, Volume 1*, L.M. Loew, Ed., CRC Press (1988) pp. 43–62; **24.** Biophys J 55, 1111 (1989); **25.** Biochemistry 29, 4380 (1990); **26.** Biochemistry 28, 5051 (1989); **27.** Biophys Chem 28, 59 (1987); **28.** Biochemistry 18, 1707 (1979); **29.** Biochim Biophys Acta 1023, 383 (1990); **30.** Biochemistry 21, 5690 (1982); **31.** Clin Chem 26, 1173 (1980); **32.** Biochemistry 16, 820 (1977); **33.** Proc Natl Acad Sci USA 77, 26 (1980); **34.** Biochemistry 33, 5791 (1994); **35.** Biophys J 50, 349 (1986); **36.** Photochem Photobiol 39, 477 (1984); **37.** Chem Phys Lipids 69, 75 (1994); **38.** Biochemistry 32, 9586 (1993); **39.** Biophys J 65, 2348 (1993); **40.** Biochim Biophys Acta 1189, 243 (1994); **41.** Chem Phys Lipids 62, 123 (1992); **42.** Biochemistry 33, 4947 (1994); **43.** Proc Natl Acad Sci USA 90, 5181 (1993); **44.** Biochemistry 33, 7670 (1994); **45.** Biophys J 65, 2493 (1993); **46.** J Biol Chem 267, 23534 (1992); **47.** J Biol Chem 270, 15076 (1995); **48.** J Biol Chem 267, 23495 (1992).

13.3 Data Table *Other Fluorescent and Spin-Labeled Fatty Acids*

Cat #	Structure	MW	Storage	Soluble	Abs	{ε × 10⁻³}	Em	Solvent	Notes
A-36	✓	505	F,L	DMSO	361	{8.0}	469	MeOH	<1>
A-37	✓	505	F,L	DMSO	361	{8.0}	467	MeOH	<1>
A-39	✓	477	F,L	DMSO	361	{8.2}	473	MeOH	<1>
A-40	✓	503	F,L,AA	DMSO	362	{6.6}	470	MeOH	<1, 2>
A-92	✓	407	F,L	DMSO	359	{7.3}	470	MeOH	<1>
A-171	✓	505	F,L	DMSO	360	{7.9}	474	MeOH	<1>
A-172	✓	505	F,L	DMSO	361	{8.5}	471	MeOH	<1>
A-176	✓	250	L	DMSO	366	{8.9}	414	MeOH	
A-241	✓	505	F,L	DMSO	362	{8.4}	467	MeOH	<1>
A-242	✓	505	F,L	DMSO	362	{8.4}	470	MeOH	<1>
A-326	✓	505	F,L	DMSO	362	{7.5}	471	MeOH	<1>
A-3880	C	~15,350	FF,L,AA	H_2O	365	{10.5}	432	H_2O	<3>
B-2602	✓	442	NC	DMSO	<300		none		
B-2603	✓	539	F,D	DMSO	<300		none		
D-94	✓	435	F,L	DMSO, EtOH	335	{4.8}	519	MeOH	
D-522	✓	385	F	DMSO, EtOH	<300		none		
D-524	✓	385	F	DMSO, EtOH	<300		none		
D-526	✓	385	F	DMSO, EtOH	<300		none		
D-531	✓	552	F	DMSO, EtOH	<300		none		
I-3881	C	~15,125	FF,AA	H_2O	280	{17}	none	pH 7	
P-1	✓	276	FF,LL,AA	EtOH	303	{76}	416	MeOH	<4>
P-2	✓	276	FF,LL,AA	EtOH	298	{85}	413	MeOH	<4>

P-1901 see P-1; P-1902 see P-2

For definitions of the contents of this data table, see "How to Use The Handbook" on page vi.
Structure: Chemical structure drawing: ✓ = shown in this section; C = not shown due to complexity.
MW: ~ indicates an approximate value.

<1> Anthroyloxy derivatives have structured excitation spectra with peaks at about 380, 360, 345 and 328 nm, with the 360 nm peak the strongest. Emission is unstructured at 445 nm in phospholipid vesicles. The fluorescence lifetime for all anthroyloxy derivatives is about 11 nsec.
<2> This product is intrinsically a liquid or an oil at room temperature.
<3> ADIFAB fatty acid indicator (A-3880) is a protein conjugate with a molecular weight of approximately 15,350 daltons. Binding of fatty acids in the micromolar range shifts the emission peak from about 432 nm to 505 nm (see Figure 20.5 in Chapter 20).
<4> Parinaric acids are very air sensitive. Use under N_2 or Ar. Butylated hydroxytoluene (BHT) may be added (1 mg/100 mL) to stock solutions to inhibit oxidative degradation. P-1 and P-2 are almost nonfluorescent in water. The fluorescence quantum yield of P-2 incorporated in DPPC model membranes is approximately 0.3 [Biochemistry 16, 819 (1977)].

13.3 Structures *Other Fluorescent and Spin-Labeled Fatty Acids*

A-37 n = 0 m = 15
A-241 n = 1 m = 14
A-242 n = 4 m = 11
A-171 n = 5 m = 10
A-172 n = 7 m = 8
A-326 n = 8 m = 7
A-36 n = 10 m = 5

A-40

A-92 n = 10
A-39 n = 15

A-176

B-2602 R = –COOH
B-2603 R = SE

D-94

D-522 n = 3 m = 12
D-524 n = 10 m = 5
D-526 n = 14 m = 1

D-531

P-1 / P-1901

P-2 / P-1902

13.3 Price List *Other Fluorescent and Spin-Labeled Fatty Acids*

Cat #	Product Name	Unit Size	Price Per Unit ($) 1–4 Units	5–24 Units
A-3880	ADIFAB fatty acid indicator	200 µg	245.00	196.00
A-176	9-anthracenepropionic acid	100 mg	38.00	30.40
A-40	12-(9-anthroyloxy)oleic acid (12-AO)	25 mg	95.00	76.00
A-39	16-(9-anthroyloxy)palmitic acid (16-AP)	100 mg	95.00	76.00
A-37	2-(9-anthroyloxy)stearic acid (2-AS)	100 mg	95.00	76.00
A-241	3-(9-anthroyloxy)stearic acid (3-AS)	25 mg	95.00	76.00
A-242	6-(9-anthroyloxy)stearic acid (6-AS)	25 mg	95.00	76.00
A-171	7-(9-anthroyloxy)stearic acid (7-AS)	25 mg	95.00	76.00
A-172	9-(9-anthroyloxy)stearic acid (9-AS)	25 mg	95.00	76.00
A-326	10-(9-anthroyloxy)stearic acid (10-AS)	25 mg	95.00	76.00
A-36	12-(9-anthroyloxy)stearic acid (12-AS)	100 mg	95.00	76.00
A-92	11-(9-anthroyloxy)undecanoic acid (11-AU)	100 mg	95.00	76.00
B-2602	12-((biotinoyl)amino)dodecanoic acid	25 mg	53.00	42.40
B-2603	12-((biotinoyl)amino)dodecanoic acid, succinimidyl ester	25 mg	48.00	38.40
D-94	11-((5-dimethylaminonaphthalene-1-sulfonyl)amino)undecanoic acid (DAUDA)	100 mg	68.00	54.40
D-531	4-(N,N-dimethyl-N-hexadecyl)ammonium-2,2,6,6-tetramethylpiperidine-1-oxyl iodide (CAT 16)	25 mg	48.00	38.40
D-526	16-doxylstearic acid	25 mg	108.00	86.40
D-522	5-doxylstearic acid	25 mg	60.00	48.00
D-524	12-doxylstearic acid	25 mg	60.00	48.00
I-3881	I-FABP fatty acid binding protein	1 mg	330.00	264.00
P-1	*cis*-parinaric acid	100 mg	155.00	124.00
P-1901	*cis*-parinaric acid *special packaging*	10x10 mg	185.00	148.00
P-2	*trans*-parinaric acid	100 mg	155.00	124.00
P-1902	*trans*-parinaric acid *special packaging*	10x10 mg	185.00	148.00

13.4 Phospholipids with Labeled Head Groups

Most of Molecular Probes' phospholipid analogs with fluorescently labeled head groups have polar fluorophores such as fluorescein, rhodamine and Texas Red attached to the terminal amino group of diacylglycerophosphoethanolamine (Table 13.6). They are generally less readily exchangeable between membranes than acyl chain–labeled analogs.[1,2] Furthermore, our extensive selection of anti-fluorophore antibodies — including antibodies directed against dansyl, fluorescein, BODIPY FL, tetramethylrhodamine and Texas Red fluorophores — provide useful tools for detecting membrane surface labels; see the Technical Note "Antibodies for Detecting Membrane Surface Labels" on page 302.

Table 13.6 Phospholipids with labeled head groups.

Head Group (Ex/Em or Application) *	Catalog Number
Dansyl (336/517)	D-57, D-3797
Pyrenesulfonyl (350/379)	P-58
NBD (463/536)	N-360
Fluorescein (496/519)	F-362
Caged fluorescein (492/517) †	D-3811
BODIPY FL (505/511)	D-3800
BODIPY 530/550 (535/552)	D-3799
Tetramethylrhodamine (540/566)	T-1391
Lissamine rhodamine (560/581)	L-1392
Texas Red (582/601)	T-1395, T-7710
Biotin (<250/none)	B-1550, B-1616
Dinitrophenyl (350/none)	D-3798
Maleimide (thiol-reactive)	M-1618
Pyridyldithio (thiol-reactive)	P-1619

* Spectral maxima (in nm) are in methanol. The spectra may be different in membranes. † Following ultraviolet photolysis.

Phospholipids with Dansyl-Labeled Head Groups

Two of our phospholipid analogs incorporate the solvent polarity–sensitive dansyl fluorophore[3] attached to the amino terminus of DHPE (D-57) or to natural diacylglycerophosphoethanolamine from egg yolk (D-3797). These are useful probes for investigating the association of protein kinase C[4-6] and other proteins[7-9] with membrane surfaces. Dansyl DHPE has been used to examine the effects of cholesterol on the accessibility of the dansyl hapten to antibodies.[10] Bilirubin (B-8446, see Section 14.6) quenches dansyl DHPE fluorescence by resonance energy transfer, a property that has been exploited to monitor its spontaneous transfer between lipid vesicles.[11]

Phospholipids with Pyrene Sulfonamide–Labeled Head Groups

Like other pyrene fluorophores, pyrenesulfonyl glycerophosphoethanolamine (pyS DHPE, P-58) forms excited-state dimers in a concentration-dependent manner, thus providing a probe suitable for investigating membrane surface phenomena.[12-15] PyS DHPE has been used in conjunction with the lipophilic nitroxide CAT 16 (D-531, see Section 13.3) and alkylamine quenchers to determine lateral diffusion coefficients in surface phospholipid monolayers.[16,17] PyS DHPE can also serve as a substrate for phospholipase C and D assays (see Section 20.5) and as a detection reagent for HPLC analysis of phospholipids.[18] Pyrenesulfonic acid (P-80, see Section 14.6), which in activated form is the precursor of pyS DHPE, has primarily been employed to probe inverted micelles.[19,20]

Phospholipids with NBD-Labeled Head Groups

The widely used membrane probe nitrobenzoxadiazole glycerophosphoethanolamine[21] (NBD-PE, N-360) has three important optical properties: photolability, which makes it suitable for photobleaching recovery measurements; concentration-dependent self-quenching; and fluorescence resonance energy transfer to rhodamine acceptors (usually rhodamine DHPE, L-1392). Spectroscopic characteristics of NBD-PE are generally similar to those described in Table 13.3 for our phospholipids with NBD-labeled acyl chains.

NBD-PE is frequently used in NBD–rhodamine fluorescence energy transfer experiments to monitor membrane fusion as described in the Technical Note "Lipid Mixing Assays of Membrane Fusion" on page 286. In addition, this methodology is applicable to monitoring intermembrane lipid transfer[22-25] and determining the transbilayer distribution of phospholipids,[26] processes that can also be followed by making use of NBD-PE fluorescence self-quenching.[27] Attachment of the NBD fluorophore to the head group makes NBD-PE resistant to transfer between vesicles.[2] NBD-PE has been used in combination with either rhodamine DHPE or Texas Red DHPE (L-1392, T-1395) for visualizing spatial relationships of lipid populations by resonance energy transfer microscopy.[28] The nitro group of NBD can be reduced with dithionite, irreversibly eliminating the dye's fluorescence. This technique can be employed to determine whether the probe is localized on the outer or inner leaflet of the cell membrane.[29-31] The argon-ion laser–excitable NBD-PE is also a frequent choice for fluorescence recovery after photobleaching (FRAP) measurements of lateral diffusion in membranes.[32-35] In addition, NBD-PE has proven to be of particular value for monitoring bilayer-to-hexagonal phase transitions because these transitions cause an increase in NBD-PE's fluorescence intensity.[36-38]

Phospholipids with Fluorescein-Labeled Head Groups

Fluorescein-derivatized dihexadecanoylglycerophosphoethanolamine (fluorescein DHPE, F-362) has been reported to be a useful cell-surface pH indicator (see Section 23.4). In a related application, fluorescein DHPE has been used to measure lateral proton conduction in large monolayers at lipid–water interfaces.[39-41] Anti-fluorescein antibody (A-889, see Section 7.3) has been employed in combination with fluorescein DHPE to investigate specific recognition interactions at membrane surfaces.[42,43]

Because of fluorescein's photolability, fluorescein DHPE is a useful reagent for measuring lateral diffusion in membranes using fluorescence photobleaching recovery methods.[44,45] Caged fluorescein DHPE (D-3811) can potentially be used for diffusion measurements using the complementary photoactivation of fluorescence (PAF) technique, in which the fluorophore is activated upon illumination rather than bleached; see Chapter 19 for a discussion of the properties and applications of caged probes. Another new technique, single-particle tracking (SPT), provides direct measurements of diffusion rates by calculating trajectories of fluorescent polystyrene beads or colloidal gold particles from time-sequential images[46,47] (see Figure 15.8 in Chapter 15). Molecular Probes' FluoSpheres® fluorescent microspheres (see Section 6.2) were labeled with streptavidin and then coupled to fluorescein DHPE using a biotinylated conjugate of anti-fluorescein monoclonal 4-4-20[46] (A-6421, see Section 7.3). Diffusion rates measured with this bridged conjugate in glass-supported phospholipid bilayers were the same as those determined with streptavidin beads coupled directly to biotin-X DHPE (B-1616). Fluorescein DHPE has also been used in conjunction with polyclonal anti-fluorescein antibody (A-889, see Section 7.3) to prepare colloidal gold probes for SPT diffusion measurements in supported phospholipid bilayers[47] and in keratocyte plasma membranes.[48]

Phospholipids with BODIPY Dye–Labeled Head Groups

Our phospholipids with BODIPY dyes attached to the head group (BODIPY FL DHPE, D-3800; BODIPY 530/550 DHPE, D-3799) have significant potential for studies of molecular recognition interactions at membrane surfaces. Spectral properties of these BODIPY probes are generally the same as those described in Table 13.2 for our phospholipids with BODIPY dye–labeled acyl chains.

Phospholipids with Rhodamine-Labeled Head Groups

The rhodamine glycerophosphoethanolamines TRITC DHPE and rhodamine DHPE (often referred to as *N*-Rh-PE) (T-1391, L-1392) do not readily transfer between separated lipid bilayers.[49,50] This property has led to the extensive use of rhodamine DHPE for membrane fusion assays based on fluorescence resonance energy transfer from NBD-PE (see the Technical Note "Lipid Mixing Assays of Membrane Fusion" on page 286). In addition, these probes are good resonance energy transfer acceptors from fluorescent lipid analogs such as the BODIPY and NBD phospholipids[51] and from protein labels such as IAF and IAEDANS[52,53] (I-3, I-14; see Sections 2.2 and 2.3). Our rhodamine-labeled phospholipids have also been used as tracers for membrane traffic during endocytosis[54,55] and for lipid processing in hepatocytes.[56]

Phospholipids with Texas Red Dye–Labeled Head Groups

Texas Red conjugates of phosphoethanolamine — Texas Red DHPE (T-1395, sometimes referred to using the generic sulforhodamine nomenclature *N*-SRh-PE) and our new Texas Red-X DHPE with an aminohexanoyl spacer ("X") (T-7710) — are principally employed as energy transfer acceptors from NBD, BODIPY and fluorescein lipid probes. The longer emission wavelength of the Texas Red dye provides superior separation of the donor and acceptor emission signals in resonance energy transfer microscopy.[29,57] This technique has enabled visualization of ATP-dependent fusion of liposomes with the Golgi apparatus.[58] Membrane flux during hemagglutinin-mediated cell–cell fusion has been visualized using Texas Red DHPE and the lipophilic carbocyanine DiI (D-282, see Section 15.4) as membrane labels.[59]

Antibodies for Detecting Membrane Surface Labels

Molecular Probes offers antibodies that recognize the dansyl (A-6398), fluorescein (A-889), BODIPY® FL (A-5770), BODIPY TR (A-6414), tetramethylrhodamine (A-6397) and Texas Red® (A-6399) fluorophores and the dinitrophenyl chromophore; see Section 7.3 for complete product information. These antibodies can be used for direct detection of labeled phospholipids and proteins via fluorescence quenching (or fluorescence enhancement in the case of the anti-dansyl antibody), as well as for ultrastructural localization using immunocytochemical techniques.[1] When used in conjunction with phospholipids with dye-labeled head groups (see Table 13.6), they are also important tools for:

- Studies of molecular recognition mechanisms at membrane surfaces[2]
- Lipid diffusion measurements[3,4]
- Quantitation of lipid internalization by endocytosis[5,6]

Because of the improved binding accessibility provided by their aminohexanoyl spacers ("X"), our Texas Red-X (T-7710) and DNP-X (D-3798) phospholipids may be particularly useful in these applications.[7,8] In addition to anti-fluorophore antibodies, we offer a selection of streptavidin (see Section 7.5) conjugates for detecting biotinylated phospholipids.

1. Cell Motility 4, 137 (1984); **2.** Angew Chem Int Ed Engl 29, 1269 (1990); **3.** J Cell Biol 120, 25 (1993); **4.** Proc Natl Acad Sci USA 88, 6274 (1991); **5.** Cell 64, 393 (1991); **6.** J Cell Biol 106, 1083 (1988); **7.** Biochim Biophys Acta 1104, 9 (1992); **8.** Biochim Biophys Acta 776, 217 (1984).

Synthetic Methods for Preparing New Phospholipid Probes

In the event that Molecular Probes' extensive selection of phospholipids does not include the particular analog of interest, several synthetic methods are available for preparing new probes.

Labeling Phospholipid Acyl Chains

Anhydrides of fluorescent or otherwise-labeled fatty acids can be used to directly acylate monoacylglycerophospholipids (lysophospholipids) in the presence of a catalyst such as 4-dimethylaminopyridine or 4-pyrrolidinopyridine.[1,2]

Labeling Head Groups

Molar equivalents of an amine-reactive succinimidyl ester, isothiocyanate or sulfonyl chloride (see Chapter 1) and diacylglycerophosphoethanolamine (e.g., D-7705) are combined in chloroform with addition of two equivalents of triethylamine. After the reaction is complete, as assessed by thin-layer chromatography, the product is purified by column chromatography on silica gel using methanol/chloroform mixtures for elution.

Converting Phosphocholine to Other Head Groups

Phosphocholine can be converted to another head group by enzymatic transphosphatidylation using phospholipase D and an appropriate alcohol (e.g., serine to convert phosphocholine to phosphoserine). This reaction can be carried out on a small scale using octyl glucoside (O-8457, see Section 16.3) to disperse the labeled diacylglycerophosphocholine substrate.[3]

1. Anal Biochem 113, 96 (1981); **2.** J Lipid Res 20, 679 (1979); **3.** J Lipid Res 31, 1719 (1990).

Phospholipids with Reactive and Hapten-Labeled Head Groups

Molecular Probes offers both reactive phospholipids and phospholipids labeled with either biotin or the dinitrophenyl hapten to facilitate binding of labeled membranes to other biomolecules. The reactive phospholipids (M-1618, P-1619) provide a means of coupling antibodies to liposomes for targeted delivery to cells [60-63] and for immunoassays and immunofluorescence labeling.[64,65] The pyridyldisulfide glycerophosphoethanolamine (PDP DHPE, P-1619) can react directly with thiol-containing biomolecules to form disulfide-linked conjugates. PDP DHPE can also be reduced to a free thiol with dithiothreitol or TCEP (D-1532, T-2556, see Section 2.1) and then conjugated to thiol-reactive dyes, phycobiliproteins [66-68] (see Section 6.4) or a biomolecule modified with one of the other crosslinking reagents in Chapter 5. The maleimide-containing phospholipid (MMCC DHPE, M-1618) forms stable crosslinks to thiolated biomolecules.[69,70]

The biotinylated glycerophosphoethanolamine derivatives (biotin DHPE, B-1550; and the longer-chain biotin-X DHPE, B-1616) can be used to couple avidin or streptavidin (see Section 7.5) to cell membranes, liposomes and lipid monolayers.[71-74] Avidin can then be employed as a bridge for antibody coupling [75] or for assembling liposomes into multiplex structures.[76] Researchers have fabricated a glucose sensor by coupling streptavidin-conjugated glucose oxidase to membranes containing biotin DHPE.[77] Interactions of biotinylated lipids with streptavidin provide a model for molecular recognition processes at membrane surfaces.[43,78,79] The phase structure of lipid assemblies incorporating biotinylated phospholipids has been studied by X-ray diffraction, ^{31}P NMR and differential scanning calorimetry.[80,81]

The 2,4-dinitrophenyl-labeled phospholipid (DNP-X DHPE, D-3798) can be used to investigate cellular and molecular recognition, including surface binding of antibodies,[82,83] molecular mechanisms of cell adhesion [84] and receptor-mediated endocytosis.[85]

Unlabeled Phospholipids

In conjunction with our extensive range of fluorescent lipid probes, Molecular Probes offers high-purity unlabeled dihexadecanoylglycerophosphocholine (DPPC, D-7704) and dihexadecanoylglycerophosphoethanolamine (DPPE, D-7705) for preparing model systems such as liposomes.[86-88] DPPE can also be used to prepare novel phospholipid probes by coupling it with amine-reactive dyes; see the Technical Note "Synthetic Methods for Preparing New Phospholipid Probes" above.

1. Kok, J.W. and Hoekstra, D. in *Fluorescent and Luminescent Probes for Biological Activity*, W.T. Mason, Ed., Academic Press, (1993) pp. 100–119; **2.** J Biol Chem 255, 5404 (1980); **3.** Biochim Biophys Acta 815, 351 (1985); **4.** Biochemistry 32, 66 (1993); **5.** Biochemistry 31, 4667 (1992); **6.** Biochemistry 31, 1125 (1992); **7.** Biochemistry 33, 13231 (1994); **8.** Biochemistry 33, 8930 (1994); **9.** Biochemistry 31, 4907 (1992); **10.** Biochim Biophys Acta 1104, 9 (1992); **11.** Biochemistry 31, 3184 (1992); **12.** J Biol Chem 267, 15184 (1992); **13.** Biophys J 57, 335 (1990); **14.** Biophys Chem 34, 163 (1989); **15.** J Phys Chem 90, 3167 (1986); **16.** Biophys J 65, 2493 (1993); **17.** J Am Chem Soc 113, 4838 (1991); **18.** J Chromatography 639, 175 (1993); **19.** J Am Chem Soc 113, 8204 (1991); **20.** Langmuir 3, 494 (1987); **21.** Chem Phys Lipids 53, 1 (1990); **22.** J Biol Chem 269, 10517 (1994); **23.** Biochemistry 29, 879 (1990); **24.** Biochim Biophys Acta 981, 178 (1989); **25.** Biochemistry 27, 3925 (1988); **26.** Biochemistry 31, 2865 (1992); **27.** Biochemistry 32, 12678 (1993); **28.** Meth Enzymol 171, 850 (1989); **29.** Biochemistry 33, 9968 (1994); **30.** Chem Phys Lipids 70, 205 (1994); **31.** Biochemistry 32, 14194 (1993); **32.** Prog Lipid Res 33, 203 (1994); **33.** J Cell Biol 122, 1253 (1993); **34.** J Cell Biol 115, 1585 (1991); **35.** J Cell Biol 115, 245 (1991); **36.** Biophys J 63, 309 (1992); **37.** Biochemistry 29, 2976 (1990); **38.** Biochemistry 27, 3947 (1988); **39.** J Am Chem Soc 113, 8818 (1991); **40.** Biochemistry 29, 59 (1990); **41.** Nature 322, 756 (1986); **42.** Biophys J 63, 823 (1992); **43.** Angew Chem Int Ed Engl 29, 1269 (1990); **44.** Biophys J 66, 25 (1994); **45.** J Cell Biol 103, 807 (1986); **46.** J Membrane Biol 135, 83 (1993); **47.** Proc Natl Acad Sci USA 88, 6274 (1991); **48.** J Cell Biol 120, 25 (1993); **49.** Biochemistry 24, 6390 (1985); **50.** Biochemistry 21, 1720 (1982); **51.** Biochemistry 20, 4093 (1981); **52.** J Biol Chem 266, 12082

(1991); **53.** Biochemistry 29, 1607 (1990); **54.** Biochim Biophys Acta 1103, 185 (1992); **55.** Eur J Cell Biol 53, 173 (1990); **56.** Biochem J 284, 259 (1992); **57.** J Cell Biol 103, 1221 (1986); **58.** Cell 55, 797 (1988); **59.** J Cell Biol 121, 543 (1993); **60.** J Immunol Meth 132, 25 (1990); **61.** Proc Natl Acad Sci USA 87, 2448 (1990); **62.** Cancer Res 48, 5237 (1988); **63.** Proc Natl Acad Sci USA 84, 246 (1987); **64.** J Immunol Meth 134, 207 (1990); **65.** J Immunol Meth 100, 59 (1987); **66.** J Fluorescence 3, 33 (1993); **67.** Cytometry 8, 562 (1987); **68.** Meth Enzymol 149, 111 (1987); **69.** Annals NY Acad Sci 446, 443 (1985); **70.** Immunology 50, 101 (1983); **71.** J Immunol Meth 158, 183 (1993); **72.** Anal Biochem 207, 341 (1992); **73.** Biophys J 59, 387 (1991); **74.** Biochim Biophys Acta 1028, 73 (1990); **75.** Biochim Biophys Acta 901, 157 (1987); **76.** Science 264, 1753 (1994); **77.** Anal Chem 65, 665 (1993); **78.** Anal Biochem 217, 128 (1994); **79.** Biophys J 65, 2160 (1993); **80.** Biophys J 66, 31 (1994); **81.** Biochemistry 32, 9960 (1993); **82.** Biophys J 63, 215 (1992); **83.** Biophys J 58, 1235 (1990); **84.** J Cell Biol 102, 2185 (1986); **85.** Proc Natl Acad Sci USA 77, 4089 (1980); **86.** Lasic, D.D., *Liposomes: From Physics to Applications*, Elsevier Science Publishers (1993); **87.** *Liposomes: A Practical Approach*, R.R.C. New, Ed., IRL Press (1990); **88.** Ann Rev Biophys Bioeng 9, 467 (1980).

13.4 Data Table *Phospholipids with Labeled Head Groups*

Cat #	Structure	MW	Storage	Soluble	Abs	{ε × 10⁻³}	Em	Solvent	Notes
B-1550	✓	1019	FF,D	<1>	<300		none		<1>
B-1616	✓	1133	FF,D	<1>	<300		none		<1>
D-57	✓	1026	FF,D,L	<1>	336	{4.5}	517	MeOH	<1>
D-3797	✓	~1080	FF,D	<1>	334	{4.5}	520	MeOH	<1>
D-3798	✓	979	FF,D,L	<1>	350	{17}	none	MeOH	<1>
D-3799	✓	1191	FF,D,L	<1>	535	{85}	552	MeOH	<1, 2>
D-3800	✓	1067	FF,D,L	<1>	505	{87}	511	MeOH	<1, 2>
D-3811	✓	1542	FF,D,LL	<1>	340	{13}	none	MeOH	<1, 3, 4>
D-7704	✓	734	FF,D	<1>	<300		none		<1>
D-7705	✓	692	FF,D	<1>	<300		none		<1>
F-362	✓	1183	FF,D,L	<1>	496	{88}	519	MeOH	<1, 5>
L-1392	✓	1334	FF,D,L	<1>	560	{95}	581	MeOH	<1>
M-1618	✓	1012	FF,D	<1>	<300		none		<1>
N-360	✓	956	FF,D,L	<1>	463	{21}	536	MeOH	<1, 6>
P-58	✓	1057	FF,D,L	<1>	350	{35}	379	MeOH	<1>
P-1619	✓	990	FF,D	<1>	281	{4.9}	none	MeOH	<1>
T-1391	✓	1237	FF,D,L	<1>	540	{93}	566	MeOH	<1>
T-1395	✓	1382	FF,D,L	<1>	582	{90}	601	MeOH	<1>
T-7710	✓	1495	FF,D,L	<1>	583	{102}	602	MeOH	<1>

For definitions of the contents of this data table, see "How to Use The Handbook" on page vi.
Structure: Chemical structure drawing: ✓ = shown in this section.
MW: ~ indicates an approximate value.

<1> Chloroform is the most generally useful solvent for preparing stock solutions of phospholipids (including sphingomyelins). Glycerophosphocholines are usually freely soluble in ethanol. Most other glycerophospholipids (phosphoethanolamines, phosphatidic acids and phosphoglycerols) are less soluble in ethanol, but solutions up to 1–2 mg/mL should be obtainable, using sonication to aid dispersion if necessary. Labeling of aqueous samples with nonmiscible phospholipid stock solutions can be accomplished by evaporating the organic solvent, followed by hydration and sonication, yielding a suspension of liposomes. Information on the solubility of natural phospholipids can be found in Marsh, D., *CRC Handbook of Lipid Bilayers*, CRC Press (1990) pp. 71–80.

<2> The absorption and fluorescence spectra of BODIPY derivatives are relatively insensitive to the solvent.

<3> All caged products are sensitive to light. They should be protected from illumination, except when photolysis is intended.

<4> This product is essentially colorless and nonfluorescent until it is converted to a derivative of fluorescein (Abs/Em = ~492/517 nm) by ultraviolet photolysis of the caging groups.

<5> Spectra of these compounds are in methanol containing a trace of KOH.

<6> NBD derivatives are almost nonfluorescent in water. The quantum yields of NBD amine derivatives increase on binding to membranes.

13.4 Structures *Phospholipids with Labeled Head Groups*

$(CH_3CH_2)_3\overset{+}{N}H$

$CH_3(CH_2)_{14}-\overset{O}{\overset{\|}{C}}-OCH_2$
$CH_3(CH_2)_{14}-\underset{O}{\underset{\|}{C}}-OCH$
$CH_2O-\overset{O}{\overset{\|}{P}}(O^-)-OCH_2CH_2-X-\overset{}{C}=O$ (biotin $(CH_2)_4$)

B-1550 X = –NH–

B-1616 X = –NH–C(=O)–$(CH_2)_5$NH–

$(CH_3CH_2)_3\overset{+}{N}H$

$CH_3(CH_2)_{14}-C(=O)-OCH_2$
$CH_3(CH_2)_{14}-C(=O)-OCH$
$CH_2O-P(=O)(O^-)-OCH_2CH_2-NH-SO_2$–naphthyl–$N(CH_3)_2$

D-57

$(CH_3CH_2)_3\overset{+}{N}H$

R–OCH_2
R–OCH
$CH_2O-P(=O)(O^-)-OCH_2CH_2NH-SO_2$–naphthyl–$N(CH_3)_2$

R = Mixed fatty acids

D-3797

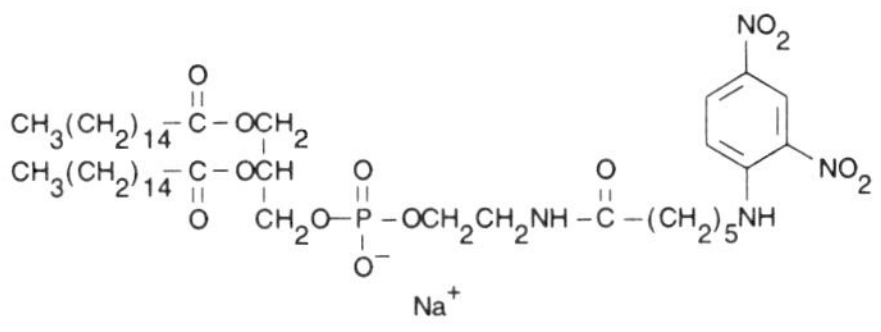

NO_2

$CH_3(CH_2)_{14}-C(=O)-OCH_2$

$CH_3(CH_2)_{14}-C(=O)-OCH$

$CH_2O-P(=O)(O^-)-OCH_2CH_2NH-C(=O)-(CH_2)_5NH$ NO_2

Na^+

D-3798

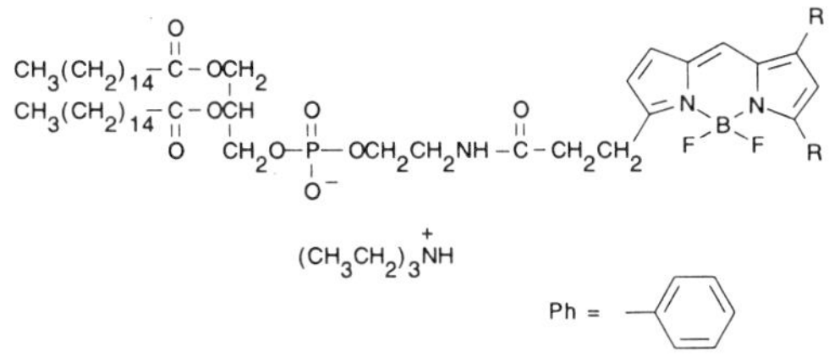

R

$CH_3(CH_2)_{14}-C(=O)-OCH_2$

$CH_3(CH_2)_{14}-C(=O)-OCH$

$CH_2O-P(=O)(O^-)-OCH_2CH_2NH-C(=O)-CH_2CH_2$ N B N F F R

$(CH_3CH_2)_3\overset{+}{N}H$

Ph = phenyl

D-3799 R = Ph

D-3800 R = $-CH_3$

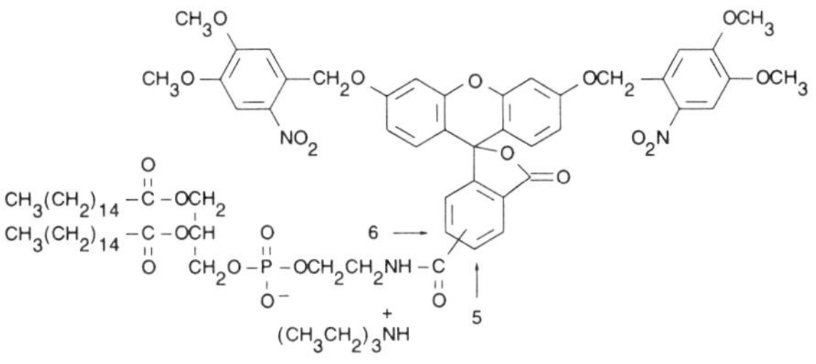

CH_3O OCH_3

CH_3O $-CH_2O$ O OCH_2- $-OCH_3$

NO_2 O_2N

O O

$CH_3(CH_2)_{14}-C(=O)-OCH_2$

$CH_3(CH_2)_{14}-C(=O)-OCH$ 6 ⟶

$CH_2O-P(=O)(O^-)-OCH_2CH_2NH-C(=O)$ 5

$(CH_3CH_2)_3\overset{+}{N}H$

D-3811

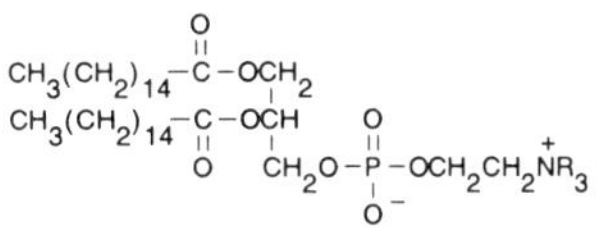

$CH_3(CH_2)_{14}-C(=O)-OCH_2$

$CH_3(CH_2)_{14}-C(=O)-OCH$

$CH_2O-P(=O)(O^-)-OCH_2CH_2\overset{+}{N}R_3$

D-7704 R = $-CH_3$

D-7705 R = H

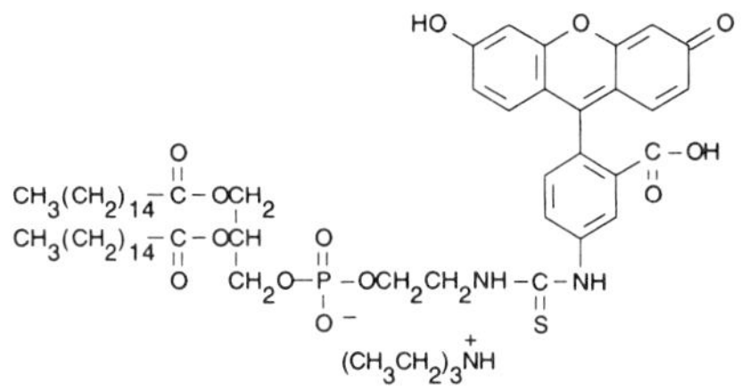

HO O O

C–OH, O

$CH_3(CH_2)_{14}-C(=O)-OCH_2$

$CH_3(CH_2)_{14}-C(=O)-OCH$

$CH_2O-P(=O)(O^-)-OCH_2CH_2NH-C(=S)-NH$

$(CH_3CH_2)_3\overset{+}{N}H$

F-362

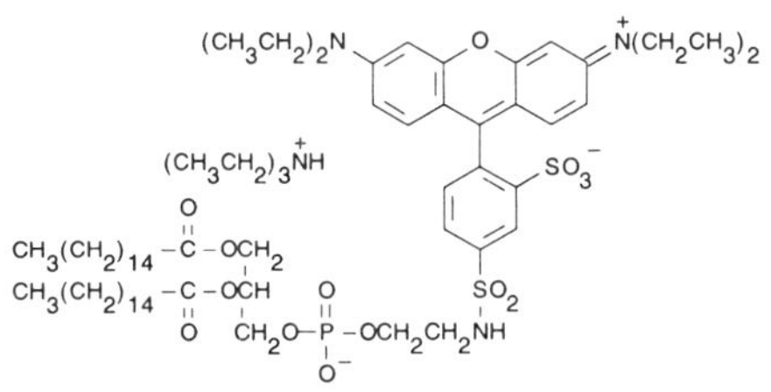

$(CH_3CH_2)_2N$ O $\overset{+}{N}(CH_2CH_3)_2$

$(CH_3CH_2)_3\overset{+}{N}H$ SO_3^-

$CH_3(CH_2)_{14}-C(=O)-OCH_2$

$CH_3(CH_2)_{14}-C(=O)-OCH$ SO_2

$CH_2O-P(=O)(O^-)-OCH_2CH_2NH$

L-1392

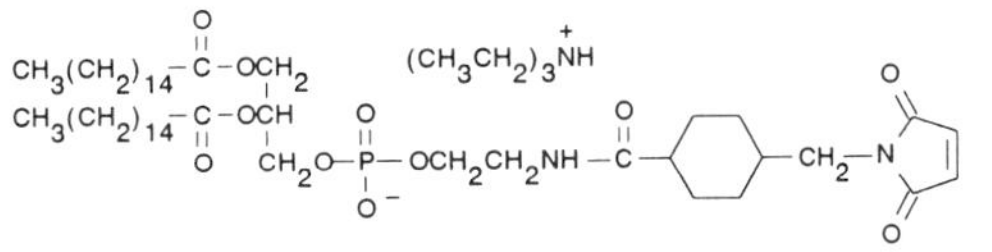

$CH_3(CH_2)_{14}-C(=O)-OCH_2$ $(CH_3CH_2)_3\overset{+}{N}H$ O

$CH_3(CH_2)_{14}-C(=O)-OCH$

$CH_2O-P(=O)(O^-)-OCH_2CH_2NH-C(=O)-$ cyclohexyl $-CH_2-N$ O

M-1618

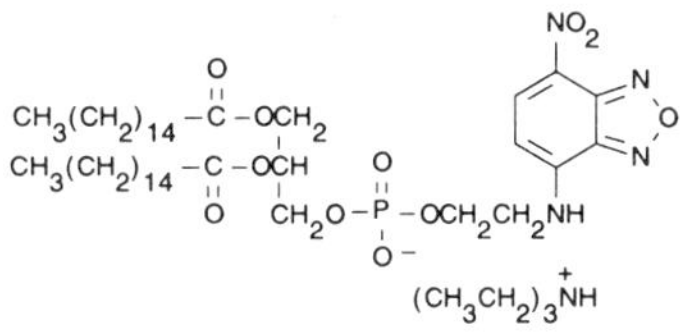

NO_2

N O N

$CH_3(CH_2)_{14}-C(=O)-OCH_2$

$CH_3(CH_2)_{14}-C(=O)-OCH$

$CH_2O-P(=O)(O^-)-OCH_2CH_2NH$

$(CH_3CH_2)_3\overset{+}{N}H$

N-360

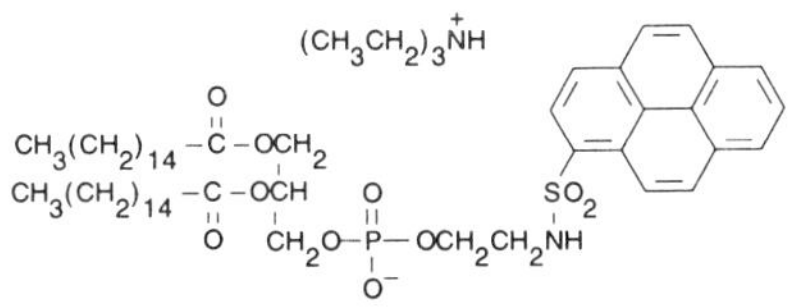

$(CH_3CH_2)_3\overset{+}{N}H$

$CH_3(CH_2)_{14}-C(=O)-OCH_2$

$CH_3(CH_2)_{14}-C(=O)-OCH$ SO_2

$CH_2O-P(=O)(O^-)-OCH_2CH_2NH$

P-58

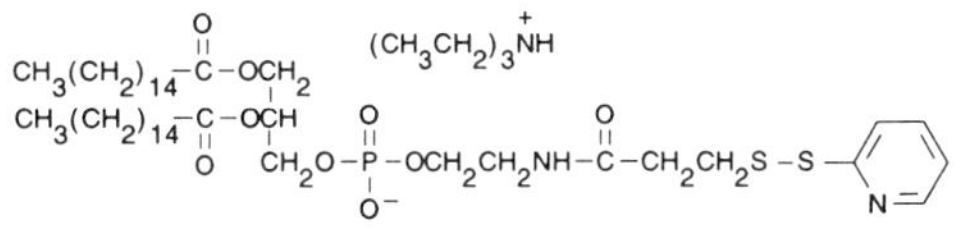

$CH_3(CH_2)_{14}-C(=O)-OCH_2$ $(CH_3CH_2)_3\overset{+}{N}H$

$CH_3(CH_2)_{14}-C(=O)-OCH$

$CH_2O-P(=O)(O^-)-OCH_2CH_2NH-C(=O)-CH_2CH_2S-S-$ N

P-1619

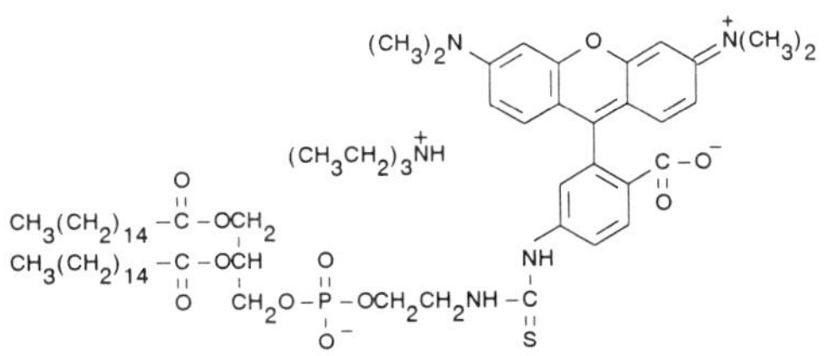

$(CH_3)_2N$ O $\overset{+}{N}(CH_3)_2$

$(CH_3CH_2)_3\overset{+}{N}H$ $C(=O)-O^-$

$CH_3(CH_2)_{14}-C(=O)-OCH_2$

$CH_3(CH_2)_{14}-C(=O)-OCH$ NH

$CH_2O-P(=O)(O^-)-OCH_2CH_2NH-C(=S)$

T-1391

$(CH_3CH_2)_3\overset{+}{N}H$ N O N+

SO_3^-

$CH_3(CH_2)_{14}-C(=O)-OCH_2$

$CH_3(CH_2)_{14}-C(=O)-OCH$ SO_2

$CH_2O-P(=O)(O^-)-OCH_2CH_2-X$

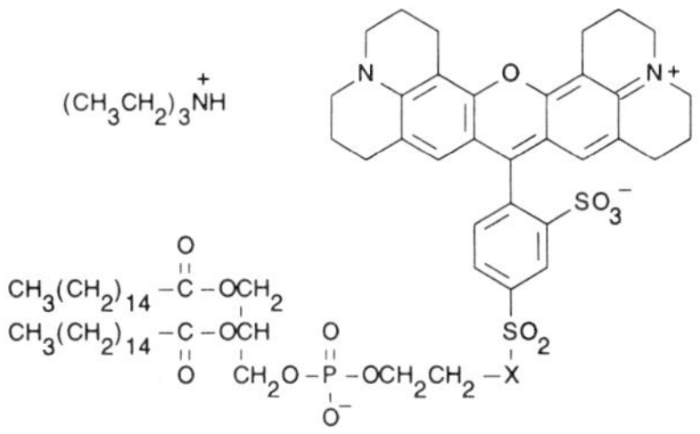

T-1395 X = $-NH-$

T-7710 X = $-NH-C(=O)-(CH_2)_5NH-$

13.4 Price List *Phospholipids with Labeled Head Groups*

Cat #		Product Name	Unit Size	Price Per Unit ($) 1–4 Units	5–24 Units
B-1616		N-((6-(biotinoyl)amino)hexanoyl)-1,2-dihexadecanoyl-*sn*-glycero-3-phosphoethanolamine, triethylammonium salt (biotin-X DHPE)	5 mg	115.00	92.00
B-1550		N-(biotinoyl)-1,2-dihexadecanoyl-*sn*-glycero-3-phosphoethanolamine, triethylammonium salt (biotin DHPE)	10 mg	115.00	92.00
D-3800		N-(4,4-difluoro-5,7-dimethyl-4-bora-3a,4a-diaza-s-indacene-3-propionyl)-1, 2-dihexadecanoyl-*sn*-glycero-3-phosphoethanolamine, triethylammonium salt (BODIPY® FL DHPE)	100 µg	75.00	60.00
D-3799		N-(4,4-difluoro-5,7-diphenyl-4-bora-3a,4a-diaza-s-indacene-3-propionyl)-1,2-dihexadecanoyl-*sn*-glycero-3-phosphoethanolamine, triethylammonium salt (BODIPY® 530/550 DHPE)	100 µg	38.00	30.40
D-7704	**New**	1,2-dihexadecanoyl-*sn*-glycero-3-phosphocholine (DPPC)	1 g	95.00	76.00
D-7705	**New**	1,2-dihexadecanoyl-*sn*-glycero-3-phosphoethanolamine (DPPE)	1 g	95.00	76.00
D-57		N-(5-dimethylaminonaphthalene-1-sulfonyl)-1,2-dihexadecanoyl-*sn*-glycero-3-phosphoethanolamine, triethylammonium salt (dansyl DHPE)	25 mg	98.00	78.40
D-3797		N-(5-dimethylaminonaphthalene-1-sulfonyl)-*sn*-glycero-3-phosphoethanolamine, triethylammonium salt *from egg PE*	25 mg	48.00	38.40
D-3798		N-((6-(2,4-dinitrophenyl)amino)hexanoyl)-1,2-dihexadecanoyl-*sn*-glycero-3-phosphoethanolamine, sodium salt (DNP-X DHPE)	10 mg	48.00	38.40
D-3811		N-(DMNB-caged fluorescein)-1,2-dihexadecanoyl-*sn*-glycero-3-phosphoethanolamine, triethylammonium salt	1 mg	38.00	30.40
F-362		N-(fluorescein-5-thiocarbamoyl)-1,2-dihexadecanoyl-*sn*-glycero-3-phosphoethanolamine, triethylammonium salt (fluorescein DHPE)	5 mg	75.00	60.00
L-1392		N-(Lissamine™ rhodamine B sulfonyl)-1,2-dihexadecanoyl-*sn*-glycero-3-phosphoethanolamine, triethylammonium salt (rhodamine DHPE)	5 mg	75.00	60.00
M-1618		N-((4-maleimidylmethyl)cyclohexane-1-carbonyl)-1,2-dihexadecanoyl-*sn*-glycero-3-phosphoethanolamine, triethylammonium salt (MMCC DHPE)	5 mg	135.00	108.00
N-360		N-(7-nitrobenz-2-oxa-1,3-diazol-4-yl)-1,2-dihexadecanoyl-*sn*-glycero-3-phosphoethanolamine, triethylammonium salt (NBD-PE)	10 mg	75.00	60.00
P-58		N-(1-pyrenesulfonyl)-1,2-dihexadecanoyl-*sn*-glycero-3-phosphoethanolamine, triethylammonium salt (pyS DHPE)	25 mg	75.00	60.00
P-1619		N-((2-pyridyldithio)propionyl)-1,2-dihexadecanoyl-*sn*-glycero-3-phosphoethanolamine, triethylammonium salt (PDP DHPE)	5 mg	135.00	108.00
T-1391		N-(6-tetramethylrhodaminethiocarbamoyl)-1,2-dihexadecanoyl-*sn*-glycero-3-phosphoethanolamine, triethylammonium salt (TRITC DHPE)	1 mg	75.00	60.00
T-1395		N-(Texas Red® sulfonyl)-1,2-dihexadecanoyl-*sn*-glycero-3-phosphoethanolamine, triethylammonium salt (Texas Red® DHPE)	1 mg	75.00	60.00
T-7710	**New**	N-(Texas Red®-X)-1,2-dihexadecanoyl-*sn*-glycero-3-phosphoethanolamine, triethylammonium salt (Texas Red®-X DHPE)	1 mg	75.00	60.00

13.5 Fluorescent Sterols, Including Cholesteryl Esters

Molecular Probes offers NBD- and pyrene-labeled cholesterol analogs in which the fluorophore replaces the terminal segment of cholesterol's flexible alkyl tail. Refer to Table 13.3 for a comparison of the spectral properties of these two fluorophores. The NBD fluorophore of the NBD cholesterol analog (N-1148) localizes in the membrane interior, unlike the anomalous positioning of NBD-labeled phospholipid acyl chains [1,2] (Figure 13.1). As with other NBD lipid analogs, this probe is useful for investigating lipid transport processes.[3-5] The pyrene cholesterol analog PMC (P-223) has been used to measure cholesterol transport rates between high-density lipoproteins [6] and phospholipid bilayer vesicles.[7]

Cholesteryl esters consist of a fatty acid esterified to the 3β-hydroxyl group of cholesterol. These highly nonpolar species are the predominant lipid components of atherosclerotic plaque and low- and high-density lipoprotein (LDL and HDL) cores. We offer cholesteryl ester analogs incorporating several different fluorophores:

- BODIPY (C-3927)
- DPH (C-7794)
- NBD (C-215, N-7709)
- *cis*-Parinaric acid (C-7789)
- Pyrene (C-212, C-213, C-7795, P-226)

With the exception of PMC oleate (P-226) and NBDEC oleate (N-7709), the fluorophore is attached to the fatty acyl tail of the cholesteryl ester. PMC oleate has been used to label acetylated LDL for following receptor-mediated endocytosis by the scavenger pathway [8] and also for selective photosensitization of cultured cells containing LDL receptors.[9] By exploiting the emission wavelength shift resulting from excimer formation, researchers have used pyrene-labeled cholesteryl esters to monitor lipid transfer among plasma lipoproteins catalyzed by specific lipid transport proteins.[10,11] Two of the pyrene cholesteryl esters (C-212, C-213) have also been characterized as fluorescent substrates for pancreatic carboxylic ester hydrolase and bile salt–activated lipase.[12]

The BODIPY FL cholesteryl ester (C-3927) can be used as a tracer of cholesterol transport and receptor-mediated endocytosis of lipoproteins by fluorescence microscopy and as a general non-exchangeable membrane marker. Researchers have also used our BODIPY FL cholesteryl ester to investigate the activity of a cholesteryl ester–transfer protein using a fluorescence microplate reader assay.[13,14] Cholesteryl esters incorporating DPH propionic acid (C-7794) and peroxidation-sensitive *cis*-parinaric acid (C-7789) have been used primarily for biophysical studies of lipoproteins and membranes.[15-17]

1. Biophys J 65, 630 (1993); **2.** Chem Phys Lipids 53, 1 (1990); **3.** J Cell Biol 129, 133 (1995); **4.** Eur J Cell Biol 59, 115 (1992); **5.** Parasitology 105, 81 (1992); **6.** J Lipid Res 27, 781 (1986); **7.** Chem Phys Lipids 63, 55 (1992); **8.** J Biol Chem 269, 21003 (1994); **9.** Proc Natl Acad Sci USA 78, 5717 (1981); **10.** Biochemistry 33, 4533 (1994); **11.** Biochemistry 32, 5029 (1993); **12.** Lipids 25, 428 (1990); **13.** Lipids 29, 811 (1994); **14.** J Lipid Res 34, 1625 (1993); **15.** Biochim Biophys Acta 735, 418 (1983); **16.** Biochem Biophys Res Comm 105, 674 (1982); **17.** Biochemistry 19, 1294 (1980).

13.5 Data Table *Fluorescent Sterols, Including Cholesteryl Esters*

Cat #	Structure	MW	Storage	Soluble	Abs	{ε × 10^{-3}}	Em	Solvent	Notes
C-212	✓	657	D,L	$CHCl_3$	342	{42}	376	MeOH	<1, 2>
C-213	✓	741	D,L	$CHCl_3$	342	{38}	378	MeOH	<1, 2>
C-215	✓	761	D,L	$CHCl_3$, MeCN	482	{21}	540	MeOH	
C-3927	✓	787	F,D,L	$CHCl_3$	505	{86}	511	MeOH	<3>
C-7789	✓	648	FF,LL,AA	$CHCl_3$	303	{68}	413	MeOH	<4>
C-7794	✓	673	F,D,L	$CHCl_3$	361	{81}	430	$CHCl_3$	<5>
C-7795	✓	685	F,D,L	$CHCl_3$	341	{39}	377	MeOH	<1, 2>
N-1148	✓	495	L	$CHCl_3$, MeCN	469	{21}	537	MeOH	
N-7709	✓	759	F,D,L	$CHCl_3$, MeCN	469	{22}	540	MeOH	
P-223	✓	561	F,D,L	$CHCl_3$, MeCN	341	{43}	376	MeOH	<1>
P-226	✓	825	FF,D,L,A	$CHCl_3$, MeCN	344	{44}	377	$CHCl_3$	<1>

For definitions of the contents of this data table, see "How to Use The Handbook" on page vi.
Structure: Chemical structure drawing: ✓ = shown in this section.

<1> Pyrene derivatives exhibit structured spectra. The absorption maximum is usually about 340 nm with a subsidiary peak at about 325 nm. There are also strong absorption peaks below 300 nm. The emission maximum is usually about 376 nm with a subsidiary peak at 396 nm. Excimer emission at about 470 nm may be observed at high concentrations.

<2> Alkylpyrene fluorescence lifetimes are up to 110 nsec and are very sensitive to oxygen.

<3> The absorption and fluorescence spectra of BODIPY derivatives are relatively insensitive to the solvent.

<4> Parinaric derivatives are very air sensitive. Use under N_2 or Ar. Butylated hydroxytoluene (BHT) may be added (1 mg/100 mL) to stock solutions to inhibit oxidative degradation.

<5> Diphenylhexatriene (DPH) and its derivatives are essentially nonfluorescent in water. Absorption and emission spectra have multiple peaks. The wavelength, resolution and relative intensity of these peaks are environment dependent. Abs and Em values are for the most intense peak in the solvent specified.

13.5 Structures *Fluorescent Sterols, Including Cholesteryl Esters*

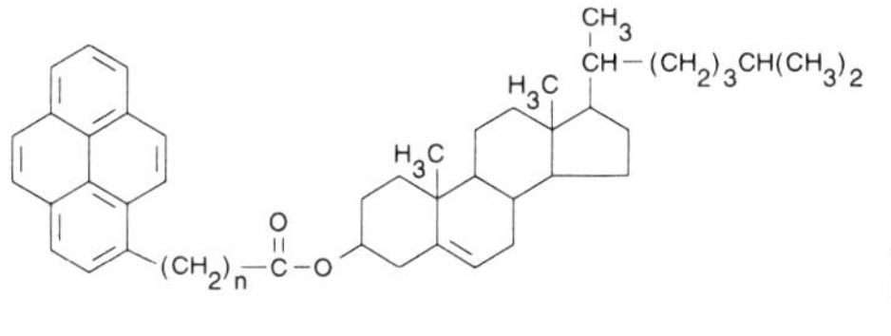

C-212 n = 3
C-7795 n = 5
C-213 n = 9

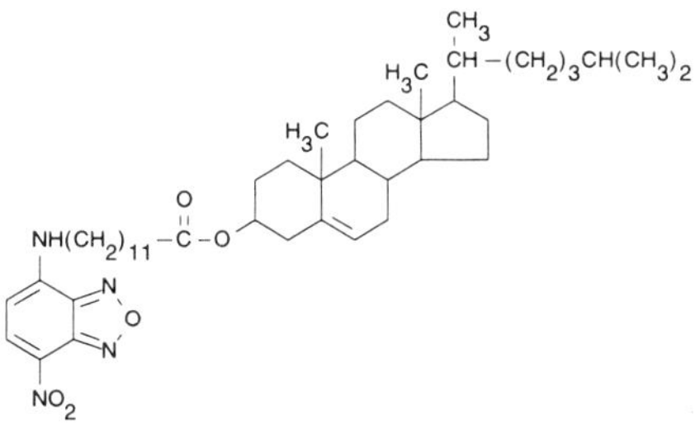

C-215

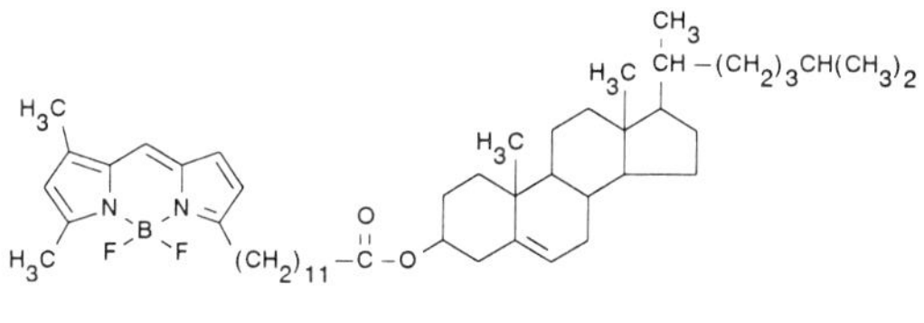

C-3927

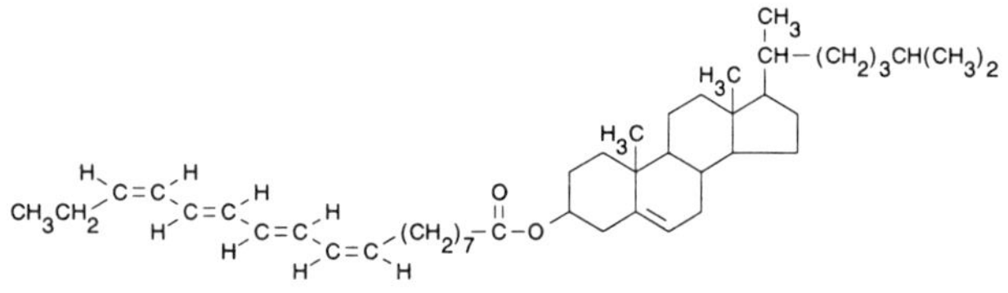

C-7789

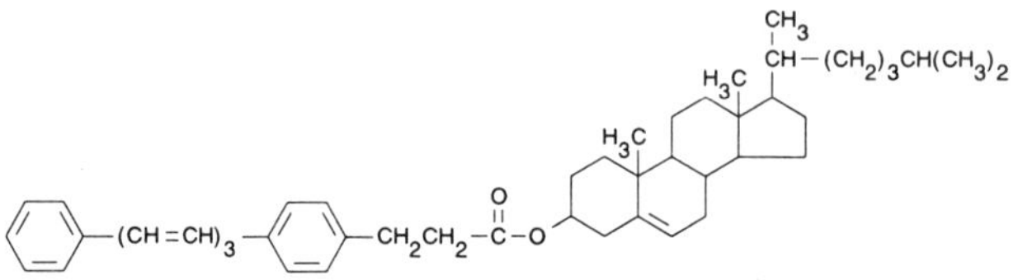

C-7794

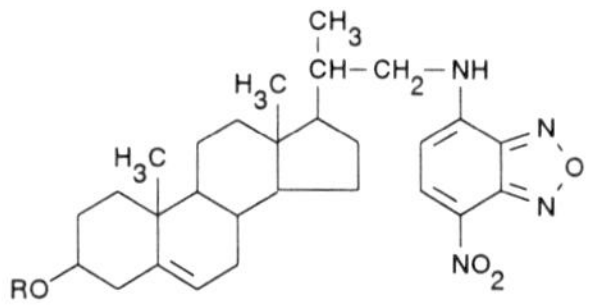

N-1148 R = H
N-7709 R = $-C(=O)-(CH_2)_7CH=CH(CH_2)_7CH_3$

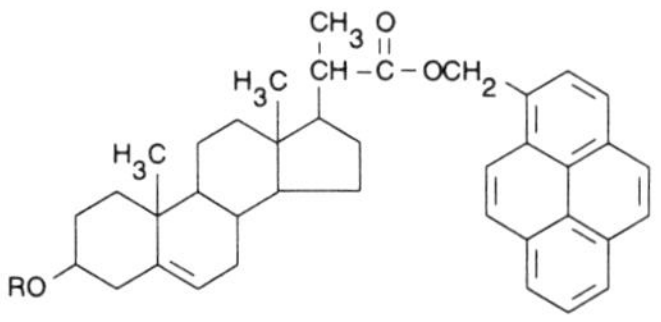

P-223 R = H
P-226 R = $-C(=O)-(CH_2)_7CH=CH(CH_2)_7CH_3$

13.5 Price List *Fluorescent Sterols, Including Cholesteryl Esters*

Cat #		Product Name	Unit Size	Price Per Unit ($) 1–4 Units	5–24 Units
C-3927		cholesteryl 4,4-difluoro-5,7-dimethyl-4-bora-3a,4a-diaza-s-indacene-3-dodecanoate (cholesteryl BODIPY® FL C_{12})	1 mg	85.00	68.00
C-215		cholesteryl 12-(N-methyl-N-(7-nitrobenz-2-oxa-1,3-diazol-4-yl)amino)dodecanoate	10 mg	50.00	40.00
C-7789	**New**	cholesteryl *cis*-parinarate	5 mg	125.00	100.00
C-7794	**New**	cholesteryl 3-((6-phenyl)-1,3,5-hexatrienyl)phenylpropionate (cholesteryl DPH propionate)	5 mg	125.00	100.00
C-212		cholesteryl 1-pyrenebutyrate	25 mg	50.00	40.00
C-213		cholesteryl 1-pyrenedecanoate	10 mg	75.00	60.00
C-7795	**New**	cholesteryl 1-pyrenehexanoate	5 mg	75.00	60.00
N-1148		22-(N-(7-nitrobenz-2-oxa-1,3-diazol-4-yl)amino)-23,24-bisnor-5-cholen-3β-ol	10 mg	135.00	108.00
N-7709	**New**	22-(N-(7-nitrobenz-2-oxa-1,3-diazol-4-yl)amino)-23,24-bisnor-5-cholen-3β-yl *cis*-9-octadecenoate	5 mg	85.00	68.00
P-223		1-pyrenemethyl 3β-hydroxy-22,23-bisnor-5-cholenate (PMC)	25 mg	125.00	100.00
P-226		1-pyrenemethyl 3β-(*cis*-9-octadecenoyloxy)-22,23-bisnor-5-cholenate (PMC oleate)	10 mg	145.00	116.00

Technical Assistance at Our Web Site (http://www.probes.com)

At Molecular Probes' Web site, we are developing an electronic version of this Handbook and other databases that should prove extremely useful for the researcher. In addition to containing all of the text from this Handbook, our Web site provides:

- **Product searches** by product name or catalog number
- **Bibliographies** for all products for which we have references
- **Keyword searches** of our entire bibliography of over 25,000 references
- **Product information sheets** for many kits and reagents
- **Technical bulletins**, including *BioProbes* newsletters and other product literature
- **Chemical structure**, **technical data** and **material safety and data sheets**
- **Color photomicrographs** that show our products in action

Visit our Web site often for new additions to our bibliography, as well as new products and upgraded search capabilities. Also look for special sales and introductory specials on some important products.

If you do not have access to the Internet or you need assistance that is not available at that site, further information on the scientific and technical background of our products can be obtained by contacting our Technical Assistance Department at the numbers listed on the inside front cover.

Chapter 14

Nonpolar and Amphiphilic Membrane Probes

Contents

Related Chapters

14.1 Introduction to Nonpolar and Amphiphilic Membrane Probes

The fluorescent membrane probes described in this chapter are either cationic or anionic amphiphiles or nonpolar neutral molecules. The positions of charged substituents relative to the fluorophore determine whether these probes are sensitive to interactions at the lipid/water interface or in the membrane interior (see Figure 13.1 in Chapter 13). In contrast to the products in Chapter 13, these probes are not structural analogs of any particular class of membrane constituents. Major applications of these products include:

- Measuring lipid diffusion (D-109, D-282; see Sections 14.2 and 14.3)
- Monitoring virus–cell fusion (O-246, see Section 14.3)
- Assessing membrane fluidity measurements (D-202, T-204; see Section 14.4)
- Investigating the molecular organization of membrane surfaces (A-47, D-250; see Section 14.5).

Most of the lipid probes in Section 14.2 are also extensively used as long-term tracers in live cells (see Section 15.4). In addition, we offer $Pd(QS)_2$ (P-1558, see Section 14.6), a reagent for hydrogenating lipids in live cells that is useful for correlating membrane composition with cellular activity.

14.2 Dialkylcarbocyanine and Dialkylaminostyryl Probes

Dialkylcarbocyanine Probes

Carbocyanines are among the most strongly absorbing dyes known and have proven to be useful tools in several different areas of research. Carbocyanines with short alkyl tails attached to the imine nitrogens are employed both as membrane-potential sensors (see Section 25.3) and as organelle stains for mitochondria and endoplasmic reticulum (see Sections 12.2 and 12.4). Those with longer alkyl tails (≥12 carbons) have an overall lipophilic character that makes them useful for neuronal tracing [1] (see Section 15.4), long-term labeling of cells in culture [2] (see Section 15.4) and noncovalent labeling of lipoproteins (see Section 17.3). This section describes the use and properties of dialkylcarbocyanines as general-purpose probes of membrane structure and dynamics.

DiI, DiO and Analogs

The most widely used carbocyanine membrane probes have been the octadecyl (C_{18}) indocarbocyanines (D-282, D-3911) and oxacarbocyanines (D-275), often referred to by the generic acronyms DiI and DiO, or more specifically as $DiIC_{18}(3)$ and $DiOC_{18}(3)$ (the subscript is the number of carbon atoms in each alkyl tail, and the bracketed numeral is the number of carbon atoms in the bridge between the indoline or benzoxazole ring systems). We also offer a large number of variations on this basic structure:

- DiI and DiO analogs with unsaturated alkyl tails (Δ^9-DiI, D-3886; *FAST* DiO™, D-3898; *FAST* DiI™, D-3899, D-7756)
- DiI and DiO analogs with shorter alkyl tails ($DiIC_{12}(3)$, D-383; $DiIC_{16}(3)$, D-384; $DiOC_{16}(3)$, D-1125)
- Long wavelength–excitable carbocyanines (DiD; D-307, D-7757)
- C_{18} thiacarbocyanine (DiS, D-1121)
- New brominated, phenyl-substituted and sulfonated derivatives of DiI and DiO (see below)

Spectral Properties of Dialkylcarbocyanines

A review of relevant spectral and physical characteristics of lipophilic carbocyanines has been compiled by Wolf.[3] The spectral properties of dialkylcarbocyanines are largely independent of the lengths of the alkyl chains, and are instead determined by the heteroatoms in the terminal ring systems and the length of the connecting bridge. The $DiIC_n(3)$ and $DiSC_n(3)$ probes have absorption and fluorescence spectra compatible with rhodamine (TRITC) optical filter sets, whereas $DiOC_n(3)$ analogs can be used with fluorescein (FITC) optical filter sets. The emission maxima of $DiIC_{18}(3)$ and $DiOC_{18}(3)$ incorporated in dioctadecenoylphosphocholine (dioleoyl PC or DOPC) liposomes (Figure 14.1) are similar to those of the dyes in methanol.

The very large molar extinction coefficients of carbocyanine fluorophores are their most outstanding spectral property. Their fluorescence quantum yields are only modest — about 0.07 for DiI in methanol [4] and about three times greater in amphiphilic solvents such as octanol;[5] their fluorescence in water is very weak. The excited-state lifetimes of carbocyanine fluorophores in lipid environments are short (~1 nanosecond), which is an advantage for flow cytometry applications because it allows more excitation/de-excita-

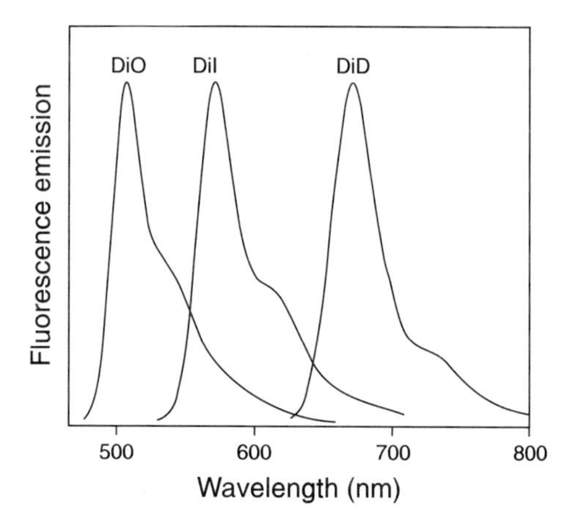

Figure 14.1 *Normalized emission spectra of DiO, DiI and DiD incorporated into dioctadecenoylglycerophosphocholine (DOPC) bilayers.*

tion cycles during flow transit;[6] the overall decay is multi-exponential.[7] Dialkylcarbocyanines are also exceptionally photostable.[8]

The red He-Ne laser–excitable indodicarbocyanines such as $DiIC_{18}(5)$ (DiD; D-307, D-7757) have very long-wavelength absorption and red emission (Figure 14.1). Their extinction coefficients are somewhat larger and fluorescence quantum yields much larger than those of carbocyanines such as $DiIC_{18}(3)$.[5] Moreover, photoexcitation of $DiIC_{18}(5)$ seems to cause less collateral damage than photoexcitation of $DiIC_{18}(3)$ in live cells.[9]

New Substituted DiI and DiO Derivatives

Molecular Probes' scientists have synthesized ten new derivatives of DiI and DiO. All of these derivatives have octadecyl (C_{18}) tails identical to those of DiI (D-282) and DiO (D-275), thereby preserving the excellent membrane retention characteristics of the parent molecules. A variety of substitutions have been made on the indoline or benzoxazole ring systems:

- Methylbenzamido (FM™-DiI, F-6999) and chloromethylbenzamido (CellTracker™ CM-DiI; C-7000, C-7001) DiI derivatives
- Brominated derivatives of DiI (D-7766) and DiO (D-7767)
- Diphenyl DiI (D-7779, D-7780) and diphenyl DiO (D-7770)
- Anionic sulfophenyl [10] (D-7777, D-7778) or sulfonate (D-7776) derivatives

Although these derivatives have primarily been developed to provide improved fixation and labeling in long-term cell tracing applications (see Section 15.4), they also offer several features that can potentially be exploited for investigations of membrane structure and dynamics. For researchers wishing to carry out comparative evaluations, our Lipophilic Tracer Sampler Kit (L-7781) provides 1 mg samples of six of the new derivatives (D-7770, D-7776 through D-7780), as well as of DiI, DiO, DiD and DiA. With the exception of the brominated derivatives, fluorescence quantum yields of these new lipophilic carbocyanines (measured in methanol) are generally two- to three-fold greater than those of DiI and DiO. In particular, we have found that the sulfophenyl derivatives (D-7777, D-7778) bound to phospholipid model membranes have approximately five-fold higher quantum yields. 5,5'-Diphenyl DiI (D-7779) exhibits a spectral shift of about 20 nm in both absorption and emission compared to DiI, making it suitable for excitation at 568 nm by the Ar-Kr mixed gas laser used in some confocal microscopes; spectra and extinction coefficients for the other substituted analogs are similar to the parent compounds.

The increased bulk of the bromo and phenyl substituents can be expected to appreciably modify lateral diffusion rates in membranes. The negative charge and greater water solubility of the new sulfonated carbocyanines should result in modified lateral and transverse distributions of these probes in lipid bilayers relative to those of DiI and DiO. The brominated derivatives of DiI (D-7766) and DiO (D-7767) have been prepared for possible use in photoconversion studies of cell membranes (see Section 15.4). The expected increase of intersystem crossing to the triplet state induced by the bromine substituents should be advantageous for measuring slow rotational diffusion processes in membranes on time scales from microseconds to seconds. These processes have been extensively studied using dialkylcarbocyanines (particularly DiI) in conjunction with polarized photobleaching recovery [11-13] and polarized fluorescence depletion [14] techniques.

DiI and DiO as Probes of Membrane Structure

The orientation of $DiIC_{18}(3)$ in membranes has been determined by fluorescence polarization microscopy.[15] The long axis of the fluorophore is parallel to the membrane surface, and the two alkyl chains protrude perpendicularly into the lipid interior (see Figure 13.1 in Chapter 13). There are conflicting reports in the literature regarding the ease of transbilayer migration ("flip-flop") of lipophilic indocarbocyanines.[16-18] The lateral partitioning behavior of dialkylindocarbocyanines in membranes has been investigated by fluorescence recovery after photobleaching [19] (FRAP), calorimetry,[20] lifetime measurements [7] and nitroxide fluorescence quenching.[21] These studies demonstrate that the probe distribution between coexisting fluid and gel phases depends on the similarity of the alkyl chain lengths of the probe and the lipid. In general, the more dissimilar the lengths, the greater the preference for fluid- over gel-phase lipids. For example, the shorter-chain $DiIC_{12}(3)$ has a substantial preference for the fluid phase (~6:1) in DOPC, whereas $DiIC_{18}(3)$ is predominantly distributed in the gel phase [21] (~1:10). Consequently, long-chain dialkylcarbocyanines are among the best probes for detecting particularly rigid gel phases.

Lipophilic carbocyanines have been used to visualize membrane fusion and cell permeabilization that occurs in response to electric fields,[22-24] as well as fusion of liposomes with planar bilayers.[25] Membrane fusion can also be measured by fluorescence resonance energy transfer to $DiIC_{18}(3)$ from dansyl- or NBD-labeled phospholipid donors [26] or by direct imaging.[27] In Langmuir–Blodgett films, excited-state energy transfer from $DiIC_{18}(3)$ to $DiIC_{18}(5)$ is exceptionally efficient because of the favorable orientations of the fluorophores.[28] Lipophilic carbocyanines have also been used to photosensitize diacetylene polymerization in multilayers,[29] to sensitize photoaffinity labeling of the viral glycoprotein hemagglutinin,[30] to image membrane domains in lipid monolayers [31] and to develop a fiber-optic potassium sensor.[32]

DiI and DiO as Probes of Membrane Dynamics

Despite their photostability, dialkylcarbocyanines are widely employed to measure lateral diffusion processes using FRAP techniques.[33] Their lateral diffusion coefficients in isolated fluid- and gel-phase bilayers are independent of the carbocyanine alkyl chain length.[19] Phase-separated populations of lipophilic carbocyanine dyes can be distinguished by their diffusion rates and can therefore be used to define lateral domains in cell membranes.[34,35] Combined lateral diffusion measurements of labeled proteins and lipids have demonstrated that transformed [36] and permeabilized [37] cells show marked changes in protein diffusion whereas lipid diffusion rates remain unchanged. In other cases, coupling of lipid and protein mobility has been identified in the form of relatively immobilized lipid domains in yeast plasma membranes [38] and around IgE receptor complexes.[39] A different photobleaching technique, which depends on the *absence* of diffusional fluorescence recovery, was employed to determine lipid flow direction in locomoting cells by following the movement of a photobleached stripe of $DiIC_{16}(3)$.[40]

Dialkylaminostyryl Probes

The lipophilic aminostyryl probes 4-Di-10-ASP (D-291), DiA (D-3883), *FAST* DiA™ (D-3897, D-7758) and DiQ (D-3885) insert in membranes with their two alkyl tails and their fluorophore oriented parallel to the phospholipid acyl chains [41] (see Figure 13.1

in Chapter 13). When these dialkylaminostyryl probes bind to membranes, they exhibit a strong fluorescence enhancement; their fluorescence in water is minimal. The interfacial solvation of the aminostyryl fluorophore causes a large blue shift of the absorption spectrum of the membrane-bound probe.[41] The absorption maximum of DiA is 456 nm when incorporated into DOPC liposomes and 490 nm when in methanol. The fluorescence emission maximum of DiA in the membrane environment is 590 nm (Figure 14.2), which is quite close to that observed for probes with shorter alkyl tails such as 4-Di-10-ASP.[41] However, its fluorescence spectrum is very broad, with appreciable intensity from about 510 nm to 690 nm. Consequently, DiA can be detected as green, orange or even red fluorescence, depending on the optical filter employed. Like the lipophilic carbocyanines, DiA is commonly used for neuronal membrane tracing (see Section 15.4). *FAST* DiA (D-3897, D-7758), the diunsaturated analog of DiA, is intended to facilitate these studies by possibly accelerating dye diffusion within the membrane.

The aminostyryl probe Di_{10}ASP-PS (S-467) is similar in structure and properties to 4-Di-10-ASP but has a sulfopropyl instead of a methyl terminal substituent. The increased charge of this substituent may retard probe internalization. Unlike Di-10-ASP and DiA, in which the alkyl chain is attached to the aminophenyl ring, our new lipophilic styrylpyridinium probe C_{22}-Di-1-ASP (D-7761) has its single C_{22}-alkyl chain attached to the pyridinium ring. C_{22}-Di-1-ASP has been used to investigate conformational transitions in inclusion complexes of the dye in amylose polymers.[42]

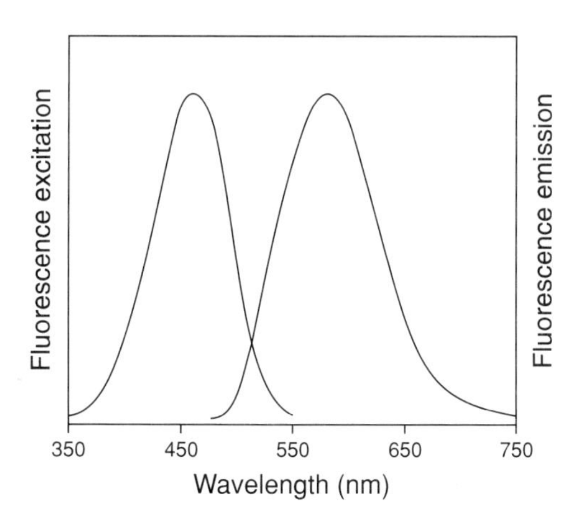

Figure 14.2 *Fluorescence excitation and emission spectra for DiA (D-3883) incorporated into dioctadecenoylglycerophosphocholine (DOPC) bilayers.*

1. Trends Neurosci 9, 333 (1989); **2.** Histochemistry 97, 329 (1992); **3.** Wolf, D.E. in *Spectroscopic Membrane Probes, Volume 1*, L.M. Loew, Ed., CRC Press (1988) p. 193–220; **4.** Tsien, R.Y. and Waggoner, A. in *Handbook of Biological Confocal Microscopy, Second Edition*, J.B. Pawley, Ed., Plenum Press (1995) pp. 267–279; **5.** Biochemistry 13, 3315 (1974); **6.** Waggoner, A.S. in *Flow Cytometry and Sorting, Second Edition*, M.R. Melamed *et al.*, Eds., Wiley-Liss (1990) pp. 209–225; **7.** Biochemistry 24, 5176 (1985); **8.** J Cell Biol 100, 1309 (1985); **9.** J Histochem Cytochem 32, 608 (1984); **10.** Biorg Medicinal Chem Lett 6, 1479 (1996); **11.** Biophys J 58, 413 (1990); **12.** Biophys J 53, 575 (1988); **13.** Biophys J 36, 73 (1981); **14.** FEBS Lett 153, 391 (1983); **15.** Biophys J 26, 557 (1979); **16.** J Cell Biol 103, 807 (1986); **17.** Biochemistry 24, 582 (1985); **18.** Nature 294, 718 (1981); **19.** Biochemistry 19, 6199 (1980); **20.** Biochim Biophys Acta 1191, 164 (1994); **21.** Biochim Biophys Acta 1023, 25 (1990); **22.** Biophys J 67, 427 (1994); **23.** Biophys J 65, 568 (1993); **24.** Biochemistry 29, 8337 (1990); **25.** J Membrane Biol 109, 221 (1989); **26.** Biochim Biophys Acta 735, 243 (1983); **27.** J Cell Biol 121, 543 (1993); **28.** Chem Phys Lett 159, 231 (1989); **29.** Ber Bunsenges Phys Chem 86, 499 (1982); **30.** J Biol Chem 269, 14614 (1994); **31.** Biophys J 65, 1019 (1993); **32.** Analyst 115, 353 (1990); **33.** Prog Lipid Res 33, 203 (1994); **34.** Chem Phys Lipids 73, 139 (1994); **35.** J Cell Biol 112, 1143 (1991); **36.** Biochim Biophys Acta 1107, 193 (1992); **37.** J Cell Physiol 158, 7 (1994); **38.** J Membrane Biol 131, 115 (1993); **39.** J Cell Biol 125, 795 (1994); **40.** Science 247, 1229 (1990); **41.** Biophys J 34, 353 (1981); **42.** Langmuir 10, 2842 (1994).

14.2 Data Table *Dialkylcarbocyanine and Dialkylaminostyryl Probes*

Cat #	Structure	MW	Storage	Soluble	Abs	{ε x 10^{-3}}	Em	Solvent	Notes
C-7001	S15.4	1052	F,D,L	DMSO, EtOH	553	{134}	570	MeOH	
D-275	S15.4	882	L	DMSO, EtOH	484	{154}	501	MeOH	
D-282	S15.4	934	L	DMSO, EtOH	549	{148}	565	MeOH	
D-291	S15.4	619	L	DMSO, EtOH	492	{53}	612	MeOH	<1>
D-307	S15.4	960	L	DMSO, EtOH	644	{195}	665	MeOH	<2>
D-383	S15.4	766	L	DMSO, EtOH	549	{144}	565	MeOH	<3>
D-384	S15.4	878	L	DMSO, EtOH	549	{148}	565	MeOH	
D-1121	✓	914	L	DMSO, EtOH	560	{150}	574	MeOH	
D-1125	S15.4	826	L	DMSO, EtOH	484	{156}	501	MeOH	
D-3883	S15.4	787	L	DMSO, EtOH	491	{52}	613	MeOH	<1>
D-3885	S15.4	837	L	DMSO, EtOH	562	{44}	681	MeOH	<1>
D-3886	S15.4	925	F,L,AA	DMSO, EtOH	549	{144}	564	MeOH	<2>
D-3897	S15.4	835	F,L,AA	DMSO, EtOH	492	{49}	612	MeOH	<1, 2>
D-3898	S15.4	874	F,L,AA	DMSO, EtOH	484	{138}	499	MeOH	
D-3899	S15.4	926	F,L,AA	DMSO, EtOH	549	{143}	564	MeOH	<2>
D-3911	S15.4	934	L	DMSO, EtOH	549	{148}	565	MeOH	
D-7756	S15.4	1018	F,L,AA	DMSO, EtOH	549	{148}	564	MeOH	
D-7757	S15.4	1052	L	DMSO, EtOH	644	{193}	663	MeOH	
D-7758	S15.4	900	F,L,AA	DMSO, EtOH	492	{41}	612	MeOH	<1>
D-7761	✓	614	L	DMSO, EtOH	479	{43}	608	MeOH	<1>
D-7766	S15.4	1092	L	DMSO, EtOH	558	{128}	575	MeOH	
D-7767	S15.4	1040	L	DMSO, EtOH	491	{117}	506	MeOH	
D-7770	S15.4	970	L	DMSO, EtOH	496	{168}	513	MeOH	

Cat #	Structure	MW	Storage	Soluble	Abs	{ε × 10^{-3}}	Em	Solvent	Notes
D-7776	S15.4	994	L	DMSO, EtOH	555	{144}	570	MeOH	
D-7777	S15.4	1146	L	DMSO, EtOH	556	{164}	573	MeOH	
D-7778	S15.4	1116	L	DMSO, EtOH	497	{175}	513	MeOH	
D-7779	S15.4	1022	L	DMSO, EtOH	576	{140}	599	MeOH	
D-7780	S15.4	1022	L	DMSO, EtOH	556	{160}	573	MeOH	
F-6999	S15.4	1017	L	DMSO, EtOH	553	{133}	570	MeOH	
S-467	✓	599	L	DMSO, EtOH	497	{56}	616	MeOH	<1>

C-7000 see C-7001

For definitions of the contents of this data table, see "How to Use the Handbook" on page vi.
Structure: Chemical structure drawing: ✓ = shown in this section; Sn.n = shown in Section number n.n.

<1> Abs and Em of styryl dyes are at shorter wavelengths in membrane environments than in reference solvents such as methanol. The difference is typically ~20 nm for absorption and ~80 nm for emission but varies considerably from one dye to another.
<2> This product is intrinsically a liquid or an oil at room temperature.
<3> This product is intrinsically a sticky gum at room temperature.

14.2 Structures *Dialkylcarbocyanine and Dialkylaminostyryl Probes*

D-1121

D-7761

S-467

Structures for all other products in this section are shown in Section 15.4.

14.2 Price List *Dialkylcarbocyanine and Dialkylaminostyryl Probes*

Cat #		Product Name	Unit Size	Price Per Unit ($) 1–4 Units	5–24 Units
C-7001	*New*	CellTracker™ CM-DiI	1 mg	110.00	88.00
C-7000	*New*	CellTracker™ CM-DiI *special packaging*	20x50 µg	110.00	88.00
D-7767	*New*	6,6′-dibromo-3,3′-dioctadecyloxacarbocyanine perchlorate (Br_2-DiOC$_{18}$(3))	5 mg	135.00	108.00
D-7766	*New*	5,5′-dibromo-1,1′-dioctadecyl-3,3,3′,3′-tetramethylindocarbocyanine perchlorate (Br_2-DiIC$_{18}$(3))	5 mg	135.00	108.00
D-291		4-(4-(didecylamino)styryl)-N-methylpyridinium iodide (4-Di-10-ASP)	25 mg	115.00	92.00
D-383		1,1′-didodecyl-3,3,3′,3′-tetramethylindocarbocyanine perchlorate (DiIC$_{12}$(3))	100 mg	135.00	108.00
D-3883		4-(4-(dihexadecylamino)styryl)-N-methylpyridinium iodide (DiA; 4-Di-16-ASP)	25 mg	115.00	92.00
D-3885		4-(4-(dihexadecylamino)styryl)-N-methylquinolinium iodide (DiQ; Di-16-ASQ)	25 mg	48.00	38.40
D-1125		3,3′-dihexadecyloxacarbocyanine perchlorate (DiOC$_{16}$(3))	25 mg	135.00	108.00
D-384		1,1′-dihexadecyl-3,3,3′,3′-tetramethylindocarbocyanine perchlorate (DiIC$_{16}$(3))	100 mg	135.00	108.00
D-7758	*New*	4-(4-(dilinoleylamino)styryl)-N-methylpyridinium, 4-chlorobenzenesulfonate (*FAST* DiA™ solid; DiΔ^{9,12}-C$_{18}$ASP, CBS)	5 mg	135.00	108.00
D-3897		4-(4-(dilinoleylamino)styryl)-N-methylpyridinium iodide (*FAST* DiA™ oil; DiΔ^{9,12}-C$_{18}$ASP, I)	5 mg	135.00	108.00
D-3898		3,3′-dilinoleyloxacarbocyanine perchlorate (*FAST* DiO™ solid; DiOΔ^{9,12}-C$_{18}$(3), ClO_4)	5 mg	135.00	108.00
D-7756	*New*	1,1′-dilinoleyl-3,3,3′,3′-tetramethylindocarbocyanine, 4-chlorobenzenesulfonate (*FAST* DiI™ solid; DiIΔ^{9,12}-C$_{18}$(3), CBS)	5 mg	135.00	108.00
D-3899		1,1′-dilinoleyl-3,3,3′,3′-tetramethylindocarbocyanine perchlorate (*FAST* DiI™ oil; DiIΔ^{9,12}-C$_{18}$(3), ClO_4)	5 mg	135.00	108.00
D-7761	*New*	4-(4-(dimethylamino)styryl)-1-docosylpyridinium bromide (C$_{22}$-Di-1-ASP)	25 mg	48.00	38.40
D-7770	*New*	3,3′-dioctadecyl-5,5′-diphenyloxacarbocyanine chloride (5,5′-Ph_2-DiOC$_{18}$(3))	5 mg	135.00	108.00
D-7779	*New*	1,1′-dioctadecyl-5,5′-diphenyl-3,3,3′,3′-tetramethylindocarbocyanine chloride (5,5′-Ph_2-DiIC$_{18}$(3))	5 mg	135.00	108.00
D-7780	*New*	1,1′-dioctadecyl-6,6′-diphenyl-3,3,3′,3′-tetramethylindocarbocyanine chloride (6,6′-Ph_2-DiIC$_{18}$(3))	5 mg	135.00	108.00
D-7778	*New*	3,3′-dioctadecyl-5,5′-di(4-sulfophenyl)-oxacarbocyanine, sodium salt (SP-DiOC$_{18}$(3))	5 mg	135.00	108.00
D-7777	*New*	1,1′-dioctadecyl-6,6′-di(4-sulfophenyl)-3,3,3′,3′-tetramethylindocarbocyanine (SP-DiIC$_{18}$(3))	5 mg	135.00	108.00
D-275		3,3′-dioctadecyloxacarbocyanine perchlorate ('DiO'; DiOC$_{18}$(3))	100 mg	135.00	108.00
D-7776	*New*	1,1′-dioctadecyl-3,3,3′,3′-tetramethylindocarbocyanine-5,5′-disulfonic acid (DiIC$_{18}$(3)-DS)	5 mg	135.00	108.00
D-282		1,1′-dioctadecyl-3,3,3′,3′-tetramethylindocarbocyanine perchlorate ('DiI'; DiIC$_{18}$(3))	100 mg	135.00	108.00

Cat #		Product Name	Unit Size	Price Per Unit ($) 1–4 Units	5–24 Units
D-3911		1,1′-dioctadecyl-3,3,3′,3′-tetramethylindocarbocyanine perchlorate *large crystals*	25 mg	100.00	80.00
D-7757	New	1,1′-dioctadecyl-3,3,3′,3′-tetramethylindodicarbocyanine, 4-chlorobenzenesulfonate salt ("DiD" solid; DiIC$_{18}$(5) solid)	10 mg	68.00	54.40
D-307		1,1′-dioctadecyl-3,3,3′,3′-tetramethylindodicarbocyanine perchlorate ("DiD" oil; DiIC$_{18}$(5) oil)	25 mg	90.00	72.00
D-1121		3,3′-dioctadecylthiacarbocyanine perchlorate ('DiS'; DiSC$_{18}$(3))	25 mg	48.00	38.40
D-3886		1,1′-dioleyl-3,3,3′,3′-tetramethylindocarbocyanine methanesulfonate (Δ^9-DiI)	25 mg	135.00	108.00
F-6999	New	FM™-DiI *special packaging*	20x50 µg	95.00	76.00
L-7781	New	Lipophilic Tracer Sampler Kit	1 kit	150.00	120.00
S-467		N-(3-sulfopropyl)-4-(4-(didecylamino)styryl)pyridinium, inner salt (Di$_{10}$ASP-PS)	100 mg	80.00	64.00

14.3 Lipophilic Derivatives of Rhodamines, Fluoresceins and Other Dyes

Octadecyl Rhodamine B

The relief of octadecyl rhodamine B (R 18, O-246) self-quenching [1] can be used to monitor membrane fusion — one of several experimental approaches developed for this application that are described in the Technical Note "Lipid Mixing Assays of Membrane Fusion" on page 286. Investigators have used octadecyl rhodamine B in conjunction with video microscopy [2-4] or digital imaging techniques [5] to monitor viral fusion processes. Membrane fusion can also be followed by monitoring fluorescence resonance energy transfer to octadecyl rhodamine B from an acylaminofluorescein donor such as 5-octadecanoylaminofluorescein [6] (O-322). Fluorescence resonance energy transfer from fluorescein or dansyl labels to octadecyl rhodamine B has been used for structural studies of the blood coagulation factor IXa, EGF receptor and receptor-bound IgE.[7-9] Octadecyl rhodamine B has also been used to stain kinesin-generated membrane tubules,[10] to characterize detergent micelles,[11] to assay for lysosomal degradation of lipoproteins [12] and to investigate the influence of proteins on lipid dynamics using time-resolved fluorescence anisotropy.[13]

Lipophilic Fluoresceins

The acylaminofluoresceins, including the eosin (tetrabromofluorescein) derivative, bind to membranes with the fluorophore at the aqueous interface and the alkyl tail protruding into the lipid interior (see Figure 13.1 in Chapter 13). 5-Dodecanoylaminofluorescein (D-109) is the hydrolysis product of our ImaGene Green™ C$_{12}$-FDG β-galactosidase substrate (D-2893, see Section 10.2). We also offer 5-hexadecanoyl- and 5-octadecanoylaminofluorescein (H-110, D-322). The octadecyl ester (F-3857) and octadecyl thiourea of fluorescein (F 18, O-3852) are both similar to 5-octadecanoylaminofluorescein but contain different linkages between the fluorophore and the alkyl tail.

Lipophilic fluorescein probes are commonly utilized for fluorescence recovery after photobleaching (FRAP) measurements of lipid lateral diffusion.[14,15] Alternatively, our DMNB-caged C$_{12}$-fluorescein (D-2512) can potentially be used for photoactivation of fluorescence (PAF) experiments, which are analogous to FRAP experiments except that the fluorophore is photoactivated upon illumination rather than bleached. Measuring the bright fluorescent signal of the photoactivated fluorophore against a dark background should be intrinsically more sensitive than measuring a dark photobleached region against a bright field. Some researchers have reported that 5-hexadecanoylaminofluorescein stays predominantly in the outer membrane leaflet of epithelia and does not pass through tight junctions, whereas the dodecanoyl derivative can "flip-flop" to the inner leaflet at 20°C (but not at <10°C) and may also pass through tight junctions.[16,17] More recent studies have indicated that the lack of tight junction penetration of 5-hexadecanoylaminofluorescein is due to probe aggregation rather than a significant difference in its transport properties.[18] The 5-hexadecanoylaminoeosin (H-193) is less pH sensitive than the fluorescein derivative and has somewhat longer-wavelength fluorescence spectra. Consequently, this probe can be employed as a membrane-bound energy-transfer acceptor from dansyl- or fluorescein-labeled proteins.[19-21] The significant phosphorescence quantum yield of eosin can be exploited to measure slow rotational motions using time-resolved phosphorescence anisotropy techniques.[22] It may also be useful for ultrastructural labeling of membranes by photoconversion of diaminobenzidine (see the Technical Note "Fluorescent Probes for Photoconversion of Diaminobenzidine" on page 414).

Alkylated Coumarins, Acridines and Resorufin

Coumarins

4-Heptadecyl-7-hydroxycoumarin (H-181) is an alkyl derivative of the pH-sensitive blue fluorophore 7-hydroxycoumarin (umbelliferone). This lipophilic coumarin is primarily useful as a pH indicator for membrane surfaces (see Section 23.4). Its pK_a in lipid assemblies is strongly dependent on the ionic composition of the membrane surface,[23,24] making it a sensitive probe of membrane-surface electrostatic potential.[25] 4-Heptadecyl-7-hydroxycoumarin has also been employed as a structural probe for the head-group region of phospholipid bilayers.[26] Our new lipophilic 7-hydroxycoumarin probe (U-6, D-7760) is a structural analog of a cationic C_{14} detergent and can potentially be used to probe electrostatic potentials and pH changes at a defined distance (nominally 6 Å) above the membrane surface.[27]

Acridines

The two acridine orange analogs (A-455, A-1372) have been widely used as mitochondrial membrane stains [28-30] (see Section 12.2) and as photosensitizers.[31,32] Their physical and spectroscopic properties are well documented.[33,34] Nonyl acridine orange (A-1372) has been utilized in a new fluorometric assay for cardiolipin phospholipids in yeast.[35]

Dodecylresorufin

Dodecylresorufin (D-7786) is the hydrolysis product of our ImaGene Red™ β-galactosidase and glucuronidase substrates (I-2906, I-2910; Section 10.2). This probe should be suitable as a long-wavelength alternative to 5-dodecanoylaminofluorescein (D-109, see above), although substantially weaker fluorescence emission can generally be expected.

1. Photochem Photobiol 60, 563 (1994); **2.** Biophys J 63, 710 (1992); **3.** J Gen Physiol 97, 1101 (1991); **4.** Proc Natl Acad Sci USA 87, 1850 (1990); **5.** FEBS Lett 250, 487 (1989); **6.** Biochim Biophys Acta 1189, 175 (1994); **7.** J Biol Chem 267, 17012 (1992); **8.** Biochemistry 30, 9125 (1991); **9.** Biochemistry 29, 8741 (1990); **10.** J Cell Biol 107, 2233 (1988); **11.** Langmuir 10, 658 (1994); **12.** Eur J Biochem 59, 116 (1992); **13.** Eur Biophys J 18, 277 (1990); **14.** Prog Lipid Res 33, 203 (1994); **15.** J Cell Sci 100, 473 (1991); **16.** Exp Cell Res 181, 375 (1989); **17.** Nature 294, 718 (1981); **18.** Biochem Biophys Res Comm 184, 160 (1992); **19.** Biophys J 66, 674 (1994); **20.** Biochemistry 32, 2409 (1993); **21.** J Membrane Biol 118 215 (1990); **22.** J Biol Chem 263, 17541 (1988); **23.** J Phys Chem 81, 1755 (1977); **24.** Biochim Biophys Acta 323, 326 (1973); **25.** Meth Enzymol 171, 376 (1989); **26.** Biochemistry 24, 573 (1985); **27.** Biochemistry 32, 10057 (1993); **28.** Cytometry 19, 304 (1995); **29.** FEBS Lett 260, 236 (1990); **30.** Histochemistry 82, 51 (1985); **31.** Photochem Photobiol 60, 503 (1994); **32.** Photochem Photobiol 29, 165 (1979); **33.** Bull Chem Soc Japan 59, 3393 (1986); **34.** Histochemistry 79, 443 (1983); **35.** Eur J Biochem 228, 113 (1995).

14.3 Data Table *Lipophilic Derivatives of Rhodamine, Fluorescein and Other Dyes*

Cat #	Structure	MW	Storage	Soluble	Abs	{ε x 10⁻³}	Em	Solvent	Notes
A-455	✓	515	L	DMSO, EtOH	495	{87}	520	MeOH	
A-1372	✓	473	L	DMSO, EtOH	495	{84}	519	MeOH	
D-109	✓	530	L	DMSO, EtOH	495	{85}	518	MeOH	<1>
D-2512	✓	920	F,D,LL	DMSO, EtOH	338	{12}	none	MeOH	<2, 3>
D-7760	S23.4	452	L	DMSO, EtOH	394	{18}	494	MeOH	<1, 4>
D-7786	✓	382	L	DMSO, EtOH	578	{69}	597	MeOH	<1>
F-3857	✓	585	L	DMSO, EtOH	504	{95}	525	MeOH	<1>
H-110	✓	586	L	DMSO, EtOH	497	{92}	519	MeOH	<1>
H-181	S23.4	401	L	DMSO, EtOH	366	{20}	453	MeOH	<1, 4>
H-193	✓	901	L	DMSO, EtOH	523	{111}	542	MeOH	
O-246	✓	731	F,DD,L	DMSO, EtOH	556	{125}	578	MeOH	<5>
O-322	✓	614	L	DMSO, EtOH	497	{84}	519	MeOH	<1>
O-3852	✓	659	L	DMSO, EtOH	497	{80}	519	MeOH	<1>

For definitions of the contents of this data table, see "How to Use the Handbook" on page vi.
Structure: Chemical structure drawing: ✓ = shown in this section; Sn.n = shown in Section number n.n.

<1> Spectra of this product are pH dependent. Data listed are for basic solutions prepared in methanol containing a trace of KOH.
<2> All caged products are sensitive to light. They should be protected from illumination, except when photolysis is intended.
<3> This product is essentially colorless and nonfluorescent until it is converted to D-109 by ultraviolet photolysis of the caging groups.
<4> Spectral data for acidic solutions are listed in Section 23.4.
<5> This product is intrinsically a sticky gum at room temperature.

14.3 Structures *Lipophilic Derivatives of Rhodamine, Fluorescein and Other Dyes*

A-455 n = 11
A-1372 n = 8

D-109 n = 10
H-110 n = 14
O-322 n = 16

D-2512

D-7760 Section 23.4

D-7786

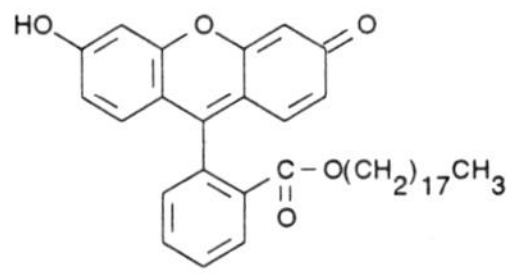

F-3857

H-110 see D-109

H-181 Section 23.4

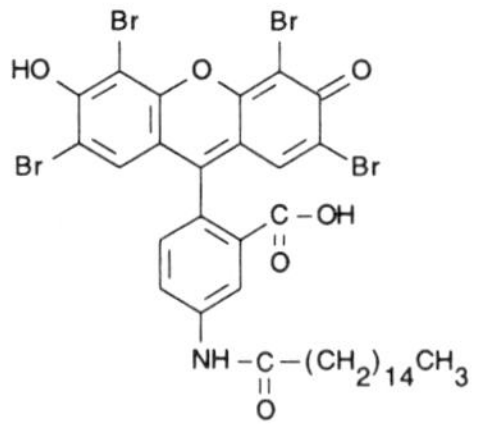

H-193

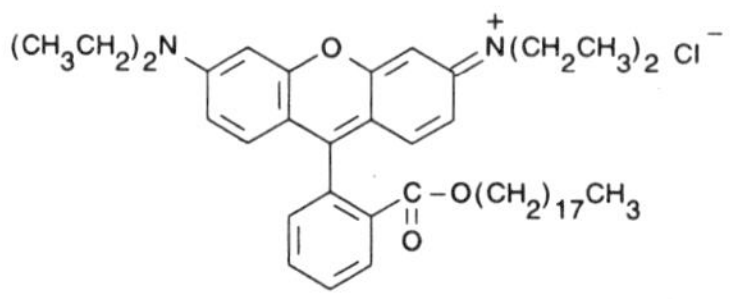

O-246

O-322 see D-109

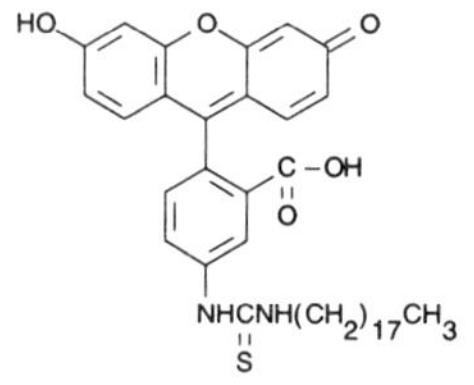

O-3852

14.3 Price List *Lipophilic Derivatives of Rhodamine, Fluorescein and Other Dyes*

Cat #		Product Name	Unit Size	Price Per Unit ($) 1–4 Units	5–24 Units
A-455		acridine orange 10-dodecyl bromide (dodecyl acridine orange)	100 mg	45.00	36.00
A-1372		acridine orange 10-nonyl bromide (nonyl acridine orange)	100 mg	45.00	36.00
D-7760	***New***	4-(N,N-dimethyl-N-tetradecylammonium)methyl-(7-hydroxycoumarin) chloride (U-6)	10 mg	48.00	38.40
D-109		5-dodecanoylaminofluorescein	100 mg	65.00	52.00
D-2512		5-dodecanoylaminofluorescein-bis-4,5-dimethoxy-2-nitrobenzyl ether (DMNB-caged C_{12}-fluorescein)	5 mg	48.00	38.40
D-7786	***New***	2-dodecylresorufin	5 mg	65.00	52.00
F-3857		fluorescein octadecyl ester	10 mg	65.00	52.00
H-181		4-heptadecyl-7-hydroxycoumarin	100 mg	65.00	52.00
H-193		5-hexadecanoylaminoeosin	100 mg	65.00	52.00
H-110		5-hexadecanoylaminofluorescein	100 mg	65.00	52.00
O-322		5-octadecanoylaminofluorescein	100 mg	65.00	52.00
O-3852		N-octadecyl-N′-(5-(fluoresceinyl))thiourea (F 18)	25 mg	65.00	52.00
O-246		octadecyl rhodamine B chloride (R 18)	10 mg	145.00	116.00

14.4 Diphenylhexatriene (DPH) and Derivatives

DPH

Membrane Labeling Properties of DPH

DPH (D-202) continues to be a popular fluorescent probe of membrane interiors, as indicated by the recent applications listed in Table 14.1. We also offer the cationic DPH derivatives TMA-DPH and TMAP-DPH (see below), as well as phospholipid and fatty acid analogs (D-476, P-459, P-1342; see Section 13.2). The orientation of DPH within lipid bilayers is loosely constrained. It is generally assumed to be oriented parallel to the lipid acyl chain axis (see Figure 13.1 in Chapter 13), but can also reside in the center of the lipid bilayer parallel to the surface, as demonstrated by time-resolved fluorescence anisotropy and polarized fluorescence measurements of oriented samples.[1-4] DPH shows no partition preference between coexisting gel- and fluid-phase phospholipids.[5] Intercalation of DPH and its derivatives into membranes is accompanied by strong enhancement of their fluorescence; their fluorescence is practically negligible in water. The fluorescence decay of DPH in lipid bilayers is complex.[6-8] Fluorescence decay data are often analyzed in terms of continuous lifetime distributions,[9-12] which are in turn interpreted as being indicative of lipid environmental heterogeneity.

Fluorescence Polarization Properties of DPH

DPH and its derivatives are cylindrically shaped molecules with absorption and fluorescence emission transition dipoles aligned approximately parallel to their long molecular axis. Consequently, their fluorescence polarization is high in the absence of rotational motion and is very sensitive to reorientation of the long axis resulting from interactions with surrounding lipids. These properties have led to their extensive use for membrane fluidity measurements (Table 14.1). The exact physical interpretation of these measurements has some contentious aspects. For instance, the probes are largely sensitive to only the angular reorientation of lipid acyl chains — a motion that does not necessarily correlate with other dynamic processes such as lateral diffusion.[13] Reviews on this subject[3,14,15,17] should be consulted for further discussion. Time-resolved fluorescence polarization measurements of lipid order are more physically rigorous because they allow the angular range of acyl chain reorientation ("lipid order") to be resolved from its rate, and considerable research has been devoted to the interpretation of these measurements.[1,9,18,19]

Table 14.1 *Selected applications of DPH and TMA-DPH (D-202, T-204).*

Measurement of Lipid Order
- Isolated or intact cell membranes [1-5]
- Model membrane systems [6-10]
- Proteins interacting with lipids: bacteriophage coat protein,[11] Ca^{2+}-ATPase,[12] protein kinase C [13,14] and signal peptides [15]
- Lipoproteins [16-19]

Agents and Conditions Influencing Membrane Fluidity
- Specific agents: alcohols,[20,21] arachidonic acid and other polyunsaturated fatty acids,[22,23] bacterial ether lipids,[24] cholesterol and other sterols,[25-27] cyclic hydrocarbons,[28] dietary lipids,[29] insecticides,[30-32] intracellular sodium,[33] mycobacterial glycopeptidolipids,[34] paclitaxel (Taxol equivalent),[35] probucol,[36] sphingosine [37] and vasodilatory drugs [38]
- Conditions: aging,[39,40] cytoskeletal modification,[41] genetic disorders,[42-44] hydration,[45] ionizing radiation,[46] lipid peroxidation [47-49] and transmembrane potential [50]

Endocytosis and Exocytosis
- Endocytosis [51-56]
- Exocytosis [53,57-60]

Analytical Applications
- Detection of gangliosides on TLC [61]
- Detection of singlet oxygen in membranes [62]
- HPLC analysis of phospholipids [63]

1. Biochim Biophys Acta 1190, 1 (1994); **2.** Biochim Biophys Acta 1147, 245 (1993); **3.** Biorheology 29, 507 (1992); **4.** Am J Physiol 260, C43 (1991); **5.** Biochim Biophys Acta 1067, 171 (1991); **6.** Biophys J 68, 1944 (1995); **7.** Chem Phys Lipids 71, 61 (1994); **8.** Biochim Biophys Acta 1149, 241 (1993); **9.** Chem Phys Lipids 64, 117 (1993); **10.** Chem Phys Lipids 64, 99 (1993); **11.** Biochemistry 32, 10720 (1993); **12.** Biochemistry 33, 4974 (1994); **13.** J Biol Chem 269, 4866 (1994); **14.** Biochemistry 31, 662 (1992); **15.** Biophys J 67, 1534 (1994); **16.** Biochemistry 33, 4879 (1994); **17.** Biochemistry 33, 2356 (1994); **18.** J Biol Chem 269, 10298 (1994); **19.** Chem Phys Lipids 60, 1 (1991); **20.** Biochemistry 34, 5945 (1995); **21.** Biochim Biophys Acta 1112, 14 (1992); **22.** Biochim Biophys Acta 1105, 285 (1992); **23.** Chem Phys Lipids 57, 87 (1991); **24.** Biochemistry 31, 12493 (1992); **25.** Biochemistry 32, 2047 (1993); **26.** Biochim Biophys Acta 1028, 82 (1990); **27.** Chem Phys Lipids 50, 43 (1989); **28.** J Biol Chem 269, 8022 (1994); **29.** Biochim Biophys Acta 1211, 198 (1994); **30.** Biochim Biophys Acta 1147, 137 (1993); **31.** Biochim Biophys Acta 1066, 124 (1991); **32.** Biochim Biophys Acta 982, 161 (1989); **33.** Life Sci 52, 1559 (1993); **34.** Biochemistry 33, 7056 (1994); **35.** Biochemistry 33, 8941 (1994); **36.** J Biol Chem 267, 12291 (1992); **37.** Chem Phys Lipids 66, 123 (1993); **38.** Biochim Biophys Acta 1153, 20 (1993); **39.** Cell Mol Biol 37, 15 (1991); **40.** J Neurochem 55, 70 (1990); **41.** J Leukocyte Biol 56, 192 (1994); **42.** Biol of the Cell 75, 135 (1992); **43.** Biochim Biophys Acta 1070, 253 (1991); **44.** Biochim Biophys Acta 1069, 151 (1991); **45.** Biochemistry 34, 6188 (1995); **46.** Int'l J Radiation Biol 61, 791 (1992); **47.** Meth Enzymol 233, 273 (1994); **48.** Physiol Res 43, 101 (1994); **49.** Am J Physiol 264, G1009 (1993); **50.** Biochim Biophys Acta 1023, 41 (1990); **51.** J Cell Biol 127, 725 (1994); **52.** J Cell Biol 125, 783 (1994); **53.** J Biol Chem 268, 6742 (1993); **54.** Biol of the Cell 71, 293 (1991); **55.** Biochim Biophys Acta 1030, 73 (1990); **56.** Cell Biophys 14, 17 (1989); **57.** Biochim Biophys Acta 1147, 194 (1993); **58.** Biochim Biophys Acta 1067, 71 (1991); **59.** Biochim Biophys Acta 901, 138 (1987); **60.** Biochemistry 25, 2149 (1986); **61.** Anal Biochem 208, 121 (1993); **62.** Biochim Biophys Acta 1082, 94 (1991); **63.** Anal Biochem 220, 172 (1994).

TMA-DPH and TMAP-DPH

Advantages of TMA-DPH and TMAP-DPH

To improve the localization of DPH in the membrane, the derivative TMA-DPH (T-204) was introduced, which contains a cationic trimethylammonium substituent that acts as a surface anchor.[20,21] More recently, we have added a new version of this probe, TMAP-DPH (T-3900), that incorporates an additional three-carbon spacer between the fluorophore and the trimethylammonium substituent. Like DPH, these derivatives readily partition from aqueous dispersions into membranes and other lipid assemblies, accompanied by strong fluorescence enhancement. The lipid–water partition coefficients (K_p) for TMA-DPH and TMAP-DPH ($K_p = 2.4 \times 10^5$ and 2.9×10^5, respectively) are lower than for DPH ($K_p = 1.3 \times 10^6$), reflecting the increased water solubility caused by their polar substituents.[22] Recent papers describe the partitioning properties of TMAP-DPH in multilamellar liposomes.[23,24] The fluorescence decay lifetime of TMA-DPH is more sensitive to changes in lipid composition and temperature than is the fluorescence decay lifetime of DPH.[25-27]

Cellular Applications of TMA-DPH and TMAP-DPH

Staining of cell membranes by TMA-DPH is much more rapid than staining by DPH. However, the duration of plasma membrane surface staining by TMA-DPH before internalization into the cytoplasm is quite prolonged.[28,29] As a consequence, TMA-DPH introduced into Madin–Darby canine kidney (MDCK) cell plasma membranes does not diffuse through tight junctions and remains in the apical domain, whereas the anionic DPH propionic acid (P-459, see Section 13.2) accumulates rapidly in intracellular membranes.[30] TMA-DPH residing in the plasma membrane can be extracted by washing with medium, thus providing a method for isolating internalized probe and monitoring endocytosis (Table 14.1, also see Section 17.2). Furthermore, because TMA-DPH is virtually nonfluorescent in water and binds in proportion to the available membrane surface,[31] its fluorescence intensity is sensitive to increases in plasma membrane surface area resulting from exocytosis (Table 14.1).

TMA-DPH fluorescence polarization measurements can be combined with video microscopy to provide spatially resolved images of phospholipid order in large liposomes and single cells.[32-35] Information regarding lipid order heterogeneity among cell populations can be obtained in a similar way using flow cytometry.[36,37]

1. Biochemistry 30, 5565 (1991); **2.** Chem Phys Lipids 57, 39 (1991); **3.** Biochimie 71, 23 (1989); **4.** Biochim Biophys Acta 859, 209 (1986); **5.** Biochim Biophys Acta 941, 102 (1988); **6.** Biophys Chem 48, 205 (1993); **7.** Biophys J 59, 466 (1991); **8.** Biophys J 56, 723 (1989); **9.** Biophys Chem 48, 337 (1994); **10.** Biochim Biophys Acta 1104, 273 (1992); **11.** Biochemistry 29, 3248 (1990); **12.** Chem Phys Lipids 50, 1 (1989); **13.** Biochim Biophys Acta 649, 471 (1981); **14.** Chem Phys Lipids 64, 117 (1993); **15.** Chem Phys Lipids 64, 93 (1993); **16.** Chem Phys Lipids 50, 171 (1989); **17.** Biochim Biophys Acta 854, 38 (1986); **18.** Chem Phys 185, 393 (1994); **19.** Chem Phys Lett 216, 559 (1993); **20.** Biochemistry 27, 7723 (1988); **21.** Biochemistry 20, 7333 (1981); **22.** Biochem Biophys Res Comm 181, 166 (1991); **23.** Chem Phys Lipids 66, 135 (1993); **24.** J Fluorescence 3, 145 (1993); **25.** Chem Phys Lipids 55, 29 (1990); **26.** Biochemistry 26, 5121 (1987); **27.** Biochemistry 26, 5113 (1987); **28.** Biochim Biophys Acta 845, 60 (1985); **29.** Cell Biophys 5, 129 (1983); **30.** Am J Physiol 255, F22 (1988); **31.** Biochemistry 25, 2149 (1986); **32.** Florine-Casteel, K., Lemasters, J.L. and Herman, B. in *Fluorescent and Luminescent Probes for Biological Activity*, W.T. Mason, Ed., Academic Press (1993) pp. 420–425; **33.** Am J Physiol 260, C1 (1991); **34.** FASEB J 5, 2078 (1991); **35.** Biophys J 57, 1199 (1990); **36.** Biochim Biophys Acta 1067, 71 (1991); **37.** Plant Physiol 94, 729 (1990).

14.4 Data Table *Diphenylhexatriene (DPH) and Derivatives*

Cat #	Structure	MW	Storage	Soluble	Abs	{ε × 10^{-3}}	Em	Solvent	Notes
D-202	✓	232	L	DMF, MeCN	350	{88}	452	MeOH	<1, 2>
P-3900	✓	504	D,L	DMF, DMSO	354	{85}	429	MeOH	<1>
T-204	✓	462	D,L	DMF, DMSO	355	{75}	430	MeOH	<1>

For definitions of the contents of this data table, see "How to Use the Handbook" on page vi.
Structure: Chemical structure drawing: ✓ = shown in this section.

<1> Diphenylhexatriene (DPH) and its derivatives are essentially nonfluorescent in water. Absorption and emission spectra have multiple peaks. The wavelength, resolution and relative intensity of these peaks are environment dependent. Abs and Em values are for the most intense peak in the solvent specified.

<2> Stock solutions of DPH (D-202) are often prepared in tetrahydrofuran (THF). Long-term storage of THF solutions is not recommended because of possible peroxide formation in that solvent.

14.4 Structures *Diphenylhexatriene (DPH) and Derivatives*

D-202

P-3900

T-204

14.4 Price List *Diphenylhexatriene (DPH) and Derivatives*

Cat #	Product Name	Unit Size	Price Per Unit ($) 1–4 Units	5–24 Units
D-202	1,6-diphenyl-1,3,5-hexatriene (DPH)	100 mg	23.00	18.40
P-3900	N-((4-(6-phenyl-1,3,5-hexatrienyl)phenyl)propyl)trimethylammonium p-toluenesulfonate (TMAP-DPH)	5 mg	148.00	118.40
T-204	1-(4-trimethylammoniumphenyl)-6-phenyl-1,3,5-hexatriene p-toluenesulfonate (TMA-DPH)	25 mg	155.00	124.00

14.5 Membrane Probes with Environment-Sensitive Spectral Shifts

Prodan, Laurdan and Patman

Prodan (P-248), introduced by Weber and Farris in 1979,[1] has both an electron-donor and an electron-acceptor substituent, resulting in a large excited-state dipole moment and extensive solvent polarity–dependent fluorescence shifts[2] (Figure 14.3). Several variants of the original probe have since been prepared, including the lipophilic derivatives laurdan (D-250) and patman (P-1387) and thiol-reactive derivatives acrylodan and badan (A-433, B-6057; see Section 2.3), which can be used to confer the environment-sensitive properties of this fluorophore on bioconjugates. When prodan or its derivatives are incorporated into membranes, their fluorescence spectra are sensitive to the physical state of the surrounding phospholipids. In membranes, prodan appears to localize at the surface,[3] although Fourier transform infrared (FTIR) measurements indicate some degree of penetration into the lipid interior.[4] Excited-state relaxation of prodan is sensitive to the nature of the linkage (ester or ether) between phospholipid hydrocarbon tails and the glycerol backbone.[5] In contrast, laurdan's excited-state relaxation

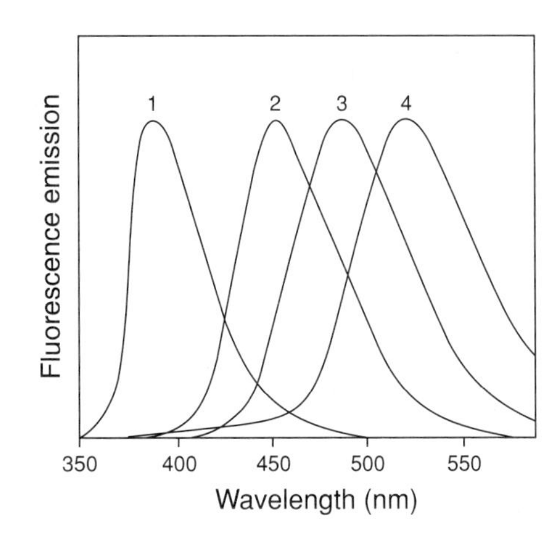

Figure 14.3 *Normalized emission spectra of prodan (P-248) in 1) cyclohexane, 2) dimethylformamide, 3) ethanol and 4) water.*

is found to be independent of head-group type, and is instead determined by water penetration into the lipid bilayer.[6,7]

Much experimental work using these probes has sought to characterize coexisting lipid domains based on their distinctive fluorescence spectra,[8-11] an approach that is intrinsically amenable to dual-wavelength ratio measurements.[7,12] Other applications include detecting nonbilayer lipid phases,[13,14] mapping changes in membrane structure induced by cholesterol and alcohols [15-17] and assessing the polarity of lipid/water interfaces.[18,19] Like ANS, prodan is also useful as a noncovalently interacting probe for proteins.[20-22]

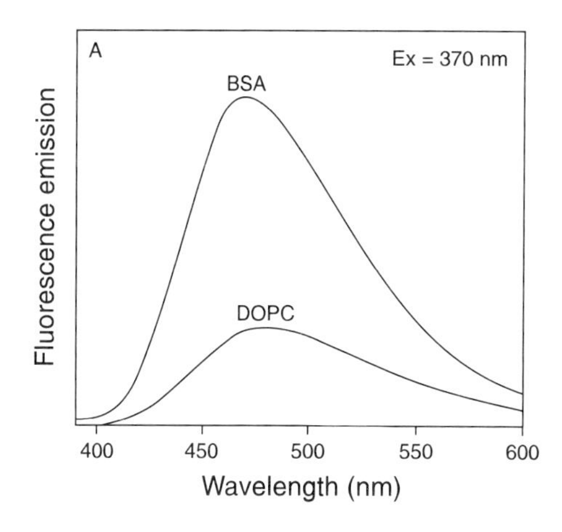

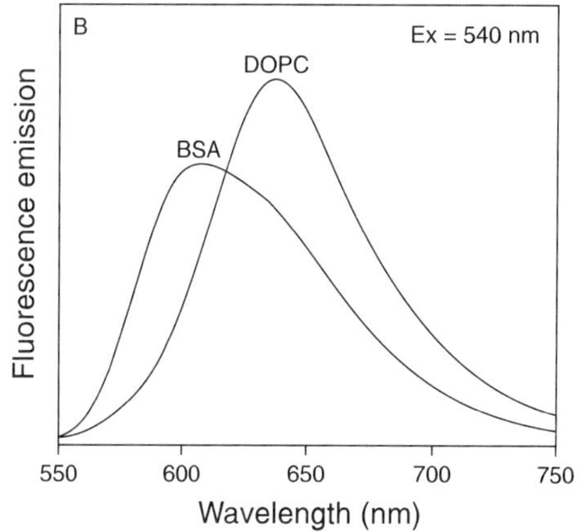

Figure 14.4 Fluorescence emission spectra of (A) 1,8-ANS (A-47) and (B) nile red (N-1142) bound to protein (BSA) and phospholipid vesicles (DOPC). Samples comprised 1 µM dye added to 20 µM bovine serum albumin (BSA) or 100 µM dioctadecenoylglycerophosphocholine (DOPC).

ANS and Related Probes

ANS

The use of anilinonaphthalene sulfonates (ANS) as fluorescent probes dates back to the pioneering work of Weber in the 1950s, and this class of probes remains valuable for studying both membrane surfaces and proteins (Table 14.2). Slavik's 1982 review of its properties is recommended reading, especially for the extensive compilation of spectral data.[23] The primary member of this class, 1,8-ANS (A-47), and its analogs 2,6-ANS (A-50) and 2,6-TNS (T-53) are all essentially nonfluorescent in water, only becoming appreciably fluorescent when bound to membranes (quantum yields ~0.25) or proteins (quantum yields ~0.7) [23-25] (Figure 14.4). This property makes them sensitive indicators of protein folding, conformational changes and other processes that modify the exposure of the probe to water (Table 14.2). Fluorescence of 2,6-ANS is also enhanced by cyclodextrins, permitting a sensitive method for separating and analyzing cyclodextrins with capillary electrophoresis.[26]

Table 14.2 Selected applications of anilinonaphthalene sulfonates (A-47, A-50, B-153, T-53).

Protein Structure and Function
- Conformational changes [1-4]
- Ligand binding [5-8]
- Phosphorylase kinase assay [9]
- Protein folding [10-15]

Membrane Surface Interactions
- Actions of bactericidal proteins (bacteriocins) [16]
- Ca^{2+} interactions with lipid bilayer surfaces [17]
- Lipid–protein interactions [18]
- Membrane surface potential [19,20]

Analytical Applications
- Cation-selective optical sensors [21,22]
- Critical micelle concentrations [23]
- Protein detection in SDS gels without destaining [24,25]
- Quantitation of lipids by TLC [26,27]
- Sensitive displacement assay for avidin–biotin interactions [28,29]

1. J Biol Chem 270, 7335 (1995); **2.** J Biol Chem 269, 17009 (1994); **3.** J Biol Chem 269, 7908 (1994); **4.** J Biol Chem 269, 7631 (1994); **5.** J Biol Chem 270, 1011 (1995); **6.** Biochemistry 33, 8991 (1994); **7.** Biochemistry 33, 7120 (1994); **8.** J Biol Chem 269, 412 (1994); **9.** Biochem Biophys Res Comm 174, 344 (1991); **10.** Biochemistry 34, 7225 (1995); **11.** Biochemistry 34, 6925 (1995); **12.** Biochemistry 34, 6815 (1995); **13.** Biochemistry 34, 5242 (1995); **14.** Biochemistry 34, 1867 (1995); **15.** Science 260, 1110 (1993); **16.** J Appl Bacteriol 76, 30 (1994); **17.** Biochim Biophys Acta 1065, 114 (1991); **18.** Biochemistry 34, 4856 (1995); **19.** Biophys J 57, 335 (1990); **20.** Biochim Biophys Acta 805, 345 (1984); **21.** Anal Chem 61, 211 (1989); **22.** Anal Chim Acta 184, 251 (1986); **23.** Anal Biochem 115, 278 (1981); **24.** Anal Biochem 165, 430 (1987); **25.** Electrophoresis 6, 527 (1985); **26.** Chromatographia 30, 414 (1990); **27.** J Chromatography 237, 522 (1982); **28.** Biochim Biophys Acta 956, 23 (1988); **29.** Anal Biochem 151, 178 (1985).

Bis-ANS

Bis-ANS (B-153) is superior to 1,8-ANS as a probe for nonpolar cavities in proteins, often binding with an affinity that is orders-of-magnitude higher.[27-30] Bis-ANS has particularly high affinity for nucleotide binding sites of proteins.[31-33] It is also useful as a structural probe for tubulin [34,35] and as an inhibitor of microtubule assembly.[36,37]

NPN

N-Phenyl-1-naphthylamine (NPN, P-65) is a nonpolar analog of ANS. It was found to be superior to DPH and nile red as a membrane stain in a comparison conducted on hyphal tips of fungi.[38] NPN is a sensitive probe of lipid mobility in lipid bilayers and monolayers and preferentially locates immediately below the phospholipid head-group region.[39,40] In addition, NPN has been found to be sensitive to redox activity in *Escherichia coli* membrane vesicles.[41,42]

DCVJ

The styrene derivative DCVJ (D-3923) is a sensitive indicator of tubulin assembly and actin polymerization.[43,44] The fluorescence quantum yield of DCVJ is strongly dependent on environmental rigidity, resulting in large fluorescence increases when the dye binds to antibodies [45] and when it is compressed in synthetic polymers or phospholipid membrane interiors.[46,47]

Dansyl Lysine

Dansyl lysine (D-372) has been used to detect lipid hexagonal phases,[14] and its fluorescence is highly sensitive to membrane cholesterol content.[48] Upon binding to anti-dansyl antibodies (A-6398, see Section 7.3), dansyl lysine exhibits increased fluorescence,[49] a feature that has been exploited for measuring the segmental flexibility of receptor-bound monoclonal antibodies.[50,51] Dansyl lysine is also a selective viability probe for cells killed by heat shock [52,53] and has been used with FITC (F-143, see Section 1.3) and Hoechst 33342 (H-1399, see Section 8.1) to simultaneously measure antigen expression and DNA content of hyperthermally sensitive and resistant cell populations.[54]

Nile Red

The phenoxazine dye nile red (N-1142) is used to localize and quantitate lipids, particularly neutral lipid droplets within cells.[55-58] It is selective for neutral lipids such as cholesteryl esters [59,60] (and therefore for lipoproteins) but is also suitable for staining lysosomal phospholipid inclusions.[61] Nile red is almost nonfluorescent in water and polar solvents but undergoes fluorescence enhancement and large absorption and emission blue shifts in nonpolar environments.[62,63] Its fluorescence enhancement upon binding to proteins is weaker than that produced by its association with lipids [63] (Figure 14.4). Ligand binding studies on tubulin [64] and tryptophan synthase [65] have exploited the environmental sensitivity of nile red's fluorescence. Nile red has also been used to detect sphingolipids on thin layer chromatograms [66] and to stain proteins after SDS-polyacrylamide gel electrophoresis.[67]

1. Biochemistry 18, 3075 (1979); **2.** Photochem Photobiol 58, 499 (1993); **3.** Biochemistry 27, 399 (1988); **4.** Biochemistry 28, 8358 (1989); **5.** Biochemistry 29, 11134 (1990); **6.** Biophys J 66, 763 (1994); **7.** Biophys J 60, 179 (1991); **8.** Biophys J 66, 120 (1994); **9.** Photochem Photobiol 57, 420 (1993); **10.** Biophys J 57, 1179 (1990); **11.** J Biol Chem 265, 20044 (1990); **12.** Photochem Photobiol 57, 403 (1993); **13.** Biophys J 57, 925 (1990); **14.** Biochemistry 31, 1550 (1992); **15.** Biophys J 68, 1895 (1995); **16.** Biophys J 65, 1404 (1993); **17.** Biochemistry 31, 9473 (1992); **18.** J Biol Chem 269, 10298 (1994); **19.** J Biol Chem 269, 7429 (1994); **20.** Biochem J 290, 411 (1993); **21.** Eur J Biochem 204, 127 (1992); **22.** Nature 319, 70 (1986); **23.** Biochim Biophys Acta 694, 1 (1982); **24.** Biochemistry 7, 3381 (1968); **25.** Biochemistry 5, 1908 (1966); **26.** J Chromatography A 680, 233 (1994); **27.** Arch Biochem Biophys 268, 239 (1989); **28.** Biochemistry 24, 3852 (1985); **29.** Biochemistry 24, 2034 (1985); **30.** Biochemistry 8, 3915 (1969); **31.** Biochemistry 31, 2982 (1992); **32.** Biochim Biophys Acta 1040, 66 (1990); **33.** Proc Natl Acad Sci USA 74, 2334 (1977); **34.** Biochemistry 33, 11900 (1994); **35.** Biochemistry 33, 11891 (1994); **36.** Biochemistry 31, 6470 (1992); **37.** J Biol Chem 259, 14647 (1984); **38.** Exp Mycol 15, 103 (1991); **39.** Biochem Cell Biol 68, 574 (1990); **40.** Biochim Biophys Acta 644, 233 (1981); **41.** Arch Biochem Biophys 282, 372 (1990); **42.** FEBS Lett 229, 127 (1987); **43.** Anal Biochem 204, 110 (1992); **44.** Biochemistry 28, 6678 (1989); **45.** Biochemistry 32, 7589 (1993); **46.** Chem Phys 169, 351 (1993); **47.** Biochemistry 25, 6114 (1986); **48.** Biophys J 42, 307 (1983); **49.** Biochemistry 21, 5738 (1982); **50.** Biochemistry 29, 4607 (1990); **51.** Biochemistry 24, 7810 (1985); **52.** Radiation Res 112, 351 (1987); **53.** Cancer Res 46, 5064 (1986); **54.** Cytometry 11, 533 (1990); **55.** Cytometry 17, 151 (1994); **56.** J Cell Biol 123, 1567 (1993); **57.** Exp Cell Res 199, 29 (1992); **58.** Brain Res 406, 215 (1987); **59.** J Cell Biol 108, 2201 (1989); **60.** J Chromatography 421, 136 (1987); **61.** Histochemistry 97, 349 (1992); **62.** Anal Chem 62, 615 (1990); **63.** Anal Biochem 167, 228 (1987); **64.** J Biol Chem 265, 14899 (1990); **65.** J Biol Chem 270, 6357 (1995); **66.** Anal Biochem 208, 121 (1994); **67.** Anal Biochem 199, 169 (1991).

14.5 Data Table *Membrane Probes with Environment-Sensitive Spectral Shifts*

Cat #	Structure	MW	Storage	Soluble	Abs	{ε × 10⁻³}	Em	Solvent	Notes
A-47	✓	299	L	pH>6, DMF	372	{7.8}	480	MeOH	<1>
A-50	✓	299	L	DMF	319	{27}	422	MeOH	<1>
B-153	✓	673	L	pH>6	395	{23}	500	MeOH	<1, 2>
D-250	✓	354	L	DMF, MeCN	364	{20}	497	MeOH	<3>
D-372	✓	379	L	pH>7	328	{4.5}	561	pH 8	<4>
D-3923	✓	249	L	DMF, DMSO	456	{61}	493	MeOH	
H-1387	✓	517	L	DMSO, EtOH	357	{17}	488	MeOH	<5>
N-1142	✓	318	L	DMF, DMSO	552	{45}	636	MeOH	<6>
P-65	✓	219	L	MeOH, DMF	337	{8.1}	426	MeOH	
P-248	✓	227	L	DMF, MeCN	361	{18}	498	MeOH	<3>
T-53	✓	335	L	DMF	318	{26}	443	MeOH	<1>

For definitions of the contents of this data table, see "How to Use the Handbook" on page vi.
Structure: Chemical structure drawing: ✓ = shown in this section.

<1> Fluorescence quantum yields of ANS and its derivatives are environment dependent and are particularly sensitive to the presence of water. QY of A-47 is about 0.4 in EtOH, 0.2 in MeOH and 0.004 in water. Em is also somewhat solvent dependent [Biochim Biophys Acta 694, 1 (1982)].

<2> B-153 is soluble in water at 0.1–1.0 mM after heating.

<3> The emission spectrum of P-248 is solvent dependent (see Figure 14.3): Em = 401 nm in cyclohexane, 440 nm in $CHCl_3$, 462 nm in MeCN, 496 nm in EtOH and 531 nm in H_2O [Biochemistry 18, 3075 (1979)]. Abs is only slightly solvent-dependent. The emission spectra of D-250 in these solvents are similar to those of P-248.

<4> Fluorescence intensity of D-372 increases at least 50-fold upon binding to phospholipid vesicles (Em = 515 nm), relative to fluorescence in H_2O [Biophys J 42, 307 (1983)].

<5> Em of H-1387 in other solvents is listed in Biochemistry 22, 5714 (1983).

<6> The absorption and fluorescence spectra and the fluorescence quantum yield of N-1142 are highly dependent on solvent [Anal Biochem 167, 228 (1987); J Lipid Res 26, 781 (1985)].

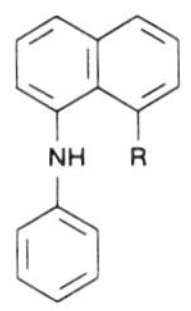

A-47 R = $-SO_3H$
P-65 R = H

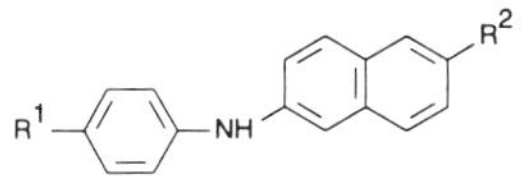

A-50 R^1 = H R^2 = $-SO_3H$
T-53 R^1 = $-CH_3$ R^2 = $-SO_3^-$ Na^+

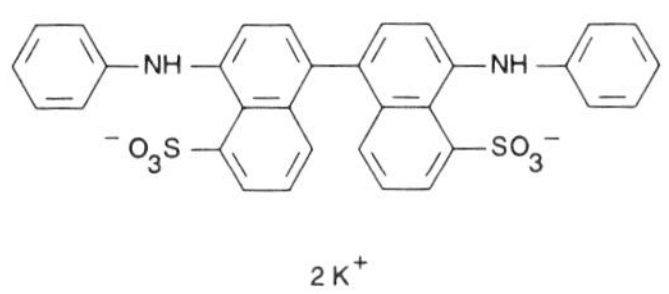

B-153

D-250 see P-248

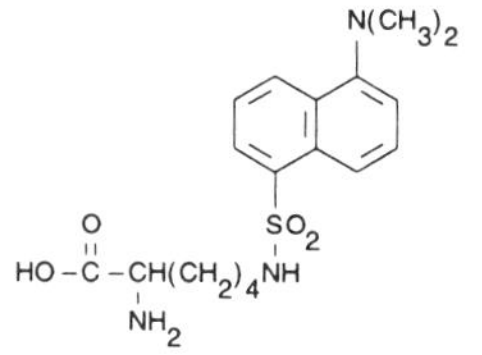

D-372

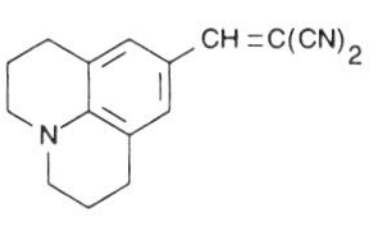

D-3923

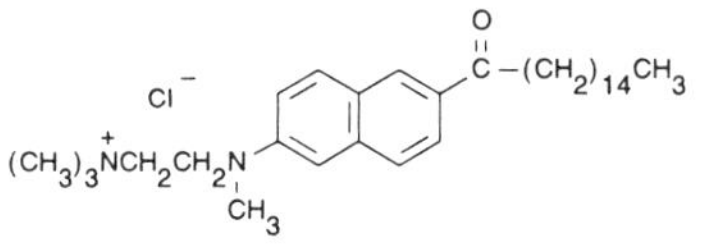

H-1387

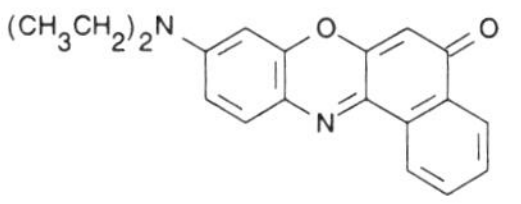

N-1142

P-65 see A-47

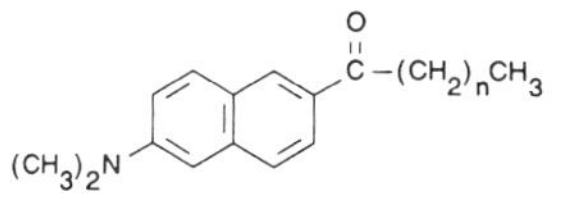

P-248 n = 1
D-250 n = 10

T-53 see A-50

14.5 Price List *Membrane Probes with Environment-Sensitive Spectral Shifts*

Cat #	Product Name	Unit Size	Price Per Unit ($) 1–4 Units	5–24 Units
A-47	1-anilinonaphthalene-8-sulfonic acid (1,8-ANS) *high purity*	100 mg	88.00	70.40
A-50	2-anilinonaphthalene-6-sulfonic acid (2,6-ANS)	100 mg	88.00	70.40
B-153	bis-ANS (4,4′-dianilino-1,1′-binaphthyl-5,5′-disulfonic acid, dipotassium salt)	10 mg	125.00	100.00
D-3923	4-(dicyanovinyl)julolidine (DCVJ)	25 mg	48.00	38.40
D-372	N-ϵ-(5-dimethylaminonaphthalene-1-sulfonyl)-L-lysine (dansyl lysine)	100 mg	48.00	38.40
D-250	6-dodecanoyl-2-dimethylaminonaphthalene (laurdan)	100 mg	135.00	108.00
H-1387	6-hexadecanoyl-2-(((2-(trimethylammonium)ethyl)methyl)amino)naphthalene chloride (patman)	5 mg	48.00	38.40
N-1142	nile red	25 mg	60.00	48.00
P-65	N-phenyl-1-naphthylamine *high purity*	1 g	48.00	38.40
P-248	6-propionyl-2-dimethylaminonaphthalene (prodan)	100 mg	135.00	108.00
T-53	2-(p-toluidinyl)naphthalene-6-sulfonic acid, sodium salt (2,6-TNS)	100 mg	48.00	38.40

14.6 Miscellaneous Membrane Probes

Pyrene and NBD Probes

We offer three bis-pyrene probes consisting of two pyrene fluorophores linked directly (B-7791) or by a C_{10} (1,10-bis-(1-pyrene)-decane, B-162) or C_3 (1,3-bis-(1-pyrene)propane, B-311) alkylene spacer. The dipyrenylalkanes are somewhat analogous to the bis-pyrenyl phospholipids (see Section 13.2) in that excimer formation (and consequently, the fluorescence emission wavelength) is controlled by intramolecular rather than bimolecular interactions. These probes are highly sensitive to constraints imposed by their environment, and can therefore be used as viscosity sensors for interior regions of lipoproteins, membranes, micelles, liquid crystals and synthetic polymers.[1-5] Excimer formation results in a spectral shift (see Figure 13.6 in Chapter 13), potentially allowing these probes to be used for ratio imaging of molecular mobility. However, pyrene fatty acids (see Section 13.2) appear to be preferable for this purpose because the uptake of 1,3-bis-(1-pyrene)-propane by cells is limited.[6] 1,3-Bis-(1-pyrene)propane's excimer formation is dependent on temperature, and this probe has been used as a fluorescence readout thermometer capable of measuring up to 400°C.[7] Uptake and biosynthetic incorporation of 1-pyrene-nonanol (P-394) by peroxisome-deficient cells are inhibited relative to those of the corresponding fatty acid. This difference, coupled with the phototoxicity of pyrene, allows selection of peroxisome-deficient mutants.[8] 1-Pyrenebutanol (P-244) may be preferable to pyrene-labeled fatty acids for photoablation of myeloid leukemia cells because it cannot be metabolized.[9] We also offer anionic pyrenesulfonic acid (P-80), which has been used to probe inverted micelles,[10,11] and cationic triphenylphosphonium (P-6921) and trimethylammonium (P-314) pyrene derivatives.[12] As with TMA-DPH, the uptake of these cationic probes by cell membranes may be faster than that of the neutral or anionic probes. Two nonpolar

NBD probes, NBD dihexadecylamine (D-69) and NBD hexadecylamine (H-429), have primarily been used for fluorescence detection of lipids separated by TLC.[13,14]

Nonpolar BODIPY Probes

Molecular Probes offers three nonpolar derivatives of the BODIPY fluorophore that offer an unusual combination of nonpolar structure and long-wavelength absorption and fluorescence. These dyes have potential applications as stains for neutral lipids and as tracers for oils and other nonpolar liquids. Staining with BODIPY 493/503 (D-3922) has been shown by flow cytometry to be more specific for cellular lipid droplets than staining with nile red [15] (N-1142). Mark Cooper at the University of Washington has found that BODIPY 505/515 (D-3921) rapidly permeates cell membranes of live zebrafish embryos, selectively staining cytoplasmic yolk platelets. This staining provides dramatic contrast enhancement of cytoplasm relative to nucleoplasm and interstitial spaces, enabling dynamic developmental processes in the embryo to be followed using time-lapse confocal microscopy. Images and video sequences generated using these techniques can be accessed at the Web site FishScope (http://weber.u.washington.edu/~fishscop). BODIPY dyes have small fluorescence Stokes shifts, extinction coefficients that are typically greater than 80,000 $cm^{-1}M^{-1}$ and high fluorescence quantum yields (0.94 for D-3921 in methanol [16]) that are not diminished in water. Their photostability is generally high; this, together with other favorable characteristics (very low triplet–triplet absorption), results in excellent performance of BODIPY 493/503 and BODIPY 505/515 as flashlamp-pumped laser dyes.[17] BODIPY 665/676 (B-3932) can be excited using the 633 nm spectral line of the He-Ne laser or the 647 nm spectral line of the Ar-Kr laser for applications requiring long-wavelength fluorescence detection.

Anthracenes and Rubrene

We offer five aryl-substituted anthracenes (B-7801, B-7802, B-7803, B-7804, B-7805) as well as bi-9-anthryl (B-7787), a directly coupled anthracene dimer. The fluorescence spectra of aryl-substituted anthracenes are highly sensitive to solvent viscosity, and they are consequently useful probes of local volume fluctuations in membranes and macromolecules.[18] Furthermore, anthracene and its derivatives dimerize upon illumination at 360 nm, yielding nonfluorescent products. Dimerization can be partially reversed by illumination at 280 nm.[19] These spectral properties can be exploited to study lateral diffusion and distribution of membrane lipids.[20] 9,10-Diphenylanthracene (D-7801) is often employed as a fluorescence quantum yield and lifetime standard.[21,22] Bis(phenylethynyl)anthracenes (B-7802, B-7803, B-7804) are also widely used as excitation energy acceptors from chemiluminescent donors such as TCPO [23] (B-1174; Section 24.3). Rubrene (R-7800) is another polyaromatic hydrocarbon that is useful as an excitation energy acceptor.[24] It is also a trap for singlet oxygen (see Section 21.3) [25] and may have applications as a visible light–excitable membrane probe.

Bilirubin

Bilirubin (B-8446) is a major product of hemoglobin decomposition that is eliminated via secretion in bile. Its conjugated tetrapyrrole chromophore produces strong absorption at about 450 nm. The fluorescence of bilirubin in fluid solutions is very weak.[26] Its absorption spectrum overlaps the emission spectrum of dansyl-labeled phospholipids (see Section 13.4), allowing quenching by fluorescence resonance energy transfer to be used to investigate bilirubin transport mechanisms.[27,28] The fluorescence of bilirubin can be measured directly in whole blood as a diagnostic for abnormal heme metabolism.[29]

$Pd(QS)_2$: A Reagent for Hydrogenating Lipids in Live Cells

Palladium disodium alizarinmonosulfonate ($Pd(QS)_2$, P-1558) is a reagent developed by Vigh and collaborators for catalytic *in situ* hydrogenation of membrane lipids.[30-33] The water-soluble $Pd(QS)_2$ complex selectively hydrogenates fatty acid double bonds in live cells, decreasing membrane fluidity and inhibiting cell development. Thus, this catalyst is useful for examining the relationship between physical properties and biological functions of cell membranes. An investigation of practical concerns such as catalyst-induced variations in reactivity and accessibility of lipid substrates within membranes and cells has been published.[34]

1. Biochim Biophys Acta 1149, 86 (1993); **2.** Biochem Biophys Res Comm 176, 356 (1991); **3.** Langmuir 6, 542 (1990); **4.** Macromolecules 20, 2241 (1987); **5.** J Am Chem Soc 106, 1045 (1984); **6.** Biochemistry 29, 1949 (1990); **7.** Appl Optics 26, 2256 (1987); **8.** Biochim Biophys Acta 1034, 132 (1990); **9.** Leukemia Res 16, 453 (1992); **10.** J Am Chem Soc 113, 8204 (1991); **11.** Langmuir 3, 494 (1987); **12.** Biochim Biophys Acta 1107, 231 (1992); **13.** Chromatographia 30, 414 (1990); **14.** J Planar Chromatography 2, 19 (1989); **15.** Cytometry 17, 151 (1994); **16.** Anal Biochem 198, 228 (1991); **17.** Optics Comm 70, 425 (1989); **18.** Biophys Chem 19, 227 (1984); **19.** J Chem Phys 43, 4175 (1965); **20.** Eur J Biochem 143, 373 (1984); **21.** Anal Biochem 148, 349 (1985); **22.** J Phys Chem 80, 969 (1976); **23.** U.S. Patent No. 3,729,426 (1973); **24.** J Am Chem Soc 89, 6515 (1967); **25.** Proc Natl Acad Sci USA 89, 11426 (1992); **26.** Angew Chem Int Ed Engl 22, 656 (1983); **27.** J Biol Chem 270, 1074 (1995); **28.** J Biol Chem 269, 19262 (1994); **29.** Anal Biochem 100, 25 (1979); **30.** Biochim Biophys Acta 937, 42 (1988); **31.** Biochim Biophys Acta 921, 167 (1987); **32.** Photobiochem Photobiophys 10, 233 (1986); **33.** Eur J Biochem 177, 477 (1985); **34.** Anal Biochem 194, 34 (1991).

14.6 Data Table *Miscellaneous Membrane Probes*

Cat #	Structure	MW	Storage	Soluble	Abs	{ε × 10^{-3}}	Em	Solvent	Notes
B-162	✓	543	L	MeCN, $CHCl_3$	345	{69}	378	$CHCl_3$	<1>
B-311	✓	445	L	MeCN, $CHCl_3$	344	{80}	378	MeOH	<1>
B-3932	✓	448	F,L	$CHCl_3$, DMSO	665	{161}	676	MeOH	<2>
B-7787	✓	354	L	$CHCl_3$	388	{25}	449	MeOH	
B-7791	✓	402	L	Toluene	349	{47}	378	Benzene	
B-7802	✓	378	L	$CHCl_3$	457	{39}	468	MeOH	<3>
B-7803	✓	407	L	$CHCl_3$	458	{37}	472	MeOH	<3>
B-7804	✓	413	L	$CHCl_3$	468	{37}	483	MeOH	<3>
B-7805	✓	453	L	$CHCl_3$	398	{11}	440	MeOH	
B-8446	✓	585	D,LL	$CHCl_3$	452	{63}	525	$CHCl_3$	<4>
D-69	✓	629	L	MeCN, DMF	486	{27}	543	MeOH	<5>
D-3921	✓	248	F,L	EtOH, DMSO	502	{98}	510	MeOH	<2>
D-3922	✓	262	F,L	EtOH, DMSO	493	{89}	504	MeOH	<2>
D-7801	✓	330	L	$CHCl_3$, MeCN	391	{13}	405	MeOH	<3>
H-429	✓	405	F,D,L	MeCN, DMF	466	{24}	538	MeOH	<5>
P-80	✓	304	L	DMF	344	{42}	376	MeOH	<6, 7>
P-244	✓	274	L	DMF, $CHCl_3$	341	{41}	376	MeOH	<7>
P-314	✓	396	L	DMF	341	{41}	377	MeOH	<7>
P-394	✓	345	L	DMF, $CHCl_3$	341	{42}	377	MeOH	<7>
P-1558	✓	789	D,L	H_2O, MeOH	ND		none		<8>
P-6921	S25.3	600	D,L	DMSO, DMF	340	{37}	377	MeOH	<7>
R-7800	S24.3	533	L	Toluene	526	{6.2}	555	Toluene	<9>

For definitions of the contents of this data table, see "How to Use the Handbook" on page vi.
Structure: Chemical structure drawing: ✓ = shown in this section; Sn.n = shown in Section number n.n.

<1> Absorption spectra of B-162 and B-311 have additional peaks at ~325 nm and <300 nm. Emission spectra include both monomer (~380 nm and ~400 nm) and excimer (~470 nm) peaks.
<2> The absorption and fluorescence spectra of BODIPY derivatives are relatively insensitive to the solvent.
<3> Fluorescence quantum yields measured in air-equilibrated benzene are 0.67 (D-7801), 0.84 (B-7802), 0.88 (B-7803) and 0.90 (B-7804). Fluorescence of D-7801 is strongly quenched by oxygen, but that of B-7802 is not [J Chem Eng Data 19, 214 (1974)].
<4> The fluorescence emission spectrum of bilirubin is excitation-wavelength dependent. The fluorescence quantum yield is very low in fluid solutions but increases when the chromophore is incorporated into a rigid matrix [Angew Chem Int Ed Engl 22, 656 (1983)].
<5> NBD derivatives are almost nonfluorescent in water. NBD derivatives of primary amines such as H-429 have a much higher quantum yield in methanol than those of secondary amines such as D-69. The quantum yields of NBD amine derivatives increase upon binding to membranes.
<6> P-80 fluorescence lifetime is 62 nsec in water, 57 nsec in phospholipid vesicles and 140 nsec in cetyltrimethylammonium micelles.
<7> Pyrene derivatives exhibit structured spectra. The absorption maximum is usually about 340 nm with a subsidiary peak at about 325 nm. There are also strong absorption peaks below 300 nm. The emission maximum is usually about 376 nm with a subsidiary peak at 396 nm. Excimer emission at about 470 nm may be observed at high concentrations.
<8> ND = not determined.
<9> Rubrene has much stronger absorption at shorter wavelengths: ε = {81} at 302 nm.

14.6 Structures *Miscellaneous Membrane Probes*

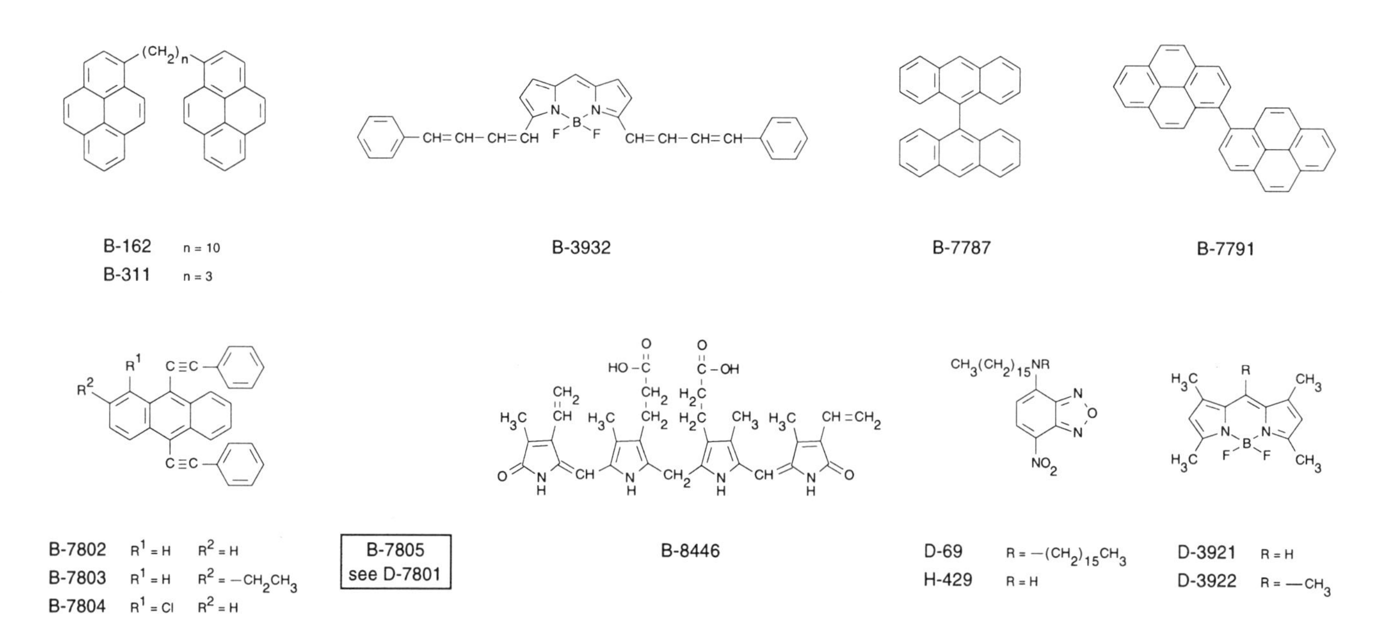

D-7801 $R^1 = H$ $R^2 = H$
B-7805 $R^1 = -OCH_2CH_3$ $R^2 = Cl$

H-429 see D-69

P-80

P-244 n = 4
P-394 n = 9

P-314

P-1558

P-6921 Section 25.3

R-7800 Section 24.3

14.6 Price List *Miscellaneous Membrane Probes*

Cat #		Product Name	Unit Size	Price Per Unit ($) 1–4 Units	5–24 Units
B-7787	*New*	bi-9-anthryl	25 mg	65.00	52.00
B-8446	*New*	bilirubin	1 g	65.00	52.00
B-7791	*New*	1,1′-bipyrene	10 mg	65.00	52.00
B-7805	*New*	9,10-bis-(4-ethoxyphenyl)-2-chloroanthracene	100 mg	18.00	14.40
B-3932		(E,E)-3,5-bis-(4-phenyl-1,3-butadienyl)-4,4-difluoro-4-bora-3a,4a-diaza-s-indacene (BODIPY® 665/676)	5 mg	75.00	60.00
B-7802	*New*	9,10-bis-(phenylethynyl)anthracene	1 g	18.00	14.40
B-7804	*New*	9,10-bis-(phenylethynyl)-1-chloroanthracene	100 mg	18.00	14.40
B-7803	*New*	9,10-bis-(phenylethynyl)-2-ethylanthracene	100 mg	18.00	14.40
B-162		1,10-bis-(1-pyrene)decane	25 mg	48.00	38.40
B-311		1,3-bis-(1-pyrenyl)propane	25 mg	165.00	132.00
D-3922		4,4-difluoro-1,3,5,7,8-pentamethyl-4-bora-3a,4a-diaza-s-indacene (BODIPY® 493/503)	10 mg	75.00	60.00
D-3921		4,4-difluoro-1,3,5,7-tetramethyl-4-bora-3a,4a-diaza-s-indacene (BODIPY® 505/515)	10 mg	75.00	60.00
D-69		4-dihexadecylamino-7-nitrobenz-2-oxa-1,3-diazole (NBD dihexadecylamine)	100 mg	48.00	38.40
D-7801	*New*	9,10-diphenylanthracene	1 g	18.00	14.40
H-429		4-hexadecylamino-7-nitrobenz-2-oxa-1,3-diazole (NBD hexadecylamine)	25 mg	48.00	38.40
P-1558		palladium disodium alizarinmonosulfonate ($Pd(QS)_2$)	100 mg	125.00	100.00
P-244		1-pyrenebutanol	100 mg	65.00	52.00
P-314		1-pyrenebutyltrimethylammonium bromide	25 mg	48.00	38.40
P-6921	*New*	1-pyrenebutyltriphenylphosphonium bromide ($PyTPP^+$)	10 mg	48.00	38.40
P-394		1-pyrenenonanol	25 mg	135.00	108.00
P-80		1-pyrenesulfonic acid, sodium salt	1 g	95.00	76.00
R-7800	*New*	rubrene	1 g	48.00	38.40

Licensing and OEM

The BODIPY®, Cascade Blue®, Oregon Green™, Rhodamine Green™, Rhodol Green™, SNAFL®, SNARF® and Texas Red® fluorophores were developed by Molecular Probes' scientists, and most are patented or have patents pending. Molecular Probes welcomes inquiries about licensing these dyes, as well as their reactive versions and conjugates, for resale or other commercial purposes. Molecular Probes also manufactures many other fluorescent dyes (e.g., FITC and TRITC; succinimidyl esters of FAM, JOE, ROX and TAMRA; reactive phycobiliproteins), crosslinking reagents (e.g., SMCC and SPDP) and biotinylation reagents, which are extensively used by other companies to prepare conjugates. We offer special discounts on almost all of our products when they are purchased in bulk quantities. Please contact our Custom and Bulk Sales Department for more specific information.

Visual Reality

Signal Amplification

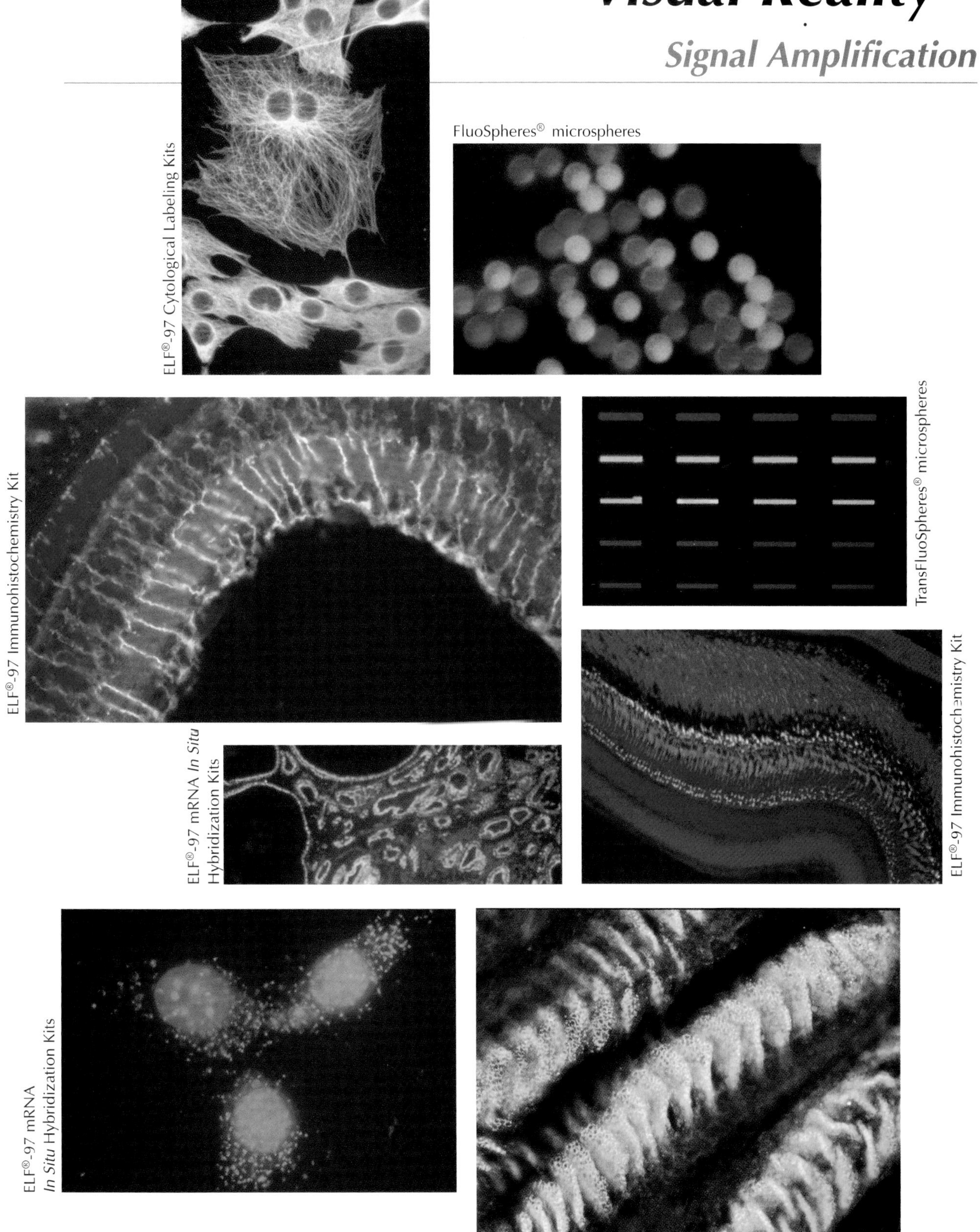

ELF®-97 Cytological Labeling Kits

FluoSpheres® microspheres

ELF®-97 Immunohistochemistry Kit

TransFluoSpheres® microspheres

ELF®-97 mRNA *In Situ* Hybridization Kits

ELF®-97 Immunohistochemistry Kit

ELF®-97 mRNA *In Situ* Hybridization Kits

ELF®-97 Endogenous Phosphatase Detection Kit

Figure captions are immediately following these Color Plates.
For a complete description of products shown, including catalog numbers and prices, see Chapter 6.

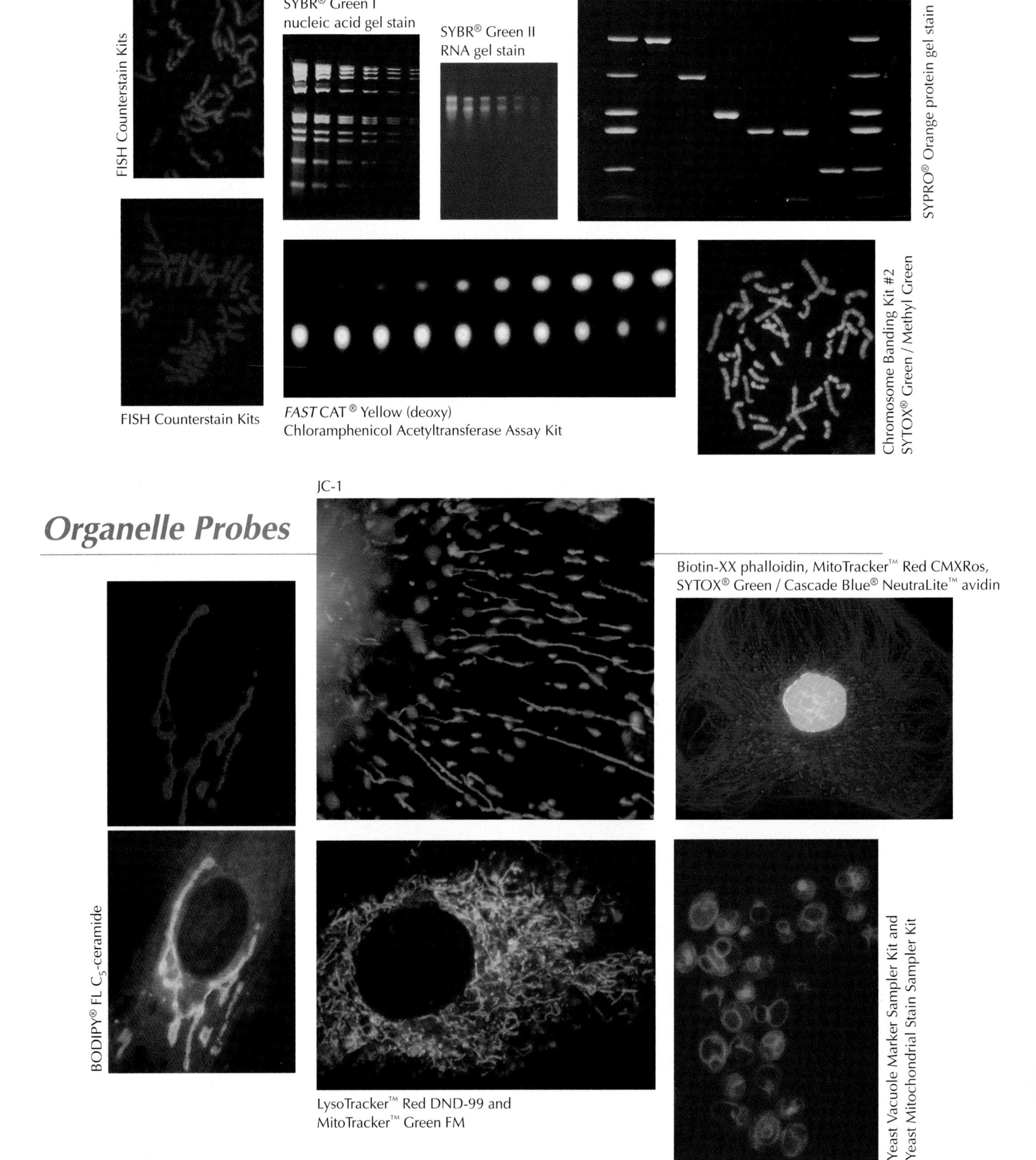

Figure captions are immediately following these Color Plates.
For a complete description of products shown, including catalog numbers and prices, see Chapters 8 through 12.

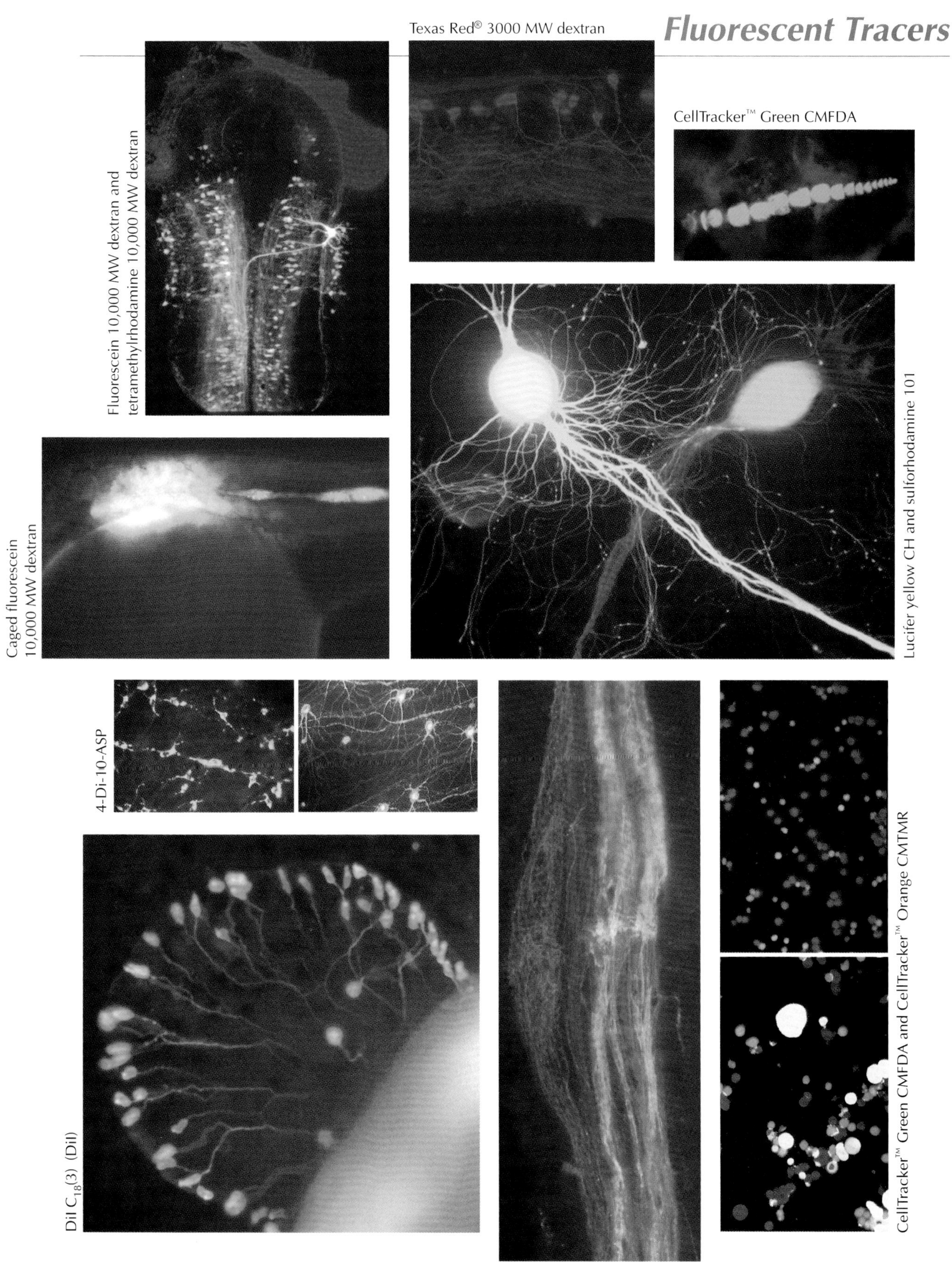

Figure captions are immediately following these Color Plates.
For a complete description of products shown, including catalog numbers and prices, see Chapter 15.

Viability Kits

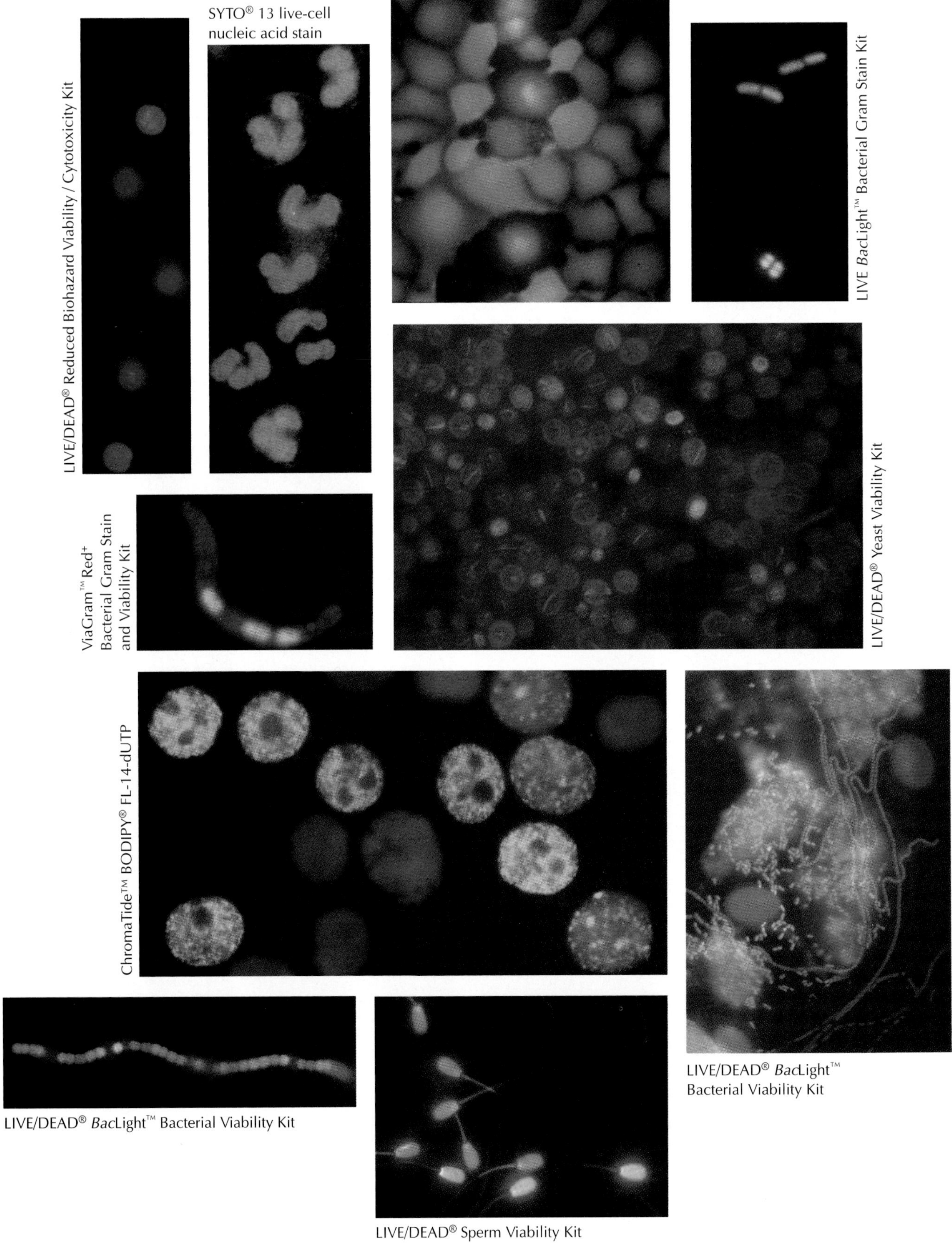

Figure captions are immediately following these Color Plates.
For a complete description of products shown, including catalog numbers and prices, see Chapter 16.

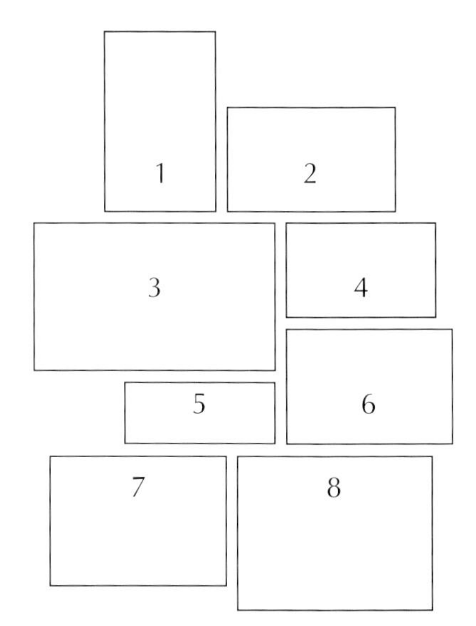

1. Microtubules labeled with mouse monoclonal anti–β-tubulin antibody in conjunction with **biotin-XX goat anti–mouse IgG antibody** (B-2763) and then developed for visualization with alkaline phosphatase–mediated techniques using our new **ELF®-97 Cytological Labeling Kit #2** (E-6603). This kit's novel ELF-97 substrate yields a yellow-green fluorescent precipitate at the site of alkaline phosphatase activity. Prior to antibody labeling, mouse fibroblasts were fixed and permeabilized in the presence of cytoskeletal stabilizing buffer and treated with **paclitaxel** (P-3456) to stabilize microtubule structures. Photo contributed by Violette Paragas, Molecular Probes, Inc.

2. A multicolored mixture of Molecular Probes' **FluoSpheres® fluorescent microspheres** photographed through a DAPI longpass optical filter set. Molecular Probes applies its proprietary fluorescent dye technology to produce a range of intensely fluorescent FluoSpheres microspheres labeled with biotin, streptavidin, NeutraLite™ avidin and protein A, providing important tools for improving the sensitivity of flow cytometry applications and immunodiagnostic assays. Photo contributed by Sam Wells, Molecular Probes, Inc.

3. A transverse section through adult zebrafish retina that has been probed with anti–glial fibrillary acidic protein (anti-GFAP, Chemicon Corp.) and then developed for visualization with alkaline phosphatase–mediated techniques using our **ELF®-97 Immunohistochemistry Kit** (E-6600). This kit's ELF-97 substrate yields an extremely photostable yellow-green signal at the site of anti-GFAP binding, clearly showing that this antibody binds to Müller cells in the zebrafish retina. The retinal section has been counterstained with **Hoechst 33342** (H-1399), which stains all nuclei blue, and with **tetramethylrhodamine wheat germ agglutinin** (W-849), which stains both the inner and outer plexiform layers as well as the photoreceptor outer segments red. Photo contributed by Karen D. Larison, Molecular Probes, Inc.

4. A positively charged membrane containing an approximately equal number of **TransFluoSpheres® fluorescent microspheres** per slot. Our series of proprietary TransFluoSpheres polystyrene beads are designed to be excited with a common wavelength and then detected at a variety of longer wavelengths with minimal spectral overlap. This nylon membrane was excited with 365 nm epi-illumination and photographed through a 400 nm longpass optical filter. Photo contributed by Vicki Singer and Sam Wells, Molecular Probes, Inc.

5. Tissue from a prostate carcinoma that has been fixed with formalin, embedded in paraffin, sectioned and hybridized with a biotinylated antisense RNA probe to gastrin-releasing peptide (GRP) receptor mRNA. Following *in situ* hybridization, the biotinylated probe was developed for visualization with alkaline phosphatase–mediated techniques using the **ELF®-97 mRNA *In Situ* Hybridization Kit #2** (E-6605). Photo contributed by Marty Bartholdi, Berlex Biosciences, Berlex Laboratories, Inc.

6. Transverse zebrafish retinal section that has been probed with FRet 43 (a monoclonal antibody that binds to double cones) and then developed for visualization with alkaline phosphatase–mediated techniques using our **ELF®-97 Immunohistochemistry Kit** (E-6600). If this tissue had been stained with only the ELF-97 alkaline phosphatase substrate, the double-cone cells would appear uniformly bright green. However, this section has also been labeled with **Hoechst 33342** (H-1399, H-3570), which stains nuclei blue, and with **tetramethylrhodamine wheat germ agglutinin** (TMR-WGA; W-849), which stains the photoreceptor outer segments and synaptic layers red. Thus, the double-cone outer segments and synaptic pedicles appear bright yellow because they are double-labeled with both the ELF-97 precipitate and TMR-WGA. Also, the ELF-97 precipitate–stained inner fibers of the double cones traverse the region occupied by the Hoechst 33342–stained rod nuclei, giving them a blue-green appearance. This image was obtained by double-exposing the film through DAPI and tetramethylrhodamine longpass optical filter sets. Photo contributed by Karen D. Larison, Molecular Probes, Inc.

7. *LacZ* mRNA in transformed mouse fibroblasts (CRE BAG 2 cells) hybridized with a singly biotinylated, complementary oligonucleotide. Hybrids were then detected by incubation with a streptavidin–alkaline phosphatase conjugate in combination with the ELF-97 alkaline phosphatase substrate, both provided in our **ELF®-97 mRNA *In Situ* Hybridization Kit** (E-6605). Cells were counterstained with **DAPI** (D-1306) and photographed using a DAPI longpass optical filter. Photo contributed by Violette Paragas, Molecular Probes, Inc.

8. A cryostat section of lightly fixed adult zebrafish gills that have been incubated with the ELF-97 substrate in our **ELF®-97 Endogenous Phosphatase Detection Kit** (E-6601). This kit's novel ELF-97 phosphatase substrate yields a yellow-green fluorescent precipitate at the site of endogenous phosphatase activity. This staining is identical to that seen when employing the conventional Gomori method for detecting phosphatase activity. Photo contributed by Karen D. Larison, Molecular Probes, Inc.

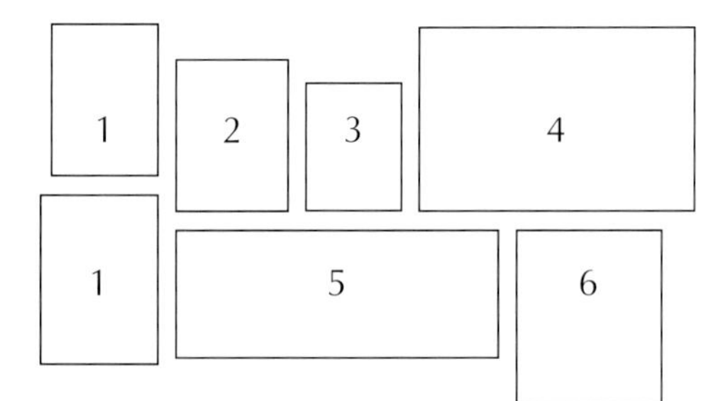

Molecular Biology Tools Color Plate 2

1. Human metaphase chromosomes stained with **DAPI** (D-1306, top panel) and **propidium iodide** (P-1304, P-3566; bottom panel). Photos contributed by Violette Paragas and Jeff Pollack, Molecular Probes, Inc.

2. A twofold dilution series of *Hae* III–digested ϕX174 RF DNA that has been electrophoresed on a 5% polyacrylamide gel and then stained with our ultrasensitive **SYBR® Green I nucleic acid gel stain** (S-7567, S-7585). Gel staining was visualized with 254 epi-illumination. Photo contributed by Xiaokui Jin and Sam Wells, Molecular Probes, Inc.

3. A twofold dilution series of *Escherichia coli* 16S and 23S ribosomal RNA (rRNA) that has been electrophoresed on a nondenaturing 1% agarose gel and then stained with our **SYBR® Green II RNA gel stain** (S-7568, S-7586). Gel staining was visualized with 254 nm epi-illumination. Photo contributed by Xiaokui Jin and Sam Wells, Molecular Probes, Inc.

4. Purified bacteriophage T4 replication proteins that have been electrophoresed on a 12% SDS-polyacrylamide gel and then stained with **SYPRO® Orange protein gel stain** (S-6650). Gel staining was visualized with 300 nm epi-illumination. T4 phage proteins were a gift from the laboratory of Peter von Hippel, Institute of Molecular Biology, University of Oregon. Photo contributed by Tom Steinberg, Jennifer Kramer and Vicki Singer, Molecular Probes, Inc.

5. Chloramphenicol acetyltransferase (CAT) assays using our ***FAST* CAT® Yellow (deoxy) Chloramphenicol Acetyltransferase Kit** (F-6617). Decreasing amounts of purified CAT enzyme (twofold dilutions) were incubated with the corresponding deoxy substrate in the presence of acetyl CoA; the reaction mixture was then separated with standard thin-layer chromatography (TLC) methods and visualized with 366 nm epi-illumination. The bottom row of fluorescent spots in each TLC represents the substrate; the top, the monoacetylated reaction product. Photo contributed by Charles Lefevre and Sam Wells, Molecular Probes, Inc.

6. Human metaphase chromosomes stained with SYTOX® Green nucleic acid stain and Methyl Green, which are provided in our **Chromosome Banding Kit #2** (C-7587), and then mounted in Cytoseal™ 60 mounting medium (Stephens Scientific). The photograph was taken using a Nikon Labophot® 2 microscope equipped with a fluorescein bandpass optical filter set. Photo contributed by Violette Paragas, Molecular Probes, Inc.

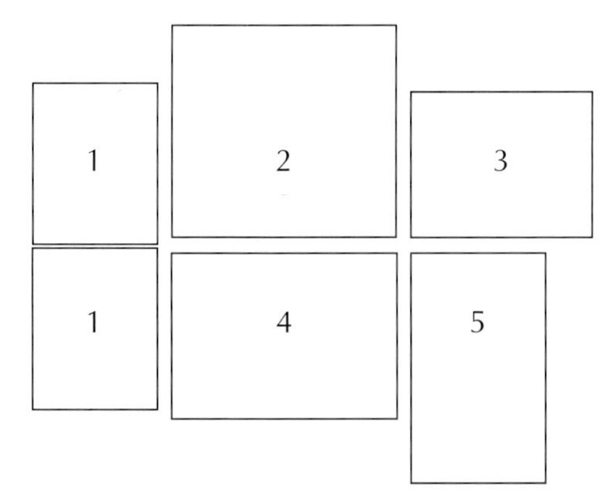

Organelle Probes Color Plate 2

1. Selective staining of the Golgi apparatus using the green fluorescent **BODIPY® FL C_5-ceramide** (D-3521) (top panel). At high concentrations, the BODIPY FL fluorophore forms excimers that can be visualized using a red longpass optical filter. The BODIPY FL C_5-ceramide accumulation in the trans-Golgi is sufficient for excimer formation (bottom panel) [J Cell Biol 113, 1267 (1991)]. Photos contributed by Richard Pagano, Mayo Foundation.

2. The potential across mitochondrial membranes, as visualized with the potential-sensitive probe **JC-1** (T-3168). The JC-1 monomer (green), which exists at low concentration or at low membrane potential, forms so-called J-aggregates (red) at higher concentrations or higher potentials. This novel membrane potential–sensitive carbocyanine has a delocalized positive charge that results in electrochromic properties, making it optically sensitive to membrane potential changes induced by ADP and metabolic inhibitors of oxidative phosphorylation [Biochemistry 30, 4480 (1991)]. Photo contributed by Lan Bo Chen, Dana Farber Cancer Institute and Harvard Medical School.

3. Bovine pulmonary arterial endothelial cells incubated with the fixable, mitochondrion-selective **MitoTracker™ Red CMXRos** (M-7512). After staining, the cells were formaldehyde-fixed, acetone-permeabilized, treated with DNase-free RNase and counterstained using SYTOX® Green nucleic acid stain from our **Cytological Nuclear Counterstain Kit** (C-7590). Microtubules were labeled with mouse monoclonal anti–β-tubulin antibody, **biotin-XX goat anti-mouse IgG (H+L) conjugate** (B-2770) and **Cascade Blue® NeutraLite™ avidin** (A-2663). This photograph was taken using multiple exposures through Texas Red®, fluorescein and DAPI bandpass optical filters using a Nikon Labophot® 2 microscope equipped with a Quadfluor™ epi-illumination system. Photo contributed by Jennifer Kramer and Vicki Singer, Molecular Probes, Inc.

4. Bovine pulmonary arterial endothelial cells incubated simultaneously with 50 nM **LysoTracker™ Red DND-99** (L-7528) and 75 nM **MitoTracker™ Green FM** (M-7514) at 37°C for 30 minutes. Both dyes showed excellent cellular retention, even after cells were fixed in 3% glutaraldehyde for 30 minutes. Photo contributed by Yu-Zhong Zhang, Molecular Probes, Inc.

5. *Saccharomyces cerevisiae* that have been stained sequentially with the red fluorescent **rhodamine B hexyl ester** (R 6, R-648), which selectively labels yeast mitochondria under these conditions, and the green fluorescent **yeast vacuole membrane marker MDY-64** (Y-7536). These probes are also provided in our **Yeast Mitochondrial Stain Sampler Kit** (Y-7530) and **Yeast Vacuole Marker Sampler Kit** (Y-7531), respectively. Stained yeast were photographed in a single exposure through an Omega® Optical triple-band filter set (O-5855), available directly from Molecular Probes. Photo contributed by Bruce Roth and Paul Millard, Molecular Probes, Inc.

Figure Captions for Color Plate 2

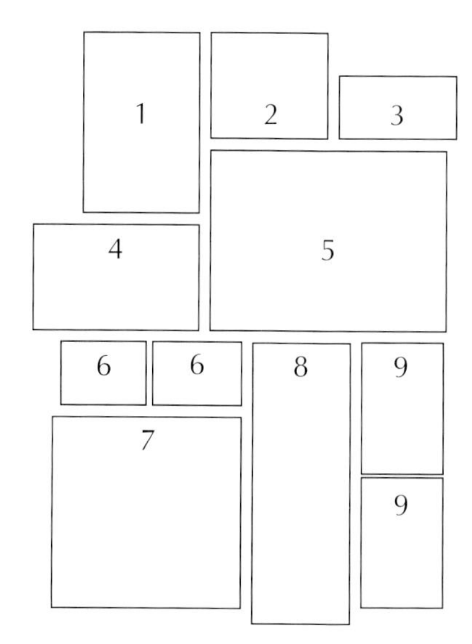

Fluorescent Tracers
Color Plate 3

1. A whole mount of the embryonic brain of *Xenopus laevis* that has been double-labeled with our **lysine-fixable fluorescein and tetramethylrhodamine 10,000 MW dextrans** (fluoro-emerald, D-1820; fluoro-ruby, D-1817). The tetramethylrhodamine dextran was used to label the neurons projecting from the retina, whereas the fluorescein dextran was applied to the transected spinal cord, thus allowing the detailed evaluation of the topological relationship of these two populations of neurons [Neurosci Lett 127, 150 (1991)]. Photo contributed by Martina Manns and Bernd Fritzsch, Department of Biomedical Sciences, Creighton University.

2. Secondary motor neurons in a spinal cord whole mount of a male western mosquitofish (*Gambusia affinis affinis*) that have been labeled with **lysine-fixable Texas Red® 3000 MW dextran** (D-3328). The dextran crystals were applied to the bipinnate inclinator muscles of the anal appendicular support fin, and the dye was transported from the axons to cell body and dendrites. Motor neurons were visualized and photographed through a Texas Red bandpass optical filter by epifluorescence microscopy. Photo contributed by E. Rosa-Molinar, Department of Cell Biology and Anatomy, University of Nebraska Medical Center, and Bernd Fritzsch, Department of Biomedical Sciences, Creighton University.

3. Detection of organisms in marine sediments by incubating an intact sediment core sample with the fixable, cell-permeant **CellTracker™ Green CMFDA** (C-2925, C-7025). The core sample was subsequently embedded, sectioned and examined for fluorescently labeled organisms. The micrograph reveals the microorganism *Leptohalysis scotti*, a marine benthic foraminifera. Photo contributed by Joan M. Bernhard, Wadsworth Center, New York State Department of Health.

4. A zebrafish embryo that was injected with a 1% solution (in 0.2 M KCl) of **caged fluorescein 10,000 MW dextran** (D-3310) at the two-cell stage and then allowed to grow for 19 hours. The posterior lateral line placode was then exposed to a five-second pulse from an epifluorescence microscope fitted with a DAPI optical filter set. After six hours of further development, the labeled primordium has migrated caudally, away from the activation site. Photo contributed by Walter K. Metcalfe, Institute of Neuroscience, University of Oregon.

5. Cultured left–upper quadrant neurons from *Aplysia californica* that have been microinjected with either **lucifer yellow CH** (L-453) or **sulforhodamine 101** (S-359). These neurons display an extensive array of overlapping processes [J Neurophysiol 66, 316 (1991)]. Photo contributed by David Kleinfeld, AT&T Bell Laboratories, and Brian Salzberg, University of Pennsylvania School of Medicine.

6. Rat retinal ganglion cells that have been allowed to regenerate for two months after transection into a grafted peripheral nerve piece and then labeled with **4-Di-10-ASP** (D-291), which is retrogradely transported within the regenerated axons and finest dendrites (right panel). Microglial cells in the retina were transcellularly labeled with 4-Di-10-ASP after phagocytosing degenerated ganglion cells (left panel). Photos contributed by Solon Thanos, Department of Ophthalmology, University of Tübingen School of Medicine.

7. The rear fin of an embryonic squid that has been labeled with **DiIC$_{18}$(3)** (DiI, D-282), showing selective staining of a population of ciliated sensory neurons whose cell bodies line the periphery of the fin. The dye was applied by soaking the embryos in a solution; no injection was used. Photo contributed by Rachel Fink, Department of Biological Sciences, Mount Holyoke College.

8. The antero-ventral and antero-dorsal lateral line nerves in an *Ambystoma mexicanum* whole-mounted brain after labeling with **fixable Rhodamine Green™ 3000 MW dextran** (D-7163) and **fixable tetramethylrhodamine 3000 MW dextran** (D-7162), respectively. In both cases, the dextran was applied to the respective cut cranial nerves. This photomicrograph shows the four segregated ventral fascicles of the mechanosensory lateral line fibers and the intermingling of the dorsal electrosensory fibers. Photo contributed by Bernd Fritzsch, Department of Biomedical Sciences, Creighton University.

9. HL60 cells that have been stained with **CellTracker™ Orange CMTMR** (C-2927) and then mixed with WEH17.1 cells stained with **CellTracker™ Green CMFDA** (C-2925, C-7025) (top panel). Several minutes after initiating cell–cell electrofusion, a CMTMR-stained HL60 cell is observed fusing with a CMFDA-stained WEH17.1 cell; cytoplasmic mixing is evident by the appearance of yellow fluorescence. After electrofusion is complete, dual-fluorescing (yellow) hybrids can be easily distinguished (bottom panel). Photos contributed by Mark J. Jaroszeski, Department of Surgery, University of South Florida.

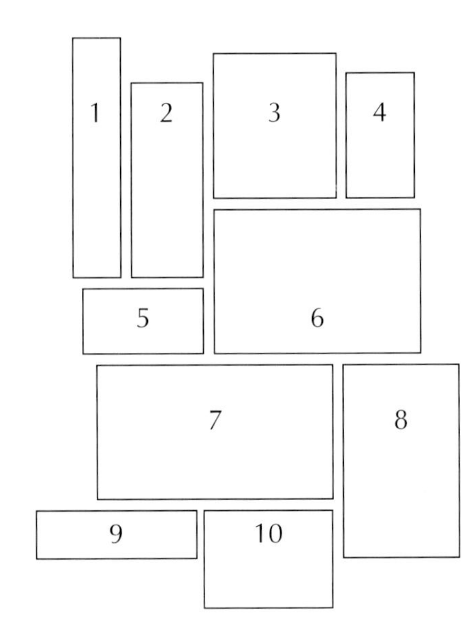

Viability Kits
Color Plate 4

1. A mixture of live and dead goat lymphocytes stained with the **LIVE/DEAD® Reduced Biohazard Viability/Cytotoxicity Kit** (L-7013) and subsequently fixed with 4% glutaraldehyde. This image was photographed in a single exposure through an Omega® Optical triple-band filter set (O-5855), available directly from Molecular Probes. Photo contributed by Bruce Roth and Paul Millard, Molecular Probes, Inc.

2. Adherent cells from human peripheral blood stained with **SYTO® 13** (S-7575), one of the five visible light–excitable cell-permeant nucleic acid stains in our new **SYTO® Live-Cell Nucleic Acid Stain Sampler Kit** (S-7572). The multilobed nuclei of these polymorphonuclear leukocytes are particularly striking in this field of view. Photo contributed by Ian Clements and Paul Millard, Molecular Probes, Inc.

3. Live and dead rat kangaroo (PtK2) cells stained with ethidium homodimer-1 and the esterase substrate calcein AM, both of which are provided in our **LIVE/DEAD® Viability/Cytotoxicity Kit** (L-3224). Live cells fluoresce a bright green, whereas dead cells with compromised membranes fluoresce orange-red. Photo contributed by Sam Wells, Molecular Probes, Inc.

4. Live *Micrococcus luteus* and *Salmonella oranienburg* bacteria stained with the **LIVE *Bac*Light™ Bacterial Gram Stain Kit** (L-7008). Gram-positive *M. luteus* cells fluoresce orange, whereas gram-negative *S. oranienburg* cells fluoresce green. This image was photographed in a single exposure through an Omega® Optical triple-band filter set (O-5855), available directly from Molecular Probes. Photo contributed by Bruce Roth and Paul Millard, Molecular Probes, Inc.

5. *Bifodobacterium sp.* bacteria stained with the **ViaGram™ Red+ Bacterial Gram Stain and Viability Kit** (V-7023). While all cells exhibit a red fluorescent surface stain (gram-positive), live cells exhibit blue fluorescent internal staining and dead cells with compromised membranes exhibit yellow-green fluorescent internal staining. This image was obtained by taking multiple exposures through DAPI, fluorescein and Texas Red® bandpass optical filter sets using a Nikon Labophot® 2 microscope equipped with a Quadfluor™ epi-illumination system. Photo contributed by Bruce Roth and Paul Millard, Molecular Probes, Inc.

6. A culture of *Saccharomyces cerevisiae* incubated in medium containing the viability indicator FUN-1™ and the counterstain Calcofluor™ White M2R, both of which are provided in our **LIVE/DEAD® Yeast Viability Kit** (L-7009). Metabolically active yeast process the FUN-1 dye, forming numerous red fluorescent cylindrical structures within their vacuoles. Calcofluor stains the cell walls fluorescent blue, regardless of the yeast's metabolic state. The yeast were photographed in a single exposure through an Omega® Optical triple-band filter set (O-5855), available directly from Molecular Probes. Photo contributed by Bruce Roth and Paul Millard, Molecular Probes, Inc.

7. Detection of BrdUrd-incorporating cells using the SBIP (strand breaks induced by photolysis) technique. Exponentially growing human promyelocytic leukemia (HL60) cells were incubated with 20 µM BrdUrd for 40 minutes and then with 20 µg/mL **Hoechst 33258** (H-1398, H-3569) in the presence of 2% DMSO for an additional 20 minutes. After this incubation, the cells were exposed to 300 nm UV light for 5 minutes to selectively photolyze DNA that contained the incorporated BrdUrd, and then fixed in 70% ethanol. Subsequent incubation of the permeabilized cells with the **ChromaTide™ BODIPY® FL-14-dUTP** (C-7614) in the presence of exogenous terminal deoxynucleotidyl transferase resulted in incorporation of the fluorophore into DNA strand breaks, thereby labeling the S phase cells. The DNA was counter-stained with the red fluorescent nucleic acid stain **propidium iodide** (P-1304, P-3566); therefore, the BODIPY FL–labeled DNA appears yellow. Photo contributed by Zbigniew Darzynkiewicz, Cancer Research Institute, New York Medical College.

8. Live and dead bacteria visualized on freshly isolated human cheek epithelial cells using our **LIVE/DEAD® *Bac*Light™ Bacterial Viability Kit** (L-7007, L-7012). When incubated with the SYTO® 9 and propidium iodide nucleic acid stains provided in this kit, live bacteria with intact cell membranes fluoresce green and dead bacteria with compromised membranes fluoresce red. This image was photographed in a single exposure through an Omega® Optical triple-band filter set (O-5855), available directly from Molecular Probes. Photo contributed by Bruce Roth and Paul Millard, Molecular Probes, Inc.

9. Use of our **LIVE/DEAD® *Bac*Light™ Bacterial Viability Kit** (L-7007, L-7012) to identify individual live and dead bacteria along a chain of *Streptococcus pyogenes*. This image was photographed in a single exposure through an Omega® Optical triple-band filter set (O-5855), available directly from Molecular Probes. Photo contributed by Bruce Roth and Paul Millard, Molecular Probes, Inc.

10. A mixture of live and dead bovine sperm cells stained with the dyes provided in our new **LIVE/DEAD® Sperm Viability Kit** (L-7011). Live sperm with intact membranes are labeled with our proprietary cell-permeant nucleic acid stain, SYBR® 14, and fluoresce green. Dead sperm, which have been killed by unprotected freeze-thawing, are labeled with propidium iodide and fluoresce orange-red. Photo contributed by Duane L. Garner, School of Veterinary Medicine, University of Nevada, Reno and Lawrence A. Johnson, USDA Agricultural Research Service.

Chapter 15

Fluorescent Tracers of Cell Morphology and Fluid Flow

Contents

Technical Notes and Product Highlights

Related Chapters

15.1 Choosing a Tracer

To serve as an effective tracer of cell morphology, a fluorescent probe or other detectable molecule must have the capacity for localized introduction into a structural entity of interest (e.g., a cell or organelle) and subsequent long-term retention within that structure. If used with live cells and tissues, then the tracer should also be biologically inert and nontoxic. When these conditions are satisfied, the fluorescence or other detectable properties of the tracer can be used to track the position of the labeled entity as a function of temporal or spatial coordinates. Fluorescent tracers can be employed to investigate flow in capillaries, to define neuronal cell connectivity and to study dye translocation through gap junctions, as well as to follow cell division, cell lysis or liposome fusion. Furthermore, they can be used to track the movements of labeled cells in culture, tissues or intact organisms.

Although the predominant tracers have been fluorescent, not all useful tracers are intrinsically detectable. For example, biotin derivatives are widely used as polar tracers, especially in neurons. However, when a biotinylated or haptenylated tracer is used in live cells, detection usually requires cell fixation and permeabilization to allow access to fluorescent- or enzyme-labeled conjugates of avidin and streptavidin (see Section 7.5) or of antibodies (see Section 7.3)

In many of these tracing applications, the physical dimensions of the tracer molecule are an important consideration. Molecular Probes offers fluorescent tracers ranging in size from small molecules about 1 nm in diameter, to polystyrene microspheres up to 15 µm in diameter. This chapter discusses our diverse selection of fluorescent tracers, as well as biotinylated, spin-labeled and other tracers:

- **Cell-permeant cytoplasmic labels** (see Section 15.2). Molecular Probes has developed several thiol-reactive CellTracker probes, which yield fluorescent products that are retained in many live cells through several generations and are not transferred to adjacent cells in a population, except possibly through gap junctions. These probes represent a significant breakthrough in the cellular retention of fluorescent dyes and are ideal long-term tracers for transplanted cells or tissues.

- **Microinjectable cytoplasmic labels** (see Section 15.3). Polar tracers such as lucifer yellow CH, Cascade Blue® hydrazide, sulforhodamine 101 and biocytin are membrane-impermeant probes that are usually introduced into cells by iontophoresis, osmotic lysis of pinocytic vesicles or comparable methods [1] (Table 15.1). These tracers are commonly used to investigate cell–cell and cell–liposome fusion, as well as membrane permeability [2] and transport through gap junctions.[3,4]

Table 15.1 Techniques for loading molecules into the cytoplasm.

Method of Plasma Membrane Breach	Size of Molecules Loaded (MW) *	References †
Chemical		
ATP	1000	1
EDTA	10,000	2
$Ca_3(PO_4)_2$	(DNA)	3
DEAE-dextran	(DNA)	4
α-Toxin of *Staphylococcus aureus*	1000	5,6
Vehicle		
Red blood cell fusion	300,000	7,8
Vesicle and liposome fusion	Very high	9,10
Mechanical		
Microinjection	Very high	11
Hypoosmotic shock	10,000	12
Osmotic lysis of pinosomes	Very high	13
Scrape loading	500,000; (DNA)	14
Agitation in cold	500,000	14
Sonication (mild)	70,000; (DNA)	15
High-velocity microprojectiles	(DNA)	16
Glass beads	150,000	17
Scratching to wound culture	150,000	18
Electrical		
Electroporation	(DNA)	19

* Molecular weight (MW) of largest molecules reported loaded (DNA is listed separately if it has been successfully introduced by a technique). † Original reports or important variations. Table adapted with permission from McNeil, P.L. in *Fluorescence Microscopy of Living Cells in Culture, Part A* (Meth Cell Biol 29) Y.-L. Wang and D.L. Taylor, Eds., Academic Press (1989) p. 153–173.

1. Biochem Biophys Res Comm 67, 1581 (1975); **2.** J Biol Chem 260, 2069 (1985); **3.** Virology 52, 456 (1973); **4.** Virology 27, 435 (1965); **5.** Meth Cell Biol 31, 63 (1989); **6.** Neuroscience 23, 1143 (1987); **7.** Cell 5, 371 (1975); **8.** J Cell Biol 101, 19 (1985); **9.** Neurosci Lett 207, 17 (1996); **10.** Meth Cell Biol 14, 33 (1976); **11.** Hoppe Seyler's Z Physiol Chem 352, 527 (1971); **12.** Science 217, 252 (1982); **13.** Cell 29, 33 (1982); **14.** J Cell Biol 98, 1556 (1984); **15.** Eur J Cell Biol 40, 242 (1986); **16.** Nature 327, 70 (1987); **17.** J Cell Sci 88, 669 (1987); **18.** Science 238, 548 (1987); **19.** Biochem Biophys Res Comm 107, 584 (1982).

- **Membrane tracers — DiI, DiO, DiA and their analogs** (see Section 15.4). Lipophilic carbocyanine and aminostyryl dyes can be introduced into membranes by direct application of a dye crystal onto a cell or by bulk loading from aqueous dispersions. Lateral diffusion of the dye within the membrane eventually stains the entire cell. These probes are widely used for neuroanatomical tracing and long-term assays of cell–cell association. Our newest DiI and DiO analogs exhibit superior solubility and brightness and, in some cases, produce a cell staining pattern that persists through aldehyde fixation and acetone permeabilization.

- **Fluorescent and biotinylated dextran conjugates** (see Section 15.5). Dextran conjugates are ideal cell-lineage tracers because they are relatively inert, exhibit low toxicity and are well retained in cells. These membrane-impermeant probes are usually loaded into cells by invasive techniques such as microinjection, scrape loading, electroporation or osmotic shock. Availability of dextrans in a range of molecular weights makes them useful as size-exclusion probes for determining pore sizes in membranes.

- **Fluorescent microspheres** (see Section 15.6). Molecular Probes' FluoSpheres and TransFluoSpheres fluorescent microspheres, which contain ~10^2 to ~10^{10} fluorescent dyes per bead, are the most intensely fluorescent tracers available (see Table 6.1 in Chapter 6). Although other multiply labeled particles such as our BioParticles® fluorescent bacteria (see Section 17.1) may be used as tracers, they are often not biologically inert nor as physically durable as fluorescent microspheres. These properties make the fluorescent beads useful as long-term markers for transplantation studies. Submicron microspheres can be injected into cells or taken up by phagocytosis. Much larger (10–15 µm) beads provide an alternative to radioactive microspheres for determination of organ blood flow, and intermediate-sized (1–5 µm) microspheres are useful for inhalation studies.

- **Proteins and Protein conjugates** (see Section 15.7). Our fluorescent protein tracers have molecular weights ranging from ~14,800 daltons (lactalbumin) to ~240,000 daltons (B- and R-phycoerythrin). Their applications are generally similar to those of the fluorescent dextrans; however, unlike the polydisperse dextrans, fluorescent protein tracers have reasonably well-defined molecular weights. Our fluorescent peroxidase conjugates are neuronal tracers that can be visualized by both fluorescence and light microscopy.

1. Meth Cell Biol 29, 153 (1989); **2.** Cytometry 21, 230 (1995); **3.** Cell 84, 381 (1996); **4.** Biochim Biophys Acta 988, 319 (1989).

15.2 Membrane-Permeant Reactive Tracers for Long-Term Cell Labeling

Thiol-Reactive CellTracker Probes

Molecular Probes' CellTracker reagents are fluorescent chloromethyl derivatives that freely diffuse through the membranes of live cells. Once inside the cell, these mildly thiol-reactive probes undergo what is believed to be a glutathione *S*-transferase–mediated reaction to produce membrane-impermeant glutathione–fluorescent dye adducts (Figure 15.1), although our experiments suggest that they may also react with other intracellular components. Regardless of the mechanism, many cell types loaded with the CellTracker probes are both fluorescent and viable for at least 24 hours after loading and often through several cell divisions. Most other cell-permeant fluorescent dyes, including the acetoxymethyl (AM) esters of calcein and BCECF (see Section 16.3), are retained in viable cells for no more than a few hours at physiological temperatures. Furthermore, unlike the free dye, the peptide–fluorescent dye adducts contain amino groups and can therefore be covalently linked to surrounding biomolecules by fixation with formaldehyde or glutaraldehyde. This property permits long-term storage of the labeled cells or tissue and, in cases where the anti-dye antibody is available (see below), amplification of the conjugate by standard immunohistochemical techniques.

CellTracker Probes in a Variety of Fluorescent Colors

Molecular Probes' CellTracker product line includes reactive chloromethyl derivatives of:

- Blue fluorescent aminocoumarin (CellTracker Blue CMAC, C-2110)
- Blue fluorescent hydroxycoumarin (CellTracker Blue CMHC, C-2111)
- Yellow-green fluorescent eosin diacetate (CellTracker Yellow-Green CMEDA, C-2926)
- Green fluorescent fluorescein diacetate (CellTracker Green CMFDA; C-2925, C-7025)
- Green fluorescent BODIPY® derivative (CellTracker Green BODIPY, C-2102)
- Orange fluorescent tetramethylrhodamine (CellTracker Orange CMTMR, C-2927)

Both CMFDA and CMEDA freely diffuse into the cell, where cytosolic esterases cleave off the acetate groups, releasing the fluorescent product. The coumarin, BODIPY and tetramethylrhodamine CellTracker probes do not require enzymatic cleavage to activate their fluorescence. Our related MitoTracker™ probes (see Section 12.2) and other mitochondrial stains may be useful for tracing actively metabolizing cells. We also offer BODIPY 493/503 methyl bromide (B-2103), a green fluorescent dye that is slightly more thiol-reactive than its chloromethyl counterpart.

Applications for CellTracker Probes

The thiol-reactive CellTracker probes are suitable for long-term cell labeling in a variety of applications, including studies of cell adhesion,[1] cell communication through gap junctions,[2] cell migration, cell tracing in mixed cultures[3,4] (see Color Plate 3 in "Visual Reality"), cytotoxicity[5] and transplantation. CellTracker Green CMFDA was used to track wild-type and myosin II mutant *Dictyostelium discoideum* cells within aggregation streams during early multicellular morphogenesis; differentiation and morphogenesis pathways were reportedly unaffected in labeled cells imaged over several days by confocal laser scanning microscopy.[6,7] CellTracker Orange CMTMR has been used to stain the cytoplasm of *engrailed*-expressing *Drosophila* cells in an *in vitro* reconstruction experiment.[8] Both CellTracker Green CMFDA and CellTracker Orange CMTMR probes were recently employed in a flow cytometry method for quantitating cell–cell electrofusion products[9,10] (see Color Plate 3 in "Visual Reality") Metabolic activity and drug-induced cytotoxicity were measured with CellTracker Blue CMAC in a fluorescence-based microplate assay.[11]

The ability to fix the intracellular products of CellTracker Green CMFDA, CellTracker Green BODIPY and CellTracker Orange CMTMR in permeabilized cells permits the stained cells to be probed with our anti-fluorescein, anti–BODIPY FL and anti-tetramethylrhodamine antibodies, respectively (see Section 7.3),

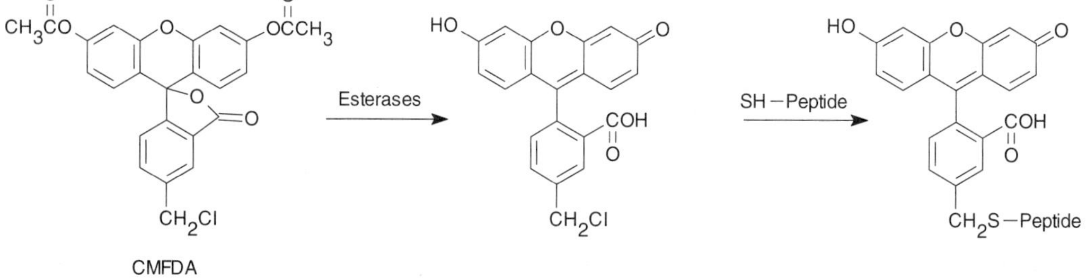

Figure 15.1 *Intracellular reactions of our fixable CellTracker Green CMFDA (5-chloromethylfluorescein diacetate, C-2925). Once this membrane-permeant probe enters a cell, esterase hydrolysis converts nonfluorescent CMFDA to fluorescent 5-chloromethylfluorescein, which can then react with thiols on proteins and peptides to form aldehyde-fixable conjugates. This probe may also react with intracellular thiol-containing biomolecules first, but the conjugate is nonfluorescent until its acetates are removed.*

and developed for electron microscopy using standard immunohistochemical techniques. Alternatively, CellTracker Yellow-Green CMEDA may be employed for correlated fluorescence and electron microscopy studies using the method reported by Deerinck and co-workers for eosin-mediated photooxidation of diaminobenzidine [12,13] (DAB); see the Technical Note "Fluorescent Probes for Photoconversion of Diaminobenzidine" on page 414. This chloromethyl eosin analog is also potentially useful for cell ablation because the eosin–glutathione conjugate is likely to be phototoxic.

Other Thiol-Reactive Tracers

Chloromethyl Derivatives of SNARF®-1 and a Dihydrofluorescein

Although they were designed for other purposes, the chloromethyl derivatives of our SNARF-1 pH indicator (C-6826) and of 2',7'-dichlorodihydrofluorescein diacetate (CM-H_2DCFDA, C-6827) possess some unique properties for tracking cells. As with the CellTracker probes, cytoplasmic enzymes hydrolyze the acetate groups from these membrane-permeant probes, and the chloromethyl moieties become conjugated to intracellular thiols. With its long Stokes shift, the SNARF derivative is the only cell tracking dye that has bright, easily distinguished red-orange fluorescence in the cytoplasm when excited at the same wavelengths used for the green fluorescent CellTracker Green CMFDA (see Figure 23.6 in Chapter 23). This property permits simultaneous tracking of two cell populations by either microscopy or flow cytometry.

As with other dihydrofluorescein derivatives (see Section 17.1), CM-H_2DCFDA requires an additional oxidation step before becoming fluorescent. This probe is potentially useful for following stimulation of oxidative activity by external agents or natural killer cells over extended periods, as well as for passively labeling cells that lack appropriate oxidative activity and then following their ingestion by scavengers such as neutrophils.

Bimanes: Blue Fluorescent Reactive Tracers

The bimane derivatives, monobromobimane (mBBr, M-1378) monochlorobimane (mBCl, M-1381) and monobromotrimethylammoniobimane, (qBBr, M-1380) are important thiol-derivatization reagents (see Section 2.4). The essentially nonfluorescent mBBr and mBCl dyes are known to passively diffuse across the plasma membrane into the cytoplasm, where they form blue fluorescent adducts with intracellular glutathione and thiol-containing proteins (see Section 16.3). Although the blue fluorescence of their thiol adducts is weaker than that of our CellTracker Blue dyes CMAC (C-2110) and CMHC (C-2111), mBBr and mBCl have been shown to serve as effective tracers of gap-junctional communication.[2,14] Likewise, the bimane qBBr is a water-soluble quaternary ammonium salt that may be suitable as a microinjectable tracer.

Fluorinated Fluorescein and Eosin Probes

In developing our new Oregon Green™ dyes (see Sections 1.4 and 1.5), we found that substitution of the "bottom" ring of fluorescein by four fluorine atoms renders the molecule susceptible to reaction with thiols. When used at 1 µM to stain live cells, the membrane-permeant diacetates of 4,5,6,7-tetrafluorofluorescein (TFFDA, T-7016) and 2',4,5,6,7,7'-hexafluorofluorescein (HFFDA, H-7019) behave like CellTracker Green CMFDA, forming a product that is well retained intracellularly.

The diacetate of 2',4',5',7'-tetrabromo-4,5,6,7-tetrafluorofluorescein (Br_4-TFFDA, T-7017) is a new polyfluorinated analog of eosin. The structural and spectral similarities of Br_4-TFFDA to eosin and CellTracker Yellow-Green CMEDA (see above) suggest that this dye may be useful for selective cell photoablation in embryo development studies or for correlated fluorescence and electron microscopy studies. We have shown that fluorination has a relatively small effect on the dye's spectra but lowers its pK_a and increases its photostability relative to that of the corresponding fluorescein and eosin dyes.

Amine-Reactive Fluorescein and Eosin Probes

The diacetates of fluorescein isothiocyanate (F-1163), carboxyfluorescein succinimidyl esters (C-1157, C-1165) and carboxyeosin succinimidyl ester (C-1164) may have applications similar to the CellTracker probes.[15] These amine-reactive reagents can passively diffuse into cells and are colorless and nonfluorescent until their acetate groups are cleaved by intracellular esterases to yield the highly fluorescent, amine-reactive fluorophores. Upon reaction with amine-containing residues of intracellular proteins, these probes form dye–protein adducts that are reportedly well retained throughout development, meiosis and *in vivo* tracing [15,16] and can survive subsequent fixation with formaldehyde or glutaraldehyde. Because it is intrinsically more reactive, the succinimidyl ester of CFDA is more likely to react at sites on the extracellular surface than is CMFDA. However, it may be possible to quench much of the fluorescence of outer-membrane fluorescein conjugates by acidifying the extracellular medium or by incubating with antifluorescein antibody [17] (A-889, see Section 7.3).

The succinimidyl ester of carboxyfluorescein diacetate (5(6)-CFDA, SE; C-1157) is reported to be particularly useful for tracing transplanted cells.[16,18] Lymphocytes labeled with the succinimidyl ester of CFDA were detected up to eight weeks after injection into mice in lymphocyte migration studies,[19,20] and viable hepatocytes similarly labeled were easily located by fluorescence microscopy even 20 days after intrahepatic transplantation.[21] The succinimidyl ester of CFDA has also been successfully employed to quantitate adhesion of neutrophils [22] and leukocytes [23] in an assay of T-cadherin–mediated cell aggregation [24] and in an *in vitro* bioassay for neurite growth.[25] Fluorescence of the intracellular hydrolysis products of the succinimidyl ester of 5-(and-6)-carboxy-2',7'-dichlorofluorescein diacetate (C-1165) is relatively insensitive to fluctuations in pH. This amine-reactive tracer was reported to be more useful than the lipophilic marker DiI (see below) in an investigation of palatal fusion in rodent embryos.[26]

1. J Cell Biol 133, 445 (1996); **2.** Cytometry 14, 747 (1993); **3.** J Cell Biol 133, 445 (1996); **4.** J Cell Biol 128, 405 (1995); **5.** Toxicol Appl Pharmacol 112, 235 (1992); **6.** Devel Biol 170, 434 (1995); **7.** J Cell Sci 108, 1105 (1995); **8.** Nature 363, 549 (1993); **9.** Anal Biochem 216, 271 (1994); **10.** Biophys J 67, 1574 (1994); **11.** Clin Chem 41, 1906 (1995); **12.** J Cell Biol 126, 901 (1994); **13.** J Cell Biol 126, 877 (1994); **14.** Biochem J 284, 631 (1993); **15.** J Cell Biol 101, 610 (1985); **16.** J Cell Biol 103, 2649 (1986); **17.** Proc Natl Acad Sci USA 78, 7540 (1981); **18.** Cell Transplantation 3, 397 (1994); **19.** Cytometry 13, 739 (1992); **20.** J Immunol Meth 133, 87 (1990); **21.** Transplantation Proc 24, 2820 (1992); **22.** J Immunol Meth 172, 25 (1994); **23.** J Immunol Meth 172, 115 (1994); **24.** J Cell Biol 119, 451 (1992); **25.** J Neurosci Meth 39, 193 (1991); **26.** Development 116, 1087 (1992).

15.2 Data Table *Membrane-Permeant Reactive Tracers for Long-Term Cell Labeling*

Cat #	Structure	MW	Storage	Soluble	Abs	{ε × 10⁻³}	Em	Solvent	Notes
B-2103	S3.4	341	F,D,L	DMSO, MeCN	515	{55}	525	MeOH	<1>
C-1157	✓	557	F,D	DMF, DMSO	<300		none		<2>
C-1164	✓	873	F,D	DMF, MeCN	<300		none		<3>
C-1165	✓	626	F,D	DMF, MeCN	<300		none		<4>
C-2102	✓	297	F,D,L	DMSO	522	{72}	529	MeOH	
C-2110	✓	210	F,D,L	DMSO	353	{14}	466	pH 9	
C-2111	✓	211	F,D,L	DMSO	372	{16}	470	pH 9	<5>
C-2925	✓	465	F,D	DMSO	<300		none		<2>
C-2926	✓	780	F,D	DMSO	<300		none		<3>
C-2927	✓	554	F,D,L	DMSO	541	{91}	565	MeOH	
C-6826	S23.2	500	F,D	DMSO	<350		none		<6>
C-6827	S21.4	536	F,D,AA	DMSO	287	{9.1}	none	MeOH	<7>
F-1163	✓	473	F,DD	DMF, DMSO	<300		none		<8>
H-7019	✓	524	F,DD	DMSO	<300		none		<9>
M-1378	S2.4	271	F,L	DMF, MeCN	398	{5.0}	<10>	pH 7	
M-1380	S2.4	409	L	H_2O*	378	{5.5}	<10>	pH 7	
M-1381	S2.4	227	F,L	DMSO	380	{6.0}	<10>	MeOH	
T-7016	✓	488	F,D	DMSO	<300		none		<9>
T-7017	✓	804	F,D	DMSO	<300		none		<11>

C-7025 see C-2925

For definitions of the contents of this data table, see "How to Use the Handbook" on page vi.
Structure: Chemical structure drawing: ✓ = shown in this section; Sn.n = shown in Section number n.n.

* Unstable in water. Use immediately.

<1> B-2103 spectra are for the unreacted reagent. The thiol adduct has Abs = 493 nm, Em = 503 nm in MeOH.

<2> C-1157 and C-2925 are converted to fluorescent products with spectra similar to C-1904 (see Section 15.3) after acetate hydrolysis.

<3> C-1164 and C-2926 are converted to fluorescent products with spectra similar to C-301, Abs = 519 nm (ε = {100}, Em = 542 nm (see Section 1.5), after acetate hydrolysis.

<4> C-1165 is converted to a fluorescent product with spectra similar to C-368, Abs = 504 nm (ε = {90}), Em = 529 nm (see Section 1.5), after acetate hydrolysis.

<5> Spectra of hydroxycoumarins are pH dependent. Below the pK_a (~7.8), Abs shifts to ~325 nm and fluorescence intensity decreases.

<6> C-6826 is converted to a fluorescent product with spectra similar to C-1270 (see Section 23.2), Abs = 576 nm (ε = {48}), Em = 635 nm at pH 10, after acetate hydrolysis.

<7> C-6827 is colorless and nonfluorescent until both the acetates are hydrolyzed and the product is subsequently oxidized to give a 2′,7′-dichlorofluorescein derivative, Abs = 504 nm (ε = {90}), Em = 529 nm. The material contains < 0.1% of oxidized derivative when initially prepared.

<8> F-1163 is converted to a fluorescent product (F-143, see Section 1.3), Abs = 494 nm (ε = {73}), Em = 519 nm, after acetate hydrolysis.

<9> T-7016 and H-7019 are converted to fluorescent products after acetate hydrolysis. T-7016 yields T-7014 (see Section 23.3), Abs = 508 nm (ε = {94}), Em = 528 nm. H-7019 yields H-7018 (see Section 23.3), Abs = 508 nm (ε = {80}), Em = 529 nm.

<10> M-1378, M-1380 and M-1381 are almost nonfluorescent until reacted with thiols. Em = 475–485 nm for thiol adducts (QY ~0.1–0.3) [Meth Enzymol 143, 76 (1987)].

<11> T-7017 is converted to a fluorescent product (T-7015, see Section 16.3), Abs = 545 nm (ε = {111}), Em = 561 nm, after acetate hydrolysis.

15.2 Structures *Membrane-Permeant Reactive Tracers for Long-Term Cell Labeling*

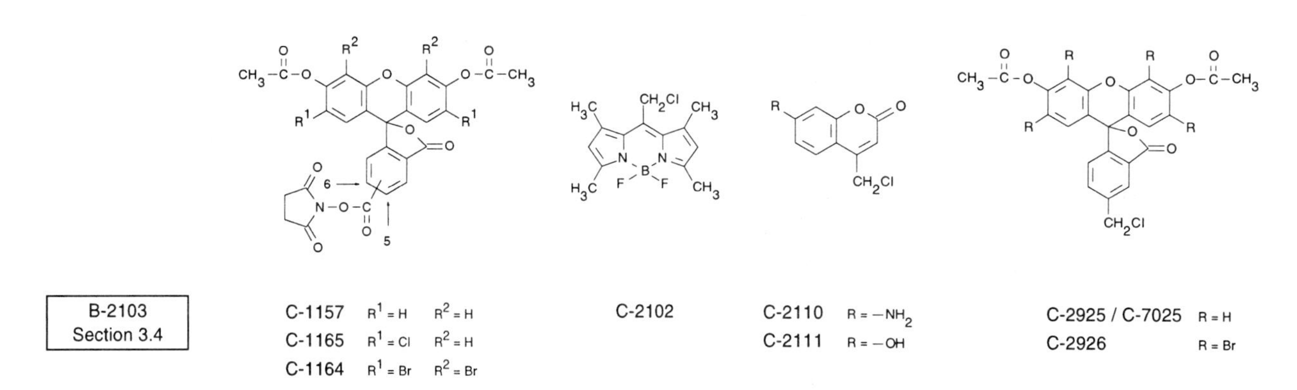

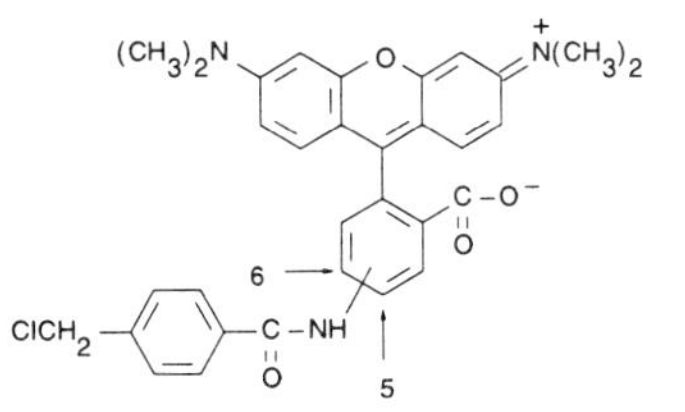

C-2927

C-6826 Section 23.2
C-6827 Section 21.4

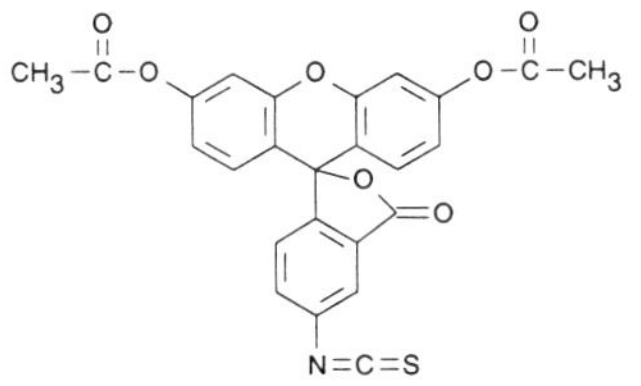

F-1163

H-7019 see T-7016
M-1378, M-1380 M-1381 Section 2.4

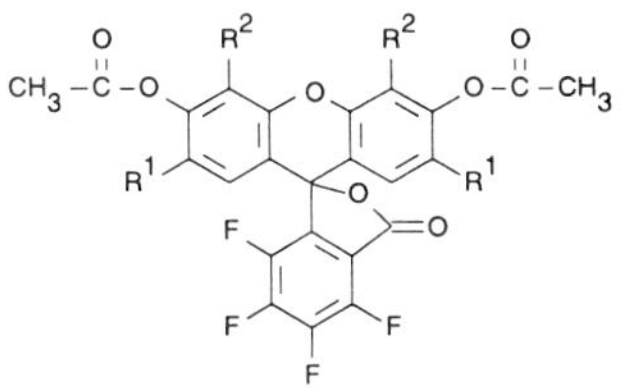

T-7016 $R^1 = H$ $R^2 = H$
H-7019 $R^1 = F$ $R^2 = H$
T-7017 $R^1 = Br$ $R^2 = Br$

15.2 Price List *Membrane-Permeant Reactive Tracers for Long-Term Cell Labeling*

Cat #		Product Name	Unit Size	Price Per Unit ($) 1–4 Units	5–24 Units
B-2103		8-bromomethyl-4,4-difluoro-1,3,5,7-tetramethyl-4-bora-3a,4a-diaza-s-indacene (BODIPY® 493/503 methyl bromide)	5 mg	95.00	76.00
C-1165		5-(and-6)-carboxy-2′,7′-dichlorofluorescein diacetate, succinimidyl ester *mixed isomers*	25 mg	115.00	92.00
C-1164		5-(and-6)-carboxyeosin diacetate, succinimidyl ester *mixed isomers*	25 mg	75.00	60.00
C-1157		5-(and-6)-carboxyfluorescein diacetate, succinimidyl ester (5(6)-CFDA, SE) *mixed isomers*	25 mg	115.00	92.00
C-2110		CellTracker™ Blue CMAC (7-amino-4-chloromethylcoumarin)	5 mg	95.00	76.00
C-2111		CellTracker™ Blue CMHC (4-chloromethyl-7-hydroxycoumarin)	5 mg	95.00	76.00
C-2102		CellTracker™ Green BODIPY® (8-chloromethyl-4,4-difluoro-1,3,5,7-tetramethyl-4-bora-3a,4a-diaza-s-indacene)	5 mg	95.00	76.00
C-2925		CellTracker™ Green CMFDA (5-chloromethylfluorescein diacetate)	1 mg	125.00	100.00
C-7025	**New**	CellTracker™ Green CMFDA (5-chloromethylfluorescein diacetate) *special packaging*	20x50 µg	145.00	116.00
C-2927		CellTracker™ Orange CMTMR (5-(and-6)-(((4-chloromethyl)benzoyl)amino)tetramethylrhodamine) *mixed isomers*	1 mg	125.00	100.00
C-2926		CellTracker™ Yellow-Green CMEDA (5-chloromethyleosin diacetate)	1 mg	125.00	100.00
C-6827	**New**	5-(and-6)-chloromethyl-2′,7′-dichlorodihydrofluorescein diacetate (CM-H_2DCFDA) *mixed isomers* *special packaging*	20x50 µg	125.00	100.00
C-6826	**New**	5-(and-6)-chloromethyl SNARF®-1, acetate *mixed isomers* *special packaging*	20x50 µg	125.00	100.00
F-1163		fluorescein 5-isothiocyanate diacetate	25 mg	115.00	92.00
H-7019	**New**	2′,4,5,6,7,7′-hexafluorofluorescein diacetate (HFFDA)	5 mg	125.00	100.00
M-1378		monobromobimane (mBBr)	25 mg	43.00	34.40
M-1380		monobromotrimethylammoniobimane bromide (qBBr)	25 mg	38.00	30.40
M-1381		monochlorobimane (mBCl)	25 mg	78.00	62.40
T-7017	**New**	2′,4′,5′,7′-tetrabromo-4,5,6,7-tetrafluorofluorescein diacetate (Br_4TFFDA)	5 mg	125.00	100.00
T-7016	**New**	4,5,6,7-tetrafluorofluorescein diacetate (TFFDA)	5 mg	125.00	100.00

15.3 Polar Tracers

Fixable Polar Tracers

Molecular Probes prepares a wide variety of highly water-soluble dyes and other detectable probes that can be used as cell tracers. Polar tracers can also be incorporated into liposomes to generate polymeric fluorescent tracers or antibody-labeling reagents.[1-5]

In most cases, these water-soluble probes are too polar to passively diffuse through cell membranes. Consequently, special methods for loading the dyes must be employed, including microinjection, pinocytosis or techniques that temporarily permeabilize the cell's membrane (Table 15.1). Permeabilization of cells with staphylococcal α-toxin or the saponin ester β-escin is reported to make the membrane of smooth muscle cells permeable to low molecular weight (<1000 daltons) molecules, while retaining high molecular weight compounds.[6-8] These reagents may facilitate the entry of many of the polar tracers in this section. Many of these tracers can also be loaded into cells noninvasively as their cell-permeant acetoxymethyl (AM) esters, which are discussed in more detail in Section 16.3.

Lucifer Yellow CH

Lucifer yellow CH (LY-CH or LY) has long been a favorite tool for studying neuronal morphology because it contains a carbohydrazide (CH) group that allows it to be covalently linked to surrounding biomolecules during aldehyde fixation.[9,10] Loading of this polar tracer and other similar impermeant dyes is usually accomplished by microinjection,[11] pinocytosis,[12] scrape loading,[13] ATP-induced permeabilization [14] or osmotic shock [15] (Table 15.1). Lucifer yellow CH localizes in the plant vacuole when taken up either through what is thought to be anion transport channels [16] or by fluid-phase endocytosis.[17] Upon injection into the epidermal cells of *Egeria densa* leaves, lucifer yellow CH moved into the cytoplasm of adjacent cells, localized in the plant vacuole or moved in and out of the nucleus.[18] The lithium salt of lucifer yellow CH [19] (L-453) is widely used for microinjection because of its high water solubility (~8%). The potassium salt (L-1177, solubility ~1%) or the ammonium salt (L-682, solubility ~6%) may be preferred in applications where lithium ions interfere with biological function.

Although its low absorbance at 488 nm (ϵ ~700 $cm^{-1}M^{-1}$) makes it inefficiently excited with the argon-ion laser, lucifer yellow CH has been used as a neuronal tracer in confocal laser scanning microscopy studies.[20-22] For electron microscopy studies, lucifer yellow can be used to photoconvert diaminobenzidine (DAB) into an insoluble, electron-dense reaction product.[23-25] Alternatively, anti–lucifer yellow antibodies may be used with enzyme-mediated immunohistochemical methods to develop a more permanent, fade-free colorimetric or electron-dense signal from dye-filled neurons that is suitable for light or electron microscopy;[26-30] see the Product Highlight "Anti–Lucifer Yellow and Anti–Cascade Blue Antibodies" on page 333.

Intracellular injection of lucifer yellow CH is extensively employed to delineate neuronal morphology in live neurons [31-33] (see Color Plate 3 in "Visual Reality") and in fixed brain slices,[34] as well as to investigate intercellular communication through gap junctions.[35-37] Lucifer yellow CH can also be used to label neurons with dye-filled electrodes during electrophysiological recording in order to correlate neuronal function with structure and connectivity.[11]

Other Lucifer Yellow Derivatives

Like lucifer yellow CH, our lucifer yellow ethylenediamine (A-1339) and lucifer yellow cadaverine (A-1340) are fixable with standard aldehyde fixatives and can be used as building blocks for new lucifer yellow derivatives.[38] The photoreactive lucifer yellow azide (lucifer yellow AB, A-629) has been used to photolabel cytoplasmic and luminal domains of Ca^{2+}-ATPase in sarcoplasmic reticulum vesicles.[39] The thiol-reactive lucifer yellow iodoacetamide (L-1338) can also be used as a microinjectable polar tracer, as well as for preparing fluorescent liposomes [1] and for detecting the accessibility of thiols in membrane-bound proteins [40,41] (see Section 9.4). In addition to these lucifer yellow derivatives, we offer lucifer yellow–conjugated dextrans (see Section 15.5).

Cascade Blue Hydrazide

Molecular Probes' Cascade Blue hydrazide is a fixable analog of the blue fluorescent tracer methoxypyrenetrisulfonic acid (MPTS, M-395).[42] All of the Cascade Blue hydrazide derivatives have good water solubility, ~1% for the sodium and potassium salts (C-687, C-3221) and ~8% for the lithium salt (C-3239). They also exhibit a higher absorbance ($\epsilon_{400\ nm}$ >28,000 $cm^{-1}M^{-1}$) and quantum yield (~0.54 in water) than lucifer yellow CH. In addition, Cascade Blue derivatives have good photostability and emissions that are well resolved from those of fluorescein and lucifer yellow CH.[43] Cascade Blue hydrazide and lucifer yellow CH can be simultaneously excited at <410 nm (Figure 15.2) for two-color detection at about

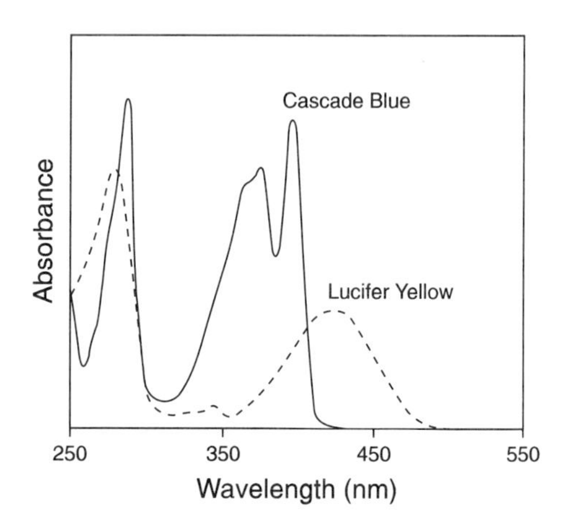

Figure 15.2 *Absorption spectra for equal concentrations of Cascade Blue hydrazide (C-687) and lucifer yellow CH (L-453) in water.*

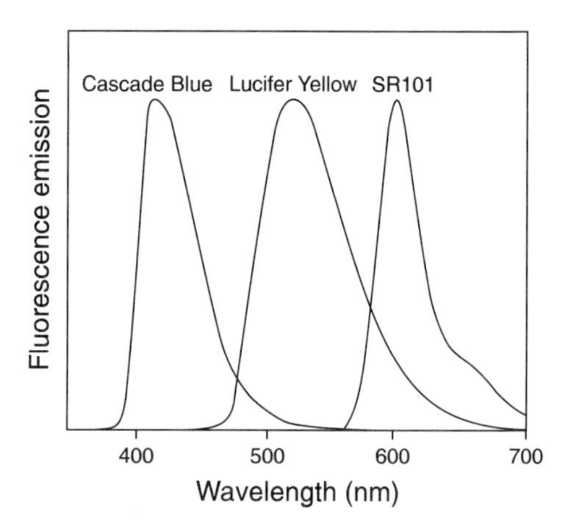

Figure 15.3 *Normalized fluorescence emission spectra for Cascade Blue hydrazide (C-687), lucifer yellow CH (L-453) and sulforhodamine 101 (S-359) in water.*

430 and 530 nm. Cascade Blue, lucifer yellow CH and sulforhodamine 101 can be used in combination for three-color mapping of neuronal processes (Figure 15.3). Molecular Probes also offers anti–Cascade Blue antibodies for localizing Cascade Blue–filled cells following fixation (see the Product Highlight "Anti–Lucifer Yellow and Anti–Cascade Blue Antibodies" on page 333). Like lucifer yellow CH, Cascade Blue hydrazide and some other polar tracers are taken up by plants and sequestered into their central vacuoles. In onion epidermal cells, this uptake is blocked by probenecid, indicating that transfer may be through anion-transport channels.[16]

Other Cascade Blue Derivatives

Cascade Blue acetyl azide (C-2284) is a water-soluble, amine-reactive tracer that can be introduced either by microinjection or by fusion of dye-filled liposomes with cells. Once inside the cell, this Cascade Blue derivative will react with the amine groups of intracellular proteins. Cascade Blue ethylenediamine and cadaverine (C-621, C-622) are aldehyde-fixable fluorophores similar to the amine derivatives of lucifer yellow (A-1339, A-1340). The photoreactive Cascade Blue azide derivative (C-625) is a photoaffinity label analogous to lucifer yellow azide (A-629) that can be microinjected into cells and subsequently fixed in place by UV illumination. Cascade Blue–labeled dextrans are also available (see Section 15.5).

Biocytin and Other Biotin Derivatives

Biocytin (ε-biotinoyl-L-lysine, B-1592) and biotin ethylenediamine (A-1593), which is structurally identical to Neurobiotin™ (a trademark of Vector Laboratories), are microinjectable anterograde and transneuronal tracers.[44-46] Retrograde transport of biocytin in neurons has also been reported.[47] These water-soluble tracers are often used to label neurons during electrophysiological measurements in order to correlate neuronal function with structure and connectivity.[48-51] Biocytin-X (B-2605) and biotin-X ethylenediamine (B-1597) contain seven-atom aminohexanoyl spacers ("X") that should enhance the accessibility of biotin to the relatively deep binding sites of avidin.[52] Biotin cadaverine (A-1594) and biotin-X cadaverine (B-1596) have slightly longer spacers than biotin ethylenediamine and biotin-X ethylenediamine, respectively.[53-55]

Biocytin, biotin ethylenediamine, biotin cadaverine, biotin-X cadaverine and N^{α}-(4-aminobenzoyl)biocytin (A-1604) all contain primary amines and can therefore be fixed in cells with formaldehyde or glutaraldehyde and subsequently detected using fluorescent- or enzyme-labeled avidin or streptavidin second-step reagents.[56] Our biocytin hydrazide (B-1603) and biocytin-X hydrazide (B-6350) can serve both as aldehyde-fixable tracers and as reactive probes for labeling glycoproteins and nucleic acids (see Section 4.2).

As with the reactive lucifer yellow and Cascade Blue derivatives discussed above, amine- or thiol-reactive biotin derivatives may be useful for some intracellular labeling applications. Biotin and biotin-X succinimidyl esters (B-1513, B-1582) have been used to trace retinal axons in avian embryos.[57] Because they are more water soluble, the *sulfo*succinimidyl esters of biotin-X and biotin-XX (B-6353, B-6352) or the thiol-reactive biocytin maleimide (M-1602) may be preferred for these applications.

Fluorescent Biotin Derivatives

Fluorescence of the finer processes of dye-filled neurons may fade rapidly or be obscured by the more intensely stained portions of the neuron, necessitating further amplification of the signal or ultrastructural detection methods. Cascade Blue biocytin (C-6949) and lucifer yellow biocytin (L-6950) incorporate a fluorophore, biotin and a fixable primary amine into a single molecule, thus enabling researchers to amplify the signals of these tracers with fluorescent or enzyme-labeled avidin or streptavidin conjugates (see Section 7.5). Although our lucifer yellow cadaverine biotin-X (L-2601) lacks a primary amine, it was reported that this tracer was well retained in aldehyde-fixed tissues, even after sectioning, extraction with detergents and several washes.[58,59] Because the lucifer yellow– and Cascade Blue–biocytin conjugates contain free primary amines, they should be even more efficiently fixed by formaldehyde or glutaraldehyde.

Fluorescent derivatives of biotin should also be useful for directly localizing biotin-binding proteins.[60] Fluorescein biotin (B-1370) is a nonfixable fluorescent biotin derivative developed by Molecular Probes as an alternative to radioactive biotin for detecting and quantitating biotin binding sites by either fluorescence or absorbance. A fluorescence polarization assay that employs competitive binding of fluorescein biotin to assess the degree of biotinylation of proteins has recently been reported.[61]

Anti–Lucifer Yellow and Anti–Cascade Blue® Antibodies

Molecular Probes' anti–lucifer yellow antibodies were specifically developed to overcome the limitations of lucifer yellow CH (L-453, L-682, L-1177), an aldehyde-fixable fluorescent cell tracer that has long been used by neuroscientists to identify patterns of gap junctional communication,[1] to assay the outgrowth of developing neurons[2] and to characterize the morphology of neurons from which electrical recordings have been made.[3] Even though the cell soma of a lucifer yellow–filled neuron may be brightly stained, its finer processes can sometimes be faint and may fade rapidly or be obscured by the more intensely stained portions of the neuron. Investigators have been able to overcome these limitations by using anti–lucifer yellow antibodies in conjunction with standard enzyme-mediated immunohistochemical methods to develop a more permanent, fade-free colorimetric signal for light microscopy.[4-7] Anti–lucifer yellow antibodies have also been used to develop tissue for electron microscopy[8] and to distinguish neurons filled with lucifer yellow CH from those injected with the lectin *Phaseolus vulgaris* leucoagglutinin[9] (PHA-L). Molecular Probes' lithium salt of Cascade Blue hydrazide (C-3239), which can also be fixed in place with aldehyde fixatives, can potentially be used as a second label with lucifer yellow CH to characterize the morphology of interacting neurons. For these applications, Molecular Probes offers unconjugated and biotinylated rabbit polyclonal anti–lucifer yellow (A-5750, A-5751) and anti–Cascade Blue (A-5760, A-5761) antibodies; see Section 7.3 for a complete description of these and other anti-fluorophore antibodies.

1. Nature 292, 17 (1981); **2.** Science 242, 700 (1988); **3.** J Neurosci Meth 36, 309 (1991); **4.** J Neurosci 14, 5267 (1994); **5.** J Neurosci Meth 41, 45 (1992); **6.** J Comp Neurol 296, 598 (1990); **7.** Devel Biol 94, 391 (1982); **8.** Circulation Res 70, 49 (1992); **9.** J Neurosci Meth 33, 207 (1990).

Cat #	Product Name	Unit Size	Price Per Unit ($) 1–4 Units	5–24 Units
A-5750	anti-lucifer yellow, rabbit IgG fraction *≥ 1 mg/mL*	0.5 mL	148.00	118.40
A-5751	anti-lucifer yellow, rabbit IgG fraction, biotin-XX conjugate *≥ 1 mg/mL*	0.5 mL	148.00	118.40
A-5760	anti-Cascade Blue®, rabbit IgG fraction *≥ 1 mg/mL*	0.5 mL	148.00	118.40
A-5761	anti-Cascade Blue®, rabbit IgG fraction, biotin-XX conjugate *≥ 1 mg/mL*	0.5 mL	148.00	118.40

Nonfixable Polar Tracers

Polar fluorescent dyes are commonly used to investigate fusion, gap-junctional communication and lysis and to detect changes in cell or liposome volume. These events are primarily monitored by following changes in a dye's fluorescence caused by interaction with nearby molecules. For example, because the fluorescence of dyes at high concentrations is quenched, various processes that result in a dilution of the dyes, such as lysis or fusion of fluorescent dye–filled cells or liposomes, can produce an increase in fluorescence, thereby providing a facile method for monitoring these events. Cell–cell and cell–liposome fusion, as well as membrane permeability and transport through gap junctions, can all be monitored using these methods. Furthermore, a fluorogenic substrate can be incorporated within a liposome or cell that lacks the enzymatic activity to generate a fluorescent product; subsequent fusion with a cell that contains the appropriate enzyme, including reporter enzymes introduced by gene transfection, will release a fluorescent product.[62] Several water-soluble substrates (e.g., fluorescein digalactoside and fluorescein diphosphate; see Chapter 10) may be useful for this application. Also, highly polar esterase substrates such as 5-sulfofluorescein diacetate (SFDA, S-1129) can be incorporated into liposomes and their uptake monitored by the increase in fluorescence following liposome fusion with cells. These polar tracers are impermeant to cell membranes and must be incorporated into cells by one of a variety of invasive techniques (Table 15.1); many of these dyes are lost from the cell upon disruption of the cell membrane.

Fluorescein Derivatives

The self-quenching of fluorescein derivatives provides a means of determining their concentration in dynamic processes such as lysis or fusion of dye-filled cells or liposomes (see the Technical Note "Assays of Volume Change, Membrane Fusion and Membrane Permeability" on page 364). Calcein (C-481) — a polyanionic fluorescein derivative that has about six negative and two positive charges at pH 7 — as well as BCECF (B-1151), carboxyfluorescein (C-194, C-1904) and fluorescein-5-(and-6)-sulfonic acid (F-1130) are all soluble in water at >100 mM at pH 7. Unlike the other fluorescein derivatives, calcein exhibits fluorescence that is essentially independent of pH between 6.5 and 12.

These four green fluorescent polar tracers are widely used for investigating:

- Cell volume changes in neurons and other cells [63-66]
- Gap junctional communication [67-69]
- Liposome formation, fusion and targeting [70-73]
- Membrane integrity and permeability [74-81]

Fluorescence of calcein (but not of carboxyfluorescein or fluorescein sulfonic acid) is strongly quenched by Fe^{3+}, Co^{2+}, Cu^{2+} and Mn^{2+} at physiological pH but not by Ca^{2+} or Mg^{2+} ions.[82,83] Monitoring the fluorescence level of cells that have been loaded with calcein (or its AM ester, see below) may provide a facile means for following uptake of Fe^{3+}, Co^{2+}, Cu^{2+}, Mn^{2+} and certain other metals through ion channels (see Tables 22.3 and 22.4 in Chapter 22). Increases in the internal volume of lipid vesicles and virus envelopes cause a decrease in Co^{2+}-induced quenching of calcein, a change that can be followed fluorometrically.[84] In addition, the Co^{2+}-quenched calcein complex is useful for both lysis and fusion assays.[85]

Molecular Probes prepares a high-purity grade of calcein (C-481) that is generally >97% pure by HPLC. Competitive commercial-grade calcein is typically <50% pure. The chemical structure assigned to "calcein" in various literature references and by commercial sources has been inconsistent;[86,87] our structure has been confirmed by NMR spectroscopy and we believe that several past assignments of other structures to calcein were incorrect. We also offer a special high-purity grade of 5-(and-6)-carboxyfluorescein (C-1904) that contains no polar or nonpolar impurities that might alter transfer rates of the dye between vesicles and cells.[88]

Cell-Permeant Fluorescein Derivatives

Cell-permeant versions of carboxyfluorescein, fluorescein sulfonic acid and calcein permit passive loading of adherent cells and cells in suspension (see Section 16.3). Acid hydrolysis of nonfluorescent carboxyfluorescein diacetate (CFDA; C-195, C-1361, C-1362; see Section 16.3) to fluorescent carboxyfluorescein has been used to detect the fusion of dye-loaded clathrin-coated vesicles with lysosomes.[89] CFDA has also been used to investigate cell–cell communication in plant cells.[90-92]

Calcein AM, but not the AM esters of BCECF or CFDA, differentially labels lymphocytes into two populations based on fluorescence intensity, only one of which is taken up by lymphoid organs. This unique property makes calcein AM a useful probe for determining the lymph node homing potential of lymphocytes.[93]

In an important new technique for studying gap junctional communication, cells are simultaneously labeled with calcein AM (C-1430, C-3099, C-3100) and DiI (D-282, see Section 15.4) and then mixed with unlabeled cells. When gap junctions are established, only the cytosolic calcein tracer (but not the DiI membrane probe) is transferred from the labeled cell to the unlabeled cell. Thus, after gap-junctional transfer, the initially unlabeled cells exhibit the green fluorescence of calcein but not the red fluorescence of DiI.[67,94,95] A similar quantitative assay that combines calcein AM and the lipid tracer PKH26 [96] to detect gap junctional communication between cells in culture has also been described.[97] In addition, calcein AM and DiI have been combined for use in following cell fusion.[98]

Fluorescein Substitutes

Sections 1.4 and 1.5 of Chapter 1 describe several new green fluorescent dyes that have exceptional optical properties. Not only do these innovative fluorescein substitutes exhibit high absorbance and quantum yields in aqueous solution, but they are significantly more photostable than fluorescein (see Figure 1.6 in Chapter 1, Figure 7.1 in Chapter 7 and Figure 11.1 in Chapter 11) and less sensitive to pH (see Figure 1.7 in Chapter 1). Their greater photostability may make them the preferred green fluorescent dyes for fluorescence microscopy. In addition to their membrane-permeant versions, which are described in Section 16.3, highly water-soluble derivatives of these fluorescein substitutes are available for use as polar tracers:

- Oregon Green™ carboxylic acids (O-6135, O-6138, O-6146, O-6148), which have low pH sensitivity at near-neutral pH
- Rhodamine Green™ carboxylic acid (5(6)-CR 110, R-6150), an exceptionally photostable green fluorescent fluorophore with pH-insensitive fluorescence in the physiological range
- Rhodol Green™ carboxylic acid (5(6)-CRF 492, R-6152), which photobleaches more slowly than fluorescein derivatives and also has low sensitivity to pH in the near-neutral range

- BODIPY 495/505 sulfonic acid (D-6961) and BODIPY 492/515 disulfonic acid (D-3238), which have high absorbance, narrow spectral bandwidths and bright green, pH-independent fluorescence
- Carboxydichlorofluorescein (C-368), which has a lower pK_a than fluorescein

In addition to its high water solubility (>5%), the lithium salt of BODIPY 495/505 sulfonic acid (D-6961) has a lower molecular weight (327 versus 457) and about 80-fold higher absorbance at the 488 nm spectral line of the argon-ion laser than lucifer yellow CH (~57,000 $cm^{-1}M^{-1}$ versus ~700 $cm^{-1}M^{-1}$). Although the dye is not expected to be fixable in cells, it should be particularly useful as a microinjectable tracer for investigating gap-junctional communication.

Stachyose Fluorescein

Stachyose fluorescein (S-6954) is a fluorescent sugar tetramer that has a molecular weight of 1146 daltons. This probe is designed to eliminate complications that arise from the fairly broad distribution of molecular weights present in dextran preparations (see Section 15.5). Stachyose fluorescein gives researchers an alternative for applications — including tracing fine neuronal projections, examining membrane exclusion properties and investigating gap-junctional communication — in which they now employ lucifer yellow CH or the polydisperse 3000 MW dextrans.

Sulforhodamines and Brilliant Sulfoflavin

Sulforhodamine 101 (S-359), sulforhodamine B (S-1307) and sulforhodamine G (S-6957) are orange to red fluorescent, water-soluble sulfonic acid tracers with strong absorbance and good photostability. Brilliant sulfoflavin (B-3245) has similarly high water solubility but green fluorescence and an unusually high Stokes shift (excitation/emission ~417/521 nm).

Sulforhodamine 101, the precursor to reactive Texas Red® derivatives, has been the preferred red fluorescent polar tracer for use in combination with lucifer yellow CH, carboxyfluorescein or calcein [99-101] (see Color Plate 3 in "Visual Reality"). Activity-dependent uptake of sulforhodamine 101 during nerve stimulation has been reported.[102-106]

Sulforhodamine B and sulforhodamine G provide low-cost alternatives to sulforhodamine 101 for investigating neuronal morphology,[107,108] preparing fluorescent liposomes,[109] studying cell–cell communications [110,111] and following phagosome–lysosome uptake and fusion.[99,112-114] Sulforhodamine G is reported to readily cross the membrane of some plant cells (wheat and legumes) when the external pH is <5.1 and to localize in the cytoplasm.[115,116] Because the fluorescence of sulforhodamine 6G is similar to that of chloroplasts, it may be useful to try the longer-wavelength sulforhodamine 101 for this application.

Sulfonerhodamine Bis-(PEG 2000)

The new polar tracer sulfonerhodamine bis-(PEG 2000) (S-6956) comprises a rhodamine dye symmetrically conjugated to two molecules of 2000-dalton poly(ethylene glycol) (PEG). With a total molecular weight of about 4500 daltons, this conjugate is less likely than other polar tracers to pass through gap junctions. Its fluorescence-to-weight ratio is higher than that typically exhibited by our low molecular weight tetramethylrhodamine dextrans (Section 15.5 below).

Polycationic Carbocyanine

The thiadicarbocyanine dye (D-1389) has three positive charges, making it a highly polar, membrane-impermeant polar tracer for cells or liposomes. To date, it has been used primarily as a probe for membrane asymmetry, in which its intense long-wavelength absorbance quenches probes, such as membrane-bound merocyanine 540 (M-299, see Section 25.3), in the outer leaflet by fluorescence resonance energy transfer [117-119] (see the Technical Note "Fluorescence Resonance Energy Transfer" on page 456). Bimolecular complexes of this polycationic dye with shorter-wavelength polyanionic dyes such as calcein are also expected to undergo fluorescence resonance energy transfer, which may be useful for following lysis or fusion.

Porphine Derivatives

Salts of tetrakis-(4-carboxyphenyl)porphine (TCPP, T-6931) and tetrakis-(4-sulfophenyl)porphine (TSPP, T-6932) offer unique possibilities as polar tracers. Both TCPP and TSPP have an intense absorption peak near 414 nm with ϵ >400,000 $cm^{-1}M^{-1}$ and several weaker absorption peaks between about 500 nm and 650 nm. Excitation at any of these absorption peak wavelengths yields emission at ~645 nm with a shoulder at about 700 nm. The broad absorption spectrum and high Stokes shift of these porphine derivatives should permit their fluorescence to be simultaneously excited but separately detected with probes such as lucifer yellow CH, calcein and sulforhodamine 101. Emission from TCPP and TSPP is pH sensitive with a pK_a of ~5.5 and is sensitive to binding of a few heavy metals (see Section 22.7). Dextran conjugates of TCPP (D-6929, D-6930) are described in Section 15.5.

Coumarins

Calcein blue (H-1426) and 7-hydroxycoumarin-3-carboxylic acid (H-185, see Section 1.7) are coumarin-based, blue fluorescent polar tracers (excitation/emission maxima of ~360/440 nm and ~388/445 nm, respectively). Their uses complement calcein and the other fluorescein-based polar tracers.[94] The membrane-permeant AM ester of calcein blue is also available (C-1429, see Section 16.3) for passive loading of cells.[120]

Polysulfonated Pyrenes

HPTS (8-hydroxypyrene-1,3,6-trisulfonic acid, also known as pyranine; H-348) is a unique pH-sensitive tracer that fluoresces blue in acidic solutions and in acidic organelles such as lysosomes, where it is taken up during endocytosis,[121-123] and fluoresces green in more basic organelles.[124] In addition to its use as a probe for proton translocation,[125-128] HPTS has been employed for intracellular labeling of neurons [129] and for delineating the cellular pathway of photosynthate transfer in the developing wheat grain.[115] HPTS forms a nonfluorescent complex with the cationic quencher DPX (X-1525), and several assays have been described that monitor the increase in HPTS fluorescence that occurs upon lysis or fusion of liposomes or cells containing this quenched complex [123,130-132] (see the Technical Note "Assays of Volume Change, Membrane Fusion and Membrane Permeability" on page 364). HPTS has also been used as a viscosity probe in unilamellar phospholipid vesicles.[133]

The pH-insensitive methyl ether of HPTS (MPTS, M-395) and 1,3,6,8-pyrenetetrasulfonic acid (P-349) are extremely soluble in water (>25%) and have been utilized as blue fluorescent tracers for detecting permeability of water.[134-136] As with HPTS, fluorescence of MPTS and 1,3,6,8-pyrenetetrasulfonic acid is quenched by

DPX [123,131,137] and by the cationic spin labels CAT 1 and TEMPO choline (see below).[138] These quenched-fluorophore complexes are useful for following lysis of cells and liposomes. DAS (D-6948) is an anthracene-based, blue fluorescent sulfonic acid tracer that may have utility similar to MPTS.

ANTS–DPX

The polyanionic dye ANTS (A-350) is often used in combination with the cationic quencher DPX (X-1525) for membrane fusion or permeability assays, including complement-mediated immune lysis [139] (see the Technical Note "Assays of Volume Change, Membrane Fusion and Membrane Permeability" on page 364). In addition, thallium (Tl^+) and cesium (Cs^+) ions quench the fluorescence of ANTS, pyrenetetrasulfonic acid and some other polyanionic fluorophores.[140,141] A review by Garcia [134] describes how this quenching effect can be utilized to determine transmembrane ion permeability. The unusually high Stokes shift of ANTS in water (>150 nm) separates its emission from much of the autofluorescence of biological samples. An approximately fourfold enhancement of the quantum yield of ANTS is induced by D_2O — a spectral characteristic that has been used to determine water permeability in red blood cell ghosts and kidney collecting tubules.[142,143] ANTS has also been employed as a neuronal tracer.[129]

Lanthanide Chelates

Terbium ion (Tb^{3+} from $TbCl_3$, T-1247) forms a chelate with dipicolinic acid (DPA, D-1248) that is 10,000-times more fluorescent than free Tb^{3+}. Fusion of vesicles that have been separately loaded with DPA and Tb^{3+} results in enhanced fluorescence, providing the basis for liposome fusion assays; see the Technical Note "Assays of Volume Change, Membrane Fusion and Membrane Permeability" on page 364. DPA can also be loaded into cells as its AM ester (DPA, AM; D-1249). The fluorescence emission spectrum of the Tb^{3+}-DPA complex exhibits two sharp peaks at 491 nm and 545 nm. Chelates of the europium ion (Eu^{3+} from $EuCl_3$, E-6960) are similar to those of Tb^{3+} but with longer-wavelength emission (peaks at 590 nm and 613 nm). Release of Eu^{3+} that has been encapsulated in liposomes is used in immunoassays,[144] and release of Eu^{3+} from cells is used in an assay for complement-mediated cytolysis.[145] With millisecond radiative lifetimes, Tb^{3+} and Eu^{3+} chelates are also useful for time-resolved detection in immunoassays.[145] Naphthoyltrifluoroacetone (NTA, T-411) and its derivatives, which require solubilization with 0.1% Triton® X-100 in aqueous systems, are generally used as Tb^{3+} and Eu^{3+} chelators.

TOTO® and YOYO® Nucleic Acid Stains

Our high affinity nucleic acid stains, including TOTO-1 and YOYO-1 (T-3600, Y-3601; see Section 8.1), form tight complexes with nucleic acids with slow off-rates for release of the dye. Consequently, nucleic acids that have been prelabeled with these dyes can be traced in cells following microinjection or during gene transfer (see photo on front cover). The cell-to-cell transport via plasmodesmata of TOTO-1–labeled RNA, single-stranded DNA and double-stranded DNA has been determined following microinjection of the labeled nucleic acids into plant cells.[146-148]

Dansyl Glucosamine

Unlike all the other probes in this section, the fluorescent carbohydrate analog, dansyl glucosamine (D-402), is an electrically *neutral* probe that has utility as a polar tracer. It has been used to follow monosaccharide transport in yeast.[149]

Isosulfan Blue

Isosulfan blue (I-6958) is a nonfluorescent, highly water-soluble dye that has very strong long-wavelength absorbance ($\epsilon_{630\ nm}$ >110,000 $cm^{-1}M^{-1}$ in methanol). Although we are not aware of its use as a polar tracer or for microinjection, its structural similarity to malachite green indicates that it may be a potent photogenerator of singlet oxygen, which could make it useful for selective photoablation of microinjected cells in cell development and differentiation studies. Isosulfan blue sulfonyl chloride (I-6204, see Section 1.6) is also available for preparing photosensitizing conjugates.

Caged Fluorescent Dye Tracers

Caged Fluorophores

UV photolysis of photoactivatable fluorescent dyes provides a means of controlling — both spatially and temporally — the release of fluorescent tracers (see also Chapter 19). Thus, caged dyes enable researchers to follow the movement of individual molecules and cellular structures,[150] as well as to study cell lineage in live organisms (see Color Plate 3 in "Visual Reality"). Caged fluorescein [151] (F-7103), caged rhodamine 110 (R-7102) and caged carboxy-Q-rhodamine (C-7100) are colorless and nonfluorescent until they are photolyzed at <365 nm to the intensely fluorescent free dyes. Movement of the liberated fluorophore from the site of photolysis can then be followed. Unlike caged fluorescein, caged 8-hydroxypyrene-1,3,6-trisulfonic acid [152] (caged HPTS, D-7060) contains only one photoremovable group, making the photolytic activation of this fluorophore more efficient. Caged HPTS is nonfluorescent and is also much more water-soluble than the other caged fluorophores. In a collaboration with Walter Lempert of Princeton University, we have shown caged HPTS and caged fluorescein dextran (D-3310, see Section 15.5 and Figure 15.6) to be excellent probes for tracing vortices in water using a technique called photoactivated nonintrusive tracking of molecular motion [153] (PHANTOMM). Our caged fluorophore–labeled dextrans are described in Section 15.5.

Reactive Caged Fluorophores

We also offer several new amine-reactive caged dyes that can be used to generate protein conjugates and other tracers for diffusion measurements using the photoactivation of fluorescence (PAF) technique.[150,154-158] PAF experiments are analogous to fluorescence recovery after photobleaching (FRAP) experiments except that the fluorophore is activated upon illumination rather than bleached. Measuring the bright fluorescent signal of the photoactivated fluorophore against a dark background should be intrinsically more sensitive than measuring a dark photobleached region against a bright field. Our amine-reactive caged dyes include CMNCBZ-caged carboxy-Q-rhodamine succinimidyl ester (C-7101), DMNB-caged Cl-NERF sulfosuccinimidyl ester (C-7079), NVOC-caged Rhodamine Green succinimidyl ester (R-7090) and the more water-soluble NVOC-caged Rhodamine Green sulfosuccinimidyl ester (R-7091). Once uncaged, these probes exhibit high extinction coefficients ($\epsilon_{537\ nm}$ ~94,000 $cm^{-1}M^{-1}$ for carboxy-Q-rhodamine, $\epsilon_{514\ nm}$ ~84,000 $cm^{-1}M^{-1}$ for Cl-NERF at pH 9 and $\epsilon_{502\ nm}$ ~89,000 $cm^{-1}M^{-1}$ for Rhodamine Green dye), good fluorescence quantum yields and excellent photostability. PAF measurements on caged fluorescein–labeled tubulin revealed microtubule migration processes in mitotic spindles.[159-161]

Fluorescent Retrograde Tracers

True Blue and Nuclear Yellow

The popular retrograde tracer true blue (T-1323) is a UV-excitable, divalent cationic dye that stains the cytoplasm with blue fluorescence.[162-166] For two-color neuronal mapping, true blue has been combined with longer-wavelength tracers such as nuclear yellow [162,167-169] (Hoechst S769121, N-6946) or diamidino yellow,[96,170-173] which primarily stain the neuronal nucleus with yellow fluorescence, or with fluorescent microspheres.[174] True blue is reported to be a less cytotoxic retrograde tracer than Fluoro-Gold™ [96,163] (a trademark of Fluorochrome, Inc.) and to be a more efficient retrograde tracer than diamidino yellow.[164] Both true blue and nuclear yellow are stable when subjected to immunohistochemical processing and can be used to photoconvert DAB into an insoluble, electron-dense reaction product.[23,24,175,176]

Hydroxystilbamidine

Hydroxystilbamidine methanesulfonate (H-7599) is an interesting probe originally developed as a trypanocide at May and Baker by the same group that developed ethidium bromide.[177,178] Hydroxystilbamidine has recently been identified by Wessendorf [179] as the active component of a dye that was named Fluoro-Gold by Schmued and Fallon [180] and later sold for retrograde tracing under that name by Fluorochrome, Inc.[181] The use of hydroxystilbamidine as a histochemical stain and as a retrograde tracer for neurons apparently goes back to the work of Snapper and collaborators in the early 1950s who, while studying the effects of hydroxystilbamidine on multiple myeloma,[182-184] showed that therapeutically administered hydroxystilbamidine gives selective staining of ganglion cells.[185,186] The comprehensive article by Wessendorf [179] also describes several other early applications of hydroxystilbamidine that do not include its therapeutic uses:

- As an AT-selective, nonintercalating nucleic acid stain for discriminating between DNA and RNA [187-189] (see Section 8.1)
- For histochemical staining of DNA, mucosubstances and elastic fibers, as well as mast cells [190]
- As a ribonuclease inhibitor [191]
- For selective identification of fungi [192]
- For lysosomal staining of live cells [193-198] (see Section 12.3)

Apparently the weakly basic properties of hydroxystilbamidine that result in its uptake by lysosomes are important for its mechanism of retrograde transport.[179,193] Hydroxystilbamidine (as its isethionate salt [96]) is reported to be nonmutagenic in *Salmonella typhimurium* by the Ames test [190] and has been used therapeutically in large doses (up to 225 mg i.v. daily).[199,200] However, the product sold as Fluoro-Gold is reported to be toxic to some cells [163] and may cause necrosis at sites of injection.[201,202]

Other Polar Tracers for Retrograde Tracing

A variety of other low molecular weight dyes have been used as fluorescent retrograde neuronal tracers.[203,204] These include propidium iodide (P-1304) [205,206] and DAPI.[207,208] Both propidium iodide and DAPI can be used to photoconvert DAB into an insoluble, electron-dense product.[23,209,210] The lactate salt of DAPI (D-3571) has much higher water solubility than the chloride salt (D-1306), making it the preferred form for microinjection.

Polar Spin Labels

The cationic spin labels CAT 1 (T-506) and TEMPO choline (D-530) are highly water soluble probes that have been used to:

- Assay complement- and drug-mediated immune lysis [211-213]
- Detect oxygen gradients and oxidation–reduction properties of cells [214-218]
- Follow transport across membranes [219]
- Quench fluorescent dyes in solutions, cells and cell membranes [220-223]
- Study liposome permeability [218,223,224]

1. J Fluorescence 3, 33 (1993); **2.** Photochem Photobiol 56, 325 (1992); **3.** J Immunol Meth 121, 1 (1989); **4.** Cytometry 8, 562 (1987); **5.** J Immunol Meth 100, 59 (1987); **6.** Ann Rev Physiol 52, 857 (1990); **7.** J Biol Chem 264, 5339 (1989); **8.** Meth Cell Biol 31, 63 (1989); **9.** Nature 292, 17 (1981); **10.** Cell 14, 741 (1978); **11.** J Neurosci Meth 36, 309 (1991); **12.** J Cell Biol 104, 1217 (1987); **13.** Exp Cell Res 168, 422 (1987); **14.** J Biol Chem 262, 8884 (1987); **15.** Cell 29, 33 (1982); **16.** J Cell Sci 99, 557 (1991); **17.** Planta 179, 257 (1989); **18.** Planta 181, 129 (1990); **19.** Lucifer yellow CH is licensed to Molecular Probes under U.S. Patent No. 4,473,693; **20.** J Neurosci Meth 52, 111 (1994); **21.** J Neurosci Meth 47, 23 (1993); **22.** Scanning Microscopy 5, 619 (1991); **23.** Lübke, J. in *Neuroscience Protocols*, F.G. Wouterlood, Ed., Elsevier Science Publishers (1993) 93-050-06-01–13; **24.** Microscopy Res Technique 24, 2 (1993); **25.** J Histochem Cytochem 36, 555 (1988); **26.** J Neurosci 14, 5267 (1994); **27.** J Neurosci Meth 41, 45 (1992); **28.** J Comp Neurol 296, 598 (1990); **29.** J Neurosci Meth 33, 207 (1990); **30.** Devel Biol 94, 391 (1982); **31.** J Neurosci 15, 1755 (1995); **32.** J Neurosci 15, 1506 (1995); **33.** J Neurosci 14, 5077 (1994); **34.** J Neurosci Meth 53, 87 (1994); **35.** Experientia 50, 125 (1994); **36.** J Neurosci 14, 999 (1994); **37.** Meth Enzymol 234, 235 (1994); **38.** Anal Biochem 211, 210 (1993); **39.** Biochim Biophys Acta 1068, 27 (1991); **40.** Biochem Soc Trans 23, 38S (1995); **41.** Biochemistry 30, 11245 (1991); **42.** Anal Biochem 198, 119 (1991); **43.** Nature 367, 69 (1994); **44.** J Neurosci 14, 3805 (1994); **45.** J Neurosci Meth 41, 31 (1992); **46.** J Neurosci Meth 37, 141 (1991); **47.** J Neurosci Meth 39, 163 (1991); **48.** Nature 377, 734 (1995); **49.** J Neurosci 14, 5578 (1994); **50.** J Neurosci 14, 4338 (1994); **51.** J Neurosci 14, 1812 (1994); **52.** Biochemistry 21, 978 (1982); **53.** J Biol Chem 269, 24596 (1994); **54.** Anal Biochem 205, 166 (1992); **55.** Biochem J 251, 935 (1988); **56.** Pilowsky, P.M. and Llewellyn-Smith, I.J. in *Neuroscience Protocols*, F.G. Wouterlood, Ed., Elsevier Science Publishers (1993) 93-050-20-01–16; **57.** Devel Biol 119, 322 (1987); **58.** J Neurosci Meth 53, 23 (1994); **59.** J Neurosci Meth 46, 59 (1993); **60.** Acta Histochem Cytochem 21, 463 (1988); **61.** Clin Chem 40, 2112 (1994); **62.** Anal Lett 18, 1847 (1985); **63.** Meth Neurosci 27, 361 (1995); **64.** Neuroscience 69, 283 (1995); **65.** J Biol Chem 267, 17658 (1992); **66.** Biochim Biophys Acta 1052, 278 (1990); **67.** BioTechniques 18, 490 (1995); **68.** Biophys J 66, 1915 (1994); **69.** J Biol Chem 268, 706 (1993); **70.** J Pharm Sci 83, 276 (1994); **71.** J Cell Biol 123, 1845 (1993); **72.** Anal Biochem 207, 109 (1992); **73.** Biochim Biophys Acta 1106, 23 (1992); **74.** Biochemistry 34, 1606 (1995); **75.** Biophys J 68, 1864 (1995); **76.** J Biol Chem 269, 14473 (1994); **77.** J Cell Biol 127, 1885 (1994); **78.** Biochemistry 31, 12424 (1992); **79.** Biochemistry 31, 9912 (1992); **80.** Anal Biochem 172, 403 (1988); **81.** Biochemistry 27, 5713 (1988); **82.** Chem Pharm Bull 39, 227 (1991); **83.** Biochim Biophys Acta 691, 332 (1982); **84.** Anal Biochem 134, 26 (1983); **85.** J Biol Chem 257, 13892 (1982); **86.** Anal Chem 35, 1035 (1963); **87.** Anal Chem 31, 456 (1959); **88.** Biochim Biophys Acta 649, 183 (1981); **89.** Cell 32, 921 (1983); **90.** J Neurochem 57, 1270 (1991); **91.** J Cell Biol 106, 715 (1988); **92.** Science 232, 525 (1986); **93.** Cytometry 13, 739 (1992); **94.** J Cell Biol 130, 987 (1995); **95.** In Vitro Cell Devel Biol 30A, 796 (1994); **96.** Not available from Molecular Probes; **97.** J Cell Biol 122, 157 (1993); **98.** J Cell Biol 131, 655 (1995); **99.** Meth Enzymol 221, 234 (1993); **100.** J Biol Chem 267, 18424 (1992); **101.** J Cell Biol 108, 2241 (1989); **102.** J Neurosci 15, 5036 (1995); **103.** J Physiol 478, 265 (1994); **104.** Exp Brain Res 97, 239 (1993); **105.** J Neurosci 12, 3187 (1992); **106.** Nature 314, 357 (1985); **107.** J Neurosci 14, 6886 (1994); **108.** J Neurosci 12, 2960 (1992); **109.** FEBS Lett 305, 185 (1992); **110.** Am J Physiol 252, H223 (1987); **111.** Proc Natl Acad Sci USA 84, 2272 (1987); **112.** J Cell Biol 104, 1749 (1987); **113.** J Leukocyte Biol 41, 111 (1987); **114.** J Leukocyte Biol 36, 273

(1984); **115.** Plant Cell Environ 17, 257 (1994); **116.** Physiol Plantarum 73, 457 (1988); **117.** J Am Chem Soc 114, 7396 (1992); **118.** Meth Enzymol 149, 281 (1987); **119.** Cell 30, 725 (1982); **120.** Science 251, 81 (1991); **121.** Proc Natl Acad Sci USA 92, 3156 (1995); **122.** J Biol Chem 268, 6742 (1993); **123.** Biochim Biophys Acta 1024, 352 (1990); **124.** Pharmaceutical Res 7, 824 (1990); **125.** J Neurochem 62, 2022 (1994); **126.** Nature 370, 379 (1994); **127.** Biochemistry 32, 7669 (1993); **128.** Biochim Biophys Acta 1111, 17 (1992); **129.** Neuron 11, 801 (1993); **130.** J Gen Virol 75, 3477 (1994); **131.** Am J Physiol 262, G30 (1992); **132.** Plant Physiol 86, 999 (1988); **133.** Arch Biochem Biophys 202, 650 (1980); **134.** Meth Enzymol 207, 501 (1992); **135.** Biophys J 54, 595 (1988); **136.** J Gen Physiol 83, 819 (1984); **137.** Biochim Biophys Acta 1142, 277 (1993); **138.** J Am Chem Soc 100, 3234 (1978); **139.** J Immunol Meth 15, 255 (1977); **140.** Proc Natl Acad Sci USA 78, 775 (1981); **141.** Proc Natl Acad Sci USA 77, 4509 (1980); **142.** Biochemistry 28, 824 (1989); **143.** Biophys J 54, 587 (1988); **144.** Clin Chem 37, 1519 (1991); **145.** J Immunol Meth 147, 13 (1992); **146.** Science 270, 1980 (1995); **147.** Virology 207, 345 (1995); **148.** Cell 76, 925 (1994); **149.** Biochem Int'l 20, 479 (1990); **150.** Ann Rev Physiol 55, 755 (1993); **151.** Exp in Fluids 21, 237 (1996); **152.** U.S. Patent No. 5,514,710; **153.** Exp in Fluids 18, 249 (1995); **154.** Biophys J 69, 1674 (1995); **155.** J Cell Biol 131, 721 (1995); **156.** J Cell Biol 122, 833 (1993); **157.** Cell 68, 923 (1992); **158.** Nature 352, 126 (1991); **159.** J Cell Biol 120, 1177 (1995); **160.** J Cell Biol 117, 105 (1992); **161.** J Cell Biol 109, 637 (1989); **162.** Neuroscience 60, 125 (1994); **163.** Neurosci Lett 128, 137 (1991); **164.** J Neurosci Meth 35, 175 (1990); **165.** J Neurosci Meth 32, 15 (1990); **166.** Meth Neurosci 3, 275 (1990); **167.** Neuroscience 28, 725 (1989); **168.** Acta Anat 122, 158 (1985); **169.** Neurosci Lett 18, 25 (1980); **170.** Brain Res 508, 289 (1990); **171.** Brain Res Bull 24, 341 (1990); **172.** J Neurosci Meth 16, 175 (1986); **173.** Exp Brain Res 51, 179 (1983); **174.** Brain Res 486, 334 (1989); **175.** J Comp Neurol 258, 230 (1987); **176.** J Neurosci Meth 14, 273 (1985); **177.** J Ultrastruct Res 37, 200 (1971); **178.** J Chem Soc 567 (1946); **179.** Brain Res 553, 135 (1991); **180.** Brain Research 377, 147 (1986); **181.** U.S. Patent No. 4,716,905; **182.** Cancer 4, 1246 (1951); **183.** Acta Hematology 3, 129 (1950); **184.** J Mt Sinai Hosp 15, 156 (1948); **185.** J Lab Clin Med 37, 562 (1951); **186.** Bull NY Acad Med 26, 269 (1950); **187.** Biochim Biophys Acta 407, 43 (1975); **188.** Biochim Biophys Acta 407, 24 (1975); **189.** Biochemistry 12, 4827 (1973); **190.** Histochemistry 74, 107 (1982); **191.** J Cell Biol 87, 292 (1980); **192.** Diagnostic Histopathol 5, 219 (1982); **193.** J Neurocytol 18, 333 (1989); **194.** J Cell Biol 90, 665 (1981); **195.** J Cell Biol 90, 656 (1981); **196.** Infection and Immunity 11, 441 (1975); **197.** Biochem Pharmacol 19, 1251 (1970); **198.** Life Sci 3, 1407 (1964); **199.** Mayo Clinic Proc 58, 223 (1983); **200.** Rev Infect Dis 2, 625 (1980); **201.** Brain Res 533, 329 (1990); **202.** Brain Res 377, 147 (1986); **203.** Proc Natl Acad Sci USA 79, 2898 (1982); **204.** J Cell Biol 89, 368 (1981); **205.** J Comp Neurol 303, 255 (1991); **206.** Neurosci Lett 117, 285 (1990); **207.** Photochem Photobiol 46, 45 (1987); **208.** Brain Res Rev 8, 99 (1984); **209.** Microscopy Res Technique 24, 2 (1993); **210.** J Neurosci Meth 45, 87 (1992); **211.** Biochem Pharmacol 33, 1851 (1984); **212.** J Immunol Meth 68, 251 (1984); **213.** J Immunol Meth 15, 147 (1977); **214.** Biochemistry 28, 2496 (1989); **215.** J Cell Physiol 140, 505 (1989); **216.** Biochim Biophys Acta 970, 270 (1988); **217.** Biochim Biophys Acta 888, 82 (1986); **218.** J Pharm Sci 75, 334 (1986); **219.** Radiation Environ Biophys 19, 275 (1981); **220.** Photochem Photobiol 59, 30 (1994); **221.** Biochemistry 32, 10826 (1993); **222.** Biochemistry 31, 703 (1992); **223.** J Membrane Biol 109, 41 (1989); **224.** Chem Phys Lipids 33, 303 (1983).

15.3 Data Table *Polar Tracers*

Cat #	Structure	MW	Storage	Soluble	Abs	{ε × 10^{-3}}	Em	Solvent	Notes
A-350	✓	427	L	H_2O	353	{7.2}	520	H_2O	
A-629	S5.3	637	F,LL	H_2O	426	{11}	534	H_2O	
A-1339	S3.3	492	L	H_2O	425	{12}	532	H_2O	
A-1340	S3.3	534	L	H_2O	426	{11}	531	H_2O	
A-1593	S4.2	367	NC	DMF, DMSO	<300		none		
A-1594	S4.2	328	NC	DMF, DMSO	<300		none		
A-1604	S4.2	606	NC	DMF, DMSO	<300		none		
B-1151	S23.2	520	L	pH>6	503	{90}	528	pH 9	<1>
B-1370	S4.3	831	L	DMF, pH>6	494	{75}	518	pH 9	
B-1513	S4.2	341	F,D	DMF, DMSO	<300		none		
B-1582	S4.2	455	F,D	DMF, DMSO	<300		none		
B-1592	S15.3	372	NC	H_2O	<300		none		
B-1596	S4.2	556	NC	DMF, DMSO	<300		none		
B-1597	S4.2	400	NC	DMF, DMSO	<300		none		
B-1603	✓	387	D	pH>6, DMF	<300		none		
B-2605	✓	486	NC	pH>6, DMF	<300		none		
B-3245	✓	404	L	H_2O	417	{16}	521	H_2O	
B-6350	✓	500	D	pH>6, DMF	<300		none		
B-6352	S4.2	670	F,D	DMF, pH>6*	<300		none		
B-6353	S4.2	557	F,D	DMF, pH>6*	<300		none		
C-194	✓	376	L	pH>6, DMF	492	{75}	517	pH 9	<1>
C-368	S1.5	445	L	pH>6, DMF	504	{90}	529	pH 8	<2>
C-481	✓	623	L	pH>5	494	{76}	517	pH 9	
C-621	S3.3	624	L	H_2O	399	{30}	423	H_2O	<3>
C-622	S3.3	667	L	H_2O	399	{29}	422	H_2O	<3>
C-625	S5.3	770	F,LL	H_2O, MeOH	400	{29}	421	H_2O	<3>
C-687	✓	596	L	H_2O†	399	{30}	421	H_2O	<3>
C-1430	S16.3	995	F,D	DMSO	<300		none		<4>
C-1904	✓	376	L	pH>6, DMF	492	{78}	517	pH 9	<1>
C-2284	S1.7	607	F,D,LL	H_2O§, MeOH	396	{29}	410	MeOH	<3>
C-3221	✓	645	L	H_2O†	399	{31}	419	H_2O	<3>
C-3239	✓	548	L	H_2O†	399	{29}	419	H_2O	<3>
C-6949	S4.3	957	D,L	H_2O	400	{31}	417	pH 8	<3>
C-7079	✓	846	FF,D,LL	H_2O*, DMSO	339	{6.1}	none	MeOH	<5, 6>
C-7100	✓	1075	F,D,LL	H_2O§, DMSO	293	{23}	none	H_2O	<5, 7>
C-7101	✓	1058	F,D,LL	DMSO	292	{27}	none	MeOH	<5, 7>
D-402	✓	412	D,L	H_2O, MeOH	340	{4.5}	520	H_2O	
D-530	✓	280	D,L	H_2O, MeOH	<300		none		
D-1248	✓	167	D	pH>5	270	{4.5}	none	pH 8	
D-1249	✓	311	F,D	DMSO	270	{4.1}	none	MeOH	

Cat #	Structure	MW	Storage	Soluble	Abs	{, ´ 10^{-3}}	Em	Solvent	Notes
D-1306	S8.1	350	L	H_2O, DMF	344	{41}	450	pH 7	
D-1389	✓	776	L	H_2O	653	{166}	674	MeOH	
D-3238	✓	484	F,D,L	H_2O	490	{97}	515	H_2O	<8>
D-3571	S8.1	457	L	H_2O, MeOH	344	{38}	450	pH 7	
D-6948	S21.3	340	L	H_2O, DMF	340	{6.3}	464	pH 7	
D-6961	✓	334	F,D,L	H_2O, MeOH	502	{89}	510	MeOH	<8>
D-7060	✓	720	FF,LL	H_2O, DMSO	402	{24}	none	H_2O	<5, 9>
E-6960	✓	366‡	D	H_2O	337	{17}	613	H_2O	<10>
F-1130	✓	478	D,L	H_2O, DMF	495	{76}	519	pH 9	<1>
F-7103	✓	827	FF,D,LL	H_2O§, DMSO	333	{12}	none	DMSO	<5, 11>
H-348	✓	524	D,L	H_2O	454	{22}	511	pH 9	<1>
H-1426	✓	321	L	pH>5	360	{16}	440	pH 9	
H-7599	S8.1	473	F,D,L	H_2O	361	{17}	536	pH 7	
I-6958	S21.2	567	D,L	H_2O	631	{111}	none	MeOH	
L-453	✓	457	L	H_2O¶	428	{12}	536	H_2O	
L-682	✓	479	L	H_2O¶	428	{12}	533	H_2O	
L-1177	✓	522	L	H_2O¶	427	{12}	535	H_2O	
L-1338	S2.4	649	F,D,L	H_2O	426	{11}	531	pH 7	<12, 13>
L-2601	✓	873	D,L	H_2O	428	{11}	531	H_2O	
L-6950	S4.3	850	D,L	H_2O	428	{11}	532	pH 7	
M-395	✓	538	L	H_2O#	404	{29}	435	pH 8	
M-1602	S4.2	524	F,D	pH>6, DMF	<300		none		
N-6946	✓	651‡	L	DMF, H_2O	347	{36}	515	MeOH	
O-6135	S1.5	448	L	pH>6, DMF	497	{84}	517	pH 9	<2>
O-6138	S1.5	512	L	pH>6, DMF	506	{88}	526	pH 9	<2>
O-6146	S1.4	412	L	pH>6, DMF	492	{85}	518	pH 9	<2>
O-6148	S1.4	412	L	pH>6, DMF	492	{88}	517	pH 9	<2>
P-349	✓	610	L	H_2O	374	{51}	403	H_2O	
P-1304	S8.1	668	L	H_2O, DMF	493	{5.9}	636	H_2O	
R-6150	S1.4	411	L	pH>6, DMF	502	{89}	524	MeOH	
R-6152	S1.4	412	DD,L	pH>6, DMF	494	{73}	521	pH 9	<2>
R-7090	✓	950	FF,D,LL	DMSO	340	{9.4}	none	MeOH	<5, 14>
R-7091	✓	1050	FF,D,LL	DMSO	336	{5.4}	none	MeOH	<5, 14>
R-7102	✓	913	F,D,LL	DMSO	284	{23}	none	DMSO	<5, 14>
S-359	✓	607	L	H_2O	586	{108}	605	H_2O	
S-1129	S16.3	518	F,D	DMSO	<300		none		<15>
S-1307	✓	559	L	H_2O	565	{84}	586	H_2O	
S-6954	✓	1146	F,D,L	H_2O	491	{60}	516	pH 7	
S-6956	✓	~4500	L	H_2O	557	{99}	581	MeOH	
S-6957	✓	553	D,L	DMSO, H_2O	529	{75}	548	MeOH	
T-411	✓	266	D,L	DMF, EtOH	337	{17}	483	MeOH	
T-506	✓	341	F,D	H_2O, MeOH	<300		none		
T-1247	✓	373‡	D	H_2O	270	{4.7}	545	H_2O	<16>
T-1323	✓	417	L	DMSO	375	{68}	403	H_2O	
T-6931	S22.7	791	D,L	pH>6, DMSO	414	{416}	645	pH 9	<17>
T-6932	S22.7	935	D,L	H_2O, DMSO	413	{472}	645	pH 9	<17>
X-1525	✓	422	D	H_2O	259	{8.8}	none	H_2O	

C-3099, C-3100 see C-1430.

For definitions of the contents of this data table, see "How to Use the Handbook" on page vi.
Structure: Chemical structure drawing: ✓ = shown in this section; Sn.n = shown in Section number n.n.
MW: ~ indicates an approximate value.

* Sulfosuccinimidyl esters are water soluble and may be dissolved in buffer at ~pH 8 for reaction with amines. Long-term storage in water is NOTrecommended due to hydrolysis.
† Maximum solubility in water is ~1% for C-687, ~1% for C-3221 and ~8% for C-3239.
‡ MW is for the hydrated form of this product.
§ Unstable in water. Use immediately.
¶ Maximum solubility in water is ~8% for L-453, ~6% for L-682 and ~1% for L-1177.
\# Maximum solubility for M-395 in water is ~25%.
<1> Spectra of this product are pH dependent (see Section 23.2).
<2> Spectra of this product are pH dependent (see Section 23.3).
<3> Cascade Blue dyes have a second absorption peak at about 376 nm with ϵ ~80% of the 395–400 nm peak.
<4> C-1430 is converted to fluorescent calcein (C-481) after acetoxymethyl ester hydrolysis.
<5> All caged products are sensitive to light. They should be protected from illumination, except when photolysis is intended.
<6> This product is colorless and nonfluorescent until converted by ultraviolet photolysis to a Cl-NERF derivative with spectral characteristics similar to C-6831 (see Section 23.3), Abs = 514 nm (ϵ = {84}), Em = 540 nm.
<7> This product is colorless and nonfluorescent until converted by ultraviolet photolysis to a carboxyrhodamine Q derivative, Abs = 537 nm (ϵ = {94}), Em = 556 nm.
<8> The absorption and fluorescence spectra of BODIPY derivatives are relatively insensitive to the solvent.
<9> Converted to H-348 by ultraviolet photolysis.
<10> Absorption and fluorescence of E-6960 are extremely weak unless it is chelated. Data are for detergent-solubilized NTA (T-411) chelate.
<11> This product is essentially colorless and nonfluorescent until converted to a derivative of fluorescein (Abs ~492 nm, Em ~517 nm) by ultraviolet photolysis of the caging groups.

<12> Spectral data of the 2-mercaptoethanol adduct.
<13> Iodoacetamides in solution undergo rapid photodecomposition to unreactive products. Minimize exposure to light prior to reaction.
<14> This product is colorless and nonfluorescent until converted by ultraviolet photolysis to a carboxyrhodamine 110 derivative with spectral characteristics similar to R-6150.
<15> S-1129 is converted to a fluorescent product (F-1130) after acetate hydrolysis.
<16> Absorption and fluorescence of T-1247 are extremely weak unless it is chelated. Data are for dipicolinic acid (D-1248) chelate. Fluorescence spectrum has secondary peak at 490 nm.
<17> Absorption spectra of this compound also contain much weaker bands in the 500–650 nm wavelength range. Absorption and fluorescence are pH dependent. At pH below 5, Abs = 440 nm for T-6931, Abs = 434 nm for T-6932.

15.3 Structures *Polar Tracers*

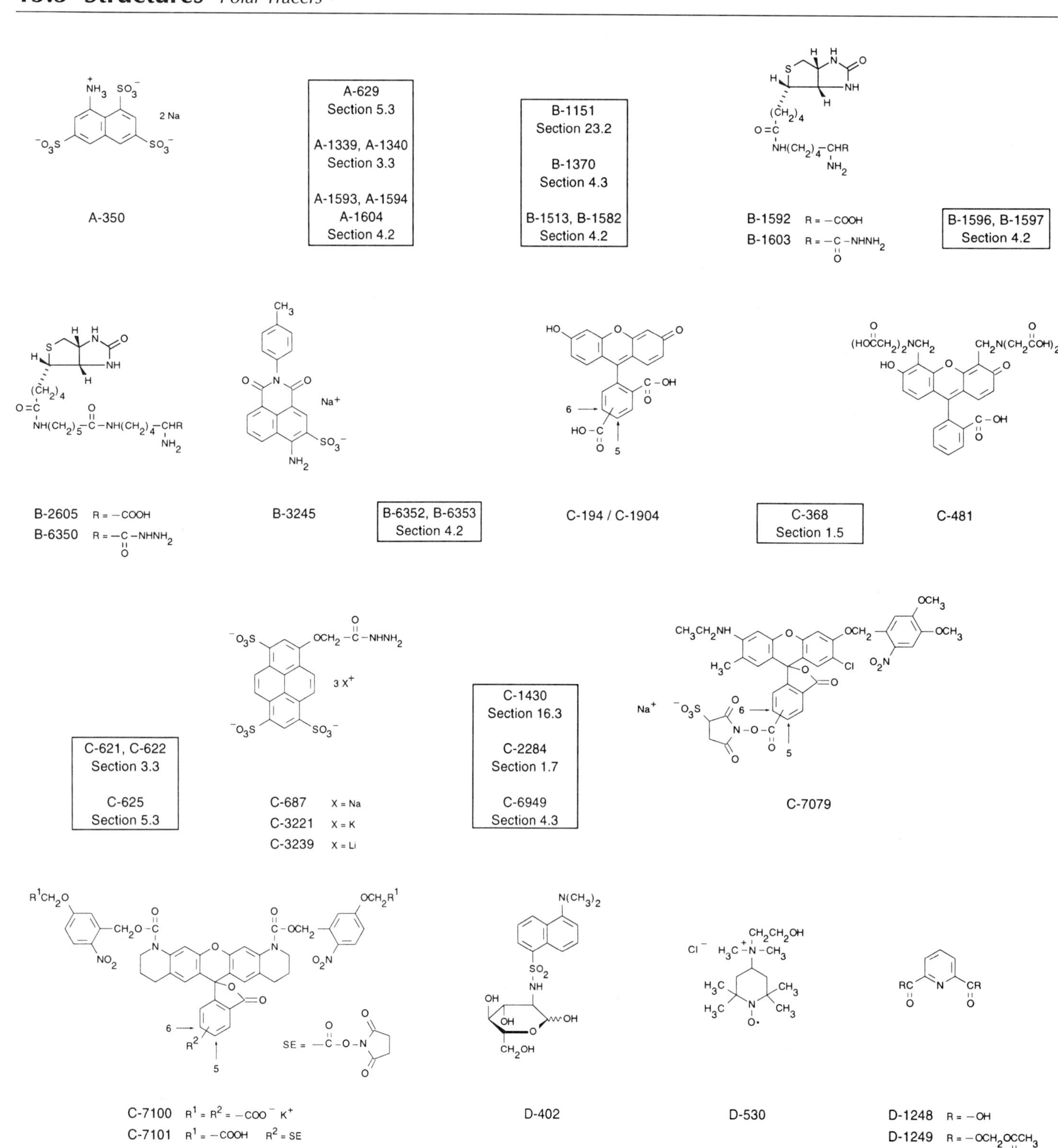

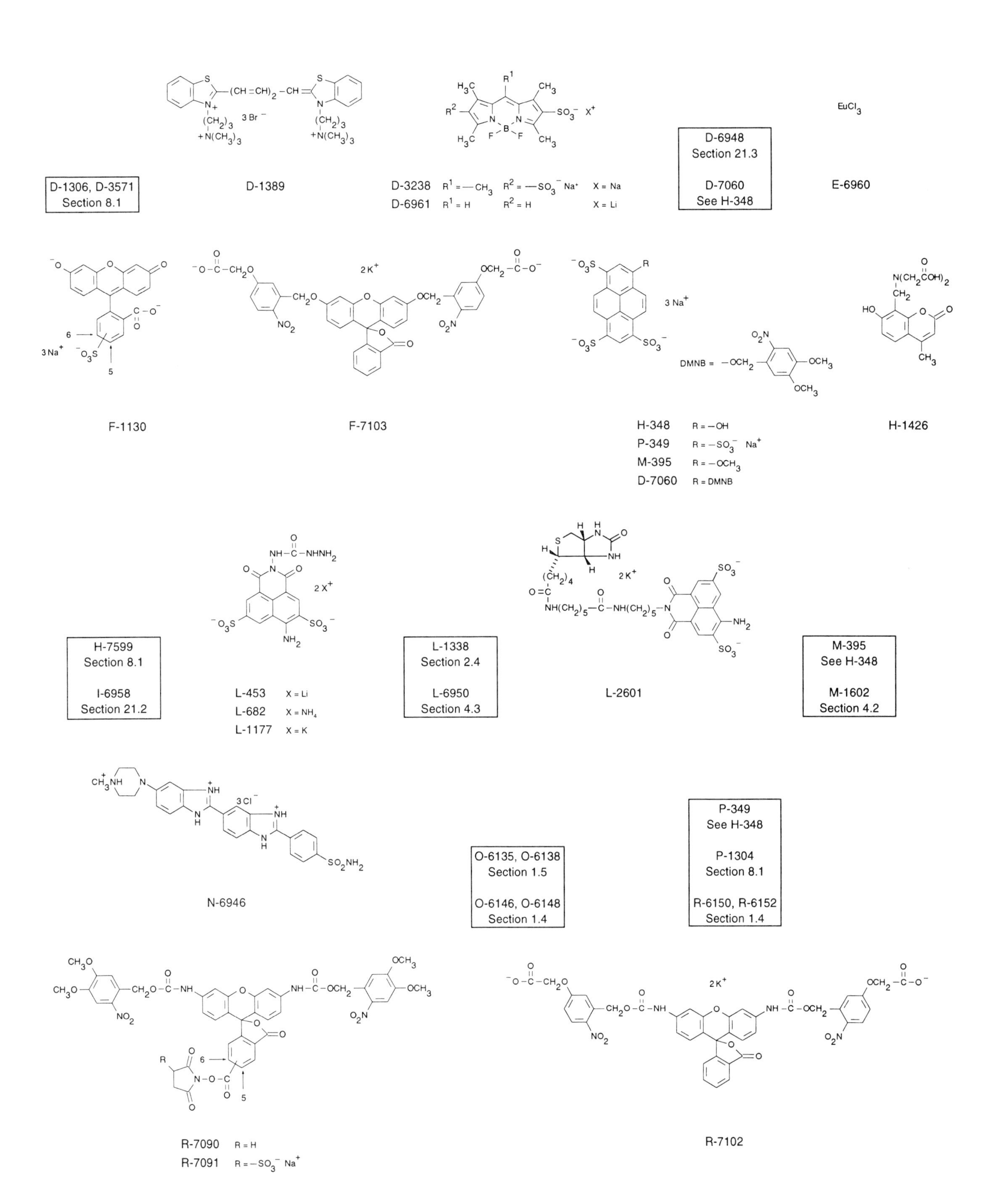
D-1306, D-3571
Section 8.1
D-1389
D-3238
D-6961
D-6948
Section 21.3
D-7060
See H-348
EuCl3
E-6960
F-1130
F-7103
DMNB
H-348
P-349
M-395
D-7060
H-1426
H-7599
Section 8.1
I-6958
Section 21.2
L-453
L-682
L-1177
L-1338
Section 2.4
L-6950
Section 4.3
L-2601
M-395
See H-348
M-1602
Section 4.2
N-6946
O-6135, O-6138
Section 1.5
O-6146, O-6148
Section 1.4
P-349
See H-348
P-1304
Section 8.1
R-6150, R-6152
Section 1.4
R-7090
R-7091
R-7102

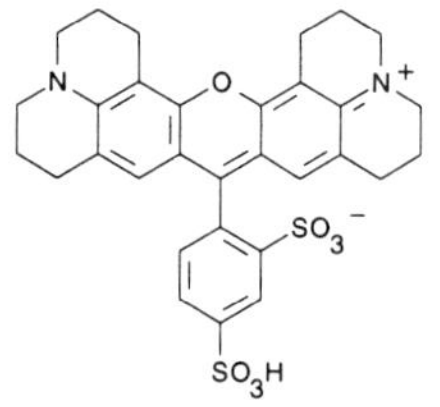

S-359

S-1129 Section 16.3

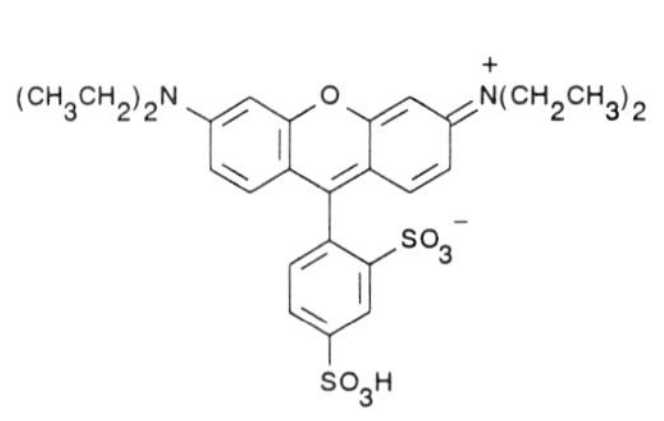

S-1307

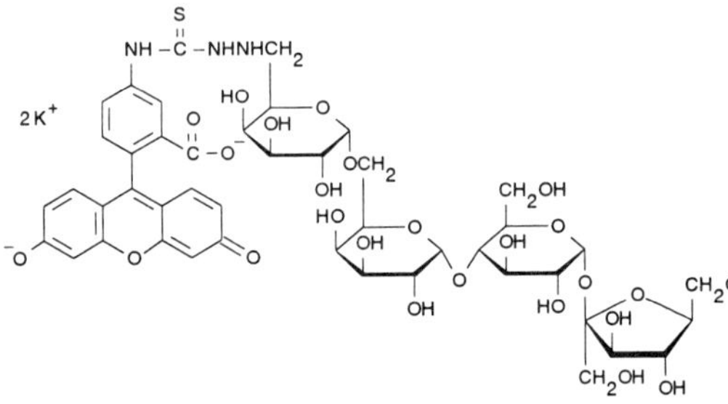

S-6954

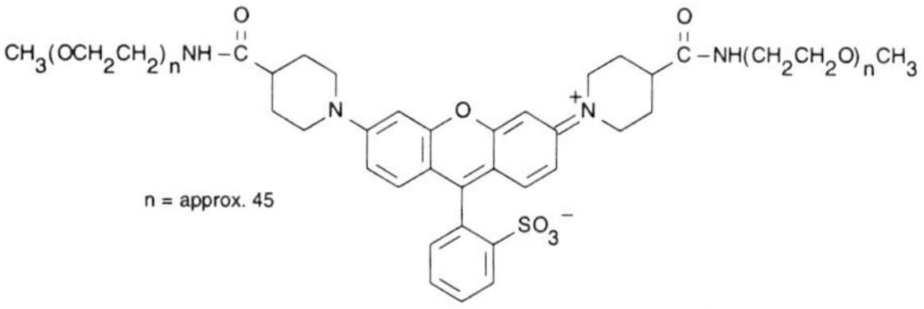

S-6956

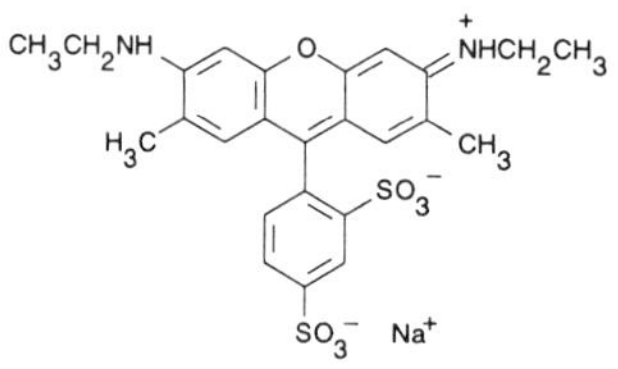

S-6957

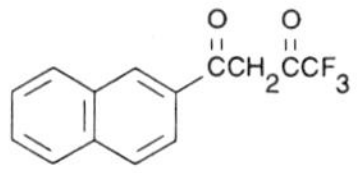

T-411

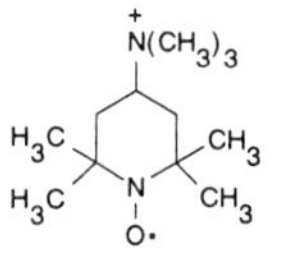

T-506

T-1247

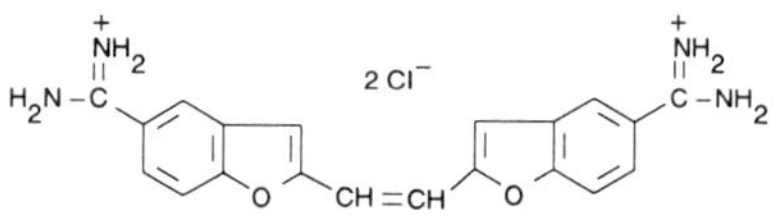

T-1323

T-6931, T-6932 Section 22.7

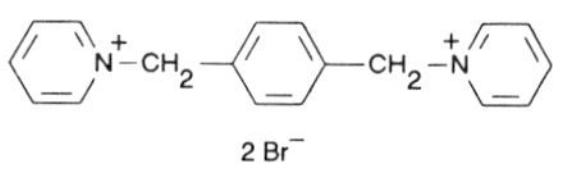

X-1525

15.3 Price List *Polar Tracers*

Cat #	Product Name	Unit Size	Price Per Unit ($) 1–4 Units	5–24 Units
A-1604	N$^{\alpha}$-(4-aminobenzoyl)biocytin, trifluoroacetate	25 mg	85.00	68.00
A-1339	N-(2-aminoethyl)-4-amino-3,6-disulfo-1,8-naphthalimide, dipotassium salt (lucifer yellow ethylenediamine)	25 mg	38.00	30.40
A-1593	N-(2-aminoethyl)biotinamide, hydrobromide (biotin ethylenediamine)	25 mg	60.00	48.00
A-350	8-aminonaphthalene-1,3,6-trisulfonic acid, disodium salt (ANTS)	1 g	88.00	70.40
A-1340	N-(5-aminopentyl)-4-amino-3,6-disulfo-1,8-naphthalimide, dipotassium salt (lucifer yellow cadaverine)	25 mg	70.00	56.00
A-1594	N-(5-aminopentyl)biotinamide (biotin cadaverine)	25 mg	85.00	68.00
A-629	N-(((4-azidobenzoyl)amino)ethyl)-4-amino-3,6-disulfo-1,8-naphthalimide, dipotassium salt (lucifer yellow AB)	10 mg	95.00	76.00
B-1592	biocytin (ε-biotinoyl-L-lysine)	100 mg	60.00	48.00
B-1603	biocytin hydrazide	25 mg	85.00	68.00
B-1582	6-((biotinoyl)amino)hexanoic acid, succinimidyl ester (biotin-X, SE; biotinamidocaproate, N-hydroxysuccinimidyl ester)	100 mg	80.00	64.00
B-6353 ***New***	6-((biotinoyl)amino)hexanoic acid, sulfosuccinimidyl ester, sodium salt (Sulfo-NHS-LC-Biotin; biotin-X, SSE)	25 mg	75.00	60.00
B-1597	2-(((N-(biotinoyl)amino)hexanoyl)amino)ethylamine (biotin-X ethylenediamine)	10 mg	75.00	60.00
B-6352 ***New***	6-((6-((biotinoyl)amino)hexanoyl)amino)hexanoic acid, sulfosuccinimidyl ester, sodium salt (biotin-XX, SSE)	25 mg	75.00	60.00
B-1596	5-(((N-(biotinoyl)amino)hexanoyl)amino)pentylamine trifluoroacetate salt (biotin-X cadaverine)	10 mg	85.00	68.00
B-1370	5-((N-(5-(N-(6-(biotinoyl)amino)hexanoyl)amino)pentyl)thioureidyl)fluorescein (fluorescein biotin)	5 mg	98.00	78.40
B-2605	ε-(6-(biotinoyl)amino)hexanoyl-L-lysine (biocytin-X)	10 mg	60.00	48.00
B-6350 ***New***	ε-(6-(biotinoyl)amino)hexanoyl-L-lysine hydrazide (biocytin-X hydrazide)	25 mg	90.00	72.00
B-1513	D-biotin, succinimidyl ester (succinimidyl D-biotin)	100 mg	53.00	42.40
B-1151	2′,7′-bis-(2-carboxyethyl)-5-(and-6)-carboxyfluorescein (BCECF acid) *mixed isomers*	1 mg	63.00	50.40
B-3245	brilliant sulfoflavin	100 mg	24.00	19.20

Cat #		Product Name	Unit Size	Price Per Unit ($) 1–4 Units	5–24 Units
C-481		calcein *high purity*	100 mg	58.00	46.40
C-1430		calcein, AM	1 mg	115.00	92.00
C-3099		calcein, AM *1 mg/mL solution in dry DMSO* *special packaging*	1 mL	120.00	96.00
C-3100		calcein, AM *special packaging*	20x50 µg	140.00	112.00
C-368		5-(and-6)-carboxy-2′,7′-dichlorofluorescein *mixed isomers*	100 mg	48.00	38.40
C-194		5-(and-6)-carboxyfluorescein *mixed isomers*	5 g	73.00	58.40
C-1904		5-(and-6)-carboxyfluorescein (5(6)-FAM) *high purity* *mixed isomers*	100 mg	55.00	44.00
C-7101	**New**	5-(and-6)-carboxy-Q-rhodamine, CMNCBZ-caged, succinimidyl ester *mixed isomers*	1 mg	68.00	54.40
C-7100	**New**	5-(and-6)-carboxy-Q-rhodamine, CMNCBZ-caged, tripotassium salt *mixed isomers*	1 mg	55.00	44.00
C-2284		Cascade Blue® acetyl azide, trisodium salt	5 mg	125.00	100.00
C-625		Cascade Blue® aminoethyl 4-azidobenzamide, trisodium salt	10 mg	95.00	76.00
C-6949	**New**	Cascade Blue® biocytin, disodium salt, fixable (biocytin Cascade Blue®)	5 mg	95.00	76.00
C-622		Cascade Blue® cadaverine, trisodium salt	10 mg	98.00	78.40
C-621		Cascade Blue® ethylenediamine, trisodium salt	10 mg	98.00	78.40
C-3239		Cascade Blue® hydrazide, trilithium salt	10 mg	95.00	76.00
C-3221		Cascade Blue® hydrazide, tripotassium salt	10 mg	95.00	76.00
C-687		Cascade Blue® hydrazide, trisodium salt	10 mg	95.00	76.00
C-7079	**New**	Cl-NERF 4,5-dimethoxy-2-nitrobenzyl ether, sulfosuccinimidyl ester, sodium salt (DMNB-caged Cl-NERF, SSE)	1 mg	48.00	38.40
D-1306		4′,6-diamidino-2-phenylindole, dihydrochloride (DAPI)	10 mg	37.00	29.60
D-3571		4′,6-diamidino-2-phenylindole, dilactate (DAPI, dilactate)	10 mg	37.00	29.60
D-3238		4,4-difluoro-1,3,5,7,8-pentamethyl-4-bora-3a,4a-diaza-s-indacene-2,6-disulfonic acid, disodium salt (BODIPY® 492/515 disulfonate)	10 mg	48.00	38.40
D-6961	**New**	4,4-difluoro-1,3,5,7-tetramethyl-4-bora-3a,4a-diaza-s-indacene-2-sulfonic acid, lithium salt (BODIPY® 495/505 sulfonate)	5 mg	75.00	60.00
D-6948	**New**	9,10-dimethoxyanthracene-2-sulfonic acid, sodium salt (DAS)	100 mg	25.00	20.00
D-7060	**New**	8-((4,5-dimethoxy-2-nitrobenzyl)oxy)pyrene-1,3,6-trisulfonic acid, trisodium salt (DMNB-caged HPTS)	5 mg	125.00	100.00
D-402		N-(5-dimethylaminonaphthalene-1-sulfonyl)glucosamine (dansyl glucosamine)	25 mg	48.00	38.40
D-530		4-(N,N-dimethyl-N-(2-hydroxyethyl))ammonium-2,2,6,6-tetramethylpiperidine-1-oxyl chloride (TEMPO choline)	25 mg	98.00	78.40
D-1248		dipicolinic acid (DPA) *cell impermeant*	1 g	17.00	13.60
D-1249		dipicolinic acid, AM (DPA, AM) *cell permeant*	25 mg	48.00	38.40
D-1389		N,N′-di(3-trimethylammoniumpropyl)thiadicarbocyanine tribromide	10 mg	80.00	64.00
E-6960	**New**	europium(III) chloride, hexahydrate *99.99%*	1 g	20.00	16.00
F-7103	**New**	fluorescein bis-(5-carboxymethoxy-2-nitrobenzyl) ether, dipotassium salt (CMNB-caged fluorescein)	5 mg	145.00	116.00
F-1130		fluorescein-5-(and-6)-sulfonic acid, trisodium salt	100 mg	78.00	62.40
H-1426		7-hydroxy-4-methylcoumarin-8-methyleneiminodiacetic acid (calcein blue)	1 g	20.00	16.00
H-348		8-hydroxypyrene-1,3,6-trisulfonic acid, trisodium salt (HPTS; pyranine)	1 g	72.00	57.60
H-7599	**New**	hydroxystilbamidine, methanesulfonate	10 mg	75.00	60.00
I-6958	**New**	isosulfan blue *>90% by HPLC*	1 g	28.00	22.40
L-6950	**New**	lucifer yellow biocytin, potassium salt, fixable (biocytin lucifer yellow)	5 mg	95.00	76.00
L-2601		lucifer yellow cadaverine biotin-X, dipotassium salt	10 mg	95.00	76.00
L-682		lucifer yellow CH, ammonium salt	25 mg	45.00	36.00
L-453		lucifer yellow CH, lithium salt	25 mg	48.00	38.40
L-1177		lucifer yellow CH, potassium salt	25 mg	45.00	36.00
L-1338		lucifer yellow iodoacetamide, dipotassium salt	25 mg	95.00	76.00
M-1602		3-(N-maleimidylpropionyl)biocytin	25 mg	68.00	54.40
M-395		8-methoxypyrene-1,3,6-trisulfonic acid, trisodium salt (MPTS)	100 mg	90.00	72.00
N-6946	**New**	nuclear yellow (Hoechst S769121)	100 mg	48.00	38.40
O-6146	**New**	Oregon Green™ 488 carboxylic acid *5-isomer*	5 mg	75.00	60.00
O-6148	**New**	Oregon Green™ 488 carboxylic acid *6-isomer*	5 mg	75.00	60.00
O-6135	**New**	Oregon Green™ 500 carboxylic acid *5-isomer*	5 mg	75.00	60.00
O-6138	**New**	Oregon Green™ 514 carboxylic acid	5 mg	75.00	60.00
P-1304		propidium iodide	100 mg	48.00	38.40
P-349		1,3,6,8-pyrenetetrasulfonic acid, tetrasodium salt	100 mg	35.00	28.00
R-7102	**New**	rhodamine 110, CMNCBZ-caged, dipotassium salt (CMNCBZ-caged rhodamine 110)	5 mg	145.00	116.00
R-7090	**New**	Rhodamine Green™ bis-(((4,5-dimethoxy-2-nitrobenzyl)oxy)carbonyl)-caged, succinimidyl ester (N-(NVOC-caged) Rhodamine Green™, SE)	5 mg	175.00	140.00
R-7091	**New**	Rhodamine Green™ bis-(((4,5-dimethoxy-2-nitrobenzyl)oxy)carbonyl)-caged, sulfosuccinimidyl ester, sodium salt (N-(NVOC-caged) Rhodamine Green™, SSE)	1 mg	68.00	54.40
R-6150	**New**	Rhodamine Green™ carboxylic acid, hydrochloride (5(6)-CR 110) *mixed isomers*	5 mg	95.00	76.00
R-6152	**New**	Rhodol Green™ carboxylic acid, hydrochloride (5(6)-CRF 492) *mixed isomers*	5 mg	75.00	60.00
S-6954	**New**	stachyose fluorescein, dipotassium salt	5 mg	95.00	76.00
S-1129		5-sulfofluorescein diacetate, sodium salt (SFDA) *single isomer*	25 mg	85.00	68.00

Cat #		Product Name	Unit Size	Price Per Unit ($) 1–4 Units	5–24 Units
S-6956	**New**	sulfonerhodamine bis-(PEG 2000)	10 mg	45.00	36.00
S-359		sulforhodamine 101	25 mg	50.00	40.00
S-1307		sulforhodamine B	5 g	78.00	62.40
S-6957	**New**	sulforhodamine G *>60% by HPLC*	5 g	78.00	62.40
T-1247		terbium(III) chloride, hexahydrate	1 g	20.00	16.00
T-6931	**New**	tetrakis-(4-carboxyphenyl)porphine (TCPP)	5 mg	48.00	38.40
T-6932	**New**	tetrakis-(4-sulfophenyl)porphine (TSPP)	5 mg	48.00	38.40
T-411		2-((trifluoroacetyl)acetyl)naphthalene (naphthoyltrifluoroacetone; NTA)	100 mg	65.00	52.00
T-506		4-trimethylammonium-2,2,6,6-tetramethylpiperidine-1-oxyl iodide (CAT 1)	100 mg	95.00	76.00
T-1323		true blue chloride	5 mg	103.00	82.40
X-1525		p-xylene-bis-pyridinium bromide (DPX)	1 g	75.00	60.00

15.4 Fluorescent Lipophilic Tracers

Cell membranes provide a convenient conduit for loading live and fixed cells with lipophilic dyes. Not only can cells tolerate a high concentration of dye, but lateral diffusion of the dye within the membrane can serve to stain the entire cell, even if the dye is applied locally. These properties have made lipophilic carbocyanine and aminostyryl dyes particularly important for anterograde and retrograde tracing in neuronal cells.[1] Lipophilic tracers are used to label cells,[2-4] organelles,[5,6] liposomes,[7] viruses [8] and lipoproteins [9] in a wide variety of long-term tracing applications, including cell transplantation, migration, adhesion and fusion studies. The distinguishing features of these carbocyanine and aminostyryl tracers are summarized in Table 15.2.

Long-Chain Carbocyanines: DiI, DiO, DiD and Analogs

DiI, DiO and DiD

The lipophilic carbocyanines DiI (DiIC_{18}(3)), DiO (DiOC_{18}(3)), and DiD (DiIC_{18}(5)) are weakly fluorescent in water but highly fluorescent and quite photostable when incorporated into membranes (see Color Plate 3 in "Visual Reality"). They have extremely high extinction coefficients (ϵ >125,000 $cm^{-1}M^{-1}$ at their longest-wavelength absorption maximum) though modest quantum yields, and short excited-state lifetimes (~1 nanosecond) in lipid environments. Once applied to cells, the dyes diffuse laterally within the plasma membrane, resulting in staining of the entire cell. Transfer of these probes between intact membranes is usually negligible. DiI and its analogs usually exhibit very low cell toxicity; however, moderate inhibition of electron transport chain activity has recently been reported for some analogs.[10]

DiI, DiO and DiD exhibit distinct orange, green and red fluorescence, respectively (see Figure 14.1 in Chapter 14), thus facilitating multicolor imaging.[11] DiO (D-275) and DiI (D-282, D-3911) can be used with standard fluorescein and rhodamine optical filter, respectively. The He-Ne laser–excitable DiD (D-307, D-7757) has much longer-wavelength excitation and emission spectra than DiI, providing a valuable alternative for labeling cells and tissues that have significant autofluorescence or in the presence of serum. DiD was used as a population marker in a flow cytometric study that also employed indo-1 to monitor intracellular Ca^{2+} mobilization.[12] Photoconversion of diaminobenzidine (DAB) by lipid tracers, including DiI and DiO derivatives, produces an insoluble, electron-dense reaction product.[13-16]

FAST DiI™ and *FAST* DiO™

Diffusion of lipophilic carbocyanine tracers from the point of their application to the terminus of a neuron can take several days to weeks.[1] The diffusion process appears to be accelerated by introducing unsaturation in the probes' alkyl tails. Molecular Probes' *FAST* DiI (D-3899, D-7756) and *FAST* DiO (D-3898) have diunsaturated linoleyl ($C_{18:2}$) tails in place of the saturated octadecyl tails ($C_{18:0}$) of DiI and DiO. Migration of the unsaturated analogs is reported to be at least 50% faster than that of DiI and DiO.[17] The perchlorate salt of *FAST* DiI does not crystallize and is a viscous oil (D-3899); however, our new *FAST* DiI 4-chlorobenzenesulfonate (D-7756) is a solid, making its crystals suitable for direct application to cells. We also offer a monounsaturated *cis*-9-octadecenyl ($C_{18:1}$) analog of DiI (Δ^9-DiI, D-3886).

Other DiO and DiI Analogs

Several other lipophilic carbocyanines are also useful tracers for labeling cell membranes. Some scientists find the slightly less lipophilic DiIC_{12}(3) (D-383), DiIC_{16}(3) (D-384) and DiOC_{16}(3) (D-1125) easier to load into cell suspensions than their C_{18} homologs. DiIC_{16}(3) has been used to measure:

- Cortical rotation during fertilization of frog eggs [18]
- Lipid diffusion in membranes by fluorescence recovery after photobleaching (FRAP) [19]
- Lipid flow during cell motility [20]
- Membrane fusion induced by electric fields [21]

Three new phenyl-substituted carbocyanine tracers — 6,6'-Ph_2-DiIC_{18}(3) (D-7780), 5,5'-Ph_2-DiIC_{18}(3) (D-7779) and 5,5'-Ph_2-DiOC_{18}(3) (D-7770) — are more lipophilic than DiI and DiO but also significantly more fluorescent, at least in methanol. 5,5'-Ph_2-DiIC_{18}(3) has spectra that fall between those of DiIC_{18}(3) and DiIC_{18}(5). These new probes can potentially be used for retrograde or anterograde neuronal tracing or other long-term tracing studies; however, they have not yet been tested in these applications.

Table 15.2 *Summary of our lipophilic carbocyanine and aminostyryl tracers.*

Cat #	Probe *	Features of Carbocyanine or Aminostyryl Tracers
DiI and Analogs		
D-282	$DiIC_{18}(3)$ "DiI"	Orange-red fluorescent lipophilic probe; widely used as a neuronal tracer
D-3911	$DiIC_{18}(3)$ "DiI"	Large-crystal form of D-282 to facilitate direct application of crystals to membranes
D-384	$DiIC_{16}(3)$	Shorter-chain DiI analog that may incorporate into membranes more easily than DiI
D-383	$DiIC_{12}(3)$	Shorter-chain DiI analog that may incorporate into membranes more easily than DiI
D-3899	*FAST* DiI oil	Unsaturated DiI analog that reportedly migrates ~50% faster than DiI within membranes
D-7756	*FAST* DiI	Solid form of D-3899 to facilitate direct application of crystals to membranes
D-3886	Δ^9-DiI oil	Unsaturated DiI analog that may migrate faster than DiI within membranes
C-7001	CellTracker CM-DiI	Chloromethylated DiI analog with enhanced solubility in culture medium; potentially retained after fixation
F-6999	FM-DiI	DiI analog with enhanced solubility in culture medium
D-7779	5,5′-Ph_2-$DiIC_{18}(3)$	More lipophilic DiI analog with a higher quantum yield and longer-wavelength spectra (red-shifted ~20 nm)
D-7780	6,6′-Ph_2-$DiIC_{18}(3)$	More lipophilic DiI analog with a higher quantum yield but similar spectra
D-7777	SP-$DiIC_{18}(3)$	Anionic DiI analog with enhanced solubility in culture medium; potentially retained after fixation
D-7776	$DiIC_{18}(3)$-DS	Anionic DiI analog with enhanced solubility in culture medium; potentially retained after fixation
D-7766	Br_2-$DiIC_{18}(3)$	Brominated DiI analog that may be useful for diaminobenzidine (DAB) photoconversion
DiD		
D-307	$DiIC_{18}(5)$ oil "DiD"	Much longer-wavelength DiI analog; useful in autofluorescent samples and as a second tracer with DiI
D-7757	$DiIC_{18}(5)$ "DiD"	Solid form of D-307 to facilitate direct application of crystals to membranes
DiO and Analogs		
D-275	$DiOC_{18}(3)$ "DiO"	Yellow-green fluorescent lipophilic probe; widely used as a second tracer with DiI
D-1125	$DiOC_{16}(3)$	Shorter-chain DiO analog that may incorporate into membranes more easily than DiO
D-3898	*FAST* DiO	Unsaturated DiO analog that reportedly migrates ~50% faster than DiO within membranes
D-7770	5,5′-Ph_2-$DiOC_{18}(3)$	More lipophilic DiO analog with a higher quantum yield and longer-wavelength spectra (red-shifted ~10 nm)
D-7778	SP-$DiOC_{18}(3)$	Anionic DiO analog with enhanced water solubility; potentially retained after fixation
D-7767	Br_2-$DiOC_{18}(3)$	Brominated DiO analog that may be useful for DAB photoconversion
DiA and Analogs		
D-3883	4-Di-16-ASP "DiA"	Yellow-green fluorescent lipophilic probe; useful as a second tracer with DiI
D-3897	*FAST* DiA oil	Unsaturated DiA analog that may migrate faster than DiA within membranes
D-7758	*FAST* DiA	Solid form of D-3897 to facilitate direct application of crystals to membranes
D-291	4-Di-10-ASP	Shorter-chain DiA analog that may incorporate into membranes more easily than DiA
DiQ		
D-3885	Di-16-ASQ "DiQ"	Orange–red fluorescent lipophilic tracer; useful as a second tracer with DiA

* Probe is in solid form unless otherwise specified.

Neuronal Tracing Studies with DiI, DiO and DiD

DiI, DiO and DiD are widely used to label neuronal projections in live and fixed tissue. The dyes insert into the outer leaflet of the plasma membrane and diffuse laterally, producing detailed labeling of fine neuronal projections. DiI diffuses about 6 mm/day in live tissue but more slowly in fixed specimens.[1] The dyes usually do not transfer from labeled to unlabeled cells, unless the membrane of the labeled cell is disrupted, and apparently do not transfer through gap junctions. Motoneurons labeled with DiI have been reported to remain viable up to four weeks in culture and one year *in vivo*.[22] Staining of neurons in fixed tissue with DiI has been reported to persist at least two years.[15,23] DiI and DiD can be used simultaneously for two-color tracing [11,24] and as a fluorescence donor–acceptor pair in excited-state energy transfer studies.[25,26] In conjunction with calcein AM (C-1430, see Section 15.3), DiI has been employed in a new technique for studying gap junctional communication in which cells simultaneously labeled with calcein AM and DiI are mixed with unlabeled cells. When gap junctions are established, only the green fluorescent calcein tracer is transferred from the labeled cell to the unlabeled cell; the red fluorescent DiI probe (or similar lipid tracer) remains associated with the membranes of the initially labeled cell.[27-29]

Cell labeling is generally performed by direct application of a dye crystal or dye-coated paper, metal or glass probe or by microinjection onto single cells from a solution in dimethylformamide or dimethylformamide/ethanol.[30,31] We also offer DiI in the form of extra large crystals (D-3911), which many researchers prefer for direct application to tissue. Detailed protocols for DiI labeling and confocal laser scanning microscopy are available.[32,33]

Long-Term Cell Tracing with Membrane Tracers

Lipophilic carbocyanine tracers like DiI are also ideal labels for long-term cell tracing *in vivo* and *in vitro*, including studies of cell migration, transplantation, adhesion and fusion.[34,35] The presence of DiI and DiO in the cell membrane does not appreciably affect cell viability, development or other basic physiological properties.[1,36] Labeling of cell suspensions for long-term tracing is typically performed from aqueous dispersions of the dyes.[37] Stock solutions of the tracers may be prepared in alcohol or dimethylsulfoxide (DMSO) and then added to cell suspensions. Several new analogues of DiI that are more soluble in culture medium should facilitate this method of labeling (see below). Alternatively, the endoplasmic reticulum (ER) has been selectively labeled by microinjecting sea urchin eggs with DiI from a solution in soybean oil, enabling researchers to follow ER reorganization during fertilization.[5]

Tissue Processing and Electron Microscopy with Long-Chain Carbocyanines

Methods have been developed to allow immunofluorescent labeling of tissue containing neurons stained with DiI.[38] Polyacrylamide has been employed to embed DiI-stained brain tissue for vibratome sectioning [39] — a method that has been reported to preserve DiI labeling better than cryostat sectioning.[40] Also, see below for a discussion of new DiI and DiO analogs, some of which may be retained through fixation and permeabilization procedures.

Like other fluorescent dyes, DiI can be used to photoconvert DAB into an insoluble, electron-dense reaction product, thus permitting examination of DiI-labeled tissue by electron microscopy.[13-16] We have recently prepared ring-brominated derivatives of DiI and DiO (Br_2-$DiIC_{18}(3)$, D-7766; Br_2-$DiOC_{18}(3)$, D-7767) that may provide even more efficient photoconversion of DAB. Bromination of dyes usually increases their capability to photosensitize production of singlet oxygen, which is believed to be responsible for the oxidative polymerization of DAB; the photoconversion efficiency has not yet been determined for these compounds. See the Technical Note "Fluorescent Probes for Photoconversion of Diaminobenzidine" on page 414 for further discussion.

New DiI and DiO Analogs, Including Membrane Tracers Retained Through Fixation

CM-DiI and FM™-DiI

CellTracker CM-DiI (C-7000, C-7001) and FM-DiI (F-6999) are two DiI derivatives that are somewhat more water soluble than $DiIC_{18}(3)$, thus facilitating the preparation of staining solutions for cell suspensions and fixed cells. In addition to its improved solubility in culture medium, CellTracker CM-DiI contains a thiol-reactive chloromethyl moiety that allows the dye to covalently bind to cellular thiols. Thus, unlike other membrane stains, the label is well retained in some cells throughout fixation and permeabilization steps [41,42] (Figure 15.4). Membrane staining with CellTracker CM-DiI reportedly persists following routine paraffin processing.[41] CellTracker CM-DiI should be useful in experiments that combine membrane labeling with subsequent immunohistochemical analysis, fluorescence *in situ* hybridization or electron microscopy. As examples, CM-DiI was used for combined neuronal tract tracing and immunohistochemistry in zebrafish (see photo on back cover), for tracking lymphocyte migration [41] and for studying the interaction of tau with the neural plasma membrane.[42]

Sulfonated Carbocyanines

Because of poor water solubility, the long-chain carbocyanine dyes such as DiI, DiO and PKH dyes are often difficult to load into cells in suspension. For example, to prevent dye precipitation, labeling with the PKH dyes requires the use of special osmolarity regulating agents and the absence of any salt in the staining

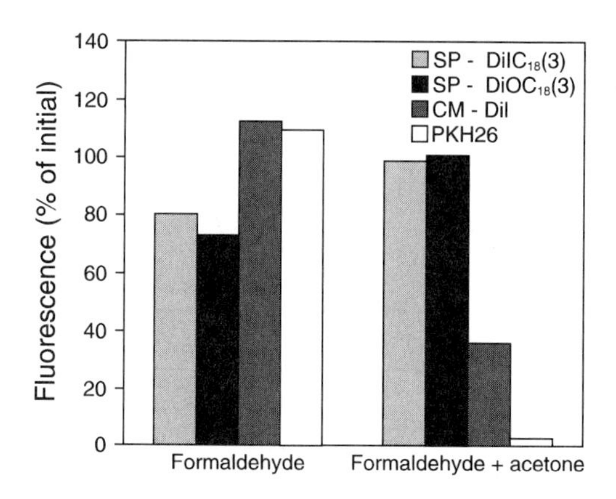

Figure 15.4 *Persistence of lipophilic tracer fluorescence following fixation. Cultured human B cells were stained with 20 µM SP-$DiOC_{18}(3)$, SP-$DiIC_{18}(3)$, CM-DiI or PKH26 and then fixed with 3.7% formaldehyde or 3.7% formaldehyde + acetone. The fixed cells were analyzed by flow cytometry (Becton Dickinson FACS® Vantage™) to generate a comparison of their fluorescence with that of the original live cell population.*

buffer.[34] To facilitate the staining of cells with long-chain carbocyanine dyes, we have developed sulfonated cyanine dyes [43] — SP-DiIC_{18}(3) (D-7777), SP-DiOC_{18}(3) (D-7778) and DiIC_{18}(3)-DS (D-7776) — that retain the 18-carbon lipophilic chains of DiI and DiO but exhibit improved solubility in culture medium. Cells can be labeled with these new dyes by simply diluting a DMSO stock solution of the dye to a labeling concentration of 1–10 µM in unmodified culture medium. We have determined that these sulfonated dyes are completely soluble at 20 µM for at least one hour in Dulbecco's phosphate-buffered saline and Hanks' balanced salt solution; aggregation begins after a few hours in solution. Uptake of the anionic carbocyanine dyes in live cells is usually slower than that of cationic carbocyanine dyes, and the resulting staining patterns also may be quite different.

The negatively charged sulfonate groups in the dyes also reduce the tendency of the dyes to aggregate in the membranes, which is a major cause of fluorescence quenching; thus, the fluorescence quantum yield of the bound dyes may be greater than that of DiI and DiO. In addition, the higher molar absorptivities of SP-DiIC_{18}(3) and especially SP-DiOC_{18}(3) relative to those of DiI and DiO make these dyes brighter. Therefore, compared with other cyanine membrane probes, these sulfonated dyes may be preferred when cells cannot tolerate excess dye or osmolarity-regulating reagents or when a subsequent fixation and lipid extraction step is required (see below). However, like DiI, DiO and the PKH dyes, these sulfonated carbocyanines also stain intracellular membranes. Also, although we have seen good retention of these anionic carbocyanine dyes in cells, their greater water solubility could result in some increase in dye transfer between cells.

Some applications require that the membrane stain remains in place following aldehyde fixation and lipid extraction. In contrast to labeling with DiI, DiO or the PKH dyes,[35] cell labeling with some of these sulfonated carbocyanines appears to be compatible with standard aldehyde fixation and acetone treatment, at least in some cell lines. Although the mechanism for cell retention of the sulfonated carbocyanines has not been determined, we have observed excellent retention of staining of human B cells with SP-DiIC_{18}(3) and SP-DiOC_{18}(3), whereas virtually all of the PKH26 dye is lost during the lipid extraction step (Figure 15.4). CM-DiI (C-7000, see above) is also well retained during this processing. Furthermore, we have observed that acetone treatment actually enhances the fluorescence of some cells stained with SP-DiIC_{18}(3) and especially with SP-DiOC_{18}(3), a phenomenon that has not been seen with other carbocyanine membrane stains. Although these properties have not yet been extensively tested, they could make SP-DiIC_{18}(3) and SP-DiOC_{18}(3) particularly useful for immunochemistry and other histochemical studies that may require tissue dehydration, clearing and membrane extraction.

Lipophilic Tracer Sampler Kit

For researchers wishing to investigate the suitability of our new membrane probes for a particular application, we have prepared a Lipophilic Tracer Sampler Kit (L-7781). This kit contains 1 mg samples of ten different membrane stains, including both new and well-established tracers:

- DiIC_{18}(3) (DiI)
- DiOC_{18}(3) (DiO)
- DiIC_{18}(5) (DiD)
- 4-Di-16-ASP (DiA)
- SP-DiIC_{18}(3)
- SP-DiOC_{18}(3)
- DiIC_{18}(3)-DS
- 6,6'-Ph_2-DiIC_{18}(3)
- 5,5'-Ph_2-DiIC_{18}(3)
- 5,5'-Ph_2-DiOC_{18}(3)

DiA, DiQ, R 18 and Octadecyl Acridine Orange

DiA and *FAST* DiA™

Reports from several researchers indicate that DiA (4-Di-16-ASP, D-3883) is a better neuronal tracer than DiO for multicolor labeling with DiI.[44-46] DiA diffuses more rapidly in membranes and is more soluble than DiO, thus facilitating cell labeling. For example, DiA and DiI have been used together to investigate the interactions between dorsal root axons and their targets [47] and axon outgrowth in the retina.[48] DiA can be excited between 440 nm and 500 nm, and its maximum emission in DOPC vesicles is ~590 nm (red-orange fluorescence; see Figure 14.2 in Chapter 14). Its fluorescence in cells, however, is usually bright green to yellow-green (depending on the optical filter set used). The polyunsaturated *FAST* DiA, offered in both oil (D-3897) and crystalline solid (D-7758) form, may diffuse faster in membranes than DiA. The C_{10} analog of DiA, 4-Di-10-ASP (D-291), has been used as a retrograde tracer to monitor injury-induced degradation of rat neurons *in vivo* and the role of microglial cells in removing debris resulting from natural and axotomy-induced neuronal cell death [49-51] (see Color Plate 3 in "Visual Reality"). 4-Di-10-ASP staining of microglia persists for at least 12 months and can be used for DAB photoconversion.[49] Axonal outgrowth of retinal ganglion cells stained with 4-Di-10-ASP has been observed.[52]

DiQ

Researchers may want to consider double-labeling cells with DiA and its analog DiQ (D-3885) because the spectra of these probes are well separated. DiA and DiQ proved suitable for tracing different afferents of the basal supraoptic nucleus.[53]

Octadecyl Rhodamine B

Octadecyl rhodamine B (R 18, O-246) is a lipophilic cation that has been extensively used as a membrane probe [54,55] (see Section 14.3). Viral particles that have been labeled with R 18 have fluorescence that is highly quenched; fusion of the particle with cell membranes relieves the quenching, making the receptor cell highly fluorescent.[55-57] Dequenching of the fluorescence of R 18 is a useful method for following a variety of other processes that involve large changes in the membrane volume accessible to the probe, including:

- Disintegration of R 18–labeled low density lipoproteins in lysosomes [58]
- Fusion of lysosomes with late endosomes [59]
- Fusion of endocytic vesicles of reticulocytes with liposomes [60]
- Receptor clustering on cell surfaces [61]

Plasma Membrane Stains

Although the lipophilic cyanine and styryl dyes described above are useful for staining cell membranes, they have been relatively difficult to apply to cells and eventually internalize, where they stain all cell membranes. Frequently, there is need to stain only the plasma membrane, without significant internalization of the label. FM™ 1-43 (T-3163), FM 4-64 (T-3166) and RH 414 (T-1111) bind rapidly and reversibly to the plasma membrane with strong fluorescence enhancement.[62-64] All three probes have large Stokes shifts and can be excited by the argon-ion laser.

FM 1-43 and FM 4-64

FM 1-43 is efficiently excited with standard fluorescein optical filters but poorly excited with standard tetramethylrhodamine optical filters. FM 1-43 has been used to outline membranes in sea urchin eggs.[65] This styryl dye has also proven extremely valuable for following synaptic recycling (see Section 17.2.).

Membranes labeled with FM 4-64 exhibit long-wavelength red fluorescence that can be distinguished from the green fluorescence of FM 1-43 staining with the proper optical filter sets, thus permitting two-color observation of membrane recycling in real time.[66,67] FM 4-64 has recently been reported to selectively stain yeast vacuolar membranes and is proving to be an important tool for visualizing vacuolar organelle morphology and dynamics and for studying the endocytic pathway in yeast[68,69] (see Section 12.3).

RH 414

Membranes stained with RH 414 exhibit orange fluorescence when observed through a longpass optical filter that permits passage of light beyond 510 nm.[70-72] In a confocal laser scanning microscopy study, the subcellular distribution of L-type Ca^{2+} channels in olfactory bulb neurons were localized using intensity ratio measurements of the green fluorescence of DM-BODIPY dihydropyridine (D-7443, see Section 18.4) to the red fluorescence of RH 414-stained plasma membranes.[73] In this method, plasma-membrane staining by RH 414 was used to control for optical artifacts and differences in membrane surface area in the optical section. RH 414 has also been used as to follow vacuolation of the transverse tubules of frog skeletal muscle[74] and to measure membrane potential (see Chapter 25.2).

Lipophilic Dextrans

Ordinarily, dextrans have low affinity for cell membranes. However, addition of a few lipophilic moieties to the dextran has been shown to make fluorescent dextrans like our lipophilic fluorescein and tetramethylrhodamine 10,000 MW dextrans (D-7169, D-7164) water-soluble tracers for labeling cell surfaces.[75] Similar lipophilic dextrans have been used to model the diffusion of receptors on muscle membrane blebs[76] and as a marker for endocytosis in chick fibroblasts.[77]

1. Trends Neurosci 12, 333 (1989); **2.** Neuron 13, 813 (1994); **3.** BioTechniques 13, 580 (1992); **4.** Development 106, 809 (1989); **5.** J Cell Biol 114, 929 (1991); **6.** J Struct Biol 105, 154 (1990); **7.** J Membrane Biol 109, 221 (1989); **8.** Cytometry 14, 16 (1993); **9.** J Cell Biol 90, 595 (1981); **10.** Biochem Pharmacol 49, 1303 (1995); **11.** J Neurosci 15, 549 (1995); **12.** Science 257, 96 (1992); **13.** J Neurosci 14, 6815 (1994); **14.** Microscopy Res Technique 24, 2 (1993); **15.** J Histochem Cytochem 38, 725 (1990); **16.** Neuroscience 28, 3 (1989); **17.** Andrea Elberger, University of Tennessee, personal communication; **18.** J Cell Biol 114, 1017 (1991); **19.** J Cell Biol 103, 1745 (1986); **20.** Science 247, 1229 (1990); **21.** Biochemistry 29, 8337 (1990); **22.** J Comp Neurol 302, 729 (1990); **23.** Exp Neurol 102, 92 (1988); **24.** J Cell Biol 126, 519 (1994); **25.** Chem Phys Lett 159, 231 (1989); **26.** Thin Solid Films 132, 55 (1985); **27.** BioTechniques 18, 490 (1995); **28.** J Cell Biol 130, 987 (1995); **29.** In Vitro Cell Devel Biol 30A, 796 (1994); **30.** Neuron 7, 819 (1991); **31.** J Neurosci 10, 3947 (1990); **32.** Mills, L.R. *et al.* in *Three Dimensional Confocal Microscopy: Investigation of Biological Specimens* (1994) pp. 325–351; **33.** Honig, M.G. in *Neuroscience Protocols*, F.G. Wouterlood, Ed., Elsevier Science Publishers (1993) 93-050-16-01–20; **34.** U.S. Patent No. 4,762,701; **35.** Meth Cell Biol 33, 469 (1990); **36.** J Cell Biol 103, 171 (1986); **37.** Histochemistry 97, 329 (1992); **38.** J Histochem Cytochem 38, 735 (1990); **39.** J Neurosci Meth 42, 65 (1992); **40.** J Neurosci Meth 42, 45 (1992); **41.** J Immunol Meth 194, 181 (1996); **42.** J Cell Biol 131, 1327 (1995); **43.** Bioorg Medicinal Chem Lett 6, 1479 (1996); **44.** J Neurosci 15, 3475 (1995); **45.** J Neurosci 15, 990 (1995); **46.** Trends Neurosci 13, 14 (1990); **47.** J Neurosci 12, 3494 (1992); **48.** Visual Neurosci 10, 117 (1993); **49.** Trends Neurosci 17, 177 (1994); **50.** J Neurosci 13, 455 (1993); **51.** Eur J Neurosci 3, 1189 (1991); **52.** J Neurosci 15, 5514 (1995); **53.** Soc Neurosci Abstr 19, 93 abstract #44.7 (1993); **54.** Biochemistry 33, 1820 (1994); **55.** Biochim Biophys Acta 1191, 375 (1994); **56.** Biochem J 300, 347 (1994); **57.** J Biol Chem 269, 5467 (1994); **58.** J Cell Biol 59, 116 (1992); **59.** J Cell Biol 126, 1173 (1994); **60.** J Biol Chem 270, 17823 (1995); **61.** Biophys J 67, 1280 (1994); **62.** EMBO J 12, 5209 (1993); **63.** J Neurosci 13, 834 (1993); **64.** Science 235, 200 (1992); **65.** J Cell Biol 131, 1183 (1995); **66.** J Cell Biol 131, 679 (1995); **67.** J Neurosci 15, 8246 (1995); **68.** Mol Biol of the Cell 7, 985 (1996); **69.** J Cell Biol 128, 779 (1995); **70.** J Neurosci 15, 6327 (1995); **71.** J Neurosci 12, 363 (1992); **72.** Science 255, 200 (1992); **73.** J Neurosci Meth 59, 183 (1995); **74.** J Muscle Res Cell Motil 16, 401 (1995); **75.** Biochemistry 19, 3893 (1980); **76.** J Cell Biol 92, 207 (1982); **77.** J Cell Biol 102, 1593 (1986).

15.4 Data Table *Fluorescent Lipophilic Tracers*

Cat #	Structure	MW	Storage	Soluble	Abs	{ε x 10⁻³}	Em	Solvent	Notes
C-7001	✓	1052	F,D,L	DMSO, EtOH	553	{134}	570	MeOH	
D-275	✓	882	L	DMSO, EtOH	484	{154}	501	MeOH	
D-282	✓	934	L	DMSO, EtOH	549	{148}	565	MeOH	
D-291	✓	619	L	DMSO, EtOH	492	{53}	612	MeOH	<1>
D-307	✓	960	L	DMSO, EtOH	644	{195}	665	MeOH	<2>
D-383	✓	766	L	DMSO, EtOH	549	{144}	565	MeOH	<3>
D-384	✓	878	L	DMSO, EtOH	549	{148}	565	MeOH	
D-1125	✓	826	L	DMSO, EtOH	484	{156}	501	MeOH	
D-3883	✓	787	L	DMSO, EtOH	491	{52}	613	MeOH	<1>
D-3885	✓	837	L	DMSO, EtOH	562	{44}	681	MeOH	<1>
D-3886	✓	925	F,L,AA	DMSO, EtOH	549	{144}	564	MeOH	<2>
D-3897	✓	835	F,L,AA	DMSO, EtOH	492	{49}	612	MeOH	<1, 2>
D-3898	✓	874	F,L,AA	DMSO, EtOH	484	{138}	499	MeOH	
D-3899	✓	926	F,L,AA	DMSO, EtOH	549	{143}	564	MeOH	<2>
D-3911	✓	934	L	DMSO, EtOH	549	{148}	565	MeOH	
D-7756	✓	1018	F,L,AA	DMSO, EtOH	549	{148}	564	MeOH	
D-7757	✓	1052	L	DMSO, EtOH	644	{193}	663	MeOH	
D-7758	✓	900	F,L,AA	DMSO, EtOH	492	{41}	612	MeOH	<1>
D-7766	✓	1092	L	DMSO, EtOH	558	{128}	575	MeOH	
D-7767	✓	1040	L	DMSO, EtOH	491	{117}	506	MeOH	

Cat #	Structure	MW	Storage	Soluble	Abs	{ε × 10⁻³}	Em	Solvent	Notes
D-7770	✓	970	L	DMSO, EtOH	496	{168}	513	MeOH	
D-7776	✓	994	L	DMSO, EtOH	555	{144}	570	MeOH	
D-7777	✓	1146	L	DMSO, EtOH	556	{164}	573	MeOH	
D-7778	✓	1116	L	DMSO, EtOH	497	{175}	513	MeOH	
D-7779	✓	1022	L	DMSO, EtOH	576	{140}	599	MeOH	
D-7780	✓	1022	L	DMSO, EtOH	556	{160}	573	MeOH	
F-6999	✓	1017	L	DMSO, EtOH	553	{133}	570	MeOH	
O-246	S14.3	731	F,DD,L	DMSO, EtOH	556	{125}	578	MeOH	<3>
T-1111	S25.2	581	D,L	DMSO, EtOH	532	{55}	716	MeOH	<1>
T-3163	S17.2	612	D,L	H_2O, DMSO	510	{56}	626	MeOH	<1, 4>
T-3166	S17.2	608	D,L	H_2O, DMSO	543	{43}	<5>	MeOH	<1>

C-7000 see C-7001.

For definitions of the contents of this data table, see "How to Use the Handbook" on page vi.
Structure: Chemical structure drawing: ✓ = shown in this section; Sn.n = shown in Section number n.n.

<1> Abs and Em of styryl dyes are at shorter wavelengths in membrane environments than in reference solvents such as methanol. The difference is typically ~20 nm for absorption and ~80 nm for emission but varies considerably from one dye to another.
<2> This product is intrinsically a liquid or an oil at room temperature.
<3> This product is intrinsically a sticky gum at room temperature.
<4> F-3163 Abs = 479 nm, Em = 598 nm bound to phospholipid vesicles (determined at Molecular Probes). Em = 565 nm bound to synaptosomal membranes [Neuron 12, 1235 (1994)].
<5> Fluorescence of T-3166 in MeOH is very weak. Excitation/emission wavelength settings at 515/640 nm are suitable for detection of yeast vacuole membrane staining with FM 4-64 [J Cell Biol 128, 779 (1995)].

15.4 Structures *Fluorescent Lipophilic Tracers*

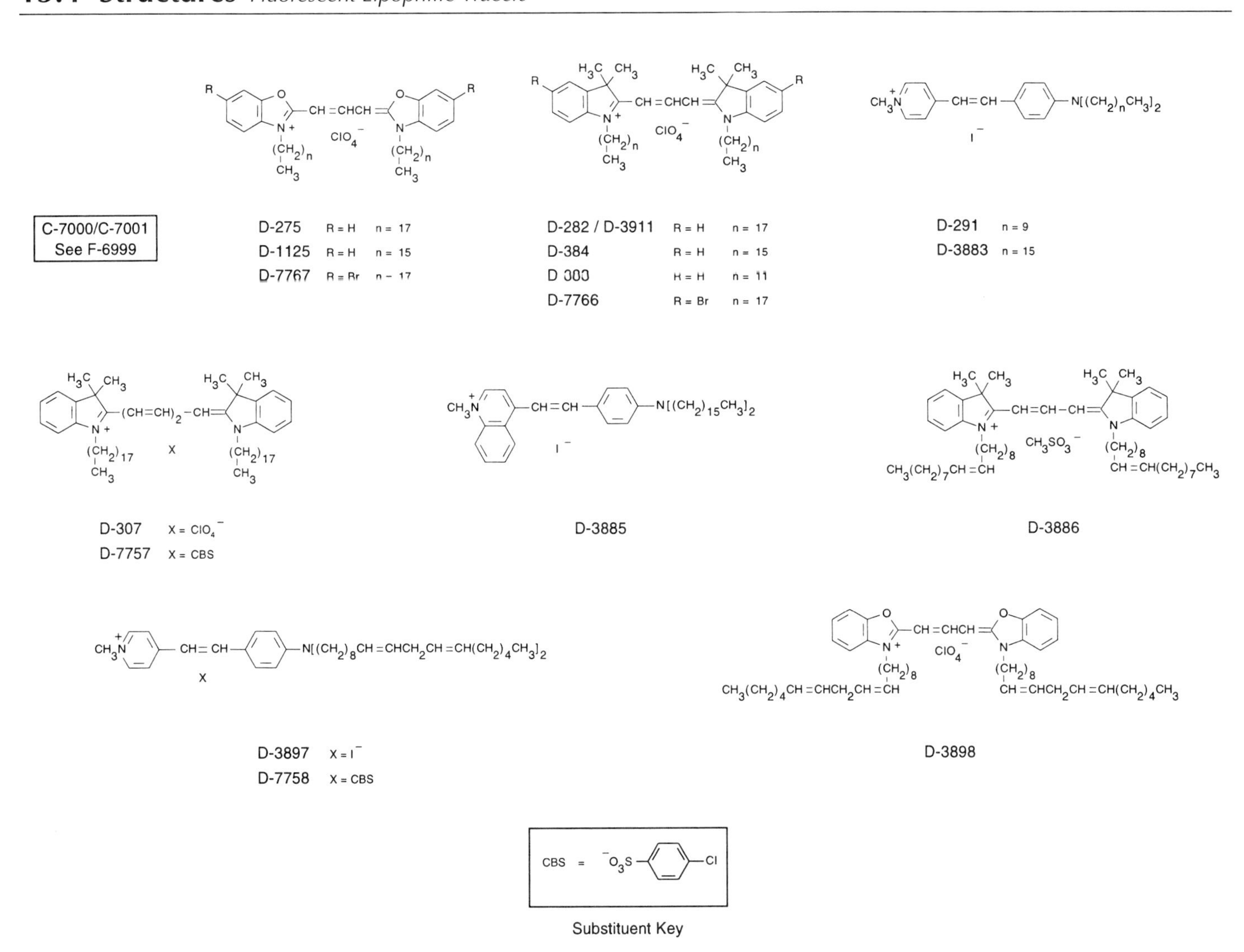

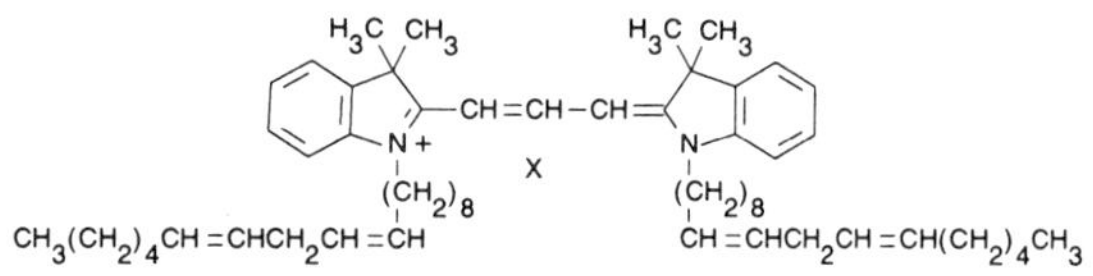

D-3899 X = ClO_4^-
D-7756 X = CBS

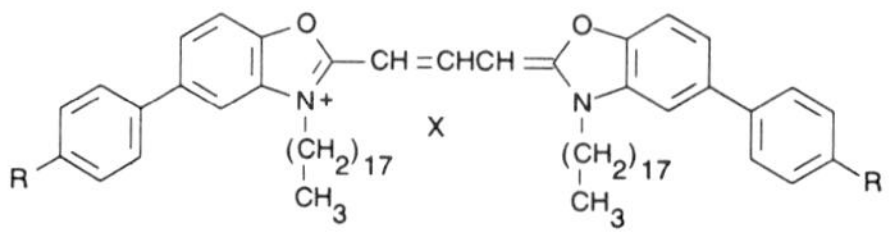

D-7770 R = H X = Cl^-
D-7778 R = —SO_3^- X = Na^+

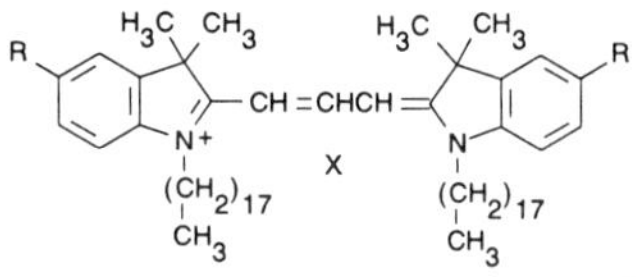

Ph = phenyl

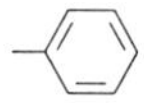

D-7776 R = —SO_3^- X = H^+
D-7779 R = Ph X = Cl^-

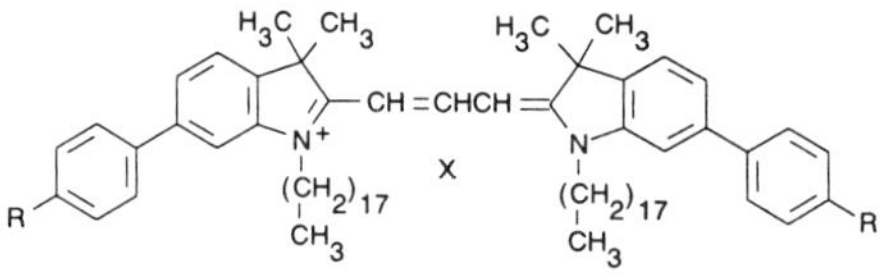

D-7780 R = H X = Cl^-
D-7777 R = —SO_3^- X = H^+

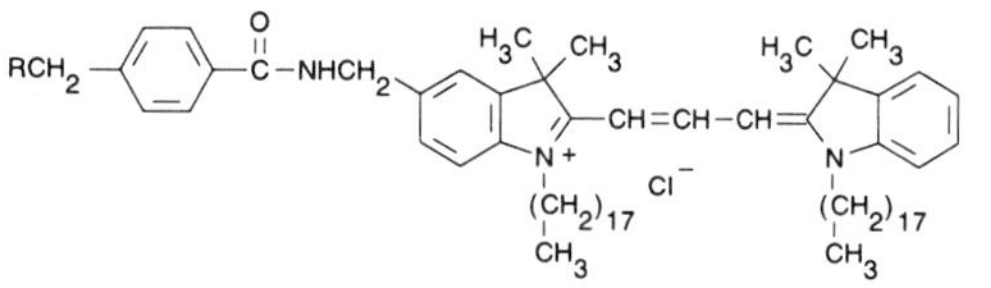

F-6999 R = H
C-7000 / C-7001 R = Cl

O-246
Section 14.3

T-1111
Section 25.2

T-3163, T-3166
Section 17.2

CBS = ^-O_3S–C_6H_4–Cl

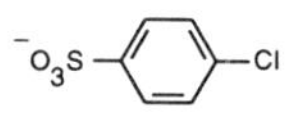

Substituent Key

15.4 Price List *Fluorescent Lipophilic Tracers*

Cat #		Product Name	Unit Size	Price Per Unit ($) 1–4 Units	Price Per Unit ($) 5–24 Units
C-7001	**New**	CellTracker™ CM-DiI	1 mg	110.00	88.00
C-7000	**New**	CellTracker™ CM-DiI *special packaging*	20x50 µg	110.00	88.00
D-7169	**New**	dextran, fluorescein, 10,000 MW, lipophilic	10 mg	95.00	76.00
D-7164	**New**	dextran, tetramethylrhodamine, 10,000 MW, lipophilic	10 mg	110.00	88.00
D-7767	**New**	6,6′-dibromo-3,3′-dioctadecyloxacarbocyanine perchlorate (Br$_2$-DiOC$_{18}$(3))	5 mg	135.00	108.00
D-7766	**New**	5,5′-dibromo-1,1′-dioctadecyl-3,3,3′,3′-tetramethylindocarbocyanine perchlorate (Br$_2$-DiIC$_{18}$(3))	5 mg	135.00	108.00
D-291		4-(4-(didecylamino)styryl)-N-methylpyridinium iodide (4-Di-10-ASP)	25 mg	115.00	92.00
D-383		1,1′-didodecyl-3,3,3′,3′-tetramethylindocarbocyanine perchlorate (DiIC$_{12}$(3))	100 mg	135.00	108.00
D-3883		4-(4-(dihexadecylamino)styryl)-N-methylpyridinium iodide (DiA; 4-Di-16-ASP)	25 mg	115.00	92.00
D-3885		4-(4-(dihexadecylamino)styryl)-N-methylquinolinium iodide (DiQ; Di-16-ASQ)	25 mg	48.00	38.40
D-1125		3,3′-dihexadecyloxacarbocyanine perchlorate (DiOC$_{16}$(3))	25 mg	135.00	108.00
D-384		1,1′-dihexadecyl-3,3,3′,3′-tetramethylindocarbocyanine perchlorate (DiIC$_{16}$(3))	100 mg	135.00	108.00
D-7758	**New**	4-(4-(dilinoleylamino)styryl)-N-methylpyridinium, 4-chlorobenzenesulfonate (*FAST* DiA™ solid; DiΔ^{9,12}-C$_{18}$ASP, CBS)	5 mg	135.00	108.00
D-3897		4-(4-(dilinoleylamino)styryl)-N-methylpyridinium iodide (*FAST* DiA™ oil; DiΔ^{9,12}-C$_{18}$ASP, I)	5 mg	135.00	108.00
D-3898		3,3′-dilinoleyloxacarbocyanine perchlorate (*FAST* DiO™ solid; DiOΔ^{9,12}-C$_{18}$(3), ClO$_4$)	5 mg	135.00	108.00
D-7756	**New**	1,1′-dilinoleyl-3,3,3′,3′-tetramethylindocarbocyanine, 4-chlorobenzenesulfonate (*FAST* DiI™ solid; DiIΔ^{9,12}-C$_{18}$(3), CBS)	5 mg	135.00	108.00
D-3899		1,1′-dilinoleyl-3,3,3′,3′-tetramethylindocarbocyanine perchlorate (*FAST* DiI™ oil; DiIΔ^{9,12}-C$_{18}$(3), ClO$_4$)	5 mg	135.00	108.00
D-7770	**New**	3,3′-dioctadecyl-5,5′-diphenyloxacarbocyanine chloride (5,5′-Ph$_2$-DiOC$_{18}$(3))	5 mg	135.00	108.00
D-7779	**New**	1,1′-dioctadecyl-5,5′-diphenyl-3,3,3′,3′-tetramethylindocarbocyanine chloride (5,5′-Ph$_2$-DiIC$_{18}$(3))	5 mg	135.00	108.00
D-7780	**New**	1,1′-dioctadecyl-6,6′-diphenyl-3,3,3′,3′-tetramethylindocarbocyanine chloride (6,6′-Ph$_2$-DiIC$_{18}$(3))	5 mg	135.00	108.00
D-7778	**New**	3,3′-dioctadecyl-5,5′-di(4-sulfophenyl)-oxacarbocyanine, sodium salt (SP-DiOC$_{18}$(3))	5 mg	135.00	108.00
D-7777	**New**	1,1′-dioctadecyl-6,6′-di(4-sulfophenyl)-3,3,3′,3′-tetramethylindocarbocyanine (SP-DiIC$_{18}$(3))	5 mg	135.00	108.00
D-275		3,3′-dioctadecyloxacarbocyanine perchlorate ('DiO'; DiOC$_{18}$(3))	100 mg	135.00	108.00
D-7776	**New**	1,1′-dioctadecyl-3,3,3′,3′-tetramethylindocarbocyanine-5,5′-disulfonic acid (DiIC$_{18}$(3)-DS)	5 mg	135.00	108.00
D-282		1,1′-dioctadecyl-3,3,3′,3′-tetramethylindocarbocyanine perchlorate ('DiI'; DiIC$_{18}$(3))	100 mg	135.00	108.00
D-3911		1,1′-dioctadecyl-3,3,3′,3′-tetramethylindocarbocyanine perchlorate *large crystals*	25 mg	100.00	80.00
D-7757	**New**	1,1′-dioctadecyl-3,3,3′,3′-tetramethylindodicarbocyanine, 4-chlorobenzenesulfonate salt ('DiD' solid; DiIC$_{18}$(5) solid)	10 mg	68.00	54.40
D-307		1,1′-dioctadecyl-3,3,3′,3′-tetramethylindodicarbocyanine perchlorate ('DiD' oil; DiIC$_{18}$(5) oil)	25 mg	90.00	72.00

Cat #		Product Name	Unit Size	Price Per Unit ($) 1–4 Units	5–24 Units
D-3886		1,1′-dioleyl-3,3,3′,3′-tetramethylindocarbocyanine methanesulfonate (Δ^9-DiI)	25 mg	135.00	108.00
F-6999	*New*	FM™-DiI *special packaging*	20x50 µg	95.00	76.00
L-7781	*New*	Lipophilic Tracer Sampler Kit	1 kit	150.00	120.00
O-246		octadecyl rhodamine B chloride (R 18)	10 mg	145.00	116.00
T-3163		N-(3-triethylammoniumpropyl)-4-(4-(dibutylamino)styryl)pyridinium dibromide (FM™ 1-43)	1 mg	165.00	132.00
T-1111		N-(3-triethylammoniumpropyl)-4-(4-(4-(diethylamino)phenyl)butadienyl)pyridinium dibromide (RH 414)	5 mg	165.00	132.00
T-3166		N-(3-triethylammoniumpropyl)-4-(6-(4-(diethylamino)phenyl)hexatrienyl)pyridinium dibromide (FM™ 4-64)	1 mg	165.00	132.00

15.5 Fluorescent and Biotinylated Dextrans

Dextrans are hydrophilic polysaccharides characterized by their moderate to high molecular weight, good water solubility and low toxicity. They are also biologically inert due to their uncommon poly-(α-D-1,6-glucose) linkages, which render them resistant to cleavage by most endogenous cellular glycosidases. Molecular Probes offers more than 100 fluorescent and biotinylated dextran conjugates in several different molecular weight ranges.

Properties of Our Dextran Conjugates

A Wide Selection of Substituents on Our Dextrans

Molecular Probes' dextrans are conjugated to a wide variety of fluorophores or to biotin (Table 15.3). In particular, we would like to highlight our dextran conjugates of the new Oregon Green, Rhodol Green and Rhodamine Green dyes, which are brighter and more photostable than most fluorescein dextrans. Also, the new dextran conjugates of tetrakis-(4-carboxyphenyl)porphine (D-6929, D-6930) have the longest-wavelength emission of any of our dextrans conjugates and can be excited at their intense absorption peak at ~415 nm or anywhere between about 500 nm and 650 nm. Dextran-conjugated fluorescent indicators for calcium and magnesium ions (see Chapter 22), pH (see Chapter 23) and sodium or chloride ions (see Chapter 24) are described with their corresponding ion indicators in other chapters.

Dextran Size

Molecular Probes' dextrans include those with nominal molecular weights (MW) of 3000, 10,000, 40,000, 70,000, 500,000 and 2,000,000 daltons, as well as a fluorescein dextran with MW of 5–40 million daltons (D-1899). Fluorescein dextran sulfates with average MW of 40,000 or 500,000 daltons (D-1890, D-1891) and a fluorescein Ficoll® of approximately 400,000 daltons (F-1965) are also available. Because unlabeled dextrans are polydisperse — and may become more so during the chemical processes required for their modification and purification — the actual molecular weights present in a particular sample may have a broad distribution. Our "3000 MW" dextran preparations contain polymers with molecular weights predominantly in the range of ~1500–3000 daltons, including the dye or other label.

Degree of Substitution of Our Dextrans

Whereas dextrans from other sources usually have a degree of substitution of 0.2 or fewer dye molecules per dextran molecule for dextrans in the 10,000 MW range, our conjugates typically contain 0.3–0.7 dyes per dextran in the 3000 MW range, 0.5–2 dyes per dextran in the 10,000 MW range, 2–4 dyes in the 40,000 MW range and 3–6 dyes in the 70,000 MW range. The actual degree of substitution is indicated on the product's label. If too many fluorophores are conjugated to the dextran molecule, quenching and undesired interactions with cellular components may occur. We have found our degree of substitution to be optimal for most applications, yielding dextrans that typically are more fluorescent than those available from other sources. It has been reported that some commercially available fluorescein isothiocyanate (FITC) dextrans yield spurious results in endocytosis studies because of the presence of free dye or metal catalysts.[1,2] To overcome this problem, Molecular Probes removes as much of the free dye as possible by a combination of precipitation, dialysis, gel filtration and other techniques. The fluorescent dextran is then assayed by thin-layer chromatography (TLC) to ensure that it is free of low molecular weight dyes. Molecular Probes prepares several unique products that have two or even three different labels, including our fluoro-ruby, mini-ruby and micro-ruby products described below. Not all individual dextran molecules of these products are expected to have all the substituents, or to be equally fixable, particularly in conjugates of the lowest molecular weight dextrans.

Dextran Net Charge and Method of Substitution

The net charge on the dextran depends on the fluorophore and the method of preparing the conjugate. Molecular Probes prepares most of its dextrans by reacting a water-soluble amino dextran (D-1860, D-1861, D-1862, D-3330, D-7144, D-7145) with the succinimidyl ester of the appropriate dye, rather than with isothiocyanate derivatives such as FITC. This method provides superior amine selectivity and yields an amide linkage, which is somewhat more stable than the corresponding thioureas formed from isothiocyanates. Except for the Rhodamine Green and Rhodol Green conjugates, once the dye has been added, the unreacted amines on the dextran are capped to yield a neutral or anionic dextran. The Cascade Blue, tetrakis-(4-carboxyphenyl)porphine, lucifer yellow, fluorescein, Oregon Green and eosin dextrans are

Table 15.3 Molecular Probes' selection of dextran conjugates.

Label(s) (Absorption/Emission Maxima) *	3000 MW	10,000 MW	40,000 MW	70,000 MW	500,000 MW	2,000,000 MW
Cascade Blue (400/420)	D-7131 D-7132 †	D-1974 D-1976 †		D-1975 D-1977 †		
Tetrakis-(4-carboxyphenyl)porphine (415/660)		D-6929		D-6930		
Lucifer yellow (428/532)		D-1825 †	D-1826 †	D-1827 †		
Oregon Green 488 (496/524)		D-7170 D-7171 †		D-7172 D-7173 †		
Fluorescein (494/518)	D-3305 D-3306 †	D-1814 D-1821 D-1820 †	D-1844 D-1845 †	D-1815 D-1823 D-1822 †	D-7136 †	D-7137 †
Fluorescein + biotin (494/518)	D-7156 †	D-7178 †		D-7165 †		
DMNB-caged fluorescein (494/518) §	D-3309	D-3310		D-3311		
DMNB-caged fluorescein + biotin (494/518) §		D-7146 †		D-7147 †		
Rhodol Green (499/525)		D-7148	D-7150			
Rhodamine Green (502/527)	D-7163	D-7152 D-7153 †	D-7154			
BODIPY FL (505/513)		D-7168 ‡				
Oregon Green 514 (511/530)		D-7174 D-7175 †		D-7176 D-7177 †		
Eosin (524/544)		D-7166 †				
Tetramethylrhodamine (555/580)	D-3307 D-3308 †	D-1816 D-1817 † D-1868 ‡	D-1842 D-1843 †	D-1819 D-1818 †		D-7139 †
Tetramethylrhodamine + biotin (555/580)	D-7162 †	D-3312 †	D-7133 †			
Rhodamine B (570/590)		D-1824	D-1840	D-1813 D-1841		
Texas Red (595/615)	D-3329 D-3328 †	D-1828 D-1863 †	D-1829	D-1830 D-1864 †		D-7141 †
Biotin (<300/none)	D-7134 D-7135 †	D-1856 D-1956 †		D-1957 †	D-7142 †	D-7143 †
Amino (<300/none)	D-3330	D-1860	D-1861	D-1862	D-7144	D-7145

* Approximate absorption and emission maxima in nm for conjugates. † "Lysine-fixable" dextrans contain lysines and can therefore be fixed in place with formaldehyde or glutaraldehyde.
‡ "Fixable" dextrans contain free amines (but not lysines) and can be fixed in place with formaldehyde or glutaraldehyde.
§ Absorption and emission maxima are for the conjugate following photolysis.

intrinsically anionic, whereas most of the dextrans labeled with the zwitterionic rhodamine B, tetramethylrhodamine and Texas Red dyes are essentially neutral. To produce more highly anionic dextrans, Molecular Probes has developed a procedure for adding negatively charged groups to the dextran carriers; these products are designated "polyanionic" dextrans. In the case of the Rhodamine Green and Rhodol Green dextrans, the unreacted amines on the dextran are not capped after dye conjugation. Thus these dextran conjugates are either neutral or cationic.

Dextran Fixability

Some applications require that the dextran tracer be subsequently treated with formaldehyde or glutaraldehyde for analysis.[3,4] For these applications, Molecular Probes offers "lysine-fixable" versions of most of our dextran conjugates of fluorophores or biotin. These dextrans have covalently bound lysine residues that permit dextran tracers to be covalently linked to surrounding biomolecules by aldehyde fixation for subsequent detection by immunohistochemical and ultrastructural techniques.

Loading Cells with Dextrans and Subsequent Tissue Processing

Unless taken up by an endocytic process, dextran conjugates are membrane impermeant and usually must be loaded by relatively invasive techniques (Table 15.1). Sterile filtration of dextran solutions before use with live cells is highly recommended.[5] A recent paper reports the use of 1.5% solution of octyl β-D-thioglucopyranoside (C_8SGlu, O-7035; see Section 16.3) to enhance the penetration of labeled dextrans through stripped skin, with penetration efficiency dependent on the dextran's molecular weight.[6] Biotin and biotinylated biomolecules with molecular weights up to >100,000 daltons are reported to be taken up by plants through an endocytic pathway.[7,8]

Our lysine-fixable dextrans can be fixed in place with formaldehyde or glutaraldehyde, allowing subsequent tissue processing such as sectioning. A protocol has been published for embedding tissues in plastic for high-resolution characterization of neurons filled with lysine-fixable fluorescent dextrans.[9] Fixation of biotinylated or fluorescent dextrans also permits the use of fluorescent- or enzyme-labeled conjugates of avidin and streptavidin (see Section 7.5) or of anti-dye antibodies (see Section 7.3), respectively, to amplify the signal, which is important for detecting fine structure in sections or for changing the detection mode. We provide antibodies to the Cascade Blue, lucifer yellow, fluorescein, BODIPY FL, tetramethylrhodamine and Texas Red fluorophores (see Section 7.3). Our antibody to fluorescein crossreacts strongly with the Oregon Green and somewhat with Rhodamine Green fluorophores, and our anti-tetramethylrhodamine and anti–Texas Red antibodies crossreact with tetramethylrhodamine, Lissamine™ rhodamine B, Rhodamine Red™ and Texas Red dyes. Molecular Probes' ELF®-97 Immunohistochemistry Labeling Kit and ELF-97 Cytological Labeling Kits (E-6600, E-6602, E-6603; see Section 6.3) may be used to further amplify the signal, yielding greater fluorescence than is possible from directly conjugated fluorescent detection reagents.

Photoconversion of neurons labeled with lysine-fixable fluorescent dextrans in the presence of diaminobenzidine (DAB) can be used to produce electron-dense products for electron microscopy.[10] Electron-dense products can also be generated from peroxidase or colloidal gold conjugates of avidin, streptavidin or anti-dye antibodies.[11]

Neuronal Tracing with Dextrans

Fluorescent and biotinylated dextrans are routinely employed to trace neuronal projections. Dextrans can function efficiently as anterograde or retrograde tracers, depending on the study method and tissue type used. Active transport of dextrans occurs only in live, not fixed tissue.[12] Comparative studies of rhodamine isothiocyanate, rhodamine B dextran (D-1824) and lysinated tetramethylrhodamine dextran (fluoro-ruby, D-1817) have shown that the dextran conjugates produce smaller injection sites and more permanent labeling than do the corresponding free dyes.[13] Dextran conjugates with molecular weights up to 70,000 daltons have been employed as neuronal tracers in a wide variety of species. The availability of fluorescent dextran conjugates with different sizes and charges permitted the analysis of direction and rate of axonal transport in the squid giant axon.[14]

Multiply Labeled Dextrans

Our fixable dextrans, most of which are lysinated dextrans (see the products marked by a single dagger † in Table 15.3), are generally preferred for neuronal tracing because they may transport more effectively and can be fixed in place with aldehydes after labeling. Molecular Probes prepares a number of multiply labeled dextrans that are fixable, including some that have acquired the distinction of unique names in various publications:

- **Fluoro-ruby** [10,13,15-19] — an orange-red fluorescent, fixable 10,000 MW dextran labeled with both tetramethylrhodamine and lysine (D-1817). 3000 MW, 40,000 MW, 70,000 MW and 2,000,000 MW versions of fluoro-ruby are also available (D-3308, D-1843, D-1818, D-7139).
- **Fluoro-emerald** [19-21] — a green fluorescent, fixable 10,000 MW dextran labeled with both fluorescein and lysine (D-1820). 3000 MW, 40,000 MW, 70,000 MW, 500,000 MW and 2,000,000 MW versions of fluoro-emerald are also available [22,23] (D-3306, D-1845, D-1822, D-7136, D-7137).
- **Micro-ruby** (D-7162) and **mini-ruby** [24,25] (D-3312) — orange-red fluorescent, fixable 3000 MW and 10,000 MW dextrans simultaneously labeled with tetramethylrhodamine, biotin and lysine. A 40,000 MW analog of this triply labeled dextran is also available (D-7133).
- **Micro-emerald** (D-7156) and **mini-emerald** (D-7178) — green fluorescent, fixable dextrans simultaneously labeled with fluorescein, biotin and lysine. A 70,000 MW analog is also available (D-7165).
- **Biotinylated dextran amine (BDA)** [5,26-33] — nonfluorescent, fixable dextrans simultaneously labeled with both biotin and lysine. These are available in several molecular weights (Table 15.3).

Fluoro-ruby and fluoro-emerald have been extensively employed for retrograde and anterograde neuronal tracing,[13] transplantation [17] and cell-lineage tracing.[16,18,19,34] Both products can be used to photoconvert DAB into an insoluble, electron-dense reaction product.[10] Like fluoro-ruby and fluoro-emerald, micro-ruby and mini-ruby are brightly fluorescent, making it easy to visualize the electrode during the injection process. In addition, because these dextrans include a covalently linked biotin, filled cells can be probed with standard enzyme-labeled avidin or streptavidin conjugates (see Section 7.5) to produce a permanent record of the experiment.[35] Mini-ruby has proven useful for intracellular filling in fixed brain slices [36] and has been reported to produce staining comparable to that achieved with lucifer yellow CH.[24] Moreover, the use of mini-ruby in conjunction with standard peroxidase-mediated avidin–biotin methods does not cause co-conversion of lipofuscin granules found in adult human brain, a common problem during photoconversion of lucifer yellow CH.[24] The lysine-fixable micro-emerald and mini-emerald — new dextrans that are triply labeled with fluorescein, biotin and lysine — provide a contrasting color that is better excited by the argon-ion laser of confocal laser scanning microscopes; they should have uses similar to micro-ruby and mini-ruby, respectively.

3000 MW Dextrans

The nominally 3000 MW dextrans offer several advantages over higher molecular weight dextrans, including faster axonal diffusion and greater access to peripheral cell processes [22] (see Color Plate 3 in "Visual Reality"). Our "3000 MW" dextran preparations contain polymers with molecular weight predominantly in the range of ~1500–3000 daltons, including the dye or other label. Our growing list of 3000 MW dextrans includes fluorescein, caged fluorescein, Rhodamine Green, tetramethylrhodamine, Texas Red and biotin conjugates. We also offer lysine-fixable 3000 MW dextrans that are simultaneously labeled with both fluorescein and biotin (micro-emerald, D-7156) or tetramethylrhodamine and biotin (micro-ruby, D-7162).

The 3000 MW fluorescein dextran and tetramethylrhodamine dextran (D-3306, D-3308) have been observed to readily undergo both anterograde and retrograde movement in live cells.[22,37] These 3000 MW dextrans appear to passively diffuse within the neuronal process, as their intracellular transport is not effectively inhibited by colchicine or nocodazole, both of which disrupt active transport by depolymerizing microtubules.[22] Moreover, these small dextrans diffuse at rates equivalent to those of smaller tracers such as sulforhodamine 101 and biocytin (~2 millimeters/hour at 22°C) and twice as fast as 10,000 MW dextrans. The relatively low molecular weight of the dextrans may result in transport of some labeled probes through gap junctions (see below).

NeuroTrace™ BDA-10,000 Neuronal Tracer Kit

Designed for both the first-time user and the experienced neuroscientist, our new NeuroTrace BDA-10,000 Neuronal Tracer Kit (N-7167) contains convenient amounts of each of the components required for neuroanatomical tracing using BDA methods, including:

- Lysine-fixable biotinylated 10,000 MW dextran amine (BDA-10,000)
- Avidin–horseradish peroxidase (avidin–HRP)
- 3,3'-Diaminobenzidine (DAB)
- Rigorously tested protocol that ensures fast, simple and inexpensive tracing experiments

The neuronal tracer BDA-10,000 is transported over long distances and fills fine processes bidirectionally, including boutons in the anterograde direction and dendritic structures in the retrograde direction.[5,30-33] Two days to two weeks after BDA-10,000 is injected into the desired region of the brain, the brain tissue is fixed and sectioned. BDA-10,000 can also be applied to cut nerves and allowed to transport. Following incubation with avidin–HRP and then DAB, the electron-dense DAB reaction product can be viewed by either light or electron microscopy[35] (Figure 15.5). The NeuroTrace BDA-10,000 labeling method can be readily combined with other anterograde or retrograde labeling methods or with immunohistochemical techniques. BDA-10,000 is available as a separate product (D-1956), as are BDA derivatives with other molecular weights — BDA-3000 (D-7135), BDA-70,000 (D-1957), BDA-500,000 (D-7142) and BDA-2,000,000 (D-7143).

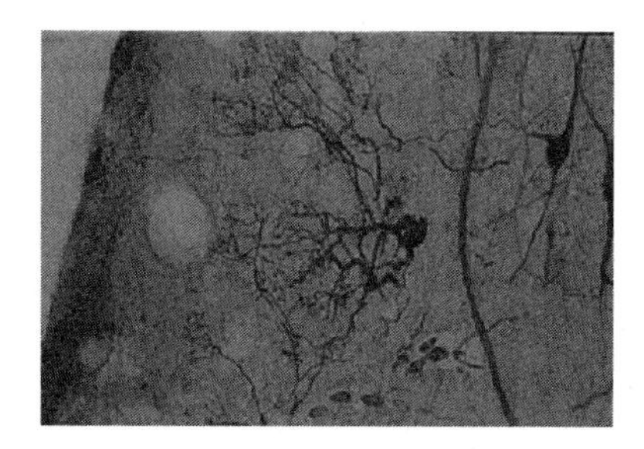

Figure 15.5 *A mitral cell in the olfactory bulb of a chinook salmon that has been retrogradely labeled using our NeuroTrace BDA-10,000 Neuronal Tracer Kit (N-7167).*

Cell Lineage Tracing with Dextrans

Fluorescent dextrans — particularly the fluorescein and rhodamine conjugates — have been used extensively for tracing cell lineage.[18,38,39] In this technique, the dextran is microinjected into a single cell of the developing embryo, and the fate of that cell and its daughters can be followed *in vivo* (see photo on back cover). Examples using this method include studies of:

- Dorsoventral axis determination in zebrafish mutants[40]
- Early cell fate commitment and lineage restrictions in developing zebrafish[41-43]
- Lineage and dopamine phenotype in tadpole hypothalamus[44]
- Migration of neural crest cells in *Xenopus*[45]
- Neural crest cell fate in zebrafish[46]
- Progeny tracing in the grasshopper neuroblast[5]

Developmental studies show that the lysinated fluorescent dextrans are also suitable for cell ablation studies, presumably through the generation of oxygen radicals.[47]

The lysine-fixable tetramethylrhodamine and Texas Red dextran conjugates (Table 15.3) are most frequently cited for lineage tracing studies; they are often preferred over other conjugates because they have bright fluorescence and are relatively photostable. As a second color, particularly in combination with the Texas Red dextrans, people have most often used our lysine-fixable fluorescein dextrans (e.g., D-3306, D-1820, D-1822).[48] However, the photostability of fluorescein conjugates is not as high as that of the tetramethylrhodamine and Texas Red conjugates. Consequently, we recommend our new green fluorescent, lysine-fixable Oregon Green 488 (D-7171, D-7173), Oregon Green 514 (D-7175, D-7177) and Rhodamine Green (D-7153, D-7163) dextran conjugates; see Figure 1.6 in Chapter 1, Figure 7.1 in Chapter 7 and Figure 11.1 in Chapter 11 for a comparison of the photostability of the various Oregon Green and Rhodamine Green dyes and fluorescein. The more photostable dextrans may also be less phototoxic in cells. While the lysine-fixable conjugates can be employed for long-term preservation of the tissue, some researchers prefer to co-inject a fluorescent, unlysinated dextran along with a nonfluorescent, lysine-fixable biotin dextran (BDA, Table 15.3). The nonfluorescent BDA can then be fixed in place with aldehyde fixatives and probed with any of our fluorescent or enzyme-labeled streptavidin and avidin conjugates described in Section 7.5.

High MW Dextran and Ficoll Conjugates

Our 500,000 and 2,000,000 MW fluorescent dextrans and 400,000 MW Ficoll conjugate (F-1965) may be particularly useful for lineage tracing at early stages of development, although these biopolymers have lower water solubility and a greater tendency to precipitate or clog microinjection needles than our lower molecular weight dextrans. Some studies suggest that lower molecular weight dextrans may leak from blastomeres, complicating the analysis. Injection of 2,000,000 MW fluorescein- and Texas Red–conjugated dextrans into separate cells of the two-cell stage zebrafish embryo allowed the construction of a fate map.[49] Our 500,000 MW and 2,000,000 MW dextrans are labeled with fluorescein, tetramethylrhodamine or Texas Red dyes or with biotin, and all contain aldehyde-fixable lysine groups. We also prepare one fluorescein dextran with a molecular weight of 5–40 million (D-1899). Our nonfluorescent 500,000 MW and 2,000,000 MW amino dextrans (D-7144, D-7145) can be conjugated with the researcher's choice of amine-reactive reagents.

Caged Fluorophore Dextrans

Dextrans with caged fluorophores are of particular interest to developmental biologists, because they can be injected early in development when the cells are large, and then later activated with UV illumination when the cells of interest may be small or buried in tissue (see Color Plate 3 in "Visual Reality"). A caged-fluorescein dextran conjugate has been used in this way to demonstrate lineage restriction boundaries in the early *Drosophila* embryo.[50] Five dextran conjugates of DMNB-caged fluorescein are available (Table 15.3). Two of these dextrans (D-7146, D-7147) have been further modified with both lysine and biotin to make them fixable, as well as detectable by avidin conjugates. See Chapter 19 for a discussion of caged probes and Section 15.3 for a description of our lower molecular weight caged tracers for microinjection.

Studying Intercellular Communication with Dextrans

The size exclusion properties of dextrans may be exploited to study connectivity between cells. Examples include studies of the passage of 3000 MW dextrans through plasmodesmata[23] and modulation of gap junctional communication by transforming growth factor–β_1 and forskolin.[51] However, the dispersion of molecular weights in our "3000 MW" dextran preparations, which contain polymers with total molecular weights predominantly in the range of ~1500–3000 daltons but may also contain molecules <1500 daltons, may complicate such analyses.

An important experimental design for identifying cells that form gap junctions makes use of simultaneous injection of the polar tracer lucifer yellow CH (~450 daltons) and a tetramethylrhodamine 10,000 MW dextran. Because low molecular weight tracers like lucifer yellow CH pass through gap junctions and dextrans do not, the injected cell exhibits red fluorescence; and cells connected through gap junctions have yellow fluorescence.[28,51] This technique has been used to follow the loss of intercellular communication in adenocarcinoma cells,[52] to show the re-establishment of communication during wound healing in *Drosophila*[53] and to investigate intercellular communication at different stages in *Xenopus* embryos.[54,55] Similar experiments could employ our lucifer yellow 10,000 MW dextran (D-1825) and a low molecular weight red fluorescent tracer such as sulforhodamine 101 (S-359, see Section 15.3) or a blue fluorescent dye such as Cascade Blue hydrazide (C-687, see Section 15.3). Fluorescein stachyose (S-6954, see Section 15.3), which has a well-defined molecular weight of 1146 daltons, is a sugar tetramer that may have the ideal size for studying the limits for passage of soluble tracers through gap junctions.

Probing Membrane Permeability with Dextrans

Labeled dextrans are often used to investigate the exclusion or transfer of macromolecules across cell membranes. For example, fluorescent dextrans have been used to monitor the effectiveness of electroporation, a technique that produces pores in the cell membrane, thus providing a convenient method for introducing materials such as exogenous DNA. Fluorescein dextrans with molecular weights ranging from 4000 to 150,000 daltons were used to determine the effect of electroporation variables — pulse size, shape and duration — on plasma-membrane pore size in chloroplasts,[56] red blood cells[57] and fibroblasts.[58] Fluorescence recovery after photobleaching (FRAP) techniques have been used to monitor nucleocytoplasmic transport of fluorescent dextrans of various molecular weights, allowing the determination of the size-exclusion limit of the nuclear pore membrane,[59-61] as well as to study the effect of epidermal growth factor and insulin on the nuclear membrane and nucleocytoplasmic transport.[62,63]

Microinjected 3000 MW fluorescent dextrans are reported to concentrate in interphase nuclei of *Drosophila* embryos, whereas 40,000 MW dextrans remain in the cytoplasm and enter the nucleus only after breakdown of the nuclear envelope during prophase. This size-exclusion phenomenon was used to follow the cyclical breakdown and reformation of the nuclear envelope during successive cell divisions.[64] Similarly, our 10,000 MW Calcium Green™ dextran conjugate (see Section 22.4) was shown to diffuse across the nuclear membrane of isolated nuclei from *Xenopus laevis* oocytes, but the 70,000 MW and 500,000 MW conjugates could not.[65] Significantly, depletion of nuclear Ca^{2+} stores by inositol 1,4,5-triphosphate (IP_3, I-3716; see Section 20.2), or by calcium chelators, blocked nuclear uptake of the 10,000 MW Calcium Green dextran conjugate but not entry of lucifer yellow CH. Fluorescent dextrans with molecular weights up to 20,000 daltons are reported to be taken up by the feeding tubes of nematodes but 40,000 MW and 70,000 MW dextrans are not.[66]

Following Endocytosis with Dextrans

Fluorescence of a number of the dyes that Molecular Probes uses to prepare its dextran conjugates is sensitive to the pH of the medium (see Chapter 23). Consequently, internalization of labeled dextrans into acidic organelles of cells can often be followed by measuring changes in the fluorescence of the dye.[67,68] Fluorescence of fluorescein-labeled dextrans is strongly quenched upon acidification (see Figure 23.2 in Chapter 23); however, fluorescein's lack of a spectral shift in acidic solution makes it difficult to discriminate between internalized probe that is quenched and the residual fluorescence of the external medium. Dextran conjugates that either shift their emission spectra, such as the SNARF and SNAFL® dextrans, or undergo significant shifts of their excitation spectra, such as BCECF, Oregon Green, HPTS, Rhodol Green, Cl-NERF and DM-NERF dextrans, are much more useful for following the internalization by ratio imaging (see the Technical Note "Loading and Calibration of Intracellular Ion Indicators" on page 549). All of our pH indicator conjugates and their optical responses are described in Section 23.4.

Discrimination of internalized fluorescent dextrans from external dextrans can be improved by adding a reagent that quenches the fluorescence of the external probe. For example, our anti-dye antibodies (see Section 7.3) usually quench the fluorescence of their conjugate dyes and may be useful for discriminating between externally bound dextrans and internalized dextrans. Although we are not aware of any publications that have used our ion indicator dextrans as probes for uptake by cells, it should be possible to follow the uptake of dextran conjugates of the fura-2, indo-1 and Calcium Green ion indicators (see Section 22.4) by using the paramagnetic Mn^{2+} ion to quench the fluorescence of the residual extracellular dextrans without affecting ingested dextrans.

Fluorescent dextrans can also be encapsulated in liposomes.[69-71] Using Texas Red– and fluorescein-labeled dextrans encapsulated in liposomes, researchers have obtained evidence that antigen processing occurs within dense lysosomes, rather than in earlier endocytic compartments.[72] Researchers have also used liposome-encapsulated fluorescent dextrans to investigate liposome fusion with isolated nuclei [73] and the effect of additives on vesicle size.[74]

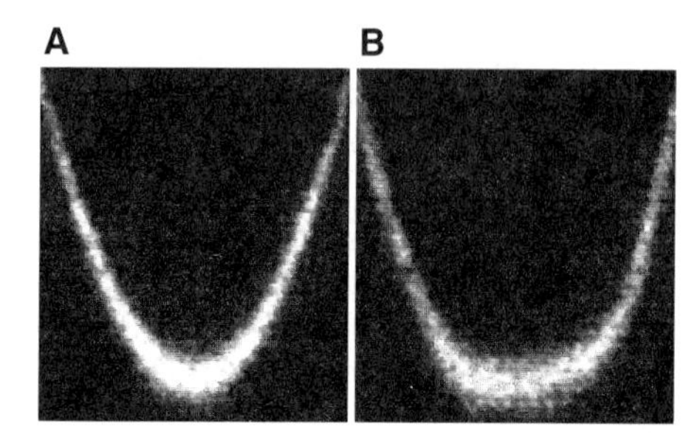

Figure 15.6 *A) Fluorescence imaging (at 490 nm excitation) of pipe flow after a thin horizontal uncaging beam (at 355 nm) was flashed into a transparent pipe containing a flowing aqueous solution of 10,000 MW dextran labeled with DMNB-caged fluorescein (D-3310). The pipe lies in the plane of the page, with the direction of flow from page top to page bottom; the UV light was focused on a slice of water in the pipe at the top of the photograph. The parabolic profile shows that water in the center of the pipe flows more rapidly than does water near the sides of the pipe. B) Same experiment but with flow rate nearly 3 times faster. Reprinted with permission from Exp in Fluids 18, 249 (1995).*

Tracing Fluid Transport with Dextrans

Fluorescent dextrans are important tools for studying the hydrodynamic properties of the cytoplasmic matrix. The intracellular mobility of these fluorescent tracers can be investigated using FRAP techniques. We offer a range of dextran sizes, thus providing a variety of hydrodynamic radii for investigating both the nature of the cytoplasmic matrix and the permeability of the surrounding membrane. Because of their solubility and biocompatibility, fluorescent dextrans have been used to monitor *in vivo* tissue permeability and flow in the uveoscleral tract,[75,76] capillaries [77,78] and proximal tubules,[79] as well as diffusion of high molecular weight substances in the brain's extracellular environment.[80] Fluorescent dextrans have also been used to assess permeability of the blood–brain barrier [81] and to monitor blood flow.[82]

DMNB-caged fluorescein dextrans (D-3309, D-3310, D-3311) and the double-labeled DMNB-caged fluorescein and biotin dextrans (D-7146, D-7147) are fluorescent only after UV photolysis,[83] enabling researchers to conduct photoactivation of fluorescence (PAF) experiments analogous to FRAP experiments in which the fluorophore is photoactivated upon illumination rather than bleached. Measuring the bright signal of the photoactivated fluorophore against a dark background should be intrinsically more sensitive than measuring a dark (photobleached) region against a bright field. In a collaboration with Walter Lempert of Princeton University, we have shown caged fluorescein dextran (D-3310) and caged HPTS (D-7060, see Section 15.3) to be excellent probes for tracing vortices in water using a technique called photoactivated nonintrusive tracking of molecular motion [84] (PHANTOMM) (Figure 15.6). Furthermore, diffusional coupling between dendritic spines and shafts was measured both by FRAP experiments with fluorescein dextran and by PAF experiments with DMNB-caged fluorescein dextran.[85] DMNB-caged fluorescein was also employed to evaluate a system that combined confocal laser scanning microscopy with focal photolysis of caged compounds.[86]

Our Dextran Bibliography

Our dextran bibliography contains over 450 references; it is available upon request from our Technical Assistance Department or through our Web site (http://www.probes.com); request bibliography D-8998. It includes references in which dextrans from several different sources were used. Because the source, molecular weight of the dextran, net charge, degree of substitution and nature of the dye may affect the application, the methods described in this section and in the references in our bibliography should be considered guides rather than definitive protocols. In most cases, however, our fluorescent dextrans are much brighter than dextrans available from other sources. Furthermore, we use rigorous methods for removing as much unconjugated dye as possible, and then assay our dextran conjugates by TLC to ensure the absence of low molecular weight dyes.

1. J Cell Sci 96, 721 (1990); **2.** J Cell Biol 105, 1981 (1987); **3.** Brain Res 526, 127 (1990); **4.** Brain Res Bull 25, 139 (1990); **5.** J Neurosci 14, 5766 (1994); **6.** J Pharm Sci 83, 1676 (1994); **7.** Plant Physiol 98, 673 (1992); **8.** Plant Physiol 93, 1492 (1990); **9.** Biotech Histochem 67, 153 (1992); **10.** J Histochem Cytochem 41, 777 (1993); **11.** Brain Res 30, 115 (1993); **12.** Trends Neurosci 13, 14 (1990); **13.** Brain Res 526, 127 (1990); **14.** Proc Natl Acad Sci USA 92, 11500 (1995); **15.** Mol Biol of the Cell 6, 1491 (1995); **16.** Nature 363, 630 (1993); **17.** Proc Natl Acad Sci USA 90, 1310 (1993); **18.** Nature 344, 431 (1990); **19.** Devel Biol 109, 509 (1985); **20.** J Cell Biol 128, 293 (1995); **21.** J Neurosci 53, 35 (1994); **22.** J Neurosci Meth 50, 95 (1993); **23.** Plant J 4, 567 (1993); **24.** J Neurosci Meth 55, 105 (1994); **25.** Brain Res 608, 78 (1993); **26.** J Neurosci 15, 5222 (1995); **27.** J Neurosci 15, 5139 (1995); **28.** J Neurosci Meth 53, 23 (1994); **29.** J Neurosci Meth 52, 153 (1994); **30.** Brain Res 607, 47 (1993); **31.** J Neurosci Meth 48, 75 (1993); **32.** J Neurosci Meth 45, 35 (1992); **33.** J Neurosci Meth 41, 239 (1992); **34.** Development 104 (suppl), 231 (1988); **35.** Reiner, A., Veenman, C.L. and Honig, M.G. in *Neuroscience Protocols*, F.G. Wouterlood, Ed., Elsevier Science Publishers (1993) 93-050-14-01–14; **36.** Brain Res 608, 78 (1993); **37.** J Neurosci Meth 53, 35 (1994); **38.** Science 252, 569 (1991); **39.** Development 108, 581 (1990); **40.** Nature 370, 468 (1994); **41.** Development 120, 483 (1994); **42.** Science 265, 517 (1994); **43.** Science 261, 109 (1993); **44.** J Neurosci 12, 1351 (1992); **45.** Development 118, 363 (1993); **46.** Development 120, 495 (1994); **47.** Devel Biol 120, 520 (1987); **48.** Devel Biol 114, 277 (1986); **49.** Nature 361, 451 (1993); **50.** Cell 68, 923 (1992); **51.** J Neurosci 15, 262 (1995); **52.** Proc Natl Acad Sci USA 85, 473 (1988); **53.** Devel Biol 127, 197 (1988); **54.** J Cell Biol 110, 115 (1990); **55.** Devel Biol 129, 265 (1988); **56.** Biophys J 58, 823 (1990); **57.** Bioelectrochem Bioenerget 20, 57 (1988); **58.** BioTechniques 6, 550 (1988); **59.** J Cell Biol 102, 1183 (1986); **60.** EMBO J 3, 1831 (1984); **61.** J Biol Chem 258, 11427 (1983); **62.** J Cell Biol 110, 559 (1990); **63.** Biochemistry 26, 1546 (1987); **64.** BioTechniques 17, 730 (1994); **65.** Science 270, 1835 (1995); **66.** Parasitology 109, 249 (1994); **67.** FASEB J 8, 573 (1994); **68.** J Cell Sci 105, 861 (1993); **69.** Agr Biol Chem 50, 399 (1986); **70.** FEBS

Lett 179, 148 (1985); **71.** J Cell Biol 32, 1069 (1983); **72.** Cell 64, 393 (1991); **73.** Biochemistry 26, 765 (1987); **74.** Biochemistry 29, 4582 (1990); **75.** Proc Natl Acad Sci USA 85, 2315 (1988); **76.** Arch Ophthalmol 105, 844 (1987); **77.** Microvascular Res 36, 172 (1988); **78.** Am J Physiol 245, H495 (1983); **79.** Am J Physiol 253, F366 (1987); **80.** Biophys J 65, 2277 (1993); **81.** Pflügers Arch 427, 86 (1994); **82.** J Cereb Blood Flow Metab 13, 359 (1993); **83.** Neuron 15, 755 (1995); **84.** Exp in Fluids 18, 249 (1995); **85.** Science 272, 716 (1996); **86.** Neuron 15, 755 (1995).

15.5 Price List *Fluorescent and Biotinylated Dextrans*

Cat #		Product Name	Unit Size	Price Per Unit ($) 1–4 Units	5–24 Units
D-3330		dextran, amino, 3000 MW	100 mg	55.00	44.00
D-1860		dextran, amino, 10,000 MW	1 g	75.00	60.00
D-1861		dextran, amino, 40,000 MW	1 g	75.00	60.00
D-1862		dextran, amino, 70,000 MW	1 g	75.00	60.00
D-7144	*New*	dextran, amino, 500,000 MW	100 mg	55.00	44.00
D-7145	*New*	dextran, amino, 2,000,000 MW	100 mg	55.00	44.00
D-7135	*New*	dextran, biotin, 3000 MW, lysine fixable (BDA-3000)	10 mg	125.00	100.00
D-7134	*New*	dextran, biotin, 3000 MW, neutral	10 mg	110.00	88.00
D-1956		dextran, biotin, 10,000 MW, lysine fixable (BDA-10,000)	25 mg	95.00	76.00
D-1856		dextran, biotin, 10,000 MW, neutral	100 mg	80.00	64.00
D-1957		dextran, biotin, 70,000 MW, lysine fixable (BDA-70,000)	25 mg	95.00	76.00
D-7142	*New*	dextran, biotin, 500,000 MW, lysine fixable (BDA-500,000)	10 mg	110.00	88.00
D-7143	*New*	dextran, biotin, 2,000,000 MW, lysine fixable (BDA-2,000,000)	10 mg	110.00	88.00
D-7168	*New*	dextran, BODIPY® FL, 10,000 MW, fixable	5 mg	110.00	88.00
D-7131	*New*	dextran, Cascade Blue®, 3000 MW, anionic	10 mg	110.00	88.00
D-7132	*New*	dextran, Cascade Blue®, 3000 MW, anionic, lysine fixable	10 mg	125.00	100.00
D-1974		dextran, Cascade Blue®, 10,000 MW, anionic	25 mg	80.00	64.00
D-1976		dextran, Cascade Blue®, 10,000 MW, anionic, lysine fixable	25 mg	95.00	76.00
D-1975		dextran, Cascade Blue®, 70,000 MW, anionic	25 mg	80.00	64.00
D-1977		dextran, Cascade Blue®, 70,000 MW, anionic, lysine fixable	25 mg	95.00	76.00
D-3309		dextran, DMNB-caged fluorescein, 3000 MW, anionic	5 mg	125.00	100.00
D-3310		dextran, DMNB-caged fluorescein, 10,000 MW, anionic	5 mg	110.00	88.00
D-3311		dextran, DMNB-caged fluorescein, 70,000 MW, anionic	5 mg	110.00	88.00
D-7146	*New*	dextran, DMNB-caged fluorescein and biotin, 10,000 MW, lysine fixable	5 mg	150.00	120.00
D-7147	*New*	dextran, DMNB-caged fluorescein and biotin, 70,000 MW, lysine fixable	5 mg	150.00	120.00
D-7166	*New*	dextran, eosin, 10,000 MW, anionic, lysine fixable	25 mg	110.00	88.00
D-3305		dextran, fluorescein, 3000 MW, anionic	10 mg	110.00	88.00
D-3306		dextran, fluorescein, 3000 MW, anionic, lysine fixable	10 mg	125.00	100.00
D-1821		dextran, fluorescein, 10,000 MW, anionic	25 mg	80.00	64.00
D-1820		dextran, fluorescein, 10,000 MW, anionic, lysine fixable (fluoro-emerald)	25 mg	95.00	76.00
D-1814		dextran, fluorescein, 10,000 MW, polyanionic	25 mg	80.00	64.00
D-1844		dextran, fluorescein, 40,000 MW, anionic	25 mg	80.00	64.00
D-1845		dextran, fluorescein, 40,000 MW, anionic, lysine fixable	25 mg	110.00	88.00
D-1823		dextran, fluorescein, 70,000 MW, anionic	25 mg	80.00	64.00
D-1822		dextran, fluorescein, 70,000 MW, anionic, lysine fixable	25 mg	110.00	88.00
D-1815		dextran, fluorescein, 70,000 MW, polyanionic	25 mg	80.00	64.00
D-7136	*New*	dextran, fluorescein, 500,000 MW, anionic, lysine fixable	10 mg	110.00	88.00
D-7137	*New*	dextran, fluorescein, 2,000,000 MW, anionic, lysine fixable	10 mg	110.00	88.00
D-1899		dextran, fluorescein, 5-40 million MW, anionic	25 mg	80.00	64.00
D-7156	*New*	dextran, fluorescein and biotin, 3000 MW, anionic, lysine fixable (micro-emerald)	5 mg	150.00	120.00
D-7178	*New*	dextran, fluorescein and biotin, 10,000 MW, anionic, lysine fixable (mini-emerald)	10 mg	150.00	120.00
D-7165	*New*	dextran, fluorescein and biotin, 70,000 MW, anionic, lysine fixable	10 mg	150.00	120.00
D-1825		dextran, lucifer yellow, 10,000 MW, anionic, lysine fixable	25 mg	80.00	64.00
D-1826		dextran, lucifer yellow, 40,000 MW, anionic, lysine fixable	25 mg	80.00	64.00
D-1827		dextran, lucifer yellow, 70,000 MW, anionic, lysine fixable	25 mg	80.00	64.00
D-7170	*New*	dextran, Oregon Green™ 488; 10,000 MW, anionic	5 mg	110.00	88.00
D-7171	*New*	dextran, Oregon Green™ 488; 10,000 MW, anionic, lysine fixable	5 mg	125.00	100.00
D-7172	*New*	dextran, Oregon Green™ 488; 70,000 MW, anionic	5 mg	110.00	88.00
D-7173	*New*	dextran, Oregon Green™ 488; 70,000 MW, anionic, lysine fixable	5 mg	125.00	100.00
D-7174	*New*	dextran, Oregon Green™ 514; 10,000 MW, anionic	5 mg	110.00	88.00
D-7175	*New*	dextran, Oregon Green™ 514; 10,000 MW, anionic, lysine fixable	5 mg	125.00	100.00
D-7176	*New*	dextran, Oregon Green™ 514; 70,000 MW, anionic	5 mg	110.00	88.00
D-7177	*New*	dextran, Oregon Green™ 514; 70,000 MW, anionic, lysine fixable	5 mg	125.00	100.00
D-1824		dextran, rhodamine B, 10,000 MW, neutral	25 mg	80.00	64.00
D-1840		dextran, rhodamine B, 40,000 MW, neutral	25 mg	80.00	64.00
D-1841		dextran, rhodamine B, 70,000 MW, neutral	25 mg	80.00	64.00

Cat #		Product Name	Unit Size	Price Per Unit ($) 1–4 Units	5–24 Units
D-1813		dextran, rhodamine B, 70,000 MW, polyanionic	25 mg	80.00	64.00
D-7163	*New*	dextran, Rhodamine Green™, 3000 MW	5 mg	125.00	100.00
D-7152	*New*	dextran, Rhodamine Green™, 10,000 MW	10 mg	110.00	88.00
D-7153	*New*	dextran, Rhodamine Green™, 10,000 MW, lysine fixable	10 mg	150.00	120.00
D-7154	*New*	dextran, Rhodamine Green™, 70,000 MW	10 mg	110.00	88.00
D-7148	*New*	dextran, Rhodol Green™, 10,000 MW	10 mg	110.00	88.00
D-7150	*New*	dextran, Rhodol Green™, 70,000 MW	10 mg	110.00	88.00
D-1890		dextran sulfate, fluorescein, 40,000 MW, anionic	25 mg	80.00	64.00
D-1891		dextran sulfate, fluorescein, 500,000 MW, anionic	25 mg	80.00	64.00
D-6929	*New*	dextran, tetrakis-(4-carboxyphenyl)porphine, 10,000 MW, anionic	10 mg	95.00	76.00
D-6930	*New*	dextran, tetrakis-(4-carboxyphenyl)porphine, 70,000 MW, anionic	10 mg	95.00	76.00
D-3307		dextran, tetramethylrhodamine, 3000 MW, anionic	10 mg	110.00	88.00
D-3308		dextran, tetramethylrhodamine, 3000 MW, anionic, lysine fixable	10 mg	125.00	100.00
D-1868		dextran, tetramethylrhodamine, 10,000 MW, anionic, fixable	25 mg	80.00	64.00
D-1817		dextran, tetramethylrhodamine, 10,000 MW, lysine fixable (fluoro-ruby)	25 mg	110.00	88.00
D-1816		dextran, tetramethylrhodamine, 10,000 MW, neutral	25 mg	80.00	64.00
D-1843		dextran, tetramethylrhodamine, 40,000 MW, lysine fixable	25 mg	110.00	88.00
D-1842		dextran, tetramethylrhodamine, 40,000 MW, neutral	25 mg	80.00	64.00
D-1818		dextran, tetramethylrhodamine, 70,000 MW, lysine fixable	25 mg	110.00	88.00
D-1819		dextran, tetramethylrhodamine, 70,000 MW, neutral	25 mg	80.00	64.00
D-7139	*New*	dextran, tetramethylrhodamine, 2,000,000 MW, lysine fixable	10 mg	110.00	88.00
D-7162	*New*	dextran, tetramethylrhodamine and biotin, 3000 MW, lysine fixable (micro-ruby)	5 mg	150.00	120.00
D-3312		dextran, tetramethylrhodamine and biotin, 10,000 MW, lysine fixable (mini-ruby)	10 mg	150.00	120.00
D-7133	*New*	dextran, tetramethylrhodamine and biotin, 40,000 MW, lysine fixable	10 mg	150.00	120.00
D-3328		dextran, Texas Red®, 3000 MW, lysine fixable	10 mg	125.00	100.00
D-3329		dextran, Texas Red®, 3000 MW, neutral	10 mg	110.00	88.00
D-1863		dextran, Texas Red®, 10,000 MW, lysine fixable	25 mg	110.00	88.00
D-1828		dextran, Texas Red®, 10,000 MW, neutral	25 mg	80.00	64.00
D-1829		dextran, Texas Red®, 40,000 MW, neutral	25 mg	80.00	64.00
D-1864		dextran, Texas Red®, 70,000 MW, lysine fixable	25 mg	110.00	88.00
D-1830		dextran, Texas Red®, 70,000 MW, neutral	25 mg	80.00	64.00
D-7141	*New*	dextran, Texas Red®, 2,000,000 MW, lysine fixable	10 mg	110.00	88.00
F-1965		Ficoll®, fluorescein, 400,000 MW, anionic	25 mg	80.00	64.00
N-7167	*New*	NeuroTrace™ BDA-10,000 Neuronal Tracer Kit	1 kit	150.00	120.00

15.6 FluoSpheres and TransFluoSpheres Fluorescent Polystyrene Microspheres for Tracing

Fluorescent and biotinylated FluoSpheres and TransFluoSpheres polystyrene microspheres satisfy several prerequisites of ideal long-term biological tracers. The fluorescence output per microsphere is significantly greater than that obtained from protein or dextran conjugates (see Table 6.2 in Chapter 6) and is relatively immune to photobleaching and other environment-dependent effects. FluoSpheres and TransFluoSpheres microspheres are also biologically inert and physically durable, and they are available from Molecular Probes with a large number of uniform sizes and surface properties. Furthermore, their spectral properties can be freely manipulated during manufacture without altering their surface properties. Our current carboxylate-modified fluorescent microspheres have much higher surface charges than were previously available, making them more hydrophilic and thus more useful as tracers. The additional carboxylate groups also make it easier to couple the microspheres to proteins and other biomolecules. See Section 6.2 for an extensive discussion of the properties of our FluoSpheres and TransFluoSpheres polystyrene beads.

Availability of intensely fluorescent, highly uniform microspheres in different colors and sizes permits diverse applications in tracking particles and cells, tracing fluid dynamics and amplifying signals. In addition, measuring the effect of various interventions on regional blood flow is an important quantitative application of fluorescent microspheres. Using a mixture of beads of different sizes, each labeled with a different fluorescent color, researchers can discriminate the size dependence of uptake or transport of microspheres *in vivo* in cells, capillaries, lung or other tissues. Our smallest microspheres can be microinjected into cells or are actively taken up by phagocytosis.

Note that we have renamed our FluoSpheres and TransFluoSpheres since publication of the Fifth Edition of this Handbook and that this has required us to renumber all products. In some cases, we have also changed the wavelengths listed in the products' names to more accurately reflect the measured wavelengths. If you have any questions about the new catalog numbers, please call our Customer Service Department and request "New Names and Catalog

Numbers for Molecular Probes' FluoSpheres® and TransFluoSpheres® Fluorescent Microspheres" (Technical Bulletin X 5000).

Fluorescent Microspheres for Regional Blood Flow Studies

Relatively large radiolabeled microspheres (10–15 µm in diameter) have long been used for regional blood flow studies in tissues. The microspheres are injected at desired locations in the circulatory system and eventually lodge in the capillaries, where they can later be counted in dissected tissue sections. To eliminate the hazards, expense and disposal problems of the radiolabeled microspheres,[1] researchers have turned to fluorescent and colored microspheres for measuring myocardial and cortical blood flow.[2-7] In addition, blood flow in brain cortex has been measured using much smaller microspheres (1.3 µm) by determining velocities of individual fluorescent beads with strobe epi-illumination.[8] Blood flow measurements using fluorescent microspheres in other organs, including the kidney,[9] lung [10] and pancreas,[11] are equally feasible. Fluorescent dextrans (see Section 15.5) also have been used to make direct measurement of capillary transit times using intravital microscopy.[12,13]

FluoSpheres Microspheres for Blood Flow Determination

Molecular Probes has used its proprietary fluorescent dye technology to produce a range of intensely fluorescent FluoSpheres microspheres specifically designed for regional blood flow determination [14] (Figure 15.7). Regional blood flow studies using our FluoSpheres polystyrene microspheres for blood flow determination, which are available in seven fluorescent colors, have been validated in three side-by-side comparisons with radioactively labeled microspheres.[4,15,16] The two methods exhibit equivalent detection sensitivity, and excellent correlation between the flow measurements has been reported. In addition, techniques have been developed to extract the microspheres and the fluorescent dyes they contain from tissue samples, allowing blood flow quantitation to be performed using readily available instrumentation such as spectrofluorometers and fluorescence microplate readers;[2] request our free technical bulletin entitled "New Sedimentation Method for Determination of Organ Blood Flow with Fluorescent Microspheres" (Technical Bulletin X 8829) from our Technical Assistance Department or through our Web site (http://www.probes.com). Our FluoSpheres microspheres for blood flow determination are also compatible with new blood flow analyzer systems for automated extraction and analysis.

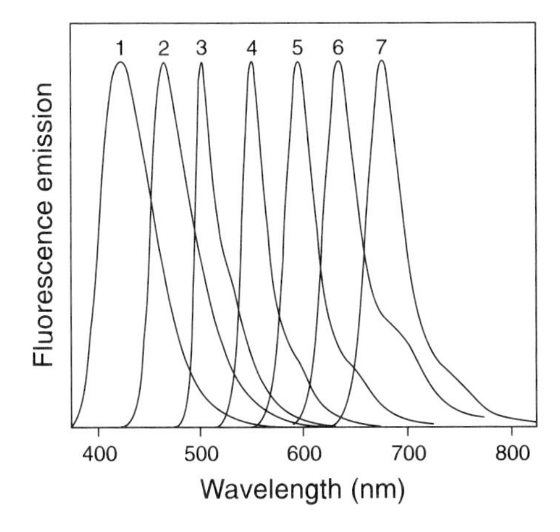

Figure 15.7 Normalized fluorescence emission spectra of the dyes contained in the polystyrene microsphere components of the FluoSpheres Blood Flow Determination Fluorescent Color Kit #2 (F-8891), after extraction into 2-ethoxyethyl acetate (Cellosolve® acetate). The seven colors of fluorescent microspheres represented are: 1) blue, 2) blue-green, 3) yellow-green, 4) orange, 5) red, 6) crimson and 7) scarlet.

FluoSpheres Color Kits for Regional Blood Flow Studies

Molecular Probes offer four different FluoSpheres Blood Flow Determination Fluorescent Color Kits:

- Kit #1 (F-8890) contains 10 mL vials of 10 µm microspheres in seven fluorescent colors (blue, blue-green, yellow-green, orange, red, crimson and scarlet).
- Kit #2 (F-8891) contains 10 mL vials of 15 µm microspheres in seven fluorescent colors (blue, blue-green, yellow-green, orange, red, crimson and scarlet).
- Kit #3 (F-8892) contains 10 mL vials of 15 µm microspheres in five fluorescent colors (blue-green, yellow-green, orange, red and crimson).
- Kit #4 (F-8893) is a sampler kit for initial testing purposes or use in smaller animals; it contains 2 mL vials of the same five fluorescent color as provided in Kit #3.

The aqueous suspensions of 10 µm and 15 µm beads contain 3.6 million and 1 million microspheres per mL, respectively. All kits include a detailed protocol for their use. Each of the colors of FluoSpheres microspheres for regional blood flow studies (Figure 15.7) can also be purchased separately (see the product list at the end of this section).

Fluorescent Microsphere Resource Center

Additional technical support, including a detailed applications manual on the use of fluorescent microspheres for blood flow determination, is available from the Fluorescent Microsphere Resource Center (FMRC) at the University of Washington, Seattle. The FMRC can be contacted by phone (206) 685–9479 or fax (206) 685–8673; information can also be obtained through a filter transfer protocol (FTP) server (fmrc.pulmcc.washington.edu), the internet (glenny@pele.pulmcc.washington.edu) or their Web site (http://fmrc.pulmcc.washington.edu/fmrc/fmrc.html).

Particle and Cell Tracking with Fluorescent Microspheres

In addition to our microspheres specially designed for blood flow studies, Molecular Probes offers a wide range of fluorescent FluoSpheres carboxylated-modified microspheres in different fluorescent colors, bead diameters and surface functional groups (see Section 6.2 for a complete list). Unlike our other fluorescent microspheres, most of which come in aqueous suspensions containing 2% solids and 2 mM sodium azide as a preservative, FluoSpheres beads with 0.04 µm diameters are now prepared as aqueous suspensions containing 5% solids without preservatives. At more than double the concentration of our standard FluoSpheres microspheres, these carboxylate-modified 0.04 µm beads are well suited to applications requiring microinjectable tracers. Yellow-green, orange, red, crimson and dark red fluorescent colors are available either individually (F-8795, F-8792, F-8793, F-8788, F-8789) or in the FluoSpheres Fluorescent Color Kit, which contains all five colors (F-8889). In addition, red-orange, far red and infrared fluorescent colors are available individually (F-8794, F-8790, F-8791). Use of the dark red, far red and infrared fluorescent FluoSpheres microspheres should permit greater light penetration through tissues. In many biological systems, the concentrated fluorescence and spherical shape of the FluoSpheres beads permit them to be detected against a relatively high but diffuse background fluorescence. However, our TransFluoSpheres microspheres, which have extremely large Stokes shifts, are preferred for some studies because their fluorescence may be better resolved from the tissue's autofluorescence (see Figure 6.3 in Chapter 6).

Transplantation and Migration Studies with Fluorescent Microspheres

Labeling cells with fluorescent microspheres prior to transplantation enables researchers to distinguish cell types and analyze graft migration in the host over long periods of time. Unlike other tracers, most of which rapidly diffuse or leach from their site of application in tissues, fluorescent microspheres tend to remain in place for periods of at least months. Their stability and biological inertness give them considerable potential for transplantation studies. Moreover, the intense fluorescence, high uniformity and low debris content of our FluoSpheres polystyrene microspheres may circumvent many of the problems encountered in use of fluorescent microspheres from other sources. Cells are generally labeled with microspheres using microinjection or other invasive methods (Table 15.1). Potential problems of sterility and unequal uptake and translocation of the beads by the cells need to be considered before using fluorescent microspheres in transplantation and other cell tracing studies.[17]

Some published transplantation applications for fluorescent microspheres obtained from various sources include studies designed to trace:

- Cholinergic neurons transplanted into an embryo [18]
- Implantation of Weaver granule cell precursors into wild-type cerebellar cortex [19]
- Microsphere-labeled myoblasts following re-injection into dystrophin-deficient mice [20]
- Migration of neutrophils from lung to tracheobronchial lymph node [21]
- Migration of transplanted neurons into regions of continuing neuronal degeneration in adult mice [22]
- Projection of efferent fibers in neural grafts [23]
- Transplantation of fetal rat brain cells into host brains [24]

Several other probes from Molecular Probes may also be useful for transplantation studies, including CellTracker dyes (see Section 15.2), the succinimidyl ester of carboxyfluorescein diacetate [25,26] (5(6)-CFDA, SE; C-1157; see Section 15.2), fluorescent dextrans [27] (see Section 15.5) and lipophilic tracers (see Section 15.4).

Neuronal Tracing with Fluorescent Microspheres

Katz was the first to use fluorescent microspheres as neuronal tracers, demonstrating that rhodamine-labeled microspheres could be retrogradely transported [28] Although the mechanism of microsphere transport is not completely understood, the process can apparently be facilitated by using high concentration of particles with small diameters (<0.05 µm) and high negative surface-charge densities.[29-31] Fluorescent microspheres are also suitable for retrograde tracing because they are not cytotoxic and persist for extraordinarily long periods in nerve cells. The fluorescence intensity of microsphere-labeled neuronal perikarya in rats was found to be undiminished one year after injection.[32]

Our red-fluorescent, carboxylate-modified FluoSpheres beads (F-8793) have been shown to be retrogradely transported in rat lumbosacral ventral root axons that had been subjected to peripheral crush injury.[33] In this study, the fluorescent beads were used to convert diaminobenzidine (DAB) into an insoluble, electron-dense reaction product in order to facilitate ultrastructural analysis by electron microscopy.[33] Because retrograde transport of fluorescent microspheres appears to be extremely sensitive to the microsphere size, surface-charge density and concentration, references for neuronal tracing in our microsphere bibliography (see below) may not always be applicable to our current products.

Following Phagocytosis with Fluorescent Microspheres

Many cell types actively ingest opsonized or nonopsonized fluorescent microspheres (see also Section 17.1). The preferred microspheres are about 0.5–2 µm in diameter. Confocal laser scanning microscopy can distinguish between ingested beads and those simply bound to the surface.[34] Flow cytometry, fluorescence microscopy and fluorescence spectrometry have been used to quantitate phagocytosis by macrophage cells [35-39] and protozoan cells.[40] Macrophage cells in primary cultures of rat cerebral cortex have been identified by their ability to selectively phagocytose fluorescent microspheres;[41] macrophage cells have also been sorted based on the absolute number of microspheres phagocytosed.[39] Fluorescent microsphere uptake has been used as a model for alveolar macrophage cell translocation and clearance of inhaled aerosols containing environmental particulates.[42-44] As with our fluorescent microspheres for blood flow determination (see above), solvent extraction of the fluorescent dye from the beads can be used to quantitate microsphere uptake by macrophage cells.[45]

Tracking the Movement of Proteins and Other Biomolecules with Fluorescent Microspheres

FluoSpheres microspheres can serve as bright, inert and extremely photostable labels for tracking particle movement and other dynamic processes over extended periods of time. The intense fluorescence of our FluoSpheres beads permits the detection and tracking of very small single particles in three dimensions.[46-48] Examples of the diverse applications of fluorescent microspheres include studies designed to track or quantitate:

- Binding of kinetochores to microtubules *in vitro* [48,49]
- Brownian motion in protein [50] and dextran solutions [51]
- Exchangeable GTP-binding site on paclitaxel-stabilized fluorescent microtubules [52]
- Direction and rate of axonal transport in the squid giant axon [53]
- Fluid dynamics and flow in brain arterioles [54] and peripheral lymph [55]
- Injection sites in tissues [56,57]
- Lateral diffusion of lipids and receptors in membranes [47,58-60]
- Movement of microinjected microspheres poleward during mitosis of sea urchin eggs [61] and sand dollar eggs [62]

- Particle flow through a model biofilm containing *Pseudomonas aeruginosa*, *P. fluorescens* and *Klebsiella pneumoniae* using confocal laser scanning microscopy [63] (Figure 15.8)
- Size dependence of particle-uptake and particle-distribution measurements in rat gastrointestinal mucosa [64]
- Three-dimensional motion of microspheres in order to assess water permeability in individual Chinese hamster ovary (CHO) cells expressing CHIP28 water channels [46]
- Uptake of seawater by starfish [65]

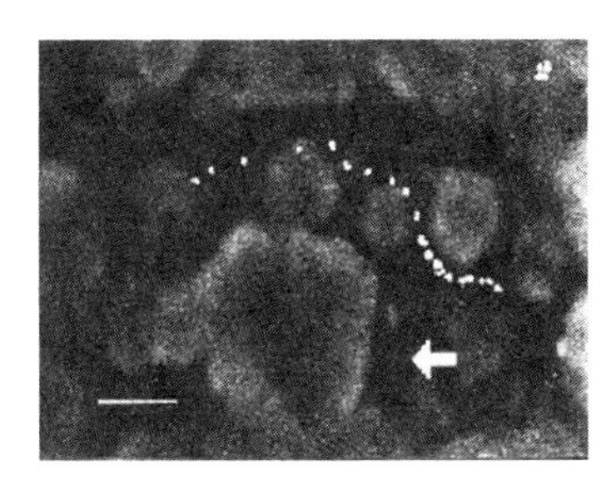

Figure 15.8 Superimposed time sequence image (2.3-second intervals), showing a single bead moving through a biofilm channel. The cell clusters autofluoresced and can be seen as lighter areas relative to the channels. The direction of bulk fluid flow is indicated by the arrow. Photo contributed by Paul Stoodley, Exeter University, and Dirk DeBeer, Max Planck Institute of Marine Biology.

Fluorescent microspheres have also been used with optical tweezers to control the movement of single myosin filaments [66] or kinesin molecules [67] and for imaging at suboptical resolution by scanning luminescence X-ray microscopy [68] In addition, our 0.1 µm red fluorescent carboxylate-modified microspheres (F-8801, see Section 6.2) have been used to model particle distribution and penetration of adenovirus-mediated gene transfer in human bronchial submucosal glands using xenografts.[69]

Our Microsphere Bibliography

Our microsphere bibliography contains over 250 references and is available upon request from our Technical Assistance Department or through our Web site (http://www.probes.com); request bibliography M-8997. It includes references in which microspheres from several different sources were used. Because the source, surface properties and size of the microspheres may affect the application, the methods described in these references should be considered guides rather than definitive protocols. In particular, the bibliography lists several references that cite the use of fluorescent microspheres for retrograde tracing, and our FluoSpheres beads have been inconsistent in this application. However, we have found our fluorescent microspheres to be significantly brighter than microspheres available from other sources (see Table 6.1 in Chapter 6).

1. Circulation Res 21, 163 (1967); **2.** Am J Physiol 269, H725 (1995); **3.** Am J Cardiovascular Pathol 4, 352 (1993); **4.** Circulatory Shock 41, 156 (1993); **5.** Circulation 83, 974 (1991); **6.** J Autonomic Nervous Sys 30, 159 (1990); **7.** Am J Physiol 251, H863 (1986); **8.** J Cereb Blood Flow Metab 13, 359 (1993); **9.** Kidney Int'l 20, 230 (1981); **10.** Arch Int'l Physiol Biochim Biophys 100, 263 (1992); **11.** Am J Physiol 265, G587 (1993); **12.** Acta Radiologica 35, 176 (1994); **13.** Science 218, 379 (1982); **14.** Molecular Probes is licensed to sell FluoSpheres microspheres for blood flow determination under patents owned by Triton Technology, Inc., including U.S. Patent Nos. 5,187,288; 5,230,343; 5,248,782 and 5,253,649; German Patent No. 40 19 025 and other U.S. and foreign patents pending; purchase of these products is accompanied by a license to use them in the practice of dye-extraction blood flow technology; **15.** Cardiovascular Res 30, 405 (1995); **16.** J Appl Physiol 74, 2585 (1993); **17.** J Neurosci Meth 44, 7 (1992); **18.** Nature 324, 569 (1986); **19.** Science 260, 367 (1993); **20.** J Histochem Cytochem 41, 1579 (1993); **21.** J Leukocyte Biol 41, 95 (1987); **22.** J Neurosci 15, 8378 (1995); **23.** Brain Res 487, 225 (1989); **24.** J Neurosci Meth 27, 121 (1989); **25.** Cell Transplantation 3, 397 (1994); **26.** J Cell Biol 103, 2649 (1986); **27.** Proc Natl Acad Sci USA 90, 1310 (1993); **28.** Nature 310, 498 (1984); **29.** Brain Res 522, 90 (1990); **30.** J Neurosci Meth 29, 1 (1989); **31.** J Neurosci Meth 24, 1 (1988); **32.** Brain Res 524, 339 (1990); **33.** Brain Res 630, 115 (1993); **34.** J Leukocyte Biol 45, 277 (1989); **35.** Cell Immunol 156, 508 (1994); **36.** Cytometry 12, 677 (1991); **37.** J Cell Biol 115, 59 (1991); **38.** J Immunol Meth 88, 175 (1986); **39.** Science 215, 64 (1982); **40.** Cytometry 11, 875 (1990); **41.** J Neurosci Res 26, 74 (1990); **42.** Cytometry 20, 23 (1995); **43.** Microscopy Res Technique 26, 437 (1993); **44.** Science 230, 1277 (1985); **45.** Anal Biochem 152, 167 (1986); **46.** Biophys J 67, 1291 (1994); **47.** J Cell Biol 127, 963 (1994); **48.** Nature 359, 533 (1992); **49.** J Cell Biol 127, 995 (1994); **50.** J Cell Biol 110, 1645 (1990); **51.** Macromolecules 24, 599 (1991); **52.** Science 261, 1044 (1993); **53.** Proc Natl Acad Sci USA 92, 11500 (1995); **54.** Microvascular Res 43, 235 (1992); **55.** Lymphology 27, 108 (1994); **56.** J Neurosci 15, 4209 (1995); **57.** J Appl Physiol 68, 1157 (1990); **58.** J Membrane Biol 144, 231 (1995); **59.** Meth Neurosci 19, 320 (1994); **60.** J Membrane Biol 135, 83 (1993); **61.** Cell Motil Cytoskeleton 8, 293 (1987); **62.** Devel Growth Differ 28, 461 (1986); **63.** Appl Environ Microbiol 60, 2711 (1994); **64.** J Pharm Pharmacol 42, 821 (1990); **65.** J Exp Zoology 255, 262 (1990); **66.** Biophys J 66, 769 (1994); **67.** Science 260, 232 (1993); **68.** J Microscopy 172, 121 (1993); **69.** Am J Physiol 268, L657 (1995).

15.6 Price List *FluoSpheres and TransFluoSpheres Fluorescent Polystyrene Microspheres for Tracing*

Cat #		Product Name	Unit Size	Price Per Unit ($) 1–4 Units	5–24 Units
F-8890		FluoSpheres® Blood Flow Determination Fluorescent Color Kit #1, polystyrene microspheres, 10 µm *seven colors, 10 mL each* *3.6x10⁶ beads/mL*	1 kit	775.00	620.00
F-8891		FluoSpheres® Blood Flow Determination Fluorescent Color Kit #2, polystyrene microspheres, 15 µm *seven colors, 10 mL each* *1.0x10⁶ beads/mL*	1 kit	775.00	620.00
F-8892	**New**	FluoSpheres® Blood Flow Determination Fluorescent Color Kit #3, polystyrene microspheres, 15 µm *five colors, 10 mL each* *1.0x10⁶ beads/mL*	1 kit	600.00	480.00
F-8893	**New**	FluoSpheres® Blood Flow Determination Fluorescent Color Kit #4, polystyrene microspheres, 15 µm *five colors, 2 mL each* *1.0x10⁶ beads/mL*	1 kit	165.00	132.00
F-8789		FluoSpheres® carboxylate-modified microspheres, 0.04 µm, dark red fluorescent (660/680) *5% solids, azide free*	1 mL	135.00	108.00
F-8790	**New**	FluoSpheres® carboxylate-modified microspheres, 0.04 µm, far red fluorescent (690/720) *5% solids, azide free*	0.4 mL	135.00	108.00

Cat #		Product Name	Unit Size	Price Per Unit ($) 1–4 Units	5–24 Units
F-8791	New	FluoSpheres® carboxylate-modified microspheres, 0.04 µm, infrared fluorescent (715/755) *5% solids, azide free*	0.4 mL	135.00	108.00
F-8792		FluoSpheres® carboxylate-modified microspheres, 0.04 µm, orange fluorescent (540/560) *5% solids, azide free*	1 mL	135.00	108.00
F-8793		FluoSpheres® carboxylate-modified microspheres, 0.04 µm, red fluorescent (580/605) *5% solids, azide free*	1 mL	135.00	108.00
F-8794	New	FluoSpheres® carboxylate-modified microspheres, 0.04 µm, red-orange fluorescent (565/580) *5% solids, azide free*	1 mL	135.00	108.00
F-8795		FluoSpheres® carboxylate-modified microspheres, 0.04 µm, yellow-green fluorescent (505/515) *5% solids, azide free*	1 mL	135.00	108.00
F-8788		FluoSpheres® carboxylate-modified microspheres, 0.04 µm, crimson fluorescent (625/645) *5% solids, azide free*	1 mL	135.00	108.00
F-8889		FluoSpheres® Fluorescent Color Kit, carboxylate-modified microspheres, 0.04 µm *five colors, 1 mL each* *5% solids, azide free*	1 kit	295.00	236.00
F-8829		FluoSpheres® polystyrene microspheres, 10 µm, blue fluorescent (365/415) *for blood flow determination* *3.6x10^6 beads/mL*	10 mL	150.00	120.00
F-8830	New	FluoSpheres® polystyrene microspheres, 10 µm, blue-green fluorescent (430/465) *for blood flow determination* *3.6x10^6 beads/mL*	10 mL	150.00	120.00
F-8831		FluoSpheres® polystyrene microspheres, 10 µm, crimson fluorescent (625/645) *for blood flow determination* *3.6x10^6 beads/mL*	10 mL	150.00	120.00
F-8833		FluoSpheres® polystyrene microspheres, 10 µm, orange fluorescent (540/560) *for blood flow determination* *3.6x10^6 beads/mL*	10 mL	150.00	120.00
F-8834		FluoSpheres® polystyrene microspheres, 10 µm, red fluorescent (580/605) *for blood flow determination* *3.6x10^6 beads/mL*	10 mL	150.00	120.00
F-8835	New	FluoSpheres® polystyrene microspheres, 10 µm, scarlet fluorescent (645/680) *for blood flow determination* *3.6x10^6 beads/mL*	10 mL	150.00	120.00
F-8836		FluoSpheres® polystyrene microspheres, 10 µm, yellow-green fluorescent (505/515) *for blood flow determination* *3.6x10^6 beads/mL*	10 mL	150.00	120.00
F-8837		FluoSpheres® polystyrene microspheres, 15 µm, blue fluorescent (365/415) *for blood flow determination* *1.0x10^6 beads/mL*	10 mL	150.00	120.00
F-8838	New	FluoSpheres® polystyrene microspheres, 15 µm, blue-green fluorescent (430/465) *for blood flow determination* *1.0x10^6 beads/mL*	10 mL	150.00	120.00
F-8839		FluoSpheres® polystyrene microspheres, 15 µm, crimson fluorescent (625/645) *for blood flow determination* *1.0x10^6 beads/mL*	10 mL	150.00	120.00
F-8841		FluoSpheres® polystyrene microspheres, 15 µm, orange fluorescent (540/560) *for blood flow determination* *1.0x10^6 beads/mL*	10 mL	150.00	120.00
F-8842		FluoSpheres® polystyrene microspheres, 15 µm, red fluorescent (580/605) *for blood flow determination* *1.0x10^6 beads/mL*	10 mL	150.00	120.00
F-8843	New	FluoSpheres® polystyrene microspheres, 15 µm, scarlet fluorescent (645/680) *for blood flow determination* *1.0x10^6 beads/mL*	10 mL	150.00	120.00
F-8844		FluoSpheres® polystyrene microspheres, 15 µm, yellow-green fluorescent (505/515) *for blood flow determination* *1.0x10^6 beads/mL*	10 mL	150.00	120.00

15.7 Protein Conjugates

Albumin Conjugates

Unlike the polydisperse dextrans, fluorescent protein tracers have reasonably well-defined molecular weights (bovine serum albumin (BSA) ~66,000 daltons; ovalbumin ~45,000 daltons; lactalbumin ~14,800 daltons). Their applications are similar to those of dextran tracers, except that the protein conjugates are more susceptible to proteolysis by enzymes. Some recent applications of fluorescent protein conjugates include:

- Quantitative studies of electroporation [1,2]
- Reconstitution of functional nuclear pores [3]
- Macromolecule trafficking to endocytic compartments [4]
- Measurement of plasma volume in rats [5]

In addition to a wide range of fluorescently labeled bovine serum albumin, lactalbumin and ovalbumin conjugates (see the product list at the end of this section), Molecular Probes prepares dinitrophenylated BSA (DNP-BSA, A-843), a reagent that is commonly used to study Fc receptor–mediated immune function [6-8] and the IgE- and IgG-mediated responses to crosslinking of DNP-specific antibodies.[9,10] For these studies, Molecular Probes offers rabbit anti-DNP antibody prepared against DNP-conjugated keyhole limpet hemocyanin (DNP-KLH) (see Section 7.3).

Casein Conjugates

BODIPY FL casein and BODIPY TR-X casein, which are components of two of our new EnzChek™ Protease Assay Kits

(E-6638, E-6639) are substrates for metallo-, serine, acid and sulfhydryl proteases, including cathepsin, chymotrypsin, elastase, papain, pepsin, thermolysin and trypsin.[11] These casein substrates are heavily labeled and therefore highly quenched conjugates; they typically exhibit <3% of the fluorescence of the corresponding free dyes. Protease-catalyzed hydrolysis relieves this quenching, yielding brightly fluorescent BODIPY FL– or BODIPY TR-X–labeled peptides. The EnzChek Protease Assay Kits provide convenient fluorescence-based assays for protease-contaminated culture media (see Section 10.4). In addition to their utility for detecting protease contamination in culture media and other experimental samples, BODIPY FL casein and BODIPY TR-X casein appear to have significant potential as:

- Fluorogenic substrates for circulating or secreted proteases in extracellular fluids
- Nontoxic, pH-insensitive markers for phagocytic cells, which will ingest and eventually cleave the quenched casein substrates to yield fluorescent BODIPY FL– or BODIPY TR-X–labeled peptides
- Microinjectable tracers to detect enhanced protease activity associated with cell activation and fusion
- Nontoxic markers for assessing various cell-loading and cell-transfection techniques, including electroporation, high-velocity microprojectiles, scrape loading and related methods

The peptide hydrolysis products of BODIPY FL casein exhibit green fluorescence that is optimally excited by the argon-ion laser, permitting flow sorting of the cells. The red fluorescent BODIPY TR-X–labeled peptides, with excitation and emission spectra similar to those of the Texas Red fluorophore, should be useful for multilabeling experiments or measurements in the presence of green autofluorescence. Following intracellular hydrolysis, some of the lower molecular weight fluorescent peptides may diffuse into other organelles or pass thorough gap junctions.

Phycobiliproteins

Phycobiliproteins make effective polar tracers because they are intensely fluorescent, very water soluble and have a low tendency to bind nonspecifically to cells. These intrinsically fluorescent proteins are stable proteins that are monodisperse and substantially larger than albumins, with molecular weights of 240,000 daltons (B- and R-phycoerythrin, P-800 and P-801) and 104,000 daltons (allophycocyanin, A-803, or crosslinked allophycocyanin, A-819). Phycobiliproteins have been employed to study nucleocytoplasmic transport of proteins,[12] permeability of reconstituted nuclear pores,[3] nuclear exclusion of nonnuclear proteins [13] and soluble cytoplasmic factors involved in protein import.[14]

Peroxidase Conjugates

Peroxidase conjugates of fluorescein or Texas Red-X dyes (P-912, P-6481) have been used in applications such as retrograde tracing, dye coupling [15] and fate mapping.[16] The fluorescent peroxidase conjugates permit visualization both during and subsequent to injection, thus ensuring that only those specimens that have been properly labeled are developed for light or electron microscopy using standard peroxidase techniques. Fluorescent peroxidase conjugates have also been used to determine aortic permeability,[17] as well as to study membrane fusion events in hormone-treated toad bladder.[18] Biotinylated peroxidase (P-917) may have utility as a retrograde tracer with detection by either its own activity or with avidin or streptavidin conjugates.

1. Biophys J 66, 1522 (1994); **2.** Biophys J 65, 414 (1993); **3.** J Biol Chem 269, 9289 (1994); **4.** J Biol Chem 270, 1443 (1995); **5.** J Appl Physiol 76, 485 (1994); **6.** Cell Regulation 2, 181 (1991); **7.** FEBS Lett 270, 115 (1990); **8.** Cell 58, 317 (1989); **9.** J Biol Chem 270, 4013 (1995); **10.** Mol Biol of the Cell 6, 825 (1995); **11.** Jones, L. *et al.*, manuscript in preparation; **12.** Biochim Biophys Acta 930, 419 (1987); **13.** J Cell Biol 103, 2091 (1986); **14.** J Cell Biol 111, 807 (1990); **15.** Brain Res 294, 9 (1984); **16.** J Exp Zool 255, 323 (1990); **17.** J Cell Biol 100, 292 (1985); **18.** J Physiol 251, C274 (1986).

15.7 Price List *Protein Conjugates*

Cat #		Product Name	Unit Size	Price Per Unit ($) 1–4 Units	5–24 Units
A-2750		albumin from bovine serum, BODIPY® FL conjugate	5 mg	35.00	28.00
A-964		albumin from bovine serum, Cascade Blue® conjugate	25 mg	98.00	78.40
A-843		albumin from bovine serum, 2,4-dinitrophenylated (DNP-BSA)	25 mg	98.00	78.40
A-841		albumin from bovine serum, fluorescein conjugate	25 mg	98.00	78.40
A-847		albumin from bovine serum, tetramethylrhodamine conjugate	25 mg	98.00	78.40
A-824		albumin from bovine serum, Texas Red® conjugate	25 mg	98.00	78.40
A-803		allophycocyanin *4 mg/mL*	0.5 mL	95.00	76.00
A-819		allophycocyanin, crosslinked (APC-XL) *4 mg/mL*	250 µL	198.00	158.40
E-6638	*New*	EnzChek™ Protease Assay Kit *green fluorescence* *100-1000 assays*	1 kit	135.00	108.00
E-6639	*New*	EnzChek™ Protease Assay Kit *red fluorescence* *100-1000 assays*	1 kit	135.00	108.00
L-838		lactalbumin, fluorescein conjugate	25 mg	75.00	60.00
O-2756		ovalbumin, BODIPY® FL conjugate	25 mg	75.00	60.00
O-835		ovalbumin, fluorescein conjugate	25 mg	75.00	60.00
O-837		ovalbumin, Texas Red® conjugate	25 mg	75.00	60.00
P-917		peroxidase from horseradish, biotin-XX conjugate	10 mg	75.00	60.00
P-912		peroxidase from horseradish, fluorescein conjugate	10 mg	75.00	60.00
P-6481	*New*	peroxidase from horseradish, Texas Red®-X conjugate	5 mg	75.00	60.00
P-800		B-phycoerythrin *4 mg/mL*	0.5 mL	95.00	76.00
P-801		R-phycoerythrin *4 mg/mL*	0.5 mL	95.00	76.00

Assays of Volume Change, Membrane Fusion and Membrane Permeability

Leakage of aqueous contents from cells or vesicles as a result of lysis, fusion or physiological permeability can be detected fluorometrically using low molecular weight soluble tracers. Assays designed to detect solution mixing often rely on a fluorophore and quencher pair, whereas assays that detect solution leakage into the external medium typically exploit the self-quenching properties of a fluorophore. Phospholipids suitable for preparing liposomes are now available from Molecular Probes (D-7704, D-7705; Section 13.4). Fusion assays that detect mixing of membrane lipids are described in the Technical Note "Lipid Mixing Assays of Membrane Fusion" on page 264.

Fluorescence Quenching Assays with ANTS/DPX

Originally developed by Smolarsky and co-workers to follow complement-mediated immune lysis,[1] the ANTS/DPX fluorescence quenching assay has since been widely used to detect membrane fusion.[2] This assay is based on the collisional quenching of the polyanionic fluorophore ANTS by the cationic quencher DPX. Separate vesicle populations are loaded with

- 25 mM ANTS (A-350; see Section 15.3), 40 mM NaCl
- 90 mM DPX (X-1525, see Section 15.3)

Vesicle fusion results in quenching of ANTS fluorescence monitored at 530 nm, with excitation at 360 nm. External leakage of vesicle contents does not cause quenching due to the accompanying dilution of ANTS and DPX. For assays of vesicle leakage, ANTS (12.5 mM) and DPX (45 mM) can be co-encapsulated into liposomes; upon dilution into surrounding medium, ANTS fluorescence will increase because quenching by DPX will be diminished.

Recent examples employing ANTS/DPX quenching include studies of the role of amino lipids [3] and phospholipid asymmetry [4] in membrane fusion; glycoprotein-mediated viral fusion with membranes;[5] and the interactions of vesicle membranes with annexins,[6] HIV fusion peptide [7] and apolipoprotein.[8]

Additional Dye/Quencher Pairs

The fluorescence of HPTS and pyrenetetrasulfonic acid is also effectively quenched by DPX.[9-11] Thallium (Tl^{+}) and cesium (Cs^{+}) ions quench the fluorescence of ANTS, pyrenetetrasulfonic acid (P-349, see Section 15.3) and some other polyanionic fluorophores. A review by Garcia [12] describes how fluorescence quenching with a variety of dye/quencher pairs can be used to determine transmembrane ion permeability.

Fluorescence Enhancement Assays with Tb^{3+}/DPA

In the Tb^{3+}/DPA assay, which was originally described by Wilschut and Papahadjopoulos,[13,14] separate vesicle populations are loaded with:

- 2.5 mM $TbCl_3$ (T-1247, see Section 15.3), 50 mM sodium citrate
- 50 mM dipicolinic acid (DPA, D-1248; see Section 15.3), 20 mM NaCl

We also offer the acetoxymethyl (AM) ester of dipicolinic acid (D-1249, see Section 15.3) for passive loading of cells. Vesicle fusion results in formation of Tb^{3+}/DPA chelates that are 10,000 times more fluorescent than free Tb^{3+}. Fluorescence of the chelates is detected at 490 nm or 545 nm, with excitation at 276 nm. Including Ca^{2+} and EDTA in the external medium inhibits formation of the complex outside the fused vesicles. The Tb^{3+}/DPA fluorescence enhancement assay has been used to investigate the role of phospholipid conformation in vesicle fusion [15] and the interaction of cardiotoxin with phospholipid vesicles.[16]

Additional Fluorescence Enhancement Pairs

As with the Tb^{3+} chelates, chelates of Eu^{3+} (E-6960, see Section 15.3) show a similar fluorescence enhancement upon formation but exhibit longer-wavelength emission (peaks at 590 nm and 613 nm). Naphthoyltrifluoroacetone (NTA, T-411; see Section 15.3) and its derivatives, which require solubilization with 0.1% Triton® X-100 for use in aqueous systems, are generally used as Eu^{3+} chelators. Tb^{3+} and Eu^{3+} chelates have millisecond radiative lifetimes, making them useful for time-resolved detection in immunoassays. An approximately fourfold enhancement of the quantum yield of ANTS is induced by D_2O; this spectral characteristic has been used to determine water permeability in kidney collecting tubules.[17]

Self-Quenching Assays with Fluorescein Derivatives

Fluorescence of carboxyfluorescein (C-194, C-1904; see Section 15.3) or calcein (C-481, see Section 15.3) is >95% self-quenched at concentrations >100 mM. Concentrated solutions of these water-soluble dyes are encapsulated in liposomes, which are then separated from any remaining free dye by gel filtration. Upon addition of a fusogen or other permeabilizing agent, dye release is accompanied by an increase fluorescence (excitation/emission maxima 490 nm/520 nm). Complete lysis of the liposomes with 0.1% Triton X-100 can be used to determine the assay end-point. Calcein may be preferred over carboxyfluorescein on the basis of higher net charge and lower pH sensitivity. Note that this assay will detect any process that causes leakage of aqueous contents, including fusion, lysis or permeabilization. In a modification of this assay designed to specifically detect vesicle fusion, a nonfluorescent Co^{2+} complex of calcein contained in a vesicle is disrupted upon fusion with a second vesicle that delivers the Co^{2+} chelator EDTA.[18,19] Recent studies employing carboxyfluorescein self-quenching include investigations of interactions of membranes with mycobacterial glycopeptidolipids [20] and with HIV glycoprotein peptide fragments.[21]

1. J Immunol Meth 15, 255 (1977); **2.** Biochemistry 24, 3099 (1985); **3.** Biochemistry 33, 12573 (1994); **4.** Biochemistry 31, 4262 (1992); **5.** Biochemistry 34, 1084 (1995); **6.** Biochemistry 32, 14194 (1993); **7.** Biochemistry 33, 3201 (1994); **8.** J Biol Chem 268, 22112 (1993); **9.** Biochim Biophys Acta 1142, 277 (1993); **10.** Am J Physiol. 262, G30 (1992); **11.** Biochim Biophys Acta 1024, 352 (1990); **12.** Meth Enzymol 207, 501 (1992); **13.** Biochemistry 19, 6011 (1980); **14.** Nature 281, 690 (1979); **15.** Biochemistry 33, 5805 (1994); **16.** J Biol Chem 269, 14473 (1994); **17.** Biophys J 54, 587 (1988); **18.** Biophys J 55, 973 (1989); **19.** J Biol Chem 257, 13892 (1982); **20.** Biochemistry 33, 7056 (1994); **21.** J Biol Chem 267, 7121 (1992).

Chapter 16

Assays for Cell Viability, Proliferation and Function

Contents

16.1 Overview of Cell Viability and Proliferation Assays

Cell viability, cell proliferation and many important live-cell functions — including programmed cell death (apoptosis), cell adhesion, chemotaxis, multidrug resistance, endocytosis (see Chapter 17) and signal transduction (see Chapter 20) — can be stimulated or monitored by application of various chemical and biological probes. Many of these processes lead to changes in intracellular radicals (see Chapter 21), free ion concentrations (see Chapters 22–24) or membrane potential (see Chapter 25) that can be detected with appropriately responsive fluorescent indicators. This chapter discusses important reagents and assays for detecting and monitoring these diverse cell processes in live cells.

Although related, viability assays and proliferation assays have fundamentally different purposes. Viability assays are used to enumerate live and dead cells in a population. In contrast, proliferation assays are designed to monitor the growth rate of a cell population or to detect daughter cells in a growing population. Fluorescence-based viability and proliferation assays are generally less hazardous and less expensive than radioisotopic techniques, more sensitive than colorimetric methods and more convenient than animal testing methods.[1] Unlike ^{51}Cr-release assays, fluorescence-based assays of cell-mediated cytotoxicity do not require large samples, which can be difficult to obtain from patients.[2] Furthermore, with trypan blue exclusion, a common colorimetric method for determining cell viability, the assay must be completed within 3–5 minutes because the number of blue-staining cells increases with time after addition of the dye.[3]

The diversity of live cells and their environments makes it impossible to devise a single viability assay applicable to all cell types. Molecular Probes' LIVE/DEAD® Viability Assay Kits offer researchers a choice of viability/cytotoxicity assays suitable for bacteria, fungi or higher eukaryotic cells (see Table 16.1 and Section 16.2 below). These kits provide the reagents and a simple protocol necessary for simultaneous two-color assessment of numbers of live and dead cells. We have also developed several proliferation assay kits that enable researchers to rapidly monitor numbers of adherent or nonadherent cells based on total nucleic acid or total protein content (see Section 16.2 below).

In addition to these kits, several of the probes for live-cell function described in Section 16.3 can be used to develop viability assays that measure a particular biochemical parameter of interest. Because viability is not easily defined in terms of a single physiological or morphological parameter, it is often desirable to combine several different measures, such as enzymatic activity, membrane permeability and redox potential. Each assay method has inherent advantages and limitations and may introduce specific biases into the experiment; thus, different applications often call for different approaches.

There is a significant overlap between probes for cell viability and probes for live-cell functions. For example, fluorogenic esterase substrates are commonly used to detect viability and proliferation, as well as to monitor cell adhesion, apoptosis and multidrug resistance. Likewise, cell-permeant and cell-impermeant nucleic acid stains are widely applicable to many live-cell function assays. We have organized Section 16.3 according to several commonly studied cell processes in order to highlight the many published applications for these probes and foster the development of new applications.

1. Burghardt, R.C. *et al.* in *Principles and Methods of Toxicology, Third Edition*, A.W. Hayes, Ed., Raven Press (1994) pp. 1231–1258; **2.** Human Immunol 57, 264 (1993); **3.** J Histochem Cytochem 33, 77 (1985).

16.2 Assay Kits for Diverse Cell Types

Viability Assay Kits for Animal Cells

LIVE/DEAD Viability/Cytotoxicity Kit for Animal Cells

Molecular Probes' LIVE/DEAD Viability/Cytotoxicity Kit (L-3224) for animal cells provides an exceptionally easy fluorescence-based method for determining viability of adherent or nonadherent cells and for assaying cytotoxicity. The kit comprises two probes: calcein AM and ethidium homodimer-1. Calcein AM is a fluorogenic esterase substrate that is hydrolyzed to a green fluorescent product (calcein); thus, green fluorescence is an indicator of cells that have esterase activity as well as an intact membrane to retain the esterase products. Ethidium homodimer-1 is a high-affinity, red fluorescent nucleic acid stain that is only able to pass through the compromised membranes of dead cells.[1] The LIVE/DEAD viability/cytotoxicity assay [2] offers several advantages:

- **Simplicity.** The reagents are simultaneously added to the cell suspension, which is then incubated for 30–45 minutes. No wash steps are required before analysis.
- **Specificity and reliability.** Green fluorescent cells are live; red fluorescent cells are dead (see Color Plate 4 in "Visual Reality").
- **Versatility.** The LIVE/DEAD viability/cytotoxicity assay is compatible with adherent cells such as astrocytes,[3] nonadherent cells and certain tissues.[4-6] Results can be analyzed by fluorescence microscopy using standard fluorescein longpass filter sets, as well as by flow cytometry (see Figure 16.1) or fluorometry. The fluorescence emission of the two probes are easily resolved (Figure 16.2).
- **Simple quantitation.** Flow cytometric measurements yield only two populations; there are rarely any doubly stained cells (Figure 16.1). Quantitative assays of bulk cells can be made using a fluorescence microplate reader or fluorometer.
- **Suitability for high-throughput screening.** The ease, reliability and low cost of the LIVE/DEAD Viability/Cytotoxicity Kit make it an economical assay for high-throughput screening of cytotoxic agents.

Table 16.1 *Summary of Molecular Probes' viability kits, which are tailored for specific organisms, as well as gram stain and counting kits for bacteria.*

Cat #	Kit Name	Kit Components	# Assays	Assay Principle
L-3224	LIVE/DEAD Viability/ Cytotoxicity Kit	• Calcein AM • Ethidium homodimer-1 • Detailed protocol	1000 microscopy assays, 1000 microplate assays or 300 flow cytometry assays	Membrane-permeant calcein AM is cleaved by esterases in live cells to yield cytoplasmic green fluorescence, and membrane-impermeant ethidium homodimer-1 labels nucleic acids of membrane-compromised cells with red fluorescence.
L-7013	LIVE/DEAD Reduced Biohazard Viability/ Cytotoxicity Kit	• SYTO 10 nucleic acid stain • DEAD Red nucleic acid stain • Detailed protocol	100 microscopy assays or 100 flow cytometry assays	Membrane-permeant SYTO 10 dye labels the nucleic acids of live cells with green fluorescence, and membrane-impermeant DEAD Red dye labels nucleic acids of membrane-compromised cells with red fluorescence. Subsequent fixation inactivates pathogens without distorting the staining pattern.
L-7010	LIVE/DEAD Cell-Mediated Cytotoxicity Kit	• $DiOC_{18}(3)$ • Propidium iodide • Detailed protocol	2000 microscopy assays or 200 flow cytometry assays	Target cells are preincubated with the green fluorescent membrane stain $DiOC_{18}(3)$ and then mixed with effector cells in the presence of the membrane-impermeant propidium iodide. Live and dead target cells retain their green fluorescent membrane stain; target and effector cells with compromised membranes exhibit red fluorescent nucleic acid staining; live effector cells are nonfluorescent.
L-7011	LIVE/DEAD Sperm Viability Kit	• SYBR 14 nucleic acid stain • Propidium iodide • Detailed protocol	1000 microscopy assays or 200 flow cytometry assays	Membrane-permeant SYBR 14 nucleic acid stain labels live sperm with green fluorescence, and membrane-impermeant propidium iodide labels the nucleic acids of membrane-compromised sperm with red fluorescence.
L-7009	LIVE/DEAD Yeast Viability Kit	• FUN-1 • Calcofluor White M2R • Detailed protocol	>1000 microscopy assays or 1000 microplate assays	Plasma membrane integrity and metabolic function of fungi are required to convert the yellow-green fluorescent intracellular staining of FUN-1 into red-orange fluorescent intravacuolar structures; Calcofluor™ White M2R labels cell-wall chitin with blue fluorescence regardless of metabolic state.
L-7012 L-7007	LIVE/DEAD *Bac*Light Bacterial Viability Kit	• SYTO 9 nucleic acid stain • Propidium iodide • *Bac*Light mounting oil • Detailed protocol	>1000 microscopy assays, 1000 microplate assays or 200 flow cytometry assays	Membrane-permeant SYBR 9 labels live bacteria with green fluorescence, membrane-impermeant propidium iodide labels membrane-compromised bacteria with red fluorescence. In Kit L-7007, stains are supplied in a mixed, two-component formulation, whereas in Kit L-7012, stains are provided as separate solutions.
V-7023	ViaGram Red⁺ Bacterial Gram Stain and Viability Kit	• DAPI • SYTOX Green nucleic acid stain • Texas Red-X conjugate of wheat germ agglutinin (WGA) • Sodium bicarbonate • *Bac*Light mounting oil • Detailed protocol	200 microscopy assays	Membrane-permeant DAPI labels live bacteria with blue fluorescence; membrane-impermeant SYTOX Green nucleic acid stain labels bacteria with compromised membranes with green fluorescence. Simultaneous Texas Red-X WGA staining produces red fluorescent surface labeling of gram-positive bacteria.
L-7005 L-7008	LIVE *Bac*Light Bacterial Gram Stain Kit	• SYTO 9 nucleic acid stain • Hexidium iodide • *Bac*Light mounting oil • Detailed protocol	>1000 microscopy assays, 1000 microplate assays or 200 flow cytometry assays	When gram-negative and gram-positive bacteria are simultaneously stained with the membrane-permeant SYTO 9 dye and hexidium iodide, gram-negative bacteria fluoresce green and gram-positive bacteria fluoresce red. In Kit L-7008, stains are supplied in a mixed, two-component formulation, whereas in Kit L-7005, stains are provided as separate solutions.
B-7277	Bacteria Counting Kit	• SYTO BC bacteria stain • Microsphere suspension standard • Detailed protocol	100 flow cytometry assays	The membrane-permeant SYTO BC stain labels both gram-positive and gram-negative bacteria with green fluorescence; the calibrated suspension of polystyrene microspheres serves as a standard for the volume of suspension analyzed.

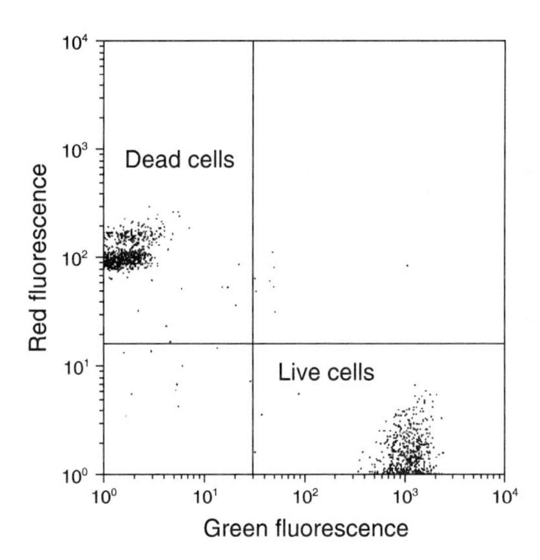

Figure 16.1 *Flow cytometric viability assay using Molecular Probes' LIVE/DEAD Viability/Cytotoxicity Kit (L-3224). A 1:1 mixture of live and ethanol-fixed human B cells was stained with calcein AM and ethidium homodimer-1 according to the kit protocol. After 5 minutes, flow cytometric analysis was carried out with excitation at 488 nm on a Becton Dickinson FACS® Vantage™ cytometer. The resulting bivariate frequency distribution shows the clear separation of the green fluorescent (530 nm) live-cell population from the red fluorescent (585 nm) dead-cell population.*

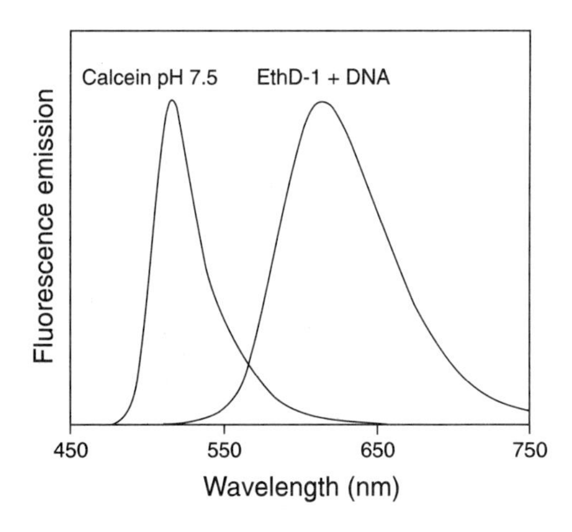

Figure 16.2 *Normalized fluorescence emission spectra of calcein and DNA-bound ethidium homodimer-1 (EthD-1), illustrating the clear spectral separation that allows simultaneous visualization of live and dead eukaryotic cells with Molecular Probes' LIVE/DEAD Viability/Cytotoxicity Kit (L-3224).*

Validity of the LIVE/DEAD viability/cytotoxicity assay for use with animal cells has been established by several laboratories. Recent published applications include measuring the toxic effects of tumor necrosis factor (TNF),[7] β-amyloid protein,[8] adenovirus E1A proteins,[9] tetrodotoxin (TTX) binding to Na^+ channels,[10] methamphetamines [11] and mitogenic sphingolipids.[12] The assay has also been adapted to quantitate lymphocyte-mediated cytotoxicity by flow cytometry,[13] cell-mediated cytotoxicity by fluorescence microscopy [14] and the viability of boar sperm by fluorescence microscopy.[15]

The LIVE/DEAD Viability/Cytotoxicity Kit is intended for use with animal cells that can be analyzed within about an hour of adding the dyes to the cells. Kit components, number of assays and assay principles are summarized in Table 16.1. This kit's two viability probes — calcein AM (C-1430, C-3099, C-3100; see Section 16.3) and ethidium homodimer-1 (E-1169, see Section 16.3) — are also available separately and may be used in combination with other probes for discrimination of live and dead cells. When assays need to be conducted over longer periods or when hazardous samples are being analyzed, we recommend our LIVE/DEAD Reduced Biohazard Viability/Cytotoxicity Kit (L-7013, see below).

LIVE/DEAD Reduced Biohazard Viability/Cytotoxicity Kit

Rigorous precautions are necessary during analysis of biohazardous specimens.[16,17] Therefore, fixation procedures that inactivate cells yet produce minimal distortion of their characteristics are highly advantageous.[18] The LIVE/DEAD Reduced Biohazard Viability/Cytotoxicity Kit (L-7013) provides a new two-color fluorescence assay for animal cell viability that is designed to reduce the risk associated with handling potential biohazards such as viral, bacterial or protozoan pathogens.[19]

Viability analysis with our LIVE/DEAD Reduced Biohazard Kit is provided by the cell-permeant, green fluorescent SYTO® 10 and the cell-impermeant, red fluorescent DEAD Red™ nucleic acid stains. The dye concentrations and their relative affinities are balanced so that a cell population exposed simultaneously to both dyes becomes differentially stained — live cells fluoresce green and dead cells fluoresce red. This assay is simple, fast and can be carried out using a fluorescence microscope, flow cytometer or fluorescence microplate reader. Moreover, the staining pattern of a cell population is retained for several hours after fixation (Figure 16.3).

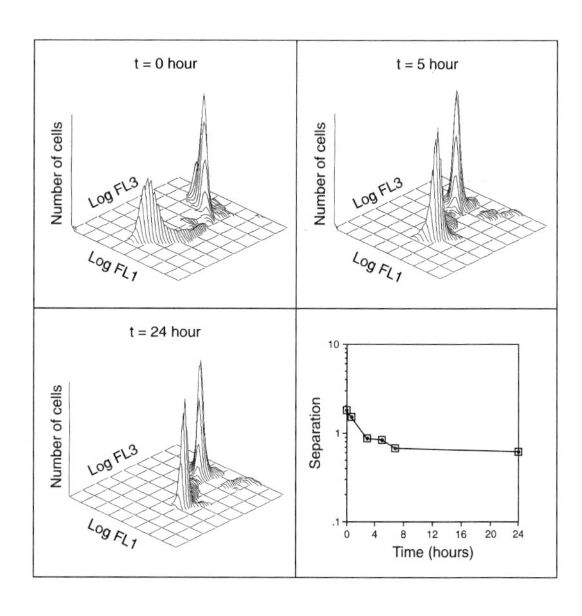

Figure 16.3 *Flow cytometric analysis of a mixed population of live and complement-treated goat lymphocytes stained using the reagents and protocols provided in our LIVE/DEAD Reduced Biohazard Viability/Cytotoxicity Kit (L-7013) and monitored over a 24-hour period. The panels (left to right, top to bottom) represent the distribution of SYTO 10 (FL1) and DEAD Red (FL3) fluorescence in lymphocytes at 0, 5 and 24 hours after fixation. The lower right panel is a plot of the separation between the live and dead population peaks as a function of time.*

The LIVE/DEAD Reduced Biohazard Viability/Cytotoxicity Kit (L-7013) has several unique features:

- **Reduced handling risks.** This kit allows viability staining to take place while the potentially pathogenic sample is well contained. Subsequent treatment with 4% glutaraldehyde (or less effectively with formaldehyde) allows for safer handling during analysis, without disrupting the distinctive staining pattern. Glutaraldehyde is known to inactivate cells and viruses, while preserving their overall morphology.[20] Also, the high sensitivity and specificity of the assay mean that sample sizes can be very small, further reducing potential biohazards.
- **Specificity and reliability.** Live cells are initially green fluorescent, and dead cells are red fluorescent. With time this discrimination is reduced but can still be detected, even after 24 hours (Figure 16.3).
- **Independence from enzymatic activity.** Because it relies on two nucleic acid stains that differ in their membrane permeability, this assay equates loss of cell viability with loss of membrane integrity. Consequently, the assay is totally independent of variations in enzymatic activity or electrical potential of the cell.
- **Versatility.** The analysis is readily quantitated with a fluorescence microscope or flow cytometer (Figure 16.3). This kit's protocol includes methods for analyzing the viability of nonadherent cells, as well as adherent cells on coverslips.
- **Convenience.** Cells can be stained and fixed at various times during the experiment, and the results can be analyzed several hours later without loss of the discrimination pattern.

Kit components, number of assays and assay principles are summarized in Table 16.1.

LIVE/DEAD Cell-Mediated Cytotoxicity Kit

Cytotoxicity triggered by a natural defense mechanism may be much slower than cell lysis triggered by a cytotoxic reagent. Our LIVE/DEAD Cell-Mediated Cytotoxicity Kit (L-7010) is intended for cytotoxicity assessments extending over time periods that are too long for effective use of cytoplasmic markers, such as calcein, which may leak out or become sequestered. This kit is based directly on procedures developed by Kroesen and colleagues for measuring natural killer (NK), lymphokine-activated killer (LAK) and T cell–mediated cytotoxicity by fluorescence microscopy.[21] The assay has also been adapted for rapid flow cytometric analysis of NK cell activity.[22,23]

Analysis of cell-mediated cytotoxicity using this kit is easy. In order to distinguish target cells, cultures are labeled overnight with $DiOC_{18}(3)$, a green fluorescent membrane stain (see Section 15.4). Target cells are then washed free of excess $DiOC_{18}(3)$ and combined in various proportions with effector cells. After a suitable incubation period, propidium iodide, a membrane-impermeant nucleic acid stain, is added. Propidium iodide labels dead effector cells, as well as dead target cells once their plasma membranes are compromised. Because the target cells retain the green fluorescent membrane stain, both live and dead effector cells and live and dead target cells can readily be discriminated with a fluorescence microscope. Dead target cells exhibit both green fluorescent membrane staining and red fluorescent nuclear staining, whereas dead effector cells show only red fluorescent nuclear staining. Live target cells have only green fluorescent membrane staining, and live effector cells are unstained. Kit components, number of assays and assay principles are summarized in Table 16.1.

LIVE/DEAD Sperm Viability Kit

The LIVE/DEAD Sperm Viability Kit (L-7011), developed in collaboration with Duane L. Garner at the University of Nevada at Reno, provides a novel fluorescence-based method for analyzing the viability of sperm in different species.[24-27] The LIVE/DEAD Sperm Viability Kit contains a patented membrane-permeant nucleic acid stain developed at Molecular Probes, SYBR® 14, along with the conventional dead-cell stain, propidium iodide. Using this combination of dyes, researchers can rapidly distinguish live and dead cells with visible-light excitation (see Color Plate 4 in "Visual Reality"), thus avoiding the harmful effects of UV exposure and allowing flow cytometric analysis of sperm viability to be performed using an argon-ion laser excitation source. When semen is incubated briefly with these two stains, live sperm with intact membranes fluoresce bright green, whereas sperm cells with damaged membranes fluoresce red. Garner and colleagues assessed bovine sperm viability with flow cytometry and with fluorescence microscopy; both techniques allowed live and dead cells to be visualized simultaneously.[27,28] Furthermore, it was reported that neither the ability to fertilize oocytes nor the development of the embryos was affected by SYBR 14 staining of porcine sperm.[24] The effect of two-photon illumination on the viability of human sperm stained with these reagents has also been analyzed.[29]

The dyes provided in the LIVE/DEAD Sperm Viability Kit stain cells more rapidly — within 5–10 minutes — than conventional stains, and both label DNA, thereby avoiding the ambiguity that may arise from targeting separate cellular components. The membrane-permeant SYBR 14 stain provided in the LIVE/DEAD Sperm Viability Kit should also provide researchers with a valuable tool for labeling and tracking live sperm, thus facilitating analysis of their motility and abundance in semen samples. Reliable viability measurements with bovine,[27] porcine, ovine, murine,[25] goat, turkey [26] and human sperm [25,29] have been published. Kit components, number of assays and assay principles are summarized in Table 16.1.

Conventional sperm viability assays have employed mixtures of two or three dyes, including fluorescein diacetate derivatives, rhodamine 123 and reduced nucleic acid stains such as dihydroethidium [26,30-33] (D-1168). Acridine orange, which fluoresces at different wavelengths when bound to DNA and RNA,[34,35] and the UV-excitable nucleic acid stains Hoechst 33258 and Hoechst 33342 [36-38] are frequently used to determine sperm viability and DNA content and to trace sperm–oocyte fusion.[39,40] These alternative viability probes are described in Section 16.3.

Proliferation Assay Kits for Animal Cells

CyQUANT™ Cell Proliferation Assay Kit

The quantity of DNA and RNA in a cell is tightly regulated. Therefore, changes in nucleic acid content can serve as a sensitive indicator of cell proliferation, as well as of cytotoxic events or pathological abnormalities that affect cell proliferation. Our new CyQUANT Cell Proliferation Assay Kit (C-7026) provides an excellent measure of proliferative activity and promises to be an important development for the rapid and quantitative screening of agents that affect cell proliferation. The CyQUANT assay is based on the use of our proprietary green fluorescent CyQUANT GR dye, which exhibits strong fluorescence enhancement when bound to cellular nucleic acids.[41] The assay is simple: the culture medium is removed (nonadherent cells require brief centrifugation); the cells are frozen, thawed and lysed by addition of buffer containing detergent and CyQUANT GR dye; and fluorescence is then measured directly in a fluorometer or fluorescence microplate reader. No washing steps, growth medium changes or long incubations are required. The CyQUANT cell proliferation assay has a number of significant advantages over other proliferation assays:

- **Sensitivity and Linearity.** The CyQUANT assay is linear from 50 or fewer cells to at least 50,000 cells in 200 µL volumes (Figure 16.4); increasing the dye concentration extends the linear range to 250,000 cells. Methods that employ Hoechst 33258 [42] (H-1398, see Section 16.3) or Hoechst 33342 [43] (H-1399, see Section 16.3) to measure cell number and proliferation are less sensitive — detection limits of 500 cells for Hoechst 33258 [42] or 2500 cells for Hoechst 33342 [43] — and have much smaller effective ranges.
- **No radioactivity.** Unlike assays that measure ^{3}H-thymidine incorporation, the CyQUANT assay does not require radioisotopes and thus does not have the hazards or the expense associated with use, storage and disposal of radioisotopes.
- **Quick and easy protocol.** The CyQUANT assay is a single-step procedure that requires no lengthy incubation steps and is complete within an hour.
- **Specificity and Reliability.** The assay is specific for total nucleic acids, with essentially no interference from other cell components. No wash steps are required because cellular growth media do not significantly interfere with CyQUANT GR fluorescence. The CyQUANT assay is reliable for cell quantitation, even without treatment to eliminate cellular RNA. However, addition of RNase or DNase permits the facile quantitation of DNA or RNA, respectively, in the sample.
- **Convenience.** Unlike assays that use tetrazolium salts, ^{3}H-thymidine, neutral red or methylene blue,[43-46] the CyQUANT procedure is not dependent on cellular metabolism. Thus, cells can be frozen and stored prior to assaying, with no reduction in signal, or they can be assayed immediately after collection. Time-course assays are simplified because data obtained from stored samples taken at widely different time intervals can be assayed together with a single standard curve determination.

We have found the CyQUANT Cell Proliferation Assay Kit to be useful for assaying widely disparate cell types, including:

- Human neonatal fibroblasts, keratinocytes, melanocytes, umbilical vein endothelial cells (HUVEC) and dermal microvascular endothelial cells (DMVEC)
- Murine fibroblasts (NIH 3T3 and CRE BAG 2 cells) and myeloma (P3X63A68) cells
- Madin–Darby canine kidney (MDCK) cells
- Chinook salmon embryo (CHSE) cells
- Rat basophilic leukemia (RBL) and glioma (C6) cells

In addition to quantitating proliferation, the CyQUANT GR reagent may supplant ^{51}Cr-release studies for monitoring T cell cytolysis and other cytolytic events.[47] Furthermore, determination of total cell number using the CyQUANT GR reagent is potentially useful for quantitating cell adhesion (see Cell Adhesion in Section 16.3) and for determining the number of cells in a tissue. Each CyQUANT Cell Proliferation Assay Kit (C-7026) includes:

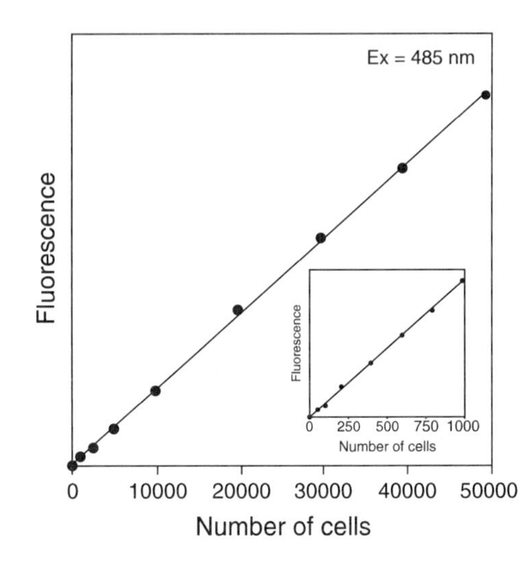

Figure 16.4 *Quantitation of NIH 3T3 fibroblasts using Molecular Probes' CyQUANT Cell Proliferation Assay Kit (C-7026). Fluorescence measurements were made using a CytoFluor® 2350 microplate reader (PerSeptive Biosystems) with excitation at 485 nm and emission detection at 530 nm. The linear range of the assay under these conditions is from 50 to 50,000 cells per 200 µL sample.*

- CyQUANT GR reagent
- Cell lysis buffer
- DNA standard for calibration
- Detailed protocol

The kit supplies sufficient materials for performing 1000 assays based on a 200 µL sample volume or a proportionately lower number of assays with a larger sample volume. The CyQUANT cell lysis buffer (20X concentrate, C-7027) is also available separately and has been formulated to produce efficient lysis, to protect nucleic acids from nuclease activity and to dissociate proteins that may interfere with dye binding to nucleic acids. It may prove generally useful in the development of other assays.

FluoReporter® Blue Fluorometric Nucleic Acid Assay Kit

The FluoReporter Blue Fluorometric dsDNA Quantitation Kit (F-2962) provides the protocols developed by Rago and colleagues [48] for analyzing cellular DNA with the blue fluorescent Hoechst 33258 nucleic acid stain. The kit enables researchers to detect ~10 ng of isolated calf thymus DNA or ~1000 mouse NIH 3T3 cells in a 200 µL sample (substantially lower levels are detectable using our new CyQUANT Cell Proliferation Assay Kit described above).

With this kit, quantitation of cellular DNA is rapid, and all manipulations can be carried out in microplate wells. The cells are lysed by freezing them in distilled water, which circumvents the requirement for extraction procedures used in other Hoechst 33258–based protocols.[45,49-52] The diluted dye solution is then added to the lysed cells, and fluorescence is measured. Kit components include:

- Hoechst 33258 in dimethylsulfoxide (DMSO)/H_2O
- TNE buffer
- Detailed protocol

Each kit provides sufficient reagents for assaying 2000 samples using a fluorescence microplate reader.

FluoReporter Fluorometric and Colorimetric Cell Protein Assay Kits

In situations where most, but not all, cells in a culture have been killed by drugs, it may be important to determine the ability of the remaining cells to proliferate in culture. Our FluoReporter Cell Protein Assay Kits are based on an electrostatic dye-binding method for measuring protein content of trichloroacetic acid (TCA)–fixed cells, a method used at the National Cancer Institute (NCI) for high-throughput screening of new chemotherapeutic reagents.[53-55] These kits enable researchers to rapidly monitor numbers of adherent or nonadherent cells based on protein content. This assay can also be used to detect small numbers of multidrug-resistant cells.[56,57] However, in some cells such as epidermal keratinocytes, the protein content of cells may increase greatly without a change in cell number.[42]

The FluoReporter Colorimetric Cell Protein Assay Kit (F-2961) contains the anionic xanthene dye sulforhodamine B, which forms an electrostatically stabilized complex with basic amino acid residues under moderately acidic conditions. The protein–dye complex (absorption maximum ~565 nm) can then be detected spectrophotometrically after removal of unbound dye from TCA-fixed cells. This FluoReporter assay is linear for determining protein content of 2000 to 200,000 mouse myeloma P3X cells, comparing favorably with tedious clonogenic assays, and is reported to have considerable advantages over tetrazolium dye–based assays.[53] Because sulforhodamine B is fluorescent, the protein–dye complex can also be measured with a fluorescence microplate reader using excitation/emission maxima of ~485/590 nm, which provides suboptimal excitation but produces a linear relationship between fluorescence intensity and cell number. The FluoReporter Colorimetric Cell Protein Assay Kit contains:

- Sulforhodamine B
- Concentrated solubilization solution
- Protocol for determining cell numbers

All manipulations are carried out in microplate wells and can be completed in less than four hours. Each kit provides sufficient reagents for ~2500 microplate-well assays.

Based on similar principles, our FluoReporter Fluorometric Cell Protein Assay Kit (F-2960) contains the red fluorescent anionic dye sulforhodamine 101 (excitation/emission maxima ~586/602 nm) and enables researchers to detect 100 to 100,000 mouse myeloma P3X cells. The FluoReporter Fluorometric Cell Protein Assay Kit contains:

- Sulforhodamine 101 in 1% acetic acid
- Concentrated counterstain solution in 1% acetic acid
- Concentrated solubilization solution
- Protocol for determining cell numbers

Each kit provides sufficient reagents for ~2000 microplate-well assays. Electrostatic dye-binding properties of sulforhodamine 101 (S-359, see Section 15.3) have also been exploited to quantitate proteins in cells by flow cytometry.[58-61] The dyes in each of these FluoReporter Cell Protein Assay Kits — sulforhodamine B (S-1307, see Section 15.3) and sulforhodamine 101 (S-359, see Section 15.3) — are available individually.

Viability Assay and Organelle Marker Kits for Yeast

LIVE/DEAD Yeast Viability Kit

Our LIVE/DEAD Yeast Viability Kit (L-7009) provides an extremely simple and sensitive assay for discriminating viable yeast and fungi in complex mixtures or in pure cultures.[62,63] This kit contains our unique two-color fluorescent viability probe, the FUN-1™ dye, which has low intrinsic fluorescence, moderate affinity for nucleic acids and exceptional membrane permeability. Also included is the UV-excitable counterstain Calcofluor™ White M2R, which labels the cell walls of yeast and fungi a fluorescent blue, regardless of the cell's metabolic state.[64-66]

The FUN-1 viability probe [67,68] displays some extraordinary spectral properties when used to stain metabolically active yeast and fungal cells, exploiting normal endogenous biochemical processing mechanisms that appear to be well conserved among different fungal species. The FUN-1 stain passively diffuses into a variety of cell types and initially stains the cytoplasm with a diffusely distributed green fluorescence. However, in several common species of yeast and fungi, subsequent processing of the dye by live cells results in the formation of distinct vacuolar structures with compact form that exhibit a striking red fluorescence, accompanied by a reduction in the green cytoplasmic fluorescence [69] (Figure 16.5). Formation of the red fluorescent intravacuolar structures requires both plasma membrane integrity and metabolic capability. Dead cells fluoresce bright yellow-green with no discernable red structures.

FUN-1 stain can be used alone or together with Calcofluor White M2R to determine the metabolic activity of single fungal cells by manual or automated fluorescence microscopy (see Color Plate 4 in "Visual Reality"). Both live and dead cells may be viewed simultaneously by fluorescence microscopy using a standard fluorescein longpass optical filter set. To visualize live and dead cells counterstained with Calcofluor White M2R, we recommend using a multiple-dye optical filter set such as the Omega®

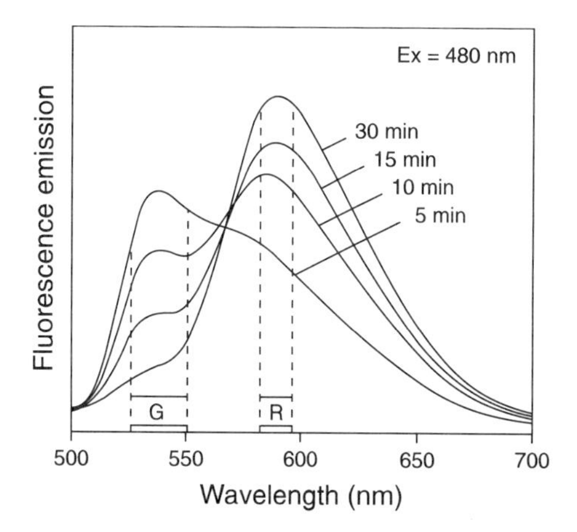

Figure 16.5 *Fluorescence emission spectra of a Saccharomyces cerevisiae suspension that has been stained with FUN-1 cell stain (F-7030). After addition of FUN-1 to the medium, its fluorescence emission spectrum (excited at 480 nm) was recorded in a spectrofluorometer at the indicated times during a 30-minute incubation period. The shift from green (G) to red (R) fluorescence reflects the processing of FUN-1 by metabolically active yeast cells.*

Optical triple-band filter set (O-5855, see Section 26.5), available directly from Molecular Probes. FUN-1 staining can also be used to assay the viability of suspensions of fungal cells using a fluorescence microplate reader or a fluorometer.

The LIVE/DEAD Yeast Viability Kit has been tested on several fungal species, including *Candida albicans*, *Candida pseudotropicalis* and several strains of *Saccharomyces cerevisiae*, under a variety of experimental conditions. Formation of the red fluorescent structures was observed not only in logarithmically growing cells but also in nonculturable cells with residual metabolic activity. The LIVE/DEAD Yeast Viability Kit should be particularly useful for detecting very low numbers of live or dead fungal cells, even in complex mixtures such as blood. Kit components, number of assays and assay principles are summarized in Table 16.1. The FUN-1 cell stain is also available separately (F-7030).

Yeast Vacuole Marker Sampler Kit

The Yeast Vacuole Marker Sampler Kit (Y-7531) contains sample quantities of a series of both well-established and novel vacuole marker probes that show promise for the study of yeast cell biology:

- 5-(and-6)-Carboxy-2',7'-dichlorofluorescein diacetate (carboxy-DCFDA)
- CellTracker™ Blue CMAC
- Aminopeptidase substrate Arg-CMAC
- Dipeptidyl peptidase substrate Ala-Pro-CMAC
- Yeast vacuole membrane marker MDY-64

Each of the components in the Yeast Vacuole Marker Sampler Kit, including the proprietary yeast vacuole membrane marker MDY-64 (Y-7536, see Section 12.3), is also available separately for those researchers who find that one particular dye is well suited for their application. See Section 12.3 for further information on the properties and uses of these components. Fluorescent dextrans (see Section 15.5) and the lipophilic styryl dye FM™ 4-64 (see Section 17.2) have also been used as vital stains for yeast vacuoles.[70,71]

Yeast Mitochondrial Stain Sampler Kit

Yeast biologists have long relied on fluorescence microscopy in the study of vital cell processes.[63,72] Molecular Probes now offers a Yeast Mitochondrial Stain Sampler Kit (Y-7530), which contains sample quantities of five different probes that have been found to selectively label yeast mitochondria. Both well-characterized and proprietary mitochondrion-selective probes are provided:

- Rhodamine 123 [73]
- Rhodamine B hexyl ester [74]
- MitoTracker™ Green FM
- SYTO 18 yeast mitochondrial stain
- $DiOC_6(3)$ [63,75,76]

The new mitochondrion-selective nucleic acid stain included in this kit — SYTO 18 yeast mitochondrial stain — exhibits a pronounced fluorescence enhancement upon binding to nucleic acids, resulting in very low background fluorescence even in the presence of dye. SYTO 18 is an effective mitochondrial stain in live yeast but neither penetrates nor stains the mitochondria of higher eukaryotic cells. Each of the components of the Yeast Mitochondrial Stain Sampler Kit is also available separately, including SYTO 18 yeast mitochondrial stain (S-7529, see Section 12.2). See Section 12.2 for further information on the properties and uses of these components.

Viability Assay and Gram Stain Kits for Bacteria

Fluorescent probes have been extensively used to assess bacterial function and viability,[62,77] as well as antibiotic susceptibility.[78] Molecular Probes' scientists have developed some important new fluorescence-based assay kits for characterizing bacteria:

- LIVE/DEAD *Bac*Light™ Bacterial Viability Kit (L-7012) for quick two-color viability assessments of both gram-negative and gram-positive bacteria
- LIVE *Bac*Light™ Bacterial Gram Stain Kit (L-7008) for rapid determination of the gram sign of live bacteria
- ViaGram™ Red+ Bacterial Gram Stain and Viability Kit (V-7023), which combines two nucleic acid stains with a lectin-based gram stain into one easy-to-use three-color fluorescence assay kit for simultaneous viability assessment and gram sign determination

All of the kits described in this section equate the presence of intact plasma membranes with viability. Viable bacteria rendered nonviable by exposure to agents that do not necessarily compromise the integrity of the plasma membrane, such as formaldehyde, usually appear viable by this criterion. In such cases, it may be preferable to use indicators of metabolic activity or membrane potential to assess bacterial viability (see Section 16.3).

The kits described here provide valuable alternatives to the insensitive and time-consuming methods used in conventional microbiological assays. Although it is impossible to predict the properties of these reagents on all bacteria under all possible environmental conditions, Molecular Probes expects that many more of the commonly encountered bacteria will stain similarly to the organisms listed in Table 16.2.

LIVE/DEAD *Bac*Light Bacterial Viability Kits

Molecular Probes' original LIVE/DEAD Viability/Cytotoxicity Kit (L-3224, see above) is a proven tool for assessing viability of animal cells but is generally not suitable for use with bacterial and yeast cells.[79] Consequently, we have developed the LIVE/DEAD *Bac*Light Bacterial Viability Kit,[67,80] which provides two different nucleic acid stains — SYTO 9 stain and propidium iodide — to rapidly distinguish live bacteria with intact plasma membranes from dead bacteria with compromised membranes.[62] This assay has several significant features:

- **Ease of use.** The reagents are simultaneously added to the bacterial suspension, which is then incubated for 5–10 minutes. No wash steps are required before analysis.
- **Specificity.** Live bacteria fluoresce green and dead bacteria fluoresce red. Live and dead bacteria can be distinguished and quantitated in minutes, even in a mixed population of bacterial species (see Color Plate 4 in "Visual Reality").
- **Reliability.** The LIVE/DEAD *Bac*Light Bacterial Viability Kits yield consistent results in tests on a variety of eubacterial genera

(Table 16.2). It can also be used to assess the viability of *Eurioplasma eurilytica* and *Mycoplasma hominus* mycoplasma and cysts of the protozoan parasite *Giardia muris*.[81]

- **Validity.** Viability measurements in fresh cultures of bacteria typically correlate well with enumeration techniques involving growth in liquid or solid media. However, variable results have been found using the LIVE/DEAD *Bac*Light reagents to assess viability in marine bacteria from environmental samples.
- **Versatility.** Bacteria can be stained in suspension or immobilized on microscope slides or filter membranes. Protocols are provided for bacterial viability analysis using a fluorescence microscope, flow cytometer, fluorometer or fluorescence microplate reader.

The intensities of the fluorescence signals produced by the SYTO 9 and propidium iodide nucleic acid stains can be adjusted by mixing different proportions of the dye solutions provided in the LIVE/DEAD *Bac*Light Kits. We have balanced the dye concentrations so that, for most bacteria, equal volumes of the two solutions provided give balanced staining of most species. The background remains virtually nonfluorescent, allowing live and dead cells to be easily differentiated in any fluorescence microscope equipped with a longpass fluorescein or comparable optical filter set. The LIVE/DEAD *Bac*Light viability assay has been used to quantitate total and viable concentrations of aerosolized *Pseudomonas fluorescens*[82] by fluorescence microscopy and to determine the antibacterial activity of quaternary ammonium compounds on *P. aeruginosa* by flow cytometry.[83]

Table 16.2 Organisms successfully stained with our LIVE/DEAD BacLight Bacterial Viability Kits (L-7007, L-7012).

Bacteria	
Gram-Positive	**Gram-Negative**
Bacillus cereus	*Agrobacterium tumefaciens*
Bacillus subtilis	*Edwardsiella ictaluri*
Clostridium perfringens	*Escherichia coli*
Lactobacillus sp.	*Klebsiella pneumoniae*
Micrococcus luteus	*Legionella pneumophila*
Mycobacterium phlei	*Pseudomonas aeruginosa*
Propionibacterium sp.	*Pseudomonas syringae*
Staphylococcus aureus	*Salmonella oranienburg*
Streptococcus pyogenes	*Serratia marcescens*
	Shigella sonnei
	Zymomonas sp.
Mycoplasma	
Eurioplasma eurilytica	
Mycoplasma hominus	
Protozoa	
Giardia muris cysts	

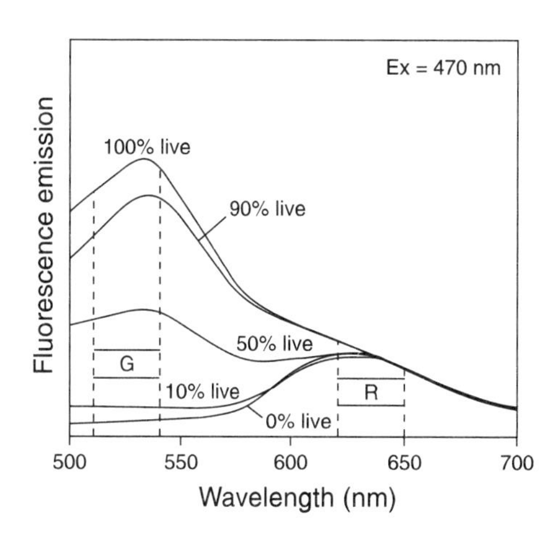

Figure 16.6 *Viability analysis of bacterial suspensions comprising various proportions of live and isopropyl alcohol–killed Escherichia coli using the reagents in Molecular Probes' LIVE/DEAD BacLight Bacterial Viability Kit (L-7007, L-7012). Live and dead bacteria are respectively stained fluorescent green (G) by SYTO 9 and fluorescent red (R) by propidium iodide. Bacterial suspensions that have been incubated in the two stains simultaneously and then excited at 470 nm exhibit a fluorescence spectral shift from red to green as the percentage of live bacteria in the sample is increased.*

Our original packaging of the LIVE/DEAD *Bac*Light Bacterial Viability Kit (L-7007), in which the dyes were mixed at different proportions in two solutions, is still available for customers who have already developed protocols using that formulation. However, we recommend use of the newer LIVE/DEAD *Bac*Light Bacterial Viability Kit (L-7012), which is more flexible because it provides separate solutions of the SYTO 9 and propidium iodide stains, thus facilitating calibration of bacterial fluorescence at each of the two emission wavelengths in quantitative assays. Kit L-7007 was designed primarily for use in fluorescence microscopy; Kit L-7012 is equally well suited for use in fluorescence microscopy and is better suited for use in quantitative analysis with a fluorometer, fluorescence microplate reader, flow cytometer (Figure 16.6) or other instrumentation. Both kits include a procedure for mounting bacteria stained with the LIVE/DEAD *Bac*Light Bacterial Viability Kit on filter membranes and a proprietary mounting oil that we have found to be useful for the direct epifluorescence filter technique (DEFT).[84]

LIVE *Bac*Light Bacterial Gram Stain Kit

Our LIVE *Bac*Light Bacterial Gram Stain Kits[67,85,86] (L-7005, L-7008) are based on differential nucleic acid staining of *live* gram-negative and gram-positive bacteria. The gram stain is one of the most important and widely used differential stains for the taxonomic classification of bacteria in both clinical and research settings. The original method involves several steps, including heat fixation of the bacteria, a two-step staining protocol, alcohol extraction and counterstaining. Over the years, several improved gram-staining techniques have been developed, but most involve cell-fixation or cell-permeabilization steps that kill the bacteria being tested. The one-step LIVE *Bac*Light Bacterial Gram Stain Kits overcome several of the problems inherent in these labor-intensive, fixation-dependent procedures.

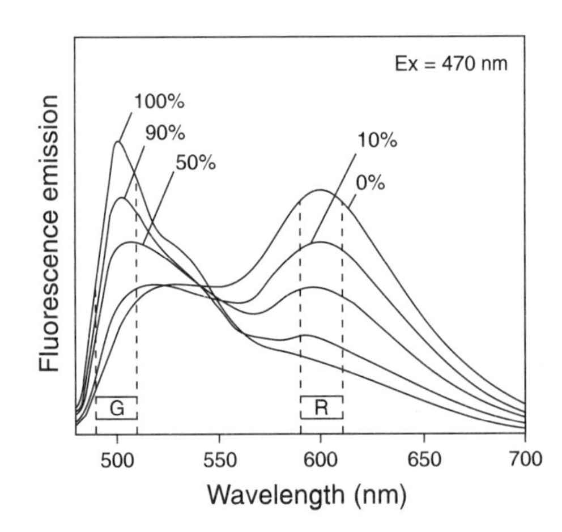

Figure 16.7 *Analysis of the percentage of gram-negative Escherichia coli in a mixed suspensions containing gram-positive Staphylococcus aureus using the reagents in Molecular Probes' LIVE BacLight Bacterial Gram Stain Kit (L-7008). Live gram-negative and gram-positive bacteria are respectively stained fluorescent green (G) by SYTO 9 and fluorescent red (R) by hexidium iodide. Bacterial suspensions that have been incubated in the two stains simultaneously and then excited at 470 nm exhibit a fluorescence spectral shift from red to green as the percentage of gram-negative bacteria in the sample is increased.*

Unlike conventional gram stain procedures, the LIVE *Bac*Light Bacterial Gram Stain Kit allows researchers to rapidly classify bacteria as either gram-negative or gram-positive in minutes using a single staining solution, no fixatives and no wash steps. The LIVE *Bac*Light Bacterial Gram Stain Kit contains our green fluorescent SYTO 9 and red fluorescent hexidium iodide nucleic acid stains. These two dyes differ both in their spectral characteristics and in their ability to label live gram-negative and gram-positive bacteria. When a mixed population of live gram-negative and gram-positive bacteria is simultaneously stained with the membrane-permeant SYTO 9 dye in combination with hexidium iodide, gram-negative bacteria fluoresce green, and the gram-positive bacteria fluoresce orange-red (see Color Plate 4 in "Visual Reality"). Dead bacteria do not exhibit predictable staining patterns. Gram-negative and gram-positive organisms can be easily differentiated in any fluorescence microscope equipped with a standard fluorescein longpass optical filter set. The assay provides a sensitive indicator system for analyzing low numbers of bacteria in the presence of background material because the unbound reagents exhibit low fluorescence when not bound to nucleic acids. The preferred use of the LIVE *Bac*Light Bacterial Gram Stain Kit may be for measuring dynamic changes in the composition of bacterial populations.

As with the LIVE/DEAD *Bac*Light Bacterial Viability Kit, our original packaging of the LIVE *Bac*Light Bacterial Gram Stain Kit (L-7008), in which the dyes were mixed at different proportions in two solutions, is still available for customers who have already developed protocols using that formulation. However, we recommend use of the newer LIVE *Bac*Light Bacterial Gram Stain Kit (L-7005), which is more flexible because it provides separate solutions of the SYTO 9 and hexidium iodide stains, thus facilitating calibration of bacterial fluorescence at each of the two emission wavelengths in quantitative assays. Kit L-7008 was designed primarily for use in fluorescence microscopy; Kit L-7005 is equally well suited for use in fluorescence microscopy and is better suited for use in quantitative analysis with a fluorometer, fluorescence microplate reader (Figure 16.7), flow cytometer or other instrumentation. Both kits include a procedure for mounting bacteria stained with the LIVE *Bac*Light Bacterial Gram Stain Kit on filter membranes and the proprietary *Bac*Light mounting oil that we have found to be useful for the direct epifluorescence filter technique (DEFT).[84] Kit components, number of assays and assay principles are summarized in Table 16.1. Hexidium iodide, the gram-positive–selective nucleic acid stain, is available separately (H-7593, see Section 16.3).

ViaGram Red+ Bacterial Gram Stain and Viability Kit

Molecular Probes' ViaGram Red+ Bacterial Gram Stain and Viability Kit (V-7023) provides an easy, three-color fluorescent staining protocol that differentially stains many gram-negative and gram-positive bacterial species and, at the same time, discriminates live from dead cells on the basis of plasma membrane integrity. The kit contains three reagents: two nucleic acid stains for viability determination — blue fluorescent cell-permeant DAPI and green fluorescent, cell-impermeant SYTOX® Green nucleic acid stain — as well as the red fluorescent Texas Red®-X wheat germ agglutinin (WGA) for gram sign determination.[87] Bacteria with intact cell membranes stain fluorescent blue with DAPI, whereas bacteria with damaged membranes stain fluorescent green with SYTOX Green nucleic acid stain. The background remains virtually nonfluorescent. The Texas Red-X WGA component selectively binds to the surface of gram-positive bacteria, providing a red fluorescent cell-surface stain that effectively distinguishes them from gram-negative bacteria, even in the presence of the viability stains. Thus, with three fluorescent colors, the four possible combinations of live or dead, gram-negative and gram-positive cells are discriminated with a fluorescence microscope (see Color Plate 4 in "Visual Reality"). This kit also includes a procedure for mounting bacteria on filter membranes and the *Bac*Light mounting oil, which we have found to be useful for the direct epifluorescence filter technique (DEFT).[84] Kit components, number of assays and assay principles are summarized in Table 16.1.

Wheat Germ Agglutinin Sampler Kit

Fluorescent lectins have proven useful in microbiology applications. Fluorescent WGA conjugates selectively stain chitin in fungal cell walls,[88] as well as the surface of gram-positive but not of gram-negative bacteria.[89,90] Fluorescent WGA has also been shown to bind to sheathed microfilariae and has been used to detect filarial infection in blood smears.[91] Our Wheat Germ Agglutinin Sampler Kit (W-7024) provides 1 mg samples of four of our brightest fluorescent WGA conjugates, spanning the spectrum from blue to red. Included in this kit are conjugates of the blue fluorescent AMCA-S, green fluorescent Oregon Green™ 488, orange fluorescent tetramethylrhodamine and red fluorescent Texas Red-X dyes. See Section 7.6 for more information on lectins, including additional WGA and concanavalin A (Con A) conjugates.

Bacteria Counting Kit

Accurate enumeration of low numbers of bacteria in samples must be performed daily in many quality-control laboratories. To facilitate this determination by flow cytometry, Molecular Probes has developed the Bacteria Counting Kit (B-7277), which provides:

- Cell-permeant, green fluorescent SYTO BC nucleic acid stain to label bacteria
- Fluorescent polystyrene microspheres to calibrate the volume of bacterial suspension analyzed
- Detailed protocol

The SYTO BC dye is a high-affinity nucleic acid stain that easily penetrates both gram-negative and gram-positive bacteria, producing an exceptionally bright green fluorescent signal. The calibrated suspension of polystyrene microspheres contains beads that exhibit a uniform density, low-level fluorescence and optimal size to clearly separate the light scattering of the microspheres from that of most bacteria.

The Bacteria Counting Kit is particularly valuable for monitoring antibiotic sensitivity because it provides a convenient and accurate means for assessing changes in a bacterial population over time. A sample of the population is simply diluted, stained briefly with the SYTO BC dye, mixed with a fixed number of microspheres and analyzed on a flow cytometer. Signals from both the stained bacteria and the beads are easily detected in the FL1 (fluorescein) channel of most flow cytometers and can be distinguished on a plot of forward scatter versus fluorescence (Figure 16.8); the density of the bacteria in the sample can be determined from the ratio of bacterial signals to microsphere signals in the cytogram. The Bacteria Counting Kit can be used with a variety of gram-negative and gram-positive species of bacteria and provides sufficient reagents for approximately 100 flow cytometry assays.

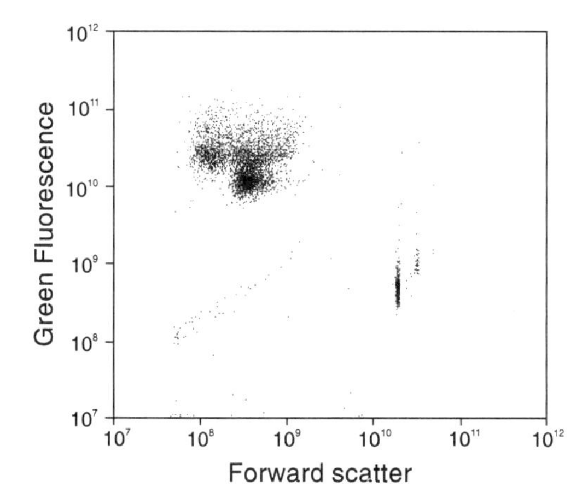

Figure 16.8 *Flow cytometric enumeration of Bacillus cereus using Molecular Probes' Bacteria Counting Kit (B-7277). Bacteria stained with SYTO BC bacterial cell stain and mixed with known concentrations of weakly fluorescent 6 µm polystyrene microsphere standards produce a bivariate frequency distribution for forward light scatter versus green fluorescence intensity that allows the bacterial population number to be determined by reference to the clearly separated microsphere standard population.*

Kits for Detecting Microorganism Contamination

Cell Culture Contamination Detection Kit

Molecular Probes' Cell Culture Contamination Detection Kit (C-7028) uses a simple and effective procedure for visually screening cell cultures for contamination by yeast (and other fungi) or by gram-negative or gram-positive bacteria. This kit not only serves to detect the contaminants, but also identifies the contaminant type, enabling the researcher to choose an appropriate course of action.

A sample of the suspected culture is subjected to two slide-staining protocols. One sample slide is stained with Calcofluor White M2R, a UV-excitable, blue fluorescent stain specific for fungal cell walls. A second slide is stained with our SYTO 9 nucleic acid stain, which identifies all bacteria irrespective of gram signature, and also with the Texas Red-X conjugate of wheat germ agglutinin (WGA), which selectively stains gram-positive bacteria.[87,89,92] On this second slide, gram-positive bacteria exhibit green fluorescent interior plus red fluorescent cell-surface staining, whereas gram-negative bacteria show only green fluorescent interior staining. Staining and examination of the slides under a fluorescence microscope can be performed in less than one hour. Each Cell Culture Contamination Detection Kit contains:

- Green fluorescent SYTO 9 nucleic acid stain
- Blue fluorescent Calcofluor White M2R fungal cell wall stain
- Red fluorescent Texas Red-X WGA, for positive identification of gram-positive bacteria
- Buffer for reconstituting Texas Red-X WGA
- Detailed protocol

This kit provides sufficient material for approximately 200 contamination detection assays.

MycoFluor™ Mycoplasma Detection Kit

Mycoplasma infections are generally difficult or impossible to detect during routine work with cultured cells because these intracellular pathogens cannot be observed by standard light microscopy. It is not surprising then that mycoplasma infections are relatively common. Estimates of contaminated cultures in the United States range from 5% to 35%, whereas the contamination rate is postulated to be much higher in those countries where systematic detection and elimination are not practiced.[93] Not only do mycoplasma cause physiological and morphological distortions that may affect experimental results, but contamination can quickly spread to other cell lines.

The MycoFluor Mycoplasma Detection Kit (M-7006) provides an extremely rapid and sensitive fluorescence microscopy–based assay for the visual identification of mycoplasma infection in laboratory cell cultures and media. In order to detect mycoplasma, the fluorescent MycoFluor reagent is added directly to the culture medium, with or without cells, and the stained sample is then examined under a fluorescence microscope. The test for the presence of mycoplasma in live or fixed cell cultures takes about 15 minutes from when the reagent is added until when the sample is viewed with a fluorescence microscope equipped with DAPI optical filters. The detection of mycoplasma in cell media requires about 30 minutes, depending on the amount of centrifugation required to concentrate potential contaminants.

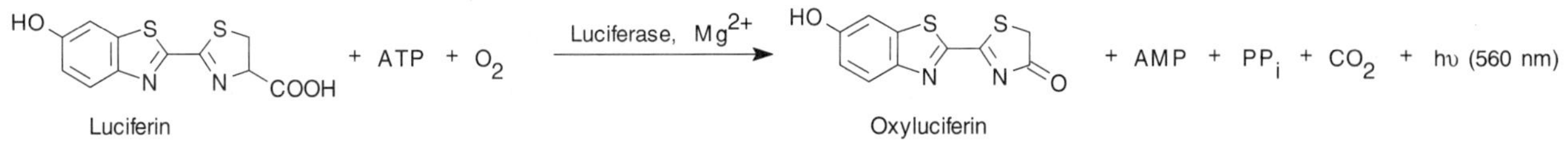

Luciferin + ATP + O_2 —(Luciferase, Mg^{2+})→ Oxyluciferin + AMP + PP_i + CO_2 + hυ (560 nm)

Figure 16.9 Reaction scheme for bioluminescence generation via luciferase-catalyzed conversion of luciferin to oxyluciferin.

Also provided with this kit are mycoplasma MORFS (Microscopic Optical Replicas for Fluorescence assays), which serve as inert positive controls that mimic the size, shape and fluorescence intensity of mycoplasma stained with the blue fluorescent MycoFluor reagent and viewed by fluorescence microscopy. The optical properties of the mycoplasma MORFS enable the researcher to discriminate between stained mycoplasma and other fluorescent material without introducing infectious biological agents into the laboratory environment. No previous experience with mycoplasma testing is required.

Each MycoFluor Mycoplasma Detection Kit supplies sufficient materials for at least 100 tests of live cells, fixed cells or culture media. Kit contents include:

- 20X Concentrated MycoFluor reagent
- Mycoplasma MORFS stock suspension
- Coverslip sealant
- Cotton swab
- Reference micrographs
- Detailed protocol

Alternatively, the LIVE/DEAD *Bac*Light Bacterial Viability Kits may be useful for detecting mycoplasma infections. Researchers have determined that the reagents in our LIVE/DEAD *Bac*Light Kits can be used for viability determinations in *Eurioplasma eurilytica* and *Mycoplasma hominus* mycoplasma.

ATP Determination Kit

Molecular Probes now offers a convenient ATP Determination Kit (A-6608) for the sensitive bioluminescence-based detection of adenosine 5'-triphosphate (ATP) with recombinant firefly luciferase and its substrate, luciferin. This assay is based on luciferase's requirement for ATP to produce light. In the presence of Mg^{2+}, luciferase catalyzes the reaction of luciferin, ATP and O_2 to form oxyluciferin, AMP, CO_2, pyrophosphate and ~560 nm light (Figure 16.9).

The luciferin–luciferase bioluminescence assay is extremely sensitive; most luminometers can detect as little as 1 picomole of pre-existing ATP or ATP as it is generated in kinetic systems. This sensitivity has led to its widespread use for detecting ATP in various enzymatic reactions, as well as for measuring viable cell number [94] and for detecting low-level bacterial contamination in samples such as blood, milk, urine, soil and sludge.[95-99] The luciferin–luciferase bioluminescence assay has also been used to determine cell proliferation and cytotoxicity in both bacterial [100,101] and mammalian cells,[102,103] and to distinguish cytostatic versus cytocidal potential of anticancer drugs on malignant cell growth.[104]

The ATP Determination Kit from Molecular Probes contains:

- Luciferin
- Luciferase
- Dithiothreitol (DTT)
- ATP
- 20X concentrated reaction buffer
- Protocol for ATP quantitation

Unlike many other commercially available ATP detection kits, our ATP Determination Kit provides the luciferase and luciferin packaged separately, which enables researchers to optimize the reaction conditions for their particular instruments and samples. Each ATP Determination Kit provides sufficient reagents to perform 200 ATP assays using 0.5 mL sample volumes or 500 ATP assays using 0.2 mL sample volumes.

1. See Burghardt, R.C. *et al.* in *Principles and Methods of Toxicology, Third Edition*, A.W. Hayes, Ed., Raven Press (1994) pp. 1231–1258 for a brief protocol for using this product for *in vitro* toxicology testing; **2.** U.S. Patent No. 5,314,805; **3.** J Neurosci 15, 5389 (1995); **4.** Biol Bull 189, 218 (1995); **5.** Opthalmologe (German) 92, 452 (1995); **6.** J Cell Sci 106, 685 (1993); **7.** Cytometry 20, 181 (1995); **8.** J Biol Chem 270, 23895 (1995); **9.** J Biol Chem 270, 7791 (1995); **10.** J Neurosci 14, 2464 (1994); **11.** J Neurosci 14, 2260 (1994); **12.** J Biol Chem 269, 6803 (1994); **13.** J Immunol Meth 177, 101 (1994); **14.** Human Immunol 57, 264 (1993); **15.** Theriogenology 43, 595 (1995); **16.** Shapiro, H.M., *Practical Flow Cytometry, Third Edition,* Wiley-Liss (1995) pp. 250–251; **17.** Meth Cell Biol 42, 359 (1994); **18.** Meth Cell Biol 42, 295 (1994); **19.** Bishop-Stewart *et al.*, manuscript in preparation; **20.** Favero, M.S. and Bond, W.W. in *Manual of Clinical Microbiology, Fifth Edition*, A. Balows, Ed., American Society for Microbiology (1991) pp. 183–200; **21.** J Immunol Meth 156, 47 (1992); **22.** J Immunol Meth 185, 209 (1995); **23.** J Immunol Meth 166, 45 (1993); **24.** Theriogenology 45, 1103 (1996); **25.** Biol Reprod 53, 276 (1995); **26.** Poultry Sci 74, 1191 (1995); **27.** J Andrology 15, 620 (1994); **28.** Theriogenology 45, 923 (1996); **29.** Nature 377, 20 (1995); **30.** Theriogenology 39, 1009 (1993); **31.** Gamete Res 22, 355 (1989); **32.** Biol Reprod 34, 127 (1986); **33.** J Histochem Cytochem 30, 279 (1982); **34.** Meth Cell Biol 33, 401 (1990); **35.** J Histochem Cytochem 25, 46 (1977); **36.** J Andrology 12, 112 (1991); **37.** Cytometry 8, 642 (1987); **38.** Cytometry 1, 132 (1980); **39.** Biol Reprod 50, 987 (1994); **40.** J Reprod Fertil 96, 581 (1992); **41.** Jones, L. *et al.*, manuscript in preparation; **42.** Anal Biochem 207, 186 (1992); **43.** J Immunol Meth 142, 199 (1991); **44.** Anal Biochem 213, 426 (1993); **45.** In Vitro Toxicol 3, 219 (1990); **46.** Exp Cell Res 124, 329 (1979); **47.** Biochemie 76, 452 (1994); **48.** Anal Biochem 191, 31 (1990); **49.** J Immunol Meth 162, 41 (1993); **50.** Cancer Res 49, 565 (1989); **51.** Anal Biochem 131, 538 (1983); **52.** Anal Biochem 102, 344 (1980); **53.** Eur J Cancer 29A, 395 (1993); **54.** J Natl Cancer Inst 82, 1113 (1990); **55.** J Natl Cancer Inst 82, 1107 (1990); **56.** Biochemistry 34, 32 (1995); **57.** J Biol Chem 270, 28145 (1995); **58.** Biotech Histochem 67, 73 (1992); **59.** Cytometry 12, 68 (1991); **60.** Histochemistry 75, 353 (1982); **61.** Stain Technol 53, 205 (1979); **62.** FEMS Microbiol Lett 133, 1 (1995); **63.** J Cell Biol 126, 1375 (1994); **64.** J Cell Biol 123, 1821 (1993); **65.** J Cell Biol 114, 111 (1991); **66.** J Cell Biol 114, 101 (1991); **67.** U.S. Patent No. 5,436,134 and patents pending; **68.** U.S. Patent No. 5,445,946; **69.** Roth, B.L. *et al.*, manuscript in preparation; **70.** J Cell Biol 128, 779 (1995); **71.** J Cell Biol 105, 1539 (1987); **72.** Meth Cell Biol 31, 357 (1989); **73.** Current Genetics 18, 265 (1990); **74.** J Cell Sci 101, 315 (1992); **75.** Cell Motil Cytoskeleton 25, 111 (1993); **76.** Biochem Int'l 2, 503 (1981); **77.** J Microbiol Meth 21, 1 (1995); **78.** J Antimicrobial Chemotherapy 34, 613 (1994); **79.** J Microbiol Meth 17, 1 (1993); **80.** U.S. Patent No. 5,534,416; **81.** R. Taghi-Kilani *et al.*, Int'l J Parasitol (1996) in press; **82.** Appl Environ Microbiol 62, 2264 (1996); **83.** Abstracts of the 95th General Meeting of the American Society for Microbiology

433, abstract #Q192 (1995); **84.** Appl Environ Microbiol 39, 423 (1980); **85.** U.S. Patent No. 5,545,535; **86.** U.S. Patent No. 5,437,980; **87.** Purchase of the ViaGram Red+ Bacterial Gram Stain and Viability Kit is accompanied by a research license under U.S. Patent No. 5,137,810 for determination of bacterial gram sign with fluorescent lectins; **88.** Investigative Ophthalmol Visual Sci 27, 500 (1986); **89.** Appl Environ Microbiol 56, 2245 (1990); **90.** U.S. Patent No. 5,137,810; **91.** Int'l J Parasitol 20, 1099 (1990); **92.** J Clin Microbiol 10, 669 (1979); **93.** Nature 339, 487 (1989); **94.** J Biolumin Chemilumin 10, 29 (1995); **95.** Anal Biochem 175, 14 (1988); **96.** Bio/Technology 6, 634 (1988); **97.** J Clin Microbiol 20, 644 (1984); **98.** J Clin Microbiol 18, 521 (1983); **99.** Meth Enzymol 57, 3 (1978); **100.** Biotech Bioeng 42, 30 (1993); **101.** J Biolumin Chemilumin 6, 193 (1991); **102.** Biochem J 295, 165 (1993); **103.** J Immunol Meth 160, 81 (1993); **104.** J Natl Cancer Inst 77, 1039 (1986).

16.2 Price List *Assay Kits for Diverse Cell Types*

Cat #		Product Name	Unit Size	Price Per Unit ($) 1–4 Units	5–24 Units
A-6608	*New*	ATP Determination Kit *200-1000 assays*	1 kit	95.00	76.00
B-7277	*New*	Bacteria Counting Kit *for flow cytometry*	1 kit	95.00	76.00
C-7028	*New*	Cell Culture Contamination Detection Kit *200 assays*	1 kit	95.00	76.00
C-7027	*New*	CyQUANT™ cell lysis buffer *20X concentrate*	50 mL	45.00	36.00
C-7026	*New*	CyQUANT™ Cell Proliferation Assay Kit *for cells in culture* *1000 assays*	1 kit	135.00	108.00
F-2962		FluoReporter® Blue Fluorometric dsDNA Quantitation Kit *200-2000 assays*	1 kit	78.00	62.40
F-2961		FluoReporter® Colorimetric Cell Protein Assay Kit	1 kit	100.00	80.00
F-2960		FluoReporter® Fluorometric Cell Protein Assay Kit	1 kit	135.00	108.00
F-7030	*New*	FUN-1™ cell stain *10 mM solution in DMSO*	100 µL	75.00	60.00
L-7008	*New*	LIVE *Bac*Light™ Bacterial Gram Stain Kit *for microscopy* *1000 assays*	1 kit	195.00	156.00
L-7005	*New*	LIVE *Bac*Light ™ Bacterial Gram Stain Kit *for microscopy and quantitative assays* *1000 assays*	1 kit	195.00	156.00
L-7007	*New*	LIVE/DEAD® *Bac*Light™ Bacterial Viability Kit *for microscopy* *1000 assays*	1 kit	195.00	156.00
L-7012	*New*	LIVE/DEAD® *Bac*Light™ Bacterial Viability Kit *for microscopy and quantitative assays* *1000 assays*	1 kit	195.00	156.00
L-7010	*New*	LIVE/DEAD® Cell-Mediated Cytotoxicity Kit *for animal cells* *2000 assays*	1 kit	95.00	76.00
L-7013	*New*	LIVE/DEAD® Reduced Biohazard Viability/Cytotoxicity Kit *for animal cells* *100 assays*	1 kit	95.00	76.00
L-7011	*New*	LIVE/DEAD® Sperm Viability Kit *200-1000 assays*	1 kit	195.00	156.00
L-3224		LIVE/DEAD® Viability/Cytotoxicity Kit *for animal cells* *1000 assays*	1 kit	195.00	156.00
L-7009	*New*	LIVE/DEAD® Yeast Viability Kit *1000 assays*	1 kit	195.00	156.00
M-7006	*New*	MycoFluor™ Mycoplasma Detection Kit	1 kit	95.00	76.00
V-7023	*New*	ViaGram™ Red+ Bacterial Gram Stain and Viability Kit *200 assays*	1 kit	95.00	76.00
W-7024	*New*	Wheat Germ Agglutinin Sampler Kit *four fluorescent conjugates, 1 mg each*	1 kit	98.00	78.40
Y-7530	*New*	Yeast Mitochondrial Stain Sampler Kit	1 kit	95.00	76.00
Y-7531	*New*	Yeast Vacuole Marker Sampler Kit	1 kit	95.00	76.00

16.3 Probes for Live-Cell Function

Viability/Cytotoxicity Assays Using Esterase Substrates

Molecular Probes prepares a wide variety of fluorogenic esterase substrates — including calcein AM, BCECF AM and fluorescein diacetate — that can be passively loaded into adherent and nonadherent cells. These cell-permeant esterase substrates serve as viability probes that measure both enzymatic activity, which is required to activate their fluorescence, and cell-membrane integrity, which is required for intracellular retention of their fluorescent products.

As electrically neutral or near-neutral molecules, the esterase substrates freely diffuse into cells. In general, cell loading of acetate or acetoxymethyl ester derivatives is accomplished by initially preparing a 1–10 mM stock solution of the dye in dimethylsulfoxide (DMSO) and then diluting the stock solution into the cell medium to a final concentration of 1–25 µM (see the Technical Note "Loading and Calibration of Intracellular Ion Indicators" on page 549). Once inside, these nonfluorescent substrates are converted by intracellular esterases into fluorescent products that are retained by cells with intact plasma membranes. In contrast, both the substrates and their products rapidly leak from dead or damaged cells with compromised membranes, even when the cells retain some residual esterase activity. Low incubation temperatures and highly charged esterase products usually favor retention, although the rate of dye loss from viable cells also depends to a large extent on cell type (see "Multidrug Resistance" in this section). For example, mast cells and epithelial cells actively secrete many polar products.[1,2] Table 16.3 lists Molecular Probes' esterase substrates that have been used for cell viability studies and compares their cell loading, retention and pH sensitivity.

Calcein AM

Of the dyes listed in Table 16.3, calcein AM (C-1430, C-3099, C-3100) stands out as the premier indicator of cell viability due to its superior cell retention and pH-insensitive fluorescence.[3-6] Calcein (C-481, see Section 15.3), which is the hydrolysis product of calcein AM, is a polyanionic fluorescein derivative that has about six negative and two positive charges at pH 7.[7] Calcein is better retained by viable cells than fluorescein, carboxyfluorescein and BCECF (Figure 16.10) and tends to have brighter fluorescence in a number of mammalian cell types. Furthermore, unlike other dyes, including BCECF AM, calcein AM does not interfere with leukocyte chemotaxis or superoxide production, nor does it affect lymphocyte–target cell conjugation.[5,8-11] Leakage of calcein from calcein AM–loaded cells has been used to measure the increase in membrane permeability that occurs above physiological temperatures,[12] as well as to assay for cytotoxic T lymphocyte activity.[4]

BCECF AM

BCECF AM (B-1150, B-1170, B-3051) is widely used for determining the ability of surviving cells to proliferate [13,14] and for detecting cytotoxicity.[15] The intracellular hydrolysis product of BCECF AM, BCECF (B-1151, see Section 23.2), has 4–5 negative charges, a property that considerably improves its cell retention in viable cells over that of fluorescein or carboxyfluorescein (Figure 16.10). However, because BCECF's emission intensity is only half maximal at pH 7.0 ($pK_a = 6.98$, see Figure 23.3 in Chapter 23) and is even further reduced in a cell's acidic compartments, BCECF's signal intensity may be less than optimal in some cell viability and cell adhesion assays.

Using monoclonal antibodies known to either enhance or inhibit NK cell function, researchers found that BCECF AM was at least as effective as ^{51}Cr for measuring NK activity. Furthermore, the fluorescence-based assay could be performed with smaller samples.[16] BCECF AM has also been used to screen for trypanocidal activity [17] and viability of islets.[18]

Fluorescein Diacetate (FDA)

Fluorescein diacetate (FDA, F-1303) was one of the first probes to be used as a fluorescent indicator of cell viability.[19-21] FDA is still occasionally used to detect cell adhesion [22] or, in combination with propidium iodide (P-1304), to determine cell viability.[23,24] However, fluorescein (F-1300, see Section 23.2) formed by intracellular hydrolysis of FDA, rapidly leaks from cells (Figure 16.10). Thus, other cell-permeant dyes such as BCECF AM and calcein AM (see above) are now preferred for cell viability assays.

Carboxyfluorescein Diacetate (CFDA) and Carboxy-2′,7′-Dichlorofluorescein Diacetate

The high leakage rate of fluorescein from cells [21,25] prompted the development of carboxyfluorescein diacetate (CFDA), which was originally used to measure intracellular pH [26] but was soon adapted for use as a cell viability indicator.[27,28] Upon hydrolysis by intracellular nonspecific esterases, CFDA forms carboxyfluorescein (5(6)-FAM; C-194, C-1904; see Section 15.3). As compared to fluorescein, carboxyfluorescein contains extra negative charges and is therefore better retained in cells [5] (Figure 16.10). CFDA is moderately permeant to most cell membranes, with uptake greater at pH 6.2 than at pH 7.4.[26]

The mixed-isomer of CFDA (5(6)-CFDA, C-195) is usually adequate for cell viability measurements; however, we also prepare high-purity single isomers of CFDA (C-1361, C-1362). In addition, we offer the electrically neutral AM ester of CFDA (5-CFDA, AM; C-1354), which facilitates cell loading. Upon hydrolysis by intracellular esterases, this AM ester also yields carboxyfluores-

Table 16.3 *Esterase substrates for cell viability studies.*

Esterase Substrate (Cat #)	Properties in Cells	pK_a of Product
BCECF AM (B-1150, B-1170, B-3051)	• Quite well retained • Released during cytolysis • pH-sensitive fluorescence	~7.0
Calcein AM (C-1430, C-3099, C-3100)	• Quite well retained • Released during cytolysis • pH-insensitive fluorescence	~5
Carboxy-2′,7′-dichlorofluorescein diacetate (carboxy-DCFDA, C-369)	• Moderately well retained • Not as pH-sensitive as CFDA	~4.8
Carboxyfluorescein diacetate (5(6)-CFDA, C-195)	• Moderately well retained • pH-sensitive fluorescence	~6.4
Carboxyfluorescein diacetate, acetoxymethyl ester (5-CFDA, AM; C-1354)	• Easier to load than CFDA yet yields the same product upon hydrolysis • pH-sensitive fluorescence	~6.4
Carboxyfluorescein diacetate, succinimidyl ester (5(6)-CFDA, SE; C-1157)	• Well retained by reaction with amines • Not completely released during cytolysis • pH-sensitive fluorescence	~6.4 *
CellTracker Green CMFDA (CMFDA; C-2925)	• Well retained by reaction with thiols • Not completely released during cytolysis • pH-sensitive fluorescence	~6.4 *
Chloromethyl SNARF-1, acetate (C-6826)	• Well retained by reaction with thiols • Not completely released during cytolysis • Long-wavelength, pH-sensitive fluorescence	~7.5 *
Fluorescein diacetate (F-1303)	• Poorly retained • pH-sensitive fluorescence • Inexpensive	~6.4
Oregon Green 488 carboxylic acid diacetate (carboxy-F_2FDA, O-6151)	• Moderately well retained • Not as pH-sensitive as CFDA	~4.7
Hexafluorofluorescein diacetate (HFFDA, H-7019) Tetrabromotetrafluorofluorescein diacetate (Br_4TFFDA, T-7017) Tetrafluorofluorescein diacetate (TFFDA, T-7016)	• Well retained by reaction with thiols • Not completely released during cytolysis • pH-insensitive fluorescence • Relatively photostable	~5.0–6.2

* This value applies to the unconjugated hydrolysis product; after conjugation to an intracellular amine or thiol, the actual pK_a value may be different.

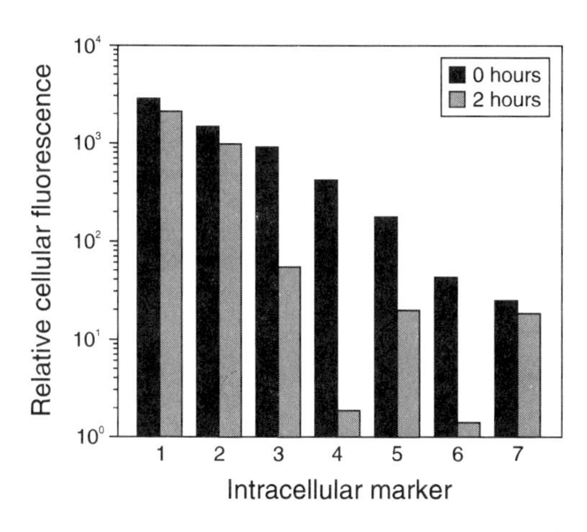

Figure 16.10 Loading and retention characteristics of intracellular marker dyes. Cells of a human lymphoid line (GePa) were loaded with the following cell-permeant acetoxymethyl ester (AM) or acetate derivatives of fluorescein: 1) calcein AM, 2) BCECF AM, 3) tetrafluorofluorescein diacetate (TFFDA), 4) fluorescein diacetate (FDA), 5) hexafluorofluorescein diacetate (HFFDA), 6) carboxyfluorescein diacetate (CFDA) and 7) chloromethylfluorescein diacetate (CMFDA). Cells were incubated in 4 µM staining solutions in Dulbecco's modified eagle medium containing 10% fetal bovine serum (DMEM+) at 37°C. After incubation for 30 minutes, cell samples were immediately analyzed on a Becton Dickinson FACS Vantage cytometer to determine the average fluorescence per cell at time zero (0 hours). Retained cell samples were subsequently washed twice by centrifugation, resuspended in DMEM+, maintained at 37°C for 2 hours and then analyzed by flow cytometry. The decrease in the average fluorescence intensity per cell in these samples relative to the time zero samples indicates the extent of intracellular dye leakage during the 2-hour incubation period.

cein.[29-31] Hemoglobin can be used to quench extracellular fluorescence due to leakage of probes or leakage of products, such as fluorescein or carboxyfluorescein.[32] Alternatively, antibodies directed against the fluorescein hapten (see Section 7.3) can be used to quench low levels of extracellular fluorescence of some fluorescein-based dyes.

CFDA has been used as a viability probe with a variety of cells, including bacteria,[33] fungi (e.g., *Saccharomyces cerevisiae*),[34] spermatozoa,[35] natural killer (NK) cells [15,16,36,37] and tumor cells [38] Cytotoxicity assays using either CFDA or 5-(and-6)-carboxy-2',7'-dichlorofluorescein diacetate (carboxy-DCFDA, C-369) show good correlation with results obtained using the radioisotopic ^{51}Cr-release method.[15,39] With its low pK_a, carboxy-DCFDA is frequently used as a selective probe for the relatively acidic yeast vacuole.[40-42] Oregon Green 488 carboxylic acid diacetate (carboxy-DFFDA, O-6151) also exhibits a low pK_a (~4.7, see Figure 23.11 in Chapter 23) and may be similarly useful as a vital stain for acidic organelles.

Sulfofluorescein Diacetate (SFDA)

Sulfofluorescein diacetate (SFDA, S-1129), which is converted by intracellular esterases to fluorescein sulfonic acid (F-1130, see Section 15.3), is more polar than CFDA and consequently may be more difficult to load into viable cells. However, SFDA's polar hydrolysis product, fluorescein sulfonic acid, is better retained in viable cells than carboxyfluorescein.[43-46] SFDA was recently used to stain live bacteria and fungi in soil; little interference from autofluorescence of soil minerals or detritus was observed.[47]

Fluorinated Fluorescein Derivatives

Molecular Probes now offers a new class of fluorescent cytoplasmic markers for cytotoxicity and cell tracing studies that includes the following fluorinated fluorescein derivatives:

- Oregon Green 488 carboxylic acid diacetate (carboxy-DFFDA, O-6151)
- 4,5,6,7-Tetrafluorofluorescein diacetate (TFFDA, T-7016)
- 2',4,5,6,7,7'-Hexafluorofluorescein diacetate (HFFDA, H-7019)
- 2',4',5',7'-Tetrabromo-4,5,6,7-tetrafluorofluorescein diacetate (Br_4TFFDA, T-7017)

Like the other fluorogenic esterase substrates, these polyfluorinated fluorescein–based cytoplasmic markers are virtually nonfluorescent until their acetate moieties are enzymatically hydrolyzed by intracellular esterases. Use of carboxy-DFFDA as a viability indicator is analogous to CFDA; however, its hydrolysis product, the 6-isomer of Oregon Green 488 carboxylic acid (O-6148, see Section 1.4), is more photostable than carboxyfluorescein (see Figure 1.6 in Chapter 1) and has a significantly lower pK_a (~4.7 versus ~6.4, see Figure 1.7 in Chapter 1).

The other cell-permeant, polyfluorinated fluorescein derivatives — TFFDA, HFFDA and Br_4TFFDA — are unique in that they react with intracellular thiols *before* their acetates are removed by esterase activity. When used at 1 µM to stain live cells, TFFDA and HFFDA behave like CellTracker Green CMFDA, forming covalent adducts that are well retained in many cell types (Figure 16.10); these adducts most likely include conjugates of glutathione and thiol-containing proteins, though this has not yet been confirmed experimentally. In contrast, if the acetate groups of TFFDA, HFFDA and Br_4TFFDA are hydrolyzed prior to thiol reaction, then these derivatives yield fluorescent products that rapidly leak from the cells. This novel class of polyfluorinated compounds should prove useful for further improving the retention of fluorescent products in applications that currently utilize BCECF AM or calcein AM, or in specialized assays for dye export (see "Multidrug Resistance" in this section).

Br_4TFFDA is converted to a tetrafluorinated eosin analog (Br_4TFF, T-7015) in viable cells. The structural and spectral similarities of Br_4TFFDA to eosin and CellTracker Yellow-Green CMEDA suggest that this dye may be useful for selective cell photoablation in embryo development studies or for correlated fluorescence and electron microscopy studies. We have shown that fluorination has a relatively small effect on Br_4TFF's spectra but lowers its pK_a and increases its photostability relative to that of the corresponding fluorescein and eosin dyes. Br_4TFF has been employed for viability testing in *Candida albicans*.[48]

CellTracker Green CMFDA and CellTracker Yellow-Green CMEDA

Molecular Probes' CellTracker dyes are thiol-reactive fluorescent dyes that are retained in many live cells through several generations (Figure 16.10) and are not transferred to adjacent cells in a population, except possibly through gap junctions. These dyes represent a significant breakthrough in the cellular retention of fluorescent probes and are ideal long-term tracers for transplanted cells or tissues (see Section 15.2).

Both CellTracker Green CMFDA (C-2925, C-7025) and CellTracker Yellow-Green CMEDA (C-2926) freely diffuse into the cell, where their weakly thiol-reactive chloromethyl moieties react with intracellular thiols and their acetate groups are cleaved by cytoplasmic esterases (see Figure 15.1 in Chapter 15), releasing the fluorescent product. The other CellTracker probes (coumarin, BODIPY® and tetramethylrhodamine derivatives, see Section 15.2) do not require enzymatic cleavage to activate their fluorescence. Because the CellTracker dyes may react with both glutathione and proteins, cells with membranes that become compromised after staining may retain some residual fluorescent conjugates. However, use of a membrane-impermeant probe such as propidium iodide (P-1304) in combination with either CellTracker Green CMFDA or CellTracker Yellow-Green CMEDA should permit relatively long-term cytotoxicity assays. CellTracker Green CMFDA and ethidium homodimer-1 (E-1169) have been used to detect viable and nonviable cells in human coronary and internal thoracic arteries sampled at autopsy.[49] As with Br_4TFFDA (see above), CellTracker Yellow-Green CMEDA is an eosin derivative that may be useful as a photosensitizer of singlet oxygen production for cell ablation.

Chloromethyl SNARF®-1 Acetate

Chloromethyl SNARF-1 acetate (C-6826) is the only cell-tracking dye (and pH indicator) that exhibits bright, easily distinguished, red cytoplasmic fluorescence when excited at the same wavelengths used to excite the green fluorescent hydrolysis product of CMFDA (see Figure 23.6 in Chapter 23). The spectral characteristics of these two dyes permit simultaneous tracking of two cell populations by either fluorescence microscopy or flow cytometry. The large Stokes shift of the SNARF fluorophore also makes chloromethyl SNARF-1 acetate useful as a viability indicator in cells that exhibit green autofluorescence when excited by the 488 nm spectral line of the argon-ion laser.

Blue Fluorescent Esterase Substrates

The new probe 2,5-dihydroxyterephthalic acid di(acetoxymethyl ester) (DHTP AM, D-7021) yields a reasonably well-retained fluorescent product with a huge Stokes shift (~200 nm) that helps to resolve the signal from cellular autofluorescence. However, DHTP appears to undergo a marked shift in its excitation spectrum with concentration, permitting its excitation by the 488 nm spectral line of the argon-ion laser as well as by UV sources. This property is potentially useful for calibrating intracellular DHTP concentration and for detecting changes in cell volume.

We also offer calcein blue AM (C-1429), a viability indicator for use with instruments optimized for the detection of blue fluorescence.[50]

Viability/Cytotoxicity Assays Using Nucleic Acid Stains

As described in Section 16.2, viability assessments of animal cells, bacteria and yeast frequently employ polar and therefore cell-impermeant nucleic acid stains, often in combination with intracellular esterase substrates (see above) or with membrane potential probes, organelle probes or other cell-permeant indicators. Although many other cell-impermeant dyes could be used to detect dead cells, the high concentrations of nucleic acids in cells, coupled with the large fluorescence enhancement exhibited by most nucleic acid stains upon binding, make cell-impermeant nucleic acid stains the logical candidates for viability probes. See Table 8.1 in Chapter 8 for a list of all of our nucleic acid stains and Section 8.1 for a general discussion of dye binding to nucleic acids.

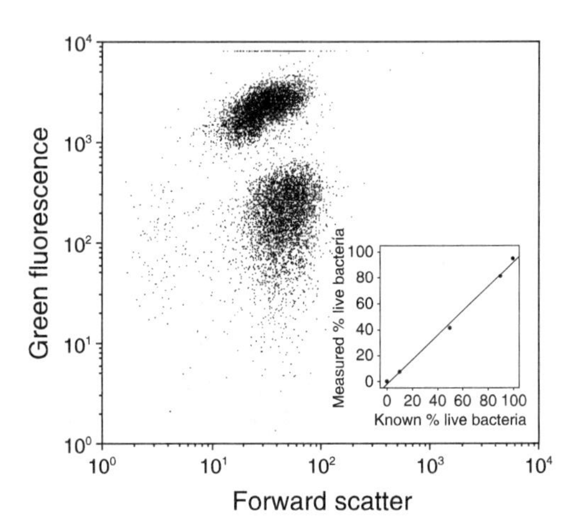

Figure 16.11 *Quantitative flow cytometric analysis of Escherichia coli viability using SYTOX Green nucleic acid stain (S-7020). A bacterial suspension containing an equal number of live and isopropyl alcohol–killed E. coli was stained with SYTOX Green and analyzed using excitation at 488 nm on a Becton Dickinson FACS Vantage cytometer. A bivariate frequency distribution for forward light scatter versus log fluorescence intensity (collected with a 510 nm longpass filter) shows two clearly distinct populations. When live and dead bacteria were mixed in varying proportions, a linear relationship between the population numbers and the actual percentage of live cells in the sample was obtained (see insert).*

SYTOX Green Nucleic Acid Stain

Many polar nucleic acid stains will enter eukaryotic cells with damaged plasma membranes yet will not stain dead bacteria with damaged plasma membranes. Our new SYTOX Green nucleic acid stain[51] (S-7020) is a high-affinity probe that easily penetrates both gram-positive and gram-negative bacteria with compromised plasma membranes, yet is completely excluded from live cells.[52] After brief incubation with SYTOX Green nucleic acid stain, dead bacteria fluoresce bright green when excited with the 488 nm spectral line of the argon-ion laser or any other 470–490 nm source. These properties, combined with its 1000-fold fluorescence enhancement upon nucleic acid binding, make our new SYTOX Green stain a simple and quantitative dead-cell indicator for use with fluorescence microscopes, fluorometers, fluorescence microplate readers or flow cytometers (Figure 16.11). We have taken advantage of the sensitivity of the SYTOX Green nucleic acid stain in our ViaGram Red$^+$ Bacterial Gram Stain and Viability Kit (V-7023, see Section 16.2).

Of course, the use of our SYTOX Green nucleic acid stain as a tool for viability assessment is not restricted to bacteria. SYTOX Green stain is also a very effective cell-impermeant counterstain in eukaryotic systems (see Section 8.4). It can be used in conjunction with blue and red fluorescent labels for multiparametric analyses in fixed cells and tissue sections (see photo on back cover). Furthermore, it may be possible to combine SYTOX Green nucleic acid stain with the membrane-permeant SYTO 17 red fluorescent nucleic acid stain (S-7579, see Section 8.1) for two-color visualization of dead and live cells.

Dimeric and Monomeric Cyanine Dyes

The 17 dimeric and monomeric cyanine dyes in the TOTO® and TO-PRO™ series (see Section 8.1) are essentially nonfluorescent unless bound to nucleic acids and have extinction coefficients 10–20 times greater than that of DNA-bound propidium iodide.

These dyes are typically impermeant to the membranes of live cells [53] but brightly stain dead cells with compromised membranes. POPO™-1 and BOBO™-1 may be useful blue fluorescent dead-cell stains, and YOYO®-3 and TOTO-3 have excitation maxima beyond 600 nm when bound to DNA. A recently developed cell viability assay utilizes YOYO-1 fluorescence before and after treatment with digitonin (D-8449, D-8450) as a measure of the dead cells and total cells, respectively, in the sample.[54]

In addition to their use as dead-cell stains, both the dimeric and monomeric cyanine dyes are also proving useful for staining viruses. YOYO-1– and POPO-1–stained viruses have been employed to identify and quantitate bacteria and cyanobacteria in marine microbial communities.[55] YO-PRO™-1 has been used to count viruses in marine and freshwater environments by epifluorescence microscopy.[56]

Our Nucleic Acid Stains Dimer and Monomer Sampler Kits (N-7565, N-7566) provide a total of 16 cyanine dyes that span the visible spectrum (the longest-wavelength dye, TO-PRO-5, is only available individually). These kits should be useful when screening dyes for their utility in viability and cytotoxicity assays.

Ethidium and Propidium Dyes

The red fluorescent, cell-impermeant ethidium and propidium dyes — ethidium bromide (E-1305, E-3565), ethidium homodimer-1 (EthD-1, E-1169) and propidium iodide (P-1304, P-3566) — can all be excited by the argon-ion laser and are therefore useful for sorting dead cells by flow cytometry.[57,58] Moreover, these dyes have large Stokes shifts and may be used in combination with fluorescein derivatives (such as calcein) for two-color applications (Figure 16.2). Both propidium iodide and ethidium bromide have been extensively used to detect dead or dying cells,[59-63] although ethidium bromide may be somewhat less reliable because it is not as highly charged. EthD-1 and propidium iodide are superior to ethidium bromide for two-color flow cytometric viability assays in which either BCECF AM or calcein AM is used as the live-cell stain because their spectra do not overlap as much with those of the green fluorescent esterase probes.[14]

With its high affinity for DNA and low membrane permeability,[64-67] EthD-1 is often the preferred red fluorescent dead-cell indicator. EthD-1 binds to nucleic acids 1000 times more tightly than does ethidium bromide and undergoes about a 40-fold enhancement of fluorescence upon binding.[66,68] Also, the high affinity of EthD-1 permits the use of very low concentrations to stain dead cells, thus avoiding the use of large quantities of the potentially hazardous ethidium bromide or propidium iodide. EthD-1, the dead-cell indicator in our LIVE/DEAD Viability/Cytotoxicity Kit (L-3224, see Section 16.2), has been used alone [69] or in combination with calcein AM [70] to detect tumor necrosis factor activity and to assay neuronal cell death.[60,71,72]

Ethidium Monoazide

Ethidium monoazide (E-1374) is a fluorescent photoaffinity label that, after photolysis, binds covalently to nucleic acids in solution and in cells with compromised membranes.[73,74] A mixed population of live and dead cells incubated with this membrane-impermeant dye can be illuminated with a visible-light source, washed, fixed and then analyzed in order to determine the viability of the cells at the time of photolysis.[74] Thus, ethidium monoazide reduces some of the hazards inherent in working with pathogenic samples because, once stained, samples can be treated with fixatives before analysis by fluorescence microscopy or flow cytometry. Immunocytochemical analyses requiring fixation are also compatible with this ethidium monoazide–based viability assay. We have developed an alternative two-color fluorescence-based assay for determining the original viability of fixed samples that employs our cell-permeant, green fluorescent SYTO 10 and cell-impermeant, red fluorescent DEAD Red nucleic acid stains; the LIVE/DEAD Reduced Biohazard Viability/Cytotoxicity Kit (L-7013) is described in Section 16.2.

Viability/Cytotoxicity Assays That Measure Oxidation or Reduction

Metabolically active cells can oxidize or reduce a variety of probes, providing a measure of cell viability and overall cell health. This measure of viability is distinct from that provided by probes designed to detect esterase activity or cell permeability.

Dihydrorhodamines and Dihydrofluoresceins

Fluorescein, rhodamine and various other dyes can be chemically reduced to colorless, nonfluorescent leuco dyes. These "dihydro" derivatives are readily oxidized back to the parent dye by some reactive oxygen species (see Section 21.4) and thus can serve as fluorogenic probes for detecting oxidative activity in cells and tissues.[75-77] Because reactive oxygen species are produced by live but not dead cells, fluorescent oxidation products that are retained in cells can be used as viability indicators for single cells or cell suspensions. Some probes that are useful for detecting oxidative activity in metabolically active cells include:

- H_2DCFDA (2',7'-dichlorodihydrofluorescein diacetate, D-399),[78] carboxy-H_2DCFDA (5-(and-6)-carboxy-2',7'-dichlorodihydrofluorescein diacetate, C-400) and the acetoxymethyl ester of H_2DCFDA (C-2938), all of which require both intracellular deacetylation and oxidation to yield green fluorescent products [79-81]
- CM-H_2DCFDA (chloromethyl-2',7'-dichlorodihydrofluorescein diacetate, C-6827), which is analogous to H_2DCFDA except that it forms a mildly thiol-reactive fluorescent product that may permit longer-term measurements
- Dihydrorhodamine 123 (D-632), dihydrorhodamine 6G (D-633) and dihydrotetramethylrosamine (D-638), which are oxidized in viable cells to the mitochondrial stains rhodamine 123,[82-85] rhodamine 6G and tetramethylrosamine,[86-88] respectively
- Dihydroethidium (D-1168), which forms the nucleic acid stain ethidium following oxidation [89,90] and has proven useful for detecting the viability of intracellular parasites [91]
- Luminol (L-8455), which is useful for chemiluminescence-based detection of oxidative events in cells rich in peroxidases, including granulocytes [92-95] and spermatozoa [96]

These probes are all described in more detail in Section 21.4, which is devoted to products for assaying oxidative activity in live cells and tissues.

Tetrazolium Salts

Tetrazolium salts are widely used for detecting redox potential of cells for viability, proliferation and cytotoxicity assays.[97-105]

Upon reduction, these water-soluble colorless compounds form uncharged, brightly colored formazans. Several of the formazans precipitate out of solution and are useful for histochemical localization of the site of reduction or, after solubilization in organic solvent, for quantitation by standard spectrophotometric techniques.

Reduction of MTT (M-6494) remains the most common assay for tetrazolium salt–based viability testing;[106-108] MTT has also been used to measure adhesion of HL60 leukemia cells onto endothelial cells.[109] Unlike MTT's purple formazan product, the extremely water-soluble, orange formazan product of XTT (X-6493) does not require solubilization prior to quantitation, thereby reducing assay time in many viability assay protocols. Moreover, sensitivity of the XTT reduction assay is reported to be similar to or better than that of the MTT reduction assay.[110] The XTT assay is particularly useful for high-throughput screening of antiviral and antitumor agents and for assessing the effect of cytokines on cell proliferation.[102,111-114]

Our other tetrazolium salts include 4-iodonitrotetrazolium violet (INT, I-6496), tetrazolium blue (T-6490), nitro blue tetrazolium (N-6495) and nitrotetrazolium violet [115] (NTV, N-6498), which yield orange, blue, deep blue and purple precipitates, respectively. Many of these salts are reduced by specific components of the electron transport chain and may be useful for determining the site of action of specific toxins.[116] Selected properties and applications for the tetrazolium salts are listed in Table 21.1 in Chapter 21.

The tetrazolium salts are available individually or as part of our Tetrazolium Salt Sampler Kit (T-6500), which contains 100 mg of INT, MTT, NBT, NTV, tetrazolium blue and XTT. This sampler kit facilitates testing of our tetrazolium salts for their suitability in a particular assay.

Other Viability/Cytotoxicity Assay Methods

A viable cell contains an ensemble of ion pumps and channels that maintain both intracellular ion concentrations and transmembrane potentials. Active maintenance of ion gradients ceases when the cell dies, and this loss of activity can be assessed using potentiometric dyes, acidotropic stains, Ca^{2+} indicators [117] (see Chapter 22) and pH indicators [117] (see Chapter 23).

Potentiometric Dyes

Molecular Probes makes a variety of dyes for detecting transmembrane potential gradients (see Chapter 25), including several cationic probes that accumulate in the mitochondria of metabolically active cells (see Section 12.2). The mitochondrion-selective rhodamine 123 [78] (R-302) has been used to assess the viability of lymphocytes,[118] human fibroblasts,[119] Simian virus–transformed human cells [120] and bacteria;[121] however, rhodamine 123 is not taken up well by gram-negative bacteria.[121] Rhodamine 123 has also been used in combination with propidium iodide (P-1304) for two-color flow cytometric viability assessment.[122]

Other potential-sensitive dyes that have proven useful in viability studies include several fast-response styryl dyes and slow-response oxonol and carbocyanine dyes. The fast-response styryl dyes such as di-4-ANEPPS (D-1199, see Section 25.2) give relatively large fluorescence response to potential changes. Di-4-ANEPPS was used for rapid measurement of toxicity in frog embryos.[123] The symmetrical bis-oxonol dyes [78] (B-413, B-438; see Section 25.3) have been used for viability assessment by flow cytometry [124-126] and imaging. These slow-response dyes have also been employed to determine antibiotic susceptibility of bacteria by flow cytometry.[127,128]

The green fluorescent cyanine dye JC-1 (T-3168) exists as a monomer at low concentrations or at low membrane potential; however, at higher concentrations (aqueous solutions above 0.1 µM) or higher potentials, JC-1 forms red fluorescent "J-aggregates" that exhibit a broad excitation spectrum and an emission maximum at ~590 nm (see Figure 25.1 in Chapter 25). JC-1 has been used to investigate apoptosis,[129-131] as well as mitochondrial poisoning, uncoupling and anoxia.[132] The ability to make ratiometric emission measurements with JC-1 makes this probe particularly useful for monitoring changes in cell health.

Acidotropic Stains

Membrane-bound proton pumps are also used to maintain low pH within the cell's acidic organelles. Our complete selection of stains for lysosomes and other acidic organelles, including LysoTracker™ and LysoSensor™ probes, are described in Section 12.3.

The lysosomal stain, neutral red (N-3246), which was first used for viability measurements by Ehrlich in 1894, has been employed in numerous cytotoxicity, cell proliferation and adhesion assays.[133-138] Although usually used as a chromophoric probe, neutral red also fluoresces at ~640 nm in viable cells and has been detected using a fluorescence microplate reader.[139] Furthermore, the fluorescence of neutral red and BCECF AM can be measured simultaneously using a single excitation wavelength of 488 nm,[139] suggesting that neutral red may be an effective probe for multicolor flow cytometric determination of cell viability. Our neutral red is highly purified to reduce contaminants that might interfere with these observations.

Acridine orange (A-1301, A-3568) concentrates in acidic organelles in a pH-dependent manner. The metachromatic green or red fluorescence of acridine orange has been used to assess islet viability [140] and bacterial spore viability [141] and to monitor physiological activity in *Escherichia coli*.[142]

Europium-Release Assay

Since their introduction by Goodman [143] in 1961, ^{51}Cr-release assays have been extensively used to measure cell-mediated cytolysis.[144] In this technique, cells are preloaded with a radioactive isotope of chromium (^{51}Cr) and then assayed for leakage of radioactivity from damaged cells. However, this assay has several disadvantages. Not only does ^{51}Cr generate radioactive waste, but it is toxic to cells, binds to cytoplasmic proteins and may not be released until the cells are completely lysed.[145,146]

In an analogous assay, cells are loaded with europium chloride ($EuCl_3$, E-6960) and diethylenetriaminepentaacetic acid (DTPA) for 15–20 minutes, washed to remove unincorporated Eu^{3+}, incubated with the cytolytic agents to be studied and centrifuged. The supernatant is then assayed by adding naphthoyltrifluoroacetone (NTA, T-411) and trioctylphosphine oxide. The resultant ternary complex (Eu^{3+}/NTA/trioctylphosphine oxide) has long-lived luminescence (100–1000 nsec) that can be distinguished from background signals by time-resolved fluorescence, as well as with a standard fluorometer or fluorescence microplate reader. The sensitivity of this Eu^{3+}-release assay for complement-mediated cytolysis using time-resolved fluorescence has been reported to be fivefold better than that of ^{51}Cr-release assays.[145] The Eu^{3+}-release assay has been used to detect cytotoxic events mediated by T lymphocytes, natural killer (NK) cells and lymphokine-activated killer (LAK) cells.[146-151]

Dansyl Lysine

Cell death that does not result in membrane disruption cannot be detected using cell-impermeant nucleic acid stains. The membrane-structure probe dansyl lysine [152] (D-372, see Section 14.5) has been used to identify cells killed by heat treatment.[153,154] Dansyl lysine–stained cells have been shown by flow cytometry to be nonviable.[155]

Cell Proliferation Assays

Reagents for quantitating cell proliferation are valuable research and diagnostic tools. Currently there is no fluorescent reagent that can be specifically incorporated into cells during cell division and directly detected on a single cell basis. Consequently, most cell proliferation assays estimate the number of cells either by measuring total nucleic acid or protein content of lysed cells or by incorporating ^{3}H-thymidine or 5-bromodeoxyuridine (BrdU, a thymidine analog) into cells during proliferation.[156,157] Our CyQUANT Cell Proliferation Assay Kit and FluoReporter Cell Protein Assay Kits, which are discussed in Section 16.2, provide reagents and tested protocols for quantitating cell proliferation using nucleic acid content measurements and protein content measurements, respectively.

Detection of Proliferating Cells Using Bromodeoxyuridine Incorporation

Cell-cycle analysis based on DNA content measurements (see "Cell-Cycle Analysis" in this section) does not discriminate proliferating cells from resting cells. In contrast, incorporation of BrdU into newly synthesized DNA permits indirect detection of rapidly proliferating cells with fluorescently labeled anti-BrdU antibodies, thereby facilitating the identification of cells that have progressed through the S-phase of the cell cycle during the BrdU labeling period.[157,158]

Because the fluorescence of the Hoechst 33258 (H-1398, H-3569) and Hoechst 33342 (H-1399, H-3570) dyes bound to DNA is *quenched* at sites where BrdU is incorporated,[156,159-161] Hoechst fluorescence can also be used to detect BrdU incorporation. This technique has been employed to quantitate the noncycling cell fraction, as well as the fraction of cells in G_1 and G_2 of two subsequent cycles.[162] The addition of ethidium bromide as a counterstain that is *insensitive* to BrdU incorporation allows the resolution of G_1, S and G_2 compartments of up to three consecutive cell cycles.[163,164]

Unlike the fluorescence of Hoechst dyes, the fluorescence of TO-PRO-3 (T-3605, see Section 8.1) and LDS 751 (L-7595) has recently been shown to be *enhanced* by the presence of bromodeoxyuridine in DNA. In conjunction with propidium iodide, these nucleic acid stains have been used to discriminate BrdU-labeled cells from nonproliferating cells by flow cytometry.[165]

Detection of Proliferating Cells Using ChromaTide™ Nucleotides

In the strand break induction by photolysis (SBIP) technique, proliferating cells that have incorporated BrdU into newly synthesized DNA are subjected to Hoechst 33258 staining followed by UV photolysis to induce DNA strand breaks.[166] Once the cells are fixed, strand breaks can be detected *in situ* using mammalian terminal deoxynucleotidyl transferase (TdT), which covalently adds labeled nucleotides to the 3'-hydroxyl ends of these DNA fragments.[167-170] Break sites have traditionally been labeled with biotinylated or haptenylated dUTP conjugates (see Section 8.2) in conjunction with fluorescently labeled streptavidin or antibody conjugates,[171,172] respectively (see Sections 7.3 and 7.5). However, a single-step procedure has been described that uses ChromaTide BODIPY FL-14-dUTP (C-7614) as a TdT substrate for detecting DNA strand breaks both in BrdU-labeled cells following SBIP and in apoptotic cells [166,173,174] (see Color Plate 4 in "Visual Reality").

The single-step BODIPY FL–based assay has several advantages over indirect detection of biotinylated or haptenylated nucleotides. With direct detection procedures, no secondary detection reagents are required; fewer protocol steps translate into less chance for error and more immediate results. Moreover, cell yields with direct detection procedures are reported to be about three times greater than those of multistep procedures employing biotin- or digoxigenin-conjugated dUTP. Both BODIPY FL– and fluorescein-labeled nucleotides can be detected with fluorescence microscopy or flow cytometry. However, BODIPY FL–labeled nucleotides provide 40% stronger signal than fluorescein-labeled nucleotides when assaying strand breaks in apoptotic versus non-apoptotic cells. Furthermore, fading of the fluorescence of the incorporated BODIPY FL dUTP is less than that of the corresponding fluorescein dUTP analog.[173]

Unlike traditional proliferation assays based on BrdU incorporation, no DNA heat- or acid-denaturation steps are required with SBIP in order to visualize the labeled strand breaks, allowing simultaneous detection of the morphology of nuclear proteins and other cellular constituents by immunocytochemical analysis. The narrow emission spectrum of the BODIPY FL–labeled nucleotides will be especially useful for multicolor labeling experiments.

Cell Adhesion Assays

The fundamental role of cell adhesion in the morphology and development of organisms, organelles and tissues has made identification of molecular mediators of cell adhesion an important research focus in cell biology and immunology.[175,176] In a typical fluorescence-based cell adhesion assay, unlabeled cell monolayers in multiwell plates are incubated with fluorescently labeled cells and then washed to separate the adherent and nonadherent populations. Cell adhesion can then be determined simply based on the correlation of retained fluorescence with cell number. An ideal fluorescent marker will retain proportionality between fluorescence and cell number and introduce minimal interference with the cell adhesion process. Because adhesion is a cell-surface phenomenon, cytoplasmic markers that can be passively loaded are preferable to compounds that label cell-surface molecules.

Cell Adhesion Assays Using Esterase Substrates

Essentially all of the esterase substrates in Table 16.3 for use in monitoring cell viability can be used for studies of cell adhesion. As with cell viability studies, calcein AM (C-1430, C-3099, C-3100) appears to best satisfy the criteria for assaying cell adhesion.[5,134,177] Calcein AM has been used in several cell adhesion assays, including those designed to measure:

- Effects of E-selectin–binding peptides [178] and integrins [179] on neutrophil adhesion
- Integrin-mediated cell adhesion in transfected K562 cells [180,181] and BSC-1 cells [182]
- Leukocyte [5,183] and neutrophil [184] adhesion to endothelial cells
- Monocyte adhesion in HIV-infected cells [185]

Other fluorogenic esterase substrates that have been used to assess cell adhesion include BCECF AM [186-191] (B-1150, B-1170, B-3051), carboxyfluorescein diacetate [5,192-194] (5(6)-CFDA, C-195), fluorescein diacetate [22] (F-1303), the succinimidyl ester of carboxyfluorescein diacetate [179] (5-CFDA, SE; C-1157) and CellTracker Green CMFDA [186] (C-2925, C-7025). The new fluorinated fluorescein derivatives — TFFDA (T-7016) and HFFDA (H-7019) — should also prove useful for cell adhesion assays, where uniform dye retention over longer periods is of paramount importance (Figure 16.10).

Cell Adhesion Assays Using the CyQUANT Cell Proliferation Assay Kit

Our CyQUANT Cell Proliferation Assay Kit (C-7026, see Section 16.2) can also serve as an important tool for both cell–cell and cell–surface adhesion. The CyQUANT assay detects total nucleic acids in cells, with a linear response between fewer than 50 to at least 50,000 cells. To quantitate cell–surface adhesion, cells are simply permitted to adhere to the surface, gently washed to remove nonadherent cells, frozen and lysed with buffer containing the CyQUANT GR reagent and then analyzed to determine total nucleic acids in the adherent cells (see Section 16.2). The ratio of fluorescence of the nucleic acids in the adherent cells to that of a matched reference containing total cells in the sample can be used to estimate the number of adherent cells. It should also be possible to extend the CyQUANT assay to studies of cell–cell adhesion by quantitating both the number of surface-adhering cells originally plated and the number of total cells after a second cell line has been introduced and allowed to adhere. A similar assay for cell adhesion based on DAPI (D-1306) has been reported.[195]

Oregon Green 488 Fibrinogen

Integrin $\alpha_{IIb}\beta_3$ (also called the glycoprotein IIb/IIIa heterodimer or GPIIb–IIIa) on the platelet membrane surface acts as a fibrinogen receptor and plays an important role in platelet adhesion and aggregation, as well as in thrombus formation.[196-198] Integrins mediate cell adhesion and migration by functionally linking extracellular structures to the cytoskeleton;[199-202] they are also known to participate in phagocytosis.[180,181,203] Integrin $\alpha_{IIb}\beta_3$ constitutes about 15% of the total membrane protein of platelets and 3% of the total platelet protein.[204]

Activation of integrin $\alpha_{IIb}\beta_3$ is required for fibrinogen binding, which leads to platelet activation, adhesion, spreading and microfilament reorganization of human endothelial cells *in vitro*.[205-207] Soluble fibrinogen binds to its receptor with a Ca^{2+}-dependent apparent K_d of 0.18 µM.[208] This binding is apparently mediated by the tripeptide sequence Arg–Gly–Asp (RGD), found in both fibrinogen and fibronectin.[198,209,210]

Our new Oregon Green 488 conjugate of human plasma fibrinogen (F-7496) should be a useful probe for detecting agonist-induced platelet activation by a variety of different fluorescence techniques, including flow cytometry and fluorescence microscopy. Use of a similar fluorescein fibrinogen for identifying activated platelets by flow cytometry has been recently described.[206,211,212] The binding of fluorescein fibrinogen to activated platelets has been shown to be saturable and can be inhibited completely by underivatized fibrinogen.[211] The preferential binding and accumulation of fluorescein fibrinogen at the endothelial border of venular blood vessels has been studied by quantitative fluorescence microscopy.[213]

Cell-Cycle Analysis

The cell-cycle distribution and proliferative state of a population of cells are important parameters when studying oncogene and tumor suppressor gene mechanisms and other live-cell functions. The cell-cycle distribution within a population can be analyzed with nucleic acid stains that, upon binding to the DNA, exhibit fluorescence emission proportional to the DNA content of the cell. Subsequent analysis by flow cytometry yields a histogram of DNA content per cell that allows determination of the fraction of cells in the G_1, S and G_2 phases of the cell cycle. This analysis is typically performed either on live cells using a cell-permeant nucleic acid stain [214,215] or on detergent-treated or fixed cells using a cell-impermeant nucleic acid stain.[159,216-218]

Cell-Cycle Analysis Using SYTOX and SYTO Nucleic Acid Stains

We have determined that SYTOX Green nucleic acid stain (S-7020) is particularly useful for cell-cycle analysis on RNase-

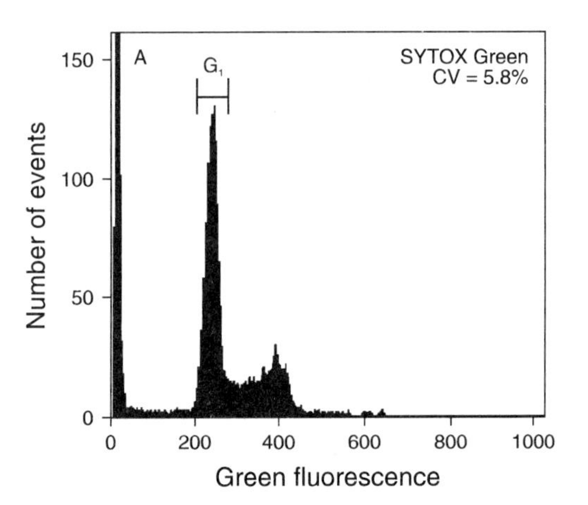

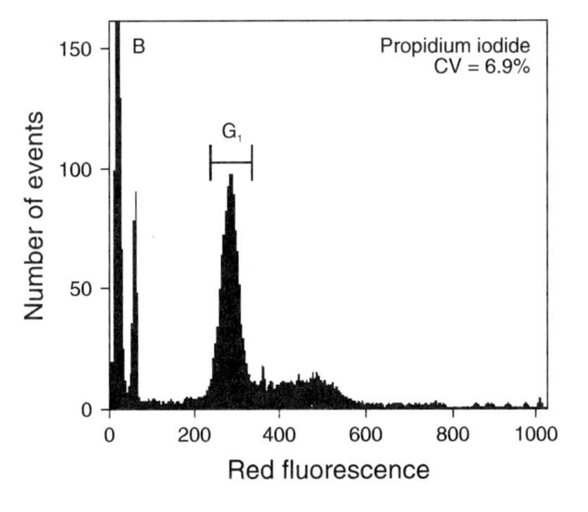

Figure 16.12 *Comparison of DNA content histograms obtained with A) SYTOX Green and B) propidium iodide. Human B cells were suspended in permeabilizing buffer (100 mM Tris pH 7.4, 154 mM NaCl, 1 mM $CaCl_2$, 0.5 mM $MgCl_2$, 0.1% Nonidet™ P-40) and then stained for 15 minutes with 0.5 µM SYTOX Green or 5 µM propidium iodide. Flow cytometric analysis of the stained cells was carried out with excitation at 488 nm on a Becton Dickinson FACS Vantage cytometer. SYTOX Green staining produces a significantly narrower G_1 phase peak, indicated by the smaller coefficient of variation (CV).*

treated fixed cells (Figure 16.12). In particular, the SYTOX Green dye produces lower coefficients of variation than propidium iodide, leading to improved resolution of cell phase. Figure 16.12 shows a comparison of DNA content histograms obtained with SYTOX Green nucleic acid stain and propidium iodide after flow cytometric analysis.

Our SYTO live-cell nucleic acid stains (see Section 8.1 and Color Plate 4 in "Visual Reality"), which are green and red fluorescent dyes that passively diffuse through cell membranes, may also be useful for cell-cycle analysis. Viable adherent cells stained with the green fluorescent SYTO 16 dye were analyzed with a new microscope-based laser scanning cytometer that measures multiple-wavelength fluorescence and light scattering for each cell on a microscope slide.[219] Although the DNA distribution was broader than that obtained with propidium iodide–stained fixed cells, data obtained with SYTO 16 dye provided sufficient resolution to distinguish mitotic cells from the G_{1a} population.

Cell-Cycle Analysis Using TOTO and YOYO Dimeric Cyanine Dyes

Most of our 17 high-sensitivity cyanine dyes in the TOTO and TO-PRO series have been reported to be useful for staining nucleic acids in fixed-cell preparations. Staining with TOTO-1 and YOYO-1 dimeric cyanine dyes allows extremely sensitive flow cytometric analysis of aldehyde-fixed marine prokaryotes,[220,221] as well as of RNase-treated nuclei and isolated human chromosomes.[222] It was reported that YOYO-1–stained nuclei exhibited over 1000 times the fluorescence signal obtained with mithramycin, and histograms of both TOTO-1– and YOYO-1–stained nuclei provided coefficients of variation that were at least as low as those found with propidium iodide or mithramycin.[159,222] Moreover, when nuclei were stained with both YOYO-1 and Hoechst 33258, the ratio of the fluorescence of these two dyes varied as a function of cell cycle, suggesting that the cyanine dyes might be useful for examining cell cycle–dependent changes that occur in chromatin structure. TOTO-1 and YOYO-1 have also been used for quantitative cytometric assessment of nuclei in formaldehyde-fixed paraffin tissue sections and reportedly produced an intense, photostable fluorescent signal that correlated well with the DNA content of diploid, tetraploid and octaploid liver nuclei.[223] The longer-wavelength TOTO-3 dimeric cyanine dye may be useful in many of the same applications as its TOTO-1 and YOYO-1 analogs. Furthermore, this dye can be excited using low-cost sources such as red He-Ne and diode lasers.[224] We offer sample sizes of all eight dimeric cyanine dyes in our Nucleic Acid Stains Dimer Sampler Kit (N-7565). Properties of the TOTO series of dimeric cyanine dyes are described in detail in Section 8.1 and Table 8.1 of Chapter 8.

Cell-Cycle Analysis Using Hoechst 33258, Hoechst 33342 and DAPI

The nucleic acid stains most frequently used for cell-cycle analysis — Hoechst 33258 (H-1398, H-3569), Hoechst 33342 (H-1399, H-3570) and DAPI (D-1306) — bind to the minor groove of DNA at AT-rich sequences. Hoechst 33342, which more rapidly permeates cells than Hoechst 33258, is commonly used for determining the DNA content of viable cells without detergent treatment or fixation.[214,215] The Hoechst dyes and DAPI can be excited with a mercury-arc lamp, the UV spectral lines of the argon-ion laser or the 325 nm spectral line of the He-Cd laser. This group of blue fluorescent nucleic acid stains preferentially bind to AT-rich sequences and also exhibit higher quantum yields when bound to AT-rich nucleic acids, thus introducing a strong bias into the measurements of nuclear DNA content.[225,226] As a consequence, data obtained with Hoechst 33342 and DAPI correlate very well with each other but less well with data obtained with propidium iodide, a red fluorescent, cell-impermeant nucleic acid stain.[227]

In addition to cell-cycle analysis, DAPI is frequently employed for DNA content–based counting of bacterial cells [228-230] and for detecting malarial infections by fluorescence microscopy.[231]

Cell-Cycle Analysis Using Ethidium Bromide and Its Analogs

Ethidium bromide (E-1305, E-3565), ethidium homodimer-1 (E-1169), propidium iodide (P-1304, P-3566) and hexidium iodide (H-7593) are phenanthridinium derivatives that intercalate between the bases of the DNA in fixed cells to yield red fluorescence. Because these base-intercalators are not specific for DNA, cell-cycle analysis usually requires treatment of fixed samples with RNase (50–100 µg/mL) to eliminate fluorescence resulting from dye binding to RNA.[159]

Cell-Cycle Analysis Using Acridine Orange

Acridine orange (A-1301, A-3568) emits green fluorescence when bound to double-stranded nucleic acids and red fluorescence when bound to single-stranded nucleic acids. This property has been exploited in methods for simultaneously analyzing the DNA and RNA content of a cell culture.[232,233] Acridine orange–based assays have enabled researchers to monitor early events during lymphocyte activation by flow cytometry,[233] as well as to study the steps leading to differentiation of keratinocytes in cell culture.[234] After staining with acridine orange, bacteria actively growing in a rich medium exhibit orange fluorescence, making them easily distinguished from bacteria in stationary phase, which have green fluorescence.[142]

Cell-Cycle Analysis Using 7-Aminoactinomycin D

7-AAD (7-aminoactinomycin D, A-1310) is a fluorescent intercalator that undergoes a spectral shift upon association with DNA; 7-AAD/DNA complexes can be excited by the argon-ion laser and emit beyond 610 nm (see Figure 8.2 in Chapter 8). This visible light–excitable nucleic acid stain is suitable for cell-cycle analysis,[235] although the precision of the distributions obtained is generally inferior to those generated with other dyes.[235-237] 7-AAD appears to be generally excluded from live cells; however, it has been reported to label the nuclear region of live cultured mouse L cells and salivary gland polytene chromosomes of *Chironomus thummi thummi* larvae.[238] Nondifferentiated cells are also reported to bind 7-AAD better than differentiated cells,[239] which may make the dye a useful marker for malignant hematopoietic cells.[240] 7-AAD binds selectively to GC regions of DNA,[241] yielding a distinct banding pattern in polytene chromosomes and chromatin.[238,242] This sequence selectivity has been exploited for chromosome banding studies [243,244] (see Section 8.4).

Apoptosis Assays

Apoptosis (programmed cell death) is the genetically controlled ablation of cells during development.[245] Cells experiencing deregulation of the cell cycle, such as tumor cells or those subjected to genetic transformation, are also prone to apoptosis.[246,247] Apoptosis is distinct from necrosis in both the biochemical and morphological changes that occur.[248,249] In contrast to necrotic cells, apoptotic cells are characterized by the relative integrity of their mitochondrial DNA [250,251] and the maintenance of membrane integrity until late in the process. Furthermore, apoptotic cells exhibit alterations in chromatin structure due to degradation of proteins [252] and nucleic acids.[253] As with cell viability, no single parameter fully defines cell death; therefore, it is often advantageous to use several different approaches when studying apoptosis. Recently, several methods have been developed to distinguish live cells from early and late apoptotic cells and from necrotic cells. See especially the review by Darzynkiewicz, Li and Gong in Methods in Cell Biology 41, 15 (1994), a book that is available from Molecular Probes (M-7887, see Section 26.6).

Assessing DNA Damage with Nucleic Acid Stains

The characteristic breakdown of the nucleus during apoptosis comprises collapse and fragmentation of the chromatin, degradation of the nuclear envelope and nuclear blebbing, resulting in the formation of micronuclei. Therefore, nucleic acid stains can be useful tools for identifying even low numbers of apoptotic cells in cell populations. Several nucleic acid stains, all of which are listed in Section 8.1, have been used to detect apoptotic cells by fluorescence imaging or flow cytometry.

- Our YO-PRO-1 (Y-3603, see Section 8.1) nucleic acid stain is the basis of an important new assay for apoptotic cells that is compatible with both fluorescence microscopy and flow cytometry. Selective uptake of YO-PRO-1 by apoptotic cells of a dexamethasone-treated population of thymocytes, an irradiated peripheral blood mononuclear cell population and a growth factor–depleted tumor B cell line was confirmed by cell sorting.[254] Unlike Hoechst 33342 staining, YO-PRO-1 staining had no effect on the ability of stained T cells to proliferate. Moreover, the visible-light absorption of the YO-PRO-1 stain eliminates the need for UV excitation capabilities in flow cytometry.
- Some of our new cell-permeant, green fluorescent SYTO dyes, including SYTO 13 and SYTO 16, are proving useful for distinguishing apoptotic neuronal cells [130] and apoptotic thymocytes.[255] Our SYTO Live-Cell Nucleic Acid Stain Sampler Kits (S-7554, S-7572; see Section 8.1) provide a total of 12 SYTO dyes that may be screened for their utility in monitoring apoptosis.
- Hoechst 33342 (H-1399, H-3570) is initially taken up by apoptotic cells, whereas cell-impermeant dyes such as propidium iodide (P-1304, P-3566) and ethidium bromide [256-258] (E-1305, E-3565) are excluded. Later stages of apoptosis are accompanied by an increase in membrane permeability, which allows propidium iodide to enter cells. Thus, a combination of Hoechst 33342 and propidium iodide has been used for simultaneous flow cytometric analysis of the stages of apoptosis and cell-cycle distribution.[256,257,259,260]
- 7-Aminoactinomycin D (7-AAD, A-1310) has been used alone or in combination with Hoechst 33342 to separate populations of live cells, early apoptotic cells and late apoptotic cells by flow cytometry.[261,262] 7-AAD's staining pattern is retained following cell fixation, and its large Stokes shift is advantageous when simultaneously staining with cell-surface labels. 7-AAD staining has also been used to detect apoptotic cells by their characteristic morphology using fluorescence microscopy.[263]
- The cell-permeant nucleic acid stain LDS 751 (L-7595) has been used to discriminate intact nucleated cells from nonnucleated cells and cells with damaged nuclei,[264,265] as well as to differentiate apoptotic cells from nonapoptotic cells.[255]
- Acridine orange (A-1301) exhibits metachromatic fluorescence that is sensitive to DNA conformation, making it a useful probe for detecting apoptotic cells.[266] When analyzed by flow cytometry, apoptotic cells stained with acridine orange show reduced green and enhanced red fluorescence in comparison to normal cells.[253]
- DAPI (D-1306) and sulforhodamine 101 (S-359, see Section 15.3) can be used together in fixed apoptotic cells to reveal concomitant breakdown of protein and DNA.[253,267,268]

DNA fragmentation can also be detected *in vitro* using electrophoresis techniques. DNA extracted from apoptotic cells, separated by gel electrophoresis and stained with ethidium bromide reveals a characteristic ladder pattern of low molecular weight DNA fragments.[267,269-272] Ethidium bromide has also been used for a dot blot assay to detect apoptotic DNA fragments.[273] Our ultrasensitive SYBR Green I nucleic acid stain (S-7567, see Section 8.3) and SYBR DX DNA blot stain (S-7550, see Section 8.3) should allow the detection of even fewer apoptotic cells in these applications.

Detecting DNA Strand Breaks with ChromaTide Nucleotides

DNA fragmentation that occurs during apoptosis produces DNA strand breaks. These DNA fragments can be end-labeled using the same methods employed in the strand break induction by photolysis (SBIP) technique for measuring cell proliferation (see above). Once the cells are fixed, DNA strand breaks can be detected *in situ* using mammalian terminal deoxynucleotidyl transferase (TdT), which covalently adds labeled nucleotides to the 3'-hydroxyl ends of these DNA fragments.[167-170] Break sites have traditionally been labeled with biotinylated or haptenylated dUTP conjugates (see Section 8.2) in conjunction with fluorophore-labeled streptavidin or antibody conjugates,[171,172] respectively (see Sections 7.3 and 7.5). Recently, however, a single-step procedure has been described that uses ChromaTide BODIPY FL-14-dUTP (C-7614) as a TdT substrate for detecting DNA strand breaks in apoptotic cells or in BrdU-labeled cells following SBIP [166,173,174] (see Color Plate 4 in "Visual Reality").

The single-step BODIPY FL–based assay has several advantages over indirect detection of biotinylated or haptenylated nucleotides, including fewer protocol steps and increased cell yields. BODIPY FL–labeled nucleotides have also proven superior to fluorescein-labeled nucleotides for detection of DNA strand breaks in apoptotic cells because they provide stronger signals, a narrower emission spectrum and less photobleaching.[173] Moreover, it has been reported that BODIPY FL-14-dUTP incorporated into the granules of the condensed chromatin structure of late-apoptotic cells — cells characterized by extensive nuclear fragmentation — exhibit yellow fluorescence, whereas uncondensed areas of the nuclei or early-apoptotic cells exhibit green fluorescence. This spectral shift, which is characteristic of the BODIPY fluorophores, is most likely a consequence of stacking of the BODIPY FL fluorophores (see Figure 13.4 in Chapter 13) and could be very useful for identifying the stages of apoptosis on a single-cell basis.

In situ DNA modifications by labeled nucleotides have been used to detect DNA fragmentation in what may be apoptotic cells in autopsy brains of Huntington's and Alzheimer's disease patients.[274]

DNA fragmentation is also associated with amyotrophic lateral sclerosis.[275] Analogous to TdT's ability to label double-strand breaks, the *E. coli* repair enzyme DNA polymerase I can be used to detect single-strand nicks,[276,277] which appear as a relatively early step in some apoptotic processes.[278-280] Because our ChromaTide BODIPY FL-14-dUTP and ChromaTide fluorescein-12-dUTP (C-7614, C-7604; see Section 8.2) are incorporated into DNA by *E. coli* DNA polymerase I, it is likely that they may also be effective for *in situ* labeling with the nick translation method.

Apoptosis Assays Using Free Radical Probes

The *bcl-2* proto-oncogene product is reported to play a role in preventing apoptosis through its antioxidant properties.[281] Following an apoptotic signal, cells sustain progressive lipid peroxidation, as detected with *cis*-parinaric acid (P-1, P-1901), that can be suppressed by *bcl-2* overexpression.[281] *cis*-Parinaric acid was also used to assess lipid peroxidation in Down's syndrome neurons, which exhibit increased levels of intracellular reactive oxygen species that lead to apoptosis.[282] The reagent diphenyl-1-pyrenylphosphine (DPPP, D-7894) is essentially nonfluorescent until it is oxidized by hydroperoxides to a phosphine oxide.[283-286] Its lipid solubility may make DPPP similarly useful for detecting hydroperoxides in the membranes of live cells.

Induction of apoptosis in human NK cells by monocytes is blocked by catalase, a scavenger of hydrogen peroxide, and by sodium azide, a myeloperoxidase inhibitor, whereas scavengers of superoxide and hydroxyl radicals do not prevent apoptosis.[287] Probes such as 10-acetyl-3,7-dihydroxyphenoxazine (A-6550) and its more lipophilic analog 10-acetyl-3,7-dihydroxy-2-dodecylphenoxazine (A-6551) react with hydrogen peroxide in the presence of a peroxidase to form red fluorescent resorufin derivatives and may therefore be useful for correlating hydrogen peroxide production in cells with apoptosis.

The peroxynitrite radical, which is generated by reaction of nitric oxide and superoxide, is able to induce apoptosis in HL-60 cells,[288] suggesting the possible utility of nitric oxide probes (see Section 21.5) in apoptosis studies. Also, because oxidative stress resulting in glutathione depletion has been reported to be a causative effect of apoptosis in mouse thymocytes,[289] our probes for glutathione (see "Thiol and Glutathione Determination in Live Cells" in this section) and reactive oxygen species (see Chapter 21) may be useful for detecting apoptotic cells.

Apoptosis Assays Using Ion Indicators

Significant changes in intracellular pH and Ca^{2+} concentrations accompany apoptosis. The role of acidification in apoptosis has been investigated using our carboxy SNARF-1 AM acetate [290,291] (C-1271, see Section 23.2) and BCECF AM (B-1150) [292] cell-permeant pH indicators.[290] Changes in intracellular Ca^{2+} levels may influence gene expression, as well as nuclease, protease and kinase activity.[293-300] Our wide selection of Ca^{2+} indicators, caged Ca^{2+} reagents, Ca^{2+} ionophores and Ca^{2+} chelators may help to sort out the mechanism of Ca^{2+} action in apoptosis.

Apoptosis Assays Using Esterase Substrates

Alterations in membrane permeability occur during apoptosis and have been monitored using nucleic acid stains (see above). These membrane changes may also affect the uptake and retention of our various general esterase substrates (see Table 16.3). Results from staining apoptotic thymocytes with esterase substrates, however, showed significant variation depending on which probe was used.[256,301] Some of this variation undoubtedly resulted from differences in the pH sensitivity of the probes; thus, calcein AM (C-1430), which has low pH sensitivity, may be the best reagent for detecting membrane permeability changes that accompany apoptosis.

Chemotaxis Assays

Chemotaxis, defined as directed cell motion toward an extracellular gradient, plays an important role during fertilization, inflammation, wound healing and hematopoiesis.[302] Measurement of chemotaxis usually has relied on the determination of the number of viable cells that have migrated through a special "chemotaxis chamber." Thus, the probes used to follow chemotaxis in live cells typically are the same esterase substrates that are useful for assaying cell viability and cell adhesion (Table 16.3). The primary esterase substrates used for this purpose are calcein AM [8,9,303] (C-1430, C-3099, C-3100) and BCECF AM [187,304] (B-1150, B-1170, B-3051). Calcein AM does not interfere with lymphocyte proliferation or with granulocyte or neutrophil chemotaxis or superoxide production,[8,304] and, unlike BCECF AM, calcein AM does not affect chemotaxis in leukocytes.[9] The tetrazolium salt MTT (M-6494) has also been used in a 96-well microplate assay to quantitate chemotaxis.[305]

Molecular Probes' products for direct detection of chemotactic peptide receptors include the fluorescein, BODIPY FL and tetramethylrhodamine conjugates of the chemotactic hexapeptide *N*-formyl-Nle-Leu-Phe-Nle-Tyr-Lys (F-1314, F-7468, F-1377; see Section 17.1), which bind to the fMLF receptor.[306,307] Molecular Probes also offers fluorescein-labeled casein (C-2990, see Section 17.1). Using fluorescein-labeled casein, researchers have demonstrated casein-specific chemotaxis receptors in human neutrophils and monocytes with flow cytometry.[308] Neutrophils activated with 4β-phorbol 12β-myristate 13α-acetate (P-3453, P-3454; see Section 20.3) have been shown to undergo a dose-dependent increase in binding of fluorescein-labeled casein.[309]

Multidrug Resistance Assays

Multidrug resistance (MDR) is a phenomenon representing a complex group of biological processes that are of growing interest in both clinical and experimental oncology.[310-312] Among the many mechanisms contributing to the acquired resistance of tumor cells to a variety of anticancer drugs are:

- Amplification of genes encoding drug-metabolizing enzymes
- Elevated levels of glutathione and glutathione-conjugating enzymes
- Mutated DNA topoisomerases
- Overexpression of plasma membrane ATP-dependent drug efflux pumps

Drug resistance in malarial parasites may share some of these same mechanisms.[313]

Based on substrate and inhibitor profiles, at least four different plasma membrane ATP-dependent drug efflux pumps have been identified.[314] The activity of the verapamil-sensitive P-glycoprotein encoded by the *MDR1* gene leads to extrusion of anthracyclins, epipodophyllotoxins, *Vinca* alkaloids and other cytostatic drugs. Many tumor cells do not express P-glycoprotein but export daunorubicin via a second, energy-dependent drug export mechanism.[315] A third, energy-dependent drug exporting mechanism has been described that is associated with a MDR-associated protein [316] (MRP), which shows high selectivity toward glutathione *S*-conjugates and is inhibited by *Vinca* alkaloids and probenecid.[317] A fourth vanadate- and verapamil-resistant but probenecid-sensitive glutathione *S*-conjugate–exporting system in mouse and rat fibroblasts has also been reported.[318]

Molecular Probes offers a variety of useful fluorescent and nonfluorescent probes for monitoring distinct aspects of the MDR phenotype.

MDR Assays Using Acetoxymethyl Esters

The discovery that acetoxymethyl (AM) ester derivatives of fluorescent calcium indicators such as indo-1 AM and fluo-3 AM (I-1203, F-1241; see Section 22.2) or other dyes such as calcein AM (C-1430, C-3099, C-3100), are rapidly extruded from cells expressing P-glycoprotein [319-322] presented a new class of highly sensitive probes for functional assays of the *MDR1*-encoded P-glycoprotein. Calcein AM (but not calcein) is an activator of P-glycoprotein in isolated membranes with a $K_a \leq 1$ µM.[321] Cells expressing P-glycoprotein rapidly remove the nonfluorescent probe calcein AM, resulting in decreased accumulation of the highly fluorescent calcein in the cytoplasmic compartment.[317,321,323] Because calcein itself is not a substrate for the P-glycoprotein, MDR can be assessed by measuring the net accumulation of intracellular fluorescence.[320] Such assays are more rapid and significantly more sensitive than conventional assays based on doxorubicin accumulation.[324]

MDR Assays Using Mitochondrial Probes

In the classical functional assay for MDR, doxorubicin efflux is measured.[325-327] The key weakness of this assay is its low sensitivity, which has stimulated the search for other fluorochromes to monitor drug efflux. MDR cells overexpressing the P-glycoprotein have been identified by using merocyanine 540 (M-299, see Section 25.3) and rhodamine 123 (R-302), as well as several other mitochondrial probes (see Section 12.2), including acridine orange 10-nonyl bromide (nonyl acridine orange, A-1372) and rhodamine 6G (R-634).[314,328-335] Furthermore, it has been reported that fluorescence excitation spectrum of rhodamine 123 is different in drug-resistant and drug-sensitive cells.[336,337]

The potential-sensitive carbocyanine dyes, including diethyl-, dipentyl-, dihexyl- and diheptyloxacarbocyanines (D-269, D-272, D-273, D-378; see Section 25.3), also have advantages over the common doxorubicin assay for MDR.[325,338] Not only are these carbocyanine dyes more fluorescent, permitting lower dye concentrations, but their fluorescence increases upon binding to cell membranes,[339] unlike the fluorescence of doxorubicin, which is significantly quenched inside cells.[340] The ratiometric, potential-sensitive di-4-ANEPPS (D-1199, see Section 25.2) has been used to demonstrate that MDR cells have decreased electrical potentials.[341]

MDR Assays Using Nucleic Acid Stains

MDR cells have also been identified by their decreased accumulation of nucleic acid–binding dyes such as Hoechst 33258, Hoechst 33342 and ethidium bromide.[335,342,343] A recent report describes the use of two new membrane-permeant nucleic acid stains developed at Molecular Probes (SY-38 and SY-3150) for improved detection of P-glycoprotein–dependent MDR by flow cytometry.[344] These dyes more closely mimic the kinetics of uptake and accumulation of many of the cytostatic drugs that are eliminated from tumor cells by the energy-dependent efflux mechanisms. Although these specific nucleic acid stains are not yet available from Molecular Probes, their chemical structures and fluorescence properties are similar to those of several of our SYTO dyes (see Section 8.1). Our SYTO Live-Cell Nucleic Acid Stain Sampler Kits (S-7554, S-7572; see Section 8.1) provide a total of 12 SYTO dyes that may be screened for their utility in detecting MDR cells.

BODIPY Verapamil and BODIPY Dihydropyridine

The Ca^{2+}-channel blocker verapamil is one of several molecules known to inhibit P-glycoprotein–mediated drug efflux.[345] Molecular Probes offers the green fluorescent BODIPY FL verapamil (B-7431) for the study of P-glycoprotein function and localization. Verapamil appears to inhibit drug efflux by acting as a substrate of the P-glycoprotein, thereby overwhelming the transporter's capacity to expel other drugs. Our BODIPY FL conjugate of verapamil, with spectral properties similar to fluorescein, also serves as a substrate of the P-glycoprotein. This fluorescent verapamil derivative preferentially accumulates in the lysosomes of normal, drug-sensitive NIH 3T3 cells but is rapidly transported out of MDR cells, as revealed by fluorescence microscopy.[346-348] The outward transport of BODIPY FL verapamil from MDR cells is inhibited by underivatized verapamil, as well as by excess vinblastine.[348] BODIPY FL verapamil should be useful for investigating P-glycoprotein–mediated multidrug resistance; however, we do not yet know its effect on L-type Ca^{2+} channels.

Like verapamil, dihydropyridines are known to inhibit drug efflux. Consequently, our new fluorescent dihydropyridines [349,350] labeled with either the green fluorescent DM-BODIPY (D-7443, D-7444; see Section 18.4) or the orange fluorescent ST-BODIPY (S-7445, S-7446; see Section 18.4) fluorophores may be useful MDR probes.

BODIPY Prazosin, BODIPY Forskolin

Photoaffinity analogs of prazosin, an α_1-adrenergic receptor antagonist,[351,352] and forskolin, an adenylate cyclase activator, have been shown to selectively photolabel isolated P-glycoprotein.[353-356] Our green fluorescent BODIPY FL prazosin (B-7433) and BODIPY FL forskolin (B-7469) and red fluorescent BODIPY 558/568 prazosin (B-7434) are potentially useful tools for probing MDR mechanisms.

MDR Assays Using Pyrenebutyltriphenylphosphonium Bromide

Lipophilic phosphonium derivatives have been shown to be substrates for P-glycoprotein, and their accumulation is blocked by verapamil.[357] Our new pyrenebutyltriphenylphosphonium bromide ($PyTPP^+$, P-6921) has not been tested for this purpose; however, its accumulation in membranes should result in formation of fluorescent excimers (see Figure 13.6 in Chapter 13). The spectral shift that results from pyrene excimer formation may permit ratiometric measurement of probe accumulation in single cells by imaging or flow cytometry.

MDR Assays Using Ion Indicators

Using fluorescent pH indicators, researchers have demonstrated that MDR cells exhibit increased intracellular pH, increased acid secretion or increased intracellular pH evoked by alkalinization of the medium.[340,341,358-361] Molecular Probes' extensive selection of fluorescent pH indicators are described in Chapter 23.

Other Probes for Multidrug Resistance

Other molecules of pharmaceutical interest available from Molecular Probes that are known to interact with P-glycoprotein or other multidrug transporters include:

- Actinomycin D[362,363] (A-7592, see Section 8.1)
- Demecolcine[362,363] (Colcemid®, D-7499; see Section 11.2)
- Gramicidin[364,365] (G-6888, see Section 24.1)
- Melittin[364] (M-98403, see Section 20.3)
- Valinomycin[364,365] (V-1644, see Section 24.1)

Nitrobenz-2-oxa-1,3-diazole–substituted phosphocholines (for example N-3786, N-3787; see Section 13.2) undergo time- and temperature-dependent translocation from the outer to inner leaflet in yeast that express the *MDR2* isoform of P-glycoproteins. This "flippase" activity can be followed by dithionite reduction of the outer leaflet of the membrane.[366]

Methotrexate Resistance and Gene Amplification

Enhanced tolerance to the cytotoxic effects of methotrexate is a common outcome of gene amplification in tumor cells.[310,367] Fluorescein methotrexate (M-1198) is useful for studying antimetabolite resistance and spontaneous gene amplification. The quantitative binding of fluorescein methotrexate to dihydrofolate reductase (DHFR), which confers methotrexate resistance, enables researchers to isolate cells based on DHFR expression.[368-371] Our fluorescein cadaverine adduct, which was originally described by Gapski and colleagues[372] appears to be equivalent to fluorescein lysine methotrexate in its applications.[368]

Glutathione Determination in Live Cells

The tripeptide glutathione (γ-L-glutamyl-L-cysteinylglycine) — the most abundant (up to 10 mM) and important nonprotein thiol in almost all mammalian cells — plays a central role in protecting cells against damage produced by free radicals, oxidants and electrophiles. A distinct mechanism of MDR involves the overexpression of energy-dependent membrane pumps dedicated to removal of glutathione *S*-conjugates from the cytoplasm by a multidrug resistance–associated protein (MRP).[373,374]

Several fluorescent reagents have been proposed for determining cellular levels of glutathione and glutathione *S*-transferase (GST), which catalyzes the formation of the glutathione *S*-conjugates, but no probe is without drawbacks in quantitative studies of live cells. The high but variable levels of intracellular glutathione make kinetic measurements under saturating substrate conditions difficult.[375,376] Isozymes of GST vary both in abundance and activity, further complicating the analysis.[377] Moreover, the fluorescent reagents designed to measure glutathione may react with intracellular thiols other than glutathione, including proteins in glutathione-depleted cells.[378] Therefore, precautions must be taken in applying the reagents mentioned here to quantitate either glutathione or GST in cells.

Glutathione Determination with Monochlorobimane

Cell-permeant monochlorobimane (mBCl, M-1381), which is essentially nonfluorescent until conjugated, has long been the preferred thiol-reactive probe for quantitating glutathione levels in cells and for measuring GST activity.[379,380] Monochlorobimane reacts more selectively with glutathione in whole cells than does monobromobimane (M-1378, see Section 2.4) and has proven useful for assaying drug resistance by flow cytometry[376,380,381] and by fluorescence microscopy.[78,382] Moreover, HPLC analysis has shown that glutathione is the only low molecular weight thiol in hepatocytes that reacts with monochlorobimane.[383] Results from glutathione determination with monochlorobimane have been shown to match those from an independent glutathione-specific assay using glutathione reductase.[384,385] However, although monochlorobimane was shown to be highly specific for glutathione in rodent cells, it failed to label human cells adequately because of its low affinity for human glutathione *S*-transferases.[376]

Monochlorobimane has been used to detect a vanadate- and verapamil-resistant but probenecid-sensitive glutathione *S*-conjugate exporting system in mouse and rat fibroblasts[318] Probenecid inhibits the ATP-dependent organic anion pump and blocks the loss of the fluorescent bimane–glutathione adduct from rat fibroblasts.[386] Monochlorobimane has also been employed to sort cells based on their expression of recombinant GST.[387]

Glutathione Determination with Visible Light–Excitable Thiol-Reactive Probes

Two of Molecular Probes' CellTracker dyes — CellTracker Green CMFDA (5-chloromethylfluorescein diacetate, C-2925) and CellTracker Yellow-Green CMEDA (5-chloromethyleosin diacetate, C-2926) — are useful alternatives to the UV-excitable monochlorobimane for determining levels of intracellular glutathione.[388] Both CellTracker Green CMFDA and CellTracker Yellow-Green CMEDA can be excited by the argon-ion laser, making them compatible with flow cytometry and confocal laser scanning microscopy applications. CellTracker Green CMFDA's enzymatic product also has a much higher absorbance and fluorescence quantum yield than that of monochlorobimane. Moreover, the specificity of CellTracker Green CMFDA was recently shown to be similar to that of monochlorobimane in determinations of mycotoxin-induced depletion of glutathione in rat granulosa cells[389] and glutathione levels in small samples from human melanoma biopsies.[390] In conjunction with Hoechst 33342 (H-1399), CellTracker Green CMFDA has also been shown to be effective for analyzing intracellular thiol levels as a function of cell cycle using flow cytometry.[391,392] Again, thiol levels determined by monochlorobimane were found to be identical to those detected using CellTracker Green CMFDA. Specificity of CellTracker Green CMFDA for glutathione (versus thiolated proteins) was shown by the isolation of >95% of the intracellular fluorescent products as a mixture of the glutathione adduct and the unconjugated hydrolysis product, chloromethylfluorescein;[376] however, the high fluorescence of unconjugated chloromethylfluorescein resulted in increased background levels.

Like CMFDA and CMEDA, chloromethyl SNARF-1 acetate (C-6826) forms adducts with intracellular thiols that are well retained in viable cells. The glutathione adduct of chloromethyl SNARF-1 can be excited by the 488 nm spectral line of the argon-ion laser yet emits beyond 630 nm (see Figure 23.6 in Chapter 23), which may prove advantageous in multicolor applications or when assaying autofluorescent samples. A number of our other CellTracker (see Section 15.2), MitoTracker (see Section 12.2) and fluorinated fluorescein derivatives (see "Viability/Cytotoxicity Assays Using Esterase Substrates" in this section) have thiol-reactive chloromethyl moieties and may be similarly useful for glutathione determination. All of these probes form glutathione *S*-conjugates that are likely to be transported from the cytoplasm by an MDR-associated protein (MRP).[373,374]

Thiol Determination With *o*-Phthaldialdehyde and Naphthalenedicarboxaldehyde

The reagent *o*-phthaldialdehyde (OPA, P-2331) reacts with both the thiol and the amine functions of glutathione, yielding a cyclic derivative that is fluorescent. The spectrum of the glutathione adduct of OPA is shifted from that of its protein adducts. This effect has occasionally been used to estimate glutathione levels in cells.[376,393] OPA has also been used as a derivatization reagent for the chromatographic determination of glutathione in cells, blood and tissues.[394,395]

The membrane-permeant naphthalenedicarboxaldehyde (NDA, N-1138) was recently used to determine glutathione levels in single cells. Cells were treated with NDA reagent and then analyzed by capillary electrophoresis;[396] glutathione labeling was reported to be complete within two minutes. NDA's glutathione adduct can be excited by the 458 nm spectral line of the argon-ion laser.

Cytotoxic and Cytostatic Agents

Natural Products and Synthetic Drugs

Molecular Probes has available several natural and synthetic products that possess cytotoxic or cytostatic activity or are useful probes for cancer research (Table 16.4).

Table 16.4 *Cytotoxic and cytostatic agents.*

Drug	Activity
Actinomycin D (A-7592, see Section 8.1)	• Nucleic acid synthesis inhibitor [1-3] • Antineoplastic antibiotic
Angelicine (isopsoralen, A-96003; see Section 8.1) Psoralen (P-96045, see Section 8.1) 4,5′,8-Trimethylpsoralen (T-1365, see Section 8.1)	• Photoreactive nucleic acid stains [4-6] • Psoralen and trimethylpsoralen photocrosslink DNA, whereas angelicine is a monofunctional photobinding reagent that does not crosslink DNA • Useful in a variety of photochemotherapies [7,8]
9-Chloro-9-(4-(diethylamino)phenyl)-9,10-dihydro-10-phenylacridine (C-8447)	• Antibacterial agent that inhibits growth of most gram-negative and gram-positive bacteria • Useful for the selective isolation of *Pseudomonas aeruginosa* [9-11]
Demecolcine (Colcemid, D-7499; see Section 11.2)	• Microtubule assembly inhibitor • Useful for synchronizing cells and blocking cells in mitosis [12,13]
2-(Dimethylaminoethyl)-3-ethyl-5-methylindole (KYR-44, D-99012)	• Cytostatic agent that affects ganglioside expression [15,16] • Serotonin and tryptamine antagonist [17]
Jasplakinolide (J-7473, see Section 11.1)	• Cytotoxic peptide from a marine sponge that induces actin polymerization [14]
Okadaic acid (O-3452, see Section 20.3)	• Protein phosphatase inhibitor • Potent tumor promoter
Paclitaxel (Taxol equivalent, T-3456; see Section 11.2)	• Microtubule assembly promoter, producing aggregates that cannot be depolymerized by microtubule-disrupting drugs [18,19] • Blocks cells in the G_2 and M phases of the cell cycle [20]
4β-Phorbol 12β-myristate 13α-acetate (PMA, TPA; P-3453, P-3454; see Section 20.3)	• Protein kinase C activator • Potent tumor promoter
Sodium polyanetholsulfonate (S-8461)	• Anticoagulant that possesses anti-complement activity and decreases the antibacterial action of blood [21]
Solasodine (S-96096)	• Steroidal alkaloid with selective toxicity for cancer cells [22]
Staurosporine (antibiotic AM-2282, S-7456; see Section 20.3)	• Potent protein kinase A and protein kinase C inhibitor [23] • Inhibits tumor cell invasion [24] • Enhances drug accumulation in MDR cells [25]
Taurocholic acid (T-8462)	• Enhances *MDR2*-mediated phosphocholine translocation in yeast [26]
Thapsigargin (T-7458, see Section 20.2)	• Endoplasmic reticulum Ca^{2+}-ATPase inhibitor that releases Ca^{2+} from intracellular stores [27] but does not directly affect plasma membrane Ca^{2+}-ATPases, IP_3 production or protein kinase C activity [28,29] • Tumor promoter

1. Biochemistry 33, 11493 (1994); **2.** Eur J Cancer Res Clin Oncol 22, 879 (1986); **3.** J Mol Biol 35, 251 (1968); **4.** Biochemistry 31, 6774 (1992); **5.** J Biol Chem 263, 11466 (1988); **6.** J Mol Biol 178, 897 (1984); **7.** Photochem Photobiol 32, 813 (1980); **8.** Ann Rev Pharmacol Toxicol 20, 235 (1980); **9.** J Clin Microbiol 30, 2728 (1992); **10.** J Clin Microbiol 26, 1910 (1988); **11.** J Clin Microbiol 26, 1901 (1988); **12.** Cell Tissue Kinetics 17, 223 (1984); **13.** Exp Cell Res 40, 660 (1965); **14.** J Biol Chem 269, 14869 (1994); **15.** Ontogenez (Moscow) 18, 413 (1987); **16.** Comp Biochem Physiol 79, 425 (1984); **17.** Ontogenez (Moscow) 17, 628 (1986); **18.** J Biol Chem 269, 23399 (1994); **19.** J Cell Biol 112, 1177 (1991); **20.** Cancer Treat Rep 62, 1219 (1978); **21.** Arch Exp Pathol Pharmacol 158, 211 (1930); **22.** Cancer Lett 55, 221 (1990); **23.** Biochem Biophys Res Comm 135, 397 (1986); **24.** J Natl Cancer Inst 82, 1753 (1990); **25.** Biochem Biophys Res Comm 173, 1252 (1990); **26.** J Biol Chem 270, 25388 (1995); **27.** J Biol Chem 270, 11731 (1995) **28.** Proc Natl Acad Sci USA 87, 2466 (1990); **29.** J Biol Chem 264, 12266 (1989).

Caged Penicillin

Molecular Probes now offers photoactivatable (caged) penicillin. The β-lactam antibiotic penicillin inflicts damage to the peptidoglycan cell walls of many bacterial species. Because the carboxylate substituent of penicillin and related antibiotics is essential for its biological activity,[397] we were able to synthesize a completely inactive, caged derivative of penicillin V (P-7061) by esterifying a DMNPE group to the carboxylate substituent at the 3-position. The release of functional penicillin V can be controlled temporally and spatially by flash photolysis with UV light (see Chapter 19 for a description of caged probes).

Cell Permeabilization Reagents

Assays of cellular cytotoxicity and viability that are based on the integrity of the plasma membrane require negative controls to correct for spontaneous cell death in the absence of the test agent and positive controls to show maximal signal of completely permeabilized cells. Cells can be completely permeabilized by using a variety of techniques involving mechanical or chemical disruption. Nonionic detergents are frequently employed for this purpose to minimize denaturation of cellular constituents and to reduce interference with fluorescent reagents. These detergents are also useful for isolating and reconstituting intrinsic membrane proteins.

Octyl Glucoside and Octyl Thioglucoside

A number of nonionic detergents can be used for solubilization of biomolecules or for membrane permeabilization with minimal perturbation of molecular structure and function. Octyl β-D-glucopyranoside (octyl glucoside, O-8457) is an important nonionic detergent that has been used extensively for the solubilization of membrane proteins in preparation for biochemical characterization [398,399] and reconstitution into artificial lipid bilayers or liposomes.[400,401] Octyl β-D-thioglucopyranoside (C_8SGlu, O-7035), which is reported to be more stable than octyl glucoside, is a mild nonionic detergent used to isolate proteins from bacteria without denaturation.[402] A recent paper reports the use of 1.5% solution of C_8SGlu to enhance the penetration of labeled dextrans through stripped skin, with penetration efficiency dependent on the dextran's molecular weight.[403]

Taurocholic Acid and Digitonin

The sodium salt of taurocholic acid [366] (T-8462), digitonin (D-8449) and a special water-soluble digitonin (D-8450) are steroid-based nonionic detergents that are thought to permeabilize cells of eukaryotic cells through their interaction with cholesterol and other unconjugated β-hydroxysteroids, which are predominantly located in the plasma membrane.[404-406] Taurocholic acid, a bile acid, extracts phospholipids out of lipid bilayers without causing leakage.[366]

1. Biochem Biophys Res Comm 172, 262 (1990); **2.** EMBO J 5, 51 (1986); **3.** J Immunol Meth 177, 101 (1994); **4.** J Immunol Meth 172, 227 (1994); **5.** J Immunol Meth 172, 115 (1994); **6.** Human Immunol 37, 264 (1993); **7.** Biophys J 18, 3 (1977); **8.** Biophys J 68, 1207 (1995); **9.** Cytometry 19, 366 (1995); **10.** Cytometry 12, 666 (1991); **11.** J Immunol Meth 139, 281 (1991); **12.** Biophys J 68, 2608 (1995); **13.** J Immunol Meth 172, 255 (1994); **14.** Cytometry 11, 244 (1990); **15.** J Immunol Meth 108, 255 (1988); **16.** J Immunol Meth 122, 15 (1989); **17.** Trop Med Parasitol 46, 45 (1995); **18.** Anal Biochem 177, 364 (1989); **19.** Cryobiology 14, 322 (1977); **20.** Vox Sanguinis 27, 13 (1974); **21.** Proc Natl Acad Sci USA 55, 134 (1966); **22.** J Immunol Meth 157, 117 (1993); **23.** Cancer Res 49, 3776 (1989); **24.** J Histochem Cytochem 33, 77 (1985); **25.** Cytometry 7, 70 (1986); **26.** Biochemistry 18, 2210 (1979); **27.** Immunol Lett 2, 187 (1980); **28.** J Immunol Meth 33, 33 (1980); **29.** Biochemistry 34, 1606 (1995); **30.** Cytometry 13, 739 (1992); **31.** J Immunol Meth 130, 251 (1990); **32.** J Immunol Meth 100, 261 (1987); **33.** Cytometry 15, 213 (1994); **34.** Appl Environ Microbiol 60, 1467 (1994); **35.** Biol Reprod 34, 127 (1986); **36.** J Immunoassay 12, 145 (1991); **37.** J Immunol Meth 86, 7 (1986); **38.** Anticancer Res 4, 927 (1994); **39.** J Immunol Meth 155, 19 (1992); **40.** J Cell Biol 128, 779 (1995); **41.** Meth Enzymol 194, 644 (1991); **42.** Meth Cell Biol 31, 357 (1989); **43.** J Cell Biol 111, 3129 (1990); **44.** J Immunol Meth 133, 87 (1990); **45.** FEBS Lett 200, 203 (1986); **46.** BioTechniques 3, 270 (1985); **47.** Appl Environ Microbiol 61, 9 (1995); **48.** Cytometry 19, 370 (1995); **49.** J Vasc Res 32, 371 (1995); **50.** Science 251, 81 (1991); **51.** U.S. Patent No. 5,436,134; **52.** Millard, P.J. *et al.*, manuscript in preparation; **53.** Appl Environ Microbiol 60, 3284 (1994); **54.** Anal Biochem 221, 78 (1994); **55.** Appl Environ Microbiol 61, 3623 (1995); **56.** Limnol Oceanogr 40, 1050 (1995); **57.** J Neurosci 6, 1492 (1986); **58.** J Immunol Meth 52, 91 (1982); **59.** Appl Environ Microbiol 61, 2521 (1995); **60.** Neuron 15, 961 (1995); **61.** Anal Biochem 220, 149 (1994); **62.** J Immunol Meth 149, 133 (1992); **63.** J Immunol Meth 134, 201 (1990); **64.** Nucleic Acids Res 23, 1215 (1995); **65.** Nucleic Acids Res 20, 2803 (1992); **66.** Biochemistry 17, 5078 (1978); **67.** Biochemistry 17, 5071 (1978); **68.** Proc Natl Acad Sci USA 87, 3851 (1990); **69.** J Immunol Meth 178, 71 (1995); **70.** Cytometry 20, 181 (1995); **71.** J Neurosci 15, 6239 (1995); **72.** J Neurosci 14, 2260 (1994); **73.** Kopp, W.C. and Wersto, R.P. in *Manual of Clinical Laboratory Immunology, Fourth Edition*, N.R. Rose *et al.*, Eds., American Society of Microbiology (1992) pp. 933–941; **74.** Cytometry 12, 133 (1991); **75.** Arch Toxicol 68, 582 (1994); **76.** Brain Res 635, 113 (1994); **77.** Chem Res Toxicol 5, 227 (1992); **78.** see Burghardt, R.C. *et al.* in *Principles and Methods of Toxicology, Third Edition*, A.W. Hayes, Ed., Raven Press (1994) pp. 1231–1258 for a brief protocol for using this product for *in vitro* toxicology testing; **79.** Biochemistry 34, 7194 (1995); **80.** Cell 75, 241 (1993); **81.** J Immunology 130, 1910 (1983); **82.** Cytometry 18, 147 (1994); **83.** Meth Enzymol 233, 539 (1994); **84.** Naturwissenschaften 75, 354 (1988); **85.** J Med Chem 30, 1757 (1987); **86.** J Euk Microbiol 41, S79 (1994); **87.** J Cell Physiol 156, 428 (1993); **88.** Biochem Biophys Res Comm 175, 387 (1991); **89.** FEMS Microbiol Lett 122, 187 (1994); **90.** J Immunol Meth 170, 117 (1994); **91.** J Immunol Meth 140, 23 (1991); **92.** J Appl Physiol 76, 539 (1994); **93.** J Leukocyte Biol 54, 300 (1993); **94.** J Biochem 106, 355 (1989); **95.** Biochem Biophys Res Comm 155, 106 (1988); **96.** J Cell Physiol 151, 466 (1992); **97.** J Immunol Meth 179, 95 (1995); **98.** J Infect Dis 172, 1153 (1995); **99.** J Appl Bacteriol 74, 433 (1993); **100.** J Immunol Meth 160, 89 (1993); **101.** J Immunol Meth 157, 203 (1993); **102.** J Immunol Meth 142, 257 (1991); **103.** Blood 76, 2327 (1990); **104.** J Natl Cancer Inst 81, 577 (1989); **105.** J Immunol Meth 65, 55 (1983); **106.** J Cell Biol 128, 201 (1995); **107.** Cancer Res 54, 3620 (1994); **108.** Cell 77, 817 (1994); **109.** J Immunol Meth 164, 255 (1993); **110.** J Infec Diseases 172, 1153 (1995); **111.** J Immunol Meth 158, 81 (1993); **112.** J Immunol Meth 147, 153 (1992); **113.** J Natl Cancer Inst 81, 577 (1989); **114.** Cancer Res 48, 4827 (1988); **115.** Histochem J 12, 619 (1980); **116.** Bitton, G. and Koopman, B. in *Toxicity Testing Using Microorganisms*, G. Bitton and B.J. Dutka, Eds., CRC Press (1986) pp. 27–55; **117.** Burghardt, R.C. *et al.* in *Principles and Methods of Toxicology, Third Edition*, A.W. Hayes, Ed., Raven Press (1994) pp. 1231–1258; **118.** Clin Bull 11, 47 (1981); **119.** J Cell Biol 91, 392 (1981); **120.** Somatic Cell Genetics 9, 375 (1983); **121.** J Appl Bacteriol 72, 410 (1992); **122.** Cancer Res 42, 799 (1982); **123.** Bull Environ Contam Toxicol 51, 557 (1993); **124.** Cytometry 15, 343 (1994); **125.** J Immunol Meth 161, 119 (1993); **126.** Jpn J Pharmacol 57, 419 (1991); **127.** J Appl Bacteriol 78, 309 (1995); **128.** J Microscopy 176, 8 (1994); **129.** J Cell Biol 130, 157 (1995); **130.** Neuron 15, 961 (1995); **131.** Exp Cell Res 214, 323 (1994); **132.** Cardiovascular Res 27, 1790 (1993); **133.** Clin Chem 41, 1906 (1995); **134.** Cell Biol Toxicol 10, 329 (1994); **135.** Anal Biochem 213, 426 (1993); **136.** Biotech Histochem 68, 29 (1993); **137.** ATLA 18, 129 (1990); **138.** Bio/Technology 8, 1248 (1990); **139.** In Vitro Toxicol 3, 219 (1990); **140.** In Vitro Cell Devel Biol 24, 266 (1988); **141.** Biotech Histochem 67, 27 (1992); **142.** J Microbiol Meth 13, 87 (1991); **143.** Nature 190, 269 (1961); **144.** Meth Enzymol 93, 233 (1983); **145.** J Immunol Meth 147, 13 (1992); **146.** J Immunol Meth 119, 45 (1989); **147.** J Immunol Meth 168, 267 (1994); **148.** J Immunol Meth 161, 135 (1993); **149.** J Immunol Meth 160, 27 (1993); **150.** Human Immunol 35, 85 (1992); **151.** J Immunol Meth 114, 191 (1988); **152.** J Fluorescence 5, 3 (1995); **153.** Cancer Res 46, 5064 (1986); **154.** J Membrane Biol 77, 115 (1984); **155.** Radiation Res 112, 351 (1987); **156.** Exp Cell Res 173, 256 (1987); **157.** Science 218, 474 (1982); **158.** Meth Cell Biol 41, 297 (1994); **159.** Meth Cell Biol 41, 195 (1994); **160.** J Histochem Cytochem 25, 913 (1977); **161.** Proc Natl Acad Sci USA 70, 3395 (1973); **162.** Proc Natl Acad Sci USA 80, 2951 (1983); **163.** Meth Cell Biol 41, 327 (1994); **164.** Exp Cell Res 174, 319

(1988); **165.** Cytometry 17, 310 (1995); **166.** Cytometry 20, 172 (1995); **167.** Cytometry 20, 257 (1995); **168.** Cytometry 20, 245 (1995); **169.** J Histochem Cytochem 41, 7 (1993); **170.** J Cell Biol 119, (1992); **171.** Cancer Res 54, 4289 (1994); **172.** J Cell Biol 119, 494 (1992); **173.** Exp Cell Res 222, 28 (1996); **174.** Cell Proliferation 28, 571 (1995); **175.** Cell 84, 345 (1996); **176.** Ann Rev Biochem 60, 155 (1991); **177.** J Immunol Meth 178, 41 (1995); **178.** J Biol Chem 270, 21129 (1995); **179.** J Immunol Meth 172, 25 (1994); **180.** J Cell Biol 130, 745 (1995); **181.** J Cell Biol 127, 1129 (1994); **182.** Proc Natl Acad Sci USA 90, 5700 (1993); **183.** J Immunol Meth 163, 181 (1993); **184.** J Biol Chem 269, 10008 (1994); **185.** J Immunology 156, 1638 (1996); **186.** J Cell Biol 133, 445 (1996); **187.** Mol Biol of the Cell 6, 661 (1995); **188.** J Biol Chem 269, 1033 (1994); **189.** J Cell Biol 125, 1395 (1994); **190.** J Biol Chem 268, 8835 (1993); **191.** J Cell Biol 123, 245 (1993); **192.** J Cell Biol 124, 609 (1994); **193.** Devel Biol 135, 133 (1989); **194.** J Cell Biol 109, 3465 (1989); **195.** J Immunol Meth 165, 93 (1992); **196.** Thrombosis Res 77, 543 (1995); **197.** Biochem J 270, 149 (1990); **198.** Biochem Pharmacol 36, 4035 (1987); **199.** Cell 84, 371 (1996); **200.** Cell 84, 359 (1996); **201.** Cell 69, 11 (1992); **202.** Nature 346, 425 (1990); **203.** J Cell Biol 131, 791 (1995); **204.** Biochemistry 33, 266 (1994); **205.** J Biol Chem 270, 28812 (1995); **206.** J Biol Chem 270, 11358 (1995); **207.** J Cell Biol 104, 1403 (1987); **208.** J Biol Chem 258, 12582 (1983); **209.** Science 231, 1559 (1986); **210.** Proc Natl Acad Sci USA 82, 8057 (1985); **211.** Cytometry 17, 287 (1994); **212.** J Lab Clin Med 123, 728 (1994); **213.** Annals NY Acad Sci 416, 426 (1983); **214.** Arch Pathol Lab Med 113, 591 (1989); **215.** J Histochem Cytochem 25, 585 (1977); **216.** Meth Cell Biol 41, 231 (1994); **217.** Meth Cell Biol 41, 218 (1994); **218.** Meth Cell Biol 41, 211 (1994); **219.** Cytometry 23, 272 (1996); **220.** Appl Environ Microbiol 62, 1649 (1996); **221.** Limnol Oceanogr 40, 1485 (1995); **222.** Cytometry 15, 129 (1994); **223.** Cytometry 17, 191 (1994); **224.** Cytometry 15, 267 (1994); **225.** Biochemistry 29, 9029 (1990); **226.** J Histochem Cytochem 24, 24 (1976); **227.** Cytometry 13, 389 (1992); **228.** J Microbiol Meth 20, 255 (1994); **229.** J Microbiol Meth 19, 89 (1994); **230.** Microbiol Rev 58, 603 (1994); **231.** J Parasitol 65, 421 (1979); **232.** Meth Cell Biol 41, 401 (1994); **233.** Proc Natl Acad Sci USA 73, 2881 (1976); **234.** J Cell Physiol 143, 279 (1990); **235.** J Immunology 136, 2769 (1986); **236.** Cytometry 13, 60 (1992); **237.** Cytometry 12, 279 (1991); **238.** Histochem J 17, 131 (1985); **239.** Cytometry 5, 355 (1984); **240.** Cancer Res 52, 5007 (1992); **241.** Biopolymers 18, 1749 (1979); **242.** Cytometry 20, 296 (1995); **243.** Chromosoma 68, 287 (1978); **244.** Proc Natl Acad Sci USA 75, 5655 (1978); **245.** J Immunology 132, 38 (1984); **246.** Blood 85, 359 (1995); **247.** Genes Dev 8, 2817 (1994); **248.** Am J Pathol 146, 3 (1995); **249.** Cell 74, 777 (1993); **250.** J Cell Biochem 52, 352 (1993); **251.** J Biol Chem 267, 10939 (1992); **252.** J Cell Biol 126, 827 (1994); **253.** Cytometry 13, 795 (1992); **254.** J Immunol Meth 185, 249 (1995); **255.** Cytometry 21, 265 (1995); **256.** Cytometry 14, 595 (1993); **257.** Anal Biochem 204, 351 (1992); **258.** Cytometry 13, 137 (1992); **259.** Cytometry 17, 59 (1994); **260.** Cytometry 16, 41 (1994); **261.** Cytometry 15, 12 (1994); **262.** J Immunol Meth 170, 145 (1994); **263.** Nature 378, 736 (1995); **264.** J Immunol Meth 123, 103 (1989); **265.** Cytometry 9, 447 (1988); **266.** Development 117, 29 (1993); **267.** Anal Biochem 218, 314 (1994); **268.** Cancer Res 44, 83 (1984); **269.** Nucleic Acids Res 22, 5506 (1994); **270.** BioTechniques 15, 1032 (1993); **271.** Cancer Res 51, 4671 (1991); **272.** Am J Pathol 136, 593 (1990); **273.** Anal Biochem 221, 431 (1994); **274.** NeuroReport 6, 1053 (1995); **275.** Acta Neuropathol 88, 207 (1994); **276.** Cancer Res 52, 1945 (1993); **277.** Int'l J Oncol 1, 639 (1992); **278.** Trends Cell Biol 5, 21 (1995); **279.** Jpn J Cancer Res 84, 566 (1993); **280.** Nucleic Acids Res 18, 4206 (1993); **281.** Cell 75, 241 (1993); **282.** Nature 378, 776 (1995); **283.** J Chromatography 628, 31 (1993); **284.** J Chromatography 622, 153 (1993); **285.** J Chromatography 596, 197 (1992); **286.** Anal Lett 21, 965 (1988); **287.** J Immunology 156, 42 (1996); **288.** J Biol Chem 270, 16487 (1995); **289.** J Immunology 155, 5133 (1995); **290.** Proc Natl Acad Sci USA 93, 654 (1996); **291.** J Biol Chem 270, 3203 (1995); **292.** J Biol Chem 270, 6235 (1995); **293.** Biochem Biophys Res Comm 214, 1130 (1995); **294.** J Neurosci 15, 1172 (1995); **295.** Biochim Biophys Acta 1223, 247 (1994); **296.** Cell Calcium 16, 279 (1994); **297.** Exp Cell Res 212, 84 (1994); **298.** FASEB J 8, 237 (1994); **299.** J Immunology 151, 5198 (1993); **300.** Exp Cell Res 197, 43 (1991); **301.** Exp Cell Res 211, 322 (1994); **302.** Ann Rev Cell Biol 4, 649 (1988); **303.** J Immunology 156, 679 (1996); **304.** J Immunol Meth 172, 115 (1994); **305.** J Immunol Meth 164, 149 (1993); **306.** Biochemistry 29, 313 (1990); **307.** J Biol Chem 265, 16725 (1990); **308.** Inflammation 7, 363 (1983); **309.** J Immunology 139, 3028 (1987); **310.** Cancer Res 54, 666 (1994); **311.** Proc Natl Acad Sci USA 91, 3497 (1994); **312.** Ann Rev Biochem 62, 385 (1993); **313.** J NIH Res 2, 28 (1990); **314.** Biochemistry 34, 32 (1995); **315.** Cancer Res 52, 17 (1992); **316.** Proc Natl Acad Sci USA 91, 13033 (1994); **317.** FEBS Lett 368, 385 (1995); **318.** Cytometry 14 (Suppl 6), 35A (1993); **319.** Cytometry 17, 343 (1994); **320.** Biochim Biophys Acta 1191, 384 (1994); **321.** J Biol Chem 268, 21493 (1993); **322.** J Natl Cancer Inst 83, 206 (1991); **323.** J Biol Chem 271, 13668 (1996); **324.** Eur J Cancer 29A, 1024 (1993); **325.** Cytometry 8, 306 (1987); **326.** Cancer Res 43, 5126 (1983); **327.** Cancer Res 40, 3895 (1980); **328.** Anticancer Res 15, 121 (1995); **329.** Cytometry 17, 50 (1994); **330.** J Biol Chem 269, 7145 (1994); **331.** Mol Pharmacol 45, 1145 (1994); **332.** Proc Natl Acad Sci USA 91, 4654 (1994); **333.** FEBS Lett 329, 63 (1993); **334.** Blood 78, 1385 (1991); **335.** Cancer Res 51, 4665 (1991); **336.** Exp Cell Res 196, 323 (1991); **337.** Cancer Commun 1, 145 (1989); **338.** J Cell Physiol 126, 266 (1986); **339.** Biochemistry 34, 3858 (1995); **340.** Biochemistry 31, 12555 (1992); **341.** Biochemistry 32, 11042 (1993); **342.** Anticancer Res 13, 1557 (1993); **343.** Cell Biochem Function 10, 9 (1992); **344.** Cytometry 20, 218 (1995); **345.** Cancer Res 47, 1421 (1987); **346.** Biochem Biophys Res Comm 212, 494 (1995); **347.** Biochim Biophys Acta 1237, 31 (1995); **348.** Mol Pharmacol 40, 490 (1991); **349.** Biochemistry 33, 11875 (1994); **350.** Proc Natl Acad Sci USA 89, 3586 (1992); **351.** Cardiology 73, 164 (1986); **352.** Life Sci 27, 1525 (1980); **353.** J Biol Chem 268, 11417 (1993); **354.** Biochem Pharmacol 43, 89 (1992); **355.** J Biol Chem 266, 20744 (1991); **356.** J Biol Chem 264, 15483 (1989); **357.** Biochemistry 31, 1992 (1992); **358.** Biochemistry 33, 7229 (1994); **359.** J Histochem Cytochem 38, 685 (1990); **360.** Br J Cancer 61, 568 (1990); **361.** J Natl Cancer Inst 81, 706 (1989); **362.** Exp Cell Res 174, 168 (1988); **363.** Hereditas 106, 97 (1987); **364.** J Biol Chem 270, 10334 (1995); **365.** Biochim Biophys Acta 1190, 72 (1994); **366.** J Biol Chem 270, 25388 (1995); **367.** Cancer Res 41, 4665 (1991); **368.** J Biol Chem 257, 14162 (1982); **369.** Cancer Res 50, 4946 (1984); **370.** Proc Natl Acad Sci USA 80, 3711 (1983); **371.** J Biol Chem 253, 5852 (1978); **372.** J Med Chem 18, 526 (1975); **373.** J Biol Chem 269, 29085 (1994); **374.** Proc Natl Acad Sci USA 91, 13033 (1994); **375.** Anal Biochem 217, 41 (1994); **376.** Cytometry 15, 349 (1994); **377.** Cancer Res 51, 1783 (1991); **378.** Cytometry 12, 366 (1991); **379.** Annals NY Acad Sci 677, 345 (1993); **380.** Cancer Res 46, 6105 (1986); **381.** J Biol Chem 263, 14107 (1988); **382.** Cytometry 19, 226 (1995); **383.** Anal Biochem 190, 212 (1990); **384.** Cancer Res 51, 4287 (1991); **385.** J Clin Lab Anal 4, 324 (1990); **386.** Cytometry 23, 78 (1996); **387.** Cytometry 12, 651 (1991); **388.** Cytometry 12, 184 (1991); **389.** Toxicol Appl Pharmacol 112, 235 (1992); **390.** Melanoma Res 5, 107 (1995); **391.** Cytometry 14, 747 (1993); **392.** Cytometry 12, 184 (1991); **393.** Exp Cell Res 163, 518 (1986); **394.** Clin Chem 41, 448 (1995); **395.** Chromatographia 36, 130 (1993); **396.** Anal Chem 67, 4261 (1995); **397.** Boyd, D.B. in *Chemistry and Biology of β-Lactam Antibiotics, Volume 1,* R.B. Morin and M. Gorman, Eds., Academic Press (1982); **398.** Biochemistry 33, 7069 (1994); **399.** Biochim Biophys Acta 382, 276 (1975); **400.** Biochemistry 20, 833 (1981); **401.** Biochemistry 19, 3088 (1980); **402.** Biochem J 222, 829 (1984); **403.** J Pharm Sci 83, 1676 (1994); **404.** Meth Enzymol 159, 193 (1988); **405.** Meth Enzymol 104, 305 (1984); **406.** Proc Natl Acad Sci USA 76, 690 (1979).

Product-Specific Bibliographies and Keyword Searches

Our Technical Assistance Department can provide you with product-specific bibliographies, as well as keyword searches of the over 25,000 literature references in our extensive bibliography database. Our bibliography database is also searchable through our Web site (http://www.probes.com).

Cat #	Structure	MW	Storage	Soluble	Abs	{ε × 10⁻³}	Em	Solvent	Notes
A-1301	S8.1	302	L	EtOH, DMF	500	{53}	526	<1>	<2>
A-1310	S8.1	1270	F,L	DMF, DMSO	546	{25}	647	<1>	
A-6550	S21.4	257	FF,D,AA	DMSO	280	{6.0}	none	pH 8	<3>
A-6551	S21.4	426	FF,D,AA	DMSO	286	{8.0}	none	MeOH	<4>
B-1150	S23.2	~615*	F,D	DMSO	<300		none		<5>
B-7431	S18.4	769	F,D,L	DMSO, EtOH	504	{74}	511	MeOH	
B-7433	S18.2	563	F,D,L	DMSO, EtOH	504	{77}	511	MeOH	
B-7434	S18.2	617	F,D,L	DMSO, EtOH	560	{76}	569	MeOH	
B-7469	S20.3	785	F,D,L	DMSO	504	{79}	511	MeOH	
C-195	✓	460	F,D	DMSO	<300		none		<6>
C-369	✓	529	F,D	DMSO	<300		none		<7>
C-1157	S15.2	557	F,D	DMF, DMSO	<300		none		<6>
C-1354	✓	532	F,D	DMSO	<300		none		<8>
C-1361	✓	460	F,D	DMSO	<300		none		<6>
C-1362	✓	460	F,D	DMSO	<300		none		<6>
C-1429	✓	465	F,D,L	DMSO	322	{14}	435	MeOH	<9>
C-1430	✓	995	F,D	DMSO	<300		none		<10>
C-2925	S15.2	465	F,D	DMSO	<300		none		<6>
C-2926	S15.2	780	F,D	DMSO	<300		none		<11>
C-6826	S23.2	500	F,D	DMSO	<350		none		<12>
C-7614	✓	~974†	FF,L	H_2O‡	504	{68}	513	pH 8	
C-8447	✓	475	F,D,L	EtOH	596	{14}	ND	MeOH	<13>
D-1306	S8.1	350	L	H_2O, DMF	358	{21}	461	<1>	
D-7021	S16.3	342	F,DD,L	DMSO	381	{5.3}	454	MeOH	
D-7894	S21.3	386	F,D,LL	MeCN	358	{29}	none	MeOH	<14>
D-8449	✓	1229	D	EtOH	<300		none		
D-8450	✓	1229	D	H_2O	<300		none		<15>
D-99012	✓	267	D	EtOH	<300		ND		<13>
E-1169	S8.1	857	F,D,L	DMSO	528	{7.0}	617	<1>	<16>
E-1305	S8.1	394	L	H_2O, DMSO	518	{5.2}	605	<1>	<17>
E-1374	S8.1	420	F,LL	DMF, EtOH	464	{5.8}	625	pH 3	<18>
E-6960	S15.3	366§	D	H_2O	337	{17}	613	H_2O	<19>
F-1303	✓	416	F,D	DMSO	<300		none		<6>
F-7496	C	~340,000	FF,D,L	H_2O	498	ND	525	pH 8	<13, 20>
H-1398	S8.1	624§	L	H_2O, DMF	352	{40}	461	<1>	
H-1399	S8.1	616§	L	H_2O, DMF	350	{45}	461	<1>	
H-7019	S15.2	524	F,DD	DMSO	<300		none		<21>
H-7593	S8.1	497	L	DMSO	518	{3.9}	600	<1>	<22>
I-6496	S21.1	500	D	H_2O, DMSO	249	{38}	none	MeOH	<23>
L-7595	S8.1	472	L	DMSO, EtOH	543	{46}	712	<1>	
M-1198	✓	979	F,L	pH>6, DMF	496	{67}	516	pH 9	
M-1381	S2.4	227	F,L	DMSO	380	{6.0}	<24>	MeOH	
M-6494	S21.4	414	D,L	H_2O, DMSO	375	{8.3}	none	MeOH	<23, 25>
N-1138	S1.8	184	L	DMF, MeCN	462	ND	520	MeOH	<13, 26>
N-3246	S12.3	289	D,L	H_2O, EtOH	541	{39}	640	<27>	
N-6495	S21.4	818	D,L	H_2O, DMSO	256	{64}	none	MeOH	<23>
N-6498	S21.4	430	D	H_2O, DMSO	246	{33}	none	MeOH	<23>
O-6151	✓	496	F,D	DMSO	<300		none		<28>
O-7035	✓	308	D	H_2O, EtOH	<300		none		
O-8457	✓	292	D	H_2O, EtOH	<300		none		
P-1	S13.3	276	FF,LL,AA	EtOH	303	{76}	416	MeOH	<29>
P-1304	S8.1	668	L	H_2O, DMF	535	{5.4}	617	<1>	<30>
P-2331	S1.8	134	L	EtOH	334	{5.7}	455	pH 9	<31>
P-6921	S25.3	600	D,L	DMSO, DMF	340	{37}	377	MeOH	<32>
P-7061	✓	560	F,D,LL	DMSO	337	{5.0}	none	MeOH	<33>
R-302	S12.2	381	F,D,L	MeOH, DMF	507	{101}	529	MeOH	
S-1129	✓	518	F,D	DMSO	<300		none		<34>
S-7020	P	~600¶	F,D,L	DMSO‡	504	{67}	523	<1>	<35>
S-8461	✓	ND	D	H_2O	<300		none		<13>
S-96096	✓	414	D	$CHCl_3$	<300		none		
T-411	S15.3	266	D,L	DMF, EtOH	337	{17}	483	MeOH	
T-3168	S25.3	652	D,L	DMSO, EtOH	514	{195}	529	MeOH	<36>
T-6490	S21.4	728	D	H_2O, DMSO	253	{53}	none	MeOH	<23>
T-7015	✓	720	L	pH>6, DMF	545	{111}	561	MeOH	
T-7016	S15.2	488	F,D	DMSO	<300		none		<21>
T-7017	S15.2	804	F,D	DMSO	<300		none		<21>
T-8462	✓	538	D	H_2O	<300		none		
X-6493	S21.4	675	D	H_2O, DMSO	286	{15}	none	MeOH	<37>

A-3568 see A-1301; B-1170, B-3051 see B-1150; C-3099, C-3100 see C-1430; C-7025 see C-2925; E-3565 see E-1305; H-3569 see H-1398; H-3570 see H-1399; P-1901 see P-1; P-3566 see P-1304.

For definitions of the contents of this data table, see "How to Use the Handbook" on page vi.
Structure: Chemical structure drawing: ✓ = shown in this section; Sn.n = shown in Section number n.n; C = not shown due to complexity; P = proprietary information.
MW: ~ indicates an approximate value.

* MW of B-1150 is approximate. See Figure 23.4 in Chapter 23 for further details.
† The molecular weight (MW) of this product is approximate because the exact degree of hydration and salt form has not been conclusively established.
‡ This product is packaged as a solution in the solvent indicated in "Soluble."
§ MW is for the hydrated form of this product.
¶ Not including counterions.
<1> Spectra represent aqueous solutions of DNA-bound dye. ϵ values are derived by comparing the absorbance of the DNA-bound dye with that of free dye in a reference solvent (H_2O or MeOH).
<2> A-1301 bound to RNA has Em ~650 nm [Cytometry 2, 201 (1982)].
<3> Peroxidase-catalyzed reaction of A-6550 with H_2O_2 produces fluorescent resorufin (R-363, see Section 10.1), Abs = 571 nm (ϵ = {54}), Em = 585 nm in pH 9 buffer.
<4> Peroxidase-catalyzed reaction of A-6551 with H_2O_2 produces fluorescent 2-dodecylresorufin (D-7786, see Section 14.3), Abs = 578 nm (ϵ = {69}), Em = 597 nm in MeOH.
<5> B-1150 is colorless and nonfluorescent until converted to B-1151 (see Section 23.2), Abs = 503 nm (ϵ = {90}), Em = 528 nm by acetoxymethyl ester hydrolysis.
<6> This compound is converted to a fluorescent product with pH-dependent spectra similar to C-1904 (see Section 23.2), Abs = 492 nm (ϵ = {78}), Em = 517 nm (see Section 23.2), after acetate hydrolysis.
<7> C-369 is converted to a fluorescent product (C-368, see Section 23.3), Abs = 504 nm (ϵ = {90}), Em = 529 nm, after acetate hydrolysis.
<8> Hydrolysis of the acetate and acetoxymethyl ester groups of C-1354 yields C-1359 (see Section 1.3), which has pH-dependent spectra similar to C-1904 (see Section 23.2), Abs = 492 nm (ϵ = {78}), Em = 517 nm.
<9> C-1429 is converted to a fluorescent product with spectra similar to H-189 (see Section 10.1), Abs = 360 nm (ϵ = {19}) Em = 449 nm in pH 9 buffer, after acetate hydrolysis.
<10> C-1430 is converted to fluorescent calcein (C-481, see Section 15.3), Abs = 494 nm (ϵ = {76}), Em = 517 nm, after acetoxymethyl ester hydrolysis.
<11> C-2926 is converted to a fluorescent product with spectra similar to C-301 (see Section 1.5), Abs = 519 nm (ϵ = {100}, Em = 542 nm (see Section 1.5), after acetate hydrolysis.
<12> C-6826 is converted to a fluorescent product with spectra similar to C-1270 (see Section 23.2), Abs = 576 nm (ϵ = {48}), Em = 635 nm at pH 10, after acetate hydrolysis.
<13> ND = not determined.
<14> Oxidation product is strongly fluorescent, Em = 379 nm. Oxidation occurs rapidly in solution when illuminated.
<15> D-8450 is soluble in water at 10 mg/mL upon heating to ~95°C. Absorbance of a 1 mg/mL aqueous solution at 280 nm is ~0.02.
<16> E-1169 in H_2O: Abs = 493 nm (ϵ = {9.1}). Very weakly fluorescent in H_2O. Fluorescence is enhanced >40-fold upon binding to DNA.
<17> E-1305 in H_2O: Abs = 480 nm (ϵ = {5.6}), Em = 620 nm (weakly fluorescent). Fluorescence is enhanced >10-fold upon binding to DNA.
<18> E-1374 spectral data are for the free dye. Fluorescence is weak, but intensity increases ~15-fold on binding to DNA. After photocrosslinking to DNA, Abs = 504 nm (ϵ ~{4.0}), Em = 600 nm [Biochemistry 19, 3221 (1980); Nucleic Acids Res 5, 4891 (1978)].
<19> Absorption and fluorescence of E-6960 are extremely weak unless it is chelated. Data are for detergent-solubilized NTA (T-411) chelate.
<20> This protein conjugate contains multiple labels. The number of labels per protein is indicated on the vial.
<21> T-7016, T-7017 and H-7019 are converted to fluorescent products after acetate hydrolysis. T-7016 yields T-7014 (see Section 23.3), Abs = 508 nm (ϵ = {94}), Em = 528 nm. T-7017 yields T-7015. H-7019 yields H-7018 (see Section 23.3), Abs = 508 nm (ϵ = {80}), Em = 529 nm.
<22> H-7593 in H_2O: Abs = 482 nm (ϵ = {5.5}), Em = 625 nm (weakly fluorescent).
<23> Enzymatic reduction products are water-insoluble formazans, with Abs = 600 nm (T-6490), 505 nm (M-6494), 605 nm (N-6495), 465 nm (I-6496) and 510 nm (N-6498) after solubilization in DMSO or DMF. See Histochemistry 76, 381 (1982) and Prog Histochem Cytochem 9, 1 (1977) for further information.
<24> M-1381 is almost nonfluorescent until reacted with thiols. Em = 475–485 nm for thiol adducts (QY ~0.1–0.3) [Meth Enzymol 251, 133 (1995); Meth Enzymol 143, 76 (1987)].
<25> M-6494 also has Abs = 242 nm (ϵ = {21}) in MeOH.
<26> Spectra of N-1138 with glycine + cyanide. Reagent Abs = 279 nm (ϵ = {5.5}), Em = 330 nm in MeOH.
<27> Spectra of N-3246 are pH dependent (pK_a ~6.7). Data reported are for 1:1 (v/v) EtOH/1% acetic acid.
<28> O-6151 is converted to a fluorescent product (O-6148, see Section 23.3), Abs = 492 nm (ϵ = {88}), Em = 517 nm, after acetate hydrolysis.
<29> P-1 is readily oxidized to nonfluorescent products. Use under N_2 or Ar except when oxidation is intended. Stock solutions should be prepared in deoxygenated solvents. Parinaric acid is appreciably fluorescent in lipid environments and organic solvents but is nonfluorescent in water.
<30> P-1304 in H_2O: Abs = 493 nm (ϵ = {5.9}), Em = 636 nm (weakly fluorescent). Fluorescence is enhanced >10-fold upon binding to DNA.
<31> Spectral data are for the reaction product of P-2331 with alanine and 2-mercaptoethanol. The spectra and the stability of the adduct depend on the amine and thiol reactants [Biochim Biophys Acta 576, 440 (1979)]. Unreacted reagent in H_2O: Abs = 257 nm (ϵ = {1.0}).
<32> Pyrene derivatives exhibit structured spectra. The absorption maximum is usually about 340 nm with a subsidiary peak at about 325 nm. There are also strong absorption peaks below 300 nm. The emission maximum is usually about 376 nm with a subsidiary peak at 396 nm. Excimer emission at about 470 nm may be observed at high concentrations.
<33> All photoactivatable probes are sensitive to light. They should be protected from illumination, except when photolysis is intended.
<34> S-1129 is converted to a fluorescent product (F-1130, see Section 23.2), Abs = 495 (ϵ = {76}), Em = 519, after acetate hydrolysis.
<35> This product is essentially nonfluorescent except when bound to DNA or RNA.
<36> T-3168 forms J-aggregates with Abs/Em = 585/590 nm at concentrations above 0.1 µM in aqueous solutions (pH 8.0) [Biochemistry 30, 4480 (1991)].
<37> Enzymatic reduction product is a water-soluble formazan, Abs = 475 nm.

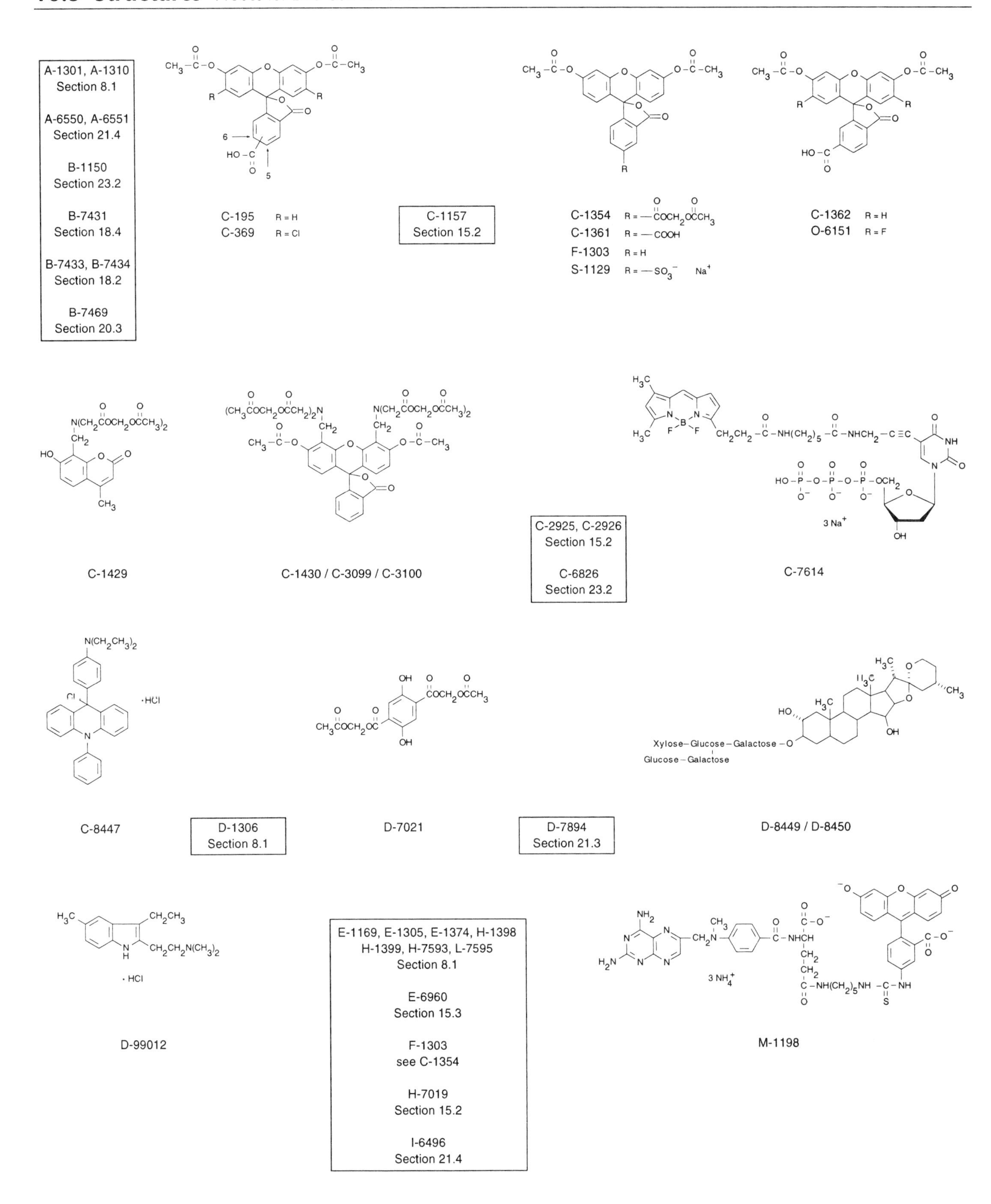
A-1301, A-1310
Section 8.1
A-6550, A-6551
Section 21.4
B-1150
Section 23.2
B-7431
Section 18.4
B-7433, B-7434
Section 18.2
B-7469
Section 20.3
C-195 R = H
C-369 R = Cl
C-1157
Section 15.2
C-1354 R = —COCH2OCCH3
C-1361 R = —COOH
F-1303 R = H
S-1129 R = —SO3⁻ Na⁺
C-1362 R = H
O-6151 R = F
C-1429
C-1430 / C-3099 / C-3100
C-2925, C-2926
Section 15.2
C-6826
Section 23.2
C-7614
C-8447
D-1306
Section 8.1
D-7021
D-7894
Section 21.3
D-8449 / D-8450
Xylose–Glucose–Galactose–O
Glucose–Galactose
D-99012
E-1169, E-1305, E-1374, H-1398
H-1399, H-7593, L-7595
Section 8.1
E-6960
Section 15.3
F-1303
see C-1354
H-7019
Section 15.2
I-6496
Section 21.4
M-1198

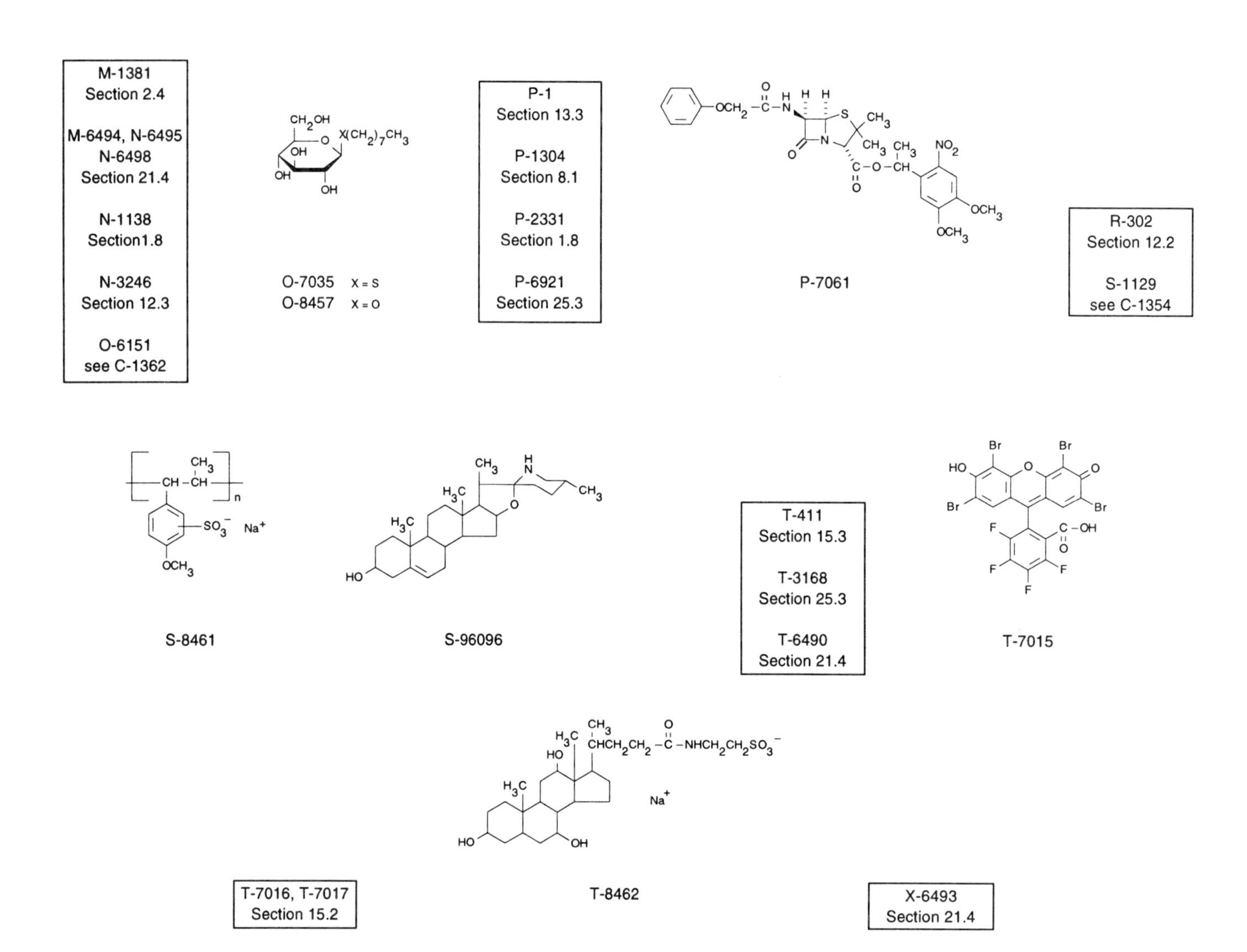

16.3 Price List *Probes for Live-Cell Function*

Cat #		Product Name	Unit Size	Price Per Unit ($) 1–4 Units	5–24 Units
A-6551	***New***	10-acetyl-3,7-dihydroxy-2-dodecylphenoxazine *special packaging*	10x100 µg	135.00	108.00
A-6550	***New***	10-acetyl-3,7-dihydroxyphenoxazine *special packaging*	10x1 mg	85.00	68.00
A-1301		acridine orange *high purity*	1 g	48.00	38.40
A-3568		acridine orange *10 mg/mL solution in water*	10 mL	28.00	22.40
A-1310		7-aminoactinomycin D (7-AAD)	1 mg	50.00	40.00
B-1150		2′,7′-bis-(2-carboxyethyl)-5-(and-6)-carboxyfluorescein, acetoxymethyl ester (BCECF, AM)	1 mg	68.00	54.40
B-1170		2′,7′-bis-(2-carboxyethyl)-5-(and-6)-carboxyfluorescein, acetoxymethyl ester (BCECF, AM) *special packaging*	20x50 µg	92.00	73.60
B-3051		2′,7′-bis-(2-carboxyethyl)-5-(and-6)-carboxyfluorescein, acetoxymethyl ester (BCECF, AM) *1 mg/mL solution in dry DMSO* *special packaging*	1 mL	73.00	58.40
B-7434	***New***	BODIPY® 558/568 prazosin	100 µg	95.00	76.00
B-7469	***New***	BODIPY® FL forskolin	100 µg	95.00	76.00
B-7433	***New***	BODIPY® FL prazosin	100 µg	95.00	76.00
B-7431	***New***	BODIPY® FL verapamil, hydrochloride	1 mg	98.00	78.40
C-1430		calcein, AM	1 mg	115.00	92.00
C-3099		calcein, AM *1 mg/mL solution in dry DMSO* *special packaging*	1 mL	120.00	96.00
C-3100		calcein, AM *special packaging*	20x50 µg	140.00	112.00
C-1429		calcein blue, AM	1 mg	50.00	40.00
C-369		5-(and-6)-carboxy-2′,7′-dichlorofluorescein diacetate (carboxy-DCFDA) *mixed isomers*	100 mg	98.00	78.40
C-1361		5-carboxyfluorescein diacetate (5-CFDA) *single isomer*	100 mg	75.00	60.00
C-1362		6-carboxyfluorescein diacetate (6-CFDA) *single isomer*	100 mg	75.00	60.00
C-195		5-(and-6)-carboxyfluorescein diacetate (5(6)-CFDA) *mixed isomers*	100 mg	75.00	60.00
C-1354		5-carboxyfluorescein diacetate, acetoxymethyl ester (5-CFDA, AM)	5 mg	68.00	54.40

Cat #		Product Name	Unit Size	Price Per Unit ($) 1–4 Units	5–24 Units
C-1157		5-(and-6)-carboxyfluorescein diacetate, succinimidyl ester (5(6)-CFDA, SE) *mixed isomers*	25 mg	115.00	92.00
C-2925		CellTracker™ Green CMFDA (5-chloromethylfluorescein diacetate)	1 mg	125.00	100.00
C-7025	**New**	CellTracker™ Green CMFDA (5-chloromethylfluorescein diacetate) *special packaging*	20x50 µg	145.00	116.00
C-2926		CellTracker™ Yellow-Green CMEDA (5-chloromethyleosin diacetate)	1 mg	125.00	100.00
C-8447	**New**	9-chloro-9-(4-(diethylamino)phenyl)-9,10-dihydro-10-phenylacridine, hydrochloride (C-390)	10 mg	75.00	60.00
C-6826	**New**	5-(and-6)-chloromethyl SNARF®-1, acetate *mixed isomers* *special packaging*	20x50 µg	125.00	100.00
C-7614	**New**	ChromaTide™ BODIPY® FL-14-dUTP (BODIPY® FL-14-dUTP) *1 mM in buffer*	25 µL	175.00	175.00
D-1306		4′,6-diamidino-2-phenylindole, dihydrochloride (DAPI)	10 mg	37.00	29.60
D-8449	**New**	digitonin *USP grade*	1 g	40.00	32.00
D-8450	**New**	digitonin *water soluble*	1 g	48.00	38.40
D-7021	**New**	2,5-dihydroxyterephthalic acid, di(acetoxymethyl ester) (DHTP, AM)	5 mg	95.00	76.00
D-99012	**New**	2-(dimethylaminoethyl)-3-ethyl-5-methylindole, hydrochloride (KYR-44)	50 µg	105.00	84.00
D-7894	**New**	diphenyl-1-pyrenylphosphine (DPPP)	5 mg	125.00	100.00
E-1305		ethidium bromide	1 g	12.00	9.60
E-3565		ethidium bromide *10 mg/mL solution in water*	10 mL	28.00	22.40
E-1169		ethidium homodimer-1 (EthD-1)	1 mg	155.00	124.00
E-1374		ethidium monoazide bromide	5 mg	115.00	92.00
E-6960	**New**	europium(III) chloride, hexahydrate *99.99%*	1 g	20.00	16.00
F-7496	**New**	fibrinogen from human plasma, Oregon Green™ 488 conjugate	5 mg	95.00	76.00
F-1303		fluorescein diacetate (FDA)	1 g	40.00	32.00
H-7019	**New**	2′,4,5,6,7,7′-hexafluorofluorescein diacetate (HFFDA)	5 mg	125.00	100.00
H-7593	**New**	hexidium iodide	5 mg	95.00	76.00
H-3569		Hoechst 33258 (bis-benzimide) *10 mg/mL solution in water*	10 mL	40.00	32.00
H-1398		Hoechst 33258, pentahydrate (bis-benzimide)	100 mg	20.00	16.00
H-3570		Hoechst 33342 *10 mg/mL solution in water*	10 mL	40.00	32.00
H-1399		Hoechst 33342, trihydrochloride, trihydrate	100 mg	20.00	16.00
I-6496	**New**	4-iodonitrotetrazolium violet (INT; 2-(4-iodophenyl)-3-(4-nitrophenyl)-5-phenyltetrazolium chloride)	1 g	40.00	32.00
L-7595	**New**	LDS 751	10 mg	65.00	52.00
M-1198		methotrexate, fluorescein, triammonium salt (fluorescein methotrexate)	1 mg	125.00	100.00
M-1381		monochlorobimane (mBCl)	25 mg	78.00	62.40
M-6494	**New**	MTT (3-(4,5-dimethylthiazol-2-yl)-2,5-diphenyltetrazolium bromide)	1 g	40.00	32.00
N-1138		naphthalene-2,3-dicarboxaldehyde (NDA)	100 mg	65.00	52.00
N-3246		neutral red *high purity*	25 mg	50.00	40.00
N-6495	**New**	nitro blue tetrazolium chloride (NBT)	1 g	65.00	52.00
N-6498	**New**	4-nitrotetrazolium violet chloride (NTV)	1 g	32.00	25.60
N-7565	**New**	Nucleic Acid Stains Dimer Sampler Kit	1 kit	125.00	100.00
N-7566	**New**	Nucleic Acid Stains Monomer Sampler Kit	1 kit	65.00	52.00
O-8457	**New**	octyl β-D-glucopyranoside (octyl glucoside)	1 g	30.00	24.00
O-7035	**New**	octyl β-D-thioglucopyranoside (C_8SGlu)	1 g	35.00	28.00
O-6151	**New**	Oregon Green™ 488 carboxylic acid diacetate (carboxy-DFFDA) *6-isomer*	5 mg	95.00	76.00
P-1		*cis*-parinaric acid	100 mg	155.00	124.00
P-1901		*cis*-parinaric acid *special packaging*	10x10 mg	185.00	148.00
P-7061	**New**	penicillin V, 1-(4,5-dimethoxy-2-nitrophenyl)ethyl ester (O-(DMNPE-caged) penicillin V)	5 mg	48.00	38.40
P-2331		o-phthaldialdehyde (OPA) *high purity*	1 g	32.00	25.60
P-1304		propidium iodide	100 mg	48.00	38.40
P-3566		propidium iodide *1.0 mg/mL solution in water*	10 mL	28.00	22.40
P-6921	**New**	1-pyrenebutyltriphenylphosphonium bromide ($PyTPP^+$)	10 mg	48.00	38.40
R-302		rhodamine 123	25 mg	30.00	24.00
S-8461	**New**	sodium polyanetholsulfonate	25 g	55.00	44.00
S-96096	**New**	solasodine	20 mg	55.00	44.00
S-1129		5-sulfofluorescein diacetate, sodium salt (SFDA) *single isomer*	25 mg	85.00	68.00
S-7572	**New**	SYTO® Live-Cell Nucleic Acid Stain Sampler Kit #1 *SYTO® dyes 11-16* *50 µL each*	1 kit	135.00	108.00
S-7554	**New**	SYTO® Live-Cell Nucleic Acid Stain Sampler Kit #2 *SYTO® dyes 20-25* *50 µL each*	1 kit	135.00	108.00
S-7020	**New**	SYTOX® Green nucleic acid stain *5 mM solution in DMSO*	250 µL	98.00	78.40
T-8462	**New**	taurocholic acid, sodium salt *high purity*	5 g	48.00	38.40
T-7015	**New**	2′,4′,5′,7′-tetrabromo-4,5,6,7-tetrafluorofluorescein (Br_4TFF)	10 mg	35.00	28.00
T-7017	**New**	2′,4′,5′,7′-tetrabromo-4,5,6,7-tetrafluorofluorescein diacetate (Br_4TFFDA)	5 mg	125.00	100.00
T-3168		5,5′,6,6′-tetrachloro-1,1′,3,3′-tetraethylbenzimidazolylcarbocyanine iodide (JC-1; CBIC$_2$(3))	5 mg	195.00	156.00
T-7016	**New**	4,5,6,7-tetrafluorofluorescein diacetate (TFFDA)	5 mg	125.00	100.00
T-6490	**New**	tetrazolium blue chloride	5 g	35.00	28.00
T-6500	**New**	Tetrazolium Salt Sampler Kit *100 mg of T-6490, X-6493, M-6494, N-6495, I-6496, N-6498*	1 kit	125.00	100.00
T-411		2-((trifluoroacetyl)acetyl)naphthalene (naphthoyltrifluoroacetone; NTA)	100 mg	65.00	52.00
X-6493	**New**	XTT (2,3-bis-(2-methoxy-4-nitro-5-sulfophenyl)-2H-tetrazolium-5-carboxanilide)	100 mg	55.00	44.00

A Sampling of Sampler Kits

Most of Molecular Probes' products are used in minute quantities, making "sample sizes" impractical. However, we have put together a number of Sampler Kits containing a set of application-specific probes, sometimes in smaller quantities and always at lower cost than the corresponding components sold separately. These Sampler Kits enable you to test our products and find the optimal probe for your particular application. Look in the designated Handbook chapter, contact our Technical Assistance Department or visit our Web site (http://www.probes.com) for more information on any of these kits.

Cat #	Product Name	Unit Size	Price Per Unit ($) 1–4 Units	5–24 Units
Chapter 7 *Protein Conjugates for Biological Detection*				
W-7024	***New*** Wheat Germ Agglutinin Sampler Kit *four fluorescent conjugates, 1 mg each*	1 kit	98.00	78.40
Chapter 8 *Nucleic Acid Detection*				
N-7565	***New*** Nucleic Acid Stains Dimer Sampler Kit	1 kit	125.00	100.00
N-7566	***New*** Nucleic Acid Stains Monomer Sampler Kit	1 kit	65.00	52.00
S-7580	***New*** SYBR® Green Nucleic Acid Gel Stain Starter Kit	1 kit	55.00	55.00
S-6655	***New*** SYPRO® Protein Gel Stain Starter Kit	1 kit	55.00	55.00
S-7572	***New*** SYTO® Live-Cell Nucleic Acid Stain Sampler Kit #1 *SYTO® dyes 11-16* *50 µL each*	1 kit	135.00	108.00
S-7554	***New*** SYTO® Live-Cell Nucleic Acid Stain Sampler Kit #2 *SYTO® dyes 20-25* *50 µL each*	1 kit	135.00	108.00
Chapter 10 *Enzymes, Enzyme Substrates and Enzyme Inhibitors*				
I-6614	***New*** Indolyl β-D-Galactopyranoside Sampler Kit *contains 5 mg each of B-1690, B-8407, I-8414, I-8420, M-8421*	1 kit	65.00	52.00
P-6548	***New*** Protease Inhibitor Sampler Kit	1 kit	195.00	156.00
R-6564	***New*** Resorufin Ether Sampler Kit	1 kit	98.00	78.40
T-6500	***New*** Tetrazolium Salt Sampler Kit *100 mg of T-6490, X-6493, M-6494, N-6495, I-6496, N-6498*	1 kit	125.00	100.00
Chapter 12 *Probes for Organelles*				
Y-7530	***New*** Yeast Mitochondrial Stain Sampler Kit	1 kit	95.00	76.00
Y-7531	***New*** Yeast Vacuole Marker Sampler Kit	1 kit	95.00	76.00
Chapter 15 *Fluorescent Tracers of Cell Morphology and Fluid Flow*				
L-7781	***New*** Lipophilic Tracer Sampler Kit	1 kit	150.00	120.00
Chapter 18 *Probes for Receptors and Ion Channels*				
B-6850	***New*** Brevetoxin Sampler Kit *5 µg of B-6851, B-6852, B-6853, B-6854, B-6855*	1 kit	135.00	108.00
Chapter 21 *Probes For Reactive Oxygen Species,Including Nitric Oxide*				
N-7925	***New*** Nitric Oxide Synthase (NOS) Inhibitor Kit	1 kit	195.00	156.00
Chapter 22 *Indicators for Ca^{2+}, Mg^{2+}, Zn^{2+} and Other Metals*				
B-6767	***New*** BAPTA Acetoxymethyl Ester Sampler Kit *2x1 mg each of B-1205, D-1207, D-1209, D-1213*	1 kit	75.00	60.00
C-6777	***New*** Coelenterazine Sampler Kit *contains 25 µg of C-2944, C-6776, C-6779, C-6780, C-6781*	1 kit	255.00	204.00
B-6768	***New*** BAPTA Buffer Kit *10 mg each of B-1204, D-1206, D-1208, D-1211*	1 kit	75.00	60.00
Chapter 26 *Tools for Fluorescence Applications*				
F-7321	***New*** Flow Cytometry Alignment Standards Sampler Kit, 2.5 µm	1 kit	55.00	44.00
F-7322	***New*** Flow Cytometry Alignment Standards Sampler Kit, 6.0 µm	1 kit	55.00	44.00
T-7284	***New*** TetraSpeck™ Fluorescent Microspheres Sampler Kit	1 kit	95.00	76.00

Chapter 17

Probes for Following Endocytosis and Exocytosis

Contents

17.1 Probes for Investigating Phagocytic Cells

Phagocytosis plays a key role in biology, providing a mechanism for defending against opportunistic infections in higher eukaryotes as well as a means of nourishment in primitive organisms. Molecular Probes offers a variety of fluorogenic and fluorescent probes for investigating phagocytosis and the cellular events associated with it. These include:

- Fc OxyBURST assay reagents, which are specifically designed to detect the oxidative burst that occurs in phagovacuoles
- Fluorogenic probes that are oxidized in cells by reactive oxygen species (see also Section 21.4)
- BioParticles fluorescent bacteria and yeast particles
- Fluorescent polystyrene microspheres of different sizes and colors (see also Section 6.2)
- Fluorescent chemotactic peptides
- Fluorescent and nonfluorescent β-glucan
- pH indicator dextrans (see also Section 23.4)

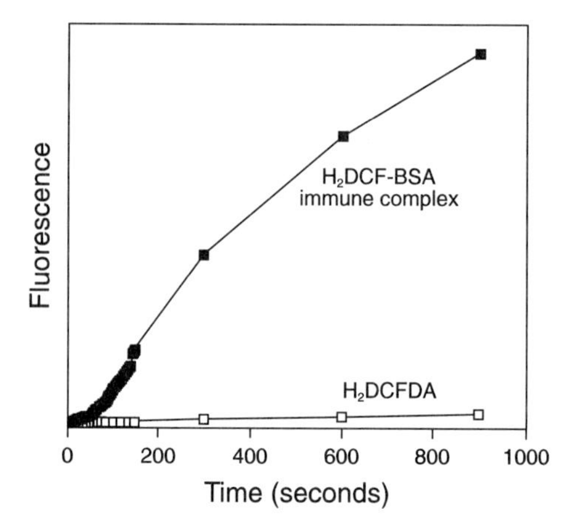

Figure 17.1 Fluorescence emission of human neutrophils challenged either with Molecular Probes' Fc OxyBURST Green assay reagent (H_2DCF-BSA immune complexes, F-2902) or with unlabeled immune complexes in the presence of dichlorodihydrofluorescein diacetate (H_2DCFDA, D-399). The Fc OxyBURST Green assay reagent generates significantly more fluorescence than does the more commonly used H_2DCFDA. Data provided by Elizabeth Simons, Boston University [J Immunol Meth 130, 223 (1990)].

Fc OxyBURST Assay Reagents and Related Probes

Fc OxyBURST Assay Reagents

When soluble or surface-bound IgG immune complexes interact with Fc receptors on phagocytic cells, a number of host defense mechanisms are activated. These include phagocytosis and activation of an NADPH oxidase–mediated oxidative burst. Dichlorodihydrofluorescein diacetate (H_2DCFDA, D-399), a cell-permeant fluorogenic probe that localizes in the cytosol, has frequently been used to monitor this oxidative burst;[1] however, its fluorescence response is limited by the diffusion rate of the reactive oxygen species from the phagovacuole where it is generated into the cytosol. In contrast, our Fc OxyBURST assay reagents permit direct measurement of the kinetics of Fc receptor–mediated internalization and the subsequent oxidative burst in the phagovacuole, yielding signals that are many times brighter than those generated by H_2DCFDA.

Molecular Probes' Fc OxyBURST Green and Orange assay reagents (F-2902, F-6691) were developed in collaboration with Elizabeth Simons of Boston University to monitor the oxidative burst in phagocytic cells using fluorescence instrumentation. The Fc OxyBURST Green assay reagent consists of bovine serum albumin (BSA) that has been covalently linked to dichlorodihydrofluorescein (H_2DCF) and then complexed with purified rabbit polyclonal anti-BSA IgG antibodies. When these immune complexes bind to Fc receptors, the nonfluorescent H_2DCF molecules are internalized within the phagovacuole and subsequently oxidized to green fluorescent dichlorofluorescein (DCF) (Figure 17.1). Unlike H_2DCFDA, the Fc OxyBURST Green assay reagent does not require intracellular esterases for activation, making this reagent particularly suitable for detecting the oxidative burst in cells with low esterase activity such as monocytes.[2] Fc OxyBURST Green assay reagent reportedly produced >8 times more fluorescence than does H_2DCFDA at 60 seconds and >20 times more at 15 minutes following internalization of the immune complex.[3]

Several recent reports describe the use of the Fc OxyBURST Green assay reagent to study the oxidative burst in phagovacuoles.

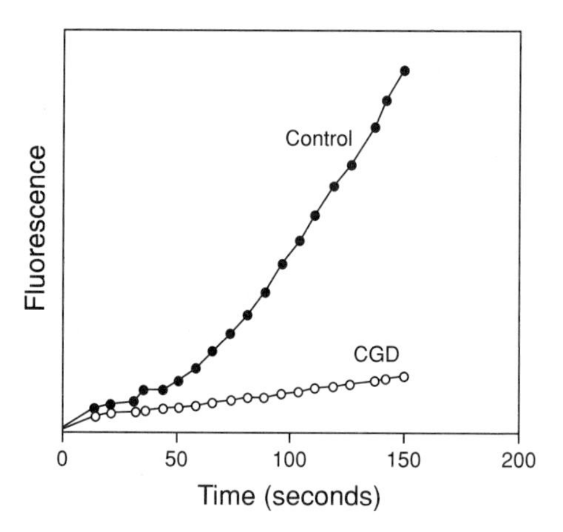

Figure 17.2 Oxidative bursts of human neutrophils from a healthy donor (control) compared to those from a patient with chronic granulomatous disease (CGD), as detected using the Fc OxyBURST Green assay reagent (F-2902). Data provided by Elizabeth Simons, Boston University [J Immunol Meth 130, 223 (1990)].

Neutrophils from patients with chronic granulomatous disease, a genetic deficiency known to disable NADPH oxidase–mediated oxidative bursts, were observed to bind but not oxidize the Fc OxyBURST Green assay reagent[3] (Figure 17.2). Using microfluorometry to detect the Fc OxyBURST Green response, researchers were able to simultaneously monitor oxidative activity and membrane currents in voltage-clamped human mononuclear cells.[4] The Fc OxyBURST Green assay reagent has also been employed

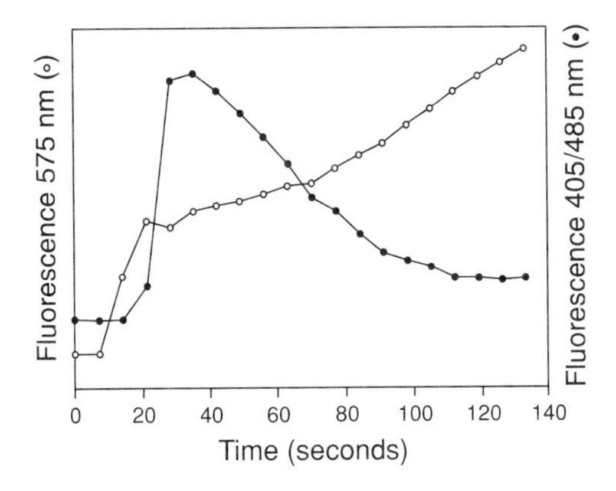

Figure 17.3 *Oxidative bursts of human polymorphonuclear leukocytes (PMNs) detected by the Fc OxyBURST Orange Assay Reagent at 575 nm (○). The stimulatory effect of the immune complex is registered by the increased intracellular Ca^{2+} levels simultaneously detected using indo-1 (●). Data provided by Elizabeth Simons, Boston University.*

to assess the effect of manganese-based superoxide dismutase mimetics on superoxide generation in human neutrophils.[5]

Our Fc OxyBURST Orange (F-6691) assay reagent is an immune complex that contains BSA labeled with the longer-wavelength tetramethylrosamine (H_2CTMRos) derivative, which exhibits excitation/emission maxima of approximately 550/574 nm upon oxidation. This reagent can be used with common rhodamine optical filters (emission 575 ± 25 nm). Simons and co-workers used the Fc OxyBURST Orange assay reagent with our indo-1 calcium indicator (see Section 22.2) to simultaneously monitor the oxidative burst and the changes in intracellular Ca^{2+} concentrations in human neutrophils (Figure 17.3).

The Fc OxyBURST immune complexes are slowly oxidized by molecular oxygen and are also susceptible to oxidation catalyzed by illumination in a fluorescence microscope. The Fc OxyBURST Green assay reagent is somewhat less prone to photooxidation than the Fc OxyBURST Orange assay reagent; of the two, only the Fc OxyBURST Green assay reagent is suitable for use in fluorescence microscopy. We have found both reagents to be reasonably stable in solution for at least six months when stored under nitrogen or argon in the dark at 4°C.

Amine-Reactive OxyBURST Reagents

As an alternative to our Fc OxyBURST assay reagents, Molecular Probes offers amine-reactive OxyBURST Green (H_2DCFDA, SE; D-2935) and OxyBURST Orange (H_2CTMRos, SE; C-2936) succinimidyl esters, which can be used to prepare oxidation-sensitive conjugates of a wide variety of biomolecules and particles, including antibodies, antigens, peptides, proteins, dextrans, bacteria, yeast and polystyrene microspheres. Following conjugation to amines, the acetates of the OxyBURST Green reagent can be removed by treatment with hydroxylamine to yield the H_2DCF conjugate. This step is not required for the OxyBURST Orange reagent. The OxyBURST Green and OxyBURST Orange conjugates are nonfluorescent until they are oxidized to the corresponding dichlorofluorescein and rosamine derivatives. Thus, like our Fc OxyBURST assay reagents, they provide a means of detecting the oxidative burst in phagocytic cells. The OxyBURST Green succinimidyl ester was recently conjugated to an antibody that binds specifically to YAC tumor cells. YAC cells opsonized with this customized OxyBURST reagent were then used in a fluorescence microscopy study to show that Fc receptor–activated neutrophils appear to deliver reactive oxygen species to the surface of their target cells.[6]

Dichlorodihydrofluorescein Diacetate and Analogs

As discussed in Section 21.4, the cell-permeant H_2DCFDA (D-399) is used to monitor oxidative activity in a wide variety of cell types. Using this fluorogenic probe, researchers have developed both flow cytometry [7,8] and fluorescence microplate assays [9] to monitor oxidative activity in neutrophils and monocytes, enabling them to examine the effects of lipopolysaccharides [10] and cytokines [11] on this important function.

H_2DCFDA passively diffuses into cells, where the acetates are cleaved by intracellular esterases. Oxidation of H_2DCF occurs almost exclusively in the cytosol and produces a fluorescence response that is much less than that of the Fc OxyBURST Green assay reagent.[3] Furthermore, the oxidation product, 2',7'-dichlorofluorescein,[12] tends to leak from cells, which may make quantitation or detection of slow oxidation difficult. Molecular Probes' 5-(and-6)-chloromethyl-2',7'-dichlorodihydrofluorescein diacetate (CM-H_2DCFDA, C-6827) should exhibit much better retention. As with CellTracker™ Green CMFDA (chloromethylfluorescein diacetate, C-2925; see Section 15.2), the mildly thiol-reactive chloromethyl groups of CM-H_2DCFDA can potentially react with intracellular thiols to form adducts that are trapped inside the cell (see Figure 15.1 in Chapter 15). In many cell types, this new chloromethyl derivative should be even better retained than our carboxylated H_2DCF (carboxy-H_2DCFDA, C-400) and its di(acetoxymethyl) ester (C-2938), both of which form the negatively charged carboxydichlorofluorescein (C-368, see Section 1.5). For further discussion of these fluorogenic probes, see Section 21.4.

Dihydrorhodamine 123

Dihydrorhodamine 123 (D-632, see also Section 21.4) is the uncharged and nonfluorescent reduction product of the mitochondrion-selective dye rhodamine 123 (R-302, see Section 12.2). Dihydrorhodamine 123 freely diffuses into most cells where it is oxidized to cationic rhodamine 123, which localizes in the mitochondria. This peroxidase substrate has been used to investigate reactive oxygen intermediates produced by human and murine phagocytes [13] and the role of the CD14 cell-surface marker in hydrogen peroxide production by human monocytes.[14] It has also been used with the Fura Red™ calcium indicator (F-3020, see Section 22.3) to simultaneously measure oxidative bursts and calcium fluxes in monocytes and granulocytes.[15] Dihydrorhodamine 123 is reportedly a more sensitive probe than H_2DCFDA for detecting granulocyte respiratory bursts.[16-18] However, dihydrorhodamine 123 also reacts with peroxynitrite,[19,20] the anion formed when nitric oxide reacts with superoxide.[21]

Dihydroethidium

Although dihydroethidium (D-1168, see also Sections 8.1 and 21.4) (also called Hydroethidine™, a trademark of Prescott Labs) is commonly used to analyze respiratory burst in phagocytes,[22-24] it has been reported that this probe undergoes significant oxidation in resting leukocytes, possibly through the uncoupling of mitochondrial oxidative phosphorylation.[25]

Chemiluminescence Probes

Luminol (L-8455) is a chemiluminescent probe that has been used to detect myeloperoxidase-mediated oxidative events in granulocytes.[26-29] The aequorin luminophore coelenterazine (C-2944) produces chemiluminescence in response to superoxide generation in phorbol ester– or chemotactic peptide–stimulated neutrophils.[30] Unlike luminol, coelenterazine exhibits luminescence that does not depend on the activity of cell-derived myeloperoxidase and is not inhibited by azide. We also offer lucigenin (L-6868), which exhibits chemiluminescence that is reported to be sensitive to the superoxide anion.[31-33] See Sections 21.3 and 21.4 for further discussion of these chemiluminescent probes.

BioParticles Fluorescent Bacteria and Yeast

Fluorescent Bacteria and Yeast Particles

Molecular Probes' BioParticles product line consists of a series of fluorescently labeled, heat- or chemically killed bacteria and yeast in a variety of sizes, shapes and natural antigenicities. These fluorescent BioParticles have been employed to study phagocytosis by fluorescence microscopy,[34] quantitative spectrofluorometry[35] and flow cytometry.[36-38] We offer *Escherichia coli* (K-12 strain), *Staphylococcus aureus* (Wood strain without protein A) and zymosan (*Saccharomyces cerevisiae*) BioParticles covalently labeled with fluorescein, BODIPY® FL, tetramethylrhodamine or Texas Red® fluorophores; special care has been taken to remove free dye after conjugation. Unlike the fluorescence of fluorescein-labeled BioParticles, which is partially quenched in acidic environments, the fluorescence of the BODIPY FL, tetramethylrhodamine and Texas Red conjugates is uniformly intense between pH 3.0 and 10.0. This property may be particularly useful in quantitating fluorescent bacteria and zymosan within acidic phagocytic vacuoles. Our BioParticles, which are freeze-dried and ready for reconstitution in the buffer of choice, come with a general protocol for measuring phagocytosis, along with a list of references. We also offer opsonizing reagents for use with each particle type as described below.

Applications for BioParticles Fluorescent Bacteria and Yeast

Fluorescent bacteria and yeast particles are proven tools for studying a variety of parameters influencing phagocytosis; for example, they have been used to:

- Detect the phagocytosis of yeast by murine peritoneal macrophage[39] and human neutrophils[35]
- Determine the effects of different opsonization procedures on the efficiency of phagocytosis of pathogenic bacteria[40] and yeast[35]
- Investigate the kinetics of phagocytosis degranulation and actin polymerization in stimulated leukocytes[35]
- Measure phagocytosis and, in conjunction with dihydroethidium, oxidative burst in leukocytes using flow cytometry[23,24]
- Quantitate the effects of anti-inflammatory drugs on phagocytosis[35]
- Show that *Dictyostelium discoideum* depleted of clathrin heavy chains are still able to undergo phagocytosis of fluorescent zymosans[41]
- Study molecular defects in phagocytic function[42]

Fluorescence of BioParticles that are bound to the surface but not internalized can be quenched by ethidium bromide,[34] trypan blue[43,44] or other quenchers. In addition to cellular applications, fluorescent BioParticles may be effective as flow cytometry calibration references when sorting bacteria and yeast mutants. These small particles may also be useful references for light scattering studies because their sizes and shapes differ in characteristic ways.

FluoReporter Phagocytosis Assay Kit

Our new FluoReporter Phagocytosis Assay Kit (F-6694) provides a convenient set of reagents for quantitating phagocytosis and assessing the effects of certain drugs or conditions on this cellular process. In this assay, cells of interest are incubated first with green fluorescent BioParticles, which are internalized by phagocytosis, and then with trypan blue, which quenches the fluorescence of extracellular BioParticles. The methodology provided by this kit was developed using the adherent murine macrophage cell line J774;[44] however, researchers can likely adapt this assay to other phagocytic cell types. Each kit provides sufficient reagents for 250 tests using a 96-well microplate format and contains:

- Fluorescein-labeled *E. coli* K-12 BioParticles
- Hanks' balanced salt solution (HBSS)
- Trypan blue
- Step-by-step instructions for performing the phagocytosis assay

Opsonizing Reagents and Nonfluorescent BioParticles

Many researchers may want to use autologous serum to opsonize their fluorescent zymosan and bacterial particles; however, we also offer special opsonizing reagents (E-2870, S-2860, Z-2850) for enhancing the uptake of each type of particle, along with a protocol for opsonization. These reagents are derived from purified rabbit polyclonal IgG antibodies that are specific for the *E. coli*, *S. aureus* or zymosan particles. Reconstitution of the lyophilized opsonizing reagents requires only the addition of water, and one unit of opsonizing reagent is sufficient to opsonize 10 mg of the corresponding BioParticles.

In addition, Molecular Probes offers nonfluorescent zymosan (Z-2849) and *S. aureus* (S-2859) BioParticles. These nonfluorescent BioParticles are useful both as controls or for custom-labeling with the reactive dye or indicator of interest.

Fluorescent Polystyrene Microspheres

Fluorescent polystyrene microspheres with diameters between 0.5 and 2.0 µm have been used to investigate phagocytic processes in rat and human neutrophils,[45,46] human trabecular meshwork cells,[47] mouse peritoneal macrophages,[48,49] ciliated protozoa[50,51] and *Dictyostelium discoideum*.[52] The phagocytosis of fluorescent microspheres has been quantitated both with image analysis[49] and with flow cytometry.[48,50,51] Section 6.2 includes a detailed description of our full line of FluoSpheres® and TransFluoSpheres® fluorescent microspheres. Because of their low nonspecific binding, carboxylate-modified microspheres appear to be best for applications involving phagocytosis. For phagocytosis experiments involving multicolor detection, we particularly recommend our new TransFluoSpheres fluorescent microspheres with Stokes shifts up to 200 nm or more. Various opsonizing reagents such as rabbit serum or fetal calf serum have been used with the microspheres to facilitate phagocytosis. Methods for coating fluorescent microspheres with the CR1 complement receptor have also been described.[53]

Fluorescent Chemotactic Peptides

A variety of white blood cells containing the formyl-Met-Leu-Phe (fMLF) receptor respond to bacterial *N*-formyl peptides by migrating to the site of bacterial invasion and then initiating an activation pathway to control the spread of infection. Activation involves calcium mobilization,[54-56] transient acidification,[57,58] actin polymerization,[59] phagocytosis [60] and production of oxidative species.[61] Molecular Probes offers fluorescein,[62-66] BODIPY FL and tetramethylrhodamine conjugates of the hexapeptide *N*-formyl-Nle-Leu-Phe-Nle-Tyr-Lys (F-1314, F-7468, F-1377), which is widely employed to investigate the fMLF receptor. These probes are well suited for instruments that employ the argon-ion laser, including confocal laser scanning microscopes and flow cytometers. The fluorescein and rhodamine conjugates have been used to study G-protein coupling and receptor structure,[67,68] expression,[69,70] distribution [71-73] and internalization.[74] For studies of fMLF receptor internalization, the fluorescence of extracellular labeled *N*-formyl hexapeptides may be quenched by adding the corresponding anti-fluorophore antibodies (see Section 7.3). Alternatively, the extracellular fluorescence of the pH-sensitive fluorescein conjugate may be reduced by dropping the pH of the external medium.

Casein and Fluorescent Casein

Using fluorescein-labeled casein (C-2990), researchers have demonstrated casein-specific chemotaxis receptors in human neutrophils and monocytes with flow cytometry.[75] Neutrophils activated with phorbol myristate acetate (P-3453, see Section 20.3) have been shown to undergo a dose-dependent increase in binding of fluorescein-labeled casein.[76] It has also been demonstrated that fluorescein casein reversibly binds to specific receptors on human monocytes but does not bind to lymphocytes *in vitro*.[75]

Our EnzChek™ Protease Assay Kits (E-6638, E-6639; see Section 10.4), which provide convenient fluorescence-based assays for protease activity, contain either green fluorescent BODIPY FL casein or red fluorescent BODIPY TR-X casein. These casein substrates are heavily labeled and therefore highly quenched conjugates; they typically exhibit <3% of the fluorescence of the corresponding free dyes. Protease-catalyzed hydrolysis relieves this quenching, yielding brightly fluorescent BODIPY FL– or BODIPY TR-X–labeled peptides (see Figure 10.11 in Chapter 10). In addition to their utility for detecting protease contamination in culture medium and other experimental samples, BODIPY FL casein and BODIPY TR-X casein appear to have significant potential as general nontoxic, pH-insensitive markers for phagocytic cells in culture. We have shown that uptake of these quenched conjugates by neutrophils is accompanied by hydrolysis of the labeled proteins by intracellular proteases and the generation of fluorescent products that are well retained in cells.[77] Once formed, small BODIPY FL– or BODIPY TR-X–labeled peptides may diffuse into other organelles or pass through gap junctions. This phagocytosis assay is readily performed in a fluorescence microplate reader or a flow cytometer; localization of the fluorescent products can be determined by fluorescence microscopy.

β-Glucan and Fluorescein β-Glucan

The polysaccharide β-glucan, a major component in fungal cell walls, has been shown to stimulate leukotriene generation [78] and to activate the alternative complement pathway [79] in human monocytes. Both mannose and β-glucan receptors are reported to be involved in phagocytosis of *S. cerevisiae* yeast.[80] In addition, it was recently demonstrated that *Pneumocystis carinii* stimulates the release of tumor necrosis factor–α from alveolar macrophages through a β-glucan–mediated pathway.[81] The avidity of Molecular Probes' β-glucan (S-2721) for the β-glucan receptor is about 30-fold higher than that of yeast-extract glucan and 300-fold better than that of algal laminarin, as determined by competitive assays. The high avidity of our β-glucan permits the use of concentrations as low as 0.1 µg/mL to enhance neutrophil phagocytosis *in vitro*. We also offer a fluorescein β-glucan conjugate (S-2720), which has been shown to bind to neutrophils in catfish anterior kidney.[82] A similar fluorescein conjugate of a soluble β-glucan was recently used to determine the carbohydrate specificity and molecular location of the β-glucan–binding lectin site of complement receptor type 3 (CD11b/CD18).[83]

1. J Immunology 130, 1910 (1983); **2.** J Leukocyte Biol 43, 304 (1988); **3.** J Immunol Meth 130, 223 (1990); **4.** J Biol Chem 270, 8328 (1995); **5.** J Biol Chem 269, 18535 (1994); **6.** J Cell Phys 156, 428 (1993); **7.** J Immunol Meth 159, 173 (1993); **8.** Cytometry 13, 615 (1992); **9.** J Immunol Meth 159, 131 (1993); **10.** Cytometry 13, 525 (1992); **11.** Exp Cell Res 209, 375 (1993); **12.** Free Rad Biol Med 16, 509 (1994); **13.** J Immunol Meth 131, 269 (1990); **14.** FEBS Lett 273, 55 (1990); **15.** Cytometry 13, 693 (1992); **16.** J Immunol Meth 178, 89 (1995); **17.** J Immunol Meth 162, 261 (1993); **18.** Naturwissenschaften 75, 354 (1988); **19.** Biochemistry 34, 3544 (1995); **20.** Free Rad Biol Med 16, 149 (1994); **21.** Free Rad Res Comm 10, 221 (1990); **22.** FEMS Microbiol Lett 122, 187 (1994); **23.** J Immunol Meth 170, 117 (1994); **24.** Cytometry 12, 687 (1991); **25.** J Leukocyte Biol 47, 440 (1990); **26.** J Appl Physiol 76, 539 (1994); **27.** J Leukocyte Biol 54, 300 (1993); **28.** J Biochem 106, 355 (1989); **29.** Biochem Biophys Res Comm 155, 106 (1988); **30.** Anal Biochem 206, 273 (1992); **31.** Blood 83, 3324 (1994); **32.** Cytometry 18, 147 (1994); **33.** Free Rad Biol Med 17, 117 (1994); **34.** J Immunol Meth 142, 31 (1991); **35.** J Immunol Meth 112, 99 (1988); **36.** J Clin Microbiol 30, 2071 (1992); **37.** J Immunol Meth 121, 203 (1989); **38.** J Immunology 133, 3303 (1984); **39.** J Immunol Meth 123, 259 (1989); **40.** J Immunol Meth 116, 235 (1989); **41.** J Cell Biol 118, 1371 (1992); **42.** Diagnostic Clin Immunol 5, 62 (1987); **43.** J Insect Physiol 40, 1045 (1994); **44.** J Immunol Meth 162, 1 (1993); **45.** J Immunol Meth 88, 175 (1986); **46.** Biochem J 266, 669 (1980); **47.** Investigative Ophthalmol Visual Sci 30, 2499 (1989); **48.** Cell Immunol 156, 508 (1994); **49.** J Leukocyte Biol 48, 403 (1990); **50.** Cytometry 13, 423 (1992); **51.** Cytometry 11, 875 (1990); **52.** J Cell Biol 126, 1433 (1994); **53.** Cytometry 12, 677 (1991); **54.** J Leukocyte Biol 49, 369 (1991); **55.** Trends Biochem Sci 15, 69 (1990); **56.** Eur J Clin Invest 19, 339 (1989); **57.** J Leukocyte Biol 53, 673 (1993); **58.** Biochem Biophys Res Comm 122, 755 (1984); **59.** Physiol Rev 67, 285 (1987); **60.** J Cell Biol 82, 517 (1979); **61.** J Biol Chem 265, 13449 (1990); **62.** Biochemistry 29, 313 (1990); **63.** J Biol Chem 265, 16725 (1990); **64.** J Cell Biol 109, 1133 (1989); **65.** Proc Natl Acad Sci USA 78, 7540 (1981); **66.** Science 205, 1412 (1979); **67.** J Biol Chem 270, 10686 (1995); **68.** J Biol Chem 269, 326 (1994); **69.** J Immunol Meth 149, 159 (1992); **70.** Biochemistry 29, 11123 (1990); **71.** J Cell Biol 121, 1281 (1993); **72.** J Cell Sci 100, 473 (1991); **73.** Biochem Soc Trans 18, 219 (1990); **74.** J Biol Chem 259, 5661 (1984); **75.** Inflammation 7, 363 (1983); **76.** J Immunology 139, 3028 (1987); **77.** Zhou, M. *et al.*, manuscript in preparation; **78.** Proc Natl Acad Sci USA 82, 2751 (1985); **79.** J Immunology 134, 2588 (1985); **80.** J Leukocyte Biol 54, 564 (1993); **81.** J Immunology 150, 3932 (1993); **82.** Vet Immunol Immunopathol 41, 141 (1994); **83.** J Immunology 156, 1235 (1996).

17.1 Data Table *Probes for Investigating Phagocytic Cells*

Cat #	Structure	MW	Storage	Soluble	Abs	{ϵ × 10⁻³}	Em	Solvent	Notes
C-400	S21.4	531	F,D,AA	pH>6, DMSO	290	{5.6}	none	MeCN	<1>
C-2936	S21.4	486	F,D,AA	DMF	234	{60}	none	MeCN	<2, 3>
C-2938	S21.4	675	F,D,AA	DMSO	291	{5.7}	none	MeOH	<1>
C-2944	S22.5	423	FF,D,LL,AA	MeOH	429	{7.5}	<4>	pH 7	
C-2990	C	~26,000	F,D,L	H_2O	496	ND	519	pH 8	<5, 6>
C-6827	S21.4	536	F,D,AA	DMSO	287	{9.1}	none	MeOH	<1>
D-399	S21.4	487	F,D,AA	pH>6, DMSO	258	{11}	none	MeOH	<1>
D-632	S21.4	346	F,D,AA	DMF, DMSO	289	{7.6}	none	MeOH	<3, 7>
D-1168	S8.1	315	FF,L,AA	DMF, DMSO	355	{14}	<8>	MeCN	<3>
D-2935	S21.4	584	F,D,AA	DMF	258	{11}	none	MeOH	<1>
F-1314	✓	1213	F,L	pH>6, DMF	494	{72}	517	pH 9	
F-1377	✓	1269	F,L	DMF, DMSO	543	{85}	571	MeOH	
F-7468	✓	1098	F,L	DMSO	504	{75}	511	MeOH	
L-6868	S24.2	511	L	H_2O	455	{7.4}	505	H_2O	<9, 10>
L-8455	S21.3	177	D,L	DMF	355	{7.5}	411	MeOH	<10>

For definitions of the contents of this data table, see "How to Use the Handbook" on page vi.
Structure: Chemical structure drawing: ✓ = shown in this section; Sn.n = shown in Section number n.n; C = not shown due to complexity.
MW: ~ indicates an approximate value.

<1> Dihydrofluorescein diacetates are colorless and nonfluorescent until both the acetates are hydrolyzed and the products are subsequently oxidized to fluorescein derivatives. The materials contain <0.1% of oxidized derivative when initially prepared. C-400, C-2938, C-6827, D-399 and D-2935 give 2′,7′-dichlorofluorescein derivatives, which have Abs = 504 nm (ϵ = {90}), Em = 529 nm.
<2> C-2936 contains <0.1% of the oxidized dye and becomes fluorescent only upon oxidation, yielding a tetramethylrosamine derivative with Abs = 550 nm (ϵ = {87}), Em = 574 nm.
<3> This compound is susceptible to oxidation, especially in solution. Store solutions under argon or nitrogen. Oxidation appears to be catalyzed by illumination.
<4> C-2944 emits chemiluminescence (Em = 466 nm) upon oxidation by superoxide [Anal Biochem 206, 273 (1992)].
<5> This product is a dye–protein conjugate. The number of dyes attached per protein is indicated on the product label. Protease hydrolysis results in fluorescence enhancement without a spectral shift.
<6> ND = not determined.
<7> D-632 is colorless and nonfluorescent until oxidized to rhodamine 123 (R-302), which exhibits Abs = 507 nm (ϵ = {101}), Em = 529 nm.
<8> D-1168 has blue fluorescence (Em ~ 420 nm) until oxidized to ethidium (E-1305), which binds to DNA with Abs = 518 nm (ϵ = {5.2}), Em = 605 nm. The reduced dye does not bind to nucleic acids [FEBS Lett 26, 169 (1972)].
<9> L-6868 has much stronger absorption at shorter wavelengths, Abs = 368 nm (ϵ = {36}).
<10> This compound emits chemiluminescence upon oxidation in basic aqueous solutions. Emission peaks are at 425 nm (L-8455) and 470 nm (L-6868).

17.1 Structures *Probes for Investigating Phagocytic Cells*

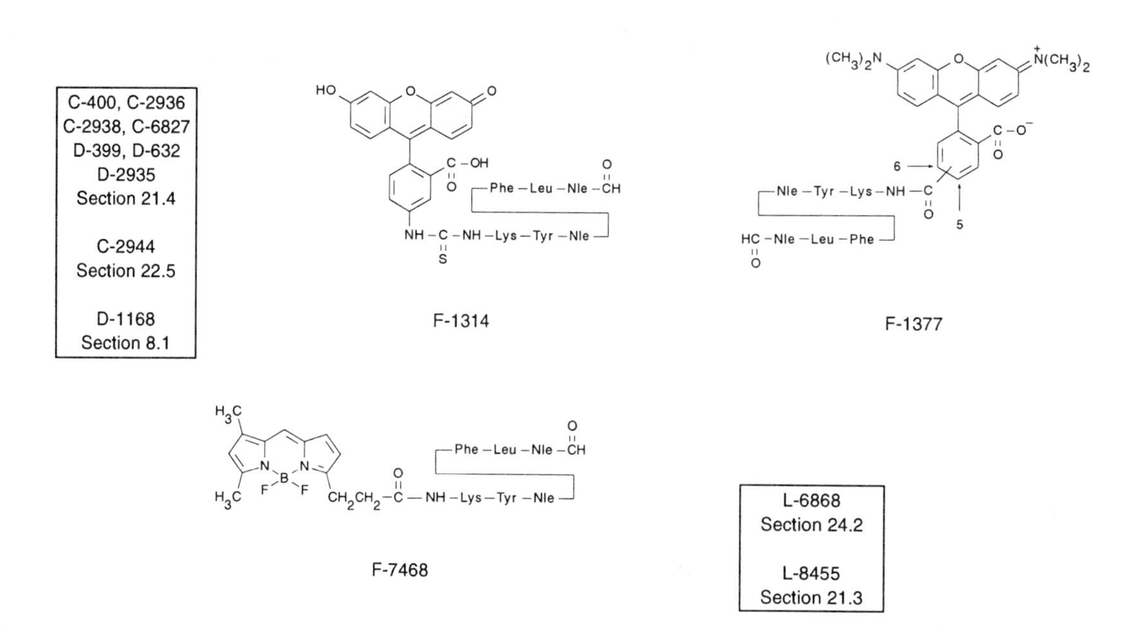

17.1 Price List *Probes for Investigating Phagocytic Cells*

Cat #		Product Name	Unit Size	Price Per Unit ($) 1–4 Units	5–24 Units
C-400		5-(and-6)-carboxy-2′,7′-dichlorodihydrofluorescein diacetate (carboxy-H_2DCFDA) *mixed isomers*	25 mg	88.00	70.40
C-2938		6-carboxy-2′,7′-dichlorodihydrofluorescein diacetate, di(acetoxymethyl ester) *single isomer*	5 mg	135.00	108.00
C-2936		4-carboxydihydrotetramethylrosamine, succinimidyl ester (OxyBURST® Orange, SE)	5 mg	55.00	44.00
C-2990		casein, fluorescein conjugate	25 mg	68.00	54.40
C-6827	**New**	5-(and-6)-chloromethyl-2′,7′-dichlorodihydrofluorescein diacetate (CM-H_2DCFDA) *mixed isomers* *special packaging*	20x50 µg	125.00	100.00
C-2944		coelenterazine	250 µg	295.00	236.00
D-399		2′,7′-dichlorodihydrofluorescein diacetate (2′,7′-dichlorofluorescin diacetate; H_2DCFDA)	100 mg	45.00	36.00
D-2935		2′,7′-dichlorodihydrofluorescein diacetate, succinimidyl ester (OxyBURST® Green, SE)	5 mg	55.00	44.00
D-1168		dihydroethidium (also called Hydroethidine - a trademark of Prescott Labs)	25 mg	98.00	78.40
D-632		dihydrorhodamine 123	10 mg	98.00	78.40
E-2870		*Escherichia coli* BioParticles® opsonizing reagent	1 U	55.00	44.00
E-2864		*Escherichia coli* (K-12 strain) BioParticles®, BODIPY® FL conjugate	10 mg	48.00	38.40
E-2861		*Escherichia coli* (K-12 strain) BioParticles®, fluorescein conjugate	10 mg	65.00	52.00
E-2862		*Escherichia coli* (K-12 strain) BioParticles®, tetramethylrhodamine conjugate	10 mg	48.00	38.40
E-2863		*Escherichia coli* (K-12 strain) BioParticles®, Texas Red® conjugate	10 mg	65.00	52.00
F-2902		Fc OxyBURST® Green assay reagent *25 assays* *3 mg/mL*	500 µL	125.00	100.00
F-6691	**New**	Fc OxyBURST® Orange assay reagent *25 assays* *3 mg/mL*	500 µL	125.00	100.00
F-6694	**New**	FluoReporter® Phagocytosis Assay Kit *250 assays*	1 kit	125.00	100.00
F-7468	**New**	formyl-Nle-Leu-Phe-Nle-Tyr-Lys, BODIPY® FL derivative	1 mg	98.00	78.40
F-1314		formyl-Nle-Leu-Phe-Nle-Tyr-Lys, fluorescein derivative	1 mg	98.00	78.40
F-1377		formyl-Nle-Leu-Phe-Nle-Tyr-Lys, tetramethylrhodamine derivative	1 mg	98.00	78.40
L-6868	**New**	lucigenin (bis-N-methylacridinium nitrate) *high purity*	10 mg	15.00	12.00
L-8455	**New**	luminol (3-aminophthalhydrazide)	25 g	75.00	60.00
S-2721		soluble β-glucan (β-glucan) *1 mg/mL*	500 µL	65.00	52.00
S-2720		soluble β-glucan, fluorescein conjugate (β-glucan, fluorescein conjugate) *1 mg/mL*	500 µL	118.00	94.40
S-2860		*Staphylococcus aureus* BioParticles® opsonizing reagent	1 U	55.00	44.00
S-2854		*Staphylococcus aureus* (Wood strain without protein A) BioParticles®, BODIPY® FL conjugate	10 mg	48.00	38.40
S-2851		*Staphylococcus aureus* (Wood strain without protein A) BioParticles®, fluorescein conjugate	10 mg	65.00	52.00
S-2852		*Staphylococcus aureus* (Wood strain without protein A) BioParticles®, tetramethylrhodamine conjugate	10 mg	48.00	38.40
S-2853		*Staphylococcus aureus* (Wood strain without protein A) BioParticles®, Texas Red® conjugate	10 mg	65.00	52.00
S-2859		*Staphylococcus aureus* (Wood strain without protein A) BioParticles®, unlabeled	100 mg	28.00	22.40
Z-2850		zymosan A BioParticles® opsonizing reagent	1 U	55.00	44.00
Z-2844		zymosan A (*S. cerevisiae*) BioParticles®, BODIPY® FL conjugate	10 mg	48.00	38.40
Z-2841		zymosan A (*S. cerevisiae*) BioParticles®, fluorescein conjugate	10 mg	65.00	52.00
Z-2842		zymosan A (*S. cerevisiae*) BioParticles®, tetramethylrhodamine conjugate	10 mg	48.00	38.40
Z-2843		zymosan A (*S. cerevisiae*) BioParticles®, Texas Red® conjugate	10 mg	65.00	52.00
Z-2849		zymosan A (*S. cerevisiae*) BioParticles®, unlabeled	100 mg	28.00	22.40

Technical Assistance at Our Web Site (http://www.probes.com)

At Molecular Probes' Web site, we are developing an electronic version of this Handbook and other databases that should prove extremely useful for the researcher. In addition to containing all of the text from this Handbook, our Web site provides:

- **Product searches** by product name or catalog number
- **Bibliographies** for all products for which we have references
- **Keyword searches** of our entire bibliography of over 25,000 references
- **Product information sheets** for many kits and reagents
- **Technical bulletins**, including *BioProbes* newsletters and other product literature
- **Chemical structure**, **technical data** and **material safety and data sheets**
- **Color photomicrographs** that show our products in action

Visit our Web site often for new additions to our bibliography, as well as new products and upgraded search capabilities. Also look for special sales and introductory specials on some important products.

If you do not have access to the Internet or you need assistance that is not available at that site, further information on the scientific and technical background of our products can be obtained by contacting our Technical Assistance Department at the numbers listed on the inside front cover.

17.2 Membrane Markers of Endocytosis and Exocytosis

FM 1-43 and Its Analogs

Molecular Probes' membrane probes FM 1-43, FM 3-25, FM 2-10, FM 1-84 and FM 4-64 are proving to be excellent reagents both for identifying actively firing neurons and for investigating the mechanisms of activity-dependent vesicle recycling in widely different species.[1] FM dyes may also be useful as general-purpose probes for investigating endocytosis. These nontoxic, water-soluble dyes, which are virtually nonfluorescent in aqueous medium, are believed to insert into the outer leaflet of the surface membrane where they become intensely fluorescent. In a neuron that is actively releasing neurotransmitter, these dyes become internalized within the recycled synaptic vesicles and the nerve terminals become brightly stained as shown in Figure 17.4. The nonspecific staining of cell-surface membranes can simply be washed off prior to viewing. The emission maxima of these FM dyes in membranes may differ considerably from those in methanol, which are listed in our data tables, and the visible color further depends on the optical filters used to isolate the emission.

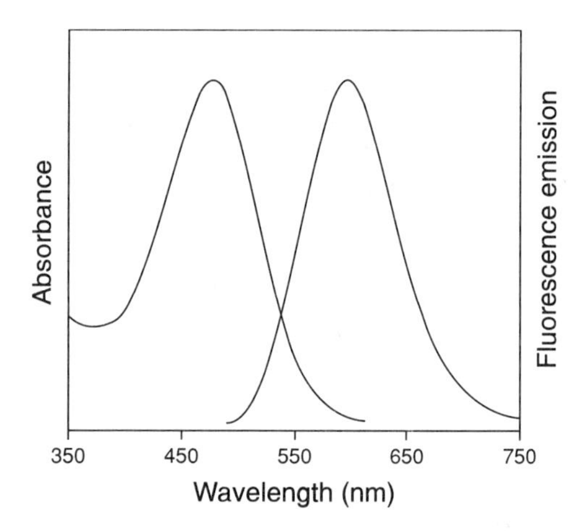

Figure 17.5 Absorption and fluorescence emission spectra of FM 1-43 (T-3163) bound to DOPC phospholipid membranes. The peaks of the spectra are at 479 nm (absorption) and 598 nm (fluorescence emission).

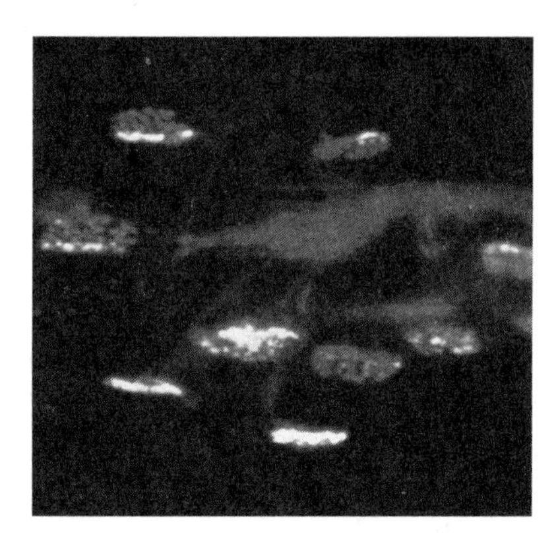

Figure 17.4 Living nerve terminals of motoneurons that innervate a rat lumbrical muscle stained with the activity-dependent dye FM 1-43 (T-3163) and observed under low magnification. The dye molecules, which insert into the outer leaflet of the surface membrane, are captured in recycled synaptic vesicles of actively firing neurons. Photo contributed by William J. Betz, University of Colorado School of Medicine.

FM 1-43

The styrylpyridinium dye FM 1-43 (T-3163) has been used to investigate synaptosomal recycling in a range of species including frog, rat and mouse.[1-7] It has also been shown that FM 1-43 can be used in *Drosophila* larvae, thereby facilitating the analysis of mutations in synaptic vesicle recycling.[8]

Betz and his colleagues, who were the first to employ FM 1-43 to observe vesicle cycling in living nerve terminals,[6,7] recently used this important probe to examine the effects of okadaic acid (O-3452, see Section 20.3) on synaptic vesicle clustering.[9] FM 1-43 was also employed in a study showing that synaptosomal endocytosis is independent of both extracellular Ca^{2+} and membrane potential in dissociated hippocampal neurons,[4] as well as in a spectrofluorometric assay demonstrating that nitric oxide–stimulated vesicle release is independent of Ca^{2+} in isolated rat hippocampal nerve terminals.[10] In addition, FM 1-43 has been used in combination with fura-2 (see Section 22.2) to simultaneously measure intracellular Ca^{2+} and membrane turnover.[11] Like most styryl dyes, the absorption and fluorescence emission spectra of FM 1-43 (absorption/emission maxima = 510/626 nm in methanol) are significantly shifted in the membrane environment (Figure 17.5); FM 1-43 is efficiently excited with standard fluorescein optical filters but poorly excited with standard tetramethylrhodamine optical filters.

Analogs of FM 1-43

A comparison of mammalian motor nerve terminals stained with either FM 1-43 or the more hydrophilic analog FM 2-10 (T-7508) revealed that FM 2-10's lower background staining and faster destaining rate may make it the preferred probe for quantitative applications.[1] However, staining with FM 2-10 requires much higher dye concentrations (100 µM compared with 2 µM for FM 1-43).[1] FM 3-25 (T-7506), FM 1-84 (T-3164), FM 4-64 (T-3166) and RH 414 (T-1111) — all more hydrophobic than FM 1-43 — may also be useful as probes for investigating endocytosis. Using time-lapse video fluorescence microscopy, Heuser and colleagues were able to follow the internalization of FM 4-64 from the *Dictyostelium* plasma membrane into the contractile vacuole and to observe up to 10 contractile vacuole cycles before the dye redistributed to the endosomes.[12] These researchers have also used FM 3-25 to stain endosomes in both protozoa and vertebrate cells.[13] FM 4-64 exhibits long-wavelength red fluorescence that can be distinguished from the green fluorescence of FM 1-43 with the proper optical filter sets, thus permitting two-color observation of membrane recycling in real time.[14,15] In addition, FM 4-64 has recently been reported to selectively stain yeast vacuolar membranes and is proving to be an important tool for visualizing vacuolar organelle morphology and dynamics and for studying the endocytic pathway in yeast [16] (see Section 12.3). Membranes stained with RH 414 exhibit orange fluorescence when observed through a longpass optical filter that permits passage of light beyond 510 nm.[6,7,17]

4-Di-2-ASP

Some cationic mitochondrial dyes such as 4-Di-2-ASP (D-289), 4-Di-1-ASP (D-288, see Section 12.2) and $DiOC_2(5)$ (D-303, see Section 12.2) stain presynaptic nerve terminals independent of neuronal activity.[18-20] The nontoxic, photostable 4-Di-2-ASP dye has been employed to stain living nerve terminals in rabbit corneal epithelium,[21] in rat epidermis [18] and at mouse, snake and frog neuromuscular junctions,[22-25] as well as to visualize the innervation of the human choroid [26] and whole mounts of the gastrointestinal tract.[27] 4-Di-2-ASP staining of neuromuscular junctions reportedly persists for several months in living mice.[25] Methods for using 4-Di-2-ASP to image neuronal cells in living animals have been described.[28] $DiOC_2(5)$ (D-303, see Section 12.2) is also sometimes used to stain living nerve terminals,[20,29] although this dye has been reported to yield high backgrounds and inconsistent staining in mammalian tissue.[19]

TMA-DPH and TMAP-DPH

Also useful as a lipid marker for endocytosis and exocytosis is the cationic linear polyene TMA-DPH (T-204), which readily incorporates in the plasma membrane of living cells.[30,31] TMA-DPH is virtually nonfluorescent in water and binds in proportion to the available membrane surface.[32] Its fluorescence intensity is therefore sensitive to physiological processes that cause a net change in membrane surface area, making it an excellent probe for monitoring events such as changes in cell volume and exocytosis.[32-35] During endocytosis, TMA-DPH progresses from the cell periphery to perinuclear regions with little loss in fluorescence intensity.[36] TMA-DPH can be extracted from plasma membranes by washing with medium, thus providing a method for isolating internalized probe and monitoring endocytosis.[37-39] To provide supporting evidence for the maturation model for endocytosis, researchers used TMA-DPH in fluorescence anisotropy assays to investigate membrane fluidity in the plasma membrane and in successive endocytic compartments of L929 cells.[36] We also offer TMAP-DPH (P-3900), which has a three-carbon spacer between the DPH fluorophore and the trimethylammonium substituent; recent papers describe the partitioning properties of TMAP-DPH probe in multilamellar liposomes.[40,41]

1. Proc R Soc Lond B 255, 61 (1994); **2.** Neuron 14, 983 (1995); **3.** Neuron 14, 773 (1995); **4.** Neuron 11, 713, (1993); **5.** J Physiol 460, 287 (1993); **6.** J Neurosci 12, 363 (1992); **7.** Science 255, 200 (1992); **8.** Neuron 13, 363 (1994); **9.** J Cell Biol 124, 843 (1994); **10.** Neuron 12, 1235 (1994); **11.** Cell Calcium 18, 440 (1995); **12.** J Cell Biol 121, 1311 (1993); **13.** John Heuser, Washington University, personal communication; **14.** J Cell Biol 131, 679 (1995); **15.** J Neurosci 15, 8246 (1995); **16.** J Cell Biol 128, 779 (1995); **17.** J Neurosci 15, 6327 (1995); **18.** Cell Tissue Res 255, 125 (1989); **19.** J Neurosci 7, 1207 (1987); **20.** Nature 310, 53 (1984); **21.** J Neurosci 13, 4511 (1993); **22.** J Neurosci 14, 5672 (1994); **23.** J Neurosci 14, 3319 (1994); **24.** J Neurosci 14, 796 (1994); **25.** J Neurosci 7, 1215 (1987); **26.** Ophthalmic Res 26, 290 (1994); **27.** J Autonomic Nervous System 38, 77 (1992); **28.** Trends Neurosci 10, 398 (1987); **29.** J Neurocytol 19, 67 (1990); **30.** J Cell Biol 127, 725 (1994); **31.** Cell Biophys 5, 129 (1983); **32.** Biochemistry 25, 2149 (1986); **33.** Biol of the Cell 79, 265 (1993); **34.** Biochim Biophys Acta 1067, 71 (1991); **35.** Biochim Biophys Acta 901, 138 (1987); **36.** J Cell Biol 125, 783 (1994); **37.** Biol of the Cell 71, 293 (1991); **38.** Biochim Biophys Acta 1030, 73 (1990); **39.** Cell Biophys 14, 17 (1989); **40.** Chem Phys Lipids 66, 135 (1993); **41.** J Fluorescence 3, 145 (1993).

17.2 Data Table *Membrane Markers of Endocytosis and Exocytosis*

Cat #	Structure	MW	Storage	Soluble	Abs	{ε × 10^-3^}	Em	Solvent	Notes
D-289	✓	394	L	H_2O, DMF	488	{48}	607	MeOH	<1>
P-3900	S14.4	504	D,L	DMF, DMSO	354	{85}	429	MeOH	<2>
T-204	S14.4	462	D,L	DMF, DMSO	355	{75}	430	MeOH	<2>
T-1111	S25.2	581	D,L	DMSO, EtOH	532	{55}	716	MeOH	<1>
T-3163	✓	612	D,L	H_2O, DMSO	510	{56}	626	MeOH	<1, 3>
T-3164	✓	640	DD,L	H_2O, DMSO	510	{52}	625	MeOH	<1>
T-3166	✓	608	D,L	H_2O, DMSO	543	{43}	<4>	MeOH	<1>
T-7506	✓	1204	D,L	DMSO, EtOH	510	{55}	624	MeOH	<1>
T-7508	✓	555	D,L	H_2O, DMSO	506	{50}	620	MeOH	<1>

For definitions of the contents of this data table, see "How to Use the Handbook" on page vi.
Structure: Chemical structure drawing: ✓ = shown in this section; Sn.n = shown in Section number n.n.

<1> Abs and Em of styryl dyes are at shorter wavelengths in membrane environments than in reference solvents such as methanol. The difference is typically ~20 nm for absorption and ~80 nm for emission but varies considerably from one dye to another.

<2> Diphenylhexatriene (DPH) and its derivatives are essentially nonfluorescent in water. Absorption and emission spectra have multiple peaks. The wavelength, resolution and relative intensity of these peaks are environment dependent. Abs and Em values are for the most intense peak in the solvent specified.

<3> Abs = 479 nm, Em = 598 nm when bound to phospholipid vesicles (determined at Molecular Probes). Em = 565 nm bound to synaptosomal membranes [Neuron 12, 1235 (1994)].

<4> Fluorescence of T-3166 in MeOH is very weak. Excitation/emission wavelength settings at 515/640 nm are suitable for detection of yeast vacuole membrane staining with this dye [J Cell Biol 128, 779 (1995)].

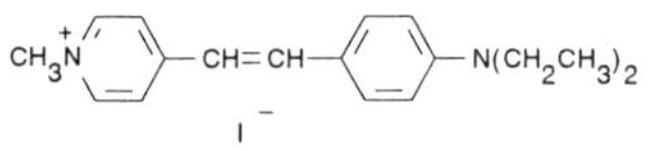

17.2 Structures *Membrane Markers of Endocytosis and Exocytosis*

D-289

P-3900, T-204
Section 14.4

T-1111
Section 25.2

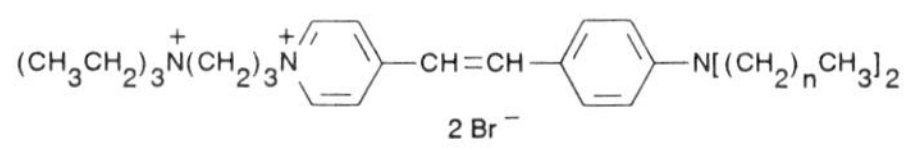

T-3164 n = 4
T-3163 n = 3
T-7508 n = 1

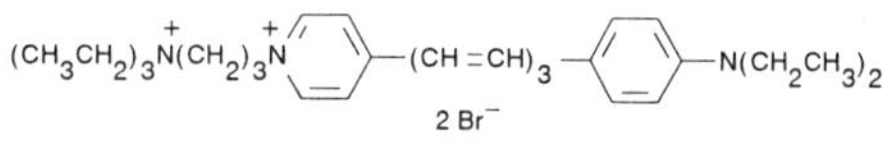

T-3166

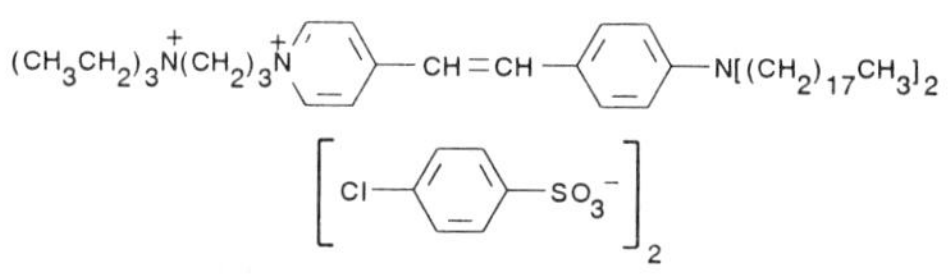

T-7506

17.2 Price List *Membrane Markers of Endocytosis and Exocytosis*

Cat #		Product Name	Unit Size	Price Per Unit ($) 1–4 Units	5–24 Units
D-289		4-(4-(diethylamino)styryl)-N-methylpyridinium iodide (4-Di-2-ASP)	1 g	75.00	60.00
P-3900		N-((4-(6-phenyl-1,3,5-hexatrienyl)phenyl)propyl)trimethylammonium p-toluenesulfonate (TMAP-DPH)	5 mg	148.00	118.40
T-3163		N-(3-triethylammoniumpropyl)-4-(4-(dibutylamino)styryl)pyridinium dibromide (FM™ 1-43)	1 mg	165.00	132.00
T-1111		N-(3-triethylammoniumpropyl)-4-(4-(4-(diethylamino)phenyl)butadienyl)pyridinium dibromide (RH 414)	5 mg	165.00	132.00
T-3166		N-(3-triethylammoniumpropyl)-4-(6-(4-(diethylamino)phenyl)hexatrienyl)pyridinium dibromide (FM™ 4-64)	1 mg	165.00	132.00
T-7508	***New***	N-(3-triethylammoniumpropyl)-4-(4-(diethylamino)styryl)pyridinium dibromide (FM™ 2-10)	5 mg	135.00	108.00
T-7506	***New***	N-(3-triethylammoniumpropyl)-4-(4-(dioctadecylamino)styryl)pyridinium, di-4-chlorobenzenesulfonate (FM™ 3-25)	5 mg	125.00	100.00
T-3164		N-(3-triethylammoniumpropyl)-4-(4-(dipentylamino)styryl)pyridinium dibromide (FM™ 1-84)	1 mg	165.00	132.00
T-204		1-(4-trimethylammoniumphenyl)-6-phenyl-1,3,5-hexatriene p-toluenesulfonate (TMA-DPH)	25 mg	155.00	124.00

17.3 Ligands for Studying Receptor-Mediated Endocytosis

Molecular Probes offers fluorescent derivatives of low-density lipoprotein (LDL), transferrin and epidermal growth factor (EGF), which are all transported into the cell by receptor-mediated endocytosis. The optical sectioning capability of the confocal laser scanning microscope makes possible direct measurements of some internalized probes. In addition, we have developed unique methods and reagents for following the internalization and processing of biomolecules that bind to cell surface receptors. These include antibodies to BODIPY FL, BODIPY TR, fluorescein, tetramethylrhodamine, Texas Red and Cascade Blue® dyes, all of which may quench the corresponding fluorophores, and an antibody to the dansyl fluorophore, which considerably enhances the fluorescence of dansylated probes; see Section 7.3 for more information.

Low-Density Lipoprotein Complexes

The human LDL complex, which delivers cholesterol to cells by receptor-mediated endocytosis, consists of a core of about 1500 molecules of cholesteryl ester and triglyceride, surrounded by a 20 Å–thick shell of phospholipids, unesterified cholesterol and a single copy of apoprotein B 100 (~500,000 daltons).[1] Once internalized, LDL dissociates from its receptor and eventually appears in lysosomes.[2] Molecular Probes offers unlabeled LDL (L-3486), which has been reported to be an effective vehicle for selectively delivering antitumor drugs to cancer cells.[3] We also offer two classes of labeled LDL probes — those containing an unmodified apoprotein, used to study the mechanisms of normal cholesterol delivery and internalization, and those with an acetylated apoprotein, used to study endothelial, microglial and other cell types that express receptors that specifically bind this modified LDL.

We prepare our fluorescent LDL products from fresh human plasma approximately every two months and ship them within two weeks of their preparation. If stored refrigerated and protected from light, our LDL products are stable for at least four to six weeks from the date of shipment; these products must not be frozen. Multiple-unit discounts will be applied to standing orders of our LDL products. Because preparation of these complexes involves several variables, some batch-to-batch variability in degree of labeling and fluorescence yield is expected.

Fluorescent LDL Complexes

Molecular Probes offers LDL labeled with either DiI (DiI LDL, L-3482) or the BODIPY FL fluorophore (BODIPY FL LDL, L-3483), highly fluorescent lipophilic dyes that diffuse into the hydrophobic portion of the LDL complex without affecting the LDL-specific binding of the apoprotein. The contrasting fluorescence of DiI LDL and fluorescein transferrin (T-2871) permits their simultaneous use to follow the lysosomally directed pathway and the recycling pathways, respectively.[4] As compared to DiI LDL, BODIPY FL LDL is more efficiently excited by the argon-ion laser, making it better suited for flow cytometry and confocal laser scanning microscopy studies. Like our BODIPY FL C_5-ceramide (D-3521, see Section 12.4), BODIPY FL LDL fluoresces somewhat in the red region, sometimes precluding its use for multicolor labeling with red fluorophores. Both the BODIPY FL LDL and DiI LDL have been used to investigate the binding specificity and partitioning of LDL throughout the *Schistosoma mansoni* parasite.[5] Fluorescent LDL complexes have also proven useful in a variety of experimental systems to:

- Count the number of cell-surface LDL receptors, analyze their motion and clustering and follow their internalization [6-8]
- Demonstrate that fibroblasts grown continuously in the presence of DiI LDL (L-3482) proliferate normally and exhibit normal morphology,[9] making DiI LDL a valuable alternative to ^{125}I-labeled LDL for quantitating LDL-receptor activity [10]
- Identify LDL receptor–deficient Chinese hamster ovary (CHO) cell mutants [11]
- Investigate the expression of LDL receptors in granulosa and luteal cells in primate and porcine ovarian follicles [12-14]
- Track the mobility of LDL receptors in an electric field [15-17]

Fluorescent Acetylated LDL Complexes

If the lysine residues of LDL's apoprotein have been acetylated, the LDL complex no longer binds to the LDL receptor,[18] but rather is taken up by macrophage and endothelial cells that possess "scavenger" receptors specific for the modified LDL.[19,20] Once the acetylated LDL (AcLDL) complexes accumulate within these cells, they assume an appearance similar to that of foam cells found in atherosclerotic plaques.[21-23] Using DiI-labeled acetylated LDL (DiI AcLDL, L-3484), researchers have discovered that the scavenger receptors on rabbit fibroblasts and smooth muscle cells appear to be up-regulated through the activation of the protein kinase C pathway.[24] Researchers have also used DiI AcLDL to show that Chinese hamster ovary (CHO) cells express AcLDL receptors that are distinct from macrophage scavenger receptors.[25,26] Ultrastructural localization of endocytic compartments that maintain a connection to the extracellular space has been achieved by photoconversion of DiI AcLDL in the presence of diaminobenzidine;[27] see the Technical Note "Fluorescent Probes for Photoconversion of Diaminobenzidine" on page 414.

It has now become routine to identify endothelial cells and microglial cells in primary cell culture by their ability to take up DiI-labeled acetylated LDL;[28-31] DiI AcLDL has also been shown to be taken up by microglia but not by astrocytes.[32] DiI AcLDL was employed in order to confirm endothelial cell identity in investigations of shear stress [33] and P-glycoprotein expression,[34] as well as to identify blood vessels in a growing murine melanoma.[35] In addition, researchers applied patch-clamp techniques to investigate membrane currents in mouse microglia, which were identified both in culture and in brain slices by their staining with DiI AcLDL.[36,37] For some applications, our BODIPY FL–labeled acetylated LDL (BODIPY FL AcLDL, L-3485) may be the preferred probe because the BODIPY dye is covalently bound to the modified apoprotein portion of the LDL complex and is therefore not extracted during subsequent manipulations of the cells.

Transferrin Conjugates

Transferrin is a monomeric serum glycoprotein (~80,000 daltons) that binds up to two Fe^{3+} atoms for delivery to vertebrate cells through receptor-mediated endocytosis.[38-40] Once iron-carrying transferrin proteins are inside endosomes, the acidic environment favors dissociation of iron from the transferrin–receptor complex. Following the release of iron, the apotransferrin is recycled to the plasma membrane, where it is released from its receptor to scavenge more iron.[41,42] Fluorescent transferrin conjugates can therefore be used with fluorescent LDL to distinguish the lysosomally directed and recycling endosomal pathways.[2,4]

Molecular Probes' fluorescent di-ferric (Fe^{3+}) human transferrin conjugates include those of fluorescein, tetramethylrhodamine, BODIPY FL and Texas Red fluorophores (T-2871, T-2872, T-2873, T-2875). Fluorescently labeled transferrin has greatly aided the investigation of endocytosis.[43-45] Because of its greater photostability, BODIPY FL–labeled transferrin was used to track early endosomes in HEp-2 cells by confocal laser scanning microscopy.[46] The pH sensitivity of fluorescein-labeled transferrin has been exploited to investigate events occurring during endosomal acidification.[47-50] Fluorescent transferrin has also been used to:

- Analyze the role of the γ-chain of type III IgG receptors in antigen–antibody complex internalization [51]
- Define the nature of several mutations that affect the endosomal pathway [52-54]
- Demonstrate that the fungal metabolite brefeldin A (B-7450, see Section 12.4) induces an increase in tubulation of transferrin receptors in BHK-21 cells [55] and in the perikaryal–dendritic region of cultured hippocampal neurons [56]
- Observe receptor trafficking in living cells by confocal laser scanning microscopy and show that recycling transferrin receptors are distributed on the surface of the leading lamella in migrating fibroblasts [57]
- Show that the endosomal compartment of living epidermoid carcinoma cells is an extensive network of tubular cisternae [58]

Epidermal Growth Factors

Epidermal growth factor (EGF) is a 53–amino acid polypeptide hormone (6045 daltons) that stimulates division of epidermal and other cells.[59-61] Molecular Probes offers unlabeled mouse submaxillary gland EGF (E-3476), as well as EGFs labeled with biotin-XX (E-3477), fluorescein (E-3478), Oregon Green™ 514 (E-7498) and tetramethylrhodamine (E-3481), all containing a single biotin or fluorophore on the N-terminal amino acid. The dissociation constant of the EGF conjugates in DMEM-HEPES medium is about

2.0 nM for human epidermoid carcinoma (A431) cells, a value that approximates that of the unlabeled EGF.

Fluorescently labeled EGFs have enabled scientists to use fluorescence resonance energy transfer techniques to assess EGF receptor–receptor and receptor–membrane interactions;[62-64] see the Technical Note "Fluorescence Resonance Energy Transfer" on page 456. Using fluorescein EGF as the donor and tetramethylrhodamine EGF as the acceptor, researchers were able to examine temperature-dependent lateral and transverse distribution of EGF receptors in A431 plasma membranes.[64] When fluorescein EGF binds to A431 cells, it apparently undergoes a biphasic quenching, which can be attributable first to changes in rotational mobility upon binding and then to receptor–ligand internalization. By monitoring this quenching in real time, the rate constants for the interaction of fluorescein EGF with its receptor were determined.[65]

Although fluorescently labeled EGF can also be used to follow lateral mobility and endocytosis of the EGF receptor,[66,67] the visualization of fluorescent EGF may require low-light imaging technology, especially in cells that express low levels of the EGF receptor.[68] Oregon Green 514–labeled EGF is one of our most fluorescent and photostable EGF conjugates. We also prepare biotinylated EGF complexed to Texas Red streptavidin (E-3480), which yields two to three times as much signal per EGF receptor as the direct conjugates. We have found that EGF receptors can easily be detected with this complex without resorting to low-light imaging technology (Figure 17.6).

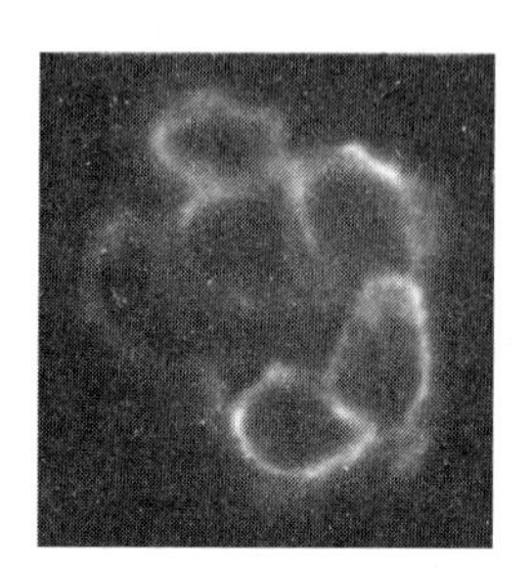

Figure 17.6 *Lightly fixed human epidermoid carcinoma cells (A431) stained with Molecular Probes' biotinylated EGF complexed to Texas Red streptavidin (E-3480). An identical cell preparation stained in the presence of 100-fold excess unlabeled EGF (E-3476) showed no fluorescent signal.*

In addition, we offer biotin-XX EGF, which contains a long spacer arm that facilitates binding of fluorescent or enzyme-conjugated streptavidins (see Section 7.5). Using biotinylated EGF and phycoerythrin-labeled secondary reagents (see Section 6.4), researchers were able to detect as few as 10,000 EGF cell-surface receptors by confocal laser scanning microscopy.[69] We have used our ELF®-97 Cytological Labeling Kits (E-6602, E-6603; see Section 6.3) in conjunction with biotinylated EGF to detect low abundance EGF receptors (see Figure 6.9C in Chapter 6).

1. Proc Natl Acad Sci USA 74, 837 (1977); **2.** J Cell Biol 121, 1257 (1993); **3.** Bioconjugate Chem 5, 105 (1994); **4.** J Cell Sci 107, 2177 (1994); **5.** Am J Pathol 138, 1173 (1991); **6.** Biophys J 66, 1301 (1994); **7.** Biophys J 67, 1280 (1994); **8.** Meth Enzymol 98, 241 (1983); **9.** In Vitro Cell Devel Biol 27A, 633 (1991); **10.** J Lipid Res 34, 325 (1993); **11.** J Biol Chem 269, 20958 (1994); **12.** Biol of Reprod 50, 204 (1994); **13.** Endocrinology 129, 3247 (1991); **14.** Biol of Reprod 47, 355 (1992); **15.** J Cell Biol 101, 148 (1985); **16.** Proc Natl Acad Sci USA 81, 5454 (1984); **17.** J Cell Biol 90, 595 (1981); **18.** J Biol Chem 253, 9053 (1978); **19.** J Supramol Struct 13, 67 (1980); **20.** J Cell Biol 82, 597 (1979); **21.** Ann Rev Biochem 52, 223 (1983); **22.** Arteriosclerosis 3, 1 (1983); **23.** Arteriosclerosis 1, 177 (1981); **24.** J Biol Chem 265, 12722 (1990); **25.** J Biol Chem 270, 1921 (1995); **26.** J Biol Chem 269, 21003 (1994); **27.** J Cell Biol 129, 133 (1995); **28.** Eur J Cell Biol 60, 48 (1993); **29.** J Cell Biol 122, 923 (1993); **30.** J Cell Biol 122, 417 (1993); **31.** J Cell Biol 99, 2034 (1984); **32.** J Neurosci 6, 2163 (1986); **33.** Exp Cell Res 198, 31 (1992); **34.** J Biol Chem 267, 20383 (1992); **35.** Histochem J 17, 1309 (1985); **36.** J Neurosci 13, 4412 (1993); **37.** J Neurosci 13, 4403 (1993); **38.** Cell 49, 423 (1987); **39.** Trends Biochem Sci 12, 350 (1987); **40.** Biochimie 68, 375 (1986); **41.** J Cell Biol 108, 1291 (1989); **42.** Cell 37, 789 (1984); **43.** J Cell Biol 125, 253 (1994); **44.** J Cell Biol 121, 61 (1993); **45.** J Biol Chem 263, 8844 (1988); **46.** J Cell Biol 128, 549 (1995); **47.** Biochemistry 31, 5820 (1992); **48.** J Bioenergetics Biomembranes 23, 147 (1991); **49.** J Biol Chem 266, 3469 (1991); **50.** J Biol Chem 265, 6688 (1990); **51.** Nature 358, 337 (1992); **52.** J Cell Biol 123, 1119 (1993); **53.** J Cell Biol 122, 1231 (1993); **54.** J Cell Biol 122, 565 (1993); **55.** J Cell Biol 118, 267 (1992); **56.** J Cell Biol 122, 1207 (1993); **57.** J Cell Biol 125, 1265 (1994); **58.** Nature 346, 335 (1990); **59.** Cell 61, 203 (1990); **60.** J Biol Chem 265, 7709 (1990); **61.** Ann Rev Biochem 56, 881 (1987); **62.** J Fluorescence 4, 295 (1994); **63.** J Biol Chem 268, 23860 (1993); **64.** J Membrane Biol 118, 215 (1990); **65.** Biochemistry 32, 12039 (1993); **66.** J Cell Biol 109, 2105 (1989); **67.** J Cell Biol 106, 1903 (1988); **68.** Proc Natl Acad Sci USA 75, 2135 (1978); **69.** J Histochem Cytochem 40, 1353 (1992).

17.3 Data Table *Ligands for Studying Receptor-Mediated Endocytosis*

Cat #	Structure	MW	Storage	Soluble	Abs	{ε × 10⁻³}	Em	Solvent	Notes
E-3476	C	~6100	FF,D	H_2O	<300		none		
E-3477	C	~6600	FF,D	H_2O	<300		none		<1>
E-3478	C	~6500	FF,D,L	H_2O	495	{84}*	517	pH 8	<1>
E-3480	C	<2>	FF,D,L	H_2O	596	ND	612	pH 7	<3>
E-3481	C	~6800	FF,D,L	H_2O	555	{85}*	581	pH 7	<1>
E-7498	C	~6600	FF,D,L	H_2O	511	{85}*	528	pH 9	<1>
L-3482	C	<4>	RR,L,AA	H_2O†	554	ND	571	H_2O†	<3, 5>
L-3483	C	<4>	RR,L,AA	H_2O†	515	ND	520	H_2O†	<3, 5>
L-3484	C	<4>	RR,L,AA	H_2O†	554	ND	571	H_2O†	<3, 5>
L-3485	C	<4>	RR,L,AA	H_2O†	510	ND	518	H_2O†	<3, 5>
L-3486	C	~3,000,000	RR,AA	H_2O†	<300		<350		<5>
T-2871	C	~80,000	FF,D,L	H_2O	496	ND	522	pH 8	<3, 6>
T-2872	C	~80,000	FF,D,L	H_2O	554	ND	578	pH 7	<3, 6>
T-2873	C	~80,000	FF,D,L	H_2O	505	ND	511	pH 7	<3, 6>
T-2875	C	~80,000	FF,D,L	H_2O	596	ND	616	pH 7	<3, 6>

For definitions of the contents of this data table, see "How to Use the Handbook" on page vi.
Structure: Chemical structure drawing: C = not shown due to complexity.
MW: ~ indicates an approximate value.

* The value of {ϵ} listed for this EGF conjugate is for the labeling dye in free solution. Use of this value for the conjugate assumes a 1:1 dye:peptide labeling ratio and no change of ϵ due to dye–peptide interactions.
† LDL complexes are packaged under argon in 10 mM Tris, 150 mM NaCl, 0.3 mM EDTA, pH 8.3 containing 2 mM sodium azide. Spectral data reported are measured in this buffer. LDL complexes must not be frozen.
<1> This EGF conjugate has approximately one label per peptide.
<2> E-3480 is a complex of E-3477 with Texas Red streptavidin (S-872, see Section 7.5), which typically contains ~3 dyes/streptavidin (MW ~60,000 daltons).
<3> ND = not determined.
<4> This LDL complex incorporates multiple fluorescent labels. The number of dyes per apoprotein B (~500,000 daltons) is indicated on the product label.
<5> LDL complexes must be stored refrigerated BUT NOT FROZEN. The maximum shelf-life under the indicated storage conditions is 4–6 weeks.
<6> This protein conjugate contains multiple labels. The number of labels per protein is indicated on the product label.

17.3 Price List *Ligands for Studying Receptor-Mediated Endocytosis*

Cat #	Product Name	Unit Size	Price Per Unit ($) 1–4 Units	5–24 Units
E-3476	epidermal growth factor (EGF) *from mouse submaxillary glands*	100 µg	73.00	58.40
E-3477	epidermal growth factor, biotin-XX conjugate (biotin EGF)	20 µg	160.00	128.00
E-3480	epidermal growth factor, biotinylated, complexed to Texas Red® streptavidin	100 µg	160.00	128.00
E-3478	epidermal growth factor, fluorescein conjugate (fluorescein EGF)	20 µg	160.00	128.00
E-7498 *New*	epidermal growth factor, Oregon Green™ 514 conjugate (Oregon Green™ 514 EGF)	20 µg	160.00	128.00
E-3481	epidermal growth factor, tetramethylrhodamine conjugate (rhodamine EGF)	20 µg	160.00	128.00
L-3486	low-density lipoprotein from human plasma (LDL) *2.5 mg/mL*	200 µL	80.00	64.00
L-3485	low-density lipoprotein from human plasma, acetylated, BODIPY® FL conjugate (BODIPY® FL AcLDL) *1 mg/mL*	200 µL	195.00	156.00
L-3484	low-density lipoprotein from human plasma, acetylated, DiI complex (DiI AcLDL) *1 mg/mL*	200 µL	195.00	156.00
L-3483	low-density lipoprotein from human plasma, BODIPY® FL complex (BODIPY® FL LDL) *1 mg/mL*	200 µL	175.00	140.00
L-3482	low-density lipoprotein from human plasma, DiI complex (DiI LDL) *1 mg/mL*	200 µL	175.00	140.00
T-2873	transferrin from human serum, BODIPY® FL conjugate	5 mg	148.00	118.40
T-2871	transferrin from human serum, fluorescein conjugate	5 mg	148.00	118.40
T-2872	transferrin from human serum, tetramethylrhodamine conjugate	5 mg	148.00	118.40
T-2875	transferrin from human serum, Texas Red® conjugate	5 mg	148.00	118.40

17.4 Miscellaneous Endocytosis and Exocytosis Probes

Fluorescent Dextrans

Fluorescent dextrans can be used to monitor the uptake and internal processing of exogenous materials by endocytosis.[1-5] Molecular Probes offers dextrans with nominal molecular weights ranging from 3000 to 2,000,000 daltons, many of which can be used as pinocytosis or phagocytosis markers (see Section 15.5 for further discussion and a complete product list). Discrimination of internalized fluorescent dextrans from external dextrans can be facilitated by adding reagents that quench the fluorescence of the external probe. For example, our anti-fluorophore antibodies (see Section 7.3) strongly quench the fluorescence of the corresponding dyes. In addition, although we are not aware of any publications that have used our Ca^{2+} indicator dextrans as probes for uptake by cells, it should be possible to follow endocytosis of dextran conjugates of fura-2, indo-1 and Calcium Green™ dyes (see Section 22.4) by using Mn^{2+} to quench the fluorescence of the residual extracellular dextrans without affecting intracellular dextrans. Using fluorescent dextran probes, researchers have investigated:

- Effects of the calmodulin antagonist W-13 on transcytosis and vesicular morphology in Madin–Darby canine kidney (MDCK) cells [6]
- Endocytic vesicles in *Plasmodium falciparum* [7,8]
- Endocytosis-defective Chinese hamster ovary (CHO) cells [9] and *Dictyostelium discoideum* [10-13]
- Fusion competence of lysosomes in parasite-infected cells [14]
- Membrane changes in prostatic adenocarcinoma [15] and basophilic leukemia cells [16,17]
- Methods to enhance receptor-mediated gene delivery [1]
- Vacuolar morphology in *Saccharomyces cerevisiae* both during the cell cycle [18-20] and in mutant cell lines [21]

Assaying Endosome Fusion with Dextrans

A novel fluorometric assay employing dextrans to follow endosome fusion has recently been described.[22] In order to pulse-label a population of endosomes, researchers incubated living baby hamster kidney (BHK) cells with a combination of rhodamine B 40,000 MW dextran (D-1840, see Section 15.5) and BODIPY FL avidin (A-2641, see Section 7.5) and then washed the cells to remove residual extracellular probe. Next, the cells were incubated with biotin 10,000 MW dextran (D-1856, see Section 15.5), biotinylated transferrin or biotinylated bovine serum albumin to label a second population of endosomes. Endosome fusion was then detected by the strong fluorescence enhancement of the BODIPY FL dye that occurs when BODIPY FL avidin complexes with biotinylated probes. Use of rhodamine B dextran as a fusion-insensitive reference enabled researchers to make ratiometric measurements of the fused organelles.

pH Indicator Dextrans

The fluorescein dextrans (pK_a ~6.4) are frequently used to investigate endocytic acidification.[23-26] Fluorescence of fluorescein-labeled dextrans is strongly quenched upon acidification (see Figure 23.2 in Chapter 23); however, fluorescein's lack of a spectral shift in acidic solution makes it difficult to discriminate between internalized probe that is quenched and residual fluorescence of the external medium. Dextran conjugates that either shift their emission spectra in acidic environments, such as the SNARF® and SNAFL® dextrans, or undergo significant shifts of their excitation spectra, such as BCECF, are preferred. We also offer dextrans labeled with the ratiometric pH-sensitive Rhodol Green™ (pK_a ~5.6), Cl-NERF (pK_a ~3.8), DM-NERF (pK_a ~5.4), Oregon Green 488 (pK_a ~4.7) and Oregon Green 514 (pK_a ~4.7) fluorophores. More information on these pH indicators can be found in Chapter 23. In addition to these pH indicator dextrans, Molecular Probes prepares a dextran that is double-labeled with fluorescein and tetramethylrhodamine (D-1950, D-1951; see Section 23.4), which has been used as a ratiometric indicator to measure endosomal acidification in Swiss 3T3 fibroblasts [27] and Hep G2 cells.[28]

Lipophilic Dextrans

In addition to the pH indicator dextrans, we also offer lipophilic 10,000 MW dextrans labeled with either fluorescein or tetramethylrhodamine [29] (D-7169, D-7164). Similar lipophilic dextrans have been used to model the diffusion of receptors on muscle membrane blebs [30] and as a marker for endocytosis in chick fibroblasts.[31]

Low Molecular Weight Markers

Hydrophilic fluorescent dyes — including sulforhodamine 101 (S-359), lucifer yellow CH (L-453), calcein (C-481), 8-hydroxypyrene-1,3,6-trisulfonic acid (HPTS, H-348) and Cascade Blue hydrazide (C-687) — are thought to be taken up by actively firing neurons through endocytosis-mediated recycling of the synaptic vesicles.[32,33] However, unlike the fluorescent FM membrane probes described in Section 17.2, the hydrophilic fluorophores appear to work for only a limited number of species in this application. Sulforhodamine 101 has been used to investigate neural activity in turtle brainstem–cerebellum,[34] neonatal rat spinal cord [35] and developing snake neuromuscular junctions.[36] The ratiometric pH indicator HPTS (see Section 23.2) has recently been employed to measure organelle acidification in the endosomal–lysosomal pathway.[37]

Staining Endocytic Vesicles for Electron Microscopy

Molecular Probes' fluorescein conjugate of cationized horse-spleen ferritin (F-948) — an electron-dense probe with a molecular weight of ~650,000 daltons and an iron content of 23% — can be used to study endocytosis by both fluorescence and electron microscopy.[38] This cationic ferritin derivative (pI >8), which has been modified to contain *N,N*-dimethyl-1,3-propanediamine groups, binds to anionic cell surfaces and other sites with high negative charges.[39-43]

Molecular Probes also offers a number of anti-fluorophore antibodies useful for detecting endocytic uptake of aldehyde-fixable fluorescent dextrans.[2] These antibodies, which are described in Section 7.3, enable researchers to develop their cell or tissue preparations for electron microscopy using standard horseradish peroxidase/diaminobenzidine [44] or colloidal gold [45] methodology.

Probes for Investigating Secretion

Intracellular calcium plays an important role in stimulus–secretion coupling, making Molecular Probes' fluorescent calcium indicators among the commonly used tools for investigating secretion. Researchers recently used our Fura Red calcium indicator to show that calcium spikes occurring in the granular areas of rat pancreatic acinar cells could be correlated with secretion.[46] Our fluorescent calcium indicators are described in detail in Chapter 22. In addition, we offer probes that modify intracellular calcium levels by a variety of mechanisms. For example, thapsigargin (T-7458, see Section 20.2), which inhibits Ca^{2+}-ATPase–mediated uptake of Ca^{2+} into the endoplasmic reticulum, has been used to further define the role of intracellular calcium in stimulus–secretion coupling in a number of cell types.[47-49] Fluorogenic and chromogenic substrates for enzymes that are released during secretion or degranulation — such as glucosaminidase, galactosidase, glucuronidase and acid phosphatase — may also be useful for detecting secretion. Many probes described in this and other chapters have been used to detect, induce or block secretion or degranulation. Among these are:

- Berberine (B-7472, Figure 17.7), a plant alkaloid that has been used to detect the Rab3AL-stimulated exocytosis of secretory granules in mast cells.[50] When used at a pH of 4 or below, berberine selectively stains heparin and mast cell granules in freeze-dried or fixed and paraffin-embedded tissue sections, producing a brilliant yellow fluorescent product that is relatively resistant to photodecomposition when exposed to UV light.[51]

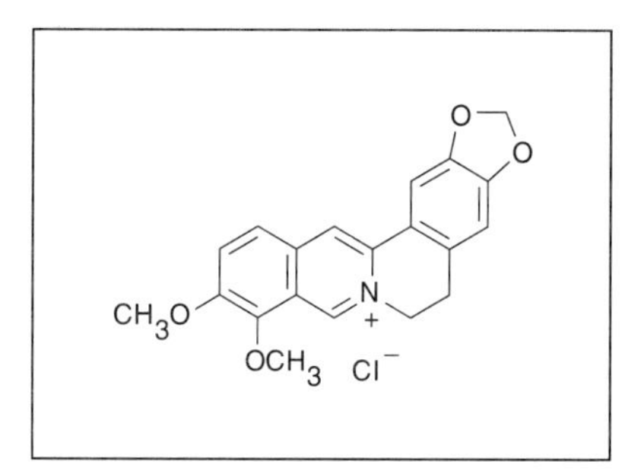

Figure 17.7 *Structure of the plant alkaloid berberine (B-7472).*

- BODIPY FL casein and BODIPY TR-X casein, substrates in our EnzChek Protease Assay Kits (E-6638, E-6639; see Section 10.4) that can potentially detect secretion of proteases from cells, including in agar plates, and may be useful for selecting enzyme-secreting mutants
- Luciferin (L-2911, see Section 10.5), which has been used to detect ATP secretion from bovine adrenal chromaffin cells [52]

- 4-Nitrophenyl *N*-acetyl-β-D-glucosaminidide (N-8426, see Section 10.2), which can detect glucosaminidase released during secretion or degranulation [53]
- TMA-DPH (T-204, see Section 17.2), which partially transfers from the plasma membrane to the mast cell granule membrane during secretion, and acridine orange (A-1301, see Section 12.3), which stains mast cell granules [54]
- Cyclic ADP-ribose (C-7619, see Section 20.2), which induces insulin secretion in pancreatic islets [55]
- Brefeldin A (B-7450, see Section 12.4), which blocks transport of NBD C_6-ceramide and BODIPY FL C_5-ceramide (N-1154, D-3521; see Section 12.4) through the Golgi apparatus [56,57]
- BODIPY FL C_5-ceramide, which is normally transported from the Golgi apparatus to the cell surface and thus permits isolation of mutant mammalian cells with defects in the secretory pathway [56]
- Anion channel blockers such as DIDS and SITS (D-337, A-339; see Section 18.5), which block secretion [58-60]
- Okadaic acid (O-3452, see Section 20.3), a protein phosphatase inhibitor that inhibits veratridine- and high K^+–induced secretion but not ionomycin-induced secretion [61]
- Nitric oxide donors, including SNAP (N-7892, see Section 21.5), which have been shown to either inhibit [62] or stimulate [63] secretion

1. J Biol Chem 269, 12918 (1994); **2.** Am J Physiol 258, C309 (1990); **3.** J Cell Physiol 131, 200 (1987); **4.** Proc Natl Acad Sci USA 82, 8523 (1985); **5.** Exp Cell Res 150, 36 (1984); **6.** J Biol Chem 269, 19005 (1994); **7.** J Cell Biol 101, 2302 (1985); **8.** Eur J Biol 34, 96 (1984); **9.** J Biol Chem 268, 25357 (1993); **10.** J Cell Biol 127, 387 (1994); **11.** J Cell Biol 126, 1433 (1994); **12.** J Cell Biol 126, 343 (1994); **13.** J Cell Biol 118, 1271 (1992); **14.** Science 249, 641 (1990); **15.** Cytometry 14, 826 (1993); **16.** J Cell Biol 101, 2156 (1985); **17.** J Cell Biol 101, 2145 (1985); **18.** J Cell Biol 107, 115 (1988); **19.** Science 241, 589 (1988); **20.** J Cell Biol 105, 1539 (1987); **21.** J Cell Biol 107, 1369 (1988); **22.** Biophys J 69, 716 (1995); **23.** FASEB J 8, 573 (1994); **24.** Proc Natl Acad Sci USA 91, 4811 (1994); **25.** J Biol Chem 268, 6742 (1993); **26.** J Cell Sci 105, 861 (1993); **27.** J Cell Biol 119, 99 (1992); **28.** J Cell Biol 130, 821 (1995); **29.** Biochemistry 19, 3893 (1980); **30.** J Cell Biol 92, 207 (1982); **31.** J Cell Biol 102, 1593 (1986); **32.** Nature 314, 357 (1985); **33.** Science 225, 851 (1984); **34.** J Neurosci 12, 3187 (1992); **35.** J Physiol 478, 265 (1994); **36.** Neuron 11, 801 (1993); **37.** Proc Natl Acad Sci USA 92, 3156 (1995); **38.** Histochemistry 98, 141 (1992); **39.** J Cell Biol 111, 2327 (1990); **40.** J Histochem Cytochem 35, 1063 (1987); **41.** J Microbiol Meth 7, 233 (1987); **42.** Proc Natl Acad Sci USA 81, 6054 (1984); **43.** Ultrastructure Res 38, 500 (1972); **44.** Histochemistry 84, 333 (1986); **45.** J Cell Biol 111, 249 (1990); **46.** EMBO J 12, 3017 (1993); **47.** Biochim Biophys Acta 1220, 199 (1994); **48.** J Neurochem 59, 2224 (1992); **49.** Cell Regulation 2, 27 (1991); **50.** Annals NY Acad Sci 710, 196, (1994); **51.** Stain Technol 55, 217 (1980); **52.** J Neurochem 49, 1266 (1987); **53.** J Cell Biol 119, 259 (1992); **54.** Biochim Biophys Acta 1067, 71 (1991); **55.** Science 259, 370 (1993); **56.** Mol Biol of the Cell 6, 135 (1995); **57.** J Biol Chem 268, 2341 (1993); **58.** Endocrinology 125, 1231 (1989); **59.** Exp Cell Res 140, 15 (1982); **60.** Proc Natl Acad Sci USA 77, 2721 (1980); **61.** Biochem Biophys Res Comm 174, 77 (1991); **62.** Biochem Biophys Res Comm 195, 1354 (1993); **63.** Am J Physiol 265, G418 (1993).

17.4 Data Table *Miscellaneous Endocytosis and Exocytosis Probes*

Cat #	Structure	MW	Storage	Soluble	Abs	{ε × 10⁻³}	Em	Solvent	Notes
B-7472	✓	408†	D,L	MeOH	429	{6.1}	none	MeOH	
C-481	S15.3	623	L	pH>5	494	{76}	517	pH 9	
C-687	S15.3	596	L	H_2O*	399	{30}	421	H_2O	<1>
H-348	S15.3	524	D,L	H_2O	454	{22}	511	pH 9	<2>
L-453	S15.3	457	L	H_2O*	428	{12}	536	H_2O	
S-359	S15.3	607	L	H_2O	586	{108}	605	H_2O	

For definitions of the contents of this data table, see "How to Use the Handbook" on page vi.
Structure: Chemical structure drawing: ✓ = shown in Figure 17.7; Sn.n = shown in Section number n.n.

* Maximum solubility in water is ~1% for C-687 and ~8% for L-453.
† MW is for the hydrated form of this product.
<1> Cascade Blue dyes have a second absorption peak at about 376 nm with ε ~80% of the 395–400 nm peak.
<2> H-348 spectra are pH dependent. See Section 23.2 for data.

17.4 Price List *Miscellaneous Endocytosis and Exocytosis Probes*

Cat #		Product Name	Unit Size	Price Per Unit ($) 1–4 Units	5–24 Units
B-7472	*New*	berberine chloride, dihydrate	1 g	15.00	12.00
C-481		calcein *high purity*	100 mg	58.00	46.40
C-687		Cascade Blue® hydrazide, trisodium salt	10 mg	95.00	76.00
D-7169	*New*	dextran, fluorescein, 10,000 MW, lipophilic	10 mg	95.00	76.00
D-7164	*New*	dextran, tetramethylrhodamine, 10,000 MW, lipophilic	10 mg	110.00	88.00
F-948		ferritin from horse spleen, cationized, fluorescein conjugate *2.5 mg/mL*	2 mL	80.00	64.00
H-348		8-hydroxypyrene-1,3,6-trisulfonic acid, trisodium salt (HPTS; pyranine)	1 g	72.00	57.60
L-453		lucifer yellow CH, lithium salt	25 mg	48.00	38.40
S-359		sulforhodamine 101	25 mg	50.00	40.00

Fluorescent Probes for Photoconversion of Diaminobenzidine

Photoconversion of Diaminobenzidine

Molecular Probes is pleased to offer a variety of fluorescent probes for photoconverting diaminobenzidine (DAB), enabling researchers to take advantage of an important development in correlated fluorescence, transmitted and electron microscopy. In 1982, Maranto first described the use of the fluorophore lucifer yellow for DAB photoconversion.[1] When a fluorophore is exposed to light of an appropriate wavelength, excitation from the electronic ground state to a higher singlet state occurs. Instead of emitting a photon, the excited state of the fluorophore may undergo intersystem crossing to the triplet state. Transfer of energy to ground state triplet oxygen ($^3O_2$) generates the toxic and highly reactive singlet oxygen ($^1O_2$), which is capable of causing damage to lipids, proteins and nucleic acids.[2] However, the reactive potential of $^1O_2$ can also be harnessed to oxidize diaminobenzidine (DAB) into an electron-opaque osmiophilic precipitate within cells. The resulting DAB reaction product exhibits exceptionally uniform, nondiffusible staining properties, making it extremely useful for subsequent electron microscopy investigation of cellular ultrastructure.[3]

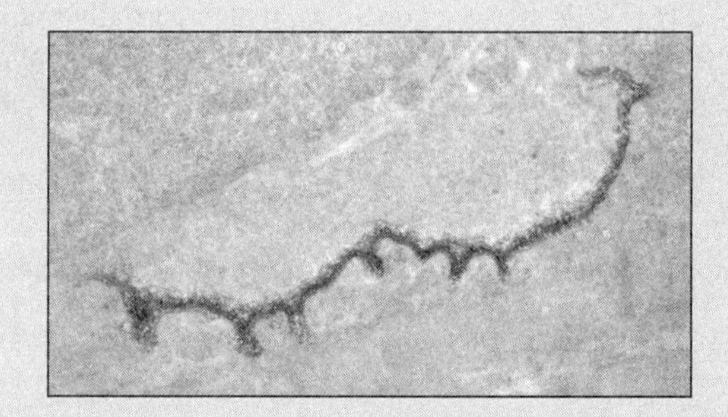

Electron micrograph of a 80 nm–thick section of formaldehyde-fixed rat solius muscle, which was first stained with eosin bungarotoxin (B-7483) and then used to photoconvert DAB into an insoluble osmiophilic polymer. Photo contributed by Thomas J. Deerinck, University of California, San Diego.

Eosin and Other Brominated Probes

In 1994, Deerinck and colleagues reported a simple method for eosin-mediated photoconversion of DAB.[4,5] Halogenated derivatives of fluorescein dyes are known to be effective photosensitizers and singlet oxygen generators.[6] Eosin is a brominated analog of fluorescein that has a $^1O_2$ yield 19 times greater than fluorescein and is an excellent dye for photoconverting DAB.[4,7] Furthermore, the small size of eosin promotes exceptional penetration into tissues resulting in increased resolution for electron microscopy. We offer amine- and thiol-reactive eosin derivatives, as well as several eosin-based conjugates, including secondary antibodies (E-6436, E-6437; see Section 7.2) and streptavidin (S-6372, see Section 7.5), that are known to photoconvert DAB.[4] We have also prepared brominated analogs of several site-selective probes, some of which appear in the product list below. Other fluorescent tracers that have been used to photoconvert DAB include:[8]

- BODIPY® FL C_5-ceramide [9] (D-3521, see Section 12.4)
- DiI [10,11] (D-282, see Section 15.4)
- Fluorescent polystyrene microspheres [12] (see Section 6.2)
- Fluoro-ruby [13] (D-1817, see Section 15.5)
- Lucifer yellow [14] (L-453, see Section 15.3)
- Propidium iodide [15] (P-1304, see Section 8.1)

1. Science 217, 953 (1982); **2.** J Photochem Photobiol 11, 241 (1991); **3.** Science 217, 3 (1982); **4.** J Cell Biol 126, 901 (1994); **5.** J Cell Biol 126, 877 (1994); **6.** Adv Photochem 18, 315 (1993); **7.** Photochem Photobiol 37, 271 (1983); **8.** Lübke, J. in *Neuroscience Protocols*, F.G. Wouterlood, Ed., Elsevier Science Publishers (1993) 93-050-06-01–13; **9.** Cell 73, 1079 (1993); **10.** J Histochem Cytochem 38, 725 (1990); **11.** Neuroscience 28, 3 (1989); **12.** Brain Res 630, 115 (1993); **13.** J Histochem Cytochem 41, 777 (1993); **14.** J Neurosci Meth 36, 309 (1991); **15.** J Neurosci Meth 45, 87 (1992).

Cat #		Product Name	Unit Size	Price Per Unit ($) 1–4 Units	5–24 Units
B-7483	*New*	α-bungarotoxin, eosin conjugate	500 µg	175.00	140.00
C-6166	*New*	5-carboxy-2′,4′,5′,7′-tetrabromosulfonefluorescein, succinimidyl ester, bis-(diisopropylethylammonium) salt	5 mg	95.00	76.00
C-2926		CellTracker™ Yellow-Green CMEDA (5-chloromethyleosin diacetate)	1 mg	125.00	100.00
D-7166	*New*	dextran, eosin, 10,000 MW, anionic, lysine fixable	25 mg	110.00	88.00
D-7546	*New*	N-(2,6-dibromo-4,4-difluoro-5,7-dimethyl-4-bora-3a,4a-diaza-s-indacene-3-pentanoyl)-sphingosine (BODIPY® FL Br_2C_5-ceramide)	250 µg	98.00	78.40
D-7767	*New*	6,6′-dibromo-3,3′-dioctadecyloxacarbocyanine perchlorate (Br_2-DiOC$_{18}$(3))	5 mg	135.00	108.00
D-7766	*New*	5,5′-dibromo-1,1′-dioctadecyl-3,3,3′,3′-tetramethylindocarbocyanine perchlorate (Br_2-DiIC$_{18}$(3))	5 mg	135.00	108.00
E-6436	*New*	eosin goat anti-mouse IgG (H+L) conjugate *2 mg/mL*	0.5 mL	75.00	60.00
E-6437	*New*	eosin goat anti-rabbit IgG (H+L) conjugate *2 mg/mL*	0.5 mL	75.00	60.00
E-99		eosin-5-iodoacetamide	100 mg	155.00	124.00
E-18		eosin-5-isothiocyanate	100 mg	98.00	78.40
E-118		eosin-5-maleimide	25 mg	155.00	124.00
E-7463	*New*	eosin phalloidin	300 U	198.00	158.40
L-7542	*New*	LysoTracker™ Green Br_2 *1 mM solution in DMSO* *special packaging*	20x50 µL	125.00	100.00
S-6372	*New*	streptavidin, eosin conjugate	1 mg	95.00	76.00
T-7539	*New*	2′,4′,5′,7′-tetrabromorhodamine 123 bromide	5 mg	48.00	38.40
T-7017	*New*	2′,4′,5′,7′-tetrabromo-4,5,6,7-tetrafluorofluorescein diacetate (Br_4TFFDA)	5 mg	125.00	100.00

Chapter 18

Probes for Receptors and Ion Channels

Contents

Related Chapters

18.1 Overview of Site-Selective Probes

Because receptor-mediated signal transduction underlies much of what occurs in cellular biochemistry and physiology,[1] fluorescent receptor ligands can provide a sensitive means of identifying and localizing some of the most pivotal molecules in cell biology. Molecular Probes offers fluorescently labeled and unlabeled ligands for cellular receptors, ion channels and carriers and many other physiologically important molecules. Our site-selective fluorescent probes, which may be used on live or fixed cells as well as in cell-free extracts, are available with a wide variety of fluorophores, spanning the spectrum from the near UV to deep red. In particular, we would like to highlight those ligands conjugated to the green fluorescent BODIPY® FL and Oregon Green™ 514 dyes and the red fluorescent Texas Red® dye, which provide extremely bright signals and superior photostability. The high sensitivity and selectivity of these fluorescent probes make them especially good candidates for measuring low-abundance receptors.

In addition, Molecular Probes has greatly expanded its selection of toxins through a collaboration with Latoxan, a distinguished producer of natural and synthetic toxins for research. Natural toxins are produced by a diverse array of plants, animals and microorganisms. Their typically high selectivity makes them important probes of the structure, function and distribution of their molecular targets.[2] The Latoxan products are grouped with our other compounds according to their biological activity; they include receptor ligands (Section 18.2), channel blockers and activators (Sections 18.4 through 18.7) and other bioactive compounds (Section 18.8). Latoxan products also appear in other chapters.

This chapter is devoted to our probes for neurotransmitter receptors and ion channels and carriers. Additional fluorescently labeled receptor ligands (including low-density lipoproteins, epidermal growth factors, transferrin conjugates and chemotactic peptides) are described in Section 17.3, along with other probes for studying receptor-mediated endocytosis. Chapter 20, entitled "Probes for Signal Transduction," focuses on reagents for investigating events — such as calcium regulation; kinase, phosphatase and phospholipase activation; and lipid trafficking — that occur downstream from the receptor–ligand interaction (see Figure 20.1 in Chapter 20). Chapters 22–24 contain our large inventory of indicators for physiologically important ions, providing a means of correlating ion channel activation with subsequent changes in intracellular ion concentration.

1. Trends Neurosci 16, 233 (1993); **2.** Meth Enzymol 207, 620 (1992).

18.2 Probes for Neurotransmitter Receptors

α-Bungarotoxin Probes for Nicotinic Acetylcholine Receptors

Fluorescent α-Bungarotoxins

Nicotinic acetylcholine receptors (nicotinic AChRs) are ligand-gated ion channels that produce an increase in Na^+ and K^+ permeability, depolarization and excitation upon activation (see Figure 20.1 in Chapter 20). α-Bungarotoxin, a 74–amino acid peptide extracted from *Bungarus multicinctus* venom, binds with high affinity to the α-subunit of the nicotinic AChR of neuromuscular junctions.[1] Molecular Probes provides an extensive selection of fluorescent α-bungarotoxin conjugates (Table 18.1) to facilitate visualization of nicotinic AChRs with a variety of instrumentation. We attach approximately one fluorophore to each molecule of α-bungarotoxin, thus retaining optimal binding specificity. The labeled bungarotoxins are then chromatographically separated from unlabeled molecules to ensure maximum labeling of the product.

Both BODIPY FL α-bungarotoxin[2] (B-3504) and Oregon Green 514 α-bungarotoxin (B-7488) have fluorescence spectra similar to those of fluorescein α-bungarotoxin (F-1176) and are therefore suitable for use with standard fluorescein optical filter sets. Oregon Green 514 conjugates exhibit bright green fluorescence and superior photostability, making them especially well suited to fluorescence and confocal laser scanning microscopy (see Figure 1.6 in Chapter 1, Figure 7.1 in Chapter 7 and Figure 11.1 in Chapter 11). Tetramethylrhodamine α-bungarotoxin (T-1175)[3,4] is currently the dominant red-orange fluorescent probe for staining the nicotinic AChR (see photo on front cover). However, our brand new Texas Red-X conjugate of α-bungarotoxin (B-7489) has a longer-wavelength emission maximum and therefore offers better spectral separation from green fluorescent dyes in multicolor experiments. Eosin α-bungarotoxin (B-7483) was designed as a probe for visualizing nicotinic AChRs either by fluorescence microscopy or by electron microscopy using the diaminobenzidine photooxidation technique described by Deerinck and colleagues[5] (see the Technical Note "Fluorescent Probes for Photoconversion of Diaminobenzidine" on page 414).

Fluorescent α-bungarotoxins have been used in a variety of informative investigations of the nicotinic AChR:

- Nicotinic AChRs tagged with fluorescently labeled α-bungarotoxin were used to correlate receptor clustering during neuromuscular development with tyrosine phosphorylation of the receptor.[6,7]
- Rhodamine-labeled α-bungarotoxin was used in conjunction with a sophisticated time-lapse video technique to document nicotinic AChR cluster formation after myoblast fusion.[8]
- Rhodamine-labeled α-bungarotoxin was used to quantitate nicotinic AChRs in a study that showed that several isotypes of agrin, a component of synaptic basal lamina, help trigger the nicotinic AChR aggregation that occurs during neuromuscular junction formation.[9]
- Studies of reinnervation of adult muscle after nerve damage have employed fluorescein- and rhodamine-labeled α-bungarotoxins to identify and visualize endplates.[10,11]

Table 18.1 *Labeled and unlabeled α-bungarotoxins.*

Cat #	Label	Ex/Em (nm)	Notes
F-1176	Fluorescein	494/518	Original green fluorescent conjugate
B-3504	BODIPY FL	504/513	Narrow emission with good photostability
B-7488	Oregon Green 514	512/530	Best photostability of the green fluorescent α-bungarotoxins
B-7483	Eosin	523/542	Potentially useful for photoconversion studies
T-1175	Tetramethyl-rhodamine	553/577	Currently the most extensively used conjugate
B-7489	Texas Red-X	593/613	Best dye to combine with green fluorescent probes
B-1196	Biotin-XX	NA	Visualized with secondary detection reagents
B-1601	Unlabeled	NA	Useful as a control, as well as for radioiodination and for preparation of new conjugates

NA = not applicable.

Biotinylated α-Bungarotoxin

Nicotinic AChRs can also be labeled with biotinylated α-bungarotoxin (B-1196), which is then localized using enzyme- or fluorophore-labeled conjugates of avidin or streptavidin[7,12-14] (see Section 7.5). In addition, the biotinylated toxin can be employed for affinity isolation of the nicotinic AChR using an avidin or streptavidin agarose (S-951, see Section 7.5) column[15] and for enzyme-linked immunosorbent assays (ELISAs) designed to detect anti–nicotinic AChR antibodies.[16]

Unlabeled α-Bungarotoxin

In addition to the fluorescent and biotinylated derivatives, we stock unlabeled α-bungarotoxin (B-1601), which has been shown to be useful for radioiodination.[1,17] Unlabeled α-bungarotoxin has also been employed for ELISA testing of nicotinic AChR binding,[18] as well as for investigating the function of the α-bungarotoxin–binding component (αBgtBC) in vertebrate neurons.[19]

β-Bungarotoxin and Related Toxins

Unlabeled β-bungarotoxin, which blocks transmission at the vertebrate neuromuscular junction via a *pre*synaptic mechanism, is discussed in Section 18.6. Toxins that are structurally similar to β-bungarotoxin but also act as nicotinic AChR antagonists include:

- α-Cobrotoxin [20-23] from *Naja naja siamensis* (C-7484), a high-affinity nicotinic AChR antagonist
- Erabutoxin a (E-98110) and erabutoxin b (E-98111), which are α-neurotoxins from the sea-snake *Laticauda semifasciata* that differ by one amino acid [21,22,24,25] (asparagine or histidine, respectively, at residue 26)
- LS III from *Laticauda semifasciata* (L-98112), which has a lower affinity than erabutoxins for the nicotinic AChR [21,26]

A recent review by Yang discusses the structure and function of α-cobrotoxin.[27] Unlabeled cobrotoxin has been used to prepare fluorescein cobrotoxin for use in visualizing membrane-bound nicotinic AChRs from *Torpedo californica*.[28,29]

Pirenzepine, ABT and Anthroylcholine Probes for Muscarinic Acetylcholine Receptors

Fluorescent Pirenzepines

Unlike nicotinic AChRs, muscarinic acetylcholine receptors (muscarinic AChRs) are G-protein–coupled receptors that produce either excitatory or inhibitory responses and are not necessarily associated with changes in ion permeability (see Figure 20.1 in Chapter 20). In collaboration with Max Cynader and colleagues at the University of British Columbia, we have developed fluorescent derivatives of pirenzepine (P-7438), a selective antagonist for the M_1 muscarinic AChR. The green fluorescent BODIPY FL (B-7436) and red fluorescent BODIPY 558/568 (B-7437) derivatives retain pirenzepine's specificity for the M_1 muscarinic receptor and exhibit similar inhibition and displacement profiles. BODIPY FL pirenzepine was employed to study the activity-dependent expression of the M_1 muscarinic receptor in cultured neurons derived from rat visual cortex.[30] Expression of the receptor increased in response to antagonist application and to chronic membrane depolarization. Conversely, receptor expression decreased when agonist or the Na^+ channel blocker tetrodotoxin (T-6913, T-6914; see Section 18.7) was applied.

Fluorescent ABT

To aid in the study of this important class of receptors, we offer 3-(2'-aminobenzhydryloxy)tropane (ABT, A-3500), a high-affinity ligand (K_d ~10^{-8} M) for muscarinic AChRs, and its green fluorescent BODIPY FL conjugate (B-7467). We expect that our new BODIPY FL ABT will selectively bind to muscarinic AChRs with high affinity because ABT similarly conjugated to agarose serves as a selective affinity matrix for purifying muscarinic AChRs.[31,32] Unlabeled ABT may be useful as an intermediate for preparing other new probes for muscarinic AChRs. We also offer benztropine methanesulfonate (benztropine mesylate, B-7477) to serve as an unlabeled control for experiments with BODIPY FL ABT.

Anthroylcholine

9-Anthroylcholine bromide (A-77) is a fluorescent ligand for muscarinic AChRs.[33,34] It acts as a competitive antagonist of acetylcholine in rat duodenum and does not appear to interact with nicotinic AChRs in *Torpedo* electric organ.[33] This fluorescent ligand also shows a large fluorescence enhancement upon binding the Ca^{2+}/calmodulin complex.[35,36]

Other Probes for Cholinergic Receptors and for Acetylcholinesterase

Unlabeled Nicotinic and Muscarinic Receptor Probes

In addition to the above-mentioned receptor ligands, Molecular Probes offers a caged carbamylcholine derivative (C-1791, see Table 18.4 in Section 18.3), as well as a large assortment of unlabeled probes for nicotinic and muscarinic AChR probes provided by the premier toxin producer Latoxan (Table 18.2). In most cases the biological activities of these probes have not been thoroughly studied.

Acetylcholinesterase Inhibitors

Through Latoxan, Molecular Probes also offers a selection of acetylcholinesterase inhibitors:

- Bomin-1, bomin-2 and bomin-3 (E-99019, P-99020, B-99021), which are selective irreversible inhibitors of carboxylesterases
- Cardiotoxin [37] from *Naja nigricollis* (C-98102)
- Desoxypeganine [38] from Zygophyllaceae (D-96009)
- Fasciculin 2 from *Dendroaspis angusticeps* (F-98107), a potent noncompetitive inhibitor of mammalian and electric eel acetylcholinesterase at picomolar concentrations [39-43] and of avian and insect acetylcholinesterases and butyryl cholinesterase at micromolar to millimolar concentrations [43,44]
- Galanthamine [45,46] from Amaryllidaceae (G-96028)
- Peganole [38,47] from Papaveraceae (P-96043), which inhibits both cholinesterases and monoamine oxidases

Prazosin Probes for α_1-Adrenergic Receptors

We have prepared the green fluorescent BODIPY FL (B-7433) and red fluorescent BODIPY 558/568 (B-7434) derivatives of prazosin, a high-affinity antagonist for the α_1-adrenergic receptor.[48,49] Research in Max Cynader's laboratory at the University of British Columbia has shown that these derivatives can be used to localize the α_1-adrenergic receptor on cultured cortical neurons.[50] BODIPY FL prazosin has been employed for flow cytometric detection of α_1-adrenergic receptors, including the α_{1a}, α_{1b} and α_{1d} receptor subtypes, on Chinese hamster ovary (CHO) cells.[51] We also offer unlabeled prazosin (P-7435) for use as a control. For information on our caged versions of several adrenergic agonists, including epinephrine (D-7057), phenylephrine (C-7083, D-7084) and norepinephrine (D-7058), see Table 18.4 in Section 18.3.

CGP 12177 Probes for β-Adrenergic Receptors

The most thoroughly studied receptor-modulated ion channel mechanism is the activation of the β-adrenergic receptor upon binding adrenaline, which causes an enhancement of Ca^{2+} current.[52] Molecular Probes now offers a green fluorescent BODIPY FL–labeled analog (B-7470) of the hydrophilic β-adrenergic receptor agonist CGP 12177.[53] A similar, but not identical, BODIPY CGP 12177 conjugate reportedly gave a strong receptor-specific signal, enabling researchers to measure on- and off-rate constants and dissociation constants that agreed well with those determined for tritiated CGP 12177.[54] For researchers studying the β-adrenergic receptor, we also offer caged versions of the agonist isoproterenol

Table 18.2 Unlabeled probes for nicotinic and muscarinic acetylcholine receptors.

Cat #	Product *Source*	Action	Ref
Nicotinic Acetylcholine Receptor (nAChR) Probes			
A-96206	14-Acetylvirescenine *Ranunculaceae*	• nAChR antagonist	1
B-1601	α-Bungarotoxin *Bungarus multicinctus*	• High-affinity nAChR antagonist • Fluorescent and biotinylated derivatives available (see Table 18.1)	2–4
C-7484	α-Cobrotoxin *Naja naja siamensis*	• High-affinity nAChR antagonist	5
C-96017	Condelphine *Ranunculaceae*	• nAChR antagonist ($IC_{50} = 3.6 \times 10^{-6}$ M) †	6
D-96052	Delcorine *Ranunculaceae*	• nAChR antagonist ($IC_{50} = 5.0 \times 10^{-5}$ M) †	7,8
D-96019	Delsoline (acomonine) *Aconitum sp. (Ranunculaceae)*	• nAChR antagonist ($IC_{50} = 1.3 \times 10^{-5}$ M) †	8,9
D-96035 ‡	*N*-Desacetyllappaconitine *Aconitum sp. (Ranunculaceae)*	• Fluorescent label for brain nAChR ($IC_{50} = 4.7 \times 10^{-6}$ M) †	10,11
E-98110	Erabutoxin a *Laticauda semifasciata*	• nAChR antagonist • Asparagine at residue 26	12–14
E-98111	Erabutoxin b *Laticauda semifasciata*	• nAChR antagonist • Histidine at residue 26	13–15
F-96026	Fluorocurarine (C-curarine III; metvine) *Vinca erecta*	• nAChR antagonist	16
K-96034	Karakoline *Ranunculaceae*	• nAChR antagonist ($IC_{50} = 1.0 \times 10^{-6}$ M) †	6,17
L-96010 ‡	Lappaconitine (allapinine) *Aconitum sp. (Ranunculaceae)*	• nAChR antagonist ($IC_{50} = 8.0 \times 10^{-5}$ M) † • Antiarrhythmic agent	10,11,18
L-98112	Laticauda semifasciata III (LS III) *Laticauda semifasciata*	• nAChR antagonist	13,19,20
L-96221	Lycaconitine *Aconitum sp.* (Ranunculaceae)	• nAChR antagonist	9
M-96204	*o*-Methylisocorydine *Corydalis sp. (Fumariaceae)*	• nAChR antagonist	21
M-96201	Methyllycaconitine (MLA) *Delphinium sp. (Ranunculaceae)*	• nAChR antagonist (nM affinity for a subtype of neuronal nAChR in brain and μM affinity for muscle nAChR)	6,7, 22–27
N-96037	Nudicauline *Ranunculaceae*	• nAChR antagonist ($IC_{50} = 5.0 \times 10^{-9}$ M) †	
Muscarinic Acetylcholine Receptor (mAChR) Probes			
B-96056	β-Belladonnine bis-(chloroethylate) *Atropa belladonna*	• M_2- and N-cholinolytic properties • Antihistaminic activity	
B-7477	Benztropine mesylate	• Useful as a control for BODIPY FL ABT (B-7467)	
I-96008 B-96099 C-96032 C-96033	Imperialine (raddeamine, sipeimine) *Petilium eduardi* 3β-Bromoimperialine 3α-Chloroimperialine 3β-Chloroimperialine	• Short-acting selective M_2 antagonist • Useful for radioiodination • Potent long-acting M_2 antagonist • Potent medium-lasting M_2 antagonist	28
I-96050	Imperialone	• Potent medium-lasting M_2 antagonist	
D-99022 D-99025	LK-66 LK-66M	• Selective M_1 antagonists	
D-99023 D-99024 D-99027 D-99028	LK-6 LK-7 LK-190 LK-252	• mAChR antagonists	
P-7438	Pirenzepine	• Selective M_1 antagonist • Fluorescent derivatives available	29–31

† IC_{50} values at [^{125}I] α-bungarotoxin–binding sites in rat brain membranes. Data provided by S. Wonnacot (University of Bath) to Latoxan. ‡ See Section 18.7. **1.** Heterocycles 12, 779 (1979); **2.** Meth Neurosci 8, 67 (1992); **3.** Biochemistry 29, 11009 (1990); **4.** Biochem Biophy Res Comm 44, 711 (1971); **5.** J Toxicol 13, 275 (1994); **6.** Can J Physiol Pharmacol 72, 104 (1994); **7.** Heterocycles 23, 2515 (1985); **8.** J Am Chem Soc 80, 4434 (1958); **9.** J Am Chem Soc 103, 6536 (1981); **10.** Eur J Pharmacol 283, 103 (1995); **11.** J Nat Prod 58, 929 (1995); **12.** J Biol Chem 270, 9362 (1995); **13.** Brain Res 211, 107 (1981); **14.** Biochem J 99, 624 (1966); **15.** Biochem Biophys Res Comm 88, 950 (1979); **16.** Pharmacol Toxicol 1, 18 (1976); **17.** Chem Nat Comp 9, 194 (1973); **18.** U.S. Patent No. 5,290,784; **19.** Biochem J 141, 389 (1974); **20.** Biochem J 141, 383 (1974); **21.** Acta Crystallography (Section C) 48, 945 (1992); **22.** FEBS Lett 365, 79 (1995); **23.** Mol Pharmacol 41, 802 (1992); **24.** FEBS Lett 270, 45 (1990); **25.** J Exp Biol 142, 215 (1989); **26.** FEBS Lett 226, 357 (1988); **27.** Experientia 42, 611 (1986); **28.** Naunyn-Schmiedeberg's Arch Pharmacol 346, 144 (1992); **29.** J Neurosci 14, 4147 (1994); **30.** J Neurosci 13, 4293 (1993); **31.** Endocrinology 131, 235 (1992).

and the antagonist propranolol (N-7066, N-7067), which are described in Table 18.4 in Section 18.3.

SCH 23390, NAPS and PPHT Probes for D_1 and D_2 Dopaminergic Receptors

There are at least six receptor subtypes for the neurotransmitter dopamine.[55] The D_1 dopaminergic receptors are coupled to G-proteins that activate adenylate cyclase, whereas D_2 dopaminergic receptors are coupled to G-proteins that inhibit adenylate cyclase activity[56] (see Figure 20.1 in Chapter 20). D_2 receptor stimulation is also reported to accelerate the turnover of phosphatidylinositol,[57] to mobilize calcium ions[58] and to increase K^+ flux.[59] Research on the D_1 and D_2 dopaminergic receptor subtypes has traditionally employed radiolabeled derivatives of the specific D_1 antagonist SCH 23390 and the corresponding D_2 antagonist NAPS.

Molecular Probes offers several fluorescent derivatives of SCH 23390, NAPS and the D_2-selective agonist 2-(*N*-phenethyl-*N*-propyl)amino-5-hydroxytetralin (PPHT) (Table 18.3). These probes retain their high affinity and specificity[60-64] and have proven useful for locating receptor subtypes in individual cells. NAPS conjugates derivatized with BODIPY FL (B-3424), tetramethylrhodamine (T-3426) and Texas Red (T-3427) were used to characterize the distribution of D_2 receptors in the retina of two different amphibian species; D_2 receptors were found on both rods and cones and appeared to be clustered in patches within the inner segment.[60] A recent paper showed that staining of D_2 receptors in the membranes of live neurons with 10 nM tetramethylrhodamine NAPS persists for about 30 minutes, whereas BODIPY FL NAPS is rapidly internalized.[65] Fluorescein-labeled SCH 23390 and NAPS (F-3421, F-3425) were employed to detect alterations in the visual cortex pattern of D_1 and D_2 receptors after operant conditioning.[66]

Table 18.3 *Fluorescent D_1 and D_2 dopaminergic receptor probes.*

Cat #	Probe	Ex/Em (nm)
D_1 Dopaminergic Receptor Antagonists		
F-3421	Fluorescein SCH 23390	495/518
B-3420	BODIPY FL SCH 23390	504/511
T-3422	Tetramethylrhodamine SCH 23390	541/568
T-3423	Texas Red SCH 23390	582/606
D_2 Dopaminergic Receptor Antagonists		
F-3425	Fluorescein NAPS	496/522
B-3424	BODIPY FL NAPS	504/511
T-3426	Tetramethylrhodamine NAPS	541/567
T-3427	Texas Red NAPS	582/600
D_2 Dopaminergic Receptor Agonists		
F-3471	Fluorescein PPHT	498/519
B-3460	BODIPY FL PPHT	504/511

BODIPY FL SCH 23390 was recently used to map D_1 receptors in vertebrate retina of several species.[67]

These dopaminergic receptor probes are available with a range of emission maxima (Table 18.3), which has made it possible to simultaneously label and distinguish receptor subtypes in populations of neurons. SCH 23390 and NAPS coupled to either the green fluorescent BODIPY FL or red fluorescent Texas Red dyes, have been used to localize D_1 and D_2 receptors in sections of rat medial prefrontal cortex.[68] In this study, D_1 and D_2 receptor subtypes occurred predominantly on different neurons.

For researchers studying dopaminergic receptors, we also offer caged versions of the agonist dopamine (C-7124, D-7064) and the antagonist haloperidol (H-7080), which are described in Table 18.4 in Section 18.3.

Diisoquinoline Alkaloid Probes for D_1 and D_2 Dopaminergic Receptors

The diisoquinoline alkaloid probes DL-tetrahydroberberine (canadine, T-96016), DL-tetrahydropalmatine (T-96038) and scoulerine (S-96203) are all antagonists of D_1 and D_2 dopaminergic receptors.[69-71] In addition, scoulerine is reported by Latoxan to be an α_1-adrenoreceptor antagonist and to have sedative and tranquilizing activity. DL-Tetrahydroberberine and DL-tetrahydropalmatine may also interact with Ca^{2+} channels and exhibit sedative, analgesic or hypnotic activity.

Naloxone and Naltrexone Probes for μ-Opioid Receptors

The μ-opioid receptors plays a critical role in analgesia. Among the antagonists that have been used to define and characterize these receptors is naltrexone, a nonaddictive drug that has been used for the treatment of opioid addiction. The fluorescent derivatives naloxone fluorescein and naltrexone fluorescein (N-1384, N-1385) have been reported to bind to the μ-opioid binding site with high affinity.[72-75] Flow cytometric analysis of the binding of fluorescein naloxone to NMDA- and μ-opioid receptors (which was displaced by NMDA and met-enkephalin, respectively) has been used to deduce the effects of operant conditioning on visual cortex receptor pattern.[66]

Conantokin-G Probes for NMDA Receptors

Conantokin-G, the 17–amino acid peptide isolated from the venom of the cone snail (*Conus geographus*), causes sleep in young mice and hyperactivity in older mice.[76] This "sleeper" peptide is a high-affinity antagonist of a distinct subclass of channel-linked glutamate receptors known as *N*-methyl-D-aspartate (NMDA) receptors.[76-80] Using BODIPY FL– and tetramethylrhodamine-labeled conantokin-G derivatives, Benke and co-workers investigated the distribution and mobility of NMDA receptors on live cortical neurons by confocal laser scanning microscopy and digital imaging.[78] BODIPY FL–labeled conantokin-G was also employed to detect alterations in the visual cortex pattern of NMDA receptors after operant conditioning.[66] Unlabeled conantokin-G (C-6909),

BODIPY FL conantokin-G (C-6904) and tetramethylrhodamine conantokin-G (C-6906) are now available from Molecular Probes.

Our selection of caged probes for the NMDA receptor, including caged NMDA (M-7069, M-7114) and caged MK-801 (D-7059), are described in Table 18.4 in Section 18.3.

Anti–NMDA Receptor Antibodies

N-methyl-D-aspartate (NMDA) receptors constitute cation channels of the central nervous system that are gated by the excitatory neurotransmitter L-glutamate.[81,82] Activation of NMDA receptors is essential for inducing long-term potentiation (LTP), a form of activity-dependent synaptic plasticity that is implicated in the learning process in animal behavioral models.[83] The biophysical properties of NMDA receptor channels contributing to LTP include Ca^{2+} permeability, voltage-dependent Mg^{2+} block and slow-gating kinetics.[84-87] NMDA receptor channel activities play a role in neuronal development and in disorders such as epilepsy and ischemic neuronal cell death. As targets for ethanol, NMDA receptors may also function in the pathology of alcoholism.[88-89]

In vitro reconstitution experiments with the cloned NMDA receptor subunit 1 and any one of the four NMDA receptor subunits 2A, 2B, 2C and 2D revealed that the physical properties of the heteromeric NMDA receptor channel appear to be imparted by the particular NMDA receptor subunit 2.[90] NMDA receptor subunits 2A and 2B are detected predominantly in the hippocampus and cortex, whereas 2C is found mainly in the cerebellum. Thus, cellular expression profiles of the NMDA receptor subunits 2A, 2B, 2C and 2D may contribute to the biophysical properties of NMDA receptors in specific central neurons.

Molecular Probes offers rabbit polyclonal antibodies to NMDA receptor subunits 2A, 2B and 2C in two forms: as affinity-purified antibodies or as unfractionated sera. The anti–NMDA receptor subunit 2A and 2B antibodies were generated against fusion proteins containing amino acid residues 1253–1391 of subunit 2A and 984–1104 of subunit 2B, respectively. These two antibodies are active against mouse, rat and human forms of the antigens and are specific for the subunit against which they were generated. In contrast, the anti–NMDA receptor subunit 2C antibody was generated against amino acid residues 25–130 of subunit 2C and recognizes the 140,000-dalton subunit 2C, as well as the 180,000-dalton subunit 2A and subunit 2B from mouse, rat and human. The affinity-purified antibodies were fractionated from sera by affinity chromatography in which NMDA receptor subunit fusion proteins were bound to a column matrix. Our affinity-purified anti–NMDA receptor subunit 2A (A-6473), 2B (A-6474) and 2C (A-6475) antibodies are suitable for immunohistochemistry, Western blots, enzyme-linked immunosorbent assays (ELISAs) and immunoprecipitations. Unfractionated rabbit sera for subunits 2A (A-6467), 2B (A-6468) and 2C (A-6469) are also available as economical alternatives for Western blot analysis and immunoprecipitation.

Philanthotoxin and Argiotoxin$_{636}$ for Glutamatergic Receptors

Philanthotoxin

Wasp philanthotoxins (PhTX) have complex actions on glutamate responses apparently mediated by a variety of receptors.[91,92] Our philanthotoxin PhTX-3.4.3 (P-7479) is a synthetic analog of a component of a wasp venom (Δ-philanthotoxin) that is a potent channel blocker of ligand-gated ion channels, including glutamate receptors and nicotinic AChRs.[93-97]

Argiotoxin$_{636}$

Argiotoxin$_{636}$ (argiopin, A-98401) potentiates the response of the quisqualate-sensitive receptor of locust muscle to L-glutamate [98,99] and is an open-channel blocker of NMDA-induced ion currents in mammalian neurons.[100-102] Argiotoxin$_{636}$ also acts on P-type Ca^{2+} channels (see Section 18.4).

Probes for Other Receptors

Probes for GABA Receptors

An extensive selection of probes for γ-aminobutyric acid (GABA) receptors can be found with our Cl^- channel and carrier probes in Section 18.5. For information on our three caged GABA derivatives (A-2506, A-7110, A-7125), see Table 18.4 in Section 18.3.

Ligands for Studying Receptor-Mediated Endocytosis

Molecular Probes offers a diverse array of fluorescent derivatives of low-density lipoprotein (LDL), transferrin, epidermal growth factor (EGF) and chemotactic peptides, which are all transported into the cell by receptor-mediated endocytosis. These probes are discussed in Chapter 17, along with other probes for following endocytosis and exocytosis.

Histamine Probe for Histamine Receptors

Molecular Probes' histamine fluorescein (H-3409) is a useful fluorescent conjugate for both confocal laser scanning microscopy and flow cytometry applications because of its strong absorption at 488 nm, which corresponds to a spectral line of the argon-ion laser. Using this probe in conjunction with fluorescence microscopy, researchers have investigated the location of cell-surface H_2 histamine receptors on neutrophils, as well as their redistribution during locomotion.[103]

Dexamethasone Probe for Glucocorticoid Receptors

Dexamethasone is a potent glucocorticoid analog that is known to induce fibrinogen gene transcription in hepatocytes.[104] Molecular Probes' fluorescein dexamethasone (D-1383) should be useful for following the internalization of steroid receptors in live cells by flow cytometry and fluorescence microscopy.[66]

1. Meth Neurosci 8, 67 (1992); **2.** Neuron 5, 773 (1990); **3.** Proc Natl Acad Sci USA 73, 4594 (1976); **4.** J Physiol 237, 385 (1974); **5.** J Cell Biol 126, 901 (1994); **6.** J Cell Biol 120, 197 (1993); **7.** J Cell Biol 120, 185 (1993); **8.** Devel Dynamics 201, 29 (1994); **9.** J Cell Biol 128, 626 (1995); **10.** J Neurosci 15, 520 (1995); **11.** J Cell Biol 124, 139 (1994); **12.** J Cell Biol 131, 441 (1995); **13.** J Biol Chem 268, 25108 (1993); **14.** Proc Natl Acad Sci USA 77, 4823 (1980); **15.** J Cell Biol 124, 661 (1994); **16.** J Immunol Meth 107, 197 (1988); **17.** Biochemistry 18, 1875 (1979); **18.** Toxicon 29, 503 (1991); **19.** Neuron 8, 353 (1992); **20.** J Toxicol 13, 275 (1994); **21.** J Protein Chem 6, 227 (1987); **22.** Brain Res 211, 107 (1981); **23.** Proc Natl Acad Sci USA 77, 2400 (1980); **24.** Biochem Biophys Res Comm 88, 950 (1979); **25.** FEBS Lett 68, 1 (1976); **26.** Biochem J 141, 389 (1974); **27.** J Toxicol, Toxin Rev 13, 275 (1994); **28.** J Biol Chem 263, 2802 (1988); **29.** J Biol Chem 257, 5632 (1982); **30.** J Neurosci 14, 4147 (1994); **31.** J Biol Chem 260, 7927 (1985); **32.** J Biol Chem 258, 13575 (1983); **33.** Gen Physiol Biophys 11, 241 (1992);

34. Biochem Soc Trans 13, 703 (1985); **35.** Biochem Biophys Res Comm 106, 1331 (1982); **36.** Biochemistry 19, 3814 (1980); **37.** Biochem J 161, 229 (1977); **38.** Chem Nat Comp 25, 729 (1989); **39.** Cell 83, 503 (1995); **40.** J Biol Chem 270, 20391 (1995); **41.** J Biol Chem 270, 19694 (1995); **42.** J Biol Chem 269, 11233 (1994); **43.** J Physiol (Paris) 79, 232 (1984); **44.** Cervenansky, C. *et al.* in *Snake Toxins*, A.L. Harvey, Ed., Pergammon Press (1991) pp. 303–321; **45.** J Pharmacol Exp Ther 265, 1474 (1993); **46.** Electroencephalography Clin Neurophysiol 82, 445 (1992); **47.** Dokl Akad Nauk SSSR 267, 469 (1982); **48.** Cardiology 73, 164 (1986); **49.** Life Sci 27, 1525 (1980); **50.** Soc Neurosci Abstr 20, 699, abstract #297.1 (1994); **51.** FEBS Lett 386, 141 (1996); **52.** Ann Rev Physiol 52, 275 (1994); **53.** J Biol Chem 258, 3496 (1983); **54.** Biochemistry 33, 9126 (1994); **55.** *Principles of Neuroscience, Third Edition*, E.R. Kandel, J.H. Schwartz, and T.M. Jessell, Eds., Elsevier (1991); **56.** Trends Neurosci 16, 233 (1993); **57.** J Biol Chem 261, 4071 (1986); **58.** Neuron 1, 27 (1988); **59.** J Physiol 392, 397 (1987); **60.** J Comp Neurol 331, 149 (1993); **61.** Brain Res 547, 208 (1991); **62.** Brain Res 547, 199 (1991); **63.** J Neurochem 52, 1641 (1989); **64.** Proc Natl Acad Sci USA 86, 8570 (1989); **65.** J Neurochem 65, 691 (1995); **66.** Biol Chem Hoppe-Seyler 376, 483 (1995); **67.** Neurochem Int'l 27, 6 (1995); **68.** J Neurosci 13, 2551 (1993); **69.** Pharmacology 43, 329 (1991); **70.** Neurochem Int'l 15, 323 (1989); **71.** Tetrahedron Lett 20, 2261 (1966); **72.** Pharmaceutical Res 3, 56 (1986); **73.** Pharmaceutical Res 6, 266 (1985); **74.** Life Sci 33, 423 (1983); **75.** Naloxone fluorescein and naltrexone fluorescein are licensed to Molecular Probes under U.S. Patent No. 4,767,718; **76.** Neurosci Lett 118, 241 (1990); **77.** J Biol Chem 268, 17173 (1993); **78.** Proc Natl Acad Sci USA 90, 7819 (1993); **79.** J Neurochem 59, 1516 (1992); **80.** J Biol Chem 265, 6025 (1990); **81.** Neuron 12, 529 (1994); **82.** Nature 354, 31 (1991); **83.** J Neurosci 9, 3040 (1989); **84.** Nature 346, 565 (1990); **85.** Nature 325, 529 (1987); **86.** Nature 321, 519 (1986); **87.** Nature 307, 462 (1984); **88.** Mol Pharmacol 45, 324 (1994); **89.** Neurosci Lett 152, 13 (1993); **90.** Science 256, 1217 (1992); **91.** Usherwood, P.N.R. in *Excitatory Amino Acids*, B.S. Meldrum, Ed., Raven Press (1991), pp. 379–395; **92.** Toxicon 28, 1333 (1990); **93.** Biochem Soc Trans 22, 888 (1994); **94.** Neurosci Lett 114, 51 (1990); **95.** J Pharmacol Exp Ther 251, 156 (1989); **96.** J Pharmacol Exp Ther 249, 123 (1989); **97.** Proc Natl Acad Sci USA 85, 4910 (1988); **98.** J Physiol 419, 569 (1989); **99.** Brain Res 459, 312 (1988); **100.** NeuroReport 6, 1037 (1995); **101.** Neurosci Lett 132, 187 (1991); **102.** Br J Pharmacol 97, 1315 (1989); **103.** Proc Natl Acad Sci USA 83, 4332 (1986); **104.** J Cell Biol 105, 1067 (1987).

18.2 Data Table *Probes for Neurotransmitter Receptors*

Cat #	Structure	MW	Storage	Soluble	Abs	{ε × 10^{-3}}	Em	Solvent	Notes
A-77	✓	388	F,D,L	MeOH, DMF	363	{6.9}	496	H_2O	
A-3500	✓	322	F	DMF	292	{2.4}	none	MeOH	
A-96206	T	466	F,DD	DMSO	<300		none		
A-98401	T	637	FF,D,A	H_2O	<300		none		
B-1196	C	~8400	FF,D	H_2O	<300		none		<1>
B-1601	C	7984	FF	H_2O	<300		<2>		
B-3420	✓	577	F,D,L	DMSO	504	{77}	511	MeOH	
B-3424	✓	789	F,D,L	DMSO	504	{89}	511	MeOH	
B-3460	✓	599	F,D,L	DMF, DMSO	504	{79}	511	MeOH	
B-3504	C	~8300	FF,D,L	H_2O	504	{83}	513	H_2O	<1, 3>
B-7433	✓	563	F,D,L	DMSO, EtOH	504	{77}	511	MeOH	
B-7434	✓	617	F,D,L	DMSO, EtOH	560	{76}	569	MeOH	
B-7436	✓	784	F,D,L	DMSO, EtOH	504	{77}	511	MeOH	
B-7437	✓	801	F,D,L	DMSO, EtOH	560	{75}	569	MeOH	
B-7467	✓	696	F,D,L	DMSO	504	{80}	511	MeOH	
B-7470	✓	651	F,D,L	DMSO	503	{65}	516	MeOH	
B-7477	✓	404	F,D	H_2O	<300		none		
B-7483	C	~8700	FF,D,L	H_2O	523	{100}	542	pH 8	<1, 3>
B-7488	C	~8500	FF,D,L	H_2O	512	{85}	530	pH 8	<1, 3>
B-7489	C	~8600	FF,D,L	H_2O	593	{95}	613	H_2O	<1, 3>
B-96056	T	672	F,D	EtOH	<300		none		
B-96099	T	493	D	EtOH	<300		none		
B-99021	T	242	F,DD	EtOH	<300		none		
C-6904	C	~2600	FF,D,L	H_2O	504	{83}	513	H_2O	<1, 4>
C-6906	C	~2700		H_2O	554	{85}	575	H_2O	<1, 4>
C-6909	C	2263	FF,D	H_2O	<300		none		
C-7484	C	~7800	FF,D	H_2O	<300		<2>		
C-96017	T	450	F,D	EtOH	<300		none		
C-96032	T	448	D	EtOH	<300		none		
C-96033	T	448	D	EtOH	<300		none		
C-98102	C	~6800	FF,D	H_2O	<300		<2>		
D-1383	✓	841	L	pH>6, DMF	494	{76}	519	pH 9	
D-96009	T	245*	D	EtOH	<300		none		
D-96019	T	468	D	DMSO	<300		none		
D-96052	T	480	D	DMSO	<300		none		
D-99022	T	417	D	H_2O	<300		none		
D-99023	T	350*	D	EtOH	<300		none		
D-99024	T	437	D	H_2O	<300		none		
D-99025	T	386	D	H_2O	<300		none		
D-99027	T	312	D	EtOH	<300		none		
D-99028	T	330	D	H_2O	<300		none		
E-98110	C	~6829	FF,D	H_2O	<300		<2>		
E-98111	C	~6852	FF,D	H_2O	<300		<2>		
E-99019	T	315	F,DD	EtOH	<300		none		
F-1176	C	~8400	FF,D,L	H_2O	494	{84}	518	pH 8	<1, 3>
F-3421	✓	726	D,L	DMF	495	{76}	518	pH 9	
F-3425	✓	938	F,D,L	DMF	496	{73}	522	pH 9	
F-3471	✓	714	L	DMSO†	498	{85}	519	MeOH‡	

Cat #	Structure	MW	Storage	Soluble	Abs	{ε × 10⁻³}	Em	Solvent	Notes
F-96026	T	343	F,D,L	H_2O	<300		none		
F-98107	C	~6735	FF,D	H_2O	<300		none		
G-96028	T	368	D	H_2O	<300		none		
H-3409	✓	544	D,L	H_2O	493	{76}	516	pH 8	
I-96008	T	430	D	EtOH	<300		none		
I-96050	T	428	NC	EtOH	<300		none		
K-96034	T	378	D	DMSO	<300		none		
L-96221	T	769	F,DD	MeOH	<300		none		
L-98112	C	~7229	FF,D	H_2O	<300		<2>		
M-96201	T	783	F,DD	MeOH	<300		none		
M-96204	T	497	D	H_2O§	<300		none		
N-1384	✓	791	D,L	EtOH, DMF	492	{79}	516	pH 9	
N-1385	✓	805	D,L	EtOH, DMF	493	{87}	516	pH 9	
N-96037	T	811	F,D	EtOH	<300		none		
P-7435	✓	420	F,D,L	DMSO, EtOH	<300		none		
P-7438	✓	424	D	DMSO, EtOH	<300		none		
P-7479	✓	545	FF,DD	H_2O	<300		none		
P-96043	T	188	D	EtOH	<300		none		
P-99020	T	228	F,DD	EtOH	<300		none		
S-96203	T	364	D,A	H_2O	<300		none		
T-1175	C	~8500	FF,D,L	H_2O	553	{85}	577	H_2O	<1, 3>
T-3422	✓	819	D,L	DMSO†	541	{85}	568	MeOH	
T-3423	✓	891	D,L	DMSO, EtOH	582	{80}	606	MeOH	
T-3426	✓	958	F,D,L	DMSO, EtOH	541	{84}	567	MeOH	
T-3427	✓	1103	F,D,L	DMSO, EtOH	582	{98}	600	MeOH	
T-96016	T	339	D,A	DMSO	<300		none		
T-96038	T	392	D,A	H_2O	<300		none		

For definitions of the contents of this data table, see “How to Use the Handbook” on page vi.
Structure: Chemical structure drawing: ✓ = shown in this section; C = not shown due to complexity; T = available on request from our Technical Assistance Department and also available through our Web site (http://www.probes.com).
MW: ~ indicates an approximate value.

- * MW is for the hydrated form of this product.
- † This product is packaged as a solution in the solvent indicated in “Soluble.”
- ‡ Spectra of this compound are in methanol containing a trace of KOH.
- § Requires heating to dissolve in H_2O.
- <1> α-Bungarotoxin and conantokin-G conjugates have approximately 1 label per peptide.
- <2> This peptide exhibits intrinsic tryptophan fluorescence (Em ~350 nm) when excited at <300 nm.
- <3> The value of {ε} listed for this α-bungarotoxin conjugate is for the labeling dye in free solution. Use of this value for the conjugate assumes a 1:1 dye:peptide labeling ratio and no change of ε due to dye–peptide interactions.
- <4> The value of {ε} listed for this conantokin-G conjugate is for the labeling dye in free solution. Use of this value for the conjugate assumes a 1:1 dye:peptide labeling ratio and no change of ε due to dye–peptide interactions.

18.2 Structures *Probes for Neurotransmitter Receptors*

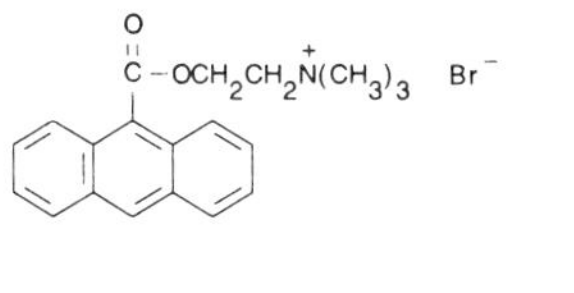

A-77

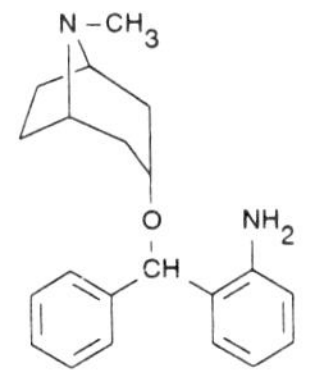

A-3500

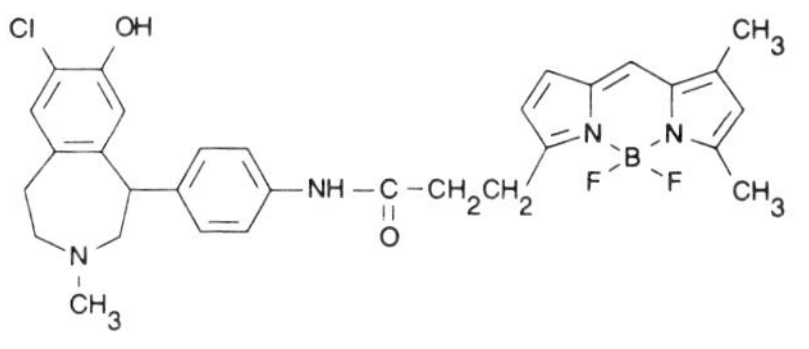

B-3420

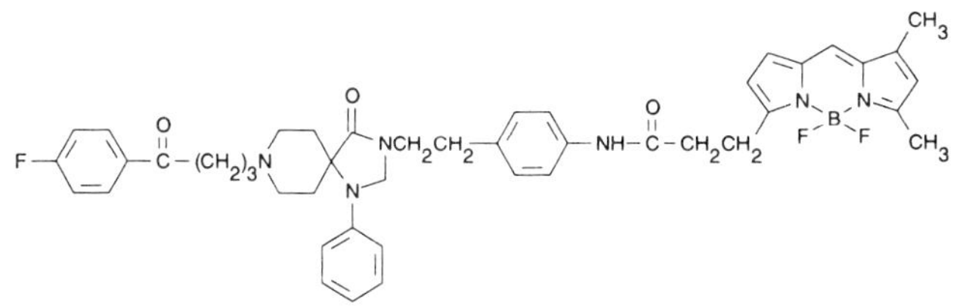

B-3424

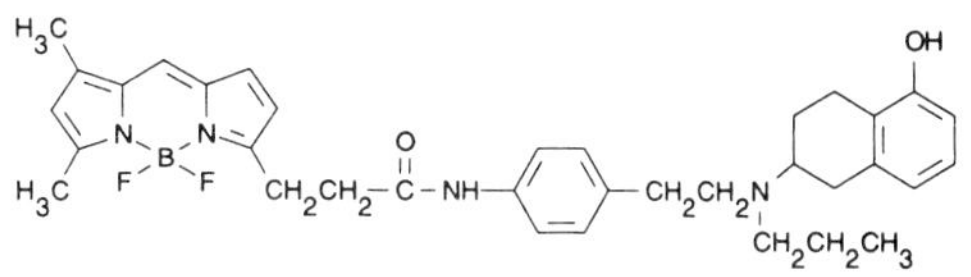

B-3460

B-7433, B-7434 see P-7435

B-7436, B-7437 see P-7438

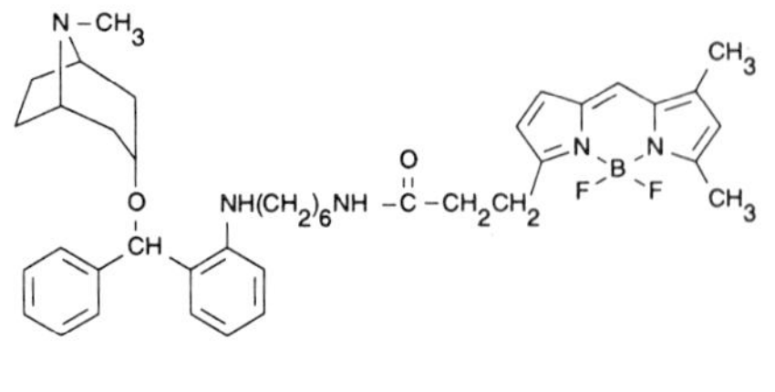

B-7467

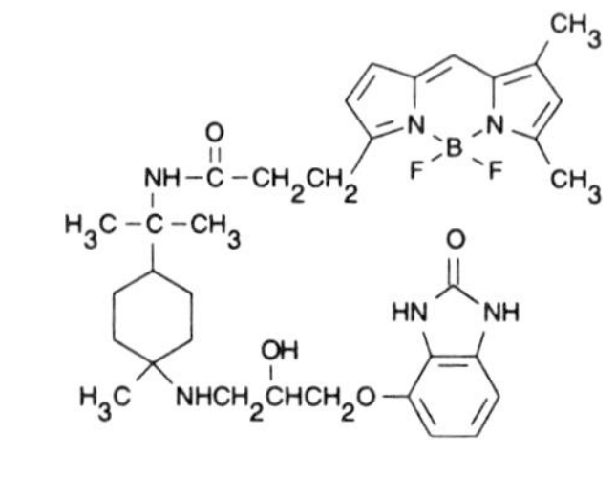

B-7470

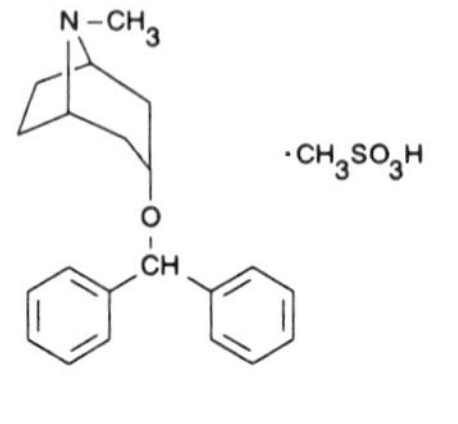

$\cdot CH_3SO_3H$

B-7477

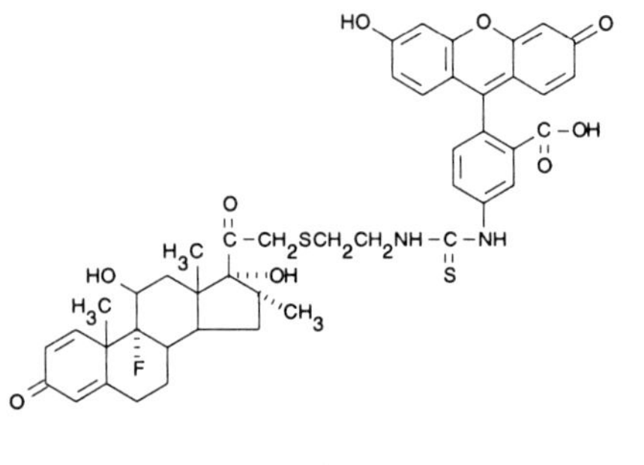

D-1383

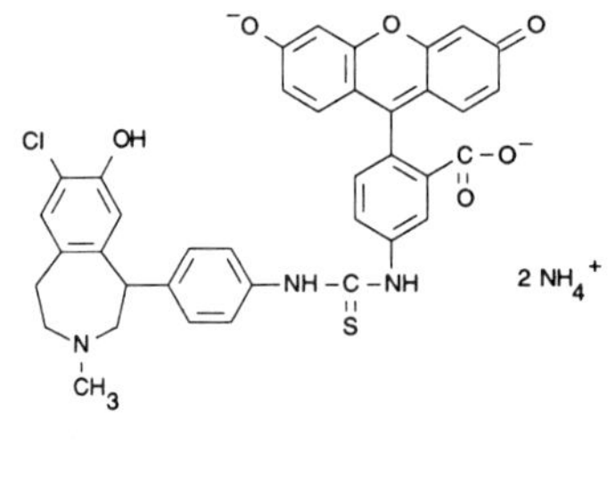

$2\ NH_4^+$

F-3421

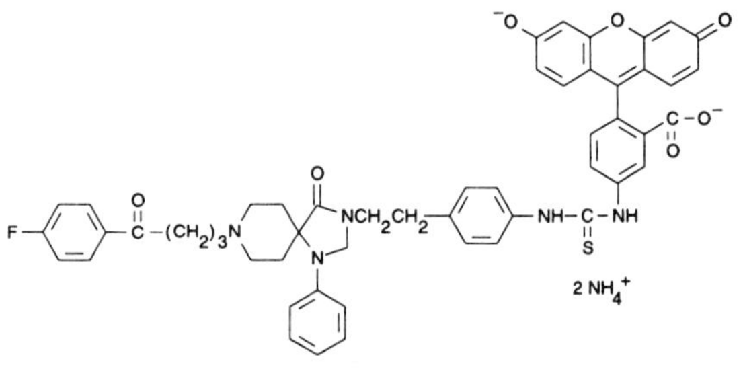

$2\ NH_4^+$

F-3425

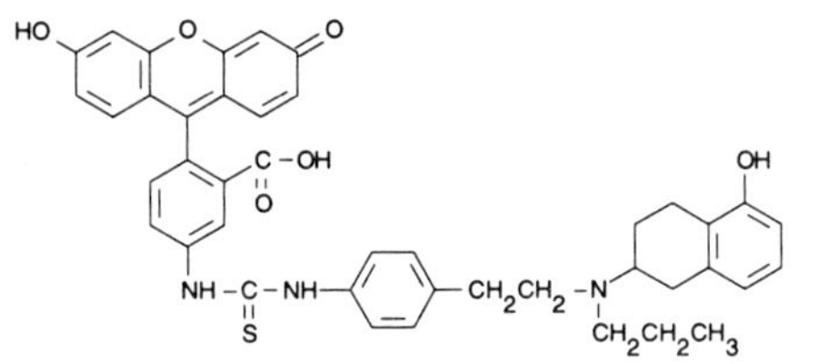

F-3471

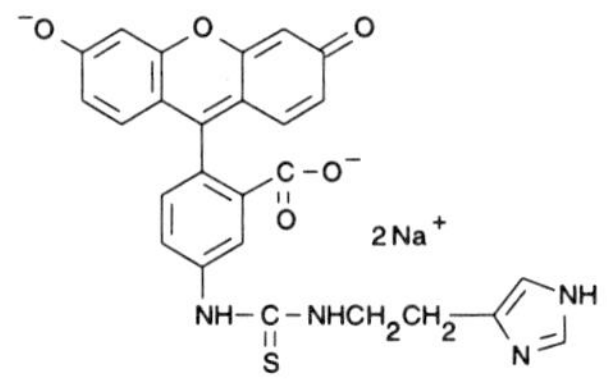

$2Na^+$

H-3409

$\cdot CH_3COOH$

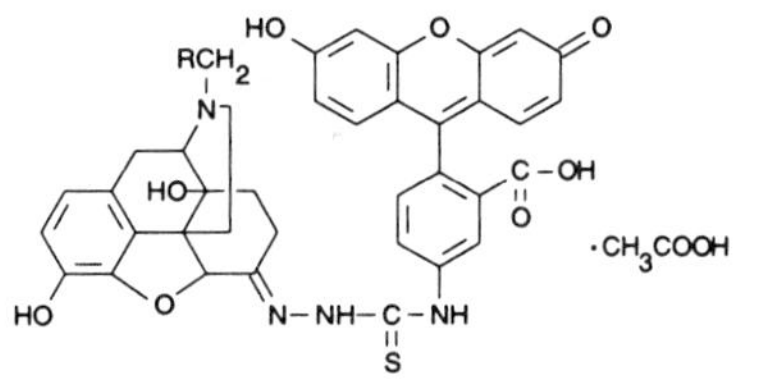

N-1384 R = $-CH=CH_2$
N-1385 R = $-CH-CH_2$ / CH_2 (cyclopropyl)

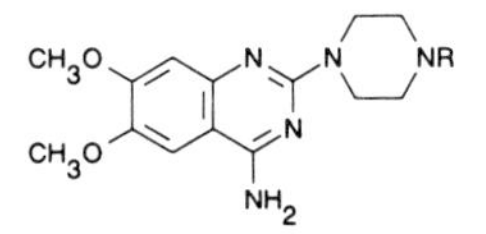

P-7435 R = $-C(=O)-$(2-furyl)
B-7433 R = BODIPY FL
B-7434 R = BODIPY 558 / 568

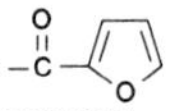

BODIPY FL =

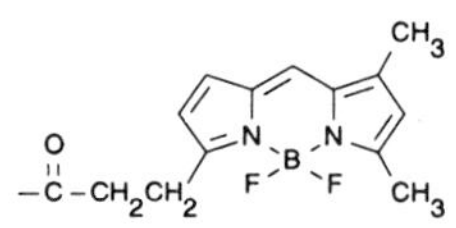

BODIPY 558 / 568 =

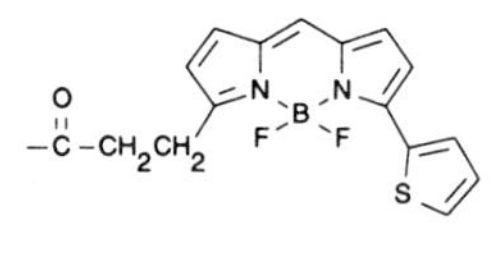

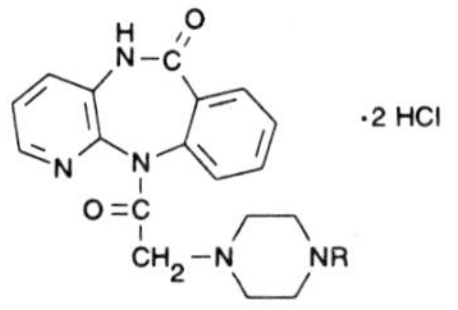

$\cdot 2\ HCl$

P-7438 R = $-CH_3$
B-7436 R = $-(CH_2)_6NH-$BODIPY FL
B-7437 R = $-(CH_2)_6NH-$BODIPY 558 / 568

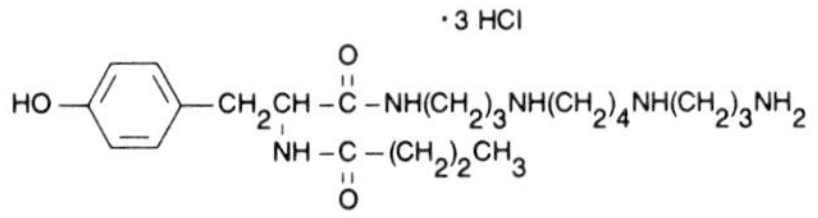

$\cdot 3\ HCl$

P-7479

$\cdot HCl$

Cl^-

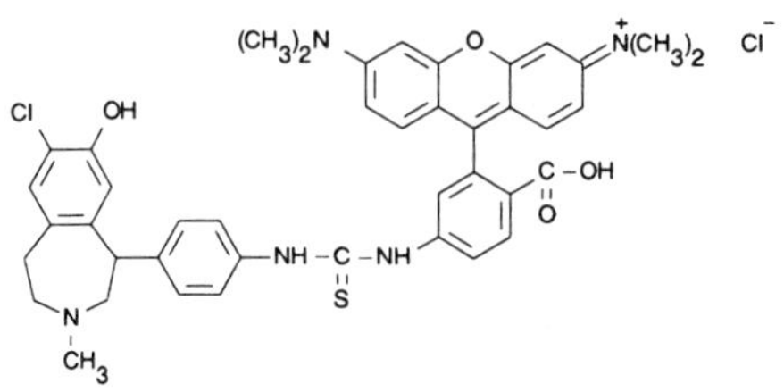

T-3422

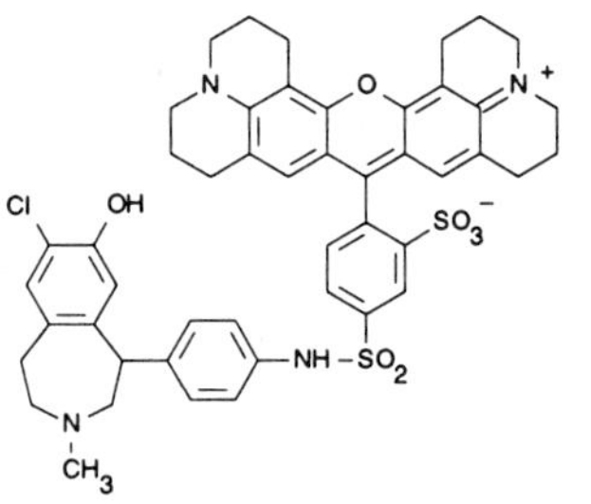

T-3423

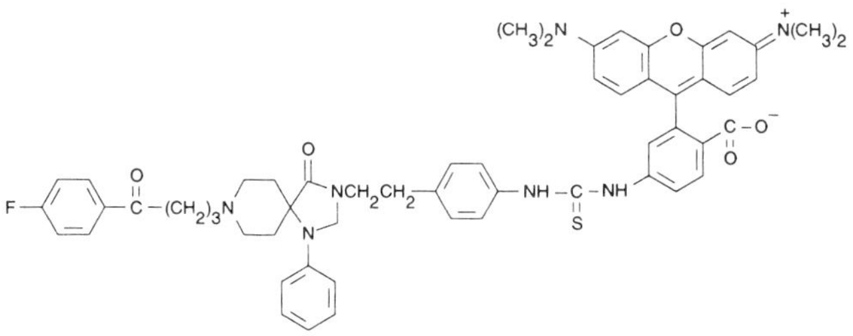

T-3426

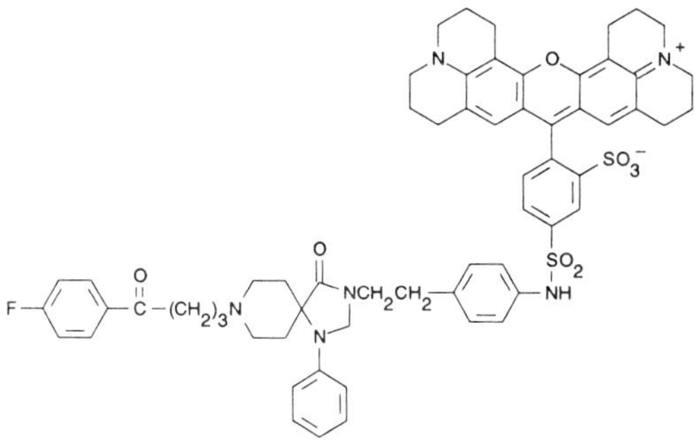

T-3427

18.2 Price List *Probes for Neurotransmitter Receptors*

Cat #		Product Name	Unit Size	Price Per Unit ($) 1–4 Units	5–24 Units
A-96206	**New**	14-acetylvirescenine	10 mg	75.00	60.00
A-3500		3-(2′-aminobenzhydryloxy)tropane (ABT)	25 mg	95.00	76.00
A-77		9-anthroylcholine bromide	100 mg	48.00	38.40
A-6473	**New**	anti-rat NMDA receptor, subunit 2A, rabbit IgG fraction *affinity purified*	10 µg	240.00	192.00
A-6467	**New**	anti-rat NMDA receptor, subunit 2A, rabbit serum	50 µL	196.00	156.80
A-6474	**New**	anti-rat NMDA receptor, subunit 2B, rabbit IgG fraction *affinity purified*	10 µg	240.00	192.00
A-6468	**New**	anti-rat NMDA receptor, subunit 2B, rabbit serum	50 µL	196.00	156.80
A-6475	**New**	anti-rat NMDA receptor, subunit 2C, rabbit IgG fraction *affinity purified*	10 µg	240.00	192.00
A-6469	**New**	anti-rat NMDA receptor, subunit 2C, rabbit serum	50 µL	176.00	140.80
A-98401	**New**	argiotoxin 636 (argiopin) *from *Argiope lobata* spider venom*	50 µg	295.00	236.00
B-96056	**New**	β-belladonnine bis-(chloroethylate)	100 mg	45.00	36.00
B-7477	**New**	benztropine methanesulfonate (benztropine mesylate)	100 mg	10.00	8.00
B-7437	**New**	BODIPY® 558/568 pirenzepine, dihydrochloride	100 µg	95.00	76.00
B-7434	**New**	BODIPY® 558/568 prazosin	100 µg	95.00	76.00
B-7467	**New**	BODIPY® FL ABT	100 µg	48.00	38.40
B-7470	**New**	BODIPY® FL (±) CGP 12177	100 µg	48.00	38.40
B-3424		BODIPY® FL NAPS	1 mg	68.00	54.40
B-7436	**New**	BODIPY® FL pirenzepine, dihydrochloride	100 µg	95.00	76.00
B-3460		BODIPY® FL PPHT	200 µg	68.00	54.40
B-7433	**New**	BODIPY® FL prazosin	100 µg	95.00	76.00
B-3420		BODIPY® FL SCH 23390	1 mg	68.00	54.40
B-96099	**New**	3β-bromoimperialine	1 mg	325.00	260.00
B-1601		α-bungarotoxin *from *Bungarus multicinctus**	1 mg	37.00	29.60
B-1196		α-bungarotoxin, biotin-XX conjugate	500 µg	175.00	140.00
B-3504		α-bungarotoxin, BODIPY® FL conjugate (BODIPY® FL α-bungarotoxin)	500 µg	175.00	140.00
B-7483	**New**	α-bungarotoxin, eosin conjugate	500 µg	175.00	140.00
B-7488	**New**	α-bungarotoxin, Oregon Green™ 514 conjugate	500 µg	195.00	156.00
B-7489	**New**	α-bungarotoxin, Texas Red®-X conjugate	500 µg	195.00	156.00
B-99021	**New**	2-butoxy-2,3-dihydro-1,4,2-benzodioxaphosphorin-2-oxide (bomin-3)	500 µg	108.00	86.40
C-98102	**New**	cardiotoxin *from *Naja nigricollis* snake venom*	1 mg	128.00	102.40
C-96032	**New**	3α-chloroimperialine	1 mg	160.00	128.00
C-96033	**New**	3β-chloroimperialine	1 mg	135.00	108.00
C-7484	**New**	α-cobrotoxin *from *Naja naja* sp.*	1 mg	21.00	16.80
C-6909	**New**	conantokin-G *purity > 95%*	25 µg	55.00	44.00
C-6904	**New**	conantokin-G, BODIPY® FL conjugate	5 µg	175.00	140.00
C-6906	**New**	conantokin-G, tetramethylrhodamine conjugate	5 µg	175.00	140.00
C-96017	**New**	condelphine	20 mg	108.00	86.40
D-96052	**New**	delcorine	20 mg	95.00	76.00
D-96019	**New**	delsoline (acomonine)	20 mg	95.00	76.00
D-96009	**New**	desoxypeganine, hydrochloride, dihydrate	20 mg	108.00	86.40
D-1383		dexamethasone fluorescein	5 mg	48.00	38.40
D-99022	**New**	1,1-dimethyl-3-(di-2-thienylmethylene)-piperidinium iodide (LK-66)	2 mg	128.00	102.40
D-99025	**New**	1,1-dimethyl-3-(di-2-thienylmethylene)-piperidinium methanesulfonate (LK-66 M)	2 mg	128.00	102.40
D-99024	**New**	1-(3,3-diphenyl-3-hydroxypropyl)-1-methylpiperidinium iodide (LK-7)	2 mg	95.00	76.00
D-99028	**New**	3-diphenylmethoxyquinuclidine, hydrochloride (LK-252)	2 mg	95.00	76.00
D-99027	**New**	3-diphenylmethylenequinuclidine, hydrochloride (LK-190)	2 mg	95.00	76.00

Cat #		Product Name	Unit Size	Price Per Unit ($) 1–4 Units	5–24 Units
D-99023	**New**	1,1-diphenyl-3-(1-piperidinyl)-1-propene, hydrochloride, dihydrate (LK-6)	2 mg	95.00	76.00
E-98110	**New**	erabutoxin a *from *Laticauda semifasciata* snake venom*	1 mg	185.00	148.00
E-98111	**New**	erabutoxin b *from *Laticauda semifasciata* snake venom*	1 mg	185.00	148.00
E-99019	**New**	2-ethoxy-2,3-dihydro-1,4,2-benzodioxaphosphorin-2-oxide (bomin-1)	500 µg	108.00	86.40
F-98107	**New**	fasciculin 2 *from *Dendroaspis angusticeps* snake venom*	100 µg	160.00	128.00
F-1176		fluorescein α-bungarotoxin (α-bungarotoxin, fluorescein conjugate)	500 µg	175.00	140.00
F-3425		fluorescein NAPS	1 mg	68.00	54.40
F-3471		fluorescein PPHT *1 mg/mL solution in dry DMSO*	200 µL	68.00	54.40
F-3421		fluorescein SCH 23390	1 mg	68.00	54.40
F-96026	**New**	fluorocurarine chloride (C-curarine III; metvine)	20 mg	128.00	102.40
G-96028	**New**	galanthamine, hydrobromide	50 mg	95.00	76.00
H-3409		histamine fluorescein, disodium salt	5 mg	68.00	54.40
I-96008	**New**	imperialine (raddeamine; sipeimine)	5 mg	135.00	108.00
I-96050	**New**	imperialone	1 mg	135.00	108.00
K-96034	**New**	karakoline	20 mg	75.00	60.00
L-98112	**New**	laticauda semifasciata III (LS III) *from *Laticauda semifasciata* snake venom*	1 mg	185.00	148.00
L-96221	**New**	lycaconitine perchlorate	10 mg	128.00	102.40
M-96204	**New**	*o*-methylisocorydine, methiodide	5 mg	95.00	76.00
M-96201	**New**	methyllycaconitine perchlorate (MLA)	5 mg	35.00	28.00
N-1384		naloxone fluorescein	5 mg	68.00	54.40
N-1385		naltrexone fluorescein	5 mg	68.00	54.40
N-96037	**New**	nudicauline perchlorate	5 mg	128.00	102.40
P-96043	**New**	peganole	20 mg	160.00	128.00
P-7479	**New**	philanthotoxin, trihydrochloride (PhTX-3.4.3)	100 µg	98.00	78.40
P-7438	**New**	pirenzepine, dihydrochloride	100 mg	10.00	8.00
P-7435	**New**	prazosin, hydrochloride	10 mg	12.00	9.60
P-99020	**New**	2-propoxy-2,3-dihydro-1,4,2-benzodioxaphosphorin-2-oxide (bomin-2)	500 µg	108.00	86.40
S-96203	**New**	scoulerine, hydrochloride	5 mg	108.00	86.40
T-96016	**New**	DL-tetrahydroberberine (canadine)	20 mg	128.00	102.40
T-96038	**New**	DL-tetrahydropalmatine, hydrochloride	5 mg	105.00	84.00
T-1175		tetramethylrhodamine α-bungarotoxin (α-bungarotoxin, tetramethylrhodamine conjugate)	500 µg	195.00	156.00
T-3426		tetramethylrhodamine NAPS	1 mg	68.00	54.40
T-3422		tetramethylrhodamine SCH 23390 *1 mg/mL solution in dry DMSO*	1 mL	68.00	54.40
T-3427		Texas Red® NAPS	1 mg	68.00	54.40
T-3423		Texas Red® SCH 23390	1 mg	68.00	54.40

18.3 Caged Neurotransmitters and Caged Drugs

When illuminated with UV light, caged neurotransmitters and drugs are converted into biologically active products that rapidly initiate or block neurotransmitter action.[1-3] Thus, these caged probes provide a means of controlling the release — both spatially and temporally — of agonist or antagonists for kinetic studies of receptor binding or channel opening. For example, *N*-(CNB-caged) carbamylcholine (C-1791) has been employed to investigate the effect of procaine on the kinetics of nicotinic acetylcholine channel opening and closing,[4] and γ-(CNB-caged) kainic acid (K-7115) has been used to study the kinetics and mechanism of activation of the glutamate receptor subtype on cultured neurons.[5]

The different caging groups confer special properties on these photoactivatable probes (see Table 19.1 in Chapter 19). We synthesize several different versions of our caged γ-aminobutyric acid (GABA) and caged L-glutamic acid probes. Our new *O*-(desyl-caged) GABA (A-7125) and γ-(desyl-caged) L-glutamic acid (G-7120) are uncaged in submicroseconds (~10^7 sec^{-1}) with a photolysis quantum yield of about 0.3, thus providing extremely rapid and efficient production of active neurotransmitter or drug upon illumination.[6,7] Our *O*-(CNB-caged) GABA (A-7110) and γ-(CNB-caged) L-glutamic acid (G-7055), which also exhibit fast uncaging rates and high photolysis quantum yields, have been used to investigate the activation kinetics of GABA receptors [8] and glutamate receptors,[9] respectively. Among our most recent additions is *N*-(CNB-caged) L-glutamic acid (C-7122), which does not hydrolyze in aqueous solution because it is caged on the amino group, enabling researchers to use very high concentrations without risk of light-independent glutamic acid production.[7,9] We have also introduced γ-(ANB-caged) L-glutamic acid (G-7121), a moderately lipophilic form of caged glutamate for stronger membrane association before photolysis.[7]

Table 18.4 summarizes our caged neurotransmitters and caged drugs, all of which are biologically inactive before photolysis. For further discussion of the properties and applications of caged probes, see Chapter 19.

We also offer RGA-30 (R-6922), which has been used as a phototransducer, eliciting action potentials in stained leech neurons in response to illumination.[10] Selective photostimulation of RGA-30–labeled neurons with a red He-Ne laser allows rapid identification of their synaptic connections.

Table 18.4 Caged neurotransmitters and caged drugs, and their biological activity after uncaging.

Cat #	Caged Probe	Biological Activity After Uncaging			References
		Agonist (+) Antagonist (–)	Natural (+) Not Natural (–)*	Receptor	
C-1791	*N*-(CNB-caged) carbamylcholine	+	–	Cholinergic	1–6
D-7057	*N*-(DMNB-caged) epinephrine	+	+	Adrenergic	
C-7083	*O*-(CNB-caged) phenylephrine	+	–	Adrenergic	7
D-7084	*N*-(DMNB-caged) phenylephrine	+	–	Adrenergic	
D-7058	*N*-(NVOC-caged) norepinephrine	+	+	Adrenergic	
N-7066	*N*-(NB-caged) isoproterenol	+	–	Adrenergic	8–10
N-7067	*N*-(NB-caged) propranolol	–	–	Adrenergic	
C-7124	*O*-(CNB-caged) dopamine	+	+	Dopaminergic	11
D-7064	*N*-(NVOC-caged) dopamine	+	+	Dopaminergic	
H-7080	*O*-(DMNB-caged) haloperidol	–	–	Dopaminergic	
A-7110	*O*-(CNB-caged) GABA	+	+	GABA	12
A-7125	*O*-(desyl-caged) GABA	+	+	GABA	13
A-2506	*O*-(DMNB-caged) GABA	+	+	GABA	14,15
G-7121	γ-(ANB-caged) L-glutamic acid	+	+	Glutamate	16
C-7122	*N*-(CNB-caged) L-glutamic acid	+	+	Glutamate	17
G-7055	γ-(CNB-caged) L-glutamic acid	+	+	Glutamate	17
G-7120	γ-(desyl-caged) L-glutamic acid	+	+	Glutamate	13,16
G-2504	α-(DMNB-caged) L-glutamic acid	+	I	Glutamate	18–20
A-2505	α-(DMNB-caged) L-aspartic acid	+	+	Glutamate	15
G-2507	*O*-(DMNB-caged) glycine	+	+	Glutamate/Glycine	15
K-7115	γ-(CNB-caged) kainic acid	+	–	Glutamate/Kainate	21
M-7114	β-(CNB-caged) NMDA	+	–	Glutamate/NMDA	22
M-7069	*N*-(NB-caged) NMDA	+	–	Glutamate/NMDA	
D-7059	*N*-(NVOC-caged) MK-801	–	–	Glutamate/NMDA	
D-7063	*N*-(NVOC-caged) serotonin	+	+	Serotonin/5-HT	

* Natural = naturally occurring neurotransmitters, for which there may be uptake mechanisms in some experimental systems; Not Natural = neurotransmitters that are not produced in biological systems.

1. J Neurosci Meth 54, 151 (1994); **2.** Proc Natl Acad Sci USA 91, 6629 (1994); **3.** Biochemistry 32, 3831 (1993); **4.** Biochemistry 32, 989 (1993); **5.** Biochemistry 31, 5507 (1992); **6.** Adv Exp Med Biol 287, 75 (1991); **7.** Biochemistry 32, 1338 (1993); **8.** J Gen Physiol 101, 337 (1996); **9.** Proc R Soc Lond B 263, 241 (1996); **10.** J Photochem Photobiol B 27, 123 (1995); **11.** J Neurosci Meth 67, 221 (1996); **12.** J Am Chem Soc 116, 8366 (1994); **13.** J Org Chem 61, 1228 (1996); **14.** Neuron 15, 755 (1995); **15.** J Org Chem 55, 1585 (1990); **16.** Soc Neurosci Abstr 21, 579, abstract #238.11 (1995); **17.** Proc Natl Acad Sci USA 91, 8752 (1994); **18.** J Neurosci Meth 54, 205 (1994); **19.** Science 265, 255 (1994); **20.** Proc Natl Acad Sci USA 90, 7661 (1993); **21.** Biochemistry 35, 2030 (1996); **22.** J Org Chem 60, 4260 (1995).

1. Ann Rev Physiol 55, 755 (1993); **2.** Corrie, J.E.T. and Trentham, D.R. in *Bioorganic Photochemistry, Volume 2*, H. Morrison, Ed., Wiley (1993) pp. 243–305; **3.** Kao, J.P.Y. and Adams, S.R. in *Optical Microscopy: Emerging Methods and Applications*, B. Herman and J.J. Lemasters, Eds., Academic Press (1993) pp. 27–85; **4.** Biochemistry 32, 3831 (1993); **5.** Biochemistry 35, 2030 (1996); **6.** J Org Chem 61, 1228 (1996); **7.** Soc Neurosci Abstr 21, 579, abstract #238.11 (1995); **8.** J Am Chem Soc 116, 8366 (1994); **9.** Proc Natl Acad Sci USA 91, 8752 (1994); **10.** Science 222, 1025 (1983).

18.3 Data Table *Caged Neurotransmitters and Caged Drugs*

Cat #	Structure	MW	Storage	Soluble	Abs	{ε × 10^{-3}}	Em	Solvent	Notes
A-2505	✓	365	F,D,LL	H_2O	347	{6.0}	none	H_2O	<1>
A-2506	✓	335	F,D,LL	H_2O	347	{5.4}	none	H_2O	<1>
A-7110	✓	396	F,D,LL	H_2O	262	{4.5}	none	pH 7	<1, 2>
A-7125	✓	411	F,D,LL	H_2O	252	{13}	none	pH 7	<1, 2>
C-1791	✓	439	F,D,LL	H_2O	264	{4.7}	none	H_2O	<1, 2>
C-7083	✓	346	F,D,LL	DMSO	266	{5.1}	none	pH 7	<1, 2>
C-7122	✓	326	F,D,LL	H_2O	266	{4.8}	none	pH 7	<1, 2>
C-7124	✓	446	F,D,LL	H_2O	275	{7.4}	none	pH 7	<1, 2>
D-7057	✓	378	F,D,LL	DMSO	340	{4.3}	none	MeOH	<1>
D-7058	✓	408	F,D,LL	DMSO	340	{6.3}	none	MeOH	<1>
D-7059	✓	460	F,D,LL	DMSO	340	{6.0}	none	MeOH	<1>
D-7063	✓	415	F,D,LL	DMSO	341	{6.4}	none	MeOH	<1>
D-7064	✓	392	F,D,LL	DMSO	340	{6.1}	none	MeOH	<1>
D-7084	✓	362	F,D,LL	DMSO	340	{4.6}	none	MeOH	<1>
G-2504	✓	379	F,D,LL	H_2O	346	{5.9}	none	H_2O	<1>
G-2507	✓	307	F,D,LL	H_2O	345	{5.7}	none	H_2O	<1>
G-7055	✓	440	F,D,LL	H_2O, DMSO	262	{5.1}	none	pH 7	<1, 2>
G-7120	✓	341	F,D,LL	MeOH	252	{12}	none	pH 7	<1, 2>
G-7121	✓	338	F,D,LL	H_2O	315	{8.2}	none	pH 7	<1>
H-7080	✓	615	F,D,LL	DMSO	335	{5.2}	none	MeOH	<1>
K-7115	✓	429	F,D,LL	H_2O	262	{2.6}	none	H_2O	<1, 2>
M-7069	✓	294	F,D,LL	H_2O	263	{3.7}	none	MeOH	<1, 2>
M-7114	✓	440	F,D,LL	H_2O	264	{5.1}	none	pH 7	<1, 2>
N-7066	✓	346	F,D,LL	DMSO	279	{4.8}	none	MeOH	<1, 2>
N-7067	✓	394	F,D,LL	DMSO	291	{7.8}	none	MeOH	<1, 2>
R-6922	S25.2	775	D,L	DMSO, EtOH	629	{83}	659	MeOH	

For definitions of the contents of this data table, see "How to Use the Handbook" on page vi.
Structure: Chemical structure drawing: ✓ = shown in this section; Sn.n = shown in Section number n.n.

<1> All photoactivatable probes are sensitive to light. They should be protected from illumination, except when photolysis is intended.
<2> This compound has weaker visible absorption at >300 nm but no discernible absorption peaks in this region.

18.3 Structures *Caged Neurotransmitters and Caged Drugs*

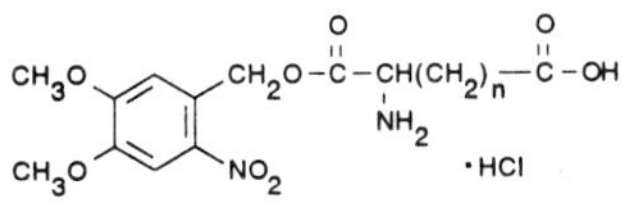

A-2505 n = 1
G-2504 n = 2

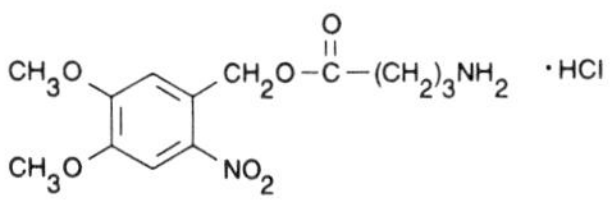

A-2506

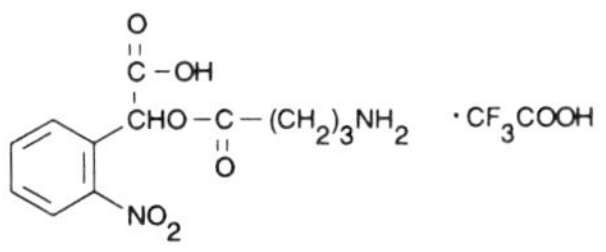

A-7110

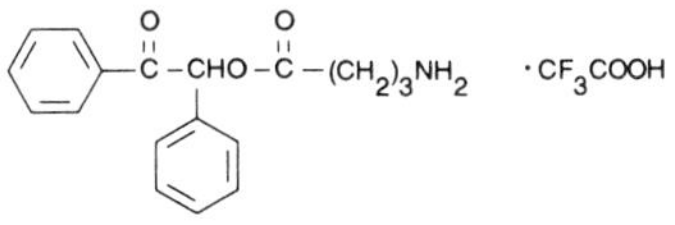

A-7125

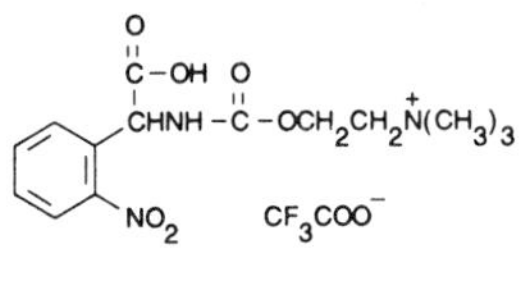

C-1791

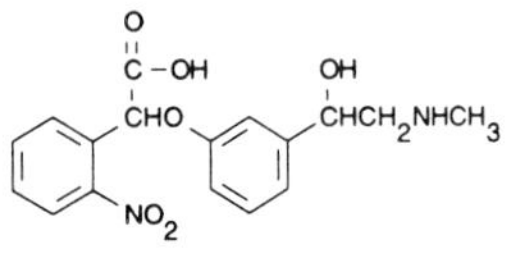

C-7083

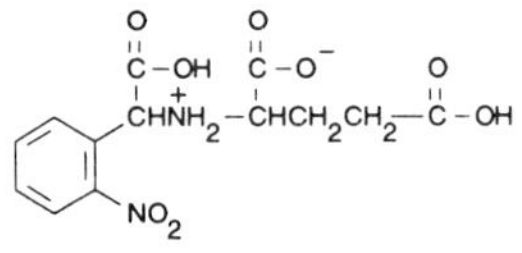

C-7122

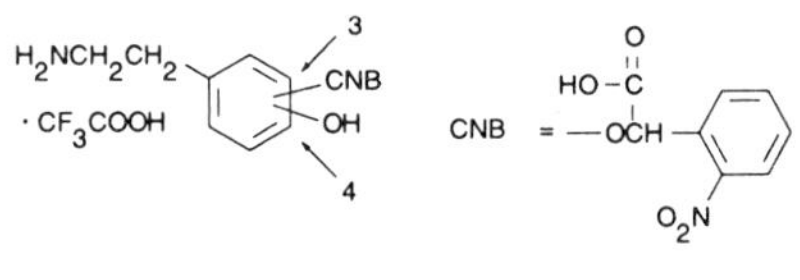

C-7124

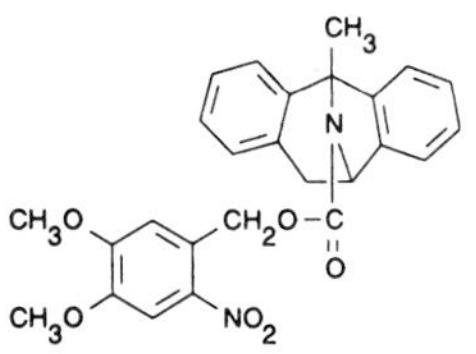

CNB =

D-7059

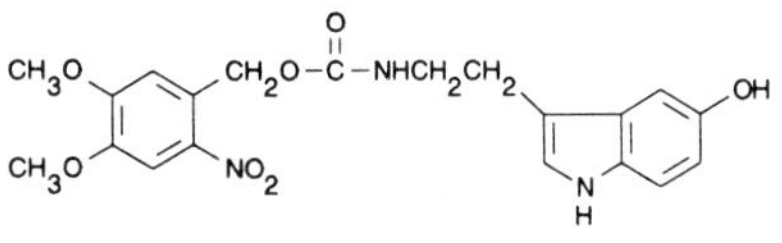

D-7063

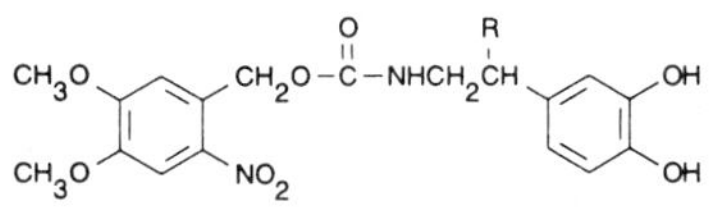

D-7064 R = H
D-7058 R = —OH

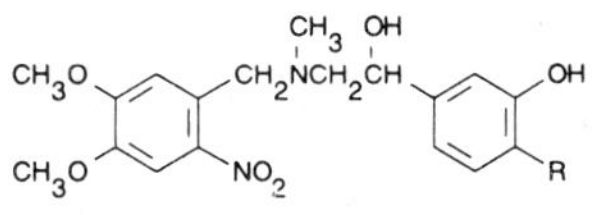

D-7084 R = H
D-7057 R = —OH

G-2504 see A-2505

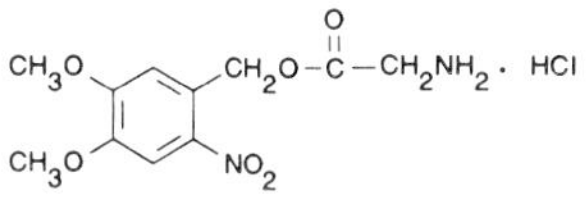

G-2507

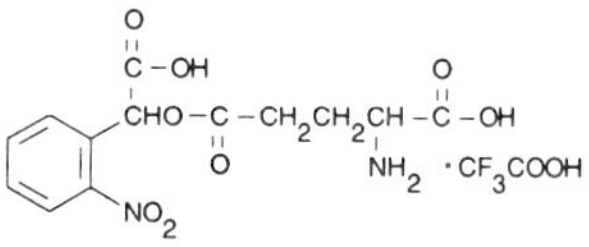

G-7055

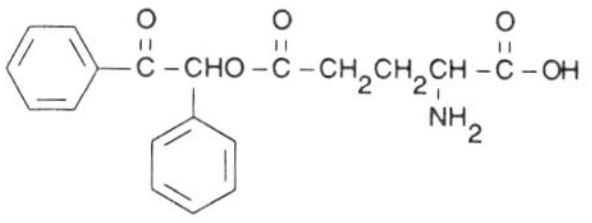

G-7120

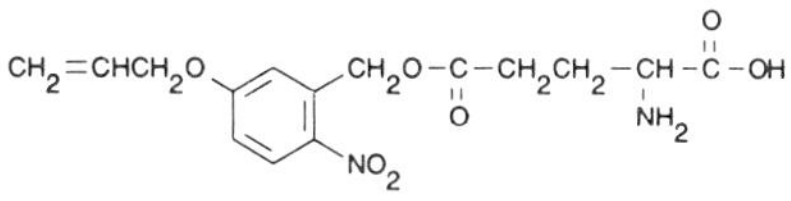

G-7121

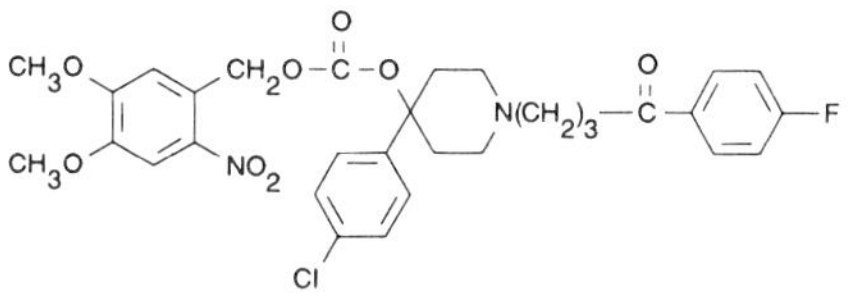

H-7080

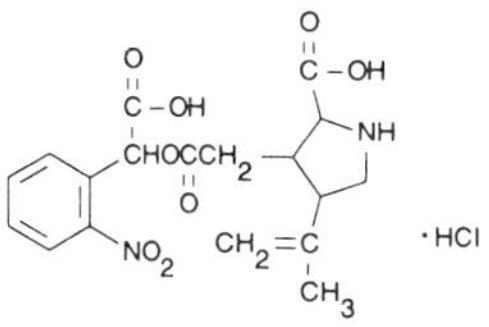

K-7115

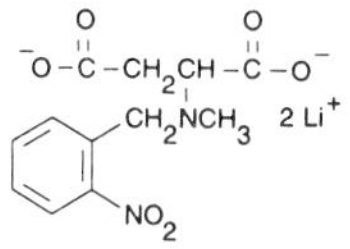

M-7069

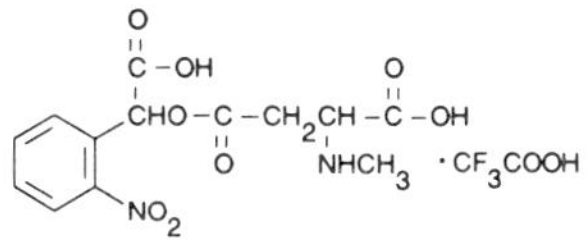

M-7114

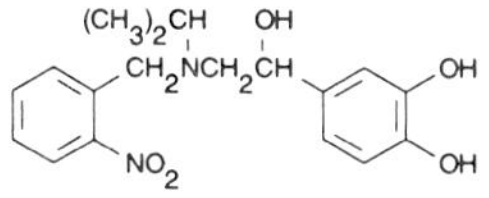

N-7066

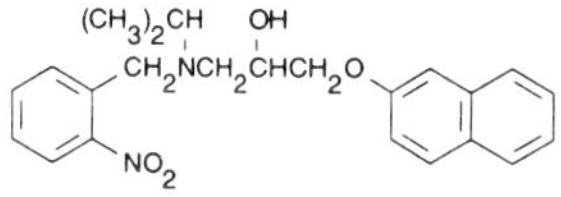

N-7067

R-6922 Section 25.2

18.3 Price List *Caged Neurotransmitters and Caged Drugs*

Cat #		Product Name	Unit Size	Price Per Unit ($) 1–4 Units	Price Per Unit ($) 5–24 Units
A-7110	**New**	γ-aminobutyric acid, α-carboxy-2-nitrobenzyl ester, trifluoroacetic acid salt (O-(CNB-caged) GABA)	5 mg	125.00	100.00
A-2506		γ-aminobutyric acid, 4,5-dimethoxy-2-nitrobenzyl ester, hydrochloride (O-(DMNB-caged) GABA)	5 mg	75.00	60.00
A-7125	**New**	γ-aminobutyric acid, (1,2-diphenyl-2-oxo)ethyl ester, trifluoroacetic acid salt (O-(desyl-caged) GABA)	5 mg	168.00	134.40
A-2505		L-aspartic acid, α-(4,5-dimethoxy-2-nitrobenzyl) ester, hydrochloride (α-(DMNB-caged) L-aspartic acid)	5 mg	75.00	60.00
C-1791		N-(α-carboxy-2-nitrobenzyl)carbamylcholine, trifluoroacetate (N-(CNB-caged) carbamylcholine)	5 mg	75.00	60.00
C-7122	**New**	N-(α-carboxy-2-nitrobenzyl)-L-glutamic acid (N-(CNB-caged) L-glutamic acid)	5 mg	168.00	134.40
C-7124	**New**	3-(and-4)-((α-carboxy-2-nitrobenzyl)oxy)-4-(and-3)-hydroxyphenethylamine, trifluoroacetic acid salt (O-(CNB-caged) dopamine) *mixed isomers*	1 mg	65.00	52.00
C-7083	**New**	O-(α-carboxy-2-nitrobenzyl)phenylephrine (O-(CNB-caged) phenylephrine)	5 mg	48.00	38.40
D-7057	**New**	N-(4,5-dimethoxy-2-nitrobenzyl)-L-epinephrine (N-(DMNB-caged) epinephrine)	5 mg	48.00	38.40
D-7064	**New**	N-(((4,5-dimethoxy-2-nitrobenzyl)oxy)carbonyl) dopamine (N-(NVOC-caged) dopamine)	5 mg	48.00	38.40
D-7059	**New**	N-(((4,5-dimethoxy-2-nitrobenzyl)oxy)carbonyl) MK-801 (N-(NVOC-caged) MK-801)	1 mg	48.00	38.40
D-7058	**New**	N-(((4,5-dimethoxy-2-nitrobenzyl)oxy)carbonyl)-L-norepinephrine (N-(NVOC-caged) norepinephrine)	5 mg	48.00	38.40
D-7063	**New**	N-(((4,5-dimethoxy-2-nitrobenzyl)oxy)carbonyl) serotonin (N-(NVOC-caged) serotonin)	5 mg	48.00	38.40
D-7084	**New**	N-(4,5-dimethoxy-2-nitrobenzyl)phenylephrine (N-(DMNB-caged) phenylephrine)	5 mg	48.00	38.40
G-7121	**New**	L-glutamic acid, γ-(5-allyloxy-2-nitrobenzyl) ester (γ-(ANB-caged) L-glutamic acid)	5 mg	168.00	134.40
G-7055	**New**	L-glutamic acid, γ-(α-carboxy-2-nitrobenzyl) ester, trifluoroacetate (γ-(CNB-caged) L-glutamic acid)	5 mg	168.00	134.40
G-2504		L-glutamic acid, α-(4,5-dimethoxy-2-nitrobenzyl) ester, hydrochloride (α-(DMNB-caged) L-glutamic acid)	5 mg	145.00	116.00
G-7120	**New**	L-glutamic acid, γ-(1,2-diphenyl-2-oxo)ethyl ester (γ-(desyl-caged) L-glutamic acid)	5 mg	168.00	134.40
G-2507		glycine, 4,5-dimethoxy-2-nitrobenzyl ester, hydrochloride (O-(DMNB-caged) glycine)	5 mg	75.00	60.00
H-7080	**New**	haloperidol, 4,5-dimethoxy-2-nitrobenzyl carbonate (O-(DMNB-caged) haloperidol)	5 mg	48.00	38.40
K-7115	**New**	kainic acid, γ-(α-carboxy-2-nitrobenzyl) ester, hydrochloride (γ-(CNB-caged) kainic acid)	1 mg	48.00	38.40
M-7114	**New**	N-methyl-D-aspartic acid, β-(α-carboxy-2-nitrobenzyl) ester, trifluoroacetic acid salt (β-(CNB-caged) NMDA)	1 mg	48.00	38.40
M-7069	**New**	N-methyl-N-(2-nitrobenzyl)-D-aspartic acid, dilithium salt (N-(NB-caged) NMDA)	5 mg	118.00	94.40
N-7066	**New**	N-(2-nitrobenzyl)-L-isoproterenol (N-(NB-caged) isoproterenol)	5 mg	48.00	38.40
N-7067	**New**	N-(2-nitrobenzyl)-DL-propranolol (N-(NB-caged) propranolol)	5 mg	48.00	38.40
R-6922	**New**	RGA-30	5 mg	118.00	94.40

18.4 Probes for Ca^{2+} Channels and Carriers

In both excitable and nonexcitable cells, intracellular Ca^{2+} levels modulate a multitude of vital cellular processes — including gene expression, cell viability, cell proliferation, cell motility and cell shape and volume regulation — thereby playing a key role in regulating cell responses to external activating agents. These dynamic changes in intracellular Ca^{2+} levels are regulated by ligand-gated and G-protein–coupled ion channels in the plasma membrane, as well as by mobilization of Ca^{2+} from intracellular stores. One of the best-studied examples of Ca^{2+}-dependent signal transduction is the depolarization of excitable cells, such as those of neuronal, cardiac, skeletal and smooth muscle tissue, which is mediated by inward Ca^{2+} and Na^+ currents. The Ca^{2+} current is attributed to the movement of ions through T-, N-, L- and P-type Ca^{2+} channels, which are defined both pharmacologically and by their biophysical properties, including conductance and voltage sensitivity. Molecular Probes offers a wide array of fluorescent ligands for N- and L-type Ca^{2+} channels, as well as a nonfluorescent probe for P-type Ca^{2+} channels. In addition, we offer ryanodine, a powerful modulator of the intracellular Ca^{2+} channels found in the sarcoplasmic reticulum and other subcellular organelles, and a green fluorescent derivative of ryanodine.

ω-Conotoxin Probes for N-Type Ca^{2+} Channels

ω-Conotoxin peptides (or ω-conopeptides) are important probes for distinguishing different types of receptors and ion channels.[1-3] These peptides were originally isolated from predatory marine snails of the genus *Conus* but are now often prepared by synthetic means.[4]

ω-Conotoxin GVIA

ω-Conotoxin GVIA (C-3117) is a 27–amino acid peptide toxin found in the venom of the cone snail *Conus geographus*.[5,6] The peptide preferentially binds to neuronal N-type Ca^{2+} channels and has little effect on L-type Ca^{2+} channels from cardiac or skeletal muscle.[7] Molecular Probes offers ω-conotoxin GVIA labeled with either fluorescein (C-6903) or tetramethylrhodamine (C-3115), both of which are biologically active.[8,9] These conotoxin GVIA analogs have dissociation constants of 10–20 nM, which compare favorably with the dissociation constant of native ω-conotoxin (0.6 nM). The tight binding of tetramethylrhodamine-labeled ω-conotoxin GVIA has permitted neuroscientists to visualize the distribution of voltage-gated Ca^{2+} channels on frog motor nerve terminals.[10] Fluorescein-labeled ω-conotoxin GVIA was used to localize N-type Ca^{2+} channels on somata, dendrites and dendritic spines of live neurons in rat hippocampal slices.[8]

ω-Conotoxin MVIIA

We offer a synthetic analog of ω-conotoxin MVIIA (C-98501), a 25–amino acid peptide blocker of N-type Ca^{2+} channels originally isolated from *Conus magus*. At very low concentrations, ω-conotoxin MVIIA and its synthetic analog (sometimes referred to as SNX-111) inhibit neurotransmitter release from synaptosomes.[5,11-15] SNX-111 has been reported to serve as a neuroprotective agent in animal models of transient global ischemia,[16] as well as of focal ischemia.[17]

Dihydropyridines, Phenylalkylamines and Calciseptine Probes for L-Type Ca^{2+} Channels

L-type Ca^{2+} channels are readily blocked by the binding of dihydropyridines and phenylalkylamines to the channel's pore-forming α_1-subunit. To facilitate the study of channel number and distribution in single cells, Molecular Probes has developed fluorescent derivatives of these two classes of Ca^{2+} channel blockers. We also offer the L-type Ca^{2+} channel blocker calciseptine.

Fluorescent Dihydropyridines

The high-affinity (–)-enantiomer and lower-affinity (+)-enantiomer of dihydropyridine are available labeled with either the green fluorescent DM-BODIPY (D-7443, D-7444) or the orange fluorescent ST-BODIPY (S-7445, S-7446) fluorophores. Knaus and colleagues have shown that all the BODIPY-conjugated dihydropyridines bind to L-type Ca^{2+} channels with high affinity and inhibit the Ca^{2+} influx in GH_3 cells.[18] For neuronal L-type Ca^{2+} channels, the (+)-enantiomers of the DM- and ST-BODIPY derivatives exhibit a K_i of 2.8 nM and 26.3 nM, respectively, whereas the (–)-enantiomers each have a K_i of 0.9 nM. Their affinities for skeletal muscle L-type Ca^{2+} channels are somewhat lower. Although the DM-BODIPY dihydropyridine exhibited a more intense fluorescence, the particularly high degree of stereoselectivity retained by the ST-BODIPY derivatives proved useful for the *in vivo* visualization of L-type Ca^{2+} channels.

DM-BODIPY dihydropyridine has been employed to investigate the molecular mechanism for dihydropyridine binding to L-type channels. Upon binding to the α_1-subunit, this ligand is reported to exhibit an increase in fluorescence quantum yield, as well as fluorescence resonance energy transfer between its fluorophore and one or more of the channel's tryptophan residues.[19] The spatial distribution and density of L-type calcium channels in cultured olfactory neurons were determined by confocal laser scanning microscopy using the ratio of the site-selective fluorescent staining produced by the (–)-enantiomer of DM-BODIPY dihydropyridine (D-7443) to the uniform fluorescent membrane staining by RH 414 (T-1111; see Section 15.4).[20] In this study, RH 414 staining served to control for optical artifacts and differences in membrane surface area. The DM-BODIPY dihydropyridine and ST-BODIPY dihydropyridine have also been used to label dihydropyridine receptors in live *Fucus* zygotes.[21]

Fluorescent Phenylalkylamines

The first fluorescently labeled phenylalkylamine drug analog, DM-BODIPY phenylalkylamine (DM-BODIPY PAA or BODIPY FL PAA, D-7441), is now available from Molecular Probes. Using DM-BODIPY PAA, Knaus and co-workers have obtained compelling evidence for the coupling between Ca^{2+} binding sites and phenylalkylamine receptors in purified skeletal muscle L-type Ca^{2+} channels.[22] Labeling of L-type Ca^{2+} channels by DM-BODIPY PAA at the nerve endplates and Z bands of muscle fibers can be blocked by preincubation with verapamil.[23] DM-BODIPY PAA is available as a racemic mixture that has a twofold lower IC_{50} for purified Ca^{2+} channels than (+/–)-desmethoxyverapamil.

The phenylalkylamine verapamil is one of several molecules known to inhibit P-glycoprotein–mediated drug efflux. The 170,000-dalton P-glycoprotein is typically overexpressed in tumor cells that have acquired resistance to a variety of anticancer drugs and is thought to mediate the ATP-dependent efflux or sequestration of structurally unrelated molecules, including actinomycin D, anthracyclines, colchicine, epipodophyllotoxins and vinblastine. Verapamil appears to inhibit drug efflux by acting as a substrate of P-glycoprotein, thereby overwhelming the transporter's capacity to expel the drugs. Molecular Probes offers a green fluorescent BODIPY FL verapamil derivative (B-7431), which also appears to serve as a substrate for P-glycoprotein. This fluorescent verapamil derivative preferentially accumulates in the lysosomes of normal, drug-sensitive NIH 3T3 cells but is rapidly transported out of multidrug-resistant cells, as revealed by fluorescence microscopy.[24] BODIPY FL verapamil should be useful for investigating P-glycoprotein–mediated multidrug resistance; however, we do not yet know its effect on L-type Ca^{2+} channels.

Calciseptine

Calciseptine (C-98109), a peptide isolated from the same snake that yields α-dendrotoxin, acts as a smooth muscle relaxant and inhibitor of cardiac contractions. Its activity resembles that of certain dihydropyridine drugs in that it blocks L-type Ca^{2+} channels but not N-type or T-type channels.[25-28]

Argiotoxin$_{636}$ for P-Type Ca^{2+} Channels

Polyamine toxins modulate the activity of P-type Ca^{2+} channels.[29] Argiotoxin$_{636}$ (argiopin, A-98401; see Section 18.2) is a polyamine toxin (636 corresponds to its molecular weight) from *Argiope labata* spider venom [30] that blocks acetylcholine release by acting on P-type Ca^{2+} channels.[31] Argiotoxin$_{636}$ is also a glutamatergic receptor antagonist; see Section 18.2 for more information.

Ryanodine Probes for Intracellular Ca^{2+} Channels

Ryanodine is a plant alkaloid that mobilizes Ca^{2+} from intracellular stores by activating a class of IP_3-insensitive Ca^{2+} channels.[32] It alters the function of the Ca^{2+} channel in a complex manner: submicromolar concentrations lock the channel in a long-lived open state, whereas micromolar or greater concentrations inhibit Ca^{2+} release.[33,34] Ryanodine can be used to modulate the Ca^{2+} concentration in sea urchin eggs [35] and in parotid acinar cells.[36] In developing skeletal muscle, ryanodine receptors localize in discrete regions of the T tubules, binding at the junctional complex between the T tubules and the sarcoplasmic reticulum.[37]

Unlabeled "ryanodine" (R-3455) from Molecular Probes is a mixture of about 47% ryanodine and 47% dehydroryanodine by HPLC. Both are equally effective with low nanomolar dissociation constants.[38,39] We also offer the purified components of this mixture: high-purity ryanodine (R-7478) and 9,21-dehydroryanodine (D-7480).

In addition to unlabeled ryanodine, we have recently prepared a monosubstituted BODIPY FL-X ryanodine derivative (B-7505), which, like the parent preparation, is a mixture of BODIPY FL-X ryanodine and BODIPY FL-X dehydroryanodine, most likely labeled at the 10 position of the ryanodine molecule. The biological activity of these derivatives has not yet been determined; however, structurally similar BODIPY ryanodines were shown to have a dissociation constant near that of ryanodine.[38,40]

Another reported label for the ryanodine receptor is the fluorogenic maleimide, CPM (D-346, see Section 2.2), which covalently modifies hyperreactive thiols of the ryanodine receptor of skeletal and cardiac sarcoplasmic reticulum.[41]

Probes for Ca^{2+}-ATPases and Ca^{2+}/H^+ Exchangers

Molecular Probes offers several unlabeled activators and inhibitors of Ca^{2+}-ATPases and Ca^{2+}/H^+ exchangers:

- The synthetic nitrodienamines M1 and M2 (D-99007, D-99008) are inhibitors of Ca^{2+}/Mg^{2+}-ATPase, whereas M3, M4 and M5 (B-99011, D-99009, D-99010) are activators of Ca^{2+}/Mg^{2+}-ATPase.
- The cardiac glycoside strophanthidin 3-acetate (acetylstrophanthidin, S-96205) is an inhibitor of a Ca^{2+}-transporting ATPase.[42,43]
- Phenylglyoxal (P-6221, see Section 18.5) has been reported to selectively block the activity of sarcoplasmic reticulum Ca^{2+}-ATPase [44] and of a red blood cell Ca^{2+}-transport protein.[45]
- The phosphonium salt adaphotris (T-99017) is reported to inhibit the $Ca^{2+}/2H^+$ exchange system in liver mitochondria.[46]

1. Meth Neurosci 8, 223 (1992); **2.** Meth Neurosci 8, 202 (1992); **3.** Meth Neurosci 8, 86 (1992); **4.** Chem Rev 93, 1923 (1993); **5.** Ann Rev Biochem 63, 823 (1994); **6.** J Biol Chem 266, 22067 (1991); **7.** Biochemistry 27, 963 (1988); **8.** J Neurosci 14, 6815 (1994); **9.** Science 244, 1189 (1989); **10.** J Neurosci 11, 1032 (1991); **11.** Biochemistry 34, 10256 (1995); **12.** Biochemistry 34, 8076 (1995); **13.** J Neurosci 14, 4882 (1994); **14.** Biochemistry 26, 2086 (1987); **15.** J Neurosci 7, 2390 (1987); **16.** Proc Natl Acad Sci USA 90, 7894 (1993); **17.** J Cereb Blood Flow Metab 14, 903 (1994); **18.** Proc Natl Acad Sci USA 89, 3586 (1992); **19.** Biochemistry 33, 11875 (1994); **20.** J Neurosci Meth 59, 183 (1995); **21.** J Cell Sci 109, 335 (1996); **22.** J Biol Chem 267, 2179 (1992); **23.** J Neurosci 16, 148 (1996); **24.** Mol Pharmacol 40, 490 (1991); **25.** Biochem Biophys Res Comm 194, 587 (1993); **26.** Peptide Res 5, 265 (1992); **27.** Proc Natl Acad Sci USA 88, 2437 (1991); **28.** Toxicon 28, 847 (1990); **29.** Trends Neurosci 16, 153 (1993); **30.** Brain Res 448, 30 (1988); **31.** J Physiol 482, 283 (1995); **32.** J Biol Chem 268, 13765 (1993); **33.** J Gen Physiol 92, 1 (1988); **34.** Am J Physiol 253, C364 (1987); **35.** Science 253, 1143 (1991); **36.** J Biol Chem 266, 14535 (1991); **37.** J Cell Biol 112, 289 (1991); **38.** J Biol Chem 269, 30243 (1994); **39.** J Labelled Compounds Radiopharmaceut 23, 215 (1985); **40.** Biochemistry 33, 6074 (1994); **41.** Mol Pharmacol 45, 189 (1994); **42.** Biochim Biophys Acta 1159, 109 (1992); **43.** J Biol Chem 260, 6843 (1985); **44.** Biochem Soc Trans 22, 381S (1994); **45.** Mol Pharmacol 44, 399 (1993); **46.** Biol Membr 6, 1278 (1989) (Russian).

Cat #	Structure	MW	Storage	Soluble	Abs	{ε x 10⁻³}	Em	Solvent	Notes
B-7431	✓	769	F,D,L	DMSO, EtOH	504	{74}	511	MeOH	
B-7505	✓	853	FF,D,L	DMSO	504	{79}	511	MeOH	
B-99011	S10.6	286	D,L	DMSO, EtOH	<300		none		
C-3115	C	~3475	FF,D,L	H_2O	555	{85}	578	H_2O	<1, 2>
C-3117	C	~3037	FF,D	H_2O	<300		none		
C-6903	C	~3425	FF,D,L	H_2O	493	{84}	518	pH 8	<1, 2>
C-98109	C	~7044	FF,D	H_2O	<300		<3>		
C-98501	C	~2639	FF,D	H_2O	<300		none		
D-7441	✓	714	F,D,L	DMSO, EtOH	504	{83}	511	MeOH	
D-7443	✓	686	F,D,L,A	DMSO, EtOH	504	{83}	511	MeOH	
D-7444	✓	686	F,D,L,A	DMSO, EtOH	504	{83}	510	MeOH	
D-7480	✓	492	F,D	MeOH, DMSO	<300		none		
D-99007	S10.6	224	D,L	DMSO, EtOH	<300		none		
D-99008	S10.6	252	D,L	DMSO, EtOH	<300		none		
D-99009	S10.6	236	D,L	DMSO, EtOH	<300		none		
D-99010	S10.6	288	D,L	DMSO, EtOH	<300		none		
R-7478	✓	494	F,D	MeOH, DMSO	<300		none		
S-7445	✓	761	F,D,L,A	DMSO, EtOH	565	{143}	570	MeOH	
S-7446	✓	761	F,D,L,A	DMSO, EtOH	565	{143}	570	MeOH	
S-96205	S10.6	447	F,D	DMSO	<300		none		
T-99017	T	295	DD	EtOH	<300		none		

R-3455 is a mixture of R-7478 and D-7480.

For definitions of the contents of this data table, see "How to Use the Handbook" on page vi.
Structure: Chemical structure drawing: ✓ = shown in this section; Sn.n = shown in Section number n.n; C = not shown due to complexity; T = available on request from our Technical Assistance Department and also through our Web site (http://www.probes.com).
MW: ~ indicates an approximate value.

<1> ω-Conotoxin conjugates have approximately 1 label per peptide.
<2> The value of {ε} listed for this ω-conotoxin conjugate is for the labeling dye in free solution. Use of this value for the conjugate assumes a 1:1 dye:peptide labeling ratio and no change of ε due to dye-peptide interactions.
<3> This peptide exhibits intrinsic tryptophan fluorescence (Em ~350 nm) when excited at <300 nm.

18.4 Structures *Probes for Ca²⁺ Channels and Carriers*

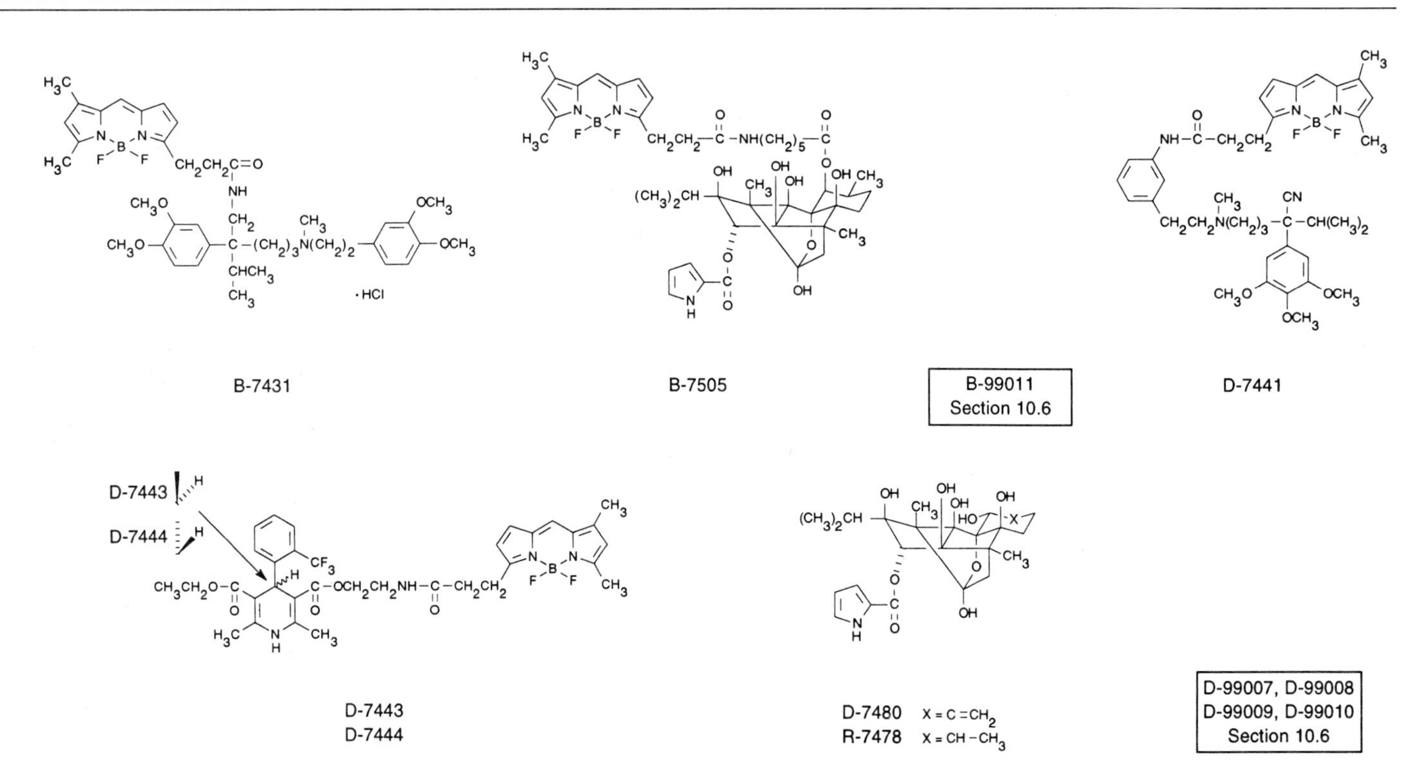

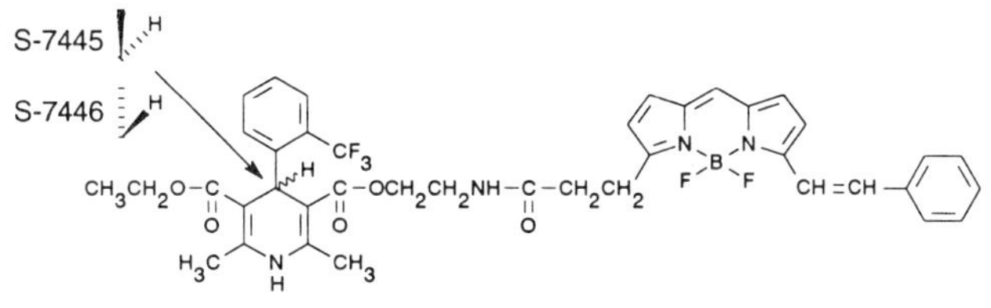

S-7445
S-7446

S-7445
S-7446

S-96205
Section 10.6

18.4 Price List *Probes for Ca^{2+} Channels and Carriers*

Cat #		Product Name	Unit Size	Price Per Unit ($) 1–4 Units	5–24 Units
B-99011	***New***	6-benzyl-1,2-dimethyl-6-ethyl-4-nitromethinyl-1,4,5,6-tetrahydropyridine (M3)	100 µg	215.00	172.00
B-7431	***New***	BODIPY® FL verapamil, hydrochloride	1 mg	98.00	78.40
B-7505	***New***	BODIPY® FL-X ryanodine	25 µg	65.00	52.00
C-98109	***New***	calciseptine *from *Dendroaspis polylepis* snake venom* *special packaging*	5x10 µg	88.00	70.40
C-3117		ω-conotoxin GVIA	5x10 µg	75.00	60.00
C-6903	***New***	ω-conotoxin GVIA, fluorescein conjugate	5 µg	295.00	236.00
C-3115		ω-conotoxin GVIA, tetramethylrhodamine conjugate	5 µg	295.00	236.00
C-98501	***New***	ω-conotoxin MVIIA *synthetic, originally from *Conus magus* molluscan venom*	100 µg	135.00	108.00
D-7480	***New***	9,21-dehydroryanodine	1 mg	38.00	30.40
D-99007	***New***	6,6-diethyl-N,2-dimethyl-4-nitromethinyl-1,4,5,6-tetrahydropyridine (M1)	100 µg	95.00	76.00
D-99008	***New***	N,2-dimethyl-6,6-dipropyl-4-nitromethinyl-1,4,5,6-tetrahydropyridine (M2)	100 µg	128.00	102.40
D-99010	***New***	1,2-dimethyl-3-nitromethylene-6-spiroadamantyl-5,6-dihydropyridine (M5)	100 µg	270.00	216.00
D-99009	***New***	1,2-dimethyl-3-nitromethylene-5-spirocyclohexyl-5,6-dihydropyridine (M4)	100 µg	215.00	172.00
D-7443	***New***	(-)-DM-BODIPY® dihydropyridine *high affinity enantiomer*	25 µg	125.00	100.00
D-7444	***New***	(+)-DM-BODIPY® dihydropyridine *low affinity enantiomer*	25 µg	125.00	100.00
D-7441	***New***	DM-BODIPY® PAA (BODIPY® FL phenylalkylamine)	25 µg	125.00	100.00
R-3455		ryanodine *mixture of ryanodine and dehydroryanodine*	10 mg	30.00	24.00
R-7478	***New***	ryanodine *≥95% by HPLC*	1 mg	38.00	30.40
S-7445	***New***	(-)-ST-BODIPY® dihydropyridine *high affinity enantiomer*	25 µg	125.00	100.00
S-7446	***New***	(+)-ST-BODIPY® dihydropyridine *low affinity enantiomer*	25 µg	125.00	100.00
S-96205	***New***	strophanthidin 3-acetate (acetylstrophanthidin)	5 mg	55.00	44.00
T-99017	***New***	P,P,P-tris-(hydroxymethyl)adamantane-1-phosphonium chloride (adaphotris)	10 mg	128.00	102.40

18.5 Probes for $GABA_A$ Receptors and Other Cl^- Channels and Carriers

Among the most ubiquitous receptors in the brain are the GABA (γ-aminobutyric acid) receptors, which are found on a wide variety of neuronal cells.[1,2] $GABA_A$ receptors are ligand-gated ion channels that produce an increase in Cl^- permeability upon GABA activation[3] (see Figure 20.1 in Chapter 20). $GABA_B$ receptors are G-protein–coupled receptors associated with modulation of K^+ channels or voltage-gated Ca^{2+} channels.[4-6]

Molecular Probes offers an extensive selection of probes for the $GABA_A$ receptor, as well as for other anion channels and carrier proteins. For information on our three caged GABA derivatives (A-2506, A-7110, A-7125), see Table 18.4 in Section 18.3.

Ro-1986 Probes for $GABA_A$ Receptors

Molecular Probes offers two fluorescent derivatives of the benzodiazepine desdiethylfluorazepam — NBD Ro-1986 and BODIPY FL Ro-1986 (N-3413, B-3414) — both of which have high affinity and selectivity for the $GABA_A$ receptor. The $GABA_A$ receptor is an oligomeric complex with an ion channel at its center; GABA, benzodiazepines and barbiturates bind cooperatively to three different sites on the receptor. Benzodiazepines are a class of CNS-depressant drugs that have strong anticonvulsant and hypnotic effects. They exert their pharmacological activity by potentiating the action of GABA, which in turn increases Cl^- channel permeability of the receptor complex. Using tritiated benzodiazepine analogs and immunohistochemical techniques, researchers have been able to localize and study $GABA_A$ receptors in nerve tissue.[7,8] Our fluorescent analogs provide a powerful alternative to radiolabeled derivatives. NBD Ro-1986 was used to study the distribution and mobility of $GABA_A$ receptors in rat spinal cord neurons.[9] BODIPY FL Ro-1986 has been employed to visualize GABA receptors in rat colon submucosa[10] and in isolated hippocampal astrocytes.[11]

$GABA_A$ Antagonists

Molecular Probes offers several noncompetitive GABA antagonists prepared by Latoxan:

- Mebicyphat (M-99003)
- Etbicyphat (E-99004)
- Propybicyphat (P-99005)
- Isobicyphat (I-99006)
- Etbicythionat (E-99015)
- Tetramethylenedisulfotetramine (TETS, T-99016)
- Cloflubicyne (C-99001)
- Flucybene (F-99002)

A comparison of the activities of these eight toxins is shown in Table 18.5.

Mebicyphat, etbicyphat, propybicyphat, isobicyphat and etbicythionat are analogs of *tert*-butylbicyclophosphorothionate [12-15] (TBPS) that block Cl^- permeability by binding to the $GABA_A$ receptor complex [14,16-18] (Table 18.5). These noncompetitive inhibitors may also affect the level of intracellular cAMP and cGMP.[15,18-20] Blockage of GABA-gated Cl^- channels by the potent convulsant TETS is reversible [21-23] and competitive with TBPS (Table 18.5). The insecticides cloflubicyne and flucybene are unusually potent noncompetitive GABA antagonists (Table 18.5).

Stilbene Disulfonates: Anion-Transport Inhibitors

Molecular Probes offers six stilbene disulfonates that have been employed to inhibit (frequently irreversibly) anion transport in a large number of mammalian cell types:[24]

- DBDS (D-675)
- DIDS (D-337)
- H_2DIDS (D-338)
- DNDS (D-673)
- NADS (A-674)
- SITS (A-339)

DNDS and SITS were among the inhibitors used to characterize three different anion exchangers in the membranes of renal brush border cells and to compare these exchangers with the band 3 anion transport protein of erythrocyte membranes.[25] Binding of DBDS to the band 3 anion transport protein is accompanied by a large increase in the probe's fluorescence that can be used to detect competitive binding of other drugs at this site.[26,27] Although binding of the photoaffinity crosslinking agent diazidostilbene disulfonic acid (DASD, D-6313; see Section 5.3) to chloride channels apparently has not been reported, this reagent may be useful as a photoactivatable inhibitor of anion channels.

These stilbene disulfonates can, in some cases, bind specifically to proteins that are not anion transporters. For example, DIDS, H_2DIDS and SITS react specifically with CD4 glycoprotein on T helper lymphocytes and macrophages, blocking HIV type-1 growth at multiple stages of the virus life cycle.[28]

Our stilbene disulfonate probes, which are 95–99% pure by HPLC, have significantly higher purity and more defined composition than available from other commercial sources. Extensive bibliographies are available from our Technical Assistance Department or through our Web site (http://www.probes.com).

Other Substrates and Inhibitors for Anion Transporters

NBD Taurine

NBD taurine (N-340) can be used as a fluorescent substrate for the anion transporter in human erythrocytes and other cells.[24,29,30] Efflux of the substrate can be monitored by measuring the quenching of NBD's fluorescence due to intracellular hemoglobin or extracellular anti-NBD antibodies [30] or by selective visualization with a fluorescence microscope.[31] NBD taurine is a competitive inhibitor of Cl^- exchange when applied to either the extracellular or intracellular membrane surface.[32]

Table 18.5 *Comparison of the activities of some Cl^- channel blockers.**

Cat #	Probe	Cl^- Channel † Blockade [IC_{100}]	GABA-Dependent $^{35}Cl^-$-Uptake Blockade [IC_{50}] ‡	GABA-Dependent $^{35}Cl^-$-Current Blockade [IC_{100}]	Inhibition of [^{3}H]-Dihydropicrotoxin Binding [IC_{50}] §	Inhibition of [^{35}S]-TBPS Binding [IC_{50}] ¶	Toxicity [LD_{50}] **
C-99001	Cloflubicyne		10 nM	5.32 nM			0.1 mg/kg
E-99004	Etbicyphat	0.05 µM			5.0 µM	5.9 µM	1 mg/kg
E-99015	Etbicythionat		0.2 nM			3.2 µM ††	1.1 mg/kg
F-99002	Flucybene		50 nM	61 nM			1 mg/kg
I-99006	Isobicyphat		0.2 nM	2.75 µM	8.0 µM	0.48 µM	0.18 mg/kg
M-99003	Mebicyphat	0.5 µM		642 µM	150 µM	100 µM	32 mg/kg
P-99005	Propybicyphat		0.5 nM	1.09 nM	8.0 µM	1.1 µM	0.38 mg/kg
T-99016	TETS					0.5 µM	0.24 mg/kg

* All data provided by Latoxan. † Pond snail (*Lymnaea stagnalis*). ‡ Rat synaptosomes. § P_2 membrane fraction from rat brain. ¶ P_2 membrane fraction from rat brain. ** Mouse, intraperitoneal. †† House fly.

$DiBAC_4(5)$ and WW 781

The membrane potential–sensing dye bis-(1,3-dibutylbarbituric acid)pentamethine oxonol ($DiBAC_4(5)$, B-436) initially inhibits Cl^- exchange with an IC_{50} of 0.146 µM; however, this inhibition increases with time to an IC_{50} of 1.05 nM, making $DiBAC_4(5)$ a more potent inhibitor than DIDS, which has an IC_{50} of 31 nM under similar conditions.[33] The potential-sensitive probe WW 781 (W-435) is also reported to be a potent inhibitor of the erythrocyte band 3 anion transporter.[34]

Aryl Glyoxals

Aryl glyoxals — including phenylglyoxal (P-6221) and 4-nitrophenylglyoxal (N-1559) — are powerful inhibitors of anion transport in red blood cells.[35] It appears that the aryl glyoxal binding site is not the same as that of DIDS and may involve modification of arginine residues.[35,36] Phenylglyoxal is also reported to modify the activity of some Ca^{2+} channels (see Section 18.4).

Eosin Maleimide

Although usually selectively reactive with thiols, eosin-5-maleimide (E-118, see Section 2.2) also reacts with a specific lysine residue of the band 3 protein in human erythrocytes, inhibiting anion exchange in these cells and providing a convenient tag for observing band 3 behavior in the membrane.[37-39] Eosin-5-isothiocyanate (E-18, see Section 1.5) has similar reactivity with band 3 proteins.[40,41]

Chlorotoxin

Chlorotoxin (C-98210) is a newly isolated 36–amino acid polypeptide toxin from the scorpion *Leiurus quinquestriatus* that inhibits small-conductance Cl^- channels when applied to the intracellular membrane surface.[42,43] Its affinity for the chloride channel in reconstituted lipid bilayers is reported to depend strongly on the membrane potential.[43]

1. Adv Exp Med Biol 368, 89 (1994); **2.** Neuromethods 4, 217 (1986); **3.** Neurochem Res 18, 365 (1993); **4.** Trends Pharmacol Sci 14, 259 (1993); **5.** Gen Pharmacol 23, 309 (1992); **6.** Trends Neurosci 15, 46 (1992); **7.** Richards, E. *et al.*, in *Neurohistochemistry: Modern Methods and Applications*, P. Panula, *et al.*, Eds., Alan R. Liss (1986) pp. 629–677; **8.** Trends Neurosci 8, 49 (1985); **9.** J Neurosci 9, 2163 (1989); **10.** Neurosci Lett 176, 32 (1994); **11.** J Neurosci 15, 2720 (1995); **12.** J Neurochem 46, 1542 (1986); **13.** Mol Pharmacol 28, 246 (1985); **14.** Mol Pharmacol 23, 326 (1985); **15.** Toxicol Appl Pharmacol 46, 411 (1978); **16.** Eur J Med Chem 13, 207 (1978); **17.** Nature 261, 601 (1976); **18.** Science 182, 1135 (1973); **19.** Biochem Pharmacol 26, 887 (1977); **20.** Nature 268, 52 (1977); **21.** Chem Res Toxicol 4, 162 (1991); **22.** Br J Pharmacol 53, 598 (1975); **23.** Br J Pharmacol 53, 422 (1975); **24.** Am J Physiol 262, C803 (1992); **25.** J Biol Chem 269, 21489 (1994); **26.** Biochim Biophys Acta 979, 193 (1989); **27.** Nature 282, 520 (1979); **28.** J Biol Chem 266, 13355 (1991); **29.** Biochemistry 31, 7301 (1992); **30.** Meth Enzymol 172, 122 (1989); **31.** Am J Physiol 249, C56 (1985); **32.** Biochem J 195, 503 (1981); **33.** Am J Physiol 269, C1073 (1995); **34.** Am J Physiol 265, C521 (1993); **35.** Biochim Biophys Acta 1026, 43 (1990); **36.** J Biosciences 15, 179 (1990); **37.** Biochemistry 34, 4880 (1995); **38.** Biophys J 66, 1726 (1994); **39.** Am J Physiol 264, C1144 (1993); **40.** Biochim Biophys Acta 897, 14 (1987); **41.** Biochim Biophys Acta 550, 328 (1979); **42.** Am J Physiol 264, C361 (1993); **43.** Toxicon 29, 1403 (1991).

18.5 Data Table *Probes for $GABA_A$ Receptors and Other Cl^- Channels and Carriers*

Cat #	Structure	MW	Storage	Soluble	Abs	{ε x 10^{-3}}	Em	Solvent	Notes
A-339	✓	498	F,DD	H_2O*	336	{47}	436	pH 7	
A-674	✓	444	L	H_2O	390	{18}	458	H_2O	
B-436	S25.3	543	L	DMSO, EtOH†	590	{160}	616	MeOH	
B-3414	✓	606	F,D,L	DMSO	504	{82}	511	MeOH	
C-98210	C	~3997	FF	H_2O	<300		none		
C-99001	T	351	DD	EtOH	<300		none		
D-337	✓	498	F,DD	H_2O*	341	{61}	415	H_2O	
D-338	✓	500	F,DD	H_2O*	286	{41}	none	MeOH	
D-673	✓	474	L	H_2O	352	{32}	none	H_2O	
D-675	✓	623	L	H_2O	343	{50}	430	MeOH	
E-99004	T	178	DD	EtOH	<300		none		
E-99015	T	194	DD	EtOH	<300		none		
F-99002	T	280	D	EtOH	<300		none		
I-99006	T	192	DD	EtOH	<300		none		
M-99003	T	164	DD	EtOH	<300		none		
N-340	✓	310	D,L	pH>6	459	{21}	534	MeOH	<1>
N-1559	S1.8	179	F	DMF, MeCN	265	{8.7}	none	MeOH	
N-3413	✓	495	F,D,L	DMSO‡	462	{18}	532	MeOH	
P-6221	S1.8	152§	F	EtOH	247	{13}	none	MeOH	
P-99005	T	192	DD	EtOH	<300		none		
T-99016	T	240	D	EtOH, DMSO	<300		none		
W-435	S25.2	757	D,L	DMSO, EtOH	605	{120}	639	MeOH	<2>

For definitions of the contents of this data table, see "How to Use the Handbook" on page vi.
Structure: Chemical structure drawing: ✓ = shown in this section; T = available on request from our Technical Assistance Department and also through our Web site (http://www.probes.com); C = not shown due to complexity; Sn.n = shown in Section number n.n.
MW: ~ indicates an approximate value.

* Isothiocyanates are unstable in water and should not be stored in aqueous solution.
† Oxonols may require addition of a base to be soluble.
‡ This product is packaged as a solution in the solvent indicated in "Soluble."
§ MW is for the hydrated form of this product.

<1> NBD taurine is sufficiently fluorescent in water to allow cellular efflux to be measured fluorometrically using excitation/emission wavelengths of 470/550 nm [Meth Enzymol 172, 122 (1989)].

<2> Fluorescence excitation/emission maxima of W-435 are 615/650 nm in cell suspensions [Biophys J 57, 835 (1990)].

18.5 Structures *Probes for $GABA_A$ Receptors and Other Cl^- Channels and Carriers*

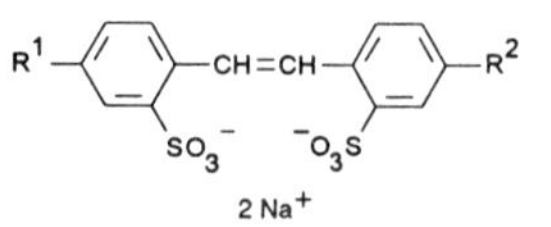

A-339 R^1 = —N=C=S R^2 = —NH–C(=O)–CH_3
A-674 R^1 = —NH_2 R^2 = —NO_2
D-337 R^1 = —N=C=S R^2 = —N=C=S
D-673 R^1 = —NO_2 R^2 = —NO_2

B-436 Section 25.3

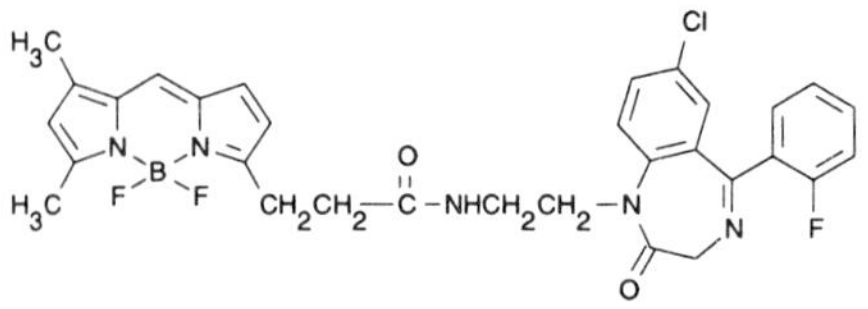

B-3414

D-337, D-673 see A-339

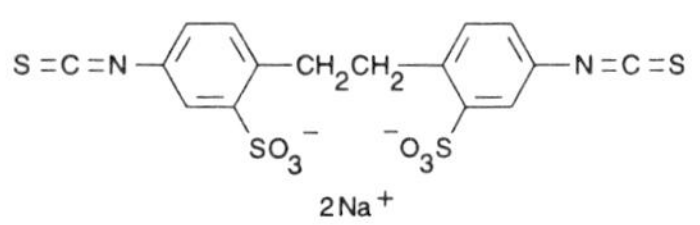

D-338

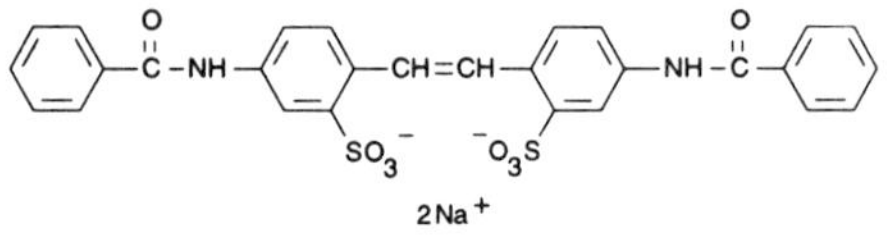

D-675

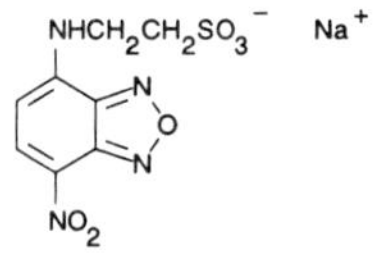

N-340

N-1559, P-6221 Section 1.8

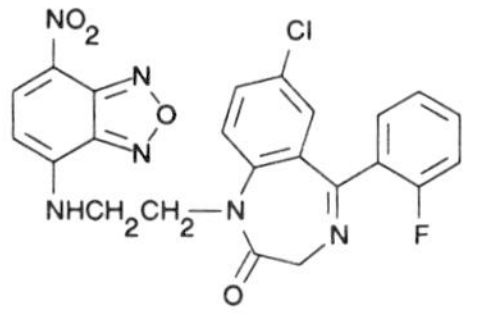

N-3413

W-435 Section 25.2

18.5 Price List *Probes for $GABA_A$ Receptors and Other Cl^- Channels and Carriers*

Cat #	Product Name	Unit Size	Price Per Unit ($) 1–4 Units	5–24 Units
A-339	4-acetamido-4′-isothiocyanatostilbene-2,2′-disulfonic acid, disodium salt (SITS)	100 mg	48.00	38.40
A-674	4-amino-4′-nitrostilbene-2,2′-disulfonic acid, disodium salt (NADS)	1 g	78.00	62.40
B-436	bis-(1,3-dibutylbarbituric acid)pentamethine oxonol ($DiBAC_4(5)$)	25 mg	65.00	52.00
B-3414	BODIPY® FL Ro-1986	1 mg	68.00	54.40
C-98210 **New**	chlorotoxin *from *Leiurus quinquestriatus* scorpion venom* *special packaging*	2x10 µg	265.00	212.00
C-99001 **New**	cloflubicyne	1 mg	205.00	164.00
D-675	4,4′-dibenzamidostilbene-2,2′-disulfonic acid, disodium salt (DBDS)	100 mg	78.00	62.40
D-338	4,4′-diisothiocyanatodihydrostilbene-2,2′-disulfonic acid, disodium salt (H_2-DIDS)	100 mg	115.00	92.00
D-337	4,4′-diisothiocyanatostilbene-2,2′-disulfonic acid, disodium salt (DIDS)	100 mg	22.00	17.60
D-673	4,4′-dinitrostilbene-2,2′-disulfonic acid, disodium salt (DNDS)	1 g	55.00	44.00
E-99004 **New**	etbicyphat	1 mg	245.00	196.00
E-99015 **New**	etbicythionat	100 µg	205.00	164.00
F-99002 **New**	flucybene	1 mg	108.00	86.40
I-99006 **New**	isobicyphat	1 mg	205.00	164.00
M-99003 **New**	mebicyphat	1 mg	135.00	108.00
N-340	2-(N-(7-nitrobenz-2-oxa-1,3-diazol-4-yl)amino)ethanesulfonic acid, sodium salt (NBD taurine)	10 mg	128.00	102.40
N-3413	N-(7-nitrobenz-2-oxa-1,3-diazol-4-yl)desdiethylfluorazepam (NBD Ro-1986) *1 mg/mL solution in dry DMSO*	1 mL	48.00	38.40
N-1559	4-nitrophenylglyoxal	100 mg	48.00	38.40
P-6221 **New**	phenylglyoxal, monohydrate	1 g	8.00	6.40
P-99005 **New**	propybicyphat	1 mg	185.00	148.00
T-99016 **New**	tetramethylenedisulfotetramide (TETS)	50 µg	155.00	124.00
W-435	WW 781, triethylammonium salt	25 mg	145.00	116.00

18.6 Probes for K^+ Channels and Carriers

Neurons and other cells contain an assortment of K^+ channels that are differentiated by their pharmacology, kinetics and voltage dependence.[1,2] High-affinity toxins from snakes, scorpions, bees and other sources have played an important role in the classification of these ion channels. Some early experiments may need to be reexamined due to lack of reliable purity criteria for the toxins, a problem that has been alleviated somewhat through characterization by HPLC and other techniques. Most of the peptide toxins from Molecular Probes have been purified by HPLC to ensure that our reagents are of the highest quality.

Glibenclamide Probes for the ATP-Dependent K^+ Channel

Glibenclamide blocks the ATP-dependent K^+ channel, thereby eliciting insulin secretion.[3] In collaboration with Gabriel Haddad of Yale University, Molecular Probes has prepared BODIPY FL glibenclamide (B-7439), a green fluorescent probe for the ATP-dependent K^+ channel. Although the binding affinity of this fluorescent probe was lower than that of the unlabeled antagonist glibenclamide (G-7440), Haddad and co-workers were able to visualize these receptors in brain sections and cultured brain cells.[4]

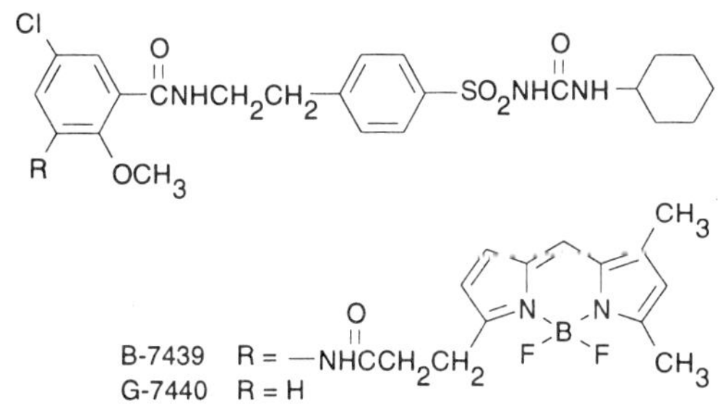

B-7439 R = —NHCCH$_2$CH$_2$ (BODIPY FL)
G-7440 R = H

Figure 18.1 Structures of BODIPY FL glibenclamide (B-7439) and glibenclamide (G-7440).

Toxin Probes for Ca^{2+}-Activated K^+ Channels and Voltage-Gated K^+ Channels

β-Bungarotoxin

β-Bungarotoxin (B-3459), a toxin extracted from *Bungarus multicinctus* venom, is composed of two subunits, each with a different physiological activity. The larger one (~13,000 daltons) exhibits moderate phospholipase A_2 activity;[5-8] the smaller one (~7000 daltons) blocks some voltage-gated K^+ channels,[9-14] inhibiting acetylcholine release at the neuromuscular junction.

Charybdotoxin

Charybdotoxin,[15] which is prepared by Latoxan both as the natural 37–amino acid peptide from *Leiurus quinquestriatus hebraeus* scorpion venom (C-98201) and as an equivalent synthetic form (C-98213), is a highly basic peptide that reversibly blocks the large-conductance apamin-insensitive Ca^{2+}-activated K^+ channel, with an apparent dissociation constant of 2.1 nM.[16-30] Charybdotoxin also exhibits high affinity for several classes of voltage-gated K^+ channels,[24,31,32] with a reported K_d of 8–14 pM (compared with 33 pM for α-dendrotoxin) for the voltage-gated K^+ channel in human and murine T lymphocytes[33] and 1.7 nM (compared with 2.8 nM for α-dendrotoxin and 185 nM for mast cell degranulating peptide) for the voltage-gated K^+ channel expressed in fibroblasts transfected with a gene from the *Shaker*-related subfamily 1.[34] The charybdotoxin–K^+ channel interaction has been investigating using both mutant ion channels[35-36] and mutants of charybdotoxin.[37-39]

Dendrotoxins

α-Dendrotoxin (D-98106), a 59–amino acid peptide from the green mamba snake *Dendroaspis angusticeps*, and dendrotoxin I (D-98113), a 60–amino acid peptide from the black mamba snake *Dendroaspis polylepis polylepis*,[40] specifically block selected voltage-gated K^+ channels.[9,41-50]

Iberiotoxin

Synthetic iberiotoxin (I-98211), which is analogous to the 37–amino acid peptide isolated from *Buthus tamulus* scorpion venom, shows 68% homology with charybdotoxin[51] and 44% homology with kaliotoxin.[52] Iberiotoxin is a highly selective blocker of large-conductance Ca^{2+}-activated K^+ channels[51,53,54] but, unlike charybdotoxin and kaliotoxin, does not affect voltage-gated K^+ channels.[33,55]

Kaliotoxins

Kaliotoxin (K-98208), a 38–amino acid peptide, and its 37–amino acid amide analog[56] (kaliotoxin (1-37) amide, K-98209) are synthetic peptides analogous to kaliotoxin originally isolated from the scorpion *Androctonus mauretanicus mauretanicus*. These peptides block large-conductance Ca^{2+}-activated K^+ channels and some voltage-gated K^+ channels.[52,57,58] In human and murine β-lymphocytes, where both K^+ channel subtypes are present, kaliotoxin acts as a selective voltage-gated K^+ channel blocker. Among the K^+ channel toxins, kaliotoxin is reportedly the most suitable probe for radioiodination because its activity appears to be the least altered after modification.[57]

Leiurotoxins

Synthetic leiurotoxin I (scyllatoxin, L-98202), a 31–amino acid peptide originally isolated from *Leiurus quinquestriatus hebraeus* scorpion venom, blocks apamin-sensitive, small-conductance Ca^{2+}-activated K^+ channels.[59-64] Although structurally unrelated to apamin, leiurotoxin I is a potent inhibitor of apamin binding to rat synaptosomal membranes, with a reported K_i = 75 pM.[65,66] An analog of leiurotoxin I, tyr^2-leiurotoxin I (T-98206), has a tyrosine residue substituted for a phenylalanine residue to facilitate radioiodination of the toxin.[64] Other leiurotoxins from *Leiurus quinquestriatus hebraeus* venom act on Na^+ channels (see Section 18.7).

Apamin

Apamin (A-98407), an 18–amino acid peptide isolated from *Apis mellifera* bee venom,[67-70] primarily blocks small-conductance

Ca^{2+}-activated K^+ channels in mammalian neurons and skeletal muscle.[2,71-75] Apamin possesses potent convulsant activity and is known to cross the blood–brain barrier.[2]

Mast Cell–Degranulating Peptide

Like apamin, mast cell–degranulating peptide (MCD peptide, M-98402) is isolated from *Apis mellifera* bee venom and possesses potent convulsant activity.[70] This 22–amino acid peptide blocks selected voltage-gated K^+ channels.[2,76] High-affinity binding sites in rat brain have been identified with radiolabeled MCD peptide, which copurified with synaptic membranes and localized primarily in the frontal cortex and the anterior hippocampal formation.[77,78] In addition, MCD peptide releases histamine from mast cells, an activity unrelated to its neurotoxicity.[78-80]

1. Annals NY Acad Sci 710, 1 (1994); **2.** J Membrane Biol 105, 95 (1988); **3.** Trends Pharmacol Sci 11, 417 (1990); **4.** G. Haddad, Yale University, personal communication; **5.** Eur J Cell Biol 43, 195 (1987); **6.** J Biochem 84, 1301 (1978); **7.** J Biochem 84, 1291 (1978); **8.** J Biochem 83, 91 (1978); **9.** Neuroscience 40, 29 (1991); **10.** Mol Pharmacol 38, 164 (1990); **11.** Eur J Biochem 178, 771 (1989); **12.** Biochemistry 27, 963 (1988); **13.** Br J Pharmacol 94, 839 (1988); **14.** Neurosci Lett 68, 141 (1986); **15.** Meth Neurosci 8, 137 (1992); **16.** Biochemistry 34, 15849 (1995); **17.** Biochemistry 34, 10771 (1995); **18.** J Biol Chem 270, 22434 (1995); **19.** Neuron 15, 5 (1995); **20.** J Biol Chem 269, 676 (1994); **21.** Biochemistry 31, 7756 (1992); **22.** J Neurosci 12, 297 (1992); **23.** Science 254, 1521 (1991); **24.** FEBS Lett 250, 433 (1989); **25.** J Gen Physiol 91, 317 (1988); **26.** J Membrane Biol 106, 243 (1988); **27.** Neurosci Lett 94, 279 (1988); **28.** Proc Natl Acad Sci USA 85, 3329 (1988); **29.** J Biol Chem 261, 14607 (1986); **30.** Nature 313, 316 (1985); **31.** Proc Natl Acad Sci USA 86, 10171 (1989); **32.** Science 239, 771 (1988); **33.** J Biol Chem 266, 3668 (1991); **34.** Neuroscience 50, 935 (1992); **35.** Neuron 5, 767 (1990); **36.** Science 245, 1382 (1989); **37.** Biochemistry 35, 6181 (1996); **38.** Biochemistry 31, 7749 (1992); **39.** Proc Natl Acad Sci USA 88, 2046 (1991); **40.** Meth Neurosci 8, 235 (1992); **41.** J Biol Chem 270, 24776 (1995); **42.** Biochemistry 33, 1617 (1994); **43.** J Physiol 464, 321 (1993); **44.** Biochemistry 31, 11084 (1992); **45.** Mol Pharmacol 34, 152 (1988); **46.** Proc Natl Acad Sci USA 85, 4919 (1988); **47.** Brain Res 377, 374 (1986); **48.** Pflügers Arch 407, 365 (1986); **49.** Proc Natl Acad Sci USA 83, 493 (1986); **50.** Nature 299, 252 (1982); **51.** J Biol Chem 265, 11083 (1990); **52.** J Biol Chem 267, 1640 (1992); **53.** Biochemistry 31, 6719 (1992); **54.** Biophys J 63, 583 (1992); **55.** Biochemistry 32, 2363 (1993); **56.** Biochemistry 33, 14256 (1994); **57.** J Biol Chem 269, 32835 (1994); **58.** J Biol Chem 268, 26302 (1993); **59.** J Membr Biol 147, 71 (1995); **60.** Int'l J Peptide Protein Res 43, 486 (1994); **61.** J Biol Chem 269, 18053 (1994); **62.** Biochemistry 31, 648 (1992); **63.** FEBS Lett 285, 271 (1991); **64.** J Biol Chem 265, 4753 (1990); **65.** Biochemistry 32, 11969 (1993); **66.** J Biol Chem 263, 10192 (1988); **67.** Biochemistry 31, 1476 (1992); **68.** Meth Neurosci 8, 15 (1992); **69.** FEBS Lett 197, 289 (1986); **70.** Science 177, 314 (1972); **71.** FEBS Lett 275, 185 (1990); **72.** Nature 323, 718 (1986); **73.** J Biol Chem 259, 1491 (1984); **74.** EMBO J 1, 1039 (1982); **75.** Proc Natl Acad Sci USA 79, 1308 (1982); **76.** J Physiol 420, 365 (1990); **77.** Brain Res 418, 235 (1987); **78.** J Biol Chem 259, 13957 (1984); **79.** Biochem J 181, 623 (1979); **80.** Br J Pharmacol 50, 383 (1974).

18.6 Data Table *Probes for K⁺ Channels and Carriers*

Cat #	Structure	MW	Storage	Soluble	Abs	{ε x 10^{-3}}	Em	Solvent
A-98407	C	~2027	FF,D	H_2O	<300		none	
B-3459	C	~20,000	FF,D	H_2O	<300		none	
B-7439	✓	783	F,D,L	DMSO, EtOH	504	{76}	511	MeOH
C-98201	C	~4296	FF,D	H_2O	<300		<1>	
D-98106	C	~7048	FF,D	H_2O	<300		<1>	
D-98113	C	~7218	FF,D	H_2O	<300		<1>	
G-7440	✓	494	D	DMSO, EtOH	299	{5.8}	none	EtOH
I-98211	C	~4230	FF,D	H_2O	<300		<1>	
K-98208	C	~4150	FF,D	H_2O	<300		none	
K-98209	C	~4021	FF,D	H_2O	<300		none	
L-98202	C	~3431	FF,D	H_2O	<300		none	
M-98402	C	~2593	FF,D	H_2O	<300		none	
T-98206	C	~3447	FF,D	H_2O	<300		none	

C-98213 see C-98201

For definitions of the contents of this data table, see "How to Use the Handbook" on page vi.
Structure: Chemical structure drawing: ✓ = see Figure 18.1; C = not shown due to complexity.
MW: ~ indicates an approximate value.

<1> This peptide exhibits intrinsic tryptophan fluorescence (Em ~350 nm) when excited at <300 nm.

18.6 Price List *Probes for K⁺ Channels and Carriers*

Cat #	Product Name	Unit Size	Price Per Unit ($) 1–4 Units	5–24 Units
A-98407 ***New***	apamin *from *Apis mellifera* bee venom* *special packaging*	2x100 µg	55.00	44.00
B-7439 ***New***	BODIPY® FL glibenclamide (BODIPY® FL glyburide)	100 µg	95.00	76.00
B-3459	β-bungarotoxin *from *Bungarus multicinctus**	1 mg	37.00	29.60
C-98201 ***New***	charybdotoxin *from *Leiurus quinquestriatus* scorpion venom*	10 µg	305.00	244.00
C-98213 ***New***	charybdotoxin *synthetic, originally from *Leiurus quinquestriatus* scorpion venom*	50 µg	135.00	108.00
D-98106 ***New***	α-dendrotoxin *from *Dendroaspis angusticeps* snake venom*	100 µg	88.00	70.40
D-98113 ***New***	dendrotoxin I (DPI) *from *Dendroaspis polylepis polylepis* snake venom* *special packaging*	2x50 µg	88.00	70.40

Cat #		Product Name	Unit Size	Price Per Unit ($) 1–4 Units	5–24 Units
G-7440	**New**	glibenclamide (glyburide)	100 mg	8.00	6.40
I-98211	**New**	iberiotoxin *synthetic, originally from *Buthus tamulus* scorpion venom*	10 µg	85.00	68.00
K-98208	**New**	kaliotoxin *synthetic, originally from *Androctonus mauretanicus mauretanicus* scorpion venom* *special packaging*	2x10 µg	128.00	102.40
K-98209	**New**	kaliotoxin (1-37) amide *special packaging*	2x10 µg	108.00	86.40
L-98202	**New**	leiurotoxin I (scyllatoxin) *synthetic, originally from *Leiurus quinquestriatus* scorpion venom*	50 µg	385.00	308.00
M-98402	**New**	MCD peptide *from *Apis mellifera* bee venom* *special packaging*	4x50 µg	108.00	86.40
T-98206	**New**	[tyr²]-leiurotoxin I ([tyr²]-scyllatoxin) *synthetic*	50 µg	385.00	308.00

18.7 Probes for Na^+ Channels and Carriers

Amiloride Analogs: Probes for the Na^+ Channel and the Na^+/H^+ Antiporter

Amiloride is a compound known to inhibit the Na^+/H^+ antiporter of vertebrate cells by acting competitively at the Na^+-binding site.[1] The antiporter extrudes protons from cells using the inward Na^+ gradient as a driving force, resulting in intracellular alkalinization. In 1967, Cragoe and co-workers reported the synthesis of amiloride and several amiloride analogs, pyrazine diuretics that inhibit the Na^+ channel in urinary epithelia.[2] Since then, about 1000 different amiloride analogs have been synthesized and many of these tested for their specificity and potency in inhibiting the Na^+ channel, Na^+/H^+ antiporter and Na^+/Ca^{2+} exchanger.[3]

Unmodified amiloride (A-3108), shown in Figure 18.2 where R^1, R^2 and R^3 are hydrogen atoms, inhibits the Na^+ channel with an IC_{50} of less than 1 µM. However, when hydrophobic substituents are added at the R^3 position, the amiloride analogs exhibit a greatly enhanced affinity and specificity for the Na^+ channel. We now offer a total of five derivatives that fall into this class of inhibitors,[3-5] including:

- Benzamil (B-6897)
- 2',4'-Dichlorobenzamil (D-6898)
- 2',4'-Dimethylbenzamil (D-6885)
- 4'-Methoxybenzamil (M-6884)
- *p*-Hydroxyphenethamil (H-6883)

These benzamil compounds are also effective inhibitors for the Na^+/Ca^{2+} exchanger, with potencies that range from 8.5 to 110 times that of amiloride.[3,4,6-9] In addition, we offer BODIPY FL amiloride (B-6905), a green fluorescent probe in which the hydrophobic BODIPY fluorophore is attached at the R^3 position.

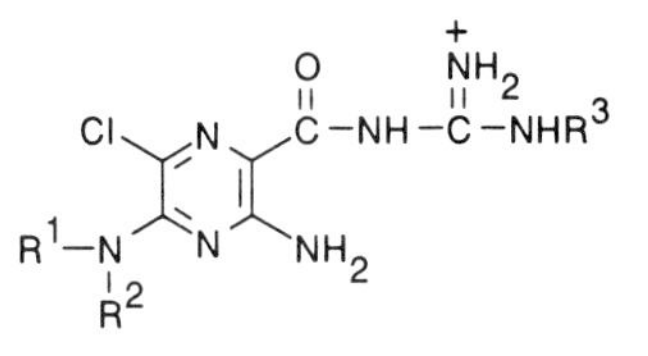

Figure 18.2 *Structure of amiloride (A-3108), when R^1, R^2 and R^3 are hydrogen atoms. Substitution at the R^1, R^2 and R^3 positions lead to differences in specificity and potency, as described in the text.*

Amiloride is also an important tool for studying the Na^+/H^+ antiporter. Structure–activity relationships have demonstrated that amiloride analogs containing hydrophobic groups at the R^1 and R^2 positions are the most potent and specific inhibitors for the Na^+/H^+ antiporter. We now offer six amiloride analogs that have been modified to increase (from twofold to over 500-fold) their relative potency for inhibiting this antiporter:[3,4,10-13]

- 5-(*N,N*-diethyl)amiloride (D-3109)
- 5-(*N,N*-dimethyl)amiloride (D-3112)
- 5-(*N*-ethyl-*N*-isopropyl)amiloride (E-3111)
- 5-(*N,N*-hexamethylene)amiloride (H-3113)
- 5-(*N*-isobutyl-*N*-methyl)amiloride (I-3110)
- 5-(*N*-isopropyl-*N*-methyl)amiloride (I-6880)

The compound 6-iodoamiloride (I-6881) has recently been described as a photoactivatable inhibitor of the Na^+/H^+ antiporter, producing irreversible inhibition upon UV illumination.[14]

Amiloride analogs are useful as both structural and functional probes in a wide variety of studies. Recent examples include:

- Regulation of intracellular pH in perfused heart,[15] human natural killer (NK) cells [16] and rat gastric cells [17]
- Topology of an amiloride-binding protein [18]
- Functional coupling of Na^+/H^+ and Na^+/Ca^{2+} exchangers in α_1-adrenoreceptor–mediated activation of hepatic functions [19]
- Glucose-stimulated medium acidification by the yeast *Schizosaccharomyces pombe* [20]

Extensive bibliographies are available for many of our amiloride analogs from our Technical Assistance Department or through our Web site (http://www.probes.com).

Tetrodotoxin, Saxitoxin and Related Probes: Blockers of Voltage-Gated Na^+ Channels

The high-affinity neurotoxins tetrodotoxin (TTX) and saxitoxin (STX) interfere with nerve transmission by selectively blocking the voltage-gated Na^+ channel.[21,22] We offer citrate-free TTX (T-6914) and citrate-containing TTX (T-6913) that are isolated in China from the pufferfish; this TTX preparation has been reported to be somewhat more active than the toxin isolated in Japan.[23] The citrate-containing TTX is more readily dissolved in water.

Saxitoxin is the most familiar member of a group of structurally similar paralytic shellfish toxins that includes the gonyautoxins GTX C1 (G-6916), GTX-1 (G-6917), GTX-IV (G-6915), GTX-VI (G-6918) and neo-STX (N-6919), which are also now available from Molecular Probes. These toxins have been employed as analytical standards in the detection of shellfish toxins using HPLC,[24-26] an automated tissue culture bioassay[27] and a solid-phase radioreceptor assay.[28] They will also likely prove useful as probes for neuroscience applications.[29]

Brevetoxins: Activators of Voltage-Gated Na^+ Channels

Brevetoxins (PbTx) are lipid-soluble cyclic polyether ladder toxins isolated from the marine dinoflagellate *Ptychodiscus brevis,* which is responsible for red tides.[30] The brevetoxins enhance cellular Na^+ influx by allosteric activation of the site 5 orphan receptor located on domain IV of the α-subunit of voltage-gated Na^+ channels.[31] The result of their binding is the opening of Na^+ channels in excitable membranes at normal resting potential and induction of persistent channel activation or prolonged open channel times.[32-36] Brevetoxins bind with high affinity to purified, reconstituted Na^+ channels.[37] Binding of brevetoxins and ciguatoxin to site 5 is competitive;[33,38,39] however, unlike cigautoxins, brevetoxins are potent icthyotoxins, with lethal doses in the nanomolar range.[38,40,41]

The chemical structures of brevetoxins fall into two classes; the more flexible brevetoxin "A" structure includes toxins PbTx-2, -3, -5, -6, -8 and -9. Brevetoxin "B" includes PbTx-1, -7 and -10.[42,43] There is some dependence of the toxicity and pharmacology of the various brevetoxins on their structure.[30] Brevetoxins PbTx-1, -2, -3, -6 and -9 in 25 µg unit sizes are available from Molecular Probes (B-6851, B-6852, B-6853, B-6854, B-6855). In addition, we provide a Brevetoxin Sampler Kit (B-6850), which contains 5 µg samples of each of our five brevetoxins. The brevetoxin analogs, which are all >95% pure by HPLC, are also useful as chromatographic standards for analysis of algal blooms.[44] Tritiated brevetoxin PbTx-3 is readily prepared by reduction of the aldehyde residue in brevetoxin PbTx-2 with sodium borotritiide.[34] Brevetoxins must not be dissolved in dimethylsulfoxide (DMSO).

Scorpion Toxins: Probes for Insect and Crustacean Na^+ Channels

Insect toxins prepared for us by Latoxan from the scorpion *Leiurus quinquestriatus hebraeus* include LqhαIT (L-98221), LTx-1 (L-98217), LTx-2 (L-98218), LTx-3 (L-92819) and LTx-4 (L-98220). LqhαIT is an α-toxin that inhibits insect and crustacean Na^+ channel inactivation;[45-47] it is also reported to have low activity on the mammalian muscle Na^+ channel.[48] The receptor site of LqhαIT on insect Na^+ channels is considered to be homologous to receptor site 3 in mammalian Na^+ channels.[49-52] The other *Leiurus* toxins from this scorpion have high structural homology with the Na^+ channel modulators neurotoxin III,[53] neurotoxin IV[54] and neurotoxin V,[55,56] but their biological activity has not been completely characterized.

Lappaconitines, Including a Fluorescent Na^+ Channel Probe

The diterpenoid plant alkaloids lappaconitine (allapinine, L-96010) and *N*-desacetyllappaconitine (D-96035) are reported to selectively block TTX-sensitive Na^+ channels without influencing the activation threshold of Na^+ channels.[57] They also have micromolar affinity for neuronal nicotinic AChR (Table 18-2). Desacetyllappaconitine is an ester of anthranilic acid and is therefore intrinsically fluorescent, making it potentially useful for labeling Na^+ channels or nicotinic AChRs.

Anthroyl Ouabain: A Probe for Na^+/K^+-ATPase

Ouabain binds to the α-subunit of Na^+/K^+-ATPase, inhibiting its transport of Na^+ across the plasma membrane. 9-Anthroyl ouabain (A-1322) — a fluorescent probe that binds to a variety of cells from different species — is useful for localizing Na^+/K^+-ATPase and for studying its membrane orientation, mobility and dynamics.[58] Anthroyl ouabain has also been employed to investigate Na^+/K^+-ATPase's active site, inhibition and conformational changes,[59-63] as well as to investigate the kinetics of cardiac glycoside binding.[64-67]

Other Probes for Na^+ Channels and Na^+/K^+-ATPase

Natural products from plants frequently exhibit biological activity that involves action at Na^+ channels. Among the Latoxan products reported in the pharmacological literature to act on Na^+ channels or Na^+/K^+-ATPase, to function as coronary dialators or sedatives or to have antiarrhythmic, hypotensive or spasmolytic activity are:

- Aconitine from *Aconitum karacolicum* (A-96001), which opens tetrodotoxin-sensitive Na^+ channels[68]
- Askendoside D from Leguminosae (A-96015), which inhibits cardiac Na^+/K^+-ATPase[69]
- Benzoylheteratisine from Ranunculaceae (B-96085), a Na^+ channel blocker[70,71]
- Deltaline from Ranunculaceae (eldeline, D-96057), a fast inward Na^+ current blocker[72]
- Diacetylkorseveriline from Liliaceae (D-96081), an antiarrhythmic
- Erysimoside from *Erysimum sp.* (Cruciferae) (E-96021), a cardiac glycoside Na^+/K^+-ATPase inhibitor[73]
- Heteratisine from Ranunculaceae (H-96061), a potent antiarrhythmic and Na^+ channel blocker in cardiomyocytes[71,74]
- Hetisine from Ranunculaceae (H-96053), a potent antiarrhythmic and local anesthetic[71]
- Nitrarine from Zygophyllaceae (N-96068), a hypotensive, spasmolytic, coronary dilator
- Severindione from Liliaceae (S-96082), an antiarrhythmic
- Veratroylzygadenine from *Veratrum sp.* (Liliaceae) (V-96077), an analog of the alkaloid veratridine that activates Na^+ channels[75]

1. J Biol Chem 258, 3503 (1983); **2.** J Med Chem 10, 66 (1967); **3.** J Membrane Biol 105, 1 (1988); **4.** Biochimie 70, 1285 (1988); **5.** Am J Physiol 242, C131 (1982); **6.** Pharmacol Toxicol 71, 95 (1992); **7.** Biochemistry 27, 2403 (1988); **8.** Biochemistry 24, 1394 (1985); **9.** Proc Natl Acad Sci USA 81, 3238 (1984); **10.** Mol Pharmacol 30, 112 (1986); **11.** Biochemistry 23, 4481 (1984); **12.** J Biol Chem 259, 4313 (1984); **13.** Mol Pharmacol 25, 131 (1984); **14.** J Biol Chem 269, 3374

(1994); **15.** Am J Physiol 265, H289 (1993); **16.** J Immunology 150, 4766 (1993); **17.** Pflügers Arch 421, 322 (1992); **18.** J Biol Chem 269, 2805 (1994); **19.** J Biol Chem 269, 860 (1994); **20.** Biochim Biophys Acta 1145, 266 (1993); **21.** Biophys J 66, 1 (1994); **22.** Hille, B., *Ionic Channels of Excitable Membranes, Second Edition*, Sinauer Associates (1992) pp. 59–62; **23.** Toxicon 23, 723 (1985); **24.** Fresenius Z Anal Chem 345, 212 (1993); **25.** J Assoc Official Anal Chem 74, 404 (1991); **26.** J Chromatography 257, 373 (1983); **27.** Toxicon 30, 1143 (1992); **28.** Anal Biochem 210, 87 (1993); **29.** *Marine Toxins: Origin, Structure, and Molecular Pharmacology (ACS Symposium Series, Volume 418)*, S. Hall and G. Strichartz, Eds., American Chemical Society (1990) pp. 29–65; **30.** Mol Brain Res 14, 64 (1992); **31.** FASEB J 3, 1807 (1989); **32.** Mol Pharmacol 40, 988 (1991); **33.** Toxicon 29, 1115 (1991); **34.** Mol Pharmacol 30, 129 (1986); **35.** J Pharmacol Exp Ther 229, 615 (1984); **36.** Mol Pharmacol 19, 345 (1981); **37.** J Biol Chem 268, 17114 (1993); **38.** Toxicon 30, 780 (1992); **39.** FEBS Lett 219, 355 (1987); **40.** J Org Chem 59, 2107 (1994); **41.** Toxicon 26, 97 (1988); **42.** J Org Chem 59, 2107 (1994); **43.** J Org Chem 59, 2101 (1994); **44.** Anal Chem 67, 1815 (1995); **45.** J Biol Chem 271, 8034 (1996); **46.** Insect Biochem Mol Biol 24, 13 (1994); **47.** Proc Natl Acad Sci USA 83, 3003 (1986); **48.** Biochemistry 29, 5941 (1990); **49.** J Biol Chem 270, 15153 (1995); **50.** Toxicol Toxin Rev 13, 25 (1994); **51.** FEBS Lett 315, 125 (1993); **52.** Biochemistry 31, 7622 (1992); **53.** Eur J Biochem 162, 589 (1987); **54.** FEBS Lett 181, 211 (1985); **55.** J Biol Chem 263, 1542 (1988); **56.** FEBS Lett 89, 54 (1978); **57.** Neurophysiologia 22, 201 (1990); **58.** Biochemistry 16, 531 (1977); **59.** Cell Biol Int'l 18, 723 (1994); **60.** Physiol Res 43, 33 (1994); **61.** Biochemistry 25, 8133 (1986); **62.** J Biol Chem 260, 14484 (1985); **63.** J Biol Chem 257, 5601 (1982); **64.** Cell Tissue Res 260, 529 (1990); **65.** J Cell Biol 103, 1473 (1986); **66.** J Biol Chem 259, 11176 (1984); **67.** Biochemistry 19, 969 (1980); **68.** Neuropharmacology 29, 567 (1990); **69.** Chem Nat Comp 19, 170 (1983); **70.** J Nat Prod 56, 2193 (1993); **71.** Phytochemistry 7, 625 (1968); **72.** J Am Chem Soc 103, 6536 (1981); **73.** Biochim Biophys Acta 937, 335 (1988); **74.** Tetrahedron 29, 3297 (1973); **75.** Naunyn-Schmiedeberg's Arch Pharmacol 342, 53 (1990).

18.7 Data Table *Probes for Na^+ Channels and Carriers*

Cat #	Structure	MW	Storage	Soluble	Abs	{ε x 10^{-3}}	Em	Solvent	Notes
A-1322	✓	789	F,D,L	DMSO	362	{7.5}	471	MeOH	
A-3108	✓	302	D,L	H_2O, MeOH	361	{18}	413	pH 7	<1>
A-96001	T	646	F,DD	DMSO	<300		none		
A-96015	T	887	D	DMSO	<300		none		
B-6851	✓	867	FF,D	DMSO*	<300		none		
B-6852	✓	895	FF,D	EtOH	<300		none		
B-6853	✓	897	FF,D	EtOH	<300		none		
B-6854	✓	911	FF,D	EtOH	<300		none		
B-6855	✓	899	FF,D	EtOH	<300		none		
B-6897	✓	356	D,L	MeOH	362	{21}	416	pH 7	<1>
B-6905	✓	611	F,D,L	DMSO, MeOH	504	{61}	511	MeOH	
B-96085	T	532	F,DD	H_2O	<300		none		
D-3109	✓	322	D,L	H_2O, MeOH	374	{24}	ND	MeOH	<1, 2>
D-3112	✓	294	D,L	H_2O, MeOH	374	{21}	420	MeOH	<1>
D-6885	✓	384	D,L	MeOH	362	{19}	415	pH 7	<1>
D-6898	✓	425	D,L	MeOH	363	{19}	415	pH 7	<1>
D-96035	T	543	F,DD	DMSO	345	{4.5}	415	MeOH	<3>
D-96057	T	508	F,DD	MeOH	<300		none		
D-96081	T	516	F,DD	EtOH	<300		none		
E-3111	✓	336	D,L	H_2O, MeOH	378	{23}	423	MeOH	<1>
E-96021	S10.6	715†	F,D	H_2O	<300		none		
G-6915	✓	411	FF,D	H_2O‡	<300		none		
G-6916	✓	475	FF,D	H_2O‡	<300		none		
G-6917	✓	411	FF,D	H_2O‡	<300		none		
G-6918	✓	396	FF,D	H_2O‡	<300		none		
H-3113	✓	348	D,L	H_2O, MeOH	374	{21}	ND	MeOH	<1, 2>
H-6883	✓	386	DD,L	MeOH	362	{16}	415	pH 7	<1>
H-96053	T	366	DD	H_2O	<300		none		
H-96061	T	392	DD	MeOH	<300		none		
I-3110	✓	336	D,L	H_2O, MeOH	376	{22}	425	MeOH	<1>
I-6880	✓	322	D,L	H_2O, MeOH	375	{24}	ND	MeOH	<1, 2>
I-6881	✓	394	F,D,L	MeOH	367	{18}	ND	pH 7	<1>
L-96010	T	666	F,DD	EtOH	<300		none		
M-6884	✓	386	D,L	MeOH	362	{20}	416	pH 7	<1>
N-6919	✓	317	FF,D	H_2O‡	<300		none		
N-96068	T	380	DD	H_2O	<300		none		
S-96082	T	428	D	EtOH	<300		none		
T-6913	✓	319	F,D	H_2O§	<300		none		
T-6914	✓	319	F,D	pH<5§	<300		none		
V-96077	T	658	F,DD	DMSO	<300		none		

For definitions of the contents of this data table, see "How to Use the Handbook" on page vi.
Structure: Chemical structure drawing: ✓ = shown in this section; Sn.n = shown in Section number n.n; T = available on request from our Technical Assistance Department and also through our Web site (http://www.probes.com).

* Brevetoxin PbTx-1 is not stable in EtOH or other alcohols.
† MW is for the hydrated form of this product.
‡ This product is supplied as a solution in 30 mM acetic acid, pH 3.
§ Tetrodotoxin is unstable in alkaline solutions.
<1> The molecular weights of A-3108 and I-6881 are for dihydrate forms. Molecular weights of other amiloride derivatives are based on the anhydrous compound.
<2> ND = not determined.
<3> Spectral data for D-96035 are estimates based on known properties of other esters of anthranilic acids; e.g., see J Biol Chem 257, 13354 (1982).

A-1322

	R^1	R^2
A-3108	$R^1 = H$	$R^2 = H$
D-3109	$R^1 = -CH_2CH_3$	$R^2 = -CH_2CH_3$
D-3112	$R^1 = -CH_3$	$R^2 = -CH_3$
E-3111	$R^1 = -CH_2CH_3$	$R^2 = -CH(CH_3)_2$
I-3110	$R^1 = -CH_3$	$R^2 = -CH_2CH(CH_3)_2$
I-6880	$R^1 = -CH_3$	$R^2 = -CH(CH_3)_2$

B-6851

B-6852 $R = -CH_2-C(=CH_2)-CHO$

B-6853 $R = -CH_2-C(=CH_2)-CH_2OH$

B-6855 $R = -CH_2CH(CH_3)-CH_2OH$

B-6854

B-6897 $R = H$

D-6885 $R = -CH_3$

D-6898 $R = Cl$

B-6905

D-3109, D-3112
E-3111
see A-3108

D-6885, D-6898
see B-6897

E-96021
Section 10.6

G-6915 $R^1 = -OSO_3^-$ $R^2 = H$

G-6917 $R^1 = H$ $R^2 = -OSO_3^-$

G-6916

G-6918 $R = -OSO_3^-$

N-6919 $R = H$

H-3113

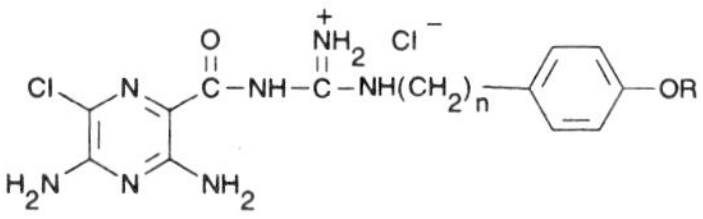

H-6883 R = H n = 2
M-6884 R = $-CH_3$ n = 1

I-3110, I-6880
see A-3108

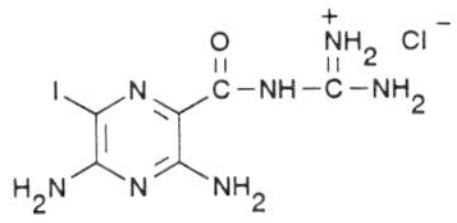

I-6881

M-6884
see H-6883

N-6919
see G-6918

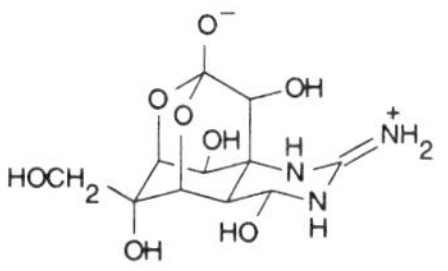

T-6913 / T-6914

18.7 Price List *Probes for Na^+ Channels and Carriers*

Cat #		Product Name	Unit Size	Price Per Unit ($) 1–4 Units	Price Per Unit ($) 5–24 Units
A-96001	***New***	aconitine	50 mg	45.00	36.00
A-3108		amiloride, hydrochloride, dihydrate	1 g	12.00	9.60
A-1322		9-anthroyl ouabain	5 mg	68.00	54.40
A-96015	***New***	askendoside D	5 mg	68.00	54.40
B-6897	***New***	benzamil, hydrochloride	25 mg	18.00	14.40
B-96085	***New***	benzoylheteratisine, hydrochloride	5 mg	75.00	60.00
B-6905	***New***	BODIPY® FL amiloride	25 µg	95.00	76.00
B-6851	***New***	brevetoxin PbTx-1 *from *Ptychodiscus brevis**	25 µg	65.00	52.00
B-6852	***New***	brevetoxin PbTx-2 *from *Ptychodiscus brevis**	25 µg	65.00	52.00
B-6853	***New***	brevetoxin PbTx-3 *from *Ptychodiscus brevis**	25 µg	65.00	52.00
B-6854	***New***	brevetoxin PbTx-6 *from *Ptychodiscus brevis**	25 µg	95.00	76.00
B-6855	***New***	brevetoxin PbTx-9 *from *Ptychodiscus brevis**	25 µg	65.00	52.00
B-6850	***New***	Brevetoxin Sampler Kit *5 µg of B-6851, B-6852, B-6853, B-6854, B-6855*	1 kit	135.00	108.00
D-96057	***New***	deltaline (eldeline)	100 mg	75.00	60.00
D-96035	***New***	N-desacetyllappaconitine	10 mg	128.00	102.40
D-96081	***New***	diacetylkorseveriline	50 mg	75.00	60.00
D-6898	***New***	2′,4′-dichlorobenzamil, hydrochloride	25 mg	38.00	30.40
D-3109		5-(N,N-diethyl)amiloride, hydrochloride	5 mg	35.00	28.00
D-3112		5-(N,N-dimethyl)amiloride, hydrochloride	5 mg	35.00	28.00
D-6885	***New***	2′,4′-dimethylbenzamil, hydrochloride	5 mg	70.00	56.00
E-96021	***New***	erysimoside, monohydrate	5 mg	60.00	48.00
E-3111		5-(N-ethyl-N-isopropyl)amiloride, hydrochloride	5 mg	70.00	56.00
G-6916	***New***	gonyautoxin C1 (epi-GTX VIII) *1 mg/mL*	100 µL	200.00	160.00
G-6917	***New***	gonyautoxin GTX-I *1 mg/mL*	100 µL	200.00	160.00
G-6915	***New***	gonyautoxin GTX-IV (neosaxitoxin 11β-sulfate) *250 µg/mL*	100 µL	200.00	160.00
G-6918	***New***	gonyautoxin GTX-VI (B2) *1 mg/mL*	100 µL	200.00	160.00
H-96061	***New***	heteratisine	10 mg	128.00	102.40
H-96053	***New***	hetisine, hydrochloride	10 mg	128.00	102.40
H-3113		5-(N,N-hexamethylene)amiloride, hydrochloride	5 mg	70.00	56.00
H-6883	***New***	p-hydroxyphenethamil, hydrochloride	5 mg	70.00	56.00
I-6881	***New***	6-iodoamiloride, hydrochloride, dihydrate	5 mg	70.00	56.00
I-3110		5-(N-isobutyl-N-methyl)amiloride, hydrochloride	5 mg	35.00	28.00
I-6880	***New***	5-(N-isopropyl-N-methyl)amiloride, hydrochloride	5 mg	70.00	56.00
L-96010	***New***	lappaconitine, hydrobromide (allapinine)	10 mg	155.00	124.00
L-98221	***New***	LqhαIT insect toxin *from *Leiurus quinquestriatus hebraeus* scorpion venom*	50 µg	415.00	332.00
L-98217	***New***	LTx-1 *from *Leiurus quinquestriatus hebraeus* scorpion venom*	50 µg	325.00	260.00
L-98218	***New***	LTx-2 *from *Leiurus quinquestriatus hebraeus* scorpion venom*	50 µg	325.00	260.00
L-98219	***New***	LTx-3 *from *Leiurus quinquestriatus hebraeus* scorpion venom*	50 µg	415.00	332.00
L-98220	***New***	LTx-4 *from *Leiurus quinquestriatus hebraeus* scorpion venom*	100 µg	525.00	420.00
M-6884	***New***	4′-methoxybenzamil, hydrochloride	5 mg	70.00	56.00
N-6919	***New***	neosaxitoxin (neo-STX) *250 µg/mL*	100 µL	200.00	160.00
N-96068	***New***	nitrarine, dihydrochloride	50 mg	95.00	76.00
S-96082	***New***	sevedindione	20 mg	105.00	84.00
T-6913	***New***	tetrodotoxin (TTX) *contains citrate buffer*	1 mg	98.00	78.40
T-6914	***New***	tetrodotoxin (TTX) *citrate free*	1 mg	98.00	78.40
V-96077	***New***	veratroylzygadenine	10 mg	55.00	44.00

18.8 Other Natural and Synthetic Toxins

Molecular Probes is pleased to supply a variety of biologically active natural products provided to us by Latoxan (Table 18.6). In many cases the research community has had limited access to these materials; consequently, most biological effects have been determined through animal studies, and detailed mechanistic information is lacking.

We expect to keep limited supplies of most of the following products in stock continuously. Other Latoxan products may be added to our current inventory. An up-to-date list of all Latoxan products currently available from Molecular Probes is available from our Technical Assistance Department or through our Web site (http://www.probes.com).

Table 18.6 *Neuroactive compounds, hormonal compounds and other bioactive alkaloids and nonalkaloids from plants.*

Cat #	Product *Source*	MW †	Store ‡	Soluble §	Activity	Ref
Bioactive Alkaloids from Plants						
C-96086	Convolamine *Convolvulaceae*	305	F,D	EtOH	• Selective sensory nerve anesthetic • Vasodilator	
C-96087	Convolvine *Convolvulaceae*	291	F,D	EtOH	• Selective sensory nerve anesthetic • Vasodilator	
C-96012	3-Cyano-4-methoxy-*N*-methyl-2-pyridone (ricinine) **Ricinus communis**	164	D	EtOH	• Convulsant • Causes respiratory depression	1
D-96020	Dubinidine *Rutaceae*	275	D	EtOH	• Anticonvulsant • Reduces motor activity	
E-96089	Evoxine (haploperine) *Rutaceae*	347	D	EtOH	• Sedative • Strychnine antagonist • Enhances effects of narcotics	2
F-96027	Foliosidine *Rutaceae*	307	D	EtOH	• Anticonvulsant • Causes hypothermia	3
G-96029	Graveoline (foliosine) *Rutaceae*	279	D	DMSO	• CNS stimulant	
H-96007	Heliotrine **Heliotropum sp.* (Boraginaceae)*	313	F,D	EtOH	• Potent liver and lung toxin • Mutagen	4–6
H-96030	Hernandezine **Thalictrum sp.* (Ranunculaceae)*	653	D	DMSO	• Inhibits platelet Aggregation • Anti-inflammatory • Analgesic • Probable Ca^{2+} antagonist	7
H-96062	Hippeastrine *Amaryllidaceae*	396	F,D	H_2O	• Hypotensive • Sedative • Antitumor agent	8
L-96011	Lycorine *Amaryllidaceae*	324	D	EtOH	• Specific inhibitor of ascorbic acid biosynthesis and of cell division in plants, yeasts, rat spermatocytes and rat fibroblasts	9–16
N-96025	Norfluorocurarine (vincanine) **Vinva erecta**	292	F,D	DMSO	• CNS stimulant • Convulsant • Probable glycinergic antagonist	
R-96095	Remerine, hydrochloride *Papaveraceae*	316	D	H_2O	• CNS stimulant • Probable dopaminergic agonist	
S-96046	Skimmianine *Rutaceae*	259	D	DMSO	• 5-Hydroxytryptamine receptor probe • Antileishmanial agent	17–20
S-96097	Songorine **Aconitum sp.* (Ranunculaceae)*	357	D	DMSO	• Adrenergic and serotonergic stimulant • Antiarrythmic	
T-96098	Tadzhaconine *Ranunculaceae*	534	F,D	EtOH	• Antiarrythmic • Local anaesthetic	
T-96049	Trichodesmine **Trichodesma incanum**	353	F,D	EtOH	• Produces encephalopathy • May be hepatotoxic	
U-96075	Ungerine **Ungernia sp.* (Amaryllidaceae)*	392	F,D	H_2O	• Sedative	
U-96076	Unsevine **Ungernia sp.* (Amaryllidaceae)*	331	F,D,A	DMSO	• Inhibits orientation reaction	

Table 18.6, continued *Neuroactive compounds, hormonal compounds and other bioactive alkaloids and nonalkaloids from plants.*

Cat #	Product *Source*	MW †	Store ‡	Soluble §	Activity	Ref
Bioactive Nonalkaloids from Plants						
A-96013	Allochroside *Acanthophyllum sp.*	~1980	F,D	H_2O	• Nonspecific immunogenic stimulant	
A-96055	Austricine *Compositae*	280	F,D	DMSO	• Hypolipidemic • Anti-inflammatory	
E-96005	Ecdysterone (β-ecdysone) *Rhaponticum sp. (Compositae)*	481	D	EtOH	• Arthropod molting hormone • Anabolic effect in mammals • Ca^{2+}-ATPase stimulant • Induces acetylcholinesterase activity in mammalian brain	21–23
E-96060	Eudesmine *Rutaceae*	386	D	DMSO	• Short-acting estrogenic activity • Sedative	24
F-96022	Ferulin *Ferula sp. (Umbelliferae)*	358	F,D	EtOH	• Estrogenic activity	25–27
F-96006	Ferutinine *Ferula sp. (Umbelliferae)*	~365	F,D	DMSO	• Estrogenic activity	28,29
G-96090	Glyrophama, dihydrate (amorphine) *Leguminosae*	741	D	pyridine	• Reduces platelet aggregation	
G-96083	Gossypol *Gossypium sp. (Malvaceae)*	519	D,AA	acetone	• Male contraceptive • Antiviral, antipsoriatic, antitumor agent	30–35
K-96063	Katacine *Polygonaceae*	~7800	D,AA	H_2O	• Antihypoxic	36
L-96065	Lapidine *Umbelliferae*	334	F,D	EtOH	• Estrogenic activity	25
L-96066	Lapiferine *Umbelliferae*	395	F,D	EtOH	• Estrogenic activity	37
L-96067	Leucomisine *Compositae*	246	F,D	DMSO	• Hypolipidemic • Anti-inflammatory	
P-96069	Pinocembrine *Leguminosae*	256	D	EtOH	• Anti-inflammatory	
S-96047	Stevioside *Stevia sp. (Compositae)*	805	F,D	H_2O	• Sweetener • Possible anti-hypertensive	38,39
T-96048	Tschimganidine *Umbelliferae*	389	F,D	EtOH	• Estrogenic activity	
T-96072	Tschimganine *Umbelliferae*	304	F,D	EtOH	• Estrogenic activity	27,29,40
T-96073	Tschimgine *Umbelliferae*	274	F,D	EtOH	• Estrogenic activity	25,27,40
U-96074	Ugaferine *Umbelliferae*	449	F,DD	EtOH	• Estrogenic activity	41

† Structures of most of these products are available from our Technical Assistance Department or through our Web site (http://www.probes.com). ‡ For definitions of storage codes, see "How to Use the Handbook" on page vi. § Suggested solvents have not been tested in all cases.

1. J Biol Chem 241, 4411 (1966); **2.** Aust J Chem 7, 87 (1954); **3.** Chem Nat Comp 2, 20 (1966); **4.** Toxicol Lett 63, 47 (1992); **5.** Mutation Res 142, 209 (1985); **6.** Cancer Res 35, 2020 (1975); **7.** Drug News and Perspectives 3, 425 (1990); **8.** J Nat Prod 56, 1423 (1993); **9.** Current Genetics 25, 80 (1994); **10.** Chem Pharm Bull 35, 1070 (1987); **11.** Current Genetics 11, 247 (1986); **12.** Exp Cell Res 150, 314 (1984); **13.** FEBS Lett 160, 129 (1983); **14.** Biochim Biophys Acta 425, 342 (1976); **15.** FEBS Lett 60, 66 (1975); **16.** Nature 256, 513 (1975); **17.** Phytochemistry 39, 1091 (1995); **18.** J Auton Pharmacol 14, 365 (1994); **19.** Antimicro Agents Chemother 37, 859 (1993); **20.** Biochem Biophys Res Comm 23, 679 (1966); **21.** J Auton Pharmacol 14, 365 (1994); **22.** Receptor 3, 203 (1993); **23.** Comp Biochem Physiol 78C, 193 (1984); **24.** Nat Products Reports 2, 195 (1985); **25.** Chem Nat Comp 17, 244 (1981); **26.** Chem Nat Comp 13, 424 (1977); **27.** Chem Nat Comp 8, 796 (1972); **28.** J Pharmacol Toxicol 4, 37 (1990) (Russian); **29.** Chem Nat Comp 8, 794 (1972); **30.** Cancer Res 54, 1707 (1994); **31.** Biochem Pharmacol 46, 251 (1993); **32.** J Animal Sci 70, 1628 (1992); **33.** J Clin Invest 79, 517 (1987); **34.** Biochim Biophys Acta 814, 405 (1985); **35.** Mol Cell Biochem 47, 65 (1982); **36.** Science 212, 927 (1981); **37.** Chem Nat Comp 19, 281 (1983); **38.** Eur J Pharmacol 183, 1822 (1990); **39.** Toxicology 72, 299 (1992); **40.** Chem Nat Comp 8, 53 (1972); **41.** Chem Nat Comp 14, 614 (1978).

18.8 Price List *Other Natural and Synthetic Toxins*

Cat #	Product Name	Unit Size	Price Per Unit ($) 1–4 Units	5–24 Units
A-96013 ***New***	allochroside *mixture of 4 glycosides from plants of the *Acanthophyllum* genus*	100 mg	95.00	76.00
A-96055 ***New***	austricine	100 mg	68.00	54.40
C-96086 ***New***	convolamine	20 mg	55.00	44.00
C-96087 ***New***	convolvine	20 mg	60.00	48.00
C-96012 ***New***	3-cyano-4-methoxy-N-methyl-2-pyridone (ricinine)	10 mg	68.00	54.40

Cat #		Product Name	Unit Size	Price Per Unit ($) 1–4 Units	5–24 Units
D-96020	***New***	dubinidine	20 mg	108.00	86.40
E-96005	***New***	ecdysterone (β-ecdysone)	10 mg	85.00	68.00
E-96060	***New***	eudesmine	10 mg	128.00	102.40
E-96089	***New***	evoxine (haploperine)	20 mg	108.00	86.40
F-96022	***New***	ferulin	10 mg	128.00	102.40
F-96006	***New***	ferutinine *mixture of ferutinine and tenuferidine*	5 mg	128.00	102.40
F-96027	***New***	foliosidine	20 mg	128.00	102.40
G-96090	***New***	glyrophama, dihydrate (amorphine)	50 mg	68.00	54.40
G-96083	***New***	gossypol	100 mg	45.00	36.00
G-96029	***New***	graveoline (foliosine)	20 mg	108.00	86.40
H-96007	***New***	heliotrine	50 mg	128.00	102.40
H-96030	***New***	hernandezine	5 mg	85.00	68.00
H-96062	***New***	hippeastrine, hydrobromide	20 mg	95.00	76.00
K-96063	***New***	katacine *mixture of polymers of proanthocyanidins*	100 mg	195.00	156.00
L-96065	***New***	lapidine	10 mg	128.00	102.40
L-96066	***New***	lapiferine	10 mg	128.00	102.40
L-96067	***New***	leucomisine	100 mg	148.00	118.40
L-96011	***New***	lycorine, hydrochloride	20 mg	75.00	60.00
N-96025	***New***	norfluorocurarine (vincanine)	20 mg	108.00	86.40
P-96069	***New***	pinocembrine	100 mg	68.00	54.40
R-96095	***New***	remerine, hydrochloride	20 mg	75.00	60.00
S-96046	***New***	skimmianine	20 mg	75.00	60.00
S-96097	***New***	songorine	5 mg	135.00	108.00
S-96047	***New***	stevioside	100 mg	45.00	36.00
T-96098	***New***	tadzhaconine	20 mg	95.00	76.00
T-96049	***New***	trichodesmine	5 mg	108.00	86.40
T-96048	***New***	tschimganidine	10 mg	95.00	76.00
T-96072	***New***	tschimganine	10 mg	75.00	60.00
T-96073	***New***	tschimgine	10 mg	108.00	86.40
U-96074	***New***	ugaferine	10 mg	108.00	86.40
U-96075	***New***	ungerine nitrate	10 mg	55.00	44.00
U-96076	***New***	unsevine	10 mg	85.00	68.00

Technical Assistance at Our Web Site (http://www.probes.com)

At Molecular Probes' Web site, we are developing an electronic version of this Handbook and other databases that should prove extremely useful for the researcher. In addition to containing all of the text from this Handbook, our Web site provides:

- **Product searches** by product name or catalog number
- **Bibliographies** for all products for which we have references
- **Keyword searches** of our entire bibliography of over 25,000 references
- **Product information sheets** for many kits and reagents
- **Technical bulletins**, including *BioProbes* newsletters and other product literature
- **Chemical structure**, **technical data** and **material safety and data sheets**
- **Color photomicrographs** that show our products in action

Visit our Web site often for new additions to our bibliography, as well as new products and upgraded search capabilities. Also look for special sales and introductory specials on some important products.

If you do not have access to the Internet or you need assistance that is not available at that site, further information on the scientific and technical background of our products can be obtained by contacting our Technical Assistance Department at the numbers listed on the inside front cover.

Chapter 19

Photoactivatable (Caged) Probes

Contents

19.1 Caging Groups and Their Photolysis

Overview of Caged Probes

Flash photolysis of photoactivatable or "caged" probes provides a means of controlling the release — both spatially and temporally — of biologically active products or other reagents of interest.[1-4] The chemical caging process may also confer membrane permeability on the caged ligand, as is the case for caged cAMP[5] and caged luciferin.[6] A variety of caged probes exists, including caged nucleotides, chelators, second messengers and neurotransmitters. These probes have tremendous potential for use with live cells. Moreover, the availability of a low-cost, high-intensity flash device for microscopes facilitates photolysis experiments in many laboratories.[7]

In addition to caged versions of biologically active molecules, this chapter describes caged fluorescent dyes that are essentially nonfluorescent until after photolysis. These caged fluorophores have proven useful for photoactivation of fluorescence (PAF) experiments, which are analogous to fluorescence recovery after photobleaching (FRAP) experiments except that the fluorophore is activated upon illumination rather than bleached. Measuring the bright fluorescent signal of the photoactivated fluorophore against a dark background should be intrinsically more sensitive than measuring a dark photobleached region against a bright field.

Properties of Different Caging Groups

The caging moiety is designed to *maximally* interfere with the binding or activity of the molecule. It is detached in microseconds to milliseconds by flash photolysis at ≤360 nm, resulting in a pulse of the active product. Uncaging can easily be accomplished with UV illumination in the fluorescence microscope or with a UV laser or flashlamp. The effects of photolytic release are frequently monitored either with fluorescent probes that measure calcium, pH, other ions or membrane potential or with electrophysiological techniques.

To date, most of the caged reagents described in the literature have been derivatives of *o*-nitrobenzylic compounds. The nitrobenzyl group is synthetically incorporated into the biologically active molecule via an ether, thioether, ester (including phosphate ester), amine or similar linkage to a hetero atom (usually O, S or N). Both the structure of the nitrobenzyl moiety and the atom to which it is attached have some effect on the efficiency and wavelength required for uncaging. Ten different caging groups are available; their properties are summarized in Table 19.1.

- Probes caged with the α-carboxy-2-nitrobenzyl (**CNB**)–caging group generally have the most advantageous properties. These include good water solubility, very fast uncaging rates in the microsecond range, high photolysis quantum yields (from 0.2–0.4) and biologically inert photo-by-products. Although the absorption maximum of the CNB-caging group is at 260 nm, its absorption spectrum tails out to approximately 360 nm, allowing successful photolysis using light with wavelengths ≤360 nm. The 1-(2-nitrophenyl)ethyl (**NPE**)–caging group, with properties similar to those of CNB, can also be photolyzed at ≤360 nm.

- As compared with CNB and NPE, the caging groups 4,5-dimethoxy-2-nitrobenzyl (**DMNB**), 1-(4,5-dimethoxy-2-nitrophenyl)-ethyl (**DMNPE**) and ((4,5-dimethoxy-2-nitrobenzyl)oxy)carbonyl (**NVOC**) have longer-wavelength absorption (absorption maximum ~355 nm) and therefore absorb 340–360 nm light

Table 19.1 *Properties of ten different caging groups.*

Probe	Uncaging Rate *	Photolysis Quantum Yield *	Inertness of the Photoproduct	Confers Water Solubility	Long-Wavelength Absorption (≥360 nm)
CNB	++++	+++++	+++++	+++++	++
NPE	+++	+++	+++	+	++
DMNB	+++	+++	++	+	+++++
NVOC	++	+++	++	+	+++++
DMNPE	+++	+++	+++	+	+++++
CMNB	+++	+++	+	++++	+++
CMNCBZ	++	+++	+	++++	+++
ANB	+++	+++	+	+	+++
NB	++	++	+	+	++
Desyl	+++++	++++	+++++	+	+++

+++++ = optimal response; + = poor response. * Both the structure of the nitrobenzyl moiety and the atom to which it is attached have some effect on the efficiency and wavelength required for uncaging.

Dichroic Mirrors for Simultaneous Photoactivation of Caged Compounds and Visualization of Fluorescent Probes in a Fluorescence Microscope

Molecular Probes now provides three dichroic mirrors (O-5804, O-5805, O-5806, see Section 26.5) that are designed by Omega® Optical specifically for use with caged compounds. All three dichroics have exceptionally high reflectance in the near UV, as well as in one or more regions of the visible spectrum (see figure). They can be used in several different configurations both for photoactivation of caged compounds and for excitation/visualization of green or orange fluorescence. These dichroic mirrors are usually used in conjunction with interference filters for discrete photoactivation of caged compounds and excitation of fluorophores. Typical configurations utilize these dichroic mirrors in the:

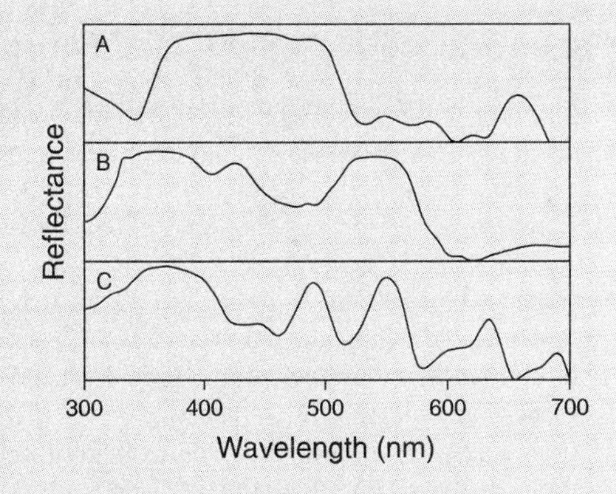

Reflectance spectra of dichroic mirrors for use with caged compounds: A) O-5804, B) O-5805 and C) O-5806 (see Section 26.5).

- **Epi-illumination path**: The dichroic mirrors may be used in conjunction with a manual or automated filter changer, scanning monochromator or dual illuminator. Because the dichroic mirror will reflect both UV light for uncaging and visible light for fluorescence excitation, the sample can be exposed to light of multiple wavelengths, either simultaneously or sequentially, without changing the dichroic mirror.

- **Transmitted light path**: The dichroic mirrors may also be mounted inside adapters used in place of the standard transmitted light condenser. A standard high numerical aperture (NA), UV-transmitting objective lens is substituted for the condenser, and UV light is directed onto the specimen via the dichroic mirror, which is placed behind this objective lens. Filtered light from a continuous arc source or flashlamp is directed through an iris onto the dichroic mirror by a light guide. Because the dichroic mirror reflects light at one or more visible wavelengths in addition to UV wavelengths, an interference filter of longer wavelength can be placed in the light path to allow visualization of the area to be photoactivated.

more efficiently. However, photolysis rates and quantum yields of DMNB-, DMNPE- and NVOC-caged probes are generally lower than those obtained for CNB-caged probes. Uncaging of NVOC-caged compounds is a two-step process, first yielding a carbamic acid derivative, which then spontaneously decarboxylates (see Figure 19.3B).

- The 5-carboxymethoxy-2-nitrobenzyl (**CMNB**)– and ((5-carboxymethoxy-2-nitrobenzyl)oxy)carbonyl (**CMNCBZ**)–caging groups provide an absorption maximum of intermediate wavelength (absorption maximum ~320 nm), while imparting significant water solubility to the caged probe. Photolysis rates and quantum yields are generally similar to those of NPE- and DMNB-caged probes. As with NVOC-caged compounds (see Figure 19.3B), CMNCBZ-caged compounds undergo a two-step uncaging process upon illumination.

- 2-Nitrobenzyl (**NB**) is one of the first groups used for caging biomolecules. Its potentially amine-reactive photo-by-product (2-nitrobenzaldehyde) may be cytotoxic.

- Desoxybenzoinyl (**desyl**)–caging groups represent an exciting new development in caging technology. Desyl-caged phosphates have been shown to be uncaged on the order of 10^8 sec^{-1}, with photolysis quantum yields >0.3.[8] The photo-by-product is expected to be biologically inert and photolysis is efficiently performed at 360 nm.

Experiments involving probes caged with all of the above caging groups except the desyl- and CNB-caging groups may require the addition of dithiothreitol (DTT, D-1532; see Section 2.1); this reducing reagent prevents the reaction between amines and the 2-nitrosobenzoyl photo-by-products.[9]

1. Current Opin Neurobiol 6, 379 (1996); **2.** Corrie, J.E.T. and Trentham, D.R. in *Bioorganic Photochemistry, Volume 2*, H. Morrison, Ed., J. Wiley (1993) pp. 243–305; **3.** Kao, J.P.Y. and Adams, S.R. in Optical Microscopy: Emerging Methods and Applications, B. Herman and J.J. Lemasters, Eds., Academic Press (1993) pp. 27–85; **4.** Ann Rev Physiol 55, 755 (1993); **5.** Nature 310, 74 (1984); **6.** BioTechniques 15, 848 (1993); **7.** Pflügers Arch 411, 200 (1988); **8.** J Am Chem Soc 115, 6001 (1993); **9.** Ann Rev Biophys Biochem 18, 239 (1989).

Product-Specific Bibliographies and Keyword Searches

Our Technical Assistance Department can provide you with product-specific bibliographies, as well as keyword searches of the over 25,000 literature references in our extensive bibliography database. Our bibliography database is also searchable through our Web site (http://www.probes.com).

19.2 Caged Probes for a Variety of Applications

Caged Nucleotides and Caged Phosphates

Our new desyl-caged ATP (A-7073) and phosphate (D-7076) are uncaged in submicroseconds and release a furan photo-by-product that is chemically inert.[1] Several of our photoactivatable nucleotides and phosphates are available with a choice of caging groups:

- Caged ATP (A-1048, A-1049, A-7073), which has been shown to release ATP in skinned muscle fibers,[2-4] sarcoplasmic reticulum vesicles,[5] submitochondrial particles [6] and membrane fragments containing Na^+/K^+-ATPase [7]
- Caged ADP (A-7056), which has been used to investigate the molecular basis of contraction of skeletal muscle fibers [8]
- Caged GTP-γ-*S* (G-1053) and GDP-β-*S* (G-3620), important probes for studying regulatory proteins such as G-proteins [9-11]
- Caged cAMP (D-1037, N-1045), which is cell-permeant and rapidly photolyzed to cAMP [12]
- Caged phosphate (H-6829, N-7065, D-7076), which can be used to produce a pulse of phosphate and, in the case of NPE-caged proton [13-15] (H-6829) and NPE-caged phosphate [16] (N-7065), to generate a photolysis-dependent proton release that results in a rapid drop in pH

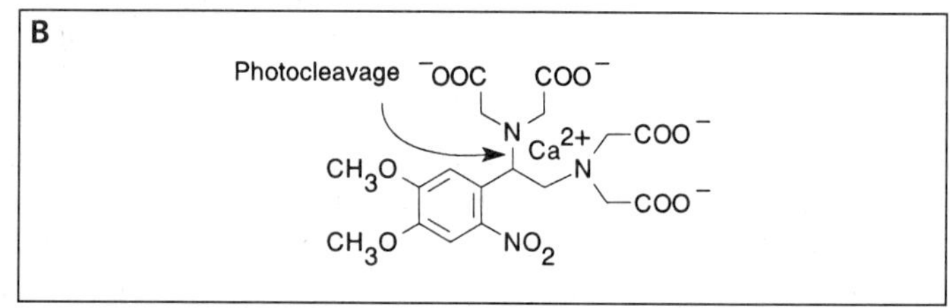

Figure 19.1 *A) NP-EGTA (N-6802) complexed with Ca^{2+}. Upon illumination, this complex is cleaved to yield free Ca^{2+} and two iminodiacetic acid photoproducts. The affinity of the photoproducts for Ca^{2+} is ~12,500-fold lower than that of NP-EGTA. B) DMNP-EDTA (D-6814) complexed with Ca^{2+}. Upon illumination, this complex is cleaved to yield free Ca^{2+} and two iminodiacetic acid photoproducts. The affinity of the photoproducts for Ca^{2+} is ~600,000-fold lower than that of DMNP-EDTA.*

Caged Ca^{2+}, Caged Ca^{2+} Chelators and Caged Ionophores

Caged ions and caged chelators can be used to influence the ionic composition of both solutions and cells, particularly for ions such as Ca^{2+} that are present at low concentrations.

Caged Ca^{2+} Reagents

Developed by Ellis-Davies and Kaplan,[17,18] nitrophenyl EGTA (NP-EGTA) is a new photolabile Ca^{2+} chelator that exhibits a high selectivity for Ca^{2+} ions, a dramatic increase in its K_d for Ca^{2+} upon illumination (from 80 nM to 1 mM) and a high photolysis quantum yield (0.23). NP-EGTA's affinity for Ca^{2+} *decreases* ~12,500-fold upon photolysis. Furthermore, its K_d for Mg^{2+} of 9 mM makes NP-EGTA essentially insensitive to physiological Mg^{2+} concentrations. We offer the tetrapotassium salt (N-6802) and the acetoxymethyl (AM) ester (N-6803) of NP-EGTA. The NP-EGTA salt can be complexed with Ca^{2+} to generate a caged Ca^{2+} reagent that will rapidly deliver Ca^{2+} upon photolysis [19-21] (Figure 19.1A). The cell-permeant AM ester of NP-EGTA does not bind Ca^{2+} unless its AM ester groups are removed. This AM ester can potentially serve as a photolabile chelator in cells because, once converted to NP-EGTA by intracellular esterases, it will bind free Ca^{2+} until photolyzed with UV light.

The first caged Ca^{2+} reagent described by Kaplan and Ellis-Davies was 1-(4,5-dimethoxy-2-nitrophenyl) EDTA (DMNP-EDTA, D-6814), which they named DM-Nitrophen™ [22-24] (now a trademark of Calbiochem-Novabiochem Corp.). Because its structure more resembles that of EDTA than EGTA, we named it as a caged EDTA derivative (Figure 19.1B). Upon illumination, DMNP-EDTA's K_d for Ca^{2+} increases from 5 nM to 3 mM. Thus, photolysis of DMNP-EDTA complexed with Ca^{2+} results in a pulse of free Ca^{2+}. Furthermore, DMNP-EDTA has significantly higher affinity for Mg^{2+} (K_d = 2.5 µM) [22] than does NP-EGTA (K_d = 9 mM),[17] making it a potentially useful caged Mg^{2+} reagent. A recent paper by Neher and Zucker discusses the uses and limitations of DMNP-EDTA.[25] In addition to the high-purity tetrapotassium salt of DMNP-EDTA (D-6814), we have prepared its AM ester (D-6815), which has not been previously available.

Caged Ca^{2+} Chelators

In contrast to NP-EGTA and DMNP-EDTA, diazo-2 is a photoactivatable Ca^{2+} scavenger. Diazo-2, which was introduced by Adams, Kao and Tsien,[26,27] exhibits a low affinity for Ca^{2+} before photolysis (K_d = 2.2 µM) that *increases* by a factor of 30 upon exposure to UV light. Diazo-2 can be microinjected as its K^+ salt (D-3034) or loaded into cells as the AM ester (D-3036). Upon illumination near 360 nm, the photolyzed diazo-2 binds cytosolic free Ca^{2+} within a few milliseconds. Diazo-2 has been used to rapidly decrease cytosolic Ca^{2+} in tensed frog muscle cells [28] and in rat fibroblasts.[26] The dipotassium salt and AM ester of diazo-3 (D-3700, D-3701) are also available.[27] Although diazo-3 binds Ca^{2+} poorly even after photolysis, it can be used as a control for diazo-2 to determine the effects of acidity and photoproducts on the cell.

Caged Ionophores

We are introducing some unique new tools for the study of Ca^{2+} regulation and ion transport. The cell-permeant DMNPE-caged ionophores derived from A-23187 (A-7109) or 4-bromo A-23187 (B-7108) can be passively loaded into cells and, at a later time, photoactivated with a UV laser or flashlamp to stimulate ion movement across the plasma membrane or to liberate Ca^{2+} from intracellular stores. We have determined that both products are inactive until uncaged with UV light. We also offer a DMNPE-caged version of the monovalent ionophore nigericin (N-7107), which generates free nigericin upon photolysis.

Caged Second Messengers and Caged Analogs of Other Biologically Active Molecules

Photoactivatable Nitric Oxide

Nitric oxide (NO) plays a critical role as a molecular mediator of a variety of physiological processes, including blood-pressure regulation and neurotransmission. However, NO's half-life is only about five seconds, thus severely limiting the types of experiments that can be carried out. Our caged inorganic NO complex, potassium nitrosylpentachlororuthenate (P-7070), provides researchers with a means of manipulating the presence of NO in experimental systems. Perfusion of hippocampal slices with this inorganic NO complex followed by photolysis is reported to release NO at concentrations sufficient to modulate NMDA receptor–mediated transmission.[29] Our new NVOC-caged SIN-1 (D-7111) is uncaged *in vitro* in microseconds after UV illumination to release the NO carrier SIN-1, which then decomposes to release NO; however, its use in live cells and tissues has not yet been reported. Other photoactivatable nitric oxide donors, including *N-tert*-butyl-α-phenylnitrone [30,31] (B-7893), 2-methyl-2-nitrosopropane dimer [32] (M-7905) and *S*-nitrosoglutathione [33] (N-7903), are discussed in Sections 21.3 and 21.5.

Caged Cyclic ADP-Ribose

Cyclic ADP-ribose (cADP-ribose) is a potent intracellular Ca^{2+} mobilizing agent that likely functions as a second messenger in an inositol-1,4,5-triphosphate (IP_3)–independent pathway.[34-36] Our new NPE-caged cADP-ribose (C-7074) induces Ca^{2+} mobilization in sea urchin egg homogenates only after photolysis, and this Ca^{2+} release is inhibited by the specific cADP-ribose antagonist 8-amino-cADP-ribose (A-7621, see Section 20.2).[37] Furthermore, when microinjected into live sea urchin eggs, NPE-caged cADP-ribose was shown to mobilize Ca^{2+} and activate cortical exocytosis after illumination with a mercury-arc lamp.[37]

Caged Luciferin

DMNPE-caged luciferin (L-7085) readily crosses cell membranes, allowing more efficient delivery of luciferin into intact cells.[38] Once the caged luciferin is inside the cell, active luciferin can be released either instantaneously by a flash of UV light, or continuously by the action of endogenous intracellular esterases found in many cell types.

Photolabile Enzyme Inhibitors

Molecular Probes offers two photolabile serine protease inhibitors — *trans-p*-(diethylamino)-*o*-hydroxy-α-methylcinnamic acid, *p*-nitrophenyl ester (D-7112) and *trans-p*-(diethylamino)-*o*-hydroxy-α-methylcinnamic acid, *p*-amidinophenyl ester (D-7113) — for investigating the active site of chymotrypsin, factor X_a and thrombin.[39] Photolabile inhibitors are important tools for effecting the photorelease of a biological catalyst and for modeling the light-dependent enzymatic activity observed in many biological systems. Molecular Probes also offers BNPA (B-6610), a reagent for caging thiols, including enzymes with active-site thiols (see below).

Caged Analogs of Other Biologically Active Molecules

We have made a strong commitment to develop unique photoactivatable probes, including caged versions of:

- Arachidonic acid (A-7062, A-7068), an important fatty acid second messenger
- Dioctanoylglycerol (D-7081), a protein kinase C activator [40]
- Paclitaxel (Taxol equivalent) (P-7082), a tubulin-assembly promoter
- Penicillin V (P-7061), an antibiotic

Caged Neurotransmitters and Caged Drugs

Once activated, caged neurotransmitters rapidly initiate or block neurotransmitter action, thus providing tools for kinetic studies of receptor binding or channel opening. We offer several caged neurotransmitters, all of which are biologically inactive before photolysis. These include our exceptional CNB-caged L-glutamic acid (G-7055), which is rapidly photolyzed with UV light [41] (Figure 19.2), as well as caged versions of:

- Aspartic acid [42] (A-2505)
- Carbamylcholine [43-50] (C-1791)
- Dopamine [51] (C-7124, D-7064)
- Epinephrine (D-7057)
- GABA [20,42,52,53] (A-2506, A-7110, A-7125)
- Glutamic acid [41,52,54-57] (C-7122, G-2504, G-7055, G-7120, G-7121)
- Glycine [42] (G-2507)
- Haloperidol (H-7080)
- Isoproterenol [58-60] (N-7066)
- Kainic acid [61] (K-7115)
- NMDA [62] (M-7069, M-7114)
- NMDA receptor antagonist MK-801 (D-7059)
- Norepinephrine (D-7058)
- Phenylephrine [63] (C-7083, D-7084)
- Propranolol (N-7067)
- Serotonin (D-7063)

For further discussion of the biological activity of our caged neurotransmitters and caged drugs after photolysis, see Section 18.3.

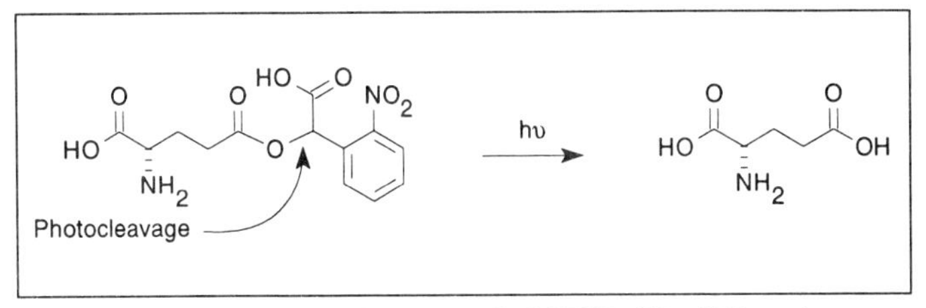

Figure 19.2 CNB-caged L-glutamic acid (G-7055). The CNB-caging group is rapidly photocleaved with UV light to release L-glutamic acid.

Caged Fluorescent Dyes

Caged Fluorescent Dyes and Their Reactive Derivatives

Photoactivatable fluorescent dyes may be one of the most useful developments in caged probe technology. In general, the caged fluorophores are colorless and nonfluorescent until photolyzed with UV light. Demonstrated and suggested applications include the study of fluid dynamics [64-66] (see Figure 15.6 in Chapter 15), hydrodynamic properties of the cytoplasmic matrix and lateral diffusion in membranes, as well as cell–cell communication and the role of gap junctions. Photoactivatable fluorophores are particularly useful for investigating cell lineage because they can be injected early in development when the cells are large, and then later activated in the growing embryo when the cells of interest may be small and buried deep within the tissue.[67] Caged fluorescent dyes have also been conjugated to proteins to follow the assembly of tubulin,[68] microtubule flux [68-70] and the dynamic behavior of actin filaments.[71-73]

Molecular Probes prepares a wide array of caged dyes and their reactive derivatives, as well as several conjugates, including:

- Polar tracers: caged fluorescein [66,74] (F-7103), caged HPTS [64,65,75] (D-7060), caged rhodamines (C-7100, R-7102)
- Reactive tracers: succinimidyl esters and sulfosuccinimidyl esters of caged carboxy-Q-rhodamine (C-7101), caged Rhodamine Green™ (R-7090, R-7091) and caged Cl-NERF (C-7079)
- Dextrans [20,76] (D-3309, D-3310, D-3311, D-7146, D-7147)
- Lipids (D-2512, D-3811)

Caged Amino Acids

Molecular Probes offers a variety of caged amino acids. The new *t*-BOC– or FMOC-protected caged amino acids (B-7077, B-7078, B-7097, B-7099, D-7098) are designed to facilitate the automated synthesis of peptides that may be biologically inactive until photolyzed. Amino acids caged on the α-amine (D-7086, D-7087, D-7088, D-7089, D-7092, D-7093, D-7094, D-7095) can block the N-terminus of the peptide during synthesis but are removable by UV illumination. See Section 9.3 for more information on our reagents for peptide synthesis.

Reagents for Caging Biomolecules

Kit for Caging Carboxylic Acids

Using organic synthesis methods, researchers can cage a diverse array of molecules. One of the preferred caging groups is the 1-(4,5-dimethoxy-2-nitrophenyl)ethyl (DMNPE) ester. Because the diazoethane precursor to DMNPE esters is unstable, we offer a kit (D-2516) for the generation of 1-(4,5-dimethoxy-2-nitrophenyl) diazoethane and the subsequent preparation of DMNPE esters.[77] This kit includes:

- 25 mg of hydrazone precursor
- MnO_2 for oxidation
- Celite® for filtration of the reaction mixture
- Detailed protocol for caging carboxylic acids

A wide range of compounds containing a weak oxy acid (with a pK_a between 3 and 7), including carboxylic acids, phenols and phosphates, should react with the diazoethane to form the DMNPE-caged analogs [10] (Figure 19.3A).

NVOC Chloride for Caging Amines

NVOC chloride (D-7118) is an amine-reactive reagent that enables researchers to cage amine-containing molecules with a DMNB group attached via an oxycarbonyl linker [78] (Figure 19.3B). It has recently been used to cage lysines of G-actin that are essential for polymerization.[79]

BNPA for Caging Thiols

The reagent α-bromo-2-nitrophenylacetic acid (BNPA, B-6610) is a synthetic precursor to caged thioethers (Figure 19.3C). This water-soluble reagent has considerable potential for caging cysteine-containing peptides and proteins,[80] as well as enzymes in which the active site is a cysteine residue.

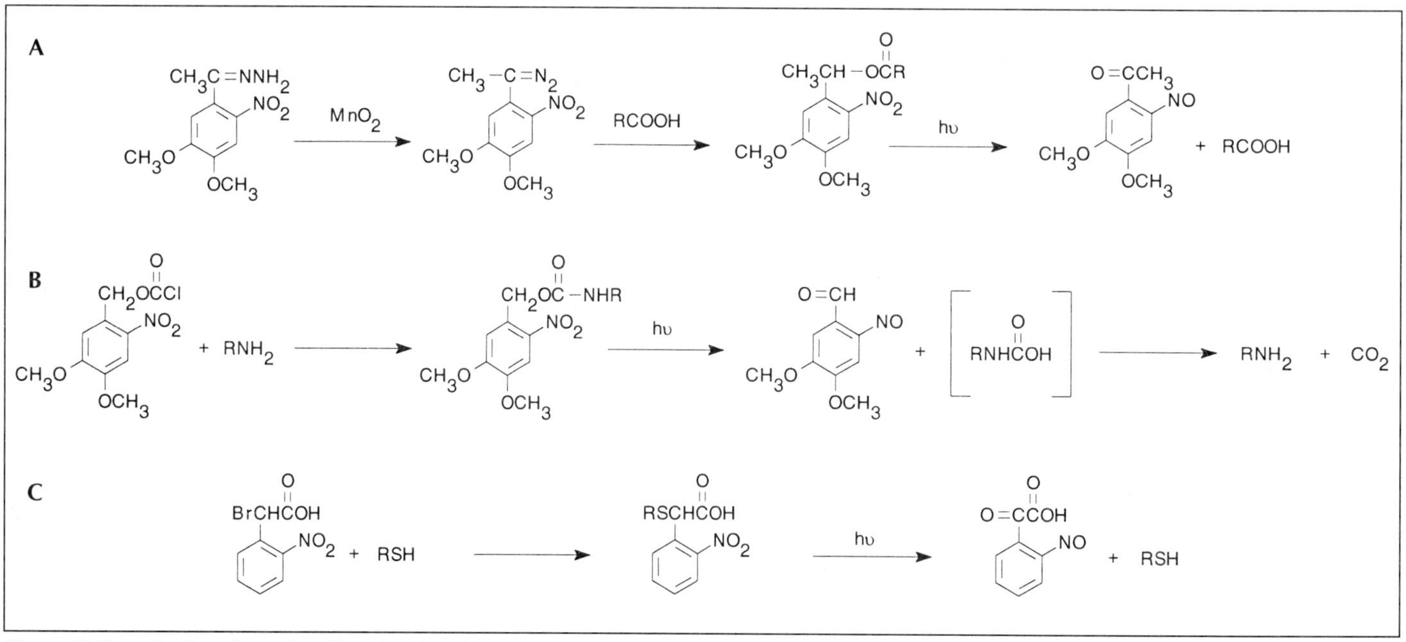

Figure 19.3 *A) Caging of a carboxylic acid using the hydrazone precursor of DMNPE, a reagent that is provided in Molecular Probes' caging kit (D-2516). B) Caging of an amine using NVOC chloride (D-7118). C) Caging of a thiol using BNPA (B-6610).*

1. J Am Chem Soc 115, 6001 (1993); **2.** Biophys J 67, 2436 (1994); **3.** Ann Rev Physiol 52, 875 (1990); **4.** Ann Rev Physiol 52, 857 (1990); **5.** Biochim Biophys Acta 1104, 207 (1992); **6.** J Biol Chem 268, 25320 (1993); **7.** Biochim Biophys Acta 939, 197 (1988); **8.** J Mol Biol 223, 185 (1992); **9.** Am J Physiol 261, 1665 (1991); **10.** Ann Rev Biophys Biophys Chem 18, 239 (1989); **11.** Pflügers Arch 411, 628 (1988); **12.** Nature 310, 74 (1984); **13.** Proc Natl Acad Sci USA 92, 9757 (1995); **14.** Biophys J 68, A364 (1995); **15.** Biophys J 65, 2368 (1993); **16.** J Mol Biol 184, 645 (1985); **17.** Proc Natl Acad Sci USA 91, 187 (1994); **18.** NP-EGTA is licensed to Molecular Probes under U.S. Patent No. 5,446,186; **19.** J Biol Chem 270, 23966 (1995); **20.** Neuron 15, 755 (1995); **21.** Science 267, 1997 (1995); **22.** Proc Natl Acad Sci USA 85, 6571 (1988); **23.** Science 241, 842 (1988); **24.** DMNP-EDTA is licensed to Molecular Probes under U.S. Patent No. 4,981,985; **25.** Neuron 10, 21 (1993); **26.** J Am Chem Soc 111, 7957 (1989); **27.** Diazo-2 and diazo-3 are licensed to Molecular Probes under U.S. Patent No. 5,141,627; **28.** FEBS Lett 255, 196 (1989); **29.** Neuropharmacology 33, 1375 (1994); **30.** Anal Chem 66, 419A (1994); **31.** J Biol Chem 268, 11520 (1993); **32.** J Org Chem 56, 5025 (1991); **33.** Photochem Photobiol 59, 463 (1994); **34.** EMBO J 13, 2038 (1994); **35.** Mol Cell Biochem 138, 229 (1994); **36.** Science 259, 370 (1993); **37.** J Biol Chem 270, 7745 (1995); **38.** BioTechniques 15, 848 (1993); **39.** J Am Chem Soc 115, 9371 (1993); **40.** Nature 308, 693 (1984); **41.** Proc Natl Acad Sci USA 91, 8752 (1994); **42.** J Org Chem 55, 1585 (1990); **43.** J Neurosci Meth 54, 151 (1994); **44.** Proc Natl Acad Sci USA 91, 6629 (1994); **45.** Biochemistry 32, 3831 (1993); **46.** Biochemistry 32, 989 (1993); **47.** Biochemistry 31, 5507 (1992); **48.** Adv Exp Med Biol 287, 75 (1991); **49.** Biochemistry 28, 49 (1989); **50.** Biochemistry 25, 1799 (1986); **51.** J Neurosci Meth 67, 221 (1996); **52.** J Org Chem 61 1228 (1996); **53.** J Am Chem Soc 116, 8366 (1994); **54.** Soc Neurosci Abstr 21, 579, abstract #238.11 (1995); **55.** J Neurosci Meth 54, 205 (1994); **56.** Science 265, 255 (1994); **57.** Proc Natl Acad Sci USA 90, 7661 (1993); **58.** J Gen Physiol 101, 337 (1996); **59.** Proc R Soc Lond B 263, 241 (1996); **60.** J Photochem Photobiol B 27, 123 (1995); **61.** Biochemistry 35, 2030 (1996); **62.** J Org Chem, 60, 4260 (1995); **63.** Biochemistry 32, 1338 (1993); **64.** AIAA Journal 34, 449 (1996); **65.** Exp in Fluids 18, 249 (1995); **66.** SPIE Proceedings 2546, abstract #2546-23, 40th Annual Optical Science, Engineering and Instrumentation International Symposium, International Society of Optical Engineering, San Diego (1995); **67.** Cell 68, 923 (1992); **68.** J Cell Biol 109, 637 (1989); **69.** J Cell Biol 126, 1455 (1994); **70.** J Cell Biol 120, 1177 (1993); **71.** J Cell Biol 122, 833 (1993); **72.** J Cell Biol 118, 367 (1992); **73.** Nature 352, 126 (1991); **74.** Exp in Fluids 21, 237 (1996); **75.** U.S. Patent No. 5,514,710; **76.** Science 272, 716 (1996); **77.** This caging kit and several of our NPE- and DMNPE-caged reagents are licensed to Molecular Probes under European Patent 0 233 403 and related patents in France, Germany, Great Britain and Italy; **78.** J Am Chem Soc 92, 6333 (1970); **79.** Biochemistry 33, 9092 (1994); **80.** Chem Biol 2, 139 (1995).

19.2 Price List *Caged Probes for a Variety of Applications*

Caged Nucleotides and Caged Phosphate *(see Chapter 20)*

Cat #		Product Name	Unit Size	Price Per Unit ($) 1–4 Units	5–24 Units
A-7056	**New**	adenosine 5′-diphosphate, P^2-(1-(2-nitrophenyl)ethyl) ester, monopotassium salt (NPE-caged ADP)	5 mg	103.00	82.40
A-1049		adenosine 5′-triphosphate, P^3-(1-(4,5-dimethoxy-2-nitrophenyl)ethyl) ester, disodium salt (DMNPE-caged ATP)	5 mg	103.00	82.40
A-7073	**New**	adenosine 5′-triphosphate, P^2-(1,2-diphenyl-2-oxo)ethyl ester, ammonium salt (desyl-caged ATP) *5 mM solution in water*	400 µL	145.00	116.00
A-1048		adenosine 5′-triphosphate, P^3-(1-(2-nitrophenyl)ethyl) ester, disodium salt (NPE-caged ATP)	5 mg	103.00	82.40
D-1037		4,5-dimethoxy-2-nitrobenzyl adenosine 3′,5′-cyclicmonophosphate (DMNB-caged cAMP)	5 mg	78.00	62.40
D-7076	**New**	(1,2-diphenyl-2-oxo)ethyl phosphate, monosodium salt, monohydrate (desyl-caged phosphate)	5 mg	95.00	76.00
G-3620		guanosine 5′-O-(2-thiodiphosphate), P$^{2(S)}$-(1-(4,5-dimethoxy-2-nitrophenyl)ethyl) ester, ammonium salt (S-(DMNPE-caged) GDP-β-S)	1 mg	155.00	124.00
G-1053		guanosine 5′-O-(3-thiotriphosphate), P$^{3(S)}$-(1-(4,5-dimethoxy-2-nitrophenyl)ethyl) ester, triammonium salt (S-(DMNPE-caged) GTP-γ-S)	1 mg	155.00	124.00
H-6829	**New**	2-hydroxyphenyl-1-(2-nitrophenyl)ethyl phosphate, sodium salt (NPE-caged proton)	25 mg	95.00	76.00
N-1045		1-(2-nitrophenyl)ethyl adenosine 3′,5′-cyclicmonophosphate (NPE-caged cAMP)	5 mg	78.00	62.40
N-7065	**New**	1-(2-nitrophenyl)ethyl phosphate, diammonium salt (NPE-caged phosphate)	5 mg	65.00	52.00

Caged Ca^{2+}, Caged Ca^{2+} Chelators and Caged Ionophores *(see Chapter 22)*

Cat #		Product Name	Unit Size	1–4 Units	5–24 Units
A-7109	**New**	A-23187, 1-(4,5-dimethoxy-2-nitrophenyl)ethyl ester (DMNPE-caged A-23187)	1 mg	98.00	78.40
B-7108	**New**	4-bromo A-23187, 1-(4,5-dimethoxy-2-nitrophenyl)ethyl ester (DMNPE-caged Br A-23187)	1 mg	98.00	78.40
D-3036		diazo-2, AM *cell permeant* *special packaging*	20x50 µg	195.00	156.00
D-3034		diazo-2, tetrapotassium salt *cell impermeant*	1 mg	195.00	156.00
D-3701		diazo-3, AM *cell permeant*	1 mg	75.00	60.00
D-3700		diazo-3, dipotassium salt *cell impermeant*	1 mg	75.00	60.00
D-6815	**New**	1-(4,5-dimethoxy-2-nitrophenyl)-1,2-diaminoethane-N,N,N′,N′-tetraacetic acid, tetra(acetoxymethyl ester) (DMNP-EDTA, AM) *cell permeant* *≥95% by HPLC* *special packaging*	20x50 µg	195.00	156.00
D-6814	**New**	1-(4,5-dimethoxy-2-nitrophenyl)-1,2-diaminoethane-N,N,N′,N′-tetraacetic acid, tetrapotassium salt (DMNP-EDTA) *cell impermeant*	5 mg	100.00	80.00
N-7107	**New**	nigericin, 1-(4,5-dimethoxy-2-nitrophenyl)ethyl ester (DMNPE-caged nigericin)	1 mg	98.00	78.40
N-6803	**New**	o-nitrophenyl EGTA, AM (NP-EGTA, AM) *cell permeant* *special packaging*	20x50 µg	195.00	156.00
N-6802	**New**	o-nitrophenyl EGTA, tetrapotassium salt (NP-EGTA) *cell impermeant* *≥95% by HPLC*	1 mg	165.00	132.00

Caged Second Messengers and Caged Analogs of Other Biologically Active Molecules *(see Chapter 20)*

Cat #		Product Name	Unit Size	Price Per Unit ($) 1–4 Units	Price Per Unit ($) 5–24 Units
A-7068	**New**	arachidonic acid, α-carboxy-2-nitrobenzyl ester (CNB-caged arachidonic acid)	5 mg	95.00	76.00
A-7062	**New**	arachidonic acid, 1-(4,5-dimethoxy-2-nitrophenyl)ethyl ester (DMNPE-caged arachidonic acid)	5 mg	95.00	76.00
C-7074	**New**	cyclic adenosine 5′-diphosphate ribose, 1-(1-(2-nitrophenyl)ethyl)ester (NPE-caged cADP-ribose) *mixed isomers*	50 µg	150.00	120.00
D-7113	**New**	*trans*-p-(diethylamino)-o-hydroxy-α-methylcinnamic acid, p-amidinophenyl ester, p-toluenesulfonic acid salt	5 mg	65.00	52.00
D-7112	**New**	*trans*-p-(diethylamino)-o-hydroxy-α-methylcinnamic acid, p-nitrophenyl ester	5 mg	65.00	52.00
D-7111	**New**	N-(((4,5-dimethoxy-2-nitrobenzyl)oxy)carbonyl)-3-morpholinosydnoneimine (NVOC-caged SIN-1)	5 mg	95.00	76.00
D-7081	**New**	1,2-dioctanoyl-3-(2-nitrobenzyl)-*sn*-glycerol (NB-caged DOG)	5 mg	95.00	76.00
L-7085	**New**	D-luciferin, 1-(4,5-dimethoxy-2-nitrophenyl)ethyl ester (DMNPE-caged luciferin)	5 mg	65.00	52.00
P-7082	**New**	paclitaxel, 2′-(4,5-dimethoxy-2-nitrobenzyl)carbonate ($O^{2'}$-(NVOC-caged) Taxol)	1 mg	95.00	76.00
P-7061	**New**	penicillin V, 1-(4,5-dimethoxy-2-nitrophenyl)ethyl ester (O-(DMNPE-caged) penicillin V)	5 mg	48.00	38.40
P-7070	**New**	potassium nitrosylpentachlororuthenate (caged nitric oxide I)	10 mg	85.00	68.00

Caged Neurotransmitters and Caged Drugs *(see Chapter 18)*

Cat #		Product Name	Unit Size	Price Per Unit ($) 1–4 Units	Price Per Unit ($) 5–24 Units
A-7110	**New**	γ-aminobutyric acid, α-carboxy-2-nitrobenzyl ester, trifluoroacetic acid salt (O-(CNB-caged) GABA)	5 mg	125.00	100.00
A-2506		γ-aminobutyric acid, 4,5-dimethoxy-2-nitrobenzyl ester, hydrochloride (O-(DMNB-caged) GABA)	5 mg	75.00	60.00
A-7125	**New**	γ-aminobutyric acid, (1,2-diphenyl-2-oxo)ethyl ester, trifluoroacetic acid salt (O-(desyl-caged) GABA)	5 mg	168.00	134.40
A-2505		L-aspartic acid, α-(4,5-dimethoxy-2-nitrobenzyl) ester, hydrochloride (α-(DMNB-caged) L-aspartic acid)	5 mg	75.00	60.00
C-1791		N-(α-carboxy-2-nitrobenzyl)carbamylcholine, trifluoroacetate (N-(CNB-caged) carbamylcholine)	5 mg	75.00	60.00
C-7122	**New**	N-(α-carboxy-2-nitrobenzyl)-L-glutamic acid (N-(CNB-caged) L-glutamic acid)	5 mg	168.00	134.40
C-7124	**New**	3-(and-4)-((α-carboxy-2-nitrobenzyl)oxy)-4-(and-3)-hydroxyphenethylamine, trifluoroacetic acid salt (O-(CNB-caged) dopamine) *mixed isomers*	1 mg	65.00	52.00
C-7083	**New**	O-(α-carboxy-2-nitrobenzyl)phenylephrine (O-(CNB-caged) phenylephrine)	5 mg	48.00	38.40
D-7057	**New**	N-(4,5-dimethoxy-2-nitrobenzyl)-L-epinephrine (N-(DMNB-caged) epinephrine)	5 mg	48.00	38.40
D-7064	**New**	N-(((4,5-dimethoxy-2-nitrobenzyl)oxy)carbonyl) dopamine (N-(NVOC-caged) dopamine)	5 mg	48.00	38.40
D-7059	**New**	N-(((4,5-dimethoxy-2-nitrobenzyl)oxy)carbonyl) MK-801 (N-(NVOC-caged) MK-801)	1 mg	48.00	38.40
D-7058	**New**	N-(((4,5-dimethoxy-2-nitrobenzyl)oxy)carbonyl)-L-norepinephrine (N-(NVOC-caged) norepinephrine)	5 mg	48.00	38.40
D-7063	**New**	N-(((4,5-dimethoxy-2-nitrobenzyl)oxy)carbonyl) serotonin (N-(NVOC-caged) serotonin)	5 mg	48.00	38.40
D-7084	**New**	N-(4,5-dimethoxy-2-nitrobenzyl)phenylephrine (N-(DMNB-caged) phenylephrine)	5 mg	48.00	38.40
G-7121	**New**	L-glutamic acid, γ-(5-allyloxy-2-nitrobenzyl) ester (γ-(ANB-caged) L-glutamic acid)	5 mg	168.00	134.40
G-7055	**New**	L-glutamic acid, γ-(α-carboxy-2-nitrobenzyl) ester, trifluoroacetate (γ-(CNB-caged) L-glutamic acid)	5 mg	168.00	134.40
G-2504		L-glutamic acid, α-(4,5-dimethoxy-2-nitrobenzyl) ester, hydrochloride (α-(DMNB-caged) L-glutamic acid)	5 mg	145.00	116.00
G-7120	**New**	L-glutamic acid, γ-(1,2-diphenyl-2-oxo)ethyl ester (γ-(desyl-caged) L-glutamic acid)	5 mg	168.00	134.40
G-2507		glycine, 4,5-dimethoxy-2-nitrobenzyl ester, hydrochloride (O-(DMNB-caged) glycine)	5 mg	75.00	60.00
H-7080	**New**	haloperidol, 4,5-dimethoxy-2-nitrobenzyl carbonate (O-(DMNB-caged) haloperidol)	5 mg	48.00	38.40
K-7115	**New**	kainic acid, γ-(α-carboxy-2-nitrobenzyl) ester, hydrochloride (γ-(CNB-caged) kainic acid)	1 mg	48.00	38.40
M-7114	**New**	N-methyl-D-aspartic acid, β-(α-carboxy-2-nitrobenzyl) ester, trifluoroacetic acid salt (β-(CNB-caged) NMDA)	1 mg	48.00	38.40
M-7069	**New**	N-methyl-N-(2-nitrobenzyl)-D-aspartic acid, dilithium salt (N-(NB-caged) NMDA)	5 mg	118.00	94.40
N-7066	**New**	N-(2-nitrobenzyl)-L-isoproterenol (N-(NB-caged) isoproterenol)	5 mg	48.00	38.40
N-7067	**New**	N-(2-nitrobenzyl)-DL-propranolol (N-(NB-caged) propranolol)	5 mg	48.00	38.40

Licensing and OEM

The BODIPY®, Cascade Blue®, Oregon Green™, Rhodamine Green™, Rhodol Green™, SNAFL®, SNARF® and Texas Red® fluorophores were developed by Molecular Probes' scientists, and most are patented or have patents pending. Molecular Probes welcomes inquiries about licensing these dyes, as well as their reactive versions and conjugates, for resale or other commercial purposes. Molecular Probes also manufactures many other fluorescent dyes (e.g., FITC and TRITC; succinimidyl esters of FAM, JOE, ROX and TAMRA; reactive phycobiliproteins), crosslinking reagents (e.g., SMCC and SPDP) and biotinylation reagents, which are extensively used by other companies to prepare conjugates. We offer special discounts on almost all of our products when they are purchased in bulk quantities. Please contact our Custom and Bulk Sales Department for more specific information.

Caged Fluorescent Dyes *(see Chapter 15)*

Cat #		Product Name	Unit Size	Price Per Unit ($) 1–4 Units	5–24 Units
C-7101	***New***	5-(and-6)-carboxy-Q-rhodamine, CMNCBZ-caged, succinimidyl ester *mixed isomers*	1 mg	68.00	54.40
C-7100	***New***	5-(and-6)-carboxy-Q-rhodamine, CMNCBZ-caged, tripotassium salt *mixed isomers*	1 mg	55.00	44.00
C-7079	***New***	Cl-NERF 4,5-dimethoxy-2-nitrobenzyl ether, sulfosuccinimidyl ester, sodium salt (DMNB-caged Cl-NERF, SSE)	1 mg	48.00	38.40
D-3309		dextran, DMNB-caged fluorescein, 3000 MW, anionic	5 mg	125.00	100.00
D-3310		dextran, DMNB-caged fluorescein, 10,000 MW, anionic	5 mg	110.00	88.00
D-3311		dextran, DMNB-caged fluorescein, 70,000 MW, anionic	5 mg	110.00	88.00
D-7146	***New***	dextran, DMNB-caged fluorescein and biotin, 10,000 MW, lysine fixable	5 mg	150.00	120.00
D-7147	***New***	dextran, DMNB-caged fluorescein and biotin, 70,000 MW, lysine fixable	5 mg	150.00	120.00
D-7060	***New***	8-((4,5-dimethoxy-2-nitrobenzyl)oxy)pyrene-1,3,6-trisulfonic acid, trisodium salt (DMNB-caged HPTS)	5 mg	125.00	100.00
D-3811		N-(DMNB-caged fluorescein)-1,2-dihexadecanoyl-*sn*-glycero-3-phosphoethanolamine, triethylammonium salt	1 mg	38.00	30.40
D-2512		5-dodecanoylaminofluorescein-bis-4,5-dimethoxy-2-nitrobenzyl ether (DMNB-caged C_{12}-fluorescein)	5 mg	48.00	38.40
F-7103	***New***	fluorescein bis-(5-carboxymethoxy-2-nitrobenzyl) ether, dipotassium salt (CMNB-caged fluorescein)	5 mg	145.00	116.00
R-7102	***New***	rhodamine 110, CMNCBZ-caged, dipotassium salt (CMNCBZ-caged rhodamine 110)	5 mg	145.00	116.00
R-7090	***New***	Rhodamine Green™ bis-(((4,5-dimethoxy-2-nitrobenzyl)oxy)carbonyl)-caged, succinimidyl ester (N-(NVOC-caged) Rhodamine Green™, SE)	5 mg	175.00	140.00
R-7091	***New***	Rhodamine Green™ bis-(((4,5-dimethoxy-2-nitrobenzyl)oxy)carbonyl)-caged, sulfosuccinimidyl ester, sodium salt (N-(NVOC-caged) Rhodamine Green™, SSE)	1 mg	68.00	54.40

Caged Amino Acids *(see Chapter 9)*

Cat #		Product Name	Unit Size	1–4 Units	5–24 Units
B-7097	***New***	N-(t-BOC)-S-(4,5-dimethoxy-2-nitrobenzyl)-L-cysteine (N-(t-BOC)-S-(DMNB-caged) L-cysteine)	100 mg	175.00	140.00
B-7099	***New***	α-(t-BOC)-ϵ-(N-(((4,5-dimethoxy-2-nitrobenzyl)oxy)carbonyl))-L-lysine (α-(t-BOC)-ϵ-(NVOC-caged) L-lysine)	100 mg	175.00	140.00
B-7077	***New***	N-(t-BOC)-O-(4,5-dimethoxy-2-nitrobenzyl)-L-tyrosine (N-(t-BOC)-O-(DMNB-caged) L-tyrosine)	100 mg	175.00	140.00
B-7078	***New***	N-(t-BOC)-L-glutamic acid γ-(4,5-dimethoxy-2-nitrobenzyl ester) (N-(t-BOC)-γ-(DMNB-caged) L-glutamic acid)	100 mg	175.00	140.00
D-7098	***New***	S-(4,5-dimethoxy-2-nitrobenzyl)-N-FMOC-L-cysteine (S-(DMNB-caged) N-FMOC-L-cysteine)	100 mg	175.00	140.00
D-7086	***New***	N-(((4,5-dimethoxy-2-nitrobenzyl)oxy)carbonyl)-L-alanine (N-(NVOC-caged) L-alanine)	25 mg	35.00	28.00
D-7087	***New***	N-(((4,5-dimethoxy-2-nitrobenzyl)oxy)carbonyl) glycine (N-(NVOC-caged) glycine)	25 mg	35.00	28.00
D-7088	***New***	N-(((4,5-dimethoxy-2-nitrobenzyl)oxy)carbonyl)-L-isoleucine (N-(NVOC-caged) L-isoleucine)	25 mg	35.00	28.00
D-7089	***New***	N-(((4,5-dimethoxy-2-nitrobenzyl)oxy)carbonyl)-L-leucine (N-(NVOC-caged) L-leucine)	25 mg	35.00	28.00
D-7092	***New***	N-(((4,5-dimethoxy-2-nitrobenzyl)oxy)carbonyl)-L-methionine (N-(NVOC-caged) L-methionine)	25 mg	35.00	28.00
D-7093	***New***	N-(((4,5-dimethoxy-2-nitrobenzyl)oxy)carbonyl)-L-phenylalanine (N-(NVOC-caged) L-phenylalanine)	25 mg	35.00	28.00
D-7094	***New***	N-(((4,5-dimethoxy-2-nitrobenzyl)oxy)carbonyl)-L-tryptophan (N-(NVOC-caged) L-tryptophan)	25 mg	35.00	28.00
D-7095	***New***	N-(((4,5-dimethoxy-2-nitrobenzyl)oxy)carbonyl)-L-valine (N-(NVOC-caged) L-valine)	25 mg	35.00	28.00

Reagents for Caging Carboxylic Acids, Amines and Thiols

Cat #		Product Name	Unit Size	1–4 Units	5–24 Units
B-6610	***New***	α-bromo-2-nitrophenylacetic acid (BNPA)	5 mg	75.00	60.00
D-7118	***New***	4,5-dimethoxy-2-nitrobenzyl chloroformate (NVOC chloride)	100 mg	48.00	38.40
D-2516		1-(4,5-Dimethoxy-2-nitrophenyl)diazoethane Generation Kit	1 kit	85.00	68.00

Fluorescence Resonance Energy Transfer

Fluorescence resonance energy transfer (FRET) is a distance-dependent interaction between the electronic excited states of two dye molecules in which excitation is transferred from a donor molecule to an acceptor molecule *without emission of a photon.* FRET is dependent on the inverse sixth power of the intermolecular separation,[1] making it useful over distances comparable with the dimensions of biological macromolecules. Thus, FRET is an important technique for investigating a variety of biological phenomena that produce changes in molecular proximity.[2-10]

Primary Conditions for FRET

- Donor and acceptor molecules must be in close proximity (typically 10–100 Å).
- Absorption spectrum of the acceptor must overlap fluorescence emission spectrum of the donor (see figure).
- Donor and acceptor transition dipole orientations must be approximately parallel.

Förster Radius

The distance at which energy transfer is 50% efficient (i.e., 50% of excited donors are deactivated by FRET) is defined by the Förster radius (R_o). The magnitude of R_o is dependent on the spectral properties of the donor and acceptor dyes:

$$R_O = [8.8 \times 10^{23} \cdot \kappa^2 \cdot n^{-4} \cdot QY_D \cdot J(\lambda)]^{1/6} \text{ Å}$$

where κ^2 = dipole orientation factor (range 0 to 4, $\kappa^2 = 2/3$ for randomly oriented donors and acceptors)

QY_D = fluorescence quantum yield of the donor in the absence of the acceptor

n = refractive index

$J(\lambda)$ = spectral overlap integral (see figure)

$= \int \epsilon_A(\lambda) \cdot F_D(\lambda) \cdot \lambda^4 d\lambda \text{ cm}^3\text{M}^{-1}$

where ϵ_A = extinction coefficient of acceptor

F_D = fluorescence emission intensity of donor as a fraction of the total integrated intensity

Donor/Acceptor Pairs

In most applications, the donor and acceptor dyes are different, in which case FRET can be detected by the appearance of sensitized fluorescence of the acceptor or by quenching of donor fluorescence. When the donor and acceptor are the same, FRET can be detected by the resulting fluorescence depolarization.[11] Some typical values of R_o are listed in the table above. Note that because the component factors of R_o (see above) are dependent on environment, the actual value observed in a specific experimental situation is somewhat variable. Extensive compilations of R_o values can be found in the literature.[4,5,7,10]

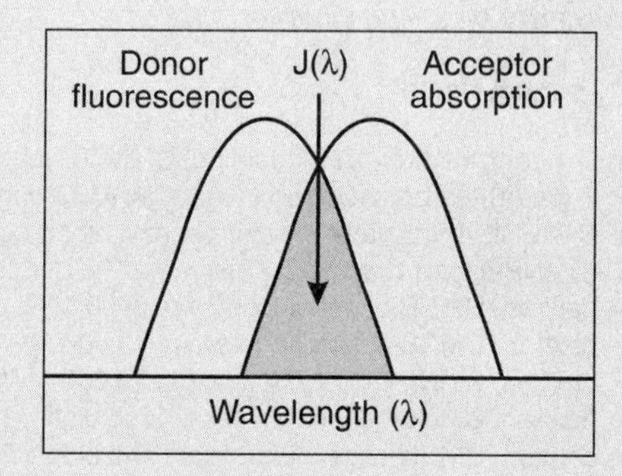

Schematic representation of the FRET spectral overlap integral.

Typical values of R_o.

Donor	Acceptor	R_o (Å)
Fluorescein	Tetramethylrhodamine	55
IAEDANS	Fluorescein	46
EDANS	DABCYL	33
Fluorescein	Fluorescein	44
BODIPY FL	BODIPY FL	57

Selected Applications of FRET

- Structure and conformation of proteins [12-17]
- Spatial distribution and assembly of protein complexes [18-22]
- Receptor/ligand interactions [23-26]
- Immunoassays [27,28]
- Structure and conformation of nucleic acids [29-34]
- Detection of nucleic acid hybridization [35-39]
- Automated DNA sequencing [40,41]
- Distribution and transport of lipids [42-45]
- Membrane fusion assays [46-49]
- Membrane potential sensing [50]
- Fluorogenic protease substrates [51-55]
- Indicators for cyclic AMP [56] and zinc [57]

1. Proc Natl Acad Sci USA 58, 719 (1967); **2.** J Struct Biol 115, 175 (1995); **3.** Meth Enzymol 246, 300 (1995); **4.** Anal Biochem 218, 1 (1994); **5.** Van der Meer, B.W. *et al., Resonance Energy Transfer Theory and Data* VCH Publishers (1994); **6.** Meth Cell Biol 30, 219 (1989); **7.** J Muscle Res Cell Motility 8, 97 (1987); **8.** Photochem Photobiol 38, 487 (1983); **9.** Ann Rev Biochem 47, 819 (1978); **10.** Meth Enzymol 48, 347 (1978); **11.** Biophys J 69, 1569 (1995); **12.** Biochemistry 35, 4795 (1996); **13.** Biochemistry 34, 8693 (1995); **14.** Biochemistry 34, 6475 (1995); **15.** Biochemistry 33, 11900 (1994); **16.** Biochemistry 33, 10171 (1994); **17.** J Biol Chem 268, 15588 (1993); **18.** Biochemistry 34, 7904 (1995); **19.** Biochemistry 33, 13102 (1994); **20.** Biochemistry 33, 5539 (1994); **21.** J Photochem Photobiol B 12, 323 (1992); **22.** J Biol Chem 264, 8699 (1989); **23.** Biochemistry 33, 11875 (1994); **24.** J Cell Physiol 159, 176 (1994); **25.** Biophys J 60, 307 (1991); **26.** J Biol Chem 259, 5717 (1984); **27.** Anal Biochem 174, 101 (1988); **28.** Anal Biochem 108, 156 (1980); **29.** Anal Biochem 221, 306 (1994); **30.** Biophys J 66, 99 (1994); **31.** Nucleic Acids Res 22, 920 (1994); **32.** Science 266, 785 (1994); **33.** Biochemistry 32, 13852 (1993); **34.** Proc Natl Acad Sci USA 90, 2994 (1993); **35.** Biochemistry 34, 285 (1995); **36.** Nucleic Acids Res 22, 662 (1994); **37.** Morrison, L.E. in *Nonisotopic DNA Probe Techniques*, L.J. Kricka, Ed., Academic Press (1992) pp. 311–352; **38.** Anal Biochem 183, 231 (1989); **39.** Proc Natl Acad Sci USA 85, 8790 (1988); **40.** Anal Chem 67, 3676 (1995); **41.** Proc Natl Acad Sci USA 92, 4347 (1995); **42.** Biochemistry 34, 4846 (1995); **43.** Biochemistry 31, 2865 (1992); **44.** Meth Cell Biol 29, 75 (1989); **45.** J Biol Chem 258, 5368 (1983); **46.** Biochim Biophys Acta 1189, 175 (1994); **47.** Biophys J 67, 1117 (1994); **48.** Meth Enzymol 221, 239 (1993); **49.** Biochemistry 20, 4093 (1981); **50.** Biophys J 69, 1272 (1995); **51.** Contillo, L.G. *et al.* in *Techniques in Protein Chemistry, Volume 5*, J. W. Crabb, Ed., Academic Press (1994) pp. 493–500; **52.** Anal Biochem 210, 351 (1993); **53.** Bioconjugate Chem 4, 537 (1993); **54.** Anal Biochem 197, 347 (1991); **55.** Science 247, 954 (1990); **56.** Nature 349, 694 (1991); **57.** J Am Chem Soc 118, 6514 (1996).

Chapter 20

Probes for Signal Transduction

Contents

Related Chapters

20.1 Introduction to Signal Transduction

Cells respond to their environment through a complex and interdependent series of signal transduction pathways that frequently begin at the cell membrane. Many cellular receptors are transmembrane proteins with extracellular domains that selectively bind a ligand. In response to ligand binding, the receptor's cytoplasmic domain may change conformation and transmit the signal across the membrane, or individual receptors may aggregate and interact with other membrane proteins in order to generate a response. Transmembrane signals trigger a cascade of events in the cell, including changes in intracellular Ca^{2+} levels, enzymatic activity and gene expression (Figure 20.1).

Molecular Probes offers several important reagents for studying signal transduction mechanisms, including Ca^{2+} regulation and the activities of other second messengers. This chapter focuses on probes for events occurring downstream from the receptor–ligand interaction. These products complement the probes for receptors and ion channels in Chapter 18, as well as the many ion indicators discussed in Chapters 22 through 24. Chapter 21 describes our new line of probes for nitric oxide research — including nitric oxide donors, nitric oxide synthase inhibitors and reagents for nitrite detection — as well as for other reactive oxygen species. Reactive oxygen species serve as second messengers in interleukin-1, tumor necrosis factor-α (TNF-α) and some growth factor signal transduction pathways [1,2] and trigger intracellular signaling transduction pathways that mediate cellular protective responses.[3-5]

1. J Biol Chem 270, 11727 (1995); **2.** BioEssays 16, 497 (1994); **3.** J Biol Chem 270, 28499 (1995); **4.** Proc Natl Acad Sci USA 92, 4527 (1995); **5.** J Biol Chem 268, 25586 (1993).

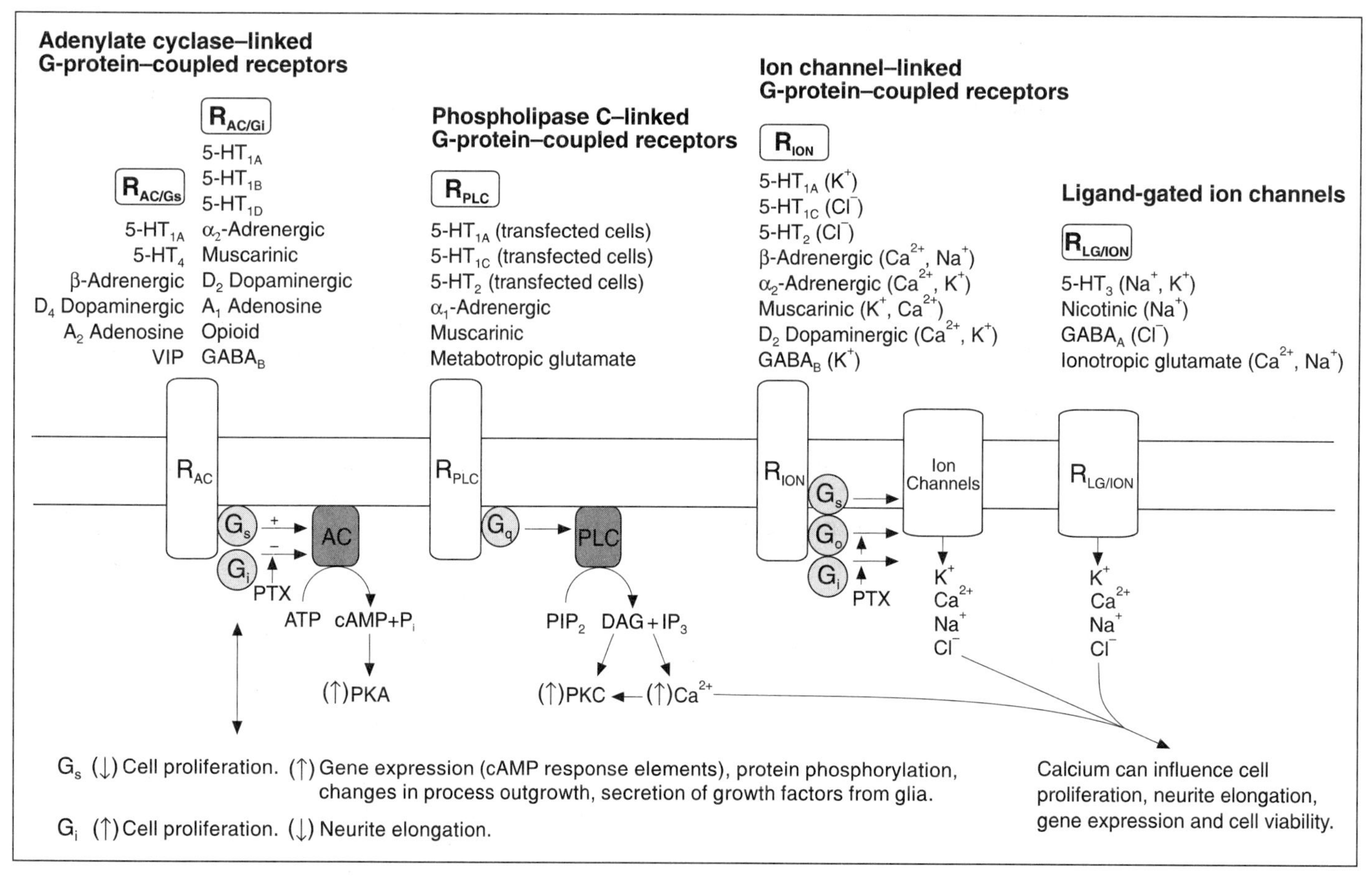

Figure 20.1 *Neurotransmitter receptors linked to second messengers mediating growth responses in neuronal and nonneuronal cells. Abbreviations:* $R_{AC/Gs}$, *receptors coupled to G-proteins that stimulate adenylate cyclase (AC) activity, leading to cAMP formation and enhanced activity of protein kinase A (PKA);* $R_{AC/Gi}$, *receptors coupled to pertussis toxin (PTX)–sensitive G-proteins that inhibit adenylate cyclase activity;* R_{PLC}, *receptors promoting the hydrolysis of phosphatidylinositol 4,5-diphosphate (PIP*$_2$*) to inositol 1,4,5-triphosphate (IP*$_3$*), which increases intracellular* Ca^{2+}*, and diacylglyercol (DAG), which activates protein kinase C (PKC);* R_{ION}, *receptors indirectly promoting ion fluxes due to coupling to various G-proteins;* $R_{LG/ION}$, *receptors that promote ion fluxes directly because they are structurally linked to ion channels (members of the superfamily of ligand-gated ion channel receptors). Stimulation of proliferation is most often associated with activation of G-proteins negatively coupled to adenylate cyclase (*G_i*), or positively coupled to phospholipase C (*G_q*) or to pertussis toxin–sensitive pathways (*G_o, G_i*). In contrast, activation of neurotransmitter receptors positively coupled to cAMP usually inhibits cell proliferation and causes changes in cell shape indicative of differentiation. Reprinted with permission from Trends Neurosci 16, 233 (1993).*

20.2 Calcium Regulation

Intracellular Ca^{2+} levels modulate a multitude of vital cellular processes — including gene expression, cell viability, cell proliferation, cell motility and cell shape and volume regulation — thereby playing a key role in regulating cell responses to external activating agents. These dynamic changes in Ca^{2+} levels are regulated by ligand-gated and G-protein–coupled ion channels in the plasma membrane and by mobilization of Ca^{2+} from intracellular stores. The generation of cytosolic Ca^{2+} spikes and oscillations typically involves the coordinated release and uptake of Ca^{2+} from these stores, mediated by intracellular Ca^{2+} channels and their response to several second messengers such as Ca^{2+} itself, cyclic ADP ribose and inositol triphosphate.[1,2] Molecular Probes is pleased to introduce several new reagents for studying Ca^{2+} regulation in live cells.

Inositol Triphosphate Pathway

D-*myo*-1,4,5-Inositol Triphosphate

Molecular Probes offers the potassium salt of D-*myo*-inositol 1,4,5-triphosphate (1,4,5-IP_3, I-3716) for researchers investigating IP_3-dependent Ca^{2+} mobilization and signal transduction mechanisms.[3] Cytoplasmic 1,4,5-IP_3 is a potent intracellular second messenger that induces Ca^{2+} release from membrane-bound stores in many tissues.[3,4] Our 1,4,5-IP_3 is at least 99% pure, as determined by paper chromatography and by ^{1}H-NMR and ^{31}P-NMR.

Fluorescent Heparin

Our new fluorescein-labeled heparin (H-7482) should be a useful tool for studying binding of this mucopolysaccharide in cells and tissues.[5] In addition to its well-known anticoagulant activity,[6] heparin binds to the IP_3 receptor and inhibits the biological cascade of events mediated by 1,4,5-IP_3.[7] Heparin exhibits a number of other biological properties, including modulation of the structure, function and metabolism of many proteins and enzymes. This mucopolysaccharide binds to low-density lipoproteins,[8] lipoprotein lipase, circulatory serine proteases and proteinase inhibitors, as well as to blood vessel–associated proteins such as fibronectin [9-11] and laminin.[12] Heparin also interacts with acidic fibroblast growth factor (HBGF) and amplifies HBGF's mutagenic and neurotrophic activity.[13] Fluorescein-labeled heparin can be prepared by a variety of methods, which may influence its applications. Therefore, fluorescein heparin from Molecular Probes may not be identical to that used in reported applications.[8-11,14-19]

Calcium-Induced Calcium Release

Cyclic ADP Ribose and Caged Cyclic ADP Ribose

Molecular Probes now offers cyclic ADP ribose (cADP-ribose, C-7619), a potent, microinjectable Ca^{2+}-mobilizing agent that likely functions as a second messenger in an inositol triphosphate–independent pathway and is involved in Ca^{2+}-induced Ca^{2+} release.[20-23] cADP-ribose is a putative physiological regulator of certain isoforms of ryanodine-activated Ca^{2+} channels [24-27] that may act through a calmodulin-mediated mechanism.[28] cADP-ribose is reported to mobilize Ca^{2+} in dorsal root ganglion cells, pituitary cells and sea urchin eggs,[29-31] where it can also act synergistically with ryanodine to release intracellular stores.[32] This Ca^{2+}-mobilizing agent is also produced in pancreatic islets by glucose stimulation [33,34] and has been shown to mediate glucose-induced insulin release in pancreatic cells [35,36] and to induce release from Ca^{2+} stores in and around the nuclear envelope.[37] Researchers have investigated the role of cADP-ribose in the agonist-evoked Ca^{2+} oscillations in pancreatic acinar cells.[38] In this study, cADP-ribose–induced Ca^{2+} spikes could be blocked with either ryanodine or heparin, implicating both ryanodine and IP_3 receptors in the Ca^{2+} spike generation.

The application of cADP-ribose in cells and tissues can be controlled spatially and temporally by flash photolysis of our new NPE-caged cADP-ribose [39] (C-7074). Photorelease of cADP-ribose from NPE-caged cADP-ribose results in initiation of Ca^{2+} release and cortical reaction in sea urchin eggs.[40]

Cyclic GDP Ribose, NAADP and Related Compounds

Cyclic ADP ribose is generated from NAD^+ by ADP-ribosyl cyclase, an enzyme found in a variety of cell types, including rat pituitary [41] and rabbit liver, brain, spleen, kidney and heart.[42] ADP-ribosyl cyclase also mediates the formation of the fluorescent nucleotide cyclic GDP ribose (cGDP-ribose, C-7620) from nicotinamide guanine dinucleotide (NGD^+), providing a convenient fluorescence assay for cyclase activity in tissue extracts or pure enzyme preparations.[43,44]

Inhibition of Ca^{2+} mobilization by cADP-ribose can be achieved specifically and reversibly with cADP-ribose antagonists,[20,23,45] including 8-amino-cADP-ribose (A-7621) and 8-bromo-cADP-ribose [46] (8-Br-cADP-ribose, B-7625). 8-Amino-cADP-ribose has been shown both to block binding of radiolabeled cADP-ribose to sea urchin egg microsomes and to inhibit cADP-ribose–mediated release of Ca^{2+} from egg homogenates, but it does not block caffeine- or ryanodine-induced Ca^{2+} release.[45]

It has recently been found that β-nicotinic acid adenine dinucleotide phosphate (NAADP, N-7622), which is generated by amine hydrolysis of NADP, stimulates Ca^{2+} mobilization from intracellular stores of sea urchin eggs by a mechanism that is distinct from that of cADP-ribose.[47] Ca^{2+} mobilization from microsomes by NAADP is markedly faster than cADP-ribose–mediated release and is not desensitized by pretreatment with cADP-ribose. In contrast to Ca^{2+} release brought about by cADP-ribose, NAADP action is not dependent on calmodulin, nor is it sensitive to the cADP-ribose antagonist, 8-amino-cADP-ribose.[47]

Ryanodine and Fluorescent Ryanodine

Ryanodine is a plant alkaloid that mobilizes Ca^{2+} from intracellular stores by activating a class of IP_3-insensitive Ca^{2+} channels.[48] It alters the function of the Ca^{2+} channel in a complex manner: submicromolar concentrations lock the channel in a long-lived open state, whereas micromolar or greater concentrations inhibit Ca^{2+} release.[49,50] Ryanodine can be used to modulate the Ca^{2+} concentration in sea urchin eggs [30] and in parotid acinar cells.[51] In developing skeletal muscle, ryanodine receptors localize in discrete regions of the T tubules, binding at the junctional complex between the T tubules and the sarcoplasmic reticulum.[52] Molecular Probes offers both unlabeled "ryanodine" and a green fluorescent derivative of ryanodine.

Unlabeled "ryanodine" (R-3455) from Molecular Probes is a mixture of about 47% ryanodine and 47% dehydroryanodine by HPLC (total >94% pure). Both are equally effective with low nanomolar dissociation constants.[53,54] We also offer the purified components of this mixture: high-purity ryanodine (R-7478) and 9,21-dehydroryanodine (D-7480).

In addition to unlabeled ryanodine, we have recently prepared a monosubstituted BODIPY® FL-X ryanodine derivative (B-7505), which, like the parent preparation, is a mixture of BODIPY FL-X ryanodine and BODIPY FL-X dehydroryanodine, most likely labeled at the 10 position of the ryanodine molecule. The biological activity of these derivatives has not yet been determined. Another reported label for the ryanodine receptor is the fluorogenic maleimide, CPM (D-346, see Section 2.2), which covalently modifies hyperreactive thiols of the ryanodine receptor of skeletal and cardiac sarcoplasmic reticulum.[55]

Caged Ca^{2+} and Caged Ca^{2+} Chelators

Caged ions and caged chelators can be used to influence the ionic composition of both solutions and cells, particularly for ions such as Ca^{2+} that are present at low concentrations.

NP-EGTA: A Caged Ca^{2+} Reagent

Ellis-Davies and Kaplan have recently developed a new photolabile chelator, *o*-nitrophenyl EGTA[56] (NP-EGTA), which exhibits a high selectivity for Ca^{2+}, a dramatic 12,500-fold *decrease* in affinity for Ca^{2+} upon UV illumination (its K_d increases from 80 nM to >1 mM) and a high photochemical quantum yield[57,58] (~0.2). Furthermore, with a K_d for Mg^{2+} of 9 mM, NP-caged EGTA does not perturb physiological levels of Mg^{2+}. We offer the potassium salt (N-6802) and the acetoxymethyl (AM) ester (N-6803) of NP-EGTA. The NP-EGTA salt can be complexed with Ca^{2+} to generate a caged calcium complex that will rapidly deliver Ca^{2+} upon photolysis (Figure 19.1A in Chapter 19). Cell-permeant NP-EGTA AM does not bind Ca^{2+} unless its AM ester groups are removed. It can potentially serve as a photolabile buffer in cells because, once converted to NP-EGTA by intracellular esterases, it will bind Ca^{2+} with high affinity until photolyzed with UV light.

DMNP-EDTA: A Caged Ca^{2+} Reagent

The first caged Ca^{2+} reagent described by Ellis-Davies and Kaplan was 1-(4,5-dimethoxy-2-nitrophenyl) EDTA (DMNP-EDTA, D-6814), which they named DM-Nitrophen™[59-61] (now a trademark of Calbiochem-Novabiochem Corp.). Because its structure more resembles that of EDTA than EGTA, we named it as a caged EDTA derivative (see Figure 19.1B in Chapter 19). Upon illumination, DMNP-EDTA's K_d for Ca^{2+} increases from 5 nM to 3 mM. Thus, photolysis of DMNP-EDTA complexed with Ca^{2+} results in a pulse of free Ca^{2+}. Furthermore, DMNP-EDTA has significantly higher affinity for Mg^{2+} (K_d = 2.5 µM) than does NP-EGTA (K_d = 9 mM).[60] Because the photolysis product's K_d for Mg^{2+} is ~3 mM, DMNP-EDTA is an effective caged Mg^{2+} source, in addition to its applications for photolytic Ca^{2+} release.[62,63] Photorelease of Ca^{2+} has been shown to occur in <180 microseconds, with even faster photorelease of Mg^{2+}.[64] A recent paper by Neher and Zucker discusses the uses and limitations of DMNP-EDTA.[63] In addition to the high-purity potassium salt of DMNP-EDTA (D-6814), we have prepared its AM ester (D-6815), which has not been previously available.

Diazo-2: A Photoactivatable Ca^{2+} Scavenger

In contrast to NP-EGTA and DMNP-EDTA, diazo-2 is a photoactivatable Ca^{2+} scavenger. Diazo-2, which was introduced by Adams, Kao and Tsien,[65-67] is a relatively weak chelator (K_d for Ca^{2+} = 2.2 µM). However, following flash photolysis at ~360 nm, cytosolic free Ca^{2+} rapidly binds to the diazo-2 photolysis product, which has a high affinity for Ca^{2+} (K_d = 73 nM). Photolysis of diazo-2 has been used to decrease cytosolic Ca^{2+} in less than two milliseconds in tensed frog muscle cells,[68] rat fibroblasts[69] and rabbit arterial smooth muscle.[70] Microinjecting a relatively low concentration of fluo-3, or one of the Calcium Green™ or Oregon Green™ 488 BAPTA indicators (see Section 22.3), along with a known quantity of diazo-2, permits the measurement of the extent of depletion of cytosolic Ca^{2+} following photolysis.[66,71,72] Diazo-2 also has been loaded into cells permeabilized with *Staphylococcus aureus* α-toxin.[70]

Diazo-2 can be microinjected into cells as its potassium salt (D-3034) or loaded as an AM ester (D-3036). Molecular Probes also offers the potassium salt and AM ester of diazo-3 (D-3700, D-3701). Diazo-3 binds Ca^{2+} with low affinity, both before and after photolysis, and therefore may serve as a control for the effects that AM ester hydrolysis and photolysis products of diazo-2 may have on the cell.

Other Probes for Calcium Regulation

Thapsigargin and BODIPY FL Thapsigargin

Thapsigargin is a naturally occurring sesquiterpene lactone isolated from the umbelliferous plant *Thapsia garganica*.[73] This tumor promoter releases Ca^{2+} from intracellular stores by specifically inhibiting the endoplasmic reticulum Ca^{2+}-ATPase;[74] it does not directly affect plasma membrane Ca^{2+}-ATPases, IP_3 production or protein kinase C activity.[75,76] Recent reports describe the effects of thapsigargin-induced Ca^{2+} signals on the transient suppression of c-*myb* mRNA levels,[77] as well as the regulation of such Ca^{2+} signals by sphingomyelinase and sphingosine.[78] Thapsigargin is available from Molecular Probes in 1 mg units (T-7458) and specially packaged in 20 vials that each contain 50 µg (T-7459). We are also pleased to introduce a membrane-permeant green fluorescent derivative of thapsigargin, BODIPY FL thapsigargin (B-7487). This derivative has spectral properties similar to those of fluorescein and may be useful for localization of thapsigargin binding sites in live cells.

Calcium Antagonist TMB-8

The Ca^{2+} antagonist TMB-8 (D-1579) has been reported to block release of Ca^{2+} from the sarcoplasmic reticulum of smooth and skeletal muscle[79] and to inhibit IP_3-induced Ca^{2+} release from vacuolar membrane vesicles of both oat roots[80] and platelet membrane vesicles.[81] TMB-8 also inhibits carbamoyl choline–stimulated amylase release in rat pancreatic acini, possibly by blocking mobilization of cellular Ca^{2+},[82] as well as the ATP-sensitive K^+ channel function in insulinoma HIT-T15 β-cells[83] and glycogen synthase activation in leukocytes.[84] In adrenal glomerulosa cells, TMB-8 strongly inhibits K^+-induced increases in intracellular Ca^{2+}

but exhibits little effect on Ca^{2+} increases induced by angiotensin II or arachidonic acid, suggesting that TMB-8 inhibits Ca^{2+} influx, rather than release, in these cells.[85] The suppression of both Ca^{2+} release and ATP secretion by TMB-8 in phorbol ester–stimulated platelets may be caused by interference with the protein kinase C pathway, and not with Ca^{2+} mobilization.[86]

Calcium Analogs

The trivalent lanthanides terbium (III) and europium (III), which are supplied by Molecular Probes as chloride salts (T-1247, E-6960), are luminescent analogs of Ca^{2+} used to study structure–function relationships in Ca^{2+}-binding proteins such as calmodulin, oncomodulin, lactalbumin and ATPases.[87-89] The long-lived luminescence of Tb^{3+} is also used to probe Ca^{2+} binding sites of alkaline phosphatase,[90] glutamine synthetase,[91] integrins,[92] protein kinase C [93] and Ca^{2+}-binding sites of ryanodine-sensitive Ca^{2+} channels.[94] Tb^{3+} reportedly binds most strongly to the I and II sites of calmodulin.[95]

Calmodulin Antagonist W-5

The calmodulin antagonist W-5 (A-1588) has a substantially lower affinity for calmodulin than its chlorinated counterpart W-7; a sixfold higher concentration of W-5 is required to effect 50% inhibition of cell proliferation in Chinese hamster ovary K_1 (CHO-K_1) cells.[96] W-5 has been reported to inhibit cell proliferation in ovarian cystadenocarcinoma cells [97] and to affect cytosolic Ca^{2+} levels in carrot cell protoplasts.[98] Because of its relatively low affinity for calmodulin, W-5 has also been used as a control for W-7 in the study of calmodulin's role in receptor-mediated endocytosis.[99,100]

1. Ann Rev Physiol 56, 297 (1994); **2.** Nature 365, 388 (1993); **3.** Nature 341, 197 (1989); **4.** Ann Rev Biochem 56, 159 (1987); **5.** Nature 372, 231 (1994); **6.** J Biol Chem 267, 8857 (1992); **7.** Biochem J 301, 155 (1994); **8.** Biochemistry 31, 5996 (1992); **9.** Biochemistry 32, 12548 (1993); **10.** Biochemistry 27, 7565 (1988); **11.** J Biol Chem 260, 7250 (1985); **12.** Anal Biochem 156, 320 (1986); **13.** Biochemistry 31, 6498 (1992); **14.** Cytometry 23, 59 (1996); **15.** Biol Pharm Bull 16, 939 (1993); **16.** Anal Biochem 160, 105 (1987); **17.** Biochim Biophys 925, 57 (1987); **18.** Anal Biochem 130, 287 (1983); **19.** Carbohydrate Res 105, 69 (1982); **20.** J Biol Chem 270, 25488 (1995); **21.** J Biol Chem 270, 9060 (1995); **22.** Cellular Signalling 6, 591 (1994); **23.** Mol Cell Biochem 138, 229 (1994); **24.** J Biol Chem 270, 17917 (1995); **25.** Pflügers Arch 429, 426 (1995); **26.** Neuron 12, 1073 (1994); **27.** Nature 364, 76 (1993); **28.** Nature 370, 307 (1994); **29.** Biochemistry 34, 2815 (1995); **30.** Science 253, 1143 (1991); **31.** J Biol Chem 264, 1608 (1989); **32.** Devel Biol 163, 1 (1994); **33.** J Biol Chem 270, 30257 (1995); **34.** J Biol Chem 270, 30045 (1995); **35.** Biochimie 77, 356 (1995); **36.** Science 259, 370 (1993); **37.** Cell 80, 439 (1995); **38.** EMBO J 13, 2038 (1994); **39.** patent pending; **40.** J Biol Chem 270, 7745 (1995); **41.** J Biol Chem 264, 11725 (1989); **42.** J Biol Chem 266, 16985 (1991); **43.** J Biol Chem 270, 30327 (1995); **44.** J Biol Chem 269, 30260 (1994); **45.** Biochim Biophys Acta 1178, 235 (1993); **46.** 8-Amino- and 8-bromo-cADP-ribose are licensed to Molecular Probes under U.S. Patent No. 5,486,604; **47.** J Biol Chem 270, 2152 (1995); **48.** J Biol Chem 268, 13765 (1993); **49.** J Gen Physiol 92, 1 (1988); **50.** Am J Physiol 253, C364 (1987); **51.** J Biol Chem 266, 14535 (1991); **52.** J Cell Biol 112, 289 (1991); **53.** J Biol Chem 269, 30243 (1994); **54.** J Labelled Compounds Radiopharmaceut 23, 215 (1985); **55.** Mol Pharmacol 45, 189 (1994); **56.** NP-EGTA is licensed to Molecular Probes under U.S. Patent No. 5,446,186; **57.** J Biol Chem 270, 23966 (1995); **58.** Proc Natl Acad Sci USA 91, 187 (1994); **59.** DMNP-EDTA is licensed to Molecular Probes under U.S. Patent No. 4,981,985; **60.** Proc Natl Acad Sci USA 85, 6571 (1988); **61.** Science 241, 842 (1988); **62.** Meth Cell Biol 40, 31 (1994); **63.** Neuron 10, 21 (1993); **64.** Biochemistry 31, 8856 (1992); **65.** Biochim Biophys Acta 1035, 378 (1990); **66.** J Am Chem Soc 111, 7957 (1989); **67.** Diazo-2 and diazo-3 are licensed to Molecular Probes under U.S. Patent No. 5,141,627; **68.** FEBS Lett 255, 1 (1989); **69.** J Am Chem Soc 111, 20 (1989); **70.** Biophys J 69, 2611 (1995); **71.** Nature 371, 603 (1994); **72.** Biophys J 65, 2537 (1993); **73.** Acta Pharm Suec 15, 133 (1978); **74.** J Biol Chem 270, 11731 (1995); **75.** Proc Natl Acad Sci USA 87, 2466 (1990); **76.** J Biol Chem 264, 12266 (1989); **77.** J Biol Chem 269, 8786 (1994); **78.** J Biol Chem 269, 5054 (1994); **79.** Br J Pharmacol 53, 279 (1975); **80.** J Biol Chem 262, 3944 (1987); **81.** Biochem Pharmacol 36, 3331 (1987); **82.** Life Sci 34, 529 (1984); **83.** Eur J Pharmacol 226, 175 (1992); **84.** Biochem Biophys Res Comm 106, 210 (1982); **85.** Biochim Biophys Acta 888, 25 (1986); **86.** FEBS Lett 176, 139 (1984); **87.** Biochemistry 33, 12238 (1994); **88.** J Biol Chem 267, 13340 (1992); **89.** Photochem Photobiol 46, 1067 (1987); **90.** J Photochem Photobiol 13, 289 (1992); **91.** Biochemistry 30, 3417 (1992); **92.** Biochemistry 29, 12238 (1994); **93.** J Biol Chem 263, 4223 (1990); **94.** J Biol Chem 269, 24864 (1994); **95.** Biochem Biophys Res Comm 138, 1243 (1986); **96.** Proc Natl Acad Sci USA 78, 4354 (1981); **97.** J Natl Cancer Inst 77, 1181 (1986); **98.** FEBS Lett 212, 133 (1987); **99.** J Cell Biochem 36, 73 (1988); **100.** J Cell Biol 105, 679 (1987).

20.2 Data Table *Calcium Regulation*

Cat #	Structure	MW	Storage	Soluble	Abs	{ε × 10⁻³}	Em	Solvent	Notes
A-1588	✓	343	NC	DMF	286	{6.8}	335	MeOH	
A-7621	✓	556	FF,D	H_2O	274	{14}	none	pH 7	
B-7487	✓	855	FF,D,L	DMSO	503	{85}	511	MeOH	
B-7505	S18.4	853	FF,D,L	DMSO	504	{79}	511	MeOH	
B-7625	✓	620	FF,D	H_2O	265	{16}	none	pH 7	
C-7074	✓	690	FF,DD,LL	H_2O	259	{16}	none	H_2O	<1, 2>
C-7619	✓	541	FF,D	H_2O	258	{9.9}	none	pH 7	
C-7620	✓	557	FF,D	H_2O	284	{5.8}	396	pH 7	
D-1579	✓	432	F,D	H_2O, DMF	268	{10}	338	$CHCl_3$	
D-3034	S22.8	711	F,D,LL	pH>6	369	{18}	none	pH 7	<1, 3>
D-3036	S22.8	847	F,D,LL	DMSO	342	{22}	none	MeOH	
D-3700	S22.8	383	F,D,LL	pH>6	374	{23}	none	MeOH	<1, 4>
D-3701	S22.8	451	F,D,LL	DMSO	344	{20}	none	MeOH	
D-6814	S22.8	626	D,LL	pH>6	348	{4.2}	none	pH 7	<1, 3>
D-6815	S22.8	762	F,D,LL	DMSO	333	{4.6}	none	MeOH	<5>
D-7480	S18.4	492	F,D	MeOH, DMSO	<300		none		
E-6960	S15.3	366*	D	H_2O	337	{17}	613	H_2O	<6>
H-7482	C	~18,000	FF,D,L	H_2O	493	ND	514	pH 8	<7, 8>
I-3716	✓	649	F,D	H_2O	<250		none		
N-6802	S22.8	654	FF,D,LL	pH>6	260	{3.5}	none	pH 7	<1, 2, 3>
N-6803	S22.8	790	FF,D,LL	DMSO	250	{4.1}	none	MeCN	<2, 5>

Cat #	Structure	MW	Storage	Soluble	Abs	{ε × 10^{-3}}	Em	Solvent	Notes
N-7622	✓	744	FF,D	H_2O	254	{17}	none	H_2O	
R-7478	S18.4	494	F,D	MeOH, DMSO	<300		none		
T-1247	S15.3	373*	D	H_2O	270	{4.7}	545	H_2O	<9>
T-7458	✓	651	F,D	DMSO, EtOH	<300		none		

R-3455 is a mixture of R-7478 and D-7480; T-7459 see T-7458

For definitions of the contents of this data table, see "How to Use the Handbook" on page vi.
Structure: Chemical structure drawing: ✓ = shown in this section; Sn.n = shown in Section number n.n; C = not shown due to complexity.
MW : ~ indicates an approximate value.

* MW is for the hydrated form of this product.
<1> All photoactivatable probes are sensitive to light. They should be protected from illumination, except when photolysis is intended.
<2> This compound has weaker visible absorption at >300 nm but no discernible absorption peaks in this region.
<3> This compound undergoes a large increase or decrease in Ca^{2+} binding affinity upon ultraviolet photolysis; see Section 22.8.
<4> Diazo-3 has negligible Ca^{2+} binding affinity before or after photolysis. It is used as a control for diazo-2 photolysis products [J Am Chem Soc, 111, 7957 (1989)].
<5> This product is intrinsically an oil at room temperature.
<6> Absorption and fluorescence of E-6960 are extremely weak unless it is chelated. Data are for detergent-solubilized NTA (T-411, see Section 15.3) chelate.
<7> ND = not determined.
<8> This product is a multiply-labeled polysaccharide conjugate. The number of labels per conjugate is indicated on the vial.
<9> Absorption and fluorescence of T-1247 are extremely weak unless it is chelated. Data are for dipicolinic acid (D-1248, see Section 15.3) chelate. Fluorescence spectrum has secondary peak at 490 nm.

20.2 Structures *Calcium Regulation*

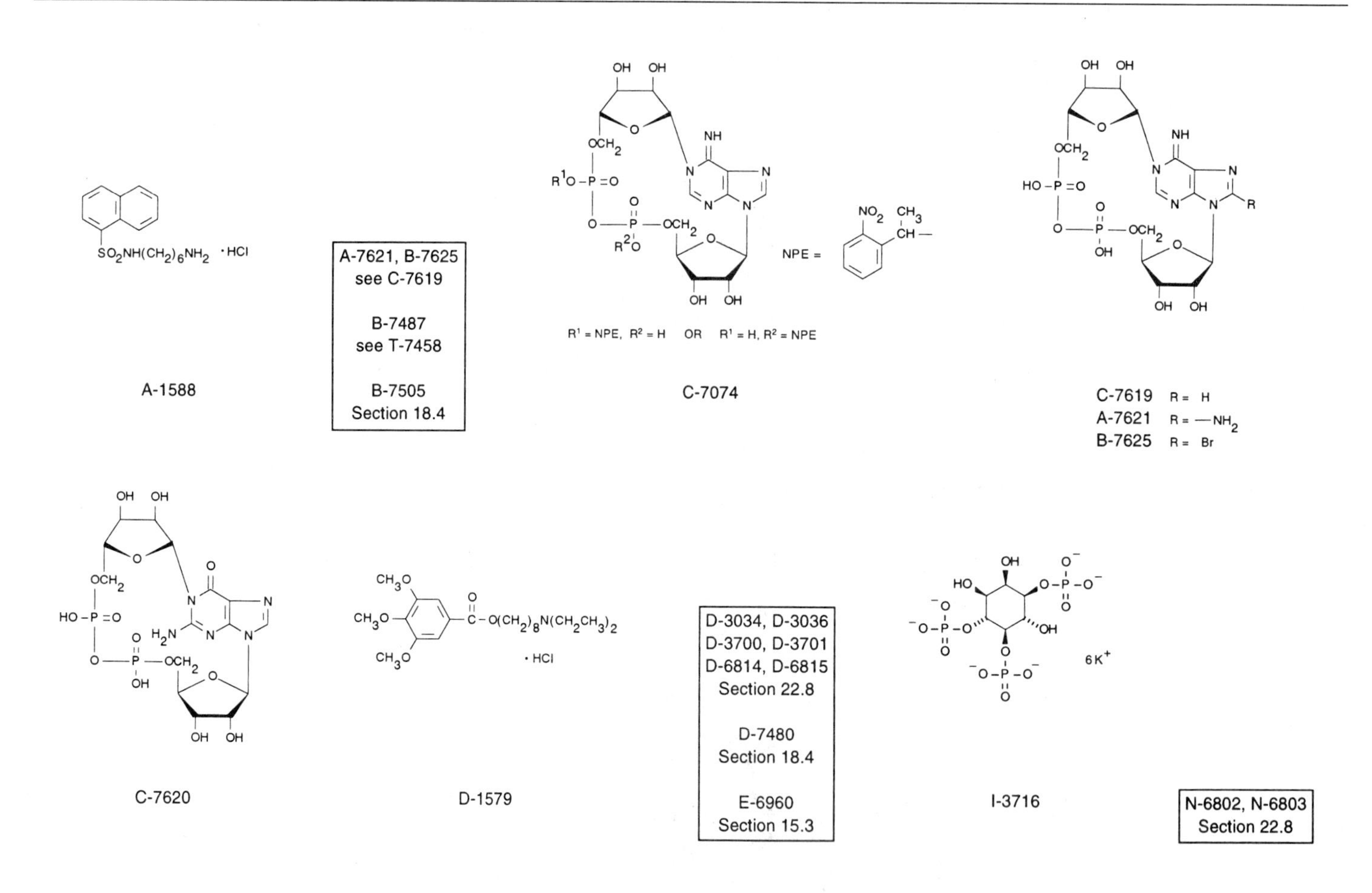

N-7622

R-7478
Section 18.4

T-1247
Section 15.3

T-7458 / T-7459 R = —C(=O)—CH2CH2CH3
B-7487 R = BODIPY FL

20.2 Price List *Calcium Regulation*

Cat #		Product Name	Unit Size	Price Per Unit ($) 1–4 Units	5–24 Units
A-7621	***New***	8-amino-cyclic adenosine 5′-diphosphate ribose (8-amino-cADP-ribose)	10 µg	100.00	80.00
A-1588		N-(6-aminohexyl)-1-naphthalenesulfonamide, hydrochloride (W-5)	100 mg	58.00	46.40
B-7487	***New***	BODIPY® FL thapsigargin	100 µg	95.00	76.00
B-7505	***New***	BODIPY® FL-X ryanodine	25 µg	65.00	52.00
B-7625	***New***	8-bromo-cyclic adenosine 5′-diphosphate ribose (8-Br-cADP-ribose)	35 µg	120.00	96.00
C-7619	***New***	cyclic adenosine 5′-diphosphate ribose (cADP-ribose)	100 µg	250.00	200.00
C-7074	***New***	cyclic adenosine 5′-diphosphate ribose, 1-(1-(2-nitrophenyl)ethyl)ester (NPE-caged cADP-ribose) *mixed isomers*	50 µg	150.00	120.00
C-7620	***New***	cyclic guanosine 5′-diphosphate ribose (cGDP-ribose)	100 µg	100.00	80.00
D-7480		9,21-dehydroryanodine	1 mg	38.00	30.40
D-3036		diazo-2, AM *cell permeant* *special packaging*	20x50 µg	195.00	156.00
D-3034		diazo-2, tetrapotassium salt *cell impermeant*	1 mg	195.00	156.00
D-3701		diazo-3, AM *cell permeant*	1 mg	75.00	60.00
D-3700		diazo-3, dipotassium salt *cell impermeant*	1 mg	75.00	60.00
D-1579		8-diethylaminooctyl 3,4,5-trimethoxybenzoate, hydrochloride (TMB-8)	100 mg	60.00	48.00
D-6815	***New***	1-(4,5-dimethoxy-2-nitrophenyl)-1,2-diaminoethane-N,N,N′,N′-tetraacetic acid, tetra-(acetoxymethyl ester) (DMNP-EDTA, AM) *cell permeant* *≥95% by HPLC* *special packaging*	20x50 µg	195.00	156.00
D-6814	***New***	1-(4,5-dimethoxy-2-nitrophenyl)-1,2-diaminoethane-N,N,N′,N′-tetraacetic acid, tetrapotassium salt (DMNP-EDTA) *cell impermeant*	5 mg	100.00	80.00
E-6960	***New***	europium(III) chloride, hexahydrate *99.99%*	1 g	20.00	16.00
H-7482	***New***	heparin, fluorescein conjugate	1 mg	95.00	76.00
I-3716		D-*myo*-inositol 1,4,5-triphosphate, hexapotassium salt (1,4,5-IP_3)	1 mg	90.00	72.00
N-7622	***New***	β-nicotinic acid adenine dinucleotide phosphate (NAADP)	10 µg	100.00	80.00
N-6803	***New***	o-nitrophenyl EGTA, AM (NP-EGTA, AM) *cell permeant* *special packaging*	20x50 µg	195.00	156.00
N-6802	***New***	o-nitrophenyl EGTA, tetrapotassium salt (NP-EGTA) *cell impermeant* *≥95% by HPLC*	1 mg	165.00	132.00
R-3455		ryanodine *mixture of ryanodine and dehydroryanodine*	10 mg	30.00	24.00
R-7478		ryanodine, *≥ 95% by HPLC*	1 mg	38.00	30.40
T-1247		terbium(III) chloride, hexahydrate	1 g	20.00	16.00
T-7458	***New***	thapsigargin	1 mg	55.00	44.00
T-7459	***New***	thapsigargin *special packaging*	20x50 µg	75.00	60.00

20.3 Probes for Protein Kinases, Protein Phosphatases and Cyclic AMP

The cascade of events that culminates in cellular response to an internal signal or environmental stimulus requires numerous molecular participants, ranging from ions to enzymes. Signal transduction pathways frequently activate specific protein kinases, leading to the phosphorylation of particular cellular proteins and subsequent initiation of a multitude of diverse cellular responses. Certain types of nucleotides have also been shown to play a major role in these activities. Molecular Probes offers a selection of native and modified biomolecules to aid the researcher in dissecting this highly complex branch of the signal transduction process.

Protein Kinases: PKC, PKA and CaM Kinase II

Protein Kinase C

Protein kinase C (PKC) is a key player in many transmembrane signal transduction systems.[1-5] This Ca^{2+}-dependent serine/threonine protein kinase is activated in the presence of certain membrane-derived lipids such as diacylglycerols and phosphatidyl serines and phosphorylates a wide variety of substrates, including ion channel proteins and cytoskeletal proteins. Molecular Probes now offers a highly purified preparation of PKC holoenzyme (P-7432), which is prepared from rat brain using ion-exchange and affinity chromatography.[6,7] These chromatographic techniques yield an enzyme preparation that is >95% pure by gel electrophoresis and contains the α-, β- and γ-isoforms (77,000–80,000 daltons). Purified PKC can be used to identify potential substrates, to determine the state of phosphorylation of PKC substrates and to serve as a positive control in PKC assays.[3,5]

We also offer the catalytic fragment of PKC (45,000–50,000 daltons, P-6461), prepared by trypsin digestion of the holoenzyme.[8] This catalytic fragment does not require Ca^{2+} or lipids for activation, yet it retains virtually the same substrate specificity and inhibitor sensitivity as the holoenzyme. The catalytic fragment is especially useful for introduction into intact cells or cell extracts because it does not require these extra factors for activation.[9,10]

Protein Kinase A

Molecular Probes now offers the catalytic subunit of cAMP-dependent protein kinase (protein kinase A or PKA, P-6460), an enzyme that mediates a variety of intracellular responses, including gene expression, carbohydrate metabolism, Ca^{2+} homeostasis and ion channel function.[11-15] Native PKA is comprised of two regulatory subunits and two catalytic subunits, both of which are released from the PKA holoenzyme upon cAMP binding. Molecular Probes' PKA catalytic subunit (~40,000 daltons) is purified from bovine heart and is active in the absence of cAMP.[15] The catalytic subunit of PKA is suitable for *in vitro* studies of PKA phosphorylation and can also be introduced directly into cells or cell extracts to modulate a number of different processes.

CaM Kinase II and Anti–Bovine Synapsin I

Ca^{2+}/calmodulin-dependent kinase II (CaM kinase II), a multifunctional kinase found in most, if not all, tissues, regulates a number of cellular functions, including neurotransmitter synthesis and release, gene expression, carbohydrate metabolism, cytoskeletal function and Ca^{2+} homeostasis.[16-21] CaM kinase II, which phosphorylates the IP_3 receptor,[22] is reported to be essential for cADP-ribose–mediated Ca^{2+} mobilization required for insulin secretion by pancreatic islet cells.[23] Molecular Probes' CaM kinase II (C-6462) is purified from rat brain and consists of the 50,000- and 59,000-dalton subunits in a ratio of about 3:1.

Synapsin I is a synaptic vesicle–associated protein that inhibits neurotransmitter release, a function that is abolished upon its phosphorylation by CaM kinase II.[24] Calmodulin also influences synapsin's function indirectly by activating CaM kinases and directly by binding to synapsin and potentiating the effects of phosphorylation.[25-27] Unphosphorylated synapsin I is able to promote actin polymerization and bundling of actin filaments in the presence of synaptic vesicles.[28,29] Upon polarization and activation of protein kinases, synapsin becomes phosphorylated, which reduces its affinity for actin and disrupts the actin network constraining the synaptic vesicles. These synaptic vesicles then migrate to the presynaptic membrane where they undergo membrane fusion and release neurotransmitter.[30,31]

Synapsin is localized exclusively to synaptic vesicles and thus serves as an excellent marker for synapses in brain and other neuronal tissue.[32,33] Antibodies directed against synapsin I have proven valuable in molecular and neurobiology research, for example to estimate synaptic density and to follow synaptogenesis.[34-36] Molecular Probes is now offering rabbit anti–bovine synapsin I, either as whole serum (A-6443) or an affinity-purified IgG fraction (A-6442). This antibody was isolated from rabbits immunized against bovine brain synapsin I but is also active against human, rat and mouse forms of the antigen; it has little or no activity against synapsin II. Affinity-purified anti–synapsin I is suitable for immunohistochemistry, Western blots, enzyme-linked immunoadsorbent assays (ELISAs) and immunoprecipitation. Unfractionated rabbit serum is provided as an economical alternative for Western blots and immunoprecipitation.

Protein Kinase C Activators

Phorbol Esters

The recognized tumor promoter 4β-phorbol 12β-myristate 13α-acetate (PMA; also called tetradecanoylphorbol 13-acetate or TPA) is a potent, nonfluorescent, cell-permeant PKC activator that is effective at nanomolar concentrations.[37,38] Molecular Probes offers 98% pure PMA both in 1 mg vials (P-3453) and as a convenient set of ten vials that each contain 100 µg (P-3454).

Diacylglycerols

Molecular Probes offers diacylglycerol analog 1,2-dioctanoyl-*sn*-glycerol (DiC_8 or DOG), another cell-permeant PKC activator.[38] Although the pure oil is relatively stable, especially when stored cold, diacylglycerols partially rearrange to inactive forms when stored in solutions of dimethylsulfoxide (DMSO), ethanol or other solvents, even in the freezer.[39] Therefore, we specially package our DOG (D-1576) into 10 separate vials that each contain 1 mg for reconstitution as needed.

Our 2-nitrobenzyl–caged dioctanoyl-*sn*-glycerol (NB-caged DOG; D-7081) should prove valuable for studying DOG-mediated activation of PKC.[38] Researchers have observed an increase in Ca^{2+} oscillation period in REF52 fibroblasts when they photolyzed a similar caged DOG probe intracellularly, rapidly stimulating the PKC pathway.[40]

Protein Kinase Inhibitors

Nonfluorescent Protein Kinase Inhibitors

Structure–activity relationship studies carried out by Toullec and colleagues led to the development of bisindolylmaleimide [41] (GF 109203X, D-7475), a competitive inhibitor of PKC with respect to ATP (K_i ~14 nM; IC_{50} = 10 nM). Bisindolylmaleimide is extremely selective for PKC over five other protein kinases.[42-44]

Molecular Probes also offers the potent, but relatively unselective, protein kinase inhibitor staurosporine [45-47] (antibiotic AM-2282, S-7456). Staurosporine competitively inhibits PKC and PKA by binding to the catalytic domains with an IC_{50} of 2.7 nM and 8.2 nM, respectively.[47] Staurosporine was recently used to investigate the function of histone phosphorylation in controlling chromosome condensation.[48]

The azepine analog ML-9 (C-1648) inhibits PKC (K_i = 54 µM), PKA (K_i = 32 µM), myosin light-chain kinase (K_i = 3.8 µM) and Ca^{2+}- and cyclic nucleotide–dependent phosphodiesterases.[49-53] ML-9 also suppresses superprecipitation of actomyosin.[54]

Melittin (M-98403), a peptide isolated from *Apis mellifera* bee venom, has a range of activity similar to ML-9. It is an inhibitor of PKC,[55,56] Ca^{2+}/calmodulin-dependent protein kinase II, myosin light-chain kinase and Na^+/K^+-ATPase from synaptosomal membranes, all with an IC_{50} of 1–4 µM [55]

Cardiotoxin (C-98102), a potent PKC-selective inhibitor peptide [55] isolated from the venom of the cobra *Naja nigricollis*, has been shown to depolarize skeletal muscle fibers *in vitro* with an IC_{50} of 1–3 µM.[57]

Fluorescent Bisindolylmaleimides: Fim-1, Rim-1 and BODIPY FL Derivatives

Bisindolylmaleimides selectively inhibit PKC by binding to the enzyme's catalytic domain. To monitor PKC activation and translocation from the cytoplasm to membranes, Chen and Poenie have developed fluorescent derivatives of bisindolylmaleimides.[58] Molecular Probes offers several fluorescent bisindolylmaleimides:

- Fim-1 (F-7462), a fluorescein-conjugated bisindolylmaleimide and Fim-1 diacetate (F-7453), its membrane-permeant derivative
- Rim-1 (R-7454), a rhodamine-conjugated bisindolylmaleimide
- Monofunctional BODIPY FL bisindolylmaleimide (B-7485), which is analogous to fim-1 and rim-1, as well as bifunctional BODIPY FL di(bisindolylmaleimide) (B-7486)

Fim-1 and rim-1 have been shown to be effective inhibitors of PKC and to exhibit 9- to 16-fold selectivity for PKC over PKA. The IC_{50} value of fim-1 for the α-isoform of PKC is 10-fold lower than that of rim-1. Enzyme kinetic analysis has demonstrated that fim-1 inhibits PKC by competing with ATP and not with phosphatidylserine or diacylglycerol, which is consistent with its binding at the catalytic domain rather than the regulatory domain.[58] Using fixed and permeabilized cells, Chen and Poenie have shown that the pattern of staining produced by both fim-1 and rim-1 was very similar to that of an anti-PKC antibody, except that the fluorescent inhibitors also appeared to stain mitochondria; this mitochondrial staining is not yet understood.[58] These fluorescent bisindolylmaleimide derivatives have significant potential for monitoring PKC activity in live cells.

Hypericin and Hypocrellin A and B

Hypericin (H-7476), a natural pigment isolated from plants of the genus *Hypericum*,[59] is a potent, selective inhibitor of PKC (IC_{50} = 1.7 µg/mL = 3.4 µM) that should be useful for probing and manipulating PKC in live cells.[60,61] When compared with other PKC probes, this aromatic polycyclic dione offers several advantages, including bright red fluorescence emission and exceptional photostability. Furthermore, hypericin exhibits light-induced inhibition of PKC,[59,60] thus providing the researcher with another level of control in manipulating and evaluating PKC activity and distribution. Hypericin has a variety of pharmacological properties, from antibacterial and antineoplastic activities to antiviral activities.[61,62] Furthermore, it is a potent photosensitizer, with a quantum yield of 0.75 for the generation of singlet oxygen.[63]

We also offer the structurally similar reagents, hypocrellin A and hypocrellin B (H-7515, H-7516), which have been shown to be selective and potent inhibitors of PKC, with IC_{50} values of 3.6 and 9.0 µM, respectively.[64,65] Like hypericin, PKC inhibition by hypocrellins is light dependent. The extremely broad absorption spectra of hypocrellins make them readily excited by a variety of light sources.[66] These chemically stable, fluorescent photosensitizers have quantum yields for singlet oxygen generation in excess of 0.7.[66,67] In addition to their utility as probes, they may serve as building blocks for preparing antiviral and anticancer agents.[65,68-70]

Protein Phosphatase Inhibitor: Okadaic Acid

Okadaic acid (O-3452, O-7457) is a polyether derivative of a 38-carbon fatty acid produced by dinoflagellates and isolated from marine sponges.[71-73] This potent tumor promoter penetrates intact cells and strongly inhibits serine/threonine protein phosphatases 1 (IC_{50} ~10–15 nM) and 2A (IC_{50} ~0.1 nM), leading to hyperphosphorylation of numerous cellular proteins.[74-76] It is much less effective in inhibiting protein phosphatase 2B (IC_{50} ~5 µM) and does not inhibit protein phosphatase 2C, tyrosine protein phosphatases, alkaline phosphatase, mitochondrial pyruvate dehydrogenase phosphatase or any of eight protein kinases tested.[72] Okadaic acid has been reported to:

- Block cell elongation of teleost rod photoreceptors [75]
- Disrupt synaptic vesicle clusters in frog motor nerve terminals [77]
- Induce reversible fragmentation of the trans-Golgi network [78]
- Inhibit collagen synthesis in cultured human fibroblasts [79]
- Modulate neutrophil activation [80]
- Potentiate heat-induced hsp 70 promoter activity [81]

Okadaic acid is available as a 0.5 mg/mL solution in DMSO (O-3452) and as the sodium salt (O-7457), which is reported to be more stable than the free acid form.

Cyclic AMP Fluorosensor

In collaboration with Atto Instruments, Inc., Molecular Probes is pleased to offer Cyclic AMP Fluorosensor (FlCRhR, C-6660), the first fluorescent probe available for the nondestructive measurement of cyclic AMP (cAMP) in live cells by digital video imaging, confocal laser scanning microscopy or microphotometry. Originally developed by Roger Tsien and co-workers,[82,83] this probe consists of cAMP-dependent protein kinase A (PKA) in which its recombinant catalytic (C) and regulatory (R) subunits are labeled with fluorescein and rhodamine, respectively. Fluorescence resonance energy transfer from fluorescein to nearby rhodamine labels occurs readily in the holoenzyme configuration (C_2R_2) but is eliminated upon subunit dissociation in response to cAMP binding (4 molecules of cAMP per C_2R_2 holoenzyme) as illustrated in Figure 20.2. Consequently, the ratio of fluorescein (~520 nm) and rhodamine (~580 nm) emission intensities excited at 488 nm can be quantitatively related to cAMP concentration.

Cyclic AMP Fluorosensor is a protein complex with an aggregate molecular weight of 172,000 daltons; therefore it must be pressure microinjected into the cytoplasm for intracellular cAMP measurements. Complete cAMP response calibration of the probe is currently only possible *in vitro*. The affinity for cAMP and the kinase activity of the probe are similar to those of the unlabeled protein kinase.[82,83] Intracellular cAMP levels can be manipulated to some extent using phosphorothioate antagonists[82,83] and cell-permeant cAMP derivatives such as dibutyryl cAMP AM[84] (D-7623, see Section 20.4). Intracellular release of cAMP can also be photolytically triggered using our membrane-permeable DMNB- and NPE-caged cAMP derivatives (D-1037, N-1045; see Section 20.4).

Because Cyclic AMP Fluorosensor has been created by modifying the subunits of a native enzyme, it is physiologically active, triggering the same downstream signaling which takes place when cAMP binds to the native kinase. Published applications for Cyclic AMP Fluorosensor by Tsien and co-workers include identification of the signaling pathway for cAMP activation of neuronal nicotinic acetylcholine receptors[85] and comparison of Ca^{2+} (measured using fura-2; see Section 22.2) and cAMP as controlling factors for intracellular vesicle motility.[86] Using simultaneous digital ratio imaging of fura-2 and Cyclic AMP Fluorosensor (Figure 20.3), researchers have visualized regulatory interactions between cAMP and Ca^{2+} at the single cell level.[87] After cAMP-dependent dissociation from the regulatory subunit, the fluorescein-labeled catalytic subunit is translocated to the nucleus (Figure 20.3).[88,89] In cAMP-dependent gene regulation processes, this translocation can be correlated with subsequent transcription factor phosphorylation using Cyclic AMP Fluorosensor.[90]

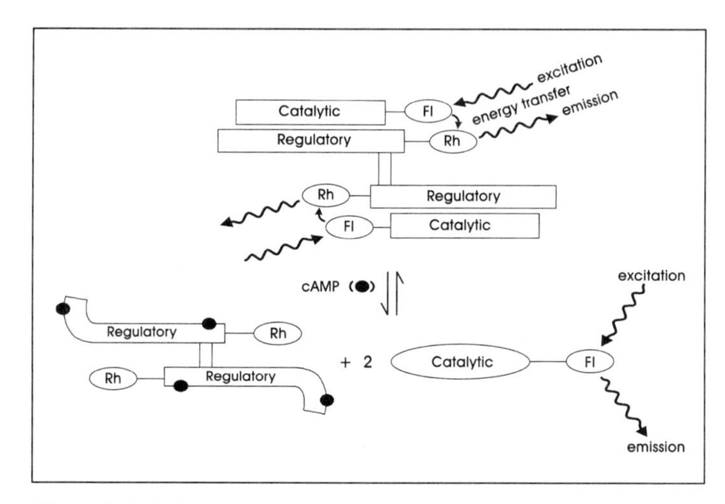

Figure 20.2 Schematic depiction of the structure and response mechanism of Cyclic AMP Fluorosensor (FlCRhR, C-6660). Fluorescence resonance energy transfer from the fluorescein-labeled (Fl) catalytic subunit to the rhodamine-labeled (Rh) regulatory subunit is eliminated upon cyclic AMP–dependent dissociation of the holoenzyme. The resulting increase of the 488 nm–excited fluorescein (520 nm) to rhodamine (580 nm) emission ratio can be used to measure intracellular cyclic AMP. Adapted with permission from Nature 349, 694 (1991).

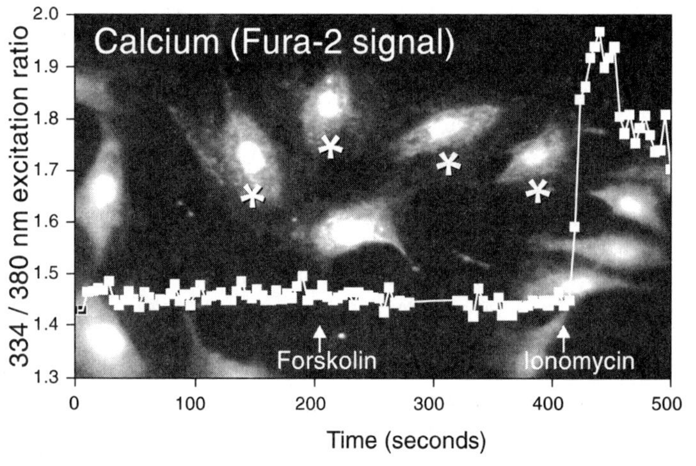

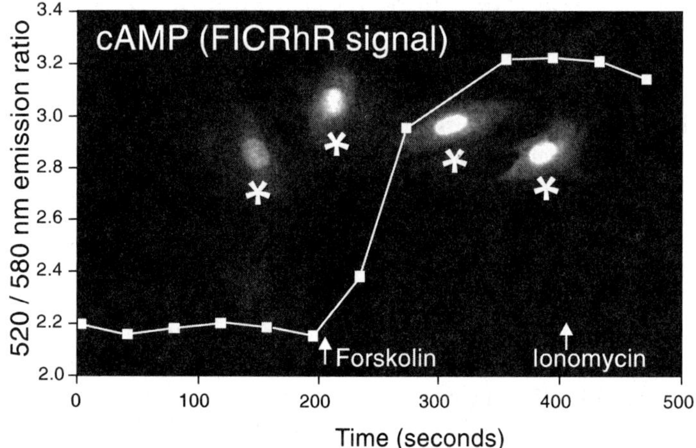

*Figure 20.3 Simultaneous imaging of Ca^{2+} and cAMP in REF 52 fibroblasts loaded with fura-2 AM, four of which (marked *) have also been microinjected with Cyclic AMP Fluorosensor (FlCRhR, C-6660).* ***Left panel:*** *the plotted photometric data represent the fura-2 334/380 nm excitation ratio (emission at 520 nm) showing a cellular Ca^{2+} response to ionomycin treatment only. The images are of ionomycin-treated cells, using 334 nm excitation only.* ***Right panel:*** *the plotted photometric data represent the FlCRhR 520/580 nm emission ratio (excitation at 488 nm) showing elevation of cAMP in response to adenylate cyclase activation by forskolin. The images (488 nm excitation, 520 nm emission), acquired about 20 minutes after forskolin treatment, show the cAMP-induced nuclear localization of the fluorescein-labeled C-subunit of FlCRhR. Images and photometric data were simultaneously acquired using the Carl Zeiss Attofluor® RatioVision® digital video imaging system with Attofluor ICCD dual emission cameras (Atto Instruments).*

Cyclic AMP Fluorosensor (C-6660) and the separate fluorescein-labeled catalytic and rhodamine-labeled regulatory subunits (C-6661, C-6662) are provided in 10 µL units at an average concentration of 20 µM (lot-to-lot variation ±5 µM) in 25 mM potassium phosphate, 1 mM EDTA, 0.5 mM β-mercaptoethanol, 2.5% glycerol, pH 7.3. This solution is ready to be microinjected into live cells. With optimal microinjection parameters, a volume of <0.5 µL of ~20 µM Cyclic AMP Fluorosensor is sufficient to load an average of >100 cells.

Adenylate Cyclase Probe: Fluorescent Forskolin

Forskolin, isolated from *Coleus forskohlii*, is a potent activator of adenylate cyclase, which catalyzes the formation of cAMP from ATP. Cervical cells stained with our new BODIPY FL forskolin (B-7469) exhibit a staining pattern identical to that seen with a fluorescent anti–adenylate cyclase antibody.[91]

1. Ann Rep Med Chem 29, 255 (1994); **2.** Eur J Biochem 208, 547 (1992); **3.** Science 258, 607 (1992); **4.** Eur J Biochem 195, 9 (1991); **5.** Ann Rev Cell Biol 2, 149 (1986); **6.** J Neurochem 52, 215 (1989); **7.** Eur J Biochem 164, 461 (1987); **8.** Biochem Biophys Res Comm 139, 320 (1986); **9.** J Biol Chem 269, 24666 (1994); **10.** J Physiol 475, 431 (1994); **11.** Biochem J 293, 413 (1993); **12.** J Neurosci 13, 5242 (1993); **13.** Nature 364, 821 (1993); **14.** Proc Natl Acad Sci USA 90, 10135 (1993); **15.** Meth Enzymol 38, 299 (1974); **16.** J Biol Chem 270, 10043 (1995); **17.** Science 268, 239 (1995); **18.** J Biol Chem 269, 20872 (1994); **19.** Trends Neurosci 17, 406 (1994); **20.** Science 260, 181 (1993); **21.** Ann Rev Biochem 61, 559 (1992); **22.** Proc Natl Acad Sci USA 88, 2232 (1991); **23.** J Biol Chem 270, 30257 (1995); **24.** Greengard, P., Benfenati, F. and Valtora F. in *Molecular and Cellular Mechanisms of Neurotransmitter Release*, edited by L. Stjarne, P. Greengard, S. Griller, T. Hökfelt and D. Ottoson, Raven Press (1994) pp. 31–45; **25.** Biochemistry 34, 1912 (1995); **26.** Biochem Soc Trans 23, 65 (1995); **27.** Eur J Biochem 224, 229 (1994); **28.** J Cell Biol 128, 905 (1995); **29.** FEBS Lett 329, 301 (1993); **30.** Science 259, 780 (1993); **31.** J Cell Biol 108, 111 (1989); **32.** Science 226, 1209 (1984); **33.** J Cell Biol 96, 1337 (1983); **34.** J Neurosci 12, 1736 (1992); **35.** J Neurosci 11, 1617 (1991); **36.** J Neurosci 9, 2151 (1989); **37.** Cell 81, 917 (1995); **38.** Nature 308, 693 (1984); **39.** Biochem Biophys Res Comm 140, 1143 (1986); **40.** Cell Calcium 12, 153 (1991); **41.** Bisindolylmaleimide is licensed to Molecular Probes under U.S. Patent No. 5,380,746; **42.** Biochem Biophys Res Comm 206, 119 (1995); **43.** J Med Chem 35, 994 (1992); **44.** J Biol Chem 266, 15771 (1991); **45.** J Immunology 149, 3894 (1992); **46.** Biochem Biophys Res Comm 158, 105 (1989); **47.** Biochem Biophys Res Comm 135, 397 (1986); **48.** J Biol Chem 269, 9568 (1994); **49.** Biochem Pharmacol 38, 2615 (1989); **50.** Biochem Biophys Res Comm 143, 1045 (1987); **51.** J Biol Chem 262, 7796 (1987); **52.** Meth Enzymol 139, 570 (1987); **53.** Biochem Biophys Res Comm 140, 280 (1986); **54.** Mol Pharmacol 33, 598 (1988); **55.** J Biol Chem 266, 2753 (1991); **56.** Brain Res 535, 131 (1990); **57.** Eur J Biochem 174, 103 (1988); **58.** J Biol Chem 268, 15812 (1993); **59.** Med Res Rev 15, 111 (1995); **60.** Biochem Biophys Res Comm 165, 1207 (1989); **61.** Proc Natl Acad Sci USA 86, 5963 (1989); **62.** Pharmacol Ther 63, 1 (1994); **63.** Photochem Photobiol 59, 529 (1994); **64.** Photochem Photobiol 61, 529 (1995); **65.** Biochem Pharmacol 47, 373 (1994); **66.** J Photochem Photobiol A: Chem. 64, 273 (1992); **67.** Photochem Photobiol 52, 609 (1990); **68.** Biochemistry 34, 15845 (1995); **69.** Photochem Photobiol 60, 253 (1994); **70.** Anti-Cancer Drug Design 8, 129 (1993); **71.** Toxicon 28, 371 (1990); **72.** Trends Biochem Sci 15, 98 (1990); **73.** J Am Chem Soc 103, 2469 (1981); **74.** J Cell Biol 131, 1291 (1995); **75.** J Neurosci 15, 6475 (1995); **76.** Eur J Biochem 193, 671 (1990); **77.** J Cell Biol 124, 843 (1994); **78.** Biochem J 301, 69 (1994); **79.** Biochem Biophys Res Comm 199, 177 (1994); **80.** Am J Physiol 262, C39 (1992); **81.** J Biol Chem 268, 1436 (1993); **82.** Nature 349, 694 (1991); **83.** Adams, S.R. *et al.*, in *Fluorescent and Luminescent Probes for Biological Activity*, W.T. Mason and G. Relf, Eds., Academic Press (1993) pp. 133–149; **84.** J Biol Chem 268, 6316 (1993); **85.** J Neurosci 14, 3540 (1994); **86.** J Cell Biol 117, 57 (1992); **87.** Proc Natl Acad Sci USA 93, 4577 (1996); **88.** Mol Biol of the Cell 4, 993 (1993); **89.** Science 260, 222 (1993); **90.** Mol Cell Biol 13, 4852 (1993); **91.** Daisy McCann, McCann Associates, Inc., personal communication.

20.3 Data Table *Probes for Protein Kinases, Protein Phosphatases and Cyclic AMP*

Cat #	Structure	MW	Storage	Soluble	Abs	$\{\epsilon \times 10^{-3}\}$	Em	Solvent	Notes
B-7469	✓	785	F,D,L	DMSO	504	{79}	511	MeOH	
B-7485	✓	717	F,D,L	DMSO, DMF	509	{78}	520	MeOH	
B-7486	✓	1097	F,D,L	DMSO, DMF	510	{93}	520	MeOH	
C-1648	✓	361	D	EtOH, H_2O	298	{8.3}	335	MeOH	
C-6660	C	~172,000	FF,L	<1>	500	ND	517	<2>	<3, 4>
C-6661	C	~40,000	FF,L	<1>	496	ND	517	<2>	<3>
C-6662	C	~100,000*	FF,L	<1>	557	ND	575	<2>	<3>
C-98102	C	~6800	FF,D	H_2O	<300		<5>		
D-1576	✓	344	FF,D	DMSO†	<300		none		<6>
D-7081	✓	480	F,D,LL	DMSO	260	{5.1}	none	MeCN	<7, 8>
D-7475	✓	412	L	DMSO, DMF	459	{8.6}	<9>	MeOH	
F-7453	✓	841	F,D,L	DMSO, DMF	464	{9.1}	none	MeOH	<10>
F-7462	✓	756	F,D,L	DMSO, DMF	494	{76}	524	pH 9	
H-7476	S21.2	504	F,D,L	DMSO, DMF	591	{37}	594	EtOH	
H-7515	S21.2	547	F,D,L	DMSO	463	{24}	600	MeOH	<11>
H-7516	S21.2	529	F,D,L	DMSO	459	{23}	616	MeOH	<11>
M-98403	C	~2847	FF,D	H_2O	<300		<5>		
O-3452	✓	805	F,D,AA	DMSO‡	<300		none		
O-7457	✓	827	F,D,AA	H_2O	<300		none		
P-3453	✓	617	F,D	EtOH, DMSO	<300		none		
R-7454	✓	939	D,L	DMSO, DMF	562	{110}	581	MeOH	
S-7456	✓	467	L	DMSO, MeOH	372	{9.4}	379	MeOH	<12>

P-3454 see P-3453

For definitions of the contents of this data table, see "How to Use the Handbook" on page vi.
Structure: Chemical structure drawing: ✓ = shown in this section; Sn.n = shown in Section number n.n; C = not shown due to complexity.
MW: ~ indicates an approximate value.

* Dimer MW.
† Isomerization of D-1576 to inactive forms occurs in solution. Store in oil form (as supplied), except when required for immediate use.
‡ This product is packaged as a solution in the solvent indicated in "Soluble."
<1> This product is supplied as a directly microinjectable 20 µM solution in 25 mM potassium phosphate, 1 mM EDTA, 0.5 mM β-mercaptoethanol, 2.5% glycerol, pH 7.3.
<2> Spectra recorded in 50 mM MOPS, pH 7.3.
<3> ND = not determined.
<4> Absorption spectrum contains a second peak at ~540 nm. Binding of cyclic AMP results in an increase of the 520/580 nm emission ratio when excited at 488 nm.
<5> This peptide exhibits intrinsic tryptophan fluorescence (Em ~350 nm) when excited at <300 nm.
<6> This product is intrinsically a liquid or an oil at room temperature.
<7> All photoactivatable probes are sensitive to light. They should be protected from illumination, except when photolysis is intended.
<8> This compound has weaker visible absorption at >300 nm but no discernible absorption peaks in this region.
<9> D-7475 is weakly fluorescent, Em = 660 nm in MeOH.
<10> Converted to F-7462 after acetate hydrolysis.
<11> H-7515 and H-7516 have weaker absorption peaks at longer wavelengths: Abs = 581 nm (ε = {12}) for H-7515, Abs = 580 nm (ε = {9}) for H-7516.
<12> S-7456 has much stronger absorption at shorter wavelengths (ε = {52} at 292 nm). Emission spectrum has additional peaks at 399 nm and 420 nm.

20.3 Structures *Probes for Protein Kinases, Protein Phosphatases and Cyclic AMP*

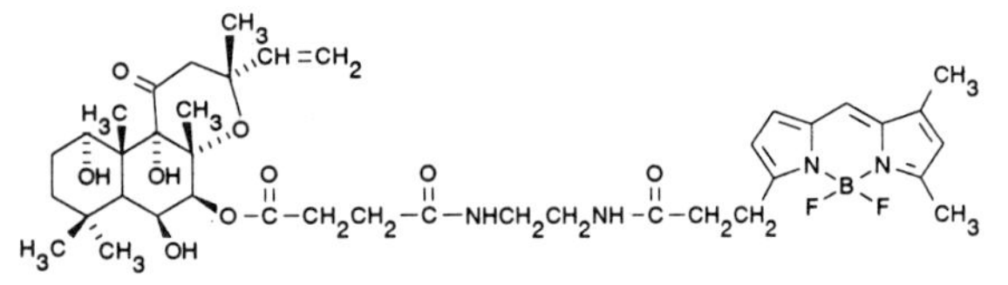

B-7469

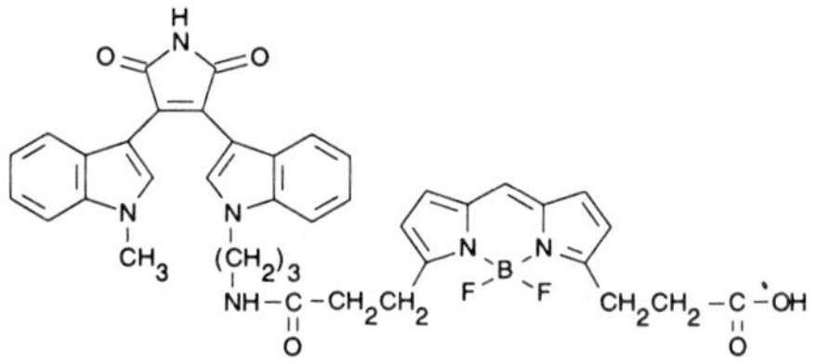

B-7485

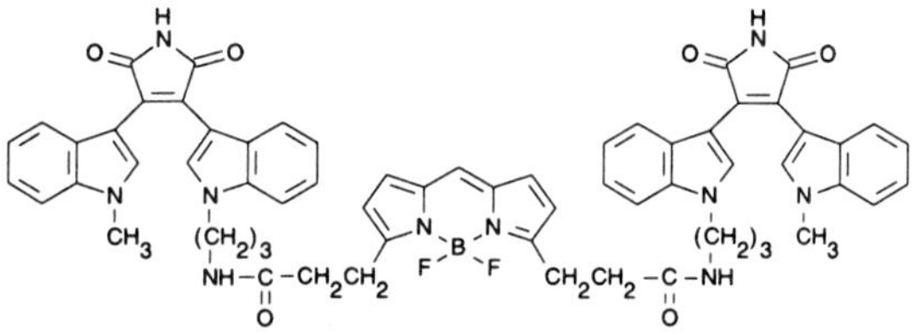

B-7486

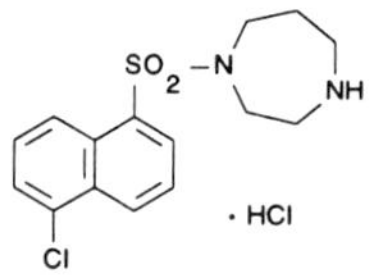

C-1648

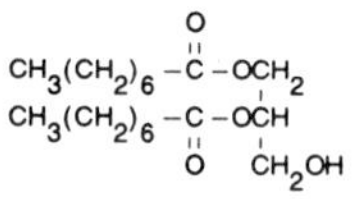

D-1576

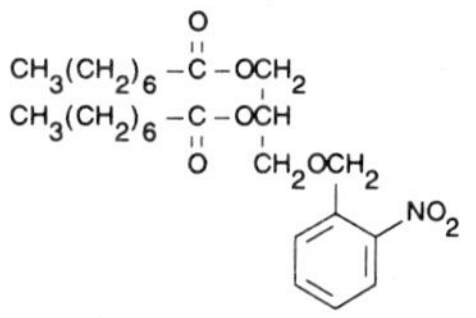

D-7081

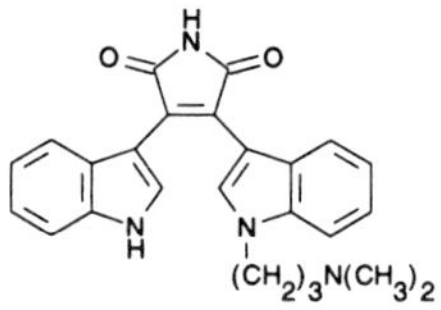

D-7475

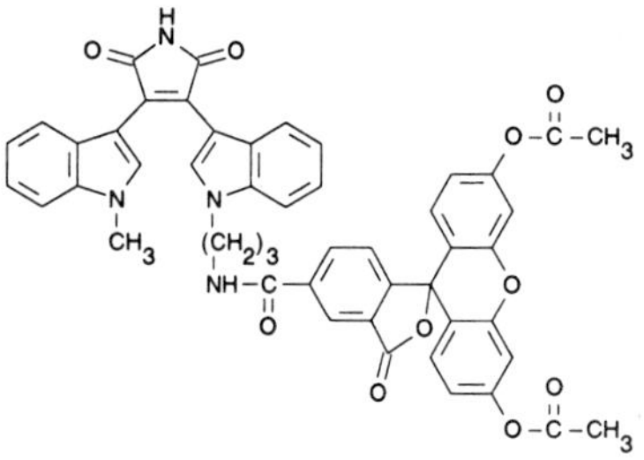

F-7453

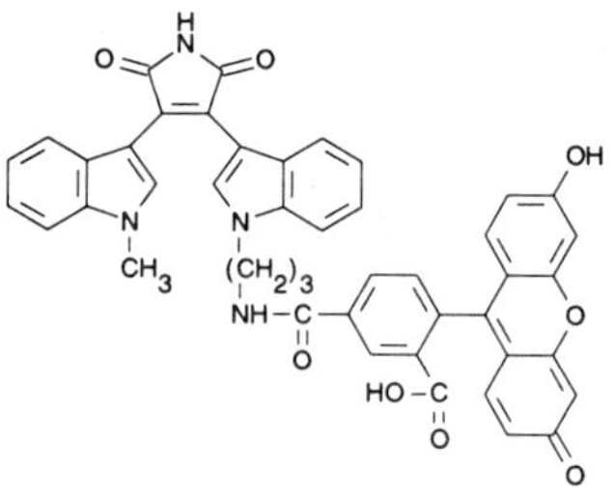

F-7462

H-7476, H-7515
H-7516
Section 21.2

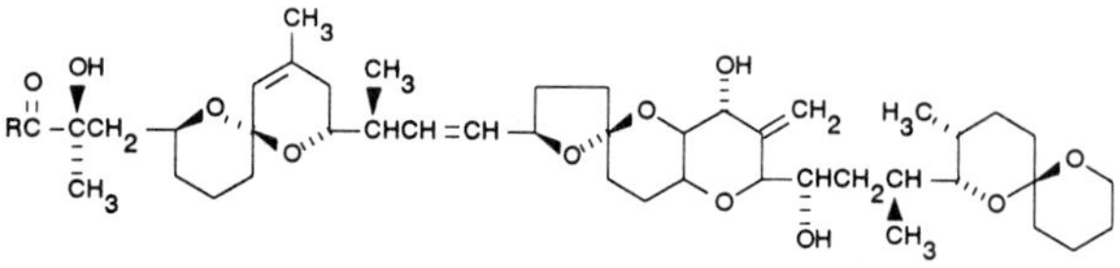

O-3452 R = —OH
O-7457 R = O^- Na^+

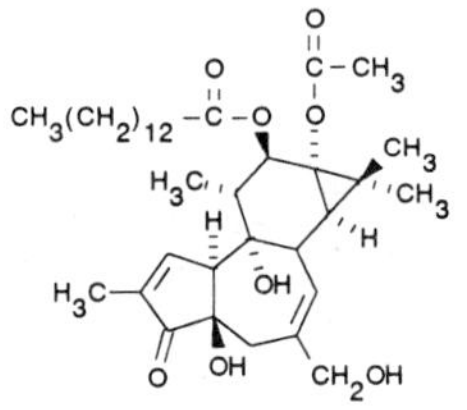

P-3453 / P-3454

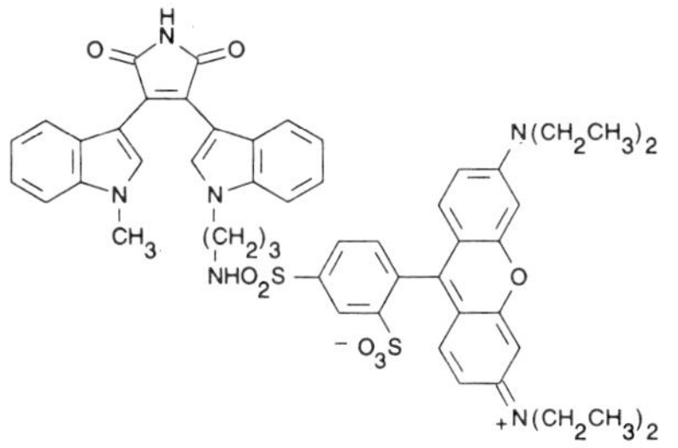

R-7454

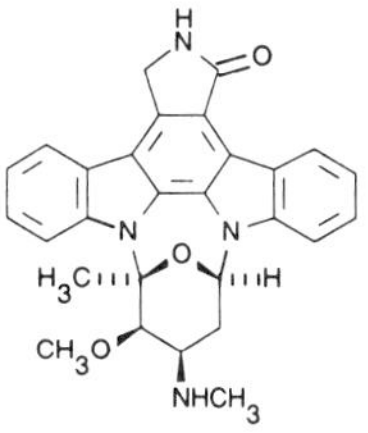

S-7456

20.3 Price List *Probes for Protein Kinases, Protein Phosphatases and Cyclic AMP*

Cat #		Product Name	Unit Size	Price Per Unit ($) 1–4 Units	5–24 Units
A-6442	***New***	anti-bovine synapsin I, rabbit IgG fraction *affinity purified* *specificity: human, bovine, rat, mouse*	10 µg	120.00	96.00
A-6443	***New***	anti-bovine synapsin I, rabbit serum *specificity: human, bovine, rat, mouse*	100 µL	98.00	78.40
B-7485	***New***	BODIPY® FL bisindolylmaleimide	250 µg	65.00	52.00
B-7486	***New***	BODIPY® FL di(bisindolylmaleimide)	250 µg	65.00	52.00
B-7469	***New***	BODIPY® FL forskolin	100 µg	95.00	76.00
C-6462	***New***	calcium/calmodulin-dependent kinase II (CaM kinase II) *from rat brain* *10 ng/µL*	100 µL	225.00	180.00
C-98102	***New***	cardiotoxin *from *Naja nigricollis* snake venom*	1 mg	128.00	102.40
C-1648		1-(5-chloronaphthalene-1-sulfonyl)-1H-hexahydroazepine, hydrochloride (ML-9)	10 mg	48.00	38.40
C-6661	***New***	cyclic AMP-dependent protein kinase, catalytic subunit, fluorescein conjugate *20±5 µM*	10 µL	295.00	295.00
C-6662	***New***	cyclic AMP-dependent protein kinase, regulatory subunit, rhodamine conjugate *20±5 µM*	10 µL	295.00	295.00
C-6660	***New***	Cyclic AMP Fluorosensor® (FlCRhR) *20±5 µM*	10 µL	495.00	495.00
D-7475	***New***	2-(1-(3-dimethylaminopropyl)-indol-3-yl)-3-(indol-3-yl)maleimide (bisindolylmaleimide; GF 109203X)	1 mg	65.00	52.00
D-1576		1,2-dioctanoyl-*sn*-glycerol (DiC_8;DOG) *special packaging*	10x1 mg	53.00	42.40
D-7081	***New***	1,2-dioctanoyl-3-(2-nitrobenzyl)-*sn*-glycerol (NB-caged DOG)	5 mg	95.00	76.00
F-7462	***New***	fim-1	1 mg	95.00	76.00
F-7453	***New***	fim-1 diacetate	1 mg	95.00	76.00
H-7476	***New***	hypericin	1 mg	48.00	38.40
H-7515	***New***	hypocrellin A	1 mg	48.00	38.40
H-7516	***New***	hypocrellin B	1 mg	48.00	38.40
M-98403	***New***	melittin *from *Apis mellifera* bee venom*	1 mg	45.00	36.00
O-3452		okadaic acid *500 µg/mL solution in DMSO*	50 µL	100.00	80.00
O-7457	***New***	okadaic acid, sodium salt	25 µg	95.00	76.00
P-3453		4β-phorbol 12β-myristate 13α-acetate (TPA; PMA)	1 mg	25.00	20.00
P-3454		4β-phorbol 12β-myristate 13α-acetate (TPA; PMA) *special packaging*	10x100 µg	35.00	28.00
P-6460	***New***	protein kinase A, catalytic subunit *from bovine heart* *20 ng/µL*	100 µL	95.00	76.00
P-7432	***New***	protein kinase C *from rat brain* *6.7 ng/µL*	2x75 µL	98.00	78.40
P-6461	***New***	protein kinase C, catalytic fragment *from rat brain* *10 ng/µL*	50 µL	225.00	180.00
R-7454	***New***	rim-1	1 mg	95.00	76.00
S-7456	***New***	staurosporine (antibiotic AM-2282)	100 µg	45.00	36.00

20.4 Nucleotides for Signal Transduction Research

Caged Nucleotides

Molecular Probes offers several "caged" nucleotides that are physiologically inactive until the caging group is cleaved by flash photolysis to release the free nucleotides (see also Chapter 19 for a description of biomolecule caging and a list of all our caged probes). Several of our photoactivatable nucleotides are available with a choice of caging groups; these include caged versions of:

- ATP (A-1048, A-1049, A-7073)
- ADP (A-7056)
- GTP-γ-*S* (G-1053)
- GDP-β-*S* (G-3620)

Caged versions of the nonhydrolyzable GTP-γ-*S* and GDP-β-*S* should prove very useful for studying regulatory proteins such as the G-proteins.[1,2]

Permeabilization of cells with staphylococcal α-toxin or with the saponin ester β-escin is reported to make the membrane of smooth muscle cells permeable to low molecular weight (<1000 daltons) molecules, while retaining high molecular weight compounds.[3,4] α-Toxin permeabilization has permitted the introduction of caged nucleotides, including caged ATP (A-1048) and caged GTP-γ-*S* (G-1053), as well as of 1,4,5-IP_3 (I-3716, see Section 20.2) into smooth muscle cells.[5,6]

Cyclic Nucleotides

Caged cAMP

Biologically active cAMP can be released by flash photolysis of DMNB-caged cAMP or NPE-caged cAMP (D-1037, N-1045). The caging process has the favorable side-effect of transforming cAMP into a membrane-permeable probe, thereby simplifying the delivery of caged cAMP into cells, where it can then be photolyzed to the natural cyclic nucleotide.[7-9]

Dibutyryl cAMP AM

Originally developed by Schultz and co-workers, the acetoxymethyl (AM) ester of $N^6,O^{2'}$-dibutyryladenosine 3',5'-cyclic monophosphate (D-7623) is an uncharged, cell-permeant derivative of cAMP that permits delivery of cAMP into cells without using invasive techniques to disrupt the cell membrane.[10] The phosphate group of cAMP is protected as an AM ester, which can be hydrolyzed by intracellular esterases (see the Technical Note "Loading and Calibration of Intracellular Ion Indicators" on page 549). To further increase the hydrophobicity of the probe, the free hydroxyl group and the free amino group of cAMP are acylated with butyryl groups, which are similarly removed by esterase activity. Extracellularly applied dibutyryl cAMP AM has been shown to elicit cellular responses that are identical to those produced by intracellular cAMP in three model systems [10] at effective concentrations two- to threefold less than required for other membrane-permeant cAMP derivatives.[10]

Brominated cAMP and cGMP

The effect of cyclic nucleotides in their natural form may be short-lived. Their activity may be prolonged markedly by halogenation of the cyclic nucleotide. Although the 8-bromo derivatives of cAMP (B-1043) and cGMP (B-1044) often induce the same physiological responses as endogenous nucleotides, these analogs are not hydrolyzed by intracellular phosphodiesterases. The cell-permeant cAMP analog 8-bromo-cAMP has been used to study phosphorylation by cAMP-dependent protein kinase [11,12] (PKA), as well as the regulation of cAMP-modulated ion channels [13-17] and of progesterone-mediated gene transcription.[18] Likewise, the cell-permeant cGMP analog 8-bromo-cGMP has been used to investigate cGMP-modulated ion channels [19-22] and the role of cGMP-dependent protein kinase in nitric oxide–mediated signaling pathways.[23,24]

MANT-cGMP

Molecular Probes' blue fluorescent methylanthraniloyl (MANT) derivative of cGMP (MANT-cGMP, M-1035) is a useful substrate for a sensitive continuous fluorescence assay of cyclic nucleotide phosphodiesterase.[25,26] Furthermore, the inhibition of calmodulin-dependent activation of cyclic nucleotide phosphodiesterase by peptide segments of HIV envelope glycoproteins has been monitored using MANT cGMP.[27]

1. Ann Rev Biophys Chem 18, 239 (1989); **2.** Pflügers Arch 411, 628 (1988); **3.** Microbiol Rev 55, 733 (1991); **4.** Meth Cell Biol 31, 63 (1989); **5.** Ann Rev Physiol 52, 857 (1990); **6.** J Biol Chem 264, 5339 (1989); **7.** J Biol Chem 267, 3242 (1992); **8.** J Mol Cell Cardiol 23, 343 (1991); **9.** Nature 310, 74 (1984); **10.** J Biol Chem 268, 6316 (1993); **11.** J Biol Chem 269, 18599 (1994); **12.** J Neurochem 56, 1723 (1991); **13.** J Biol Chem 270, 17749 (1995); **14.** J Neurosci 15, 1025 (1995); **15.** J Neurosci 12, 290 (1992); **16.** Proc Natl Acad Sci USA 88, 6946 (1991); **17.** Neurosci Lett 105, 137 (1989); **18.** Science 250, 1740 (1990); **19.** Biophys J 69, 409 (1995); **20.** J Neurosci 15, 7196 (1995); **21.** Biochemistry 32, 10089 (1993); **22.** Biochemistry 27, 4396 (1988); **23.** J Biol Chem 270 22277 (1995); **24.** J Neurosci 15, 7653 (1995); **25.** Anal Biochem 162, 291 (1987); **26.** J Biol Chem 257, 13354 (1982); **27.** J Biol Chem 268, 22895 (1993).

20.4 Data Table *Nucleotides for Signal Transduction Research*

Cat #	Structure	MW	Storage	Soluble	Abs	{ε x 10^{-3}}	Em	Solvent	Notes
A-1048	S11.3	~700	FF,D,LL	H_2O	259	{18}	none	MeOH	<1, 2, 3>
A-1049	S11.3	~760	FF,D,LL	H_2O	351	{4.4}	none	H_2O	<1, 2>
A-7056	S11.3	~614	FF,D,LL	H_2O	259	{15}	none	MeOH	<1, 2, 3>
A-7073	S11.3	~753	FF,LL	H_2O*	254	{24}	none	pH 7	<1, 2, 3>
B-1043	✓	~408	F,D	pH>6	264	{16}	none	pH 8	
B-1044	✓	~424	F,D	pH>6	263	{17}	none	pH 8	
D-1037	✓	~524	F,D,LL	DMSO	338	{6.1}	none	MeOH	<1, 2>
D-7623	✓	~541	FF,D	DMSO	271	{17}	none	MeOH	
G-1053	✓	~800	FF,D,LL	H_2O	352	{3.7}	none	H_2O	<1, 2>
G-3620	✓	~703	FF,D,LL	H_2O	352	{3.1}	none	H_2O	<1, 2>
M-1035	✓	~500	F,D,L	H_2O	359	{5.2}	448	pH 7	
N-1045	✓	~478	FF,D,LL	DMSO	258	{19}	none	MeOH	<1, 2, 3>

For definitions of the contents of this data table, see "How to Use the Handbook" on page vi.
Structure: Chemical structure drawing: ✓ = shown in this section; Sn.n = shown in Section number n.n.
MW: ~ indicates that the molecular weight (MW) of this product is approximate because the exact degree of hydration and salt form have not been conclusively established.

* This product is packaged as a solution in the solvent indicated in "Soluble."
<1> Caged nucleotide esters are free of contaminating free nucleotides when initially prepared. However, some decomposition may occur during storage.
<2> All photoactivatable probes are sensitive to light. They should be protected from illumination, except when photolysis is intended.
<3> This compound has weaker visible absorption at >300 nm but no discernible absorption peaks in this region.

20.4 Structures *Nucleotides for Signal Transduction Research*

A-1048, A-1049
A-7056, A-7073
Section 11.3

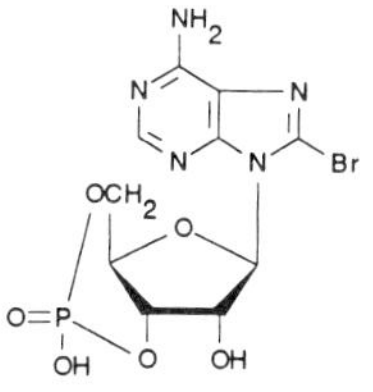

B-1043

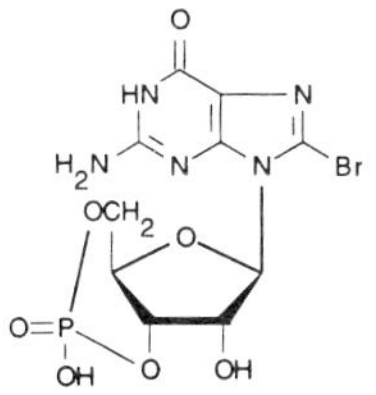

B-1044

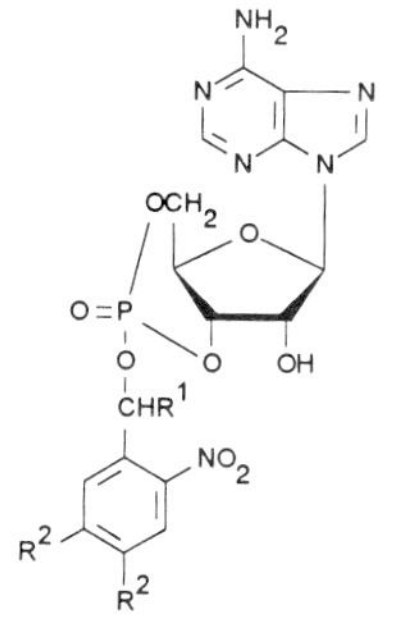

D-1037 R¹ = H R² = —OCH_3
N-1045 R¹ = —CH_3 R² = H

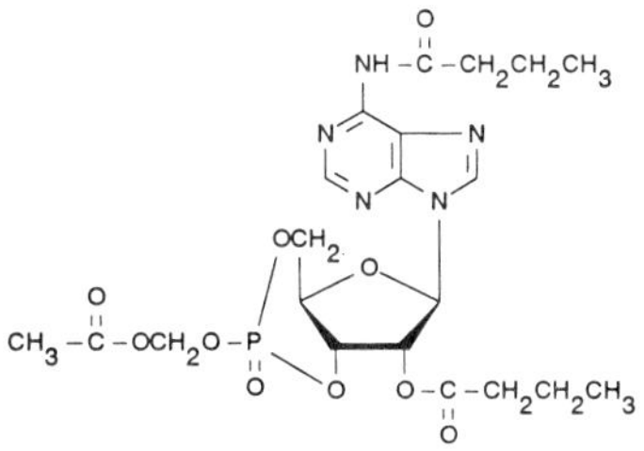

D-7623

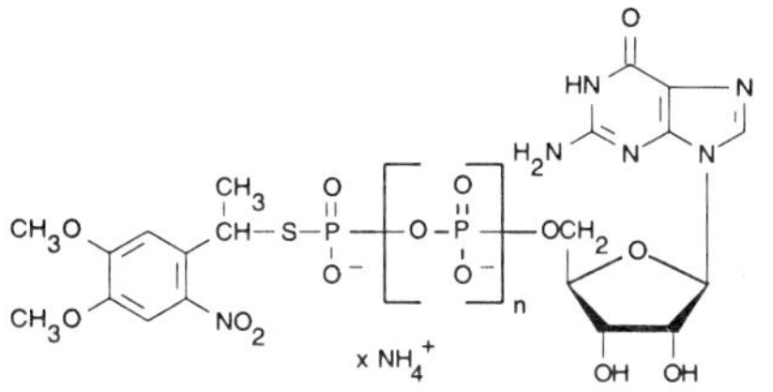

G-1053 n = 2 x = 3
G-3620 n = 1 x = 2

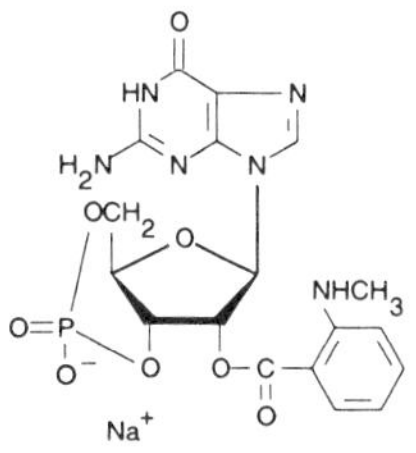

M-1035

N-1045
see D-1037

20.4 Price List *Nucleotides for Signal Transduction Research*

Cat #		Product Name	Unit Size	Price Per Unit ($) 1–4 Units	5–24 Units
A-7056	**New**	adenosine 5′-diphosphate, P²-(1-(2-nitrophenyl)ethyl) ester, monopotassium salt (NPE-caged ADP)	5 mg	103.00	82.40
A-1049		adenosine 5′-triphosphate, P³-(1-(4,5-dimethoxy-2-nitrophenyl)ethyl) ester, disodium salt (DMNPE-caged ATP)	5 mg	103.00	82.40
A-7073	**New**	adenosine 5′-triphosphate, P²-(1,2-diphenyl-2-oxo)ethyl ester, ammonium salt (desyl-caged ATP) *5 mM solution in water*	400 µL	145.00	116.00
A-1048		adenosine 5′-triphosphate, P³-(1-(2-nitrophenyl)ethyl) ester, disodium salt (NPE-caged ATP)	5 mg	103.00	82.40
B-1043		8-bromoadenosine 3′,5′-cyclicmonophosphate, free acid	100 mg	90.00	72.00
B-1044		8-bromoguanosine 3′,5′-cyclicmonophosphate, free acid	25 mg	42.00	33.60
D-7623	**New**	N⁶,O²′-dibutyryladenosine 3′,5′-cyclic monophosphate, acetoxymethyl ester *cell permeant* *special packaging*	10x100 µg	75.00	60.00
D-1037		4,5-dimethoxy-2-nitrobenzyl adenosine 3′,5′-cyclicmonophosphate (DMNB-caged cAMP)	5 mg	78.00	62.40
G-3620		guanosine 5′-O-(2-thiodiphosphate), P²⁽ˢ⁾-(1-(4,5-dimethoxy-2-nitrophenyl)ethyl) ester, ammonium salt (S-(DMNPE-caged) GDP-β-S)	1 mg	155.00	124.00
G-1053		guanosine 5′-O-(3-thiotriphosphate), P³⁽ˢ⁾-(1-(4,5-dimethoxy-2-nitrophenyl)ethyl) ester, triammonium salt (S-(DMNPE-caged) GTP-γ-S)	1 mg	155.00	124.00
M-1035		2′-(N-methylanthraniloyl)guanosine 3′,5′-cyclicmonophosphate, sodium salt (MANT-cGMP)	10 mg	95.00	76.00
N-1045		1-(2-nitrophenyl)ethyl adenosine 3′,5′-cyclicmonophosphate (NPE-caged cAMP)	5 mg	78.00	62.40

20.5 Detection of Phospholipase Activity

Phospholipases play an important role in cellular signaling processes via the generation of second messengers such as diacylglycerols, arachidonate and inositol triphosphate.[1-4] In addition, phospholipase A_2 activation is a key step in inflammation processes, making these enzymes important therapeutic targets.[5] Phospholipases are classified according to the cleavage site on the phospholipid substrate (Figure 20.4). There are two types of fluorescence-based phospholipase detection methods:[6]

- **Continuous methods**, which permit direct fluorometric monitoring of enzymatic activity using self-quenching or excimer-forming probes
- **Discontinuous methods**, which require resolution of fluorescent substrates and products by TLC, HPLC or other separation techniques.

Table 20.1 summarizes Molecular Probes' products for fluorescence-based phospholipase assays. Other applications for our wide range of fluorescent phospholipids are described in Chapter 13.

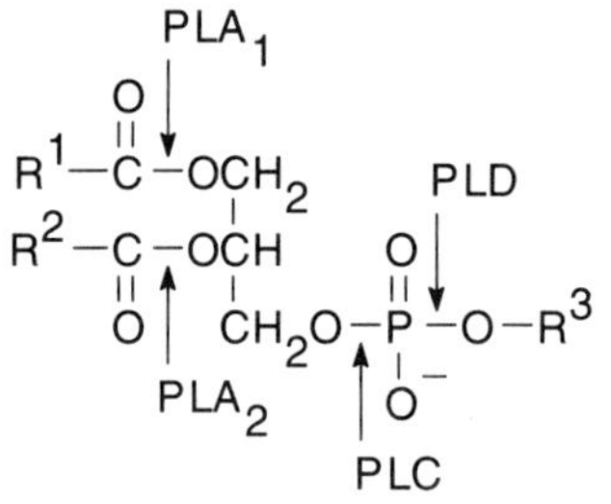

Figure 20.4 *Cleavage specificities of phospholipases.*

Phospholipases A_1 and A_2

Bis-BODIPY-C_{11}-PC

Our phospholipase A substrate — bis-BODIPY-glycerophosphocholine (bis-BODIPY FL C_{11}-PC, B-7701) — has been specifically designed to allow continuous monitoring of phospholipase A action

Table 20.1 *Fluorescence-based phospholipase assays.*

PL *	Probes	Assay Principle	Detection Method	References
A_1, A_2	B-7701	Intramolecular self-quenching	Fluorescence increase at ~515 nm	1
A_1, A_2	B-3781, B-3782	Intramolecular excimer formation	Emission ratio 380/470 nm	2–4
A_1, A_2	B-3760, D-3761	Liberation of free thiol groups	Fluorogenic or chromogenic thiol reagents	5
A_1, A_2	A-3880 †	Free fatty acid sensor	Emission ratio 432/505 nm	6–8
A_1, A_2	D-94 and I-3881	Displacement of protein-bound fluorescent fatty acid	Fluorescence decrease at ~500 nm	9,10
A_2	N-3786, N-3787	Intermolecular self-quenching	Fluorescence increase at ~530 nm	11,12
A_2	H-361, H-3743 H-3810, H-3818	Intermolecular excimer formation	Emission ratio 380/470 nm	13–15
A_2	O-7703	Release of MeOH/H_2O–soluble fatty acid	Phase separation of fluorescent product (P-3840, see Section 13.2)	16,17
A_2, C	H-7708	Intramolecular quenching by trinitrophenyl group	Fluorescence increase	18,19
A_2, C, D	D-3771, D-7707	Formation of fluorescent *O*-alkylglycerol derivative	TLC or HPLC	20,21
A_2, C, D	H-361, P-58	Quenching by disulfide-polymerized lipid matrix	Fluorescence increase at ~380 nm	22,23
C	M-2885	Hydrolysis of nonfluorescent phosphoester	Fluorescent product (H-189, see Section 10.1)	24
PI-C	B-7706, P-3764	Release of low MW fluorescent alcohol	TLC or HPLC	25,26

* Phospholipase specificity: A_1, A_2, C or D (see Figure 20.4 for definitions); PI-C = phosphatidyl inositol–specific phospholipase C (bacterial enzyme only). † See Section 20.6
1. J Biol Chem 267, 21465 (1992); **2.** Biochemistry 32, 583 (1993); **3.** Anal Biochem 116, 553 (1981); **4.** Biochim Biophys Acta 1192, 132 (1994); **5.** J Biol Chem 259, 5734 (1984); **6.** Anal Biochem 229, 256 (1995); **7.** Biochem J 298, 23 (1994); **8.** J Biol Chem 267, 23495 (1992); **9.** Anal Biochem 212, 65 (1993); **10.** Biochem J 266, 435 (1990); **11.** Lipids 24, 691 (1989); **12.** Biochem Biophys Res Comm 118, 894 (1984); **13.** Biochim Biophys Acta 917, 411 (1987); **14.** Chem Phys Lipids 53, 129 (1990); **15.** Anal Biochem 177, 103 (1989); **16.** Clin Chem 31, 714 (1985); **17.** Biochemistry 32, 5373 (1993); **18.** Chem Phys Lipids 59, 69 (1991); **19.** Anal Biochem 170, 248 (1988); **20.** Anal Biochem 218, 136 (1994); **21.** Biochem J 307, 799 (1995); **22.** Anal Biochem 221, 152 (1994); **23.** J Biol Chem 270, 263 (1995); **24.** Can J Biochem Cell Biol 63, 272 (1985); **25.** Biochemistry 31, 12169 (1992); **26.** Bioorg Med Chem Lett 1, 619 (1991).

and to be spectrally compatible with argon-ion laser excitation sources. When this probe is incorporated into cell membranes, the proximity of the BODIPY FL fluorophores on adjacent phospholipid acyl chains causes fluorescence self-quenching. Separation of the fluorophores upon hydrolytic cleavage of one of the acyl chains by phospholipase A_1 or A_2 results in increased fluorescence. Bis-BODIPY-C_{11}-PC has been developed in collaboration with Elizabeth Simons, who has successfully employed it for flow cytometric detection of phospholipase A activity in neutrophils. Using bis-BODIPY-C_{11}-PC and indo-1 simultaneously, researchers in the Simons lab at Boston University have demonstrated that a rise in intracellular Ca^{2+} precedes phospholipase A activation in immune complex–stimulated cells.[7] Plasma membrane loading of bis-BODIPY-C_{11}-PC was accomplished by incubation of cells with liposomes comprising a 1:10 molar ratio of bis-BODIPY-C_{11}-PC to unlabeled phosphatidylserine.

Bis-Pyrenyl Phospholipase A Probes

Our bis-pyrenyl phospholipase A probes (B-3781, B-3782) both emit at ~470 nm, indicating that their adjacent pyrene fluorophores form excited-state dimers (see Figure 13.6 in Chapter 13). Phospholipase-mediated hydrolysis separates the fluorophores, which then emit as monomers at ~380 nm.[8] These substrates have proven to be effective phospholipase A_2 substrates in model membrane systems (Table 20.1); however, it has been reported that 1,2-bis-(1-pyrenebutanoyl)-*sn*-glycero-3-phosphocholine (B-3781) is highly resistant to degradation by phospholipases in human skin fibroblasts.[9] 1,2-Bis-(1-pyrenebutanoyl)-*sn*-glycero-3-phosphocholine has been used in a sensitive, continuous assay for lecithin: cholesterol acyl transferase [10] (LCAT). These probes have several other reported applications, including investigations of protein kinase C (PKC) interactions with lipids,[11] DNA binding to liposomes [12,13] and lipid dynamics.[14,15]

Singly Labeled Fluorescent Phospholipids

Phospholipase A_2 activity has also been measured using phospholipids labeled with a single pyrene (H-361, H-3743, H-3810, H-3818) or NBD (N-3786, N-3787) fluorophore (Table 20.1). Because only the *sn*-2 phospholipid acyl chain is labeled, these probes can discriminate between phospholipase A_2 and phospholipase A_1 activity. To obtain a direct fluorescence response to enzymatic cleavage, sufficient phospholipid must be loaded to cause either intermolecular self-quenching (NBD-acyl phospholipids) or excimer formation (pyreneacyl phospholipids). This initial condition may be difficult to achieve when labeling live cells; however, labeling via biosynthetic incorporation of the corresponding fluorescent fatty acids (see Section 13.2) may be more efficient.[16] Acidic pyrene-labeled phospholipids — particularly the phosphomethanol derivatives [17-21] (H-3810, H-3743) — are preferred as substrates by pancreatic and intestinal phospholipase A_2, whereas phosphocholines (H-361, H-3818) are preferred by phospholipase A_2 from snake venom.[21] Pyrene-labeled phospholipids have been inserted into liposomes consisting of disulfide-linked polymerized phospholipids, which form a stable host matrix that is resistant to degradation by phospholipase A_2.[22,23] Pyrene fluorescence is quenched by interactions with the lipid matrix and increases upon cleavage of the probe by phospholipase. This versatile approach can be adapted for detection of phospholipase A_2 by using β-py-C_{10}-HPC (H-361) or of phospholipases C and D by using pyS DHPE [24] (P-58).

O-Alkyl and *S*-Acetyl Phospholipids

Specificity for phospholipase A_2 versus phospholipase A_1 can be obtained using phospholipids with nonhydrolyzable, ether-linked alkyl chains in the *sn*-1 position. 1-Octacosanyl-2-(1-pyrenehexanoyl)-*sn*-glycero-3-phosphomethanol (C_{28}-*O*-PHPM; O-7703) was developed for monitoring phospholipase A_2 in serum because the bis-pyrenyl probes (see above) are not specific for phospholipase A_2 and can also produce false indications of activity due to interactions with serum proteins. The cleavage product of C_{28}-*O*-PHPM hydrolysis by phospholipase A_2 — 1-pyrenehexanoic acid (P-3840, see Section 13.2) — cannot be directly distinguished from the substrate on the basis of fluorescence. However, this product can be readily resolved by extraction into aqueous methanol.[25] 1-*O*-Alkyl–substituted phospholipids containing the BODIPY FL (D-3771, D-7707), NBD (D-3768) and dansyl (D-3765) fluorophores are also useful substrates for phospholipase A_2–specific chromatographic assays.[26,27]

Phospholipase A_1–specific assays can be achieved using the thioacyl substrate 1-(*S*-decanoyl)-2-decanoyl-1-thio-*sn*-glycero-3-phosphocholine (D-3761). Hydrolysis of the thioester linkage by phospholipase A_1 liberates a free thiol group, which can be detected spectrophotometrically using reagents such as dithiopyridine or DTNB (D-8451, see Section 5.2) [28] or, potentially with much higher sensitivity, using our new Thiol and Sulfide Quantitation Kit (T-6060, see Section 5.2). 1,2-Bis-(*S*-decanoyl)-1,2-dithio-*sn*-glycero-3-phosphocholine (B-3760) operates on the same principle, but, like the bis-pyrenyl phospholipase A probes (B-3781, B-3782), is sensitive to hydrolysis by both phospholipase A_1 and phospholipase A_2.

ADIFAB: A Different View of Phospholipase A Activity

ADIFAB fatty acid indicator (A-3880, see Section 20.6) functions as a fluorescent sensor for phospholipase A cleavage products (free fatty acids). Therefore, it does not require membrane loading and can be used to monitor hydrolysis of natural (rather than synthetic) substrates. Assaying lysophospholipase activity with ADIFAB yields sensitivity comparable to radioisotopic methods.[29] Richieri and Kleinfeld have recently described methodology for using ADIFAB to measure the activity of phospholipase A_2 on cell and lipid-vesicle membranes. Their assay is capable of detecting hydrolysis rates as low as 10^{-12} mole/minute.[30] Displacement of dansylundecanoic acid (DAUDA, D-94) from fatty acid binding proteins (I-3881, see Section 20.6) is the basis of another phospholipase A_2 assay method (Table 20.1) that is conceptually similar to the detection mechanism of ADIFAB.[31-34]

Enzymes and Inhibitors

In addition to our probes for detection of phospholipase activity, we also offer phospholipase A_2 isolated from *Apis mellifera* bee venom [35] (P-98408) and several neurotoxins with phospholipase A_2 activity, including β-bungarotoxin [36,37] (B-3459), ammodytoxin [38] (A-98105) and notexin [39] (N-98103, N-98104). 4-Bromophenacyl bromide (B-6286, see Section 10.6) is a widely used phospholipase A_2 inhibitor [40-42] that reacts with a histidine residue in the enzyme's catalytic site.[43]

Phospholipases C and D

4-Methylumbelliferyl Phosphocholine

The nonfluorescent 4-methylumbelliferyl phosphocholine (M-2885) is cleaved by phospholipase C phosphodiesterases, yielding blue fluorescent 7-hydroxy-4-methylcoumarin (H-189, see Section 10.1). Hydrolysis of 4-methylumbelliferyl phosphocholine by sphingomyelinase — which usually hydrolyzes sphingomyelin to a ceramide plus phosphocholine — has also been demonstrated.[44] Sphingomyelinase activity is deficient in Niemann–Pick disease, a lysosomal lipid-storage disorder. The biological importance of ceramides is discussed in Section 20.6.

Trinitrophenylated Phospholipid

Phospholipids incorporating both a pyrene-labeled acyl chain and a trinitrophenyl (TNP)-labeled head group are internally quenched fluorogenic substrates for the *continuous* assay of phospholipase A_2 [45] or phospholipase C activity.[46] Cleavage by either enzyme results in separation of the fluorophore and quencher and an increase in the fluorescence of the pyrene label. We now offer TNP β-py-C_6-HPE (H-7708), a TNP- and pyrene-labeled phospholipid analog originally prepared by Kinnunen and co-workers.[45]

D 609: A Selective Phospholipase C Inhibitor

D 609 (tricyclodecan-9-yl xanthogenate, T-6615) is a selective inhibitor of phosphatidylcholine-specific phospholipase C.[47] D 609 also is reported to have antiviral [48-50] and antitumor [51-53] activity.

Pyrene and BODIPY PI-PLC Substrates

Phosphatidylinositol-specific phospholipase C (PI-PLC, EC 3.1.4.10) from *Bacillus cereus* cleaves phosphatidylinositol (PI), yielding water-soluble D-*myo*-inositol 1,2-cyclic monophosphate and lipid-soluble diacylglycerol (DAG).[54] This enzyme also functions to release enzymes that are linked to glycosyl phosphatidylinositol (GPI) membrane anchors.[55] Enzymatic hydrolysis of our fluorescent PI-PLC substrate 1-pyrenebutyl *myo*-inositol-1-phosphate (P-3764) yields 1-pyrenebutanol (P-244, see Section 14.6), which can be quantitated by TLC or HPLC.[56] Although the maximum activity of *Bacillus cereus* PI-PLC is about 20-fold lower with this fluorescent analog than with its natural substrates, it is still substantially better than with other existing substrate analogs. For example, the use of 1-pyrenebutyl *myo*-inositol-1-phosphate allows the detection of about 0.2 ng of *Bacillus cereus* PI-PLC.[57] However, this analog is not an effective substrate for detecting PI-PLC activity of eukaryotic cells that generate IP_3 and DAG in the PKC signal transduction pathway.[58] Our BODIPY FL C_5 *myo*-inositol 1-phosphate (B-7706) has similar characteristics to 1-pyrenebutyl *myo*-inositol-1-phosphate as a PI-PLC substrate; its longer-wavelength spectral characteristics and high absorbance and fluorescence should facilitate analysis using fluorescence scanners that employ the argon-ion laser.

Bacillus cereus PI-PLC and anti–PI-PLC

Molecular Probes now offers highly purified *Bacillus cereus* PI-PLC (P-6466), which has been used in studies of PI synthesis and export across the plasma membrane.[59] PI-PLC generates diacylglycerols for PKC-linked signal transduction studies [60] and provides an efficient means of releasing most GPI-anchored proteins from cell surfaces under conditions in which the cells remain viable.[61,62]

To complement this enzyme, we also offer an inhibitory mouse monoclonal antibody (A-6400) that is suitable for immunological detection of *Bacillus cereus* PI-PLC in Western blots or enzyme-linked immunosorbent assays (ELISAs) and also for structure–function studies of the active enzyme. This mouse monoclonal A72–24 antibody crossreacts with PI-PLC from the closely related *Bacillus thuringiensis* by ELISA and Western blotting.[63] Activity of PI-PLC for both PI hydrolysis and GPI-linked acetylcholinesterase release is inhibited by the antibody.

Phospholipase D

The products of phospholipase A_2, C and D cleavage of fluorescent 1-*O*-alkyl-2-decanoyl-*sn*-glycero-3-phosphocholines (D-3771, D-3768, D-3765) can be separated and independently quantitated based on their different extents of migration on TLC or HPLC.[26] The BODIPY analog (D-3771) is preferred for this application because it is relatively photostable and the fluorescence properties of its different enzymatic products are all very similar.[64] Researchers have taken advantage of these features to assay for phospholipase D activation in vascular smooth muscle cells [64,65] and in yeast.[66] In these experiments, this BODIPY analog was dispersed with octyl glucoside (O-8457, see Section 16.3) for incubation with cell and tissue extracts. Andrew Gewirtz and Elizabeth Simons at Boston University have found that biosynthetic incorporation of the corresponding lysophospholipid (BODIPY FL C_{11}-lyso PAF, D-3772; see Section 20.6) — which is readily dispersed in culture medium at concentrations up to 1 mM — is an efficient method for *in situ* generation of the acylated BODIPY phospholipid in neutrophils. The new 2-hexadecanoyl BODIPY phospholipid (D-7707) is designed as a chromatographic standard for the biosynthetically incorporated product. Using these methods, researchers have identified phospholipase D activation as a key process controlling the degranulation of neutrophils in response to phagocytic stimuli.[67] TNP β-py-C_6-HPE (H-7708), discussed above, can potentially allow for the continuous assay of phospholipase activity.

1. Cell 80, 269 (1995); **2.** Biochim Biophys Acta 1212, 26 (1994); **3.** Cell 77, 329 (1994); **4.** FASEB J 5, 2068 (1991); **5.** FASEB J 8, 916 (1994); **6.** Anal Biochem 219, 1 (1994); **7.** J Biol Chem 267, 21465 (1992); **8.** Anal Biochem 116, 553 (1981); **9.** Biochemistry 34, 2049 (1995); **10.** J Lipid Res 33, 1863 (1992); **11.** J Biol Chem 270, 1254 (1995); **12.** Chem Phys Lipids 72, 77 (1994); **13.** Chem Phys Lipids 66, 75 (1993); **14.** Biophys J 66, 1981 (1994); **15.** Biophys J 64, 137 (1993); **16.** Biochemistry 32, 11330 (1993); **17.** J Biol Chem 270, 17327 (1995); **18.** Biochemistry 30, 8074 (1991); **19.** Biochem Biophys Res Comm 168, 644 (1990); **20.** Anal Biochem 177, 103 (1989); **21.** Biochim Biophys Acta 917, 411 (1987); **22.** Biochemistry 32, 13902 (1993); **23.** J Am Chem Soc 108, 7789 (1986); **24.** Anal Biochem 221, 152 (1994); **25.** Clin Chem 31, 714 (1985); **26.** Anal Biochem 185, 80 (1990); **27.** Anal Biochem 174, 477 (1988); **28.** Chem Phys Lipids 53, 115 (1990); **29.** Biochem J 298, 23 (1994); **30.** Anal Biochem 229, 256 (1995); **31.** Anal Biochem 212, 65 (1993); **32.** Biochem J 278, 843 (1991); **33.** Biochem J 276, 129 (1991); **34.** Biochem J 266, 435 (1990); **35.** Eur J Biochem 213, 1193 (1993); **36.** Biochem Pharmacol 44, 1073 (1992); **37.** Toxicon 28, 1423 (1990); **38.** Biochim Biophys Acta 828, 306 (1985); **39.** Eur J Biochem 185, 263 (1989); **40.** Eur J Biochem 219, 401 (1994); **41.** Brain Res 628, 340 (1993); **42.** Biochemistry 29, 6082 (1990); **43.** Biochemistry 19, 745 (1980); **44.** Can J Biochem 63, 272 (1985); **45.** Anal Biochem 170, 248 (1988); **46.** Chem Phys Lipids 59, 69 (1991); **47.** Cell 71, 765 (1992); **48.** Virology 181, 101 (1991); **49.** AIDS Res Human Retroviruses 4, 71 (1988); **50.** Proc Natl Acad Sci USA 81, 3263 (1984); **51.** Cancer Lett 46, 143 (1989); **52.** Int'l J Cancer 43, 508 (1989); **53.** Cancer Lett 35, 237 (1987); **54.** Meth Enzymol 197, 493 (1991); **55.** Science 239, 268 (1988); **56.** Bioorg Med Chem Lett 1, 619 (1991); **57.** Biochemistry 31, 12169 (1992); **58.** Science 244, 546 (1989); **59.** Biochim Biophys Acta 1224, 247 (1994); **60.** J Biol Chem 269,

4098 (1994); **61.** J Biol Chem 266, 1926 (1991); **62.** J Neurochem 57, 67 (1991); **63.** Biochim Biophys Acta 1047, 47 (1990); **64.** Anal Biochem 218, 136 (1994); **65.** J Biol Chem 269, 23790 (1994); **66.** Biochem J 307, 799 (1995); **67.** Mol Biol of the Cell 6, 225a, abstract #1305 (1995).

20.5 Data Table *Detection of Phospholipase Activity*

Cat #	Structure	MW	Storage	Soluble	Abs	{$\epsilon \times 10^{-3}$}	Em	Solvent	Notes
B-3760	✓	598	FF,D	<1>	<300		none		
B-3781	✓	798	FF,D,L	<1>	342	{75}	471	EtOH	<2>
B-3782	✓	966	FF,D,L	<1>	340	{62}	473	EtOH	<2>
B-7701	✓	1030	FF,D,L	<1>	505	{123}	512	MeOH	<3>
B-7706	✓	554	FF,D,L	MeOH	505	{80}	511	MeOH	
D-94	S13.3	435	F,L	DMSO, EtOH	335	{4.8}	519	MeOH	
D-3761	✓	582	FF,D	<1>	<300		none		
D-3765	✓	814	FF,D,L	<1>	335	{3.5}	520	EtOH	
D-3768	✓	758	FF,D,L	<1>	465	{18}	540	EtOH	<4>
D-3771	✓	855	FF,D,L	<1>	506	{71}	512	EtOH	
D-7707	✓	939	FF,D,L	<1>	506	{73}	512	EtOH	
H-361	S13.2	850	FF,D,L	<1>	342	{37}	376	MeOH	<5, 6, 7>
H-3743	S13.2	745	FF,D,L	<1>	341	{37}	377	MeOH	<5, 6, 7>
H-3810	S13.2	801	FF,D,L	<1>	341	{40}	376	MeOH	<5, 6, 7>
H-3818	S13.2	794	FF,D,L	<1>	341	{37}	377	MeOH	<5, 6, 7>
H-7708	✓	985	FF,D,L	<1>	342	{40}	376	MeOH	<6, 7>
M-2885	✓	341	F,D	DMSO, MeOH	313	{12}	370	pH 9	<8>
N-3786	S13.2	772	FF,D,L	<1>	465	{21}	533	EtOH	<4, 9>
N-3787	S13.2	856	FF,D,L	<1>	465	{22}	534	EtOH	<4, 9>
O-7703	✓	894	FF,D,L	$CHCl_3$, EtOH	341	{37}	376	MeOH	<6, 7>
P-58	S13.4	1057	FF,D,L	<1>	350	{35}	379	MeOH	
P-3764	✓	522	FF,D,L	MeOH	341	{44}	376	MeOH	<10>
T-6615	✓	266	F,D	DMSO, DMF	<300		none		

For definitions of the contents of this data table, see "How to Use the Handbook" on page vi.
Structure: Chemical structure drawing: ✓ = shown in this section; Sn.n = shown in Section number n.n.

<1> Chloroform is the most generally useful solvent for preparing stock solutions of phospholipids. Glycerophosphocholines are usually freely soluble in ethanol. Most other glycerophospholipids are less soluble in ethanol, but solutions up to 1-2 mg/mL should be obtainable, using sonication to aid dispersion if necessary. Labeling of aqueous samples with nonmiscible phospholipid stock solutions can be accomplished by evaporating the organic solvent, followed by hydration and sonication, yielding a suspension of liposomes. Information on solubility of natural phospholipids can be found in Marsh, D., CRC Handbook of Lipid Bilayers, CRC Press (1990) pp. 71-80.
<2> The intact compound exhibits pyrene excimer fluorescence. Lipase hydrolysis yields pyrene monomers with spectra similar to P-31 or P-32 (see Section 13.2).
<3> Phospholipase A cleavage results in increased fluorescence with essentially no wavelength shift. The cleavage products are D-3862 (see Section 13.2) and a fluorescent lysophospholipid.
<4> NBD derivatives are almost nonfluorescent in water. The quantum yields of NBD phospholipid derivatives increase upon binding to membranes.
<5> Alkylpyrene fluorescence lifetimes are up to 110 nsec and are very sensitive to oxygen.
<6> Pyrene derivatives exhibit structured spectra. The absorption maximum is usually about 340 nm with a subsidiary peak at about 325 nm. There are also strong absorption peaks below 300 nm. The emission maximum is usually about 376 nm with a subsidiary peak at 396 nm. Excimer emission at about 470 nm may be observed at high concentrations.
<7> Phospholipase A_2 hydrolysis releases a fluorescent fatty acid; P-31 (see Section 13.2) from H-361 and H-3810 or P-3840 (see Section 13.2) from H-3743, H-3818, O-7703 and H-7708.
<8> Enzymatic hydrolysis of this substrate yields 7-hydroxy-4-methylcoumarin (H-189, see Section 10.1), which has Abs = 360 nm (ϵ = {19}), Em = 449 nm in pH 9 buffer.
<9> Phospholipase A_2 hydrolysis releases a fluorescent fatty acid; N-316 (see Section 13.2) from N-3786 or N-678 (see Section 13.2) from N-3787.
<10> Cleavage by PI-specific phospholipase C (P-6466) yields P-244 (see Section 14.6).

Product-Specific Bibliographies and Keyword Searches

Our Technical Assistance Department can provide you with product-specific bibliographies, as well as keyword searches of the over 25,000 literature references in our extensive bibliography database. Our bibliography database is also searchable through our Web site (http://www.probes.com).

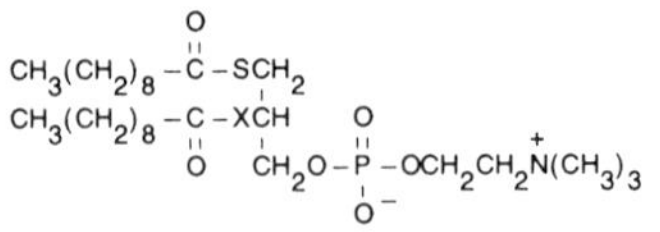

B-3760 X = S
D-3761 X = O

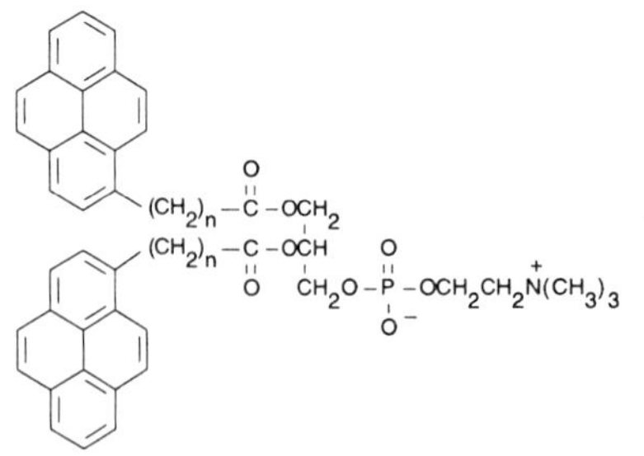

B-3781 n = 3
B-3782 n = 9

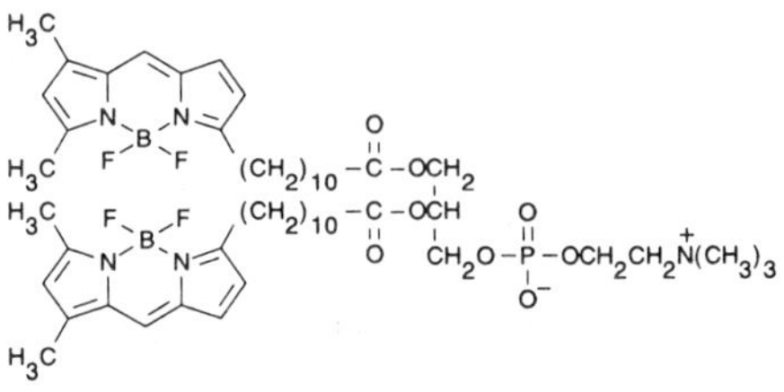

B-7701

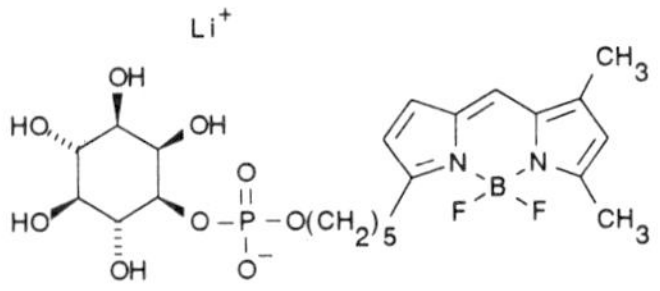

B-7706

D-94
Section 13.3

D-3761
see B-3760

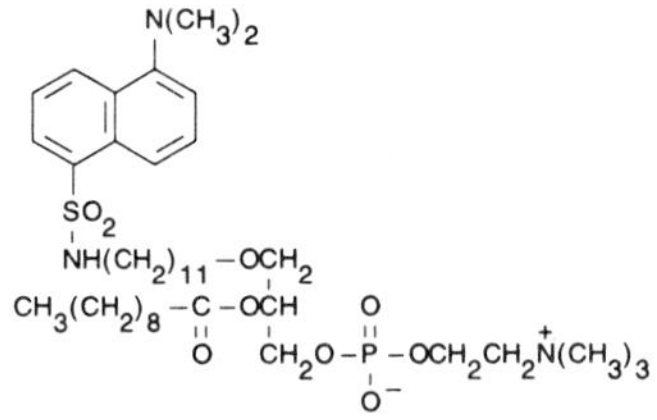

D-3765

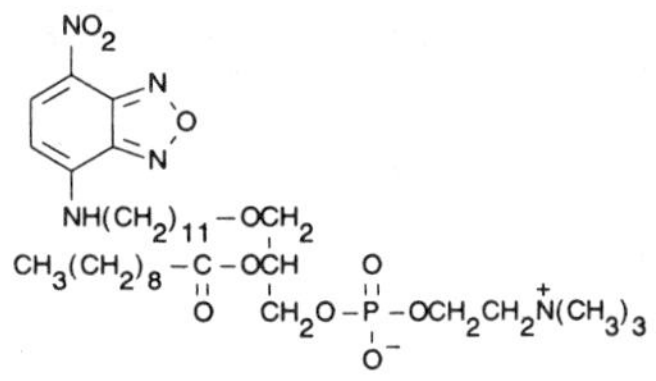

D-3768

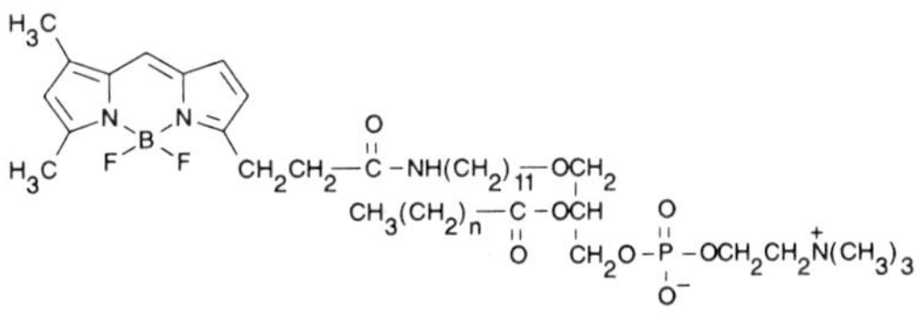

D-3771 n = 8
D-7707 n = 14

H-361, H-3743
H-3810, H-3818
Section 13.2

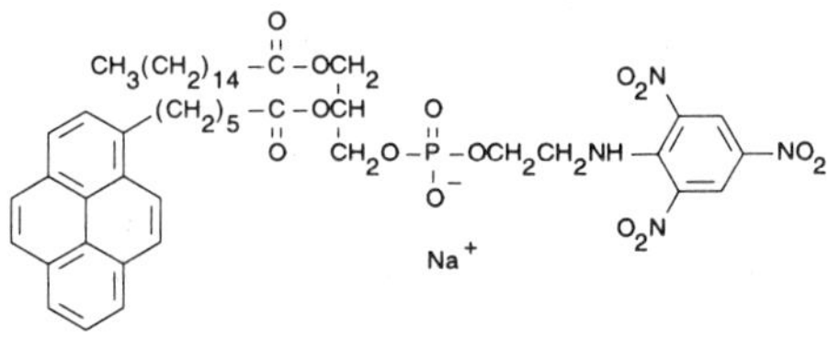

H-7708

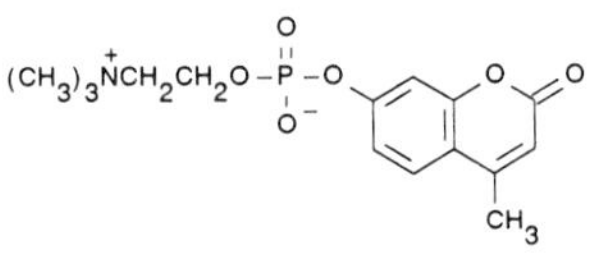

M-2885

N-3786, N-3787
Section 13.2

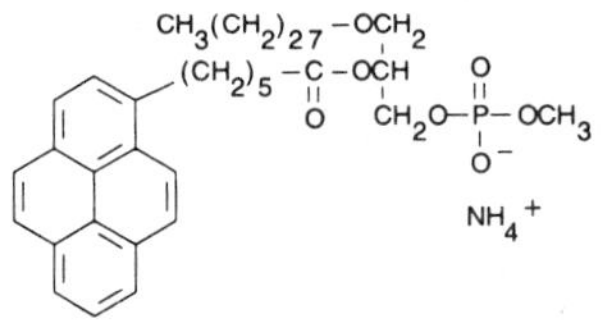

O-7703

P-58
Section 13.4

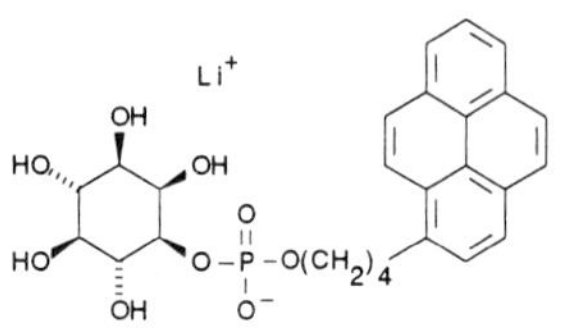

P-3764

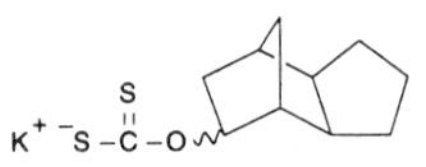

T-6615

Cat #	Product Name	Unit Size	Price Per Unit ($) 1–4 Units	5–24 Units
A-98105 **New**	ammodytoxin *phospholipase A_2 from *Vipera ammodytes* snake venom*	250 µg	95.00	76.00
A-6400 **New**	anti-*Bacillus cereus* phosphatidylinositol-specific phospholipase C, mouse monoclonal A72-24	100 µg	98.00	78.40
B-3760	1,2-bis-(S-decanoyl)-1,2-dithio-*sn*-glycero-3-phosphocholine	1 mg	60.00	48.00
B-7701 **New**	1,2-bis-(4,4-difluoro-5,7-dimethyl-4-bora-3a,4a-diaza-s-indacene-3-undecanoyl)-*sn*-glycero-3-phosphocholine (bis-BODIPY® FL C_{11}-PC)	100 µg	95.00	76.00
B-3781	1,2-bis-(1-pyrenebutanoyl)-*sn*-glycero-3-phosphocholine	1 mg	85.00	68.00
B-3782	1,2-bis-(1-pyrenedecanoyl)-*sn*-glycero-3-phosphocholine	1 mg	85.00	68.00
B-7706 **New**	BODIPY® FL C_5 *myo*-inositol 1-phosphate, lithium salt	100 µg	95.00	76.00
B-3459	β-bungarotoxin *from *Bungarus multicinctus**	1 mg	37.00	29.60
D-3761	1-(S-decanoyl)-2-decanoyl-1-thio-*sn*-glycero-3-phosphocholine	1 mg	45.00	36.00
D-3771	2-decanoyl-1-(O-(11-(4,4-difluoro-5,7-dimethyl-4-bora-3a,4a-diaza-s-indacene-3-propionyl)amino)-undecyl)-*sn*-glycero-3-phosphocholine	1 mg	98.00	78.40
D-3768	2-decanoyl-1-(O-(N-(7-nitrobenz-2-oxa-1,3-diazol-4-yl)amino)undecyl)-*sn*-glycero-3-phosphocholine	1 mg	68.00	54.40
D-7707 **New**	1-(O-(11-(4,4-difluoro-5,7-dimethyl-4-bora-3a,4a-diaza-s-indacene-3-propionyl)amino)undecyl)-2-hexadecanoyl-*sn*-glycero-3-phosphocholine	1 mg	98.00	78.40
D-94	11-((5-dimethylaminonaphthalene-1-sulfonyl)amino)undecanoic acid (DAUDA)	100 mg	68.00	54.40
D-3765	1-(O-(11-(5-dimethylaminonaphthalene-1-sulfonyl)amino)undecyl)-2-decanoyl-*sn*-glycero-3-phosphocholine	1 mg	68.00	54.40
H-361	1-hexadecanoyl-2-(1-pyrenedecanoyl)-*sn*-glycero-3-phosphocholine (β-py-C_{10}-HPC)	1 mg	65.00	52.00
H-3810	1-hexadecanoyl-2-(1-pyrenedecanoyl)-*sn*-glycero-3-phosphomethanol, sodium salt (β-py-C_{10}-HPM)	1 mg	65.00	52.00
H-3818	1-hexadecanoyl-2-(1-pyrenehexanoyl)-*sn*-glycero-3-phosphocholine (β-py-C_6-HPC)	1 mg	65.00	52.00
H-3743	1-hexadecanoyl-2-(1-pyrenehexanoyl)-*sn*-glycero-3-phosphomethanol, sodium salt (β-py-C_6-HPM)	1 mg	65.00	52.00
H-7708 **New**	1-hexadecanoyl-2-(1-pyrenehexanoyl)-N-(2,4,6-trinitrophenyl)-*sn*-glycero-3-phosphoethanolamine, sodium salt (TNP β-py-C_6-HPE)	1 mg	85.00	68.00
M-2885	4-methylumbelliferyl phosphocholine	10 mg	85.00	68.00
N-3787	2-(12-(7-nitrobenz-2-oxa-1,3-diazol-4-yl)amino)dodecanoyl-1-hexadecanoyl-*sn*-glycero-3-phosphocholine (NBD-C_{12}-HPC)	5 mg	50.00	40.00
N-3786	2-(6-(7-nitrobenz-2-oxa-1,3-diazol-4-yl)amino)hexanoyl-1-hexadecanoyl-*sn*-glycero-3-phosphocholine (NBD-C_6-HPC)	5 mg	50.00	40.00
N-98103 **New**	notexin fraction *phospholipase A_2 from *Notechis scutatus* snake venom*	1 mg	245.00	196.00
N-98104 **New**	notexin Np *phospholipase A_2 from *Notechis scutatus* snake venom*	100 µg	215.00	172.00
O-7703 **New**	1-octacosanyl-2-(1-pyrenehexanoyl)-*sn*-glycero-3-phosphomethanol, ammonium salt (C_{28}-O-PHPM)	250 µg	200.00	160.00
P-98408 **New**	phospholipase A_2 *from *Apis mellifera* bee venom*	1 mg	38.00	30.40
P-6466 **New**	phospholipase C, phosphatidylinositol-specific *from *Bacillus cereus** *100 U/mL*	50 µl	40.00	32.00
P-3764	1-pyrenebutyl *myo*-inositol 1-phosphate, lithium salt	500 µg	160.00	128.00
P-58	N-(1-pyrenesulfonyl)-1,2-dihexadecanoyl-*sn*-glycero-3-phosphoethanolamine, triethylammonium salt (pyS DHPE)	25 mg	75.00	60.00
T-6615 **New**	tricyclodecan-9-yl xanthogenate, potassium salt (D 609)	5 mg	48.00	38.40

20.6 Probes for Lipid Metabolism and Signaling

Lipid metabolites produced by phospholipase action and subsequent intracellular modifications are extensively involved in cellular signaling processes [1-3] and produce a wide range of physiological and pathological effects including apoptosis, inflammatory responses, cell differentiation and proliferation. To support research in these areas, Molecular Probes offers:

- Fluorescent BODIPY and NBD ceramides, sphingomyelins and glycosylceramide synthase inhibitors
- Reagents for elevating intracellular ceramide levels
- Fluorescent analogs of platelet-activating factor (PAF)
- ADIFAB, a fluorescent sensor for detection of fatty acid metabolites (see also Section 20.5)

Sphingolipids

BODIPY Sphingolipids

Ceramides (*N*-acylsphingosines), like diacylglycerols, are lipid second messengers that function in signal transduction processes.[4] The concentration-dependent spectral properties of Molecular Probes' BODIPY FL C_5-ceramide (D-3521) and sphingomyelin (D-3522) make them particularly suitable for investigating sphingolipid transport and metabolism, in addition to their applications as structural markers for the Golgi complex (see Section 12.4).[5] BODIPY FL C_5-ceramide can be visualized by fluorescence microscopy [6,7] or by electron microscopy following diaminobenzidine (DAB) photoconversion to an electron-dense product.[8]

Our new dibrominated BODIPY FL C_5-ceramide (BODIPY FL Br_2C_5-ceramide, D-7546) is designed to increase photoconversion efficiency and thus provide more intense staining for electron microscopy (see the Technical Note "Fluorescent Probes for Photoconversion of Diaminobenzidine" on page 414). Our range of BODIPY sphingolipids also includes the long wavelength–excitable BODIPY TR ceramide (D-7540) and three BODIPY FL cerebrosides (glycosylated ceramides; D-7519, D-7547, D-7548). All of Molecular Probes' ceramides are prepared from D-*erythro*-sphingosine and therefore have the same stereochemical conformation as natural biologically active sphingolipids.

BODIPY FL C_5-ceramide has been used to investigate the linkage of sphingolipid metabolism to protein secretory pathways [9-11] and neuronal growth.[12] Internalization of BODIPY FL C_5-sphingomyelin (D-3522) from the plasma membrane of human skin fibroblasts results in a mixed population of labeled endosomes that can be distinguished based on the concentration-dependent green (~515 nm) or red (~620 nm) emission of the probe.[13] Our new BODIPY FL cerebrosides should be useful tools for the study of glycosphingolipid transport and signaling pathways in cells [14,15] and for diagnosis of storage disorders — for example, Krabbe disease and Gaucher disease — that result in the intracellular accumulation of cerebrosides.[16-21] Studies by Martin and Pagano [22] have shown that the internalization routes for BODIPY FL C_5-glucocerebroside (D-7548) follow both endocytic and nonendocytic pathways and are quite different from those for BODIPY FL C_5-sphingomyelin.

NBD Sphingolipids

NBD C_6-ceramide (N-1154) and NBD C_6-sphingomyelin (N-3524) analogs predate their BODIPY counterparts and have been extensively used for following sphingolipid metabolism in cells [14,23,24] and in multicellular organisms.[25] Koval and Pagano have prepared NBD analogs of both the naturally occurring D-*erythro* and the nonnatural L-*threo* stereoisomers of sphingomyelin and have compared their intracellular transport behavior in Chinese hamster ovary (CHO) fibroblasts.[26] NBD C_6-ceramide lacks the useful concentration-dependent optical properties of the BODIPY FL analog and is less photostable. The photobleaching kinetics of NBD C_6-ceramide were shown to be a useful property for estimating cholesterol levels in the Golgi apparatus.[27]

Cell-Permeant Acetyl Ceramide and Hexanoyl Ceramide

The cell-permeant probes sphingosine derivatives, acetyl ceramide (A-7509) and hexanoyl ceramide (H-7524), have enabled researchers to assess the function of ceramides in processes such as transcription-factor activation,[28] cell-growth inhibition [29] and apoptosis.[30] These ceramides have several important effects on cellular processes, including:

- Activation of stress-activated protein kinases [31] and protein phosphatase 2A [32]
- Induction of DNA fragmentation [33] and effects on the signal transduction pathway that leads to apoptosis [30,34]
- Inhibition of protein kinase C–α translocation [35]
- Inhibition of superoxide formation in neutrophils [36,37]
- Reversal of brefeldin A (BFA) resistance in BFA-resistant cell lines [38]
- Modulation of endocytosis [39]
- Stimulation of cell proliferation in Swiss 3T3 fibroblasts [40]

Generation of ceramide by hydrolysis of sphingomyelin appears to play a role in mediating the effects of exposure to tumor necrosis factor–α [41] (TNF-α), γ-interferon and several other agents, all of which induce an apoptosis-like cell death.[33] Molecular Probes' wide array of reagents for following the diverse morphological and biochemical changes that occur during apoptosis are described in Section 16.3.

Glucosylceramide Synthase Inhibitors

The glucosylceramide synthase inhibitors, PDMP (D-7522) and the more potent PPMP (H-7523), elevate intracellular ceramide levels by blocking the metabolism of ceramide to glucosylceramide in the Golgi complex.[42-46] Using both PDMP and hexanoyl ceramide (H-7524), researchers have investigated the effects of increasing ceramide levels on sphingolipid-induced mitogenesis.[47,48] PDMP-induced increases in ceramide levels arrest cell-cycle progression and inhibit growth factor–stimulated cell proliferation.[42] Our new BODIPY FL and BODIPY TMR derivatives of these glucosylceramide synthase inhibitors (D-7520, D-7521) are potentially useful probes for studying the subcellular localization of glucosylceramide synthase [14] and its abundance in tissues.[49] A BODIPY analog of PDMP synthesized by Rosenwald and Pagano localized in lysosomes of cultured CHO cells and induced changes in their morphology.[50]

Fluorescent Analogs of Platelet-Activating Factor

Platelet-activating factors (PAF) are ether phospholipids — specifically 1-*O*-alkyl-2-acetyl-*sn*-glycero-3-phosphocholine — that activate a variety of intracellular signaling pathways via G-protein–linked membrane receptors.[51] Their biological effects are numerous, including platelet aggregation, blood-pressure regulation, leukocyte stimulation, uterine implantation of fertilized eggs, anaphylaxis and inflammation.[51,52] Molecular Probes' fluorescent PAF analogs, which were developed in collaboration with Stewart and Betty Hendrickson, incorporate the dansyl, NBD and BODIPY FL fluorophores attached to the terminus of the *sn*-1 chain (A-3767, A-3770, A-3773). Dansyl and BODIPY FL analogs of deacetylated (lyso) PAF are also available (D-3766, D-3772). Enzymatic acetylation of dansyl-C_{11}-lyso PAF (D-3766) to the PAF analog (A-3767) has been followed chromatographically.[53] The NBD and BODIPY PAF analogs are designed to allow imaging of PAF activation and metabolic trafficking. The optical properties of these probes are similar to those of the corresponding diacylphospholipids described in Section 13.2.

Free Fatty Acids and Their Analogs

ADIFAB: A Unique Free Fatty Acid Indicator

The elevated levels of free fatty acids (FFA) — which are associated with multiple pathological states, including cancer and diabetes — are generated by inflammatory responses, phospholipase A activity and cytotoxic phenomena.[54] Sensitive techniques are required to detect and quantitate free fatty acids because these important metabolites have low aqueous solubility and usually exist in complexed form. ADIFAB (A-3880) is a dual-wavelength fluorescent FFA indicator that consists of a polarity-sensitive fluorescent probe (acrylodan, A-433; see Section 2.3) conjugated to intes-

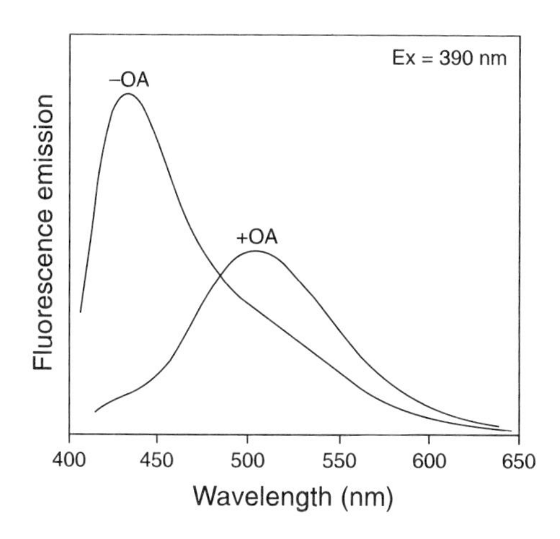

Figure 20.5 *The free fatty acid–dependent spectral shift of ADIFAB (A-3880). Spectra shown represent 0.2 µM ADIFAB in pH 8.0 buffer with (+OA) and without (-OA) addition of 4.7 µM cis-9-octadecenoic (oleic) acid (OA). The ratio of fluorescence emission intensities at 505 nm and 432 nm can be quantitatively related to free fatty acid concentrations.*

tinal fatty acid–binding protein (I-FABP fatty acid binding protein, I-3881), a low molecular weight (15,000 daltons) protein with high binding affinity for FFA.[55-57]

As shown in Figure 20.5, titration of ADIFAB with oleic acid results in a shift of the fluorescence maximum from ~432 nm to 505 nm. The ratio (R) of these signals (505 nm/432 nm) can be converted to a FFA concentration using the FFA dissociation constant (K_d), by employing analysis procedures similar to those developed for Ca^{2+} indicators [58] (see Chapter 22). Values of K_d vary considerably for different fatty acids; a typical value is 0.28 µM for oleic acid [57] (determined at 37°C). There is little, if any, interference from bile acids, glycerides, sterols or bilirubin. With appropriate precautions, which are described in the product information sheet included with the material, ADIFAB can be used to determine FFA concentrations in the range 1 nM to >20 µM.

ADIFAB was used to investigate the physical basis of *cis*-unsaturated fatty acid inhibition of cytotoxic T cells.[59] Recently, this effect has been found to be due to inhibition of a specific tyrosine phosphorylation event that normally accompanies antigen stimulation.[60,61] Measurements using ADIFAB have also revealed previously undetected differences in FFA binding affinities among fatty acid–binding proteins from different tissues [62] and have enabled quantitation of FFA levels in human serum as a potential diagnostic tool.[63]

Caged Arachidonic Acid and Anandamide

Molecular Probes' chemists have synthesized a caged arachidonic acid derivative (A-7062) by esterification of a DMNPE group to the carboxylate of this important fatty acid second messenger species.[64,65] We also prepare arachidonic acid caged with a carboxynitrobenzyl (CNB) group (A-7068). The carboxyl moiety of the latter probe improves its water solubility and should increase its rate of uncaging (see Section 19.1).

In addition to caged arachidonic acid, Molecular Probes offers anandamide (H-7492), which is metabolically derived from arachidonic acid in brain tissue [66] and is an endogenous ligand for cannabinoid receptors, eliciting inhibition of N-type Ca^{2+} channel opening and adenylate cyclase.[67] Anandamide activates inwardly rectifying potassium conductance.[68]

Oleamide

Oleamide (O-7541), the amide of oleic acid, has recently been reported to be a natural lipid present in cerebrospinal fluid.[69] Oleamide and other fatty amides apparently act by a previously unrecognized signaling pathway to induce sleep when injected into rats. Oleamide from Molecular Probes is prepared from high-purity oleic acid.

1. Cell 80, 269 (1995); **2.** Cell 77, 329 (1994); **3.** *Lipid Metabolism in Signaling Systems (Meth Neurosci 18)*, J.N. Fain, Ed., Academic Press (1993); **4.** J Biol Chem 269, 3125 (1994); **5.** J Cell Biol 113, 1267 (1991); **6.** Cytometry 14, 251 (1993); **7.** J Cell Biol 120, 399 (1993); **8.** Eur J Cell Biol 58, 214 (1992); **9.** Mol Biol of the Cell 6, 135 (1995); **10.** J Biol Chem 268, 4577 (1993); **11.** Biochemistry 31, 3581 (1992); **12.** J Biol Chem 268, 14476 (1993); **13.** Mol Biol of the Cell 6, 231a, abstract #1339 (1995); **14.** Biochim Biophys Acta 1113, 277 (1992); **15.** Brain Research 597, 108 (1992); **16.** J Biol Chem 268, 14861 (1993); **17.** Biochim Biophys Acta 915, 87 (1987); **18.** Anal Biochem 136, 223 (1984); **19.** Clin Chem Acta 142, 313 (1984); **20.** Clin Chem Acta 124, 123 (1982); **21.** Biochem Biophys Res Comm 18, 221 (1965); **22.** J Cell Biol 125, 769 (1994); **23.** Adv Cell Mol Biol Membranes 1, 199 (1993); **24.** Biochim Biophys Acta 1082, 113 (1991); **25.** Parasitology 105, 81 (1992); **26.** J Cell Biol 108, 2169 (1989); **27.** Proc Natl Acad Sci USA 90 2661 (1993); **28.** J Biol Chem 269, 8455 (1994); **29.** J Biol Chem 269, 8937 (1994); **30.** J Biol Chem 270, 27326 (1995); **31.** J Biol Chem 270, 22689 (1995); **32.** J Biol Chem 268, 15523 (1993); **33.** Science 259, 1769 (1993); **34.** Science 259, 1771 (1993); **35.** J Biol Chem 270, 5007 (1995); **36.** J Biol Chem 270, 3056 (1995); **37.** J Biol Chem 269, 18384 (1994); **38.** J Biol Chem 270, 4088 (1995); **39.** J Biol Chem 270, 13291 (1995); **40.** J Biol Chem 267, 18483 (1992); **41.** J Biol Chem 268, 17762 (1993); **42.** J Biol Chem 270, 2859 (1995); **43.** Adv Lipid Res 26, 183 (1993); **44.** J Biochem 111, 191 (1992); **45.** J Lipid Res 28, 565 (1987); **46.** Chem Phys Lipids 26, 265 (1980); **47.** J Biol Chem 269, 6803 (1994); **48.** J Cell Biol 121, 1385 (1993); **49.** Am J Physiol 262, F24 (1992); **50.** J Lipid Res 35, 1232 (1994); **51.** Biochem J 292, 617 (1993); **52.** J NIH Res 3, 80 (1991); **53.** Anal Biochem 174, 477 (1988); **54.** J Immunology 147, 2809 (1991); **55.** J Biol Chem 270, 15076 (1995); **56.** Biochemistry 32, 7574 (1993); **57.** J Biol Chem 267, 23495 (1992); **58.** J Biol Chem 260, 3440 (1985); **59.** Biochemistry 32, 530 (1993); **60.** J Biol Chem 269, 9506 (1994); **61.** J Biol Chem 268, 17578 (1993); **62.** J Biol Chem 269, 23918 (1994); **63.** J Lipid Res 36, 229 (1995); **64.** Science 251, 204 (1991); **65.** Prog Lipid Res 26, 125 (1987); **66.** J Biol Chem 270, 23823 (1995); **67.** Proc Natl Acad Sci USA 91, 6698 (1994); **68.** J Neurosci 15, 6552 (1995); **69.** Science 268, 1506 (1995).

20.6 Data Table *Probes for Lipid Metabolism and Signaling*

Cat #	Structure	MW	Storage	Soluble	Abs	{ε × 10^{-3}}	Em	Solvent	Notes
A-3767	✓	702	FF,D,L	<1>	337	{3.2}	525	EtOH	
A-3770	✓	632	FF,D,L	<1>	465	{18}	540	EtOH	<2>
A-3773	✓	743	FF,D,L	<1>	506	{76}	512	EtOH	
A-3880	C	~15,350	FF,L,AA	H_2O	365	{10.5}	432	H_2O	<3>
A-7062	✓	514	FF,D,LL,AA	DMSO	339	{5.2}	none	MeOH	<4>
A-7068	✓	484	FF,D,LL,AA	DMSO	<5>	<5>	none	MeOH	<4, 6, 7>

Cat #	Structure	MW	Storage	Soluble	Abs	{ε × 10^{-3}}	Em	Solvent	Notes
A-7509	✓	342	F	DMSO, EtOH	<300		none		
D-3521	S12.4	602	FF,D,L	$CHCl_3$, DMSO	505	{91}	511	MeOH	<8>
D-3522	S13.2	767	FF,D,L	<1>	505	{77}	512	MeOH	
D-3766	✓	660	FF,D,L	<1>	335	{3.5}	525	EtOH	
D-3772	✓	701	FF,D,L	<1>	506	{75}	512	EtOH	
D-7519	✓	862	F,D,L	DMSO, EtOH	505	{85}	511	MeOH	
D-7520	✓	637	F,D,L	DMSO, EtOH	505	{85}	511	MeOH	
D-7521	✓	715	F,D,L	DMSO, EtOH	542	{60}	563	MeOH	
D-7522	✓	427	F,D	DMSO, EtOH	<300		none		
D-7540	S12.4	706	FF,D,L	$CHCl_3$, DMSO	589	{60}	617	MeOH	
D-7546	S12.4	759	FF,D,L	$CHCl_3$, DMSO	533	{80}	545	MeOH	
D-7547	✓	862	F,D,L	DMSO, EtOH	505	{85}	511	MeOH	
D-7548	✓	764	F,D,L	DMSO, EtOH	505	{85}	511	MeOH	
H-7492	✓	348	FF,D,L,AA	EtOH	<300		none		<7>
H-7523	✓	511	F,D	DMSO, EtOH	<300		none		
H-7524	✓	398	FF,D	DMSO, EtOH	<300		none		
I-3881	C	~15,125	FF,AA	H_2O	280	{17}	none	pH 7	
N-1154	S12.4	576	FF,D,L	$CHCl_3$, DMSO	466	{22}	536	MeOH	<2>
N-3524	S13.2	741	FF,D,L	<1>	466	{22}	536	MeOH	<2>
O-7541	✓	281	F,A	EtOH	<300		none		

For definitions of the contents of this data table, see "How to Use the Handbook" on page vi.
Structure: Chemical structure drawing: ✓ = shown in this section; Sn.n = shown in Section number n.n.
MW: ~ indicates an approximate value.

<1> Chloroform is the most generally useful solvent for preparing stock solutions of phospholipids (including sphingomyelins). Glycerophosphocholines are usually freely soluble in ethanol. Information on solubility of natural phospholipids can be found in Marsh, D., *CRC Handbook of Lipid Bilayers*, CRC Press (1990) pp. 71–80.
<2> NBD derivatives are almost nonfluorescent in water. The quantum yields of NBD phospholipid derivatives increase upon binding to membranes.
<3> ADIFAB fatty acid indicator (A-3880) is a protein conjugate with a molecular weight of approximately 15,350 daltons. Binding of fatty acids in the micromolar range shifts the emission peak from about 432 nm to 505 nm.
<4> All photoactivatable probes are sensitive to light. They should be protected from illumination except when photolysis is intended.
<5> No resolved absorption peaks >250 nm. ε ~{5} at 260 nm.
<6> This compound has weak visible absorption at >300 nm but no discernible absorption peaks in this region.
<7> This product is intrinsically a liquid or an oil at room temperature.
<8> Em for D-3521 shifts to ~620 nm when high concentrations of the probe (>5 mol %) are incorporated in lipid mixtures [J Cell Biol 113, 1267 (1991)].

20.6 Structures *Probes for Lipid Metabolism and Signaling*

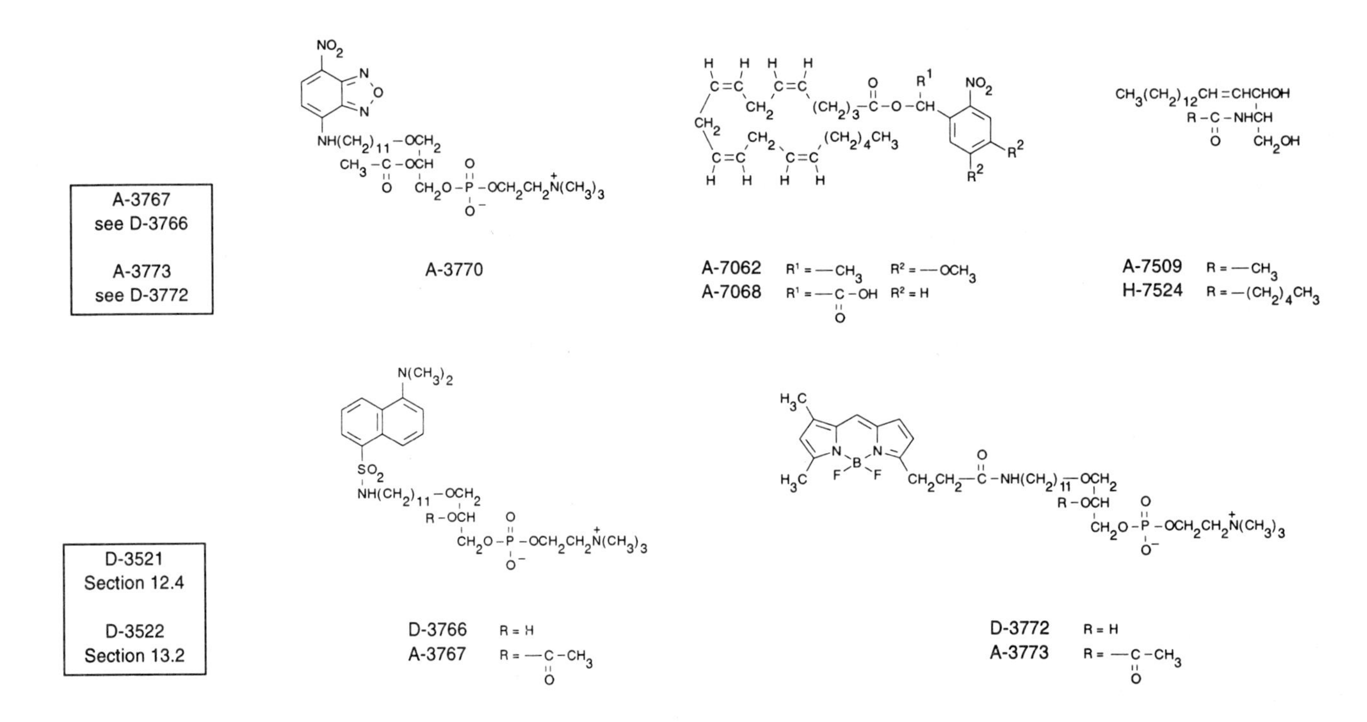

D-7519

D-7520

D-7521

D-7522 n = 8
H-7523 n = 14

D-7540, D-7546
Section 12.4

D-7548 n = 4
D-7547 n = 11

H-7492

H-7523
see D-7522

H-7524
see A-7509

N-1154
Section 12.4

N-3524
Section 13.2

O-7541

20.6 Price List *Probes for Lipid Metabolism and Signaling*

Cat #		Product Name	Unit Size	Price Per Unit ($) 1–4 Units	5–24 Units
A-3773		2-acetyl-1-(O-(11-(4,4-difluoro-5,7-dimethyl-4-bora-3a,4a-diaza-s-indacene-3-propionyl)amino)-undecyl)-*sn*-glycero-3-phosphocholine (BODIPY® FL C_{11}-PAF)	500 µg	68.00	54.40
A-3767		2-acetyl-1-(O-(11-(5-dimethylaminonaphthalene-1-sulfonyl)amino)undecyl)-*sn*-glycero-3-phosphocholine (dansyl-C_{11}-PAF)	500 µg	68.00	54.40
A-7509	**New**	N-acetyl-D-erythro-sphingosine (acetyl ceramide; C_2-ceramide)	5 mg	95.00	76.00
A-3770		2-acetyl-1-(O-(11-(7-nitrobenz-2-oxa-1,3-diazol-4-yl)amino)undecyl)-*sn*-glycero-3-phosphocholine (NBD-C_{11}-PAF)	500 µg	68.00	54.40
A-3880		ADIFAB fatty acid indicator	200 µg	245.00	196.00
A-7068	**New**	arachidonic acid, α-carboxy-2-nitrobenzyl ester (CNB-caged arachidonic acid)	5 mg	95.00	76.00
A-7062	**New**	arachidonic acid, 1-(4,5-dimethoxy-2-nitrophenyl)ethyl ester (DMNPE-caged arachidonic acid)	5 mg	95.00	76.00
D-7522	**New**	DL-*threo*-2-decanoylamino-3-morpholino-1-phenyl-1-propanol, hydrochloride (PDMP)	5 mg	48.00	38.40
D-7546	**New**	N-(2,6-dibromo-4,4-difluoro-5,7-dimethyl-4-bora-3a,4a-diaza-s-indacene-3-pentanoyl)sphingosine (BODIPY® FL Br_2C_5-ceramide)	250 µg	98.00	78.40
D-7520	**New**	DL-*threo*-2-(4,4-difluoro-5,7-dimethyl-4-bora-3a,4a-diaza-s-indacene-3-dodecanoyl)amino-3-morpholino-1-phenyl-1-propanol, hydrochloride (BODIPY® FL C_{12}-MPP)	25 µg	95.00	76.00
D-7519	**New**	N-(4,4-difluoro-5,7-dimethyl-4-bora-3a,4a-diaza-s-indacene-3-dodecanoyl)sphingosyl 1-β-D-galactopyranoside (BODIPY® FL C_{12}-galactocerebroside)	25 µg	95.00	76.00
D-7547	**New**	N-(4,4-difluoro-5,7-dimethyl-4-bora-3a,4a-diaza-s-indacene-3-dodecanoyl)sphingosyl 1-β-D-glucopyranoside (BODIPY® FL C_{12}-glucocerebroside)	250 µg	98.00	78.40
D-3521		N-(4,4-difluoro-5,7-dimethyl-4-bora-3a,4a-diaza-s-indacene-3-pentanoyl)sphingosine (BODIPY® FL C_5-ceramide)	250 µg	125.00	100.00
D-7548	**New**	N-(4,4-difluoro-5,7-dimethyl-4-bora-3a,4a-diaza-s-indacene-3-pentanoyl)sphingosyl 1-β-D-glucopyranoside (BODIPY® FL C_5-glucocerebroside)	250 µg	98.00	78.40
D-3522		N-(4,4-difluoro-5,7-dimethyl-4-bora-3a,4a-diaza-s-indacene-3-pentanoyl)sphingosyl phosphocholine (BODIPY® FL C_5-sphingomyelin)	250 µg	125.00	100.00
D-3772		1-(O-(11-(4,4-difluoro-5,7-dimethyl-4-bora-3a,4a-diaza-s-indacene-3-propionyl)amino)undecyl)-*sn*-glycero-3-phosphocholine (BODIPY® FL C_{11}-lyso PAF)	500 µg	90.00	72.00
D-7521	**New**	DL-*threo*-2-(4,4-difluoro-5-(4-methoxyphenyl)-4-bora-3a,4a-diaza-s-indacene-3-dodecanoyl)amino-3-morpholino-1-phenyl-1-propanol (BODIPY® TMR C_{12}-MPP)	25 µg	95.00	76.00
D-7540	**New**	N-((4-(4,4-difluoro-5-(2-thienyl)-4-bora-3a,4a-diaza-s-indacene-3-yl)phenoxy)acetyl)sphingosine (BODIPY® TR ceramide)	250 µg	98.00	78.40
D-3766		1-(O-(11-(5-dimethylaminonaphthalene-1-sulfonyl)amino)undecyl)-*sn*-glycero-3-phosphocholine (dansyl-C_{11}-lyso PAF)	500 µg	68.00	54.40
H-7523	**New**	DL-*threo*-2-hexadecanoylamino-3-morpholino-1-phenyl-1-propanol, hydrochloride (PPMP)	5 mg	48.00	38.40

Cat #		Product Name	Unit Size	Price Per Unit ($) 1–4 Units	5–24 Units
H-7524	***New***	N-hexanoyl-D-erythro-sphingosine (hexanoyl ceramide; C_6-ceramide)	5 mg	95.00	76.00
H-7492	***New***	N-(2-hydroxyethyl)-*cis,cis,cis,cis*-5,8,11,14-eicosatetraenoic acid amide (anandamide) *special packaging*	10x1 mg	48.00	38.40
I-3881		I-FABP fatty acid binding protein	1 mg	330.00	264.00
N-1154		6-((N-(7-nitrobenz-2-oxa-1,3-diazol-4-yl)amino)hexanoyl)sphingosine (NBD C_6-ceramide)	1 mg	125.00	100.00
N-3524		6-((N-(7-nitrobenz-2-oxa-1,3-diazol-4-yl)amino)hexanoyl)sphingosyl phosphocholine (NBD C_6-sphingomyelin)	1 mg	125.00	100.00
O-7541	***New***	*cis*-9-octadecenamide (oleamide) *≥98%*	100 mg	68.00	54.40

Hints for Finding the Product of Interest

This Handbook includes products released by Molecular Probes before September 1996. Additional products and search capabilities are available through our Web site (http://www.probes.com). All information, including price and availability, is subject to change without notice.

- **If you know the name of the product you want:** Look up the product by name in the **Master Price List/Product Index** at the back of the book. The Master Price List/Product Index is arranged alphabetically, by both product name and alternative name, using the alphabetization rules outlined on page vi. This index will refer you directly to all the Handbook sections that discuss the product.

- **If you DO NOT know the name of the product you want:** Search the **Keyword Index** for the words that best describe the applications in which you are interested. The Keyword Index will refer you directly to page numbers on which the listed topics are discussed. Alternatively, search the **Table of Contents** for the chapter or section title that most closely relates to your application.

- **If you only know the alpha-numeric catalog number of the product you want but not its name:** Look for the catalog number in the **Master Price List/Product Index** at the back of the book. First locate the group of catalog numbers with the same alpha identifier as the catalog number of interest and then scan this group for the correct numeric identifier. The Master Price List/Product Index will refer you directly to all the Handbook sections that discuss this product. Alternatively, you may use the search features available through our Web site (http://www.probes.com) to obtain the product name of any of our current products given only the catalog number (or numeric identifier).

- **If you only know part of a product name or are searching for a product introduced after publication of this Handbook:** Search Molecular Probes' up-to-date product list through our **Web site (http://www.probes.com)**. The product search feature on our home page allows you to search our entire product list, including products that have been added since the completion of this Handbook, using a partial product name, complete product name or catalog number. Product-specific bibliographies can also be obtained from our Web site once the product of interest has been located.

- **If you still cannot find the product that you are looking for or are seeking advice in choosing a product:** Please contact our Technical Assistance Department by phone, e-mail, fax or mail.

Chapter 21

Probes For Reactive Oxygen Species, Including Nitric Oxide

Contents

21.1 Introduction to Reactive Oxygen Species

Molecular Probes has developed several probes that either generate or detect various reactive oxygen species, including singlet oxygen ($^1O_2$), superoxide ($O_2^{\bar{\bullet}}$), hydroxyl radical (OH•) and various peroxides (ROOR′) and hydroperoxides (ROOH). Produced during a number of physiological processes, these activated oxygen species react with a large variety of easily oxidizable cellular components, including NADH, NADPH, dopa, ascorbic acid, histidine, tryptophan, tyrosine, cysteine, glutathione, proteins and nucleic acids.[1-5] Reactive oxygen species can also oxidize cholesterol and unsaturated fatty acids, causing membrane lipid peroxidation.[6,7] Several recent reviews discuss the chemistry of the different reactive oxygen species and their lipid peroxidation products.[8-12]

The importance of the nitric oxide radical (abbreviated NO• or NO) and other reactive oxygen species as biological messengers has been increasingly recognized since publication of the Fifth Edition of our Handbook.[13-19] Consequently, almost all of the probes described in this chapter are new to our product line. We are continuing to develop new probes for generating and detecting nitric oxide, as well as new inhibitors for nitric oxide synthase.

1. Ann Rev Biochem 64, 97 (1995); **2.** Meth Enzymol 234 (1994); **3.** Meth Enzymol 233 (1994); **4.** Meth Enzymol 186 (1990); **5.** Meth Enzymol 105 (1984); **6.** J Biol Chem 268, 18502 (1993); **7.** Pharmacol Ther 53, 375 (1992); **8.** Clin Chem 41, 1819 (1995); **9.** Free Rad Biol Med 19, 77 (1995); **10.** Free Rad Biol Med 18, 1033 (1995); **11.** Free Rad Biol Med 18, 775 (1995); **12.** Biochimie 76, 355 (1994); **13.** Chem Ind 828 (1995); **14.** Current Biology 2, 437 (1995); **15.** J Med Chem 38, 4342 (1995); **16.** Ann Rev Biochem 63, 175 (1994); **17.** Cell 78, 919 (1994); **18.** Am J Physiol 262, G379 (1992); **19.** Science 257, 494 (1992).

21.2 Generating Reactive Oxygen Species

Singlet Oxygen

Singlet oxygen is responsible for much of the physiological damage caused by reactive oxygen species, including nucleic acid modification through selective reaction with deoxyguanosine.[1] The lifetime of singlet oxygen is sufficiently long (4.4 microseconds in water [2]) to permit significant diffusion in cells and tissues.[3] In the laboratory singlet oxygen is usually generated in one of three ways: photochemically from dioxygen ($^3O_2$) using a photosensitizing dye [4] or chemically either by thermal decomposition of a peroxide or dioxetane or by microwave discharge through an oxygen stream. Singlet oxygen can be detected by its characteristic weak chemiluminescence at 1268 nm [5] or at 634 and 703 nm.[6]

Hypocrellins and Hypericin

Among the most efficient reagents for generating singlet oxygen are the photosensitizers hypocrellin A (H-7515), hypocrellin B (H-7516) and hypericin (H-7476). These heat-stable dyes exhibit quantum yields for singlet oxygen generation in excess of 0.7, as well as high photostability, making them important agents for both anticancer and antiviral therapy.[7-12] Hypocrellins are also specific and potent inhibitors of protein kinase C [13,14] (see Section 20.3), whereas hypericin is an effective inhibitor of both protein kinase C and tyrosine protein kinase [15] with antiretroviral activity.[16]

Because their chemical reactivities are well characterized,[17,18] hypocrellins and hypericin are amenable to conjugation to a variety of primary and secondary detection reagents. Not only do these photosensitizing dyes efficiently oxidize diaminobenzidine (DAB) to form an insoluble, electron-dense DAB oxidation product (see the Technical Note "Fluorescent Probes for Photoconversion of Diaminobenzidine" on page 414), but they exhibit modest fluorescence quantum yields and broad UV and visible spectra. Thus, hypocrellin- and hypericin-labeled detection reagents are compatible with fluorescence, light or electron microscopy applications.[19] Hypocrellins A and B have been used as building blocks to prepare various bioconjugates for both immunoassays and immunotherapeutics.[20,21]

Halogenated Fluoresceins and Rhodamines, Including Rose Bengal

Halogenated derivatives of fluorescein dyes are known to be effective photosensitizers and singlet oxygen generators.[22] Bioconjugates of eosin, erythrosin and our new halogenated BODIPY® fluorophores can be prepared using the reactive derivatives described in Chapters 1–3. The practical use of eosin and other conjugates to improve ultrastructural resolution through singlet oxygen generation is described in the Technical Note "Fluorescent Probes for Photoconversion of Diaminobenzidine" on page 414. The generation of singlet oxygen can be targeted to the mitochondria by several cationic dyes [23] such as tetrabromorhodamine 123 (T-7539), which has a quantum yield for singlet oxygen generation of 0.65–0.7.[23,24]

Rose bengal (4,5,6,7-tetrachloro-2',4',5',7'-tetraiodofluorescein) has a quantum yield for singlet oxygen generation of 0.75 in water, as compared to 0.63 for erythrosin, 0.57 for eosin and 0.03 for fluorescein.[22] Rose bengal agarose (R-7932) comprises rose bengal noncovalently immobilized on amino agarose.[25] Immobilization permits rapid removal of the dye from photooxidized proteins, nucleic acids or cells, resulting in better control of the modification.[25] Photolysis of immobilized rose bengal can also be used to generate singlet oxygen in the gas phase, which has the advantage of providing a relatively constant concentration of singlet oxygen.[26]

Endoperoxides

The water-soluble endoperoxides of anthracenedipropionic acid and naphthalenedipropionic acid have also been demonstrated to efficiently release singlet oxygen under physiological conditions, producing strand breaks in DNA.[27-29] These endoperoxides can be readily prepared from anthracene-9,10-dipropionic acid (A-7896) and naphthalene-1,4-dipropionic acid (N-7897) using hypocrellins as photosensitizers [10] (see Figure 21.1 in Section 21.3).

Merocyanine 540

Photolysis of merocyanine 540 (M-299) is reported to produce both singlet oxygen and other reactive oxygen species, including oxygen radicals.[30-33] Merocyanine 540 is commonly used as a photosensitizer in photodynamic therapy.[34-43]

Hydroxyl, Alkoxy and Superoxide Radicals

The hydroxyl radical is a very reactive oxygen species that has a lifetime of about 2 nanoseconds in aqueous solution and a radius of diffusion of about 20 Å. Thus it induces peroxidation only when it is generated in close proximity to its target. The hydroxyl radical can be derived from superoxide in a reaction catalyzed by Fe^{2+} or other transition metals, as well as by the effect of ionizing radiation on dioxygen. Molecular Probes prepares several probes that generate hydroxyl or alkoxy radicals. Superoxide is most effectively generated from a hypoxanthine–xanthine oxidase generating system.[44-46]

FotoFenton™ Reagents

Hydroxyl radicals are conveniently produced by photolysis of our FotoFenton dyes — 2-mercaptopyridine *N*-oxide[47-49] (FotoFenton 1, M-7904) and 2-hydroxyacetophenone oxime[50] (FotoFenton 2, H-7895). When activated by UV illumination, these hydroxyl radical generators produce fewer side products than does Fenton's reagent (Fe^{2+}/H_2O_2), the hydroxyl radical generator currently used by most researchers. Hydroxyl radical generation from FotoFenton 1 is not dependent on metals.[48]

Malachite Green and Isosulfan Blue Derivatives

Malachite green is a nonfluorescent photosensitizer that absorbs at long wavelengths (~630 nm). Its photosensitizing action can be targeted to particular cellular sites by conjugating malachite green isothiocyanate (M-689) to specific antibodies. Enzymes and other proteins within ~10 Å of the binding site of the malachite green–labeled antibody can then be selectively destroyed upon irradiation with long-wavelength light. Recent studies by Jay and colleagues have demonstrated that this photoinduced destruction of enzymes in the immediate vicinity of the chromophore is apparently the result of localized production of hydroxyl radicals, which have short lifetimes that limit their diffusion from the site of their generation.[51]

Isosulfan blue (I-6958) is a nonfluorescent, highly water-soluble dye that has very strong long-wavelength absorbance (ϵ >110,000 $cm^{-1}M^{-1}$ in methanol). Although we are not aware of its use as a photosensitizer, its structural similarity to malachite green indicates that it may be a potent photogenerator of singlet oxygen, which could make it useful for selective photoablation of microinjected cells in cell development and differentiation studies. Our new reactive sulfonyl chloride of isosulfan blue (I-6204) is potentially a lower cost alternative to malachite green isothiocyanate for the preparation of photosensitizing bioconjugates.

Phenanthrolines

Conjugation of the bromoacetamide or iodoacetamide of 1,10-phenanthroline (P-6878, P-6879) to thiol-containing ligands confers the metal-binding properties of this important complexing agent on the ligand. For example, the covalent copper–phenanthroline complex of oligonucleotides or nucleic acid–binding molecules in combination with hydrogen peroxide acts as a chemical nuclease to selectively cleave DNA or RNA.[52-57] Hydroxyl radicals or other reactive oxygen species appear to be involved in this cleavage.[58,59]

Artemisinins

The antimalarial drugs artemisinin (qinghaosu, A-7455) and dihydroartemisinin (D-96058) produce alkoxy radicals in a reaction catalyzed by intracellular Fe^{2+}.[60,61] Radical formation appears to be an important step in the mechanism of their action.[62] A recent paper also presents evidence for the involvement of a high-valent Fe(IV)=O species in the mechanism of artemisinin's action.[63]

Pyrenealkanols

1-Pyrenebutanol (P-244) and 1-pyrenenonanol (P-394) are reported to be generators of reactive oxygen intermediates useful for UV-mediated photoablation.[64-66] 1-Pyrenebutanol has been used as a selective nonmetabolizable photosensitizer for human myeloid leukemia cells.[65]

5-Aminolevulinic Acid

5-Aminolevulinic acid (ALA, A-8445) is a natural heme precursor that accumulates as a result of acute intermittent porphyria and lead poisoning.[67] ALA undergoes iron-catalyzed generation of reactive oxygen species, apparently including both superoxide and hydroxyl radicals, that results in lipid peroxidation and DNA scission.[68-70] Superoxide generated by ALA may be important in the mobilization of iron from the endoplasmic reticulum,[71] as well as from ferritin.[68]

1. J Photochem Photobiol 11, 241 (1991); **2.** J Am Chem Soc 104, 5541 (1982); **3.** J Chem Soc Faraday Trans I 82, 1627 (1986); **4.** Meth Enzymol 186, 635 (1990); **5.** Proc Natl Acad Sci USA 76, 6047 (1979); **6.** J Am Chem Soc 111, 2909 (1989); **7.** Photochem Photobiol 61, 632 (1995); **8.** Photochem Photobiol 61, 529 (1995); **9.** Pharmacol Ther 63, 1 (1994); **10.** J Photochem Photobiol A: Chem 64, 273 (1992); **11.** Photochem Photobiol 54, 95 (1991); **12.** Photochem Photobiol 53, 169 (1991); **13.** Biochem Pharmacol 47, 373 (1994); **14.** Biochem Biophys Res Comm 165, 1207 (1989); **15.** Biochem Pharmacol 46, 1929 (1993); **16.** Antiviral Res 17, 63 (1992); **17.** Photochem Photobiol 52, 609 (1990); **18.** Photochem Photobiol 43, 677 (1986); **19.** J Cell Biol 126, 901 (1994); **20.** Photochem Photobiol 60, 253 (1994); **21.** Anti-Cancer Drug Design 8, 129 (1993); **22.** Adv Photochem 18, 315 (1993); **23.** Photochem Photobiol 55, 81 (1992); **24.** Proc SPIE 997, 48 (1988); **25.** FEBS Lett 62, 334 (1976); **26.** Photochem Photobiol 56, 441 (1992); **27.** J Biol Chem 268, 18502 (1993); **28.** Proc Natl Acad Sci USA 89, 11428 (1992); **29.** Proc Natl Acad Sci USA 89, 11426 (1992); **30.** Biochim Biophys Acta 1105, 333 (1992); **31.** Arch Biochem Biophys 291, 43 (1991); **32.** J Photochem Photobiol 58, 339 (1991); **33.** J Photochem Photobiol 1, 437 (1988); **34.** J Immunol Meth 168, 245 (1994); **35.** Pharm Ther 63, 1 (1994); **36.** Arch Biochem Biophys 300, 714 (1993); **37.** Bone Marrow Transplantation 12, 191 (1993); **38.** Cancer Res 53, 806 (1993); **39.** Free Rad Biol Med 12, 389 (1992); **40.** Biochem Biophys Acta 1075, 28 (1991); **41.** J Infectious Diseases 163, 1312 (1991); **42.** Photochem Photobiol 53, 1 (1991); **43.** Cancer Res 49, 3637 (1989); **44.** Arch Biochem Biophys 314, 284 (1994); **45.** J Reprod Fert 97, 441 (1993); **46.** Meth Enzymol 12, 5 (1967); **47.** Photochem Photobiol 61, 269 (1995); **48.** Anal Biochem 206, 309 (1992); **49.** Tetrahedron Lett 31, 6869 (1990); **50.** Tetrahedron Lett 42, 3707 (1974); **51.** Proc Natl Acad Sci USA 91, 2659 (1994); **52.** Nucleic Acids Res 22, 4789 (1994); **53.** Bioconjugate Chem 4, 69 (1993); **54.** Meth Enzymol 208, 414 (1991); **55.** Ann Rev Biochem 59, 207 (1990); **56.** J Am Chem Soc 111, 4941 (1989); **57.** Proc Natl Acad Sci USA 86, 9702 (1989); **58.** Biochemistry 29, 8447 (1990); **59.** J Am Chem Soc 109, 1990 (1987); **60.** Antimicro Agents Chemother 37, 1108 (1993); **61.** Med Res Rev 7, 29 (1987); **62.** Prog Clin Biol Res 313, 95 (1989); **63.** J Am Chem Soc 117, 5885 (1995); **64.** J Biol Chem 267, 8299 (1992); **65.** Leukemia Res 16, 453 (1992); **66.** Biochim Biophys Acta 1034, 132 (1990); **67.** Clin Chem 32, 1255 (1986); **68.** Arch Biochem Biophys 316, 607 (1995); **69.** Biochim Biophys Acta 1225, 259 (1994); **70.** Biochim Biophys Acta 881, 100 (1986); **71.** Arch Biochem Biophys 297, 189 (1992).

21.2 Data Table *Generating Reactive Oxygen Species*

Cat #	Structure	MW	Storage	Soluble	Abs	{ϵ x 10^{-3}}	Em	Solvent	Notes
A-7455	✓	282	F,D	DMSO	<300		none		
A-7896	✓	366	D,L	H_2O	400	{8.5}	429	pH 7	
A-8445	✓	168	F,D	H_2O	<300		none		
D-96058	✓	284	F,D	DMSO	<300		none		
H-7476	✓	504	F,D,L	DMSO, DMF	591	{37}	594	EtOH	
H-7515	✓	547	F,D,L	DMSO	463	{24}	600	MeOH	<1>
H-7516	✓	529	F,D,L	DMSO	459	{23}	616	MeOH	<1>
H-7895	✓	151	L	MeOH	303	{4.0}	none	MeOH	
I-6204	✓	563	F,DD,L	DMF, MeCN	646	{87}	none	MeOH	
I-6958	✓	567	D,L	H_2O	631	{111}	none	MeOH	
M-299	S25.3	570	D,L	DMSO, EtOH	555	{160}	578	MeOH	
M-689	S1.6	486	F,DD,L	DMF, DMSO	628	{76}	none	MeOH	<2>
M-7904	✓	149	F,D,L	H_2O	345	{3.3}	none	MeOH	<3>
N-7897	✓	316	D,L	H_2O	289	{8.1}	337	MeOH	
P-244	S14.6	274	L	DMF, $CHCl_3$	341	{41}	376	MeOH	<4>
P-394	S14.6	345	L	DMF, $CHCl_3$	341	{42}	377	MeOH	<4>
P-6878	S2.4	316	F,D,L	DMSO	270	{29}	none	$CHCl_3$	
P-6879	S2.4	363	F,D,L	DMSO	270	{28}	none	$CHCl_3$	<5>
T-7539	S12.2	741	F,D,L	MeOH, DMF	524	{91}	550	MeOH	

For definitions of the contents of this data table, see "How to Use the Handbook" on page vi.
Structure: Chemical structure drawing: ✓ = shown in this section; Sn.n = shown in Section number n.n.

<1> H-7515 and H-7516 have weaker absorption peaks at longer wavelengths: Abs = 581 nm (ϵ = {12}) for H-7515, Abs = 580 nm (ϵ = {9}) for H-7516. Their quantum yields for singlet oxygen production (0.83 for H-7515, 0.76 for H-7516) are essentially independent of the irradiation wavelength between 430 and 580 nm [J Photochem Photobiol A: Chem 64, 273 (1992)].

<2> Isothiocyanates are unstable in water and should not be stored in aqueous solution.

<3> M-7904 generates hydroxyl radicals (OH•) in aqueous solutions upon irradiation with visible light. Organic buffers such as Tris react readily with OH• and should be avoided; borate buffers are recommended [Anal Biochem 206, 309 (1992)].

<4> Pyrene derivatives exhibit structured spectra. The absorption maximum is usually about 340 nm with a subsidiary peak at about 325 nm. There are also strong absorption peaks below 300 nm. The emission maximum is usually about 376 nm with a subsidiary peak at 396 nm. Excimer emission at about 470 nm may be observed at high concentrations.

<5> Iodoacetamides in solution undergo rapid photodecomposition to unreactive products. Minimize exposure to light prior to reaction.

21.2 Structures *Generating Reactive Oxygen Species*

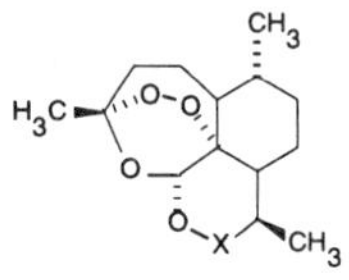

A-7455 X = C=O
D-96058 X = CHOH

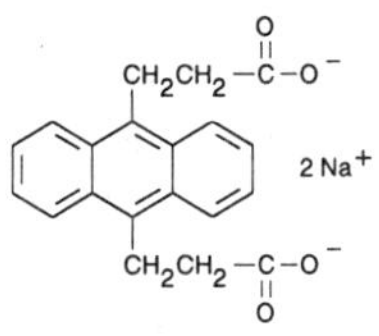

A-7896

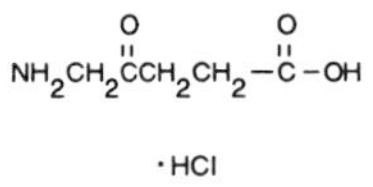

A-8445

D-96058
see A-7455

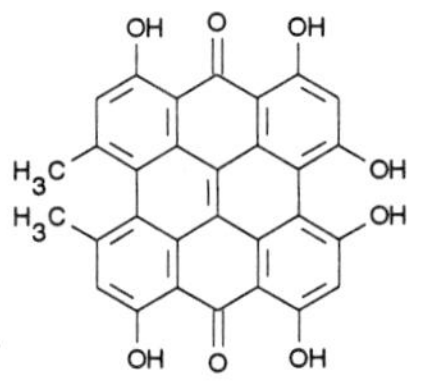

H-7476

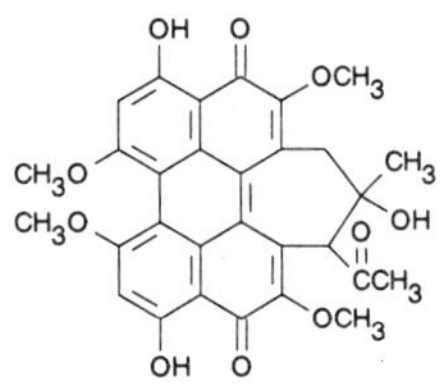

H-7515

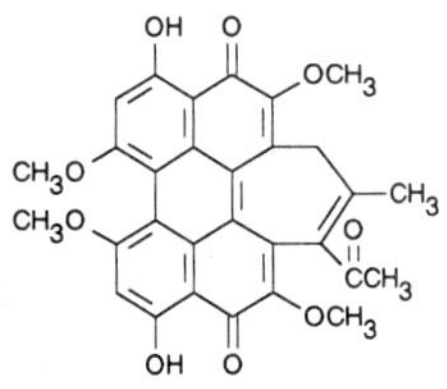

H-7516

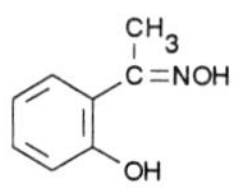

H-7895

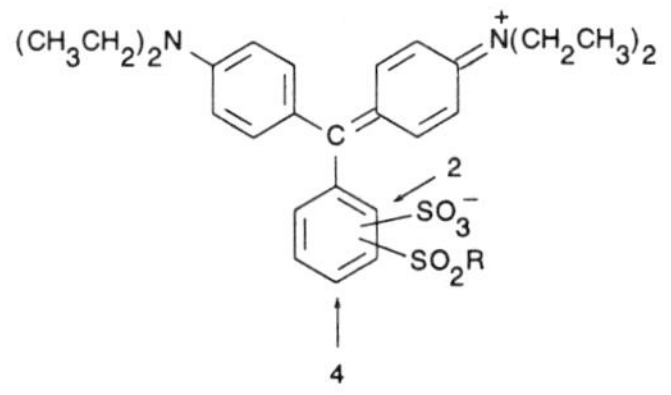

I-6958 R = O^- Na^+
I-6204 R = Cl

M-299
Section 25.3

M-689
Section 1.6

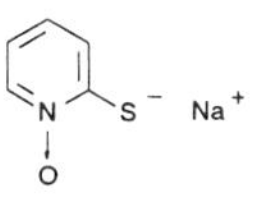

M-7904

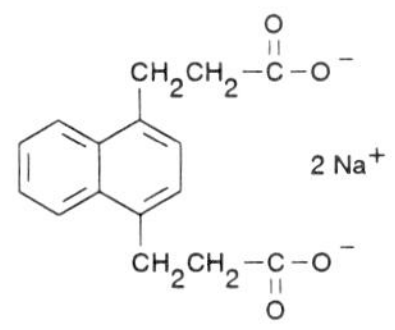

N-7897

P-244, P
Section 14.6

P-6878, P-6879
Section 2.4

T-7539
Section 12.2

21.2 Price List *Generating Reactive Oxygen Species*

Cat #		Product Name	Unit Size	Price Per Unit ($) 1–4 Units	5–24 Units
A-8445	**New**	5-aminolevulinic acid, hydrochloride (ALA)	1 g	55.00	44.00
A-7896	**New**	anthracene-9,10-dipropionic acid, disodium salt	25 mg	65.00	52.00
A-7455	**New**	artemisinin (qinghaosu)	100 mg	20.00	16.00
D-96058	**New**	dihydroartemisinin	50 mg	128.00	102.40
H-7895	**New**	2-hydroxyacetophenone oxime (FotoFenton™ 2)	1 g	35.00	28.00
H-7476	**New**	hypericin	1 mg	48.00	38.40
H-7515	**New**	hypocrellin A	1 mg	48.00	38.40
H-7516	**New**	hypocrellin B	1 mg	48.00	38.40
I-6958	**New**	isosulfan blue *>90% by HPLC*	1 g	28.00	22.40
I-6204	**New**	isosulfan blue sulfonyl chloride *mixed isomers*	5 mg	78.00	62.40
M-689		malachite green isothiocyanate	10 mg	135.00	108.00
M-7904	**New**	2-mercaptopyridine N-oxide, sodium salt (FotoFenton™ 1)	100 mg	35.00	28.00
M-299		merocyanine 540	100 mg	30.00	24.00
N-7897	**New**	naphthalene-1,4-dipropionic acid, disodium salt	25 mg	65.00	52.00
P-6878	**New**	N-(1,10-phenanthrolin-5-yl)bromoacetamide	5 mg	75.00	60.00
P-6879	**New**	N-(1,10-phenanthrolin-5-yl)iodoacetamide	5 mg	75.00	60.00
P-244		1-pyrenebutanol	100 mg	65.00	52.00
P-394		1-pyrenenonanol	25 mg	135.00	108.00
R-7932	**New**	rose bengal agarose *sedimented bead suspension*	10 mL	95.00	76.00
T-7539	**New**	2′,4′,5′,7′-tetrabromorhodamine 123 bromide	5 mg	48.00	38.40

21.3 Detecting Reactive Oxygen Species

Although there are no equilibrium sensors that continuously monitor the level of reactive oxygen species, there exist a number of probes that trap or otherwise react with singlet oxygen, hydroxyl radicals or superoxide. The optical or electron spin properties of the resulting products are in some way a measure of the presence or quantity of the reactive oxygen species. In this section we discuss the assays most useful for detecting reactive oxygen species that either are in solution or have leaked from cells. Section 21.4 includes the reagents most useful for measuring reactive oxygen molecules formed within living cells and tissues.

Singlet Oxygen

Anthracene and Naphthalene Derivatives

Many compounds quench singlet oxygen without necessarily being useful for detecting singlet oxygen. However, singlet oxygen is known to bleach the fluorescence of a variety of aromatic compounds,[1] including anthracene-9,10-dipropionic acid and naphthalene-1,4-dipropionic acid (A-7896, N-7897; see Section 21.2). The quantum yields and kinetics of singlet oxygen generation can therefore be determined by measuring the decrease in fluorescence of such reagents.[2] These two compounds also have been used to trap singlet oxygen in various chemical and biological processes[3-5] (Figure 21.1). Our 9,10-dimethoxyanthracene-2-sulfonic acid sodium salt (DAS, D-6948) is a water-soluble derivative that can also potentially be used for monitoring singlet oxygen generation.

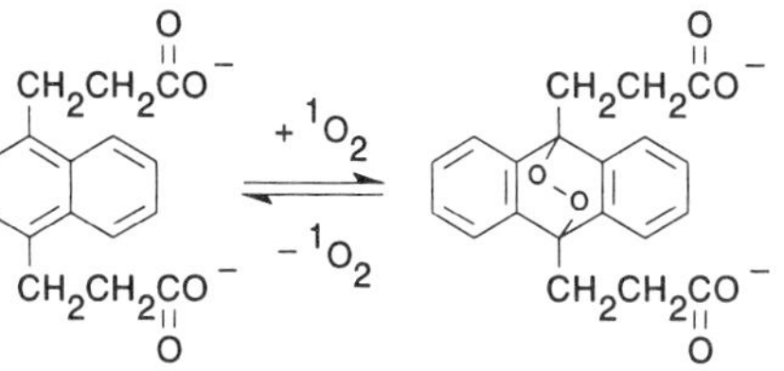

anthracene-9,10-dipropionic acid (fluorescent) $\underset{-{}^1O_2}{\overset{+{}^1O_2}{\rightleftharpoons}}$ endoperoxide (nonfluorescent)

Figure 21.1 *Singlet oxygen ($^1O_2$) trapping by anthracene-9,10-dipropionic acid (A-7896), yielding the nonfluorescent endoperoxide.*

enylisobenzofuran, Rubrene and Diphenylanthracene

The lipophilic dye 1,3-diphenylisobenzofuran (D-7919) is an fficient trapping agent for singlet oxygen formed in model membranes [6] and in cells.[7] Its intrinsic fluorescence is quenched upon endoperoxide formation.[8] The polycyclic hydrocarbon rubrene (R-7800, see Section 24.3) reacts rapidly with singlet oxygen, quenching its bright red fluorescence.[9] Similarly, the blue fluorescent hydrocarbon 9,10-diphenylanthracene (D-7801) forms a colorless, nonfluorescent endoperoxide with $^1O_2$.[10,11]

trans-1-(2'-Methoxyvinyl)pyrene

trans-1-(2'-Methoxyvinyl)pyrene (M-7913) can be used to detect picomole quantities of singlet oxygen in chemical and biological systems (Figure 21.2), making this compound perhaps the most sensitive singlet oxygen probe currently available. Furthermore, this highly selective chemiluminescent probe does not react with other activated oxygen species such as hydroxyl radical, superoxide or hydrogen peroxide.[12,13]

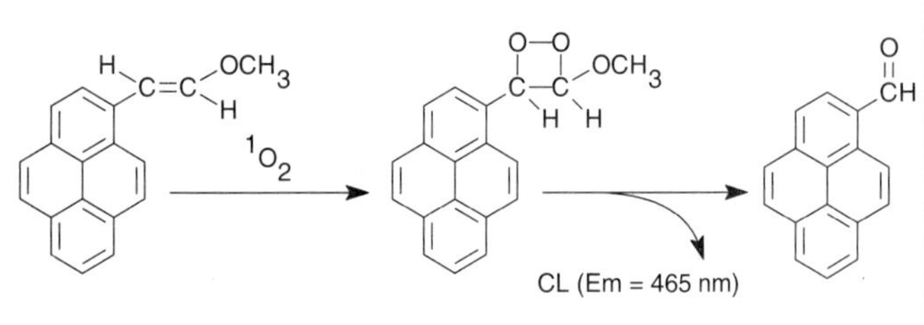

Figure 21.2 *Reaction of trans-1-(2'-methoxyvinyl)pyrene (M-7913) with singlet oxygen ($^1O_2$), yielding a dioxetane intermediate that generates chemiluminescence (CL) upon decomposition to 1-pyrenecarboxaldehyde.*

Hydroperoxides

Peroxidation of unsaturated lipids plays an important role in cell membrane properties,[14] signal transduction pathways,[15-17] apoptosis [18] and the deterioration of foods and other biological compounds. Lipid oxidation may also be responsible for aging, as well as pathological processes such as drug-induced phototoxicity and atherosclerosis. Lipid hydroperoxides have been reported to accumulate in oxidatively stressed individuals, including HIV-infected patients.[19] To assess the extent of lipid peroxidation, researchers either measure the amount of lipid hydroperoxides directly or detect the presence of secondary reaction products (e.g., 4-hydroxy-2-nonenal or malonaldehyde; see below).

Diphenyl-1-Pyrenylphosphine

The direct fluorometric detection of hydroperoxides in solution may now be possible using the reagent diphenyl-1-pyrenylphosphine (DPPP, D-7894). DPPP is essentially nonfluorescent until oxidized to a phosphine oxide by peroxides. DPPP has previously been used to detect picomole levels of hydroperoxides by HPLC.[20-23] Its solubility in lipids may make it quite useful for detecting hydroperoxides in the membranes of living cells, as well as for studying superoxide dismutase (SOD), which catalyzes the conversion of superoxide to hydrogen peroxide.

Luminol

Phospholipid hydroperoxides have been determined directly by chemiluminescence-detected HPLC with luminol (L-8455) [24,25] or a combination of luminol and cytochrome *c*.[26] This chemiluminescent probe has also been employed as a chemical sensor for dioxygen and nitrogen dioxide.[27] The use of luminol to detect peroxidase- and metal ion–mediated oxidative events in cells is discussed in Section 21.4.

Detecting 4-Hydroxy-2-Nonenal

4-Hydroxy-2-nonenal is a major cause of lipid peroxidation–induced liver toxicity. Several reagents were described in Section 3.2 for the direct fluorometric detection of aldehydes. The modification of 4-hydroxy-2-nonenal or malonaldehyde with a fluorescent or chromophoric hydrazine reagent coupled with the separation and detection of the reaction product by a chromatography-based technique has rarely been reported but appears to be one of the most promising approaches for detecting these lipid peroxidation products.[28-30]

Hydroxyl and Superoxide Radicals

In recent years, hydroxyl and superoxide radicals have been implicated in a number of pathological conditions, including ischemia, reperfusion and aging. It is also thought that the superoxide anion may play a role in regulating normal vascular function.[31-34]

Spin Traps, Including New Fluorogenic Spin Traps

Hydroxyl radicals have usually been detected after reaction with spin traps such as *N-tert*-butyl-α-phenylnitrone (PBN, B-7893) and 2-methyl-2-nitrosopropane dimer (M-7905), both of which form relatively stable free radicals that can be detected by their electron paramagnetic resonance.[35-38] The spin-trapping techniques predominantly used to detect these free radical species provide little information on the reactivity or precise site of formation of the hydroxyl radical or superoxide. We now offer TEMPO-9-AC and proxyl fluorescamine, two fluorogenic probes for detecting hydroxyl radicals and superoxide. Each of these molecules contains a nitroxide moiety that effectively quenches its fluorescence.[39,40] However, once TEMPO-9-AC or proxyl fluorescamine encounters a hydroxyl radical or superoxide, its fluorescence is restored and the radical's spin signal is destroyed, making these probes useful for detecting radicals either by fluorescence or by electron spin resonance spectroscopy. Radical-specific scavengers — such as the superoxide-specific *p*-benzoquinone and superoxide dismutase or the hydroxyl radical–specific mannitol and dimethylsulfoxide (DMSO) — can be used to identify the detected species.

Hydroxylation of Aromatic Hydrocarbons

Hydroxyl radicals (but not superoxide) are sufficiently reactive to hydroxylate certain aromatic hydrocarbons [41] such as phenylalanine.[42] The succinimidyl ester of coumarin-3-carboxylic acid (SECCA, C-7933) is a useful new fluorogenic reagent for detecting hydroxyl radicals produced either by ionizing radiation or in a Cu^{2+}-mediated reaction. This amine-reactive reagent — which can be coupled to a wide variety of biomolecules to create a probe selective for cells, organelles, proteins and nucleic acids — is converted by hydroxyl radicals to a highly fluorescent 7-hydroxycoumarin adduct.[43-45]

Phenoxazone (P-6549) can be hydroxylated to highly fluorescent resorufin by aryl hydrocarbon hydroxylase.[46,47] Phenoxazone's structural similarity to SECCA suggests that it may also be useful for detecting hydroxyl radicals, although its use for that purpose has apparently not been reported.

An unusual reaction has been reported for detecting superoxide production in heat-shocked cells of *Neurospora crassa* that may also involve hydroxylation of an aromatic hydrocarbon with expulsion of a sugar. In this assay, the carbohydrate ether of 4-methylumbelliferyl β-D-glucopyranoside (M-8423, see Section 10.2) is cleaved, with liberation of blue fluorescent 7-hydroxy-4-methylcoumarin (H-189, see Section 10.1). The generality of this reaction needs to be established; however, other glycosidase substrates in Section 10.2 may provide enhanced detectability at longer wavelengths.

Detection of Abasic Sites in Nucleic Acids

The effect of radiation-induced damage of nucleic acids by hydroxyl radicals is often cleavage of the base without scission of the nucleic acid chain.[48] Formation of apurinic/apyrimidinic sites is thought to be an important step in carcinogenesis.[49-51] The aldehydic product formed by irradiation of DNA has been detected using the biotinylated hydroxylamine derivative ARP (A-6346).[50] Once the aldehyde group in an abasic site is modified with ARP, the resulting biotinylated DNA can be quantitated with avidin conjugates (see Section 7.5). Fluorescent hydrazines discussed in Section 3.2 may be similarly useful for detecting these aldehyde groups.

Other Reagents for Detecting Reactive Oxygen Species

Some other reagents from Molecular Probes for detecting superoxide or peroxy radicals include:

- Coelenterazine (C-2944), which, in the absence of apoaequorin, produces chemiluminescence when oxidized by superoxide — a reaction reported to be quite specific for superoxide[52]
- Lucigenin[53-58] (L-6868), a chemiluminescent probe reported to be sensitive to the superoxide anion
- Nitro blue tetrazolium salt (NBT) (N-6495, see Section 21.4) and other tetrazolium salts (see also Section 16.3), chromogenic probes useful for superoxide determination[59,60]
- R-phycoerythrin (P-801, see Section 6.4), which has been used to detect and measure total plasma antioxidant capacity, including peroxy radicals[61-64]

1. Yagi, K., *Active Oxygens, Lipid Peroxides, and Antioxidants*, Lewis Publisher (1993); **2.** Photochem Photobiol 37, 271 (1983); **3.** Arch Pharmacol 325, 219 (1992); **4.** J Biol Chem 260, 5358 (1985); **5.** J Am Chem Soc 102, 5590 (1980); **6.** J Biochem Biophys Meth 27, 143 (1993); **7.** Biochem Mol Biol Int'l 32, 1093 1994); **8.** Biochim Biophys Acta 108, 94 (1991); **9.** Proc Natl Acad Sci USA 89, 11426 (1992); **10.** J Am Chem Soc 117, 9159 (1995); **11.** Bioorg Med Chem Lett 2, 1137 (1992); **12.** Meth Enzymol 133, 569 (1986); **13.** Biochem Biophys Res Comm 123, 869 (1984); **14.** Meth Enzymol 233, 459 (1994); **15.** Free Rad Biol Med 15, 365 (1993); **16.** Free Rad Biol Med 11, 81 (1991); **17.** Biochem Pharmacol 38, 543 (1987); **18.** Cell 75, 241 (1993); **19.** J Biol Chem 269, 798 (1994); **20.** J Chromatography 628, 31 (1993); **21.** J Chromatography 622, 153 (1993); **22.** J Chromatography 596, 197 (1992); **23.** Anal Lett 21, 965 (1988); **24.** Biomed Chromatography 4, 131 (1990); **25.** Free Rad Biol Med 7, 209 (1989); **26.** Meth Enzymol 233, 324 (1994); **27.** Anal Chem 67, 2224 (1995); **28.** Anal Chem 67, 1603 (1995); **29.** Anal Biochem 220, 391 (1994); **30.** Am J Pathol 142, 1353 (1993); **31.** Am J Physiol 264, H859 (1993); **32.** Am J Physiol 263, H537 (1992); **33.** J Appl Physiol 67, 1903 (1989); **34.** J Appl Physiol 66, 167 (1989); **35.** Anal Biochem 217, 76 (1994); **36.** Meth Enzymol 233, 112 (1994); **37.** Meth Enzymol 233, 105 (1994); **38.** Inorganic Chem 31, 2333 (1992); **39.** FASEB J 9, 1085 (1995); **40.** Anal Biochem 212, 85 (1993); **41.** Meth Enzymol 233, 67 (1994); **42.** Anal Biochem 220, 11 (1994); **43.** Free Rad Biol Med 18, 669 (1995); **44.** Radiation Res 138, 177 (1994); **45.** Int'l J Radiation Biol 63, 445 (1993); **46.** Chem-Biol Interact 45, 243 (1983); **47.** Biochem Pharmacol 34, 3337 1985); **48.** Free Rad Biol Med 18, 1033 (1995); **49.** Biochemistry 32, 8276 (1993); **50.** Biochemistry 31, 3703 (1992); **51.** Biochemistry 11, 3610 (1972); **52.** Anal Biochem 206, 273 (1992); **53.** Am J Physiol 267, L815 (1994); **54.** Blood 83, 3324 (1994); **55.** Cytometry 18, 147 (1994); **56.** Free Rad Biol Med 17, 117 (1994); **57.** Free Rad Biol Med 14, 457 (1993); **58.** J Immunol Meth 97, 209 (1987); **59.** Chem Soc Rev 6, 195 (1977); **60.** Can J Biochem 51, 158 (1973); **61.** Clin Chem 41, 1738 (1995); **62.** Free Rad Biol Med 18, 29 (1995); **63.** Anal Biochem 177, 300 (1989); **64.** FASEB J 2, 2487 (1988).

21.3 Data Table *Detecting Reactive Oxygen Species*

Cat #	Structure	MW	Storage	Soluble	Abs	{ε x 10^{-3}}	Em	Solvent	Notes
A-6346	S4.2	331	F,D	DMF, DMSO	<300		none		
A-7923	✓	376	F,D,L	DMSO	358	{11}	424	MeOH	<1>
B-7893	✓	177	F,D	EtOH	292	{20}	none	MeOH	
C-2944	S22.5	423	FF,D,LL,AA	MeOH	429	{7.5}	<2>	pH 7	
C-7924	✓	488	F,D,L	DMSO, H_2O	385	{5.8}	485	pH 7	<1>
C-7933	✓	287	F,D,L	DMSO	299	{16}	431	MeOH	<3>
D-6948	✓	340	L	H_2O, DMF	340	{6.3}	464	pH 7	
D-7801	S14.6	330	L	$CHCl_3$, MeCN	391	{13}	405	MeOH	
D-7894	✓	386	F,D,LL	MeCN	358	{29}	none	MeOH	<4>
D-7919	✓	270	L	DMSO	409	{24}	476	MeOH	
L-6868	S24.2	511	L	H_2O	455	{7.4}	505	H_2O	<5, 6>
L-8455	✓	177	D,L	DMF	355	{7.5}	411	MeOH	<6>
M-7905	✓	174	FF,D	H_2O	287	{8.3}	none	H_2O	<7>
M-7913	✓	258	F,L	DMF, DMSO	352	{30}	401	MeOH	<8>
P-6549	✓	125	L	DMSO	445	{10}	none	MeOH	<9>

For definitions of the contents of this data table, see "How to Use the Handbook" on page vi.
Structure: Chemical structure drawing: ✓ = shown in this section; Sn.n = shown in Section number n.n.

<1> Fluorescence of A-7923 and C-7924 is weak. Reduction of nitroxide by superoxide in the presence of reducing agents results in increased fluorescence without a spectral shift [Anal Biochem 212, 85 (1993)].
<2> C-2944 emits chemiluminescence (Em = 466 nm) upon oxidation by superoxide [Anal Biochem 206, 273 (1992)].
<3> Enzymatic hydroxylation yields a 7-hydroxycoumarin derivative with spectroscopic properties similar to those of H-185 (see Section 10.1).
<4> Oxidation product is strongly fluorescent with Em = 379 nm. Oxidation occurs rapidly in solution when illuminated.

<5> L-6868 has much stronger absorption at shorter wavelengths: Abs = 368 nm (ε = {36}).
<6> This compound emits chemiluminescence upon oxidation in basic aqueous solutions. Emission peaks are at 425 nm (L-8455) and 470 nm (L-6868).
<7> Spontaneously decomposes in solution.
<8> Generates chemiluminescence (Em = 465 nm in 0.1 M SDS) upon reaction with singlet oxygen [J Am Chem Soc 108, 4498 (1986)].
<9> Enzymatic hydroxylation yields resorufin (R-363, see Section 10.1).

21.3 Structures *Detecting Reactive Oxygen Species*

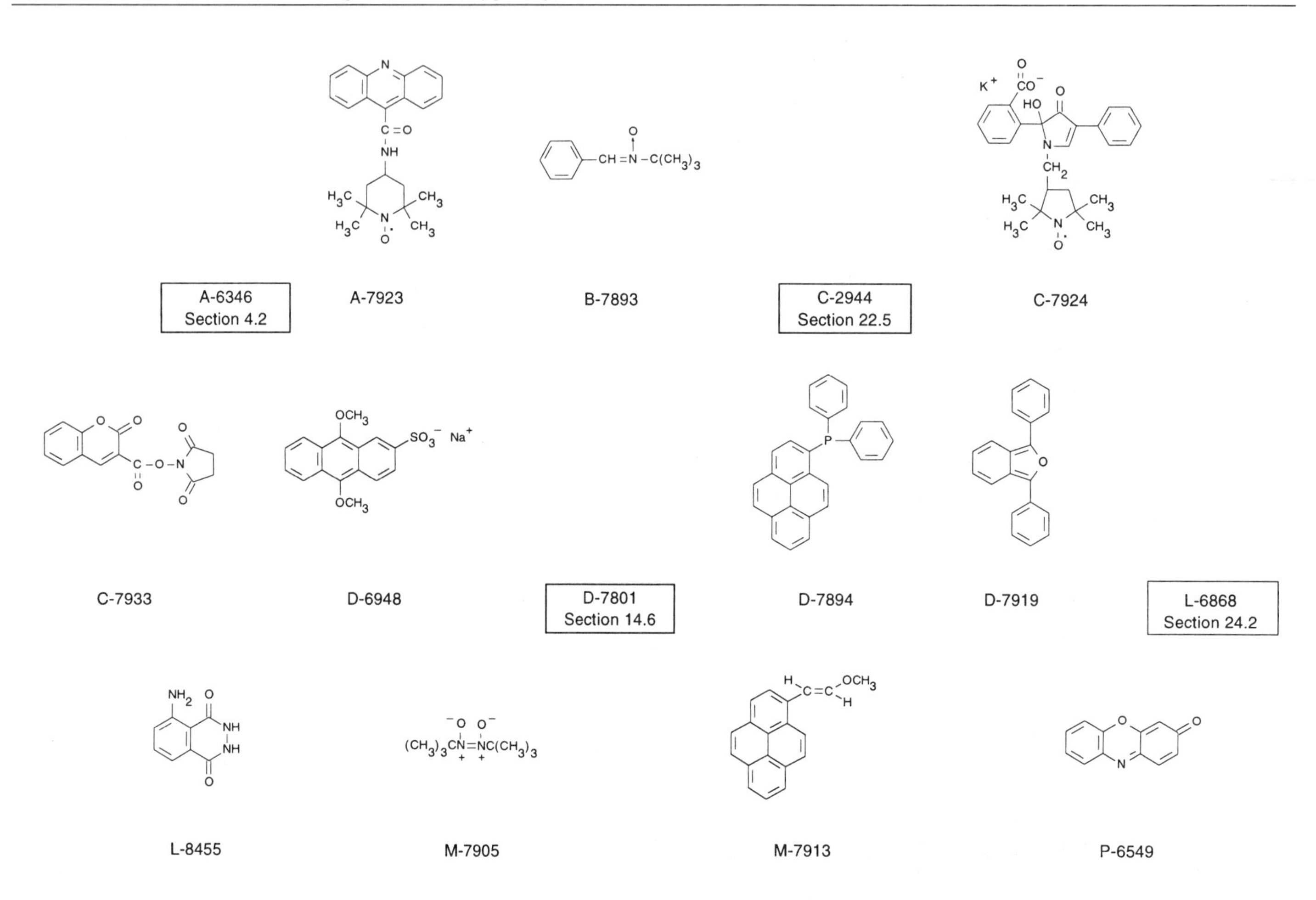

21.3 Price List *Detecting Reactive Oxygen Species*

Cat #		Product Name	Unit Size	Price Per Unit ($) 1–4 Units	5–24 Units
A-7923	**New**	4-((9-acridinecarbonyl)amino)-2,2,6,6-tetramethylpiperidin-1-oxyl, free radical (TEMPO-9-AC) *>90% by HPLC*	5 mg	65.00	52.00
A-6346	**New**	N-(aminooxyacetyl)-N'-(D-biotinoyl) hydrazine (ARP)	10 mg	98.00	78.40
B-7893	**New**	N-*tert*-butyl-α-phenylnitrone (PBN)	1 g	25.00	20.00
C-7924	**New**	5-(2-carboxyphenyl)-5-hydroxy-1-((2,2,5,5-tetramethyl-1-oxypyrrolidin-3-yl)methyl)-3-phenyl-2-pyrrolin-4-one, potassium salt (proxyl fluorescamine)	5 mg	65.00	52.00
C-2944		coelenterazine	250 µg	295.00	236.00
C-7933	**New**	coumarin-3-carboxylic acid, succinimidyl ester (SECCA)	25 mg	65.00	52.00
D-6948	**New**	9,10-dimethoxyanthracene-2-sulfonic acid, sodium salt (DAS)	100 mg	25.00	20.00
D-7801	**New**	9,10-diphenylanthracene	1 g	18.00	14.40
D-7919	**New**	1,3-diphenylisobenzofuran	1 g	28.00	22.40
D-7894	**New**	diphenyl-1-pyrenylphosphine (DPPP)	5 mg	125.00	100.00
L-6868	**New**	lucigenin (bis-N-methylacridinium nitrate) *high purity*	10 mg	15.00	12.00
L-8455	**New**	luminol (3-aminophthalhydrazide)	25 g	75.00	60.00
M-7913	**New**	*trans*-1-(2′-methoxyvinyl)pyrene	1 mg	75.00	60.00
M-7905	**New**	2-methyl-2-nitrosopropane dimer	1 g	45.00	36.00
P-6549	**New**	phenoxazone	10 mg	98.00	78.40

21.4 Assaying Oxidative Activity in Live Cells and Tissue

Assaying oxidative activity in living cells with fluorogenic, chemiluminescent or chromogenic probes is complicated by the possibility of having multiple forms of reactive oxygen in the same cell. In addition, the nitric oxide radical (see Section 21.5) may produce the same changes in the optical properties of the probe as do other reactive oxygen molecules. Blocking agents and enzyme inhibitors can sometimes help to sort out the species responsible for the probe's optical response. Quantitative analysis is also difficult because of: 1) the high intracellular concentration of glutathione, which can form thiyl or sulfinyl radicals or otherwise trap or reduce oxygen species;[1] 2) the variable concentration of metals, which can either catalyze or inhibit radical reactions; and 3) the presence of other free radical–quenching agents such as spermine.[2]

Fluorescein, rhodamine and various other dyes can be chemically reduced to colorless, nonfluorescent leuco dyes. These "dihydro" derivatives are readily oxidized back to the parent dye by some reactive oxygen species and thus can serve as fluorogenic probes for detecting oxidative activity in cells and tissues;[3-5] however, their oxidation may not easily discriminate between the various reactive oxygen species. It has been reported that dihydroethidium, dichlorodihydrofluorescein (H_2DCF) and dihydrorhodamine 123 react with intracellular hydrogen peroxide[6,7] — a reaction mediated by peroxidase, cytochrome *c* or Fe^{2+}.[5-8] The leuco dyes also serve as fluorogenic substrates for peroxidase enzymes (see Section 10.5). All these reagents are slowly oxidized by air back to the parent fluorescent dyes, and in some cases light appears to accelerate their oxidation.

Dichlorodihydrofluorescein Diacetate and Derivatives

Dichlorodihydrofluorescein Diacetate

Dichlorodihydrofluorescein diacetate (H_2DCFDA, D-399), also known as dichlorofluores*cin* diacetate, is commonly used to detect the generation of reactive oxygen intermediates in neutrophils and macrophages[9-13] (see also Section 17.1). Recent investigations suggest that the cell-permeant H_2DCFDA may also be extremely useful for assessing the overall oxidative stress in toxicological phenomena.[4,5] In a recent review, Tsuchiya and colleagues outlined methods for visualizing the generation of oxidative species in whole animals. For example, they suggest using propidium iodide (P-1304, see Section 8.1) with H_2DCFDA to simultaneously monitor oxidant production and cell injury.[14] H_2DCFDA has been used to visualize oxidative changes in carbon tetrachloride–perfused rat liver[15] and in venular endothelium during neutrophil activation,[16] as well as to examine the effect of ischemia and reperfusion in lung and heart tissue.[17,18] Using H_2DCFDA, researchers characterized hypoxia-dependent peroxide production in *Saccharomyces cerevisiae* as a possible model for ischemic tissue destruction.[19] A variety of toxicological phenomena in cultured cells have also been investigated with H_2DCFDA, including:

- Amyloid β protein–mediated increases in hydrogen peroxide in PC12 cells[20]
- Effects of calcium antagonists on oxidative metabolism in dissociated rat cerebellar and cortical neurons[21]
- Methamphetamine-induced oxidative stress in dopaminergic neurons[22]
- Effect of lipopolysaccharides on the level of oxygen metabolites in rat liver Kupffer cells[23]
- Nephrotoxin-induced oxidative stress in isolated proximal tubular cells[24]
- Nickel-induced increases in oxidant levels in Chinese hamster ovary (CHO) cells[25]
- Effect of transforming growth factor–β1 on the overall oxidized state of mouse osteoblastic cells[26]

In neutrophils, H_2DCFDA has proven useful for flow cytometric analysis of nitric oxide, forming a product that has spectral properties identical to those produced when it reacts with hydrogen peroxide.[27] In this study, H_2DCFDA's reaction with nitric oxide was blocked by adding the arginine analog N^G-methyl-L-arginine (L-NMMA, M-7898; see Section 21.5) to the cell suspension.[27] It has been reported that H_2DCF is not directly oxidized by either superoxide or free hydroxyl radical in aqueous solution.[5]

Improved Versions of H_2DCFDA

Intracellular oxidation of H_2DCF tends to be accompanied by leakage of the product, 2',7'-dichlorofluorescein,[28] which may make quantitation or detection of slow oxidation difficult. To enhance retention of the fluorescent product, Molecular Probes offers the carboxylated H_2DCFDA analog (carboxy-H_2DCFDA, C-400), which has two negative charges at physiological pH, and its di(acetoxymethyl ester) (C-2938), which should more easily pass through membranes during cell loading. Upon oxidation and cleavage of the acetate and ester groups by intracellular esterases, both analogs form carboxydichlorofluorescein (C-368, see Section 23.3), with additional negative charges that should impede its leakage out of the cell. Carboxy-H_2DCFDA (C-400) has been used to assess the oxidative process in isolated perfused rat heart tissue[18] and in transfected *cos*-1 cells expressing native or mutagenized prostaglandin endoperoxide H synthase.[29] Its di(acetoxymethyl ester) (C-2938) has been employed to investigate the role of the *bcl-2* proto-oncogene product in preventing apoptosis through its antioxidant properties.[30]

Molecular Probes has also recently introduced 5-(and-6)-chloromethyl-2',7'-dichlorodihydrofluorescein diacetate (CM-H_2DCFDA, C-6827) — a chloromethyl derivative of H_2DCFDA that should exhibit much better retention in live cells. As with our other chloromethyl derivatives (see the description of our CellTracker™ probes in Section 15.2), we believe that CM-H_2DCFDA passively diffuses into cells, where its acetate groups are cleaved by intracellular esterases and its thiol-reactive chloromethyl group reacts with intracellular glutathione and other thiols. Subsequent oxidation yields a fluorescent adduct that is trapped inside the cell, thus facilitating long-term studies.

The diacetate derivatives of the dichlorodihydrofluoresceins are quite stable. When used in intracellular applications, the acetates are cleaved by endogenous esterases, releasing the corresponding dichlorodihydrofluorescein derivative. If, however, these nonfluorescent diacetate derivatives are used in *in vitro* assays, they must first be hydrolyzed with mild base to form the colorless probe. For further discussion of the utility of these reagents as probes for investigating phagocytic cells, see Section 17.1.

Dihydrorhodamine and Dihydrorosamine Derivatives

Dihydrorhodamine 123

Dihydrorhodamine 123 (D-632, see also Section 17.1) is the uncharged and nonfluorescent reduction product of the mitochondrion-selective dye rhodamine 123 (R-302, see Section 12.2). This leuco dye passively diffuses across most cell membranes and is oxidized to cationic rhodamine 123, which localizes in mitochondria. Dihydrorhodamine 123 has been reported to be more sensitive than H_2DCFDA for detecting granulocyte respiratory burst activity.[31,32] Dihydrorhodamine 123 has also been used to investigate the production of reactive oxygen species in activated rat mast cells[33] and cultured endothelial cells.[8] Like H_2DCF, dihydrorhodamine 123 does not directly detect superoxide,[7] but rather reacts with hydrogen peroxide in the presence of peroxidase,[7] cytochrome *c* or Fe^{2+}.[8]

It has recently been reported that dihydrorhodamine 123 can also be used to detect peroxynitrite,[34,35] the anion formed when nitric oxide reacts with superoxide or hydrogen peroxide.[34,36] Peroxynitrite, which may play a role in many pathological conditions,[35,37] has been shown to react with sulfhydryl groups,[38] DNA[39] and membrane phospholipids,[40] as well as with tyrosine[41] and other phenolic compounds.[42]

Longer-Wavelength Reduced Rhodamines and Rosamines

Intracellular oxidation of dihydrorhodamine 6G and dihydrotetramethylrosamine (D-633, D-638) yields rhodamine 6G and tetramethylrosamine (R-634, T-639), respectively, both of which localize in the mitochondria of living cells (see Section 12.2). Their cationic oxidation products have longer-wavelength spectra than those of dihydrorhodamine 123, making them especially useful for multicolor applications[43] and in autofluorescent cells and tissues. Dihydrotetramethylrosamine was used to demonstrate that reactive oxygen metabolites cross the membranes of target cells during neutrophil-mediated, antibody-dependent cellular toxicity.[43] This oxidative probe has also been used to stain phagocytic vacuoles and mitochondria in living amoeba.[44]

Reduced MitoTracker™ Probes

Two of our new MitoTracker probes — MitoTracker Orange CM-H_2TMRos and MitoTracker Red CM-H_2XRos (M-7511, M-7513) — are chemically reactive reduced rosamines. Unlike MitoTracker Orange CMTMRos and MitoTracker Red CMXRos (M-7510, M-7512; see Section 12.2), the reduced versions of these probes do not fluoresce until they enter an actively respiring cell, where they are oxidized by reactive oxygen species to the fluorescent mitochondrion-selective probe and then sequestered in the mitochondria. Both MitoTracker Orange CMTMRos and the reduced MitoTracker Orange CM-H_2TMRos have been used to investigate the metabolic state of *Pneumocystis carinii* mitochondria.[45]

Chemically Reactive OxyBURST® Probes

Molecular Probes' amine-reactive OxyBURST Green and OxyBURST Orange succinimidyl esters (2',7'-dichlorodihydrofluorescein diacetate, succinimidyl ester, D-2935; 4'-carboxydihydrotetramethylrosamine, succinimidyl ester, C-2936) can be used to prepare oxidation-sensitive conjugates of a wide variety of other biomolecules.[43,46] Following conjugation to amines, the acetates of the OxyBURST Green reagent can be removed by treatment with hydroxylamine to yield the H_2DCF conjugate. This step is not required for the OxyBURST Orange reagent, which yields a reduced rosamine (H_2CTMRos) conjugate. The H_2DCF and H_2CTMRos conjugates are nonfluorescent until they are oxidized by reactive oxygen species to the corresponding dichlorofluorescein and rosamine derivatives. Thus, they enable researchers to follow the phagocytosis and resultant oxidative burst of labeled antibodies, antigens, peptides, proteins, dextrans, bacteria, yeast and polystyrene microspheres.[43] See Section 17.1 for further discussion of these reactive OxyBURST reagents, as well as of our Fc OxyBURST immune complexes.

Dihyroethidium

Although dihydroethidium (D-1168, see also Sections 8.1 and 17.1) (also called Hydroethidine™, a trademark of Prescott Labs) is commonly used to analyze respiratory burst activity in phagocytes,[47-49] it has been reported that this probe undergoes significant oxidation in resting leukocytes, possibly through the uncoupling of mitochondrial oxidative phosphorylation.[6] Cytosolic dihydroethidium exhibits blue fluorescence; however, once this probe is oxidized to ethidium, it intercalates within the cell's DNA, staining its nucleus a bright fluorescent red.[50-52] The mechanism of dihydroethidium's interaction with lysosomes and DNA has been described.[53] Dihydroethidium has been used to:

- Detect multidrug-resistant cancer cells[54]
- Follow phagocytosis and oxidative bursts by phagocytic blood cells[47,48,55,56]
- Investigate spermatozoal viability[57,58]
- Quantitate killer cell–target cell conjugates by flow cytometry methods[59-61]

Lipid Probes for Detecting Hydroperoxides

cis-Parinaric Acid

Fluorescence quenching of the fatty acid analog *cis*-parinaric acid (P-1, P-1901) has been used in several lipid peroxidation assays, including quantitative determinations in living cells.[62-67] In a study investigating the membrane antioxidant properties of the *bcl-2* proto-oncogene product, researchers used *cis*-parinaric acid to detect lipid hydroperoxides together with 6-carboxy-2',7'-dichlorodihydrofluorescein diacetate, di(acetoxymethyl ester) (C-2938) to detect cytosolic reactive oxygen molecules.[30] Parinaric acid's extensive unsaturation makes it quite susceptible to oxidation if not rigorously protected from air.[68] Consequently, we offer *cis*-parinaric acid in a 100 mg unit size (P-1) and specially packaged in 10 ampules sealed under argon, each containing 10 mg (P-1901).

During experiments, we advise handling parinaric acid samples under inert gas and preparing solutions using degassed buffers and solvents. Parinaric acid is also somewhat photolabile and undergoes photodimerization when exposed to intense illumination, resulting in loss of fluorescence.[69]

Resorufin Derivatives

Oxidation of 10-acetyl-3,7-dihydroxyphenoxazine (A-6550) forms resorufin, which exhibits strong red fluorescence. This resorufin derivative has greater stability, yields less background and produces a product that is more readily detected than the similar chromogenic reduced methylene blue derivatives commonly used to determine lipid peroxides in plasma, sera, cell extracts and a variety of membrane systems.[70-72] Using 10-acetyl-3,7-dihydroxyphenoxazine and horseradish peroxidase, we have recently found that release of hydrogen peroxide to the medium by as few as 2000 phorbol ester–stimulated neutrophils can be detected in a fluorescence microplate reader. The new reagent 10-acetyl-3,7-dihydroxy-2-dodecylphenoxazine (A-6551) is based on the lipophilic resorufin derivative 2-dodecylresorufin (D-7786; see Section 14.3), which also forms the basis of our ImaGene Red™ glycosidase substrates (see Section 10.2). Therefore, we anticipate that oxidation of this new peroxidase substrate by reactive oxygen species, including hydroperoxides, will form a fluorescent product that is well retained in living cells by virtue of its lipophilic tail.

Chemiluminescent Probes for Detecting Superoxide and Other Radicals

Lucigenin

The chemiluminescent probe lucigenin (L-6868, see Section 21.3) has been shown to be useful for detecting the superoxide anion.[73,74] Lucigenin has been employed to investigate superoxide generation in spermatozoa,[75] chondrocytes contained within the matrix of living cartilage tissue [76] and mitochondria of alveolar macrophages [77] and L929 cells.[78] It has also been used to examine the role of the superoxide anion in reoxygenation injury in isolated rat hepatocytes.[79,80] We have purified our lucigenin to remove a bright blue fluorescent impurity that is found in some commercial samples.

Luminol

Although luminol (L-8455, see Section 21.3) is not useful for detecting superoxide in living cells,[73] it is commonly employed to detect peroxidase- or metal ion–mediated oxidative events.[81-83] Used alone, luminol can detect oxidative events in cells rich in peroxidases, including granulocytes [84-87] and spermatozoa.[75] This probe has also been used in conjunction with horseradish peroxidase (HRP) to investigate reoxygenation injury in rat hepatocytes.[79,88] In these experiments, it is thought that the primary species being detected is hydrogen peroxide. In addition, luminol has been employed to detect peroxynitrite,[89-91] a molecule thought to be generated in a variety of pathological conditions.[37] It has been reported that luminol's chemiluminescence response to oxidative species may be competitively inhibited by biomolecules containing sulfhydryl and thioether groups.[92-94]

Coelenterazine

In the absence of apoaequorin, the luminophore coelenterazine (C-2944, see Section 21.3) produces chemiluminescence in response to superoxide generation in phorbol ester– or chemotactic peptide–stimulated neutrophils.[95] Unlike luminol, coelenterazine exhibits luminescence that is not dependent on the activity of cell-derived myeloperoxidase or inhibited by azide.

Superior Antifade Reagents for a Multitude of Applications

The main purpose of any antifade reagent is to sustain dye fluorescence, which is usually accomplished by inhibiting the generation and diffusion of reactive oxygen species. Molecular Probes offers a variety of antifade reagents to meet the diverse needs of the research community. The antifade formulation in our original *SlowFade*™ Antifade Kit (S-2828) reduces the fading rate of fluorescein to almost zero, making this antifade reagent especially useful for quantitative measurements. However, this *SlowFade* formulation does initially quench fluorescein's fluorescence. Our new *SlowFade Light* Antifade Kit (S-7461) contains an antifade formulation that slows fluorescein's photobleaching by about fivefold with little or no quenching of the fluorescent signal, thereby dramatically increasing the signal-to-noise ratio in photomicroscopy. The latest addition to our antifade arsenal — ProLong™ Antifade Kit (P-7481) — outperforms all other commercially available reagents, with little or no quenching of the fluorescent signal. In addition to inhibiting the fading of fluorescein, tetramethylrhodamine and Texas Red® dyes, the ProLong antifade reagent retards the fading of DNA-bound nucleic dyes such as DAPI, propidium iodide and YOYO®-1. Its compatibility with a multitude of dyes and dye complexes makes the Prolong Antifade Kit an especially valuable tool for multiparameter analyses such as fluorescence *in situ* hybridization (FISH). For more information on these antifade reagents, see Section 26.1.

Cat #	Product Name	Unit Size	Price Per Unit ($) 1–4 Units	5–24 Units
P-7481	***New*** ProLong™ Antifade Kit	1 kit	98.00	78.40
S-2828	*SlowFade*™ Antifade Kit	1 kit	55.00	44.00
S-7461	***New*** *SlowFade*™ *Light* Antifade Kit	1 kit	55.00	44.00

Table 21.1 Tetrazolium salts for detecting redox potential in living cells and tissues.

Cat #	Tetrazolium Salt	Color of Formazan	Water Solubility of Formazan	Applications
I-6496 (INT)	4-Iodonitrotetrazolium violet	orange	no	• Cytochrome *b*–ubiquinone complex activity [1] • Succinic dehydrogenase histochemistry [2] • Heavy metal toxicity in yeast [1]
M-6494 (MTT)	3-(4,5-Dimethylthiazol-2-yl)-2,5-diphenyltetrazolium bromide	purple	no	• Superoxide generation by fumarate reductase [3] and nitric oxide synthase [4] • Mitochondrial dehydrogenase activity [5] • Cell viability and proliferation [6-10] • Neuronal cell death [11] • Platelet activation [12] • Tumor cell adhesion [13] and invasion [14] • Multidrug resistance [15] • *In vitro* toxicity testing [16-18]
N-6495 (NBT)	Nitro blue tetrazolium chloride	deep blue	no	• Superoxide generation by xanthine oxidase [19] • Superoxide release from single pulmonary alveolar macrophages [20] • Neutrophil oxidative metabolism [21,22] • Ascorbate peroxidase activity [23] • NADPH diaphorase activity [24-26] • Succinic dehydrogenase histochemistry [2]
N-6498 (NTV)	4-Nitrotetrazolium violet chloride	purple	no	• Catalytic properties of xanthine dehydrogenases [27] and oxidoreductases in organelles [28]
T-6490	Tetrazolium blue chloride	blue	no	• Detection of impaired electron transport in *Chlamydomonas* [29] • Succinic dehydrogenase histochemistry [2]
X-6493 (XTT)	2,3-Bis-(2-methoxy-4-nitro-5-sulfophenyl)-2*H*-tetrazolium-5-carboxanilide	orange	yes	• Antifungal susceptibility [30] • Drug sensitivity of cells [31] • Parasitic nematode viability [32] • Tumor cell cytotoxicity [33]

1. Bitton, G. and Koopman, B. in *Toxicity Testing Using Microorganisms*, G. Bitton and B.J. Dutka, Eds., CRC Press (1986) pp. 27–55; **2.** Histochemistry 76, 381 (1982); **3.** J Biol Chem 270, 19767 (1995); **4.** J Biol Chem 269, 12589 (1994); **5.** Cytometry 13, 532 (1992); **6.** J Immunol Meth 168, 253 (1994); **7.** Anal Biochem 214, 190 (1993); **8.** J Immunol Meth 164, 149 (1993); **9.** Anal Biochem 205, 8 (1992); **10.** J Immunol Meth 89, 271 (1986); **11.** J Cell Biol 128, 201 (1995); **12.** J Immunol Meth 159, 253 (1993); **13.** J Immunol Meth 164, 255 1993); **14.** Cancer Res 54, 3620 (1994); **15.** Leukemia Res 16, 1165 (1992); **16.** Biosci Biotech Biochem 56, 1472 (1992); **17.** J Immunol Meth 144, 141 (1991); **18.** J Immunol Meth 131, 165 (1990); **19.** J Reprod Fert 97, 441 (1993); **20.** Am J Physiol 252, C677 (1987); **21.** Clin Chim Acta 221, 197 (1993); **22.** J Leukocyte Biol 53, 404 (1993); **23.** Anal Biochem 212, 540 (1993); **24.** Neurosci Lett 155, 61 (1993); **25.** Proc Natl Acad Sci USA 88, 7797 (1991); **26.** Proc Natl Acad Sci USA 88, 2811 (1991); **27.** ICRS Med Sci 14, 887 (1986); **28.** Anal Biochem 131, 404 (1983); **29.** Biosci Rep 3, 367 (1983); **30.** Antimicro Agents Chemother 36, 1619 (1992); **31.** Cancer Res 48, 4827 (1988); **32.** Parasitology 107, 175 (1993); **33.** J Immunol Meth 147, 153 (1992).

Tetrazolium Salts — Chromogenic Redox Indicators

Tetrazolium salts are widely used for detecting redox potential of cells for viability, proliferation and cytotoxicity assays. Upon reduction, these water-soluble colorless compounds form uncharged, brightly colored formazans. Several of the formazans precipitate out of solution and are useful for histochemical localization of the site of reduction or, after solubilization in organic solvent, for quantitation by standard spectrophotometric techniques. The extremely water-soluble formazan product of XTT does not require solubilization prior to quantitation. Many of these salts are reduced by specific components of the electron transport chain and may be useful for determining the site of action of specific toxins.[96] Selected applications of the tetrazolium salts are listed in Table 21.1. Also see Section 16.3 for applications of tetrazolium salts in cells. The tetrazolium salts are available individually or as part of our Tetrazolium Salt Sampler Kit (T-6500), which contains 100 mg of each of the six tetrazolium salts in Table 21.1.

1. Arch Biochem Biophys 314, 282 (1994); **2.** Proc Natl Acad Sci USA 89, 11426 (1992); **3.** Arch Toxicol 68, 582 (1994); **4.** Brain Res 635, 113 (1994); **5.** Chem Res Toxicol 5, 227 (1992); **6.** J Leukocyte Biol 47, 440 (1990); **7.** Eur J Biochem 217, 973 (1993); **8.** Arch Biochem Biophys 302, 348 (1993); **9.** Exp Cell Res 209, 375 (1993); **10.** J Immunol Meth 159, 173 (1993); **11.** J Immunol Meth 159, 131 (1993); **12.** Cytometry 13, 615 (1992); **13.** Cytometry 13, 525 (1992); **14.** Meth Enzymol 233, 128 (1994); **15.** Lab Invest 64, 167 (1991); **16.** Am J Physiol 264, H881 (1993); **17.** Lab Invest 70, 579 (1994); **18.** Free Rad Res Comm 16, 217 (1992); **19.** Cytometry 14, 287 (1993); **20.** Cell 77, 817 (1994); **21.** Brain Res 610, 172 (1993); **22.** J Neurosci 14, 2260 (1994); **23.** Eur J Cell Biol 65, 200 (1994); **24.** J Biol Chem 269, 14546 (1994); **25.** Toxicol Appl Pharmacol 120, 29 (1993); **26.** J Cell Biol 126, 1079 (1994); **27.** J Leukocyte Biol 51, 496 (1992); **28.** Free Rad Biol Med 16, 509 (1994); **29.** Biochemistry 34, 7194 (1995); **30.** Cell 75, 241 (1993); **31.** J Immunol Meth 178, 89 (1995); **32.** Naturwissenschaften 75, 354 (1988); **33.** APMIS 102, 474 (1994); **34.** Biochemistry 34, 3544 (1995); **35.** Free Rad Biol Med 16, 149 (1994); **36.** Free Rad Res Comm 10, 221 (1990); **37.** Proc Natl Acad Sci USA 87, 1620 (1990); **38.** J Biol Chem 266, 4244 (1991); **39.** J Am Chem Soc 114, 5430 (1992); **40.** Arch Biochem Biophys 288, 481 (1991); **41.** Arch Biochem Biophys 298, 431 (1992); **42.** Arch Biochem Biophys 298, 438 (1992); **43.** J Cell Physiol 156, 428 (1993); **44.** Biochem Biophys Res Comm 175, 387 (1991); **45.** J Euk Microbiol 41, S79 (1994); **46.** J Immunol Meth 130, 223 (1990); **47.** FEMS Microbiol Lett 122, 187 (1994); **48.** J Immunol Meth 170, 117 (1994); **49.** Cytometry 12, 687 (1991); **50.** J Leukocyte Biol 55, 253 (1994); **51.** J Histochem Cytochem 34, 1109 (1986); **52.** BioTechniques 3, 270 (1985); **53.** Histochemistry 94, 205 (1990); **54.** Exp Cell Res 190, 69 (1990); **55.** Cytometry 17, 294 (1994); **56.** J Immunology 151, 1463 (1993); **57.** J Histochem Cytochem 39, 485 (1991); **58.** Gamete Res 22, 355 (1989); **59.** Cytometry 12, 666 (1991); **60.** J Immunol Meth 139, 281 (1991); **61.** J Immunol Meth 130, 251 (1990); **62.** Proc Natl Acad Sci USA 91, 1183 (1994); **63.** Arch Biochem Biophys 297, 147 (1992); **64.** Cytometry 13, 686 (1992); **65.** Anal Biochem 196, 443 (1991); **66.** Chem Phys Lipids 53, 309 (1990); **67.** Biochim Biophys Acta 944, 29 (1988); **68.** Biochemistry 16, 820 (1977); **69.** Proc Natl Acad Sci USA 77, 26 (1980); **70.** Proc Soc Exp Biol Med 206, 53 (1994); **71.** Free Rad Biol Med 12, 389 (1992); **72.** Biochem Int'l 10, 205 (1985); **73.** Free Rad Biol Med 15, 447 (1993); **74.** J Immunol Meth 97, 209 (1987); **75.** J Cell Physiol 151, 466 (1992); **76.** Free Rad Biol Med 15, 143 (1993); **77.** Free Rad Biol Med 17, 117 (1994); **78.** Biochem J 289, 587 (1993); **79.** Am J Physiol 266, G799 (1994); **80.** Life Sci 55, 1427 (1994); **81.** Biochem Mol Biol Int'l 33, 1179 (1994); **82.** Free Rad Biol Med 6, 623 (1989); **83.** J Immunology 129, 1589 (1982); **84.** J

Appl Physiol 76, 539 (1994); **85.** J Leukocyte Biol 54, 300 (1993); **86.** J Biochem 106, 355 (1989); **87.** Biochem Biophys Res Comm 155, 106 (1988); **88.** Am J Physiol 262, G1015 (1992); **89.** Arch Biochem Biophys 310, 352 (1994); **90.** Anal Chem 65, 1794 (1993); **91.** Biochem J 290, 51 (1993); **92.** Biochim Biophys Acta 1097, 145 (1991); **93.** Biomed Biochim Acta 50, 967 (1991); **94.** Biomed Biochim Acta 49, 991 (1990); **95.** Anal Biochem 206, 273 (1992); **96.** Bitton, G. and Koopman, B. in *Toxicity Testing Using Microorganisms*, G. Bitton and B.J. Dutka, Eds., CRC Press (1986) pp. 27–55.

21.4 Data Table *Assaying Oxidative Activity in Live Cells and Tissue*

Cat #	Structure	MW	Storage	Soluble	Abs	{ε × 10⁻³}	Em	Solvent	Notes
A-6550	✓	257	FF,D,AA	DMSO	280	{6.0}	none	pH 8	<1>
A-6551	✓	426	FF,D,AA	DMSO	286	{8.0}	none	MeOH	<2>
C-400	✓	531	F,D,AA	pH>6, DMSO	290	{5.6}	none	MeCN	<3>
C-2936	✓	486	F,D,AA	DMF	234	{60}	none	MeCN	<4, 5>
C-2938	✓	675	F,D,AA	DMSO	291	{5.7}	none	MeOH	<3>
C-6827	✓	536	F,D,AA	DMSO	287	{9.1}	none	MeOH	<3>
D-399	✓	487	F,D,AA	pH>6, DMSO	258	{11}	none	MeOH	<3>
D-632	✓	346	F,D,AA	DMF, DMSO	289	{7.6}	none	MeOH	<5, 6>
D-633	✓	445	F,D,AA	DMF, DMSO	296	{11}	none	MeOH	<5, 6>
D-638	✓	344	F,D,AA	THF, DMSO	235	{58}	none	MeOH	<5, 6, 7>
D-1168	S8.1	315	FF,L,AA	DMF, DMSO	355	{14}	<8>	MeCN	<5>
D-2935	✓	584	F,D,AA	DMF	258	{11}	none	MeOH	<3>
I-6496	✓	506	D	H_2O, DMSO	249	{38}	none	MeOH	<9>
M-6494	✓	414	D,L	H_2O, DMSO	375	{8.3}	none	MeOH	<9, 10>
M-7511	S12.2	393	F,D,AA	DMSO	235	{57}	none	MeOH	<5, 6>
M-7513	S12.2	497	F,D,AA	DMSO	245	{45}	none	MeOH	<5, 6>
N-6495	✓	818	D,L	H_2O, DMSO	256	{64}	none	MeOH	<9>
N-6498	✓	430	D	H_2O, DMSO	246	{33}	none	MeOH	<9>
P-1	S13.3	276	FF,LL,AA	EtOH	303	{76}	416	MeOH	<11>
T-6490	✓	728	D	H_2O, DMSO	253	{53}	none	MeOH	<9>
X-6493	✓	675	D	H_2O, DMSO	286	{15}	none	MeOH	<12>

P-1901 see P-1.

For definitions of the contents of this data table, see "How to Use the Handbook" on page vi.
Structure: Chemical structure drawing: ✓ = shown in this section; Sn.n = shown in Section number n.n.

<1> Reaction of A-6550 with H_2O_2 produces fluorescent resorufin (R-363, see Section 10.1), Abs = 571 nm (ε = {54}), Em = 585 nm in pH 9 buffer.
<2> Reaction of A-6551 with H_2O_2 produces fluorescent 2-dodecylresorufin (D-7786, see Section 14.3), Abs = 578 nm (ε = {69}), Em = 597 nm in MeOH.
<3> Dihydrofluorescein diacetates are colorless and nonfluorescent until both the acetates are hydrolyzed and the products are subsequently oxidized to fluorescein derivatives. The materials contain <0.1% of oxidized derivative when initially prepared. C-400, C-2938, C-6827, D-399 and D-2935 yield 2′,7′-dichlorofluorescein derivatives, which have Abs = 504 nm (ε = {90}), Em = 529 nm.
<4> C-2936 contains <0.1% of the oxidized dye and becomes fluorescent only upon oxidation, yielding a tetramethylrosamine derivative with Abs = 550 nm (ε = {87}), Em = 574 nm.
<5> This compound is susceptible to oxidation, especially in solution. Store solutions under argon or nitrogen. Oxidation appears to be catalyzed by illumination.
<6> These compounds are essentially colorless and nonfluorescent until oxidized. Oxidation products (in parentheses) are as follows: D-632 (R-302); D-633 (R-634); D-638 (T-639); M-7511 (M-7510); M-7513 (M-7512). Spectral properties of the oxidation products are listed in Section 12.2.
<7> Prepare a 4–5 mM stock solution of D-638 by initially dissolving in THF and diluting 1:5 with DMSO [Biochem Biophys Res Comm 175, 387 (1991)]. Long-term storage of THF solutions should be avoided due to tendency for peroxide formation.
<8> D-1168 has blue fluorescence (Em ~420 nm) until oxidized to ethidium (E-1305), which binds to DNA with Abs = 518 nm (ε = {5.2}), Em = 605 nm. The reduced dye does not bind to nucleic acids [FEBS Lett 26, 169 (1972)].
<9> Enzymatic reduction products are water-insoluble formazans, with Abs = 600 nm (T-6490), 505 nm (M-6494), 605 nm (N-6495), 465 nm (I-6496) and 510 nm (N-6498) after solubilization in DMSO or DMF. See Histochemistry 76, 381 (1982) and Prog Histochem Cytochem 9, 1 (1977) for further information.
<10> M-6494 also has Abs = 242 nm (ε = {21}) in MeOH.
<11> P-1 is readily oxidized to nonfluorescent products. Use under N_2 or Ar except when oxidation is intended. Stock solutions should be prepared in deoxygenated solvents. Parinaric acid is appreciably fluorescent in lipid environments and organic solvents but is nonfluorescent in water.
<12> Enzymatic reduction product is a water-soluble formazan, Abs = 475 nm.

Product-Specific Bibliographies and Keyword Searches

Our Technical Assistance Department can provide you with product-specific bibliographies, as well as keyword searches of the over 25,000 literature references in our extensive bibliography database. Our bibliography database is also searchable through our Web site (http://www.probes.com).

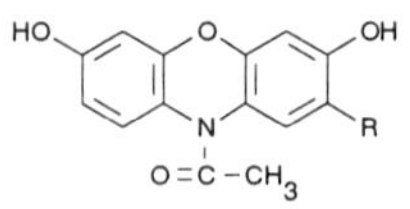

A-6550 R = H
A-6551 R = $-(CH_2)_{11}CH_3$

C-400, C-6827
see D-399

C-2936
see D-638

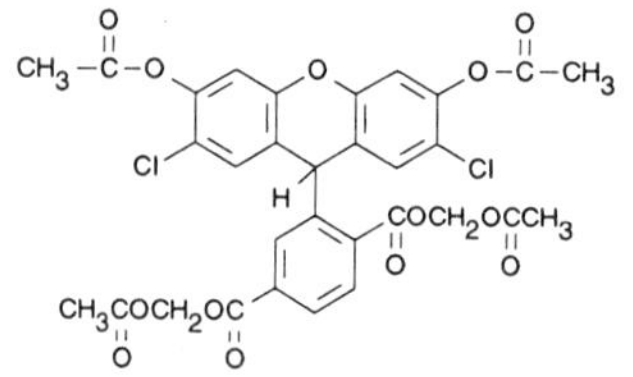

C-2938

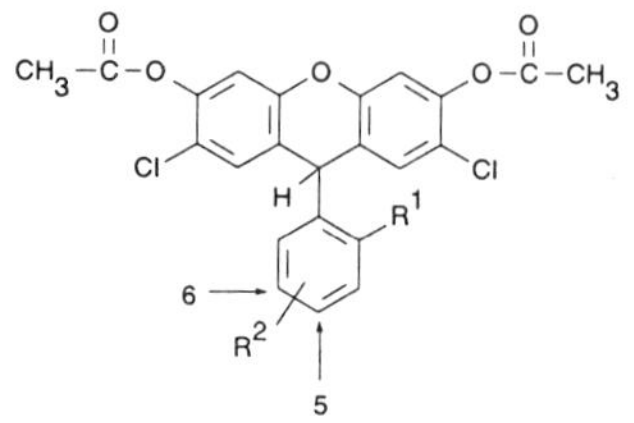

	R^1	R^2
D-399	$R^1 = -COOH$	$R^2 = H$
C-400	$R^1 = -COOH$	$R^2 = -COOH$
D-2935	R^1 = SE	$R^2 = H$
C-6827	$R^1 = -COOH$	$R^2 = -CH_2Cl$

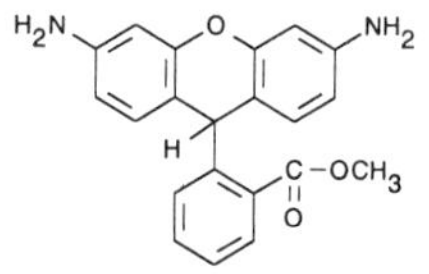

D-632

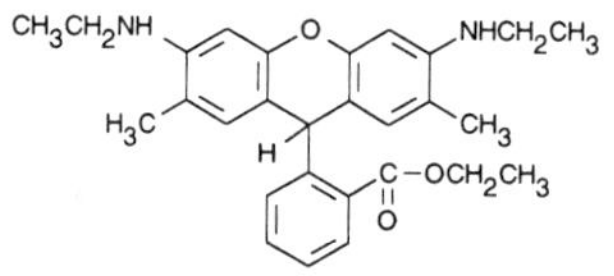

D-633

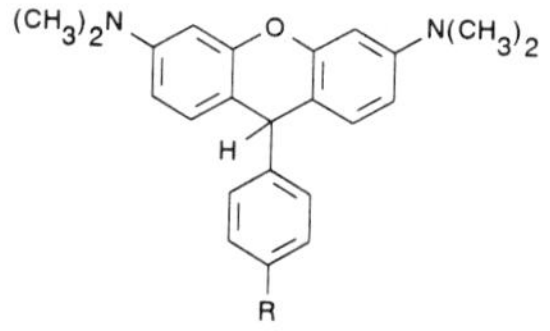

D-638 R = H
C-2936 R = SE

D-1168
Section 8.1

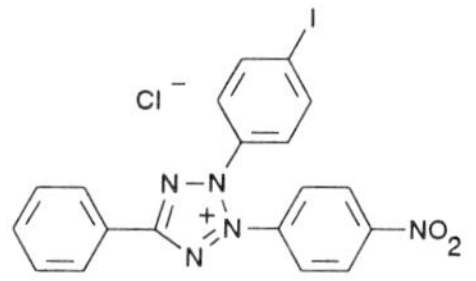

I-6496

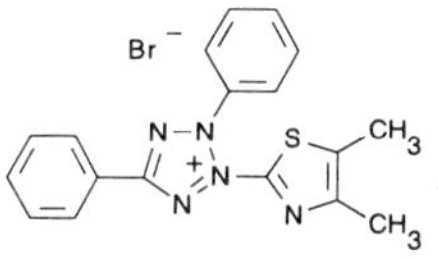

M-6494

M-7511, M-7513
Section 12.2

N-6495
see T-6490

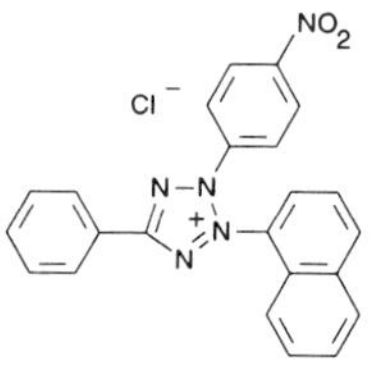

N-6498

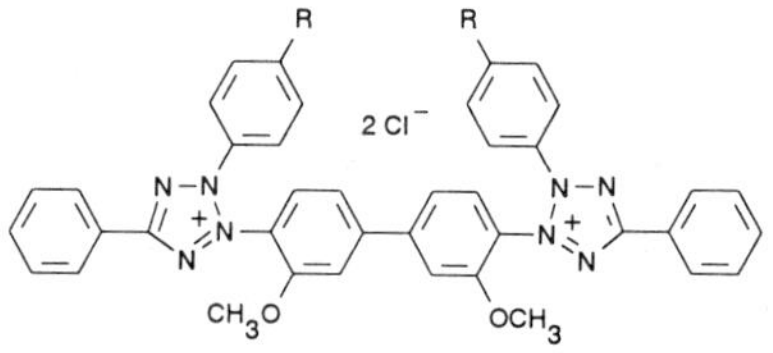

P-1
Section 13.3

T-6490 R = H
N-6495 R = $-NO_2$

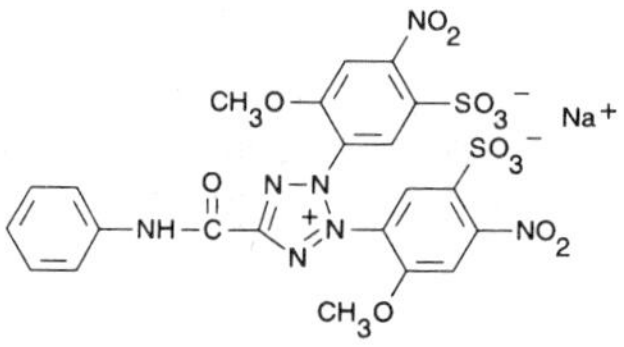

X-6493

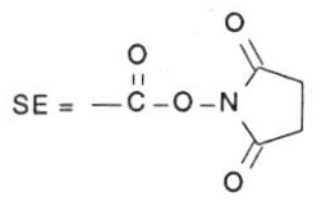

SE = $-\overset{O}{\overset{\|}{C}}-O-N$ (succinimide)

Substituent Key

Custom Synthesis Services

Molecular Probes' custom synthesis services can provide an almost unlimited variety of organic compounds and bioconjugates, including labeled antibodies, immunogens and analogs of biologically active small molecules. We can prepare known compounds using published procedures, or if these are not available, we can apply our expertise to design and synthesize reagents. We also offer quantity discounts for multiple-unit purchases of most of our catalog products. Please contact our Custom and Bulk Sales Department to discuss your requirements.

Cat #		Product Name	Unit Size	Price Per Unit ($) 1–4 Units	5–24 Units
A-6551	**New**	10-acetyl-3,7-dihydroxy-2-dodecylphenoxazine *special packaging*	10x100 µg	135.00	108.00
A-6550	**New**	10-acetyl-3,7-dihydroxyphenoxazine *special packaging*	10x1 mg	85.00	68.00
C-400		5-(and-6)-carboxy-2′,7′-dichlorodihydrofluorescein diacetate (carboxy-H_2DCFDA) *mixed isomers*	25 mg	88.00	70.40
C-2938		6-carboxy-2′,7′-dichlorodihydrofluorescein diacetate, di(acetoxymethyl ester) *single isomer*	5 mg	135.00	108.00
C-2936		4-carboxydihydrotetramethylrosamine, succinimidyl ester (OxyBURST® Orange, SE)	5 mg	55.00	44.00
C-6827	**New**	5-(and-6)-chloromethyl-2′,7′-dichlorodihydrofluorescein diacetate (CM-H_2DCFDA) *mixed isomers* *special packaging*	20x50 µg	125.00	100.00
D-399		2′,7′-dichlorodihydrofluorescein diacetate (2′,7′-dichlorofluorescin diacetate; H_2DCFDA)	100 mg	45.00	36.00
D-2935		2′,7′-dichlorodihydrofluorescein diacetate, succinimidyl ester (OxyBURST® Green, SE)	5 mg	55.00	44.00
D-1168		dihydroethidium (also called Hydroethidine - a trademark of Prescott Labs)	25 mg	98.00	78.40
D-632		dihydrorhodamine 123	10 mg	98.00	78.40
D-633		dihydrorhodamine 6G	25 mg	98.00	78.40
D-638		dihydrotetramethylrosamine	10 mg	65.00	52.00
I-6496	**New**	4-iodonitrotetrazolium violet (INT; 2-(4-iodophenyl)-3-(4-nitrophenyl)-5-phenyltetrazolium chloride)	1 g	40.00	32.00
M-7511	**New**	MitoTracker™ Orange CM-H_2TMRos *special packaging*	20x50 µg	125.00	100.00
M-7513	**New**	MitoTracker™ Red CM-H_2XRos *special packaging*	20x50 µg	125.00	100.00
M-6494	**New**	MTT (3-(4,5-dimethylthiazol-2-yl)-2,5-diphenyltetrazolium bromide)	1 g	40.00	32.00
N-6495	**New**	nitro blue tetrazolium chloride (NBT)	1 g	65.00	52.00
N-6498	**New**	4-nitrotetrazolium violet chloride (NTV)	1 g	32.00	25.60
P-1		*cis*-parinaric acid	100 mg	155.00	124.00
P-1901		*cis*-parinaric acid *special packaging*	10x10 mg	185.00	148.00
T-6490	**New**	tetrazolium blue chloride	5 g	35.00	28.00
T-6500	**New**	Tetrazolium Salt Sampler Kit *100 mg of T-6490, X-6493, M-6494, N-6495, I-6496, N-6498*	1 kit	125.00	100.00
X-6493	**New**	XTT (2,3-bis-(2-methoxy-4-nitro-5-sulfophenyl)-2H-tetrazolium-5-carboxanilide)	100 mg	55.00	44.00

21.5 Probes for Nitric Oxide Research

Nitric oxide (NO) plays a critical role as a molecular mediator of a variety of physiological processes, including blood-pressure regulation and neurotransmission.[1-7] However, because free NO is a transient species with a half-life of about five seconds, investigations of this gaseous molecule have relied largely on studies of nitric oxide synthases (NOS). NO is a highly reactive species that in the presence of oxygen is also reported to release superoxide anion. Thus, preparation of solutions containing NO requires special precautions to achieve reproducibility.[8]

Photoactivatable Nitric Oxide Donors

The potential spatial and temporal control of nitric oxide release made possible by photolysis of NO precursors makes this an attractive approach for generating NO in experimental systems. Molecular Probes' selection of photoactivatable NO donors now includes:

- Caged inorganic NO complex [9] (potassium nitrosylpentachlororuthenate, P-7070)
- *N-tert*-Butyl-α-phenylnitrone [10,11] (B-7893, see Section 21.3)
- 2-Methyl-2-nitrosopropane dimer (M-7905, see Section 21.3) [12]
- NVOC-caged SIN-1 (D-7111)
- *S*-Nitrosoglutathione [13] (N-7903)

Perfusion of hippocampal slices with the caged inorganic NO complex (P-7070) followed by photolysis is reported to release NO at concentrations sufficient to modulate NMDA receptor–mediated transmission.[9] Illumination of the spin trap *N-tert*-butyl-α-phenylnitrone with UV or even ambient light releases NO in aqueous solutions.[11] Our new NVOC-caged SIN-1 is uncaged *in vitro* in microseconds after UV illumination to release the NO carrier SIN-1, which then decomposes to release NO; however, its use in living cells and tissues has not yet been reported.

Spontaneous Nitric Oxide Donors

As an alternative to our photoactivatable NO probes, Molecular Probes offers several NO carriers that release NO spontaneously under physiological conditions (Figure 21.3), including:

- DEANO [14] (diethylamine nitric oxide, D-7915) (Figure 21.3A)
- Spermine NONOate [14,15] (S-7916) (Figure 21.3A)
- *S*-Nitrosoglutathione [13,16,17] (N-7903) (Figure 21.3B)
- SNAP [18-20] (*S*-nitroso-*N*-acetylpenicillamine; N-7892, N-7927) (Figure 21.3B)
- SIN-1 [21,22] (3-morpholinosydnonimine, hydrochloride; M-7891, M-7926) (Figure 21.3C)

DEANO and Spermine NONOate

The DEANO and spermine NONOate solids provide a means of preparing aqueous NO solutions. When these solids are dissolved in buffer, cell culture medium or blood, they dissociate to form two molecules of NO and one molecule of the corresponding amine [14] (Figure 21.3A). The delivery of NO can easily be controlled by

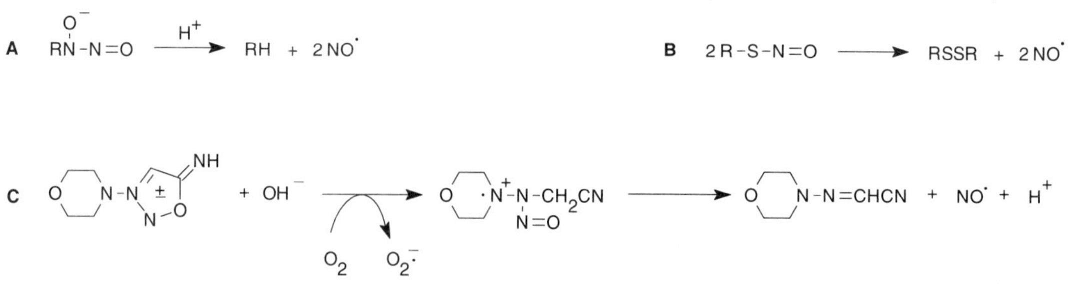

A $RN(O^-)\text{-}N{=}O \xrightarrow{H^+} RH + 2NO^{\cdot}$

B $2R\text{-}S\text{-}N{=}O \longrightarrow RSSR + 2NO^{\cdot}$

C SIN-1 + OH^- ($O_2 \rightarrow O_2^{-\cdot}$) ⟶ $O\ \cdot N^+\text{-}N(N{=}O)\text{-}CH_2CN$ ⟶ $O\ N\text{-}N{=}CHCN + NO^{\cdot} + H^+$

Figure 21.3 Mechanisms of spontaneous NO release by: A) DEANO (D-7915) or spermine NONOate (S-7916); B) S-nitrosoglutathione (N-7903) or SNAP (N-7892, N-7927); and C) SIN-1 (M-7891, M-7926).

preparing basic solutions of these NONOates and then lowering the pH to initiate NO generation.

DEANO has a half-life of two minutes in pH 7.4 phosphate buffer at 37°C, releasing two molecules of NO and one molecule of diethylamine.[14,23,24] DEANO therefore allows more sustained delivery of NO than is possible using NO-saturated aqueous solutions (Figure 21.4). The data in Figure 21.4 are from a study of cytochrome P-450 inhibition by NO in which DEANO was used in conjunction with fluorogenic microsomal dealkylase substrates (R-441 and R-352, see Section 10.5).[25] DEANO has also been used as a NO donor to stimulate ADP-ribosylation of various proteins.[26]

With a half-life of 39 minutes at 37°C in pH 7.4 buffer, spermine NONOate releases NO much more slowly than DEANO, making it suitable for whole animal infusions and experiments with long incubations.[14] Also, spermine, the by-product of spermine NONOate dissociation, is less likely to be deleterious in living systems and may also demonstrate biological activity of its own.[27] Both DEANO and spermine NONOate are reported to inhibit *in vitro* proliferation of A375 melanoma cells.[28]

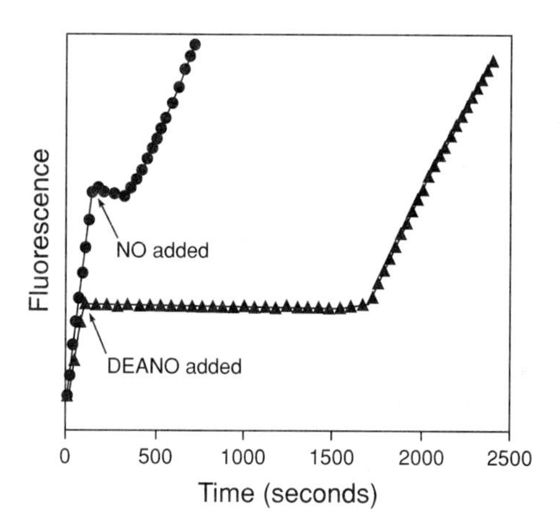

Figure 21.4 Inhibition of phenobarbital-induced rat hepatic microsomal dealkylase activity by 100 µM NO (●, added from a saturated aqueous stock solution) or 50 µM DEANO (▲, D-7915). Enzymatic activity was monitored using the fluorogenic substrate benzyloxyresorufin (R-441, see Section 10.5). Reprinted with permission from Arch Biochem Biophys 300, 115 (1993).

SNAP and SIN-1

Molecular Probes also offers the NO donors SNAP (N-7892, N-7927) and SIN-1 (M-7891, M-7926), which spontaneously release NO (and superoxide in the case of SIN-1) under physiological conditions (Figures 21.3B and 21.3C), thereby stimulating cyclic GMP production.[21,29-32] SNAP and SIN-1 have been shown to be potent vasodilators *in vivo* and *in vitro* and to inhibit smooth muscle cell mitogenesis and proliferation.[19,20,22,33] The relationship between NO generated from SNAP and SIN-1 and intracellular Ca^{2+} has been studied using fluorescent Ca^{2+} indicators [34-36] (see Chapter 22). It has also been reported that NO released from SNAP stimulates Ca^{2+}-independent synaptic vesicle release,[36] which can be detected with FM 1-43 (T-3163, see Section 17.2). Because of potential instability of SNAP and SIN-1 in solution, we also offer each product in a specially packaged set of 20 vials, each containing 1 mg (N-7927, M-7926).

Carboxy PTIO: A Nitric Oxide Antagonist

Molecular Probes now offers carboxy-PTIO (C-7912), a water-soluble and stable free radical molecule that reacts stoichiometrically with NO.[16,37,38] Carboxy-PTIO can be used *in vivo* to inhibit the physiological effects mediated by NO,[16,38,39] or *in vitro* to quantitate NO levels by ESR spectrometry.[40,41]

Nitric Oxide Synthase Probes

NOS Inhibitors and NOS Inhibitor Kit

Nitric oxide synthase catalyzes the NADPH- and O_2-dependent oxidation of L-arginine to NO and L-citrulline.[42-44] To help unravel the biological effects of NO, we offer a selection of NOS inhibitors. These inhibitors differ in their effectiveness and selectivity for the inhibition of different NOS isoforms. Our inhibitors include:

- *S*-Methyl-L-thiocitrulline chloride (M-7914), a citrulline analog that is reported to be more potent than N^G-methyl-L-arginine (L-NMMA) for inhibiting the constitutive brain and inducible smooth muscle NOS isoforms [45]
- L-NIO (N^5-(1-iminoethyl)-L-ornithine, I-7899), a potent, irreversible inhibitor of neutrophil NOS [46-48]
- NVOC-caged L-NIO (D-7917), a photoactivatable NOS inhibitor that has significant potential for *in vivo* experiments, though it has not yet been tested in living cells
- L-NIL (N^6-(1-iminoethyl)-L-lysine, I-7922), a more potent inhibitor of mouse inducible NOS than of rat brain constitutive NOS [49]
- L-NMMA (N^G-methyl-L-arginine, M-7898), which competitively inhibits NOS in a wide variety of cells,[50-53] and N^G-methyl-D-arginine (D-NMMA, M-7906), the enantiomer of

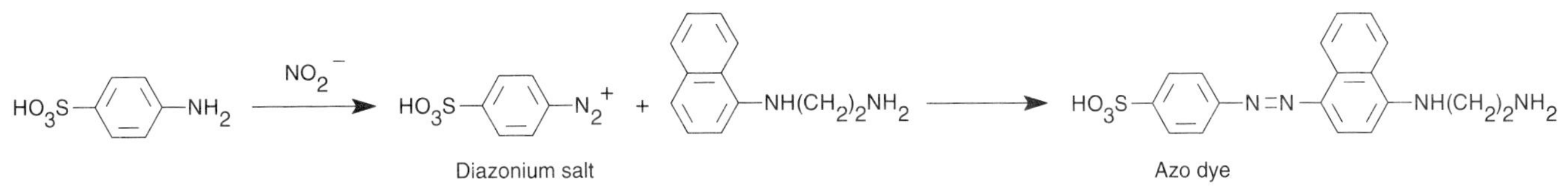

$HO_3S-C_6H_4-NH_2 \xrightarrow{NO_2^-} HO_3S-C_6H_4-N_2^+ + C_{10}H_7-NH(CH_2)_2NH_2 \longrightarrow HO_3S-C_6H_4-N{=}N-C_{10}H_6-NH(CH_2)_2NH_2$

Diazonium salt

Azo dye

Figure 21.5 *Principle of nitrite quantitation using the Griess Reagent Kit (G-7921). Formation of the azo dye is detected via its absorbance at 548 nm.*

L-NMMA, which does not inhibit NOS and can thus serve as a control for L-NMMA's nonspecific effects [48]

- L-NAME [48] (N^G-nitro-L-arginine methyl ester, N-7907)
- 7-NI (7-nitroindazole, N-7902) and its water-soluble sodium salt [54] (N-7928), which are inhibitors of brain NOS; 7-NI does not show cardiovascular side-effects in mouse and rabbit *in vivo* studies, indicating that it does not inhibit endothelial NOS activity in these species [55-58]
- 3-Bromo-7-nitroindazole (B-7929), a brominated analog that is reported to be a more potent inhibitor of NOS than 7-NI [59]
- Diphenyleneiodonium chloride (D-7909) and diphenyleneiodonium *p*-toluenesulfonate (D-7911), which are potent irreversible inhibitors of NOS, as well as of other flavoproteins such as NADPH oxidase [60-63]

Our Nitric Oxide Synthase (NOS) Inhibitor Kit (N-7925) contains seven of the most important NOS inhibitors — *S*-methyl-L-thiocitrulline (5 mg), L-NAME (100 mg), L-NMMA (25 mg), L-NIL (5 mg), L-NIO (5 mg), 7-NI (25 mg) and diphenyleneiodonium chloride (25 mg) — at less than half the component prices.

It was recently reported that L-NAME and several other nitro-containing molecules spontaneously release a small but detectable amount of NO.[64] Using an extremely sensitive NO assay, researchers have demonstrated that the photochemical generation of NO was dependent on both light intensity and solute concentration, with UV light stimulating the greatest NO production.

NOS Substrate and Cofactor

Oxidation of the biosynthetic intermediate N^G-hydroxy-L-arginine (H-7910) by NOS generates NO and citrulline.[65-68]

6*R*-BH_4 ((6*R*)-5,6,7,8-tetrahydro-L-biopterin, T-7920) is a cofactor for all NOS isozymes described so far,[3,24,69,70] including the constitutive neuronal NOS and endothelial NOS and the inducible NOS. 6*R*-BH_4 also serves as a cofactor of phenylalanine, tyrosine and tryptophan hydroxylases, which are central enzymes in the biosynthesis of melanin and catecholamines.[71,72]

NADPH Diaphorase Detection

Histochemical staining for NADPH diaphorase activity has been found to be a reliable technique for identifying NOS-containing cells in a number of vertebrate species,[7,73,74] including rat,[75] mouse,[76] rabbit,[77] guinea pig,[78] dog [79] and carp.[80] When NADPH diaphorase was purified from rat brain to homogeneity, researchers found that NOS activity co-purified with their preparation and that the NADPH substrate nitro blue tetrazolium (NBT, N-6495; see Section 21.4) competitively inhibited NOS.[74] However, the tight correlation between NADPH diaphorase activity, as measured by the reduction of nitro blue tetrazolium (NBT) to the blue formazan precipitate, and NOS activity is reportedly only observed after tissue fixation.[81] For dehydrogenase assays and viability measurements, Molecular Probes offers a wide array of tetrazolium salts, including INT, tetrazolium blue, NBT, NVT, MTT and the extremely water-soluble XTT (I-6496, T-6490, N-6495, N-6498, M-6494, X-6493; see Section 21.4).

Detecting Nitric Oxide, Nitrite and Nitrate

Griess Reagent Kit

Under physiological conditions, NO is oxidized to nitrite and nitrate. The Griess reagent provides a simple and well-characterized colorimetric assay for nitrites, and nitrates that have been reduced to nitrites, with a detection limit of about 100 nM.[82,83] Nitrites react with the Griess reagent to form a purple azo derivative that can be monitored by absorbance at 548 nm (Figure 21.5). Our new Griess Reagent Kit (G-7921) contains all of the reagents required for nitrite quantitation, including:

- *N*-(1-Naphthyl)ethylenediamine dihydrochloride
- Sulfanilic acid in 5% H_3PO_4
- Concentrated nitrite standard for generating calibration curves

Both the *N*-(1-naphthyl)ethylenediamine dihydrochloride and the sulfanilic acid in 5% H_3PO_4 are provided in convenient dropper bottles for easy preparation of the Griess reagent. Sample pretreatment with nitrate reductase and glucose-6-phosphate dehydrogenase is reported to reduce nitrate without producing excess NADPH, which can interfere with the Griess reaction.[84]

2,3-Diaminonaphthalene

We also now offer 2,3-diaminonaphthalene (D-7918), which reacts with nitrite to form the fluorescent product 1*H*-naphthotriazole. Using 2,3-diaminonaphthalene, researchers have developed a rapid, quantitative fluorometric assay that can detect from 10 nM to 10 µM nitrite and is compatible with a 96-well microplate format.[85] Neutral red (N-3246, see Section 12.3) is reported to be useful for an absorptimetric assay of nitrite with a limit of detection of 14 ng/mL.[86]

Other Probes for Detecting NO

Continuous fluorometric detection of NO in the presence of other reactive oxygen species is not yet feasible. The dihydrofluoresceins, dihydrorhodamines and dihydrorosamines that are extensively used to detect various forms of oxygen are known to react with NO, yielding the same oxidation products.[87] NO also reacts with superoxide or hydrogen peroxide to produce the reactive peroxynitrite anion, which oxidizes dihydrorhodamine 123 (D-632, see Section 21.4).[88,89]

Continuous chemiluminescent detection of NO production by perfused organs using luminol (L-8455, see Section 21.3) has been reported.[90] The species actually detected in this assay is peroxynitrite.

1. Chem Ind 828 (1995); **2.** Current Biology 2, 437 (1995); **3.** J Med Chem 38, 4343 (1995); **4.** Ann Rev Biochem 63, 175 (1994); **5.** Cell 78, 919 (1994); **6.** Am J Physiol 262, G379 (1992); **7.** Science 257, 494 (1992); **8.** J Cardiovascular Pharmacol 17 (Suppl 3), S25 (1991); **9.** Neuropharmacology 33, 1375 (1994); **10.** Anal Chem 66, 419A (1994); **11.** J Biol Chem 268, 11520 (1993); **12.** J Org Chem 56, 5025 (1991); **13.** Photochem Photobiol 59, 463 (1994); **14.** J Med Chem 34, 3242 (1991); **15.** Spermine NONOate is licensed to Molecular Probes under U.S. Patent No. 5,155,137; **16.** Biochemistry 34, 7177 (1995); **17.** Br J Pharmacol 107, 745 (1992); **18.** J Biol Chem 268, 27180 (1993); **19.** J Pharmacol Exp Ther 260, 286 (1992); **20.** Eur J Pharmacol 144, 379 (1987); **21.** Nature 364, 626 (1993); **22.** J Pharmacol Exp Ther 248, 762 (1989); **23.** J Biol Chem 270, 17355 (1995); **24.** J Biol Chem 270, 655 (1995); **25.** Arch Biochem Biophys 300, 115 (1993); **26.** J Leukocyte Biol 57, 152 (1995); **27.** *The Physiology of Polyamines, Volumes I and II*, U. Bachrach and Y.M. Heimer, Eds., CRC Press (1989); **28.** Cancer Res 53, 564 (1993); **29.** Nature 375, 68 (1995); **30.** Brain Res 619, 344 (1993); **31.** FEBS Lett 315, 139 (1993); **32.** Thrombosis Res 70, 405 (1993); **33.** J Clin Invest 83, 1774 (1989); **34.** Am J Physiol 266, L9 (1994); **35.** Life Sci 54, 1449 (1994); **36.** Neuron 12, 1235 (1994); **37.** Biochem Biophys Res Comm 202, 923 (1994); **38.** Biochemistry 32, 827 (1993); **39.** Infection and Immunity 61, 3552 (1993); **40.** Photochem Photobiol 61, 325 (1995); **41.** Life Sci 54, PL-185 (1994); **42.** Biochem J 298, 249 (1994); **43.** Cell 78, 915 (1994); **44.** J Med Chem 37, 1899 (1994); **45.** J Med Chem 37, 885 (1994); **46.** Br J Pharmacol 107, 1159 (1992); **47.** Br J Pharmacol 102, 234 (1991); **48.** Br J Pharmacol 101, 746 (1990); **49.** J Med Chem 37, 3886 (1994); **50.** Biochem Biophys Res Comm 199, 147 (1994); **51.** Biochemistry 33, 14784 (1994); **52.** J Biol Chem 269, 1674 (1994); **53.** J Neurosci 14, 1985 (1994); **54.** Br J Pharmacol 114, 257 (1995); **55.** J Neurochem 64, 936 (1995); **56.** Br J Pharmacol 110, 225 (1993); **57.** Br J Pharmacol 110, 219 (1993); **58.** Br J Pharmacol 108, 296 (1993); **59.** Br J Pharmacol 112, 351P (1994); **60.** J Biol Chem 270, 11727 (1995); **61.** J Biol Chem 270, 8328 (1995); **62.** Biochem J 290, 41 (1993); **63.** FASEB J 5, 98 (1991); **64.** Life Sci 54, PL-1 (1993); **65.** Biochemistry 34, 1930 (1995); **66.** J Biol Chem 269, 17776 (1994); **67.** Biochem Biophys Res Comm 192, 53 (1993); **68.** J Biol Chem 268, 14781 (1993); **69.** J Biol Chem 269, 13861 (1994); **70.** J Biol Chem 269, 13725 (1994); **71.** Brain Res 635, 59 (1994); **72.** Science 263, 1444 (1994); **73.** Proc Natl Acad Sci USA 88, 7797 (1991); **74.** Proc Natl Acad Sci USA 88, 2811 (1991); **75.** Neurosci Lett 161, 49 (1993); **76.** Acta Histochem 93, 397 (1992); **77.** Cell Tissue Res 274, 539 (1993); **78.** Neurosci Lett 143, 65 (1992); **79.** Am J Physiol 263, G277 (1992); **80.** Neurosci Lett 158, 151 (1993); **81.** Neurosci Lett 155, 61 (1993); **82.** FASEB J 7, 349 (1993); **83.** Anal Biochem 126, 131 (1982); **84.** Anal Biochem 224, 502 (1995); **85.** Anal Biochem 214, 11 (1993); **86.** Anal Lett 27, 991 (1994); **87.** J Leukocyte Biol 51, 496 (1992); **88.** Biochemistry 34, 3544 (1995); **89.** Free Rad Biol Med 16, 149 (1994); **90.** Anal Chem 65, 1794 (1993).

21.5 Data Table *Probes for Nitric Oxide Research*

Cat #	Structure	MW	Storage	Soluble	Abs	{ε × 10⁻³}	Em	Solvent	Notes
B-7929	✓	242	D,L	DMSO	360	{8.7}	none	MeOH	
C-7912	✓	315	FF,D	H_2O	367	{9.3}	none	MeOH	
D-7111	✓	409	F,D,LL	DMSO, MeCN	316	{22}	none	MeOH	<1>
D-7909	✓	315	F,D,L	DMSO, DMF	263	{17}	none	MeOH	
D-7911	✓	450	F,D,L	DMSO, DMF	262	{18}	none	MeOH	
D-7915	✓	155	FF,DD,A	H_2O, DMSO	248	{8.0}	none	pH 12	<2>
D-7917	✓	412	F,D,LL	DMSO, H_2O	348	{5.5}	none	pH 7	<1>
D-7918	✓	158	L	DMSO, MeOH	340	{5.1}	377	MeOH	
H-7910	✓	250	D	H_2O	<300		none		
I-7899	✓	210	DD	H_2O	<300		none		
I-7922	✓	224	DD	H_2O	<300		none		
M-7891	✓	207	FF,D,LL	DMSO, H_2O	291	{11}	none	pH 7	<3>
M-7898	✓	248	D	H_2O	<300		none		
M-7906	✓	248	D	H_2O	<300		none		
M-7914	✓	278	F,D	H_2O	<300		none		
N-7892	✓	220	FF,D,LL	DMSO, H_2O	342	{0.7}	none	MeOH	<3>
N-7902	✓	163	D,L	DMSO	356	{8.3}	none	MeOH	
N-7903	✓	336	FF,D,LL	H_2O	336	{0.9}	none	pH 7	<3>
N-7907	✓	270	F,D	H_2O	<300		none		
N-7928	✓	171	DD,L	H_2O	362	{8.9}	none	pH 7	
P-7070	✓	387	D,LL	H_2O	516	{0.05}	none	H_2O	
S-7916	✓	262	FF,DD,A	H_2O, DMSO	248	{6.7}	none	pH 12	<2>
T-7920	✓	314	F,D,L,A	DMSO	297	{8.4}	none	pH 7	

M-7926 see M-7891; N-7927 see N-7892.

For definitions of the contents of this data table, see "How to Use the Handbook" on page vi.
Structure: Chemical structure drawing: ✓ = shown in this section.

<1> All caged products are sensitive to light. They should be protected from illumination, except when photolysis is intended.
<2> Releases nitric oxide upon acid-catalyzed dissociation in solution (see Figure 21.3). Stable in alkaline solutions [Meth Enzymol 268, 281 (1996)].
<3> Spontaneously decomposes in solution (see Figure 21.3).

21.5 Structures *Probes for Nitric Oxide Research*

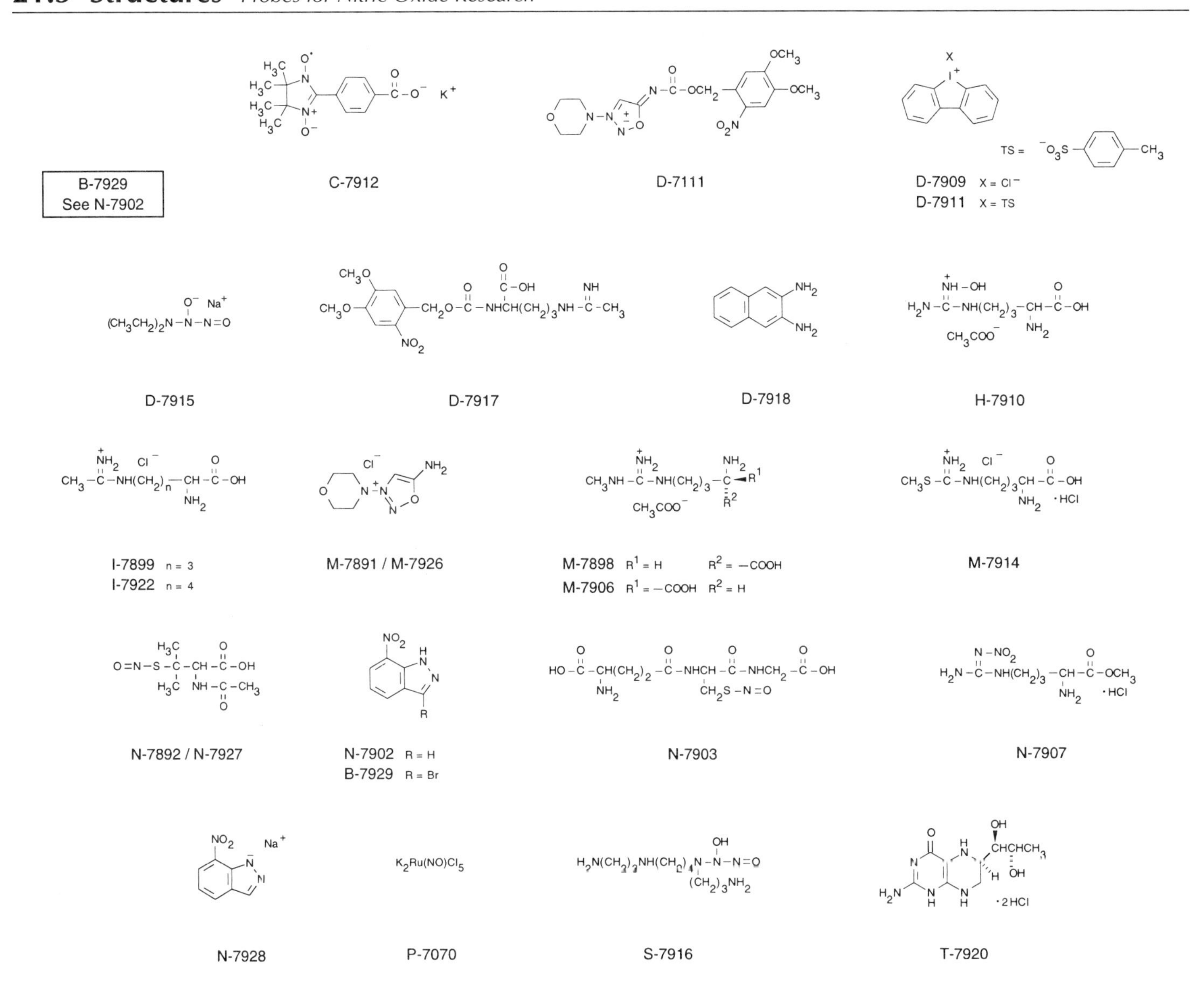

21.5 Price List *Probes for Nitric Oxide Research*

Cat #		Product Name	Unit Size	Price Per Unit ($) 1–4 Units	5–24 Units
B-7929	***New***	3-bromo-7-nitroindazole	25 mg	45.00	36.00
C-7912	***New***	2-(4-carboxyphenyl)-4,4,5,5-tetramethylimidazoline-1-oxyl-3-oxide, potassium salt (carboxy-PTIO)	25 mg	45.00	36.00
D-7918	***New***	2,3-diaminonaphthalene	100 mg	18.00	14.40
D-7915	***New***	diethylamine nitric oxide, sodium salt (DEANO)	10 mg	48.00	38.40
D-7917	***New***	N^1-(((4,5-dimethoxy-2-nitrobenzyl)oxy)carbonyl)-N^5-(1-iminoethyl)-L-ornithine, hydrate (NVOC-caged L-NIO)	5 mg	95.00	76.00
D-7111	***New***	N-(((4,5-dimethoxy-2-nitrobenzyl)oxy)carbonyl)-3-morpholinosydnoneimine (NVOC-caged SIN-1)	5 mg	95.00	76.00
D-7909	***New***	diphenyleneiodonium chloride	25 mg	35.00	28.00
D-7911	***New***	diphenyleneiodonium p-toluenesulfonate	25 mg	35.00	28.00
G-7921	***New***	Griess Reagent Kit *for nitrite quantitation*	1 kit	55.00	44.00
H-7910	***New***	N^G-hydroxy-L-arginine, monoacetate	5 mg	48.00	38.40
I-7922	***New***	N^6-(1-iminoethyl)-L-lysine, hydrochloride (L-NIL)	5 mg	85.00	68.00
I-7899	***New***	N^5-(1-iminoethyl)-L-ornithine, hydrochloride (L-NIO)	5 mg	85.00	68.00
M-7906	***New***	N^G-methyl-D-arginine, acetate salt (D-NMMA)	25 mg	55.00	44.00
M-7898	***New***	N^G-methyl-L-arginine, acetate salt (L-NMMA)	25 mg	45.00	36.00
M-7914	***New***	S-methyl-L-thiocitrulline chloride, hydrochloride	5 mg	95.00	76.00

Cat #		Product Name	Unit Size	Price Per Unit ($) 1–4 Units	5–24 Units
M-7891	***New***	3-morpholinosydnonimine, hydrochloride (SIN-1)	25 mg	45.00	36.00
M-7926	***New***	3-morpholinosydnonimine, hydrochloride (SIN-1) *special packaging*	20x1 mg	65.00	52.00
N-7925	***New***	Nitric Oxide Synthase (NOS) Inhibitor Kit	1 kit	195.00	156.00
N-7907	***New***	N^G-nitro-L-arginine methyl ester, hydrochloride (L-NAME)	1 g	9.00	7.20
N-7902	***New***	7-nitroindazole	25 mg	45.00	36.00
N-7928	***New***	7-nitroindazole, sodium salt	25 mg	45.00	36.00
N-7892	***New***	S-nitroso-N-acetylpenicillamine (SNAP)	25 mg	45.00	36.00
N-7927	***New***	S-nitroso-N-acetylpenicillamine (SNAP) *special packaging*	20x1 mg	65.00	52.00
N-7903	***New***	S-nitrosoglutathione	25 mg	45.00	36.00
P-7070	***New***	potassium nitrosylpentachlororuthenate (caged nitric oxide I)	10 mg	85.00	68.00
S-7916	***New***	spermine NONOate	10 mg	48.00	38.40
T-7920	***New***	(6R)-5,6,7,8-tetrahydro-L-biopterin, dihydrochloride (6R-BH_4)	25 mg	48.00	38.40

Technical Assistance at Our Web Site (http://www.probes.com)

At Molecular Probes' Web site, we are developing an electronic version of this Handbook and other databases that should prove extremely useful for the researcher. In addition to containing all of the text from this Handbook, our Web site provides:

- **Product searches** by product name or catalog number
- **Bibliographies** for all products for which we have references
- **Keyword searches** of our entire bibliography of over 25,000 references
- **Product information sheets** for many kits and reagents
- **Technical bulletins**, including *BioProbes* newsletters and other product literature
- **Chemical structure**, **technical data** and **material safety and data sheets**
- **Color photomicrographs** that show our products in action

Visit our Web site often for new additions to our bibliography, as well as new products and upgraded search capabilities. Also look for special sales and introductory specials on some important products.

If you do not have access to the Internet or you need assistance that is not available at that site, further information on the scientific and technical background of our products can be obtained by contacting our Technical Assistance Department at the numbers listed on the inside front cover.

Chapter 22

Indicators for Ca^{2+}, Mg^{2+}, Zn^{2+} and Other Metals

Contents

22.8 Chelators, Calibration Buffers and Ionophores 540

Technical Notes and Product Highlights

Related Chapters

22.1 Introduction to Ca^{2+} Measurements with Fluorescent Indicators

Fluorescent probes that show a spectral response upon binding Ca^{2+} have enabled researchers to investigate changes in intracellular free Ca^{2+} concentrations using fluorescence microscopy, flow cytometry and fluorescence spectroscopy.[1] These fluorescent indicators, most of which are variations of the nonfluorescent Ca^{2+} chelators EGTA and BAPTA,[2] have evolved largely through the efforts of Roger Tsien and his colleagues and, more recently, through those of scientists at Molecular Probes.

Selection Criteria for Fluorescent Ca^{2+} Indicators

Molecular Probes offers the widest available selection of fluorescent Ca^{2+} indicators for detecting changes in intracellular Ca^{2+} over the range of <50 nM to >50 µM (Table 22.1). Not only are we the primary supplier of fura-2, indo-1, quin-2, fluo-3 and rhod-2, but we exclusively offer a number of other indicators for intracellular Ca^{2+}. These include the new bis-fura-2 indicator and the visible light–excitable Oregon Green BAPTA, Calcium Green, Calcium Orange, Calcium Crimson and Fura Red indicators. In addition to a wide range of affinities and selectivities, these unique indicators offer increased brightness, reduced phototoxicity and faster responses. We also offer indicators that are conjugated to low and high molecular weight dextrans for improved cell retention and less compartmentalization, as well as lipophilic Ca^{2+} indicators for possible use in studying near-membrane Ca^{2+} (see Section 22.4).

A number of factors should be considered when selecting a fluorescent Ca^{2+} indicator, some of which are summarized in Table 22.1 and include:

- **Indicator form** (salt, AM ester or dextran), which influences the cell loading method and affects the indicator's intracellular distribution and retention. The salt and dextran forms are typically microinjected or scrape loaded into cells (see Table 15.1 in Chapter 15). In contrast, the cell-permeant acetoxymethyl (AM) esters can be passively loaded into cells, where they are cleaved to cell-impermeant products by intracellular esterases. For a discussion of ratiometric methods and AM ester loading, see the Technical Note "Loading and Calibration of Intracellular Ion Indicators" on page 549.
- **Measurement mode**, which is dictated by whether qualitative or quantitative ion concentration data is required. Ion indicators that exhibit spectral shifts upon ion binding can be used for ratiometric measurements of Ca^{2+} concentration, which are essentially independent of uneven dye loading, cell thickness, photobleaching and dye leakage. Excitation and emission wavelength preferences depend on the type of instrumentation being used, as well as on sample autofluorescence and on the presence of other fluorescent or photoactivatable probes in the experiment.
- **Dissociation constant (K_d)**, which must be compatible with the Ca^{2+} concentration range of interest. Indicators have a detectable response in the concentration range from approximately $0.1 \times K_d$ to $10 \times K_d$. The K_d of Ca^{2+} indicators is dependent on many factors, including pH, temperature, ionic strength, viscosity, protein binding and the presence of Mg^{2+} and other ions.

Table 22.1 *Summary of fluorescent Ca^{2+} indicators available from Molecular Probes.*

Ca^{2+} Indicator	Salt *	AM Ester †	Dextran ‡	Mode §	K_d (nM) ¶	Notes
Bis-fura	B-6810			Ex 340/380	370	1
BTC	B-6790	B-6791		Ex 400/480	7000	2
Calcium Green-1	C-3010	C-3011, C-3012	C-6765, C-3713, C-3714, C-6766	Em 530	190	3,4
Calcium Green-2	C-3730	C-3732		Em 535	550	3,5
Calcium Green-5N	C-3737	C-3739		Em 530	14,000	3
Calcium Orange	C-3013	C-3015		Em 575	185	2
Calcium Orange-5N	C-6770	C-6771		Em 580	20,000	2
Calcium Crimson	C-3016	C-3018	C-6824, C-6825	Em 615	185	2
Fluo-3	F-1240, F-3715	F-1241, F-1242		Em 525	390	3,4
Fura 2	F-1200, F-6799	F-1201, F-1221, F-1225	F-6764, F-3029, F-3030	Ex 340/380	145	2
Fura Red	F-3019	F-3020, F-3021		Ex 420/480	140	2,6,7
Indo-1	I-1202	I-1203, I-1223, I-1226	I-3032, I-3033	Em 405/485	230	2
Mag-fura-2	M-1290	M-1291, M-1292		Ex 340/380	25,000	2
Mag-fura-5	M-3103	M-3105		Ex 340/380	28,000	2
Mag-indo-1	M-1293	M-1295	M-6907, M-6908	Em 405/485	35,000	2,8
Magnesium Green	M-3733	M-3735		Em 530	6000	3
Oregon Green 488 BAPTA-1	O-6806	O-6807	O-6798, O-6797	Em 520	170	3
Oregon Green 488 BAPTA-2	O-6808	O-6809		Em 520	580	3,9
Oregon Green 488 BAPTA-5N	O-6812	O-6813		Em 520	20,000	3
Quin-2	Q-1287	Q-1288, Q-1289		Em 495	60	2,10
Rhod-2	R-1243	R-1244, R-1245		Em 570	570	2
Texas Red–Calcium Green			C-6800	Em 535/615	370	11

* Catalog number for cell-impermeant salt. † Catalog number for cell-permeant AM ester. ‡ Catalog number for dextran conjugate. § Measurement wavelengths (in nm), where Ex = Fluorescence excitation and Em = Fluorescence emission. Indicators for which a pair of wavelengths are listed have dual-wavelength ratio-measurement capability. ¶ Ca^{2+} dissociation constant; measured at Molecular Probes *in vitro* at 22°C in 100 mM KCl, 10 mM MOPS pH 7.2, unless otherwise noted. K_d values depend on temperature, ionic strength, pH and other factors, and are usually higher *in vivo*. Because indicator dextrans are intrinsically polydisperse and have variable degrees of substitution, lot-specific K_d values are printed on the vial in most cases.

1. Similar Ca^{2+}-dependent fluorescence response to fura-2 but ~75% greater molar absorptivity; **2.** AM ester form is fluorescent (a major potential source of error in Ca^{2+} measurements); **3.** AM ester form is nonfluorescent; **4.** Calcium Green-1 is more fluorescent than fluo-3 in both Ca^{2+}-bound and Ca^{2+}-free forms. Magnitude of Ca^{2+}-dependent fluorescence increase is greater for fluo-3; **5.** Larger Ca^{2+}-dependent fluorescence increase than Calcium Green-1; **6.** Can also be used in combination with fluo-3 for dual-wavelength ratio measurements, Ex = 488 nm, Em = 530/670 nm [Cytometry 17, 135 (1994); Cell Calcium 14, 359 (1993)]; **7.** Mag-Fura Red has similar spectral properties, with K_d for Ca^{2+} = 17 µM; **8.** K_d determined in 100 mM KCl, 40 mM HEPES, pH 7.0 at 22°C [Biochem Biophys Res Comm 177, 184 (1991)]; **9.** Larger Ca^{2+}-dependent fluorescence increase than Oregon Green 488 BAPTA-1; **10.** K_d determined in 120 mM KCl, 20 mM NaCl, pH 7.05 at 37°C [Meth Enzymol 172, 230 (1989)]; **11.** This indicator consists of Ca^{2+}-sensitive Calcium Green-1 and Ca^{2+}-insensitive Texas Red dyes linked to the same dextran.

High-Purity Indicators from Molecular Probes

Molecular Probes strives to provide the highest-purity indicators available anywhere. The AM esters of most of our indicators are certified to be at least 95% pure by both UV and visible HPLC analysis, although purity often exceeds 98%. Furthermore, the AM esters of many of the Ca^{2+} and Mg^{2+} indicators are available in special packaging for more convenient handling and for reduced risk of deterioration during storage.

Intracellular calibration of Ca^{2+} indicators may be achieved either by manipulating Ca^{2+} levels inside cells using an ionophore or by releasing the indicator into the surrounding medium of known Ca^{2+} concentration via detergent lysis of the cells. We offer several control compounds and buffers for measuring and manipulating intracellular and extracellular Ca^{2+}, which are discussed in Section 22.8. These include caged Ca^{2+} reagents and caged chelators (NP-EGTA, DMNP-EDTA and diazo-2), as well as Calcium Calibration Buffer Kits, BAPTA-derived buffers, ion-selective chelating polymers (Calcium Sponge™ products) and the Ca^{2+} ionophores A-23187 and its nonfluorescent analog 4-bromo A-23187. Our reagents for probing Ca^{2+} regulation and second messenger activity are described in more detail in Chapter 20.

Reference Guides for Using Fluorescent Ca^{2+} Indicators

In order to meet the needs of researchers new to this technology, Molecular Probes has recently begun to offer selected books that provide a survey of fluorescent probes and their techniques.

- *Methods in Cell Biology, Volume 40: A Practical Guide to the Study of Calcium in Living Cells* (M-7890), edited by Nuccitelli, is an indispensable guide for all researchers using fluorescent ion indicators.
- *Fluorescent and Luminescent Probes for Biological Activity. A Practical Guide to Technology for Quantitative Real-Time Analysis* (F-7889), edited by Mason, is a comprehensive survey of optical probe techniques, including fluorescent ion indicators.
- A recent review by June and Rabinowicz in *Methods of Cell Biology, Volume 41: Flow Cytometry, Part A* (M-7887) describes flow cytometric measurements of intracellular Ca^{2+} using indo-1, fluo-3 and the combination of fluo-3 and Fura Red.

Other recent reviews of these indicators include those by Thomas,[3] Blinks[4] and Kuhn.[5] Several earlier reviews on ion indicators also contain useful technical information.[1,6-10]

1. Biochem J 248, 313 (1987); **2.** Biochemistry 19, 2396 (1980); **3.** Meth Toxicol 1B, 287 (1994); **4.** Blinks, J.R. in *The Heart and Cardiovascular System, Second Edition*, H.A. Fozzard, Ed., Raven Press (1992) pp. 1171–1201; **5.** Kuhn, M. in *Fluorescent Chemosensors for Ion and Molecule Recognition*, A.W. Czarnik, Ed., American Chemical Society (1993) pp. 147–161; **6.** Ann Rev Physiol 52, 467 (1990); **7.** Ann Rev Physiol 52, 431 (1990); **8.** Meth Enzymol 192, 38 (1990); **9.** Ann Rev Neurosci 12, 227 (1989); **10.** Meth Cell Biol 30, 127 (1989).

22.1 Price List *Introduction to Ca^{2+} Measurements with Fluorescent Indicators*

Cat #		Product Name	Unit Size	Price Per Unit ($) 1–4 Units	5–24 Units
F-7889	*New*	Fluorescent and Luminescent Probes for Biological Activity. A Practical Guide to Technology for Quantitative Real-Time Analysis. W.T. Mason, ed. Academic Press (1993); 433 pages, comb bound	each	65.00	65.00
M-7890	*New*	Methods in Cell Biology, Volume 40: A Practical Guide to the Study of Calcium in Living Cells. R. Nuccitelli, ed. Academic Press (1994); 342 pages, comb bound	each	45.00	45.00
M-7887	*New*	Methods in Cell Biology, Volume 41: Flow Cytometry, Part A. Z. Darzynkiewicz, J.P. Robinson, H.A. Crissman, eds. Academic Press (1994); 591 pages, comb bound	each	59.00	59.00

22.2 Fluorescent Ca^{2+} Indicators Excited by Ultraviolet Light

Fura-2, Indo-1 and Derivatives

Fura-2 and Indo-1

Fura-2 and indo-1 are UV-excitable, ratiometric Ca^{2+} indicators [1] that are generally considered to be interchangeable in most experiments. Fura-2 has become the dye of choice for ratio-imaging microscopy, in which it is more practical to change excitation wavelengths than emission wavelengths.[2] Upon binding Ca^{2+}, fura-2 exhibits an absorption shift that can be observed by scanning the excitation spectrum between 300 and 400 nm, while monitoring the emission at ~510 nm (Figure 22.1). In contrast, indo-1 is the preferred dye for flow cytometry, where it is more practical to use a single laser for excitation — usually the 351–356 nm spectral lines of the argon-ion laser — and monitor two emissions.[3] The emission maximum of indo-1 shifts from ~475 nm in Ca^{2+}-free medium to ~400 nm when the dye is saturated with Ca^{2+} (Figure 22.2). Indo-1 does not require special quartz optics for microscopy and may be less subject to compartmentalization than fura-2.[4] However, fura-2 is more resistant to photobleaching than indo-1.[5,6]

The potassium salt of fura-2 (F-1200), new sodium salt of fura-2 (F-6799) and potassium salt of indo-1 (I-1202) are cell-impermeant probes that can be microinjected into cells, thus avoiding the presence of residual, unhydrolyzed AM esters, which can complicate analysis. Free acids of fura-2 and indo-1 can also be loaded into some plant cells at pH 4–5.[7-11] In addition, these salts are useful as standards for calibrating Ca^{2+} measurements.

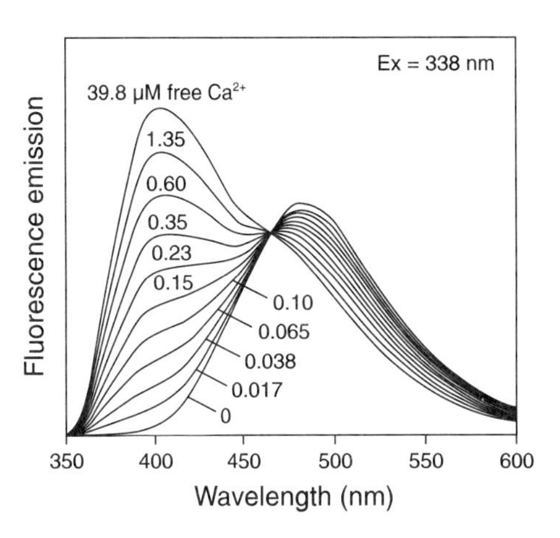

Figure 22.2 Fluorescence emission spectra of indo-1 (I-1202) in solutions containing zero to 39.8 µM free Ca^{2+}.

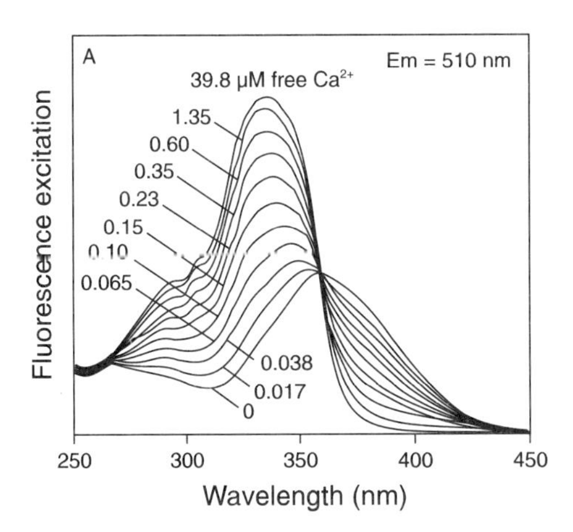

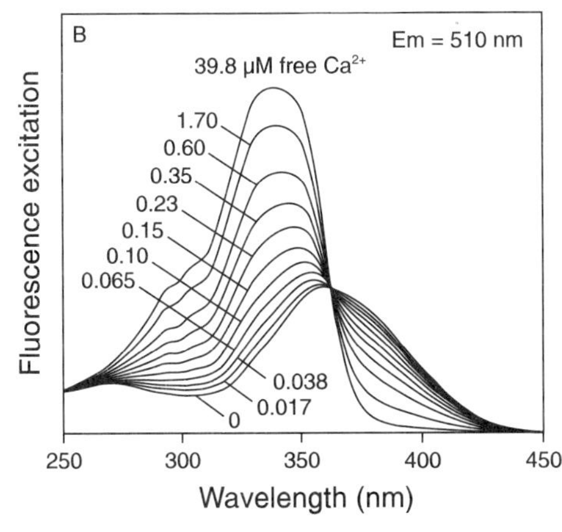

Figure 22.1 Fluorescence excitation spectra of A) fura-2 (F-1200) and B) bis-fura-2 (B-6810) in solutions containing zero to 39.8 µM free Ca^{2+}.

Unlike the salt forms, the AM esters of fura-2 and indo-1 can passively diffuse across cell membranes, enabling researchers to avoid the use of invasive loading techniques. Once inside the cell, these esters are cleaved by intracellular esterases to yield cell-impermeant fluorescent indicators; see "Loading and Calibration of Intracellular Ion Indicators" on page 549. Molecular Probes offers fura-2 AM and indo-1 AM in 1 mg vials (F-1201, I-1203) or specially packaged in 20 vials of 50 µg each (F-1221, I-1223); the special packaging is recommended when small quantities of the dyes are to be used over a long period of time. We also provide stock solutions of fura-2 AM and indo-1 AM in anhydrous DMSO at ~1 mg/mL (1 mM; F-1225, I-1226).

Dextran conjugates of fura and indo, which can be loaded by microinjection, and a lipid analog of fura for measuring near-membrane Ca^{2+} are described in Section 22.4. Fura-2 and indo-1 also exhibit high affinities for other divalent cations such as Zn^{2+} and Hg^{2+}, a property that is discussed further in Section 22.7.

Extensive bibliographies are available for these products upon request from our Technical Assistance Department or through our Web site (http://www.probes.com). Our bibliographic database currently contains over 2000 publications that cite the use of fura-2 for measuring intracellular free Ca^{2+} and over 500 publications for indo-1.

Bis-Fura-2: Brighter Signal with Lower Affinity for Ca^{2+}

An indicator that accurately senses higher concentrations of Ca^{2+} — for instance spikes up to or exceeding 1 µM — may be required when studying some Ca^{2+}-mediated signal transduction mechanisms. As discussed in Section 22.1, ion measurements with any indicator are most sensitive in the narrow concentration range near the indicator's K_d. For example, the K_d for Ca^{2+} of fura-2 is ~135 nM in Mg^{2+}-free Ca^{2+} buffers and ~224 nM in the presence of 1 mM Mg^{2+}.[2,12] Although it may accurately indicate peaks of

Ca^{2+} up to 500 nM in cells, fura-2 exhibits limited sensitivity above 500 nM Ca^{2+} (Figure 22.1A).

By linking two fura fluorophores with one Ca^{2+} binding site, we have produced a new indicator that we call bis-fura-2, which exhibits twice the absorbance of fura-2 and a K_d for Ca^{2+} of ~370 nM and ~525 nM in the absence and presence of 1 mM Mg^{2+}, respectively.[13] In other aspects, bis-fura-2's spectral response to Ca^{2+} (Figure 22.1B) and quantum yield are virtually identical to those of fura-2. Although the difference between the K_d of fura-2 and bis-fura-2 for Ca^{2+} is small, the change in excitation ratio for bis-fura-2 in response to Ca^{2+} concentrations >500 nM is larger than that of fura-2 (Figure 22.1B); this difference could improve the dynamic range for Ca^{2+} measurements in cells. Other potential advantages of bis-fura-2 include:

- Higher fluorescence output per indicator, which may allow the use of lower dye concentrations
- Lower affinity for Ca^{2+}, which should decrease the buffering of intracellular Ca^{2+} and produce a faster response to Ca^{2+} spikes
- Additional negative charge, which facilitates dye retention

The hexapotassium salt of bis-fura-2 (B-6810) is now available for microinjection and other loading techniques (see Table 15.1 in Chapter 15); we do not currently offer a membrane-permeant form of bis-fura-2.

Quin-2 and Derivatives

Quin-2 and Quin-2 AM

Quin-2 belongs to the first generation of Ca^{2+} indicators developed by Tsien.[14] Quin-2 has a lower absorbance and quantum yield than fura-2, indo-1, fluo-3 and Calcium Green and thus requires higher loading concentrations. The resulting high intracellular concentration of the indicator may buffer intracellular Ca^{2+} transients.[15] Quin-2 AM has been used to intentionally deplete cytosolic free Ca^{2+} [16] and to ensure unidirectional Ca^{2+} influx.[17] Measurement of cytosolic free Ca^{2+} with quin-2 has been thoroughly reviewed by Tsien and Pozzan.[18]

Molecular Probes prepares quin-2 as a high-purity cell-impermeant free acid (Q-1287) and as its cell-permeant AM ester (Q-1288, Q-1289).

Methoxyquin MF

A fluorinated derivative of quin-2, methoxyquin MF [19] has been used for ^{19}F NMR studies of insulin-dependent increases in cytosolic free Ca^{2+} concentrations, from 20 nM to 60 nM, in human red blood cells.[20] Methoxyquin MF has also been employed as a fluorescent Ca^{2+} indicator in a stopped-flow experiment to determine the kinetics of Ca^{2+} dissociation from calmodulin.[21] In addition to the cell-impermeant free acid (M-6794), we offer the AM ester of methoxyquin MF (M-6795) for noninvasive cell loading.

Low-Affinity Calcium Indicators

Mag-Fura-2, Mag-Fura-5 and Mag-Indo-1

Mag-fura-2 (also called furaptra), mag-fura-5 and mag-indo-1 were originally designed to report intracellular Mg^{2+} levels (see Section 22.6); however, these indicators actually have much higher affinity for Ca^{2+} than for Mg^{2+}. Although Ca^{2+} binding by these indicators may complicate analysis when they are employed to measure intracellular Mg^{2+},[22] their increased effective range and improved linearity for Ca^{2+} measurements has been exploited for measuring intracellular Ca^{2+} levels between 1 µM and 100 µM.[23,24]

The spectral shifts of mag-fura-2, mag-fura-5 and mag-indo-1 are very similar to those of fura-2 and indo-1 but occur at higher Ca^{2+} concentrations. Because the off-rates for Ca^{2+} binding of these indicators are faster than those of fura-2 and indo-1, these dyes have been used to image action potentials in skeletal muscle with little or no kinetic delay in the contractile response of the fibers, thus permitting simultaneous measurements of Ca^{2+}, tension and stiffness.[25-27] A recent paper compares the spectral properties, kinetics and selectivity of several of our low-affinity indicators and their utility in skeletal muscle fibers.[28]

The moderate Ca^{2+} affinity of mag-fura-2 and the tendency of its acetoxymethyl (AM) ester to accumulate in subcellular compartments has proven useful for *in situ* monitoring of IP_3-sensitive Ca^{2+} stores.[29] Mag-fura-2 has also been employed to follow Ca^{2+} transients in presynaptic nerve terminals,[30] gastric epithelial cells [31] and cultured myocytes.[32] Imaging of mag-fura-2 using a single excitation wavelength (420 nm) is reported to improve the detection of high-level Ca^{2+} transients in various cells, including Purkinje neurons and frog muscle.[23,27] Mag-indo-1 has been used to detect gonadotropin-releasing hormone–induced Ca^{2+} oscillations in gonadotropes [33] and to investigate low-affinity Ca^{2+}-binding sites in barnacle troponins.[34]

Mag-fura-2, mag-fura-5 and mag-indo-1 are available as cell-impermeant potassium salts (M-1290, M-3103, M-1293) or as cell-permeant AM esters (M-1291, M-1292, M-3105, M-1295).

BTC

Molecular Probes has collaborated with Haralambos Katerinopoulos of the University of Crete in the development of the coumarin benzothiazole–based indicator BTC (B-6790) and its cell-permeant derivative BTC AM [35] (B-6791). This Ca^{2+} indicator exhibits a shift in excitation maximum from about 480 nm to 400 nm upon binding Ca^{2+} (Figure 22.3), permitting ratiometric measurements that are essentially independent of uneven dye loading, cell thickness, photobleaching and dye leakage. Its high selectivity and moderate affinity for Ca^{2+} (K_d ~7 µM) should allow

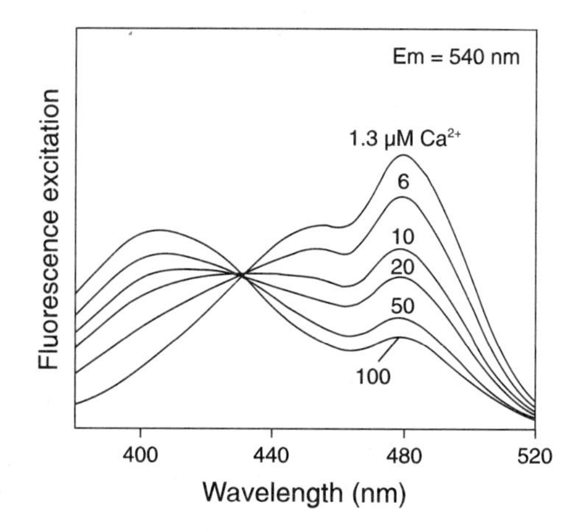

Figure 22.3 *Fluorescence excitation spectra of BTC (B-6790) in solutions containing 1.3 to 100 µM free Ca^{2+}.*

accurate quantitation of high intracellular Ca^{2+} levels, which is not currently possible with either fura-2 or indo-1. Furthermore, because BTC is excited at longer wavelengths than the ratioable fura-2 and indo-1 indicators, cellular photodamage and autofluorescence may be less of a problem. When loaded as its AM ester, BTC exhibited little compartmentalization and no toxicity in three cell types.[36]

A recent paper compares the low-affinity Ca^{2+} indicators BTC and mag-fura-2 with the higher-affinity indicators fura-2 and Calcium Green-2 and demonstrates that the Ca^{2+} transients in neurons measured with the low-affinity indicators are significantly more rapid than when measured with the higher-affinity indicators.[30] Furthermore, cells labeled with BTC and mag-fura-2 are reportedly brighter than those labeled with fura-2. Another recent paper describes the unusual use of the tetrapotassium salt of BTC to localize atherosclerotic plaque.[37]

1. Fura-2 and indo-1 are licensed to Molecular Probes under U.S. Patent No. 4,603,209 and related patents; **2.** J Biol Chem 260, 3440 (1985); **3.** Meth Cell Biol 41, 150 (1994); **4.** Cell Calcium 11, 487 (1990); **5.** Am J Physiol 258, C533 (1990); **6.** Arch Biochem Biophys 261, 91 (1988); **7.** Proc Natl Acad Sci USA 89, 3591 (1992); **8.** Plant Physiol 93, 841 (1990); **9.** Plant Sci 67, 125 (1990); **10.** Cell Calcium 8, 455 (1988); **11.** Eur J Cell Biol 46, 466 (1988); **12.** Measured at 37°C in 100 mM KCl, 10 mM MOPS, pH 7.0; **13.** K_d values for Ca^{2+} are determined at Molecular Probes at ~22°C using our Calcium Calibration Buffer Kits (see Section 22.8); **14.** Biochemistry 19, 2396 (1980); **15.** J Biol Chem 258, 4876 (1983); **16.** Brain Res 528, 48 (1990); **17.** Biochemistry 26, 6995 (1987); **18.** Meth Enzymol 172, 230 (1989); **19.** Methoxyquin MF was formerly called quin-2 MF by Molecular Probes; **20.** Diabetologia 36, 146 (1993); **21.** J Biol Chem 271, 62 (1996); **22.** Am J Physiol 263, C300 (1992); **23.** Pflügers Arch 429, 587 (1995); **24.** Neuron 10, 21 (1993); **25.** J Physiol 475, 319 (1994); **26.** Biochem Biophys Res Comm 177, 184 (1991); **27.** J Gen Physiol 97, 271 (1991); **28.** Biophys J 70, 896 (1996); **29.** Proc Natl Acad Sci USA 90, 2598 (1993); **30.** Biophys J 68, 2156 (1995); **31.** Am J Physiol 267, G442 (1994); **32.** Am J Physiol 264, C1259 (1993); **33.** Proc Natl Acad Sci USA 91, 9750 (1994); **34.** Biochemistry 30, 702 (1991); **35.** U.S. Patent No. 5,501,980; **36.** Cell Calcium 15, 190 (1994); **37.** J Photochem Photobiol 27, 81 (1995).

22.2 Data Table *Fluorescent Ca^{2+} Indicators Excited by Ultraviolet Light*

					Low Ca^{2+}			High Ca^{2+}					
Cat #	**Structure**	**MW**	**Storage**	**Soluble**	**Abs**	**{ε × 10⁻³}**	**Em**	**Abs**	**{ε × 10⁻³}**	**Em**	**Solvent**	**Cleaved***	**K_d†**	**Notes**
B-6790	✓	844	L	pH>6	464	{29}	533	401	{20}	529	<1>		7.0 µM	<2>
B-6791	✓	980	F,D,L	DMSO	433	{39}	504				MeOH	B-6790		
B-6810	✓	1007	D,L	pH>6	366	{49}	511	338	{59}	504	<1>		370 nM	<2, 3>
F-1200	✓	832	D,L	pH>6	363	{28}	512	335	{34}	505	<1>		145 nM	<2, 3>
F-1201	✓	1002	F,D,L	DMSO	370	{31}	476				EtOAc	F-1200		
F-6799	✓	751	DD,L	pH>6	363	{28}	512	335	{34}	505	<1>		145 nM	<2, 3>
I-1202	✓	840	D,L	pH>6	346	{33}	475	330	{33}	401	<1>		230 nM	<2, 3>
I-1203	✓	1010	F,D,L	DMSO	356	{38}	478				MeOH	I-1202		
M-1290	S22.6	587	L	pH>6	369	{22}	511	329	{26}	508	<1>		25 µM	<4>
M-1291	S22.6	723	F,D,L	DMSO	366	{31}	475				EtOAc	M-1290		
M-1293	S22.6	595	L	pH>6	349	{38}	480	328	{35}	390	<1>		35 µM	<5, 6>
M-1295	S22.6	731	F,D,L	DMSO	354	{37}	472				MeOH	M-1293		
M-3103	S22.6	601	L	pH>6	369	{23}	505	330	{25}	500	<1>		28 µM	<4>
M-3105	S22.6	737	F,D,L	DMSO	365	{31}	461				EtOAc	M-3103		
M-6794	✓	560	D,L	pH>6	353	{4.0}	495	333	{3.9}	495	<1>		65 nM	<2>
M-6795	✓	848	F,D,L	DMSO	348	{4.0}	446				MeOH	M-6794		
Q-1287	✓	542	D,L	pH>6	353	{4.0}	495	333	{3.9}	495	<1>		60 nM	<7>
Q-1288	✓	830	F,D,L	DMSO	348	{4.0}	446				MeOH	Q-1287		

F-1221, F-1225 see F-1201; I-1223, I-1226 see I-1203; M-1292 see M-1291; Q-1289 see Q-1288.

For definitions of the contents of this data table, see "How to Use the Handbook" on page vi.
Structure: Chemical structure drawing: ✓ = shown in this section; Sn.n = shown in Section number n.n.

* Catalog numbers listed in this column indicate the cleavage product generated by enzymatic or chemical hydrolysis of the acetoxymethyl ester groups of the parent compound.

† Data on dissociation constants come from several different sources. These values are known to vary considerably depending on the temperature, pH, ionic strength, viscosity, protein binding, presence of other ions (especially polyvalent ions), instrument setup and other factors. It is strongly recommended that these values be specifically determined under your experimental conditions by using our Calcium Calibration Buffer Kits (see Section 22.8).

<1> Spectra measured in aqueous buffers containing 10 mM EGTA (low Ca^{2+}) or a >10-fold excess of free Ca^{2+} relative to the K_d (high Ca^{2+}).

<2> Dissociation constant determined at Molecular Probes by fluorescence measurements in 100 mM KCl, 10 mM MOPS, pH 7.2, 0 to 10 mM CaEGTA at 22°C.

<3> K_d (Ca^{2+}) for F-1200 and I-1202 from the original reference by Grynkiewicz, Pozzan and Tsien [J Biol Chem 260, 3440 (1985)] are 224 nM and 250 nM, respectively, measured in 1 mM EGTA, 100 mM KCl, 1 mM free Mg^{2+}, 10 mM MOPS, pH 7.0 at 37°C. For B-6810, K_d (Ca^{2+}) in presence of Mg^{2+} is 525 nM (determined at Molecular Probes in 100 mM KCl, 10 mM MOPS, pH 7.2, 1 mM Mg^{2+} at 22°C).

<4> Dissociation constant determined at Molecular Probes by fluorescence measurements in 100 mM KCl, 10 mM MOPS, pH 7.2, 0 to 1 mM free Ca^{2+} at 22°C.

<5> Dissociation constant determined in 100 mM KCl, 40 mM HEPES, pH 7.0 at 22°C [Biochem Biophys Res Comm 177, 184 (1991)].

<6> The emission spectrum of Ca^{2+}-bound M-1293 excited at 340 nm has approximately equal peak intensities at ~390 nm and ~480 nm [Biochemistry 30, 702 (1991)].

<7> K_d (Ca^{2+}) for quin-2 was measured in 120 mM KCl, 20 mM NaCl, pH 7.05 at 37°C. Under the same conditions with addition of 1 mM Mg^{2+}, K_d = 115 nM [Meth Enzymol 172, 230 (1989)].

22.2 Structures *Fluorescent Ca^{2+} Indicators Excited by Ultraviolet Light*

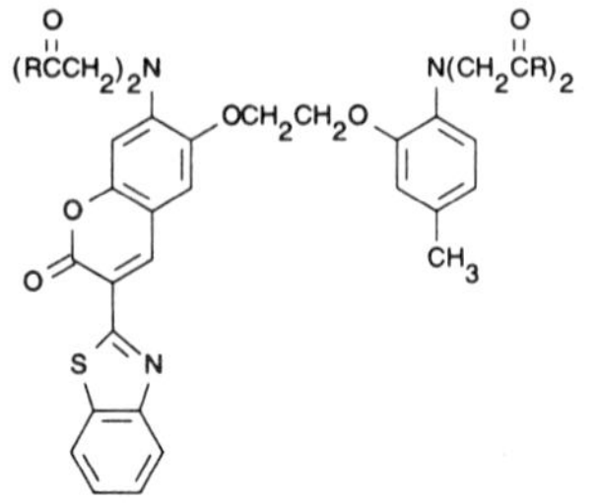

B-6790 R = $O^- K^+$
B-6791 R = $-OCH_2OCCH_3$ (C=O)

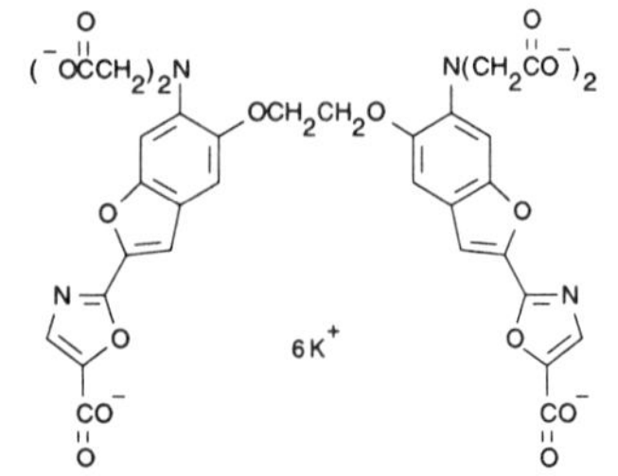

B-6810

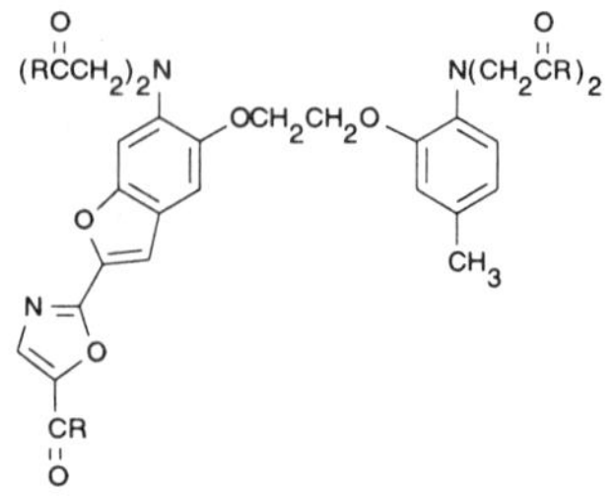

F-1200 R = $O^- K^+$
F-6799 R = $O^- Na^+$
F-1201 / F-1221 / F-1225 R = $-OCH_2OCCH_3$ (C=O)

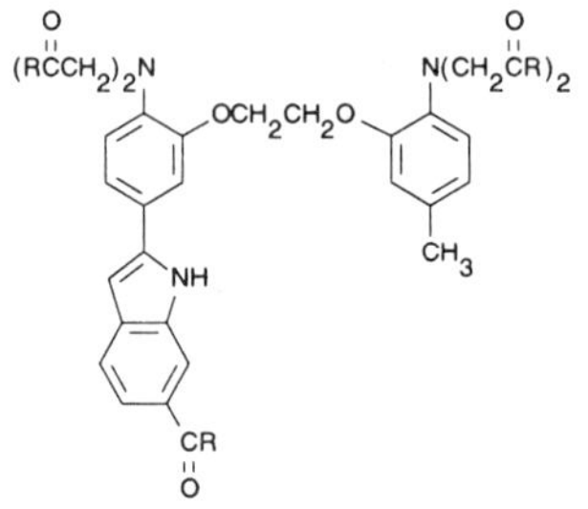

I-1202 R = $O^- K^+$
I-1203 / I-1223 / I-1226 R = $-OCH_2OCCH_3$ (C=O)

M-1290, M-1291
M-1293, M-1295
M-3103, M-3105
Section 22.6

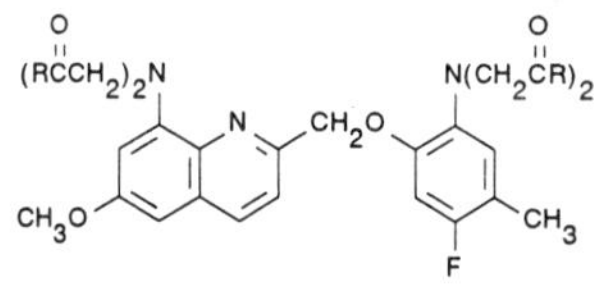

M-6794 R = $-OH$
M-6795 R = $-OCH_2OCCH_3$ (C=O)

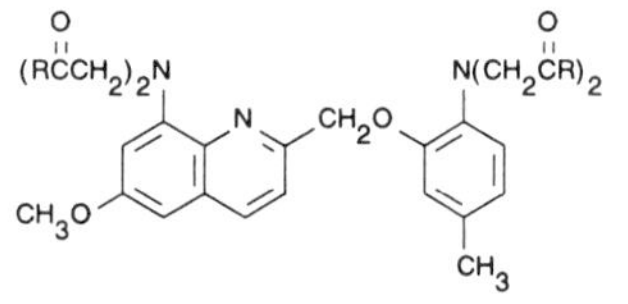

Q-1287 R = $-OH$
Q-1288 / Q-1289 R = $-OCH_2OCCH_3$ (C=O)

22.2 Price List *Fluorescent Ca^{2+} Indicators Excited by Ultraviolet Light*

Cat #		Product Name	Unit Size	Price Per Unit ($) 1–4 Units	5–24 Units
B-6810	***New***	bis-fura-2, hexapotassium salt *cell impermeant*	1 mg	98.00	78.40
B-6791	***New***	BTC, AM *cell permeant* *≥95% by HPLC* *special packaging*	20x50 µg	110.00	88.00
B-6790	***New***	BTC, tetrapotassium salt *cell impermeant*	1 mg	85.00	68.00
F-1201		fura-2, AM *cell permeant* *≥95% by HPLC*	1 mg	100.00	80.00
F-1221		fura-2, AM *cell permeant* *≥95% by HPLC* *special packaging*	20x50 µg	138.00	110.40
F-1225		fura-2, AM *1 mM solution in dry DMSO* *cell permeant* *special packaging*	1 mL	125.00	100.00
F-1200		fura-2, pentapotassium salt *cell impermeant* *≥95% by HPLC*	1 mg	83.00	66.40
F-6799	***New***	fura-2, pentasodium salt *cell impermeant*	1 mg	83.00	66.40
I-1203		indo-1, AM *cell permeant*	1 mg	90.00	72.00
I-1223		indo-1, AM *cell permeant* *special packaging*	20x50 µg	128.00	102.40
I-1226		indo-1, AM *1 mM solution in dry DMSO* *cell permeant* *special packaging*	1 mL	115.00	92.00
I-1202		indo-1, pentapotassium salt *cell impermeant* *≥95% by HPLC*	1 mg	70.00	56.00
M-1291		mag-fura-2, AM *cell permeant* *≥95% by HPLC*	1 mg	168.00	134.40
M-1292		mag-fura-2, AM *cell permeant* *≥95% by HPLC* *special packaging*	20x50 µg	198.00	158.40
M-1290		mag-fura-2, tetrapotassium salt *cell impermeant* *≥95% by HPLC*	1 mg	148.00	118.40
M-3105		mag-fura-5, AM *cell permeant* *≥95% by HPLC* *special packaging*	20x50 µg	198.00	158.40
M-3103		mag-fura-5, tetrapotassium salt *cell impermeant* *≥95% by HPLC*	1 mg	148.00	118.40
M-1295		mag-indo-1, AM *cell permeant* *≥95% by HPLC* *special packaging*	20x50 µg	198.00	158.40
M-1293		mag-indo-1, tetrapotassium salt *cell impermeant*	1 mg	148.00	118.40
M-6795	***New***	methoxyquin MF, AM *cell permeant*	5 mg	95.00	76.00
M-6794	***New***	methoxyquin MF, free acid *cell impermeant*	10 mg	135.00	108.00
Q-1288		quin-2, AM *cell permeant* *≥95% by HPLC*	25 mg	115.00	92.00
Q-1289		quin-2, AM *cell permeant* *≥95% by HPLC* *special packaging*	10x1 mg	73.00	58.40
Q-1287		quin-2, free acid *cell impermeant*	25 mg	63.00	50.40

22.3 Fluorescent Ca^{2+} Indicators Excited by Visible Light

Visible light–excitable Ca^{2+} indicators offer several advantages over UV-excitable indicators:

- Efficient excitation with most laser-based instrumentation, including confocal laser scanning microscopes and flow cytometers
- Reduced interference from sample autofluorescence
- Less cellular photodamage and scatter
- Higher absorbance, which may permit the use of lower dye concentrations and therefore lower phototoxicity to live cells
- Compatibility with photoactivatable ("caged") probes and other UV-absorbing reagents, increasing options for multiparameter measurements

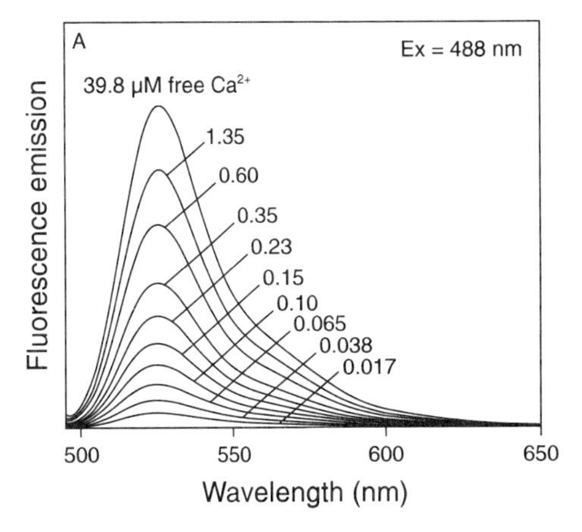

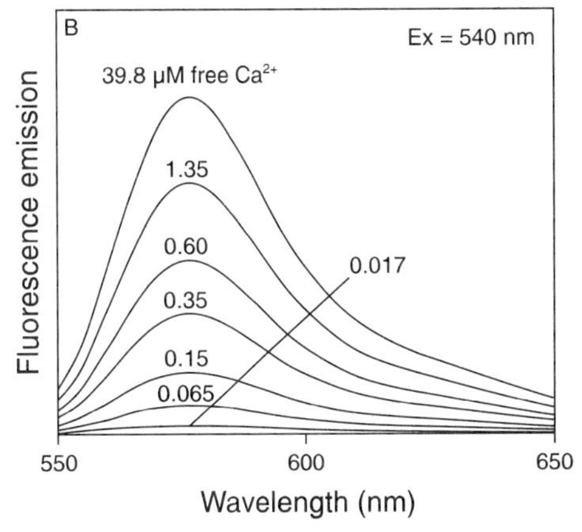

Figure 22.4 *Ca^{2+}-dependent fluorescence emission spectra of A) fluo-3 (F-1240) and B) rhod-2 (R-1243). The spectra for Ca^{2+}-free solutions are indistinguishable from the baseline in both cases.*

Fluo-3, Rhod-2 and Derivatives

Fluo-3 and Rhod-2

The Ca^{2+} indicators fluo-3 and rhod-2 were developed by Tsien and colleagues for use with visible-light excitation sources,[1] thereby providing tools for laser-based instrumentation.[2] Fluo-3 can be used with standard fluorescein optical filters, whereas rhod-2 is compatible with standard rhodamine optical filters. Rhod-2 is much less efficiently excited by the argon-ion laser than fluo-3, although it was reported to give more consistent results than fluo-3 when measuring Ca^{2+} in synaptic boutons by confocal laser scanning microscopy.[3]

Fluo-3 is essentially nonfluorescent unless bound to Ca^{2+} and exhibits a quantum yield at saturating Ca^{2+} of ~0.18.[2] As prepared by Minta, Kao and Tsien,[2] fluo-3 was originally reported to undergo an ~40-fold increase in fluorescence upon binding Ca^{2+}. However, during 1990 we improved our purification of fluo-3 so that the enhancement is usually at least 100-fold, and may even exceed 200-fold[4] (Figure 22.4). Between normal resting cytosolic free Ca^{2+} concentrations and indicator saturation, the enhancement is generally between 5- and 10-fold, making fluo-3 particularly useful for measuring the kinetics of Ca^{2+} transients. In a careful study of the spectral properties of highly purified fluo-3, Harkins, Kurebayashi and Baylor characterized the effects of pH and viscosity on Ca^{2+} measurements with fluo-2 and demonstrated that binding of the indicator to proteins has a significant effect on its K_d for Ca^{2+}.[4]

Rhod-2 was originally reported to exhibit only a 3.4-fold enhancement of fluorescence upon binding Ca^{2+}.[2] Our chemists have also improved the purification of rhod-2, yielding a highly purified preparation that shows greater than 100-fold enhancement in fluorescence upon binding Ca^{2+} (Figure 22.4). Note that, as a consequence of this increased purity, the K_d for Ca^{2+} of rhod-2 in the absence of Mg^{2+} has now been determined to be 570 nM,[5] which is considerably lower than that cited in the Fifth Edition of our *Handbook of Fluorescent Probes and Research Chemicals* and in the original paper on rhod-2.[2]

Neither fluo-3 nor rhod-2 undergoes a significant shift in emission or excitation wavelength upon binding to Ca^{2+}, which precludes the use of ratiometric measurements with these dyes alone (see the Technical Note "Loading and Calibration of Intracellular Ion Indicators" on page 549). Simultaneous loading of cells with fluo-3 and Fura Red, which exhibit reciprocal shifts in fluorescence intensity upon binding Ca^{2+}, has enabled researchers to make ratiometric measurements of intracellular Ca^{2+} (Figure 22.5) using confocal laser scanning microscopy[6-11] or flow cytometry.[12] For ratiometric measurements, Fluo-3 may also be co-loaded into cells with a Ca^{2+}-insensitive dye. Carboxy SNARF® AM acetate (C-1271, see Section 23.2), which can be excited along with fluo-3 but is detected at much longer wavelengths (see Figure 23.6 in Chapter 23), can serve as the Ca^{2+}-insensitive dye, provided that the pH within the cells remains constant during the experiment.[13-16] Co-loading of fluo-3 and carboxy SNARF also permits the simultaneous imaging of Ca^{2+} transients and intracellular pH in experiments in which the concentrations of both ions are changing.[17,18]

Fluo-3 is available as a cell-impermeant potassium (F-3715) or ammonium salt (F-1240) or as a cell-permeant AM ester (F-1241, F-1242). Unlike the AM esters of fura-2 and indo-1, fluo-3 AM is not fluorescent until hydrolyzed inside the cell. Fluo-3 has also been loaded into plant cell protoplasts as its free acid.[19] Rhod-2 is available as the cell-impermeant ammonium salt (R-1243) or as the cell-permeant AM ester (R-1244, R-1245). The AM esters are

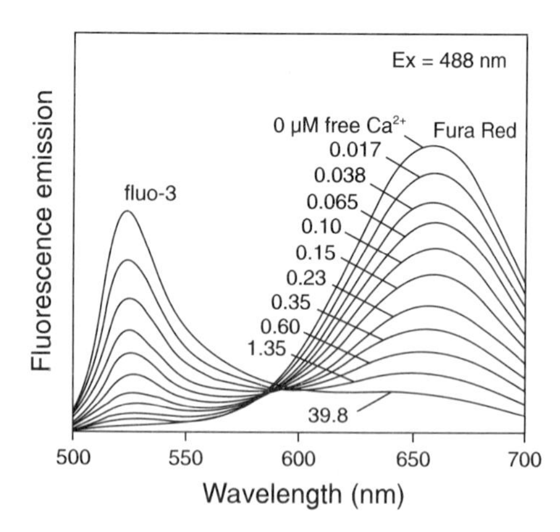

Figure 22.5 Fluorescence emission spectra of a 1:10 mole:mole mixture of fluo-3 (F-1240) and Fura Red (F-3019) indicators, simultaneously excited at 488 nm, in solutions containing zero to 39.8 μM free Ca^{2+}.

available specially packaged as sets of 10 vials, each containing 50 μg.

Indicators of Mitochondrial Ca^{2+} Transients

The AM ester of rhod-2 (R-1244, R-1245) is the only cell-permeant Ca^{2+} indicator that has a net positive charge. This property promotes its sequestration into mitochondria in some cells, most likely via membrane potential–driven uptake. By reducing rhod-2 AM to the colorless, nonfluorescent dihydrorhod-2 AM, the discrimination between cytosolic and mitochondrially localized dye can be further improved.[20] The AM ester of dihydrorhod-2 exhibits Ca^{2+}-dependent fluorescence only after it is oxidized and its AM esters are cleaved to yield the rhod-2 indicator, processes that occur rapidly in the mitochondrial environment. A detailed protocol for reducing rhod-2 AM to generate dihydrorhod-2 AM is provided with every purchase of rhod-2 AM.

Loading of cells with the AM ester of fluo-3 is also reported to yield significant mitochondrial staining, as demonstrated with confocal laser scanning microscopy by the colocalization of fluo-3 fluorescence with rhodamine 123 fluorescence.[21] Use of the mitochondrial probe tetramethylrhodamine methyl ester (TMRM, T-668; see Section 12.2) in combination with fluo-3 enabled researchers to simultaneously measure changes in mitochondrial and cytosolic Ca^{2+} levels.[22]

Calcium Green, Calcium Orange and Calcium Crimson Indicators

Calcium Green-1 and Calcium Green-2 Indicators

Calcium Green-1 and Calcium Green-2, as well as Calcium Orange and Calcium Crimson described below, are visible light–excitable indicators developed at Molecular Probes.[23] Like fluo-3, the Calcium Green indicators exhibit an increase in fluorescence emission intensity upon binding Ca^{2+} with little shift in wavelength[24] (Figure 22.6A and 22.6B); the fluorescence spectra of the Calcium Green indicators are almost identical to those of fluo-3. Further comparison of the Calcium Green indicators and fluo-3 reveals that, at high Ca^{2+} levels, Calcium Green-1 and Calcium Green-2 are several times brighter than fluo-3; Calcium Green-1 has a quantum yield at saturating Ca^{2+} of 0.75[19] as compared to about 0.18 for fluo-3.[2]

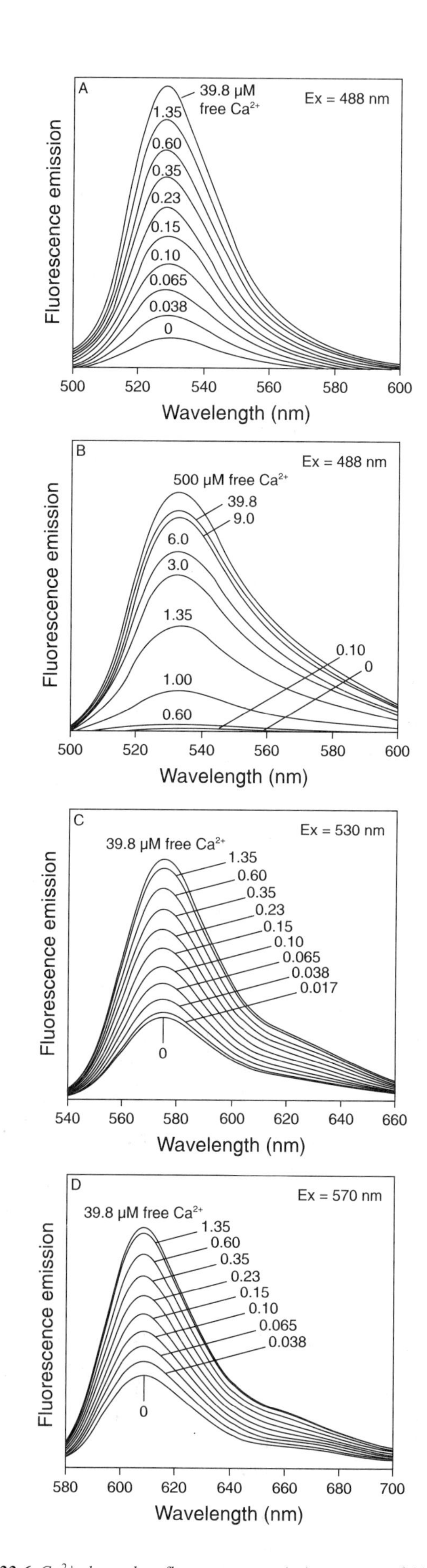

Figure 22.6 Ca^{2+}-dependent fluorescence emission spectra of A) Calcium Green-1 (C-3010), B) Calcium Green-2 (C-3730), C) Calcium Orange (C-3013) and D) Calcium Crimson (C-3016) indicators.

The Calcium Green indicators have several other important features:

- Calcium Green-1 is more fluorescent in resting cells than fluo-3, which increases the visibility of unstimulated cells, facilitates the determination of baseline fluorescence and makes calculations of intracellular Ca^{2+} concentrations more reliable.
- Calcium Green-1 is the only visible light–excitable indicator that is useful for measurements of intracellular Ca^{2+} by fluorescence lifetime imaging.[25,26]
- The Ca^{2+} affinity of Calcium Green-1 in the absence of Mg^{2+} (K_d = 190 nM) is higher than that of fluo-3 (K_d = 390 nM) or Calcium Green-2 (K_d = 550 nM).[5]
- Like fluo-3, Calcium Green-2 is essentially nonfluorescent in the absence of Ca^{2+} and exhibits an approximately 100-fold increase in emission intensity upon Ca^{2+} binding, which leads to a very large dynamic range.
- In contrast to Calcium Green-1, Calcium Green-2 has two fluorescent reporter groups, which are believed to quench one another in the absence of Ca^{2+}. Calcium Green-2 undergoes a much larger increase in fluorescence emission upon Ca^{2+} binding than does Calcium Green-1, and its lower affinity for Ca^{2+} makes it particularly well suited to the measurement of relatively high spikes of Ca^{2+} (up to ~10 µM).
- Like fluo-3 AM, the AM esters of the Calcium Green indicators are nonfluorescent.

Furthermore, the Calcium Green indicators are reported to be less phototoxic to cells than fluo-3.[27,28] This observation stems at least in part from the fact that the Calcium Green indicators are intrinsically more fluorescent than fluo-3, thus permitting less illumination and lower dye concentrations to achieve the same signal.[29] Calcium Green-1 has been used to image spiral Ca^{2+} waves [30-32] and to measure Ca^{2+} release invoked by photolysis of caged IP_3.[33] Simultaneous loading of Calcium Green-2 and carboxy SNARF-1 enabled researchers to make ratiometric measurements of intracellular Ca^{2+} in cardiac myocytes.[29]

Calcium Green-1 and Calcium Green-2 are available as cell-impermeant potassium salts (C-3010, C-3730) or as cell-permeant AM esters (C-3011, C-3012, C-3732).

Calcium Orange and Calcium Crimson Indicators

Like the Calcium Green indicators, Calcium Orange (C-3013) and Calcium Crimson (C-3016) exhibit an increase in fluorescence emission intensity upon binding to Ca^{2+} with little shift in wavelength (Figure 22.6C and 22.6D) and can be loaded into cells as their acetoxymethyl (AM) esters (C-3015, C-3018). Both indicators are more photostable than either fluo-3 or the Calcium Green indicators.

Calcium Orange has an excitation maximum near 550 nm and is compatible with standard rhodamine optical filters. Calcium Orange has been used to monitor Ca^{2+} in intact photoreceptors containing a genetically altered rhodopsin pigment,[34] as well as to follow calcium influx and release in hippocampal astrocytes.[35]

Calcium Crimson's excitation maximum (~590 nm) is similar to that of our Texas Red® dyes (~595 nm), making it our longest-wavelength Ca^{2+} indicator. Calcium Crimson's bright red fluorescence may help to eliminate interference by cellular autofluorescence.[36] Because this indicator tends to compartmentalize in some cell types, we now offer 10,000 MW and 70,000 MW dextran conjugates of Calcium Crimson (see Section 22.4).

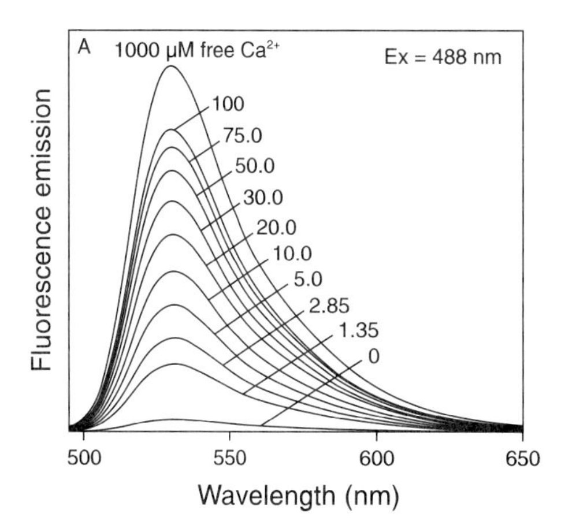

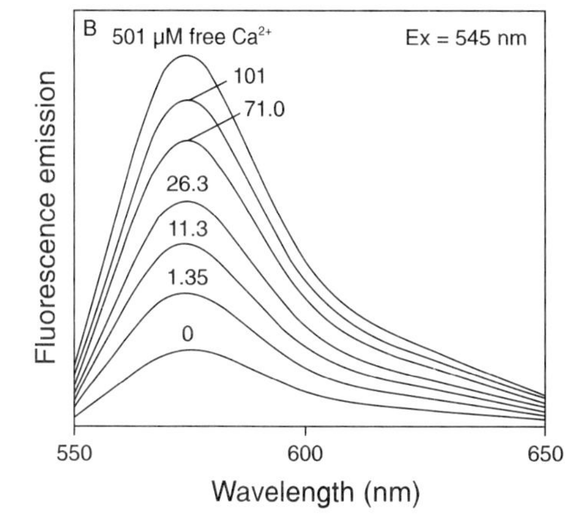

Figure 22.7 *Ca^{2+}-dependent fluorescence emission spectra of A) Calcium Green-5N (C-3737) and B) Calcium Orange-5N (C-6770) indicators.*

Calcium Green-5N and Calcium Orange-5N: Low-Affinity Ca^{2+} Indicators

The Ca^{2+} indicators Calcium Green-5N and Calcium Orange-5N, as well as the Mg^{2+} indicator Magnesium Green, have K_ds for Ca^{2+} in the absence of Mg^{2+} of ~14 µM, 20 µM and 6 µM, respectively[5] (Figure 22.7A and 22.7B). These low-affinity Ca^{2+} indicators exhibit relatively little fluorescence in all cells except those undergoing exceptionally high pulses of intracellular Ca^{2+}, which may occur during Ca^{2+}-induced Ca^{2+} release.[37,38] Calcium Green-5N, Calcium Orange-5N and Magnesium Green™ are not expected to be affected by normal levels of intracellular Mg^{2+}, and they buffer intracellular Ca^{2+} to a lesser extent than do the higher-affinity Ca^{2+} indicators. Furthermore, the high Ca^{2+} dissociation rates of these indicators are advantageous for tracking rapid Ca^{2+}-release kinetics.[39,40]

Use of the low-affinity Ca^{2+} indicator Calcium Green-5N in combination with the higher-affinity indicator Calcium Green-2 in the same experimental protocol can give an indication of the absolute magnitude of Ca^{2+} spikes.[41] Furthermore, coinjection of Ca^{2+}-sensitive Calcium Green-5N and Ca^{2+}-insensitive 8-aminonaphthalene-1,3,6-trisulfonic acid (ANTS, A-350; see Section 15.3) into *Limulus* ventral nerve photoreceptors permitted ratiometric measurement of Ca^{2+} flux.[42] Calcium Green-5N has also been shown to be effective for:

- Confocal imaging of Ca^{2+} microdomains in turtle hair cells [43]
- Following changes in IP_3-mediated Ca^{2+} distribution in *Xenopus* oocytes [39,44,45]
- Measuring rapid transient Ca^{2+} in dendrites and somata of Purkinje neurons [46]
- Monitoring increases in neuronal free Ca^{2+} associated with glutamate excitotoxicity [47,48]
- Studying light-stimulated Ca^{2+} flux in *Drosophila* photoreceptors [49]
- Visualizing Ca^{2+} transients in a single sarcomere with pulsed laser excitation and confocal spot detection [50]

Calcium Green-5N and Calcium Orange-5N are available as cell-impermeant potassium salts (C-3737, C-6770) or as cell-permeant AM esters (C-3739, C-6771). Magnesium Green, which is also discussed in Section 22.6 with the other Mg^{2+} indicators, is available as a cell-impermeant potassium salt (M-3733) or as a cell-permeant AM ester (M-3735).

Oregon Green 488 BAPTA Indicators

Our newest visible light–excitable Ca^{2+} indicators are Oregon Green 488 BAPTA-1, Oregon Green 488 BAPTA-2 and Oregon Green 488 BAPTA-5N. Based on our outstanding Oregon Green 488 fluorescein substitutes (see Section 1.4), these indicators have spectra virtually identical to those of fluorescein (excitation/emission maxima ~492/517 nm). The Oregon Green 488 BAPTA indicators are more efficiently excited by the 488 nm spectral line of the argon-ion laser than are the fluo-3 and Calcium Green indicators, both of which are based on the 2',7'-dichlorofluorescein chromophore (excitation/emission maxima ~504/529 nm).

Oregon Green 488 BAPTA-1 and Oregon Green BAPTA-2

The spectral properties of the Oregon Green BAPTA indicators may permit the use of lower dye concentrations when using the argon-ion laser for excitation, making the Oregon Green 488 BAPTA indicators the preferred reagents for intracellular Ca^{2+} measurements by confocal scanning laser microscopy. The absorbance of Oregon Green 488 BAPTA-1 at 488 nm is ~93% of the absorbance at its absorption wavelength maxima, whereas the absorbance of fluo-3 and Calcium Green at 488 nm is only ~45% of their maxima. Furthermore, the quantum yields of the Ca^{2+} complexes of Oregon Green BAPTA-1 and Calcium Green-1 are ~0.7, as compared to only ~0.18 for fluo-3.

As with Calcium Green-1 (see Figure 22.6A), Oregon Green 488 BAPTA-1 is moderately fluorescent in Ca^{2+}-free solution, and its fluorescence is enhanced about 14-fold at saturating Ca^{2+} (Figure 22.8A). Oregon Green 488 BAPTA-1 has a K_d for Ca^{2+} in the absence of Mg^{2+} of about 170 nM.[5] Oregon Green 488 BAPTA-2 is similar to Calcium Green-2 in that it contains two dye molecules per BAPTA chelator and exhibits very low fluorescence in the absence of Ca^{2+}. The fluorescence of Oregon Green 488 BAPTA-2 is enhanced at least 37-fold at saturating Ca^{2+}, and it has a K_d for Ca^{2+} in the absence of Mg^{2+} of ~580 nM [5] (Figure 22.8B). Other properties of the Oregon Green indicators essentially mimic those of the Calcium Green-1 and Calcium Green-2 indicators.

Oregon Green 488 BAPTA-1 and Oregon Green BAPTA-2 are available as cell-impermeant potassium salts (O-6806, O-6808) or as cell-permeant AM esters (O-6807, O-6809), which are specially packaged as sets of 10 vials, each containing 50 µg.

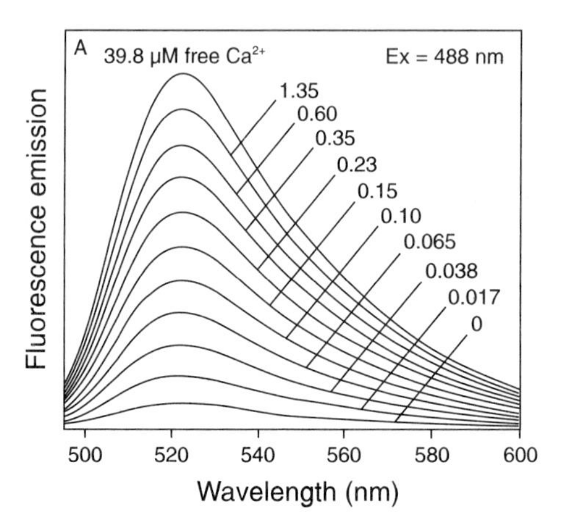

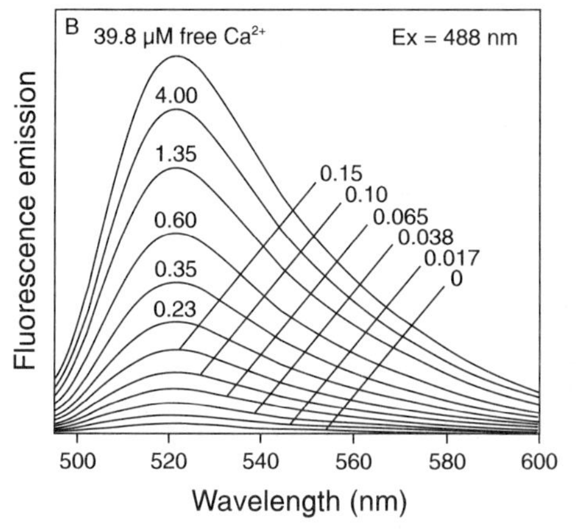

Figure 22.8 *Ca^{2+}-dependent fluorescence emission spectra of A) Oregon Green 488 BAPTA-1 (O-6806) and B) Oregon Green 488 BAPTA-2 (O-6808).*

Oregon Green 488 BAPTA-5N: A Low-Affinity Ca^{2+} Indicator

As with Calcium Green-5N (see above), our new Oregon Green BAPTA-5N is a lower-affinity Ca^{2+} indicator (K_d in the absence of Mg^{2+} ~20 µM) [5] that is expected to be useful for measuring Ca^{2+} spikes above 1 µM. Saturation of the indicator with Ca^{2+} increases its fluorescence at 521 nm by >50-fold. Both the cell-impermeant potassium salt (O-6812) and cell-permeant AM ester (O-6813) of Oregon Green BAPTA-5N are available.

Fura Red Indicator

Our visible light–excitable fura-2 analog, Fura Red [51] offers unique possibilities for ratiometric measurement of Ca^{2+} in single cells by microphotometry, imaging or flow cytometry. The visible-wavelength excitation (450–500 nm) and very long-wavelength emission maximum (~660 nm) of Fura Red eliminate interference from autofluorescence in most cells. Fluorescence of Fura Red excited at 488 nm *decreases* once the indicator binds Ca^{2+} (see Figure 22.5). Even in the absence of Ca^{2+}, the fluorescence of Fura Red is much weaker than that of the other visible light–excitable Ca^{2+} indicators, necessitating the use of higher concentrations of the indicator in cells to produce equivalent fluorescence.

Ratiometric measurements of Ca^{2+} levels in frog skeletal muscle fibers with Fura Red have been made using excitation wavelengths of 420 nm and 480 nm.[52] A simultaneous assay for Ca^{2+} uptake and ATP hydrolysis by sarcoplasmic reticulum has been developed that uses the large *absorbance* change of Fura Red upon Ca^{2+} binding.[53] This assay, which can measure Ca^{2+} uptake — and probably uptake of heavy metal ions through channels (see Table 22.3 in Section 22.7) — from the medium in real time, avoids the need for radioactive Ca^{2+} and should be generally useful for measuring Ca^{2+} uptake by cells. In several cell types, simultaneous labeling with Fura Red and fluo-3 has enabled researchers to use ratiometric measurements for estimating intracellular Ca^{2+} levels using confocal laser scanning microscopy [6-9,11,54] or flow cytometry.[12]

Furthermore, the huge Stokes shift of Fura Red permits multicolor analysis of Fura Red in combination with fluorescein or fluorescein-like dyes using only a single excitation wavelength (Figure 22.5). For example, researchers have been able to simultaneously measure Ca^{2+} fluxes and oxidative bursts in monocytes and granulocytes by simultaneously measuring the fluorescence of Fura Red and rhodamine 123 — the oxidation product of the probe dihydrorhodamine 123 (D-632, see Section 17.1).[55]

Fura Red is available as either a cell-impermeant tetraammonium salt (F-3019) or as a cell-permeant AM ester (F-3020, F-3021).

Calcein and Calcein Blue

Calcein (C-481, see Section 22.7) and calcein blue (H-1426, see Section 22.7) are relatively low-affinity Ca^{2+} chelators. The K_ds for both the Ca^{2+} and Mg^{2+} complexes of calcein at physiological pH are about 10^{-3} to 10^{-4} M.[56] As derivatives of iminodiacetic acid, these dyes exhibit an ion affinity that increases considerably at higher pH, and neither dye is particularly useful for measuring Ca^{2+} or Mg^{2+} in cells. Calcein is nonfluorescent above pH 12 but forms fluorescent complexes with Ca^{2+}, Mg^{2+} and several other metals [57-60] (see Table 22.3 in Section 22.7).

Calcein has been used to assay for Ca^{2+} in serum,[61,62] to measure Ca^{2+} binding to sarcoplasmic reticulum [56] and to determine intracellular Fe^{3+} [63] (see Section 22.7). Calcein blue has also been used for fluorometric determination of Ca^{2+} in serum.[61,62]

1. Fluo-3 and rhod-2 are licensed to Molecular Probes under U.S. Patent No. 5,049,673 and related patents; **2.** J Biol Chem 264, 8171 (1989); **3.** J Neurosci 13, 632 (1993); **4.** Biophys J 65, 865 (1993); **5.** K_d values for Ca^{2+} are determined at Molecular Probes at ~22°C using our Calcium Calibration Buffer Kits (see Section 22.8); **6.** Cell Calcium 19, 255 (1996); **7.** Cell Calcium 19, 3 (1996); **8.** Cell Calcium 18, 5 (1995); **9.** Cell Calcium 16, 279 (1994); **10.** Cell Calcium 15, 341 (1994); **11.** Cell Calcium 14, 359 (1993); **12.** Cytometry 17, 135 (1994); **13.** Exp Cell Res 217, 410 (1995); **14.** J Biol Chem 270, 29781 (1995); **15.** J Biol Chem 269, 30636 (1994); **16.** Cytometry 11, 923 (1990); **17.** Cytometry 24, 99 (1996); **18.** Cytometry 14, 257 (1993); **19.** Proc Natl Acad Sci USA 89, 3591 (1992); **20.** Cell 82, 415 (1995); **21.** Mills, L. in *Three Dimensional Confocal Microscopy: Volume Investigation of Biological Specimens* J.K. Stevens, L.R. Mills and J.E. Trogadis, Eds. Academic Press (1994) pp. 253–280; **22.** FEBS Lett 382, 31 (1996); **23.** U.S. Patent No. 5,453,517 and patents pending; **24.** Biochem Biophys Res Comm 180, 209 (1991); **25.** J Fluorescence 4, 291 (1994); **26.** SPIE Proc 1640, 390 (1992); **27.** Stricker, S.A. in *Three Dimensional Confocal Microscopy: Volume Investigation of Biological Specimens*, J.K. Stevens, L.R. Mills and J.E. Trogadis, Eds. pp. 281–300 Academic Press (1994); **28.** Devel Biol 149, 370 (1992); **29.** Pflügers Arch 430, 529 (1995); **30.** Clapham, D.E. and Sneyd, J. in *Advances in Second Messengers and Phosphoprotein Research*, A.R. Means, Ed., Raven Press pp. 1–24 (1995); **31.** Science 260, 229 (1993); **32.** Biophys J 61, 509 (1992); **33.** J Biol Chem 267, 17722 (1992); **34.** Neuron 13, 837 (1994); **35.** J Neurosci 16, 71 (1996); **36.** J Neurosci 15, 5535 (1995); **37.** Proc Natl Acad Sci USA 87, 1461 (1990); **38.** J Theor Biol 93, 1009 (1981); **39.** Biophys J 70, 1006 (1996); **40.** FEBS Lett 364, 335 (1995); **41.** J Neurosci 15, 4209 (1995); **42.** J Gen Physiol 105, 95 (1995); **43.** Neuron 15, 1323 (1995); **44.** Cell Calcium 15, 276 (1994); **45.** J Physiol 476, 17 (1994); **46.** Proc Natl Acad Sci USA 92, 10272 (1995); **47.** Neuron 11, 751 (1993); **48.** Neurosci Lett 162, 149 (1993); **49.** Neuron 12, 1257 (1994); **50.** Nature 367, 739 (1994); **51.** Fura Red is licensed to Molecular Probes under U.S. Patent No. 4,849,362; **52.** Biophys J 64, 1934 (1993); **53.** Anal Biochem 227, 328 (1995); **54.** Cell Calcium 15, 341 (1994) (1994); **55.** Cytometry 13, 693 (1992); **56.** Biophys J 18, 3 (1977); **57.** Anal Chem 46, 2036 (1974); **58.** Anal Chem 35, 1035 (1963); **59.** Anal Chem 31, 456 (1959); **60.** Analyst 82, 284 (1957); **61.** Analyst 113, 251 (1988); **62.** Anal Chem 35, 1238 (1963); **63.** J Biol Chem 270, 24209 (1995).

22.3 Data Table *Fluorescent Ca^{2+} Indicators Excited by Visible Light*

Cat #	Structure	MW	Storage	Soluble	Low Ca^{2+} Abs	Low Ca^{2+} {ε × 10^{-3}}	Low Ca^{2+} Em	High Ca^{2+} Abs	High Ca^{2+} {ε × 10^{-3}}	High Ca^{2+} Em	Solvent	Cleaved*	K_d†	Notes
C-3010	✓	1147	D,L	pH>6	506	{81}	531	506	{82}	531	<1>		190 nM	<2, 3>
C-3011	✓	1291	F,D	DMSO	302	{17}	none				MeOH	C-3010		
C-3013	✓	1087	D,L	pH>6	549	{80}	575	549	{80}	576	<1>		185 nM	<2, 3>
C-3015	✓	1223	F,D,L	DMF, DMSO	540	{94}	566				MeOH	C-3013		
C-3016	✓	1233	D,L	pH>6	590	{96}	615	589	{92}	615	<1>		185 nM	<2, 3>
C-3018	✓	1368	F,D,L	DMSO	583	{113}	602				MeOH	C-3016		
C-3730	✓	1666	D,L	pH>6	506	{95}	536	503	{147}	536	<1>		550 nM	<2, 3>
C-3732	✓	1817	F,D	DMSO	302	{29}	none				MeOH	C-3730		
C-3737	✓	1192	D,L	pH>6	506	{83}	532	506	{82}	532	<1>		14 µM	<3, 4>
C-3739	✓	1336	F,D	DMSO	361	{15}	none				EtOAc	C-3737		
C-6770	✓	1132	D,L	pH>6	549	{69}	582	549	{72}	582	<1>		20 µM	<3, 4>
C-6771	✓	1268	F,D,L	DMSO	540	{85}	566				MeOH	C-6770		
F-1240	✓	855	D,L	pH>6	503	{90}	<5>	506	{100}	526	<1>		390 nM	<2>
F-1241	✓	1130	F,D,L	DMSO	464	{26}	<6>				MeOH	F-1240		
F-3019	✓	725	D,L	pH>6	472	{29}	657	436	{41}	637	<1>		140 nM	<2, 7>
F-3020	✓	1089	F,D,L	DMSO	458	{43}	597				MeOH	F-3019		
F-3715	✓	960	D,L	pH>6	506	{90}	<5>	506	{100}	526	<1>		390 nM	<2>
M-3733	S22.6	916	L	pH>6	506	{77}	531	506	{77}	531	<1>		6 µM	<3, 4>
M-3735	S22.6	1026	F,D	DMSO	302	{16}	none				MeOH	M-3733		

Cat #	Structure	MW	Storage	Soluble	Abs	Low Ca^{2+} {ε × 10^{-3}}	Em	Abs	High Ca^{2+} {ε × 10^{-3}}	Em	Solvent	Cleaved*	K_d†	Notes
O-6806	✓	1114	D,L	pH>6	494	{76}	523	494	{78}	523	<1>		170 nM	<2, 3>
O-6807	✓	1258	F,D	DMSO	299	{19}	none				EtOAc	O-6806		
O-6808	✓	1600	D,L	pH>6	494	{105}	523	494	{140}	523	<1>		580 nM	<2, 3>
O-6809	✓	1751	F,D,L	DMSO	299	{31}	none				MeOH	O-6808		
O-6812	✓	1159	DD,L	pH>6	494	{72}	521	494	{76}	521	<1>		20 µM	<3, 4>
O-6813	✓	1303	F,D	DMSO	300	{17}	none				MeOH	O-6812		
R-1243	✓	806	D,L	pH>6	549	{79}	<5>	552	{82}	581	<1>		570 nM	<2>
R-1244	✓	1124	F,D,L	DMSO	550	{125}	571				<8>	R-1243		

C-3012 see C-3011; F-1242 see F-1241; F-3021 see F-3020; R-1245 see R-1244.

For definitions of the contents of this data table, see "How to Use the Handbook" on page vi.
Structure: Chemical structure drawing: ✓ = shown in this section; Sn.n = shown in Section number n.n.

* Catalog numbers listed in this column indicate the cleavage product generated by enzymatic or chemical hydrolysis of the acetoxymethyl ester groups of the parent compound.

† Data on dissociation constants come from several different sources. These values are known to vary considerably depending on the temperature, pH, ionic strength, viscosity, protein binding, presence of other ions (especially polyvalent ions), instrument setup and other factors. It is strongly recommended that these values be specifically determined under your experimental conditions by using our Calcium Calibration Buffer Kits (see Section 22.8).

<1> Spectra measured in aqueous buffers containing 10 mM EGTA (low Ca^{2+}) or a >10-fold excess of free Ca^{2+} relative to the K_d (high Ca^{2+}).

<2> Dissociation constant determined at Molecular Probes by fluorescence measurements in 100 mM KCl, 10 mM MOPS, pH 7.2, 0 to 10 mM CaEGTA at 22°C.

<3> This indicator exhibits fluorescence enhancement in response to ion binding, with essentially no change in absorption or emission wavelengths.

<4> Dissociation constant determined at Molecular Probes by fluorescence measurements in 100 mM KCl, 10 mM MOPS, pH 7.2, 0 to 1 mM free Ca^{2+} at 22°C.

<5> Fluorescence of unbound fluo-3 and rhod-2 is very weak and is enhanced >100-fold on binding Ca^{2+}.

<6> F-1241 fluorescence is very weak. Fluorescence is enhanced only after complete hydrolysis to F-1240 followed by binding of Ca^{2+}.

<7> The fluorescence quantum yield of Fura Red is low (~0.013 in Ca^{2+}-free solution [Meth Cell Biol 40, 155 (1994)]).

<8> Spectra measured in 90:10 (v/v) $CHCl_3$:MeOH.

22.3 Structures *Fluorescent Ca^{2+} Indicators Excited by Visible Light*

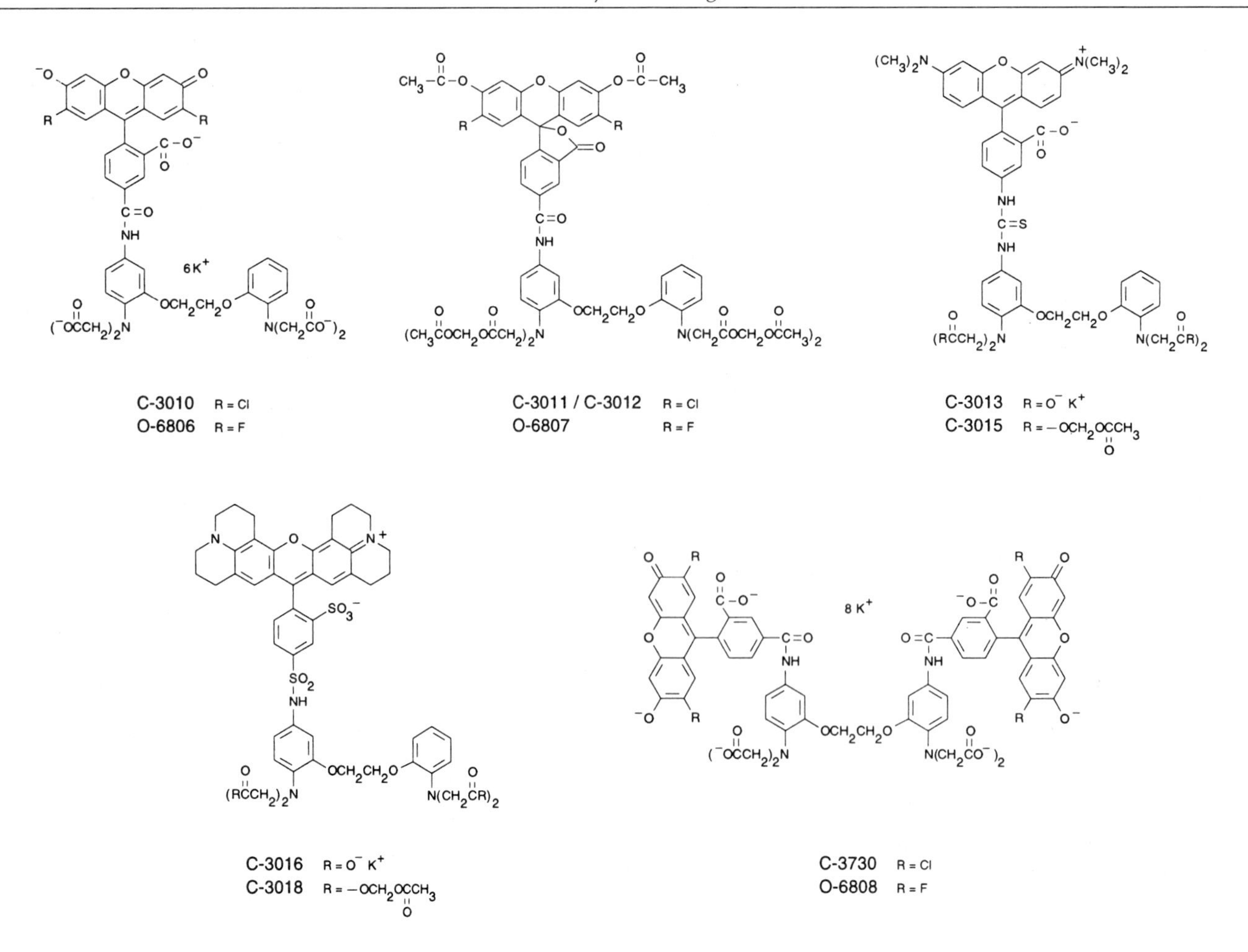

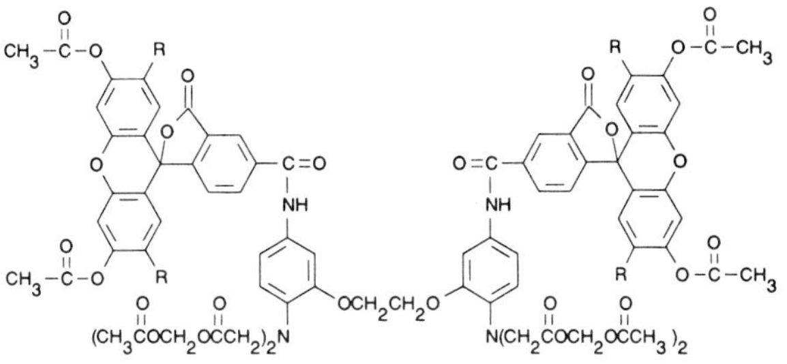

$CH_3-C(=O)-O$ R O $O-C(=O)-CH_3$ R O
O C=O O=C O
NH NH
$CH_3-C(=O)-O$ R R $O-C(=O)-CH_3$
$(CH_3COCH_2OCCH_2)_2N$ OCH_2CH_2O $N(CH_2COCH_2OCCH_3)_2$

C-3732 R = Cl
O-6809 R = F

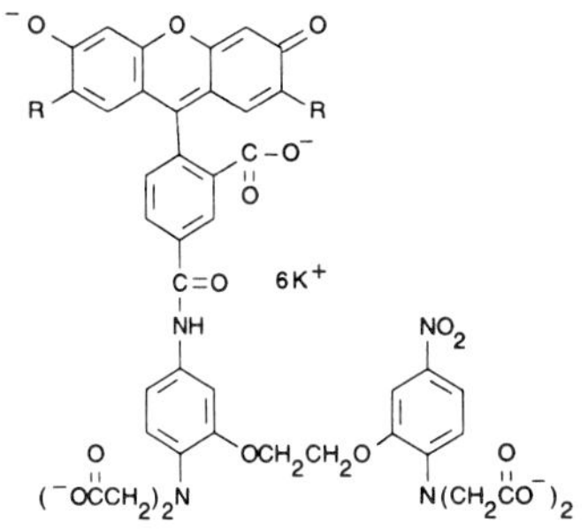

^-O O O R R
$C-O^-$ (C=O)
C=O $6K^+$
NH NO_2
$(^-OCCH_2)_2N$ OCH_2CH_2O $N(CH_2CO^-)_2$

C-3737 R = Cl
O-6812 R = F

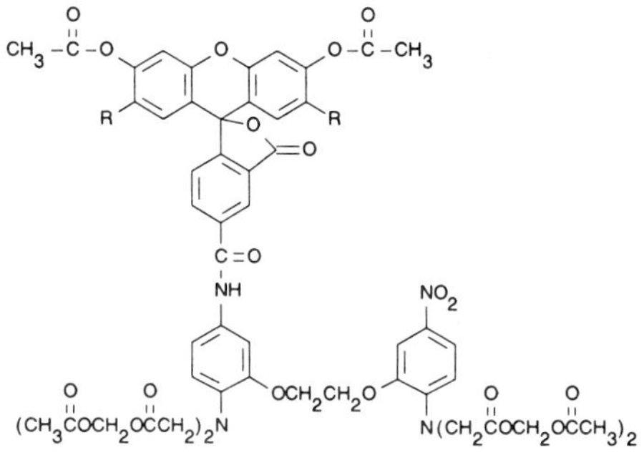

$CH_3-C(=O)-O$ O $O-C(=O)-CH_3$
R O R
O
C=O
NH NO_2
$(CH_3COCH_2OCCH_2)_2N$ OCH_2CH_2O $N(CH_2COCH_2OCCH_3)_2$

C-3739 R = Cl
O-6813 R = F

$(CH_3)_2N$ O $\overset{+}{N}(CH_3)_2$
$C-O^-$ (C=O)
NH
C=S
NH NO_2
$4K^+$
$(RCCH_2)_2N$ OCH_2CH_2O $N(CH_2CR)_2$

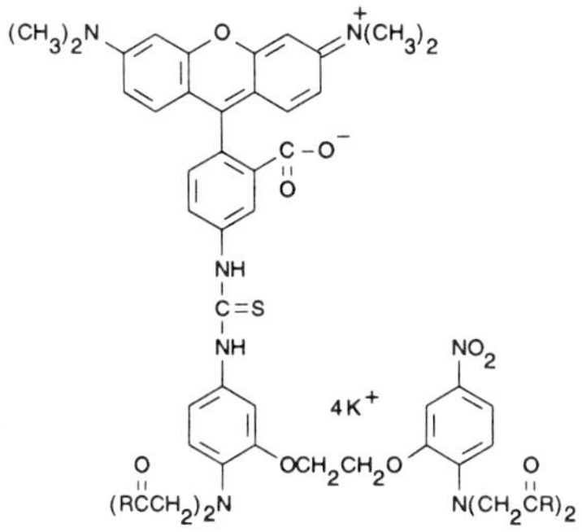

C-6770 R = O^- K^+
C-6771 R = $-OCH_2OCCH_3$ (C=O)

R O O
Cl Cl CH_3
$(RCCH_2)_2N$ OCH_2CH_2O $N(CH_2CR)_2$

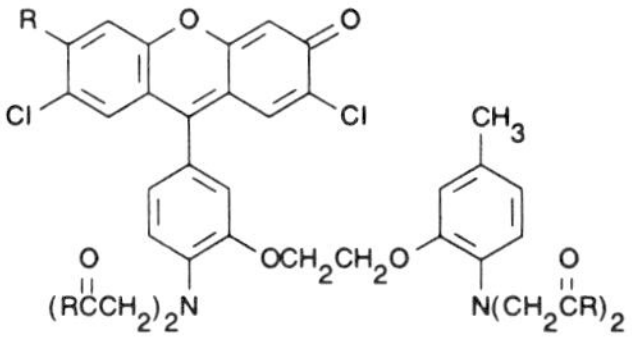

F-1240 R = O^- NH_4^+
F-3715 R = O^- K^+
F-1241 / F-1242 R = $-OCH_2OCCH_3$ (C=O)

$(^-OCCH_2)_2N$ $N(CH_2CO^-)_2$
OCH_2CH_2O
O $4NH_4^+$ CH_3
HC
HN O
S NH

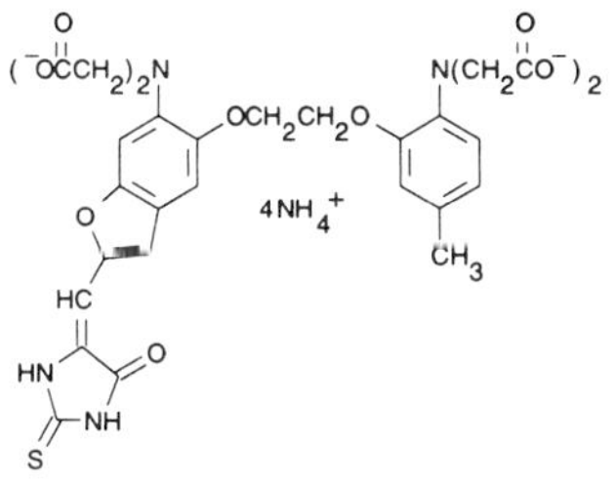

F-3019

$(CH_3COCH_2OCCH_2)_2N$ $N(CH_2COCH_2OCCH_3)_2$
OCH_2CH_2O
O CH_3
HC
CH_3COCH_2-N O
S N CH_2OCCH_3

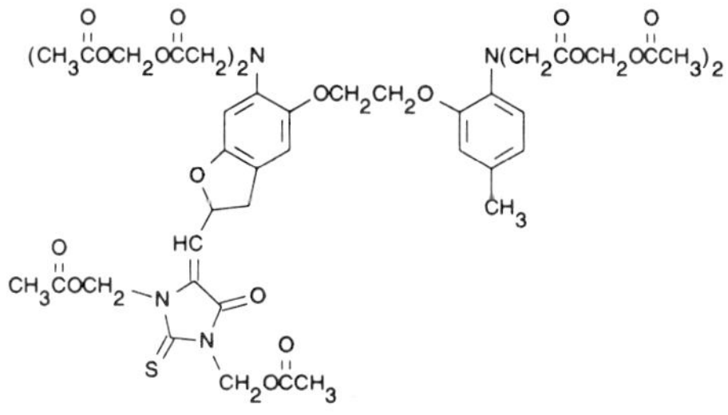

F-3020 / F-3021

M-3733, M-3735
Section 22.6

O-6806	see C-3010
O-6807	see C-3011
O-6808	see C-3730
O-6809	see C-3732
O-6812	see C-3737
O-6813	see C-3739

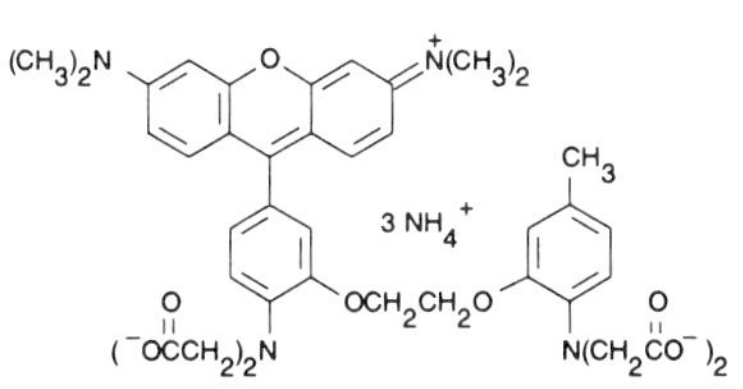

$(CH_3)_2N$ O $\overset{+}{N}(CH_3)_2$
CH_3
$3\ NH_4^+$
$(^-OCCH_2)_2N$ OCH_2CH_2O $N(CH_2CO^-)_2$

R-1243

$(CH_3)_2N$ O $\overset{+}{N}(CH_3)_2$ Br^-
CH_3
$(CH_3COCH_2OCCH_2)_2N$ OCH_2CH_2O $N(CH_2COCH_2OCCH_3)_2$

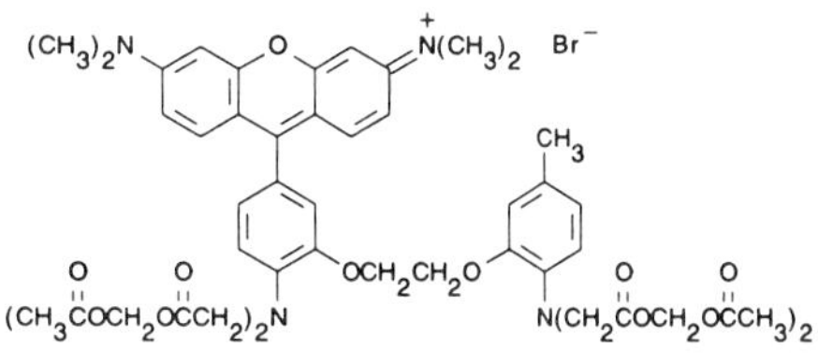

R-1244 / R-1245

Cat #		Product Name	Unit Size	Price Per Unit ($) 1–4 Units	5–24 Units
C-3018		Calcium Crimson™, AM *cell permeant* *special packaging*	10x50 µg	95.00	76.00
C-3016		Calcium Crimson™, tetrapotassium salt *cell impermeant*	500 µg	68.00	54.40
C-3011		Calcium Green™-1, AM *cell permeant* *≥97% by HPLC*	500 µg	95.00	76.00
C-3012		Calcium Green™-1, AM *cell permeant* *≥97% by HPLC* *special packaging*	10x50 µg	110.00	88.00
C-3010		Calcium Green™-1, hexapotassium salt *cell impermeant*	500 µg	85.00	68.00
C-3732		Calcium Green™-2, AM *cell permeant* *special packaging*	10x50 µg	95.00	76.00
C-3730		Calcium Green™-2, octapotassium salt *cell impermeant*	500 µg	68.00	54.40
C-3739		Calcium Green™-5N, AM *cell permeant* *≥95% by HPLC* *special packaging*	10x50 µg	110.00	88.00
C-3737		Calcium Green™-5N, hexapotassium salt *cell impermeant*	500 µg	85.00	68.00
C-3015		Calcium Orange™, AM *cell permeant* *≥95% by HPLC* *special packaging*	10x50 µg	95.00	76.00
C-3013		Calcium Orange™, tetrapotassium salt *cell impermeant*	500 µg	68.00	54.40
C-6771	**New**	Calcium Orange™-5N, AM *cell permeant* *≥95% by HPLC* *special packaging*	10x50 µg	110.00	88.00
C-6770	**New**	Calcium Orange™-5N, tetrapotassium salt *cell impermeant*	500 µg	85.00	68.00
F-1241		fluo-3, AM *cell permeant*	1 mg	165.00	132.00
F-1242		fluo-3, AM *cell permeant* *special packaging*	20x50 µg	195.00	156.00
F-1240		fluo-3, pentaammonium salt *cell impermeant*	1 mg	135.00	108.00
F-3715		fluo-3, pentapotassium salt *cell impermeant*	1 mg	155.00	124.00
F-3020		Fura Red™, AM *cell permeant* *≥95% by HPLC*	500 µg	155.00	124.00
F-3021		Fura Red™, AM *cell permeant* *≥95% by HPLC* *special packaging*	10x50 µg	175.00	140.00
F-3019		Fura Red™, tetraammonium salt *cell impermeant*	500 µg	125.00	100.00
M-3735		Magnesium Green™, AM *cell permeant* *≥95% by HPLC* *special packaging*	20x50 µg	165.00	132.00
M-3733		Magnesium Green™, pentapotassium salt *cell impermeant*	1 mg	100.00	80.00
O-6807	**New**	Oregon Green™ 488 BAPTA-1, AM *cell permeant* *≥95% by HPLC* *special packaging*	10x50 µg	125.00	100.00
O-6806	**New**	Oregon Green™ 488 BAPTA-1, hexapotassium salt *cell impermeant* *≥95% by HPLC*	500 µg	95.00	76.00
O-6809	**New**	Oregon Green™ 488 BAPTA-2, AM *cell permeant* *special packaging*	10x50 µg	125.00	100.00
O-6808	**New**	Oregon Green™ 488 BAPTA-2, octapotassium salt *cell impermeant*	500 µg	95.00	76.00
O-6813	**New**	Oregon Green™ 488 BAPTA-5N, AM *cell permeant* *≥95% by HPLC* *special packaging*	10x50 µg	125.00	100.00
O-6812	**New**	Oregon Green™ 488 BAPTA-5N, hexapotassium salt *cell impermeant* *≥95% by HPLC*	500 µg	95.00	76.00
R-1244		rhod-2, AM *cell permeant*	1 mg	165.00	132.00
R-1245		rhod-2, AM *cell permeant* *special packaging*	20x50 µg	195.00	156.00
R-1243		rhod-2, triammonium salt *cell impermeant*	1 mg	135.00	108.00

22.4 Fluorescent Ca^{2+} Indicator Conjugates

Dextran Conjugates

When ion indicators are loaded into cells as their acetoxymethyl (AM) esters, they may translocate to intracellular compartments, where they are still fluorescent but no longer respond to changes in cytosolic ion levels.[1,2] This problem frequently limits the experiment's duration because sequestration of the indicator into organelles will cause errors in the estimated cytosolic ion levels. Furthermore, Ca^{2+} indicators such as fura-2 and indo-1 may bind to cellular proteins,[1-5] which can markedly alter the indicator's response to Ca^{2+}. In one case, up to 85% of fura-2 was reported to be bound to proteins, resulting in a threefold change in its K_d for Ca^{2+}.[6]

To overcome these limitations, Molecular Probes prepares dextran conjugates of ion indicators. Dextrans are hydrophilic polysaccharides characterized by their moderate to high molecular weight, good water solubility and low toxicity. They are also biologically inert due to their uncommon poly-(α-D-1,6-glucose) linkages, which render them resistant to cleavage by most endogenous cellular glycosidases. Indicator dextrans must be loaded into cells by microinjection, patch-clamp methods, scrape loading, endocytosis, liposome fusion or comparable techniques (see Table 15.1 in Chapter 15). However, once loaded, dextran conjugates are well retained in viable cells, will not pass through gap junctions and are less likely to become compartmentalized.[7,8] Also, fluorescence photobleaching measurements have shown that, as compared to low molecular weight dyes, dextran conjugates are much less likely to bind to proteins.[9] Because dextran conjugates are intrinsically polydisperse and their degree of substitution may vary with the production lot, the K_d of each lot of these indicators should be calibrated independently.

Fura and Indo Dextrans

Molecular Probes' dextran conjugates of fura and indo [10] tend to remain in the cytosol without compartmentalization or leakage and are less likely to bind to cellular proteins, making them useful for long-term Ca^{2+} measurements.[11,12] Although the spectral response curves of the conjugates are very similar to those of the free dyes, their affinity for Ca^{2+} is somewhat weaker. The K_ds for Ca^{2+} of fura and indo dextrans in the absence of Mg^{2+} vary between 200 nM and 400 nM,[13] depending on the molecular weight of the dextran and individual batch characteristics.

Dye compartmentalization has been especially problematic for measurements of ions in plant cells, where the dye is frequently transported out of the cytosol in minutes. Unlike microinjected fura-2 salt, fura dextran is retained for hours in the cytosol of stamen hair cells and *Lilium* pollen tubes.[14] A comparison of the dextran conjugates of fura and Calcium Green with conventional indicators for imaging Ca^{2+} levels in plant and fungal cells has been published.[15] Fura dextran has been used to monitor Ca^{2+} levels in a wide variety of cells, including:

- *Dictyostelium discoideum*, where it was used for long-term Ca^{2+} measurements [12]
- Fertilized bovine eggs,[16] rabbit eggs,[17] chicken eggs [18] and *Fucus* eggs [19]
- Giant presynaptic terminals of the chick ciliary ganglion [20]
- Growing unicells of green algae [21]
- Isolated liver nuclei, where fura dextran is uniformly distributed in the nucleoplasm [22]
- Mouse neuroblastoma cells, where fura dextran has been used to determine the effect of Ca^{2+} buffers on the propagation of Ca^{2+} waves [23]
- Plant and fungal cells [15]
- Retrograde-labeled chick embryo neurons [18,24]

Indo dextran has been used to measure the Ca^{2+} feedback signal in the phototransduction cascade of vertebrate rods.[25]

Fura is available conjugated to 3000, 10,000 and 70,000 MW dextrans (F-6764, F-3029, F-3030); indo is available conjugated to 10,000 and 70,000 MW dextrans (I-3032, I-3033).

Calcium Green, Oregon Green 488 BAPTA and Calcium Crimson Dextran

Molecular Probes offers 3000, 10,000, 70,000 and 500,000 MW dextran conjugates of our Calcium Green-1 indicator (C-6765, C-3713, C-3714, C-6766) and 10,000 and 70,000 MW dextran conjugates of our Oregon Green 488 BAPTA-1 indicator [10] (O-6798, O-6797). A review by Read and co-workers compares the Calcium Green-1 and fura dextrans with conventional indicators for imaging Ca^{2+} levels in plant and fungal cells.[15] Spectra of the new Oregon Green 488 BAPTA-1 dextrans match the 488 nm spectral line of the argon-ion laser and the standard fluorescein optical filters better than do the Calcium Green-1 indicator dextrans, which should permit the use of lower probe concentrations to achieve the same signal. The visible light–excitable Calcium Green-1 indicator dextrans have been used to examine:

- Ca^{2+} levels for up to four days in retrogradely labeled spinal cord and brainstem neurons of the lamprey [26]
- Ca^{2+} levels in neurons of live zebrafish [27] and in fertilized starfish eggs (Figure 22.9)
- Changes in Ca^{2+} levels during mitosis in one-cell mouse embryos [28] and Ca^{2+} oscillations that occur during the growth and maturation of mouse oocytes [29]
- Ca^{2+}-regulated exocytosis [30] and caffeine-induced Ca^{2+} release [31] in sea urchin eggs and embryos
- Changes in intraciliary Ca^{2+}-controlling ciliary motility in the ctenophore *Mnemiopsis leidyi* [32]
- Cytosolic free Ca^{2+} during *Fucus* egg activation [19]
- Effects of antidromic stimulation on Ca^{2+} levels in retrogradely labeled motoneurons [18,24]
- Increase in intracellular free Ca^{2+} in serum-stimulated fibroblasts [33]

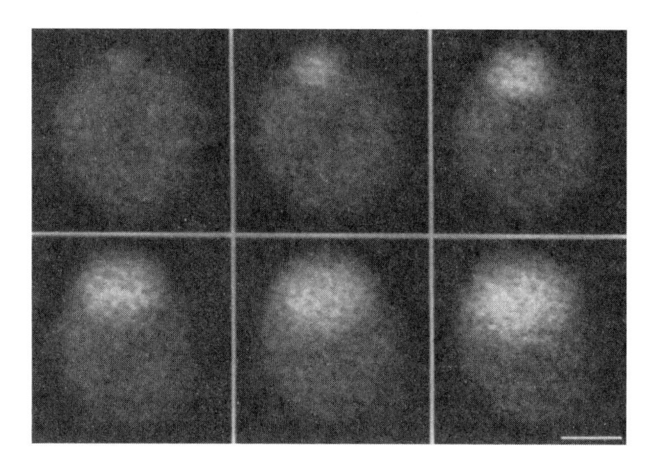

Figure 22.9 Propagation of a fertilization-induced Ca^{2+} wave in a starfish (Asterina miniata) egg visualized at one-second intervals using microinjected Calcium Green-1 dextran (C-3713). Photo contributed by Mark Terasaki, National Institutes of Health.

- Regulation of size-specific entry of molecules into the *Xenopus* oocyte nucleus by the nuclear Ca^{2+} store [34]
- Role of gibberellic and abscisic acids on Ca^{2+}-modulated secretion in barley aleurone protoplasts [35]
- Synchronized Ca^{2+} waves in *Xenopus laevis* oocytes [36]
- Transmission of changes in cytosolic Ca^{2+} levels to the nucleus by the nuclear membrane [37]
- Variation in intracellular Ca^{2+} levels in zebrafish embryos during early cell cleavage [38,39]

The combination of Calcium Green dextran and Fura Red (F-3020, see Section 22.3) has been used to simultaneously monitor intracellular and extracellular Ca^{2+} levels.[40] Calcium Green-1 dextran has also been coinjected into starfish embryos with a Ca^{2+}-insensitive rhodamine dextran (D-1824, see Section 15.5) to permit dual-channel confocal ratioing.[41] Our Calcium Green-1 and Oregon Green 488 BAPTA-1 dextran conjugates should also be useful for monitoring the changes in intracellular free Ca^{2+} that accompany photolysis of diazo-2, caged EGTA, caged EDTA or other caged Ca^{2+} probes (see Section 22.8).

Our newest indicator dextrans are the Calcium Crimson 10,000 MW and 70,000 MW dextrans (C-6824, C-6825). Although the fluorescence increase of these conjugates upon binding is small — about 2.5–3 fold between zero and saturating Ca^{2+} — the Calcium Crimson dextran has the longest-wavelength fluorescence spectra (excitation maximum ~593 nm) of all the available Ca^{2+} indicators, thus providing a probe with good spectral separation from the fluorescence of fluorescein-like dyes.

Calcium Green-1 and Texas Red Dextran: A Ratioable Indicator Dextran

A shortcoming of the visible light–excitable Ca^{2+} indicators has been their inability to report signals that may be quantitated ratiometrically. The ion-dependent spectral shifts of indicators such as fura-2 and indo-1 allow them to be used ratiometrically, making the Ca^{2+} measurements essentially independent of differences in dye distribution, cell thickness, photobleaching and other factors. Our new 70,000 MW dextran simultaneously conjugated to Calcium Green-1 and Texas Red dyes (C-6800) provides a Ca^{2+} indicator

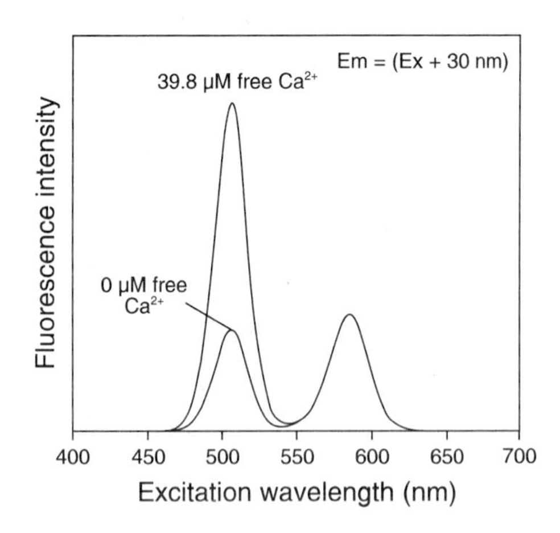

Figure 22.10 *Ca^{2+}-dependent fluorescence spectra of Calcium Green-1 and Texas Red 70,000 MW dextran (C-6800) in solutions containing zero and 39.8 µM free Ca^{2+}. Excitation and emission wavelengths were scanned synchronously with a 30 nm offset. The magnitude of this response may vary among different batches of indicator.*

that is excited with visible wavelengths and simultaneously emits Ca^{2+}-sensitive green and Ca^{2+}-insensitive red fluorescent signals (Figure 22.10). Because these signals originate from structurally linked fluorophores, the dyes must have equivalent distribution patterns, unlike the uneven distribution that may result from co-injection of Fura Red and fluo-3.[42] The double-labeled Calcium Green and Texas Red dextran should be useful for ratiometric Ca^{2+} measurements using confocal laser scanning microscopes equipped with the Ar-Kr mixed gas laser, which provides both 488 nm (for Ca^{2+}-dependent Calcium Green excitation) and 568 nm (for Ca^{2+}-independent Texas Red excitation) excitation. As with all of our indicator dextrans, the K_d of each lot of this dextran conjugate should be calibrated independently because it may vary with the production lot.

Mixtures of Calcium Green dextran and Ca^{2+}-insensitive dextrans (such as tetramethylrhodamine or Texas Red dextrans, see Section 15.5) have been coinjected into cells for use in ratio-imaging microscopy.[35,38,39,41] Tetrakis-(4-carboxyphenyl)porphine (TCPP, T-6931; see Section 22.7), tetrakis-(4-sulfophenyl)porphine (TSPP, T-6932; see Section 22.7) and our porphine dextrans (D-6929, D-6930; see Section 22.7) may also be useful as Ca^{2+}- and Mg^{2+}-insensitive standards (see Table 22.4 in Section 22.7) for dual-emission ratio-imaging microscopy. TCPP and TSPP have extremely strong absorption near 414 nm (ϵ >400,000 $cm^{-1}M^{-1}$) and several weaker absorption peaks between 500 nm and 650 nm; excitation at any of these absorption peak wavelengths yields emission at ~645 nm with a shoulder at about 700 nm.

Mag-Indo Dextrans: Low-Affinity Ca^{2+} Indicators

Mag-indo is a dual-emission indicator that has a higher affinity for Ca^{2+} than for Mg^{2+}. When conjugated to a 10,000 or 70,000 MW dextran (M-6907, M-6908), the indicator exhibits a K_d for Ca^{2+} of ~35 µM [13] and K_d for Mg^{2+} of about 2.5 mM.[43] These conjugates may have specialized uses for measuring Ca^{2+} levels in the 1–100 µM range and for measuring fast transients.

Protein Conjugates

Molecular Probes occasionally receives requests for fura conjugated to biomolecules other than dextrans for site-selective targeting. Our chemists have developed an amine-reactive fura derivative [10] that we will covalently attach to the biomolecule of your choice. One example of such a conjugate is our new fura BSA (A-6796), which may be useful as a microinjectable probe for intracellular Ca^{2+} measurements. To discuss your specific research requirements, please contact our Custom and Bulk Sales Department.

Lipophilic Derivatives for Detecting Calcium Near Membrane Surfaces

Fura C_{18} and Calcium Green C_{18}

Changes in intracellular Ca^{2+} levels often involve the movement of Ca^{2+} ions through channels and transporters in the plasma membrane.[44,45] Thus, Ca^{2+} concentrations at the inner-membrane surface may, at least transiently, reach levels markedly different from those in the cytosol. In a recent paper, Etter and colleagues describe the use of our fura-C_{18} (F-6792) [10] to investigate near-membrane Ca^{2+} changes in isolated smooth muscle cells upon membrane depolarization.[46] Their results suggest that this lipophilic Ca^{2+} indicator may be used to resolve localized changes in Ca^{2+} concentration that cannot be detected with indicators distributed throughout the cytosol. Fura-C_{18} was recently used to measure time courses of Ca^{2+} in microinjected *Xenopus* oocytes.[47]

Molecular Probes now offers Calcium Green-C_{18} (C-6804),[10] a lipophilic indicator that preferentially associates with cell membranes. Spectral properties are similar to those of Calcium Green-1 (see Figure 22.6A), but the K_d for Ca^{2+} of the lipophilic indicators may be strongly affected by membrane binding. Calcium Green-C_{18} has recently been used to image efflux of Ca^{2+} from osteoblasts by detecting *extracellular* near-membrane Ca^{2+} levels.[48]

Fura-Indoline-C_{18}

Fura-indoline-C_{18} (F-6805), a hybrid of fura and the widely used membrane probe DiI (DiIC_{18}(3), D-282; see Section 15.4), has a weak fluorescence in water that increases when it is inserted into membranes. The fluorescence excitation maximum of fura-indoline-C_{18} shifts from ~610 nm to ~500 nm upon binding Ca^{2+}, which may permit its use for ratiometric measurements, especially for confocal laser scanning microscopes that have the Ar-Kr mixed-gas laser. Fura-indoline-C_{18} is also a sensitive colorimetric indicator that undergoes a large absorbance change from deep blue to light pink upon binding Ca^{2+}.

1. Biophys J 58, 1491 (1990); **2.** Cell Calcium 11, 63 (1990); **3.** Biophys J 67, 1646 (1994); **4.** Biophys J 65, 865 (1993); **5.** Cell Calcium 13, 59 (1992); **6.** Biochem J 54, 1089 (1988); **7.** J Exp Biol 196, 419 (1994); **8.** Meth Cell Biol 29, 59 (1989); **9.** J Cell Physiol 141, 410 (1989); **10.** U.S. Patent No. 5,453,517 and patents pending; **11.** J Cell Biol 58, 172 (1992); **12.** Eur J Cell Biol 58, 172 (1992); **13.** K_d values for Ca^{2+} are determined at Molecular Probes at ~22°C using our Calcium Calibration Buffer Kits (see Section 22.8); **14.** J Cell Sci 101, 7 (1992); **15.** J Microscopy 166, 57 (1992); **16.** Biol of Reprod 47, 960 (1992); **17.** Devel Biol 166, 634 (1994); **18.** J Neurosci 14, 6354 (1994); **19.** Development 120, 155 (1994); **20.** Nature 365, 256 (1993); **21.** Eur J Cell Biol 67, 363 (1995); **22.** Cell 80, 439 (1995); **23.** Biophys J 69, 1683 (1995); **24.** J Neurosci Meth 46, 91 (1993); **25.** Neuron 13, 849 (1994); **26.** Brain Res 663, 61 (1994); **27.** J Neurophysiol 73, 399 (1995); **28.** J Cell Biol 132, 915 (1996); **29.** Development 120, 3507 (1994); **30.** J Cell Biol 131, 1747 (1995); **31.** Devel Biol 161, 370 (1994); **32.** J Cell Biol 125, 1127

(1994); **33.** Nature 359, 736 (1992); **34.** Science 270, 1835 (1995); **35.** Proc Natl Acad Sci USA 89, 3591 (1992); **36.** Nature 377, 438 (1995); **37.** Nature 367, 745 (1994); **38.** J Cell Biol 131, 1539 (1995); **39.** Biol Bull 187, 234 (1994); **40.** J Biol Chem 271, 7615 (1996); **41.** Devel Biol 170, 496 (1995); **42.** Cell Calcium 14, 359 (1993); **43.** Determined at Molecular Probes using our Magnesium Calibration Standard Kit (M-3120, see Section 22.8); **44.** Endocrinology 132, 2176 (1993); **45.** J Immunology 150, 2620 (1993); **46.** J Biol Chem 269, 10141 (1994); **47.** Pflügers Arch 431, 379 (1996); **48.** J Biol Chem 270, 22445 (1995).

22.4 Data Table *Fluorescent Ca^{2+} Indicator Conjugates*

Cat #	Structure	MW	Storage	Soluble	Abs	Low Ca^{2+} {ε × 10^{-3}}	Em	Abs	High Ca^{2+} {ε × 10^{-3}}	Em	Solvent	K_d*	Notes
A-6796	C	~66,000	FF,D,L	H_2O	360	ND	491	339	ND	488	<1>	165 nM	<2, 3, 4>
C-3713	C	<5>	F,D,L	H_2O	508	ND	533	508	ND	533	<1>	260 nM	<2, 6, 7>
C-3714	C	<5>	F,D,L	H_2O	510	ND	535	510	ND	535	<1>	240 nM	<2, 6, 7>
C-6765	C	<5>	F,D,L	H_2O	510	ND	535	510	ND	535	<1>	540 nM	<2, 6, 7>
C-6766	C	<5>	F,D,L	H_2O	509	ND	534	509	ND	534	<1>	310 nM	<2, 6, 7>
C-6800	C	<5>	F,D,L	H_2O	<8>	ND	<8>	<8>	ND	<8>		370 nM	<2, 6>
C-6804	✓	1429	D,L	DMSO	509	{75}	530	509	ND	530	<1>	280 nM	<2, 7, 9>
C-6824	C	<5>	F,D,L	H_2O	592	ND	611	592	ND	611	<1>	335 nM	<2, 6, 7>
C-6825	C	<5>	F,D,L	H_2O	592	ND	611	592	ND	611	<1>	440 nM	<2, 6, 7>
F-3029	C	<5>	F,D,L	H_2O	364	ND	501	338	ND	494	<1>	240 nM	<2, 6>
F-3030	C	<5>	F,D,L	H_2O	366	ND	501	337	ND	495	<1>	205 nM	<2, 6>
F-6764	C	<5>	F,D,L	H_2O	366	ND	503	337	ND	497	<1>	265 nM	<2, 6>
F-6792	✓	1113	D,L	pH>6	365	{25}	501	338	{30}	494	<1>	150 nM	<10>
F-6805	✓	952	D,L	DMSO, MeOH	612	{31}	711	498	ND	694	<1>	260 nM	<11>
I-3032	C	<5>	F,D,L	H_2O	356	ND	466	341	ND	408	<1>	320 nM	<2, 6>
I-3033	C	<5>	F,D,L	H_2O	352	ND	464	340	ND	408	<1>	360 nM	<2, 6>
M-6907	C	<5>	F,D,L	H_2O	355	ND	466	340	ND	435	<1>	<12>	
M-6908	C	<5>	F,D,L	H_2O	355	ND	466	340	ND	438	<1>	<12>	
O-6797	C	<5>	F,D,L	H_2O	496	ND	524	498	ND	524	<1>	225 nM	<2, 6, 7>
O-6798	C	<5>	F,D,L	H_2O	496	ND	524	497	ND	524	<1>	265 nM	<2, 6, 7>

For definitions of the contents of this data table, see "How to Use the Handbook" on page vi.
Structure: Chemical structure drawing: ✓ = shown in this section; C = not shown due to complexity.
MW: ~ indicates an approximate value.
ND = not determined.

* Data on dissociation constants come from several different sources. These values are known to vary considerably depending on the temperature, pH, ionic strength, viscosity, protein binding, presence of other ions (especially polyvalent ions), instrument setup and other factors. It is strongly recommended that these values be specifically determined under your experimental conditions by using our Calcium Calibration Buffer Kits (see Section 22.8).

<1> Spectra measured in aqueous buffers containing 10 mM EGTA (low Ca^{2+}) or a >10-fold excess of free Ca^{2+} relative to the K_d (high Ca^{2+}).
<2> Dissociation constant determined at Molecular Probes by fluorescence measurements in 100 mM KCl, 10 mM MOPS, pH 7.2, 0 to 10 mM CaEGTA at 22°C.
<3> This protein conjugate contains multiple labels. The number of labels per protein is indicated on the vial.
<4> Dissociation constant and spectral parameters may vary between different lots of this product. Lot-specific K_d values are indicated on the vial.
<5> Molecular weight is nominally as specified in the product name but may have a broad distribution.
<6> Because indicator dextrans are polydisperse both in molecular weight and degree of substitution, dissociation constants and spectra may vary between batches. Lot-specific K_d values are indicated on the vial.
<7> This indicator exhibits fluorescence enhancement in response to ion binding, with essentially no change in absorption or emission wavelengths.
<8> This dextran conjugate incorporates both Ca^{2+}-sensitive Calcium Green-1 (Abs = 510 nm, Em = 534 nm) and Ca^{2+}-insensitive Texas Red (Abs = 599 nm, Em = 615 nm) fluorophores. The 535/615 nm emission ratio (excited at 488 nm) is sensitive to Ca^{2+} concentration.
<9> K_d values for C-6804 reported in J Biol Chem 270, 22445 (1995) are 230 nM in aqueous solution (pH 7.2, 0.1 M ionic strength, 22°C) and 62 nM in the presence of phospholipid vesicles.
<10> K_d value for F-6792 bound to smooth muscle cell membranes [J Biol Chem 269, 10141 (1994)].
<11> K_d value for F-6805 determined at Molecular Probes in 100 mM KCl, 10 mM MOPS, pH 7.2, 0 to 10 mM CaEGTA at 22°C using absorbance measurements at 610 nm.
<12> K_d (Ca^{2+}) values for M-6907 and M-6908 have not been determined. Values are expected to be similar to that of M-1293 (see Section 22.2).

C-6804 F-6792 F-6805

22.4 Price List *Fluorescent Ca^{2+} Indicator Conjugates*

Cat #		Product Name	Unit Size	Price Per Unit ($) 1–4 Units	5–24 Units
A-6796	***New***	albumin from bovine serum, fura conjugate (fura BSA)	5 mg	125.00	100.00
C-6824	***New***	Calcium Crimson™ dextran, 10,000 MW, anionic	5 mg	150.00	120.00
C-6825	***New***	Calcium Crimson™ dextran, 70,000 MW, anionic	5 mg	150.00	120.00
C-6804	***New***	Calcium Green™-C_{18}, hexapotassium salt	1 mg	150.00	120.00
C-6765	***New***	Calcium Green™-1 dextran, potassium salt, 3000 MW, anionic	5 mg	175.00	140.00
C-3713		Calcium Green™-1 dextran, potassium salt, 10,000 MW, anionic	5 mg	150.00	120.00
C-3714		Calcium Green™-1 dextran, potassium salt, 70,000 MW, anionic	5 mg	150.00	120.00
C-6766	***New***	Calcium Green™-1 dextran, potassium salt, 500,000 MW, anionic	5 mg	150.00	120.00
C-6800	***New***	Calcium Green™-1 and Texas Red® dextran, 70,000 MW, anionic	5 mg	150.00	120.00
F-6792	***New***	fura-C_{18}, pentapotassium salt	1 mg	150.00	120.00
F-6764	***New***	fura dextran, potassium salt, 3000 MW, anionic	5 mg	275.00	220.00
F-3029		fura dextran, potassium salt, 10,000 MW, anionic	5 mg	248.00	198.40
F-3030		fura dextran, potassium salt, 70,000 MW, anionic	5 mg	195.00	156.00
F-6805	***New***	fura-indoline-C_{18}	1 mg	85.00	68.00
I-3032		indo dextran, potassium salt, 10,000 MW, anionic	5 mg	248.00	198.40
I-3033		indo dextran, potassium salt, 70,000 MW, anionic	5 mg	195.00	156.00
M-6907	***New***	mag-indo dextran, potassium salt, 10,000 MW, anionic	5 mg	148.00	118.40
M-6908	***New***	mag-indo dextran, potassium salt, 70,000 MW, anionic	5 mg	148.00	118.40
O-6798	***New***	Oregon Green™ 488 BAPTA-1 dextran, potassium salt, 10,000 MW, anionic	5 mg	150.00	120.00
O-6797	***New***	Oregon Green™ 488 BAPTA-1 dextran, potassium salt, 70,000 MW, anionic	5 mg	150.00	120.00

22.5 Aequorin: A Bioluminescent Ca^{2+} Indicator

Bioluminescence is defined as the production of light by biological organisms. Because light is produced by a chemical reaction of specific photoproteins within the organism and does not require illumination, bioluminescence-based assays can be extremely sensitive and free of background. However, the intensity of light produced by bioluminescent cells is often very low, necessitating use of image enhancement to obtain sufficient signals. A useful practical overview of bioluminescence and fluorescence methods for monitoring intracellular Ca^{2+} has been published.[1]

Properties and Applications of Aequorin

Molecular Probes is pleased to offer a full line of reagents for quantitative Ca^{2+} measurements with aequorin, a photoprotein originally isolated from luminescent jellyfish and other marine organisms. The aequorin complex consists of a 22,000 MW apoaequorin protein, molecular oxygen and the luminophore coelenterazine.[2-5] When three Ca^{2+} ions bind to this complex, coelenterazine is oxidized to coelenteramide, with a concomitant release of carbon dioxide and blue light (emission maximum ~466 nm, Figure 22.11).[6,7] The approximately third-power dependence of aequorin's bioluminescence on Ca^{2+} concentration gives it a broad detection range, allowing the measurement of Ca^{2+} concentrations from ~0.1 µM to >100 µM.[8,9]

Unlike fluorescent Ca^{2+} indicators, Ca^{2+}-bound aequorin can be detected without illuminating the sample, thereby eliminating interference from autofluorescence and allowing simultaneous labeling with caged probes.[10] Moreover, aequorin that has been microinjected into eggs usually reports higher wave amplitudes (3–30 µM)

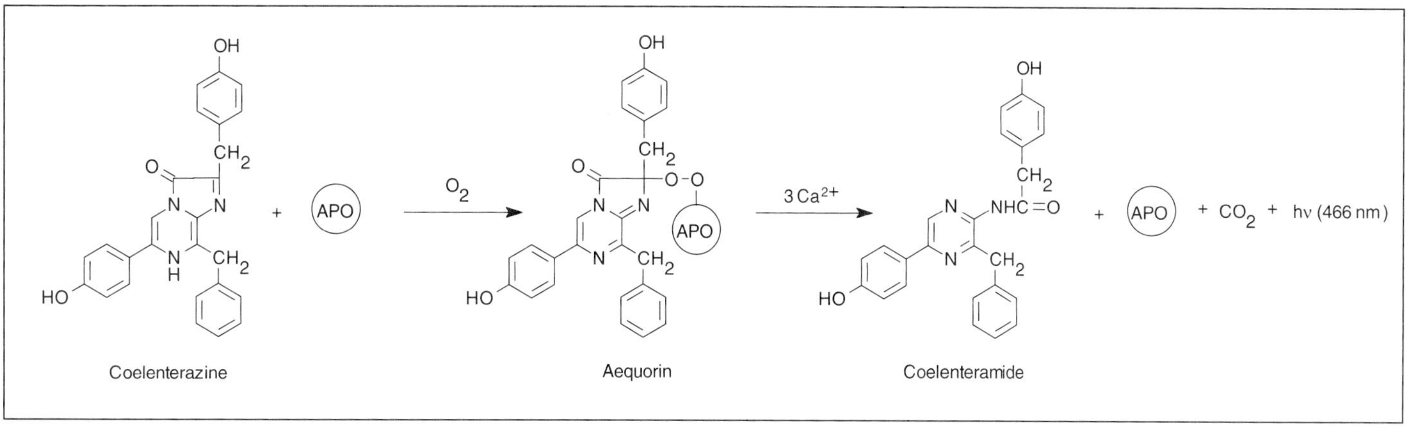

Figure 22.11 Ca^{2+}*-dependent generation of luminescence by the aequorin complex, which contains apoaequorin (APO) and coelenterazine.*

than do fluorescent ion indicators.[11-14] Aequorin is not exported or secreted, nor is it compartmentalized or sequestered within cells; thus, aequorin measurements can be used to detect Ca^{2+} changes that occur over relatively long periods. In several experimental systems, aequorin's luminescence was found to be detectable many hours to days after cell loading.[8,15,16] Aequorin also reportedly does not disrupt cell functions or embryo development.[8]

Recombinant Aequorin

Conventional purification of aequorin from the jellyfish *Aequorea victoria* requires laborious extraction procedures and has sometimes yielded preparations that are substantially heterogeneous or that are toxic to the organisms under study.[17,18] Two tons of jellyfish typically yield 125 mg of purified photoprotein.[19] In contrast, Molecular Probes' recombinant AquaLite® aequorin (A-6785) is produced by purifying apoaequorin from genetically engineered *Escherichia coli*, followed by reconstitution of the aequorin complex *in vitro* with pure coelenterazine.[20] This method of preparation yields pure, nontoxic, fully charged aequorin complex that is suitable for measuring intracellular Ca^{2+} by microinjection or other loading techniques, as well as for calibrating aequorin assays. Pressure injection is a commonly cited loading method, despite the fact that only large cells can be loaded in this way. Pressure injection has been employed to study the effects of caffeine on mouse diaphragm muscle fibers [21] and the role of Ca^{2+} in the fertilization of sea urchin eggs.[22] Alternatively, human platelets have been transiently permeabilized to the aequorin complex with DMSO [23] and monkey kidney cells have been loaded by hypoosmotic shock.[24]

Because of its Ca^{2+}-dependent emission, the aequorin complex has been extensively used as an intracellular Ca^{2+} indicator. *Aequorea victoria* aequorin has been used to:

- Analyze the secretion response of single adrenal chromaffin cells to nicotinic cholinergic agonists [25]
- Clarify the role of Ca^{2+} release in heart muscle damage [26]
- Demonstrate the massive release of Ca^{2+} during fertilization [27]
- Study the regulation of the sarcoplasmic reticulum Ca^{2+} pump expression in developing chick myoblasts [28]

Molecular Probes' biotin and streptavidin conjugates of AquaLite recombinant aequorin (A-6786, A-6787; see Section 7.5) are extremely versatile bioluminescent probes for detecting antigens, nucleic acids and other biological targets.[29] The extremely low background and broad detection range of aequorin makes these bioluminescence-based assays more sensitive than chemiluminescence-based, fluorometric and colorimetric enzyme-mediated techniques.[30,31]

Aequorin Expression Vectors

Production of Aequorin in Transfected or Transformed Cells

The cloning of apoaequorin cDNA [20,32] allows many new uses of this photoprotein, including the ability to transiently or stably express apoaequorin in transfected cells and to target apoaequorin to specific cell organelles. Once cells or tissues have been transfected with an "aequorin expression vector," which contains the apoaequorin structural gene, they are incubated in a medium containing cell-permeant coelenterazine (C-2944) or one of its analogs (see below) in order to reconstitute the aequorin complex. After formation of the active aequorin complex, intracellular Ca^{2+} can be measured by assaying cells for light production using a luminometer or a low level–light imaging detector. In collaboration with several university research laboratories, Molecular Probes now offers aequorin expression vectors suitable for use in mammalian and plant cells for transient and stable cell transfections.[33] Our mammalian expression vectors were constructed and characterized by R. Rizzuto, T. Pozzan and colleagues at the University of Padova and by A.K. Campbell and colleagues at the University of Wales College of Medicine. Our plant expression vectors were constructed and characterized by M. Knight and colleagues at Oxford University. Detailed plasmid maps are available from our Technical Assistance Department and through our Web site (http://www.probes.com).

Cytoplasmic- and Organelle-Targeting Aequorin Expression Vectors for Mammalian Cells

The identification of organelle-specific protein targeting sequences allows researchers to use apoaequorin expression vectors to measure Ca^{2+} concentrations at the subcellular level.[34,35] Molecular Probes offers a choice of several organelle-targeting aequorin expression vectors, as well as two cytoplasmic-targeting vectors:

- Cytoplasmic-targeting aequorin expression vector pSVAEQN (A-6782)
- Cytoplasmic-targeting aequorin expression vector cytAEQ/pcDNAI (A-6821)
- Mitochondrial-targeting aequorin expression vector mtAEQ/pMT2 (A-6788)
- Mitochondrial-targeting aequorin expression vector mtAEQ/pcDNAI (A-6823)
- Nuclear-targeting aequorin expression vector nu/cytAEQ/pcDNAI (A-6818)
- Nuclear-targeting aequorin expression vector pSVNPA (A-6783)
- Endoplasmic reticulum–targeting aequorin expression vector erAEQ/pcDNAI (A-6822)
- Endoplasmic reticulum–targeting aequorin expression vector pSVAEQERK (A-6784)

The **cytoplasmic-targeting pSVAEQN**, which contains the apoaequorin structural gene in the expression vector pSV7d, has been used to compare Ca^{2+} levels in the cytoplasm and endoplasmic reticulum of COS7 cells.[36] The **cytoplasmic-targeting cytAEQ/pcDNAI** contains sequences encoding the HA1 hemagglutinin epitope fused to the apoaequorin structural gene in the expression vector pcDNAI. The presence of the HA1 epitope confers the additional possibility of immunolocalization of the recombinant protein.[37] Brini and colleagues have published a detailed evaluation of cytosolic Ca^{2+} measurements made with aequorin, which was expressed in HeLa cells and primary cultures of rat embryonic cortex neurons transfected with cytAEQ/pcDNAI and then incubated with coelenterazine.[38] These researchers report that, unlike fluorescent ion indicators, recombinantly expressed aequorin exhibits no leakage or sequestration into organelles, negligible Ca^{2+} buffering and an extremely large detection range, allowing measurement of large changes in neuronal free Ca^{2+} concentration. Furthermore, in HeLa cells cotransfected with cytAEQ/pcDNAI and a plasmid encoding an α_1-adrenergic receptor (coupled to inositol 1,4,5-triphosphate generation), aequorin-based measurements enabled these researchers to monitor the subpopulation of cells expressing the receptor. In contrast, the Ca^{2+} change detected by the classical Ca^{2+} indicator fura-2 was dominated by the untransfected cells. The cytAEQ/pcDNAI vector was also used in combination with the mtAEQ/pMT2 vector (see below) to measure cytosolic free Ca^{2+} independently from mitochondrial free Ca^{2+}.[39]

The **mitochondrial-targeting mtAEQ/pMT2** contains sequences encoding the mitochondrial-targeting peptide from subunit VIII of human cytochrome *c* oxidase fused to the apoaequorin structural gene in the expression vector pMT2. Using this plasmid, researchers observed rapid, transient rises in mitochondrial Ca^{2+} in transfected mammalian cells, including bovine aortic endothelial, HeLa and Chinese hamster ovary (CHO) cells, as well as primary cultures of cardiac myocytes and neurons.[34,35,39-41] Ca^{2+} measurements with mitochondrial-targeted aequorin have been compared to those with the positively charged rhod-2, which also localizes in the mitochondria.[40] The mtAEQ/pMT2 vector was also used to show that mitochondria can detect the presence of high Ca^{2+} in their immediate vicinity.[42] Similarly, the **mitochondrial-targeting mtAEQ/pcDNAI** encodes an HA1 epitope–tagged version of this fusion protein in the expression vector pcDNAI to facilitate immunolocalization of the recombinant protein.[43]

The **inducible nuclear-targeting nu/cytAEQ/pcDNAI** contains sequences encoding the nuclear localization signal from the rat glucocorticoid receptor (GR) fused to the HA1 epitope–tagged apoaequorin structural gene in the expression vector pcDNAI. The GR domain of the fusion protein includes a binding site for glucocorticoid hormones. In the absence of hormone, the fusion protein remains in the cytoplasm; however, when hormone is added, the fusion protein is targeted to the nucleus.[44,45] The **nuclear-targeting pSVNPA** contains sequences encoding nucleoplasmin — a structural nuclear protein from *Xenopus laevis* — fused to the apoaequorin structural gene in the expression vector pSV7d. In experiments measuring the effects of elevated cytosolic Ca^{2+} on nuclear Ca^{2+}, this chimeric fusion protein was shown to be localized to the nucleus of transfected COS7 cells.[46]

The **endoplasmic reticulum–targeting erAEQ/pcDNAI** contains sequences encoding the peptide leader sequence and the VDJ and CH1 domains of mouse immunoglobulin γ2b heavy chain fused to the HA1 epitope–tagged apoaequorin structural gene in the expression vector pcDNAI. Localization of the aequorin fusion protein to the endoplasmic reticulum depends upon the presence of the CH1 domain, which is selectively retained in the endoplasmic reticulum in the absence of the immunoglobulin light chain.[47] The point mutation (Asp^{119} to Ala) in the apoaequorin coding region results in the production of aequorin with reduced affinity for Ca^{2+}, thereby allowing accurate quantitation of the relatively high levels of Ca^{2+} present in the endoplasmic reticulum.[48] This construct has been used to investigate dynamic changes in the free Ca^{2+} concentration in the endoplasmic reticulum of HeLa cells.[49] The **endoplasmic reticulum–targeting pSVAEQERK** encodes a tripartite apoaequorin fusion protein — which comprises the endoplasmic reticulum–targeting signal sequence of rabbit calreticulin fused to the N-terminus of apoaequorin and the endoplasmic reticulum retention signal sequence KDEL fused to the C-terminus of apoaequorin — in the expression vector pSV7d. This construct was used to detect changes in free Ca^{2+} in the endoplasmic reticulum of COS7 cells.[36,50]

Aequorin Expression Vectors for Plant Cells

Many of the techniques developed for measuring intracellular Ca^{2+} in mammalian cells cannot be readily applied to plant cells.[51] Because of the rigid cell wall, wall-associated esterase activities and turgor pressure of plant cells, conventional methods for introducing probes into the cytoplasm are ineffective. In addition, several popular fluorescent Ca^{2+} indicators are rapidly sequestered into the plant vacuole, resulting both in loss of probe from the cytoplasm as well as in erroneous measurements. Autofluorescence of the plant's cell walls and chloroplasts can also severely limit the detectability of fluorescent probes.

With the advent of reliable plant transformation procedures and the isolation of apoaequorin cDNA,[20,52] apoaequorin can be transiently or stably expressed in plant cells. Functional aequorin can then be reconstituted by incubating apoaequorin-expressing cells with coelenterazine or one of its analogs, which passively diffuses through plant cell walls and membranes.

To facilitate aequorin-based Ca^{2+} measurements in plants, Molecular Probes offers two apoaequorin expression vectors:

- Cytoplasm-targeting aequorin expression vector pMAQ2 (A-6793)
- Chloroplast-targeting aequorin expression vector pMAQ6 (A-6819)

The **cytoplasm-targeting pMAQ2** contains the apoaequorin structural gene fused to the constitutive cauliflower mosaic virus (CaMV) 35S promoter.[53] In addition, the vector carries the *Agrobacterium tumefaciens* plasmid sequences required for plant transformation. Tobacco plants transformed with this vector have been used to investigate changes in cytosolic Ca^{2+} that occur in response to wind-induced plant motion,[54] as well as in response to touch, cold-shock and fungal elicitors.[53,55]

The **chloroplast-targeting pMAQ6** contains the coding sequence of the transit peptide from the small subunit of ribulose-1,5-bisphosphate carboxylase–oxygenase fused to the 5'-terminus of the apoaequorin structural gene. This construct has been used in combination with pMAQ2 to monitor the circadian oscillations of cytosolic and chloroplastic free Ca^{2+} in tobacco and *Arabidopsis*.[56]

Table 22.2 *Coelenterazines and their properties.*

Cat #	Coelenterazine Analog	Em * (nm)	RLC †	Relative Intensity ‡	Half-rise Time (msec) §
C-2944	native	466	1.00	1	6–30
C-6779	*f*	472	0.80	20	6–30
C-6780	*h*	466	0.75	16	6–30
C-6781	*hcp*	445	0.65	500	2–5
C-6776	*n*	468	0.25	0.15	6–30

* Emission maxima. † Relative luminescence capacity = total time-integrated emission of aequorin in saturating Ca^{2+} relative to native aequorin = 1.0. ‡ Relative intensity at 100 nM Ca^{2+}. § Half-rise time = time for the luminescence signal to reach 50% of the maximum after addition of 1 mM Ca^{2+} to a standard of aequorin reconstituted with the coelenterazine analog of interest. The effective rise time in cells and tissues will probably be longer because Mg^{2+} reduces the initial rate of rise [Blinks, J.R. and Moore, E.D.W. in *Soc Gen Physiol Ser 40 (Opt Methods Cell Physiol)* (1986) pp. 229–238]. All data from Cell Calcium 14, 373 (1993).

Coelenterazine and Its Synthetic Analogs

Molecular Probes offers coelenterazine and several synthetic coelenterazine analogs for reconstituting aequorin in cells that have been transfected with apoaequorin cDNA. Coelenterazine is also required for generating the bioluminescent aequorin complex when using chimeric aequorin constructs.[57] Coelenterazine has been shown to be readily cell permeant in organisms as diverse as *E. coli*,[58] yeast,[59,60] *Dictyostelium* cells,[61] mammalian cells[41,62,63] and plants.[10,53,55]

In addition to native coelenterazine (C-2944), we have synthesized four derivatives of coelenterazine that confer different Ca^{2+} affinities and spectral properties to the aequorin complex[64-66] (Table 22.2). Like native coelenterazine, these four novel derivatives can be used to reconstitute the aequorin complex both *in vivo* and *in vitro*. Recombinant apoaequorin reconstituted with coelenterazine *hcp* (C-6781) is reported to have the best luminescence overall, with both a high quantum yield and a fast response time[67] (Table 22.2). However, generation of the apoaequorin/coelenterazine *hcp* complex can be relatively slow in cells.[65] Aequorins containing the *f* or *h* form of coelenterazine (C-6779, C-6780) exhibit relative intensities that are 10–20 times that of apoaequorin reconstituted with native coelenterazine.[10,66,68] Coelenterazine *n* (C-6776) is reportedly the most useful low-sensitivity coelenterazine, producing an apoaequorin/coelenterazine *n* complex that exhibits 10,000-fold lower luminescent intensity than the apoaequorin/coelenterazine *hcp* complex.[67] To facilitate the evaluation of the various coelenterazines, we offer a Coelenterazine Sampler Kit (C-6777), which contains 25 µg samples of all four of our coelenterazine derivatives (coelenterazine *f*, *h*, *hcp* and *n*) as well as 25 µg of native coelenterazine.

1. *Cellular Calcium: A Practical Approach*, J.G. McCormack and P.H. Cobbold, Eds., IRL Press (1991); **2.** J Biochem 105, 473 (1989); **3.** J Chem Soc Chem Commun 21, 1566 (1986); **4.** Meth Enzymol 57, 271 (1978); **5.** Symp Soc Exp Biol 30, 41 (1976); **6.** J Cell Comp Physiol 62, 1 (1962); **7.** J Cell Comp Physiol 59, 223 (1962); **8.** Meth Cell Biol 40, 305 (1994); **9.** Meth Enzymol 172, 164 (1989); **10.** Cell Biol Int'l 17, 111 (1993); **11.** Cell Calcium 14, 736 (1993); **12.** Ann NY Acad Sci 639, 112 (1991); **13.** Devel Biol 135, 182 (1989); **14.** Devel Biol 118, 259 (1986); **15.** Cobbold, P.H. and Lee, J.A.C. in *Cellular Calcium: A Practical Approach*, J.G. McCormack and P.H. Cobbold, Eds., IRL Press at Oxford University Press, New York (1991) pp. 1–54; **16.** J Cell Biol 115, 1259 (1991); **17.** Biochem J 270, 309 (1990); **18.** J Gen Physiol 85, 189 (1985); **19.** Biochemistry 11, 1602 (1972); **20.** Biochem Biophys Res Comm 126, 1259 (1985); **21.** Neurosci Lett 127, 28 (1991); **22.** J Cell Biol 100, 1522 (1985); **23.** Biochem Biophys Res Comm 177, 888 (1991); **24.** Am J Physiol 247, C396 (1984); **25.** FEBS Lett 211, 44 (1987); **26.** Nature 312, 444 (1984); **27.** Proc Natl Acad Sci USA 74, 623 (1977); **28.** Am J Physiol 251, C512 (1986); **29.** Stults, N.L. *et al.* in *Bioluminescence and Chemiluminescence: Current Status*, P. Stanley and L. Kricka, Eds., Wiley (1991) pp. 533–536; **30.** Biochemistry 31, 1433 (1992); **31.** Smith, D.F. *et al.* in *Bioluminescence and Chemiluminescence: Current Status*, P. Stanley and L. Kricka, Eds., John Wiley & Sons (1991) pp. 529–532; **32.** Proc Natl Acad Sci USA 82, 3154 (1985); **33.** Aequorin expression vectors are licensed to Molecular Probes under U.S. Patent Nos. 4,237,224; 4,656,134; 4,468,464; 4,740,470; 5,162,227; **34.** J Biolumin Chemilumin 9, 177 (1994); **35.** Meth Cell Biol 40, 339 (1994); **36.** Biochem Biophys Res Comm 189, 1008 (1992); **37.** Mol Cell Biol 8, 2159 (1988); **38.** J Biol Chem 270, 9896 (1995); **39.** J Biol Chem 268, 22385 (1993); **40.** Proc Natl Acad Sci USA 93, 5489 (1996); **41.** Nature 358, 325 (1992); **42.** Science 262, 744 (1993); **43.** Meth Enzymol 260, 417 (1995); **44.** Cell Calcium 16, 259 (1994); **45.** EMBO J 12, 4813 (1993); **46.** Exp Cell Res 216, 236 (1995); **47.** Cell 60, 781 (1990); **48.** Biochem Biophys Res Comm 187, 1091 (1992); **49.** EMBO J 14, 5467 (1995); **50.** Anal Biochem 221, 173 (1994); **51.** Callaham, D.A. and Hepler, P.K. in *Cellular Calcium: A Practical Approach*, J.G. McCormack and P.H. Cobbold, Eds., IRL Press at Oxford University Press, New York (1991) pp. 383–410; **52.** Proc Natl Acad Sci USA 82, 3154 (1985); **53.** Nature 352, 524 (1991); **54.** Proc Natl Acad Sci USA 89, 4967 (1992); **55.** J Cell Biol 121, 83 (1993); **56.** Science 269, 1863 (1995); **57.** J Biolumin Chemilumin 4, 346 (1989); **58.** FEBS Lett 282, 405 (1991); **59.** Biochem Biophys Res Comm 174, 115 (1991); **60.** Proc Natl Acad Sci USA 88, 6878 (1991); **61.** FEBS Lett 337, 43 (1994); **62.** Anal Biochem 209, 343 (1993); **63.** Cell Calcium 14, 663 (1993); **64.** Biochem J 306, 537 (1995); **65.** Biochem J 296, 549 (1993); **66.** Biochem J 261, 913 (1989); **67.** Cell Calcium 14, 373 (1993); **68.** Cell Calcium 12, 635 (1991).

22.5 Data Table *Aequorin: A Bioluminescent Ca^{2+} Indicator*

Cat #	Structure	MW	Storage	Soluble	Abs	{ε x 10^{-3}}	Em	Solvent
C-2944	✓	423	FF,D,LL,AA	MeOH	429	{7.5}	<1>	pH 7
C-6776	✓	458	FF,D,LL,AA	MeOH	431	{9.0}	<1>	MeOH
C-6779	✓	425	FF,D,LL,AA	MeOH	437	{8.7}	<1>	MeOH
C-6780	✓	407	FF,D,LL,AA	MeOH	437	{9.5}	<1>	MeOH
C-6781	✓	415	FF,D,LL,AA	MeOH	430	{7.0}	<1>	MeOH

For definitions of the contents of this data table, see "How to Use the Handbook" on page vi.
Structure: Chemical structure drawing: ✓ = shown in this section.

<1> Coelenterazine complexes with aequorin emit calcium-dependent bioluminescence. Bioluminescence emission maxima (relative intensity at 100 nM Ca^{2+}) are as follows: C-2944, 466 nm (1); C-6776, 468 nm (0.15); C-6779, 472 nm (20); C-6780, 466 nm (16); C-6781, 445 nm (500). Data from Cell Calcium 14, 373 (1993).

22.5 Structures *Aequorin: A Bioluminescent Ca^{2+} Indicator*

C-2944 R = –OH
C-6779 R = F
C-6780 R = H

C-6776

C-6781

22.5 Price List *Aequorin: A Bioluminescent Ca^{2+} Indicator*

Cat #		Product Name	Unit Size	Price Per Unit ($) 1–4 Units	5–24 Units
A-6821	***New***	aequorin expression vector cytAEQ/pcDNAI, mammalian, cytoplasmic targeting *500 µg/mL*	50 µL	295.00	236.00
A-6822	***New***	aequorin expression vector erAEQ/pcDNAI, mammalian, endoplasmic reticulum targeting *500 µg/mL*	50 µL	295.00	236.00
A-6823	***New***	aequorin expression vector mtAEQ/pcDNAI, mammalian, mitochondrial targeting *500 µg/mL*	50 µL	295.00	236.00
A-6788	***New***	aequorin expression vector mtAEQ/pMT2, mammalian, mitochondrial targeting *500 µg/mL*	50 µL	295.00	236.00
A-6818	***New***	aequorin expression vector nu/cytAEQ/pcDNAI, mammalian, inducible nuclear targeting *500 µg/mL*	50 µL	295.00	236.00
A-6793	***New***	aequorin expression vector pMAQ2, plant, cytoplasmic targeting *500 µg/mL*	50 µL	295.00	236.00
A-6819	***New***	aequorin expression vector pMAQ6, plant, chloroplast targeting *500 µg/mL*	50 µL	295.00	236.00
A-6784	***New***	aequorin expression vector pSVAEQERK, mammalian, endoplasmic reticulum targeting *500 µg/mL*	50 µL	295.00	236.00
A-6782	***New***	aequorin expression vector pSVAEQN, mammalian, cytoplasmic targeting *500 µg/mL*	50 µL	295.00	236.00
A-6783	***New***	aequorin expression vector pSVNPA, mammalian, nuclear targeting *500 µg/mL*	50 µL	295.00	236.00
A-6785	***New***	AquaLite® aequorin (aequorin) *recombinant*	25 µg	98.00	78.40
C-2944		coelenterazine	250 µg	295.00	236.00
C-6779	***New***	coelenterazine f	250 µg	295.00	236.00
C-6780	***New***	coelenterazine h	250 µg	295.00	236.00
C-6781	***New***	coelenterazine hcp	250 µg	295.00	236.00
C-6776	***New***	coelenterazine n	250 µg	295.00	236.00
C-6777	***New***	Coelenterazine Sampler Kit *contains 25 µg of C-2944, C-6776, C-6779, C-6780, C-6781*	1 kit	255.00	204.00

22.6 Fluorescent Mg^{2+} Indicators

Intracellular Mg^{2+} is important for mediating enzymatic reactions, DNA synthesis, hormonal secretion and muscular contraction. To facilitate the investigation of magnesium's role in these and other cellular functions, Molecular Probes offers several different fluorescent indicators for measuring intracellular Mg^{2+} concentration. They include furaptra,[1,2] which we refer to as mag-fura-2 to denote the similarity of its structure and spectral response with the Ca^{2+} indicator fura-2; mag-indo-1, with a structure and spectral response similar to that of indo-1; and mag-fura-5. For applications such as confocal laser scanning microscopy and flow cytometry, we offer a series of visible light–excitable Mg^{2+} indicators, including Magnesium Green, Magnesium Orange™ and Mag-Fura Red™. Our mag-indo dextran conjugates should be especially useful for applications requiring long-term cell retention.

Mg^{2+} indicators are generally designed to maximally respond to the Mg^{2+} concentrations commonly found in cells, typically ranging from about 0.1 mM to 6 mM.[3,4] Intracellular free Mg^{2+} levels have been reported to be ~0.3 mM in synaptosomes,[4] 0.37 mM in hepatocytes[5] and 0.5–1.2 mM in cardiac cells,[6] whereas the concentration of Mg^{2+} in normal serum is ~0.44–1.5 mM.[7] Measurements using fluorescent Mg^{2+} indicators are somewhat more demanding than intracellular Ca^{2+} determinations because physiological changes in Mg^{2+} concentration are relatively small.[8] Mg^{2+} indicators also bind Ca^{2+}; however, typical physiological Ca^{2+} concentrations (10 nM–1 µM) do not usually interfere with Mg^{2+} measurements because the affinity of these indicators for Ca^{2+} is low. Although Ca^{2+} binding by Mg^{2+} indicators can be a complicating factor in Mg^{2+} measurements,[9] this property can also be exploited for measuring high Ca^{2+} concentrations[10-12] (1–100 µM); see Sections 22.2 and 22.3 for examples.

For calibrating Mg^{2+} indicators, Molecular Probes offers the Magnesium Calibration Standard Kit (M-3120, see Section 22.8), which contains solutions with precisely defined Mg^{2+} concentrations. Also available are the ionophores A-23187 and the nonfluorescent 4-bromo A-23187 (A-1493, B-1494; see Section 22.8), which are preferred over ionomycin for Mg^{2+} calibrations because they transport Mg^{2+} more effectively.[2] Solutions used to calibrate Mg^{2+} indicators should be initially free of heavy metals such as Mn^{2+} that can interact with the indicators. These metals can be removed by treating the solution with the divalent cation chelator TPEN (T-1210, see Section 22.8).

Magnesium Indicators Excited by Ultraviolet Light

Mag-Fura-2, Mag-Fura-5 and Mag-Indo-1

The K_d for Mg^{2+} of mag-fura-5 and mag-indo-1 are 2.3 mM and 2.7 mM, respectively,[13] slightly higher than that of mag-fura-2, which is 1.9 mM. Mag-fura-2 was first used to detect Mg^{2+} fluctuations in embryonic chicken heart cells[14] and rat liver cells.[2] The lower-affinity mag-fura-5 and mag-indo-1 are sensitive to somewhat higher spikes in intracellular Mg^{2+}.[8,15] The affinities of mag-fura-2 and mag-indo-1 for Mg^{2+} are reported to be unchanged at pH values between 5.5 and 7.4 and at temperatures between 22°C and 37°C.[16]

As with their Ca^{2+}-indicating analogs, mag-fura-2 undergoes an appreciable shift in excitation wavelength upon Mg^{2+} binding (Figure 22.12), and mag-indo-1 exhibits shifts in both excitation and emission wavelengths (Figure 22.13). Equipment, optical filters and calibration methods are very similar to those required for the Ca^{2+} indicators. The excitation-ratioable mag-fura-2 and mag-fura-5 are most useful for fluorescence microscopy, whereas the emission-ratioable mag-indo-1 is preferred for flow cytometry.

A comparison of intracellular and solution K_ds for mag-fura-2 has been published by Hurley and co-workers.[9] Researchers have used mag-fura-2 to measure intracellular Mg^{2+} in a wide variety of cells and tissues:

- Swiss 3T3 cells[17]
- BC_3-H-1 cells[18]
- Jurkatt cells[19]
- Rat adipocytes[20]
- Vascular smooth muscle[21]
- Cortical neurons[22]
- Lymphocytes[23]

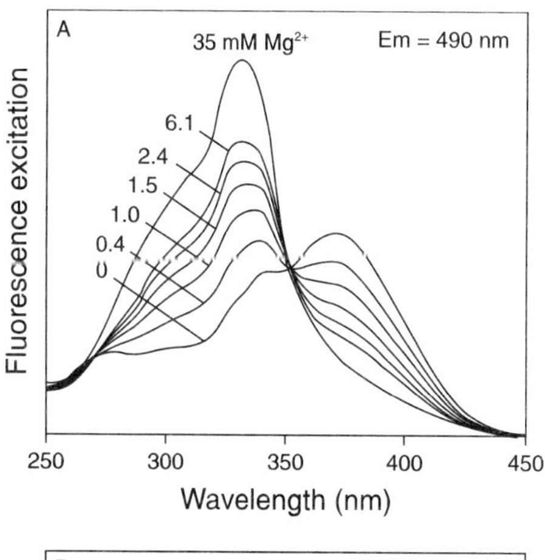

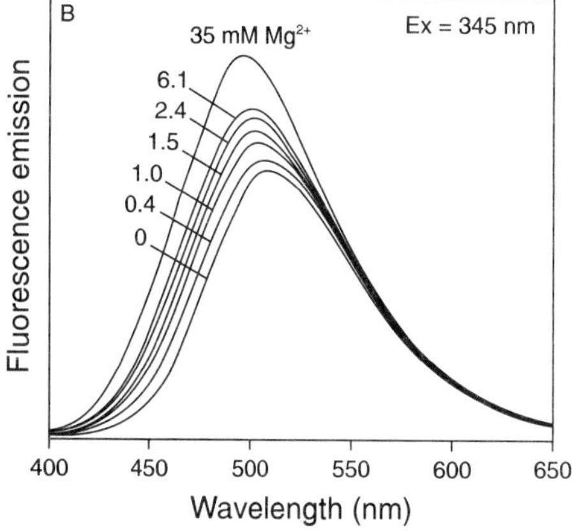

Figure 22.12 *A) Fluorescence excitation and B) fluorescence emission spectra of mag-fura-2 (M-1290) in solutions containing zero to 35 mM* Mg^{2+}.

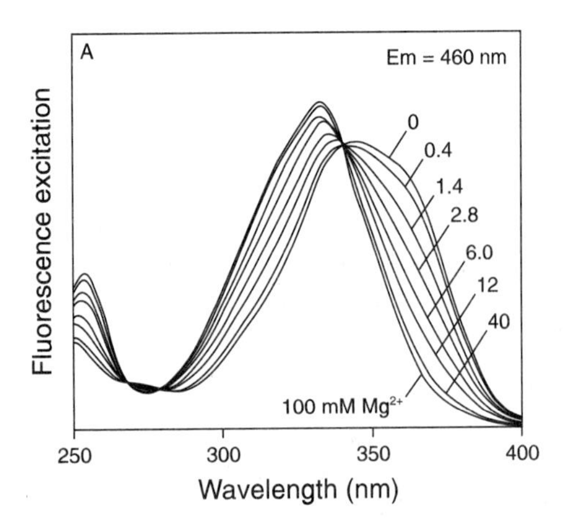

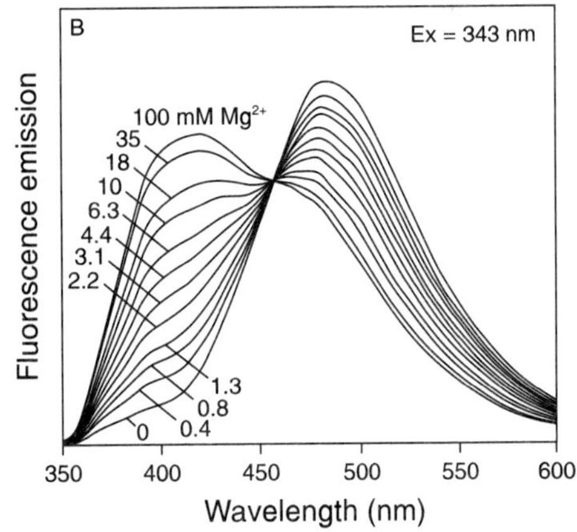

Figure 22.13 *A) Fluorescence excitation and B) fluorescence emission spectra of mag-indo-1 (M-1293) in solutions containing zero to 100 mM* Mg^{2+}.

Simultaneous flow cytometric measurements of Ca^{2+} and Mg^{2+} using fluo-3 and mag-indo-1 have been made.[24] Discussion of various methods for measuring intracellular Mg^{2+} can be found in two reviews.[1,6]

In addition to the cell-impermeant potassium salts (M-1290, M-3103, M-1293), we offer the cell-permeant AM esters of mag-fura-2, mag-fura-5 and mag-indo-1 as sets of 20 vials, each containing 50 µg (M-1292, M-3105, M-1295). The special packaging is recommended when small quantities of the dyes are to be used over a long period of time. Mag-fura-2 AM is also available in a single vial containing 1 mg (M-1291).

Mag-Indo Dextrans

For improved cell retention and reduced compartmentalization, we offer mag-indo conjugated to 10,000 MW and 70,000 MW dextrans (M-6907, M-6908), which exhibit a K_d for Mg^{2+} of about 2.7 mM and 2.3 mM, respectively.[13] Physiological concentrations of Ca^{2+} below ~10 µM are not expected to interfere with Mg^{2+} measurements made with these dextrans.

Magnesium Indicators Excited by Visible Light

Molecular Probes offers several visible light–excitable Mg^{2+} indicators, including Magnesium Green, Magnesium Orange and Mag-Fura Red indicators. As with mag-fura-2, mag-fura-5 and mag-indo-1, these visible light–excitable Mg^{2+} indicators can also be used as low-affinity Ca^{2+} indicators (see Section 22.3). A recent paper describes the properties of several of our low-affinity Ca^{2+} indicators, including Calcium Green-5N, Calcium Orange-5N, Magnesium Green, Magnesium Orange, mag-fura-2, mag-indo-1 and Mag-Fura Red, and compares their utility in skeletal muscle fibers.[25]

Magnesium Green Indicator

Magnesium Green exhibits a higher affinity for Mg^{2+} (K_d ~1.0 mM)[13] than does mag-fura-2 (K_d ~1.9 mM) or mag-indo-1 (K_d ~2.7 mM);[13] this indicator also binds Ca^{2+} with moderate affinity (K_d for Ca^{2+} in the absence of Mg^{2+} ~6 µM).[26] The spectral properties of Magnesium Green are similar to those of the Calcium Green indicators. Upon binding Mg^{2+}, Magnesium Green exhibits an increase in fluorescence emission intensity without a shift in wavelength (Figure 22.14). This Mg^{2+} indicator has been used to measure the binding of free Mg^{2+} by the SecA protein.[27] Magnesium Green is available as a cell-impermeant potassium salt (M-3733) or as a cell-permeant AM ester (M-3735).

Magnesium Orange Indicator

The fluorescence response of Magnesium Orange in increasing concentrations of Mg^{2+} is shown in Figure 22.14. Its K_d for Mg^{2+} at ~22°C is ~3.9 mM,[13] whereas its K_d for Ca^{2+} is ~12 µM.[26] Magnesium Orange is available as a cell-impermeant potassium salt (M-6890) or as a cell-permeant AM ester (M-6891).

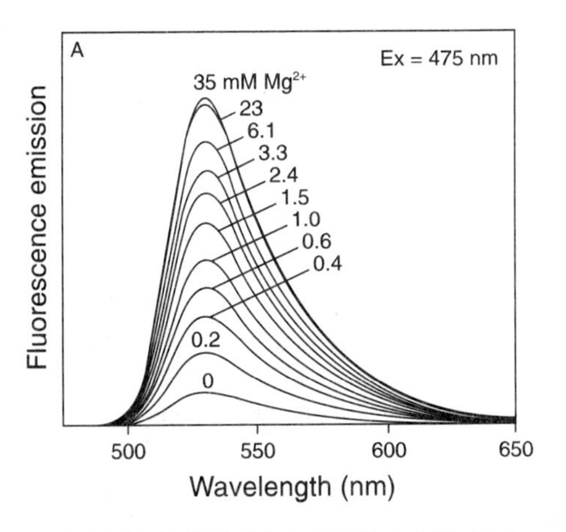

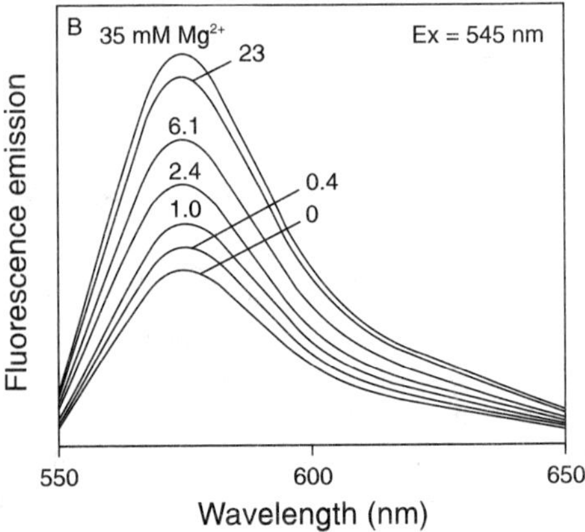

Figure 22.14 Mg^{2+}*-dependent fluorescence emission spectra of A) Magnesium Green (M-3733) and B) Magnesium Orange (M-6890) indicators.*

Mag-Fura Red Indicator

The affinity of Mag-Fura Red for Mg^{2+} is slightly higher than that of Magnesium Orange; its K_d for Mg^{2+} at ~22°C is 2.5 mM, whereas its K_d for Ca^{2+} is 17 µM. Like Fura Red (see Figure 22.5), the fluorescence of Mag-Fura Red *decreases* when the concentration of Mg^{2+} increases, with only a small shift of its 650 nm emission peak. Mag-Fura Red is available as a cell-impermeant potassium salt (M-6892) or as a cell-permeant AM ester (M-6893).

Analytical Reagents for Mg^{2+}

N,N'-bis-(salicylidene)-2,3-diaminobenzofuran (SABF, B-6874) forms a fluorescent complex with Mg^{2+} (excitation/emission maxima ~485/545 nm) that can be extracted into organic solvents. This reagent has been used to detect Mg^{2+} in serum [28] and in renal tubular fluid.[29] We also offer *N,N'*-bis-(salicylidene)ethylenediamine (B-6875), which forms a similar complex with shorter-wavelength excitation/emission maxima (355/440 nm); the Mg^{2+} detection limit of this reagent is about 0.17 parts per billion.[30]

A fluorometric method that uses 2,2'-dihydroxyazobenzene (D-6873, see Section 22.7) permits the detection of 2–25 µg of Mg^{2+} in the presence of up to 5 mg Ca^{2+}.[31]

1. Ann Rev Physiol 53, 241 (1991); **2.** Am J Physiol 256, C540 (1989); **3.** Pflügers Arch 420, 347 (1992); **4.** Biochim Biophys Acta 898, 331 (1987); **5.** J Biol Chem 261, 2567 (1986); **6.** Ann Rev Physiol 53, 273 (1991); **7.** Clin Chem 35, 1492 (1989); **8.** Pflügers Arch 422, 179 (1992); **9.** Am J Physiol 263, C300 (1992); **10.** Neuron 10, 21 (1993); **11.** Proc Natl Acad Sci USA 90, 2598 (1993); **12.** J Gen Physiol 97, 271 (1991); **13.** K_d values for Mg^{2+} determined at Molecular Probes at 22°C using our Magnesium Calibration Standard Kit (M-3120, see Section 22.8); **14.** Proc Natl Acad Sci USA 86, 2981 (1989); **15.** J Physiol 475, 319 (1994); **16.** Biochem Biophys Res Comm 177, 184 (1991); **17.** J Biochem 115, 730 (1994); **18.** Mol Cell Biochem 136, 11 (1994); **19.** Shock 1, 213 (1994); **20.** Biochem Biophys Res Comm 202, 416 (1994); **21.** Am J Physiol 265, H281 (1993); **22.** Neuron 11, 751 (1993); **23.** Clin Sci 80, 539 (1991); **24.** Biochem J 289, 373 (1993); **25.** Biophys J 70, 896 (1996); **26.** K_d values for Ca^{2+} are determined at Molecular Probes at ~22°C using our Calcium Calibration Buffer Kits (see Section 22.8); **27.** J Biol Chem 270, 18975 (1995); **28.** Analyst 92, 20 (1967); **29.** Anal Biochem 65, 79 (1975); **30.** Anal Chem 31, 2083 (1959); **31.** Anal Chem 35, 1144 (1963).

22.6 Data Table *Fluorescent Mg^{2+} Indicators*

Cat #	Structure	MW	Storage	Soluble	Abs	Low Mg^{2+} {ε × 10^{-3}}	Em	Abs	High Mg^{2+} {ε × 10^{-3}}	Em	Solvent	Cleaved*	K_d†	Notes
B-6874	✓	356	D,L	DMSO	430	{27}	550	<1>		<1>	MeOH			
B-6875	✓	268	D,L	DMSO	316	{7.6}	none	<2>		<2>	MeOH			
M-1290	✓	587	L	pH>6	369	{22}	511	330	{24}	491	<3>		1.9 mM	<4, 5>
M-1291	✓	723	F,D,L	DMSO	366	{31}	475				EtOAc	M-1290		
M-1293	✓	595	L	pH>6	349	{38}	480	330	{33}	417	<3>		2.7 mM	<4, 5>
M-1295	✓	731	F,D,L	DMSO	354	{37}	472				MeOH	M-1293		
M-3103	✓	601	L	pH>6	369	{23}	505	332	{25}	482	<3>		2.3 mM	<4, 5>
M-3105	✓	737	F,D,L	DMSO	365	{31}	401				EtOAc	M-3103		
M-3733	✓	916	L	pH>6	506	{77}	531	506	{75}	531	<3>		1.0 mM	<4, 5, 6>
M-3735	✓	1026	F,D	DMSO	302	{16}	none				MeOH	M-3733		
M-6890	✓	856	L	pH>6	550	{76}	575	550	{74}	575	<3>		3.9 mM	<4, 5, 6>
M-6891	✓	958	F,D,L	DMSO	540	{90}	566				MeOH	M-6890		
M-6892	✓	564	L	pH>6	483	{23}	659	427	{29}	631	<3>		2.5 mM	<4, 5>
M-6893	✓	810	F,D,L	DMSO	450	{40}	650				MeOH	M-6892		
M-6907	C	<7>	F,D,L	H_2O	355	ND	466	341	ND	412	<3>		2.7 mM	<8>
M-6908	C	<7>	F,D,L	H_2O	355	ND	466	341	ND	412	<3>		2.3 mM	<8>

M-1292 see M-1291

For definitions of the contents of this data table, see "How to Use the Handbook" on page vi.
Structure: Chemical structure drawing: ✓ = shown in this section; C = not shown due to complexity.

* Catalog numbers listed in this column indicate the cleavage product generated by enzymatic or chemical hydrolysis of the acetoxymethyl ester groups of the parent compound.

† Data on dissociation constants come from several different sources. These values are known to vary considerably depending on the temperature, pH, ionic strength, viscosity, protein binding, presence of other ions (especially polyvalent ions), instrument setup and other factors. It is strongly recommended that these values be specifically determined under your experimental conditions by using our Magnesium Calibration Standard Kit (see Section 22.8).

<1> Fluorescence of B-6874 is enhanced upon binding Mg^{2+}. Complex with Mg^{2+} has Abs ~485 nm, Em ~545 nm in 50% aqueous MeOH [Analyst 92, 20 (1967)].

<2> B-6875 forms a fluorescent complex with Mg^{2+}, Abs ~355 nm, Em ~440 nm in DMF [Anal Chem 31, 2083 (1959)].

<3> Spectra measured in aqueous buffers containing zero or 35 mM Mg^{2+}, indicated as low and high Mg^{2+}, respectively.

<4> This indicator binds Ca^{2+} with higher affinity than Mg^{2+}, producing a similar spectral response (see Section 22.2 for M-1290, M-1293 and M-3103, see Section 22.3 for M-3733). K_d (Ca^{2+}) values for M-6890 and M-6892 are 12 µM and 17 µM, respectively.

<5> K_d (Mg^{2+}) values have been determined at Molecular Probes in 115 mM KCl, 20 mM NaCl, 10 mM Tris, pH 7.05, 0 to 35 mM Mg^{2+} at 22°C.

<6> This indicator exhibits fluorescence enhancement in response to ion binding, with essentially no change in absorption or emission wavelengths.

<7> Molecular weight is nominally as specified in the product name but may have a broad distribution.

<8> Because indicator dextrans are polydisperse both in molecular weight and degree of substitution, dissociation constants and spectra may vary between batches.

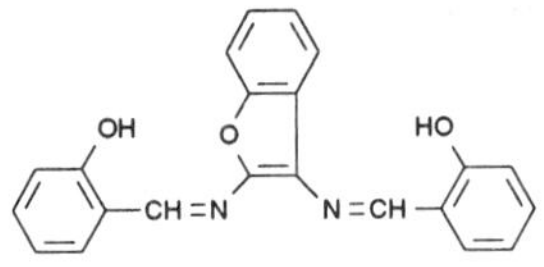

22.6 Structures *Fluorescent Mg^{2+} Indicators*

B-6874

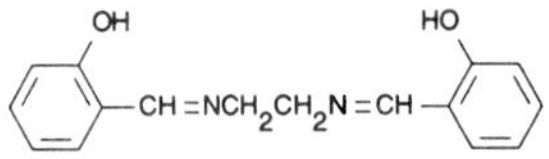

B-6875

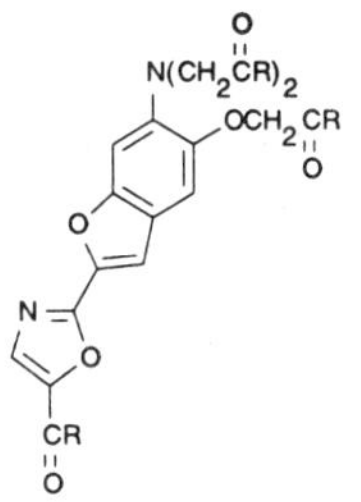

M-1290 R = $O^- K^+$
M-1291 / M-1292 R = $-OCH_2OCCH_3$ (C=O)

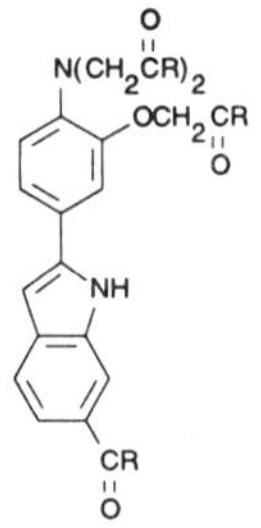

M-1293 R = $O^- K^+$
M-1295 R = $-OCH_2OCCH_3$ (C=O)

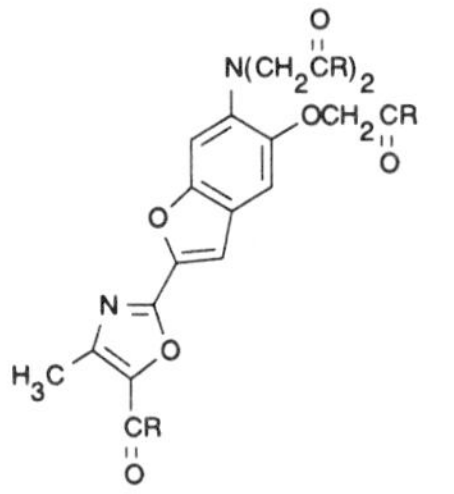

M-3103 R = $O^- K^+$
M-3105 R = $-OCH_2OCCH_3$ (C=O)

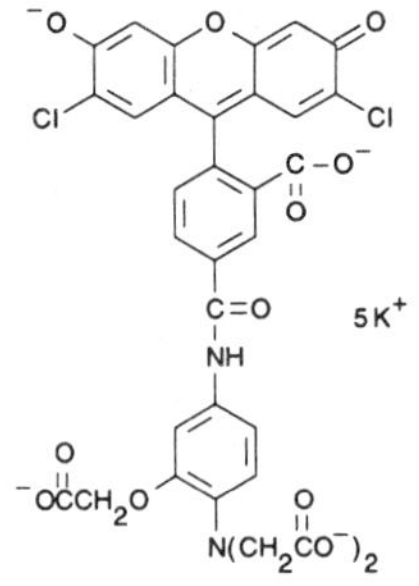

M-3733

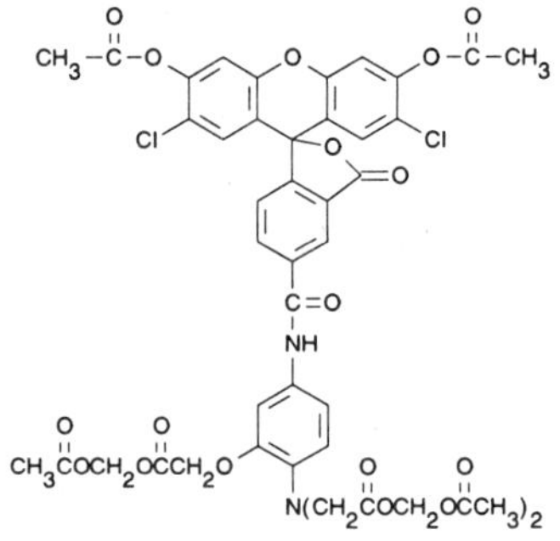

M-3735

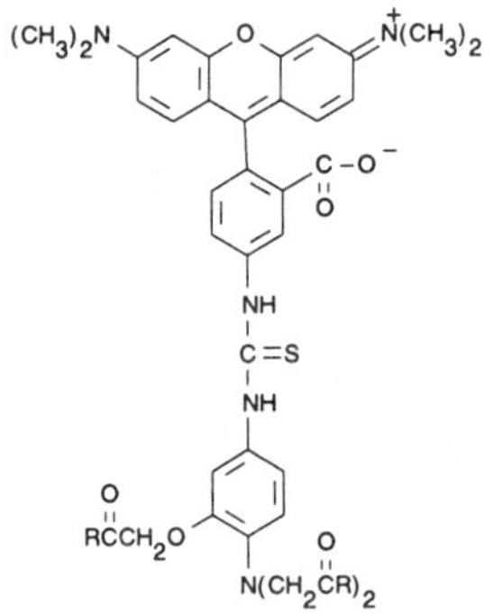

M-6890 R = $O^- K^+$
M-6891 R = $-OCH_2OCCH_3$ (C=O)

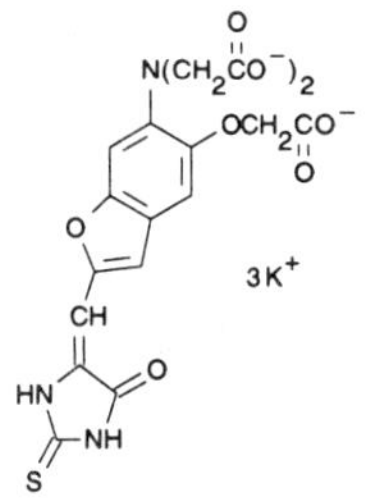

M-6892

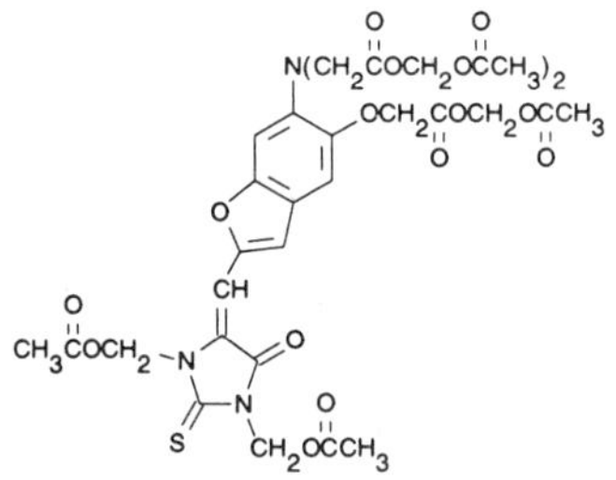

M-6893

22.6 Price List *Fluorescent Mg^{2+} Indicators*

Cat #		Product Name	Unit Size	Price Per Unit ($) 1–4 Units	5–24 Units
B-6874	***New***	N,N′-bis-(salicylidene)-2,3-diaminobenzofuran (SABF)	100 mg	45.00	36.00
B-6875	***New***	N,N′-bis-(salicylidene)ethylenediamine	5 g	24.00	19.20
M-1291		mag-fura-2, AM *cell permeant* *≥95% by HPLC*	1 mg	168.00	134.40
M-1292		mag-fura-2, AM *cell permeant* *≥95% by HPLC* *special packaging*	20x50 µg	198.00	158.40
M-1290		mag-fura-2, tetrapotassium salt *cell impermeant* *≥95% by HPLC*	1 mg	148.00	118.40
M-3105		mag-fura-5, AM *cell permeant* *≥95% by HPLC* *special packaging*	20x50 µg	198.00	158.40
M-3103		mag-fura-5, tetrapotassium salt *cell impermeant* *≥95% by HPLC*	1 mg	148.00	118.40
M-6893	***New***	Mag-Fura Red™, AM *cell permeant* *≥95% by HPLC* *special packaging*	20x50 µg	165.00	132.00

Cat #		Product Name	Unit Size	Price Per Unit ($) 1–4 Units	5–24 Units
M-6892	New	Mag-Fura Red™, tripotassium salt *cell impermeant* *≥95% by HPLC*	1 mg	125.00	100.00
M-1295		mag-indo-1, AM *cell permeant* *≥95% by HPLC* *special packaging*	20x50 µg	198.00	158.40
M-1293		mag-indo-1, tetrapotassium salt *cell impermeant*	1 mg	148.00	118.40
M-6907	New	mag-indo dextran, potassium salt, 10,000 MW, anionic	5 mg	148.00	118.40
M-6908	New	mag-indo dextran, potassium salt, 70,000 MW, anionic	5 mg	148.00	118.40
M-3735		Magnesium Green™, AM *cell permeant* *≥95% by HPLC* *special packaging*	20x50 µg	165.00	132.00
M-3733		Magnesium Green™, pentapotassium salt *cell impermeant*	1 mg	100.00	80.00
M-6891	New	Magnesium Orange™, AM *cell permeant* *≥95% by HPLC* *special packaging*	20x50 µg	155.00	124.00
M-6890	New	Magnesium Orange™, tripotassium salt *cell impermeant*	1 mg	95.00	76.00

22.7 Fluorescent Indicators for Zn^{2+} and Other Metals

Not only do certain metals play an important role in biological structure and activity, but they can also serve as useful probes of biological processes, including ion transport through Ca^{2+} channels.[1,2] Metal contamination, however, can cause ecological problems and present significant risks to health. For example, Hg^{2+} at submicromolar concentrations can cause a rapid and sustained increase in intracellular Ca^{2+} levels in rat T lymphocytes [3] and modify the depolarization- and agonist-stimulated Ca^{2+} signals in neuroadrenergic PC12 cells.[4] Rat astrocytes exposed to Pb^{2+} show significant increases in intracellular inositol triphosphate levels.[5] Furthermore, Pb^{2+} can mimic Ca^{2+} in important cellular processes and has recently been shown to activate protein kinase C from rat brain.[6]

Measuring heavy metal ion concentrations in cells and environmental samples with traditional ion indicators has been hampered by competitive binding of far more abundant cations, such as Ca^{2+} and Mg^{2+}. Molecular Probes has made a strong commitment to the development of fluorescence methods for the ultrasensitive detection of a wide variety of soluble metal ions. Our recent focus has been on the synthesis of new analytical reagents for detecting biologically relevant metals, as well as for sensing low concentrations of heavy metal ions in environmental samples. The goal of this research is to develop new sensor molecules that will selectively respond to submicromolar concentrations of metal ions such as Zn^{2+}, Cu^{2+}/Cu^{+}, Fe^{3+}/Fe^{2+}, Ni^{2+}, Hg^{2+}, Pb^{2+} and Cd^{2+} in the presence of relatively high concentrations of Ca^{2+}, Mg^{2+}, Na^{+} and K^{+}. The response of some fluorescent probes to the monovalent cations Rb^{+} and Cs^{+} is described in Chapter 24. New detection methods rely both on novel fluorescent ion sensors specifically designed for metal detection, as well as on new applications for indicators originally designed for detection of Ca^{2+} and Mg^{2+} (Table 22.3).

Existing spectrophotometric methods for detecting metal ions in environmental samples have usually required sample pretreatment, separation or concentration.[7] With our novel fluorescent indicators, a solution of the indicating dye is simply added to an aqueous sample, and the resulting fluorescence intensity is compared to that of a target-ion–free control. Several of our fluorescent indicators can be used to selectively determine polyvalent cation concentrations inside cells or to follow metal ion transport through ion channels. Other indicators are primarily useful for measurements in solutions or in extracts of environmental samples. In most cases, the high affinity of the indicators for metal ions allows interference from other compounds to be minimized by diluting the sample with deionized water. Most of the spectral responses and ion dissociation constants (K_d) reported in this section have been measured in 5.0 mM MOPS buffer, pH 7.0. These responses may vary significantly in more complex environments such as seawater. As with all ion indicators, the K_d is dependent on many factors, including pH, temperature, viscosity, protein binding and the presence of Mg^{2+} and other ions.

Ca^{2+} and Mg^{2+} Indicators as Reagents for Zn^{2+} and Other Metals

We have tested the responses of the Ca^{2+} and Mg^{2+} indicators (described in Sections 22.2 through 22.6) to a series of polyvalent metal ions and were surprised to find that most of the BAPTA-based indicators, including the Calcium Green dyes and fluo-3, exhibit the highest emission intensities upon binding Tb^{3+}, Hg^{2+} and Cd^{2+} [8] (Table 22.3). For example, Tb^{3+} enhances the fluorescence of fluo-3 more than 200-fold and has about 2.5 times the fluorescence of the Ca^{2+} complex of fluo-3. Fluorescence of the Calcium Green and Oregon Green 488 BAPTA-1 indicators is particularly sensitive to low levels of the environmental pollutants Cd^{2+}, Hg^{2+}, Ni^{2+} and Ba^{2+}.[8] Note that most of the measurements reported in Table 22.3 were made using 5 µM metal ion, which does not saturate the response for all indicators, and that the intensities reported in the table are relative to the intensity observed under the most fluorescent conditions for that indicator. The relative intensities for the different indicators in this table should not be compared because the indicators may have considerably different fluorescence quantum yields and extinction coefficients. Furthermore, our selection of excitation and emission wavelengths for the measurements affect the relative magnitude of changes observed.

It should be possible to use several of these indicators to monitor the permeability of Ca^{2+} channels to Ba^{2+}, Mn^{2+}, Co^{2+}, Cd^{2+}, Pb^{2+}, La^{3+} and other metals.[1,2] For example, the fluorescence quenching of Ca^{2+} indicators by Mn^{2+} has been used to:

- Calibrate fluo-3's spectral response [9]
- Determine the background contributed by cell autofluorescence and unhydrolyzed dye [10]
- Indirectly measure Ca^{2+} influx through ion channels [11-13]
- Quench indicators that have leaked from the cell in cell suspension measurements [14]

Table 22.3 *Ca^{2+} and Mg^{2+} indicator responses to 5 µM solutions of metal ions (unless otherwise indicated).*

Indicator	EGTA (10 mM)	Ca^{2+} (100 µM)	Mg^{2+} (10 mM)	Zn^{2+}	Cu^{2+}	Cu^{+}	Cd^{2+}	Hg^{2+}	Ni^{2+}	Co^{2+}	Ba^{2+}	Pb^{2+}	As^{3+}	Al^{3+}	Tb^{3+}	La^{3+}
Calcein	100	84	92	92	<1	<1	73	45	<1	<1	75	79	78	65	<1	84
Calcium Green-1	7	100	14	52	15	12	100	100	91	14	72	16	18	16	100	18
Calcium Green-2	~1	100	18	54	22	4	100	80	19	45	45	20	19	20	53	25
Calcium Green-5N	~3	72	12	48	14	9	96	100	31	40	46	14	14	24	90	39
Calcium Orange	31	96	31	84	34	19	100	55	71	41	81	37	46	32	89	33
Calcium Crimson	28	84	31	87	31	~3	100	100	16	12	76	33	47	30	95	32
Fluo-3	<1	41	5	25	4	<1	35	43	12	4	6	6	4	4	100	7
Fura Red	100	20	88	50	76	<1	9	<1	12	5	40	81	82	82	8	32
Oregon Green BAPTA-1	7	98	57	54	44	20	97	100	84	50	71	48	45	41	99	56
Magnesium Green	5	88	27	94	29	62	100	50	75	53	48	25	28	24	93	50
Magnesium Orange	73	78	70	100	64	66	64	35	80	74	93	65	66	65	66	65
Mag-Fura Red	100	46	70	<1	72	~1	<1	<1	~2	6	80	70	18	82	7	67

The indicator response values listed in this table represent the relative fluorescence of each indicator at 22°C in 5 mM MOPS buffer, pH 7.0, containing 5 µM metal ion (unless otherwise indicated); indicator concentrations were typically 1–5 µM. Metal binding by indicators may not be saturating under these conditions. The samples were excited at the absorption maximum of the uncomplexed dye, and the emission spectrum of the indicator/metal ion complex was scanned using a Hitachi F-4500 fluorescence spectrofluorometer with spectral correction. Although some indicators exhibit a shift in the emission maximum depending on the metal ion, the fluorescence intensities of all the indicator/metal ion complexes formed with a given indicator were measured at a single emission wavelength, chosen as the peak emission wavelength for the majority of complexes formed with this indicator. For each indicator, the relative fluorescence emission intensity of the indicator/metal ion complex is expressed relative to the intensity (arbitrarily set at 100) of the most fluorescent metal ion complex formed with this indicator. This method of data analysis normalizes the responses for differences in quantum yields and extinction coefficients between individual indicators. The indicator responses listed in this table represent relative fluorescence intensities in the presence of the designated pure metal ions at neutral pH and will vary with pH, ionic strength, temperature and presence of other ions. Data are provided only as a guideline for future research.

Fluorescence of calcein (C-481) is strongly quenched by Co^{2+}, Ni^{2+}, Cu^{+}, Cu^{2+}, Mn^{2+} and Tb^{3+} at physiological pH (Table 22.3) but not by Ca^{2+} or Mg^{2+} ions.[15,16] Monitoring the fluorescence level of cells that have been microinjected with calcein or loaded using calcein AM may provide a means for following uptake of toxic metals through the cell's ion channels.

Indicators for Zinc

Zinc is an important divalent cation in biological systems,[17] influencing DNA synthesis, microtubule polymerization, apoptosis,[18] gene expression [19] and protein structure and function.[20,21] Zn^{2+} is also implicated in the formation of amyloid plaques during the onset of Alzheimer's disease.[22-25] In addition, Zn^{2+} is reported to activate protein kinase C, and this activation is blocked by the heavy metal complexing agent TPEN [26,27] (T-1210, see Section 22.8). The intracellular concentration of free Zn^{2+} is extremely low in most cells (<1 nM),[28-30] with the remainder being bound to proteins or nucleic acids. Zn^{2+} binds to a variety of sites in cells that exhibit different affinities for the ion.[31] These sites produce several labile pools of Zn^{2+}, each with its own biological role and availability to complexing agents and chelators.

Newport Green™ and Newport Green Diacetate

Our Newport Green indicator (N-7990) exhibits an increase in fluorescence emission intensity upon binding Zn^{2+} (also Ni^{2+}, Co^{2+} and to a lesser degree Cd^{2+}) with little shift in wavelength (Table 22.4). Zn^{2+} binding to Newport Green in solution is slightly stronger than Ni^{2+} binding (K_d = 1.0 µM for Zn^{2+} and 1.5 µM for Ni^{2+}) but produces a less fluorescent complex. Relatively high levels of Ca^{2+} and Mg^{2+} are unlikely to interfere with Zn^{2+} detection with Newport Green (Table 22.4). We have observed that the weak cytosolic fluorescence of NIH 3T3 fibroblasts and rat basophilic leukemia (RBL) cells loaded with the cell-permeant Newport Green diacetate (N-7991) is substantially enhanced when the cells are exposed to 1 to 5 µM extracellular Zn^{2+} (without an ionophore). In these cells, the dye's initially uniform cytosolic staining changes to a primarily nuclear localization, with particularly prominent staining of nucleoli. We did not observe this change when either Ni^{2+} or Co^{2+} was added extracellularly.

TSQ

Use of the membrane-permeant probe *N*-(6-methoxy-8-quinolyl)-*p*-toluenesulfonamide (TSQ, M-688) in cells was first described by Fredrickson.[32] TSQ is selective for Zn^{2+} in the presence of physiological concentrations of Ca^{2+} and Mg^{2+} ions. The complex of TSQ with free Zn^{2+} apparently has a stoichiometry of two dye molecules per metal atom,[33] but a 1:1 complex may be formed with metalloproteins. The intracellular Zn^{2+} chelator dithizone blocks TSQ binding of Zn^{2+}.[34]

Several reports suggest that TSQ can be used to localize Zn^{2+} pools in the central nervous system. Zn^{2+} moves from presynaptic nerve terminals into postsynaptic nerve terminals when blood flow is constricted in the brain. This translocation is reported to correlate with ischemia-caused neurodegeneration, as determined by

Table 22.4 *Indicator responses to 25 µM solutions of metal ions (unless otherwise indicated).*

Indicator	EGTA (10 mM)	Zn^{2+}	Cu^{+}	Cu^{2+}	Cd^{2+}	Hg^{2+}	Ni^{2+}	Co^{2+}	Pb^{2+}	Fe^{2+}	Fe^{3+}	Ca^{2+} (100 µM)	Mg^{2+} (35 mM)
APTRA-BTC	100	46	17	42	64	96	48	74	87	100	63	89	98
Bis-BTC	~2	40	<1	~1	100	35	4	~2	5	~2	~2	4	~2
BTC-5N	~1	46	<1	~1	100	40	4	~2	7	~2	5	5	~1
Calcein	84	100	<1	<1	100	64	<1	<1	68	48	54	84	92
FITC-Gly-His	100	82	16	62	89	7	84	84	76	70	62	81	89
FITC-Gly-Gly-His	100	82	13	64	83	9	80	79	85	62	50	86	89
FL-DFO	100	84	~3	13	87	50	79	73	67	12	16	86	89
Newport Green	6	63	~1	~2	37	11	100	61	8	7	9	9	7
Phen Green	100	40	<1	<1	5	<1	47	24	26	7	34	60	70
TCPP	74	90	2	73	82	48	94	95	95	90	100	92	68
TSPP	89	38	<1	<1	74	21	100	92	92	92	100	100	100

The indicator response values listed in this table represent the relative fluorescence of each indicator at 22°C in 5 mM MOPS buffer, pH 7.0, containing 25 µM metal ion (unless otherwise indicated); indicator concentrations were typically 1–5 µM. Metal binding by indicators may not be saturating under these conditions. The samples were excited at the absorption maximum of the uncomplexed dye, and the emission spectrum of the indicator/metal ion complex was scanned using a Hitachi F-4500 fluorescence spectrofluorometer with spectral correction. Although some indicators, including calcein and BTC, exhibit a shift in the emission maximum depending on the metal ion, the fluorescence intensities of all the indicator/metal ion complexes formed with a given indicator were measured at a single emission wavelength, chosen as the peak emission wavelength for the majority of complexes formed with this indicator. For each indicator, the relative fluorescence emission intensity of the indicator/metal ion complex is expressed relative to the intensity (arbitrarily set at 100) of the most fluorescent metal ion complex formed with this indicator. This method of data analysis normalizes the responses for differences in quantum yields and extinction coefficients between individual indicators. The indicator responses listed in this table represent relative fluorescence intensities in the presence of the designated pure metal ions at neutral pH and will vary with pH, ionic strength, temperature and presence of other ions. Data are provided only as a guideline for future research.

the fluorescence of TSQ.[34] TSQ has also been used to study zinc release from axonal boutons of hippocampal mossy fibers [35] and changes in Zn^{2+} distribution in the rat hippocampus and amygdala [36] during and after kainic acid–induced seizures. In addition, TSQ is a selective nontoxic stain for pancreatic islet cells, which have a high content of Zn^{2+}, and may be useful for their flow cytometric isolation.[37,38]

TSQ-based assays for Zn^{2+} in seawater and other biological systems exhibit a detection limit of 0.1 nM.[39,40] The simultaneous determination of Zn^{2+} and Cd^{2+} by spectrofluorometry using TSQ in an SDS micelle has also been reported.[33] TSQ has recently been used to measure Zn^{2+} levels in artificial lipid vesicles and live sperm cells by flow cytometry.[41] In this study, the fluorescence yield of the TSQ-Zn^{2+} complex was shown to be much higher when bound to lipids than in aqueous solution, indicating that quantitative cell assays for Zn^{2+} based on the fluorescence intensity of TSQ may not be accurate because of uncertainty in the quantum yield of the dye when bound to membranes.

APTRA-BTC: Ratiometric Determination of Zn^{2+} and Cd^{2+}

We have modified the structure of our coumarin benzothiazole–based Ca^{2+} indicator BTC (see Section 22.2) to create APTRA-BTC (A-6895), a ratioable fluorescent indicator for Zn^{2+} and Cd^{2+}. Upon binding either of these metal ions, APTRA-BTC exhibits a shift in its excitation maximum, with an isosbestic point near 420 nm (Figure 22.15). Using the ratio of excitation intensities at 380 nm and 460 nm, we have determined the K_d for Zn^{2+} and Cd^{2+} to be ~1.4 µM and ~1.0 µM, respectively, at pH 7.0 and 22°C. Fluorescence of APTRA-BTC is only minimally affected by 100 µM Ca^{2+} or 35 mM Mg^{2+} (Table 22.4), and is not significantly affected by 25 µM Al^{3+}, Ba^{2+}, Bi^{3+}, Ca^{2+}, Co^{2+}, Cu^{2+}, Eu^{3+}, Hg^{2+}, Mg^{2+}, Ni^{2+} or Pb^{2+}. Because of the relatively weak binding of Zn^{2+} and Cd^{2+} to APTRA-BTC, this indicator may be more suitable for measuring the metals in environmental samples than in cells;[39] however, we have prepared the AM ester of APTRA-BTC (A-6896) for potential use in monitoring uptake of these ions by cells.

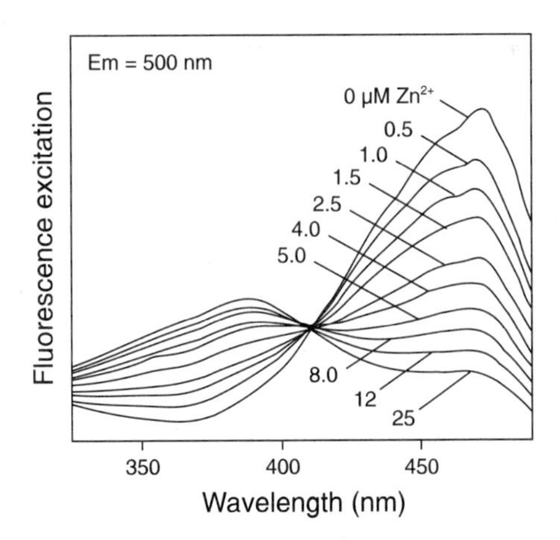

Figure 22.15 *Fluorescence excitation spectra of APTRA-BTC (A-6895) in solutions containing zero to 25 µM Zn^{2+}.*

Traditional Ca^{2+} and Mg^{2+} Indicators as Zn^{2+} Indicators

Zn^{2+} binds tightly to most traditional Ca^{2+} indicators with a higher affinity than Ca^{2+}. For example, fura-2 (F-1200, see Section 22.2) exhibits a K_d for Zn^{2+} in the absence of Ca^{2+} of 3 nM;[42] the spectral responses of the fura-2 complexes of Zn^{2+} and Ca^{2+} are virtually the same.[43] The lack of saturation of fura-2 fluorescence in cells is a further indication of the low intracellular concentration of free Zn^{2+}. Fura-2 remains sensitive to nanomolar Zn^{2+} levels in the presence of 25–100 nM free Ca^{2+}, allowing detection of intracellular Zn^{2+} influx via voltage-gated Ca^{2+} channels.[42]

Mag-fura-2 (M-1290, see Section 22.6) exhibits slightly altered spectral characteristics upon binding Zn^{2+} (K_d ~20 nM at pH 7.0–7.8 and 37°C), allowing Zn^{2+} to be measured in the presence of Ca^{2+}.[44] Similar measurements may be feasible with Mag-Fura Red (M-6892, see Section 22.6), which exhibits fluorescence that is strongly quenched by low concentrations of Zn^{2+} (K_d ~20 nM) but is relatively unaffected by 5 µM Ca^{2+} (see Table 22.3). The emission intensity of Magnesium Green (M-3733, see Section 22.6) is strongly enhanced by Zn^{2+} binding (K_d of ~20 nM).[8]

Displacement of Co^{2+} by Zn^{2+} from the essentially nonfluorescent Co^{2+}-calcein (C-481) complex provides an easy assay for Zn^{2+} in solution that has a micromolar sensitivity.[45]

TPEN (T-1210, see Section 22.8) is a strong complexing agent of Zn^{2+} that does not bind either Ca^{2+} or Mg^{2+};[46] it can be used to indirectly determine whether Zn^{2+} may be affecting the fluorescence of ion indicators and other dyes.

Indicators for Copper

Copper is third in abundance (after Fe^{3+} and Zn^{2+}) among the essential heavy metals in the human body. Dietary copper is required for normal hemoglobin synthesis, for prevention of anemia and for some enzyme activity. It may also play a role in dexamethasone- and etoposide-induced apoptosis.[47] Reduction of Cu^{2+} to Cu^{+} is catalyzed by the β-amyloid precursor protein, which is converted to plaques that are characteristic of Alzheimer's disease.[48]

Peptide Sensors for Cu^{+}

We have recently developed some novel indicators with high selectivity and sensitivity for Cu^{+} ions. The tripeptide, Gly-Gly-His, commonly called copper-binding peptide, has been used to mimic the Cu^{+}-binding site of human serum albumin.[49] More recently, the Ni^{2+} complex of this peptide has been examined as a catalyst for protein crosslinking.[50] Labeling this peptide and its shorter version (Gly-His) on the N-terminal amine with fluorescein yields two brightly fluorescent sensors — FITC-Gly-Gly-His (F-6865) and FITC-Gly-His (F-6864) — that are strongly quenched by submicromolar concentrations of Cu^{+} and Hg^{2+}. Exposure of these indicators to 5 µM concentrations of sixteen other metal ions produces little decrease in fluorescence. The excellent selectivity of these indicators may make them particularly useful for detecting the reduction or oxidation of copper and mercury ions in solution.

Phen Green for Cu^{+}

Our new phenanthroline-based Phen Green™ indicator (P-6801) is an excellent general-purpose heavy metal sensor capable of detecting a broad range of metal ions, including both Cu^{2+} and Cu^{+}.[8] The use of Phen Green for detecting Fe^{2+}, Hg^{2+}, Pb^{2+}, Cd^{2+} and Ni^{2+} at submicromolar concentrations is described below. Uncomplexed Phen Green is brightly fluorescent, with a fluorescence quantum yield of ~0.8. Binding of certain heavy metal ions is registered by strong fluorescence quenching (Table 22.4). The emission intensity of Phen Green depends both on metal ion concentration and on the indicator's concentration. Phen Green diacetate (P-6763) may be useful for loading this indicator in cells, although the response of intracellular Phen Green to copper ions has not yet been tested.

Porphine Indicators for Cu^{+}

The new tetrakis-(4-carboxyphenyl)porphine (TCPP, T-6931), its 10,000 MW and 70,000 MW dextran conjugates (D-6929, D-6930) and tetrakis-(4-sulfophenyl)porphine (TSPP, T-6932) are unique heavy metal indicators. TCPP and TSPP absorb intensely near 414 nm with ϵ >400,000 $cm^{-1}M^{-1}$ and several weaker absorption peaks between about 500 and 650 nm. Excitation at any of these absorption peak wavelengths yields emission at ~645 nm with a shoulder at about 700 nm. Binding of submicromolar levels of Cu^{+} (but not to Cu^{2+}) to either TCPP or its dextran conjugate results in strong fluorescence quenching with a K_d of ~0.5 µM. With the exception of Hg^{2+}, fluorescence of TCPP is either insensitive or slightly enhanced by metal binding. Thus, TCPP and its dextran conjugates may be useful for following reduction of Cu^{2+} to Cu^{+} and for detecting low levels of Cu^{+} in solutions. TCPP's use for ratiometric detection of Cd^{2+} is described below. In contrast to TCPP, TSPP exhibits fluorescence that is strongly quenched by both Cu^{+} and Cu^{2+}; however, we have also observed quenching by Zn^{2+} and Cd^{2+} at pH 7 (Table 22.4).

Traditional Ca^{2+} and Mg^{2+} Indicators as Copper Indicators

Few fluorescent indicators have been described for fluorometric determination of either Cu^{+} or Cu^{2+}. The data in Table 22.3 indicates that the fluorescence of Magnesium Green undergoes an approximately 12-fold increase upon binding monovalent Cu^{+} and 6-fold increase upon binding divalent Cu^{2+}, the fluorescence of Calcium Green-2 is enhanced about 5-fold by Cu^{+} and about 22-fold by Cu^{2+}, and the fluorescence of Fura Red and Mag-Fura Red is almost totally quenched by Cu^{+} but is relatively unaffected by Cu^{2+} (Table 22.3). Thus, the Fura Red and Mag-Fura Red indicators may provide a convenient means of following the reduction of Cu^{2+} to Cu^{+}, which occurs when using Cu^{2+} for protein quantitation.[51]

Indicators for Iron

Phen Green for Fe^{2+}

Our new general-purpose heavy metal sensor, Phen Green indicator (P-6801), is capable of detecting Fe^{2+}, as well as Cu^{2+}, Cu^{+}, Hg^{2+}, Pb^{2+}, Cd^{2+} and Ni^{2+}.[8] The fluorescence of Phen Green is strongly quenched upon binding Fe^{2+}, exhibiting over a 10-fold decrease in intensity (Table 22.4). The emission intensity of Phen Green depends both on metal ion concentration and on the indicator's concentration. Phen Green diacetate (P-6763) may be useful for loading this indicator in cells, although its use as an intracellular indicator has not yet been tested.

Calcein

Calcein (C-481) and calcein blue (H-1426) are rarely used now as fluorometric Ca^{2+} or Mg^{2+} indicators (see Section 22.3) because they require strongly alkaline pH.[45,52-56] Above pH ~12, calcein becomes nonfluorescent, but some of its metal complexes remain fluorescent.[45,57] A recent paper describes the use of fluorescence quenching of calcein (see Tables 22.3 and 22.4) at neutral pH to measure cytosolic Fe^{2+} concentrations from about 0.1 µM to 1.0 µM.[58] In this study, the AM ester of calcein (C-1430, C-3099, C-3100) was used to passively load the cells in order to follow intracellular release of iron from transferrin.

Fluorescein Desferrioxamine

The fluoresceinated derivative of the naturally occurring microbial siderophore[59,60] desferrioxamine B (FL-DFO, F-6877) is a potential probe for use in studies of iron transport, storage and metabolism. Binding of low levels of either Fe^{2+} or Fe^{3+} strongly quenches the 515 nm fluorescence of FL-DFO. Fluorescence of FL-DFO is insensitive to 0.1 mM Ca^{2+} and 35 mM Mg^{2+} and exhibits little or no quenching by several divalent heavy metal cations, including 25 µM Zn^{2+}, Cd^{2+}, Ni^{2+} and Cu^{2+} (Table 22.4). The specificity of this response is similar to that of an NBD-labeled DFO described by Lytton and co-workers.[61] NBD-DFO has been used to inhibit intraerythrocytic *Plasmodium falciparum* growth,[61] as well as to identify regions of iron uptake by cotton and maize root membranes and demonstrate the involvement of rhizosphere microorganisms in this process.[60]

Indicators for Mercury, Lead and Cadmium

Phen Green

Our new phenanthroline-based Phen Green indicator (P-6801) is an excellent general-purpose heavy metal sensor capable of detecting a broad range of metal ions, including Cu^{2+}, Cu^{+} and Fe^{2+} (see above), as well as micromolar concentrations of Hg^{2+}, Pb^{2+}, Cd^{2+}, Zn^{2+} and Ni^{2+}.[8] The fluorescence of 1 µM Phen Green is >99% quenched by 5 µM Hg^{2+} (Table 22.4). Like phenanthroline itself, the stoichiometry of metal binding to Phen Green is not always 1:1, permitting the detection range to be tuned by varying the indicator's concentration. For Hg^{2+}, the metal concentration that yields the half-maximal fluorescence response is 0.02 µM when the Phen Green concentration is 0.12 µM and increases to 1.2 µM when the Phen Green concentration is 5 µM. Concentrations of soluble Pb^{2+} as low as ~2 µM can be determined using ~1 µM Phen Green. Reducing the Phen Green concentration to <1 µM further improves the sensitivity for detection of Pb^{2+}.

We have used this versatile sensor to detect metal ions in a variety of matrices, including seawater and various contaminated solids such as paint sludge and soil. In such samples, Phen Green detects only the readily soluble (bioavailable) fraction of the total metal ions. Phen Green is well suited for initial field testing of metal ion contamination in aqueous samples; the large fluorescence changes produced by submicromolar ion concentrations of these heavy metals (Table 22.4) are easily visible upon illuminating the sample with a hand-held UV light source. Phen Green diacetate (P-6763) may be useful for loading this indicator in cells, although the response of intracellular Phen Green to metal ions has not yet been tested.

Bis-BTC and BTC-5N

The fluorescence of BTC-5N and bis-BTC excited at 405–415 nm increases upon binding Cd^{2+} and Zn^{2+} but is not significantly affected by millimolar levels of Ca^{2+}, Mg^{2+}, Na^{+} or K^{+}, or by 25 µM levels of Al^{3+}, Bi^{3+}, Co^{2+}, Cu^{+}, Cu^{2+}, Ni^{2+} or Pb^{2+}.[62] We have determined the affinity of BTC-5N for several metals:

- K_d for Cd^{2+} = 0.1 µM
- K_d for Zn^{2+} = 0.2 µM
- K_d for Hg^{2+} = 0.4 µM
- K_d for Pb^{2+} = 6.3 µM
- K_d for Ba^{2+} = 68 µM

As with most fluorescent indicators, complex ionic environments result in somewhat lower detection sensitivity; for example, the K_d of BTC-5N for Cd^{2+} in seawater is ~1.4 µM. A unique combination of Cd^{2+}-sensitive BTC-5N and the Cd^{2+}-insensitive porphine dye TCPP (T-6931) permits ratiometric measurement of Cd^{2+} (Figure 22.16).

Uncomplexed bis-BTC is virtually nonfluorescent; Cd^{2+}-dependent fluorescence enhancement in the 50 nM to 1.0 µM range (K_d = 140 nM) is readily detectable at 505 nm when excited at 410 nm (Figure 22.17).

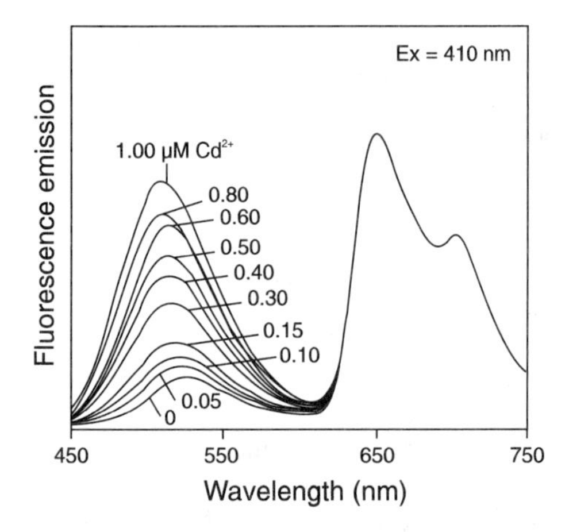

Figure 22.16 Fluorescence emission spectra of a mixture of 1 µM BTC-5N (B-6845) and 1 µM TCPP (T-6931) in solutions containing zero to 1.0 µM Cd^{2+}.

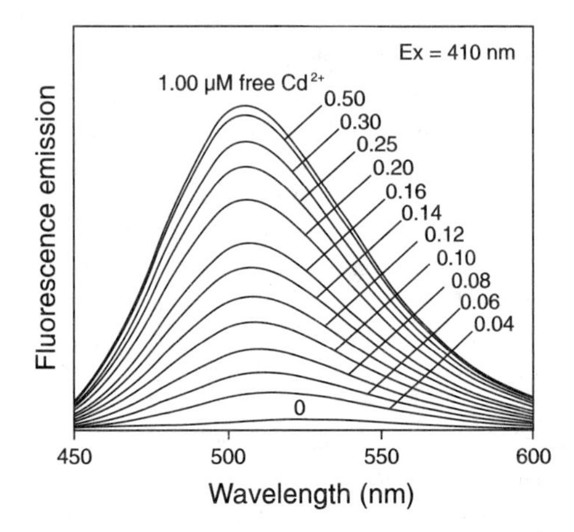

Figure 22.17 Fluorescence emission spectra of bis-BTC (B-6847) in solutions containing zero to 1.0 µM Cd^{2+}.

BTC-5N and bis-BTC are available as cell-impermeant potassium and ammonium salts (B-6845, B-6847), respectively, and as cell-permeant AM esters (B-6846, B-6848). The AM esters should be useful for measuring cellular Cd^{2+} permeability and Cd^{2+} transport through ion channels.

Traditional Ca^{2+} and Mg^{2+} Indicators as Cd^{2+}, Hg^{2+} and Pb^{2+} Indicators

The fluorescence of several of our traditional Ca^{2+} and Mg^{2+} indicators is strongly affected by binding of Cd^{2+}, Hg^{2+} and Pb^{2+} (Table 22.3). Fura-2 and quin-2 bind Cd^{2+} with extremely high affinity. The excitation response of fura-2 to Cd^{2+} — almost an exact match to that of fura-2 to Ca^{2+} (see Figure 22.1A) — has been used to monitor Cd^{2+} uptake by cells and to image intracellular free Cd^{2+}.[63] The response is reversed by TPEN (T-1210, see Section 22.8), which complexes heavy metals but not Ca^{2+} or Mg^{2+}.[46] Mag-fura-2 (M-1290, see Section 22.6) has also proven to be a useful intracellular Cd^{2+} indicator,[64] and indo-1 can be used to simultaneously determine the intracellular concentrations of Ca^{2+} and Cd^{2+} or Ca^{2+} and Ba^{2+}.[1] Fluorometric detection of Cd^{2+} at pH 13.3 using calcein (C-481) has been reported.[57]

Pb^{2+} entry into cells has been monitored using fura-2 as an intracellular indicator.[65,66] ^{19}F NMR can also be used to detect Pb^{2+} uptake by platelets using fluorinated BAPTA derivatives[67] (see Section 22.8).

Fura Red and Mag-Fura Red fluorescence is almost totally quenched by extremely low levels of Hg^{2+}, as shown in Table 22.3.

Indicators for Nickel and Cobalt

Newport Green: A Fluorescent Indicator for Nickel

Much of the recent interest in Ni^{2+} comes from its use in the isolation of poly-histidine–containing proteins by metal-chelate affinity chromatography.[68,69] We have developed the Newport Green indicator (N-7990), which we find to be an exceptionally sensitive probe for Ni^{2+} and Zn^{2+} in solution (K_d = 1.0 µM for Zn^{2+} and 1.5 µM for Ni^{2+}). 25 µM Ni^{2+} enhances the fluorescence of Newport Green approximately 16-fold without a spectral shift; Zn^{2+} and Cd^{2+} enhance this indicator's fluorescence to a lesser extent (Table 22.4). Ni^{2+} strongly enhances the fluorescence of several other ion indicators and significantly quenches the fluorescence of Mag-Fura Red (Table 22.3).

Traditional Ca^{2+} and Mg^{2+} Indicators as Ni^{2+} and Co^{2+} Indicators

Co^{2+} enhances the fluorescence of Calcium Green-2 at least 45-fold and strongly quenches both Fura Red and Mag-Fura Red (Table 22.3). Co^{2+} and Ni^{2+} (as well as Cu^{+}, Cu^{2+}, Tb^{3+} and other polycations; see Tables 22.3 and 22.4) totally quench the fluorescence of calcein, even at pH 7;[15,16,70] thus, it should be possible to follow the kinetics of uptake of these ions into cells loaded with calcein AM (C-1430, C-3099, C-3100). Note that the 1:1 stoichiometry of calcein–metal binding will require the use of low loading levels of this probe to achieve significant quenching by limited amounts of metal transport. The quenching of calcein fluorescence by Co^{2+} or Ni^{2+} has been used to detect liposome fusion[15,16] (see Section the Technical Note "Assays of Volume Change, Membrane Fusion and Membrane Permeability" on page 549).

Indicators for Aluminum and Gallium

Equimolar complexes of Al^{3+} and Ga^{3+} with several azo dyes, including 2,2'-dihydroxyazobenzene (2,2'-azodiphenol, D-6873), are fluorescent, particularly at slightly acidic pH values.[71,72] With the exception of Calcium Green-2 and Oregon Green BAPTA-2 indicators, Al^{3+} binding has little effect on the fluorescence of most of the traditional Ca^{2+} and Mg^{2+} indicators shown in Table 22.3. However, Al^{3+} is reported to selectively form a fluorescent complex with calcein (C-481) at acidic pH that can be detected with micromolar sensitivity.[45]

Indicators for Lanthanides

NTA and DPA

Europium ion (Eu^{3+} from $EuCl_3$; E-6960) forms a highly fluorescent complex with naphthoyltrifluoroacetone (NTA, T-411). The resulting chelate has an extremely long lifetime and is widely used in highly sensitive time-resolved fluoroimmunoassays [73] (TR-FIA). Dipicolinic acid (DPA, D-1248) also forms highly fluorescent complexes with other lanthanides; it has been used to detect terbium ion (Tb^{3+} from $TbCl_3$; T-1247) and Eu^{3+} with sensitivities down to ~4 ng/mL.[74] In addition to the cell-impermeant acid, we offer the cell-permeant AM ester of DPA (DPA, AM; D-1249) to facilitate passive loading of cells.

Traditional Ca^{2+} and Mg^{2+} Indicators as Lanthanide Indicators

Tb^{3+} has a strong effect on the fluorescence of a wide variety of traditional Ca^{2+} and Mg^{2+} indicators (Table 22.3). For example, the fluorescence yield of fluo-3 upon binding Tb^{3+} increases about 200-fold and is almost threefold that of the Ca^{2+}-saturated indicator (Table 22.3). However, low levels of Tb^{3+} totally quench the fluorescence of calcein, even at pH 7. Long-lived luminescence of Tb^{3+} is also used to probe Ca^{2+} and other metal binding sites of alkaline phosphatase,[75] glutamine synthetase,[76] integrins,[77] protein kinase C [78] and other proteins.

Fluorescence of calcein blue (H-1426) is strongly and selectively quenched by lanthanides, including Tb^{3+}, Eu^{3+}, Sm^{3+} and Dy^{3+}.[79] Sensitivity of the assay is reported to be about 0.01–0.02 µg/mL, depending on the metal ion.

1. Mol Cell Biochem 151, 91 (1995); **2.** Endocrinology 131, 1936 (1992); **3.** J Toxicol Environ Health 38, 159 (1993); **4.** FASEB J 7, 1507 (1993); **5.** Brain Res 618, 9 (1993); **6.** Nature 269, 834 (1994); **7.** *Photometric Determination of Trace Metals Volume 3, Part IIA*, E.B. Sandel and H. Onishi, Eds., John Wiley and Sons (1986); **8.** SPIE Proc 2388, 238 (1995); **9.** J Biol Chem 264, 8171 (1989); **10.** J Biol Chem 258, 4876 (1983); **11.** J Physiol 421, 55 (1990); **12.** Cell Calcium 10, 477 (1989); **13.** J Biol Chem 264, 18349 (1989); **14.** Cell Calcium 10, 171 (1989); **15.** Chem Pharm Bull 39, 227 (1991); **16.** Biochim Biophys Acta 691, 332 (1982); **17.** Acct Chem Res 26, 543 (1993); **18.** Biochem J 296, 403 (1993); **19.** J Biol Chem 265, 6513 (1990); **20.** Biochem J 303, 781 (1994); **21.** Ann Rev Biochem 61, 897 (1992); **22.** Science 268, 1920 (1995); **23.** J Biol Chem 269, 12152 (1994); **24.** Science 265, 1464 (1994); **25.** Science 265, 1365 (1994); **26.** Biochim Biophys Acta 1053, 113 (1990); **27.** J Biol Chem 263, 6487 (1988); **28.** J Membrane Biol 123, 61 (1991); **29.** J Biol Chem 262, 11140 (1987); **30.** J Biol Chem 246, 1160 (1971); **31.** Acta Histochem 89, 107 (1990); **32.** J Neurosci Meth 20, 91 (1987); **33.** J Fluorescence 1, 267 (1991); **34.** Science 272, 1013 (1996); **35.** Brain Res 446, 383 (1988); **36.** Brain Res 480, 317 (1989); **37.** Biotechnic Histochem 4, 196 (1993); **38.** Transplantation 56, 1282 (1993); **39.** Anal Chem 66, 2732 (1994); **40.** Biol Res 27, 49 (1994); **41.** Cytometry 21, 153 (1995); **42.** J Biol Chem 270, 2473 (1995); **43.** J Biol Chem 260, 3440 (1985); **44.** J Biochem Biophys Meth 27, 25 (1993); **45.** Anal Chem 35, 1035 (1963); **46.** J Biol Chem 260, 2719 (1985); **47.** FEBS Lett 352, 58 (1994); **48.** Science 271, 1406 (1996); **49.** J Biol Chem 248, 5878 (1974); **50.** Biochemistry 34, 4733 (1995); **51.** Anal Biochem 150, 76 (1985); **52.** Analyst 113, 251 (1988); **53.** Biophys J 18, 3 (1977); **54.** Clin Chem 18, 1411 (1972); **55.** Analyst 83, 188 (1958); **56.** Analyst 82, 284 (1957); **57.** Anal Chem 46, 2036 (1974); **58.** J Biol Chem 270, 24209 (1995); **59.** J Biol Chem 270, 26723 (1995); **60.** Plant Physiol 99, 1329 (1992); **61.** Mol Pharmacol 40, 584 (1991); **62.** U.S. Patent No. 5,459,276; **63.** J Biol Chem 267, 25553 (1992); **64.** Kidney Int'l 41, 1237 (1992); **65.** Biochim Biophys Acta 1069, 197 (1991); **66.** Am J Physiol 259, C762 (1990); **67.** Biochim Biophys Acta 1092, 341 (1991); **68.** Bio/Technology 6, 1321 (1988); **69.** J Biol Chem 258, 5994 (1983); **70.** Anal Chem 56, 810 (1984); **71.** Anal Chim Acta 32, 398 (1965); **72.** Anal Chim Acta 26, 458 (1962); **73.** Analyst 117, 1879 (1992); **74.** Gaodeng Xuexiao Huaxue Xuebao 4, 115 (1983) cited in *Practical Fluorescence, Second Edition*, G. Guilbault, Ed., Marcel Dekker, (1990) pp. 198 and 215; **75.** J Photochem Photobiol 13, 289 (1992); **76.** Biochemistry 30, 3417 (1992); **77.** Biochemistry 29, 12238 (1994); **78.** J Biol Chem 263, 4223 (1990); **79.** Anal Chem 59, 1122 (1987).

22.7 Data Table *Fluorescent Indicators for Zn^{2+} and Other Metals*

Cat #	Structure	MW	Storage	Soluble	Abs	Low Ion {ε × 10⁻³}	Em	Abs	High Ion {ε × 10⁻³}	Em	Solvent	Cleaved*	K_d†	Notes
A-6895	✓	599	L	pH>6	466	{31}	520	380	{16}	530	H_2O/Zn^{2+}		1.4 µM	<1, 2, 3>
A-6896	✓	701	F,D,L	DMSO	429	{41}	496				MeOH	A-6895		
B-6845	✓	875	L	pH>6	459	{40}	517	417	{25}	532	H_2O/Zn^{2+}		210 nM	<1, 2, 4>
B-6846	✓	1011	F,D,L	DMSO	428	{42}	482				EtOAc	B-6845		
B-6847	✓	947	L	pH>6	455	{53}	529	405	{42}	505	H_2O/Cd^{2+}		140 nM	<1, 2>
B-6848	✓	1167	F,D,L	DMSO	429	{77}	495				EtOAc	B-6847		
C-481	S15.3	623	L	pH>5	494	{76}	517				pH 9			<5>
C-1430	S16.3	995	F,D	DMSO	<300		none					C-481		
D-1248	S15.3	167	D	pH>5	270	{4.5}	none				pH 8			
D-1249	S15.3	311	F,D	DMSO	270	{4.1}	none				MeOH	D-1248		
D-6873	✓	214	D,L	DMSO, DMF	394	{13}	none	<6>		<6>	MeOH			
D-6929	C	<7>	F,D,L	H_2O	418	ND	660			<8>	pH 7			<9>
D-6930	C	<7>	F,D,L	H_2O	415	ND	662			<8>	pH 7			<9>
E-6960	S15.3	366‡	D	H_2O				337	{17}	613	H_2O			<10>
F-6864	✓	678	F,D,L	H_2O	493	{80}	517			<11>	pH 9			
F-6865	✓	735	F,D,L	H_2O	493	{84}	517			<11>	pH 9			
F-6877	✓	1035	F,D,L	H_2O	493	{73}	515				pH 9			
H-1426	S15.3	321	L	pH>5	360	{16}	440				pH 9			
M-688	✓	328	L	EtOH	334	{4.2}	385			<12>	MeOH			

Cat #	Structure	MW	Storage	Soluble	Low Ion Abs	Low Ion {ε × 10⁻³}	Low Ion Em	High Ion Abs	High Ion {ε × 10⁻³}	High Ion Em	Solvent	Cleaved*	K_d†	Notes
N-7990	✓	794	D,L	pH>6	506	{82}	535	506	{82}	535	H_2O/Ni^{2+}		1.5 µM	<1, 2>
N-7991	✓	802	F,D	DMSO	302	{15}	none				MeOH	N-7990		
P-6763	✓	669	F,D	DMSO	<300		none					P-6801		
P-6801	✓	661	D,L	pH>6	492	{65}	517			<13>	pH 9			
T-411	S15.3	266	D,L	DMF, EtOH	337	{17}	483				MeOH			
T-1247	S15.3	373‡	D	H_2O				270	{4.7}	545	H_2O			<14>
T-6931	✓	791	D,L	pH>6, DMSO	414	{416}	645			<8>	pH 9			<15>
T-6932	✓	935	D,L	H_2O, DMSO	413	{472}	645			<16>	pH 9			<15>

C-3099, C-3100 see C-1430

For definitions of the contents of this data table, see "How to Use the Handbook" on page vi.
Structure: Chemical structure drawing: ✓ = shown in this section; Sn.n = shown in Section number n.n; C = not shown due to complexity.
ND = not determined.

* Catalog numbers listed in this column indicate the cleavage product generated by enzymatic or chemical hydrolysis of the acetate or acetoxymethyl ester groups of the parent compound.

† Dissociation constants determined at Molecular Probes by fluorescence measurements in 5 mM MOPS, pH 7.0 at 22°C. These values are known to vary considerably depending on the temperature, pH, ionic strength, viscosity, protein binding, presence of other ions (especially polyvalent ions), instrument setup and other factors. It is strongly recommended that these values be specifically determined under your experimental conditions.

‡ MW is for the hydrated form of this product.

<1> For "Low Ion" spectra, the concentration of the ion indicated in the column headed "Solvent" is zero. For "High Ion" spectra, the concentration of this ion is above that required to saturate the response of the indicator.

<2> Dissociation constant is for ion indicated in the column headed "Solvent."

<3> A-6895 also binds Cd^{2+} with similar affinity (K_d ~1 µM) and spectral response.

<4> Fluorescence excited at 415 nm increases 20-fold upon binding Zn^{2+}. Fluorescence enhancement, emission maximum and K_d for other ions (Cd^{2+}, Hg^{2+}, Pb^{2+}, Ba^{2+}) is variable.

<5> C-481 fluorescence is strongly quenched by micromolar concentrations of Fe^{2+}, Fe^{3+}, Co^{2+}, Ni^{2+}, Cu^{+}, Cu^{2+}, Mn^{2+} and Tb^{3+} at pH 7.

<6> D-6873 forms a fluorescent complex with Al^{3+}, Abs = 480 nm, Em = 573 nm in aqueous solutions at pH 6.3 containing 35% EtOH [Anal Chim Acta 32, 398 (1965)].

<7> Molecular weight is nominally as specified in the product name but may have a broad distribution.

<8> Binding of Cu^{+} (K_d ~0.5 µM) results in fluorescence quenching with no change in emission wavelength.

<9> Because indicator dextrans are polydisperse both in molecular weight and degree of substitution, dissociation constants and spectra may vary between batches.

<10> Absorption and fluorescence of E-6960 are extremely weak unless it is chelated. Data are for the detergent-solubilized NTA (T-411) chelate.

<11> Fluorescence of F-6864 and F-6865 is quenched by micromolar concentrations of Cu^{+} and Hg^{2+} with no change in emission wavelength.

<12> Fluorescence of M-688 is very sensitive to solvent polarity. Em = 495 nm in aqueous buffer (pH 7.4). Fluorescence is enhanced upon binding Zn^{2+} (0.01–10 µM) with no change in emission wavelength [Cytometry 21, 153 (1995)].

<13> Fluorescence of P-6801 is quenched by Cu^{+}, Cu^{2+}, Cd^{2+}, Fe^{2+}, Hg^{2+}, Ni^{2+} and Pb^{2+} with no change in emission wavelength. The sensitivity of this response is dependent on the concentration of indicator in the sample.

<14> Absorption and fluorescence of T-1247 are extremely weak unless it is chelated. Data are for dipicolinic acid (D-1248) chelate. Fluorescence spectrum has secondary peak at 490 nm.

<15> Absorption spectra of this compound also contain much weaker bands in the 500–650 nm wavelength range. Absorption and fluorescence are pH dependent. At pH below 5, Abs = 440 nm for T-6931, Abs = 434 nm for T-6932.

<16> T-6932 forms a fluorescent complex (Em = 604 nm) with Zn^{2+} at pH >8. Complex with Cd^{2+} (Em = 633 nm) is strongly quenched [Anal Chim Acta 281, 347 (1993)].

22.7 Structures *Fluorescent Indicators for Zn^{2+} and Other Metals*

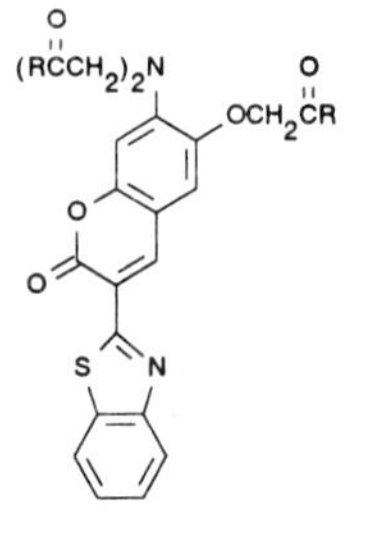

A-6895 R = O^- K^+
A-6896 R = $-OCH_2OC(O)CH_3$

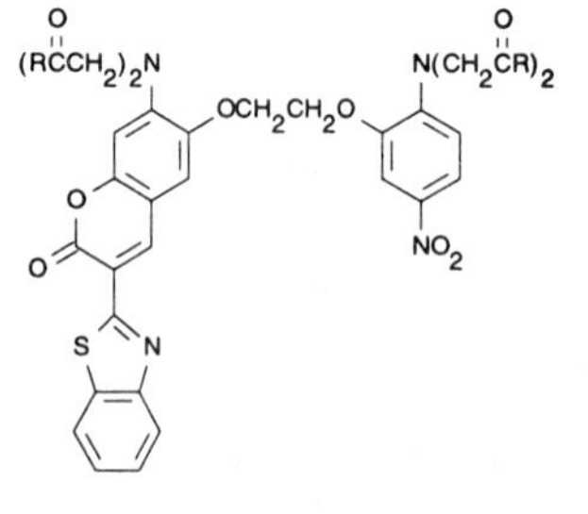

B-6845 R = O^- K^+
B-6846 R = $-OCH_2OC(O)CH_3$

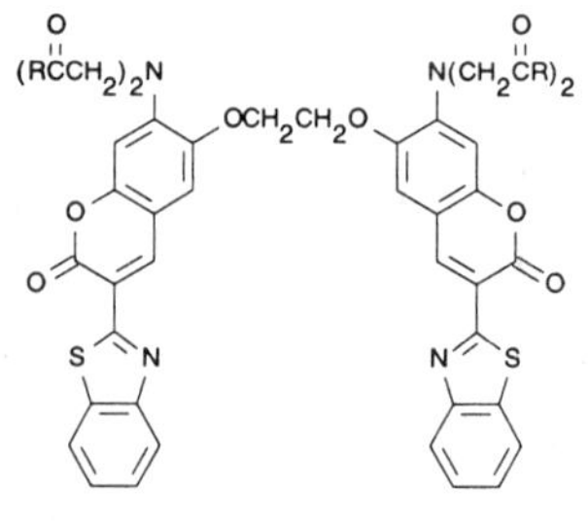

B-6847 R = O^- NH_4^+
B-6848 R = $-OCH_2OC(O)CH_3$

C-481, D-1248
D-1249
Section 15.3
C-1430
Section 16.3

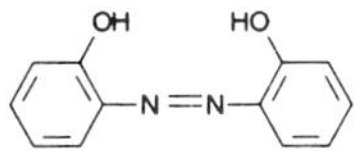

D-6873

E-6960
Section 15.3

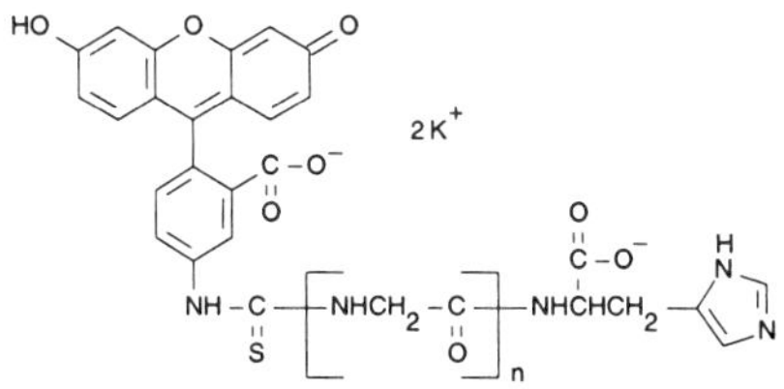

F-6864 n = 1
F-6865 n = 2

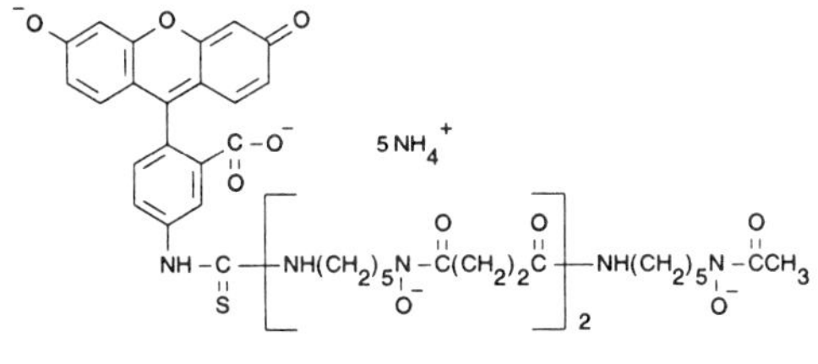

F-6877

H-1426
Section 15.3

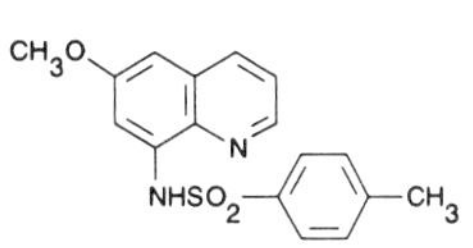

M-688

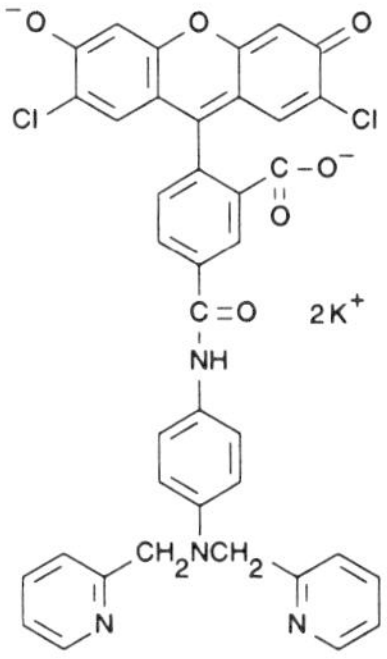

N-7990

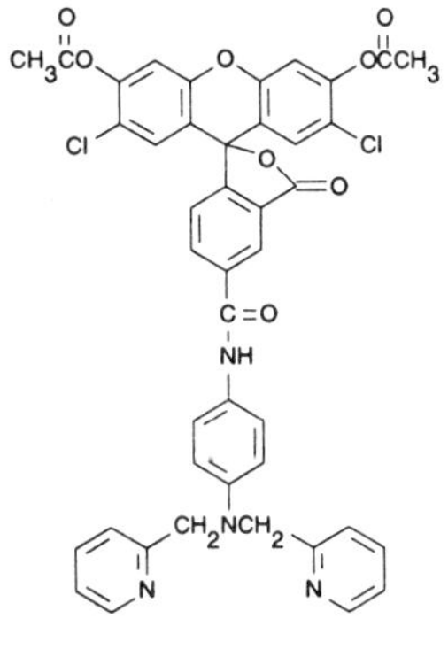

N-7991

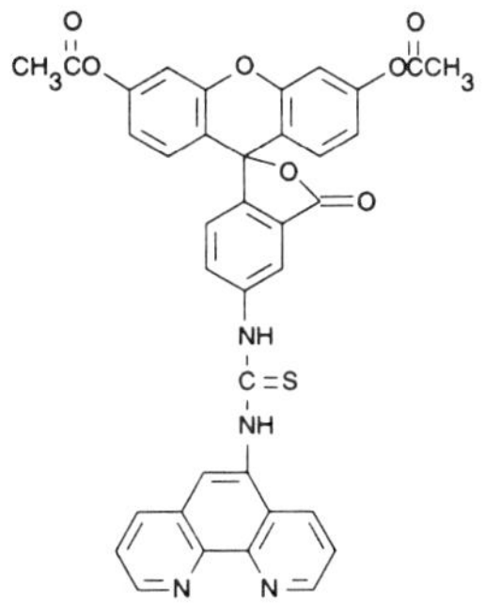

P-6763

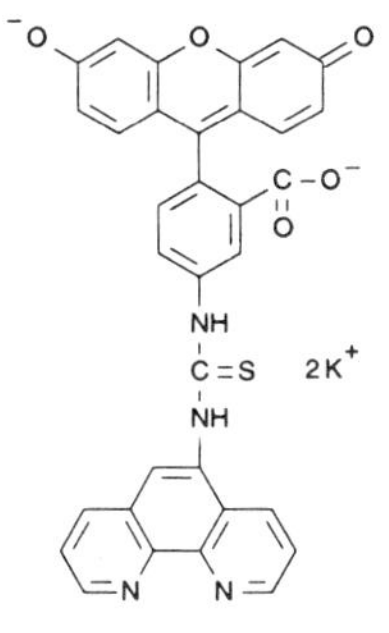

P-6801

T-411, T-1247
Section 15.3

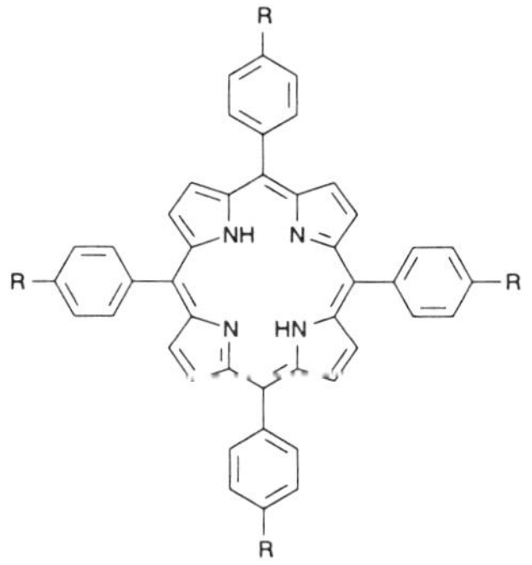

T-6931 R = —COOH
T-6932 R = —SO_3H

22.7 Price List *Fluorescent Indicators for Zn^{2+} and Other Metals*

Cat #		Product Name	Unit Size	Price Per Unit ($) 1–4 Units	5–24 Units
A-6896	***New***	APTRA-BTC, AM *cell permeant* *≥95% by HPLC*	20x50 µg	98.00	78.40
A-6895	***New***	APTRA-BTC, tripotassium salt *cell impermeant*	1 mg	80.00	64.00
B-6848	***New***	bis-BTC, AM *cell permeant* *≥95% by HPLC* *special packaging*	20x50 µg	110.00	88.00
B-6847	***New***	bis-BTC, tetraammonium salt *cell impermeant*	1 mg	98.00	78.40
B-6846	***New***	BTC-5N, AM *cell permeant* *≥95% by HPLC* *special packaging*	20x50 µg	125.00	100.00
B-6845	***New***	BTC-5N, tetrapotassium salt *cell impermeant*	1 mg	95.00	76.00
C-481		calcein *high purity*	100 mg	58.00	46.40
C-1430		calcein, AM	1 mg	115.00	92.00
C-3099		calcein, AM *1 mg/mL solution in dry DMSO* *special packaging*	1 mL	120.00	96.00
C-3100		calcein, AM *special packaging*	20x50 µg	140.00	112.00
D-6929	***New***	dextran, tetrakis-(4-carboxyphenyl)porphine, 10,000 MW, anionic	10 mg	95.00	76.00
D-6930	***New***	dextran, tetrakis-(4-carboxyphenyl)porphine, 70,000 MW, anionic	10 mg	95.00	76.00
D-6873	***New***	2,2′-dihydroxyazobenzene (2,2′-azodiphenol)	100 mg	18.00	14.40

Cat #		Product Name	Unit Size	Price Per Unit ($) 1–4 Units	5–24 Units
D-1248		dipicolinic acid (DPA) *cell impermeant*	1 g	17.00	13.60
D-1249		dipicolinic acid, AM (DPA, AM) *cell permeant*	25 mg	48.00	38.40
E-6960	*New*	europium(III) chloride, hexahydrate *99.99%*	1 g	20.00	16.00
F-6877	*New*	N-(fluorescein-5-thiocarbamoyl)desferrioxamine, pentaammonium salt (FL-DFO) *special packaging*	20x50 µg	95.00	76.00
F-6865	*New*	(fluorescein-5-thioureidyl)glycylglycyl-L-histidine, dipotassium salt (FITC-Gly-Gly-His)	1 mg	65.00	52.00
F-6864	*New*	(fluorescein-5-thioureidyl)glycyl-L-histidine, dipotassium salt (FITC-Gly-His)	1 mg	65.00	52.00
H-1426		7-hydroxy-4-methylcoumarin-8-methyleneiminodiacetic acid (calcein blue)	1 g	20.00	16.00
M-688		N-(6-methoxy-8-quinolyl)-p-toluenesulfonamide (TSQ)	25 mg	85.00	68.00
N-7991	*New*	Newport Green™ diacetate *cell permeant*	1 mg	90.00	72.00
N-7990	*New*	Newport Green™, dipotassium salt *cell impermeant*	1 mg	80.00	64.00
P-6763	*New*	Phen Green™ diacetate *cell permeant*	1 mg	90.00	72.00
P-6801	*New*	Phen Green™, dipotassium salt	1 mg	80.00	64.00
T-1247		terbium(III) chloride, hexahydrate	1 g	20.00	16.00
T-6931	*New*	tetrakis-(4-carboxyphenyl)porphine (TCPP)	5 mg	48.00	38.40
T-6932	*New*	tetrakis-(4-sulfophenyl)porphine (TSPP)	5 mg	48.00	38.40
T-411		2-((trifluoroacetyl)acetyl)naphthalene (naphthoyltrifluoroacetone; NTA)	100 mg	65.00	52.00

22.8 Chelators, Calibration Buffers and Ionophores

Caged Calcium and Caged Calcium Chelators

Nitrophenyl EGTA: A Superior Caged Calcium Reagent

As an alternative to solely monitoring Ca^{2+} changes using fluorescent indicators, scientists may want to rapidly raise or lower the intracellular Ca^{2+} concentration and study the physiological response that results. Ellis-Davies and Kaplan have recently developed a new photolabile chelator, *o*-nitrophenyl EGTA [1] (NP-EGTA), which exhibits a high selectivity for Ca^{2+}, a dramatic 12,500-fold decrease in affinity for Ca^{2+} upon UV illumination (its K_d increases from 80 nM to >1 mM) and a high photochemical quantum yield (~0.2).[2-4] Photolysis of NP-EGTA is slightly faster than that of DMNP-EDTA,[5] which is frequently called DM-Nitrophen™ (see below). Furthermore, with a K_d for Mg^{2+} of 9 mM, NP-caged EGTA does not bind physiological levels of Mg^{2+} and thus reduces interference from this abundant cation.[6,7] Skinned muscle fibers equilibrated with the caged probe were shown to contract maximally upon irradiation with a single flash from a frequency-doubled ruby laser (347 nm illumination).[4] Using a system that combined confocal laser scanning microscopy with focal photolysis, researchers have selectively photolyzed Ca^{2+}-activated K^+ channels in the soma of a single NP-EGTA–loaded Purkinje cell.[8]

We offer the tetrapotassium salt (N-6802) and the acetoxymethyl (AM) ester (N-6803) of NP-EGTA. The NP-EGTA salt can be complexed with Ca^{2+} to generate a caged Ca^{2+} reagent that will rapidly deliver Ca^{2+} upon photolysis (Figure 19.1A in Chapter 19). The cell-permeant of NP-EGTA AM does not bind Ca^{2+} unless its AM ester groups are removed. This AM ester can potentially serve as a photolabile buffer in cells because, once converted to NP-EGTA by intracellular esterases, it will bind Ca^{2+} with high affinity until photolyzed with UV light.

DMNP-EDTA

The first caged Ca^{2+} reagent to be described by Kaplan and Ellis-Davies was 1-(4,5-dimethoxy-2-nitrophenyl) EDTA (DMNP-EDTA, D-6814), which they named DM-Nitrophen [9-11] (now a trademark of Calbiochem-Novabiochem Corp.). Because its structure more resembles that of EDTA than EGTA, we named it as a caged EDTA derivative (Figure 19.1B in Chapter 19). Upon illumination, DMNP-EDTA's K_d for Ca^{2+} increases from 5 nM to 3 mM. Thus, photolysis of DMNP-EDTA complexed with Ca^{2+} results in a pulse of free Ca^{2+}. Furthermore, DMNP-EDTA has significantly higher affinity for Mg^{2+} (K_d = 2.5 µM) than does NP-EGTA (K_d = 9 mM).[9] Because the photolysis product's K_d for Mg^{2+} is ~3 mM,

Product-Specific Bibliographies and Keyword Searches

Our Technical Assistance Department can provide you with product-specific bibliographies, as well as keyword searches of the over 25,000 literature references in our extensive bibliography database. Our bibliography database is also searchable through our Web site (http://www.probes.com).

DMNP-EDTA is an effective caged Mg^{2+} source, in addition to its applications for photolytic Ca^{2+} release.[12,13] Photorelease of Ca^{2+} has been shown to occur in <180 µsec, with even faster photorelease of Mg^{2+}.[14] Moreover, based on properties of the parent chelator EDTA, we expect DMNP-EDTA to also be useful for photolytic release of divalent cations such as Hg^{2+}, Cu^{2+}, Cd^{2+} and Zn^{2+}.

Photolysis of DMNP-EDTA has been used extensively to release Ca^{2+} in a variety of systems. This caged Ca^{2+} reagent has been employed to:

- Evoke secretion in bovine chromaffin cells [13]
- Investigate temperature modulation of Ca^{2+}-activated neurotransmitter release from squid neurons [15]
- Measure Ca^{2+} pump kinetics in erythrocyte ghosts using Calcium Green-2 and Calcium Green-5N [16]
- Stimulate neurotransmitter release in nerve terminals [17]
- Study cardiac and skeletal muscle [18]

A recent paper by Neher and Zucker discusses the uses and limitations of DMNP-EDTA.[13]

In addition to the tetrapotassium salt of DMNP-EDTA (D-6814), we have prepared the AM ester of this probe (D-6815), which has not been previously available.

Photoactivatable Calcium Scavengers

In contrast to NP-EGTA and DMNP-EDTA, diazo-2 is a photoactivatable Ca^{2+} scavenger. Diazo-2, which was introduced by Adams, Kao and Tsien,[19-21] is a relatively weak chelator (K_d for Ca^{2+} = 2.2 µM). However, following flash photolysis at ~360 nm, cytosolic free Ca^{2+} rapidly binds to the high-affinity diazo-2 photolysis product (K_d = 73 nM). Photolysis of diazo-2 has been used to decrease cytosolic Ca^{2+} in less than three milliseconds in tensed frog muscle cells,[22] rat fibroblasts [20] and rabbit arterial smooth muscle.[23] Trapping of Ca^{2+} following photolysis of diazo-2 results in relaxation of skinned cardiac or skeletal muscle [24] and rapid depletion of Ca^{2+} in ventricular myocytes.[25] Microinjecting a relatively low concentration of a visible light–excitable Ca^{2+} indicator — such as fluo-3 or one of our Calcium Green or Oregon Green 488 BAPTA indicators — along with a known quantity of diazo-2 permits the measurement of the extent of depletion of cytosolic Ca^{2+} following photolysis.[20,26,27]

Diazo-2 can be microinjected into larger cells [7] or electroporated into smaller cells [28] as its potassium salt (D-3034). Diazo-2 also has been loaded into cells permeabilized by *Staphylococcus aureus* α-toxin.[23] Alternatively, diazo-2 can be passively loaded into cells as its AM ester (D-3036).

In addition to diazo-2, Molecular Probes offers the potassium salt and AM ester of diazo-3 (D-3700, D-3701). Diazo-3 binds Ca^{2+} with low affinity, both before and after photolysis, and therefore may serve as a control for the effects that AM ester hydrolysis and photolysis products of diazo-2 may have on the cell.[28]

Nonfluorescent Chelators

BAPTA and BAPTA AM

The BAPTA buffers of Tsien [29] are highly selective for Ca^{2+} over Mg^{2+} and can be used to control the level of both intracellular and extracellular Ca^{2+} (Table 20.4 and Figure 22.18). The BAPTA buffers are more selective for Ca^{2+} than EDTA and EGTA, and their metal binding is also much less pH sensitive. Furthermore, BAPTA buffers bind and release Ca^{2+} ions about 50–400 times faster than EGTA. Both BAPTA and its membrane-permeant AM ester are extensively used to clamp intracellular Ca^{2+} concentrations, providing insights on the role of free cytosolic Ca^{2+} in a number of important cell systems. Extensive bibliographies are available for each of these products from our Technical Assistance Department or through our Web site (http://www.probes.com).

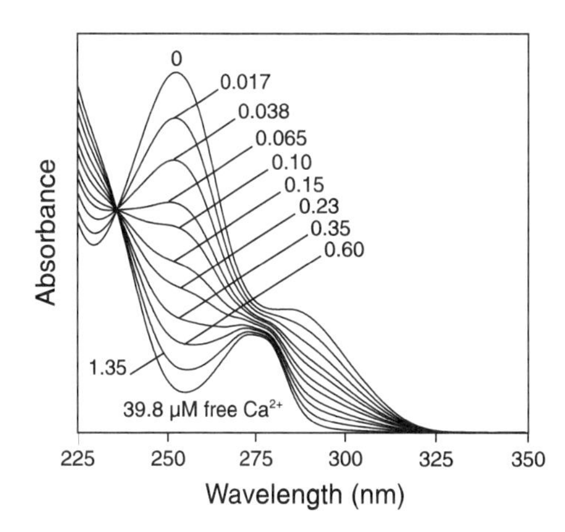

Figure 22.18 *Absorption spectra of BAPTA (B-1204) in solutions containing zero to 39.8 µM free Ca^{2+}.*

BAPTA is available as a cell-impermeant potassium, cesium [30] or sodium salt (B-1204, B-1212, B-1214); the Cs^+ salt of BAPTA has frequently been used for patch-clamp experiments.[30] In addition, we offer the cell-permeant BAPTA AM ester (B-1205). For the convenience of customers, we also make BAPTA AM available in a special packaged set of 20 vials, each containing 1 mg (B-6769). BAPTA and BAPTA AM are included in our new BAPTA and BAPTA Acetoxymethyl Ester Sampler Kits (see below). In addition to these products, we offer 10,000 and 70,000 MW dextran conjugates of BAPTA, which we call Calcium Sponge D10 and D70 (C-3040, C-3041), respectively; see below for a complete description.

Other BAPTA Derivatives

Other BAPTA derivatives available from Molecular Probes are listed in Table 20.4, along with their K_ds for Ca^{2+}. The most powerful Ca^{2+} chelator among these is 5,5'-dimethyl BAPTA (D-1206).[31-34] BAPTA derivatives with intermediate affinity for Ca^{2+} have been extensively used to study Ca^{2+} mobilization, spatial Ca^{2+} buffering and Ca^{2+} shuttling in a variety of cells, including *Xenopus* eggs,[35-38] fucoid eggs,[39] plants [40-42] and hair cells.[43] It was reported that 4,4'-difluoro BAPTA (D-1215) and 5,5'-dibromo BAPTA (D-1211) were effective at blocking fucoid egg development without killing the eggs, whereas both the stronger Ca^{2+} buffers BAPTA such as 5,5'-dimethyl BAPTA and the weaker buffers 5-nitro BAPTA (N-3006) and 5,5'-dinitro BAPTA (D-3001) required higher levels to realize the same inhibition.[39,44,45] 5,5'-Dibromo BAPTA has also been used to block nuclear vesicle fusion in *Xenopus* egg extracts,[46] to determine the affinity of a Ca^{2+}-binding protein [47] and to load Ca^{2+} into stamenal hairs of *Setcreasea purpurea* in a study of cytoplasmic streaming.[48] 5,5'-Dibromo BAPTA and lower-affinity chelators protect neurons against excitotoxic and ischemic injury, without markedly attenuating intracellular Ca^{2+} levels.[49] Loading of neuroblastoma cells with BAPTA derivatives slows Ca^{2+} waves.[50]

Fluorinated BAPTA derivatives (D-1208, D-1209, D-1215, D-1216, T-6816, T-6817) have been employed for optical imaging studies [51,52] but are most widely used for NMR analysis of Ca^{2+} in live cells and tissues,[53] including kidney,[54-56] heart,[57-61] brain,[62] erythrocytes [63] and platelets.[64] The ^{19}F NMR shifts of the fluorinated BAPTA buffers have been reported to correlate with intracellular Ca^{2+} in BALB/c thymocytes,[65] normal [66] and sickle [67] erythrocytes and ferret hearts.[68,69] ^{19}F NMR has also been used to detect Pb^{2+} uptake by platelets.[64]

BAPTA and BAPTA Acetoxymethyl Ester Sampler Kits

Molecular Probes offers eight different BAPTA salts (Table 22.5), as well as seven different BAPTA AM esters, including:

- BAPTA AM (B-1205, B-6769)
- 5,5'-dibromo BAPTA AM (D-1213)
- 5,5'-dimethyl BAPTA AM (D-1207)
- 5,5'-difluoro BAPTA AM (D-1209)
- 5,5'-dinitro BAPTA AM (D-3005)
- 5'-nitro BAPTA AM (N-3007)
- 5,5',6,6'-tetrafluoro BAPTA AM (T-6817).

In addition, our BAPTA Buffer Kit (B-6768) contains 10 mg samples of four BAPTA buffers with a range of affinities for Ca^{2+} (BAPTA, 5,5'-dimethyl BAPTA, 5,5'-difluoro BAPTA and 5,5'-dibromo BAPTA), and our BAPTA Acetoxymethyl Ester Sampler Kit (B-6767) contains two 1 mg samples of each of the corresponding AM esters.

EDTA AM and EGTA AM

Molecular Probes' AM esters of EDTA and EGTA (E-1218, E-1219) can be passively loaded into cells and used to generate intracellular EDTA and EGTA. The slower on-rate of EDTA and EGTA relative to the BAPTA-based buffers reduces their ability to inhibit Ca^{2+} diffusion in cells.[50,70] Because Ca^{2+} binding by intracellular EDTA or EGTA is slow, it is possible to distinguish between buffering of rapid Ca^{2+} transients, which can occur with BAPTA-derived buffers, and the slower effects of general Ca^{2+} buffering.[31,49,71] EDTA is known to chelate many metals; thus, the cell-permeant EDTA AM may be useful for studying metal cytotoxicity, as well as for dissecting the physiological role of metals, for example in zinc-finger proteins and other metalloproteins.[72] EGTA AM has been reported to:

- Block oxidative stress caused by hydroperoxide formation in kidney proximal tubular cells [73]
- Clamp the Ca^{2+} concentration in macrophages [74]
- Confer neuroprotective activity [49]
- Inhibit the substance K–induced Ca^{2+} rise in glioma cells [75]

Calcium Sponge Polymers

Molecular Probes offers biologically compatible conjugates of the BAPTA chelator to selectively remove specific polyvalent ions from solution, as well as from the binding sites of indicators, proteins and polynucleotides. These BAPTA conjugates — Calcium

Table 22.5 Ca^{2+} affinities of BAPTA chelators.

Catalog #	Chelator	R^1	R^2	R^3	R^4	K_d for Ca^{2+}
B-1204 B-1212 B-1214	BAPTA	H	H	H	H	• No Mg^{2+}: 160 nM * • 1 mM Mg^{2+}: 700 nM †
D-1211	5,5′-Dibromo BAPTA	H	Br	H	Br	• No Mg^{2+}: 1.6 µM ‡
D-1215	4,4′-Difluoro BAPTA	F	H	H	H	• No Mg^{2+}: 4.6 µM †
D-1208	5,5′-Difluoro BAPTA	H	F	H	F	• No Mg^{2+}: 635 nM § • 1 mM Mg^{2+}: 705 nM §
D-1206	5,5′-Dimethyl BAPTA	H	CH_3	H	CH_3	• No Mg^{2+}: 40 nM ‡
D-3001	5,5′-Dinitro BAPTA	H	NO_2	H	NO_2	• No Mg^{2+}: 7.4 mM *
N-3006	5-Nitro BAPTA	H	NO_2	H	H	• No Mg^{2+}: 42 µM *
T-6816	5,5′,6,6′-Tetrafluoro BAPTA	H	F	F	F	• No Mg^{2+}: 65 µM ¶

* Measured at Molecular Probes in 10 mM MOPS, 100 mM KCl, pH 7.2 at 22°C.
† Measured in 10 mM HEPES, 300 mM KCl, pH 7.0 at 22°C [Cell Calcium 10, 491 (1989)].
‡ Measured in 10 mM MOPS, 100 mM KCl, pH 7.3 at 22°C [Biochemistry 19, 2396 (1980)].
§ Measured in 10 mM HEPES, 115 mM KCl, 20 mM NaCl, pH 7.05 [Am J Physiol 252, C441 (1987)].
¶ Measured in 10 mM HEPES, 115 mM KCl, 20 mM NaCl, pH 7.2 at 37°C [Am J Physiol 266, C1313 (1994)].

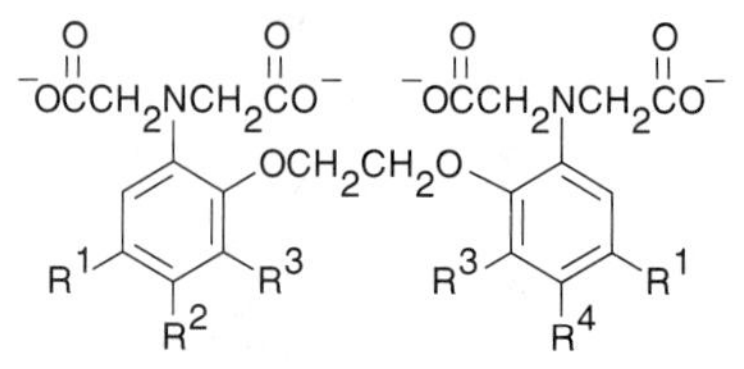

Sponge S (polystyrene) and Calcium Sponge D (dextran) — are selective for Ca^{2+} and certain other ions, including Zn^{2+} and some heavy metals, in the presence of relatively high levels of Mg^{2+}.

Many contaminating polycations can be selectively removed from aqueous solutions simply by stirring a solution with the water insoluble Calcium Sponge S polymer (BAPTA polystyrene, C-3047). For example, free Ca^{2+} can be reduced to less than 40 nM (measured with fura-2) by passing 3 mL of a 100 µM $CaCl_2$ solution through one gram of Calcium Sponge S. The polymer can be regenerated several times by washing it with pH 4 buffer then readjusting to neutral pH with base.

The water-soluble Calcium Sponge D10 and D70 dextran conjugates (BAPTA 10,000 MW dextran, C-3040; BAPTA 70,000 MW dextran, C-3041) may be particularly useful for buffering or chelating polyvalent ions inside or outside cells. Like the fluorescent indicator dextrans, these polymeric conjugates are less likely than the free chelators to become compartmentalized or bind to intracellular proteins and should not pass through gap junctions.

TPEN

Because TPEN (T-1210) binds heavy metals without disturbing Ca^{2+} and Mg^{2+} concentrations,[76] it is ideal for identifying artifacts that may arise from heavy metal binding by BAPTA-derived buffers and indicators. Heavy metal chelation by TPEN has been used to show that quin-2's effects on inositol phosphate formation are due to specific buffering of Ca^{2+} [77] and to prove that BAPTA's effects on mitotic progression in Swiss 3T3 fibroblasts are caused specifically by Ca^{2+} buffering and not by binding of essential heavy metal ions.[78] Likewise, Zn^{2+} chelation by TPEN has been shown to prevent translocation of protein kinase C induced by phorbol esters and antigen activation [79] and to block the phorbol ester–induced reorganization of the actin cytoskeleton.[80,81] The complex of TPEN with Fe^{2+} has been reported to serve as a superoxide dismutase mimic and as a Fenton catalyst.[82]

Calcium Calibration Buffer Kits

Calibration of fluorescent Ca^{2+} indicators is a prerequisite for accurate Ca^{2+} measurements. Molecular Probes offers a number of kits designed to facilitate this calibration using a laboratory fluorometer or quantitative imaging system. All of these kits contain buffers and detailed protocols — including methods for calculation of K_d, a sample response curve and tables to aid in the determination of the exact concentration of free Ca^{2+} under conditions of varying pH, temperature and ionic strength. A discussion of methods to correct the fura-2 K_d for differences in temperature and ionic strength has been published.[83]

Calcium Calibration Buffer Kits for High- to Moderate-Affinity Ca^{2+} Indicators

Because cells contain very low levels of *free* Ca^{2+}, it is essential to use calibrated Ca^{2+} buffers such as EGTA to precisely calibrate the Ca^{2+} indicators in your own equipment under your experimental conditions. When the concentrations of Ca^{2+} and EGTA are very close to each other, the only free Ca^{2+} available is that which is in equilibrium with EGTA. Thus, the free Ca^{2+} concentration is determined by the K_d of CaEGTA at a controlled pH, temperature and ionic strength. In order to attain Ca^{2+} and EGTA concentrations sufficiently close to each other, our Calcium Calibration Buffer Kits (C-3008, C-3009, C-3721, C-3722) contain CaEGTA solutions that have been accurately prepared using Roger Tsien's "pH-metric" method, which makes use of the fact that the ion binding of EGTA causes an acidification of the solution.[84] These kits can be used to obtain calibration curves and K_ds for all of our high- to moderate-affinity Ca^{2+} indicators, including fura-2, indo-1, fluo-3 and the Calcium Green-1, Calcium Green-2, Calcium Orange, Calcium Crimson, Oregon Green 488 BAPTA-1, Oregon Green 488 BAPTA-2 and Fura Red indicators. A recent review by Neher discusses calibration methods for fura-2 in cells.[85]

Calcium Calibration Buffer Kit #1 (C-3008) contains:

- 10 mM K_2EGTA buffered solution (zero free Ca^{2+})
- 10 mM CaEGTA buffered solution (40 µM free Ca^{2+} using the protocol provided)
- Detailed protocol for calibrating Ca^{2+} indicators

When used according to the protocol provided, each kit provides sufficient reagents for five complete calibrations using 2 mL samples and a standard fluorometer cuvette. Many more calibrations can be done by digital imaging microscopy.[86,87] This kit employs a reciprocal dilution method — an equal amount of dye is added to a portion of the zero and 40 µM free Ca^{2+} solutions, and the two are then crossdiluted to give a series of solutions — which minimizes indicator concentration errors but requires more effort in making the dilutions than does Kit #2 (see below). With ratiometric indicators, this method yields a series of curves that exhibit an accurate isosbestic point; it is the method regularly used at Molecular Probes to determine Ca^{2+} affinities. Our Calcium Calibration Buffer Kit #1 was used as a source of known free Ca^{2+} concentrations in an experiment that examined the effect of Ca^{2+} binding on the fluorescence of calretinin.[88]

A concentrated form of Calcium Calibration Buffer Kit #1 (Calcium Calibration Buffer Concentrate Kit, C-3723) is also offered. This kit allows the researcher to dilute the Ca^{2+} concentrates into buffered solutions of any composition in order to more closely mimic the pH, ionic strength or viscosity of the cytoplasmic environment or of solutions with unusual ion levels such as sea water.

In Calcium Calibration Buffer Kit #2 (C-3009), 11 prediluted K_2EGTA/CaEGTA solutions containing from 0 to 10 mM CaEGTA (0 to 40 µM free Ca^{2+}) are bottled separately, making the calibrations faster, especially for cases in which all the dilutions are not required. This kit is intrinsically less accurate than Kit #1 because it requires precise addition of the same amount of fluorescent indicator to each solution. Calcium Calibration Buffer Kit #2 is preferred for calibrating the response of Ca^{2+} imaging systems, in which single drops of the fura-2–containing buffer are placed on a microscope slide.[86,87] Calcium Calibration Buffer Kit #2 was used to calibrate a new fiber-optic Ca^{2+} sensor constructed by covalent immobilization of a Calcium Green derivative.[89]

Calcium Calibration Buffer Kits with Magnesium

We also offer two Calcium Calibration Buffer Kits with Magnesium (C-3721, C-3722), which are analogous to the Calcium Calibration Buffer Kits #1 and #2 except that the buffers contain 1 mM Mg^{2+}. This approximately physiological Mg^{2+} level may strongly affect the K_d for Ca^{2+} of the BAPTA-derived indicators. Thus, these kits are suitable for simulating the intracellular Mg^{2+} environment during the calibration procedure and for determining the effect of this ion on the K_d of the Ca^{2+} indicator. Such effects can be significant; for example, Tsien reports that the K_d of fura-2 increases from 135 nM in the absence of Mg^{2+} to 224 nM in the presence of 1 mM Mg^{2+}.[90]

New Calcium Calibration Buffer Kit for Low-Affinity Ca^{2+} Indicators

The Calcium Calibration Buffer Kit #3 (C-6775) provides 11 ready-to-use Ca^{2+} solutions containing from 1 µM to 1 mM free Ca^{2+}. The K_2EGTA/CaEGTA buffering system is employed to generate the two lowest Ca^{2+} concentrations; solutions with higher Ca^{2+} concentrations do not have a buffered Ca^{2+} level and are verified using an ion-selective electrode. This kit is ideal for calibrating our low-affinity Ca^{2+} indicators, including Calcium Green-5N, Calcium Orange-5N, Oregon Green BAPTA-5N, mag-fura-2, mag-fura-5 and our ratioable Ca^{2+} indicator BTC.

Fura-2 Calcium Imaging Calibration Kit

The new Fura-2 Calcium Imaging Calibration Kit (F-6774), which is designed to facilitate rapid calibration and standardization of digital imaging microscopes,[86,87] contains the same 11 prediluted buffers as in our Calcium Calibration Buffer Kit #2. However, in this kit the buffers also include 50 µM fura-2, as well as 15 µm unstained polystyrene microspheres to act both as spacers that ensure uniform separation between the slide and the coverslip and as focusing aids. We also provide a twelfth buffer, identical to the 10 mM CaEGTA standard but lacking fura-2, that serves as a control for background fluorescence.

Magnesium Calibration Standard Kit

Molecular Probes' Magnesium Calibration Standard Kit (M-3120) provides a simple and convenient approach to determination of the K_d of fluorescent Mg^{2+} indicators in solution. Using the Magnesium Calibration Standard Kit and a fluorometer or quantitative imaging system, a researcher can calculate the K_d of a Mg^{2+} indicator from a plot of the indicator's fluorescence in the presence of several different known Mg^{2+} concentrations, generated using a reciprocal dilution method. Each kit contains:

- Zero Mg^{2+} buffered solution
- 35 mM Mg^{2+} buffered solution
- Detailed protocol for calibrating Mg^{2+} indicators

Sufficient reagents are provided for at least five dilution series using 2 mL samples in a standard fluorometer cuvette; many more calibrations can be done by digital imaging microscopy.

Ionophores

A-23187 and 4-Bromo A-23187

The Ca^{2+} ionophore A-23187 (A-1493) is commonly used for *in situ* calibrations of fluorescent Ca^{2+} indicators, to equilibrate intracellular and extracellular Ca^{2+} concentrations and to permit Mn^{2+} to enter the cell to quench intracellular dye fluorescence. Although the intrinsic fluorescence of A-23187 is too high for use with fura-2, indo-1 and quin-2, it is suitable for use with the visible light–excitable indicators, including Calcium Green, Calcium Orange, Calcium Crimson, Oregon Green 488 BAPTA, fluo-3, rhod-2 and Fura Red. Brominated A-23187 (4-bromo A-23187, B-1494), which is essentially nonfluorescent, is the best ionophore for use with fura-2, indo-1 and quin-2. Like A-23187, it rapidly transports both Ca^{2+} and Mn^{2+} into cells.[91] Both A-23187 and 4-bromo A-23187 can also be used to equilibrate intracellular and extracellular Mg^{2+} concentrations, making them useful for calibrating Mg^{2+} indicators.

Caged A-23187 and Caged 4-Bromo A-23187

We are introducing some unique new tools for the study of ion transport and Ca^{2+} regulation. Caged ionophores derived from A-23187 (A-7109) or 4-bromo A-23187 (B-7108) enter cells passively, where they subsequently may be photoactivated with a UV flashlamp or laser pulse to stimulate ion movement across the plasma membrane or to liberate Ca^{2+} from intracellular stores.[92] Both products are inactive until uncaged with UV light. See Chapter 19 for further discussion of caged probes.

Pluronic F-127: A Useful Dispersing Reagent

Because AM esters have low aqueous solubility, dispersing agents, such as fetal calf serum, bovine serum albumin or Pluronic F-127, are often used to facilitate cell loading.[93-95] Molecular Probes provides Pluronic F-127 in three forms, all of which have low UV absorbance ($OD_{280\ nm}$ <0.02 at 10 mg/mL):

- Powder (P-6867)
- 10% Sterile solution of Pluronic F-127 in water (P-6866), for use in tissue culture and other applications requiring sterile reagents
- 20% Solution in DMSO (P-3000)

Digitonin for Cell Lysis

It is frequently necessary to lyse cells in order to calibrate the ion indicators, especially when the cells are not easily permeabilized by ionophores.[96] Probably the most common lysing agent is digitonin, which we have available with low UV contamination ($OD_{280\ nm}$ <0.02 at 1 mg/mL) as either USP grade (D-8449) or in a more water-soluble preparation (D-8450). Lysis into buffers containing Mn^{2+} quenches almost all of the fluorescence of the indicators and is a good test for the completeness of AM ester hydrolysis.[97,98] Triton® X-100 is also used to lyse cells for calibration of Ca^{2+} and Mg^{2+} indicators.[99-103]

Reagents for Investigating Calcium Modulation and Second Messenger Activity

Molecular Probes prepares several important reagents to investigate signal transduction, second messenger activity, ion channels and endogenous modulators of intracellular Ca^{2+} concentrations.

Probes for Ca^{2+} Regulation (see Section 20.2)

- cADP ribose, caged cADP ribose, cGDP ribose and NAADP
- *myo*-inositol triphosphate (IP_3)
- Ryanodine and BODIPY FL-X ryanodine
- Thapsigargin and BODIPY FL thapsigargin
- TMB-8, a calcium antagonist

Probes for Protein Kinases, Protein Phosphatases and Adenylate Cyclase (see Section 20.3)

- Bisindolylmaleimides, including GF 109203X, BODIPY FL bisindolylmaleimide, fim-1 and rim-1
- BODIPY FL forskolin
- Ca^{2+}/calmodulin-dependent kinase II (CaM kinase II)
- Cyclic AMP Fluorosensor® (FlCRhR) — an indicator of intracellular cAMP levels
- 1,2-Dioctanoyl-*sn*-glycerol and 4β-phorbol 12β-myristate 13α-acetate (PMA)
- Okadaic acid
- Protein kinase A catalytic subunit
- Protein kinase C

Probes for Ca^{2+} Channels and Carriers (see Section 18.4)

- ω-Conotoxin GVIA, ω-conotoxin MVIIA and fluorescent conotoxins for N-type Ca^{2+} channels
- Dihydropyridines, phenylalkylamines and calciseptine for L-type Ca^{2+} channels
- $Argiotoxin_{636}$ for P-type Ca^{2+} channels
- Ryanodine probes for intracellular Ca^{2+} channels
- Probes for Ca^{2+}-ATPases and the Ca^{2+}/H^+ exchange

See also our caged nucleotides (Section 20.4) and our caged receptor probes (Section 18.3), some of which modulate the Ca^{2+} flux through ion channels.

1. NP-EGTA is licensed to Molecular Probes under U.S. Patent No. 5,446,186; **2.** J Biol Chem 270, 23966 (1995); **3.** Science 267, 1997 (1995); **4.** Proc Natl Acad Sci USA 91, 187 (1994); **5.** Biophys J 70, 1006 (1996); **6.** Cell Calcium 19, 185 (1996); **7.** Cell Calcium 19, 3 (1996); **8.** Neuron 15, 755 (1995); **9.** Proc Natl Acad Sci USA 85, 6571 (1988); **10.** Science 241, 842 (1988); **11.** DMNP-EDTA is licensed to Molecular Probes under U.S. Patent No. 4,981,985; **12.** Meth Cell Biol 40, 31 (1994); **13.** Neuron 10, 21 (1993); **14.** Biochemistry 31, 8856 (1992); **15.** J Physiol 426, 473 (1990); **16.** Biophys J 69, 30 (1995); **17.** J Physiol 462, 243 (1993); **18.** Biophys J 66, 1879 (1994); **19.** Biochim Biophys Acta 1035, 378 (1990); **20.** J Am Chem Soc 111, 7957 (1989); **21.** Diazo-2 and diazo-3 are licensed to Molecular Probes under U.S. Patent No. 5,141,627; **22.** FEBS Lett 255, 196 (1989); **23.** Biophys J 69, 2611 (1995); **24.** Pflügers Arch 425, 175 (1993); **25.** Cell Calcium 19, 255 (1996); **26.** Nature 371, 603 (1994); **27.** Biophys J 65, 2537 (1993); **28.** Proc Natl Acad Sci USA 89, 11804 (1992); **29.** Biochemistry 19, 2396 (1980); **30.** J Neurosci 15, 903 (1995); **31.** J Neurosci 15, 2867 (1995); **32.** Exp Physiol 79, 269 (1994); **33.** J Biol Chem 268, 6511 (1994); **34.** Mol Biol of the Cell 4, 293 (1993); **35.** Biol Bull 183, 368 (1993); **36.** Devel Biol 158, 200 (1993); **37.** J Cell Biol 122, 387 (1993); **38.** J Cell Sci 106, 523 (1993); **39.** Proc Natl Acad Sci USA 86, 6607 (1989); **40.** Cell Calcium 16, 322 (1994); **41.** Plant Cell 6, 1815 (1994); **42.** J Cell Sci 101, 7 (1992); **43.** Nature 363, 74 (1993); **44.** Development 120, 155 (1994); **45.** Cell Calcium 10, 491 (1989); **46.** Cell 73, 1411 (1993); **47.** J Biol Chem 265, 9838 (1990); **48.** Planta 182, 34 (1990); **49.** J Neurophysiol 72, 1973 (1994); **50.** Biophys J 69, 1683 (1995); **51.** Neuron 11, 221 (1993); **52.** Proc Natl Acad Sci USA 86, 6179 (1986); **53.** Proc Natl Acad Sci USA 80, 7178 (1983); **54.** Biochim Biophys Acta 1226, 83 (1994); **55.** J Biol Chem 268, 991 (1993); **56.** J Biol Chem 267, 3637 (1992); **57.** J Biol Chem 271, 7398 (1996); **58.** Am J Physiol 266, C1323 (1994); **59.** Biophys J 65, 2547 (1993); **60.** J Biol Chem 265, 1394 (1990); **61.** Proc Natl Acad Sci USA 84, 6005 (1987); **62.** J Neurochem 55, 878 (1990); **63.** Biochim Biophys Acta 1136, 155 (1992); **64.** Biochim Biophys Acta 1092, 341 (1991); **65.** Proc Natl Acad Sci USA 80, 7178 (1983); **66.** Am J Physiol 252, C441 (1987); **67.** Blood 69, 1469 (1987); **68.** Am J Physiol 258, H9 (1990); **69.** Proc Natl Acad Sci USA 84, 6005 (1987); **70.** J Physiol 86, 129 (1992); **71.** J Neurosci 14, 523 (1994); **72.** J Biol Chem 265, 6513 (1990); **73.** J Biol Chem 269, 14546 (1994); **74.** J Cell Biol 113, 757 (1991); **75.** Proc Natl Acad Sci USA 88, 6878 (1991); **76.** J Biol Chem 260, 2719 (1985); **77.** J Biol Chem 263, 7581 (1988); **78.** J Cell Biol 111, 183 (1990); **79.** J Biol Chem 263, 6487 (1988); **80.** J Cell Physiol 158, 337 (1994); **81.** Cell Regulation 2, 1067 (1991); **82.** Arch Biochem Biophys 293, 153 (1992); **83.** Cell Calcium 12, 279 (1991); **84.** Meth Enzymol 172, 230 (1989); **85.** Neuropharmacology 34, 1423 (1995); **86.** J Neurochem 62, 890 (1994); **87.** Cell Calcium 11, 75 (1990); **88.** Biochemistry 34, 15389 (1995); **89.** Anal Chem 68, 1414 (1996); **90.** J Biol Chem 260, 3440 (1985); **91.** Anal Biochem 146, 349 (1985); **92.** J Biol Chem 262, 12801 (1987); **93.** Ishihara, A. *et al.*, manuscript submitted; **94.** Proc Natl Acad Sci USA 84, 7793 (1987); **95.** Science 233, 886 (1986); **96.** J Biol Chem 266, 13646 (1991); **97.** Cell Calcium 10, 171 (1989); **98.** J Biol Chem 258, 4876 (1983); **99.** Am J Physiol 256, C300 (1992); **100.** Biochem J 273, 541 (1991); **101.** *Cellular Calcium: A Practical Approach*, J.G. McCormack and P.H. Cobbold, Eds., IRL Press (1991); **102.** Biochem Biophys Res Comm 173, 396 (1990); **103.** Am J Physiol 256, C540 (1989).

22.8 Data Table *Chelators, Calibration Buffers and Ionophores*

Cat #	Structure	MW	Storage	Soluble	Abs	{ε x 10^{-3}}	Em	Solvent	Cleaved*	K_d†	Notes
A-1493	✓	524	F,L	DMSO, EtOH	378	{8.9}	438	MeOH			
A-7109	✓	733	FF,D,LL	DMSO, EtOH	370	{10}	none	MeOH			<1>
B-1204	✓	629	D	pH>6	284	{5.1}	<2>	<3>		160 nM	<4, 5>
B-1205	✓	765	F,D	DMSO	287	{5.9}	ND	$CHCl_3$	B-1204		
B-1212	✓	1004	D	pH>6	285	{5.2}	<2>	<3>		160 nM	<4, 5>
B-1214	✓	564	D	pH>6	285	{5.1}	<2>	<3>		160 nM	<4, 5>
B-1494	✓	603	F,D	DMSO, EtOH	289	{20}	none	MeOH			<6>
B-7108	✓	812	FF,D,LL	DMSO, EtOH	337	{11}	none	MeOH			<1>
C-3040	C	<7>	D	H_2O	252	ND	<2>	<3>		175 nM	<4, 5, 8>
C-3041	C	<7>	D	H_2O	245	ND	<2>	<3>		115 nM	<4, 5, 8>
D-1206	✓	657	D	pH>6	290	{5.1}	ND	<3>		40 nM	<5, 9>
D-1207	✓	793	F,D	DMSO	291	{5.9}	ND	$CHCl_3$	D-1206		
D-1208	✓	665	D	pH>6	289	{5.1}	ND	<3>		635 nM	<5, 10>
D-1209	✓	801	F,D	DMSO	290	{5.7}	ND	EtOAc	D-1208		
D-1211	✓	787	D	pH>6	263	{18}	ND	<3>		1.6 µM	<5, 9>
D-1213	✓	922	F,D	DMSO	296	{6.9}	ND	EtOAc	D-1211		

Cat #	Structure	MW	Storage	Soluble	Abs	{ε x 10^{-3}}	Em	Solvent	Cleaved*	Kd†	Notes
D-1215	✓	665	D	pH>6	292	{7.6}	ND	<3>		4.6 µM	<5, 11>
D-1216	✓	801	F,D	DMSO	291	{8.7}	ND	MeOH	D-1215		
D-3001	✓	566	D,L	pH>6	431	{30}	ND	<3>		7.4 mM	<4, 5>
D-3005	✓	855	F,D,L	DMSO	373	{26}	ND	EtOAc	D-3001		
D-3034	✓	711	F,D,LL	pH>6	369	{18}	none	<3>		2.2 µM	<1, 12>
D-3036	✓	847	F,D,LL	DMSO	342	{22}	none	MeOH	D-3034		
D-3700	✓	383	F,D,LL	pH>6	374	{23}	none	MeOH		<13>	<1>
D-3701	✓	451	F,D,LL	DMSO	344	{20}	none	MeOH	D-3700		
D-6814	✓	626	D,LL	pH>6	348	{4.2}	none	<3>		5 nM	<1, 14>
D-6815	✓	762	F,D,LL	DMSO	333	{4.6}	none	MeOH	D-6814		<15>
D-8449	S16.3	1229	D	EtOH	<300		none				
D-8450	S16.3	1229	D	H_2O	<300		none				<16>
E-1218	✓	580	F,D	DMSO	<300		none				
E-1219	✓	669	F,D	DMSO	<300		none				<15>
N-3006	✓	521	D,L	pH>6	430	{12}	ND	<3>		42 µM	<4, 5>
N-3007	✓	810	F,D,L	DMSO	373	{14	ND	EtOAc	N-3006		
N-6802	✓	654	FF,D,LL	pH>6	260	{3.5}	none	<3>		80 nM	<1, 6, 17>
N-6803	✓	790	FF,D,LL	DMSO	250	{4.1}	none	MeCN	N-6802		<15>
T-1210	✓	425	D	EtOH	261	{14}	ND	MeOH			
T-6816	✓	701	D	pH>6	294	{3.7}	ND	<3>		65 µM	<5, 10>
T-6817	✓	837	F,D	DMSO	284	{2.7}	ND	EtOAc	T-6816		<15>

B-6769 see B-1205

For definitions of the contents of this data table, see "How to Use the Handbook" on page vi.
Structure: Chemical structure drawing: ✓ = shown in this section; Sn.n = shown in Section number n.n.
ND = not determined.

* Catalog numbers listed in this column indicate the cleavage product generated by enzymatic or chemical hydrolysis of the acetoxymethyl ester groups of the parent compound.

† Data on Ca^{2+} dissociation constants come from several different sources. These values are known to vary considerably depending on the temperature, pH, ionic strength, viscosity, protein binding, presence of other ions (especially polyvalent ions), instrument setup and other factors. It is strongly recommended that these values be specifically determined under your experimental conditions by using our Calcium Calibration Buffer Kits.

<1> All photoactivatable probes are sensitive to light. They should be protected from illumination, except when photolysis is intended.

<2> BAPTA is weakly fluorescent in aqueous solutions (Em = 363 nm, QY = 0.03) [Biochemistry 19, 2396 (1980)].

<3> Abs and ε values determined in Ca^{2+}-free solution (100 mM KCl, 10 mM EGTA, 10mM MOPS, pH 7.2).

<4> Dissociation constant determined at Molecular Probes from absorption measurements in 100 mM KCl, 10 mM MOPS, pH 7.2 at 22°C. Ca^{2+} concentrations below 5 µM were controlled using CaEGTA buffering (Calcium Calibration Buffer Kit #1 (C-3008)).

<5> Absorption spectra of BAPTA and its derivatives are dependent on Ca^{2+} concentration (see Figure 22.18).

<6> This compound has weaker visible absorption at >300 nm but no discernible absorption peaks in this region.

<7> Molecular weight is nominally as specified in the product name but may have a broad distribution.

<8> Because BAPTA dextran conjugates are polydisperse both in molecular weight and degree of substitution, dissociation constants and spectra may vary between batches.

<9> Dissociation constant determined in CaEGTA buffers in 100 mM KCl, 10 mM MOPS, pH 7.3 [Biochemistry 19, 2396 (1980)].

<10> Dissociation constant determined in 115 mM KCl, 20 mM NaCl, 10 mM HEPES, pH 7.05 (D-1208) or 7.2 (T-6816) [Am J Physiol 252, C441 (1987); Am J Physiol 266, C1313 (1994)].

<11> Dissociation constant determined in 300 mM KCl, 10 mM HEPES pH 7.0 [Cell Calcium 10, 491 (1989)].

<12> K_d (Ca^{2+}) decreases to 73 nM after ultraviolet photolysis. The absorption spectrum of the photolysis product is similar to that of B-1204 [J Am Chem Soc 111, 7957 (1989)].

<13> Diazo-3 has negligible Ca^{2+} binding affinity before or after photolysis. It is used as a control for diazo-2 photolysis products [J Am Chem Soc, 111, 7957 (1989)].

<14> K_d (Ca^{2+}) increases to 3 mM after ultraviolet photolysis. K_d values determined in 130 mM KCl, 10 mM HEPES, pH 7.1 [Proc Natl Acad Sci USA 85, 6571 (1988)].

<15> This product is intrinsically an oil at room temperature.

<16> D-8450 is soluble in water at 10 mg/mL upon heating to ~95°C. Absorbance of a 1 mg/mL aqueous solution at 280 nm is ~0.02.

<17> K_d (Ca^{2+}) increases to 1 mM after ultraviolet photolysis. K_d values determined in 100 mM KCl, 40 mM HEPES, pH 7.2 [Proc Natl Acad Sci USA 91, 187 (1994)].

22.8 Structures *Chelators, Calibration Buffers and Ionophores*

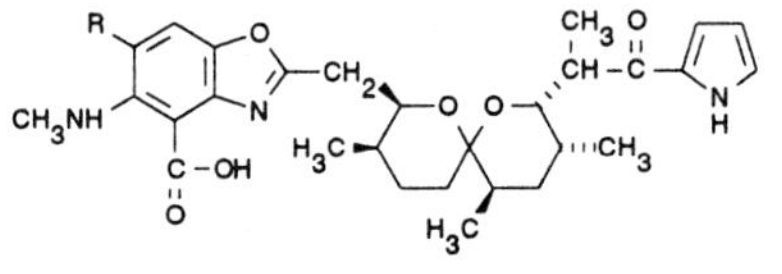

A-1493 R = H
B-1494 R = Br

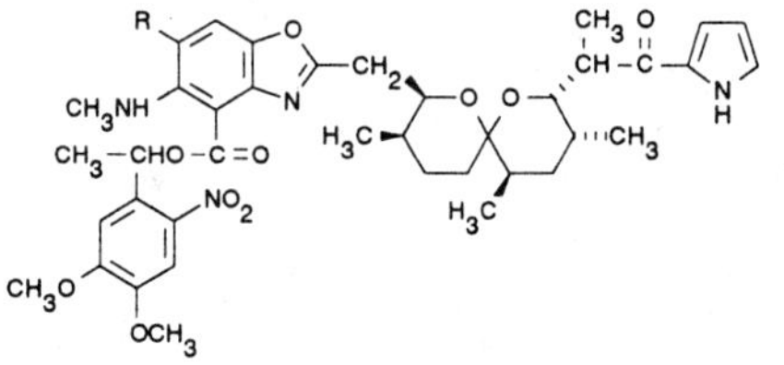

A-7109 R = H
B-7108 R = Br

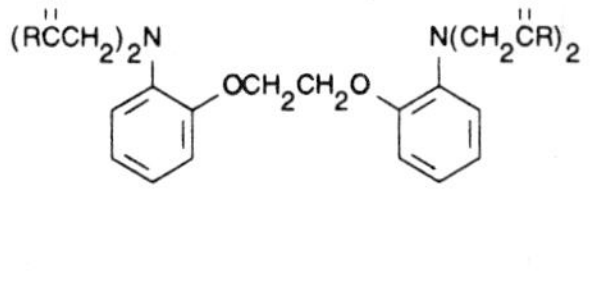

B-1204 $R = O^- K^+$
B-1214 $R = O^- Na^+$
B-1212 $R = O^- Cs^+$
B-1205 / B-6769 $R = -OCH_2OCCH_3$ (C=O)

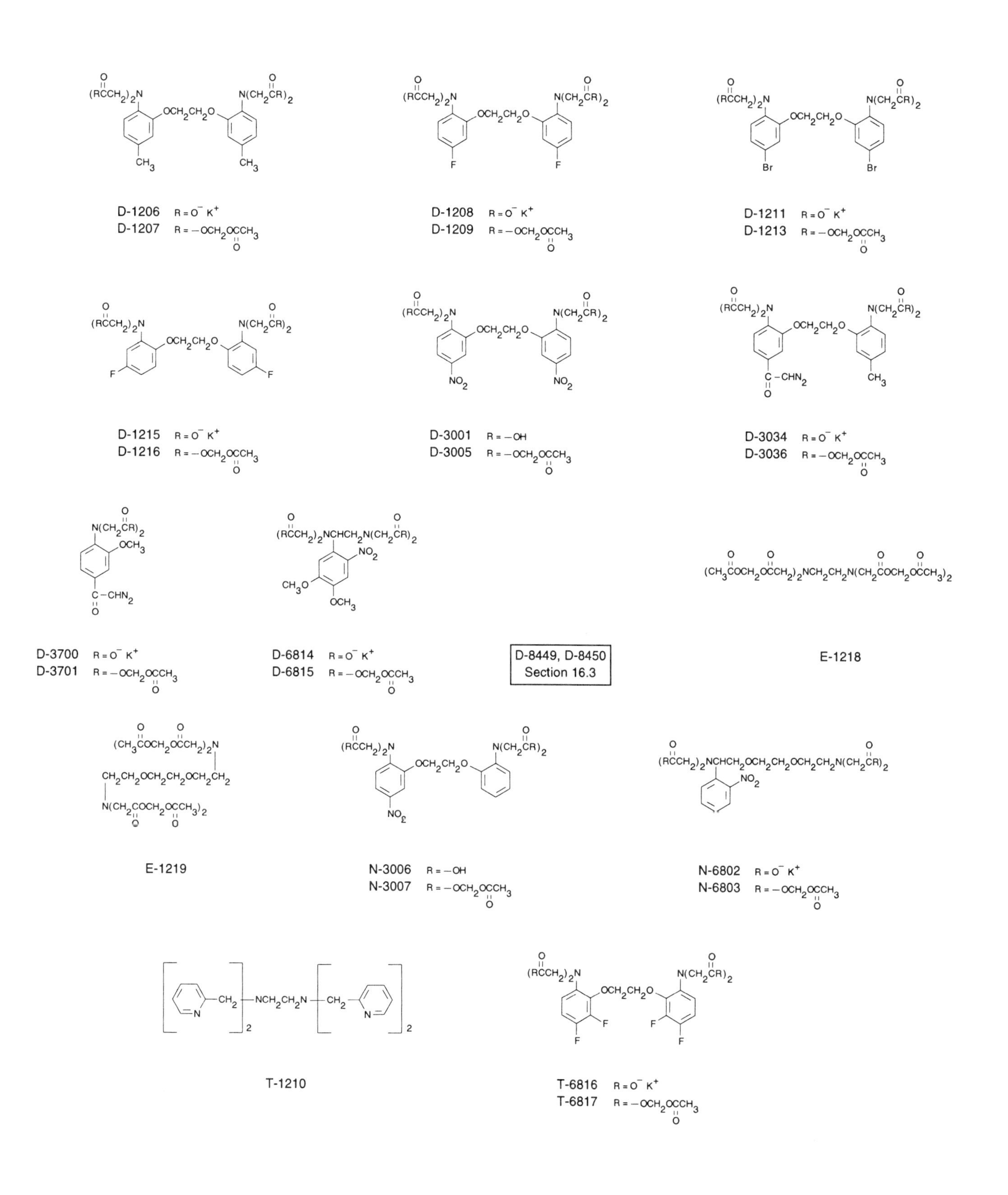
D-1206 R = O⁻ K⁺
D-1207 R = −OCH2OCCH3
D-1208 R = O⁻ K⁺
D-1209 R = −OCH2OCCH3
D-1211 R = O⁻ K⁺
D-1213 R = −OCH2OCCH3
D-1215 R = O⁻ K⁺
D-1216 R = −OCH2OCCH3
D-3001 R = −OH
D-3005 R = −OCH2OCCH3
D-3034 R = O⁻ K⁺
D-3036 R = −OCH2OCCH3
D-3700 R = O⁻ K⁺
D-3701 R = −OCH2OCCH3
D-6814 R = O⁻ K⁺
D-6815 R = −OCH2OCCH3
D-8449, D-8450
Section 16.3
E-1218
E-1219
N-3006 R = −OH
N-3007 R = −OCH2OCCH3
N-6802 R = O⁻ K⁺
N-6803 R = −OCH2OCCH3
T-1210
T-6816 R = O⁻ K⁺
T-6817 R = −OCH2OCCH3

Cat #		Product Name	Unit Size	Price Per Unit ($) 1–4 Units	5–24 Units
A-7109	**New**	A-23187, 1-(4,5-dimethoxy-2-nitrophenyl)ethyl ester (DMNPE-caged A-23187)	1 mg	98.00	78.40
A-1493		A-23187 free acid (calcimycin)	10 mg	68.00	54.40
B-6767	**New**	BAPTA Acetoxymethyl Ester Sampler Kit *2x1 mg each of B-1205, D-1207, D-1209, D-1213*	1 kit	75.00	60.00
B-1205		BAPTA, AM *cell permeant* *≥95% by HPLC*	25 mg	85.00	68.00
B-6769	**New**	BAPTA, AM *cell permeant* *special packaging*	20x1 mg	95.00	76.00
B-6768	**New**	BAPTA Buffer Kit *10 mg each of B-1204, D-1206, D-1208, D-1211*	1 kit	75.00	60.00
B-1212		BAPTA, tetracesium salt *cell impermeant* *≥95% by HPLC*	1 g	128.00	102.40
B-1204		BAPTA, tetrapotassium salt *cell impermeant* *≥95% by HPLC*	1 g	73.00	58.40
B-1214		BAPTA, tetrasodium salt *cell impermeant*	1 g	73.00	58.40
B-7108	**New**	4-bromo A-23187, 1-(4,5-dimethoxy-2-nitrophenyl)ethyl ester (DMNPE-caged Br A-23187)	1 mg	98.00	78.40
B-1494		4-bromo A-23187, free acid	1 mg	78.00	62.40
C-3723		Calcium Calibration Buffer Concentrate Kit *zero and 100 mM CaEGTA (2 x 5 mL)*	1 kit	95.00	76.00
C-3008		Calcium Calibration Buffer Kit #1 *zero and 10 mM CaEGTA (2 x 50 mL)*	1 kit	95.00	76.00
C-3009		Calcium Calibration Buffer Kit #2 *zero to 10 mM CaEGTA (11 x 10 mL)*	1 kit	95.00	76.00
C-6775	**New**	Calcium Calibration Buffer Kit #3 *1 µM to 1 mM range (11 x 10 mL)*	1 kit	95.00	76.00
C-3721		Calcium Calibration Buffer Kit with Magnesium #1 *zero and 10 mM CaEGTA with 1 mM Mg^{2+} (2 x 50 mL)*	1 kit	95.00	76.00
C-3722		Calcium Calibration Buffer Kit with Magnesium #2 *zero to 10 mM CaEGTA with 1 mM Mg^{2+} (11 x 10 mL)*	1 kit	95.00	76.00
C-3040		Calcium Sponge™ D10 (BAPTA dextran, 10,000 MW, anionic)	10 mg	48.00	38.40
C-3041		Calcium Sponge™ D70 (BAPTA dextran, 70,000 MW, anionic)	10 mg	48.00	38.40
C-3047		Calcium Sponge™ S (BAPTA polystyrene)	1 g	98.00	78.40
D-3036		diazo-2, AM *cell permeant* *special packaging*	20x50 µg	195.00	156.00
D-3034		diazo-2, tetrapotassium salt *cell impermeant*	1 mg	195.00	156.00
D-3701		diazo-3, AM *cell permeant*	1 mg	75.00	60.00
D-3700		diazo-3, dipotassium salt *cell impermeant*	1 mg	75.00	60.00
D-1213		5,5′-dibromo BAPTA, AM *cell permeant* *≥95% by HPLC*	25 mg	48.00	38.40
D-1211		5,5′-dibromo BAPTA, tetrapotassium salt *cell impermeant*	100 mg	73.00	58.40
D-1216		4,4′-difluoro BAPTA, AM *cell permeant*	25 mg	48.00	38.40
D-1215		4,4′-difluoro BAPTA, tetrapotassium salt *cell impermeant*	100 mg	48.00	38.40
D-1209		5,5′-difluoro BAPTA, AM *cell permeant* *≥95% by HPLC*	25 mg	125.00	100.00
D-1208		5,5′-difluoro BAPTA, tetrapotassium salt *cell impermeant*	100 mg	98.00	78.40
D-8449	**New**	digitonin *USP grade*	1 g	40.00	32.00
D-8450	**New**	digitonin *water soluble*	1 g	48.00	38.40
D-6815	**New**	1-(4,5-dimethoxy-2-nitrophenyl)-1,2-diaminoethane-N,N,N′,N′-tetraacetic acid, tetra(acetoxymethyl ester) (DMNP-EDTA, AM) *cell permeant* *≥95% by HPLC* *special packaging*	20x50 µg	195.00	156.00
D-6814	**New**	1-(4,5-dimethoxy-2-nitrophenyl)-1,2-diaminoethane-N,N,N′,N′-tetraacetic acid, tetrapotassium salt (DMNP-EDTA) *cell impermeant*	5 mg	100.00	80.00
D-1207		5,5′-dimethyl BAPTA, AM *cell permeant* *≥95% by HPLC*	25 mg	95.00	76.00
D-1206		5,5′-dimethyl BAPTA, tetrapotassium salt *cell impermeant*	100 mg	48.00	38.40
D-3005		5,5′-dinitro BAPTA, AM *cell permeant* *≥95% by HPLC*	10 mg	48.00	38.40
D-3001		5,5′-dinitro BAPTA, free acid *cell impermeant*	25 mg	73.00	58.40
E-1218		EDTA, tetra(acetoxymethyl ester) (EDTA, AM)	25 mg	85.00	68.00
E-1219		EGTA, tetra(acetoxymethyl ester) (EGTA, AM)	10 mg	125.00	100.00
F-6774	**New**	Fura-2 Calcium Imaging Calibration Kit *zero to 10 mM CaEGTA, 50 µM fura-2 (11 x 1 mL)*	1 kit	95.00	76.00
M-3120		Magnesium Calibration Standard Kit *zero and 35 mM Mg^{2+} (2 x 50 mL)*	1 kit	48.00	38.40
N-3007		5-nitro BAPTA, AM *cell permeant* *≥95% by HPLC*	5 mg	85.00	68.00
N-3006		5-nitro BAPTA, free acid *cell impermeant*	10 mg	73.00	58.40
N-6803	**New**	o-nitrophenyl EGTA, AM (NP-EGTA, AM) *cell permeant* *special packaging*	20x50 µg	195.00	156.00
N-6802	**New**	o-nitrophenyl EGTA, tetrapotassium salt (NP-EGTA) *cell impermeant* *≥95% by HPLC*	1 mg	165.00	132.00
P-6867	**New**	Pluronic® F-127 *low UV absorbance*	2 g	8.00	6.40
P-3000		Pluronic® F-127 *20% solution in DMSO*	1 mL	15.00	12.00
P-6866	**New**	Pluronic® F-127 *sterile 10% solution in water*	30 mL	35.00	28.00
T-6817	**New**	5,5′,6,6′-tetrafluoro BAPTA, AM *cell permeant* *≥95% by HPLC*	10 mg	65.00	52.00
T-6816	**New**	5,5′,6,6′-tetrafluoro BAPTA, tetrapotassium salt *cell impermeant*	25 mg	65.00	52.00
T-1210		tetrakis-(2-pyridylmethyl)ethylenediamine (TPEN)	100 mg	50.00	40.00

Loading and Calibration of Intracellular Ion Indicators

There are two major prerequisites for measuring intracellular ion concentrations using fluorescent indicators:

- **Loading**: The indicator must be localized in the region (most commonly the cytosol) where the ion concentration is to be measured.
- **Calibration**: The fluorescence of the indicator must be quantitatively related to the concentration of the free ion.

This technical note focuses on these prerequisites. Further information on the practical aspects of ion measurements using fluorescent indicators can be found in several reviews.[1-5]

Loading

Cell loading methods can be divided into two groups. Bulk loading procedures are applicable to large populations of cells and include:

- Acetoxymethyl (AM) ester loading [6]
- Acid loading [7,8] (particularly applicable to plant cells)
- ATP-induced permeabilization [9]
- Cationic liposome delivery [10]
- Electroporation [11]
- Hypoosmotic shock
- Scrape loading

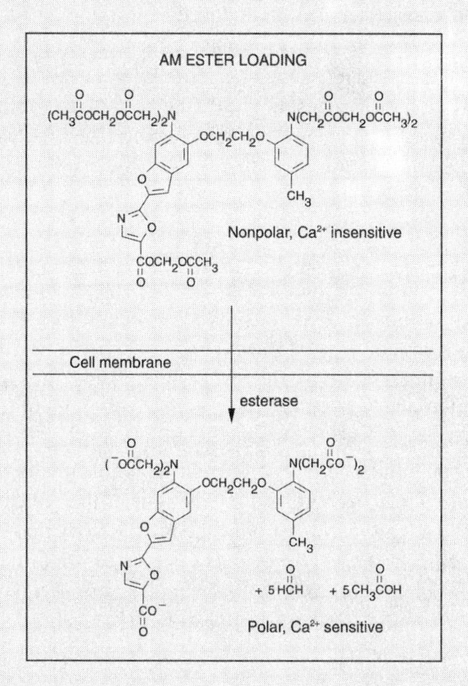

Figure 1 *Schematic diagram of the processes involved in loading cells using membrane-permeant acetoxymethyl (AM) ester derivatives of fluorescent indicators, in this case fura-2. Note the generation of potentially toxic by-products (formaldehyde and acetic acid).*

Single cell procedures such as microinjection and patch pipette perfusion must be carried out one cell at a time. Reviews of some of these techniques have been published;[12,13] see also Table 15.1 in Chapter 15.

The AM Ester Loading Technique

The noninvasive and technically straightforward AM ester technique is by far the most popular method for loading fluorescent ion indicators (Figure 1). The carboxylate groups of indicators for Ca^{2+} and other cations and the phenolic hydroxyl groups of pH indicators are derivatized as acetoxymethyl or acetate esters, respectively, rendering the indicator permeant to membranes and insensitive to ions. Once inside the cell, these derivatized indicators are hydrolyzed by intracellular esterases, releasing the ion-sensitive, polyanionic indicator.

In practice, a 1–10 mM stock solution of the ester probe in *anhydrous* dimethylsulfoxide (DMSO) is prepared and divided into appropriately sized aliquots that can be stored desiccated at -20°C. This procedure will curtail the spontaneous ester hydrolysis that can occur in moist environments. Before loading, the DMSO stock solution should be diluted at least 1:200 in serum-free culture medium to a final concentration of about 1–10 µM. The nonionic and nondenaturing detergent Pluronic® F-127 (P-3000, P-6866, P-6867; see Section 22.8) is frequently added to help disperse the indicator in the loading medium.[14] After incubation at 20–37°C for 15–60 minutes, the cells should be washed 2–3 times with fresh serum-free culture medium (serum may contain esterase activity). The loading medium should also be free of amino acids or buffers containing primary or secondary amines because aliphatic amines may cleave the AM esters and prevent loading.

Problems with AM Ester Loading

Compartmentalization. For calibration purposes (see below), it is usually assumed that fluorescent indicators are homogeneously distributed in the cytosol and equally responsive to variations of intracellular ion concentration. However, AM esters are capable of accumulating in any membrane-enclosed structure within the cell. In addition, indicators in polyanionic form may be sequestered within organelles via active transport processes.[15] Compartmentalization is usually more pronounced at higher loading temperatures and is particularly acute in plant and fungal cells.[16,17] The extent of compartmentalization can be assessed by image analysis, as well as fluorometrically using membrane permeabilization reagents, such as digitonin (D-8449, D-8450; see Section 22.8) and Triton® X-100.[14]

Incomplete AM ester hydrolysis: Residual unhydrolyzed AM esters may be present extracellularly due to incomplete removal by washing. Inside the cell, low levels of intracellular esterase activity, which can vary considerably from one cell type to another, may produce only partial AM ester hydrolysis.[18-20] Because even partially hydrolyzed AM esters are Ca^{2+}-insensitive, detection of their fluorescence as part of the total signal leads to an underestimation of the Ca^{2+} concentration.[21,22] Fluorescence quenching by Mn^{2+}, which only binds to completely de-esterified indicators, can be used to quantitate these effects. Note that although some indicators are fluorescent in the AM ester form, others are not; see Table 22.1 in Chapter 22.

Leakage: Extrusion of anionic indicators from cells by organic ion transporters can be reduced by cooling the sample or by applying inhibitors such as probenecid and sulfinpyrazone.[15] AM esters are extruded by the P-glycoprotein multidrug transporter.[23] Ratiometric measurements (see below) help to minimize the impact of indicator leakage on experimental data.

Calibration

Ion Dissociation Constants

The dissociation constant (K_d) is the key conversion parameter linking fluorescence signals to ion concentrations. For pH indicators, K_d is conventionally expressed as the negative log (pK_a). The concentration range over which an indicator produces an observable response is approximately $0.1 \times K_d$ to $10 \times K_d$. For BAPTA-based Ca^{2+} indicators in particular, the K_d is very sensitive to a number of environmental factors, including temperature, pH, ionic strength and interactions of the indicator with proteins.[24-27] Examination of published data shows that values of K_d determined *in situ* within cells can be up to fivefold higher than values determined *in vitro*,[24,28,29] underscoring the importance of performing calibrations to determine K_d directly in the system under study.

Calibration Methodology

Calibration procedures basically consist of recording fluorescence signals corresponding to a series of precisely manipulated ion concentrations. The resulting sigmoidal titration curve is either linearized by means of a Hill plot or analyzed directly by nonlinear regression to yield K_d. For *in vitro* calibrations of Ca^{2+} indicators, EGTA buffering is widely used to produce defined Ca^{2+} concentrations that can be calculated from the K_d of the Ca^{2+}-EGTA complex.[5,14,30] This technique is utilized in Molecular Probes' Calcium Calibration Buffer Kits (see Section 22.8). *In situ* calibrations of intracellular indicators generally utilize an ionophore to equilibrate the controlled external ion concentration with the ion concentration within the cell. Commonly used ionophores include:

- A-23187 or 4-bromo A-23187 (A-1493, B-1494; see Section 22.8) for Ca^{2+} and Mg^{2+}
- Nigericin (N-1495, see Sections 23.2 and 24.2) for H^+ and Cl^-
- Gramicidin (G-6888, see Section 24.1.) for Na^+
- Valinomycin (V-1644, see Section 24.1) for K^+

Ratiometric Calibration

Indicators that show an excitation or emission spectral shift upon ion binding can be calibrated using a **ratio** of the fluorescence intensities measured at two different wavelengths, resulting in the cancellation of artifactual variations in the fluorescence signal that might otherwise be misinterpreted as changes in ion concentration (Figure 2). Note that background levels must be subtracted from the component fluorescence intensities *before* calculation of the ratio. Examples of indicators exhibiting ion-dependent spectral shifts can be found in Figures 22.1 and 22.2 in Chapter 22 and Figures 23.3 and 23.6 in Chapter 23. The ratio of two intensities with opposite ion-sensitive responses (for example, 340 nm/380 nm in Figure 22.1) gives the largest possible dynamic range of ratio signals for a given indicator. Alternatively, the ratio of an ion-sensitive intensity to an ion-insensitive intensity (measured at a spectral isosbestic point, e.g., 360 nm in Figure 22.1) can be used (Figure 2). Ratiometric measurements reduce or eliminate variations of several determining factors in the measured fluorescence intensity, including indicator concentration, excitation pathlength, excitation intensity and detection efficiency.[31,32] Artifacts that are eliminated include photobleaching and leakage of the indicator, variable cell thickness, and nonuniform indicator distribution within cells (due to compartmentalization) or among populations of cells (due to loading efficacy variations).

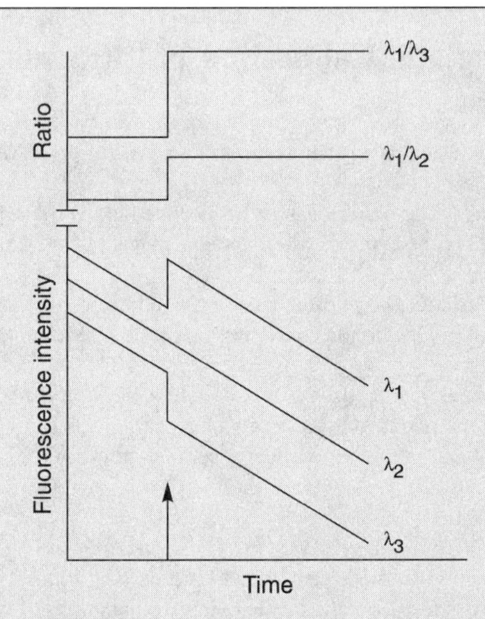

***Figure 2** Simulated data demonstrating the practical importance of ratiometric fluorescence techniques. The figure represents an ion indicator that exhibits a fluorescence intensity increase in response to ion binding at wavelength λ_1 and a corresponding decrease at λ_3. Fluorescence measured at an isosbestic point (λ_2) is independent of ion concentration. The intracellular indicator concentration diminishes rapidly due to photobleaching, leakage (assuming the extracellular indicator is not detectable) or some other process. The change of intracellular ion concentration due to a stimulus applied at the time indicated by the arrow is unambiguously identified by recording the fluorescence intensity ratios λ_1/λ_3 or λ_1/λ_2.*

1. Meth Cell Biol 40, 155 (1994) **available from Molecular Probes as catalog #M-7890**; **2.** Meth Toxicology 1B, 287 (1994); **3.** *Cellular Calcium: A Practical Approach*, J.G. McCormack and P.H. Cobbold, Eds., pp. 1–54, Oxford University Press (1991); **4.** Meth Enzymol 192, 38 (1990); **5.** Meth Enzymol 172, 230 (1989); **6.** Nature 290, 527 (1981); **7.** Plant Physiol 93, 841 (1990); **8.** Cell Calcium 8, 455 (1987); **9.** J Biol Chem 262, 8884 (1987); **10.** Neurosci Lett 207, 17 (1996); **11.** *Guide to Electroporation and Electrofusion*, D.C. Chang *et al.*, Eds., Academic Press (1992); **12.** Cytometry 14, 265 (1993); **13.** Meth Cell Biol 29, 153 (1989); **14.** Meth Cell Biol 40, 155 (1994); **15.** Cell Calcium 11, 57 (1990); **16.** J Exp Biol 196, 419 (1994); **17.** J Microscopy 166, 57 (1992); **18.** Cell Calcium 11, 63 (1990); **19.** Am J Physiol 255, C304 (1988); **20.** Anal Biochem 169, 159 (1988); **21.** Biophys J 67, 476 (1994); **22.** Biophys J 65, 561 (1993); **23.** J Biol Chem 268, 21493 (1993); **24.** Biophys J 67, 1646 (1994); **25.** Biophys J 63, 89 (1992); **26.** Biochem Biophys Res Comm 180, 209 (1991); **27.** Biochem Biophys Res Comm 177, 184 (1991); **28.** Biophys J 68, 1453 (1995); **29.** Biophys J 65, 865 (1993); **30.** Cell Calcium 12, 279 (1991); **31.** Meth Cell Biol 30, 157 (1989); **32.** J Biol Chem 260, 3440 (1985).

Chapter 23

pH Indicators

Contents

Technical Notes and Product Highlights

Related Chapters

23.1 Overview of pH Indicators

The ability of dyes — notably litmus, phenolphthalein and phenol red — to change their color in response to a pH change has found widespread application in research and industry. However, only fluorescent dyes can provide the greater sensitivity required for optical pH measurements *inside* live cells. The demand for sensitive intracellular pH indicators has spurred the search for improved fluorescent dyes that can sense pH changes within physiological ranges. Of course, many of the same fluorescent pH indicators can also be used as pH sensors in cell-free media.

To quantitatively measure pH, it is essential to match the indicator's pK_a to the pH of the experimental system. Consequently, the following two sections of this chapter are divided into pH indicators for use in environments with near-neutral pH (see Section 23.2) and pH indicators for use in acidic environments (see Section 23.3). Intracellular pH is generally between ~6.8 and 7.4 in the cytosol and ~4.5 and 6.0 in the cell's acidic organelles. Unlike intracellular free Ca^{2+} concentrations, which can rapidly change by perhaps 100-fold, the pH inside a cell varies by only fractions of a pH unit, and such changes may be quite slow. Although the optical change of even the best fluorescent pH probes is usually relatively small, they have proven to be effective tools for investigating the role of intracellular pH in diverse physiological and pathological processes, including cell proliferation,[1] apoptosis,[2,3] fertilization,[4] malignancy,[5] multidrug resistance,[6-8] ion transport,[9-11] lysosomal storage disorders and Alzheimer's disease.[12]

Molecular Probes offers a variety of fluorescent pH indicators, their conjugates and other reagents for pH measurements in biological systems. Over the last several years, we have introduced a number of innovative pH indicators:

- Our visible light–excitable SNARF and SNAFL pH indicators enable researchers to determine intracellular pH in the physiological range using dual-emission or dual-excitation ratiometric techniques (see Section 23.2), thus providing important tools for confocal laser scanning microscopy and flow cytometry.
- Our new LysoSensor probes, as well as indicators based on the Oregon Green and rhodol fluorophores, can be used to estimate pH in a cell's acidic organelles (see Section 23.3).
- We also offer a number of fluorescent pH indicators coupled to dextrans (see Section 23.4). Indicator dextrans are extremely well retained, do not bind to cellular proteins and have a reduced tendency to compartmentalize.[13]

Families of pH indicators available from Molecular Probes are listed in Table 23.1 in approximate order of decreasing pK_a values.

Table 23.1 *Molecular Probes' pH indicator families, in order of decreasing* pK_a.

Parent Fluorophore	pH Range	Typical Measurement
SNAFL indicators	7.2–8.2	Excitation ratio 490/540 nm or emission ratio 540/630 nm
SNARF indicators	7.0–8.0	Emission ratio 580/640 nm
HPTS (pyranine)	7.0–8.0	Excitation ratio 450/405 nm
BCECF	6.5–7.5	Excitation ratio 490/440 nm
Fluoresceins and carboxyfluoresceins	6.0–7.2	Excitation ratio 490/450 nm
Oregon Green dyes	4.2–5.7	Excitation ratio 510/450 nm or excitation ratio 490/440 nm
Rhodols (including NERF dyes)	4.0–6.0	Excitation ratio 514/488 nm or excitation ratio 500/450 nm
LysoSensor probes	3.5–8.0 *	Excitation ratio 340/380 nm †

* Depends on pK_a of selected probe; see Table 23.2 for pK_a of each LysoSensor probe. † Applies to L-7545 only. Other LysoSensor probes allow single excitation and emission measurements only; see Table 23.2 for wavelengths.

1. Cell Physiol Biochem 2, 159 (1992); **2.** J Biol Chem 270, 6235 (1995); **3.** J Biol Chem 270, 3203 (1995); **4.** Biophys J 68, 739 (1995); **5.** Cancer Res 54, 5670 (1994); **6.** Biophys J 69, 883 (1995); **7.** Proc Natl Acad Sci USA 91, 1128 (1994); **8.** Biochemistry 32, 11042 (1993); **9.** J Biol Chem 270, 19599 (1995); **10.** J Biol Chem 270, 13716 (1995); **11.** J Biol Chem 270, 7915 (1995); **12.** Biochem Biophys Res Comm 194, 537 (1993); **13.** Meth Cell Biol 29, 59 (1989).

23.2 Probes Useful at Near-Neutral pH

Fluorescein and Fluorescein Derivatives

Fluorescein and many of its derivatives exhibit multiple, pH-dependent ionic equilibria.[1-4] Both the phenol and carboxylic acid functional groups of fluorescein are almost totally ionized in aqueous solutions above pH 9 (Figure 23.1). Acidification of the fluorescein dianion first protonates the phenol (pK_a ~6.4) to yield the fluorescein monoanion, then the carboxylic acid (pK_a <5) to produce the neutral species of fluorescein. Further acidification generates a fluorescein cation (pK_a ~2.1). Only the monoanion and dianion of fluorescein are fluorescent, with quantum yields of 0.93 and 0.37, respectively. However, excitation of either the neutral or cationic species is reported to produce emission from the anion with effective quantum yields of 0.31 and 0.18.[1] A further equilibrium involves formation of a colorless, nonfluorescent lactone (Figure 23.1). The lactone is not formed in aqueous solution above pH 5 but may be the dominant form of neutral fluorescein in solvents such as acetone. The pH-dependent absorption spectra of fluorescein (Figure 23.2A) clearly show the blue shift and decreased absorptivity indicative of the formation of protonated species.

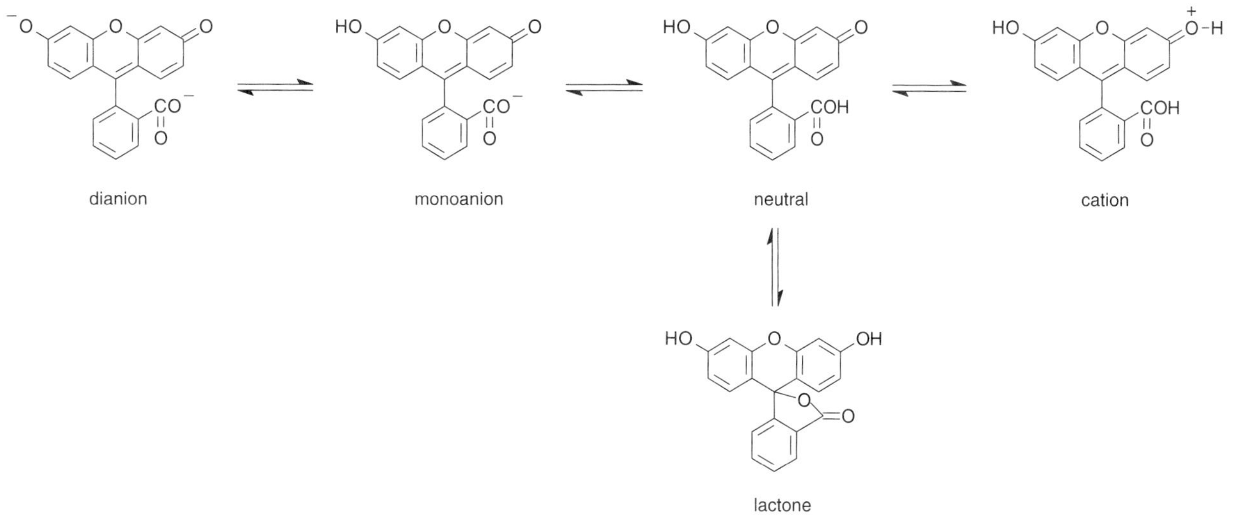

Figure 23.1 *Ionization equilibria of fluorescein.*

However, the fluorescence emission spectrum of most fluorescein derivatives, even in acidic solution, is dominated by the dianion, with only small contributions from the monoanion. Consequently, the wavelength and shape of the emission spectra resulting from excitation close to the dianion absorption peak at 490 nm are relatively independent of pH, but the fluorescence intensity is dramatically reduced at acidic pH (Figure 23.2B).

Molecular Probes offers a broad variety of fluorescein-derived reagents and fluoresceinated probes that can serve as sensitive fluorescent pH indicators in a wide range of applications. Chemical substitutions of fluorescein may shift absorption and fluorescence maxima and change the pK_a of the dye; however, the effects of acidification on the spectral characteristics illustrated in Figure 23.2 are generally maintained in fluorescein derivatives.

Fluorescein and Its Diacetate

The cell-permeant fluorescein diacetate (FDA, F-1303) is still occasionally used to measure intracellular pH,[5] as well as to study cell adhesion [6] or, in combination with propidium iodide (P-1304, see Section 8.1), to determine cell viability.[7,8] However, fluorescein (F-1300) formed by intracellular hydrolysis of FDA rapidly leaks from cells (see Figure 16.10 in Chapter 16). Thus, other cell-permeant dyes such as the acetoxymethyl (AM) esters of BCECF and calcein are now preferred for intracellular pH measurements and cell viability assays.

Carboxyfluorescein and Its Cell-Permeant Esters

The high leakage rate of fluorescein from cells makes it very difficult to quantitate intracellular pH because the decrease in the cell's fluorescence due to dye leakage cannot be easily distinguished from that due to acidification. The use of carboxyfluorescein diacetate (5(6)-CFDA, C-195) for intracellular pH measurements addresses this problem.[9] CFDA is moderately permeant to most cell membranes and, upon hydrolysis by intracellular nonspecific esterases, forms carboxyfluorescein (5(6)-FAM; C-194, C-1904), which has a pH-dependent spectral response very similar to that of fluorescein. As compared to fluorescein, carboxyfluorescein contains extra negative charges and is therefore better retained in cells [10] (see Figure 16.10 in Chapter 16). The mixed-isomer preparation of CFDA (C-195) is usually adequate for intracellular pH measurements because the single isomers of carboxyfluorescein exhibit essentially identical pH-dependent spectra with a pK_a ~6.5. For experiments requiring a pure isomer, the single-isomer preparations of carboxyfluorescein (C-1359, C-1360; see Section 1.3) and CFDA (C-1361, C-1362; see Section 16.3) are available. In addi-

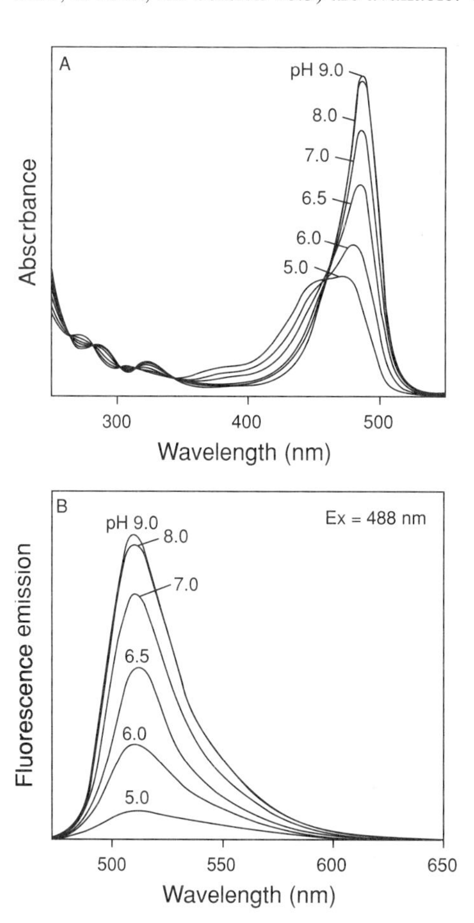

Figure 23.2 *The pH-dependent spectra of fluorescein: A) absorption spectra, B) emission spectra.*

tion, we offer the AM ester of CFDA (5-CFDA, AM; C-1354), which is electrically neutral and facilitates cell loading. Upon hydrolysis by intracellular esterases, this AM ester also yields carboxyfluorescein.[11-13]

Intracellular pH measurements have been made using carboxyfluorescein,[14,15] although the spectral and pK_a properties of SNARF, SNAFL and BCECF (see below) make these indicators superior probes for most pH studies. Carboxyfluorescein is also commonly employed as a polar tracer (see Section 15.3), and CFDA and its AM ester are used for monitoring viability,[13,16] apoptosis [17] and cell adhesion [10,18] (see Section 16.3).

BCECF and Its AM Ester

Although carboxyfluorescein is better retained in cells than fluorescein, its pK_a of ~6.5 is lower than the cytosolic pH of most cells (pH ~6.8–7.4). Consequently, its fluorescence change is less than optimal for detecting small pH changes above pH 7. Since its introduction by Roger Tsien,[19] the polar fluorescein derivative BCECF (B-1151) and its AM ester (B-1150, B-1170, B-3051) have become the most widely used fluorescent indicators for estimating intracellular pH. BCECF's four to five negative charges at pH 7–8 improve its retention in cells (see Figure 16.10 in Chapter 16), and its pK_a of 6.98 is ideal for typical intracellular pH measurements.

As with fluorescein and carboxyfluorescein, absorption of the phenolate anion (basic) form of BCECF is red-shifted and has increased molar absorptivity relative to the protonated (acidic) form (Figure 23.3A); there is little, if any, pH-dependent shift in the fluorescence emission spectrum of BCECF upon excitation at 505 nm (Figure 23.3B). BCECF is typically used as a dual-excitation ratiometric pH indicator. Signal errors caused by variations in concentration, pathlength, leakage and photobleaching are greatly reduced with ratiometric methods (see the Technical Note "Loading and Calibration of Intracellular Ion Indicators" on page 549). Intracellular pH measurements with BCECF are made by determining the pH-dependent ratio of emission intensity (at 535 nm) when the dye is excited at 505 nm versus the emission intensity when excited at its isosbestic point of 439 nm (Figure 23.3C). Because BCECF absorption at 439 nm is quite weak (Figure 23.3A), increasing the denominator wavelength to ~450 nm provides improved signal-to-noise characteristics for ratio imaging applications.[20] A few flow cytometry studies have also employed BCECF as a dual-emission ratiometric pH indicator, using an emission ratio of 525/640 nm.[21] As with other intracellular pH indicators, *in situ* calibration of BCECF's fluorescence response is usually accomplished using 10–50 µM nigericin (N-1495, see below) in the presence of 100–150 mM K^+ to equilibrate internal and external pH.[9,22] Alternative calibration methods have also been reported.[23,24]

Loading of live cells for measurement of intracellular pH is readily accomplished by incubating cell suspensions or adherent cells in a 1–10 µM solution of the AM ester of BCECF. Three different molecular species can be obtained in synthetic preparations of the AM ester of BCECF; however, all three forms shown in Figure 23.4 appear to be converted to the same product — BCECF acid (B-1151) — by intracellular esterase hydrolysis. Although we can readily prepare the pure triacetoxymethyl ester form (Form I in Figure 23.4), some researchers have found that its solubility in a mostly aqueous medium is lower than that of the mixed product containing Forms II and III and that cell loading with the mixed product is more efficient. Consequently, we now produce BCECF AM as a mixture of Forms II and III with a typical percentage composition ratio of 45:55, as determined by HPLC and NMR. Because the molecular weights of the different species are not equal, the effective molecular weight for each production lot is reported on the product's label. The AM ester of BCECF is available in a single 1 mg vial (B-1150), specially packaged as a set of 20 vials that each contain 50 µg (B-1170) and as a 1 mg/mL solution (~1.6 mM) in anhydrous dimethylsulfoxide (DMSO) (B-3051). We highly recommend purchasing the set of 20 vials in order to reduce the potential for product deterioration caused by exposure to moisture.

Our bibliography on BCECF AM — which is available upon request from our Technical Assistance Department or through our Web site (http://www.probes.com) — lists over 500 journal citations, including recent references for the use of BCECF AM to investigate:

- Cl^-/HCO_3^- exchange [25-27]
- K^+/H^+ exchange [28]
- Lactate transport [29,30]
- Na^+/H^+ exchange [31,32]

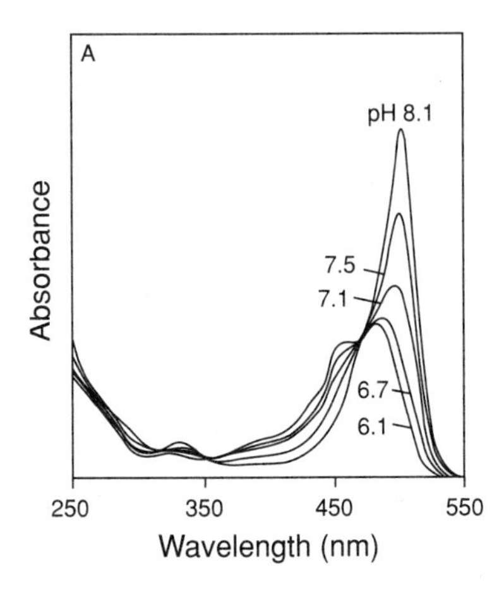

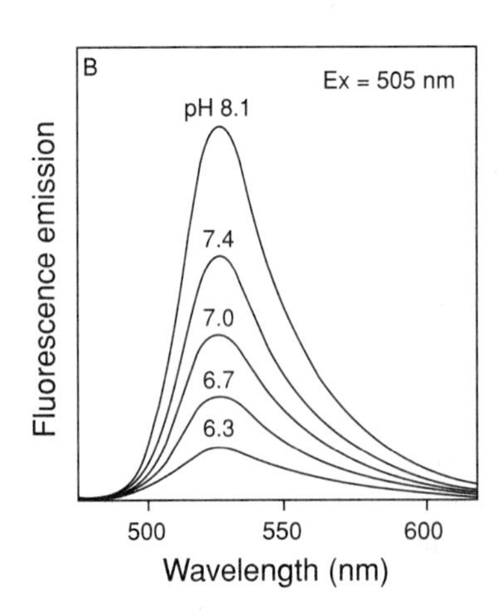

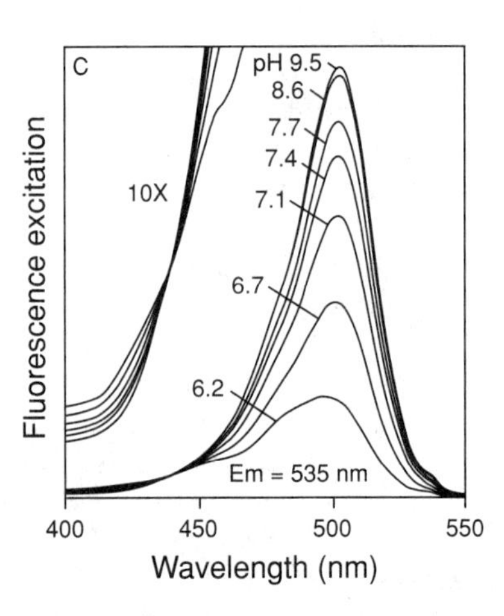

Figure 23.3 *The pH-dependent spectra of BCECF (B-1151): A) absorption spectra, B) emission spectra and C) excitation spectra. The fluorescence excitation spectra have been enlarged on the left to reveal BCECF's 439 nm isosbestic point. Note that the isosbestic point of the excitation spectra of BCECF is different from that of the absorption spectra (compare panels A and C).*

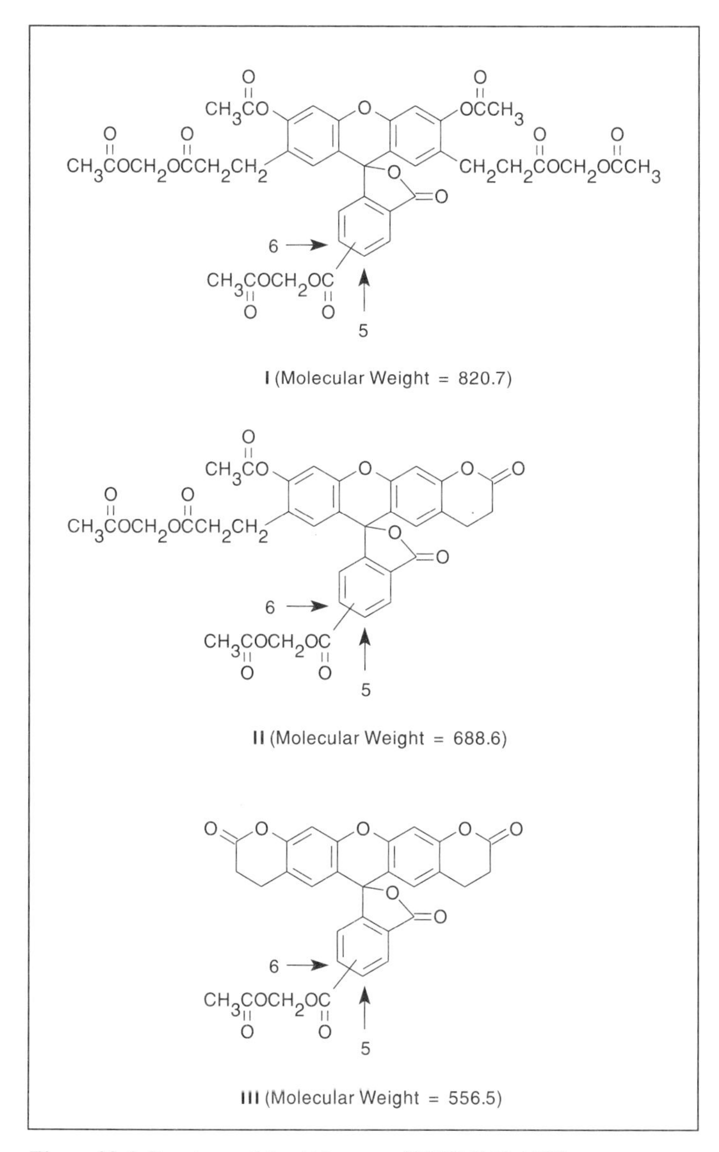

Figure 23.4 Structures of the AM esters of BCECF (B-1150).

- Na^+/Ca^{2+} exchange [33,34]
- NH_4^+ transport [35]
- Apoptosis [36-38]
- Cytotoxicity [39,40]
- Multidrug resistance [25,41-45]
- Cell volume changes [46,47]
- Cytosolic pH regulation in osteoblasts [48] and osteoclasts [49]
- pH in lateral intercellular spaces of epithelial cell monolayers [50] and interstitial spaces of normal and neoplastic tissue [51]
- Plant vacuoles [52]

The cell-impermeant BCECF acid (B-1151) is useful both for calibration and for loading cells by microinjection, electroporation and comparable techniques. The free acid of BCECF has been loaded into bacterial cells by brief incubation at pH ~2.[53] Molecular Probes also offers dextran conjugates of BCECF (see Section 23.4).

Carboxy-4′,5′-Dimethylfluorescein and Its Diacetate

The introduction of electron-releasing groups such as methyl groups into fluorescein dyes decreases the acidity of the indicator. The methylated derivative 5-(and-6)-carboxy-4',5'-dimethylfluorescein (C-366) exhibits a pK_a of ~7.0. Ratiometric measurements using a excitation ratio of 490/450 nm are possible with this indicator.[54] Its cell-permeant diacetate (C-367) has been used for pH measurements in smooth muscle,[55] soybean cells,[56] rabbit proximal tubules [57,58] and various other cell types.

Fluorescein Sulfonic Acid and Its Diacetate

The fluorescein-5-(and-6)-sulfonic acid (F-1130) is much more polar than carboxyfluorescein. Consequently, once inside cells or liposomes, it is relatively well retained. Some cells can be loaded directly with 5-sulfofluorescein diacetate [59-62] (SFDA, S-1129). Direct ratiometric measurement of the pH in the trans-Golgi of live human fibroblasts was achieved by microinjecting liposomes loaded with both fluorescein sulfonic acid and sulforhodamine 101 [63] (S-359, see Section 15.3).

Chemically Reactive Fluorescein Diacetates

One means for overcoming the cell leakage problem common to the above pH indicators, including BCECF, is to trap the indicator inside the cell via conjugation to intracellular constituents. The chloromethyl derivatives CellTracker™ Green CMFDA (C-2925, C-7025) and chloromethyl SNARF-1 (C-6826, see below) probably have the greatest potential for long-term cell tracing and pH studies. In many cell types, the weakly thiol-reactive chloromethyl moiety of CMFDA reacts with intracellular thiols, including glutathione and proteins, to yield well-retained products (see Figure 16.10 in Chapter 16). Cleavage of the acetate groups of the CMFDA conjugate by intracellular esterases yields a conjugate that retains the pH-dependent spectral properties of fluorescein.

Similarly, the amine-reactive succinimidyl ester of CFDA (5(6)-CFDA, SE; C-1157) can potentially be used for long-term pH studies of live cells, producing a conjugate with the pH-sensitive properties of carboxyfluorescein. Because it is intrinsically more reactive, the succinimidyl ester of CFDA is more likely to react at sites on the extracellular surface than is CMFDA. However, it may be possible to quench much of the fluorescence of outer-membrane fluorescein conjugates by acidifying the extracellular medium or by incubating with anti-fluorescein antibody [64] (A-889, see Section 7.3).

Naphthofluorescein

Naphthofluorescein (N-650) and its carboxy derivative (C-652) have pH-dependent red fluorescence [65] with a relatively high pK_a of ~7.6. The long-wavelength pH-dependent spectra of carboxynaphthofluorescein have been exploited in the construction of a photodiode-based fiber optic pH sensor.[65]

SNARF and SNAFL pH Indicators

The seminaphthorhodafluors (SNARF dyes) and seminaphthofluoresceins (SNAFL dyes) are visible light–excitable fluorescent pH indicators developed at Molecular Probes.[66,67] The SNARF and SNAFL indicators have both dual-emission and dual-excitation properties, making them useful for confocal laser scanning microscopy,[68] flow cytometry [21,69-71] and microspectrofluorometry.[72] The dual-emission properties of SNARF dyes may make these the preferred dyes for use in fiber-optic pH sensors.[73,74] The fluorophores can be excited by the 488 or 514 nm spectral lines of the argon-ion laser and are sensitive to pHs within the physiological range. Dextran conjugates of the SNARF and SNAFL dyes are described in Section 23.4.

Carboxy SNARF-1 and Its Cell-Permeant Ester

Carboxy SNARF-1 (C-1270), which is easily loaded into cells as its cell-permeant AM ester acetate (C-1271, C-1272), has a pK_a of about 7.5 at room temperature and 7.3–7.4 at 37°C. Thus, carboxy SNARF-1 is useful for measuring pH changes between pH ~7 and 8. Like fluorescein and BCECF, the absorption spectrum of carboxy SNARF-1 undergoes a red-shift upon deprotonation of its phenolic substituent (Figure 23.5). In contrast to the fluorescein-based indicators, however, carboxy SNARF-1 also exhibits a significant pH-dependent emission shift from yellow-orange to deep red fluorescence under acidic and basic conditions, respectively (Figure 23.6). This pH dependence allows the ratio of the fluorescence intensities from the dye at two emission wavelengths — typically 580 nm and 640 nm — to be used for quantitative determinations of pH (see the Technical Note "Loading and Calibration of Intracellular Ion Indicators" on page 549). For practical purposes, it is often desirable to bias the detection of carboxy SNARF-1 fluorescence towards the less-fluorescent acidic form by using an excitation wavelength between 488 nm and the excitation isosbestic point at ~530 nm, yielding balanced signals for the two emission-ratio components (Figure 23.6). When excited at 488 nm, carboxy SNARF-1 exhibits an emission isosbestic point of ~610 nm and a lower fluorescent signal than obtained with 514 nm excitation.[68] Alternatively, when excited by the 568 nm spectral line of the Ar-Kr laser found in some confocal laser scanning microscopes, carboxy SNARF-1 exhibits a fluorescence increase at 640 nm as the pH increases and an emission isosbestic point at 585 nm.[68] As with other ion indicators, intracellular environments may cause significant modifications of both the spectral properties and pK_a of carboxy SNARF-1,[75-77] and the indicator should always be calibrated in the system under study.

The spectra of carboxy SNARF-1 are well resolved from those of the fura-2,[78,79] fluo-3,[70] Calcium Green™ and Oregon Green 488 BAPTA indicators for calcium (see Section 22.3), permitting simultaneous measurements of intracellular pH and Ca^{2+}. Carboxy SNARF-1 has also been used in combination with the Na^+ sodium indicator SBFI (S-1263, see Section 24.1) to simultaneously detect pH and Na^+ changes in heart mitochondria.[80] The relatively long-wavelength excitation and emission characteristics of carboxy SNARF-1 permit studies that employ the anion-transport inhibitor DIDS[81,82] (D-337, see Section 18.5), amiloride derivatives[81] (see Section 18.7), caged probes (see Chapter 19) and other modifiers of cell function that are excited with UV light. In addition, carboxy SNARF-1 has served as a Ca^{2+}-*insensitive* standard in order to make ratiometric measurements of intracellular Ca^{2+} with the non-ratiometric Ca^{2+} indicators fluo-3[83-86] and Calcium Green-2.[87]

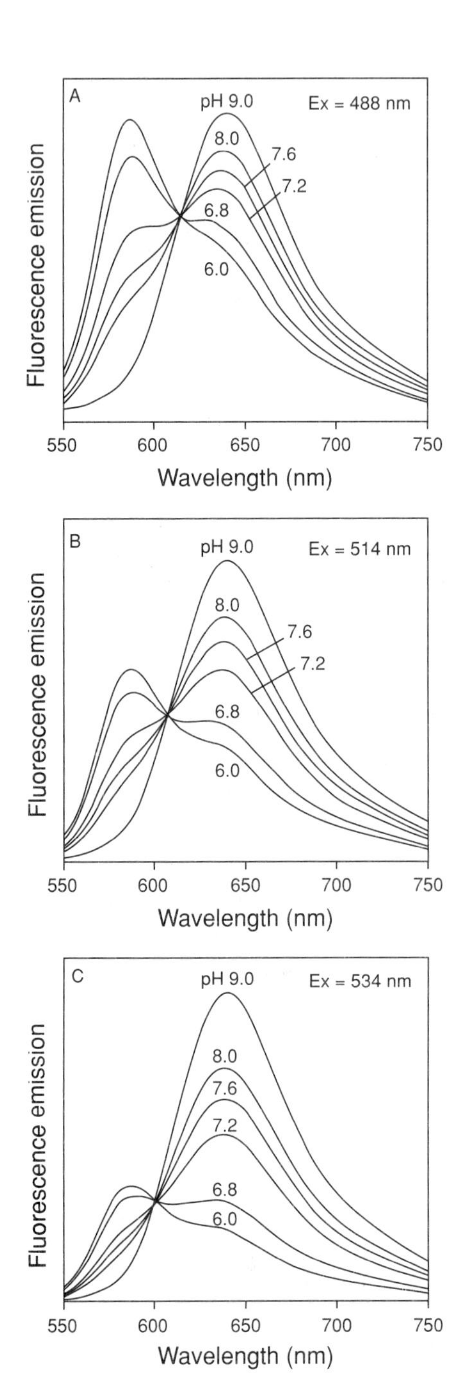

Figure 23.6 The pH-dependent emission spectra of carboxy SNARF-1 (C-1270) when it is excited at A) 488 nm, B) 514 nm and C) 534 nm.

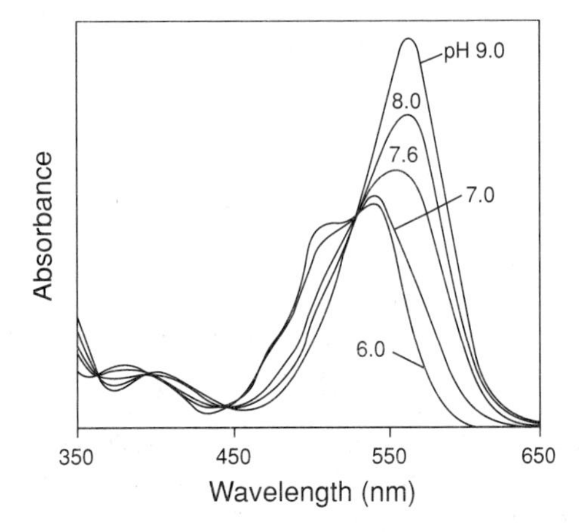

Figure 23.5 The pH-dependent absorption spectra of carboxy SNARF-1 (C-1270).

Chloromethyl SNARF-1 Acetate

Our new 5-(and-6)-chloromethyl SNARF-1 acetate (C-6826) contains a chloromethyl group that is mildly reactive with intracellular thiols, forming adducts that should improve cellular retention of the SNARF fluorophore. As with CellTracker Green CMFDA (see above), improved retention of this conjugate in cells may permit monitoring of intracellular pH over longer time periods than is possible with other intracellular pH indicators.

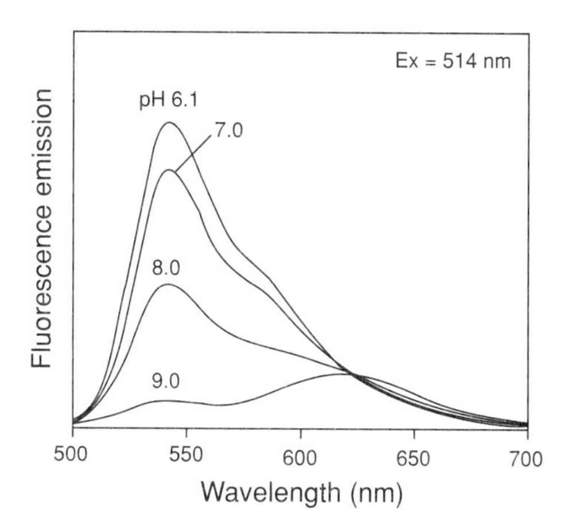

Figure 23.7 *The pH-dependent emission spectra of carboxy SNAFL-1 (C-1255).*

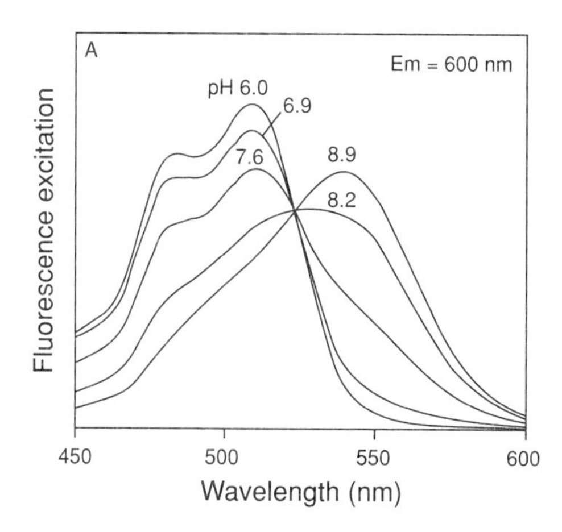

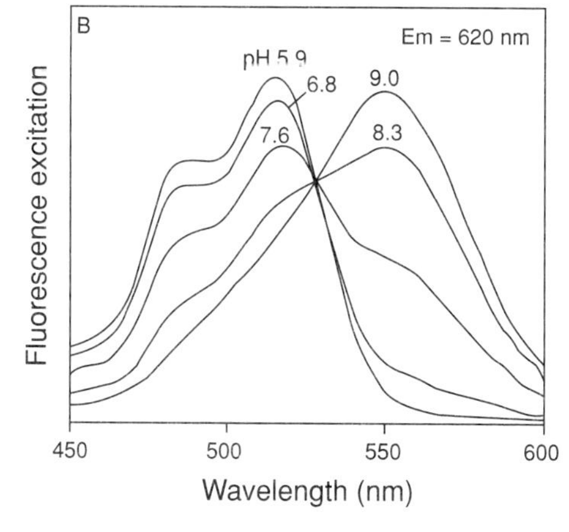

Figure 23.8 *The pH-dependent excitation spectra of A) carboxy SNAFL-1 (C-1255) with emission monitored at 600 nm and B) carboxy SNAFL-2 (C-1260) with emission monitored at 620 nm.*

Carboxy SNAFL-1 and Carboxy SNAFL-2

Carboxy SNAFL-1 and carboxy SNAFL-2, which are available as the free acids (C-1255, C-1260) and as their cell-permeant diacetates (C-1256, C-1261), have higher fluorescence quantum yields as acids than as bases. Excitation near the absorption maxima of their acidic forms (~514 nm) results in strong emission from the acid and relatively weak emission from the base [66] (Figure 23.7). These indicators can also be used as dual-excitation probes (Figure 23.8) and as fluorescence lifetime–based pH sensors.[88,89]

SNARF and SNAFL Calceins

Molecular Probes' polar SNARF calcein (S-3055) and SNAFL calcein (S-3052), also available as the cell-permeant AM esters (S-3057, S-3054), are dual-emission pH indicators [90] that incorporate hydrophilic iminodiacetic acid groups such as those found in calcein. These hydrophilic groups may endow SNARF and SNAFL calceins with the enhanced intracellular retention properties of calcein (see Figure 16.10 in Chapter 16). The pH-dependent fluorescence spectra of these indicators are quite complex and do not exhibit clear isosbestic points, suggesting the existence of multiple emitting species. However, the absorption spectra of both SNARF calcein and SNAFL calcein closely resemble those of carboxy SNARF-1 and carboxy SNAFL-1 and indicate pK_a values of ~7. Moreover, the ratio of emission intensities at 535 nm and 625 nm is clearly pH dependent between pH 6 and 7.5;[90] another emission starts to appear in these indicators above pH ~8, apparently as the result of ionization of the iminodiacetic acid group.

SNAFL calcein has been used to measure intracellular pH during transdifferentiation of retinal pigment epithelium into neuroepithelium and under conditions of reduced partial O_2 pressure.[91,92] This probe has also been employed for pH measurements in Madin–Darby canine kidney (MDCK) cells by confocal laser scanning microscopy.[90]

8-Hydroxypyrene-1,3,6-Trisulfonic Acid (HPTS)

8-Hydroxypyrene-1,3,6-trisulfonic acid (HPTS, also known as pyranine; H-348) is a highly water-soluble, membrane-impermeant pH indicator with a pK_a ~7.3 in aqueous buffers.[93] Unlike SNARF-, SNAFL- and fluorescein-based indicators, there is no membrane-permeant form of HPTS available. Consequently, HPTS must be introduced into cells by microinjection, patch-pipet techniques, liposomes,[94] acidification,[95] endocytosis [96] or scrape loading [97] (see Table 15.1 in Chapter 15). The pK_a of HPTS is reported to rise to 7.5–7.8 in the cytosol of some cells.[97] HPTS exhibits a pH-dependent absorption shift (Figure 23.9), allowing ratiometric measurements using an excitation ratio of 450/405 nm.[98]

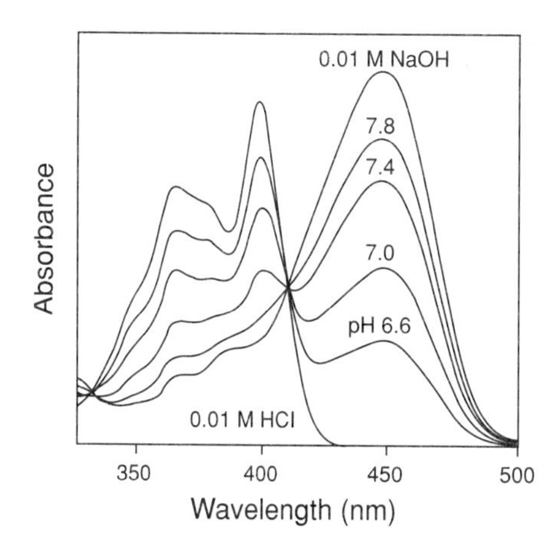

Figure 23.9 *The pH-dependent absorption spectra of HPTS (H-348).*

The unique pH-dependent spectral properties, high water solubility and low cost of HPTS make its applications numerous. They include:

- Detecting proton permeability of liposomes [99,100]
- Fiber-optic sensing of oxygen and carbon dioxide,[101,102] ammonia [103] and enzymatic activity [104]
- Following proton release from cells [105]
- Liposome tracking and fusion [94,96,106-109]
- Measuring acidity of lysosomes and other organelles [98,110,111]
- Monitoring lateral proton diffusion and transfer in membranes [112-115]
- Studying lipid transbilayer transport in cell membranes [116-118]

We also offer a caged form of HPTS (H-7060, see Section 15.3) that is nonfluorescent until photolyzed by UV light, as well as a new dextran conjugate of HPTS (D-7179, see Section 23.4).

Auxiliary Probes for pH Studies

In addition to the fluorescent pH indicators described above and in the next sections, Molecular Probes provides nigericin, which is widely used for calibrating intracellular pH indicators, as well as some unique caged compounds that can be used for localized generation of protons by UV photolysis.

Nigericin and Caged Nigericin

Intracellular calibration of the fluorescence response of cytosolic pH indicators is typically performed using the ionophore nigericin (N-1495), which causes an exchange of internal K^+ for external H^+ [9,22] (see the Technical Note "Loading and Calibration of Intracellular Ion Indicators" on page 549). Photolysis of our new DMNPE-caged nigericin (N-7107) can potentially provide spatial and temporal control of K^+/H^+ exchange in cells; see Chapter 19 for further discussion of caged probes.

Caged Protons

Molecular Probes prepares three caged probes that can photolytically liberate protons. The NPE-caged proton (H-6829) generates a photolysis-dependent proton release that results in a pH drop of up to several orders of magnitude in nanoseconds to microseconds.[119] The rapid proton release during this photolysis has been detected by enhancement of the fluorescence of carboxy SNAFL-1 [120,121] (C-1255). Photolysis of either NPE-caged phosphate [122] (N-7065) or desyl-caged phosphate (D-7076) liberates inorganic phosphate, which rapidly ionizes to release a proton. See Chapter 19 for a more complete discussion of the properties of caged probes.

Amiloride Derivatives

Molecular Probes offers amiloride and a wide selection of amiloride derivatives (see Section 18.7) for manipulating intracellular pH. Amiloride and related pyrazines are efficient inhibitors of the Na^+/H^+ antiporter in mammalian cells.[123-125] The compound 6-iodoamiloride (I-6881) has recently been described as a photoactivatable inhibitor of the Na^+/H^+ antiporter, producing irreversible inhibition upon UV illumination.[126] The Na^+/H^+ antiporter extrudes protons from cells using the inward Na^+ gradient as a driving force, resulting in intracellular alkalinization.

1. Spectrochim Acta (Part A) 51, L7 (1995); **2.** Photochem Photobiol 60, 435 (1994); **3.** J Luminescence 10, 381 (1975); **4.** J Phys Chem 75, 245 (1971); **5.** FEBS Lett 341, 125 (1994); **6.** J Immunol Meth 157, 117 (1993); **7.** Cancer Res 49, 3776 (1989); **8.** Histochem Cytochem 33, 77 (1985); **9.** Biochemistry 18, 2210 (1979); **10.** J Immunol Meth 172, 115 (1994); **11.** Biochemistry 34, 1606 (1995); **12.** Cytometry 13, 739 (1992); **13.** J Immunol Meth 130, 251 (1990); **14.** Cytometry 19, 235 (1995); **15.** Photochem Photobiol 60, 274 (1994); **16.** Anticancer Res 14, 927 (1994); **17.** Exp Cell Res 211, 322 (1994); **18.** J Cell Biol 124, 609 (1994); **19.** J Cell Biol 95, 189 (1982); **20.** Meth Cell Biol 30, 157 (1989); **21.** Meth Cell Biol 41, 135 (1994); **22.** Meth Enzymol 192, 38 (1990); **23.** J Cell Physiol 151, 596 (1992); **24.** J Fluorescence 2, 191 (1992); **25.** Biochemistry 33, 7239 (1994); **26.** J Biol Chem 269, 4116 (1994); **27.** J Physiol 475, 59 (1994); **28.** Biochemistry 34, 15157 (1995); **29.** J Biol Chem 271, 861 (1996); **30.** Biochem J 304, 751 (1994); **31.** J Biol Chem 270, 13716 (1995); **32.** J Biol Chem 270, 11051 (1995); **33.** J Biol Chem 270, 9137 (1995); **34.** J Biol Chem 269, 860 (1994); **35.** J Biol Chem 269, 21962 (1994); **36.** J Biol Chem 270, 6235 (1995); **37.** J Biol Chem 269, 12084 (1994); **38.** Cytometry 14, 595 (1993); **39.** Cytometry 20, 281 (1995); **40.** J Immunol Meth 172, 255 (1994); **41.** Biophys J 69, 883 (1995); **42.** Biochemistry 33, 11008 (1994); **43.** Cytometry 17, 343 (1994); **44.** Biochemistry 32, 11042 (1993); **45.** J Biol Chem 268, 21493 (1993); **46.** Meth Neurosci 27, 361 (1995); **47.** J Biol Chem 267, 17658 (1992); **48.** Mineral Electrolyte Metab 20, 16 (1994); **49.** J Biol Chem 270, 2203 (1995); **50.** Am J Physiol 266, C73 (1994); **51.** Cancer Res 54, 5670 (1994); **52.** J Plant Physiol 145, 57 (1995); **53.** Biochim Biophys Acta 1115, 75 (1991); **54.** Anal Biochem 156, 202 (1986); **55.** Jpn J Physiol 43, 103 (1993); **56.** Plant Physiol 98, 680 (1992); **57.** J Gen Physiol 92, 395 (1988); **58.** J Gen Physiol 92, 369 (1988); **59.** J Cell Biol 111, 3129 (1990); **60.** J Immunol Meth 133, 87 (1990); **61.** FEBS Lett 200, 203 (1986); **62.** BioTechniques 3, 270 (1985); **63.** J Biol Chem 270, 4967 (1995); **64.** Proc Natl Acad Sci USA 78, 7540 (1981); **65.** Mikrochim Acta 108, 133 (1992); **66.** Anal Biochem 194, 330 (1991); **67.** U.S. Patent No. 4,945,171; **68.** Biophys J 66, 942 (1994); **69.** Cytometry 14, 916 (1993); **70.** Cytometry 14, 257 (1993); **71.** Cytometry 12, 127 (1991); **72.** Anal Biochem 193, 49 (1991); **73.** Anal Chem 67, 2264 (1995); **74.** Anal Chem 65, 2329 (1993); **75.** Pflügers Arch 427, 332 (1994); **76.** Anal Biochem 204, 65 (1992); **77.** J Fluorescence 2, 75 (1992); **78.** J Cell Physiol 161, 129 (1994); **79.** Am J Physiol 260, C297 (1991); **80.** J Biol Chem 270, 672 (1995); **81.** J Biol Chem 270, 1315 (1995); **82.** Pflügers Arch 417, 234 (1990); **83.** J Biol Chem 270, 29781 (1995); **84.** J Biol Chem 269, 30636 (1994); **85.** Biochem J 289, 373 (1993); **86.** Cytometry 11, 923 (1990); **87.** Pflügers Arch 430, 529 (1995); **88.** Anal Chem 65, 1668 (1993); **89.** Anal Chem 65, 853 (1993); **90.** J Cell Physiol 164, 9 (1995); **91.** Biochem Cell Biol 72, 257 (1994); **92.** Cell Calcium 16, 404 (1994); **93.** Fresenius Z Anal Chem 314, 119 (1983); **94.** Biochemistry 29, 4929 (1990); **95.** Biochim Biophys Acta 1142, 277 (1993); **96.** Biochim Biophys Acta 1024, 352 (1990); **97.** Anal Biochem 167, 362 (1987); **98.** Proc Natl Acad Sci USA 92 3156 (1995); **99.** J Neurochem 62, 2022 (1994); **100.** Biochim Biophys Acta 1111, 17 (1992); **101.** Analyst 118, 839 (1993); **102.** Anal Chem 60, 2028 (1988); **103.** Anal Chim Acta 185, 321 (1986); **104.** Anal Chem 58, 2874 (1986); **105.** Biophys J 68, 1518 (1995); **106.** Biochemistry 33, 4562 (1994); **107.** Biochim Biophys Acta 1146, 87 (1993); **108.** Biochim Biophys Acta 1103, 185 (1992); **109.** Cancer Res 48, 5237 (1988); **110.** J Biol Chem 268, 6742 (1993); **111.** Am J Physiol 262, G30 (1992); **112.** J Biol Chem 270, 4368 (1995); **113.** Proc Natl Acad Sci USA 92, 372 (1995); **114.** Biochemistry 33, 873 (1994); **115.** Nature 370, 379 (1994); **116.** Biochemistry 32, 11074 (1993); **117.** Proc Natl Acad Sci USA 89, 11367 (1992); **118.** Biochemistry 29, 3046 (1990); **119.** Biophys J 68, A364 (1995); **120.** Proc Natl Acad Sci USA 92, 9757 (1995); **121.** Biophys J 65, 2368 (1993); **122.** J Mol Biol 184, 645 (1985); **123.** J Membrane Biol 105, 1 (1988); **124.** J Biol Chem 258, 3503 (1983); **125.** J Med Chem 10, 66 (1967); **126.** J Biol Chem 269, 3374 (1994).

23.2 Data Table *Probes Useful at Near-Neutral pH*

Cat #	Structure	MW	Storage	Soluble	Acidic/Neutral Solution			Basic Solution						
					Abs	{ε × 10^{-3}}	Em	Abs	{ε × 10^{-3}}	Em	Solvent	Cleaved*	pK_a†	Notes
B-1150	✓	~615‡	F,D	DMSO	<300		none					B-1151		
B-1151	✓	520	L	pH>6	482	{35}	520	503	{90}	528	<1>		~7.0	
C-194	S15.3	376	L	pH>6, DMF	475	{28}	517	492	{75}	517	<1>		~6.4	
C-195	S16.3	460	F,D	DMSO	<300		none					C-194		
C-366	✓	404	L	pH>6	500	{21}	537	507	{62}	537	<1>		~7.0	
C-367	✓	488	F,D	DMSO	<300		none					C-366		
C-652	S1.6	476	L	pH>6, DMF	512	{11}	563	598	{49}	668	<1>		~7.6	<2>
C-1157	S15.2	557	F,D	DMF, DMSO	<300		none					<3>		
C-1255	✓	426	L	pH>6	508	{29}	543	540	{52}	623	<1>		~7.8	<4>
C-1256	✓	510	F,D	DMSO	<350		none					C-1255		
C-1260	✓	461	L	pH>6	514	{31}	546	543	{50}	630	<1>		~7.7	<4>
C-1261	✓	545	F,D	DMSO	<350		none					C-1260		
C-1270	✓	453	L	pH>6	548	{27}	587	576	{48}	635	<1>		~7.5	<4>
C-1271	✓	568	F,D	DMSO	<350		none					C-1270		
C-1354	S16.3	532	F,D	DMSO	<300		none					<5>		
C-1904	S15.3	376	L	pH>6, DMF	475	{29}	517	492	{78}	517	<1>		~6.4	
C-2925	S15.2	465	F,D	DMSO	<300		none					<3>		
C-6826	✓	500	F,D	DMSO	<350		none					<6>		
D-7076	✓	332§	F,D,LL	H_2O	250	{12}	none				H_2O			<7, 8>
F-1130	S15.3	478	D,L	H_2O, DMF	476	{31}	519	495	{76}	519	<1>		~6.4	
F-1300	S10.1	332	L	pH>6, DMF	473	{32}	514	490	{88}	514	<1>		~6.4	
F-1303	S16.3	416	F,D	DMSO	<300		none					F-1300		
H-348	S15.3	524	D,L	H_2O	403	{20}	511	454	{22}	511	<1>		~7.3	<9>
H-6829	✓	361	F,D,LL	pH>6	269	{5.7}	none				MeOH			<7, 8>
N-650	✓	432	L	pH>8, DMF	510	{10}	561	594	{46}	663	<1>		~8.0	
N-1495	✓	725	F,D	MeOH	<300		none							
N-7065	✓	281	F,D,LL	H_2O	259	{5.7}	none				MeOH			<7, 8>
N-7107	✓	934	FF,D,LL	MeOH	338	{11}	none				MeOH			<7>
S-1129	S16.3	518	F,D	DMSO	<300		none					F-1130		
S-3052	✓	758	D,L	pH>6	506	{27}	535	535	{50}	620	<1>		~7.2	<10, 11>
S-3054	✓	1045	F,D	DMSO	<350		none					S-3052		
S-3055	✓	599	D,L	pH>6	552	{27}	~590	574	{45}	629	<1>		~6.9	<10, 11>
S-3057	✓	857	F,D	DMSO	<350		none					S-3055		

B-1170, B-3051 see B-1150; C-1272 see C-1271; C-7025 see C-2925

For definitions of the contents of this data table, see "How to Use the Handbook" on page vi.
Structure: Chemical structure drawing: ✓ = shown in this section; Sn.n = shown in Section number n.n.

* Catalog numbers and footnotes listed in this column indicate the cleavage product generated by enzymatic or chemical hydrolysis of the acetate and/or acetoxymethyl ester groups of the parent compound.
† pK_a values may vary considerably depending on the temperature, ionic strength, viscosity, protein binding and other factors. Unless otherwise noted, values listed have been determined from pH-dependent fluorescence measurements at 22°C.
‡ MW of B-1150 is approximate. See Figure 23.4 for further details.
§ MW is for the hydrated form of this product.
<1> Spectra are in aqueous buffers adjusted to >1 pH unit above (basic solution) or below (acidic/neutral solution) the pK_a.
<2> Data on pH dependence of C-652 spectra obtained at Molecular Probes. For additional data, see Mikrochim Acta 108, 133 (1992).
<3> C-1157 and C-2925 are converted to fluorescent products with spectra similar to C-1904 after acetate hydrolysis.
<4> Values of pK_a for SNAFL and SNARF indicators are as reported in Anal Biochem 194, 330 (1991).
<5> Hydrolysis of the acetate and acetoxymethyl ester groups of this compound yields C-1359 (see Section 1.3), which has pH-dependent spectra similar to C-1904.
<6> C-6826 is converted to a fluorescent product with spectra similar to C-1270 after acetate hydrolysis.
<7> All caged products are sensitive to light. They should be protected from illumination, except when photolysis is intended.
<8> This compound has weaker visible absorption at >300 nm but no discernible absorption peaks in this region.
<9> The pK_a for H-348 was determined in 0.066 M phosphate buffers at 22°C [Fresenius Z Anal Chem 314, 119 (1983)].
<10> The pK_a value for this product is determined from the pH-dependent variation of the absorption spectrum.
<11> The fluorescence excitation and emission spectra of this compound are pH dependent but do not show clear isosbestic points.

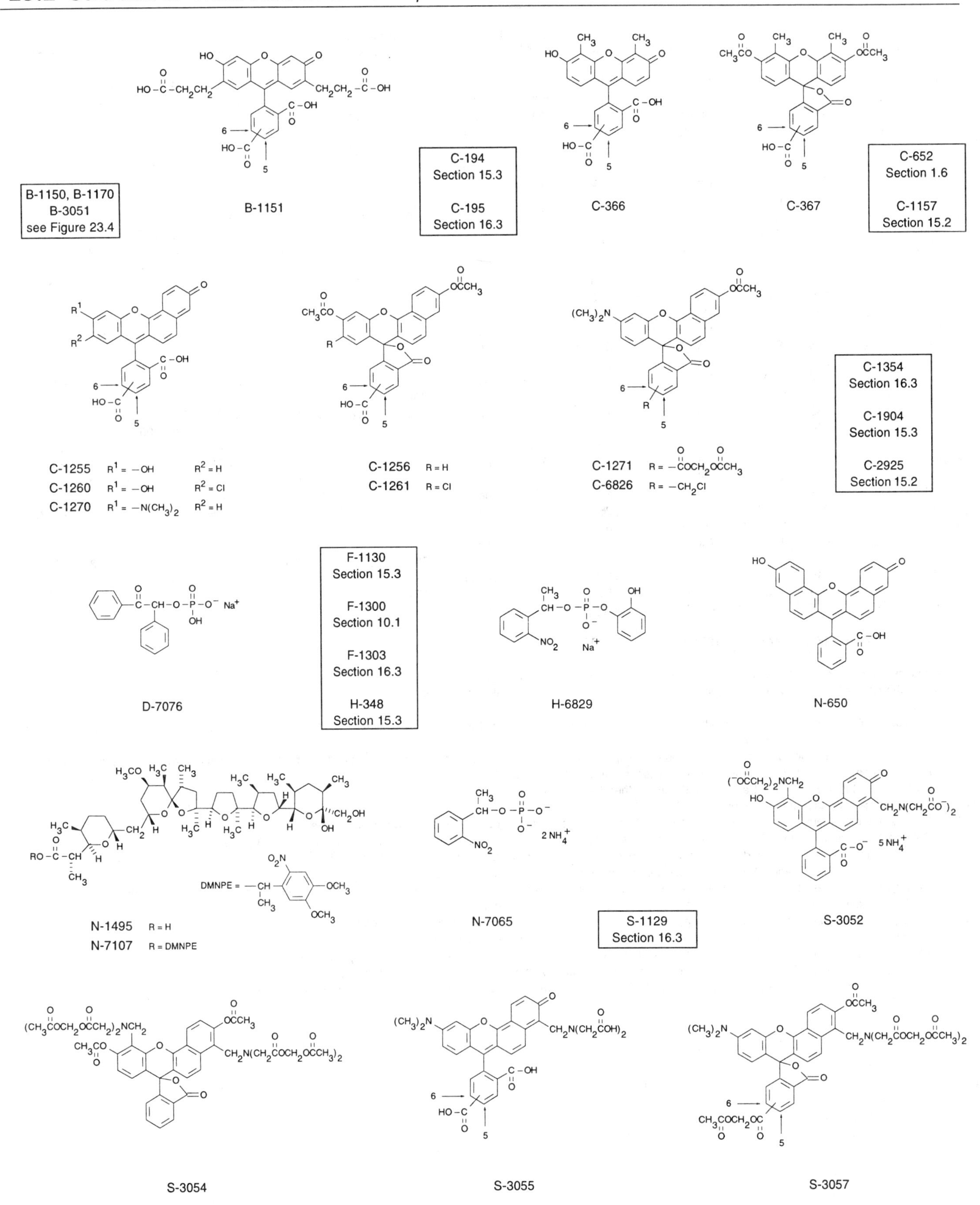

B-1150, B-1170
B-3051
see Figure 23.4
B-1151
C-194
Section 15.3
C-195
Section 16.3
C-366
C-367
C-652
Section 1.6
C-1157
Section 15.2
C-1255 R1 = —OH R2 = H
C-1260 R1 = —OH R2 = Cl
C-1270 R1 = —N(CH3)2 R2 = H
C-1256 R = H
C-1261 R = Cl
C-1271
C-6826 R = —CH2Cl
C-1354
Section 16.3
C-1904
Section 15.3
C-2925
Section 15.2
D-7076
F-1130
Section 15.3
F-1300
Section 10.1
F-1303
Section 16.3
H-348
Section 15.3
H-6829
N-650
N-1495 R = H
N-7107 R = DMNPE
N-7065
S-1129
Section 16.3
S-3052
S-3054
S-3055
S-3057

23.2 Price List *Probes Useful at Near-Neutral pH*

Cat #		Product Name	Unit Size	Price Per Unit ($) 1–4 Units	5–24 Units
B-1151		2′,7′-bis-(2-carboxyethyl)-5-(and-6)-carboxyfluorescein (BCECF acid) *mixed isomers*	1 mg	63.00	50.40
B-1150		2′,7′-bis-(2-carboxyethyl)-5-(and-6)-carboxyfluorescein, acetoxymethyl ester (BCECF, AM)	1 mg	68.00	54.40
B-1170		2′,7′-bis-(2-carboxyethyl)-5-(and-6)-carboxyfluorescein, acetoxymethyl ester (BCECF, AM) *special packaging*	20x50 µg	92.00	73.60
B-3051		2′,7′-bis-(2-carboxyethyl)-5-(and-6)-carboxyfluorescein, acetoxymethyl ester (BCECF, AM) *1 mg/mL in dry DMSO* *special packaging*	1 mL	73.00	58.40
C-366		5-(and-6)-carboxy-4′,5′-dimethylfluorescein *mixed isomers*	100 mg	48.00	38.40
C-367		5-(and-6)-carboxy-4′,5′-dimethylfluorescein diacetate *mixed isomers*	100 mg	98.00	78.40
C-194		5-(and-6)-carboxyfluorescein *mixed isomers*	5 g	73.00	58.40
C-1904		5-(and-6)-carboxyfluorescein (5(6)-FAM) *high purity* *mixed isomers*	100 mg	55.00	44.00
C-195		5-(and-6)-carboxyfluorescein diacetate (5(6)-CFDA) *mixed isomers*	100 mg	75.00	60.00
C-1354		5-carboxyfluorescein diacetate, acetoxymethyl ester (CFDA, AM)	5 mg	68.00	54.40
C-1157		5-(and-6)-carboxyfluorescein diacetate, succinimidyl ester (CFDA, SE) *mixed isomers*	25 mg	115.00	92.00
C-652		5-(and-6)-carboxynaphthofluorescein *mixed isomers*	100 mg	45.00	36.00
C-1255		5-(and-6)-carboxy SNAFL®-1	1 mg	90.00	72.00
C-1256		5-(and-6)-carboxy SNAFL®-1, diacetate	1 mg	118.00	94.40
C-1260		5-(and-6)-carboxy SNAFL®-2	1 mg	90.00	72.00
C-1261		5-(and-6)-carboxy SNAFL®-2, diacetate	1 mg	118.00	94.40
C-1270		5-(and-6)-carboxy SNARF®-1	1 mg	125.00	100.00
C-1271		5-(and-6)-carboxy SNARF®-1, acetoxymethyl ester, acetate	1 mg	145.00	116.00
C-1272		5-(and-6)-carboxy SNARF®-1, acetoxymethyl ester, acetate *special packaging*	20x50 µg	175.00	140.00
C-2925		CellTracker™ Green CMFDA (5-chloromethylfluorescein diacetate)	1 mg	125.00	100.00
C-7025	**New**	CellTracker™ Green CMFDA (5-chloromethylfluorescein diacetate) *special packaging*	20x50 µg	145.00	116.00
C-6826	**New**	5-(and-6)-chloromethyl SNARF®-1, acetate *mixed isomers* *special packaging*	20x50 µg	125.00	100.00
D-7076	**New**	(1,2-diphenyl-2-oxo)ethyl phosphate, monosodium salt, monohydrate (desyl-caged phosphate)	5 mg	95.00	76.00
F-1300		fluorescein *reference standard*	1 g	40.00	32.00
F-1303		fluorescein diacetate (FDA)	1 g	40.00	32.00
F-1130		fluorescein-5-(and-6)-sulfonic acid, trisodium salt	100 mg	78.00	62.40
H-6829	**New**	2-hydroxyphenyl-1-(2-nitrophenyl)ethyl phosphate, sodium salt (NPE-caged proton)	25 mg	95.00	76.00
H-348		8-hydroxypyrene-1,3,6-trisulfonic acid, trisodium salt (HPTS, pyranine)	1 g	72.00	57.60
N-650		naphthofluorescein	100 mg	18.00	14.40
N-7107	**New**	nigericin, 1-(4,5-dimethoxy-2-nitrophenyl)ethyl ester (DMNPE-caged nigericin)	1 mg	98.00	78.40
N-1495		nigericin, free acid	10 mg	70.00	56.00
N-7065	**New**	1-(2-nitrophenyl)ethyl phosphate, diammonium salt (NPE-caged phosphate)	5 mg	65.00	52.00
S-3054		SNAFL® calcein, acetoxymethyl ester *cell permeant* *special packaging*	20x50 µg	68.00	54.40
S-3052		SNAFL® calcein, ammonium salt *cell impermeant*	1 mg	42.00	33.60
S-3055		SNARF® calcein *cell impermeant*	1 mg	42.00	33.60
S-3057		SNARF® calcein, acetoxymethyl ester *cell permeant* *special packaging*	20x50 µg	68.00	54.40
S-1129		5-sulfofluorescein diacetate, sodium salt (SFDA) *single isomer*	25 mg	85.00	68.00

23.3 Probes Useful at Acidic pH

The cellular infrastructure is composed of compartments with different degrees of acidity. For example, biomolecules brought into cells by receptor-mediated endocytosis or phagocytosis are initially processed through organelles of decreasing pH, and specialized organelles such as plant vacuoles and the acrosome of spermatozoa are intrinsically acidic. A low intracompartmental pH activates enzymes and other protein functions, such as iron release from transferrin, that would be too slow at neutral pH, thereby facilitating cellular metabolism.

The fluorescent pH indicators used to detect acidic organelles and to follow trafficking through acidic organelles must have a different fluorescence response — including a lower pK_a — than those described in Section 23.2. Also, unlike most pH indicators for cytosolic measurements, pH indicators for acidic organelles need not be intrinsically permeant to membranes. More often they are covalently attached to large biomolecules that are actively taken up and processed through acidic organelles by the cell's own mechanisms (see Chapter 17 and Section 23.4).

LysoSensor Probes

Our new LysoSensor probes are weak bases that are selectively concentrated in acidic organelles as a result of protonation. This protonation also relieves the fluorescence quenching of the dye by its weak-base side chain. Thus, unlike most other pH indicators in this chapter, the LysoSensor dyes become *more* fluorescent in acidic environments (Figure 23.10). LysoSensor Yellow/Blue (L-7545) undergoes a pH-dependent emission shift to longer wavelengths in acidic environments when illuminated near its excitation

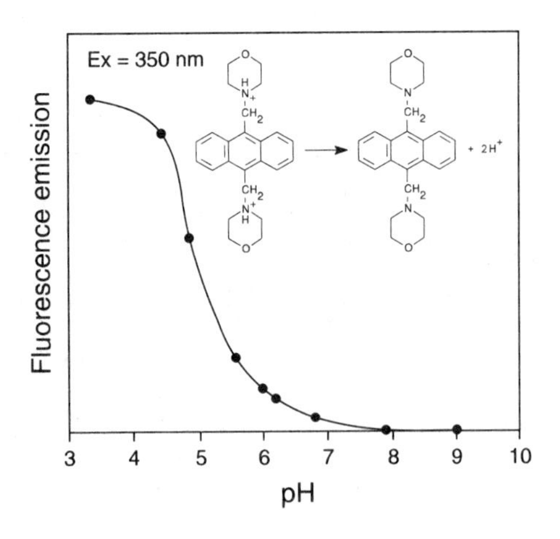

Figure 23.10 *The pH titration curve of LysoSensor Blue DND-167 probe (L-7533), which exhibits a* pK_a *~5.1.*

isosbestic point (~360 nm), as well as a pH-dependent excitation shift when detected near its emission isosbestic point (~490 nm) (see Figure 12.5 in Chapter 12). Although these spectra may change inside cells, ratiometric pH measurements in acidic organelles may be possible, as well as color-differentiated staining of acidic versus neutral organelles.

The blue fluorescent LysoSensor Blue and green fluorescent LysoSensor Green probes are available with optimal pH sensitivity in either the acidic or neutral range (pK_a ~5.2 or ~7.5). With their low pK_a values, LysoSensor Blue DND-167 (L-7533) and LysoSensor Green DND-189 (L-7535) are almost nonfluorescent except when inside acidic compartments, whereas LysoSensor Blue DND-192 (L-7532) and LysoSensor Green DND-153 (L-7534) are brightly fluorescent at neutral pH. Because the LysoSensor probes may localize in the membranes of the acidic organelle rather than in an aqueous medium, it is probable that the pK_a values listed in Table 23.2 will be different in cells and that only qualitative or semiquantitative comparisons of organelle pH may be possible.

LysoSensor probes can be used either singly or potentially in combination with the less pH-sensitive LysoTracker™ dyes

Table 23.2 *Summary of the pH response of our LysoSensor probes.*

Cat #	LysoSensor Probe	Abs/Em * (nm)	pK_a †	Useful pH Range †
L-7532	LysoSensor Blue DND-192	374/424	7.5	6.5–8.0
L-7533	LysoSensor Blue DND-167	373/425	5.1	4.5–6.0
L-7534	LysoSensor Green DND-153	442/505	7.5	6.5–8.0
L-7535	LysoSensor Green DND-189	443/505	5.2	4.5–6.0
L-7545	LysoSensor Yellow/Blue DND-160	384/540 ‡ 329/440 §	4.2	3.0–5.0

* Absorption (Abs) and fluorescence emission (Em) maxima, at pH 5; values may vary somewhat in cellular environments. † All pK_a values were determined *in vitro*; values are likely to be different in cells. ‡ at pH 3; § at pH 7.

described in Section 12.3 to investigate the acidification of lysosomes and alterations of lysosomal function or trafficking that occur in some cells. For example, lysosomes in some tumor cells have a lower pH than normal lysosomes,[1] whereas other tumor cells contain lysosomes with higher pH.[2] In addition, cystic fibrosis and other diseases result in defects in the acidification of some intracellular organelles,[3] and the LysoSensor probes may prove useful in studying these aberrations.

Oregon Green and Dichlorofluorescein Derivatives

Introduction of electron-withdrawing groups into fluorescein dyes lowers the pK_a of the phenolic group to 5 or below. We have used this property to virtually eliminate the fluorescence sensitivity at neutral pH of some of our new amine- and thiol-reactive Oregon Green dyes (see Sections 1.4, 1.5 and 2.2). However, these fluorinated fluorescein dyes are still pH sensitive in moderately acidic solutions, with pK_a values of ~4.7 (Figure 23.11). With the exception of their lower pK_a values, the pH-dependent spectral characteristics of the Oregon Green dyes closely parallel those of other fluorescein-based dyes (Figures 23.2 and 23.3), allowing dual-excitation ratio measurements with the same general configuration used for BCECF (see Section 23.2).

The Oregon Green carboxylic acids (O-6135, O-6138, O-6146, O-6148), polyfluorinated fluoresceins (Cl_2TFF, D-6612; HFF, H-7018; TFF, T-7014) and 5-(and-6)-carboxy-2',7'-dichlorofluores-

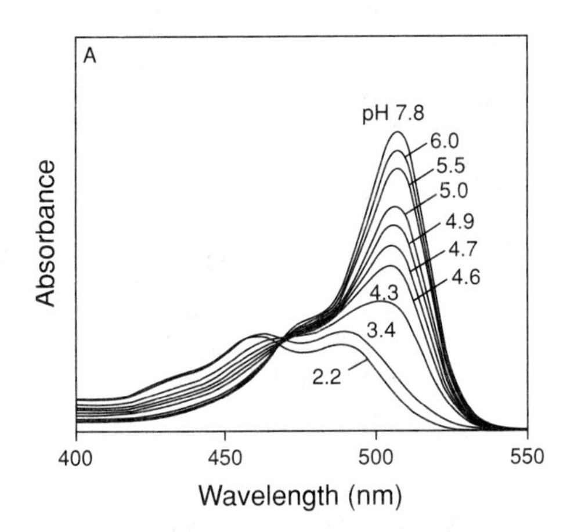

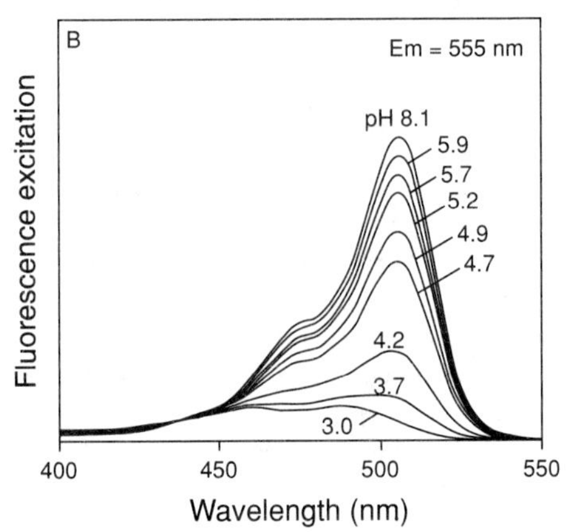

Figure 23.11 *The pH-dependent spectra of Oregon Green 514 carboxylic acid (O-6138): A) absorption spectra and B) excitation spectra.*

cein (C-368) do not readily enter cells but may be useful as fluid phase markers for endocytosis. These fluorinated and chlorinated fluorescein derivatives, as well as the rhodol derivatives described below, will probably be most useful in the form of conjugates that are endocytosed and processed through acidic organelles (see Section 23.4); dextran conjugates of Oregon Green 488 and Oregon Green 514 dyes are described in Section 23.4. Cell-permeant versions of many of these fluorophores are also available (carboxy-DCFDA, C-369; HFFDA, H-7019; carboxy-DFFDA, O-6151; TFFDA, T-7016; see Section 16.3). Carboxy-DCFDA has been used to measure pH in acidic organelles,[4] as well as in the cytosol and vacuoles of plants and yeast.[5,6]

Rhodol Derivatives

Rhodol Green™ carboxylic acid, DM-NERF and Cl-NERF[7,8] (R-6152, D-6830, C-6831) can be considered structural hybrids of fluorescein and rhodamine dyes (see Sections 1.4 and 1.5). These rhodol derivatives retain the pH-sensitive properties conferred by a phenolic hydroxyl group, but their amine substituents reduce the pK_a values to 5.6 for Rhodol Green carboxylic acid, 5.4 for DM-NERF and 3.8 for Cl-NERF (Figure 23.12). Rhodol dyes, which are all more photostable than fluorescein (see Figure 1.6 in Chapter 1), exhibit high extinction coefficients (>70,000 $cm^{-1}M^{-1}$) and high quantum yields (~0.8–0.9) in aqueous solution, with absorption maxima between ~490 nm and >520 nm and emission maxima between ~520 and 545 nm.[8] As with fluorescein-based dyes, the fluorescence excitation profiles of rhodol-based dyes are pH dependent (Figures 23.12 and 23.15), allowing ratiometric pH determinations. In addition, excitation of Rhodol Green carboxylic acid at 450 nm reveals pH-dependent changes in its emission profile (Figure 23.13). Similar pH-dependent changes in the emission spectra are observed for other fluorescein and rhodol dyes but the stronger absorptivity and higher fluorescence quantum yields of the basic forms obscure the emission shifts when the dyes are excited in the 490–520 nm range[9] (Figures 23.2B and 23.3B).

Cl-NERF and DM-NERF are very useful dyes for ratiometric pH measurements, especially with instruments that employ the argon-ion laser. In addition to 510/450 nm excitation ratio configurations, these NERF derivatives permit ratiometric pH measurements using the two principal visible lines of the argon-ion laser (514 nm and 488 nm).[8] Cell uptake of the relatively polar rhodol carboxylic acids into acidic organelles may make these derivatives useful as fluid phase markers for endocytosis; however, it is more common to use the Rhodol Green, DM-NERF and Cl-NERF dyes in the form of conjugates (see Section 23.4) to detect the uptake of labeled probes into acidic organelles.

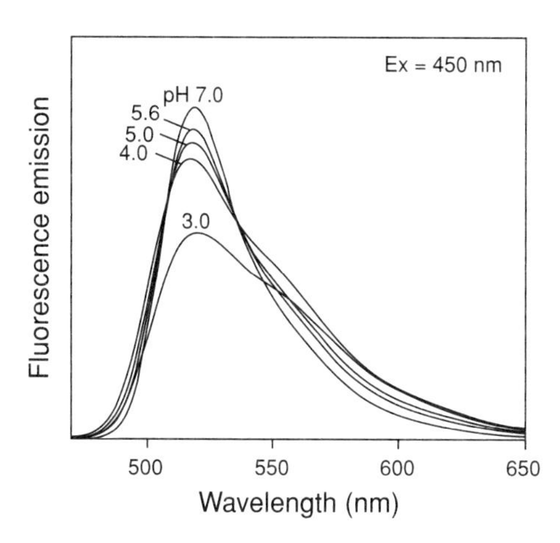

Figure 23.13 The pH-dependent emission spectra of Rhodol Green carboxylic acid (R-6152) excited at 450 nm. The spectrum at pH 3 compared to those between pH 7 and pH 4 indicates the onset of a second protolytic equilibrium.

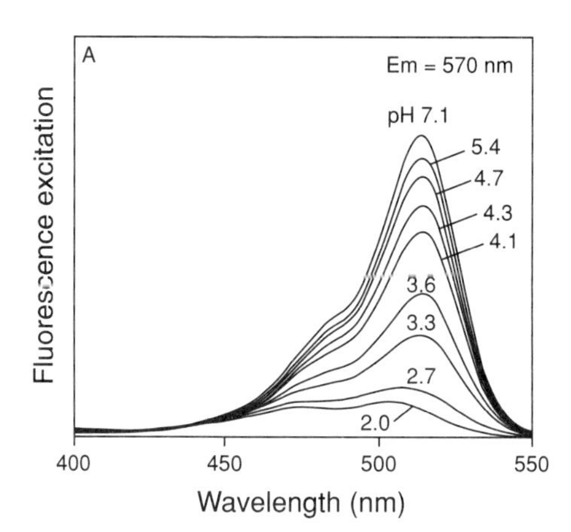

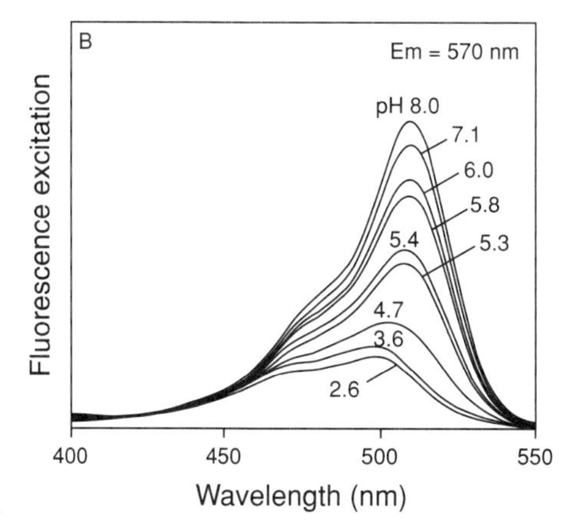

Figure 23.12 The pH-dependent excitation spectra of A) Cl-NERF (C-6831) and B) DM-NERF (D-6830).

Other pH Indicators for Use in Acidic Environments

9-Amino-6-Chloro-2-Methoxyacridine

The nucleic acid stain 9-amino-6-chloro-2-methoxyacridine (ACMA, A-1324) apparently binds to membranes in the energized state and becomes quenched if a pH gradient forms.[10] Mechanistically, this probe resembles the membrane potential–sensitive carbocyanines (see Section 25.3) more than the other probes in this chapter. ACMA has been employed to follow cation and anion movement across membranes of chromatophores,[11] chloroplasts,[12] bacterial photosynthetic membranes[13] and submitochondrial particles,[10] as well as to study the proton-pumping activity of various membrane ATPases.[14,15]

Carboxy SNARF and Carboxy SNAFL Indicators

Although fluorescence of the carboxy SNARF (C-1270; see Section 23.2) and carboxy SNAFL indicators (C-1255, C-1256; see Section 23.2) is not very pH sensitive below pH 6 (see Figures 23.6 and 23.7 in Section 23.2 above), they may still be useful as tracers for uptake into endosomes and lysosomes. At pH <6, carboxy SNARF-1 will be fully in its acidic form, which exhibits yellow-orange fluorescence (Figure 23.6) that is easily distinguished from the red or red-orange fluorescence for the same dye in less acidic organelles. Unlike most pH indicators, the fluorescence quantum yield of the acidic form of the carboxy SNAFL dyes is greater than that of the basic form; thus, these dyes will exhibit more fluorescence and shorter-wavelength spectra in acidic organelles than in basic organelles. Retinal outer segments labeled with amine-reactive forms of either carboxy SNAFL-2 (see Section 23.4) or 2',7'-dichlorodihydrofluorescein (see Section 17.1) have been used to follow phagocytosis by human retinal pigment epithelial cells.[16,17] A chemotactic peptide prepared from the succinimidyl ester of carboxy SNAFL-2 (C-3062, see Section 23.4) has been used to follow pH changes that occur during receptor binding and internalization into acidic organelles.[18]

1. Wilman, D.E.V. and Connors, T.A., *Molecular Aspects of Anticancer Drug Action*, S. Neidle and M.J. Waring, Eds., Macmillian (1983) pp. 233–282; **2.** J Biol Chem 265, 4775 (1990); **3.** Nature 352, 70 (1991); **4.** Anal Biochem 187, 109 (1990); **5.** Plant Physiol 98, 680 (1992); **6.** Meth Enzymol 194, 644 (1991); **7.** U.S. Patent Nos. 5,227,487; 5,442,045; and other patents pending; **8.** Anal Biochem 207, 267 (1992); **9.** Meth Cell Biol 30, 127 (1989); **10.** Biochim Biophys Acta 722, 107 (1983); **11.** Eur Biophys J 19, 189 (1991); **12.** Biochemistry 19, 1922 (1980); **13.** Biochim Biophys Acta 1143, 215 (1993); **14.** J Biol Chem 269, 10221 (1994); **15.** Biochim Biophys Acta 1183, 161 (1993); **16.** Exp Cell Res 214, 242 (1994); **17.** Exp Eye Res 59, 271 (1994); **18.** Cytometry 15, 148 (1994).

23.3 Data Table *Probes Useful at Acidic pH*

					Acidic/Neutral Solution			Basic Solution					
Cat #	Structure	MW	Storage	Soluble	Abs	{ε × 10^{-3}}	Em	Abs	{ε × 10^{-3}}	Em	Solvent	pK_a*	Notes
A-1324	S8.1	259	L	DMF, DMSO	412	{8.2}	471	<1>	<1>	<1>	MeOH	~8.6	<2>
C-368	S1.5	445	L	pH>6, DMF	488	{17}	529	504	{90}	529	<3>	~4.8	
C-6831	✓	452	D,L	pH >6	504	{39}	540	514	{84}	540	<3>	~3.8	
D-6612	✓	473	L	pH>6, DMF	502	{26}	540	520	{93}	540	<3>	~4.4	
D-6830	✓	431	D,L	pH >6	497	{42}	527	510	{76}	536	<3>	~5.4	
H-7018	✓	440	L	pH>6, DMF	490	{24}	529	508	{80}	529	<3>	~5.0	
L-7532	S12.3	292	L	DMSO†	374	{11}	424	374	{11}	424	<3>	~7.5	<4>
L-7533	S12.3	376	L	DMSO†	373	{11}	425	373	{11}	425	<3>	~5.1	<4>
L-7534	S12.3	356	L	DMSO†	442	{17}	505	442	{17}	505	<3>	~7.5	<4>
L-7535	S12.3	398	L	DMSO†	443	{16}	505	443	{16}	505	<3>	~5.2	<4>
L-7545	S12.3	366	L	DMSO†	384	{21}	540	329	{23}	440	<3>	~4.2	<5>
O-6135	S1.5	448	L	pH>6, DMF	481	{29}	517	497	{84}	517	<3>	~4.7	
O-6138	S1.5	512	L	pH>6, DMF	489	{27}	526	506	{88}	526	<3>	~4.7	
O-6146	S1.4	412	L	pH>6, DMF	478	{27}	518	492	{85}	518	<3>	~4.7	
O-6148	S1.4	412	L	pH>6, DMF	478	{28}	517	492	{88}	517	<3>	~4.7	
R-6152	S1.4	412	DD,L	pH>6, DMF	480	{42}	521	494	{73}	521	<3>	~5.6	
T-7014	✓	404	L	pH>6, DMF	490	{29}	528	508	{94}	528	<3>	~6.2	

For definitions of the contents of this data table, see "How to Use the Handbook" on page vi.
Structure: Chemical structure drawing: ✓ = shown in this section; Sn.n = shown in Section number n.n.

* pK_a values may vary considerably depending on the temperature, ionic strength, viscosity, protein binding and other factors. Unless otherwise noted, values listed have been determined from pH-dependent fluorescence measurements at 22°C.

† This product is packaged as a solution in the solvent indicated in "Soluble."

<1> Absorption and fluorescence spectra of the protonated and deprotonated forms of A-1324 are quite similar. This observation and the listed pK_a value are reported in Eur Biophys J 13, 251 (1986). Accumulation of the probe in the presence of transmembrane pH gradients results in fluorescence quenching [Biochim Biophys Acta 722, 107 (1983)].

<2> Spectra of this compound are in methanol acidified with a trace of HCl.

<3> Spectra are in aqueous buffers adjusted to >1 pH unit above (basic solution) or below (acidic/neutral solution) the pK_a.

<4> This LysoSensor dye exhibits increasing fluorescence as pH decreases with no spectral shift. Absorption spectra are essentially pH independent within 2 units of pK_a. L-7532 and L-7533 have additional peaks at Abs = 394 nm and Em = 401 nm.

<5> The pK_a value for this product is determined from the pH-dependent variation of the absorption spectrum.

23.3 Structures *Probes Useful at Acidic pH*

A-1324
Section 8.1

C-368
Section 1.5

C-6831 R = Cl
D-6830 R = $-CH_3$

D-6612 R = Cl
H-7018 R = F
T-7014 R = H

L-7532, L-7533
L-7534, L-7535
L-7545
Section 12.3

O-6135, O-6138
Section 1.5

O-6146, O-6148
R-6152
Section 1.4

23.3 Price List *Probes Useful at Acidic pH*

Cat #		Product Name	Unit Size	Price Per Unit ($) 1–4 Units	5–24 Units
A-1324		9-amino-6-chloro-2-methoxyacridine (ACMA)	100 mg	78.00	62.40
C-368		5-(and-6)-carboxy-2′,7′-dichlorofluorescein *mixed isomers*	100 mg	48.00	38.40
C-6831	***New***	Cl-NERF	5 mg	95.00	76.00
D-6612	***New***	2′,7′-dichloro-4,5,6,7-tetrafluorofluorescein (Cl_2TFF)	5 mg	48.00	38.40
D-6830	***New***	DM-NERF	5 mg	95.00	76.00
H-7018	***New***	2′,4,5,6,7,7′-hexafluorofluorescein (HFF)	10 mg	75.00	60.00
L-7533	***New***	LysoSensor™ Blue DND-167 *1 mM solution in DMSO* *special packaging*	20x50 µL	125.00	100.00
L-7532	***New***	LysoSensor™ Blue DND-192 *1 mM solution in DMSO* *special packaging*	20x50 µL	125.00	100.00
L-7534	***New***	LysoSensor™ Green DND-153 *1 mM solution in DMSO* *special packaging*	20x50 µL	125.00	100.00
L-7535	***New***	LysoSensor™ Green DND-189 *1 mM solution in DMSO* *special packaging*	20x50 µL	125.00	100.00
L-7545	***New***	LysoSensor™ Yellow/Blue DND-160 *1 mM solution in DMSO* *special packaging*	20x50 µL	125.00	100.00
O-6146	***New***	Oregon Green™ 488 carboxylic acid *5-isomer*	5 mg	75.00	60.00
O-6148	***New***	Oregon Green™ 488 carboxylic acid *6-isomer*	5 mg	75.00	60.00
O-6135	***New***	Oregon Green™ 500 carboxylic acid *5-isomer*	5 mg	75.00	60.00
O-6138	***New***	Oregon Green™ 514 carboxylic acid	5 mg	75.00	60.00
R-6152	***New***	Rhodol Green™ carboxylic acid, hydrochloride (5(6)-CRF 492) *mixed isomers*	5 mg	75.00	60.00
T-7014	***New***	4,5,6,7-tetrafluorofluorescein (TFF)	10 mg	35.00	20.00

Technical Assistance at Our Web Site (http://www.probes.com)

At Molecular Probes' Web site, we are developing an electronic version of this Handbook and other databases that should prove extremely useful for the researcher. In addition to containing all of the text from this Handbook, our Web site provides:

- **Product searches** by product name or catalog number
- **Bibliographies** for all products for which we have references
- **Keyword searches** of our entire bibliography of over 25,000 references
- **Product information sheets** for many kits and reagents
- **Technical bulletins**, including *BioProbes* newsletters and other product literature
- **Chemical structure**, **technical data** and **material safety and data sheets**
- **Color photomicrographs** that show our products in action

If you do not have access to the Internet or you need assistance that is not available at that site, further information on the scientific and technical background of our products can be obtained by contacting our Technical Assistance Department at the numbers listed on the inside front cover.

23.4 pH Indicator Conjugates

This section includes our selection of pH indicators conjugated to dextrans and lipids, as well as our chemically reactive pH indicators for preparing new pH-sensitive conjugates.

pH Indicator Dextrans

The pH-sensitive properties of the pH indicators described in Sections 23.2 and 23.3 are usually not significantly affected upon conjugation to dextrans. However, coupling of pH indicators to these relatively inert polysaccharides changes several other properties of the dyes:

- The conjugates have high water solubility and must be loaded into cells by microinjection, patch-clamp techniques, scrape loading, endocytosis, liposome fusion or comparable techniques (see Table 15.1 in Chapter 15).
- Once loaded, the dextrans are well retained in viable cells and will not pass through gap junctions.
- Attachment to a dextran significantly decreases the likelihood that the indicator will become compartmentalized, thereby avoiding a substantial problem associated with cell-permeant acetoxymethyl (AM) ester derivatives.[1]

The properties of some of the most useful pH indicator dextrans available from Molecular Probes are listed in Table 23.3 in approximate order of decreasing pK_a values.

Indicator Dextrans for Measuring Near-Neutral pH

BCECF dextrans are important dual-excitation pH indicator conjugates for pH measurements near pH 7.0.[2] As compared to Swiss 3T3 cells labeled with the AM ester of BCECF, cells labeled with BCECF dextran reportedly showed much more stable fluorescent signals, reduced probe compartmentalization and 10-fold more

Table 23.3 *Molecular Probes' pH indicator dextrans, in order of decreasing pK_a.*

Dye	Cat #	pK_a *	Measurement Wavelengths	Application Notes
SNAFL	D-3301, D-3302	~7.8	Excitation ratio 490/540 nm detected at 580 nm Emission ratio 540/630 nm excited at 514 nm	• Acidic form has the higher quantum yield (see Figure 23.7) • Fluorescence increases in acidic organelles.
SNARF	D-3303, D-3304	~7.5	Emission ratio 580/640 nm excited at 514 or 488 nm	• Best conjugate for ratiometric emission measurements, with spectra similar to carboxy SNARF-1 (see Figure 23.6)
HPTS	D-7179	~7.0	Excitation ratio 470/380 nm detected at 530 nm	• Spectra of dextran conjugate are significantly shifted (~20 nm) relative to free dye
BCECF	D-1878, D-1879, D-1880	~7.0	Excitation ratio 490/440 nm detected at 530 nm	• Best conjugate for ratiometric excitation measurements, with spectra similar to BCECF (see Figure 23.3)
Fluorescein	D-1821, D-1823, D-1844, D-1899, D-3305	~6.4	Excitation ratio 490/450 nm detected at 520 nm	• Fluorescence is strongly quenched upon uptake into acidic organelles (see Figure 23.2B)
Fluorescein and tetramethylrhodamine	D-1950, D-1951	~6.4	Excitation ratio 495/555 nm detected at 580 nm †	• Conjugate incorporating both pH-sensitive and pH-insensitive fluorescent dyes (see Figure 23.14)
Rhodol Green	D-7148, D-7150	~5.6	Excitation ratio 500/450 nm detected at 530 nm	• High photostability • Most useful below pH 6 (see Figure 23.15)
DM-NERF	D-3319, D-3320	~5.4	Excitation ratio 510/450 nm detected at 540 nm	• Useful at a higher pH than Cl-NERF dextrans (see Figure 23.12)
Oregon Green 488	D-7170, D-7172	~4.7	Excitation ratio 490/440 nm detected at 520 nm	• Good photostability • Optimum pH sensitivity range between that of DM-NERF and Cl-NERF
Oregon Green 514	D-7174, D-7176	~4.7	Excitation ratio 510/450 nm detected at 530 nm	• Excellent photostability • Optimum pH sensitivity range between that of DM-NERF and Cl-NERF (see Figure 23.11)
Cl-NERF	D-3324, D-3325	~3.8	Excitation ratio 510/450 nm detected at 540 nm	• Useful at a lower pH than DM-NERF dextrans (see Figure 23.12)

* pK_a values are those determined for the free dyes. Actual values for dextran conjugates may differ by up to +/− 0.3 pH units and may vary between production lots. † Ratiometric emission measurements at 520/580 nm (with excitation at 495 nm) are also possible in principle; however, the response may be complicated by fluorescence resonance energy transfer.

resistance to light-induced damage.[2] Our 40,000 MW BCECF dextran (D-1879) was employed to monitor pH changes during fertilization and activation in hydrozoan eggs,[3] and our 10,000 MW BCECF dextran (D-1878) was used with patch-pipette techniques to measure nuclear and cytosolic pH in frog neurons.[4]

SNARF-1 dextran (D-3304) has been microinjected into rhizoid cells of the alga *Pelvetia fastigata* and examined by ratiometric imaging in order to measure pH gradients associated with polar tip growth.[5] SNARF-1 dextran has also been used to follow the alkalinization of the cytosol that occurs during activation of the multidrug transporter.[6]

The new HPTS dextran (D-7179) has its absorption and emission spectra shifted to longer wavelengths by about 20 nm relative to those of free HPTS (Figure 23.9); note, however, that the absorption maximum of the acidic form of the conjugated dye (~375 nm) is actually shorter than that of the free dye (~403 nm), due to a change in the shape of the spectrum. HPTS dextran retains the ratiometric capabilities of the parent fluorophore discussed above (Section 23.2) and exhibits a pK_a of almost exactly 7.0.

Indicator Dextrans for Measuring Acidic pH

Although the fluorescein, BCECF, HPTS, SNARF and SNAFL dextrans are intended for pH measurements between pH ~6 and 8, these dextrans are equally useful for detecting uptake into acidic organelles, such as occurs during endocytosis. In particular, when these indicator dextrans enter moderately acidic compartments (pH <5.5):

- Fluorescence of the fluorescein, BCECF and HPTS dextrans is strongly quenched[7-9] (Figures 23.2 and 23.3).
- In the double-labeled fluorescein–tetramethylrhodamine dextrans (D-1950, D-1951), fluorescein's fluorescence is quenched, while tetramethylrhodamine's fluorescence is unchanged[10] (Figure 23.14).
- SNARF dextrans are completely converted to their yellow orange fluorescent acidic forms, rather than the red fluorescent basic form (Figure 23.6).
- SNAFL dextrans exhibit blue-shifted emission spectra and a strong increase in fluorescence upon acidification (Figure 23.7).

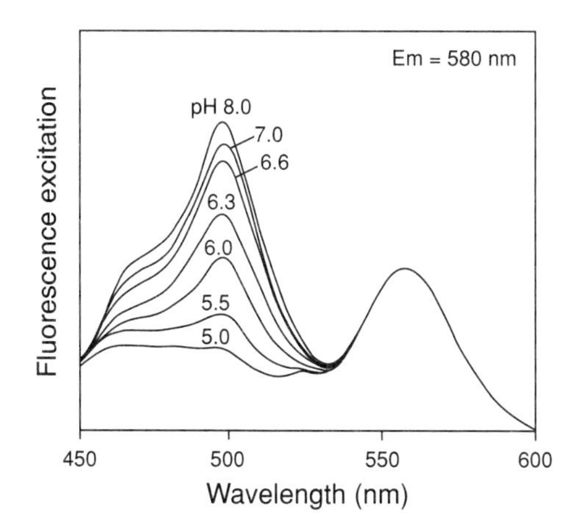

Figure 23.14 *The excitation spectra of double-labeled fluorescein–tetramethylrhodamine dextran (D-1950, D-1951), which contains pH-dependent (fluorescein) and pH-independent (tetramethylrhodamine) dyes.*

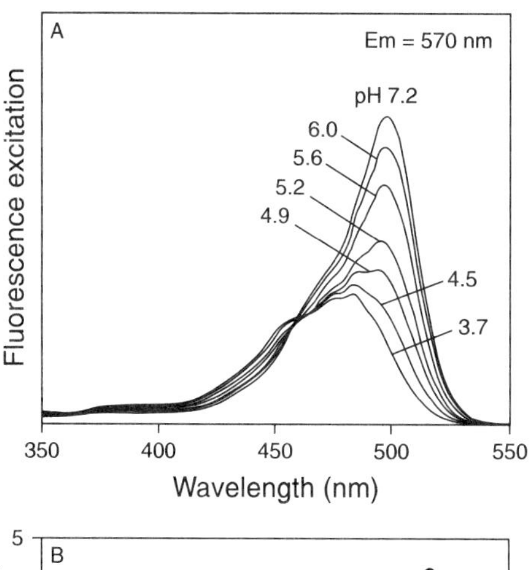

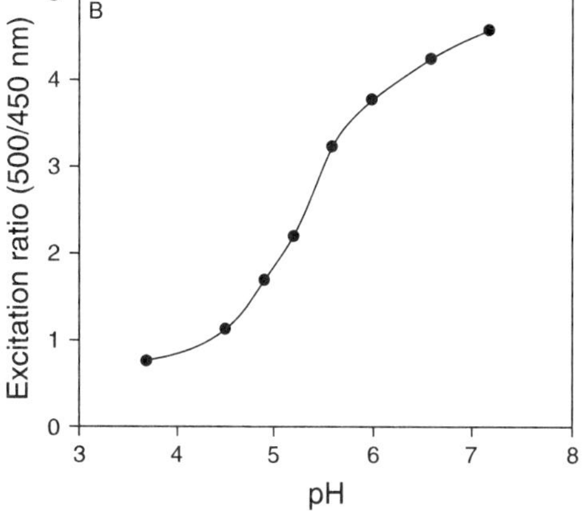

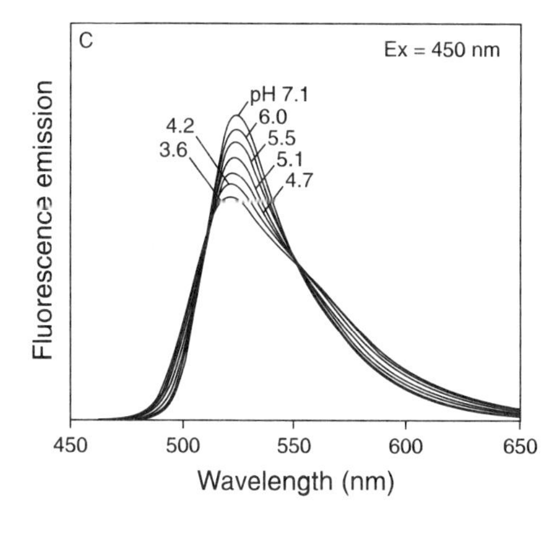

Figure 23.15 *The pH-dependent spectra of Rhodol Green dextran (D-7148, D-7150): A) excitation spectra, B) excitation ratios (500/450 nm), C) emission spectra (excited at 450 nm).*

Although the indicator dextrans discussed above are useful for detecting translocation into acidic compartments, the relative insensitivity of their fluorescence below pH ~6 limits quantitative pH estimation. The lower pK_a values of our Rhodol Green, DM-NERF, Cl-NERF and Oregon Green dextran conjugates (see Table 23.3) make them more suitable indicators for estimating the pH of relatively acidic lysosomal environments. Moreover, the shift in their excitation spectra in acidic medium permits ratiometric pH measurements (Figure 23.15). The membrane-impermeant DM-NERF and Cl-NERF dextrans (D-3319, D-3324) have been shown to be taken up from the extracellular medium by endocytic processes, eventually localizing in lysosomes.[11]

Lipophilic pH Indicators

Measurement of the pH adjacent to membrane surfaces is complicated by electrostatic charge and solvation effects on the pK_a of surface-bound indicators.[12,13] Consequently, lipophilic pH indicators are more useful for detecting changes rather than absolute values of pH at membrane–water interfaces.

7-Hydroxycoumarin Derivatives

The anionic forms of 7-hydroxycoumarins (umbelliferones) have absorption maxima at ~360 nm and bright blue fluorescence emissions at ~450 nm, whereas the neutral phenolic forms are weakly fluorescent with blue-shifted absorptions.[14] The pK_a of 7-hydroxycoumarins in aqueous solution is usually ~7.8.[15] However, the pK_a of 4-heptadecyl-7-hydroxycoumarin (H-181) — a lipophilic 7-hydroxycoumarin derivative — varies from 6.35 in the cationic detergent CTAB to 11.15 in the anionic detergent sodium dodecyl sulfate (SDS), as measured by its fluorescence response.[13] 4-Heptadecyl-7-hydroxycoumarin has been used to measure pH differences at membrane interfaces in isolated plasma membranes of normal and multidrug-resistant murine leukemia cells.[16,17] Our newest lipophilic 7-hydroxycoumarin probe (U-6, D-7760) is derived from a cationic C_{14} detergent; it has been used to probe interfacial potentials and pH changes at a defined distance, nominally 6Å, above the membrane surface.[18]

Fluorescein Conjugates

When inserted in the outer monolayer of egg phosphatidylcholine vesicles, fluorescein DHPE (F-362) has a pK_a of ~6.2, close to that of free fluorescein.[12] Researchers have used the pH-dependent fluorescence of fluorescein DHPE to measure lateral proton conduction along lipid monolayers.[19-22] This fluorescein-labeled phospholipid has also been used to follow proton translocation from internal compartments in phospholipid vesicles.[23] Our other fluorescein lipid analogs — 5-dodecanoylaminofluorescein, 5-hexadecanoylaminofluorescein and 5-octadecanoylaminofluorescein (D-109, H-110, O-322; see Section 14.3) — may have similar applications.

Table 23.4 *Reactive pH indicator dyes.*

pH Indicator	Preferred Reactive Form
BCECF	BCECF (B-1151, see Section 23.2) *
Carboxyfluorescein	5-(and-6)-carboxyfluorescein, succinimidyl ester (C-1311, see Section 1.3)
Cl-NERF	Cl-NERF (C-6831, see Section 23.3) *
Dichlorofluorescein	2′,7′-dichlorofluorescein-5-isothiocyanate (D-6078, see Section 1.5)
Dimethylfluorescein	5-(and-6)-carboxy-4′,5′-dimethylfluorescein (C-366, see Section 23.2) *
DM-NERF	DM-NERF (D-6830, see Section 23.3) *
Naphthofluorescein	5-(and-6)-carboxynaphthofluorescein, succinimidyl ester (C-653)
Oregon Green 488	Oregon Green 488 carboxylic acid, succinimidyl ester (O-6147, O-6149)
Oregon Green 500	Oregon Green 500 carboxylic acid, succinimidyl ester (O-6136)
Oregon Green 514	Oregon Green 514 carboxylic acid, succinimidyl ester (O-6139)
Rhodol Green	Rhodol Green carboxylic acid, succinimidyl ester (R-6108)
SNAFL-1	5-(and-6)-carboxy SNAFL-1, succinimidyl ester (C-3061)
SNAFL-2	5-(and-6)-carboxy SNAFL-2, succinimidyl ester (C-3062)
SNARF-1	5-(and-6)-carboxy SNARF-1 (C-1270, see Section 23.2) *

* Carboxylic acids require activation with EDAC/NHSS before reaction with amines (see Section1.1).

Reactive Dyes for Preparing pH-Sensitive Conjugates

Most of the pH indicators described in Sections 23.2 and 23.3 can be conjugated to biological molecules in order to generate pH-sensitive derivatives. The resulting conjugates can then be used as tracers to follow receptor-mediated internalization, organelle trafficking and other processes. Conjugates of SNARF and SNAFL indicators with peptides, bacteria, transferrin, oligonucleotides and other biomolecules exhibit shorter-wavelength emission spectra (and a strong increase in fluorescence in the case of SNAFL conjugates) upon internalization and acidification by endocytosis or phagocytosis. Retinal outer segments labeled with amine-reactive forms of either carboxy SNAFL-2 (see Section 23.4) or 2',7'-dichlorodihydrofluorescein (see Section 17.1) have been used to follow phagocytosis by human retinal pigment epithelial cells.[24,25] In this study, the fluorescence increase at 550–610 nm exhibited by the carboxy SNAFL-2 conjugate in the acidic phagosomes allowed internalized conjugate to be distinguished from bound conjugate. A chemotactic peptide prepared from the succinimidyl ester of carboxy SNAFL-2 (C-3062) has been used to follow pH changes that occur upon receptor binding and internalization into acidic organelles.[26] SNARF and SNAFL conjugates are also useful pH indicators for environments with near-neutral pH such as cytosol.

Most of the reagents or methods required to prepare conjugates have been described in Chapter 1. Most common is the modification of amines by succinimidyl esters or isothiocyanates of the pH indicator. When the amine-reactive pH indicator is not available, sulfosuccinimidyl esters can generally be prepared *in situ* simply by dissolving the carboxylic acid dye in a buffer that contains *N*-hydroxysulfosuccinimide and 1-ethyl-3-(3-dimethylaminopropyl)carbodiimide (NHSS, H-2249; EDAC, E-2247; see Section 3.3). Addition of NHSS to the buffer has been shown to enhance the yield of carbodiimide-mediated conjugations.[27] Suitable amine-reactive pH indicators or dyes that can be made reactive using EDAC/NHSS are listed in Table 23.4.

1. J Exp Biol 196, 419 (1994); **2.** J Cell Physiol 141, 410 (1989); **3.** Devel Biol 156, 176 (1993); **4.** J Neurosci 14, 5741 (1994); **5.** Science 263, 1419 (1994); **6.** J Histochem Cytochem 38, 685 (1990); **7.** Proc Natl Acad Sci USA 91, 4811 (1994); **8.** J Biol Chem 268, 25320 (1993); **9.** J Cell Sci 105, 861 (1993); **10.** J Cell Biol 821, 130 (1995); **11.** FASEB J 8, 573 (1994); **12.** Biochim Biophys Acta 939, 289 (1988); **13.** J Phys Chem 81, 1755 (1977); **14.** Meth Cell Biol 30, 127 (1989); **15.** Bull Chem Soc Jpn 56, 6 (1983); **16.** Biochemistry 29, 7275 (1990); **17.** Biochim Biophys Acta 729, 185 (1983); **18.** Biochemistry 32, 10057 (1993); **19.** Biochemistry 29, 59 (1990); **20.** J Am Chem Soc 113, 8818 (1990); **21.** Eur J Biochem 162, 379 (1987); **22.** Nature 322, 756 (1986); **23.** Biochim Biophys Acta 766, 161 (1984); **24.** Exp Cell Res 214, 242 (1994); **25.** Exp Eye Res 59, 271 (1994); **26.** Cytometry 15, 148 (1994); **27.** Anal Biochem 156, 220 (1986).

23.4 Data Table *pH Indicator Conjugates*

					Acidic/Neutral Solution			Basic Solution					
Cat #	Structure	MW	Storage	Soluble	Abs	{ε x 10^{-3}}	Em	Abs	{ε x 10^{-3}}	Em	Solvent	pK_a*	Notes
C-653	S1.6	574	F,D,L	DMF, DMSO	515	{10}	565	602	{42}	672	<1>	~7.6	<2>
C-3061	✓	523	F,D,L	DMSO	510	{26}	545	542	{47}	625	<1>	~7.8	<2>
C-3062	✓	558	F,D,L	DMSO	520	{29}	545	550	{47}	632	<1>	~7.7	<2, 3>
D-1821, D-1823, D-1844, D-1899, D-3305	C	<4>	F,L	H_2O	473	ND	514	490	ND	514	<1>	~6.4	<5, 6>
D-1878, D-1879, D-1880	C	<4>	F,L	H_2O	482	ND	520	503	ND	528	<1>	~7.0	<5, 6>
D-1950, D-1951	C	<4>	F,L	H_2O	<7>	ND	<7>	<7>	ND	<7>			<6>
D-3301, D-3302	C	<4>	F,L	H_2O	514	ND	546	543	ND	630	<1>	~7.7	<5, 6>
D-3303, D-3304	C	<4>	F,L	H_2O	548	ND	587	576	ND	635	<1>	~7.5	<5, 6>
D-3319, D-3320	C	<4>	F,L	H_2O	497	ND	527	510	ND	536	<1>	~5.4	<5, 6>
D-3324, D-3325	C	<4>	F,L	H_2O	504	ND	540	514	ND	540	<1>	~3.8	<5, 6>
D-7148, D-7150	C	<4>	F,L	H_2O	480	ND	521	494	ND	521	<1>	~5.6	<5, 6>
D-7170, D-7172	C	<4>	F,L	H_2O	478	ND	518	492	ND	518	<1>	~4.7	<5, 6>
D-7174, D-7176	C	<4>	F,L	H_2O	489	ND	526	506	ND	526	<1>	~4.7	<5, 6>
D-7179	C	<4>	F,L	H_2O	375	ND	536	469	ND	532	<1>	~7.0	<6>
D-7760	✓	452	L	DMSO, EtOH	337	{14}	417	394	{18}	494	MeOH	~7.1	<8>
F-362	S13.4	1183	FF,D,L	<9>	476	{32}	519	476	{32}	519	MeOH	~6.2	<8>
H-181	✓	401	L	DMSO, EtOH	325	{15}	385	366	{20}	453	MeOH	~8.9	<8>
O-6136	S1.5	647	F,D,L	DMF, DMSO	483	{27}	519	499	{78}	519	<1>	~4.7	<2>
O-6139	S1.5	609	F,D,L	DMF, DMSO	489	{26}	526	506	{85}	526	<1>	~4.7	<2>
O-6147	S1.4	509	F,D,L	DMF, DMSO	480	{24}	521	495	{76}	521	<1>	~4.7	<2>
O-6149	S1.4	509	F,D,L	DMF, DMSO	480	{26}	516	496	{82}	516	<1>	~4.7	<2>
R-6108	S1.4	472	F,D,L	DMF, DMSO	482	{36}	523	496	{63}	523	<1>	~5.6	<2>

For definitions of the contents of this data table, see "How to Use the Handbook" on page vi.
Structure: Chemical structure drawing: ✓ = shown in this section; Sn.n = shown in Section number n.n.; C = not shown due to complexity.

* pK_a values may vary considerably depending on the temperature, ionic strength, viscosity, protein binding and other factors. Unless otherwise noted, values listed have been determined from pH-dependent fluorescence measurements at 22°C.

<1> Spectra are in aqueous buffers adjusted to >1 pH unit above (basic solution) or below (acidic/neutral solution) the pK_a.

<2> Spectral data for this product represents the unreacted succinimidyl ester. The pK_a value and the spectral data for acidic solutions have been estimated based on the spectra of the parent carboxylic acid.

<3> For examples of pH-dependent spectra of C-3062 conjugated to peptides, see Cytometry 15, 148 (1994).

<4> Molecular weights are nominally as specified in the product name but may have a broad distribution.

<5> Abs, Em and pK_a values listed for this dextran conjugate are those obtained for the free dye. Values for the conjugates are typically very similar, with slight variations between different production lots.

<6> ND = not determined.

<7> These conjugates contain both pH-sensitive fluorescein (Abs = 495 nm, Em = 520 nm) and pH-insensitive tetramethylrhodamine (Abs = 555 nm, Em = 575 nm) fluorophores. Resulting pH-dependent fluorescence excitation spectra are shown in Figure 23.14.

<8> The pK_a values of these lipophilic pH indicators may vary considerably with the surface charge of the membrane [Biochemistry 32, 10057 (1993); Biochim Biophys Acta 939, 289 (1988); J Phys Chem 81, 1755 (1977)]. The pK_a values listed are for neutrally charged liposomes or micelles, as reported in these publications. Spectra are in MeOH containing a trace of HCl for acidic spectra or a trace of KOH for basic spectra.

<9> Chloroform is the most generally useful solvent for preparing stock solutions of phospholipids. Solutions of glycerophosphoethanolamines in ethanol up to 1–2 mg/mL should be obtainable, using sonication to aid dispersion if necessary. Information on solubility of natural phospholipids can be found in Marsh, D., *CRC Handbook of Lipid Bilayers*, CRC Press (1990) pp. 71–80.

Product-Specific Bibliographies and Keyword Searches

Our Technical Assistance Department can provide you with product-specific bibliographies, as well as keyword searches of the over 25,000 literature references in our extensive bibliography database. Our bibliography database is also searchable through our Web site (http://www.probes.com).

C-653 Section 1.6

C-3061 R = H
C-3062 R = Cl

D-7760

F-362 Section 13.4

H-181

O-6136, O-6139 Section 1.5

O-6147, O-6149 R-6108 Section 1.4

23.4 Price List *pH Indicator Conjugates*

Cat #		Product Name	Unit Size	Price Per Unit ($) 1–4 Units	5–24 Units
C-653		5-(and-6)-carboxynaphthofluorescein, succinimidyl ester *mixed isomers*	25 mg	48.00	38.40
C-3061		5-(and-6)-carboxy SNAFL®-1, succinimidyl ester *mixed isomers*	1 mg	35.00	28.00
C-3062		5-(and-6)-carboxy SNAFL®-2, succinimidyl ester *mixed isomers*	1 mg	35.00	28.00
D-1878		dextran, BCECF, 10,000 MW, anionic	10 mg	80.00	64.00
D-1879		dextran, BCECF, 40,000 MW, anionic	10 mg	80.00	64.00
D-1880		dextran, BCECF, 70,000 MW, anionic	10 mg	80.00	64.00
D-3324		dextran, Cl-NERF, 10,000 MW, anionic	5 mg	80.00	64.00
D-3325		dextran, Cl-NERF, 70,000 MW, anionic	5 mg	80.00	64.00
D-3319		dextran, DM-NERF, 10,000 MW, anionic	5 mg	80.00	64.00
D-3320		dextran, DM-NERF, 70,000 MW, anionic	5 mg	80.00	64.00
D-3305		dextran, fluorescein, 3000 MW, anionic	10 mg	110.00	88.00
D-1821		dextran, fluorescein, 10,000 MW, anionic	25 mg	80.00	64.00
D-1844		dextran, fluorescein, 40,000 MW, anionic	25 mg	80.00	64.00
D-1823		dextran, fluorescein, 70,000 MW, anionic	25 mg	80.00	64.00
D-1899		dextran, fluorescein, 5-40 million MW, anionic	25 mg	80.00	64.00
D-1950		dextran, fluorescein and tetramethylrhodamine, 10,000 MW, anionic	10 mg	95.00	76.00
D-1951		dextran, fluorescein and tetramethylrhodamine, 70,000 MW, anionic	10 mg	95.00	76.00
D-7179	**New**	dextran, 8-hydroxypyrene-1,3,6-trisulfonic acid, 10,000 MW, anionic (HPTS dextran)	5 mg	80.00	64.00
D-7170	**New**	dextran, Oregon Green™ 488; 10,000 MW, anionic	5 mg	110.00	88.00
D-7172	**New**	dextran, Oregon Green™ 488; 70,000 MW, anionic	5 mg	110.00	88.00
D-7174	**New**	dextran, Oregon Green™ 514; 10,000 MW, anionic	5 mg	110.00	88.00
D-7176	**New**	dextran, Oregon Green™ 514; 70,000 MW, anionic	5 mg	110.00	88.00
D-7148	**New**	dextran, Rhodol Green™, 10,000 MW	10 mg	110.00	88.00
D-7150	**New**	dextran, Rhodol Green™, 70,000 MW	10 mg	110.00	88.00
D-3301		dextran, SNAFL®-2, 10,000 MW, anionic	5 mg	80.00	64.00
D-3302		dextran, SNAFL®-2, 70,000 MW, anionic	5 mg	80.00	64.00
D-3303		dextran, SNARF®-1, 10,000 MW, anionic	5 mg	80.00	64.00
D-3304		dextran, SNARF®-1, 70,000 MW, anionic	5 mg	80.00	64.00
D-7760	**New**	4-(N,N-dimethyl-N-tetradecylammonium)methyl-(7-hydroxycoumarin) chloride (U-6)	10 mg	48.00	38.40
F-362		N-(fluorescein-5-thiocarbamoyl)-1,2-dihexadecanoyl-*sn*-glycero-3-phosphoethanolamine, triethylammonium salt (fluorescein DHPE)	5 mg	75.00	60.00
H-181		4-heptadecyl-7-hydroxycoumarin	100 mg	65.00	52.00
O-6147	**New**	Oregon Green™ 488 carboxylic acid, succinimidyl ester *5-isomer*	5 mg	95.00	76.00
O-6149	**New**	Oregon Green™ 488 carboxylic acid, succinimidyl ester *6-isomer*	5 mg	95.00	76.00
O-6136	**New**	Oregon Green™ 500 carboxylic acid, succinimidyl ester, triethylammonium salt *5-isomer*	5 mg	95.00	76.00
O-6139	**New**	Oregon Green™ 514 carboxylic acid, succinimidyl ester	5 mg	95.00	76.00
R-6108	**New**	Rhodol Green™ carboxylic acid, succinimidyl ester (5(6)-CRF 492, SE) *mixed isomers*	5 mg	145.00	116.00

Chapter 24

Indicators for Na^+, K^+, Cl^- and Other Inorganic Ions

Contents

24.1 Fluorescent Na^+ and K^+ Indicators

SBFI, SBFO and PBFI

Properties of SBFI, SBFO and PBFI

The sodium-sensitive dyes SBFI and SBFO [1] (S-1262, S-6899) and the potassium-sensitive dye PBFI [1,2] (P-1265) are useful for the fluorometric determination of physiological levels of Na^+ and K^+, respectively. Developed by Roger Tsien, these benzofuranyl crown ethers and their cell-permeant acetoxymethyl (AM) esters are sufficiently selective to be used in the presence of physiological concentrations of other monovalent cations.[1] Furthermore, their spectral responses upon ion binding permit excitation ratio measurements (see the Technical Note "Loading and Calibration of Intracellular Ion Indicators" on page 549), and they can be used with the same optical filters and equipment as used for fura-2.[3]

These Na^+ and K^+ indicators consist of fluorophores linked to the nitrogens of a crown ether with a cavity size that confers selectivity for the respective ligand. When an ion binds to SBFI, SBFO or PBFI, the indicator's fluorescence quantum yield increases, its excitation peak narrows and its excitation maximum shifts to shorter wavelengths, causing a significant change in the ratio of fluorescence intensities excited at 340/380 nm (Figures 24.1 and 24.2). This fluorescence is relatively unaffected by changes in pH between 6.5 and 7.5, though it is strongly affected by ionic strength [4] and increases with viscosity.[1]

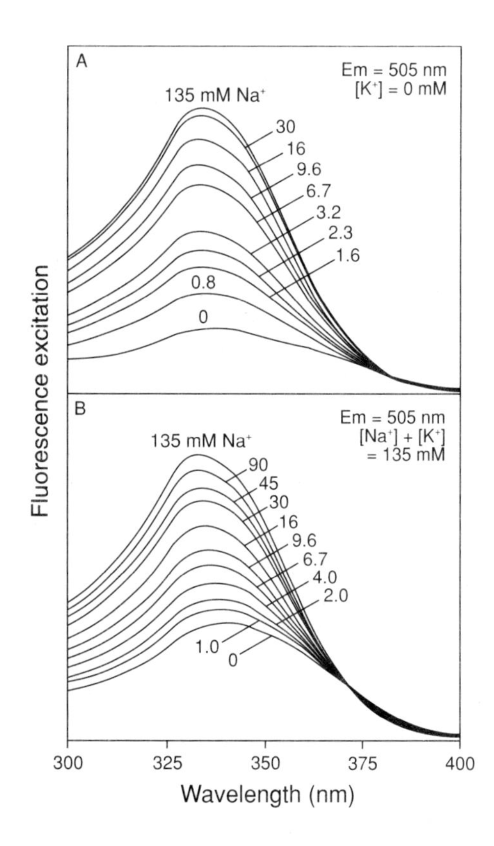

Figure 24.1 SBFI's excitation spectral response to Na^+: A) in K^+-free solution and B) in solutions containing K^+ with the combined Na^+ and K^+ concentration equal to 135 mM. The scale on the vertical axis is the same for both panels.

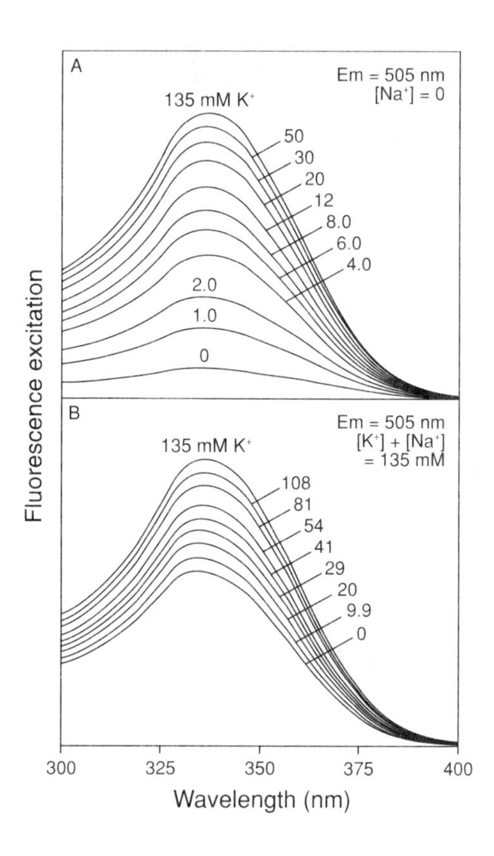

Figure 24.2 PBFI's excitation spectral response to K^+: A) in Na^+-free solution and B) in solutions containing Na^+ with the combined K^+ and Na^+ concentration equal to 135 mM. The scale on the vertical axis is the same for both panels.

Although SBFI is quite selective for Na^+, there is some effect of K^+ on its Na^+ affinity (Figure 24.1). The dissociation constant (K_d) of SBFI for Na^+ is 3.8 mM in the absence of K^+ and 11.3 mM in solutions with combined Na^+ and K^+ concentration of 135 mM, approximating physiological ionic strength. SBFI is ~18-fold more selective for Na^+ than for K^+. As compared to SBFI, SBFO exhibits a lower affinity for Na^+ and a greater fluorescence quantum yield. The K_d of SBFO for Na^+ is 31 mM in the absence of K^+ and 34 mM in solutions with combined Na^+ and K^+ concentrations of 135 mM.

Likewise, the K_d of PBFI for K^+ is strongly dependent on whether Na^+ is present (Figure 24.2), with a value of 5.1 mM in the absence of Na^+ and 44 mM in solutions with combined Na^+ and K^+ concentrations of 135 mM. In buffers in which the Na^+ is replaced by tetramethylammonium chloride, the K_d of PBFI for K^+ is 11 mM; choline chloride and *N*-methylglucamine are two other possible replacements for Na^+ in the medium. PBFI's lower K_d in Na^+-free solutions may permit researchers to continuously monitor the rate of K^+ efflux from cells, supplementing or potentially replacing ^{86}Rb efflux measurements. Furthermore, although PBFI is only 1.5-fold more selective for K^+ than for Na^+, this selectivity is often sufficient because there is normally about 10 times more K^+ than Na^+ in cells.

The K_d of all ion indicators depends on factors such as pH, temperature, ionic strength, levels of other cations and protein concentration. The crown-ether Na^+ and K^+ indicators, including SBFI,

PBFI and Sodium Green, tend to bind to intracellular proteins, which may further alter their dissociation constants and spectral responses. Because the K_d of these indicators may be different in cells than in solution, intracellular SBFI and SBFO should be calibrated using the pore-forming antibiotic gramicidin [3] (G-6888), and intracellular PBFI should be calibrated using the K^+ ionophore valinomycin [5] (V-1644).

Cell Loading with SBFI and PBFI

SBFI and PBFI are available both as cell-impermeant acid salts (S-1262, P-1265) and as cell-permeant AM esters (S-1263, P-1266). At present, SBFO (S-6899) is available only as its cell-impermeant diammonium salt. Because the AM esters are somewhat labile, we also offer these products specially packaged in sets of 20 separate vials (S-1264, P-1267), each containing 50 µg for reconstitution in dimethylsulfoxide (DMSO) as required. Several researchers have found that Pluronic® F-127 facilitates loading of the AM esters of SBFI and PBFI and that relatively long loading times — up to 4 hours — may be required (see the Technical Note "Loading and Calibration of Intracellular Ion Indicators" on page 549). Pluronic F-127 is available as a 20% solution in DMSO (P-3000), as a sterile 10% solution in water (P-6866) and as a solid (P-6867). Somewhat higher working concentrations of PBFI and SBFI than those used for fura-2 may be required because of the lower quantum yield of these indicators. It has been reported that compartmentalization of SBFI is greater when the indicator is loaded at 37°C than when it is loaded at lower temperatures.[6] Other practical aspects of loading and calibrating SBFI have been reviewed by Negulescu and Machen.[3]

Applications of SBFI and SBFO

SBFI has been employed to estimate Na^+ gradients in isolated mitochondria,[7,8] as well as to measure intracellular Na^+ levels or Na^+ efflux in cells from a variety of tissues:

- Blood — platelets,[9,10] fibroblasts and lymphocytes [11] and macrophages [12]
- Brain — astrocytes,[13] hippocampal neurons,[14,15] Purkinje cells [16] and synaptosomes [17]
- Muscle — heart muscle [18,19] and smooth muscle [20,21]

SBFI has also been used in combination with other ion indicators to correlate changes in intracellular Na^+ and Ca^{2+} concentrations [6,18,22,23] or intracellular Na^+ concentrations and pH.[24,25]

A recent paper describes the use of SBFO and video microscopy to measure the Na^+ concentration in intercellular spaces of confluent Madin–Darby canine kidney (MDCK) cells, where this indicator's high K_d for Na^+ is ideal.[26] When 1% bovine serum albumin (BSA) was present, SBFO's isosbestic point was observed to shift from ~376 nm to ~360 nm. Common components of the extracellular fluid, including K^+ (0–20 mM), Ca^{2+} (0–10 mM) and Mg^{2+} (0–5 mM), as well as pH (6–8), did not affect the spectral response of SBFO to 100 mM Na^+.[26]

Applications of PBFI

PBFI is useful for following K^+ transport across membranes [27,28] and in plant vacuoles and protoplasts.[29,30] It has been employed to measure K^+ transport in liposomes reconstituted with a mitochondrial K^+/H^+ antiporter [5] and with a glibenclamide-sensitive, ATP-dependent K^+ channel [31] from beef heart and rat liver. The AM ester of PBFI (P-1266) was loaded into rat glomerular mesangial cells for ratiometric (340 nm/380 nm excitation) K^+ measurements.[32] This cell-permeant derivative has also been used recently to assess intramitochondrial free K^+ in a study of Ca^{2+}-dependent cytotoxicity in rat hepatocytes.[33]

CD 222: A New Probe for Extracellular K^+

Movement of K^+ across the plasma membrane is closely associated with a variety of cellular responses and a number of distinct K^+ channels have now been described. Most excitable cells contain voltage-dependent K^+ channels that modulate such processes as sperm motility,[34] neurotransmitter release from presynaptic nerve terminals [35] and synaptosomes [36] and regulation of membrane potential in resting and stimulated T lymphocytes.[37] The typical K^+ concentration in the extracellular medium is about 5 mM.

CD 222 (coumarin diacid cryptand [2.2.2], C-6870) is a new fluorescent probe developed by Crossley, Goolamali and Sammes for measuring K^+ efflux.[38] In Na^+-free medium, CD 222 has ratioable excitation properties with maximal emission near 465 nm (Figure 24.3). We have determined that its K_d for K^+ in the absence of Na^+ is 0.9 mM. However, Cd 222's K_d for K^+ is strongly affected by Na^+, rising to 58 mM when measured in the presence of 135 mM Na^+. Low levels of K^+ efflux from cells can potentially be detected with CD 222 in Na^+-free solutions with physiological ionic strength by replacing extracellular Na^+ with tetramethylammonium ion. Unlike Na^+, tetramethylammonium ion does not significantly affect CD 222's K_d for K^+ or its spectral response. Thus, this approach may permit the continuous real-time quantitation of stimulated K^+ efflux from cells and has the potential to supplement or replace ^{86}Rb efflux measurements. The fluorescence intensity of CD 222 is also relatively unaffected by 1 mM Ca^{2+}, 1 mM Mg^{2+}, 135 mM choline chloride and 135 mM *N*-methylglucamine hydrochloride and by changes in pH between 6 and 8.

In addition to its ratioable K^+ response, CD 222 undergoes a shift in its excitation spectrum in response to the binding of rubidium ion (Rb^+), with an apparent K_d in the absence of K^+ of about

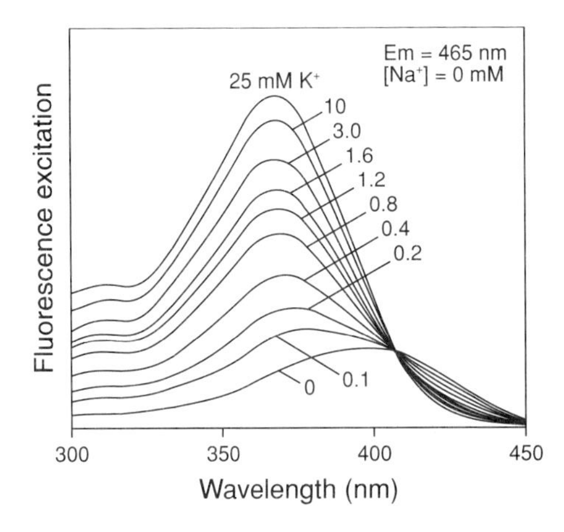

Figure 24.3 *Fluorescence excitation spectra of CD 222 in increasing concentrations of K^+.*

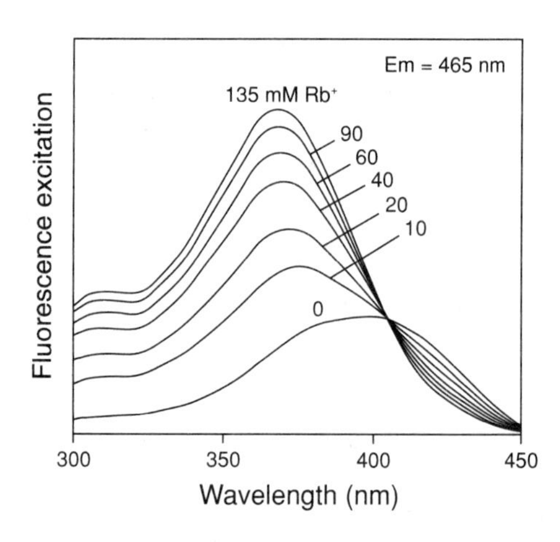

Figure 24.4 *Fluorescence excitation spectra of CD 222 in increasing concentrations of Rb^+.*

10 mM (Figure 24.4). This property is not observed with PBFI. CD 222's spectral shift upon binding Rb^+ may permit ratiometric measurements of Rb^+ concentration under certain circumstances.

Sodium Green and Sodium Green Dextran

Molecular Probes' new Sodium Green indicator is a Na^+-sensitive dye that can be excited with visible light. This Na^+ indicator provides a valuable alternative to the UV-excitable SBFI, especially for instruments that employ the argon-ion laser.[39] We offer the cell-impermeant tetra(tetramethylammonium) salt of Sodium Green (S-6900), as well as its cell-permeant tetraacetate (S-6901). Some researchers have reported that Sodium Green may bind to intracellular proteins, altering its K_d and possibly losing its response to Na^+. Consequently, we have conjugated Sodium Green to 10,000 MW dextran (S-6911) to produce a Na^+ indicator that is expected to exhibit improved cellular retention and decreased protein binding in viable cells.

Sodium Green consists of a fluorescein analog linked to each of the nitrogens of a crown ether with a cavity size that confers selectivity for the Na^+ ion. Upon binding Na^+, Sodium Green exhibits an increase in fluorescence emission intensity with little shift in wavelength (Figure 24.5). As compared to SBFI, Sodium Green shows greater selectivity for Na^+ than K^+ (~41-fold versus ~18-fold) and displays a much higher quantum yield (0.2 versus 0.08) in Na^+-containing solutions. The K_d of Sodium Green for Na^+ is about 6 mM in K^+-free solution and about 21 mM in solutions with combined Na^+ and K^+ concentration of 135 mM, approximating physiological ionic strength. Because its K_d may be different in cells than in solution, intracellular Sodium Green should be calibrated using the pore-forming antibiotic gramicidin [3] (G-6888).

The visible-wavelength spectral characteristics of Sodium Green offer several potential advantages over SBFI. For example, its peak excitation/emission wavelengths of ~507/535 nm:

- Are compatible with instruments that employ the argon-ion laser such as flow cytometers and confocal microscopes
- Are in regions of the spectrum where cellular autofluorescence and scattering backgrounds are less of a problem
- Reduce the potential for photodamage to the cell because the energy of the excitation light is lower than that of UV light
- Do not overlap with wavelengths required for photoactivation of caged probes
- Enable researchers to simultaneously monitor Na^+ concentrations along with other cell parameters — such as Ca^{2+} concentration, pH and membrane potential — that can be detected with shorter- or longer-wavelength dyes

The requirement of Sodium Green for visible-light excitation has proven useful for flow cytometric analysis of intracellular Na^+ concentrations,[40] as well as for studying the effects of amiloride, which emits strongly when excited with UV light, on Na^+ transport in submandibular salivary ducts.[41]

It should be noted that calculations of free Na^+ levels with the single-wavelength indicator Sodium Green are inherently less accurate than those measured with a dual-wavelength ratiometric indicator such as SBFI. Potentially, dual-emission measurements of Na^+ should be possible by simultaneous excitation of Sodium Green and the carboxy SNARF®-1 pH indicator (C-1270, see Section 23.2), using the pH-independent emission isosbestic point of carboxy SNARF-1 at ~610 nm. This strategy was shown to be effective for making ratioable Ca^{2+} measurements with the single-wavelength indicator fluo-3.[42] Co-injecting Sodium Green dextran with a Na^+-insensitive dextran such as Texas Red® dextran may also permit dual-emission ratiometric Na^+ measurements that are analogous to the ratiometric Ca^{2+} measurements obtained by co-injecting Calcium Green™ dextran and Cascade Blue® dextran.[43]

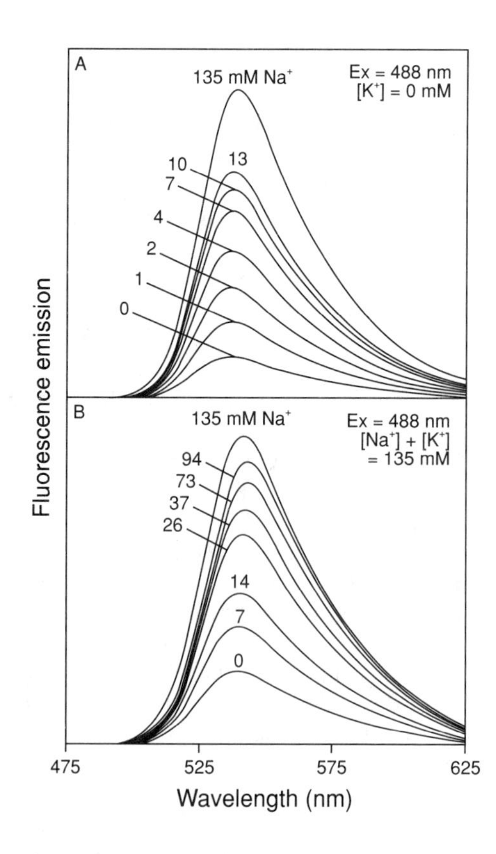

Figure 24.5 *Excitation spectral response of the Sodium Green indicator to Na^+: A) in K^+-free solution and B) in solutions containing K^+ with the combined Na^+ and K^+ concentration equal to 135 mM. The scale on the vertical axis is the same for both panels.*

Alternative Fluorescence Techniques for Measuring Na^+ and K^+

Several lipophilic dyes have been incorporated into membranes or polymers and used in combination with a selective ion carrier to detect Na^+ or K^+ in solution. Energy transfer from the rhodamine glycerophosphoethanolamine (L-1392, see Section 13.4) to a ternary complex of valinomycin (V-1644), K^+ and MEDPIN (a dye that shows a strong spectral shift upon conversion to the K^+ salt) has been used as an alternative fluorescence-based method for determining aqueous K^+ concentrations.[44] Once incorporated within membranes, octadecyl rhodamine B (O-246, see Section 14.3) can be used as an optical sensor for K^+.[45] Similarly, membrane-incorporated 1,8-ANS (A-47, see Section 14.5) has been used as an optical sensor for Na^+.[46] Researchers have also described fiber optic systems for detecting K^+ that use either $DiIC_{18}(5)$[45,47] (D-307, see Section 14.2) or dodecyl acridine orange[48] (A-455, see Section 14.3). A unique optical fiber ion sensor for K^+ that is sensitive over the range of 1 µM to 1 mM employs our red fluorescent FluoSpheres® microspheres (see Section 6.2) in combination with a pH-sensitive dye.[49]

1. J Biol Chem 264, 19449 (1989); **2.** Biophys J 68, 2469 (1995); **3.** Meth Enzymol 192, 38 (1990); **4.** Cell Regulation 1, 259 (1990); **5.** J Biol Chem 265, 10522 (1990); **6.** Science 254, 1014 (1991); **7.** J Biol Chem 270, 672 (1995); **8.** Am J Physiol 262, C1047 (1992); **9.** Biochim Biophys Acta 1220, 248 (1994); **10.** J Biol Chem 265, 19543 (1990); **11.** J Biol Chem 264, 19458 (1989); **12.** J Biol Chem 268, 18640 (1993); **13.** J Neurosci 14, 2464 (1994); **14.** Mol Biol Int'l 32, 475 (1994); **15.** Nature 357, 244 (1992); **16.** J Neurochem 60, 1236 (1993); **17.** J Neurochem 61, 818 (1993); **18.** Biochem Biophys Res Comm 203, 1050 (1994); **19.** Biochem Biophys Acta 1137, 135 (1992); **20.** Am J Physiol 267, H2214 (1994); **21.** Proc Natl Acad Sci USA 90, 8058 (1993); **22.** Am J Physiol 267, H568 (1994); **23.** Am J Physiol 266, G601 (1994); **24.** J Biol Chem 268, 22408 (1993); **25.** Am J Physiol 262, C348 (1992); **26.** J Membrane Biol 144, 11 (1995); **27.** Biochemistry 33, 13769 (1994); **28.** Biochim Biophys Acta 1146, 87 (1993); **29.** J Biol Chem 270, 4368 (1995); **30.** Planta 195, 525 (1995); **31.** J Biol Chem 267, 26062 (1992); **32.** Am J Physiol 262, F462 (1992); **33.** Biochem J 299, 539 (1994); **34.** Mol Reprod Devel 39, 409 (1994); **35.** J Neurosci 12, 297 (1992); **36.** J Physiol 361, 419 (1985); **37.** J Cell Physiol 125, 72 (1985); **38.** J Chem Soc Perkin Trans II 1615 (1994); **39.** U.S. Patent Nos. 5,405,975; 5,516,864; and patent pending; **40.** Cytometry 21, 248 (1995); **41.** J Biol Chem 270, 19606 (1995); **42.** Cytometry 11, 923 (1990); **43.** Nature 359, 736 (1992); **44.** Biophys Chem 33, 295 (1989); **45.** Analyst 113, 693 (1988); **46.** Anal Chim Acta 184, 251 (1986); **47.** Analyst 115, 353 (1990); **48.** Anal Chem 62, 1528 (1990); **49.** Anal Chem 65, 123 (1993).

24.1 Data Table *Fluorescent Na^+ and K^+ Indicators*

Cat#	Structure	MW	Storage	Soluble	Abs	Low Ion* {ε × 10^{-3}}	Em	Abs	High Ion* {ε × 10^{-3}}	Em	Solvent	K_d	Notes
C-6870	✓	697	F,D,L	DMSO	396	{32}	480	363	{21}	467	H_2O/K^+	0.9 mM	<1, 2>
G-6888	C	~1880	D	MeOH	<300		none						
P-1265	✓	951	L	MeOH	336	{33}	557	338	{41}	507	H_2O/K^+	5.1 mM	<1, 3>
P-1266	✓	1171	F,D,L	DMSO	369	{37}	<4>				MeOH		
S-1262	✓	907	L	pH>8	339	{45}	565	333	{52}	539	H_2O/Na^+	3.8 mM	<1, 5>
S-1263	✓	1127	F,D,L	DMSO	379	{32}	<4>				MeOH		
S-6899	✓	735	L	pH >6	354	{50}	515	343	{49}	500	H_2O/Na^+	31 mM	<1, 6>
S-6900	✓	1668	L	pH >6	506	{92}	535	507	{100}	535	H_2O/Na^+	6.0 mM	<1, 5, 7>
S-6901	✓	1543	F,DD,L	DMSO	302	{21}	none				MeOH		
S-6911	C	<8>	F,D,L	H_2O	509	ND	533	509	ND	533	H_2O/Na^+	20 mM	<1, 5, 9, 10>
V-1644	C	1111	F,L	EtOH	<300		none				MeCN		

P-1267 see P-1266; S-1264 see S-1263.

For definitions of the contents of this data table, see "How to Use The Handbook" on page vi.
Structure: Chemical structure drawing: ✓ = shown in this section; C = not shown due to complexity.
MW: ~ indicates an approximate value.

* For "Low Ion" spectra, the concentration of the ion indicated in the column headed "solvent" is zero. For "High Ion" spectra, the concentration of this ion is above that required to saturate the response of the indicator.

<1> Dissociation constant values vary considerably depending on presence of other ions, temperature, pH, ionic strength, viscosity, protein binding and other factors. It is essential that the spectral response of the probe be calibrated in your system.

<2> $K_d(K^+)$ determined at Molecular Probes in 10 mM MOPS, pH 7.0 (adjusted with tetramethylammonium hydroxide) at 22°C. $K_d(K^+)$ is strongly dependent on Na^+ ($K_d(K^+)$ = 58 mM in presence of constant 135 mM Na^+), but is essentially unaffected by tetramethylammonium ions.

<3> $K_d(K^+)$ determined at Molecular Probes in 10 mM MOPS, pH 7.0 (adjusted with tetramethylammonium hydroxide) at 22°C. $K_d(K^+)$ is strongly dependent on Na^+ concentration. In solutions with $[Na^+] + [K^+]$ = 135 mM, $K_d(K^+)$ = 44 mM.

<4> Fluorescence of S-1263 and P-1266 is very weak.

<5> $K_d(Na^+)$ determined at Molecular Probes in 10 mM MOPS, pH 7.0 (adjusted with tetramethylammonium hydroxide) at 22°C. Na^+ dissociation constants for these indicators are dependent on K^+ concentration. In solutions with $[Na^+] + [K^+]$ = 135 mM, $K_d(Na^+)$ = 11.3 mM (S-1262) and 21 mM (S-6900).

<6> $K_d(Na^+)$ for SBFO (S-6899) determined at Molecular Probes in 10 mM MOPS, pH 7.05 at 22°C. In solutions with $[Na^+] + [K^+]$ = 135 mM, $K_d(Na^+)$ = 34 mM.

<7> Sodium Green (S-6900) exhibits fluorescence enhancement in response to ion binding with essentially no change in absorption or emission wavelengths.

<8> Molecular weight is nominally as specified in the product name but may have a broad distribution.

<9> Dissociation constant (K_d) and spectral parameters may vary between different lots of this product.

<10> ND = not determined.

24.1 Structures *Fluorescent Na^+ and K^+ Indicators*

C-6870

P-1265 R = $O^- \; NH_4^+$

P-1266 / P-1267 R = $-OCH_2OCCH_3$ (C=O)

S-1262 R = $O^- \; NH_4^+$

S-1263 / S-1264 R = $-OCH_2OCCH_3$ (C=O)

S-6899 ($2\,NH_4^+$)

S-6900 ($4\,(CH_3)_4N^+$)

S-6901

24.1 Price List *Fluorescent Na^+ and K^+ Indicators*

Cat #		Product Name	Unit Size	Price Per Unit ($) 1–4 Units	5–24 Units
C-6870	***New***	CD 222 potassium indicator	1 mg	95.00	76.00
G-6888	***New***	gramicidin	100 mg	15.00	12.00
P-1266		PBFI, AM *cell permeant*	1 mg	245.00	196.00
P-1267		PBFI, AM *cell permeant* *special packaging*	20x50 µg	275.00	220.00
P-1265		PBFI, tetraammonium salt *cell impermeant*	1 mg	198.00	158.40

Cat #		Product Name	Unit Size	Price Per Unit ($) 1–4 Units	5–24 Units
P-6867	New	Pluronic® F-127 *low UV absorbance*	2 g	8.00	6.40
P-3000		Pluronic® F-127 *20% solution in DMSO*	1 mL	15.00	12.00
P-6866	New	Pluronic® F-127 *sterile 10% solution in water*	30 mL	35.00	28.00
S-1263		SBFI, AM *cell permeant*	1 mg	245.00	196.00
S-1264		SBFI, AM *cell permeant* *special packaging*	20x50 µg	275.00	220.00
S-1262		SBFI, tetraammonium salt *cell impermeant*	1 mg	198.00	158.40
S-6899	New	SBFO, diammonium salt *cell impermeant*	1 mg	125.00	100.00
S-6911	New	Sodium Green™ dextran, 10,000 MW, anionic	5 mg	135.00	108.00
S-6901	New	Sodium Green™ tetraacetate *cell permeant* *special packaging*	20x50 µg	138.00	110.40
S-6900	New	Sodium Green™, tetra(tetramethylammonium) salt *cell impermeant*	1 mg	98.00	78.40
V-1644		valinomycin	25 mg	42.00	33.60

24.2 Fluorescent Cl⁻ Indicators

Measurement of intracellular Cl⁻ concentrations and the study of Cl⁻ channels have been stimulated by the discovery that the genetic effects of cystic fibrosis are manifested by changes in Cl⁻ transport.[1] Probes that act directly on the Cl⁻ channel are described in Section 18.5. However, most current techniques for investigating the kinetics of Cl⁻ transport in cells are seriously limited. Methods based on measurement of ^{36}Cl have poor sensitivity because of the low specific activity of this isotope, and microelectrodes have poor Cl⁻ selectivity and can be used only in large cells. Although ratiometric Cl⁻ indicators are not yet available, recently developed Cl⁻-sensitive dyes are proving to be more sensitive and selective than conventional methods and in some cases can be loaded into cells noninvasively. We offer these fluorescent Cl⁻ indicators:

- 6-methoxy-*N*-(3-sulfopropyl)quinolinium (SPQ, M-440)
- *N*-(3-sulfopropyl)acridinium (SPA, S-460)
- *N*-(ethoxycarbonylmethyl)-6-methoxyquinolinium bromide (MQAE, E-3101)
- *N*-(carboxymethyl)-6-methoxyquinolinium bromide (MQAA, C-3106)
- 6-methoxy-*N*-ethylquinolinium iodide (MEQ, M-6886)
- *N*-(4-aminobutyl)-6-methoxyquinolinium chloride (ABQ, A-6912) and ABQ dextran (D-6887)
- lucigenin (L-6868)

All of the above dyes are collisionally quenched by halide ions,[2,3] resulting in an ion concentration–dependent fluorescence decrease without a shift in wavelength (Figure 24.6). The efficiency of the quenching process is represented by the Stern–Volmer quenching constant (K_{SV}), which is the reciprocal of the ion concentration that produces 50% of maximum quenching. Quenching by other halides, such as Br⁻ and I⁻, and other anions, such as thiocyanate, is more efficient than Cl⁻ quenching. Physiological concentrations of nonchloride ions and physiological pH do not significantly affect the fluorescence of SPQ and analogous Cl⁻ indicators; however, pH sensitivity of the fluorescence of SPQ in certain buffers has recently been reported.[4]

Methods for measuring Cl⁻ transport in vesicles, liposomes and live cells, including a discussion on SPQ and MQAE, have been reviewed by Verkman.[5] We currently recommend MQAE or dihydro-MEQ, which is prepared *in situ* from our reagent MEQ, as the preferred indicators for Cl⁻ in cells unless they can be microinjected with ABQ dextran. Because the Cl⁻-dependent quenching of these indicators may be different in cells than in solution, intracellular indicators should be calibrated using high-K^+ buffers and the K^+/H^+ ionophore nigericin (N-1495) in conjunction with tributyltin chloride, an organometallic compound that acts as a Cl⁻/OH⁻ antiporter.[5-7]

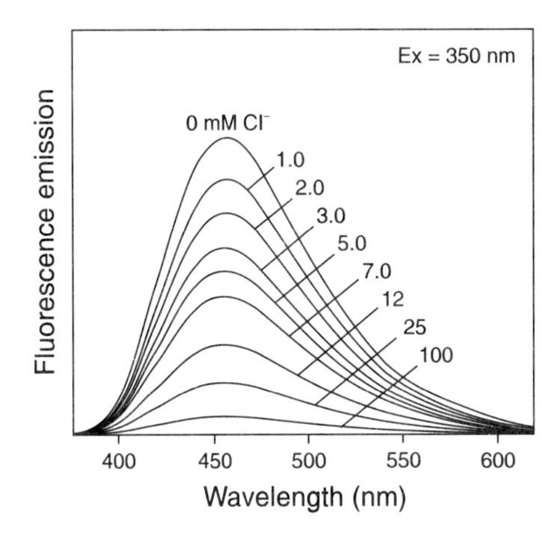

Figure 24.6 Fluorescence emission spectra of MQAE in increasing concentrations of Cl⁻.

Permeability of Cl⁻ has been measured in single cells by exposing cells to a medium containing nitrate (NO_3^-), which is readily permeable through Cl⁻ channels but does not quench the indicator's fluorescence. Intracellular Cl⁻ is rapidly and reversibly lost. Subsequent rapid exchange of the extracellular medium to a Cl⁻-containing buffer permits quantitation of unidirectional Cl⁻ flux into cells. Both SPQ [8,9] and MQAE [10] have been used in this assay; other Cl⁻ indicators listed above should behave similarly.

SPQ

SPQ (M-440), which is usually loaded into cells by brief hypotonic shock, enables researchers to follow Cl⁻ transport with millisecond time resolution in a variety of experimental systems. However, this indicator is poorly retained in cells, with a half-life of only 8–9 minutes at 38°C, and quenching of SPQ fluorescence by Cl^- is less effective inside cells (K_{SV} = 12 M^{-1}) than in aqueous solution (K_{SV} = 118 M^{-1}).[11] SPQ has been employed to investigate Cl^- levels in epithelial cells,[12] fibroblasts [7] and normal and cystic fibrosis lymphocytes,[13] as well as in membrane vesicles from the renal brush border,[14] proximal tubules [15] and plants.[16,17]

MQAE and MQAA

Developed by Verkman and colleagues, the improved Cl^- indicator MQAE (E-3101) has greater sensitivity to Cl^- (K_{SV} = 200 M^{-1}) than SPQ (K_{SV} = 118 M^{-1}) [5,6] and a higher fluorescence quantum yield. In cells, K_{SV} for MQAE is 64 M^{-1}, making it about five times as sensitive to intracellular Cl^- as SPQ.[18] In a study of Cl^- transport in liposomes and LLC-PK1 cells, MQAE was found to undergo slow leakage from cells (<20% per hour at 37°C).[6] Its fluorescence was unaffected by pH or by bicarbonate, borate, nitrate, or sulfate anions,[6] though some quenching by $H_2PO_4^-$ and carboxylates has been reported.[19] The ester group of MQAE may slowly hydrolyze inside cells, resulting in a change in its fluorescence response.[19] The zwitterionic MQAA (C-3106) is not as cell-permeant as MQAE. Recent papers describe the use of MQAE in reconstituted membranes containing the $GABA_A$ receptor,[20] synaptoneurosomes [18] and salivary glands.[10]

MEQ and Cell-Permeant DiH-MEQ

The Cl^- indicator 6-methoxy-*N*-ethylquinolinium iodide (MEQ) can be rendered cell-permeant by masking its positively charged nitrogen to create a lipophilic, Cl^--insensitive compound, 6-methoxy-*N*-ethyl-1,2-dihydroquinoline (diH-MEQ).[21,22] This reduced quinoline derivative can then be loaded noninvasively into cells, where it is rapidly reoxidized to the cell-impermeant, Cl^--sensitive MEQ. Because diH-MEQ is susceptible to spontaneous oxidation upon storage, we offer the diH-MEQ precursor MEQ (M-6886), along with an simple protocol for reducing MEQ to the cell-permeant derivative with sodium borohydride (not supplied) just prior to cell loading. Using this protocol, researchers may find that MEQ provides an improvement over poorly retained SPQ and MQAE for investigating intracellular Cl^- levels. Quenching of MEQ fluorescence by Cl^- has a K_{SV} of 19 M^{-1} in cells, a value that is slightly higher than that reported for SPQ in fibroblasts.

ABQ and ABQ Dextran

ABQ (A-6912) has spectral properties similar to SPQ (excitation/emission maxima ~355/450 nm) but is much more sensitive to Cl^- (K_{SV} = 237 M^{-1} at pH 7.5).[23] ABQ can be used directly with cells and may be useful for generating new Cl^- indicators. It is the intermediate we use to prepare ABQ dextran (D-6887), a microinjectable Cl^- indicator for applications requiring long-term retention in cells. ABQ dextran is also potentially valuable for studying localized Cl^- concentrations in endocytic vesicles and interstitial compartments. The Cl^- sensitivity of ABQ dextran *in vitro* (K_{SV} = 145 M^{-1} at pH 7.5) is slightly greater than that of SPQ.[23] Dextran conjugates (10,000 MW) of ABQ and the Cl^--insensitive Cl-NERF dye (D-3324, see Section 23.4) have been used simultaneously to measure changes in the Cl^- concentration of intercellular spaces of MDCK monolayers grown on glass coverslips.[24] Excitation of both dyes at 325 nm produces a ratioable Cl^--sensitive emission that can be monitored at ~450 nm and ~540 nm. However, the fluorescence of ABQ is strongly affected by proteins in the medium, whereas that of Cl-NERF is not.[24]

Lucigenin

The fluorescence of lucigenin (L-6868) is quantitatively quenched by Cl^- with a reported K_{SV} = 390 M^{-1}.[25] Lucigenin absorbs maximally at both 368 nm (ϵ = 36,000 $cm^{-1}M^{-1}$) and 455 nm (ϵ = 7400 $cm^{-1}M^{-1}$), with an emission maximum at 505 nm. Its fluorescence emission has a quantum yield of ~0.6 and is insensitive to nitrate, phosphate and sulfate. Lucigenin is a useful Cl^- indicator in liposomes and reconstituted membrane vesicles; however, because its fluorescence is reported to be unstable in cytoplasm, it is not likely to be suitable for determining intracellular Cl^-.[25] Lucigenin is also a potentially valuable tool for measuring Cl^- efflux from cells in which the extracellular medium is suddenly made Cl^--free by superfusion. Lucigenin from Molecular Probes has been highly purified to remove a bright blue fluorescent contaminant that is found in some commercial samples and can interfere with these measurements.

1. Nature 358, 536 (1992); **2.** Fresenius Z Anal Chem 314, 577 (1983); **3.** J Heterocyclic Chem 19, 841 (1982); **4.** Am J Physiol 264, C27 (1993); **5.** Am J Physiol 259, C375 (1990); **6.** Anal Biochem 178, 355 (1989); **7.** Biophys J 56, 1071 (1989); **8.** Biochim Biophys Acta 1152, 83 (1993); **9.** Proc Natl Acad Sci USA 88, 7500 (1991); **10.** Pflügers Arch 427, 24 (1994); **11.** Biophys J 53, 955 (1988); **12.** Science 251, 679 (1991); **13.** Science 248, 1416 (1990); **14.** Am J Physiol 254, F114 (1988); **15.** Biophys J 55, 1251 (1989); **16.** J Membrane Biol 116, 129 (1990); **17.** Planta 181, 406 (1990); **18.** Biochim Biophys Acta 1153, 262 (1993); **19.** Am J Physiol 267, H2114 (1994); **20.** Biochemistry 33, 755 (1994); **21.** Br J Pharmacol 108, 469 (1993); **22.** Biochemistry 30, 7879 (1991); **23.** Am J Physiol 262, C243 (1992); **24.** J Membrane Biol 144, 21 (1995); **25.** Anal Biochem 219, 139 (1994).

24.2 Data Table *Fluorescent Cl^- Indicators*

Cat#	Structure	MW	Storage	Soluble	Abs	{ε x 10^{-3}}	Em	Solvent	K_{SV}*	Notes
A-6912	✓	303	D,L	H_2O	344	{3.5}	445	H_2O	237 M^{-1}	<1, 2, 3>
C-3106	✓	298	D,L	H_2O	346	{2.9}	445	H_2O	113 M^{-1}	<1, 2, 3>
D-6887	C	<5>	F,D,L	H_2O	345	ND	447	H_2O	145 M^{-1}	<1, 2, 4, 6>
E-3101	✓	326	F,D,L	H_2O	350	{2.8}	460	H_2O	200 M^{-1}	<1, 2, 3>
L-6868	✓	511	L	H_2O	455	{7.4}	505	H_2O	390 M^{-1}	<2, 3, 7, 8>

Cat#	Structure	MW	Storage	Soluble	Abs	{ε × 10^{-3}}	Em	Solvent	K_{SV}*	Notes
M-440	✓	281	L	H_2O	344	{3.7}	443	H_2O	118 M^{-1}	<1, 2, 3>
M-6886	✓	315	L	H_2O	344	{3.9}	442	H_2O	145 M^{-1}	<1, 2, 3, 9>
N-1495	S23.2	725	F,D	MeOH	<300		none			
S-460	✓	301	L	H_2O	417	{4.2}	489	H_2O	5 M^{-1}	<2, 3, 7>

For definitions of the contents of this data table, see "How to Use The Handbook" on page vi.
Structure: Chemical structure drawing: ✓ = shown in this section; Sn.n = shown in section number n.n; C = not shown due to complexity.

* K_{SV} is the Stern–Volmer quenching constant for Cl^- ions, representing the reciprocal of the ion concentration that produces 50% fluorescence quenching. The quenching constant is very dependent on viscosity and is usually significantly lower in cells. These indicators are quenched more effectively by bromide, iodide and certain other anions.

<1> These quinolinium dyes also have slightly stronger (~50%) absorption peaks 25–30 nm shorter than the listed Abs wavelength.

<2> This product undergoes Cl^--dependent fluorescence quenching with essentially no change in absorption or emission wavelengths.

<3> Values of K_{SV} are taken from literature sources: Anal Biochem 219, 139 (1994); Am J Physiol 262, C243 (1992); Biochemistry 30, 7879 (1991); Anal Biochem 178, 355 (1989).

<4> Spectral parameters may vary between different lots of this product.

<5> Molecular weight is nominally as specified in the product name but may have a broad distribution.

<6> ND = not determined.

<7> S-460 and L-6868 have much stronger absorption at shorter wavelengths: Abs = 358 nm (ε = {21}) for S-460; Abs = 368 nm (ε = {36}) for L-6868.

<8> L-6868 emits chemiluminescence upon oxidation in basic aqueous solutions (Em = 470 nm).

<9> M-6886 may be chemically reduced to cell-permeant diH-MEQ [Biochemistry 30, 7879 (1990)].

24.2 Structures *Fluorescent Cl⁻ Indicators*

A-6912 See M-440

C-3106 R = H
E-3101 R = $-CH_2CH_3$

L-6868

M-440 R = $-(CH_2)_3SO_3^-$
M-6886 R = $-CH_2CH_3$ I^-
A-6912 R = $-(CH_2)_4NH_2$ Cl^- ·HCl

N-1495 Section 23.2

S-460

24.2 Price List *Fluorescent Cl⁻ Indicators*

Cat #		Product Name	Unit Size	Price Per Unit ($) 1–4 Units	5–24 Units
A-6912	**New**	N-(4-aminobutyl)-6-methoxyquinolinium chloride, hydrochloride (ABQ)	25 mg	65.00	52.00
C-3106		N-(carboxymethyl)-6-methoxyquinolinium bromide (MQAA)	100 mg	48.00	38.40
D-6887	**New**	dextran, ABQ, chloride-indicating, 10,000 MW, cationic	5 mg	85.00	68.00
E-3101		N-(ethoxycarbonylmethyl)-6-methoxyquinolinium bromide (MQAE)	100 mg	78.00	62.40
L-6868	**New**	lucigenin (bis-N-methylacridinium nitrate) *high purity*	10 mg	15.00	12.00
M-6886	**New**	6-methoxy-N-ethylquinolinium iodide (MEQ)	100 mg	45.00	36.00
M-440		6-methoxy-N-(3-sulfopropyl)quinolinium, inner salt (SPQ)	100 mg	63.00	50.40
N-1495		nigericin, free acid	10 mg	70.00	56.00
S-460		N-(3-sulfopropyl)acridinium, inner salt (SPA)	100 mg	80.00	64.00

Product-Specific Bibliographies and Keyword Searches

Our Technical Assistance Department can provide you with product-specific bibliographies, as well as keyword searches of the over 25,000 literature references in our extensive bibliography database. Our bibliography database is also searchable through our Web site (http://www.probes.com).

24.3 Probes for Other Inorganic Ions

Sections 24.1 and 24.2 focused on cell-impermeant and cell-permeant fluorescent indicators for Na^+, K^+ and Cl^-. This section describes probes and analytical methods for direct or indirect quantitation of other inorganic anions — including bromide, iodide, hypochlorite, cyanide, sulfide, sulfite, nitrite, nitrate and phosphate — as well as the cation cesium.[1] Molecular Probes anticipates that these dyes will both facilitate and stimulate research on the role of these ions in biological systems and in the environment.

Chloride, Bromide and Iodide

As mentioned above, fluorescence of the Cl^- indicators, including SPQ and the acridinium salt SPA (M-440, S-460; see Section 24.2), is quenched by collision with a variety of anions, including (in order of increasing quenching efficiency) Cl^-, Br^-, I^- and thiocyanate [2] (SCN^-). For example, SPQ is quenched by the anionic TES (*N*-tris(hydroxymethyl)methyl-2-aminoethanesulfonic acid) but not by the protonated TES zwitterion, a property that has been exploited to measure proton efflux from proteoliposomes.[3] Anion detectability using diffusional fluorescence quenching of these fluorophores is typically limited to the millimolar range. I^- quenches many other fluorophores and is commonly used to determine the accessibility of fluorophores to quenching in proteins and membranes.[4,5]

Nanogram quantities of halides, sulfite, nitrite and other readily oxidizable compounds can be detected by their effect on the chemiluminescence of aromatic hydrocarbons, such as rubrene (R-7800) and 9,10-diphenylanthracene (D-7801), which is induced by TCPO [6] (B-1174). Halides can also be oxidized to hypohalites (ClO^-, BrO^-, IO^-), which react with rhodamine 6G (R-634) to produce chemiluminescent products.[7,8] The reaction of ClO^- with fluorescein conjugates is the basis of a sensitive chemiluminescent immunoassay.[9] ClO^- also reacts with fluorescein (F-1300) to produce fluorescent products,[10] permitting analysis of ClO^- levels in water.

Cyanide

The homologous aromatic dialdehydes, *o*-phthaldialdehyde [11] (OPA, P-2331), naphthalene-2,3-dicarboxaldehyde [12] (NDA, N-1138) and anthracene-2,3-dicarboxaldehyde (ADA, A-1139), are essentially nonfluorescent until reacted with a primary amine in the presence of excess cyanide or a thiol, such as 2-mercaptoethanol, 3-mercaptopropionic acid or the less obnoxious sulfite,[13] to yield a fluorescent isoindole (see Figure 1.13 in Chapter 1). Modified protocols that use an excess of an amine and limiting amounts of other nucleophiles permit the determination of cyanide in blood, urine and other samples.[14-17]

Molecular Probes also offers ATTO-TAG™ CBQCA (A-6222), which is similar to OPA and NDA in that it reacts with primary amines in the presence of cyanide or thiols to form highly fluorescent isoindoles [18-26] (see Figure 1.13D in Chapter 1). ATTO-TAG CBQCA should also be useful for detecting cyanide in a variety of biological samples.

We have recently found that our Thiol and Sulfide Quantitation Kit (T-6060) also provides an ultrasensitive enzymatic assay for cyanide with a detection limit of ~5 nanomoles. Interference would be expected from thiols, sulfides, sulfites and other reducing agents. See below for a complete description of this kit's contents.

Sulfide, Sulfite, Sulfate and Thiosulfate

Thiol and Sulfide Quantitation Kit

Colorimetric and fluorometric detection and quantitation of thiols, sulfites and related environmental pollutants are now possible using a variety of reagents available from Molecular Probes. In particular, our Thiol and Sulfide Quantitation Kit (T-6060) provides an ultrasensitive colorimetric assay for thiols (also called mercaptans or sulfhydryls) and inorganic sulfides (or hydrogen sulfide). In this assay, which is based on a method reported by Singh,[27,28] thiols or sulfides reduce a disulfide-inhibited derivative of papain, stoichiometrically releasing the active enzyme. Activity of the enzyme is then measured using the chromogenic papain substrate L-BAPNA. Although thiols and sulfides can also be quantitated using 5,5'-dithiobis-(2-nitrobenzoic acid) (DTNB or Ellman's reagent, D-8451), the enzymatic amplification step in our quantitation kit enables researchers to detect as little as 0.2 nanomoles of thiols or sulfides — a sensitivity that is about 100-fold better than that achieved with DTNB. Furthermore, incorporation of the disulfide cystamine into the solution permits quantitation of accessible thiols on proteins and potentially other high molecular weight molecules. The Thiol and Sulfide Quantitation Kit contains:

- Papain-$SSCH_3$, the disulfide-inhibited papain derivative
- L-BAPNA, a chromogenic papain substrate
- DTNB (Ellman's reagent), for calibrating the assay
- Cystamine
- L-Cysteine, a thiol standard
- Buffer
- Protocols for measuring thiols, inorganic sulfides and maleimides

Sufficient reagents are provided for approximately 50 assays using 1 mL assay volumes and standard cuvettes or 250 assays using a microplate format. This kit can also be used to detect phosphines, sulfites and cyanides with detection limits of about 0.5, 1 and 5 nanomoles, respectively. We expect that this kit can be adapted to directly detect other disulfide-reducing reagents and to indirectly detect certain heavy metals that are quantitatively precipitated by low levels of sulfide such as Hg^{2+}, Pb^{2+} and Cd^{2+}.

Other Reagents

A number of reagents that react with thiols have been described in Chapter 2. 5,5'-Dithiobis-(2-nitrobenzoic acid) (DTNB or Ellman's reagent, D-8451) is, by far, the most common reagent for colorimetric thiol quantitation (see Section 2.4). Among the reagents for fluorometric analysis, the nonfluorescent reagent monobromobimane (M-1378) is widely used to detect inorganic sulfide, sulfite and thiosulfate, including levels dissolved in plasma, by conversion to fluorescent products followed by chromatographic separation.[29] Sulfite, and indirectly sulfur dioxide, can be quantitated by making it the limiting reagent in combination with excess amine and *o*-phthaldialdehyde (P-2331) or naphthalene-2,3-dicarboxaldehyde (N-1138).[13,30,31] Good fluorescence-based assays for sulfate have not been reported. However, we propose that a combination of an indicator whose fluorescence is altered by binding Ba^{2+} such as Calcium Green™-1, Calcium Green-2 or BTC (see Table 22.3 in Chapter 22) can be used to indirectly detect sulfate, which will quantitatively precipitate Ba^{2+} from the solution.

Nitrite and Nitrate

Until recently, the primary application for nitrite (NO_2^-) detection was in the testing of processed meats. However, with the discovery of the role of nitric oxide in signal transduction (see Section 21.5), assays for NO_2^- have assumed new importance. Molecular Probes has several reagents for NO_2^- quantitation, including our Griess Reagent Kit (G-7921), which provides the reagents required to spectrophotometrically detect NO_2^- with the Griess diazotization reaction.[32-35] We also offer 2,3-diaminonaphthalene (D-7918), which reacts with NO_2^- to form the fluorescent product 1*H*-naphthotriazole. A rapid, quantitative fluorometric assay that employs 2,3-diaminonaphthalene can reportedly detect from 10 nM to 10 µM NO_2^- and is compatible with a 96-well microplate format.[36] Nitrate (NO_3^-) does not interfere with this assay; however, NO_3^- can be reduced to NO_2^- by hydrazine and then detected using the same reagent.[37]

Phosphate and Pyrophosphate

Our EnzChek™ Phosphate Assay Kit (E-6646), which is based on a method originally described by Webb,[38,39] provides a facile enzymatic assay for detecting inorganic phosphate from a wide variety of sources through formation of a chromophoric product (see Figure 10.5 in Chapter 10). Although this kit is usually used to determine the inorganic phosphate produced by enzymes such as ATPases, kinases and phosphatases, it can also be used to specifically quantitate inorganic phosphate with a sensitivity of ~2 µM (~0.2 µg/mL). Moreover, this colorimetric assay has proven useful for determining inorganic phosphate levels in the presence of high concentrations of acid-labile phosphates using a microplate reader.[40] Each EnzChek Phosphate Assay Kit contains:

- 2-Amino-6-mercapto-7-methylpurine ribonucleoside (MESG)
- Purine nucleoside phosphorylase (PNP)
- Concentrated reaction buffer
- KH_2PO_4 standard
- Protocol for detecting and quantitating phosphate

In the presence of inorganic phosphate, MESG is enzymatically converted by PNP to ribose-1-phosphate and the chromophoric product 2-amino-6-mercapto-7-methylpurine. This kit contains sufficient reagents for about 100 phosphate assays using 1 mL assay volumes and standard cuvettes.

Furthermore, we have adapted the method provided in our EnzChek Phosphate Assay Kit to permit the sensitive spectrophotometric detection of pyrophosphate, which is split by the enzyme pyrophosphatase to inorganic phosphate.[41] The EnzChek Pyrophosphate Assay Kit (E-6645) enables researchers to detect as little as ~1 µM pyrophosphate (~0.2 µg/mL) and can potentially be adapted to continuously detect enzymes that liberate pyrophosphate such as DNA and RNA polymerases, adenylate cyclase and acyl-CoA synthetase. Each EnzChek Pyrophosphate Assay Kit contains:

- Inorganic pyrophosphatase
- 2-Amino-6-mercapto-7-methylpurine ribonucleoside (MESG)
- Purine nucleoside phosphorylase (PNP)
- Concentrated reaction buffer
- $Na_2P_2O_7$ standard
- Protocol for detecting and quantitating pyrophosphate

The kit contains sufficient reagents for about 100 pyrophosphate assays using standard 1 mL assay volumes and standard cuvettes.

Selenium

Selenium, hydrogen selenide and selenium-containing compounds such as selenomethionine are readily assayed by their specific formation of a fluorescent heterocycle (excitation/emission maxima ~375–380 nm/520–580 nm) with 2,3-diaminonaphthalene (D-7918). Sensitivity is reported to be in the low nanogram range.[42-46] This assay has been adapted to permit detection of selenium-containing compounds on TLC sheets.[47]

Cesium as a Probe for Cation Flux

Transport of cesium ion (Cs^+) in a neuronal cell line has been used as a measure of cation flux by following its quenching of anthracene-1,5-dicarboxylic acid [48,49] (ADC, A-683). Our new AM ester of ADC (A-6871) permits passive loading of cells with this dye.

1. Analyst 107, 465 (1982); **2.** Fresenius Z Anal Chem 314, 577 (1983); **3.** J Biol Chem 269, 7435 (1994); **4.** Chem Phys Lipids 60, 127 (1991); **5.** Anal Biochem 114, 199 (1981); **6.** Biomed Chromatography 4, 92 (1990); **7.** Analyst 114, 1275 (1989); **8.** Anal Lett 21, 1887 (1988); **9.** U.S. Patent No. 4,491,634; **10.** Chem Pharm Bull 32, 3702 (1984); **11.** Proc Natl Acad Sci USA 72, 619 (1975); **12.** Anal Chem 59, 1096 (1987); **13.** J Chromatography A 668, 323 (1994); **14.** J Chromatography 582, 131 (1992); **15.** Anal Chim Acta 225, 351 (1989); **16.** Biomed Chromatography 3, 209 (1989); **17.** Anal Sci 2, 491 (1986); **18.** Anal Chem 66, 3512 (1994); **19.** Anal Chem 66, 3477 (1994); **20.** Electrophoresis 14, 373 (1993); **21.** Anal Chem 64, 973 (1992); **22.** Anal Chem 63, 413 (1991); **23.** Anal Chem 63, 408 (1991); **24.** J Chromatography 559, 223 (1991); **25.** Proc Natl Acad Sci USA 88, 2302 (1991); **26.** J Chromatography 499, 579 (1990); **27.** Bioconjugate Chem 5, 348 (1994); **28.** Anal Biochem 213, 49 (1993); **29.** Chem Pharm Bull 40, 3000 (1992); **30.** J Chromatography 564, 258 (1991); **31.** Anal Chem 61, 408 (1989); **32.** J Pharmacol Exp Ther 252, 922 (1990); **33.** Biochem Biophys Res Comm 161, 420 (1989); **34.** J Exp Med 169, 1543 (1989); **35.** Proc Natl Acad Sci USA 84, 9265 (1987); **36.** Anal Biochem 214, 11 (1993); **37.** Anal Lett 4, 761 (1971); **38.** Anal Biochem 218, 449 (1994); **39.** Proc Natl Acad Sci 89, 4884 (1992); **40.** Anal Biochem 230, 173 (1995); **41.** Upson, R.H., Haugland, R.P., Malekzadeh, N. and Haugland, R.P., Anal Biochem, in press; **42.** Anal Sci 6, 475 (1990); **43.** Talanta 29, 1117 (1982); **44.** Analyst 104, 778 (1979); **45.** Anal Chim Acta 89, 29 (1977); **46.** Analyst 87, 558 (1962); **47.** J Chromatography 537, 397 (1991); **48.** Anal Biochem 157, 353 (1986); **49.** Arch Biochem Biophys 225, 500 (1983).

24.3 Data Table *Probes for Other Inorganic Ions*

Cat #	Structure	MW	Storage	Soluble	Abs	{ε x 10⁻³}	Em	Solvent	Notes
A-683	✓	266	L	pH>6	368	{7.3}	435	pH 8	
A-1139	S1.8	234	L	EtOH	546	ND	570	MeOH	<1, 2>
A-6222	S1.8	305	F,D,L	MeOH	465	ND	560	MeOH	<2, 3, 4>
A-6871	✓	410	F,D,L	DMSO, MeCN	386	{8.2}	448	MeCN	
B-1174	✓	449	F,D	DMF, MeCN	296	{3.8}	none	MeOH	<5>
D-7801	S14.6	330	L	$CHCl_3$, MeCN	391	{13}	405	MeOH	
D-7918	S21.5	158	L	DMSO, MeOH	340	{5.1}	377	MeOH	
D-8451	S5.2	396	D	pH>6	324	{11}	none	pH 8	<6>
F-1300	S10.1	332	L	pH>6, DMF	490	{88}	514	pH 9	
M-1378	S2.4	271	F,L	DMF, MeCN	398	{5.0}	<7>	pH 7	
N-1138	S1.8	184	L	DMF, MeCN	462	ND	520	MeOH	<2, 8>
P-2331	S1.8	134	L	EtOH	334	{5.7}	455	pH 9	<9>
R-634	S12.2	479	F,D,L	EtOH	528	{100}	551	MeOH	
R-7800	✓	533	L	Toluene	526	{6.2}	555	Toluene	<10>

For definitions of the contents of this data table, see "How to Use The Handbook" on page vi.
Structure: Chemical structure drawing: ✓ = shown in this section; Sn.n = shown in section number n.n.

<1> Spectra of A-1139 with butylamine + cyanide. Absorption spectrum has multiple peaks. Unreacted reagent Abs = 361 nm (ε = {5.4}), Em = 410 nm in MeOH.
<2> ND = not determined.
<3> Spectra of A-6222 with glycine + cyanide. Unreacted reagent: Abs = 254 nm (ε = {46}), nonfluorescent in MeOH.
<4> Solubility in methanol is improved by addition of base (e.g., 1–5% (v/v) 0.2 M KOH).
<5> B-1174 is most effective at pH 6–8 [Anal Chim Acta 177, 103 (1985)].
<6> D-8451 reaction product with thiols has Abs = 410 nm (ε = {14}) [Meth Enzymol 233, 380 (1994)].
<7> M-1378 is almost nonfluorescent until reacted with thiols. Em = 475–485 nm for thiol adducts (QY 0.1–0.3) [Meth Enzymol 143, 76 (1987)].
<8> Spectra of N-1138 with glycine + cyanide. Reagent Abs = 279 nm (ε = {5.5}), Em = 330 nm in MeOH.
<9> Spectral data are for the reaction product of P-2331 with alanine and 2-mercaptoethanol. The spectra and the stability of the adduct depend on the amine and thiol reactants [Biochim Biophys Acta 576, 440 (1979)]. Unreacted P-2331 has Abs = 257 nm (ε = {1.0}) in H_2O.
<10> Rubrene has much stronger absorption at shorter wavelengths: ε = {81} at 302 nm.

24.3 Structures *Probes for Other Inorganic Ions*

A-683 R = –OH
A-6871 R = $-OCH_2OCCH_3$ (C=O)

A-1139, A-6222
Section 1.8

B-1174

D-7801
Section 14.6

D-7918
Section 21.5

D-8451
Section 5.2

F-1300
Section 10.1

M-1378
Section 2.4

N-1138, P-2331
Section 1.8

R-634
Section 12.2

R-7800

24.3 Price List *Probes for Other Inorganic Ions*

Cat #		Product Name	Unit Size	Price Per Unit ($) 1–4 Units	5–24 Units
A-1139		anthracene-2,3-dicarboxaldehyde (ADA)	25 mg	40.00	32.00
A-683		anthracene-1,5-dicarboxylic acid (ADC)	100 mg	48.00	38.40
A-6871	*New*	anthracene-1,5-dicarboxylic acid, di(acetoxymethyl ester) (ADC, AM)	5 mg	68.00	54.40
A-6222	*New*	ATTO-TAG™ CBQCA derivatization reagent (CBQCA; 3-(4-carboxybenzoyl)quinoline-2-carboxaldehyde)	10 mg	110.00	88.00
B-1174		bis-(2,4,6-trichlorophenyl) oxalate (TCPO)	1 g	25.00	20.00
D-7918	*New*	2,3-diaminonaphthalene	100 mg	18.00	14.40
D-7801	*New*	9,10-diphenylanthracene	1 g	18.00	14.40
D-8451	*New*	5,5′-dithiobis-(2-nitrobenzoic acid) (DTNB; Ellman's reagent)	10 g	45.00	36.00
E-6646	*New*	EnzChek™ Phosphate Assay Kit *100 assays*	1 kit	135.00	108.00
E-6645	*New*	EnzChek™ Pyrophosphate Assay Kit *100 assays*	1 kit	155.00	124.00
F-1300		fluorescein *reference standard*	1 g	40.00	32.00
G-7921	*New*	Griess Reagent Kit *for nitrite quantitation*	1 kit	55.00	44.00
M-1378		monobromobimane (mBBr)	25 mg	43.00	34.40
N-1138		naphthalene-2,3-dicarboxaldehyde (NDA)	100 mg	65.00	52.00
P-2331		o-phthaldialdehyde (OPA) *high purity*	1 g	32.00	25.60
R-634		rhodamine 6G chloride	1 g	18.00	14.40
R-7800	*New*	rubrene	1 g	48.00	38.40
T-6060	*New*	Thiol and Sulfide Quantitation Kit *50-250 assays*	1 kit	95.00	76.00

Technical Assistance at Our Web Site (http://www.probes.com)

At Molecular Probes' Web site, we are developing an electronic version of this Handbook and other databases that should prove extremely useful for the researcher. In addition to containing all of the text from this Handbook, our Web site provides:

- **Product searches** by product name or catalog number
- **Bibliographies** for all products for which we have references
- **Keyword searches** of our entire bibliography of over 25,000 references
- **Product information sheets** for many kits and reagents
- **Technical bulletins**, including *BioProbes* newsletters and other product literature
- **Chemical structure**, **technical data** and **material safety and data sheets**
- **Color photomicrographs** that show our products in action

Visit our Web site often for new additions to our bibliography, as well as new products and upgraded search capabilities. Also look for special sales and introductory specials on some important products.

If you do not have access to the Internet or you need assistance that is not available at that site, further information on the scientific and technical background of our products can be obtained by contacting our Technical Assistance Department at the numbers listed on the inside front cover.

A Sampling of Sampler Kits

Most of Molecular Probes' products are used in minute quantities, making "sample sizes" impractical. However, we have put together a number of Sampler Kits containing a set of application-specific probes, sometimes in smaller quantities and always at lower cost than the corresponding components sold separately. These Sampler Kits enable you to test our products and find the optimal probe for your particular application. Look in the designated Handbook chapter, contact our Technical Assistance Department or visit our Web site (http://www.probes.com) for more information on any of these kits.

Cat #		Product Name	Unit Size	Price Per Unit ($) 1–4 Units	5–24 Units
Chapter 7 *Protein Conjugates for Biological Detection*					
W-7024	***New***	Wheat Germ Agglutinin Sampler Kit *four fluorescent conjugates, 1 mg each*	1 kit	98.00	78.40
Chapter 8 *Nucleic Acid Detection*					
N-7565	***New***	Nucleic Acid Stains Dimer Sampler Kit	1 kit	125.00	100.00
N-7566	***New***	Nucleic Acid Stains Monomer Sampler Kit	1 kit	65.00	52.00
S-7580	***New***	SYBR® Green Nucleic Acid Gel Stain Starter Kit	1 kit	55.00	55.00
S-6655	***New***	SYPRO® Protein Gel Stain Starter Kit	1 kit	55.00	55.00
S-7572	***New***	SYTO® Live-Cell Nucleic Acid Stain Sampler Kit #1 *SYTO® dyes 11-16* *50 µL each*	1 kit	135.00	108.00
S-7554	***New***	SYTO® Live-Cell Nucleic Acid Stain Sampler Kit #2 *SYTO® dyes 20-25* *50 µL each*	1 kit	135.00	108.00
Chapter 10 *Enzymes, Enzyme Substrates and Enzyme Inhibitors*					
I-6614	***New***	Indolyl β-D-Galactopyranoside Sampler Kit *contains 5 mg each of B-1690, B-8407, I-8414, I-8420, M-8421*	1 kit	65.00	52.00
P-6548	***New***	Protease Inhibitor Sampler Kit	1 kit	195.00	156.00
R-6564	***New***	Resorufin Ether Sampler Kit	1 kit	98.00	78.40
T-6500	***New***	Tetrazolium Salt Sampler Kit *100 mg of T-6490, X-6493, M-6494, N-6495, I-6496, N-6498*	1 kit	125.00	100.00
Chapter 12 *Probes for Organelles*					
Y-7530	***New***	Yeast Mitochondrial Stain Sampler Kit	1 kit	95.00	76.00
Y-7531	***New***	Yeast Vacuole Marker Sampler Kit	1 kit	95.00	76.00
Chapter 15 *Fluorescent Tracers of Cell Morphology and Fluid Flow*					
L-7781	***New***	Lipophilic Tracer Sampler Kit	1 kit	150.00	120.00
Chapter 18 *Probes for Receptors and Ion Channels*					
B-6850	***New***	Brevetoxin Sampler Kit *5 µg of B-6851, B-6852, B-6853, B-6854, B-6855*	1 kit	135.00	108.00
Chapter 21 *Probes For Reactive Oxygen Species,Including Nitric Oxide*					
N-7925	***New***	Nitric Oxide Synthase (NOS) Inhibitor Kit	1 kit	195.00	156.00
Chapter 22 *Indicators for Ca^{2+}, Mg^{2+}, Zn^{2+} and Other Metals*					
B-6767	***New***	BAPTA Acetoxymethyl Ester Sampler Kit *2x1 mg each of B-1205, D-1207, D-1209, D-1213*	1 kit	75.00	60.00
C-6777	***New***	Coelenterazine Sampler Kit *contains 25 µg of C-2944, C-6776, C-6779, C-6780, C-6781*	1 kit	255.00	204.00
B-6768	***New***	BAPTA Buffer Kit *10 mg each of B-1204, D-1206, D-1208, D-1211*	1 kit	75.00	60.00
Chapter 26 *Tools for Fluorescence Applications*					
F-7321	***New***	Flow Cytometry Alignment Standards Sampler Kit, 2.5 µm	1 kit	55.00	44.00
F-7322	***New***	Flow Cytometry Alignment Standards Sampler Kit, 6.0 µm	1 kit	55.00	44.00
T-7284	***New***	TetraSpeck™ Fluorescent Microspheres Sampler Kit	1 kit	95.00	76.00

Chapter 25

Probes for Membrane Potential

Contents

Related Chapters

25.1 Introduction to Potentiometric Probes

Potentiometric optical probes enable researchers to perform membrane potential measurements in organelles and in cells that are too small to allow the use of microelectrodes. Moreover, in conjunction with imaging techniques, these probes can be employed to map variations in membrane potential along neurons [1] and among cell populations with spatial resolution and sampling frequency that are difficult to achieve using microelectrodes.

Common Applications for Potentiometric Probes

The plasma membrane of a cell typically has a transmembrane potential of approximately –70 mV (negative inside) as a consequence of K^+, Na^+ and Cl^- concentration gradients that are maintained by active transport processes. Potentiometric probes offer an indirect method of detecting the translocation of these ions, whereas the fluorescent ion indicators discussed in Chapter 24 can be used to directly measure changes in specific ion concentrations. Increases and decreases in membrane potential — referred to as membrane hyperpolarization and depolarization, respectively — play a central role in many physiological processes, including nerve-impulse propagation, muscle contraction, cell signaling and ion-channel gating. Potentiometric probes are important for studying these processes, as well as for visualizing mitochondria (which exhibit transmembrane potentials of approximately –150 mV, negative inside matrix), for assaying cell viability and for monitoring synaptic activity; see Sections 12.2, 16.3 and 17.2, respectively for more information on these applications. Potentiometric probes include the cationic or zwitterionic styryl dyes, the cationic carbocyanines and rhodamines, the anionic oxonols and hybrid oxonols and merocyanine 540. The class of dye determines factors such as accumulation in cells, response mechanism and toxicity. Surveys of techniques and applications using membrane potential probes can be found in several reviews.[2-8]

Selecting a Potentiometric Probe

Selecting the best potentiometric probe for a particular application can be complicated by the substantial variations in their optical responses, phototoxicity and interactions with other molecules. Probes can be divided into two categories based on their response mechanism:

- **Fast-response probes** (usually styrylpyridinium dyes) operate by means of a change in their electronic structure, and consequently their fluorescence properties, in response to a change in the surrounding electric field. Their optical response is sufficiently fast to detect transient (millisecond) potential changes in excitable cells, including single neurons, cardiac cells and intact brains. However, the magnitude of their potential-dependent fluorescence change is often small; fast-response probes typically show a 2–10% fluorescence change per 100 mV.

- **Slow-response probes** exhibit potential-dependent changes in their transmembrane distribution that are accompanied by a fluorescence change. The magnitude of their optical responses is usually much larger than that of fast-response probes. Slow-response probes, which include cationic carbocyanines and rhodamines and anionic oxonols, are suitable for detecting changes in average membrane potentials of nonexcitable cells caused by respiratory activity, ion channels, drugs and other factors.

Calibration of potentiometric probes can be accomplished by imposing a transmembrane potential using valinomycin (V-1644; see Section 24.1) in conjunction with externally applied K^+ solutions.[9,10] The descriptions in this chapter focus on the response characteristics of the various types of potential-sensitive dyes offered by Molecular Probes.

1. Biophys J 67, 1301 (1994); **2.** J Photochem Photobiol B 33, 101 (1996); **3.** Adv Chem Ser 235, 151 (1994); **4.** Loew, L.M. in *Fluorescent and Luminescent Probes for Biological Activity*, W.T. Mason, Ed., Academic Press (1994) pp. 150–160; **5.** Wu, J.-Y. and Cohen, L.B. in *Fluorescent and Luminescent Probes for Biological Activity*, W.T. Mason, Ed., Academic Press (1994) pp. 389–404; **6.** Biochim Biophys Acta 1016, 1 (1990); **7.** Meth Cell Biol 30, 193 (1989); **8.** Meth Enzymol 172, 102 (1989); **9.** Biochemistry 28, 4536 (1989); **10.** Ann Rev Biophys Bioeng 8, 47 (1979).

25.2 Fast-Response Probes

Fast-response probes are listed in Table 25.1, along with their charge, optical response and selected applications.

Styryl Dyes

Di-4-ANEPPS and Di-8-ANEPPS

The ANEP (aminonaphthylethenylpyridinium) dyes developed by Leslie Loew and colleagues [1] exhibit the most consistently sensitive responses of the fast-response probes. Di-4-ANEPPS (D-1199) has a fairly uniform 10% per 100 mV change in fluorescence intensity in a variety of tissue, cell and model membrane systems.[2] However, di-4-ANEPPS is internalized in the cell rather rapidly, precluding its use in all but very short-term experiments. In contrast, di-8-ANEPPS (D-3167) is better retained in the outer leaflet of the plasma membrane. In addition, although both ANEP dyes exhibit good photostability and low toxicity, di-8-ANEPPS is reported to be slightly more photostable and significantly less phototoxic than di-4-ANEPPS.[3,4]

Like other styryl dyes, the ANEP dyes are essentially nonfluorescent in aqueous solutions and exhibit spectral properties that are strongly dependent on their environment.[5] When bound to phos-

Table 25.1 Characteristics and selected applications of Molecular Probes' fast-response probes.

Dyes (Cat #)	Structure (Charge)	Optical Response	Selected Applications
Di-4-ANEPPS (D-1199) Di-8-ANEPPS (D-3167) Di-2-ANEPQ (D-6923) Di-8-ANEPPQ (D-6925) Di-12-ANEPPQ (D-6927) Di-18:2-ANEPPS (D-6928)	Styryl (cationic or zwitterionic)	FAST; fluorescence excitation ratio 440/505 nm decreases upon membrane hyperpolarization	• Mapping of membrane potentials along neurons [1-4] and muscle fibers [5-8] • Combined potentiometric and pH measurements [9,10] • Energy transduction [11-14]
RH 160 (S-1107) RH 237 (S-1109) RH 414 (T-1111) RH 421 (S-1108) RH 461 (T-1113) RH 704 (T-6693) RH 773 (D-6695) RH 795 (R-649)	Styryl (cationic or zwitterionic)	FAST; generally similar to ANEP dyes with excitation red shift upon membrane hyperpolarization	• Imaging of membrane potentials evoked by visual [15,16] and olfactory [17,18] stimuli • Functional tracing of neurons [19-22] • Detection of synaptic activity [23-25] and ion channel activity [26,27]
RH 155 (R-1114) WW 781 (W-435)	Hybrid oxonol (anionic)	FAST; RH 155 absorbance (at ~720 nm) changes and WW 781 fluorescence decreases upon membrane hyperpolarization	• Invertebrate neurons [28,29] • Membrane potentials in skeletal and cardiac muscle [30-32] • Voltage-sensitive calcium channels [33-34] • Red blood cells [35-36]

1. J Neurosci 15, 1392 (1995); **2.** Biophys J 67, 1301 (1994); **3.** Neuron 13, 1187 (1994); **4.** Neuron 9, 393 (1992); **5.** Biophys J 66, 719 (1994); **6.** Biol Bull 183, 342 (1993); **7.** Nature 355, 349 (1992); **8.** Biophys J 56, 623 (1989); **9.** Biochemistry 33, 7239 (1994); **10.** Biochemistry 32, 11042 (1993); **11.** J Biol Chem 269, 21576 (1994); **12.** Am J Physiol 265, H453 (1993); **13.** Am J Physiol 265, H445 (1993); **14.** Proc Natl Acad Sci USA 89, 5996 (1992); **15.** J Neurosci 14, 2545 (1994); **16.** J Neurosci Meth 42, 195 (1992); **17.** Ann Rev Neurosci 15, 321 (1992); **18.** J Neurosci 12, 976 (1992); **19.** J Neurosci Meth 54, 151 (1994); **20.** Proc Natl Acad Sci USA 91, 4604 (1994); **21.** Brain Res 94, 371 (1993); **22.** J Neurosci 12, 3054 (1992); **23.** J Neurosci 14, 3319 (1994); **24.** Nature 364, 635 (1993); **25.** Biophys J 51, 643 (1987); **26.** J Biol Chem 269, 21620 (1994); **27.** Nature 367, 739 (1994); **28.** J Neurosci 14, 1366 (1994); **29.** Biophys J 56, 213 (1989); **30.** Circulation 85, 1865 (1992); **31.** Circulation Res 69, 842 (1991); **32.** J Gen Physiol 95, 147 (1990); **33.** Proc Natl Acad Sci USA 85, 4075 (1988); **34.** J Neurosci 7, 3877 (1987); **35.** Am J Physiol 265, C521 (1993); **36.** Biophys J 57, 835 (1990).

pholipid vesicles, di-8-ANEPPS has absorption/emission maxima of ~488/605 nm, as compared to ~498/713 nm in methanol. The fluorescence excitation/emission maxima of di-4-ANEPPS bound to neuronal membranes are ~475/617 nm.[6] Both di-4-ANEPPS and di-8-ANEPPS respond to increases in membrane potential (hyperpolarization) with a decrease in fluorescence excited at approximately 440 nm and an increase in fluorescence excited at 530 nm.[5,7] These spectral shifts permit the use of ratiometric methods (see the Technical Note "Loading and Calibration of Intracellular Ion Indicators" on page 549) to correlate the change in fluorescence signal with membrane potential. Using di-8-ANEPPS, Loew and colleagues were able to define differences between transmembrane potentials of neurites and somata [8] and to follow changes in membrane potential along the surface of a single mouse neuroblastoma cell in their study of the mechanisms underlying cathode-directed neurite elongation.[9] Other applications are listed in Table 25.1.

New ANEP Dyes

In collaboration with Leslie Loew and Joe Wuskell of the University of Connecticut, Molecular Probes is pleased to offer a series of four new potential-sensitive ANEP dyes.[10] The very water-soluble di-2-ANEPEQ (D-6923) is specifically designed to be microinjected into cells, a mode of delivery that intensifies the staining of remote neuronal processes. Microinjection of di-2-ANEPEQ (also referred to as JPW 1114) into neurons in ganglia of the snail *Helix aspersa* produced an approximately 50-fold improvement in voltage-sensitive signals from distal processes over that obtained with conventional absorption and fluorescence staining methods.[11] Di-8-ANEPPQ (D-6925) and di-12-ANEPPQ (D-6927) are useful for potential-sensitive retrograde labeling of neurons [10] using techniques similar to those used for lipophilic carbocyanine and aminostyryl tracers (see Section 15.4). The unsaturated analog di-$C_{18:2}$-ANEPPS (D-6928) is structurally similar to *FAST* DiA™ (D-3897, D-7758; Section 15.4) and may be similarly useful for neuronal labeling.

RH Dyes

Originally synthesized by Rina Hildesheim, the RH dyes include an extensive series of dialkylaminophenylpolyenylpyridinium dyes that are principally used for functional imaging of neurons (Table 25.1). The existence of numerous RH dye analogs reflects the observation that no single dye provides the optimal response under all experimental conditions.[3,12,13] Currently, the most widely used RH dyes are RH 414 (T-1111) and RH 795 (R-649). RH 421 (S-1108) has yielded the most sensitive response recorded for a fast potentiometric probe, exhibiting >20% fluorescence change per 100 mV on neuroblastoma cells.[14] Physiological effects of staining with different analogs are not equivalent. For example, staining of the cortex with RH 414 causes arterial constriction, whereas staining with RH 795 does not.[15] RH 795 produced negligible side effects when tested *in vitro* using hippocampal slices and *in vivo* using single-unit recordings in cat and monkey visual cortex.[16]

Like the ANEP dyes, the RH dyes exhibit varying degrees of fluorescence excitation and emission spectral shifts in response to membrane potential changes.[6] Their absorption and fluorescence spectra are also strongly dependent on the environment.[17] Spectra of RH 414 bound to phospholipid vesicles are similar to those obtained on neuronal plasma membranes.[6] The RH dyes belong to the same structural class as our FM™ dyes (see Section 17.2) and can likewise be used to detect activity-dependent synaptosomal recycling at live nerve terminals.[18,19]

Impermeant Oxonols: WW 781, RH 155 and RGA-30

WW 781 (W-435) and RH 155 (R-1114) are oxonol dyes with phenylsulfonate substituents that tend to make them impermeant to cell membranes. Consequently, their potentiometric responses are faster but of smaller magnitude than those of other oxonol dyes [20] (see Section 25.3). Because they are anionic, adsorption of these dyes to the membrane increases upon depolarization. WW 781 exhibits a fluorescence increase accompanied by an absorption red shift as the membrane potential decreases (depolarization).[21] Like stilbene disulfonates (see Section 18.5), WW 781 is a potent inhibitor of the erythrocyte band 3 anion transporter.[22] RH 155 is nonfluorescent; its potentiometric response is detected using absorbance changes at approximately 720 nm.[23,24]

RGA-30 (R-6922) is a thio analog of WW 781 that acts as a phototransducer, eliciting action potentials in stained neurons in response to illumination. Selective laser photostimulation of RGA-30–stained neurons allows rapid identification of their synaptic connections. However, it is reported that RGA-30 is not useful as a potentiometric probe.[25]

1. Biochemistry 24, 5749 (1985); **2.** J Membrane Biol 130, 1 (1992); **3.** Biophys J 67, 1301 (1994); **4.** Pflügers Arch 426, 548 (1994); **5.** Biochemistry 28, 4536 (1989); **6.** Biochim Biophys Acta 1150, 111 (1993); **7.** Biophys J 67, 208 (1994); **8.** Neuron 13, 1187 (1994); **9.** Neuron 9, 393 (1992); **10.** Biophys J 68, A291 (1995); **11.** J Neurosci 15, 1392 (1995); **12.** Physiol Rev 68, 1285 (1988); **13.** Ann Rev Neurosci 8, 263 (1985); **14.** Biophys J 42, 195 (1983); **15.** Nature 324, 361 (1986); **16.** J Neurosci 14, 2545 (1994); **17.** Biophys J 68, 1406 (1995); **18.** J Neurosci 12, 363 (1992); **19.** Science 255, 200 (1992); **20.** Chem Phys Lett 129, 225 (1986); **21.** Biophys J 57, 835 (1990); **22.** Am J Physiol 265, C521 (1993); **23.** Ann Rev Physiol 51, 507 (1989); **24.** J Neurosci 7, 649 (1987); **25.** Science 222, 1025 (1983).

25.2 Data Table *Fast-Response Probes*

Cat #	Structure	MW	Storage	Soluble	Abs	{ε x 10⁻³}	Em	Solvent	Notes
D-1199	✓	481	D,L	DMSO, EtOH	496	{39}	705	MeOH	<1>
D-3167	✓	593	DD,L	DMSO, EtOH	498	{35}	713	MeOH	<1>
D-6923	✓	549	D,L	DMSO, EtOH	517	{36}	721	EtOH	<1>
D-6925	✓	732	D,L	DMSO, EtOH	516	{36}	721	EtOH	<1>
D-6927	✓	844	D,L	DMSO, EtOH	519	{36}	719	EtOH	<1>
D-6928	✓	865	D,L	DMSO, EtOH	501	{36}	705	EtOH	<1>
R-649	✓	544	D,L	DMSO, EtOH	530	{33}	712	MeOH	<1>
R-1114	✓	950	D,L	DMSO, EtOH	650	{120}	none	MeOH	
R-6695	✓	569	D,L	DMSO, EtOH	532	{49}	716	MeOH	<1>
R-6922	✓	775	D,L	DMSO, EtOH	629	{83}	659	MeOH	
S-1107	✓	471	D,L	DMSO, EtOH	518	{50}	705	MeOH	<1, 2>
S-1108	✓	499	D,L	DMSO, EtOH	515	{50}	704	MeOH	<1, 2>
S-1109	✓	497	D,L	DMSO, EtOH	528	{53}	782	MeOH	<1, 2>
T-1111	✓	581	D,L	DMSO, EtOH	532	{55}	716	MeOH	<1>
T-1113	✓	539	D,L	DMSO, EtOH	533	{51}	716	MeOH	<1>
T-6693	✓	596	D,L	DMSO, EtOH	526	{47}	711	MeOH	<1>
W-435	✓	757	D,L	DMSO, EtOH	605	{120}	639	MeOH	<3>

For definitions of the contents of this data table, see "How to Use The Handbook" on page vi.
Structure: Chemical structure drawing: ✓ = shown in this section.

<1> The absorption and emission maxima of styryl dyes are at shorter wavelengths in membrane environments than in reference solvents such as methanol. The difference is typically ~20 nm for absorption and ~80 nm for emission but varies considerably from one dye to another.

<2> Excitation and emission maxima for these dyes adsorbed on neuronal plasma membranes are 498/631 nm (S-1107), 493/638 nm (S-1108) and 506/687 nm (S-1109) [Biochim Biophys Acta 1150, 111 (1993)].

<3> Fluorescence excitation/emission maxima of W-435 are 615/650 nm in cell suspensions [Biophys J 57, 835 (1990)].

25.2 Structures *Fast-Response Probes*

$^{-}O_3S(CH_2)_3\overset{+}{N}$ ... CH = CH ... NR_2

D-1199 R = $-(CH_2)_3CH_3$
D-3167 R = $-(CH_2)_7CH_3$
D-6928 R = $-(CH_2)_8CH{=}CHCH_2CH{=}CH(CH_2)_4CH_3$

$(CH_3)_3\overset{+}{N}(CH_2)_n\overset{+}{N}$ ... CH = CH ... $N[(CH_2)_mCH_3]_2$
2 Br^-

D-6923 n = 2 m = 1
D-6925 n = 3 m = 7
D-6927 n = 3 m = 11

$HOCH_2CH_2\overset{+}{N}(CH_3)_2CH_2CH(OH)CH_2\overset{+}{N}$ ... CH = CH − CH = CH ... $N[(CH_2)_3CH_3]_2$
2 (CH_3COO^-)

R-649

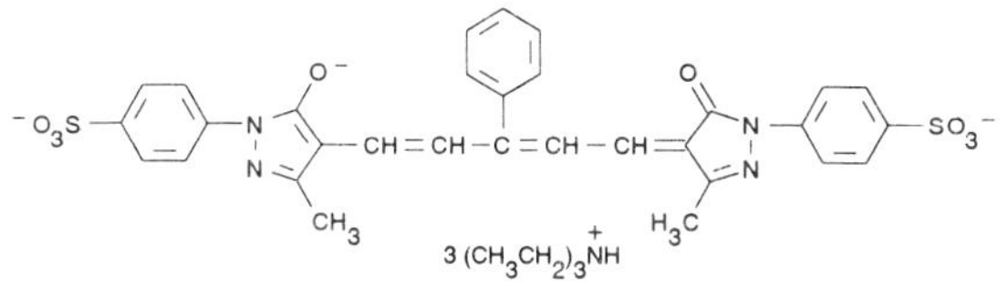

R-1114

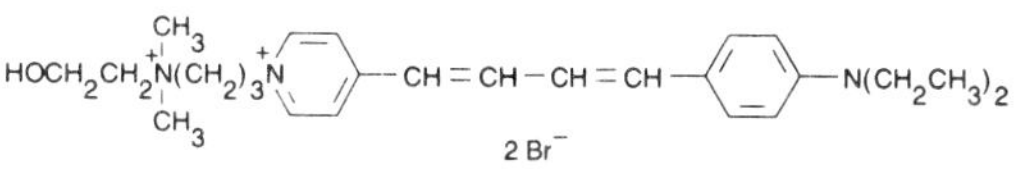

R-6695

R-6922 see W-435

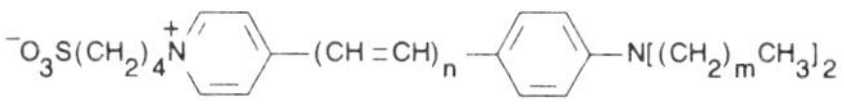

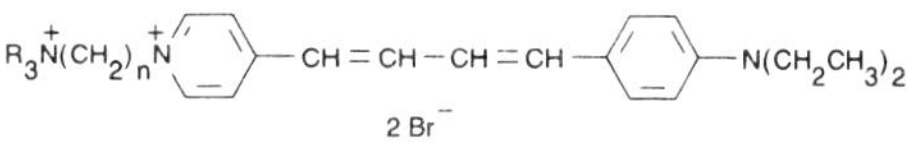

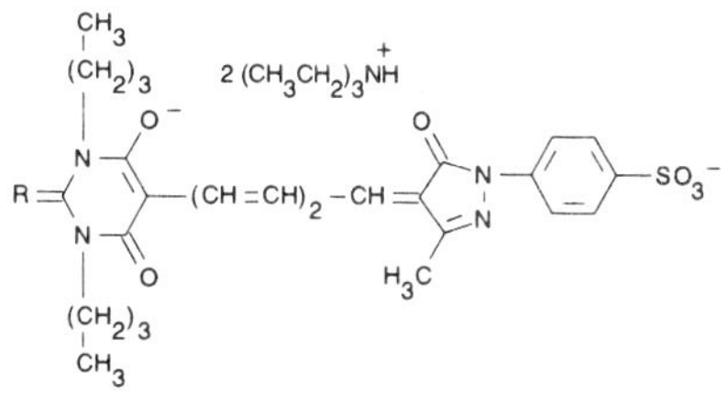

	n	m
S-1107	n = 2	m = 3
S-1108	n = 2	m = 4
S-1109	n = 3	m = 3

	R	n
T-1111	R = $-CH_2CH_3$	n = 3
T-1113	R = $-CH_3$	n = 3
T-6693	R = $-CH_2CH_3$	n = 4

	R
W-435	R = O
R-6922	R = S

25.2 Price List *Fast-Response Probes*

Cat #		Product Name	Unit Size	Price Per Unit ($) 1–4 Units	5–24 Units
D-6923	***New***	di-2-ANEPEQ (JPW 1114)	5 mg	175.00	140.00
D-6925	***New***	di-8-ANEPPQ	5 mg	175.00	140.00
D-6927	***New***	di-12-ANEPPQ	5 mg	175.00	140.00
D-1199		di-4-ANEPPS	5 mg	155.00	124.00
D-3167		di-8-ANEPPS	5 mg	195.00	156.00
D-6928	***New***	di-18:2-ANEPPS	5 mg	175.00	140.00
R-6922	***New***	RGA-30	5 mg	118.00	94.40
R-1114		RH 155	25 mg	145.00	116.00
R-6695	***New***	RH 773	5 mg	118.00	94.40
R-649		RH 795	1 mg	165.00	132.00
S-1107		N-(4-sulfobutyl)-4-(4-(4-(dibutylamino)phenyl)butadienyl)pyridinium, inner salt (RH 160)	25 mg	75.00	60.00
S-1109		N-(4-sulfobutyl)-4-(6-(4-(dibutylamino)phenyl)hexatrienyl)pyridinium, inner salt (RH 237)	5 mg	75.00	60.00
S-1108		N-(4-sulfobutyl)-4-(4-(4-(dipentylamino)phenyl)butadienyl)pyridinium, inner salt (RH 421)	25 mg	165.00	132.00
T-6693	***New***	N-(4-triethylammoniumbutyl)-4-(4-(4-(diethylamino)phenyl)butadienyl)pyridinium dibromide (RH 704)	5 mg	145.00	116.00
T-1111		N-(3-triethylammoniumpropyl)-4-(4-(4-(diethylamino)phenyl)butadienyl)pyridinium dibromide (RH 414)	5 mg	165.00	132.00
T-1113		N-(3-trimethylammoniumpropyl)-4-(4-(4-(diethylamino)phenyl)butadienyl)pyridinium dibromide (RH 461)	5 mg	75.00	60.00
W-435		WW 781, triethylammonium salt	25 mg	145.00	116.00

25.3 Slow-Response Dyes

Slow-response probes are listed in Table 25.2, along with their charge, optical response and selected applications.

Carbocyanines

DiI, DiS and DiO Derivatives

Indo- (DiI), thia- (DiS) and oxa- (DiO) carbocyanines with short alkyl tails (between one and seven carbon atoms) were among the first potentiometric fluorescent probes developed.[1] These cationic dyes accumulate on hyperpolarized membranes and are translocated into the lipid bilayer.[2] Aggregation within the confined membrane interior usually results in decreased fluorescence and absorption shifts,[3] although the magnitude and even the direction of the fluorescence response is strongly dependent on the concentration of the dye and its structural characteristics.[1,4] Relatively large fluorescence changes are obtainable;[1] for example, the thiadicarbocyanine DiSC$_2$(5) (D-304) is quenched up to 98% in response to a 100 mV polarizing potential.[5] DiSC$_3$(5) (D-306) is the most widely used carbocyanine dye for membrane potential measurements,[6] followed closely by DiOC$_5$(3) (D-272); see Table 25.2 for selected references. Carbocyanine dyes selectively stain mitochondrial membranes,[7] causing respiratory inhibition,[4,8] and are therefore often relatively toxic to cells.[9]

JC-1 and Analogs

The green fluorescent JC-1 (5,5',6,6'-tetrachloro-1,1',3,3'-tetraethylbenzimidazolylcarbocyanine iodide; T-3168) exists as a monomer at low concentrations or at low membrane potential. However, at higher concentrations (aqueous solutions above 0.1 µM) or higher potentials, JC-1 forms red fluorescent "J-aggregates," which exhibit a broad excitation spectrum and a very narrow emission spectrum (Figure 25.1). Because J-aggregate formation increases linearly with applied membrane potential over the range of 30–180 mV, this phenomenon can be exploited for potentiometric measurements[10,11] (Table 25.2). Various types of ratio measurements are possible by combining signals from the green fluorescent JC-1 monomer (absorption/emission maxima ~510/527 nm) and the red fluorescent J-aggregate (absorption/emission maximum ~585/590 nm), which can be effectively excited anywhere between

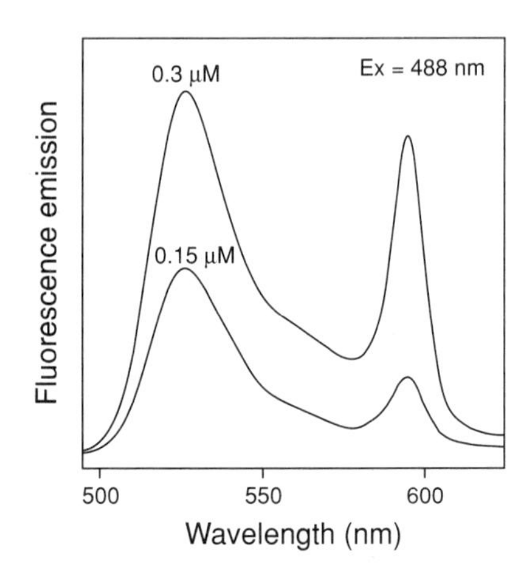

Figure 25.1 *Fluorescence emission spectra of 0.15 and 0.3 µM JC-1 (T-3168) in 50 mM Tris-HCl pH 8.2 containing 1% DMSO, showing the concentration-dependent increase of long-wavelength J-aggregate emission at 590 nm.*

Table 25.2 *Characteristics and selected applications of Molecular Probes' slow-response probes.*

Dyes (Cat #)	Structure (Charge)	Optical Response	Selected Applications
$DiOC_2(3)$ (D-269) $DiOC_2(5)$ (D-303) $DiOC_5(3)$ (D-272) $DiOC_6(3)$ (D-273) $DiOC_7(3)$ (D-378) $DiSC_2(5)$ (D-304) $DiSC_3(5)$ (D-306) $DiIC_1(3)$ (H-379)	Carbocyanine (cationic)	SLOW; fluorescence generally decreases upon membrane hyperpolarization.	• Calcium channels [1,2] and other ion transport systems [3-5] • Mitochondrial activity [6-9] • Neurons and brain tissue [10-12] • Flow cytometry [13,14]
JC-1 (T-3168)	Carbocyanine (cationic)	SLOW; fluorescence emission ratio 585/520 nm increases upon membrane hyperpolarization.	• Ratio imaging of membrane potential [15-18] • Dual-emission flow cytometry [19,20]
Tetramethylrhodamine methyl and ethyl esters (T-668, T-669)	Rhodamine (cationic)	SLOW; used to obtain unbiased images of potential-dependent dye distribution.	• Membrane potentials for single mitochondria in living cells [21] • Simultaneous mitochondrial potential and intracellular calcium [22,23] • Confocal imaging of membrane potential [24-26]
Oxonol V (O-266) Oxonol VI (O-267)	Oxonol (anionic)	SLOW; fluorescence decreases upon membrane hyperpolarization.	• Plant vacuoles [27-29] • Ion channels and electrogenic pumps [30-34] • Liposomes [35-37]
$DiBAC_4(3)$ (B-438) $DiBAC_4(5)$ (B-436) $DiSBAC_2(3)$ (B-413)	Oxonol (anionic)	SLOW; fluorescence decreases upon membrane hyperpolarization.	• Calcium and membrane potential [38-44] • Stimulus–secretion coupling [45,46] • Flow cytometry assays of cell viability [47-49] and membrane potential [50-53] • Hypertonic stress [54]
Merocyanine 540 (M-299)	Merocyanine	FAST/SLOW (biphasic response)	• Membrane potentials in mitochondria [55,56] and skeletal muscle [57] • Structure of membrane surfaces [58-60] • Membrane lipid asymmetry [61,62] • Photodynamic therapy

1. Am J Physiol 266, C67 (1994); **2.** J Biol Chem 268, 18151 (1993); **3.** J Biol Chem 269, 29509 (1994); **4.** J Biol Chem 269, 21962 (1994); **5.** J Biol Chem 268, 206 (1993); **6.** J Biol Chem 270, 3788 (1995); **7.** Biophys J 56, 979 (1989); **8.** Plant Physiol 84, 1385 (1987); **9.** J Cell Biol 88, 526 (1981); **10.** Brain Res 595, 79 (1992); **11.** Glia 4, 611 (1991); **12.** J Neurosci Meth 13, 199 (1985); **13.** Meth Cell Biol 41, 121 (1994); **14.** J Immunol Meth 107, 129 (1988); **15.** Biophys J 65, 1767 (1993); **16.** Cardiovascular Res 27, 1790 (1993); **17.** Biochemistry 30, 4480 (1991); **18.** Proc Natl Acad Sci USA 88, 3671 (1991); **19.** Exp Cell Res 214, 323 (1994); **20.** Biochem Biophys Res Comm 197, 40 (1993); **21.** Biophys J 65, 2396 (1993); **22.** Biophys J 66, 942 (1994); **23.** Proc Natl Acad Sci USA 91, 12579 (1994); **24.** Loew, L.M. in *Cell Biology: A Laboratory Handbook, Volume 2*, J.E. Celis, Ed., Academic Press (1994) pp. 399–403; **25.** Meth Cell Biol 38, 195 (1993); **26.** Biophys J 56, 1053 (1989); **27.** J Biol Chem 267, 21850 (1992); **28.** Plant J 2, 97 (1992); **29.** J Biol Chem 265, 9617 (1990); **30.** J Biol Chem 268, 23122 (1993); **31.** Biochemistry 29, 3859 (1990); **32.** Biochim Biophys Acta 1023, 81 (1990); **33.** Biochim Biophys Acta 1017, 221 (1990); **34.** Biochim Biophys Acta 980, 139 (1989); **35.** Biochim Biophys Acta 1146, 87 (1993); **36.** Biochem Biophys Res Comm 173, 1008 (1990); **37.** Biophys Chem 34, 225 (1989); **38.** Am J Physiol 266, H1416 (1994); **39.** J Biol Chem 269, 29451 (1994); **40.** J Biol Chem 269, 23597 (1994); **41.** J Cell Biol 120, 1003 (1993); **42.** Am J Physiol 263, C1302 (1992); **43.** Proc Natl Acad Sci USA 89, 1690 (1992); **44.** J Biol Chem 265, 15003 (1990); **45.** J Cell Physiol 158, 309 (1994); **46.** Biochem J 287, 59 (1992); **47.** Cytometry 15, 343 (1994); **48.** J Microscopy 176, 8 (1994); **49.** J Immunol Meth 161, 119 (1993); **50.** Cytometry 14, 59 (1993); **51.** Cytometry 13, 545 (1992); **52.** Devel Neurosci 13, 11 (1991); **53.** J Neurosci Meth 22, 203 (1988); **54.** J Biol Chem 269, 10485 (1994); **55.** J Biol Chem 266, 803 (1991); **56.** J Membrane Biol 123, 23 (1991); **57.** J Gen Physiol 95, 147 (1990); **58.** Biochim Biophys Acta 1146, 169 (1993); **59.** Biochim Biophys Acta 1107, 245 (1992); **60.** J Cell Physiol 138, 61 (1989); **61.** Biochim Biophys Acta 1062, 24 (1991); **62.** Proc Natl Acad Sci USA 83, 3311 (1986).

485 nm and its absorption maximum. Optical filters designed for fluorescein and tetramethylrhodamine (see Section 26.5) can be used to separately visualize the monomer and J-aggregate forms, respectively. Alternatively, both forms can be observed simultaneously using a fluorescein longpass optical filter set (see Section 26.5). A recent review by Chen and Smiley describes the properties of JC-1 and its use for investigating mitochondrial potentials in live cells.[12]

In addition to JC-1, we also offer two sulfonated analogs, TDBC-3 (B-6952) and TDBC-4 (B-6953), that form J-aggregates with spectral properties similar to those of JC-1.[13,14] Use of these compounds as potentiometric probes has not yet been reported; however, it seems likely that their potentiometric response will be smaller than that of JC-1 due to reduced membrane permeability. Another sulfonated analog, the nonfluorescent 3,3'-bis-(3-sulfopropyl)-5,5'-dichloro-9-ethylthiacarbocyanine (B-6951) exhibits a five-fold decrease in its 625/550 nm absorbance ratio between 25°C and 35°C and therefore may be a useful ratiometric temperature sensor.[15]

Rhodamine and Pyrene Probes

Rhodamine 123 and Tetramethylrhodamine Methyl and Ethyl Esters

Rhodamine 123 (R-302) is widely used as a structural marker for mitochondria (see Section 12.2) and as an indicator of mitochondrial activity (see Section 16.3). Specific potential-dependent staining of mitochondria is obtained by setting the extracellular K^+ concentration close to intracellular values (~137 mM), thereby depolarizing the plasma membrane.[16] Closely related to rhodamine 123, the methyl and ethyl esters of tetramethylrhodamine (TMRM and TMRE; T-668 and T-669, respectively) are currently the preferred dyes for determining membrane potential by quantitative imaging [17,18] (Table 25.2). Quantitative membrane potential measurements with the Nernst equation require that the transmembrane distribution of the dye depends only on the membrane potential and that other processes such as dye aggregation and potential-independent interactions with intracellular components contribute minimally. TMRM and TMRE fulfill these requirements in several respects.[19,20] They are more membrane-permeant than rhodamine 123, and their strong fluorescence means that they can be used at low concentrations, thus avoiding aggregation. Because their fluorescence is relatively insensitive to environment, spatially resolved fluorescence of TMRM and TMRE presents an unbiased profile of their transmembrane distribution that can be directly related to the membrane potential via the Nernst equation.[17,21]

Triphenylphosphonium Analog

Potential-dependent uptake of triphenylmethylphosphonium ($TPMP^+$) or tetraphenylphosphonium (TPP^+) ions by cells can be quantitatively determined using tritiated analogs or selective ion electrodes.[22-25] Our new pyrenebutyltriphenylphosphonium bromide ($PyTPP^+$, P-6921) is expected to show similar uptake behavior and may also exhibit a potential-dependent fluorescence emission shift due to pyrene excimer formation (see Figure 13.6 in Chapter 13). A structurally related spin-labeled triphenylphosphonium has been used to investigate internal electrostatic potentials in model phospholipid bilayer membranes.[26]

Oxonols

Oxonol V and Oxonol VI

The anionic bis-isoxazolone oxonols (O-266, O-267) accumulate in the cytoplasm of depolarized cells by a Nernst equilibrium–dependent uptake from the extracellular solution.[27] Their voltage-dependent partitioning between water and membranes is often measured by absorption rather than fluorescence. Of the oxonols studied by Smith and Chance,[28] oxonol VI (O-267) gave the largest spectral shifts, with an isosbestic point at 603 nm. In addition, oxonol VI responds to potential changes more rapidly than oxonol V and is therefore considered to be a better probe for fast potential changes.[29]

DiBAC Dyes

The three bis-barbituric acid oxonols, often referred to as DiBAC dyes, form a family of spectrally distinct potentiometric probes with excitation maxima at approximately 490 nm ($DiBAC_4(3)$, B-438), 530 nm ($DiSBAC_2(3)$, B-413) and 590 nm ($DiBAC_4(5)$, B-436). The dyes enter depolarized cells where they bind to intracellular proteins or membranes and exhibit enhanced fluorescence and red spectral shifts.[30] Hyperpolarization results in extrusion of the anionic dye and thus a decrease in fluorescence. $DiBAC_4(3)$ reportedly has the highest voltage sensitivity.[31] The long-wavelength $DiSBAC_2(3)$ has frequently been used in combination with the UV-excitable Ca^{2+} indicators indo-1 or fura-2 for simultaneous measurements of membrane potential and Ca^{2+} concentrations (Table 25.2). Interactions between anionic oxonols and the cationic K^+-valinomycin complex complicate the use of this ionophore to calibrate potentiometric responses.[32,33] Like the bis-isoxazolone oxonols, the DiBAC dyes are excluded from mitochondria on account of their overall negative charge, making them superior to carbocyanines for measuring plasma membrane potentials by flow cytometry (Table 25.2).

Merocyanine 540

Although merocyanine 540 (M-299) was among the first fluorescent dyes to be used as a potentiometric probe,[34] its use for this application has declined with the advent of superior probes. A significant disadvantage of merocyanine 540 is its extreme phototoxicity; consequently it is now more commonly used as a photosensitizer.[35-44] Merocyanine 540 exhibits a biphasic kinetic response to membrane polarization changes. It binds to the surface of polarized membranes in a perpendicular orientation, reorienting as the membrane depolarizes to form nonfluorescent dimers with altered absorption spectra.[45,46] This fast (microseconds) reorientation is followed by a slower response caused by an increased dye uptake.

Merocyanine 540 is also a useful probe of lipid packing because it binds preferentially to membranes with highly disordered lipids.[47,48] Like dansyl lysine (D-372, see Section 14.5), merocyanine 540 is sensitive to heat-induced changes in the organization of membrane lipids.[49] In addition, merocyanine 540 is a probe of transmembrane lipid asymmetry when used in conjunction with a membrane-impermeant carbocyanine (D-1389, see Section 15.3), which reportedly quenches merocyanine 540 only in the outer monolayer.[50]

1. Biochemistry 13, 3315 (1974); **2.** J Membrane Biol 92, 171 (1986); **3.** J Membrane Biol 59, 1 (1981); **4.** Biophys J 56, 979 (1989); **5.** Biochemistry 24, 2101 (1985); **6.** Biochim Biophys Acta 1196, 181 (1994); **7.** J Cell Biol 88, 526 (1981); **8.** J Biol Chem 256 1108 (1981); **9.** Biochem Pharmacol 45, 691 (1993); **10.** Biochemistry 30, 4480 (1991); **11.** Proc Natl Acad Sci USA 88, 3671 (1991); **12.** Chen, L.B. and Smiley, S.T. in *Fluorescent and Luminescent Probes for Biological Activity*, W.T. Mason, Ed., Academic Press (1994) pp. 124–132; **13.** J Fluorescence 4, 57 (1994); **14.** Bull Chem Soc Jpn 53, 3120 (1980); **15.** J Imaging Science and Technology 37, 585 (1993); **16.** Meth Cell Biol 29, 103 (1989); **17.** Meth Cell Biol 38, 195 (1993); **18.** Meth Cell Biol 30, 193 (1989); **19.** Biophys J 56, 1053 (1989); **20.** Biophys J 53, 785 (1988); **21.** Loew, L.M. in *Cell Biology: A Laboratory Handbook, Volume 2*, J.E. Celis, Ed., Academic Press (1994) pp. 399–403; **22.** Biochem J 255, 357 (1988); **23.** Biochim Biophys Acta 893, 69 (1987); **24.** J Biol Chem 260, 2869 (1985); **25.** J Membrane Biol 56, 191 (1980); **26.** Biophys J 65, 289 (1993); **27.** Biochim Biophys Acta 903, 480 (1987); **28.** J Membrane Biol 46, 255 (1979); **29.** Biophys Chem 34, 225 (1989); **30.** Chem Phys Lipids 69, 137 (1994); **31.** Biochim Biophys Acta 771, 208 (1984); **32.** Biophys J 57, 835 (1990); **33.** Biochim Biophys Acta 809, 228 (1985); **34.** Ann Rev Biophys Bioeng 8, 47 (1979); **35.** J Immunol Meth 168, 245 (1994); **36.** Pharm Ther 63, 1 (1994); **37.** Arch Biochem Biophys 300, 714 (1993); **38.** Bone Marrow Transplantation 12, 191 (1993); **39.** Cancer Res 53, 806 (1993); **40.** Free Rad Biol Med 12, 389 (1992); **41.** Biochem Biophys Acta 1075, 28 (1991); **42.** J Infectious Diseases 163, 1312 (1991); **43.** Photochem Photobiol 53, 1 (1991); **44.** Cancer Res 49, 3637 (1989); **45.** Biochemistry 24, 7117 (1985); **46.** Biochemistry 17, 5228 (1978); **47.** Biochim Biophys Acta 1146, 136 (1993); **48.** Biochim Biophys Acta 732, 387 (1983); **49.** Biochim Biophys Acta 1030, 269 (1990); **50.** Meth Enzymol 149, 281 (1987).

25.3 Data Table *Slow-Response Dyes*

Cat #	Structure	MW	Storage	Soluble	Abs	{$\epsilon \times 10^{-3}$}	Em	Solvent	Notes
B-413	✓	436	L	DMSO, EtOH*	535	{170}	560	MeOH	
B-436	✓	543	L	DMSO, EtOH*	590	{160}	616	MeOH	
B-438	✓	517	L	DMSO, EtOH*	493	{123}	516	MeOH	
B-6951	✓	672	D,L	DMSO, EtOH	571	{106}	none	H_2O	<1>
B-6952	✓	751	D,L	DMSO	525	{153}	542	DMSO	<2>
B-6953	✓	779	D,L	DMSO	525	{159}	542	DMSO	<2>
D-269	✓	460	D,L	DMSO	482	{153}	500	MeOH	
D-272	✓	544	D,L	DMSO	484	{155}	500	MeOH	
D-273	✓	573	D,L	DMSO	484	{154}	501	MeOH	
D-303	✓	486	D,L	DMSO	579	{238}	601	MeOH	
D-304	✓	518	D,L	DMSO	651	{260}	671	MeOH	
D-306	✓	547	D,L	DMSO	651	{258}	675	MeOH	
D-378	✓	601	D,L	DMSO	482	{148}	504	MeOH	
H-379	✓	484	D,L	DMSO	539	{142}	564	MeOH	
M-299	✓	570	D,L	DMSO, EtOH	555	{160}	578	MeOH	
O-266	✓	384	L	DMSO, EtOH*	610	{135}	639	MeOH	
O-267	✓	316	L	DMSO, EtOH*	599	{136}	634	MeOH	
P-6921	✓	600	D,L	DMSO	340	{37}	377	MeOH	<3>
R-302	S12.2	381	F,D,L	MeOH, DMF	507	{101}	529	MeOH	
T-668	✓	501	F,D,L	DMSO, MeOH	548	{110}	573	MeOH	
T-669	✓	515	F,D,L	DMSO, EtOH	549	{109}	574	MeOH	
T-3168	✓	652	D,L	DMSO, EtOH	514	{195}	529	MeOH	<4>

For definitions of the contents of this data table, see "How to Use The Handbook" on page vi.
Structure: Chemical structure drawing: ✓ = shown in this section; Sn.n = shown in Section number n.n.

* Oxonols may require addition of a base to be soluble.
<1> ϵ determined for 2 x 10^{-6} M solution in 1:9 MeOH:H_2O. Absorption spectrum is variable due to concentration-dependent dye aggregation.
<2> Absorption spectrum in water is sensitive to concentration-dependent dye aggregation.
<3> Pyrene derivatives exhibit structured spectra. The absorption maximum is usually about 340 nm with a subsidiary peak at about 325 nm. There are also strong absorption peaks below 300 nm. The emission maximum is usually about 376 nm with a subsidiary peak at 396 nm. Excimer emission at about 470 nm may be observed at high concentrations.
<4> T-3168 forms J-aggregates with absorption/fluorescence emission maxima at 585/590 nm at concentrations above 0.1 micromolar in aqueous solutions (pH 8.0) (Figure 25.1; Biochemistry 30, 4480 (1991)).

25.3 Structures *Slow-Response Dyes*

B-413

B-438 n = 1
B-436 n = 2

B-6951

B-6952, B-6953 see T-3168

D-269 R = $-CH_2CH_3$ n = 1
D-303 R = $-CH_2CH_3$ n = 2
D-272 R = $-(CH_2)_4CH_3$ n = 1
D-273 R = $-(CH_2)_5CH_3$ n = 1
D-378 R = $-(CH_2)_6CH_3$ n = 1

D-304 R = $-CH_2CH_3$
D-306 R = $-(CH_2)_2CH_3$

H-379

M-299

O-266 R = Ph
O-267 R = $-(CH_2)_2CH_3$

Ph = phenyl

P-6921

R-302 Section 12.2

T-668 R = $-OCH_3$
T-669 R = $-OCH_2CH_3$

T-3168 R = $-CH_2CH_3$ X = I^-
B-6952 R = $-(CH_2)_3SO_3^-$ X = K^+
B-6953 R = $-(CH_2)_4SO_3^-$ X = K^+

25.3 Price List *Slow-Response Dyes*

Cat #		Product Name	Unit Size	Price Per Unit ($) 1–4 Units	Price Per Unit ($) 5–24 Units
B-436		bis-(1,3-dibutylbarbituric acid)pentamethine oxonol (DiBAC$_4$(5))	25 mg	65.00	52.00
B-438		bis-(1,3-dibutylbarbituric acid)trimethine oxonol (DiBAC$_4$(3))	25 mg	115.00	92.00
B-413		bis-(1,3-diethylthiobarbituric acid)trimethine oxonol (DiSBAC$_2$(3))	100 mg	115.00	92.00
B-6953	**New**	3,3′-bis-(4-sulfobutyl)-1,1′-diethyl-5,5′,6,6′-tetrachlorobenzimidazolylcarbocyanine, potassium salt (TDBC-4)	5 mg	48.00	38.40
B-6951	**New**	3,3′-bis-(3-sulfopropyl)-5,5′-dichloro-9-ethylthiacarbocyanine, sodium salt	5 mg	48.00	38.40
B-6952	**New**	3,3′-bis-(3-sulfopropyl)-1,1′-diethyl-5,5′,6,6′-tetrachlorobenzimidazolylcarbocyanine, potassium salt (TDBC-3)	5 mg	48.00	38.40
D-269		3,3′-diethyloxacarbocyanine iodide (DiOC$_2$(3))	100 mg	24.00	19.20
D-303		3,3′-diethyloxadicarbocyanine iodide (DiOC$_2$(5))	1 g	35.00	28.00
D-304		3,3′-diethylthiadicarbocyanine iodide (DiSC$_2$(5))	1 g	35.00	28.00
D-378		3,3′-diheptyloxacarbocyanine iodide (DiOC$_7$(3))	100 mg	68.00	54.40
D-273		3,3′-dihexyloxacarbocyanine iodide (DiOC$_6$(3))	100 mg	85.00	68.00
D-272		3,3′-dipentyloxacarbocyanine iodide (DiOC$_5$(3))	100 mg	85.00	68.00
D-306		3,3′-dipropylthiadicarbocyanine iodide (DiSC$_3$(5))	100 mg	145.00	116.00
H-379		1,1′,3,3,3′,3′-hexamethylindocarbocyanine iodide (DiIC$_1$(3))	100 mg	35.00	28.00
M-299		merocyanine 540	100 mg	30.00	24.00

Cat #		Product Name	Unit Size	Price Per Unit ($) 1–4 Units	5–24 Units
O-266		oxonol V (bis-(3-phenyl-5-oxoisoxazol-4-yl)pentamethine oxonol)	100 mg	145.00	116.00
O-267		oxonol VI (bis-(3-propyl-5-oxoisoxazol-4-yl)pentamethine oxonol)	100 mg	145.00	116.00
P-6921	*New*	1-pyrenebutyltriphenylphosphonium bromide (PyTPP+)	10 mg	48.00	38.40
R-302		rhodamine 123	25 mg	30.00	24.00
T-3168		5,5′,6,6′-tetrachloro-1,1′,3,3′-tetraethylbenzimidazolylcarbocyanine iodide (JC-1; $CBIC_2(3)$)	5 mg	195.00	156.00
T-669		tetramethylrhodamine, ethyl ester, perchlorate (TMRE)	25 mg	72.00	57.60
T-668		tetramethylrhodamine, methyl ester, perchlorate (TMRM)	25 mg	72.00	57.60

Hints for Finding the Product of Interest

This Handbook includes products released by Molecular Probes before September 1996. Additional products and search capabilities are available through our Web site (http://www.probes.com). All information, including price and availability, is subject to change without notice.

- **If you know the name of the product you want:** Look up the product by name in the **Master Price List/Product Index** at the back of the book. The Master Price List/Product Index is arranged alphabetically, by both product name and alternative name, using the alphabetization rules outlined on page vi. This index will refer you directly to all the Handbook sections that discuss the product.

- **If you DO NOT know the name of the product you want:** Search the **Keyword Index** for the words that best describe the applications in which you are interested. The Keyword Index will refer you directly to page numbers on which the listed topics are discussed. Alternatively, search the **Table of Contents** for the chapter or section title that most closely relates to your application.

- **If you only know the alpha-numeric catalog number of the product you want but not its name:** Look for the catalog number in the **Master Price List/Product Index** at the back of the book. First locate the group of catalog numbers with the same alpha identifier as the catalog number of interest and then scan this group for the correct numeric identifier. The Master Price List/Product Index will refer you directly to all the Handbook sections that discuss this product. Alternatively, you may use the search features available through our Web site (http://www.probes.com) to obtain the product name of any of our current products given only the catalog number (or numeric identifier).

- **If you only know part of a product name or are searching for a product introduced after publication of this Handbook:** Search Molecular Probes' up-to-date product list through our **Web site (http://www.probes.com)**. The product search feature on our home page allows you to search our entire product list, including products that have been added since the completion of this Handbook, using a partial product name, complete product name or catalog number. Product-specific bibliographies can also be obtained from our Web site once the product of interest has been located.

- **If you still cannot find the product that you are looking for or are seeking advice in choosing a product:** Please contact our Technical Assistance Department by phone, e-mail, fax or mail.

Chapter 26

Tools for Fluorescence Applications

Contents

Related Chapters

26.1 Fluorescence Microscopy Reference Standards and Antifade Reagents

To obtain accurate and reproducible results from fluorescence imaging applications, it is essential to maximize the performance of the optical system. Careful calibration and instrumentation adjustment are required for high-precision imaging of fluorescent probes, particularly in multicolor applications that involve multiple exposures, repetitive scans or three-dimensional sectioning. Molecular Probes offers a variety of microsphere reference standards designed to facilitate adjustment and calibration of both conventional fluorescence microscopes and confocal laser scanning microscopes (Table 26.1). In addition, because imaging sensitivity depends not only on a well-adjusted instrument but also on the intensity and stability of the fluorescence signal, Molecular Probes has developed several antifade reagents to minimize photobleaching of fluorescently labeled specimens.

FocalCheck Fluorescent Microspheres

Our FocalCheck fluorescent microspheres are specifically designed for examining the alignment, sensitivity and stability of confocal laser scanning microscopes. We expect that they will be particularly useful for confirming the optical sectioning thickness (Z-resolution) in three-dimensional imaging applications. These 15 µm polystyrene beads have been treated by a proprietary method in which a fluorescent dye is allowed to stain only the outermost portion of the microsphere. The resulting beads have a well-defined dye layer, which, when viewed in cross section in the confocal laser scanning microscope, appears as a fluorescent ring of varying dimensions depending on the focal plane (Figure 26.1). We refer to this proprietary staining procedure as ring staining, in order to differentiate it from routine staining throughout the bead.

The excitation/emission maxima exhibited by the different stains in our FocalCheck microspheres — blue (365/430 nm), green (505/515 nm), orange (560/580 nm) and dark red (660/680 nm) — are well matched to the laser sources and optical filters commonly used in confocal laser scanning microscopy. Moreover, because the dyes are localized *within* the bead and therefore protected from environmental factors, the FocalCheck microspheres are brighter and more photostable than conventional surface-stained beads.

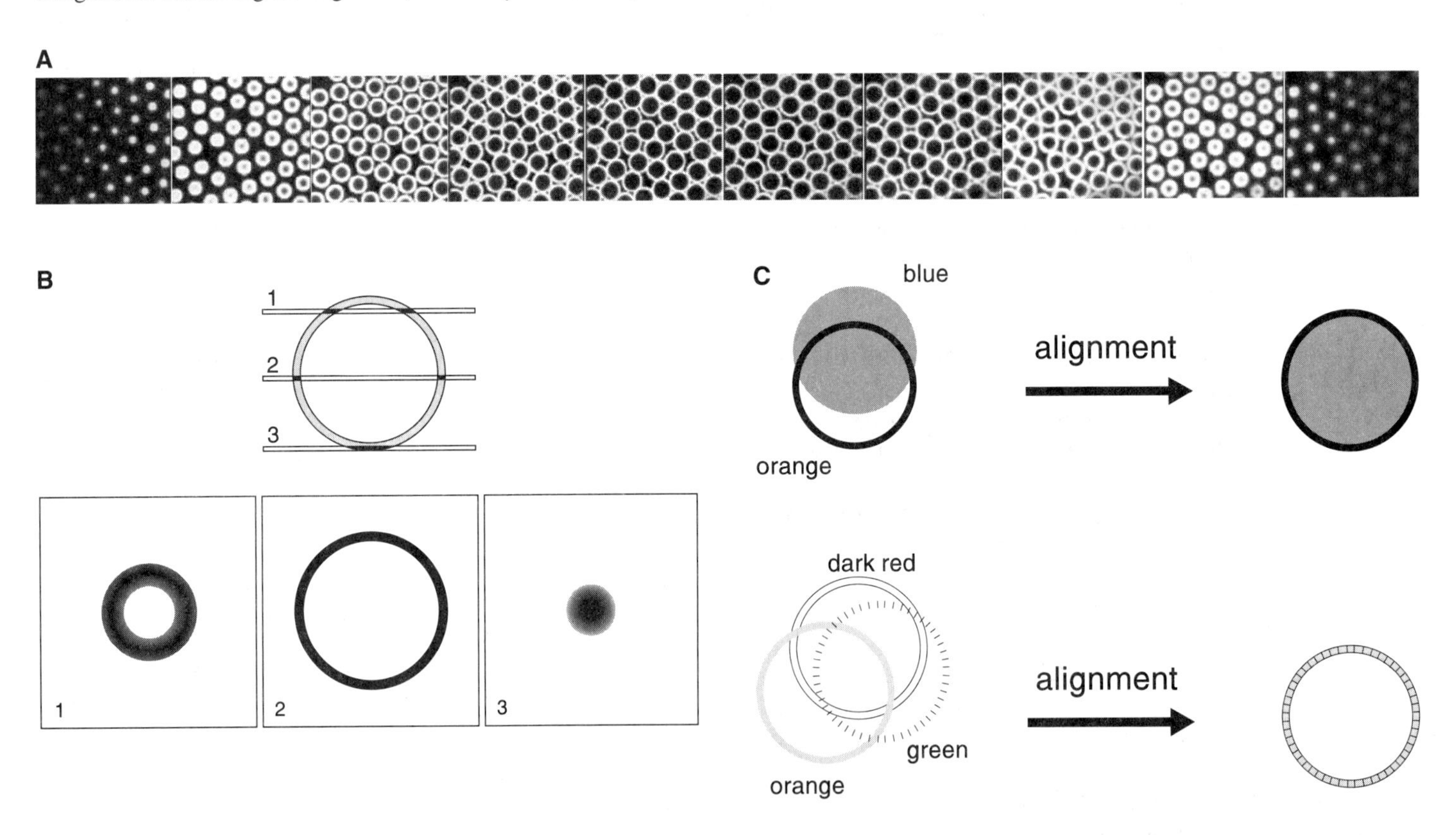

Figure 26.1 *Confocal laser scanning microscope optical cross sectioning and alignment with FocalCheck microspheres. A) Serial optical sectioning from top to bottom along the Z-axis of ring-stained microspheres reveals a continuous pattern of disc-to-ring-to-disc images. B) The diameter of the fluorescent ring (or disc) seen is dependent on the depth of the optical focal plane. C) In the confocal laser scanning microscope, separate light paths exist for UV and visible wavelengths. Also, fluorescence emitted is detected by different photomultiplier detectors. Proper optical alignment may be obtained with either of two types of FocalCheck microspheres. For example, the orange ring stain/blue throughout microspheres allow UV/visible wavelength alignment in three dimensions upon aligning the orange ring with the blue disc. Focal alignment is also possible simultaneously in three colors by aligning the green, orange and dark red rings of the FocalCheck microspheres containing fluorescent green/orange/dark red ring stains.*

Table 26.1 Summary of Molecular Probes' fluorescence microscopy reference standards.

Product Name (Cat #)	Contents	Features and Applications
FocalCheck Microspheres • Fluorescent blue/orange ring stains (F-7234) • Fluorescent green/dark red ring stains (F-7240) • Fluorescent green/orange/dark red ring stains (F-7235) • Fluorescent green ring stain/blue throughout (F-7237) • Fluorescent green ring stain/dark red throughout (F-7238) • Fluorescent orange ring stain/blue throughout (F-7236) • Fluorescent dark red ring stain/green throughout (F-7239)	FocalCheck microspheres are available with ring stains of two or three different fluorescent dyes or with a ring stain of one fluorescent color combined with a stain of a second fluorescent color throughout the bead. Sufficient material is provided for ~100 slides.	FocalCheck microspheres are specifically designed for examining the alignment, sensitivity and stability of confocal laser scanning microscopes. These 15 µm beads have been stained to allow the dye(s) to penetrate only a short distance below the bead surface, such that in optical cross section the beads display a well-defined fluorescent ring(s). The excitation/emission maxima exhibited by the different stains in our FocalCheck microspheres — blue (365/430 nm), green (505/515 nm), orange (560/580 nm) and dark red (660/680 nm) — are well matched to the laser sources and optical filters commonly used in confocal laser scanning microscopy.
MultiSpeck Multispectral Fluorescence Microscopy Standards Kits • Prepared slides (M-7900) • In suspension (M-7901)	Kit M-7900 includes three prepared slides, each with two distinct zones: • MultiSpeck beads and dim MultiSpeck beads • MultiSpeck beads and singly labeled beads • MultiSpeck beads in both a 2D and 3D array Kit M-7901 includes: • MultiSpeck beads • Mixture of singly labeled beads • Mounting medium Sufficient material is provided for ~50 slides of each bead type.	MultiSpeck microspheres serve as multicolored reference particles for use in fluorescence microscopy and confocal laser scanning microscopy. These 4 µm beads have been stained with multiple fluorescent dyes, yielding beads that exhibit *three* distinct excitation and emission bands that are compatible with standard DAPI, fluorescein and tetramethylrhodamine (or Texas Red) optical filter sets. A mixture of beads singly labeled with the blue, green and red dyes of the multicolored beads is also included.
TetraSpeck Fluorescent Microspheres • 0.1 µm (T-7279) • 0.2 µm (T-7280) • 0.5 µm (T-7281) • 1.0 µm (T-7282) • 4.0 µm (T-7283) **TetraSpeck Fluorescent Microspheres Sampler Kit** (T-7284)	Each TetraSpeck product provides a suspension of four-color (blue/green/orange/dark red) fluorescent microspheres. Sufficient material is provided for ~100 slides. The TetraSpeck Sampler Kit includes a sample suspension of: • 0.1 µm TetraSpeck microspheres • 0.5 µm TetraSpeck microspheres • 4.0 µm TetraSpeck microspheres Sufficient material is provided for ~20 slides of each bead type.	TetraSpeck microspheres, available in five different bead diameters, serve as multicolored reference particles for use in fluorescence microscopy and confocal laser scanning microscopy. TetraSpeck microspheres have each been stained with four different fluorescent dyes, yielding beads that exhibit *four* well-separated excitation and emission peaks: blue (365/430 nm), green (505/515 nm), orange (560/580 nm) and dark red (660/680 nm). The TetraSpeck Sampler Kit provides a convenient and cost-effective way to experiment with TetraSpeck beads of different diameters.
PS-Speck Microscope Point Source Kit (P-7220)	This kit includes: • Blue fluorescent beads (365/415 nm) • Green fluorescent beads (505/515 nm) • Orange fluorescent beads (540/560 nm) • Deep red fluorescent beads (633/660 nm) • Mounting medium Sufficient material is provided for ~100 slides of each bead type.	PS-Speck microspheres serve as subresolution (0.175 µm in diameter) fluorescent point sources for calibrating instrument optics, especially in three-dimensional imaging applications.
InSpeck Microscope Image Intensity Calibration Kits • Blue (350/440 nm) (I-7221) • Green (505/515 nm) (I-7219) • Orange (540/560 nm) (I-7223) • Red (580/605 nm) (I-7224) • Deep Red (633/660 nm) (I-7225)	Each kit includes: • 100% relative fluorescence intensity beads • 30% relative fluorescence intensity beads • 10% relative fluorescence intensity beads • 3% relative fluorescence intensity beads • 1% relative fluorescence intensity beads • 0.5% relative fluorescence intensity beads • Unlabeled control beads • Mounting medium Sufficient material is provided for ~100 slides of each bead type.	InSpeck Kits supply suspensions of 2.5 µm microspheres that can be used as calibrated intensity standards for fluorescence microscopy. The seven bead suspensions supplied in each kit exhibit fluorescence intensities that cover the range of intensities commonly encountered in microscopy applications. We offer microscopists a choice of five InSpeck Kits, which provide fluorescent microspheres in five different fluorescent colors.

FocalCheck products are currently available in seven different configurations, including three suspensions that contain microspheres exhibiting ring stains of two or three different fluorescent colors:

- Fluorescent blue/orange ring stains (F-7234)
- Fluorescent green/dark red ring stains (F-7240)
- Fluorescent green/orange/dark red ring stains (F-7235)

and four suspensions that contain microspheres exhibiting a ring stain of one fluorescent color combined with a stain of a second fluorescent color throughout the bead:

- Fluorescent green ring stain/blue throughout (F-7237)
- Fluorescent green ring stain/dark red throughout (F-7238)
- Fluorescent orange ring stain/blue throughout (F-7236)
- Fluorescent dark red ring stain/green throughout (F-7239)

The sharp ring stains exhibited by the FocalCheck microspheres produce a striking visual representation of instrument misalignment or other aberrations, making them ideal as reference standards for confocal laser scanning microscopy. Correct image registration is indicated when the multiple ring images of the ring-stained FocalCheck beads (or the ring and disk images of the combination ring-stained and stained-throughout FocalCheck beads) are perfectly coincident in all dimensions (Figure 26.1). Furthermore, because the FocalCheck beads are available in a number of multicolor options, they are especially useful in testing and aligning confocal laser scanning microscopes with multiple laser lines and detection channels.

MultiSpeck and TetraSpeck Fluorescent Microspheres

Our unique MultiSpeck and TetraSpeck fluorescent microspheres promise to greatly facilitate the adjustment and calibration of conventional fluorescence microscopes, confocal laser scanning microscopes and associated image-processing equipment in both scientific and commercial imaging, especially for multicolor applications. These uniform, multiply stained microspheres are particularly useful for verifying the ability of instrumentation to co-localize and resolve objects emitting different wavelengths of light in the same optical plane.

MultiSpeck Multispectral Fluorescence Microscopy Standards Kits

The 4 µm MultiSpeck microspheres in our MultiSpeck Multispectral Fluorescence Microscopy Standards Kits (M-7900, M-7901) exhibit three relatively distinct emission bands — blue, green and red — throughout every particle. The spectral characteristics are compatible with optical filter sets designed for commonly used blue, green and red fluorophores (e.g., DAPI, fluorescein and rhodamine or Texas Red® dyes). The MultiSpeck beads can be used as external references for comparing images collected with different optics, on different instruments and in different laboratories, as well as for monitoring routine day-to-day variations in instrument performance. Furthermore, because a single multispectral microsphere will appear as different colors depending on the optical filters used for observation, these beads can be used to assess image registration in two and three dimensions, allowing the researcher to accurately determine the spatial relationships of different labels in a multiparameter experiment.

Molecular Probes' original MultiSpeck Multispectral Fluorescence Microscopy Standards Kit (M-7900) provides three ready-to-use microscope slides, each with two clearly delineated zones. Each of the three slides has a zone of MultiSpeck microspheres, containing a monolayer of these multispectral beads immobilized in mounting medium. The second zone in each of three slides contains either:

- Low-intensity MultiSpeck microspheres (Dim Sample) for testing instrument sensitivity
- Mixture of singly labeled red, green and blue beads (RGB Sample), made with the same dyes but in *separate* particles, for assessing the performance of multiband optical filter sets
- Three-dimensional array of the MultiSpeck microspheres (3D Sample) to allow calibration of Z-axis registration of each of the three colors

For those researchers who would prefer to customize the use of the MultiSpeck microspheres, we have also introduced a MultiSpeck Multispectral Fluorescence Microscopy Standards Kit (M-7901) that contains the multispectral beads in suspension rather than mounted on slides. In addition to the suspension of MultiSpeck microspheres, each kit contains a mixed suspension of individually stained blue, green and red fluorescent microspheres and a vial of mounting medium. The kit supplies material sufficient for ~50 slide preparations using either bead suspension.

TetraSpeck Fluorescent Microspheres and Sampler Kits

Our new TetraSpeck fluorescent microspheres expand the multispectral strategy introduced with the MultiSpeck beads in two important ways. First, the TetraSpeck beads have been stained throughout with four different fluorescent dyes, yielding beads that each display four well-separated excitation and emission peaks — 365/430 nm (blue), 505/515 nm (green), 560/580 nm (orange) and 660/680 nm (dark red) — with little or no fluorescence resonance energy transfer between dyes. Second, these microspheres are available in five nominal sizes (actual bead diameters are indicated on product labels), spanning the range from subresolution to nearly cell-size particles:

- 0.1 µm (T-7279)
- 0.2 µm (T-7280)
- 0.5 µm (T-7281)
- 1.0 µm (T-7282)
- 4.0 µm (T-7283)

For the convenience of first-time users, we offer the TetraSpeck Fluorescent Microspheres Sampler Kit (T-7284), containing separate samples of our 0.1 µm, 0.5 µm and 4.0 µm TetraSpeck beads.

PS-Speck Microscope Point Source Kit

The fluorescent microspheres in Molecular Probes' PS-Speck Microscope Point Source Kit (P-7220) have a diameter of 0.175 ± 0.005 µm; thus, they are ideal as subresolution fluorescent sources for calibrating instrument optics, especially in three-dimensional applications. This kit's four ready-to-use suspensions contain bright, monodisperse particles in the following fluorescent colors (and excitation/emission peaks):

- Blue (365/415 nm)
- Green (505/515 nm)
- Orange (540/560 nm)
- Deep red (633/660 nm)

The kit also includes mounting medium and a mounting protocol for the user's convenience. Each suspension provides sufficient material to mount about 100 slides.

InSpeck Microscopy Image Intensity Calibration Kits

InSpeck Microscope Image Intensity Calibration Kits provide fluorescence microscopists with intensity references for generating calibration curves and evaluating sample brightness. The kits are offered in a choice of five different fluorescent colors, allowing calibrations over a wide range of excitation and emission wavelengths:

- InSpeck Blue (350/440 nm) Kit (I-7221)
- InSpeck Green (505/515 nm) Kit (I-7219)
- InSpeck Orange (540/560 nm) Kit (I-7223)
- InSpeck Red (580/605 nm) Kit (I-7224)
- InSpeck Deep Red (633/660 nm) Kit (I-7225)

Each kit includes six separate suspensions of InSpeck fluorescent microspheres with fluorescence intensities covering the range of intensities commonly encountered in microscopy applications:

- 100% relative fluorescence intensity
- 30% relative fluorescence intensity
- 10% relative fluorescence intensity
- 3% relative fluorescence intensity
- 1% relative fluorescence intensity
- 0.5% relative fluorescence intensity

Unstained control beads and mounting medium are also supplied. The aqueous suspensions of microspheres may be applied directly to the sample for calibrating fluorescence intensities or mounted separately in an adjacent well or on another slide. Each suspension provides enough material to prepare about 100 slides.

ProLong and *SlowFade* Antifade Kits

Loss of fluorescence through the irreversible photobleaching process can lead to a significant reduction in sensitivity, particularly when target molecules are of low abundance or when excitation light is of high intensity or long duration. To minimize photobleaching of experimental samples, we have developed the ProLong, *SlowFade* and *SlowFade Light* Antifade Kits, which have been proven to increase the photostability of many of our fluorophores in fixed cells, fixed tissues and cell-free preparations. The primary function of any antifade reagent is to sustain dye fluorescence, usually by inhibiting the generation and diffusion of reactive oxygen species. Strategies for further maximizing the fluorescence signal in both living and nonliving specimens include minimizing the exposure of fluorescently labeled samples to excitation light with neutral density filters, as well as using high–numerical aperture objectives, relatively low magnification, high-quality optical filters and high-speed film or high-efficiency detectors.

ProLong Antifade Kit

The latest addition to our antifade arsenal — ProLong Antifade Kit (P-7481) — outperforms all other commercially available reagents with little or no quenching of the fluorescent signal (Figure 26.2). Each ProLong Antifade Kit contains:

- ProLong antifade reagent powder
- ProLong mounting medium
- Protocol for mounting samples

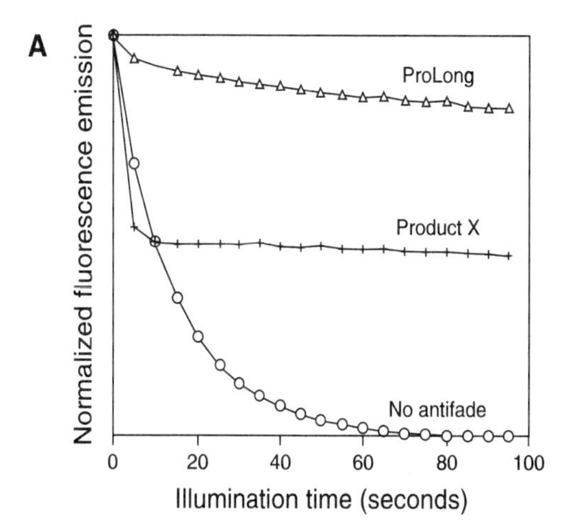

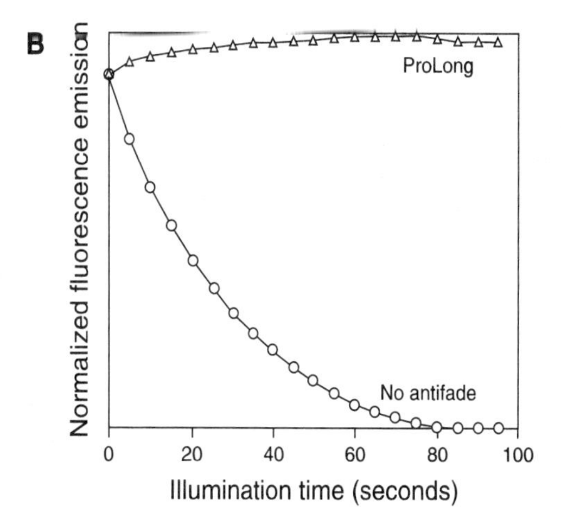

Figure 26.2 *Bleaching profiles of A) fluorescein and B) Texas Red dyes in cell samples. In these photobleaching experiments, human epithelium (HEp-2) cells were probed with human anti-nuclear antibodies and then developed for visualization with fluorophore-labeled secondary reagents. The data were collected through a 40× objective with a Star 1™ cooled CCD camera (Photometrics Ltd.) and analyzed with an Image-1®/AT image processor (Universal Imaging). Identical samples were mounted in ProLong antifade reagent (△), Product X (+) or medium containing no antifade reagent (○). Although these data were normalized, we observed little or no quenching of samples mounted with the ProLong mounting medium.*

Figure 26.2 shows that the ProLong Antifade Kit provides more fluorescence output than a popular *p*-phenylenediamine–containing antifade reagent [1] when used to mount fluorescein-stained HEp-2 cells. Furthermore, not only is the Texas Red dye unquenched by the addition of the ProLong reagents, but its signal becomes noticeably brighter when viewed through the Texas Red optical filter set. The ProLong Antifade Kit also inhibits the fading of tetramethylrhodamine, as well as fading of DNA-bound nucleic dyes such as DAPI, propidium iodide and YOYO®-1, (D-1306, P-1304, Y-3601; see Section 8.1), again without significantly quenching these dyes. The compatibility of the ProLong Kit with a multitude of dyes and dye complexes makes it an especially valuable tool for multiparameter analyses such as multicolor fluorescence *in situ* hybridization.

SlowFade and *SlowFade Light* Antifade Kits

Our original *SlowFade* formulation (S-2828) was designed to reduce the fading rate of fluorescein to almost zero. Because it provides a nearly constant emission intensity from fluorescein, the original *SlowFade* reagent is especially useful for quantitative measurements and applications that employ a confocal laser scanning microscope, in which the excitation intensities can be extreme and prolonged. The *SlowFade* reagent can extend the useful fluorescence emission of fluorescein more than 50-fold and can preserve the signal in cell and tissue mounts for up to two years. However, the original *SlowFade* formulation does substantially quench fluorescein's fluorescence and almost completely quenches that of the Cascade Blue® and AMCA fluorophores.

To overcome this limitation, Molecular Probes' researchers have developed the *SlowFade Light* Antifade Kit (S-7461). The antifade formulation in our *SlowFade Light* Antifade Kit slows fluorescein's fading rate by about fivefold without significantly reducing fluorescein's initial fluorescence intensity (Figure 26.3), thereby dramatically increasing the signal-to-noise ratio in photomicroscopy. Moreover, the quenching of Cascade Blue, AMCA, tetramethylrhodamine and Texas Red dyes is minimal. In fact, the *SlowFade Light* antifade reagent reduces the fading rate of the Cascade Blue fluorophore to almost zero, while decreasing its emission intensity by only about 30%.

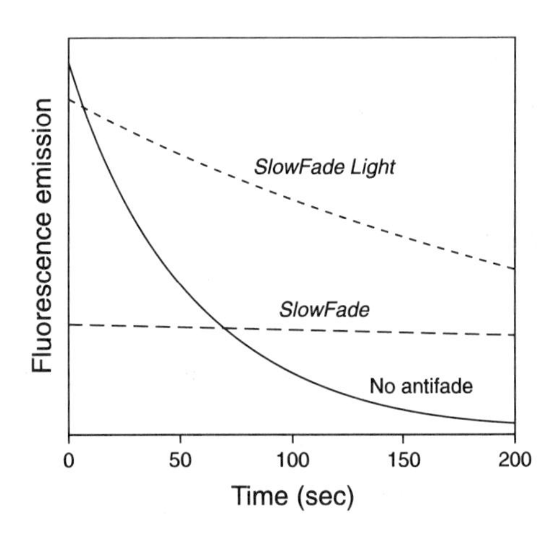

Figure 26.3 *The fluorescence intensity of fluorescein as a function of illumination time, under the following conditions: in the presence of the SlowFade Light antifade reagent ("SlowFade Light"), in the presence of the SlowFade antifade reagent ("SlowFade") and in the absence of an antifade reagent ("no antifade"). In these experiments, we added free fluorescein directly to a solution and then examined the mixture in a capillary tube using a 20× / 0.4 lens. In real samples such as cells and tissues, we find that the local environment influences the bleaching rates, yielding results that are sometimes different from those shown here.*

Each *SlowFade* or *SlowFade Light* Antifade Kit contains:

- *SlowFade* or *SlowFade Light* antifade reagent in 50% (v/v) glycerol, ready to use and sufficient for at least 200 coverslip-size experiments
- 2X Concentrated *SlowFade* or *SlowFade Light* antifade reagent solution, provided for those applications in which glycerol may not be compatible

Attofluor® Cell Chamber

The Attofluor cell chamber (A-7816) is a durable and practical coverslip holder designed for viewing live cell specimens on upright or inverted microscopes (see figure). Features include:

- Surgical stainless steel construction
- Autoclavable, allowing cells to be grown directly in the chamber
- O-ring seal design that prevents sample contamination by oil and leakage of media from the coverslip.
- Accepts 25 mm–diameter round coverslips and mounts in a standard 35 mm–diameter stage holder.
- Thin 0.5 mm base, allowing clearance for the objective when focusing.

For detailed specifications, please contact our Technical Assistance Department or see our Web site at http://www.probes.com.

Attofluor cell chamber.

Cat #	Product Name	Unit Size	Price Per Unit ($) 1–4 Units	5–24 Units
A-7816	***New*** Attofluor® cell chamber *for microscopy*	each	158.00	158.00

- Equilibration buffer, which raises the pH of the sample, increasing the protection afforded by both *SlowFade* formulations
- Protocol for mounting samples

The *SlowFade* Antifade Kits are especially designed for those who want a quick, no-fuss solution to photobleaching. Unlike the ProLong Antifade Kit, the *SlowFade* Kits require no mixing. Simply pre-equilibrate the sample in the wash buffer provided and then apply the mounting medium directly from our convenient applicator bottle.

Liumogen

Liumogen (2,2'-dihydroxy-1,1'-naphthaldiazine, D-7792) is used as an optical coating in the production of CCD-imaging detectors [2] and can be sublimed onto glass substrates such as microscope slides to create a uniform fluorophore coating (excitation/emission maxima = 406/509 nm) for use in fluorescence microscopy.[3]

1. J Histochem Cytochem 41, 1833 (1993); **2.** Appl Optics 23, 146 (1984); **3.** J Histochem Cytochem 41, 1833 (1993).

26.1 Price List *Fluorescence Microscopy Reference Standards and Antifade Reagents*

Cat #		Product Name	Unit Size	Price Per Unit ($) 1–4 Units	5–24 Units
A-7816	***New***	Attofluor® cell chamber *for microscopy*	each	158.00	158.00
D-7792	***New***	2,2′-dihydroxy-1,1′-naphthaldiazine (liumogen)	100 mg	65.00	52.00
F-7234	***New***	FocalCheck™ microspheres, 15 µm, fluorescent blue/orange ring stains	0.5 mL	135.00	108.00
F-7239	***New***	FocalCheck™ microspheres, 15 µm, fluorescent dark red ring stain/green throughout	0.5 mL	135.00	108.00
F-7240	***New***	FocalCheck™ microspheres, 15 µm, fluorescent green/dark red ring stains	0.5 mL	135.00	108.00
F-7235	***New***	FocalCheck™ microspheres, 15 µm, fluorescent green/orange/dark red ring stains	0.5 mL	135.00	108.00
F-7238	***New***	FocalCheck™ microspheres, 15 µm, fluorescent green ring stain/dark red throughout	0.5 mL	135.00	108.00
F-7237	***New***	FocalCheck™ microspheres, 15 µm, fluorescent green ring stain/blue throughout	0.5 mL	135.00	108.00
F-7236	***New***	FocalCheck™ microspheres, 15 µm, fluorescent orange ring stain/blue throughout	0.5 mL	135.00	108.00
I-7221	***New***	InSpeck™ Blue (350/440) Microscope Image Intensity Calibration Kit	1 kit	80.00	64.00
I-7225	***New***	InSpeck™ Deep Red (633/660) Microscope Image Intensity Calibration Kit	1 kit	80.00	64.00
I-7219	***New***	InSpeck™ Green (505/515) Microscope Image Intensity Calibration Kit	1 kit	80.00	64.00
I-7223	***New***	InSpeck™ Orange (540/560) Microscope Image Intensity Calibration Kit	1 kit	80.00	64.00
I-7224	***New***	InSpeck™ Red (580/605) Microscope Image Intensity Calibration Kit	1 kit	80.00	64.00
M-7901	***New***	MultiSpeck™ Multispectral Fluorescence Microscopy Standards Kit *in suspension*	1 kit	110.00	88.00
M-7900	***New***	MultiSpeck™ Multispectral Fluorescence Microscopy Standards Kit *prepared slides*	1 kit	125.00	100.00
P-7481	***New***	ProLong™ Antifade Kit	1 kit	98.00	78.40
P-7220	***New***	PS-Speck™ Microscope Point Source Kit *blue, green, orange and deep red fluorescent beads*	1 kit	80.00	64.00
S-2828		*SlowFade*™ Antifade Kit	1 kit	55.00	44.00
S-7461	***New***	*SlowFade*™ *Light* Antifade Kit	1 kit	55.00	44.00
T-7284	***New***	TetraSpeck™ Fluorescent Microspheres Sampler Kit	1 kit	95.00	76.00
T-7279	***New***	TetraSpeck™ microspheres, 0.1 µm, fluorescent blue/green/orange/dark red	0.5 mL	95.00	76.00
T-7280	***New***	TetraSpeck™ microspheres, 0.2 µm, fluorescent blue/green/orange/dark red	0.5 mL	95.00	76.00
T-7281	***New***	TetraSpeck™ microspheres, 0.5 µm, fluorescent blue/green/orange/dark red	0.5 mL	95.00	76.00
T-7282	***New***	TetraSpeck™ microspheres, 1.0 µm, fluorescent blue/green/orange/dark red	0.5 mL	95.00	76.00
T-7283	***New***	TetraSpeck™ microspheres, 4.0 µm, fluorescent blue/green/orange/dark red	0.5 mL	95.00	76.00

Product-Specific Bibliographies and Keyword Searches

Our Technical Assistance Department can provide you with product-specific bibliographies, as well as keyword searches of the over 25,000 literature references in our extensive bibliography database. Our bibliography database is also searchable through our Web site (http://www.probes.com).

26.2 Flow Cytometry Reference Standards

Flow cytometers are designed to perform quantitative measurements on individual cells and other particles with high precision, speed and accuracy. As with all high-performance instrumentation, flow cytometers must be calibrated frequently to ensure accuracy and reliability. The stability, uniformity and reproducibility of our fluorescent microsphere products make them ideal reference standards for flow cytometry (Table 26.2). Our flow cytometry reference standards exhibit superior stability and greater uniformity than other commercially available microparticles, and they are accompanied by a full one-year warranty (see page 603).

AlignFlow and AlignFlow Plus Flow Cytometry Alignment Beads

Due to recent advances in flow cytometry and the development of new fluorescent probes, a scientist can now perform multiparameter analyses on biological samples. In order to ensure accurate and reproducible results, flow cytometers should be checked daily for proper performance. Molecular Probes' 2.5 µm AlignFlow (A-7302, A-7304, A-7312) and 6.0 µm AlignFlow Plus (A-7303, A-7305, A-7313) flow cytometry alignment beads are reliable references for aligning, focusing and calibrating flow cytometers. These beads are highly uniform polystyrene microspheres that have been stained with stable fluorescent dyes. When excited with the 488 nm spectral line of the argon-ion laser, the AlignFlow beads emit broadly from 515 nm to 660 nm — covering the spectral range of fluorophores commonly used to label cells. The AlignFlow beads for UV (350–360 nm) excitation emit between 400 nm and 470 nm, and the AlignFlow beads for 633 nm excitation emit between 645 nm and 680 nm.

Using AlignFlow or AlignFlow Plus beads, a flow cytometrist can reproduce adjustment parameters crucial to flow cytometry before an experiment is started. The beads approximate the size, emission wavelength and intensity of many biological samples and permit the calibration of a flow cytometer's laser, optics and stream flow without wasting valuable and sensitive experimental material. Because the dyes are contained inside the microspheres, instead of merely on the bead surfaces, AlignFlow beads have superior signal stability and can serve as daily reference standards by which an instrument's performance and reliability are evaluated.

DAPI-Like Flow Cytometry Reference Beads

Our inexpensive DAPI-like flow cytometry reference beads simulate DAPI- or Hoechst-stained cells. These 2.5 µm (F-7310) or 6.0 µm (F-7311) polystyrene beads are similar to our AlignFlow and AlignFlow Plus beads for UV excitation except that they are slightly less intense (~70% of the intensity of AlignFlow beads) and have a greater bead-to-bead variation in intensity. These DAPI-like beads exhibit intensity variations comparable to those commonly encountered with stained cells and therefore can substitute for stained cells in preliminary instrument setup and calibration.

CompenFlow Flow Cytometry Compensation Kit

Using multicolor analysis with a modern flow cytometer, a scientist can resolve subsets of cells that are differentially labeled with fluorescent probes. However, the emission spectra of the dyes typically used for cell labeling are broader than the light collection windows generated by a flow cytometer's optical filters. Thus, the fluorescence signal of a dye may be received in more than one fluorescence detector. The emission overlap can be corrected using electronic processing called "color compensation." Molecular Probes' CompenFlow Flow Cytometry Compensation Kit (C-7301) is designed to help flow cytometer operators set up compensation circuits that properly remove unwanted signals from secondary channels.

The CompenFlow beads have several unique properties and show a nearly perfect spectral match to labeled cells. Each CompenFlow Kit contains three suspensions of 6.0 µm polystyrene microspheres:

- Fluorescein-like beads
- R-phycoerythrin–like beads
- Control beads, lightly stained to mimic autofluorescence of unstained lymphocytes

When excited at 488 nm by an argon-ion laser, the fluorescein-like polystyrene beads emit yellow-green light that matches the spectrum of fluorescein-labeled cells (emission maximum of approximately 515 nm), and the R-phycoerythrin–like polystyrene beads emit red-orange light that matches the spectrum of R-phycoerythrin–labeled cells (emission maximum of approximately 575 nm, see Figure 26.4).

Moreover, because the dye is contained inside each CompenFlow microsphere, instead of merely on the bead surface, it is

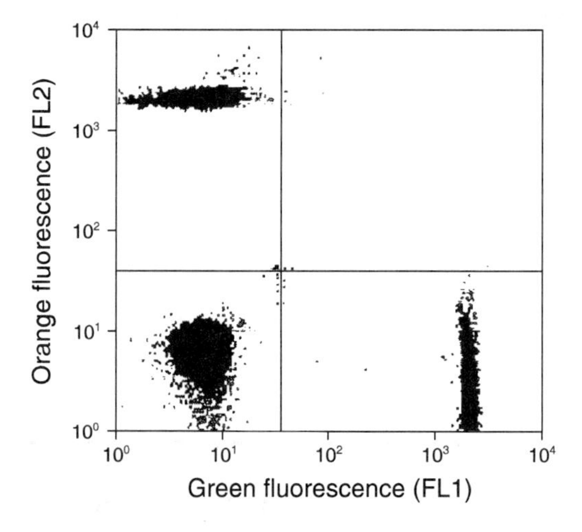

Figure 26.4 *Flow cytometric analysis of the polystyrene beads supplied in our CompenFlow Flow Cytometry Compensation Kit demonstrates the clear separation of fluorescein-like (lower right), R-phycoerythrin–like (upper left) and control (lower left) populations. Analysis was performed on a Becton Dickinson FACScan™ flow cytometer using excitation at 488 nm.*

Table 26.2 Summary of Molecular Probes' flow cytometry reference standards.

Product Name (Cat #)	Contents	Features and Applications
AlignFlow Flow Cytometry Alignment Beads • For UV (350–360 nm) excitation (A-7304) • For 488 nm excitation (A-7302) • For 633 nm excitation (A-7312)	Each product provides a suspension of AlignFlow microspheres in dropper vials for convenient application.	AlignFlow beads are 2.5 µm reference standards for aligning, focusing and calibrating flow cytometers. The beads are available for UV (350–360 nm) excitation (emission range 400–470 nm), for 488 nm excitation (emission range 515–660 nm) and for 633 nm excitation (emission range 645–680 nm).
AlignFlow Plus Flow Cytometry Alignment Beads • For UV (350–360 nm) excitation (A-7305) • For 488 nm excitation (A-7303) • For 633 nm excitation (A-7313)	Each product provides a suspension of AlignFlow Plus microsphere in dropper vials for convenient application.	AlignFlow Plus beads are 6.0 reference standards, otherwise identical to AlignFlow beads.
DAPI-Like Flow Cytometry Reference Beads • 2.5 µm beads (F-7310) • 6.0 µm beads (F-7311)	Each product provides a suspension of DAPI-like microspheres in dropper vials for convenient application.	The 2.5 µm and 6.0 µm DAPI-like flow cytometry reference beads simulate DAPI-stained cells and can be used for preliminary instrument setup and calibration. Their fluorescence is slightly less intense than that of our AlignFlow products, and they show more bead-to-bead variation.
CompenFlow Flow Cytometry Compensation Kit (C-7301)	This kit includes three microsphere suspensions in dropper vials for convenient application: • Fluorescein-like beads • R-Phycoerythrin–like beads • Control beads	The CompenFlow Kit provides reference standards for setting up the compensation circuits of a flow cytometer. These 6.0 µm microspheres show a nearly perfect spectral match to cells labeled with either fluorescein or R-phycoerythrin, and the control beads mimic the autofluorescence of unstained cells.
LinearFlow Flow Cytometry Intensity Calibration Kits • Green (505/515 nm) (L-7306) • Orange (540/560 nm) (L-7307) • Carmine (580/620 nm) (L-7308) • Deep Red (633/660 nm) (L-7314)	Each kit includes four microsphere suspensions in dropper vials for convenient application: • 100% relative fluorescence intensity • 10% relative fluorescence intensity • 2.0% relative fluorescence intensity • 0.4% relative fluorescence intensity	LinearFlow Kits supply suspensions of 2.5 µm microspheres that can be used as calibrated intensity standards for generating calibration curves and evaluating sample brightness in flow cytometry experiments. The kits are available in four different fluorescent colors.
Flow Cytometry Alignment Standards Sampler Kits • 2.5 µm beads (F-7321) • 6.0 µm beads (F-7322)	Each kit includes samples of three AlignFlow or AlignFlow Plus microsphere suspensions: • Beads for UV excitation • Beads for 488 nm excitation • Beads for 633 nm excitation	These sampler kits provide a convenient and cost-effective way to try our AlignFlow or AlignFlow Plus beads.
Flow Cytometry Compensation and Intensity Standards Kit (F-7320)	This kit includes sample suspensions of: • CompenFlow fluorescein-like beads • CompenFlow R-phycoerythrin–like beads • CompenFlow control beads • LinearFlow Green beads, 100% intensity • LinearFlow Green beads, 10% intensity • LinearFlow Green beads, 2.0% intensity • LinearFlow Green beads, 0.4% intensity	This standards kit provides an inexpensive way to try our CompenFlow and LinearFlow Green products.

Warranty on Molecular Probes' Reference Standards for Flow Cytometry

Molecular Probes warrants that its flow cytometry reference standards will remain suitable for calibration of flow cytometers for one year from the date of purchase if stored protected from light, either refrigerated or at room temperature. We will replace or refund the purchase price of any product returned to us that we find to be out of compliance with our specifications. Please contact our Customer Service Department for more information.

shielded from many environmental factors that cause quenching of exposed fluorophores. Consequently, the CompenFlow beads demonstrate a constant intensity for much longer than other commercially available beads that are merely surface-labeled. Our stability tests indicate that after dilution into sheath fluid, our bead suspensions are stable for at least a week; beads from other sources fade after only one day. Also, because our fluorescein-like beads are stained with a dye that is pH insensitive, their emission spectrum will not shift with changes in pH, as with conventional fluorescein labeled beads.

LinearFlow Flow Cytometry Intensity Calibration Kits

LinearFlow Flow Cytometry Intensity Calibration Kits provide flow cytometry operators with intensity references for generating calibration curves and evaluating sample brightness. Each kit contains four vials of 2.5 µm polystyrene microspheres that are fluorescently labeled to provide four precisely determined intensity levels when excited in a flow cytometer (Figure 26.5):

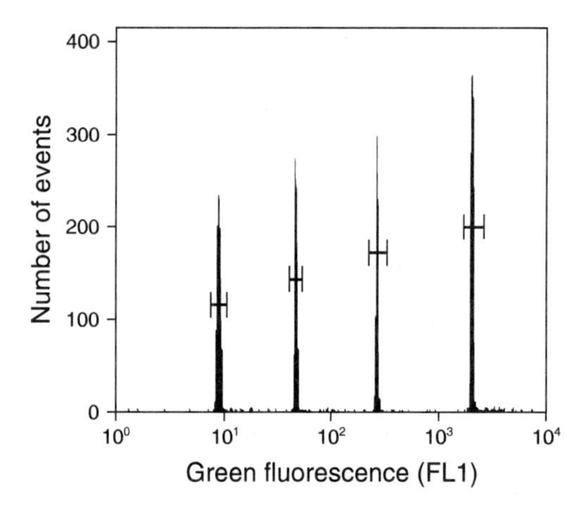

Figure 26.5 *Fluorescence intensity histogram of the four polystyrene bead samples supplied in our LinearFlow Green (505/515) Flow Cytometry Intensity Calibration Kit. Fluorescence measurements were performed on a Becton Dickinson FACScan flow cytometer using excitation at 488 nm.*

- 100% relative fluorescence intensity
- 10% relative fluorescence intensity
- 2.0% relative fluorescence intensity
- 0.4% relative fluorescence intensity

The LinearFlow Flow Cytometry Intensity Calibration Kits are available in four different fluorescent colors, which cover the spectral range commonly encountered in flow cytometry:

- Green (505/515) (L-7306)
- Orange (540/560) (L-7307)
- Carmine (580/620) (L-7308)
- Deep Red (633/660) (L-7314)

The LinearFlow Green Kit contains microspheres with a fluorescence excitation maximum of 505 nm and emission maximum of 515 nm, closely matching the fluorescence of fluorescein, BODIPY® FL and SYTOX® Green dyes. This kit is ideal for calibrating the green (FL1) channel of a flow cytometer. Microspheres in the LinearFlow Orange Kit are spectrally similar to phycoerythrins and tetramethylrhodamine conjugates, making this kit useful for calibrating the orange (FL2) channel. Microspheres in the LinearFlow Carmine Kit exhibit excitation and emission spectra similar to those of propidium iodide/DNA complexes and Texas Red conjugates, whereas those in the LinearFlow Deep Red Kit match Cy™5 and allophycocyanin conjugates; both kits are suitable for calibrating the red (FL3) channel.

Sampler Kits for Flow Cytometry Reference Standards

Flow Cytometry Alignment Standards Sampler Kits

We are pleased to offer two convenient and cost-effective Flow Cytometry Alignment Standards Sampler Kits (F-7321, F-7322) that contain introductory sizes (one-tenth volumes) of either the three AlignFlow 2.5 µm beads or the three AlignFlow Plus 6.0 µm beads for UV, 488 nm and 633 nm excitation.

Flow Cytometry Compensation and Intensity Standards Kit

The Flow Cytometry Compensation and Intensity Standards Kit (F-7320) features introductory sizes (one-tenth volumes) of both the CompenFlow Flow Cytometry Compensation Kit and the LinearFlow Green (505/515) Flow Cytometry Intensity Calibration Kit, providing scientists with an inexpensive way to try several of our new flow cytometry reference standards.

26.2 Price List *Flow Cytometry Reference Standards*

Cat #		Product Name	Unit Size	Price Per Unit ($) 1–4 Units	5–24 Units
A-7304	***New***	AlignFlow™ flow cytometry alignment beads, 2.5 µm *for UV excitation*	3 mL	120.00	96.00
A-7302	***New***	AlignFlow™ flow cytometry alignment beads, 2.5 µm *for 488 nm excitation*	3 mL	120.00	96.00
A-7312	***New***	AlignFlow™ flow cytometry alignment beads, 2.5 µm *for 633 nm excitation*	3 mL	120.00	96.00
A-7305	***New***	AlignFlow™ Plus flow cytometry alignment beads, 6.0 µm *for UV excitation*	3 mL	120.00	96.00
A-7303	***New***	AlignFlow™ Plus flow cytometry alignment beads, 6.0 µm *for 488 nm excitation*	3 mL	120.00	96.00
A-7313	***New***	AlignFlow™ Plus flow cytometry alignment beads, 6.0 µm *for 633 nm excitation*	3 mL	120.00	96.00
C-7301	***New***	CompenFlow™ Flow Cytometry Compensation Kit	1 kit	140.00	112.00
F-7321	***New***	Flow Cytometry Alignment Standards Sampler Kit, 2.5 µm	1 kit	55.00	44.00

Cat #		Product Name	Unit Size	Price Per Unit ($) 1–4 Units	5–24 Units
F-7322	***New***	Flow Cytometry Alignment Standards Sampler Kit, 6.0 µm	1 kit	55.00	44.00
F-7320	***New***	Flow Cytometry Compensation and Intensity Standards Kit	1 kit	45.00	36.00
F-7310	***New***	flow cytometry reference beads, DAPI-like, 2.5 µm	3 mL	85.00	68.00
F-7311	***New***	flow cytometry reference beads, DAPI-like, 6.0 µm	3 mL	85.00	68.00
L-7308	***New***	LinearFlow™ Carmine (580/620) Flow Cytometry Intensity Calibration Kit	1 kit	140.00	112.00
L-7314	***New***	LinearFlow™ Deep Red (633/660) Flow Cytometry Intensity Calibration Kit	1 kit	140.00	112.00
L-7306	***New***	LinearFlow™ Green (505/515) Flow Cytometry Intensity Calibration Kit	1 kit	140.00	112.00
L-7307	***New***	LinearFlow™ Orange (540/560) Flow Cytometry Intensity Calibration Kit	1 kit	140.00	112.00

26.3 Minifluorometer and Cuvettes

Molecular Probes is pleased to offer researchers what we consider to be one of the finest compact fluorometers in the marketplace today. We also provide an array of fluorometer accessories and cuvettes in order to meet the requirements for precise and accurate fluorometric measurement.

Turner Designs TD-700 Minifluorometer

In collaboration with Turner Designs, Inc., Molecular Probes offers the TD-700 Laboratory Fluorometer (F-8000) (Figure 26.6 and Table 26.3). Our scientists have found that, in quantitative assays utilizing a single pair of excitation/emission wavelengths, the quality of data produced by the TD-700 equals that obtainable with much more costly spectrofluorometers. This instrument therefore provides an ideal platform for researchers wishing to gain access to the sensitivity and convenience offered by fluorometric assays, without major expenditure on equipment. The TD-700 accommodates various combinations of light sources and optical filters depending on the wavelength range required for a particular application. For example, the TD-700 equipped with quartz–halogen lamp (Q-8010) and fluorescein-range optical filter set (F-8020) produces excellent results with the following products and assays:

- CyQUANT™ Cell Proliferation Assay Kit (see Section 16.2)
- PicoGreen® dsDNA quantitation reagent (see Section 8.3)
- OliGreen™ ssDNA quantitation reagent (see Section 8.3)
- NanoOrange™ Protein Quantitation Kit (see Section 9.1)
- EnzChek™ Protease Assay Kits (see Section 10.4)

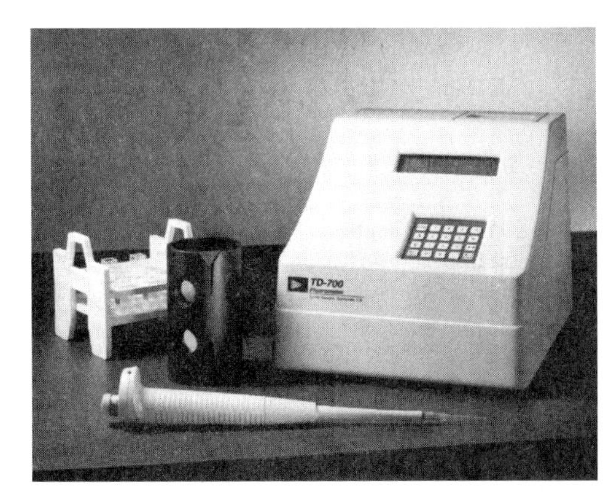

Figure 26.6 *Turner Designs TD-700 minifluorometer (F-8000).*

Table 26.3 *Specifications for the Turner minifluorometer (F-8000).*

Power	External power supply, 100–240 VAC, 30 watts maximum
Weight	5.9 kg (13 lbs)
Operating Temperature	60–95°F (15–36°C)
Display	16 × 2 character LCD (3.86" × 0.86"; 9.8 cm × 2.18 cm)
Keypad	4 × 5 keys (3" × 2.7"; 7.62 cm × 6.86 cm)
Data Output	100% ASCII format through a 9-pin RS-232 serial cable at 9600 baud
Sample Adapters	Includes adapter for 10 mm × 10 mm × 45 mm cuvettes (S-8060). Sample chamber accommodates optional adapters for 25 mm × 150 mm (S-8063) and 13 mm × 100 mm test tubes (S-8062). Optional adapter with short (15 mm × 7 mm) aperture (S-8064) for measurements on small (2 mL) samples in 10 mm × 10 mm × 45 mm cuvettes.
Filter Cylinder	Accommodates eight, 1-inch round filters (four excitation and four emission)
Detector	Factory-installed photomultiplier tube with wavelength range of 300–650 nm
Lamp	Quartz–halogen lamp (Q-8010) (20 watts; lamp life = 2000 hours) or low-pressure mercury-vapor lamp (M-8111, M-8112) (4 watts; lamp life = 8000 hours)
Readout	Fluorescence intensity or calibrated concentration
Calibration	Multipoint calibration for concentration measurement or single point fluorescence intensity calibration
Blank	Reads and subtracts blank data
Discrete Sample Averaging	Sample readings can be averaged to improve accuracy (7-second delay; 12-second signal averaging; 5-second display readout)
Printer	Optional Star® DP 8340 printer (P-8051)

Many other applications are feasible, including detection and quantitation of β-galactosidase, alkaline phosphatase and other enzymes (see Chapter 10); bacterial viability analysis using SYTOX Green nucleic acid stain (see Chapter 16); and analysis of metal ions in environmental samples with fluorescent ion indicators (see Chapter 22). Some dual-wavelength measurement applications, including bacterial viability assays with our LIVE/DEAD® *Bac*Light™ Kits (see Section 16.2) and fatty acid quantitation with our ADIFAB fatty acid indicator (see Section 20.6), have also been successfully implemented on the TD-700. Suitable optical filter combinations for dual-wavelength measurements can be assembled from Molecular Probes' range of Omega® Optical filters (see Section 26.5); for assistance with selection, please contact our Technical Assistance Department.

Features of the TD-700

- Compact footprint: 9.25" width × 11" depth × 8.25" height (23.5 cm width × 28 cm depth × 21 cm height)
- Cylindrical optical filter holder providing easy interchange of excitation/emission wavelengths
- Sample compartment accommodating cuvette or test tube sample configurations
- Automatic range optimization, ensuring correct selection of photometric gain and measurement range for each set of samples
- Multipoint calibration, allowing up to five calibration points to be stored for quantitative analysis applications
- ASCII data output that interfaces to a PC or serial printer

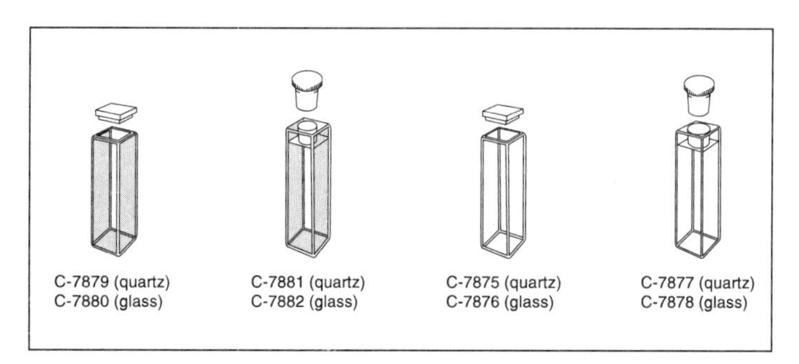

Figure 26.7 *Cuvettes for absorption and fluorescence spectroscopy. Usable wavelength ranges: quartz, 170 nm to 2700 nm; glass, 334 nm to 2500 nm. Cuvette dimensions are 10 mm × 10 mm × 45 mm. Absorption cuvettes are offered with Teflon® covers (C-7879, C-7880) or Teflon stoppers (C-7881, C-7882). Fluorescence cuvettes are offered with Teflon covers (C-7875, C-7876) or Teflon stoppers (C-7877, C-7878).*

Absorption and Fluorescence Cuvettes

We offer 10 mm × 10 mm × 45 mm sample cuvettes for fluorometric and spectrophotometric applications. These cuvettes are available in both optical-quality glass and quartz, with either a Teflon cover or Teflon stopper (Figure 26.7). The glass cuvettes have a usable range of 334 nm to 2500 nm, whereas the quartz cuvettes have a usable range of 170 nm to 2700 nm.

26.3 Price List *Minifluorometer and Cuvettes*

Ordering Information

To obtain a complete TD-700 package you must order three separate items: fluorometer, lamp and optical filter set (see the product list at the end of this section). Lamp and optical filter set selection depends on the application for which the instrument is to be used. Please contact our Technical Assistance Department if you have questions about the most suitable options for a particular application.

At present the TD-700 is available directly from Molecular Probes in the United States only. We invite inquiries from other countries through our Technical Assistance Department; we will endeavor to put interested customers in contact with a local supplier. The instrument is backed by Turner Designs' one-year warranty.

Cat #		Product Name	Unit Size	Price Per Unit ($) 1–4 Units	5–24 Units
		Fluorometer			
F-8000	***New***	fluorometer, Turner Designs TD-700, with standard photomultiplier *includes S-8060 sample adaptor*	each	3765.00	3765.00
		Lamps			
M-8112	***New***	mercury vapor lamp for TD-700 fluorometer, ultraviolet range *coumarin excitation*	each	45.00	45.00
M-8111	***New***	mercury vapor lamp for TD-700 fluorometer, visible range *rhodamine excitation*	each	55.00	55.00
Q-8010	***New***	quartz-halogen lamp assembly for TD-700 fluorometer *fluorescein excitation*	each	405.00	405.00
		Filter Sets			
F-8020	***New***	filter set for TD-700 fluorometer, fluorescein range *excitation 481-491 nm; emission 510-700 nm*	each	295.00	295.00
F-8021	***New***	filter set for TD-700 fluorometer, rhodamine range *excitation 545-555 nm; emission 570-700 nm*	each	365.00	365.00
F-8022	***New***	filter set for TD-700 fluorometer, ultraviolet range *excitation 300-400 nm; emission 410-610 nm*	each	215.00	215.00
		Accessories			
F-8072	***New***	filter O-rings for TD-700 fluorometer, set of 4	each	10.00	10.00
P-8071	***New***	paper for Star® DP 8340 printer, set of 5 rolls	each	35.00	35.00
P-8051	***New***	printer, Star® DP 8340, with power supply and serial cable	each	410.00	410.00
Q-8070	***New***	quartz-halogen replacement bulb for TD-700 fluorometer	each	50.00	50.00

Cat #		Product Name	Unit Size	Price Per Unit ($) 1–4 Units	5–24 Units
S-8064	*New*	sample adaptor for TD-700 fluorometer, 15 mm x 7 mm aperture, for 10 mm x 10 mm x 45 mm cuvettes	each	190.00	190.00
S-8060	*New*	sample adaptor for TD-700 fluorometer, 25 mm x 6 mm aperture, for 10 mm x 10 mm x 45 mm cuvettes	each	190.00	190.00
S-8062	*New*	sample adaptor for TD-700 fluorometer, fixed aperture, for 13 mm diameter test tubes	each	95.00	95.00
S-8063	*New*	sample adaptor for TD-700 fluorometer, fixed aperture, for 25 mm diameter test tubes	each	95.00	95.00
		Cuvettes			
C-7880	*New*	cuvette for absorption, optical glass, 10 mm x 10 mm, with Teflon® cover *useable range 334 to 2500 nm*	each	25.00	25.00
C-7882	*New*	cuvette for absorption, optical glass, 10 mm x 10 mm, with Teflon® stopper *useable range 334 to 2500 nm*	each	65.00	65.00
C-7879	*New*	cuvette for absorption, quartz, 10 mm x 10 mm, with Teflon® cover *useable range 170 to 2700 nm*	each	75.00	75.00
C-7881	*New*	cuvette for absorption, quartz, 10 mm x 10 mm, with Teflon® stopper *useable range 170 to 2700 nm*	each	98.00	98.00
C-7876	*New*	cuvette for fluorescence, optical glass, 10 mm x 10 mm, with Teflon® cover *useable range 334 to 2500 nm*	each	55.00	55.00
C-7878	*New*	cuvette for fluorescence, optical glass, 10 mm x 10 mm, with Teflon® stopper *useable range 334 to 2500 nm*	each	78.00	78.00
C-7875	*New*	cuvette for fluorescence, quartz, 10 mm x 10 mm, with Teflon® cover *useable range 170 to 2700 nm*	each	110.00	110.00
C-7877	*New*	cuvette for fluorescence, quartz, 10 mm x 10 mm, with Teflon® stopper *useable range 170 to 2700 nm*	each	135.00	135.00

26.4 Photographic Filters for Gel Electrophoresis

Molecular Probes offers several unique fluorescent reagents for staining nucleic acids and proteins that have been separated by gel electrophoresis. Pre-eminent among these gel stains are our SYBR Green nucleic acid gel stains (see Section 8.3) and SYPRO protein gel stains (see Section 9.2). To photograph such gels with optimal sensitivity, it is essential to select a photographic filter with spectral properties that closely match those of the fluorescent dye used. Molecular Probes provides three filters optimized for black-and-white photography of gels stained with fluorescent reagents. These 75 mm × 75 mm gelatin filters may be used with our gel photographic filter holder (G-6657), a metal frame that protects the filter and keeps it flat during photography. The photographic filters fit securely into the filter carrier of most camera setups used in molecular biology laboratories, or they may be cut to fit specialized holders.

SYBR Green Gel Stain Photographic Filter

For optimal sensitivity with Polaroid® 667 black-and-white print film and UV illumination, DNA or RNA gels stained with our proprietary SYBR Green I nucleic acid gel stain or SYBR Green II RNA gel stain should be photographed through Molecular Probes' SYBR Green gel stain photographic filter (S-7569).

Ethidium Bromide Gel Stain Photographic Filter

For nucleic acids stained conventionally with ethidium bromide, Molecular Probes offers the ethidium bromide gel stain photographic filter (E-7591) for use with black-and-white photography.

SYPRO Orange/Red Protein Gel Stain Photographic Filter

Protein gels stained with our new SYPRO Orange or SYPRO Red protein gel stains are optimally photographed using Polaroid 667 black-and-white print film through our SYPRO Orange/Red protein gel stain photographic filter (S-6656).

26.4 Price List *Photographic Filters for Gel Electrophoresis*

Cat #		Product Name	Unit Size	Price Per Unit ($) 1–4 Units	5–24 Units
E-7591	*New*	ethidium bromide gel stain photographic filter	each	29.00	29.00
G-6657	*New*	gel photographic filter holder	each	18.00	18.00
S-7569	*New*	SYBR® Green gel stain photographic filter	each	29.00	29.00
S-6656	*New*	SYPRO® Orange/Red protein gel stain photographic filter	each	35.00	35.00

26.5 Optical Filters for Fluorescence Microscopy

Molecular Probes offers high-quality Omega Optical filter sets for fluorescence microscopy (see Tables 26.5, 26.6 and 26.7). In addition to designing new optical filters, Omega Optical continually refines their existing filters to meet new requirements of fluorescence microscopists. Molecular Probes can assist you in choosing the best optical filter set for your application, as well as the best probes for your current instrumentation. The process of choosing a particular filter combination may involve a straightforward recommendation or a complex analysis of the spectral relationships of dyes and optical filters. Molecular Probes works closely with Omega Optical to make certain that optical filter sets are optimized to match the spectral characteristics of our fluorescent dyes. We also strive to offer both bandpass and longpass optical filter sets for our most common single emission dyes, providing researchers with the greatest flexibility in designing their experiments.

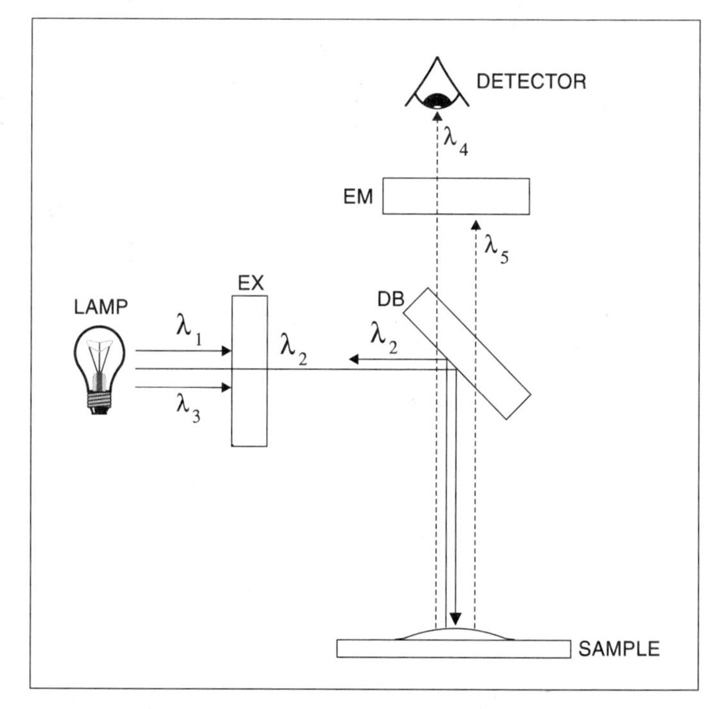

Figure 26.8 *Functions of fluorescence microscope filter set components. The desired excitation wavelength (λ_2) is selected from the spectral output of the lamp by the excitation filter (EX) and directed to the sample via the dichroic beamsplitter (DB). The beamsplitter separates emitted fluorescence (---) from scattered excitation light (—). The emission filter (EM) selectively transmits a portion of the sample's fluorescence emission (λ_4) for detection and blocks other emission components (λ_5).*

Optical Filter Definition and Design

Fluorescence is one of the most sensitive, yet demanding, areas in microscopy. The ability to selectively excite and detect dye fluorescence in the presence of extraneous light is crucial, given that potential background sources are typically tens of thousands of times greater than emission intensities. In addition, it is important to collect as much of the emission as possible because the total fluorescence of a dye is limited by its photobleaching rate.

The Optical Filter Set

A set of optical filters for selective excitation and detection of fluorescence typically consists of a minimum of three components: an excitation filter, a dichroic mirror (aka dichroic beamsplitter) and an emission filter (aka barrier filter) (Figure 26.8). The excitation filter selectively transmits a portion of the emission from the light source (Table 26.4). The dichroic mirror then reflects the selected light, directing it to the sample. When the sample fluoresces, the light is transmitted back through the dichroic mirror to the emission filter. The emission filter efficiently blocks any residual excitation light that may be reflected from glass or metal surfaces.

Table 26.4 *Fluorescence excitation sources.*

Source	Principal Lines (nm)
Mercury-arc lamp	366, 405, 436, 546, 578 *
Xenon-arc lamp	250–1000 *
Tungsten–halogen lamp	350–1000 *
Cr:LiSAF laser	430 †
Helium–cadmium laser	325, 442
Argon-ion laser	488, 514
Nd:YAG laser	532 †
Helium–neon laser	543, 594, 633
Krypton-ion laser	568, 647

* Continuous white-light source. † Frequency-doubled principal line output.

The Trade-Off in Optical Filter Set Design

For optimal fluorescence detection, the excitation and emission filters should be centered on the dye's absorption and emission peaks. To maximize the signal, you may choose excitation and emission filters with wide bandwidths. However, this strategy may result in unacceptable overlap of the emission signal with the excitation signal, resulting in poor resolution. To minimize spectral overlap, you may instead choose excitation and emission filters that are narrow in bandwidth and well-separated spectrally to increase signal isolation. This approach will reduce optical noise but may also reduce the signal strength to unacceptable levels. Omega Optical carefully considers this trade-off between signal isolation and signal intensity in designing its optical filter sets and seeks to find the optimal balance between these two parameters. In addition, Omega Optical filters are comprised of materials with minimal intrinsic fluorescence and are coated using advanced optical fabrication techniques.

Selecting an Optical Filter Set

The optical filter sets presented here have been selected to optimize the fluorescence excitation and emission of a number of

Table 26.5 Bandpass and longpass filter sets.

Cat #	Excitation *	Dichroic *	Emission *	Cat #	Excitation *	Dichroic *	Emission *
O-5700	254 ± 12.5	290	330 ± 30	O-5836 †	500 ± 12.5	525	545 ± 17.5
O-5701	330 ± 40	400	450 ± 32.5	O-5837 †	500 ± 12.5	525	≥530
O-5702	330 ± 40	400	≥400	O-5719	510 ± 11.5	550	605 ± 25
O-5704	365 ± 12.5	400	450 ± 32.5	O-5749	510 ± 11.5	540	≥550
O-5706	365 ± 12.5	420	535 ± 17.5	O-5721	515 ± 1.0	565	≥590
O-5703	365 ± 12.5	390	≥400	O-5834 †	525 ± 22.5	557	645 ± 37.5
O-5705	365 ± 12.5	390	≥515	O-5835 †	525 ± 22.5	557	≥565
O-5707	380 ± 7.5	410	450 ± 32.5	O-5723	535 ± 17.5	575	590 ± 17.5
O-5708	400 ± 7.5	420	450 ± 32.5	O-5724	535 ± 17.5	570	605 ± 25
O-5801	405 ± 20	450	520 ± 20	O-5725	535 ± 17.5	575	635 ± 27.5
O-5839	405 ± 5	420	≥435	O-5726	535 ± 17.5	585	660 ± 16
O-5748	405 ± 20	450	≥450	O-5722	535 ± 17.5	565	≥590
O-5709	425 ± 22.5	475	535 ± 27.5	O-5728	546 ± 5	560	580 ± 15
O-5710	425 ± 22.5	475	≥515	O-5729	546 ± 5	564	≥590
O-5803	445 ± 15	475	≥480	O-5727	550 ± 15	575	615 ± 22.5
O-5720	450 ± 27.5	495	535 ± 17.5	O-5731	560 ± 20	595	610 ± 10
O-5802	450 ± 22.5	495	≥500	O-5732	560 ± 20	595	635 ± 27.5
O-5711	455 ± 35	505	≥515	O-5859 †	560 ± 27.5	595	645 ± 37.5
O-5713	470 ± 17.5	515	560 ± 7.5	O-5730	560 ± 20	590	≥600
O-5712	470 ± 17.5	505	≥515	O-5733	580 ± 13.5	600	630 ± 15
O-5833 †	475 ± 20	505	535 ± 22.5	O-5734	590 ± 22.5	630	660 ± 16
O-5714	480 ± 30	560	635 ± 27.5	O-5735	610 ± 10	645	670 ± 20
O-5715	485 ± 11	505	530 ± 15	O-5838	610 ± 35	660	690 ± 20
O-5716	485 ± 11	505	535 ± 17.5	O-5736	633 ± 1.5	645	670 ± 20
O-5717	485 ± 11	505	≥530	O-5737	640 ± 10	670	682 ± 11
O-5718	490 ± 10	550	660 ± 25	O-5738	670 ± 10	690	≥700
				O-5739	740 ± 12.5	770	≥780

* Measured in nm. A 530 nm bandpass filter with a 30 nm bandwidth is specified by 530 ± 15 nm in the Emission column. A 530 nm longpass filter is specified by ≥ 530 in the Emission column. † VIVID™ single dye filter sets are especially designed for applications that yield low signals. These sets pass maximum excitation light to the sample and capture as much emission light as possible.

Molecular Probes' dyes. Some dyes can be used with more than one optical filter set, as indicated in Table 26.8 on page 612. Thus, optimal excitation — both with different light sources and for multiple-dye applications — can be easily achieved. Excitation filters in some optical filter sets do not overlap precisely with the dye's excitation peak, but instead with a peak corresponding to a principal spectral line of the excitation source (Table 26.4). The spectrum of the xenon-arc lamp is almost continuous, allowing the user to select an optical filter that exhibits optimum overlap of transmittance with a dye's excitation peak. Some excitation filters are specifically designed to be used interchangeably with either a mercury-arc lamp or an alternate excitation source.

Emission filters can provide either longpass or bandpass wavelength transmission. For example, a typical longpass emission filter

Table 26.6 *Filter sets for ratioable indicators.*

Cat #	Excitation #1 *	Excitation #2 *	Dichroic #1 *	Dichroic #2 *	Emission #1 *	Emission #2 *
O-5740	340 ± 7.5	380 ± 7.5	430		510 ± 20	
O-5741	440 ± 10	490 ± 10	515		535 ± 12.5	
O-5742	440 ± 10	490 ± 10	515		660 ± 25	
O-5743	510 ± 11.5	555 ± 13.5	585		620 ± 17.5	
O-5744	355 ± 7.5		390	455	405 ± 21.5	495 ± 10
O-5745	485 ± 11		505		530 ± 15	≥590
O-5746	510 ± 11.5		540	610	580 ± 15	640 ± 16.5
O-5747	515 ± 1		540	610	580 ± 15	640 ± 16.5

* Measured in nm.

might transmit all wavelengths ≥530 nm, whereas a typical bandpass filter might transmit only wavelengths between 515 and 545 nm. Longpass filters should be used when the application requires maximum emission collection or when spectral discrimination is not an issue, a circumstance that sometimes occurs when only a single emitting species is involved or when multiple emitting species are well-separated spectrally. Such is the case for the reagents included in the LIVE/DEAD Viability/Cytotoxicity Kit (see Section 16.2).

Bandpass filters should be used when spectral discrimination is necessary. Relatively broadband filter sets are designed to maximize image brightness and clarity while maintaining high signal isolation. These sets are ideal for direct or video-enhanced fluorescence microscopy. The narrowband filter sets are designed to maximize the signal-to-noise ratio for applications where wavelength isolation is more important than image brightness. Some applications may require the use of sensitive photomultiplier (PMT) detectors, as in confocal laser scanning microscopy. Alternatively, a linear photometric charge-coupled device (CCD), diode array or intensified video camera may be employed for quantitative imaging or microspectrofluorometry. The spectral sensitivity of the detection system should also be considered in order to achieve optimum detector signal-to-noise or accurate color rendition.

Table 26.5 lists our longpass and bandpass optical filter sets for single dyes, along with the specifications for their associated excitation filter, dichroic mirror and emission filter. Optical filter sets have also been designed for use with Molecular Probes' ratiometric dyes, including several of our pH and ion indicators (Table 26.6). Multiband filter sets enable microscopists to *simultaneously* excite and detect two, three or even four fluorophores. Molecular Probes and Omega Optical offer several choices of high-performance multiband filter sets for use in multiple-dye applications (Table 26.7).

Omega Optical's center wavelengths (CWLs) and full bandwidths at half-maximum transmissions (FWHMs) are carefully selected to optimize each dye's signal-to-noise ratio. By comparing the filter specifications with the excitation and emission peaks of the fluorophore, you can be assured of selecting the best optical filters for your application. These comparisons should be made with care because some dyes may have significantly different spectral properties in your application from those reported for the dye in solution. For example, the spectral characteristics of many nucleic acid stains depend on whether the dyes are in aqueous solution or bound to DNA or RNA. Similarly, styryl dyes have emission maxima that depend on whether they are dissolved in solvent or associated with membranes. To help the researcher choose the correct optical filter set, Molecular Probes has provided a list of representative dyes and associated filter sets for fluorescence microscopy (Table 26.8). This table includes spectral characteristics for the most common environment in which the dye would be found in most experimental applications. We welcome consultation with our Technical Assistance Department for help in selecting the correct optical filter for your application.

Dichroic Mirrors for Photoactivation of Caged Compounds

In addition to a wide range of filter sets, Molecular Probes provides three dichroic mirrors (O-5804, O-5805, O-5806) designed by Omega Optical specifically for use with caged compounds. All

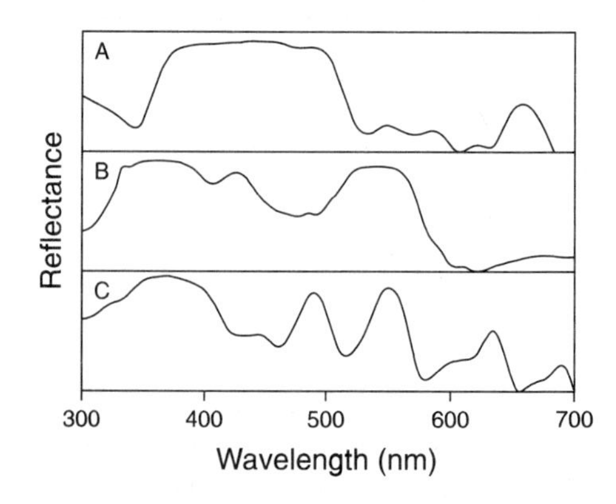

Figure 26.9 *Reflectance spectra of dichroic mirrors for use with caged compounds: A) O-5804, B) O-5805 and C) O-5806.*

Table 26.7 *Multiple-dye filter sets for two, three and four dyes.*

Cat #	Dye Combination
Two-Dye Filter Sets	
O-5850	DAPI/fluorescein
O-5840	DAPI/ tetramethylrhodamine
O-5852	Fluorescein/tetramethylrhodamine
O-5854	Fluorescein/Texas Red
O-5853	Fluorescein/allophycocyanin
Three-Dye Filter Sets	
O-5857	DAPI/fluorescein/tetramethylrhodamine
O-5848 *	DAPI/fluorescein/tetramethylrhodamine
O-5846 †	DAPI/fluorescein/tetramethylrhodamine
O-5855	DAPI/fluorescein/Texas Red
O-5844 ‡	DAPI/fluorescein/Texas Red
O-5847 §	DAPI/fluorescein/Texas Red
O-5858	DAPI/fluorescein/propidium iodide
Four-Dye Filter Set	
O-5856	DAPI/fluorescein/tetramethylrhodamine/Cy™5

* Similar to O-5846 but has higher red transmission. † Yields a bright orange signal and works well with propidium iodide. ‡ Contains very narrow excitation bands that are optimized for the mercury-arc lamp and result in reduced sample photobleaching. § Similar to O-5855 but has higher red transmission.

three dichroics have exceptionally high reflectance in the near UV, as well as in one or more regions of the visible spectrum (Figure 26.9). They can be used in several different configurations, both for photoactivation of caged compounds and for excitation/visualization of green or orange fluorescence. These dichroic mirrors are usually used in conjunction with interference filters for discrete photoactivation of caged compounds and excitation of fluorophores. The dichroic mirrors are typically configured in either the epi-illumination or the transmitted light path:

- **Epi-illumination path**. The dichroic mirrors may be used in conjunction with a manual or automated filter changer, scanning monochromator or dual illuminator. Because the dichroic mirror will reflect both UV light for uncaging and visible light for fluorescence excitation, the sample can be exposed to light of multiple wavelengths, either simultaneously or sequentially, without changing the dichroic mirror.

- **Transmitted light path**. The dichroic mirrors may also be mounted inside adapters used in place of the standard transmitted light condenser. A standard high numerical aperture, UV-transmitting objective lens is substituted for the condenser, and UV light is directed onto the specimen via the dichroic mirror, which is placed behind this objective lens. Filtered light from a continuous arc source or flashlamp is directed through an iris onto the dichroic mirror by a light guide. Because the dichroic mirror reflects light at one or more visible wavelengths in addition to UV wavelengths, an interference filter of longer wavelength can be placed in the light path to allow visualization of the area to be photoactivated.

Optical Filter Sets and Individual Filters

Optical Filter Sets

Molecular Probes offers Omega Optical filters configured for Leitz-Leica, Nikon, Olympus and Zeiss microscopes. If you have requirements for optical filters compatible with other types of instruments, please contact our Technical Assistance Department.

These optical filters may be purchased as unmounted optical filter *sets* (containing excitation filter(s), dichroic mirror(s) and emission filter(s)), as well as mounted and aligned in the appropriate filter holder. Molecular Probes offers a wide assortment of filter holders for Leitz-Leica, Nikon, Olympus and Zeiss microscopes; just specify the make and model of your instrument when placing your order. For Nikon microscopes, filter sets are *only* offered mounted and aligned in a Nikon filter holder. We also offer individual unmounted excitation filters, dichroic mirrors and emission filters.

Technical Assistance

Optimal selection of optical filter sets for fluorescence applications requires consideration of several variables. We invite you to contact Molecular Probes' Technical Assistance Department for advice. When calling, please be prepared to describe the dye(s), instrumentation and method of detection you are using. A Technical Assistance Representative will then advise you on the most effective filter configuration for your purposes. Requests for special filter mounting and alignment services and for filters with custom center wavelengths, bandwidths and sizes will be forwarded to the engineering staff at Omega Optical. The feasibility of such requests must be assessed prior to pricing and ordering.

Ordering Information

To place your order for Omega Optical filters, call Molecular Probes' Customer Service Department. Please be prepared to give the following information to our representatives:

- Molecular Probes' catalog number (see Tables 26.5 through 26.6)
- Make and model of your instrument
- Your choice of a complete optical filter *set* or an *individual* excitation filter, dichroic mirror or emission filter
- Your choice of a *mounted* or *unmounted* optical filter set

Table 26.8 *Spectral characteristics and suggested filter sets for Molecular Probes' dyes.*

Dye	Absorption Maximum (nm)	Emission Maximum (nm)	Bandpass Filter Sets	Longpass Filter Sets
Acridine orange	500 (+DNA) 460 (+RNA)	526 (+DNA) 650 (+RNA)	O-5715 (DNA) O-5714 (RNA)	O-5712 (DNA + RNA)
Allophycocyanin	650	660	O-5735, O-5737	
7-Aminoactinomycin D (7-AAD) *	546	647	O-5725, O-5726	O-5722
AMCA-S, AMC †	345	445	O-5701, O-5704	O-5702, O-5703
BCECF	482 (low pH) 503 (high pH)	520 (low pH) 528 (high pH)	O-5741 ‡	
BOBO-1, BO-PRO-1	462	481	O-5709	O-5803
BODIPY FL	505	513	O-5715, O- 5716,	O-5717
BODIPY TMR	542	574	O-5723, O-5724, O-5728	O-5722, O-5729
BODIPY TR	589	617	O-5732, O-5733, O-5734	O-5730
Calcein	494	517	O-5715, O-5716,	O-5717
Calcium Crimson indicator	590	615	O-5733	O-5730
Calcium Green indicators	506	533	O-5716, O-5836	O-5717, O-5749
Calcium Orange indicator	549	576	O-5728	O-5722
6-Carboxyrhodamine 6G	525	555	O-5836	O-5837
5-(and-6-)-Carboxy SNARF-1 indicator	548 (low pH) 576 (high pH)	587 (low pH) 635 (high pH)	O-5746 ‡	
Cascade Blue fluorophore	400	420	O-5704, O-5707, O-5708	O-5748
Cl-NERF	504 (low pH) 514 (high pH)	540 (low pH) 540 (high pH)	O-5741 ‡	
DAPI *	358	461	O-5701, O-5704	O-5702, O-5703
Di-8-ANEPPS §	488	605	O-5742 ‡	
DiA §	456	590	O-5714	O-5711
DiD ($DiIC_{18}(5)$)	644	665	O-5735, O-5737	
DiI ($DiIC_{18}(3)$)	549	565	O-5723, O-5724, O-5728	O-5729
DiO ($DiOC_{18}(3)$)	484	501	O-5715, O-5716, O-5720	O-5717
DM-NERF	497 (low pH) 510 (high pH)	527 (low pH) 536 (high pH)	O-5741 ‡	
ELF-97 alcohol	345	530	O-5706	O-5702, O-5703, O-5705
Eosin	524	544	O-5836	O-5749, O-5837
EthD-1 *	528	617	O-5725	O-5717, O-5722

Table 26.8, continued *Spectral characteristics and suggested filter sets for Molecular Probes' dyes.*

Dye	Absorption Maximum (nm)	Emission Maximum (nm)	Bandpass Filter Sets	Longpass Filter Sets
Ethidium bromide *	518	605	O-5725	O-5722
Fluorescein	494	518	O-5715, O-5716	O-5717
Fluo-3	506	526	O-5716, O-5836	O-5717, O-5837
FM 1-43 §	479	598	O-5714	O-5717
Fura-2 indicator	363 (low [Ca^{2+}]) 335 (high [Ca^{2+}])	512 (low [Ca^{2+}]) 505 (high [Ca^{2+}])	O-5740 ‡	
Fura Red indicator	472 (low [Ca^{2+}]) 436 (high [Ca^{2+}])	657 (low [Ca^{2+}]) 637 (high [Ca^{2+}])	O-5742 ‡	
Hoechst 33258, Hoechst 33342 *	352	461	O-5701, O-5704	O-5702, O-5703
7-Hydroxy-4-methylcoumarin	360	455	O-5701	O-5702
Indo-1	346 (low [Ca^{2+}]) 330 (high [Ca^{2+}])	475 (low [Ca^{2+}]) 401 (high [Ca^{2+}])	O-5744 ‡	
Lissamine rhodamine B	570	590	O-5731	O-5730
Lucifer yellow CH	428	536	O-5709	O-5710
LysoSensor Blue DND-192, DND-167	374	425	O-5704, O-5707	O-5703
LysoSensor Green DND-153, DND-189	442	505	O-5720	O-5710
LysoSensor Yellow/Blue	384 (low pH) 329 (high pH)	540 (low pH) 440 (high pH)	O-5740 ‡, O-5744 ‡	
LysoTracker Green	504	511	O-5715, O-5716	O-5717
LysoTracker Yellow	534	551	O-5723	O-5722
LysoTracker Red	577	592	O-5731	O-5730
Magnesium Green indicator	506	531	O-5715, O-5716, O-5836	O-5717, O-5837
Magnesium Orange indicator	550	575	O-5723, O-5724, O-5728	O-5722, O-5729
MitoTracker Green FM	490	516	O-5715, O-5716	O-5717
MitoTracker Orange CMTMRos	551	576	O-5723, O-5728	O-5722, O-5729
MitoTracker Red CMXRos	578	599	O-5731	O-5730
NBD	465	535	O-5713, O-5716	O-5712, O-5717
Nile red §	549	628	O-5725, O-5727	O-5722
Oregon Green 488 fluorophore	496	524	O-5715, O-5716	O-5717
Oregon Green 500 fluorophore	503	522	O-5715, O-5716	O-5717
Oregon Green 514 fluorophore	511	530	O-5716, O-5836	O-5717, O-5837
POPO-1, PO-PRO-1 *	434	456	O-5708	O-5748, O-5839
Propidium iodide *	536	617	O-5725	O-5722
Rhodamine 110	496	520	O-5715, O-5716	O-5717

Table 26.8, continued *Spectral characteristics and suggested filter sets for Molecular Probes' dyes.*

Dye	Absorption Maximum (nm)	Emission Maximum (nm)	Bandpass Filter Sets	Longpass Filter Sets
Rhodamine Red	570	590	O-5731	O-5730
R-phycoerythrin	565	575	O-5723, O-5728	O-5722, O-5749
Resorufin	570	585	O-5727, O-5728	O-5729, O-5730
RH 414 §	500	635	O-5714	O-5717, O-5749
Rhodamine Green fluorophore	502	527	O-5715, O-5716	O-5717
Rhodol Green fluorophore	499	525	O-5715, O-5716	O-5717
Rhodamine 123	507	529	O-5716, O-5836	O-5717, O-5837
Sodium Green indicator	507	535	O-5716, O-5836	O-5717, O-5837
SYTOX Green nucleic acid stain *	504	523	O-5716, O-5836	O-5717, O-5837
SYTO green fluorescent nucleic acid stains 12, 13, 16, 21, 23, 24 *	494 ± 6	515 ± 7	O-5715, O-5716	O-5717
SYTO green fluorescent nucleic acid stains 11, 14, 15, 20, 22, 25 *	515 ± 7	543 ± 13	O-5716, O-5836	O-5717, O-5837
SYTO 17 red fluorescent nucleic acid stain *	621	634	O-5733	
Tetramethylrhodamine, Rhodamine B	555	580	O-5723, O-5724, O-5728	O-5722, O-5729
Texas Red	595	615	O-5732, O-5733, O-5734	O-5730
TOTO-1, TO-PRO-1 *	514	533	O-5716, O-5836	O-5717, O-5837
TOTO-3, TO-PRO-3 *	642	660	O-5735, O-5737	
X-Rhodamine	580	605	O-5731	O-5730
YOYO-1, YO-PRO-1 *	491	509	O-5715, O-5716	O-5717
YOYO-3, YO-PRO-3 *	612	631	O-5733	

* Excitation and emission peaks listed for dye/DNA complex. † AMC = 7-amino-4-methylcoumarin. ‡ For filter sets for ratiometric dyes, see Table 26.6. § Excitation and emission peaks listed for lipid-bound dye.

Custom Synthesis Services

Molecular Probes' custom synthesis services can provide an almost unlimited variety of organic compounds and bioconjugates, including labeled antibodies, immunogens and analogs of biologically active small molecules. We can prepare known compounds using published procedures, or if these are not available, we can apply our expertise to design and synthesize reagents. We also offer quantity discounts for multiple-unit purchases of most of our catalog products. Please contact our Custom and Bulk Sales Department to discuss your requirements.

26.6 Books and Videotape

Books from Molecular Probes

To augment our selection of fluorescent probes, we are pleased to offer several valuable reference books for scientists utilizing fluorescence techniques for their research applications. We recommend these recent editions to anyone interested in up-to-date guides on fluorescence methods.

A Practical Guide to the Study of Calcium in Living Cells

Volume 40 of *Methods in Cell Biology*, entitled *A Practical Guide to the Study of Calcium in Living Cells* (M-7890), describes popular techniques, along with helpful tips for optimizing intracellular Ca^{2+} measurements — one of the largest single applications of fluorescence in the biosciences. This laboratory guide, edited by Nuccitelli, is designed for researchers, graduate students and technicians in a wide variety of disciplines.

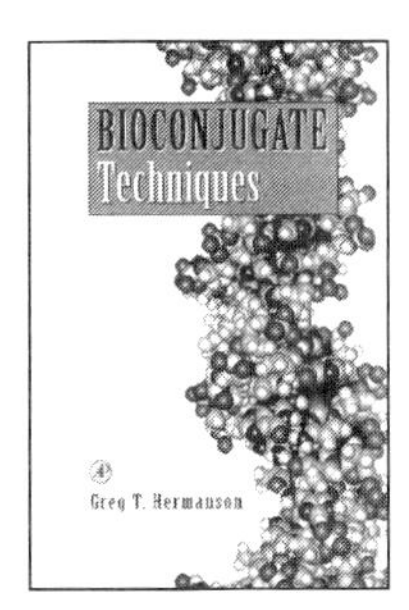

Bioconjugate Techniques

Bioconjugate Techniques (B-7884) by Hermanson is an essential guide that captures the entire field of bioconjugate chemistry in a single volume. This well-illustrated and highly referenced book provides researchers and clinicians with easy-to-follow protocols describing the preparation and use of bioconjugates for detecting, quantitating and analyzing biomolecules.

Fluorescent and Luminescent Probes for Biological Activity

The development of optical probes for biological activity has contributed greatly to important technical and experimental advances in the biomedical sciences. Fluorescent and luminescent probes have in many instances displaced radioisotopes as standard research tools, providing significant improvements in speed and ease of detection. *Fluorescent and Luminescent Probes for Biological Activity* (F-7889), a comb-bound guide edited by Mason, provides a comprehensive survey of the current scope of optical probe techniques. Also presented are detailed discussions of practical biological questions, underlying principles of optical and microelectronic instrumentation and future developments of optical probe technology.

Flow Cytometry, Second Edition

Flow Cytometry, comprising volumes 41 (Part A, M-7887) and 42 (Part B, M-7888) of *Methods in Cell Biology*, is a comprehensive comb-bound laboratory guide edited by Darzynkiewicz, Robinson and Crissman. This invaluable reference describes the essential flow cytometry techniques using a step-by-step format, supplemented by explanations and troubleshooting tips — a must for researchers, clinicians and students.

Handbook of Fluorescent Probes and Research Chemicals, Sixth Edition

Molecular Probes' *Handbook of Fluorescent Probes and Research Chemicals, Sixth Edition* (H-8999) is a comprehensive reference book of reagents for fluorescence applications. With 27 chapters organized according to application, this Handbook provides researchers with a thorough characterization of nearly 3000 reagents — including spectral information, structures and literature citations — as well as a survey of recent and potential applications. The first copy of this Handbook is sent free to our customers. Subsequent copies are available for purchase; multiple-copy discounts are also available.

CELLebration! Videotape

The CELLebration! videotape (C-7886), edited by Fink, celebrates the dynamic nature of life at the cellular level. CELLebration's contributors explore life at the intercellular and intracellular level using state-of-the-art imaging techniques. Sponsored by Molecular Probes and the American Society for Cell Biology, this NSTC-formatted videotape presentation demonstrates a number of fluorescence imaging applications, including Molecular Probes' MitoTracker™ mitochondrion-selective probes, fluo-3 Ca^{2+} indicator and fluorescently labeled tubulin. This video, offered in VHS format, is of special interest to educators, and a portion of the proceeds from its sale will go to the American Society for Cell Biology to support future educational initiatives.

26.6 Price List *Books and Videotape*

Cat #		Product Name	Unit Size	Price Per Unit ($) 1–4 Units	5–24 Units
B-7884	*New*	Bioconjugate Techniques. G.T. Hermanson, Academic Press (1996); 785 pages, soft cover	each	60.00	60.00
C-7886	*New*	CELLebration: The ASCB Educational Video Project. R. Fink, ed. Sinauer Associates (1995) *VHS format*	each	85.00	85.00
F-7889	*New*	Fluorescent and Luminescent Probes for Biological Activity. A Practical Guide to Technology for Quantitative Real-Time Analysis. W.T. Mason, ed. Academic Press (1993); 433 pages, comb bound	each	65.00	65.00
H-8999	*New*	Handbook of Fluorescent Probes and Research Chemicals, Sixth Edition. R.P. Haugland, Molecular Probes, Inc. (1996); 700 pages, soft cover	each	25.00	20.00
M-7890	*New*	Methods in Cell Biology, Volume 40: A Practical Guide to the Study of Calcium in Living Cells. R. Nuccitelli, ed. Academic Press (1994); 342 pages, comb bound	each	45.00	45.00
M-7887	*New*	Methods in Cell Biology, Volume 41: Flow Cytometry, Part A. Z. Darzynkiewicz, J.P. Robinson, H.A. Crissman, eds. Academic Press (1994); 591 pages, comb bound	each	59.00	59.00
M-7888	*New*	Methods in Cell Biology, Volume 42: Flow Cytometry, Part B. Z. Darzynkiewicz, J.P. Robinson, H.A. Crissman, eds. Academic Press (1994); 697 pages, comb bound	each	59.00	59.00

Hints for Finding the Product of Interest

This Handbook includes products released by Molecular Probes before September 1996. Additional products and search capabilities are available through our Web site (http://www.probes.com). All information, including price and availability, is subject to change without notice.

- **If you know the name of the product you want:** Look up the product by name in the **Master Price List/Product Index** at the back of the book. The Master Price List/Product Index is arranged alphabetically, by both product name and alternative name, using the alphabetization rules outlined on page vi. This index will refer you directly to all the Handbook sections that discuss the product.

- **If you DO NOT know the name of the product you want:** Search the **Keyword Index** for the words that best describe the applications in which you are interested. The Keyword Index will refer you directly to page numbers on which the listed topics are discussed. Alternatively, search the **Table of Contents** for the chapter or section title that most closely relates to your application.

- **If you only know the alpha-numeric catalog number of the product you want but not its name:** Look for the catalog number in the **Master Price List/Product Index** at the back of the book. First locate the group of catalog numbers with the same alpha identifier as the catalog number of interest and then scan this group for the correct numeric identifier. The Master Price List/Product Index will refer you directly to all the Handbook sections that discuss this product. Alternatively, you may use the search features available through our Web site (http://www.probes.com) to obtain the product name of any of our current products given only the catalog number (or numeric identifier).

- **If you only know part of a product name or are searching for a product introduced after publication of this Handbook:** Search Molecular Probes' up-to-date product list through our **Web site (http://www.probes.com)**. The product search feature on our home page allows you to search our entire product list, including products that have been added since the completion of this Handbook, using a partial product name, complete product name or catalog number. Product-specific bibliographies can also be obtained from our Web site once the product of interest has been located.

- **If you still cannot find the product that you are looking for or are seeking advice in choosing a product:** Please contact our Technical Assistance Department by phone, e-mail, fax or mail.

Chapter 27

Handbook Addendum

27.1 New Products from Molecular Probes

Molecular Probes continues to add innovative reagents to our current selection of nearly 3000 products. The products listed below were added to our inventory too late for inclusion into the appropriate Handbook chapter. The section(s) of this Handbook in which they belong are indicated in the far right column. Further technical information on these products is available from our Technical Assistance Department or through our Web site (http://www.probes.com). Also look for descriptions of these products in upcoming editions of *BioProbes*, our periodic technical publication. Once the electronic version of our Handbook debuts at our Web site, we will be continuously integrating new products into the appropriate chapters and sections. Visit Molecular Probes' home page often for the most up-to-date product listing and technical information.

Cat #		Product Name	Unit Size	Price Per Unit ($) 1–4 Units	Price Per Unit ($) 5–24 Units	Section #
A-6033	***New***	N-(2-aminoethyl)maleimide, trifluoroacetic acid salt	10 mg	48.00	38.40	2.4, 3.3, 5.2
A-6032	***New***	N-(5-aminopentyl)maleimide, trifluoroacetic acid salt	10 mg	48.00	38.40	2.4, 3.3, 5.2
A-6257	***New***	8-aminopyrene-1,3,6-trisulfonic acid, trisodium salt (APTS)	10 mg	95.00	76.00	3.2, 15.3
B-6187	***New***	6-(4-benzoylbenzamido)hexanoic acid, succinimidyl ester (BzBz-X, SE)	25 mg	45.00	36.00	5.3
B-6351	***New***	ε-biotinoyl-α-(9-fluorenylmethoxycarbonyl)-L-lysine (FMOC biocytin)	100 mg	175.00	140.00	4.2, 9.3
B-6357	***New***	S-biotinylhomocysteine	5 mg	55.00	44.00	4.3, 15.3
D-6226	***New***	2,4-dichloro-6-(4-(N,N-diethylamino)phenyl)-1,3,5-triazine	10 mg	78.00	62.40	1.8
D-7711	***New***	N-(4,4-difluoro-5,7-dimethyl-4-bora-3a,4a-diaza-s-indacene-3-dodecanoyl) sphingosyl phosphocholine (BODIPY® FL C_{12}-sphingomyelin) *1 mg/mL in DMSO*	250 µL	125.00	100.00	12.4, 20.6
D-6012	***New***	N-(5-((4,4-difluoro-1,3-dimethyl-5-(4-methoxyphenyl)-4-bora-3a,4a-diaza-s-indacene-2-propionyl)amino)pentyl)iodoacetamide (BODIPY® TMR cadaverine IA)	5 mg	95.00	76.00	2.2
D-6186	***New***	6-((4,4-difluoro-5-phenyl-4-bora-3a,4a-diaza-s-indacene-3-propionyl)amino) hexanoic acid, succinimidyl ester (BODIPY® R6G-X, SE)	5 mg	145.00	116.00	1.2
D-6011	***New***	N-(5-(((4-(4,4-difluoro-5-(2-thienyl)-4-bora-3a,4a-diaza-s-indacene-3-yl)phenoxy) acetyl)amino)pentyl)iodoacetamide (BODIPY® TR cadaverine IA)	5 mg	125.00	100.00	2.2
D-7713	***New***	1,2-dihexadecanoyl-N-(eosin-5-thiocarbamoyl)-*sn*-glycero-3-phosphoethanolamine, triethylammonium salt (eosin DHPE)	1 mg	75.00	60.00	13.4
D-7117	***New***	4,5-dimethoxy-2-nitrobenzyl bromide (BMNDMB)	100 mg	98.00	78.40	10.6, 19.2
D-7119	***New***	*trans*-2-(4,5-dimethoxy-2-nitrophenyl)nitroethene	5 mg	38.00	30.40	10.6, 19.2
D-6563	***New***	1,2-dioleoyl-3-(1-pyrenedecanoyl)-*rac*-glycerol	1 mg	65.00	52.00	13.5
D-6562	***New***	1,2-dioleoyl-3-(1-pyrenedodecanoyl)-*rac*-glycerol	1 mg	65.00	52.00	13.5
G-7493	***New***	globotriose GB_3, fluorescein conjugate	10 µg	175.00	140.00	17.3
M-6205	***New***	8-methoxypyrene-1,3,6-trisulfonyl chloride	10 mg	45.00	36.00	1.7
O-7490	***New***	Oregon Green™ 488 bisindolylmaleimide	250 µg	65.00	52.00	20.3
O-6034	***New***	Oregon Green™ 488 maleimide	5 mg	95.00	76.00	2.2
O-6185	***New***	Oregon Green™ 488-X, succinimidyl ester *6-isomer*	5 mg	95.00	76.00	1.4
P-7071	***New***	(µ-peroxo) (µ-hydroxo)bis(bis(bipyridyl)cobalt(III)) perchlorate (caged oxygen)	25 mg	95.00	76.00	19.2, 21.2
P-6169	***New***	N-(1-pyrenebutanoyl)cysteic acid, potassium salt	5 mg	65.00	52.00	1.7, 13.3
R-6029	***New***	Rhodamine Red™ C_2 maleimide	5 mg	95.00	76.00	2.2
R-7712	***New***	Rhodamine Red™-X 1,2-dihexadecanoyl-*sn*-glycero-3-phosphoethanolamine, triethylammonium salt (Rhodamine Red™-X DHPE)	1 mg	75.00	60.00	13.4
S-6902	***New***	Sulfite Blue™ coumarin	10 mg	95.00	76.00	24.3
T-7106	***New***	4,5,6,7-tetrafluorofluorescein bis-(5-carboxymethoxy-2-nitrobenzyl) ether, dipotassium salt (CMNB-caged TFF)	5 mg	155.00	124.00	15.3, 19.2
T-6568	***New***	1-(2-trifluoromethylphenyl)imidazole (TRIM)	100 mg	65.00	52.00	21.5

27.2 Sale Products from Molecular Probes

Molecular Probes has reduced the price of approximately 300 products listed below by up to 90%, with further discounts on purchases of multiple units. Because these products are scheduled to be discontinued once current stocks are depleted, we recommend that you purchase enough product for all planned experiments. A limited number of these products will continue to be available by custom synthesis; call our Custom and Bulk Sales Department for more information. For most of these products, technical information, including chemical structures, technical data, applications and references, is still available in the Fifth Edition of our *Handbook of Fluorescent Probes and Research Chemicals* or through our Web site (http://www.probes.com). These products are warranted to conform to their chemical descriptions or we will refund your purchase price; however, we cannot replace the materials with new batches or perform additional quality control on these products due to their limited availability. To help you locate products of interest, we have divided the sale products into categories corresponding to the chapters of this Handbook and then alphabetized them within these categories.

Chapter 1 — *Fluorophores and Their Amine-Reactive Derivatives*

Cat #	Product Name	Unit Size	Price Per Unit ($) 1–4 Units	Price Per Unit ($) 5–24 Units
A-1422	acridone-2-acetic acid	25 mg	20.00	16.00
A-1425	acridone-10-acetic acid	25 mg	20.00	16.00
A-2335	ATTO-TAG™ BQ Amine Derivatization Kit	1 kit	20.00	16.00
C-1423	7-carboxymethoxy-4-methylcoumarin	100 mg	20.00	16.00
C-1145	5-(3-carboxypropionyl)aminofluorescein (fluorescein succinamide)	100 mg	20.00	16.00
D-1915	5-dimethylaminonaphthalene-1-sulfonyl chloride *10% adsorbed on Celite®*	5 g	20.00	16.00
D-1557	4-(2,4-dinitroanilino)benzenesulfonyl chloride	100 mg	20.00	16.00
D-1620	N$^{\alpha}$-(2,4-dinitro-5-fluorophenyl)-L-alanine amide (Marfey's reagent; FDAA)	100 mg	20.00	16.00
M-188	7-methoxycoumarin-4-acetic acid	100 mg	20.00	16.00
S-1637	succinimidyl acrylate	100 mg	20.00	16.00
S-131	succinimidyl 9-anthracenecarboxylate	100 mg	20.00	16.00
S-412	succinimidyl 3-(9-carbazole)propionate	100 mg	20.00	16.00
S-330	succinimidyl 1-pyreneacrylate	25 mg	20.00	16.00
S-328	succinimidyl 1-pyrenenonanoate	25 mg	20.00	16.00

Chapter 2 — *Thiol-Reactive Probes*

Cat #	Product Name	Unit Size	Price Per Unit ($) 1–4 Units	Price Per Unit ($) 5–24 Units
A-408	2-anthraceneiodoacetamide	100 mg	20.00	16.00
A-98	anthraniloyl aziridine	100 mg	20.00	16.00
D-82	2,5-dimethoxystilbene-4′-maleimide	100 mg	20.00	16.00
E-568	4,4-epoxymethano-2,2,6,6-tetramethylpiperidine-1-oxyl	25 mg	20.00	16.00
N-197	N-(1-naphthyl)maleimide	100 mg	20.00	16.00
P-1512	4-phenylazophenylmaleimide	1 g	20.00	16.00
T-536	2,2,6,6-tetramethylpiperidin-1-oxyl-4-yl methanesulfonate	100 mg	20.00	16.00

Chapter 3 — *Reagents for Reaction with Groups Other than Thiols or Amines*

Cat #	Product Name	Unit Size	Price Per Unit ($) 1–4 Units	Price Per Unit ($) 5–24 Units
A-1462	7-acetoxycoumarin-4-carboxaldehyde	100 mg	20.00	16.00
A-457	5-((((2-aminoethyl)thio)acetyl)amino)fluorescein	25 mg	20.00	16.00
A-180	9-anthracenecarboxaldehyde	1 g	20.00	16.00
A-102	anthracene-9-carboxaldehyde carbohydrazone	100 mg	20.00	16.00
A-178	9-anthracenemethanol	1 g	20.00	16.00
B-1352	4′,5′-bis-(aminomethyl)fluorescein, dihydrochloride	25 mg	20.00	16.00
B-1457	4-(bromomethyl)-7,8-benzcoumarin	100 mg	20.00	16.00
C-357	5-(((2-(carbohydrazino)methyl)thio)acetyl)aminoeosin	25 mg	20.00	16.00
D-1403	4-(diazomethyl)-7-hydroxycoumarin	25 mg	20.00	16.00
D-450	2-diphenylacetyl-1,3-indandione-1-hydrazone	100 mg	20.00	16.00
M-1698	malachite green acrylamide	10 mg	20.00	16.00
M-1460	7-methoxycoumarin-4-carboxaldehyde	100 mg	20.00	16.00
M-1464	2-methoxy-1-naphthaldehyde	1 g	20.00	16.00
N-318	2-(7-nitrobenz-2-oxa-1,3-diazol-4-yl)methylaminoacetaldehyde	25 mg	20.00	16.00

Chapter 3 — *Reagents for Reaction with Groups Other than Thiols or Amines, continued*

Cat #	Product Name	Unit Size	Price Per Unit ($) 1–4 Units	5–24 Units
P-1466	2-(pentafluorophenyl)malondialdehyde	100 mg	20.00	16.00
P-105	1-pyrenecarboxaldehyde carbohydrazone	100 mg	20.00	16.00
P-410	1-pyreneglyoxal	100 mg	20.00	16.00
P-163	1-pyrenemethanol	1 g	20.00	16.00

Chapter 4 — *Biotins and Haptens*

Cat #	Product Name	Unit Size	1–4 Units	5–24 Units
N-1524	4-nitrophenyl D-biotin	100 mg	20.00	16.00

Chapter 5 — *Crosslinkers and Photoreactive Reagents*

Cat #	Product Name	Unit Size	1–4 Units	5–24 Units
A-1519	N-(2-aminoethyl)-4-azido-2-nitroaniline	100 mg	20.00	16.00
A-1502	4-azidobenzoic acid	1 g	20.00	16.00
N-1504	4-nitrophenyl bromoacetate	1 g	20.00	16.00
S-1581	succinimidyl 4-formylphenoxyacetate (SFPA)	100 mg	20.00	16.00
S-1667	succinimidyl 4-(((iodoacetyl)amino)methyl)cyclohexane-1-carboxylate (SIAC)	100 mg	20.00	16.00

Chapter 6 — *Fluorescence Detection Methods, Including FluoSpheres® and ELF® Technologies*

Cat #	Product Name	Unit Size	1–4 Units	5–24 Units
L-5343	latex, FluoSpheres®, amidine, 0.05 µm, 2% solids, red fluorescent (580/605)	10 mL	20.00	16.00
L-5340	latex, FluoSpheres®, amidine, 0.05 µm, 2% solids, blue fluorescent (365/415)	10 mL	20.00	16.00
L-5360	latex, FluoSpheres®, amidine, 0.5 µm, 2% solids, blue fluorescent (365/415)	10 mL	20.00	16.00
L-5341	latex, FluoSpheres®, amidine, 0.05 µm, 2% solids, yellow-green fluorescent (505/515)	10 mL	20.00	16.00
L-5217	latex, FluoSpheres®, carboxylate-modified, biotin-labeled, 0.03 µm, 1% solids, nonfluorescent	0.4 mL	20.00	16.00
L-5253	latex, FluoSpheres®, carboxylate-modified, biotin-labeled, 1.0 µm, 1% solids, red fluorescent (580/605)	0.4 mL	20.00	16.00
L-5269	latex, FluoSpheres®, carboxylate-modified, 0.5 µm, 2% solids, crimson fluorescent (625/645)	2 mL	20.00	16.00
L-5264	latex, FluoSpheres®, carboxylate-modified, 0.5 µm, 2% solids, dark red fluorescent (660/680)	2 mL	20.00	16.00
L-5284	latex, FluoSpheres®, carboxylate-modified, 1.0 µm, 2% solids, dark red fluorescent (660/680)	2 mL	20.00	16.00
L-5304	latex, FluoSpheres®, carboxylate-modified, 2.0 µm, 2% solids, dark red fluorescent (660/680)	2 mL	20.00	16.00
L-5186	latex, FluoSpheres®, carboxyl, 1.0 µm, 2% solids, Mango Red-colored (470-570 nm)	10 mL	20.00	16.00
L-5187	latex, FluoSpheres®, carboxyl, 1.0 µm, 2% solids, Royal Blue-colored (640-690 nm)	10 mL	20.00	16.00
L-5140	latex, FluoSpheres®, carboxyl, 0.2 µm, 2% solids, blue fluorescent (365/415)	10 mL	20.00	16.00
L-5144	latex, FluoSpheres®, carboxyl, 0.2 µm, 2% solids, dark red fluorescent (660/680)	2 mL	20.00	16.00
L-5119	latex, FluoSpheres®, polystyrene, 15 µm, 2% solids, crimson fluorescent (625/645)	1 mL	20.00	16.00
L-5113	latex, FluoSpheres®, polystyrene, 15 µm, 2% solids, red fluorescent (580/605)	1 mL	20.00	16.00
L-5133	latex, FluoSpheres®, polystyrene, 10 µm, 2% solids, red fluorescent (580/605)	1 mL	20.00	16.00
L-5110	latex, FluoSpheres®, polystyrene, 15 µm, 2% solids, blue fluorescent (365/415)	1 mL	20.00	16.00
L-5138	latex, FluoSpheres®, polystyrene, 10 µm, 2% solids, nile red fluorescent (535/575)	1 mL	20.00	16.00
L-5112	latex, FluoSpheres®, polystyrene, 15 µm, 2% solids, orange fluorescent (540/560)	1 mL	20.00	16.00
L-5045	latex, FluoSpheres®, sulfate, 0.2 µm, 2% solids, Canary Yellow-colored (400-470 nm)	10 mL	20.00	16.00
L-5085	latex, FluoSpheres®, sulfate, 1.0 µm, 2% solids, Canary Yellow-colored (400-470 nm)	10 mL	20.00	16.00
L-5009	latex, FluoSpheres®, sulfate, 0.02 µm, 2% solids, crimson fluorescent (625/645)	2 mL	20.00	16.00
L-5029	latex, FluoSpheres®, sulfate, 0.1 µm, 2% solids, crimson fluorescent (625/645)	2 mL	20.00	16.00
L-5049	latex, FluoSpheres®, sulfate, 0.2 µm, 2% solids, crimson fluorescent (625/645)	2 mL	20.00	16.00
L-5089	latex, FluoSpheres®, sulfate, 1.0 µm, 2% solids, crimson fluorescent (625/645)	2 mL	20.00	16.00
L-5109	latex, FluoSpheres®, sulfate, 2.0 µm, 2% solids, crimson fluorescent (625/645)	2 mL	20.00	16.00
L-5103	latex, FluoSpheres®, sulfate, 2.0 µm, 2% solids, red fluorescent (580/605)	2 mL	20.00	16.00
L-5100	latex, FluoSpheres®, sulfate, 2.0 µm, 2% solids, blue fluorescent (365/415)	2 mL	20.00	16.00
L-5044	latex, FluoSpheres®, sulfate, 0.2 µm, 2% solids, dark red fluorescent (660/680)	2 mL	20.00	16.00

Chapter 7 — *Protein Conjugates for Biological Detection, Including Immunoreagents and Avidin and Lectin Conjugates*

Cat #	Product Name	Unit Size	1–4 Units	5–24 Units
P-844	protein A, tetramethylrhodamine conjugate	1 mg	20.00	16.00

Chapter 8 — *Nucleic Acid Detection*

Cat #	Product Name	Unit Size	Price Per Unit ($) 1–4 Units	5–24 Units
A-1116	9-azidoacridine	100 mg	20.00	16.00
C-137	9-(4′-carboxyanilino)-6-chloro-2-methoxyacridine	100 mg	20.00	16.00
F-3623	FluoroTide™ M13/pUC (-21) primer, 5′-BODIPY® 558/568 labeled, X d(TGTAAAACGACGGCCAGT)	0.1 OD	20.00	16.00
F-3626	FluoroTide™ M13/pUC (-21) primer, 5′-(2,4-dinitrophenyl) labeled, X d(TGTAAAACGACGGCCAGT)	0.1 OD	20.00	16.00
F-3635	FluoroTide™ M13/pUC reverse primer, 5′-biotin labeled, X d(CAGGAAACAGCTATGACC)	0.1 OD	20.00	16.00
F-3632	FluoroTide™ M13/pUC reverse primer, 5′-BODIPY® FL labeled, X d(CAGGAAACAGCTATGACC)	0.1 OD	20.00	16.00
F-3634	FluoroTide™ M13/pUC reverse primer, 5′-BODIPY® 591/618 labeled, X d(CAGGAAACAGCTATGACC)	0.1 OD	20.00	16.00
F-3636	FluoroTide™ M13/pUC reverse primer, 5′-(2,4-dinitrophenyl) labeled, X d(CAGGAAACAGCTATGACC)	0.1 OD	20.00	16.00
F-3631	FluoroTide™ M13/pUC reverse primer, 5′-fluorescein labeled, X d(CAGGAAACAGCTATGACC)	0.1 OD	20.00	16.00
F-3655	FluoroTide™ SP6 primer, 5′-biotin labeled, X d(CGATTTAGGTGACACTATAG)	0.1 OD	20.00	16.00
F-3652	FluoroTide™ SP6 primer, 5′-BODIPY® FL labeled, X d(CGATTTAGGTGACACTATAG)	0.1 OD	20.00	16.00
F-3653	FluoroTide™ SP6 primer, 5′-BODIPY® 558/568 labeled, X d(CGATTTAGGTGACACTATAG)	0.1 OD	20.00	16.00
F-3654	FluoroTide™ SP6 primer, 5′-BODIPY® 591/618 labeled, X d(CGATTTAGGTGACACTATAG)	0.1 OD	20.00	16.00
F-3656	FluoroTide™ SP6 primer, 5′-(2,4-dinitrophenyl) labeled, X d(CGATTTAGGTGACACTATAG)	0.1 OD	20.00	16.00
F-3651	FluoroTide™ SP6 primer, 5′-fluorescein labeled, X d(CGATTTAGGTGACACTATAG)	0.1 OD	20.00	16.00
F-3642	FluoroTide™ T7 primer, 5′-BODIPY® FL labeled, X d(TAATACGACTCACTATAGGG)	0.1 OD	20.00	16.00
F-3643	FluoroTide™ T7 primer, 5′-BODIPY® 558/568 labeled, X d(TAATACGACTCACTATAGGG)	0.1 OD	20.00	16.00
F-3641	FluoroTide™ T7 primer, 5′-fluorescein labeled, X d(TAATACGACTCACTATAGGG)	0.1 OD	20.00	16.00

Chapter 9 — *Peptide and Protein Detection, Analysis and Synthesis*

Cat #	Product Name	Unit Size	1–4 Units	5–24 Units
D-665	1,9-dimethylmethylene blue chloride *high purity*	100 mg	20.00	16.00
N-1540	4-nitrophenyl ester of 6-(phosphocholine)hexanoic acid	25 mg	20.00	16.00
P-1541	6-(phosphocholine)hexanoic acid	25 mg	20.00	16.00

Chapter 10 — *Enzymes, Enzyme Substrates and Enzyme Inhibitors*

Cat #	Product Name	Unit Size	1–4 Units	5–24 Units
A-1436	7-acetoxy-3-cyanocoumarin	100 mg	20.00	16.00
A-396	8-acetoxypyrene-1,3,6-trisulfonic acid, trisodium salt (HPTS acetate)	100 mg	20.00	16.00
A-2895	5-acetylaminofluorescein di-β-D-galactopyranoside (C_2FDG)	5 mg	20.00	16.00
A-6478	2-amino-4-methoxynaphthalene p-toluenesulfonic acid salt (MNA) *reference standard*	10 mg	20.00	16.00
C-640	coenzyme A, 12-(N-methyl-N-(7-nitrobenz-2-oxa-1,3-diazol-4-yl)amino)dodecanoic acid ester	1 mg	20.00	16.00
C-1347	coenzyme A, 6-((7-nitrobenz-2-oxa-1,3-diazol-4-yl)amino)hexanoic acid ester	1 mg	20.00	16.00
C-1348	coenzyme A, 1-pyrenebutanoic acid ester	1 mg	20.00	16.00
D-419	6,7-diacetoxy-4-methylcoumarin	100 mg	20.00	16.00
D-2898	2′,7′-dichlorofluorescein di-β-D-galactopyranoside (DCFDG)	5 mg	20.00	16.00
D-431	2-dimethylaminonaphthalene-5-sulfonamide	25 mg	20.00	16.00
E-187	7-ethoxy-4-methylcoumarin	100 mg	20.00	16.00
E-184	ethyl 7-hydroxycoumarin-3-carboxylate	100 mg	20.00	16.00
F-482	fluorescein monomethyl ether methyl ester (MFME)	100 mg	20.00	16.00
M-483	5-(and-6)-methoxycarbonylfluorescein methyl ether methyl ester *mixed isomers*	25 mg	20.00	16.00
M-310	7-methoxy-4-methylcoumarin	100 mg	20.00	16.00
N-651	naphthofluorescein diacetate	100 mg	20.00	16.00
N-656	naphthofluorescein di-β-D-galactopyranoside	5 mg	20.00	16.00
N-5	p-nitrophenyl anthranilate	1 g	20.00	16.00
N-6454	5-nitrosalicylaldehyde (5-NSA)	1 g	20.00	16.00
P-2941	1-pyrenemethyl laurate	25 mg	20.00	16.00
P-166	1-pyrenesulfonamide	100 mg	20.00	16.00
R-365	resazurin acetate	25 mg	20.00	16.00

Chapter 11 — *Probes for Actin, Tubulin and Nucleotide-Binding Proteins*

Cat #	Product Name	Unit Size	1–4 Units	5–24 Units
C-1639	colchicine biotin-XX	5 mg	20.00	16.00
E-1020	1,N^6-ethenoadenosine, hydrochloride (ε-AD)	25 mg	20.00	16.00
E-1026	1,N^6-etheno-2-azaadenosine	10 mg	20.00	16.00
T-1040	8-(tetramethylrhodamine)thioadenosine 3′,5′-cyclicmonophosphate	5 mg	20.00	16.00

Chapter 12 — *Probes for Organelles*

Cat #	Product Name	Unit Size	Price Per Unit ($) 1–4 Units	5–24 Units
D-280	1,1′-dihexyl-3,3,3′,3′-tetramethylindocarbocyanine iodide (DiIC$_6$(3))	100 mg	20.00	16.00
D-324	1,1′-dihexyl-3,3,3′,3′-tetramethylindodicarbocyanine iodide (DiIC$_6$(5))	100 mg	20.00	16.00
D-287	3,3′-dihexylthiacarbocyanine iodide (DiSC$_6$(3))	100 mg	20.00	16.00
D-292	4-((4-(dimethylamino)phenyl)butadienyl)-N-methylpyridinium iodide	100 mg	20.00	16.00
D-297	4-(4-(dimethylamino)styryl)-N-methylquinolinium iodide	100 mg	20.00	16.00
D-279	1,1′-dipentyl-3,3,3′,3′-tetramethylindocarbocyanine iodide (DiIC$_5$(3))	100 mg	20.00	16.00
D-286	3,3′-dipentylthiacarbocyanine iodide (DiSC$_5$(3))	100 mg	20.00	16.00

Chapter 13 — *Fluorescent Phospholipids, Fatty Acids and Sterols*

Cat #	Product Name	Unit Size	1–4 Units	5–24 Units
A-229	9-anthracenemethyl 3β-acetoxy-22,23-bisnor-5-cholenate (AMC acetate)	10 mg	20.00	16.00
A-231	9-anthracenemethyl 3β-(*cis*-9-octadecenoyloxy)-22,23-bisnor-5-cholenate (AMC oleate)	10 mg	20.00	16.00
A-387	21-(9-anthroyloxy)deoxycorticosterone	5 mg	20.00	16.00
D-527	3-doxyl-5α-cholestane	25 mg	20.00	16.00
N-233	NBD-methylaminoethyl 3β-acetoxy-22,23-bisnor-5-cholenate (NBDEC acetate)	10 mg	20.00	16.00
P-1529	15-phenylpentadecanoic acid	100 mg	20.00	16.00
P-225	1-pyrenemethyl 3β-octadecanoyloxy-22,23-bisnor-5-cholenate (PMC stearate)	10 mg	20.00	16.00

Chapter 14 — *Nonpolar and Amphiphilic Membrane Probes*

Cat #	Product Name	Unit Size	1–4 Units	5–24 Units
A-154	1-acetylpyrene	100 mg	20.00	16.00
A-46	1-anilinonaphthalene-2-sulfonic acid (1,2-ANS)	100 mg	20.00	16.00
A-1373	N-(9-anthracene)propyl trimethylammonium bromide	100 mg	20.00	16.00
A-207	1-(9-anthracenyl)-4-phenyl-1,3-butadiene	100 mg	20.00	16.00
A-2502	11-(5-azido-1-naphthoxy)undecanoic acid	10 mg	20.00	16.00
A-2503	11-(5-azido-1-naphthoxy)undecanoic acid, succinimidyl ester	10 mg	20.00	16.00
B-252	3-butyryl-7-hydroxycoumarin	100 mg	20.00	16.00
D-392	1-decanoylpyrene	100 mg	20.00	16.00
D-41	2-(N-decyl)aminonaphthalene-6-sulfonic acid, sodium salt	100 mg	20.00	16.00
D-1118	4-diethylaminobenzonitrile *high purity*	1 g	20.00	16.00
D-443	4-(dimethylaminobenzylidene)malononitrile	100 mg	20.00	16.00
D-256	1-dimethylaminomethylpyrene, hydrochloride	100 mg	20.00	16.00
D-68	N-(5-dimethylaminonaphthalene-1-sulfonyl)dihexadecylamine	100 mg	20.00	16.00
D-67	N-(5-dimethylaminonaphthalene-1-sulfonyl)hexadecylamine	100 mg	20.00	16.00
D-203	1-(4-dimethylaminophenyl)-6-phenylhexatriene (DMA-DPH)	25 mg	20.00	16.00
D-554	4-(N,N-dimethyl-N-decyl)ammonium-2,2,6,6-tetramethylpiperidine-1-oxyl iodide (CAT 10)	25 mg	20.00	16.00
D-553	4-(N,N-dimethyl-N-octyl)ammonium-2,2,6,6-tetramethylpiperidine-1-oxyl iodide (CAT 8)	25 mg	20.00	16.00
D-1132	4-(N,N-dimethyl-N-(1-pyrenemethyl)ammonium)butanesulfonate, inner salt	25 mg	20.00	16.00
D-1131	3-(N,N-dimethyl-N-(1-pyrenemethyl)ammonium)propanesulfonate, inner salt	25 mg	20.00	16.00
D-70	4-dioctylamino-7-nitrobenz-2-oxa-1,3-diazole (NBD dioctylamine)	100 mg	20.00	16.00
D-495	3,3′-ditetradecylthiadicarbocyanine perchlorate (DiSC$_{14}$(5))	25 mg	20.00	16.00
D-158	1-dodecanoylpyrene	100 mg	20.00	16.00
D-141	6-dodecyl-7-hydroxy-4-methylcoumarin	100 mg	20.00	16.00
D-528	doxylcyclohexane	100 mg	20.00	16.00
D-529	10-doxylnonadecane (10N19)	25 mg	20.00	16.00
D-585	4-doxylpentanoic acid, ethyl ester	25 mg	20.00	16.00
D-545	10-doxylstearic acid, methyl ester	25 mg	20.00	16.00
E-155	1-ethylpyrene	100 mg	20.00	16.00
H-1325	5-hexadecanoylaminofluorescein diacetate	25 mg	20.00	16.00
H-255	3-hexadecanoyl-7-dimethylaminocoumarin	10 mg	20.00	16.00
H-182	3-hexadecanoyl-7-hydroxycoumarin	25 mg	20.00	16.00
H-44	2-(N-hexadecyl)aminonaphthalene-6-sulfonic acid, sodium salt	100 mg	20.00	16.00
H-161	1-hexadecylpyrene	100 mg	20.00	16.00
H-446	4-hydroxybenzylidenemalononitrile	100 mg	20.00	16.00
M-54	2-(N-methylanilino)naphthalene-6-sulfonic acid, sodium salt (2,6-MANS)	100 mg	20.00	16.00
M-319	9-methylanthracene	100 mg	20.00	16.00
M-201	methyl 1-pyrenebutyrate	100 mg	20.00	16.00
M-451	methyl 1-pyrenenonanoate	25 mg	20.00	16.00
M-264	N-methyl-4-(1-pyrene)vinylpyridinium iodide	100 mg	20.00	16.00
O-175	octadecyl anthracene-9-carboxylate	100 mg	20.00	16.00
O-156	1-octanoylpyrene	100 mg	20.00	16.00
P-391	1-pentylpyrene	25 mg	20.00	16.00

Chapter 14 — *Nonpolar and Amphiphilic Membrane Probes, continued*

Cat #	Product Name	Unit Size	Price Per Unit ($) 1–4 Units	5–24 Units
P-66	N-phenyl-2-naphthylamine *high purity*	1 g	20.00	16.00
P-208	1-phenyl-4-(1-pyrenyl)-1,3-butadiene	25 mg	20.00	16.00
P-1189	1-phenyl-6-((4-trifluoromethyl)phenyl)-1,3,5-hexatriene (CF_3-DPH)	25 mg	20.00	16.00
P-249	6-propionyl-2-methoxynaphthalene (promen)	100 mg	20.00	16.00
P-315	1-pyrenenonanoic acid	25 mg	20.00	16.00
P-95	11-((1-pyrenesulfonyl)amino)undecanoic acid	100 mg	20.00	16.00
S-474	N-(4-sulfobutyl)-4-(4-(didecylamino)styryl)pyridinium, inner salt (Di_{10}ASP-BS)	100 mg	20.00	16.00
T-43	2-(N-tetradecyl)aminonaphthalene-6-sulfonic acid, sodium salt	100 mg	20.00	16.00
T-500	2,2,6,6-tetramethylpiperidine-1-oxyl (TEMPO)	1 g	20.00	16.00
T-551	2,2,6,6-tetramethylpiperidin-1-oxyl-4-yl *cis*-9-octadecenoate (TEMPO oleate)	25 mg	20.00	16.00
T-575	2,2,6,6-tetramethylpiperidin-1-oxyl-4-yl decanoate	25 mg	20.00	16.00
T-573	2,2,6,6-tetramethylpiperidin-1-oxyl-4-yl heptanoate	25 mg	20.00	16.00
T-572	2,2,6,6-tetramethylpiperidin-1-oxyl-4-yl hexanoate	25 mg	20.00	16.00
T-574	2,2,6,6-tetramethylpiperidin-1-oxyl-4-yl nonanoate	25 mg	20.00	16.00
T-550	2,2,6,6-tetramethylpiperidin-1-oxyl-4-yl octadecanoate (TEMPO stearate)	25 mg	20.00	16.00
T-510	2,2,6,6-tetramethylpiperidin-1-oxyl-4-yl octanoate	25 mg	20.00	16.00
V-390	1-valeroylpyrene	100 mg	20.00	16.00

Chapter 15 — *Fluorescent Tracers of Cell Morphology and Fluid Flow*

Cat #	Product Name	Unit Size	1–4 Units	5–24 Units
B-3225	α,α′-bis-(N-nicotinamidium)-p-xylene dibromide (DNX)	100 mg	20.00	16.00
B-3242	4,4′-bis-(p-sulfophenyl)stilbene, dipotassium salt	25 mg	20.00	16.00
B-3243	4,4′-bis-(o-sulfostyryl)biphenyl, disodium salt	25 mg	20.00	16.00
C-375	5-(and-6)-carboxyeosin diacetate *mixed isomers*	25 mg	20.00	16.00
D-1953	dextran, carboxymethyl, 70,000 MW	1 g	20.00	16.00
D-1866	dextran, coumarinamino, 40,000 MW, neutral	25 mg	20.00	16.00
D-1867	dextran, coumarinamino, 70,000 MW, neutral	25 mg	20.00	16.00
D-1835	dextran, dansyl, 40,000 MW, neutral	25 mg	20.00	16.00
D-1836	dextran, dansyl, 70,000 MW, neutral	25 mg	20.00	16.00
D-1807	dextran, eosin, 10,000 MW, polyanionic, fixable	25 mg	20.00	16.00
D-1837	dextran, N-methylanthraniloyl, 10,000 MW, neutral	25 mg	20.00	16.00
D-1838	dextran, N-methylanthraniloyl, 40,000 MW, neutral	25 mg	20.00	16.00
D-1839	dextran, N-methylanthraniloyl, 70,000 MW, neutral	25 mg	20.00	16.00
D-1876	dextran, β-methylumbelliferone, 10,000 MW, neutral	25 mg	20.00	16.00
D-1877	dextran, β-methylumbelliferone, 70,000 MW, neutral	25 mg	20.00	16.00
D-1831	dextran, stilbene, 10,000 MW, anionic	25 mg	20.00	16.00
D-1832	dextran, stilbene, 40,000 MW, anionic	25 mg	20.00	16.00
D-1833	dextran, stilbene, 70,000 MW, anionic	25 mg	20.00	16.00
D-1869	dextran, tetramethylrhodamine, 40,000 MW, anionic, fixable	25 mg	20.00	16.00
D-1895	dextran, tetramethylrhodamine and biotin, 10,000 MW, neutral	10 mg	20.00	16.00
D-3235	4,4-difluoro-1,3,5,7-tetraphenyl-4-bora-3a,4a-diaza-s-indacene-2,6-disulfonic acid, disodium salt (BODIPY® 522/549 disulfonate)	10 mg	20.00	16.00
D-196	4-(N,N-di(2-hydroxyethyl)amino)-7-nitrobenz-2-oxa-1,3-diazole	25 mg	20.00	16.00
D-72	5-dimethylaminonaphthalene-1-sulfonic acid	1 g	20.00	16.00
D-73	1-dimethylaminonaphthalene-7-sulfonic acid	100 mg	20.00	16.00
D-3247	N-ϵ-2,4-dinitrophenyl-L-lysine, hydrochloride salt (ϵ-DNP lysine)	100 mg	20.00	16.00
D-1611	N,N′-di-(3-sulfopropyl)-4,4′-bipyridinium, inner salt	1 g	20.00	16.00
E-1613	ethane bis-(2,2′-bipyridinium) dibromide	1 g	20.00	16.00
M-1341	N-(methoxycarbonylamino)-4-amino-3,6-disulfo-1,8-naphthalimide, dipotassium salt	25 mg	20.00	16.00
N-401	N-(7-nitrobenz-2-oxa-1,3-diazol-4-yl)-N-methylglucamine	25 mg	20.00	16.00
S-1366	N-(3-sulfopropyl)acridone, sodium salt	100 mg	20.00	16.00
T-3227	α,α′,α″,α‴-tetrakis-(N-nicotinamidium)durene tetrabromide (TND)	100 mg	20.00	16.00
T-3226	α,α′,α″,α‴-tetrakis-(N-pyridinium)durene tetrabromide (TPD)	100 mg	20.00	16.00

Chapter 17 — *Probes for Following Endocytosis and Exocytosis*

Cat #	Product Name	Unit Size	Price Per Unit ($) 1–4 Units	5–24 Units
E-2865	*Escherichia coli* (K-12 strain) BioParticles®, Cascade Blue® conjugate	10 mg	20.00	16.00
E-2712	*Escherichia coli* (K-12 strain) BioParticles®, Cl-NERF conjugate	2 mg	20.00	16.00
E-2711	*Escherichia coli* (K-12 strain) BioParticles®, DM-NERF conjugate	2 mg	20.00	16.00
E-2713	*Escherichia coli* (K-12 strain) BioParticles®, SNARF® conjugate	2 mg	20.00	16.00
E-2869	*Escherichia coli* (K-12 strain) BioParticles®, unlabeled	100 mg	20.00	16.00
T-2874	transferrin from human serum, Cascade Blue® conjugate	5 mg	20.00	16.00
Z-2716	ZYMOCEL™ BioParticles®, BODIPY® FL conjugate	2 mg	20.00	16.00
Z-2715	ZYMOCEL™ BioParticles®, fluorescein conjugate	2 mg	20.00	16.00
Z-2717	ZYMOCEL™ BioParticles®, Texas Red® conjugate	2 mg	20.00	16.00
Z-2718	ZYMOCEL™ BioParticles®, unlabeled	2 mg	20.00	16.00
Z-2845	zymosan A (*S. cerevisiae*) BioParticles®, Cascade Blue® conjugate	10 mg	20.00	16.00
Z-2701	zymosan A (*S. cerevisiae*) BioParticles®, DM-NERF conjugate	2 mg	20.00	16.00
Z-2703	zymosan A (*S. cerevisiae*) BioParticles®, SNARF® conjugate	2 mg	20.00	16.00

Chapter 18 — *Probes for Receptors and Ion Channels*

Cat #	Product Name	Unit Size	1–4 Units	5–24 Units
A-508	4-(N-(2-acetoxyethyl)-N,N-dimethyl)ammonium-2,2,6,6-tetramethylpiperidine-1-oxyl bromide	25 mg	20.00	16.00
A-3499	2-aminostrychnine	5 mg	20.00	16.00
B-1527	3,3′-bis-trimethylammoniummethylazobenzene dibromide (*trans*-Bis-Q)	100 mg	20.00	16.00
D-1626	dexamethasone	1 g	20.00	16.00
D-1629	dexamethasone 21-(2-(aminoethyl)thio) derivative, trifluoroacetate	25 mg	20.00	16.00
D-1628	dexamethasone 21-((((6-(biotinamido)hexanoyl)amine)ethyl)thio) derivative	5 mg	20.00	16.00
D-604	dexamethasone cadaverine fluorescein	5 mg	20.00	16.00
D-1627	dexamethasone 21-methanesulfonate	25 mg	20.00	16.00
D-60	5-dimethylaminonaphthalene-1-sulfonamidopropyltrimethylammonium iodide	100 mg	20.00	16.00
D-134	5-(((5-dimethylamino-1-naphthalene)sulfonyl)amino)-1,3-benzodioxole	100 mg	20.00	16.00
D-79	3-(5-dimethylaminonaphthalene-1-sulfonyl)aminoquinuclidine methiodide	100 mg	20.00	16.00
H-261	7-hydroxy-N-methylquinolinium iodide	100 mg	20.00	16.00
M-135	N-methylacridinium iodide	100 mg	20.00	16.00
M-260	N-methyl-5,6-benzoquinolinium iodide	100 mg	20.00	16.00
M-259	N-methylphenanthridium iodide	100 mg	20.00	16.00
P-76	1-pyrenebutyrylcholine bromide	100 mg	20.00	16.00

Chapter 19 — *Photoactivatable (Caged) Probes*

Cat #	Product Name	Unit Size	1–4 Units	5–24 Units
G-1051	guanosine 5′-O-(3-thiotriphosphate), $P^{3(S)}$-(1-(2-nitrophenyl)ethyl) ester, triammonium salt (NPE-caged GTP-γ-S)	1 mg	20.00	16.00
N-1790	N-(1-(2-nitrophenyl)ethyl)carbamoyl choline iodide (NPE-caged carbamoyl choline)	10 mg	20.00	16.00

Chapter 20 — *Probes for Signal Transduction*

Cat #	Product Name	Unit Size	1–4 Units	5–24 Units
E-1566	ethylene glycol dioctanoate	100 mg	20.00	16.00
E-1567	ethylene glycol dioleate	100 mg	20.00	16.00
P-1571	1,2-propyleneglycol dioleate	100 mg	20.00	16.00

Chapter 22 — *Indicators for Ca^{2+}, Mg^{2+}, Zn^{2+} and Other Metals*

Cat #	Product Name	Unit Size	1–4 Units	5–24 Units
A-3127	2-amino-5-fluoro-4-methylphenol-N,N,O-triacetic acid, tri(acetoxymethyl ester) (5-fluoro-4-methyl APTRA, AM)	5 mg	20.00	16.00
A-3126	2-amino-5-fluoro-4-methylphenol-N,N,O-triacetic acid, tripotassium salt (5-fluoro-4-methyl APTRA)	10 mg	20.00	16.00
A-3125	2-amino-5-fluorophenol-N,N,O-triacetic acid, tri(acetoxymethyl ester) (5-fluoro APTRA, AM)	5 mg	20.00	16.00
A-3129	2-amino-4-fluorophenol-N,N,O-triacetic acid, tri(acetoxymethyl ester) (4-fluoro APTRA, AM)	5 mg	20.00	16.00
A-3128	2-amino-4-fluorophenol-N,N,O-triacetic acid, tripotassium salt (4-fluoro APTRA)	10 mg	20.00	16.00
A-3123	2-aminophenol-N,N,O-triacetic acid, tri(acetoxymethyl ester) (APTRA, AM)	25 mg	20.00	16.00
D-6869	2,3-dihydroxynaphthalene (DHN)	5 g	20.00	16.00
M-1296	mag-quin-1 *cell impermeant*	10 mg	20.00	16.00
M-1297	mag-quin-1, AM *cell permeant*	5 mg	20.00	16.00
M-1298	mag-quin-2 *cell impermeant*	10 mg	20.00	16.00
M-3727	N-(2-methoxyphenyl)iminodiacetic acid ('half BAPTA')	10 mg	20.00	16.00
M-3728	N-(2-methoxyphenyl)iminodiacetic acid, AM ('half BAPTA, AM')	5 mg	20.00	16.00

Chapter 23 — *pH Indicators*

Cat #	Product Name	Unit Size	Price Per Unit ($) 1–4 Units	5–24 Units
C-1275	5-(and-6)-carboxy SNARF®-2	1 mg	20.00	16.00
C-1285	5-(and-6)-carboxy SNARF®-2, acetoxymethyl ester, acetate	1 mg	20.00	16.00
C-1281	5-(and-6)-carboxy SNARF®-X	1 mg	20.00	16.00
D-3327	dextran, Cl-NERF, 10,000 MW, anionic, lysine fixable	5 mg	20.00	16.00
D-3322	dextran, DM-NERF, 10,000 MW, anionic, lysine fixable	5 mg	20.00	16.00
D-370	1,4-dihydroxyphthalonitrile *high purity*	100 mg	20.00	16.00
M-3223	1-methyl-8-oxyquinolinium, inner salt	100 mg	20.00	16.00
S-1250	SNAFL®-1	1 mg	20.00	16.00
S-1251	SNAFL®-1, diacetate	1 mg	20.00	16.00

Chapter 25 — *Probes for Membrane Potential*

Cat #	Product Name	Unit Size	1–4 Units	5–24 Units
D-278	1,1′-dibutyl-3,3,3′,3′-tetramethylindocarbocyanine iodide ($DiIC_4(3)$)	100 mg	20.00	16.00
D-285	3,3′-dibutylthiacarbocyanine iodide ($DiSC_4(3)$)	100 mg	20.00	16.00
D-276	1,1′-diethyl-3,3,3′,3′-tetramethylindocarbocyanine iodide ($DiIC_2(3)$)	100 mg	20.00	16.00
D-1124	1,1′-diethyl-3,3,3′,3′-tetramethylindodicarbocyanine iodide ($DiIC_2(5)$)	100 mg	20.00	16.00
D-283	3,3′-diethylthiacarbocyanine iodide ($DiSC_2(3)$)	100 mg	20.00	16.00
D-305	3,3′-diethylthiatricarbocyanine iodide ($DiSC_2(7)$)	100 mg	20.00	16.00
D-377	3,3′-dimethylthiacarbocyanine iodide ($DiSC_1(3)$)	100 mg	20.00	16.00
D-277	1,1′-dipropyl-3,3,3′,3′-tetramethylindocarbocyanine iodide ($DiIC_3(3)$)	100 mg	20.00	16.00
D-284	3,3′-dipropylthiacarbocyanine iodide ($DiSC_3(3)$)	100 mg	20.00	16.00
H-380	1,1′,3,3,3′,3′-hexamethylindodicarbocyanine iodide ($DiIC_1(5)$)	100 mg	20.00	16.00
H-381	1,1′,3,3,3′,3′-hexamethylindotricarbocyanine perchlorate ($DiIC_1(7)$)	100 mg	20.00	16.00
M-1329	N-methyl-4-(pyrrolidinyl)styrylpyridinium iodide	100 mg	20.00	16.00
S-469	N-(4-sulfobutyl)-4-(4-(diethylamino)styryl)pyridinium, inner salt (Di_2ASP-BS)	100 mg	20.00	16.00
S-468	N-(4-sulfobutyl)-4-(4-(dimethylamino)styryl)pyridinium, inner salt (Di_1ASP-BS; RH 369)	100 mg	20.00	16.00
S-463	N-(3-sulfopropyl)-4-(4-(dibutylamino)styryl)pyridinium, inner salt (Di_4ASP-PS)	100 mg	20.00	16.00
S-462	N-(3-sulfopropyl)-4-(4-(diethylamino)styryl)pyridinium, inner salt (Di_2ASP-PS)	100 mg	20.00	16.00
S-461	N-(3-sulfopropyl)-4-(4-(dimethylamino)styryl)pyridinium, inner salt (Di_1ASP-PS)	100 mg	20.00	16.00

Customer Service and Technical Assistance

Molecular Probes has offices in Eugene, Oregon, USA and in Leiden, The Netherlands. **Molecular Probes, Inc.** in Eugene, Oregon serves customers from North America, Central America, South America, Asia, Australia, New Zealand, Pakistan and islands in the Pacific. **Molecular Probes Europe BV** in Leiden, The Netherlands handles all accounts in Europe, Africa, the Middle East, Russia and the Commonwealth of Independent States; orders from these areas must be placed with the Leiden office.

Molecular Probes, Inc.
4849 Pitchford Avenue
Eugene, OR 97402-9165 USA

PO Box 22010
Eugene, OR 97402-0469 USA
Phone: (541) 465-8300
Fax: (541) 344-6504
Web: http://www.probes.com

Customer Service Department
Hours: 7:00 am to 5:00 pm (Pacific Time)
Phone: (541) 465-8338
Fax: (541) 344-6504
E-mail: order@probes.com

For US and Canada
Toll Free Order Phone: (800) 438-2209
Toll Free Order Fax: (800) 438-0228

Technical Assistance Department
Hours: 8:00 am to 4:00 pm
Phone: (541) 465-8353
Fax: (541) 465-4593
E-mail: tech@probes.com

Custom and Bulk Sales
Phone: (541) 465-8390
Fax: (541) 984-5658
E-mail: custom@probes.com

Molecular Probes Europe BV
PoortGebouw
Rijnsburgerweg 10
2333 AA Leiden
The Netherlands
Phone: +31-71-5233378
Fax: +31-71-5233419
Web: http://www.probes.com

Customer Service Department
Hours: 9:00 to 16:30 (Central European Time)
Phone: +31-71-5236850
Fax: +31-71-5233419
E-mail: eurorder@probes.nl

For Germany, Finland,
France and Switzerland
Toll Free Order Phone: +31-800-5550
Toll Free Order Fax: +31-800-5551

Technical Assistance Department
Hours: 9:00 to 16:30
Phone: +31-71-5233431
Fax: +31-71-5233419
E-mail: eurotech@probes.nl

Custom and Bulk Sales
Phone: +31-71-5236850
Fax: +31-71-5233419
E-mail: eurocustom@probes.nl

Please type or print clearly

Last Name ....

First Name/Middle Initial ....

Department ....

Institution ....

Mailing Address ....

City/State ....

Zip/Postal Code .... **Country** ....

Telephone # .... **Fax #** ....

E-mail ....

Check as many as apply

Specializations

__ 2. bioanalytical chemistry
__ 3. biochemistry
__ 4. biophysics
__ 5. biotechnology
__ 6. cell biology
__ 7. clinical chemistry
__ 8. cytogenetics
__ 9. developmental biology
__ 154. forensics
__ 10. hematology
__ 11. histochemistry
__ 12. immunology
__ 25. instrumentation
__ 27. marine biology
__ 13. medicine
__ 142. microbiology
__ 14. molecular biology
__ 15. neuroscience
__ 16. organic chemistry
__ 17. pathology
__ 168. pharmacology
__ 19. physiology
__ 1. plant biology
__ 20. polymer chemistry
__ 21. toxicology

Instrumentation

__ 160. capillary electrophoresis
__ 65. confocal microscopy
__ 38. electron microscopy
__ 42. flow cytometry
__ 159. fluorescence microplate readers
__ 66. fluorometry
__ 51. HPLC
__ 37. lasers
__ 55. microscopy/imaging
__ 138. optical filters

Custom services used

__ 76. antibody production
__ 187. assay development
__ 78. biotin conjugation
__ 74. diagnostic reagents
__ 77. fluorophore conjugation
__ 75. immunogen preparation
__ 72. oligonucleotide synthesis
__ 70. organic synthesis
__ 71. peptide synthesis
__ 79. protein crosslinking

Interests

__ 110. actin
__ 140. aequorin
__ 30. affinity chromatography
__ 169. apoptosis
__ 99. avidin/biotin reagents
__ 143. bacteria
__ 41. biosensors
__ 157. blood flow
__ 31. caged probes
__ 113. carbohydrates
__ 170. cell adhesion
__ 129. chelators
__ 32. chemiluminescence
__ 171. chemotaxis
__ 105. colloidal gold
__ 96. crosslinking reagents
__ 172. cyclic nucleotides
__ 108. detergents
__ 33. DNA detection
__ 35. DNA sequencing
__ 36. DNA synthesis
__ 162. electrophysiology
__ 100. enzyme conjugates
__ 131. enzyme substrates
__ 174. • peroxidases
__ 139. • phospholipases
__ 150. • proteases
__ 173. • protein kinases/ phosphatases
__ 132. fatty acids
__ 101. ferritin conjugates
__ 45. fluorescence-based ELISA
__ 43. fluorescence energy transfer
__ 44. fluorescence polarization
__ 98. fluorescent antibodies
__ 93. fluorescent dextrans
__ 39. gel electrophoresis/blotting
__ 151. • proteins
__ 152. • nucleic acids
__ 144. • Western blots
__ 175. • Southern blots
__ 176. • Northern blots
__ 177. glutathione
__ 158. high-throughput screening
__ 97. HPLC derivatization
__ 52. immunofluorescence
__ 88. inositol phosphates
__ 67. *in situ* hybridization
__ 89. ion channels
__ 80. ion indicators
__ 81. • pH
__ 83. • sodium
__ 84. • potassium
__ 86. • chloride
__ 82. • calcium
__ 85. • magnesium
__ 148. • zinc
__ 149. • heavy metals
__ 104. lectins
__ 53. liposomes
__ 146. long-term cell tracing
__ 106. membrane probes
__ 54. microinjection
__ 134. microspheres
__ 178. multidrug resistance
__ 161. near IR dyes
__ 179. neurotransmitters
__ 155. nitric oxide
__ 121. nucleic acids
__ 122. nucleotide analogs
__ 135. oligonucleotides
__ 188. organelle probes
__ 153. • endoplasmic reticulum
__ 118. • Golgi
__ 180. • lysosomes/endosomes
__ 120. • mitochondria
__ 181. • nucleus/chromosomes
__ 182. PCR
__ 57. peptide sequencing
__ 56. peptide synthesis
__ 94. phagocytosis/endocytosis
__ 90. phorbol esters/DAG
__ 107. phospholipids
__ 69. phosphorescence
__ 59. photoaffinity reagents
__ 58. photobleaching
__ 102. phycobiliproteins
__ 92. polar tracers
__ 91. potential-sensitive probes
__ 183. protein quantitation
__ 147. reactive oxygen species
__ 95. reactive probes
__ 109. receptor probes
__ 111. • adrenergic
__ 112. • benzodiazepine
__ 114. • cholinergic
__ 115. • dopaminergic
__ 116. • drugs of abuse
__ 117. • fluorescent drugs
__ 130. • growth factors/ interleukins
__ 119. • insulin
__ 133. • lipoproteins
__ 123. • peptides
__ 124. • steroid/hormone
__ 125. • toxins
__ 126. • transferrin
__ 46. reporter genes
__ 47. • chloramphenicol acetyltransferase
__ 48. • galactosidase
__ 49. • glucuronidase
__ 184. • green fluorescent protein
__ 141. • luciferase
__ 61. RNA detection
__ 185. secretion/exocytosis
__ 163. signal transduction
__ 186. transplantation
__ 127. tubulin/taxol
__ 63. viability/cytotoxicity
__ 145. yeast

HB6

Please type or print clearly

Last Name ....

First Name/Middle Initial ....

Department ....

Institution ....

Mailing Address ....

City/State ....

Zip/Postal Code .... **Country** ....

Telephone # .... **Fax #** ....

E-mail ....

Check as many as apply

Specializations

__ 2. bioanalytical chemistry
__ 3. biochemistry
__ 4. biophysics
__ 5. biotechnology
__ 6. cell biology
__ 7. clinical chemistry
__ 8. cytogenetics
__ 9. developmental biology
__ 154. forensics
__ 10. hematology
__ 11. histochemistry
__ 12. immunology
__ 25. instrumentation
__ 27. marine biology
__ 13. medicine
__ 142. microbiology
__ 14. molecular biology
__ 15. neuroscience
__ 16. organic chemistry
__ 17. pathology
__ 168. pharmacology
__ 19. physiology
__ 1. plant biology
__ 20. polymer chemistry
__ 21. toxicology

Instrumentation

__ 160. capillary electrophoresis
__ 65. confocal microscopy
__ 38. electron microscopy
__ 42. flow cytometry
__ 159. fluorescence microplate readers
__ 66. fluorometry
__ 51. HPLC
__ 37. lasers
__ 55. microscopy/imaging
__ 138. optical filters

Custom services used

__ 76. antibody production
__ 187. assay development
__ 78. biotin conjugation
__ 74. diagnostic reagents
__ 77. fluorophore conjugation
__ 75. immunogen preparation
__ 72. oligonucleotide synthesis
__ 70. organic synthesis
__ 71. peptide synthesis
__ 79. protein crosslinking

Interests

__ 110. actin
__ 140. aequorin
__ 30. affinity chromatography
__ 169. apoptosis
__ 99. avidin/biotin reagents
__ 143. bacteria
__ 41. biosensors
__ 157. blood flow
__ 31. caged probes
__ 113. carbohydrates
__ 170. cell adhesion
__ 129. chelators
__ 32. chemiluminescence
__ 171. chemotaxis
__ 105. colloidal gold
__ 96. crosslinking reagents
__ 172. cyclic nucleotides
__ 108. detergents
__ 33. DNA detection
__ 35. DNA sequencing
__ 36. DNA synthesis
__ 162. electrophysiology
__ 100. enzyme conjugates
__ 131. enzyme substrates
__ 174. • peroxidases
__ 139. • phospholipases
__ 150. • proteases
__ 173. • protein kinases/ phosphatases
__ 132. fatty acids
__ 101. ferritin conjugates
__ 45. fluorescence-based ELISA
__ 43. fluorescence energy transfer
__ 44. fluorescence polarization
__ 98. fluorescent antibodies
__ 93. fluorescent dextrans
__ 39. gel electrophoresis/blotting
__ 151. • proteins
__ 152. • nucleic acids
__ 144. • Western blots
__ 175. • Southern blots
__ 176. • Northern blots
__ 177. glutathione
__ 158. high-throughput screening
__ 97. HPLC derivatization
__ 52. immunofluorescence
__ 88. inositol phosphates
__ 67. *in situ* hybridization
__ 89. ion channels
__ 80. ion indicators
__ 81. • pH
__ 83. • sodium
__ 84. • potassium
__ 86. • chloride
__ 82. • calcium
__ 85. • magnesium
__ 148. • zinc
__ 149. • heavy metals
__ 104. lectins
__ 53. liposomes
__ 146. long-term cell tracing
__ 106. membrane probes
__ 54. microinjection
__ 134. microspheres
__ 178. multidrug resistance
__ 161. near IR dyes
__ 179. neurotransmitters
__ 155. nitric oxide
__ 121. nucleic acids
__ 122. nucleotide analogs
__ 135. oligonucleotides
__ 188. organelle probes
__ 153. • endoplasmic reticulum
__ 118. • Golgi
__ 180. • lysosomes/endosomes
__ 120. • mitochondria
__ 181. • nucleus/chromosomes
__ 182. PCR
__ 57. peptide sequencing
__ 56. peptide synthesis
__ 94. phagocytosis/endocytosis
__ 90. phorbol esters/DAG
__ 107. phospholipids
__ 69. phosphorescence
__ 59. photoaffinity reagents
__ 58. photobleaching
__ 102. phycobiliproteins
__ 92. polar tracers
__ 91. potential-sensitive probes
__ 183. protein quantitation
__ 147. reactive oxygen species
__ 95. reactive probes
__ 109. receptor probes
__ 111. • adrenergic
__ 112. • benzodiazepine
__ 114. • cholinergic
__ 115. • dopaminergic
__ 116. • drugs of abuse
__ 117. • fluorescent drugs
__ 130. • growth factors/ interleukins
__ 119. • insulin
__ 133. • lipoproteins
__ 123. • peptides
__ 124. • steroid/hormone
__ 125. • toxins
__ 126. • transferrin
__ 46. reporter genes
__ 47. • chloramphenicol acetyltransferase
__ 48. • galactosidase
__ 49. • glucuronidase
__ 184. • green fluorescent protein
__ 141. • luciferase
__ 61. RNA detection
__ 185. secretion/exocytosis
__ 163. signal transduction
__ 186. transplantation
__ 127. tubulin/taxol
__ 63. viability/cytotoxicity
__ 145. yeast

HB6

Please
Affix
Postage

Molecular Probes, Inc.
PO Box 22010
Eugene, OR 97402-0469
USA

Please
Affix
Postage

Molecular Probes, Inc.
PO Box 22010
Eugene, OR 97402-0469
USA

Master Price List/Product Index

Molecular Probes' *Handbook of Fluorescent Probes and Research Chemicals*

The Master Price List/Product Index includes all products available from Molecular Probes at the time of printing, with the exception of dextrans (Section 15.5), FluoSpheres® polystyrene microspheres (Section 6.2) and optical filters (Section 26.5). As in the section price lists, products are arranged in alphabetical order of name. Many products that have alternative names are listed under both the principal and alternative name to aid in locating a product of interest. The last column of the Master Price List/Product Index serves as a product index by indicating the Handbook section(s) in which technical information on the product can be found.

Cat #		Product Name	Unit Size	Price Per Unit ($) 1–4 Units	5–24 Units	Section #
A-7109	**New**	A-23187, 1-(4,5-dimethoxy-2-nitrophenyl)ethyl ester (DMNPE-caged A-23187)	1 mg	98.00	78.40	19.2, 22.8
A-1493		A-23187 free acid (calcimycin)	10 mg	68.00	54.40	22.8
A-1310		7-AAD (7-aminoactinomycin D)	1 mg	50.00	40.00	8.1, 16.3
F-6053	**New**	ABD-F (7-fluorobenz-2-oxa-1,3-diazole-4-sulfonamide)	10 mg	55.00	44.00	2.3
A-6912	**New**	ABQ (N-(4-aminobutyl)-6-methoxyquinolinium chloride, hydrochloride)	25 mg	65.00	52.00	24.2
A-3500		ABT (3-(2′-aminobenzhydryloxy)tropane)	25 mg	95.00	76.00	18.2
A-484		4-acetamido-4′-((iodoacetyl)amino)stilbene-2,2′-disulfonic acid, disodium salt	25 mg	95.00	76.00	2.4, 9.4
A-339		4-acetamido-4′-isothiocyanatostilbene-2,2′-disulfonic acid, disodium salt (SITS)	100 mg	48.00	38.40	18.5
A-485		4-acetamido-4′-maleimidylstilbene-2,2′-disulfonic acid, disodium salt	25 mg	95.00	76.00	2.4, 9.4
A-1642		2-acetamido-4-mercaptobutanoic acid hydrazide (AMBH)	100 mg	40.00	32.00	3.2, 5.2
A-1434		7-acetoxy-N-methylquinolinium iodide	25 mg	35.00	28.00	10.5
A-7509	**New**	acetyl ceramide; C_2-ceramide (N-acetyl-D-erythro-sphingosine)	5 mg	95.00	76.00	20.6
A-3773		2-acetyl-1-(O-(11-(4,4-difluoro-5,7-dimethyl-4-bora-3a,4a-diaza-s-indacene-3-propionyl)-amino)undecyl)-*sn*-glycero-3-phosphocholine (BODIPY® FL C_{11}-PAF)	500 µg	68.00	54.40	20.6
A-6551	**New**	10-acetyl-3,7-dihydroxy-2-dodecylphenoxazine *special packaging*	10x100 µg	135.00	108.00	10.5, 16.3, 21.4
A-6550	**New**	10-acetyl-3,7-dihydroxyphenoxazine *special packaging*	10x1 mg	85.00	68.00	10.5, 16.3, 21.4
A-3767		2-acetyl-1-(O-(11-(5-dimethylaminonaphthalene-1-sulfonyl)amino)undecyl)-*sn*-glycero-3-phosphocholine (dansyl-C_{11}-PAF)	500 µg	68.00	54.40	20.6
A-7509	**New**	N-acetyl-D-erythro-sphingosine (acetyl ceramide; C_2-ceramide)	5 mg	95.00	76.00	20.6
A-685		5-((2-(and-3)-S-(acetylmercapto)succinoyl)amino)fluorescein (SAMSA fluorescein) *mixed isomers*	25 mg	95.00	76.00	5.2
A-8444	**New**	N-acetylneuraminic acid (sialic acid)	100 mg	42.00	33.60	10.6
A-3770		2-acetyl-1-(O-(11-(7-nitrobenz-2-oxa-1,3-diazol-4-yl)amino)undecyl)-*sn*-glycero-3-phosphocholine (NBD-C_{11}-PAF)	500 µg	68.00	54.40	20.6
A-6609	**New**	6-(N-acetyl-L-phenylalanyl)amino-6-deoxyluciferin	1 mg	95.00	76.00	10.4
A-96206	**New**	14-acetylvirescenine	10 mg	75.00	60.00	18.2
A-1324		ACMA (9-amino-6-chloro-2-methoxyacridine)	100 mg	78.00	62.40	8.1, 23.3
A-96001	**New**	aconitine	50 mg	45.00	36.00	18.7
A-7923	**New**	4-((9-acridinecarbonyl)amino)-2,2,6,6-tetramethylpiperidin-1-oxyl, free radical (TEMPO-9-AC) *>90% by HPLC*	5 mg	65.00	52.00	21.3
A-1127		acridine-9-carboxylic acid, succinimidyl ester (9-AC, SE)	100 mg	78.00	62.40	1.8
A-666		acridine homodimer (bis-(6-chloro-2-methoxy-9-acridinyl)spermine)	10 mg	98.00	78.40	8.1
A-417		9-acridineisothiocyanate	100 mg	35.00	28.00	1.8
A-1301		acridine orange *high purity*	1 g	48.00	38.40	8.1, 12.3, 16.3
A-3568		acridine orange *10 mg/mL solution in water*	10 mL	28.00	22.40	8.1, 12.3, 16.3
A-455		acridine orange 10-dodecyl bromide (dodecyl acridine orange)	100 mg	45.00	36.00	14.3
A-1372		acridine orange 10-nonyl bromide (nonyl acridine orange)	100 mg	45.00	36.00	12.2, 14.3
A-433		acrylodan (6-acryloyl-2-dimethylaminonaphthalene)	25 mg	155.00	124.00	2.3
A-433		6-acryloyl-2-dimethylaminonaphthalene (acrylodan)	25 mg	155.00	124.00	2.3
A-1127		9-AC, SE (acridine-9-carboxylic acid, succinimidyl ester)	100 mg	78.00	62.40	1.8
A-7592	**New**	actinomycin D *≥98% by HPLC*	10 mg	55.00	44.00	8.1
A-1400		ADAM (9-anthryldiazomethane)	25 mg	118.00	94.40	3.4
T-99017	**New**	adaphotris (*P,P,P*-tris-(hydroxymethyl)adamantane-1-phosphonium chloride)	10 mg	128.00	12.40	18.4
A-683		ADC (anthracene-1,5-dicarboxylic acid)	100 mg	48.00	38.40	24.3
A-6871	**New**	ADC, AM (anthracene-1,5-dicarboxylic acid, di(acetoxymethyl ester))	5 mg	68.00	54.40	24.3
A-7056	**New**	adenosine 5′-diphosphate, P^2-(1-(2-nitrophenyl)ethyl) ester, monopotassium salt (NPE-caged ADP)	5 mg	103.00	82.40	11.3, 19.2, 20.4
A-1049		adenosine 5′-triphosphate, P^3-(1-(4,5-dimethoxy-2-nitrophenyl)ethyl) ester, disodium salt (DMNPE-caged ATP)	5 mg	103.00	82.40	11.3, 19.2, 20.4

Cat #		Product Name	Unit Size	Price Per Unit ($) 1–4 Units	5–24 Units	Section #
A-7073	**New**	adenosine 5′-triphosphate, P^2-(1,2-diphenyl-2-oxo)ethyl ester, ammonium salt (desyl-caged ATP) *5 mM solution in water*	400 µL	145.00	116.00	11.3, 19.2, 20.4
A-1048		adenosine 5′-triphosphate, P^3-(1-(2-nitrophenyl)ethyl) ester, disodium salt (NPE-caged ATP)	5 mg	103.00	82.40	11.3, 19.2, 20.4
A-3880		ADIFAB fatty acid indicator	200 µg	245.00	196.00	13.3, 20.6
E-1022		ϵ-ADP (1,N^6-ethenoadenosine 5′-diphosphate)	25 mg	65.00	52.00	11.3
A-6202	**New**	AEBSF (4-(2-aminoethyl)benzenesulfonyl fluoride, hydrochloride)	25 mg	28.00	22.40	10.6
A-6785	**New**	aequorin (AquaLite® aequorin) *recombinant*	25 µg	98.00	78.40	22.5
A-6821	**New**	aequorin expression vector cytAEQ/pcDNAI, mammalian, cytoplasmic targeting *500 µg/mL*	50 µL	295.00	236.00	22.5
A-6822	**New**	aequorin expression vector erAEQ/pcDNAI, mammalian, endoplasmic reticulum targeting *500 µg/mL*	50 µL	295.00	236.00	22.5
A-6823	**New**	aequorin expression vector mtAEQ/pcDNAI, mammalian, mitochondrial targeting *500 µg/mL*	50 µL	295.00	236.00	22.5
A-6788	**New**	aequorin expression vector mtAEQ/pMT2, mammalian, mitochondrial targeting *500 µg/mL*	50 µL	295.00	236.00	22.5
A-6818	**New**	aequorin expression vector nu/cytAEQ/pcDNAI, mammalian, inducible nuclear targeting *500 µg/mL*	50 µL	295.00	236.00	22.5
A-6793	**New**	aequorin expression vector pMAQ2, plant, cytoplasmic targeting *500 µg/mL*	50 µL	295.00	236.00	22.5
A-6819	**New**	aequorin expression vector pMAQ6, plant, chloroplast targeting *500 µg/mL*	50 µL	295.00	236.00	22.5
A-6784	**New**	aequorin expression vector pSVAEQERK, mammalian, endoplasmic reticulum targeting *500 µg/mL*	50 µL	295.00	236.00	22.5
A-6782	**New**	aequorin expression vector pSVAEQN, mammalian, cytoplasmic targeting *500 µg/mL*	50 µL	295.00	236.00	22.5
A-6783	**New**	aequorin expression vector pSVNPA, mammalian, nuclear targeting *500 µg/mL*	50 µL	295.00	236.00	22.5
A-6663	**New**	Albumin Blue 580, potassium salt	10 mg	90.00	72.00	9.1
A-2750		albumin from bovine serum, BODIPY® FL conjugate	5 mg	35.00	28.00	15.7
A-964		albumin from bovine serum, Cascade Blue® conjugate	25 mg	98.00	78.40	15.7
A-843		albumin from bovine serum, 2,4-dinitrophenylated (DNP-BSA)	25 mg	98.00	78.40	15.7
A-841		albumin from bovine serum, fluorescein conjugate	25 mg	98.00	78.40	15.7
A-6796	**New**	albumin from bovine serum, fura conjugate (fura BSA)	5 mg	125.00	100.00	22.4
A-847		albumin from bovine serum, tetramethylrhodamine conjugate	25 mg	98.00	78.40	15.7
A-824		albumin from bovine serum, Texas Red® conjugate	25 mg	98.00	78.40	15.7
A-7304	**New**	AlignFlow™ flow cytometry alignment beads, 2.5 µm *for UV excitation*	3 mL	120.00	96.00	26.2
A-7302	**New**	AlignFlow™ flow cytometry alignment beads, 2.5 µm *for 488 nm excitation*	3 mL	120.00	96.00	26.2
A-7312	**New**	AlignFlow™ flow cytometry alignment beads, 2.5 µm *for 633 nm excitation*	3 mL	120.00	96.00	26.2
A-7305	**New**	AlignFlow™ Plus flow cytometry alignment beads, 6.0 µm *for UV excitation*	3 mL	120.00	96.00	26.2
A-7303	**New**	AlignFlow™ Plus flow cytometry alignment beads, 6.0 µm *for 488 nm excitation*	3 mL	120.00	96.00	26.2
A-7313	**New**	AlignFlow™ Plus flow cytometry alignment beads, 6.0 µm *for 633 nm excitation*	3 mL	120.00	96.00	26.2
A-927		alkaline phosphatase, biotin-XX conjugate *5 mg/mL*	200 µL	95.00	76.00	4.3, 6.3, 7.5
A-96013	**New**	allochroside *mixture of 4 glycosides from plants of the *Acanthophyllum* genus*	100 mg	95.00	76.00	18.8
A-803		allophycocyanin *4 mg/mL*	0.5 mL	95.00	76.00	6.4, 15.7
A-819		allophycocyanin, crosslinked (APC-XL) *4 mg/mL*	250 µL	198.00	158.40	6.4, 15.7
A-865		allophycocyanin, crosslinked, goat anti-mouse IgG (H+L) conjugate *1 mg/mL*	0.5 mL	195.00	156.00	6.4, 7.2
A-6920	**New**	α-amanitin	1 mg	90.00	72.00	8.3, 10.6
A-6471	**New**	AMCA-S goat anti-mouse IgG (H+L) conjugate *2 mg/mL*	0.5 mL	75.00	60.00	7.2
A-6472	**New**	AMCA-S goat anti-rabbit IgG (H+L) conjugate *2 mg/mL*	0.5 mL	75.00	60.00	7.2
A-6120	**New**	AMCA-S, SE (7-amino-3-((((succinimidyl)oxy)carbonyl)methyl)-4-methylcoumarin-6-sulfonic acid)	5 mg	95.00	76.00	1.7
A-6440	**New**	AMCA-X goat anti-mouse IgG (H+L) conjugate *2 mg/mL*	0.5 mL	75.00	60.00	7.2
A-6441	**New**	AMCA-X goat anti-rabbit IgG (H+L) conjugate *2 mg/mL*	0.5 mL	75.00	60.00	7.2
A-6118	**New**	AMCA-X, SE (6-((7-amino-4-methylcoumarin-3-acetyl)amino)hexanoic acid, succinimidyl ester)	10 mg	98.00	78.40	1.7
A-3108		amiloride, hydrochloride, dihydrate	1 g	12.00	9.60	18.7
A-1363		5-(aminoacetamido)fluorescein (fluoresceinyl glycine amide)	10 mg	98.00	78.40	3.3
A-1364		4′-((aminoacetamido)methyl)fluorescein	10 mg	98.00	78.40	3.3
A-6289	**New**	2-aminoacridone, hydrochloride	25 mg	75.00	60.00	3.2
A-1310		7-aminoactinomycin D (7-AAD)	1 mg	50.00	40.00	8.1, 16.3
A-6480	**New**	4-aminobenzamidine, dihydrochloride	1 g	18.00	14.40	10.6
A-3500		3-(2′-aminobenzhydryloxy)tropane (ABT)	25 mg	95.00	76.00	18.2
A-1604		N^{α}-(4-aminobenzoyl)biocytin, trifluoroacetate	25 mg	85.00	68.00	4.2, 9.4, 15.3
A-6912	**New**	N-(4-aminobutyl)-6-methoxyquinolinium chloride, hydrochloride (ABQ)	25 mg	65.00	52.00	24.2
A-7110	**New**	γ-aminobutyric acid, α-carboxy-2-nitrobenzyl ester, trifluoroacetic acid salt (O-(CNB-caged) GABA)	5 mg	125.00	100.00	18.3, 19.2
A-2506		γ-aminobutyric acid, 4,5-dimethoxy-2-nitrobenzyl ester, hydrochloride (O-(DMNB-caged) GABA)	5 mg	75.00	60.00	18.3, 19.2
A-7125	**New**	γ-aminobutyric acid, (1,2-diphenyl-2-oxo)ethyl ester, trifluoroacetic acid salt (O-(desyl-caged) GABA)	5 mg	168.00	134.40	18.3, 19.2
A-1324		9-amino-6-chloro-2-methoxyacridine (ACMA)	100 mg	78.00	62.40	8.1, 23.3

Cat #		Product Name	Unit Size	Price Per Unit ($) 1–4 Units	5–24 Units	Section #
C-2110		7-amino-4-chloromethylcoumarin (CellTracker™ Blue CMAC)	5 mg	95.00	76.00	10.1, 15.2
A-6527	New	7-amino-4-chloromethylcoumarin, L-alanine amide, hydrochloride (CMAC, Ala)	5 mg	95.00	76.00	10.4
A-6524	New	7-amino-4-chloromethylcoumarin, L-alanyl-L-proline amide, hydrochloride (CMAC, Ala-Pro)	5 mg	95.00	76.00	10.4
A-6528	New	7-amino-4-chloromethylcoumarin, L-arginine amide, dihydrochloride (CMAC, Arg)	5 mg	95.00	76.00	10.4
A-6520	New	7-amino-4-chloromethylcoumarin, t-BOC-L-leucyl-L-methionine amide (CMAC, t-BOC-Leu-Met)	5 mg	95.00	76.00	10.4
A-6575	New	7-amino-4-chloromethylcoumarin, CBZ-L-arginine amide, hydrochloride (CMAC, CBZ-Arg)	5 mg	95.00	76.00	10.4
A-6576	New	7-amino-4-chloromethylcoumarin, CBZ-L-isoleucyl-L-prolyl-L-arginine amide, hydrochloride (CMAC, CBZ-Ile-Pro-Arg)	5 mg	95.00	76.00	10.4
A-6522	New	7-amino-4-chloromethylcoumarin, CBZ-L-phenylalanyl-L-arginine amide, hydrochloride (CMAC, CBZ-Phe-Arg)	5 mg	95.00	76.00	10.4
A-6523	New	7-amino-4-chloromethylcoumarin, CBZ-L-valyl-L-leucyl-L-lysine amide, hydrochloride (CMAC, CBZ-Val-Leu-Lys)	5 mg	95.00	76.00	10.4
A-6526	New	7-amino-4-chloromethylcoumarin, glycine amide, hydrochloride (CMAC, Gly)	5 mg	95.00	76.00	10.4
A-6573	New	7-amino-4-chloromethylcoumarin, glycyl-L-arginine amide, dihydrochloride (CMAC, Gly-Arg)	5 mg	95.00	76.00	10.4
A-6574	New	7-amino-4-chloromethylcoumarin, glycyl-L-proline amide, hydrochloride (CMAC, Gly-Pro)	5 mg	95.00	76.00	10.4
A-6525	New	7-amino-4-chloromethylcoumarin, L-leucine amide, hydrochloride (CMAC, Leu)	5 mg	95.00	76.00	10.4
A-6570	New	7-amino-4-chloromethylcoumarin, L-lysine amide, dihydrochloride (CMAC, Lys)	5 mg	95.00	76.00	10.4
A-6571	New	7-amino-4-chloromethylcoumarin, L-phenylalanyl-L-proline amide, hydrochloride (CMAC, Phe-Pro)	5 mg	95.00	76.00	10.4
A-7621	New	8-amino-cyclic adenosine 5′-diphosphate ribose (8-amino-cADP-ribose)	10 µg	100.00	80.00	20.2
A-2952		3-amino-3-deoxydigoxigenin hemisuccinamide, succinimidyl ester	5 mg	95.00	76.00	4.2
A-6611	New	6-amino-6-deoxyluciferin (ADL)	5 mg	95.00	76.00	10.5
A-117		5-aminoeosin	100 mg	70.00	56.00	3.3
A-1339		N-(2-aminoethyl)-4-amino-3,6-disulfo-1,8-naphthalimide, dipotassium salt (lucifer yellow ethylenediamine)	25 mg	38.00	30.40	3.3, 15.3
A-91		5-((2-aminoethyl)amino)naphthalene-1-sulfonic acid, sodium salt (EDANS)	1 g	75.00	60.00	3.3
A-6202	New	4-(2-aminoethyl)benzenesulfonyl fluoride, hydrochloride (AEBSF)	25 mg	28.00	22.40	10.6
A-1593		N-(2-aminoethyl)biotinamide, hydrobromide (biotin ethylenediamine)	25 mg	60.00	48.00	4.2, 15.3
A-6033	New	N-(2-aminoethyl)maleimide, trifluoroacetic acid salt	10 mg	48.00	38.40	27.1
A-6255	New	5-aminofluorescein *single isomer*	100 mg	18.00	14.40	3.3
A-6565	New	7-amino-4-(S-glutathionyl)methylcoumarin, L-lysine amide	5 mg	135.00	108.00	10.4
A-1588		N-(6-aminohexyl)-1-naphthalenesulfonamide, hydrochloride (W-5)	100 mg	58.00	46.40	20.2
A-8445	New	5-aminolevulinic acid, hydrochloride (ALA)	1 g	55.00	44.00	21.2
A-191		7-amino-4-methylcoumarin *reference standard*	100 mg	45.00	36.00	3.3, 10.1
A-6118	New	6-((7-amino-4-methylcoumarin-3-acetyl)amino)hexanoic acid, succinimidyl ester (AMCA-X, SE)	10 mg	98.00	78.40	1.7
A-8437	New	7-amino-4-methylcoumarin, L-alanine amide, trifluoroacetate salt	100 mg	65.00	52.00	10.4
A-6519	New	7-amino-4-methylcoumarin, CBZ-L-arginyl-L-arginine amide, dihydrochloride	25 mg	95.00	76.00	10.4
A-6521	New	7-amino-4-methylcoumarin, CBZ-L-phenylalanyl-L-arginine amide, hydrochloride	25 mg	95.00	76.00	10.4
A-8438	New	7-amino-4-methylcoumarin, L-citrulline amide, hydrobromide	10 mg	75.00	60.00	10.4
A-8439	New	7-amino-4-methylcoumarin, glycine amide, hydrobromide	100 mg	75.00	60.00	10.4
A-8440	New	7-amino-4-methylcoumarin, L-leucine amide, hydrochloride	25 mg	38.00	30.40	10.4
A-8441	New	7-amino-4-methylcoumarin, L-methionine amide, trifluoroacetate salt	25 mg	38.00	30.40	10.4
A-8442	New	7-amino-4-methylcoumarin, L-phenylalanine amide, trifluoroacetate salt	25 mg	38.00	30.40	10.4
A-8443	New	7-amino-4-methylcoumarin, L-pyroglutamic acid amide	5 mg	38.00	30.40	10.4
A-1351		4′-(aminomethyl)fluorescein, hydrochloride	25 mg	98.00	78.40	3.3
A-1353		5-(aminomethyl)fluorescein, hydrochloride	10 mg	98.00	78.40	3.3
P-2421		1-aminomethylpyrene, hydrochloride (1-pyrenemethylamine, hydrochloride)	100 mg	58.00	46.40	3.3
A-350		8-aminonaphthalene-1,3,6-trisulfonic acid, disodium salt (ANTS)	1 g	88.00	70.40	3.2, 15.3
A-674		4-amino-4′-nitrostilbene-2,2′-disulfonic acid, disodium salt (NADS)	1 g	78.00	62.40	18.5
A-6346	New	N-(aminooxyacetyl)-N′-(D-biotinoyl) hydrazine (ARP)	10 mg	98.00	78.40	4.2, 8.2, 21.3
A-6250	New	4-amino-3-pentene-2-one (fluoral P)	1 g	38.00	30.40	3.2
A-1318		5-(and-6)-((N-(5-aminopentyl)amino)carbonyl)tetramethylrhodamine (tetramethylrhodamine cadaverine) *mixed isomers*	10 mg	98.00	78.40	3.3
A-1340		N-(5-aminopentyl)-4-amino-3,6-disulfo-1,8-naphthalimide, dipotassium salt (lucifer yellow cadaverine)	25 mg	70.00	56.00	3.3, 15.3
A-1594		N-(5-aminopentyl)biotinamide (biotin cadaverine)	25 mg	85.00	68.00	4.2, 15.3
A-6032	New	N-(5-aminopentyl)maleimide, trifluoroacetic acid salt	10 mg	48.00	38.40	27.1
A-434		5-((5-aminopentyl)thioureidyl)eosin, hydrochloride (eosin cadaverine)	25 mg	48.00	38.40	3.3
A-456		5-((5-aminopentyl)thioureidyl)fluorescein (fluorescein cadaverine)	25 mg	138.00	110.40	3.3
A-7494	New	p-aminophenyl phosphoryl choline	100 mg	18.00	14.40	9.1
L-8455	New	3-aminophthalhydrazide (luminol)	25 g	75.00	60.00	10.5, 17.1, 21.3
A-6257	New	8-aminopyrene-1,3,6-trisulfonic acid, trisodium salt (APTS)	10 mg	95.00	76.00	27.1
A-6252	New	2-aminopyridine	5 g	16.00	12.80	3.2

Cat #		Product Name	Unit Size	Price Per Unit ($) 1–4 Units	Price Per Unit ($) 5–24 Units	Section #
A-497		6-aminoquinoline (6-AQ) *reference standard*	1 g	24.00	19.20	3.3, 10.1
A-6591	**New**	6-aminoquinoline, L-alanine amide, di(trifluoroacetate) salt	10 mg	45.00	36.00	10.4
A-6590	**New**	6-aminoquinoline, L-glycine amide, di(trifluoroacetate) salt	10 mg	45.00	36.00	10.4
A-6597	**New**	6-aminoquinoline, L-hydroxyproline amide, di(trifluoroacetate) salt	10 mg	45.00	36.00	10.4
A-6595	**New**	6-aminoquinoline, L-isoleucine amide, di(trifluoroacetate) salt	10 mg	45.00	36.00	10.4
A-6598	**New**	6-aminoquinoline, L-lysine amide, tri(trifluoroacetate) salt	10 mg	45.00	36.00	10.4
A-6594	**New**	6-aminoquinoline, L-methionine amide, di(trifluoroacetate) salt	10 mg	45.00	36.00	10.4
A-6593	**New**	6-aminoquinoline, L-phenylalanine amide, di(trifluoroacetate) salt	10 mg	45.00	36.00	10.4
A-6596	**New**	6-aminoquinoline, L-proline amide, di(trifluoroacetate) salt	10 mg	45.00	36.00	10.4
A-6592	**New**	6-aminoquinoline, L-valine amide, di(trifluoroacetate) salt	10 mg	45.00	36.00	10.4
A-6120	**New**	7-amino-3-((((succinimidyl)oxy)carbonyl)methyl)-4-methylcoumarin-6-sulfonic acid (AMCA-S, SE)	5 mg	95.00	76.00	1.7
A-6249	**New**	7-amino-4-trifluoromethylcoumarin (AFC) *reference standard*	100 mg	18.00	14.40	3.3, 10.1
A-98105	**New**	ammodytoxin *phospholipase A_2 from *Vipera ammodytes* snake venom*	250 µg	95.00	76.00	20.5
H-7492	**New**	anandamide (N-(2-hydroxyethyl)-*cis,cis,cis,cis*-5,8,11,14-eicosatetraenoic acid amide) *special packaging*	10x1 mg	48.00	38.40	20.6
G-7121	**New**	γ-(ANB-caged) L-glutamic acid (L-glutamic acid, γ-(5-allyloxy-2-nitrobenzyl) ester)	5 mg	168.00	134.40	18.3, 19.2
A-96003	**New**	angelicine (isopsoralen)	20 mg	55.00	44.00	8.1
A-47		1-anilinonaphthalene-8-sulfonic acid (1,8-ANS) *high purity*	100 mg	88.00	70.40	14.5
A-50		2-anilinonaphthalene-6-sulfonic acid (2,6-ANS)	100 mg	88.00	70.40	14.5
A-47		1,8-ANS (1-anilinonaphthalene-8-sulfonic acid) *high purity*	100 mg	88.00	70.40	14.5
A-50		2,6-ANS (2-anilinonaphthalene-6-sulfonic acid)	100 mg	88.00	70.40	14.5
A-1139		anthracene-2,3-dicarboxaldehyde (ADA)	25 mg	40.00	32.00	1.8, 24.3
A-683		anthracene-1,5-dicarboxylic acid (ADC)	100 mg	48.00	38.40	24.3
A-6871	**New**	anthracene-1,5-dicarboxylic acid, di(acetoxymethyl ester) (ADC, AM)	5 mg	68.00	54.40	24.3
A-7896	**New**	anthracene-9,10-dipropionic acid, disodium salt	25 mg	65.00	52.00	21.2
A-409		2-anthraceneisothiocyanate	100 mg	95.00	76.00	1.8
A-176		9-anthracenepropionic acid	100 mg	38.00	30.40	1.8, 13.3
A-448		2-anthracenesulfonyl chloride	100 mg	48.00	38.40	1.8
A-77		9-anthroylcholine bromide	100 mg	48.00	38.40	18.2
A-1440		9-anthroylnitrile	25 mg	55.00	44.00	3.1
A-1322		9-anthroyl ouabain	5 mg	68.00	54.40	18.7
A-40		12-(9-anthroyloxy)oleic acid (12-AO)	25 mg	95.00	76.00	13.3
A-39		16-(9-anthroyloxy)palmitic acid (16-AP)	100 mg	95.00	76.00	13.3
A-1122		4-(9-anthroyloxy)phenacyl bromide (panacyl bromide)	100 mg	118.00	94.40	3.4
A-37		2-(9-anthroyloxy)stearic acid (2-AS)	100 mg	95.00	76.00	13.3
A-241		3-(9-anthroyloxy)stearic acid (3-AS)	25 mg	95.00	76.00	13.3
A-242		6-(9-anthroyloxy)stearic acid (6-AS)	25 mg	95.00	76.00	13.3
A-171		7-(9-anthroyloxy)stearic acid (7-AS)	25 mg	95.00	76.00	13.3
A-172		9-(9-anthroyloxy)stearic acid (9-AS)	25 mg	95.00	76.00	13.3
A-326		10-(9-anthroyloxy)stearic acid (10-AS)	25 mg	95.00	76.00	13.3
A-36		12-(9-anthroyloxy)stearic acid (12-AS)	100 mg	95.00	76.00	13.3
A-92		11-(9-anthroyloxy)undecanoic acid (11-AU)	100 mg	95.00	76.00	13.3
A-1400		9-anthryldiazomethane (ADAM)	25 mg	118.00	94.40	3.4
		antibodies — primary antibodies are listed under anti-, and secondary antibody conjugates are listed under the name of the fluorophore or hapten label; see also Sections 7.2, 7.3 and 7.4.				
A-6455	**New**	anti-*Aequorea victoria* green fluorescent protein, rabbit serum (anti-GFP)	100 µL	98.00	78.40	7.4
A-6400	**New**	anti-*Bacillus cereus* phosphatidylinositol-specific phospholipase C, mouse monoclonal A72-24	100 µg	98.00	78.40	7.4, 20.5
S-7456	**New**	antibiotic AM-2282 (staurosporine)	100 µg	45.00	36.00	20.3
A-5770		anti-BODIPY® FL, rabbit IgG fraction *≥ 1 mg/mL*	0.5 mL	148.00	118.40	7.3
A-5771		anti-BODIPY® FL, rabbit IgG fraction, biotin-XX conjugate *≥ 1 mg/mL*	0.5 mL	148.00	118.40	4.3, 7.3
A-6420	**New**	anti-BODIPY® FL, rabbit IgG fraction, fluorescein conjugate *2 mg/mL*	0.5 mL	148.00	118.40	7.3
A-6414	**New**	anti-BODIPY® TR, rabbit IgG fraction *≥ 1 mg/mL*	0.5 mL	148.00	118.40	7.3
A-6409	**New**	anti-bovine cytochrome oxidase subunit IV, mouse monoclonal 10G8-D12-C12 *specificity: human, bovine, chicken*	250 µg	195.00	156.00	7.4, 12.5
A-6431	**New**	anti-bovine cytochrome oxidase subunit IV, mouse monoclonal 20E8-C12 *specificity: human, bovine, rat, mouse*	250 µg	195.00	156.00	7.4, 12.5
A-6456	**New**	anti-bovine cytochrome oxidase subunit Vb, mouse monoclonal 16H12-H9 *specificity: human, bovine, rat, mouse, chicken*	250 µg	195.00	156.00	7.4, 12.5
A-6410	**New**	anti-bovine cytochrome oxidase subunit VIa-H, mouse monoclonal 4H2-A5	250 µg	195.00	156.00	7.4, 12.5
A-6411	**New**	anti-bovine cytochrome oxidase subunit VIa-L, mouse monoclonal 14A3-AD2-BH4	250 µg	195.00	156.00	7.4, 12.5
A-6401	**New**	anti-bovine cytochrome oxidase subunit VIc, mouse monoclonal 3G5-F7-G3 *specificity: human, bovine, rat*	100 µg	195.00	156.00	7.4, 12.5
A-6442	**New**	anti-bovine synapsin I, rabbit IgG fraction *affinity purified* *specificity: human, bovine, rat, mouse*	10 µg	120.00	96.00	7.4, 20.3
A-6443	**New**	anti-bovine synapsin I, rabbit serum *specificity: human, bovine, rat, mouse*	100 µL	98.00	78.40	7.4, 20.3
A-5760		anti-Cascade Blue®, rabbit IgG fraction *≥ 1 mg/mL*	0.5 mL	148.00	118.40	7.3

Cat #		Product Name	Unit Size	Price Per Unit ($) 1–4 Units	5–24 Units	Section #
A-5761		anti-Cascade Blue®, rabbit IgG fraction, biotin-XX conjugate *≥ 1 mg/mL*	0.5 mL	148.00	118.40	4.3, 7.3
A-6398	**New**	anti-dansyl, rabbit IgG fraction *≥ 1 mg/mL*	0.5 mL	148.00	118.40	7.3
A-6430	**New**	anti-dinitrophenyl-KLH, rabbit IgG fraction	1 mg	148.00	118.40	7.3
A-6435	**New**	anti-dinitrophenyl-KLH, rabbit IgG fraction, biotin-XX conjugate *2 mg/mL*	0.5 mL	148.00	118.40	4.3, 7.3
A-6423	**New**	anti-dinitrophenyl-KLH, rabbit IgG fraction, fluorescein conjugate *2 mg/mL*	0.5 mL	148.00	118.40	7.3
A-6424	**New**	anti-dinitrophenyl-KLH, rabbit IgG fraction, tetramethylrhodamine conjugate *2 mg/mL*	0.5 mL	148.00	118.40	7.3
A-6396	**New**	anti-dinitrophenyl-KLH, rabbit IgG fraction, Texas Red®-X conjugate *2 mg/mL*	0.5 mL	148.00	118.40	7.3
		antifade reagents — see *SlowFade*™ and ProLong™ products; see also Section 26.1.				
A-6421	**New**	anti-fluorescein, mouse monoclonal 4-4-20	0.5 mg	195.00	156.00	7.3
A-6413	**New**	anti-fluorescein, rabbit IgG Fab fragments *≥ 1 mg/mL*	0.5 mL	195.00	156.00	7.3
A-889		anti-fluorescein, rabbit IgG fraction *≥ 1 mg/mL*	0.5 mL	148.00	118.40	7.3
A-982		anti-fluorescein, rabbit IgG fraction, biotin-XX conjugate *≥ 1 mg/mL*	0.5 mL	148.00	118.40	4.3, 7.3
A-6418	**New**	anti-fluorescein, rabbit IgG fraction, BODIPY® FL-X conjugate *2 mg/mL*	0.5 mL	148.00	118.40	7.3
A-981		anti-fluorescein, rabbit IgG fraction, Texas Red® conjugate *≥ 1 mg/mL*	0.5 mL	148.00	118.40	7.3
A-6455	**New**	anti-GFP (anti-*Aequorea victoria* green fluorescent protein, rabbit serum)	100 µL	98.00	78.40	7.4
A-5790		anti-β-glucuronidase, rabbit IgG fraction *2 mg/mL*	0.5 mL	148.00	118.40	7.4, 10.2
A-5800		anti-glutathione S-transferase, rabbit IgG fraction *3 mg/mL*	0.5 mL	155.00	124.00	7.4, 9.1
A-6403	**New**	anti-human cytochrome oxidase subunit I, mouse monoclonal 1D6-E1-A8	100 µg	195.00	156.00	7.4, 12.5
A-6402	**New**	anti-human cytochrome oxidase subunit I, mouse monoclonal 5D4-F5	100 µg	195.00	156.00	7.4, 12.5
A-6404	**New**	anti-human cytochrome oxidase subunit II, mouse monoclonal 12C4-F12	100 µg	195.00	156.00	7.4, 12.5
A-5750		anti-lucifer yellow, rabbit IgG fraction *≥ 1 mg/mL*	0.5 mL	148.00	118.40	7.3
A-5751		anti-lucifer yellow, rabbit IgG fraction, biotin-XX conjugate *≥ 1 mg/mL*	0.5 mL	148.00	118.40	4.3, 7.3
A-6473	**New**	anti-rat NMDA receptor, subunit 2A, rabbit IgG fraction *affinity purified*	10 µg	240.00	192.00	7.4, 18.2
A-6467	**New**	anti-rat NMDA receptor, subunit 2A, rabbit serum	50 µL	196.00	156.80	7.4, 18.2
A-6474	**New**	anti-rat NMDA receptor, subunit 2B, rabbit IgG fraction *affinity purified*	10 µg	240.00	192.00	7.4, 18.2
A-6468	**New**	anti-rat NMDA receptor, subunit 2B, rabbit serum	50 µL	196.00	156.80	7.4, 18.2
A-6475	**New**	anti-rat NMDA receptor, subunit 2C, rabbit IgG fraction *affinity purified*	10 µg	240.00	192.00	7.4, 18.2
A-6469	**New**	anti-rat NMDA receptor, subunit 2C, rabbit serum	50 µL	176.00	140.80	7.4, 18.2
A-6397	**New**	anti-tetramethylrhodamine, rabbit IgG fraction *≥ 1 mg/mL*	0.5 mL	148.00	118.40	7.3
A-6399	**New**	anti-Texas Red®, rabbit IgG fraction *≥ 1 mg/mL*	0.5 mL	148.00	118.40	7.3
A-6405	**New**	anti-yeast cytochrome oxidase subunit I, mouse monoclonal 11D8-B7	100 µg	195.00	156.00	7.4, 12.5
A-6407	**New**	anti-yeast cytochrome oxidase subunit II, mouse monoclonal 4B12-A5	250 µg	195.00	156.00	7.4, 12.5
A-6408	**New**	anti-yeast cytochrome oxidase subunit III, mouse monoclonal DA5	250 µg	195.00	156.00	7.4, 12.5
A-6432	**New**	anti-yeast cytochrome oxidase subunit IV, mouse monoclonal 1A12-A12	250 µg	195.00	156.00	7.4, 12.5
A-6429	**New**	anti-yeast dolichol phosphate mannose synthase, mouse monoclonal 5C5-A7	250 µg	195.00	156.00	7.4, 12.5
A-6449	**New**	anti-yeast mitochondrial porin, mouse monoclonal 16G9-E6	250 µg	195.00	156.00	7.4, 12.5
A-6457	**New**	anti-yeast 3-phosphoglycerate kinase (PGK), mouse monoclonal 22C5-D8	250 µg	195.00	156.00	7.4, 12.5
A-6458	**New**	anti-yeast vacuolar alkaline phosphatase, mouse monoclonal 1D3-A10	250 µg	195.00	156.00	7.4, 12.5
A-6428	**New**	anti-yeast vacuolar carboxypeptidase Y (CPY), mouse monoclonal 10A5-B5	250 µg	195.00	156.00	7.4, 12.5
A-6427	**New**	anti-yeast vacuolar H^+-ATPase 60 kDa subunit, mouse monoclonal 13D11-B2	250 µg	195.00	156.00	7.4, 12.5
A-6422	**New**	anti-yeast vacuolar H^+-ATPase 69 kDa subunit, mouse monoclonal 8B1-F3	250 µg	195.00	156.00	7.4, 12.5
A-6426	**New**	anti-yeast vacuolar H^+-ATPase 100 kDa subunit, mouse monoclonal 10D7-A7-B2	250 µg	195.00	156.00	7.4, 12.5
A-350		ANTS (8-aminonaphthalene-1,3,6-trisulfonic acid, disodium salt)	1 g	88.00	70.40	3.2, 15.3
A-98407	**New**	apamin *from *Apis mellifera* bee venom* *special packaging*	2x100 µg	55.00	44.00	18.6
A-6896	**New**	APTRA-BTC, AM *cell permeant* *≥95% by HPLC*	20x50 µg	98.00	78.40	22.7
A-6895	**New**	APTRA-BTC, tripotassium salt *cell impermeant*	1 mg	80.00	64.00	22.7
A-6257	**New**	APTS (8-aminopyrene-1,3,6-trisulfonic acid, trisodium salt)	10 mg	95.00	76.00	27.1
A-6785	**New**	AquaLite® aequorin (aequorin) *recombinant*	25 µg	98.00	78.40	22.5
A-6786	**New**	AquaLite® aequorin, biotinylated *recombinant*	25 µg	175.00	140.00	4.3, 7.5
A-6787	**New**	AquaLite® aequorin, streptavidin *recombinant*	1 unit	240.00	192.00	7.5
A-7068	**New**	arachidonic acid, α-carboxy-2-nitrobenzyl ester (CNB-caged arachidonic acid)	5 mg	95.00	76.00	19.2, 20.6
A-7062	**New**	arachidonic acid, 1-(4,5-dimethoxy-2-nitrophenyl)ethyl ester (DMNPE-caged arachidonic acid)	5 mg	95.00	76.00	19.2, 20.6
A-98401	**New**	argiotoxin 636 (argiopin) *from *Argiope lobata* spider venom*	50 µg	295.00	236.00	18.2
A-6346	**New**	ARP (N-(aminooxyacetyl)-N′-(D-biotinoyl) hydrazine)	10 mg	98.00	78.40	4.2, 8.2, 21.3
A-7455	**New**	artemisinin (qinghaosu)	100 mg	20.00	16.00	21.2
A-36		12-AS (12-(9-anthroyloxy)stearic acid)	100 mg	95.00	76.00	13.3
A-96015	**New**	askendoside D	5 mg	68.00	54.40	10.6, 18.7
A-2505		L-aspartic acid, α-(4,5-dimethoxy-2-nitrobenzyl) ester, hydrochloride (α-(DMNB-caged) L-aspartic acid)	5 mg	75.00	60.00	18.3, 19.2
E-1023		ε-ATP (1,N^6-ethenoadenosine 5′-triphosphate)	25 mg	118.00	94.40	11.3
A-6608	**New**	ATP Determination Kit *200-1000 assays*	1 kit	95.00	76.00	10.3, 11.3, 16.2
A-7816	**New**	Attofluor® cell chamber *for microscopy*	each	158.00	158.00	26.1
A-2333		ATTO-TAG™ CBQCA Amine Derivatization Kit	1 kit	98.00	78.40	1.8
A-6222	**New**	ATTO-TAG™ CBQCA derivatization reagent (CBQCA; 3-(4-carboxybenzoyl)-quinoline-2-carboxaldehyde)	10 mg	110.00	88.00	1.8, 9.1, 24.3
A-2334		ATTO-TAG™ FQ Amine Derivatization Kit	1 kit	98.00	78.40	1.8

Cat #		Product Name	Unit Size	Price Per Unit ($) 1–4 Units	5–24 Units	Section #
A-96055	New	austricine	100 mg	68.00	54.40	18.8
A-2641		avidin, BODIPY® FL conjugate	5 mg	75.00	60.00	7.5
A-887		avidin, egg white	100 mg	195.00	156.00	7.5
A-2667		avidin, egg white	5 mg	18.00	14.40	7.5
A-821		avidin, fluorescein conjugate	5 mg	75.00	60.00	7.5
A-2666		avidin, NeutraLite™	5 mg	78.00	62.40	7.5
A-6379	New	avidin, NeutraLite™, AMCA-S conjugate	1 mg	65.00	52.00	7.5
A-2661		avidin, NeutraLite™, BODIPY® FL conjugate	1 mg	65.00	52.00	7.5
A-2663		avidin, NeutraLite™, Cascade Blue® conjugate	1 mg	65.00	52.00	7.5
A-2662		avidin, NeutraLite™, fluorescein conjugate	1 mg	65.00	52.00	7.5
A-2664		avidin, NeutraLite™, horseradish peroxidase conjugate	1 mg	65.00	52.00	7.5
A-6374	New	avidin, NeutraLite™, Oregon Green™ 488 conjugate	1 mg	65.00	52.00	7.5
A-6376	New	avidin, NeutraLite™, Oregon Green™ 500 conjugate	1 mg	65.00	52.00	7.5
A-6375	New	avidin, NeutraLite™, Oregon Green™ 514 conjugate	1 mg	65.00	52.00	7.5
A-6378	New	avidin, NeutraLite™, Rhodamine Red™-X conjugate	1 mg	65.00	52.00	7.5
A-2660		avidin, NeutraLite™, R-phycoerythrin conjugate *1 mg/mL*	1 mL	145.00	116.00	6.4, 7.5
A-6373	New	avidin, NeutraLite™, tetramethylrhodamine conjugate	1 mg	65.00	52.00	7.5
A-2665		avidin, NeutraLite™, Texas Red® conjugate	1 mg	65.00	52.00	7.5
A-820		avidin, Texas Red® conjugate	5 mg	75.00	60.00	7.5
A-1548		4-azidobenzoic acid, succinimidyl ester	100 mg	38.00	30.40	5.3
A-629		N-(((4-azidobenzoyl)amino)ethyl)-4-amino-3,6-disulfo-1,8-naphthalimide, dipotassium salt (lucifer yellow AB)	10 mg	95.00	76.00	5.3, 9.4, 15.3
A-1156		5-azidonaphthalene-1-sulfonic acid, sodium salt	100 mg	48.00	38.40	5.3
A-6310	New	5-azidonaphthalene-1-sulfonyl chloride	10 mg	68.00	54.40	5.3
A-1506		4-azidophenyl disulfide (4,4′-dithiobis-(phenylazide))	100 mg	45.00	36.00	5.3
A-6312	New	N-(4-azidophenylthio)phthalimide	25 mg	28.00	22.40	5.3
A-6307	New	4-azidosalicylic acid, succinimidyl ester	25 mg	38.00	30.40	5.3
A-2521		4-azido-2,3,5,6-tetrafluorobenzoic acid	100 mg	53.00	42.40	5.3
A-2522		4-azido-2,3,5,6-tetrafluorobenzoic acid, succinimidyl ester	25 mg	85.00	68.00	5.3
A-6311	New	N-(2-((2-(((4-azido-2,3,5,6-tetrafluoro)benzoyl)amino)ethyl)dithio)ethyl)maleimide (TFPAM-SS1)	5 mg	98.00	78.40	5.3, 9.4
A-6308	New	4-azido-2,3,5,6-tetrafluorobenzyl amine	25 mg	85.00	68.00	5.3
A-6315	New	N-(4-azido-2,3,5,6-tetrafluorobenzyl)-6-maleimidylhexanamide (TFPAM-6)	5 mg	95.00	76.00	5.3
A-6309	New	N-(4-azido-2,3,5,6-tetrafluorobenzyl)-3-maleimidylpropionamide (TFPAM-3)	5 mg	95.00	76.00	5.3
A-6320	New	6-azido-4,5,7-trifluorofluorescein (6-ATFF)	10 mg	95.00	76.00	5.3
A-6499	New	2,2′-azino-bis-(3-ethylbenzothiazine-6-sulfonic acid), diammonium salt	1 g	18.00	14.40	10.5
D-6873	New	2,2′-azodiphenol (2,2′-dihydroxyazobenzene)	100 mg	18.00	14.40	22.7
		*Bac*Light™ Reagents — see LIVE/DEAD® *Bac*Light™ Bacterial Viability Kits and LIVE *Bac*Light™ Bacterial Gram Stain Kits; see also Section 16.2.				
B-7277	New	Bacteria Counting Kit *for flow cytometry*	1 kit	95.00	76.00	16.2
B-6057	New	badan (6-bromoacetyl-2-dimethylaminonaphthalene)	10 mg	75.00	60.00	2.3, 3.4
B-6767	New	BAPTA Acetoxymethyl Ester Sampler Kit *2x1 mg each of B-1205, D-1207, D-1209, D-1213*	1 kit	75.00	60.00	22.8
B-1205		BAPTA, AM *cell permeant* *≥95% by HPLC*	25 mg	85.00	68.00	22.8
B-6769	New	BAPTA, AM *cell permeant* *special packaging*	20x1 mg	95.00	76.00	22.8
B-6768	New	BAPTA Buffer Kit *10 mg each of B-1204, D-1206, D-1208, D-1211*	1 kit	75.00	60.00	22.8
C-3040		BAPTA dextran, 10,000 MW, anionic (Calcium Sponge™ D10)	10 mg	48.00	38.40	22.8
C-3041		BAPTA dextran, 70,000 MW, anionic (Calcium Sponge™ D70)	10 mg	48.00	38.40	22.8
C-3047		BAPTA polystyrene (Calcium Sponge™ S)	1 g	98.00	78.40	22.8
B-1212		BAPTA, tetracesium salt *cell impermeant* *≥95% by HPLC*	1 g	128.00	102.40	22.8
B-1204		BAPTA, tetrapotassium salt *cell impermeant* *≥95% by HPLC*	1 g	73.00	58.40	22.8
B-1214		BAPTA, tetrasodium salt *cell impermeant*	1 g	73.00	58.40	22.8
B-1151		BCECF acid (2′,7′-bis-(2-carboxyethyl)-5-(and-6)-carboxyfluorescein) *mixed isomers*	1 mg	63.00	50.40	15.3, 23.2
B-1150		BCECF, AM (2′,7′-bis-(2-carboxyethyl)-5-(and-6)-carboxyfluorescein, acetoxymethyl ester)	1 mg	68.00	54.40	16.3, 23.2
B-1170		BCECF, AM (2′,7′-bis-(2-carboxyethyl)-5-(and-6)-carboxyfluorescein, acetoxymethyl ester) *special packaging*	20x50 µg	92.00	73.60	16.3, 23.2
B-3051		BCECF, AM (2′,7′-bis-(2-carboxyethyl)-5-(and-6)-carboxyfluorescein, acetoxymethyl ester) *1 mg/mL solution in dry DMSO* *special packaging*	1 mL	73.00	58.40	16.3, 23.2
B-6492	New	BCIP, Na (5-bromo-4-chloro-3-indolyl phosphate, disodium salt)	1 g	75.00	60.00	9.2, 10.3
B-8405	New	BCIP, toluidine (5-bromo-4-chloro-3-indolyl phosphate, p-toluidine salt)	1 g	110.00	88.00	9.2, 10.3
B-96056	New	β-belladonnine bis-(chloroethylate)	100 mg	45.00	36.00	18.2
B-6546	New	benzamidine, hydrochloride, monohydrate	5 g	18.00	14.40	3.2, 10.6
B-6897	New	benzamil, hydrochloride	25 mg	18.00	14.40	18.7
B-1509		benzophenone-4-iodoacetamide	100 mg	85.00	68.00	5.3
B-1526		benzophenone-4-isothiocyanate	100 mg	85.00	68.00	5.3
B-1508		benzophenone-4-maleimide	100 mg	85.00	68.00	5.3

Cat #		Product Name	Unit Size	Price Per Unit ($) 1–4 Units	5–24 Units	Section #
B-681		benzotriazol-1-yl 4-aminobenzoate	100 mg	48.00	38.40	1.8
B-680		benzotriazol-1-yl anthranilate	100 mg	48.00	38.40	1.8
B-6947	**New**	benzoxanthene yellow (Hoechst 2495)	100 mg	18.00	14.40	9.2
B-6553	**New**	N-benzoyl-DL-arginine, p-nitroanilide, hydrochloride (DL-BAPNA)	1 g	16.00	12.80	10.4
B-6552	**New**	N-benzoyl-L-arginine, p-nitroanilide, hydrochloride (L-BAPNA)	100 mg	32.00	25.60	10.4
B-6187	**New**	6-(4-benzoylbenzamido)hexanoic acid, succinimidyl ester (BzBz-X, SE)	25 mg	45.00	36.00	27.1
B-1577		4-benzoylbenzoic acid, succinimidyl ester	100 mg	35.00	28.00	5.3
B-96085	**New**	benzoylheteratisine, hydrochloride	5 mg	75.00	60.00	18.7
B-1584		p-benzoyl-DL-phenylalanine (DL-Bpa)	25 mg	24.00	19.20	5.3, 9.3
B-7477	**New**	benztropine methanesulfonate (benztropine mesylate)	100 mg	10.00	8.00	18.2
B-99011	**New**	6-benzyl-1,2-dimethyl-6-ethyl-4-nitromethinyl-1,4,5,6-tetrahydropyridine (M3)	100 µg	215.00	172.00	10.6, 18.4
R-441		benzyloxyresorufin (resorufin benzyl ether)	10 mg	98.00	78.40	10.5
B-7472	**New**	berberine chloride, dihydrate	1 g	15.00	12.00	17.4
B-6529	**New**	betaine aldehyde chloride, monohydrate	100 mg	68.00	54.40	10.6
B-7787	**New**	bi-9-anthryl	25 mg	65.00	52.00	14.6
B-8446	**New**	bilirubin	1 g	65.00	52.00	14.6
		bimanes — see Section 1.7.				
B-7884	**New**	Bioconjugate Techniques. G.T. Hermanson, Academic Press (1996); 785 pages, soft cover	each	60.00	60.00	26.6
B-1592		biocytin (ε-biotinoyl-L-lysine)	100 mg	60.00	48.00	4.2, 15.3
C-6949	**New**	biocytin Cascade Blue® (Cascade Blue® biocytin, disodium salt, fixable)	5 mg	95.00	76.00	4.3, 15.3
B-1603		biocytin hydrazide	25 mg	85.00	68.00	4.2, 15.3
L-6950	**New**	biocytin lucifer yellow (lucifer yellow biocytin, potassium salt, fixable)	5 mg	95.00	76.00	4.3, 15.3
B-2605		biocytin-X (ε-(6-(biotinoyl)amino)hexanoyl-L-lysine)	10 mg	60.00	48.00	4.2, 15.3
B-6350	**New**	biocytin-X hydrazide (ε-(6-(biotinoyl)amino)hexanoyl-L-lysine, hydrazide)	25 mg	90.00	72.00	4.2, 15.3
		BioParticles® labeled bacteria — products are listed under *Escherichia coli, Staphylococcus aureus* and zymosan A; see also Section 17.1 for a complete list.				
B-1595		D(+)-biotin	1 g	32.00	25.60	4.2
B-1582		biotinamidocaproate, N-hydroxysuccinimidyl ester; biotin-X, SE (6-((biotinoyl)-amino) hexanoic acid, succinimidyl ester)	100 mg	80.00	64.00	4.2, 15.3
A-1594		biotin cadaverine (N-(5-aminopentyl)biotinamide)	25 mg	85.00	68.00	4.2, 15.3
B-1550		biotin DHPE (N-(biotinoyl)-1,2-dihexadecanoyl-*sn*-glycero-3-phosphoethanolamine, triethylammonium salt)	10 mg	115.00	92.00	4.3, 5.2, 13.4
E-3477		biotin EGF (epidermal growth factor, biotin-XX conjugate)	20 µg	160.00	128.00	4.3, 17.3
A-1593		biotin ethylenediamine (N-(2-aminoethyl)biotinamide, hydrobromide)	25 mg	60.00	48.00	4.2, 15.3
B-1549		biotin hydrazide	100 mg	24.00	19.20	4.2
B-2602		12-((biotinoyl)amino)dodecanoic acid	25 mg	53.00	42.40	4.3, 13.3
B-2603		12-((biotinoyl)amino)dodecanoic acid, succinimidyl ester	25 mg	48.00	38.40	4.3, 13.3
B-1598		6-((biotinoyl)amino)hexanoic acid (biotin-X)	100 mg	48.00	38.40	4.2
B-1600		6-((biotinoyl)amino)hexanoic acid, hydrazide (biotin-X hydrazide; BACH)	50 mg	48.00	38.40	4.2, 9.4
B-1582		6-((biotinoyl)amino)hexanoic acid, succinimidyl ester (biotin-X, SE; biotinamidocaproate, N-hydroxysuccinimidyl ester)	100 mg	80.00	64.00	4.2, 15.3
B-6353	**New**	6-((biotinoyl)amino)hexanoic acid, sulfosuccinimidyl ester, sodium salt (Sulfo-NHS-LC-Biotin; biotin-X, SSE)	25 mg	75.00	60.00	4.2, 9.4, 15.3
B-1597		2-(((N-(biotinoyl)amino)hexanoyl)amino)ethylamine (biotin-X ethylenediamine)	10 mg	75.00	60.00	4.2, 15.3
B-2600		6-((6-((biotinoyl)amino)hexanoyl)amino)hexanoic acid, hydrazide (biotin-XX hydrazide)	25 mg	80.00	64.00	4.2
B-1606		6-((6-((biotinoyl)amino)hexanoyl)amino)hexanoic acid, succinimidyl ester (biotin-XX, SE)	100 mg	135.00	108.00	4.2
B-6352	**New**	6-((6-((biotinoyl)amino)hexanoyl)amino)hexanoic acid, sulfosuccinimidyl ester, sodium salt (biotin-XX, SSE)	25 mg	75.00	60.00	4.2, 9.4, 15.3
B-1596		5-(((N-(biotinoyl)amino)hexanoyl)amino)pentylamine trifluoroacetate salt (biotin-X cadaverine)	10 mg	85.00	68.00	4.2, 15.3
B-1370		5-((N-(5-(N-(6-(biotinoyl)amino)hexanoyl)amino)pentyl)thioureidyl)fluorescein (fluorescein biotin)	5 mg	98.00	78.40	4.3, 15.3
B-1616		N-((6-(biotinoyl)amino)hexanoyl)-1,2-dihexadecanoyl-*sn*-glycero-3-phosphoethanolamine, triethylammonium salt (biotin-X DHPE)	5 mg	115.00	92.00	4.3, 5.2, 13.4
B-2605		ε-(6-(biotinoyl)amino)hexanoyl-L-lysine (biocytin-X)	10 mg	60.00	48.00	4.2, 15.3
B-6350	**New**	ε-(6-(biotinoyl)amino)hexanoyl-L-lysine, hydrazide (biocytin-X hydrazide)	25 mg	90.00	72.00	4.2, 15.3
B-6349	**New**	ε-biotinoyl-α-*tert*-butoxycarbonyl-L-lysine (α-(t-BOC) biocytin)	100 mg	95.00	76.00	4.2, 9.3
B-1550		N-(biotinoyl)-1,2-dihexadecanoyl-*sn*-glycero-3-phosphoethanolamine, triethylammonium salt (biotin DHPE)	10 mg	115.00	92.00	4.3, 5.2, 13.4
B-6351	**New**	ε-biotinoyl-α-(9-fluorenylmethoxycarbonyl)-L-lysine (FMOC biocytin)	100 mg	175.00	140.00	27.1
B-1591		N-(biotinoyl)-N′-(iodoacetyl)ethylenediamine	25 mg	88.00	70.40	4.2
B-1592		ε-biotinoyl-L-lysine (biocytin)	100 mg	60.00	48.00	4.2, 15.3
B-1513		D-biotin, succinimidyl ester (succinimidyl D-biotin)	100 mg	53.00	42.40	4.2, 15.3
B-1598		biotin-X (6-((biotinoyl)amino)hexanoic acid)	100 mg	48.00	38.40	4.2

Cat #		Product Name	Unit Size	Price Per Unit ($) 1–4 Units	5–24 Units	Section #
B-1596		biotin-X cadaverine (5-(((N-(biotinoyl)amino)hexanoyl)amino)pentylamine trifluoroacetate salt)	10 mg	85.00	68.00	4.2, 15.3
B-1616		biotin-X DHPE (N-((6-(biotinoyl)amino)hexanoyl)-1,2-dihexadecanoyl-*sn*-glycero-3-phosphoethanolamine, triethylammonium salt)	5 mg	115.00	92.00	4.3, 5.2, 13.4
B-2604		biotin-X 2,4-dinitrophenyl-X-L-lysine, succinimidyl ester (DNP-X-biocytin-X, SE)	5 mg	145.00	116.00	4.2
B-1597		biotin-X ethylenediamine (2-(((N-(biotinoyl)amino)hexanoyl)amino)ethylamine)	10 mg	75.00	60.00	4.2, 15.3
B-1600		biotin-X hydrazide; BACH (6-((biotinoyl)amino)hexanoic acid, hydrazide)	50 mg	48.00	38.40	4.2, 9.4
B-1582		biotin-X, SE; biotinamidocaproate, N-hydroxysuccinimidyl ester (6-((biotinoyl)amino)hexanoic acid, succinimidyl ester)	100 mg	80.00	64.00	4.2, 15.3
B-6353	**New**	biotin-X, SSE (6-((biotinoyl)amino)hexanoic acid, sulfosuccinimidyl ester, sodium salt)	25 mg	75.00	60.00	4.2, 9.4, 15.3
B-6453	**New**	biotin-XX donkey anti-sheep IgG (H+L) conjugate *2 mg/mL*	0.5 mL	75.00	60.00	7.2
B-6452	**New**	biotin-XX goat anti-human IgG (H+L) conjugate *2 mg/mL*	0.5 mL	75.00	60.00	7.2
B-2763		biotin-XX goat anti-mouse IgG (H+L) conjugate *2 mg/mL*	0.5 mL	75.00	60.00	4.3, 7.2
B-2770		biotin-XX goat anti-rabbit IgG (H+L) conjugate *2 mg/mL*	0.5 mL	75.00	60.00	4.3, 7.2
B-2600		biotin-XX hydrazide (6-((6-((biotinoyl)amino)hexanoyl)amino)hexanoic acid, hydrazide)	25 mg	80.00	64.00	4.2
B-7474	**New**	biotin-XX phalloidin	50 U	198.00	158.40	4.3, 11.1
B-1606		biotin-XX, SE (6-((6-((biotinoyl)amino)hexanoyl)amino)hexanoic acid, succinimidyl ester)	100 mg	135.00	108.00	4.2
B-6352	**New**	biotin-XX, SSE (6-((6-((biotinoyl)amino)hexanoyl)amino)hexanoic acid, sulfosuccinimidyl ester, sodium salt)	25 mg	75.00	60.00	4.2, 9.4, 15.3

biotinylation reagents and biotin conjugates — see Sections 4.2 and 4.3, respectively, for complete lists of biotinylation reagents and biotin conjugates, all of which are also listed in this Master Price List/Product Index.

Cat #		Product Name	Unit Size	1–4 Units	5–24 Units	Section #
B-6357	**New**	S-biotinylhomocysteine	5 mg	55.00	44.00	27.1
B-7791	**New**	1,1′-bipyrene	10 mg	65.00	52.00	14.6
B-153		bis-ANS (4,4′-dianilino-1,1′-binaphthyl-5,5′-disulfonic acid, dipotassium salt)	10 mg	125.00	100.00	11.2, 14.5
H-1398		bis-benzimide (Hoechst 33258, pentahydrate)	100 mg	20.00	16.00	8.1, 16.3
B-6848	**New**	bis-BTC, AM *cell permeant* *≥95% by HPLC* *special packaging*	20x50 µg	110.00	88.00	22.7
B-6847	**New**	bis-BTC, tetraammonium salt *cell impermeant*	1 mg	98.00	78.40	22.7
B-1151		2′,7′-bis-(2-carboxyethyl)-5-(and-6)-carboxyfluorescein (BCECF acid) *mixed isomers*	1 mg	63.00	50.40	15.3, 23.2
B-1150		2′,7′-bis-(2-carboxyethyl)-5-(and-6)-carboxyfluorescein, acetoxymethyl ester (BCECF, AM)	1 mg	68.00	54.40	16.3, 23.2
B-1170		2′,7′-bis-(2-carboxyethyl)-5-(and-6)-carboxyfluorescein, acetoxymethyl ester (BCECF, AM) *special packaging*	20x50 µg	92.00	73.60	16.3, 23.2
B-3051		2′,7′-bis-(2-carboxyethyl)-5-(and-6)-carboxyfluorescein, acetoxymethyl ester (BCECF, AM) *1 mg/mL solution in dry DMSO* *special packaging*	1 mL	73.00	58.40	16.3, 23.2
B-3760		1,2-bis-(S-decanoyl)-1,2-dithio-*sn*-glycero-3-phosphocholine	1 mg	60.00	48.00	20.5
B-436		bis-(1,3-dibutylbarbituric acid)pentamethine oxonol (DiBAC$_4$(5))	25 mg	65.00	52.00	18.5, 25.3
B-438		bis-(1,3-dibutylbarbituric acid)trimethine oxonol (DiBAC$_4$(3))	25 mg	115.00	92.00	25.3
B-413		bis-(1,3-diethylthiobarbituric acid)trimethine oxonol (DiSBAC$_2$(3))	100 mg	115.00	92.00	25.3
B-7701	**New**	1,2-bis-(4,4-difluoro-5,7-dimethyl-4-bora-3a,4a-diaza-s-indacene-3-undecanoyl)-*sn*-glycero-3-phosphocholine (bis-BODIPY® FL C$_{11}$-PC)	100 µg	95.00	76.00	20.5
B-7805	**New**	9,10-bis-(4-ethoxyphenyl)-2-chloroanthracene	100 mg	18.00	14.40	14.6
B-6810	**New**	bis-fura-2, hexapotassium salt *cell impermeant*	1 mg	98.00	78.40	22.2
D-7475	**New**	bisindolylmaleimide; GF 109203X (2-(1-(3-dimethylaminopropyl)-indol-3-yl)-3-(indol-3-yl)maleimide)	1 mg	65.00	52.00	20.3
B-2282		bis-(4-methoxyphenyl)-1-(pyrenyl)methyl chloride (BMPM chloride)	25 mg	35.00	28.00	3.1
L-6868	**New**	bis-N-methylacridinium nitrate (lucigenin) *high purity*	10 mg	15.00	12.00	12.2, 17.1, 21.3, 24.2

bis-oxonol — B-413, B-436 and B-438 (see Section 25.3) have all been called "bis-oxonol" in the literature.

Cat #		Product Name	Unit Size	1–4 Units	5–24 Units	Section #
B-3932		(E,E)-3,5-bis-(4-phenyl-1,3-butadienyl)-4,4-difluoro-4-bora-3a,4a-diaza-s-indacene (BODIPY® 665/676)	5 mg	75.00	60.00	14.6
B-7802	**New**	9,10-bis-(phenylethynyl)anthracene	1 g	18.00	14.40	14.6
B-7804	**New**	9,10-bis-(phenylethynyl)-1-chloroanthracene	100 mg	18.00	14.40	14.6
B-7803	**New**	9,10-bis-(phenylethynyl)-2-ethylanthracene	100 mg	18.00	14.40	14.6
O-266		bis-(3-phenyl-5-oxoisoxazol-4-yl)pentamethine oxonol (oxonol V)	100 mg	145.00	116.00	25.3
O-267		bis-(3-propyl-5-oxoisoxazol-4-yl)pentamethine oxonol (oxonol VI)	100 mg	145.00	116.00	25.3
B-3781		1,2-bis-(1-pyrenebutanoyl)-*sn*-glycero-3-phosphocholine	1 mg	85.00	68.00	20.5
B-162		1,10-bis-(1-pyrene)decane	25 mg	48.00	38.40	14.6
B-3782		1,2-bis-(1-pyrenedecanoyl)-*sn*-glycero-3-phosphocholine	1 mg	85.00	68.00	20.5
B-311		1,3-bis-(1-pyrenyl)propane	25 mg	165.00	132.00	14.6
B-6874	**New**	N,N′-bis-(salicylidene)-2,3-diaminobenzofuran (SABF)	100 mg	45.00	36.00	22.6
B-6875	**New**	N,N′-bis-(salicylidene)ethylenediamine	5 g	24.00	19.20	22.6
B-6953	**New**	3,3′-bis-(4-sulfobutyl)-1,1′-diethyl-5,5′,6,6′-tetrachlorobenzimidazolylcarbocyanine, potassium salt (TDBC-4)	5 mg	48.00	38.40	25.3

Cat #		Product Name	Unit Size	Price Per Unit ($) 1–4 Units	5–24 Units	Section #
B-6951	*New*	3,3′-bis-(3-sulfopropyl)-5,5′-dichloro-9-ethylthiacarbocyanine, sodium salt	5 mg	48.00	38.40	25.3
B-6952	*New*	3,3′-bis-(3-sulfopropyl)-1,1′-diethyl-5,5′,6,6′-tetrachlorobenzimidazolylcarbocyanine, potassium salt (TDBC-3)	5 mg	48.00	38.40	25.3
B-1174		bis-(2,4,6-trichlorophenyl) oxalate (TCPO)	1 g	25.00	20.00	1.8, 24.3
		blood flow microspheres — see FluoSpheres® polystyrene microspheres in Section 15.6.				
B-3582		BOBO™-1 iodide (462/481) *1 mM solution in DMSO*	200 µL	225.00	180.00	8.1
B-3586		BOBO™-3 iodide (570/602) *1 mM solution in DMSO*	200 µL	225.00	180.00	8.1
B-6349	*New*	α-(t-BOC) biocytin (ε-biotinoyl-α-*tert*-butoxycarbonyl-L-lysine)	100 mg	95.00	76.00	4.2, 9.3
B-7097	*New*	N-(t-BOC)-S-(4,5-dimethoxy-2-nitrobenzyl)-L-cysteine (N-(t-BOC)-S-(DMNB-caged) L-cysteine)	100 mg	175.00	140.00	9.3, 19.2
B-7099	*New*	α-(t-BOC)-ε-(N-(((4,5-dimethoxy-2-nitrobenzyl)oxy)carbonyl))-L-lysine (α-(t-BOC)-ε-(NVOC-caged) L-lysine)	100 mg	175.00	140.00	9.3, 19.2
B-7077	*New*	N-(t-BOC)-O-(4,5-dimethoxy-2-nitrobenzyl)-L-tyrosine (N-(t-BOC)-O-(DMNB-caged) L-tyrosine)	100 mg	175.00	140.00	9.3, 19.2
B-6142	*New*	α-(t-BOC)-ε-(5-dimethylaminonaphthalene-1-sulfonyl)-L-lysine (α-(t-BOC)-ε-dansyl-L-lysine)	100 mg	175.00	140.00	9.3
B-6217	*New*	α-(t-BOC)-ε-(4-dimethylaminophenylazobenzoyl)-L-lysine (α-(t-BOC)-ε-dabcyl-L-lysine)	100 mg	175.00	140.00	9.3
B-7078	*New*	N-(t-BOC)-L-glutamic acid γ-(4,5-dimethoxy-2-nitrobenzyl ester) (N-(t-BOC)-γ-(DMNB-caged) L-glutamic acid)	100 mg	175.00	140.00	9.3, 19.2
B-6215	*New*	5-((2-(t-BOC)-γ-glutamylaminoethyl)amino)naphthalene-1-sulfonic acid	100 mg	175.00	140.00	9.3
		BODIPY® dyes — the names of many of our BODIPY® products start with 4,4-difluoro...; see Sections 1.2, 13.2 and 13.3 for examples.				
B-3475		BODIPY® 558/568 phalloidin	300 U	198.00	158.40	11.1
B-7437	*New*	BODIPY® 558/568 pirenzepine, dihydrochloride	100 µg	95.00	76.00	18.2
B-7434	*New*	BODIPY® 558/568 prazosin	100 µg	95.00	76.00	16.3, 18.2
P-7501	*New*	BODIPY® 564/570 Taxol (paclitaxel, BODIPY® 564/570 conjugate)	10 µg	168.00	134.40	11.2
B-3416		BODIPY® 581/591 phalloidin	300 U	198.00	158.40	11.1
C-7614	*New*	BODIPY® FL-14-dUTP (ChromaTide™ BODIPY® FL-14-dUTP) *1 mM in buffer*	25 µL	175.00	175.00	8.2, 16.3
C-7613	*New*	BODIPY® FL-14-UTP (ChromaTide™ BODIPY® FL-14-UTP) *1 mM in buffer*	25 µL	175.00	175.00	8.2
B-7467	*New*	BODIPY® FL ABT	100 µg	48.00	38.40	18.2
D-6201	*New*	BODIPY® FL AEBSF (4-(2-(4,4-difluoro-5,7-dimethyl-4-bora-3a,4a-diaza-s-indacene-3-propionyl)aminoethyl)benzenesulfonyl fluoride)	1 mg	48.00	38.40	9.4, 10.6
B-6905	*New*	BODIPY® FL amiloride	25 µg	95.00	76.00	18.7
B-7485	*New*	BODIPY® FL bisindolylmaleimide	250 µg	65.00	52.00	20.3
D-7546	*New*	BODIPY® FL Br_2C_5-ceramide (N-(2,6-dibromo-4,4-difluoro-5,7-dimethyl-4-bora-3a,4a-diaza-s-indacene-3-pentanoyl)sphingosine)	250 µg	98.00	78.40	12.4, 20.6
B-3504		BODIPY® FL α-bungarotoxin (α-bungarotoxin, BODIPY® FL conjugate)	500 µg	175.00	140.00	18.2
D-3521		BODIPY® FL C_5-ceramide (N-(4,4-difluoro-5,7-dimethyl-4-bora-3a,4a-diaza-s-indacene-3-pentanoyl)sphingosine)	250 µg	125.00	100.00	12.4, 20.6
D-7548	*New*	BODIPY® FL C_5-glucocerebroside (N-(4,4-difluoro-5,7-dimethyl-4-bora-3a,4a-diaza-s-indacene-3-pentanoyl)sphingosyl 1-β-D-glucopyranoside)	250 µg	98.00	78.40	20.6
B-7706	*New*	BODIPY® FL C_5 *myo*-inositol 1-phosphate, lithium salt	100 µg	95.00	76.00	20.5
D-3522		BODIPY® FL C_5-sphingomyelin (N-(4,4-difluoro-5,7-dimethyl-4-bora-3a,4a-diaza-s-indacene-3-pentanoyl)sphingosyl phosphocholine)	250 µg	125.00	100.00	12.4, 13.2, 20.6
D-7519	*New*	BODIPY® FL C_{12}-galactocerebroside (N-(4,4-difluoro-5,7-dimethyl-4-bora-3a,4a-diaza-s-indacene-3-dodecanoyl)sphingosyl 1-β-D-galactopyranoside)	25 µg	95.00	76.00	20.6
D-7547	*New*	BODIPY® FL C_{12}-glucocerebroside (N-(4,4-difluoro-5,7-dimethyl-4-bora-3a,4a-diaza-s-indacene-3-dodecanoyl)sphingosyl 1-β-D-glucopyranoside)	250 µg	98.00	78.40	20.6
D-7520	*New*	BODIPY® FL C_{12}-MPP (DL-*threo*-2-(4,4-difluoro-5,7-dimethyl-4-bora-3a,4a-diaza-s-indacene-3-dodecanoyl)amino-3-morpholino-1-phenyl-1-propanol, hydrochloride)	25 µg	95.00	76.00	20.6
D-7711	*New*	BODIPY® FL C_{12}-sphingomyelin (N-(4,4-difluoro-5,7-dimethyl-4-bora-3a,4a-diaza-s-indacene-3-dodecanoyl)sphingosyl phosphocholine) *1 mg/mL in DMSO*	250 µL	125.00	100.00	27.1
B-7470	*New*	BODIPY® FL (±) CGP 12177	100 µg	48.00	38.40	18.2
B-7486	*New*	BODIPY® FL di(bisindolylmaleimide)	250 µg	65.00	52.00	20.3
B-7469	*New*	BODIPY® FL forskolin	100 µg	95.00	76.00	16.3, 20.3
B-7439	*New*	BODIPY® FL glibenclamide (BODIPY® FL glyburide)	100 µg	95.00	76.00	18.6
B-7439	*New*	BODIPY® FL glyburide (BODIPY® FL glibenclamide)	100 µg	95.00	76.00	18.6
B-2752		BODIPY® FL goat anti-mouse IgG (H+L) conjugate	1 mg	75.00	60.00	7.2
B-2766		BODIPY® FL goat anti-rabbit IgG (H+L) conjugate	1 mg	75.00	60.00	7.2
L-3483		BODIPY® FL LDL (low-density lipoprotein from human plasma, BODIPY® FL complex) *1 mg/mL*	200 µL	175.00	140.00	17.3
B-3424		BODIPY® FL NAPS	1 mg	68.00	54.40	18.2
B-607		BODIPY® FL phallacidin	300 U	198.00	158.40	11.1
D-7441	*New*	BODIPY® FL phenylalkylamine (DM-BODIPY® PAA)	25 µg	125.00	100.00	18.4
B-7436	*New*	BODIPY® FL pirenzepine, dihydrochloride	100 µg	95.00	76.00	18.2
B-3460		BODIPY® FL PPHT	200 µg	68.00	54.40	18.2

Cat #		Product Name	Unit Size	Price Per Unit ($) 1–4 Units	5–24 Units	Section #
B-7433	**New**	BODIPY® FL prazosin	100 µg	95.00	76.00	16.3, 18.2
D-6300	**New**	BODIPY® FL propanol (4,4-difluoro-5,7-dimethyl-4-bora-3a,4a-diaza-s-indacene-3-propanol)	10 mg	95.00	76.00	3.4
B-3414		BODIPY® FL Ro-1986	1 mg	68.00	54.40	18.5
B-3420		BODIPY® FL SCH 23390	1 mg	68.00	54.40	18.2
D-2184		BODIPY® FL, SE (4,4-difluoro-5,7-dimethyl-4-bora-3a,4a-diaza-s-indacene-3-propionic acid, succinimidyl ester)	5 mg	145.00	116.00	1.2
D-6140	**New**	BODIPY® FL, SSE (4,4-difluoro-5,7-dimethyl-4-bora-3a,4a-diaza-s-indacene-3-propionic acid, sulfosuccinimidyl ester, sodium salt)	5 mg	145.00	116.00	1.2, 9.4
P-7500	**New**	BODIPY® FL Taxol (paclitaxel, BODIPY® FL conjugate)	10 µg	168.00	134.40	11.2
B-7487	**New**	BODIPY® FL thapsigargin	100 µg	95.00	76.00	20.2
B-7431	**New**	BODIPY® FL verapamil, hydrochloride	1 mg	98.00	78.40	16.3, 18.4
B-7505	**New**	BODIPY® FL-X ryanodine	25 µg	65.00	52.00	18.4, 20.2
D-6102	**New**	BODIPY® FL-X, SE (6-((4,4-difluoro-5,7-dimethyl-4-bora-3a,4a-diaza-s-indacene-3-propionyl)amino)hexanoic acid, succinimidyl ester)	5 mg	145.00	116.00	1.2, 4.2
B-7491	**New**	BODIPY® R6G phalloidin	300 U	198.00	158.40	11.1
D-6180	**New**	BODIPY® R6G, SE (4,4-difluoro-5-phenyl-4-bora-3a,4a-diaza-s-indacene-3-propionic acid, succinimidyl ester)	5 mg	145.00	116.00	1.2
D-6186	**New**	BODIPY® R6G-X, SE (6-((4,4-difluoro-5-phenyl-4-bora-3a,4a-diaza-s-indacene-3-propionyl)amino)hexanoic acid, succinimidyl ester)	5 mg	145.00	116.00	27.1
C-7616	**New**	BODIPY® TMR-14-dUTP (ChromaTide™ BODIPY® TMR-14-dUTP) *1 mM in buffer*	25 µL	175.00	175.00	8.2
C-7615	**New**	BODIPY® TMR-14-UTP (ChromaTide™ BODIPY® TMR-14-UTP) *1 mM in buffer*	25 µL	175.00	175.00	8.2
D-7521	**New**	BODIPY® TMR C_{12}-MPP (DL-*threo*-2-(4,4-difluoro-5-(4-methoxyphenyl)-4-bora-3a,4a-diaza-s-indacene-3-dodecanoyl)amino-3-morpholino-1-phenyl-1-propanol)	25 µg	95.00	76.00	20.6
D-6012	**New**	BODIPY® TMR cadaverine IA (N-(5-((4,4-difluoro-1,3-dimethyl-5-(4-methoxyphenyl)-4-bora-3a,4a-diaza-s-indacene-2-propionyl)amino)pentyl)iodoacetamide)	5 mg	95.00	76.00	27.1
B-6450	**New**	BODIPY® TMR-X goat anti-mouse IgG (H+L) conjugate	1 mg	75.00	60.00	7.2
D-6117	**New**	BODIPY® TMR-X, SE (6-((4,4-difluoro-1,3-dimethyl-5-(4-methoxyphenyl)-4-bora-3a,4a-diaza-s-indacene-2-propionyl)amino)hexanoic acid, succinimidyl ester)	5 mg	145.00	116.00	1.2
C-7618	**New**	BODIPY® TR-14-dUTP (ChromaTide™ BODIPY® TR-14-dUTP) *1 mM in buffer*	25 µL	175.00	175.00	8.2
C-7617	**New**	BODIPY® TR-14-UTP (ChromaTide™ BODIPY® TR-14-UTP) *1 mM in buffer*	25 µL	175.00	175.00	8.2
B-7464	**New**	BODIPY® TR-X phallacidin	300 U	198.00	158.40	11.1
D-6116	**New**	BODIPY® TR-X, SE (6-(((4-(4,4-difluoro-5-(2-thienyl)-4-bora-3a,4a-diaza-s-indacene-3-yl)phenoxy)acetyl)amino)hexanoic acid, succinimidyl ester)	5 mg	145.00	116.00	1.2, 4.2
H-1586		Bolton-Hunter reagent (3-(4-hydroxyphenyl)propionic acid, succinimidyl ester)	1 g	38.00	30.40	1.8, 9.3
E-99019	**New**	bomin-1 (2-ethoxy-2,3-dihydro-1,4,2-benzodioxaphosphorin-2-oxide)	500 µg	108.00	86.40	18.2
P-99020	**New**	bomin-2 (2-propoxy-2,3-dihydro-1,4,2-benzodioxaphosphorin-2-oxide)	500 µg	108.00	86.40	18.2
B-99021	**New**	bomin-3 (2-butoxy-2,3-dihydro-1,4,2-benzodioxaphosphorin-2-oxide)	500 µg	108.00	86.40	18.2
B-3583		BO-PRO™-1 iodide (462/481) *1 mM solution in DMSO*	1 mL	135.00	108.00	8.1
B-3587		BO-PRO™-3 iodide (575/599) *1 mM solution in DMSO*	1 mL	135.00	108.00	8.1
B-7450	**New**	brefeldin A *from *Penicillium brefeldianum**	5 mg	42.00	33.60	12.4
B-7449	**New**	brefeldin A, BODIPY® 558/568 conjugate *isomer 1*	25 µg	95.00	76.00	12.4
B-7447	**New**	brefeldin A, BODIPY® FL conjugate *isomer 1*	25 µg	95.00	76.00	12.4
B-6851	**New**	brevetoxin PbTx-1 *from *Ptychodiscus brevis**	25 µg	65.00	52.00	18.7
B-6852	**New**	brevetoxin PbTx-2 *from *Ptychodiscus brevis**	25 µg	65.00	52.00	18.7
B-6853	**New**	brevetoxin PbTx-3 *from *Ptychodiscus brevis**	25 µg	65.00	52.00	18.7
B-6854	**New**	brevetoxin PbTx-6 *from *Ptychodiscus brevis**	25 µg	95.00	76.00	18.7
B-6855	**New**	brevetoxin PbTx-9 *from *Ptychodiscus brevis**	25 µg	65.00	52.00	18.7
B-6850	**New**	Brevetoxin Sampler Kit *5 µg of B-6851, B-6852, B-6853, B-6854, B-6855*	1 kit	135.00	108.00	18.7
B-3245		brilliant sulfoflavin	100 mg	24.00	19.20	15.3
B-1450		3-bromoacetyl-7-diethylaminocoumarin	25 mg	60.00	48.00	3.4
B-6057	**New**	6-bromoacetyl-2-dimethylaminonaphthalene (badan)	10 mg	75.00	60.00	2.3, 3.4
B-1115		1-(bromoacetyl)pyrene	100 mg	48.00	38.40	3.4
B-1043		8-bromoadenosine 3′,5′-cyclicmonophosphate, free acid	100 mg	90.00	72.00	20.4
B-7108	**New**	4-bromo A-23187, 1-(4,5-dimethoxy-2-nitrophenyl)ethyl ester (DMNPE-caged Br A-23187)	1 mg	98.00	78.40	19.2, 22.8
B-1494		4-bromo A-23187, free acid	1 mg	78.00	62.40	22.8
B-8400	**New**	5-bromo-4-chloro-3-indolyl N-acetyl-β-D-galactosaminide	25 mg	95.00	76.00	10.2
B-8401	**New**	5-bromo-4-chloro-3-indolyl N-acetyl-β-D-glucosaminide	25 mg	48.00	38.40	10.2
B-8402	**New**	5-bromo-4-chloro-3-indolyl butyrate	25 mg	48.00	38.40	10.5
B-8403	**New**	5-bromo-4-chloro-3-indolyl caprylate	25 mg	48.00	38.40	10.5
B-1690		5-bromo-4-chloro-3-indolyl β-D-galactopyranoside (X-Gal)	1 g	55.00	44.00	9.2, 10.2
B-1689		5-bromo-4-chloro-3-indolyl β-D-glucopyranoside (X-Glu)	100 mg	98.00	78.40	10.2
B-1691		5-bromo-4-chloro-3-indolyl β-D-glucuronide, cyclohexylammonium salt (X-GlcU, CHA)	100 mg	115.00	92.00	9.2, 10.2
B-8404	**New**	5-bromo-4-chloro-3-indolyl β-D-glucuronide, sodium salt (X-GlcU, Na)	50 mg	120.00	96.00	10.2
B-6492	**New**	5-bromo-4-chloro-3-indolyl phosphate, disodium salt (BCIP, Na)	1 g	75.00	60.00	9.2, 10.3

Cat #		Product Name	Unit Size	Price Per Unit ($) 1–4 Units	Price Per Unit ($) 5–24 Units	Section #
B-8405	New	5-bromo-4-chloro-3-indolyl phosphate, p-toluidine salt (BCIP, toluidine)	1 g	110.00	88.00	9.2, 10.3
B-8406	New	5-bromo-4-chloro-3-indolyl sulfate, potassium salt (X-sulfate)	100 mg	35.00	28.00	10.5
B-8407	New	5-bromo-6-chloro-3-indolyl β-D-galactopyranoside	100 mg	65.00	52.00	10.2
B-8408	New	5-bromo-6-chloro-3-indolyl β-D-glucuronide, cyclohexylammonium salt	25 mg	95.00	76.00	10.2
B-8409	New	5-bromo-6-chloro-3-indolyl phosphate, p-toluidine salt	100 mg	48.00	38.40	10.3
B-8410	New	5-bromo-6-chloro-3-indolyl sulfate, potassium salt	100 mg	48.00	38.40	10.5
B-7625	New	8-bromo-cyclic adenosine 5′-diphosphate ribose (8-Br-cADP-ribose)	35 µg	120.00	96.00	20.2
B-1044		8-bromoguanosine 3′,5′-cyclicmonophosphate, free acid	25 mg	42.00	33.60	20.4
B-96099	New	3β-bromoimperialine	1 mg	325.00	260.00	18.2
B-2103		8-bromomethyl-4,4-difluoro-1,3,5,7-tetramethyl-4-bora-3a,4a-diaza-s-indacene (BODIPY® 493/503 methyl bromide)	5 mg	95.00	76.00	2.2, 3.4, 15.2
B-1456		4-bromomethyl-6,7-dimethoxycoumarin	100 mg	24.00	19.20	3.4
B-1458		3-bromomethyl-6,7-dimethoxy-1-methyl-2(1H)-quinoxazolinone	25 mg	48.00	38.40	3.4
B-1355		5-(bromomethyl)fluorescein	10 mg	125.00	100.00	2.2, 3.4
B-344		4-bromomethyl-7-methoxycoumarin (BMB)	100 mg	38.00	30.40	3.4
B-7929	New	3-bromo-7-nitroindazole	25 mg	45.00	36.00	21.5
B-6610	New	α-bromo-2-nitrophenylacetic acid (BNPA)	5 mg	75.00	60.00	10.6, 19.2
B-6286	New	4-bromophenacyl bromide (2,4′-dibromoacetophenone)	5 g	18.00	14.40	3.4, 10.6
B-6287	New	4-bromophenacyl trifluoromethanesulfonate	25 mg	48.00	38.40	3.4
B-6058	New	3-bromopyruvic acid	1 g	8.00	6.40	2.4, 3.2
		BSA (bovine serum albumin) conjugates — products are listed under albumin from bovine serum; see also Section 15.7.				
B-6791	New	BTC, AM *cell permeant* *≥95% by HPLC* *special packaging*	20x50 µg	110.00	88.00	22.2
B-6790	New	BTC, tetrapotassium salt *cell impermeant*	1 mg	85.00	68.00	22.2
B-6846	New	BTC-5N, AM *cell permeant* *≥95% by HPLC* *special packaging*	20x50 µg	125.00	100.00	22.7
B-6845	New	BTC-5N, tetrapotassium salt *cell impermeant*	1 mg	95.00	76.00	22.7
B-1601		α-bungarotoxin *from *Bungarus multicinctus**	1 mg	37.00	29.60	18.2
B-3459		β-bungarotoxin *from *Bungarus multicinctus**	1 mg	37.00	29.60	18.6, 20.5
B-1196		α-bungarotoxin, biotin-XX conjugate	500 µg	175.00	140.00	4.3, 18.2
B-3504		α-bungarotoxin, BODIPY® FL conjugate (BODIPY® FL α-bungarotoxin)	500 µg	175.00	140.00	18.2
B-7483	New	α-bungarotoxin, eosin conjugate	500 µg	175.00	140.00	18.2
F-1176		α-bungarotoxin, fluorescein conjugate (fluorescein α-bungarotoxin)	500 µg	175.00	140.00	18.2
B-7488	New	α-bungarotoxin, Oregon Green™ 514 conjugate	500 µg	195.00	156.00	18.2
T-1175		α-bungarotoxin, tetramethylrhodamine conjugate (tetramethylrhodamine α-bungarotoxin)	500 µg	195.00	156.00	18.2
B-7489	New	α-bungarotoxin, Texas Red®-X conjugate	500 µg	195.00	156.00	18.2
B-99021	New	2-butoxy-2,3-dihydro-1,4,2-benzodioxaphosphorin-2-oxide (bomin-3)	500 µg	108.00	86.40	18.2
B-3824		5-butyl-4,4-difluoro-4-bora-3a,4a-diaza-s-indacene-3-nonanoic acid (C_4-BODIPY® 500/510 C_9)	1 mg	38.00	30.40	13.2
B-3794		2-(5-butyl-4,4-difluoro-4-bora-3a,4a-diaza-s-indacene-3-nonanoyl)-1-hexadecanoyl-*sn*-glycero-3-phosphocholine (β-C_4-BODIPY® 500/510 C_9-HPC)	100 µg	38.00	30.40	13.2
B-7893	New	N-*tert*-butyl-α-phenylnitrone (PBN)	1 g	25.00	20.00	21.3
D-2893		C_{12}FDG (5-dodecanoylaminofluorescein di-β-D-galactopyranoside)	5 mg	195.00	156.00	10.2
C-7619	New	cADP-ribose (cyclic adenosine 5′-diphosphate ribose)	100 µg	250.00	200.00	20.2
		caged ATP — A-1048, A-1049, A-7073 (see Section 19.2) have all been called "caged ATP" in the literature.				
		caged calcium — see NP-EGTA and DMNP-EDTA (N-6802, D-6814); see also Sections 19.2, 20.2 and 22.8.				
P-7070	New	caged nitric oxide I (potassium nitrosylpentachlororuthenate)	10 mg	85.00	68.00	19.2, 21.5
P-7071	New	caged oxygen ((µ-peroxo) (µ-hydroxo)bis(bis(bipyridyl)cobalt(III)) perchlorate)	25 mg	95.00	76.00	27.1
		caged probes — see Section 19.2 for a complete list of caged probes, all of which are also listed in this Master Price List/Product Index.				
C-481		calcein *high purity*	100 mg	58.00	46.40	15.3, 17.4, 22.7
C-1430		calcein, AM	1 mg	115.00	92.00	15.3, 16.3, 22.7
C-3099		calcein, AM *1 mg/mL solution in dry DMSO* *special packaging*	1 mL	120.00	96.00	15.3, 16.3, 22.7
C-3100		calcein, AM *special packaging*	20x50 µg	140.00	112.00	15.3, 16.3, 22.7
H-1426		calcein blue (7-hydroxy-4-methylcoumarin-8-methyleneiminodiacetic acid)	1 g	20.00	16.00	15.3, 22.7
C-1429		calcein blue, AM	1 mg	50.00	40.00	16.3
A-1493		calcimycin (A-23187 free acid)	10 mg	68.00	54.40	22.8
C-98109	New	calciseptine *from *Dendroaspis polylepis* snake venom* *special packaging*	5x10 µg	88.00	70.40	18.4
C-3723		Calcium Calibration Buffer Concentrate Kit *zero and 100 mM CaEGTA (2 x 5 mL)*	1 kit	95.00	76.00	22.8
C-3008		Calcium Calibration Buffer Kit #1 *zero and 10 mM CaEGTA (2 x 50 mL)*	1 kit	95.00	76.00	22.8
C-3009		Calcium Calibration Buffer Kit #2 *zero to 10 mM CaEGTA (11 x 10 mL)*	1 kit	95.00	76.00	22.8
C-6775	New	Calcium Calibration Buffer Kit #3 *1 µM to 1 mM range (11 x 10 mL)*	1 kit	95.00	76.00	22.8
C-3721		Calcium Calibration Buffer Kit with Magnesium #1 *zero and 10 mM CaEGTA with 1 mM Mg^{2+} (2 x 50 mL)*	1 kit	95.00	76.00	22.8
C-3722		Calcium Calibration Buffer Kit with Magnesium #2 *zero to 10 mM CaEGTA with 1 mM Mg^{2+} (11 x 10 mL)*	1 kit	95.00	76.00	22.8

Cat #		Product Name	Unit Size	Price Per Unit ($) 1–4 Units	5–24 Units	Section #
C-6462	*New*	calcium/calmodulin-dependent kinase II (CaM kinase II) *from rat brain* *10 ng/µL*	100 µL	225.00	180.00	20.3
C-3018		Calcium Crimson™, AM *cell permeant* *special packaging*	10x50 µg	95.00	76.00	22.3
C-3016		Calcium Crimson™, tetrapotassium salt *cell impermeant*	500 µg	68.00	54.40	22.3
C-6824	*New*	Calcium Crimson™ dextran, 10,000 MW, anionic	5 mg	150.00	120.00	22.4
C-6825	*New*	Calcium Crimson™ dextran, 70,000 MW, anionic	5 mg	150.00	120.00	22.4
C-3011		Calcium Green™-1, AM *cell permeant* *≥97% by HPLC*	500 µg	95.00	76.00	22.3
C-3012		Calcium Green™-1, AM *cell permeant* *≥97% by HPLC* *special packaging*	10x50 µg	110.00	88.00	22.3
C-3010		Calcium Green™-1, hexapotassium salt *cell impermeant*	500 µg	85.00	68.00	22.3
C-3732		Calcium Green™-2, AM *cell permeant* *special packaging*	10x50 µg	95.00	76.00	22.3
C-3730		Calcium Green™-2, octapotassium salt *cell impermeant*	500 µg	68.00	54.40	22.3
C-3739		Calcium Green™-5N, AM *cell permeant* *≥95% by HPLC* *special packaging*	10x50 µg	110.00	88.00	22.3
C-3737		Calcium Green™-5N, hexapotassium salt *cell impermeant*	500 µg	85.00	68.00	22.3
C-6804	*New*	Calcium Green™-C_{18}, hexapotassium salt	1 mg	150.00	120.00	22.4
C-6765	*New*	Calcium Green™-1 dextran, potassium salt, 3000 MW, anionic	5 mg	175.00	140.00	22.4
C-3713		Calcium Green™-1 dextran, potassium salt, 10,000 MW, anionic	5 mg	150.00	120.00	22.4
C-3714		Calcium Green™-1 dextran, potassium salt, 70,000 MW, anionic	5 mg	150.00	120.00	22.4
C-6766	*New*	Calcium Green™-1 dextran, potassium salt, 500,000 MW, anionic	5 mg	150.00	120.00	22.4
C-6800	*New*	Calcium Green™-1 and Texas Red® dextran, 70,000 MW, anionic	5 mg	150.00	120.00	22.4
C-3015		Calcium Orange™, AM *cell permeant* *≥95% by HPLC* *special packaging*	10x50 µg	95.00	76.00	22.3
C-3013		Calcium Orange™, tetrapotassium salt *cell impermeant*	500 µg	68.00	54.40	22.3
C-6771	*New*	Calcium Orange™-5N, AM *cell permeant* *≥95% by HPLC* *special packaging*	10x50 µg	110.00	88.00	22.3
C-6770	*New*	Calcium Orange™-5N, tetrapotassium salt *cell impermeant*	500 µg	85.00	68.00	22.3
C-3040		Calcium Sponge™ D10 (BAPTA dextran, 10,000 MW, anionic)	10 mg	48.00	38.40	22.8
C-3041		Calcium Sponge™ D70 (BAPTA dextran, 70,000 MW, anionic)	10 mg	48.00	38.40	22.8
C-3047		Calcium Sponge™ S (BAPTA polystyrene)	1 g	98.00	78.40	22.8
C-6462	*New*	CaM kinase II (calcium/calmodulin-dependent kinase II) *from rat brain* *10 ng/µL*	100 µL	225.00	180.00	20.3
T-96016	*New*	canadine (DL-tetrahydroberberine)	20 mg	128.00	102.40	18.2
C-356		5-(((2-(carbohydrazino)methyl)thio)acetyl)aminofluorescein	25 mg	138.00	110.40	3.2
C-400		5-(and-6)-carboxy-2′,7′-dichlorodihydrofluorescein diacetate (carboxy-H_2DCFDA) *mixed isomers*	25 mg	88.00	70.40	10.5, 17.1, 21.4
C-2938		6-carboxy-2′,7′-dichlorodihydrofluorescein diacetate, di(acetoxymethyl ester) *single isomer*	5 mg	135.00	108.00	17.1, 21.4
C-6170	*New*	6-carboxy-4′,5′-dichloro-2′,7′-dimethoxyfluorescein (6-JOE)	5 mg	75.00	60.00	1.5
C-6171	*New*	6-carboxy-4′,5′-dichloro-2′,7′-dimethoxyfluorescein, succinimidyl ester (6-JOE, SE)	5 mg	125.00	100.00	1.5
C-368		5-(and-6)-carboxy-2′,7′-dichlorofluorescein *mixed isomers*	100 mg	48.00	38.40	1.5, 15.3, 23.3
C-369		5-(and-6)-carboxy-2′,7′-dichlorofluorescein diacetate (carboxy-DCFDA) *mixed isomers*	100 mg	98.00	78.40	16.3
C-1165		5-(and-6)-carboxy-2′,7′-dichlorofluorescein diacetate, succinimidyl ester *mixed isomers*	25 mg	115.00	92.00	15.2
C-2936		4-carboxydihydrotetramethylrosamine, succinimidyl ester (OxyBURST® Orange, SE)	5 mg	55.00	44.00	17.1, 21.4
C-366		5-(and-6)-carboxy-4′,5′-dimethylfluorescein *mixed isomers*	100 mg	48.00	38.40	23.2
C-367		5-(and-6)-carboxy-4′,5′-dimethylfluorescein diacetate *mixed isomers*	100 mg	98.00	78.40	23.2
C-301		5-(and-6)-carboxyeosin *mixed isomers*	100 mg	38.00	30.40	1.5
C-1164		5-(and-6)-carboxyeosin diacetate, succinimidyl ester *mixed isomers*	25 mg	75.00	60.00	15.2
C-194		5-(and-6)-carboxyfluorescein *mixed isomers*	5 g	73.00	58.40	15.3, 23.2
C-1359		5-carboxyfluorescein (5-FAM) *single isomer*	100 mg	55.00	44.00	1.3
C-1360		6-carboxyfluorescein (6-FAM) *single isomer*	100 mg	55.00	44.00	1.3
C-1904		5-(and-6)-carboxyfluorescein (5(6)-FAM) *high purity* *mixed isomers*	100 mg	55.00	44.00	1.3, 15.3, 23.2
C-1361		5-carboxyfluorescein diacetate (5-CFDA) *single isomer*	100 mg	75.00	60.00	16.3
C-1362		6-carboxyfluorescein diacetate (6-CFDA) *single isomer*	100 mg	75.00	60.00	16.3
C-195		5-(and-6)-carboxyfluorescein diacetate (5(6)-CFDA) *mixed isomers*	100 mg	75.00	60.00	16.3, 23.2
C-1354		5-carboxyfluorescein diacetate, acetoxymethyl ester (5-CFDA, AM)	5 mg	68.00	54.40	16.3, 23.2
C-1157		5-(and-6)-carboxyfluorescein diacetate, succinimidyl ester (5(6)-CFDA, SE) *mixed isomers*	25 mg	115.00	92.00	15.2, 16.3, 23.2
C-2210		5-carboxyfluorescein, succinimidyl ester (5-FAM, SE) *single isomer*	10 mg	95.00	76.00	1.3
C-6164	*New*	6-carboxyfluorescein, succinimidyl ester (6-FAM, SE) *single isomer*	10 mg	95.00	76.00	1.3
C-1311		5-(and-6)-carboxyfluorescein, succinimidyl ester (5(6)-FAM, SE) *mixed isomers*	100 mg	75.00	60.00	1.3
C-400		carboxy-H_2DCFDA (5-(and-6)-carboxy-2′,7′-dichlorodihydrofluorescein diacetate) *mixed isomers*	25 mg	88.00	70.40	10.5, 17.1, 21.4
C-6541	*New*	6′-(O-(carboxymethyl))-2′,7′-dichlorofluorescein 3′-(O-(N-acetyl-β-D-glucosaminide)) (CM-DCF-NAG)	5 mg	95.00	76.00	10.2
C-3106		N-(carboxymethyl)-6-methoxyquinolinium bromide (MQAA)	100 mg	48.00	38.40	24.2
C-6167	*New*	6-carboxymethylthio-2′,4′,5′,7′-tetrabromo-4,5,7-trifluorofluorescein (eosin F_3S)	5 mg	75.00	60.00	1.5
C-652		5-(and-6)-carboxynaphthofluorescein *mixed isomers*	100 mg	45.00	36.00	1.6, 23.2
C-653		5-(and-6)-carboxynaphthofluorescein, succinimidyl ester *mixed isomers*	25 mg	48.00	38.40	1.6, 23.4

Cat #		Product Name	Unit Size	Price Per Unit ($) 1–4 Units	5–24 Units	Section #
C-1791		N-(α-carboxy-2-nitrobenzyl)carbamylcholine, trifluoroacetate (N-(CNB-caged) carbamylcholine)	5 mg	75.00	60.00	18.3, 19.2
C-7122	New	N-(α-carboxy-2-nitrobenzyl)-L-glutamic acid (N-(CNB-caged) L-glutamic acid)	5 mg	168.00	134.40	18.3, 19.2
C-7124	New	3-(and-4)-((α-carboxy-2-nitrobenzyl)oxy)-4-(and-3)-hydroxyphenethylamine, trifluoroacetic acid salt (O-(CNB-caged) dopamine) *mixed isomers*	1 mg	65.00	52.00	18.3, 19.2
C-7083	New	O-(α-carboxy-2-nitrobenzyl)phenylephrine (O-(CNB-caged) phenylephrine)	5 mg	48.00	38.40	18.3, 19.2
C-7924	New	5-(2-carboxyphenyl)-5-hydroxy-1-((2,2,5,5-tetramethyl-1-oxypyrrolidin-3-yl)-methyl)-3-phenyl-2-pyrrolin-4-one, potassium salt (proxyl fluorescamine)	5 mg	65.00	52.00	21.3
C-7912	New	2-(4-carboxyphenyl)-4,4,5,5-tetramethylimidazoline-1-oxyl-3-oxide, potassium salt (carboxy-PTIO)	25 mg	45.00	36.00	21.5
C-7912	New	carboxy-PTIO (2-(4-carboxyphenyl)-4,4,5,5-tetramethylimidazoline-1-oxyl-3-oxide, potassium salt)	25 mg	45.00	36.00	21.5
C-7101	New	5-(and-6)-carboxy-Q-rhodamine, CMNCBZ-caged, succinimidyl ester *mixed isomers*	1 mg	68.00	54.40	15.3, 19.2
C-7100	New	5-(and-6)-carboxy-Q-rhodamine, CMNCBZ-caged, tripotassium salt *mixed isomers*	1 mg	55.00	44.00	15.3, 19.2
C-6109	New	5-carboxyrhodamine 6G, hydrochloride (5-CR 6G) *single isomer*	5 mg	75.00	60.00	1.5
C-2213		6-carboxyrhodamine 6G, hydrochloride (6-CR 6G) *single isomer*	5 mg	75.00	60.00	1.5
C-6127	New	5-carboxyrhodamine 6G, succinimidyl ester (5-CR 6G, SE) *single isomer*	5 mg	125.00	100.00	1.5
C-6128	New	6-carboxyrhodamine 6G, succinimidyl ester (6-CR 6G, SE) *single isomer*	5 mg	125.00	100.00	1.5
C-6157	New	5-(and-6)-carboxyrhodamine 6G, succinimidyl ester (5(6)-CR 6G, SE) *mixed isomers*	5 mg	75.00	60.00	1.5
C-1255		5-(and-6)-carboxy SNAFL®-1	1 mg	90.00	72.00	23.2
C-1256		5-(and-6)-carboxy SNAFL®-1, diacetate	1 mg	118.00	94.40	23.2
C-1260		5-(and-6)-carboxy SNAFL®-2	1 mg	90.00	72.00	23.2
C-1261		5-(and-6)-carboxy SNAFL®-2, diacetate	1 mg	118.00	94.40	23.2
C-3061		5-(and-6)-carboxy SNAFL®-1, succinimidyl ester *mixed isomers*	1 mg	35.00	28.00	23.4
C-3062		5-(and-6)-carboxy SNAFL®-2, succinimidyl ester *mixed isomers*	1 mg	35.00	28.00	23.4
C-1270		5-(and-6)-carboxy SNARF®-1	1 mg	125.00	100.00	23.2
C-1271		5-(and-6)-carboxy SNARF®-1, acetoxymethyl ester, acetate	1 mg	145.00	116.00	23.2
C-1272		5-(and-6)-carboxy SNARF®-1, acetoxymethyl ester, acetate *special packaging*	20x50 µg	175.00	140.00	23.2
C-6165	New	5-carboxy-2′,4′,5′,7′-tetrabromosulfonefluorescein	5 mg	75.00	60.00	1.5
C-6166	New	5-carboxy-2′,4′,5′,7′-tetrabromosulfonefluorescein, succinimidyl ester, bis-(diisopropylethylammonium) salt	5 mg	95.00	76.00	1.5
C-6121	New	5-carboxytetramethylrhodamine (5-TAMRA) *single isomer*	10 mg	95.00	76.00	1.6
C-6122	New	6-carboxytetramethylrhodamine (6-TAMRA) *single isomer*	10 mg	95.00	76.00	1.6
C-300		5-(and-6)-carboxytetramethylrhodamine (5(6)-TAMRA) *mixed isomers*	100 mg	80.00	64.00	1.6
C-2211		5-carboxytetramethylrhodamine, succinimidyl ester (5-TAMRA, SE) *single isomer*	5 mg	148.00	118.40	1.6
C-6123	New	6-carboxytetramethylrhodamine, succinimidyl ester (6-TAMRA, SE) *single isomer*	5 mg	135.00	108.00	1.6
C-1171		5-(and-6)-carboxytetramethylrhodamine, succinimidyl ester (5(6)-TAMRA, SE) *mixed isomers*	25 mg	118.00	94.40	1.6
C-1488		3-carboxyumbelliferyl β-D-galactopyranoside (CUG)	10 mg	98.00	78.40	10.2
C-1492		3-carboxyumbelliferyl β-D-glucuronide (CUGlcU)	10 mg	48.00	38.40	10.2
C-6124	New	5-carboxy-X-rhodamine (5-ROX) *single isomer*	10 mg	95.00	76.00	1.6
C-6156	New	6-carboxy-X-rhodamine (6-ROX) *single isomer*	10 mg	95.00	76.00	1.6
C-1308		5-(and-6)-carboxy-X-rhodamine (5(6)-ROX) *mixed isomers*	25 mg	60.00	48.00	1.6
C-6125	New	5-carboxy-X-rhodamine, succinimidyl ester (5-ROX, SE) *single isomer*	5 mg	135.00	108.00	1.6
C-6126	New	6-carboxy-X-rhodamine, succinimidyl ester (6-ROX, SE) *single isomer*	5 mg	135.00	108.00	1.6
C-1309		5-(and-6)-carboxy-X-rhodamine, succinimidyl ester (5(6)-ROX, SE) *mixed isomers*	25 mg	118.00	94.40	1.6
C-98102	New	cardiotoxin *from *Naja nigricollis* snake venom*	1 mg	128.00	102.40	18.2, 20.3
C-7612	New	Cascade Blue®-7-dUTP (ChromaTide™ Cascade Blue®-7-dUTP) *1 mM in buffer*	25 µL	175.00	175.00	8.2
C-7611	New	Cascade Blue®-7-UTP (ChromaTide™ Cascade Blue®-7-UTP) *1 mM in buffer*	25 µL	175.00	175.00	8.2
C-2284		Cascade Blue® acetyl azide, trisodium salt	5 mg	125.00	100.00	1.7, 4.2, 9.4, 15.3
C-625		Cascade Blue® aminoethyl 4-azidobenzamide, trisodium salt	10 mg	95.00	76.00	5.3, 9.4, 15.3
C-6949	New	Cascade Blue® biocytin, disodium salt, fixable (biocytin Cascade Blue®)	5 mg	95.00	76.00	4.3, 15.3
C-622		Cascade Blue® cadaverine, trisodium salt	10 mg	98.00	78.40	3.3, 15.3
C-621		Cascade Blue® ethylenediamine, trisodium salt	10 mg	98.00	78.40	3.3, 15.3
C-962		Cascade Blue® goat anti-mouse IgG (H+L) conjugate *2 mg/mL*	0.5 mL	75.00	60.00	7.2
C-2764		Cascade Blue® goat anti-rabbit IgG (H+L) conjugate *2 mg/mL*	0.5 mL	75.00	60.00	7.2
C-2794		Cascade Blue® goat anti-rat IgG (H+L) conjugate *2 mg/mL*	0.5 mL	75.00	60.00	7.2
C-3239		Cascade Blue® hydrazide, trilithium salt	10 mg	95.00	76.00	15.3
C-3221		Cascade Blue® hydrazide, tripotassium salt	10 mg	95.00	76.00	15.3
C-687		Cascade Blue® hydrazide, trisodium salt	10 mg	95.00	76.00	3.2, 9.4, 15.3, 17.4
C-2990		casein, fluorescein conjugate	25 mg	68.00	54.40	10.4, 17.1
T-506		CAT 1 (4-trimethylammonium-2,2,6,6-tetramethylpiperidine-1-oxyl iodide)	100 mg	95.00	76.00	15.3

Cat #		Product Name	Unit Size	Price Per Unit ($) 1–4 Units	5–24 Units	Section #
D-531		CAT 16 (4-(N,N-dimethyl-N-hexadecyl)ammonium-2,2,6,6-tetramethylpiperidine-1-oxyl iodide)	25 mg	48.00	38.40	13.3
A-6222	**New**	CBQCA; 3-(4-carboxybenzoyl)quinoline-2-carboxaldehyde (ATTO-TAG™ CBQCA derivatization reagent)	10 mg	110.00	88.00	1.8, 9.1, 24.3
C-6667	**New**	CBQCA Protein Quantitation Kit *300-800 assays*	1 kit	98.00	78.40	9.1
C-6485	**New**	N-CBZ-L-leucyl-L-leucyl-L-tyrosine, diazomethyl ketone	10 mg	95.00	76.00	10.6
C-6870	**New**	CD 222 potassium indicator	1 mg	95.00	76.00	24.1
C-7028	**New**	Cell Culture Contamination Detection Kit *200 assays*	1 kit	95.00	76.00	16.2
C-7886	**New**	CELLebration: The ASCB Educational Video Project. R. Fink, ed. Sinauer Associates (1995) *VHS format*	each	85.00	85.00	26.6
		cell proliferation assay — see CyQUANT™ Cell Proliferation Assay Kit; see also Section 16.2.				
C-2110		CellTracker™ Blue CMAC (7-amino-4-chloromethylcoumarin)	5 mg	95.00	76.00	10.1, 15.2
C-2111		CellTracker™ Blue CMHC (4-chloromethyl-7-hydroxycoumarin)	5 mg	95.00	76.00	15.2
C-7001	**New**	CellTracker™ CM-DiI	1 mg	110.00	88.00	14.2, 15.4
C-7000	**New**	CellTracker™ CM-DiI *special packaging*	20x50 µg	110.00	88.00	14.2, 15.4
C-2102		CellTracker™ Green BODIPY® (8-chloromethyl-4,4-difluoro-1,3,5,7-tetramethyl-4-bora-3a,4a-diaza-s-indacene)	5 mg	95.00	76.00	15.2
C-2925		CellTracker™ Green CMFDA (5-chloromethylfluorescein diacetate)	1 mg	125.00	100.00	15.2, 16.3, 23.2
C-7025	**New**	CellTracker™ Green CMFDA (5-chloromethylfluorescein diacetate) *special packaging*	20x50 µg	145.00	116.00	15.2, 16.3, 23.2
C-2927		CellTracker™ Orange CMTMR (5-(and-6)-(((4-chloromethyl)benzoyl)amino)-tetramethylrhodamine) *mixed isomers*	1 mg	125.00	100.00	15.2
C-2926		CellTracker™ Yellow-Green CMEDA (5-chloromethyleosin diacetate)	1 mg	125.00	100.00	15.2, 16.3
		ceramides — most of the ceramides are named as derivatives of sphingosine; see Section 20.6 for a complete list.				
A-7509	**New**	C_2-ceramide; acetyl ceramide (N-acetyl-D-erythro-sphingosine)	5 mg	95.00	76.00	20.6
H-7524	**New**	C_6-ceramide; hexanoyl ceramide; (N-hexanoyl-D-erythro-sphingosine)	5 mg	95.00	76.00	20.6
C-1361		5-CFDA (5-carboxyfluorescein diacetate) *single isomer*	100 mg	75.00	60.00	16.3
C-1362		6-CFDA (6-carboxyfluorescein diacetate) *single isomer*	100 mg	75.00	60.00	16.3
C-195		5(6)-CFDA (5-(and-6)-carboxyfluorescein diacetate) *mixed isomers*	100 mg	75.00	60.00	16.3, 23.2
C-1354		5-CFDA, AM (5-carboxyfluorescein diacetate, acetoxymethyl ester)	5 mg	68.00	54.40	16.3, 23.2
C-1157		5(6)-CFDA, SE (5-(and-6)-carboxyfluorescein diacetate, succinimidyl ester) *mixed isomers*	25 mg	115.00	92.00	15.2, 16.3, 23.2
C-7620	**New**	cGDP-ribose (cyclic guanosine 5′-diphosphate ribose)	100 µg	100.00	80.00	10.3, 20.2
C-98201	**New**	charybdotoxin *from *Leiurus quinquestriatus* scorpion venom*	10 µg	305.00	244.00	18.6
C-98213	**New**	charybdotoxin *synthetic, originally from *Leiurus quinquestriatus* scorpion venom*	50 µg	135.00	108.00	18.6
C-6056	**New**	7-chlorobenz-2-oxa-1,3-diazole-4-sulfonyl chloride	25 mg	45.00	36.00	5.2
C-6055	**New**	N-(7-chlorobenz-2-oxa-1,3-diazol-4-yl)sulfonyl morpholine	10 mg	45.00	36.00	2.3
C-8447	**New**	9-chloro-9-(4-(diethylamino)phenyl)-9,10-dihydro-10-phenylacridine, hydrochloride (C-390)	10 mg	75.00	60.00	16.3
C-96032	**New**	3α-chloroimperialine	1 mg	160.00	128.00	18.2
C-96033	**New**	3β-chloroimperialine	1 mg	135.00	108.00	18.2
C-8411	**New**	6-chloro-3-indolyl β-D-galactopyranoside	100 mg	inquire	inquire	10.2
C-8412	**New**	6-chloro-3-indolyl β-D-glucuronide, cyclohexylammonium salt	25 mg	inquire	inquire	10.2
C-8413	**New**	6-chloro-3-indolyl phosphate, p-toluidine salt	100 mg	inquire	inquire	10.3
C-2927		5-(and-6)-(((4-chloromethyl)benzoyl)amino)tetramethylrhodamine (CellTracker™ Orange CMTMR) *mixed isomers*	1 mg	125.00	100.00	15.2
C-6827	**New**	5-(and-6)-chloromethyl-2′,7′-dichlorodihydrofluorescein diacetate (CM-H_2DCFDA) *mixed isomers* *special packaging*	20x50 µg	125.00	100.00	15.2, 17.1, 21.4
C-2102		8-chloromethyl-4,4-difluoro-1,3,5,7-tetramethyl-4-bora-3a,4a-diaza-s-indacene (CellTracker™ Green BODIPY®)	5 mg	95.00	76.00	15.2
C-2926		5-chloromethyleosin diacetate (CellTracker™ Yellow-Green CMEDA)	1 mg	125.00	100.00	15.2, 16.3
C-6532	**New**	4-chloromethyl-7-ethoxycoumarin	10 mg	48.00	38.40	10.5
C-2925		5-chloromethylfluorescein diacetate (CellTracker™ Green CMFDA)	1 mg	125.00	100.00	15.2, 16.3, 23.2
C-7025	**New**	5-chloromethylfluorescein diacetate (CellTracker™ Green CMFDA) *special packaging*	20x50 µg	145.00	116.00	15.2, 16.3, 23.2
C-6533	**New**	5-(and-6)-chloromethylfluorescein diethyl ether *mixed isomers*	5 mg	95.00	76.00	10.5
C-2111		4-chloromethyl-7-hydroxycoumarin (CellTracker™ Blue CMHC)	5 mg	95.00	76.00	15.2
C-6569	**New**	5-(and-6)-chloromethylrhodamine 110, bis-(CBZ-L-alanyl-L-alanine amide) (CM-R110, $(CBZ\text{-}Ala\text{-}Ala)_2$) *mixed isomers*	1 mg	75.00	60.00	10.4
C-6826	**New**	5-(and-6)-chloromethyl SNARF®-1, acetate *mixed isomers* *special packaging*	20x50 µg	125.00	100.00	15.2, 16.3, 23.2
C-1648		1-(5-chloronaphthalene-1-sulfonyl)-1H-hexahydroazepine, hydrochloride (ML-9)	10 mg	48.00	38.40	20.3
C-10		4-chloro-7-nitrobenz-2-oxa-1,3-diazole (NBD chloride; 4-chloro-7-nitrobenzofurazan)	1 g	24.00	19.20	1.8, 2.3
C-98210	**New**	chlorotoxin *from *Leiurus quinquestriatus* scorpion venom* *special packaging*	2x10 µg	265.00	212.00	18.5
C-3927		cholesteryl BODIPY® FL C_{12} (cholesteryl 4,4-difluoro-5,7-dimethyl-4-bora-3a,4a-diaza-s-indacene-3-dodecanoate)	1 mg	85.00	68.00	13.5

Cat #		Product Name	Unit Size	Price Per Unit ($) 1–4 Units	5–24 Units	Section #
C-3927		cholesteryl 4,4-difluoro-5,7-dimethyl-4-bora-3a,4a-diaza-s-indacene-3-dodecanoate (cholesteryl BODIPY® FL C_{12})	1 mg	85.00	68.00	13.5
C-7794	**New**	cholesteryl DPH propionate (cholesteryl 3-((6-phenyl)-1,3,5-hexatrienyl)-phenylpropionate)	5 mg	125.00	100.00	13.5
C-215		cholesteryl 12-(N-methyl-N-(7-nitrobenz-2-oxa-1,3-diazol-4-yl)amino)dodecanoate	10 mg	50.00	40.00	13.5
C-7789	**New**	cholesteryl *cis*-parinarate	5 mg	125.00	100.00	13.5
C-7794	**New**	cholesteryl 3-((6-phenyl)-1,3,5-hexatrienyl)phenylpropionate (cholesteryl DPH propionate)	5 mg	125.00	100.00	13.5
C-212		cholesteryl 1-pyrenebutyrate	25 mg	50.00	40.00	13.5
C-213		cholesteryl 1-pyrenedecanoate	10 mg	75.00	60.00	13.5
C-7795	**New**	cholesteryl 1-pyrenehexanoate	5 mg	75.00	60.00	13.5
C-7614	**New**	ChromaTide™ BODIPY® FL-14-dUTP (BODIPY® FL-14-dUTP) *1 mM in buffer*	25 µL	175.00	175.00	8.2, 16.3
C-7613	**New**	ChromaTide™ BODIPY® FL-14-UTP (BODIPY® FL-14-UTP) *1 mM in buffer*	25 µL	175.00	175.00	8.2
C-7616	**New**	ChromaTide™ BODIPY® TMR-14-dUTP (BODIPY® TMR-14-dUTP) *1 mM in buffer*	25 µL	175.00	175.00	8.2
C-7615	**New**	ChromaTide™ BODIPY® TMR-14-UTP (BODIPY® TMR-14-UTP) *1 mM in buffer*	25 µL	175.00	175.00	8.2
C-7618	**New**	ChromaTide™ BODIPY® TR-14-dUTP (BODIPY® TR-14-dUTP) *1 mM in buffer*	25 µL	175.00	175.00	8.2
C-7617	**New**	ChromaTide™ BODIPY® TR-14-UTP (BODIPY® TR-14-UTP) *1 mM in buffer*	25 µL	175.00	175.00	8.2
C-7612	**New**	ChromaTide™ Cascade Blue®-7-dUTP (Cascade Blue®-7-dUTP) *1 mM in buffer*	25 µL	175.00	175.00	8.2
C-7611	**New**	ChromaTide™ Cascade Blue®-7-UTP (Cascade Blue®-7-UTP) *1 mM in buffer*	25 µL	175.00	175.00	8.2
C-7604	**New**	ChromaTide™ fluorescein-12-dUTP (fluorescein-12-dUTP) *1 mM in buffer*	25 µL	175.00	175.00	8.2
C-7603	**New**	ChromaTide™ fluorescein-12-UTP (fluorescein-12-UTP) *1 mM in buffer*	25 µL	175.00	175.00	8.2
C-7630	**New**	ChromaTide™ Oregon Green™ 488-5-dUTP (Oregon Green™ 488-5-dUTP) *1 mM in buffer*	25 µL	175.00	175.00	8.2
C-7629	**New**	ChromaTide™ Rhodamine Green™-5-dUTP (Rhodamine Green™-5-dUTP) *1 mM in buffer*	25 µL	175.00	175.00	8.2
C-7628	**New**	ChromaTide™ Rhodamine Green™-5-UTP (Rhodamine Green™-5-UTP) *1 mM in buffer*	25 µL	175.00	175.00	8.2
C-7606	**New**	ChromaTide™ tetramethylrhodamine-5-dUTP (TMR-5-dUTP) *1 mM in buffer*	25 µL	175.00	175.00	8.2
C-7605	**New**	ChromaTide™ tetramethylrhodamine-5-UTP (TMR-5-UTP) *1 mM in buffer*	25 µL	175.00	175.00	8.2
C-7608	**New**	ChromaTide™ Texas Red®-5-dUTP (Texas Red®-5-dUTP) *1 mM in buffer*	25 µL	175.00	175.00	8.2
C-7631	**New**	ChromaTide™ Texas Red®-12-dUTP (Texas Red®-12-dUTP) *1 mM in buffer*	25 µL	175.00	175.00	8.2
C-7607	**New**	ChromaTide™ Texas Red®-5-UTP (Texas Red®-5-UTP) *1 mM in buffer*	25 µL	175.00	175.00	8.2
C-7584	**New**	Chromosome Banding Kit #1 *YOYO®-1/Methyl Green* *200 slides*	1 kit	95.00	76.00	8.4
C-7587	**New**	Chromosome Banding Kit #2 *SYTOX® Green/Methyl Green* *200 slides*	1 kit	95.00	76.00	8.4
C-6831	**New**	Cl-NERF	5 mg	95.00	76.00	1.5, 23.3
C-7079	**New**	Cl-NERF 4,5-dimethoxy-2-nitrobenzyl ether, sulfosuccinimidyl ester, sodium salt (DMNB-caged Cl-NERF, SSE)	1 mg	48.00	38.40	15.3, 19.2
C-99001	**New**	cloflubicyne	1 mg	205.00	164.00	18.5
C-7484	**New**	α-cobrotoxin *from *Naja naja* sp.*	1 mg	21.00	16.80	18.2
C-2944		coelenterazine	250 µg	295.00	236.00	17.1, 21.3, 22.5
C-6779	**New**	coelenterazine f	250 µg	295.00	236.00	22.5
C-6780	**New**	coelenterazine h	250 µg	295.00	236.00	22.5
C-6781	**New**	coelenterazine hcp	250 µg	295.00	236.00	22.5
C-6776	**New**	coelenterazine n	250 µg	295.00	236.00	22.5
C-6777	**New**	Coelenterazine Sampler Kit *contains 25 µg of C-2944, C-6776, C-6779, C-6780, C-6781*	1 kit	255.00	204.00	22.5
D-7499	**New**	Colcemid®; N-desacetyl-N-methylcolchicine (demecolcine)	10 mg	38.00	30.40	11.2
C-662		colchicine fluorescein	5 mg	125.00	100.00	11.2
C-7301	**New**	CompenFlow™ Flow Cytometry Compensation Kit	1 kit	140.00	112.00	26.2
C-6909	**New**	conantokin-G *purity > 95%*	25 µg	55.00	44.00	18.2
C-6904	**New**	conantokin-G, BODIPY® FL conjugate	5 µg	175.00	140.00	18.2
C-6906	**New**	conantokin-G, tetramethylrhodamine conjugate	5 µg	175.00	140.00	18.2
C-6749	**New**	concanavalin A, AMCA-S conjugate	5 mg	48.00	38.40	7.6
C-827		concanavalin A, fluorescein conjugate	10 mg	48.00	38.40	7.6
C-6741	**New**	concanavalin A, Oregon Green™ 488 conjugate	5 mg	48.00	38.40	7.6
C-6742	**New**	concanavalin A, Oregon Green™ 514 conjugate	5 mg	48.00	38.40	7.6
C-860		concanavalin A, tetramethylrhodamine conjugate	10 mg	48.00	38.40	7.6
C-825		concanavalin A, Texas Red® conjugate	10 mg	48.00	38.40	7.6
C-96017	**New**	condelphine	20 mg	108.00	86.40	18.2
C-3117		ω-conotoxin GVIA	5x10 µg	75.00	60.00	18.4
C-6903	**New**	ω-conotoxin GVIA, fluorescein conjugate	5 µg	295.00	236.00	18.4
C-3115		ω-conotoxin GVIA, tetramethylrhodamine conjugate	5 µg	295.00	236.00	18.4
C-98501	**New**	ω-conotoxin MVIIA *synthetic, originally from *Conus magus* molluscan venom*	100 µg	135.00	108.00	18.4
C-96086	**New**	convolamine	20 mg	55.00	44.00	18.8
C-96087	**New**	convolvine	20 mg	60.00	48.00	18.8
C-7933	**New**	coumarin-3-carboxylic acid, succinimidyl ester (SECCA)	25 mg	65.00	52.00	10.5, 21.3
C-606		coumarin phallacidin	300 U	198.00	158.40	11.1
D-346		CPM (7-diethylamino-3-(4′-maleimidylphenyl)-4-methylcoumarin)	25 mg	158.00	126.40	2.2, 9.4

Cat #		Product Name	Unit Size	Price Per Unit ($) 1–4 Units	Price Per Unit ($) 5–24 Units	Section #
C-8448	***New***	creatine phosphate, disodium salt, tetrahydrate (phosphocreatine)	5 g	44.00	35.20	10.3
C-1488		CUG (3-carboxyumbelliferyl β-D-galactopyranoside)	10 mg	98.00	78.40	10.2
C-7880	***New***	cuvette for absorption, optical glass, 10 mm x 10 mm, with Teflon® cover *useable range 334 to 2500 nm*	each	25.00	25.00	26.3
C-7882	***New***	cuvette for absorption, optical glass, 10 mm x 10 mm, with Teflon® stopper *useable range 334 to 2500 nm*	each	65.00	65.00	26.3
C-7879	***New***	cuvette for absorption, quartz, 10 mm x 10 mm, with Teflon® cover *useable range 170 to 2700 nm*	each	75.00	75.00	26.3
C-7881	***New***	cuvette for absorption, quartz, 10 mm x 10 mm, with Teflon® stopper *useable range 170 to 2700 nm*	each	98.00	98.00	26.3
C-7876	***New***	cuvette for fluorescence, optical glass, 10 mm x 10 mm, with Teflon® cover *useable range 334 to 2500 nm*	each	55.00	55.00	26.3
C-7878	***New***	cuvette for fluorescence, optical glass, 10 mm x 10 mm, with Teflon® stopper *useable range 334 to 2500 nm*	each	78.00	78.00	26.3
C-7875	***New***	cuvette for fluorescence, quartz, 10 mm x 10 mm, with Teflon® cover *useable range 170 to 2700 nm*	each	110.00	110.00	26.3
C-7877	***New***	cuvette for fluorescence, quartz, 10 mm x 10 mm, with Teflon® stopper *useable range 170 to 2700 nm*	each	135.00	135.00	26.3
C-684		3-cyano-7-ethoxycoumarin	10 mg	85.00	68.00	10.5
C-183		3-cyano-7-hydroxycoumarin	100 mg	35.00	28.00	10.1
C-7619	***New***	cyclic adenosine 5′-diphosphate ribose (cADP-ribose)	100 µg	250.00	200.00	20.2
C-7074	***New***	cyclic adenosine 5′-diphosphate ribose, 1-(1-(2-nitrophenyl)ethyl)ester (NPE-caged cADP-ribose) *mixed isomers*	50 µg	150.00	120.00	19.2, 20.2
C-6661	***New***	cyclic AMP-dependent protein kinase, catalytic subunit, fluorescein conjugate *20±5 µM*	10 µL	295.00	295.00	20.3
C-6662	***New***	cyclic AMP-dependent protein kinase, regulatory subunit, rhodamine conjugate *20±5 µM*	10 µL	295.00	295.00	20.3
C-6660	***New***	Cyclic AMP Fluorosensor® (FlCRhR) *20±5 µM*	10 µL	495.00	495.00	20.3
C-7620	***New***	cyclic guanosine 5′-diphosphate ribose (cGDP-ribose)	100 µg	100.00	80.00	10.3, 20.2
C-428		N-cyclohexyl-N′-(4-(dimethylamino)naphthyl)carbodiimide (NCD-4)	25 mg	138.00	110.40	3.3, 9.4
C-442		N-cyclohexyl-N′-(1-pyrenyl)carbodiimide	25 mg	65.00	52.00	3.3
C-7027	***New***	CyQUANT™ cell lysis buffer *20X concentrate*	50 mL	45.00	36.00	16.2
C-7026	***New***	CyQUANT™ Cell Proliferation Assay Kit *for cells in culture* *1000 assays*	1 kit	135.00	108.00	16.2
C-7590	***New***	Cytological Nuclear Counterstain Kit *DAPI/SYTOX® Green/PI* *300 slides*	1 kit	75.00	60.00	8.4
T-6615	***New***	D 609 (tricyclodecan-9-yl xanthogenate, potassium salt)	5 mg	48.00	38.40	20.5
D-6216	***New***	ε-dabcyl-α-FMOC-L-lysine (ε-(4-((4-(dimethylamino)phenyl)azo)benzoyl)-α-9-fluorenylmethoxycarbonyl-L-lysine)	100 mg	175.00	140.00	9.3
D-2245		dabcyl, SE (4-((4-(dimethylamino)phenyl)azo)benzoic acid, succinimidyl ester)	100 mg	68.00	54.40	1.8, 9.3
D-1537		dabsyl chloride (4-dimethylaminoazobenzene-4′-sulfonyl chloride)	100 mg	24.00	19.20	1.8, 9.3
D-1552		DAMP (N-(3-((2,4-dinitrophenyl)amino)propyl)-N-(3-aminopropyl)methylamine, dihydrochloride)	100 mg	98.00	78.40	12.3
D-2281		m-dansylaminophenylboronic acid	100 mg	48.00	38.40	3.1
D-151		dansyl aziridine (5-dimethylaminonaphthalene-1-sulfonyl aziridine)	100 mg	68.00	54.40	2.3
D-113		dansyl cadaverine (5-dimethylaminonaphthalene-1-(N-(5-aminopentyl))-sulfonamide)	100 mg	78.00	62.40	3.3, 9.4, 12.3
D-21		dansyl chloride (5-dimethylaminonaphthalene-1-sulfonyl chloride)	1 g	40.00	32.00	1.8, 9.3
D-112		dansyl ethylenediamine (5-dimethylaminonaphthalene-1-(N-(2-aminoethyl))-sulfonamide)	100 mg	78.00	62.40	3.3
D-11		dansyl fluoride (5-dimethylaminonaphthalene-1-sulfonyl fluoride)	100 mg	18.00	14.40	9.4, 10.6
D-100		dansyl hydrazine (5-dimethylaminonaphthalene-1-sulfonyl hydrazine)	100 mg	27.00	21.60	3.2, 9.3
D-1306		DAPI (4′,6-diamidino-2-phenylindole, dihydrochloride)	10 mg	37.00	29.60	8.1, 11.2, 15.3 16.3
D-3571		DAPI, dilactate (4′,6-diamidino-2-phenylindole, dilactate)	10 mg	37.00	29.60	8.1, 15.3
D-426		DASPEI (2-(4-(dimethylamino)styryl)-N-ethylpyridinium iodide)	1 g	75.00	60.00	12.2
D-94		DAUDA (11-((5-dimethylaminonaphthalene-1-sulfonyl)amino)undecanoic acid)	100 mg	68.00	54.40	13.3, 20.5
H-6482	***New***	DDAO (7-hydroxy-9H-(1,3-dichloro-9,9-dimethylacridin-2-one)) *reference standard*	10 mg	45.00	36.00	10.1
D-6488	***New***	DDAO galactoside (9H-(1,3-dichloro-9,9-dimethylacridin-2-one-7-yl) β-D-galactopyranoside)	5 mg	135.00	18.00	10.2
D-6486	***New***	DDAO GlcU (9H-(1,3-dichloro-9,9-dimethylacridin-2-one-7-yl) β-D-glucuronide)	5 mg	135.00	108.00	10.2
D-6487	***New***	DDAO phosphate (9H-(1,3-dichloro-9,9-dimethylacridin-2-one-7-yl) phosphate, diammonium salt)	5 mg	135.00	108.00	10.3
D-7915	***New***	DEANO (diethylamine nitric oxide, sodium salt)	25 mg	48.00	38.40	21.5
D-7522	***New***	DL-*threo*-2-decanoylamino-3-morpholino-1-phenyl-1-propanol, hydrochloride (PDMP)	5 mg	48.00	38.40	20.6
D-3761		1-(S-decanoyl)-2-decanoyl-1-thio-*sn*-glycero-3-phosphocholine	1 mg	45.00	36.00	20.5
D-3771		2-decanoyl-1-(O-(11-(4,4-difluoro-5,7-dimethyl-4-bora-3a,4a-diaza-s-indacene-3-propionyl)amino)undecyl)-*sn*-glycero-3-phosphocholine	1 mg	98.00	78.40	20.5

Cat #	Product Name	Unit Size	Price Per Unit ($) 1–4 Units	5–24 Units	Section #
D-3768	2-decanoyl-1-(O-(N-(7-nitrobenz-2-oxa-1,3-diazol-4-yl)amino)undecyl)-*sn*-glycero-3-phosphocholine	1 mg	68.00	54.40	20.5
D-3826	5-decyl-4,4-difluoro-4-bora-3a,4a-diaza-s-indacene-3-propionic acid (C_{10}-BODIPY® 500/510 C_3)	1 mg	38.00	30.40	13.2
D-7480 **New**	9,21-dehydroryanodine	1 mg	38.00	30.40	18.4, 20.2
D-96052 **New**	delcorine	20 mg	95.00	76.00	18.2
D-96019 **New**	delsoline (acomonine)	20 mg	95.00	76.00	18.2
D-96057 **New**	deltaline (eldeline)	100 mg	75.00	60.00	18.7
D-7499 **New**	demecolcine (Colcemid®; N-desacetyl-N-methylcolchicine)	10 mg	38.00	30.40	11.2
D-98106 **New**	α-dendrotoxin *from *Dendroaspis angusticeps* snake venom*	100 µg	88.00	70.40	18.6
D-98113 **New**	dendrotoxin I (DPI) *from *Dendroaspis polylepis polylepis* snake venom* *special packaging*	2x50 µg	88.00	70.40	18.6
D-6620 **New**	1-deoxychloramphenicol (L-(2-dichloroacetylamino)-3-(4-nitrophenyl)propanol)	5 mg	98.00	78.40	10.5
D-970	deoxyribonuclease I, fluorescein conjugate	5 mg	135.00	108.00	11.1
D-7497 **New**	deoxyribonuclease I, Oregon Green™ 488 conjugate	5 mg	135.00	108.00	11.1
D-971	deoxyribonuclease I, tetramethylrhodamine conjugate	5 mg	135.00	108.00	11.1
D-972	deoxyribonuclease I, Texas Red® conjugate	5 mg	135.00	108.00	11.1
D-1638	desacetyl colchicine	10 mg	48.00	38.40	11.2
D-96035 **New**	N-desacetyllappaconitine	10 mg	128.00	102.40	18.7
D-96009 **New**	desoxypeganine, hydrochloride, dihydrate	20 mg	108.00	86.40	18.2
A-7073 **New**	desyl-caged ATP (adenosine 5′-triphosphate, P^2-(1,2-diphenyl-2-oxo)ethyl ester, ammonium salt) *5 mM solution in water*	400 µL	145.00	116.00	11.3, 19.2, 20.4
A-7125 **New**	O-(desyl-caged) GABA (γ-aminobutyric acid, (1,2-diphenyl-2-oxo)ethyl ester, trifluoroacetic acid salt)	5 mg	168.00	134.40	18.3, 19.2
G-7120 **New**	γ-(desyl-caged) L-glutamic acid (L-glutamic acid, γ-(1,2-diphenyl-2-oxo)ethyl ester)	5 mg	168.00	134.40	18.3, 19.2
D-7076 **New**	desyl-caged phosphate ((1,2-diphenyl-2-oxo)ethyl phosphate, monosodium salt, monohydrate)	5 mg	95.00	76.00	19.2, 23.2
D-2921	DetectaGene™ Blue CMCG *lacZ* Gene Expression Kit	1 kit	148.00	118.40	10.2
D-2920	DetectaGene™ Green CMFDG *lacZ* Gene Expression Kit	1 kit	295.00	236.00	10.2
D-1383	dexamethasone fluorescein	5 mg	48.00	38.40	18.2
	dextrans — see Section 15.5 for dextran conjugates and Sections 22.4, 22.6, 23.4, 24.1 and 24.2 for indicator dextrans.				
D-3883	DiA; 4-Di-16-ASP (4-(4-(dihexadecylamino)styryl)-N-methylpyridinium iodide)	25 mg	115.00	92.00	14.2, 15.4
D-96081 **New**	diacetylkorseveriline	50 mg	75.00	60.00	18.7
D-1306	4′,6-diamidino-2-phenylindole, dihydrochloride (DAPI)	10 mg	37.00	29.60	8.1, 11.2, 15.3 16.3
D-3571	4′,6-diamidino-2-phenylindole, dilactate (DAPI, dilactate)	10 mg	37.00	29.60	8.1, 15.3
D-1463	1,2-diamino-4,5-dimethoxybenzene, dihydrochloride (DDB)	100 mg	73.00	58.40	3.2
D-7918 **New**	2,3-diaminonaphthalene	100 mg	18.00	14.40	21.5, 24.3
D-6923 **New**	di-2-ANEPEQ (JPW 1114)	5 mg	175.00	140.00	25.2
D-6925 **New**	di-8-ANEPPQ	5 mg	175.00	140.00	25.2
D-6927 **New**	di-12-ANEPPQ	5 mg	175.00	140.00	25.2
D-1199	di-4-ANEPPS	5 mg	155.00	124.00	25.2
D-3167	di-8-ANEPPS	5 mg	195.00	156.00	25.2
D-6928 **New**	di-18:2-ANEPPS	5 mg	175.00	140.00	25.2
B-153	4,4′-dianilino-1,1′-binaphthyl-5,5′-disulfonic acid, dipotassium salt (bis-ANS)	10 mg	125.00	100.00	11.2, 14.5
D-6313 **New**	4,4′-diazidostilbene-2,2′-disulfonic acid, disodium salt, tetrahydrate (DASD)	1 g	35.00	28.00	5.3
D-3036	diazo-2, AM *cell permeant* *special packaging*	20x50 µg	195.00	156.00	19.2, 20.2, 22.8
D-3034	diazo-2, tetrapotassium salt *cell impermeant*	1 mg	195.00	156.00	19.2, 20.2, 22.8
D-3701	diazo-3, AM *cell permeant*	1 mg	75.00	60.00	19.2, 20.2, 22.8
D-3700	diazo-3, dipotassium salt *cell impermeant*	1 mg	75.00	60.00	19.2, 20.2, 22.8
D-1402	4-diazomethyl-7-diethylaminocoumarin	10 mg	88.00	70.40	3.4, 9.4
D-1401	4-diazomethyl-7-methoxycoumarin	25 mg	88.00	70.40	3.4
B-438	$DiBAC_4(3)$ (bis-(1,3-dibutylbarbituric acid)trimethine oxonol)	25 mg	115.00	92.00	25.3
B-436	$DiBAC_4(5)$ (bis-(1,3-dibutylbarbituric acid)pentamethine oxonol)	25 mg	65.00	52.00	18.5, 25.3
D-675	4,4′-dibenzamidostilbene-2,2′-disulfonic acid, disodium salt (DBDS)	100 mg	78.00	62.40	18.5
D-1213	5,5′-dibromo BAPTA, AM *cell permeant* *≥95% by HPLC*	25 mg	48.00	38.40	22.8
D-1211	5,5′-dibromo BAPTA, tetrapotassium salt *cell impermeant*	100 mg	73.00	58.40	22.8
D-1379	dibromobimane (bBBr)	25 mg	35.00	28.00	2.4, 5.2
D-7546 **New**	N-(2,6-dibromo-4,4-difluoro-5,7-dimethyl-4-bora-3a,4a-diaza-s-indacene-3-pentanoyl)sphingosine (BODIPY® FL Br_2C_5-ceramide)	250 µg	98.00	78.40	12.4, 20.6
D-6143 **New**	2,6-dibromo-4,4-difluoro-5,7-dimethyl-4-bora-3a,4a-diaza-s-indacene-3-propionic acid (BODIPY® FL Br_2)	5 mg	95.00	76.00	1.2
D-6144 **New**	2,6-dibromo-4,4-difluoro-5,7-dimethyl-4-bora-3a,4a-diaza-s-indacene-3-propionic acid, succinimidyl ester (BODIPY® FL Br_2, SE)	5 mg	145.00	116.00	1.2
D-7767 **New**	6,6′-dibromo-3,3′-dioctadecyloxacarbocyanine perchlorate (Br_2-$DiOC_{18}(3)$)	5 mg	135.00	108.00	14.2, 15.4
D-7766 **New**	5,5′-dibromo-1,1′-dioctadecyl-3,3,3′,3′-tetramethylindocarbocyanine perchlorate (Br_2-$DiIC_{18}(3)$)	5 mg	135.00	108.00	14.2, 15.4

Cat #		Product Name	Unit Size	Price Per Unit ($) 1–4 Units	5–24 Units	Section #
D-7623	***New***	N^6,$O^{2'}$-dibutyryladenosine 3′,5′-cyclic monophosphate, acetoxymethyl ester *cell permeant* *special packaging*	10x100 µg	75.00	60.00	20.4
D-6898	***New***	2′,4′-dichlorobenzamil, hydrochloride	25 mg	38.00	30.40	18.7
D-6226	***New***	2,4-dichloro-6-(4-(N,N-diethylamino)phenyl)-1,3,5-triazine	10 mg	78.00	62.40	27.1
D-399		2′,7′-dichlorodihydrofluorescein diacetate (2′,7′-dichlorofluorescin diacetate; H_2DCFDA)	100 mg	45.00	36.00	10.5, 17.1, 21.4
D-2935		2′,7′-dichlorodihydrofluorescein diacetate, succinimidyl ester (OxyBURST® Green, SE)	5 mg	55.00	44.00	17.1, 21.4
D-6488	***New***	9H-(1,3-dichloro-9,9-dimethylacridin-2-one-7-yl) β-D-galactopyranoside (DDAO galactoside)	5 mg	135.00	108.00	10.2
D-6486	***New***	9H-(1,3-dichloro-9,9-dimethylacridin-2-one-7-yl) β-D-glucuronide (DDAO GlcU)	5 mg	135.00	108.00	10.2
D-6487	***New***	9H-(1,3-dichloro-9,9-dimethylacridin-2-one-7-yl) phosphate, diammonium salt (DDAO phosphate)	5 mg	135.00	108.00	10.3

dichlorofluorescein diacetate (DCF-DA) usually refers to the colorless, nonfluorescent "dihydro" derivative 2′,7′-dichlorodihydrofluorescein diacetate (2′,7′-dichlorofluores*cin* diacetate, H_2DCFDA; D-399), which is also abbreviated DCDFDA; see also Sections 10.5, 17.1 and 21.4.

Cat #		Product Name	Unit Size	Price Per Unit ($) 1–4 Units	5–24 Units	Section #
D-6078	***New***	2′,7′-dichlorofluorescein-5-isothiocyanate (Cl_2FITC)	25 mg	55.00	44.00	1.5
D-399		2′,7′-dichlorofluorescin diacetate; H_2DCFDA (2′,7′-dichlorodihydrofluorescein diacetate)	100 mg	45.00	36.00	10.5, 17.1, 21.4
D-6612	***New***	2′,7′-dichloro-4,5,6,7-tetrafluorofluorescein (Cl_2TFF)	5 mg	48.00	38.40	23.3
D-16		5-(4,6-dichlorotriazinyl)aminofluorescein (5-DTAF) *single isomer*	100 mg	75.00	60.00	1.3, 3.1
D-17		6-(4,6-dichlorotriazinyl)aminofluorescein (6-DTAF) *single isomer*	100 mg	75.00	60.00	1.3, 3.1
D-3923		4-(dicyanovinyl)julolidine (DCVJ)	25 mg	48.00	38.40	11.2, 14.5
D-307		'DiD' oil; $DiIC_{18}(5)$ oil (1,1′-dioctadecyl-3,3,3′,3′-tetramethylindodicarbocyanine perchlorate)	25 mg	90.00	72.00	14.2, 15.4
D-7757	***New***	'DiD' solid; $DiIC_{18}(5)$ solid (1,1′-dioctadecyl-3,3,3′,3′-tetramethylindodicarbocyanine, 4-chlorobenzenesulfonate salt)	10 mg	68.00	54.40	14.2, 15.4
D-146		didansyl-L-cystine	100 mg	48.00	38.40	2.3
D-291		4-(4-(didecylamino)styryl)-N-methylpyridinium iodide (4-Di-10-ASP)	25 mg	115.00	92.00	14.2, 15.4
D-383		1,1′-didodecyl-3,3,3′,3′-tetramethylindocarbocyanine perchlorate ($DiIC_{12}(3)$)	100 mg	135.00	108.00	14.2, 15.4
D-337		DIDS (4,4′-diisothiocyanatostilbene-2,2′-disulfonic acid, disodium salt)	100 mg	22.00	17.60	18.5
D-3109		5-(N,N-diethyl)amiloride, hydrochloride	5 mg	35.00	28.00	18.7
D-7915	***New***	diethylamine nitric oxide, sodium salt (DEANO)	25 mg	48.00	38.40	21.5
D-1446		7-diethylaminocoumarin-3-carbonyl azide	25 mg	88.00	70.40	3.1
D-1421		7-diethylaminocoumarin-3-carboxylic acid	100 mg	38.00	30.40	1.7
D-355		7-diethylaminocoumarin-3-carboxylic acid, hydrazide (DCCH)	25 mg	90.00	72.00	3.2
D-1412		7-diethylaminocoumarin-3-carboxylic acid, succinimidyl ester	25 mg	75.00	60.00	1.7
D-7113	***New***	*trans*-p-(diethylamino)-o-hydroxy-α-methylcinnamic acid, p-amidinophenyl ester, p-toluenesulfonic acid salt	5 mg	65.00	52.00	10.6, 19.2
D-7112	***New***	*trans*-p-(diethylamino)-o-hydroxy-α-methylcinnamic acid, p-nitrophenyl ester	5 mg	65.00	52.00	10.6, 19.2
D-404		7-diethylamino-3-((4′-(iodoacetyl)amino)phenyl)-4-methylcoumarin (DCIA)	25 mg	158.00	126.40	2.2
D-347		7-diethylamino-3-(4′-isothiocyanatophenyl)-4-methylcoumarin (CPI)	25 mg	98.00	78.40	1.7
D-346		7-diethylamino-3-(4′-maleimidylphenyl)-4-methylcoumarin (CPM)	25 mg	158.00	126.40	2.2, 9.4
D-1579		8-diethylaminooctyl 3,4,5-trimethoxybenzoate, hydrochloride (TMB-8)	100 mg	60.00	48.00	20.2
D-289		4-(4-(diethylamino)styryl)-N-methylpyridinium iodide (4-Di-2-ASP)	1 g	75.00	60.00	17.2
D-99007	***New***	6,6-diethyl-N,2-dimethyl-4-nitromethinyl-1,4,5,6-tetrahydropyridine (M1)	100 µg	95.00	76.00	10.6, 18.4
D-269		3,3′-diethyloxacarbocyanine iodide ($DiOC_2(3)$)	100 mg	24.00	19.20	25.3
D-303		3,3′-diethyloxadicarbocyanine iodide ($DiOC_2(5)$)	1 g	35.00	28.00	12.2, 25.3
D-304		3,3′-diethylthiadicarbocyanine iodide ($DiSC_2(5)$)	1 g	35.00	28.00	25.3
D-1216		4,4′-difluoro BAPTA, AM *cell permeant*	25 mg	48.00	38.40	22.8
D-1215		4,4′-difluoro BAPTA, tetrapotassium salt *cell impermeant*	100 mg	48.00	38.40	22.8
D-1209		5,5′-difluoro BAPTA, AM *cell permeant* *≥95% by HPLC*	25 mg	125.00	100.00	22.8
D-1208		5,5′-difluoro BAPTA, tetrapotassium salt *cell impermeant*	100 mg	98.00	78.40	22.8
D-6103	***New***	4,4-difluoro-4-bora-3a,4a-diaza-s-indacene-3,5-dipropionic acid (BODIPY® 500/510)	5 mg	95.00	76.00	1.2
D-3829		4,4-difluoro-5,7-dimethyl-4-bora-3a,4a-diaza-s-indacene-3-decyl sulfate, sodium salt (BODIPY® FL C_{10}-sulfate)	1 mg	38.00	30.40	13.2
D-3822		4,4-difluoro-5,7-dimethyl-4-bora-3a,4a-diaza-s-indacene-3-dodecanoic acid (BODIPY® FL C_{12})	1 mg	38.00	30.40	13.2
D-7520	***New***	DL-*threo*-2-(4,4-difluoro-5,7-dimethyl-4-bora-3a,4a-diaza-s-indacene-3-dodecanoyl)amino-3-morpholino-1-phenyl-1-propanol, hydrochloride (BODIPY® FL C_{12}-MPP)	25 µg	95.00	76.00	20.6
D-3792		2-(4,4-difluoro-5,7-dimethyl-4-bora-3a,4a-diaza-s-indacene-3-dodecanoyl)-1-hexadecanoyl-*sn*-glycero-3-phosphocholine (β-BODIPY® FL C_{12}-HPC)	100 µg	38.00	30.40	13.2
D-3802		2-(4,4-difluoro-5,7-dimethyl-4-bora-3a,4a-diaza-s-indacene-3-dodecanoyl)-1-hexadecanoyl-*sn*-glycero-3-phosphoethanolamine (β-BODIPY® FL C_{12}-HPE)	100 µg	38.00	30.40	13.2

Cat #		Product Name	Unit Size	Price Per Unit ($) 1–4 Units	5–24 Units	Section #
D-7519	*New*	N-(4,4-difluoro-5,7-dimethyl-4-bora-3a,4a-diaza-s-indacene-3-dodecanoyl)-sphingosyl 1-β-D-galactopyranoside (BODIPY® FL C_{12}-galactocerebroside)	25 µg	95.00	76.00	20.6
D-7547	*New*	N-(4,4-difluoro-5,7-dimethyl-4-bora-3a,4a-diaza-s-indacene-3-dodecanoyl)-sphingosyl 1-β-D-glucopyranoside (BODIPY® FL C_{12}-glucocerebroside)	250 µg	98.00	78.40	20.6
D-7711	*New*	N-(4,4-difluoro-5,7-dimethyl-4-bora-3a,4a-diaza-s-indacene-3-dodecanoyl)-sphingosyl phosphocholine (BODIPY® FL C_{12}-sphingomyelin) *1 mg/mL in DMSO*	250 µL	125.00	100.00	27.1
D-3821		4,4-difluoro-5,7-dimethyl-4-bora-3a,4a-diaza-s-indacene-3-hexadecanoic acid (BODIPY® FL C_{16})	1 mg	38.00	30.40	13.2
D-3834		4,4-difluoro-5,7-dimethyl-4-bora-3a,4a-diaza-s-indacene-3-pentanoic acid (BODIPY® FL C_5)	1 mg	38.00	30.40	1.2, 13.2
D-6184	*New*	4,4-difluoro-5,7-dimethyl-4-bora-3a,4a-diaza-s-indacene-3-pentanoic acid, succinimidyl ester (BODIPY® FL C_5, SE)	5 mg	145.00	116.00	1.2
D-3805		2-(4,4-difluoro-5,7-dimethyl-4-bora-3a,4a-diaza-s-indacene-3-pentanoyl)-1-hexadecanoyl-*sn*-glycero-3-phosphate, diammonium salt (β-BODIPY® FL C_5-HPA)	100 µg	38.00	30.40	13.2
D-3803		2-(4,4-difluoro-5,7-dimethyl-4-bora-3a,4a-diaza-s-indacene-3-pentanoyl)-1-hexadecanoyl-*sn*-glycero-3-phosphocholine (β-BODIPY® FL C_5-HPC)	100 µg	38.00	30.40	13.2
D-3804		2-(4,4-difluoro-5,7-dimethyl-4-bora-3a,4a-diaza-s-indacene-3-pentanoyl)-1-hexadecanoyl-*sn*-glycero-3-phosphoethanolamine (β-BODIPY® FL C_5-HPE)	100 µg	38.00	30.40	13.2
D-3521		N-(4,4-difluoro-5,7-dimethyl-4-bora-3a,4a-diaza-s-indacene-3-pentanoyl)-sphingosine (BODIPY® FL C_5-ceramide)	250 µg	125.00	100.00	12.4, 20.6
D-7548	*New*	N-(4,4-difluoro-5,7-dimethyl-4-bora-3a,4a-diaza-s-indacene-3-pentanoyl)-sphingosyl 1-β-D-glucopyranoside (BODIPY® FL C_5-glucocerebroside)	250 µg	98.00	78.40	20.6
D-3522		N-(4,4-difluoro-5,7-dimethyl-4-bora-3a,4a-diaza-s-indacene-3-pentanoyl)-sphingosyl phosphocholine (BODIPY® FL C_5-sphingomyelin)	250 µg	125.00	100.00	12.4, 13.2, 20.6
D-6300	*New*	4,4-difluoro-5,7-dimethyl-4-bora-3a,4a-diaza-s-indacene-3-propanol (BODIPY® FL propanol)	10 mg	95.00	76.00	3.4
D-2183		4,4-difluoro-5,7-dimethyl-4-bora-3a,4a-diaza-s-indacene-3-propionic acid (BODIPY® FL)	5 mg	95.00	76.00	1.2
D-2371		4,4-difluoro-5,7-dimethyl-4-bora-3a,4a-diaza-s-indacene-3-propionic acid, hydrazide (BODIPY® FL hydrazide)	5 mg	95.00	76.00	3.2
D-2184		4,4-difluoro-5,7-dimethyl-4-bora-3a,4a-diaza-s-indacene-3-propionic acid, succinimidyl ester (BODIPY® FL, SE)	5 mg	145.00	116.00	1.2
D-6140	*New*	4,4-difluoro-5,7-dimethyl-4-bora-3a,4a-diaza-s-indacene-3-propionic acid, sulfosuccinimidyl ester, sodium salt (BODIPY® FL, SSE)	5 mg	145.00	116.00	1.2, 9.4
D-6201	*New*	4-(2-(4,4-difluoro-5,7-dimethyl-4-bora-3a,4a-diaza-s-indacene-3-propionyl)-aminoethyl)benzenesulfonyl fluoride (BODIPY® FL AEBSF)	1 mg	48.00	38.40	9.4, 10.6
D-6101	*New*	6-((4,4-difluoro-5,7-dimethyl-4-bora-3a,4a-diaza-s-indacene-3-propionyl)amino)-hexanoic acid (BODIPY® FL-X)	5 mg	95.00	76.00	1.2
D-6102	*New*	6-((4,4-difluoro-5,7-dimethyl-4-bora-3a,4a-diaza-s-indacene-3-propionyl)amino)-hexanoic acid, succinimidyl ester (BODIPY® FL-X, SE)	5 mg	145.00	116.00	1.2, 4.2
D-3772		1-(O-(11-(4,4-difluoro-5,7-dimethyl-4-bora-3a,4a-diaza-s-indacene-3-propionyl)-amino)undecyl)-*sn*-glycero-3-phosphocholine (BODIPY® FL C_{11}-lyso PAF)	500 µg	90.00	72.00	20.6
D-7707	*New*	1-(O-(11-(4,4-difluoro-5,7-dimethyl-4-bora-3a,4a-diaza-s-indacene-3-propionyl)-amino)undecyl)-2-hexadecanoyl-*sn*-glycero-3-phosphocholine	1 mg	98.00	78.40	20.5
D-6141	*New*	N-(4,4-difluoro-5,7-dimethyl-4-bora-3a,4a-diaza-s-indacene-3-propionyl)-cysteic acid, succinimidyl ester, triethylammonium salt (BODIPY® FL, CASE)	5 mg	125.00	100.00	1.2, 9.4
D-3800		N-(4,4-difluoro-5,7-dimethyl-4-bora-3a,4a-diaza-s-indacene-3-propionyl)-1,2-dihexadecanoyl-*sn*-glycero-3-phosphoethanolamine, triethylammonium salt (BODIPY® FL DHPE)	100 µg	75.00	60.00	13.4
D-2390		4,4-difluoro-5,7-dimethyl-4-bora-3a,4a-diaza-s-indacene-3-propionyl ethylenediamine, hydrochloride (BODIPY® FL EDA)	5 mg	145.00	116.00	3.3
D-2005		N-(4,4-difluoro-5,7-dimethyl-4-bora-3a,4a-diaza-s-indacene-3-propionyl)-N′-iodoacetyl ethylenediamine (BODIPY® FL IA)	5 mg	95.00	76.00	2.2
D-3862		4,4-difluoro-5,7-dimethyl-4-bora-3a,4a-diaza-s-indacene-3-undecanoic acid (BODIPY® FL C_{11})	1 mg	38.00	30.40	13.2
D-6003	*New*	N-(4,4-difluoro-5,7-dimethyl-4-bora-3a,4a-diaza-s-indacene-3-yl)methyl)-iodoacetamide (BODIPY® FL C_1-IA)	5 mg	95.00	76.00	2.2
D-6117	*New*	6-((4,4-difluoro-1,3-dimethyl-5-(4-methoxyphenyl)-4-bora-3a,4a-diaza-s-indacene-2-propionyl)amino)hexanoic acid, succinimidyl ester (BODIPY® TMR-X, SE)	5 mg	145.00	116.00	1.2
D-6012	*New*	N-(5-((4,4-difluoro-1,3-dimethyl-5-(4-methoxyphenyl)-4-bora-3a,4a-diaza-s-indacene-2-propionyl)amino)pentyl)iodoacetamide (BODIPY® TMR cadaverine IA)	5 mg	95.00	76.00	27.1
D-3832		4,4-difluoro-5,7-diphenyl-4-bora-3a,4a-diaza-s-indacene-3-dodecanoic acid (BODIPY® 530/550 C_{12})	1 mg	38.00	30.40	13.2
D-3812		2-(4,4-difluoro-5,7-diphenyl-4-bora-3a,4a-diaza-s-indacene-3-dodecanoyl)-1-hexadecanoyl-*sn*-glycero-3-phosphocholine (β-BODIPY® 530/550 C_{12}-HPC)	100 µg	38.00	30.40	13.2
D-3813		2-(4,4-difluoro-5,7-diphenyl-4-bora-3a,4a-diaza-s-indacene-3-dodecanoyl)-1-hexadecanoyl-*sn*-glycero-3-phosphoethanolamine (β-BODIPY® 530/550 C_{12}-HPE)	100 µg	38.00	30.40	13.2

Cat #		Product Name	Unit Size	Price Per Unit ($) 1–4 Units	5–24 Units	Section #
D-3833		4,4-difluoro-5,7-diphenyl-4-bora-3a,4a-diaza-s-indacene-3-pentanoic acid (BODIPY® 530/550 C_5)	1 mg	38.00	30.40	13.2
D-3815		2-(4,4-difluoro-5,7-diphenyl-4-bora-3a,4a-diaza-s-indacene-3-pentanoyl)-1-hexadecanoyl-*sn*-glycero-3-phosphocholine (β-BODIPY® 530/550 C_5-HPC)	100 µg	38.00	30.40	13.2
D-2186		4,4-difluoro-5,7-diphenyl-4-bora-3a,4a-diaza-s-indacene-3-propionic acid (BODIPY® 530/550)	5 mg	95.00	76.00	1.2
D-2372		4,4-difluoro-5,7-diphenyl-4-bora-3a,4a-diaza-s-indacene-3-propionic acid, hydrazide (BODIPY® 530/550 hydrazide)	5 mg	95.00	76.00	3.2
D-2187		4,4-difluoro-5,7-diphenyl-4-bora-3a,4a-diaza-s-indacene-3-propionic acid, succinimidyl ester (BODIPY® 530/550, SE)	5 mg	145.00	116.00	1.2
D-3799		N-(4,4-difluoro-5,7-diphenyl-4-bora-3a,4a-diaza-s-indacene-3-propionyl)-1,2-dihexadecanoyl-*sn*-glycero-3-phosphoethanolamine, triethylammonium salt (BODIPY® 530/550 DHPE)	100 µg	38.00	30.40	13.4
D-2391		4,4-difluoro-5,7-diphenyl-4-bora-3a,4a-diaza-s-indacene-3-propionyl ethylenediamine, hydrochloride (BODIPY® 530/550 EDA)	5 mg	95.00	76.00	3.3
D-2006		N-(4,4-difluoro-5,7-diphenyl-4-bora-3a,4a-diaza-s-indacene-3-propionyl)-N′-iodoacetylethylenediamine (BODIPY® 530/550 IA)	5 mg	95.00	76.00	2.2
D-6145	**New**	2′,7′-difluorofluorescein (Oregon Green™ 488)	10 mg	20.00	16.00	1.4
D-6566	**New**	6,8-difluoro-7-hydroxy-4-methylcoumarin *reference standard*	10 mg	45.00	36.00	10.1
D-7521	**New**	DL-*threo*-2-(4,4-difluoro-5-(4-methoxyphenyl)-4-bora-3a,4a-diaza-s-indacene-3-dodecanoyl)amino-3-morpholino-1-phenyl-1-propanol (BODIPY® TMR C_{12}-MPP)	25 µg	95.00	76.00	20.6
D-3836		4,4-difluoro-5-(4-methoxyphenyl)-4-bora-3a,4a-diaza-s-indacene-3-pentanoic acid (BODIPY® 542/563 C_5)	1 mg	38.00	30.40	13.2
D-3837		4,4-difluoro-5-(4-methoxyphenyl)-4-bora-3a,4a-diaza-s-indacene-3-undecanoic acid (BODIPY® 542/563 C_{11})	1 mg	38.00	30.40	13.2
D-3823		4,4-difluoro-5-methyl-4-bora-3a,4a-diaza-s-indacene-3-dodecanoic acid (C_1-BODIPY® 500/510 C_{12})	1 mg	38.00	30.40	13.2
D-3793		2-(4,4-difluoro-5-methyl-4-bora-3a,4a-diaza-s-indacene-3-dodecanoyl)-1-hexadecanoyl-*sn*-glycero-3-phosphocholine (β-BODIPY® 500/510 C_{12}-HPC)	100 µg	38.00	30.40	13.2
D-6567	**New**	6,8-difluoro-4-methylumbelliferyl phosphate, ammonium salt (DiFMUP)	5 mg	38.00	30.40	10.3
D-3825		4,4-difluoro-5-octyl-4-bora-3a,4a-diaza-s-indacene-3-pentanoic acid (C_8-BODIPY® 500/510 C_5)	1 mg	38.00	30.40	13.2
D-3795		2-(4,4-difluoro-5-octyl-4-bora-3a,4a-diaza-s-indacene-3-pentanoyl)-1-hexadecanoyl-*sn*-glycero-3-phosphocholine (β-C_8-BODIPY® 500/510 C_5-HPC)	100 µg	38.00	30.40	13.2
D-3922		4,4-difluoro-1,3,5,7,8-pentamethyl-4-bora-3a,4a-diaza-s-indacene (BODIPY® 493/503)	10 mg	75.00	60.00	14.6
D-3238		4,4-difluoro-1,3,5,7,8-pentamethyl-4-bora-3a,4a-diaza-s-indacene-2,6-disulfonic acid, disodium salt (BODIPY® 492/515 disulfonate)	10 mg	48.00	38.40	15.3
D-6180	**New**	4,4-difluoro-5-phenyl-4-bora-3a,4a-diaza-s-indacene-3-propionic acid, succinimidyl ester (BODIPY® R6G, SE)	5 mg	145.00	116.00	1.2
D-6186	**New**	6-((4,4-difluoro-5-phenyl-4-bora-3a,4a-diaza-s-indacene-3-propionyl)amino)-hexanoic acid, succinimidyl ester (BODIPY® R6G-X, SE)	5 mg	145.00	116.00	27.1
D-3860		4,4-difluoro-5-(4-phenyl-1,3-butadienyl)-4-bora-3a,4a-diaza-s-indacene-3-pentanoic acid (BODIPY® 581/591 C_5)	1 mg	38.00	30.40	13.2
D-3806		2-(4,4-difluoro-5-(4-phenyl-1,3-butadienyl)-4-bora-3a,4a-diaza-s-indacene-3-pentanoyl)-1-hexadecanoyl-*sn*-glycero-3-phosphocholine (β-BODIPY® 581/591 C_5-HPC)	100 µg	38.00	30.40	13.2
D-2227		4,4-difluoro-5-(4-phenyl-1,3-butadienyl)-4-bora-3a,4a-diaza-s-indacene-3-propionic acid (BODIPY® 581/591)	5 mg	95.00	76.00	1.2
D-2228		4,4-difluoro-5-(4-phenyl-1,3-butadienyl)-4-bora-3a,4a-diaza-s-indacene-3-propionic acid, succinimidyl ester (BODIPY® 581/591, SE)	5 mg	145.00	116.00	1.2
D-3861		4,4-difluoro-5-(4-phenyl-1,3-butadienyl)-4-bora-3a,4a-diaza-s-indacene-3-undecanoic acid (BODIPY® 581/591 C_{11})	1 mg	38.00	30.40	13.2
D-3858		4,4-difluoro-5-(2-pyrrolyl)-4-bora-3a,4a-diaza-s-indacene-3-pentanoic acid (BODIPY® 576/589 C_5)	1 mg	38.00	30.40	13.2
D-2224		4,4-difluoro-5-(2-pyrrolyl)-4-bora-3a,4a-diaza-s-indacene-3-propionic acid (BODIPY® 576/589)	5 mg	95.00	76.00	1.2
D-2225		4,4-difluoro-5-(2-pyrrolyl)-4-bora-3a,4a-diaza-s-indacene-3-propionic acid, succinimidyl ester (BODIPY® 576/589, SE)	5 mg	145.00	116.00	1.2
D-3859		4,4-difluoro-5-(2-pyrrolyl)-4-bora-3a,4a-diaza-s-indacene-3-undecanoic acid (BODIPY® 576/589 C_{11})	1 mg	38.00	30.40	13.2
D-3838		4,4-difluoro-5-styryl-4-bora-3a,4a-diaza-s-indacene-3-pentanoic acid (BODIPY® 564/570 C_5)	1 mg	38.00	30.40	13.2
D-2221		4,4-difluoro-5-styryl-4-bora-3a,4a-diaza-s-indacene-3-propionic acid (BODIPY® 564/570)	5 mg	95.00	76.00	1.2
D-2222		4,4-difluoro-5-styryl-4-bora-3a,4a-diaza-s-indacene-3-propionic acid, succinimidyl ester (BODIPY® 564/570, SE)	5 mg	145.00	116.00	1.2
D-3839		4,4-difluoro-5-styryl-4-bora-3a,4a-diaza-s-indacene-3-undecanoic acid (BODIPY® 564/570 C_{11})	1 mg	38.00	30.40	13.2

Cat #		Product Name	Unit Size	Price Per Unit ($) 1–4 Units	Price Per Unit ($) 5–24 Units	Section #
D-3921		4,4-difluoro-1,3,5,7-tetramethyl-4-bora-3a,4a-diaza-s-indacene (BODIPY® 505/515)	10 mg	75.00	60.00	14.6
D-3827		4,4-difluoro-1,3,5,7-tetramethyl-4-bora-3a,4a-diaza-s-indacene-8-nonanoic acid (BODIPY® 493/503 C_9)	1 mg	38.00	30.40	13.2
D-2190		4,4-difluoro-1,3,5,7-tetramethyl-4-bora-3a,4a-diaza-s-indacene-8-propionic acid (BODIPY® 493/503)	5 mg	95.00	76.00	1.2
D-2191		4,4-difluoro-1,3,5,7-tetramethyl-4-bora-3a,4a-diaza-s-indacene-8-propionic acid, succinimidyl ester (BODIPY® 493/503, SE)	5 mg	145.00	116.00	1.2
D-6961	***New***	4,4-difluoro-1,3,5,7-tetramethyl-4-bora-3a,4a-diaza-s-indacene-2-sulfonic acid, lithium salt (BODIPY® 495/505 sulfonate)	5 mg	75.00	60.00	15.3
D-6004	***New***	N-(4,4-difluoro-1,3,5,7-tetramethyl-4-bora-3a,4a-diaza-s-indacene-2-yl)-iodoacetamide (BODIPY® 507/545 IA)	5 mg	95.00	76.00	2.2
D-3835		4,4-difluoro-5-(2-thienyl)-4-bora-3a,4a-diaza-s-indacene-3-dodecanoic acid (BODIPY® 558/568 C_{12})	1 mg	38.00	30.40	13.2
D-2218		4,4-difluoro-5-(2-thienyl)-4-bora-3a,4a-diaza-s-indacene-3-propionic acid (BODIPY® 558/568)	5 mg	95.00	76.00	1.2
D-2219		4,4-difluoro-5-(2-thienyl)-4-bora-3a,4a-diaza-s-indacene-3-propionic acid, succinimidyl ester (BODIPY® 558/568, SE)	5 mg	145.00	116.00	1.2
D-6116	***New***	6-(((4-(4,4-difluoro-5-(2-thienyl)-4-bora-3a,4a-diaza-s-indacene-3-yl)phenoxy)-acetyl)amino)hexanoic acid, succinimidyl ester (BODIPY® TR-X, SE)	5 mg	145.00	116.00	1.2, 4.2
D-6251	***New***	5-(((4-(4,4-difluoro-5-(2-thienyl)-4-bora-3a,4a-diaza-s-indacene-3-yl)phenoxy)-acetyl)amino)pentylamine, hydrochloride (BODIPY® TR cadaverine)	5 mg	95.00	76.00	3.3
D-6011	***New***	N-(5-(((4-(4,4-difluoro-5-(2-thienyl)-4-bora-3a,4a-diaza-s-indacene-3-yl)phenoxy)-acetyl)amino)pentyl)iodoacetamide (BODIPY® TR cadaverine IA)	5 mg	125.00	100.00	27.1
D-7540	***New***	N-((4-(4,4-difluoro-5-(2-thienyl)-4-bora-3a,4a-diaza-s-indacene-3-yl)phenoxy)-acetyl)sphingosine (BODIPY® TR ceramide)	250 µg	98.00	78.40	12.4, 20.6
D-6567	***New***	DiFMUP (6,8-difluoro-4-methylumbelliferyl phosphate, ammonium salt)	5 mg	38.00	30.40	10.3
D-8449	***New***	digitonin *USP grade*	1 g	40.00	32.00	16.3, 22.8
D-8450	***New***	digitonin *water soluble*	1 g	48.00	38.40	16.3, 22.8
D-378		3,3′-diheptyloxacarbocyanine iodide (DiOC$_7$(3))	100 mg	68.00	54.40	12.2, 25.3
D-7713	***New***	1,2-dihexadecanoyl-N-(eosin-5-thiocarbamoyl)-*sn*-glycero-3-phospho-ethanolamine, triethylammonium salt (eosin DHPE)	1 mg	75.00	60.00	27.1
D-7704	***New***	1,2-dihexadecanoyl-*sn*-glycero-3-phosphocholine (DPPC)	1 g	95.00	76.00	5.2, 13.4
D-7705	***New***	1,2-dihexadecanoyl-*sn*-glycero-3-phosphoethanolamine (DPPE)	1 g	95.00	76.00	5.2, 13.4
D-69		4-dihexadecylamino-7-nitrobenz-2-oxa-1,3-diazole (NBD dihexadecylamine)	100 mg	48.00	38.40	14.6
D-3883		4-(4-(dihexadecylamino)styryl)-N-methylpyridinium iodide (DiA; 4-Di-16-ASP)	25 mg	115.00	92.00	14.2, 15.4
D-3885		4-(4-(dihexadecylamino)styryl)-N-methylquinolinium iodide (DiQ; Di-16-ASQ)	25 mg	48.00	38.40	14.2, 15.4
D-1125		3,3′-dihexadecyloxacarbocyanine perchlorate (DiOC$_{16}$(3))	25 mg	135.00	108.00	14.2, 15.4
D-384		1,1′-dihexadecyl-3,3,3′,3′-tetramethylindocarbocyanine perchlorate (DiIC$_{16}$(3))	100 mg	135.00	108.00	12.4, 14.2, 15.4
D-273		3,3′-dihexyloxacarbocyanine iodide (DiOC$_6$(3))	100 mg	85.00	68.00	12.2, 25.3
D-96058	***New***	dihydroartemisinin	50 mg	128.00	102.40	21.2
D-1168		dihydroethidium (also called Hydroethidine - a trademark of Prescott Labs)	25 mg	98.00	78.40	8.1, 17.1, 21.4
D-1194		dihydrofluorescein diacetate (H_2FDA)	1 g	58.00	46.40	10.5
D-632		dihydrorhodamine 123	10 mg	98.00	78.40	12.2, 17.1, 21.4
D-633		dihydrorhodamine 6G	25 mg	98.00	78.40	12.2, 21.4
D-638		dihydrotetramethylrosamine	10 mg	65.00	52.00	12.2, 21.4
D-6873	***New***	2,2′-dihydroxyazobenzene (2,2′-azodiphenol)	100 mg	18.00	14.40	22.7
D-7792	***New***	2,2′-dihydroxy-1,1′-naphthaldiazine (liumogen)	100 mg	65.00	52.00	26.1
D-7021	***New***	2,5-dihydroxyterephthalic acid, di(acetoxymethyl ester) (DHTP, AM)	5 mg	95.00	76.00	16.3
D-282		'DiI'; DiIC$_{18}$(3) (1,1′-dioctadecyl-3,3,3′,3′-tetramethylindocarbocyanine perchlorate)	100 mg	135.00	108.00	12.4, 14.2, 15.4
L-3484		DiI AcLDL (low-density lipoprotein from human plasma, acetylated, DiI complex) *1 mg/mL*	200 µL	195.00	156.00	17.3
D-383		DiIC$_{12}$(3) (1,1′-didodecyl-3,3,3′,3′-tetramethylindocarbocyanine perchlorate)	100 mg	135.00	108.00	14.2, 15.4
D-384		DiIC$_{16}$(3) (1,1′-dihexadecyl-3,3,3′,3′-tetramethylindocarbocyanine perchlorate)	100 mg	135.00	108.00	12.4, 14.2, 15.4
D-282		DiIC$_{18}$(3); 'DiI' (1,1′-dioctadecyl-3,3,3′,3′-tetramethylindocarbocyanine perchlorate)	100 mg	135.00	108.00	12.4, 14.2, 15.4
D-7776	***New***	DiIC$_{18}$(3)-DS (1,1′-dioctadecyl-3,3,3′,3′-tetramethylindocarbocyanine-5,5′-disulfonic acid)	5 mg	135.00	108.00	14.2, 15.4
L-3482		DiI LDL (low-density lipoprotein from human plasma, DiI complex) *1 mg/mL*	200 µL	175.00	140.00	17.3
D-7594	***New***	4′,6-(diimidazolin-2-yl)-2-phenylindole, dihydrochloride (DIPI)	10 mg	65.00	52.00	8.1
D-338		4,4′-diisothiocyanatodihydrostilbene-2,2′-disulfonic acid, disodium salt (H_2-DIDS)	100 mg	115.00	92.00	18.5
D-337		4,4′-diisothiocyanatostilbene-2,2′-disulfonic acid, disodium salt (DIDS)	100 mg	22.00	17.60	18.5
D-7758	***New***	4-(4-(dilinoleylamino)styryl)-N-methylpyridinium, 4-chlorobenzenesulfonate (*FAST* DiA™ solid; Di$\Delta^{9,12}$-C$_{18}$ASP, CBS)	5 mg	135.00	108.00	14.2, 15.4
D-3897		4-(4-(dilinoleylamino)styryl)-N-methylpyridinium iodide (*FAST* DiA™ oil; Di$\Delta^{9,12}$-C$_{18}$ASP, I)	5 mg	135.00	108.00	14.2, 15.4
D-3898		3,3′-dilinoleyloxacarbocyanine perchlorate (*FAST* DiO™ solid; DiO$\Delta^{9,12}$-C$_{18}$(3), ClO_4)	5 mg	135.00	108.00	14.2, 15.4

Cat #		Product Name	Unit Size	Price Per Unit ($)		Section #
				1–4 Units	5–24 Units	
D-7756	***New***	1,1′-dilinoleyl-3,3,3′,3′-tetramethylindocarbocyanine, 4-chlorobenzenesulfonate (*FAST* DiI™ solid; $DiI\Delta^{9,12}$-C_{18}(3), CBS)	5 mg	135.00	108.00	14.2, 15.4
D-3899		1,1′-dilinoleyl-3,3,3′,3′-tetramethylindocarbocyanine perchlorate (*FAST* DiI™ oil; $DiI\Delta^{9,12}$-C_{18}(3), ClO_4)	5 mg	135.00	108.00	14.2, 15.4
D-6948	***New***	9,10-dimethoxyanthracene-2-sulfonic acid, sodium salt (DAS)	100 mg	25.00	20.00	1.8, 15.3, 21.3
D-1037		4,5-dimethoxy-2-nitrobenzyl adenosine 3′,5′-cyclicmonophosphate (DMNB-caged cAMP)	5 mg	78.00	62.40	11.3, 19.2, 20.4
D-7117	***New***	4,5-dimethoxy-2-nitrobenzyl bromide (BMNDMB)	100 mg	98.00	78.40	27.1
D-7118	***New***	4,5-dimethoxy-2-nitrobenzyl chloroformate (NVOC chloride)	100 mg	48.00	38.40	19.2
D-7057	***New***	N-(4,5-dimethoxy-2-nitrobenzyl)-L-epinephrine (N-(DMNB-caged) epinephrine)	5 mg	48.00	38.40	18.3, 19.2
D-7098	***New***	S-(4,5-dimethoxy-2-nitrobenzyl)-N-FMOC-L-cysteine (S-(DMNB-caged) N-FMOC-L-cysteine)	100 mg	175.00	140.00	9.3, 19.2
D-7086	***New***	N-(((4,5-dimethoxy-2-nitrobenzyl)oxy)carbonyl)-L-alanine (N-(NVOC-caged) L-alanine)	25 mg	35.00	28.00	9.3, 19.2
D-7064	***New***	N-(((4,5-dimethoxy-2-nitrobenzyl)oxy)carbonyl) dopamine (N-(NVOC-caged) dopamine)	5 mg	48.00	38.40	18.3, 19.2
D-7087	***New***	N-(((4,5-dimethoxy-2-nitrobenzyl)oxy)carbonyl) glycine (N-(NVOC-caged) glycine)	25 mg	35.00	28.00	9.3, 19.2
D-7917	***New***	N^1-(((4,5-dimethoxy-2-nitrobenzyl)oxy)carbonyl)-N^5-(1-iminoethyl)-L-ornithine, hydrate (NVOC-caged L-NIO)	5 mg	95.00	76.00	21.5
D-7088	***New***	N-(((4,5-dimethoxy-2-nitrobenzyl)oxy)carbonyl)-L-isoleucine (N-(NVOC-caged) L-isoleucine)	25 mg	35.00	28.00	9.3, 19.2
D-7089	***New***	N-(((4,5-dimethoxy-2-nitrobenzyl)oxy)carbonyl)-L-leucine (N-(NVOC-caged) L-leucine)	25 mg	35.00	28.00	9.3, 19.2
D-7092	***New***	N-(((4,5-dimethoxy-2-nitrobenzyl)oxy)carbonyl)-L-methionine (N-(NVOC-caged) L-methionine)	25 mg	35.00	28.00	9.3, 19.2
D-7059	***New***	N-(((4,5-dimethoxy-2-nitrobenzyl)oxy)carbonyl) MK-801 (N-(NVOC-caged) MK-801)	1 mg	48.00	38.40	18.3, 19.2
D-7111	***New***	N-(((4,5-dimethoxy-2-nitrobenzyl)oxy)carbonyl)-3-morpholinosydnoneimine (NVOC-caged SIN-1)	5 mg	95.00	76.00	19.2, 21.5
D-7058	***New***	N-(((4,5-dimethoxy-2-nitrobenzyl)oxy)carbonyl)-L-norepinephrine (N-(NVOC-caged) norepinephrine)	5 mg	48.00	38.40	18.3, 19.2
D-7093	***New***	N-(((4,5-dimethoxy-2-nitrobenzyl)oxy)carbonyl)-L-phenylalanine (N-(NVOC-caged) L-phenylalanine)	25 mg	35.00	28.00	9.3, 19.2
D-7063	***New***	N-(((4,5-dimethoxy-2-nitrobenzyl)oxy)carbonyl) serotonin (N-(NVOC-caged) serotonin)	5 mg	48.00	38.40	18.3, 19.2
D-7094	***New***	N-(((4,5-dimethoxy-2-nitrobenzyl)oxy)carbonyl)-L-tryptophan (N-(NVOC-caged) L-tryptophan)	25 mg	35.00	28.00	9.3, 19.2
D-7095	***New***	N-(((4,5-dimethoxy-2-nitrobenzyl)oxy)carbonyl)-L-valine (N-(NVOC-caged) L-valine)	25 mg	35.00	28.00	9.3, 19.2
D-7060	***New***	8-((4,5-dimethoxy-2-nitrobenzyl)oxy)pyrene-1,3,6-trisulfonic acid, trisodium salt (DMNB-caged HPTS)	5 mg	125.00	100.00	15.3, 19.2
D-7084	***New***	N-(4,5-dimethoxy-2-nitrobenzyl)phenylephrine (N-(DMNB-caged) phenylephrine)	5 mg	48.00	38.40	18.3, 19.2
D-6815	***New***	1-(4,5-dimethoxy-2-nitrophenyl)-1,2-diaminoethane-N,N,N′,N′-tetraacetic acid, tetra(acetoxymethyl ester) (DMNP-EDTA, AM) *cell permeant* *≥95% by HPLC* *special packaging*	20x50 µg	195.00	156.00	19.2, 20.2, 22.8
D-6814	***New***	1-(4,5-dimethoxy-2-nitrophenyl)-1,2-diaminoethane-N,N,N′,N′-tetraacetic acid, tetrapotassium salt (DMNP-EDTA) *cell impermeant*	5 mg	100.00	80.00	19.2, 20.2, 22.8
D-2516		1-(4,5-Dimethoxy-2-nitrophenyl)diazoethane Generation Kit	1 kit	85.00	68.00	19.2
D-7119	***New***	*trans*-2-(4,5-dimethoxy-2-nitrophenyl)nitroethene	5 mg	38.00	30.40	27.1
D-3112		5-(N,N-dimethyl)amiloride, hydrochloride	5 mg	35.00	28.00	18.7
D-1537		4-dimethylaminoazobenzene-4′-sulfonyl chloride (dabsyl chloride)	100 mg	24.00	19.20	1.8, 9.3
D-1332		N-(4-(6-dimethylamino-2-benzofuranyl)phenylisothiocyanate	25 mg	48.00	38.40	1.8
D-1333		N-(4-(6-dimethylamino-2-benzofuranyl)phenyl)maleimide	25 mg	38.00	30.40	2.4
D-126		7-dimethylaminocoumarin-4-acetic acid (DMACA)	100 mg	88.00	70.40	1.7
D-374		7-dimethylaminocoumarin-4-acetic acid, succinimidyl ester (DMACA, SE)	25 mg	115.00	92.00	1.7
D-99012	***New***	2-(dimethylaminoethyl)-3-ethyl-5-methylindole, hydrochloride (KYR-44)	50 µg	105.00	84.00	16.3
D-112		5-dimethylaminonaphthalene-1-(N-(2-aminoethyl))sulfonamide (dansyl ethylenediamine)	100 mg	78.00	62.40	3.3
D-113		5-dimethylaminonaphthalene-1-(N-(5-aminopentyl))sulfonamide (dansyl cadaverine)	100 mg	78.00	62.40	3.3, 9.4, 12.3
D-6104	***New***	6-((5-dimethylaminonaphthalene-1-sulfonyl)amino)hexanoic acid, succinimidyl ester (dansyl-X, SE)	25 mg	48.00	38.40	1.7, 4.2
D-1302		4-((5-dimethylaminonaphthalene-1-sulfonyl)amino)phenylisothiocyanate	100 mg	38.00	30.40	1.8, 9.3
D-94		11-((5-dimethylaminonaphthalene-1-sulfonyl)amino)undecanoic acid (DAUDA)	100 mg	68.00	54.40	13.3, 20.5
D-3765		1-(O-(11-(5-dimethylaminonaphthalene-1-sulfonyl)amino)undecyl)-2-decanoyl-*sn*-glycero-3-phosphocholine	1 mg	68.00	54.40	20.5
D-3766		1-(O-(11-(5-dimethylaminonaphthalene-1-sulfonyl)amino)undecyl)-*sn*-glycero-3-phosphocholine (dansyl-C_{11}-lyso PAF)	500 µg	68.00	54.40	20.6

Cat #		Product Name	Unit Size	Price Per Unit ($) 1–4 Units	Price Per Unit ($) 5–24 Units	Section #
D-151		5-dimethylaminonaphthalene-1-sulfonyl aziridine (dansyl aziridine)	100 mg	68.00	54.40	2.3
D-21		5-dimethylaminonaphthalene-1-sulfonyl chloride (dansyl chloride)	1 g	40.00	32.00	1.8, 9.3
D-22		2-dimethylaminonaphthalene-5-sulfonyl chloride	100 mg	125.00	100.00	1.7
D-23		2-dimethylaminonaphthalene-6-sulfonyl chloride	100 mg	125.00	100.00	1.7
D-57		N-(5-dimethylaminonaphthalene-1-sulfonyl)-1,2-dihexadecanoyl-*sn*-glycero-3-phosphoethanolamine, triethylammonium salt (dansyl DHPE)	25 mg	98.00	78.40	13.4
D-11		5-dimethylaminonaphthalene-1-sulfonyl fluoride (dansyl fluoride)	100 mg	18.00	14.40	9.4, 10.6
D-402		N-(5-dimethylaminonaphthalene-1-sulfonyl)glucosamine (dansyl glucosamine)	25 mg	48.00	38.40	15.3
D-3797		N-(5-dimethylaminonaphthalene-1-sulfonyl)-*sn*-glycero-3-phosphoethanolamine, triethylammonium salt *from egg PE*	25 mg	48.00	38.40	13.4
D-100		5-dimethylaminonaphthalene-1-sulfonyl hydrazine (dansyl hydrazine)	100 mg	27.00	21.60	3.2, 9.3
D-372		N-ϵ-(5-dimethylaminonaphthalene-1-sulfonyl)-L-lysine (dansyl lysine)	100 mg	48.00	38.40	14.5
D-427		4-dimethylaminonaphthyl-1-isothiocyanate	100 mg	28.00	22.40	1.8
D-2245		4-((4-(dimethylamino)phenyl)azo)benzoic acid, succinimidyl ester (dabcyl, SE)	100 mg	68.00	54.40	1.8, 9.3
D-6216	***New***	ϵ-(4-((4-(dimethylamino)phenyl)azo)benzoyl)-α-9-fluorenylmethoxycarbonyl-L-lysine (ϵ-dabcyl-α-FMOC-L-lysine)	100 mg	175.00	140.00	9.3
D-1521		4-dimethylaminophenylazophenyl-4′-maleimide (DABMI)	100 mg	55.00	44.00	2.4
D-7475	***New***	2-(1-(3-dimethylaminopropyl)-indol-3-yl)-3-(indol-3-yl)maleimide (bisindolylmaleimide; GF 109203X)	1 mg	65.00	52.00	20.3
D-7761	***New***	4-(4-(dimethylamino)styryl)-1-docosylpyridinium bromide (C_{22}-Di-1-ASP)	25 mg	48.00	38.40	14.2
D-426		2-(4-(dimethylamino)styryl)-N-ethylpyridinium iodide (DASPEI)	1 g	75.00	60.00	12.2
D-288		4-(4-(dimethylamino)styryl)-N-methylpyridinium iodide (4-Di-1-ASP)	1 g	75.00	60.00	12.2
D-308		2-(4-(dimethylamino)styryl)-N-methylpyridinium iodide (2-Di-1-ASP)	1 g	75.00	60.00	12.2
D-7517	***New***	4-dimethylaminotetramethylrosamine chloride	5 mg	45.00	36.00	12.2
D-1207		5,5′-dimethyl BAPTA, AM *cell permeant* *≥95% by HPLC*	25 mg	95.00	76.00	22.8
D-1206		5,5′-dimethyl BAPTA, tetrapotassium salt *cell impermeant*	100 mg	48.00	38.40	22.8
D-6885	***New***	2′,4′-dimethylbenzamil, hydrochloride	5 mg	70.00	56.00	18.7
D-99008	***New***	N,2-dimethyl-6,6-dipropyl-4-nitromethinyl-1,4,5,6-tetrahydropyridine (M2)	100 µg	128.00	102.40	10.6, 18.4
D-99022	***New***	1,1-dimethyl-3-(di-2-thienylmethylene)-piperidinium iodide (LK-66)	2 mg	128.00	102.40	18.2
D-99025	***New***	1,1-dimethyl-3-(di-2-thienylmethylene)-piperidinium methanesulfonate (LK-66 M)	2 mg	128.00	102.40	18.2
D-531		4-(N,N-dimethyl-N-hexadecyl)ammonium-2,2,6,6-tetramethylpiperidine-1-oxyl iodide (CAT 16)	25 mg	48.00	38.40	13.3
D-530		4-(N,N-dimethyl-N-(2-hydroxyethyl))ammonium-2,2,6,6-tetramethylpiperidine-1-oxyl chloride (TEMPO choline)	25 mg	98.00	78.40	15.3
D-2004		N,N′-dimethyl-N-(iodoacetyl)-N′-(7-nitrobenz-2-oxa-1,3-diazol-4-yl)-ethylenediamine (IANBD amide)	25 mg	95.00	76.00	2.3
D-99010	***New***	1,2-dimethyl-3-nitromethylene-6-spiroadamantyl-5,6-dihydropyridine (M5)	100 µg	270.00	216.00	10.6, 18.4
D-99009	***New***	1,2-dimethyl-3-nitromethylene-5-spirocyclohexyl-5,6-dihydropyridine (M4)	100 µg	215.00	172.00	10.6, 18.4
D-7700	***New***	4-(N,N-dimethyl-N-tetradecylammonium)methyl-(7-hydroxycoumarin) chloride (U-6)	10 mg	48.00	38.40	14.3, 23.4
M-6494	***New***	3-(4,5-dimethylthiazol-2-yl)-2,5-diphenyltetrazolium bromide (MTT)	1 g	40.00	32.00	16.3, 21.4
D-3005		5,5′-dinitro BAPTA, AM *cell permeant* *≥95% by HPLC*	10 mg	48.00	38.40	22.8
D-3001		5,5′-dinitro BAPTA, free acid *cell impermeant*	25 mg	73.00	58.40	22.8
D-2248		6-(2,4-dinitrophenyl)aminohexanoic acid, succinimidyl ester (DNP-X, SE)	25 mg	75.00	60.00	4.2
D-3798		N-((6-(2,4-dinitrophenyl)amino)hexanoyl)-1,2-dihexadecanoyl-*sn*-glycero-3-phosphoethanolamine, sodium salt (DNP-X DHPE)	10 mg	48.00	38.40	13.4
D-1552		N-(3-((2,4-dinitrophenyl)amino)propyl)-N-(3-aminopropyl)methylamine, dihydrochloride (DAMP)	100 mg	98.00	78.40	12.3
D-673		4,4′-dinitrostilbene-2,2′-disulfonic acid, disodium salt (DNDS)	1 g	55.00	44.00	18.5
D-275		'DiO'; $DiOC_{18}(3)$ (3,3′-dioctadecyloxacarbocyanine perchlorate)	100 mg	135.00	108.00	14.2, 15.4
D-272		$DiOC_5(3)$ (3,3′-dipentyloxacarbocyanine iodide)	100 mg	85.00	68.00	9.2, 12.4, 25.3
D-275		$DiOC_{18}(3)$; 'DiO' (3,3′-dioctadecyloxacarbocyanine perchlorate)	100 mg	135.00	108.00	14.2, 15.4
D-7770	***New***	3,3′-dioctadecyl-5,5′-diphenyloxacarbocyanine chloride (5,5′-Ph_2-$DiOC_{18}(3)$)	5 mg	135.00	108.00	14.2, 15.4
D-7779	***New***	1,1′-dioctadecyl-5,5′-diphenyl-3,3,3′,3′-tetramethylindocarbocyanine chloride (5,5′-Ph_2-$DiIC_{18}(3)$)	5 mg	135.00	108.00	14.2, 15.4
D-7780	***New***	1,1′-dioctadecyl-6,6′-diphenyl-3,3,3′,3′-tetramethylindocarbocyanine chloride (6,6′-Ph_2-$DiIC_{18}(3)$)	5 mg	135.00	108.00	14.2, 15.4
D-7778	***New***	3,3′-dioctadecyl-5,5′-di(4-sulfophenyl)-oxacarbocyanine, sodium salt (SP-$DiOC_{18}(3)$)	5 mg	135.00	108.00	14.2, 15.4
D-7777	***New***	1,1′-dioctadecyl-6,6′-di(4-sulfophenyl)-3,3,3′,3′-tetramethylindocarbocyanine (SP-$DiIC_{18}(3)$)	5 mg	135.00	108.00	14.2, 15.4
D-275		3,3′-dioctadecyloxacarbocyanine perchlorate ('DiO'; $DiOC_{18}(3)$)	100 mg	135.00	108.00	14.2, 15.4
D-7776	***New***	1,1′-dioctadecyl-3,3,3′,3′-tetramethylindocarbocyanine-5,5′-disulfonic acid ($DiIC_{18}(3)$-DS)	5 mg	135.00	108.00	14.2, 15.4
D-282		1,1′-dioctadecyl-3,3,3′,3′-tetramethylindocarbocyanine perchlorate ('DiI'; $DiIC_{18}(3)$)	100 mg	135.00	108.00	12.4, 14.2, 15.4
D-3911		1,1′-dioctadecyl-3,3,3′,3′-tetramethylindocarbocyanine perchlorate *large crystals*	25 mg	100.00	80.00	14.2, 15.4
D-7757	***New***	1,1′-dioctadecyl-3,3,3′,3′-tetramethylindodicarbocyanine, 4-chlorobenzene-sulfonate salt ('DiD' solid; $DiIC_{18}(5)$ solid)	10 mg	68.00	54.40	14.2, 15.4

Cat #		Product Name	Unit Size	Price Per Unit ($) 1–4 Units	5–24 Units	Section #
D-307		1,1′-dioctadecyl-3,3,3′,3′-tetramethylindodicarbocyanine perchlorate ('DiD' oil; DiIC$_{18}$(5) oil)	25 mg	90.00	72.00	14.2, 15.4
D-1121		3,3′-dioctadecylthiacarbocyanine perchlorate ('DiS'; DiSC$_{18}$(3))	25 mg	48.00	38.40	14.2
D-1576		1,2-dioctanoyl-*sn*-glycerol (DiC$_8$;DOG) *special packaging*	10x1 mg	53.00	42.40	20.3
D-7081	**New**	1,2-dioctanoyl-3-(2-nitrobenzyl)-*sn*-glycerol (NB-caged DOG)	5 mg	95.00	76.00	19.2, 20.3
D-6563	**New**	1,2-dioleoyl-3-(1-pyrenedecanoyl)-*rac*-glycerol	1 mg	65.00	52.00	27.1
D-6562	**New**	1,2-dioleoyl-3-(1-pyrenedodecanoyl)-*rac*-glycerol	1 mg	65.00	52.00	27.1
D-3886		1,1′-dioleyl-3,3,3′,3′-tetramethylindocarbocyanine methanesulfonate (Δ^9-DiI)	25 mg	135.00	108.00	14.2, 15.4
D-6137	**New**	3,6-dioxaheptanoic acid, succinimidyl ester	25 mg	48.00	38.40	1.8
D-272		3,3′-dipentyloxacarbocyanine iodide (DiOC$_5$(3))	100 mg	85.00	68.00	9.2, 12.4, 25.3
D-7801	**New**	9,10-diphenylanthracene	1 g	18.00	14.40	14.6, 21.3, 24.3
D-7909	**New**	diphenyleneiodonium chloride	25 mg	35.00	28.00	21.5
D-7911	**New**	diphenyleneiodonium p-toluenesulfonate	25 mg	35.00	28.00	21.5
D-202		1,6-diphenyl-1,3,5-hexatriene (DPH)	100 mg	23.00	18.40	14.4
D-476		2-(3-(diphenylhexatrienyl)propanoyl)-1-hexadecanoyl-*sn*-glycero-3-phosphocholine (β-DPH HPC)	1 mg	48.00	38.40	13.2
D-99024	**New**	1-(3,3-diphenyl-3-hydroxypropyl)-1-methylpiperidinium iodide (LK-7)	2 mg	95.00	76.00	18.2
D-7919	**New**	1,3-diphenylisobenzofuran	1 g	28.00	22.40	21.3
D-99028	**New**	3-diphenylmethoxyquinuclidine, hydrochloride (LK-252)	2 mg	95.00	76.00	18.2
D-99027	**New**	3-diphenylmethylenequinuclidine, hydrochloride (LK-190)	2 mg	95.00	76.00	18.2
D-7076	**New**	(1,2-diphenyl-2-oxo)ethyl phosphate, monosodium salt, monohydrate (desyl-caged phosphate)	5 mg	95.00	76.00	19.2, 23.2
D-99023	**New**	1,1-diphenyl-3-(1-piperidinyl)-1-propene, hydrochloride, dihydrate (LK-6)	2 mg	95.00	76.00	18.2
D-7894	**New**	diphenyl-1-pyrenylphosphine (DPPP)	5 mg	125.00	100.00	16.3, 21.3
D-7594	**New**	DIPI (4′,6-(diimidazolin-2-yl)-2-phenylindole, dihydrochloride)	10 mg	65.00	52.00	8.1
D-1248		dipicolinic acid (DPA) *cell impermeant*	1 g	17.00	13.60	15.3, 22.7
D-1249		dipicolinic acid, AM (DPA, AM) *cell permeant*	25 mg	48.00	38.40	15.3, 22.7
D-306		3,3′-dipropylthiadicarbocyanine iodide (DiSC$_3$(5))	100 mg	145.00	116.00	25.3
D-3885		DiQ; Di-16-ASQ (4-(4-(dihexadecylamino)styryl)-N-methylquinolinium iodide)	25 mg	48.00	38.40	14.2, 15.4
B-413		DiSBAC$_2$(3) (bis-(1,3-diethylthiobarbituric acid)trimethine oxonol)	100 mg	115.00	92.00	25.3
D-306		DiSC$_3$(5) (3,3′-dipropylthiadicarbocyanine iodide)	100 mg	145.00	116.00	25.3
D-1121		'DiS'; DiSC$_{18}$(3) (3,3′-dioctadecylthiacarbocyanine perchlorate)	25 mg	48.00	38.40	14.2
D-8451	**New**	5,5′-dithiobis-(2-nitrobenzoic acid) (DTNB; Ellman's reagent)	10 g	45.00	36.00	2.4, 5.2, 24.3
D-6316	**New**	5,5′-dithiobis-(2-nitrobenzoic acid, succinimidyl ester) (DTNB, SE)	25 mg	65.00	52.00	5.2, 9.4
D-8452	**New**	1,4-dithioerythritol (DTE)	1 g	21.00	16.80	2.1, 5.2
D-1532		dithiothreitol (DTT)	1 g	18.00	14.40	2.1, 5.2
D-1389		N,N′-di(3-trimethylammoniumpropyl)thiadicarbocyanine tribromide	10 mg	80.00	64.00	15.3
D-7443	**New**	(-)-DM-BODIPY® dihydropyridine *high affinity enantiomer*	25 µg	125.00	100.00	18.4
D-7444	**New**	(+)-DM-BODIPY® dihydropyridine *low affinity enantiomer*	25 µg	125.00	100.00	18.4
D-7441	**New**	DM-BODIPY® PAA (BODIPY® FL phenylalkylamine)	25 µg	125.00	100.00	18.4
D-6830	**New**	DM-NERF	5 mg	95.00	76.00	1.5, 23.3
		DM-Nitrophen™ — a structurally identical product is listed under (4,5-dimethoxy-2-nitrophenyl)-1,2-diaminoethane-N,N,N′,N′-tetraacetic acid (DMNP-EDTA, D-6814; see Sections 19.2, 20.2 and 22.8).				
D-7562	**New**	DNA gel-loading solution with 15% Ficoll® *6X concentrate*	1 mL	24.00	19.20	8.3
D-7561	**New**	DNA gel-loading solution with 40% sucrose *6X concentrate*	1 mL	24.00	19.20	8.3
		DNase I conjugates — products are listed under deoxyribonuclease I; see also Section 11.1.				
D-109		5-dodecanoylaminofluorescein	100 mg	65.00	52.00	14.3
D-2512		5-dodecanoylaminofluorescein-bis-4,5-dimethoxy-2-nitrobenzyl ether (DMNB-caged C$_{12}$-fluorescein)	5 mg	48.00	38.40	14.3, 19.2
D-6534	**New**	5-dodecanoylaminofluorescein diethyl ether	5 mg	95.00	76.00	10.5
D-2893		5-dodecanoylaminofluorescein di-β-D-galactopyranoside (C$_{12}$FDG)	5 mg	195.00	156.00	10.2
D-250		6-dodecanoyl-2-dimethylaminonaphthalene (laurdan)	100 mg	135.00	108.00	14.5
A-455		dodecyl acridine orange (acridine orange 10-dodecyl bromide)	100 mg	45.00	36.00	14.3
D-6530	**New**	2-dodecyl-6-ethoxyphenoxazin-3-one (C$_{12}$-ethoxyresorufin)	5 mg	95.00	76.00	10.5
D-7786	**New**	2-dodecylresorufin	5 mg	65.00	52.00	14.3
D-526		16-doxylstearic acid	25 mg	108.00	86.40	13.3
D-522		5-doxylstearic acid	25 mg	60.00	48.00	13.3
D-524		12-doxylstearic acid	25 mg	60.00	48.00	13.3
D-202		DPH (1,6-diphenyl-1,3,5-hexatriene)	100 mg	23.00	18.40	14.4
P-459		DPH propionic acid (3-(4-(6-phenyl)-1,3,5-hexatrienyl)phenylpropionic acid)	25 mg	148.00	118.40	13.2
D-7704	**New**	DPPC (1,2-dihexadecanoyl-*sn*-glycero-3-phosphocholine)	1 g	95.00	76.00	5.2, 13.4
D-7705	**New**	DPPE (1,2-dihexadecanoyl-*sn*-glycero-3-phosphoethanolamine)	1 g	95.00	76.00	5.2, 13.4
X-1525		DPX (p-xylene-bis-pyridinium bromide)	1 g	75.00	60.00	15.3
D-16		5-DTAF (5-(4,6-dichlorotriazinyl)aminofluorescein) *single isomer*	100 mg	75.00	60.00	1.3, 3.1
D-8452	**New**	DTE (1,4-dithioerythritol)	1 g	21.00	16.80	2.1, 5.2
D-8451	**New**	DTNB; Ellman's reagent (5,5′-dithiobis-(2-nitrobenzoic acid))	10 g	45.00	36.00	2.4, 5.2, 24.3
D-1532		DTT (dithiothreitol)	1 g	18.00	14.40	2.1, 5.2

Cat #	Product Name	Unit Size	Price Per Unit ($) 1–4 Units	5–24 Units	Section #
D-96020 ***New***	dubinidine	20 mg	108.00	86.40	18.8
E-6545 ***New***	E-64 (*trans*-epoxysuccinyl-L-leucylamido-(4-guanidino)butane)	10 mg	65.00	52.00	10.6
E-96005 ***New***	ecdysterone (β-ecdysone)	10 mg	85.00	68.00	18.8
E-2247	EDAC (1-ethyl-3-(3-dimethylaminopropyl)carbodiimide, hydrochloride)	100 mg	15.00	12.00	3.3, 5.2
A-91	EDANS (5-((2-aminoethyl)amino)naphthalene-1-sulfonic acid, sodium salt)	1 g	75.00	60.00	3.3
E-1218	EDTA, tetra(acetoxymethyl ester) (EDTA, AM)	25 mg	85.00	68.00	22.8
E-6288 ***New***	EEDQ (2-ethoxy-1-ethoxycarbonyl-1,2-dihydroquinoline)	5 g	16.00	12.80	3.3, 5.2
E-1219	EGTA, tetra(acetoxymethyl ester) (EGTA, AM)	10 mg	125.00	100.00	22.8
E-6580 ***New***	ELF®-97 acetate (ELF®-97 esterase substrate)	5 mg	95.00	76.00	10.5
E-6578 ***New***	ELF®-97 alcohol *1 mM solution in DMSO*	1 mL	24.00	19.20	6.3, 10.1
E-6602 ***New***	ELF®-97 Cytological Labeling Kit #1 *50 assays*	1 kit	148.00	118.40	6.3, 10.3
E-6603 ***New***	ELF®-97 Cytological Labeling Kit #2 *with streptavidin, alkaline phosphatase conjugate* *50 assays*	1 kit	198.00	158.40	6.3, 10.3
E-6601 ***New***	ELF®-97 Endogenous Phosphatase Detection Kit	1 kit	135.00	108.00	6.3, 10.3
E-6581 ***New***	ELF®-97 ethyl ether (ELF®-97 microsomal dealkylase substrate)	5 mg	95.00	76.00	10.5
E-6587 ***New***	ELF®-97 β-D-glucuronide (ELF®-97 β-D-glucuronidase substrate)	5 mg	175.00	140.00	10.2
E-6585 ***New***	ELF®-97 guanidinobenzoate (ELF®-97 guanidinobenzoatase substrate) *>70%*	5 mg	125.00	100.00	10.5
E-6600 ***New***	ELF®-97 Immunohistochemistry Kit	1 kit	198.00	158.40	6.3, 10.3
E-6604 ***New***	ELF®-97 mRNA *In Situ* Hybridization Kit #1 *50 assays*	1 kit	175.00	140.00	6.3, 8.4, 10.3
E-6605 ***New***	ELF®-97 mRNA *In Situ* Hybridization Kit #2 *with streptavidin, alkaline phosphatase conjugate* *50 assays*	1 kit	225.00	180.00	6.3, 8.4, 10.3
E-6583 ***New***	ELF®-97 palmitate (ELF®-97 lipase substrate)	5 mg	95.00	76.00	10.5
E-6589 ***New***	ELF®-97 phosphate (ELF®-97 phosphatase substrate) *5 mM solution in water* *contains 2 mM azide*	1 mL	150.00	120.00	6.3, 10.3
E-6588 ***New***	ELF®-97 phosphate (ELF®-97 phosphatase substrate) *5 mM solution in water* *sterile filtered*	1 mL	160.00	128.00	6.3, 10.3
E-6579 ***New***	ELF®-97 sulfate (ELF®-97 sulfatase substrate)	5 mg	125.00	100.00	10.5
E-6606 ***New***	ELF® spin filters *20 filters*	1 box	48.00	38.40	6.3, 8.4, 10.3
E-6644 ***New***	EnzChek™ Dextranase Assay Kit *100-1000 assays*	1 kit	135.00	108.00	10.2
E-6636 ***New***	EnzChek™ DNase Gel Assay Kit *100 assays*	1 kit	135.00	108.00	8.3, 10.3
E-6637 ***New***	EnzChek™ DNase Solution Assay Kit *25 assays*	1 kit	135.00	108.00	8.3, 10.3
E-6643 ***New***	EnzChek™ 5′-Nucleotidase Assay Kit *100 assays*	1 kit	95.00	76.00	10.3
E-6646 ***New***	EnzChek™ Phosphate Assay Kit *100 assays*	1 kit	135.00	108.00	10.3, 11.3, 24.3
E-6658 ***New***	EnzChek™ Polarization Assay Kit for Proteases *green fluorescence* *100-1000 assays*	1 kit	135.00	108.00	10.4
E-6659 ***New***	EnzChek™ Polarization Assay Kit for Proteases *red fluorescence* *100-1000 assays*	1 kit	135.00	108.00	10.4
E-6638 ***New***	EnzChek™ Protease Assay Kit *green fluorescence* *100-1000 assays*	1 kit	135.00	108.00	10.4, 15.7
E-6639 ***New***	EnzChek™ Protease Assay Kit *red fluorescence* *100-1000 assays*	1 kit	135.00	108.00	10.4, 15.7
E-6645 ***New***	EnzChek™ Pyrophosphate Assay Kit *100 assays*	1 kit	155.00	124.00	10.3, 11.3, 24.3
E-6641 ***New***	EnzChek™ RNase Gel Assay Kit *100 assays*	1 kit	135.00	108.00	8.3, 10.3
E-6436 ***New***	eosin goat anti-mouse IgG (H+L) conjugate *2 mg/mL*	0.5 mL	75.00	60.00	7.2
E-6437 ***New***	eosin goat anti-rabbit IgG (H+L) conjugate *2 mg/mL*	0.5 mL	75.00	60.00	7.2
E-99	eosin-5-iodoacetamide	100 mg	155.00	124.00	2.2
E-18	eosin-5-isothiocyanate	100 mg	98.00	78.40	1.5
E-118	eosin-5-maleimide	25 mg	155.00	124.00	2.2
E-7463 ***New***	eosin phalloidin	300 U	198.00	158.40	11.1
E-3476	epidermal growth factor (EGF) *from mouse submaxillary glands*	100 µg	73.00	58.40	17.3
E-3477	epidermal growth factor, biotin-XX conjugate (biotin EGF)	20 µg	160.00	128.00	4.3, 17.3
E-3480	epidermal growth factor, biotinylated, complexed to Texas Red® streptavidin	100 µg	160.00	128.00	4.3, 17.3
E-3478	epidermal growth factor, fluorescein conjugate (fluorescein EGF)	20 µg	160.00	128.00	17.3
E-7498 ***New***	epidermal growth factor, Oregon Green™ 514 conjugate (Oregon Green™ 514 EGF)	20 µg	160.00	128.00	17.3
E-3481	epidermal growth factor, tetramethylrhodamine conjugate (rhodamine EGF)	20 µg	160.00	128.00	17.3
E-6538 ***New***	1,2-epoxy-3-(4-nitrophenoxy)propane (EPNP) *high purity*	100 mg	24.00	19.20	10.6
E-6051 ***New***	1-(2,3-epoxypropyl)-4-(5-(4-methoxyphenyl)oxazol-2-yl)pyridinium trifluoromethanesulfonate (PyMPO epoxide)	5 mg	98.00	78.40	2.2
E-6545 ***New***	*trans*-epoxysuccinyl-L-leucylamido-(4-guanidino)butane (E-64)	10 mg	65.00	52.00	10.6
E-98110 ***New***	erabutoxin a *from *Laticauda semifasciata* snake venom*	1 mg	185.00	148.00	18.2
E-98111 ***New***	erabutoxin b *from *Laticauda semifasciata* snake venom*	1 mg	185.00	148.00	18.2
E-96021 ***New***	erysimoside, monohydrate	5 mg	60.00	48.00	10.6, 18.7
E-333	erythrosin-5-iodoacetamide	25 mg	155.00	124.00	2.2
E-332	erythrosin-5-isothiocyanate	25 mg	185.00	148.00	1.5
E-2870	*Escherichia coli* BioParticles® opsonizing reagent	1 U	55.00	44.00	17.1
E-2864	*Escherichia coli* (K-12 strain) BioParticles®, BODIPY® FL conjugate	10 mg	48.00	38.40	17.1
E-2861	*Escherichia coli* (K-12 strain) BioParticles®, fluorescein conjugate	10 mg	65.00	52.00	17.1
E-2862	*Escherichia coli* (K-12 strain) BioParticles®, tetramethylrhodamine conjugate	10 mg	48.00	38.40	17.1
E-2863	*Escherichia coli* (K-12 strain) BioParticles®, Texas Red® conjugate	10 mg	65.00	52.00	17.1

Cat #	Product Name	Unit Size	Price Per Unit ($) 1–4 Units	Price Per Unit ($) 5–24 Units	Section #
E-99004 ***New***	etbicyphat	1 mg	245.00	196.00	18.5
E-99015 ***New***	etbicythionat	100 µg	205.00	164.00	18.5
E-1022	1,N^6-ethenoadenosine 5′-diphosphate (ε-ADP)	25 mg	65.00	52.00	11.3
E-1023	1,N^6-ethenoadenosine 5′-triphosphate (ε-ATP)	25 mg	118.00	94.40	11.3
E-667	ethidium-acridine heterodimer	1 mg	95.00	76.00	8.1
E-1305	ethidium bromide	1 g	12.00	9.60	8.1, 16.3
E-3565	ethidium bromide *10 mg/mL solution in water*	10 mL	28.00	22.40	8.1, 16.3
E-7591 ***New***	ethidium bromide gel stain photographic filter	each	29.00	29.00	8.3, 26.4
E-3561	ethidium diazide chloride	5 mg	95.00	76.00	5.3, 8.1
E-1169	ethidium homodimer-1 (EthD-1)	1 mg	155.00	124.00	8.1, 16.3
E-3599	ethidium homodimer-2 (EthD-2) *1 mM solution in DMSO*	200 µL	138.00	110.40	8.1
E-1374	ethidium monoazide bromide	5 mg	115.00	92.00	5.3, 8.1, 16.3
E-3101	N-(ethoxycarbonylmethyl)-6-methoxyquinolinium bromide (MQAE)	100 mg	78.00	62.40	24.2
E-186	7-ethoxycoumarin	100 mg	24.00	19.20	10.5
E-99019 ***New***	2-ethoxy-2,3-dihydro-1,4,2-benzodioxaphosphorin-2-oxide (bomin-1)	500 µg	108.00	86.40	18.2
E-6288 ***New***	2-ethoxy-1-ethoxycarbonyl-1,2-dihydroquinoline (EEDQ)	5 g	16.00	12.80	3.3, 5.2
E-6531 ***New***	7-ethoxy-4-heptadecylcoumarin	10 mg	55.00	44.00	10.5
R-352	ethoxyresorufin (resorufin ethyl ether)	5 mg	98.00	78.40	10.5
E-2882	7-ethoxy-4-trifluoromethylcoumarin	25 mg	60.00	48.00	10.5
E-2247	1-ethyl-3-(3-dimethylaminopropyl)carbodiimide, hydrochloride (EDAC)	100 mg	15.00	12.00	3.3, 5.2
E-6306 ***New***	ethylene glycol bis-(succinic acid), bis-(succinimidyl ester) (EGS)	100 mg	18.00	14.40	5.2, 9.4
E-3111	5-(N-ethyl-N-isopropyl)amiloride, hydrochloride	5 mg	70.00	56.00	18.7
E-96060 ***New***	eudesmine	10 mg	128.00	102.40	18.8
E-6960 ***New***	europium(III) chloride, hexahydrate *99.99%*	1 g	20.00	16.00	9.2, 15.3, 16.3 20.2, 22.7
E-96089 ***New***	evoxine (haploperine)	20 mg	108.00	86.40	18.8
O-3852	F18 (N-octadecyl-N′-(5-(fluoresceinyl))thiourea)	25 mg	65.00	52.00	14.3
C-1359	5-FAM (5-carboxyfluorescein) *single isomer*	100 mg	55.00	44.00	1.3
C-2210	5-FAM, SE (5-carboxyfluorescein, succinimidyl ester) *single isomer*	10 mg	95.00	76.00	1.3
C-1311	5(6)-FAM, SE (5-(and-6)-carboxyfluorescein, succinimidyl ester) *mixed isomers*	100 mg	75.00	60.00	1.3
F-98107 ***New***	fasciculin 2 *from *Dendroaspis angusticeps* snake venom*	100 µg	160.00	128.00	18.2
F-2900	*FAST* CAT® Chloramphenicol Acetyltransferase Assay Kit *100 assays*	1 kit	95.00	76.00	10.5
F-6616 ***New***	*FAST* CAT® Green (deoxy) Chloramphenicol Acetyltransferase Assay Kit *100 assays*	1 kit	125.00	100.00	10.5
F-6617 ***New***	*FAST* CAT® Yellow (deoxy) Chloramphenicol Acetyltransferase Assay Kit *100 assays*	1 kit	125.00	100.00	10.5
D-3897	*FAST* DiA™ oil; DiΔ^{9,12}-C$_{18}$ASP, I (4-(4-(dilinoleylamino)styryl)-N-methylpyridinium iodide)	5 mg	135.00	108.00	14.2, 15.4
D-7758 ***New***	*FAST* DiA™ solid; DiΔ^{9,12}-C$_{18}$ASP, CBS (4-(4-(dilinoleylamino)styryl)-N-methylpyridinium, 4-chlorobenzenesulfonate)	5 mg	135.00	108.00	14.2, 15.4
D-3899	*FAST* DiI™ oil; DiIΔ^{9,12}-C$_{18}$(3), ClO$_4$ (1,1′-dilinoleyl-3,3,3′,3′-tetramethylindocarbocyanine perchlorate)	5 mg	135.00	108.00	14.2, 15.4
D-7756 ***New***	*FAST* DiI™ solid; DiIΔ^{9,12}-C$_{18}$(3), CBS (1,1′-dilinoleyl-3,3,3′,3′-tetramethylindocarbocyanine, 4-chlorobenzenesulfonate)	5 mg	135.00	108.00	14.2, 15.4
D-3898	*FAST* DiO™ solid; DiOΔ^{9,12}-C$_{18}$(3), ClO$_4$ (3,3′-dilinoleyloxacarbocyanine perchlorate)	5 mg	135.00	108.00	14.2, 15.4
F-2902	Fc OxyBURST® Green assay reagent *25 assays* *3 mg/mL*	500 µL	125.00	100.00	17.1
F-6691 ***New***	Fc OxyBURST® Orange assay reagent *25 assays* *3 mg/mL*	500 µL	125.00	100.00	17.1
F-1179	FDG (fluorescein di-β-D-galactopyranoside)	5 mg	115.00	92.00	10.2
F-2915	FDGlcU (fluorescein di-β-D-glucuronide)	5 mg	148.00	118.40	10.2
F-2999	FDP (fluorescein diphosphate, tetraammonium salt)	5 mg	165.00	132.00	10.3
F-948	ferritin from horse spleen, cationized, fluorescein conjugate *2.5 mg/mL*	2 mL	80.00	64.00	17.4
F-96022 ***New***	ferulin	10 mg	128.00	102.40	18.8
F-96006 ***New***	ferutinine *mixture of ferutinine and tenuferidine*	5 mg	128.00	102.40	18.8
O-6080 ***New***	F$_2$FITC (Oregon Green™ 488 isothiocyanate) *mixed isomers*	5 mg	95.00	76.00	1.4
F-7496 ***New***	fibrinogen from human plasma, Oregon Green™ 488 conjugate	5 mg	95.00	76.00	16.3
F-1965	Ficoll®, fluorescein, 400,000 MW, anionic	25 mg	80.00	64.00	15.5
F-8072 ***New***	filter O-rings for TD-700 fluorometer, set of 4	each	10.00	10.00	26.3
F-8020 ***New***	filter set for TD-700 fluorometer, fluorescein range *excitation 481-491 nm; emission 510-700 nm*	each	295.00	295.00	26.3
F-8021 ***New***	filter set for TD-700 fluorometer, rhodamine range *excitation 545-555 nm; emission 570-700 nm*	each	365.00	365.00	26.3
F-8022 ***New***	filter set for TD-700 fluorometer, ultraviolet range *excitation 300-400 nm; emission 410-610 nm*	each	215.00	215.00	26.3
F-7462 ***New***	fim-1	1 mg	95.00	76.00	20.3
F-7453 ***New***	fim-1 diacetate	1 mg	95.00	76.00	20.3
F-7583 ***New***	FISH Counterstain Kit #1 *DAPI/YOYO®-1/PI* *100 slides*	1 kit	95.00	76.00	8.4
F-7588 ***New***	FISH Counterstain Kit #2 *DAPI/SYTOX® Green/PI* *100 slides*	1 kit	95.00	76.00	8.4

Cat #		Product Name	Unit Size	Price Per Unit ($) 1–4 Units	5–24 Units	Section #
F-99013	***New***	fisonate	10 mg	60.00	48.00	10.6
F-6865	***New***	FITC-Gly-Gly-His ((fluorescein-5-thioureidyl)glycylglycyl-L-histidine, dipotassium salt)	1 mg	65.00	52.00	22.7
F-6864	***New***	FITC-Gly-His ((fluorescein-5-thioureidyl)glycyl-L-histidine, dipotassium salt)	1 mg	65.00	52.00	22.7
F-143		FITC 'Isomer I' (fluorescein-5-isothiocyanate) *≥95% by HPLC*	1 g	98.00	78.40	1.3, 9.3
F-1906		FITC 'Isomer I' (fluorescein-5-isothiocyanate) *≥95% by HPLC* *special packaging*	10x10 mg	48.00	38.40	1.3, 9.3
F-1907		FITC 'Isomer I' (fluorescein-5-isothiocyanate) *≥95% by HPLC* *special packaging*	10x100 mg	125.00	100.00	1.3, 9.3
F-144		FITC 'Isomer II' (fluorescein-6-isothiocyanate)	1 g	98.00	78.40	1.3
C-6660	***New***	FlCRhR (Cyclic AMP Fluorosensor®) *20±5 µM*	10 µL	495.00	495.00	20.3
F-6877	***New***	FL-DFO (N-(fluorescein-5-thiocarbamoyl)desferrioxamine, pentaammonium salt) *special packaging*	20x50 µg	95.00	76.00	22.7
F-99014	***New***	flodiphoside	10 mg	60.00	48.00	10.6
F-7321	***New***	Flow Cytometry Alignment Standards Sampler Kit, 2.5 µm	1 kit	55.00	44.00	26.2
F-7322	***New***	Flow Cytometry Alignment Standards Sampler Kit, 6.0 µm	1 kit	55.00	44.00	26.2
F-7320	***New***	Flow Cytometry Compensation and Intensity Standards Kit	1 kit	45.00	36.00	26.2
F-7310	***New***	flow cytometry reference beads, DAPI-like, 2.5 µm	3 mL	85.00	68.00	26.2
F-7311	***New***	flow cytometry reference beads, DAPI-like, 6.0 µm	3 mL	85.00	68.00	26.2
		flow cytometry standards — see Section 26.2 for a complete list.				
F-99002	***New***	flucybene	1 mg	108.00	86.40	18.5
F-1241		fluo-3, AM *cell permeant*	1 mg	165.00	132.00	22.3
F-1242		fluo-3, AM *cell permeant* *special packaging*	20x50 µg	195.00	156.00	22.3
F-1240		fluo-3, pentaammonium salt *cell impermeant*	1 mg	135.00	108.00	22.3
F-3715		fluo-3, pentapotassium salt *cell impermeant*	1 mg	155.00	124.00	22.3
A-6250	***New***	fluoral P (4-amino-3-pentene-2-one)	1 g	38.00	30.40	3.2
F-6290	***New***	N-(9-fluorenylmethoxycarbonyl) hydrazine (FMOC hydrazine)	25 mg	68.00	54.40	3.2
F-7452	***New***	FluoReporter® Actin Assay Kit	1 kit	195.00	156.00	11.1
F-6348	***New***	FluoReporter® Biotin/DNP Protein Labeling Kit *5-10 labelings*	1 kit	245.00	196.00	4.2
F-6095	***New***	FluoReporter® Biotin-X-C_5 Oligonucleotide Phosphate Labeling Kit *5 labelings*	1 kit	175.00	140.00	4.2, 8.2
F-6081	***New***	FluoReporter® Biotin-XX Oligonucleotide Amine Labeling Kit *5 labelings*	1 kit	175.00	140.00	4.2, 8.2
F-2610		FluoReporter® Biotin-XX Protein Labeling Kit *5 labelings of 5-20 mg protein each*	1 kit	165.00	132.00	4.2
F-2962		FluoReporter® Blue Fluorometric dsDNA Quantitation Kit *200-2000 assays*	1 kit	78.00	62.40	16.2
F-6096	***New***	FluoReporter® BODIPY® FL-C_5 Oligonucleotide Phosphate Labeling Kit *5 labelings*	1 kit	175.00	140.00	8.2
F-6079	***New***	FluoReporter® BODIPY® FL Oligonucleotide Amine Labeling Kit *5 labelings*	1 kit	175.00	140.00	1.2, 8.2
F-6082	***New***	FluoReporter® BODIPY® FL-X Oligonucleotide Amine Labeling Kit *5 labelings*	1 kit	175.00	140.00	1.2, 8.2
F-6092	***New***	FluoReporter® BODIPY® R6G Oligonucleotide Amine Labeling Kit *5 labelings*	1 kit	175.00	140.00	1.2, 8.2
F-6097	***New***	FluoReporter® BODIPY® TMR-C_5 Oligonucleotide Phosphate Labeling Kit *5 labelings*	1 kit	175.00	140.00	8.2
F-6083	***New***	FluoReporter® BODIPY® TMR-X Oligonucleotide Amine Labeling Kit *5 labelings*	1 kit	175.00	140.00	1.2, 8.2
F-6084	***New***	FluoReporter® BODIPY® TR-X Oligonucleotide Amine Labeling Kit *5 labelings*	1 kit	175.00	140.00	1.2, 8.2
F-6093	***New***	FluoReporter® BODIPY® 564/570 Oligonucleotide Amine Labeling Kit *5 labelings*	1 kit	175.00	140.00	1.2, 8.2
F-6094	***New***	FluoReporter® BODIPY® 581/591 Oligonucleotide Amine Labeling Kit *5 labelings*	1 kit	175.00	140.00	1.2, 8.2
F-2961		FluoReporter® Colorimetric Cell Protein Assay Kit	1 kit	100.00	80.00	9.1, 16.2
F-6085	***New***	FluoReporter® Dinitrophenyl-X (DNP-X) Oligonucleotide Amine Labeling Kit *5 labelings*	1 kit	175.00	140.00	4.2, 8.2
F-6434	***New***	FluoReporter® FITC Protein Labeling Kit *5-10 labelings*	1 kit	195.00	156.00	1.3
F-6433	***New***	FluoReporter® Fluorescein-EX Protein Labeling Kit *5-10 labelings*	1 kit	195.00	156.00	1.3
F-6086	***New***	FluoReporter® Fluorescein-X Oligonucleotide Amine Labeling Kit *5 labelings*	1 kit	175.00	140.00	1.3, 8.2
F-2960		FluoReporter® Fluorometric Cell Protein Assay Kit	1 kit	135.00	108.00	9.1, 16.2
F-6100	***New***	FluoReporter® Labeled Oligonucleotide Purification Kit *five spin columns plus buffers*	1 kit	50.00	40.00	4.2, 8.2
F-1930		FluoReporter® *lacZ* Flow Cytometry Kit *50 assays*	1 kit	178.00	142.40	10.2
F-1931		FluoReporter® *lacZ* Flow Cytometry Kit *250 assays*	1 kit	725.00	580.00	10.2
F-2905		FluoReporter® *lacZ*/Galactosidase Quantitation Kit *300 assays*	1 kit	98.00	78.40	10.2
F-6347	***New***	FluoReporter® Mini-biotin-XX Protein Labeling Kit *5 labelings of 0.1-3 mg protein each*	1 kit	165.00	132.00	4.2
F-6087	***New***	FluoReporter® Oregon Green™ 488 Oligonucleotide Amine Labeling Kit *5 labelings*	1 kit	175.00	140.00	1.4, 8.2
F-6153	***New***	FluoReporter® Oregon Green™ 488 Protein Labeling Kit *5-10 labelings*	1 kit	195.00	156.00	1.4
F-6154	***New***	FluoReporter® Oregon Green™ 500 Protein Labeling Kit *5-10 labelings*	1 kit	195.00	156.00	1.5
F-6155	***New***	FluoReporter® Oregon Green™ 514 Protein Labeling Kit *5-10 labelings*	1 kit	195.00	156.00	1.5
F-6694	***New***	FluoReporter® Phagocytosis Assay Kit *250 assays*	1 kit	125.00	100.00	17.1
F-6088	***New***	FluoReporter® Rhodamine Green™-X Oligonucleotide Amine Labeling Kit *5 labelings*	1 kit	175.00	140.00	1.4, 8.2

Cat #		Product Name	Unit Size	Price Per Unit ($) 1–4 Units	5–24 Units	Section #
F-6098	*New*	FluoReporter® Rhodamine Red™-C_2 Oligonucleotide Phosphate Labeling Kit *5 labelings*	1 kit	175.00	140.00	8.2
F-6089	*New*	FluoReporter® Rhodamine Red™-X Oligonucleotide Amine Labeling Kit *5 labelings*	1 kit	175.00	140.00	1.6, 8.2
F-6161	*New*	FluoReporter® Rhodamine Red™-X Protein Labeling Kit *5-10 labelings*	1 kit	195.00	156.00	1.6
F-6163	*New*	FluoReporter® Tetramethylrhodamine Protein Labeling Kit *5-10 labelings*	1 kit	195.00	156.00	1.6
F-6090	*New*	FluoReporter® Tetramethylrhodamine (5-TAMRA) Oligonucleotide Amine Labeling Kit *5 labelings*	1 kit	175.00	140.00	1.6, 8.2
F-6099	*New*	FluoReporter® Texas Red®-C_5 Oligonucleotide Phosphate Labeling Kit *5 labelings*	1 kit	175.00	140.00	8.2
F-6091	*New*	FluoReporter® Texas Red®-X Oligonucleotide Amine Labeling Kit *5 labelings*	1 kit	175.00	140.00	1.6, 8.2
F-6162	*New*	FluoReporter® Texas Red®-X Protein Labeling Kit *5-10 labelings*	1 kit	195.00	156.00	1.6
F-2332		fluorescamine	100 mg	20.00	16.00	1.8, 9.1
F-1300		fluorescein *reference standard*	1 g	40.00	32.00	1.3, 10.1, 23.2 24.3
C-7604	*New*	fluorescein-12-dUTP (ChromaTide™ fluorescein-12-dUTP) *1 mM in buffer*	25 µL	175.00	175.00	8.2
C-7603	*New*	fluorescein-12-UTP (ChromaTide™ fluorescein-12-UTP) *1 mM in buffer*	25 µL	175.00	175.00	8.2
B-1370		fluorescein biotin (5-((N-(5-(N-(6-(biotinoyl)amino)hexanoyl)amino)pentyl)-thioureidyl)fluorescein)	5 mg	98.00	78.40	4.3, 15.3
F-7103	*New*	fluorescein bis-(5-carboxymethoxy-2-nitrobenzyl) ether, dipotassium salt (CMNB-caged fluorescein)	5 mg	145.00	116.00	15.3, 19.2
F-1176		fluorescein α-bungarotoxin (α-bungarotoxin, fluorescein conjugate)	500 µg	175.00	140.00	18.2
A-456		fluorescein cadaverine (5-((5-aminopentyl)thioureidyl)fluorescein)	25 mg	138.00	110.40	3.3
F-6218	*New*	fluorescein-5-carbonyl azide, diacetate	10 mg	78.00	62.40	3.1
F-6106	*New*	6-(fluorescein-5-carboxamido)hexanoic acid, succinimidyl ester (5-SFX) *single isomer*	5 mg	65.00	52.00	1.3
F-2181		6-(fluorescein-5-(and-6)-carboxamido)hexanoic acid, succinimidyl ester (5(6)-SFX) *mixed isomers*	10 mg	65.00	52.00	1.3, 4.2
F-6129	*New*	6-(fluorescein-5-(and-6)-carboxamido)hexanoic acid, succinimidyl ester (5(6)-SFX) *mixed isomers* *special packaging*	10x1 mg	78.00	62.40	1.3
F-362		fluorescein DHPE (N-(fluorescein-5-thiocarbamoyl)-1,2-dihexadecanoyl-*sn*-glycero-3-phosphoethanolamine, triethylammonium salt)	5 mg	75.00	60.00	13.4, 23.4
F-1303		fluorescein diacetate (FDA)	1 g	40.00	32.00	16.3, 23.2
F-1179		fluorescein di-β-D-galactopyranoside (FDG)	5 mg	115.00	92.00	10.2
F-2881		fluorescein di-β-D-glucopyranoside (FDGlu)	5 mg	70.00	56.00	10.2
F-2915		fluorescein di-β-D-glucuronide (FDGlcU)	5 mg	148.00	118.40	10.2
F-2890		fluorescein di-(4-guanidinobenzoate), dihydrochloride	5 mg	75.00	60.00	10.5
F-2999		fluorescein diphosphate, tetraammonium salt (FDP)	5 mg	165.00	132.00	10.3
F-2810		fluorescein donkey anti-sheep IgG (H+L) conjugate *2 mg/mL*	0.5 mL	75.00	60.00	7.2
E-3478		fluorescein EGF (epidermal growth factor, fluorescein conjugate)	20 µg	160.00	128.00	17.3
F-6130	*New*	fluorescein-5-EX, succinimidyl ester *single isomer*	10 mg	65.00	52.00	1.3, 4.2
F-2780		fluorescein goat anti-human IgG (H+L) conjugate *2 mg/mL*	0.5 mL	75.00	60.00	7.2
F-2761		fluorescein goat anti-mouse IgG (H+L) conjugate *2 mg/mL*	0.5 mL	75.00	60.00	7.2
F-2765		fluorescein goat anti-rabbit IgG (H+L) conjugate *2 mg/mL*	0.5 mL	75.00	60.00	7.2
F-143		fluorescein-5-isothiocyanate (FITC 'Isomer I') *≥95% by HPLC*	1 g	98.00	78.40	1.3, 9.3
F-1906		fluorescein-5-isothiocyanate (FITC 'Isomer I') *≥95% by HPLC* *special packaging*	10x10 mg	48.00	38.40	1.3, 9.3
F-1907		fluorescein-5-isothiocyanate (FITC 'Isomer I') *≥95% by HPLC* *special packaging*	10x100 mg	125.00	100.00	1.3, 9.3
F-1919		fluorescein-5-isothiocyanate *10% adsorbed on Celite®*	1 g	45.00	36.00	1.3
F-144		fluorescein-6-isothiocyanate (FITC 'Isomer II')	1 g	98.00	78.40	1.3
F-1163		fluorescein-5-isothiocyanate diacetate	25 mg	115.00	92.00	15.2
F-150		fluorescein-5-maleimide	25 mg	155.00	124.00	2.2, 9.4
M-1198		fluorescein methotrexate (methotrexate, fluorescein, triammonium salt)	1 mg	125.00	100.00	16.3
F-6489	*New*	fluorescein mono-(4-guanidinobenzoate), hydrochloride	5 mg	95.00	76.00	10.5
F-3425		fluorescein NAPS	1 mg	68.00	54.40	18.2
F-3857		fluorescein octadecyl ester	10 mg	65.00	52.00	14.3
F-432		fluorescein phalloidin	300 U	198.00	158.40	11.1
F-3471		fluorescein PPHT *1 mg/mL solution in dry DMSO*	200 µL	68.00	54.40	18.2
F-3421		fluorescein SCH 23390	1 mg	68.00	54.40	18.2
F-1130		fluorescein-5-(and-6)-sulfonic acid, trisodium salt	100 mg	78.00	62.40	15.3, 23.2
F-6877	*New*	N-(fluorescein-5-thiocarbamoyl)desferrioxamine, pentaammonium salt (FL-DFO) *special packaging*	20x50 µg	95.00	76.00	22.7
F-362		N-(fluorescein-5-thiocarbamoyl)-1,2-dihexadecanoyl-*sn*-glycero-3-phospho-ethanolamine, triethylammonium salt (fluorescein DHPE)	5 mg	75.00	60.00	13.4, 23.4
F-121		fluorescein-5-thiosemicarbazide	100 mg	95.00	76.00	3.2, 9.4
F-6865	*New*	(fluorescein-5-thioureidyl)glycylglycyl-L-histidine, dipotassium salt (FITC-Gly-Gly-His)	1 mg	65.00	52.00	22.7

Cat #		Product Name	Unit Size	Price Per Unit ($) 1–4 Units	5–24 Units	Section #
F-6864	New	(fluorescein-5-thioureidyl)glycyl-L-histidine, dipotassium salt (FITC-Gly-His)	1 mg	65.00	52.00	22.7
A-1363		fluoresceinyl glycine amide (5-(aminoacetamido)fluorescein)	10 mg	98.00	78.40	3.3
F-7889	New	Fluorescent and Luminescent Probes for Biological Activity. A Practical Guide to Technology for Quantitative Real-Time Analysis. W.T. Mason, ed. Academic Press (1993); 433 pages, comb bound	each	65.00	65.00	22.1, 26.6
F-6053	New	7-fluorobenz-2-oxa-1,3-diazole-4-sulfonamide (ABD-F)	10 mg	55.00	44.00	2.3
F-96026	New	fluorocurarine chloride (C-curarine III; metvine)	20 mg	128.00	102.40	18.2
D-1820		fluoro-emerald (dextran, fluorescein, 10,000 MW, anionic, lysine fixable)	25 mg	95.00	76.00	15.5
F-8000	New	fluorometer, Turner Designs TD-700, with standard photomultiplier *includes S-8060 sample adaptor*	each	3765.00	3765.00	26.3
F-486		4-fluoro-7-nitrobenz-2-oxa-1,3-diazole (NBD fluoride; 4-fluoro-7-nitrobenzofurazan)	25 mg	95.00	76.00	1.8, 2.3
F-1501		4-fluoro-3-nitrophenyl azide (FNPA)	100 mg	24.00	19.20	5.3
D-1817		fluoro-ruby (dextran, tetramethylrhodamine, 10,000 MW, lysine fixable)	25 mg	110.00	88.00	15.5
F-3622		FluoroTide™ M13/pUC (-21) primer, 5′-BODIPY® FL labeled, X d(TGTAAAACGACGGCCAGT)	0.1 OD	95.00	76.00	8.2
F-7632	New	FluoroTide™ M13/pUC (-21) primer, 5′-BODIPY® R6G labeled, X d(TGTAAAACGACGGCCAGT)	0.1 OD	95.00	76.00	8.2
F-7633	New	FluoroTide™ M13/pUC (-21) primer, 5′-BODIPY® TMR labeled, X d(TGTAAAACGACGGCCAGT)	0.1 OD	95.00	76.00	8.2
F-7634	New	FluoroTide™ M13/pUC (-21) primer, 5′-BODIPY® TR labeled, X d(TGTAAAACGACGGCCAGT)	0.1 OD	95.00	76.00	8.2
F-3621		FluoroTide™ M13/pUC (-21) primer, 5′-fluorescein labeled, X d(TGTAAAACGACGGCCAGT)	0.1 OD	95.00	76.00	8.2
F-6677	New	FluoroTide™ M13/pUC (-21) primer, 5′-Oregon Green™ 488 labeled, X d(TGTAAAACGACGGCCAGT)	0.1 OD	95.00	76.00	8.2
F-6676	New	FluoroTide™ M13/pUC (-21) primer, 5′-Texas Red®-X labeled, X d(TGTAAAACGACGGCCAGT)	0.1 OD	95.00	76.00	8.2
		FluoSpheres® polystyrene microspheres — see Section 6.3 for a complete list of FluoSpheres® and TransFluoSpheres® polystyrene microspheres.				
B-6351	New	FMOC biocytin (ϵ-biotinoyl-α-(9-fluorenylmethoxycarbonyl)-L-lysine)	100 mg	175.00	140.00	27.1
F-6290	New	FMOC hydrazine (N-(9-fluorenylmethoxycarbonyl) hydrazine)	25 mg	68.00	54.40	3.2
T-3163		FM™ 1-43 (N-(3-triethylammoniumpropyl)-4-(4-(dibutylamino)styryl)pyridinium dibromide)	1 mg	165.00	132.00	15.4, 17.2
T-3164		FM™ 1-84 (N-(3-triethylammoniumpropyl)-4-(4-(dipentylamino)styryl)pyridinium dibromide)	1 mg	165.00	132.00	17.2
T-7508	New	FM™ 2-10 (N-(3-triethylammoniumpropyl)-4-(4-(diethylamino)styryl)pyridinium dibromide)	5 mg	135.00	108.00	17.2
T-7506	New	FM™ 3-25 (N-(3-triethylammoniumpropyl)-4-(4-(dioctadecylamino)styryl)-pyridinium, di-4-chlorobenzenesulfonate)	5 mg	125.00	100.00	17.2
T-3166		FM™ 4-64 (N-(3-triethylammoniumpropyl)-4-(6-(4-(diethylamino)phenyl)-hexatrienyl)pyridinium dibromide)	1 mg	165.00	132.00	12.3, 15.4, 17.2
F-6999	New	FM™ DiI *special packaging*	20x50 µg	95.00	76.00	14.2, 15.4
F-7234	New	FocalCheck™ microspheres, 15 µm, fluorescent blue/orange ring stains	0.5 mL	135.00	108.00	26.1
F-7239	New	FocalCheck™ microspheres, 15 µm, fluorescent dark red ring stain/green throughout	0.5 mL	135.00	108.00	26.1
F-7240	New	FocalCheck™ microspheres, 15 µm, fluorescent green/dark red ring stains	0.5 mL	135.00	108.00	26.1
F-7235	New	FocalCheck™ microspheres, 15 µm, fluorescent green/orange/dark red ring stains	0.5 mL	135.00	108.00	26.1
F-7237	New	FocalCheck™ microspheres, 15 µm, fluorescent green ring stain/blue throughout	0.5 mL	135.00	108.00	26.1
F-7238	New	FocalCheck™ microspheres, 15 µm, fluorescent green ring stain/dark red throughout	0.5 mL	135.00	108.00	26.1
F-7236	New	FocalCheck™ microspheres, 15 µm, fluorescent orange ring stain/blue throughout	0.5 mL	135.00	108.00	26.1
F-96027	New	foliosidine	20 mg	128.00	102.40	18.8
F-7468	New	formyl-Nle-Leu-Phe-Nle-Tyr-Lys, BODIPY® FL derivative	1 mg	98.00	78.40	17.1
F-1314		formyl-Nle-Leu-Phe-Nle-Tyr-Lys, fluorescein derivative	1 mg	98.00	78.40	17.1
F-1377		formyl-Nle-Leu-Phe-Nle-Tyr-Lys, tetramethylrhodamine derivative	1 mg	98.00	78.40	17.1
M-7904	New	FotoFenton™ 1 (2-mercaptopyridine N-oxide, sodium salt)	100 mg	35.00	28.00	21.2
F-7030	New	FUN-1™ cell stain *10 mM solution in DMSO*	100 µL	75.00	60.00	12.3, 16.2
		furaptra — see mag-fura-2 (M-1290); see also Sections 22.2 and 22.6.				
F-1201		fura-2, AM *cell permeant* *≥95% by HPLC*	1 mg	100.00	80.00	22.2
F-1221		fura-2, AM *cell permeant* *≥95% by HPLC* *special packaging*	20x50 µg	138.00	110.40	22.2
F-1225		fura-2, AM *1 mM solution in dry DMSO* *cell permeant* *special packaging*	1 mL	125.00	100.00	22.2
F-6774	New	Fura-2 Calcium Imaging Calibration Kit *zero to 10 mM CaEGTA, 50 µM fura-2 (11 x 1 mL)*	1 kit	95.00	76.00	22.8
F-1200		fura-2, pentapotassium salt *cell impermeant* *≥95% by HPLC*	1 mg	83.00	66.40	22.2
F-6799	New	fura-2, pentasodium salt *cell impermeant*	1 mg	83.00	66.40	22.2
A-6796	New	fura BSA (albumin from bovine serum, fura conjugate)	5 mg	125.00	100.00	22.4

Cat #		Product Name	Unit Size	Price Per Unit ($) 1–4 Units	5–24 Units	Section #
F-6792	**New**	fura-C_{18}, pentapotassium salt	1 mg	150.00	120.00	22.4
F-6764	**New**	fura dextran, potassium salt, 3000 MW, anionic	5 mg	275.00	220.00	22.4
F-3029		fura dextran, potassium salt, 10,000 MW, anionic	5 mg	248.00	198.40	22.4
F-3030		fura dextran, potassium salt, 70,000 MW, anionic	5 mg	195.00	156.00	22.4
F-6805	**New**	fura-indoline-C_{18}	1 mg	85.00	68.00	22.4
F-3020		Fura Red™, AM *cell permeant* *≥95% by HPLC*	500 µg	155.00	124.00	22.3
F-3021		Fura Red™, AM *cell permeant* *≥95% by HPLC* *special packaging*	10x50 µg	175.00	140.00	22.3
F-3019		Fura Red™, tetraammonium salt *cell impermeant*	500 µg	125.00	100.00	22.3
G-8453	**New**	D-(+)-galactosamine, hydrochloride	1 g	25.00	20.00	10.6
G-8454	**New**	α-D-galactose 1-phosphate, dipotassium salt, pentahydrate	100 mg	38.00	30.40	10.6
G-6463	**New**	β-galactosidase *from *Escherichia coli**	5 mg	95.00	76.00	10.2
G-96028	**New**	galanthamine, hydrobromide	50 mg	95.00	76.00	18.2
G-6657	**New**	gel photographic filter holder	each	18.00	18.00	8.3, 9.2, 26.4
G-7440	**New**	glibenclamide (glyburide)	100 mg	8.00	6.40	18.6
G-7493	**New**	globotriose GB_3, fluorescein conjugate	10 µg	175.00	140.00	27.1
S-2721		β-glucan (soluble β-glucan) *1 mg/mL*	500 µL	65.00	52.00	17.1
S-2720		β-glucan, fluorescein conjugate (soluble β-glucan, fluorescein conjugate) *1 mg/mL*	500 µL	118.00	94.40	17.1
G-7121	**New**	L-glutamic acid, γ-(5-allyloxy-2-nitrobenzyl) ester (γ-(ANB-caged) L-glutamic acid)	5 mg	168.00	134.40	18.3, 19.2
G-7055	**New**	L-glutamic acid, γ-(α-carboxy-2-nitrobenzyl) ester, trifluoroacetate (γ-(CNB-caged) L-glutamic acid)	5 mg	168.00	134.40	18.3, 19.2
G-2504		L-glutamic acid, α-(4,5-dimethoxy-2-nitrobenzyl) ester, hydrochloride (α-(DMNB-caged) L-glutamic acid)	5 mg	145.00	116.00	18.3, 19.2
G-7120	**New**	L-glutamic acid, γ-(1,2-diphenyl-2-oxo)ethyl ester (γ-(desyl-caged) L-glutamic acid)	5 mg	168.00	134.40	18.3, 19.2
G-2879		glutathione agarose, linked through sulfur *sedimented bead suspension*	10 mL	90.00	72.00	9.1
G-6664	**New**	glutathione agarose, linked through sulfur *eight 1.0 mL prepacked columns*	1 set	125.00	100.00	9.1
G-6665	**New**	glutathione agarose, linked through sulfur *five 2.0 mL prepacked columns*	1 set	125.00	100.00	9.1
G-7440	**New**	glyburide (glibenclamide)	100 mg	8.00	6.40	18.6
G-7702	**New**	glycerol tris-(1-pyrenebutyrate)	5 mg	48.00	38.40	10.5
G-2507		glycine, 4,5-dimethoxy-2-nitrobenzyl ester, hydrochloride (O-(DMNB-caged) glycine)	5 mg	75.00	60.00	18.3, 19.2
G-96090	**New**	glyrophama, dihydrate (amorphine)	50 mg	68.00	54.40	18.8
G-6448	**New**	goat anti-mouse IgG (H+L) *affinity adsorbed against human IgG and human serum* *5 mg/mL*	1 mL	75.00	60.00	7.2
G-6446	**New**	goat anti-rabbit IgG (H+L) *affinity adsorbed against human, mouse and bovine serum* *5 mg/mL*	1 mL	95.00	76.00	7.2
G-6447	**New**	goat anti-rat IgG (H+L) *affinity adsorbed against mouse IgG and mouse serum* *5 mg/mL*	1 mL	125.00	100.00	7.2
G-6916	**New**	gonyautoxin C1 (epi-GTX VIII) *1 mg/mL*	100 µL	200.00	160.00	18.7
G-6917	**New**	gonyautoxin GTX-I *1 mg/mL*	100 µL	200.00	160.00	18.7
G-6915	**New**	gonyautoxin GTX-IV (neosaxitoxin 11β-sulfate) *250 µg/mL*	100 µL	200.00	160.00	18.7
G-6918	**New**	gonyautoxin GTX-VI (B2) *1 mg/mL*	100 µL	200.00	160.00	18.7
G-96083	**New**	gossypol	100 mg	45.00	36.00	18.8
G-6888	**New**	gramicidin	100 mg	15.00	12.00	24.1
G-96029	**New**	graveoline (foliosine)	20 mg	108.00	86.40	18.8
G-7921	**New**	Griess Reagent Kit *for nitrite quantitation*	1 kit	55.00	44.00	21.5, 24.3
G-3620		guanosine 5′-O-(2-thiodiphosphate), $P^{2(S)}$-(1-(4,5-dimethoxy-2-nitrophenyl)ethyl) ester, ammonium salt (S-(DMNPE-caged) GDP-β-S)	1 mg	155.00	124.00	11.3, 19.2, 20.4
G-1053		guanosine 5′-O-(3-thiotriphosphate), $P^{3(S)}$-(1-(4,5-dimethoxy-2-nitrophenyl)ethyl) ester, triammonium salt (S-(DMNPE-caged) GTP-γ-S)	1 mg	155.00	124.00	11.3, 19.2, 20.4
D-338		H_2-DIDS (4,4′-diisothiocyanatodihydrostilbene-2,2′-disulfonic acid, disodium salt)	100 mg	115.00	92.00	18.5
H-7080	**New**	haloperidol, 4,5-dimethoxy-2-nitrobenzyl carbonate (O-(DMNB-caged) haloperidol)	5 mg	48.00	38.40	18.3, 19.2
H-8999		Handbook of Fluorescent Probes and Research Chemicals, Sixth Edition. R.P. Haugland, Molecular Probes, Inc. (1996) 700 pages, soft cover	each	25.00	20.00	26.6
H-96007	**New**	heliotrine	50 mg	128.00	102.40	18.8
H-7482	**New**	heparin, fluorescein conjugate	1 mg	95.00	76.00	20.2
H-181		4-heptadecyl-7-hydroxycoumarin	100 mg	65.00	52.00	14.3, 23.4
H-96030	**New**	hernandezine	5 mg	85.00	68.00	18.8
H-96061	**New**	heteratisine	10 mg	128.00	102.40	18.7
H-96053	**New**	hetisine, hydrochloride	10 mg	128.00	102.40	18.7
H-193		5-hexadecanoylaminoeosin	100 mg	65.00	52.00	14.3
H-110		5-hexadecanoylaminofluorescein	100 mg	65.00	52.00	14.3
H-7523	**New**	DL-*threo*-2-hexadecanoylamino-3-morpholino-1-phenyl-1-propanol, hydrochloride (PPMP)	5 mg	48.00	38.40	20.6
H-3790		1-hexadecanoyl-2-(3-perylenedodecanoyl)-*sn*-glycero-3-phosphocholine	1 mg	38.00	30.40	13.2
H-361		1-hexadecanoyl-2-(1-pyrenedecanoyl)-*sn*-glycero-3-phosphocholine (β-py-C_{10}-HPC)	1 mg	65.00	52.00	13.2, 20.5
H-3784		1-hexadecanoyl-2-(1-pyrenedecanoyl)-*sn*-glycero-3-phosphoethanolamine (β-py-C_{10}-HPE)	1 mg	65.00	52.00	13.2

Cat #	Product Name	Unit Size	Price Per Unit ($) 1–4 Units	Price Per Unit ($) 5–24 Units	Section #
H-3809	1-hexadecanoyl-2-(1-pyrenedecanoyl)-*sn*-glycero-3-phosphoglycerol, ammonium salt (β-py-C_{10}-PG)	1 mg	65.00	52.00	13.2
H-3810	1-hexadecanoyl-2-(1-pyrenedecanoyl)-*sn*-glycero-3-phosphomethanol, sodium salt (β-py-C_{10}-HPM)	1 mg	65.00	52.00	13.2, 20.5
H-3818	1-hexadecanoyl-2-(1-pyrenehexanoyl)-*sn*-glycero-3-phosphocholine (β-py-C_6-HPC)	1 mg	65.00	52.00	13.2, 20.5
H-3743	1-hexadecanoyl-2-(1-pyrenehexanoyl)-*sn*-glycero-3-phosphomethanol, sodium salt (β-py-C_6-HPM)	1 mg	65.00	52.00	13.2, 20.5
H-7708 ***New***	1-hexadecanoyl-2-(1-pyrenehexanoyl)-N-(2,4,6-trinitrophenyl)-*sn*-glycero-3-phosphoethanolamine, sodium salt (TNP β-py-C_6-HPE)	1 mg	85.00	68.00	20.5
H-1387	6-hexadecanoyl-2-(((2-(trimethylammonium)ethyl)methyl)amino)naphthalene chloride (patman)	5 mg	48.00	38.40	14.5
H-429	4-hexadecylamino-7-nitrobenz-2-oxa-1,3-diazole (NBD hexadecylamine)	25 mg	48.00	38.40	14.6
H-7018 ***New***	2′,4,5,6,7,7′-hexafluorofluorescein (HFF)	10 mg	75.00	60.00	23.3
H-7019 ***New***	2′,4,5,6,7,7′-hexafluorofluorescein diacetate (HFFDA)	5 mg	125.00	100.00	15.2, 16.3
H-3113	5-(N,N-hexamethylene)amiloride, hydrochloride	5 mg	70.00	56.00	18.7
H-379	1,1′,3,3,3′,3′-hexamethylindocarbocyanine iodide (DiIC_1(3))	100 mg	35.00	28.00	25.3
H-7524 ***New***	hexanoyl ceramide; C_6-ceramide (N-hexanoyl-D-erythro-sphingosine)	5 mg	95.00	76.00	20.6
H-7524 ***New***	N-hexanoyl-D-erythro-sphingosine (hexanoyl ceramide; C_6-ceramide)	5 mg	95.00	76.00	20.6
H-7593 ***New***	hexidium iodide	5 mg	95.00	76.00	8.1, 16.3
H-96062 ***New***	hippeastrine, hydrobromide	20 mg	95.00	76.00	18.8
H-3409	histamine fluorescein, disodium salt	5 mg	68.00	54.40	18.2
H-2930	HIV Protease Substrate 1 (Arg-Glu(EDANS)-Ser-Gln-Asn-Tyr-Pro-Ile-Val-Gln-Lys (DABCYL)-Arg)	1 mg	150.00	120.00	10.4
B-6947 ***New***	Hoechst 2495 (benzoxanthene yellow)	100 mg	18.00	14.40	9.2
H-3569	Hoechst 33258 (bis-benzimide) *10 mg/mL solution in water*	10 mL	40.00	32.00	8.1, 16.3
H-1398	Hoechst 33258, pentahydrate (bis-benzimide)	100 mg	20.00	16.00	8.1, 16.3
H-3570	Hoechst 33342 *10 mg/mL solution in water*	10 mL	40.00	32.00	8.1, 16.3
H-1399	Hoechst 33342, trihydrochloride, trihydrate	100 mg	20.00	16.00	8.1, 16.3
N-6946 ***New***	Hoechst S769121 (nuclear yellow)	100 mg	48.00	38.40	15.3
	horseradish peroxidase (HRP) — products are listed under peroxidase from horseradish; see also Section 15.7.				
H-348	HPTS; pyranine (8-hydroxypyrene-1,3,6-trisulfonic acid, trisodium salt)	1 g	72.00	57.60	15.3, 17.4, 23.2
	Hydroethidine™ — a structurally identical product is listed under dihydroethidium (D-1168, see Sections 8.1, 17.1 and 21.4).				
H-7895 ***New***	2-hydroxyacetophenone oxime (FotoFenton™ 2)	1 g	35.00	28.00	21.2
H-7910 ***New***	N^G-hydroxy-L-arginine, monoacetate	5 mg	48.00	38.40	21.5
H-185	7-hydroxycoumarin-3-carboxylic acid *reference standard*	100 mg	95.00	76.00	1.7, 10.1
H-1103	7-hydroxycoumarin-3-carboxylic acid, succinimidyl ester	25 mg	78.00	62.40	1.7
H-6482 ***New***	7-hydroxy-9H-(1,3-dichloro-9,9-dimethylacridin-2-one) (DDAO) *reference standard*	10 mg	45.00	36.00	10.1
H-7492 ***New***	N-(2-hydroxyethyl)-*cis,cis,cis,cis*-5,8,11,14-eicosatetraenoic acid amide (anandamide) *special packaging*	10x1 mg	48.00	38.40	20.6
H-189	7-hydroxy-4-methylcoumarin *reference standard*	1 g	12.00	9.60	10.1
H-1428	7-hydroxy-4-methylcoumarin-3-acetic acid	100 mg	60.00	48.00	1.7
H-1411	7-hydroxy-4-methylcoumarin-3-acetic acid, succinimidyl ester	25 mg	78.00	62.40	1.7, 9.3
H-1426	7-hydroxy-4-methylcoumarin-8-methyleneiminodiacetic acid (calcein blue)	1 g	20.00	16.00	15.3, 22.7
H-6883 ***New***	p-hydroxyphenethamil, hydrochloride	5 mg	70.00	56.00	18.7
H-6829 ***New***	2-hydroxyphenyl-1-(2-nitrophenyl)ethyl phosphate, sodium salt (NPE-caged proton)	25 mg	95.00	76.00	19.2, 23.2
H-1586	3-(4-hydroxyphenyl)propionic acid, succinimidyl ester (Bolton-Hunter reagent)	1 g	38.00	30.40	1.8, 9.3
H-348	8-hydroxypyrene-1,3,6-trisulfonic acid, trisodium salt (HPTS; pyranine)	1 g	72.00	57.60	15.3, 17.4, 23.2
H-7599 ***New***	hydroxystilbamidine, methanesulfonate	10 mg	75.00	60.00	8.1, 10.6, 12.3 15.3
H-2249	N-hydroxysulfosuccinimide, sodium salt (NHSS)	100 mg	25.00	20.00	3.3, 5.2
T-659	7-hydroxy-4-trifluoromethylcoumarin (β-trifluoromethylumbelliferone) *reference standard*	100 mg	18.00	14.40	10.1
H-7476 ***New***	hypericin	1 mg	48.00	38.40	20.3, 21.2
H-7515 ***New***	hypocrellin A	1 mg	48.00	38.40	20.3, 21.2
H-7516 ***New***	hypocrellin B	1 mg	48.00	38.40	20.3, 21.2
I-7	IAANS (2-(4′-(iodoacetamido)anilino)naphthalene-6-sulfonic acid, sodium salt)	100 mg	95.00	76.00	2.3
I-14	1,5-IAEDANS (5-((((2-iodoacetyl)amino)ethyl)amino)naphthalene-1-sulfonic acid)	100 mg	35.00	28.00	2.3, 9.4
I-3	5-IAF (5-iodoacetamidofluorescein)	100 mg	138.00	110.40	2.2, 9.4
D-2004	IANBD amide (N,N′-dimethyl-N-(iodoacetyl)-N′-(7-nitrobenz-2-oxa-1,3-diazol-4-yl)ethylenediamine)	25 mg	95.00	76.00	2.3
I-9	IANBD ester (N-((2-(iodoacetoxy)ethyl)-N-methyl)amino-7-nitrobenz-2-oxa-1,3-diazole)	100 mg	95.00	76.00	2.3, 9.4
I-98211 ***New***	iberiotoxin *synthetic, originally from *Buthus tamulus* scorpion venom*	10 µg	85.00	68.00	18.6

Cat #		Product Name	Unit Size	Price Per Unit ($) 1–4 Units	5–24 Units	Section #
I-3881		I-FABP fatty acid binding protein	1 mg	330.00	264.00	13.3, 20.6
I-2904		ImaGene Green™ C_{12}FDG *lacZ* Gene Expression Kit	1 kit	225.00	180.00	10.2
I-2908		ImaGene Green™ C_{12}FDGlcU GUS Gene Expression Kit	1 kit	225.00	180.00	10.2
I-2906		ImaGene Red™ C_{12}RG *lacZ* Gene Expression Kit	1 kit	195.00	156.00	10.2
I-2910		ImaGene Red™ C_{12}RGlcU GUS Gene Expression Kit	1 kit	195.00	156.00	10.2
		imaging standards — see Section 26.1 for a complete list.				
I-7922	**New**	N^6-(1-iminoethyl)-L-lysine, hydrochloride (L-NIL)	5 mg	85.00	68.00	21.5
I-7899	**New**	N^5-(1-iminoethyl)-L-ornithine, hydrochloride (L-NIO)	5 mg	85.00	68.00	21.5
I-2553		2-iminothiolane, hydrochloride	100 mg	15.00	12.00	5.2
I-96008	**New**	imperialine (raddeamine; sipeimine)	5 mg	135.00	108.00	18.2
I-96050	**New**	imperialone	1 mg	135.00	108.00	18.2
I-1203		indo-1, AM *cell permeant*	1 mg	90.00	72.00	22.2
I-1223		indo-1, AM *cell permeant* *special packaging*	20x50 µg	128.00	102.40	22.2
I-1226		indo-1, AM *1 mM solution in dry DMSO* *cell permeant* *special packaging*	1 mL	115.00	92.00	22.2
I-1202		indo-1, pentapotassium salt *cell impermeant* *≥95% by HPLC*	1 mg	70.00	56.00	22.2
I-3032		indo dextran, potassium salt, 10,000 MW, anionic	5 mg	248.00	198.40	22.4
I-3033		indo dextran, potassium salt, 70,000 MW, anionic	5 mg	195.00	156.00	22.4
I-6614	**New**	Indolyl β-D-Galactopyranoside Sampler Kit *contains 5 mg each of B-1690, B-8407, I-8414, I-8420, M-8421*	1 kit	65.00	52.00	10.2
I-8414	**New**	3-indoxyl β-D-galactopyranoside	100 mg	85.00	68.00	10.2
I-8415	**New**	3-indoxyl β-D-glucopyranoside, trihydrate	100 mg	35.00	28.00	10.2
I-8416	**New**	3-indoxyl β-D-glucuronide, cyclohexylammonium salt	25 mg	38.00	30.40	10.2
I-8417	**New**	3-indoxyl phosphate, di(2-amino-2-methyl-1,3-propanediol) salt	1 g	75.00	60.00	10.3
I-8418	**New**	3-indoxyl phosphate, disodium salt	1 g	75.00	60.00	10.3
I-8419	**New**	3-indoxyl phosphate, p-toluidine salt	1 g	75.00	60.00	10.3
I-3716		D-*myo*-inositol 1,4,5-triphosphate, hexapotassium salt (1,4,5-IP$_3$)	1 mg	90.00	72.00	20.2
I-7221	**New**	InSpeck™ Blue (350/440) Microscope Image Intensity Calibration Kit	1 kit	80.00	64.00	26.1
I-7225	**New**	InSpeck™ Deep Red (633/660) Microscope Image Intensity Calibration Kit	1 kit	80.00	64.00	26.1
I-7219	**New**	InSpeck™ Green (505/515) Microscope Image Intensity Calibration Kit	1 kit	80.00	64.00	26.1
I-7223	**New**	InSpeck™ Orange (540/560) Microscope Image Intensity Calibration Kit	1 kit	80.00	64.00	26.1
I-7224	**New**	InSpeck™ Red (580/605) Microscope Image Intensity Calibration Kit	1 kit	80.00	64.00	26.1
I-6496	**New**	INT; 2-(4-iodophenyl)-3-(4-nitrophenyl)-5-phenyltetrazolium chloride (4-iodonitrotetrazolium violet)	1 g	40.00	32.00	16.3, 21.4
I-7		2-(4′-(iodoacetamido)anilino)naphthalene-6-sulfonic acid, sodium salt (IAANS)	100 mg	95.00	76.00	2.3
I-3		5-iodoacetamidofluorescein (5-IAF)	100 mg	138.00	110.40	2.2, 9.4
I-15		6-iodoacetamidofluorescein (6-IAF)	100 mg	138.00	110.40	2.2
I-106		4-iodoacetamidosalicylic acid	100 mg	48.00	38.40	2.4
I-6001	**New**	iodoacetic acid, succinimidyl ester (succinimidyl iodoacetate)	100 mg	65.00	52.00	5.2
I-9		N-((2-(iodoacetoxy)ethyl)-N-methyl)amino-7-nitrobenz-2-oxa-1,3-diazole (IANBD ester)	100 mg	95.00	76.00	2.3, 9.4
I-2953		3-iodoacetylamino-3-deoxydigoxigenin	5 mg	95.00	76.00	4.2
I-14		5-((((2-iodoacetyl)amino)ethyl)amino)naphthalene-1-sulfonic acid (1,5-IAEDANS)	100 mg	35.00	28.00	2.3, 9.4
I-591		N-(4-(iodoacetyl)amino)-2,2,6,6-tetramethylpiperidinyl-1-oxyl (TEMPO iodoacetamide)	25 mg	55.00	44.00	2.4
I-6881	**New**	6-iodoamiloride, hydrochloride, dihydrate	5 mg	70.00	56.00	18.7
I-1536		N-(2-iodoethyl)trifluoroacetamide	1 g	35.00	28.00	2.4, 9.3
I-8420	**New**	5-iodo-3-indolyl β-D-galactopyranoside	25 mg	48.00	38.40	10.2
I-6496	**New**	4-iodonitrotetrazolium violet (INT; 2-(4-iodophenyl)-3-(4-nitrophenyl)-5-phenyltetrazolium chloride)	1 g	40.00	32.00	16.3, 21.4
I-3716		1,4,5-IP$_3$ (D-*myo*-inositol 1,4,5-triphosphate, hexapotassium salt)	1 mg	90.00	72.00	20.2
I-99006	**New**	isobicyphat	1 mg	205.00	164.00	18.5
I-3110		5-(N-isobutyl-N-methyl)amiloride, hydrochloride	5 mg	35.00	28.00	18.7
I-1195		isoluminol isothiocyanate (ILITC)	100 mg	48.00	38.40	1.8
I-6002	**New**	N-isopropyl iodoacetamide (NIPIA)	100 mg	45.00	36.00	2.4, 9.3
I-6880	**New**	5-(N-isopropyl-N-methyl)amiloride, hydrochloride	5 mg	70.00	56.00	18.7
I-6621	**New**	isopropyl β-D-thiogalactopyranoside (IPTG) *dioxane free*	1 g	24.00	19.20	10.6
A-96003	**New**	isopsoralen (angelicine)	20 mg	55.00	44.00	8.1
I-6958	**New**	isosulfan blue *>90% by HPLC*	1 g	28.00	22.40	15.3, 21.2
I-6204	**New**	isosulfan blue sulfonyl chloride *mixed isomers*	5 mg	78.00	62.40	1.6, 21.2
I-6076	**New**	1-(3-isothiocyanatophenyl)-4-(5-(4-methoxyphenyl)oxazol-2-yl)pyridinium bromide (PyMPO, ITC)	5 mg	98.00	78.40	1.7
P-6542	**New**	iso-valeryl-L-Val-L-Val-Sta-L-Ala-Sta (pepstatin A)	25 mg	55.00	44.00	10.6
J-7473	**New**	jasplakinolide	100 µg	198.00	158.40	11.1
T-3168		JC-1; CBIC$_2$(3) (5,5′,6,6′-tetrachloro-1,1′,3,3′-tetraethylbenzimidazolylcarbocyanine iodide)	5 mg	195.00	156.00	12.2, 16.3, 25.3
C-6170	**New**	6-JOE (6-carboxy-4′,5′-dichloro-2′,7′-dimethoxyfluorescein)	5 mg	75.00	60.00	1.5
C-6171	**New**	6-JOE, SE (6-carboxy-4′,5′-dichloro-2′,7′-dimethoxyfluorescein, succinimidyl ester)	5 mg	125.00	100.00	1.5

Cat #		Product Name	Unit Size	Price Per Unit ($) 1–4 Units	5–24 Units	Section #
D-6923	New	JPW 1114 (di-2-ANEPEQ)	5 mg	175.00	140.00	25.2
K-7115	New	kainic acid, γ-(α-carboxy-2-nitrobenzyl) ester, hydrochloride (γ-(CNB-caged) kainic acid)	1 mg	48.00	38.40	18.3, 19.2
K-98208	New	kaliotoxin *synthetic, originally from *Androctonus mauretanicus mauretanicus* scorpion venom* *special packaging*	2x10 µg	128.00	102.40	18.6
K-98209	New	kaliotoxin (1-37) amide *special packaging*	2x10 µg	108.00	86.40	18.6
K-96034	New	karakoline	20 mg	75.00	60.00	18.2
K-96063	New	katacine *mixture of polymers of proanthocyanidins*	100 mg	195.00	156.00	18.8
L-838		lactalbumin, fluorescein conjugate	25 mg	75.00	60.00	15.7
L-96065	New	lapidine	10 mg	128.00	102.40	18.8
L-96066	New	lapiferine	10 mg	128.00	102.40	18.8
L-96010	New	lappaconitine, hydrobromide (allapinine)	10 mg	155.00	124.00	18.7
		latex — see Section 6.2 for a complete list of FluoSpheres® and TransFluoSpheres® polystyrene microspheres.				
L-98112	New	laticauda semifasciata III (LS III) *from *Laticauda semifasciata* snake venom*	1 mg	185.00	148.00	18.2
D-250		laurdan (6-dodecanoyl-2-dimethylaminonaphthalene)	100 mg	135.00	108.00	14.5
L-3486		LDL (low-density lipoprotein from human plasma) *2.5 mg/mL*	200 µL	80.00	64.00	17.3
L-7595	New	LDS 751	10 mg	65.00	52.00	8.1, 16.3
L-98202	New	leiurotoxin I (scyllatoxin) *synthetic, originally from *Leiurus quinquestriatus* scorpion venom*	50 µg	385.00	308.00	18.6
L-96067	New	leucomisine	100 mg	148.00	118.40	18.8
L-6543	New	leupeptin hemisulfate	10 mg	45.00	36.00	10.6
L-7308	New	LinearFlow™ Carmine (580/620) Flow Cytometry Intensity Calibration Kit	1 kit	140.00	112.00	26.2
L-7314	New	LinearFlow™ Deep Red (633/660) Flow Cytometry Intensity Calibration Kit	1 kit	140.00	112.00	26.2
L-7306	New	LinearFlow™ Green (505/515) Flow Cytometry Intensity Calibration Kit	1 kit	140.00	112.00	26.2
L-7307	New	LinearFlow™ Orange (540/560) Flow Cytometry Intensity Calibration Kit	1 kit	140.00	112.00	26.2
L-7781	New	Lipophilic Tracer Sampler Kit	1 kit	150.00	120.00	14.2, 15.4
L-2424		Lissamine™ rhodamine B ethylenediamine	10 mg	115.00	92.00	3.3
L-20		Lissamine™ rhodamine B sulfonyl chloride *mixed isomers*	1 g	135.00	108.00	1.6
L-1908		Lissamine™ rhodamine B sulfonyl chloride *mixed isomers* *special packaging*	10x10 mg	65.00	52.00	1.6
L-1392		N-(Lissamine™ rhodamine B sulfonyl)-1,2-dihexadecanoyl-*sn*-glycero-3-phosphoethanolamine, triethylammonium salt (rhodamine DHPE)	5 mg	75.00	60.00	13.4
D-7792	New	liumogen (2,2′-dihydroxy-1,1′-naphthaldiazine)	100 mg	65.00	52.00	26.1
L-7008	New	LIVE *Bac*Light™ Bacterial Gram Stain Kit *for microscopy* *1000 assays*	1 kit	195.00	156.00	16.2
L-7005	New	LIVE *Bac*Light™ Bacterial Gram Stain Kit *for microscopy and quantitative assays* *1000 assays*	1 kit	195.00	156.00	16.2
L-7007	New	LIVE/DEAD® *Bac*Light™ Bacterial Viability Kit *for microscopy* *1000 assays*	1 kit	195.00	156.00	16.2
L-7012	New	LIVE/DEAD® *Bac*Light™ Bacterial Viability Kit *for microscopy and quantitative assays* *1000 assays*	1 kit	195.00	156.00	16.2
L-7010	New	LIVE/DEAD® Cell-Mediated Cytotoxicity Kit *for animal cells* *2000 assays*	1 kit	95.00	76.00	16.2
L-7013	New	LIVE/DEAD® Reduced Biohazard Viability/Cytotoxicity Kit *for animal cells* *100 assays*	1 kit	95.00	76.00	16.2
L-7011	New	LIVE/DEAD® Sperm Viability Kit *200-1000 assays*	1 kit	195.00	156.00	16.2
L-3224		LIVE/DEAD® Viability/Cytotoxicity Kit *for animal cells* *1000 assays*	1 kit	195.00	156.00	16.2
L-7009	New	LIVE/DEAD® Yeast Viability Kit *1000 assays*	1 kit	195.00	156.00	16.2
L-3486		low-density lipoprotein from human plasma (LDL) *2.5 mg/mL*	200 µL	80.00	64.00	17.3
L-3485		low-density lipoprotein from human plasma, acetylated, BODIPY® FL conjugate (BODIPY® FL AcLDL) *1 mg/mL*	200 µL	195.00	156.00	17.3
L-3484		low-density lipoprotein from human plasma, acetylated, DiI complex (DiI AcLDL) *1 mg/mL*	200 µL	195.00	156.00	17.3
L-3483		low-density lipoprotein from human plasma, BODIPY® FL complex (BODIPY® FL LDL) *1 mg/mL*	200 µL	175.00	140.00	17.3
L-3482		low-density lipoprotein from human plasma, DiI complex (DiI LDL) *1 mg/mL*	200 µL	175.00	140.00	17.3
L-98221	New	LqhαIT insect toxin *from *Leiurus quinquestriatus hebraeus* scorpion venom*	50 µg	415.00	332.00	18.7
L-98112	New	LS III (laticauda semifasciata III) *from *Laticauda semifasciata* snake venom*	1 mg	185.00	148.00	18.2
L-98217	New	LTx-1 *from *Leiurus quinquestriatus hebraeus* scorpion venom*	50 µg	325.00	260.00	18.7
L-98218	New	LTx-2 *from *Leiurus quinquestriatus hebraeus* scorpion venom*	50 µg	325.00	260.00	18.7
L-98219	New	LTx-3 *from *Leiurus quinquestriatus hebraeus* scorpion venom*	50 µg	415.00	332.00	18.7
L-98220	New	LTx-4 *from *Leiurus quinquestriatus hebraeus* scorpion venom*	100 µg	525.00	420.00	18.7
L-7105	New	luciferase, firefly *recombinant* *in solution*	1 mg	80.00	64.00	10.5
L-7085	New	D-luciferin, 1-(4,5-dimethoxy-2-nitrophenyl)ethyl ester (DMNPE-caged luciferin)	5 mg	65.00	52.00	10.5, 19.2
L-2911		D-luciferin, free acid	25 mg	80.00	64.00	10.5
L-6619	New	luciferin methyl ether	5 mg	85.00	68.00	10.5
L-2916		D-luciferin, potassium salt	25 mg	88.00	70.40	10.5
L-2912		D-luciferin, sodium salt	25 mg	88.00	70.40	10.5
A-629		lucifer yellow AB (N-(((4-azidobenzoyl)amino)ethyl)-4-amino-3,6-disulfo-1,8-naphthalimide, dipotassium salt)	10 mg	95.00	76.00	5.3, 9.4, 15.3
L-6950	New	lucifer yellow biocytin, potassium salt, fixable (biocytin lucifer yellow)	5 mg	95.00	76.00	4.3, 15.3

Cat #		Product Name	Unit Size	Price Per Unit ($) 1–4 Units	5–24 Units	Section #
A-1340		lucifer yellow cadaverine (N-(5-aminopentyl)-4-amino-3,6-disulfo-1,8-naphthalimide, dipotassium salt)	25 mg	70.00	56.00	3.3, 15.3
L-2601		lucifer yellow cadaverine biotin-X, dipotassium salt	10 mg	95.00	76.00	4.3, 15.3
L-682		lucifer yellow CH, ammonium salt	25 mg	45.00	36.00	15.3
L-453		lucifer yellow CH, lithium salt	25 mg	48.00	38.40	3.2, 9.4, 15.3, 17.4
L-1177		lucifer yellow CH, potassium salt	25 mg	45.00	36.00	15.3
A-1339		lucifer yellow ethylenediamine (N-(2-aminoethyl)-4-amino-3,6-disulfo-1,8-naphthalimide, dipotassium salt)	25 mg	38.00	30.40	3.3, 15.3
L-1338		lucifer yellow iodoacetamide, dipotassium salt	25 mg	95.00	76.00	2.4, 4.2, 9.4, 15.3
L-6868	***New***	lucigenin (bis-N-methylacridinium nitrate) *high purity*	10 mg	15.00	12.00	12.2, 17.1, 21.3, 24.2
L-8455	***New***	luminol (3-aminophthalhydrazide)	25 g	75.00	60.00	10.5, 17.1, 21.3
L-96221	***New***	lycaconitine perchlorate	10 mg	128.00	102.40	18.2
L-96011	***New***	lycorine, hydrochloride	20 mg	75.00	60.00	18.8
L-7533	***New***	LysoSensor™ Blue DND-167 *1 mM solution in DMSO* *special packaging*	20x50 µL	125.00	100.00	12.3, 23.3
L-7532	***New***	LysoSensor™ Blue DND-192 *1 mM solution in DMSO* *special packaging*	20x50 µL	125.00	100.00	12.3, 23.3
L-7534	***New***	LysoSensor™ Green DND-153 *1 mM solution in DMSO* *special packaging*	20x50 µL	125.00	100.00	12.3, 23.3
L-7535	***New***	LysoSensor™ Green DND-189 *1 mM solution in DMSO* *special packaging*	20x50 µL	125.00	100.00	12.3, 23.3
L-7545	***New***	LysoSensor™ Yellow/Blue DND-160 *1 mM solution in DMSO* *special packaging*	20x50 µL	125.00	100.00	12.3, 23.3
L-7537	***New***	LysoTracker™ biotin *4 mM solution in DMSO* *special packaging*	10x50 µL	125.00	100.00	4.3, 12.3
L-7538	***New***	LysoTracker™ biotin/DNP *4 mM solution in DMSO* *special packaging*	10x50 µL	125.00	100.00	4.3, 12.3
L-7525	***New***	LysoTracker™ Blue DND-22 *1 mM solution in DMSO* *special packaging*	20x50 µL	125.00	100.00	12.3
L-7542	***New***	LysoTracker™ Green Br_2 *1 mM solution in DMSO* *special packaging*	20x50 µL	125.00	100.00	12.3
L-7526	***New***	LysoTracker™ Green DND-26 *1 mM solution in DMSO* *special packaging*	20x50 µL	125.00	100.00	12.3
L-7528	***New***	LysoTracker™ Red DND-99 *1 mM solution in DMSO* *special packaging*	20x50 µL	125.00	100.00	12.3
L-7527	***New***	LysoTracker™ Yellow DND-68 *1 mM solution in DMSO* *special packaging*	20x50 µL	125.00	100.00	12.3
M-1291		mag-fura-2, AM *cell permeant* *≥95% by HPLC*	1 mg	168.00	134.40	22.2, 22.6
M-1292		mag-fura-2, AM *cell permeant* *≥95% by HPLC* *special packaging*	20x50 µg	198.00	158.40	22.2, 22.6
M-1290		mag-fura-2, tetrapotassium salt *cell impermeant* *≥95% by HPLC*	1 mg	148.00	118.40	22.2, 22.6
M-3105		mag-fura-5, AM *cell permeant* *≥95% by HPLC* *special packaging*	20x50 µg	198.00	158.40	22.2, 22.6
M-3103		mag-fura-5, tetrapotassium salt *cell impermeant* *≥95% by HPLC*	1 mg	148.00	118.40	22.2, 22.6
M-6893	***New***	Mag-Fura Red™, AM *cell permeant* *≥95% by HPLC* *special packaging*	20x50 µg	165.00	132.00	22.6
M-6892	***New***	Mag-Fura Red™, tripotassium salt *cell impermeant* *≥95% by HPLC*	1 mg	125.00	100.00	22.6
M-1295		mag-indo-1, AM *cell permeant* *≥95% by HPLC* *special packaging*	20x50 µg	198.00	158.40	22.2, 22.6
M-1293		mag-indo-1, tetrapotassium salt *cell impermeant*	1 mg	148.00	118.40	22.2, 22.6
M-6907	***New***	mag-indo dextran, potassium salt, 10,000 MW, anionic	5 mg	148.00	118.40	22.4, 22.6
M-6908	***New***	mag-indo dextran, potassium salt, 70,000 MW, anionic	5 mg	148.00	118.40	22.4, 22.6
M-3120		Magnesium Calibration Standard Kit *zero and 35 mM Mg^{2+} (2 x 50 mL)*	1 kit	48.00	38.40	22.8
M-3735		Magnesium Green™, AM *cell permeant* *≥95% by HPLC* *special packaging*	20x50 µg	165.00	132.00	22.3, 22.6
M-3733		Magnesium Green™, pentapotassium salt *cell impermeant*	1 mg	100.00	80.00	22.3, 22.6
M-6891	***New***	Magnesium Orange™, AM *cell permeant* *≥95% by HPLC* *special packaging*	20x50 µg	155.00	124.00	22.6
M-6890	***New***	Magnesium Orange™, tripotassium salt *cell impermeant*	1 mg	95.00	76.00	22.6
M-689		malachite green isothiocyanate	10 mg	135.00	108.00	1.6, 21.2
M-8		2-(4′-maleimidylanilino)naphthalene-6-sulfonic acid, sodium salt (MIANS)	100 mg	95.00	76.00	2.3
M-6026	***New***	1-(2-maleimidylethyl)-4-(5-(4-methoxyphenyl)oxazol-2-yl)pyridinium methanesulfonate (PyMPO maleimide)	5 mg	98.00	78.40	2.2
M-1618		N-((4-maleimidylmethyl)cyclohexane-1-carbonyl)-1,2-dihexadecanoyl-*sn*-glycero-3-phosphoethanolamine, triethylammonium salt (MMCC DHPE)	5 mg	135.00	108.00	5.2, 13.4
M-1602		3-(N-maleimidylpropionyl)biocytin	25 mg	68.00	54.40	4.2, 15.3
M-84		4-maleimidylsalicylic acid	100 mg	48.00	38.40	2.4
M-405		mansyl chloride (6-(N-methylanilino)naphthalene-2-sulfonyl chloride)	25 mg	135.00	108.00	1.7
		MAPTAM — see BAPTA (B-1204); see also Section 22.8.				
M-1378		mBBr (monobromobimane)	25 mg	43.00	34.40	2.4, 9.4, 15.2, 24.3
M-1381		mBCl (monochlorobimane)	25 mg	78.00	62.40	2.4, 15.2, 16.3
M-98402	***New***	MCD peptide *from *Apis mellifera* bee venom* *special packaging*	4x50 µg	108.00	86.40	18.6
M-99003	***New***	mebicyphat	1 mg	135.00	108.00	18.5
M-98403	***New***	melittin *from *Apis mellifera* bee venom*	1 mg	45.00	36.00	10.6, 20.3
M-6886	***New***	MEQ (6-methoxy-N-ethylquinolinium iodide)	100 mg	45.00	36.00	24.2
M-7904	***New***	2-mercaptopyridine N-oxide, sodium salt (FotoFenton™ 1)	100 mg	35.00	28.00	21.2
M-8112	***New***	mercury vapor lamp for TD-700 fluorometer, ultraviolet range *coumarin excitation*	each	45.00	45.00	26.3
M-8111	***New***	mercury vapor lamp for TD-700 fluorometer, visible range *rhodamine excitation*	each	55.00	55.00	26.3
M-299		merocyanine 540	100 mg	30.00	24.00	21.2, 25.3
M-7890	***New***	Methods in Cell Biology, Volume 40: A Practical Guide to the Study of Calcium in Living Cells. R. Nuccitelli, ed. Academic Press (1994); 342 pages, comb bound	each	45.00	45.00	22.1, 26.6

Cat #		Product Name	Unit Size	Price Per Unit ($)		Section #
				1–4 Units	5–24 Units	
M-7887	**New**	Methods in Cell Biology, Volume 41: Flow Cytometry, Part A. Z. Darzynkiewicz, J.P. Robinson, H.A. Crissman, eds. Academic Press (1994); 591 pages, comb bound	each	59.00	59.00	22.1, 26.6
M-7888	**New**	Methods in Cell Biology, Volume 42: Flow Cytometry, Part B. Z. Darzynkiewicz, J.P. Robinson, H.A. Crissman, eds. Academic Press (1994); 697 pages, comb bound	each	59.00	59.00	26.6
M-1198		methotrexate, fluorescein, triammonium salt (fluorescein methotrexate)	1 mg	125.00	100.00	16.3
M-6884	**New**	4′-methoxybenzamil, hydrochloride	5 mg	70.00	56.00	18.7
M-1445		7-methoxycoumarin-3-carbonyl azide	25 mg	88.00	70.40	3.1
M-1420		7-methoxycoumarin-3-carboxylic acid	100 mg	48.00	38.40	1.7
M-1410		7-methoxycoumarin-3-carboxylic acid, succinimidyl ester	25 mg	65.00	52.00	1.7
M-6886	**New**	6-methoxy-N-ethylquinolinium iodide (MEQ)	100 mg	45.00	36.00	24.2
M-440		6-methoxy-N-(3-sulfopropyl)quinolinium, inner salt (SPQ)	100 mg	63.00	50.40	24.2
M-395		8-methoxypyrene-1,3,6-trisulfonic acid, trisodium salt (MPTS)	100 mg	90.00	72.00	15.3
M-6205	**New**	8-methoxypyrene-1,3,6-trisulfonyl chloride	10 mg	45.00	36.00	27.1
M-6795	**New**	methoxyquin MF, AM *cell permeant*	5 mg	95.00	76.00	22.2
M-6794	**New**	methoxyquin MF, free acid *cell impermeant*	10 mg	135.00	108.00	22.2
M-688		N-(6-methoxy-8-quinolyl)-p-toluenesulfonamide (TSQ)	25 mg	85.00	68.00	22.7
R-351		methoxyresorufin (resorufin methyl ether)	5 mg	98.00	78.40	10.5
M-7913	**New**	*trans*-1-(2′-methoxyvinyl)pyrene	1 mg	75.00	60.00	21.3
M-405		6-(N-methylanilino)naphthalene-2-sulfonyl chloride (mansyl chloride)	25 mg	135.00	108.00	1.7
M-1035		2′-(N-methylanthraniloyl)guanosine 3′,5′-cyclicmonophosphate, sodium salt (MANT-cGMP)	10 mg	95.00	76.00	10.3, 20.4
M-7906	**New**	N^G-methyl-D-arginine, acetate salt (D-NMMA)	25 mg	55.00	44.00	21.5
M-7898	**New**	N^G-methyl-L-arginine, acetate salt (L-NMMA)	25 mg	45.00	36.00	21.5
M-7114	**New**	N-methyl-D-aspartic acid, β-(α-carboxy-2-nitrobenzyl) ester, trifluoroacetic acid salt (β-(CNB-caged) NMDA)	1 mg	48.00	38.40	18.3, 19.2
M-8456	**New**	methyl β-D-glucuronide, sodium salt	100 mg	35.00	28.00	10.6
M-8421	**New**	N-methyl-3-indolyl β-D-galactopyranoside	25 mg	48.00	38.40	10.2
M-25		N-methylisatoic anhydride *high purity*	1 g	75.00	60.00	1.8, 3.1
M-96204	**New**	*o*-methylisocorydine, methiodide	5 mg	95.00	76.00	18.2
M-96201	**New**	methyllycaconitine perchlorate (MLA)	5 mg	35.00	28.00	18.2
M-6539	**New**	methyl methanethiolsulfonate (MMTS) *special packaging*	10x10 mg	35.00	28.00	2.4, 10.6
M-1119		12-(N-methyl)-N-((7-nitrobenz-2-oxa-1,3-diazol-4-yl)amino)octadecanoic acid (12-NBD stearic acid)	5 mg	65.00	52.00	13.2
M-7069	**New**	N-methyl-N-(2-nitrobenzyl)-D-aspartic acid, dilithium salt (N-(NB-caged) NMDA)	5 mg	118.00	94.40	18.3, 19.2
M-7905	**New**	2-methyl-2-nitrosopropane dimer	1 g	45.00	36.00	21.3
M-7914	**New**	S-methyl-L-thiocitrulline chloride, hydrochloride	5 mg	95.00	76.00	21.5
M-1489		4-methylumbelliferyl β-D-galactopyranoside (MUG)	1 g	25.00	20.00	10.2
M-8422	**New**	4-methylumbelliferyl α-D-glucopyranoside	100 mg	38.00	30.40	10.2
M-8423	**New**	4-methylumbelliferyl β-D-glucopyranoside (MUGlu)	1 g	38.00	30.40	10.2
M-1490		4-methylumbelliferyl β-D-glucuronide (MUGlcU)	100 mg	48.00	38.40	10.2
M-8424	**New**	4-methylumbelliferyl palmitate	100 mg	20.00	16.00	10.5
M-8425	**New**	4-methylumbelliferyl phosphate, dicyclohexylammonium salt, trihydrate	1 g	50.00	40.00	10.3
M-6491	**New**	4-methylumbelliferyl phosphate, free acid	1 g	38.00	30.40	10.3
M-2885		4-methylumbelliferyl phosphocholine	10 mg	85.00	68.00	20.5
M-8		MIANS (2-(4′-maleimidylanilino)naphthalene-6-sulfonic acid, sodium salt)	100 mg	95.00	76.00	2.3
		microscopy standards — see Section 26.1 for a complete list.				
D-7156	**New**	micro-emerald (dextran, fluorescein and biotin, 3000 MW, anionic, lysine fixable)	5 mg	150.00	120.00	4.3, 15.5
D-7162	**New**	micro-ruby (dextran, tetramethylrhodamine and biotin, 3000 MW, lysine fixable)	5 mg	150.00	120.00	4.3, 15.5
D-7178	**New**	mini-emerald (dextran, fluorescein and biotin, 10,000 MW, anionic, lysine fixable)	10 mg	150.00	120.00	4.3, 15.5
D-3312		mini-ruby (dextran, tetramethylrhodamine and biotin, 10,000 MW, lysine fixable)	10 mg	150.00	120.00	4.3, 15.5
M-7502	**New**	MitoFluor™ Green	1 mg	98.00	78.40	12.2
M-7514	**New**	MitoTracker™ Green FM *special packaging*	20x50 µg	98.00	78.40	12.2
M-7511	**New**	MitoTracker™ Orange CM-H$_2$TMRos *special packaging*	20x50 µg	125.00	100.00	12.2, 21.4
M-7510	**New**	MitoTracker™ Orange CMTMRos *special packaging*	20x50 µg	98.00	78.40	12.2
M-7513	**New**	MitoTracker™ Red CM-H$_2$XRos *special packaging*	20x50 µg	125.00	100.00	12.2, 21.4
M-7512	**New**	MitoTracker™ Red CMXRos *special packaging*	20x50 µg	98.00	78.40	12.2
M-6246	**New**	mono-N-(t-BOC)-cadaverine	100 mg	35.00	28.00	3.3
M-6247	**New**	mono-N-(t-BOC)-ethylenediamine	1 g	35.00	28.00	3.3
M-6248	**New**	mono-N-(t-BOC)-propylenediamine	1 g	35.00	28.00	3.3
M-1378		monobromobimane (mBBr)	25 mg	43.00	34.40	2.4, 9.4, 15.2, 24.3
M-1380		monobromotrimethylammoniobimane bromide (qBBr)	25 mg	38.00	30.40	2.4, 9.4, 15.2
M-1381		monochlorobimane (mBCl)	25 mg	78.00	62.40	2.4, 15.2, 16.3
M-7891	**New**	3-morpholinosydnonimine, hydrochloride (SIN-1)	25 mg	45.00	36.00	21.5
M-7926	**New**	3-morpholinosydnonimine, hydrochloride (SIN-1) *special packaging*	20x1 mg	65.00	52.00	21.5

Cat #		Product Name	Unit Size	Price Per Unit ($)		Section #
				1–4 Units	5–24 Units	
C-3106		MQAA (N-(carboxymethyl)-6-methoxyquinolinium bromide)	100 mg	48.00	38.40	24.2
E-3101		MQAE (N-(ethoxycarbonylmethyl)-6-methoxyquinolinium bromide)	100 mg	78.00	62.40	24.2
M-6494	**New**	MTT (3-(4,5-dimethylthiazol-2-yl)-2,5-diphenyltetrazolium bromide)	1 g	40.00	32.00	16.3, 21.4
M-1489		MUG (4-methylumbelliferyl β-D-galactopyranoside)	1 g	25.00	20.00	10.2
M-1490		MUGlcU (4-methylumbelliferyl β-D-glucuronide)	100 mg	48.00	38.40	10.2
M-7901	**New**	MultiSpeck™ Multispectral Fluorescence Microscopy Standards Kit *in suspension*	1 kit	110.00	88.00	26.1
M-7900	**New**	MultiSpeck™ Multispectral Fluorescence Microscopy Standards Kit *prepared slides*	1 kit	125.00	100.00	26.1
M-7006	**New**	MycoFluor™ Mycoplasma Detection Kit	1 kit	95.00	76.00	16.2
N-7622	**New**	NAADP (β-nicotinic acid adenine dinucleotide phosphate)	10 µg	100.00	80.00	20.2
N-1384		naloxone fluorescein	5 mg	68.00	54.40	18.2
N-1385		naltrexone fluorescein	5 mg	68.00	54.40	18.2
N-7907	**New**	L-NAME (N^G-nitro-L-arginine methyl ester, hydrochloride)	1 g	9.00	7.20	21.5
N-6666	**New**	NanoOrange™ Protein Quantitation Kit *200-2000 assays*	1 kit	135.00	108.00	9.1
N-1138		naphthalene-2,3-dicarboxaldehyde (NDA)	100 mg	65.00	52.00	1.8, 16.3, 24.3
N-7897	**New**	naphthalene-1,4-dipropionic acid, disodium salt	25 mg	65.00	52.00	21.2
N-2461		2-(2,3-naphthalimino)ethyl trifluoromethanesulfonate	100 mg	53.00	42.40	3.4
N-650		naphthofluorescein	100 mg	18.00	14.40	10.1, 23.2
T-411		naphthoyltrifluoroacetone; NTA (2-((trifluoroacetyl)acetyl)naphthalene)	100 mg	65.00	52.00	15.3, 16.3, 22.7
C-10		NBD chloride; 4-chloro-7-nitrobenzofurazan (4-chloro-7-nitrobenz-2-oxa-1,3-diazole)	1 g	24.00	19.20	1.8, 2.3
F-486		NBD fluoride; 4-fluoro-7-nitrobenzofurazan (4-fluoro-7-nitrobenz-2-oxa-1,3-diazole)	25 mg	95.00	76.00	1.8, 2.3
N-1154		NBD C_6-ceramide (6-((N-(7-nitrobenz-2-oxa-1,3-diazol-4-yl)amino)hexanoyl)-sphingosine)	1 mg	125.00	100.00	12.4, 20.6
N-360		NBD-PE (N-(7-nitrobenz-2-oxa-1,3-diazol-4-yl)-1,2-dihexadecanoyl-*sn*-glycero-3-phosphoethanolamine, triethylammonium salt)	10 mg	75.00	60.00	13.4
N-354		NBD phallacidin (N-(7-nitrobenz-2-oxa-1,3-diazol-4-yl)phallacidin)	300 U	198.00	158.40	11.1
N-3413		NBD Ro-1986 (N-(7-nitrobenz-2-oxa-1,3-diazol-4-yl)desdiethylfluorazepam) *1 mg/mL solution in dry DMSO*	1 mL	48.00	38.40	18.5
N-3524		NBD C_6-sphingomyelin (6-((N-(7-nitrobenz-2-oxa-1,3-diazol-4-yl)amino)-hexanoyl)sphingosyl phosphocholine)	1 mg	125.00	100.00	12.4, 13.2, 20.6
M-1119		12-NBD stearic acid (12-(N-methyl)-N-((7-nitrobenz-2-oxa-1,3-diazol-4-yl)amino)-octadecanoic acid)	5 mg	65.00	52.00	13.2
N-340		NBD taurine (2-(N-(7-nitrobenz-2-oxa-1,3-diazol-4-yl)amino)ethanesulfonic acid, sodium salt)	10 mg	128.00	102.40	18.5
N-6495	**New**	NBT (nitro blue tetrazolium chloride)	1 g	65.00	52.00	9.2, 16.3, 21.4
N-6547	**New**	NBT/BCIP Reagent Kit	1 kit	135.00	108.00	9.2, 10.3
N-1138		NDA (naphthalene-2,3-dicarboxaldehyde)	100 mg	65.00	52.00	1.8, 16.3, 24.3
N-6919	**New**	neosaxitoxin (neo-STX) *250 µg/mL*	100 µL	200.00	160.00	18.7
G-6915	**New**	neosaxitoxin 11β-sulfate (gonyautoxin GTX-IV) *250 µg/mL*	100 µL	200.00	160.00	18.7
N-7167	**New**	NeuroTrace™ BDA-10,000 Neuronal Tracer Kit	1 kit	150.00	120.00	15.5
N-3246		neutral red *high purity*	25 mg	50.00	40.00	12.3, 16.3
N-7991	**New**	Newport Green™ diacetate *cell permeant*	1 mg	90.00	72.00	22.7
N-7990	**New**	Newport Green™, dipotassium salt *cell impermeant*	1 mg	80.00	64.00	22.7
H-2249		NHSS (N-hydroxysulfosuccinimide, sodium salt)	100 mg	25.00	20.00	3.3, 5.2
N-7622	**New**	β-nicotinic acid adenine dinucleotide phosphate (NAADP)	10 µg	100.00	80.00	20.2
N-7107	**New**	nigericin, 1-(4,5-dimethoxy-2-nitrophenyl)ethyl ester (DMNPE-caged nigericin)	1 mg	98.00	78.40	19.2, 23.2
N-1495		nigericin, free acid	10 mg	70.00	56.00	23.2, 24.2
I-7922	**New**	L-NIL (N^6-(1-iminoethyl)-L-lysine, hydrochloride)	5 mg	85.00	68.00	21.5
N-1142		nile red	25 mg	60.00	48.00	11.2, 14.5
N-96068	**New**	nitrarine, dihydrochloride	50 mg	95.00	76.00	18.7
N-7925	**New**	Nitric Oxide Synthase (NOS) Inhibitor Kit	1 kit	195.00	156.00	21.5
N-7907	**New**	N^G-nitro-L-arginine methyl ester, hydrochloride (L-NAME)	1 g	9.00	7.20	21.5
N-3007		5-nitro BAPTA, AM *cell permeant* *≥95% by HPLC*	5 mg	85.00	68.00	22.8
N-3006		5-nitro BAPTA, free acid *cell impermeant*	10 mg	73.00	58.40	22.8
N-1148		22-(N-(7-nitrobenz-2-oxa-1,3-diazol-4-yl)amino)-23,24-bisnor-5-cholen-3β-ol	10 mg	135.00	108.00	13.5
N-7709	**New**	22-(N-(7-nitrobenz-2-oxa-1,3-diazol-4-yl)amino)-23,24-bisnor-5-cholen-3β-yl *cis*-9-octadecenoate	5 mg	85.00	68.00	13.5
N-678		12-(N-(7-nitrobenz-2-oxa-1,3-diazol-4-yl)amino)dodecanoic acid	100 mg	148.00	118.40	13.2
N-3787		2-(12-(7-nitrobenz-2-oxa-1,3-diazol-4-yl)amino)dodecanoyl-1-hexadecanoyl-*sn*-glycero-3-phosphocholine (NBD-C_{12}-HPC)	5 mg	50.00	40.00	13.2, 20.5
N-340		2-(N-(7-nitrobenz-2-oxa-1,3-diazol-4-yl)amino)ethanesulfonic acid, sodium salt (NBD taurine)	10 mg	128.00	102.40	18.5
N-316		6-(N-(7-nitrobenz-2-oxa-1,3-diazol-4-yl)amino)hexanoic acid (NBD-X)	100 mg	50.00	40.00	1.8, 13.2
N-3786		2-(6-(7-nitrobenz-2-oxa-1,3-diazol-4-yl)amino)hexanoyl-1-hexadecanoyl-*sn*-glycero-3-phosphocholine (NBD-C_6-HPC)	5 mg	50.00	40.00	13.2, 20.5

Cat #		Product Name	Unit Size	Price Per Unit ($) 1–4 Units	5–24 Units	Section #
N-1154		6-((N-(7-nitrobenz-2-oxa-1,3-diazol-4-yl)amino)hexanoyl)sphingosine (NBD C_6-ceramide)	1 mg	125.00	100.00	12.4, 20.6
N-3524		6-((N-(7-nitrobenz-2-oxa-1,3-diazol-4-yl)amino)hexanoyl)sphingosyl phosphocholine (NBD C_6-sphingomyelin)	1 mg	125.00	100.00	12.4, 13.2, 20.6
N-3413		N-(7-nitrobenz-2-oxa-1,3-diazol-4-yl)desdiethylfluorazepam (NBD Ro-1986) *1 mg/mL solution in dry DMSO*	1 mL	48.00	38.40	18.5
N-360		N-(7-nitrobenz-2-oxa-1,3-diazol-4-yl)-1,2-dihexadecanoyl-*sn*-glycero-3-phosphoethanolamine, triethylammonium salt (NBD-PE)	10 mg	75.00	60.00	13.4
N-354		N-(7-nitrobenz-2-oxa-1,3-diazol-4-yl)phallacidin (NBD phallacidin)	300 U	198.00	158.40	11.1
N-7066	***New***	N-(2-nitrobenzyl)-L-isoproterenol (N-(NB-caged) isoproterenol)	5 mg	48.00	38.40	18.3, 19.2
N-7067	***New***	N-(2-nitrobenzyl)-DL-propranolol (N-(NB-caged) propranolol)	5 mg	48.00	38.40	18.3, 19.2
N-6495	***New***	nitro blue tetrazolium chloride (NBT)	1 g	65.00	52.00	9.2, 16.3, 21.4
N-7902	***New***	7-nitroindazole	25 mg	45.00	36.00	21.5
N-7928	***New***	7-nitroindazole, sodium salt	25 mg	45.00	36.00	21.5
N-6476	***New***	2-nitrophenol (o-nitrophenol) *reference standard*	1 g	8.00	6.40	10.1
N-6477	***New***	4-nitrophenol (p-nitrophenol) *reference standard*	1 g	8.00	6.40	10.1
N-8426	***New***	4-nitrophenyl N-acetyl-β-D-galactosaminide	100 mg	35.00	28.00	10.2
N-8427	***New***	4-nitrophenyl N-acetyl-β-D-glucosaminide	1 g	75.00	60.00	10.2
N-8428	***New***	4-nitrophenyl β-D-cellobioside	100 mg	48.00	38.40	10.2
N-2551		4-nitrophenyl 3-diazopyruvate	25 mg	45.00	36.00	5.3
N-6803	***New***	o-nitrophenyl EGTA, AM (NP-EGTA, AM) *cell permeant* *special packaging*	20x50 µg	195.00	156.00	19.2, 20.2, 22.8
N-6802	***New***	o-nitrophenyl EGTA, tetrapotassium salt (NP-EGTA) *cell impermeant* *≥95% by HPLC*	1 mg	165.00	132.00	19.2, 20.2, 22.8
N-1045		1-(2-nitrophenyl)ethyl adenosine 3′,5′-cyclicmonophosphate (NPE-caged cAMP)	5 mg	78.00	62.40	11.3, 19.2, 20.4
N-7065	***New***	1-(2-nitrophenyl)ethyl phosphate, diammonium salt (NPE-caged phosphate)	5 mg	65.00	52.00	19.2, 23.2
N-8429	***New***	4-nitrophenyl β-D-fucopyranoside	250 mg	55.00	44.00	10.2
N-8430	***New***	4-nitrophenyl α-D-galactopyranoside	1 g	35.00	28.00	10.2
N-8431	***New***	2-nitrophenyl β-D-galactopyranoside (ONPG)	5 g	35.00	28.00	10.2
N-8432	***New***	4-nitrophenyl β-D-galactopyranoside	1 g	24.00	19.20	10.2
N-8433	***New***	4-nitrophenyl α-D-glucopyranoside	1 g	20.00	16.00	10.2
N-8434	***New***	4-nitrophenyl β-D-glucopyranoside	1 g	35.00	28.00	10.2
N-8435	***New***	4-nitrophenyl β-D-glucuronide	100 mg	25.00	20.00	10.2
N-1559		4-nitrophenylglyoxal	100 mg	48.00	38.40	1.8, 18.5
N-1505		4-nitrophenyl iodoacetate (NPIA)	1 g	48.00	38.40	5.2
N-8436	***New***	4-nitrophenyl α-D-mannopyranoside	1 g	48.00	38.40	10.2
N-6613	***New***	4-nitrophenyl phosphate, disodium salt, hexahydrate (PNPP)	5 g	75.00	60.00	10.3
N-7892	***New***	S-nitroso-N-acetylpenicillamine (SNAP)	25 mg	45.00	36.00	21.5
N-7927	***New***	S-nitroso-N-acetylpenicillamine (SNAP) *special packaging*	20x1 mg	65.00	52.00	21.5
N-7903	***New***	S-nitrosoglutathione	25 mg	45.00	36.00	21.5
N-6498	***New***	4-nitrotetrazolium violet chloride (NTV)	1 g	32.00	25.60	16.3, 21.4
M-7906	***New***	D-NMMA (N^G-methyl-D-arginine, acetate salt)	25 mg	55.00	44.00	21.5
M-7898	***New***	L-NMMA (N^G-methyl-L-arginine, acetate salt)	25 mg	45.00	36.00	21.5
A-1372		nonyl acridine orange (acridine orange 10-nonyl bromide)	100 mg	45.00	36.00	12.2, 14.3
N-6356	***New***	norbiotinamine, hydrochloride	10 mg	65.00	52.00	4.2
N-96025	***New***	norfluorocurarine (vincanine)	20 mg	108.00	86.40	18.8
N-98103	***New***	notexin fraction *phospholipase A_2 from *Notechis scutatus* snake venom*	1 mg	245.00	196.00	20.5
N-98104	***New***	notexin Np *phospholipase A_2 from *Notechis scutatus* snake venom*	100 µg	215.00	172.00	20.5
N-6802	***New***	NP-EGTA (o-nitrophenyl EGTA, tetrapotassium salt) *cell impermeant* *≥95% by HPLC*	1 mg	165.00	132.00	19.2, 20.2, 22.8
N-6803	***New***	NP-EGTA, AM (o-nitrophenyl EGTA, AM) *cell permeant* *special packaging*	20x50 µg	195.00	156.00	19.2, 20.2, 22.8
N-6946	***New***	nuclear yellow (Hoechst S769121)	100 mg	48.00	38.40	15.3
N-7565	***New***	Nucleic Acid Stains Dimer Sampler Kit	1 kit	125.00	100.00	8.1, 16.3
N-7566	***New***	Nucleic Acid Stains Monomer Sampler Kit	1 kit	65.00	52.00	8.1, 16.3
		nucleotides — see especially Chromatide™ nucleotides and in Section 8.2; other nucleotides are listed in Sections 11.3 and 20.4				
N-96037	***New***	nudicauline perchlorate	5 mg	128.00	102.40	18.2
O-7703	***New***	1-octacosanyl-2-(1-pyrenehexanoyl)-*sn*-glycero-3-phosphomethanol, ammonium salt (C_{28}-O-PHPM)	250 µg	200.00	160.00	20.5
O-322		5-octadecanoylaminofluorescein	100 mg	65.00	52.00	14.3
O-7541	***New***	*cis*-9-octadecenamide (oleamide) *≥98%*	100 mg	68.00	54.40	20.6
O-3852		N-octadecyl-N′-(5-(fluoresceinyl))thiourea (F18)	25 mg	65.00	52.00	14.3
O-246		octadecyl rhodamine B chloride (R 18)	10 mg	145.00	116.00	14.3, 15.4
O-2892		5-octanoylaminofluorescein di-β-D-galactopyranoside (C_8FDG)	5 mg	135.00	108.00	10.2
O-8457	***New***	octyl glucoside (octyl β-D-glucopyranoside)	1 g	30.00	24.00	16.3
O-7035	***New***	octyl β-D-thioglucopyranoside (C_8SGlu)	1 g	35.00	28.00	16.3
O-3452		okadaic acid *500 µg/mL solution in DMSO*	50 µL	100.00	80.00	10.6, 20.3
O-7457	***New***	okadaic acid, sodium salt	25 µg	95.00	76.00	10.6, 20.3
O-7541	***New***	oleamide (*cis*-9-octadecenamide) *≥98%*	100 mg	68.00	54.40	20.6

Cat #		Product Name	Unit Size	Price Per Unit ($) 1–4 Units	5–24 Units	Section #
O-6682	***New***	oligonucleotide molecular weight standards *50 gel lanes*	500 µL	50.00	40.00	8.3
O-7582	***New***	OliGreen™ ssDNA quantitation reagent *for ssDNA and oligonucleotides* *200-2000 assays*	1 mL	175.00	140.00	8.1, 8.3
		Omega® Optical filters — see Section 26.5 for a complete list.				
		ONLY™ Labeling Kits — see FluoReporter® Oligonucleotide and Protein Labeling Kits; for a complete list, see page 18.				
N-8431	***New***	ONPG (2-nitrophenyl β-D-galactopyranoside)	5 g	35.00	28.00	10.2
P-2331		OPA (o-phthaldialdehyde) *high purity*	1 g	32.00	25.60	1.8, 9.1, 16.3, 24.3
D-6145	***New***	Oregon Green™ 488 (2′,7′-difluorofluorescein)	10 mg	20.00	16.00	1.4
O-6807	***New***	Oregon Green™ 488 BAPTA-1, AM *cell permeant* *≥95% by HPLC* *special packaging*	10x50 µg	125.00	100.00	22.3
O-6798	***New***	Oregon Green™ 488 BAPTA-1 dextran, potassium salt, 10,000 MW, anionic	5 mg	150.00	120.00	22.4
O-6797	***New***	Oregon Green™ 488 BAPTA-1 dextran, potassium salt, 70,000 MW, anionic	5 mg	150.00	120.00	22.4
O-6806	***New***	Oregon Green™ 488 BAPTA-1, hexapotassium salt *cell impermeant* *≥95% by HPLC*	500 µg	95.00	76.00	22.3
O-6809	***New***	Oregon Green™ 488 BAPTA-2, AM *cell permeant* *special packaging*	10x50 µg	125.00	100.00	22.3
O-6808	***New***	Oregon Green™ 488 BAPTA-2, octapotassium salt *cell impermeant*	500 µg	95.00	76.00	22.3
O-6813	***New***	Oregon Green™ 488 BAPTA-5N, AM *cell permeant* *≥95% by HPLC* *special packaging*	10x50 µg	125.00	100.00	22.3
O-6812	***New***	Oregon Green™ 488 BAPTA-5N, hexapotassium salt *cell impermeant* *≥95% by HPLC*	500 µg	95.00	76.00	22.3
O-7490	***New***	Oregon Green™ 488 bisindolylmaleimide	250 µg	65.00	52.00	27.1
O-6146	***New***	Oregon Green™ 488 carboxylic acid *5-isomer*	5 mg	75.00	60.00	1.4, 15.3, 23.3
O-6148	***New***	Oregon Green™ 488 carboxylic acid *6-isomer*	5 mg	75.00	60.00	1.4, 15.3, 23.3
O-6151	***New***	Oregon Green™ 488 carboxylic acid diacetate (carboxy-DFFDA) *6-isomer*	5 mg	95.00	76.00	16.3
O-6147	***New***	Oregon Green™ 488 carboxylic acid, succinimidyl ester *5-isomer*	5 mg	95.00	76.00	1.4, 23.4
O-6149	***New***	Oregon Green™ 488 carboxylic acid, succinimidyl ester *6-isomer*	5 mg	95.00	76.00	1.4, 23.4
C-7630	***New***	Oregon Green™ 488-5-dUTP (ChromaTide™ Oregon Green™ 488-5-dUTP) *1 mM in buffer*	25 µL	175.00	175.00	8.2
O-6380	***New***	Oregon Green™ 488 goat anti-mouse IgG (H+L) conjugate *2 mg/mL*	0.5 mL	75.00	60.00	7.2
O-6381	***New***	Oregon Green™ 488 goat anti-rabbit IgG (H+L) conjugate *2 mg/mL*	0.5 mL	75.00	60.00	7.2
O-6382	***New***	Oregon Green™ 488 goat anti-rat IgG (H+L) conjugate *2 mg/mL*	0.5 mL	75.00	60.00	7.2
O-6010	***New***	Oregon Green™ 488 iodoacetamide *mixed isomers*	5 mg	95.00	76.00	2.2
O-6080	***New***	Oregon Green™ 488 isothiocyanate (F_2FITC) *mixed isomers*	5 mg	95.00	76.00	1.4
O-6034	***New***	Oregon Green™ 488 maleimide	5 mg	95.00	76.00	27.1
O-7466	***New***	Oregon Green™ 488 phalloidin	300 U	198.00	158.40	11.1
O-6185	***New***	Oregon Green™ 488-X, succinimidyl ester *6-isomer*	5 mg	95.00	76.00	27.1
O-6135	***New***	Oregon Green™ 500 carboxylic acid *5-isomer*	5 mg	75.00	60.00	1.5, 15.3, 23.3
O-6136	***New***	Oregon Green™ 500 carboxylic acid, succinimidyl ester, triethylammonium salt *5-isomer*	5 mg	95.00	76.00	1.5, 23.4
O-6386	***New***	Oregon Green™ 500 goat anti-mouse IgG (H+L) conjugate *2 mg/mL*	0.5 mL	75.00	60.00	7.2
O-6387	***New***	Oregon Green™ 500 goat anti-rabbit IgG (H+L) conjugate *2 mg/mL*	0.5 mL	75.00	60.00	7.2
O-6138	***New***	Oregon Green™ 514 carboxylic acid	5 mg	75.00	60.00	1.5, 15.3, 23.3
O-6139	***New***	Oregon Green™ 514 carboxylic acid, succinimidyl ester	5 mg	95.00	76.00	1.5, 23.4
E-7498	***New***	Oregon Green™ 514 EGF (epidermal growth factor, Oregon Green™ 514 conjugate)	20 µg	160.00	128.00	17.3
O-6383	***New***	Oregon Green™ 514 goat anti-mouse IgG (H+L) conjugate *2 mg/mL*	0.5 mL	75.00	60.00	7.2
O-6384	***New***	Oregon Green™ 514 goat anti-rabbit IgG (H+L) conjugate *2 mg/mL*	0.5 mL	75.00	60.00	7.2
O-7465	***New***	Oregon Green™ 514 phalloidin	300 U	198.00	158.40	11.1
O-2756		ovalbumin, BODIPY® FL conjugate	25 mg	75.00	60.00	15.7
O-835		ovalbumin, fluorescein conjugate	25 mg	75.00	60.00	15.7
O-837		ovalbumin, Texas Red® conjugate	25 mg	75.00	60.00	15.7
O-266		oxonol V (bis-(3-phenyl-5-oxoisoxazol-4-yl)pentamethine oxonol)	100 mg	145.00	116.00	25.3
O-267		oxonol VI (bis-(3-propyl-5-oxoisoxazol-4-yl)pentamethine oxonol)	100 mg	145.00	116.00	25.3
D-2935		OxyBURST® Green, SE (2′,7′-dichlorodihydrofluorescein diacetate, succinimidyl ester)	5 mg	55.00	44.00	17.1, 21.4
C-2936		OxyBURST® Orange, SE (4-carboxydihydrotetramethylrosamine, succinimidyl ester)	5 mg	55.00	44.00	17.1, 21.4
P-3456		paclitaxel (Taxol equivalent) *for use in research only*	5 mg	40.00	32.00	11.2
P-7501	***New***	paclitaxel, BODIPY® 564/570 conjugate (BODIPY® 564/570 Taxol)	10 µg	168.00	134.40	11.2
P-7500	***New***	paclitaxel, BODIPY® FL conjugate (BODIPY® FL Taxol)	10 µg	168.00	134.40	11.2
P-7082	***New***	paclitaxel, 2′-(4,5-dimethoxy-2-nitrobenzyl)carbonate ($O^{2'}$-(NVOC-caged) Taxol)	1 mg	95.00	76.00	11.2, 19.2
P-1558		palladium disodium alizarinmonosulfonate ($Pd(QS)_2$)	100 mg	125.00	100.00	14.6
A-1122		panacyl bromide (4-(9-anthroyloxy)phenacyl bromide)	100 mg	118.00	94.40	3.4
P-8071	***New***	paper for Star® DP 8340 printer, set of 5 rolls	each	35.00	35.00	26.3
P-1		*cis*-parinaric acid	100 mg	155.00	124.00	13.3, 16.3, 21.4
P-1901		*cis*-parinaric acid *special packaging*	10x10 mg	185.00	148.00	13.3, 16.3, 21.4

Cat #		Product Name	Unit Size	Price Per Unit ($) 1–4 Units	Price Per Unit ($) 5–24 Units	Section #
P-2		*trans*-parinaric acid	100 mg	155.00	124.00	13.3
P-1902		*trans*-parinaric acid *special packaging*	10x10 mg	185.00	148.00	13.3
H-1387		patman (6-hexadecanoyl-2-(((2-(trimethylammonium)ethyl)methyl)amino)-naphthalene chloride)	5 mg	48.00	38.40	14.5
P-1266		PBFI, AM *cell permeant*	1 mg	245.00	196.00	24.1
P-1267		PBFI, AM *cell permeant* *special packaging*	20x50 µg	275.00	220.00	24.1
P-1265		PBFI, tetraammonium salt *cell impermeant*	1 mg	198.00	158.40	24.1
B-7893	**New**	PBN (N-*tert*-butyl-α-phenylnitrone)	1 g	25.00	20.00	21.3
P-1405		PDAM (1-pyrenyldiazomethane)	25 mg	88.00	70.40	3.4, 9.4
D-7522	**New**	PDMP (DL-*threo*-2-decanoylamino-3-morpholino-1-phenyl-1-propanol, hydrochloride)	5 mg	48.00	38.40	20.6
P-6317	**New**	PEAS; AET (N-((2-pyridyldithio)ethyl)-4-azidosalicylamide)	10 mg	65.00	52.00	5.3, 9.4
P-96043	**New**	peganole	20 mg	160.00	128.00	18.2
P-7061	**New**	penicillin V, 1-(4,5-dimethoxy-2-nitrophenyl)ethyl ester (O-(DMNPE-caged) penicillin V)	5 mg	48.00	38.40	16.3, 19.2
P-96054	**New**	pentazolone, hydrochloride	100 mg	38.00	30.40	10.6
R-1147		pentoxyresorufin (resorufin pentyl ether)	5 mg	98.00	78.40	10.5
P-6542	**New**	pepstatin A (iso-valeryl-L-Val-L-Val-Sta-L-Ala-Sta)	25 mg	55.00	44.00	10.6
P-917		peroxidase from horseradish, biotin-XX conjugate	10 mg	75.00	60.00	4.3, 7.5, 15.7
P-912		peroxidase from horseradish, fluorescein conjugate	10 mg	75.00	60.00	15.7
P-6481	**New**	peroxidase from horseradish, Texas Red®-X conjugate	5 mg	75.00	60.00	15.7
P-7071	**New**	(µ-peroxo) (µ-hydroxo)bis(bis(bipyridyl)cobalt(III)) perchlorate (caged oxygen)	25 mg	95.00	76.00	27.1
P-646		3-perylenedodecanoic acid	10 mg	148.00	118.40	13.2
P-1692		PETG (phenylethyl β-D-thiogalactopyranoside)	10 mg	43.00	34.40	10.6
P-3458		phallacidin	1 mg	80.00	64.00	11.1
P-3457		phalloidin	1 mg	80.00	64.00	11.1
P-6544	**New**	1,10-phenanthroline, monohydrate *≥99%*	5 g	18.00	14.40	10.6
P-6878	**New**	N-(1,10-phenanthrolin-5-yl)bromoacetamide	5 mg	75.00	60.00	2.4, 21.2
P-6879	**New**	N-(1,10-phenanthrolin-5-yl)iodoacetamide	5 mg	75.00	60.00	2.4, 21.2
P-6763	**New**	Phen Green™ diacetate *cell permeant*	1 mg	90.00	72.00	22.7
P-6801	**New**	Phen Green™, dipotassium salt	1 mg	80.00	64.00	22.7
P-6549	**New**	phenoxazone	10 mg	98.00	78.40	10.5, 21.3
P-1692		phenylethyl β-D-thiogalactopyranoside (PETG)	10 mg	43.00	34.40	10.6
P-8458	**New**	phenyl β-D-galactopyranoside	1 g	22.00	17.60	10.2
P-8459	**New**	phenyl β-D-glucuronide, monohydrate	100 mg	35.00	28.00	10.2
P-6221	**New**	phenylglyoxal, monohydrate	1 g	8.00	6.40	1.8, 18.5
P-1342		4-((6-phenyl)-1,3,5-hexatrienyl)benzoic acid (DPH carboxylic acid)	25 mg	38.00	30.40	13.2
P-459		3-(4-(6-phenyl)-1,3,5-hexatrienyl)phenylpropionic acid (DPH propionic acid)	25 mg	148.00	118.40	13.2
P-3900		N-((4-(6-phenyl-1,3,5-hexatrienyl)phenyl)propyl)trimethylammonium p-toluenesulfonate (TMAP-DPH)	5 mg	148.00	118.40	14.4, 17.2
P-6203	**New**	phenylmethanesulfonyl fluoride (PMSF) *high purity*	5 g	32.00	25.60	10.6
P-65		N-phenyl-1-naphthylamine *high purity*	1 g	48.00	38.40	14.5
P-8460	**New**	phenyl β-D-thioglucuronide	100 mg	35.00	28.00	10.6
P-7479	**New**	philanthotoxin, trihydrochloride (PhTX-3.4.3)	100 µg	98.00	78.40	18.2
P-3453		4β-phorbol 12β-myristate 13α-acetate (TPA; PMA)	1 mg	25.00	20.00	20.3
P-3454		4β-phorbol 12β-myristate 13α-acetate (TPA; PMA) *special packaging*	10x100 µg	35.00	28.00	20.3
C-8448	**New**	phosphocreatine (creatine phosphate, disodium salt, tetrahydrate)	5 g	44.00	35.20	10.3
P-98408	**New**	phospholipase A_2 *from *Apis mellifera* bee venom*	1 mg	38.00	30.40	20.5
P-6466	**New**	phospholipase C, phosphatidylinositol-specific *from *Bacillus cereus** *100 U/mL*	50 µL	40.00	32.00	20.5
		phospholipids — see Sections 13.2 and 13.4 for a discussion of our naming conventions and complete lists.				
P-2331		o-phthaldialdehyde (OPA) *high purity*	1 g	32.00	25.60	1.8, 9.1, 16.3, 24.3
P-800		B-phycoerythrin *4 mg/mL*	0.5 mL	95.00	76.00	6.4, 15.7
P-801		R-phycoerythrin *4 mg/mL*	0.5 mL	95.00	76.00	6.4, 15.7
P-811		R-phycoerythrin, biotin-XX conjugate *4 mg/mL*	0.5 mL	135.00	108.00	4.3, 6.4
P-852		R-phycoerythrin goat anti-mouse IgG (H+L) conjugate *1 mg/mL*	1 mL	165.00	132.00	6.4, 7.2
P-2771		R-phycoerythrin goat anti-rabbit IgG (H+L) conjugate *1 mg/mL*	0.5 mL	95.00	76.00	6.4, 7.2
P-806		R-phycoerythrin, pyridyldisulfide derivative *2 mg/mL*	1 mL	135.00	108.00	6.4
P-7589	**New**	PicoGreen® dsDNA Quantitation Kit *200-2000 assays*	1 kit	175.00	140.00	8.1, 8.3
P-7581	**New**	PicoGreen® dsDNA quantitation reagent *200-2000 assays*	1 mL	150.00	120.00	8.1, 8.3
P-96069	**New**	pinocembrine	100 mg	68.00	54.40	18.8
P-7438	**New**	pirenzepine, dihydrochloride	100 mg	10.00	8.00	18.2
P-6867	**New**	Pluronic® F-127 *low UV absorbance*	2 g	8.00	6.40	22.8, 24.1
P-3000		Pluronic® F-127 *20% solution in DMSO*	1 mL	15.00	12.00	22.8, 24.1
P-6866	**New**	Pluronic® F-127 *sterile 10% solution in water*	30 mL	35.00	28.00	22.8, 24.1
P-6203	**New**	PMSF (phenylmethanesulfonyl fluoride) *high purity*	5 g	32.00	25.60	18.2
P-6283	**New**	poly(ethylene glycol) methyl ether, amine-terminated, average MW 550	100 mg	38.00	30.40	3.3

Cat #		Product Name	Unit Size	Price Per Unit ($) 1–4 Units	5–24 Units	Section #
P-6282	**New**	poly(ethylene glycol) methyl ether, amine-terminated, average MW 750	100 mg	38.00	30.40	3.3
P-6285	**New**	poly(ethylene glycol) methyl ether, amine-terminated, average MW 2000	1 g	95.00	76.00	3.3
P-6281	**New**	poly(ethylene glycol) methyl ether, amine-terminated, average MW 5000	1 g	95.00	76.00	3.3
P-3580		POPO™-1 iodide (434/456) *1 mM solution in DMSO*	200 µL	225.00	180.00	8.1
P-3584		POPO™-3 iodide (534/570) *1 mM solution in DMSO*	200 µL	225.00	180.00	8.1
P-3581		PO-PRO™-1 iodide (435/455) *1 mM solution in DMSO*	1 mL	135.00	108.00	8.1
P-3585		PO-PRO™-3 iodide (539/567) *1 mM solution in DMSO*	1 mL	135.00	108.00	8.1
P-7070	**New**	potassium nitrosylpentachlororuthenate (caged nitric oxide I)	10 mg	85.00	68.00	19.2, 21.5
H-7523	**New**	PPMP (DL-*threo*-2-hexadecanoylamino-3-morpholino-1-phenyl-1-propanol, hydrochloride)	5 mg	48.00	38.40	20.6
P-7435	**New**	prazosin, hydrochloride	10 mg	12.00	9.60	18.2
P-8051	**New**	printer, Star® DP 8340, with power supply and serial cable	each	410.00	410.00	26.3
P-248		prodan (6-propionyl-2-dimethylaminonaphthalene)	100 mg	135.00	108.00	11.2, 14.5
P-7481	**New**	ProLong™ Antifade Kit	1 kit	98.00	78.40	26.1
P-1304		propidium iodide	100 mg	48.00	38.40	8.1, 15.3, 16.3
P-3566		propidium iodide *1.0 mg/mL solution in water*	10 mL	28.00	22.40	8.1, 16.3
P-248		6-propionyl-2-dimethylaminonaphthalene (prodan)	100 mg	135.00	108.00	11.2, 14.5
P-99020	**New**	2-propoxy-2,3-dihydro-1,4,2-benzodioxaphosphorin-2-oxide (bomin-2)	500 µg	108.00	86.40	18.2
P-99005	**New**	propybicyphat	1 mg	185.00	148.00	18.5
P-6548	**New**	Protease Inhibitor Sampler Kit	1 kit	195.00	156.00	10.6
P-6445	**New**	protein A *recombinant*	5 mg	45.00	36.00	7.2
P-2757		protein A, biotin-XX conjugate	1 mg	75.00	60.00	4.3, 7.2
P-963		protein A, Cascade Blue® conjugate	1 mg	75.00	60.00	7.2
P-846		protein A, fluorescein conjugate	1 mg	75.00	60.00	7.2
P-6395	**New**	protein A, Oregon Green™ 488 conjugate	1 mg	75.00	60.00	7.2
P-826		protein A, Texas Red® conjugate	1 mg	75.00	60.00	7.2
P-6444	**New**	protein G *recombinant*	5 mg	45.00	36.00	7.2
P-6460	**New**	protein kinase A, catalytic subunit *from bovine heart* *20 ng/µL*	100 µL	95.00	76.00	20.3
P-7432	**New**	protein kinase C *from rat brain* *6.7 ng/µL*	2x75 µL	98.00	78.40	20.3
P-6461	**New**	protein kinase C, catalytic fragment *from rat brain* *10 ng/µL*	50 µL	225.00	180.00	20.3
P-6649	**New**	protein molecular weight standards *broad range* *200 gel lanes*	400 µL	65.00	52.00	9.2
P-6647	**New**	protein molecular weight standards *high range* *200 gel lanes*	400 µL	45.00	36.00	9.2
P-6648	**New**	protein molecular weight standards *low range* *200 gel lanes*	400 µL	45.00	36.00	9.2
P-6305	**New**	Protein-Protein Crosslinking Kit *3 conjugations*	1 kit	175.00	140.00	5.2
C-7924	**New**	proxyl fluorescamine (5-(2-carboxyphenyl)-5-hydroxy-1-((2,2,5,5-tetramethyl-1-oxypyrrolidin-3-yl)methyl)-3-phenyl-2-pyrrolin-4-one, potassium salt)	5 mg	65.00	52.00	21.3
P-96045	**New**	psoralen	100 mg	95.00	76.00	8.1
P-7220	**New**	PS-Speck™ Microscope Point Source Kit *blue, green, orange and deep red fluorescent beads*	1 kit	80.00	64.00	26.1
P-6115	**New**	1-pyreneacetic acid, succinimidyl ester	25 mg	48.00	38.40	1.7
P-32		1-pyrenebutanoic acid (pyrenebutyric acid)	1 g	68.00	54.40	13.2
P-1903		1-pyrenebutanoic acid *high purity*	100 mg	68.00	54.40	13.2
P-101		1-pyrenebutanoic acid, hydrazide	100 mg	95.00	76.00	3.2, 9.4
P-130		1-pyrenebutanoic acid, succinimidyl ester	100 mg	115.00	92.00	1.7
P-244		1-pyrenebutanol	100 mg	65.00	52.00	3.4, 14.6, 21.2
P-6169	**New**	N-(1-pyrenebutanoyl)cysteic acid, potassium salt	5 mg	65.00	52.00	27.1
P-6114	**New**	N-(1-pyrenebutanoyl)cysteic acid, succinimidyl ester, sodium salt	5 mg	75.00	60.00	1.7
P-3764		1-pyrenebutyl *myo*-inositol 1-phosphate, lithium salt	500 µg	160.00	128.00	20.5
P-314		1-pyrenebutyltrimethylammonium bromide	25 mg	48.00	38.40	14.6
P-6921	**New**	1-pyrenebutyltriphenylphosphonium bromide (PyTPP+)	10 mg	48.00	38.40	14.6, 16.3, 25.3
P-32		pyrenebutyric acid (1-pyrenebutanoic acid)	1 g	68.00	54.40	13.2
P-164		1-pyrenecarboxylic acid	100 mg	48.00	38.40	1.8, 13.2
P-31		1-pyrenedecanoic acid	25 mg	68.00	54.40	13.2
P-3527		N-(1-pyrenedecanoyl)sphingosylphosphocholine (py-C_{10}-sphingomyelin)	1 mg	115.00	92.00	13.2
P-96		1-pyrenedodecanoic acid	25 mg	68.00	54.40	13.2
P-3779		12-(1-pyrenedodecyl)phosphocholine	1 mg	48.00	38.40	13.2
P-6253	**New**	1-pyreneethylamine, hydrochloride	5 mg	48.00	38.40	3.3
P-6030	**New**	N-(1-pyreneethyl)iodoacetamide	5 mg	48.00	38.40	2.3
P-243		1-pyrenehexadecanoic acid	5 mg	98.00	78.40	13.2
P-3840		1-pyrenehexanoic acid	100 mg	68.00	54.40	13.2
P-29		N-(1-pyrene)iodoacetamide	100 mg	115.00	92.00	2.3, 9.4, 11.1
P-331		1-pyreneisothiocyanate	100 mg	95.00	76.00	1.8
P-28		N-(1-pyrene)maleimide	100 mg	75.00	60.00	2.3, 9.4
P-2421		1-pyrenemethylamine, hydrochloride (1-aminomethylpyrene, hydrochloride)	100 mg	58.00	46.40	3.3
P-223		1-pyrenemethyl 3β-hydroxy-22,23-bisnor-5-cholenate (PMC)	25 mg	125.00	100.00	13.5
P-2007		N-(1-pyrenemethyl)iodoacetamide (PMIA amide)	25 mg	95.00	76.00	2.3
P-4		1-pyrenemethyl iodoacetate (PMIA ester)	100 mg	95.00	76.00	2.3

Cat #		Product Name	Unit Size	Price Per Unit ($) 1–4 Units	5–24 Units	Section #
P-226		1-pyrenemethyl 3β-(*cis*-9-octadecenoyloxy)-22,23-bisnor-5-cholenate (PMC oleate)	10 mg	145.00	116.00	13.5
P-394		1-pyrenenonanol	25 mg	135.00	108.00	14.6, 21.2
P-6254	**New**	1-pyrenepropylamine, hydrochloride	5 mg	48.00	38.40	3.3
P-6031	**New**	N-(1-pyrenepropyl)iodoacetamide	10 mg	65.00	52.00	2.3
P-80		1-pyrenesulfonic acid, sodium salt	1 g	95.00	76.00	14.6
P-24		1-pyrenesulfonyl chloride	100 mg	78.00	62.40	1.8
P-58		N-(1-pyrenesulfonyl)-1,2-dihexadecanoyl-*sn*-glycero-3-phosphoethanolamine, triethylammonium salt (pyS DHPE)	25 mg	75.00	60.00	13.4, 20.5
P-349		1,3,6,8-pyrenetetrasulfonic acid, tetrasodium salt	100 mg	35.00	28.00	15.3
P-1405		1-pyrenyldiazomethane (PDAM)	25 mg	88.00	70.40	3.4, 9.4
P-6317	**New**	N-((2-pyridyldithio)ethyl)-4-azidosalicylamide (PEAS; AET)	10 mg	65.00	52.00	5.3, 9.4
P-1619		N-((2-pyridyldithio)propionyl)-1,2-dihexadecanoyl-*sn*-glycero-3-phosphoethanolamine, triethylammonium salt (PDP DHPE)	5 mg	135.00	108.00	5.2, 13.4
Q-8010	**New**	quartz-halogen lamp assembly for TD-700 fluorometer *fluorescein excitation*	each	405.00	405.00	26.3
Q-8070	**New**	quartz-halogen replacement bulb for TD-700 fluorometer	each	50.00	50.00	26.3
Q-1288		quin-2, AM *cell permeant* *≥95% by HPLC*	25 mg	115.00	92.00	22.2
Q-1289		quin-2, AM *cell permeant* *≥95% by HPLC* *special packaging*	10x1 mg	73.00	58.40	22.2
Q-1287		quin-2, free acid *cell impermeant*	25 mg	63.00	50.40	22.2
Q-1590		quinuclidine	100 mg	8.00	6.40	3.1
R-6479	**New**	R110 (rhodamine 110) *reference standard*	25 mg	12.00	9.60	10.1
R-96095	**New**	remerine, hydrochloride	20 mg	75.00	60.00	18.8
R-2931		Renin Substrate 1 (Arg-Glu(EDANS)-Ile-His-Pro-Phe-His-Leu-Val-Ile-His-Thr-Lys(DABCYL)-Arg)	1 mg	150.00	120.00	10.4
R-6540	**New**	resorufin N-acetyl-β-D-glucosaminide (RNAG)	5 mg	128.00	102.40	10.2
R-441		resorufin benzyl ether (benzyloxyresorufin)	10 mg	98.00	78.40	10.5
R-6564	**New**	Resorufin Ether Sampler Kit	1 kit	98.00	78.40	10.5
R-352		resorufin ethyl ether (ethoxyresorufin)	5 mg	98.00	78.40	10.5
R-1159		resorufin β-D-galactopyranoside	25 mg	103.00	82.40	10.2
R-1160		resorufin β-D-glucopyranoside	25 mg	103.00	82.40	10.2
R-1161		resorufin β-D-glucuronide, potassium salt	5 mg	128.00	102.40	10.2
R-6535	**New**	resorufin 3β-hydroxy-22,23-bisnor-5-cholenyl ether	1 mg	125.00	100.00	10.5
R-351		resorufin methyl ether (methoxyresorufin)	5 mg	98.00	78.40	10.5
R-1147		resorufin pentyl ether (pentoxyresorufin)	5 mg	98.00	78.40	10.5
R-363		resorufin, sodium salt *reference standard*	100 mg	18.00	14.40	10.1
R-6922	**New**	RGA-30	5 mg	118.00	94.40	18.3, 25.2
R-1114		RH 155	25 mg	145.00	116.00	25.2
S-1107		RH 160 (N-(4-sulfobutyl)-4-(4-(4-(dibutylamino)phenyl)butadienyl)pyridinium, inner salt)	25 mg	75.00	60.00	25.2
S-1109		RH 237 (N-(4-sulfobutyl)-4-(6-(4-(dibutylamino)phenyl)hexatrienyl)pyridinium, inner salt)	5 mg	75.00	60.00	25.2
T-1111		RH 414 (N-(3-triethylammoniumpropyl)-4-(4-(4-(diethylamino)phenyl)butadienyl)-pyridinium dibromide)	5 mg	165.00	132.00	15.4, 17.2, 25.2
S-1108		RH 421 (N-(4-sulfobutyl)-4-(4-(4-(dipentylamino)phenyl)butadienyl)pyridinium, inner salt)	25 mg	165.00	132.00	25.2
T-1113		RH 461 (N-(3-trimethylammoniumpropyl)-4-(4-(4-(diethylamino)phenyl)-butadienyl)pyridinium dibromide)	5 mg	75.00	60.00	25.2
T-6693	**New**	RH 704 (N-(4-triethylammoniumbutyl)-4-(4-(4-(diethylamino)phenyl)-butadienyl)pyridinium dibromide)	5 mg	145.00	116.00	25.2
R-6695	**New**	RH 773	5 mg	118.00	94.40	25.2
R-649		RH 795	1 mg	165.00	132.00	25.2
R-1244		rhod-2, AM *cell permeant*	1 mg	165.00	132.00	22.3
R-1245		rhod-2, AM *cell permeant* *special packaging*	20x50 µg	195.00	156.00	22.3
R-1243		rhod-2, triammonium salt *cell impermeant*	1 mg	135.00	108.00	22.3
R-302		rhodamine 123	25 mg	30.00	24.00	12.2, 16.3, 25.3
R-6479	**New**	rhodamine 110 (R110) *reference standard*	25 mg	12.00	9.60	10.1
R-634		rhodamine 6G chloride	1 g	18.00	14.40	12.2, 24.3
R-648		rhodamine B, hexyl ester, chloride (R 6)	10 mg	85.00	68.00	12.2
R-6515	**New**	rhodamine 110, bis-(t-BOC-L-leucylglycyl-L-arginine amide), dihydrochloride	5 mg	75.00	60.00	10.4
R-6513	**New**	rhodamine 110, bis-(t-BOC-L-leucyl-L-methionine amide)	5 mg	75.00	60.00	10.4
R-6504	**New**	rhodamine 110, bis-(CBZ-L-alanine amide)	5 mg	75.00	60.00	10.4
R-6506	**New**	rhodamine 110, bis-(CBZ-L-alanyl-L-alanyl-L-alanyl-L-alanine amide)	5 mg	75.00	60.00	10.4
R-6508	**New**	rhodamine 110, bis-(CBZ-L-alanyl-L-arginine amide), dihydrochloride	5 mg	75.00	60.00	10.4
R-6501	**New**	rhodamine 110, bis-(CBZ-L-arginine amide), dihydrochloride (BZAR)	5 mg	75.00	60.00	10.4
R-6514	**New**	rhodamine 110, bis-(CBZ-glycylglycyl-L-arginine amide), dihydrochloride	5 mg	75.00	60.00	10.4
R-6517	**New**	rhodamine 110, bis-(CBZ-glycyl-L-prolyl-L-arginine amide), dihydrochloride	5 mg	75.00	60.00	10.4
R-6505	**New**	rhodamine 110, bis-(CBZ-L-isoleucyl-L-prolyl-L-arginine amide), dihydrochloride (BZiPAR)	5 mg	75.00	60.00	10.4

Cat #		Product Name	Unit Size	Price Per Unit ($) 1–4 Units	5–24 Units	Section #
R-6502	***New***	rhodamine 110, bis-(CBZ-L-phenylalanyl-L-arginine amide), dihydrochloride	5 mg	75.00	60.00	10.4
R-6507	***New***	rhodamine 110, bis-(CBZ-L-prolyl-L-arginine amide), dihydrochloride	5 mg	75.00	60.00	10.4
R-6512	***New***	rhodamine 110, bis-(CBZ-L-valyl-L-lysyl-L-methionine amide), dihydrochloride	5 mg	75.00	60.00	10.4
R-6516	***New***	rhodamine 110, bis-(CBZ-L-valyl-L-prolyl-L-arginine amide), dihydrochloride	5 mg	75.00	60.00	10.4
R-6510	***New***	rhodamine 110, bis-(glutarylglycyl-L-arginine amide), dihydrochloride	5 mg	75.00	60.00	10.4
R-6509	***New***	rhodamine 110, bis-(L-leucine amide), dihydrochloride	5 mg	75.00	60.00	10.4
R-6511	***New***	rhodamine 110, bis-(L-leucyl-L-arginylglycylglycine amide), tetrahydrochloride	5 mg	75.00	60.00	10.4
R-6577	***New***	rhodamine 110, bis-(L-phenylalanine amide), di(trifluoroacetic acid) salt	5 mg	75.00	60.00	10.4
R-6518	***New***	rhodamine 110, bis-(L-prolyl-L-phenylalanyl-L-arginine amide), tetrahydrochloride	5 mg	75.00	60.00	10.4
R-6559	***New***	rhodamine 110, 4-(chloromethyl)benzoyl amide, L-alanine amide, hydrochloride (CMB-R110-Ala)	1 mg	75.00	60.00	10.4
R-6555	***New***	rhodamine 110, 4-(chloromethyl)benzoyl amide, L-alanyl-L-proline amide, hydrochloride (CMB-R110-Ala-Pro)	1 mg	75.00	60.00	10.4
R-6560	***New***	rhodamine 110, 4-(chloromethyl)benzoyl amide, CBZ-L-arginine amide, hydrochloride (CMB-R110-Arg)	1 mg	75.00	60.00	10.4
R-6556	***New***	rhodamine 110, 4-(chloromethyl)benzoyl amide, glycine amide, hydrochloride (CMB-R110-Gly)	1 mg	75.00	60.00	10.4
R-6561	***New***	rhodamine 110, 4-(chloromethyl)benzoyl amide, glycyl-L-proline amide, hydrochloride (CMB-R110-Gly-Pro)	1 mg	75.00	60.00	10.4
R-6557	***New***	rhodamine 110, 4-(chloromethyl)benzoyl amide, L-leucine amide, hydrochloride (CMB-R110-Leu)	1 mg	75.00	60.00	10.4
R-6558	***New***	rhodamine 110, 4-(chloromethyl)benzoyl amide, L-phenylalanyl-L-proline amide, hydrochloride (CMB-R110-Phe-Pro)	1 mg	75.00	60.00	10.4
R-7102	***New***	rhodamine 110, CMNCBZ-caged, dipotassium salt (CMNCBZ-caged rhodamine 110)	5 mg	145.00	116.00	15.3, 19.2
L-1392		rhodamine DHPE (N-(Lissamine™ rhodamine B sulfonyl)-1,2-dihexadecanoyl-*sn*-glycero-3-phosphoethanolamine, triethylammonium salt)	5 mg	75.00	60.00	13.4
C-7629	***New***	Rhodamine Green™-5-dUTP (ChromaTide™ Rhodamine Green™-5-dUTP) *1 mM in buffer*	25 µL	175.00	175.00	8.2
C-7628	***New***	Rhodamine Green™-5-UTP (ChromaTide™ Rhodamine Green™-5-UTP) *1 mM in buffer*	25 µL	175.00	175.00	8.2
R-7090	***New***	Rhodamine Green™ bis-(((4,5-dimethoxy-2-nitrobenzyl)oxy)carbonyl)-caged, succinimidyl ester (N-(NVOC-caged) Rhodamine Green™, SE)	5 mg	175.00	140.00	15.3, 19.2
R-7091	***New***	Rhodamine Green™ bis-(((4,5-dimethoxy-2-nitrobenzyl)oxy)carbonyl)-caged, sulfosuccinimidyl ester, sodium salt (N-(NVOC-caged) Rhodamine Green™, SSE)	1 mg	68.00	54.40	15.3, 19.2
R-6150	***New***	Rhodamine Green™ carboxylic acid, hydrochloride (5(6)-CR 110) *mixed isomers*	5 mg	95.00	76.00	1.4, 15.3
R-6107	***New***	Rhodamine Green™ carboxylic acid, succinimidyl ester, hydrochloride (5(6)-CR 110, SE) *mixed isomers*	5 mg	135.00	108.00	1.4
R-6112	***New***	Rhodamine Green™ carboxylic acid, trifluoroacetamide, succinimidyl ester (5(6)-CR 110 TFA, SE) *mixed isomers*	5 mg	135.00	108.00	1.4
R-6113	***New***	Rhodamine Green™-X, succinimidyl ester, hydrochloride *mixed isomers*	5 mg	135.00	108.00	1.4
R-415		rhodamine phalloidin	300 U	198.00	158.40	11.1
R-6029	***New***	Rhodamine Red™ C_2 maleimide	5 mg	95.00	76.00	27.1
R-7712	***New***	Rhodamine Red™-X DHPE (Rhodamine Red™-X 1,2-dihexadecanoyl-*sn*-glycero-3-phosphoethanolamine, triethylammonium salt)	1 mg	75.00	60.00	27.1
R-7712	***New***	Rhodamine Red™-X 1,2-dihexadecanoyl-*sn*-glycero-3-phosphoethanolamine, triethylammonium salt (Rhodamine Red™-X DHPE)	1 mg	75.00	60.00	27.1
R-6393	***New***	Rhodamine Red™-X goat anti-mouse IgG (H+L) conjugate *2 mg/mL*	0.5 mL	75.00	60.00	7.2
R-6394	***New***	Rhodamine Red™-X goat anti-rabbit IgG (H+L) conjugate *2 mg/mL*	0.5 mL	75.00	60.00	7.2
R-6160	***New***	Rhodamine Red™-X, succinimidyl ester *mixed isomers*	5 mg	75.00	60.00	1.6, 4.2
R-6111	***New***	Rhodol Green™ carboxylic acid, N,O-bis-(trifluoroacetyl), succinimidyl ester (5(6)-CRF 492 TFA, SE) *mixed isomers*	5 mg	135.00	108.00	1.4
R-6152	***New***	Rhodol Green™ carboxylic acid, hydrochloride (5(6)-CRF 492) *mixed isomers*	5 mg	75.00	60.00	1.4, 15.3, 23.3
R-6108	***New***	Rhodol Green™ carboxylic acid, succinimidyl ester (5(6)-CRF 492, SE) *mixed isomers*	5 mg	145.00	116.00	1.4, 23.4
R-6412	***New***	Rhodol Green™ goat anti-mouse IgG (H+L) conjugate *2 mg/mL*	0.5 mL	75.00	60.00	7.2
R-6470	***New***	Rhodol Green™ goat anti-rabbit IgG (H+L) conjugate *2 mg/mL*	0.5 mL	75.00	60.00	7.2
C-96012	***New***	ricinine (3-cyano-4-methoxy-N-methyl-2-pyridone)	10 mg	68.00	54.40	18.8
R-7454	***New***	rim-1	1 mg	95.00	76.00	20.3
R-6540	***New***	RNAG (resorufin N-acetyl-β-D-glucosaminide)	5 mg	128.00	102.40	10.2
R-7932	***New***	rose bengal agarose *sedimented bead suspension*	10 mL	95.00	76.00	21.2
C-6125	***New***	5-ROX, SE (5-carboxy-X-rhodamine, succinimidyl ester) *single isomer*	5 mg	135.00	108.00	1.6
C-6126	***New***	6-ROX, SE (6-carboxy-X-rhodamine, succinimidyl ester) *single isomer*	5 mg	135.00	108.00	1.6
C-1309		5(6)-ROX, SE (5-(and-6)-carboxy-X-rhodamine, succinimidyl ester) *mixed isomers*	25 mg	118.00	94.40	1.6
R-7800	***New***	rubrene	1 g	48.00	38.40	14.6, 24.3
R-3455		ryanodine *mixture of ryanodine and dehydroryanodine*	10 mg	30.00	24.00	18.4, 20.2

Cat #		Product Name	Unit Size	Price Per Unit ($) 1–4 Units	5–24 Units	Section #
R-7478	**New**	ryanodine *≥95% by HPLC*	1 mg	38.00	30.40	18.4, 20.2
S-8064	**New**	sample adaptor for TD-700 fluorometer, 15 mm x 7 mm aperture, for 10 mm x 10 mm x 45 mm cuvettes	each	370.00	370.00	26.3
S-8060	**New**	sample adaptor for TD-700 fluorometer, 25 mm x 6 mm aperture, for 10 mm x 10 mm x 45 cuvettes	each	190.00	190.00	26.3
S-8062	**New**	sample adaptor for TD-700 fluorometer, fixed aperture, for 13 mm diameter test tubes	each	95.00	95.00	26.3
S-8063	**New**	sample adaptor for TD-700 fluorometer, fixed aperture, for 25 mm diameter test tubes	each	95.00	95.00	26.3
A-685		SAMSA fluorescein (5-((2-(and-3)-S-(acetylmercapto)succinoyl)amino)-fluorescein) *mixed isomers*	25 mg	95.00	76.00	5.2
S-1553		SATA (succinimidyl acetylthioacetate)	100 mg	48.00	38.40	5.2
S-1573		SATP (succinimidyl acetylthiopropionate)	100 mg	48.00	38.40	5.2
S-1263		SBFI, AM *cell permeant*	1 mg	245.00	196.00	24.1
S-1264		SBFI, AM *cell permeant* *special packaging*	20x50 µg	275.00	220.00	24.1
S-1262		SBFI, tetraammonium salt *cell impermeant*	1 mg	198.00	158.40	24.1
S-6899	**New**	SBFO, diammonium salt *cell impermeant*	1 mg	125.00	100.00	24.1
S-96203	**New**	scoulerine, hydrochloride	5 mg	108.00	86.40	18.2
L-98202	**New**	scyllatoxin (leiurotoxin I) *synthetic, originally from *Leiurus quinquestriatus* scorpion venom*	50 µg	385.00	308.00	18.6
S-96082	**New**	sevedindione	20 mg	105.00	84.00	18.7
F-6106	**New**	5-SFX (6-(fluorescein-5-carboxamido)hexanoic acid, succinimidyl ester) *single isomer*	5 mg	65.00	52.00	1.3
F-2181		5(6)-SFX (6-(fluorescein-5-(and-6)-carboxamido)hexanoic acid, succinimidyl ester) *mixed isomers*	10 mg	65.00	52.00	1.3, 4.2
F-6129	**New**	5(6)-SFX (6-(fluorescein-5-(and-6)-carboxamido)hexanoic acid, succinimidyl ester) *mixed isomers* *special packaging*	10x1 mg	78.00	62.40	1.3
M-7926	**New**	SIN-1 (3-morpholinosydnonimine, hydrochloride) *special packaging*	20x1 mg	65.00	52.00	21.5
A-339		SITS (4-acetamido-4′-isothiocyanatostilbene-2,2′-disulfonic acid, disodium salt)	100 mg	48.00	38.40	18.5
S-96046	**New**	skimmianine	20 mg	75.00	60.00	18.8
S-2828		*SlowFade*™ Antifade Kit	1 kit	55.00	44.00	26.1
S-7461	**New**	*SlowFade*™ *Light* Antifade Kit	1 kit	55.00	44.00	26.1
S-1534		SMCC (succinimidyl *trans*-4-(maleimidylmethyl)cyclohexane-1-carboxylate)	100 mg	98.00	78.40	5.2
S-3054		SNAFL® calcein, acetoxymethyl ester *cell permeant* *special packaging*	20x50 µg	68.00	54.40	23.2
S-3052		SNAFL® calcein, ammonium salt *cell impermeant*	1 mg	42.00	33.60	23.2
N-7892	**New**	SNAP (S-nitroso-N-acetylpenicillamine)	25 mg	45.00	36.00	21.5
N-7927	**New**	SNAP (S-nitroso-N-acetylpenicillamine) *special packaging*	20x1 mg	65.00	52.00	21.5
S-3055		SNARF® calcein *cell impermeant*	1 mg	42.00	33.60	23.2
S-3057		SNARF® calcein, acetoxymethyl ester *cell permeant* *special packaging*	20x50 µg	68.00	54.40	23.2
		SNARF® indicators — see carboxy SNARF products; see also Section 23.2.				
S-6911	**New**	Sodium Green™ dextran, 10,000 MW, anionic	5 mg	135.00	108.00	24.1
S-6901	**New**	Sodium Green™ tetraacetate *cell permeant* *special packaging*	20x50 µg	138.00	110.40	24.1
S-6900	**New**	Sodium Green™, tetra(tetramethylammonium) salt *cell impermeant*	1 mg	98.00	78.40	24.1
S-8461	**New**	sodium polyanetholsulfonate	25 g	55.00	44.00	16.3
S-96096	**New**	solasodine	20 mg	55.00	44.00	16.3
S-2721		soluble β-glucan (β-glucan) *1 mg/mL*	500 µL	65.00	52.00	17.1
S-2720		soluble β-glucan, fluorescein conjugate (β-glucan, fluorescein conjugate) *1 mg/mL*	500 µL	118.00	94.40	17.1
S-96097	**New**	songorine	5 mg	135.00	108.00	18.8
S-1531		SPDP (succinimidyl 3-(2-pyridyldithio)propionate)	100 mg	115.00	92.00	5.2
S-7916	**New**	spermine NONOate	10 mg	48.00	38.40	21.5
M-440		SPQ (6-methoxy-N-(3-sulfopropyl)quinolinium, inner salt)	100 mg	63.00	50.40	24.2
S-6954	**New**	stachyose fluorescein, dipotassium salt	5 mg	95.00	76.00	15.3
S-2860		*Staphylococcus aureus* BioParticles® opsonizing reagent	1 U	55.00	44.00	17.1
S-2854		*Staphylococcus aureus* (Wood strain without protein A) BioParticles®, BODIPY® FL conjugate	10 mg	48.00	38.40	17.1
S-2851		*Staphylococcus aureus* (Wood strain without protein A) BioParticles®, fluorescein conjugate	10 mg	65.00	52.00	17.1
S-2852		*Staphylococcus aureus* (Wood strain without protein A) BioParticles®, tetramethylrhodamine conjugate	10 mg	48.00	38.40	17.1
S-2853		*Staphylococcus aureus* (Wood strain without protein A) BioParticles®, Texas Red® conjugate	10 mg	65.00	52.00	17.1
S-2859		*Staphylococcus aureus* (Wood strain without protein A) BioParticles®, unlabeled	100 mg	28.00	22.40	17.1
S-7456	**New**	staurosporine (antibiotic AM-2282)	100 µg	45.00	36.00	20.3
S-7445	**New**	(-)-ST-BODIPY® dihydropyridine *high affinity enantiomer*	25 µg	125.00	100.00	18.4
S-7446	**New**	(+)-ST-BODIPY® dihydropyridine *low affinity enantiomer*	25 µg	125.00	100.00	18.4
S-96047	**New**	stevioside	100 mg	45.00	36.00	18.8

Cat #		Product Name	Unit Size	Price Per Unit ($) 1–4 Units	5–24 Units	Section #
S-888		streptavidin	5 mg	135.00	108.00	7.5
S-2669		streptavidin *special packaging*	10x100 µg	48.00	38.40	7.5
S-951		streptavidin agarose *sedimented bead suspension*	5 mL	125.00	100.00	7.5, 9.1
S-921		streptavidin, alkaline phosphatase conjugate *2 mg/mL*	0.5 mL	155.00	124.00	6.3, 7.5, 9.2, 10.3
S-868		streptavidin, allophycocyanin, crosslinked, conjugate *1 mg/mL*	0.5 mL	285.00	228.00	6.4, 7.5
S-6364	**New**	streptavidin, AMCA-S conjugate	1 mg	95.00	76.00	7.5
S-2642		streptavidin, BODIPY® FL conjugate	1 mg	95.00	76.00	7.5
S-6372	**New**	streptavidin, eosin conjugate	1 mg	95.00	76.00	7.5
S-869		streptavidin, fluorescein conjugate	1 mg	95.00	76.00	7.5
S-931		streptavidin, β-galactosidase conjugate	1 mg	125.00	100.00	7.5, 9.2, 10.2
S-884		streptavidin, lucifer yellow conjugate	1 mg	95.00	76.00	7.5
S-6368	**New**	streptavidin, Oregon Green™ 488 conjugate	1 mg	95.00	76.00	7.5
S-6367	**New**	streptavidin, Oregon Green™ 500 conjugate	1 mg	95.00	76.00	7.5
S-6369	**New**	streptavidin, Oregon Green™ 514 conjugate	1 mg	95.00	76.00	7.5
S-911		streptavidin, peroxidase from horseradish conjugate	1 mg	95.00	76.00	7.5, 9.2, 10.5
S-867		streptavidin, B-phycoerythrin conjugate *1 mg/mL*	1 mL	145.00	116.00	6.4, 7.5
S-866		streptavidin, R-phycoerythrin conjugate *1 mg/mL*	1 mL	145.00	116.00	6.4, 7.5
S-871		streptavidin, rhodamine B conjugate	1 mg	95.00	76.00	7.5
S-6366	**New**	streptavidin, Rhodamine Red™-X conjugate	1 mg	95.00	76.00	7.5
S-6371	**New**	streptavidin, Rhodol Green™ conjugate	1 mg	95.00	76.00	7.5
S-870		streptavidin, tetramethylrhodamine conjugate	1 mg	95.00	76.00	7.5
S-872		streptavidin, Texas Red® conjugate	1 mg	95.00	76.00	7.5
S-6370	**New**	streptavidin, Texas Red®-X conjugate	1 mg	95.00	76.00	7.5
S-96205	**New**	strophanthidin 3-acetate (acetylstrophanthidin)	5 mg	55.00	44.00	10.6, 18.4
S-1553		succinimidyl acetylthioacetate (SATA)	100 mg	48.00	38.40	5.2
S-1573		succinimidyl acetylthiopropionate (SATP)	100 mg	48.00	38.40	5.2
S-1144		succinimidyl 9-anthracenepropionate	100 mg	38.00	30.40	1.8
B-1513		succinimidyl D-biotin (D-biotin, succinimidyl ester)	100 mg	53.00	42.40	4.2, 15.3
S-2182		succinimidyl 4-O-(4,4′-dimethoxytrityl)oxybutyrate	25 mg	45.00	36.00	1.8
S-1580		succinimidyl 4-formylbenzoate (SFB)	100 mg	20.00	16.00	3.2
I-6001	**New**	succinimidyl iodoacetate (iodoacetic acid, succinimidyl ester)	100 mg	65.00	52.00	5.2
S-1666		succinimidyl 6-((iodoacetyl)amino)hexanoate (SIAX)	100 mg	85.00	68.00	5.2
S-1668		succinimidyl 6-(6-(((iodoacetyl)amino)hexanoyl)amino)hexanoate (SIAXX)	25 mg	85.00	68.00	5.2
S-1535		succinimidyl 3-maleimidylbenzoate (SMB)	100 mg	42.00	33.60	5.2
S-1563		succinimidyl 6-maleimidylhexanoate (EMCS)	25 mg	48.00	38.40	5.2
S-1534		succinimidyl *trans*-4-(maleimidylmethyl)cyclohexane-1-carboxylate (SMCC)	100 mg	98.00	78.40	5.2
S-128		succinimidyl N-methylanthranilate	100 mg	98.00	78.40	1.8
S-1167		succinimidyl 6-(N-(7-nitrobenz-2-oxa-1,3-diazol-4-yl)amino)hexanoate (NBD-X, SE)	25 mg	98.00	78.40	1.8
S-6110	**New**	1-(3-(succinimidyloxycarbonyl)benzyl)-4-(5-(4-methoxyphenyl)oxazol-2-yl)pyridinium bromide (PyMPO, SE)	5 mg	98.00	78.40	1.7
S-1531		succinimidyl 3-(2-pyridyldithio)propionate (SPDP)	100 mg	115.00	92.00	5.2
S-520		succinimidyl 2,2,5,5-tetramethyl-3-pyrroline-1-oxyl-3-carboxylate	100 mg	55.00	44.00	1.8
S-6902	**New**	Sulfite Blue™ coumarin	10 mg	95.00	76.00	27.1
S-1107		N-(4-sulfobutyl)-4-(4-(4-(dibutylamino)phenyl)butadienyl)pyridinium, inner salt (RH 160)	25 mg	75.00	60.00	25.2
S-1109		N-(4-sulfobutyl)-4-(6-(4-(dibutylamino)phenyl)hexatrienyl)pyridinium, inner salt (RH 237)	5 mg	75.00	60.00	25.2
S-1108		N-(4-sulfobutyl)-4-(4-(4-(dipentylamino)phenyl)butadienyl)pyridinium, inner salt (RH 421)	25 mg	165.00	132.00	25.2
S-1129		5-sulfofluorescein diacetate, sodium salt (SFDA) *single isomer*	25 mg	85.00	68.00	15.3, 16.3, 23.2
S-6956	**New**	sulfonerhodamine bis-(PEG 2000)	10 mg	45.00	36.00	15.3
S-460		N-(3-sulfopropyl)acridinium, inner salt (SPA)	100 mg	80.00	64.00	24.2
S-467		N-(3-sulfopropyl)-4-(4-(didecylamino)styryl)pyridinium, inner salt (Di_{10}ASP-PS)	100 mg	80.00	64.00	14.2
S-359		sulforhodamine 101	25 mg	50.00	40.00	15.3, 17.4
S-1307		sulforhodamine B	5 g	78.00	62.40	9.2, 15.3
S-6957	**New**	sulforhodamine G *>60% by HPLC*	5 g	78.00	62.40	15.3
S-6680	**New**	SYBR® DNA molecular weight standards *low range* *100 gel lanes*	1 mL	50.00	40.00	8.3
S-6681	**New**	SYBR® DNA molecular weight standards *high range* *100 gel lanes*	1 mL	50.00	40.00	8.3
S-7550	**New**	SYBR® DX DNA blot stain *1000X concentrate in DMSO*	1 mL	175.00	140.00	8.1, 8.3
S-7563	**New**	SYBR® Green I nucleic acid gel stain *10,000X concentrate in DMSO*	500 µL	125.00	100.00	8.1, 8.3
S-7567	**New**	SYBR® Green I nucleic acid gel stain *10,000X concentrate in DMSO*	1 mL	235.00	188.00	8.1, 8.3
S-7585	**New**	SYBR® Green I nucleic acid gel stain *10,000X concentrate in DMSO* *special packaging*	20x50 µL	265.00	212.00	8.1, 8.3
S-7564	**New**	SYBR® Green II RNA gel stain *10,000X concentrate in DMSO*	500 µL	125.00	100.00	8.1, 8.3
S-7568	**New**	SYBR® Green II RNA gel stain *10,000X concentrate in DMSO*	1 mL	235.00	188.00	8.1, 8.3
S-7586	**New**	SYBR® Green II RNA gel stain *10,000X concentrate in DMSO* *special packaging*	20x50 µL	265.00	212.00	8.1, 8.3

Cat #		Product Name	Unit Size	Price Per Unit ($) 1–4 Units	5–24 Units	Section #
S-7569	*New*	SYBR® Green gel stain photographic filter	each	29.00	29.00	8.3, 26.4
S-7580	*New*	SYBR® Green Nucleic Acid Gel Stain Starter Kit	1 kit	55.00	55.00	8.1, 8.3
S-6650	*New*	SYPRO® Orange protein gel stain *5000X concentrate in DMSO*	500 µL	95.00	76.00	9.2
S-6651	*New*	SYPRO® Orange protein gel stain *5000X concentrate in DMSO* *special packaging*	10x50 µL	110.00	88.00	9.2
S-6656	*New*	SYPRO® Orange/Red protein gel stain photographic filter	each	35.00	35.00	9.2, 26.4
S-6655	*New*	SYPRO® Protein Gel Stain Starter Kit	1 kit	55.00	55.00	9.2
S-6653	*New*	SYPRO® Red protein gel stain *5000X concentrate in DMSO*	500 µL	95.00	76.00	9.2
S-6654	*New*	SYPRO® Red protein gel stain *5000X concentrate in DMSO* *special packaging*	10x50 µL	110.00	88.00	9.2
S-7573	*New*	SYTO® 11 live-cell nucleic acid stain *5 mM solution in DMSO*	250 µL	98.00	78.40	8.1
S-7574	*New*	SYTO® 12 live-cell nucleic acid stain *5 mM solution in DMSO*	250 µL	98.00	78.40	8.1
S-7575	*New*	SYTO® 13 live-cell nucleic acid stain *5 mM solution in DMSO*	250 µL	98.00	78.40	8.1
S-7576	*New*	SYTO® 14 live-cell nucleic acid stain *5 mM solution in DMSO*	250 µL	98.00	78.40	8.1
S-7577	*New*	SYTO® 15 live-cell nucleic acid stain *5 mM solution in DMSO*	250 µL	98.00	78.40	8.1
S-7578	*New*	SYTO® 16 live-cell nucleic acid stain *1 mM solution in DMSO*	250 µL	98.00	78.40	8.1
S-7555	*New*	SYTO® 20 live-cell nucleic acid stain *1 mM solution in DMSO*	250 µL	98.00	78.40	8.1
S-7556	*New*	SYTO® 21 live-cell nucleic acid stain *5 mM solution in DMSO*	250 µL	98.00	78.40	8.1
S-7557	*New*	SYTO® 22 live-cell nucleic acid stain *5 mM solution in DMSO*	250 µL	98.00	78.40	8.1
S-7558	*New*	SYTO® 23 live-cell nucleic acid stain *5 mM solution in DMSO*	250 µL	98.00	78.40	8.1
S-7559	*New*	SYTO® 24 live-cell nucleic acid stain *5 mM solution in DMSO*	250 µL	98.00	78.40	8.1
S-7560	*New*	SYTO® 25 live-cell nucleic acid stain *5 mM solution in DMSO*	250 µL	98.00	78.40	8.1
S-7572	*New*	SYTO® Live-Cell Nucleic Acid Stain Sampler Kit #1 *SYTO® dyes 11-16* *50 µL each*	1 kit	135.00	108.00	8.1, 16.3
S-7554	*New*	SYTO® Live-Cell Nucleic Acid Stain Sampler Kit #2 *SYTO® dyes 20-25* *50 µL each*	1 kit	135.00	108.00	8.1, 16.3
S-7579	*New*	SYTO® 17 red fluorescent nucleic acid stain *5 mM solution in DMSO*	250 µL	98.00	78.40	8.1
S-7529	*New*	SYTO® 18 yeast mitochondrial stain *5 mM solution in DMSO*	250 µL	98.00	78.40	12.2
S-7020	*New*	SYTOX® Green nucleic acid stain *5 mM solution in DMSO*	250 µL	98.00	78.40	8.1, 16.3
T-96098	*New*	tadzhaconine	20 mg	95.00	76.00	18.8
T-6536	*New*	TAME (N^{α}-p-tosyl-L-arginine methyl ester, hydrochloride)	5 g	18.00	14.40	10.4
C-6121	*New*	5-TAMRA (5-carboxytetramethylrhodamine) *single isomer*	10 mg	95.00	76.00	1.6
C-6122	*New*	6-TAMRA (6-carboxytetramethylrhodamine) *single isomer*	10 mg	95.00	76.00	1.6
C-300		5(6)-TAMRA (5-(and-6)-carboxytetramethylrhodamine) *mixed isomers*	100 mg	80.00	64.00	1.6
C-2211		5-TAMRA, SE (5-carboxytetramethylrhodamine, succinimidyl ester) *single isomer*	5 mg	148.00	118.40	1.6
C-6123	*New*	6-TAMRA, SE (6-carboxytetramethylrhodamine, succinimidyl ester) *single isomer*	5 mg	135.00	108.00	1.6
C-1171		5(6)-TAMRA, SE (5-(and-6)-carboxytetramethylrhodamine, succinimidyl ester) *mixed isomers*	25 mg	118.00	94.40	1.6
T-6105	*New*	5(6)-TAMRA-X, SE (6-(tetramethylrhodamine-5-(and-6)-carboxamido)hexanoic acid, succinimidyl ester) *mixed isomers*	10 mg	118.00	94.40	1.6, 4.2
T-8462	*New*	taurocholic acid, sodium salt *high purity*	5 g	48.00	38.40	16.3
P-3456		Taxol equivalent (paclitaxel) *for use in research only*	5 mg	40.00	32.00	11.2
T-2556		TCEP (tris-(2-carboxyethyl)phosphine, hydrochloride)	1 g	25.00	20.00	2.1, 5.2, 9.4
B-1174		TCPO (bis-(2,4,6-trichlorophenyl) oxalate)	1 g	25.00	20.00	1.8, 24.3
D-530		TEMPO choline (4-(N,N-dimethyl-N-(2-hydroxyethyl))ammonium-2,2,6,6-tetramethylpiperidine-1-oxyl chloride)	25 mg	98.00	78.40	15.3
I-591		TEMPO iodoacetamide (N-(4-(iodoacetyl)amino)-2,2,6,6-tetramethyl-piperidinyl-1-oxyl)	25 mg	55.00	44.00	2.4
T-1247		terbium(III) chloride, hexahydrate	1 g	20.00	16.00	9.2, 15.3, 20.2, 22.7
T-7539	*New*	2′,4′,5′,7′-tetrabromorhodamine 123 bromide	5 mg	48.00	38.40	12.2, 21.2
T-7015	*New*	2′,4′,5′,7′-tetrabromo-4,5,6,7-tetrafluorofluorescein (Br_4TFF)	10 mg	35.00	28.00	16.3
T-7017	*New*	2′,4′,5′,7′-tetrabromo-4,5,6,7-tetrafluorofluorescein diacetate (Br_4TFFDA)	5 mg	125.00	100.00	15.2, 16.3
T-3168		5,5′,6,6′-tetrachloro-1,1′,3,3′-tetraethylbenzimidazolylcarbocyanine iodide (JC-1; CBIC$_2$(3))	5 mg	195.00	156.00	12.2, 16.3, 25.3
T-6817	*New*	5,5′,6,6′-tetrafluoro BAPTA, AM *cell permeant* *≥95% by HPLC*	10 mg	65.00	52.00	22.8
T-6816	*New*	5,5′,6,6′-tetrafluoro BAPTA, tetrapotassium salt *cell impermeant*	25 mg	65.00	52.00	22.8
T-7014	*New*	4,5,6,7-tetrafluorofluorescein (TFF)	10 mg	35.00	28.00	23.3
T-7106	*New*	4,5,6,7-tetrafluorofluorescein bis-(5-carboxymethoxy-2-nitrobenzyl) ether, dipotassium salt (CMNB-caged TFF)	5 mg	155.00	124.00	27.1
T-7016	*New*	4,5,6,7-tetrafluorofluorescein diacetate (TFFDA)	5 mg	125.00	100.00	15.2, 16.3
T-96016	*New*	DL-tetrahydroberberine (canadine)	20 mg	128.00	102.40	18.2
T-7920	*New*	(6R)-5,6,7,8-tetrahydro-L-biopterin, dihydrochloride (6R-BH_4)	25 mg	48.00	38.40	21.5
T-96038	*New*	DL-tetrahydropalmatine, hydrochloride	5 mg	105.00	84.00	18.2
T-6931	*New*	tetrakis-(4-carboxyphenyl)porphine (TCPP)	5 mg	48.00	38.40	15.3, 22.7
T-1210		tetrakis-(2-pyridylmethyl)ethylenediamine (TPEN)	100 mg	50.00	40.00	22.8

Cat #		Product Name	Unit Size	Price Per Unit ($) 1–4 Units	5–24 Units	Section #
T-6932	***New***	tetrakis-(4-sulfophenyl)porphine (TSPP)	5 mg	48.00	38.40	15.3, 22.7
T-8464	***New***	3,3′,5,5′-tetramethylbenzidine, dihydrochloride (TMB, HCl)	5 g	95.00	76.00	9.2, 10.5
T-8463	***New***	3,3′,5,5′-tetramethylbenzidine, free base (TMB)	5 g	75.00	60.00	9.2, 10.5
T-99016	***New***	tetramethylenedisulfotetramide (TETS)	50 µg	155.00	124.00	18.5
T-6105	***New***	6-(tetramethylrhodamine-5-(and-6)-carboxamido)hexanoic acid, succinimidyl ester (5(6)-TAMRA-X, SE) *mixed isomers*	10 mg	118.00	94.40	1.6, 4.2
T-1175		tetramethylrhodamine α-bungarotoxin (α-bungarotoxin, tetramethylrhodamine conjugate)	500 µg	195.00	156.00	18.2
A-1318		tetramethylrhodamine cadaverine (5-(and-6)-((N-(5-aminopentyl)amino)-carbonyl)tetramethylrhodamine) *mixed isomers*	10 mg	98.00	78.40	3.3
T-6219	***New***	tetramethylrhodamine-5-carbonyl azide	5 mg	98.00	78.40	3.1
T-669		tetramethylrhodamine, ethyl ester, perchlorate (TMRE)	25 mg	72.00	57.60	12.2, 25.3
T-2762		tetramethylrhodamine goat anti-mouse IgG (H+L) conjugate *2 mg/mL*	0.5 mL	75.00	60.00	7.2
T-2769		tetramethylrhodamine goat anti-rabbit IgG (H+L) conjugate *2 mg/mL*	0.5 mL	75.00	60.00	7.2
T-6006	***New***	tetramethylrhodamine-5-iodoacetamide (5-TMRIA) *single isomer*	5 mg	175.00	140.00	2.2, 9.4
T-1480		tetramethylrhodamine-5-isothiocyanate (5-TRITC; G isomer)	5 mg	75.00	60.00	1.6
T-1481		tetramethylrhodamine-6-isothiocyanate (6-TRITC; R isomer)	5 mg	75.00	60.00	1.6
T-490		tetramethylrhodamine-5-(and-6)-isothiocyanate (5(6)-TRITC) *mixed isomers*	10 mg	85.00	68.00	1.6
T-6027	***New***	tetramethylrhodamine-5-maleimide *single isomer*	5 mg	125.00	100.00	2.2, 9.4
T-6028	***New***	tetramethylrhodamine-6-maleimide *single isomer*	5 mg	125.00	100.00	2.2, 9.4
T-668		tetramethylrhodamine, methyl ester, perchlorate (TMRM)	25 mg	72.00	57.60	12.2, 25.3
T-3426		tetramethylrhodamine NAPS	1 mg	68.00	54.40	18.2
T-3422		tetramethylrhodamine SCH 23390 *1 mg/mL solution in dry DMSO*	1 mL	68.00	54.40	18.2
T-1391		N-(6-tetramethylrhodaminethiocarbamoyl)-1,2-dihexadecanoyl-*sn*-glycero-3-phosphoethanolamine, triethylammonium salt (TRITC DHPE)	1 mg	75.00	60.00	13.4
T-639		tetramethylrosamine chloride	25 mg	72.00	57.60	12.2
T-7284	***New***	TetraSpeck™ Fluorescent Microspheres Sampler Kit	1 kit	95.00	76.00	26.1
T-7279	***New***	TetraSpeck™ microspheres, 0.1 µm, fluorescent blue/green/orange/dark red	0.5 mL	95.00	76.00	26.1
T-7280	***New***	TetraSpeck™ microspheres, 0.2 µm, fluorescent blue/green/orange/dark red	0.5 mL	95.00	76.00	26.1
T-7281	***New***	TetraSpeck™ microspheres, 0.5 µm, fluorescent blue/green/orange/dark red	0.5 mL	95.00	76.00	26.1
T-7282	***New***	TetraSpeck™ microspheres, 1.0 µm, fluorescent blue/green/orange/dark red	0.5 mL	95.00	76.00	26.1
T-7283	***New***	TetraSpeck™ microspheres, 4.0 µm, fluorescent blue/green/orange/dark red	0.5 mL	95.00	76.00	26.1
T-6490	***New***	tetrazolium blue chloride	5 g	35.00	28.00	16.3, 21.4
T-6500	***New***	Tetrazolium Salt Sampler Kit *100 mg of T-6490, X-6493, M-6494, N-6495, I-6496, N-6498*	1 kit	125.00	100.00	16.3, 21.4
T-6913	***New***	tetrodotoxin (TTX) *contains citrate buffer*	1 mg	98.00	78.40	18.7
T-6914	***New***	tetrodotoxin (TTX) *citrate free*	1 mg	98.00	78.40	18.7
T-99016	***New***	TETS (tetramethylenedisulfotetramide)	50 µg	155.00	124.00	18.5
C-7608	***New***	Texas Red®-5-dUTP (ChromaTide™ Texas Red®-5-dUTP) *1 mM in buffer*	25 µL	175.00	175.00	8.2
C-7607	***New***	Texas Red®-5-UTP (ChromaTide™ Texas Red®-5-UTP) *1 mM in buffer*	25 µL	175.00	175.00	8.2
T-6009	***New***	Texas Red® C_5 bromoacetamide	5 mg	95.00	76.00	2.2
T-6008	***New***	Texas Red® C_2 maleimide	5 mg	95.00	76.00	2.2
T-2425		Texas Red® cadaverine (Texas Red® C_5)	5 mg	125.00	100.00	3.3
T-1395		Texas Red® DHPE (N-(Texas Red® sulfonyl)-1,2-dihexadecanoyl-*sn*-glycero-3-phosphoethanolamine, triethylammonium salt)	1 mg	75.00	60.00	13.4
C-7631	***New***	Texas Red®-12-dUTP (ChromaTide™ Texas Red®-12-dUTP) *1 mM in buffer*	25 µL	175.00	175.00	8.2
T-862		Texas Red® goat anti-mouse IgG (H+L) conjugate *2 mg/mL*	0.5 mL	75.00	60.00	7.2
T-2767		Texas Red® goat anti-rabbit IgG (H+L) conjugate *2 mg/mL*	0.5 mL	75.00	60.00	7.2
T-6256	***New***	Texas Red® hydrazide *>90% single isomer*	5 mg	95.00	76.00	3.2, 9.4
T-3427		Texas Red® NAPS	1 mg	68.00	54.40	18.2
T-3423		Texas Red® SCH 23390	1 mg	68.00	54.40	18.2
T-353		Texas Red® sulfonyl chloride *mixed isomers*	10 mg	98.00	78.40	1.6
T-1905		Texas Red® sulfonyl chloride *mixed isomers* *special packaging*	10x~1 mg	118.00	94.40	1.6
T-1395		N-(Texas Red® sulfonyl)-1,2-dihexadecanoyl-*sn*-glycero-3-phosphoethanolamine, triethylammonium salt (Texas Red® DHPE)	1 mg	75.00	60.00	13.4
T-7710	***New***	N-(Texas Red®-X)-1,2-dihexadecanoyl-*sn*-glycero-3-phosphoethanolamine, triethylammonium salt (Texas Red®-X DHPE)	1 mg	75.00	60.00	13.4
T-6390	***New***	Texas Red®-X goat anti-mouse IgG (H+L) conjugate *2 mg/mL*	0.5 mL	75.00	60.00	7.2
T-6391	***New***	Texas Red®-X goat anti-rabbit IgG (H+L) conjugate *2 mg/mL*	0.5 mL	75.00	60.00	7.2
T-6392	***New***	Texas Red®-X goat anti-rat IgG (H+L) conjugate *2 mg/mL*	0.5 mL	75.00	60.00	7.2
T-7471	***New***	Texas Red®-X phalloidin	300 U	198.00	158.40	11.1
T-6134	***New***	Texas Red®-X, succinimidyl ester *mixed isomers*	5 mg	95.00	76.00	1.6, 4.2
T-7458	***New***	thapsigargin	1 mg	55.00	44.00	20.2
T-7459	***New***	thapsigargin *special packaging*	20x50 µg	75.00	60.00	20.2
T-6060	***New***	Thiol and Sulfide Quantitation Kit *50-250 assays*	1 kit	95.00	76.00	2.4, 5.2, 24.3
T-6483	***New***	TLCK (N^{α}-tosyl-L-lysine chloromethyl ketone, hydrochloride)	100 mg	25.00	20.00	10.6
T-204		TMA-DPH (1-(4-trimethylammoniumphenyl)-6-phenyl-1,3,5-hexatriene p-toluenesulfonate)	25 mg	155.00	124.00	14.4, 17.2

Cat #		Product Name	Unit Size	Price Per Unit ($) 1–4 Units	5–24 Units	Section #
P-3900		TMAP-DPH (N-((4-(6-phenyl-1,3,5-hexatrienyl)phenyl)propyl)trimethylammonium p-toluenesulfonate)	5 mg	148.00	118.40	14.4, 17.2
D-1579		TMB-8 (8-diethylaminooctyl 3,4,5-trimethoxybenzoate, hydrochloride)	100 mg	60.00	48.00	20.2
C-7606	New	TMR-5-dUTP (ChromaTide™ tetramethylrhodamine-5-dUTP) *1 mM in buffer*	25 µL	175.00	175.00	8.2
C-7605	New	TMR-5-UTP (ChromaTide™ tetramethylrhodamine-5-UTP) *1 mM in buffer*	25 µL	175.00	175.00	8.2
T-6006	New	5-TMRIA (tetramethylrhodamine-5-iodoacetamide) *single isomer*	5 mg	175.00	140.00	2.2, 9.4
T-7601	New	TNP-ADP (2′-(or-3′)-O-(trinitrophenyl)adenosine 5′-diphosphate, disodium salt) *5 mg/mL in water*	2 mL	160.00	128.00	11.3
T-7624	New	TNP-AMP (2′-(or-3′)-O-(trinitrophenyl)adenosine 5′-monophosphate, sodium salt) *5 mg/mL in water*	2 mL	48.00	38.40	11.3
T-7602	New	TNP-ATP (2′-(or-3′)-O-(trinitrophenyl)adenosine 5′-triphosphate, trisodium salt) *5 mg/mL in 0.1 M Tris buffer pH 9*	2 mL	160.00	128.00	11.3
T-7600	New	TNP-GTP (2′-(or-3′)-O-(trinitrophenyl)guanosine 5′-triphosphate, trisodium salt) *5 mg/mL in water*	1 mL	160.00	128.00	11.3
T-53		2,6-TNS (2-(p-toluidinyl)naphthalene-6-sulfonic acid, sodium salt)	100 mg	48.00	38.40	14.5
T-406		TNS chloride (6-(p-toluidinyl)naphthalene-2-sulfonyl chloride)	25 mg	135.00	108.00	1.7
T-53		2-(p-toluidinyl)naphthalene-6-sulfonic acid, sodium salt (2,6-TNS)	100 mg	48.00	38.40	14.5
T-406		6-(p-toluidinyl)naphthalene-2-sulfonyl chloride (TNS chloride)	25 mg	135.00	108.00	1.7
T-3602		TO-PRO™-1 iodide (515/531) *1 mM solution in DMSO*	1 mL	135.00	108.00	8.1
T-3605		TO-PRO™-3 iodide (642/661) *1 mM solution in DMSO*	1 mL	135.00	108.00	8.1
T-7596	New	TO-PRO™-5 iodide (745/770) *1 mM solution in DMSO*	1 mL	135.00	108.00	8.1
T-6536	New	N^{α}-p-tosyl-L-arginine methyl ester, hydrochloride (TAME)	5 g	18.00	14.40	10.4
T-6483	New	N^{α}-tosyl-L-lysine chloromethyl ketone, hydrochloride (TLCK)	100 mg	25.00	20.00	10.6
T-6484	New	N-tosyl-L-phenylalanine chloromethyl ketone (TPCK)	1 g	42.00	33.60	10.6
T-3600		TOTO®-1 iodide (514/533) *1 mM solution in DMSO*	200 µL	225.00	180.00	8.1
T-3604		TOTO®-3 iodide (642/660) *1 mM solution in DMSO*	200 µL	225.00	180.00	8.1
P-3453		TPA; PMA (4β-phorbol 12β-myristate 13α-acetate)	1 mg	25.00	20.00	20.3
P-3454		TPA; PMA (4β-phorbol 12β-myristate 13α-acetate) *special packaging*	10x100 µg	35.00	28.00	20.3
T-6484	New	TPCK (N-tosyl-L-phenylalanine chloromethyl ketone)	1 g	42.00	33.60	10.6
T-1210		TPEN (tetrakis-(2-pyridylmethyl)ethylenediamine)	100 mg	50.00	40.00	22.8
T-2873		transferrin from human serum, BODIPY® FL conjugate	5 mg	148.00	118.40	17.3
T-2871		transferrin from human serum, fluorescein conjugate	5 mg	148.00	118.40	17.3
T-2872		transferrin from human serum, tetramethylrhodamine conjugate	5 mg	148.00	118.40	17.3
T-2875		transferrin from human serum, Texas Red® conjugate	5 mg	148.00	118.40	17.3
		TransFluoSpheres® polystyrene microspheres — see Section 6.3 for a complete list of FluoSpheres® and TransFluoSpheres® polystyrene microspheres.				
T-96049	New	trichodesmine	5 mg	108.00	86.40	18.8
T-6615	New	tricyclodecan-9-yl xanthogenate, potassium salt (D 609)	5 mg	48.00	38.40	20.5
T-6693	New	N-(4-triethylammoniumbutyl)-4-(4-(4-(diethylamino)phenyl)butadienyl)pyridinium dibromide (RH 704)	5 mg	145.00	116.00	25.2
T-3163		N-(3-triethylammoniumpropyl)-4-(4-(dibutylamino)styryl)pyridinium dibromide (FM™ 1-43)	1 mg	165.00	132.00	15.4, 17.2
T-1111		N-(3-triethylammoniumpropyl)-4-(4-(4-(diethylamino)phenyl)butadienyl)pyridinium dibromide (RH 414)	5 mg	165.00	132.00	15.4, 17.2, 25.2
T-3166		N-(3-triethylammoniumpropyl)-4-(6-(4-(diethylamino)phenyl)hexatrienyl)pyridinium dibromide (FM™ 4-64)	1 mg	165.00	132.00	12.3, 15.4, 17.2
T-7508	New	N-(3-triethylammoniumpropyl)-4-(4-(diethylamino)styryl)pyridinium dibromide (FM™ 2-10)	5 mg	135.00	108.00	17.2
T-7506	New	N-(3-triethylammoniumpropyl)-4-(4-(dioctadecylamino)styryl)pyridinium, di-4-chlorobenzenesulfonate (FM™ 3-25)	5 mg	125.00	100.00	17.2
T-3164		N-(3-triethylammoniumpropyl)-4-(4-(dipentylamino)styryl)pyridinium dibromide (FM™ 1-84)	1 mg	165.00	132.00	17.2
T-411		2-((trifluoroacetyl)acetyl)naphthalene (naphthoyltrifluoroacetone; NTA)	100 mg	65.00	52.00	15.3, 16.3, 22.7
T-6537	New	trifluoromethanesulfonamide	100 mg	38.00	30.40	10.6
T-6568	New	1-(2-trifluoromethylphenyl)imidazole (TRIM)	100 mg	65.00	52.00	27.1
T-659		β-trifluoromethylumbelliferone (7-hydroxy-4-trifluoromethylcoumarin) *reference standard*	100 mg	18.00	14.40	10.1
T-657		β-trifluoromethylumbelliferyl β-D-galactopyranoside	100 mg	68.00	54.40	10.2
T-658		β-trifluoromethylumbelliferyl β-D-glucuronide	25 mg	125.00	100.00	10.2
T-6568	New	TRIM (1-(2-trifluoromethylphenyl)imidazole)	100 mg	65.00	52.00	27.1
T-204		1-(4-trimethylammoniumphenyl)-6-phenyl-1,3,5-hexatriene p-toluenesulfonate (TMA-DPH)	25 mg	155.00	124.00	14.4, 17.2
T-1113		N-(3-trimethylammoniumpropyl)-4-(4-(4-(diethylamino)phenyl)butadienyl)-pyridinium dibromide (RH 461)	5 mg	75.00	60.00	25.2
T-506		4-trimethylammonium-2,2,6,6-tetramethylpiperidine-1-oxyl iodide (CAT 1)	100 mg	95.00	76.00	15.3
T-1365		4,5′,8-trimethylpsoralen (trioxsalen)	100 mg	45.00	36.00	8.1

Cat #		Product Name	Unit Size	Price Per Unit ($) 1–4 Units	5–24 Units	Section #
T-7601	New	2′-(or-3′)-O-(trinitrophenyl)adenosine 5′-diphosphate, disodium salt (TNP-ADP) *5 mg/mL in water*	2 mL	160.00	128.00	11.3
T-7624	New	2′-(or-3′)-O-(trinitrophenyl)adenosine 5′-monophosphate, sodium salt (TNP-AMP) *5 mg/mL in water*	2 mL	48.00	38.40	11.3
T-7602	New	2′-(or-3′)-O-(trinitrophenyl)adenosine 5′-triphosphate, trisodium salt (TNP-ATP) *5 mg/mL in 0.1 M Tris buffer pH 9*	2 mL	160.00	128.00	11.3
T-7600	New	2′-(or-3′)-O-(trinitrophenyl)guanosine 5′-triphosphate, trisodium salt (TNP-GTP) *5 mg/mL in water*	1 mL	160.00	128.00	11.3
T-1365		trioxsalen (4,5′,8-trimethylpsoralen)	100 mg	45.00	36.00	8.1
T-2556		tris-(2-carboxyethyl)phosphine, hydrochloride (TCEP)	1 g	25.00	20.00	2.1, 5.2, 9.4
T-6052	New	tris-(2-cyanoethyl)phosphine	1 g	24.00	19.20	2.1, 5.2, 9.4
T-99017	New	*P,P,P*-tris-(hydroxymethyl)adamantane-1-phosphonium chloride (adaphotris)	10 mg	128.00	102.40	18.4
T-1480		5-TRITC; G isomer (tetramethylrhodamine-5-isothiocyanate)	5 mg	75.00	60.00	1.6
T-1481		6-TRITC; R isomer (tetramethylrhodamine-6-isothiocyanate)	5 mg	75.00	60.00	1.6
T-490		5(6)-TRITC (tetramethylrhodamine-5-(and-6)-isothiocyanate) *mixed isomers*	10 mg	85.00	68.00	1.6
T-1391		TRITC DHPE (N-(6-tetramethylrhodaminethiocarbamoyl)-1,2-dihexadecanoyl-*sn*-glycero-3-phosphoethanolamine, triethylammonium salt)	1 mg	75.00	60.00	13.4
T-1323		true blue chloride	5 mg	103.00	82.40	15.3
T-96048	New	tschimganidine	10 mg	95.00	76.00	18.8
T-96072	New	tschimganine	10 mg	75.00	60.00	18.8
T-96073	New	tschimgine	10 mg	108.00	86.40	18.8
M-688		TSQ (N-(6-methoxy-8-quinolyl)-p-toluenesulfonamide)	25 mg	85.00	68.00	22.7
T-6913	New	TTX (tetrodotoxin) *contains citrate buffer*	1 mg	98.00	78.40	18.7
T-6914	New	TTX (tetrodotoxin) *citrate free*	1 mg	98.00	78.40	18.7
T-7451	New	tubulin from bovine brain *10 mg/mL*	100 µL	125.00	100.00	11.2
T-7460	New	tubulin from bovine brain, tetramethylrhodamine conjugate *10 mg/mL*	10 µL	145.00	116.00	11.2
		Turner fluorometer — see fluorometer, Turner Designs TD-700; Section 26.3.				
T-98206	New	[tyr²]-leiurotoxin I ([tyr²]-scyllatoxin) *synthetic*	50 µg	385.00	308.00	18.6
D-7760	New	U-6 (4-(N,N-dimethyl-N-tetradecylammonium)methyl-(7-hydroxycoumarin) chloride)	10 mg	48.00	38.40	14.3, 23.4
U-96074	New	ugaferine	10 mg	108.00	86.40	18.8
U-96075	New	ungerine nitrate	10 mg	55.00	44.00	18.8
U-6059	New	UniBlue A vinyl sulfone, sodium salt *>80% by HPLC*	100 mg	18.00	14.40	2.4
U-96076	New	unsevine	10 mg	85.00	68.00	18.8
V-1644		valinomycin	25 mg	42.00	33.60	24.1
V-96077	New	veratroylzygadenine	10 mg	55.00	44.00	18.7
V-7023	New	ViaGram™ Red⁺ Bacterial Gram Stain and Viability Kit *200 assays*	1 kit	95.00	76.00	16.2
N-96025	New	vincanine (norfluorocurarine)	20 mg	108.00	86.40	18.8
W-6745	New	wheat germ agglutinin, AMCA-S conjugate	5 mg	98.00	78.40	7.6
W-966		wheat germ agglutinin, Cascade Blue® conjugate	5 mg	98.00	78.40	7.6
W-834		wheat germ agglutinin, fluorescein conjugate	5 mg	98.00	78.40	7.6
W-6748	New	wheat germ agglutinin, Oregon Green™ 488 conjugate	5 mg	98.00	78.40	7.6
W-7024	New	Wheat Germ Agglutinin Sampler Kit *four fluorescent conjugates, 1 mg each*	1 kit	98.00	78.40	7.6, 16.2
W-849		wheat germ agglutinin, tetramethylrhodamine conjugate	5 mg	98.00	78.40	7.6
W-6746	New	wheat germ agglutinin, Texas Red®-X conjugate	5 mg	98.00	78.40	7.6
W-435		WW 781, triethylammonium salt	25 mg	145.00	116.00	18.5, 25.2
B-1690		X-Gal (5-bromo-4-chloro-3-indolyl β-D-galactopyranoside)	1 g	55.00	44.00	9.2, 10.2
B-1691		X-GlcU, CHA (5-bromo-4-chloro-3-indolyl β-D-glucuronide, cyclohexylammonium salt)	100 mg	115.00	92.00	9.2, 10.2
B-8404	New	X-GlcU, Na (5-bromo-4-chloro-3-indolyl β-D-glucuronide, sodium salt)	50 mg	120.00	96.00	10.2
X-491		X-rhodamine-5-(and-6)-isothiocyanate (5(6)-XRITC) *mixed isomers*	10 mg	95.00	76.00	1.6
X-491		5(6)-XRITC (X-rhodamine-5-(and-6)-isothiocyanate) *mixed isomers*	10 mg	95.00	76.00	1.6
X-6493	New	XTT (2,3-bis-(2-methoxy-4-nitro-5-sulfophenyl)-2H-tetrazolium-5-carboxanilide)	100 mg	55.00	44.00	16.3, 21.4
X-1525		p-xylene-bis-pyridinium bromide (DPX)	1 g	75.00	60.00	15.3
Y-7530	New	Yeast Mitochondrial Stain Sampler Kit	1 kit	95.00	76.00	12.2, 16.2
Y-7531	New	Yeast Vacuole Marker Sampler Kit	1 kit	95.00	76.00	12.3, 16.2
Y-7536	New	yeast vacuole membrane marker MDY-64	1 mg	48.00	38.40	12.3
Y-3603		YO-PRO™-1 iodide (491/509) *1 mM solution in DMSO*	1 mL	135.00	108.00	8.1
Y-3607		YO-PRO™-3 iodide (612/631) *1 mM solution in DMSO*	1 mL	135.00	108.00	8.1
Y-3601		YOYO®-1 iodide (491/509) *1 mM solution in DMSO*	200 µL	225.00	180.00	8.1
Y-3606		YOYO®-3 iodide (612/631) *1 mM solution in DMSO*	200 µL	225.00	180.00	8.1
Z-2850		zymosan A BioParticles® opsonizing reagent	1 U	55.00	44.00	17.1
Z-2844		zymosan A (*S. cerevisiae*) BioParticles®, BODIPY® FL conjugate	10 mg	48.00	38.40	17.1
Z-2841		zymosan A (*S. cerevisiae*) BioParticles®, fluorescein conjugate	10 mg	65.00	52.00	17.1
Z-2842		zymosan A (*S. cerevisiae*) BioParticles®, tetramethylrhodamine conjugate	10 mg	48.00	38.40	17.1
Z-2843		zymosan A (*S. cerevisiae*) BioParticles®, Texas Red® conjugate	10 mg	65.00	52.00	17.1
Z-2849		zymosan A (*S. cerevisiae*) BioParticles®, unlabeled	100 mg	28.00	22.40	17.1

Index

A

B

C

K

L

M

N

O

P

Q

R

Trademark and Patent Information

Trademark Information

Molecular Probes has achieved worldwide recognition of its trademarks. Please respect our trademarks, as we will vigorously protect their proper usage. The following are trademarks of Molecular Probes, Inc.:

AlignFlow™
ATTO-TAG™
*Bac*Light™
BioParticles®
BOBO™
BODIPY®
BO-PRO™
Calcium Crimson™
Calcium Green™
Calcium Orange™
Calcium Sponge™
Cascade Blue®
CellTracker™
ChromaTide™
CompenFlow™
CyQUANT™
DEAD Red™
DetectaGene™
ELF®
EnzChek™
FAST CAT®
FAST DiA™
FAST DiI™
FAST DiO™
FluoReporter®
FluoroTide™
FluoSpheres®
FM™
FocalCheck™
FotoFenton™
FUN-1™
Fura Red™
ImaGene™
ImaGene Green™
ImaGene Red™
InSpeck™
LinearFlow™
LIVE/DEAD®
LysoSensor™
LysoTracker™
Mag-Fura Red™
Magnesium Green™
Magnesium Orange™
MitoFluor™
MitoTracker™
MultiSpeck™
MycoFluor™
NanoOrange™
NeuroTrace™
Newport Green™
OliGreen™
Oregon Green™
OxyBURST®
Phen Green™
PicoGreen®
POPO™
PO-PRO™
ProLong™
PS-Speck™
Rhodamine Green™
Rhodamine Red™
Rhodol Green™
SlowFade™
SNAFL®
SNARF®
Sodium Green™
Sulfite Blue™
SYBR®
SYPRO®
SYTO®
SYTOX®
TetraSpeck™
Texas Red®
TO-PRO™
TOTO®
TransFluoSpheres®
ViaGram™
YO-PRO™
YOYO®

Trademark of other companies that are referred to in our product literature include:

ABTS™ (Boehringer Mannheim GmbH)
AquaLite® (SeaLite Sciences, Inc.)
Attofluor® (Atto Instruments, Inc.)
Calcofluor™ (American Cyanamid)
Celite® (Celite Corp.)
Cellosolve® (Union Carbide Corp.)
ChemWindow® (SoftShell International, Ltd.)
Colcemid® (Ciba, Ltd.)
Coomassie® (Imperial Chemical Industries, Ltd.)
Cy™ (Amersham International PLC)
CytoFluor® (PerSeptive Biosystems, Inc.)
Cytoseal™ (Stephens Scientific)
DM-Nitrophen™ (Calbiochem-Novabiochem Corp.)
FACS® (Becton Dickinson and Co.)
FACScan™ (Becton Dickinson and Co.)
Ficoll® (Pharmacia Fine Chemicals, Inc.)
FluorImager™ (Molecular Dynamics, Inc.)
Fluoro-Gold™ (Fluorochrome, Inc.)
Fluorosensor® (Atto Instruments, Inc.)
FMBIO® (Hitachi Software Engineering Co., Ltd.)
Haema-Line® (Serono Diagnostics, Inc.)
Hybond® (Amersham International PLC)
Hydroethidine™ (Prescott Labs)
Image-1® (Universal Imaging Corp.)
Labophot® (Nikon, Inc.)
Lissamine™ (Imperial Chemical Industries, Ltd.)
Neurobiotin™ (Vector Laboratories)
NeutraLite™ (Belovo Chemicals)
Nonidet™ (Shell International Petroleum Co., Ltd.)
Omega® (Omega Optical, Inc.)
Parafilm® (American National Can Co.)
Pluronic® (BASF Corp.)
Polaroid® (Polaroid Corp.)
Quadfluor™ (Nikon, Inc)
RatioVision® (Atto Instruments, Inc.)
Sephadex® (Pharmacia, Inc.)
SoftShell® (SoftShell International, Ltd.)
SPF-500™ (SLM-Aminco)
Star® (Star Micronics America, Inc.)
Star 1™ (Photometrics)
Storm™ (Molecular Dynamics, Inc.)
Taxol® (Bristol-Meyers Squibb Co.)
Teflon® (E.I. DuPont de Nemours and Co.)
Threshold® (Molecular Devices Corp.)
Triton® (Union Carbide Corp.)
Tween® (ICI Americas, Inc.)
Vantage™ (Becton Dickinson and Co.)
VIVID™ (Omega Optical, Inc.)
ZYMOCEL™ (Alpha-Beta Technology, Inc.)

All names containing the designation ® are registered with the U.S. Patent and Trademark Office.

Licensing and Patent Summary

Products and applications discussed in Molecular Probes' product literature may be covered by one or more U.S. or foreign patents, or patents pending, owned by Molecular Probes, Inc.:

4,213,904	5,227,487	5,316,906	5,405,975	5,445,946	5,516,864
4,774,339	5,242,805	5,321,130	5,410,030	5,451,663	5,534,416
4,945,171	5,248,782	5,326,692	5,433,896	5,453,517	5,545,535
5,132,432	5,262,545	5,338,854	5,436,134	5,459,268	5,573,909
5,187,288	5,274,113	5,362,628	5,437,980	5,459,276	5,576,424
5,208,148	5,314,805	5,364,764	5,442,045	5,501,980	EP 0 641 351
			5,443,986	5,514,710	CA 2,119,126

or licensed by Molecular Probes, Inc.:

4,226,978	4,557,862	4,810,636	5,110,725	5,182,214	DE 40 19 025
4,237,224	4,603,209	4,849,362	5,134,232	5,230,343	EP 0 233 403
4,468,464	4,640,893	4,859,582	5,137,810	5,253,649	EP 0 314 480
4,473,693	4,656,134	4,981,985	5,141,627	5,380,746	
4,489,001	4,656,252	5,049,673	5,155,137	5,446,186	
4,499,284	4,740,470	5,055,556	5,162,227	5,459,272	
4,520,110	4,767,718	5,070,012	5,180,828	5,486,604	

Some restrictions may apply. All products are offered for research purposes only and are not available for resale or other commercial uses without a specific agreement from Molecular Probes, Inc. We welcome inquiries about licensing our products and applications for nonresearch uses.